S0-BLM-134

PROTOSTARS & PLANETS II

PROTOSTARS & PLANETS II

Edited by

David C. Black
Mildred Shapley Matthews

With 67 collaborating authors

THE UNIVERSITY OF ARIZONA PRESS
TUCSON, ARIZONA

SPACE SCIENCE SERIES

Tom Gehrels, General Editor

PLANETS, STARS AND NEBULAE, STUDIED WITH PHOTOPOLARIMETRY, T. Gehrels, ed., 1974, 1133 pp.
JUPITER, T. Gehrels, ed., 1976, 1254 pp.
PLANETARY SATELLITES, J. A. Burns, ed., 1977, 598 pp.
PROTOSTARS AND PLANETS, T. Gehrels, ed., 1978, 756 pp.
ASTEROIDS, T. Gehrels, ed., 1979, 1181 pp.
COMETS, L. L. Wilkening, ed., 1982, 766 pp.
THE SATELLITES OF JUPITER, D. Morrison, ed., 1982, 972 pp.
VENUS, D. M. Hunten et al., eds., 1983, 1143 pp.
SATURN, T. Gehrels & M. Matthews, eds., 1984, 968 pp.
PLANETARY RINGS, R. Greenberg & A. Brahic, eds., 1984, 784 pp.
PROTOSTARS & PLANETS II, D. C. Black & M. S. Matthews, eds., 1985, 1293 pp.

About the cover. A cocoon nebula, perhaps the primordial solar nebula, is depicted on the cover. Dust particles have condensed in the nebular disk and begun accreting into small planetesimals. Inhomogeneities in dust distribution block starlight from the inner nebula in some regions, but allow light to escape in certain directions. Painting by William K. Hartmann.

THE UNIVERSITY OF ARIZONA PRESS

This book was set in 10/12 Linotron 202 Times Roman
Manufactured in the U.S.A.

Library of Congress Cataloging-in-Publication Data
Main entry under title:

Protostars & planets II.

(Space science series)
Bibliography: p.
Includes index.
1. Protostars—Congresses. 2. Planets—Origin—Congresses. 3. Solar System—Origin—Congresses. 4. Cosmochemistry—Congresses. I. Black, David C. II. Matthews, Mildred Shapley. III. Title: Protostars and planets II. IV. Series.
QB806.P77 1985 521'.58 85-11223

ISBN 0-8165-0950-6 (alk. paper)

To
Bart J. Bok

CONTENTS

COLLABORATING AUTHORS

FOREWORD

On 5 August 1983, the international community of astronomers lost one of its most beloved and influential members. Bart J. Bok died in his Tucson home at the age of 77. He was a truly remarkable individual, a grand pundit of Milky Way research, a complete and model scientist who made equally outstanding contributions to astronomical research, administration, teaching, and public education.

A member of the National Academy of Sciences, Bok was also a former president of the American Astronomical Society, and former vice-president of the International Astronomical Union. He served as Associate Director of Harvard College Observatory, Director of Mt. Stromlo Observatory and Director of Steward Observatory. He was as well a recipient of the Henry Norris Russell Prize, American astronomy's highest honor.

Although primarily concerned with galactic structure, Bok had a lifelong interest in star formation; he contributed a number of important works in this area and had planned to be a contributor to this volume. It is particularly fitting, therefore, that this book be dedicated to Bart J. Bok. Briefly, I wish to focus on his contributions to star-formation research, a major emphasis of this book.

Bart Bok was born on 28 April 1906 in a Dutch Army warehouse in the city of Hoorn, Holland; he was the son of an army sergeant major. He became interested in astronomy at the ripe old age of 12 while a boy scout in Holland. In 1924 he entered the University of Leiden to begin a serious and dedicated pursuit of astronomical studies. Incidentally, when Bok arrived at the Leiden train station to begin his college career, the first person he met was fellow incoming freshman astronomy student G. P. Kuiper, who became a pioneer in planetary studies, and also advocated interdisciplinary scientific endeavors like those exemplified by the first (Gehrels 1978) and the present Protostars and Planets books.

When Bok began his studies of astronomy, the concept of star formation ocurring in the present epoch of galactic history was not yet established. However, his classic 1934 work on the stability of moving clusters laid the foundation for one of the major proofs of the existence of recent star formation in our Galaxy. He found that open clusters were not necessarily dynamically stable over long periods of time. Galactic tidal forces and encounters with other clusters would destabilize most clusters causing them to disintegrate in a time considerably less than the age of the universe. This implied that moving clusters were stellar systems which formed long after the Galaxy came into existence. In 1947, the Armenian astronomer Ambartsumian, using Bok's criteria for stability, showed that some systems, known as OB associations, were so unstable that they could not have existed for more than ten million years, a mere instant in the life of our Galaxy.

Ever since his earliest university days, Bok realized that if star formation occurred in the Galaxy, it had to happen within interstellar clouds of gas and dust. As a young student he became interested in the Great Carina emission nebula and in the dark obscuring nebulae, observed by E. E. Barnard, the well-known Barnard objects. However, his early interest in such matters was not shared by all his colleagues. He enjoyed relating the story of how his thesis advisor, the master of statistical star counting, P. J. van Rhijn, admonished him not to waste valuable time on such objects because they interferred with star counts and therefore got in the way of useful astronomy. Needless to say, Bok did not take this warning very seriously. In the early 1950s he made perhaps his most important contribution to astronomy and star formation research. Between 1952 and 1956 he assembled a group of promising young physics students to undertake the construction of one of the first radio-telescope facilities in the United States dedicated to the study of interstellar atomic hydrogen emission at 21-cm wavelength. In doing so, he was instrumental in training students who became some of the most influential and prominent radio astronomers in the United States. With their new instrument Bok and his students made careful studies of hydrogen clouds in the Galaxy and made one of the early determinations of the gas-to-dust ratios in interstellar clouds. In 1947 and 1948 he published papers calling attention to dark dust globules (now called Bok Globules) as probable sites of star formation and early stellar evolution. However, in the early 1950s when they turned their radio telescope to measure the expected interstellar atomic hydrogen in these and other dark nebulae, they were astonished to find very little interstellar atomic gas. As early as 1955, they were the first to realize and suggest that such dark nebulae primarily contained molecular hydrogen. They anticipated the importance of molecular clouds nearly eight years before the first radio-molecular line was actually discovered.

In 1978, Bart Bok and I sat together for the opening talks at the first Protostars and Planets meeting. I remember having lively conversations with him about sequential star formation and supernova-induced star formation,

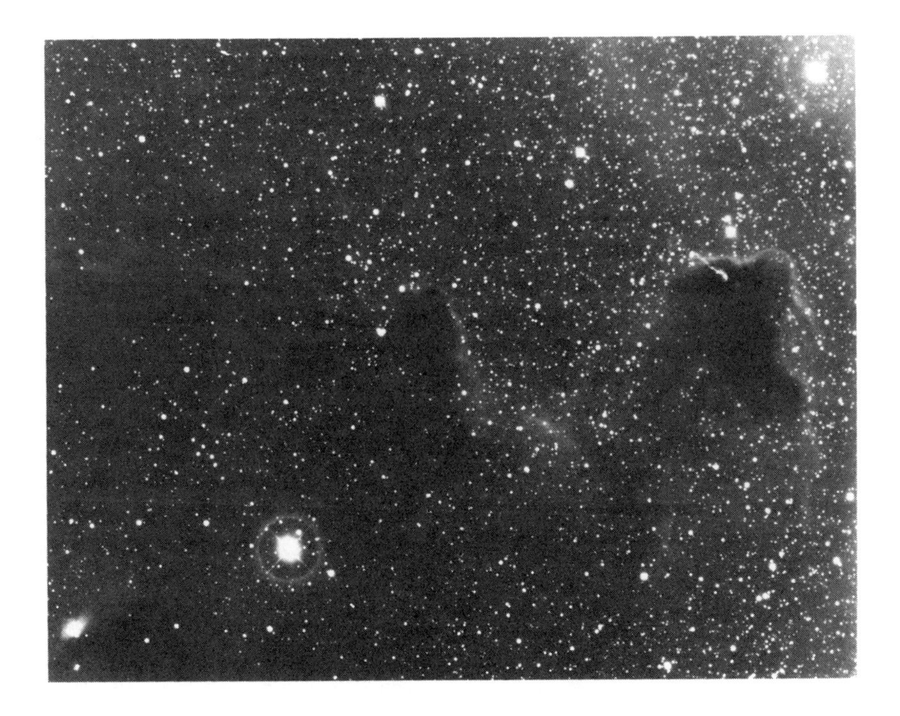

topics much discussed at that meeting. Bok also talked about globules and their relation to star formation and presented a paper on that subject. Most people at that meeting were skeptical about the role of globules as star-formation sites. However, this did not deter Bok; later that year, in February, he went to Chile to photograph some globules where he expected star formation to be occurring. It was to be his last observing run; on Valentine's night in 1978 he took this photograph of a globule in the Gum Nebula (see above). It turned out to be an interesting globule; it contains a young embedded star indicating that star formation could take place in a Bok Globule. Moreover, the globule's location at the edge of the Gum Nebula suggested that it might have been compressed by the expansion of that nebula; perhaps this was a case of triggered star formation. Finally, and most interestingly, the young object was shown to be associated with a string of Herbig-Haro objects which were later found to be a highly collimated, high-velocity jet of interstellar gas ejected by the star. This discovery of jets around young stars has stimulated a tremendous amount of new research in star formation studies and their significance has been a major topic of discussion at the 1984 meeting from which this book evolves. Bok would have been very pleased.

Although impressive by any standards, Bok's research accomplishments were only part of his prodigious contribution to astronomy. Above all, he was a great teacher and inspirational leader. He had a strong interest in the education and promotion of young researchers. He directed over 60 Ph.D. dissertations. Many of his students have become prominent astronomers such as J. L. Greenstein, I. R. King, M. K. V. Bappu, S. McCuskey, E. M. Lindsay, to name a few. He had an unabashed love for astronomy and was never too busy to discuss an astronomical result, no matter how minor, with an aspiring student. He encouraged and promoted the careers of countless young astronomers and his infectious enthusiasm was an inspiration for three generations of scientists.

Bok's interests did not stop with the work of professional astronomers; he was also greatly concerned and involved with amateur astronomers and the lay public. During the last eight years of his life he embarked on an exhaustive schedule of public lectures and writing. He accompanied numerous nonprofessional eclipse expeditions around the world (India, the Soviet Union, and Indonesia, to name a few recent locations), in the capacity of lecturer and scientific advisor. It is perhaps a measure of this man and his love for astronomy that he could contribute so profoundly to so many areas of endeavor. His legacy to astronomy is great. He will be missed.

CHARLES J. LADA

PREFACE

In January of 1978 a meeting was convened under the leadership of T. Gehrels with the intent of bringing together a diverse set of scientists working on the problems of star formation and the formation of the solar system. That "Protostars and Planets" meeting and the book of the same title resulting from the meeting were by almost any measure an immense success. The book, with 39 chapters and 51 contributing authors, has served as a valuable source of information and ideas for researchers in a variety of fields. Recognition that many significant experimental, observational, and theoretical advances had been made in virtually all facets of the study of star formation and the origin of the solar system led in 1983 to the notion that it was time to convene once again the researchers working on these problems, both in an effort to make all researchers concerned with these problems aware of recent advances, and to create the opportunity for as much cross-fertilization of ideas as possible. Gehrels anticipated the need for a second "Protostars and Planets" meeting when he noted in his introduction to *Protostars and Planets:* "We propose to produce the next volume on cosmogony of stars and planets with a meeting the first week of 1984 in Tucson." "Protostars and Planets II" was indeed held in Tucson in the first week of 1984, and this book presents the written thoughts of the principle speakers (and their colleagues) at that 1984 meeting.

A consensus has been held in the scientific community for several decades that the formation of the solar system is but one example of a phenomenon that occurs frequently in nature, and that that phenomenon is intimately related to, if not a natural consequence of, the formation of stars. This consensus underlies the synergistic relationship that exists between planetary science and stellar astronomy/astrophysics. To date, flow of information has been mainly in the sense that studies of solar system objects have affected models of important astrophysical phenomena (e.g., nucleosynthesis); however, there has also been information flow in the other direction (e.g., accretion disk theory). It is interesting to look once again at Gehrels' introduction in *Protostars and Planets* where he attempts to codify this synergistic activity as a "new discipline." He offered such terms as "Originary Sciences," "Astrogony," and "Cosmogony" to describe this new discipline.

While interactions of the type discussed above have been useful and productive, there has not yet been a major advance in our collective understanding of the processes involved in the formation of stars and planetary systems.

Part of what makes this book and the meeting from which it derives special in this regard is that now it appears that an intellectual and technical threshold has been reached on several fronts, and as a consequence there will be a burst of insight and knowledge into these fundamental problems. What is the nature of this threshold and why can we expect a quantum leap in our understanding of star and planetary system formation?

In the area of star formation we will see major advances following analysis of data from the Infrared Astronomical Satellite, the Hubble Space Telescope, the Space Infrared Telescope Facility, and a variety of groundbased millimeter interferometric systems. The Hubble Space Telescope, along with the Space Infrared Telescope Facility, will permit limited direct imaging of substellar companions to nearby stars. All of these instruments will establish a data base that should contain information to characterize the evolution of material from interstellar cloud to main sequence star, including the time history of circumstellar disks—the presumed birthplace of planetary companions to other stars.

In a complementary manner we see what is now planetary science stretching beyond the solar system to include the detection and study of other planetary systems. This extension of planetary science into the observational study of other stars and their environs is proper and natural; the *only* way we will ever understand the origin of the solar system is to test our models against nature. Will we find, for example, that planetary systems are much rarer than most of us would expect? Will we find that the solar system is reproduced in its gross structure around every single star or binary star? Is there something fundamental or prototypical of planetary systems in the character of the solar system, and if so, what? The answers to these questions go beyond discovering another planetary system, a fundamental step that remains to be taken. They require a comprehensive statistical survey of the nearest several hundred stars to a level of accuracy that is beyond the capability of existing telescopes. This level of performance is, however, not beyond our present technology, and plans are under consideration for a space-based astrometric telescope of sufficient accuracy to detect objects as small as Uranus around any star within a few tens of parsecs distant from the Sun.

The elements of Gehrels' "new discipline" are now beginning to emerge and their interrelationships are being defined. Pursuing his attempt to name this new discipline, it is perhaps appropriate to designate this area of overlap between planetary science and stellar astronomy/astrophysics *planetary systems science,* as the full scope of any union between these current disciplines is realized in the planetary systems which are, if our current theoretical construct is correct, the long-lived remnants of the birth of both stars and planets. An appreciation of this new era in research is gained from many of the discussions contained in this book, and we predict that, when "Protostars and Planets III" is held in 1990, this new discipline will be well into adolescence.

A few remarks are in order concerning the structure of the meeting and this book. There were some 200 attendees at the January 1984 meeting, and we received over 100 requests to present contributed talks. However, in light of the highly interdisciplinary character of the meeting, a decision was made to have no contributed (i.e., 5–10 minute) talks. Rather, we decided to have fewer and therefore longer (30 minute) talks on topics which seemed to the organizing committee to have undergone significant advances in recent years. Speakers were asked to summarize the key developments in their areas without providing a historical overview, indicate the major outstanding problems, and suggest possible lines of research over the next five years. These topical discussions were grouped into half-day sessions. Each session was initiated by a 45 minute general overview of the subject of the session. The intent was that the general overview would permit nonexperts in the audience to gain some perspective on the material to be covered in the session. A great deal of emphasis was placed on time for discussion during the meeting; each talk was structured so that two-thirds of the allotted time was for formal presentation and the remaining time was for discussion of that particular talk. A half-hour slot was preserved at the end of each half-day session for discussion of the entire session. A twenty minute break was taken during each session to permit casual interaction over coffee. However, we recognized the intellectual value of contributed papers; all contributed papers were given as evening "poster talks" so that they would not conflict with the main discussions during the day. Many people were concerned before the meeting that this format would not work because no one would come to the poster session after dinner. However, the success of this format exceeded our greatest expectations; the three hour evening poster talks were heavily attended and the forum proved an effective one to not only discuss the posters, but to discuss all topics related to the meeting. We strongly recommend this format for future meetings of this type.

The book is organized along the lines of the formal presentations during the meeting. There are some chapters which do not follow the general structure of the majority of the contributions. Summary talks from a planetary perspective by E. H. Levy and from an astronomical perspective by S. E. Strom, which brought the Tucson 1984 meeting to a close, are the lead chapters in the book. The other chapters which are of a different flavor are those representing the four principal schools of thought on models of solar system formation. Neither C. Hayashi nor V. S. Safronov were able to attend the meeting but each contributed a chapter for the book which described their current thinking on the origin of the solar system. The chapters by these authors, together with those of A. G. W. Cameron and of D. N. C. Lin and J. Papaloizou, provide in one book the opportunity for students of the subject to compare and contrast these current schools of thought.

Many people are necessary if endeavors such as "Protostars and Planets II" are to happen. We cannot here mention by name everyone who played a role in the meeting and this publication; however, in the back of the book we list the many individuals who contributed in one way or another to this effort. We would especially like to thank the authors, the reviewers, and the members of the scientific organizing committee for giving much of their time and effort to *Protostars and Planets II*. Finally we wish to acknowledge the efforts of M. A. Matthews during the meeting and of T. S. Mullin who kept both the meeting and the early work for the book running smoothly, as well as those of M. Magisos who has since stepped in to carry a substantial load in the preparation of this book. All these people deserve credit for any of the success associated with both the meeting and this book.

Astronomy in general, and the subject of this book in particular, lost a great and inspirational force when Bart J. Bok died a few months before the Protostars and Planets II meeting. He recognized and enthusiastically embraced the dawning of this new era in the study of stars and planetary systems, which is why we chose to dedicate this book to his pioneering spirit.

David C. Black
Mildred Shapley Matthews

PART I
Introduction

PROTOSTARS AND PLANETS: OVERVIEW FROM A PLANETARY PERSPECTIVE

E. H. LEVY
University of Arizona

Today we understand the simultaneous formation of a star and planetary system to be natural and coupled consequences of the collapse of an interstellar cloud of molecular gas and dust. Ideas put forward to account for the existence of our own solar system seem generally applicable to star formation throughout the Universe. If this is indeed the case, then planetary systems should be present in large numbers throughout our Universe. Even the gross structural features of our planetary system seem largely to be the result of nearly deterministic processes, rather than the result of chance alone; many planetary systems throughout our Universe may resemble this one. This chapter summarizes and examines some of the implications of ideas about the formation of stars and planetary systems.

I. INTRODUCTION

In the past, conferences on topics such as this one often began with a review of several diverse, competing theories for the origin of the solar system. However, recent conferences, including the one that inspires this book, are marked by the absence of any such listing of a large array of competing cosmogonical theories. The absence of such an open-minded review in this book and the dispatch with which these chapters get down to the business of elucidating our present ideas and identifying specific observational and theoretical problems, reflects the fact that today, to a first approximation, there exist no competing theories for the origin of our solar system. There is remarkable agreement about the general character of the events and the physical processes which produced this Sun and these planets. Moreover, the events and processes themselves are thought to be unremarkable.

To be sure, numerous mysteries remain; unsolved problems and uncertainties still exceed confident solutions and certainties although our ideas about the origin of this solar system—and of stars more generally—become better understood as time passes. Studies since the mid 1960s trace a plausible evolutionary track from interstellar gas to stars and planets. Increasingly sophisticated theoretical studies, together with laboratory measurements of astonishing virtuosity on primitive meteorites, and astronomical studies of the interstellar medium and protostars, are painting a picture with increasing detail about physical conditions associated with the formation of stars, protoplanets, and planets.

Today we understand the nearly simultaneous formation of a star and of a planetary system to be natural, and possibly common, consequences of the collapse of interstellar gas into a compact disk nebula. The disk-shaped nebula forms by virtue of the angular momentum carried by the infalling gas, which allows the gas to accumulate more easily along the axis of rotation than toward it. In ways barely understood, this nebula organizes itself into a dissipative structure that sheds angular momentum at the peripheries, allowing a star to accumulate at the center. At the same time, the distribution of solid matter is governed almost deterministically by evaporation and condensation in the nebula's radially varying thermodynamic and physical conditions. The aggregation of this condensed material produces planets whose general compositions, and even relative masses, are in turn a nearly inevitable result of a nebula's structure. The inevitability of this line of reasoning is indeed intoxicating; we might as well go on with it, if only to see where the most extreme view will carry us.

The scenario recited above is thought to be representative of star formation in general. One cannot help but be impressed by the evidence reviewed in this book (see chapters by Harvey and by Rydgren and Cohen), and by the new results of the IRAS investigations, that disks, perhaps a hundred or a thousand astronomical units across, are common, and maybe universal, features of protostars. It is not ridiculous then to speculate that planetary systems, similar to our own, are common in the universe. Of course, there are some obvious things left out of such a precipitate speculation. It glosses over the fact that most stars are in bound multiple systems, and the formation of multiple stars we understand even less than the formation of single stars. This speculation also leaves out the possible effects of varying initial conditions in the collapsing gas. However, a protostellar disk is a dissipative structure, evolving irreversibly and losing memory of its earlier states, but for a few conserved integrals; it is clear that much work needs to be done to define the effects of varying initial conditions on the course of star formation.

Despite these uncertainties, our present skeletal picture of star formation leads naturally to the idea that planetary systems much like our own may be common, at least around single stars that resemble the Sun. Even hospitable, terrestrial-type planets can emerge as commonplace. Although this seductive idea may be true, it may be only a manifestation of our continuing ignorance.

If we continue to follow the extreme view to which our ideas can lead—a view to which I admit at least an emotional attachment—it is interesting to note how close we are to the ultimate realization of the Copernican program. Benign planetary environments occur widely, in numbers which, while uncertain, are sure to be large. We are left only one or two short steps away from completely removing man from any privileged position in the Universe. According to modern textbook descriptions, the appearance of self-replicating entities seems almost as inevitable as the allegedly ubiquitous existence of thermodynamically and compositionally hospitable planets. Subsequent evolution proceeds inevitably in response to selective pressures in a competitive environment with limited resources. Again the result appears almost inevitable: advanced states of sentient, mobile, and manipulative beings, characterized by opposing thumbs and introspective temperaments, disposed to contemplate the events which led to their own existence. And thus, here we are today—it seems almost inescapable.

There is reason beyond idle amusement to contemplate these lengths to which our present ideas about the formation of stars and planetary systems carry us if we merely nudge them along. It helps to point out the fact that, for all the scope of our ideas on the subject, we are unsure whether planetary systems are common or rare in the Universe. We do not know if the one star and planetary system that we know best is representative of the many, or if our very existence here, on planets, demands that this system be idiosyncratic and different from most of the other stars in the Universe. This knowledge is certainly fundamental to our understanding of planet formation; furthermore, by virtue of our present picture of the linked contemporaneous formation of the Sun and planets, it must also influence our understanding of star formation. We human beings are dependent for our existence on the particular events which produced our particular planet. To understand how common are such events, and how common are such planets, will cut to the core of our perceptions about our very existence.

II. BEFORE THE BEGINNING OF THE SOLAR SYSTEM

While the framework of our ideas about star and planet formation seems to be growing more secure, vast areas of essential ignorance still stand between us and understanding. Only a few can be alluded to below.

One of the things that we teach our graduate students is how to do back-of-the-envelope, order-of-magnitude calculations; and we become irritated if they fail to get the right answer. For example, stars form from the collapse of presumably gravitationally bound, giant molecular clouds; and such clouds, if they were to collapse, have dynamical collapse times of about two million years. The Galaxy has roughly two billion solar masses invested in such molecular clouds. It is easy to see at once (on the basis of arguments which are dimensional, and therefore irrefutable) that the Galaxy should be putting some thousand solar masses a year into new stars. It is irritating, therefore, that our Galaxy fails to get the correct answer; the argument fails and overestimates the stellar birth rate by three orders of magnitude (chapter by Solomon and Sanders).

It is clear that essential elements of the equilibrium and dynamical behavior of the interstellar medium are not understood. Consider a few of the basic questions which remain unanswered:

—What is the general character of the virial equilibrium of interstellar clouds? That is to say: How much internal energy do these clouds contain, how is it distributed, what are its forms, how does it behave under compression during gravitational collapse?
—What is the role of rotation in the equilibria and the dynamical behaviors of these clouds? How strong are the magnetic fields, how do they evolve in collapsing clouds, and what are their effects?
—What role does turbulence play in these clouds? How do the turbulence and the magnetic fields interact?
—What is the ionization fraction in cloud material and how does the coupling of the gas and magnetic field behave during collapse?
—At what point does free-fall collapse of cloud gas occur? To what extent is the transition from diffuse interstellar gas to star mediated by a sequence of quasi-static, near-equilibrium states?
—What are the effects of the onset of star formation on the subsequent states and evolution of a molecular cloud? In what forms is energy liberated and how does the released energy interact with the surrounding cloud material?

Knowing the distribution and evolution of internal energy and angular momentum in protostellar clouds is fundamental to our understanding of the entire star formation process. Clouds are not in free-fall collapse despite the apparent absence of sufficient thermal pressure to withstand gravity (see the chapter by Scalo). Other forms of internal energy which can inhibit collapse include, for example, fluid turbulence, rotation, and magnetic fields; however, our detailed understanding of how these behave in real collapsing clouds is insecure.

The degree to which a collapsing cloud retains mechanical connection with the surrounding medium, through magnetic fields which allow it to shed angular momentum and which may provide some additional, tensional support against gravity, is unknown (chapter by Mestel). We do not know when, during stellar collapse, the bulk of initial angular momentum is lost. This question looms large in our attempts to understand the conditions which give rise to single, as opposed to multiple, stars (Mouschovias 1978).

It is worth recalling that we do not know what determines the main sequence masses of stars. What, in the formation of even an idealized single star, ultimately truncates the inflow of material? Is the distribution of final masses determined primarily by fragmentation of the larger cloud complex, or early in the collapse of the fragments, so that the final stellar masses result from an actual depletion of infalling matter, or, does some later-stage behavior of the protostar halt the inflow of what otherwise would be an essentially unlimited accretion to much higher mass?

Consider, in this same context, the question of the existence of external triggers of star formation. In the first *Protostars and Planets* book (Gehrels 1978), you could barely make a star without a supernova to trigger cloud collapse. It was widely conjectured that the material from which our own solar system formed had been inoculated with fresh supernova material within a few million years of its formation. This conjecture was driven, to a large extent, by the apparent presence of live ^{26}Al, a short-lived, *r*-process radionuclide, in meteorites assembled during the solar system's birth (Lee et al. 1976*c*). Recent work, however, calls this conjecture into question. The apparent absence of the medium-lived radioisotope ^{247}Cm in the early solar system suggests (chapter by Wasserburg) that supernova nucleosynthesis, contemporaneous with the formation of the solar system, might not have been responsible for the ^{26}Al. Surely these questions remain open; the immediate prehistory and very early history of the solar system still hide many important events and processes, evidence for which is probably hidden in primitive meteorites and other primordial materials.

Astronomical evidence for the general role of external star formation triggers (e.g., supernova blast waves or spiral density-wave shocks) is not clear-cut (chapter by Strom). There is a variety of conceivable star formation triggers (chapter by Cameron); however, such generational relationships in star formation seem still unclear. Several open questions are tied up with the general issue of star formation triggers. The most obvious is whether some especially violent dynamical perturbation plays an important role in precipitating star formation, or whether stars just form spontaneously and quiescently in undisturbed interstellar clouds, perhaps controlled by a gentle rain of neutral gas settling out of magnetized clouds through ambipolar diffusion (chapter by Cassen et al.).

A significant question, related to the role of star formation triggers, involves the frequency with which stars form from material that is different in composition from that of average cosmic background. Even if most star formation does occur in the vicinity of massive stars (chapter by Elmegreen), it remains to be established what causal relationships occur, and what role newly synthesized radioactive stellar fallout plays as an energy source in the evolution of cool, protoplanetary matter. And, on a related point which continues to confuse attempts to read solar system history from meteorites, we do not know how much intact prenebular solid material was incorporated into solar system objects (chapters by Kerridge and Chang and by Clayton et al.).

The matter in molecular cloud cores apparently undergoes substantial chemical processing. A considerable amount of isotopic fractionation can occur in these low-temperature chemical reactors (chapter by Irvine et al.), especially in relatively low-mass species like deuterium and carbon. How such fractionations may survive in the nebula and be related to observed isotopic heterogeneity of solar system material is not well understood (chapter by Draine). Chemical reactions in cloud cores probably have important influences on the clouds' continuing dynamical evolution (chapters by Glassgold and by Herbst). The extent of ionization in cloud cores influences the degree to which the matter remains coupled to the magnetic field; cooling rates, which exert significant control on cloud collapse, depend on the chemical state of the matter.

III. DURING THE BEGINNING OF THE SOLAR SYSTEM

During the transition from diffuse gas to a compact assemblage, large amounts of energy and angular momentum were shed from a protostellar system. When faced with such excesses of resources, nature continually surprises us with its imaginative means of expenditure. To take an example close at hand in the solar system, it would be easy to imagine our Sun simply emitting photons and neutrinos from a gently glowing, quiescent surface. Yet, faced with a little energy and angular momentum to spend, nature goes on a rampage and confounds us with a vast array of solar-active phenomena: an oscillatory magnetic field, explosive flares, energetic particles, radio outbursts, a hot corona, and a supersonic solar wind which spreads this confusion throughout the entire solar system.

A similarly rich menagerie of behaviors is likely to characterize protostellar and protoplanetary systems, which have much energy and angular momentum to spend in such a short time. Perhaps nothing so clearly reveals protostars as scenes of unforeseen dynamical activity and vigor as do the discoveries of collimated, jet-like, bipolar outflows and Herbig-Haro objects, which seem to be associated with protostellar systems. Prodigious amounts of energy are radiated away from young stellar objects through this nonthermal, mechanical mode. In one observed object (chapter by Schwartz), the mechanical luminosity can exceed several tenths of a solar luminosity ($L_\odot$). This apparently corresponds to at least one percent of the central star's total luminosity. By way of comparison, it is interesting to note that our own Sun's solar wind carries away only about $10^{-7}\ L_\odot$. The very nature of Herbig-Haro objects remains a mystery, and their relation to young star outflows is still unclear: Are they stellar ejecta or are they obstacles caught in the flow?

It seems likely that such channeled outflows arise primarily because of the rotation of the young protostar. Perhaps, as many have suggested (in this context, as well as in some attempts to explain bipolar collimated outflows in extragalactic radio sources), the highly collimated flows arise from otherwise spherical supersonic winds which are truncated laterally by material in a surrounding disk. However, one probably should not completely discard the possibility that external control is exerted on some of these flows; in this regard the existence—alluded to during the discussion—of fields of several such flows all seeming to be aligned along a superposed magnetic field is intriguing. It is probably fair to say that we have no real understanding of the dynamical origin and character of these collimated flows.

The accretion disk remains a vague and difficult object to perceive from both observational and theoretical perspectives. The accretion disk, more than any other single entity, ties together the two aspects of this book into a single subject. On the one hand, the disk serves as a transitory dissipative object in which protostellar matter seems able to shed large amounts of angular momentum during its transition from cloud gas to the main sequence; on the other hand, the disk seems to be the realm of planet formation.

During the formation of a star, the intermediate accretion disk seems to behave as a dissipative structure (see, e.g., Prigogine 1980) which transports excess angular momentum away from the center of accretion, thereby permitting the accretion to proceed at a rate far faster than could otherwise be the case. The microscopic mechanical transport coefficients of protostellar matter are too small to carry angular momentum rapidly through the fluid. The enhanced rate of transport and dissipation is produced by macroscopic, collective effects; among the several collective phenomena which may be important in the evolution of protostars and protoplanetary disks are convection (turbulence), larger-scale fluid circulations, magnetic fields, and spiral density waves induced at orbital resonances.

Turbulent convection substitutes a large coefficient of friction, derived from the mixing of momentum by macroscopic convective motions, for the negligibly small momentum transport produced by molecular motions. Several different driving mechanisms have been invoked to account for the presence of the turbulence, including agitation by the still infalling gas or unstable meridional circulations (Cameron 1978*c*) and thermal instability driven by the energy liberated in the viscous evolution of a dusty, high-opacity disk (chapter by Lin and Papaloizou).

In this field of study, as in others, our understanding remains clouded by the absence of a deductive theory of turbulence and turbulent transport; we are confined by estimates based on relatively simple mixing-length theory. While this approach probably carries our calculations toward reality, significant uncertainty remains. For example, at least in the inner parts of a protoplanetary accretion disk, the angular period of orbital motion is likely to be smaller than the turnover time of a convective eddy. Under this condition the Coriolis force becomes significant in the behavior of the fluid; the effect of the Coriolis force on such convection is to enforce a kind of two-dimensionality on the flow and to diminish transport in the radial direction. One early attempt to deal with this problem (Canuto et al. 1984), based on a linearized calculation, leads to results which seem not fundamentally changed from the estimates based on isotropic turbulence (chapter by Cameron); surely more will be heard on this subject as time goes by.

Meteorites show unambiguous evidence that they were assembled in the presence of a surprisingly strong magnetic field, perhaps as intense as one Gauss or more (Levy and Sonett 1978). Such a strong, nebular-scale magnetic field would have had profound effects on the nebula's evolution and could have been generated by the nebula itself (Levy 1978). It is possible, for example, that such a field could have provided a dominating contribution to momentum transport in the very inner nebula, where the Coriolis force might have inhibited convective radial transport. It is interesting to note that such a strong magnetic field could be expected to generate local flaring activity (especially above the disk in a nebular corona) in a manner similar to solar flares or geomagnetic tail storms. Nebular flares would produce local transient energy release, and could be responsible for such unexplained phenomena as chondrule formation in meteorites and oxygen isotope separations.

One of the inevitable by-products of turbulent transport of angular momentum is that the turbulence equally well transports small bits of solid matter around the solar system. The confrontation of this fact with the observed character of meteorites raises interesting questions. On the one hand, the large implied radial mixing seems capable of accounting for the complicated histories of some meteorite particles (chapter by Morfill et al.). On the other hand, extreme isotopic heterogeneity of solar system material (chapter by Clayton et al.) and the absence of calcium-aluminum-rich inclusions (CAI) in ordinary chondrites (chapter by Boynton) argues against such extreme homogenization of nebular material.

Primitive meteorites remain perhaps our most sensitive tracers of conditions in the protoplanetary nebula. The challenges posed by recent analyses of meteorites loom very large. The extreme isotopic heterogeneity of meteoritic material suggests the existence of several distinct and unhomogenized reservoirs of protoplanetary matter, with different isotopic compositions (chapter by Clayton et al.). The genesis of these reservoirs of matter, and the course of events which allowed them to remain unhomogenized, are not well understood in a manner that is demonstrably consistent with other facts about the solar system. The temperature history of meteorite components eludes detailed understanding. Evidence seems to suggest that some material at several AU from the Sun was exposed to at least transiently high temperatures, on the order of 1200 to 1400 K (chapter by Boynton).

It seems to me that the most critical issues in understanding the evolution of accretion disks are the nature and influences of the collective phenomena which dominate the transport processes, and thereby drive the disk's evolution, and which may play an important role in the accumulation of planets. It is not clear, in looking at the whole disk and over the whole of its evolution, that a single phenomenon dominates. It may be that, at different times and in different parts of the disk, different effects (such as turbulence, magnetic fields, density waves, etc.) predominated.

The ring systems of the outer planets have revealed important new information about the behavior of dissipative Keplerian disk systems (chapter by Lissauer and Cuzzi). The newly detailed pictures of Saturn's rings, made available by the Voyager spacecraft, are stimulating major advances in our understanding of the evolution of disks, and show them to have a rich repertory of dynamical behaviors. Spiral density waves are observed to occur, induced by the presence of moderately massive bodies, which may play a significant role in transporting angular momentum (and therefore mass) in the rings. Similar collective effects may have been important in some regions of the protoplanetary nebula. Planetary rings are thought to be confined to narrow bands by so-called shepherding satellites. Such satellites open and maintain gaps in the distribution of ring material. A similar clearing of gaps around large protoplanets may have had important effects on planetary accretion processes. More speculatively, other nonlinear effects in the disk may have worked on the formation of planets, with the formation of one object resonantly influencing the formation of others, possibly accounting for what some believe to be the unexpectedly regular spacings of the planetary orbits.

The ultimate fate of the disk gas still eludes us. The processes and the timing of its eventual dissipation are not known. This is particularly important for our understanding of the accretion of solid matter, because the orbital dynamics of small solid particles will depend on whether ponderable amounts of gas are present. One outstanding question is what relationship is there, if any, between the dissipation of disks and the observed energetic outflows.

IV. PLANETS AND PLANETESIMALS

Surprisingly little has been written in this book on the subject of planet formation. Two basic processes dominate our thinking about the aggregation of planets: solid-matter accretion and hydrodynamic-collapse instabilities in the nebular gas. Considering that the planets had to start off as dispersed matter and end up as dense accumulations, these two processes seem to have the most reasonable possibilities fairly well surrounded. Indeed, during recent years, virtually every combination of these processes has been invoked to account for the planets.

Planetary compositions of condensible species seem to track a condensation sequence, as a function of distance from the Sun. This fact argues strongly in favor of a dominant role in planet formation for the accretion of solid matter which had been processed through regimes of evaporation and condensation in a radially varying nebular temperature. Even Jupiter and Saturn, which most closely resemble the Sun in composition, have excessive amounts of condensible matter (chapters by Podolak and Reynolds, and by Gautier and Owen). The similarity in total masses of the condensible fraction of the four giant planets is commonly held to be evidence for the accretion of solid matter in the formation of these planets, prior to the induced hydrodynamical collapse of a nebular gas envelope.

According to this prevalent accretionary line of thought, the inner planets formed almost entirely from the accretion of solid particles of rock and metal to successively larger bodies, ultimately aggregating to the few terrestrial planets; the giant, gas-rich planets resulted from subsequent gravitationally-induced hydrodynamic collapse of nebular gas onto large ice and rock cores which formed through accretion. Presumably, during the gas collapse phase, a small viscous accretion disk formed, from which the regular satellite systems were born.

Although this crude picture seems able to account for the gross compositional and structural features of the planetary system, important unsolved problems remain. Perhaps most significant is the problem associated with the time scales of planet formation (chapter by Pollack). The problem arises because, in the outer parts of the nebula, the space density of planetesimals was probably not very much greater (and perhaps smaller) than in the inner nebula. Moreover, the dynamical times, which scale with the orbital periods, are much longer. Consequently, the rate of giant planet accretion through two-body encounters is expected to be lower than for the terrestrial bodies. This difficulty may be sharpest for Uranus and Neptune, because it is not yet clear that such an accretion model can account for formation of these objects in the entire age of the solar system. Even for Jupiter, the requirement, albeit hypothetical and arguable, that that planet form early enough to truncate the accretion of Mars and to abort planetary accretion in the asteroid belt does not seem satisfied.

Altogether, the compositions of the planets argue for the accretion of condensible matter in planet formation; the apparent rapidity of planet formation argues a role for collective phenomena, of which hydrodynamic collapse to giant gaseous protoplanets (Cameron 1978*c*) is perhaps the most extreme, and most rapid, example. More complete knowledge of the interior structures of the giant planets and a sharper understanding of their evolutions could advance our ideas about their formation (chapter by Pollack). For example, if the excess condensible matter should largely be confined to a core, and if such segregation is unlikely after formation, then this would argue for giant planet formation by late gas collapse onto a preexisting core. But these issues are not finally resolved. Critical information pertaining to the origin of the giant planets, including elemental and isotopic abundances, remains to be gathered by future investigations, especially by probes penetrating into the planets and analyzing the composition of the matter.

The terrestrial planets suffered even more neglect than did their Jovian cousins. However, it is worth pointing out that significant questions plague our understanding of the terrestrial planets as well. For example, the genesis of the volatile inventory of the terrestrial planets remains unknown. Some volatiles may have been trapped in condensed rock; some of the volatiles may have been caught later by accretion of ice-rich planetesimals (comets) formed farther out in the nebula. The early dynamical mixing of planetesimals between the inner and outer solar system is not fully understood.

The origin of the Earth's Moon is a major source of annoyance. On the one hand, its composition is sufficiently different from that of the Earth itself that there are few satisfactory ways of accounting for its existence through simple accretion near the Earth. On the other hand, making it elsewhere and later bringing it here, to be captured as an intact body, is dynamically difficult. Neither of these simple scenarios seems adequate to account for the Moon. Earth and Moon may once have been a highly interacting composite system, perhaps the result of a large collision (W. K. Hartmann and D. R. Davis 1975; Ward and Cameron 1978) in which matter was chemically and physically partitioned between the two to produce the present peculiar compositional differences. Probably there is a lot to be learned about planet formation through a satisfactory explanation of the Moon's existence. However, the Moon problem is sufficiently perplexing that at least one worker has found solace in its small size; he noted that the Moon is of sufficiently small mass that "to a first approximation, it does not exist," and, therefore, we can ignore it for a while (Wetherill, personal communication).

Regular satellite systems of the giant planets, of which Jupiter's four Galilean satellites are the best behaved example, probably formed in a circumplanetary disk-shaped nebula, in a manner analogous to the solar system's formation (see chapter by Lissauer and Cuzzi). The increase in volatile fraction, with distance from Jupiter, among the Galilean satellites is reminiscent of the regular compositional variation of the planets with distance from the Sun. This hints at a similar nebular origin for the Galilean satellites, although compositional alteration by subsequent evolutionary effects cannot yet be excluded for these satellites. Indeed, Io has undergone an extreme measure of sustained evolution as a result of tidal heating. Altogether, the solar system's formation seems to have involved at least a two-level hierarchy of disk systems, the primary disk giving rise to the solar system itself, while perhaps several smaller disks were involved in the formation of the giant planets and at least some of their satellites.

In this respect, the early evolutionary history of the giant planets is important to our understanding of their satellite systems. The luminosity of the nascent giant planets (chapter by Bodenheimer), and the energy liberated in their accretion disks, may have dominated physical conditions in their vicinities.

The irregular satellites, orbiting near the peripheries of the giant planet systems, were most likely solar nebula planetesimals captured by gas drag as they passed through the circumplanetary gas. The survival of these relatively small orbiting bodies against drag-induced evolution into the central planet apparently argues for rapid dissipation of the giant planet nebulae. Thus, the present-day orbits of these small, irregular moons may communicate important clues about the evolution of the giant planet nebulae.

Comets and asteroids are expected to store unique clues about conditions associated with the formation of the solar system. This is a result of the fact that, because of their small size, many did not achieve the high temperatures and pressures which sustained the extensive evolutionary alteration that occurred on planets. Thus, comets and asteroids are primitive in comparison with other solar system bodies.

Comets are thought to be the most primitive and fully representative of condensible protoplanetary nebula material. The presence of abundant and very volatile ice in contemporary comets testifies to the low-temperature regimes of their formation and to the compositional integrity that was maintained by their long-term cold storage in the distant Oort cloud. Future detailed study of comets which will be made possible by spacecraft investigations is expected to produce important advances in our knowledge of protosolar nebular condensates.

A few plausible suggestions have been put forward for the origin of comets (see the chapter by Weissman, which includes in addition discussions of several implausible suggestions). Plausible suggestions include formation near Uranus and Neptune, with subsequent ejection by gravitational interactions with the outer planets, formation in the far outer solar system with subsequent orbit expansion induced by mass loss from the Sun and surrounding nebula, and formation in distant subnebulae, well outside the realm of the planets. However, there is, at present, little to distinguish definitively among the several possibilities. Because comets formed so far from the Sun, at very low temperatures, it is likely that a large part of their material consists of unprocessed dust and molecules from the solar system's parent interstellar cloud. Detailed study of cometary material can be expected to forge an important link in our understanding of the transition from a molecular cloud core to a star- and planet-producing nebula.

Astronomical investigations of asteroids suggest that they are not unaltered nebular condensates, but rather that many asteroids have undergone substantial alteration processes. Indeed many main belt asteroids seem to be remnants of larger bodies which had undergone major differentiation processes before being disrupted. Presumably, gravitational perturbations from Jupiter scrambled asteroidal orbits to the extent that they were prevented from continuing accretion; since the early birth of Jupiter, the evolution of the asteroid population is thought to have been dominated collisional fragmentation.

It is generally believed, although not finally proven, that asteroids were never incorporated into objects large enough to have been melted by the low-level radioactive heat sources that drive evolution on the larger planets. Thus, additional sources of energy seem to have been at work, and the asteroids may be remnants of early heating processes which are no longer at work in the solar system. Among the possible sources of energy are electrical induction in a vigorous, early solar wind (Sonett et al. 1968) or heating by now extinct radionuclides. In this respect, asteroids are the products of evolutionary processes no longer at work in the solar system, and they must hide significant clues about conditions during the solar system's formation.

The main belt asteroid population, taken as a whole, exhibits important regularities when examined spectroscopically. The asteroids can be fitted into spectral classes, which seem to correspond, at least partially and crudely, to the compositional classes of meteorites. Moreover, different groups predominate at different heliocentric radii, with presumably more volatile-rich species predominating at greater distances from the Sun. This suggests that the asteroid belt is a compositionally structured assemblage of bodies, whose distribution may retain memory of physical conditions in the nebula from which they were born. Fortuitously, but apparently not accidentally, this preserved information spans the region of transition between the inner part of the solar system dominated by relatively refractory condensates and the outer part dominated by relatively volatile condensates. Altogether, detailed study of the asteroids promises to reveal important information about the protoplanetary nebula and protoplanetary collision and accretion processes.

V. CONCLUSION

Substantial advances in our understanding of the origin and evolution of the solar system and mechanisms of star formation, since the mid 1960s, resulted from a happy interaction of detailed experimental research on primitive solar system material, astronomical research, a much deeper understanding of the planets brought about largely by deep space missions, and impressive theoretical advances. It is worth asking: What is the goal of such cosmogonical research? Surely we can have no expectation of uncovering all of the detailed history of events which propelled some 10^{57} atoms along the evolutionary path from a diffuse interstellar cloud to the present peculiar assemblage of Sun, planets, and people. But we can hope to put together observations, experiments, and calculations to draw a picture of the general physical nature of events that make stars and planets, and to ascertain their prevalence and general characteristics in our Universe.

One of the compelling new lines of further research in this area would be the discovery and subsequent detailed analysis of other planetary systems. The scientific technology for this search seems now to be within our reach (D. C. Black 1980; Gatewood et al. 1980; R. S. McMillan et al. 1984). A remarkable manifestation of our present ideas about the formation of stars and planets is that most of us would be astounded if planetary systems at least crudely similar to our own should turn out to be uncommon. But, common or uncommon, whatever the answer should turn out to be, the impact on our understanding of the Universe in which we live will be most remarkable.

In conclusion, consider what should be the potentially observable signatures of distant planetary systems (as opposed to, say, double or triple stars) if our basic ideas about the formation of our own system are correct and more generally applicable. The structure intermediate between a collapsing cloud and the final star-planet system is the dissipative disk, and the basic morphological features of the solar system derive from the disk's behavior. The evidence of our own solar system suggests that a high degree of dissipation persisted to rather late stages in the system's formation. The planetary orbits are all relaxed to a relatively thin disk, orbiting in the same direction and near the Sun's equatorial plane; the orbits are largely circular. The compositions of the planets follow a condensation sequence. The small rock and metal objects reside near the center, where temperatures were too high for the more abundant volatile ices to condense or to remain solid; the much more massive ice- and gas-rich planets dwell farther out, where the much more abundant ices could condense and remain in solid form. The division between these two regimes seems likely to coincide with the boundary between those nebular temperatures at which water ice condenses and those temperatures at which water remains vapor.

Thus, as a tentative working definition, the signature of a planetary system should contain two components: evidence that the orbit morphology was produced by a highly dissipative dynamical evolution (concentric, circular, coplanar orbits) and evidence of a condensation sequence preserved in the radial distribution of planetary masses and compositions. The discovery of systems with this signature would lend strong support to our current ideas; the subsequent detailed study of the discovered systems would vastly expand our understanding planet-system and star formation. To establish the contrary, that such systems are absent or very uncommon, would challenge fundamental aspects of our present thinking.

Acknowledgment. This work was supported in part by the National Aeronautics and Space Administration.

PROTOSTARS AND PLANETS: OVERVIEW FROM AN ASTRONOMICAL PERSPECTIVE

STEPHEN E. STROM
University of Massachusetts

Current problems in understanding the formation and early evolution of stars are reviewed. First discussed are observational studies of external galaxies which may provide insight into the factors controlling star-forming efficiency and the initial mass function. Advances in the classification of molecular cloud properties in the Milky Way are summarized. Considerable emphasis is given to a discussion of recent observations of mass outflows from young stellar objects and the possibility that such outflows may provide a guide to the role played by magnetic fields during the prestellar collapse phases. Such flows also appear to serve as signposts for the discovery of circumstellar disks. Recent observations of possible disk structures surrounding young stellar objects are presented along with a picture of one such disk, that surrounding the T Tauri star, HL Tau. The formation of disks may be a fairly frequent outcome of the star-forming process. However, a direct connection between these disks and the formation of planetary systems has yet to be established.

For much of the past four decades, astronomers have chosen to study the star-formation process by observing and analyzing regions of recent stellar birth within the Milky Way. That this "microscopic" approach continues to be fruitful is shown by the wealth of new insights gained from infrared and millimeter-line investigations of such regions. Many contributions to this book reflect the vigor of such research.

However, in my view, studies of star formation would benefit from a greater investment in attempts to understand this process on a "macroscopic" level. Actively star-forming galaxies provide laboratories in which we can witness the orchestration of star formation on a galactic scale and can evaluate the influence of gas density, metal abundance, and other parameters on the efficiency of star formation and on the stellar initial mass function. The efficacy of proposed triggering mechanisms in inducing star formation can be tested as well.

Several chapters suggest the growing interest in synthesizing insights gained from both the microscopic and macroscopic approaches (see, e.g., chapters by Elmegreen, by Solomon and Sanders, by Goldsmith and Arquilla, by Evans, by Scalo, by Wilking and Lada, and by Harvey). As preparation for future protostars and planets research, I offer here some further thoughts on the macroscopic approach.

I. STAR-FORMATION EFFICIENCY IN DISK GALAXIES

Star formation efficiency in disk galaxies can be evaluated in two ways: (1) directly, by observing the current rate of star formation over the galactic disk, and (2) indirectly, by using relative gas content as a gauge of the past conversion rate of gas into stars.

Kennicutt and coworkers (Kennicutt 1983; Kennicutt and Kent 1983) have used the observed equivalent width of H α combined with galaxy colors to compute the ratio of newly formed, massive stars to the number of stars formed in the past. The integrated emission of a galaxy correlates with morphological (Hubble) type (Searle et al. 1973; J. G. Cohen 1976) although the dispersion among galaxies of a given type is large. For late-type, disk-dominated systems for which Kennicutt's estimates should be fairly accurate, the rate of star formation is nearly constant with time. The current epoch OB star formation rate appears to be linked only weakly to the H I content and the galaxy luminosity. On the other hand, Bothun (1984) finds a strong relation between neutral hydrogen content and galaxy luminosity; low-luminosity systems have significantly higher H I mass to 1.6 μm luminosity than do their more luminous counterparts. Hence, the rate of star formation, averaged over a Hubble time (the presumed age of the universe), appears to be higher in higher-luminosity systems. As sample sizes increase, we may expect further elucidation of the global dependence of star formation efficiency on galaxy type and luminosity.

A major advance in the study of bulk properties of galaxies has come from observations of their molecular gas content (see Young 1983 for recent summary). In the inner regions of some galaxies, the molecular hydrogen mass, $m(H_2)$, appears to exceed the neutral hydrogen mass, m(H I), by a considerable factor. At present, however, evaluation of the molecular hydrogen abundance depends on the adopted conversion between the observed strength of CO emission and H_2 column density. Not only do observers disagree on the conversion factor for the solar neighborhood, but the effects of metal abundance variations and differential heating of molecular clouds have yet to be included in current discussions. Subject to these uncertainties, it appears as if the molecular gas content to luminosity ratio is approximately constant over a factor of 100 variation in galaxy luminosity (Young and Scoville 1982). If so, the ratio $m(H_2)/m(\text{H I})$ decreases with decreasing galaxy luminosity. Perhaps the relative decrease in molecular hydrogen content is related in some way to the lower yield of new stars (averaged over a Hubble time) in low-luminosity systems. Future work aimed at providing H_2/H I ratios for a wider range of galactic types and luminosities will be critical to discussions of differential star formation rates in disk galaxies.

In addition to attempts to chart the integrated gas content and star-forming activity for disk galaxies, several investigators have attempted detailed studies of individual galaxies. In relatively nearby ($d<10$ Mpc) systems, it is possible to study the H I and H_2 content and current epoch star-forming activity as a function of position in the disk. For example, Jensen et al. (1981) mapped the star-forming activity in M83. Their study yields both the current epoch star formation rate for OB stars, from observation of H α, and the rate averaged over 10^8 yr from careful study of ultraviolet (0.35 μm), blue (0.42 μm), and red (0.65 μm) surface photometry. They compare these rates with the local gas surface density and conclude that the yield of new stars per H_2 nucleon is approximately constant. DeGioia-Eastwood et al. (1984) and Young and Scoville (1982*b*) also argue that the yield of new stars depends primarily on the mass of H_2 available locally (however, see Kennicutt [1983] who concludes that no relation between H I content and current epoch star-forming activity can be found). The estimated gas-consumption time is relatively short ($\approx$2 Gyr) if the average star formation rate is equal to its current epoch value.

Jensen et al. (1981) and Kaufman and coworkers (see, e.g., Rumstay and Kaufman 1983) have used observations of H II regions to evaluate proposed triggering mechanisms for initiating star-forming events. They conclude that star formation in the giant H II regions defining the spiral arms is initiated by passage of gas through the density wave pattern present in old disk stars (W. W. Roberts et al. 1975). However, a significant fraction ($\lesssim$50%) of star-forming activity takes place outside the arms and appears not to require a trigger.

The detailed study of star-forming activity in nearby galaxies should be further pursued. From such study, we may learn whether the formation of molecular clouds is both necessary and sufficient to account for all newly formed stars, or whether triggers, related to the dynamical properties of the galaxy, come into play as well (e.g., the rotation properties, arm amplitudes, density wave pattern speed, etc.). We may learn as well, from detailed mapping of the behavior of CO to H I emission around various annuli within a galaxy, what factors may control the rate of conversion of H I to H_2, and whether these factors depend on global (galaxy dynamics) or local (e.g., metal abundance, number of massive stars, and consequent supernova rate) properties. Also, the disagreement concerning the yield of new stars per gas nucleon and the relatively short gas-depletion times computed from estimates of the current epoch star formation rate must be understood.

II. THE INITIAL MASS FUNCTION

In the early 1970s, Searle et al. (1973) used the observed integrated colors of galaxies to deduce the shape of the initial mass function (IMF) (see the chapter by Scalo) as well as the rate of star formation as a function of time. More recently, Kennicutt (1983) has made use of the integrated H α emission, along with galaxy colors, to estimate the shape of the IMF averaged over the lifetime of a galaxy. For a sample of nearly 200 galaxies, the best fit to the observations is found for models of galaxy evolution characterized by an IMF slope ($n = 2.35$) close to Salpeter's original value and in accord with the original conclusions of Searle et al. These conclusions are most trustworthy in the case of spiral galaxies of late Hubble type, in which the contribution of the galactic bulge to the total light output is negligible. Kennicutt's (1983) conclusion that gas-depletion times are typically $\leq$3 Gyr leads one to suspect:

1. IMF in late-type galaxies includes relatively few low-mass stars;
2. High-mass and low-mass stars are formed in different regions and at rates that are little correlated;
3. Gas replenishment from external sources, such as infall from halos or capture of nearby dwarf galaxies, is an important effect during the lifetime of a galaxy;
4. Star formation in disk galaxies is sporadic or cyclical or;
5. We are living at a time of relatively rapid galaxy evolution.

Shields and Tinsley (1976) used the observed equivalent width of H β (W(H β)) to study variations in the IMF as a function of position within the Sc galaxy M101. W(H β) measures the ratio of ultraviolet quanta arising from very high-mass stars in the association responsible for producing the H II region to the number of visible quanta arising from association members of all masses; hence a rough estimate of the slope of the IMF can be made. Shields and Tinsley concluded that few stars of mass in excess of 30 $M_\odot$ can be found in regions of high metal abundance; such objects are seen preferentially in parts of the galaxy characterized by low metallicity. Their study, while a landmark attempt to evaluate a possible underlying cause for variations in the IMF, suffers from an important observational selection effect; only bright, rather than representative H II regions were included in their analysis. A far more detailed study of a more representative sample of H II regions in M101 and other galaxies will be necessary before the link between metallicity and upper mass limit can be considered firm. Proper application of the Shields and Tinsley method to other galaxies offers the hope of charting the dependence of the IMF on metallicity and other factors.

Another hold on the IMF and its variation with time is provided by observation of the metal content of galaxies. For closed systems, Searle and Sargent (1973) showed that the current epoch metal-to-hydrogen ratio Z depends upon the relative gas content $M_{\rm gas}/M_{\rm stars}$:

$$Z = (\text{yield}) \times \log \left(\frac{1 + M_{\rm gas}}{M_*} \right) \tag{1}$$

where the "yield" is the number of heavy elements returned to the interstellar medium per generation of stars. The yield depends critically on the IMF as metals come primarily from high-mass stars while low-mass stars consume gas, returning no heavy elements. Hence, by observing both Z, which derives from analysis of H II region spectra, and the H I and H_2 mass, the yield and hence the time averaged IMF can be deduced. Application of this technique has been attempted primarily in dwarf disk systems, where the assumption of "closed" evolution seems most plausible. Preliminary conclusions (Lequeux et al. 1979) suggest that the yield is uniform. Many more systems must be analyzed before this conclusion can be accepted.

Creative application of metallicity and gas content observations in combination with the Shields and Tinsley (1976) technique offers the possibility of significant advance in our understanding of the IMF and its dependence on varying physical conditions. Groundbased and Space Telescope studies of the luminous stellar population in nearby galaxies (Humphreys and McElroy 1984) offer an independent check on the conclusions drawn from these less direct methods.

III. MILKY WAY STUDIES

Charting the gas distribution and star-forming activity in the Milky Way has proven to be a major challenge. By all accounts, our galaxy is not a picture book spiral system. Embedded as we are in its disk, it is difficult to discern spiral features either in the distribution of H I or molecular gas. At this time, it is probably wisest to focus efforts to understand spiral structure, and its relation to star formation and the cycling of atomic and molecular gas, in other galaxies where the patterns are more plainly defined. The following issues are of specific interest here:

1. What defines a star-forming molecular cloud complex?
2. What are the statistical properties of such complexes and how are they distributed within the galaxy?
3. Do we observe hierarchical structures in clouds (e.g., complexes, individual clouds, cloud fragments or cores)?
4. If so, what are the physical properties of these structures?

Attempts to map the temperature, density and velocity fields in large numbers of cloud complexes, while valuable, would benefit enormously from improvements in spatial resolution and in sensitivity to emission from species which probe the domain of densities $>10^4$ cm^{-3}. Extension of current work seems likely to provide a detailed empirical picture of the fragmentation of cloud complexes.

While it is fashionable to talk about the role of magnetic fields, very few observational studies have been aimed at assessing its importance in the process of cloud fragmentation and prestellar collapse. At minimum, the field bears faithful witness to events which lead to cloud formation and fragmentation (Vrba 1977) and almost certainly plays a significant, if not the dominant, role in transferring angular momentum from prestellar cores to the surrounding cloud medium (Young et al. 1982). Advances in descriptions of cloud structure demand a comparable effort to relate this new information to the geometry and strength of the magnetic field. Near-infrared and infrared polarimetric studies (Monetti et al. 1984) offer the opportunity to map the field geometry from the periphery of cloud complexes to the periphery of condensed cores. In combination with density maps and descriptions of the cloud velocity field, observation of magnetic field geometry should provide a probe of the fragmentation and core formation processes.

IV. THE STAR-FORMING HISTORY OF INDIVIDUAL CLOUD COMPLEXES

Once molecular clouds form, does the star-forming process proceed:

1. At once throughout a cloud on a time scale comparable to a free-fall time?
2. Continuously over time scales long compared to a free-fall time?
3. At random at all stellar masses simultaneously, or sequentially, from low to high or high to low masses?
4. Spontaneously or with external triggers?
5. At special places or uniformly within the cloud?

Infrared maps of embedded stellar populations and optical studies of young cluster HR diagrams provide our soundest answers to these questions. Our present best guess is that the time scale for star formation in a typical cloud is at least several times 10^7 yr and perhaps as long as 10^8 yr (chapter by Harvey; M. T. Adams et al. 1983; Stauffer 1984); these times are much longer than the nominal free-fall times computed for cloud complexes of typical total mass 10^4 $M_\odot$ and extent of several pc. Infrared pictures of the Taurus, Chameleon and Ophiuchus cloud complexes suggest that low-mass ($M < 1.5$ $M_\odot$) stars form continuously and uniformly throughout a cloud.

These low-mass stars are the first to form; formation of higher-mass stars is favored later in the star-forming history of a complex (M. T. Adams et al. 1983; Iben and Talbot 1966; Herbig 1962). If this sequence of star formation proves commonplace, it suggests a uniformity in the fragmentation process which merits close scrutiny by theorists (Silk 1985; C. A. Norman and J. Silk 1980).

However, time scales for pre-main sequence evolution, and hence our conclusions regarding the sequence of star formation, are determined by computed evolutionary tracks. As yet, no challenging test of these tracks has been made. Computed tracks certainly blanket the domain of the HR diagram occupied by pre-main sequence stars. However, we have not been able to check whether a track for a star of a specified mass indeed passes through the locus of points occupied by stars, differing in age, but having the same mass. By adding a surface gravity measurement to our estimate of luminosity and effective temperature, we can, in principle, make that check. Mould and Wallis (1977) established a methodology for surface gravity determinations of young stellar objects; their lead should be followed.

The question of where massive stars form within cloud complexes has been addressed recently by Waller (1984). By combining the galactic plane survey of the University of Massachusetts and SUNY Stony Brook consortium with the Altenhoff et al. (1978) radio continuum and Downes et al. (1980) recombination line surveys, Waller is able to locate sites of massive star formation within 35 molecular clouds. He finds that massive stars are more likely to form near cloud centers, rather than at their peripheries. If this proves correct, Waller's work suggests that the role of external triggers may not be essential to high-mass star formation except insofar as proposed triggers act to gather molecular material. A similarly motivated approach making use of Infrared Astronomical Satellite (IRAS) pictures of isolated clouds may also prove fruitful.

V. THE ROLE OF EMBEDDED STELLAR POPULATIONS

We now know that it is difficult, if not impossible, to make sense of the observed temperature, density, and velocity-field data for molecular clouds without knowledge of the character and number of embedded young stellar objects. Such objects have the following characteristics:

1. Winds that alter local velocity fields and contribute to overall cloud support by increasing the internal energy of the cloud;
2. Moderate-to-high photospheric and envelope luminosities that can heat the cloud and dissociate nearby molecular material;
3. Possibly trigger star formation in post-shock regions associated with wind-cloud interaction (C. A. Norman and J. Silk 1980) or ionization front propagation (B. G. Elmegreen and C. J. Lada 1977);
4. Possibly chemically contaminate clouds, particularly if the time scale for star formation is long compared with the stellar evolution time scale for stars massive enough to become supernovae.

Detailed study of the stellar content of molecular clouds began in the early 1970s with the advent of infrared detectors of sensitivity sufficient to carry out extensive mapping at 2.2 μm (K. M. Strom et al. 1976). Later work refined methods of sorting between embedded and background stellar populations (Elias 1978; Hyland et al. 1982) as well as probing to fainter limiting magnitudes (Hyland et al. 1982). Recent studies (see, e.g., the chapter by Wilking and Lada) have begun to yield insight regarding the efficiency of star formation and the mechanisms leading to the formation of bound clusters and unbound associations.

In some regions (e.g., the dense core in the Ophiuchus complex studied by Wilking and Lada [see their chapter]), the density of newly-formed stars is sufficiently high to provide support of the cloud against further collapse (C. A. Norman and J. Silk 1980) if the winds from these embedded young stellar objects are similar to those observed in other environments.

It should be noted that current detector sensitivity permits mapping to limits 10 to 20 times fainter than those achieved to date. Deep maps should be of particular value in discussing (1) cloud support from low-mass star winds, and (2) luminosity functions, particularly toward the low luminosity end. For the latter issue, it is especially important to realize that a census of embedded populations, however complete, may provide only a snapshot of a still actively star-forming cloud.

Several contributors (see, e.g., the chapter by Harvey) emphasize the importance of seeking pictures, as opposed to source detections, in understanding embedded stellar populations. It is now generally accepted that some fraction of early point source detections are actually (1) patches of infrared reflection nebulae, or (2) shock-excited or photoionized gas. Multi-wavelength pictures can provide guidance regarding the true nature of a source both from its morphology and from its spectral energy distribution. Polarization measurements are also useful in this regard. Infrared pictures should prove valuable in locating dense structures in the vicinity of young stellar objects, a feature of particular value in seeking an understanding of collimating mechanisms for mass flows emanating from these objects.

The high-resolution infrared spectroscopic studies discussed in the chapter by Scoville provide a high spatial resolution probe of molecular cloud densities and temperatures along a ray to an embedded young stellar object. The range of physical conditions that can be examined is truly impressive. From early work both outflows emanating from massive young stellar objects, and infalling molecular material from an as yet undispersed protostellar cloud have been discovered. These studies, which require significant allocation of time on the largest groundbased telescopes, may make unique and fundamental contributions to the study of embedded stars and their environs. It is particularly important to search more thoroughly for evidence of infall (see chapter by Scoville; Moran 1983).

VI. OUTFLOWS FROM YOUNG STELLAR OBJECTS

Mass outflows appear to be a common characteristic of young stars of all masses. These outflows have been known for some time (1) from analysis of optical spectra, and (2) from their effect on nearby molecular cloud material; both broad-line CO sources and optical Herbig-Haro objects (see the chapter by Schwartz) result from wind-cloud interactions. Recent studies have shown that many outflows are well-collimated (opening angles typically far smaller than 45°) and bipolar (Snell et al. 1980; Bally and Lada 1983; Snell and Edwards 1983,1984). These molecular flows extend over distances of several parsecs. From their linear extent and an estimate of the characteristic velocity of the flow ($\approx$200 km s^{-1}), a typical flow time scale of several times 10^4 yr can be deduced.

Studies of mass outflows from young stellar objects have benefited recently from narrowband imaging surveys carried out at optical wavelengths (Mundt and Fried 1983; K. M. Strom et al. 1983; S. E. Strom 1983; Morgan et al. 1984). Very sensitive maps of wind-cloud interactions can be made by photographing regions surrounding Herbig-Haro and other young stellar objects through filters admitting spectral features sensitive to shock-excited gas emission. These maps reveal highly collimated jets as well as sinuous (precessing?) outflows; the length of these optical features ranges from 10^3 to 10^5 AU. Study of the excitation conditions and velocity fields characterizing these features is just beginning.

One unexpected outcome from the narrowband imaging survey is a remarkable tendency (K. M. Strom et al. 1983; Morgan et al. 1984) for outflows from individual young stellar objects embedded within a cloud complex to be aligned (a) one with another; (b) along or perpendicular to the direction of the long axis of the cloud; and (c) along the direction of the magnetic field as judged by the observed orientation of the electric-vector for stars viewed through the periphery of the cloud complexes. The best illustrations of alignment are found in the southwest region of the Orion complex (flows associated with Herbig-Haro 1-3, Haro 4-249, Haro 4-255, Haro 13a, Haro 14a and Herbig-Haro 33 and 40 appear to be parallel to within $\pm 6°$ over a projected distance in excess of 20 pc) and in the Taurus clouds (where the flows from DG Tau, Haro 6-5c, Haro 6-10, and Haro 6-13 are aligned with the direction of the magnetic field (see Monetti et al. 1984). The mechanism(s) responsible for collimation of the flows are apparently related to one or more large-scale characteristics (rotation, magnetic fields?) of the cloud. Do more mature young stars preserve a record of these mechanisms, for example, in the orientation of their rotation axes?

In several cases (the Cohen-Schwartz star [Cohen and Schwartz 1979], HH 12/107 [K. M. Strom et al. 1983], HH 101 [Morgan et al. 1984], and the HL/XZ Tau region [Mundt and Fried 1983; Morgan et al. 1984*a*]), the object apparently responsible for a collimated mass outflow and knots of HH nebulosity is not heavily obscured. Hence, it may be possible to catalog the optical properties of these objects and to compare them with other young stellar objects as well as to relate the present-day mass loss activity with the mass loss averaged over several hundred to several times 10^4 yr (if optical spectra of sufficient signal-to-noise and spectral resolution can be obtained for these faint stars).

Study of the molecular outflows provides an estimate of the wind luminosity. Often, this mechanical luminosity represents a significant fraction of the star's photon luminosity (between 1 and 10%). What accounts for this high mechanical luminosity? How is the wind fed and collimated? Are winds steady or sporadic? What is the duration of the strong wind phase? Answers to these questions are not only important in themselves but to understanding the role played by mass outflows from young stellar objects in controlling the evolution of the host molecular cloud.

VII. RECENT STUDIES OF INDIVIDUAL YOUNG STELLAR OBJECTS

Rydgren (see his chapter) has given an excellent review of progress in charting the physical properties of young stellar objects and their envelopes. Among the highlights of current research are:

1. Recognition that large spots and active regions are characteristic of intermediate and low-mass young stellar objects (T Tauri stars). These regions appear to be responsible for modulating the light output from these stars and suggest the possibility of strong sub-photospheric magnetic fields;
2. Charting of the chromospheric and coronal emission from T Tauri stars. The chromospheric temperature rise in these stars may be responsible for the majority of emission-line activity and spectral veiling observed in these objects;
3. Modeling of the dust envelopes surrounding T Tauri and Herbig Ae and Be stars. Rydgren and collaborators (see Rydgren et al. 1982) have succeeded in matching the infrared spectral energy distributions of these objects to composite photosphere, spherical dust envelope models;
4. Observation of extended gas envelopes surrounding T Tauri stars. Jankovics et al. (1983) have begun to make use of the forbidden-line spectra of T Tauri stars to study the outermost parts of their mass outflows. Mapping the geometry of T Tauri mass outflows may be possible from these spectroscopic studies in combination with high spatial resolution imaging in the forbidden lines.

Among the outstanding uncertainties regarding the interpretation of visible young stellar objects, I would list:

1. The large amplitude ($>$2.5 mag) light variations;
2. The large envelope luminosity. The combined optical and near-ultraviolet excess line and continuum radiation is in some cases comparable to the photospheric luminosity;
3. The large mechanical luminosity of the winds;
4. The extraordinary collimation of mass outflows, at distances $\leq$100 AU from the stellar surface.

It is possible that all these characteristics are in some way related to a combination of stellar rotation and mass accretion from a circumstellar disk. An accretion disk (Lynden-Bell and Pringle 1974) can provide significant luminosity if mass inflow rates are sufficiently high. Large amplitude variations can result from irregular inflow. Either inflow of material from a disk onto a rotating star (F. Shu 1983, personal communication) or evaporation of disk material (Choe 1984) appears capable of driving and collimating a stellar wind. Unfortunately, very few investigators (however, see Thompson et al. 1977) have chosen to carry out investigations aimed at establishing the presence or absence of such hot accretion disks. Surely, this is a topic meriting closer attention.

VIII. CIRCUMSTELLAR DISKS

Several chapters (see, e.g., chapters by Harvey and by Cassen et al.) provide strong support for the view that circumstellar disks form naturally and quite commonly during the early phases of stellar evolution. Three classes of observed disk structures can be outlined as follows:

1. Elongated structures (of dimension 1000 to 10^5 AU and average density 10^4 cm^{-3}) located in star-forming molecular cloud complexes. In the case of the region surrounding L 1551-IRS 5, recent high spatial resolution (beam size of 33″) CS observations (Kaifu et al. 1984) suggest the presence of a dense, rotating disk with its rotation axis directed parallel to the molecular outflow (Snell et al. 1980);
2. Far-infrared emission disks (in which the radiation from cool [$T \approx 50$ K] dust renders the disk visible) of dimension 300 to 3000 AU. M. Cohen (1983) presents evidence that a disk of this size may surround the young T Tauri star HL Tau, while the IRAS science team has suggested that disks (dimension 30 to 300 AU and total mass in dust on the order of 1 $M_\oplus$) of large particles surround stars of intermediate age;
3. Near-infrared scattering disks (in which dust grains near a young stellar object scatter light from these objects in the observer's direction) of dimension $\leq$300 AU. Beckwith et al. (1984) and Grasdalen et al. (1984) have detected such disks at wavelengths of 1 to 3 μm from application of speckle interferometric and maximum entropy reconstruction techniques, respectively. In some cases, scattering disks may define the inner regions of more extended disks manifesting themselves at far-infrared wavelengths.

As an example of a scattering disk, we show in Fig. 1 a 1.6 μm image of HL Tau derived from application of a maximum entropy algorithm to heavily oversampled data obtained at the Infrared Telescope Facility (IRTF) during January, 1984 (Grasdalen et al. 1984). A single channel InSb photometer was used to carry out observations of HL Tau and nearby, unresolved standard stars through a 2″ circular aperture. In Fig. 2, we show images of this star

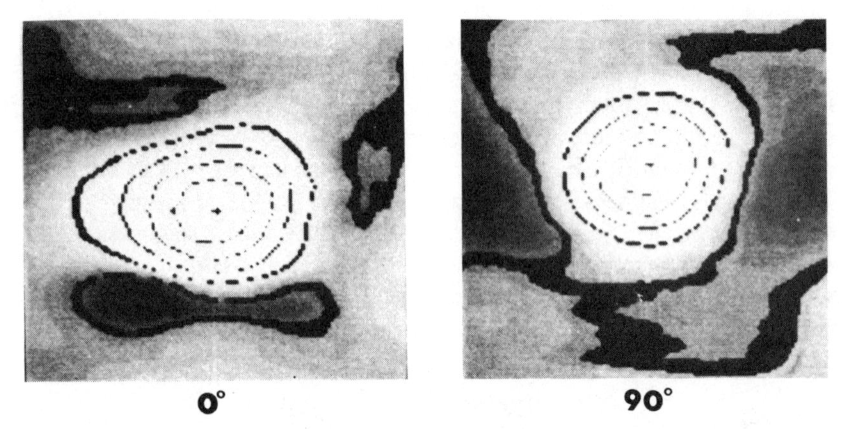

Fig. 1. A reconstruction of a 1.6 μm image of HL Tau. This is the most successful reconstruction of a set of 8 (as judged by the x^2 test). Several others of the set show the same general features. The reconstructed image of the unresolved star is shown to the same scale in the lower right inset. The contour levels have a constant (logarithmic) spacing of 0.5 dex. North is at the top and east at the left. The images are 12″8 on a side.

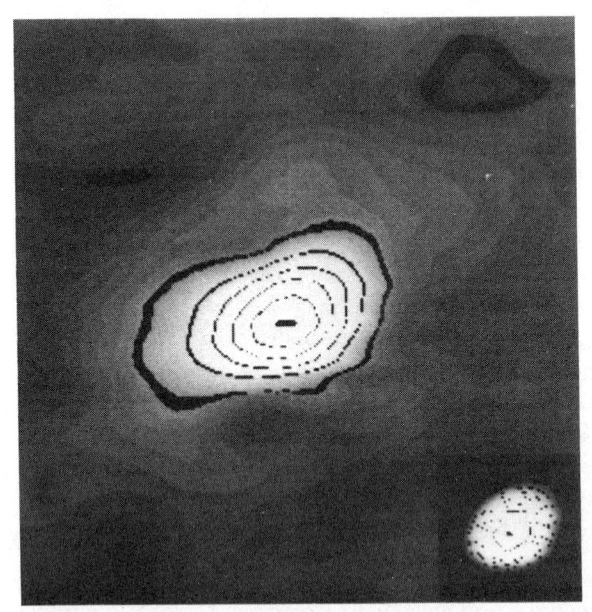

Fig. 2. Reconstructions of 2.2 μm images of HL Tau taken with a linear polarization analyzer in the beam. The images are 6″4 on a side. The contour levels have a constant (logarithmetic) spacing of 0.5 dex. North is at the top and east at the left. Note that the image of HL Tau is extended east-west for the analyzer position angle of 0°; it is unresolved with the analyzer in position angle 90°. If the extended structure is produced by starlight scattered by dust in a circumstellar disk, we would expect this pattern.

taken at 2.2 μm through a polarizer oriented at position angles 0° (left-hand panel) and 90° (right-hand). If the nearly east-west extension seen in Fig. 1 arises from scattering from dust grains in a disk, then the light from the extended structure should be strongly polarized with an electric-vector position angle of 0°; the extended structure should disappear at an analyzer position angle of 90°, in accord with Fig. 2. The approximate diameter of the disk is 270 AU, while the mass in scattering grains is estimated to be 5×10^{-7} $M_{\odot}$. Similar conclusions regarding the HL Tau disk were reached contemporaneously by Beckwith et al. (1984).

A critical next step will be to determine whether these disks represent tracers of remnant molecular cloud material or of more advanced, relatively gas-free aggregates containing larger solid bodies.

Young stellar objects located in nearby molecular cloud complexes at distances of ≈150 pc are amenable to study with both maximum entropy and speckle techniques. Speckle techniques should be able to discern disks to distances up to 1 Kpc. High spatial resolution observation at longer wavelengths (20 μm) should place constraints on the outer extent of these disks.

Lunar occultation observations and optical speckle observations of young stellar objects in forbidden lines might also prove to be effective tools for detecting disk-like structures and mapping gas outflows close to stellar surfaces.

IX. FUTURE GOALS

Gains in understanding the process of star and planet formation since 1978 have been significant. New results from IRAS and from sensitive groundbased optical, infrared, and millimeter telescopes have wrought impressive changes in our thinking. The next major steps in understanding star-forming process must come from very high spatial and spectral resolution observations at infrared, sub-millimeter and millimeter wavelengths and from detailed study of emission from very cool objects. New tools will be needed such as:

1. Space Infrared Telescope Facility, a cryogenically cooled infrared telescope: to find and study the youngest stars and protostars and to locate star-forming regions in external galaxies;
2. A 15-meter Class Optical and Infrared Telescope: to provide the collecting area and spatial resolution for high-resolution infrared spectroscopy and for speckle interferometric studies of young stars and their environs;
3. The Large Deployable Reflector and a groundbased millimeter wave interferometer: to study the chemistry and physics of fragmenting clouds and to chart the route from prestellar clump to protostellar object.

The investment in these new tools will be formidable. The problems they are designed to address, however, go to the heart of the quest to understand our origins and place in the universe.

PART II
Molecular Clouds and Star Formation

MOLECULAR CLOUDS AND STAR FORMATION: AN OVERVIEW

BRUCE G. ELMEGREEN
Columbia University

Theoretical work on star formation has progressed rapidly in the last decade because molecular-line observations have provided detailed information about star-forming clouds. This overview will summarize the most general aspects of these clouds, such as why they are molecular, what observations of the molecules have revealed about star formation, what observations of cloud masses and the distribution of clouds in the Galaxy suggest about star formation on a large scale and about the evolution and lifetime of molecular clouds, and why future observations of extremely high densities and clumping in cloud cores are essential to our understanding of the star formation mechanism.

I. HISTORICAL INTRODUCTION

Molecules have been observed in diffuse clouds since the late 1930s, when interstellar absorption lines from CH, CH^+ and CN were discovered in the visible spectra of several stars (Dunham 1937,1939,1941; Dunham and Adams 1937; W. S. Adams 1941,1943; Swings and Rosenfeld 1937; McKellar 1940; Douglas and Herzberg 1941). Molecular hydrogen was detected much later, using ultraviolet spectrometers above the atmosphere (Carruthers 1970). In the 1970s, the relative abundance and temperature of H_2 was determined for over 50 nearby diffuse clouds from absorption-line observations by the Copernicus satellite (Spitzer et al. 1973,1974).

Molecular hydrogen is not so easily observed in dense star-forming clouds. The clouds are usually too opaque for ultraviolet absorption line analysis, and H_2 does not radiate well from its lower rotational level because it has no permanent electronic dipole moment. Cold molecular hydrogen in star-forming clouds has been recognized only by its collisional excitation of other molecules, and by the relative decrease in 21-cm emission or absorption line strength from the depletion of atomic hydrogen. For example, thirty years ago, Bok (1955) noted that the ratio of the H I column density to the dust extinction is much lower in local dark clouds than it is in diffuse clouds, and he conjectured that the missing hydrogen is in the form of undetected molecules. Similar studies by Mezaros (1968), Heiles (1969), Quiroga and Varsavsky (1970), Sancisi and Wesselius (1970), Knapp (1972) and others also found a depletion of H I in dust clouds.

Direct observations of trace molecules in star-forming regions began with the detection of OH (Heiles 1968), NH_3 (A. C. Cheung et al. 1968) and H_2CO (Palmer et al. 1969) in dust clouds at centimeter wavelengths, and of CO near H II regions (R. W. Wilson et al. 1970) and dust clouds (Penzias et al. 1972) at millimeter wavelengths. H_2 was immediately identified as the dominant component of these clouds, even though it was not detected directly, because a total gas density of 10^3 cm^{-3} or more is required for collisional excitation of CO, and densities exceeding 10^4 cm^{-3} are required for excitation of CS (Penzias et al. 1971). Only H_2 could be so dense and escape detection at 21 cm.

The inferred H_2 densities in star-forming clouds are much higher than the H I densities measured from 21-cm observations. 21-cm observations of Orion OB 1 (Menon 1958) led Bok (1955) to conclude that the associated cloud mass is 60,000 $M_\odot$ and the average hydrogen density is 5 atoms cm^{-3}. Bok (1955) also estimated from star counts on the Palomar Observatory Sky Survey that the extinction in the center of the ρ Oph cloud "runs as high as 8 magnitudes." Recent molecular-line observations indicate that the average density in the Orion cloud is larger than 100 molecules cm^{-3} (Tucker et al. 1973; Liszt et al. 1974; Kutner et al. 1977), and that the visual extinction through the core of the Ophiuchus cloud exceeds 100 mag (Lada and Wilking 1980).

21-cm and extinction observations of star-forming regions often underestimated the cloud masses and gas densities. This led to erroneous conclusions about star formation. For example, low-density (10–20 cm^{-3}) atomic clouds were found surrounding many optical H II regions (see, e.g., Riegel 1967). The H I density was usually less than the density in the ionized gas, and the total H I mass was usually low (e.g., 10^3 $M_\odot$). This seemed to imply that star formation follows the isolated collapse of a cloud core, and that the resulting H II region expands spherically into a uniform, low-density medium, completely ionizing the initial cloud and producing a "Stromgren sphere." The high densities implied by the mere detection of CO completely changed this picture; most massive star formation is now thought to occur near the periphery of molecular cloud complexes, and most H II regions expand like blisters away from the denser parts of this periphery (Zuckerman 1973; Israel

1978; Gilmore 1980). What was formerly thought to be an initially spherical expansion of the H II region is now viewed as an asymmetric "champagne flow" from the dense cloud core to the surrounding low-density medium (Tenorio-Tagle 1979). Such asymmetric expansion pushes the neutral cloud to one side and ionizes only a small fraction of the initial gas. This leaves a substantial amount of neutral matter nearby for further star formation (B. G. Elmegreen and C. J. Lada 1977).

This example, and others given here, illustrate why molecular-line observations are essential for determining the mechanisms of star formation. In many cases, even the most qualitative ideas about star formation have been critically dependent on the quantitative properties of molecular clouds. The purpose of this overview is to summarize the most fundamental properties of molecular clouds, and to discuss how these properties have influenced our concept of star formation. Section II explains why star-forming clouds are usually cold and molecular. Section III summarizes what molecular observations of cloud densities, velocity dispersions, and ionization fractions have revealed about star formation. Section IV discusses observations of the cloud mass function, the distribution of clouds in the Galaxy, and what these observations imply about cloud formation, evolution, and lifetime. Section V illustrates what more we could learn about star formation, if we had better observations of cloud densities and internal clump structures.

II. WHY STAR-FORMING CLOUDS ARE USUALLY MOLECULAR

Star formation occurs when part of an interstellar cloud collapses under the force of its own gravity. Star-forming clouds are, therefore, strongly self-gravitating, in the sense that the binding force from gravity in the cloud core exceeds the binding force from pressure in the remote external medium. Such a gravitational force threshold corresponds to a minimum value of the average cloud mass column density μ that is proportional to the square root of the external pressure P

$$\mu > 1.6\left(\frac{P}{G}\right)^{1/2} \tag{1}$$

where G is the gravitational constant. The numerical factor in this relationship is based on the Bonner (1956)-Ebert (1955) condition for the critically stable mass M and radius R of a spherical isothermal cloud immersed in a medium of constant pressure (i.e., $\mu = M/\pi R^2$). The fact that a threshold in *column* density must be exceeded can be seen most easily by comparing the self-gravitational energy density in a critically stable cloud, GM^2/R^4, to the external pressure. This gives $M/R^2 \sim (P/G)^{1/2}$, as in the exact solution.

The local value of the dust-to-gas ratio in interstellar space makes μ equal to 0.005 g cm^{-2} times the visual extinction A_V in magnitudes (Jenkins and Savage 1974). Thus, the condition for strong internal gravity is approximately

$$A_V > 0.8\left(\frac{P}{3000\ k}\right)^{1/2} \tag{2}$$

for a cloud in a typical environment with a pressure of $\sim$3000 k for Boltzmann's constant k.

Why are such self-gravitating clouds also molecular? Molecules form where the opacity from dust is large enough to absorb most of the background stellar ultraviolet radiation; the transition from atomic to molecular gas is usually abrupt because of molecular self-shielding (Hollenbach et al. 1971; Spitzer and Jenkins 1975; Bally and Langer 1982). H_2 and CO, for example, have been found to occur in clouds where A_V exceeds approximately 0.5 mag (Spitzer and Jenkins 1975; Bally and Langer 1982). Each isotope of CO appears when its column density exceeds between 1 and 2×10^{14} cm^{-2} (Frerking et al. 1982). Coincidently, this self-shielding threshold is the same as the mass column density threshold that makes a cloud strongly self-gravitating in the local environment. Thus, the local star-forming clouds are also molecular.

The exclusion of background starlight also removes the heat input to a cloud from the photoelectric effect and from ultraviolet ionization of weakly bound atoms. The cloud temperature then drops from a typical value of $\sim$80 K in a transparent cloud (Spitzer 1978) to $\lesssim$10 K in an opaque cloud. This temperature drop occurs even though the cooling rate per gram of molecular material is typically a factor of $\sim$20 less than the cooling rate per gram of atomic material at the same pressure (based on a cooling rate of $2 \times 10^{-27}\ n^2$ erg cm^3 s^{-1} for a diffuse cloud at T = 80 K and n = 40 cm^{-3} [Dalgarno and McCray 1972], and a cooling rate of $6 \times 10^{-28}\ n$ erg s^{-1} for a molecular cloud at T = 10 K and n = 320 cm^{-3} [P. F. Goldsmith and W. D. Langer 1978]).

Not all of a star-forming cloud will be cold and molecular. The above discussion applies only to the strongly self-gravitating *part* of a cloud (i.e., the cloud core). Star-forming clouds should also have warm atomic envelopes, where the ultraviolet radiation from external starlight photodissociates molecules and heats the gas. Molecular clouds should not be viewed as isolated clouds, but only as the opaque *regions* of more extended cloud complexes (Wannier et al. 1983)

The fraction of a cloud's total mass that is in the form of molecules can be small. As an example, consider a cloud to be defined as a region where the density equals or exceeds 40 atoms cm^{-3}, the typical density in a diffuse cloud. This is the density where gas heated to 80 K by ultraviolet radiation can be in pressure equilibrium with the warm intercloud medium, so it may be used to define the cloud boundary. Then consider, as above, that molecules form in the inner part of this cloud where the visual extinction to the cloud surface exceeds 0.5 mag. If the cloud is spherical, and has an internal density variation of the form

$$n(R) = 40\ (R/R_{edge})^{-\alpha}\ \mathrm{cm}^{-3} \tag{3}$$

then the ratio of the molecular mass to the total cloud mass is easily found to be

$$M_{MOLECULAR}/M_{TOTAL} = (1 + 5\ \mathrm{parsecs}/R_{edge})^{\alpha-3}. \tag{4}$$

The column density of hydrogen nuclei corresponding to a visual extinction of $A_V = 0.5$ mag has been assumed to equal 1.2×10^{21} cm^{-2} from Jenkins and Savage (1974).

Figure 1 shows the molecular mass fraction as a function of R_{edge}. For $\alpha = 2$, as in a pressure-supported isothermal cloud, more than half of the cloud's hydrogen will be in atomic form if $R_{edge} < 5$ pc. A more uniform cloud ($\alpha < 2$), or a nonspherical cloud, contains an even higher fraction of atomic hydrogen.

The molecular fraction of a star-forming cloud depends on at least three properties of the overall environment:

1. The *heavy element abundance,* because of (a) the role of dust opacity in shielding the molecules from ultraviolet radiation; (b) the dependence of

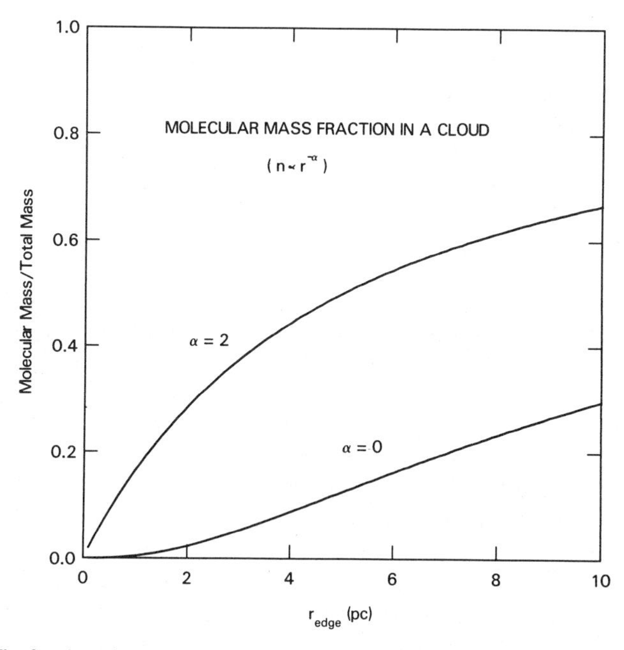

Fig. 1. The fraction of a spherical cloud's mass that is in molecular form plotted as a function of the cloud radius. α is the power in the power-law dependence of density on radius. The cloud is defined as a region where the density > 40 atoms cm^{-3}, and the molecular part of the cloud is defined as where the visual extinction to the surface > 0.5 mag.

the molecule formation rates on the relative densities of the constituents (e.g., for CO) or on the total surface area of dust grains (for H_2); (c) the dependence of the molecule formation rates on the thermal temperature, which depends on the heavy element cooling rate; and (d) the dependence of the external radiation field, and therefore the photodissociation rate, on the heavy element abundance, because of the likely dependence of the initial stellar mass function on this abundance, and because of the dependence of the ultraviolet flux from stars on their atmospheric opacities.

2. The molecular fraction depends on the *local star density,* which determines the flux of ultraviolet radiation incident on the cloud.
3. The molecular fraction also depends on the *cosmic ray flux,* because cosmic rays produce ionization in a cloud, which determines the formation rate of molecules by ion-molecule reactions, and because cosmic rays are an important heat source in the cloud. This dependence on the cosmic ray flux introduces additional dependences on the local supernova rate and the magnetic field strength.

Evidently the molecular content of a star-forming cloud is a complicated function of a variety of environmental factors. Variations in the abundances of dust and heavy elements can make the molecular contents and temperatures different from the local values. Magellanic irregular galaxies, for example, have relatively low metal and dust abundances, and the star-forming clouds in these galaxies have relatively weak CO emission (B. G. Elmegreen et al. 1980; Gordon et al. 1982; Young et al. 1984; Tacconi and Young 1984). Such variations can make the determination of cloud properties difficult or impossible, unless all of the factors that determine the abundance and excitation of the molecules used for cloud diagnostics are known for each region. A comprehensive theory of star formation must be based on observations of both molecular and atomic gas in clouds that are found in different types of galaxies and in different environments of our own Galaxy.

III. WHAT MOLECULAR OBSERVATIONS REVEAL ABOUT STAR FORMATION

Molecular-line observations are the most versatile probe of the internal properties of star-forming clouds. Molecular excitation is usually by collision, so the relative strengths of different emission-line transitions can be used to determine the local density of H_2 and the thermal temperature. Once the excitation is understood, the total emission in a spectral line gives the column density of the emitting species. The line profile gives the mean cloud velocity and velocity dispersion, and a map of the emission reveals the internal cloud structure and molecular boundary. Other cloud properties, such as the ionization fraction, can be determined from the relative abundances of ionized species once the gas-phase chemistry is understood.

What do observations of cloud densities, thermal temperatures, emission-line profiles, and ionization fractions reveal about star formation? The most fundamental information probably comes from the gas density; the relative importance of thermal and turbulent velocities during star formation is not well understood, and the ionization fraction has only limited use for dynamics studies in the absence of direct magnetic field observations. Nevertheless, observations of molecular clouds have revealed a number of new properties of star formation, as summarized in the following sections.

A. Observations of Cloud Densities

Most star-forming clouds have regions with densities in excess of 10^4 hydrogen molecules cm^{-3} (R. N. Martin and A. H. Barrett 1975; Evans and Kutner 1976; Liszt and Leung 1977; Linke and Goldsmith 1980; Wootten et al. 1980b; Loren 1981; Wootten 1981; Snell 1981; Vanden Bout et al. 1983). The overall profile of mean cloud density is often close to a power law, where the density decreases with radius to a power between 1.5 and 2 (Westbrook et al. 1976; L. H. Cheung et al. 1980). The structure sometimes appears clumped, however, so the densest regions are probably immersed in a lower density interclump medium (Zuckerman and Evans 1974; Townes 1976; Evans et al. 1979; Blitz and Shu 1980; Norman and Silk 1980 [and references therein]). Very young stars tend to appear in the densest clumps (Beichman, personal communication). Unfortunately, the star formation process inside a cloud clump, and the interaction between clumps inside a cloud are not well understood (cf. Sec. V.C).

The recognition that the density in the star-forming part of a cloud exceeds 10^3 or 10^4 molecules cm^{-3}, instead of only ~ 10 cm^{-3}, as formerly inferred from 21-cm observations, has led to a number of new concepts given below.

(1) H II regions have densities comparable to or lower than the molecular density in a cloud core, so most visible H II regions contain freshly ionized material that is expanding away from an adjoining cloud. H II regions are often like blisters on the surface of a cloud. The first blister model was for Orion (Zuckerman 1973), but subsequent observations revealed that many H II regions have this structure (Israel 1978).

(2) The time scale for the self-gravitational collapse of a cloud core depends primarily on the gas density. For a density of 10^4 cm^{-3}, the initial free-fall time is only 0.6×10^6 yr. This time is short enough to explain the simultaneous appearance of several short-lived O-type stars in a single, Trapezium-like cluster. It is also short enough to allow star formation to propagate inside a cloud, thus explaining the sequential appearance of subgroups in OB associations (Blaauw 1964). Theory predicts that the time interval between star formation epochs should be comparable to the gravitational free-fall time in the ambient cloud (B. G. Elmegreen and C. J. Lada 1977). If the cloud density is too low, then the time for a second generation of stars to form in a

compressed region will be longer than the maximum age of the stars that drive the compression; massive stars will still drive shocks into the cloud, but the shocks will disperse before a new generation of stars can form. This was the objection raised by Dibai (1958) to Oort's (1954) theory of such propagating star formation. Dibai argued that the low value of the density thought to be present in star-forming clouds delayed the pressurized triggering of gravitational instabilities beyond the time when the pressure was available. Dibai thought that star formation propagated by the direct squeezing of preexisting dense globules (which also seems possible—see, e.g., Dibai and Kaplan [1965]; Dyson [1968]; Kahn [1969]; Tenorio-Tagle [1977]; Whitworth [1981]; Sandford et al. [1982]; R. I. Klein et al. [1983]; LaRosa [1983]), but not by the formation of new globules or clumps in gravitational instabilities that arise in the shocked gas. Dibai's objection to Oort's theory disappeared when cloud densities were observed to be high.

(3) High densities also imply that star formation in OB associations is very inefficient. For the observed mass of young stars in a typical OB association, the observed value of the gas density implies that the ratio of the star mass to the gas mass is low. Inefficient star formation implies that most of a molecular cloud does not form stars at all. The clouds are either dispersed before much of the mass can turn into stars, or they are prevented or delayed from collapsing for much longer than a free-fall time by internal pressures, such as turbulence or magnetic fields. Zuckerman and Evans (1974) were the first to point out that the mass of dense cloudy material in the Galaxy is so high that the observed star formation rate can be explained only if most of this material is somehow prevented from forming stars in a free-fall time. The solution to this problem appears to be a combination of two effects: (a) low star formation efficiency resulting from effective cloud dispersal (efficiencies of 0.1% to 1% are likely for OB associations) (Duerr et al. 1982); and (b) cloud support that delays the free-fall collapse in the core for perhaps 3 free-fall times. This factor of 3 comes from the observation that molecular self-absorption in optically thick line profiles is sometimes redshifted by about 30% of the line half-width (i.e., implying contraction of a cool envelope onto a warm core [Snell and Loren 1977; Loren et al. 1981; see also, Leung and Brown 1977]). Because line half-widths are typically virial theorem velocities for the cloud cores (see, e.g., R. B. Larson 1981), the turbulent crossing time equals approximately the free-fall time. Thus, the observed redshift implies a contraction time of about $1/0.3 = 3$ free-fall times. Other clouds studied by Myers (1980) appear to be contracting at the free-fall rate onto thermal-pressure supported cores. Giant cloud envelopes could possibly resist self-gravitational collapse for a much longer time because of support from magnetic pressure and Alfvén waves. Such support would last for the magnetic diffusion time (cf. Sec. III.C).

(4) High densities and low efficiencies in star-forming regions may explain the expansion of OB associations. When a giant cloud disperses after

inefficient star formation, very little of the total mass remains behind to bind the embedded star cluster together (Hills 1980; Duerr et al. 1982; B. G. Elmegreen 1983*c*). The cluster then expands at a velocity comparable to the velocity dispersion in the former cloud. Moving clumps inside a cloud will also scatter the embedded stars during disruption (McCrea 1955).

(5) On a galactic scale, the high densities in molecular clouds imply that at least half of the interstellar medium inside the solar circle is molecular, and that the mean density of interstellar matter is between 2 and 5 times the value formerly obtained from 21-cm observations (Scoville and Solomon 1975; Burton et al. 1975; Bash and Peters 1976; Gordon and Burton 1976; Burton and Gordon 1976; Scoville et al. 1976; R. S. Cohen and P. Thaddeus 1977; Solomon et al. 1979*b*). Such high mean densities allow large-scale gravitational instabilities to grow in only $\sim 10^{7.5}$ yr (B. G. Elmegreen 1979*a*; Cowie 1981; B. G. Elmegreen and D. M. Elmegreen 1983; Jog and Solomon 1984*a*), a time that is short compared to the flow-through time in a galactic spiral density wave. High mean densities also imply that self-gravitational forces exceed magnetic pressure forces from the pure Parker instability (B. G. Elmegreen 1982*a*). Such gravitational instabilities may, therefore, explain how cloud and star formation is triggered in the spiral arms of galaxies that contain global density waves.

B. Observations of Gas Motions

Observations of thermal temperatures and velocity dispersions in star-forming clouds have been slightly less revealing than observations of the density, because no one yet understands what all the different types of gas motions imply. Theories of star formation often introduce a Jeans mass, which is essentially the cube of the Jeans length, $c/(G\rho)^{1/2}$ multiplied by the density ρ, but the appropriate value of the rms velocity c is not really known; i.e., should c be the thermal speed or the turbulent speed? Perhaps each speed corresponds to a different interpretation of the Jeans length, the first being the scale of a small, thermal-pressure supported condensation or clump in a cloud, and the second being the scale of a whole cloud. Furthermore, no one knows if the velocity dispersion is the result of systematic motions, such as contraction, expansion, or oscillation of a whole cloud, or if the dispersion is from convection, turbulence, or orbital motions of clumps. The thermal temperature can usually be explained using the known sources of heating and cooling in a cloud (see, e.g., P. F. Goldsmith and W. D. Langer 1978; Evans et al. 1982*a*; Vanden Bout and Evans 1982), but the origin of the macroscopic motions, which are often supersonic for the observed thermal temperatures, is unknown. Internal cloud dynamics, and the dynamics of star formation, cannot be understood until molecular line profiles are explained.

One apparently successful application of the thermal temperature in a discussion involving the Jeans mass has been made by Stahler (1983*a*), who points out that the upper boundary to the location of stars on the pre-main sequence track in an HR diagram is coincident with a theoretically predicted birthline, based on a mass accretion rate for protostars that is given by the expression c^3/G (which is the Jeans mass divided by the Jeans instability time) for thermal sound speed c. This reinforces the common notion that gravitational collapse to stars occurs within a Jeans size clump inside a cloud of many clumps (cf. Sec. V.C).

One of the most important results to come from molecular-line profiles is the inference that nearly all stars have strong winds at an early stage. Emission lines from the part of the cloud surrounding an embedded star often show broad pedestals from high-velocity flows (Zuckerman et al. 1976). The stellar winds are apparently pushing the ambient molecular gas away from the star (see Snell et al. 1980; Bally and Lada 1983).

C. Observations of the Ionization Fraction

Another important observation is of HCO^+ and DCO^+, which provides an estimate of the electron density in the core of a molecular cloud (Guelin et al. 1977; Wootten et al. 1979). The electron density in a molecular cloud indicates how well the magnetic field is coupled to the neutral gas. The magnetic field responds directly only to the charged particles in a cloud, such as molecular and atomic ions, electrons, and charged grains. Collisions between these charged particles and neutral atoms and molecules allow the field to exert a force on the bulk of the cloud, which is neutral. Such magnetic forces can be important in transferring angular momentum from a cloud core to the cloud envelope, thereby allowing the core to contract into a star without conserving the core's angular momentum (see chapter by Mestel). Magnetic forces can also play an important role in the overall balance between pressure and gravity inside a cloud (see reviews by Mouschovias [1978] and by Nakano [1984]). The magnetic lines of force that emerge from a cloud can also transfer momentum between the cloud and the external gas. Such magnetic connections may influence the clouds' translational motion (B. G. Elmegreen 1981*a*).

The ionization fractions inferred for dense molecular clouds are on the order of 10^{-7} (Langer 1984). These fractions are high enough to allow magnetic fields to exert a significant force on the neutral matter in a cloud, yet low enough to drive cloud evolution as the field gradually slips away. The diffusion time is given by the approximate expression

$$\tau_{\text{diff}} \simeq 4 \times 10^6 R^{1/2} x_7{}^{1/2} \text{ yr} \tag{5}$$

for cloud radius R in pc and ionization fraction x (in units of 10^{-7}). This expression was derived for a cloud with supersonic ion-neutral slip (corresponding to $R > 1.2x_7$ pc), and a field strength that is large enough to support the cloud against self-gravity (B. G. Elmegreen 1979*b*). Magnetic diffusion could delay the collapse of a large cloud for $\geq 10^7$ yr, which is long enough to explain the total duration of star formation in a typical OB association.

IV. WHAT OBSERVATIONS OF CLOUD MASSES AND THE DISTRIBUTION OF CLOUDS IN THE GALAXY HAVE REVEALED ABOUT STAR FORMATION AND CLOUD EVOLUTION

A. Small Molecular Clouds are More Common than Large Molecular Clouds

The mass distribution function for molecular clouds appears to vary with cloud mass approximately as $M^{-1.5\pm0.1}$ (Dame 1983; see also chapter by Solomon). After normalizing this function to the total molecular density in the solar neighborhood ($0.5 m_H cm^{-3}$) for clouds in the mass range of 10^2 to 10^6 $M_\odot$ and a scale height of 100 pc, the number density of molecular clouds becomes

$$n(M)dM = 10^{3.2\pm0.5} M^{-1.5\pm0.1} dM \tag{6}$$

for M in $M_\odot$ and $n(M)$ given in kpc^{-2}.

This distribution function implies that small molecular clouds are more common than large molecular clouds. Thus, the clouds closest to the Sun, which, from probability alone, will be the most common clouds, are also among the smallest clouds. The expected distance in pc to the nearest cloud in a logarithmic mass interval around the mass M is

$$D = 25\ (M/M_\odot)^{0.25}. \tag{7}$$

This equation was obtained from the inverse square root of $Mn(M)$. It implies that the nearest clouds of mass 10^3 $M_\odot$, 10^4 $M_\odot$, and 10^5 $M_\odot$ should be at distances of 140 pc, 250 pc and 440 pc, respectively, which is not unreasonable, considering the distances to the Ophiuchus and Taurus clouds at the low-mass end, and the Perseus and Orion clouds at the high-mass end.

Because molecular clouds have large extinctions (Sec. II), and nearby clouds have few foreground stars, the nearby molecular clouds will appear dark in projection against the background starfield. If molecular clouds are also randomly located with respect to mass, then most dark clouds must be molecular clouds at the low-mass end of the mass spectrum.

The term "dark cloud" is not a good name for a cloud. It can refer to different objects depending on the quality, depth, and color sensitivity of the photographs used. Barnard's (1927) photographs show the local dust clouds better than the Palomar Observatory Sky Survey because Barnard's plate sensitivity was too low to see many faint stars behind the clouds. Red-sensitive plates usually show nearby molecular clouds better than blue-sensitive plates (even though the extinction through the cloud is less in the red), because there are more faint red stars than blue stars in the background field. Perhaps a better name for a dark cloud is a *dwarf molecular cloud.* This term emphasizes the small size of the cloud and de-emphasizes any special circumstance of the cloud's location relative to the Sun. It is also an appropriate antonym for the other term commonly used, "giant molecular cloud." To be precise, we might refer to clouds with masses $\lesssim 10^4$ $M_\odot$ as dwarf molecular clouds (abbreviated DMC). This mass threshold appears to be important because clouds this small seldom form O-type stars, cloud disruption is seldom rapid (because there are no O stars), and bound clusters occasionally appear (because cloud disruption is slow). Clouds with masses $>10^4$ $M_\odot$ are giant molecular clouds (GMCs); they differ in a significant way from DMCs because GMCs usually form disruptive O stars, and the embedded clusters or associations usually end up unbound (because of the rapid cloud dispersal).

B. Giant Molecular Clouds Contain Most of the Molecular Mass

From the mass distribution function given above, the total mass density of local clouds in the mass range M_1 to M_2, expressed in units of $M_\odot$ kpc^{-2}, equals

$$\int_{M_1}^{M_2} 10^{3.2} M^{-1.5} M \, dM = 3200(M_2^{1/2} - M_1^{1/2}). \quad (8)$$

This total mass density increases with M_1 and M_2, so the largest clouds contain most of the mass (Solomon and Sanders 1980). This implies that most star formation occurs in giant molecular clouds. For a total cloud mass range of 10^2 $M_\odot$ to 10^7 $M_\odot$, the Sun had a 97% chance of forming in a cloud with a mass $>10^4$ $M_\odot$, and a 90% chance of forming in a cloud with a mass $>10^5$ $M_\odot$.

Because GMCs have more total mass than DMCs, they are more likely to form (rare) massive stars than DMCs. This implies the following:

1. Most star formation occurs in the same clouds where massive stars form. Along with massive stars come strong winds, H II regions, supernova explosions, and possible isotopic anomalies.
2. Most star formation may occur slightly before the cloud core is disrupted. Core disruption presumably follows the formation of O-type stars, so these stars usually form last out of a group of stars (Herbig (1962*b*). Other stellar groups and O stars may form elsewhere in the same cloud complex (Blaauw 1964).
3. Most star formation leaves no bound cluster behind when the gas leaves (i.e., no open or galactic cluster). This follows because rapid cloud disruption halts star formation in a cloud core before the star formation efficiency can become high.

High star formation efficiencies are essential for a bound cluster to remain when the gas leaves. If the upper mass limit for a cloud or cloud core that forms a bound cluster is 10^4 $M_\odot$ (see, e.g., B. G. Elmegreen 1983*c*), and the cloud mass function is as given above, then the Sun had a 3% chance of forming in a bound cluster. The Sun probably formed in the same way as most other stars, in an unbound subgroup in an OB association, or on the periphery of an OB association.

The largest clouds (10^6–10^7 $M_\odot$) probably form star complexes (Efremov 1979). Clouds with intermediate mass (10^5–10^6 $M_\odot$) form OB associations and occasionally bound clusters, and smaller clouds ($<10^5$ $M_\odot$) probably form loose stellar aggregates, and an occasional bound cluster.

C. Carbon Monoxide Spiral Structure in the Galaxy

The largest molecular clouds in the first quadrant of the Galaxy lie in the Sagittarius and Scutum spiral arms (Dame et al. 1984). The only large interarm feature is the Aquila spur, previously recognized by its 21-cm emission from H I (Weaver 1974). Figure 2 shows the distribution of the largest clouds from the perspective of an observer located 2 kpc above the position of the Sun (insert), and from the CO maps of the sky distribution, clipped to show only CO antenna temperatures above 2 K (Dame et al. 1984). Cloud distances were determined from the distances to the associated H II regions (Georgelin and Georgelin 1976; Lockman 1979; Downes et al. 1980).

Smaller molecular clouds may also be concentrated in these spiral arms, but such a distribution has not yet been demonstrated unambiguously. A problem with calibrating the distances to small clouds is that the associated H II regions are difficult to recognize or isolate at a distance of 5 to 10 kpc. Some may not even have H II regions. Small molecular clouds are known to be present in the interarm regions, because the terminal velocity curve on a longitude-velocity diagram of CO emission is almost continuous, and because the interarm gaps in other regions of this diagram are observed to contain such small clouds (see, e.g., Stark 1979).

Carbon monoxide emission in other spiral galaxies is occasionally observed to be concentrated in spiral arms. This occurs for M31 (Boulanger et al. 1981), NGC 4321 and NGC 1097 (D. M. Elmegreen and B. G. Elmegreen 1982; Scoville et al. 1983*b*), which are among the few galaxies with spiral arms that have been mapped with high enough angular resolution to see the CO arm structure. This observation of a spiral enhancement in the total CO emission implies not only that the large clouds are positioned in the arms, but also that a large fraction of the total CO mass is in the arms as well.

Galaxies with long and symmetric spiral arms are often called grand design galaxies. One reason why the CO intensity or concentration of large clouds is larger in the arms than in the interarms of such galaxies is that grand design spiral arms are usually density waves. This density-wave nature of grand design spirals has been established on the basis of the following observations:

1. Galactic surface photometry indicates that the arm/interarm intensity ratio is about the same for the blue and red or near-infrared passbands, so the arms have about the same colors as the interarms (Schweizer 1976; S. E.

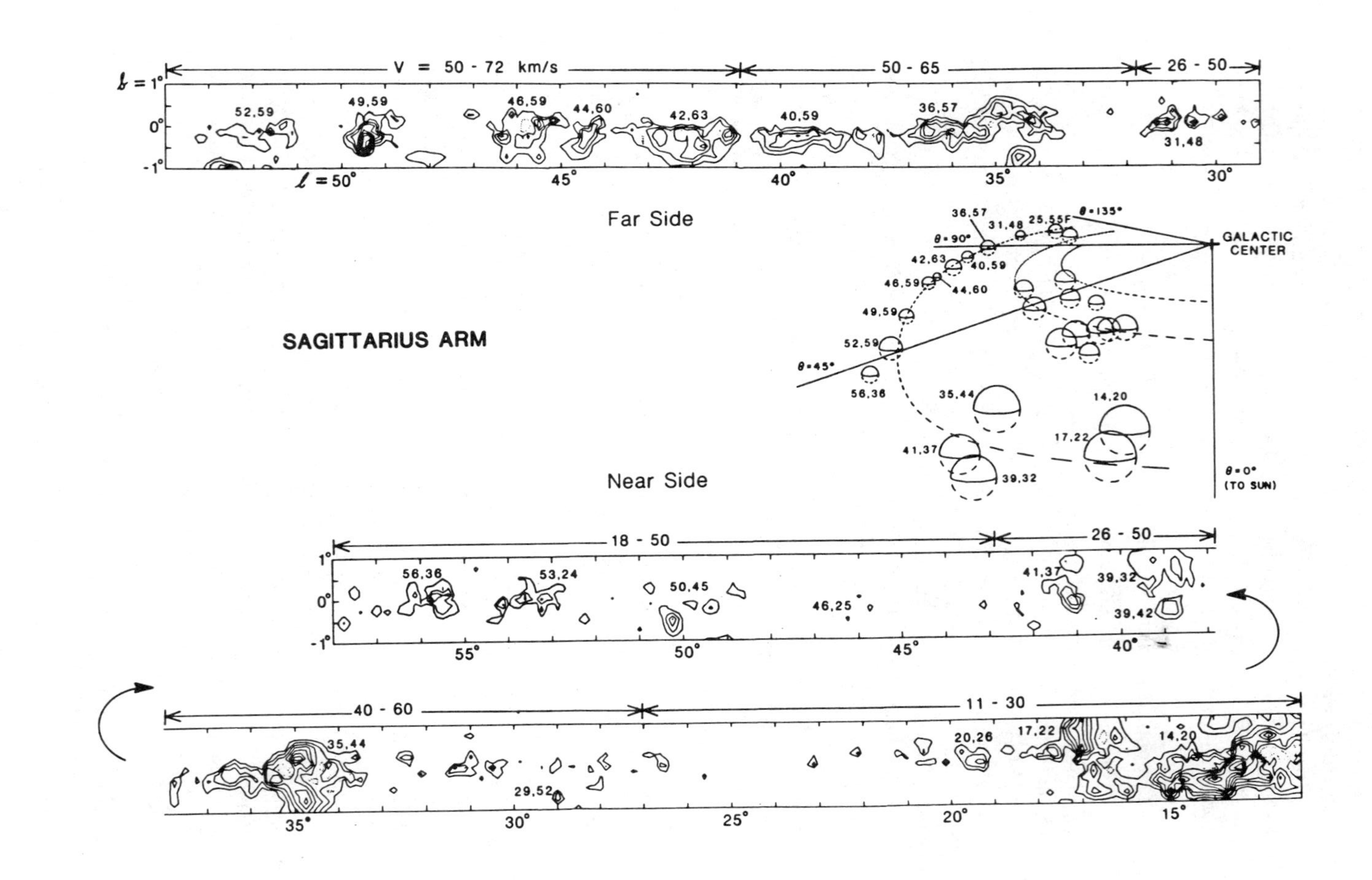
V = 50 - 72 km/s
50 - 65
26 - 50
$b = 1°$
0°
-1°
52,59
49,59
46,59
44,60
42,63
40,59
36,57
31,48
$l = 50°$
45°
40°
35°
30°
Far Side
SAGITTARIUS ARM
Near Side
θ = 135°
θ = 90°
θ = 45°
θ = 0° (TO SUN)
GALACTIC CENTER
25,55F
56,36
35,44
14,20
17,22
41,37
39,32
18 - 50
56,36
53,24
50,45
46,25
41,37
39,32
39,42
55°
40 - 60
11 - 30
35,44
29,52
20,26
17,22
14,20
25°
20°
15°

Strom et al. 1976; D. M. Elmegreen and B. G. Elmegreen 1984). Thus grand design arms are almost pure density enhancements compared to the interarms; star formation in such arms contributes only a small amount to the total intensity of the arms in either the blue or near-infrared passbands.

2. Grand design spiral arms show streaming motions of the gas (Rots 1975; Bosma 1978; Visser 1980; Newton 1980). This indicates that the arms contain a mass enhancement that is large enough to deflect the otherwise circular flow around the galaxy.
3. Magnetic compression in theoretically predicted spiral arm shocks (W. W. Roberts 1969; W. W. Roberts and C. Yuan 1970) has been inferred from radio continuum observations (Mathewson et al. 1972).

Each component of a galactic disk should be concentrated in the spiral arms if a density wave causes the arms. The density enhancement in the arms should depend on the velocity dispersion of the component, smaller dispersions giving larger enhancements. The arm/interarm density contrast for objects formed randomly in a grand design galaxy is essentially the ratio of the time the material spends in the arms to the time it spends in the interarm regions. The material spends a longer time in the arms than in the interarms because the gravitational force of the arms causes this material to flow nearly parallel to the spiral pattern within the arms and somewhat perpendicular to the spiral pattern between the arms (W. W. Roberts 1969).

If a galaxy has irregular spiral structure and no density waves, then the arms may be sheared patches of star formation (Seiden and Gerola 1982; D. M. Elmegreen and B. G. Elmegreen 1984). Molecular clouds should be concentrated in the arms whether these arms are from density waves or pure star formation.

This expected correlation between spiral arms and CO emission implies that carbon monoxide spiral structure should reveal little about cloud lifetimes. CO spiral structure does not imply that clouds are young, for example, as suggested by R. S. Cohen et al. (1980), and a lack of obvious CO spiral

---

Fig. 2. Spatial maps of the large clouds in the Sagittarius spiral arm shown by three longitude strips (from Dame et al. 1984). The maps were produced by setting all spectral channels with $T < 2$ K equal to zero before integrating over velocity. The velocity integration limits change in discrete steps along the maps in order to follow the velocity of the arms; the limits are indicated directly above each map. The contour interval in all of the maps is 9.8 K km s^{-1}. The figure between the maps shows the positions of the molecular clouds in the galactic plane, as viewed from the perspective of an observer 2 kpc above the Sun. The circle diameters are proportional to the cube roots of the cloud masses and the inverses of the cloud distances, and all the clouds are assumed to lie in the galactic plane. Straight lines from the galactic center have galactocentric longitudes of 0°, 45°, 90°, and 135°. Six clouds with masses $< 10^6$ $M_\odot$ are not shown in the central figure but are shown and labeled in the spatial maps. The dotted curves are logarithmic spirals shown in perspective. For the Sagittarius arm, the spiral is a least-squares fit through the positions of the clouds that outline the arm.

structure does not imply that clouds are old, as suggested by Scoville and Hersch (1979) and Solomon and Sanders (1980). Clouds can be very old and still appear in density-wave spiral arms (as in the case of H I gas, for example, which can be as old as the Galaxy). Conversely, molecular clouds can be very young but not appear in any obvious spirals if these spirals are irregular and hard to recognize.

The total fraction of the gas that is in molecular form (Scoville and Hersh 1979; Solomon and Sanders 1980) also reveals little about individual cloud lifetimes. This is because the arm/interarm surface brightness contrasts in grand design galaxies are so large (4 to 1, to 12 to 1 in many cases; see Schweizer [1976]; D. H. Elmegreen and B. G. Elmegreen [1984]) that a high fraction of the gas can be molecular (80% or more) and still the molecular clouds can form in the arms and be destroyed in the interarms.

The previous conclusion that molecular clouds are long-lived because (a) they lack spiral structure, and (b) the total molecular mass fraction may be large in the inner galaxy (Scoville and Hersh 1979; Solomon and Sanders 1980) is only partly true. The small clouds in the interarm region could certainly be longer-lived than the large clouds, but because these large clouds do not appear in the interarms, their ages must be less than the arm crossing time, which is between 10^8 and 2×10^8 yr, depending on galactocentric radius. Some smaller clouds may not survive the interarm transit either, but this is more difficult to determine. Similarly, the previous conclusion that molecular clouds are short-lived because (a) CO spirals are present at some intensity level and (b) the molecular mass fraction may, in fact, be small in the inner galaxy (R. S. Cohen et al. 1980), is also only partly true. Clouds could, in principle, live forever and still show strong spiral structure if the density-wave enhancement of Population I objects is strong.

D. Cloud Ages

Cloud ages may be obtained from the ages of associated star clusters (Bash et al. 1977) and from the statistics of cloud counts. The age of the associated star cluster gives the time *after* star formation began, and the cloud counts for a large sample give the mean time *before* star formation begins in clouds of that type. The average age of a cloud of a particular type may be defined from the expression

$$<\mathrm{AGE}_{\mathrm{cloud}}> = \mathrm{AGE}_{\mathrm{cluster}}\,[1 + (N_{\mathrm{without}}/N_{\mathrm{with}})] \quad (9)$$

where N_{without} is the number density of similar clouds without star formation and N_{with} is the number density of similar clouds with star formation. For example, if a typical cloud spends 10^8 yr without star formation and 10^7 yr with star formation, then, on average, there will be 10 times as many of these clouds without star formation as with star formation. The mean age of such a cloud is, therefore, the maximum age of a typical cluster in one of the clouds

TABLE I
Cloud Lifetimes

Mass ($M_\odot$)	**Cloud Formation Time ($\times$ 10^6 yr)**	**Star Formation Time ($\times$ 10^6 yr)**
10^6–10^7	50–200 (*L*/v, spiral structure)[a]	20–100 (ages of star complexes)
10^4–10^6	20–40 (*L*/v for isolated cloud) 20–200 (if part of a larger cloud)	10–20 (ages of OB associations)
10^1–10^4	1–10 (molecule formation; *L*/v) 20–200 (if part of a larger cloud)	0.1–10 (ages of T Tauri associations) or 1–200? (if stars do not destroy it)

[a]Parenthetical notes indicate methods for limiting the ages. *L*/v represents a typical formation time from the ratio of the cloud size to the rms velocity of the gas out of which the cloud forms. CO spiral structure gives a maximum cloud age for the largest clouds. Other notes suggest that some clouds form as pieces of larger clouds, and may remain for a long time without much star formation, if their density is low.

currently forming stars, multiplied by 11. Actually, the observed ratio $N_{\mathrm{without}}/N_{\mathrm{with}}$ is probably ≤ 1, since not many clouds without star formation have been found. Thus the total age of a molecular cloud is probably within a factor of 2 or 3 of the age of the stars that similar clouds contain.

Table I gives an estimate for the average ages of clouds of various masses, and it gives the methods or constraints used to obtain these ages. The age of a molecular cloud probably ranges between 10^5 yr (the molecule formation time) and several times 10^8 yr, depending on the cloud's size, among other things. Constraints on the cloud formation time come from CO observations of spiral structure (for the largest cloud) and from the ratio of the cloud size to the rms velocity of the gas out of which the cloud forms (denoted by *L*/v in the table). The largest clouds, which measure over 100 pc in length, may take more than 50×10^6 yr to form. Smaller clouds, such as the Orion cloud complex, probably form in only 20 to 40×10^6 yr. In the case of Orion, the formation process probably included a 40×10^6 yr episode of gas accumulation into a giant shell (the "Lindblad expanding ring"), followed by gravitational collapse along the shell periphery (B. G. Elmegreen 1982*b*). Cloud destruction times come from the maximum ages of star complexes, OB associations, T Tauri-star or reflection nebula associations, or from the age spread in galactic clusters, depending on the cloud mass. Small clouds could, in principle, last for a very long time if the small stars they form cannot destroy them, but there is no direct evidence for such long lives at the present time.

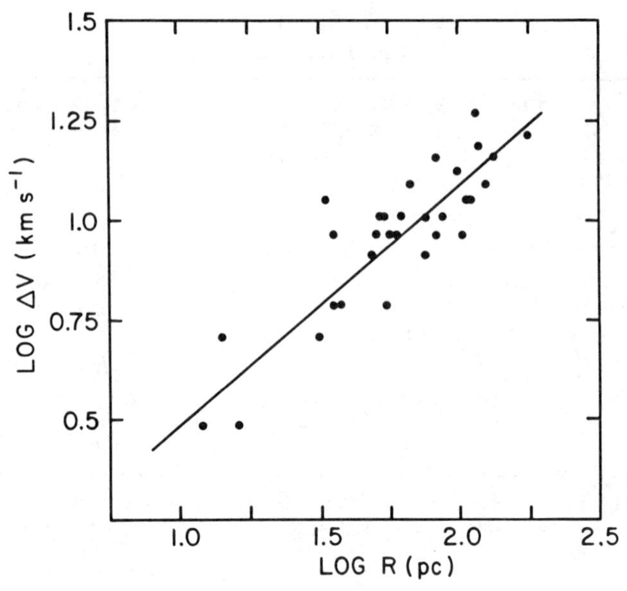

Fig. 3. The logarithm of the observed linewidth Δv (FWHM) of each complex listed in Dame et al. (1984) plotted against the logarithm of the radius R. The straight line is a least-squares fit.

E. New Way to Map Carbon Monoxide Clouds in the Galaxy

Observations of a large number of molecular clouds have shown that the diameters of such clouds are a function of the linewidths, although there is a considerable amount of scatter about a mean relation. Figure 3 shows such a function for CO clouds in the survey of Dame et al. (1984). The relation is

$$d = 3.2\ \Delta v^{5/3} \tag{10}$$

for cloud diameter d measured in pc and defined to be $(4A/\pi)^{1/2}$ for cloud area A, and for full width at half maximum of the CO profile Δv, in km s^{-1}. This function should not necessarily be viewed as a universal relationship. It may depend on the telescope used and on the method for identifying clouds in the data (see R. B. Larson 1981; chapter by Solomon). The distance (in kpc) to a cloud whose angular diameter is θ (in arcmin) is therefore

$$D = 11\ \Delta v^{5/3}/\theta. \tag{11}$$

One way to map the mean distribution of CO in the Galaxy is to use this relationship to determine the distances to a large number of clouds whose velocity dispersions and angular sizes can be derived from galactic CO surveys. Statistical corrections for blending will have to be applied to account for clouds that have the same sky positions and the same velocities but occur at the near and far kinematic distances. The positions of clouds on a Galaxy map should then indicate the overall distribution of CO. Spiral arms may not appear because the errors in the distances to individual clouds may be too large. Nevertheless, a mean radial distribution could be obtained.

One potentially important result would be the discovery of an azimuthal asymmetry in the CO distribution. The survey of large clouds by Dame et al. (1984) suggested that the CO is concentrated in the near side of the Galaxy at the distance of the Norma-Scutum spiral arm. If true, this would imply that the estimated total molecular mass in the 5 kpc ring would have to be reduced (by a factor of 2, perhaps) from that derived under the assumption that the gas is uniformly distributed around the galactic center (Dame 1983). The same may be true of the total emissivity and star formation rate derived from infrared and radio continuum surveys of the inner Galaxy.

The rotation curve of the Galaxy could also be determined from the azimuthal velocity that corresponds to the observed radial velocity of each cloud at its estimated distance. Velocity streaming motions may be detected by plotting on a map of the Galaxy the difference between the inferred azimuthal velocity component for each cloud and the average rotational velocity at the radius of that cloud.

V. WHY FUTURE OBSERVATIONS OF CLOUD DENSITIES AND CLUMPING IN CLOUD CORES ARE ESSENTIAL FOR THE CONTINUED PROGRESS OF STAR FORMATION THEORY

A. Density Diagnostics

Gas density is an important property of a molecular cloud. It is related to almost every aspect of a cloud's equilibrium state and evolution, and it determines the basic time and length scales for the formation of stars within the cloud (Sec. III.A). Gas density distributions in star-forming clouds should be mapped fully, and the densities at which rapid magnetic diffusion, gravitational collapse, and various other physical processes become important should be determined (Table II).

Emission line ratios of CS, NH_3 or H_2CO are used to determine gas densities between 10^4 and 10^6 cm^{-3} (see references in Sec. III.A). Lower densities can be measured from carbon fine-structure transitions in the ultraviolet for the transparent parts of the cloud (de Boer and Morton 1974), or from the amount of optical extinction per unit line-of-sight depth. At higher densities (10^7–10^{10} cm^{-3}), infrared line emission from vibrationally excited CO molecules may be useful (Scoville et al. 1980). More density calibrators are needed for high densities.

What more can we learn about star formation with better knowledge of gas densities? One of the most important things to determine now is the size, distribution, and motion of the clumps or fragments inside a giant molecular cloud. Even if the clumps are unresolved, they may still be studied in several

TABLE II
Density Thresholds for Processes during Star Formation

Density (cm^{-3})	Diagnostic[a]	Region or Physical Process
10^0	A_V/L	Atomic cloud envelopes, galactic spiral arms.
10^1	A_V/L, H I, C(UV)	Cloud envelopes, cloud formation.
10^2	N(CO)/L	Self-gravity important, molecules form, temperature begins to drop to 10 K.
10^3	N(CO)/L	Cloud contraction, magnetic braking of spin(?), important interactions between clumps.
10^4	CS, H_2CO, NH_3	Thermal clumps, begin collapse of cloud clumps to stars(?).
10^5–10^6	CS, H_2CO, NH_3	Begin rapid magnetic diffusion(?), end of magnetic braking of rotation(?).
10^7–10^{12}	CO vibrational transitions + ??	Formation phase for protostellar disks(?), accretion shocks on protostars(?), opaque protostellar cores(?).
$>10^{12}$	??	Begin formation of planets & binary stars.
10^{25}	—	Begin hydrogen nuclear fusion in star.

[a]Density diagnostics include the extinction or H I column density per unit length for the low-density regions, carbon fine-structure excitation for diffuse clouds, molecular column densities per unit length for intermediate-density molecular clouds, and molecular excitation analyses of various types for the highest-density regions.

ways. One way is to determine the radial distribution of the mass of gas at a particular density n, $M_n(R)$. This quantity could be derived from molecular transitions that are sensitive to a small density range around n. For example, suppose that a cloud contains a large number of clumps of identical mass, each of which is supported against self-gravity by thermal pressure at a uniform temperature. Then the gas density n inside each clump will be approximately proportional to the inverse square of the radial position r inside that clump (see, e.g., Shu 1977)

$$n(r) = n_0 \frac{r_0^2}{r^2}. \tag{12}$$

The density and radius at the edge of the clump are denoted by n_0 and r_0. In addition, suppose that the ensemble of clumps in the cloud also has an isothermal distribution, in the sense that the number density N of clumps in the cloud is proportional to the inverse square of the radial position R inside the cloud

$$N(R) = N_0 \frac{R_0^2}{R^2} \tag{13}$$

for clump density N_0 at the edge of the cloud of radius R_0 (cf. Sec. III.A for the average density distribution). Finally, suppose that the clumps are separated by an interclump medium of constant temperature, which is in pressure equilibrium with each clump, and in hydrostatic balance with the gravitational force in the cloud. Then the pressure in the interclump medium will vary as R^{-2}. Because the clumps are assumed to be in pressure equilibrium with this medium, the density at the edge of each clump n_o will vary as R^{-2}, and because all the clumps are assumed to have the same mass, the clump radius will vary as $r_0 \propto n_0^{-1/3} \sim R^{2/3}$.

What is the mass of gas at density n that is enclosed inside the radius R? For each clump the mass of gas that has a density between n and $n + \delta n$, is given by the expression

$$\delta m_n = 4\pi r^2\, n(r)\, \delta r = 4\pi n_o\, r_o^2\, \delta r \tag{14}$$

for a radial interval derived from the $n(r)$ distribution

$$\delta r = -0.5 n_o^{1/2}\, r_o\, n^{-3/2}\, \delta n. \tag{15}$$

Thus, δm_n varies with cloud radius as

$$\delta m_n = -2\pi n_o^{3/2}\, r_o^3\, n^{-3/2}\, \delta n \propto n_o^{3/2}\, r_o^3\, \delta n \propto R^{-1}\, \delta n. \tag{16}$$

The total mass of gas having a density between n and $n + \delta n$ from all clumps within the radius R is given by

$$\begin{aligned} \delta M_n (R) &= \int \delta m_n (R)\, N(R)\, 4\pi R^2\, \mathrm{d}R \propto \delta n \int R^{-1}\, R^{-2}\, R^2\, \mathrm{d}R \\ &\propto \delta n \int R^{-1}\, \mathrm{d}R \propto \ln(R)\, \delta n. \end{aligned} \tag{17}$$

This radial dependence should be compared to that of the total mass of gas at any density, which increases linearly with R for this example. Thus, the mass distribution obtained from observations of a molecular transition sensitive to a specific density interval may differ from the total mass distribution in a cloud if the cloud is clumpy. A comparison of mass distributions for different densities could determine the clump properties even if the clumps are unresolved.

A similar exercise would be to plot the distribution of mass with respect to density at a particular radius R, $M_R\,(n)$. This may indicate the spectrum of density fluctuations in the cloud. If the mass spectrum decreases with density, for example, then the clumps that are present may be related to Jeans unstable fragments, for which the clump mass is proportional to the inverse square root of density. If $M_R\,(n)$ increases with clump density, then the clump size may be relatively constant from clump to clump at a particular R.

The velocity structure inside a cloud might be obtained from maps of isolated spherical clouds, such as isolated Bok globules, in several different density-sensitive transitions (see, Loren 1977*a*; Martin and Barrett 1978). A plot of $\Delta v^2 (R)$ versus $n(R)R^2$ for linewidth Δv and density n would be interesting because (a) if the cloud is isothermal and in turbulent/magnetic pressure balance, then all radii in the cloud will have about the same values of $n(R)R^2$ and Δv, and the plot for different R would occupy a small range of values; alternatively, (b) if the cloud is not isothermal but is still supported by turbulent or magnetic pressure, then $\Delta v(R)$ should increase with increasing nR^2 because, for virial theorem velocities, $\Delta v^2 \propto nR^2$; or, (c) if the cloud is collapsing in a steady state, then $\Delta v(r)$ should decrease with increasing nR^2, because the inflow flux, $\Delta v n R^2$ will be constant with radius. Such simplified distinctions between steady-state models of clouds illustrate how observations of densities and linewidths might be used to probe the internal velocity field. More detailed analyses of such equilibria are necessary.

B. Internal Cloud Structure: Remnant Diffuse Clouds and Thermal Clumps

A giant molecular cloud must be assembled from smaller clouds, because the interstellar medium is cloudy on small scales. The assemblage is probably not random, because that takes a long time ($>10^8$ yr; see Kwan [1979] and Scoville and Hersch [1979]). A number of forces are present that can cause interstellar diffuse clouds to assemble into giant molecular clouds faster than by random coagulation. Mutual gravity between small clouds is one such force. The evidence that self-gravity in the ambient interstellar medium is responsible for the formation of some giant molecular clouds has been summarized by B. G. Elmegreen and D. M. Elmegreen (1983). Another force comes from deformations in the ambient interstellar magnetic field, as in the E. N. Parker (1966) instability (see Mouschovias et al. [1974]; Blitz and Shu [1980]; B. G. Elmegreen [1982*a*] for application to molecular clouds). Some clouds may be the gravitationally collapsed parts of giant shells that were swept up from the ambient interstellar gas (B. G. Elmegreen 1982*b*).

Whatever the cause of a molecular cloud's formation, the diffuse clouds that were present in the gas before the molecular cloud formed may still be present as clumps inside the molecular cloud after it formed. Since the average mass of a diffuse cloud is $\sim 10^3$ $M_\odot$, each 10^5 $M_\odot$ molecular cloud should contain ~ 100 former diffuse clouds. At the pressures inside a giant molecular cloud, such remnant diffuse clouds may be about 1 pc in diameter and have an average density slightly in excess of 10^3 cm^{-3}; the temperature should be low (6–10 K) and the gas should be molecular (cf. Sec. II). Because the magnetic diffusion time in an isolated diffuse cloud is much longer than the molecular cloud formation time (i.e., $\tau_{diff} >> 10^8$ yr), the magnetic field that was present in each diffuse cloud should still be present in the molecular cloud. Remnant diffuse clouds in giant molecular clouds should be interconnected by remnant magnetic flux tubes.

The existence of remnant diffuse clouds in giant molecular clouds may have little direct impact on star formation, although cloud clumping in general may influence the molecular cloud's rate of energy dissipation and evolution. Remnant diffuse clouds are important to observe because they may illustrate the mechanism and time scale for cloud formation. Molecular clouds with well-separated remnant diffuse clouds should be young compared to the coalescence time, and molecular clouds with only a few clumps, or with a core-halo structure to the clump distribution, should be old enough to have had significant interactions between the various cloud components.

Another type of internal structure expected for molecular clouds is a clump that has the size of a thermal pressure scale length, which is a Jeans length ($=c/[G\rho]^{1/2}$ for thermal speed c and mass density ρ). This size is about 0.2 pc for a gas temperature of 10 K and an H_2 density of 10^4 cm^{-3}; each clump will contain about 1 $M_\odot$. All molecular clouds should contain Jeans size clumps, or "thermal clumps," where the thermal pressure balances local gravity.

Figure 4 shows a chain of 5 apparently similar globules (S. Schneider and B. G. Elmegreen 1979), of which only one (to the south) has a uniform background illumination. This one shows considerable substructure on the thermal pressure scale length of 0.1 pc (subtending 1 millimeter on the Palomar Observatory Sky Survey). The others do not show this structure, but presumably only because their background illumination is not smooth. All molecular clouds larger than a Jeans mass should contain substructure on a thermal pressure scale length because of local gravitational forces.

C. Clump Theories

Many theories of star formation require better observations of the clumpy substructure inside molecular clouds (see, e.g., Zuckerman and Evans 1974; Norman and Silk 1980; Blitz and Shu 1980; Scalo and Pumphrey 1982). One interesting implication of these theories is that a cloud with clumps can be *unstable* to form stars on a *small* scale, but *stable,* long-lived and resistant to overall collapse on a *large* scale.

Consider the following example of how individual cloud clumps could form stars in 10^5 yr bursts while interacting so weakly that the whole cloud, which is an ensemble of clumps, remains supported for a much longer time (e.g., 10^7 yr). This particular model assumes that the clumps are on magnetic flux tubes inside a cloud, and that clump-clump interactions are controlled by magnetic forces. Cloud clumps may seldom collide physically; clump interactions could be confined entirely to the bending and stretching of field lines, not to direct physical contact and shock formation between mass elements (P. Clifford and B. G. Elmegreen 1983). Clump collisions would then dissipate little energy, even at superthermal velocities. In this model, the energy lost per

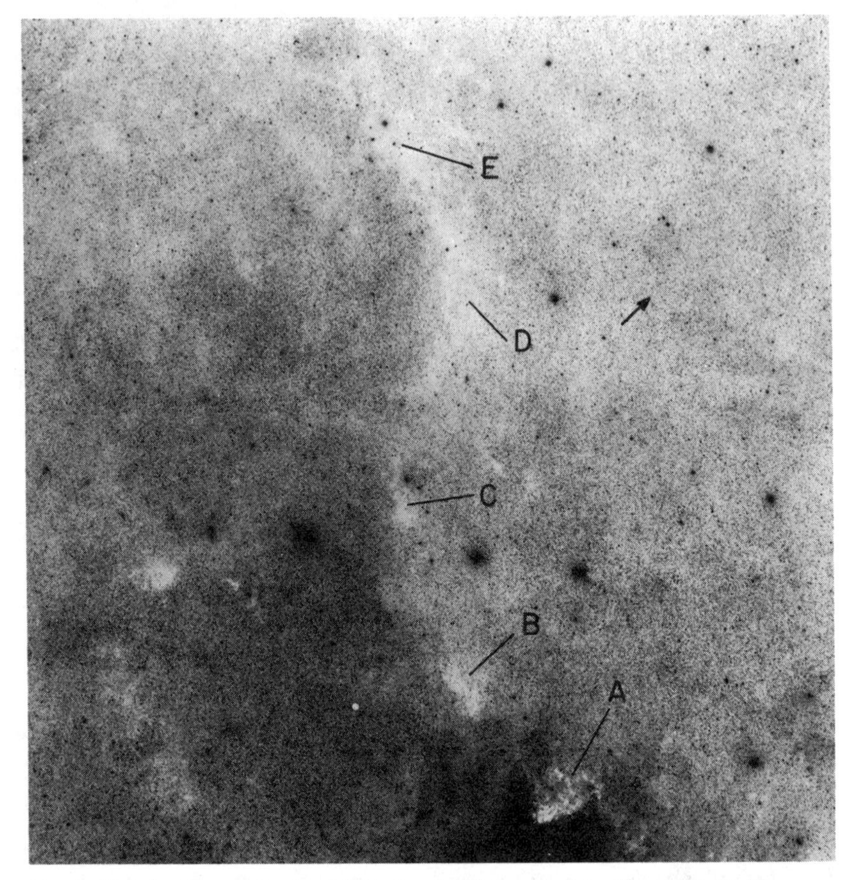

Fig. 4. Globular filament Number 7 ($20^h\ 58^m$, $+48°$) from S. Schneider and B. G. Elmegreen (1979) containing five distinct globules (A–E) that appear to be similar. The internal structure inside Globule A is revealed by a uniform background light source. The clumps in this photograph measure 0.1 pc in diameter and contain approximately 1 $M_{\odot}$ of gas (estimated from the extinction). They appear to be "thermal clumps," in the sense that their size is the thermal pressure scale height in the cloud. Star formation may occur in a clump when accretion from the surrounding interclump medium increases the clump's mass above the critical value for self-gravitational instability.

clump interaction will be determined by ion-neutral particle collisions in the gas. When field-line entanglements cause a clump to slow down and start up again in the other direction, the neutral molecules, which are not connected to the field directly, will at first continue on in the same direction; only the ions will be pulled back directly by the field. The ions will at first move through the neutrals and create a viscous force that accelerates the neutrals; this force will pull the neutrals back along with the ions and the magnetic field. The loss of energy to heat and radiation that results from this viscous force may deter-

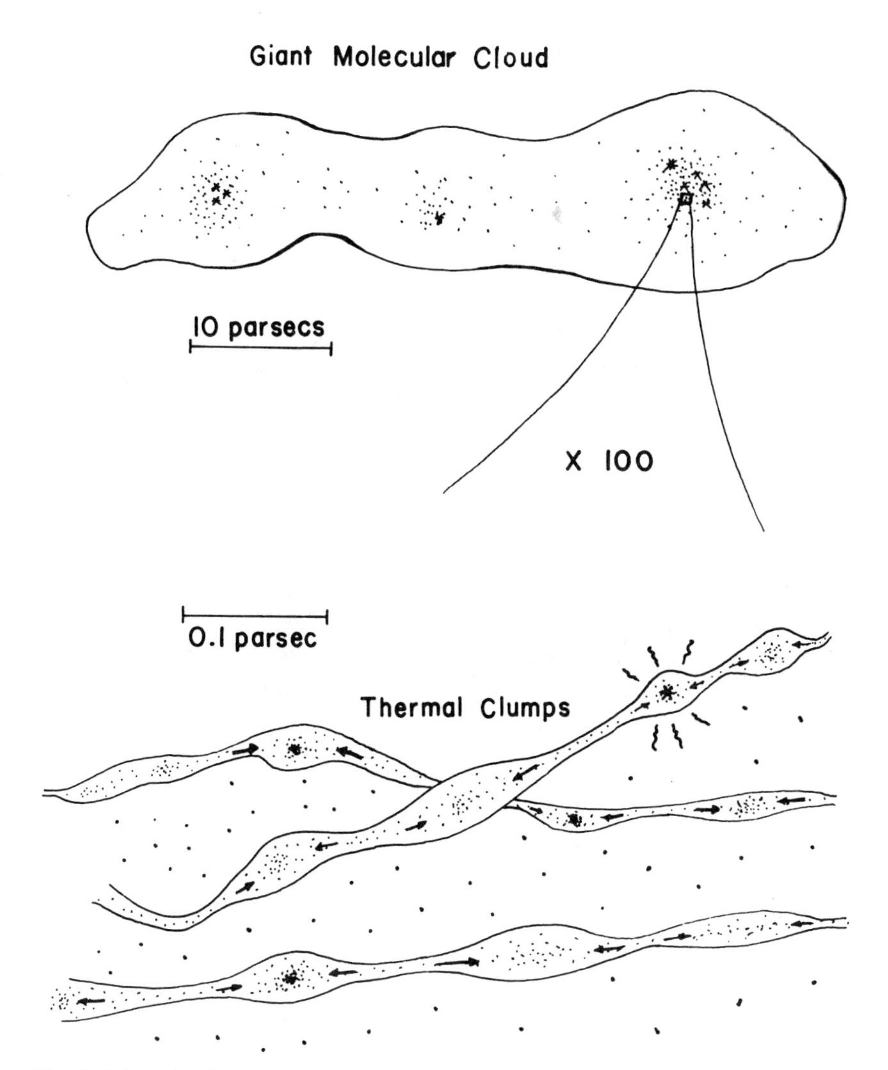

Fig. 5. Schematic diagram showing the possible configuration of Jeans-mass size "thermal clumps" inside a giant molecular cloud complex. The size of the whole cloud is taken to be ~50 pc, and the size of a typical clump is ~0.1 pc. Clumps may be strung out along the ambient magnetic flux tubes, as indicated. The inevitable accretion of gas from the flux tubes to the clump cores (driven by gravity) may slowly trigger the formation of stars in one clump after another, until a sufficiently massive star disrupts the local ensemble. Young stars are denoted by an "⋆" symbol.

mine the time scale for cloud evolution during the contraction phase. This time scale could be much longer than the turbulent dissipation time in the absence of magnetic fields.

The mechanism of star formation inside a clumpy cloud may be very different from that inside a uniform cloud. Norman and Silk (1980), R. B. Larson (1981), and others suggest that clumps coalesce until they reach a critical size, and then they collapse to form stars. If collisions are limited by magnetic fields, however, then such coalescence can be rare. Another possibility is that the clumps gradually accrete matter from the intervening magnetic flux tubes (Fig. 5). The gas that lies on the magnetic field lines between the clumps will slowly fall toward the nearest clump because of gravitational forces.

This process of filament slip may actually be visible in the Taurus filaments at the present time. Each filament in Taurus contains condensations that are more or less equally spaced. This is an expected result of gravitational condensation in filamentary geometries (Chandrasekhar and Fermi 1953). As time proceeds, each condensation accretes more and more of the surrounding filamentary gas. One might suppose that eventually the condensations pick up so much mass that they can no longer support themselves by thermal pressure. Then, they will collapse and make a star or a small stellar system. Star formation may always result from rapid and small-scale clump instabilities inside a complex of clumps that contracts slowly as it loses turbulent energy through ion-neutral viscosity.

STAR FORMATION IN A GALACTIC CONTEXT: THE LOCATION AND PROPERTIES OF MOLECULAR CLOUDS

P. M. SOLOMON
State University of New York, Stony Brook

and

D. B. SANDERS
University of Massachusetts

Giant molecular clouds, the site of star formation, are the largest and most massive objects in the Galaxy. Their physical properties and distribution are best traced by millimeter-wave observations of CO. We review the overall galactic distribution of molecular clouds and compare it with other galaxies showing that the surface density of molecular hydrogen and of molecular clouds is less than that in the strongest molecular galaxies such as M51, comparable to that in M101 and much greater than that in M31. Within the Galaxy, molecular clouds are the dominant component of the interstellar medium in the inner half of the disk at $R < 0.8\ R_0$. *Most of the molecular gas is in clouds at the high end of the mass spectrum with mass* $> 10^5 M_\odot$ *with over half the mass in clouds with* $M > 4 \times 10^5 M_\odot$. *We present CO maps of the northern galactic plane from the high-resolution Massachusetts–Stony Brook CO Galactic Plane Survey. Analysis of this survey shows the presence of two populations of molecular clouds. Over two thousand molecular cloud cores have been identified in the region between galactic longitude 20° and 50°. They are divided into two populations based on the intensity of CO emission which reflects the kinetic temperature. The warm molecular cloud cores exhibit a nonaxisymmetric galactic distribution, are clearly associated with H II regions, appear to be clustered, and are a spiral arm population. The cold molecular cloud cores are a widespread disk population located both in and out of spiral arms. Thus dense molecular clouds are not confined to spiral arms and star formation is taking place throughout the disk.*

The largest clusters of warm clouds have a scale size between 100 and 150 pc and contain about 5 × $10^6 M_\odot$. Giant molecular clouds do not fit into a two- or three-phase pressure-equilibrium picture of interstellar matter. They are gravitationally bound and their internal pressure, dominated by chaotic motions, is 2 orders of magnitude greater than the standard interstellar medium pressure. A likely origin for these clouds, particularly for the large clusters, is gravitational instability.

Millimeter-wave observations of the CO molecule have proved over the past decade to be one of the more important probes of interstellar matter in the Galaxy. CO is the most abundant molecule with a permanent dipole moment in dense interstellar clouds. Molecular hydrogen, H_2, has no dipole moment or radio transitions and is largely unobservable directly except in those regions that are shock heated (to temperatures > 1000 K) to produce quadrupole vibration-rotation emission.

The fundamental CO transition frequency ($J = 1 - 0$) of 115,271.2 MHz corresponding to $h\nu/K = 5.5$ K is excited by collisions with molecular hydrogen even in clouds with very low kinetic temperature. The minimum local H_2 density which produces excitation significantly above the microwave background is about $n_{H_2} > 100$ to 300 cm^{-3} (see, e.g., Goldreich and Kwan 1974; Scoville and Solomon 1974; Solomon 1978; Liszt et al. 1981). CO emission is the best available tracer of hydrogen in all interstellar clouds with a total hydrogen density $(n_H + 2n_{H_2}) > 100$ cm^{-3} because this corresponds to the effective threshold for H_2 self shielding against photodissociation in the Lyman bands (Solomon and Wickramasinghe 1969; Hollenbach et al. 1971); above these densities there is expected to be an almost complete conversion of H I into H_2. Thus the dense star-forming interstellar clouds are molecular clouds with most of the primary molecule H_2 unobservable at any wavelength.

Over the past decade several surveys of ^{12}CO and ^{13}CO emission on a galactic scale have revealed the overall distribution and quantity of molecular gas in the Galaxy (Scoville and Solomon 1975; Burton et al. 1975; R. S. Cohen and P. Thaddeus 1977; Burton and Gordon 1978; Solomon et al. 1979*a*; R. S. Cohen et al. 1980; D. B. Sanders et al. 1984*b*) and the concentration of molecular gas into giant molecular clouds with characteristic mass $> 10^5$ $M_\odot$ (Solomon et al. 1979*a*; Solomon and Sanders 1980; Stark 1979; Liszt et al. 1981; Dame 1984; Sanders et al. 1985).

Giant molecular clouds are the largest and most massive objects in our Galaxy and the primary site of star formation. Their physical properties and distribution are fundamental to studies of the interstellar medium, star formation, the structure of the Galaxy, and the dynamics of the galactic disk. In this chapter we summarize the main results of the CO survey carried out jointly by the State University of New York (SUNY) at Stony Brook and the University of Massachusetts at Amherst. This survey utilized the largest available antennas in the U.S., the National Radio Astronomy Observatory (NRAO) 10 meter during 1977–1980 and the Five Colleges Radio Astronomy Observatory (FCRAO) 14-meter during 1981. (Many of the results of the Goddard Institute for Space Studies (GISS) Survey with the 1.2-meter minitelescope are discussed in the chapter by Elmegreen.) In Sec. I the axisymmetric distribution of molecular clouds in the galactic disk is presented showing that most of the molecular gas in our Galaxy is inside a region between 4 and 8 kpc from the galactic center; thus the most active star formation in the Galaxy is in the inner half of the disk, a view drastically different from that obtained by H I observations. Section II includes the Australian observations of the southern galactic plane with those from the north, discusses the differences and presents the combined distribution. The identification of giant molecular clouds from CO observations is discussed in Sec. III which also summarizes their physical properties and mass spectrum. All of the results in Secs. I through III are based on about 3000 CO spectra obtained in latitude or longitude strips. Section IV introduces some preliminary results from the new comprehensive Massachusetts—Stony Brook CO galactic plane survey with 40,000 spectra mapping almost the entire first quadrant of the Galaxy primarily with 3 arcmin spacing. This survey obtained during 1981–1983 gives the first high-resolution picture of virtually all molecular clouds in the inner Galaxy $R < R_0$. Pictures and contour diagrams are shown of CO emission from huge clusters of molecular clouds with mass $\sim 5 \times 10^6$ $M_\odot$. In Sec. V evidence is presented for two populations of molecular clouds: a spiral arm population of warm cloud cores which is confined to restricted locations in longitude-velocity space coincident with H II regions and a disk population of cold molecular cloud cores distributed widely in the Galaxy and not confined to spiral arms. Section VI points out that giant molecular clouds do not fit into a standard two- or three-phase interstellar medium and their origin is likely to involve self gravity as the formation mechanism rather than pressure equilibrium.

I. AXISYMMETRIC DISTRIBUTION OF MOLECULAR CLOUDS

Due to differential galactic rotation and the resulting Doppler shift, it is possible to observe clouds across the Galaxy even if individual clouds are optically thick. Figure 1 shows a sample line profile of CO emission at 2.6 mm taken in the galactic plane at latitude $b = 0°$ and longitude $\ell = 30°$. Five distinct clouds are visible at different radial velocities ranging from 6 to 98 km s^{-1}. This spectrum demonstrates one of the chief characteristics of CO emission; unlike H I most of the emission along the line of sight is confined to discrete clouds with a contrast in and out of the cloud of one to two orders of magnitude. Each radial velocity can be associated with a corresponding distance from the galactic center, using a galactic rotation law, and either a near or far distance along the line of sight. The higher velocities originate from the region along the line of sight nearest to the galactic center. Thus the very

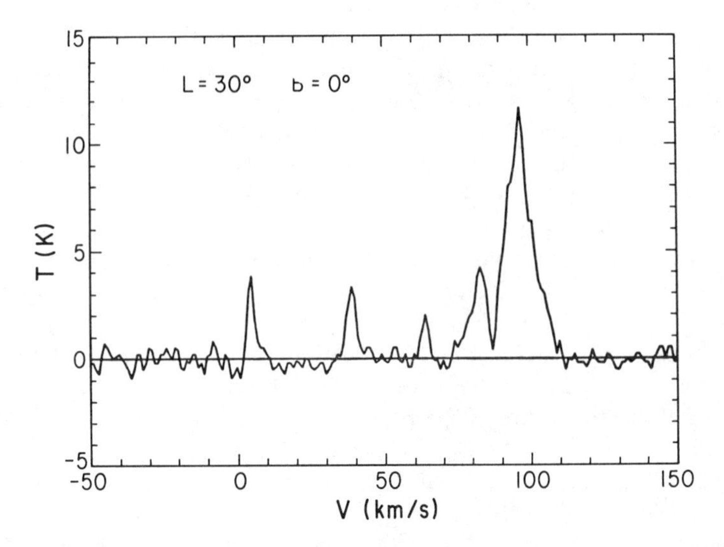

Fig. 1. CO spectra of emission from the galactic plane at longitude $\ell = 30°$ and latitude $b = 0°$. Emission from five discrete clouds at different radial velocities along the line of sight can be seen. A longitude-velocity diagram of the type presented in Color Plates 1 and 2 is made from a series of spectra like this.

strong emission at v = 98 km s^{-1} is from a cloud at R = 5 kpc (actually R = 0.5 R_0, we assume here R_0 = 10 kpc).

The most prominent characteristic of the distribution of CO emission in the Galaxy is the concentration to a ring-like morphology with a peak between 4 and 8 kpc (Scoville and Solomon 1975; Burton et al. 1975; R. S. Cohen and P. Thaddeus 1977). There is also a very strong peak in emission in the inner few hundred parsecs of the Galaxy. Figure 2 shows the axisymmetric distribution of total CO emission from the disk of the Milky Way as it would be seen if viewed face-on compared to that of 7 other galaxies (D. B. Sanders et al. 1984*b*). The interpretation of CO emission in terms of total molecular hydrogen by Scoville and Solomon (1975) was the first indication that there was more interstellar matter in molecular hydrogen than in H I in the inner Galaxy and more importantly that it was distributed differently.

The axisymmetric picture is a simplification but it does serve to place the molecular cloud distribution in a context of the Galaxy as a whole. A recent analysis of the distribution (D. B. Sanders et al. 1984*b*) leads to the following conclusions:

H_2 is the dominant form of the interstellar medium at $R < R_0$. At $R < 16$ kpc the total masses of H_2 (3.5×10^9 $M_\odot$) and H I (3×10^9 $M_\odot$) are approximately equal, but 90% of H_2 is found inside the solar circle, while only 30% of H I is found there. Within the region R = 4 to 8 kpc, the H_2 mass exceeds the mass of H I by a factor of 4. (The uncertainty in the mass of H_2 is due to variations in the measured conversion factor for translating observed

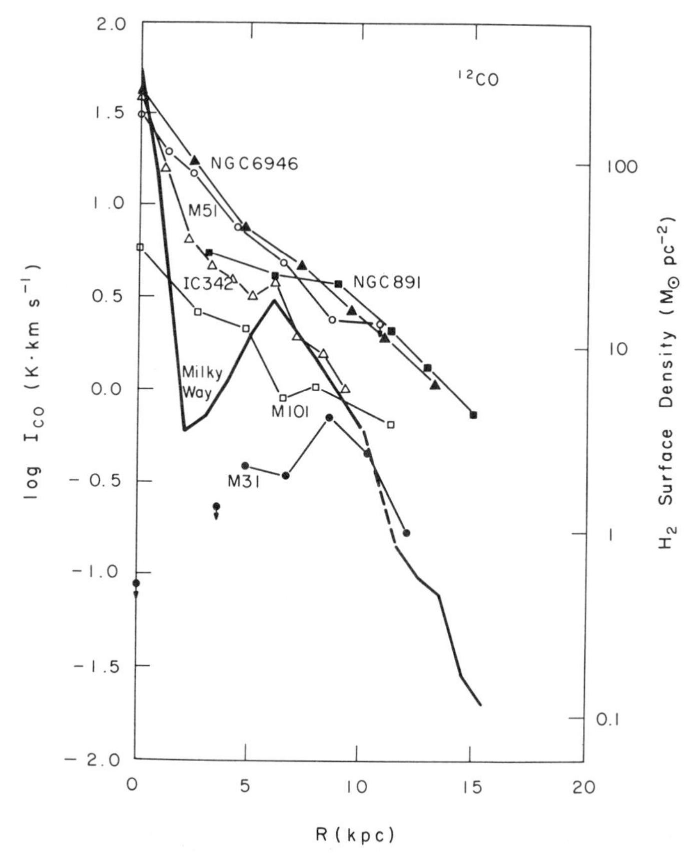

Fig. 2. Radial distribution of CO integrated intensity I_{CO} in the Milky Way compared with six galaxies for which radial distribution data are currently available: IC342, NGC6946 (Young and Scoville 1982); M51 (Scoville and Young 1983); M101 (Solomon et al. 1983*a*); NGC891 (Solomon 1981); M31 (Stark 1979).

CO integrated intensity to H_2 column density. Existing empirical measures of this factor range from 2 to 5×10^{20} cm^{-2} K^{-1} km^{-1} s based on calibration by ^{13}CO/Ag, galactic gamma rays and by comparison with masses determined from the virial theorem. We have adopted 3.6×10^{20}.)

The radial distribution of molecules as measured from CO surface brightness shows a strong peak in the galactic center at $R < 1.5$ kpc, a minimum between 2 and 4 kpc, a rise to a secondary maximum at between $R = 5$ and 7 kpc, and a sharp fall toward larger radius. There is very little emission beyond $R = 15$ kpc. The fall in CO surface brightness between $R = 6$ and 15 kpc is nearly a factor of 100. Thus, for a constant (with R) conversion factor of CO emission to H_2 there is a change of two orders of magnitude in the surface density of H_2 and consequently molecular clouds between 6 and 15 kpc galactocentric radius, while H I is virtually constant over this range.

The half-width at half-maximum thickness of the disk $z_{1/2}$ for CO emission is in the range from 40 to 75 pc between R = 3 and 10 kpc. In the solar neighborhood the scale heights of gas, as measured by CO emission, and young stars are approximately equal.

In comparison with other galaxies, as can be seen from Fig. 2, the Milky Way is a moderate CO emitter with a surface density of H_2 less than that of the strongest galaxies such as NGC 6946, and comparable to M101 although neither M101 or NGC 6946, for example, have rings. In sharp contrast, the nearby galaxy M31 does not show a primarily molecular inner disk. The CO emission is weak throughout the disk in comparison with the Milky Way and appears closely correlated with H I. The fraction of the interstellar medium which is molecular in M31 appears to be small ($<40\%$) at all galactocentric distances. The total surface density of all interstellar matter in M31 is substantially less than in the Milky Way, particularly for $R \leq$ 8 kpc. Thus the star-forming environment may be quite different in M31 from that in the Milky Way.

The existance of very rapid falloff in (molecular cloud) H_2 surface density, beyond the solar circle, was disputed by Kutner and Meade (1981) who concluded that the mass of H_2 in the outer Galaxy ($R >$ 10 kpc) was comparable to that at 4 to 8 kpc. However, their observations in the outer Galaxy appear to be erroneous. Solomon et al. (1982) using two separate antennas at FCRAO and Bell Laboratories could not verify most of the CO emission features claimed to exist by Kutner and Meade. The rapid decline of H_2 in the outer part of a galaxy is a common feature of many galaxies in contrast to H I which often continues beyond the optical disk.

II. COMBINED DISTRIBUTION IN NORTHERN AND SOUTHERN GALACTIC PLANE

The longitude-velocity distribution for the CO emission, including both northern and southern hemisphere data (Robinson et al. 1984), is shown in Fig. 3. The southern hemisphere data, obtained by the CSIRO group in Australia, has been combined with northern data from the comprehensive, high-resolution Massachusetts-Stony Brook survey including over 40,000 spectra. In Fig. 3 lines of constant galactic radius are superimposed based upon galactic rotation. The galactic center shows extremely strong emission with large velocity spread. Beyond 6° longitude most of the emission in both hemispheres is concentrated between the lines corresponding to 0.45 R_0 to 0.85 R_0. The molecular gas content is approximately symmetric between both hemispheres with the total CO emissivity, weighted by the area at the different radii, being about 20% greater in the south.

Plate 1 in the Color Section shows the same longitude-velocity diagram in color. The strong concentrations in yellow and red represent some of the largest cloud complexes (clusters) in the Galaxy. As shown in Sec. V, they are primarily associated with giant H II regions.

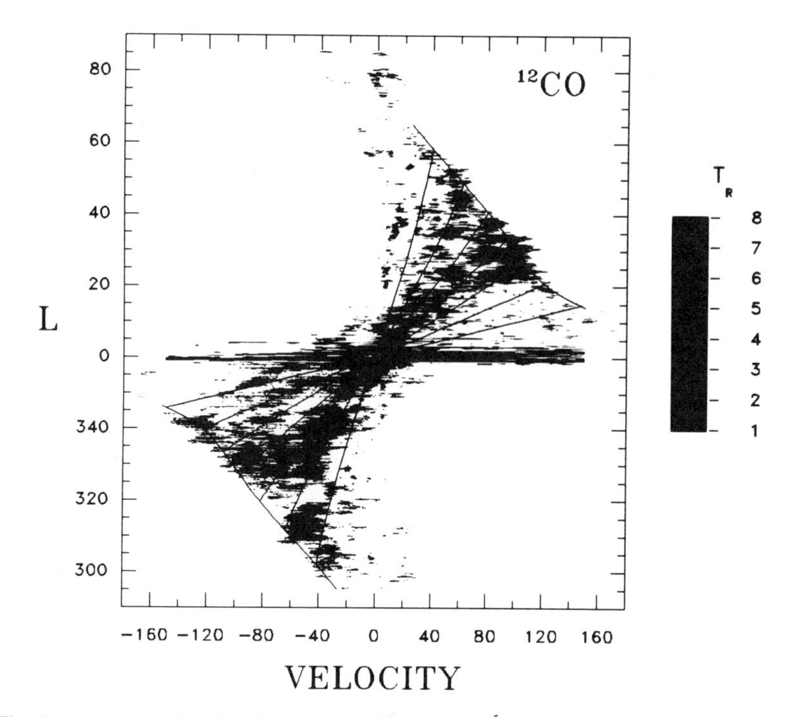

Fig. 3. A representation of the intensity of millimeter-wave emission from CO at a wavelength of 2.6 mm along the inner galactic plane between 290° and +90° longitude (L). The velocity of the emission indicates the distance from the galactic center, with the highest velocity at each longitude (or lowest velocity at negative longitude) corresponding to the closest approach of that line of sight to the galactic center. The intensities are represented by a gray scale in units of T_A^* with a saturation level of 6 K and a lower cutoff at 1.0 K. Superimposed lines of constant galactocentric radius at 0.1 R_0 intervals between 0.25 R_0 and 0.95 R_0 and the terminal velocity were computed using the H I galactic rotation model of Burton and Gordon (1978).

Figure 4 shows a comparison between the CO emissivity distribution in the northern and southern hemispheres. The detailed shape of the molecular ring in each hemisphere is different. The southern distribution is much wider (0.6 R_0 full-width at half-maximum [FWHM] compared to 0.3 R_0), and may possibly have two peaks near 0.3 and 0.7 R_0. The standard deviation of the mean J (emissivity), determined in each radius bin in Fig. 4, is typically $<$ 5%. Significant differences between the northern and southern distributions occur in four radial bins. Within three of these (0.3, 0.5, and 0.8 R_0) the differences exceed twice the internal dispersion of 10° longitude subsets of data from the individual hemispheres. This suggests that structures, probably spiral arms or arm segments that are $>$ 10° in length, are responsible for the differences. In part, these differences may also result from the presence of clusters of giant molecular clouds which may contribute up to 50% of the emissivity in a given annulus.

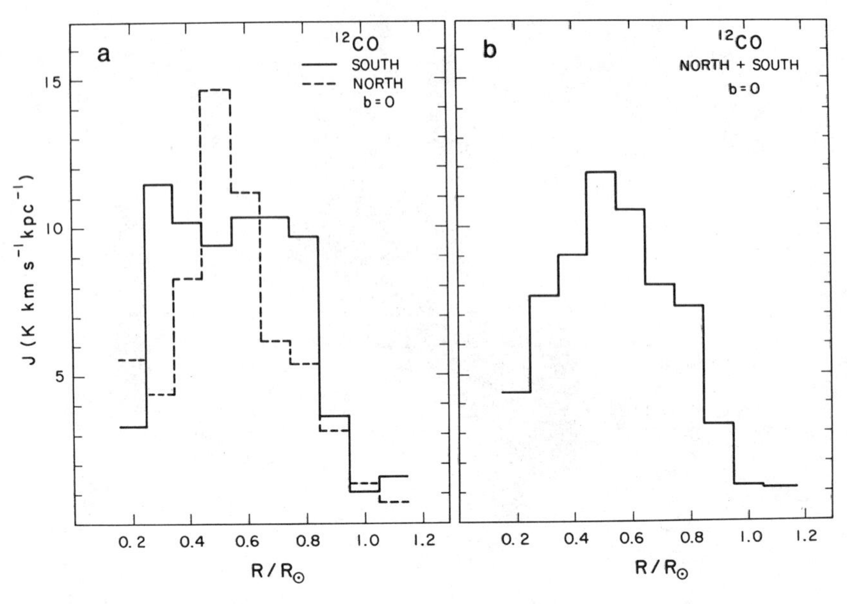

Fig. 4. (a) The radial distribution of CO emissivity in the southern and northern hemispheres at $b = 0°$ between galactocentric radii 0.2 R_0 and R_0. (b) The combined (north and south) radial distribution of CO emission at $b = 0°$ for $\ell = 294°$ to $70°$.

III. IDENTIFICATION OF GIANT MOLECULAR CLOUDS AS THE DOMINANT COMPONENT OF INTERSTELLAR MATTER IN THE RING

The identification of giant molecular clouds as responsible for much of the CO emission in the galactic disk is evident from the concentrations of CO emission in longitude-velocity space (Solomon et al. 1979*a*; Solomon and Sanders 1980) in data obtained with high spatial resolution. The breakup of the emission from the galactic plane at $b = 0$ and $\ell = 23°$ to $30°$ showed 38 clouds at $R = 4$ to 8 kpc into clumps with typical length of 20 to 80 pc (see Color Plate 2). Based on these observations, we showed that a substantial part of all CO emission and most of the mass was contained in giant clouds with mass between 10^5 and 3×10^6 $M_\odot$ with a cloud mass spectrum showing most of the molecular ISM in the most massive clouds. A similar study by Liszt et al. (1981) using ^{13}CO emission came to a similar conclusion but with an upper limit for cloud size of about 45 pc. In an analysis of 80 clouds closer than 6 kpc from the Sun, Sanders et al. (1985*a*) utilized the virial theorem rather than column density measurements to determine cloud mass. The mass spectrum of clouds (see Fig. 5) expressed as the fraction of mass in clouds of mass m per logarithmic mass interval was found to be

$$g(m) \propto m^{0.42} \tag{1}$$

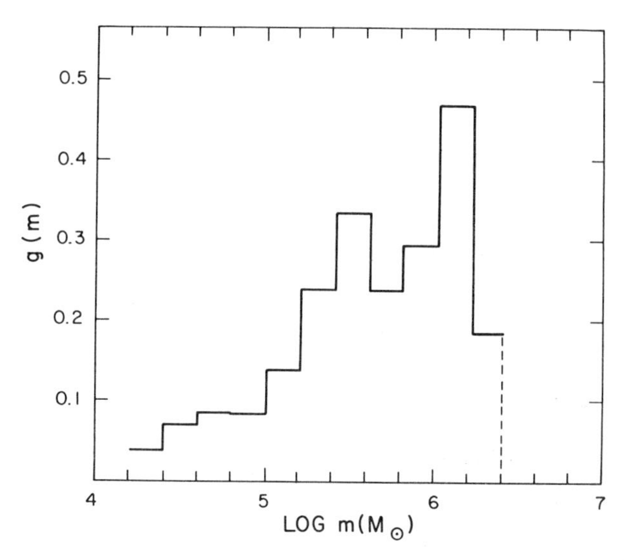

Fig. 5. Fraction of molecular cloud mass in clouds of mass m per logarithmic mass interval.

between cloud mass of about 4×10^4 and 2×10^6 $M_\odot$. The positive exponent indicates that most of the mass in molecular clouds is in the high-mass end of the spectrum. Over half of the total mass is in clouds with mass $> 4 \times 10^5$ $M_\odot$ and diameter > 48 pc; 90% of the mass is in clouds > 20 pc and mass > 8×10^4 $M_\odot$. All measurements of the mass spectrum (Solomon et al. 1979*a*; Solomon and Sanders 1980; Liszt et al. 1981; Dame 1983; Sanders et al. 1985*b*) yield a mass distribution where > 80% of the molecular mass is in clouds with $m > 10^5$ $M_\odot$. Thus giant molecular clouds contain most of the mass in dense (molecular) interstellar clouds. A basic characteristic of star formation is the fact that the parent clouds are not only very much more massive than individual stars; they are also much more massive than galactic star clusters.

The measurement of the velocity widths of the CO emission shows a linewidth-size relationship that is clearly different from that expected from pure turbulence as suggested by R. B. Larson (1981). Figure 6 shows the data along with a fit for the velocity full-width at half-maximum of

$$\Delta v = 0.88\, D^{0.62} \text{ km s}^{-1}. \qquad (2)$$

Although a power law relationship is not proven by this data, the exponent derived is almost twice that expected for a single turbulent spectrum.

The above result leads to a determination of mean cloud density. For a cloud with a $1/r$ density law, the mean number density in cm^{-3} of H_2 is

$$n(H_2) = 290\, (D/20 \text{ pc})^{-0.75}. \qquad (3)$$

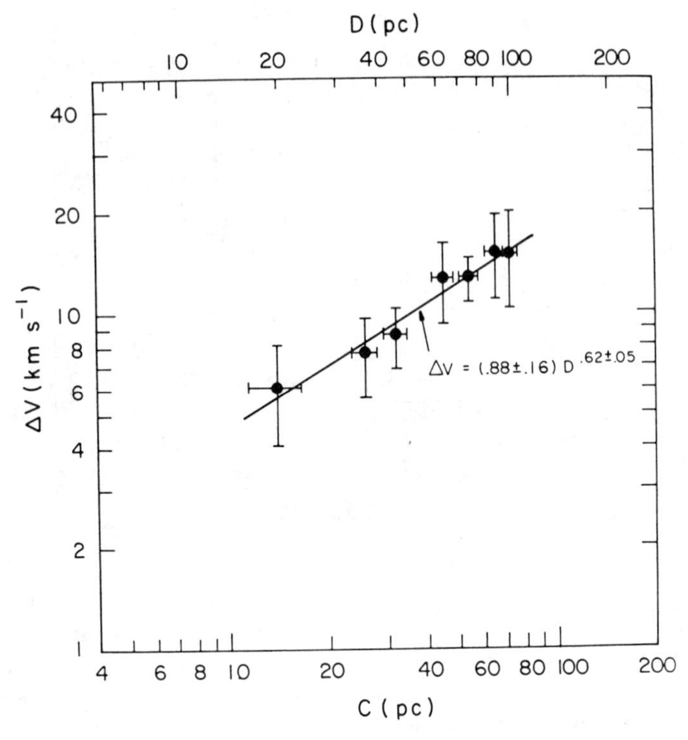

Fig. 6. Velocity full width at half maximum as a function of cloud diameter (top axis) for 80 clouds from Sanders et al. (1985).

For clouds with a diameter of 20 pc the mean density is 290 cm^{-3} and the mass is 8.1×10^4 $M_\odot$; for clouds with $D = 80$ pc, the mean density is $\sim$ 100 cm^{-3} and the mass is 1.9×10^6 $M_\odot$.

IV. THE MASSACHUSETTS–STONY BROOK HIGH-RESOLUTION CO SURVEY

The joint University of Massachusetts–SUNY, Stony Brook CO survey of the galactic disk consists of 40,572 spectral-line observations of CO emission at $\lambda = 2.6$ mm carried out with the FCRAO 14-m antenna with a beamwidth of 47″. The observations cover the northern galactic plane between longitudes 8° to 90° and latitudes $-1\overset{\circ}{.}05$ to $+1\overset{\circ}{.}00$ with spacing every 3′ in ℓ and b in the region $18° < \ell < 54°$ and every 6′ at $8° < \ell < 18°$ and $54° < \ell < 90°$. The full data are presented in the form of latitude-velocity diagrams (D. B. Sanders et al. 1985*a*), longitude-velocity diagrams (Clemens et al. 1985) and spatial, latitude-longitude diagrams of intensity in 5 km s^{-1} bins (Solomon et al. 1985).

The survey spacing of 3′ was chosen to allow observation of essentially all molecular clouds or cloud components in the 4 to 8 kpc molecular ring with a size > 10 pc. For example, at a distance of 10 kpc from the Sun, the beam of 47″ subtends 2.3 pc and the beam spacing of 3′ corresponds to a projected separation of 8.7 pc. Even at the farthest distances from the Sun within the molecular ring, the beam spacing is only 12 pc and beam size is 3 pc. The previous two large-scale CO surveys have not been able to resolve most individual clouds within the ring. The GISS minitelescope (R. S. Cohen et al. 1980) has a beamwidth of 9′, much too large to resolve many individual clouds or cloud components, while the first Stony Brook–Massachusetts survey carried out primarily with the National Radio Astronomy Observatory (NRAO) 36-foot antenna (D. B. Sanders et al. 1984*c*, 1985*a*), although it utilized a small beam of 65″, it had a spacing of 12′ in latitude. High-resolution data was obtained only in one spatial dimension along $b = 0°$.

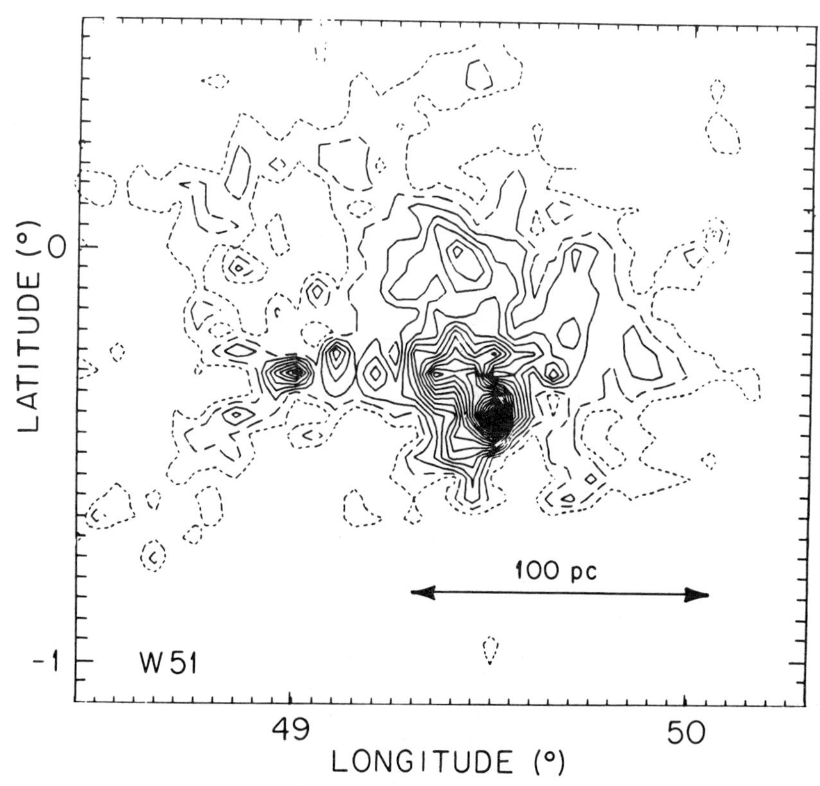

Fig. 7. Longitude-latitude plot of CO emission contours from the molecular cloud cluster associated with W51. Contour intervals are 1,1.5,2, . . . K, antenna temperature averaged over a 15 km s^{-1} velocity interval centered at the 63 km s^{-1} mean velocity of the cluster. Giant H II regions are located near each of the strong CO maxima.

Color Plate 3 shows 12 high-resolution pictures of CO emission from the galactic plane between longitude 20° and 50°; each picture represents the CO emission intensity at 5 km s^{-1} intervals. Many hundreds of individual giant molecular clouds contribute to the emission. Some of the most prominent features represent huge cloud clusters such as the region at $\ell = 34°$ to 36° and including velocities from 30 to 60 km s^{-1}, associated with the supernova remnant W44, at a distance from the Sun of about 3 kpc (see Color Plate 4). The extent of this region and other large clusters is about 100 pc (D. B. Sanders et al. 1985*b*); other examples include the region at $\ell = 28°$, v = 70 to 85 km s^{-1}, $\ell = 48.°5$ to 50.°0 and v = 45 to 70 km s^{-1} (the W51 region), ℓ 22.°5 to 23.°3 and v = 60 to 80 and $\ell = 23.°2$ to 24°, v = 75 to 90 km s^{-1}. As an example, Fig. 7 shows a contour diagram of the cloud cluster associated with the radio continuum source W51 at a distance of 8 kpc from the Sun. This region clearly has many components yet is a coherent structure over an extent of about 1° square corresponding to a linear dimension of 140 pc. The mass determined both by utilizing measured velocity dispersion to obtain a virial

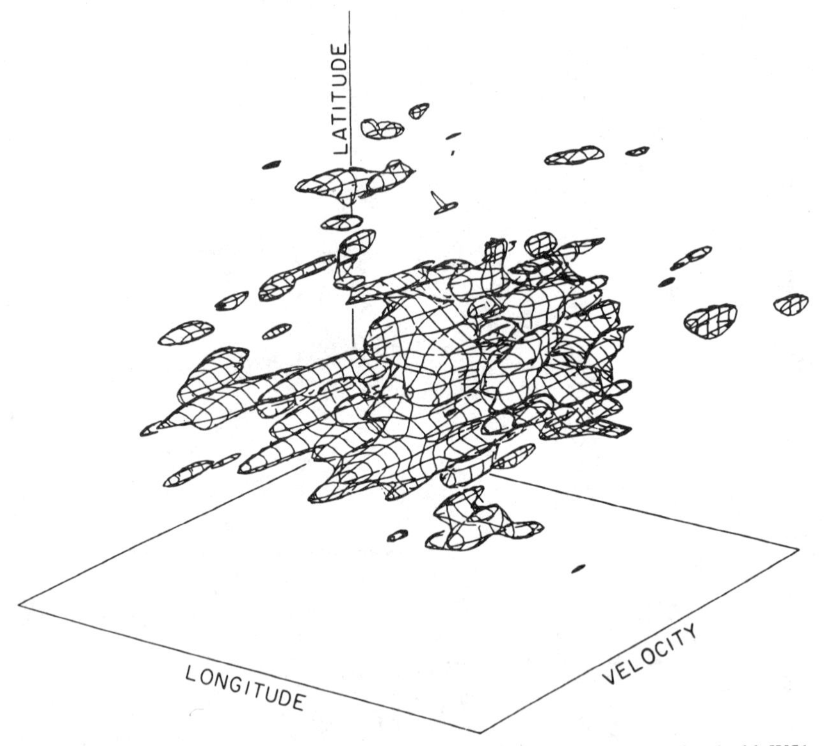

Fig. 8. Three-dimensional perspective view of the molecular cloud cluster associated with W51. This picture shows the full velocity extent of the cluster (25 km s^{-1}). The contour surface represents a gas kinetic temperature of 11 K. The range of longitude along the axis is from 48.°5 to 50.°0, latitude from −1° to +1° and velocity from 45 to 75 km s^{-1}.

mass, and an H_2 mass determined from CO luminosity is $\sim 5 \times 10^6$ $M_\odot$. Figure 8 shows a three-dimensional representation in ℓ-b-v space showing a full velocity extent of nearly 30 km s^{-1} for this object.

Also apparent in the Color Plate 3 is the rich detail of emission from the far side of the Galaxy at distances of typically 10 to 18 kpc. Two-thirds of the disk is on the far side of the tangent point and no previous survey has had sufficient angular resolution and coverage to resolve clouds at this distance. For example, the emission from longitude 23° to 32° and velocity 35 to 50 km s^{-1} with a very narrow latitude extent at a distance of 13 kpc consists of at least 30 separate clouds larger than 20 pc (giant molecular clouds with mass > 10^5 $M_\odot$) (see Color Plate 5). This chain of clouds may be a spiral arm segment.

V. DISK AND SPIRAL ARM MOLECULAR CLOUD POPULATION

In the region between $\ell = 20°$ and 50° at all positive velocities approximately 2000 CO emission centers have been identified from contour diagrams spaced every 5 km s^{-1} (Solomon et al. 1984*c*). The quantity which was mapped is the integrated CO intensity over 5 km s^{-1} or I_5; $I_5/5$ is essentially the mean antenna temperature averaged over 5 km s^{-1}. The sources were divided into two populations, cold and warm, based on the intensity I_5. Figures 9, 10,11 and 12 show examples of contour diagrams of I_5.

By utilizing the emission centers rather than just total CO emission of clouds defined by outer boundaries, it is possible to pinpoint a source location in ℓ, b, and v without ambiguity. Observationally the local maxima observed at high resolution are well-defined entities and their distribution represents the distribution of star-forming molecular cloud cores. This high-resolution survey thus makes it possible to study the galactic distribution of star formation regions, rather than more amorphous emission, and to examine their degree of confinement to spiral arms. Figures 13 and 14 show the distribution of cold and warm cloud cores, respectively, on the longitude-velocity plane. (A small amount of jitter in velocity of ± 2.5 km s^{-1} has been added since the catalog data were binned every 5 km s^{-1}.) Superposed on the figures are histograms showing the counts in the respective longitude bins. The rotation curve maximum velocity is shown by the smooth lines at the upper right.

The cold molecular cloud cores (Fig. 12) show a fairly smooth distribution in ℓ, with a gradual decline from ~90 clouds per degree at $\ell = 22°$ to 28° to 40 clouds per degree at $\ell = 40°$ to 50°. This is roughly what is expected from a ring-like galactic structure between galactocentric radii of 4 to 8 kpc. There is no evidence of confinement in ℓ-v space, or consequently in the galactic plane, to spiral arm patterns. These cold cores account for three-quarters of the total population and about half the CO emission of molecular cloud cores. They represent molecular clouds with local densities $n(H_2) >$ 300 cm^{-3} (see, e.g., Solomon and Sanders 1980; Liszt et al. 1981) but with

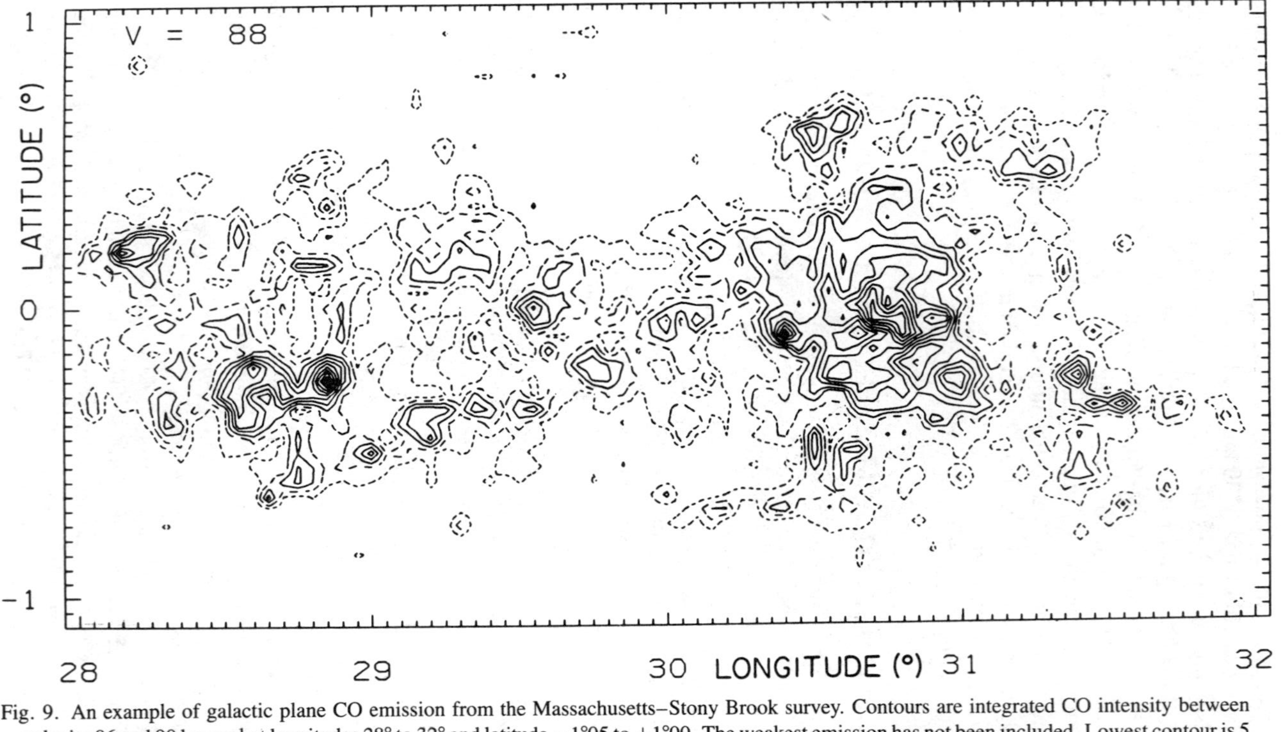

Fig. 9. An example of galactic plane CO emission from the Massachusetts–Stony Brook survey. Contours are integrated CO intensity between velocity 86 and 90 km s^{-1} at longitudes 28° to 32° and latitude $-1\overset{\circ}{.}05$ to $+1\overset{\circ}{.}00$. The weakest emission has not been included. Lowest contour is 5 K km s^{-1} and spacing is 5 K km s^{-1}. Note the breakup into discrete sources even in this region of the galactic plane which is in the heart of the molecular ring at $R = 5$ kpc, above the third contour levels. Contours may also be regarded as mean antenna temperatures over a full width of 5 km s^{-1} beginning at $T_R^* = 1, 2, 3 \ldots$ K.

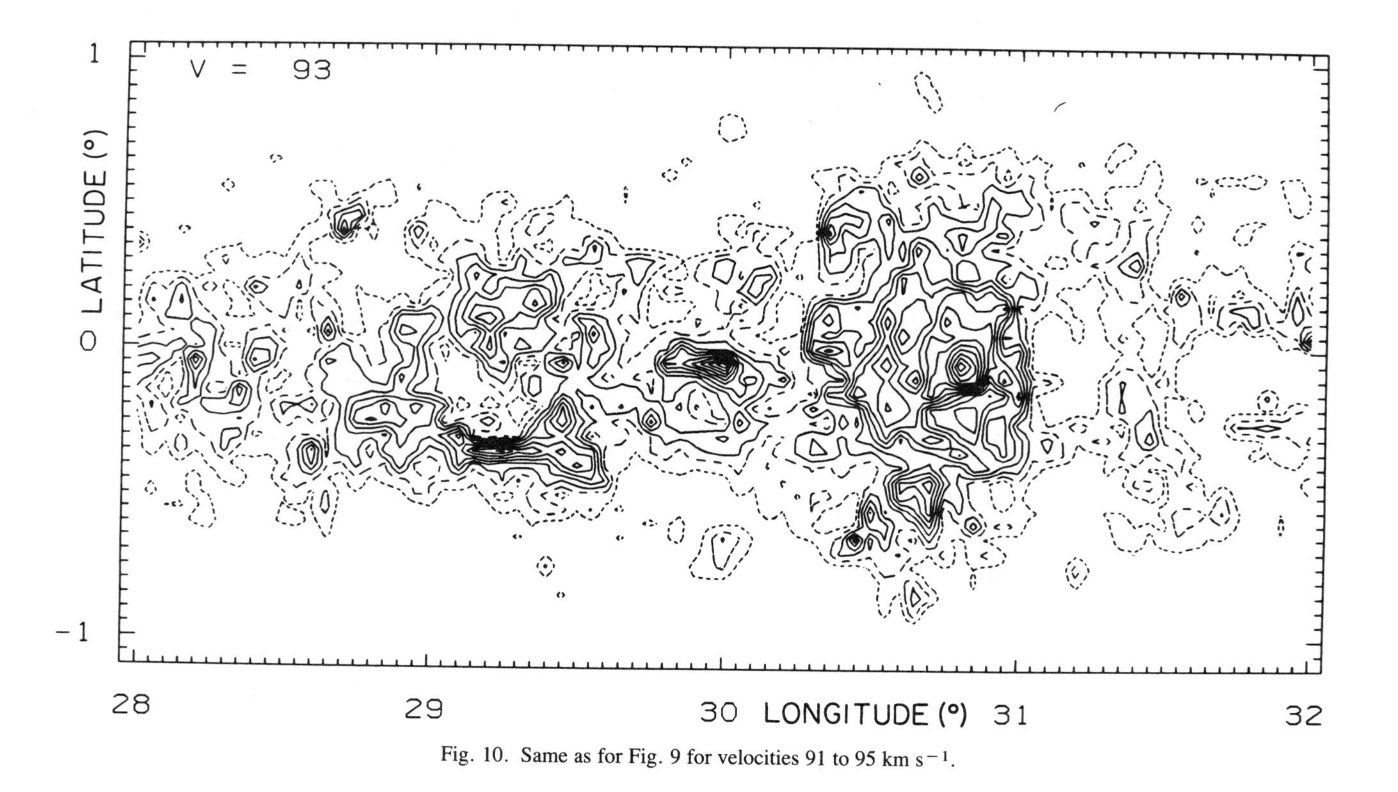

Fig. 10. Same as for Fig. 9 for velocities 91 to 95 km s^{-1}.

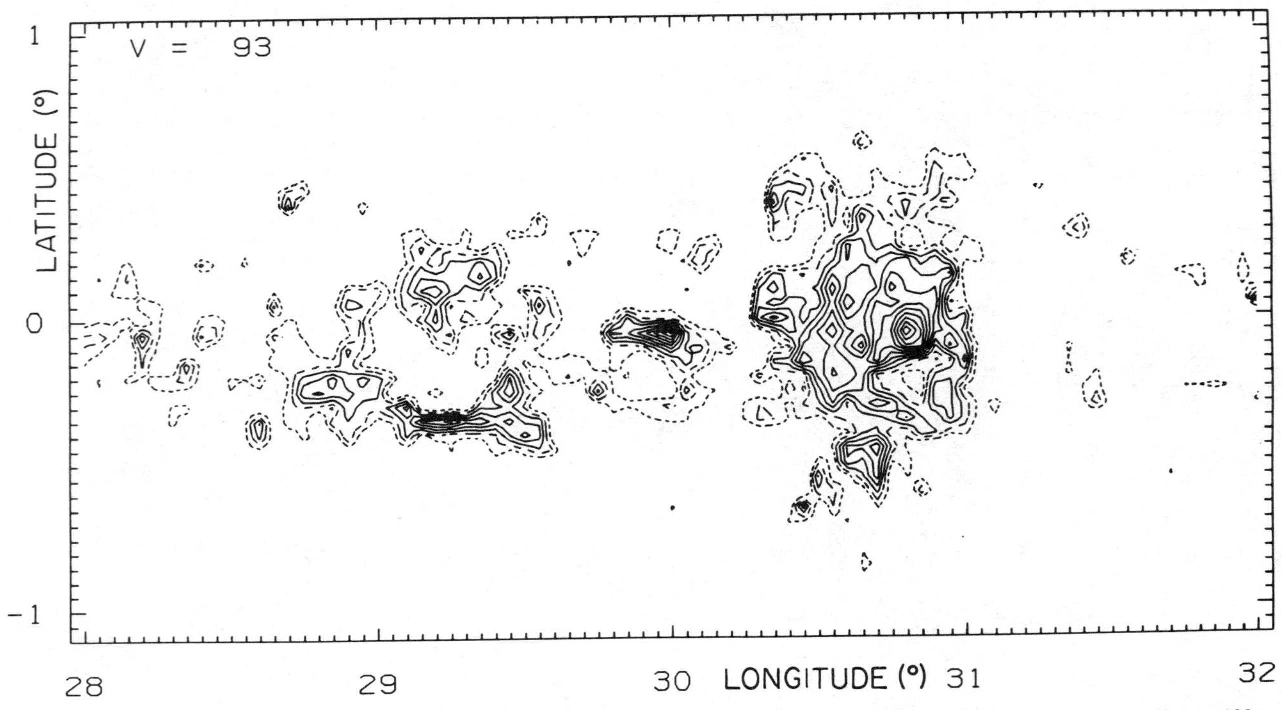

Fig. 11. Same as Fig. 10 but with 2 lowest contours missing. Note the large cloud cluster at longitudes 30°.2 to 30°.9 which has a diameter of about 120 pc and mass about 5×10^6 $M_\odot$.

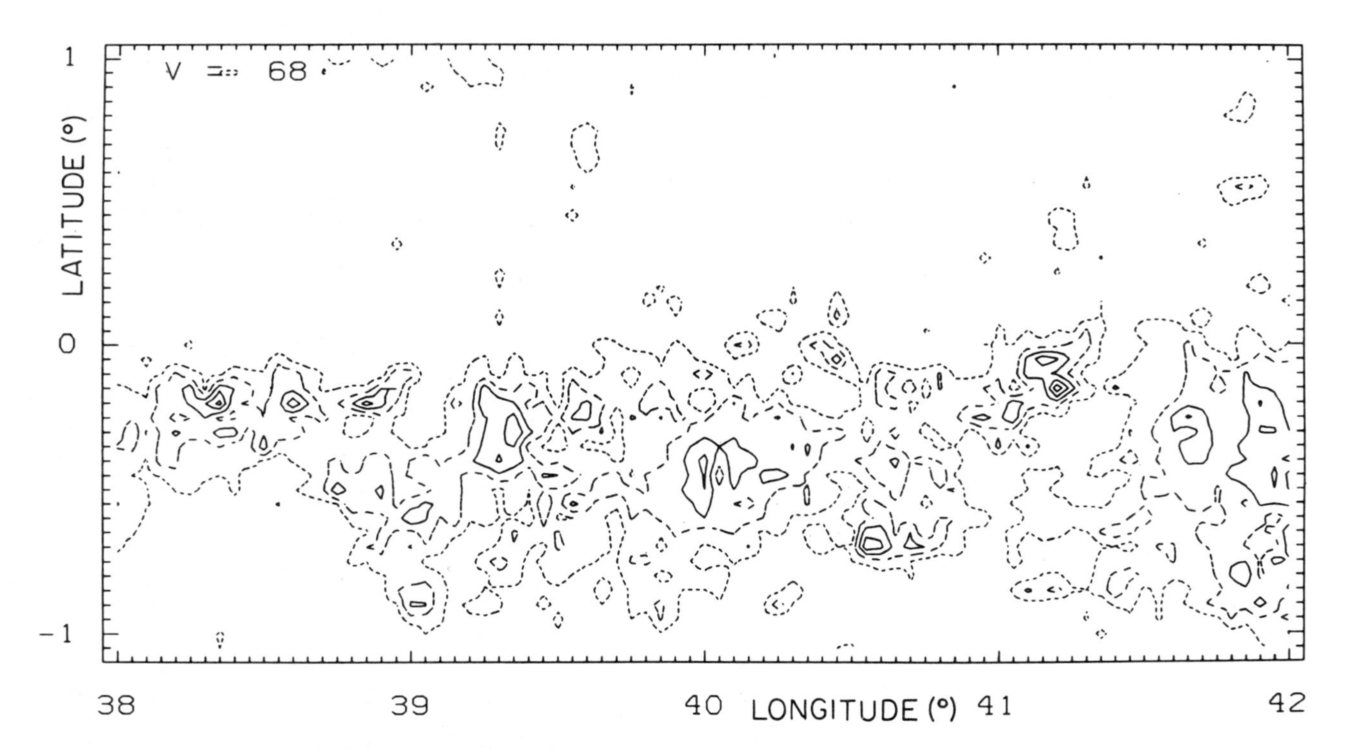

Fig. 12. A contour diagram showing a region of cold CO emission between velocities 65 and 70 km s^{-1}. The contour levels are the same as in Fig. 9. Most of the emission centers in this figure belong to the cold population.

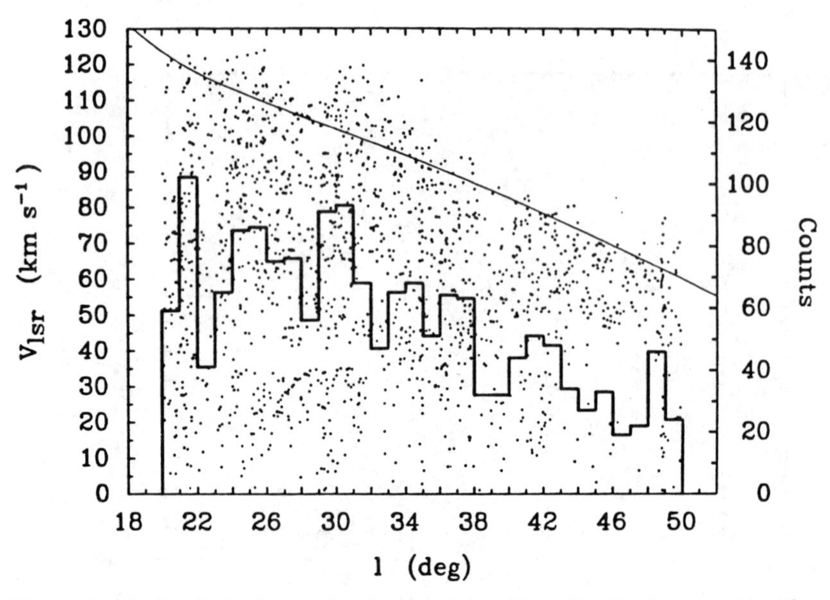

Fig. 13. Longitude-velocity diagram showing the location of the cold molecular cores with $-1° < b < 0.°4$. The histogram is a number count as a function of longitude in bins spaced every 1°.

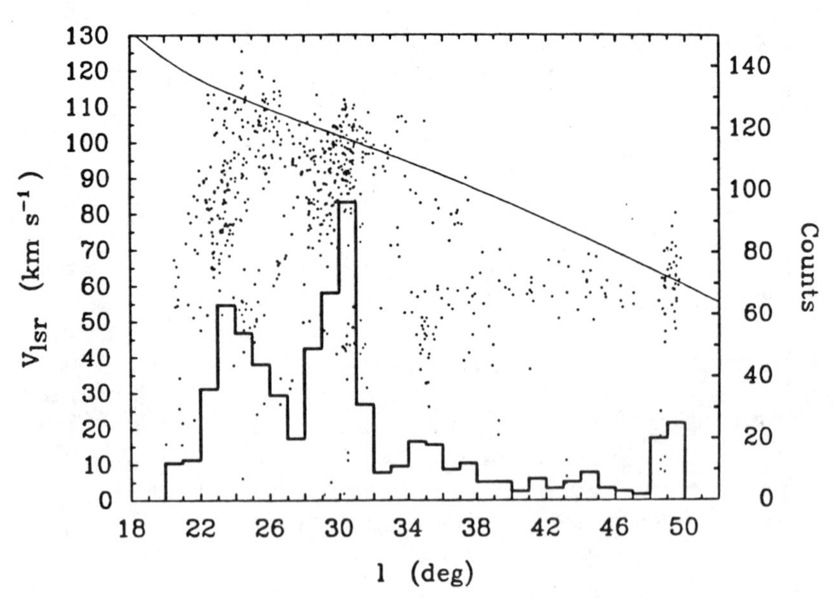

Fig. 14. Longitude-velocity diagram showing the location of the warm molecular cores.

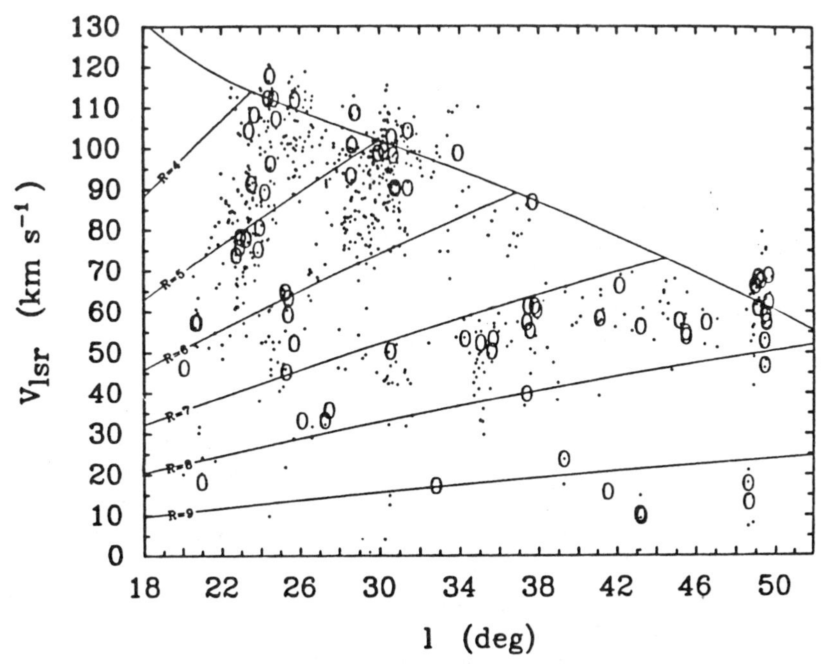

Fig. 15. Longitude-velocity diagram of radio H II regions indicated by ovals, adapted from Downes et al. (1980), superimposed on warm molecular cloud cores indicated by dots.

kinetic temperature significantly lower than those of the warm cores. They are widely distributed in ℓ-v space and hence in the galactic disk.

In contrast the warm cores (Fig. 13) show a strikingly different distribution with two sharp peaks at $\ell = 24°$ and $\ell = 30°$ and a smaller one at $\ell = 50°$. These 3 peaks, each a few degrees wide in ℓ, contain about one-half of all warm cloud cores. The distribution of the warm molecular cloud cores in the ℓ-v plane exhibits a very close similarity to that of radio H II regions. Figure 15 shows the distribution of warm cores in the ℓ-v plane superposed on the distribution of known H II regions, with available recombination line velocities, taken from Downes et al. (1980). It is clear that clusters of warm cloud cores are coincident with H II regions. This must represent that portion of emitting molecular clouds within the galactic spiral arms. H II regions are often the defining feature of spiral arms on photographs of galaxies. The hot young stars responsible for the H II regions are also the likely heat sources for the warm molecular clouds. Significant heating may even be taking place due to embedded massive stars with dust-limited or very small H II regions.

The cold cloud cores represent dense interstellar clouds without substantial heating. Because a typical dimension subtented by the survey beam is about 3 pc, the lack of substantial heating refers to regions of this size or greater, and a velocity width of 5 km s^{-1}. Heating on a much smaller scale may still be occurring, but the effects of beam dilution prevent its observation. The characteristic kinetic temperature over the beam size is obtained by converting the antenna temperature, $T_5 = I_5/5$ into a Planck brightness temperature which gives a gas kinetic temperature of < 10 K for the cold cores (see, e.g., Eq. 20 in Solomon et al. [1979*b*]). At a velocity resolution of 1 km s^{-1} the upper limit temperature is about 15 K. Their widespread distribution in ℓ-v space, smooth radial distribution, and moderate temperatures suggest that the cold cores represent dense molecular clouds widely distributed throughout the galactic disk. The most massive stars forming in these clouds are probably late B stars.

The warm cores represent molecular gas with temperatures > 11 K (15 K at 1 km s^{-1} resolution) including many cores with kinetic temperatures greater than 20 K. The association of warm cores with H II regions and also with supernova remnants clearly points to heating by recently formed or presently forming O and B stars. The high kinetic temperatures of the warm cores, their confinement to restricted regions of ℓ-v space, and the breakup of the northern molecular ring into two peaks at $R = 5$ and $R = 7.5$ kpc clearly point to this subset of molecular clouds as the spiral arm population.

The existence of the disk population proves that molecular clouds cannot be confined to spiral arms since they are located at all longitudes and velocities permitted by galactic rotation within the 4 to 8 kpc ring, and exhibit a smooth number count as a function of longitude. Our finding of the disk population contradicts the interpretation of R. S. Cohen et al. (1980) and P. Thaddeus (1981) (GISS survey) that in the inner Galaxy as in the outer Galaxy ". . . molecular clouds as commonly defined by CO observations—objects larger than 5 pc with CO temperatures greater than 1-2K—do not exist between the [spiral] arms, or are extremely rare." Their evidence for the deduced absence of clouds between spiral arms are holes or missing emission in ℓ-v space. Our survey data shows that there are indeed large holes in ℓ-v space when only the warm clouds are counted. However, it also shows hundreds of cold CO emission centers (cold cloud cores) within the specific holes. For example, Thaddeus and Dame (1984) discussing a cloud-free area state "the most telling such region is a gap at $\ell = 25°$, v = 75 km s^{-1} in the heart of the so-called molecular ring." Their picture shows this hole from $\ell = 24°$ to 27°, v = 70 km s^{-1} to 85 km s^{-1}. We find *46 CO emission centers in this hole, 42 from the cold population and 4 from the warm population*; within an area in ℓ-v space of 45 deg·km s^{-1} the expected number for a random distribution over all ℓ-v space between $R = 4$ to 8 kpc would be about 48 $\pm$ 7 and 18 $\pm$ 4, respectively, for the hot and cold population. Thus there is no absence of cold molecular clouds here but there is a strong minimum in the warm population.

It appears that the GISS survey is not detecting the presence of the cold population. Their small 1.2-meter antenna bandwidth (HPBW = 8′ at 115 GHz) subtends a region of 30 pc at 12 kpc distance. Thus they dilute the intensity of all but the nearby cloud cores. Most of the cold cores will be diluted to an intensity $< T_R^* = 2$ K. Two-thirds of the ring is beyond 8 kpc distance and one-half beyond 11 kpc. Entire clouds of size 30 pc will be unresolved and appear as a confused background. Their methods of analysis including the adding together of spectra between latitudes $\pm 1°$ (R. S. Cohen et al. 1980; Thaddeus and Dame 1984) and for cloud identification, subtracting off all emission $< T_R^* = 2$ K as observed by their antenna (Dame 1983; Thaddeus and Dame 1984), emphasizes large nearby clouds, dilutes or misses most of the far side of the galactic plane, and subtracts almost all of the cold molecular clouds.

The preliminary analysis of the Massachusetts–Stony Brook CO survey shows that the Galaxy has two populations of molecular clouds characterized by:

1. Warm molecular cloud cores with one-fourth of the population and about one-half of the emission. They exhibit a nonaxisymmetric galactic distribution; are clearly associated with H II regions; appear to be clustered; and are a spiral arm population.
2. Cold molecular cloud cores containing three-fourths of the total number. They are a widespread disk population located both in and out of spiral arms.

An analysis of the spatial correlation function of the CO cloud cores shows that the warm cloud cores are clustered on scales up to about 150 pc (Rivolo, Solomon and Sanders, in preparation). Combined with a measured core one-dimensional radial velocity dispersion of 7 km s^{-1}, a characteristic mass for the clustering scale length of 150 pc diameter is found to be about 5×10^6 $M_\odot$.

VI. CONCLUSIONS

The association of young stars and protostars, or at least stars very recently formed, with molecular clouds has been demonstrated over the past 10 or 15 years both on a local and galactic scale. Sign posts of star formation such as imbedded infrared sources, H_2O maser sources, compact H II regions, T Tauri stars, and outflows with molecular composition all occur in molecular clouds. There is little doubt that the important population I of the Galaxy and all galaxies is the molecular component and not H I. This is a drastic change in our concept of star formation from that prevalent in the recent past.

The molecular clouds described here with mass greater than 10^5 $M_\odot$ are dense; in particular their density is far above that needed to satisfy the Jeans criterion for a thermal gas. These clouds are gravitationally bound, their pressure, which is due primarily to large-scale chaotic motions with velocities (FWHM) of 8 to 15 km s^{-1} at average densities of ~ 150 cm^{-3} (corresponds to $n\,T \approx 10^6$ K cm^{-3}) is two or more orders of magnitude higher than the intercloud medium. They do not fit into the two- or three-phase pictures of the interstellar medium in pressure equilibrium.

The existence of very massive clusters of clouds with 5×10^6 $M_\odot$ which we have discussed here strongly points to gravitational instabilities as a mechanism for cloud formation for the most massive objects. Application of the Toomre (1964) criterion for gravitational instabilities in the disk to the gas component alone, including H I and H_2 (Jog and Solomon 1984*b*, their Appendix A), shows that the gas alone within the disk (from $R = 4$ to $R = 10$ kpc) is within a factor of two of instability. Alternatively, gravitational instabilities induced in the gas within a two-fluid system (stars + gas) (Jog and Solomon 1984*b*) will produce regions of 10^7 $M_\odot$.

The star formation process is itself affected by the feedback from star formation. Even the sign of this feedback is not always clear. While shock fronts may compress regions, outflow may inject sufficient energy into the clouds to stop collapse. The support of the clouds must be at least in part due to star formation. What is clear from the observations which show 3×10^9 $M_\odot$ in gravitationally contained clouds in the Galaxy is that star formation in the Galaxy is an inefficient process. The rate of star formation is controlled by the mechanisms that govern the growth, support and disruption of molecular clouds. These processes occur both in and outside spiral arms. Formation of the most massive stars may require the most massive clouds in spiral arms but star formation is occurring throughout the galactic disk inside molecular clouds.

MOLECULAR CLOUD CORES

PHILIP C. MYERS
Harvard-Smithsonian Center for Astrophysics

We describe recent progress in observations and interpretations of molecular cloud cores, i.e. the dense ($n \gtrsim 10^4$ cm^{-3}) *parts of molecular clouds. Cores can be divided into two groups. Low-mass cores contain* ~ 1 $M_\odot$, *have a temperature of 10 K, have subsonic turbulence, and are found near low-mass T Tauri stars. Massive cores are more diverse, but contain 10 to* 10^3 $M_\odot$, *have temperatures in the range 30 to 100 K, have supersonic turbulence, and are found near massive OB stars. Cores with embedded stars are likely to have CO outflow, and several such cores have an elongated shape suggestive of a disk or toroid. Some of these have velocity gradients* > 10 $km\ s^{-1}\ pc^{-1}$, *indicating a relatively fast rotation compared to that of larger cloud regions. Cores and larger cloud regions appear to be near virial equilibrium, to obey a "condensation" law* $n \propto L^{-1}$ *between density and size, and to obey a "turbulence" law* $\sigma \propto L^{0.3-0.6}$ *between velocity dispersion and size. Most regions in virial equilibrium have supersonic turbulence, which must contribute to cloud support and which must be replenished rapidly. Some cores contain young stars, as indicated by coincidence between infrared sources and core maps, and by interaction of outflow sources and cores. Many tens of core-star associations are known, suggesting that cores form stars. Core evolution is uncertain. IRAS results suggest that low-mass cores typically form low-mass stars within* $\sim 10^6$ *yr. Stellar outflows appear capable of dispersing cores, although outflow duration and mechanisms remain unclear.*

Since 1978 there has been considerable progress in our understanding of molecular cloud cores, the dense parts of molecular clouds. Reviews in *Protostars and Planets* (Gehrels 1978) by Evans, by Larson, and by Field give good summaries of what was known at that time. Since then, many more regions have been well studied in density-sensitive lines, with single telescopes and with interferometers. The basic physical properties, i.e., sizes, temperatures, densities, and velocity dispersions, are better known than before, and interesting relationships among these quantities are now emerging. The role of molecular cloud cores in star formation is more evident than before, although many important questions remain open. Finally, some steps toward a picture of core evolution can now be taken.

In this review we discuss molecular cloud cores from the viewpoint of recent observations, especially observations of regions of low-mass star formation. Some recent reviews that may be complementary are those by Evans (see his chapter) and Wynn-Williams (1982), primarily on young, massive stars and their environments; by Ho and Townes (1983) on studies of dense regions using lines of NH_3; and by Elmegreen (see his chapter) and Silk (1985*b*) on molecular clouds and star formation from a theoretical viewpoint.

I. DEFINITIONS AND BASIC PROPERTIES

We take a molecular cloud core to be a subregion of a molecular cloud, having mean gas density $n \gtrsim 10^4$ cm^{-3}. Molecular clouds are most commonly defined by maps in the 2.6 mm J = 1-0 rotational line of CO, which generally requires $n \gtrsim 300$ cm^{-3} for excitation. Such clouds are typically 3 to 100 pc in size, and nearby ones (within ~ 1 kpc) coincide reasonably well with regions of dust extinction visible on optical photographs. Molecular cloud cores are 10 to 100 times smaller than CO clouds, and are not easily identified from CO maps unless they are appreciably hotter than surrounding gas. In the nearest molecular clouds, cores can be recognized as regions of enhanced visual obscuration (Figs. 1 and 2). A line or combination of lines sensitive to $n \gtrsim 10^4$ cm^{-3} is needed to study cloud cores, and several such cm- and mm-wavelength lines have been used successfully. These include rotational lines of CS (Linke and Goldsmith 1980) and HC_3N (Avery 1980), the K-doublet and rotational lines of H_2CO (Evans and Kutner 1976), and several inversion lines of NH_3 (Ho and Townes 1983).

Each spectral line differs somewhat from the next in its sensitivity to the distribution of gas density, temperature, column density, and emitter abundance in the telescope beam, so it is important to specify the line(s) used when discussing measurements of cloud properties. Furthermore, it is desirable to base summarizing statements about cloud properties (e.g., cloud size) on measurements made with the same line and with the same spectral and angular resolution, when possible. To illustrate the strong dependence of map size on the spectral line used, Fig. 3 shows maps of the molecular cloud associated with S 252 in lines of ^{12}CO, ^{13}CO, and CS (Lada and Wooden 1979). The FWHM (full width at half maximum) map size in the north-south direction decreases by a factor ~ 5 from the ^{12}CO map, which defines the cloud, to the CS map, which defines the core. This is due mainly to the higher gas density needed to excite the CS line than the ^{12}CO line, and to the prevalence of

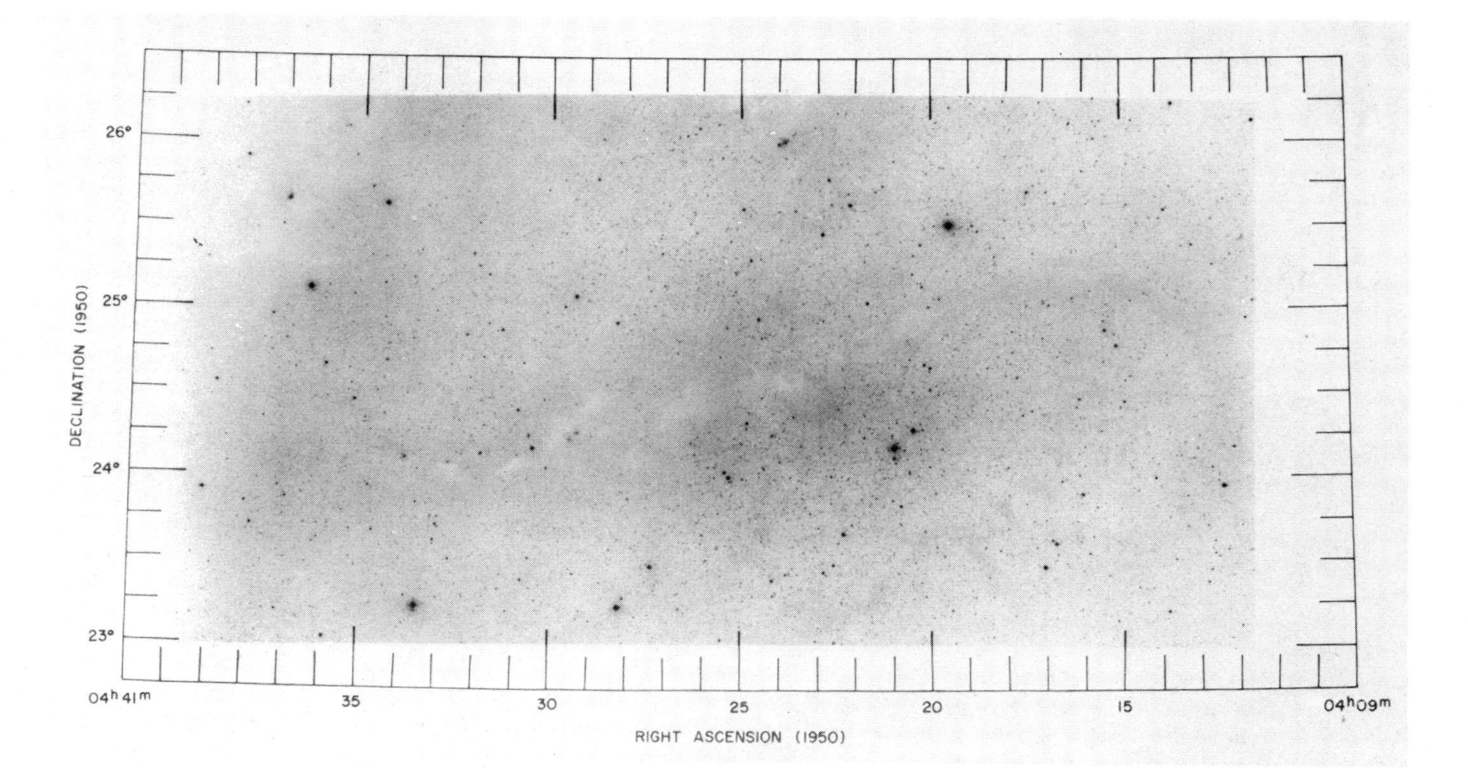

Fig. 1. Reproduction of Palomar Sky Atlas red print, centered on the Barnard 18 region in Taurus (Myers 1982). Numerous opaque condensations are visible; most contain low-mass cores as described in Sec. I. Scale: 1 deg = 2.4 pc.

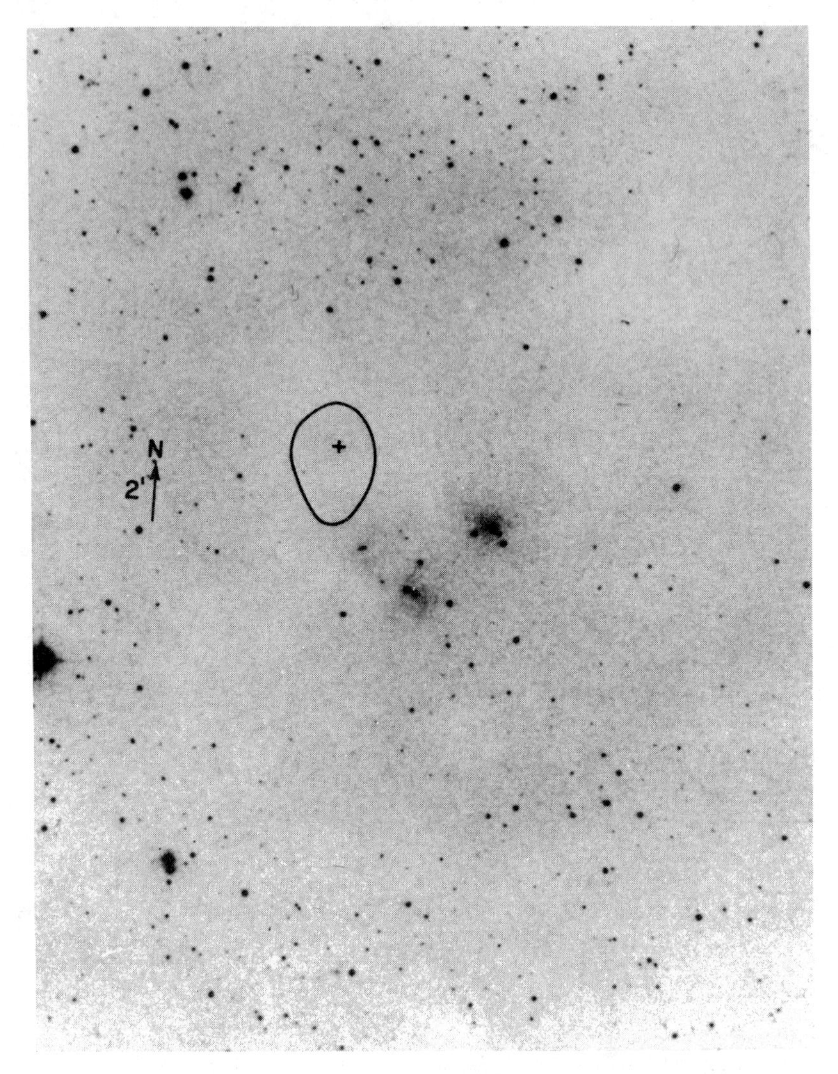

Fig. 2. Detail of Barnard 18 region, showing the half-maximum contour of NH_3 $(J,K) = (1,1)$ line emission that defines the low-mass core TMC-2. Several associated low-mass stars are visible. Scale: 2 arcmin = 0.08 pc.

denser gas in smaller regions in molecular clouds. Therefore, molecular clouds differ from planets in that measurements of cloud size can vary by large factors, depending on the probe in use.

Because the deduced size of a cloud region depends on the gas density to which the line observations are sensitive, one can expect that, conversely, the density deduced from line observations will vary with the line map size, from line to line in a given cloud and from cloud to cloud in a given line. In Sec. II

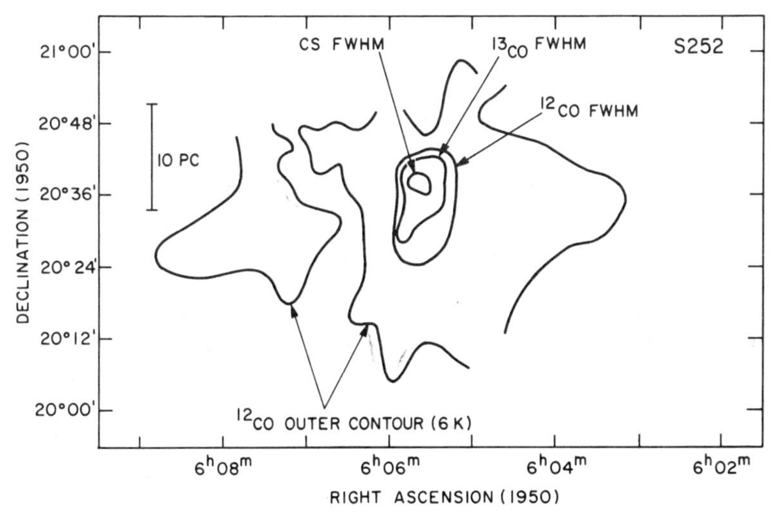

Fig. 3. Maps of the molecular cloud region S 252 in the J = 1-0 lines of ^{12}CO, ^{13}CO, and CS, showing strong dependence of map size on the line used to make the map (Lada and Wooden 1979).

we show that this variation of density with size is significant, as is a probably related variation of velocity dispersion with cloud size. Therefore, it is important to refer density, velocity dispersion, and other quantities derived from them (e.g., mass) to the size scale being measured. With these considerations in mind, we base a summary of core properties mainly on observations of the NH_3 (J,K) = (1,1) and (2,2) lines, and divide cloud cores into the two groups in Table I.

TABLE I
Properties of Molecular Cloud Cores

	L[c] (pc)	log n[d] (cm^{-3})	T[e] (K)	Δv[f] (km s^{-1})	M[g] ($M_\odot$)	log τ_{ff}[h] (yr)	Nearby Stars
Low-mass[a]	0.05–0.2	4–5	9–12	0.2–0.4	0.3–10	5.0–5.3	T Tauri
Massive[b]	0.1–3	4–6	30–100	1–3	10–10^3	4.3–5.3	OB

[a]Numbers in this row are based primarily on NH_3 observations reported by Myers and Benson (1983).
[b]Numbers in this row are based primarily on NH_3 observations reported by Ho et al. (1981) and others cited in Sec. I.
[c]L is the FWHM map diameter.
[d]n is the mean density of collision partners (H_2 molecules and He atoms) needed to excite the observed transition.
[e]T is the kinetic temperature of the emitting gas.
[f]Δv is the FWHM velocity width of the NH_3 emission, corrected for hyperfine broadening.
[g]M is the mass enclosed by a sphere of diameter L and mean density n. [It should be noted that the masses in Myers' and Benson's (1983) Table 1 are too high by a factor of 3, due to a calculation error.]
[h]τ_{ff} is the free-fall time for density n.

Low-Mass Cores

Low-mass cores have been found in nearby dark clouds by detecting molecular-line emission from regions of high visual obscuration, a few arcmin in extent. A fairly typical core is TMC-2 (Lynds #1529), shown in Fig. 2. First studied in the 6-cm H_2CO line by Kutner (1973), it has proven to be representative of the ~ 30 regions of high visual obscuration studied in the NH_3 (1,1) and (2,2) lines by Myers and Benson (1983) and Benson (1983). In the NH_3 (1,1) line TMC-2 has FWHM map diameter $L \sim 0.1$ pc and mean density required for excitation $n \sim 3 \times 10^4$ cm^{-3}, giving ~ 1 $M_\odot$ of gas within the FWHM map diameter. The kinetic temperature is 10 K, and the FWHM velocity width is 0.3 km s^{-1}, implying that the typical H_2 molecule in the core has a turbulent speed of only about half of its thermal speed. Thus low-mass cores have subsonic turbulence[a] (Myers 1983). Several low-mass T Tauri stars are seen within a few times 0.1 pc in the more extended, less dense region of obscuration Barnard 18 (Myers 1982). A region with the density of TMC-2 has a free-fall time of 2×10^5 yr, comparable to the estimated ages of the youngest T Tauri stars (Cohen and Kuhi 1979; Stahler 1983). We shall see that optically invisible, low-luminosity ($\lesssim 10\ L_\odot$) IRAS (Infrared Astronomical Satellite) sources are also associated with a significant number of low-mass cores (see Sec. III) (Beichman 1984).

Massive Cores

Massive cores have greater size, density, temperature, velocity dispersion, and mass than low-mass cores, and are found near young stars of earlier spectral type. They appear to have a much greater range in their properties than low-mass cores have. Their size, density, and temperature each vary by factors of 3 to 10 or more. Part of this apparent increase in diversity occurs because of resolution effects. Low-mass cores are plentiful at close distances, within ~ 200 pc, so that single-antenna beam widths, typically 1 arcmin, resolve most of the emission from a core. In contrast, nearly all known massive cores are at least 1 kpc distant, so most present-day single-antenna radio observations cannot resolve structure on scales smaller than ~ 0.4 pc. Consequently massive cores observed with single antennas have more ambiguous estimates of density, temperature, and mass than do low-mass cores. Among ten cores near ultracompact H II regions and/or luminous ($\gtrsim 10^4\ L_\odot$) infrared sources, the median gas density and mass are 3×10^3 cm^{-3} and 300 $M_\odot$, respectively, if clumping is negligible; and 1×10^5 cm^{-3} and 1.5×10^3 $M_\odot$, respectively, if the emission arises from a single source that occupies 0.1 of the projected beam area (Ho et al. 1981). Thus, these cores have mass greater than the low-mass cores by two to three orders of magnitude. It may be signif-

---

[a]Here, turbulence refers to motions having neither a thermal nor a simple systematic character, but which must be present in order to account for the observed line broadening. In this context, turbulence is a general term and does not necessarily imply that the observed gas has the properties of subsonic incompressible fluids.

icant that the giant molecular clouds where such massive cores are found are more massive, by factors of $\sim$ 100, than the molecular clouds where low-mass cores are found. However, very little is known about the prevalence of low-mass cores in giant clouds.

Some of the single-antenna resolution limitations described above are now being overcome by aperture synthesis observations. In Orion, single-dish observations (Batrla et al. 1983*a*; Ziurys et al. 1981; Ho et al. 1979) have mapped the spike emission region ($n \sim 10^6$ cm^{-3}; $L \gtrsim 0.1$ pc; $T \sim 70$ K; $M \sim 100$ M$_\odot$) around the infrared Kleinmann-Low Nebula (KL) and a cooler source 4′ to the north, among others in the Orion region. VLA (Very Large Array) observations have resolved a hot core within KL, associated with the Becklin-Neugebauer source and other 20-μm sources, which has $n \gtrsim 10^7$ cm^{-3}, $L \sim 0.01$ pc, $T \sim 200$ K, and $M \sim 10$ M$_\odot$ (Pauls et al. 1983; Genzel et al. 1982; see also Plambeck et al. 1982; Townes et al. 1983). In the source 4′ north of KL, there are two rapidly rotating clumps, each with $n \sim 10^5$ to 10^6 cm^{-3}, $L \sim 0.05$ pc, $T \sim 20$ K, and $M \sim 0.2$ to 10 M$_\odot$, but with no known infrared source (Harris et al. 1983). It is not yet known how typical the Orion results are; however, VLA observations of W 51 show three NH_3 emission knots, each with $L \sim 0.1$ pc and $T \sim 100$ K, and each coincident with a compact H II region; these resemble the KL region of Orion (Ho et al. 1983).

In summary, the low-mass cores appear to form a relatively homogeneous group with $L \sim 0.1$ pc, $n \sim 10^4$ cm^{-3}, $T \sim 10$ K, subsonic turbulence, and ~ 1 M$_\odot$. They are found near T Tauri stars and low-luminosity ($\lesssim 10$ L$_\odot$) infrared stars. The massive cores are more diverse, but are greater in size L by factors 1 to 30, in n by factors 1 to 100, in T by factors 3 to 10, in Δv by factors 3 to 10, and in M by factors 10 to 10^3. They have highly supersonic turbulence and are found near compact H II regions and luminous ($\gtrsim 10^4$ L$_\odot$) infrared stars.

Magnetic Fields

The properties in Table I are suitable for a quantitative summary. In addition, three aspects of cloud cores more difficult to quantify deserve mention: magnetic fields, rotation, and chemical complexity. At present, no direct measurements of magnetic field strength are available at the size and density scale of cloud cores, and measurements in other regions are few; thus, inferences about fields in cores are very uncertain. Possible ways to measure magnetic fields in molecular clouds are discussed by B. G. Elmegreen (1978*a*) in regard to Zeeman splitting and by Goldreich and Kylafis (1981, 1982) concerning linear polarization.

Zeeman splitting of the 18-cm OH maser lines indicates fields B of a few mGauss on scales 10^{15} to 10^{16} cm near the compact H II region W 3 (OH) (Moran et al. 1978; Reid et al. 1981), and similar results are found in other OH maser sources. Zeeman splitting of the 21-cm H line in a CO cloud south of OMC-1 indicates $B \sim 10$ μGauss, probably on a scale $\sim 10^{18}$ cm (Heiles and Troland 1982). Heiles and Troland estimate the volume density of this cloud to be $n \sim 400$ cm^{-3}. Application of the flux-freezing relation $B \propto n^K$, $1/3 \lesssim K \lesssim 1/2$ (Mouschovias 1978), to the results of Heiles and Troland indicates $B \sim 40$ to 160 μGauss in low-mass cores and 40 to 500 μGauss in massive cores.

To be dynamically important, core magnetic fields should be comparable to $B_{support} = 3\pi\sqrt{G/5}\,\mu n L$, where G is the gravitational constant and μ is the mean mass of a cloud particle (B. G. Elmegreen 1978*a*). If $B = B_{support}$, then mid-range values of n and L in Table I require $K_{support} \sim 0.3$ for low-mass cores and $K_{support} \sim 0.9$ for massive cores. If, as is often assumed, $K = 0.5$, these estimates suggest that frozen-in fields would be too weak in massive cores and too strong in low-mass cores for balance against gravity. As low-mass cores show no evidence of expansion, this latter inference implies that they may typically have less magnetic flux than the frozen value. Such a loss of flux has been suggested as a mechanism for forming low-mass cores (Shu 1983; see also the chapter by Cassen et al.).

The role played by rotational motions in cloud cores is discussed by Goldsmith and Arquilla (see their chapter). Observations of cores in the $J = 1 - 0$ line of ^{13}CO, summarized by Goldsmith and Arquilla, and in the $(J,K) = (1,1)$ line of NH_3, described by Ungerechts et al. (1982) and Myers and Benson (1983), suggest that many cores have velocity gradients ~ 0.1 to 1 km s^{-1} pc^{-1}. These indications of rotational motion generally are insufficient, by themselves, to account for the nonthermal part of the observed line broadening, and the rotational contribution to the velocity dispersion rarely exceeds the turbulent contribution. On the other hand, recent observations, generally at finer angular resolution than those described above, and within ~ 0.1 pc of young stars, suggest that rotational motions may play a more prominent role. In Orion, observations of NH_3 emission north of the Kleinmann-Low nebula reveals two clumps with 20 to 30 km s^{-1} pc^{-1}, in VLA maps (Harris et al. 1983). Velocity gradients of similar magnitude have been reported in NH_3 emission in ON-1 (Zheng et al. 1984), in CS emission in the outflow sources NGC 2071 (Takano et al. 1984), and L 1551 (Kaifu et al. 1984). However, such gradients need to be confirmed; in L 1551, NH_3 observations with angular and spectral resolution nearly identical to those of Kaifu et al. show no detectable velocity gradient (Menten and Walmsley 1984).

As discussed by Irvine et al. (see their chapter), the relative abundances of molecules from cloud to cloud, and from core to core, are generally similar, with a number of distinct exceptions. We note that this number of exceptions has increased rapidly in recent years. This trend tends to challenge the familiar assumption that molecular abundances are essentially constant across a map. Variations in abundance of ^{13}CO from dark cloud centers to edges by factors $\gtrsim 3$ are indicated by observations of Dickman et al. (1979), McCutcheon et al. (1980), and Langer et al. (1980). These have been attributed to increased fractionation of ^{13}CO at cloud edges, due to enhanced photoionization (W. D. Watson et al. 1976). A closely related variation in ^{12}CO and ^{13}CO abundance, relative to that of C^{18}O, has been reported at a molecular cloud edge near the H II region S 68 (Bally and Langer 1982); this is attributed to both ^{13}CO fractionation and isotope-selective photodestruction.

Variations by factors as great as $\sim$ 10 in the abundance of the long carbon-chain molecules HC_3N, HC_5N, etc. have been noted from cloud to cloud (Avery 1980; Benson and Myers 1983). Similar variations within TMC-1 have been suggested in the abundance of long-chain molecules relative to that of NH_3 (Little et al. 1979*b*; Toelle et al. 1981) although Bujarrabal et al. (1981) claim constant abundance. In TMC-1, the map peaks of NH_3 and HC_3N emission are displaced by several arcmin. However, no such shift is seen in either L 1544 or TMC-1C, the latter just 10′ north of TMC-1 (Benson and Myers 1983). Significant differences in map shape, difficult to explain with excitation or radiative transfer arguments, are also evident in maps of L 183 in the NH_3 (1,1) line (Ungerechts et al. 1980), in the SO_2 $3_{13} - 2_{02}$ line (Irvine et al. 1983), and in the CS $J = 1 - 0$ line (Snell et al. 1982). Similar discrepancies are seen in maps of OMC-1 in lines of HCO^+, HCN, and SO (Plambeck 1984). These observations suggest that there may be new opportunities to study chemical processes in cloud cores, but they raise the caution that interpretation of maps on the basis of uniform molecular abundance may sometimes be misleading.

A development having potential importance for physical and chemical properties of cloud cores is the recognition that core lifetimes may be much shorter than cloud lifetimes. Boland and de Jong (1982) have estimated that molecules containing heavy elements will condense onto grains with high efficiency if they remain in dense cores for more than a depletion time $\sim 2 \times 10^5$ yr. To avoid this, they propose turbulent circulation of dense gas and dust between cloud interior and cloud edge, where ultraviolet photons strike grains and thereby return molecules to the gas phase. An alternative explanation might be that some cores lose their identity within a depletion time, due to star formation and subsequent winds and outflows. In either case, a core lifetime much shorter than the cloud lifetime implies that molecular abundances in cores may in part reflect molecular formation and destruction processes at times before the core was recognizable as such, and under physical conditions of lower density, higher temperature, and greater ultraviolet flux than in present-day cores.

II. CORE STRUCTURE AND PHYSICS

Knowledge of the internal density, temperature, and velocity structure of cloud cores is of great importance for understanding the main physical processes in cores, and core evolution. In this section, we first discuss structural information based on observations of individual cores and then consider trends among properties of cores and larger cloud regions. At present, knowledge of core structure is still meager. At small scales, cores with associated stars are better studied and better known than cores without stars; we distinguish these two groups in this discussion.

Density Structure

In relation to surrounding gas, cores are local maxima in density, as indicated by comparison of maps such as in Figs. 2 and 3 in lines sensitive to distinctly different density. Because cores form stars, it is likely that some cores have considerable internal structure at high densities and small sizes compared to those that define the core. However, the details of this structure are just beginning to be known, and we need more observations of high-density tracers at high angular resolution. A crude method of estimating the dependence of mean density on size is to plot the density required to excite a line against the map extent in that line. In low-mass cores, such plots give $n \propto r^{-p}$, $1 \lesssim p \lesssim 3$ over ~ 0.1 to 0.5 pc (R. N. Martin and A. H. Barrett 1978; Snell 1981). Studies of isolated globules based on maps of visual extinction derived from reddening of background stars (Jones et al. 1984) and ^{13}CO line area (Dickman and Clemens 1983) suggest a similar degree of central condensation on the scale of 0.1 pc. In massive cores, a more complex situation may be present. Mundy (1984) finds that massive cores near M 17, S 140, and NGC 2024 have CS map sizes that do not decrease significantly with increasing rotational quantum number J, as expected if $n \propto r^{-p}$, $p > 0$. A possible explanation is that these massive cores have significant clumpy structure unresolved by the telescope beam. Accurate determination of radial density profiles is of interest as a possible dynamical probe; if an isothermal gas sphere is in hydrostatic equilibrium, it has $n \propto r^{-2}$ and if in free-fall collapse, it has $n \propto r^{-3/2}$. However, at present the prospect for measurements of sufficient accuracy to distinguish these cases seems remote.

The most significant recent development in understanding the density structure of cores with stars is the discovery of dense elongated features associated with stars having gas outflow. Such structures have been widely interpreted as disks or toroids (Plambeck et al. 1982: Orion; Bally et al. 1983: S 106; Kaifu et al. 1984: L 1551; Güsten et al. 1984: Cepheus A; Torrelles et al. 1983: numerous sources). These are generally $\lesssim 0.1$ pc in diameter, denser than 10^4 cm^{-3}, closely associated with the outflows, and in some cases have their long dimension roughly perpendicular to the outflow direction. The causal relationship of the dense structures and the outflow is unclear: the dense gas may serve to collimate the outflow on interstellar scales (Canto et al. 1981); or the collimation may occur on stellar (L. Hartmann and K. B. MacGregor 1982*b*) or circumstellar (Snell et al. 1980) scales. In these latter cases the shape of the intersteller disk might be more a result of the collimation of the outflow than its cause. Recently, evidence for a circumstellar disk ~ 30 times smaller than the molecular objects considered above has been found on the basis of mid- and near-infrared observations of HL Tauri (Cohen 1983; Grasdalen et al. 1984; Beckwith et al. 1984).

Temperature Structure

The temperature structure of cores is poorly known; a notable exception is the detailed study of near-infrared emission and absorption lines from the

Becklin-Neugebauer object (Scoville et al. 1983*a*). This indicates an ionized wind with temperature $T = 10^4$ K out to radius $R = 20$ AU from the central star; a possible shock front with $T = 3500$ K at $R \geq 3$ AU; a dust envelope around the ionized region with $T = 900$ K at $R = 20 - 25$ AU; and an ambient molecular cloud with $T = 600$ K at $R \geq 25$ AU. However, thermal structure is not known in such detail for any core for scales $R \sim 100$ AU to $\sim 10^4$ AU. Generally, cores containing OBA stars have higher CO temperatures than their surroundings, while cores containing low-mass stars, and low-mass cores without stars, show little or no such increase. On size scales of a few pc, studies based on mm-wavelength CO lines and the infrared continuum from massive cores around nearby H II regions give temperatures ~ 40 K at ~ 1 pc from the peak, declining to ~ 20 K at ~ 5 pc (Evans et al. 1981,1982).

Systematic Velocity Structure

The velocity structure of cores can be divided into four categories: outflows, rotation, infall, and random motions. The prevalence, structure, and energetics of outflows are reviewed by Lada (1985; see also the chapter by Evans). Here, we note that the presence of CO outflow from young low-mass stars is highly correlated with the presence of a dense core around the star. More than half of the known low-mass cores with embedded stars have CO outflows (Myers et al. 1985, in preparation) while ~ 0.1 of T Tauri stars (which generally lack cores) surveyed by Edwards and Snell (1982) have outflows. Thus, the presence of a dense core around a star may be about as good an indicator of stellar youth as is stellar outflow. Evidence for the presence of rotation in cores has been summarized in Sec. I, but detailed structural information is still lacking. Evidence for infall of circumstellar gas onto a star has been presented by Scoville et al. (1983*a*) for infall of CO toward the Becklin-Neugebauer object, and by Reid et al. (1980) and Garay et al. (1984) for infall of OH masers toward compact H II regions. However, maps of infall motions are not yet available, and it is not entirely clear whether these cases of apparent infall are gravitational in origin. The structure of random motions in cores and surrounding regions has been the subject of several comparisons of line widths. These and related trends are discussed in the remaining part of this section.

Structure of Random Motions

The core properties described in Sec. I suggest many questions about the physical mechanisms responsible for the density structure and internal motions of cores and their environments. Significant clues concerning these questions have come from the demonstration of three interrelated trends among size L, mean density n, and velocity dispersion σ. [In this section it is desirable to refer to σ, the three-dimensional velocity dispersion of the particle of mean mass μ. This more fundamental quantity σ is related to Δv, the observed FWHM of the distribution of line-of-sight velocities of the emitter of mass m, referred to in Section II, by $\sigma = [3\,(8 \ln 2)^{-1}\,(\Delta v)^2 + kT\,(\mu^{-1} - m^{-1})]^{1/2}$, where k is Boltzmann's constant (Myers 1983). For the largest

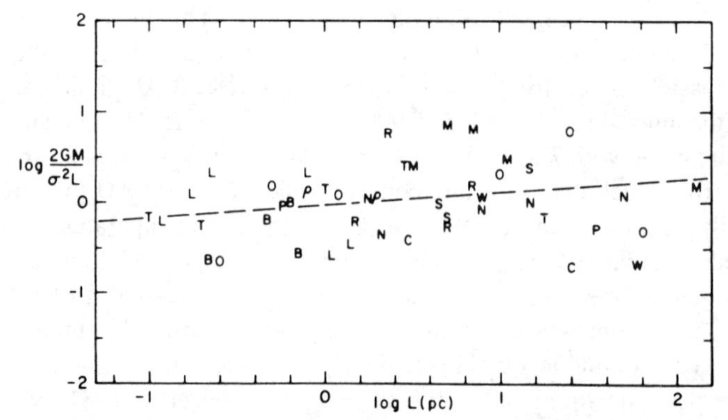

Fig. 4. Plot of the virial ratio 2 $GM/\sigma^2 L$ vs. region size L for 46 cloud regions, showing that most regions appear close to virial equilibrium, over three decades in L (Larson 1981).

clouds, an additional term representing systematic motions may be important, and should be included (Larson 1981).]

Larson (1981) showed that over 50 recent molecular cloud measurements, mostly in the 2.7 mm $J = 1\text{-}0$ line of ^{13}CO, satisfy approximately three laws as illustrated by Eqs. (1)–(3) and Figs. 4–6:

(1) Virial equilibrium,

$$\sigma = \left[\frac{\pi \mu G}{3}\right]^{\frac{1}{2}} n^{\frac{1}{2}} L \tag{1a}$$

or, assuming $\mu = 2.33$ amu

$$\sigma = 0.016\, n^{1/2}\, L \text{ km s}^{-1} \tag{1b}$$

where n is in cm^{-3} and L is in pc (see Fig. 4);

(2) Condensation law, whose best fit is

$$n = 3900\, L^{-1.2} \text{ cm}^{-3} \tag{2}$$

(see Fig. 5);

(3) "Turbulence" law, whose best fit is

$$\sigma = 1.2\, L^{0.3} \text{ km s}^{-1} \tag{3}$$

(see Fig. 6). Equations (2) and (3) differ slightly from Larson's (1984) equations (5) and (1), respectively, because they are linear least-squares fits to log-log plots, while Larson's equation (1) is eye-fitted and his equation (5) is derived from his equations (1) and (2). Also our Equation (2) is expressed in terms of n rather than Larson's $\langle n(H_2) \rangle$.

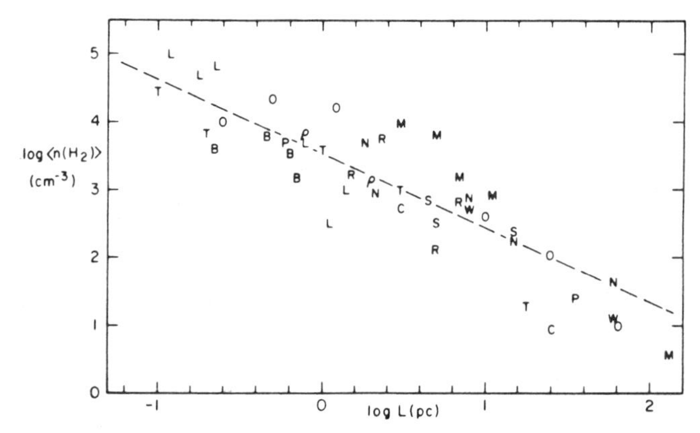

Fig. 5. Plot of mean H_2 density vs. region size L, showing that the mean density varies with L approximately as L^{-1} (Larson 1981).

Each of these relations has a correlation coefficient of ~ 0.9 and holds over three decades in cloud size L. The typical uncertainty in the exponent of L is 0.1. Each of Eqs. (1)–(3) is consistent, within its uncertainties, with that derived from the other two. Larson (1981) also showed that progressively larger subregions within clouds support the turbulence law, Eq. (3), but with less data and more scatter (see Fig. 7).

The relations, Eqs. (1)–(3) are interesting and important, and several authors have sought to verify, criticize, or interpret them. Leung et al. (1982), Myers (1983), and Torrelles et al. (1983) investigated 49 clouds small compared to Larson's (L = 0.05 pc to 24 pc) in lines of ^{12}CO, ^{13}CO, and NH_3; Dame et al. (1985) studied 34 clouds large compared to Larson's (L = 12 pc to 384 pc) in the ^{12}CO line. In each case, the data have significant correlations that appear consistent with relations (1)–(3), within factors of ~ 3 in the coefficients and within ~ 0.3 in the exponents. Because of noise,

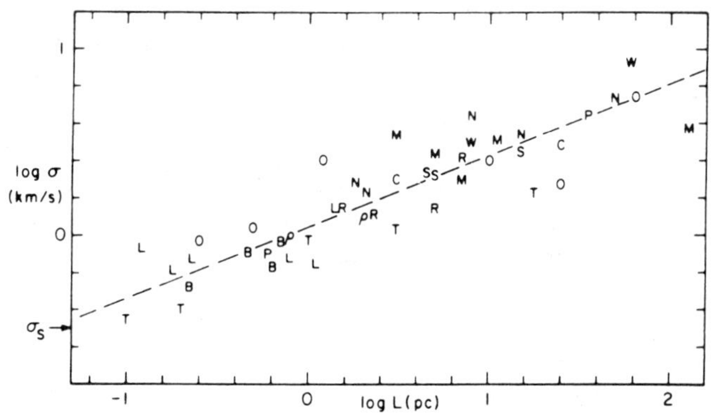

Fig. 6. Plot of velocity dispersion σ vs. region size L, showing that σ varies approximately as $L^{0.3}$ (Larson 1981).

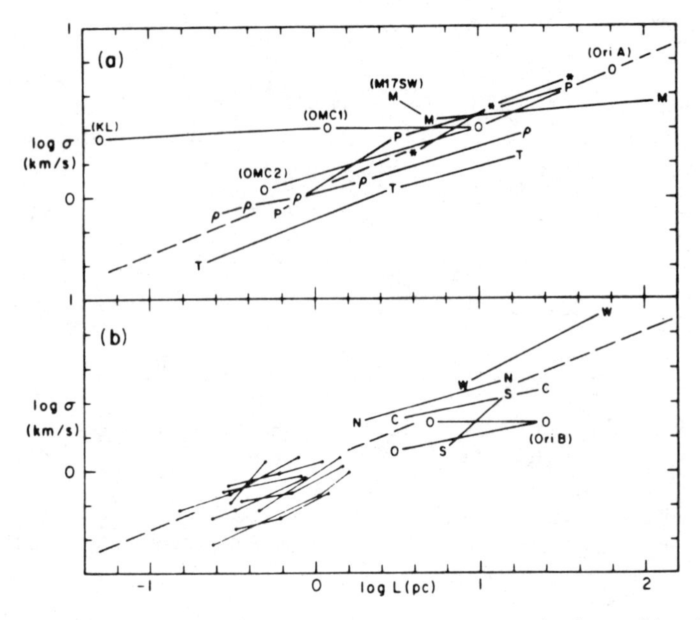

Fig. 7. Plot similar to Fig. 6 but with regions belonging to the same cloud, some hierarchically nested, connected by straight lines (Larson 1981). The data are grouped in upper and lower panels to increase clarity. The data in the left part of panel (b) are from Snell (1979).

selection effects, differing resolutions, and other reasons, it is difficult to determine whether each of these trends is statistically consistent with each of the others. However, each is statistically significant, and there seems little doubt that Eqs. (1)–(3) are approximately satisfied by a wide variety of interstellar clouds, and probably also by progressively larger regions within individual clouds.

The physical significance of these trends is clearest for Eq. (1), virial equilibrium. Because cloud ages deduced from ages of embedded stars usually exceed cloud free-fall times by factors $\gtrsim 10$, the appearance of equilibrium is not surprising. If a cloud consists only of relatively stable, independently moving clumps, then the predominance of equilibrium velocities can be understood to mean that clumps with discrepant velocities have escaped, collapsed, or interacted with the other clumps so as to drive their velocities toward equilibrium values. The constraints on these interactions appear to be easily satisfied in simulated collisions (Scalo and Pumphrey 1982). If instead, a cloud is a relatively continuous fluid, then the appearance of virial equilibrium requires that collisions between supersonically moving fluid elements exert a supporting pressure, as do collisions between subsonic molecules in an idealized self-gravitating cloud in equilibrium. The continuous-fluid picture is probably more accurate, at least in well-resolved nearby clouds, in that the dense clumps that are present contain relatively little mass, while the most widespread low-density gas contains most of the cloud mass. Therefore, the appearance of virial equilibrium in many clouds and cloud regions suggests that nonthermal turbulent motions play a central role in supporting and maintaining molecular clouds. This role has been suggested for more than a decade (see, e.g., Zuckerman and Evans 1974), but the evidence is now extensive for size scales 0.1 pc to 100 pc.

The appearance of equilibrium does not exclude the possibility that some regions are collapsing, especially on small scales, because collapse speeds are only slightly different from virial speeds. However, if clouds are supported globally by turbulent pressure, then the turbulent motions must be continually supplied with fresh energy, because the free decay time of supersonic turbulence is much shorter than typical cloud lifetimes (Fleck 1981, and references therein).

Interpretation of the condensation and turbulence laws, Eqs. (2) and (3), has been more difficult and controversial than interpretation of virial equilibrium. These trends are probably not independent, since within their uncertainties each can be obtained from the other in combination with the virial equilibrium law. Thus, either one might be fundamental and the other derivative. Larson (1981) noted that the condensation law can result from selection, if the observations are sensitive to a relative range of column density $N = nL$ small compared to the relative range of L. If the condensation law is real, it has the unexplained consequence that cloud column densities lie in a relatively narrow range, a factor ~ 10.

The turbulence law might be real and fundamental, but might arise for statistical reasons, giving little insight into the mechanisms of cloud turbulence (Scalo 1984). Scalo noted that longer path lengths L can be expected to contain larger numbers of turbulent velocity fluctuations; if their central velocities are randomly distributed, then the velocity dispersion along L may vary as $L^{1/2}$, close to that which is observed. In an alternate view, the turbulence law may arise because each cloud region is at a point of critical stability against collapse due to gravity and external pressure P; in which case $\sigma \propto P^{1/4} L^{1/2}$ (Chieze 1985). A view which may have fundamental implications for cloud physics, if correct, was proposed by Larson (1981) and elaborated by Fleck (1981,1983). In this picture, the turbulence law reflects an energy cascade from larger to smaller scales, similar to the process in idealized incompressible fluids, where fully developed subsonic turbulence obeys the Kolmogorov law $\sigma \propto L^{1/3}$. While suggestive, the analogy to Kolmogorov turbulence cannot be exact: because of the density structure and compressibility of interstellar clouds; because of significant gravitational forces; because of supersonic motions in all but the smallest cloud regions; and because of the probable importance of magnetic fields in cloud dynamics, at least at low densities.

It will be some time before the many interpretations of the trends discussed here can be sorted out. More rigorous approaches to analysis of cloud turbulence, based on mean velocities rather than line widths, have been presented by Scalo (1984), and Dickman (see his chapter); these may prove useful. However, at this point it seems clear that:

1. Virial equilibrium is a widespread principle governing the motions and densities in clouds and cloud cores, from ~ 0.1 pc to ~ 100 pc;
2. Supersonic turbulent motions, visible in all cloud regions except low-mass cores, are probably supporting motions;
3. From cloud to cloud, mean densities n, sizes L, and velocity dispersions σ obey the trends $n \propto L^{-1}$ and $\sigma \propto L^{0.3-0.6}$;
4. Only one of these two trends, $n(L)$ and $\sigma(L)$, need be independent, and it is possible that one is fundamental and the other derivative.

The trends and implications described here are based partly on measurements of cores and partly on measurements of less dense, more extended regions. Nonetheless, they raise at least 4 questions about cloud cores that can be studied with new observations: (1) How well do trends of $n(L)$ and $\sigma(L)$, deduced from maps of different lines in the same cloud (Fig. 7), agree with those deduced from maps of the same line in different clouds (Fig. 6)? These two types of trend may conceivably have different explanations. The main sources of multiple-line data, as in Fig. 7, are the studies of R. N. Martin and A. H. Barrett (1978) and Snell (1981) in dark clouds. These can now be extended significantly because of improvements in receiver sensitivity and telescope resolution. (2) What is the internal density structure of cores, especially over length scales where σ changes significantly? If virial equilibrium holds on scales smaller than the core size, changes on these scales in σ imply changes in n. For example, if a low-mass core is isothermal and has negligible turbulence, but is surrounded by gas satisfying $\sigma \propto L^{1/2}$, then a possible phenomenological model might be $\sigma^2 \propto T\,(1 + L/L_o)$, where L_o is a characteristic core size. This model and virial equilibrium imply

$$n \propto [(L/L_o)^{-2} + (L/L_o)^{-1}]. \quad (4)$$

Thus, the dependence of n on L might change significantly from $L > L_o$ to $L < L_o$ (Myers 1983), and such a change might be easier to detect than the difference between $n \propto r^{-2}$ (isothermal equilibrium) and $n \propto r^{-3/2}$ (free fall), discussed earlier in this section. If a massive core has internal sources of heating and turbulence, similar considerations suggest that the density gradient will again become steeper at smaller length scales in the core. High-resolution observations in lines sensitive to high densities should clarify this picture. (3) Do the trends $n(L)$ and $\sigma(L)$ differ between samples containing only massive cores and only low-mass cores? If they do, the details may give clues as to how these apparently different core types are supplied with turbulent energy. (4) Do the trends $n(L)$ and $\sigma(L)$ differ between samples of cores with associated stars and cores without associated stars? It will be important to

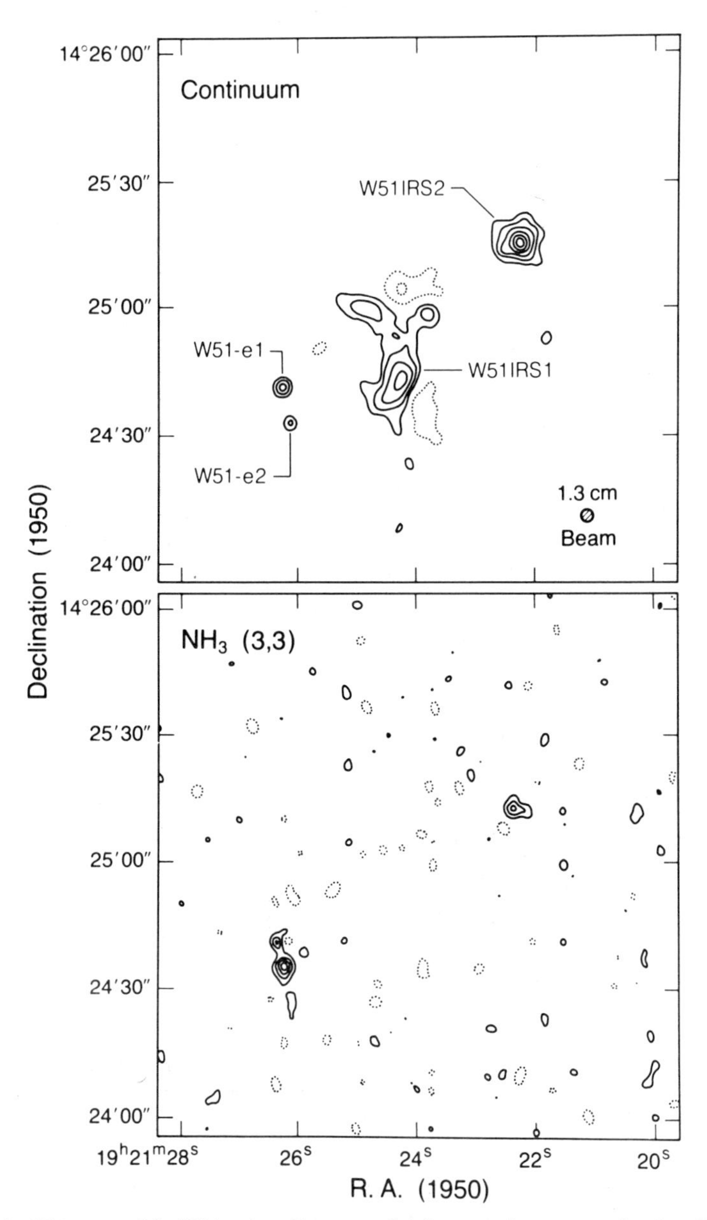

Fig. 8. VLA maps of the W51 region. *Upper panel:* 1.3-cm continuum map, showing ultracompact H II regions of size ~ 0.1 pc. *Lower panel:* 1.3-cm map of integrated NH_3 $(J,K) = (3,3)$ line emission, showing close coincidence of NH_3 emission (massive cores) with ultracompact H II regions. Scale: 30″ = 1 pc (Ho et al. 1983).

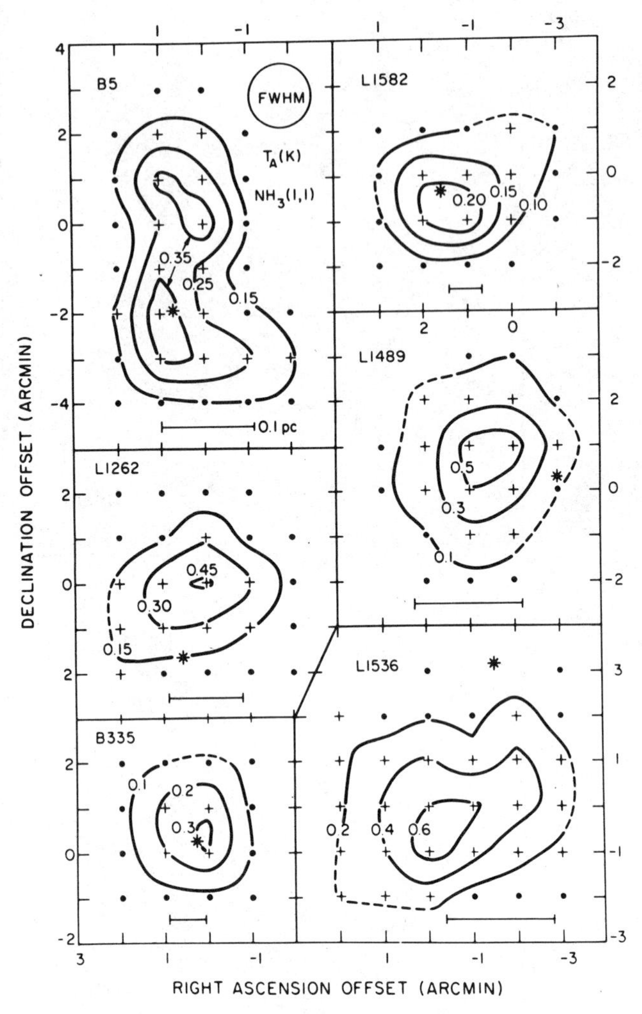

Fig. 9. 1.3-cm maps of six low-mass cores in the NH_3 (J,K) = (1,1) line, showing relative locations of 2-μm sources and the far-infrared source in B335 (*asterisks*). The horizontal line in each panel indicates 0.1 pc (Benson et al. 1984).

learn the size scale in a core where these trends break down or change, perhaps as a consequence of stellar luminosity and winds.

III. CORES AND STAR FORMATION

There are now three lines of evidence that the youngest stars are found in cores: (1) For massive stars, luminous infrared sources and/or compact H II regions have long been known to be associated with hot, dense gas, but the last few years have seen a great increase in the number of such regions with maps of high-density tracers (Wynn-Williams 1982) (Table I). Some of these maps have sufficiently high-density sensitivity and angular resolution to isolate dense regions only about as massive as their associated stars (see, e.g., Ho et al. 1983) (Fig. 8). Others show groupings of compact H II regions, infrared sources, and multiple cores, suggestive of clusters of OB stars in the process of formation.

(2) For low-mass stars, there have been significant increases in the number of known low-mass cores, as described in Sec. I. Many have associated low-luminosity ($\lesssim 10\ L_{\odot}$) sources, found in the far infrared (Keene et al. 1983; J. Davidson and B. Jaffe 1984) or the near infrared (Benson et al. 1984) (Fig. 9). The most comprehensive census of stars in low-mass cores will come from the IRAS survey. Preliminary results indicate that about one-third of the cores studied by Myers et al. (1983) have associated low-luminosity IRAS sources (Beichman 1984).

(3) For both massive and low-mass stars, there is now an easily identified and widely detectable signature of star-core interaction: the presence of spatially displaced maps of CO line emission in the blue- and red-line wings. This CO bipolar flow detected around many infrared sources (Bally and Lada

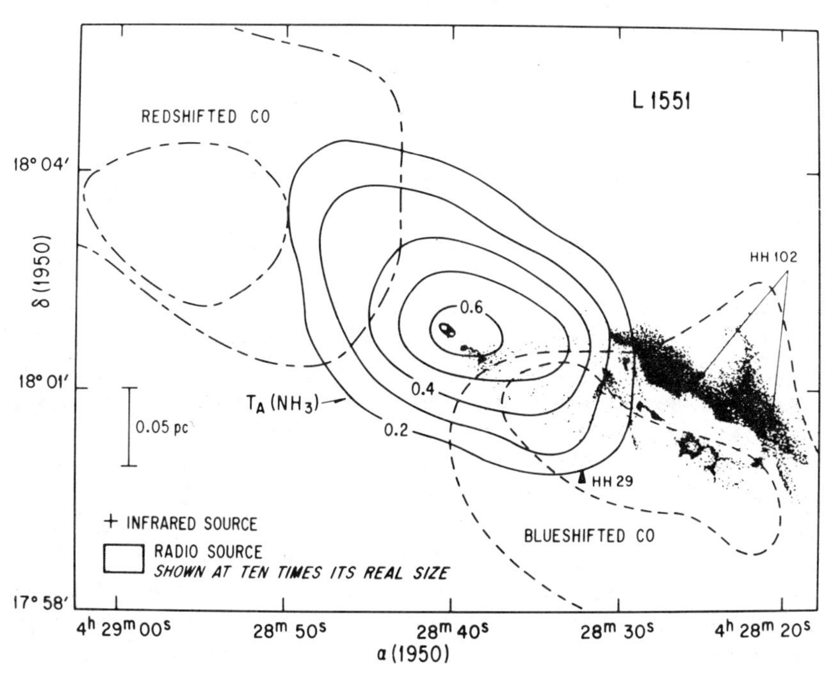

Fig. 10. 1.3-cm map of NH_3 $(J,K) = (1,1)$ emission from the low-mass core and outflow source in Lynds 1551, showing location of infrared source IRS 5, radio continuum source, and CO outflow lobes (Torrelles et al. 1983).

1983; Lada 1985) signals that dense gas is still associated with the stellar source, an association confirmed in some cases by CS and NH_3 maps (Torrelles et al. 1983) (Fig. 10). The flow-core interaction evident in the CO line wings provides better evidence of star-core association than the mere projection of the star on the core map.

Therefore, there are now many tens of cases known where extremely young stars are associated with cores of dense gas, with core mass $\gtrsim$ star mass. It appears fair to describe these cores as star-forming, in that the stars probably formed in the cores where we now see them. Acquisition of these many cases is an extremely important development. It implies that at least some star-forming cores last long enough, perhaps $\sim 10^5$ yr, after their stars have formed to be identified and studied, and that we now have practical techniques to find more such regions. Areas that can now be investigated are the distribution and multiplicity of stars in cores, and the star formation efficiency (stellar mass/stellar-plus-core-gas mass) of those cores with stars.

IV. CORE EVOLUTION

How do molecular cloud cores form; interact with their parent clouds and with each other; produce stars; and die? In this section we sketch a highly tentative picture, emphasizing where more information is needed.

Core Formation

It is important to underscore the point in Sec. I that core properties depend critically on the probe used to measure them. The core sizes and masses summarized in Table I refer to NH_3 line observations sensitive to density $\gtrsim 10^4$ cm^{-3}. In the ^{13}CO $J = 1$-0 line, the same cores would appear larger, less dense, and more massive. Thus, models of core formation and evolution need to account not only for core properties at a convenient fiducial density at the time of observation, but also for core properties at the density where the core became distinguishable from background at the time of formation. The density at which cores are distinguishable above background is difficult to estimate, but probably lies near $\sim 10^3$ cm^{-3} for low-mass cores.

Several schemes have been suggested for forming cores, but it seems premature to try to distinguish among them. Larson (1981) pictures cores as structures that arise from turbulent motions and that survive because they happen to be self-gravitating. Shu (1983; see also chapter by Cassen et al.) proposes that cores condense as interior cloud regions lose their magnetic support due to ambipolar diffusion, on a time scale $\sim 10^7$ yr. Cameron (see his chapter) suggests that cores can acquire the isotopic radioactivity properties of the solar system if they form as a result of random traversals of molecular clouds by asymptotic-giant-branch stars.

Core Interaction

How often, and with what outcome, do cores interact with each other and with their background clouds? For massive cores, the scarcity of high-resolution data makes realistic answers difficult. Nonetheless, the evidence for clumpy structure within dense regions described in Sec. I, and CO observations of relatively massive clouds (see, e.g., Blitz and Thaddeus 1980), suggest models of clumps moving with respect to each other at virial speeds, with relatively little background. Scalo and Pumphrey (1982), Pumphrey and Scalo (1983), and Gilden (1984) have simulated interactions among clumps of this type. They find that collisions are frequent and can inhibit or promote collapse, depending on the circumstances of the collision. For low-mass cores, a model of the Barnard 18 region suggests that the mass of cores is small compared to low-density ($n \lesssim 10^3$ cm^{-3}) background gas (Myers 1982). Furthermore, velocity differences between core gas and background gas in Taurus are negligible (Baudry et al. 1981). Hence, there is little basis for significant interaction of low-mass cores with each other, or with background gas.

Core Destruction

Evidently some cores form stars, as discussed in Sec. III, and evidently many stars lose their parent cores by the time they reach the main sequence. There is abundant evidence that outflows and winds from young stars have the momentum and energy to disperse significant amounts of core mass, and the prevalence of outflows suggests that they are a common, if not ubiquitous, feature of early stellar evolution. However, the outflow mechanisms and time scales are still unclear, as are the relations between outflows and winds. The time for a steady T Tauri-like wind to drive a shell to typical low-mass core radius is $\sim 4 \times 10^5$ yr (Chevalier 1983). The dynamical lifetime of bipolar flows is $\sim 10^4$ yr, based on CO spectra and maps (see, e.g., Bally and Lada 1983), but the flow duration, and the time for a flow to disperse a core, could be longer. If the dispersal time were as short as 10^4 yr, cases where two or three outflows are observed in a single cloud (Calvet et al. 1983: L 1551; Goldsmith et al. 1984: L 1455) would seem to require an implausibly high degree of synchronization. If the dispersal time were $\sim 10^5$ yr, then the similarity to the free-fall time suggests that in some cases the dispersing outflow can have significant overlap in time with the late stages of the infall that formed the star.

For massive cores, outflows are more energetic ($\sim 10^{46}$ erg) than for low-mass cores ($\sim 10^{44}$ ergs) and are energetically capable of significant massive core dispersal (Lada 1985; Goldsmith et al. 1984). However, young massive stars are more luminous than young low-mass stars by factors $\gtrsim 10^3$, and this great luminosity increase allows several other mechanisms of core dispersal that are unlikely for low-mass stars. The roles of supernovae and expanding H II regions are well summarized by Silk (1985). Luminous stars may also significantly alter nearby cores by photoionization. Vidal (1982) suggests that "nebular condensations" near the Trapezium stars may be partly ionized by Θ_1 C Orionis. Reipurth (1983) proposes that photoionization due to O stars destroys some dense cores and may remove the lower-density gas around others. In the latter case, the surviving core would have the isolated appearance and steep density gradients characteristic of Bok Globules.

V. PROSPECTS

The study of cloud cores, and their relationship to associated young stars, appears likely to benefit greatly from the many high-resolution telescopes coming into use at millimeter, submillimeter, and infrared wavelengths in the near future. Some of the more interesting areas to be studied are discussed below.

Map Geometry

Much of our understanding of bipolar flows has come from interpretation of the shape and geometry of maps, including (a) the displaced maps of CO red and blue wings first found by Snell et al. (1980); (b) the orientation of dense gas structures perpendicular to the CO flow direction (Schwartz et al. 1983; Torrelles et al. 1983); and (c) the progression of CS line velocity along the dense gas direction in Orion (Hasegawa et al. 1984) and L 1551 (Kaifu et al. 1984), and others cited in Sec. I suggesting the presence of rotating disks. It will be important in future studies to further refine the geometrical relation of cores to their stars; for example, do young stars without outflow appear more centered in their core maps than young stars with outflow?

Velocity Dispersions

The velocity dispersions of low-mass cores are among the narrowest known in interstellar clouds, consistent with the suggestions of Larson (1981), Leung et al. (1982), and Myers (1983) that cores form low-mass stars after their nonthermal support has largely dissipated. For massive cores, velocity dispersions are an order of magnitude greater than for low-mass cores (an increase often attributed solely to the greater luminous and mechanical energy provided by newly formed massive stars). However, it is possible that much of the massive-core velocity dispersion precedes star formation and simply satisfies the requirements of virial support of the core. It may be possible to test this question with detailed comparison of core velocity dispersions among regions forming stars of low, intermediate, and high mass.

Core and Star Masses

Similar comparisons of core mass and star mass may also be useful. Ho et al. (1981) and Larson (1982) have presented data showing that the mass of the most massive star in a cloud-embedded cluster tends to correlate with the cloud mass. Further studies involving cores that are more specific to their young stars may help to determine the relative importance of core properties in influencing the properties of the stars they produce. Complete surveys of cloud complexes for dense cores should reveal the distribution of core masses in each complex and the variation in clumpiness from complex to complex. Are the low-mass and massive cores described here distinctly different, or are they part of a continuous spectrum of core masses?

Gravitational Collapse

The case that dense cores form stars, while very strong, is still circumstantial. Detailed observational evidence for gravitational motions of star-forming gas has not yet been presented, for at least two reasons: (1) good candidate regions may still be lacking. The young stellar objects found at optical and near-infrared wavelengths may be too evolved, i.e., they may have completed their accretion; (2) instrumental sensitivity and resolution may still be inadequate. Detection of free-fall motion with speed greater than a few km s^{-1} onto a few-solar-mass star in Taurus requires arcsec resolution in a sufficiently strong emission line. However, each of these deficiencies may soon be reduced. The IRAS survey has significantly increased the number of young star candidates in nearby molecular cloud complexes, to 10^3–10^4 sources (Rowan-Robinson et al. 1984). New instruments of high spectral and spatial resolution are now operating or being developed, including the VLA at centimeter wavelengths; the Berkeley and California Institute of Technology (Caltech) arrays, and others being planned, at millimeter wavelengths; the United Kingdom/Netherlands, Caltech, and Arizona/Max-Planck-Institute for Radioastronomy groundbased submillimeter telescopes; and the NASA SIRTF (Space Infrared Telescope Facility) and LDR (Large Deployable Reflector) spaceborne submillimeter and infrared telescopes. These developments offer some basis for optimism that cores may be detected in the process of forming stars in the relatively near future.

Acknowledgments. I thank P. Thaddeus of the NASA Goddard Institute for Space Studies and I. I. Shapiro of the Harvard-Smithsonian Center for Astrophysics for their support, D. C. Black and M. S. Matthews for their editorial encouragement and patience; C. Barrett for excellent preparation of the manuscript; A. Goodman for assistance in analyzing data; numerous colleagues for preprints and discussions, and the National Academy of Sciences-National Research Council for a Research Associateship.

PHYSICAL CONDITIONS IN ISOLATED DARK GLOBULES

CHUN MING LEUNG
Rensselaer Polytechnic Institute

Isolated dark globules, also known as Bok globules or Barnard objects, are nearby dense interstellar clouds of gas and dust with no significant internal heat source. They appear visibly as well-defined patches of obscuration against the general background of stars and are isolated from bright nebulosities and H II regions. The current status of observational and theoretical research on the physical conditions in these objects is reviewed. In particular, the physics of the gas and dust components is discussed. Observations of both line and continuum radiation at optical, infrared and millimeter wavelengths are shown to provide important constraints on their physical structure, dynamical state and evolutionary status. While most globules appear to be gravitationally bound and in virial equilibrium, recent infrared and radio observations suggest that low-mass star formation may be taking place in some globules, a well-studied example being B335. Finally, suggestions for future theoretical and observational studies to determine their physical origin and their relationship to star formation are outlined.

I. HISTORICAL BACKGROUND

"It is not difficult to estimate that there must be about 25,000 globules near the central plane of our Galaxy. How did these originate? To be perfectly frank, I do not know how the isolated globules were formed. All I can say is that I am glad that the Good Lord put them there and that Edward Emerson Barnard looked systematically for them and prepared his very useful list of Barnard Objects."

Bart Bok (1977)

Since the eighteenth century, astronomers have noted the absence of stars, or the presence of dark patches, in certain regions of the sky. It was not until the turn of this century, however, that these dark regions were recognized as absorbing clouds of dust situated between the background star field and the observer, and were not merely voids in the general background of stars. A strong argument against these dark regions being holes in the sky was that many of them have well-defined edges. If these were merely holes their age must be of the order of 10^8 yr in which time the random motions of stars in the neighborhood would long since have eliminated the sharp boundaries, if not the hole itself. The study of external galaxies, with their dark lanes of obscuring dust, provided another strong argument for the existence of dark dust clouds in our own Galaxy.

With the advent of photography in astronomy, it became evident that in our Galaxy there are many dark regions of different sizes and some have complex structural features. In 1919, based on his own photographic survey of the Milky Way, Barnard published the first catalog of dark nebulae (Barnard 1919), and later extended the list to include a total of 349 objects (Barnard 1927). Since Barnard's pioneering work, several extensive catalogs and/or atlases of dark nebulae have appeared.

1. Lundmark and Melotte (Lundmark 1926) conducted a search for dark nebulae using the Franklin-Adams photographic survey of the whole sky. The Lundmark catalog, which gives the positions, sizes and descriptions of 1550 objects, is the only search to date which covers the entire sky.
2. Using both the Ross-Calvert photographic *Atlas of the Milky Way* (F. E. Ross and M. R. Calvert 1934) and Hayden's *Photographic Atlas of the Southern Milky Way* (1952), Khavtassi made a survey of dark nebulae and published it as a catalog (Khavtassi 1955) and as an atlas (Khavtassi 1960). His catalog lists 797 nebulae over the whole sky between latitudes $-20° \leq b \leq +20°$, together with their coordinates, areas and estimates of opacity.
3. Schoenberg (1964), using both the Ross-Calvert atlas (1934) and the Lick photographs of the Milky Way (Barnard 1913), compiled a list of 1456 dark nebulae, together with their coordinates, areas, shapes and estimates of opacity.
4. Lynds (1962) made a complete survey of dark nebulae in the northern hemisphere using the *National Geographic Society—Palomar Observatory Sky Survey.* In her catalog, the coordinates, surface area and visual estimate of the opacity (on a scale of 1 to 6) are tabulated for each of the 1802 clouds, most of which are condensations within the general obscuration of extensive dark cloud complexes.
5. Sim (1968), using the Palomar Observatory Sky Survey prints, made a detailed catalog of dark condensations observed in and near 66 OB associations and young star clusters. She found 63 certain and 63 probable

objects; their positions, shapes, orientations and angular sizes were tabulated.

6. Recently Feitzinger and Stuwe (1984), using photographic plates from the European Southern Observatory and the United Kingdom Schmidt Telescope, completed the first comprehensive catalog of dark nebulae and globules found in the southern sky. This survey is similar to the work of Lynds (1962) but for dark nebulae in the southern hemisphere. In the catalog, visual estimates of opacity (on a scale of 1 to 6) are tabulated for all dark clouds greater than 0.01 square degree between galactic longitudes 240° and 360°. A total of 437 clouds are found in the southern sky, covering 264 square degrees.

In general, the ability to detect a dark cloud not only depends on the limiting magnitude of the survey but also on the wavelength sensitivity of the photographic plate. Furthermore, clouds of greatest opacity increase in number with the resolving limit of the survey. Due to the presence of interstellar extinction, most of the dark nebulae identified (in visible wavelengths) to date are nearby clouds which lie within 1 kpc or so of the solar neighborhood. Most of these dark clouds lie along and are confined to within 20° of the galactic plane. Their angular sizes vary in extent from many degrees to less than one minute of arc, and the obscuration can vary by many magnitudes from one cloud to another as well as within the same cloud. The number of dark nebulae decreases rapidly with increasing apparent sizes, and small clouds ($\leq$ 1 square degree in surface area) constitute more than 70% of all known nebulae.

II. CLASSIFICATION OF GLOBULES BY MORPHOLOGY

The nearby dark nebulae may be roughly divided into large dark cloud complexes and globules. The large dust complexes are often several degrees in size and many have embedded in them significant numbers of low-luminosity stars. The ρ Ophiuchi cloud, many dark nebulae in Taurus, and the Southern Coalsack are the best-known examples of this class of dark clouds. With typical masses ranging from 10^3 to 10^4 $M_\odot$, dark cloud complexes may play an important role in the formation of low-mass stars. On the other hand, the term globules, first introduced by Bok and Reilly (1947), has been used to describe a variety of dense interstellar clouds of gas and dust, specifically those which appear visibly as well-defined patches of obscuration against a general background of stars, nebulosities or H II regions. On the basis of their optical morphology, four types of globules can be distinguished: (a) elephant-trunk and speck globules, (b) cometary globules, (c) globular filaments, and (d) isolated dark globules. Because these various types of globules may be related and may represent different evolutionary stages of small dark nebulae, below we describe in some detail their structure and possible formation mechanisms.

Fig. 1. Elephant-trunk and speck globules in the Rosette nebula (figure from Herbig 1974).

A. Elephant-trunk and Speck Globules

Elephant-trunk globules are long tongues of obscuring neutral gas and dust, often observed in projection against the luminous background of H II regions. The elephant trunks generally point toward the exciting star(s) in the H II region and are often surrounded by bright rims. They frequently appear in chains. Their sizes range between 0.01 and 0.5 pc. The larger ones have masses of 5 to 10 $M_\odot$ with densities on the order of 10^3 to 10^4 cm^{-3}. On the other hand, speck globules, which have shapes varying from teardrops to round and compact spheres, represent the smallest-unit clouds that we can observe directly. They are often seen in groups surrounding the elephant trunks. The physical association of speck globules and elephant trunks strongly suggests that the former may simply be fragments of the latter (Herbig 1974). Examples of elephant-trunk and speck globules can be found in M8, M16, IC1396 and NGC 2244 (the Rosette nebula). Note that not all H II regions have these objects associated with them. The Orion complex seems to be devoid of these features. Fig. 1 shows a picture of the Rosette nebula with the associated elephant-trunk and speck globules.

Recently, Schneps et al. (1980) studied the CO emission from the elephant-trunk globules in the Rosette nebula. They found that the globules appear to form a chain-like structure which is expanding away from the star cluster at the same supersonic speeds as the ionized gas, suggesting that the neutral globules along with the ionized gas are swept out by stellar winds from the central cavity. Furthermore, there is a radial velocity gradient along the lengths of the globules, increasing toward the star cluster. They interpreted this motion as the stretching action responsible for the elongated appearance of the globules. The time scale for stretching appears to be comparable to the dynamical (expansion) age of the nebula. The large globules also appear to be splintering into small teardrop like fragments (speck globules). These authors concluded that the globules are remnants of a swept-up stellar wind shell in an expanding H II region. The dense shell swept up by these winds can lead to elephant-trunk formation either through density inhomogeneities in the ambient medium or through a Rayleigh-Taylor instability. In the first process, studied by Pikelner and Sorockenko (1974), the globules form when the shell sweeps up a denser patch of gas in the medium outside the shell. The denser section of the shell, having a greater inertia, expands at a slower rate and is eventually left behind, producing an elongated structure which points toward the center of the sphere. However, this picture does not explain why the globules have roughly the same velocity gradient which implies a common formation time for all the elephant trunks. The second process is considerably more complex and works as follows. Stellar winds from an early-type star sweep out a cavity in the molecular cloud, driving the supersonic expansion of a thin dense shell. When a portion of the shell reaches the edge of the molecular cloud and continues to expand into a region of lower density, the lower-density medium offers less resistance against the expansion, causing this portion of the shell to accelerate outward. This acceleration, driven by the pressure inside the shell, is Rayleigh-Taylor unstable, i.e., when the low-density gas is unable to support the denser shell, the shell will burst. As the gas inside bubbles out through the shell, long tongues of dense shell material stretch inward, forming elephant-trunk globules. In this way an entire section of shell material becomes unstable at once and all the globules form at roughly the same time. The globules continue to grow as the remnants of the shell expand past the cloud boundary. Eventually the unstable shell adjusts to the lower-density environment and the shell once again decelerates as it continues to sweep up the low-density gas, halting further instability. Consequently, a new shell will form which expands at a rate different from that of the shell still contained within the molecular cloud. Thus a Rayleigh-Taylor instability in a stellar-wind shell could explain the presence of elephant-trunk globules. Once formed, the globules, which are subjected to the intense ionizing radiation of the cluster O stars, will form H II layers on their outer surfaces which are visible as bright rims. While the recombination in this layer shield the rest of the globule from total ionization, the globule may be slowly etched away by a flow of ionized gas which evaporates into the surrounding low-density medium. It appears that while fragmentation can occur in the large globules, evaporation is the dominant destruction mechanism for elephant-trunk globules because the evaporation rate, (5 to 10) $\times$ 10^{-5} $M_{\odot}$ yr^{-1}, is 5 to 10 times greater than the fragmentation rate. The stability against evaporation of globules embedded in an H II region was studied in some detail by Tenorio-Tagle (1977). Recently Sandford et al. (1982*a,* 1984) performed radation hydrodynamics calculations to study the formation and stability of dusty globules illuminated by an external radiation source. They demonstrated the importance of radiation-driven implosion to star formation in globules. Their results indicate that an optically thick globule can resist evaporation (through shielding by dust) and survive long enough for self gravity to act, resulting in the formation of either low-mass stars or long-lived, stable dark globules.

B. Cometary Globules

Cometary globules (CG's), which were first noted by Hawarden and Brand (1976), are comet like clouds with a head-tail appearance. A typical CG has a compact, dusty, optically-opaque head which sometimes contains embedded stars. Surrounded on one side of the head is a narrow bright rim while extended from the other side is a long, faintly-luminous tail. On the basis of optical measurements, the length of the tail varies from 0.2 pc to over 7 pc while the size of the head ranges from 0.1 pc up to 1 pc. The average ratio of head width to tail length is 0.3 ± 0.2. Typical mass of a CG varies from $< 1\ M_\odot$ to several $M_\odot$. The characteristic gas densities of CG's are similar to those of elephant-trunk globules. Some CG's have several heads with one combined tail. CG's are generally isolated and not physically associated with any neighboring nebulosities. Fig. 2 shows an example of a cometary globule found in the Gum-Vela region.

Zealey et al. (1983) have tabulated the positions, orientations and angular tail lengths of most of the known CG's. Of the 38 known, 29 are associated with the Gum-Vela nebula which is a large region of ionized gas with an angular radius of $\sim 18°$ (at its estimated distance of 400 pc, this corresponds to a physical radius of 125 pc). The tails of all the CG's in the Gum-Vela region appear to be directed from a common central region in the nebula where the young Vela pulsar, the highly energetic O4If star ζ Puppis, and the binary star (WC8+O9) γ^2 Velorum are located. In addition, there are at least 13 stars of type B1.5 or earlier in the region. Several suggestions have been made about the nature of the Gum-Vela nebula: a fossil sphere of ionized hydrogen gas, an evolved H II region, an old supernova remnant or a stellar wind bubble. The formation mechanism for CG's in the Gum-Vela region is unknown. Two different scenarios have been proposed to describe the formation of CG's in general and the cometary globule (CG1) in the Gum-Vela region in particular. Brand (1981) and Brand et al. (1983), who considered the association of a young variable star (Bernes 135) with CGI, argued that CG's are shocked clouds, formed when a blast wave from a supernova remnant collided with an initially spherical cloud. The implosion produced a forward-moving shock and a reverse shock (a rarefaction wave) within the cloud. These together resulted in gravitationally unstable material in a dense clump (the head) and downstream-ejected cloud material (the tail) which is considerably denser than the intercloud medium. A single star or group of stars may be formed in the head. In the case of CG1, the illuminating star, which was formed in the

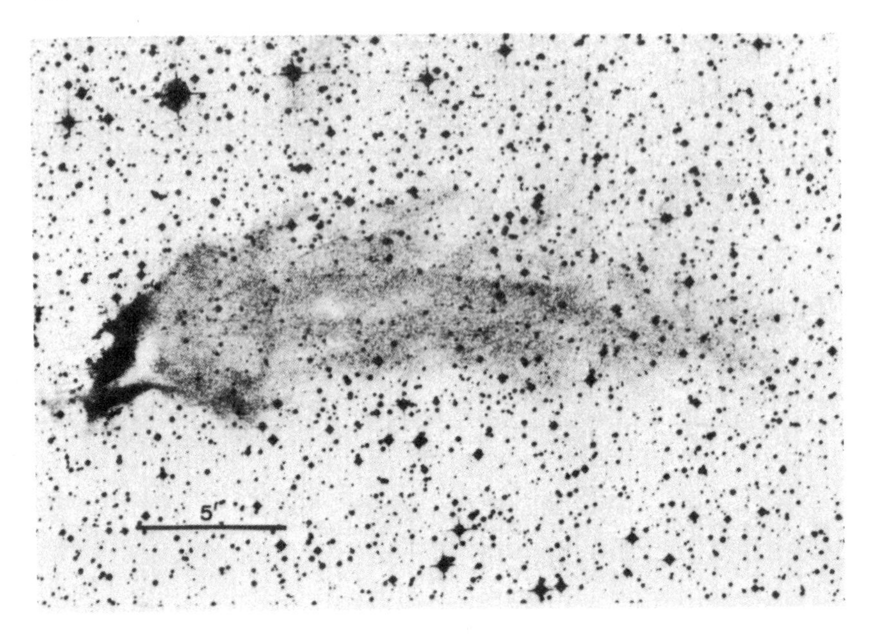

Fig. 2. A cometary globule in the Gum-Vela nebula (figure from Reipurth 1983).

head after the implosion, has since drifted to the front edge. This picture assumes that many spatially distinct clouds were originally present before the supernova explosion and that each cloud would give rise to a cometary globule. It also suggests that CG's may be the most probable configuration for interstellar clouds whose dynamics are dominated by supernova explosions.

The second scenario, described by Reipurth (1983), suggested that when a massive O star like ζ Puppis is ignited in the neighborhood of small clumpy molecular clouds, the ultraviolet radiation impinging on a neutral cloud creates an ionization front which is seen as a bright-rim structure around the dense core. Subsequent evaporation by the radiation and disruption by the associated shocks separate the less dense material from the cloud cores. Thus the exposed and compressed cloud cores will for a time show bright rims and faintly luminous tails (which are basically eroded material from the core) in their shadow regions. After the disappearance of the short-lived O star, clouds that have not evaporated or been disrupted (by internal star formation due to external compression) will survive for some time as isolated dark globules (discussed below). Low-mass star formation may also occur in and around CG's at various stages of this process. The recent calculations at Sandford et al. (1982*a*, 1984) tend to support the plausibility of this scenario.

C. Globular Filaments

Globular filaments are dark nebulae which have elongated and filamentary structure. They often show signs of internal fragmentation or condensa-

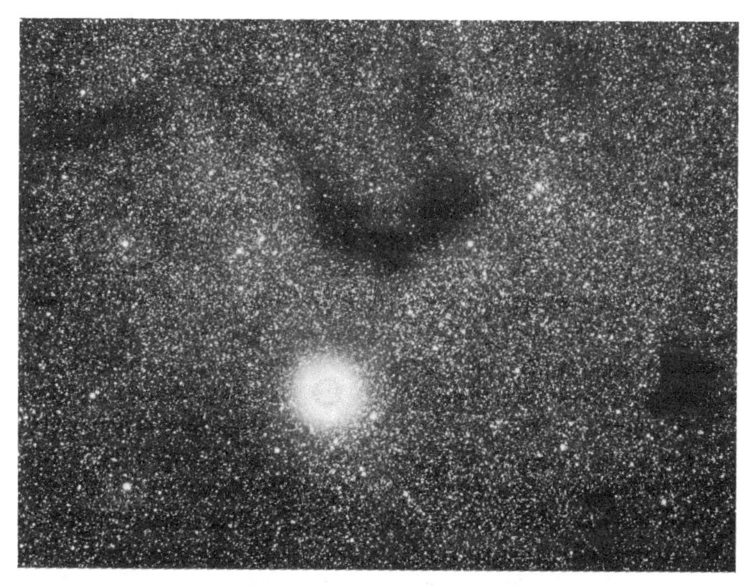

Fig. 3. An example of a globular filament, the Snake (L66 or B72) (figure from Bok 1977).

tion into small, discrete, globule-like beads. The angular length varies from 0°.5 to 9° and the angular width ranges from 2′ to 40′. Examples are those found in the Taurus cloud complex. Fig. 3 shows an example of a globular filament, the Snake (L66 or B72), which is located near the Ophiuchus cloud. Using the *Palomar Observatory Sky Survey,* Schneider and Elmegreen (1979) compiled a list of 23 globular filaments. For each filament, they gave the position, angular size and description of the hierarchies of condensations.

One characteristic feature observed among all globular filaments is the regularity of their segmentation; the average ratio of fragment separation S to fragment diameter D is $S/D \sim 3 \pm 1$. The uniformity of this ratio within any one filament and for different filaments strongly suggests that there is a characteristic length for fragmentation from an initially uniform filament. Another characteristic feature is the appearance of externally applied forces in a transverse direction (to the length of the filament) giving these clouds a windswept or asymmetric appearance. Further evidence is shown by a star-count differential across the filament or the presence of sharper edges along one side of the filament. As proposed by Schneider and Elmegreen (1979), the overall structure and asymmetric appearance of globular filaments would be best explained if an initially plane-parallel accumulation of interstellar gas first fragmented into parallel filaments which align along the direction of the interstellar magnetic field. Some external forces then swept up matter perpendicular to the filaments, giving rise to a windswept appearance. This was followed by a gravitationally-induced condensation into individual globules along the fila-

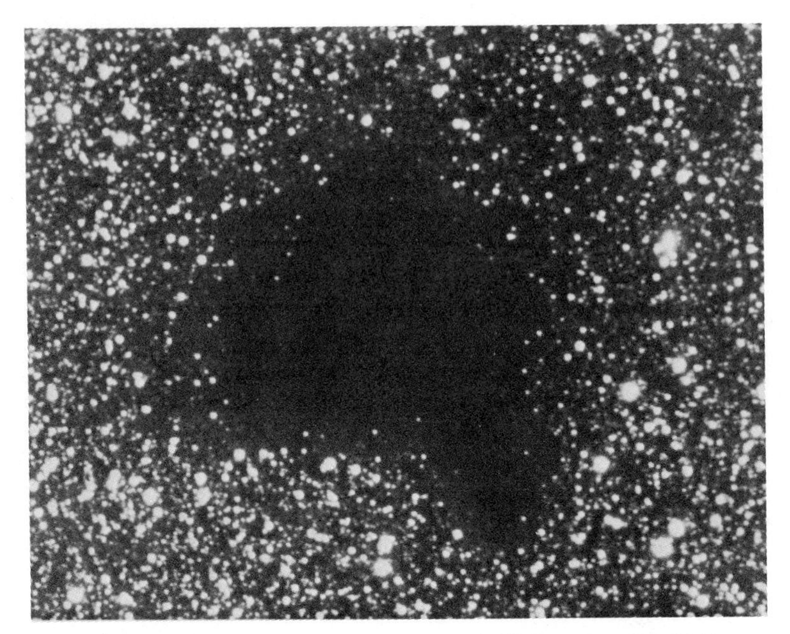

Fig. 4. An example of an isolated dark globule, B68 (figure from Bok 1977).

ment, with the internal fragmentation proceeding by hierarchies, where fragments contain subfragments. According to this picture, globular filaments may be an evolutionary link between unfragmented filaments like B44 in Ophiuchus and the commonly studied, isolated globules like B68.

D. Isolated Dark Globules

These are dense clouds which display strikingly simple structure (usually round) with relatively sharp boundaries. They are also known as Bok globules or Barnard objects. They are isolated from bright nebulosities and H II regions and often appear visibly as well-defined patches of obscuration against the general background of stars. Some examples of isolated globules are B68, B227, B335 and B361. Fig. 4 is a picture of an isolated dark globule, B68. While isolated dark globules were first studied by Barnard (1919), it was Bok and Reilly (1947) who first called attention to their possible role in star formation. Subsequently, Bok and his coworkers have cataloged a number of these globules and deduced some of their properties from optical observations (see, e.g., Bok et al. 1971; Bok and Cordwell 1973; Bok and McCarthy 1974; Bok 1977, 1978).

It is important to point out that while isolated globules have physical conditions and chemical composition similar in many ways to dark clouds, they differ in that dark clouds generally do not appear as isolated clouds but are usually associated with or physically attached to other dark clouds or dark cloud complexes, as noted by Snell (1981). The dark clouds and dark cloud complexes are probably condensations in extended, low-density, atomic-hydrogen clouds. The first comprehensive review on dark clouds (of which globules form a subset) was given by Heiles (1971). Recent reviews on isolated dark globules include those (at a popular level) by Dickman (1977), Reipurth (1984), and (at a technical level) Hyland (1981). In this review we concentrate on discussion of their physical conditions.

As a result of their simple environment and geometry, the range of physical processes which can occur in these objects is likely to be fairly limited, and one expects only minor variations among globules. This means that even a relatively limited study of dark globules may yield a substantial insight into their general characteristics. Furthermore, their simple geometry and environment also make them excellent candidates for theoretical studies and detailed modeling. This is in marked contrast to the complexity and diversity among giant molecular clouds with internal heat sources. Table I gives a summary of the typical physical parameters, both observed and derived, of isolated dark globules. The values for the parameters were compiled from a variety of sources in the literature, as discussed in detail throughout the rest of this chapter. To facilitate later discussions on the physical conditions inside these objects, we present in Fig. 5 a schematic overview of the physical processes involved in an isolated dark globule without significant internal heat source. Continuum radiation from the ambient interstellar radiation field (ISRF) is absorbed mostly by dust grains (processes 2 and 4 in the figure) which then

TABLE I
Physical Parameters of Isolated Dark Globules

Observed quantities		
Visible angular size	$\theta(')$	3–30
Dust extinction	A_v(mag)	1–25
CO line width	ΔV(km s^{-1})	1–3
CO intensity	T_R^*(K)	6–15
Far-infrared flux	F_ν(Jy)	10–100
Distance	D(pc)	150–600
Location	$\|b\|$	$<15°$
Molecular species		CO, CS, H_2CO, NH_3
Examples		B227, B335
Derived quantities		
Radius	R(pc)	0.1–2.0
Gas density	n_{H_2}(cm^{-3})	10^3–10^4
Gas temperature	T_k(K)	8–20
Dust temperature	T_d(K)	10–20
Mass	$M(M_\odot)$	5–500
Jean's mass	$M_J(M_\odot)$	2–20
Jean's length	L_J(pc)	0.2–1.0
Free-fall time	t_{ff}(yr)	$\lesssim 10^6$

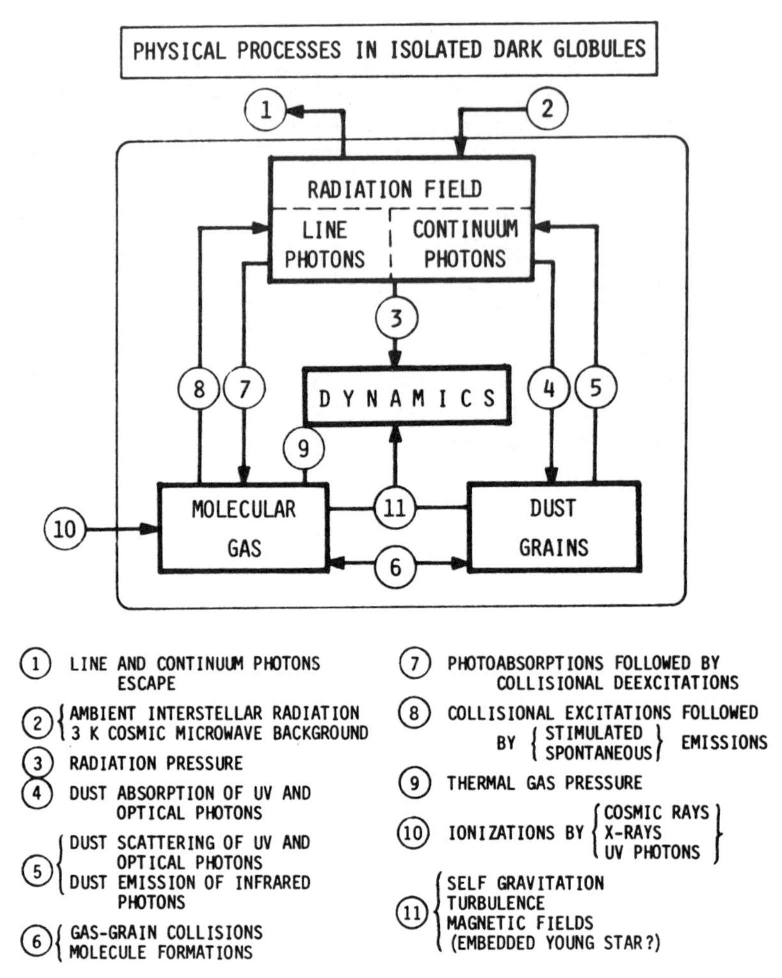

Fig. 5. Schematic diagram showing the major physical processes which can occur in isolated dark globules.

emit infrared radiation (5). The temperature of the dust grains is established by the thermal balance between heating (absorption of the ISRF) and cooling (infrared emission). Because only a small part of the thermal energy stored in the dust grains is transmitted to the gas as kinetic energy through collisions (6), the dust temperature determination is nearly independent of gas-dust interactions, except when the gas density exceeds 10^6 cm^{-3}. The molecular gas is heated mostly by cosmic ray protons (10) and is cooled (through emission of line protons) when the net energy absorbed in collisional excitations with trace species (8) exceeds the net energy delivered in collisional deexcitations (7). The velocity structure and dynamical state of the cloud are governed by the interactions among gravitation (11), thermal gas pressure (9), radiation pressure (3), and other sources such as turbulence and magnetic fields. For globules with embedded protostars, their dynamics may also be affected. In Secs. III and IV we discuss in more detail the physics of the dust component and observations from which the physical conditions in isolated dark globules are determined.

III. PHYSICS OF THE DUST COMPONENT

A. Basic Physical Concepts

Before discussing the observations pertaining to the dust component in dark globules, it is useful to review the basic physical processes which are involved in the transport of radiation through a medium of dust particles. When radiation passes through a region of space containing dust grains, the photons can be either absorbed, with conversion of radiation into heat, or scattered, with a change in the photon's direction of travel. The probabilities of absorbing and scattering photons are given, respectively, by the absorption and scattering cross sections, the sum of which is called the extinction cross section. The ratio of the scattering cross section to the total extinction cross section is called the albedo. These cross sections depend on the composition, shape and size of the dust particles, as well as on the wavelength of the photon. In particular, they depend on the ratio of the particle size to the photon wavelength. In general, these cross sections are a complicated function of wavelength and must be calculated from the so-called Mie theory of scattering. One of the outstanding problems in astronomy is to determine the chemical composition, shape and size of interstellar grains. The current view is that they may be particles of graphite, silicate or ice, or a combination of these. Typical grain sizes are on the order of 0.1 μm, which is determined from the wavelength dependence of interstellar extinction. For these grain types and size, both the extinction cross section and the albedo decrease as the wavelength increases. In particular, the albedo approaches unity in the ultraviolet and visible, while in the infrared, there is very little scattering and the albedo essentially goes to zero.

An interstellar grain is heated by the absorption of starlight in the ultraviolet and visible where the absorption cross section is large. On the other hand, it is cooled when energy is radiated away, usually in the infrared. In a steady state, the temperature of a single interstellar grain is determined by the energy balance between the absorption of dilute radiation from stars and the reemission of photons in the infrared. The temperature of most interstellar grains varies from 15 to 30 K, depending on the grain material. In the presence of many grains, such as inside an isolated dark globule, the equilibrium grain temperature is determined not only by the absorption and scattering properties of the grain and the spectral distribution of the incident radiation field, but also by the density distribution of the dust grains within the cloud as well as the total amount of dust present, which is essentially measured by the total dust opacity of the cloud. For optically thin clouds and in the outer parts of opaque clouds, the heating is mainly by the attenuated and scattered components of starlight in the ultraviolet and visible. In the interior of moderately opaque clouds, where very few of the heating photons in the ultraviolet and visible can penetrate, the grains are heated mainly either by the near-infrared component of starlight or by the thermal radiation emitted by grains in regions near the heat source. Because the absorption cross section is much smaller in the infrared than in the visible, the result of grain shielding is that as one moves toward the cloud interior, i.e., as the dust column density increases, the grain temperature will decrease. This means that the higher the opacity, the lower the temperature is. In general, both the emergent intensity and spectrum in the infrared essentially depend on the distribution of grain temperature inside the cloud.

The thermal energy stored in dust grains can also be transmitted to the gas as kinetic energy through collisions. Depending on whether the dust grains are hotter or colder than the gas kinetic temperature, they may be either cooled or heated by the gas. However, this thermal coupling between the gas and dust components is important only when the gas density exceeds 10^6 cm^{-3}. Hence, except in the cloud core where the gas density is high, the temperature distribution of dust grains is mostly determined by the ambient ISRF.

B. Interstellar Radiation Field

The ambient ISRF incident on a dark globule depends on the height of the cloud above the galactic plane and on its proximity to hot stars. Recently Mathis et al. (1983) modeled in detail the ISRF in the solar neighborhood and its variation with galactocentric distance, taking into account the latest observational results and their interpretations. In general, the ISRF can be divided into three distinct spectral components.

1. Ultraviolet, visible and near-infrared ($0.0912 \leq \lambda \leq 8$ μm). Radiation in this region peaks at ~ 1 μm and is dominated by stellar radiation from early-type (O/B) stars, disk (A/F) stars and late-type (M giant) stars.
2. Middle and far-infrared ($8 \leq \lambda \leq 500$ μm). Radiation in this region peaks at ~ 100 μm and is dominated by thermal radiation from interstellar dust. About 80% of this radiation comes from dust embedded in extended low-density H II regions and from dust in diffuse intercloud gas. The remaining 20% comes from dust associated with compact H II regions. The contribution from dust in quiescent molecular clouds (i.e., clouds with no significant internal source of heating) is negligible.
3. Submillimeter and millimeter ($\lambda > 500$ μm). Radiation in this region is dominated by the 2.8 K cosmic microwave background which peaks at ~ 1000 μm.

In the solar neighborhood the total energy density of the first two components corresponds to that of a blackbody at $T \simeq 3.2$ K, of which 80% comes from the first component.

The ultraviolet component ($0.0912 \leq 0.25$ μm) of the ISRF is dominated by emission from early-type stars and starlight scattered by interstellar dust (diffuse galactic light). This ultraviolet radiation is important in the energetics of dark globules. It affects the thermal structure of the gas (e.g., heating of the gas through ionization and photoelectric emission from dust grains) and the dust grains. It also affects the chemical composition of the gas and the lifetime of atomic and molecular species through photoionization and photodissociation. By shielding the interior of the cloud from the ambient ultraviolet radiation, dust grains create an environment in the cloud interior which favors low temperatures and molecule formation. The problem of ultraviolet radiation transport in dust in dark globules has been solved in different ways by a number of authors (see, e.g., Sandell and Mattila 1975; Leung 1975*a*; Whitworth 1976; Bernes and Sandqvist 1977; Sandell 1978; Flannery et al. 1980). The radiation density depends sensitively on the grain albedo and the asymmetry parameter which describes the forward-scattering phase function. It is found that with increasing optical depth, the molecular lifetimes become very sensitive to the degree of anisotropic scattering. For forward scattering, photodissociation is a significant destruction mechanism for molecules deep in molecular clouds.

Before extensive observations of the interstellar ultraviolet radiation became available, several authors have made quantitative estimates of this radiation (see, e.g., Habing 1968; Witt and Johnson 1973; Jura 1974; Gondhalekar and Wilson 1975; Henry 1977). These computations depend on the form of the particular galactic model employed and on the choice of model parameters. The intensities predicted by these authors are typically within factors of 2 to 3 of each other. Recently observations of the galactic ultraviolet radiation have been made by a number of authors (see, e.g., Kurt and Sunyaev 1968; Hayakawa et al. 1969; Henry 1973; Henry et al. 1974, 1977; Lillie and Witt 1976; R. C. Anderson et al. 1977; Morgan et al. 1978; Paresce et al. 1979; Pitz et al. 1979; Gondhalekar et al. 1980). These observations, which were taken over different regions of the sky and cover different wavelength ranges, indicate significant variation of the intensity with both galactic latitude and longitude. These variations should be taken into account when comparing observations taken over different regions of the sky. The uncertainties in the observations are somewhat higher compared to those in the theoretical calculations.

C. Observational and Theoretical Studies

There are essentially three observational methods for determining the physical properties (e.g., grain properties, opacity, density distribution and temperature) of the dust component in dark globules: (1) measurement of surface brightness profiles in the visible, (2) observations in the visible or near-infrared of background field stars behind a globule, and (3) observation in the far-infrared of thermal emission by dust in globules. Hildebrand (1983) has given a tutorial review on the physical principles by which dust masses and total masses of interstellar clouds and certain characteristics of interstellar dust grains can be derived from observations of far-infrared and submillimeter thermal emission. Below we briefly discuss the observational results and associated theoretical studies of these three approaches.

Observational results indicate that the visible surface brightness of many isolated dark globules is not uniform. They exhibit dark cores and rims which appear to be luminous, although they are still opaque to any background starlight. Pioneer work in this field was done by Struve and Elvey (1936). It was recognized by Struve (1937) that this limb-brightening effect implies the presence of strongly forward-scattering dust particles within the globules. Lynds (1967,1968), who reviewed this subject, called these "bright dark nebulae." The luminous rim is produced by anisotropic dust scattering of starlight and diffuse galactic light. Monte Carlo simulations of this radiation scattering problem have been considered by Mattila (1970) and by Witt and Stephens (1974). They determined the surface brightness distribution of clouds having a range of albedo, scattering asymmetry and optical depth. They found that the surface brightness profiles of dark globules with large optical depth depend on the density distribution of the cloud as well as on the scattering properties of the grains. Dust grains with albedo between 0.2 and 0.7 and scattering asymmetry parameter of 0.7 to 0.8 are implied. They also found that for clouds with centrally condensed density distribution, the relative bright rim thickness to cloud radius is a useful probe of maximum cloud opacity. Thus the analysis of surface brightness profiles can be used to determine the mass distribution and central densities of dark globules.

Various star-count techniques have been used by Bok and his coworkers in the optical study of isolated globules (see Bok and Cordwell 1973, for a review) to determine their dust opacity. Schmidt (1975) applied these techniques to determine the dust opacity and density distribution in globule B361. Tomita et al. (1979) examined by means of star-count techniques the dust density distributions in 14 globules. They found that the projected radial density distributions can be approximated by power laws with exponents varying from -3 to -5. Dickman (1978) also used star-counting techniques to determine the dust extinction in a number of globules. More recently, I. P. Williams and H. C. Bhatt (1982) considered the formation of centrally condensed dust distributions in isolated globules, taking into account the effects of both gravitation and radiation pressure.

Because the extinction due to dust is much lower at wavelengths of ~ 1 μm than in the visible, near-infrared surveys provide an extremely useful probe of globules. T. J. Jones et al. (1984*a*) conducted near-infrared photometry, spectroscopy and polarimetry of background field stars behind globule 2 in the Southern Coalsack. In addition to determining the dust density distribution, they found that there are definitely aligned grains and that there is weak evidence for structure in the polarization map indicative of a mildly compressed magnetic field.

Keene (1981) has presented observations of far-infrared emission from nine globules. The observed intensity and the remarkable uniformity of the emission argue against the presence of internal heat source and confirm that the dominant heat source for the grains is the ISRF. The far-infrared intensities of those globules are roughly equal: the brightest and faintest differ by only a factor of ~ 5. If the heat source is the ISRF, the observed dispersion can easily be explained by differences in cloud geometries or by local variation on the ISRF. The observed spectra, which peak at ~ 225 μm, are consistent with optically thin thermal emission from dust with temperatures 13 to 16 K (slightly higher but comparable to the gas temperatures derived from molecular-line observations). The far-infrared spectra also imply that the emissivity of the grains must fall at least as fast as λ^{-2} from 500 μm to 1 mm. In general, the observed spectra are in reasonable agreement with the theoretical predictions of Spencer and Leung (1978).

An interesting discrepency appears to exist between the observation and theoretical prediction. Spencer and Leung (1978) predicted that for globules of moderate extinction ($A_v \geq 10$) and without an internal heat source, the thermal structure is similar to that of a cold core surrounded by a warm envelope so that significant infrared limb-brightening occurs at wavelengths of grain emission ($20 \leq \lambda \leq 600$ μm). The infrared limb-brightening is most pronounced in uniform density models. On the other hand, the observations of Keene (1981) indicated that the surface brightness in the far-infrared for the two globules, B361 and B335, which were mapped is either fairly uniform (in B361) or shows limb-darkening (in B335). The discrepency between the observation and the model could be explained (1) by the presence of an infrared component in the ISRF which can penetrate the globule interior and heats the entire cloud to produce a more uniform dust temperature distribution; (2) by the presence of a centrally condensed density distribution; or (3) by the presence of an embedded infrared source. Indeed B361 is known to have a centrally condensed dust distribution with a steep density gradient (Schmidt 1975) while B335 was later found to have an embedded far-infrared source.

To summarize, on the basis of optical and infrared observations as well as theoretical studies, the dust density distribution in isolated globules is centrally condensed and the dust temperature (about 15 K) increases radially outward from the cloud core. The dust particles exhibit strongly forward-scattering properties and have a fairly high albedo in the visible (between 0.2 and 0.7). The far-infrared emissivity of the dust appears to fall roughly as λ^{-2}. There is also some evidence for aligned grains in isolated globules, which is indicative of a slightly compressed magnetic field.

IV. PHYSICS OF THE GAS COMPONENT

A. Density Structure

The column density and space density of H_2 gas in isolated dark globules are generally inferred indirectly from observations of molecular lines of the trace species. Martin and Barrett (1978) discussed in some detail the standard procedure for such determinations, with emphasis on applications to isolated dark globules. In general transitions with widely different thermalization properties will be formed in different density regions of a cloud (Leung 1978). Hence by observing a variety of molecular lines, one can infer the density and temperature in different parts of a cloud. In particular, observations of CO and ^{13}CO are useful for the study of the cloud envelope where the gas density is not high. On the other hand, transitions of CS, HCN and H_2CO will serve as useful density probes for the cloud core where the gas density is high. For example, Snell et al. (1982) have determined the density structure in several dark clouds and globules from CS observations. In an effort to study the detailed gas density distribution in dark clouds and globules, Fulkerson and Clark (1984) have constructed models for the observed surface brightness of selected transitions of H_2CO for several clouds. Comparison with observations indicates that the gas is centrally condensed with a density gradient which closely approximates an inverse square law. A similar conclusion has been drawn by Snell (1981) in his studies of several dark clouds. In general, these studies indicate that most isolated dark globules have a core-envelope density structure. The physical significance of an inverse square density law in cloud cores has been investigated by Terebey et al. (1984*a*). It appears that the relative gas density distribution in a globule can be determined simply from the intensity contours of either optically thin transitions or optically thick transitions which are formed only in high-density regions. Hence molecular transitions from CS, H_2CO, and $C^{18}O$ can be used as tracers for the gas density distribution in a globule as demonstrated by the modeling results shown in Fig. 6. Frerking and Langer (1984) have determined the density distribution in the cloud core of B335 from observations of $C^{18}O$.

While observations of molecular species are useful in probing the inner regions of dark globules, observations of the 21-cm line of atomic hydrogen can provide information on the physical conditions in the outer envelope. Observations (see, e.g., Arnal and Gergely 1977; Bowers et al. 1980; Levinson and Brown 1980) of H I self-absorption lines indicate that the outer envelope of a globule is likely to be made up of atomic hydrogen. While the gas density is lower in the envelope, the gas temperature appears to be higher. The velocity dispersion also seems to be larger.

B. Energetics and Temperature Structure

Observationally the gas kinetic temperature is usually determined from CO line transitions. The standard derivation assumes that the $J = 1 - 0$

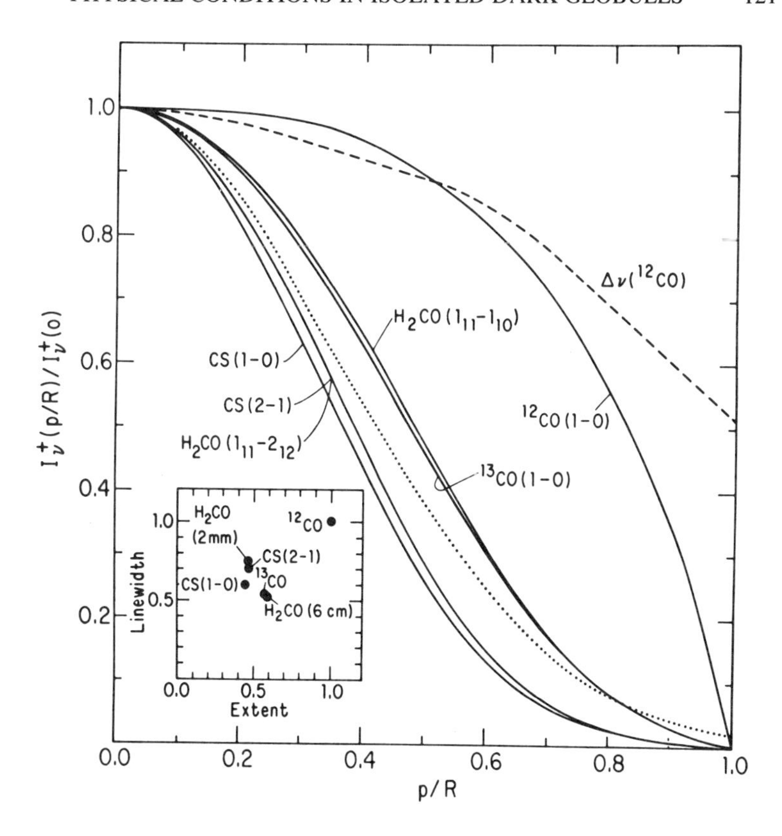

Fig. 6. The spatial variation across the cloud surface of the normalized emergent intensities of the observed transition of CO, ^{13}CO, CS and H_2CO. Also plotted are the variation of the linewidth of CO (dashed line) and the gas density distribution (dotted line). The intensity profiles of either optically thin transitions (^{13}CO) or optically thick transitions with long thermalization lengths (CS, H_2CO) follow closely the gas density distribution. The insert shows the relationship between predicted linewidths and spatial extents, as determined by the half-maximum intensity contours. All quantities are normalized to those of CO. The results are from a cloud model in which the gas temperature, relative abundance of molecules and the turbulent velocity are all uniform.

transition of CO is optically thick and thermalized so that the observed intensity, which then measures the excitation temperature of the transition, is equal to the gas kinetic temperature of the cloud. An upper limit to the gas temperature can also be estimated by measuring the rotational temperature of the hyperfine transitions of NH_3 (see Martin and Barrett [1978] for a discussion). Results indicate that isolated dark globules have a rather narrow range in gas temperature, i.e., from 8 to 15 K.

Theoretically the temperature distribution in the gas component is determined by the thermal balance between various heating and cooling processes. The gas cools by the emission of line photons. In general, the gas cooling rate is determined by the difference between the kinetic energy loss through collisional excitations and the kinetic energy gain through collisional deexcitations. For isothermal and homogeneous clouds, the cooling rate decreases toward the cloud interior where radiation trapping is most significant. Because cooling results from a binary process (collisions), we expect the cooling rate to depend quadratically on the gas density under typical conditions in molecular clouds. In extreme cases of very high densities and optical depth, the cooling rate is roughly independent of density, because most of the cooling lines are saturated and thermalized. The cooling rate increases with temperature because more cooling species are excited collisionally to higher levels. Because photons from higher transitions not only carry more energy but also are created more frequently, the net cooling rate is increased. In general, the line transitions contributing most to the total cooling rate are those in which the energy for excitation is comparable to the thermal energy of the gas.

Various heating mechanisms have been proposed (cf. Goldsmith and Langer 1978) for supplying energy to the gas in molecular clouds. Below we list the more important ones:

1. Cosmic ray ionization;
2. Gravitational contraction;
3. Dissipation of turbulence;
4. Photoionization of trace elements;
5. Photoelectrons from dust grains;
6. H_2 formation on grain surfaces;
7. Photodissociation of H_2;
8. Energy exchange between gas and grains through collisions;
9. Magnetic ion-neutral slip.

In the interior of dark globules where ultraviolet radiation cannot penetrate, the dominant heating mechanism is likely to be cosmic ray ionization. Gravitational contraction and dissipation of turbulence may also be important while the other mechanisms are operative in the outer regions of a cloud. Goldsmith and Langer (1978) analysed in detail the cooling produced by line emission from a variety of molecular and atomic species, including those observed as well as those theoretically expected in dense molecular clouds. They have also evaluated the contribution of various heating mechanisms. Under the physical conditions in dark globules, cosmic ray heating alone may be sufficient to balance the gas cooling by CO line emission. CO is the dominant coolant simply because it is the most abundant molecule next to molecular hydrogen. They found that the gas temperature throughout most of the volume of a dark globule would lie in the range 8 to 12 K if cosmic rays are the only source of heating. This result, which is relatively independent of density, is in remarkable agreement with the nearly constant temperature measured in dark globules. For example, Dickman (1975) conducted a CO survey of 68 dark

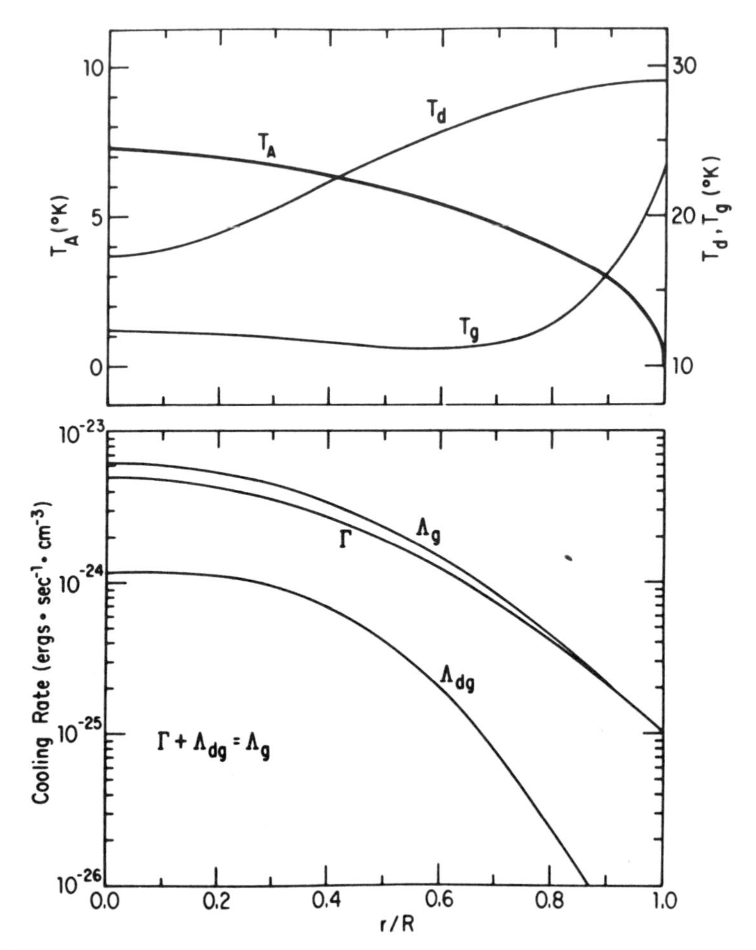

Fig. 7. Effect of density gradient on the gas temperature distribution in a globule. The upper panel shows the gas and dust temperature distributions (T_g, T_d). Also plotted is the variation across the cloud surface of the emergent antenna temperature (T_A) for the $J = 1 - 0$ transition of CO. T_A falls off toward the cloud edge despite the increase of T_g toward the surface, i.e., surface heating. The lower panel compares the different cooling and heating rates for the gas and grain components. Surface heating occurs even though the heating rates increase toward the center (figure from Leung 1978).

clouds from the Lynds (1962) catalog. He found that the peak CO emissions are remarkably uniform. Because the CO transition is optically thick and thermalized, the constancy of the peak CO emission indicates a rough equality of gas kinetic temperature in the sources as well.

The temperature distribution in a globule depends sensitively on the presence of density gradients and the density dependence of the heating and cooling rates. Leung (1978) modeled the thermal structure of dark globules by considering the following processes: heating by cosmic rays, cooling by CO and thermal coupling between gas and grains through collisions. He found that even though the heating rate per unit volume is higher at the cloud center, the equilibrium gas temperature decreases toward the interior because of a density-gradient effect (see Fig. 7). Other effects, such as heating by various photolytic processes at the cloud surface, may also contribute to a higher gas temperature toward the cloud boundary. Analysis of CO observations of the globule B5 by Young et al. (1982) indicates that the gas temperature appears to increase toward the cloud edges. Observations (see, e.g., Bowers et al. 1980; Levinson and Brown 1980) of H I self-absorption lines also indicate that the outer envelope of a globule is likely to be made up of atomic hydrogen at a higher temperature.

C. Internal Velocity Structure

One of the most important questions about dark globules is their velocity structure which governs their evolutionary process. In particular, because on the basis of simple analysis most isolated globules should be gravitationally unstable. (cf. Table I; Field 1978), it is crucial to determine the nature of the velocity field (e.g., turbulence, expansion, contraction and rotation) in dark globules. It is well known that while the temperature and density of these objects may be inferred from molecular line intensities, the velocity structure manifests itself in the details of the line profiles. Indeed CO line profiles can provide quite specific and decipherable information on the structure and dynamics of the emitting region. A detailed discussion of how one can extract information on the velocity structure from various properties of the line profile is given by Leung (1978). In Table II we summarize the various types of observations of the gas component in dark globules and their interpretations, with particular emphasis on the velocity field.

In an attempt to study the internal velocity structure and dynamical state of dark globules, Leung et al. (1982) have obtained high-resolution CO observations of over a dozen isolated dark globules. Their results indicate that a power law correlation exists between the internal velocity dispersion and cloud size (approximately of the form $\Delta v \propto R^{\frac{1}{2}}$). The power law relation is consistent with the presence of supersonic turbulence which is similar to that found earlier by R. B. Larson (1981) in other diffuse clouds and molecular cloud complexes. This strongly suggests that there may be an evolutionary link between globules and other molecular clouds and that the observed motions are all part of a common hierarchy of interstellar turbulent motions.[a] R. B. Larson (1981) attributed this power law correlation to a turbulent energy

[a]In hydrodynamics, turbulence is a well-defined phenomenon and is a state of motion with high Reynolds numbers, in which the inertial terms in the equations of motion dominate the viscous dissipation. On the other hand, while the term turbulence in radiative transfer and here could be identical with hydrodynamic turbulence, it could also be a field of shock waves, or even a velocity field of more regular structure. Here the term refers to motions having neither a thermal nor systematic character, but which must be present to account for the observed line width.

TABLE II
Studies of the Gas Component

Observational Results	Implications/Interpretations
Density Structure	
21-cm self-absorption lines	Presence of atomic hydrogen envelope
Varying spatial extent of emission of different molecular species	Centrally condensed density distribution
Temperature Structure	
Uniform CO intensities among many different sources	Insignificant or no internal heat source within globule; similar physical environment
Velocity Structure	
Average gas density correlates with cloud size	Role of gravitational contraction
Velocity dispersion correlates with cloud size	Velocity field dominated by random mass motions or "turbulence"
Linewidth increases with spatial extent of emission for different molecular species	"Turbulent" velocity decreases toward cloud interior; presence of density gradient
General lack of self-absorption profiles in CO	Finite "turbulence" scales in outer cloud regions
Self-absorption profiles and smaller linewidths for lines formed in dense regions	Subsonic "turbulent" velocity in quiescent cloud core
Asymmetric line profiles in optically thick lines; symmetric profiles in optically thin lines	Presence of systematic motions due to rotation or contraction
Systematic shifts in radial velocity across cloud surface	Presence of cloud rotation
Distinct spatial distribution of red-shifted and blue-shifted line wing intensities	Bipolar flow structure of embedded energy source

cascade phenomenon in which large-scale motions decay as a result of various instabilities into motions of smaller and smaller scales which thus derive their energy from the cascade. This leads to a systematic increase of velocity dispersion with length scale or cloud size. On the other hand, the observed power law correlation could arise from the tendency of clouds to evolve toward virial equilibrium ($\Delta v \propto n^{\frac{1}{2}}R$) and from the tendency of clouds to obey $n \propto R^{-1}$ where n is the gas density. The difficulty with the latter explanation is that one must explain the origin of the density-size relation from cloud to cloud and within clouds (see chapter by Myers; Myers 1983). Recently, Canto et al. (1984) have attributed the power law correlation to the tendency of clouds to evolve toward virial equilibrium in which the dissipation of random mass motions is balanced by radiative cooling. This scenario suggests that the dissipation of chaotic bulk motions is the dominant heating mechanism for dark globules. Finally, Henriksen and Turner (1984) attributed the power law correlation to the existence of supersonic turbulence that is gravitationally driven

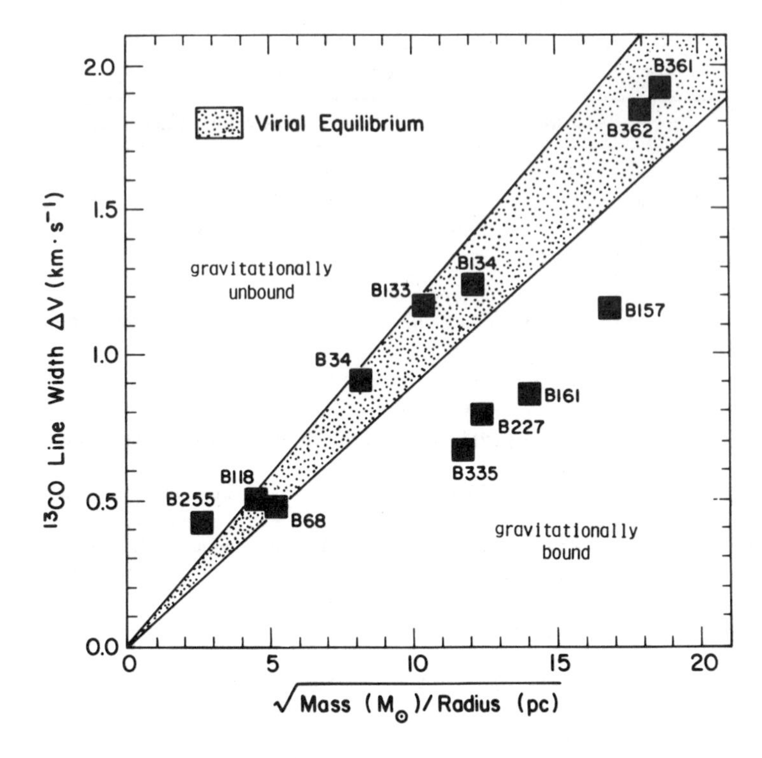

Fig. 8. The observed ^{13}CO line width (Δv) plotted as a function of the derived quantity $(\text{mass/radius})^{\frac{1}{2}}$ for a sample of globules. Also plotted are the predicted values based on models of virial equilibrium. The width of the shaded band takes into account the effects of line saturation and density gradients on Δv. Objects lying below the equilibrium band should be gravitationally bound and may have undergone significant gravitational contraction (figure adopted from Leung et al. 1982).

and limited by angular momentum constraints. While each of the proposed scenarios has its own problems upon close scrutiny, it is possible that the observed power law correlation is a combined manifestation of several phenomena. More detailed discussion on this is given by Myers (1983; also his chapter). The question on the maintenance and dissipation of supersonic turbulence in molecular clouds has been addressed by Scalo and Pumphrey (1982).

Leung et al. (1982) also have found that most of the dark globules studied appeared to be in virial equilibrium and a few may have even undergone significant gravitational contraction[a] (see Fig. 8). The average gas density

---

[a]The term gravitational contraction implies a slow shrinking process in the presence of other resisting forces such as magnetic field, thermal gas pressure, rotation and turbulence. On the other hand, the term gravitational collapse usually implies a rapid infalling process under mostly pressure-free conditions, e.g., free-fall collapse.

seems to decrease with increasing cloud size, suggesting that gravitational contraction may account for the difference in gas density. The effects of gravitational contraction, which tends to increase the central gas density, coupled with the more rapid dissipation of turbulent motions in high-density regions may account for the observed difference in velocity dispersion in different parts of a cloud. In particular, Martin and Barrett (1978) found that the velocity dispersion determined from transitions (such as those of CS and HCN) which are formed in the inner high-density regions are smaller than those

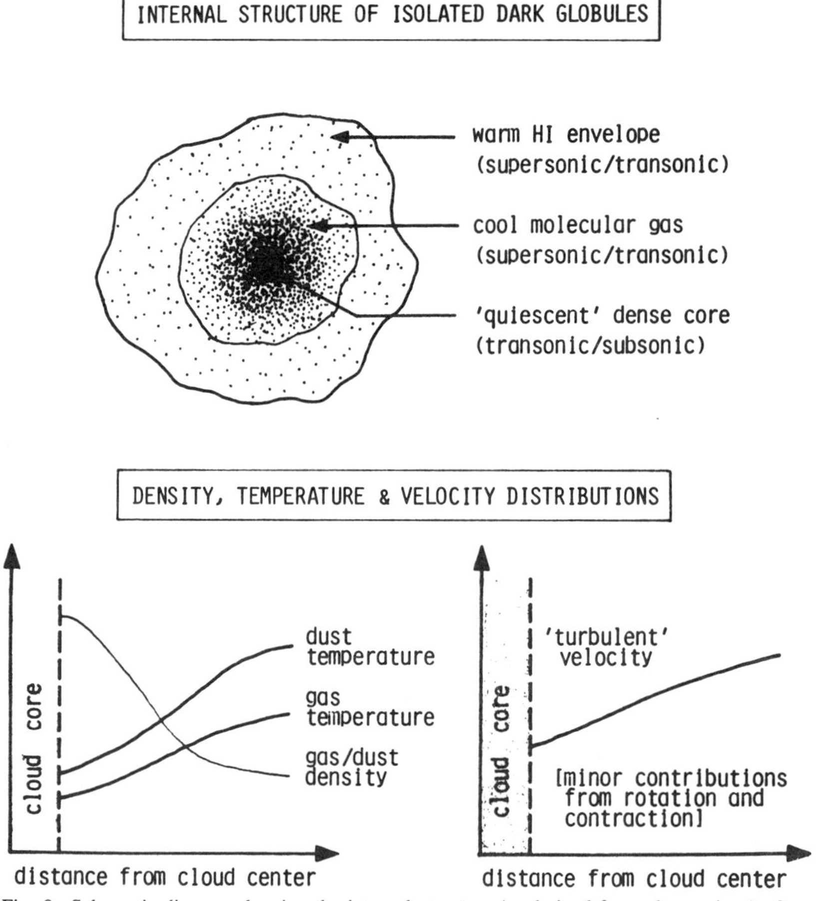

Fig. 9. Schematic diagram showing the internal structure (as derived from observations) of an isolated dark globule. Typically, a warm and diffuse envelope of atomic hydrogen gas surrounds a cloud of cooler molecular hydrogen gas in which is embedded a cold, dense, quiescent core. The lower panel shows qualitatively the general behavior of the distributions in density and temperature of the gas and dust components and the velocity structure. The structure of the cloud core is not well determined.

derived from transitions of CO which are formed in the outer regions. They interpreted this as evidence for gravitational collapse with a radial velocity law which increases with radius. However, such an interpretation is inconsistent with the observed power law correlation between velocity dispersion and cloud size. Hence an interpretation consistent with various observations would be that the magnitude of the velocity field due to random mass motions increases with the radial distance from the cloud center.

While rotation is detected in a few globules, e.g., B361 (Milman 1977) and B5 (Young et al. 1982), it does not seem to play a significant role in the dynamics of isolated globules. Recently, from a sample of 16 dark clouds (a few of which are globules) which show evidence of rotation, Goldsmith and Arquilla (see their chapter) found that the shifts in velocity produced by rotation follow a power law similar to that found for the variation of linewidth with cloud size. In addition, their results indicate a low degree of alignment of cloud angular momentum vectors relative to the angular momentum vector of the Galaxy. They concluded that rotation in dark clouds may result from fluctuations in the random velocity field that is present (as first proposed by Fleck and Clark [1981]) and that rotation alone cannot support these clouds in virial equilibrium.

To summarize, while there is observational evidence for the presence of rotation and random mass motions in isolated globules, there is as yet no direct and conclusive evidence for gravitational contraction or collapse. Furthermore, most isolated globules appear to be in virial equilibrium. Based on existing observations and their interpretations which are summarized in Table II, Fig. 9 is a schematic diagram showing the internal structure of a typical isolated globule and its density, temperature and velocity structure. Note that because thermal velocity depends on the mass of a molecule, a cloud region may have subsonic turbulence for some species but supersonic turbulence for other species. Due to the finite spatial resolution of existing radio telescopes, detailed structure of the cloud core is not well resolved. Because arc-second resolution is required for such studies, the density and temperature distributions as well as the velocity field in the core region of dark globules are largely undetermined. An overview of the physics of molecular cloud cores is given in the chapter by Myers.

V. THEORETICAL MODELS

There are two different approaches in the theoretical modeling of molecular clouds in general and dark globules in particular. In the first approach, which may be called a case-study approach, the model parameters are adjusted to maximize the agreement between model results and observations of one particular source. In the second approach, which may be called an ensemble approach, agreement with the model results is sought only for the *average* properties of a large number of sources of similar nature. The second approach is normally used when a general and preliminary understanding of the physical processes involved in the model is sought.

In constructing realistic models for the structure and evolution of dark globules, a large number of physical processes must be considered. These include: (a) various chemical processes for the formation and destruction of atomic and molecular species to model the chemical composition and its evolution; (b) various heating and cooling mechanisms for the gas and dust components to determine the gas and dust temperature distributions; (c) radiation transport in the line and continuum radiation to determine the emergent energy spectra and line profiles for comparison with observations; (d) various dynamical processes such as gravitational collapse, rotation, turbulence, and magnetic field to determine the gas density distribution. Because the simultaneous inclusion of all conceivable processes in a self-consistent treatment would be an almost impossible computational task, most authors have taken a piece-meal approach in the theoretical modeling of dark globules, i.e., model one or two processes at a time while ignoring all others or assuming other processes are known or can be decoupled. Below we review the more important work on theoretical modeling of the physical conditions in dark globules, starting with models in which steady state condition is assumed. We mostly include empirical models that are particularly relevant to the discussion in Secs. III and IV. Theoretical models which emphasize only the chemical evolution will not be discussed here; they are reviewed in the chapter by Herbst.

Leung (1975*a,b*) solved in detail the problem of radiation transport in line and continuum radiation for the gas and dust components in dark globules. The gas and dust temperature distributions are also calculated for clouds of given opacity and density distribution. Spherically symmetric geometry is assumed. Spencer and Leung (1978) also calculated the infrared spectrum from dark globules. Kenyon and Starrfield (1979) contructed polytropic models for dark globules and used the calculated density distribution to obtain theoretical curves of visual extinction versus cloud radius. When comparing these curves with observations of B361 and Coalsack globule 2, they found that B361 resembles $n = 1.75$ polytrope while globule 2 resembles an $n = 2.5$ polytrope. Clavel et al. (1978) modeled the chemical and thermal structure of dark clouds that are externally heated. Plane-parallel geometry and uniform density are assumed. Various heating and cooling processes for the gas component are included and the radiative transfer equation for the cooling line photons is solved using an escape probability approximation. The temperature of the dust component is not solved. De Jong et al. (1980) performed a similar calculation but used a more extensive chemical network. The gas density is calculated assuming the cloud is in hydrostatic equilibrium supported by a constant turbulent pressure. This work has recently been updated and extended to spherical cloud geometry (Boland and de Jong 1984). In addition to considering the chemical and thermal structure of dark clouds, Arshutkin and Kolesnik (1981) also studied the effect of turbulent motion on the heating and equilibrium state and spherically symmetric clouds. They found that a turbulent velocity which decreases toward the cloud center is necessary for maintaining the equilibrium state of molecular clouds with a core-envelope structure, a conclusion in apparent agreement with observations.

To interpret the radio and optical observation of an isolated globule which appears to have subsonic turbulence, Dickman and Clemens (1983) constructed a polytropic model which is pressure-bounded, centrally-condensed, hydrostatic and has a gas temperature increasing outward. The model results yield good agreement with the observed cloud radius, central visual extinction and inferred cloud mass. The hydrostatic model was then used to construct CO line profiles which agree reasonably well with observations. Finally, Krugel et al. (1983) have constructed a comprehensive model for the physical structure (i.e., density, temperature, velocity field and molecular abundances) of B335 using the available line and continuum observations as constraints.

While the above theoretical studies have considered in some detail the chemical and thermal structure of globules as well as the problem of radiation transport, all hydrodynamical processes are ignored. On the other hand, although extensive theoretical studies have been made on the dynamical evolution of interstellar clouds in general, these calculations emphasize only the role played by gravitation, rotation and magnetic field on the stability against cloud collapse and fragmentation. They completely ignore the effects of chemistry, thermal structure and radiation transport and they deemphasize observational comparisons. A recent review and tutorial on these models is given by Tohline (1982). Here we discuss only models which also emphasize observational comparisons. Villere and Black (1980,1982) simulated the dynamic evolution of dark globules using a two-dimensional hydrodynamic code in which the details of chemical and thermal processes are ignored. Thus they considered the isothermal evolution of rotating and collapsing clouds which initially have uniform temperature, density, angular velocity and spherical geometry. By comparing the properties of the cloud models with available observations of several globules, they concluded that most of the globules are undergoing gravitational collapse and that their ages are smaller than their free-fall times. However, as pointed out by these authors, because turbulence and magnetic fields were not taken into account and comparisons were made using derived rather than directly observable quantities, these conclusions should be viewed with caution and be reexamined with improvements in both observations and theoretical models. Gerola and Glassgold (1978) have constructed an integrated and self-consistent one-dimensional hydrodynamic model of an externally heated cloud of mass 2×10^4 $M_\odot$. While the mass of the cloud model is too large for dark globules, the physical processes should be similar. A variety of heating and cooling mechanisms were included along with important chemical reaction schemes. Their results indicate that the chemical evolution of the cloud is governed by the ratios of the chemical time scales to the dynamical time scale. Most recently Tarafdar et al. (1984) constructed detailed models to study the dynamical and chemical evolution of initially diffuse, externally heated, interstellar clouds of both low and high masses. Instead of determining the gas temperatures self-consistently by including various heating and cooling mechanisms, they assumed a temperature distribution which depends on extinction and density and which is semi-empirically determined. Compared to the work of Gerola and Glassgold (1978), the chemical network employed is more extensive. In comparing their model results with observations for both diffuse and dark clouds, they found general agreement in the dependence of abundance on both gas density (for dark clouds) and visual extinction (for diffuse clouds) for various molecules and ions. They concluded that diffuse and dark clouds may represent different stages in the chemical evolution of interstellar clouds.

VI. STAR FORMATION IN ISOLATED DARK GLOBULES

Bok and Reilly (1947) first called attention to the isolated globules as precursors of star formation and hypothesized that they are in a state of gravitational collapse, ultimately leading to the formation of stars. Despite the lack of direct observational evidence to support such an assumption, this view has persisted in the literature but remained somewhat controversial. For example, van den Bergh (1972) concluded that "there is at present no firm evidence that any stars are formed from globules." The weight of recent infrared and radio observations suggests that stars form in associations from fragmentation of much more massive molecular clouds. Using radio and microwave observations, detailed studies of the physical, chemical and dynamical properties of dark clouds have been made by Nachman (1979) and Snell (1981). Frerking and Langer (1982) made a survey of 180 dark clouds to search for broad CO-line wings; they detected wings with characteristic dispersions of 3 to 10 km-s^{-1} in several sources. They suggested that these features probably arise from stellar winds associated with embedded low-mass stars and that the statistics are consistent with the formation rate of low-mass stars in the solar neighborhood. Recent infrared observations of the Infrared Astronomical Satellite (IRAS) also confirmed that many dark clouds have internal heat sources (Beichman 1984). An example of a dark cloud which has been extensively studied in both the radio and infrared is B5 (Young et al. 1982; Beichman et al. 1984) that contains several internal heat sources. The characters of these sources range from a fully-formed star embedded within the cloud core, to a dense clump of warm material at the position of a local density maximum, to stars separate from the cloud but enshrouded in luminous dust shells. B5 is located near one end of a large molecular complex which includes both NGC 1333 and the young open cluster IC 348. Recently in an effort to detect young stars associated with dark globules, Ogura and Hasegawa (1983) made a survey for Hα-emission stars of the T Tau type in the vicinity of 15 globules; they found a total of 60 emission-line stars, indicating that the surface density of emission-line stars in the vicinity of globules is higher than that in the ordinary field by a factor of 2. Because star formation is known to occur in dark clouds and dark clouds do not usually appear as isolated clouds but are generally associated with or physically attached to other dark cloud complexes, it is not clear whether star formation can take place in isolated dark globules. However, recent observations of the isolated globule B335 may partially support the original conjecture of Bok and Reilly in that star formation is possible even in an isolated globule. Below we summarize the observational results of B335.

B335, an outstanding example of an isolated dark globule, is the best studied representative of its class. It has a highly opaque core with a more transparent envelope. The visually opaque region is roughly elliptical in shape, with an angular size of approximately 3′ E-W × 4′ N-S. The central 40″ region has a visual extinction of at least 30 mag. The absence of foreground stars places an upper limit on its distance of 400 pc (Bok and McCarthy 1974) while star count results indicate its distance to be 250 pc (Tomita et al. 1979).

Radio observations of molecular-line emission from B335 have been made and there now exist line maps, cross scans or center position measurements from different molecules and transitions, e.g., CO (Martin and Barrett 1978; Wilson et al. 1981; Frerking and Langer 1982; Goldsmith et al. 1984), CS (Martin and Barrett 1978; Snell et al. 1982), NH_3 (Martin and Barrett 1978; Ungerechts et al. 1982; Menten et al. 1984), H_2CO (Rickard et al. 1977; Martin and Barrett 1978), OH (Martin and Barrett 1978) and HCO^+. CO observations indicate a total cloud mass (core and envelope) of 100 $M_\odot$, and average hydrogen density $\langle n(H_2)\rangle \sim 1.3 \times 10^3$ cm^{-3} (Leung et al. 1982), and a gas temperature of $T_k \sim 10$ K (Martin and Barrett 1978). From CS lines, the central hydrogen density is estimated to be $\geq 10^4$ cm^{-3} (Snell et al. 1982). From estimates of its mass, radius and velocity dispersion, it is concluded (Leung et al. 1982) that B335 must have undergone significant gravitational contraction (see Fig. 6), unlike most other isolated globules which seem to be in viral equilibrium. Furthermore, there is strong observational evidence that high-velocity gas motions exist in B335 as indicated by the presence of broad and asymmetric CO line wings, $\Delta v(\text{wing}) \geq 4$ km s^{-1}, which suggests the possible presence of stellar winds associated with a newly formed star (Frerking and Langer 1982; Leung et al. 1982). In addition, as shown in Fig. 10 the spatial distributions of the blue- and red-shifted gas indicate a bipolar flow structure (Frerking and Langer 1982; Goldsmith et al. 1984), similar to many bipolar-flow sources associated with protostars and young stellar objects (Bally and Lada 1983). The presence of rotation in the cloud core is also suggested by both CO and NH_3 observations (Frerking and Langer 1982; Menten et al. 1984).

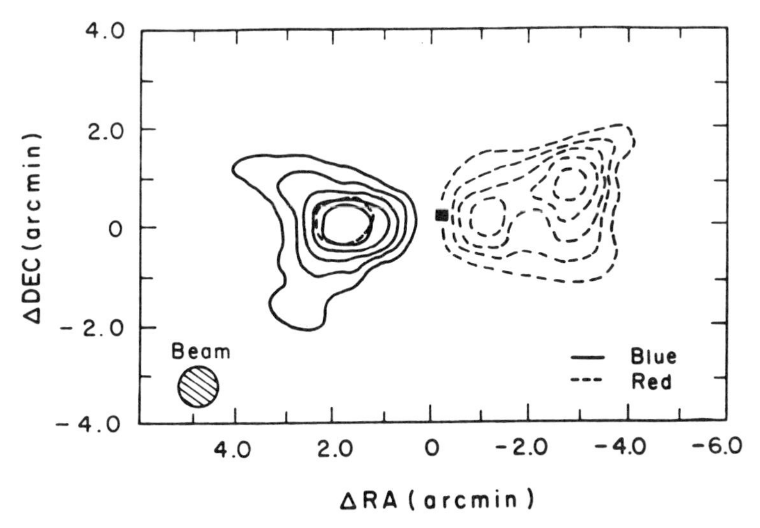

Fig. 10. A map of B335 showing the spatial distribution of the integrated intensity of the $J = 1-0$ transition of CO over the velocity intervals 3 to 7.5 km s^{-1} (blue) and 9.5 to 14 km s^{-1} (red). The contours (units: K-km-s^{-1}) extend from 1.25 to 3.25 with a contour spacing of 0.5. The solid square denotes the position of the far-infrared source detected by Keene et al. (1983) (figure from Goldsmith et al. 1984).

In addition to radio-line observations, far-infrared emission by dust (heated by the ambient ISRF) has also been detected in B335 (Keene et al. 1980). Keene et al. (1983), using broad-band photometry in the far-infrared (60 to 1000 μm), detected a compact source at the center of B335 (see Fig. 11). Both the 200 and 400 μm observations are consistent with a diameter for the compact source of $\leq 30''$. The bolometric luminosity of the globule within a 90″ beam is 3.0 $L_\odot$ $(D/250 \text{ pc})^2$ of which $\geq 70\%$ probably comes from an embedded source and the rest from the ISRF. The density distribution of the globule is sharply peaked toward the center. The mass of the central core derived from the 400 μm flux density is 2.5 $M_\odot (D/250 \text{ pc})^2$. For a uniform spherical concentration $\leq 30''$ diameter, this implies a H_2 density of $\geq 1.6 \times 10^6$ cm$^{-3} (D/250 \text{ pc})^{-1}$. While there appears to be no radio continuum counterpart for the embedded source, Krugel et al. (1983) have detected a near-infrared (at 2.2 μm) source at the same position using near-infrared (1.25 to 3.5 μm) photometry in the Johnson *J, H, K, L* bands. The positions of the bipolar-flow source, the far-infrared source and the near-infrared point source coincide to within the different beam sizes. However, the observations of the 2.2 μm source remain to be confirmed. Finally, Krugel et al. (1983) have constructed a model for the physical structure of B335 using the available line and continuum observations.

The above observations, when taken together, appear to support the notion that star formation is taking place in B335.

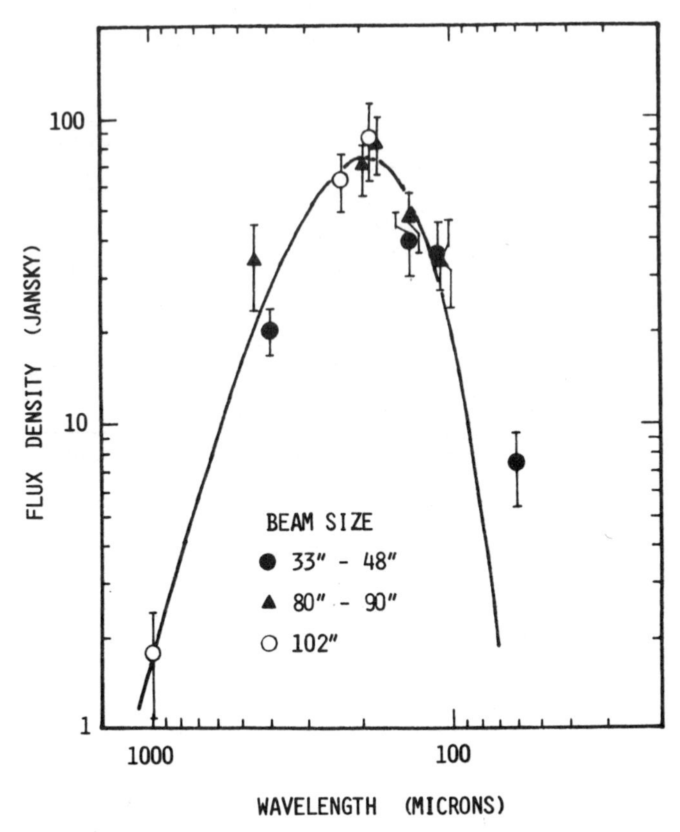

Fig. 11. Far-infrared spectrum of B335. The solid line, which represents a spectrum of the form $\nu^2 B(\nu, 15K)$, is a satisfactory fit to the observed large beam fluxes (80″ to 102″) (figure adopted from Keene et al. 1983).

1. The relationship among the mass, radius and velocity dispersion indicates that B335 is not in virial equilibrium but has undergone significant gravitational contraction;
2. The presence of broad, asymmetric CO-line wings at the globule center indicates a bipolar-flow structure which resembles those associated with what are believed to be pre-main sequence objects undergoing mass loss;
3. The presence of a compact, far-infrared source of low luminosity with a near-infrared counterpart at the position of bipolar-flow source implies that this embedded source may power the bipolar flow.

Thus B335 may represent an outstanding example of an isolated dark globule in which a low-mass, low-luminosity protostar is being formed in a relatively simple environment. It is clear that high angular resolution observations of both line and continuum emission at different wavelengths will be crucial in further studies of low-mass star formation in isolated dark globules.

VI. SOME UNSOLVED PROBLEMS

In this survey we have attempted to review some recent progress, both observational and theoretical, in the study of isolated dark globules. While a good many general characteristics of the gas and dust components of dark globules have been determined observationally and are fairly well understood theoretically, there still remain many unanswered questions, some of which can only be tackled by a combination of observational work and theoretical modeling. Below we outline some of these problems.

What are the relationships, if any, among the different kinds of globules, i.e., isolated dark globules, globular filaments, cometary globules and elephant-trunk globules? While it is likely that cometary and elephant-trunk globules may represent globules formed by similar physical processes but are at different evolutionary stages, it is not clear whether isolated dark globules are simply remnants of cometary globules (Reipurth and Bouchet 1984), or represent condensed fragments from nearby filamentary cloud complexes, the formation of which may depend strongly on the properties of interstellar turbulence (Schneider and Elmegreen 1979; Leung et al. 1982). If the latter scenario is favored, what would be the role of interstellar turbulence in the formation of globular filaments and isolated dark globules? The question on the evolution of globules associated with young OB clusters has been considered by Ott and Sanders (1980) who conclude that they expand with age, possibly after rapid compression to some minimum size. A statistical analysis of the correlations in physical conditions, chemical composition and spatial distributions among the different types of globules using detailed and comprehensive observational data may shed light on this question.

Why do most isolated dark globules not collapse but appear to be in virial equilibrium? What roles do gravitation and magnetic fields play in their dynamical evolution? Observational determinations of (a) the form of systematic velocity law due to rotation and contraction (if they are present), (b) the precise form of the gas density distribution in the cloud core, and (c) the magnetic field strength are crucial in shedding light on these questions. Interferometric observations which provide high spatial resolution will be required.

While the presence of turbulence is suggested from molecular line observations of dark globules, the effects of fluid flow instabilities and turbulence have so far not been considered in hydrodynamical modeling. In order to understand the role played by turbulence in the evolution of dark globules, one may need to include the turbulent velocity field as an initial condition in future modeling of cloud dynamics and consider the coupling between gravitational contraction and turbulence dissipation as well as maintenance. Ignoring these effects will likely lead to erroneous conclusions.

In conclusion, it is fair to say that despite the apparent simplicity of the physical environment and in the appearance of isolated dark globules, much remains to be learned about these "holes in the heavens" and their role in the formation of low-mass stars such as our Sun.

Acknowledgments. The author's research on dark globules was partially supported by a grant from the Research Corporation. A major portion of this review chapter was written while the author was a visiting staff member at the Los Alamos National Laboratory (LANL). He is grateful to W. F. Huebner of the Theoretical Division at LANL for hospitality and financial support.

ROTATION IN DARK CLOUDS

PAUL F. GOLDSMITH AND RICHARD ARQUILLA
University of Massachusetts

Recent observations of dark interstellar clouds are analyzed with regard to the role of rotation in determining the structure and energetics of these objects. Observational techniques are discussed. A sample of sixteen objects or condensations which show evidence for rotation have been obtained. The mean densities and sizes of the clouds are strongly inversely correlated, with n α*(size)*$^{-1.5}$*, while the cloud angular velocities* ω *vary as (size)*$^{-0.6}$*. The velocity shifts due to rotation of a given cloud are proportional to (size)*$^{0.4}$*, which is close to the linewidth-size relationship found for a large sample of clouds not specifically thought to be rotating. The implication is that rotation is closely linked to the processes responsible for random motions in interstellar clouds, although rotational motion alone does not appear to support these clouds in virial equilibrium.*

Rotation in interstellar molecular clouds can have a significant effect on the structure of these objects, as well as play an important role in determining their evolution in terms of contraction, fragmentation, and ultimately the formation of stars. Systems of widely varying size (galactic, stellar, and planetary) include rotational motion as an important component of their dynamical equilibrium. In the case of molecular clouds, their formation from a differentially rotating galactic disk provides a possible origin for rotation, and if angular momentum is conserved, rotation should be easily detectable when densities have reached $\sim 10^3$ cm^{-3}. Studying the role of rotation in molecular clouds is not only hampered by their irregularity and the presence of effects producing line broadening and shifts in the centroid of spectral profiles, but by the paucity of observations of the high quality required; conclusions concerning the importance of rotation have been necessarily tentative (cf. Field 1978). In this review we analyze recent molecular-line observations of a number of dark clouds and assess the role played by rotation. In Sec. I we discuss some of the different techniques available for discerning the presence of rotation. From a sample of dark clouds with detectable rotation, we are able in Sec. II to establish some correlations between angular momentum and other cloud properties. The role of rotation in the energetics of dark clouds is discussed in Sec. III.

I. OBSERVATIONAL APPROACH

Information about molecular cloud motions is limited to the component of their velocity field along the line of sight; the intensity of radiation reaching the observer at a particular frequency can also be affected by radiative transfer effects. To determine the velocity field within a cloud in a relatively unbiased way, it is preferable to use spectral lines of low or moderate opacity; those most widely employed are the rotational transitions of ^{13}CO and $C^{18}O$, and the inversion transitions of NH_3. It must be recognized, however, that features with relatively low column density can be seen only in ^{12}CO and will be missed by this approach, an example being the high-velocity bipolar outflow features which have been found in a number of dark clouds (Frerking and Langer 1982; P. F. Goldsmith et al. 1984). In a simple model of a rotating cloud, the ordered aspect of the velocity field provides a distinctive feature which in principle allows us to discriminate between this form of motion and other forms like velocity perturbations due to shear, peculiar motion of subcondensations, asymmetric collapse or expansion, and due to a possible turbulent component of the velocity field having significant correlation length. In practice an unambiguous rotational signature is difficult to observe and in many cases the presence of rotation is only marginally certain.

In trying to define a possible rotational component of the velocity field, we must also recognize that the imperfect isolation and lack of symmetry of interstellar molecular clouds suggest that the pattern of the velocity field will likely deviate from that of an idealized isolated model cloud. As an example, we show in Fig. 1 the visual extinction in the region of the dark cloud L 1257 (Arquilla 1984). The condensation within the SE promontory of obscuring material was previously found on the basis of ^{12}CO observations to be rotating in a fairly rigid-body-like fashion (Snell 1981), with the rotation axis oriented NE-SW. The visual extinction data show that this kinematic fragment is a part of a much larger region. The situation of a dense condensation (or condensations) being appended to an extended region of relatively low density is found to prevail in the majority of the clouds discussed in this chapter. In maps of ^{13}CO, the regions which appear to be rotating are relatively isolated, but their behavior is very likely affected by coupling to the much larger, nearby quiescent regions.

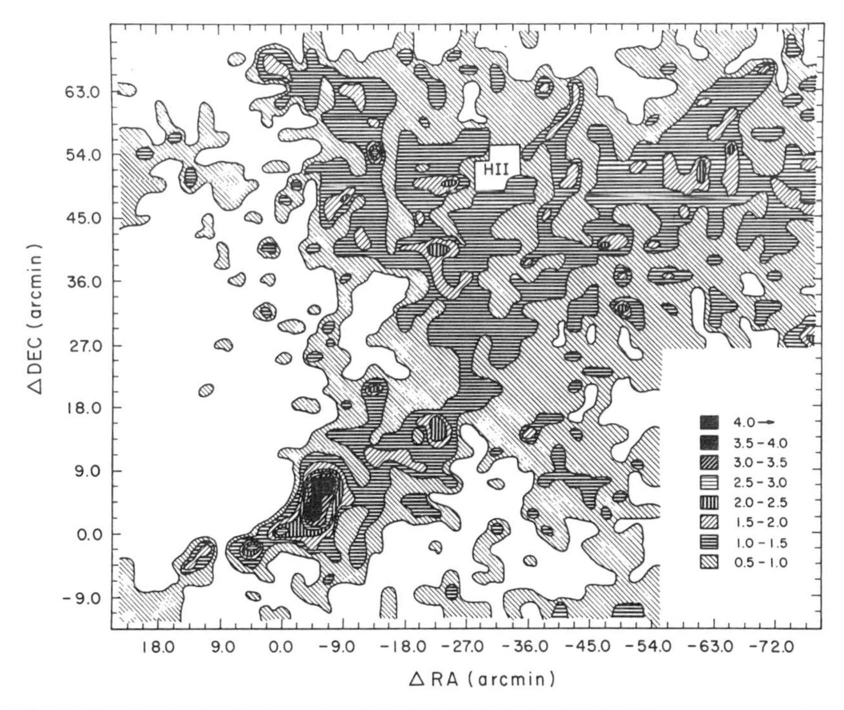

Fig. 1. Visual extinction in the region of the L 1257 dark cloud derived from star counts of a Palomar Observatory Sky Survey red print.

The rotation of a cloud is generally established by observation of systematic shifts in the velocity of maximum emission which correspond to the pattern expected from a rotating object; the shifts of the line centroid must be a significant fraction of the linewidth in order to identify the rotation. One exception to this methodological approach is the analysis of the Taurus complex by Kleiner and Dickman (1984) in which an autocorrelation analysis was used to determine the presence of a large-scale gradient. A method widely used to examine the systematic behavior is to compare spectra along a cut through the cloud; these can conveniently be analyzed in the form of a spatial-velocity diagram. An example of such a presentation of data is the NW-SE cut through B361 shown in Fig. 2 (Arquilla 1984). In order for the cloud to be considered as rotating, such maps taken at different position angles must show the appropriate behavior of the velocity variation. The data for B361 satisfies this requirement fairly well, while also clearly showing evidence for nonrigid-body behavior. Spatial-velocity maps are especially useful for distinguishing between rotation and other velocity fields which exhibit similar patterns. Bipolar outflows, for example, can produce systematic shifts in the mean emission velocity of ^{12}CO. However, the relatively small column densities in these flows reduce their emission in ^{13}CO to quite low levels compared to ^{12}CO and

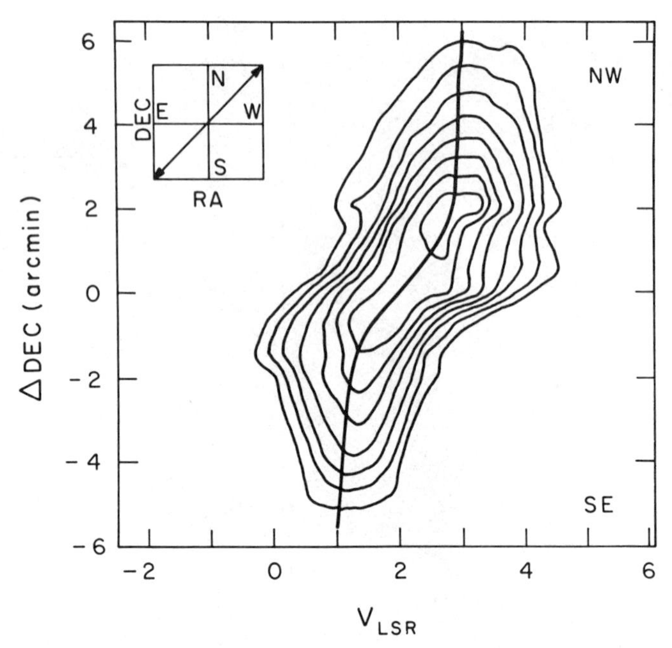

Fig. 2. Spatial-velocity map of ^{13}CO emission along a cut through the center of B361. The offset along the NW-SE axis is presented in terms of the declination offset from the cloud center. The solid curve represents the equilibrium velocity profile produced by a fitted density distribution (from Arquilla 1984) which comprises a uniform density core and a r^{-2} density variation in the envelope.

to present-day detection thresholds. Also, the velocity of maximum emission, if one considers the quiescent and perturbed gas together, is not affected by the presence of the outflow, and hence the spike in a spatial-velocity map is not affected. GL 437, a molecular cloud previously thought to be rotating (Schneps et al. 1978), was found on closer study to contain a bipolar outflow (Arquilla and Goldsmith 1984). While we do not feel that this type of perturbation is likely to be responsible for the velocity shifts seen in the data presented below, nonrotational contributions to the velocity field cannot be eliminated.

An alternative method which presents a more limited view of the velocity field, is to display the velocity of peak emission as a function of two spatial coordinates. This approach is useful if the cloud contains localized perturbed regions which can make individual spatial-velocity cuts look highly irregular. If these areas are isolated in a two-dimensional presentation, the overall form of the velocity field can still be ascertained. The mean velocity of the ^{13}CO emission can be used to confirm behavior seen in the map of velocity of maximum emission. The advantage of improved accuracy in the velocity determination is partially offset by the increased sensitivity to line-wing asymmetries. It is also possible to make contours of equal emission intensity within different velocity intervals (single-channel maps). This approach appears to be the least satisfactory in that it emphasizes irregularities in the column density and the degree of molecular excitation, rather than the form of the velocity field.

II. RESULTS: PROPERTIES OF ROTATING CLOUDS

In Table I we indicate the sixteen entities comprising the data set used in this study. These are not all distinct; several clouds are fragments of the Taurus complex. This data set is more selective than the compilations given by Field (1978) and Fleck and Clark (1981) in that only dark clouds (those without evident high-mass star formation) are included; the increased quantity of data presently available reflects the increased capability of current millimeter systems. Much of these data have been obtained from the literature; the mass and size estimates are obtained from ^{13}CO observations except where noted. The cloud sizes are obtained from inspection of maps preferably of column density; dimensions to half-maximum intensity are used and the size given in Table I is the full diameter if the cloud appears symmetric, or the geometric mean of the extreme perpendicular diameters for those clouds having an elongated appearance. The cloud masses have either been given in the references cited, or are calculated from the column densities and cloud dimensions. Angular velocities are computed by assuming rigid-body rotation without correction for projection effects onto the plane of the sky (correction considered at the

TABLE I
Properties of 16 Rotating Clouds

Object	Mass ($M_\odot$)	Size (pc)	J ($M_\odot$ km s^{-1} pc)[a]	J/M (km s^{-1} pc)[b]	ω (km s^{-1} pc^{-1})[c]	References[d]
Taurus complex	5600	17	5.4×10^4	9.7	0.2	1
B18	1100	10	2200	2.0	0.24	2
B5	500	4.0	300	0.6	0.67	3
TMC-1 ring	200	1.2	46	0.23	1.0	4
L 183	150	1.1	13	0.09	1.2	5
B361	98	1.2	22	0.23	1.7	6
L 1544	42	0.6	2.5	0.06	1.7	7
L 134	30	0.5	0.75	0.025	1.0	8
L 1535	20	0.5	1.3	0.064	1.3	9
L 1257 (M)	15	0.6	0.90	0.063	1.8	6
B213 NW	14	0.2	0.24	0.02	5.7	10
B163	8	0.5	0.38	0.05	1.1	6
L 1253	7	0.4	0.2	0.03	1.8	6
B163 SW	4	0.5	0.29	0.07	2.9	6
B68	1.1	0.24	0.013	0.012	2.0	11
L 1535 (core)	0.7	0.1	0.003	0.004	4.0	12

[a] 1 $M_\odot$ km s^{-1} pc = 6.1×10^{56} g cm^2 s^{-1}; [b] 1 km s^{-1} pc = 3.1×10^{23} cm^2 s^{-1}; [c] 1 km s^{-1} pc^{-1} = 3.2×10^{-14} s^{-1}.
[d] 1. Kleiner and Dickman (1984): correlation analysis; 2. Baudry et al. (1981): different molecules; 3. Basic data from Young et al (1981) refer to bulk of cloud but revised distance of 300 pc has been used; 4. Schloerb and Snell (1982,1984); 5. Clark and Johnson (1981); 6. Arquilla (1984); 7. Snell 1981 and personal communication; 8. Mattila et al. (1979): OH data referring to core of cloud; 9. P. F. Goldsmith and M. Sernyak (1984); 10. Clark and Johnson (1978): 5.5 km s^{-1} fragment in H_2CO; 11. R. N. Martin and A. H. Barrett (1978); 12. Ungerechts et al. (1982): NH_3.

end of Sec. III), and by defining $\omega = \Delta V/L$ where ΔV is the maximum edge-to-edge velocity shift and L the extent of the cloud along the direction of the velocity gradient.

The angular momentum is obtained by assuming uniform density, so that $J = 2/5\, MR^2\omega$ where R is one half of the size defined above, and M is the cloud mass. The specific angular momentum J/M in this analysis is given approximately by $L\Delta V/10$, and so depends linearly on the distance to the cloud while $\omega = \Delta V/L$ varies inversely as the distance to the cloud. The mass of the cloud determined from column densities is proportional to the square of its distance.

These objects fall into the category of dark clouds, defined as having appreciable visual extinction (greater than a few magnitudes) but many contain embedded low-luminosity sources. The clouds B5 (Beichman et al. 1984) and B335 (Keene et al. 1983) have been studied in the far infrared and contain sources with total luminosities of a few $L_\odot$. It is probable that this is the case for the other clouds, which are also the sites of low-mass star formation. It is not yet clear whether it is useful to make a distinction between those clouds with and without luminous energy sources; over most of the area of the clouds under consideration here, the kinetic temperature of the gas is between 10 K and 15 K.

The estimates of cloud properties are subject to a large number of possible errors, with the distance to the cloud being probably the most important. The relatively consistent method in which the data of these clouds have been analyzed will result in preserving well-defined trends in the results, even though values of the different quantities may be significantly in error.

From the cloud size and mass, we calculate a mean hydrogen density defined by $\bar{n} = 3M/4\pi R^3 m_{H_2}$ where R is half the size L of the cloud, and m_{H_2} is the mass of the H_2 molecule. A comparison of $\bar{n}$ with L is shown in Fig. 3. An unweighted least-squares fit, with $\bar{n}$ in cm^{-3} and L in pc, yields

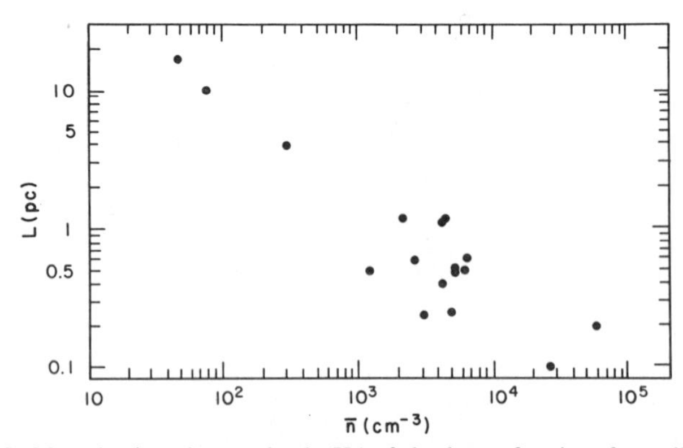

Fig. 3. Mean density $\bar{n}$ (assumed to be H_2) of clouds as a function of overall size L.

$$\bar{n} = 2000\,L^{-1.3}. \quad (1)$$

This is a somewhat steeper relationship than found by R. G. Larson (1981), $n = 3400\ L^{-1.1}$. Larson's sample included clouds of widely differing characteristics and with properties determined by different methods; the significance (if any) of the difference in exponents is not at present clear; the standard deviation of the fitted slope is 0.15.

The distribution of the angular velocities for the clouds in this study is given in Fig. 4. It is appropriate at this point to recall that, (1) this study is biased towards clouds suspected to be rotating, and (2) no correction has been made for projection effects. There are readily observable correlations of specific angular momentum and angular velocity with cloud size. For the former, an unweighted least-squares fit yields with specific angular momentum in units of km s^{-1} pc^{-1} and cloud size in parsecs:

$$J/M = 0.14\,L^{1.4} \quad (2)$$

with a correlation coefficient of 0.93 (Bevington 1969) and a standard deviation of the slope of 0.11. The data are shown in Fig. 5. The issue of angular momentum-size correlation is discussed further by Arquilla (1984). The analogous information expressed in terms of the angular velocity (in units of km s^{-1} pc^{-1}) and cloud size in parsecs yields:

$$\omega = 1.2\,L^{-0.6}. \quad (3)$$

This relationship, with a standard deviation of slope equal to 0.07, was determined directly by least-squares fitting, which is why the coefficient is slightly different than that obtained from the relationship $J/M = \omega L^2/10$ for rigidly rotating, uniform-density spheres.

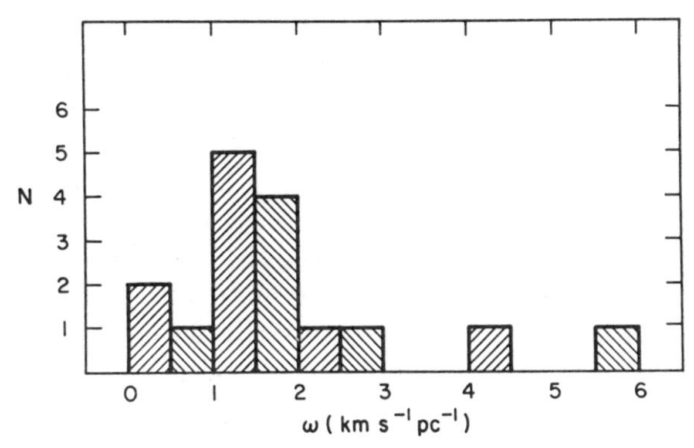

Fig. 4. Distribution of angular velocities of the 16 clouds in this study. 1 km s^{-1} pc^{-1} = 3.2×10^{-14} s^{-1}.

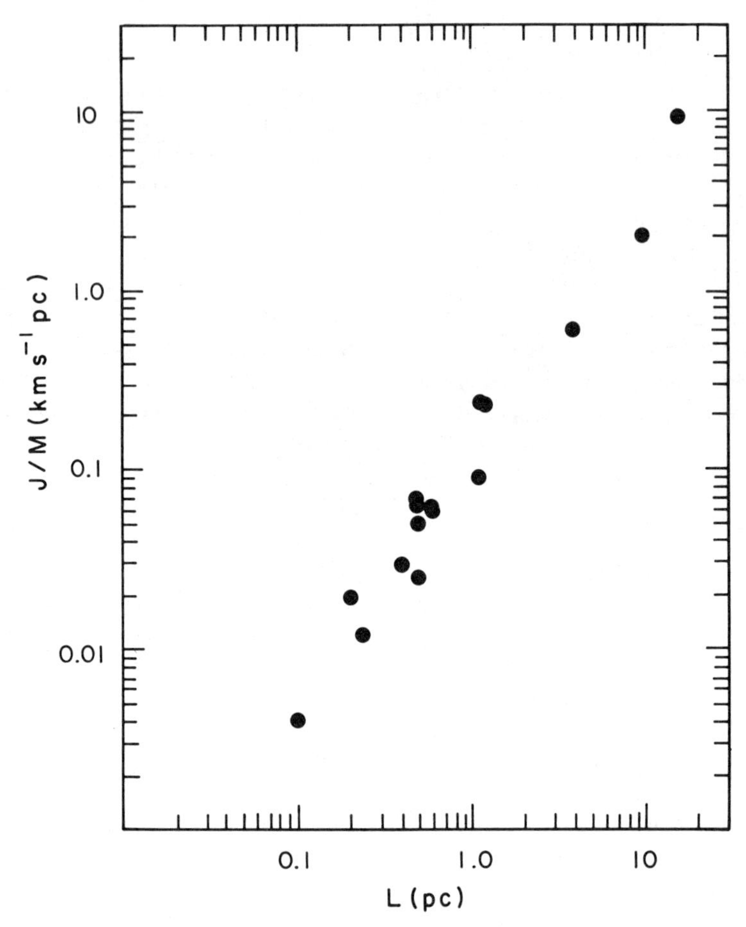

Fig. 5. Specific angular momentum plotted as a function of cloud size. 1 km s^{-1} pc = 3.1 × 10^{23} cm^2 s^{-1}.

The correlation between specific angular momentum and cloud size must be regarded with caution inasmuch as clouds with significantly lower rotational velocities may not be recognized as rotating. Thus, the line which fits the data in Fig. 5 should probably be regarded as an upper envelope of rotation in dark clouds; this is discussed further below. If we consider the evolution from lower-density to higher-density phases by a combination of contraction and fragmentation, it is clear that considerable rotational angular momentum is lost in the process. The most obvious sink is the orbital motion of the fragments; this is consistent with the results of Bodenheimer et al. (1980*b*) who found in numerical studies that the ratio of spin to orbital angular momentum for a fragment was 0.2. Alternative scenarios such as those involving magnetic braking can, with proper choice of parameters, explain the observations. Detailed modeling is necessary to evaluate accurately this effect, especially with regard to the coupling of the magnetic field and the gas.

While the physical significance of the correlation of specific angular momentum and cloud size is not yet clear, a connection between the velocity shifts due to rotation and the linewidths observed in clouds without any indication of rotation can be seen. If we express the angular velocity–cloud size relationship found in Eq. (3) in terms of the velocity shift, we obtain

$$\Delta V = 1.2\, L^{0.4} \tag{4}$$

where ΔV is given in km s^{-1} and L in pc. This has a close similarity to the relationship between the velocity dispersion of profiles for various molecules from different types of clouds found by R. G. Larson (1981), namely $\sigma = 1.1\, L^{0.38}$. It is also similar to the relationships obtained by Leung et al. (1982) for the turbulent velocity in globules, $V_{\rm turb} \simeq R^{0.3\text{–}0.5}$, with the different exponents for high and low categories of cloud temperature. The present study thus offers an empirical suggestion that the processes that create and sustain turbulence in interstellar clouds are also responsible for producing rotational motions; in this interpretation, rotation may arise from a somewhat unusual fluctuation in the velocity field. This interpretation was previously advocated by Fleck and Clark (1981) who also suggested that galactic differential rotation could sustain the turbulent velocity field in molecular clouds.

The picture of rotation as a component or outgrowth of the essentially random velocity field in dark clouds is consistent with the orientations of the directions of the inferred angular momentum vectors of various regions in the Taurus cloud complex. As discussed by P. F. Goldsmith and M. Sernyak (1984), there is no clear correlation between the projected cloud rotation axes and the galactic plane (or the axis of galactic rotation). Magnetic braking (cf. Mouschovias and Paleologou 1980; see also chapter by Mestel), cloud collisions (Horedt 1982), and possibly other processes could randomize the direction of cloud angular momentum initially parallel or antiparallel (cf. Mestel 1966*b*) to the angular momentum of the Galaxy. It appears that these processes dominate the situation, or that the angular momentum vectors are initially created with a largely random distribution of directions. The lack of correlation of the axes of cloud rotation with that of the Galaxy implies that galactic differential rotation does not directly impart angular momentum to molecular clouds which is subsequently transferred to stars forming within them—consistent with the conclusions of Wolff et al. (1982).

III. ROTATION AND CLOUD ENERGETICS

Calculation of the form and stability of a rotating interstellar cloud is a formidable undertaking, as witnessed, for example, by the recent efforts of Stahler (1983*b*,*c*), Tohline (1984), Miyama et al. (1984), and Terebey et al.

(1984). The spectral line profiles expected from a particular theoretical model depend not only on the velocity and density along the line of sight through the cloud, but also on the gas temperature and the fractional abundance of the molecular species being considered. The necessary quantity of high-quality observational data required for comparison with models is becoming available and analysis is at present underway (Arquilla 1984). Here, it seems appropriate only to address the simplest question, namely, whether the rotation observed in this (biased) sample of clouds is likely playing a significant role in their energetics. In this discussion we ignore the role of internal and external pressure. The relatively small difference in linewidth between those clouds identifiable as rotating and those without this behavior, supports the view that the observed linewidths are not, in fact, primarily due to rotational motion. The effective internal pressure due to turbulence may greatly exceed the thermal pressure; magnetic fields may also be playing a role.

For simplicity, we will ignore all contributions other than rotation and gravity, and model only uniformly rotating spherical clouds. We take the density profile to be given by

$$\rho(r) = \rho_0\left(\frac{r_0}{r}\right)^{\alpha} \tag{5}$$

where ρ_0 is the density at radius r_0, and $\alpha < 3$. The kinetic energy is given by $T = \frac{1}{2}I\omega^2$, with the moment of inertia

$$I = \frac{2}{3}MR^2\left(\frac{3 - \alpha}{5 - \alpha}\right) \tag{6}$$

where M is the mass and R the radius of the cloud. The potential energy is given by

$$|U| = \frac{GM^2}{R}\left(\frac{3 - \alpha}{5 - 2\alpha}\right) \tag{7}$$

The degree of central condensation is seen to have a major effect inasmuch as the moment of inertia is decreased and the magnitude of the gravitational energy is increased for steeper density distributions. The angular momentum and specific angular momentum as calculated in Table I can also be corrected for the effect of central condensation by multiplying by the factors $\frac{5}{6}$ and $\frac{5}{9}$ for $\alpha = 1$ and 2, respectively. Substituting the kinetic energy of rotation and the gravitational potential energy into the virial theorem yields, for equilibrium

$$\omega_c = \left[2\pi G\bar{\rho}\left(\frac{5 - \alpha}{5 - 2\alpha}\right)\right]^{1/2} \tag{8}$$

where $\bar{\rho}$ is the mean mass density. We can write

$$\omega_c = \beta\omega_{CO} \quad (9)$$

where ω_{CO} is the required equilibrium angular velocity for a uniform cloud of given density, ω_{CO} (km s^{-1}pc^{-1}) = 0.037 $\bar{n}$(cm^{-3})$^{0.5}$ assuming a mean molecular weight of 2 atomic mass units. For $\alpha = 0$, $\beta = 1$; for $\alpha = 1$, $\beta = 1.15$; and for $\alpha = 2$, $\beta = 1.73$. The observational data show that many isolated dark clouds are centrally condensed (Snell 1981; Dickman and Clemens 1983; Arquilla 1984; Arquilla and Goldsmith 1984) with $\alpha \sim 2$.

A least-squares fit to the data yields the relationship (with ω in km s^{-1} pc^{-1} and $\bar{n}$ in cm^{-3})

$$\omega = 0.053\ \bar{n}^{0.4}$$

with a correlation coefficient of 0.8 and a standard deviation of the slope equal to 0.06. In Fig. 6 we show the dependence of ω on $\bar{n}$ determined from the observations, together with theoretical equilibrium curves for $\alpha = 0$ and $\alpha = 2$. We see that, while the observed ω increase somewhat more slowly with increasing $\bar{n}$ than required to satisfy virial equilibrium, the magnitudes do not differ significantly from those required for uniform density clouds [$\alpha = 0$]. The presence of a strong degree of central condensation demands angular velocities considerably greater than those observed, if rotation is to have a dominant role in the energetics of these dark clouds.

An effect possibly causing us to underestimate the angular momentum of these clouds is the inclination of the rotation axis. We observe only the projection of the angular velocity (or J, or J/M) perpendicular to the line of sight with the relationship between the observed and the true angular velocity given by

$$\omega_{OBS} = \omega_{TRUE} \sin i \quad (11)$$

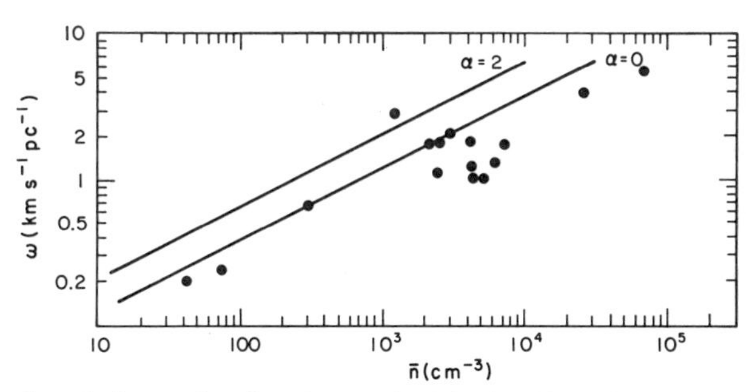

Fig. 6. Angular velocity as a function of mean cloud density. The solid lines represent the densities required for virial equilibrium (see Eq. 8) in the case of uniform density ($\alpha = 0$) and of strong central condensation $n(r) \sim r^{-2}$ ($\alpha = 2$).

where i is the angle between the axis of rotation and the line of sight. This is, of course, the same relationship as for the velocities of rotating stars. In the case of stellar rotation, the distribution of true velocities can be obtained from that of observed velocities if, for example, the plausible assumption of random orientation in space is made (cf. Tassoul 1978). In the case of the clouds studied here, there is a clear dependence of ω on the density, with the result that we must consider clouds of a particular density or those within a density interval as distinct samples. Then we find (analogously with stellar rotation) that the mean value of the observed angular velocity (at a particular density) is related to the mean value of the true angular velocity by

$$\langle \omega_{OBS} \rangle = \frac{\pi}{4} \langle \omega_{TRUE} \rangle \tag{12}$$

(Tassoul 1978), independent of the distribution of $|\omega|$. For clouds having a particular density, then, $\langle \omega_{TRUE} \rangle = 1.27 \langle \omega_{OBS} \rangle$.

This correction obviously helps bring the data and theoretical virial equilibrium curves into better agreement. However, this type of analysis must be viewed with some caution, because observational selection already has presumably eliminated clouds having an angular velocity below some threshold from consideration. Since clouds with large $|\omega_{TRUE}|$ but very small i are not identified as rotating and hence are not entering into the determination of $\langle \omega_{OBS} \rangle$, we should not correct for their presence. The fact that the correlation of the peak velocity shift in rotating clouds with cloud size (Eq. 4) is similar to the correlation of their dispersion with size, means that there is no scale size at which rotation is relatively more or less easily discernable. In the present situation, it seems difficult to justify any significant correction for projection, and we thus conclude that on the whole rotation does not play a dominant role in cloud energetics, being by itself generally insufficient to support the clouds against gravitational collapse. This view is supported by the lack of significant flattening in a manner consistent with rotational support against gravitational contraction.

IV. SUMMARY

We have used a sample of 16 interstellar molecular clouds which show evidence for rotation, to study the correlation of angular momentum with other cloud properties and to investigate the role of rotation in the energetics of these objects. The angular velocity and density are inversely correlated with cloud size, while the specific angular momentum is highly correlated with the size of the region under consideration. To explain the observed correlation, we must assume either that rotational motion is introduced over a wide range of scale sizes, or that a sequential fragmentation process takes place in which the majority of rotational angular momentum is lost in the breakup of condensations into smaller fragments, with the bulk of the angular momentum going into orbital motions. The shifts in velocity produced by rotation follow a law similar to that found for the variation of linewidth with cloud size. This agreement, together with the low degree of alignment of cloud angular momentum vectors relative to the angular momentum vector of the Galaxy, suggests that rotation in molecular clouds may result from fluctuations in the chaotic velocity field that is present. Although subject to considerable uncertainty due to possible projection effects, it does not appear that rotation alone can provide virial equilibrium for these clouds.

Acknowledgments. We wish to express our gratitude to R. Dickman. R. Snell, P. Schloerb and S. Kleiner for useful and stimulating discussions as well as generous sharing of data, and to D. Plosia and K. Lewis for expert assistance in preparing the manuscript. P. Myers, D. Black, and F. Shu contributed by careful reading and helpful suggestions for improving the manuscript. P. F. G. thanks the Observatoire de Paris, Meudon, for support and hospitality during part of the time that this chapter was being prepared. Support for astronomy research at Five College Radio Astronomy Observatory by the National Science Foundation is gratefully acknowledged.

TURBULENCE IN MOLECULAR CLOUDS

ROBERT L. DICKMAN
University of Massachusetts

It has long been appreciated that turbulence may play a crucial role in the formation of stars and planetary systems. Molecular clouds have been recognized as primary stellar birthsites for nearly fifteen years, and speculation that they may harbor widespread gas motions so chaotic as to merit the designation turbulence extends over basically the same period. Only during the past several years, however, have observational and analytical techniques matured to the point where the systematic study of stochastic motions in molecular clouds is a realistic goal. The basic aim of this chapter is to offer a primer of basic concepts and methods of analysis for observationally-oriented individuals who wish to work in this rapidly developing area. First the difficulties which beset early attempts to determine the nature of gas motions within molecular clouds are reviewed. Some aspects of turbulence as a hydrodynamic phenomenon are considered next along with an introduction to the statistical vocabulary of the subject which is required to understand the methods for analyzing observational data. A simple and useful approximation for estimating the velocity correlation length of a molecular cloud is also described. The chapter concludes with a final perspective, which considers the extent to which size-velocity dispersion correlations can serve as a probe of chaotic velocity fields in molecular clouds.

I. INTRODUCTION

Despite intensive observational efforts, star formation remains an event as yet glimpsed only by computers. An evolutionary path of uncertain duration, which encompasses the clear onset of gravitational instability in a prestellar mass to a point quite possibly far beyond the thermonuclear ignition stage (chapter by Scoville), lies largely devoid of observational data. Pending further progress in instrumentation, study of the formation of stars and planetary systems will likely remain for some time an essentially theoretical arena whose predictions can be tested only indirectly. The observational study of molecular clouds therefore possesses an importance beyond its own innate interest; it provides both initial and boundary conditions for the theoretical star formation problem.

For present purposes, we may consider the term turbulence to denote a spatially irregular, eddying state of fluid motion which is characterized by an underlying statistical order. (A more complete description is given in Sec. III.) Turbulence is never completely chaotic; on spatial scales described by the correlation length, turbulent motions begin to become coherent and shed their seemingly random, fluctuating behavior. Thus, a turbulent velocity field will appear very different on scales much larger and much smaller than the correlation length. As a consequence, the fluid stress associated with a turbulent flow has an intrinsically scale-dependent character. This fact, along with the essentially nondeterministic elements which turbulence may introduce into the gravitational instability problem, implies that if turbulence is present in molecular clouds, it may affect seriously the support and fragmentation mechanisms that govern star formation (chapter by Scalo). From a theoretical point of view, there are thus two related issues of importance: does turbulence comprise a significant component of the velocity field in molecular clouds, and if so, what does its velocity fluctuation spectrum look like, i.e., what are the important correlation lengths of the motions?

Gas density, temperature, and velocity all represent key state parameters for molecular clouds, and each poses special problems of observational determination and interpretation. But it has proven especially difficult to establish even a broadly qualitative picture of gas motions in a typical molecular cloud. Do they possess a high degree of order, or are they in fact so spatially disorganized (except possibly in a statistical sense) as to merit the designation turbulence? Until recently, convincing observational evidence on behalf of either scenario was lacking. However, since the first Protostars and Planets conference (Gehrels 1978), there have been increasing indications that turbulence may need to be seriously reckoned with in the study of molecular clouds. Analytical methods which enable one to quantify statistically the role which spatially disordered gas motions play in the interstellar medium have been known for some thirty or more years. Observational sensitivity has now improved to the point where obtaining the large bodies of spectroscopic data required for a statistical velocity-field analysis is not particularly painful; this has stimulated several groups to begin intensive velocity-field studies of molecular clouds aimed specifically at clarifying the role which turbulence may play within them (Scalo 1984; chapter by Scalo; Kleiner and Dickman 1984; Stenholm 1984; Kleiner 1985; Kleiner and Dickman 1985*a,b*).

There has also been a discernible shift in theoretical perspective over the past few years. A basic conceptual stumbling block has always existed for turbulent models of the dense intestellar medium, in the sense that at least mildly supersonic motions appear to be the rule rather than the exception in molecular clouds. As a result, attempts to attribute a highly disordered character to these motions had also to provide a physical mechanism capable of replenishing the substantial energy losses which would inevitably accompany the turbulence. While it would be an exaggeration to maintain that clear resolution of the energy problem has occurred, the present picture is much less bleak than it appeared at the time of the first Protostars and Planets conference. Possibly as a result theoretical interest in the role of turbulence in dense interstellar clouds has seen something of a renaissance in the intervening period.

All these facts suggest that a summary of current viewpoints on turbulence in molecular clouds may be useful. My basic goal is to provide a primer of concepts and methods for observationally oriented individuals who will work on this newly emerging subject. Analysis of observations is emphasized; this chapter is not intended as a comprehensive theoretical review of turbulence and its potential applications to the dynamics of star formation. (See Scalo's chapter which covers a related area—fragmentation and hierarchical structure in the interstellar medium—from just such a perspective.)

This chapter begins by reviewing the difficulties which beset early attempts to determine the nature of gas motions within molecular clouds (Sec. II). The fundamental problems here were the lack of an unambiguous observational signature for turbulence, coupled with conceptual uncertainties about the viability of such flows. In Sec. III some aspects of turbulence as a hydrodynamic phenomenon are reviewed. Distillation of so broad a topic may be of value to observers who find they need to deal with the subject; acquaintance with the statistical vocabulary of the subject is in any case essential in order to understand the analytical methods and to evaluate their shortcomings. Section IV delves into the velocity-correlation methods which can be used to study turbulence in molecular clouds. Some simple approximations are noted, as well as some potential pitfalls. It should be emphasized that the methods described make no presuppositions concerning the origin or nature of the turbulence to be analyzed; they are as capable of revealing the spectrum of supersonic magnetohydrodynamic turbulence as they are of uncovering the structure of a Kolmogorov-Obukhov flow. A final perspective, which considers the extent to which size-velocity dispersion correlations can serve as a probe of chaotic velocity fields in molecular clouds, is given in Sec. V.

II. OBSERVATIONS AND THE ISSUE OF TURBULENCE

With the exception of a handful of clearly rotating globules (Arquilla 1984), and a growing number of bidirectional jets associated with the vagaries of stellar adolescence (Snell 1983), remarkably little is known about the nature of velocity fields in molecular clouds. That there was cause for puzzlement became evident at about the same time as molecular clouds were recognized to be basic units of the interstellar medium, i.e., during the early 1970s. Almost without exception, spectroscopic observations showed emission lines far broader (from about 0.5 to 2 km s^{-1} for dark clouds, and from about 2 to $>$ 20 km s^{-1} for giant molecular clouds) than could be accounted for by thermal motions. Pressure broadening, Zeeman splitting and other more exotic mechanisms could be rather straightforwardly discarded as sources of these linewidths, and it seemed clear that the Doppler effect must be responsible. This immediately raised questions concerning the nature of the mass motions producing the line broadening. Were they highly ordered and systematic, largely chaotic and turbulent, or were they a hybrid of the two?

Molecular clouds tend to be localized objects, in sharp contrast to the much more pervasive distributions of neutral interstellar hydrogen which exist in the Galaxy. The relative compactness of molecular clouds appeared to lead to serious energy difficulties if their generally supersonic internal motions were largely attributed to turbulence. In many instances, the energy density of the postulated velocity fluctuations was so high as to produce serious doubts that the clouds in question could remain gravitationally bound. Furthermore, while all turbulent flows are dissipative (Sec. III), if the supersonic motions indicated by the observations were presumed to reflect a purely random velocity field, the inevitable radiative shock losses would quickly deplete the clouds of turbulent energy unless sources to replenish the lost energy were postulated. The power requirements for these sources appeared in many cases to be prohibitive if the coherence length for the turbulent motions was assumed to be much smaller than the cloud size. This follows from simple dimensional arguments, because in a cloud of mass M, turbulent velocity dispersion v, with coherence length L, a typical collision time for gas elements is of order L/v; one thus expects an energy loss rate of order $\sim M\mathrm{v}^3/L$. Of course, one can reduce this power loss by assuming appropriately larger values of the coherence scale (up to the limit imposed by the cloud size, at any rate), but it was widely appreciated that if such large velocity correlation lengths were present in molecular clouds, they should have been readily identified by mapping observations—and generally speaking, this was not the case. As a consequence, it was anticipated that a very robust mechanism would be required to maintain the observed velocity dispersions in molecular clouds if they were due to turbulence.

It was also occasionally noted that if turbulence were actually widespread in molecular clouds, the fact that observational evidence appeared to point toward relatively small correlation lengths led to a variety of problems in understanding molecular line profiles, particularly those of the widely employed λ2.6 mm carbon monoxide (CO) line. These problems all arose in the context of microturbulent line-formation models, in which the correlation length of turbulent motions is assumed to be much smaller than all compositional, density and source function gradients along the line of sight. While the validity of this assumption could not be tested observationally (no measurements of turbulent correlation lengths were available), it was suspected but not conclusively demonstrated that its somewhat extreme nature might be responsible for the widely predicted, but almost never observed, self-reversed line shapes characteristic of the microturbulent model (see, e.g., Leung and Lizst 1976). Additional problems occurred in understanding how the very optically thick CO line could reflect the presence of the hot, and deeply embedded stellar birthsites that it appeared to trace so well (see, e.g., Goldreich and Kwan 1975); a related issue was why cloud maps made in optically thin spectral transitions appeared so similar to those made in the 2.6 mm CO line. In both cases, it was commonly asserted that the optically thick CO could probe only the surface layers of a turbulent cloud.

Only slightly more appealing were schemes postulating velocity fields dominated by highly ordered flows. Clearly, the presence of supersonic velocities in a cloud need not entail the energy dissipation problems encountered in the turbulent scenarios if regions with large relative velocity differences are physically separated; indeed, if source geometry and flow pattern are suitably chosen, such regions need never interact.

Dynamical models of this sort are attractive in other ways. Spectral line formation becomes a basically local problem, in sharp contrast to the microturbulent case (Goldreich and Kwan 1975; see also Rybicki and Hummer 1978), and rarely produces the nonphysical line profiles which plague the microturbulent models. Even a decade ago, it was recognized that linewidths toward the active cores of giant molecular clouds were totally incompatible with rotation as a global line-broadening mechanism (Penzias 1975). Although considerably smaller widths are seen toward dark clouds and globules, a fact which led Field (1978) to propose rotation as the most satisfactory spectral line-broadening mechanism in dense clouds, observations imply that a more isotropic type of motion is responsible (Arquilla 1984). This essentially leaves one with expansion, collapse, or oscillation as coherent sources of line broadening (see, e.g., Villere and Black 1982). The first and second both define time scales $t \sim$ (cloud size)/(velocity) which are considerably smaller than 10^7 yr; this is an uncomfortable result given the relative compactness of molecular clouds and the apparently large mass fraction of molecular material in the Galaxy (see, e.g., Scoville and Hersh 1979). Collapse scenarios have additional difficulties in avoiding observationally unacceptable high star formation rates on a galaxy-wide basis (Zuckerman and Palmer 1974; but see Wilking and Lada 1983), and in accounting for the subsequent dissipation of unprocessed material left deep in the gravitational well of an evolved cloud. Oscillations, possibly regulated by an embedded magnetic field (Arons and Max 1975), have been suggested but no detailed models posed. Finally, all models which posit simple spatial-velocity relationships for molecular clouds must come to grips with the visibly irregular geometries of most dark nebulae; on the largest scales at least, these appear to be utterly incompatible with simple motions such as radial collapse.

In reviewing these same issues almost ten years ago, Penzias (1975) remarked that despite the apparently severe difficulties faced by turbulent models, the lack of a clearly favored dynamical picture for molecular clouds probably indicated a need for careful review of the reasons for skeptism of turbulent scenarios. To a considerable extent this is precisely what has occurred in the intervening decade, and the early 1980s have seen many, if not all, of the problems associated with turbulence in molecular clouds substantially alleviated or eliminated.

Let us review some of these changes in outlook. While it is by no means true that a satisfactory theory of turbulent line formation is now available, the situation has become less troublesome from several standpoints. The assertion that one cannot "see" deeply into turbulent media in a high-opacity spectral line such as the λ2.6 mm transition of CO, for example, was shown by White (1977) to be a misconception. In addition, line-formation studies were performed that relaxed the microturbulent assumption of a vanishingly small velocity correlation length (Dickman and Kleiner 1981); these revealed that even model molecular clouds with steep core-edge excitation temperature gradients could produce realistic CO line profiles (Fig. 1) when the correlation

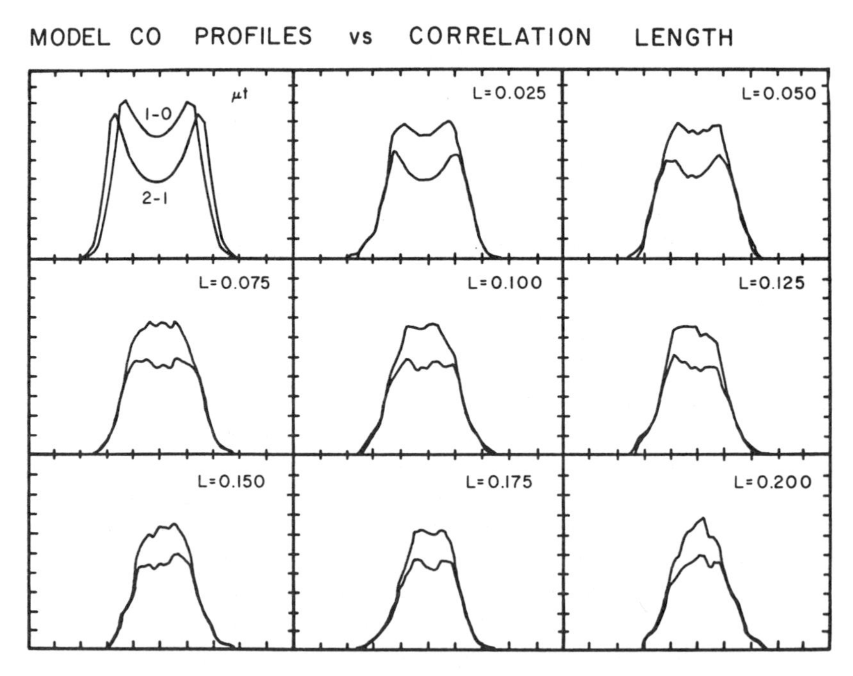

Fig. 1. Theoretical line profiles of the two lowest rotational transitions of CO in a series of turbulent giant molecular cloud models. Each panel contains spectra of each transition (the lower-frequency line being the stronger in each case), and the sole distinction between the panels is the correlation length of the model turbulence; this is given in units of the cloud diameter in the upper right-hand corner of each panel. The upper left-hand panel is the microturbulent limit of zero correlation length, and one can see clearly the deep, central self-reversals in the model spectra; features of this sort are rarely observed. The noise in the profiles is due to the Monte Carlo technique used to calculate the models.

length grew sufficiently large ($\sim$10% of the cloud size in the model shown in the figure). Models of this sort cannot as yet be considered entirely satisfactory, in part because they still tend to predict the presence of self-reversed lines when the clouds are examined at sufficiently high spatial resolution, an effect for which there is no observational support.

Another approach to avoiding line-profile pathologies is to postulate a turbulent cloud whose spectral lines are produced by numerous, small clumps; intensity contrasts in such a model are largely achieved by gradients in the number of clumps along each line of sight, rather than by direct variations in excitation conditions. While a self-consistent theoretical model of this sort has yet to be attempted in a rigorous line formation context (but see Snell et al. 1984), the model of Zuckerman and Evans (1974) serves as a prototype, both from a line formation as well as a dynamical standpoint. The extreme energy loss rate associated with small-scale turbulence is avoided in this observationally motivated picture by postulating that molecular clouds consist of many small, dense clumps, each possessing a subsonic internal velocity field, and each moving through a considerably less dense buffer medium. If the clumps are to be in pressure equilibrium with their surroundings, the buffer clearly must be much warmer than the condensations themselves. This could even allow the clumps to move subsonically relative to the ambient gas; however, the essential point of the model is that clumps must not fill the entire body of the parent cloud, so that the rate of collisional dissipation is reduced.

Penzias discussed this picture critically in 1975, and noted two basic faults: (1) dissipation may still be significant, and (2) individual clumps cannot be gravitationally bound, nor kept together by ambient pressure. With the hindsight of the past ten years, we can inquire whether these remain truly fatal problems. Only recently has the issue of dissipation been examined in detail for clump models (Scalo and Pumphrey 1982; Gilden 1984), with serious attention paid to radiative (shock) loss mechanisms and hydrodynamics. Energy loss rates are found to be substantial, but hardly insupportable (see below). The second problem, i.e., that the clumps are ephemeral, is hardly troubling at all if we recognize that this is precisely what one expects for strongly turbulent flow in a highly compressible medium; islands of enhanced density continually coalesce and then dissolve as they participate in the irregular motion of the fluid. Without the action of gravity to inhibit this evanescence, the only stable structure possessed by the fluid will be statistical in character. Indeed, it is inconceivable that supersonic turbulence could exist in a molecular cloud without dramatic density fluctuations; even for subsonic motion, a rule-of-thumb for estimating the rms amplitude of density fluctuations is $[\langle\delta\rho^2\rangle^{1/2}/\langle\rho\rangle] \sim M^2$, where M is the Mach number of the turbulence (Jones 1976).

Finally, let us consider briefly the replenishment of turbulent energy losses in molecular clouds. Collapse and star formation may themselves lead to the generation of gas velocities so chaotic as to warrant the designation turbulence. Several years ago Buff et al. (1979) showed that even the highly idealized radial collapse of an initially perfect sphere leads to the stochastic formation of unstable subcondensations. Even so, because of computational grid resolution limits, it is unclear to what extent current finite-difference hydrodynamic simulations of the gravitational collapse problem can properly follow tendencies toward chaotic motion. It is noteworthy that N-body simulations of cloud collapse (Benz 1984) develop disordered velocity field components which have appreciable spatial correlation scales, in times on the order of the free-fall time. It is also now widely accepted that most pre-main sequence stars go through an energetic mass loss phase (Snell 1983). It has been suggested that stellar winds from young T Tauri stars could regenerate otherwise decaying turbulent motions in dark clouds (Norman and Silk 1980; Dickman et al. 1980). Various external sources of energy potentially capable of exciting stochastic motions in interstellar clouds have also been proposed. Perhaps the most fundamental of these is the shear of galactic differential rotation (Fleck 1980,1981), but other sources such as OB stellar winds or supernova blast waves could also introduce mechanical energy (Cox 1979) in the form of randomized acoustic or sound waves (Spitzer 1982; Dickman and Clemens 1983).

A particularly interesting mechanism for sustaining turbulence in interstellar clouds is that recently proposed by Kegel and Völk (1983). Treating stars and interstellar gas as gravitationally coupled fluids (see also Jog and Solomon 1984*a,b*), Kegel and Völk study the effect of stellar dynamical friction on the gas. Random stellar motions lead naturally to the excitation of stochastic modes in the gas; these can include supersonic velocity fluctuations, as well as significant density fluctuations. Even more importantly, the velocity and density correlation functions (Sec. III.B) induced in a gaseous medium can be calculated from the model. Although this has not yet been done except for simplified, illustrative cases, this new formalism should be actively pursued, so that comparisons can be made with the emerging observational estimates for these quantities (Sec. IV).

III. PROPERTIES OF TURBULENCE

A satisfactory terse definition of turbulence is difficult to provide. Nonetheless, elaborating the working definition adopted in Sec. I is useful for several reasons. First, the richness and generality of turbulence as a hydrodynamic phenomenon is best clarified by reviewing its qualitative attributes. Second, familiarity with the statistical measures commonly used to characterize turbulent flow is essential, because the observational analysis of interstellar gas motions (Sec. IV) must be framed in these same terms. Third, it is important, particularly for those who would interpret their observations, to appreciate the lack of self-consistency and the essentially phenomenological orientation possessed by most theories of turbulence. The success and utility of the Kolmogorov-Obukhov theory (Sec. III.C), for example, should not blind one to the fundamental incompatibility of its basic postulates to the compressible, supersonic motions believed common in the interstellar medium; definitive proof that the Kolmogorov 2/3 law applies to gas flow in molecular clouds would almost surely necessitate a substantial revision of the picture of these objects described in the previous section.

Numerous standard works on turbulence exist, among them those by Batchelor (1956), Landau and Lifshitz (1959, ch. XVI), Tennekes and Lumley (1972), Monin and Yaglom (1975), and Leslie (1973). These should be consulted for further information on this vast area of physics. The collection of historical papers edited by Friedlander and Topper (1961) is both fascinating and valuable; significant early contributions to the subject are reprinted in English, some (like the papers of Kolmogorov [1941*a,b,c*]) very difficult to obtain otherwise.

A. Qualitative Aspects

Turbulence arises in hydrodynamic systems when the inertial forces associated with the fluid motions become so large relative to viscous restoring forces as to permit the strongly nonlinear growth of small velocity fluctuations. Irrespective of any initial order (laminar flow is perhaps the most commonly adduced example), the fluid then begins to exhibit motions over a very wide range of degrees of freedom. Mathematically, the path of the system through a generalized configuration space becomes chaotic; states initially separated by arbitrarily small intervals diverge from one another at an exponential rate (see, e.g., Swinney and Gollub 1978; E. Ott 1981; Kadanoff 1983; Ford 1983). Because a transition to turbulence can be viewed as the runaway growth of fluctuations on a large range of flow scales, the ensuing loss of any initial order is, in a practical sense, nondeterministic. This suggests that the use of statistical measures to characterize the resulting fluid state may be highly profitable.

Turbulence embraces a remarkably wide class of fluid motions, and in its most general guise involves both velocity and density fields. It is inherently compressible and three dimensional. While aerodynamicists have long known that gas compressibility can largely be disregarded at Mach numbers below $\sim$ 0.8 (see, e.g., Roberson and Crowe 1980), it is difficult to specify general circumstances in which the dimensionality of turbulence can be similarly ignored. As a result, all dimensionally simplified models and approximations should be applied with caution (von Neumann 1949). (Kraichnan and Montgomery [1980], for example, show that when the dimensionality of turbulent motion changes from two to three, the time evolution of velocity correlations alters dramatically.) The rotational character of turbulence, which is perhaps its most striking visible aspect, and which underlies the somewhat idealized picture of turbulent eddies cannot, of course, be accommodated except in more than one dimension.

Turbulence ultimately originates in the nonlinearity of fluid motion. This makes for extraordinary mathematical difficulties and unsurprisingly, workers frequently resort to dimensional analysis as a simplifying tool. Dimensional approaches can be extremely rewarding (see, e.g., Frisch et al. 1978), but as shown very beautifully by Lighthill (1955), they may also contain dangerous pitfalls which result from the very nonlinearities which motivate their use in the first place. As in all nonlinear applications, a healthy skepticism concerning the use of dimensional approximations should be retained.

Turbulence is intrinsically dissipative. While conservative dynamical systems can also exhibit chaotic behavior, the lossiness of fluid flow is responsible for much of the particular character of turbulence (Eckmann 1981; E. Ott 1981). Any nonideal fluid in motion eventually comes to rest, its translational energy lost to heat through viscosity. Thus, any turbulent system possesses an energy sink on a viscous length scale more or less equal to its molecular mean free path. (This fundamental fact serves as an important normalization constraint in certain theories, such as that of Heisenberg [1948; see also Canuto et al. 1984].) In systems where supersonic flow exists, the correlation structure of the velocity field may permit fluid elements to collide fairly frequently at relative speeds exceeding that of sound. The ensuing shock waves represent additional, direct loss modes for the medium. Supersonic turbulent motion (Sec. I) may be the rule, rather than the exception, in molecular clouds.

The literature of turbulence abounds with a great variety of dimensionless ratios, each germane to a particular class of hydrodynamic system and each useful in elucidating the onset of stochastic motions in them. The Reynolds number (Re) is perhaps the best-known of these; it serves to estimate the ratio of the inertial forces in a fluid (which tend to promote nonlinear motion) to restoring viscous forces:

$$\mathrm{Re} = VL/\nu. \tag{1}$$

Here V is the fluid velocity, L a characteristic dimension associated with the perturbing element driving the onset of turbulent instability in the flow, and ν the kinematic viscosity of the fluid (see, e.g., Tennekes and Lumley 1972). For a dilute isothermal gas with mean free path λ and sound speed c, ν can be written (see, e.g., Condon and Odishaw 1967)

$$\nu = \lambda c. \tag{2}$$

The Reynolds number is widely used to describe the onset of turbulence in initially laminar flows. This occurs for $\mathrm{Re} >> 1$ where the scale L is associated with the size of the element disrupting the smooth flow (a grid in a wind tunnel, for example).

As a general matter, dimensionless ratios like the Reynolds number should be used with care in astrophysical applications. It is obvious, for example, that because $V \simeq c$ in molecular clouds (Sec. I), the Reynolds number for these objects is essentially just the length-scale ratio L/λ. Because the molecular mean free path is enormously smaller than any macroscopic structures which one might associate with interstellar clouds, it is not surprising that one finds very large values of Re ($\sim 10^4$ to 10^6) associated with such objects. However, even for laboratory flows, the critical Reynolds number at which turbulence is initiated can range over some two orders of magnitude in different systems (von Neumann 1949). Furthermore, while viscosity and gas velocity can be estimated reliably from astronomical observation, it is unclear what value L should be assigned in evaluating the Reynolds criterion for a molecular cloud. For that matter, is it even appropriate to consider laminar flow as a reasonable starting point for the growth of disordered motions in such a system? Most dimensionless turbulence parameters like the Reynolds number have their genesis in phenomenological descriptions of fluid systems studied in the laboratory; at best, these are highly imperfect analogs for astrophysical conditions.

B. Mathematical Vocabulary

That turbulence could be analyzed as a statistical phenomenon was first appreciated by Taylor (1921) during the first quarter of this century. The use of statistical measures to characterize turbulence is now usually regarded as a nearly indispensible strategy (see Deissler [1984] for another point of view, however). We have already noted that in a practical sense, the loss of determinacy associated with the onset of turbulent flow implies that in two virtually identical systems very different patterns of turbulent motion may develop. Nonetheless, the patterns emerge from the operation of the same physical processes, subject to essentially indistinguishable initial and boundary conditions. Statistical measures emphasize this unity, extracting underlying regularities of the flow and rejecting incidental details.

There are certain difficulties associated with statistical (or stochastic) characterizations of turbulence in interstellar clouds. Defining statistical moments of the hydrodynamic variables, for example, is not entirely straightforward. While introducing an ensemble of clouds, each containing a realization of the turbulent flow is a conceptually satisfactory way to accomplish this, it is of little use in the actual computation of the moments themselves. Usually one is forced to consider various subregions of a cloud as statistically equivalent to one another (the ergodicity assumption), so that ensemble averages can be calculated as spatial averages. Furthermore, in astronomical applications, unless one is willing to assume that all nebulae of a given type are subject to basically identical physical processes, there will always be additional difficulties in interpreting cloud structure on scales comparable in size to the objects themselves. This occurs because on large scales any individual system can provide only a small number of statistically independent samples of itself; one may never then be certain if one is dealing with statistical aberrations or with large-scale properties common to the cloud ensemble.

A second and quite general limitation to statistical descriptions of turbulence is deeply rooted in the nonlinearity of hydrodynamics and is usually referred to as the closure problem. From a stochastic point of view, the actual density and velocity fields characterizing a particular region in a turbulent fluid are of limited and secondary use. Instead, their statistical moments, which can be represented symbolically as $<v^n>$, $<\rho^n>$, etc., are of primary concern (see, e.g., Batchelor 1956). Equations for these moments can be derived easily from the fluid equations. However, it is then found that the equation for $<v^2>$ depends upon $<v^3>$, that for $<v^4>$ upon $<v^5>$, and so on (see, e.g., Leslie 1973). Any attempt to solve for the density and velocity moments which are central to a statistical treatment of turbulence must first truncate this open-ended progression, and close the system of governing differential equations. Arguably, the difficulty in doing so has been the basic issue in the theoretical study of turbulent flow for at least the last forty years (see, e.g., Heisenberg 1948; Chandrasekhar 1951*a,b,*1955,1956; Leslie 1973; Canuto et al. 1984).

Perhaps the most widely employed statistical measures for describing turbulent flows are the *n*-point correlations of the fluctuating fluid fields, which express the degree of mutual coherence exhibited by the turbulence at two or more points in space and time. In general, such correlations may involve both the velocity and density, as well as their cross correlations (see, e.g., Chandrasekhar 1951*a,b,*1955); in certain cases additional dynamical variables must also be considered. For simplicity, however, we shall consider the velocity field alone in what follows. The most basic measure which one can introduce in this case is the two-point covariance tensor, C_{ij}, first used extensively by Taylor (1935). If the subscripts i and j are used to denote Cartesian velocity components, and the symbols $<>$ used to denote ensemble average, then

$$C_{ij}(\mathbf{x},t;\mathbf{x}',t') \equiv$$
$$<[v_i(\mathbf{x}',t')-<v_i(\mathbf{x}',t')>][v_j(\mathbf{x}'+\mathbf{x},t'+t)-<v_j(\mathbf{x}'+\mathbf{x},t'+t>]>. \quad (3)$$

As is common, C_{ij} has been defined here in terms of fluctuations about the mean flow, rather than in terms of the flow itself. C_{ij} specifies the average degree of mutual coherence that can be expected between the ith component of velocity fluctuations at $\mathbf{x}'$ and time t', and the jth fluctuation component lying at a vector distance (or lag) $\mathbf{x}$ away, at later time $(t'+t)$. Provided that the ensemble averages exist, C_{ij} transforms like a second-rank tensor. Although the notation involved becomes quite unwieldy, higher-order tensors can be defined in analogous fashion. Once the (infinite) set of *n*-point correlation tensors appropriate to a particular type of fluid system have been specified in a region of space and time of interest, the system can be considered solved. Even for the simplest hydrodynamic systems, this has never been acomplished in a theoretically self-consistent fashion.

It is reasonable to introduce some simplifications into our conceptual picture at this point. Temporal correlations are of course inaccessible to direct astronomical measurement and can be ignored hereafter. Furthermore, we may also disregard the issue of how velocity correlations evolve with time; this is reasonable from an observational point of view because velocity fields in the interstellar medium do not usually change appreciably on the scale of human lifetimes. Finally, we shall assume that we are dealing with spatially homogeneous turbulence, so that all ensemble-averaged properties depend not upon position, but only upon *differences* in position. (Mathematically, this amounts to the assumption that the flow is "wide-sense stationary"; see, e.g., Leslie [1973].) With these assumptions, the two-point covariance tensor C_{ij} is now much simpler:

$$C_{ij}(\mathbf{x}) = <[\mathrm{v}_i(\mathbf{x}')-<\mathrm{v}_i>][\mathrm{v}_j(\mathbf{x}'+\mathbf{x})-<\mathrm{v}_j>]>. \tag{4}$$

Clearly, C_{ij} is both symmetric and reflection-invariant with respect to vector lag:

$$C_{ij} = C_{ji}; \quad C_{ij}(\mathbf{x}) = C_{ij}(-\mathbf{x}). \tag{5}$$

The autocorrelation tensor, $\alpha_{ij}(\mathbf{x})$, is a commonly used normalized form of C_{ij}:

$$\alpha_{ij}(\mathbf{x}) \equiv C_{ij}(\mathbf{x})/C_{ij}(0). \tag{6}$$

It has the intuitively appealing property that $\alpha_{ij}(0) \equiv 1$; even the most chaotic systems possess perfect self-coherence.

The autocovariance tensor is a basic descriptive tool in studies of turbulent flow. In an incompressible fluid, for example, the influence of velocity fluctuations on the mean flow is described by the Reynolds stress tensor, $\rho C_{ij}(0)$ (Tennekes and Lumley 1972), which is simply proportional to the autocovariance tensor at zero lag. Furthermore, the trace of the stress tensor is simply the turbulent energy density of the fluid. In the case of an isotropic system, where rotational invariance demands that $C_{ij}(\mathbf{x}) = C_{ij}(|\mathbf{x}|)$ and all diagonal components be equal [e.g., $C_{11}(x) = C_{22}(x)$, etc.], this last point is particularly easy to see, because

$$C_{\mathrm{ii}}(0) = <[\mathrm{v}_i(x) - <\mathrm{v}_i>]^2> = <\mathrm{v}_i^2> - <\mathrm{v}_i>^2. \tag{7}$$

This is just the squared turbulent velocity dispersion along any one of three (statistically independent) Cartesian axes of the system.

In a turbulent fluid, covariance measures like $C_{ij}(\mathbf{x})$ must gradually decline in magnitude at nonzero lags. This is plausible on purely physical grounds, because very distant points in a turbulent system can be expected to lose eventually any deterministic links. This does not imply, however, that the fluid's coherence must decline monotonically. In complex systems, particularly ones with multiple sources and sinks of energy on different length scales, there is no fundamental mechanism which prohibits a partial recorrelation of fluctuations at intermediate lags. Thus, velocity fluctuations may display a somewhat enhanced coherence on large scales while showing much less organization on smaller scales; density fluctuations may exhibit similar behavior (see Kleiner and Dickman [1984] for an astronomical example).

These considerations lead naturally to the concept of correlation scales for turbulent systems. These are generally defined as lags at which the coherence of a hydrodynamic variable declines to $(1/e)$ of its zero-lag value. From the definition of the velocity autocovariance given above, it is clear that, in general, a correlation scale may depend upon the Cartesian axes (i and j) which have been chosen for a coordinate system, as well as upon the directions in which correlations are evaluated. In the case of isotropic turbulence, one can speak of the correlation *lengths* of a velocity field, because simple scalars then describe the statistical coherence of the system; but again there may be more than one of these. In the idealized case of isotropic turbulence characterized by a single velocity correlation length, it is frequently asserted that the correlation length corresponds approximately to the size of the largest turbulent eddies in the system.

Alternatives to the covariance tensors, of which the two-point velocity measure introduced above is the simplest example, can also be defined; the covariance measures are not unique. One alternative to the two-point covariance which has found frequent use in astronomy and related fields is the structure tensor (see, e.g., Münch 1958; Kaplan and Klimishin 1964; Scalo 1984; Tatarskii 1961; Peebles 1980). For a homogeneous, time-independent flow the structure tensor is defined:

$$S_{ij}(\mathbf{x}) \equiv \langle [\delta v_i(\mathbf{x}') - \delta v_j(\mathbf{x}' + \mathbf{x})]^2 \rangle \tag{8}$$

where

$$\delta v_i \equiv v_i - \langle v_i \rangle . \tag{9}$$

Hence

$$S_{ij}(\mathbf{x}) = \sigma_i^2 + \sigma_j^2 - 2C_{ij}(\mathbf{x}) \tag{10}$$

where

$$\sigma_i^2 \equiv <[v_i(\mathbf{x}) - <v_i>]^2> \equiv <v_i^2> - <v_i>^2. \tag{11}$$

For the especially simple case of isotropic turbulence, the structure tensor simply reduces to the structure function, related to the velocity autocorrelation function by:

$$S(x) = 2\sigma^2[1 - \alpha(x)]. \tag{12}$$

While the structural and covariance tensors of a turbulent fluid can be put into complete one-to-one correspondence theoretically, as can be seen from Eq. (10), they differ in several important respects when applied to data sets of finite extent (see, e.g., Peebles 1980; Scalo 1984; Kleiner and Dickman 1985*a*). Most notable of these distinctions is the tendency of structural measures to suppress linear trends across the data; this is evident from the definition (Eq. 8). In Sec. IV we briefly discuss some consequences of this bias.

C. Limitations of Turbulence Theory

A complete theory of turbulence would deduce, *ab initio* from the hydrodynamic equations, the correlation tensors for a fluid with specific initial and boundary conditions, or at least would provide compact algorithms for their computation. That such a theory does not exist in general terms is not surprising, but it is rather remarkable that *no* deductive theories whatever exist, even for what might be judged very simple model systems. For example, no derivation of the two-point velocity correlation function of isotropic, fully-developed, incompressible turbulent flow has ever been given, although approximations, some exceedingly serviceable, abound.

In the absence of any exact theory of turbulence, it is useful to examine briefly what is perhaps the best-known approximate theory of turbulence, namely that due to Kolmogorov (1941*a,b,c*) and, independently, to Obukhov (1941). Of special interest here is the extent to which it can constitute a viable benchmark for astrophysical turbulence, particularly in the dense interstellar medium.

The Kolmogorov-Obukhov theory is pertinent to incompressible, isotropic, homogeneous and fully-developed turbulence. Its major prediction (and triumph) is the so-called Kolmogorov 2/3 law which asserts that typical squared velocity differences in the model fluid (essentially just the structure function discussed above in Sec. III.B) grow as the two-thirds power of the distance separating the sampled regions. Symbolically, for a particular Cartesian velocity component v_i:

$$<[v_i(\mathbf{x}_1) - v_i(\mathbf{x}_2)]^2> \sim (|\mathbf{x}_1 - \mathbf{x}_2|/L)^{2/3}. \tag{13}$$

Here the scale L is fundamental to the physical system being modeled, and can be related simply to the correlation length of the turbulence (see, e.g., Panchev 1971). Although the original derivations are more elaborate, simple physical arguments serve readily to establish Eq. (13) if one assumes a uniform energy density per Fourier mode for the fluid (see, e.g., Kaplan 1966; Kaplan and Pikel'ner 1970); this is essentially the same as assuming that once energy is injected into the system (at scales $> L$), it is simply transferred without loss to progressively smaller scales until dissipated by viscosity. Notice that regardless of its merits, this sort of energy cascade is patently inapplicable when direct energy sinks, such as radiative shock losses, exist in addition to viscosity.

The utility of the Kolmogorov model as a practical, analytical tool is impressive in certain disciplines. For example, in studies of propagation phenomena such as scintillation or radiance fluctuations driven by turbulence in the terrestrial atmosphere, the two-thirds law appears to be closely adhered to (see, e.g., Tatarskii 1961; Baars 1970; Mavrokoukoulakis et al. 1978). Certainly, the basic model requirements of incompressibility (i.e., Mach number $<< 1$) and inertial-range scales (far from energy-deposition and viscosity-related effects) are well satisfied by the meterology underlying the studies above; however, it is not entirely clear why boundary-layer anisotropies and influences like wind-shear effects do not mitigate the model's applicability more strongly.

It has been known for a number of years that the Kolmogorov-Obukhov model is both flawed and highly limited in the range of length scales which it describes; this is particularly so at high Reynolds numbers (Townsend and Stewart 1951). It also fails to correspond well with experiment in more subtle ways. For example, the model fails to replicate satisfactorily observed velocity correlations among three or more points, a problem which may stem from a breakdown in the model's incompressibility assumption. In a pictorial sense, what happens is that smaller turbulent eddies fail to populate the stratum of fluid fluctuations as uniformly as their larger counterparts, an effect termed "intermittency." Ways of modifying the Kolmogorov-Obukhov theory to account for it are reviewed by Frisch and his collaborators (1978), who themselves present a basically dimensional approach to the problem which they call the beta model.

How suitable is the Kolmogorov-Obukhov theory as a conceptual framework for the study of astrophysical turbulence? Strongly in its favor as a theoretical tool is its innate simplicity and its success in describing certain classes of phenomena such as those noted above in connection with the Earth's atmosphere. Moreover, although there are relatively few published determinations of the spectrum of turbulent gas motions in the interstellar medium (or of associated phenomena, such as nebular brightness fluctuations), it is not unusual to find note taken of a closeness to, or consistency with, the Kolmogorov-Obukhov structural index of 2/3 (Armstrong and Rickett 1981;

and Klimishin 1964; Kaplan 1966; Chandrasekhar and Münch 1952; 1958; R. B. Larson 1981). Recent efforts by Ferrini and coworkers et al. 1982,1983*a*) to apply the beta model of Frisch et al. (1978) essentially accept this apparent concordance, and attempt to ascertain whether intermittency corrections lead to a more realistic picture of the dense interstellar medium, in which molecular clouds are regarded as rather small turbulent structures.

Nonetheless, the Kolmogorov-Obukhov theory is rooted largely in the phenomenology of essentially incompressible, subsonic dynamical systems which in their simplest form contain but one source of turbulent energy, injected on the outer scale appropriate to the system, and but one sink, viscosity. The basic suitability of this sort of model to describe the generally supersonic, compressible motions of an interstellar medium possibly subject to energy deposition on several scales at once, should be viewed with caution if not with outright skepticism. Especially troubling insofar as molecular clouds are concerned is how and where gravity fits in. Molecular clouds are manifestly dominated by gravity. Any phenomenological description of turbulence which relies on the statistical characterization of velocity fluctuations as a function of region or scale size, and which at the same time ignores the interplay of gravity with thermal and scale-dependent turbulent pressure (see, e.g., Chandrasekhar 1951*c*), is severely handicapped from the outset. It should also be emphasized strongly that determining a structural index near the Kolmogorov 2/3 value does not necessarily imply that the fluid studied is in a state near that predicted by the theory.

Observation, of course, is the ultimate arbiter of any physical theory. At present, despite early indications that it might be relevant in describing stochastic velocity structure in molecular clouds, the Kolmogorov-Obukhov theory appears to be inapplicable to molecular clouds (Scalo 1984; Kleiner and Dickman 1985*a,b*). Because other credible trial theories pertinent to interstellar turbulence which are capable of concise test are simply nonexistent, it appears essential at this point to assemble, as far as possible, a purely observational phenomenology of the subject which can then be subjected to *ex post facto* theoretical analysis. How this may actually be accomplished is discussed in Sec. IV.

IV. OBSERVATIONAL TECHNIQUES

Despite the fact that an infinite set of density and velocity correlation tensors is formally necessary for a complete description of a compressible, stationary turbulent flow, the two-point velocity correlation can be regarded as a basic starting point in the study of turbulence in molecular clouds (Sec. I). Given the daunting prospects for a reasonably accurate *ab initio* theory of turbulence appropriate to these objects, we have argued that the determination of this quantity should be approached observationally. In this section we consider how and to what extent this can be done using spectroscopy.

Some limitations must be accepted at the outset. Spectral lines provide only collective probes of the line formation path. We have previously emphasized, however, that turbulence is a fundamentally three-dimensional phenomenon. Thus, although an observer may have the ability to make a high-resolution map of an interstellar cloud, relations along the line of sight can never be certain, except possibly in a statistical sense. This means that directly determining the behavior of a particular correlation tensor with respect to three-dimensional vector lag is impossible unless assumptions are made regarding the tensor's symmetries; invariably, all one can do is to determine average correlations along the line of sight. Instrumental noise imposes additional limits to the estimation of velocity correlations; so also does the finite extent of molecular clouds, particularly if a source happens to be comparable in extent to the size of a suspected correlation scale. The number of statistically independent data samples available is then irrevocably limited, and as a consequence, so is the signal-to-noise ratio of any deduced correlations (Kleiner and Dickman 1984,1985*a*).

A. Basic Concepts

In the 1950s it was first pointed out that correlation properties of interstellar turbulence could be probed by spectroscopic observations (Kampé de Fériet 1955). The essential idea is that by mapping an interstellar cloud in an optically thin spectral line, one statistically samples the velocity field of the source. Studies of this type in the optical part of the spectrum have been used to examine the behavior of turbulence in diffuse nebulae (see, e.g., Münch 1958; Kaplan and Klimishin 1964); radio investigations of neutral atomic clouds in the 21 cm line of H I have also been carried out (see, e.g., Baker 1973; Chiezé and Lazareff 1980). Until recently, however, molecular clouds were not subjected to a similar scrutiny. This lapse is largely attributable to the fact that as in any statistical approach, fairly large data sets are required for a meaningful study; only recently have millimeter-wave receivers become sensitive enough to acquire the necessary data bases in reasonable amounts of time.

Let us consider a hypothetical interstellar cloud which we shall suppose to be sampled at high spatial and spectral resolution in an optically thin resonance line of some atom or molecule. We shall also assume the source to be so large and statistically uniform that meaningful ensemble averages can be constructed by spatially averaging the data (ergodicity assumption). For the present, we shall ignore the smoothing produced over the cloud face by the finite spatial resolution of the telescope (e.g., diffraction and error patterns in a radio telescope, or slit size/detector sensitivity compromises in an optical telescope). The impact of instrumental noise on the analysis which follows is also ignored here for clarity (but is treated in detail by Dickman and Kleiner [1985]).

Let us denote by $I(u,\boldsymbol{\xi})$ the specific intensity of radiation observed at position $\boldsymbol{\xi}$ on the face of the cloud at a frequency corresponding to Doppler-shift velocity $u = [(\nu - \nu_0)/\nu_0]c$ with respect to the rest velocity of the source. For simplicity, and without loss of generality, the cloud velocity may be assumed to be zero. The velocity centroid of the spectral-line profile at $\boldsymbol{\xi}$ is defined:

$$\mathrm{v}_c(\boldsymbol{\xi}) \equiv \int \mathrm{d}u\, uI(u,\boldsymbol{\xi}) \,/\, \int \mathrm{d}u\, I(u,\boldsymbol{\xi}) \tag{14}$$

where the integrations extend over the profile.

We shall adopt a z-axis orthogonal to $\hat{\boldsymbol{\xi}}$ for the purpose of measuring distances into the model cloud, and we shall denote by $\mathrm{v}(z,\boldsymbol{\xi})$ the line-of-sight component of the velocity field at position $(z,\boldsymbol{\xi})$. One can then show (Dickman and Kleiner 1985) that for optically thin radiation, the measured centroid is related to the gas motions which produce the line profile via:

$$\mathrm{v}_c(\boldsymbol{\xi}) = \int \mathrm{d}z\, \mathrm{v}(z,\boldsymbol{\xi})\, S_\nu(z,\boldsymbol{\xi})\, \kappa_0(z,\boldsymbol{\xi}) / \int \mathrm{d}z\, S_\nu(z,\boldsymbol{\xi})\, \kappa_0(z,\boldsymbol{\xi}). \tag{15}$$

Here the symbols S_ν and κ_0 denote, respectively, the source function and line-center absorption coefficient at location $(z,\boldsymbol{\xi})$ in the cloud. Both integrals extend from 0 (at the front of the cloud), to a depth Z.

Equation (15) provides a link between velocity field and spectral profile. In order to forge it into a workable tool, it is necessary to assume that the cloud is essentially uniform along each sampled line of sight.[a] Letting Z denote the depth of the cloud, we then have the especially simple relationship:

$$\mathrm{v}_c(\boldsymbol{\xi}) = Z^{-1} \int_0^Z \mathrm{d}z\; \mathrm{v}(z,\boldsymbol{\xi}). \tag{16}$$

This result is quite intuitive. Photons in an optically thin spectral line are emitted without subsequent interaction with their parent medium. If the medium is excited uniformly along each line of sight, the photon spectrum replicates perfectly the velocity distribution of the emitting particles, and the line-profile centroid coincides perfectly with the average line-of-sight velocity of the gas.

[a]Because the source function and absorption coefficients appear in both the numerator and denominator above, to lowest order, departures from constancy will leave the results which follow essentially unchanged. The impact of deviations from the obviously unrealistic ideal of uniform line-of-sight depth in the formalism are more difficult to characterize. Generally, variations in cloud depth lead to spurious, nonstationary trends in the velocity correlations. However, these primarily contaminate the estimation process at lags on the order of the depth variations. These are apt to be rather large, whereas the correlation techniques we discuss are most valuable in explicating structure on moderate-to-small cloud scales.

Let us consider how the quantity $v_c(\xi)$ might be expected to fluctuate in a map of our hypothetical cloud if turbulence is present. The amplitude of the fluctuations is most simply described by their variance, and this can depend upon only two basic parameters: (1) the overall magnitude of turbulent velocities in the cloud; and (2) the ratio of the cloud depth Z to the largest correlation scale L possessed by the velocity field along the line of sight. This ratio is the only dimensionless ratio in the problem if opacity effects are unimportant, and it determines the number of statistically independent velocity-field samples which typically occur along a sight line: $N \sim Z/L$. When this number is small, the statistical character of the velocity field governs the way in which independent flow cells combine along the individual sight lines to produce particular values of $v_c(\xi)$; measurements of centroid velocities then exhibit substantial positional fluctuations. On the other hand, when the correlation length is much smaller than the depth of the cloud, a typical line formation path contains so many turbulent cells that their average velocity closely approaches that of the whole cloud; thus, when N becomes very large, all line-profile centroids tend to the same fixed value.

To the extent that the N turbulent cells along a typical line of sight can be treated as statistically independent regions whose barycentric velocities reflect the statistics of the underlying velocity field, the foregoing considerations can be made quantitative by using simple statistical arguments. Let us denote the line-of-sight velocity of a turbulent cell by u. By assumption, the statistics of the cell velocities are identical to those of the gas in the model cloud. Letting the variance of the turbulent velocity fluctuations in the cloud be σ^2, we then have:

$$\begin{aligned} <u> &= 0 \\ <u^2> &= \sigma^2. \end{aligned} \tag{17}$$

Now suppose that an average of N cells populates each line of sight where a centroid velocity v_c is measured. Then

$$v_c \cong N^{-1} \sum_i u_i \tag{18}$$

where the sum extends over the N cells. The average and variance of the centroids are then easily calculated:

$$\langle v_c \rangle = \left\langle N^{-1} \sum_i u_i \right\rangle = N^{-1} \sum_i \langle u_i \rangle = 0$$

$$\langle v_c^2 \rangle = N^{-2} \sum_i \sum_j \langle u_i u_j \rangle. \tag{19}$$

Because the turbulent cells are assumed to be statistically independent, $<u_i u_j> = <u^2> \delta_{ij}$, so that

$$<v_c^2> = N^{-1} <u^2>. \quad (20)$$

Thus, the variance in line-of-sight centroid velocity of the N cells which comprise a typical observation path should be given approximately by:

$$\sigma^2_c \sim \sigma^2/N = \sigma^2/(Z/L). \quad (21)$$

Despite these somewhat simplistic arguments, Eq. (21) turns out to be a remarkably robust approximation to the more exact result which can be derived using the autocorrelation function (Dickman and Kleiner 1985). This makes it useful for estimating the (largest) correlation scale in a molecular cloud from a set of observations without the need for elaborate analysis. Once σ^2_c has been ascertained, and Z has been suitably estimated (usually by assuming a depth comparable to the mean transverse extent of the source), it is only necessary to determine the value of σ^2 in order to apply Eq. (21). This can be done in several ways. The simplest way is to average together the spectral profiles which comprise the data set. If there are no strongly nonstationary trends in the velocity field, such as rotation or bipolar flow, the resulting composite spectrum usually will be nearly Gaussian with variance σ^2. [The Gaussian distribution is basically a consequence of the central limit theorem (statistical law of large numbers; see, e.g., Papoulis 1965) and is considerably less informative about the turbulence than one might at first suppose (Leslie 1973).]

It may happen, however, that one does not have available the spectral profiles (which may occur in dealing with published data), or that one wishes to examine a rotating cloud. In this case, one should, if possible, first remove (by least-squares fitting) any clearly systematic effects in the data, and recalculate σ^2_c. (This will avoid any possibility of obtaining a length scale characteristic of the systematic trends which could mask other statistical scales in the data.) Provided that one can assume that individual spectral line profiles can be adequately represented by Gaussians, it is then possible to use the following simple relation to determine σ^2 (Dickman and Kleiner 1985):

$$\sigma^2 = \sigma^2_c + <\Delta v>^2/8 \ln 2 \quad (22)$$

where $<\Delta v>$ is the average full width at half intensity of the individual spectra. Thus, the largest correlation scales of a turbulent molecular cloud can be probed in an approximate fashion by applying the simple formulae above.

The foregoing arguments rest basically upon the fact that the observability of turbulent velocity fluctuations in an interstellar source is degraded by a factor $\sim (L/Z)$ due to cloud depth effects. A similar loss of sensitivity is produced by resolution-smoothing in observational data, such as that due to the finite beamwidth of a radio telescope. If the angular resolution of the instrument is ϕ radians, and one studies a source d parsecs distant with velocity correlation length L, then irrespective of the path-length dilution, there will be an additional decrease in sensitivity to turbulent velocity fluctuations of order $\sim L/(\phi d)$. The lesson here is clear: nearby molecular clouds are the best candidates for the study of turbulence, and for this reason the Taurus and Ophiuchus nebular complexes ($d \cong 140$ and 160 pc, respectively) merit special attention.

B. Two-Dimensional Analysis

In order to obtain the fullest picture of the velocity field of a turbulent interstellar source, including estimates for any secondary coherence lengths and the actual form of the autocorrelation function, a more elaborate treatment of the data is necessary; one must form the two-dimensional autocorrelation function of the velocity centroid map. A theoretical framework for interpreting the correlation surfaces obtained with this procedure is straightforwardly derived from the definition (Eq. 16) of the spectral line centroid and need not be repeated here (Dickman and Kleiner 1985); techniques for maximizing the signal-to-noise ratio of the derived correlations at various lags are summarized by Kleiner (1985) and Kleiner and Dickman (1984,1985*a*). The basic relation between the calculated autocorrelation function of spectral velocity centroids, and the true autocorrelation of the line-of-sight velocity field α is the following:

$$\frac{\langle v_c(\boldsymbol{\xi}) v_c(\boldsymbol{\xi} + \mathbf{x})\rangle}{\langle v_c(\boldsymbol{\xi}) v_c(\boldsymbol{\xi})\rangle} = \frac{\int [1 - (z/Z)] \alpha[(z^2 + \mathbf{x}^2)^{1/2}]\, dz}{\int [1 - (z/Z)] \alpha(z)\, dz}\,. \tag{23}$$

Kleiner (1985) discusses the impact of the line-of-sight averaging which occurs in Eq. (23). He points out that because the correlation function is usually a decreasing function of lag, the triangular weighting tends to insure that spacings near the observed lag $\mathbf{x}$ contribute most heavily to the integrals. Thus, although $\alpha(z)$ cannot be reproduced exactly by correlating a map of spectral profile centroids, a reasonably good estimate, which may tend to overstate slightly the scale of any correlations (cf. Kleiner and Dickman 1985 *a*), can nonetheless be made.

It should be emphasized that the analysis to be carried out here is inherently two-dimensional. Even should one wish to assume that one is dealing with isotropic turbulence, the cloud in question may nonetheless reveal initial anisotropies; these may be the result of systematic motions (like rotation), or may simply be due to unavoidably impoverished statistical sampling (such as for correlated motions on scales an appreciable fraction of the cloud size). As a practical matter, it is therefore essential to begin the study of any interstellar cloud in a fully two-dimensional manner. The interpretation and removal of nonisotropic motions (by high-pass filtering, for example) can then be carried out, and any remaining rotationally invariant correlations studied.

The necessity of performing a fully two-dimensional evaluation of the correlation function (Eq. 23) suggests that fast Fourier transform (FFT) techniques be used in implementing the schemes described above. Essentially, one simply calculates the power spectrum of the data set using two-dimensional FFT's (Brenner 1976), and performs a final fast transform to obtain the desired two-dimensional correlation. For large data bases, the resultant savings in computation time made possible by this somewhat circuitous approach makes for a highly appealing strategy (cf. Kleiner 1985). [For N data points, use of a FFT in place of a conventional slow Fourier transform entails a fractional time saving of roughly $(N\log_2 N/N^2)$.] However, several imperatives must be adhered to in its use. The most important of these is that an astronomer must begin with a spatially rectilinear data set. Unevenly spaced data must first be interpolated into regular grid form before the FFT's can be carried out; this inevitably introduces an uncertainty into the derived correlations whose magnitude is difficult to estimate.

Practically speaking, the inherent attractiveness of an FFT approach to cloud correlation studies is mitigated to whatever extent pre- or post-processing of the data is necessary. High- and low-pass filtering fall into this category. Accordingly, there are circumstances in which an approach to turbulent velocity fields might be made as expeditiously by using two-dimensional structure measures. This approach has been lucidly discussed by Scalo (1984). Probably the most distinctive attribute of the structural method is its relative insensitivity to linear data trends already noted in Sec. III. If such trends can be legitimately regarded as spurious, small-scale correlations which might otherwise be lost in a naive correlation analysis, will be highly visible when a structural analysis is performed. This is a two-edged sword, however; the ability to discern certain features of potential physical significance in one's data is also sacrificed. My own preference is to use correlation, rather than structural functions (Kleiner and Dickman 1984,1985*a,b*), although as computing costs continue to decline, there may be little computational incentive to prefer one technique to the other.

V. PERSPECTIVES

The observational study of turbulence in molecular clouds is a subject still in its infancy. Far too small a sample of objects has been subjected to a systematic two-dimensional velocity analysis[a] for there to be even preliminary hints

[a]As of this writing, we are aware of only three efforts in this area: (a) Kleiner's studies of the Taurus complex using the λ2.7 mm ^{13}CO line, made jointly with the author (1985*a,b*); (b) Scalo's (1984) structure function study of the ρ Oph core region, using the λ2.7 mm $C^{18}O$ line, and (c) Falgarone and Puget's in-progress study of cold condensations in the Cygnus region, using both the above molecular transitions (E. Falgarone, personal communication).

regarding which, if any, turbulent properties may be common to the velocity fields of molecular clouds. In this connection, it is important to place in perspective the size-linewidth relations for interstellar clouds which were first intensively explored by R. B. Larson (1979).

R. B. Larson (1981) pointed out that over a range of length scales between about 0.5 and 60 pc in the interstellar medium, molecular cloud size appeared to correlate closely with internal velocity dispersion. A reasonably good fit to the data was obtained with the relation $\Delta v \propto R^b$, with $b = 0.38$. This result was consistent with an earlier study (R. B. Larson 1979) which also embraced neutral hydrogen clouds, and which contained data on lengths scales ranging up to nearly 1 kpc; a power-law index $b = 0.33$ was inferred from the earlier data set. Subsequent studies (Leung et al. 1982; Myers 1983) have apparently confirmed the general existence of such a correlation for molecular clouds, although the scatter in the data is appreciable and the value of the index b therefore ill-determined (in part, this is because uncertainties in cloud distances make reliable size estimates correspondingly difficult).

Does this correlation mirror the presence of a global, scale-invariant turbulent process within the Galaxy? In originally making this suggestion, Larson (1981) noted the closeness of his derived power-law index to the Kolmogorov-Obukhov value of ⅓ which characterizes an incompressible, dissipationless turbulent cascade (recall that the "2/3 law" applies to *squared* velocity differences (Sec. III.C) whereas the correlation studies in question here involve velocity *dispersions*). Regardless of any hesitation one may have in accepting the applicability of such a theory to interstellar velocity fields (cf. Sec. III.C), it is not difficult to demonstrate rigorously that, on average, the velocity dispersion within regions of size R in a Kolmogorov-Obukhov flow does scale as $R^{1/3}$ (Kleiner and Dickman 1985*a*).

However, if the interstellar size-velocity dispersion correlation is due to a turbulent cascade, the outer scale (Sec. III.C) of the process must exceed considerably the size of the largest region in which the correlation is observed to apply. Moreover, it is easily shown that the correlation length associated with Kolmogorov turbulence is ~½ the outer scale (Kleiner and Dickman 1985*a*), so that a value for the former quantity of at least some tens of parsecs must be expected in the dense interstellar medium on the basis of Larson's hypothesis. This is so large that highly correlated, turbulent motions should be easily detectable in molecular clouds using the techniques reviewed in this chapter.

Attempts to do so have failed in both the central regions of the ρ Oph cloud, and in the Taurus complex of dark nebulae (Scalo 1984; Kleiner and Dickman 1985*a*). Furthermore, in Heiles' Cloud 2, where the velocity field has been resolved into a rotational component carrying about ⅓ the kinetic energy of the gas, and an isotropic turbulent component, carrying the remaining ⅔ of the energy, the correlation structure of the turbulence does not at all resemble the Kolmogorov-Obukhov model (Kleiner and Dickman 1985*b*); instead, the correlation length of ~ 0.05 pc in Cloud 2 marks a relatively sharp break in the velocity spectrum, beyond which the turbulent flow rapidly becomes incoherent. While a recent study of correlations in the dark cloud Barnard 5 by Stenholm (1984) has reported some evidence for a ⅔ law, the volume of data was limited and studied with one-, not two-dimensional techniques; we are therefore inclined to be cautious about Stenholm's result until a larger data base for this cloud has been rigorously analyzed. In summary, we are at present aware of no compelling direct evidence that Kolmogorov-Obukhov turbulence is operative in molecular clouds.

The foregoing considerations certainly do not preclude a fundamentally dynamical origin for Larson's linewidth-cloud size relation. However, regardless of its physical origin, its applicability as a probe of turbulence in molecular clouds rests on the postulate that the gas motions in all such objects are part of a single, homogeneous velocity field within the Galaxy. Given the profoundly differentiated character of the interstellar medium, this seems difficult to justify. As a result, the correlation methods reviewed in this chapter, or their structural variants, presently appear to be the best available methods for probing the properties of turbulent motion in dense clouds.

Further refinements of the techniques are clearly called for. It seems ludicrous to discard the overwhelming bulk of information contained in a spectral line profile, and to employ only its first velocity moment as turbulence diagnostic. Related to this is the need to establish a formalism for estimating velocity-density cross correlations. Until this is possible, observations will not be truly capable of painting a complete picture of molecular clouds, and we shall lack observational keys to understand the influence of compressibility in the dense interstellar medium. Finally, it is essential that error estimation capabilities keep pace with the unavoidable imperfections of the observations, as well as with the possibly less-avoidable simplifying assumptions which underlie the interpretations of the data. If they do not, the creation of statistical sand castles which lack solid foundations seems an all too worrisome possibility.

Acknowledgments. I am pleased to acknowledge many useful conversations with my coworker S. Kleiner on much of the material in this chapter. I am also grateful to R. Fleck for first stimulating my interest in the question of whether properties of interstellar turbulence might be observable, and for numerous instructive conversations on the topic. I am grateful to him and to D. Black for their constructive comments in the cause of improving an early version of this chapter. I would also like to acknowledge the comments of an anonymous referee on the same early draft. Finally, I am deeply grateful to D. Black and M. Matthews for their extraordinary editorial patience with this manuscript.

STAR FORMATION: AN OVERVIEW

NEAL J. EVANS II
The University of Texas at Austin

Studies of clusters, associations, and field stars provide a kind of fossil record of the star formation process. Recent studies of clusters indicate that star formation may proceed over a very long time, with lower-mass stars forming first and higher-mass stars only later. Studies of regions of current star formation provide snapshots of the process and a knowledge of the conditions which precede star formation. Infrared observations have identified regions of star formation, and molecular line studies are characterizing their properties. Infrared mapping and photometry have revealed many compact sources, often embedded in more extensive warm dust associated with a molecular cloud core. More detailed study of these objects is now beginning, and traditional interpretations are being questioned. Some compact sources are now thought to be density enhancements which are not self-luminous. Infrared excesses around young stars may not always be caused by circumstellar dust; speckle measurements have shown that at least some of the excess toward T Tauri is caused by an infrared companion. Consequently, interpretation of infrared excesses in terms of disks should be viewed with caution. Spectroscopic studies of the dense, star-forming cores and of the compact objects themselves have uncovered a wealth of new phenomena, including the widespread occurrence of energetic outflows. New discoveries with the Infrared Astronomical Satellite and with other planned infrared telescopes will continue to advance this field.

I. THE FOSSIL RECORD

The astronomer contemplating the study of stellar evolution faces problems similar to those of the biologist who attempts to reconstruct the process of biological evolution because the process generally occurs on time scales much longer than those encompassed by scientific investigation. The data at our disposal consists of fragmented and incomplete fossil records, together with locations where evolution is unusually rapid or otherwise clear cut, from which we hope to piece together the general outlines of the process. In the particular case of star formation, the H-R diagrams of open clusters and associations provide a kind of fossil record of individual realizations of star-formation process. Like the biological fossil record, they are incomplete and plagued by various selection effects: the high-mass stars may have evolved off the main sequence before the low-mass stars reach it; mass segregation may cause our census to undercount the low-mass stars; and some theory of how the main sequence is reached is necessary for a valid interpretation. If these problems can be overcome, we would hope to learn the distribution of masses which formed (the initial mass function) and the time scales of formation (e.g., was it sudden or spread over a long period?). Further, do all clusters give the same answers? Either way, theories of star formation must account for either the similarities or the differences between clusters.

Since most stars apparently form, not in clusters, but in looser aggregations, perhaps associations (M. S. Roberts 1957; G. E. Miller and J. M. Scalo 1978), a thorough understanding of star formation would include a study of the initial mass functions of associations and field stars. Because stellar associations are not gravitationally bound, they are very poorly preserved fossils, and one is readily misled by selection effects. The field stars, on the other hand, are like a great jumble of bones washed up by a river, representing the remains of star formation at a wide range of times in galactic history and locations in the galaxy (Scalo 1985*a*). Thus, while field stars represent the majority of stars, their evolutionary history is extremely tangled. Since chapters by Scalo and by Wilking and Lada will discuss these topics in more detail, I will restrict myself to mentioning a few of the general conclusions of these studies and to pointing out several outstanding puzzles.

First, the studies of open clusters seem to indicate that initial mass functions vary from cluster to cluster, but the kinds of incompleteness mentioned above make this conclusion uncertain (Scalo 1985*a*). Second, the initial mass function for associations is consistent with the composite cluster IMF and the field star IMF over the range of 3 to 10 $M_\odot$, where the data are reasonably complete (Scalo 1985*a*). Third, the time scale of star formation appears to vary from cluster to cluster, as measured by the length of the main sequence. Since conventional theory of stellar evolution indicates that massive stars will evolve off the main sequence before low-mass stars can evolve *to* it, only a certain range of masses can be on the main sequence simultaneously if all stars formed at the same time. Cluster HR diagrams indicate that this range of masses is exceeded in many cases and the length of the main sequence gives a lower limit to the time interval over which stars have formed. These time intervals vary from cluster to cluster and are extremely long in some cases: the most dramatic example is the Pleiades, where the age of the high-mass stars is

$\sim 7 \times 10^7$ yr and the apparent age of the lower main sequence is $\sim 2.5 \times 10^8$ yr (Stauffer 1980). While other clusters are less extreme, many indicate a range of stellar ages (Iben and Talbot 1966). An immediate conclusion is that clusters form over a range of time which is not short compared to other relevant evolutionary time scales—star formation is not coeval. Faced with an earlier and less extreme form of this problem, Herbig (1962*a*) suggested that low-mass stars form first and high-mass stars form only near the end of cluster formation. This phenomenon could in principle result from a stochastic process in which a constant mass distribution function is sampled randomly, because the average time required to sample the high-mass end will be greater; hence high-mass stars will not, on average, form as early in the history of the cluster. This stochastic model does however predict that low-mass stars will continue to form, and some should appear above the main sequence. The deficiency of such stars in some clusters cannot be explained by a stochastic process and requires that low-mass star formation ceased before high-mass star formation began. In fact, stars seem to have formed in a temporal sequence of increasing mass in NGC 2264 (M. T. Adams et al. 1983). This peculiar result may lead one to question conventional time scales for pre-main sequence evolution. These time scales are also suspect because many regions of current star formation show an apparent decline in star formation rate during the last 10^5 yr (Stahler 1983*a*). The latter decline may be resolved by new considerations of where stars make their first appearance on the HR diagram—their birthline—according to Stahler (1983*a*), but the decline of low-mass (0.1 to 0.5 $M_{\odot}$) star formation in NGC 2264 appears to have occurred over 3×10^6 yr ago. Unless theoretical time scales are further modified, we must find reasons for clouds to form stars sequentially in mass. If the first stars heat the cloud or inject turbulent energy, the Jeans mass will increase ($M_J \sim T^{3/2}$, or $V^3_{\rm turb}$ if turbulent energy is equated to an effective temperature). If, on the other hand, fragmentation processes determine the initial mass function, a minimum mass for fragments exists because small fragments become so opaque that the radiative cooling time becomes longer than the free-fall time. This minimum mass is a weak ($\sim T^{1/4}$) function of temperature (Spitzer 1978). More detailed examination of these questions are necessary, but at the moment, these observations appear to present a major puzzle.

II. REGIONS OF CURRENT STAR FORMATION

An alternative to examining the fossil record is to study regions of rapid evolution. With a few exceptions, like FU Orionis events and flare activity, the time scales for early stellar evolution are still long compared to the span of recorded observations. On the other hand, we have in regions of current star formation a series of snapshots of different epochs in the star formation process. If all star formation regions follow roughly the same temporal sequence, we may attempt to construct a history of the average star formation process by assembling enough snapshots into a temporal sequence. This is a major advantage we have over the biologists since terrestrial evolution does not repeat itself. Even in the astronomical case, the evidence cited above for variations from cluster to cluster in initial mass function and time interval of star formation must make us cautious about the resulting average history. Further, we have the problem of assigning the correct times to the various snapshots. Finally, we have only a few examples of regions which appear to be forming clusters, although we should be able to find more (see chapter by Wilking and Lada).

Another purpose for studying regions of current star formation is to describe the environment for star formation. The conditions in regions of star formation must be determined so that theoretical models can be made to reflect realistic situations. Let us turn then to the question of how we discover and characterize regions of current star formation. Infrared observations have been the primary means used to search for stars in the process of formation. At first these infrared searches were concentrated in the vicinity of compact H II regions, based on the premise that still younger objects might be associated with the young hot stars that ionize the gas in H II regions (see, e.g., Wynn-Williams and Becklin 1974). Later, the advent of millimeter-wavelength astronomy provided a larger frame for the picture. It is now clear that nearly all reasonably compact H II regions are associated with molecular clouds which are invariably much larger and more massive than either the H II region or the aggregate of the stars ionizing it. It is also clear that the dark clouds that have long been known to be associated with T Tauri stars are merely molecular clouds which are near enough to us to produce a substantial dark area on the sky. Maps of these molecular clouds in the spectral lines of carbon monoxide (CO) revealed regions of elevated temperature which have turned out to be excellent hunting grounds for higher-mass young stars.

The result of a decade of this work is that we now have a large list of regions of star formation (cf. Wynn-Williams 1982) and detailed studies of many of them. The searches have, however, been biased by application of search techniques already proven successful. Because many of these techniques involved searching where known stars had already formed, we may not have found many regions where star formation is just beginning. This situation will soon be rectified; in regions where it is not limited by confusion, the Infrared Astronomical Satellite (IRAS) survey will include essentially all stellar and protostellar objects surrounded by significant amounts of dust out to a distance of $\sim$1 kpc for a luminosity of 1 $L_{\odot}$ (Werner 1982). If we can separate these sources from other kinds of infrared sources, a much less biased and a more complete picture of star formation will result.

With the prospect of such a picture soon to emerge, it may be foolish to attempt a review of star formation now. If, however, we attempt to summarize the *current* picture, something like the following emerges. Star formation appears to occur invariably in molecular clouds. This idea has considerable appeal. After a period of controversy over whether molecular clouds are (Bash and Peters 1976; R. S. Cohen et al. 1980) or are not (Scoville et al. 1979) confined to spiral arms, a consensus has emerged that molecular clouds are at least enhanced in spiral arms, while not being totally confined to them. This distribution is reasonably consistent with the traditional view, based largely on the blue colors of spiral arms in other galaxies, that massive stars are born in spiral arms. Furthermore, molecular clouds are cold and massive objects, conditions which should favor star formation if simple Jeans mass arguments have any validity.

If we try to develop a more detailed scenario for star formation, we quickly encounter inconsistencies and puzzles. Unlike diffuse atomic clouds, which are confined by pressure from the hotter intercloud medium, essentially all molecular clouds are gravitationally bound. A simple analysis leads to the conclusion that nearly all molecular clouds should be collapsing at the free-fall rate. Even at the lowest densities thought to exist in molecular clouds ($n \sim 100$ cm^{-3}), the time for free-fall collapse is $t_{ff} = 4 \times 10^7 n^{-1/2} = 4 \times 10^6$ yr. Since estimates for the total mass of molecular clouds in the galactic disk range upward from 10^9 M$_\odot$, a naive interpretation would suggest a star formation rate of ≥ 250 M$_\odot$ yr^{-1}, which is much larger than conventional estimates. Furthermore, most of the molecular gas is concentrated into very large clouds or complexes with $M = 10^5$–10^6 M$_\odot$, much bigger than open clusters or associations, which have typically 10^3 M$_\odot$ of stars. Evidently, the process of star formation on large scales is inefficient in some sense; only a small fraction (0.1 to 1%) of a molecular cloud forms stars.[a] This formal solution, low efficiency of star formation, is not satisfactory because it also requires explanation: what prevents the overall collapse of massive molecular clouds? Fragmentation can channel the collapse into multiple star formation, avoiding the formation of a supermassive object, but how is the gravitational binding energy overcome? How does a gravitationally bound object of 10^6 M$_\odot$ become a 10^3 M$_\odot$ cluster or association of stars and 10^6 M$_\odot$ of unbound gas? While stellar winds, H II regions, and supernova explosions have all been proposed as means to disperse the unused gas, they have also been suggested as triggers to further star formation. The complexity of hydrodynamical processes may preclude a clear theoretical analysis of cloud dispersal through these processes. Observational results do present a clear challenge to theory; CO observations of clusters indicate that the molecular cloud is dispersed by the time the last O9 star leaves the main sequence, corresponding to a duration of 20–40 $\times$ 10^6 yr for the cloud, once *massive* stars have formed (Bash et al. 1977).

---

[a]The efficiency appears to be much higher on smaller scales: the portion of a cloud forming a bound cluster must have a higher efficiency under certain assumption; and the efficiency in forming an individual star also appears to be reasonably high.

III. FORMATION OF A STAR

Let us now consider the process of formation of individual stars. The basic assumption in most theoretical calculations of star formation is that one can isolate the formation of a single star from the effects of its environment, that is, one begins with a mass of gas which collapses to form into a star. Based largely on the pioneering work of R. B. Larson (1969), several theorists developed models for hydrodynamical collapse. These spherically symmetric models inevitably found that a core forms when the density reaches a point where infrared photons are trapped. The core then contracts more slowly than the free-fall velocity toward the main sequence. Spherical accretion and, eventually, nuclear reactions provide a luminosity source embedded in the accreting envelope for a considerable length of time. Such an object would appear only as an infrared source until the envelope cleared, revealing a visible object. The term "protostar" is often used to describe those objects in which nuclear reactions have not yet begun. After nuclear reactions begin, massive stars are expected to achieve surface temperatures sufficient to produce many photons capable of ionizing hydrogen. Massive stars may then pass through a phase which can be described as a "cocoon star" (K. Davidson and M. Harwit 1967). Such a star, on or near the main sequence, and its ionized gas are surrounded by a dust shell. The pressure inside the dust shell eventually disperses the shell, leaving a visible star with a compact H II region.

Since compact H II regions are well-known objects, the observational challenge contained in the theoretical calculations was to find objects in the earlier stages of the process: collapsing masses of gas, protostars, and cocoon stars. Attempts to identify collapsing portions of molecular clouds have generally been frustrated by the difficulty in extracting dynamical information from the shapes of molecular lines. The collapse phase of an individual star formation event will also be difficult to identify by its dust emission because it is not expected to be very luminous nor hot until after a core has formed. Some regions of nearby globular molecular clouds have been found to contain cold ($T_D < 25$ K), low luminosity ($< 10\ L_\odot$) sources of far-infrared emission (Keene et al. 1983; Beichman et al. 1984). While these may be examples of very early stages of gravitational collapse, alternative explanations are also possible (Beichman et al. 1984). On the whole, I doubt that any unambiguous examples of the collapse phase are known.

Calculations (see, e.g., Yorke and Shustov 1981) of the predicted energy distribution of protostars agree roughly with those of many objects found in molecular clouds: generally a peak in flux density around 5 to 20 μm, with broad spectral depressions at wavelengths corresponding to dust grain resonances (e.g., the 9.7-μm silicate feature). Consequently, many protostar candidates exist (cf. Wynn-Williams 1982); the problem is that other objects could look quite similar to protostars. Cocoon stars, or even post-main sequence stars which have ejected a dust shell, can have energy distributions very similar to those of protostar models. Indeed sensitive searches for radio continuum emission or infrared recombination lines have revealed that many protostar candidates have ionized regions inside dust shells and might better be called cocoon stars. Given the rules of the game, massive infrared objects can only lay provisional claim to the title of protostar; if observations with better sensitivity reveal ionized gas, the object falls into the less exciting category of cocoon stars. If we also require evidence for accretion, then it cannot be said that we have found any unambiguous massive protostars.

The evolution of low-mass ($M \lesssim 5\ M_\odot$) objects is rather different. The time between formation of a core and the initiation of nuclear reactions can become so long that all the envelope can accrete, leaving a visible object, which may even wander outside the molecular cloud. These objects are sometimes called protostars, but if ongoing substantial accretion is required for protostellar status, they do not qualify. They are usually referred to as pre-main sequence stars (PMS) or young stellar objects (YSOs), and they include such objects as T Tauri stars.

This section has presented a very simplistic picture of individual star formation as we might have viewed it at the time of the last protostars and planets meeting (Gehrels 1978). The purpose has been to motivate some of the subsequent observational developments. These developments will in turn indicate the need for substantial modifications to this picture.

IV. RECENT OBSERVATIONAL DEVELOPMENTS

High spatial resolution studies of infrared emission at 10 to 20 μm have revealed that many emission regions break up into several sources (Beichman et al. 1979). Thus, it appeared that formation of very compact clusters was a common phenomenon. In a review, Wynn-Williams (1982) has found that more than half of the protostars he catalogued were members of double or multiple groups with sizes less than 0.25 pc. In this situation, the picture of a single, isolated star forming may not be realistic (Wynn-Williams 1982). However,the prevalence of compact clusters has been called into question recently by observations in the first-discovered and prototypical infrared cluster, the Kleinmann-Low nebula in the Orion molecular cloud. Polarization studies at 3.8 μm have shown that the polarization vectors form halos around two of the compact sources, IRc 2 and the BN object (Werner et al. 1983). The high degree of polarization ($> 30\%$) identifies scattered light as the primary source of polarization, rather than dichroic absorption. The polarization is high at the location of IRc 3 and IRc 4, two prominent sources at 20 μm, but low toward IRc 2 and BN. These results suggest that IRc 3 and IRc 4 are not self-luminous. This possibility has recently been confirmed by high spatial resolution observations at 2 to 30 μm (Wynn-Williams et al. 1984). These maps reveal that the color temperatures determined from 8 to 12.5 μm and from 12.5 to 20 μm peak only on BN and IRc 2 and are constant over the rest of the Kleinmann-Low nebula, including IRc 3 and IRc 4. The 20 and 30 μm emission peaks, in contrast, coincide with the regions of strong scattering of 3.8 μm emission. Wynn-Williams et al. interpret these peaks as resulting from heating by the self-luminous objects, principally IRc 2. If this situation obtains in other regions where multiple sources have been found, we may have to revise the notion that massive stars commonly form in very tight clusters.

An analogous technique has been applied to compact H II regions, for which radio continuum maps with high spatial resolution often reveal multiple sources. These sources could be self-excited H II regions, each with its own star, or dense clumps. Observations of ionization structure through infrared fine-structure lines (Lacy et al. 1982) have shown that some of the sources are clumps which may be the ionized skins of neutral condensations, remnant from the molecular cloud. Such condensations appear very dramatically in VLA (Very Large Array) maps of M17 by Felli et al. (1984), who suggest clump parameters (size $\sim$ 0.24 pc, density $\geq 10^5$ cm^{-3}, mass $\sim$ 36 $M_\odot$) which are in reasonable agreement with those deduced for the adjacent molecular cloud (Snell et al. 1984). Further high spatial resolution studies of clumps in molecular clouds should be valuable in constraining the initial conditions for star formation with parameters more realistic than those derived by neglecting the small-scale structure.

Meanwhile, observations with still higher spatial resolution have shown that a number of compact infrared sources are in fact double sources (Dyck and Howell 1982; Howell et al. 1981; Neugebauer et al. 1982) with spatial separations of 700 to 3000 AU. In the case of Mon R2 IRS 3, McCarthy (1982) finds evidence for a third, more extended (2″), component, whose colors indicate that it is scattered light from an envelope of dust centered on the double. The most dramatic example of this type is the speckle observation showing T Tauri to be a double star, with the cool (600 to 800 K) companion perhaps accounting for T Tauri's infrared excess between 2 and 20 μm (Dyck et al. 1982). The separation is only 80 to 150 AU. The nature of the infrared companion is as yet unclear, but Hanson et al. (1983) have suggested that it may be a giant gaseous protoplanet which is accreting matter from a disk around T Tauri. Regardless of the exact nature of the T Tauri infrared companion, its discovery calls into question the traditional interpretations of infrared excesses around young, optically visible stars. In particular, interpretations involving disks should be viewed with considerable caution until direct evidence is available on the geometry. Interferometric infrared techniques with substantial baselines will be essential for studying the nature of infrared emission regions around young stars; these techniques promise to be among the most important avenues for further exploration of star formation over the next decade.

Just as observations with high spatial resolution in the infrared have challenged traditional pictures of star formation, so have observations with high spectral resolution opened new avenues of exploration and, by indirect means, allowed study of the environment of young stars at spatial resolutions still higher than those obtainable by direct means. Given the picture of protostar evolution presented in Sec. III, the game plan seemed clear: find a protostar candidate with no radio continuum emission or infrared recombination lines, and find evidence for infall. The latter problem could be attacked by high-resolution infrared spectroscopy; using the infrared source as a background lamp, one should be able to identify infalling gas by virtue of its redshift relative to the rest of the molecular cloud, the velocity of which can be accurately determined from radio spectroscopy.

When this plan was followed in the case of the BN object (Scoville et al. 1983*a* and references therein), the results were dramatic and totally contrary to expectations based on the picture discussed above. While a CO rovibrational system at 9 km s^{-1} (the ambient cloud velocity) is present, there are also distinct systems at -18 and -3 km s^{-1}, indicative of gas which is blueshifted, rather than redshifted relative to the ambient cloud. Also present in the spectrum are broad ($\Delta v \sim 200$ km s^{-1}) hydrogen recombination lines and CO overtone bandheads in emission at $\lambda \sim 2.3$ μm. These fascinating results will be presented in more detail in the chapter by Scoville.

No other protostar candidate has been studied in as much detail as BN, which is the brightest of these objects, but a similar study has now been made for NGC 2024 IRS 2 (J. H. Black and S. P. Willner 1984). In this case, only the ambient molecular cloud was seen in the CO rovibrational absorption lines; the velocity and width of these lines agree with measurements of emission lines at radio and millimeter wavelengths. As in BN, a wide ($\Delta v = 137$ km s^{-1}) Br γ line was seen.

The dominance of outflow over infall toward BN is less surprising when placed in the context of other developments. Indeed, Grasdalen (1976) had already detected a hydrogen recombination line from BN, and radio continuum emission has now also been detected (Moran et al. 1983). Consequently, one would have suspected that BN might have passed beyond the protostellar phase. Furthermore, millimeter CO lines revealed broad ($\Delta v \sim$ 150 km s^{-1}) wings superposed on the narrow spike caused by the ambient cloud (Zuckerman et al. 1976; Kwan and Scoville 1976). Finally molecular H_2 quadrupole rovibrational emission was also seen in this region (T. Gautier et al. 1976). Further studies of this emission have shown that it arises in hot ($T \sim 1000$ to 3000 K) gas which has been shocked (cf. Beckwith et al. 1983 and references therein). Also arising from this gas are high J (up to 34) rotational lines of CO which appear throughout the far infrared (D. M. Watson 1982). All these data point toward a powerful outflow in the core of the Orion molecular cloud. The dominant outflow appears to be driven not by BN but by the more deeply embedded and highly luminous ($L \sim 10^5$ $L_{\odot}$) IRc 2 (Downes et al. 1981). This picture of outflow has been combined with the polarization and color temperature results described above to synthesize a new model of the Orion core as a clumpy cavity centered on IRc 2; most of the other infrared sources are interpreted as irregularities in the material defining the edge of the cavity (Wynn-Williams et al. 1984). An alternative view (Werner 1982; Werner et al. 1983) is that the other infrared sources may be colder cloud fragments, perhaps even the long-sought protostars. Persuasive evidence of infall in these cold objects, probably based on a narrow, long-wavelength, spectral line will be necessary to substantiate this conjecture. IRc 4 is especially interesting in this regard because of its association with a dense region revealed by highly excited radio NH_3 emission (Zuckerman et al. 1981) and recently detected far-infrared NH_3 emission (Townes et al. 1983).

V. THE OUTFLOW PHENOMENON

While the Orion region has received the most detailed scrutiny, the phenomenon of outflow around young stars has been increasingly recognized as extremely common. Bally and Lada (1983) have found evidence for such outflows in about 35 regions of star formation. While most of the flows studied by Bally and Lada are driven by quite luminous stars, some are not. Beckwith et al. (1983) argue that low-mass stars are likely to produce the bulk of the flows. Indeed, a CO $J = 2 \rightarrow 1$ survey of relatively lower-mass pre-main sequence stars (T Tauri and related objects) in molecular clouds has found that $\sim$ 30 to 50% show evidence of outflow (Levreault 1983). Thus, outflow seems to be a nearly ubiquitous phase of early stellar evolution (cf. Genzel and Downes 1983). The majority of the flows show evidence for bipolarity, as a redshifted lobe in one direction is matched by a blueshifted lobe in the opposite direction. The classic example of a bipolar outflow is found in the dark cloud L 1551 (Snell et al. 1980). Apparently driven by an embedded infrared source of modest (30 $L_\odot$) luminosity (Beichman and Harris 1981), the redshifted and blueshifted lobes are very distinct and well-collimated. Several blueshifted Herbig-Haro objects are associated with the blueshifted CO lobe and proper motion studies indicate a transverse motion away from the central infrared source at velocities of $\sim$ 150 km s^{-1} (Cudworth and Herbig 1979). Finally, M. Cohen et al. (1982) have detected a compact ($\sim 2''$) radio continuum structure which is also elongated and aligned with the much larger CO flow structure.

While the velocity and other properties of the outflow vary considerably from source to source, systematic patterns suggest a common phenomenon as the ultimate source of the flows. For example, the mechanical luminosity in the high velocity flows ($L_{HVF} = ½\, MV^2/\tau$, where M is the swept-up mass, V is the velocity, and τ is the lifetime of the phenomenon), is an increasing function of L_*, the total stellar luminosity. While it is reassuring that $L_{HVF} <$ L_* in all cases, the mechanism for accelerating the flow is probably not radiation pressure. The required driving force ($F = \dot{M}V$) invariably exceeds that available from radiation pressure (L_*/c) by 10^2 to 10^3. Unless photons can be scattered 10^2 to 10^3 times before escaping, an alternative driving force must be found. While optical depths in some molecular clouds are high enough (cf. Phillips and Beckman 1980), the albedo is not sufficiently high to prevent degradation of optical and ultraviolet photons to longer wavelengths where they will quickly escape. The same arguments apply to those sources of outflow which have been identified by virtue of their 2 μm H_2 emission (cf. Beckwith 1981). Consequently, most outflows are attributed to stellar winds which are not driven by radiation pressure. The bipolarity may be explained as an intrinsic property of the wind or as an initially isotropic wind which is channeled by circumstellar (Snell et al. 1980) or interstellar (Cantó et al. 1981; Königl 1982) disks. Even models for intrinsically bipolar flows make use of nonspherical geometry in the form of an accretion disk (Torbett 1983).

Leaving aside the details, it is quite clear that the bipolarity of the flows challenges the relevance of any spherically symmetric models of star formation. Rotation and magnetic fields are obvious choices for producing a favored axis, and the increases in magnetic field strength and rotational velocity that might accompany collapse make them likely candidates for the energy reservoir which drives the outflow. A model of rotationally driven winds has been proposed by L. Hartmann and K. B. MacGregor (1982*a*), while a model using magnetic bubbles produced by rotation to drive the flow has been suggested by Draine (1983). Further theoretical effort can be expected; detailed predictions will be necessary for comparison with the increasing sophistication of the observational probes. More generally, the current situation indicates that rotation and magnetic fields will have to be included in any realistic hydrodynamical calculation of protostar formation, a daunting prospect given current limitations on computational power.

The apparent ubiquity of outflow from young objects may also have implications for some of the questions about large-scale star formation. C. A. Norman and J. Silk (1980) proposed that stellar winds from stars in the T Tauri phase may provide sufficient energy to support a molecular cloud against overall collapse. They argue that the stellar winds stimulate the formation of other low-mass stars and a feedback process produces a slow and steady rate of low-mass star formation which avoids the extremes of cloud disruption or overall collapse. A criticism of this model is that it requires a very high density of T Tauri stars ($\gtrsim 10$ pc^{-3}), but the much higher mass-loss rates and forces ($\dot{M}V \sim 10^{-5}$ to 10^{-1} M$_\odot$ yr^{-1} km s^{-1} as opposed to 3×10^{-6}, as assumed by Norman and Silk) associated with the outflows revealed by CO and H_2 studies will decrease the required density. Bally and Lada (1983) find that these outflows may be sufficiently numerous to support molecular clouds, depending on various uncertain assumptions. Taking a particular model for rotationally driven winds to provide support, Franco and Cox (1983) find that consistency with total galactic star formation rates of 5 to 10 M$_\odot$ yr^{-1} (see, e.g., L. F. Smith et al. 1978) can be achieved if only 2×10^8 M$_\odot$ of the galactic molecular material ($\sim$ 10%) is contained in structures with a density $\sim 10^3$ cm^{-3}.

VI. SUMMARY

Recent observational advances have quite dramatically changed our picture of star formation. The process is much more complex and violent than we imagined in the late 1970s. The search for evidence of infall has been rewarded with nearly ubiquitous evidence of outflow. How do we reconcile this evidence with the generally held belief that stars form from contracting regions of molecular clouds? Perhaps the answer lies in the aspherical geometry implied in most of the outflow regions. The new picture of star formation may require outflow along one axis, accompanied and perhaps driven by infall in a plane perpendicular to that axis. In this case, one may be justified in saying that "outflow means infall," but the Orwellian overtones of such statements render them particularly suspicious at the present time. Such a combined infall/outflow model may also be related to the way in which a star sheds its angular momentum. If a small amount of material can be ejected to large distances, a large amount of angular momentum can be carried off thus allowing infall to continue.

More detailed calculations involving rotation and magnetic fields are required to provide firm predictions of the observational consequences of such models. Another question which needs examination is the role of the environment; how does the presence of nearby young stars, or of very inhomogeneous clouds and radiation fields, affect the formation of an individual star? The formation of binary stars should also be calculated, since high spatial resolution studies are beginning to provide observational data on this subject. Finally, the time scales for pre-main sequence evolution should be reevaluated in order to clarify the implications of the open cluster observations discussed in Sec. I.

In the observational area, two major avenues for further investigation are evident. Essentially, they represent further application of the techniques of high spatial and spectral resolution discussed above. To advance both these efforts, large telescopes are important. Infrared speckle interferometry on a new generation of large groundbased telescopes (diameters from 7 to 15 m), which are planned for the mid 1990s, will allow us to probe still smaller scales and assess the importance of binary systems in star formation. An alternative approach involves the use of several separated, preferably mobile, telescopes used as an interferometer. Large groundbased telescopes will also allow high spectral resolution studies of more protostar candidates to assess the generality of the phenomena associated with BN. The combination of high spectral resolution with speckle techniques (Dyck et al. 1983), when applied to the star formation problem, can provide a probe of protostellar structure in temperature, density, and velocity.

While much useful work can be done from the ground, infrared telescopes in space will be essential in developing a full understanding of star formation. The Space Infrared Telescope Facility (SIRTF), being planned for the early 1990s, will provide still more sensitive photometric studies of star-forming regions and, with appropriate instrumentation, detailed spectroscopic study of many more star-forming regions like Orion. Another exciting prospect is the Large Deployable Reflector (LDR), a large ($\sim$ 20 m) space telescope being planned for the late 1990s. LDR would achieve $1''$ resolution at 100 μm, bringing a wholly new clarity to the picture of star formation in our own and in other galaxies. Together with European efforts like the Infrared Space Observatory (ISO) and the Far-Infrared Space Telescope (FIRST), these space telescopes will revolutionize our understanding of star formation.

Acknowledgments. I would like to thank my colleagues, especially H. Dinerstein, B. Wilking, R. Levreault, C. Beichman, and L. Blitz, for suggestions and comments. This work was supported in part by grants from the National Science Foundation and the National Aeronautical and Space Administration.

INFRARED SPECTROSCOPY OF PROTOSTARS AND YOUNG STELLAR OBJECTS

N. Z. SCOVILLE
California Institute of Technology

Near-infrared spectroscopy offers a unique opportunity to observe young stellar objects (possibly protostars) and their circumstellar nebula. The atmospheric windows at $\lambda = 2$ *to 5* μm *enable the study of molecular rotation-vibration bands of CO and* H_2 *in addition to recombination lines of H I. Recent observations giving kinematic resolution of 6 to 20 km* s^{-1} *are reviewed for the Becklin-Neugebauer (BN) object in Orion. The observations suggest lingering accretion in a neutral shell* (R $\approx$ *25 AU) surrounding the star and its outflowing ionized wind.*

A major theme of infrared astronomy has been the study of protostars and young stellar objects (YSOs). Maps of dust clouds in the near infrared have revealed numerous compact sources with no apparent optical counterparts. These continuum data at $\lambda = 2$ to 20 μm serve to pinpoint the location of embedded "stars" recently formed within the massive dust clouds. Often these sources are clustered on a scale of 10 to 30 arcsec (Beichmann et al. 1979). Their spectra often have a low color temperature (200–600 K) suggesting that the luminosity from a central hot source, if such exists in the interior, must be degraded into the near infrared via absorption and reemission in the surrounding dust shroud. From the observed fluxes and color temperatures the implied radii of the dust photospheres are in the range 20 to 100 AU.

Observations at longer infrared wavelengths ($\lambda = 50$–350 μm) have been most useful in estimating the total luminosity within a given star formation region. Although generally of too low spatial resolution to pinpoint accurately the particular near-infrared object, which is the dominant source of luminosity, these observations provide a critical constraint (the total luminosity) needed to model the overall evolution of such star formation regions. These data have unfortunately, provided few clues regarding the evolutionary status of the individual infrared sources. Whether the observed luminosity sources derive their energy from nuclear reactions, or whether they are truly protostellar in the sense that the observed luminosity has a gravitational origin is never clear from photometric studies alone.

It is the development of near-infrared spectroscopy which holds forth the greatest promise for understanding the central embedded sources, their spectral type and luminosity. Spectroscopy also enables study of the immediate circumstellar gas: its chemical makeup, ionization state, temperature, and density. Measurements of the gas dynamics are of course critical to pin down whether the central star is still accreting gas or is expelling its protostellar shroud.

In this chapter I will review recent results in the field of near-infrared spectroscopy of protostars, drawing heavily upon observations of the Orion nebula. Though our studies here are in a rather formative stage, they underscore the potential of spectroscopy to probe young stellar objects. Observations of the infrared continuum in Orion are covered in the chapter by Evans in this book, and the far-infrared spectroscopy of the larger cloud-core region has been summarized in a review by D. M. Watson (1984).

I. THE ORION CLOUD CORE

Due to its proximity, the core of the Orion molecular cloud has become a prototype for the study of massive star formation. A most fascinating aspect of the many studies there is the great diversity of phenomena. Within a region of about 40 arcsec in size ($R \leq 2 \times 10^{17}$ cm), one sees a clustering of luminous sources, highly supersonic molecular gas flows, and shock-heated molecular hydrogen emission. The relationship of the energetic phenomena seen in the cloud core to recently formed stars has become a central theme of studies in this region.

The primary spectroscopic probes available in the near infrared include the recombination lines of atomic hydrogen (e.g., Br α and γ lines) at 4.05 and 2.16 μm and rotation-vibration lines of H_2 and CO. The recombination lines originate from extremely localized H II regions near two of the infrared sources: BN and IRS 2 (Hall et al. 1978; Scoville et al. 1983*a*). The molecular lines arise from gas more widely distributed in the cloud core. The H_2 and CO emissions originate from highly excited ($T > 600$ K) neutral gas and the absorption in CO from generally low-excitation molecular gas in the ambient molecular cloud along the line of sight to the continuum sources. A most impressive aspect of near-infrared spectroscopy is the ability to probe diverse regions ranging from circumstellar H II regions, presumably at 10^4K to molecular gas in shock fronts at 2000 K and in the ambient cloud at 100 K.

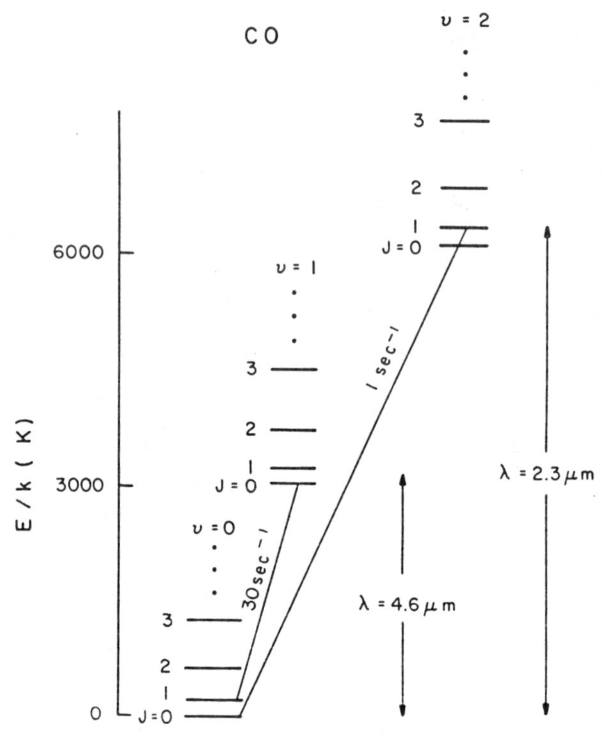

Fig. 1. Energy levels and radiative decay rates for the first three vibrational states of CO. The decay rates include the sum of R and P branch transitions in the near infrared. Rotational energies are not drawn to scale.

To illustrate the types of molecular transitions observed in the near infrared, Fig. 1 shows the energy levels of CO in the first 3 vibrational states. Within each vibrational band (e.g., v = 0) there exists a series of rotational energy levels, $J = 0, 1, 2, \ldots$, separated by increasing energy as a function of increasing J. Permitted electric dipole transitions correspond to $\Delta J = \pm 1$ and the first pure rotational transition ($J = 1 - 0$) with an energy-different equivalent to 5.5 K occurs at a wavelength of 2.6 mm. The transitions observed in the near infrared correspond to vibrational quanta change by $\Delta v = 1$ or 2 (e.g., v = 0 − 1 or v = 0 − 2). The former, occurring at a wavelength of 4.6 μm, is designated the fundamental band; the latter, at a wavelength of 2.35 μm, is designated the first overtone band.

Also shown in Fig. 1 are the spontaneous decay rates for the vibrational transitions. These decay rates are approximately a factor of 10^8 faster than those encountered in the pure rotational transitions. Thus, since the vibrational energy quanta correspond to $E/k = 3000$ K, the observation of significant emission arising from either of the CO v = 1 and v = 2 states necessarily

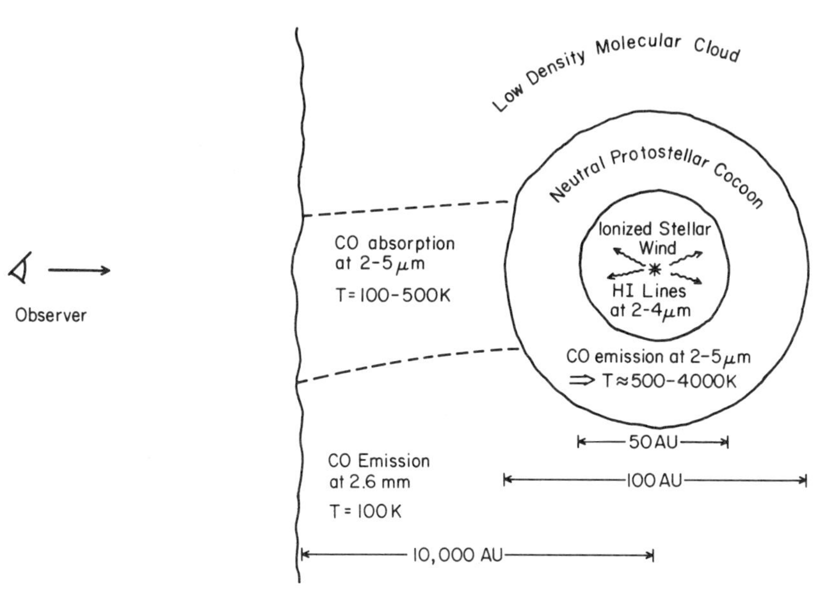

Fig. 2. A schematic view of the types of regions probed by different spectroscopic transitions in a typical star formation region such as the Orion molecular cloud.

implies a high temperature for the emission region and if comparable amounts of emission are seen from *both* states, a high density ($N_{H_2} > 10^{10}$ cm^{-3}) is also implied. On the other hand, absorption in CO transitions out of low J states in v = 0 can occur from low temperature, low density gas along the line of sight to continuum sources. A brief review of excitation criteria for the near-infrared molecular transitions is given by Scoville (1984). A schematic view of the various regions probed by different spectroscopic tracers is given in Fig. 2.

Though early continuum maps of the Orion cloud core indicated a cluster of about a dozen sources, it now appears that possibly only three of these represent true sources of luminosity. The basis for selecting these three (BN, IRc-2 and IRS 2) is the observation of steep temperature gradients in the surrounding dust, as would be expected if there were a central luminosity source (Wynn-Williams et al. 1984). Many of the remaining condensations in the emission exhibit high near-infrared polarization, suggesting that their radiation is probably reflected light (Werner et al. 1983).

To date, only the BN object has been extensively studied by near-infrared spectroscopy (Hall et al. 1978; Scoville et al. 1983*a*). In this source the Bracket and Pfund series lines of H I have been observed in emission and the CO fundamental and overtone bands have been seen in both emission and absorption. Molecular hydrogen emission is also seen, but this probably arises from the extensive shock fronts in the cloud core well away from BN. Extensive studies of the near-infrared hydrogen lines in the embedded objects of other cloud cores have been made by Thompson and Tokunaga (1978, 1980), Thompson (1982), T. Simon et al. (1979), and M. Simon et al. (1981*b*). Most recently McGregor et al. (1984) have obtained optical spectra (λ = 6000–10,000 Å) of many of these objects. Studies of the near-infrared CO lines similar to these observed in BN have been made for about five other young stellar objects by Kleinmann et al. (1984). Because of the more complete nature of the BN observations, we concentrate here on those results. Observations of the infrared and optical H II lines are reviewed in the chapter by Thompson.

A. Near-Infrared Spectroscopy of BN

Figures 3 and 4 show the near infrared spectra of BN in the vicinity of the CO fundamental and overtone bands (Scoville et al. 1983*a*). At λ = 4.6 μm (Fig. 3) one sees strong absorption by line-of-sight gas in the ambient molecular cloud up to approximately J = 15 on the R and P branches. (The gaps appearing in the 4.6 μm band adjacent to each of the absorption lines are due to telluric CO absorption; our ability to observe the interstellar CO is made possible by the differential Doppler shift of the source out of the telluric lines.) In addition to the strong absorption by ^{12}CO, one also sees weaker components due to ^{13}CO and an emission feature on the wings of the ^{12}CO absorption lines. In the overtone band at λ = 2.35 μm, Fig. 4 shows absorption in both the R and P branch up to J = 10. Due to the lower line strength in the overtone transitions, the absorption lines in the overtone band are considerably less saturated than those in the fundamental band. Also shown in Fig. 4 are four broad emission features corresponding to the overtone bandheads (v = 5 $\rightarrow$ 3, 4 $\rightarrow$ 2, 3 $\rightarrow$ 1, and 2 $\rightarrow$ 0). Because these emission features arise from high-energy states with a nearly thermal distribution of population, the source of this emission must be an extremely dense, hot medium.

The kinematic profiles obtained by adding up the separate rotation vibration lines in the ^{13}CO fundamental, the ^{12}CO overtone and fundamental bands are shown in Fig. 5. Taking account of the differing band strengths and isotopic abundances, the three profiles provide a sequence in increasing opacity with the CO v = 0 $\rightarrow$ 2 band having the lowest opacity and the CO v = 0 $\rightarrow$ 1 being the most saturated.

Clearly seen in all three bands are two distinct kinematic components at velocities with respect to local standard of rest V_{LSR} = +9 and -17 km s^{-1}. Also evident in the more opaque bands (the fundamental bands of CO and ^{13}CO) are an emission component at V_{LSR} = +20 km s^{-1} and a weak absorption component at V_{LSR} = +30 km s^{-1}. The latter two features originate from considerably hotter gas than the former two. This fact may be seen in Fig. 6 where averages of small groups of rotation-vibration lines in the CO

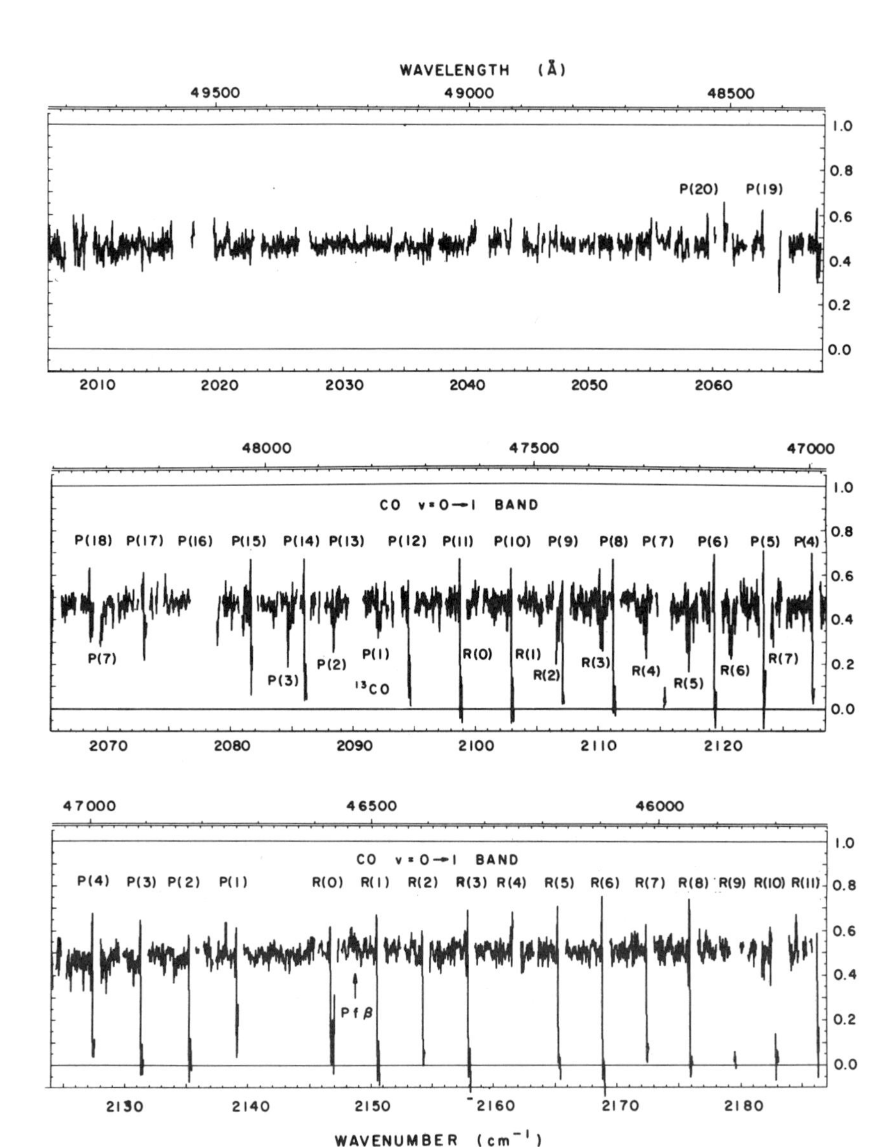

Fig. 3. The λ = 4.6 spectra of BN (Scoville et al. 1983*a*). To correct for telluric absorption features the raw spectrum of BN must be ratioed to a comparison star, in this case Vega. In this figure only the ratio spectrum is shown. (The gaps correspond to regions of telluric absorption > 50%.) Note the strong absorption in the low *J* lines of the CO fundamental band at 4.6 μm in addition to the ^{13}CO isotopic band at 4.8 μm.

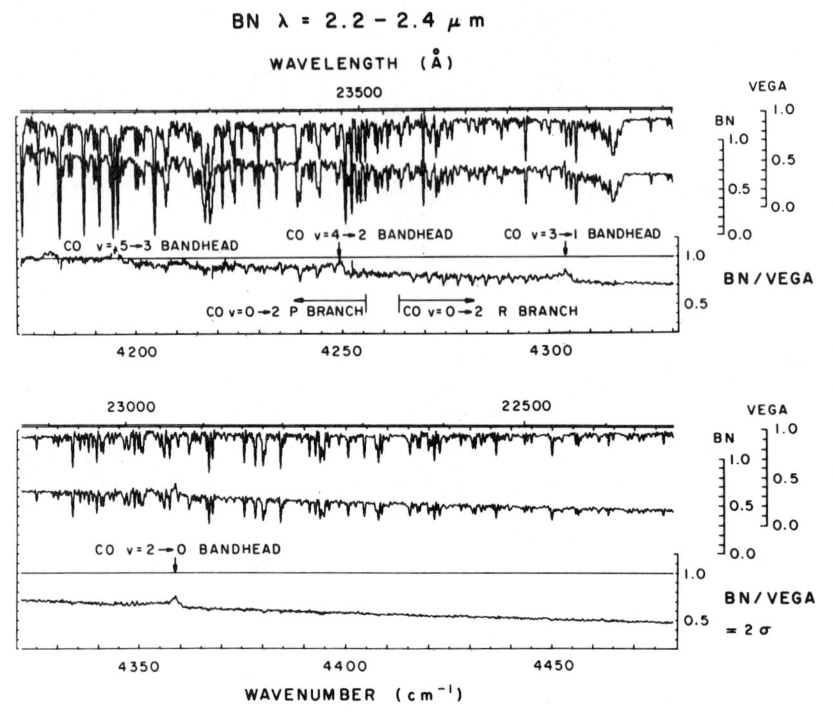

Fig. 4. The 2.3 μm spectra of BN (Scoville et al. 1983*a*). As seen in Fig. 3, to correct for telluric absorption features the raw spectrum of BN must be ratioed to a comparison star, in this case Vega. Both the spectra of Vega and BN may be seen in addition to the ratioed spectrum. In this figure one sees absorption at low J in the first overtone band (v $= 0 \rightarrow 2$) in addition to broad emission features at the four overtone bandheads.

fundamental band are plotted. Viewing this figure, it is clear that the absorption at +9 and −17 km s^{-1} dies out long before the emission at +20 km s^{-1} despite the much greater opacity in the absorption features. From a fit to the strengths of the separate rotation-vibration lines, the rotational temperature is found to be 150 K for the +9 and −17 km s^{-1} gas and 600 K for the +20 km s^{-1} gas.

On the basis of its velocity and low rotational temperature, it is clear that the +9 km s^{-1} gas originates from the column of gas in the OMC 1 cloud along the line of sight to BN. The −17 km s^{-1} absorption also has a low rotational temperature which should probably be attributed to a part of the expanding plateau source in front of and along the line of sight to BN. The fact that this feature is seen in absorption at negative velocity indicates definitely that the high-velocity gas in the plateau feature must be expanding out of the cloud core.

Perhaps most relevant to understanding the evolutionary status of BN itself are the 4.6 μm features at +20 and +30 km s^{-1} and the 2.3 μm band-

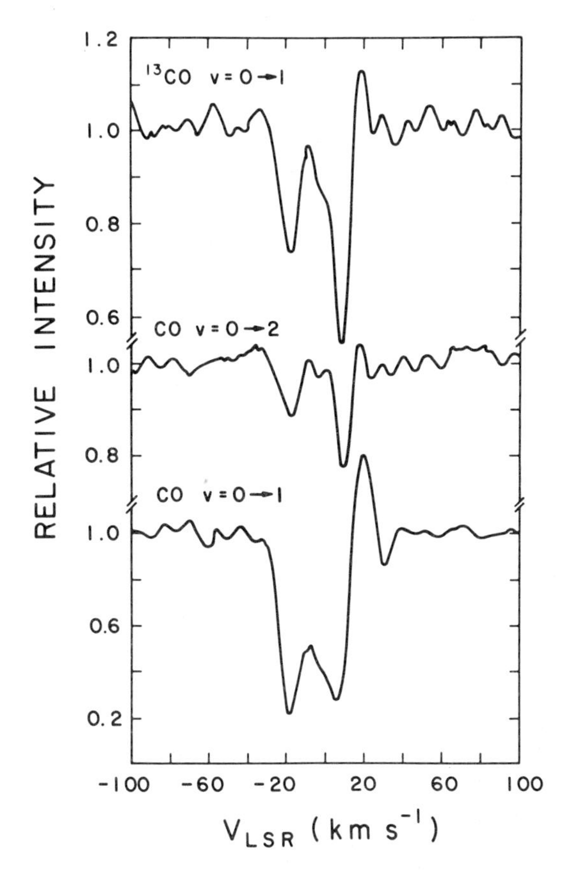

Fig. 5. Kinematic profiles for the CO and ^{13}CO bands obtained by averaging the rotation-vibration lines of all detected transitions. In general, the lines in the CO fundamental band have the highest optical depth followed by the ^{13}CO fundamental band lines and lastly the CO overtone bands. (Quantitative estimates of the relative opacities in separate bands cannot be obtained from this figure since the profiles were obtained by averaging different rotational states for the different bands.) Two strong absorption features are seen at +9 and −17.8 km s^{-1} corresponding to line-of-sight gas in OMC 1 and the plateau source, respectively. Both features are formed in gas with rotational temperature T_R = 150 K. The emission feature at +20 km s^{-1} and the weaker absorption at +30 km s^{-1} probably originate in higher-temperature gas within 25 AU of BN.

head emission. The components of neutral gas most closely associated with BN, i.e., the CO bandhead source and the +20 to 30 km s^{-1} molecular gas, probably originate in two distinct regions with very different temperatures, 3500 K and 600 K, respectively.

The bandhead emission is thought to originate in a compact region $< 2/3$ of an AU in size with density exceeding 10^{10} cm^{-3} (Scoville et al. 1983*a*). However, based on the narrow width of this feature ($\Delta V < 75$ km s^{-1}), it is

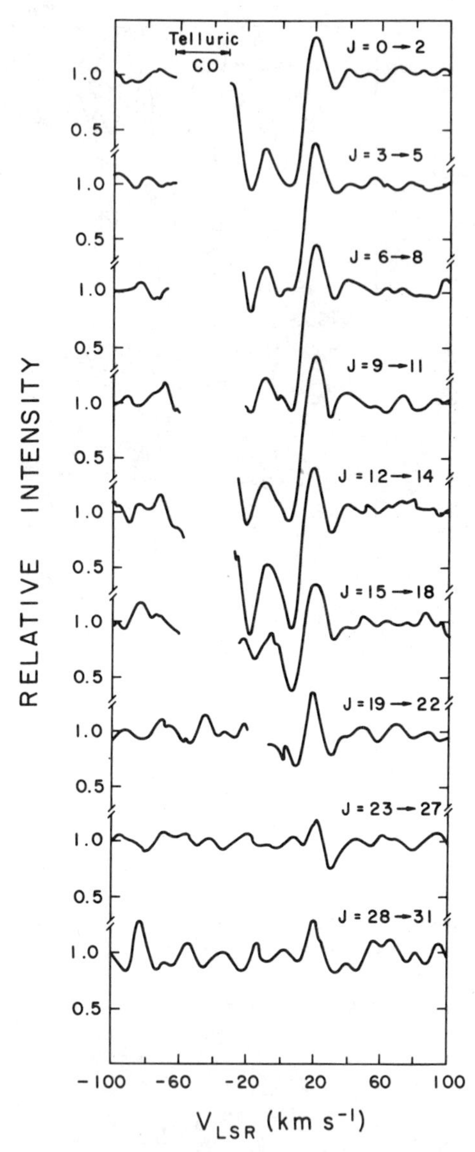

Fig. 6. Averages of subsets of the CO fundamental-band lines indicating relative temperatures of the absorbing gas clouds. Note that the +9 and −17 km s^{-1} features disappear at J lower than the +20 km s^{-1} feature despite their larger opacity in the low J lines. The former must arise from gas at 150 K, the latter at 600 K.

difficult to see how this emission could originate with a radius of 1 AU, where the orbital velocities should be several hundred km s^{-1}.

On the other hand, the +20 and 30 km s^{-1} CO features seen at 4.6 μm probably originate in a shell of gas immediately outside the dust photosphere at $R \sim 20$ AU. This situation is suggested by the moderate rotational temperature (~600 K) for these features and the fact that one of them is seen in absorption of the 4.6 μm continuum. Assuming the standard CO-to-H_2 abundance ratio (corresponding to 10% of the available C in CO), the measured CO column density implies a hydrogen density of $\sim 10^7$ cm^{-3} in this shell.

Critical to understanding the zone in which the CO lines are produced is a determination of the velocity of the central star. If the two features at +30 and +20 km s^{-1} are attributed to a single zone near BN, then the observed velocity profile (cf. Fig. 5) can be modeled as a shell of gas infalling at 10 km s^{-1}, probably the last, lingering phases of accretion flow onto the recently formed star. Based on the H II lines, it is believed that the center of mass velocity for the star is 20 to 21 km s^{-1}. Thus, a circumstellar shell, infalling at a relative velocity of 10 km s^{-1} will produce an absorption line at +30 km s^{-1} in the portion of the shell between us and the central star. The CO emission at +20 km s^{-1} would arise from gas at the shell periphery, infalling perpendicular to our line of sight.

The ionization requirements of the H II region seen in the Brackett lines could be supplied by a B 0.5 main sequence star (T_{eff} = 28000 K, $L \sim 2 \times 10^4 L_\odot$). An important feature of the Brackett lines are the broad wings extending over 300 km s^{-1} for which the line shape can be modeled by an optically thin, radial outflowing wind with $V \propto R^{-2/3}$. The implied mass loss rate is $\sim 2 \times 10^{-6} M_\odot$ yr^{-1}. Very long array (VLA) measurements of the radio emission from BN indicate that the size of the H II region producing the near-infrared hydrogen lines must be < 20 AU in radius (Moran et al. 1983).

The approximate properties and conditions in the various zones surrounding the Becklin-Neugebauer (BN) object are summarized in Table I. If one attempts to unify the observations of BN into a single model, such a model must involve both infall (evidenced by the inverse P-Cygni profile) and outflow (as evidenced by the Brackett lines). Because the source of ionization for the Brackett lines is probably the ultraviolet continuum of the central star, it is natural to place the outflowing ionized gas wind immediately adjacent to the star. The CO observed at λ = 4.6 μm would then originate in a neutral cocoon surrounding the H II region. The size estimate obtained for the molecular gas is entirely consistent with that of the ionized region estimated from the VLA measurements. This model is shown schematically in Fig. 7.

A major uncertainty in our modeling of BN is the placement of the CO bandhead emission source. The high temperature suggests that this emission could arise in a region close to the star, possibly a circumstellar disk. On the other hand, the small velocity dispersion of the emission provides rather compelling evidence that this region cannot be within 1 AU of the central star. If it

TABLE I
Emission Regions in the Becklin-Neugebauer Objects

Zone	Possible Model	Mass ($M_\odot$)	Radius (AU)	Temperature (K)	Kinematics
Star	B 0.5 main sequence	18	0.025	28,000	$V_{LSR} \simeq 21$ km s^{-1}, $V_{escape} = 1000$ km s^{-1}
H II	Outflowing wind	10^{-6}	out to 20	10,000	$\bar{V} = 21$ km s^{-1}, $V \propto R^{-2/3}$
CO bandhead source	Shock front?	10^{-5}	size $\leq 2/3$ AU at $R \geq 3$ AU	3,500	$\bar{V} = 20$ km s^{-1}, $\Delta V \leq 75$ km s^{-1}
Dust photosphere	Envelope surrounding H II region	—	20–25	600	—
CO 4.6 μm emission	Accreting ambient cloud	10^{-4}	25	600	$\bar{V} \simeq 21$ km s^{-1}, infall at ~ 10 km s^{-1}

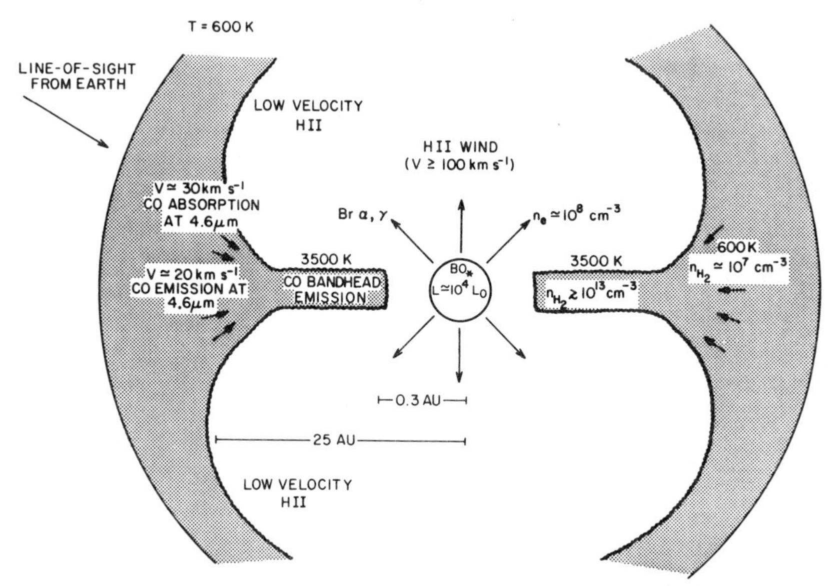

Fig. 7. Schematic model for the separate zones surrounding BN. In the center is a star with ionizing flux equivalent to a B 0.5 main sequence star. An outflowing wind seen in the ionized gas has an estimated mass loss rate of $2 \times 10^{-6} M_\odot$ yr^{-1}. The velocity law providing the best fit to the line shapes ($V \alpha R^{-2/3}$ suggests that significant deceleration of this outflying gas must occur. The outer size of the H II region is determined by radio measurements (Moran et al. 1983). In the model, the H II region is bounded at its outside by a dust shell. The molecular gas at +20 and +30 km s^{-1} occurs just outside this dust photosphere. This exterior gas appears to be infalling at 10 km s^{-1}. Most uncertain is the placement of the CO bandhead emission region for which the high temperature and density (T_K = 3500 K, $n_H \geq 10^{10}$ cm^{-3} would suggest proximity to the central star, yet the small velocity dispersion ($\Delta V < 75$ km s^{-1}) requires a radius > 1 AU from the star. Whether the bandhead emission arises in a disk within the H II region (as shown in the figure) or farther out in a shock front is unclear at present.

is produced at a larger radius, then it is difficult to understand why its size corresponds to < ⅔ AU. Critical to defining the location of this emission will be future measurements of the kinematics of the bandhead emission as a function of the band energy. The possibility that BN is, in fact, a more complex source (consisting of an early-type star needed to ionize the H II plus a late-type star with the circumstellar bandhead emission source) cannot be ruled out at present.

B. Other Embedded Objects

The richness of the phenomena observed in the molecular lines associated with BN underscore the need for more extensive infrared spectroscopy of other embedded sources. At present, the only instrument capable of operation in the 4.6 μm CO band with the resolution required to separate the telluric and interstellar CO lines at 4.6 μm is the Fourier Transform Spectroscope (FTS) on the 4 m telescope at Kitt Peak National Observatory. Several of the other sources observed with this instrument include UOA 27 (GL 2591), LkHα101, and MWC 297 (Kleinmann et al. 1984). In the former source, the CO rotational temperature is found to be comparable with that observed in OMC 1, but in the latter two sources the CO appears to be at much lower temperatures ~10 K. An interesting correlation is found between the column density of CO as indicated by the absorption lines and the dust extinction as indicated by the hydrogen line ratios in these sources. In the hotter sources, (UOA 27 and BN) approximately 100% of the available carbon must be in carbon monoxide while in the two colder sources only 10% is required. These measurements suggest that the CO to H_2 ratio may be a variable from cloud to cloud, especially between cloud cores and the cloud envelopes. It is possible that in the colder clouds a large fraction of the CO is depleted onto dust grains but in the warmer regions where the dust is above 50 K, the CO sublimates back into the gas.

II. PROTOSTARS

Despite the great yearnings of infrared astronomers to detect and observe protostars, probably none of the objects observed in the near-infrared are truly protostellar in the sense that they derive their energy from gravitational collapse (Wynn-Williams 1982). The reason for this is easily seen considering the expected properties of such objects. Not only will they have extremely low temperatures ($T < 500$ K), but they will also be heavily dust enshrouded. For example, a nebula containing 1 $M_{\odot}$ within a radius of 100 AU will have a mean density $\sim 4 \times 10^{10}$ cm^{-3} and the hydrogen column density from the outside of this nebula to the central object is a staggering 6×10^{25} cm^{-2}. For a standard gas-to-dust ratio, this corresponds to a visual extinction of 6×10^4 magnitudes. Clearly, such an object would be totally obscured to both near- and far-infrared observations, becoming transparent only at millimeter wavelengths. With the advent of millimeter-wave interferometers and sub-millimeter telescopes, we may at last begin to see such objects via their long wavelength continuum emission. We can, therefore, anticipate that the next ten years will be an exciting period for the discovery and observation of the elusive protostars.

Acknowledgment. Some of this work was done while the author was on the staff of the Five College Radio Astronomy Observatory, University of Massachusetts, Amherst.

FRAGMENTATION AND HIERARCHICAL STRUCTURE IN THE INTERSTELLAR MEDIUM

JOHN M. SCALO
The University of Texas at Austin

Observational and theoretical work related to the fragmentation of interstellar structures is reviewed. Comparison of information obtained using different techniques shows that interstellar cloud structure is hierarchical. Each major complex, with size of $\sim$ 1 kpc and mass of $\sim 10^7 M_{\odot}$, contains internal structure with at least 4 or 5 hierarchical levels, over a range in size scale from $\sim$ 1 kpc to below 0.1 pc. The evidence consists of direct mapping using a variety of techniques for the nearby complexes and direct and more indirect arguments for distant complexes. The hierarchical structure of the Taurus complex is discussed in detail. The dominant structure at each level of the hierarchy is typically an elongated, probably prolate, cloud containing a few bound fragments. Evidence for filamentary and shell structure in lower-column density regions is summarized. A similarity with extragalactic structures is pointed out. The empirical correlations between density, internal velocity dispersion, and length scale are summarized, and various empirical estimates of the cloud size and mass spectrum are reviewed. These data are used to estimate the parameters of the interstellar hierarchy, assuming it is strictly self-similar. The results favor a small ($\sim$ 2 to 5) number of fragments per fragmentation stage and a small ($\lesssim$ 0.1) mass efficiency fragmentation. Numerical collapse calculations are reviewed. The fact that their spatial resolution is presently insufficient to reveal highly nonlinear developments like multiple fragmentation stages or any sort of turbulence is emphasized. The calculations do show that fragmentation may occur through purely gravitational effects, possibly requiring elongated geometry or sizable initial velocity and/or density fluctuations, or may be controlled by rotation and/or magnetic fields, depending almost entirely on initial conditions. This situation is a fundamental reflection of the severe practical restrictions on spatial resolution of current numerical techniques. Generally, however, all the numerical calculations agree roughly with observations in the sense that the *calculations predict a small number of fragments per fragmentation stage and a small mass efficiency of fragmentation. It may be possible to better distinguish the models on the basis of the velocity field. Fragmentation may also be induced by stars. The processes reviewed include fragmentation associated with several instabilities in expanding shells around early-type stars, radiatively-induced implosion or disruption of preexisting clumps, and the ability of field stars to induce large density and velocity fluctuations both through the effects of winds and H II regions as well as through fluctuations in the field star gravitational potential. There is no dearth of sources for density and velocity fluctuations, suggesting that the appropriate initial conditions for numerical collapse studies must incorporate significant inhomogeneity, a practice which has not been common in most published work. The role of turbulence in fragmentation is also reviewed. In the general sense of stochastic/ordered structure induced by nonlinear processes over a range of scales, cloud structure is turbulent, but there is no general agreement on the dominant effects or source of energy and the situation remains thoroughly confused. Several conceptual models as well as more quantitative studies based on statistical turbulence theories are critically reviewed. A discussion is given of the origin of the largest-scale cloud structures, including gravitational instability, thermal instability of a cloud fluid, and collisional buildup of structure through collisional coalescence, and several problems are pointed out. Finally, observational and theoretical considerations relevant to large shell structures are outlined. These shells may be causally related to the turbulent energy injection and fragmentation structure.*

I. INTRODUCTION

"The difficulty of form is its essential intranslatability on the one hand and its self-revealing obviousness on the other."

Thomas M. Messer

"Order is established by means of its opposite."

Yagyu (c. 1630)

One of the more fundamental and longest-known observational clues to the physical processes which control star formation is the fact that nearly all stars form in groups within large massive complexes of gas and dust. Combined with the fact that the gas associated with star formation is very inhomogeneous over a large range of size scales, one is forced to conclude that stars form by the fragmentation of large gas complexes, an idea now almost universally accepted. Because the nature of star formation in galaxies is observed to depend on galactic properties and environment, fragmentation is the process which couples galactic and stellar scales, a bridge between galactic evolution and star formation.

The beginnings of intensive study of fragmentation can be traced to a number of papers which appeared in the early 1950s, and these can in turn be traced to earlier discussions of the optical appearance of the interstellar medium by Barnard, Bok, McCuskey, and others. The discussion of turbulent cosmogony by von Weizäcker and of gravitational fragmentation by Hoyle provided a theoretical framework in which the evolution from gas cloud to protostar could be envisioned. Linear perturbation analyses by Chandrasekhar and others showed how magnetic fields and rotation might affect gravitational instability. Statistical studies by Ambartsumian, and Chandrasekhar and Münch, showed how a simple cloud model could be quantitatively compared with observations to derive estimates of cloud properties. The introduction of the frequency distribution of stellar masses at birth by Salpeter suggested a direct observational connection between star formation and fragmentation.

The major highlights of the subsequent thirty years include: a growing understanding of the dominant heating and cooling processes in interstellar gas; efforts at a quantified analytical treatment of gravitational fragmentation on the basis of perturbation analysis and the critical assessment of the viability of gravitational fragmentation by Layzer; the advent of radio astronomy, which led to extensive studies of interstellar structure using continuum, H I, and molecular emission; a growing awareness of the possible roles of magnetic fields, rotation, turbulence, chemistry, fragment interactions, and stellar energy inputs; and the use of numerical techniques to solve the hydrodynamical equations governing the behavior of self-gravitating clouds.

These advances provided few definite answers to the fundamental questions related to fragmentation, but introduced a host of new and more specific questions to which much of current research is directed, and have led to a stage of development in which the enormous complexity of the problem and the need for coordinated activity between various observational and theoretical areas of research is recognized. We now see that cloud complexes which spawn stars are extremely complicated systems in which nonlinear gravitational, rotational, thermal, chemical, magnetic, and turbulent effects may be important, and different processes may dominate or compete at different stages of fragmentation. In addition, fragments may interact with each other by direct collisions or tidal encounters, and newly-formed stars within the cloud may significantly affect the fragmentation and star formation through the injection of mass, radiation, and mechanical energy.

Despite this challenging richness of phenomena, there has been significant recent progress in delineating the possible effects of particular processes and an increasing wealth of observational data on the internal structure of clouds which can be used to constrain the theoretical ideas. This chapter is an attempt to summarize our present observational knowledge of the morphology of fragmentation, correlations between fragment properties, and the frequency distribution of interstellar cloud masses, and to outline some of the many theoretical ideas and results on fragmentation which can be related to these observations. The literature in this area is vast; I concentrate here on recent work. Previous reviews of fragmentation theories with extensive references to earlier work can be found in Mestel (1977), Silk (1978,1981), G. E. Miller and J. M. Scalo (1979), Zinnecker (1981) and Tohline (1983). A more general review of star formation processes has been given by Silk (1985*b*). There do not appear to be any previous general reviews on observations of hierarchical fragmentation or the mass distribution of interstellar clouds. A brief but well-balanced discussion of some aspects of turbulence in the interstellar medium is given by Dickey (1984). Two topics which are omitted because of space limitations are the stellar initial mass function (Scalo 1985*a,b*) and the use of two-point spatial statistics such as correlation and structure functions to characterize interstellar structure (Dickman's chapter, and Houlahan and Scalo, in preparation).

Also left out for lack of space are discussions of thermal, chemical and acoustic instabilities and the effects of these processes on gravitational instability (see Yoneyama 1973; Kegel and Traving 1976; Glassgold and Langer 1976; Giaretta 1979,1980; Ibanez 1981; Flannery and Press 1979; S. L. W. McMillan et al. 1980), the possibility that fragmentation may be viewed as a phase transition (Ferrini et al. 1983*b*; Tohline 1985), and fragmentation in the early universe (see Tohline 1980*b*; Palla et al. 1983; Silk 1983).

I have also omitted most of the so-called standard material on linearized perturbation analyses of the fluid equations. Linear perturbation results can yield useful characteristic length and time scales for cloud formation by various mechanisms, especially at the largest scales, but it seems clear to me both from observations of cloud structure and numerical hydrodynamic calculations that the essence of fragmentation is intrinsically nonlinear. From the observational point of view, one need only examine the density contrasts and velocity fluctuation Mach numbers to see that the structure and evolution of the interstellar medium is highly nonlinear in the sense of compressibility, and there may be additional equally important nonlinear aspects of the problem which are not so clearly perceived. Theoretically, most numerical (nonlinear) studies of cloud evolution, as well as more conceptual models, variously suggest strong nonlinearities in density, velocity, and/or magnetic field variations. In addition, the essential results of numerical calculations often have little to do with the predictions of linearized analyses. For these reasons, this chapter concentrates on nonlinear processes, although in some cases linearized analyses are reviewed when no other treatments of potentially important processes are available. This emphasis is also motivated partly by the existence of recent reviews of the linearized perturbation results (see R. B. Larson [1985] and B. G. Elmegreen [1985] for complementary presentations and motivations, although both of these papers also review nonlinear aspects of the problem). The theoretical part of the present text reflects the fact that nonlinear phenomena must for the most part be approached numerically or conceptually,[a] with little middle ground. Analytical mathematics has not so

---

[a]It seems important to clarify the nature of these "conceptual" discussions. A more accurate description would be "intuitive" without its negative connotations. When results of order of magnitude calculations are presented, the essential idea is intuitive, and is merely developed by calculation. Nevertheless, I will use the more conventional term "conceptual" in what follows without attempting any further definition.

far yielded much information on the nonlinear behavior of fluids or other nonlinear systems. The conceptual approaches to the fragmentation problem are equally important as the numerical studies because, although they are speculative, they are nevertheless capable of examining a large range, both in amplitude and nature, of nonlinear phenomena.

The present review places a major emphasis on the hierarchical nature of fragmentation, especially when discussing observations, because such structure is clearly indicated by the available data and seems fundamental to an understanding of fragmentation. That hierarchical structure has not been discussed much in previous work is a result of the fact that the severe practical limitations to the spatial resolution of both observational mapping and numerical studies have obscured the range of scales involved, and because there has been little coordination or even communication among workers who employ different observational techniques.

The goal of fragmentation theory is an understanding of the evolutionary process responsible for the observed spatial structure of the interstellar medium. Section II is an attempt to clarify the essential features of cool interstellar structures related to star formation. Section III shows one way in which observations can be used to quantify the structure. The numerous proposed physical processes which may lead to fragmentation are reviewed in Sec. IV, while theories for the origin of the largest-scale structures are discussed in Sec. V.

II. MORPHOLOGY OF FRAGMENTATION

The existence of star clusters and associations shows that clouds fragment to produce stars, but observations of the mass spectrum and other properties of the stars themselves only tell us about the end result of fragmentation, and are extremely difficult to relate to the various effects which may be important during evolution from cloud to star. Direct information on the earlier stages of fragmentation must come from studies of the internal structure of cloud complexes, where fragmentation and star formation are now occurring. There are a number of techniques available for structural investigations, including extinction, reddening, polarization, optical, H I, and molecular lines, and infrared mapping.

A. Practical Considerations

Despite the large amount of data which has accumulated in these areas in the last decade, it has proven surprisingly difficult to relate the observations to the physics of fragmentation. One problem is that the various types of observations are useful only in different specific ranges of column density, and so each type of data gives information over a narrow range of fragment properties, or possibly range of evolutionary states. For example, reddening and optical lines can only probe relatively small column densities. There has been little effort in relating these different types of observational data. Secondly, an observational study of fragmentation requires fully sampled data covering a large range of scales. To illustrate the severity of the problem, consider the Taurus complex, which has an extent of $\gtrsim 20°$ and contains substructure on all scales down to ≤ 1 arcmin. In order to quantitatively describe the velocity and column density structure using, say, the ^{13}CO line would require that at least 10^4 positions be sampled, and more than 10^6 positions would be needed to cover the entire range of scales. Considering the relatively small line strengths and the high signal-to-noise ratio and good velocity resolution needed for such a study, the total required observing time is discouragingly large. Work in this direction is proceeding, but it will be some time before accurate and homogeneous data covering a factor of 100 or so in scale are available. Similar considerations apply to other methods, such as extinction, in which an enormous number of stars would need to be counted to resolve the relevant range of scales. Automated star counting to very faint apparent magnitudes will be required for studies of fragmentation. Techniques for such a study are now available (see, e.g., Jarvis and Tyson 1981), but have not yet been applied to interstellar structure. In addition, it is important to realize that any single observational method is biased toward a particular range of cloud parameters, such as density, column density, or temperature, so selection effects are important. It is also important to note that the highest angular resolution ordinarily attainable by any method ($\sim$ 1 arcmin, although much better resolution can be obtained using the Very Large Array [VLA] and automated deep star counts) means that we can only hope to study a large range of fragmentation scales in the nearby complexes.

For these reasons, and others, most observational studies have concentrated on the properties of a particular class of object on a particular scale (e.g., giant molecular clouds, dark-cloud cores, etc.), and it is difficult to combine and compare the results in order to infer the entire spectrum of fragmentation. Thus, it must be understood that the description of interstellar structure outlined below is biased by selection effects, especially those concerning column density. These selection effects are specified below. Nevertheless, an attempt is made to consolidate the available observational data into a coherent picture.

B. Basic Observational Approaches to Fragmentation

There are at least three basic approaches by which observational data can be used to study fragmentation processes. One of these approaches involves anatomical studies which construct a map of a given region using well-sampled observations of some quantity such as extinction or intensity in a spectral line. These studies can yield the large-scale properties of the region, like density, temperature, total mass, molecular abundances, density and velocity gradients, etc., as well as the properties of any subcondensations. The general morphology which emerges from these studies is reviewed in this section.

Another approach involves surveys of a large number of objects with qualitatively similar properties, like small dark clouds, which can be used to generate a catalog of properties for a given class of structures. The frequency distribution of, and correlations among, these properties, such as size, shape, column density, and velocity dispersion, can provide some basic constraints on theories of fragmentation. The major problem here is that a comparison with theory requires a quantitative understanding of the selection effects and identification criteria which enter the catalog construction. A discussion of the frequency distribution of cloud masses inferred from catalogs and its implications for hierarchical fragmentation models is given in Sec. III.

A third approach is a quantitative characterization of the spatial structure of interstellar clouds which can be obtained from a statistical description of the spatial relation between fluctuations in the amplitudes of observed variables at different positions on a well-sampled spatial array, using, for example, moments of a multipoint probability distribution. This approach, which was pioneered by Chandrasekhar and Münch (1952), has proven useful in studies of galaxy clustering, but has not been applied systematically to interstellar cloud complexes, primarily because it requires that a very large number of positions ($\gtrsim$ 1000) be sampled. In addition, the method involves some very dangerous assumptions, especially stochastic homogeneity and ergodicity, which are difficult to justify, and it is not yet clear that the statistics (e.g., the autocorrelation function or the structure function) can distinguish effectively between different possible structural models (Houlahan and Scalo 1985). Interest in these methods as a probe of fragmentation has increased greatly in the past two years, but most of the work has not yet been published (see, e.g., Kleiner and Dickman 1985; Lichten personal communication, 1984; Perault et al. 1985*b*; Houlahan and Scalo, in preparation) or does not yield readily to interpretation (see, e.g., Crovisier and Dickey 1983; Scalo 1984*b*), and so will not be reviewed here. However, structural statistics may become an increasingly prominent and useful tool in the near future, especially if fragmentation can be viewed as a stochastic process.

C. Disorder

Interstellar matter possesses two apparent structural characteristics which probably contain the keys to the physics of fragmentation. The first property is the existence of an ordered, hierarchical component which appears over a wide range of size scales and densities. It is the evidence for this component which is emphasized below. At the same time, the structure seems to possess a disordered, or chaotic component. The irregular appearance of interstellar gas and dust is perhaps its most striking and longest-known property. Column density fluctuations, usually seen on all scales larger than the spatial resolution of the observing equipment, encompass a great variety of sizes, shapes, and relative configurations. As an example, Fig. 1 shows part of Lynds' (1962) map of the distribution of extinction based on visual inspection of

Fig. 1. Spatial distribution of dark cloud structures for galactic longitudes from 350° through 70°, as determined from visual inspection of POSS plates by Lynds (1962). Over 900 individually cataloged clouds are represented. The five shadings correspond to Lynds opacity classes 1,2,3,4, and 5 + 6, from lightest to darkest.

Palomar Optical Sky Survey (POSS) photographs. Studies of optical and molecular spectral lines often reveal spatial fluctuations of the mean line-of-sight radial velocity and the presence of multiple velocity components along individual lines of sight, indicating that the velocity field also possesses a disordered component. The observed velocity differences and line widths imply that the density fluctuations move at supersonic, but perhaps sub-Alfvénic, speeds.

While the present review will discuss only the evidence for hierarchical structure, it is important to recognize the implications of the chaotic component. The simultaneous existence of chaotic and organized structure, randomness and order, is quite typical of many nonlinear systems, and is commonly referred to as turbulence, regardless of the physical nature of the system (see, e.g., May 1976; Kadanoff 1984, for popular accounts). This definition of turbulence as the characteristic tendency of nonlinear systems to exhibit strongly fluctuating substructure extending over a range of scales is relevant to fluids, chemically reacting systems, populations, biological activity, and a wide variety of other systems. Given the structural appearance of the interstellar medium and the probable importance of such nonlinear processes as gravity, shocks, and fluid advection, there should be no hesitation in using the term turbulence to describe its structure. However, it must be emphasized that this description does not necessarily imply any close connection with the incompressible fluid turbulence studied in terrestrial environments and laboratories, which is only one manifestation of a more general type of behavior. By discussing only the hierarchical structure we are here examining only one side of the coin; a full understanding of fragmentation must certainly involve a description of its stochastic elements. A discussion of some aspects of turbulent behavior in clouds is given in chapters by Dickman and by Myers, and in Sec. IV below.

D. Hierarchical Order

Interstellar structure contains an ordered, hierarchical component covering at least four orders of magnitude in density and length scale. An attempt at an anatomical review of the components of this hierarchy is given here. At scales on the order of 1 kpc the gas appears to be gathered into rather diffuse structures which are most easily seen in large-scale reddening studies. Figures 2 and 3 show two different determinations of the spatial distribution of color excess per unit length in the galactic plane within 2 or 3 kpc from the Sun, taken from Lucke (1978) and Neckel and Klare (1981). Although the details vary, both studies, along with earlier work using smaller stellar samples (see, e.g., FitzGerald 1968), clearly show the presence of structures with sizes ranging from $\sim$ 0.5 to 2 kpc. The existence of such large structures is also known from H I emission studies of our Galaxy (McGee and Murray 1961; McGee et al. 1963; McGee and Milton 1964), the Large Magellanic Cloud (see, e.g., McGee and Milton 1966), and M101 (Allen and Goss 1979; Vialla-

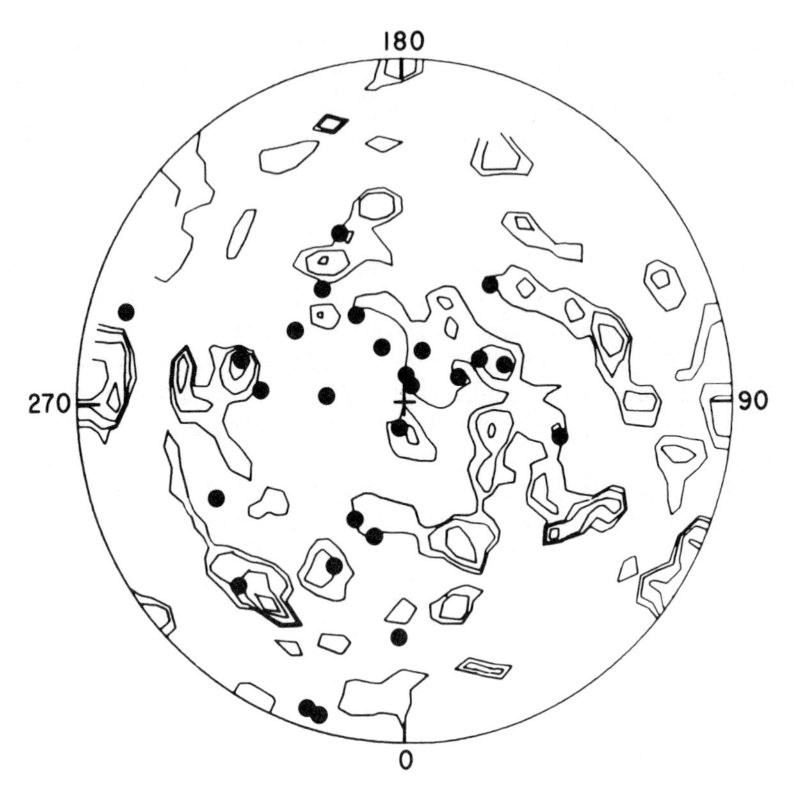

Fig.2. Contours of equal color excess per unit length for stars out to a distance of 2 kpc and within 200 pc of the galactic plane. Contour intervals are 0.5, 1.0, 2.0, and 4.0 mag kpc^{-1} in $E(B\text{-}V)$, and the smoothing scale is 200 pc. Solid circles are positions of R associations. (Figure from Lucke 1978.)

fond et al. 1981). The inferred H I masses of these large structures are of order $10^7 M_\odot$. It is interesting to note that the line widths are in all cases close to the value required for internal support against gravitational collapse, a property which seems to be shared by smaller structures (see Sec. III).

These largest coherent structures (that is, besides spiral arms) have also been recognized by their fossilized stellar imprints in the form of large star complexes with sizes of ~ 500 pc (see Efremov 1979; Stal'bovski and Schevchenco 1981), as discussed in B. G. Elmegreen and D. M. Elmegreen (1983). The Sun may even be immersed in such a star complex (Efremov 1979; Eggen 1982). In other spiral galaxies these large star clouds are called superassociations by Wray and de Vaucoulers (1980). The geometry of these largest gas structures is not clear; they may be extreme examples of the H I shells studied by Heiles (1980), although most of Heiles' shells are much smaller. B. G. Elmegreen and D. M. Elmegreen (1983) have emphasized the existence of

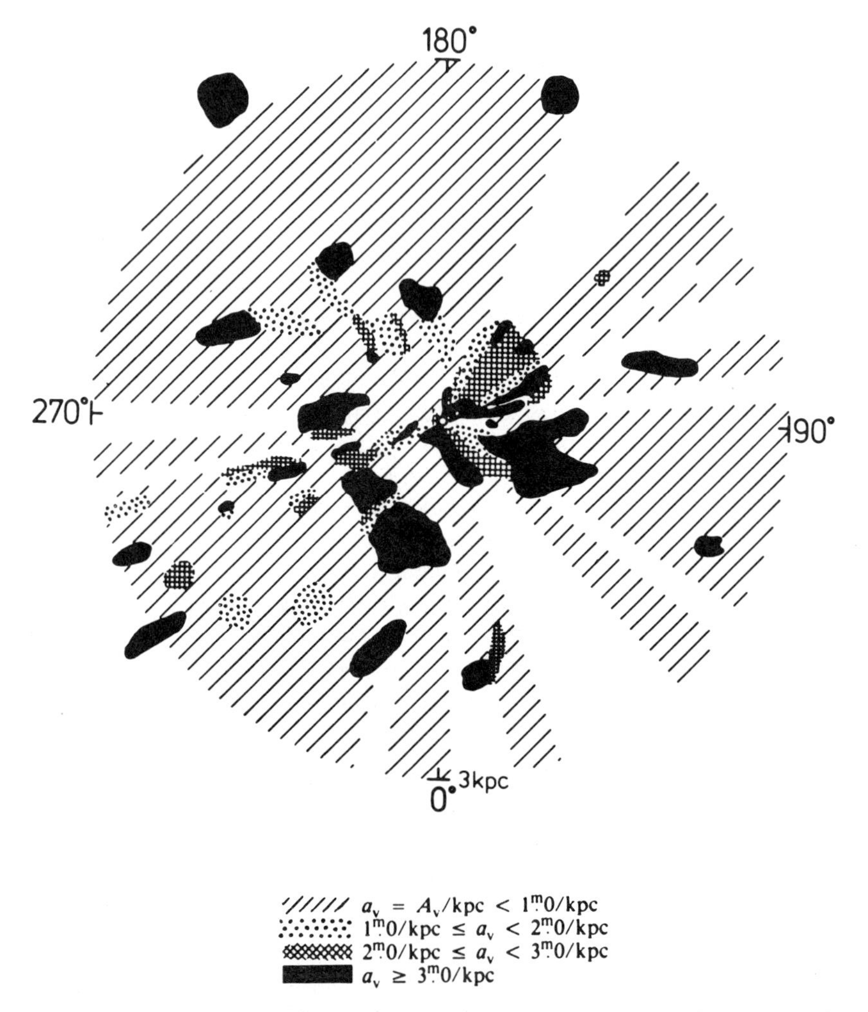

Fig. 3. Sketch of contours of equal color excess per unit length, as indicated, for stars out to 3 kpc. (Figure from Neckel and Klare 1981.)

these kpc-scale structures, and refer to them as superclouds, a designation which will be used in what follows.

These large reddening-H I superclouds typically contain a number of denser entities with scales ranging from ~ 20 pc to 100 pc, which are known as extinction complexes (see Lynds 1968) or giant molecular clouds (GMC's) (Solomon and Sanders 1980; Kutner and Mead 1981; Liszt et al. 1981; Sanders et al. 1984*a*), depending on how they are observed. These complexes are usually elongated, often along the galactic plane, and have mean internal densities of ~ 10^2 cm^{-3}. Local examples are found in Taurus-Auriga, Scorpius-

Ophiuchus, the Southern Coalsack, and Orion. These local complexes, as well as the larger structures in which they reside, can also be seen in gamma-ray emission arising from interactions of cosmic rays with molecular gas (Mayer-Hasselwander et al. 1982), in H I emission (McGee and Murray 1961; Heiles and Jenkins 1976), and even in optical interstellar lines (see Heiles 1974). For example, the existence of the Orion complex in optical interstellar-line spatial structure has been recognized for over 30 years from W. S. Adams' (1949) catalog. That the extinction-molecular complexes are only the denser cores of the reddening-H I superclouds has been emphasized by B. G. Elmegreen and D. M. Elmegreen (1983). Additional support for this idea comes from the work of Sanders et al. (1984*a*) and Rivolo et al. (1985), who find that CO clouds with sizes from $\sim$ 20 to 60 pc are clustered on a scale of 100 to 300 pc. Similar scales are indicated for external galaxies, for example by the combined near-infrared, Hα, and radio continuum study of the NGC 604 complex in M33 by Israel et al. (1982).

It is at the scale of extinction-molecular complexes that the association with star formation becomes clearest, because these complexes often contain young clusters and associations, H II regions, and ^{12}CO hot spots and infrared sources which can be identified with protostellar heating (see the review by Rowan-Robinson 1979). The most interesting differences between these complexes is the level of star formation activity, with some, like the Southern Coalsack, almost devoid of signs of star formation, and others, like Orion and most giant molecular clouds (GMC's), showing intense formation of stars, at least stars of high mass.

The presence of more than one young star cluster within some individual extinction complexes and the tendency of OB associations to exhibit subgroups with differing ages (Blaauw 1964) indicates the next level of fragmentation into a number of subcondensations. More directly, it is seen that the gas in all the local complexes exhibits internal clumping down to the limits of resolution. At scales of $\sim$ 1 to 10 pc the internal structure can be seen in CO mapping (see, e.g., Blitz 1980), resolved in both position and velocity, and large-scale extinction studies (see, e.g., Rossano 1978*a,b*) of complexes. Rossano's (1978*a*) extinction map of the Corona Australis complex is shown in Fig. 4; the hierarchical nature of the cloud structure is apparent, as emphasized by Rossano.

With increasing spatial resolution one sees similar structures on smaller spatial scales. For the local complexes, these are most of the dark clouds which are seen in photographs of the Milky Way, and which have been cataloged by Lynds (1962) and others. Most of the dark clouds with Lynds opacity class ($\approx A_V$) $\geqq$ 2 to 3 are associated with the larger extinction-molecular complexes (Lynds 1968) and there are very few truly isolated dark clouds. For example, Clark and Johnson (1981) interpret the clouds L 134, L 183, and L 1778 as a system of associated fragments within a larger cloud. The sizes of these dark clouds range from $\sim$ 20 pc down to $\sim$ 0.05 pc or

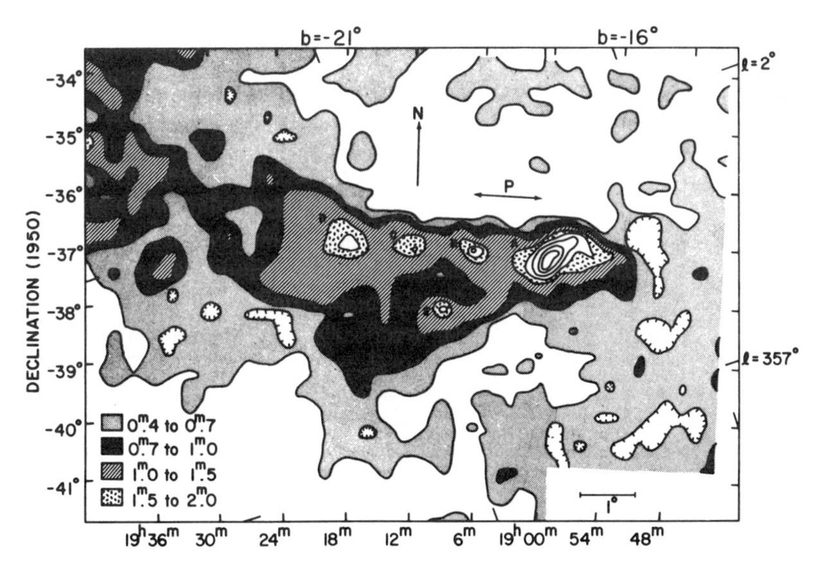

Fig. 4. Distribution of extinction in the Corona Australis dark-cloud complex from Rossano (1978*a*). Contours are for extinction of 0.4, 0.7, 1.0, 1.5, 2.0, 2.5, 3.0, and 3.5 mag. The first four contour levels are shaded.

smaller, with typical densities increasing at each smaller scale, reaching 10^4 to 10^5 cm^{-3} at length scale $L \sim 0.1$ pc. The densities and sizes scale roughly as $n \propto r^{-1}$ (see the chapter by Myers, and Sec. III below).

Usually the dark clouds with $L \sim 1$ to 10 pc are further fragmented, often in a hierarchical manner. The study of structured filamentary dark clouds by Schneider and Elmegreen (1979) shows that many filaments with lengths of $\sim$ 1 to 5 pc contain one or more subcondensations rather regularly spaced along the filament, and that some of these subcondensations are themselves fragmented in a similar fashion. In a few cases the fragmentation can be traced through three internal levels of hierarchy within a single filament, which in turn is contained within a larger, lower-density complex of clouds. The smallest subcondensations have dimensions of ~ 0.02 pc, a limit set essentially by the $\sim$ 1 arcmin resolution limit of dark-cloud identification imposed by the background star density. Sizes and other properties at this level have also been derived from NH_3 observations (Ungerechts et al. 1982; Myers and Benson 1983; Gaida et al. 1984). Additional evidence for very small-scale structure is summarized below. (Evidence related to the *very* small-scale fluctuations in electron density, not discussed here, can be found in Armstrong and Rickett [1981].) Since the masses of the smallest dark clouds derived from molecular lines and extinction are on the order of stellar masses, these clumps probably represent the stage of fragmentation at which star formation actually occurs (see chapter by Myers).

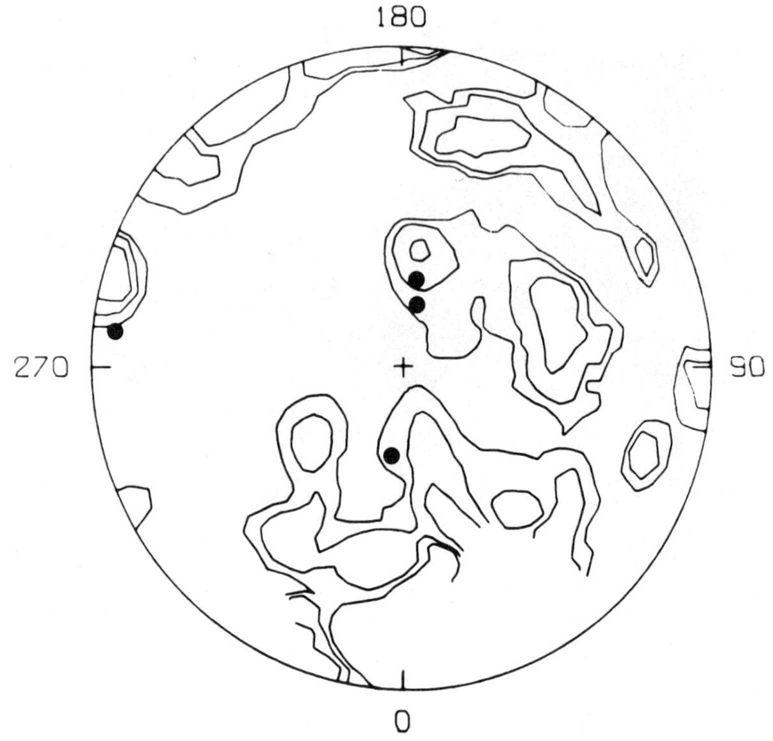

Fig. 5. Contours of equal-color excess per unit length for stars out to 500 pc and within 50 pc of the galactic plane. Contour intervals are 0.5, 1.0, 2.0, and 4.0 mag kpc^{-1} in $E(B\text{-}V)$, and the smoothing scale is 100 pc. Solid circles are positions of R associations. The local complexes in Scorpius, Ophiuchus, Taurus, Perseus, and part of Orion can be seen. (Figure from Lucke 1978.)

E. An Example of Hierarchical Structure

The most accessible examples of the similarity of fragmented structure observed over a large range of scale are the nearby Sco-Oph and Taurus complexes. The Taurus region is discussed in detail here; first, a brief summary for Sco-Oph is given.

For the Sco-Oph region, the hierarchical structure may be followed from the large reddening complex toward the galactic center (Lucke 1978; see Fig. 5), to the Sco-Oph extinction complex (extent $\geqq$ 100 pc) which is the most prominant of the higher-density fragments within the reddening complex, to the two, or possibly three further levels of fragmentation at scales of $\sim$ 2 to 20 pc, in the extinction map of Rossano (1978*b*) and also in a plot of positions and areas of Lynds' (1962) clouds in this region, down to the higher extinction core of the ρ Oph cloud (size $\sim$ 1 pc) which is itself broken up into several fragments with sizes of a few tenths of a parsec (see Myers et al. 1978; Wilking and Lada 1984). At this scale the resolution limit of the radio observations is approached, but the separations of embedded protostars shows that fragmentation to smaller scales has in fact occurred. Similar substructure from 10 pc to 0.1 pc in other subregions of the Sco-Oph complex can be seen by examination of the positions, sizes, and opacity classes of the over 200 Lynds dark clouds which fall within the complex. Recent maps of parts of the complex can be found in Lebrun and Huang (1984: 370 square degrees; ^{12}CO) and Wouterloot (1981,1984: 48 square degrees; OH).

The best-studied example of hierarchical fragmentation is the Taurus extinction complex. This nearby ($\sim$ 140 pc) complex, toward $\ell \approx 170°$, b $\approx -15°$, is part of a more extensive and lower-density structure which can be seen in the reddening map of Fig. 5, and also in a more recent reddening study of Perry and Johnston (1982). The gas in the older Perseus complex, containing the association Per OB2, at a distance of $\sim$ 300 pc at slightly smaller longitudes in Fig. 5, is also a part of this structure, and it seems likely that the Taurus and Perseus complexes are physically related (see Wouterloot 1981). It is also worth noting that the Pleiades cluster lies in the same general direction at about the same distance, as pointed out by Jones and Herbig (1982).

Figure 6 shows McCuskey's (1938) extinction map of most of the entire Taurus complex, and two hierarchical levels of internal fragmentation can be seen. At a distance of 140 pc, the Taurus complex is seen to cover a region at least as large as 100 pc by 40 pc. (The often-quoted ^{12}CO size is smaller because the stellar heating sources are concentrated toward the central regions; see the map in Herbig [1977].) Kleiner and Dickman (1985) have mapped the central region of the complex in ^{13}CO at 15 arcmin resolution,

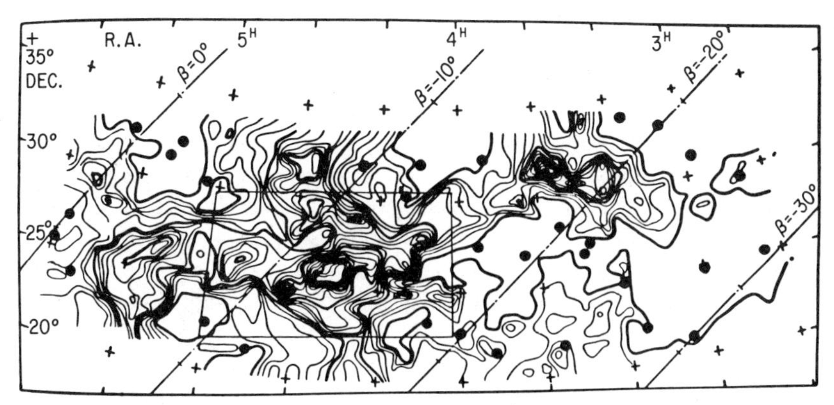

Fig. 6. Contours of extinction in the Taurus complex, with contour interval 0.2 mag. Black circles are areas selected as unobscured. (Figure from McCuskey 1938.) The rectangle refers to the region mapped in Fig. 7.

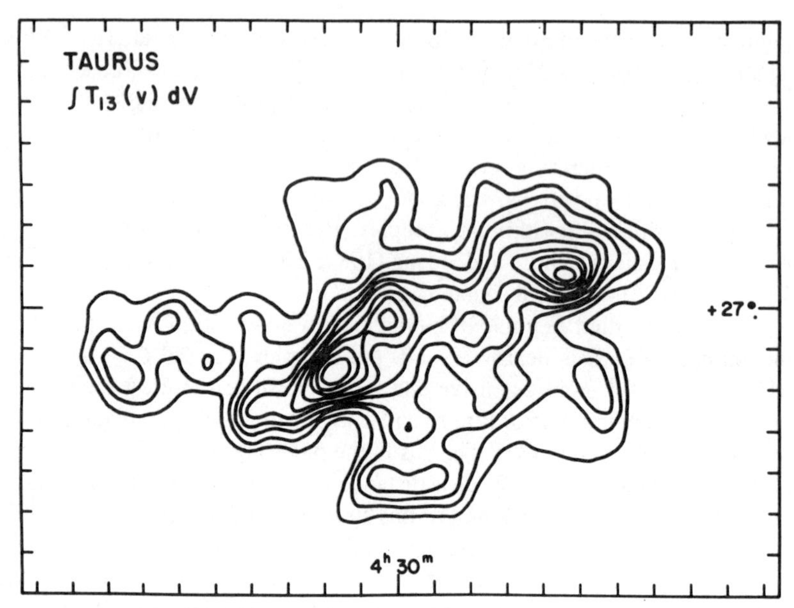

Fig. 7. Contours of equal integrated ^{13}CO emission in the Taurus complex. (Figure from Kleiner and Dickman 1984.) Boundaries of this map are shown in Fig. 6.

and their map, which corresponds to the rectangle of Fig. 6, is shown in Fig. 7. (For a ^{12}CO map of a larger area see Baran and Thaddeus [1977].) Notice the excellent agreement between the ^{13}CO and extinction maps. The Taurus complex is elongated, and a velocity gradient of 0.07 km s^{-1} pc^{-1} across the *minor* axis has been found by Kleiner and Dickman. While this apparent rotation is dynamically insignificant, its orientation suggests that the complex is prolate.

Within the Taurus complex are seen numerous smaller and denser dark clouds (see, e.g., Lynds 1962) with sizes from 10 pc down to 0.1 pc or smaller. The clouds in Taurus are not simply superimposed along the line of sight (Straizys and Meistas 1980). These dark clouds are often internally fragmented in a hierarchical manner. Most observational work has concentrated on the densest central region of the complex, especially the roundish dark cloud Heiles Cloud 2, and the filaments west of Heiles Cloud 2. A sketch of the extinction in this region based on POSS plates from Walmsley et al. (1980) is shown in Fig. 8. Cloud 2 is the region near $\alpha = 4^h\ 38^m$, $\delta = 26°$. The region to the east of Cloud 2 has apparently not been studied in molecular lines.

When mapped at high resolution, Cloud 2 is found to consist of a number of fragments located within a ring structure. The recent observations of the Cloud 2 structure by Schloerb and Snell (1984) are represented in Fig. 9, which shows a map of the subcondensations observed in $C^{18}O$. (See also

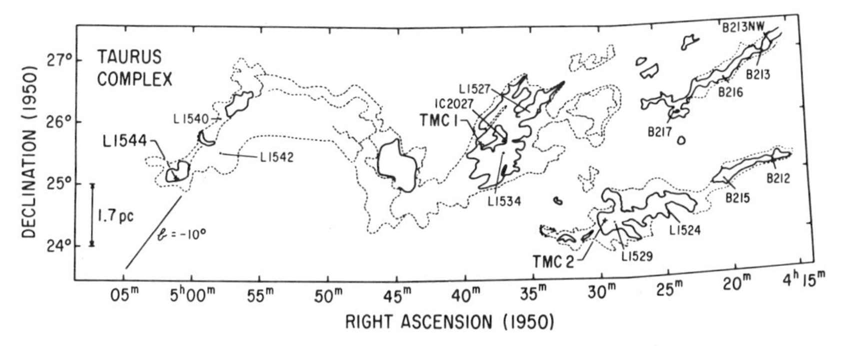

Fig. 8. Sketch of the cloud structure in the central region of the Taurus complex from Walmsley et al. (1980), based on POSS prints. Heavy lines indicate outline of dark clouds, dashed lines indicate connecting dust lanes. The dark cloud at $\alpha = 4^h\,37^m$ is Heiles Cloud 2, while the three filamentary clouds to the west are, clockwise from top, GF13, GF14, and GF16.

Cernicharo et al. [1984] for an extinction map.) The dotted contours represent CS observations of the denser clump TMC 1 (Snell et al. 1982; see also Avery et al. 1982; Myers and Benson 1983); more clumps like TMC 1 probably exist in the region (see Cernicharo et al. 1984) but would not be detected in CO. Schloerb and Snell find that the ring, of diameter ~ 1.5 pc, is rotating at about

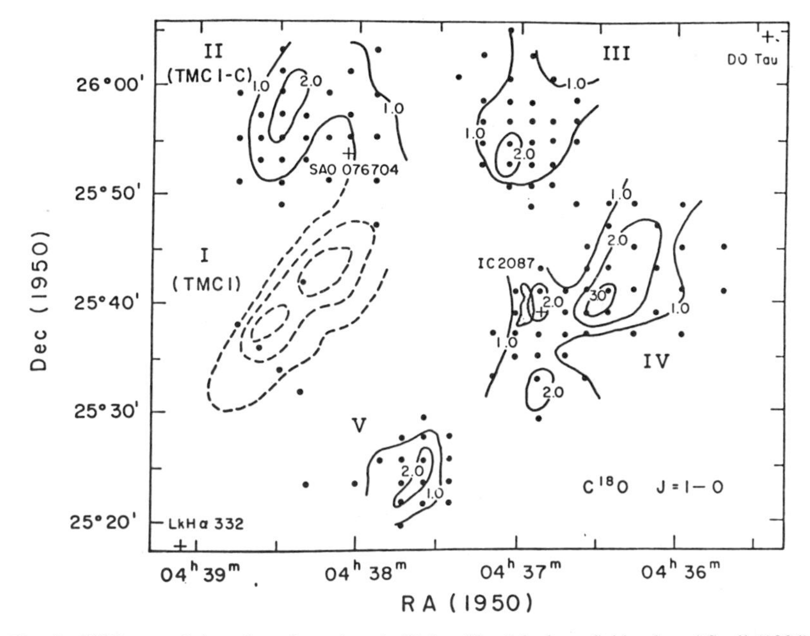

Fig. 9. $C^{18}O$ map of the subcondensations in Heiles Cloud 2, from Schloerb and Snell (1984). Dots represent the locations of individual $C^{18}O$ observations. The dashed curve represents the denser clump TMC 1, based on CS observations.

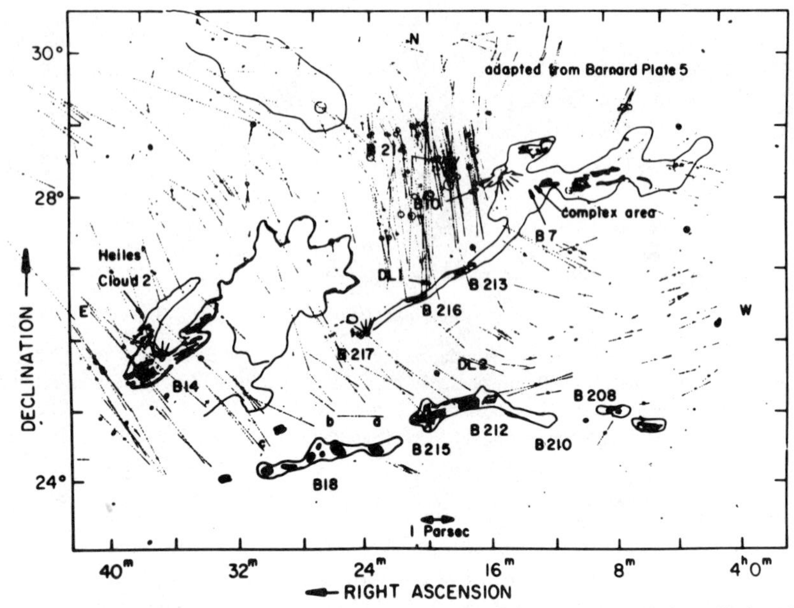

Fig. 10. Sketch of cloud structure around Heiles Cloud 2 and GF13, 14, and 16, from Clark et al. (1977), along with polarization vectors from Hsu (1984).

1 km s^{-1}, and that the properties of this fragmented rotating ring are in impressively good agreement with multidimensional numerical calculations of rotating nonmagnetic cloud collapse (see Sec. IV). The linear polarization near Cloud 2 has been recently studied by Moneti et. al. (1984) and Hsu (1984).

The filaments to the west of Cloud 2 in Fig. 8, with lengths of $\sim$ 5 pc, have also been studied extensively. A sketch of the filaments adapted from Barnard's (1927) study by Clark et al. (1977), is reproduced in Fig. 10 and superimposed on polarization vectors from Hsu's (1984) study. The filaments are obviously fragmented in extinction. Barnard (1927) catalogued a number of condensations in both the northern and southern filaments, spaced rather regularly along their lengths and indicated in Fig. 10. Schneider and Elmegreen (1979) discussed the extinction structure of these filaments and others exhibited on POSS plates. The northern filament, called GF13, contains three main subcondensations, and one of these is split into five subcondensations. The southern filament is actually two filaments which may or may not be connected at their ends; Schneider and Elmegreen label these GF14 and GF16. The western filament GF14 shows four subcondensations in extinction, while GF16 has five, and one of them (GF16A) is further fragmented into five subcondensations. Figure 11 shows extinction maps of GF14 (upper) and GF16 (lower) based on star counts from red POSS plates by Baudry et al.

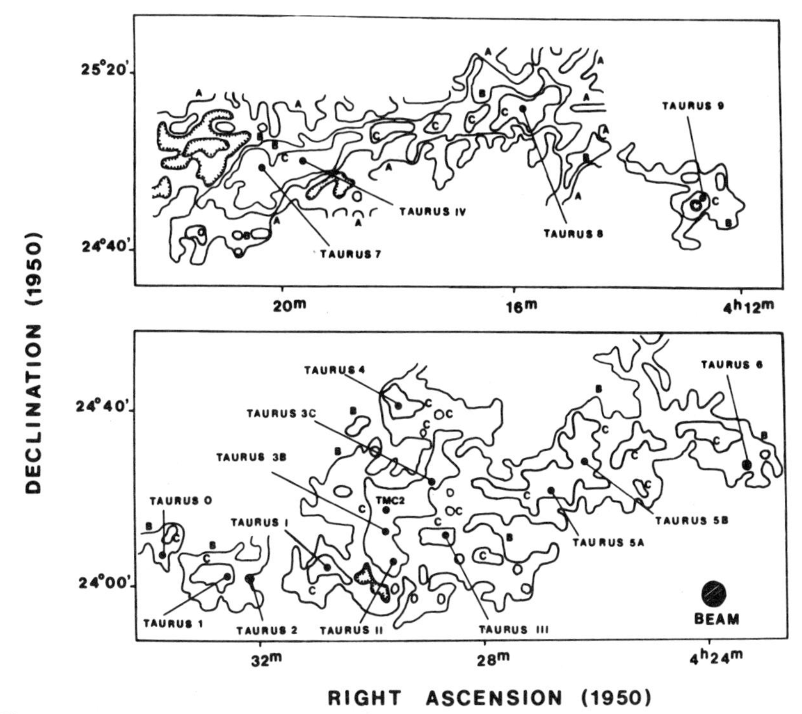

Fig. 11. Contours of visual extinction in the filamentary clouds GF14 (top) and GF16 (bottom), from Baudry et al. (1981). The contours labeled A, B, and C correspond to $A_V > 1.5$ to 2, $A_V > 2.5$ to 3.5, and $A_V > 4$ mag, respectively. The very small unlabeled contours correspond to $A_V > 5$ mag. The hatched areas outline holes in the visual extinction. The solid circles represent positions observed in HCO^+ by Baudry et al.

(1981). An independent extinction map of GF16 can be found in Batrla et al. (1981); more recent extinction maps of the Taurus filaments are given by Gaida et al. (1984). The resolution corresponds to about 0.1 pc, and several condensations of this size can be seen. A study of the subcondensations in GF16 in H_2CO has been given by Pöppel et al. (1983). The fragmentation of these filaments has also been inferred from the existence of multiple velocity components. For example, Clark et al. (1977) observed ^{12}CO and ^{13}CO in eight of the clumps within the filaments, and interpreted the multiple velocity components found in three of the clumps as evidence for the orbital motion of subcondensations. An extensive discussion of observations of CO, NH_3, H_2CO, OH, and HCO^+ in GF14 and GF16 is given by Baudry et al. (1981). The dense clump called TMC 2, a source of emission by heavy carbon chain molecules, is located near the eastern end of GF16. The velocity dispersion in GF16 is larger than any systematic velocity gradient (Baudry et al. 1981; Pöppel et al. 1983), but there is some evidence for a systematic velocity shift of ~ 1 km s^{-1} along the length of GF14 (Baudry et al. 1981; Fig. 12). A recent discussion of the properties of the condensation near the eastern end of GF16 has been given by P. F. Goldsmith and M. J. Sernyak (1984), who also give a useful summary of the velocity field in the Taurus complex at various scales.

Because of the regular spacing of the subcondensations in these and other globular filaments, Schneider and Elmegreen (1979) argued that it is unlikely that the filaments represent rings or disks viewed edge-on. The filaments seem to be elongated roughly perpendicular to the polarization vectors in this region (Hsu 1984; see Fig. 10), a configuration which is difficult to reconcile with fragmentation theories based on the dominance of magnetic effects if the filaments are actually prolate. However, the recent study of McDavid (1984) of a much larger ($\sim$ 50 pc) filament, GF7, not in the Taurus complex, shows that the direction of the field does coincide with the long axis of the field, and that the filament and its five internal fragments can be understood in terms of the dominance of the magnetic field.

Prolate structures appear common in the dense interstellar medium. Three important examples are the streamers or filaments in Ophiuchus, the strings of H II regions in Cep A (Hughes and Wouterloot 1984), and the filamentary H I structures mapped by Verschuur (1974*a,b;* see Sec. II.G below). However, the geometry of elongated dark clouds is still not completely clear. The statistical study of Hopper and Disney (1974) and Disney and Hopper (1975) indicated that most of these clouds must be flattened structures, not prolate, and that they are aligned in the galactic plane but not in the magnetic field direction. As mentioned earlier, the Taurus complex itself is probably prolate. In large-scale H I maps both shells and filaments appear (see Sec. II.G for small-scale filaments and Sec. IV for a discussion of large shell structures).

If GF14 and GF16 are physically related, if not connected, then this southern filament presents evidence for at least three levels of hierarchical fragmentation covering sizes from $\sim$ 5 pc down to below 0.1 pc. The study of Schneider and Elmegreen (1979) shows that such structures are typical of dense filamentary clouds. The above discussion shows that cloud structure in Taurus can be traced from large scales ($\gtrsim$ 300 pc) down to at least 0.1 pc, the resolution limit of the observations. Furthermore, these structures appear to be hierarchical.

F. More Distant Regions

Although fragmentation at the smallest-size scales can only be observed readily in the local ($\lesssim$ 300 pc) complexes, there is substantial evidence that such structure exists in more distant regions. The regions which have been most studied are selected by their association with giant H II regions and OB associations; because of the stellar heat input to the dust and the existence of dust-gas energy exchange, the gas is warm enough ($\gtrsim$ 30 K) to emit strong CO lines. It is doubtful whether complexes like Taurus and Sco-Oph would even be noticed by most surveys at distances $\gtrsim$ 1 kpc, except for their densest portions.

At the largest-size scales, it is difficult to say whether the distant warm cloud complexes are fragments of superclouds because the reddening statistics are uncertain at large distances, as are the distances of the complexes, but the suggestion of large H I envelopes around giant molecular clouds (see, e.g., Bridle and Kesteven 1970; Blitz and Thaddeus 1980), the observation of superclouds (see B. G. Elmegreen and D. M. Elmegreen 1983) and superassociations (Wray and de Vaucoulers 1980) in other spiral galaxies, and the association of local complexes within the larger reddening superclouds support the idea. CO observations of the more distant warm molecular complexes certainly show internal fragmentation at the limit of resolution (~5 pc) similar to the local complexes; examples are the Cep OB3 molecular cloud (Sargent 1977; J. Carr 1984), the M17 cloud (Lada and Elmegreen 1978; H. M. Martin et al. 1984), and the Carina Nebula (de Graauw et al. 1981).

Several arguments give indirect evidence for smaller-scale clumping. For example, the star density in typical star clusters implies fragmentation on scales on the order of $\leq$ 0.1 pc. The subthermal excitation of NH_3 is commonly interpreted as due to small beam filling factors of the dense NH_3 gas, with suggested clump scales of ~ 0.05 to 0.2 pc (Schwartz et al. 1977; Matsakis et al. 1977; Barrett et al. 1977; Little et al. 1980), because the densities are large enough to ensure thermalization of the level populations. Similar results have been suggested on the basis of studies of excitation and self-absorption of H_2CO, HCO^+, and HC_3N (see, e.g., Morris et al. 1977; Evans et al. 1979); the different behavior of CS and CO line strength ratios (Kwan 1978); the result that column density increases without a corresponding increase in density inferred from CS and H_2CO (Mundy 1984); and the similarity of line profiles from different CO transitions (H. M. Martin et al. 1984). Wilson and Jaffe (1981) interpreted the differing shapes of H_2CO emission and absorption lines in molecular complexes in terms of clumping on scales of ~ 0.1 pc. Batrla et al. (1983*b*) have interpreted their H_2CO observations of the cloud toward Cas A in terms of a large number of clumps with sizes of ~ 0.4 pc and number densities $n \sim 10^4$ cm^{-3}. The large line widths observed in the warm complexes can also be interpreted as arising from the random velocities of fragments smaller than the beam, although other explanations involving large-scale organized motions such as collapse, expansion, and rotation are also possible. In general, all these indirect arguments overwhelmingly suggest small-scale ($\lesssim$ 0.1 pc) clumping in a number of regions.

There are some direct observations of clumping at small scales in the more distant complexes. At intermediate distances, the detailed analysis of ^{13}CO observations of a complex at a distance of ~ 750 pc by Perault et al. (1985*a*) indicates that the complex contains a large number of condensations on scales of 1 to 3 pc which are often themselves internally fragmented. At larger distance, the direct indications of small-scale structure include VLA radio continuum observations of the M17 H II region, which show clumping on scales of ~0.2 pc (H. M. Martin et al. 1984; Selli et al. 1984); Westerbork H_2CO observations of the giant H II regions DR 21 and W3, which reveal structure on scales of ~ 0.2 pc (see, e.g., Forster et al. 1981; Arnal 1982); and VLA observations of DR 21 in NH_3 (Matsakis et al. 1981) and in H_2CO (Dickel et al. 1983) and of W3 (OH) in H_2CO (Dickel et al. 1984), which suggest density structure on scales as small as 0.01 pc. Hughes and Wouterloot (1984) have presented Westerbork and VLA observations of radio continuum emission in Cep A and find two strings of very young (~ 10^3 yr) H II regions, each containing around 14 O stars; the length of each string is ~ 0.1 pc. In these cases, we can trace the hierarchical structure from ~ 0.01 pc all the way up to supercloud dimensions of ~ 1 kpc. A good example, pointed out by B. G. Elmegreen and D. M. Elmegreen (1983), is that the W3 giant H II region-OB association, itself internally fragmented, as discussed above, is part of a complex containing the clusters h and χ Persei, the H II region-OB associations W3, W4, and W5, and two giant molecular clouds. It should be pointed out that not all studies suggest structure on scales of $\lesssim 0.2$ pc. Examples are the NH_3 observations of NGC 2071 in the Orion complex (Calamai et al. 1982) and CS observations of the core of the Orion Nebula (P. F. Goldsmith et al. 1980). Nevertheless, most of the evidence points strongly to nested structure from ~ 1 kpc down to $\lesssim 0.1$ pc, just as for the local complexes.

G. Structure at Smaller Column Density

The discussion so far has been directed from large rarefied structures to progressively smaller and denser objects. However, this progression is misleading because of major selection effects. Historically, the smaller-scale structure was first recognized in extinction, and most subsequent studies, especially in molecular lines, have been directed towards these regions of large column density. The tendency to study regions of recent star formation leads to a similar bias. One should not make the mistake of concluding that these are the only or even the most common structures. Studies of H I emission and absorption, reddening, and optical interstellar lines reveal that a large fraction of the sky is filled with lower-column density clouds, shells, and filaments covering a wide range of size scales. These can be seen in H I emission from the large Hat Creek, Parkes, and Argentine surveys (see Colomb et al. 1980, and references therein), with resolution of ~0°.5, Heiles' (1980) analysis of

---

Fig. 12. Small scale H I emission structure in Verschuur's (1974*a*) region B. Top left: Integrated column density map for radial velocities between -7 km s^{-1} and $+17$ km s^{-1}; contour unit is 2.2×10^{19} cm^{-2} and maximum column density is 5.5×10^{20} cm^{-2}. Top right: Schematic representation of filamentary structures; velocities (km s^{-1}) within the filaments are indicated at various points in the maps. Lower nine panels: H I antenna temperature contours for selected radial velocities, with a velocity integration interval of 1.0 km s^{-1}.

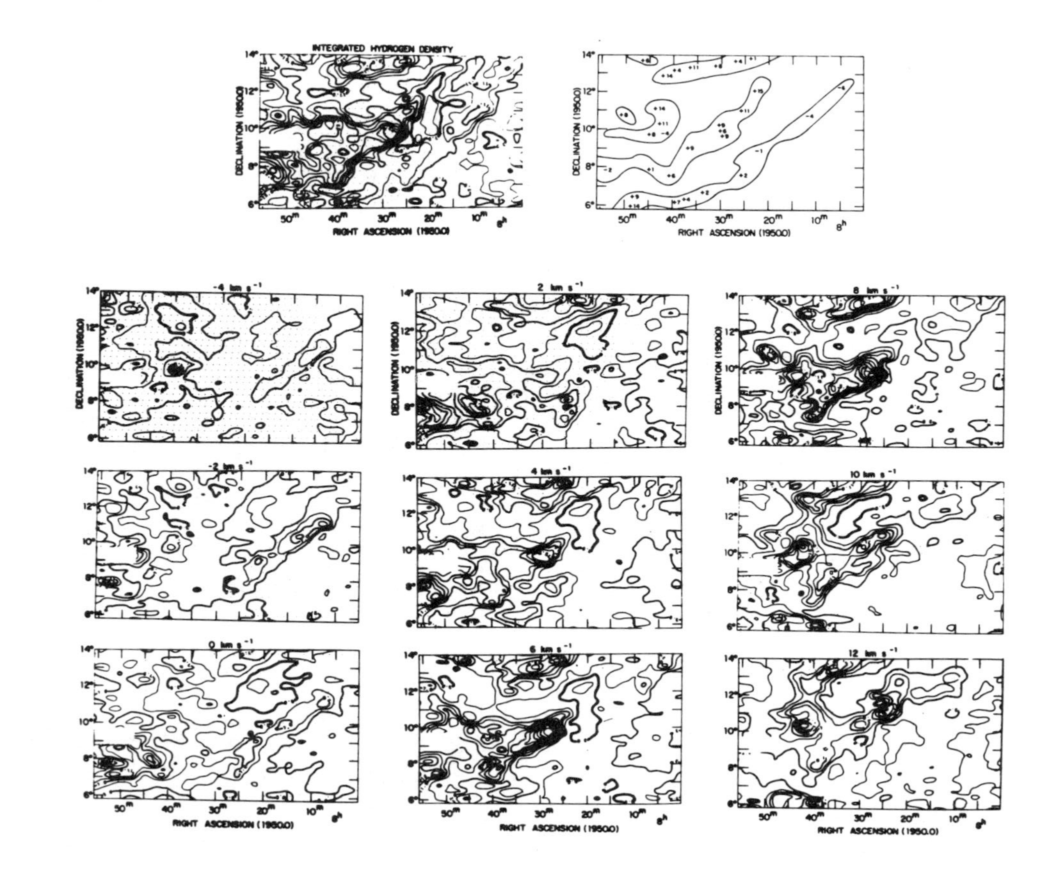

INTEGRATED HYDROGEN DENSITY
DECLINATION (1950.0)
RIGHT ASCENSION (1950.0)
-4 km s⁻¹
2 km s⁻¹
8 km s⁻¹
-2 km s⁻¹
4 km s⁻¹
10 km s⁻¹
0 km s⁻¹
6 km s⁻¹
12 km s⁻¹

the low-latitude Weaver-Williams survey, and the high-resolution mapping of selected small-scale regions by Heiles (1967) and Verschuur (1974*a,b*).

Verschuur's (1974*a*) maps of a roughly $\sim$ 100 square degree region toward $\ell = 215$, $b = +27$ (his region B) are shown in Fig. 12. The top two figures show the integrated column density map and Verschuur's sketch of the four major filaments in the region, while the lower maps show the cloud structure at 9 different radial velocities. Similar structure was found in Verschuur's region A, but with less pronounced filamentary structure. In two additional regions, Verschuur (1974*b*) found no clouds, just filaments.

Space does not permit a discussion of the average and statistical properties of structures observed in H I, or the large body of work on reddening and optical lines (some results for the mass distribution are given in Sec. III). The important point is that these structures have small column densities ($A_V \lesssim 0.5$) and internal densities ($n \lesssim$ 20–50 cm^{-3}), yet cover a range of scales from $\sim$ 0.5 pc (resolution limit) to $\gtrsim$10 pc.

The smallest scales of diffuse H I structure appear to be $\sim$ 0.2 pc, judging from a recent study of H I absorption towards radio sources by Dickey et al. (1984). It is not clear whether the variations on these small scales actually represent separate clouds, or just variations within larger structures.

An important issue that has not been resolved (or even discussed much) concerns whether the low column density structure represents the outer, transparent parts of cloud complexes, or a complicated system of density and velocity fluctuations which pervade most of the interior volume of superclouds. If the latter is correct, then one can easily imagine a broad spectrum of density fluctuations within superclouds, of which only a small fraction are sufficiently large and dense to remain gravitationally bound. It is these bound fluctuations that would evolve to the cloud complexes and smaller-scale substructure discussed earlier. Indeed the dense structures nearly all seem to be balanced on the verge of collapse by some internal support mechanism, as discussed in more detail below. In this case, the observed hierarchy described above must be viewed as the bound condensations that gravity has filtered out of the general fluctuation spectrum. This distinction is of major importance if we are to interpret the observed structure in terms of fragmentation. For this reason we shall refer to the structure discussed earlier as the bound hierarchy.

H. Implications of a Simple Hierarchical Model

Taken together, the observations suggest five or more levels of bound hierarchical fragmentation, starting with the reddening structures or superclouds, and continuing down to dense ($n \gtrsim 10^4$ cm^{-3}) clumps with sizes of $\lesssim$ 0.1 pc. This hierarchy, also emphasized by Rossano (1978*a,b*) for a smaller range of scales, is qualitatively very similar to that originally suggested by von Weizäcker and Hoyle, although the physical processes responsible for the structure may be very different from the turbulent cascade or spontaneous gravitational fragmentation conceptualized by these authors. The density at

TABLE I
Bound Interstellar Hierarchical Structures

Level	Size Scale (pc)	Mean Density (cm^{-3})	Mass ($M_{\odot}$)	Designation
I	10^3	1	2×10^7	reddening—H I superclouds
II	10^2	100	3×10^6	extinction—molecular complexes, GMC's
III	10	500	1×10^4	major fragments in complexes
IV	1	3×10^3	1×10^2	dark clouds
V	10^{-1}	3×10^4	1	dark-cloud cores, globules

each level scales (roughly) inversely with the size, as emphasized by R. B. Larson (1981), and each level typically appears to consist of $N_f \sim 2$ to 10 fragments, with a tendency for elongated or irregularly shaped fragments.

Table I gives a schematic summary of the properties of the bound hierarchy at each level. This table is meant only as an illustrative guide, and should not be taken too literally. The properties at each level show a significant dispersion among different regions or within a given region and the hierarchy is probably not so discrete, because the discreteness indicated in Table I is partly a result of the different observational techniques used at each size and column density. In addition, this characterization of the structure does not incorporate the lower column-density fluctuations, as discussed in Sec. II.G above.

As an example, if we adopt five levels of fragmentation from 10^3 pc to 0.1 pc, N_f fragments at each level, and assume that density is related to size by $n \propto r^p$ with $-1.5 \lesssim p \lesssim -1$ (see Sec. III below), then the ratio of the masses contained in consecutive levels is $10^{-3+p} N_f$, which is about 0.03 ($p = -1$, $N_f = 3$) to 0.3 ($p = -1.5$, $N_f = 10$). This suggests that the fragmentation only involves a small fraction ($\sim 10\%$) of the mass at each stage; most of the mass remains in the larger interfragment structures. This inefficiency agrees marginally with the numerical collapse calculations reviewed in Sec. IV. The differential mass distribution of the fragments among the five levels is $f(m) \sim m^{\gamma}$, with $\gamma \approx -1.3$ to -1.6 for $p = -1.0$ to -1.5; however, the mass spectrum within each fragmentation stage may be very different, because it is determined by the detailed physics of the fragmentation process at that stage. This result is consistent with the observational estimates of the fragment mass spectrum presented in the next section.

If only the fifth level represents clouds which collapse to form stars, without further fragmentation, then the fraction of gas which eventually becomes stars, or the "efficiency of star formation," is $10^{-4(3+p)} N_f^4$. Within a single extinction-molecular complex (level II), the efficiency is $10^{-3(3+p)} N_f^3$. For example, if $p = -1.3$ and $N_f = 5$, about 0.1% of the mass in a molecular complex would be converted to stars. These estimates are extremely crude, and are only meant to point out the low star formation efficiency which is implied by such a hierarchical model. This discussion does suggest, however, that it is more meaningful to speak of the inefficiency of fragmentation rather than the inefficiency of star formation.

Herbig (1978) has argued that star formation is not a hierarchical process, using the fact that T Tauri and related stars, especially in Taurus, do not reside in small clusters but instead are distributed rather homogeneously. However, the apparent absence of very tight groups like those seen in some star-forming regions containing OB stars might be due to OB stars forming preferentially in regions of largest gas density, or might just imply that the clusters which formed were not tightly bound and so were easily disrupted; a velocity dispersion of 1 km s^{-1} would be sufficient to erase the substructure on scales $\lesssim 1$ pc in the latter case. Furthermore, the spatial distribution of the stars in Taurus presented by Herbig (1977*b*) does appear clustered on scales of 1 to 3 pc; the number of stars is too small to say anything about clustering on other scales, especially considering the possible importance of projection effects. In any case, the gas distribution discussed earlier shows overwhelmingly that the structure is hierarchical. The stars that form from this structure cannot be easily used as an imprint because of their small numbers and significant random velocities.

I. Similarity to Extragalactic Structure?

It is difficult not to notice the resemblance of this hierarchical structure to the arrangement of galaxies in the universe. According to a recent study by Bahcall and Soniera (1984; see also Oort 1983), most superclusters contain 2 to 15 rich clusters, and the larger superclusters appear elongated. At large density enhancements, some superclusters are cores of even larger superclusters. The galaxy clusters within the superclusters may possess internal structures (see, e.g., Geller and Beers 1982) which has survived cluster dynamical evolution; the Local Group is often considered as such a subcondensation within the Virgo cluster. Beyond that level we see galaxies, often in pairs or higher-multiplicity groupings. One cannot but be struck by the similarity of this structure to the interstellar hierarchy described above. The similarity is even more obvious if we consider the sequence star cloud, OB association, subgroups, open star clusters, and multiple stars. Even the relative sizes and density enhancements are in rough agreement between the two hierarchies.

The similarity goes beyond structural appearance. R. B. Larson (1979) has emphasized that the velocity dispersion-size relation on subgalactic scales (stellar and interstellar; see R. B. Larson [1981]) extends up to the sizes of galaxy clusters, with the same normalization, and Fleck (1982*a*) and others have noted other properties, like angular momentum, that seem to obey scaling relations which hold over an extremely large range of size scales. One may also point out the rough similarity between the estimated mass spectra of stars, globular clusters, and galaxies. It is as if the universe is a self-similar hierarchy covering over nine orders of magnitude in scale, but with two special length scales, corresponding to galaxies and stars. The first ($\sim$ 100 kpc) may be established because of the nature of cooling processes (Rees and Ostriker 1977; Silk 1977*b*) while the second may be related to the mass below which a contracting cloud becomes opaque to its gravitational energy release (see, e.g., Low and Lynden-Bell 1976). Attempts to understand the processes responsible for extragalactic structure currently center on distinguishing between models in which structure builds from small scales through gravitational clustering, and models in which the structure arises through fragmentation beginning at the largest scales. The debate is very similar to the question of whether interstellar structures build up through cloud collisions or cascade down in size through gravitational, magnetic, turbulent, or other instabilities, a subject which is reviewed in Sec. V. One may also note the recent finding by Byrd and Valtonen (1985) that galaxy groups are expanding as an analog to the expansion of OB associations. The idea that the similarity between subgalactic and extragalactic structure reflects a common origin is very speculative at present, but its philosophical significance and far-reaching implications suggest that a more quantified comparison is desirable.

III. STATISTICAL PROPERTIES

The anatomical studies discussed above strongly suggest that fragmentation is hierarchical, with at least five (rather ill-defined) levels leading to star formation, but do not give a quantitative description of this hierarchy because the observations generally refer to specific regions (which have been selected according to various criteria, like regions of known recent star formation). The objects studied may not be a fair sample of the ensemble of structures whose average properties we desire. Certainly some quantification of the observations is required to infer more information concerning the spectrum and origin of density and velocity fluctuations which are involved in fragmentation. Several statistical approaches to quantitative information are possible:

1. Correlations between fragment properties;
2. The frequency distribution of fragment masses and sizes, which can yield estimates of parameters characterizing a self-similar hierarchy, as shown in Sec. III.C;
3. The frequency distribution of stellar masses, or initial mass function (IMF);
4. Two-point measures of spatial structure, such as the autocorrelation function.

Most of these lines of evidence cannot presently be compared with theory, except in a very limited manner, because the theories are based on either numerical calculations with insufficient resolution, linear stability analyses, or heuristic arguments. The present discussion will be confined to the first two lines of evidence listed above.

A. Correlations between Fragment Properties

R. B. Larson (1981) has emphasized that densities and velocity dispersions inferred from molecular-line observations scale with size of region in a manner which seems to hold over nearly 3 orders of magnitude in size. Since these relations are discussed in the chapter by Myers, only a brief discussion of their relevance to the fragmentation problem is given here.

A roughly inverse relation between density and size appears to hold over the entire bound hierarchy of Table I, and also among the individual clouds at the smallest scales. This relation may contain information on the geometry of the fragmentation process, because compression or accumulation of material in one, two, or three dimensions would yield $n \sim r^p$ with $p = -1, -2$, or -3, respectively. p appears to be in the range -1 to -1.5 and so seems to rule out a spherically symmetric process. A value of $p \approx -1$ would suggest the formation of oblate pancake-shaped fragments, as might arise from supersonic cloud collisions, contraction along magnetic field lines, or the importance of rotation. Recent evidence for a $\sim$ 0.2 pc rotating disk around the Orion-KL object has been presented by Hasegawa et al. (1984). However cloud structure often appears filamentary, and, as emphasized by Schneider and Elmegreen (1979), the regular spacing of subcondensations within these filaments implies that they are prolate structures, so we would expect $p \approx -2$. The observationally-determined value of p may be somewhat smaller than -1 for small dark clouds, and so may reflect the contributions of both prolate and oblate fragments. For example, it is obvious from inspection in Lynds' (1962) dark-cloud catalog that the smaller clouds have larger column densities on the average, so $p < -1$. In addition, it is also true that the empirical estimates at small scales are biased because roughly spherical-looking condensations are usually chosen for study. D. B. Sanders et al. (1984*a*) find $p \approx -0.75$ for GMC's but their densities are derived by assuming virial equilibrium. Prolate structures appear common; three specific examples cited earlier are the streamers in Ophiuchus, the strings of H II regions in Cep A (Hughes and Wouterloot 1984), and the filamentary H I structure found by Verschuur (1974 *a,b*) in at least three of four directions.

More likely, the scaling relation may represent effects besides geometry. It would be quite interesting to estimate the relation between density and size separately for globular filaments and more spherically symmetric dark clouds. For a self-similar hierarchy, the value of p, when combined with the mass distribution, can give information on the mean number of fragments per level of the hierarchy, as outlined in Sec. III.C.

While the following discussion will assume that density scales roughly inversely with size scale ($p \sim -1$), it must be emphasized that this result is only known to apply to bound condensations and may also suffer from various selection effects. As an extreme example, note that Stenholm (1984) suggested that column density *increases* with size scale, based upon his analysis of the spatial power spectrum of CO peak antenna temperatures in the dark cloud B5. However, this result is uncertain because of the presence of inhomogeneities on scales comparable to the image size, i.e. the data does not represent a stationary process.

There also exists a scaling relation between internal velocity dispersion, as measured by molecular line widths, and size of the form $\Delta v \sim L^q$, with $q \approx$ 0.4 to 0.6 (R. B. Larson 1981; Leung et al. 1983; Myers 1983). Dickey (1984) gives a review of the data. A similar relation has been found for GMC's, with $q \approx 0.6$, from the CO survey of D. B. Sanders et al. (1984*b*). Quiroga (1983) suggests a similar relation for diffuse H I clouds.

There are a number of possible interpretations of this result. R. B. Larson (1981) assumed that the relation is a measure of the structure function (mean square velocity difference between pairs of positions as a function of separation) used to characterize stochastic fields, and suggested that the result $q \approx 0.4$ implies a similarity with incompressible turbulence, for which $q = 1/3$ would be expected (the Kolmogorov-Obhukov spectrum). This point of view is also taken by Quiroga (1983) in his discussion of diffuse H I clouds. This identification of the line width-size relation with the structure function is not at all clear; direct measurements of the structure function using well-sampled radial velocities are quite difficult with existing data (Scalo 1984). Other possible explanations include a purely statistical effect in which the velocity dispersion of fragments naturally increases with the depth (size) of the regions studied (Scalo 1984), a simple radial gradient in turbulent velocity dispersion, and the velocity dispersion-size relation expected for clouds in virial equilibrium between gravity and fragment random kinetic energy (R. B. Larson 1981; Dickey 1984; Myers 1983; Scalo 1984*b*; see Sec. III.C). Both effects would give $\alpha \approx 0.5$ if $p \approx -1$. Additional considerations can be found in Dickey (1984) and in the chapter by Myers. Much additional observational work on the velocity fields of clouds will be needed before such data can be used to constrain the fragmentation process.

A possibly more important clue relevant to the fragmentation question is the fact that the relation between velocity dispersion, mass, and size approximately satisfies the virial relation, $\Delta v \sim (GM/L)^{1/2}$, over a large range of structure sizes (R. B. Larson 1981). Because there is little evidence for dynamical collapse, at least on larger scales, most clouds therefore appear to be in quasi-virial equilibrium between gravity and some internal support mechanism. A recent discussion of the situation for small dark clouds is given by Myers (1983), who finds strong evidence for virial equilibrium in these clouds. As mentioned in Sec. II, H I studies of superclouds with sizes of ~ 1 kpc also suggest virial equilibrium.

This conclusion is in agreement with a number of indirect arguments suggesting cloud lifetimes in excess of the free-fall time. Briefly, the main lines of evidence are based on comparisons of the galactic star formation rate, mean surface density of H_2, and estimated star formation efficiencies (Silk 1981); ages of OB associations in GMC's and studies of the association of CO with clusters of various ages (Bash et al. 1977); comparison of a ballistic CO-cloud model with observations (Bash 1979); chemical ages of clouds (Knapp 1974; Allen and Robinson 1977; Stahler 1984); requirements of the Oort model for cloud evolution (Kwan 1979); and arguments based on an interpretation of the correlation of stellar velocity dispersion with stellar age as a result of cloud-star gravitational scattering (Icke 1982). Upper limits to cloud lifetimes have also been estimated based on the erasure of substructure by coalescence (Blitz and Shu 1980), confinement to spiral arms (see the chapter by Solomon and Sanders), and the predicted effects of dynamical friction between field stars and large clouds on cloud velocity dispersions (Hausman, personal communication). These lines of evidence are individually quite weak, but taken together they suggest $10^7 \lesssim \tau \lesssim 10^8$ yr for extinction molecular complexes, while the free-fall times are $\lesssim 3 \times 10^6$ yr. Something is holding up the collapse over a wide range of scales, but whether it involves magnetic fields, rotation, turbulence, stellar winds, or something else is still a matter of debate which will not be discussed here. Of particular importance, however, would seem to be the fact that whatever process or processes are involved must operate over a wide range of size scales, because the available evidence suggests that the bound hierarchy is virialized at all levels.

The facts that the density-size and line width-size relations hold within objects of a single level in Table I, and that structures as different in scale as GMC's and small dark clouds seem to follow the same line width-size relation, suggests that the hierarchy described schematically in Table 1 is actually more continuous, in the sense that there is a significant variation in properties among the members of each level even though the process of star formation may always require around five or six levels of fragmentation on the average. As an extreme case, Fig. 13 depicts the variation of a number of physically interesting properties of interstellar structures as a function of size scale, assuming only that, whatever the physics of fragmentation, the density and size (diameter) scale as $n = n_o (L/L_o)^p$ with uncertain normalization $n_o = 10^4$ cm^{-3} at $L_o = 1$ pc and p a parameter with a likely value around -1.0 to -1.5. The quantities plotted are mass $m = 0.02\, nL^3\, M_\odot$; visual extinction A_V calculated from $A_V = 1.1 \times 10^{-21} N$ mag, where N is the column density; ratio of size L to thermal Jeans length $L_J = 8.5\, T^{1/2}/n^{1/2}$ pc; free-fall time $\tau_{ff} = 4 \times 10^7/n^{1/2}$ yr; free-fall velocity $v_{ff} \approx (2GM/L)^{1/2} = 0.021\, n^{1/2} L$ km s^{-1}; critical rotational velocity gradient at which rotation balances gravity perpendicular to the rotational axis $(dV/dr)_{cr} = 0.04\, n^{1/2}$ km s^{-1} pc^{-1} (see Lequeux 1977; Field 1978); and the ratio of magnetic field strength scaled according to $B = B_o(n/n_o)^k$, with $B_o = 3$ μGauss at $n_o = 1$ cm^{-3} and $k = 1/2$ or $1/3$, to the critical field strength at which the field balances gravity perpendicular to the field $B_{cr} = 0.06\, n\ell$ μGauss (see, e.g., Mestel 1977). The velocity v_{ff} is also about equal to the critical rotational velocity for support

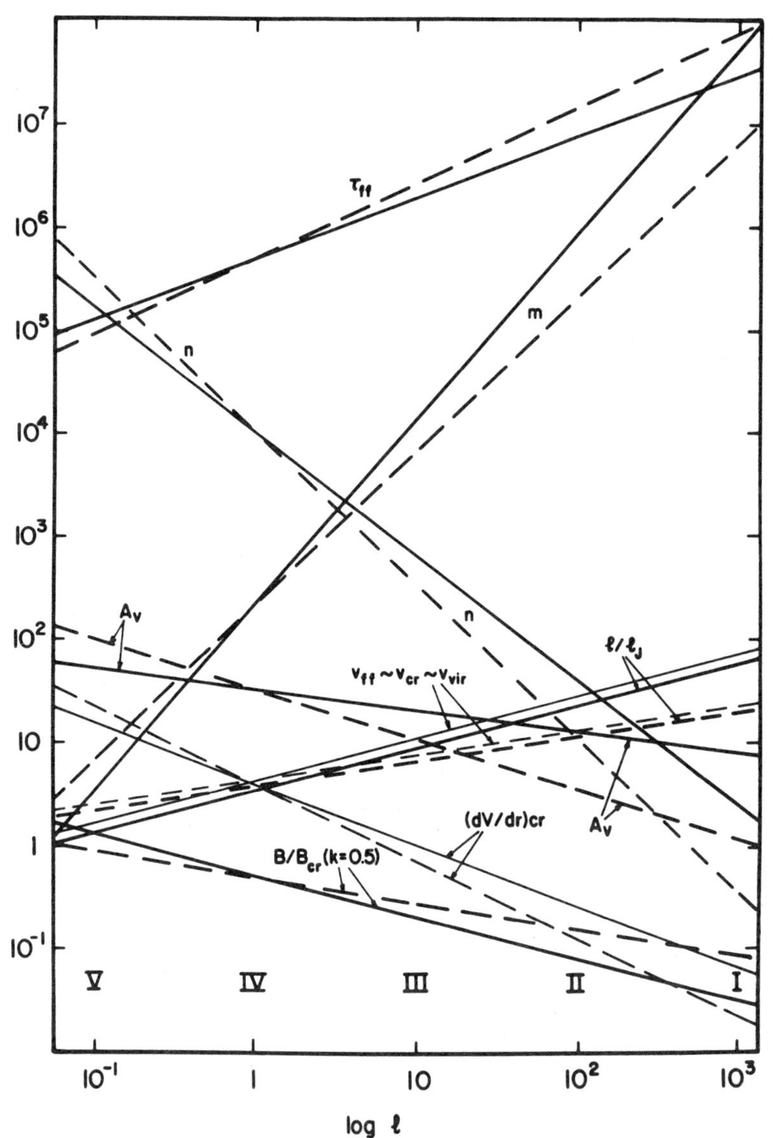

Fig. 13. Size scale dependence of physical quantities for clouds following a power law density-size relation $n \propto n_o (\ell/\ell_o)^p$, where ℓ is the diameter in parsecs, for $p = -1.2$ (solid lines) and -1.5 (dashed lines). The adopted normalization is $n_o = 10^4$ cm^{-3} at $\ell_o = 1$ pc.

against gravity and the characteristic velocity to which subcondensations within that scale will be accelerated by gravity (virial velocity). The solid and dashed lines correspond to $p = -1.2$ and -1.5, respectively. The approximate scales of the five hierarchical levels discussed earlier are indicated, but it must be remembered that these levels may be observationally selected.

With the above parameterization several interesting features emerge. First, the structure is unstable to gravitational collapse against thermal pressure ($L/L_J > 1$, $v_{ff}/c_s > 1$) for scales larger than about 0.1 pc. The fact that thermal pressure can balance gravity at $L \sim 0.1$ pc, $m \sim 1\ M_\odot$ has been noted by several authors in various contexts, especially as the scale below which fragmentation is not possible without assistance by, for example, rotation, and as the scale below which the random cloud relative velocities, or turbulence, becomes subsonic. Second, the available observed velocity gradients across the faces of clouds give a range of values at a given scale which, when compared with $(dV/dr)_{cr}$, indicates that rotational effects are unimportant in more massive clouds (see, e.g., Kutner et al. 1976; Loren 1977*a,b;* Kleiner and Dickman 1985), but are important in a small but significant fraction of small very opaque clouds (see Fleck and Clark 1981; R. N. Martin and A. H. Barrett 1978; Snell 1981; Ungerechts et al. 1980; Myers and Benson 1983). Fleck (1981) and others have pointed out that at the largest scales the critical rotational angular velocity $\omega_{cr} = 2 \times 10^{-15}\, n^{1/2}\ s^{-1}$ is comparable to the angular velocity expected from differential galactic rotation, so that galactic rotational energy could serve as the large-scale source of kinetic energy for turbulence, assuming that some mechanism exists for coupling the shear on this scale with the smaller sizes. Third, the magnetic field estimate, while extremely uncertain, suggests that fields play a minor role if $B \propto n^k$ with $k \lesssim 0.5$, although the conclusion is still arguable.

B. Frequency Distribution of Cloud Masses

The frequency distribution of cloud masses is much more difficult to determine than the stellar mass spectrum, essentially because there is no analogue of the mass-luminosity relation for clouds, i.e. there is no readily observable quantity which can be reliably related to mass. The only "direct" method consists of the following steps:

1. Devise a suitable working definition of a cloud in terms of an observable property (e.g., a line strength, reddening, etc.) which shows correlated behavior over some interval of space or velocity.
2. Measure the angular sizes of a large number of such clouds and somehow estimate the distance to convert to linear sizes.
3. Determine internal densities for each cloud either from column density and size, or from arguments involving excitation of molecular rotational lines.

In practice it is extremely difficult to carry out this program. In many cases, the best we can do is estimate a size distribution, $f_r(r)$, and then make some hopefully reasonable assumptions concerning the internal densities n. For example, if $f_r(r) \propto r^\alpha$ and n is related to r by $n \propto r^p$, we have $f_m(m) = f_r(r)$ $|dr/dm| \propto m^\gamma$, with $\gamma = (\alpha - p - 2)/(3 + p)$.

Estimation of size and mass distributions by means of identification and counting of individual clouds also suffers from the necessarily subjective definition of a cloud which must be made. That hierarchical complexes of clouds are chaotic and irregular and may contain density fluctuations covering a large range in amplitudes implies that all such studies will contain selection effects.

1. "Direct" Estimates. In some cases both angular size and column density can be measured, which gives cloud masses directly for an assumed cloud shape if the distances are known. Such data for diffuse clouds using H I emission have been presented in the extensive studies by Heiles (1967) and Vershuur (1974*a,b*). Heiles (1967) mapped an area of about 160 square degrees near $\ell \sim 120$, b ~ 15, and identified over 800 small clouds embedded within two large sheet-like structures which are probably associated. Average column densities, radii, particle densities, and masses are 2×10^{19} cm^{-2}, 2 pc, 3 cm^{-3}, and 2 M$_\odot$, respectively, using a distance of 300 pc on the basis of arguments given by Ames and Heiles (1970). Heiles also identified 13 larger concentrations within the sheets. It is impossible to estimate reliable power law size and mass distributions from Heiles' data (although such results are occasionally quoted in the literature), partly because incompleteness sets in at ~ 1 pc, and because the distribution is not smooth at larger sizes. For $r > 1$ pc, I find $\alpha \sim -4$ and $\gamma \sim -1.5$, but these are uncertain by at least ± 0.5.

Verschuur (1974*a*) mapped two regions, referred to as regions A and B, around $\ell \sim 220$, b ~ 35, covering a total of 420 square degrees. Spatial contour maps were constructed for every 1 km s^{-1} velocity interval, and clouds were defined as sets of closed contours recognizable in at least 2 adjacent velocity intervals (see Fig. 10). The column densities of these clouds were in the range 1–8 $\times 10^{20}$ cm^{-2}, and typical sizes, densities, and masses in region A(B) were 3(2) pc, 30(70) cm^{-3}, and 17(7) M$_\odot$, for an assumed distance of 100 pc. Many of the clouds in region B were located in four major filaments. Frequency distributions of radii and masses can be constructed from the data of Verschuur, but, as with Heiles' data, they are incomplete because of resolution for $r \lesssim 0.5$ pc. At larger sizes the histograms are not very smooth; a power law fit gives $\alpha \sim -4.5$, $\gamma \sim -2$ for region A, but these are uncertain by at least ± 0.5. These H I results mostly refer to diffuse clouds which are not associated with cloud complexes and star formation, so their relevance to the type of fragmentation of interest here is not obvious.

A related method for directly estimating masses was used by Knude (1979) in a study of local reddening statistics. Knude identified a cloud as an angular area over which the reddening appeared visually correlated. This is of course a subjective criterion, but does allow an estimate of sizes, which can be combined with the derived reddenings (proportional to column density) to obtain volume densities, and hence masses. Dimensions and masses of 94 clouds ($A_V \lesssim 0.4$) were obtained in this way, and for the 68 clouds with $7.5 < m < 78$ (the lower limit is imposed by incompleteness at small sizes), Knude finds $f(m) \propto m^{-1.4}$. Again the result may not be directly relevant to regions of star formation.

For molecular clouds, the recent ^{13}CO study of Casoli et al. (1984) provides a direct estimate of the cloud mass spectrum over a fairly broad range of masses. Casoli et al. surveyed regions in the second galactic quadrant around $\ell \approx 115°$, avoiding the distance ambiguities and blending problems associated with the first quadrant. The regions included 12 square degrees in the Perseus arm and 6 square degrees in the Orion arm. The 8 arcmin sampling corresponds to 8 pc in Perseus and 2 pc in Orion. The criterion for counting clouds was that adjacent (in space) spectra have an emission maximum within a specified velocity interval, either 1 km s^{-1} or 4 km s^{-1}; the larger interval identifies larger clouds. Masses were obtained by combining column densities derived from ^{13}CO integrated emission, for an optical depth of 0.2, with a size defined by the 0.3 K peak antenna temperature contour. With the 1 km s^{-1} bins, 165 clouds in Perseus and 63 in Orion were used to determine maximum likelihood estimates of the index of a power law mass spectrum. The result was $\gamma = -1.4$ to -1.6 in both regions, depending on the velocity bin size and the adopted distance to the Perseus complex ($\sim$ 3 to 5 kpc). This result refers to masses in the range 400 to 2×10^5 M$_\odot$ in Perseus, and smaller masses in Orion.

Bhatt et al. (1984) have recently used the same approach to estimate mass spectra for dark clouds in Lynds' (1962) catalog, obtaining sizes from tabulated angular areas and column densities assuming that the opacity class is equal to the visual extinction; actually the choice $A_V = 6$ for opacity class 6 clouds is a lower limit. Bhatt et al. find $\gamma \approx -1$ for 28 clouds in Orion (30 to 2×10^4 M$_\odot$) and 27 clouds in the ρ Oph cloud (10 to 2000 M$_\odot$). For 42 clouds (2 to 1000 M$_\odot$) in Taurus, the mass spectrum is not a power law, but flattens at smaller masses. For $m \gtrsim 40$ M$_\odot$ a power law fit gives $\gamma \approx -1.5$, but for 2 to 40 M$_\odot$ $\gamma \sim -1$. Assuming that the clouds in Lynds' catalog are distributed randomly in space, Bhatt et al. also constructed the mass spectrum of 1697 clouds in the mass range 3 to 10^5 M$_\odot$; the result is very similar to Taurus, with $\gamma \sim -1.5$ for $m \gtrsim 100$ M$_\odot$ but flatter at smaller masses.

An estimate of the mass spectrum of the smallest fragments, which might represent a near-final preprotostellar fragmentation stage, is interesting for comparison with the stellar initial mass spectrum. Differences might be related to the mass changes that occur during the protostellar stage due to accretion, winds, or other effects (see Ebert and Zinnecker 1981; Smith 1985). The only published data which allow a direct estimate of the mass spectrum for these fragments within complexes is the extensive study of dark-cloud cores by Myers et al. (1983), who provide upper limits to the masses of 48 small dark ($A_V \gtrsim 5$) clouds with estimated distances. Most are in Taurus, Ophiu-

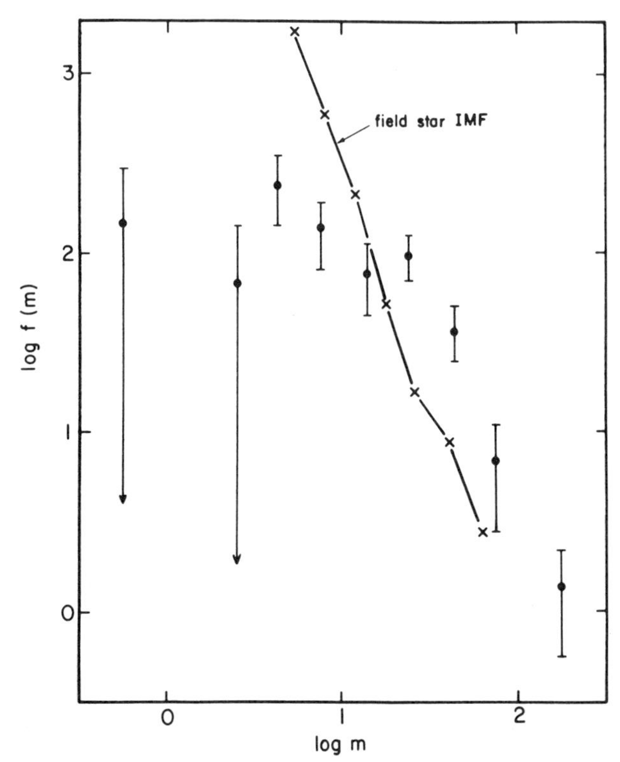

Fig. 14. Frequency distributions of masses of dark-cloud cores studied by Myers et al. (1983). Field star initial mass function (Scalo 1985*a*) is shown for comparison.

chus, or Orion. The masses were obtained by combining the $C^{18}O$ column density, an adopted conversion from $C^{18}O$ column density to total column density based on extinction, and a size based on visual appearance on POSS plates. It is difficult to assign uncertainties, but a factor of two seems conservative. The masses are upper limits to the masses of the $C^{18}O$-emitting regions because the total column density from extinction must refer to a larger path length than the very opaque core to which the $C^{18}O$ column density refers; however, the correction to total mass is probably nearly constant among the clouds, so this should not affect the shape of the mass spectrum. The resulting frequency distribution of masses (arbitrary normalization) is shown in Fig. 14, along with $N^{1/2}$ counting uncertainties. Omitting the two clouds with the smallest masses, a power law fit of the form $f(m) \propto m^{\gamma}$ would give γ in the range of -1.0 to -1.5 for $m \gtrsim 4$. However, the distribution is steeper for $m \gtrsim 20$, with $\gamma \approx -2$.

A problem with this sample is that it is restricted to clouds with a narrow range of linear sizes, because the clouds were chosen such that their angular sizes were $\lesssim 5$ arcmin, while the resolution limit is about 1 arcmin and most of the clouds are at about the same distance ($\sim$ 150 pc). Thus, most clouds with diameters outside the range of 0.05 to 0.3 pc are excluded. In addition, the sample excludes clouds with $A_V \lesssim 5$, while these lower-opacity clouds are the most numerous in Lynds' (1962) catalog. It is difficult to guess how the mass spectrum for a less restricted sample would differ; some evidence related to this point is discussed below.

For illustrative purposes, the field star mass spectrum (Scalo 1985*a*) is also shown (with arbitrary normalization). At masses of $\gtrsim 20$ $M_\odot$ the shapes are very similar. (The slope of the stellar mass spectrum for $m \gtrsim 10$ is very uncertain, however.) The cloud mass spectrum extends to higher masses, but the absolute scale of the cloud masses (and stellar masses) is uncertain. The cloud masses are overestimates for the $C^{18}O$-emitting regions, but if the less opaque surrounding material is considered a part of the cloud then they may be underestimates, because the appropriate linear sizes would be larger. The stellar masses may be too small if convective overshoot brightening is not as important as was assumed in deriving the stellar IMF. For masses below ~ 20 $M_\odot$ the cloud mass spectrum is much flatter, but this might be an artifact caused by incompleteness, because the lower limit on detectable dark-cloud angular sizes ($\sim$1 arcmin) and the restriction of the sample to clouds with large column densities ($A_V \gtrsim 5$) both bias the sample against lower-mass clouds.

Although a valid comparison of fragment and stellar mass spectra cannot yet be performed, it is important to realize that there are reasons why we should expect differences, and that these differences would provide valuable clues concerning the physical processes at work. For example, higher-mass protostars probably blow off a significant fraction of their original (fragment) mass before they settle onto the main sequence, causing the mass spectrum to steepen. Protostars may also be able to accrete surrounding gas, which flattens the mass spectrum if the accretion efficiency is an increasing function of protostellar mass (Zinnecker 1981; see also Ebert and Zinnecker 1981). Reliable comparisons of fragment, protostellar, and stellar mass spectra could tell us much about these finishing touches. Also, the fragment and stellar mass spectra may be time-dependent in a given region, so comparisons of mass spectra for a sequence of regions like the Southern Coalsack, Taurus, Sco-Oph, and Orion might allow us to understand the evolution of the mass spectrum.

2. *Size Distributions.* In many cases we are forced to examine the size distribution and make some assumptions about the internal densities in order to convert to a mass distribution. Size distributions can be determined from either angular sizes and distances of individual clouds, or from a frequency distribution of angular sizes and an assumption concerning the spatial distribution of clouds. The frequency distribution of column densities, which measures the size distribution if we make some assumptions about the internal densities, has also been used for diffuse clouds (Penston et al. 1969; Hobbs

1974; McKee and Ostriker 1977; Crovisier 1978,1981), and could also be used for Lynds' dark clouds. For example, if the column density distribution is a power law with index θ, then the mass spectrum is a power law with index $[\theta(p+1) - 4] / (3+p)$ if the spatial sampling is random (as for optical lines and H I absorption against radio sources), or $[\theta(p+1) - 2] / (3+p)$ if the spatial sampling is not random (as for Lynds' clouds). However, the uncertainties in this approach are very large, and will be omitted from the following discussion.

It is usually assumed, often for simplicity, that internal density is unrelated to size for diffuse clouds. Certainly the densities and sizes given by Heiles (1967) and Verschuur (1974*a*) show no correlation although there are large variations. For cloud complexes, the study of local GMC's by Blitz (1980) and the larger-scale surveys mentioned in Sec. II suggest that the mean internal densities cover a rather narrow range, and we might again assume constant density, independent of size. However, for the smaller dark clouds of most interest to fragmentation, there is strong evidence that density increases with decreasing size. As discussed above, the available observations suggest $n \propto r^p$ with $p \sim -1$. As mentioned earlier, the fact that the average sizes of dark clouds in Lynds' (1962) catalog decrease with increasing opacity class shows that p must be somewhat smaller than -1. Myers (1983) finds $p = -1.3$ when 27 dark-cloud cores observed in NH_3 are combined with 16 clouds mapped in ^{13}CO by Leung et al. (1982), and $p = -1.0$ (with a much smaller correlation coefficient) for the NH_3 clouds alone. In what follows, power law fits to size distributions will be summarized, and the mass spectrum will be estimated assuming constant density for diffuse clouds and complexes, and $-1.5 \leq p \leq -1.0$ for dark clouds.

At the large scale, several studies have attempted to determine the size distribution of large molecular cloud complexes. Liszt et al. (1981) used the ^{13}CO line, which is a better probe of size and column density than ^{12}CO because it is optically thinner and suffers less from blending in space and velocity, which was a problem in earlier CO surveys. Liszt et al. identified 375 clouds in the first galactic quadrant visually from longitude-velocity maps by requiring that each feature have a peak central intensity > 0.4 K and that it occur in more than one spectrum. The linear size was calculated from the angular extent and the kinematic distance, which is probably the largest uncertainty. There are also problems with blending and the ambiguity between near and far kinematic distances, but by using reasonable assumptions a size distribution could be determined. The result showed that a delta function, Gaussian, or flat distribution can be rejected. An exponential of the form $f(r) \propto \exp[-(r-8)/6]$ (r in pc) provides a good fit. A power law with a slope of -3.3 ($9\ \mathrm{pc} \lesssim r\ \$\ 30\ \mathrm{pc}$) fits almost as well. Using the power law, if all the complexes have the same mean internal density, the mass distribution would be $f(m) \propto m^{-1.8}$. The mass spectrum would be steeper if density and size are anticorrelated, e.g., $\gamma = -2.2$ if $p = -1$. The mass spectrum found by Casoli et al. (1984) using sizes and individual column densities to obtain masses is flatter, with $\gamma \approx -1.5$; Casoli et al. ascribe the discrepancy to the distance ambiguities and blending at the tangent point in the Liszt et al. (1981) data.

A new attempt to estimate the mass spectrum from the size distribution of GMC's has been reported by D. B. Sanders et al. (1985*a*), who use a ^{12}CO survey of the first galactic quadrant sampled every 1° in longitude and 12′ in latitude. Taking care to avoid the distance ambiguities by using 80 features on the near side of the tangent point, Sanders et al. construct the distribution of latitude chord lengths and then convert to a diameter distribution by correcting for a selection effect associated with the sampling technique. The resulting size distribution can be fit by a power law with index of -2.5 ± 0.25 for sizes in the range of 15 to 80 pc. Using densities derived from the virial theorem, the size distribution was converted into a mass spectrum, which is a power law of index $\gamma = -1.6$ for the mass range 8×10^4 to 2×10^6 $M_\odot$. This result agrees with the direct ^{13}CO estimate of Casoli et al. (1984), although it should be noted that the Casoli et al. result refers to smaller masses and also that they found densities roughly independent of size.

Also of relevance to the fragmentation problem are the smaller and denser clouds which form the substructure of molecular cloud complexes. A large amount of information on these clouds is contained in the Lynds' (1962) dark-cloud catalog, which gives positions, areas, and visually-estimated opacities for 1802 objects. The major problem lies in the unknown distances to most of these clouds. However, if we only examine clouds which are observed toward local complexes, such as Taurus, Sco-Oph, and Orion, it may reasonably be assumed that they all have the same distance, so the shape of the size distribution follows directly from the angular sizes. Rowan-Robinson (1979) applied this technique to the dark clouds in the Taurus complex and found that a power law size distribution would have an index $\gamma \approx -1$. He suggested that data for a smaller number of clouds toward ρ Oph (16 clouds) and Orion (23 clouds) were roughly consistent with this result.

Mass distributions for the dark clouds from Lynds' catalog within three complexes, derived from the angular area distribution, are shown in Fig. 15 (Scalo 1985*c*). The adopted boundaries of the regions were rectangular in galactic coordinates, and cover 300 square degrees for Taurus (roughly McCuskey's [1938] extinction region and Wouterloot's [1981] complex A), 120 square degrees for Ophiuchus (about a quarter of the area of the Sco-Oph complex mapped in extinction by Rossano [1978*b*]), and 150 square degrees for Orion. The number of clouds in each region are indicated in the figure. The adopted distances were 140 pc for Taurus, 175 pc for Sco-Oph, and 450 pc for Orion. In converting from mass to size, it was assumed that $n = n_o(r/r_o)^p$ with a rather uncertain normalization of $n_o = 10^3$ cm^{-3} at $r_o = 1$ pc. Results for $p = -1.0$ and $p = -1.5$ are shown; the mass distribution from the data of Myers et al. (1983) for small opaque clouds, discussed earlier, is shown at the bottom for comparison. The number of clouds used in Oph is much larger than that used by Bhatt et al. (1984) because they concentrated on

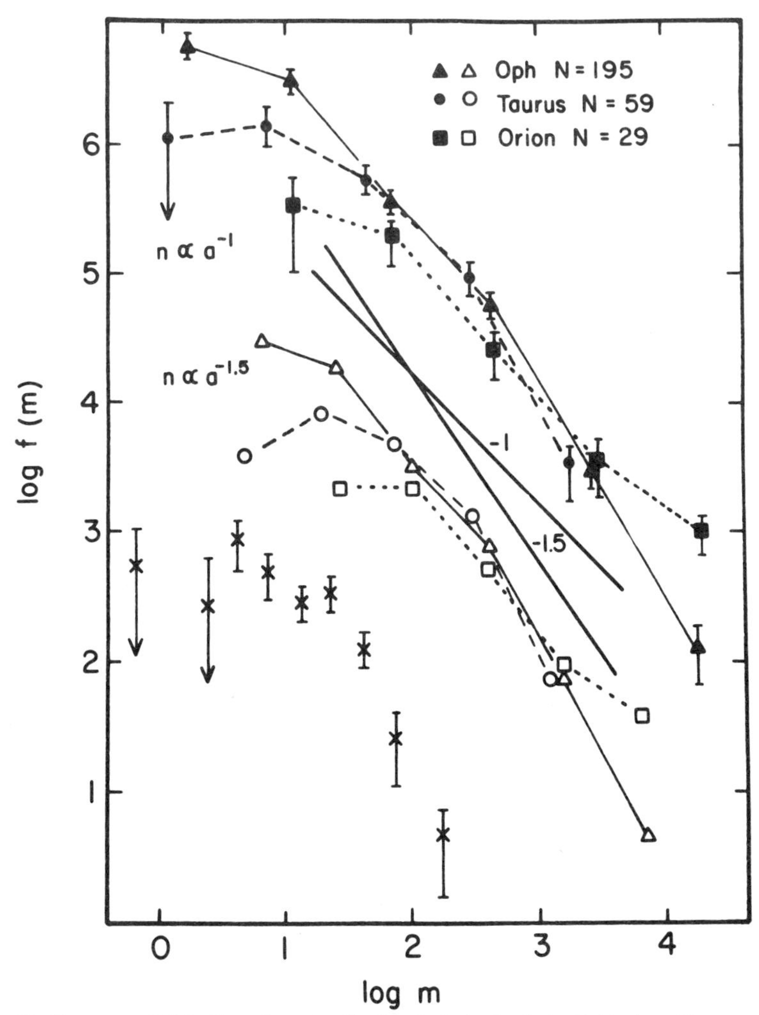

Fig. 15. Frequency distribution of masses for Lynds dark clouds in Ophiuchus, Taurus, and Orion, derived from the angular area distribution and an assumed relation between number density and radius of the form $n = n_o\,(r/r_o)^{-p}$ with normalization $n_o = 10^3\ \mathrm{cm}^{-3}$ at $r = 1$ pc and $p = 1.0$ and 1.5, from Scalo (1985*b*). The number of clouds used in each region is indicated in the upper right. Crosses represent the mass spectrum of small dark clouds studied by Myers et al. (1983), also shown in Fig. 14.

the region near ρ Oph. In general, the results of both studies are consistent, even though the mass distributions were derived in different manners.

Some differences between the mass distributions of the three regions are apparent, but should not be overinterpreted because of selection effects. In particular, the turnover in the Orion distribution is almost certainly due to incompleteness at small angular sizes. The turnover in the Taurus distribution and flattening in Oph may also be due to incompleteness, but because these two regions are at about the same distance, the Taurus complex may actually be deficient in low-mass clouds compared with Oph. That the Orion distribution is somewhat flatter and extends to larger sizes is consistent with intuitive expectations, because Orion contains stars with larger masses than the other regions, and agrees with R. B. Larson's (1982) finding that the mass of the most massive clouds and stars in star-forming regions are correlated. However, it should be noted that this result must be at least partially due to the poorer linear resolution at the distance of Orion; one large cloud in Orion might be identified as two or more smaller clouds at the distance of Taurus or Ophiuchus.

At larger masses, all three mass spectra are consistent with a power law of index $\gamma \approx -1.0$ to -1.5. Although the mass scales are somewhat uncertain, the Myers et al. mass spectrum (mostly in Taurus and Oph) is certainly deficient at large masses ($\gamma \approx -2$) compared to the other mass spectra in Fig. 15. The discrepancy is due to the fact that the Myers et al. clouds were all very small and most had opacity classes 5 or 6. A more detailed study shows that the derived mass distributions of Lynds' clouds steepens and shifts to smaller masses with increasing opacity class, and the result for classes 5 and 6 is consistent with the Myers et al. distribution (Scalo 1985*b*). Unfortunately, it is not possible to say how much these results are influenced by incompleteness at small angular sizes and small opacity classes.

In order to use a larger statistical sample, we can examine the angular area distribution for all the clouds in Lynds' catalog. Assuming that these clouds are distributed uniformly in space out to a limiting distance D in a disk of thickness $2H$, the angular area distribution can be evaluated for a given distance from the assumed size distribution and then integrated over all distances $< D$. The resulting integral is complicated, but one can show that for a power law $f_r\,(r)$ with exponent α, the result is $h(\sigma) \propto \sigma^{(\alpha-1)/2}$. This result allows us to infer a size, and hence a mass distribution from the observed distribution of angular areas.

Figure 16 shows the distribution of angular surface areas for all clouds in the dark-cloud catalogs of Lynds (1962), Khavtassi (1955,1960), and Schoenberg (1967) as given in Lynds (1968). The Lynds distribution is obviously not a power law over the entire range of areas, but $h(\sigma) \propto \sigma^{-1.2}$ is a reasonable fit between about 0.04 and 10 square degrees. This would suggest $f_r(r) \propto r^{-1.4}$. Khavtassi's (1955) area distribution agrees well with Lynds for areas $\gtrsim 1$ square degree but is flatter for smaller areas, probably because of the brighter limiting magnitude of the plates used by Khavtassi, making it more difficult to identify small clouds. If we take $\alpha = -1.4 \pm 0.2$, then the mass spectrum has $\gamma = -1.2 \pm 0.3$, for $-1.5 \leq p \leq -1.0$.

However, Schoenberg's area distribution is much steeper than Lynds', which suggests that selection effects may be quite significant in all the catalogs. For example, clouds with angular sizes $\lesssim 1$ arcmin cannot be detected

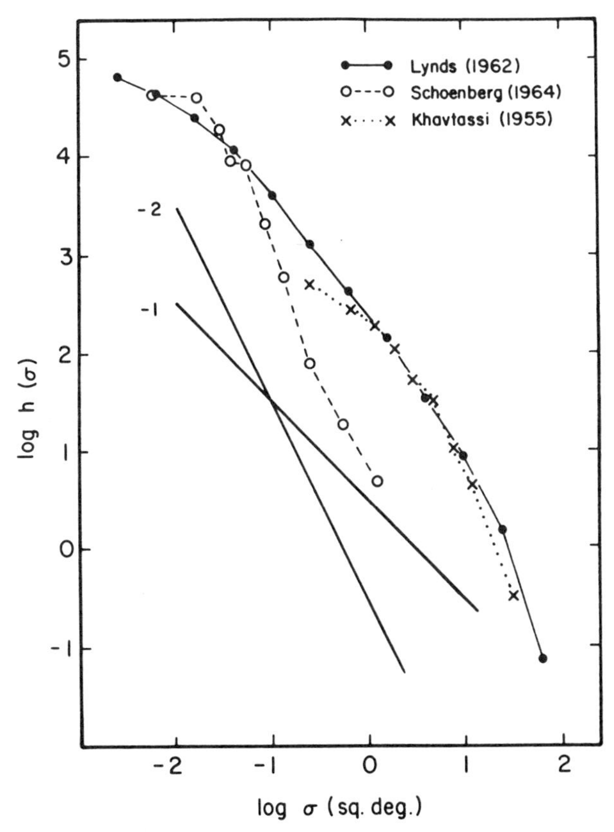

Fig. 16. Frequency distribution of dark-cloud angular areas from catalogs by Khavtassi (1955), Lynds (1962), and Schoenberg (1964) as given by Lynds (1968).

on Palomar Schmidt plates because the number of background stars is too small; this corresponds to a minimum detectable cloud size which increases with distance. In Taurus the limit is ~ 0.04 pc, while in Orion it is about 0.12 pc; incompleteness sets in at sizes 3 to 4 times larger. It becomes increasingly difficult to detect clouds more distant than several hundred parsecs because of the increasing number of foreground stars. It is also more difficult to identify and measure clouds with low opacities ($A_V \lesssim 1$ to 2 mag), which may bias the distribution toward smaller clouds. A complicated (and unknown) selection function would therefore need to be included in the integral mentioned above. In addition, there must always be personal selection effects in dark-cloud identification and area estimates which cannot be quantified. Another potentially serious problem is that the observed distribution of extinction is a projection of the true spatial distribution of dark clouds, and it is likely that nearby clouds obscure more distant clouds, that some small clouds located in

TABLE II
Estimates of the Mass Spectrum Index γ

Type of Object	γ	Sources of Data[a]
From sizes and column densities:		
Diffuse clouds, H I emission	-1.5: to -2:	1, 2
Diffuse clouds, reddening	-1.4	3
Molecular clouds, ^{13}CO	-1.4 to -1.6	4
Dark clouds, extinction	-1.0 to -1.5	5
Dark-cloud cores, $C^{18}O$	-1.0 to -1.5, $m \geq 4$ $M_\odot$ -2, $m \geq 20$ $M_\odot$	6
From size distribution and density assumption:		
Molecular clouds, ^{13}CO	-1.8	7
Molecular clouds, ^{12}CO	-1.6	8
Dark clouds, extinction	-1.2 ± 0.3	9
From column density distribution and density assumption:		
Diffuse clouds, Ca II lines	-1.5	10
Diffuse clouds, K I lines	-2.0	11
Diffuse clouds, H I absorption	-1.9	12
Diffuse clouds, reddening	-2 for $m \leq 5 \times 10^3$ $M_\odot$	13
Lynds' clouds	-1 to -1.3	(see text)

[a] 1: Heiles 1967; 2: Verschuur 1974*a*; 3: Knude 1979; 4: Casoli et al. 1984; 5: Bhatt et al. 1984; 6: Myers et al. 1983; 7: Liszt et al. 1981; 8: Sanders et al. 1985; 9: Scalo 1985*c*; 10: Penston et al. 1969; 11: Hobbs 1974; 12: Crovisier 1978, 1981; 13: Scheffler 1967.

front of larger clouds are not detected, and so on. An analytic correction for these projection effects appears formidable. An attempt to circumvent some of these problems, using digitized Schmidt plates with automated cloud identification procedures and comparisons with simulations to correct for projection and quantify selection effects, is in progress (Houlahan and Scalo, in preparation).

Bania and Lockman (1984) have recently presented the apparent size (chord) distribution of 177 H I self-absorption features found in their high-resolution (4 arcmin) latitude scans, but they emphasize that the selection effects involved in the catalog construction are severe, and probably preclude a useful estimate of the mass spectrum, especially considering that nothing is known about the relation between density and size for these clouds.

The results discussed above are listed in Table II. Results based on column density distributions are added for completeness. The column density distribution result for Lynds' clouds was derived assuming a uniform distribution of column densities, as there are roughly an equal number of clouds in opacity classes 1 through 5. It is seen that all the estimates of γ fall in the range of -1 to -2; excluding the diffuse clouds, nearly all the results are in the range of -1.0 to -1.6. The exceptions are the steeper distributions found for GMC's by Liszt et al. (1981), which Casoli et al. (1984) attribute to distance uncertainties and blends at the tangent point, and for dark-cloud cores with $m \gtrsim 20\ M_\odot$ in the study of Myers et al. (1983) which appear to be a selection effect due to the exclusion of clouds with smaller column density. There is also some evidence that γ increases (mass spectrum flattens) at smaller masses. A recent discussion by Drapatz and Zinnecker (1984) reaches the same conclusion based on size distributions from a smaller set of studies.

C. Parameters of Hierarchical Fragmentation Implied by Statistical Results

The observed correlations between fragment properties and observed estimates of the frequency distributions of cloud masses and sizes obviously contain a great deal of information concerning the physics of fragmentation. Unfortunately theoretical studies of fragmentation (discussed in Sec. IV) are not yet developed to the point where a direct comparison is possible, primarily because of the practical limitations on spatial resolution of the numerical experiments. However, the observations can yield estimates of the gross properties of fragmentation, such as its mass efficiency and number of condensations per fragmentation stage, if we assume that fragmentation is a strictly self-similar hierarchical process, an assumption that is supported by the observations discussed earlier.

We can characterize a self-similar hierarchy by three parameters, which can be chosen for ease of comparison with numerical simulations. One parameter is the number of fragments formed at each level of the fragmentation hierarchy, denoted by η. The second is the mass efficiency of fragmentation

$$f \equiv \eta m_i / m_{i-1} \tag{1}$$

where m_{i-1} and m_i refer to the average masses of individual fragments at levels $i-1$ and i of the hierarchy. The third parameter is chosen as the shrinkage factor

$$\theta \equiv r_i / r_{i-1} \tag{2}$$

which measures the relative scales of fragments at consecutive levels. The parameter θ is not very well defined for comparison with theory because the process is time-dependent. Instead of θ, the third parameter could be chosen as the volume-filling factor $\epsilon = \eta(r_i/r_{i-1})^3 = \eta\theta^3$ or the mean density contrast, defined as the ratio of the density of fragments at level i to the mean density at level $i-1$, including the fragments it contains. This mean density contrast can be expressed as $\delta = f/\epsilon = f/\eta\theta^3$. In what follows, we use the shrinkage factor θ because it is easiest to estimate from published numerical simulations. The hierarchy is self-similar if η, f, and θ are constants, having the same value at every scale.

For a self-similar hierarchy, the density-size relation, the differential mass spectrum, and the differential size spectrum are power laws: $\rho \propto r^p$, $f(m) \propto m^\gamma$, $f(r) \propto r^\alpha$. The indices p, γ, α, are related to the parameters η, f, θ by (Scalo 1985*c*)

$$\begin{aligned} p &= \log(f/\eta\theta^3)/\log\theta \\ \gamma &= \log(\eta^2/f)/\log(f/\eta) \\ \alpha &= \log(\eta/\theta)/\log\theta. \end{aligned} \tag{3}$$

If the fragments are assumed virialized at each level of the hierarchy, the velocity-size relation is also a power law $v \propto r^q$, with

$$q = 1/2 \log(f/\eta\theta)/\log\theta = (p+2)/2. \tag{4}$$

Note that if $p = -1$, $q = 0.5$, close to the empirical determinations.

The empirical estimates of p, γ, and α can be combined with theoretical calculations to test the consistency of the theoretical results with the empirical constraints. For example, Fig. 17 shows the relations $\gamma(\eta,f)$ and $\alpha(\eta,\theta)$ for the

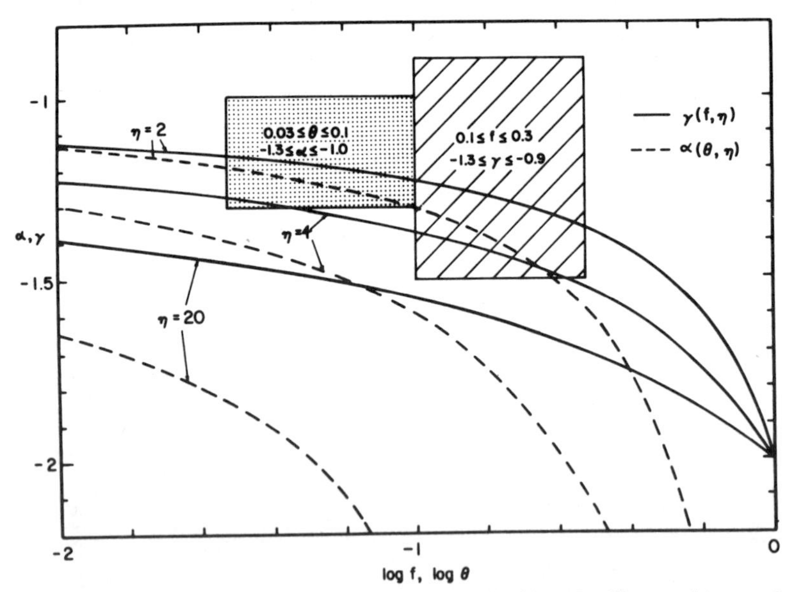

Fig. 17. Relations between parameters of a strictly self-similar hierarchy. The quantities α and γ are the indices of power law size and mass frequency distributions, which can be estimated from observations. The mass efficiency of fragmentation at each level of the hierarchy is f, θ is the shrinkage factor at each level, and η is the mean number of fragments formed per level. Solid lines refer to $\gamma(f,\eta)$, dashed lines to $\alpha(\theta,\eta)$. Boxes show the allowed ranges of α and γ from observations, and of θ and f taken from a number of numerical calculations of rotating cloud collapse. Both comparisons suggest consistency of theory and observations with $\eta \approx 2$ to 4.

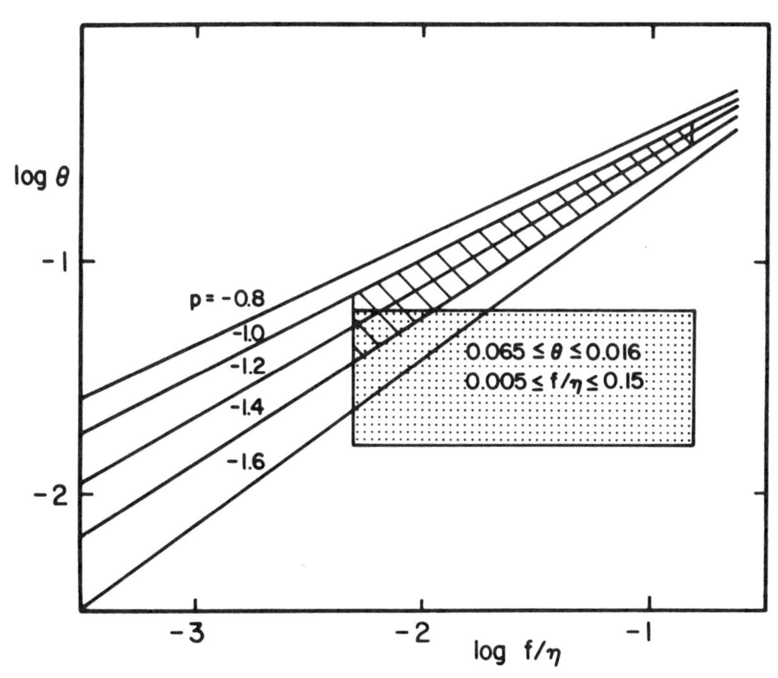

Fig. 18. Relation between shrinkage factor θ and the quantity f/η, where f is the mass efficiency of fragmentation and η is the mean number of fragments per fragmentation stage, for a strictly self-similar hierarchy, for different values of the exponent p in an assumed power law relation between density and size. The box indicates the ranges of θ and f/η inferred from numerical collapse calculations. This comparison, combined with that shown in Fig. 17, suggests that the fragmentation efficiency may be smaller than predicted by nonmagnetic collapse calculations.

self-similar hierarchical model. Empirical estimates of γ and α are mostly in the range $-1.3 \lesssim \alpha \lesssim -1.0$, $-1.5 \lesssim \gamma \lesssim -0.9$, while numerical simulations (see Sec. IV) suggest $0.03 \lesssim \theta \lesssim 0.1$, $0.1 \lesssim f \lesssim 0.3$. These limits are shown as the hatched and stipled boxes in Fig. 17. As can be seen, the $\gamma(f,\eta)$ relation requires $\eta \lesssim 4$, with $\eta \approx 2$ favored. The $\alpha(\theta,\eta)$ comparison yields the same result. These constraints on the mean number of fragments per level are in excellent agreement with the morphological appearance described in Sec. II and the numerical simulations, as reviewed in Sec. IV, which all suggest $2 \leq \eta \lesssim 5$.

A further comparison is possible using the density-size relation. Figure 18 shows the predicted relation between the shrinkage factor θ and the ratio f/η for various values of the index of the density-size relation p. The hatched region corresponds to $-1.4 \leqq p \leqq -1.0$ as suggested by observations. The stipled region is bounded in θ by $0.03 \leqq \theta \leqq 0.1$, estimated from published numerical simulations, and $0.005 \leqq f/\eta \leqq 0.15$, using the rather generous limits of $0.05 \leqq f \leqq 0.3$ and $2 \leqq \eta \leqq 10$. The overlap of these regions requires $\theta \approx 0.05$ and $f/\eta \sim 0.007$.

The θ constraint is marginally consistent with theoretical results, but the f/η constraint, $f \approx 0.007\ \eta$, suggests mass efficiencies smaller than most numerical predictions if $\eta \sim 2$ to 5. Some of this discrepancy may reflect the lack of spatial resolution in the numerical calculations. In Sec. IV it is suggested that fragmentation models dominated by magnetic effects might predict significantly different efficiencies, so the present approach might be used to distinguish between them.

These comparisons are only meant to suggest the manner in which theory and observations can be compared, and to motivate further work. While the comparisons suggest encouragingly good consistency between observations and most of the numerical collapse-fragmentation studies reviewed below, they cannot as yet be used to rule out models in which other effects, such as magnetic fields, are important, because these other models may give similar results for the number of fragments, efficiency, and shrinkage factor. Ultimately, the numerical calculations must be able to account *directly* for the morphology, correlations, and frequency distributions, but such a confrontation of theory with observation must await an improvement of *at least* an order of magnitude in the spatial resolution of numerical experiments.

IV. THEORIES OF FRAGMENTATION

A. Numerical Calculations

The only rigorous theoretical approach to the fragmentation problem is through numerical solutions of the hydrodynamical equations in three dimensions. Although there has been encouraging progress in this area in the last decade, a realistic calculation including all the relevant effects lies in the distant future. The major problem is that a proper resolution of the large range of relevant spatial scales is beyond the capabilities of present-day computers. Current calculations are limited to $\sim 10^4$ grid points, at least 10^3 to 10^4 short of the number required to study hierarchical fragmentation. In this respect the numerical problem is very similar to that encountered in simulations of incompressible turbulence at large Reynolds numbers, and is analogous to the observational problem of mapping the full range of structure in cloud complexes. Because small-scale behavior cannot be studied, chaotic nonlinear behavior such as the transition to turbulence cannot be revealed by existing numerical codes, just as transition cannot be studied in incompressible flows (except for very low Reynolds numbers). That such nonlinear behavior might be expected is suggested by the sensitivity of the calculated evolution to initial conditions (Buff et al. 1980; Boss 1980), a typical symptom of chaotic behavior. Even with existing codes, the behavior at the smallest resolvable scales is uncertain because of numerical diffusion, which smears out small-scale structure. Some of these problems will become less severe with advances in numerical methods (e.g., deformable or hierarchical grids and methods which allow variable time steps in different spatial regions) and computer capabilities.

For these reasons numerical calculations cannot presently predict a clump mass spectrum or a spatial autocorrelation function for a fragmenting cloud, but they can shed some light on how fragmentation occurs and the importance of various processes; therefore a brief summary is given here. Comprehensive reviews of earlier work in this area are given in Bodenheimer (1981) and Tohline (1983). All the calculations referred to here assume an isothermal gas, which is a good approximation for the structures of interest, but an assumption which suppresses any potential thermal or chemical instabilities. A recent calculation of multidimensional nonisothermal collapse including a treatment of the radiation field has been presented by Boss (1984*b*); these calculations are relevant only to the last, preprotostellar phases of fragmentation, which occur at very small size scales and are not included in this review.

1. Nonrotating, Nonmagnetic Cloud Collapse. The first question involves the ability of a self-gravitating but nonrotating and nonmagnetic cloud to undergo fragmentation. This problem may be relevant to the largest structural scales, where rotation can possibly be ignored. As originally envisioned by Hoyle (1953), the contraction of a marginally Jeans-unstable isothermal cloud would lead to a reduction of the critical mass for instability ($\propto \rho^{-1/2}$) and hence contraction of smaller regions within the original cloud. The process would continue on smaller scales within the contracting fragments, leading to a hierarchical arrangement of subcondensations, and would terminate when clumps become so dense that they are opaque to the radiation generated by their collapse. This conceptual prediction resembles the observed structure reviewed in Sec. II. While subsequent perturbation analyses for a pressure-free gas supported this idea, Layzer (1963) argued that the fragments could not survive the collapse of the larger structure, and that tidal forces would tear apart the fragments as they fell to the cloud center. More recent numerical investigations have generally supported Layzer's contention, although the main effect is the rapid buildup of a pressure gradient within the collapsing cloud that smears out density fluctuations which might otherwise grow.

Although this result seems commonly accepted by most workers, it should be emphasized that it applies only to very special initial conditions. Fragments *can* survive cloud collapse *if* the cloud mass is much larger than the Jeans mass *and* the cloud initially contains either sizable ($\delta\rho/\rho \gtrsim 1$) density fluctuations (Boss and Bodenheimer 1979; Bodenheimer et al. 1980; Tohline 1980; Rozyczka 1983; see Fig. 19) or random subsonic velocity fluctuations (Rozyczka et al. 1980*b*; see Fig. 20). It is quite interesting to note that the fragments in Rozyczka's (1983) calculation developed spin angular momentum through tidal interactions. Because there are a number of mechanisms which may excite internal density fluctuations in clouds (for example, fluctuations induced by the velocity fluctuations of field stars, as discussed below) and because large clouds are observed to contain large-amplitude den-

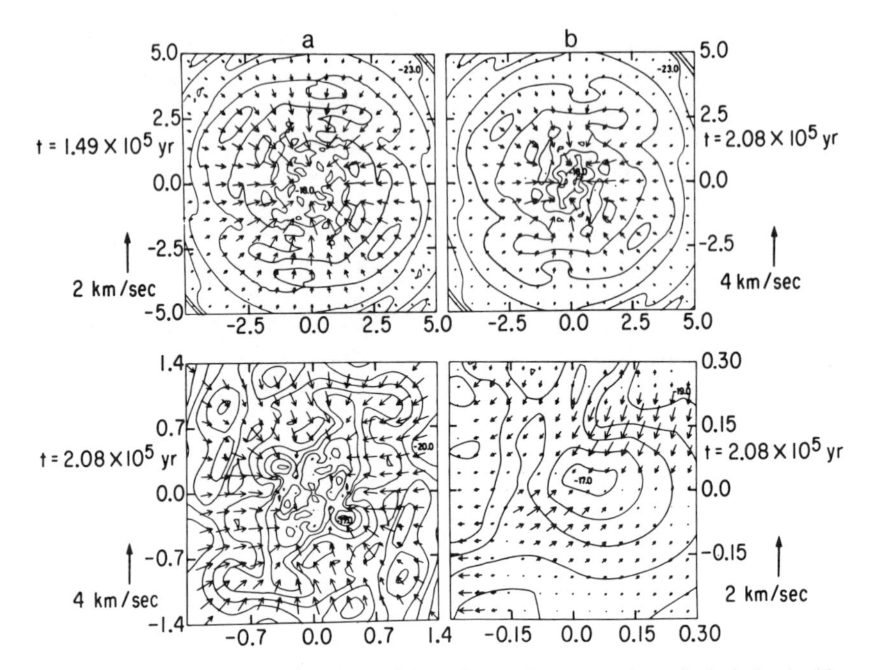

Fig. 19. Three-dimensional calculations of the collapse of a nonrotating spherical cloud with random initial density perturbations of amplitude up to $\delta\rho/\rho = 3$ (Rozyczka 1983). Isodensity contours and velocity field (arrows) in the equatorial cross section of the cloud are shown. The highest and lowest density contours are labeled by log ρ, and the arrow near the lower corner of each frame gives the scale of the velocity vectors. The length unit is 10^{18} cm. Frames *a* and *b* show the structure at 0.78 and 1.1 initial free-fall times, while *c* and *d* are magnified versions of *b*. Frame *d* is centered at the blob seen to the right of and below the center of frame *c*, and the velocity vectors in *d* represent the component generated by tidal forces. This calculation demonstrates that fragmentation can occur during collapse in the presence of initial density fluctuations of amplitude $\delta\rho/\rho \sim 1$.

sity fluctuations, it seems that the case against Hoyle's proposition is not as strong as usually supposed, at least on large scales when rotation is unimportant and $m >> m_J$. One must also consider R. B. Larson's (1978*b*) unconventional particle simulation of three-dimensional collapse, which might represent extremely inhomogeneous initial conditions. Larson found that a cloud evolved into a number of accreting cores, with the number of fragments roughly equal to the initial number of Jeans masses contained in the original cloud, and hierarchical structures were sometimes found. Larson's numerical method has been widely criticized as a representation of the hydrodynamical equations, but it remains true that his method allowed much better spatial resolution than in Eulerian finite-difference calculations. Because finite-difference calculations for clouds with $m >> m_J$ and sizable initial perturbations do yield fragmentation, the fact that Larson's models became increasingly irregular and clumpy as m/m_J increased may still be valid.

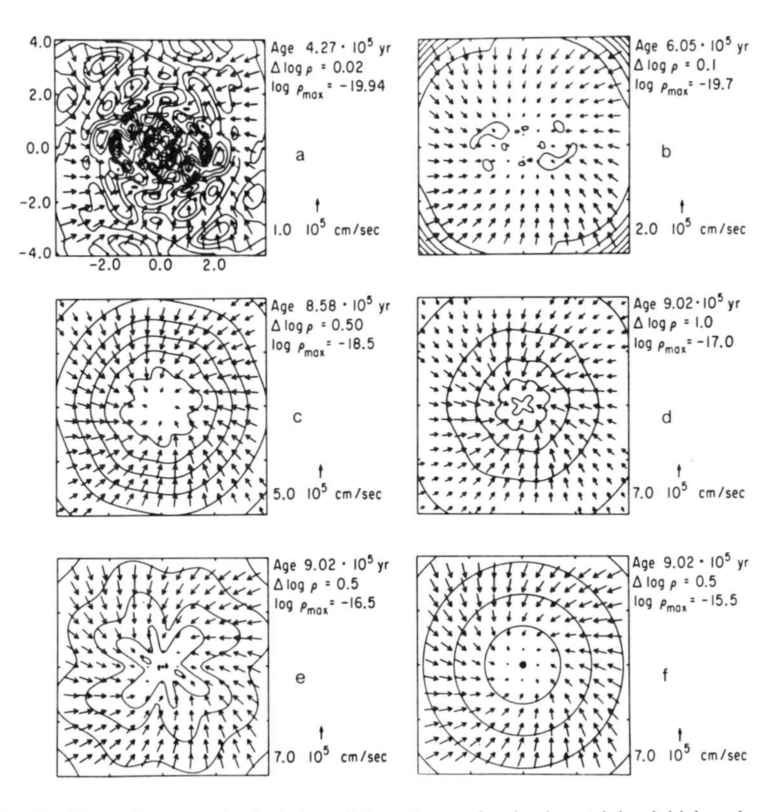

Fig. 20. Three-dimensional calculation of the collapse of a cloud containing initial random velocity fluctuations uniformly distributed between 0 and 30% of the sound speed (for $T = 10$ K), from Rozyczka et al. (1980*b*). The parameters of the calculation were mass = 5000 $M_\odot$, initial density = 5.6×10^{-21} g cm^{-3}, and initial uniform angular velocity = 7.8×10^{-15} s^{-1}. The initial ratios of thermal and rotational energies to gravitational energy were $\alpha = 0.02$, $\beta = 0.01$. Each panel shows contours of equal density and velocity vectors, with scale indicated to the right. At 4×10^5 yr $\approx 0.5\ \tau_{ff}$ (panel *a*), the velocity fluctuations have generated a chaotic density structure, a feature which also appears in other models with initial velocity fluctuations but larger α. At later times (panels *b* and *c*) the density structure disappears in the outer regions but survives near the center in the form of several independent subcondensations whose density increases with time (panel *e;* note change in scale). At least 2 of these subcondensations possess spin angular momentum. Panel *f* is the same model except that no initial velocity perturbations were imposed, showing that the fragmentation in panel *e* was induced by the turbulent velocity field.

The ability to fragment may also depend on the initial geometry of the cloud. Silk (1982) suggested that collapse to a flattened structure would enhance the growth of fluctuations, but this basically geometrical argument is not supported by Rozyczka's (1983) three-dimensional numerical calculations of nonrotating clouds with different degrees of initial flattening. Bastien

(1983) has presented two-dimensional calculations of collapsing nonrotating cylinders and describes how their fragmentation depends on their elongation and the number of initial Jeans masses in the cylinders. These calculations may be quite relevant to interstellar clouds, because, as pointed out in Sec. II, there is a tendency for elongated structures which are probably prolate. A useful summary of the criteria for instability of self-gravitating sheets and filaments, including the effects of rotation and magnetic fields, has recently been given by R. B. Larson (1984).

Additional insight into the influence of geometry on fragmentation may be obtained by examination of Villumsen's (1984) *N*-body simulations of dissipationless collapse of protogalaxies. These calculations are not exactly comparable with gas dynamical collapse because, even though the stellar velocity dispersion is equivalent to a gas pressure, the influence of long-range gravitational scattering has no analogue for the gaseous fragmentation process. Nevertheless, it is significant that Villumsen's collapses which started from disk or bar structures give much more substructure than collapses from an initially spherical configuration. For example, in the disk collapses, small-scale disturbances grow rapidly, destroying the symmetry and causing the development of strong asymmetries in the gravitational potential, which enhance fragmentation. For low-temperature bars, the fragmentation is similar to that found in the gas dynamical calculations of Bastien (1983). These results not only emphasize the importance of initial geometry for fragmentation, but also suggest again the close connection between fragmentation on extragalactic and interstellar scales.

None of this discussion directly addresses Layzer's (1963) specific suggestion that fragments will be destroyed through encounters. *N*-body simulations of systems of fragments by Arny and Weismann (1963) and Pumphrey and Scalo (1984*a*) indicate that fragments can survive tidal encounters and direct collisions, respectively, and numerical work by Boss (1981*b*) suggests that tidal interactions can actually induce further fragmentation. However, coalescence of fragments does occur in several numerical studies (see, e.g., Gingold and Monaghan 1983; Villumsen 1984). All the above results suggest that purely gravitational fragmentation can occur during the collapse of massive ($m >> m_J$) clouds if a source of initial perturbations is present, and that fragmentation is probably easier in nonspherical clouds.

2. *Rotational Fragmentation.* Most numerical work has centered on rotating clouds. The major attraction of rotation is that rotationally-induced fragmentation can convert spin angular momentum into orbital angular momentum, thereby alleviating the angular momentum problem which has been central to discussions of star formation for over 20 years. Indeed, all published numerical calculations agree that the specific angular momenta of fragments formed in a rotating cloud can be reduced by an order of magnitude or more, and Bodenheimer (1978) has presented a hierarchical scheme which

shows how spin-orbit angular momentum transfer can account for the observed rotational properties of single stars and binaries. Although there are no known cases of large ($\gtrsim$ 10 pc) clouds or complexes with dynamically significant rotation, there are several smaller dark clouds in which rotation appears important (see summary in Fleck and Clark 1981), and one would expect rotation to become important during protostellar collapse. Therefore these numerical calculations are primarily relevant to the smallest fragmentation scales.

Without discussing the detailed differences between the various calculations, the following general statements can be made. Fragmentation does not occur during the dynamical collapse phase (assuming uniform initial conditions), but must await the establishment of a quasi-static central ring, disk, or bar structure. Whether the intermediary structure is ring-like, disk-like, or bar-like depends on adopted initial conditions, the numerical method, considerations of spurious numerical angular momentum diffusion (see, e.g., Narita et al. 1983, and references therein), and the possible existence of a physical turbulent viscosity (Deissler 1976; Tscharnuter 1978; Regev and Shaviv 1981), which may lead to solid-body rotation and a concentration of mass near the center instead of in a ring. The important point is that the quasi-static structures are in most cases unstable to the growth of nonaxisymmetric density

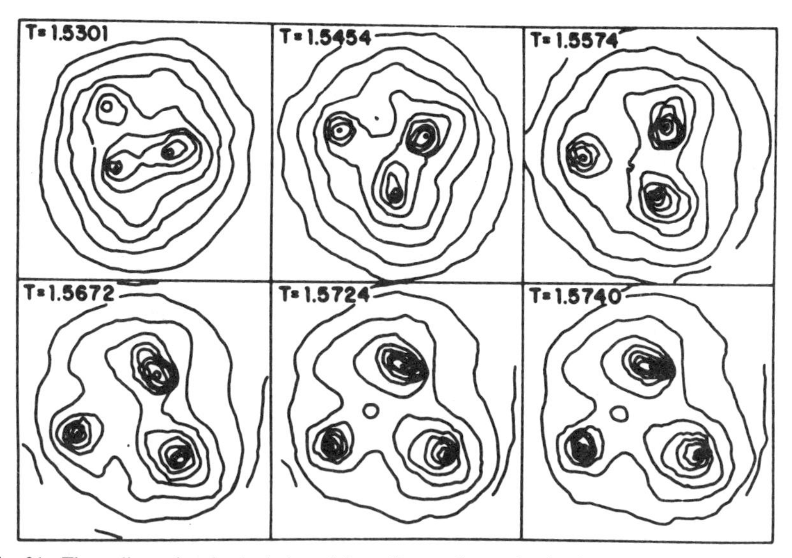

Fig. 21. Three-dimensional calculation of the collapse of a rotating isothermal cloud, from Wood (1982). Each isodensity contour (in the equatorial plane viewed along the rotational axis) represents a factor of 2 increase in density, and T is the time in units of the initial free-fall time. In this example fragmentation occurred directly from a ring wave set up by a competition between centrifugal and gravitational forces in a collapsing disk structure. With larger α, fragmentation was found to develop from a bar structure which was distorted by the ring wave.

perturbations, and tend to fragment into a few (2 to 8) clumps which contain a relatively small ($\sim$ 20 to 50%) fraction of the mass and which have had their specific angular momenta reduced by a factor of 10 to 20 relative to the initial configuration.

The mass efficiency of the process and the number of fragments formed depend on the factors named above. Finite-difference calculations suggested that usually two fragments should result, but fluid element or particle methods (see, e.g., Wood 1982) often find a larger number of fragments. Figure 21 shows the evolution of one of Wood's (1982) models. Gingold and Monaghan (1983) have pointed out that the character of fragmentation in a rotating cloud depends on the assumed initial distribution of angular velocity. In particular, a differentially rotating cloud is more susceptible to fragmentation than a uniformly rotating cloud. Gingold and Monaghan presented numerical calculations using a particle method which showed that in the case of differential rotation, more fragments ($\sim$ 4 or 5) are formed although these subsequently coalesce[a] and the fraction of mass in the fragments is increased relative to the uniformly rotating case. The evolution is not sensitive to the number of initial Jeans masses in the cloud. The evolution of one of their differentially rotating models, viewed perpendicular and parallel to the rotation axis, is shown in Fig. 22. It should be noted that the existence of turbulence in clouds is expected to damp differential rotation.

Miyama et al. (1984) have recently presented a number of three-dimensional collapse calculations, using the smoothed-particle method, for a wide range of initial values of α and β. The models assume that the initial clouds are rigidly rotating uniform density spheres with only small ($< 5\%$) density fluctuations. Their results show that the character of fragmentation depends on the product of $\alpha\beta$, at least for their chosen initial conditions. Fragmentation occurs for $\alpha\beta \lesssim 0.12$, with the number of fragments increasing with decreasing $\alpha\beta$, for sub-Jeans initial perturbations. The number of fragments ranged from 3 to 8 for the range of conditions examined. The fragmentation mass efficiencies were in the range of 0.3 to 0.4 and the specific angular momentum was reduced by factors of 18 to 40 by conversion into orbital angular momentum, in good agreement with previous work. An important point is that each fragment has small values of α and β, suggesting that a hierarchical fragmentation sequence will occur.

A still more recent set of calculations using the particle method has been reported by Benz (1985), who finds that collapsing uniformly rotating isothermal clouds develop into a central massive disk surrounded by many less massive collapsing fragments, and that the structure is due to Rayleigh-Taylor instabilities arising from small random initial perturbations.

---

[a]Fragmentation into several subunits followed by coalescence is also found in *N*-body simulations of dissipationless protogalaxy collapse (see McGlynn 1984; Villumsen 1984).

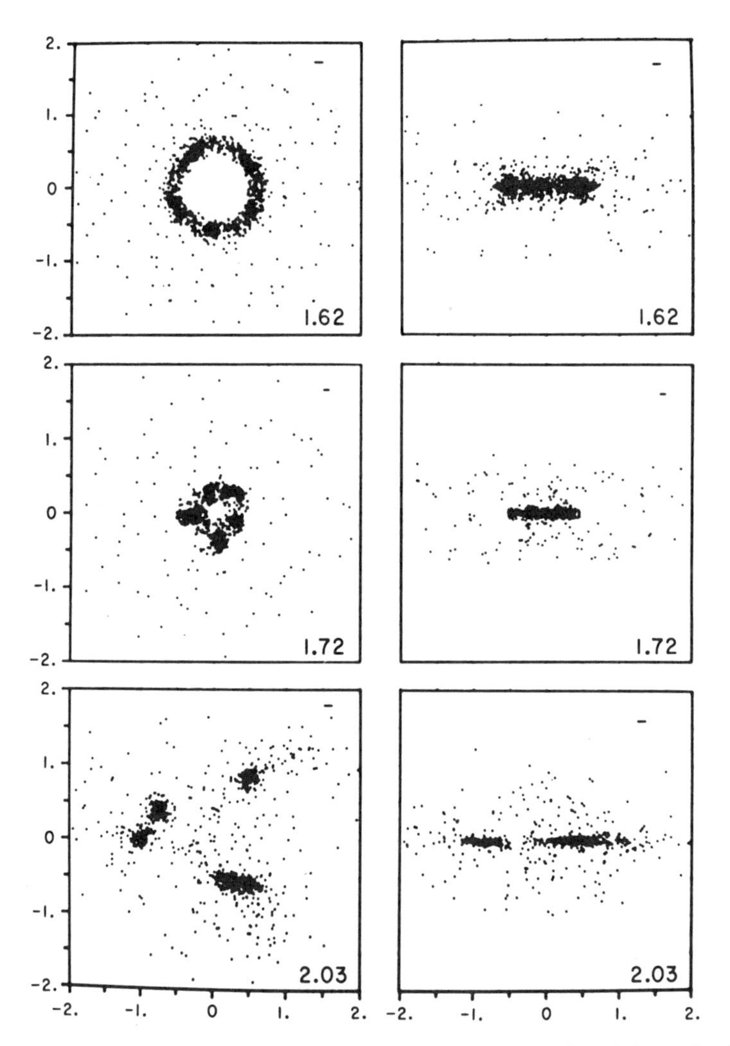

Fig. 22. Three-dimensional calculation of the fragmentation of a differentially rotating isothermal cloud ($\alpha = 0.24$, $\beta = 0.17$), from Gingold and Monaghan (1983). Figures show the density distribution viewed parallel (left) and perpendicular (right) to the rotation axis. Times in units of initial free-fall time are given in the lower right corner of each frame. A ring fragments into five subcondensations, and the system of fragments then expands. Subsequent evolution (not shown) resulted in the coalescence of the five fragments; by 3 τ_{ff} there was only a main amorphous fragment and a weaker satellite. This calculation suggests that differential rotation may lead to a larger number of fragments, at least before coalescence, with a larger fraction of the original mass, compared to solid body rotation.

While all of the above calculations help elucidate the conditions for and results of fragmentation in rotating clouds, they clearly point to the importance of initial conditions. When one considers that these calculations assume initially spherical clouds with only small density and velocity fluctuations, the sensitivity to initial conditions becomes even more apparent, because the calculations discussed in Sec. IV.A.1 above indicate that geometry and initial fluctuations can greatly affect the outcome of collapse.

The observational results of Schloerb and Snell (1984; see Fig. 9), which indicate that Heiles Cloud 2 is a fragmented rotating ring, provide evidence that rotationally induced fragmentation does occur in the manner predicted by some numerical calculations in some clouds. It is possible that VLA mapping of the fragments in NH_3 and H_2CO (if their core densities are large enough to excite these lines) can reveal their rotational properties and allow stronger constraints to be placed on numerical models. Rotation is thought to be important in a number of other dark clouds (e.g., B163 and B163W, L 1257), but molecular-line studies of small, roundish, very dark (Lynds opacity class 5 or 6) clouds (globules) show that rotation is not dynamically significant in most cases (see, e.g., R. N. Martin and A. H. Barrett 1978; Snell 1981; Ungerechts et al. 1980; Myers and Benson 1983). The implication is that some other mechanism, such as magnetic braking or turbulent viscosity, has already been effective in transferring angular momentum out of the clouds. It is also possible, but unlikely, that we happen to observe these clouds just after they have fragmented out of larger structures but before they undergo significant collapse themselves, so that their rotational velocity gradients are undetectable at the present time. Because magnetic braking should become ineffective at very large densities if plasma drift decouples the field from the bulk of the gas, it seems inevitable that rotation must become important during the later, near-protostellar stages of fragmentation, and that rotationally-induced fragmentation is a dominant process in the formation of binary and multiple protostellar systems. It should be pointed out at this point that, while rotational or magnetic loss of angular momentum is probably important at some stages of fragmentation, additional mechanisms may be required to account for the observed rotational properties of stars. The small values of v sin i found by Wolff et al. (1982) in a significant fraction of their sample of 306 B stars led them to conclude that angular momentum loss must occur *after* star formation is essentially complete, and they suggested gravitational encounters as a mechanism.

3. Magnetic Clouds. The possible effects of magnetic fields on the collapse and fragmentation of interstellar clouds have received much discussion beginning with the paper by Mestel and Spitzer (1956). Reviews of the theoretical possibilities have been given by Mestel (1965*a,b,*1977; see also his chapter), Mestel and Paris (1984), Mouschovias (1978,1981), and Nakano (1985); only a brief summary is given here. More recent work on related topics include a detailed study of magnetohydrodynamic (MHD) shocks in clouds (Draine et al. 1983); an interpretation of high-velocity protostellar outflows and driven turbulence in terms of magnetic bubbles (Draine 1983); the effect of magnetic fields on cloud collision cross sections (Clifford and Elmegreen 1983); MHD waves as the cause of suprathermal line widths in clouds (Zweibel and Josafatsson 1983; see Sec. IV.C.6 below); a numerical study of the effects of plasma drift (ambipolar diffusion) on cloud collapse (D. C. Black and E. H. Scott 1982; see Scott [1984] for analytic solutions and B. G. Elmegreen [1979*b*] for a discussion of the physical processes); Nakano's (1985, and references therein) discussion of the possibility that fragmentation is controlled by gradual reduction of the magnetic Jeans mass through plasma drift; and the first three-dimensional numerical calculation of the evolution of a magnetic, rotating, self-gravitating cloud (Dorfi 1982).

Dorfi's (1982) calculations are especially relevant to the present discussion. Using a hierarchical numerical grid to adequately resolve the nonuniform structures and large gradients which develop in the central regions, Dorfi was able to follow the evolution for over two initial free-fall times. The initial conditions were a uniform spherical cloud with $n = 10\ \mathrm{cm}^{-3}$, $R = 20$ pc, $m = 10^4\ M_{\odot}$, and $T = 100$ K, permeated by a uniform 3 μGauss magnetic field and uniformly rotating with an angular velocity of $10^{-15}\ \mathrm{s}^{-1}$. With these conditions the ratio of rotational to gravitational energies is only 0.01, while the ratio of magnetic to gravitational energies is 3.4. Thus, only the inner regions can collapse along the field, while the flow is stabilized across the field. Even though rotation was initially unimportant, the interaction between the rotational flow and the field resulted in some interesting structures.

Dorfi studied two cases, one with the field perpendicular and the other parallel to the rotational axis. In the perpendicular case, gas initially flows in along the field and out across the field, forming a sheet-like structure (Dorfi calls it a bar when referring to two-dimensional projections, but in three dimensions it is more like a sheet). The sheet is tilted with respect to a plane perpendicular to the field because of rotation, which twists the field into an *s*-shaped structure when viewed along the rotational axis. Magnetic braking is extremely effective on times as short as 0.1 τ_{ff}. By $2\tau_{ff}$ about half the mass has escaped, leaving a central bound region with maximum density contrast of 300. A central condensation forms within the sheet; it is oblate, with short axis parallel to the field and maximum extent of ~ 10 pc. This central condensation breaks up into two blobs lined up along the rotational axis. They are formed by the gas interactions which rotationally twist the field into the *s*-shape when observed along the rotational axis. The mass of the final collapsing core region is only 4% of the original mass. In this sense magnetic collapse and fragmentation is inefficient, as in the rotational calculations of fragmentation described above, and as required by the observations discussed earlier in Sec. III. If the mass efficiency of fragmentation f is really smaller in the magnetic field-dominated case ($f \approx 0.04$) than in the rotation-dominated case ($f \approx 0.2$ to 0.5 for most of the calculations reviewed in Sec. IV.A.2 above), then it may be possible to choose between these models on the basis of observations. For example, the discussion of Sec. III.C suggested that very small values of f may be required.

In the case of parallel rotation and field, the gas collapses into a disk, as expected, with density contrast of ~ 100 at $t = 0.6\,\tau_{ff}$, at which time magnetic braking is essentially complete. By $2\tau_{ff}$ a ring forms at $r \approx 1$ pc because of the dominance of gravity interior and field pressure exterior to this point. The ring is surrounded by a nearly poloidal field which disconnects from the overall field because of numerical diffusion, which Dorfi speculates might simulate physical diffusion. The ring rapidly breaks up, and the subsequent evolution of the central region is complex and caused at least partially by numerical diffusion.

Although Dorfi's work is a pilot study, it does suggest that collapse and possibly fragmentation can occur in a strongly magnetic cloud, that they are mass-inefficient, and that magnetic braking of rotation is as efficient as found by earlier analytical treatments. A much larger set of calculations with improved spatial resolution will be required before any definite predictions can be made concerning fragmentation, but it appears that it will be difficult to distinguish observationally between magnetic and rotational fragmentation on the basis of predicted density structures. While it may be possible to use the difference in predicted fragmentation efficiency for this purpose (Scalo 1985*b*), the most promising discriminant would seem to be the predicted velocity fields.

A fundamental problem is that observations provide very little guide to the appropriate initial conditions for the magnetic field. Dorfi's results were largely dictated by the assumed dominance of the field in the initial cloud. If magnetic fields could be shown to be dynamically insignificant on a certain size scale, then it is unlikely that they could be important on smaller scales. Alternately, if fields are important, then the efficiency of magnetic braking indicates that rotation will be unimportant during subsequent phases. Unfortunately, the necessary observations are very difficult and their interpretation often ambiguous, as discussed at length in the reviews by Chaisson and Vrba (1978) and Verschuur (1979). More recent work on the large-scale galactic field (Simard-Normandin and Kronberg 1980; Inoue and Tabara 1981; Phillips et al. 1981; Vallee 1983; see also Beck [1982,1983] on M31); 21-cm Zeeman observations (see, e.g., Troland and Heiles 1982*a*,*b*,1982); OH Zeeman observations (Crutcher et al. 1981); optical polarization in M17 (Chesterman et al. 1982); GF7 (McDavid 1984); Taurus (Moneti et al. 1984; Hsu 1984); and Zeeman splitting in OH masers (Wouterloot et al. 1980; Hansen 1982) have not yet resolved the basic question of the dynamical importance of magnetic fields in regions of different sizes and densities (see Mouschovias 1981; also Brown and Chang 1983; Fleck 1983*b*). Preliminary results of a program to measure Zeeman splitting in OH toward molecular clouds has been reported by Crutcher (1983), and this work should greatly improve the constraints on theoretical models. The recent work on the structure function of interstellar magnetic field fluctuations by Simonetti et al. (1984), who used rotation measure and by Hsu (1984), for optical polarization in Taurus, offers another method for quantifying the magnetic field structure which can someday be compared with models. See Ichimar (1976) for a theoretical calculation of field fluctuation functions. Despite these problems, it is still very encouraging that theoretical work is approaching the level of sophistication at which the effects of a magnetic field of assumed initial importance can be quantified, and the resulting structures can be presented in sufficient detail to allow future comparisons with observations.

In conclusion, numerical studies to date leave open the possibilities that fragmentation may occur through purely gravitational effects, possibly requiring elongated geometry or sizable initial velocity and density fluctuations, or may be controlled by rotation and/or magnetic fields. Fundamentally, the choice is dictated by the assumed initial conditions. Because purely gravitational fragmentation occurs during initial collapse while rotational and magnetic fragmentation must wait longer, it would be interesting to examine the evolution of a cloud which is initially extremely inhomogeneous in density with associated random velocity fluctuations (i.e., a hybrid of the models presented by Roczycska et al. (1980*b*); and Roczycska (1983) but rotating rapidly enough so that rotational fragmentation would otherwise occur. Such a calculation would tax presently attainable numerical resolution, but could demonstrate whether one effect damps or amplifies the other.

B. Stellar-Induced Fragmentation

One of the most interesting developments of the past few years has been the recognition that stellar explosions, winds, and ionization fronts may play a major role in controlling fragmentation. Much of the work in this area has been directed toward showing how expansions around hot stars can directly trigger star formation, as suggested long ago by Öpik, Bok, Kossacki, and others. The evidence for this view is summarized in Lada et al. (1978) and Herbst and Assousa (1978). The most compelling evidence is that OB associations often contain spatially distinct subgroups which lie roughly in a sequence of decreasing age, implying that star formation has propagated rather discretely through the parent cloud complex.

B. G. Elmegreen and C. J. Lada (1977) suggested that the accumulation of compressed ambient gas between the shock front and ionization front around an early-type star would eventually lead to gravitational instability and star formation. Massive stars formed within this shell would then initiate a new ionization-shock (I-S) front, and so on. A schematic illustration of the process is shown in Fig. 23. Elmegreen and Lada showed that such sequential star formation was consistent with observations of gas and young stars in several regions. A recent study of W5 by Wilking et al. (1984*b*) generally

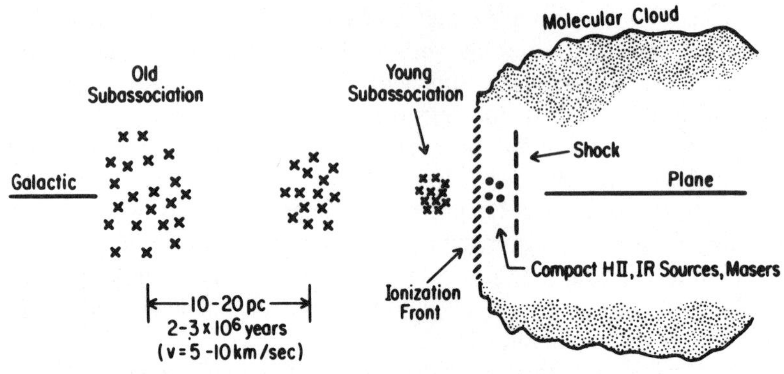

Fig. 23. Schematic illustration of sequential formation of OB subgroups by gravitational instability in an I-S front. Figure from Thaddeus (1977), after Elmegreen and Lada (1977).

supports this idea. B. G. Elmegreen and W.-H. Chiang (1982) and B. G. Elmegreen (1983*b*) discuss how the same type of process may occur on larger scales, due to shells blown away from entire OB associations (see Sec. IV.D).

The question of how such a shell fragments has been addressed in a number of papers. B. G. Elmegreen and D. M. Elmegreen (1978) performed a linear perturbation analysis on a self-gravitating layer bounded by equal external pressures. The mass column density σ increases approximately linearly with time, and Elmegreen and Elmegreen found that the wavelength and

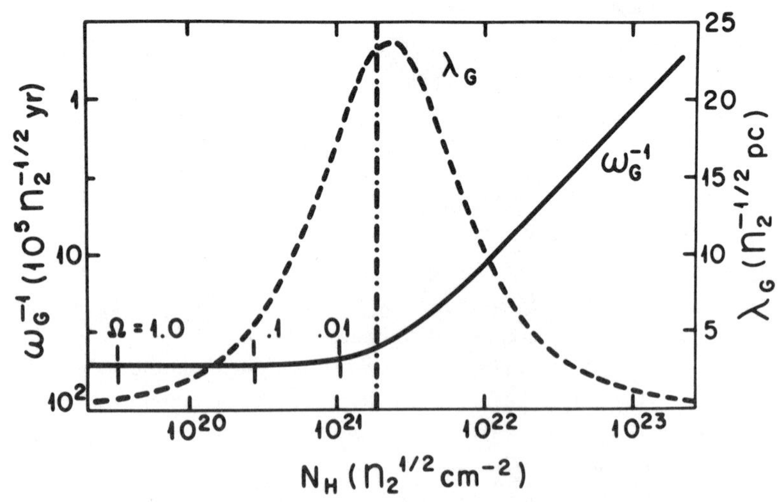

Fig. 24. The *e*-folding time ω_G^{-1} of the fastest-growing wavelength λ_G for the gravitational instability of an I-S front as a function of its column density N_H, from Giuliani (1980). The quantity n_2 is the number density behind the I front. Rapid star formation is expected to the right of the dot-dashed line.

growth rate of the fastest-growing wavelength depends basically on the ratio of $\pi G\sigma^2/2$ to the external pressure. Giuliani (1980) has presented a useful discussion of the behavior of these quantities, which are reproduced in Fig. 24 as a function of the slab column density N_H in units of $n_2^{1/2}$ cm^{-2}, where n_2 is the number density in the ionized gas just behind the ionization front. Since N_H increases with time, this graph shows how the wavelength and time scale of the fastest growing mode vary as the front propagates. Note that the most unstable wavelength peaks at an intermediate column density and that the growth time of this mode decreases with increasing N_H past this point, for a given n_2. The results are only valid if the evolutionary time of the slab is much greater than the growth time; the tick marks in Fig. 24 show the values of N_H below which this condition does not hold, for three values of $\Omega = \rho_2/\rho_0$, the density ratio between the ionized gas behind the ionization front and the neutral cloud gas ahead of the shock front.

Even though the slab may be gravitationally unstable, collapsing perturbations may become stable after reaching a spherical shape. Elmegreen and Elmegreen showed that the mass of a cylinder of diameter $\lambda_G/2$ exceeds the critical mass for collapse of an isothermal pressure-bounded sphere when the column density exceeds a critical value, indicated by the vertical dot-dash line in Fig. 24. To the right of this line the instability should lead to collapse of the fragments, and prompt star formation if no further fragmentation occurs. It should be possible to derive a qualitative predicted IMF for this process by assuming that at any time the spectrum of growing modes is a delta function at the fastest-growing wavelength. The fraction of stars at a given mass will then be related to the fraction of time spent at each interval of critical wavelength, obtained from the column-density evolution and the dependence of wavelength on column density. When the delta-function assumption is dropped, an interesting aspect of this problem is that the maximum mass that can form in the shell is an increasing function of time, even though the dominant mass decreases with time (see Fig. 24). Elmegreen and Elmegreen make the plausible assumption that the maximum mass is related to the wavelength whose growth time is just smaller than the shock duration τ, and find a maximum mass which varies as $\tau^4\sigma^3$.

Welter and Schmid-Burgk (1981) and Welter (1982) showed how to modify the analysis to include sphericity and to treat the compressed layer as moving and shock-bounded instead of stationary and pressure-bounded, and found that the differences in the resulting dispersion relation are relatively small. These authors did not treat the early evolution when the shell is thin. Vishniac (1983) finds that in this regime the boundary conditions do make a difference, and that the Elmegreen and Elmegreen dispersion relation is incorrect in this case, as can be seen by noting that in the thin-shell limit the dispersion relation does not involve the sound speed. According to Vishniac's analysis, when the appropriate boundary conditions are employed, the shell becomes gravitationally unstable at a later time than found by Elmegreen and

Elmegreen (by a factor of the square root of the Mach number), and at this time the pancake-like perturbations are themselves unstable. These effects were not noticed by Welter and Schmid-Burgk because they only considered the evolution after the shell was thick, in the sense that the gravitational scale height is smaller than the shell thickness, while Vishniac treats the thin-shell case. The situation regarding gravitational instability of a shock-bounded layer therefore remains unclear, at least to this author.

The fact that the observed spatial separation between subgroups seems to decrease with decreasing age was explained by Elmegreen and Lada (1976) as a consequence of a positive density gradient ahead of the I-S front. Bedijn and Tenorio-Tagle (1980) subsequently studied the problem. Starting with a stationary subgroup with zero velocity, they follow the motion of the shell until the critical column density is reached, and then give the next subgroup the velocity of the front at this time, continuing the sequence. Bedijn and Tenorio-Tagle follow a sample sequence of five subgroups in this way, for a prefront density of 10^3 cm^{-3} and an initial Strömgren radius of 0.1 pc. The total time interval is 1.6×10^7 yr. The time intervals between formation of subgroups and the distances between subgroups increase toward the younger subgroups, but taking into account the motion of the subgroups since their formation gives subgroup separations and age differences which are largest for the oldest subgroups. There is thus no necessity to introduce density gradients, whose presence can either enhance or counteract the effect. The subgroup separations found by Bedijn and Tenorio-Tagle vary from 60 pc to 6 pc, with a total extent of 100 pc, in good agreement with observations. They compare their calculations with the properties of the four subgroups in Orion (Ia Ori to Id Ori = Trapezium), or five subgroups if the BN object and KL nebula are considered a newly-formed subgroup. The calculated velocity differences between subgroups change by 3.5 km s^{-1}, so accurate radial velocity studies of associations could test the model. Another possible discriminant between this model and the density gradient model is that, if the cloud has a constant density but moving subgroups, the age differences increase toward younger subgroups, by a factor of 1.5 for each subgroup for the parameters chosen, but a positive density gradient alone would give a decrease in age difference from the first to last subgroup, with a total decrease of a factor of ~ 5.

One problem with the Elmegreen and Lada model, emphasized by Woodward (1980), is that it requires that the undisturbed gas must be somehow supported against collapse for a time long enough for the unstable compressed layer to accumulate in the shell. Welter and Nepveu (1982) also recognized this problem, and suggested that the preshock gas is supported by turbulence which becomes suppressed by the shock passage. The reader is referred to Welter and Nepveu (1982) for details.

It is also not clear whether other instabilities can occur which will overwhelm the gravitational instability. Giuliani (1979) showed that I-S fronts are subject to a photon-driven slab (PDS) instability, which might explain the existence of "elephant trunk" structures observed around H II regions. The PDS instability grows from a rippling of the I-S front, and is physically and geometrically distinct from the gravitational instability. The physical interpretation of this instability, however, is unclear. Giuliani suggests that the I-S front can be viewed as a forced harmonic oscillator subject to damping by accretion of ambient gas in the perturbed gas, and that the instability arises when the forcing terms, related to variations in the emission measure of the H II region, become out of phase with the natural oscillation frequency of the shell. If the PDS instability dominates, a fragment that would otherwise be subject to collapse may be disrupted. The competition was studied in detail by Giuliani (1980), who showed that the PDS instability dominates the gravitational instability and that the fastest-growing wavelength is smaller for the PDS instability during the early stages of I-S front propagation, which should delay star formation by exciting internal turbulence. Giuliani's (1979) Fig. 4 shows how this delay time can easily exceed the lifetime of the exciting stars ($< 10^7$ yr) for plausible ranges of the parameters.

Vishniac (1983) demonstrated that another potentially important instability in shells is purely dynamical. Because the thermal pressure interior to the shell always acts in a direction normal to the shell while the external ram pressure acts normal to the local direction of shell motion, an undulation in the shell can grow. Vishniac shows that a geometrically thin shell is dynamically unstable if the adiabatic index $\gamma < 1.3$, as would occur in shells with efficient cooling, and the growth time is about equal to the sound crossing time of the shell. However, this instability will not produce gravitationally bound fragments unless they were bound in the initial cloud; this is a different aspect of the problem mentioned above, that a source of support is required prior to shock passage. Vishniac suggests that the end result of dynamical instability is more likely to be turbulence at roughly sonic velocities. If this dynamical instability also occurs in the thick shells relevant to the Elmegreen and Lada model, then it, like the PDS instability, may foil the attempted growth of nascent bound fragments. Of course what is needed at this point is a numerical simulation of the nonlinear evolution of a self-gravitating propagating I-S front. Although three dimensions are required even in the plane-parallel case, it appears that current collapse codes are capable of sufficient spatial resolution to shed some light on these issues.

The effect of a magnetic field on the slab gravitational instability is still a matter of speculation. Elmegreen and Lada had discussed the possibility that the field could affect or control the direction of propagation of star-forming I-S fronts. They argued that if an I-S front propagates across the field, the compressed field will substantially increase the critical mass for gravitational instability, inhibiting star formation. Then star formation will only occur for fronts which happen to propagate nearly in the field direction. In this way magnetic fields, although probably not strong enough to directly guide the front, can still impose preferred directions on propagating star formation, and would, if the above argument is correct, give an additional complicated but significant twist to the predicted IMF, as well as variations depending on fluctuations in the mean field strength over scales from about 10 pc to 100 pc (the 100 pc fluctuations are well known and have a significant amplitude). Welter and Nepveu (1982) have examined the effect of a magnetic field in more detail, and argue that, for a range of possible physical conditions, one can obtain gravitational instability with about the same time scale both for propagation along and across the field. They suggest that OB subgroup alignment is mostly a result of the observed elongation of many molecular cloud complexes roughly parallel to the galactic plane.

A very different idea, the magnetically-induced fragmentation of shells, has been proposed by Baierlein et al. (1981) and Baierlein (1983). The basic idea is that an expanding decelerating shell experiences an effective gravitational field and is supported against this field by the swept-up ambient magnetic field. A perturbation in the intially uniform field configuration leads to a Lorentz force which pushes gas from the "hills" of the magnetic field into the "valleys", i.e., the shell is unstable to the Parker instability. If $s \equiv v_A^2/c_s^2$ is the ratio of magnetic and gas pressures (v_A = Alfvén speed), then the ratio of the fastest growing wavelength in the horizontal direction to the shell scale height H is about $\pi s^{-1/4}$, and the growth time for this mode is about $(H/c_s)s^{-1/2}$. When radiative cooling is effective in the shell, the quantity s will be large, and so the shell can fragment into blobs with sizes $\lesssim H$ on times much shorter than the sound crossing time of the shell, because the flow velocity is limited by the Alfvén velocity. For this reason the growth can be much more rapid than gravitational instability. Numerical calculations reported by Baierlein (1983) indicate that the fastest-growing density perturbations reach an amplitude $\approx 2\pi v_A/c_s$ in a time of about $5H/v_A$, and that the gas flows into the fragment with a maximum speed $\sim(c_s v_A)^{1/2}$. The nonlinear evolution is of course still very uncertain. A major question is whether the compressed magnetic field will be effective in opposing the self-gravitation of the fragments; if not, then the fragmentation could lead to star formation in the shell.

Another way in which stars can influence the internal structure of clouds is through the compression of preexisting density fluctuations. Bok and Oort had both suggested that expanding H II regions could compress clouds and lead to star formation. One-dimensional calculations by Tenorio-Tagle (1977) showed that this implosion process may be viable for clumps which are sufficiently dense and massive to escape destruction by ionization and evaporation. Sandford et al. (1984, and earlier references therein; see also chapter by Klein et al.) have studied this problem in more detail using a two-dimensional numerical method which includes a realistic coupling between radiative transfer and gas dynamics and separate treatments of gas and dust. These calculations show how an early-type star can cause the radiatively-driven implosion of a nearby dense clump by means of a convergent shock. Although self-gravity was not included in the calculation, Sandford et al. suggest that stellar-mass clumps can survive long enough for gravity to take over, leading to star formation. Klein et al. (1983) found a similar result with a different geometry, in which a clump is located between two early-type stars, and sketched a scenario in which a chain reaction to star formation induced by radiatively-driven implosions might occur. It should be noted that because the process requires the prior existence of dense clumps, it is not a fragmentation mechanism but a means for driving stable fragments to star formation. Perhaps an equally important consequence is the destruction of low-mass, low-density fragments which might have otherwise formed stars. As pointed out by Klein et al., the process may in this way effectively terminate the formation of low-mass stars by wiping out the fragmentation structure on small scales.

This suggestion is supported by La Rosa (1983), who studied a simplified model for the radiatively-induced implosion of clumps which illustrates several significant features. The most important aspect of La Rosa's calculations is that the relation between internal velocity dispersion and size given by R. B. Larson (1981) was used for the initial clumps, with the dispersion treated as a source of internal pressure. Because internal velocity dispersion increases with size, a shock cannot penetrate as easily into larger clumps, and it becomes difficult to implode clumps larger than about 1 pc. Smaller clumps are susceptible to implosion, but will most likely be evaporated by ionization if their radius is $\lesssim 0.5$ pc and their internal density is $\lesssim 10^3$ cm^{-3}. I conclude that, given the density structure observed in complexes which do not yet contain O stars, implosion may be rare, and the dominant effect of H II region-clump interactions is probably the termination of low-mass star formation. This could partially explain why inferred star formation rates as a function of time for low-mass stars often declines as the rate for higher-mass stars increases (see, e.g., Cohen and Kuhi 1979; M. T. Adams et al. 1983).

A number of papers have also examined the effect of a supernova blast wave on nearby clouds (see, e.g., Nittmann et al. 1982; Heathcote and Brand 1983, and references therein) and the possible disruption of the cloud complex in which it was born (Wheeler et al. 1980; Shull 1980). A two-dimensional calculation of the compression of a cloud by a nearby supernova, including self-gravitation, has been recently carried out by Krebs and Hillebrandt (1984). They find that a supernova can induce gravitational collapse, but only if the original cloud mass is already near the Jeans mass.

All of the calculations discussed so far were primarily concerned with triggering star formation, and do not directly address the question of how stellar energy sources may actually cause or control fragmentation. A study more relevant to the fragmentation problem has been made by Bania and Lyon (1980), who calculate the effects of randomly positioned OB stars on an initially uniform gas. The OB stars are allowed to ionize and push on the gas on a two-dimensional grid for the duration of their main sequence lifetimes. The hydrodynamics is done on a 40×40 grid of 4.5 pc square cells using an

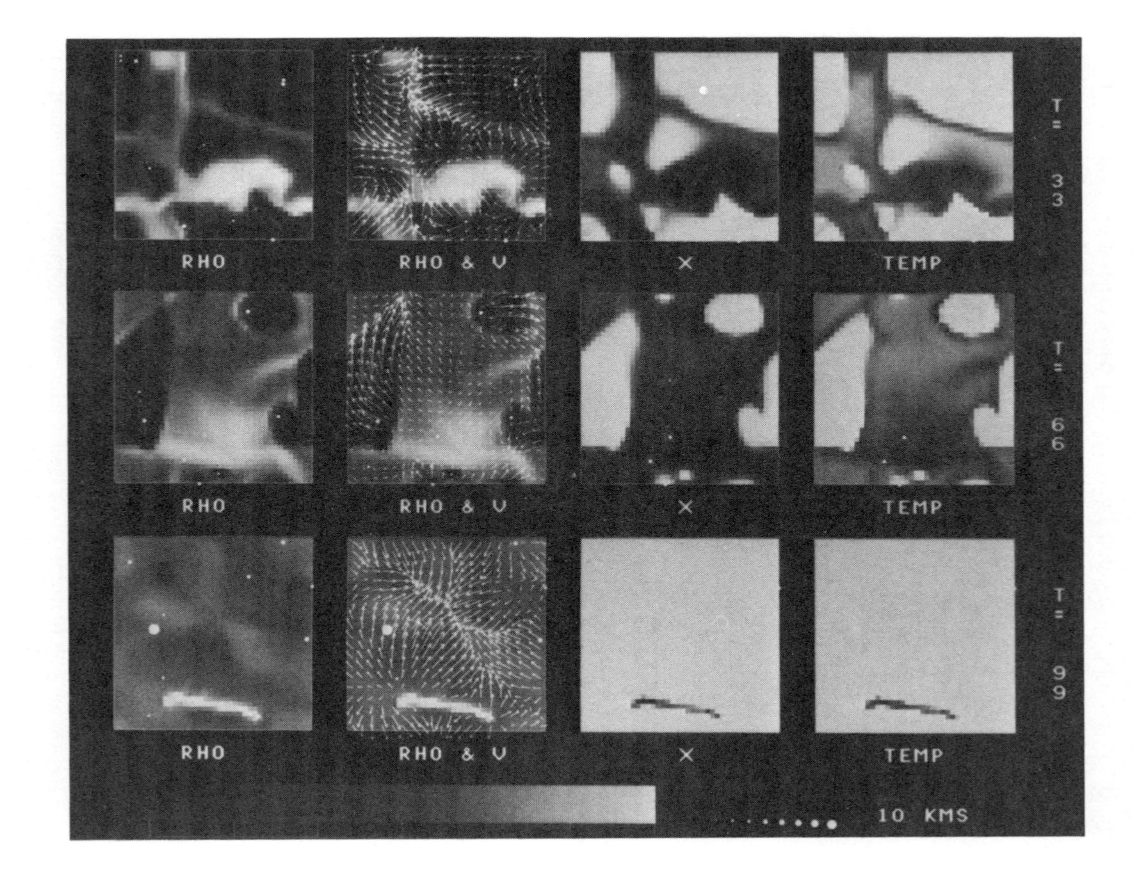
T = 33
RHO
RHO & V
X
TEMP
T = 66
RHO
RHO & V
X
TEMP
T = 99
RHO
RHO & V
X
TEMP
10 KMS

explicit Eulerian technique, and the ionization, heating and cooling, and radiative transfer (on-the-spot approximation) are applied after each hydrodynamic time step. Although the calculations were meant to simulate the large-scale interstellar medium, with an original uniform density of ~ 1 cm^{-3}, the results are also relevant to OB stars within denser structures. The general behavior is that the H II regions create low-density voids surrounded by shell-like structures which eventually refill the cavities after the stars die and the H II regions recombine. The dense structures are accelerated by the momentum from the expanding H II regions and the "rocket effect." The gas evolves to an interesting and complex structure, examples of which are shown in Fig. 25.

By defining a cloud as a transition from H II to H I at an ionization fraction of 0.01 and analyzing the simulations at 10^6 yr intervals, Bania and Lyon were able to determine the ensemble average statistical properties of the cloud structure. The comparisons of the models with observations of H I emission and other data are very extensive and cannot be reviewed properly in the space available here, except to say that the agreement is impressive. The most important results for the fragmentation problem are:

1. The time average of the clumpiness measure $<n^2>/<n>^2$ was 6 to 10 for the five models examined, with a maximum value ranging from about 30 to 90.
2. A wide range of cloud masses is produced. The mass spectrum had an index $\gamma \approx -1.5$ to -2 at $m \gtrsim 10^4$, and was flatter at smaller masses. The mean cloud mass ranged from a few times 10^3 $M_\odot$ to 10^5 $M_\odot$ in the different models, and a substantial fraction of the clouds should be gravitationally unstable.
3. The frequency distribution of cloud sizes r varied roughly as r^{-1} and the distribution of column densities N as N^{-2}. The size of the largest structures is determined basically by the mean distance between the OB stars.
4. Because the clouds are inertially confined, the stable clouds have lifetimes comparable to the time to establish large local pressure changes, which is related to the main sequence lifetimes of the OB stars, and may be as large as a few times 10^7 yr.
5. The cloud random velocities do not depend on cloud mass, in good agreement with more recent observational results (see Hausman [1982] for a review of this problem). However, the magnitude of the cloud-to-cloud velocity dispersion (~ 1 km s^{-1}) is smaller than observed, probably because O star winds and supernova explosions were not included in the calculations.

Fig. 25. Grey-scale images of logarithms of density, ionization fraction, and temperature, along with velocity field, at the three times (in 10^6 yr) indicated to the right, for one of the two-dimensional simulations of the effects of OB star ionization fronts on an initially uniform gas by Bania and Lyon (1980). Stellar positions are shown as white circles with size proportional to ionizing flux.

Bania (1981) later presented more detailed models in which the OB stars were distributed with a realistic initial mass function and the star formation rate was allowed to depend on different powers of the local gas density. Although the general results are similar to the earlier models, it is interesting to note that the introduction of a dependence of the OB star birthrate on the gas density led to sequential star formation in 14 spatially distinct regions during the 2.6×10^8 yr time span modeled by the calculation, suggestive of the subgroups found in OB associations. Each region produced 8 ± 4 OB stars from an average gas mass of around 4×10^4 $M_\odot$. The average separation between sequential star formation sites was about 90 pc, with relative velocities around 10 km s^{-1}.

These simulations suggest a plausible means by which fragmented structure and consequent gravitational instability can arise, requiring only the presence of stars (field stars or newly-formed stars within a cloud) which can provide the necessary momentum input. It is not clear whether the process can account for the entire range of interstellar cloud structure, and in particular its hierarchical appearance. One can imagine an originally smooth medium containing only field stars whose large mean separation gives rise to large structures which form new stars on smaller scales, producing smaller and denser structures in a hierarchical manner.

The calculations of Bania and Lyon also give support to a modified version of the model of C. A. Norman and J. Silk (1980) for the self-sustaining fragmentation and star formation induced by T Tauri winds. Norman and Silk imagined that the dense shells driven by these winds would collide and fragment into small, sub-Jeans fragments if the star density is large enough, and that these clumps could be confined long enough by ram pressure so that coalescing clump collisions would bring them above the Jeans mass, leading to further star formation and fragmentation. There are two basic problems with this scenario. First, it is not clear how the shells fragment due to collisions. Gravitational instability is very inefficient in one-dimensional compressions, as shown by the dispersion relation derived by M. E. Stone (1970) and the two-dimensional hydrodynamic calculations of Gilden (1984); also the shells are confined inertially from both sides so dynamical instability as discussed by Vishniac (1983) cannot occur. Because the shells will generally collide obliquely, it is possible that the collision will cause shell breakup, but it is difficult to imagine the formation of the small well-defined clumps envisioned by Norman and Silk. The possibility of gravitational instability in a T Tauri wind shell *à la* B. G. Elmegreen and C. J. Lada (1977) has apparently not been discussed. The second problem involves the coalescence of sub-Jeans mass clumps. As shown by the two-dimensional hydrodynamic calculations of Gilden (1984), efficient shock cooling does not ensure coalescence. For the collision of identical clumps, the criterion for coalescence is that the velocity of reexpansion after the initial compression must be smaller than the escape velocity from the combined system. Since the reexpansion is a rarefaction with velocity on the order of the sound speed, and because the escape velocity is smaller than the sound speed for sub-Jeans clumps, it is unlikely that coalescence of identical stable clumps can occur. For nonidentical stable clumps, Gilden argues from momentum conservation that nondisruptive coalescence can only occur if the mass ratio of the colliding clumps exceeds the Mach number of their relative collision velocity. Because the relative clump velocities in the Norman and Silk model are fairly large, significant coalescence would require a flat mass spectrum extending over a large range of masses (in order to get a significant probability of large mass ratio collisions), unlike the rather narrow and steep mass spectrum preferred by Norman and Silk.

Despite these difficulties, the general idea of the Norman and Silk model is very attractive, especially since more recent observations have emphasized the prevalence of protostellar outflows. The simulations of Bania and Lyon suggest a resolution to these difficulties, because the wind-driven shells around randomly positioned T Tauri stars are qualitatively similar to the situation studied by Bania and Lyon. The major modification would then be the way in which the fragmentation structure is generated and evolves. Clouds with a wide range of masses would be produced without the necessity of coalescence and some fraction of these would be massive enough to collapse and produce new T Tauri stars in the manner suggested by Norman and Silk. Norman and Silk's arguments that a mass spectrum flatter than $\gamma = -2$ would lead to bursts of star formation while a steeper mass spectrum leads to long-lived self-sustaining star formation may still be valid. Bania's (1981) simulation of a similar scenario using OB stars suggests that the statistical properties to the structure and the star formation rate can remain stable for long periods of time.

Another interesting way in which stars can induce structure in the interstellar gas is through purely gravitational interactions. Stars with a dispersion in velocity will induce velocity and density fluctuations in the gas, and if the stellar and gaseous components have a relative velocity, the criterion for gravitational stability is significantly altered. This problem has been discussed by Niimi (1970) and more recently by Kegel and Völk (1983). Kegel and Völk consider the behavior of linearized perturbations using the Vlasov equation for the stars, the equations of continuity and momentum conservation for the gas, the Poisson equation for the gravitational potential, and a polytropic equation of state. For plane waves of wavelength λ, if the drift velocity between gas and stars in the direction of wave propagation is u_o, then the condition for gravitational stability is

$$\lambda / a_g \, [a_g^2 / a_s^2 + c_s^2 / (c_s^2 - u_o^2)]^{1/2} \leq 1 \qquad (5)$$

where a_g and a_s are the Jeans lengths for the gas and stars alone and c_s is the adiabatic sound speed in the gas. This result shows that if the drift velocity approaches the sound speed, wavelengths much smaller than the Jeans length can be unstable.

By considering a superposition of stable modes, Kegel and Völk are able to derive expressions for the spatial autocorrelation functions of the induced velocity and density fluctuations, assuming a cutoff at a wavelength equal to the mean stellar separation, because the stellar distribution function is discontinuous for smaller wavelengths. The correlation length for velocity fluctuations turns out to be approximately the mean stellar separation, while for the density fluctuations it is roughly the smaller of the mean stellar separation or the Jeans length in the gas. For no drift between gas and stars, the rms-induced fluctuations in velocity relative to c_s and in density relative to the unperturbed gas density are both given, to order of magnitude, by

$$\delta v / c_s \approx \delta\rho / \rho_o \approx (\rho_s / \rho_o)^{1/2} \, \sigma / c_s \tag{6}$$

where ρ_s is the unperturbed mass density of stars and σ is the stellar velocity dispersion (a Maxwellian velocity distribution was assumed). When the drift velocity is comparable to c_s the fluctuation amplitudes are much larger, but then most wavelengths will be gravitationally unstable.

The above expression shows that significant fluctuations can occur in cloud complexes. If we consider a 10 K cloud which has converted 10% of its mass into protostars with a velocity dispersion of 1 km s^{-1}, the velocity fluctuations should be mildly supersonic and the relative density fluctuations are > 1. Larger fluctuations can be induced by the field stars which happen to be moving within a cloud complex, because even though ρ_s/ρ_o is only 10^{-2} to 10^{-3} in that case, the velocity dispersion is $\sim$ 30 km s^{-1}. This process may be an important cause of density and velocity fluctuations within clouds, and deserves additional study. In particular, it may provide a long-lasting source of supersonic velocity dispersions and cloud support even in the presence of dissipation.

In summary, it appears likely that interactions with stars are an important source of fragmentation structure, but the manner in which this occurs is still quite uncertain. The I-S front sequential star formation model of B. G. Elmegreen and C. J. Lada (1977) and B. G. Elmegreen and D. M. Elmegreen (1978) is still an appealing source of induced star formation if other instabilities do not intervene. A numerical study of the nonlinear evolution is needed at this point. The magnetic shell instability discussed by Baierlein et al. (1981) may be just as important if the enhanced magnetic field does not prevent the survival of the fragments. Star-driven implosions as in Sandford et al. (1982*a*) may also cause star formation, but require preexisting dense clouds, and so do not actually cause fragmentation. The calculations may, however, provide a means for destroying low-mass clumps and halting the formation of low-mass stars once massive stars begin to form. The most significant sources of fragmentation structure seem to be related to the effects of groups of stars on the gas: H II regions as in Bania and Lyon (1980), T Tauri winds as in C. A. Norman and J. Silk (1980), and gravitationally-induced velocity and density fluctuations as in Kegel and Völk (1983). Further work on these models should be able to predict a mass spectrum for the fragments and the spatial correlation function of the velocity and density structure.

C. Turbulence and Fragmentation

It was remarked earlier that the evolution of numerical collapse models may be extremely sensitive to initial conditions, as was demonstrated for spherically symmetric collapse by Buff et al. (1979) and was indicated by Boss' (1980*b*) calculations of fragmentation in rotating clouds. This strongly suggests that we can expect the nonlinear stages of gravitational instability, whether or not in the presence of rotation (and probably magnetic fields), to exhibit a transition to chaotic behavior or turbulence. The idea that such a transition occurs and that the interstellar medium should be describable as a turbulent process was apparent in early discussions by Hoyle, von Weizäcker, Chandrasekhar and Münch, and others. In their discussion of the collapse of pregalactic protoglobular clusters, Peebles and Dicke (1968,p. 902) remarked intuitively: "It is not in fact very realistic to carry the integration much beyond the point at which the center achieves free fall, because the free fall surely generates turbulence." While this evolution may be unpredictable, it is still meaningful to discuss the resultant spatial structure of the density and velocity fields and the probability distribution of these and other variables in terms of ensemble averages over a number of independent realizations of the process.

Observations have also led to increasingly common interpretations in terms of some sort of turbulence (see the chapter by Dickman, and below). Although there is not even any agreement on what is meant by the word turbulence as it applies to the interstellar medium, it is becoming clear that the concept of fragmentation and star formation as a stochastic process may provide the basis for the next stage in our understanding of these phenomena. For these reasons it is worthwhile to review some of the suggestions which have been made recently concerning the nature of cloud turbulence. It should be emphasized that all these arguments are extremely crude and speculative, either assuming an analogy with incompressible turbulence or some conceptual model for the structure of turbulent self-gravitating clouds. Remembering that the turbulent behavior of even very simple nonlinear systems is not understood, we should not expect that any quick intuitive arguments will be able to represent the complicated processes which may occur in interstellar clouds. However, the following discussion may contain some of the essential features characterizing cloud turbulence, and will hopefully motivate more complete and quantitative models. (I emphasize that the term turbulence is used in what follows in a general sense, referring to chaotic nonlinear behavior which may or may not be related to incompressible fluid turbulence.) Some discussion of analytical weakly nonlinear calculations is given in Sec. IV.C.7. Ultimately the answers will come from large series' of numerical models with a range of initial conditions and improved spatial resolution.

1. Historical Perspective. Active discussions of the role of turbulence in fragmentation and star formation have only regained prominence during the last several years, after a long dormancy from the time of the classic paper by von Weizäcker and several intermittent attempts at resurrection. There were a number of reasons for the hesitation to involve turbulence. The supersonic line widths observed in both atomic and molecular gas were sometimes interpreted in terms of small-scale random motions, but it was difficult to see how such motions could persist for a significant time in the face of efficient shock dissipation (see, e.g., Field 1978). Also, the line widths seemed compatible with free-fall collapse, at least for the molecular clouds, so it was not clear that another process was required. Perhaps more fundamentally, there was always the underlying knowledge that, once we admit that the interstellar medium is turbulent, there is very little theoretical basis on which to proceed. It remains true that, apart from essentially phenomenological arguments, there is no quantitative theory which can account for the properties of even fully-developed homogeneous isotropic incompressible turbulence, despite the great variety of approaches which have been pursued. All published statistical theories (see, e.g., Heisenberg, Kraichnen, etc.; for a good survey see Beran [1968]) must assume a particular form for the nonlinear interaction terms in order to obtain even rough agreement with the most basic empirical results (like the Kolmogorov constant), and are incapable of predicting more detailed properties. Numerical calculations with subgrid modeling for the smaller scales have become increasingly successful in some respects, but there are very few calculations which can reveal the actual transition to turbulence, and these are all for small Reynolds numbers.

A number of developments have contributed to the renewed interest in turbulent models for interstellar cloud structure.

a. Attempts to fit molecular-line profiles, line-intensity ratios, and the radial distribution of line width, mean velocity, and density inferred from molecular-line observations in clouds using pure collapse models were generally (with some exceptions) unsuccessful, and macroturbulent models or combined turbulent core-collapsing envelope models were proposed as an alternative (see, e.g., Scoville and Wannier 1979, and references therein). Line transfer is quite a difficult problem for such models, and simplified computations of line profiles for a turbulent model with finite correlation length are only now becoming available (see the chapter by Dickman).

b. Observations continued to reveal density and velocity structure on all scales (see Sec. II) giving support to the idea that small-scale clumping within the beam is an important factor in controlling the line widths and line profiles.

c. It has become increasingly evident since the paper by Zuckerman and Evans (1974) that cloud lifetimes probably exceed their free-fall times by a significant margin, as outlined earlier (Sec. III), so some sort of support in excess of thermal pressure seemed necessary, and random motions of clumps (loosely "turbulent pressure") were a prime candidate.

d. It also became recognized that the dissipation problem may not be as disastrous for turbulence theories as was once thought, both because the dissipation is relatively inefficient compared to early estimates and because fresh sources of random kinetic energy, in the form of stellar winds and explosions, collisions with other clouds, galactic rotation, and the gravitational potential of the cloud, are available.

e. Finally, recent work in a number of diverse nonastronomical fields has demonstrated that chaotic, or turbulent, behavior can be expected to occur in a wide variety of nonlinear systems, suggesting the plausibility of some sort of turbulence in the highly nonlinear interstellar medium. The vigorous current activity in the study of chaotic behavior cannot but cause some optimism that an understanding of generalized turbulent behavior lies in the near future.

In addition, turbulence has been advocated as an explanation for the random orientations of angular momentum vectors in stars, binaries, and clouds (Fleck and Clark 1981), and as a means to explain observed carbon abundances in clouds by cycling material through the outer relatively transparent layers where ultraviolet radiation can penetrate (Boland and de Jong 1982, 1984). The renewed interest in turbulent clouds has resulted in a number of heuristic models which try to account for certain observed features.

2. *Kolmogorov Revisited.* The conceptual approaches to turbulent behavior in the interstellar medium which have been discussed in the literature can basically be divided into a few viewpoints. The first, tracing back to von Weizäcker, pictures interstellar turbulence as resembling incompressible turbulence, and then uses phenomenological scaling arguments based on Kolmogorov's ideas concerning the existence of a dissipationless energy cascade, or inertial subrange, modified to account for compressibility and other effects, to derive density-size and velocity-size relations and even an initial mass function (Fleck 1980,1981,1982*a,b*,1983*a;* also Arny 1971). The large Reynolds numbers ($\sim 10^7$ to 10^9) and very small Kolmogorov microscales ($\lesssim 10^{12}$ cm) at which viscous dissipation is important in the interstellar medium are suggestive of an inertial subrange, and the hierarchical structure of the interstellar medium is certainly reminiscent of descriptions of incompressible turbulence as "whirls within whirls" or a "self-similar cascade." Moreover, differential galactic rotation may provide a long-lasting source of energy at the largest size scales (see Fleck 1980,1981).

In this picture, the ultimate source of the turbulence is a shear instability associated with differential galactic rotation, which injects energy at large sizes. These large eddies are assumed to generate smaller-scale structure through some sort of dissipationless vorticity stretching, as is often invoked for incompressible turbulence. In this model, gravity and dissipation are only important insofar as they affect the compressibility of the motion. Fleck and Clark (1981) point out that in this model one expects random directions for the angular momentum vectors of clouds on smaller scales, and present a number of lines of evidence concerning the distribution of observed cloud and stellar rotational axes and binary star orbital axes which agree with this expectation. It should be pointed out, however, that the same sort of random orientations could be produced by tidal torques among fragments, for which there is some numerical evidence (Roczycska 1983), or direct collisions between fragments, and might be a general feature of any turbulent model.

This Kolmogorov-type picture tries to account for the observed scaling relations by assuming virial equilibrium (the support mechanism is unspecified) in the form $L/v_L \sim \rho^{-1/2}$, and constant kinetic energy density transfer rate at every scale, $\rho v_L{}^3/L$ = constant, giving $v_L \sim L^{3/5}$, $\rho \propto L^{-4/5}$, close to that which is observed. One could also abandon the virial assumption and just introduce a parameter such as the density enhancement at each scale, as was done long ago by von Weizäcker and Von Hoerner (see Fleck 1980,1981). Fleck (1983*b*) pointed out that the constancy of the volume rate of kinetic energy transfer of $\sim \rho v^3/L$ leads to $v \propto (L/\rho)^{1/2}$, so an inverse relation between density and size could give a reasonable velocity-size relation without invoking virial equilibrium. A similar model that explicitly treats the hierarchical nature of this scenario is given in Ferrini et al. (1984). However, these models do not offer any physical explanation of just how the energy is supposed to be transferred to smaller scales; why the energy transfer time scale is supposed to be L/v; why the structure is in virial equilibrium at all scales; or how the smaller-scale structure can remain supersonic without shock dissipation, unless magnetic fields are important, in which case the scaling arguments are inappropriate. The time scale discussed by Fleck (1981) is just the time over which the energy injection at the largest scale would slow down the rotation of the entire galaxy, and is not relevant to the dissipation of the smaller-scale structure. With dissipation, the smaller structure would possess essentially no random motion even if galactic rotation continually pumps in new kinetic energy at the large scales. Nevertheless, the availability of galactic rotational energy remains an attractive source for the initial large-scale instability, although a more quantitative examination of the problem along the lines of Goldreich and Lynden-Bell (1965*a,b*) and J. H. Hunter and T. Horak (1983) is lacking.

3. Rotational Collapse Revisited. Henriksen and Turner (1984) have recently discussed how the tendency for gravitational contraction to be counteracted by rotation might lead to a hierarchical structure with properties like those observed. At the largest scale, the energy source is assumed to be galactic differential rotation, as proposed by Fleck. At each smaller scale the two forces attempt to achieve balance, leading to virial equilibrium and a transfer of angular momentum flux to larger scales. The plausibility of such gravitational torquing has been recently emphasized by Larson (1984) and Boss

(1985), and may be due to tidal effects or the growth of nonaxisymmetric trailing perturbations. Assuming a constant rate of change of angular momentum density, which is $\rho_L v_L L/\tau_{ff}$ in this model, and virial equilibrium, one obtains $v \sim L^{1/2}$, $\rho_L \sim \ell^{-1}$, again close to the observed relations. (This model was first used by Arny [1971] in a discussion of the IMF produced by turbulence.) Henriksen and Turner also give some interesting implications for the termination of the self-similar process at stellar masses and the stellar IMF.

As discussed in Sec. IV.A, numerical collapse calculations are nearly unanimous in concluding that initially-uniform rotating clouds do evolve to a quasi-equilibirum disk, ring, or bar in which rotation tries to balance gravity, and that these configurations rapidly break up into a few fragments. Furthermore, the observations of Heiles Cloud 2 by Schloerb and Snell (1984) strongly suggest that some clouds do evolve in this manner. Although Henriksen and Turner make no mention of how fragmentation occurs in their model, it seems that if it is rotationally induced, then the model is just basically the hierarchical model discussed by Bodenheimer (1978). Henriksen and Turner have therefore essentially extended Bodenheimer's model to show that, if the angular momentum source rate is constant at all levels of the hierarchy, then the observed density-size and velocity-size relations follow directly. Without fragmentation, the model would require rotational support at every scale, definitely not an observed feature of interstellar structure. A possible problem, as with Bodenheimer's scenario, is that the model implies a common rotational axis on all scales, which is not observed. However, tidal interactions and direct collisions may be capable of randomizing the angular momentum vectors on each scale.

One may justifiably question the use of the term turbulence for the model just described. Bodenheimer's model is certainly not usually considered as a turbulent model. Henriksen and Turner apparently use the term to refer to any self-similar hierarchical structure, not necessarily a stochastic process. In this sense all theories of hierarchical nonlinear fragmentation (and hierarchical galaxy clustering) should be considered turbulence theories; I believe this is a useful point of view. In addition, the process may possess a sensitivity to initial conditions which is similar to turbulent processes.

Besides this semantic consideration, the model encounters practical difficulties when confronted with observations of cloud structure. There is very little evidence that the virialized state of most observed structures is related to rotation, although rotation seems definitely important in some small fraction of small dark clouds. Also, the prevalence of filamentary structure, the fact that the Taurus complex is probably prolate and rotating at a dynamically insignificant speed about the long axis, and the arguments of Schneider and Elmegreen that most dark filaments are prolate and easily fragmented do not seem to fit in with the picture evoked by the rotational model. (The same remarks apply to Fleck's model.) The model also predicts that a significant fraction of clouds on all scales should be in nearly free-fall collapse, because the time to break up the quasi-equilibrium structure into fragments is not long compared to the initial cloud free-fall time. These criticisms are meant to motivate more quantitative studies of gravitational-rotational hierarchies which can address these questions.

4. Gravitational Virialization. A very different approach to conceptual turbulence models was motivated more by the question of energy dissipation than any analogies with incompressible flows. Here it is recognized from the outset that interstellar turbulence is unlikely to resemble Kolmogorov's picture of freely-decaying incompressible turbulence because of the importance of dissipation and energy injection over a range of different size scales, and the fact that the observed density fluctuations are extremely nonlinear, with $\delta\rho/\rho \gtrsim 10$. Because the line widths are supersonic, and cannot be accounted for solely by collapse or rotation, it is difficult to see how dissipation can be avoided unless the magnetic field is dynamically important. The importance of dissipation was recognized in the apparently never-completed program of Von Hoerner's group (see Von Hoerner 1958) to model cloud turbulence as a field of interacting shock waves. This was the major argument against turbulent models, and cannot be easily dismissed. The problem, then, is: How does a system of supersonic density fluctuations in a cloud maintain its random kinetic energy over a time scale longer than the dynamical time scale of the cloud?

One suggestion (Scalo and Pumphrey 1982) is that the fragments continually extract energy from the gravitational field of their parent cloud. As the parent cloud collapses, the change in potential energy can be channeled into the random motions of density fluctuations until a quasi-virialized state is attained in which the fragments move at the circular velocity. The system slowly contracts at a rate set by the rate of dissipation due to clump collisions and drag. This idea of gravitational virialization by direct collisions was independently suggested long ago by Hoyle, Odgers and Stewart, and Woolfsen and later mentioned by Larson, McCrea, and Mestel. Scalo and Pumphrey suggested that, by accounting for the inefficiency of noncentral collisions in dissipating energy, the dissipation and hence contraction time can exceed the free-fall time by a significant margin, especially if the fragments are centrally condensed. *N*-cloud simulations (Pumphrey and Scalo 1984) show that such a quasi-virialized state can be attained in an unstable cloud, but only if the clump collisions involve sufficient transverse momentum transfer to impart some randomization to the clump velocity vectors, and if the clumps are internally supported for a time in excess of their own free-fall times. The randomization requirement may not be too severe, but improved multidimensional cloud collision calculations are needed to test this point. However, the internal support requirement seems contrived; if rotational support is invoked, then there arises the same inconsistency with observations encountered above. A system of fragments in a cloud may acquire and maintain some part of their random velocities through interaction with the gravitational potential, but some additional source of energy, or of new density and velocity fluctuations, seems necessary to maintain the motions for times in excess of the free-fall time if the fragments themselves are not internally supported.

A recently reported numerical calculation of rotating cloud collapse using a particle method by Benz (1985) apparently supports the idea of gravitational virialization. Benz finds that turbulent motions are generated from the input of gravitational energy, and that the turbulence greatly slows down the collapse.

5. Energy Sources. There are a number of potentially important sources of turbulent energy. As discussed in Sec. IV, OB star winds and ionization fronts can generate considerable density and velocity structure and perhaps trigger star formation through I-S gravitational instability and shock compression of dense preexisting clumps. However, this process can only occur in complexes where OB stars have already formed, unless field OB stars are numerous enough, which seems unlikely. The calculations of Kegel and Völk (1983) suggest that the gravitational field of passing field stars alone may provide a continual supply of density and velocity fluctuations in cloud complexes, as described in Sec. IV.B.

Bash et al. (1982) have argued that energy is supplied to large clouds through random collisions with smaller clouds. The problem here lies in the existence of these dense small clouds *outside* of complexes and the question of why the collision rate should be such as to just keep the complex supported against gravity for longer than a dynamical time scale; if the collision rate is too large (small), the large cloud will expand (contract) and the increased (decreased) cross section will further accelerate its dispersal (collapse) unless the large cloud encounters variations in the small-cloud number density on time scales of $\lesssim 10^6$ yr. Nevertheless, fragments within complexes probably encounter and coalesce with smaller fragments, and if the mean collision time is less than the fragment free-fall time, the collisions might provide a means of preventing collapse (see Pumphrey and Scalo 1984). However, numerical simulations of such collisions (Gilden 1984) show that most of the kinetic energy lost by the small cloud as it plows through a larger cloud is radiated away through a bow shock. Support of a large cloud by intrusions of small clouds might therefore require that many small clouds be in the process of penetration at all times.

On small scales, protostellar winds are probably capable of sustaining the observed line widths for long periods of time (Norman and Silk 1980; see Sec. IV.B). In this model the turbulence consists of randomly moving interacting clumps which are continually generated by the collision and breakup of wind-driven shells. The rapidly accumulating evidence for high-velocity molecular outflows around protostellar objects has provided strong support for this idea, although the details may differ somewhat from the original Norman and Silk model (see Sec. IV.B). One potential problem is that the required mass loss rates and/or protostellar number densities may be larger than allowed by observation.

Silk (1985) has shown how the breakup of momentum-conserving shells can account for the observed density-size scale relation if virial equilibrium is assumed, but the cause of the internal support is left unspecified.

In a variation of this model, Franco (1983) has argued that the dissipation of excess protostellar rotational energy by rotationally-driven winds (see also L. Hartmann and K. B. MacGregor [1982*a*] and, for a more detailed model, Pudritz and Norman [1983]) can be an important source of turbulence and heating in clouds. Unlike the Norman and Silk model, this model requires a rather small number density of protostars. Another possibility is that the winds are generated by the transformation of the protostellar rotational energy into magnetic energy, and hence pressure, by the twisting of field lines (Draine 1983). A recent discussion of these various mechanisms is given by Franco (1984). Evidently there is no paucity of models to explain the observed protostellar energy input; unfortunately this diversity means that we cannot confidently estimate the turbulent energy injection rate as a function of time, or its effect on further star formation.

If internal stellar objects are important sources of turbulent energy in cloud complexes, then it is obvious that the turbulence cannot possibly be modeled by analogy with the idea of a Kolmogorov cascade. Instead of injection at large scales with cascade to smaller scales, the process would be powered by essentially microscopic explosions (Dickey 1984) with dissipation important at intermediate scales. This description may suggest some useful phenomenological approaches to the problem of cloud turbulence powered by internal stellar sources. It may also be relevant to the development of turbulent structure in clusters and superclusters of protogalaxies powered by protogalactic winds.

Although protostellar winds undoubtedly do inject energy into clouds, it is still unclear whether they are the dominant source of turbulent motions in most clouds. One test is provided by observations of complexes which contain very few protostars. The best-known example is the Southern Coalsack. The line widths of H I, CO, and H_2CO are all definitely supersonic even though no protostellar sources have been found within the complex (see T. J. Jones et al. 1980; Bowers et al. 1980, and references therein) and there is no evidence for collapse (T. J. Jones et al. 1984*a*). A similar result is obtained for the northern complex Khav 141, in which there is little evidence for star formation (see Saito et al. 1981). This indicates that, whatever the nature of the turbulence, it can persist without the presence of protostars.

In reality, the true nature of turbulence in cloud structures probably involves, to various unknown degrees, injection of energy by rotational shear at large scales; gravitational fragmentation influenced by rotation and magnetic fields leading to a sort of cascade (*not* dissipationless); the constant excitation

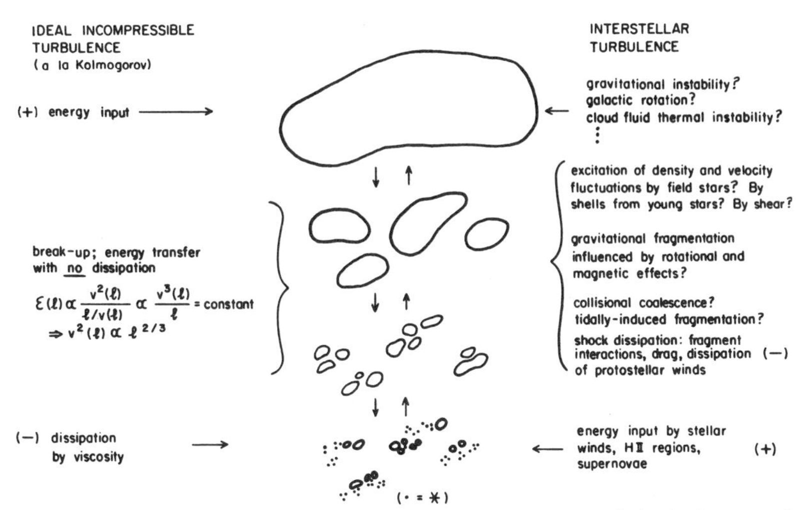

Fig. 26. Schematic illustration of the contrast between the Kolmogorov dissipationless cascade model and the processes suspected of being important in real interstellar turbulence.

of density and velocity fluctuations by field stars; energy injection at the microscopic scale by internal star formation activity; and dissipation at all scales due to the interaction of protostellar winds with the ambient medium, fragment interactions, and the motion of fragments through the interfragment medium. The contrast with the Kolmogorov cascade for incompressible turbulence is illustrated schematically in Fig. 26. Evidently, if we are to understand interstellar turbulence by means of a phenomenological model, the essential feature will be a description of the balances and transfers of mass, momentum, energy, and angular momentum between these various processes. This statement also suggests that models which concentrate on a single process and/or ignore the range of scales involved (this includes nearly all the studies reviewed here) should be viewed with caution.

6. Magnetohydrodynamic Waves. Another possibility is that the clouds of all sizes are supported by a magnetic field and that turbulence which broadens lines is a spectrum of MHD waves. A recent discussion of this problem by Zweibel and Josaffetson (1983) indicates that the damping time may be reasonably large. These linear wave modes are expected to be incompressible, so the observed clumpiness would probably have to be attributed to gravity, although the nonlinear wave interactions might also lead to substantial density fluctuations. Because the detailed evolution of the wave spectrum and the resulting velocity field are unknown, it is not possible to compare this model directly with observations. However, the observed line widths imply virial balance at all size scales, and, if the line widths represent the Alfvén speed, it may be quite difficult to construct a model which involves a hierarchy of magnetically limited gravitational fragmentation (see Mestel 1977; chapter by Mestel; Nakano 1985) and which at the same time reproduces the observed line width-size relation. Related to this problem is the possibility that fluid turbulence can be the driving force for MHD waves, as discussed by Eilik and Henriksen (1984). A detailed discussion of MHD turbulence and its possible relevance to the small-scale structure of the interstellar medium has been recently given by Higdon (1985; see also Grappin et al. 1982).

It is also possible that the observed supersonic line widths in clouds are due not to MHD waves but complex nonradial oscillations of the clouds. This topic has not been addressed in the literature (see, however, S. V. Weber 1976), and so will not be discussed further here; it merely emphasizes the wide range of problems which still require investigation before we can claim any understanding of turbulence in the interstellar medium.

The above discussion does not exhaust the possible sources of energy or initial fluctuations for fragmentation. For example, Fleck (1984) has interpreted the four main fragments in the Corona Australis complex (see Fig. 4) as due to a Kelvin-Helmholtz instability at the boundary layer of the complex produced as the complex fell into the galactic plane. While this process is not strictly an energy-injection mechanism, it is certainly well known that the nonlinear stages of the Kelvin-Helmholtz instability are turbulent so we may add another possible process to our list of processes which may trigger turbulence.

7. Studies of Weak Nonlinearities. There is a middle way between the order-of-magnitude conceptual approaches to the nonlinear fragmentation-turbulence problem, which may be too simplistic or founded on linearized results, and numerical calculations which in principle could solve the complete problem but which in practice are stymied by their lack of spatial resolution and the problem of understanding the results. This intermediate approach is the realm of quantitative analytical treatments of turbulence. There are a staggering number of papers and approaches in this area in the fluid dynamics literature, but so far none of these has attained predictive power, and most have not been able to account, even after the fact, for fundamental empirical results, like the Kolmogorov constant, for isotropic homogeneous incompressible turbulence, without a disturbing amount of twiddling. Most of these approaches are statistical in the sense that they only attempt to follow the low-order moments of the two-point probability distribution of fluctuations, such as the autocorrelation function. As with all such problems, there is a severe difficulty with closure of the moment equations, and the various theories differ mostly in the nature of their closure assumptions. In addition, these theories often assume that the turbulence is weak in some sense, i.e., not too nonlinear.

There have only been a few attempts to apply these techniques to astrophysical problems relevant to fragmentation, and even these have been largely ignored in the astronomical literature, perhaps because of the formidable mathematical appearance of the theories. However, the apparent complexity usually masks some very simple fundamental assumptions, and the results, even though based on questionable closure hypotheses, can still shed light on the physical nature of the problem, and for that reason a few of these papers are reviewed briefly here. The discussion is confined to studies which are directly relevant to the problem of hierarchical fragmentation.

Olson and Sachs (1973) extended the classic work of Proudman and Reid on the statistical evolution of vorticity to include expanding or contracting background states. Although they were concerned with the evolution of turbulence in the early expanding universe, their basic result is relevant to contracting interstellar clouds. A differential equation for the mean-square vorticity can be derived by using the "fourth-order cumulant discard" closure assumption, which basically assumes that fourth-order velocity correlations can be decomposed to products of second-order correlations as if the fluctuations were a joint-Gaussian process. The equations also assume that the turbulence is incompressible, isotropic, and homogeneous. Their calculations indicate that vorticity increases during contraction of a turbulent cloud; neglecting molecular viscosity, the vorticity becomes infinite in a finite time. While the fourth-order cumulant discard closure method is now known to give nonphysical results for long times (for a review of a fairly successful modification including magnetic field, see Poquet [1979]), this basic conclusion may still be valid. One may be tempted to interpret the vorticity increase in a contracting medium as an obvious consequence of angular momentum conservation. However, the important point is that if vorticity is an essential feature of interstellar turbulence and fragmentation, this turbulence should become stronger during gravitational contraction, a result which may be related to the support of the bound interstellar structures by turbulent pressure. Because the calculation was for an incompressible medium, it remains to be seen how the result may be altered by the nonlinear interaction of density fluctuations (compressible modes) and vorticity fluctuations (solenoidal modes).

The work by J. H. Hunter and K. S. Schweiker (1981) and J. H. Hunter and T. Horak (1983) on the behavior of rotational and irrotational perturbations in a shear flow is related to the above problem. These authors find that in many cases the vorticity modes grow while the density modes damp, suggesting a possible origin for the vorticity fluctuations which would be amplified during contraction.

An important but neglected paper by Sasao (1973) investigates quantitatively the nature of density fluctuations excited by purely vortical turbulence in an isothermal self-gravitating medium. The source of the initial rotational velocity fluctuations is not specified. A number of possible sources of fluctuations associated with newly-formed or field stars were discussed in Sec. IV.B, but these are unlikely to be purely rotational. For this reason Sasao's work should be viewed simply as an illustration of how the rotational part of the velocity field can generate density fluctuations. J. H. Hunter and K. S. Schweiker (1981) have shown how vorticity modes may grow in the presence of shear.

The coupling between the compressible and vortical modes is intrinsically nonlinear. Sasao obtains and solves the equations for correlation functions by an expansion to first order in the fluctuations in density and potential velocity, and to second order in the rotational velocity component. He closes the moment equations by use of the fourth-order cumulant discard assumption. A unique feature of Sasao's treatment is that the moment equations are derived from the equation for the characteristic functional (essentially the joint probability function of all Fourier components of the density and velocity fluctuations; see Beran [1968]), including compressibility and self-gravitation. The final equations describe the time evolution of the wavenumber spectrum of density and potential velocity fluctuations induced by the rotational incompressible turbulence, and the evolution of the nonlinear source terms which feed the irrotational (compressible) modes. Sasao solves these equations for three assumed forms of the irrotational energy spectrum: a delta function in wavenumber k, a k^4 spectrum cutoff by a Gaussian, and a Kolmogorov spectrum.

While these calculations are only valid for weakly nonlinear turbulence, several interesting results emerge which may be relevant to an understanding of the manner in which the observed very nonlinear structure arose. First, the behavior of the induced density fluctuations is in some ways very similar to the classical Jeans result; that is, for wavelengths much smaller than the initial Jeans length, the induced fluctuations quickly become sound waves with rms density amplitude $\delta\rho/\rho \propto (u/c_s)^2$ (where u is the rms solenoidal velocity fluctuation amplitude and c_s is the sound speed), while wavelengths larger than the Jeans wavelength grow rapidly. The second result concerns the dominant wavelength of the growing modes. If the initial velocity perturbations span a range of wavelengths, as for the Kolmogorov energy spectrum, then the peak of the spectrum of induced density fluctuations is comparable to the Jeans wavelength, whether or not the cloud is contracting. This supports and extends R. B. Larson's (1984) contention that the ordinary (thermal) Jeans length plays a fundamental role in fragmentation, even when rotation and magnetic fields are significant. However, an initial turbulent energy spectrum which is sharply peaked at some wavelength produces a density spectrum with a peak at the same wavelength, at least for noncontracting clouds.

Third, the shape of the density fluctuations spectrum is the same as that of the initial turbulent spectrum. This prediction could be tested by estimating power spectra for densities and velocities in real interstellar clouds. Finally, the growth of the relative rms density fluctuation is found to be very rapid in a contracting cloud. Because the induced density fluctuations depend on the square of the vortical velocity component, the result of Olson and Sachs (1973) that vorticity increases in a contracting cloud further accentuates the growth of density fluctuations.

It is often claimed that turbulence can control the effective Jeans mass. This conclusion stems from the inclusion of a turbulent pressure, for example, in the virial theorem. Such a turbulent pressure would not only affect the critical mass for gravitational instability, but would inhibit the growth of density fluctuations in much the same way that ordinary pressure is found to inhibit fragmentation in numerical collapse calculations (Sec. IV.A). However, it is important to realize that all published estimates of the effects of turbulent pressure on gravitational instability which essentially compare turbulent and gravitational energies completely neglect third-order correlations (e.g., $<\delta\rho u_i u_j>$), and these correlations *enhance* fragmentation through the coupling of rotational and compressible modes. As explained by Sasao, it is likely that the growth by third-order correlations dominates the damping by turbulent pressure; this is why the dominant unstable wavelength is closer to the thermal Jeans mass than the turbulent Jeans mass. Turbulent pressure may support a cloud against collapse without preventing its fragmentation. For this reason, arguments that rely on a turbulent Jeans mass, such as the estimates of mass distributions by Arny (1971) and Fleck (1982*b*) must be considered tentative at best.

The work of Olson and Sachs (1973) and Sasao (1973) cannot tell us anything about the strongly nonlinear effects involved in fragmentation. However, together they suggest a physical picture in which turbulence in a self-gravitating interstellar cloud promotes fragmentation because of the coupling between vorticity and density fluctuations, both of which increase with time in a contracting turbulent cloud. It is easy to imagine that these processes may result in hierarchical fragmentation, but such a speculation cannot yet be discussed profitably because the theoretical work can only treat the effects of weak nonlinearities. Nevertheless, it is interesting to note that this picture essentially represents a synthesis of the early ideas of Hoyle (1953) and von Weizäcker (1951). Hoyle ignored vortical turbulence and thus considered that density fluctuations grew rapidly through gravitational effects alone. Von Weizäcker, on the other hand, emphasized turbulent cascade by analogy with incompressible turbulence, but only considered the effects of compressibility on fragmentation in a very qualitative way, based on earlier work by von Hoerner, and suggested that large density fluctuations arise because of shocks, not because of ordinary gravitational instability. In general terms, the idea that the interaction between gravitational and rotational effects is a dominant process in fragmentation is similar to several more recent theoretical studies which were discussed earlier.

Space precludes discussion of additional studies which may be relevant to turbulence in the interstellar medium. For example, the work of S. S. Aggarwal and G. L. Kalra (1984) suggests that nonlinear wave-wave interactions may trigger gravitational instability at wavelengths smaller than the Jeans length. Vithal and Vats (1983*a*) show that wave-wave interactions lead to cascading of energy to smaller wavelengths in a compressible medium, although dissipation and self-gravitation are not included in the analysis (see Bhatia and Kalra [1980] and Vithal and Vats [1983*b*]).

Further quantitative calculations of the statistical properties of turbulence in contracting clouds along the lines of Olson and Sachs (1973) and Sasao (1973) are highly desirable at the present time. The quasi-normal closure assumption made by these authors is known to give nonphysical results for large Reynolds numbers and/or late stages of the flow development, but these deficiencies are now understood in terms of excessive memory in the model system. A remedy for this problem is the eddy-damped quasi-normal Markov approximation, which has been applied with some success even for MHD turbulence (see Poquet 1979). An application of this method to an astrophysical problem involving turbulent amplification of magnetic fields in radio sources can be found in DeYoung (1980). The method must be extended to the compressible case.

V. FORMATION OF LARGE-SCALE (0.1 TO 1 kpc) STRUCTURES

A. Gravitational Instability

The most straightforward mechanism for the formation of the kpc-scale reddening - H I structures is gravitational instability of the galactic disk. The classical Jeans criterion for instability in a gas with ratio of specific heats equal to γ is

$$\gamma c^2 k^2 - 4\pi G\rho > 0 \tag{7}$$

where c is the velocity dispersion, k the wavenumber, and ρ the average density. If we consider the gaseous disk to consist of clouds with velocity dispersion c_{10} in units of 10 km s^{-1} and smeared out particle number density n_o, the growth time scale (in yr) and critical wavelength (in kpc) can be expressed as

$$\tau_J \approx 3 \times 10^7 \, n_o^{-1/2} \tag{8}$$

$$\lambda_J \approx 1 \, n_o^{-1/2} \, c_{10} \tag{9}$$

and so λ_J is comparable to the sizes of the observed structures. However, this result assumes an infinite medium (Jeans swindle), and neglects rotation, magnetic fields, dissipation, and coalescence.

The pure gravitational instability of a differentially rotating isothermal disk to the growth of cylindrically symmetric perturbations can only occur if (Toomre 1964; Goldreich and Lynden-Bell 1965*a,b*)

$$Q \equiv \kappa \frac{C}{3.36 G\mu} \leq 1 \tag{10}$$

where $\kappa = [4B(B-A)]^{1/2}$ is the epicyclic frequency ($\sim 3 \times 10^8$ yr^{-1} in the solar neighborhood), C is the gas velocity dispersion, and μ is the gaseous surface density. As emphasized by Jog and Solomon (1984*a*), this criterion predicts that the gas disk of our Galaxy should be stable, although by a small margin. Jog and Solomon (1984*a,b;* see also Niimi 1970) show that when the disk is treated as a two-fluid system, the gravitational interaction between the gas fluid and star fluid can lead to instability, even when each fluid would be stable by itself. The relative contribution of each component to the instability is proportional to its surface density weighted by the inverse of its velocity dispersion; because the gas velocity dispersion is much smaller than for the stars, the gas component makes a substantial contribution to the instability even if its surface density is much smaller than that of the stars. By using values of the relevant parameters appropriate to our Galaxy, Jog and Solomon find that the wavelength and growth time of the fastest growing mode are typically $\lambda \sim 2$ to 3 kpc, $\tau \sim 2$ to 4×10^7 yr, containing a mass of 4 to 10×10^7 $M_\odot$. Jog and Solomon (1984*b*) suggest that the growth of this two-fluid instability will increase the gas density above the limit required for gravitational instability of the gas fluid alone, giving clusters of large structures with sizes of $\sim$ 400 pc and masses of $\sim 10^7$ $M_\odot$. This suggestion is rather speculative because it depends on the nonlinear behavior of the original two-fluid instability. It is also not clear why the density increase required for the one-fluid instability could not be obtained simply by compression in spiral density wave shocks.

Thus in this picture a two-fluid gravitational instability initiates the evolution of interstellar cloud fragmentation with the growth of very large structures, and the density increase within these structures promotes a one-fluid gravitational fragmentation.

It is important to note that Jog and Solomon's two-fluid analysis was partially motivated by the apparent result that the one-fluid interstellar medium is marginally stable according to the Q criterion given by Eq. (10). However, as pointed out by B. G. Elmegreen and D. M. Elmegreen (1983), the presence of a small azimuthal galactic magnetic field results in a stability criterion for a rotating medium which is nearly the same as that for a nonrotating, nonmagnetic medium (Lynden-Bell 1966), independent of Q. This occurs because magnetic forces oppose the Coriolis forces (Chandrasekhar 1954; see Stephenson [1961] for an extension including viscosity). The value of Q only affects the growth time scale, not the unstable wavelength. If this is correct, then the arguments of Jog and Solomon, including their stability analysis for the two-fluid case, would require modification. The two-fluid case would then boil down to the criteria presented by Niimi (1970) and Kegel and Völk (1983) discussed in Sec. IV.B above. For no relative velocity between the stars and gas, the unstable wavelength is just the ordinary Jeans wavelength for the gas multiplied by a factor (< 1) which depends on the ratio of velocity dispersions and mass densities of stars and gas; for our Galaxy this factor is not much less than unity, so the one-fluid Jeans wavelength is recovered even in the two-fluid case. However, the analysis also shows that the existence of a drift velocity between gas and star fluids, an effect not considered by Jog and Solomon, can reduce the critical wavelength much below the Jeans wavelength because the drift velocity is comparable to the gas sound speed. For these reasons it is not at all clear that the two-fluid critical wavelengths and masses derived by Jog and Solomon are relevant to the real interstellar medium, nor that they are comparable to the scales of observed superclouds.

B. G. Elmegreen (1982*a,c*) has presented a very detailed analysis of the linearized perturbation equations for the combined Parker-Jeans instability of a self-gravitating, isothermal exponential gas layer, neglecting rotation. Elmegreen shows that at lower densities, $n_o \lesssim 1$ cm^{-3}, the Parker instability dominates gravitational instability, and the dominant cloud mass decreases with decreasing density. For example, at $n_o = 0.3$ cm^{-3} the cloud mass is only about 2×10^4 $M_\odot$. These masses are comparable to those estimated for clouds in which optical polarization suggests the operation of Parker instability (see Appenzeller 1974; Vrba 1977; Vrba et al. 1981). At larger initial densities, however, Parker instability is unimportant and gravitational instability produces clouds with final masses of $\sim 10^6$ $M_\odot$, approximately independent of density, with growth time essentially equal to τ_J. Note that this final mass for the fastest growing wavelength is much smaller than found by Jog and Solomon, even for the one-fluid case at enhanced densities. This discrepancy only reflects the fact that estimated masses associated with unstable wavelengths depends on the assumed geometry; for different assumed elongations, Elmegreen's (1982*c*) masses can exceed 10^7 $M_\odot$ (Elmegreen, personal communication). Elmegreen suggests that the essential feature promoting the compression required for the gravitational formation of these clouds is a spiral density wave, explaining why large cloud complexes and giant H II regions appear in the spiral arms of galaxies (see also B. G. Elmegreen [1979*a*] for a more general discussion of the collapse and fragmentation of dust lanes). We see that Elmegreen's picture of disk instability is much different than that proposed by Jog and Solomon. Jog and Solomon neglected the effect of a magnetic field in opposing the Coriolis force while Elmegreen neglected the effect of the presence of stars.

Both of the studies discussed above assume an isothermal cloud fluid, and so it is interesting to consider the effects of dissipation. Struck-Marcell and Scalo (1984) have examined the linear stability of a cloud system when cloud collisional dissipation, coalescence, and energy input due to star formation are included. However, the effects of stellar gravitation are not included. The results show that the classical Jeans mode can be suppressed when the cloud collision time scale is smaller than τ_J. However, an unstable modified isentropic Jeans mode always exists for wavelengths $\lambda \gtrsim \lambda_J/2$; when collisions are important the growth time scale is somewhat larger than τ_J, but the effect is not large. It seems plausible, therefore, that the isentropic Jeans mode may preserve the general conclusions of Elmegreen and Jog and Solomon, even when dissipation and coalescence are important in suppressing the ordinary Jeans instability (although this suggestion needs to be examined in detail).

B. Thermal Instability of a Cloud Fluid

If we consider the interstellar medium to be composed of a large number of clouds which interact collisionally and which acquire energy through stellar explosions, winds, etc., then the system of clouds may be subject to a thermal instability if the energy injection rate does not depend on the number density of clouds, n_{cl}.

Cowie (1980) pointed out that the effective pressure-density relation in an equilibrium cloud fluid can be obtained by equating the collisional dissipation rate of $\sim n^2c^3$ to the energy input. If the energy input is constant, then the pressure $nc^2/3$ will vary as $(nc)^{-1}$, which decreases during compression. A compression therefore leads to a pressure deficit, which causes collapse. However, Cowie neglected the effect of coalescence, which reduces n_{cl} during the compression. A linear perturbation analysis of model cloud fluid equations (Struck-Marcell and Scalo 1984) confirms Cowie's suggestion: An isobaric thermal instability occurs as long as the efficiency of dissipation is not much smaller than that of coalescence, which is plausible. The instability is entirely analogous to the isobaric mode encountered in studies of a nonadiabatic chemically reacting interstellar gas. The critical wavenumber is given by $k_c = (5\tau^2c^2/9)^{-1/2}$, where τ is the mean cloud collision time scale. The corresponding wavelength, which is about 9 times the cloud mean free path, can be written (in kpc)

$$\lambda_c \approx 4\,(\bar{m}/10^5\,M_\odot)^{1/3}\,(n_{in}/100\,\mathrm{cm}^{-3})^{2/3}\,n_o^{-1} \tag{11}$$

where $\bar{m}$ is the mean cloud mass, n_{in} is the average internal particle number density of the clouds, and n_o is the smeared-out particle number density. The unstable wavelengths must be smaller than λ_c so that collapse, communicated by pressure waves in the cloud fluid, can proceed more quickly than dissipation.

The growth time for this instability is essentially the cloud collision time τ. Relative to the classical Jeans growth time τ_J one finds

$$\tau/\tau_J = 3\,(n_{in}/100)^{2/3}\,(\bar{m}/10^5)^{1/3}\,n_o^{-1/2}c_{10}^{-1}. \tag{12}$$

This expression shows that the thermal instability will proceed faster than gravitational instability for small cloud masses and/or internal densities, or in shock-compressed regions of the interstellar medium where n_0 and c are large. In particular, this thermal instability could dominate gravitational instability in regions which have passed through a spiral density wave. The instability would lead to a two-phase cloud fluid: a low-density, large-velocity dispersion phase, and a high-density, low-velocity dispersion phase; the latter phase would correspond to the gathering of small clouds into large complexes with sizes of ~ 0.2 to 2 kpc, depending on the parameters. The model requires the prior existence of relatively low-mass clouds; perhaps these are formed by the Parker instability as found by Elmegreen (1982*c*). The instability also depends crucially on the assumption that the energy injection rate is independent of the number density of clouds. For example, the instability does not occur in the standard Oort model in which energy input is due to disruption of massive clouds (Struck-Marcell and Scalo 1984); energy injection due to winds and explosions of field stars, as in the model of Bania and Lyon (1980) is required.

C. Collisional Buildup: The Oort Model

The idea that massive ($\sim 10^5$ to 10^6 $M_\odot$) star-forming interstellar clouds form by means of collisional coalescence of smaller clouds has a long and venerable history. The most commonly discussed model, the Oort model (Oort 1954), assumes that small clouds are formed when massive clouds are disrupted by internal stellar activity such as expanding H II regions, winds, or supernova explosions. These small clouds are supposed to collide and coalesce, eventually creating a new generation of massive star-forming clouds which disrupt, and so on. The Oort model is appealing because estimates for the mean collision time scale for diffuse clouds, based on statistics of H I emission and absorption, optical interstellar lines, and reddening, is estimated to be only around 10^7 yr, assuming that clouds are randomly distributed, and because the efficiency of radiative shock dissipation in cloud collisions suggests that collisions should lead to coalescence.

The basic kinetic equation governing the evolution of the mass distribution has been expounded in a large number of papers, and the reader is referred to Kwan (1979), Cowie (1980), and Pumphrey and Scalo (1983) for earlier references. Refinements include an allowance for a finite time lag between the formation of massive star-forming clouds and their disruption, the evaporation of clouds and other processes related to interactions with supernova remnants (Chieze and Lazareff 1980), and the effects of spiral density waves. The kinetic equation can only be solved analytically for a very restricted set of assumptions concerning the form of the collision cross section and the cloud velocity distribution, but more recent *N*-body simulations have included more realism (see, e.g., Hausman 1982). The linear stability of a hydrodynamic description of the Oort model (and other models for interstellar cloud evolution including the Oort model) is discussed in Struck-Marcel and Scalo (1984*a*). The Oort model is very stable to both linear and nonlinear perturbations when time lags are ignored, but a finite lag between massive cloud formation and cloud disruption leads to limit cycles, chaotic behavior, and strong bursts of star formation (Scalo and Struck-Marcell 1985).

There are two basic problems with the Oort model which have received recent attention. The first is that it may be difficult to form clouds as massive as GMC's ($\sim 10^5$ to 10^6 $M_\odot$) unless these clouds have lifetimes at least as long as a few times 10^8 yr (Scoville and Hersh 1979; Kwan 1979), which now seems unlikely. Furthermore, the time required to build massive clouds is so long that it is difficult to understand why star formation occurs preferentially in spiral arms. Casoli and Combes (1982) showed that, if the density enhancement of clouds in spiral arms is a factor of 5 or more, the resulting increase in the collision rate can result in GMC's concentrated in spiral arms, even if the massive cloud lifetime is as small as 4×10^7 yr. However, the density enhancement was an assumed parameter. Kwan and Valdez (1983) and Tomisaka (1984) have presented *N*-body simulations of the Oort model which self-consistently follow the cloud collisional evolution in a spiral gravitational potential. Kwan and Valdez (1983) show that the rate of cloud growth is increased by a factor of 3 to 6 in the arms because of the effect of the potential minimum in increasing the space density of clouds. However, the model still required $\sim 10^8$ yr to form massive clouds. Tomisaka (1984) performed similar calculations, but finds that the model can give a sizable percentage of the total gas mass in the form of GMC's with masses $> 10^5$ $M_\odot$ even if the massive cloud lifetime is $\sim 4 \times 10^7$ yr. The massive cloud lifetime which is required to explain the observed (and uncertain) CO cloud mass distribution is still too large ($\sim 2 \times 10^8$ yr) if the internal densities are as large as 300 cm^{-3}; however, Tomisaka suggests that lifetimes $\lesssim 4 \times 10^7$ yr can be obtained if the internal densities are about a factor of 5 smaller, or if the observed mass distribution refers mostly to the molecular ring at a galactocentric radius of $\sim$ 5 kpc. It thus appears that the Oort model may be consistent with GMC lifetimes of $\sim 3 \times 10^7$ yr when the effects of a spiral density wave are considered. However, none of these calculations include the effects of collisional fragmentation, which may be quite important according to Hausman (1981, 1982).

The second problem is that a collisional model which assumes momentum conservation in collisions predicts a velocity dispersion-mass relation of the form $v(m) \propto m^{-1/2}$, which is not observed, as pointed out by Kwan and Valdez (1983). A recent study of the kinematics of local molecular clouds by Stark (1984) indicates a velocity dispersion of about 9.0 km s^{-1} for low-mass (10^2 to 10^4 $M_\odot$) clouds and 6.6 km s^{-1} for moderate-mass (10^4 to $10^{5.5}$ $M_\odot$) clouds. Kinetic energy is not constant for $m \lesssim 10^5$ $M_\odot$, although there is a tendency for equipartition at larger masses.

A discussion of this problem was given by Hausman (1982), who finds from Monte-Carlo simulations that any model which gives a roughly constant $v(m)$ as observed, gives a cloud mass spectrum which is too steep compared to observations; this result is due to the effects of collisional fragmentation. Casoli and Combes (1982) find, however, that $v(m) \propto m^{-0.1}$ or so, with a satisfactory mass distribution, even though collisional fragmentation was included. They claim that the flat $v(m)$ occurs because there is not enough time to establish the $m^{-1/2}$ relation. The *N*-body calculations of Tomisaka (1984) give a similarly flat $v(m)$, but in this work the larger velocity dispersion of massive clouds is apparently due to their acceleration and deceleration near the potential minimum; however, collisional fragmentation was again not included. It therefore seems possible, though not firmly established, that the observed relatively flat v(m) may not be a devastating blow to the Oort model, but the physical processes involved still need to be clarified.

There are additional, perhaps more fundamental, issues which need to be addressed. First, there is as yet only weak observational evidence that massive clouds disrupt into a number of small clouds which can remain as coherent objects at least until they collide with other clouds. A recent study of CO towards a large number of young clusters by Leisawitz (1985, in preparation) gives support to this picture. It is suggested below that the optical and H I shell structures observed in our Galaxy and the Large Magellanic Cloud may break up into the clouds required by the Oort model.

Secondly, it is not at all clear that collisional growth can yield the type of hierarchical spatial structure which was described in Sec. II. The observed hierarchical structure at mass m would have to be interpreted as a frozen history of all the cloud collisions which occurred during the buildup of a cloud of mass m. The smaller cloud involved in a coalescent collision can probably only retain its identity as a fragment within the larger cloud for a time on the order of L/c_s, where L is the size of the smaller cloud and c_s is the sound speed, as suggested by the two-dimensional cloud collision calculations of Gilden (1984). This time must exceed the growth time of the most massive clouds if we are to observe hierarchical internal structure. The model is furthermore restricted by the fact that the observed structure seems to contain perhaps 3 to 5 fragments per level of hierarchy, so that at least 3 to 5 collisions with smaller clouds must occur before the cloud itself meets a larger cloud, because once it is swallowed there is little chance that it could later consume a still smaller cloud. This should impose a constraint on the cloud mass spectrum. However, even then it is difficult to reconcile this idea with the rather subjective impression that fragments within any scale seem to have similar sizes and regular spacings. (Similar remarks apply to the GMC model of Bash et al. (1981) which requires collisions with small clouds to power the observed supersonic internal motions and support the GMC's.)

A possible answer to these objections is simply that the internal structure resulting from cloud collisions is in fact erased on a short time scale, so that the observed structure represents fragmentation processes which occur subsequent to the buildup of a GMC. In this case the collisional buildup determines the appropriate initial conditions for true fragmentation processes. Also, one might expect very small rotational velocities for the GMC's because the angular momenta of the clouds involved in a coalescent collision should be randomized and because rotational energy may be efficiently converted into internal energy, in analogy with numerical calculations of galaxy mergers.

Another problem concerns the existence of the reddening-H I complexes with masses of $\sim 10^7$ $M_\odot$. It is not possible for collisions to build up such a large structure in a reasonable amount of time. If extinction-molecule complexes are actually fragments within the larger diffuse complexes (discussed earlier), then one must appeal to a separate mechanism, like gravitational instability of the disk, to form the largest structures. This suggests an interesting variant of the Oort model in which the galactic disk is replaced by the interior of a reddening-H I complex as the volume in which the coalescence-disruption cycle operates. The idea is attractive because within the diffuse complex the cloud number density and total mass density are larger than in the galactic disk scenario, so that less time is required to build up successive generations of GMC-sized clouds.

One must also question the assumption that cloud collisions will generally lead to coalescence. While it is true that shock radiation dissipates relative kinetic energy efficiently, there is still a large amount of momentum deposited, which may be sufficient to disrupt the large clump. This point is discussed quantitatively by Hausman (1982), Pumphrey and Scalo (1984), and Gilden (1984). Gilden shows that in order for the collision of two Jeans-stable clouds to resist disruption, the mass ratio must exceed the Mach number of the collision. Because the appropriate Mach number is large ($\gtrsim 10$), a system of interacting Jeans-stable clouds is likely to resist coalescence unless the cloud mass distribution is very flat, so that collisions with large mass ratios can occur with sufficient probability. One can show that for a power law differential mass spectrum of the form $f(m) \propto m^{-\gamma}$, the fraction of collisions in which the mass ratio m_2/m_1 ($\leqq 1$) exceeds a given value is approximately equal to $(m_2/m_1)^{\gamma-1}$, for $\gamma > 1$, and for a lower-mass limit much smaller than the upper-mass limit. If $\gamma \approx 1.7$ as found by many investigations of the Oort model, then only about 20% of Mach 10 collisions should lead to coalescence.

Gilden (1984) also points out that, in a collision of two clouds with different sizes, the larger cloud can entrap the smaller cloud, giving coalescence, only if the column density of the large cloud exceeds that of the small cloud. However, as pointed out earlier, observations of dark clouds indicate that column density decreases somewhat with increasing size, making coalescence unlikely. It is possible that the inelasticity of cloud collisions can still lead to the clustering of clouds into large structures within a spiral gravitational potential, even without coalescence, as shown by the recent *N*-body simulations of W. W. Roberts and M. A. Hausman (1984) and Hausman and Roberts (1984).

More generally, the entire kinetic equation approach to the cloud evolution problem may be questioned because of its overly simplistic parameterization of the outcome of collisions, even when these parametrized treatments appear complex, as in Hausman (1983) and Pumphrey and Scalo (1984). Lattanzio et al. (1985) have emphasized this problem in their detailed three-dimensional simulations of interstellar cloud collisions.

Finally, if interstellar clouds are linked magnetically to the external medi-

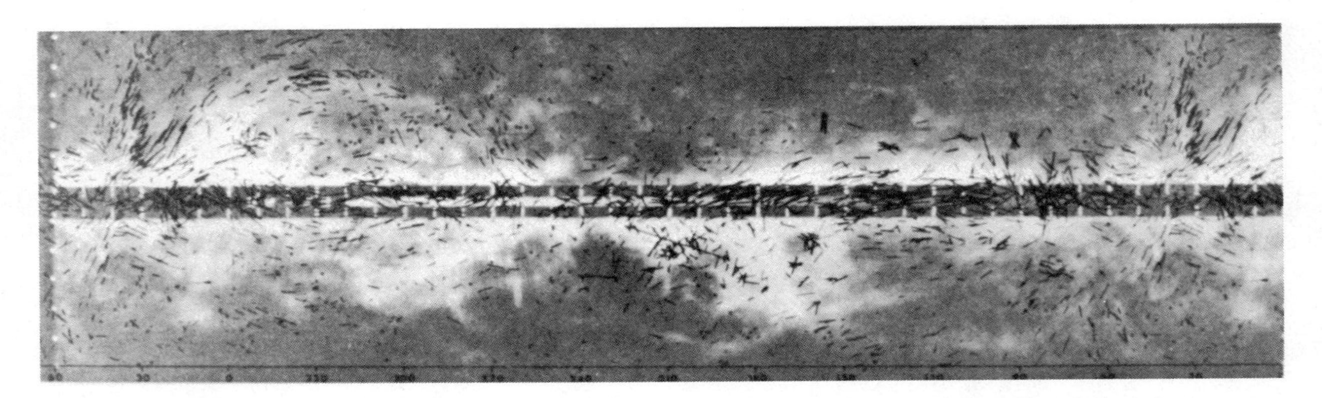

Fig. 27. Large-scale H I emission map in galactic coordinates showing filamentary and shell structure, along with optical polarization vectors. Figure from Colomb et al. (1980).

um, it may be that clouds rarely undergo physical collisions at all (Clifford and Elmegreen 1983).

We have not discussed the possibility that cloud collisions can induce internal fragmentation and star formation. Some such considerations are discussed in Gilden (1984). If fragmentation and star formation are viewed as a phase transition occurring within clouds, then it is interesting to note the similarity to recent models for intense rainfall induced by the collisional coalescence of cumulus clouds (see Bennetts et al. 1982).

D. Expanding Shells

The morphology of interstellar structures provides valuable clues to the physical processes responsible for their formation and internal substructure. There is substantial evidence that at least some part of the large-scale structure is in the form of expanding shells. For example, the local extinction-molecule complexes such as Taurus and Orion are part of a large ($\sim$ kpc) expanding ring of material, probably associated with Gould's belt (Lindblad 1967), suggesting that shell fragmentation is an important process at large scales. Cowie et al. (1981) found shells with sizes of $\sim$ 100 pc around 2 of 13 OB associations using International Ultraviolet Explorer (IUE) data. Shell structures can be clearly seen in intermediate-latitude H I emission surveys of our Galaxy (an example is shown in Fig. 27). The fact that some shells are so large ($\sim 40°$) that they could be seen at large distances led Heiles (1979; see also Hu 1981) to examine the Weaver and Williams (1973) low-latitude H I survey for the presence of shells. Heiles tabulates estimated parameters for some 63 H I shells with a large range in sizes ($\sim$ 10 pc to 1.2 kpc) and masses ($\sim 10^3$ to 10^8 $M_{\odot}$). Heiles (1984) has recently given a new discussion of the H I shells and reemphasizes that the subjective nature of the shell identifications makes these lists of shell structures unsuitable for statistical purposes. Heiles also points out that there appears to be no unique relationship between shells and any particular astronomical object. While some shells can be associated with supernovae or young clusters, not all supernovae and clusters have shells, and most shells are not associated with any known astronomical object.

Similar shells, mostly around OB associations, are seen optically in the Large Magellanic Cloud (LMC) (see, e.g., Meaburn 1980; Braunsfurth and Feitzinger 1983, and references therein) and M31 (Brinks 1981; Brinks and Bajaja 1983). Two photographs of the LMC structure in Hα + [NII] are presented in Figs. 28 and 29. Georgelin et al. (1983) have investigated the variations of the [SII]6716/Hα line ratio in and among 43 LMC shell structures with sizes $\lesssim$ 200 pc. They concluded that the large shells appear intermediate between classical H II regions and supernova remnants, both in their morphology and their internal motions and line ratios. The large internal velocities found in many of these shells may be taken as some support for cyclic interstellar medium models like the Oort model.

Interpretations of these shells include the combined effects of H II regions, stellar winds, and supernova explosions in a star cluster (see, e.g.,

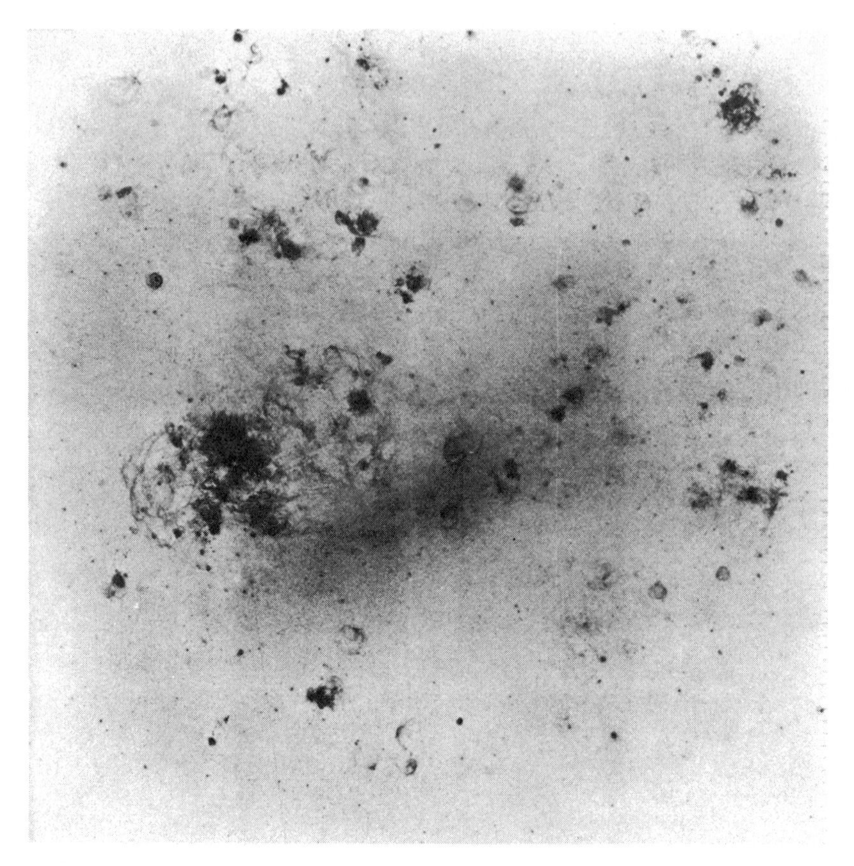

Fig. 28. Hα+[N II] photograph of the LMC showing shell structures (figure from Davies et al. 1976). The largest structures are the 30 Doradus supergiant shells with a diameter of 1.3 kpc; the core of 30 Dor has a size of about 300 pc.

Weaver 1979; Bruhweiler et al. 1980; Elmegreen 1981*b*), large high-velocity clouds falling through the galactic disk (Tenorio-Tagle 1980,1981), and runaway expansion of shells driven by radiation pressure from field stars (B. G. Elmegreen and W.-H. Chiang 1982). A good review of the attributes and failings of these models is given by Heiles (1984). It may also be possible to explain some of these shells as resulting from Parker instability, as in the case of Barnard's Loop in Orion (Appenteller 1974). Because many (but not all) of the observed galactic shells seem to be expanding, we can probably reject the idea that they are the result of a coalescence event, the currently popular explanation for shells around some elliptical galaxies (see Schweizer 1983; Quinn 1983).

An appealing theory for kpc-sized shells, which may be related to the reddening - H I superclouds discussed in Sec. II, is the radiation-pressure mechanism examined by Elmegreen and Chiang (1982). The basic idea is that

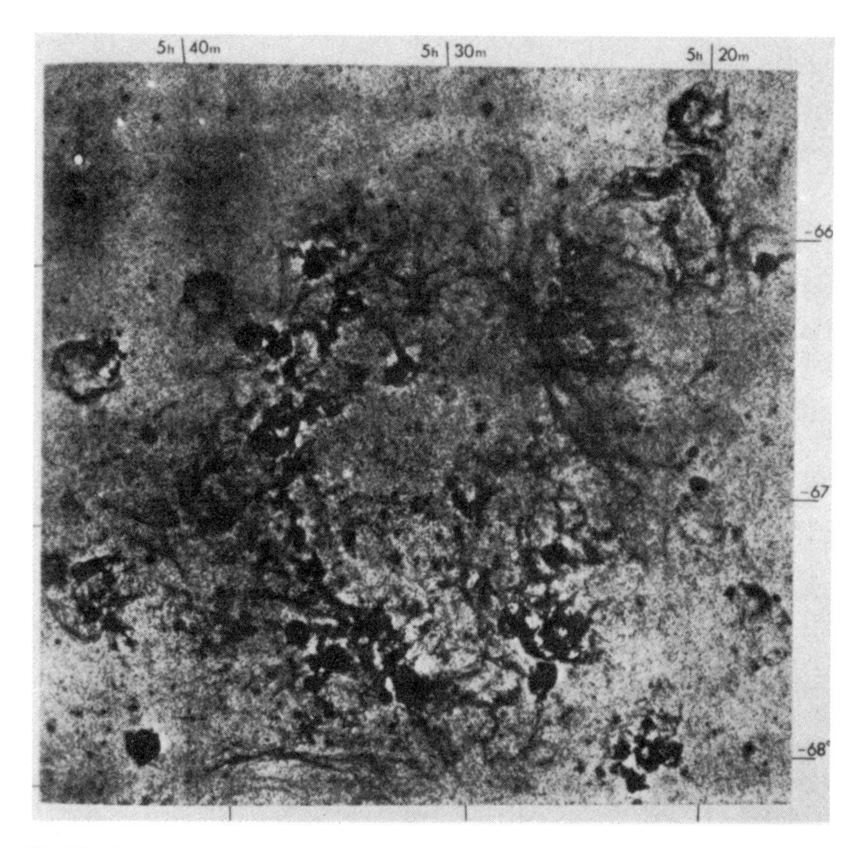

Fig. 29. A portion of the plate shown in Fig. 28 (near the top of Fig. 28 north of the 30 Dor complex), enhanced by unsharp masking and high contrast printing, revealing new supergiant shells comparable in size to the 30 Dor shells (figure from Meaburn 1980).

if an initial event, such as stellar winds or explosions, clears a small cavity, the lack of absorption in the cavity makes the net radiation flux from *field* stars anisotropic. External pressure attempts to keep the shell from expanding, but the net radiation force of the field stars and the cluster radiation force can overcome this confinement for a reasonable value of the relevant parameters. The radiation force from the cluster decreases as R^{-2} as the shell expands, but the field star contribution depends on $n_s/R^2 \propto R$, where n_s is the number density of field stars, and so increases as the shell expands, leading to runaway expansion. Elmegreen and Chiang show that a shell can grow from 10^2 pc to 10^3 pc in about 10^8 yr, and suggest several possible mechanisms for the eventual breakup of the shell. First, galactic shear may distort the shell into spiral-like pieces with sizes of $\sim$ 1 kpc. Second, when the shell size exceeds the Jeans length in the galactic disk ($\sim$ 1 kpc), the shell may collapse and fragment. Finally, when the optical depth through the shell exceeds unity, the radiation force is only exerted on the inner surface, leading to breakup by the Rayleigh-Taylor instability. These large-shell breakup mechanisms deserve a detailed investigation.

A problem with the B. G. Elmegreen and W.-H. Chiang model (Heiles 1984) is that the runaway expansion can only occur for a preexisting shell having a radius and expansion velocity greater than some critical value, which may be so large as to make it difficult to understand energetically how such a preexisting shell could form. A possible way out of this difficulty, as pointed out by Heiles, is to assume that the ambient material swept up by the shell is in the form of clouds, so that the effective external pressure would be very small. Elmegreen and Chiang show that in such a case runaway expansion should occur for any shell within radius $>$ 100 to 300 pc, which may not be unreasonable.

Another attractive idea for explaining the extremely large energies inferred for some supershells is collisions of high-velocity clouds with the galactic disk, as modeled by Tenorio-Tagle (1980,1981). Heiles (1984) gives three arguments, based on his observational study of H I shells, which support this model:

1. The large energy requirements for supershells preclude a supernova/OB cluster wind source, while the infalling cloud model can account for the inferred energies;
2. The association of some supershells with high-velocity gas, and especially;
3. The fact that, for the most energetic expanding supershells, and most of all the shells, only one hemisphere is visible, a direct prediction of the theory.

If such expanding supershells can lead to star formation, perhaps as in the local Lindbland ring, and if the infalling clouds are identified with the Magellanic Stream, then the above remarks suggest one way in which tidal encounters between galaxies can induce star formation, a phenomenon which is well established observationally.

Combining these ideas with theories for sequential star formation by ionization-shock fronts discussed earlier (Sec. IV.B), one can imagine a whole hierarchy of fragmented shell structures covering the range of 10 to 1000 pc. A similar bootstrapping of shell structures has been proposed as a model for galaxy formation (Ostriker and Cowie 1981). For the interstellar medium it is clear that shells on the requisite scales do exist, along with plausible, though rather qualitative, theories for their formation. However, it remains to be established whether these shells are the dominant structures initiating fragmentation and star formation on large scales, in view of the number of other mechanisms which have been proposed. It is also important to study the internal density and velocity structure of the smaller shells, not only because they show how gas is ejected from star-forming regions, but also because they may provide constraints on the masses and injection velocities of the small clouds postulated in the Oort model.

VI. CONCLUDING REMARKS

This review has emphasized two major themes: the fact that the observed structure of the interstellar medium is hierarchical, and the diversity of physical processes which may be responsible for this structure. On the observational side, there is a need for quantification of the structural characteristics over a broad range of spatial scales which will only be achieved by a coordinated deployment of the data available from observational techniques sensitive to different ranges of column density. The most effective future programs for this purpose are likely to be large-scale extinction and reddening studies. Wide-field CCD images can yield accurate star counts down to at least 26th magnitude, and can probe regions with visual extinctions between about 1 and 10 mag with an angular resolution approaching 1 arcsec. While a large fraction of the sky cannot be studied in this fashion, the nearby complexes could in principal be structurally analyzed in column density over a range of $\sim 10^4$ to 10^5 in spatial scales. Reddening statistics can be used at smaller column densities although the angular resolution will be much poorer because of the brighter limiting magnitude imposed by the necessity of spectroscopic observations.

Neither extinction nor reddening give any information on the velocity structure; this must come from molecular lines (^{13}CO or ^{18}CO) at large column densities and H I or optical interstellar lines at smaller column densities. The important point is that, if we are to quantify interstellar velocity structure, the observations must sample a very large number ($\gtrsim 10^4$) of positions; because good signal-to-noise and velocity resolution are essential in such a study, this will be a laborious feat with current-generation equipment. However, determination of velocity structure over a range of scales is necessary to distinguish between various fragmentation mechanisms, such as gravitational contraction controlled by either rotational or magnetic effects.

Theoretically, the major problems again involve the ability to examine the coupling between processes which occur over a range of spatial scales. In particular, current numerical hydrodynamic techniques are incapable of following more than one level of fragmentation. Furthermore, the results of the available hydrodynamic studies reviewed above are highly dependent on the initial conditions and other assumptions. Not only are the calculations sensitive to the initial partitioning of gravitational, thermal, rotational, and magnetic energies, but to the presence of initial fluctuations in density and velocity, which are likely to be significant, and the possible importance of turbulent viscosity and other poorly understood phenomena. In addition, the dynamical range of current numerical calculations is too small to reveal what may prove to be the most important feature of the problem: the tendency of nonlinear systems to exhibit chaotic, or turbulent, behavior. The associated sensitivity of the evolution to initial conditions, which is already suggested by existing numerical experiments, remains to be explored.

Finally, a good deal of theoretical and observational work strongly suggests that the formation of stars affects the subsequent evolution of the cloud structure, whether by wind-driven turbulence, fragmentation of expanding shells, or implosion or disruption of preexisting fragments. It will not be possible to include these effects rigorously in hydrodynamic calculations in the forseeable future, but it certainly seems worthwhile to consider an approximate modeling of the stellar effects within the framework of a hydrodynamic calculation, in much the same way that small-scale processes are included in subgrid modeling techniques common in numerical studies of incompressible turbulence studies which suffer from the same deficiency in range. The simulations of Bania and Lyon (1980) may be considered as prototypes of this approach.

These comments should not be taken as some sort of vilification of previous theoretical and observational work, which has pushed fruitfully against existing computational and instrumental limitations, but rather to underline the exigency for coordinated activity between observers and theoreticians, in preparation for the rapid expansion in computational and instrumental capabilities which should occur in the next few years.

In summary, the study of fragmentation must be regarded as an attempt to understand relationships between the scales of stars and galaxies. Indeed, one of the only clear impressions which emerges from the mass of existing theoretical and observational work is that the entire range of scales is coupled in a nonlinear manner. It is in this fundamental sense that fragmentation is a form of turbulence, and, just as in all such studies of turbulent nonlinear systems, the ultimate goal is an understanding of the wonderful order which appears amidst a welter of forms.

THE FORMATION OF BOUND STELLAR CLUSTERS

BRUCE A. WILKING
University of Texas at Austin

and

CHARLES J. LADA
University of Arizona

Recent theoretical and observational studies of the formation of bound and unbound stellar systems are reviewed. Current observational and theoretical evidence appears to favor the hypothesis that both gravitationally unbound associations and bound clusters form initially in bound molecular clouds. Whether or not a stellar system will remain bound after formation depends primarily on the efficiency of conversion of gas to stars in a cloud, the manner in which gas not used up in star formation is dispersed from the system and, to a lesser extent, on the initial mass density of the protostellar cloud. Associations are a result of the global star formation process in a giant molecular cloud (GMC). Observations indicate that the overall efficiency of star formation in GMCs is small; only about 0.2 to 5% of the total gaseous mass is ever converted to stars in such clouds. Consequently, the inevitable destruction of a GMC by H II regions, stellar winds, and supernovae which accompany the formation of new stars, results in the dispersal of the majority of the initial binding mass of the star-forming complex. Therefore, an unbound system of stars is left behind, expanding into the field with a velocity on the order of a few km s^{-1}, characteristic of the velocity dispersion in the original molecular cloud. Open clusters appear to form as a result of enhanced star formation efficiency in a small, localized region or core of high gas density within a GMC. Theoretical considerations suggest that star formation efficiencies of at least 30% are required to produce clusters similar to the Pleiades from typical molecular cloud cores. It is not yet clear from observations whether the galactic star formation process can *produce efficiencies as high as 50%, but if such efficiencies cannot be attained then gas removal from protocluster systems must occur over periods of at least a few million years. Observational techniques used in the identification of young dust-embedded clusters and the results of recent studies of young star-forming regions are discussed in terms of their likelihood of forming bound systems. Finally, the utility of infrared observations to determine the luminosity functions of dust-embedded stellar populations is discussed in the context of recent observations of the* ρ *Ophiuchi dark cloud.*

The observed presence of long-lived open clusters in the disk of our Galaxy is clear evidence that stars form in gravitationally bound systems. However, we understand very little about the processes by which these bound clusters are formed. Only about 10% of the field stars in the solar neighborhood can be accounted for from the dissipation of bound stellar systems and the remaining 90% are believed to have formed in unbound associations (M. S. Roberts 1957; G. E. Miller and J. M. Scalo 1978). Therefore, the star formation process in our Galaxy would appear to favor the formation of unbound associations. What special conditions in molecular clouds lead to the formation of bound clusters as opposed to unbound associations? What observational procedures are best suited for locating and studying these dense stellar aggregates?

The identification of young dust-embedded clusters is critical for our understanding of low-mass star formation. These high-density stellar groups provide a unique laboratory in which to investigate the formation and early evolution of solar-mass type stars. Another major motivation for studying young embedded clusters is the unique opportunity to explore how the mass function of a cluster evolves from its earliest stages of formation. Despite the basic similarity of the initial mass functions of open clusters for $M > 3$ $M_{\odot}$ (see, e.g., Scalo 1985*a*), there is good evidence that not only do cluster-forming clouds produce stars over a significant fraction of the lifetime of clusters (see, e.g., Herbig 1962*a*; Landolt 1979; Stauffer 1980; W. Herbst and G. E. Miller 1982) but this star formation process may proceed sequentially in mass (e.g., NGC 2264 [M. T. Adams et al. 1983]). Hence, investigations into how the cluster initial mass function (IMF) evolves will give us insight into the actual process by which protostellar fragments are created and how they produce an IMF whose form is apparently common to all clusters.

In this chapter, we will review the properties of visible associations and open clusters and discuss what these properties infer about the processes by which these stellar aggregates are formed. To assist in the identification of young embedded clusters, we will discuss the required efficiency of conversion of gas into stars and the molecular cloud environment necessary to produce bound clusters. Quantitative estimates for the final star formation efficiency in bound systems will be presented from both analytic and numerical studies. Observational techniques which are used to identify regions forming groups of young stellar objects (YSOs) will be reviewed. Existing observations of young stellar groups will be presented and compared to the proposed scenarios for association and cluster formation. A preliminary study of the luminosity function of the young dust-embedded cluster in the ρ Ophiuchi dark cloud will be presented.

I. FORMATION OF CLUSTERS AND ASSOCIATIONS

A. Review of Optical Properties

Associations are low-density stellar aggregates consisting of stars of the same physical type which are enhanced relative to the field population; see Blaauw (1964) for detailed review. Associations can be collections of massive stars (OB), reflection nebulae (R) or emission-line stars (T). Stellar densities in associations are usually < 0.1 $M_{\odot}$ pc^{-3} making them unstable to disruption by galactic tides ($\rho_{crit} > 0.1$ $M_{\odot}$ pc^{-3} for stability [Bok 1934; Mineur 1939]) and passing interstellar clouds ($\rho_{crit} > 1$ $M_{\odot}$ pc^{-3} for stability over average cluster lifetime of 200×10^6 yr [Spitzer 1958]). Therefore, it is not surprising to find that such low-density collections of stars are young ($< 10 \times 10^6$ yr old).

Open clusters are high-density stellar aggregates composed of stars over a wide range of mass. This mass distribution varies little from cluster to cluster and closely resembles the field star IMF (see Scalo 1985*a*). Stellar densities in open clusters are typically > 1 $M_{\odot}$ pc^{-3} allowing them to be relatively stable against external disruptive forces. Open clusters are found to be long-lived; 50% of observed open clusters reach an age of 200×10^6 yr (Wielen 1971).

From these observed properties of associations and open clusters, it appears that associations are gravitationally unbound. Supporting their unbound nature is the observation of the low efficiency with which associations form from molecular clouds. A comparison of the stellar mass of OB associations with the gaseous mass of the giant molecular clouds (GMCs) from which they formed suggests that only 0.2% to 5.0% of the total mass of a GMC is ever converted into stars (Blitz 1978; Duerr et al. 1982). Thus, as first suggested by von Hoerner (1968), the dispersal of an association-forming cloud will remove the majority of the initial binding mass of the system and result in an unbound, expanding association. Conversely, open clusters owe their longevity to their formation as gravitationally bound systems which are stable against disruption from external forces. Apparently, the conversion of gas into stars during cluster formation is sufficiently great such that the new stellar system retains a substantial fraction of the initial binding mass of the cloud.

B. Models for Association Formation

Observations of unbound associations first led astronomers to the assumption that these associations, unlike open clusters, formed in unbound molecular clouds. This initial expansion of molecular clouds from which associations were formed was proposed to have been driven by a variety of mechanisms. Öpik (1953) suggested that expanding shells from a supernova explosion could sweep up and compress molecular material from which new stars are formed. These newly formed stars would retain the outward motion of the supernova shell and form an unbound association which was expanding away from the original site of the supernova explosion.

Oort (1954) proposed a similar scenario for association formation based upon the expansion of the H II region into neutral cloud material. The observed concentration of young stellar groups within regions of ionized gas led Ambartsumian (1955) to propose an alternative model in which associations arise from the breakup and expansion of a single, massive protostar which subsequently forms a trapezium-like system.

Observations of the velocity dispersions in molecular clouds and young associations do not support the concept of associations forming in unbound molecular clouds. First, the velocity dispersions observed in molecular clouds are consistent with the clouds being gravitationally bound and approximately in virial equilibrium (R. B. Larson 1981). This observation holds for clouds over a wide range of mass ($1-10^5$ $M_\odot$) and includes clouds forming massive OB stars. Second, young stars in associations are observed to acquire the same velocity dispersion as the molecular gas from which they form. Hence, the new system of stars appears to form in equilibrium with the bound molecular cloud. For example, the velocity dispersion of emission-line stars in the Taurus-Auriga T association is measured to be 1 to 2 km s^{-1} and resembles the velocity dispersion of the Taurus molecular cloud (Dieter 1976; B. F. Jones and G. H. Herbig 1979). The velocity dispersion for members of the λ Ori OB T association is also about 2 km s^{-1} (Mathieu and Latham, preprint). Even though this association is no longer confined by the gas from which it formed, it displays motions characteristic of the molecular gas in the nearby B 30 and B 35 clouds which border the association (Lada and Black 1976; Deurr et al. 1982).

In summary, recent study of the velocity dispersions in molecular clouds and of their associated stars have shown that stars, both in associations and in open clusters, must form in bound molecular clouds. Upon their formation, these stars acquire the velocity dispersion of their associated molecular gas. If the efficiency of conversion of gas to stars is low, then the removal of the gas from the system will also remove the majority of the initial binding mass. As a result, the velocity dispersion of the association will be greater than the escape velocity from the naked system of stars, and the association will diffuse into the general background population in $< 10^7$ yr. Conversely, if the star formation process in clouds results in a high efficiency of conversion of gas into stars, then the dispersal of the cloud will remove only a small fraction of the initial binding mass of the system. In this case, the velocity dispersion of this bare stellar group will be much less than the escape velocity from the system and allow the group to remain gravitationally bound until evaporation and external forces disperse the cluster after several hundred million years.

C. Star Formation Efficiency

In order to quantify the efficiency with which gas must be converted into stars to produce gravitationally bound systems, the parameter η, called the star formation efficiency, has been introduced. This parameter is defined as

$$\eta = M_{\mathrm{stars}}/(M_{\mathrm{stars}} + M_{\mathrm{gas}}) \tag{1}$$

and is a time-dependent quantity related to the total mass of stars present (M_{stars}) as compared to the initial cloud mass ($M_{\mathrm{stars}} + M_{\mathrm{gas}}$). This quantity η will slowly increase over the star-forming lifetime of a molecular cloud.

Predictions for the final value of η in a bound cluster depend critically on how fast the gas is removed from the system of stars and gas. The dependence on η on the gas release time has been considered by analytic studies for two limiting cases; the gas release time is rapid (much shorter) or slow (much longer) compared to the crossing time for a system in virial equilibrium. Numerical studies of the star formation efficiency in bound clusters have the advantage of being able to consider the entire range of possible gas release times and attach realistic values to the time scales for the slow (adiabactic) removal of gas.

1. Analytic Studies. Various methods have been used to investigate the final star formation efficiency necessary to produce a bound cluster in the limiting cases of rapid and slow gas release. In the case of rapid gas release, the gas is removed in much less than a crossing time such that the gas-free cluster initially has the same radius R_i and velocity dispersion as the system of gas and stars. However, the sudden drop in cluster binding mass results in a large number of stars having great enough velocities to escape the system. The remaining cluster expands and revirializes reaching a final radius R_f given by

$$\frac{R_f}{R_i} = \frac{\eta}{2\eta - 1} \quad (\eta > 0.5). \tag{2}$$

(see, e.g., Hills 1980). In the case of slow gas release, the gas mass loss from the system occurs on a time scale much greater than the crossing time and allows the system to readjust and remain in virial equilibrium. The final cluster radius is given by the adiabactic invariant

$$\frac{R_f}{R_i} = \frac{1}{\eta} \tag{3}$$

(see, e.g., Hills 1980). The initial and final cluster radii can be expressed in terms of either velocity dispersions (using the virial theorem) or densities (assuming a simple geometry). By comparing the observed properties of visible (gas-free) open clusters (cluster radius, stellar density, velocity dispersion of stars) to conditions present in cluster-forming molecular clouds (cloud radius, gas density, velocity dispersion of gas), we can infer the final values of η which were necessary to form a bound stellar system.

B. G. Elmegreen (1983*c*) has compared the observed stellar densities in open clusters to typical gas spatial densities in star-forming cloud cores [$n(H_2) = 10^4$ cm^{-3}]. From this comparison he derives values for the star formation efficiency of 27% or 59% depending upon whether the cloud disruption is slow or rapid. Mathieu (1983) used three independent techniques to derive values for η in open clusters. He considered the changes in the velocity dispersion, mass density, and radius of clusters as the gas from these systems is released. All three techniques gave similar results; the star formation efficiency in open clusters must reach 30% or 55% for the slow or rapid dispersal of gas, respectively.

While these analytic methods have the advantage of requiring knowledge of only the initial and final states of young clusters, they cannot reveal the dependence of the star formation on the gas release time apart from two limiting cases. In addition, these analytic studies cannot properly account for the relaxation and evaporation of the cluster as the cloud is disrupted. Realistic time scales for the adiabactic removal of gas are not available from analytic methods. Many of the shortcomings of analytic studies of the star formation efficiency are remediated by numerical calculations of the evolution of young clusters.

2. *Numerical Studies.* Numerical simulations of the dynamical evolution of clusters as the residual star-forming gas is removed from the system have been performed by Lada et al. (1984). Their models were performed for protocluster systems of 50 or 100 stars for various values of the star formation efficiency (just prior to gas dispersal) and the gas release time. By varying these parameters, it is possible to explore what combinations of the star formation efficiency and gas release time produce bound clusters.

The protocluster systems were modeled with a *N*-body code in which stars were simulated as point masses and the gas was represented by an additional term in gravitational potential function determining stellar motion. The stellar mass distribution was chosen to mimic the field star IMF (which also closely resembles the composite cluster IMF) which can be approximated as a power law in the mass with a spectral index of 2.5 (G. E. Miller and J. A. Scalo 1979). The gas was assumed to be initially in virial equilibrium and in a state where the kinetic energy of turbulent clumps of gas support the molecular cloud against gravitational collapse.

An example of one such model is shown in Fig. 1 for a cluster of 50 stars with a star formation efficiency of 40% and a gas release time of 3 crossing times (about 3×10^6 yr). As characteristic of all models, the cluster expands in response to the release of gas and a significant number of stars escape from the protocluster (as indicated by circles). Following the dynamical evolution of this cluster to 40 initial crossing times results in a bound cluster comprised of about 80% of the original protocluster.

The results of these *N*-body calculations are summarized in Fig. 2 in a graph of the gas release time (in units of the crossing time) versus the star formation efficiency for an initially virialized protocluster. The stability criterion for the formation of a bound cluster requires the final stellar density of the cluster to exceed 0.1 $M_\odot$ pc^{-3}, i.e., that density which is stable against shear from galactic tides. For a gas release time which is rapid ($\tau_R < 10^6$ yr), a star formation efficiency of at least 50% is necessary to form a bound cluster. For a gas release time in excess of 4 to 5 crossing times (corresponding to 4 to 5×10^6 yr), the models are reduced to the analytic models for adiabactic gas release and efficiencies of only 20 to 30% are required to result in bound stellar systems. Thus, as shown in Fig. 2, these model calculations are in agreement with the aforementioned analytic models in the limits of the slow and rapid removal of gas and permit the accurate interpolation of star formation efficiencies for bound systems for intermediate gas release times.

D. Molecular Cloud Environment for Bound Clusters

The molecular material which is found associated with OB associations is usually in the form of giant molecular clouds. These GMCs have masses of $\sim 10^5$ $M_\odot$, average spatial densities of 300 cm^{-3} and often extend for over 100 pc (see, e.g., Blitz 1978). Similarly, T associations are found to lie in large extended molecular cloud complexes with no well-defined cloud center such as the Taurus-Auriga dark cloud complex (see, e.g., B. F. Jones and G. H. Herbig 1979).

From simple considerations of the observed properties of open clusters and the final values for the star formation efficiency necessary to produce bound clusters, we can determine the initial properties of the molecular clouds or cloud cores which bear these clusters. The radius of an open cluster increases slowly over time as the cluster emerges from its cloud and relaxes (see, e.g., Spitzer and Härm 1958). Typical radii of 2 to 3 pc for relaxed clusters imply that cluster-forming cloud cores should be fairly compact ($r \lesssim$ 2 pc) and much smaller than association-forming complexes. The total stellar mass of an open cluster is usually about 500 $M_\odot$ when the mass in stars which have escaped from the cluster or remain undetected are accounted for. For example, the well-studied Pleiades cluster has about 460 $M_\odot$ in stars (B. F. Jones 1970) and the Hyades cluster about 300 $M_\odot$ (Pels et al. 1975). These cluster masses imply an initial cloud mass on the order of 1000 $M_\odot$ if the cluster formed with $\eta \sim 50\%$.

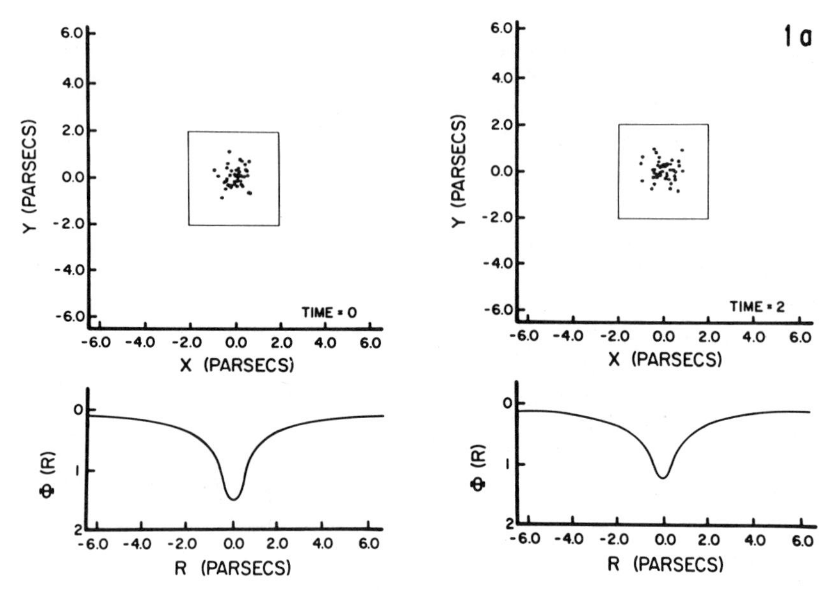

Fig. 1 a–f. The dynamical evolution of a young cluster as gas is released from the system. From one of the models computed by Lada et al. (1984), snapshots are presented of the x–y projection of stellar positions for a cluster of 50 stars which has a star formation efficiency of 40% and a gas release time of three crossing times ($\sim 3 \times 10^6$ yr). Plots a and b depict cluster evolution during gas removal. The time-varying gravitational potential for the molecular gas is indicated. Plots c–f depict cluster evolution after gas removal is complete. The box near the center of the cluster is shown to illustrate the scale. Unbound stars are enclosed by circles.

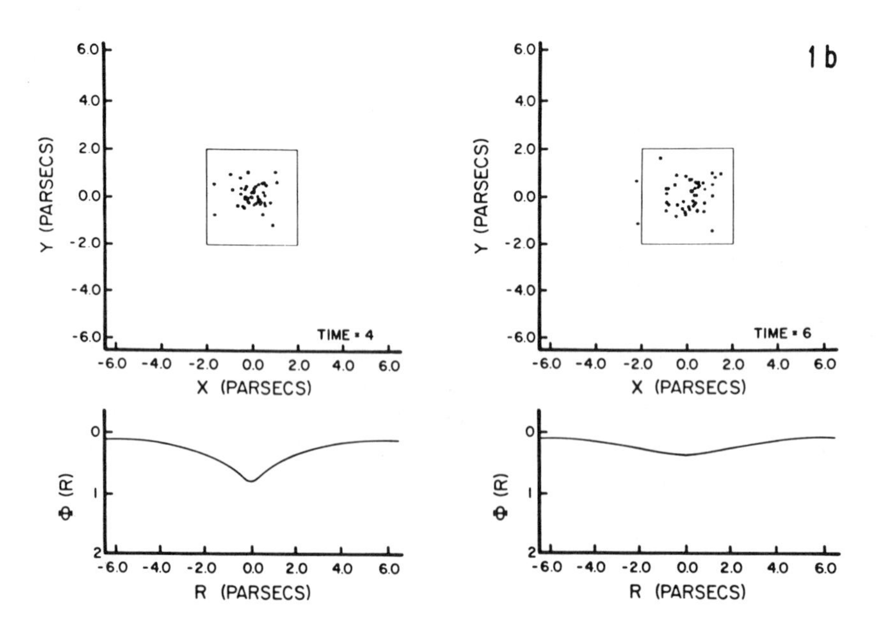

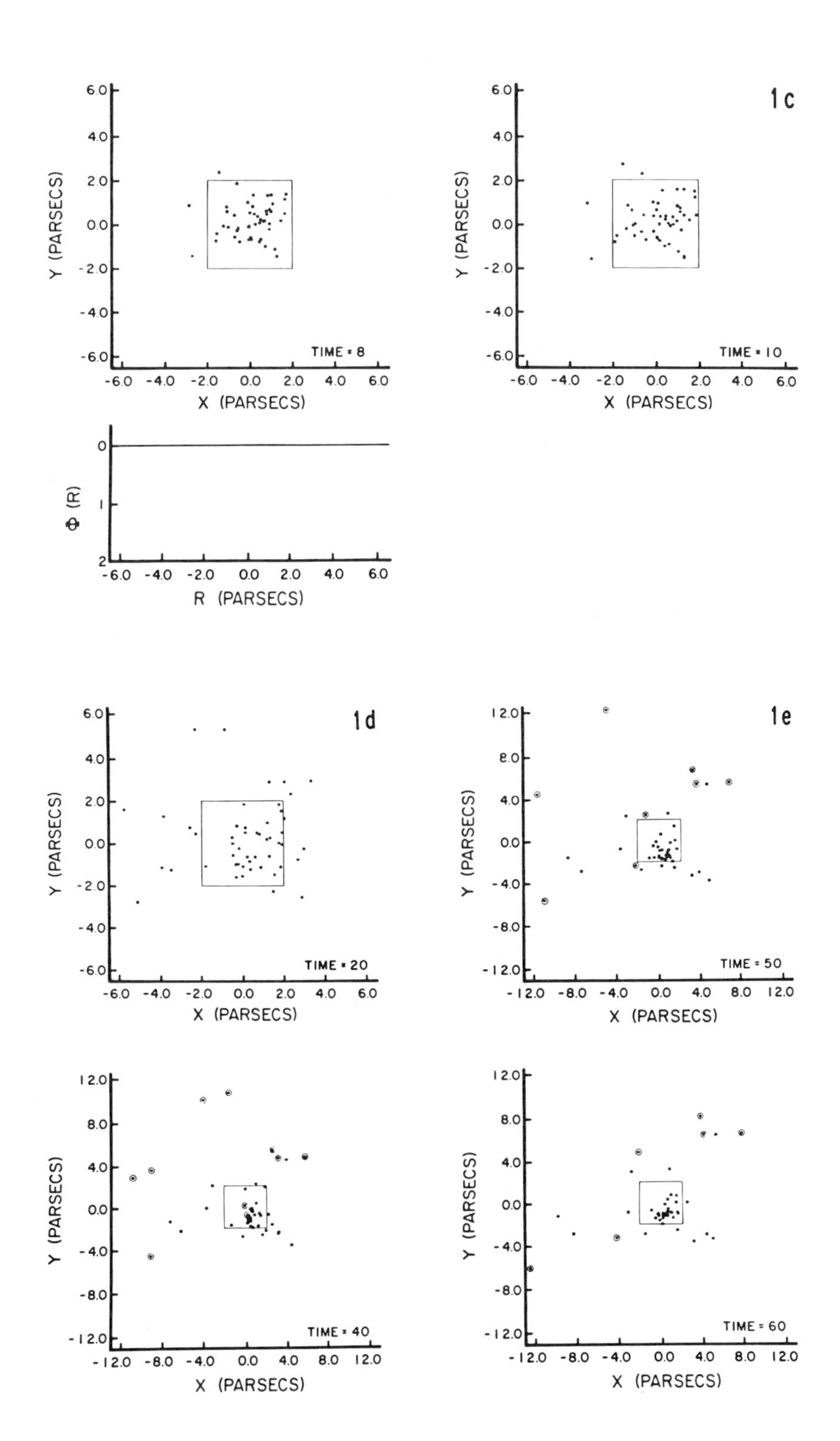

1c
TIME = 8
TIME = 10
Y (PARSECS)
X (PARSECS)
Φ (R)
R (PARSECS)
1d
1e
TIME = 20
TIME = 50
TIME = 40
TIME = 60

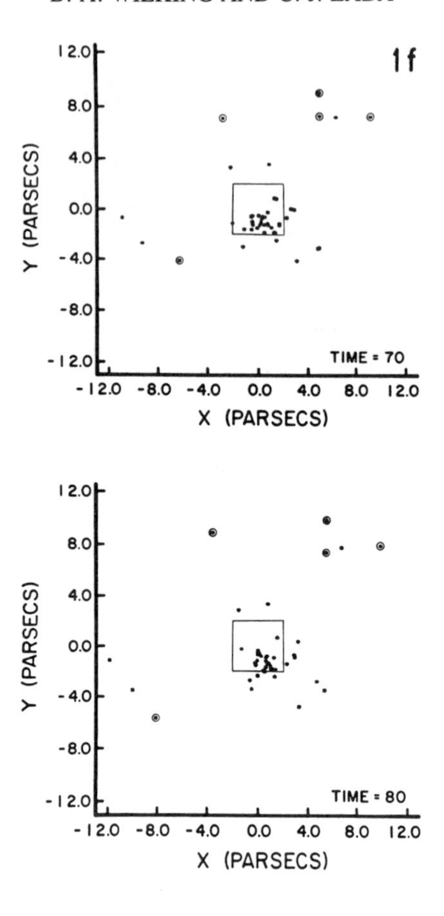

More precise limits for the initial conditions present in clouds forming bound clusters are available from theoretical studies by B. G. Elmegreen (1983*b,c*) and Lada et al. (1984). Elmegreen has modeled the star formation process in clouds as one where the stars appear at random time intervals with a frequency weighted by their relative distribution given by the IMF (see, e.g., Scalo 1985*a*). Hence, low-mass stars have the highest probability of formation and are the predominant stellar component in a low-mass molecular cloud. Considering only radiation pressure from newly formed stars as the disruptive force in these clouds, Elmegreen's models predict the maximum mass for a molecular cloud or cloud core which forms a bound cluster to be a few times 10^3 $M_\odot$. Later models of cluster formation and cloud disruption described by B. G. Elmegreen (1983*b*) show that the majority of these lower-mass clouds may in fact form unbound stellar systems. The *N*-body code of Lada et al. (1984) can be used to place constraints on the size and density of cloud cores

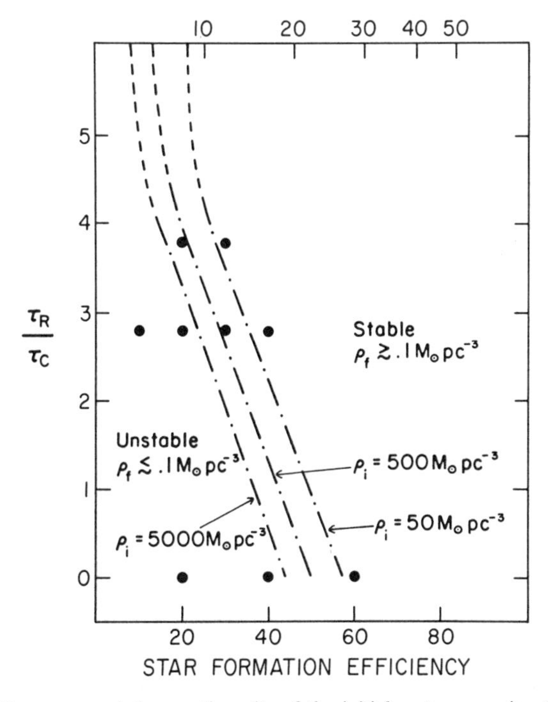

Fig. 2. Plot of gas removal time τ_r (in units of the initial system crossing time) versus star formation efficiency. This plot represents the full range of models run by Lada et al. (1984) (shown as black dots) which bridges the gap between analytic studies which can only consider either rapid or slow gas release. This plot is separated into two regions by the dot-dashed lines. To the right of the lines, final clusters have densities > 0.1 M$_\odot$ pc^{-3} and are consequently stable to the shear resulting from galactic tides. To the left of the lines, the clusters are unstable to this shear. Since the final cluster densities are dependent upon how the models are scaled, the lines dividing the two regions are presented for a range of initial system densities which span two orders of magnitude. In the limit of very long gas removal, the lines asymptote to the analytic results of Mathieu (1983). For $\tau_r < 4\tau_c$, the lines are placed on the plot by approximate interpolation between models. As shown, the models are relatively insensitive to the initial cluster density.

which form bound clusters. By scaling the final mass densities produced by their models to those typical of open clusters such as the Pleiades (~2 M$_\odot$ pc^{-3} [B. F. Jones 1970]), they find that the initial mass densities can range from $n(H_2) = 10^3$ to 10^6 cm^{-3} for initial half-mass core radii of 0.7 to 0.1 pc.

The molecular cloud environment most conducive to the formation of a bound cluster is one in which massive stars are absent. From his model for cluster formation, B. G. Elmegreen (1983*c*) views this absence as a result of long delay times for the formation of massive stars in low-mass clouds. The absence of massive stars insures that the continuous production of low-mass stars can proceed uninterrupted and the release of gas from the system is slow. As a result, the efficiency of conversion of gas to stars need not be as high as in the presence of massive stars forming bound clusters (see Sec. I.C). Therefore, cool, quiescent clouds displaying low-velocity dispersions in the molecular gas (1–2 km s^{-1}) are most favorable for the formation of the bound stellar systems. Clearly, at some point in the formation of a cluster, massive stars may form and, through their disruptive influence, actually mark the end of cluster formation and the beginning of the cluster's emergence from the cloud as suggested by Herbig (1962*a*). But due to the rapid dispersal of gas which is likely to occur with H II regions and stellar winds associated with massive stars (see, e.g., Mazurek 1980), such regions should display very high star formation efficiencies ($\eta > 50\%$) if they are to remain bound.

II. OBSERVATIONAL STUDIES OF YOUNG CLUSTERS AND ASSOCIATIONS

In the preceding sections, we have outlined specific characteristics of young star-forming regions which may point to their participation in the formation of a bound cluster such as high star formation efficiencies, slow gas release times, and cool, quiescent, and compact molecular clouds or cloud cores. Additional considerations might include a cloud's association with high concentrations of infrared sources, emission-line stars, or reflection nebulae. In the following section, we discuss the observational procedures available for the identification of young dust-embedded clusters and review the properties of several candidate clouds. The observational studies presented will favor the study of clouds within several hundred parsecs of the Sun which allow the detection of the most deeply embedded or lowest luminosity objects in the cloud. In addition, nearby clouds have the advantage of a minimum of foreground star confusion and of high spatial resolution in the cloud.

A. Observational Procedures

The observational procedures chosen must be designed to include cluster stars over a wide range of luminosity and evolutionary states. Thus, the broadest wavelength coverage possible (i.e., optical to far-infrared) is desired to study these young stars from the unobscured outer-cloud regions as well as from the dense cloud cores. However, it should be emphasized that because of the high stellar densities involved, high spatial resolution must be preserved in a multi-wavelength study of individual sources.

1. Cluster Membership. Distinguishing between background field stars and those stars associated with the molecular cloud is of the greatest importance in establishing the existence of an embedded cluster. This was first seriously considered for embedded infrared sources by Elias (1978*a,b,c*) who compared the observed density of 2 μm point sources to those in nearby unobscured regions and to models for the galactic stellar distribution (see, e.g., Elias 1978*a;* T. J. Jones et al. 1981). In addition, Elias used the fact that the majority of field stars were K and M giants and performed narrowband photometry to search for the 2.3 μm CO absorption feature which would most likely be absent or obscured by dust emission in the young cluster stars.

A variety of methods have been used to detect the presence of circumstellar gas and dust which is expected to be found around young embedded stars. Hydrogen emission lines in the optical and infrared are used as indicators of circumstellar gas (see, e.g., Hyland et al. 1982). Continuum emission from circumstellar gas is often evident through the presence of an ultraviolet excess or veiling of optical absorption lines in the blue region of the spectrum (Walker 1972). Broadband near-infrared colors can be used to indicate the existence of excess emission association with hot circumstellar dust (see, e.g., M. Cohen and L. V. Kuhi 1979). Near-infrared color-color diagrams (i.e., *J-H* vs. *H-K*) have been used effectively to identify sources with excess emission (see Hyland [1981] for a review of this method).

Other methods used to differentiate cluster stars from field stars toward star-forming clouds include: radial velocity or proper motion surveys (see, e.g., Walker 1983), variability (M. T. Adams et al. 1983; Sellgren 1983), the superposition of sources on regions of high column density (Wilking and Lada 1983), and the orientation of polarization position angles with respect to the general field direction (Wilking et al. 1979). More recently, it has become possible to use X-ray emission and far-infrared emission to locate the positions of young low-mass stars; however, these lower-resolution observations are often difficult to compare with the higher-resolution optical and near-infrared observations (see, e.g., Montmerle et al. 1983; Wilking et al. 1984*a*). Clearly, one should use as many of the available methods as possible to obtain an unbiased sample of cluster member stars and to serve as a consistency check between methods.

2. *Star Formation Efficiency.* The star formation efficiency is a difficult quantity to compute because of the large uncertainties involved in deriving stellar masses and the coexistent mass of molecular gas. If the cluster members are visible, then it is possible to measure both the bolometric luminosity and effective temperature of each star and estimate their mass from their position on the HR diagram. If the stars lie above the main sequence, these mass estimates will rely on the validity of the pre-main sequence evolutionary tracks which are adopted (see M. Cohen and L. V. Kuhi 1979). If the cluster members are still deeply embedded, then only their bolometric luminosities are accessable and only upper limits to the star's mass can be derived because, for stars with $M > 1\ M_{\odot}$, they approach the main sequence from a constant or higher-luminosity regime (Iben 1965; Stahler et al. 1980*a*).

The mass of star-forming gas is most directly computed from observations of optically thin molecular emission lines (usually ^{13}CO or C^{18}O) and the gas excitation temperature (as derived from ^{12}CO emission lines). From these we can calculate a column density assuming local thermodynamic equilibrium (LTE) and convert this to a hydrogen column density via a relation, for example, between ^{13}CO column density and visual extinction and a gas-to-dust ratio (Dickman 1978). However, summing the hydrogen column densities over the areal extent of the cloud yields an estimate for the mass which inherently carries a factor of two uncertainty. In addition, the relations necessary to convert ^{13}CO column density to hydrogen column density are derived for cool clouds with low visual extinction ($A_v < 5$ mag) regions. Similar relations in regions of higher extinction ($A_v \sim 12$–20 mag) have been determined for the Taurus and ρ Oph clouds (Frerking et al. 1982). One advantage of computing the cloud mass in this manner is that it does not require the knowledge of the line-of-sight extent of the cloud.

Since the cluster population is usually not completely sampled throughout the entire depth of the cloud, some correction must be applied to the cluster mass to account for this unseen population. Alternatively, if one can estimate to what depth into the cloud cluster members have been well sampled in terms of the maximum visual extinction, then we can simply compute the mass of gas and the star formation efficiency in this star-forming layer. In this case, the mass of gas is given directly through the gas-to-dust ratio and without the need to convert ^{13}CO column densities to visual extinction (see, e.g., Wilking and Lada 1983).

B. Observational Results

In the following section, we review recent observations of young star-forming regions in which there is a high density of young stars. The cloud morphologies, stellar densities, and star formation efficiencies in these regions will be compared to our expectations for young association-forming or cluster-forming complexes (see Sec. I). From these comparisons, we can more readily identify regions which are in the process of forming bound clusters. The observational results from each cloud are summarized in Table I. The investigation of the evolution of the cluster IMF in the ρ Ophiuchi cloud is discussed in Sec. II.B.4.

1. The λ Ori OB and T Association. The λ Ori OB and T association is dominated by the O8 III star λ^1 Ori which has excited a large H II region within which most of the association members now reside. Bordering the H II region is a large shell of neutral gas and dust ($M \sim 10^5$ $M_\odot$) which appears to be the last remnants of the giant molecular cloud from which the association formed (see Fig. 3). This shell is clearly visible in the color image of Orion (Color Plate 6). The total stellar mass of this association which is comprised of 83 emission-line stars and 12 OB stars is about 200 to 300 $M_\odot$ (Deurr et al. 1982). When the mass of this relatively unobscured stellar population is com-

TABLE I
Star Formation Efficiencies and Stellar Densities in Several Dark Cloud Complexes

Cloud Complex	Distance (pc)	Most Massive Star	Estimated Cluster Mass ($M_\odot$)	η(%)	$M_\odot$ pc^{-3}	References[a]
λ Ori (OB, T)	400	O8 III	200–300	0.2–0.3	0.03	(1)
Tau-Aur (T)	130	<B5	176[b]	2	0.5	(2),(3)
Chamaeleon (T)	<200	B9 V	95	12	8	(4)
NGC 7023	440	B3 III-IVe	10–23	30–50	6–14[d]	(5)
NCG 2023	500	B1.5 V	10–15	2–3	10–15[d]	(5)
NGC 2068	500	B3 V	30–70	10–20	30–70[d]	(5)
ρ Oph	160	B2 V	96[c]	25	145	(6),(7)

(1) Duerr et al. 1982; (2) B. F. Jones and G. H. Herbig 1979; (3) Baran 1975; (4) Hyland et al. 1982; (5) Sellgren 1983; (6) Wilking and Lada 1983; (7) Lada and Wilking 1984.
Emission-line stars only.
Estimate for core region where $A_v \geqq 50$ mag.
Assumes plane-parallel cloud geometry.

pared to the mass of the surrounding gas, very low star formation efficiencies are implied: η = 0.2 to 0.3%.

The star-forming history for the λ Ori region (which is outlined in Duerr et al. [1982]) is similar to that proposed for association-forming complexes in Sec. I. The OB and T association has formed from a giant molecular cloud which occupied a region now populated by the association. The conversion efficiency of gas to stars was extremely low such that as the OB association and λ^1 Ori now clear away the molecular cloud material, the majority of the original binding mass of the cluster is also being removed. The escape velocity for a star within the association is calculated to be only 0.2 km s^{-1} (Duerr et al. 1982) which is small compared to the observed velocity dispersions of 2 km s^{-1} calculated for 18 association members by Mathieu and Latham (preprint). Therefore, as a result of the low star formation efficiency, the fate of the λ Ori OB and T association is that of an unbound, expanding association.

2. *The Taurus-Auriga and Chamaeleon Dark Cloud Complexes.* Both the Tau-Aur and Chamaeleon dark cloud complexes have been the object of extensive multi-wavelength studies to investigate their young stellar and protostellar populations. In the Tau-Aur cloud, emission-line studies, proper motion surveys and infrared surveys have been performed and the young stellar population appears dominated by visible emission-line stars (B. F. Jones and G. H. Herbig 1979; M. Cohen and L. V. Kuhi 1979; Elias 1978*c*; Benson et al. 1984). B. F. Jones and G. H. Herbig (1979) estimate that there are probably about 176 emission-line stars throughout the entire complex. Comparing this to estimates for the mass of the molecular gas in the complex of 8000 $M_\odot$ (Baran 1975) results in a low star formation efficiency and a low stellar density for the entire complex: $\eta \sim 2\%$ and 0.5 $M_\odot$ pc^{-3}. As pointed out by M.

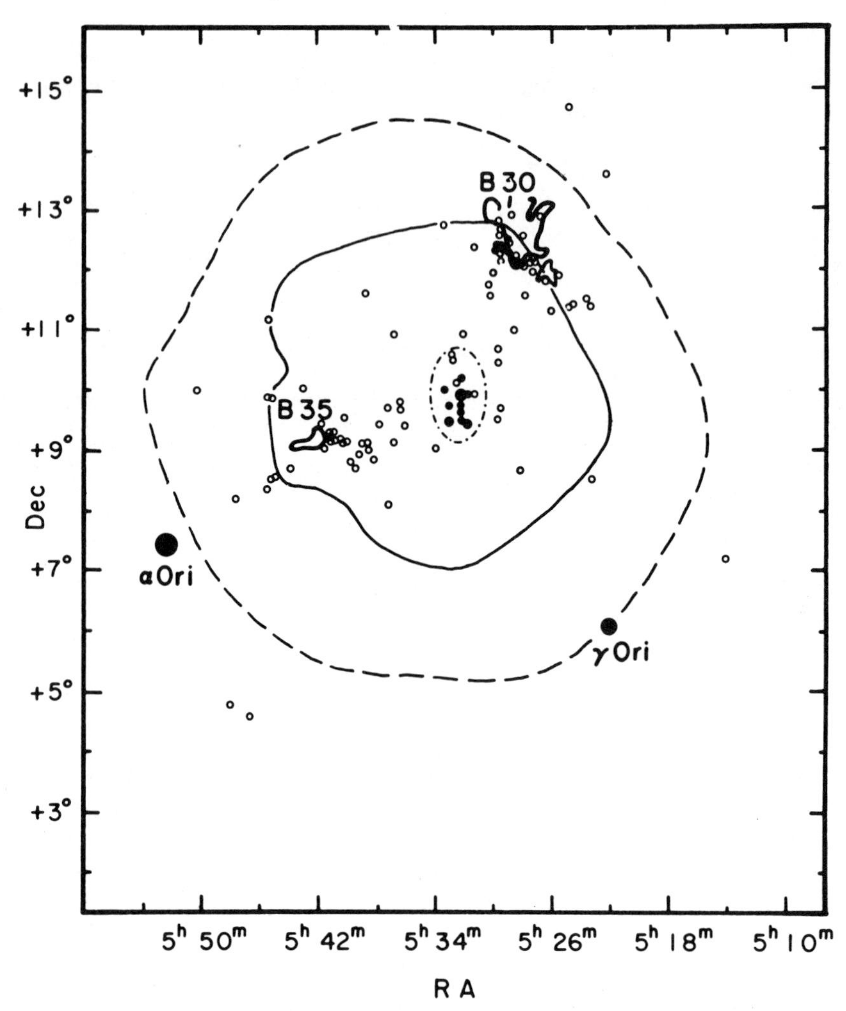

Fig. 3. The morphology of the λ Ori region as reproduced from Duerr et al. (1982). Open circles represent known Hα emission objects while small filled circles denote the positions of some of the known OB association members. The dot-dashed oval represents the extent of the OB association. Heavy solid lines mark the extent of the molecular emission in the B30 and B35 complexes. The medium solid line marks the boundaries of the H II region while the dashed line delineates the boundaries of the dark cloud material surrounding the λ Ori region.

Cohen and L. V. Kuhi (1979), higher efficiencies ($\eta \lesssim 10\%$) and densities ($\lesssim$ 30 $M_{\odot}$ pc^3) result when considering localized concentrations of stars within the complex.

Emission-line and infrared surveys of the Chamaeleon dark cloud complex have revealed 43 association members (Heinze and Mendoza 1973; Hyland et al. 1982). Accounting for embedded stars which may have been fainter than the limits of their survey or which lie in unsurveyed regions, Hyland et al. (1982) estimate that the total stellar population in the complex totals about 95 stars. Computing the cloud mass from their extinction map, Hyland et al. (1982) estimate the star formation efficiency and stellar density for the entire complex to be η = 12% and 8 $M_{\odot}$ pc^{-3}, respectively. More recently, T. J. Jones et al. (1984*b*) have used a deeper infrared survey to reveal a region within the Chamaeleon complex with a high density of embedded stars which potentially could form a bound subgroup with a dozen or fewer members.

Unlike the molecular cloud morphology predicted for cluster-forming clouds (see Sec. I.D), both the Tau-Aur and Chamaeleon dark cloud complexes, and the star formation associated with them, extend over tens of parsecs with no well-defined center (see, e.g., Baudry et al. 1981). The current values for the star formation efficiency are quite low in these complexes. Thus, it would appear unlikely that either region would form a bound cluster although it is possible that small subgroups of stars could form bound. It has been suggested by R. B. Larson (1982*a*) that filamentary cloud complexes such as these eventually collect into progressively more massive and condensed cores and proceed to form progressively more massive stars. Both Tau-Aur and the Chamaeleon clouds are relatively young (there has been no cloud disruption by massive stars) and display high enough stellar densities that (as suggested by Hyland et al. [1982] for the Chamaeleon) T Tauri stellar winds could be important in sustaining low-mass star formation in the manner proposed by C. A. Norman and J. Silk (1980). Hence, it is conceivable that future star formation in these complexes could lead to high enough star formation efficiencies to form bound clusters. It should be noted that R. B. Larson's (1982*a*) model for the cloud evolution of star-forming complexes is based on a supposed deficiency of low-mass stars in Orion and NGC 2264 which could be a selection effect (see, e.g., M. T. Adams et al. (1983).

3. NGC 7023, 2023, and 2068. Sellgren (1983) has made extensive near-infrared surveys of three dark clouds associated with reflection nebulae. Her study has revealed numerous sources associated with each reflection nebula as established through their properties of excess infrared emission, variability, or hydrogen emission lines. The molecular-line data available in these regions show that these clusters lie in the cores of extended clouds (see, e.g., D. M. Elmegreen and B. G. Elmegreen 1978; Milman et al. 1975). As summarized in Table I, the star formation efficiencies in the young, undisturbed cloud cores of NGC 7023 and 2068 are greater than $\eta \geq 10\%$. Both of these clouds are candidates for the formation of bound clusters.

4. The Rho Ophiuchi Dark Cloud. The ρ Oph dark cloud complex lies at a distance of 160 pc from the Sun at the edge of the Sco-Cen OB association. An IRAS image of this region is shown in Color Plate 7. Using optically-thin $C^{18}O$ emission lines, Wilking and Lada (1983) have traced out a 1 pc × 2 pc ridge of gas that forms the centrally condensed core of the ρ Oph cloud (see Fig. 4). Extremely large gas and dust column densities prevail in this core: $N^{18}{}_{LTE} = 1.4$ to 2.9×10^{16} cm^{-2} and $A_v = 50$ to 100 mag. A new and sensitive near-infrared survey of the central 10 × 10 square arcmin region of the core, which was previously thought to be devoid of star formation, revealed 20 sources (16 previously unknown). On the basis of comparison of the infrared cluster density with star background counts, near-infrared excesses, and the superposition of infrared sources on areas of high visual extinction (as inferred from the $C^{18}O$ column densities), all 20 sources were shown to be associated with the cloud. Combining these new infrared data with existing studies of emission-line stars (Rydgren et al. 1976) and of embedded infrared sources (Vrba et al. 1975; Elias 1978*b*), Wilking and Lada were able to establish a population of 44 cluster members in the core region of the ρ Oph cloud.

The best estimates for the star formation efficiency in ρ Oph are for the well-studied core region where $A_v \gtrsim 50$ mag (see Fig. 4). Star formation efficiencies of about 25% are obtained within a star-forming layer of depth $A_v = 40$ mag; this is the maximum depth to which the survey of Wilking and Lada could detect a typical solar-luminosity T Tauri star (Lada and Wilking 1984). This implies a total cluster mass of $\sim$96 $M_\odot$ for the high column density core and stellar densities of 145 $M_\odot$ pc^{-3}. Such a high density of embedded stars in the cool, compact cloud core of the ρ Oph cloud led Wilking and Lada to conclude that the formation of a bound cluster in this cloud is imminent. The ρ Oph core and infrared cluster are the prototypical region for clouds in the process of forming bound stellar systems. Underlining the uniqueness of the ρ Oph cluster is the degree to which it is centrally condensed. Although the Chamaeleon cloud complex has been well studied by sensitive near-infrared surveys over an area of $\sim$1 pc × 5 pc (Hyland et al. 1982; T. J. Jones et al. 1984*b*), it is estimated to contain the same number of cluster members as the ρ Oph cluster, but over an area which is $\sim$12 times larger than the ρ Oph core region.

Upon the realization that the ρ Oph cloud will most likely form a gas-free bound cluster, it is important to begin studying the distribution of luminosity of these young stars. The luminosity function of such an embedded cluster, once corrected for selection effects, can be directly compared to those of open clusters. To this end, Lada and Wilking (1984) have obtained 5 to 20 μm photometry for many of the known cluster members in ρ Oph. This photome-

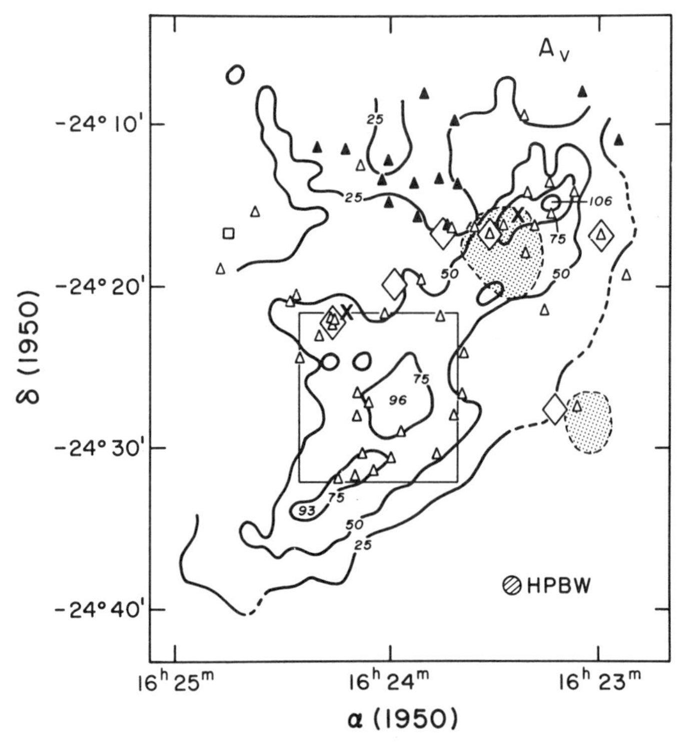

Fig. 4. The core region of the ρ Ophiuchi dark cloud as determined from the $C^{18}O$ observations of Wilking and Lada (1983). Column densities of $C^{18}O$ have been converted to visual extinction and these contours are marked in increments of 25 mag with the peak values given for the three largest enhancements. Open triangles represent 42 of the 44 sources which have been determined to be embedded in the ρ Oph cloud while solid triangles represent unclassified sources. The sole field star which has been identified in the core region is shown by an open square. Also displayed in this figure are: (1) two of the 78 μm sources observed by Fazio et al. (1976), represented by dot patterns; (2) six sources of radio continuum radiation believed to be compact H II regions, indicated by large diamonds, as identified by R. L. Brown and B. Zuckerman (1975) and Falgarone and Gilmore (1981); and (3) two areas of high spatial density ($n(H_2) > 10^6$ cm^{-3}), marked by the X's, as found by Loren et al. (1983).

try revealed that most of these sources are low-luminosity objects which radiate the bulk of their luminosity in the 2 to 20μm spectral region, and which are probably still contracting toward the main sequence. This enabled good estimates for the bolometric luminosities to be obtained (within a factor of 2) by simply integrating the energy distribution of each source over the observed wavelength range. The resulting luminosity function for 37 stars which have well-determined luminosities is shown in Fig. 5a. For comparison, the luminosity function corresponding to a log-normal IMF (G. E. Miller and J. M.

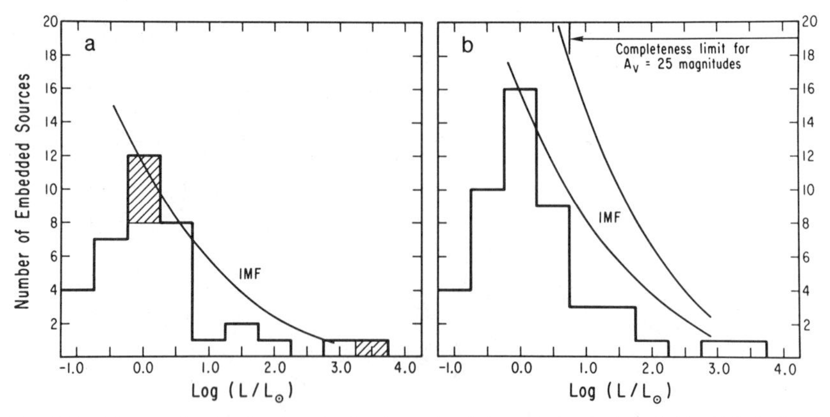

Fig. 5 (a). Observed luminosity function for the embedded population in the ρ Ophiuchi dark cloud from Lada and Wilking (1984). The hatched areas of the histogram represent sources whose luminosities were determined from either M. Cohen and L. V. Kuhi (1979) or Garrison (1967). The luminosity function corresponding to a log-normal initial mass function (IMF) (G. E. Miller and J. M. Scalo 1979) is drawn for comparison. It has been normalized to reproduce the observed number of stars with $-0.25 \leqq \log L/L_\odot \leqq 0.75$. (b) Observed luminosity function for the embedded stars including the 11 remaining known members whose luminosities are not well determined but are represented by approximate upper limits in the histogram (Lada and Wilking 1984). The luminosity function corresponding to a log-normal IMF is presented for comparison with two different normalizations. The lowermost IMF curve has been normalized in the same manner as that in Fig. 5a. The uppermost IMF curve has been normalized to produce the correct number of stars in the range $-0.25 \leqq \log L/L_\odot \leqq 0.75$ after observational selection effects are taken into account.

Scalo 1979) is also plotted and normalized to the observed low-luminosity population. The deviation of the observed luminosity function from that of the IMF at low luminosities reflects the sensitivity limits of infrared surveys and there are no other statistically significant deviations.

Before a meaningful comparison of the ρ Oph cluster luminosity function can be made to that from the IMF, all known selection effects must be considered. Fig. 5b shows the observed distribution of luminosity of all known cluster members (48 stars); eleven member stars which have not been observed at 10 or 20 μm and which have more uncertain estimates for their bolometric luminosities have been included. Once again there are no statistically significant deviations of this distribution from that of the IMF.

However, accounting for two remaining selection effects results in a significant departure of the ρ Oph cluster luminosity function from that of the IMF at intermediate-to-high luminosities ($L > 5\ L_\odot$). The first of these effects is a result of the inability of infrared surveys to sample the embedded population throughout the entire depth of the dense ρ Oph core and the fact that the depth to which stars are completely sampled varies with luminosity. Simply stated, if stars are evenly distributed throughout the ρ Oph cloud, then infrared surveys have sampled the higher luminosity stars to a depth which is roughly twice that to which lower luminosity stars have been completely sampled. Hence, to form a luminosity function for the ρ Oph cluster which represents all stars expected within a star-forming layer, as defined by the higher-luminosity stars, it is necessary to increase the low-luminosity population by approximately a factor of two. As shown in Fig. 5b, if the luminosity function derived from the IMF were normalized to this corrected low-luminosity population, a significant deficiency of intermediate luminosity stars is suggested.

A second effect, the subsequent evolution of a large pre-main sequence population, must be considered before comparing the observed luminosity function with that calculated from the mass distribution of stars on the zero-age main sequence (i.e., the IMF). The predominance of pre-main sequence stars in the ρ Oph cluster suggests that the sample as a whole is most likely more luminous than one would expect if all stars were to lie on the main sequence. Thus, if star formation in the ρ Oph cloud were to end abruptly, then the evolution of stars to the main sequence would shift the observed luminosity function to the left, thereby increasing the deficit of intermediate-to-high mass stars ($L > 5\ L_{\odot}$).

As mentioned earlier, the end result of star formation in open clusters appears to be a mass distribution which closely resembles the field star IMF (see Scalo [1985*a*] for a review). If subsequent observations confirm the apparent deficiency of intermediate-luminosity stars in ρ Oph, then it appears that future star formation in this cloud would favor the formation of more massive stars. Indeed, this type of evolution for the cluster IMF in NGC 2264 is observed by M. T. Adams et al. (1983) where stars appear to have formed sequentially in mass and yet produced a mass spectrum consistent with the IMF.

Further study of the ρ Oph cluster luminosity function could have important implications for the basic processes by which protostellar fragments are produced. While a general deficiency of stars with $L > 5\ L_{\odot}$ is suggested, two very luminous B stars, Source 1 ($L = 1500\ L_{\odot}$) and HD 147889 ($L = 5000\ L_{\odot}$) have also formed in the core region. Their presence is reminiscent of a bimodal mass distribution predicted by models for the growth of protostellar fragments by accretion (Zinnecker 1982) and by collisional coaelesence (Pumphrey and Scalo 1984). In their models, these growth mechanisms lead to a runaway growth of the most massive fragment and a gap in the resulting mass function. A much more extensive study of the ρ Oph cluster would be necessary to yield such important information on the growth processes of protostellar fragments.

Regardless of the final form or interpretation of the ρ Oph luminosity function, the great utility of broadband infrared photometry to compute the bolometric luminosity of embedded cluster members has been shown. This calorimetric technique, when combined with lower-resolution far-infrared surveys, can be used as a powerful tool with which to study the evolution of the mass functions of young dust-embedded clusters.

III. SUMMARY AND FUTURE WORK

The star formation process in our Galaxy appears to be very inefficient. The majority ($\sim$ 90%) of stars in the solar neighborhood are produced in bound giant molecular clouds which convert only a small fraction ($\eta \lesssim 5\%$) of their molecular gas into stars. Upon the dispersal of the GMC, the resulting stellar association becomes unbound and diffuses into the general field population.

Under special circumstances, however, bound molecular clouds will convert gas into stars with a high efficiency ($\eta \gtrsim 20$–50%) and produce a cluster which can remain gravitationally bound after the gas is removed from the system. Insight into what special requirements in the star formation process lead to a bound cluster have been provided by analytic and numerical studies of the evolution of young embedded clusters and by the observed properties of open clusters. The final star formation efficiency necessary to produce a bound cluster is a sensitive function of the gas release time. For the rapid removal of gas ($\tau_R < 10^6$ yr), efficiencies must exceed about 50%. However, if the gas is released more slowly over 4 to 5 $\times$ 10^6 yr, then the final star formation efficiency need be only 20 to 30%. Clearly, the formation of bound clusters will be more favorable in clouds which have yet to form massive, disruptive stars and where the gas release times will be longest. From a study of the masses and radii of open clusters, we predict that bound clusters will most likely be formed in compact, quiescent clouds or cloud cores with radii $<$1 pc and masses of $\sim$1000 $M_\odot$.

Theoretical models for the dynamical evolution of clusters as their residual gas is removed can be used to describe a broad continuum of initial cloud conditions, star formation efficiencies, and gas release times. It is, therefore, of the greatest importance to begin collecting the observational data in young star-forming regions which can be compared with theory to identify areas in the process of forming bound clusters. Observational techniques which are used to establish the membership of young dust-embedded clusters and to compute star formation efficiencies have been reviewed and emphasize the need for the broadest wavelength coverage possible.

The results of several well-sampled surveys of young star-forming regions have been presented and compared to our expectations for association and cluster formation. The λ Ori OB and T association is a prime example of a cluster that formed with a very low star formation efficiency ($\eta \sim$ 0.2–0.3%) and, as a consequence, is unbound and expanding. The prototypical region for clouds in the process of forming bound clusters is the centrally condensed core of the ρ Oph dark cloud. Star formation efficiencies of $\sim$25% and the absence of massive stars insure the formation of a bound stellar system in this cloud.

Upon the identification of regions in the process of forming bound clusters, we can begin investigating how the cluster mass function evolves over time. The results of a calorimetric technique for determining the luminosity function of cluster members in the ρ Oph cloud have been presented. The value of such studies lie in their potential to infer the manner by which protostellar fragments grow into stars and the mass dependence of the early stages of star formation. More accurate determinations of the luminosity function of young embedded clusters will rely heavily on longer wavelength surveys of dark clouds such as those by the Infrared Astronomical Satellite (IRAS) and on the high resolution capabilities of the Space Infrared Telescope Facility (SIRTF) at wavelengths beyond 20 μm.

Acknowledgments. We would like to thank N. Evans and J. Scalo for enlightening discussions and their comments on the early version of this manuscript. R. Mathieu and M. Margulis are gratefully acknowledged for data supplied in advance of publication. This work was supported in part by a National Science Foundation grant to The University of Texas at Austin. C. J. L. is a Alfred P. Sloan Foundation Fellow.

MAGNETIC FIELDS

L. MESTEL
University of Sussex

The principal topics discussed are: (1) the dynamical and observational consequences of the anisotropic collapse of cool gas clouds permeated by the local galactic magnetic field; (2) magneto-gravitational equilibria of such clouds with sub-critical mass-flux ratios, especially in the thin disk approximation; (3) magnetic braking of both subcritical and supercritical masses; and (4) the consequences of flux leakage during the molecular cloud phase, including the effect on field topology.

I. INTRODUCTION: ANISOTROPIC COLLAPSE

This review concentrates on the role of the galactic magnetic field in the early stages of star formation. It is easy to be convinced that the observed field is strong enough to modify profoundly the dynamics of gravitational contraction, in particular by introducing an essential anisotropy into the problem. To fix ideas, consider a simple picture of the formation of a magnetic cloud, in which gas flows along the local galactic field $\mathbf{B}_o$, triggered, e.g., by the passage of the galactic spiral shock, or by an expanding supernova shell. As gas accumulates, assisted by increased cooling at higher densities, the growing self-gravitation of the cloud causes motion of gas plus inductively-coupled field lines in the directions perpendicular to $\mathbf{B}_o$, so generating magnetic forces that oppose further lateral contraction. There is ultimately no substitute for precise studies of this difficult gas-dynamical problem (cf. Sec. II); however, judicious application of the virial theorems to plausible simulations of the mass and field distributions enables one easily to isolate a critical mass $\mathcal{M}_c$, defined essentially by the magnetic flux $\mathcal{F}$ trapped in the cloud:

$$G\mathcal{M}_c^2 = \frac{4k}{3\pi^2}\mathcal{F}^2 \tag{1}$$

where k is a nondimensional factor which decreases from near unity for a sphere to about ¼ for a flattened body (Chandrasekhar and Fermi 1953; Mestel and Spitzer 1956; Mestel 1965,1969; Strittmatter 1966; Mestel and Paris 1979,1984). If the cloud mass $\mathcal{M} > \mathcal{M}_c$, the magnetic forces alone are never strong enough to prevent indefinite gravitational contraction; on the other hand, if $\mathcal{M} < \mathcal{M}_c$, the cloud will settle into equilibrium with approximate magneto-gravitational balance holding across the field and hydrostatic balance along the field.

At this point, we immediately run up against the observational fact that cool molecular clouds are nevertheless held up against self-gravitation in all three dimensions, i.e., along as well as perpendicular to the direction of the inferred large-scale galactic magnetic field. The most plausible explanation is that mass motions (described by the all-embracing term turbulence) give the cloud an effective dynamical temperature that is much greater than that due to microscopic motions. To offset their spontaneous decay, such motions require energy input, e.g., from winds emitted from newly formed stars (C. A. Norman and J. Silk 1980), and it may be that a cutoff in this supply is the necessary condition for the initiation of new star formation. For simplicity, we continue to speak of the whole cloud settling into a flattened structure, but much of the discussion is applicable to a more realistic picture in which flow along the field following the decline in turbulent pressure occurs locally within the cloud.

It is convenient to relate the cloud parameters to those pertaining to an idealization of the galactic background as the warm interstellar medium, of uniform density ρ_o, high temperature T_o, and permeated by the uniform magnetic field $\mathbf{B}_o$. Thus the magnetic flux $\mathcal{F}$ trapped within the cloud is written as

$$\mathcal{F} = \pi B_o R_o^2 = \pi \bar{B} \bar{R}^2 \tag{2}$$

so that R_o is the radius of a background flux tube of strength F, and $\bar{B}$ is the mean internal field strength when the cloud has the trans-field radius $\bar{R}$ (see Fig. 1). The cloud mass, written as

$$\mathcal{M} = K\left(\frac{4\pi}{3}\rho_0 R_0^3\right) = \frac{2\pi}{3}\rho_0 R_0^2 L_0, \qquad L_0 = 2KR_0 \tag{3}$$

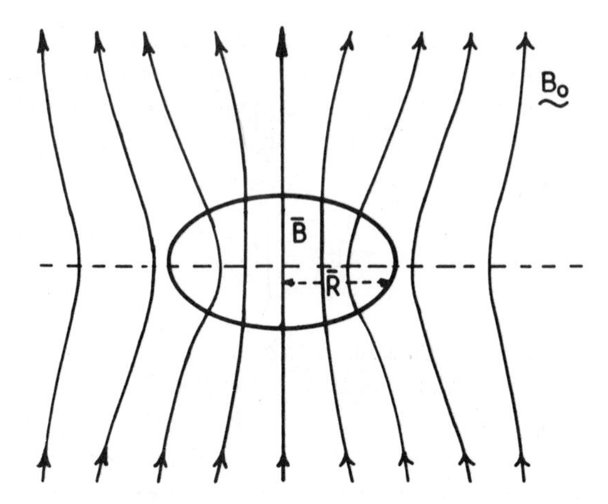

Fig. 1. Schematic picture of a magnetic cloud.

is K times the mass of a background sphere R_0, or equivalently the mass of a background prolate spheroid of semiminor and semimajor axes R_0, KR_0 (anticipating that in the interesting cases $K >> 1$). If the cloud is approximated as an oblate spheroid of density $\bar{\rho}$, and minor and major axes $2\bar{z}$ and $2\bar{R}$, then

$$\mathcal{M} = \frac{4\pi}{3}\bar{\rho}\bar{z}\bar{R}^2 = \frac{2\pi}{3}\bar{M}\bar{R}^2 \tag{4}$$

with $\bar{M} = 2\bar{\rho}\bar{z}$ the projected mass per unit area on the axis. It is then instructive to write the condition for gravitational collapse in the form (Field 1970; B. G. Elmegreen 1981*b*; Nakano 1981; Mestel and Paris 1984)

$$\bar{M} > \bar{M}_c = \left(\frac{3k}{\pi^2 G}\right)^{1/2}\bar{B}. \tag{5}$$

This is equivalent to setting a lower limit L_c on the *accumulation length* L_0 defined in Eq. (3):

$$L_0 > L_c = \frac{B_0}{\pi\rho_0}\left(\frac{3k}{G}\right)^{1/2} \simeq 1200\frac{(B_0/3 \times 10^{-6})}{(n_0/1)} \tag{6}$$

in parsecs, where k is taken as unity and standard values are inserted for B_o and the number density n_o. One is immediately impressed by the very large value for L_c. Maybe, as pointed out by L. Spitzer and F. Shu (personal communications), one should not be embarrassed, as such lengths are predicted by the Rayleigh-Taylor-Parker instability in the interstellar medium; however, one is led to speculate that flux freezing may not have held during cloud formation, but that instead an effective turbulent resistivity has enabled substantial trans-field diffusion to occur (cf. Sweet 1950). If cloud formation is violent, following a local input of kinetic energy density far exceeding $B_o^2/8\pi$, then one can picture eddies tangling the field into loops which can break off from the main field because of finite resistivity acting on the currents at the neck. A disconnected loop will itself be spontaneously unstable and may disappear comparatively quickly, leaving the system with ρ/B much increased, but without there having been the enormous flow down the field implied by L_c, which becomes a notional length.

In any case, it is certainly important to keep distinct the problem of cloud *formation* from that of cloud *equilibrium* and *collapse*. Let us suppose, as an example, that a supercritical spherical cloud has formed (whether or not through temporary breakdown in field freezing), so that Eq. (5) holds with $k = 1$. If the cloud is cool, as it contracts laterally it will simultaneously flatten into a roughly spheroidal structure, but although $\bar{M}$ and $\bar{B}$ increase, the ratio $\bar{M}/\bar{B}$ remains constant; thus when the cloud semi-thickness is $\bar{z}$, the sphere of radius $\bar{z}$ and density $\bar{\rho} = \bar{M}/2\bar{z}$ is again supercritical. One can regard Eq. (5) as defining a minimum accumulation length $\bar{L}_c$ corresponding to $\bar{B}$ and the instantaneous density $\bar{\rho}$; with $k = 1$

$$\bar{L}_c = \frac{\bar{B}}{\pi\bar{\rho}}\left(\frac{3}{G}\right)^{1/2} \tag{7}$$

but as the sphere flattens, both the semi-thickness $\bar{z}$ and $\bar{L}_c$ decrease as $1/\bar{\rho}$. This leads at once to the idea of magnetic fragmentation; if the quasi-spherical cloud is supercritical, then the virial condition allows the possibility that roughly spherical subcondensations can separate out of the flattened cloud (Mestel 1965*b*). Writing the condition in the form of Eq. (7) makes this manifest, because no length-scale perpendicular to the field enters (purely lateral contraction leaving $\bar{B}/\bar{\rho}$ invariant), while distance along the field enters multiplied by the density.

Showing that fragmentation is allowed by the virial condition does not of course prove that it will happen. If centrifugal forces are weak, then the supercritical flattened cloud will collapse, albeit anisotropically, and the same queries arise as in the simpler nonmagnetic problem. The virial condition is a censor—a necessary but by no means sufficient condition for the spontaneous gravitational amplification of a density perturbation against a background that is itself growing in density. The important points to emphasize are: (1) without flux leakage, a subcritical cloud can neither collapse nor fragment; and (2) in a supercritical cloud, although flux leakage will clearly make fragmentation easier (cf. Sec. IV), it is not absolutely essential, because spontaneous anisotropic flow can yield fragments that satisfy the virial censor (Eq. 5). (If, in fact, centrifugal forces are maintained comparable with gravity [cf. Sec. III], then the fragmentation problem in a flattened system becomes closer in spirit to the classical Jeans problem of the spontaneous gravitational amplification of a density perturbation in a noncollapsing medium.)

However, one should note that fragments forming in this way from a magnetically strong cloud will themselves also have a magnetic energy comparable with their gravitational energy. This is in marked contrast to all the evidence from main sequence stars. Even for those with the strongest surface magnetic fields, such as Babcock's 35 kGauss star, no plausible inward extrapolation yields a total magnetic energy that is more than a very small fraction of the gravitational; for a typical magnetic star, the ratio is likely to be $\sim 10^{-5}$, corresponding to an internal field strength of $\sim 10^6$ Gauss. It cannot be stated too strongly that although flux-freezing plus anisotropic flow can allow the formation of gravitationally-bound, strongly magnetic subcondensations, it cannot account for the lowness of the flux retained by stars. It is easy to extend the previous discussion (cf. B. G. Elmegreen 1981*b;* Nakano 1983*a;* Mestel and Paris 1984) to show that to produce the very small inferred ratios of $\mathcal{F}/G^{1/2}\mathcal{M}$, the necessary accumulation length would need to be absurdly long (typically 40 kpc). Alternatively, if the accumulation length is kept reasonable, then $\mathcal{F}/G^{1/2}\mathcal{M}$ remains high, and a body forming in this way, which would reach a main sequence density $\rho \simeq 1$ g cm^{-3} with an interior frozen-in field strength of $\simeq 10^6$ Gauss, would have a mass of only $\simeq 10^{-4}$ $M_\odot$. Thus, there is no doubt that the bulk of the magnetic flux threading the gas from which a protostar forms must have been lost *at some stage* (cf. Sec. IV). Equally, the later the epoch of flux loss, the longer the magnetic field remains dynamically important, e.g., for dealing with the angular momentum problem (Sec. III).

II. MAGNETO-THERMO-GRAVITATIONAL EQUILIBRIA

Our discussion of the consequences of spontaneous flow along the field is relevant to a cool, weakly turbulent cloud, which can reach a balance between pressure and gravity along the field with the semi-thickness given in terms of the isothermal sound speed $\bar{a}$ by (see, e.g., McCrea 1957)

$$\bar{z} \simeq \left(\frac{\bar{a}^2}{2\pi G\bar{\rho}}\right)^{1/2} \tag{8}$$

and with mean values $\bar{B}$ and $\bar{\rho}$ then related by

$$\frac{\bar{B}}{\bar{\rho}^{1/2}} \simeq \frac{\mathcal{F}}{G^{1/2}\mathcal{M}}\left(\frac{8\bar{a}^2}{9\pi}\right)^{1/2} \simeq 4\left(\frac{2\pi}{15}\right)^{1/2}\bar{a}\left(\frac{\mathcal{M}_c}{\mathcal{M}}\right) \tag{9}$$

where the value $\mathcal{M}_c \simeq (\sqrt{15}/6\pi)\ \mathcal{F}/G^{1/2}$ appropriate to a highly flattened body has been adopted. If $\mathcal{M} > \mathcal{M}_c$, the cloud can contract laterally, with Eq. (8) continuing to hold, so that if $\mathcal{F}$ and $\mathcal{M}$ both remain constant, Eq. (9) predicts $\bar{B} \propto \bar{\rho}^{1/2}$, a result implicit in earlier theoretical studies and verified in numerical studies by Scott and Black (1980) (which follow the detailed evolution of the ($\mathbf{B}$, ρ) fields). Note, however, that the constant of proportionality must not be fixed by insertion of the background values (B_o, ρ_o); the temperature of the cool cloud enters crucially through $\bar{a}^2$, because it defines the minimum density increase the cloud gas would suffer in condensing from the hot intercloud medium even before any lateral contraction had increased $\bar{B}$ above B_o.

If $\mathcal{M} < \mathcal{M}_c$, then the lateral contraction of the cloud is limited by the magnetic forces. In the spirit of the discussion of Sec. I, we consider first a model with the cool cloud idealized as an oblate spheroid $\bar{S}$ (Mestel and Paris 1984), which defines a class of confocal spheroids. Between $\bar{S}$ and a limiting spheroid S_o, the field lines lie on orthogonal hyperboloids. Beyond S_o the field reduces to $\mathbf{B}_o$; the normal components of $\mathbf{B}$ on $\bar{S}$ and S_o are continuous, but surface currents maintain discontinuities in tangential components. (When the eccentricity $\bar{e}$ of $\bar{S}$ is zero, the model reverts to an earlier one [Gillis et al. 1974,1979; Mestel and Paris 1979] with $\bar{S}$ and S_o both spheres and with radial field lines in between.) Proper use of the tensorial virial theorem (including the Maxwell stress integral over a surface surrounding S_o) yields for the semi-major axis of the spheroid

$$\frac{\bar{R}}{R_0} \simeq 1 - \left(\frac{\mathcal{M}}{\mathcal{M}_c}\right)^2. \tag{10}$$

This indicates that unless $\mathcal{M}$ is close to $\mathcal{M}_c$, $\bar{R} \simeq R_o$ and the gravitational distortion of the field will be weak; but, if $\mathcal{M} \simeq \mathcal{M}_c$, then $\bar{B}/B_o$ from Eq. (2) can be well above unity. This model is the basis of the interpretation (Mestel 1969; Mouschovias 1976*b*) of the locally strong fields discovered by the Zeeman effect on the 21-cm line.

The virial theorems are clearly very useful tools, able to yield quickly the essential qualitative features following insertion of simple but intuitively reasonable representations of the magnetic field structure. Thus Eq. (10) shows clearly the transition from the magneto-gravitational equilibrium, achieved by subcritical clouds, to the inevitable gravitational collapse when $\mathcal{M} > \mathcal{M}_c$. The limitations of the virial treatment are equally patent; there is ultimately no substitute for the construction of models with force balance holding at each point rather than just through the virial integrals. Both supercritical and subcritical clouds are of interest for the theory of star formation, but it is clearly easier to construct models in equilibrium. We are interested in getting some feel for the detailed distribution of flux with respect to mass within different equilibria, and in knowing how much the critical value of the parameter $f \equiv \mathcal{F}/\pi^2 G^{1/2} \mathcal{M}$ can vary. The spontaneous evolution of such clouds may give one clues towards the understanding of at least some modes of star formation. Besides the obvious processes of mass accretion and slow flux leakage, there is the question of the detachment of the cloud field from the background field, its connection with the breakdown of magneto-gravitational equilibrium, and the effect on the efficiency of magnetic braking (cf. Sec. III).

There are two paths to the construction of equilibrium models. The first starts with a hypothetical parent cloud in an idealized background interstellar medium. Thus, in the most extensive set of computations to date, Mouschovias (1976*a,b*) takes as the parent cloud a spherical region in a warm isothermal background of uniform density ρ_o and permeated by a uniform magnetic field $\mathbf{B}_o$, so fixing not only the total flux $\mathcal{F}$ and the total mass $\mathcal{M}$, but also the detailed mass-flux relation. One can then postulate that the parent cloud cools and collapses with *strict flux-freezing holding*. Then, if $\mathcal{M}$ is subcritical, the cloud cannot contract indefinitely; after dissipation of the kinetic energy fed from the gravitational field, the cloud settles into a state of thermo-magneto-gravitational equilibrium *with the same mass-flux relation as in the parent cloud.* The numerical scheme required is difficult, and Mouschovias's successful construction of a large class of models with finite temperatures, and without any tangential field discontinuities (which would imply unbalanced stresses) is all the more impressive. However, there is inevitably some arbitrariness, both in the simple picture of the interstellar medium and even more so in the adoption of a spherical parent cloud. Once at least a tentative doubt has been cast also on the constraint of strict flux-freezing during cloud formation, one is led naturally to the alternative path pioneered by D. A. Parker (1973,1974), in which the precise mass-flux relations are not prescribed in advance of the solution, but instead emerge from the equilibrium models, constructed because of their mathematical simplicity and their intrinsic interest. Once a model has been built, then a convenient way of representing the mass-flux relation does, in fact, come from performing the unscrambling exercise; with strict flux-freezing supposed to have held, one can construct the shape of the hypothetical parent cloud in the assumed ($\mathbf{B}_o$, ρ_o) background. All methods have some degree of arbitrariness; the second approach may offer scope for studying a wider variety of models with fewer technical difficulties.

Some recent computations (Mestel and Ray 1984) follow D. A. Parker (1974) in making the further simplification of a disk-like structure. We have noted that the most plausible explanation of the apparent long lifetimes of cool, giant molecular clouds is that they are held up in the direction along the field by a strong dynamical pressure, which, however, will need maintenance against decay, e.g., by input of energy from previously formed stars (C. A. Norman and J. Silk 1980). Such clouds can be modeled by giving the gas a high temperature to represent the turbulent energy. Equally, if gravitationally bound subcondensations are to form within giant molecular clouds, it is either through significant local trans-field motion (cf. Sec. IV), or because sufficient flow along the field lines does occur locally, presumably through a decline in the energy supply to the turbulence. This is part of the motivation for concentrating on disk-like equilibria; but in any case, there is a strong case for studying the problems of magneto-gravitational equilibrium in their simplest form.

The first class of models studied by Mestel and Ray (1984) have in $0 \leq \tilde{\omega} \leq R$ the mass $M(\tilde{\omega})$ per unit area which generates the gravitational field

$$g(\tilde{\omega}) = \frac{V^2 \tilde{\omega}}{(\tilde{\omega}^2 + \ell^2 R^2)}, \qquad V = \text{constant}. \tag{11}$$

The parameter ℓ is always < 1, and in the most interesting cases is taken small (0.1–0.3), corresponding to a moderately centrally condensed mass distribution. Within $x \equiv \tilde{\omega}/R = \ell$, $g \propto \tilde{\omega}$ approximately, but beyond $x = \ell$, $g \simeq V^2/\tilde{\omega}$, the familiar form that in a disk-like galaxy is balanced by the centrifugal acceleration of a flat rotation law $\Omega\tilde{\omega} = V$. The models are, in fact, generalizations of the point singularity form with $\ell = 0$, for which $M(\tilde{\omega}) = (V^2/\pi^2 G\tilde{\omega}) \cos^{-1}(\tilde{\omega}/R)$. When $\ell \neq 0$, the form for $M(\tilde{\omega})$ may be found from standard theory (see, e.g., Mestel 1963); within $x = \ell$, $M(\tilde{\omega})$ is approximately uniform, and beyond $x = \ell$, $M(\tilde{\omega}) \propto 1/\tilde{\omega}$, approximately. Equilibrium is maintained by the Lorentz force per unit area that acts on the sheet current within the disk balancing the gravitational force per unit area, equivalent to

$$M(\tilde{\omega})g(\tilde{\omega}) = \frac{B_{\tilde{\omega}} B_z}{2\pi} \tag{12}$$

where $B_{\tilde{\omega}}$ is the component at $z = 0+$, and B_z is the sum of the field due to the currents in the disk and the background field B_o. The hot, low-density medium outside feels negligible gravity and can be taken as having uniform pressure and with a curl-free field. The equations are solved by iteration. Within $x = \ell$, B_z is approximately uniform, $B_{\tilde{\omega}} \propto \tilde{\omega}$; beyond $x = \ell$, B_z and $B_{\tilde{\omega}}$ both behave approximately as $1/\tilde{\omega}$. As the parameter $f \equiv \mathcal{F}/\pi^2 G^{1/2} \mathcal{M}$ decreases, the ratio $B_z(\tilde{\omega} = 0)/B_o$ increases; there is less flux, but it is compressed so as to yield the stronger magnetic force required to balance the simultaneously increased gravitational force. As expected, there is a limiting value f_c of the parameter f below which no equilibria exist (for the value of ℓ chosen); it is found that the value of $f_c \simeq 0.74$ is very insensitive to the choice of ℓ. As an example, it is found that for $\ell = 0.1$, $B_z(0) \simeq 14\,B_o$ in the limiting model, with the area density at the origin $M(0) \simeq 12\mathcal{M}/\pi R^2$.

A qualitatively new feature that these detailed equilibria yield is that, in the limiting equilibrium state, there is a cusp-like neutral point at the edge of the disk. This is a not unexpected consequence of the curl-free condition on the external field. For example, in the simplest possible approximate model—a sphere of radius $\bar{R}$ with a uniform field $\bar{B}$ that is a frozen-in amplification of the background field B_o—the curl-free external field develops a neutral point at the cloud surface when $\bar{B}/B_o = 2$. It is also understandable that this must be the limiting equilibrium model; a field in which the cusp has evolved into a pair of O- and X-type neutral points (with the O-point necessarily within the disk) would clearly fail to satisfy force balance at the O-point.

Of these "moderately centrally condensed" models, those with the parameter f near the critical value have a mass/flux distribution that is not too dissimilar from that in a uniform prolate spheroid with a uniform field $\mathbf{B}_o$. Another class of models constructed by Mestel and Ray (1984) has the more centrally condensed density $M(\tilde{\omega})$ that generates the gravitational field

$$g(\tilde{\omega}) = \frac{V^2\tilde{\omega}\ell R}{(\tilde{\omega}^3 + \ell^3R^3)}, \qquad 0 < \ell < 1 \tag{13}$$

which beyond $\tilde{\omega} = \ell R$ approximates to a point-source field. For $\ell \geq 0.6$, it was possible to construct the equilibrium models all the way to the limit, where again there is a cusp-like neutral point at the edge. For smaller ℓ, numerical difficulties seem responsible for impeding computation of models all the way to the limit, but there does not seem any reason to expect qualitatively different behavior. The value of the parameter in the expected limiting model can be found by extrapolation. It is remarkable that for the class of model defined by Eq. (13), again f_c (whether found directly or by extrapolation) is near the value 0.74.

A still more striking conclusion emerges from comparison of some limiting solutions (for different ℓ values) of the classes defined by Eqs. (11) and (13). It is found that although these models have very different density and magnetic field distributions in physical space, the mass-flux relations, emerging, e.g., from the unscrambling exercise, are remarkably close. Equivalently, for models with f close to 0.74, there is a remarkable sensitivity of the spatial M and B_z distributions to small variations in the mass-flux relation.

In both classes of model the field within the disk is only slightly distorted from $\mathbf{B}_o$ if $\mathcal{M} << \mathcal{M}_c$ (cf. Eqs. 2 and 10). As $\mathcal{M}$ approaches $\mathcal{M}_c$, then the amplified field within $\tilde{\omega} = \ell R$ stays nearly parallel to $\mathbf{B}_o$, but beyond $\tilde{\omega} = \ell R$, $B_{\tilde{\omega}}$ and B_z are comparable, and the Maxwell stresses exert strong pinching forces on the disk. This effect was illustrated in an early study (Mestel 1966*a*) of an initially uniform field subject to a strong nonhomologous radial motion; the possible consequences for the topology of the field will be noted in Sec. IV.

III. MAGNETIC BRAKING

So far, the magnetic field has appeared as an impediment to collapse and fragmentation. We now discuss the possibly decisive role it may play in the early phases of star formation through its ability to redistribute angular momentum. For a fuller account on this subject, see Mestel and Paris (1984).

The physics of magnetic braking is a simple application of basic magnetohydrodynamics. Consider a cloud with at least some of its magnetic field lines extending far into the surrounding medium, which is supposed to have for simplicity a uniform angular velocity Ω_o. If the cloud has an angular velocity $\bar{\Omega} > \Omega_o$, the shear generates a toroidal component $\mathbf{B}_t$ from the initial poloidal component $\mathbf{B}_p$. The component $(\nabla \times \mathbf{B}_t) \times \mathbf{B}_p/4\pi$ of the magnetic force density exerts a torque which removes angular momentum from the cloud and transfers it to the surroundings; angular momentum is transported outwards by torsional Alfvén waves, so that a noncontracting cloud will asymptotically be corotating with the surroundings. It is at once clear that the actual evolution of the cloud depends on whether the cloud mass is subcritical or supercritical. The contraction of a subcritical cloud is limited by the magnetic forces that grow as the field is distorted, as well as by centrifugal forces. Some contraction occurs following angular momentum removal, but ultimately the cloud will reach equilibrium with gravity balanced by magnetic force assisted by thermal pressure and the centrifugal force of corotation. By contrast, a cloud with supercritical mass cannot be held up by magnetic forces alone. If initially centrifugal force and thermal and turbulent pressure are weak, the cloud contracts at the magnetically-diluted free-fall rate, spins up and so generates magnetic torques. The question then is whether, in spite of the contraction, magnetic braking can nevertheless be efficient enough to keep the cloud in near corotation with the background, or at least keep the centrifugal forces weak compared with gravity. If so, then collapse will continue at roughly the free-fall rate, accompanied by quite efficient braking; but if not, then we may expect the cloud to approach thermo-magneto-centrifugo-gravitational balance, with its subsequent contraction rate determined by the rate of braking rather than by gravity.

The issue clearly depends on time scales. We anticipate the answer to the query just posed, and estimate a lower limit to the characteristic time of braking by systematically exaggerating the rate at which angular momentum is transported outwards. We assume that all field lines remain infinite and so contribute to the braking, as shown in Fig. 1. We consider first the axisymmetric problem, with the rotation and magnetic axes coinciding. Along the part of the field with approximately radial field lines the angular velocity obeys the standard one-dimensional wave equation, but with the wave speed $B_r/(4\pi\rho)^{1/2}$ decreasing with r. Then in the simplest illustrative example, with a discontinuity in Ω imposed at the cloud surface at $t = 0$, a wave front advances with the local Alfvén speed, but the angular velocity gradient behind the wave front is negative (in contrast to the cylindrical problem [Ebert et al. 1960], for which all waves travel undistorted). Thus we compute an upper limit to the efficiency of braking if we replace the actual solution by one in which Ω does stay constant behind the wave front. A lower limit to the characteristic time of braking τ_b can then be easily estimated by requiring that the moment of inertia of the cloud be equal to that of the external gas defined by the position of the Alfvén wave fronts at time τ_b:

$$\tau_b = \frac{\sqrt{\pi}}{2}\frac{1}{\rho_0^{1/2}}\frac{\mathcal{M}}{\mathcal{F}}\left(\frac{\bar{R}}{R_0}\right)^{6/5}\frac{1}{K^{2/5}} \tag{14}$$

where K is again defined by $\mathcal{M} = (4\pi\rho_o R_o^3/3)K$. This expression shows the dependence on the two cloud invariants $\mathcal{M}$ and $\mathcal{F}$, and on the density ρ_o of the medium that is absorbing the angular momentum. The shape of the field enters through the nontrivial factor $(\bar{R}/R_o)^{6/5}$, which makes the time shorter than for a cylindrical model.

Equation (14) can be written in the equivalent forms

$$\tau_b = \frac{\sqrt{2}}{3}\left(\frac{\bar{\rho}}{\rho_0}\right)^{1/2}\frac{\bar{a}}{\bar{B}G^{1/2}}\left(\frac{\bar{R}}{R_0}\right)^{6/5}\frac{1}{K^{2/5}} \tag{15}$$

in terms of cloud quantities $\bar{B}$, $\bar{\rho}$; or as

$$\tau_b = \frac{2\sqrt{\pi}}{3}\frac{\rho_0^{1/2}R_0}{B_0}\left(\frac{\bar{R}}{R_0}\right)^{6/5}K^{3/5} \tag{16}$$

in terms of background values B_0, ρ_0. It is clear from Eqs. (14) and (16) that τ_b does not depend on $\bar{a}$, which, however, fixes the flattening of the cloud and so also the time of travel of Alfvén waves through the cloud. This is normally found to be less than τ_b, so justifying the implicit assumption that the braking time is fixed essentially by the external travel time. (The factor $\bar{a}$ in Eq. [15] is exactly compensated for a cloud of given mass by $\bar{\rho}^{1/2}$.)

We now compare the estimate given by Eq. (14) with the free-fall time, finding

$$\frac{\tau_b}{\tau_{ff}} \simeq \frac{2}{3}\left(\frac{\mathcal{M}}{\mathcal{M}_c}\right)\left(\frac{R_0}{\bar{R}}\right)^{3/10}K^{1/10} \tag{17}$$

where $\mathcal{M}_c = 2\sqrt{3}\mathcal{F}/3\pi\sqrt{G}$ is the standard virial theorem form (Eq. 1) for the critical mass of a spherical cloud. Ignoring the weak dependence of Eq. (17) on K and on $(R_o/\bar{R})$, we see that the ratio $\mathcal{M}/\mathcal{M}_c$ plays a dual role in the theory; it is the parameter that decides whether the cloud has enough mass for indefinite gravitational contraction, and also whether the braking time is longer or shorter than the free-fall time. Thus if $\mathcal{M}/\mathcal{M}_c < 1$, $\tau_b < \tau_{ff}$, but collapse is in any case limited by the magnetic forces. Corotation is established quite quickly, but the amount of angular momentum lost is limited, simply because the magnetic forces limit the degree of contraction. Recall that if $\mathcal{M}$ is close to though less than $\mathcal{M}_c$, the cloud radius $\bar{R} << R_o$ (cf. Eq. 10), so that with $\bar{\Omega} = \Omega_o$, $\bar{\Omega}^2\bar{R} <<$

$G\mathcal{M}/\bar{R}^2$. On the other hand, if $\mathcal{M}/\mathcal{M}_c > 1$, then magnetic forces alone cannot stop the contraction, and since $\tau_b > \tau_{ff}$, an initially slowly rotating cloud will begin its collapse effectively conserving its angular momentum; but once it has spun up sufficiently it will indeed settle into a state of approximate magneto-thermo-centrifugo-gravitational equilibrium, normally with $\bar{\Omega}^2\bar{R} \simeq G\mathcal{M}/\bar{R}^2$. Contraction will continue, but now at a rate fixed by magnetic braking, and so in the time scale τ_b which is longer than τ_{ff}.

These conclusions are confirmed by detailed numerical treatment. In the latest study (Mestel and Paris 1984) of the axisymmetric problem, supercritical clouds are modeled by uniform oblate spheroids; the virial theorem is used to describe the contraction in both dimensions, and the upper limit to the rate of removal of angular momentum is computed by the approximation outlined above. It is found that the contraction path of the cloud oscillates about the curve defined by force balance, as anticipated.

Arguments similar to those used above suggest that when the magnetic and rotation axes are perpendicular, the quasi-radial field yields a braking efficiency of the same order, being again significantly larger than for an aligned cylindrical field. This is to some extent confirmed by the preliminary study in Mestel and Paris (1984) and in numerical studies by Dorfi (1982). An exact treatment (Mouschovias and Paleologou 1979) of an illustrative example for the perpendicular case—the braking of a stationary cylindrical cloud with a basic field directed everywhere along cylindrical radii—yields again a markedly shorter braking time than for the case of a cylinder with the basic field parallel to the axis. The lesson to be drawn is that estimates of the braking time are to some extent sensitive to the model adopted for the basic field. Again, if the mass of the cloud is subcritical, the perpendicular rotator will ultimately corotate with the background. During the approach to the asymptotic state, the centrifugal forces may cause outward motion of the surrounding gas, so that the problem will have some features in common with that of a magnetic stellar wind, and with the transport of angular momentum by waves limited by the existence of an "Alfvénic surface." An interesting prediction by Mouschovias and Paleologou is the recurrent appearance of retrograde rotation within their perpendicular cylindrical model. Some observational support comes from Young et al. (1981), who find that the core of a particular dark cloud rotates in the opposite sense to the bulk. When the basic magnetic field is inclined to the rotation axis, a precessional torque is generated as well as the braking torque. A preliminary study (Paris 1971) indicated that this would act to rotate the angular momentum vector of the cloud towards the direction of the field.

All these studies adopt idealized models for the basic magnetic field structure. A cloud of mass less than but close to $\mathcal{M}_c$ will ultimately feel only a weak centrifugal force, so that one or other of the models discussed in Sec. II is appropriate, as long as field detachment has not occurred. When $\mathcal{M} > \mathcal{M}_c$, the cloud continues to contract, and the field structure is determined by the gas motion; but computer programs to construct the evolving magnetic, rotation, and velocity fields are often bedevilled by side effects of the artificial viscosity introduced to ensure numerical convergence. The consequent unphysical diffusion may be misleading; but equally, an adequate physical theory of diffusion is all the more essential, especially since field line disconnection from the background field can clearly affect the efficiency of magnetic braking (cf. Sec. IV).

In spite of the patent limitations of the studies to date, it appears that the rotational evolution of a cloud depends rather sensitively on whether it begins its life with a subcritical or supercritical mass. With $\mathcal{M}$ subcritical, the magnetic stresses are able to enforce corotation with the background of at least those parts of the cloud which have retained infinite field lines. A generalization of the models of Sec. II could have magneto-gravitational balance holding in the central regions and approximate centrifugo-gravitational balance in the outer parts with detached field lines; however, for illustration, we temporarily assume that no field disconnection has occurred. Then if the ratio $f \equiv \mathcal{F}/\pi^2 G^{1/2}\mathcal{M}$ decreases steadily, due either to accretion of gas along the field lines, or more probably to slow but significant flux leakage, but in a time longer than the magnetic braking time, then the picture is of a slow contraction with corotation being maintained. As f approaches f_c (we have already noted the apparent insensitivity of f_c to the precise mass-flux distribution), the centrifugal force of corotation felt by most of the gas will become small because of the growing central condensation; contraction with Ω maintained constant corresponds to a very large reduction in the cloud angular momentum. Once the cloud has become supercritical in mass (e.g., because of accelerated flux leakage at the dense molecular cloud phase), then we are left with a slowly rotating cloud which will then collapse gravitationally and perhaps also fragment, conserving its remaining angular momentum; but the consequent spin-up of a fragment will not yield a centrifugal acceleration again comparable with gravity until densities are reached well into the opaque phases (Mestel 1965*b*; Mouschovias 1977; Mestel and Paris 1979). By contrast, we have seen that a cloud formed with a supercritical mass will steadily contract at a rate fixed by magnetic braking, and with centrifugal forces normally comparable with gravity at all stages. Fragmentation following flattening along the field may take place, but contraction of masses of stellar order to high densities depends on a continuing redistribution of angular momentum. If a substantial magnetic flux is retained right up to the opaque phases, then magnetic torques may remain the dominant process. One can hardly expect the field of a fragment to remain indefinitely a distorted part of the background galactic field; however, magnetic torques may be able, for example, to convert much of a fragment's spin angular momentum into that in the mutual orbit of a binary pair. But if most of the flux is lost at the dense molecular cloud phase, then an initially supercritical cloud will find itself weakly magnetic but retain a high angular momentum. The subsequent history of such a cloud remains conjectural. It may fragment into subcondensations of stellar order, but again it is difficult to see how the fragments can reach pre-main sequence densities without some redistribution of angular momentum. If a fragment remains in a more or less stable state of thermo-centrifugo-gravitational equilibrium, then the dust grains may settle into a thin disk, which is subsequently the locale for the formation of cometesimals by gravitational instability (Goldreich and Ward 1973; L. Biermann and K. W. Michel 1978). Another possibility is that the cloud evolves through redistribution of angular momentum by gravitational torques (J. E. Pringle, personal communication; R. B. Larson 1984). For the moment, we emphasize that a quite modest variation in the parameter f can make the difference between a supercritical and subcritical case, with radically different subsequent evolution. Observational selection is likely to favor the discovery of clouds of subcritical mass, but supercritical clouds or fragments should also occur and are of cosmogonical interest.

IV. FLUX LEAKAGE

We have already noted the profound qualitative changes that result from relaxation of the severe flux-freezing constraint. Flux leakage in these early stages of star formation occurs through the process often referred to as ambipolar diffusion. The electrons and ions making up the small charged fraction of the gas (for brevity referred to as the plasma) remain inductively coupled to the magnetic field lines; the magnetic force drives the plasma plus field through the neutral gas, at a rate fixed by the balance of the magnetic force with the frictional force acting on the ions through mutual collisions with the neutral particles (Mestel and Spitzer 1956). Thus, e.g., in each of the equilibrium models of Sec. II, the gravitationally distorted field lines are steadily trying to straighten themselves, so causing both flux leakage from the cloud as a whole and flux redistribution within the cloud. The cloud adjusts itself to a new equilibrium corresponding to the reduced parameter $f = \mathcal{F}/\pi^2 G^{1/2}\mathcal{M}$ and the associated change in the mass-flux distribution, as long as an equilibrium state exists; but ultimately, either the cloud as a whole or a section of it (e.g., the central core) will find itself supercritical in mass and will begin to contract.

The frictional force density acting on the ions is conveniently written as

$$\alpha\left(\frac{n_i}{n_n}\right)\rho^2(\mathbf{v}_n - \mathbf{v}_i) \tag{18}$$

where n_i, $\mathbf{v}_i$, n_n, $\mathbf{v}_n$ are the number density and mean velocity of the ions and neutrals, respectively, and $\alpha = \langle\sigma v\rangle/2m_H$, with the rate coefficient for ion-H_2 collisions $\langle\sigma v\rangle = 1.9 \times 10^{-9}\ cm^3 s^{-1}$ (Osterbrock 1961), yielding $\alpha = 5.7 \times$

10^{14}. The crucial question is of course that of the characteristic time for significant flux leakage, and this depends both on the geometry of the body considered and on the ionization fraction n_i/n_n, which is determined by the complicated details of the microphysics. In a body that is roughly an oblate spheroid about the direction of the magnetic axis, and with $f \equiv \mathcal{F}/\pi^2 G^{1/2}\mathcal{M}$ close to the critical value f_c, then the characteristic time τ_{ad} for substantial flux leakage is

$$\begin{aligned}\tau_{ad} &\simeq \frac{R}{|\mathbf{v}_n - \mathbf{v}_i|} \simeq \alpha \frac{(n_i/n_n)\rho^2 R}{|(\nabla \times \mathbf{B}) \times \mathbf{B}/4\pi|} \\ &\simeq \alpha \left(\frac{n_i}{n_n}\right)\left(\frac{4\pi\rho R^2 z}{3}\right)^2 \frac{(9/16\pi^2 R^3 z^2)}{(\mathbf{B}^2/4\pi z)} \\ &\simeq \alpha \left(\frac{n_i}{n_n}\right)\left(\frac{\mathcal{M}^2}{\mathcal{F}^2}\right)\left(\frac{9\pi}{4}\right)\left(\frac{R}{z}\right) \simeq 10^{14}\left(\frac{n_i}{n_n}\right)\left(\frac{R}{z}\right)\end{aligned} \tag{19}$$

in years. (Note that the flattening along the field increases the density by R/z, but the increased curvature of the field lines introduces a factor R/z into the magnetic force density, so that the time is increased by only one factor R/z.) It is at once clear that only in molecular clouds with very low n_i/n_n will τ_{ad} be short enough for the process to be of global interest.

Detailed studies must take account of the variation of (n_i/n_n) with the varying density through a cloud. The rate of ionization depends on the background galactic ultraviolet, X-ray, and cosmic ray fluxes, and on their attenuation in passage through dense nearly neutral clouds. Silk and Norman (1983) have pointed out that locally-produced X-rays from newly-formed pre-main sequence stars can increase markedly the ionization rate: a star formation model that is critically dependent on flux leakage could be self-regulating. The destruction of plasma may be significantly accelerated if most of the positive charges are attached to molecular ions which are neutralized by the comparatively fast chemical process of dissociative recombination. There is the opposing effect of charge-exchange reactions which can neutralize molecular ions by producing metallic (atomic) ions, which recombine with electrons at the much slower radiative rate, but this in turn depends on a sufficient fraction of the metallic atoms remaining free rather than being locked up in dust grains. A further complication enters at high cloud densities, when the corresponding ion density becomes so low that the magnetic field becomes inductively coupled to the negatively charged dust grains rather than to the ions (Nakano and Umebayashi 1980).

A complete treatment of the microphysics is complicated, and bedevilled by uncertainties. However, it is normally the case that n_i/n_n decreases with increasing density, simply because ionization rates vary as ρ and recombination rates roughly as ρ^2. This is the basic cause of an instability analysed in papers by Nakano (1976,1983*b*) which do, in fact, discuss the microphysical details. A region of a subcritical cloud with lower magnetic flux than the average will compensate by becoming denser so as to build up the magnetic force to the level required to balance gravity and pressure; but at higher densities the ratio n_i/n_n decreases, so allowing faster flux leakage, etc. until the region becomes locally supercritical in mass and can condense gravitationally out of the cloud. Probably the most important consequence of flux leakage is this transformation of the cloud as a whole or a local region (such as the core) from having a subcritical to a supercritical $G^{1/2}\mathcal{M}/\mathcal{F}$; and as already noted, if the transition is moderately slow, corotation with the surroundings may be maintained, corresponding to an enormous loss of angular momentum.

However, one should note that what happens is the reduction of flux to below the level where it can prevent contraction; this does not imply that the magnetic field has become dynamically insignificant, still less that the flux has become as low as that inferred to be within the main sequence stars with even the strongest observed magnetic fields. This is again brought out by the two-dimensional computations of D. C. Black and E. H. Scott (1982), who parametrize the fractional ionization n_i/n_n by kn_n^{-q}, where k and q are adjustable constants. These authors find that if $q < 0.8$, the field stays nearly frozen into the collapsing cloud, and even if $q \simeq 1$, the field is still amplified during collapse in spite of the very low values of n_i/n_n. Asymptotic flux freezing holds if the drift velocity v_D of the plasma relative to the neutral gas becomes small compared with the overall contraction velocity. If the magnetic field strength inside the cloud is increasing as ρ^x, then the flux tube that originally threaded the cloud may be thought of as having a radius $R' \propto \rho^{-x/2}$. The magnetic force density F_B will be at least $B^2/4\pi R' \propto \rho^{5x/2}$, and as the radius of curvature of the field is likely to be smaller than R', a somewhat stronger law may be appropriate (Black and Scott suggest ρ^{3x}). The drift velocity $v_D \propto F_B/\rho^2(n_i/n_n) \propto \rho^y$, where $y = (5x/2 - 2 + q)$ at least, and possibly $(3x - 2 + q)$. A supercritical cloud freely falling in two dimensions and satisfying the hydrostatic balance condition (Eq. 8) in the third dimension (along the field) has an inward velocity $\propto \rho^{1/8}$. Then if the appropriate flux-freezing law $B \propto \rho^{1/2}$ (cf. Sec. II) is to hold, with $F_B \propto \rho^{5x/2}$, we require $q < 7/8$, and with $F_B \propto \rho^{3x}$, $q < 5/8$; in particular, the case $q = 0.5$ (valid when a linear ionization rate balances a quadratic recombination rate) satisfies either criterion.

The discussion assumes implicitly that the constant k in the assumed relation $n_i/n_n = kn_n^{-q}$ has not undergone any sudden reduction. If, on the contrary, n_i/n_n were reduced by an order of magnitude or more in a time short compared with the free-fall time, then the flux within the cloud or blob would clearly decrease sharply, reducing the magnetic force to a value which does yield a drift velocity less than the free-fall velocity; but again, if q is not too large, the remaining flux would be approximately trapped in the condensing mass.

A number of authors (Shu 1983; Paleologou and Mouschovias 1983; Scott 1984) have studied the analogous problem in the simpler plane-parallel geometry, with the magnetic field having just the component $B_x(z)$. The magnetic and thermal pressure gradients balance the z-component of gravity; the neutral particles slowly settle, with again the mutual ion-neutral friction the means of transferring the magnetic force to act on the neutrals, which feel virtually all the gravitational force density. The fundamental difference between this problem and that of Black and Scott is that in plane-parallel geometry the gravitational field saturates at the value $4\pi G$ (mass/unit area), so that there is an asymptotic equilibrium state in which the magnetic field plus plasma has drifted out and gravity is balanced by thermal pressure alone. Shu has solved the problem for the case $q = 0.5$; he discovers a universal asymptotic self-similar form for B_x as the neutral gas evolves towards the ultimate state of thermo-gravitational equilibrium. We shall see below that there is one important problem for which this model is an acceptable approximation over a significant time scale; but in general, the very different properties of both the gravitational and magnetic fields in this geometry suggest that detailed conclusions should be applied with caution.

As already emphasized, the basic magnetic impediment to collapse and fragmentation is, in principle, resolved once flux diffusion has transformed a subcritical mass into a supercritical one. It is nevertheless of interest and importance to inquire (see, e.g., Paleologou and Mouschovias 1983) whether ambipolar diffusion in the diffuse phases may sometimes be so efficient (due to a catastrophic reduction in n_i/n_n) that the remnant flux is as low as is inferred from observations of even the strongest main sequence magnetic stars. On the other hand, we should note that excess flux may be lost by other processes at later phases, with possible striking observational consequences. For example, once a fragment has become opaque, then its contraction rate will be fixed by heat loss rather than by either (magnetically-diluted) gravitation or by magnetic braking. It is very likely that over this more leisurely time scale, excess magnetic flux may float to the surface via magnetic buoyancy (a secular instability, dependent on heat exchange), to be dissipated in the low-density atmosphere. The suggestion has been made that the T Tauri phenomenon (in which sometimes as much as 10% of the luminosity is emerging as mechanical flux) may have as its source the energy of the remnant primeval flux.

A sizeable magnetic flux retained by a supercritical mass can continue to affect the subsequent diffuse phase of dynamically controlled contraction. Consider for the moment the collapse of a cool, nonmagnetic, slowly rotating gas cloud. In principle this can fragment into systematically smaller masses, with the gravitational energy released and randomized available ultimately to maintain the system of fragments as a protostar cluster. An absolute lower limit to the masses that can form in this way is fixed by the steady growth in optical depth, which ensures that ultimately the heat of compression generated by a contracting fragment is trapped so that isothermality ceases to hold. All estimates of this opacity-limited minimum mass are embarrassingly low. It may be that even in the nonmagnetic problem, the opacity argument merely supplies an extreme lower limit, and that it is instead the dynamics of the later stages of star formation that fixes both the mean protostellar mass and the mass-spectrum; if so, this encourages one to look more closely at the effect of a nearly frozen-in flux on the dynamics of a cool cloud. The field will clearly increase the minimum mass that can form at any epoch, since the sound speed in the Jeans mass has to be replaced by the magneto-sonic speed $(a^2 + B^2/4\pi\rho)^{1/2}$; but as already noted in Sec. I, flow along the field lines (such as spontaneous flattening) allows systematically smaller masses to form within an already supercritical cloud. In a rotating magnetic cloud but with the rotation and magnetic vectors nearly parallel, the centrifugal forces do not interfere with the flattening. The minimum mass that can form is again fixed by the breakdown in isothermality due to the increase in optical depth, and it is not clear that this minimum will be greatly increased by the presence of the field. By contrast, a rapidly rotating cloud or fragment with the rotation and magnetic axes inclined at a large angle may achieve a dynamically stable equilibrium state as a quasi-ellipsoid with its longest axis parallel to the magnetic axis (Mestel 1965*b*). If the cloud is slightly supercritical, then as magnetic braking removes angular momentum it will contract approximately *isotropically,* with the gravitational energy released being converted into magnetic energy and kinetic energy of rotation, and so will not fragment ($B/\rho^{2/3}$ stays constant instead of decreasing). The crucial difference is that the centrifugal forces of spin now prevent flattening along the field. If the steady contraction through angular momentum loss were to continue indefinitely, the cloud would reach the opaque phase as one rapidly rotating body; the gravitational energy required to supply the random kinetic energy of fragments would instead have been largely converted into magnetic energy, to be subsequently dissipated. More plausibly, the evolution of such a cloud would be a competition between angular momentum loss and slow flux leakage. As the cloud contracts, the steady detachment of field lines from the background would reduce the rate of angular momentum loss; meanwhile, flux leakage (assisted by hydromagnetic instabilities near *O*-type neutral points) would cause the cloud to flatten, approaching a structure which could be approximated by the plane-parallel models just mentioned. However, once a little more than half the flux has been lost, the rotating cloud can in principle fission into a magnetically linked binary pair, each of which can contract to much higher densities through magnetic conversion of spin angular momentum into orbital. Subsequent further flux leakage from each binary component may allow the fission process to be repeated. The model needs to be studied in detail, in order to estimate the masses that ultimately reach the opaque phase, but they are virtually certain to be larger than masses estimated from the opacity limit. The essential points are that in the diffuse phase, fragmentation occurs only because of the reduction in $B/\rho^{2/3}$ by flux leakage, whereas the increase of density, and so the approach to optically thick conditions, occurs also through angular momentum loss, which causes isotropic contraction with $B/\rho^{2/3}$ remaining constant. The ultimate growth in opacity again causes isothermality to break down and so halts gravitational break-up, but the final masses are fixed by the previous history of competition between flux leakage and magnetic braking.

Finally, we note the role of ambipolar diffusion in the subtle question of magnetic field line detachment. This is illustrated by referring back to the disk-like equilibrium models of Sec. II. We then noted that the equilibrium states with limiting parameters f_c have a cusp-like neutral point just outside; we shall now see that the equilibrium of subcritical models may break down through spontaneous changes in field topology. Outside the central regions, the field has comparable components $B_{\tilde{\omega}}$ and B_z, and so in addition to the force $(B_{\tilde{\omega}})_+ B_z/2\pi$ per unit area which balances gravity in the $\tilde{\omega}$-direction, there is now a magnetic pinching stress $B_{\tilde{\omega}}^2/8\pi$ acting on both faces. Note the contrast with the previously cited models where the magnetic field is symmetric rather than antisymmetric in the equator and so exerts forces opposing rather than assisting gravity. The models in reality have a finite thickness, fixed initially by the balance between the self-gravitation, the pressure gradient (essentially of the neutral gas) and the magnetic pressure gradient. Again this magnetic force is felt directly by the plasma, which, therefore, drifts towards the equator, with the magnetic force being transferred to the neutrals via the mutual friction. The process stops when the plasma forms a magnetically pinched thin layer at the equator and the neutral gas conforms to thermo-gravitational balance. The time scale for this is $\simeq \alpha(n_i/n_n)\, 2\pi M^2/B_{\tilde{\omega}}^2$ (depending on the area density M, rather than the volume density ρ, because the thickness $\bar{z}$ enters multiplying ρ). In a strongly centrally condensed model the ratio $M/B_{\tilde{\omega}}$ decreases sharply outside $\tilde{\omega} = \ell R$, so that this time for ambipolar diffusion in the z-direction is locally much shorter than the time for flux to leak out of the cloud by diffusion in the $\tilde{\omega}$-direction. Note that this process has not led to any change in field topology, because the field lines have been assumed to remain tied to the plasma. However, the ions and electrons are now compressed to densities that are at least 10^4 times larger, and so will largely recombine; the neutralized former plasma will drift across the field lines, leaving again only a small fraction in pressure balance with the magnetic pressure $B_{\tilde{\omega}}^2/8\pi$. At the same time, the steady reduction in the scale of variation of $B_{\tilde{\omega}}$ in the z-direction increases the local ratio of the radial magnetic force density to the gravitational; the magnetic field lines will tend to straighten, so dragging out the remaining charges in a dynamical time scale, and reducing still further the separation of the pinching field lines. Thus, the freezing of the field into the charged component of the gas must break down; Ohmic diffusion yields first a cusp-like and then O- and X-type neutral points (Mestel and Strittmatter 1967). Rough estimates suggest that for the disk-like

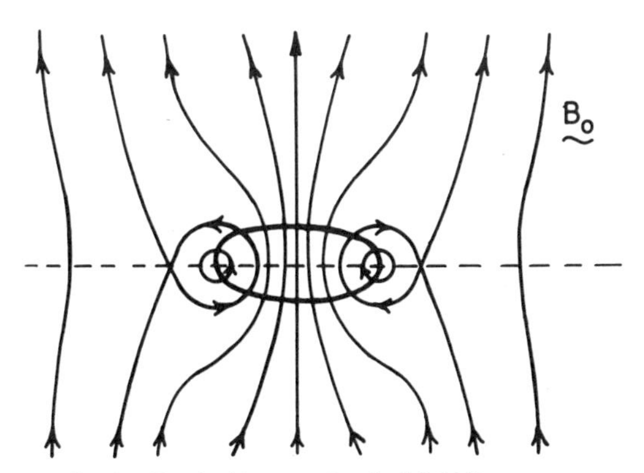

Fig. 2. Cloud with some detached field lines.

models discussed in Sec. II, the time scale of the whole process is essentially that of the initial equatorial concentration of plasma, and that for strongly centrally condensed bodies, most of the field lines may have time to disconnect from the background field. The disconnected part of the field will itself be subject to topological hydromagnetic instabilities near *O*-type neutral points, which may cause locally accelerated flux decay.

These changes in the field structure are clearly significant for the magnetic braking problem. Most studies have implicitly or explicitly maximized the efficiency of braking by adopting field structures which link all parts of a cloud with the surrounding gas that absorbs the angular momentum; whereas in a cloud as shown in Fig. 2, the inner parts will suffer continuous braking, while the outer parts may effectively conserve the angular momentum they have at the epoch of field disconnection. The same picture will be applicable to strongly magnetic, supercritical fragments that have condensed out of a massive cloud. In Sec. III we noted a suggestion that a cold, rapidly rotating, weakly magnetic fragment could be the locale for the formation of cometesimals. Figure 2 suggests a unitary picture in which the inner regions suffer so much more angular momentum loss that they can contract, for example, to the orbit of Pluto to become the protosolar nebula, while the high angular momentum, outer regions are the locale of the Oort cometary cloud.

PROCESSES AND PROBLEMS IN SECONDARY STAR FORMATION

RICHARD I. KLEIN
University of California at Berkeley

RODNEY W. WHITAKER AND MAXWELL T. SANDFORD II
Los Alamos National Laboratory

Recent developments relating the conditions in molecular clouds to star formation triggered by a prior stellar generation are reviewed. Primary processes are those that lead to the formation of a first stellar generation. The secondary processes that produce stars in response to effects caused by existing stars are compared and evaluated in terms of observational data presently available. We discuss the role of turbulence to produce clumpy cloud structures and introduce new work on colliding intercloud gas flows leading to nonlinear inhomogeneous cloud structures in an initially smooth cloud. This clumpy morphology has important consequences for secondary formation. The triggering processes of supernovae, stellar winds, and H II regions are discussed with emphasis on the consequences for radiation-driven implosion as a promising secondary star formation mechanism. Detailed two-dimensional, radiation-hydrodynamic calculations of radiation-driven implosion are discussed. This mechanism is shown to be highly efficient in synchronizing the formation of new stars in $\sim 1\text{–}3 \times 10^4$ *yr and could account for the recent evidence for new massive star formation in several ultracompact H II regions. It is concluded that, while no single theory adequately explains the variety of star formation observed, a uniform description of star formation is likely to involve several secondary processes. Advances in the theory of star formation will require multi-dimensional calculations of coupled processes. Important nonlinear interactions include hydrodynamics, radiation transport, and magnetic fields.*

Theories that explain the formation of stars lie at the foundation of astronomy because our view of nearly all structure in the universe depends upon their existence. Progress in recent years has been possible because of instrumentation advances that have allowed observations over a wider portion of the electromagnetic spectrum. The series of steps whereby a diffuse cloud of molecular gas undergoes initial compression, eventually leading to a first generation of protostellar objects and subsequently giving rise to future generations of new stars, is slowly becoming unraveled. The early 1980s have brought radio wavelength observations at high resolution with the Very Large Array (VLA), a new infrared survey by the Infrared Astronomical Satellite (IRAS), and supercomputers which have made possible more realistic theoretical models. This chapter reviews recent theoretical and observational work bearing on processes that form stars as a consequence of interactions between existing stars and surrounding cloud material. The processes are referred to as *secondary* or *induced* star formation as opposed to *spontaneous* star formation. Because secondary star formation processes depend on molecular cloud morphology, we also briefly review theories for the development of the physical conditions within clouds.

We distinguish between primary and secondary star formation by following B. G. Elmegreen's (1980) definitions. Spontaneous (primary) star formation follows from the gravitational collapse of part or all of a molecular cloud, and secondary star formation includes processes triggered by the first generation of stars, or by occurrences external to the cloud. Suitable triggers might include cloud collisions, supernova detonations, H II region expansions, stellar winds, or spiral density wave shocks.

New observations confirm the view that star formation is ongoing in our Galaxy and the newest data surprisingly favor low-mass ($\leq 5\ M_{\odot}$) objects, due to the short life and apparent rarity of O stars. It is now known that stars form deep within dense cloud cores as well as near their edges, in isolated dark globules, and perhaps within elephant trunks and cometary nebulae. The initial mass function (IMF) determined from field and cluster stars (see Scalo 1985*b*) appears to have a certain universal character, although there are variations at low mass for cluster stars. At least one young cluster (NGC 2264) studied by M. T. Adams et al. (1983) shows no deficiency of low-mass stars when compared with the field. Meusinger (1983) studied the present-day mass function and found that temporal variation of the IMF must depend upon stellar mass. An IMF that decreases exponentially in time, over the entire mass range, fails to produce the observed present-day mass function. This result is independent of the star formation rate used; therefore, it makes a strong case for the continuous formation of different masses. Several different primary and secondary processes probably contribute to the star formation we observe and their relative importance is likely to depend upon the cloud structure. The complexity of molecular cloud evolution undoubtedly supports processes that depend intimately upon one another, but once initiated they appear to compete for dominance in the birth of stars. It is, therefore, improbable that a single, complex theory will prove comprehensive. The IMF is determined from observations of the accessible regions and it consequently represents an agglomeration of the "bones" left from all star formation and does not necessarily represent a single process (see chapter by Leung).

This review concentrates on recent developments of secondary star formation processes. We emphasize processes that depend on interactions of existing stars with the surrounding cloud morphology. In Sec. I we discuss the development of an inhomogeneous (clumpy) cloud environment that appears to influence secondary star formation. We briefly discuss the roles of turbulence and cloud fragmentation, and we introduce new work on gas flow interactions. In Sec. II we review secondary star formation mechanisms including those driven by supernovae, stellar winds, and expanding H II regions. We discuss new work on a theory for the implosion of cloud clumps by radiation from O and B stars. This chapter introduces a new generation of numerical models based upon the solution of the equations for two-dimensional, time-dependent, radiation hydrodynamics. In Sec. III we consider possible directions for future research.

I. DEVELOPMENT OF THE CLOUD ENVIRONMENT

A. Does Turbulence Play a Significant Role in Star Formation?

It is becoming clear that molecular cloud evolution leading to star formation involves interdependent processes that combine to yield an initial stellar mass function having considerable variation (Freeman 1977). The internal structure of dense clouds may strongly influence the processes of star formation operating within the cloud. Cloud cores appear to have internal, supersonic, random motions and normally display a clumpy structure. Supersonic clumps could partake in a hierarchical cascade of energy dissipation which would affect the medium surrounding the primary generation of new stars. The stars formed in this primary generation in turn influence the next generation born within the clumpy cloud environment. Recent observations (Sargent 1979; see also chapter by Dickman) reveal that most clouds contain inhomogeneities (clumps) down to very small scales. Radio wavelength linewidths can be interpreted as due to supersonic internal motions which could easily result in density inhomogeneities. These inhomogeneities may dominate the structure of molecular clouds and set the scale for the mass spectrum of collapsing objects. Supersonic motion of density clumps in a lower-density ($\cong 10^3$ cm^{-3}) cloud medium suggests turbulent flow. If turbulent flow is indeed present in molecular clouds and plays a significant role in the dynamics, the conventional picture of fragmentation requires major revision.

Considerations such as these, and observations of disordered filamentary structure in the Taurus dark cloud led R. B. Larson (1979,1981) to study the velocity-size spectrum for a wide variety of interstellar regions that included isolated clouds, cloud complexes, and regions within giant clouds. In order to determine if motions within these regions follow a systematic flow suggestive of turbulence, he determined the three-dimensional velocity dispersion due to large- and small-scale variations, and thermal motions, and correlated this dispersion with region size. R. B. Larson (1981) found a relationship between velocity dispersion σ (km s^{-1}) and the region size L (pc) that can be represented by

$$\sigma = 1.1\, L^{0.38} \tag{1}$$

for size scales $0.1 \leq L \leq 1000$ pc. Because this power-law correlation extends over several orders of magnitude and includes very small scales, he suggested that the observed motions are components in a hierarchy of interstellar turbulent motion.

More recent studies (Myers 1983; Dame et al. 1984) essentially confirmed the velocity dispersion correlation with region size found by Larson, but gave a steeper exponential dependence $\sigma \propto L^{0.5-0.6}$ which resulted in a different constant coefficient. These differences may be attributed to both the larger sample used in current work, and to the specific determination of the dispersive velocity component. If most clouds are in virial equilibrium as R. B. Larson (1981) suggests, then the observed clumpiness of molecular clouds may result from processes other than gravitational collapse. The strong correlation between velocity dispersion and region size and the observation of irregular substructure in clouds may indicate the presence of turbulent flow.

Additional observational evidence supporting the presence of turbulence comes from studies of the rotational properties of molecular clouds. Fleck and Clark (1981) noted that if angular momentum transport is responsible for cloud rotation during cloud collapse, then the rotational angular velocity (ω) scales as $\omega \propto n_o^{-2/3}$, where n_o is the density required for gravitational collapse to occur. The Jeans criterion implies that low-mass clouds form in regions of high n_o, and momentum conservation therefore requires that they rotate more slowly than high-mass clouds. Observations indicate the opposite: small clouds rotate more rapidly than large ones. Additionally, if stars derived their angular momenta from galactic rotation, one would expect their angular momentum vectors to be aligned perpendicular to the galactic plane. Observations indicate, however, that early-type stars and field Ap stars have randomly distributed rotation axes (S. S. Huang and O. Struve 1954; Abt et al. 1972). Also, the orientation of eclipsing binary orbital planes displays no evidence for a preferred galactic distribution. Taken together, these observations suggest that the rotation of clouds does not result from the ordered motion of centrifugal balance, but from a more random process; perhaps the motions are turbulent.

Myers and Benson (1983) surveyed $\sim$ 100 dark clouds and found significant correlations between the velocity dispersion of CO and NH_3 molecular lines and the region size, as well as between mean density and size. These observations extended the work of R. B. Larson (1981) into the subsonic regime for dense cloud cores and confirmed the earlier correlations found for larger clouds. Myers and Benson (1983) found the velocity dispersion $\sigma \propto L^{0.5}$ and the cloud density $n \propto L^{-1.3}$, and pointed out that the correlation between the velocity dispersion and region size is not evidence for subsonic turbulence because it can occur as a consequence of the tendency for the clouds studied to be in virial equilibrium (which gives $\sigma \propto n^{1/2}L$), and to have nearly constant column density $N = nL \propto L^{-0.3}$. Unfortunately these correlations have equal significance, so it is unclear which of the relationships are fundamental and which are simply derivative. One cannot therefore conclude that turbulence is a fundamental process in dense cores until observations can be made with enough resolution to show significant differences in the correlations.

Some evidence for turbulent dissipation in dense cores was found in vigorous star-forming regions studied by Myers (1983). He noted that the detection of dense cores in the Taurus-Auriga and Ophiuchus regions, which are known to contain many T Tauri stars, is several times the rate in Aquila which is a region with few low-mass stars. Myers proposed that some cloud cores collapse rapidly and are therefore cut off from the source of turbulent energy, if this energy is supplied as in the Scalo and Pumphrey (1982) picture which attributes turbulence to collisions and drag by the clumps. Consequently, a possible conclusion is that some dense cores are on the verge of collapse because they dissipate their *initial* turbulent energy on a decay time scale $L/u \cong 5 \times 10^5$ yr (where u is characteristic fluid velocity) which is comparable to the core free-fall time. A possible problem is apparent if one considers the presence of magnetic fields. Assuming interstellar magnetic fields to be significant in molecular cloud regions, several workers have determined the leakage of these fields through the neutrals in a medium of low fractional ionization (Mestel and Spitzer 1956; Mouschovias 1981; Shu 1983). Shu and Tereby (1984) demonstrated that, for reasonable ratios (1 to 10) of the initial magnetic pressure to gas pressure, the time scale for an isothermal, self-gravitating slab to lose a significant part of its initial magnetic flux is $\sim$ 10 to 10^2 sound crossing times. For the cores observed by Myers (1983) this corresponds to $\sim 10^6$ to 10^7 yr leakage time. The time scale for collisional interactions and clump accretion to mediate turbulent energy transfer via the picture of Scalo and Pumphrey (1982) is also 10^6 to 10^7 yr, assuming a $\sim$ 1 pc cloud radius and $\sim$ 0.3 pc diameter clumps. The magnetic field leakage time is comparable to the turbulent energy transfer time, indicating that magnetic fields may indeed support clumps long enough for turbulence to affect the collapse of dense cores. It therefore appears that turbulent interactions may involve magnetic fields even down to size scales of dense cores. Cassen et al. (see their chapter) argue that the fundamental process operating in the creation of clumps and dense cores is ambipolar diffusion, but most existing theories avoid the question of how a cloud gets its original clumpy structure and it may be that clumps are the remains of a once turbulent magnetic eddy flow.

Additional evidence for the presence of turbulence and its importance in the star formation process may come from inspection of the low-mass end of the IMF. At low masses the IMF flattens and turns over for some open clusters (G. E. Miller and J. M. Scalo 1979, 1985*b*). This observation led J. H. Hunter and R. C. Fleck (1982) to investigate the influence of large-scale flows in clouds on the IMF. They assumed that imposed turbulent velocity fields, which affect gravitational instability, take the form of the Gaussian distribution suggested by observations (Dickman et al. 1980). After deriving the form of the Jeans mass with such a velocity distribution, they showed that a flattening or turnover of the IMF occurs for low-mass stars at the expected place in the spectrum. Moreover, they argued that the observed IMF variations for masses $\leq 1\ M_\odot$ (often taken as evidence against a universal IMF) are due to the expected variations in local turbulence within the primary clouds. Similar variations in the magnetic field strength in primary clouds could also be related to variations in turbulent structure.

Observations of the flow morphology in certain cloud systems such as the Taurus dark clouds offers compelling evidence for turbulent motion. Tajima and Leboeuf (1980) performed detailed calculations of the nonlinear Kelvin-Helmholtz instability in a compressible, supersonic flow with and without an external magnetic field parallel to the shear flow direction. These calculations used a magnetohydrodynamic particle code and the results (Tajima and Leboeuf 1980, see their Fig. 1) show the development of wavy filaments and turbulence strikingly similar to the patterns observed in molecular clouds. A possible conclusion from this calculation is that the morphology observed in clouds can indeed originate from fluid instabilities.

We conclude that the presence of some form of turbulent flow is likely present in many molecular clouds, that this motion can cause clumping and filamentary structure on many size scales, and that stars form in these regions. The important role of cloud inhomogeneities for secondary star formation processes is discussed in Sec. II.

B. Fragmentation in Molecular Clouds

The existence of gas fragments within molecular clouds is no longer in doubt. While fragmentation is clearly an important aspect of star formation, we cannot overemphasize that fragmentation theory does not appear to be the only explanation for the reason stars form, and against which observations should be compared. Since publication of the first *Protostars and Planets* book (Gehrels 1978) many calculations have appeared in the literature, analytic work has continued, and a few new approaches to fragmentation have been made. Tohline (1982) presented a comprehensive review of hydrodynamic collapse including a summary of earlier pressure-free calculations and comparisons of recent analytical results to three-dimensional (3D) collapse calculations. The reader is encouraged to consult this review for references to individual calculations and for specific numerical detail. A subsequent

(Tohline 1984) analytic approach to the collapse and fragmentation of rotating clouds clarified much of the numerical work. This work provides a new, unified framework in which to understand the variety of published numerical results.

The observational evidence in support of collapse and fragmentation primarily consists of clusters and of binary and multiple systems. The clumpy nature of molecular clouds (see chapter by Myers) is suggestive and is supported by observations of multiple infrared sources in dense cores (Beichman et al. 1979). However, Evans (see his chapter) discusses the possibility that scattering from clumps may be the origin of some multiple infrared sources.

We agree with Tohline's (1982) view that with presently available computers, 3D calculations can only provide information on the first stage of the fragmentation process. Physically detailed, high-resolution, 3D computations which use the present algorithms, require computational capability beyond the next generation of computers. For example, the current 3D hydrodynamic models require improvement by a factor of at least ten in each spatial dimension. Because the time step is related to the spatial resolution, the existing methods require $\sim 10 \times 10^3$ times more computing power; but the best factor that is likely to be available with the new class VII machines is only ~ 20. Thus, intrinsic fragmentation of molecular clouds must still be approached analytically or from an empirical, statistical viewpoint.

C. Fragment Interactions

Theoretical models for the IMF (see Scalo [1978, 1985*b*] for reviews of the IMF) are based upon the premise that molecular cloud cores fragment into clumps, and that the distribution of clump masses represents the mass distribution of the stars that form. These theories calculate the asymptotic evolution of the fragment masses from an initial distribution. Processes that increase the density in a clump such as molecular line cooling, clump collisions, and accretion all act to reduce the Jeans mass, thereby favoring star formation. The fragment interactions and physical conditions within the cloud are thus considered to be connected with the observed IMF. The fragmented cloud picture is supported by observations of small-scale intensity variations when clouds are viewed in CO radio lines and by suprathermal linewidths interpreted as due to space velocity variations between unresolved clumps. Massive stars are not found in certain quiescent star-forming regions (Myers and Benson 1983), and the formation process in these clouds may be a primary one. Thus, in some clouds, a study of fragment interactions may adequately describe the star formation process. In other clouds, stars form from clumps as a consequence of their interaction with ionizing radiation from O and B stars. The mass distribution and presence of clumps is therefore important in secondary star formation and we find it necessary to review the most recent theoretical work on the evolution of the fragment mass distribution.

Silk and Takahashi (1979) describe the interaction between clumps by a coagulation equation for the evolution of the number density $N(m,t)$ of clumps between mass m and $m + dm$ from the time of their formation to time t. Coagulation models exclude explicit consideration of the clump density (size), thermodynamics, and dynamical interactions with the ambient medium. Silk and Takahashi (1979) found analytic solutions for a coagulation equation which includes the effect of accretion. Their asymptotic solution gives a deficiency in the number of clumps of low mass due to the effect of accretion on the higher-mass clumps. This result means that the low-mass stars observed in some clusters either originate from the field, or possibly form as a result of evaporative implosions of the higher-mass clumps by hot stars.

A more detailed theory for fragment interactions was formulated by Pumphrey and Scalo (1983) in order to study the clump velocity distribution and to incorporate drag forces. Their formulation explicitly included dependence upon the parent cloud total mass, radius, and density. They wrote a kinetic equation for a clump distribution function that depends upon a clump size parameter (defined in terms of the individual fragment mass and density), and upon the fragment mass, position, and velocity vectors. Their equation is unfortunately too general to admit solutions for real systems, but they demonstrated that averaging in position and velocity space recovers the condensation-coalescence equation. The kinetic equation can be directly simulated with a numerical model which solves equations of motion for an ensemble of statistical particles. Pumphrey and Scalo (1983) contended that avoiding an explicit consideration of thermodynamic and hydrodynamic processes is an advantage because these difficult details are parameterized. Processes fundamental to clump interactions such as shocks, molecular cooling, and density compressions are incorporated only to the extent that their effects can be included in stochastic and collision terms. This is in fact a severe limitation because hydrodynamic calculations by Gilden (1984) showed that speeds for both clump collisions and drag effects are transonic and result in complicated hydrodynamic flows. For example, the conditions under which clump collisions lead to coalescence, fragmentation, or gravitational instability are more complex than assumed by Pumphrey and Scalo (1983). Thus, the additional detail provided by the kinetic equation in comparison with the coagulation theory does not seem justified. We conclude from Gilden's (1984) results, and from our own recent work (J. H. Hunter et al. 1984), that details of the clump interactions cannot be considered as strictly external processes. A rigorous simulation must start from microscopic kinetic equations from which the hydrodynamic equations can be recovered as an approximation (Gail and Sedlmayer 1979). It has also been demonstrated that the internal hydrodynamic evolution of clumps, e.g., their thermodynamic and dynamical condition, is fundamental to determining the time scale for star formation (Stahler 1983*a*; J. H. Hunter 1979).

J. H. Hunter et al. (1984) have investigated the interaction of large-scale ($\sim$ 1 pc) gas flows in molecular clouds. These flows may result from the action of forces external to the cloud: for example, expanding H II regions, spiral arm shocks, and supernovae detonations. Colliding gas flows form cool compressed regions. Gas flows of initial atomic hydrogen density 600 cm^{-3}, colliding at 10 km s^{-1} relative velocity initially form a disk with density $\sim 8 \times 10^4$ cm^{-3}. M. E. Stone (1970) studied colliding flows and concluded that the dense layers are dynamically unstable. B. G. Elmegreen and D. M. Elmegreen (1978) performed a linear stability analysis of gravitating layers and concluded that they fragment and form stars. Our results in Fig. 1 show the velocity vectors and atomic hydrogen density contours that result from the collision of two clumpy, 5 km s^{-1} flows, each of $\sim$ 1 pc length. The collision forms a self-gravitating disk that becomes unstable, fragments, and results in the formation of a collapsing protostellar cloud and several smaller compressed regions. The fragments that form are not well represented by a distribution of discrete clumps and their evolution is therefore not properly treated by coagulation and kinetic models. In particular, the central condensation that becomes Jeans unstable grows by accretion of gas moving at subsonic speed between two irregular shocks. The calculations show no evidence for fragment growth or dissipation due to clump collisions. In the more general case of three-dimensional gas flows, numerous clumps of many sizes are expected to form and their evolution is likely to be determined by the velocity field existing in gas decelerated through irregular shocks. It is thus possible that the observed CO linewidths and the clumpy structure in cloud cores result from fluid instabilities that are the consequences of the interaction of laminar gas flows. The models based on statistical interactions in a preexisting clump population seem too simple to be useful.

To summarize, neither coagulation nor kinetic clump interaction theories appear capable of describing completely the physical conditions within molecular cloud cores leading to star formation. Bastien (1981) critically examined fragment interaction models and showed that clump collisions are important only for masses $\geq 0.8\ M_\odot$. Low-mass stars therefore may form with the mass distribution determined by the initial fragmentation process, or else their number is determined by secondary star formation processes. A complete understanding of cloud clump morphology is beyond our present means. Until the phenomenology leading to star formation from the cloud clumps is better understood it seems inappropriate to pursue incomplete theoretical models that attempt to produce the universal IMF.

II. SECONDARY STAR FORMATION

Observations of young T Tauri associations forming from the ρ Ophiuchus and Taurus dark cloud material support a picture in which fragmentation

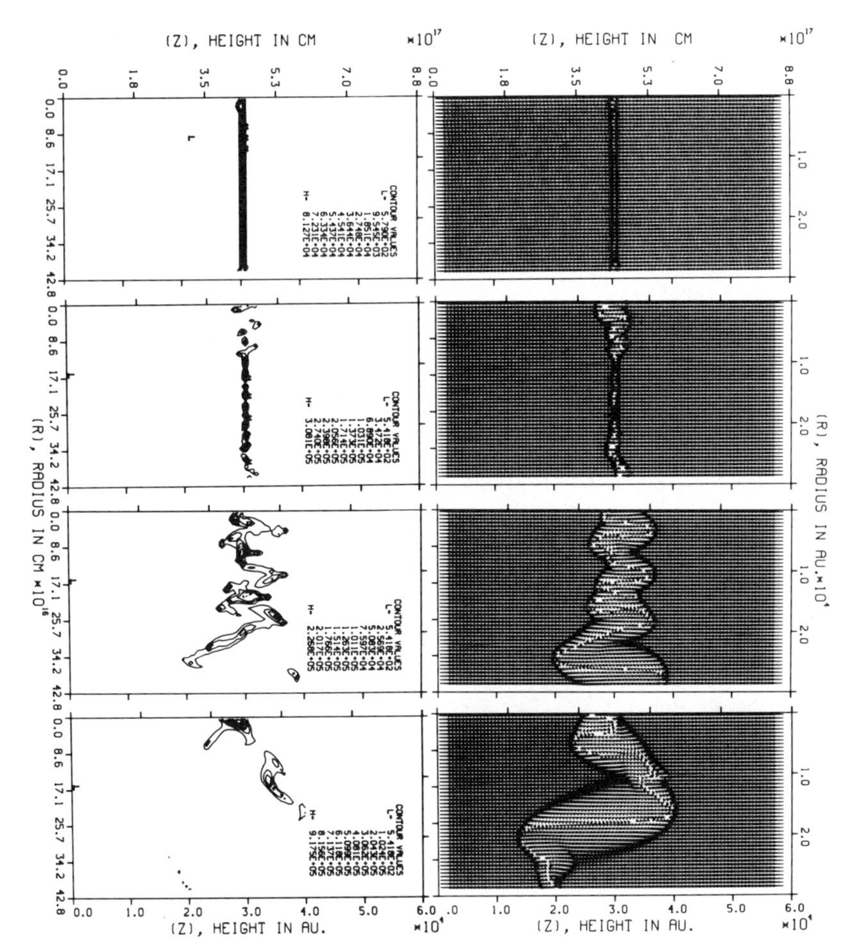

Fig. 1. Time evolution of colliding gas flows at $t = 1.24 \times 10^{12}$, 2.90×10^{12}, 6.44×10^{12}, and 1.06×10^{13} s. A continuous flow of clumpy gas at $n_{\rm H} \cong 500$ cm^{-3} ($\mp$ 10 %) enters the top and bottom of the computing grid at 5.0 km s^{-1}. The collision of the flows results in the formation of an unstable, self-gravitating disk at $t \cong 1.24 \times 10^{12}$ s (*top panel*). The disk fragments (*center panels*) and forms a centrally condensed, Jeans unstable object (*bottom panel*) at $t \cong 1.06 \times 10^{13}$ s. Contour values in number density are cm^{-3} units, the high (H) and low (L) levels are marked, and the fifth contour level is dashed. Gas velocity vectors are shown on the right.

quiescently forms stars. Solomon and Sanders (see their chapter) have reviewed data showing that while the total number of observed, bound molecular clouds fills the available space in the galactic disk, the "strong" source ($T_A > 9$ K, $T_{\rm kinetic} \cong 15 - 20$ K) clouds which all contain H II regions are found only in spiral arms. When the mass of all clouds is considered, the observed star formation rate (~ 3 M$_\odot$ yr^{-1}) is much smaller than the rate expected ($\sim 10^3$ M$_\odot$ yr^{-1}) which infers that star formation is inhibited outside the spiral arms, and that the process is not particularly efficient. The newest data on the velocity structure within star-forming clouds seems to indicate energy is input to the gas (Fleck 1983). Gravitating clouds appear to be internally supported against collapse (see the chapter by Goldsmith and Arquilla); yet stars form in their cores, and other young objects seem to have been born in low-mass, stable clumps near cloud edges.

Lada (1980) has presented a body of observational evidence supporting the idea that massive stars form primarily at molecular cloud edges. It was separately suggested that their formation requires an external triggering mechanism (Lada et al. 1978). Several mechanisms external to the cloud can account for the compression necessary to drive star formation at either the molecular cloud surface or interior. In particular, suitable compressions result from spiral density wave shocks (Woodward 1976), from ionization-shock fronts around sequentially formed OB subgroups (B. G. Elmegreen and C. J. Lada 1977), from stellar winds (Castor et al 1975*a*), from cloud-cloud collisions (Loren 1976), and from supernova explosions (W. Herbst and G. Assousa 1978).

In this section we concentrate on theories that incorporate a triggering mechanism, emphasizing processes in which secondary star formation is either caused or accelerated by interactions between the energy output from existing stars and the cloud material. We confine most of our discussion to mechanisms that produce radiatively driven shocks because this theory is more quantitatively developed and because the other mechanisms have been included in several previous reviews. In particular, we discuss the consequence for star formation of the interaction of O and B star radiation with a clumpy cloud environment.

Massive stars are short-lived and may be supernovae at the end of their evolution, propagating strong shock waves in the interstellar medium which can compress clouds to instability (Woodward 1976). W. Herbst and G. Assousa (1978) compared the efficiency of density-wave and supernova-shock compressions and concluded that supernovae shocks are the most important triggers; however, Lada et al. (1978) found that the effects of expanding H II regions are much more efficient because they provide a steady source of pressure, while supernovae are impulsive sources. Only supernovae that ignite early ($< 10^6$ yr) during the evolution of an OB subgroup affect the dynamics of the star-forming layer.

Kossacki (1968) considered the gravitational instability of condensations behind a spherically converging shock and determined a criterion for the masses of the condensations. Kossacki's work predates the related study of B. G. Elmegreen and C. J. Lada (1977) who proposed that the process by which the ionization-shock front from an OB subgroup propagates into a plane-parallel molecular cloud and creates a cool, dense, star-forming layer is a sequential one. They termed the dense layer the "cooled-post-shock" layer. The plane-parallel cooled-post-shock layer is subject to gravitational instability and it fragments to form members of a new OB star subgroup that continue the process. Sequential star formation is in superficial agreement with observations that some of the youngest OB subgroups formed near the edges of clouds, and that subgroups within an association are in temporal sequence. If one accepts the premise that external effects confine primary star formation to spiral arms, then the reason that secondary, sequential star formation fails to propagate into the interarm clouds must be explained. A possible explanation may be found in the work by Cox (1983) who concluded that the star formation rate during the formation of the Galaxy was much higher and that star formation in our present Galaxy is inhibited.

For the Cep OB3 region, where star formation efficiency is high and a new subgroup is forming, Sargent (1979) concluded that the *onset* of star formation was brought on by conditions different from the sequential star formation mechanism that now seems to operate within the association. Isobe and Sasaki (1982) studied the Orion association and reached a similar conclusion. Harvey and Gatley (1983) concluded that in NGC 6334, which contains the greatest number of protostellar OB stars of any region yet surveyed, the sequential triggering mechanism is not viable. Triggering by a spiral density wave was suggested as a plausible explanation for the widespread O and B star formation in NGC 6334. Jaffe and Fazio (1982) investigated the O and B star formation mechanism in the M 17 SW molecular cloud using the first high-resolution, high-sensitivity, far-infrared data. Their results favor triggering by a spiral shock, and they discarded the sequential star formation hypothesis because the data do not show evidence for a temporal sequence. Thus, while there is evidence that the sequential process envisioned by B. G. Elmegreen and C. J. Lada (1977) may have been very important in the prehistory of the Galaxy, and that it may operate within some young associations, many secondary star formation regions seem to be the result of a different type of interaction.

The importance of star formation near H II regions has been well documented by several observational studies (see, e.g., Blitz and Stark 1982). Blitz (1980) emphasized the effects of stellar winds from O and B stars (Castor et al. 1975*a*) as a viable mechanism for massive star formation. In this picture, winds from hot stars create ionized bubbles in the interstellar medium and compress the gas in the bubble volume into thin, dense shells. Wind-driven shells are similar to the shells produced by a supernova explosion. They form a cool, dense, star-forming layer analogous to the cooled post-shocked layer in the B. G. Elmegreen and C. J. Lada (1977) sequential star formation mechanism. This layer may become gravitationally unstable leading to star formation. Stellar winds may also be effective in triggering massive star formation because they stir the ambient cloud and increase the rate of clump aggregation (Silk 1984), which in turn is proportional to the rate of star formation. Blitz (1980) pointed out that it is not obvious how one differentiates between the stellar wind star formation mechanism and the ionization-shock front mechanism proposed by B. G. Elmegreen and C. J. Lada (1977). Both expanding H II regions and strong wind outflows emanate from newly formed OB subgroups, making discrimination between these two secondary star formation mechanisms difficult. This problem was addressed by Hughes (1982) with the conclusion that an isolated B star in IC 1805 formed as a result of wind-driven shock compression.

Hydrodynamic models calculated by Sandford and Whitaker (1983) showed that a hot wind encountering a cloud fragment forms a bow shock around the denser, neutral gas. Their numerical results were confirmed by a one-dimensional, analytic analysis and hot shocks were found to be rather ineffective as a means to compress the clump. Cool, neutral winds could transfer momentum into cloud clumps supporting the Norman and Silk (1980) proposal that T Tauri winds initiate subsequent, secondary star formation. Thus, Blitz's (1980) proposal and Hughes' (1982) conclusion that stellar winds of hot OB subgroups initiate the compressions leading to massive star formation requires further study.

A major goal for observers of star-forming regions must be the discrimination between the possible secondary star formation mechanisms. Improvement in the calculations of the interaction of stars with their surroundings will of course provide new observational tests. Several suggestions for observational tests have already emerged. Lada (1980) proposed looking for correlations between the propagation direction of spiral density waves and the location of stellar birth complexes at the edge of giant molecular clouds. This test assumes that the edge of the giant cloud which first encounters the spiral wave will be the edge where stars form first. This is not necessarily the case. If the cloud structure is sufficiently smooth near the edge, star formation may be more efficient in the clumpy interior. A more promising test would be a correlation between the temporal sequence of young subgroups and the direction of galactic rotation inside the corotation radius (Lada 1980).

The importance of supernovae as a triggering mechanism would benefit by the discovery of massive star formation within a supernova remnant. The prospect of an observation that does not show an ionization-shock front due to the O star trigger is small (B. G. Elmegreen 1980) because such ionizing radiation always preceeds the explosion. The CMa R1 region cited by W. Herbst and G. Assousa (1977) as evidence of supernova-triggered star formation has an ionization front in the vicinity of the expected compression direction. Thus, without additional evidence for the origin of this front, one does not know if it results from the supernova or from the original O star. An interesting case where secondary star formation could potentially result from several mechanisms is currently being studied by Ho (personal communication, 1984). Ho observed a compact H II region (G34.1), $\sim$ 40 pc projected distance from W44, which contains a supernova remnant. The H II region is cometary in appearance, surrounds a dense neutral globule, and the ionization appears to point precisely to the location of the remnant center. Ho estimates the amount of energy necessary to deform the neutral globule is $\sim 10^{50}$ ergs (close to the energy of a supernova explosion). The remnant and shock have not yet reached the compact H II region so the direct action of the supernova is not responsible for the ionization in G34.1. Another possibility is that the H II region is a result of prompt X-ray radiation from the supernova. Calculations of this effect (R. I. Klein and R. A. Chevalier 1978) show that during the interval of peak X-ray luminosity ($\sim 10^3$ s) only $\sim 10^{48}$ erg is generated. The 40 pc distance would dilute and absorb the ionizing flux of the original O star, excluding this as a possible source. The most probable remaining possibility is that the H II region results from the stellar wind of the original O star. Ho and collaborators are continuing to study this interesting region.

Calculations that follow the evolution of an ionization-shock front in a molecular cloud at the time a supernova precursor explodes and augments the clump compression are badly needed. These calculations could help clarify the relative importance of each process while providing signatures of their combined effects. In the following subsections we discuss recent work that investigates the consequences of ionization-shock fronts generated by O and B star radiation propagating into an inhomogeneous cloud environment. These calculations provide the background for future more complex calculations involving additional impulsive forces from stellar winds and supernovae.

A. Radiation-Driven Implosions

Massive stars are such strong sources of energetic radiation that they have an important, immediate influence on their local environment (G. H. Smith 1982). Their radiative output generates enormous H II regions in low-density gas, they create compact H II regions in dense cloud cores, and they can evaporate existing cloud clumps (Whitworth 1979). R. I. Klein et al. (1980) have shown that hot stars drive convergent ionization-shock fronts into cloud clumps, imploding them to gravitationally unstable densities. This extended the previous work by Oort (1954) and Dibai (1958,1960). LaRosa (1983) examined a simplified radiative implosion model and analytically derived estimates of the conditions for clump implosion or erosion by radiation.

In a series of papers investigating the effects of ionization-shock fronts in a clumpy cloud, R. I. Klein et al. (1980,1983) and Sandford et al. (1982*a,b,* 1984) have presented results of calculations with a new 2D, two-phase flow, radiation hydrodynamics code. We used an implicit, multiphase hydrodynamics method (Harlow and Amsden 1975) which is coupled to a 2D, time-dependent, discrete-ordinate technique for radiative transfer due to Lathrop and Brinkley (1973). Thus, for the first time, the frequency and angle dependence of the multidimensional equations of transfer and gas dynamics were considered. Our cylindrically-symmetric, Eulerian-mesh radiation hydrodynamics method included a second material phase (dust) moving relative to the compressible gas which consisted of hydrogen ions, neutrals, and electrons. We solved the set of hydrodynamic conservation equations for the gas and dust phases simultaneously with a combined equation of transfer that includes three frequency groups for ionizing radiation absorbed by gas and attenuated by dust extinction, and with the appropriate ionization rate and constitutive equations. The dust and gas fluid components are connected through phase changes, if they occur, and by appropriate heat and momentum interchange functions.

B. Implosions by Single Stars

The results of Klein et al. (1980) showed that ionizing stellar radiation which irradiates only one face of a cloud clump can effectively compress the clump with the shock which precedes the ionization front. In Sandford et al. (1982*a*) the adiabatic results by Klein et al. (1980) were extended by including a molecular cooling model. Results were compared for the implosion evolution of a low-mass ($\sim 1 \times 10^{-5}$ $M_\odot$) globule and one of higher mass ($\sim$ 2 $M_\odot$). The physical environment modeled by our calculations treated the effect of ionizing blackbody radiation from an O9 star ($T_* = 30{,}000$ K) approximately 8 pc distant from a molecular cloud clump.

Two cases from Sandford et al. (1982*a*) which indicate the important results of the radiative implosion model will be summarized. Both calculations were performed on a 25 × 50 zone, Eulerian cylindrical computational mesh, and the diluted nongrey O star radiation was incident on a clump having initial atomic hydrogen density $n_H \cong 600$ cm^{-3}. The neutral clump and surrounding ionized gas were initially in approximate pressure equilibrium. The temperature in the neutral gas clump was $T_g \cong 15$ K, and the surrounding ionized medium had an electron density $n_e \cong 0.24$ cm^{-3} and $T_g \cong 15{,}000$ K. The low-mass case had a grid size of $\Delta r = \Delta z = 1 \times 10^{15}$ cm, and the higher-mass case had a grid size 50 times larger. In the low-mass case the computing cells were optically thin and the ionization front was resolved, covering about two mesh cells, but the ionization structure was not resolved in the higher-mass case. In neither calculation was the shock resolved. The simple scheme used for molecular cooling gave isothermal shocks near 100 K.

In each case the radiation flowed around the curved clump face generating an ionization-shock front which surrounded most of it, and in which motions were focused toward the symmetry axis. The interaction of the radiation with the neutral clump also caused hot ionized gas to expand outward into the intercloud medium. The ionization-shock front progressed into the clump and soon a convergent shock separated from the ionization-front and traveled into the interior of the clump. By 1300 yr, for the low-mass case, the convergent shock drove neutral gas toward a point on the symmetry axis and produced a centrally condensed globule with a compression ratio $\sim$ 70. Woodward (1976) found similar motions toward the symmetry axis, resulting from a shock wave striking a clump. Figure 2 presents, for the low-mass case, contours of H I number density and H II number density at $t \sim 1300$ yr. The

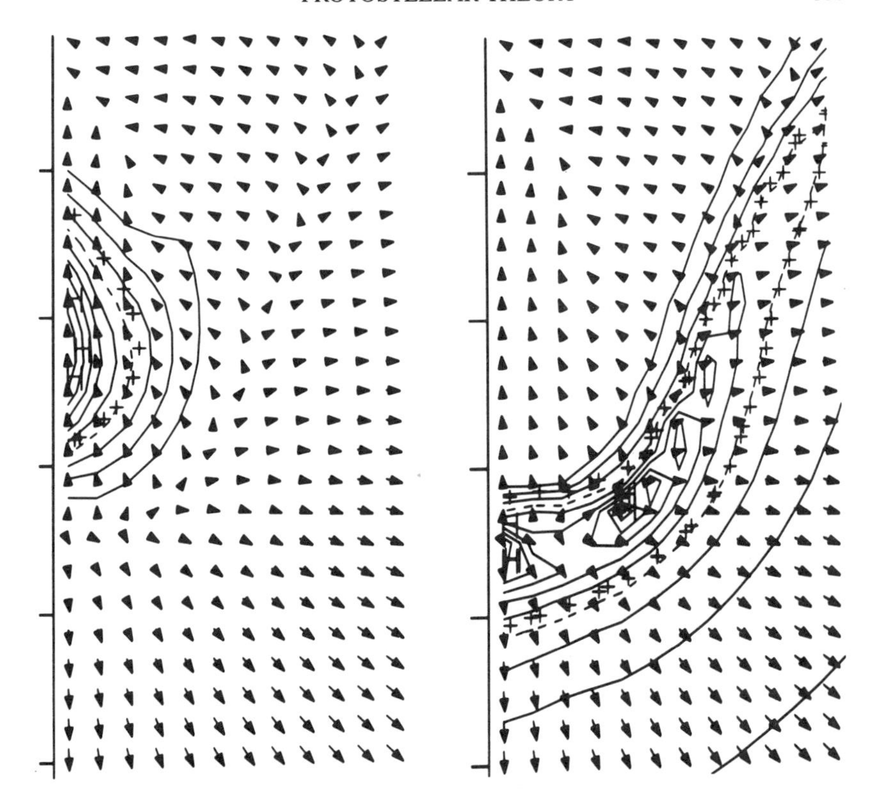

Fig. 2. Contours of neutral (*left*) and ionized (*right*) hydrogen and the gas velocity field for a low-mass globule compressed by radiation from an O9 star located 8 pc below the computing grid (Cloud 10 model, by Sandford [1982*a*]). The maximum velocity is $\sim$ 15 km s^{-1}. The inner (H) and outer contours of neutral hydrogen density are 3.16×10^4 and 3.51×10^3 cm^{-3}, and the contour spacing is 3.5×10^3 cm^{-3}. The closed (H) contour on the axis in the ionized hydrogen plot is 179 cm^{-3}, the lowest contour is 20 cm^{-3}, and contour spacing is 20 cm^{-3}.

contours are superimposed on the hydrogen gas velocity vectors at the same time. The separation of the shocked neutral material from the ionization front is evident as well as the higher ionization of the cloud material facing the distant O9 star. It is seen that the flow of radiation around the original clump surface was effective in driving the convergent shock wave that produced a large compression in the neutral globule. The convergent flow in the clump and the outward flow in the H II region are well illustrated by the velocity vectors.

For the higher-mass case, the evolution proceeds in a similar fashion and at the end of the calculation ($\sim 6 \times 10^4$ yr), a toroidal globule formed near the symmetry axis. At this time, the compression ratio is nearly 19 and the mass in the toroidal condensation is $\sim$ 0.8 $M_\odot$. In both models, about 40% of the initial cloud gas is compressed into the globule, with the remaining mass ablated back into the intercloud medium.

While neither globule was Jeans unstable at the end of the calculation, the process of radiation-driven implosion was effective in producing small dense structures. Two important considerations are worthy of comment. First, the simple molecular cooling law used gave minimum temperatures of about 100 K. Cooling by CO molecules would reduce the minimum temperature to much lower values, yielding cooler, denser structures of smaller Jeans mass. Second, J. H. Hunter (1979) analyzed the effect of an inwardly directed velocity flow on the Jeans mass and found (for velocities of 1 km s^{-1} at 10 K) reductions of the Jeans mass by factors of 5 to 20. These effects augment the radiation driven implosion, and the process may be sufficient to drive even small globules to Jeans instability provided that they are not evaporated by the ionizing radiation from the star.

Some recent observations tend to support the radiative implosion model. Felli et al. (1980) found a point source near an arc-like ridge of ionization in radio continuum observations of M17; they concluded that the object may be an example of star formation induced by shock focusing. More recently, McCutcheon (1982) reported on CO observations in the vicinity of IC 5146 in which they found three regions of enhanced emission, all on the edge of the Sharpless region S 125. They discussed the need for a source to maintain the emission and concluded that an embedded protostar (about B1) is a likely candidate for the most intense region. They also proposed that *all three regions* were formed by an interaction with the H II region generated by the central star of S 125. The three compressed regions could have formed as a result of radiative implosions driven by the central star. Reipurth (1983) observed bright rimmed globules in the Gum Nebula and concluded that some are forming stars as a consequence of compression by converging ionization-shock fronts.

C. Implosions by Multiple Sources

Our first calculations (Klein et al. 1980; Sandford et al 1982*a*) investigated the mechanism of radiation-driven implosion by single O stars and resulted in highly compressed globules nearly massive enough to gravitationally collapse to form new stars. These results led us to the conclusion that radiation-driven implosion is a promising star formation mechanism and that the morphological structure surrounding a young cluster of O stars may significantly influence the subsequent star formation in the cloud interior. The mass spectrum of young stellar objects would depend upon the size scale of the initial inhomogeneities.

Radio observations of ultracompact H II regions by Ho and Haschick (1981) and Dreher et al. (1983) discovered multiple sources having size scales comparable to their separation distances (≤ 0.1 pc), and formation time-scale differences less than a few tens of thousands of years. These small-scale structures were interpreted as the result of the most recent episode of star formation that produced new members of an OB star cluster. Further radio continuum and line observations by Haschick and Ho (1983) of O and B star formation in the W33 complex show that the most recent episode of star formation resulted in massive stars confined to a 1 pc core within a dense molecular gas. The appearance of a highly clumped H II region and evidence for several sites of star formation suggests cluster formation *within* the core of W33. These observations challenge the more traditional view that massive star formation is confined to the surface of molecular clouds (Lada 1980) and demonstrate that they may form throughout the interiors of cloud complexes as well. Optical observations are largely responsible for the suggestion that the birthplace of O stars is at the surface of molecular clouds. This could be a selection effect because optical identification of O stars is normally restricted to cloud surfaces due to interior obscuration. In addition, observations based on giant H II regions could reflect the disruption of cloud material by O stars born in their interior. These stars would now have the appearance of being born on the cloud edges. This opens the possibility for new secondary star formation mechanisms that operate in cloud *interiors* and depend on the cloud morphology for their efficiency.

Motivated by these observations, and by the compelling evidence that widespread clumping and inhomogeneities exist in molecular clouds down to scales ≤ 0.1 pc throughout star-forming regions, we calculated (Klein et al. 1983) the detailed time-dependent evolution of an inhomogeneity initially containing 84 $M_{\odot}$ that is embedded between and irradiated by two O7 stars, each at a distance of 0.5 pc and from the clump. The inhomogeneity was represented by a sphere of neutral hydrogen at a temperature $T = 30$ K and with uniform density $n_{\rm H} = 3000$ cm^{-3}, and was surrounded by an intercloud medium fully ionized with density $n_{\rm H} = 0.24$ cm^{-3}. The objective of the calculation was to follow the evolution of a radiatively compressed inhomogeneity embedded in the environment produced by an earlier generation of O and B stars, and to determine the consequences of this evolution for new star formation. Our calculations were performed with the two-dimensional, implicit, Eulerian, radiation-hydrodynamics code described previously (Sandford et al. 1982*a*) and did not include the effects of dust and self-gravity. The embedded inhomogeneity was placed on the cylindrical symmetry axis colinear with the ionizing O stars.

The results showed that within 10^3 yr the ionizing ultraviolet radiation ablates a layer of plasma which flows from the surface of the neutral clump, resulting in the propagation of two-dimensional shockwaves moving into the neutral gas. The time evolution of this clump is illustrated in Fig. 3 at 1×10^4, 1.5×10^4, and 2.0×10^4 yr. Convergent, strong shock waves propagate within the clump and produce density enhancements (compressions) of 4.4, 7.6, and 24.0 over the ambient density at the three respective times. Dynamical evolution suggests that individual local sites of high density, initially present at low density due to the "staircase" effect of discrete zoning, are amplified by the shock convergence. At 2×10^4 yr the converging shocks coalesce and drive the individual enhancements into a disk. The possibility exists that local density or pressure fluctuations could amplify during the compression

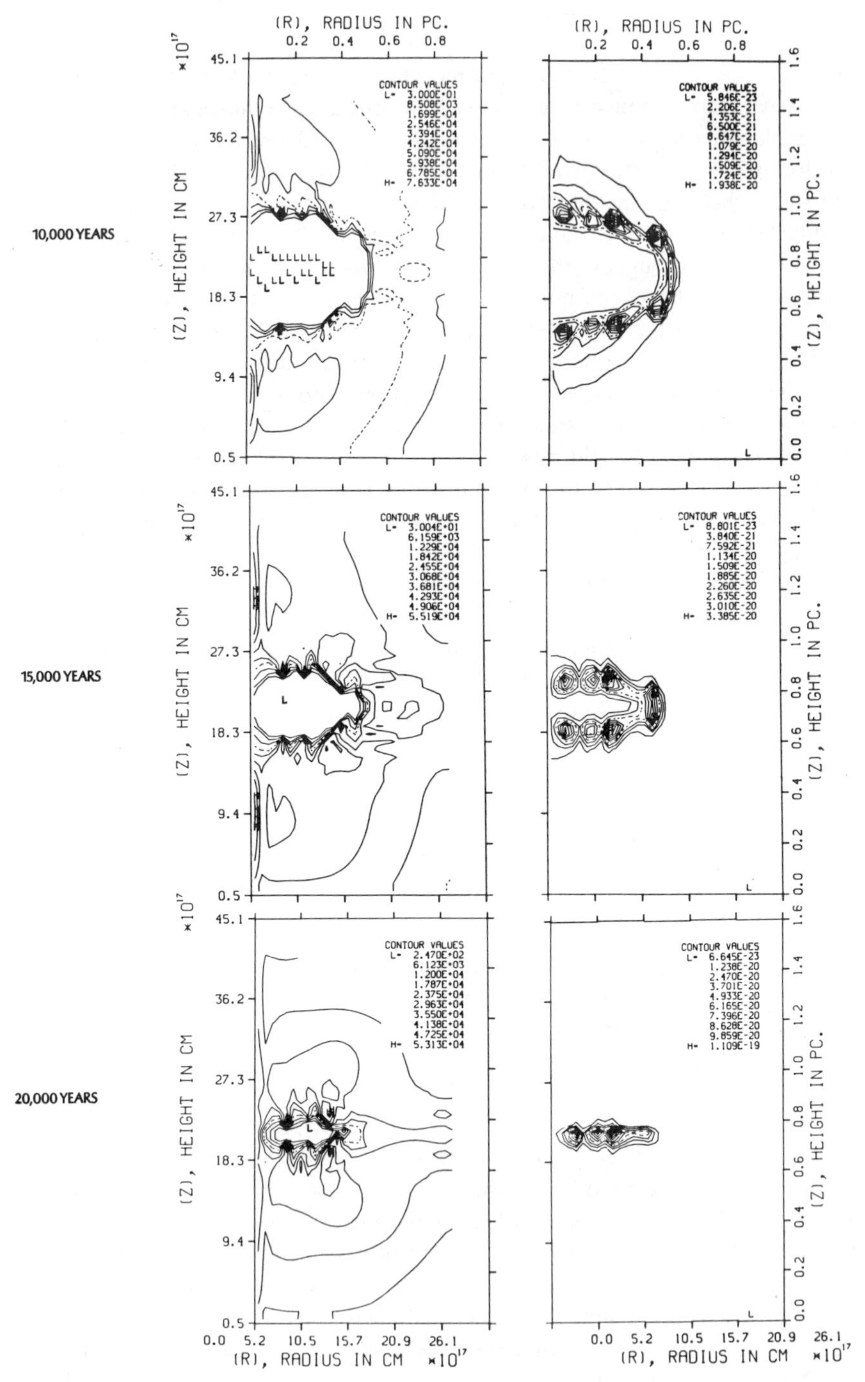

Fig. 3. Gas density (*right*) and absolute temperature (*left*) contours for the radiation implosion driven by multiple sources. Contour values are in cm^{-3} and K units, the fifth contour level is dashed, and the high (H) and low (L) levels are marked (from Klein et al. 1983).

and later fragment to become individual compact objects. At 3×10^4 yr, the convergent shocks surrounding the clump interact nonlinearly in the interior producing a compression factor of 170 over the initial density. The final density is $n_H \cong 5 \times 10^5$ cm^{-3}. The globule that forms has a radius < 0.1 pc, $T_g \cong$ 200 K, and an ablation outflow speed > 30 km s^{-1}. The calculation was terminated because the grid lacks sufficient resolution at the longest times. Of the initial 84 $M_\odot$, 40 $M_\odot$ remained intact and 44 $M_\odot$ evaporated from the object into the interclump medium. The compressed globule was found to be a factor of 3 smaller than the static Jeans mass at 3×10^4 yr, but cooling due to molecular species other than H_2 could drive the clump temperature to $T_g <$ 100 K and would cause gravitational collapse in a free-fall time $< 6 \times 10^4$ yr. An estimate of the mass lost due to the continued irradiation of the imploded clump was made by Klein et al. (1983). Their estimate was made by combining mass conservation with analytic expressions due to Kahn (1969) relating the Lyman continuum flux at the clump to the particle density and temperature, and to the flow velocity in the ablated, ionized gas. By equating the derived mass loss over a given time interval with the mass inside a shell of given radius and thickness, Klein et al. (1983) obtained an estimate for the evaporative lifetime of the compressed clump. For an illuminating star O7 or earlier, the lifetime is $\sim 3 \times 10^5$ yr, but for an O9 or later star, the lifetime is $\sim$ 1–2 $\times 10^6$ yr. The inclusion of dust effects (see Sec. III.D below) would increase the lifetime. Hence, if the object is gravitationally unstable at the time of maximum compression by the convergent shocks, it would likely collapse in $\sim 10^4$ yr.

The two-dimensional symmetry of these calculations leads to the formation of a disk and later to a toroidal structure, either of which may fragment before gravitational collapse. In a real object, the radiation driven implosion would occur in three-dimensional geometry. Some indication of the stability of the toroid resulting from two-dimensional calculations may be obtained by considering the effects of nonaxisymmetric perturbations on a rotating, self-gravitating torroidal object. Cook (1977) considered this question by performing time dependent, three-dimensional, hydrodynamic calculations of collapsing, adiabatic torroids. The Virial theorem relates the gravitational, internal, and rotational energies of an object in equilibrium and, when combined with an approximate criterion for the torroidal fragmentation, it yields a stability condition. Using Cook's (1977) three-dimensional results for the toroid resulting from our (Klein et al. 1983) calculation, we find that it would fragment into $\sim$ 2 masses if rotating at 2×10^{-14} s^{-1}. As the temperature of the torroid drops, the thermal pressure which resists the action of a perturbation is reduced, and the theory predicts fragmentation into more objects. Fragmentation might produce several objects, each of 6 to 7 $M_\odot$ and could therefore lead to low-mass star formation. Alternatively, inhomogeneities with initial masses > 84 $M_\odot$ driven to high compression by shock focusing could result in massive star formation even for unstable torroids. On the other hand, an initial core-halo density distribution might keep the collapse on the axis leading to larger-mass compressed clumps. In three dimensions, irradiation from several surrounding stars might surpress torroidal growth and could lead directly to rapid massive star formation.

The key point made by Klein et al. (1983) is that *shock focusing,* which occurs in either two- or three-dimensional geometry, is a consequence of radiation-driven implosion that substantially increases the compression and is expected to drive clumps to gravitational instability on time scales short compared to evaporation times. This provides a mechanism for forming additional stars within an OB subgroup that is born in a molecular cloud core. Because the multiply driven implosion mechanism is efficient (40 to 50% of the clump mass remains after evaporation), it is possible for a few O and B star triggers to implode many embedded clumps to form a new generation of stars in a few times 10^4 yr (Fig. 4). If this process gives birth to O stars, the additional imploded clumps could form yet another generation of stars. Thus, a region 1 pc in size could contain several newly formed O and B stars and ultracompact H II regions that appear to be coeval within a few tens of thousands of years. The densities and time scales found in our calculations agree well with the observations of Ho and Haschick (1981) lending support to multiply driven radiation implosions as the mechanism for synchronizing the formation of OB subgroup stars.

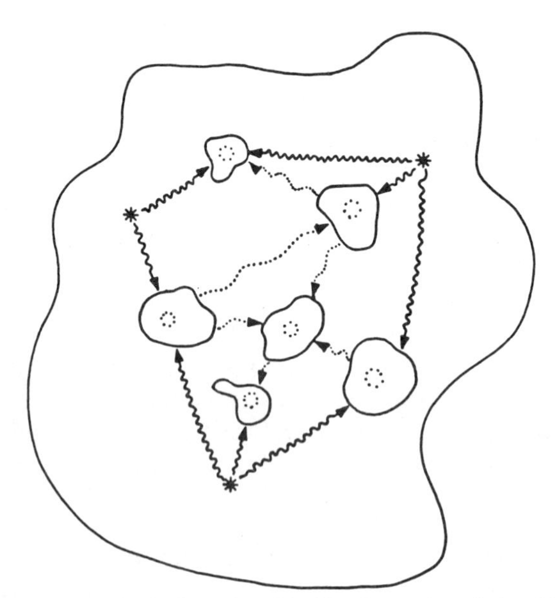

Fig. 4. Schematic geometry for the formation of an OB subgroup by radiation-driven implosions (from Klein et al. 1983). Primary generation O and B stars (*asterisks*) implode cloud clumps which subsequently form new OB stars (*dashed circles*). These implode the remaining clumps, as indicated by the wavy lines of interaction.

An alternative explanation for the observations was offered by Shu (personal communication, 1983). If the individual stars have peculiar velocities on the order of the sound speed, their successive Stromgren spheres are left behind to recombine in one crossing time. Hence the star never stays long enough in any one place to build up a large H II region, and can be older than the apparent age. These alternative theories can be tested by a high-sensitivity continuum study of the low-level emission in ultracompact H II regions, in which one looks for the interaction between H II regions and surrounding neutral material or for evidence of low-level emission trailing the condensations.

Clearly, more theoretical work remains to explore the consequences of multiply driven radiation implosions. As mentioned above, it is important to determine the efficiency of the implosion mechanism for O star triggers that are also blowing stellar winds, or for stars that enter a supernova phase while contributing large amounts of ionizing radiation to imploding embedded inhomogeneities. It also remains to be shown how structure within clumps affects the convergent shock and to determine if the masses that can be compressed will survive evaporation to form low-mass stars.

D. Effects of Dust

When dust is included as a component of the interclump medium, radiation pressure from distant O stars accelerates grains with respect to the gas and the ionization-shock front becomes dust-bounded as it converges into a clump (Sandford et al. 1984). The ionization front speed slows to that of the dust grains which undergo drag interactions with the gas (~ 2 km s^{-1}). The time scale to reach the same compression is longer than in the absence of dust, but it remains short compared with the free-fall time. For ionization front densities ~ 200 cm^{-3} and with dilute radiation fields, sputtering and vaporization effects are not important. Grains survive and accumulate in the ionization front and the dust-to-gas mass ratio increases from 0.02 to 0.04–0.13. Gravity becomes an effective collapse force at about the time the ionization front becomes dust-bounded and its effect on the subsequent dust and gas distribution remains to be determined. The evolution of a globule in these circumstances is shown schematically in Fig. 5. The lifetimes of cloud clumps are found to be longer than estimated by LaRosa (1983) because globules become centrally condensed and are less easily eroded. Dust in the ionization front attenuates ionizing radiation and reduces the ablation rate, and incident radiation is also absorbed by the evaporated (ablated) gas to maintain its ionization. Low-mass (< 1 M$_\odot$) globules can be confined by a dilute radiation field long enough to become gravitationally unstable. It is thus proposed that isolated, stable dark globules were formed by radiative implosions of small cloud fragments at the distant edges of an ancient H II region.

When effects of dust are included, low-mass globules are stable long enough to contract and to form a star if they are in the presence of a weak,

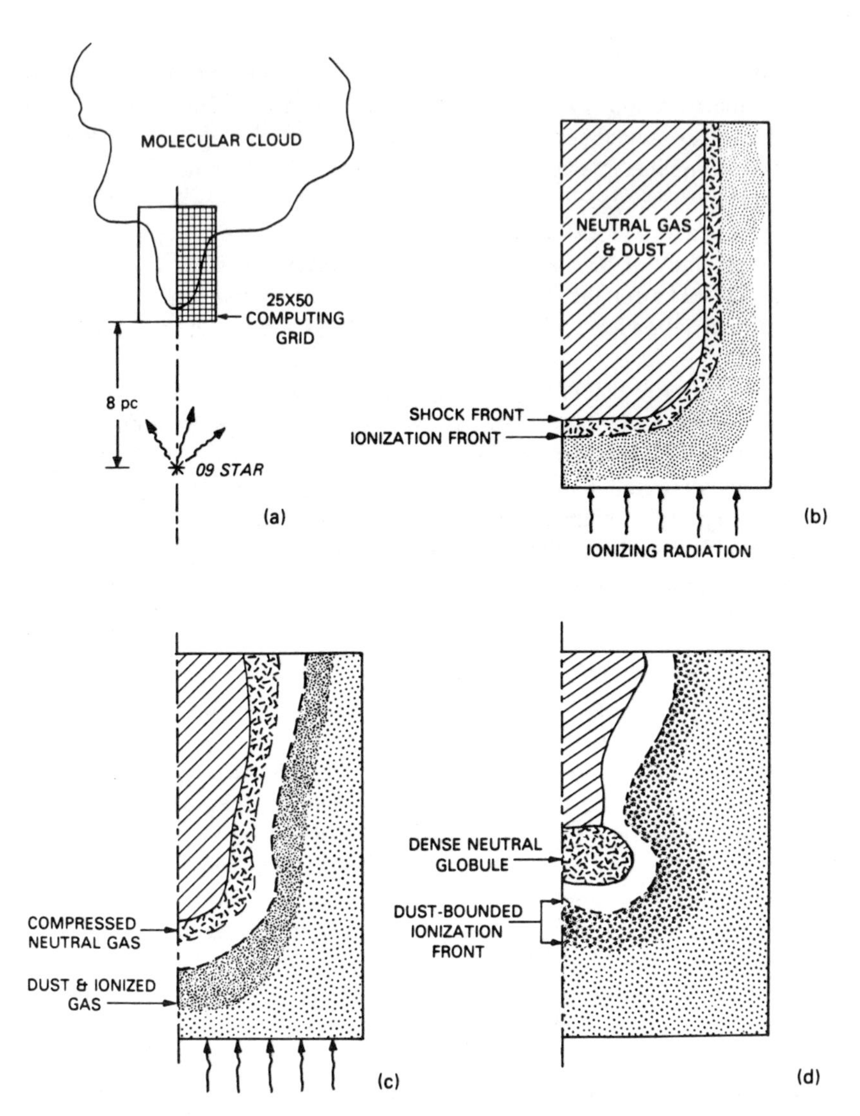

Fig. 5. Schematic diagram of a radiatively driven dust-bounded implosion (from Sandford et al. 1984). In (a), an O9 star illuminates a dusty molecular cloud. In (b), an ionization-shock front propagates a shock into the neutral gas and dust (*diagonal lines*) and compresses gas behind the shock (*scattered lines*). Ionized gas (*shading*) ablates from the neutral cloud edge at the ionization front (*dashed line*). In (c), the shock converges toward the symmetry axis and dust accumulates in the ionization front (*dark shading*). In (d), a globule of neutral gas and dust is formed and is surrounded by a dust-bounded ionization front (*dark shading*) within an H II region (*light shading*).

external, ionizing radiation field. Observational evidence for star formation in dark globules is increasing. Krügel et al. (1983) and Keene et al. (1983) independently concluded that the Bok Globule B335 is forming, or has recently formed, a star of ~ 1 to 2 $M_{\odot}$. T. J. Jones et al. (1980) and T. J. Jones et al. (1984*a*) made extensive observations of Bok Globule 2 in the Southern Coalsack. This ~ 12 $M_{\odot}$ object shows no evidence for a central density enhancement, as would be expected for a gravitationally contracting globule. The stability of the globule is interpreted as being due to some form (turbulent or magnetic) of internal support. Our calculations (Sandford et al. 1984) indicated that at an early stage in the evolution of a globule confined by an ionization-shock front, the density profile becomes flat and the internal motions are acoustic. Additional calculations are needed to determine the effect of the demise of the radiation source, leading to the collapse of the ionization-shock front. It is possible that reasonably uniform, stable globules containing sonic (turbulent?) gas dynamical flows could form. A theoretical prediction of the probable lifetime and eventual fate of such globules would be of great observational interest.

Globules compressed by radiation-driven implosions are predicted to contain relatively more dust than ones that gravitated from cloud fragments, and second-generation protostellar objects would therefore be born with far-infrared signatures different than those from objects formed by another mechanism. For NGC 2264 which contains hot stars capable of imploding clumps, T. M. Adams et al. (1983) concluded that the star formation history extends over ~ 10^7 yr, and that the youngest, low-mass, pre-main sequence stars show circumstellar dust emission. Wesselius et al. (1984) studied young objects in the Chameleon I dark cloud (which contains no O and B stars) and found two pre-main sequence A0 stars embedded in large, extended (~ 0.2 pc) dust complexes. These observations, and those of Baud et al. (1984) showing that G- and K-type young stellar objects have relatively little dust, are consistent with the conjecture that primary star formation from cloud fragments forms extended dust cocoons but secondary formation in imploded globules results in compact, circumstellar emission.

The stellar material originating from the central condensation of an imploded globule contains the ambient cloud abundances; however, the envelope, atmosphere, and accretion disk or shell forms from dustier gas and is therefore predicted to be metal-rich. Main sequence stars showing evidence for a far-infrared excess, such as Vega (Aumann et al. 1984), could therefore represent a population formed by radiation-driven implosions and their statistical frequency of occurrence would measure the efficiency of the process. The formation of low-mass, isolated, main sequence stars would be relatively rare if the process is 1 to 2% efficient (Reipurth 1983). Dark globules may provide the best examples of isolated, quiescent star formation. New observations to determine the conditions in a large number of globules are needed. The evolutionary history of dark globules could perhaps be determined if their internal structures are found to form a progression, and if the relationship between the isolated globules and those embedded in H II regions can be established.

E. Importance of Radiation-Driven Implosions

The efficiency of radiation-driven implosions for stars forming from cometary nebulae in the Gum Nebula was estimated to be $\sim$ 1 to 2% (Reipurth 1983). The star formation efficiency required to produce a bound cluster is high ($>$ 50%) according to Wilking and Lada (1983), but the overall formation efficiency for field stars is apparently quite low ($\sim$ 0.1%). Because hot stars tend to disrupt their parent clouds, and in the process implode some of the fragments, one expects that this secondary star formation may be more efficient than the overall process, but perhaps less efficient than required to form bound clusters. Near the edges of an expanding H II region, radiative implosions of individual clumps may tend to form lower-mass objects because (1) compressions from single sources are weaker than within an OB subgroup, (2) time scales are short compared to the lifetime of the source star, and (3) the clumps driven to instability are small ones that could not otherwise contract. O and B type stars are observed to be forming in dense cloud cores and our calculations show that radiative implosions in these regions may lead to a higher efficiency of massive star formation on a short time scale. Thus, we believe that radiatively driven implosions may stimulate star formation over a wide mass range, in different cloud morphologies, and with correspondingly different formation efficiencies. Observers are encouraged to test this view.

As an example of the possible importance of secondary star formation, possibly driven by radiative implosions or stellar winds, we review observations of the NGC 2264 region. T. M. Adams et al. (1983) employed new, unsharp-masking methods to locate and include faint, low-mass stars in this star-forming region. No evidence for a deficiency in the low-mass IMF was found. On the other hand, Scalo (1978, 1985*b*) presented evidence for a real deficiency in the numbers of low-mass stars for a variety of galactic clusters, and he concluded (in agreement with coagulation theory results) that the low-mass stars found in some clusters may form via a different mechanism. This mechanism may be radiation-driven implosion. NGC 2264 appears to be an excellent region in which to study the star formation process at low mass.

A generally accepted suggestion that the star formation rate is mass dependent followed Herbig's (1954) study of NGC 2264, and is based upon the difference between nuclear ages of clusters and the (greater) ages of low-mass members. The conclusion is drawn that low-mass stars form first, but this result is not explained by coagulation models (primary star formation) or by radiative implosions (secondary star formation). Ages for low-mass, pre-main sequence stars are determined by comparing their positions in the HR diagram with *theoretical* isochrones. Based on the older, widely-used isochrones, T. M. Adams et al. (1983) concluded, in agreement with Herbig's (1954) result, that the star formation rate in NGC 2264 peaked at different times in the past, and that the lowest masses began to form first. Stahler et al. (1980*a,b*,1981) presented a new, efficient, more general approach to pre-main sequence evolution modeling and calculated evolutionary tracks that differ from the old results. Stahler (1983*a*) examined the age determination problem in view of the new calculations and identified a "birthline" in the HR diagram for T Tauri stars. The new stellar birthline is in good agreement with a large number of observations for several clusters and necessitates a revison of the ages inferred for low-mass stars. Corrected ages result in star formation rates that increase from the past to the present for all masses, and the rate peaks found by T. M. Adams et al. (1983) are thus removed. The increasing star formation rate may be due to secondary star formation processes, such as radiation-driven implosion.

New far-infrared observations of $C^{18}O$ sources within condensations in some cloud cores just beginning to form stars reveal the youngest objects to be B stars (Jaffe et al. 1984). A similar result, but for O stars, is found in new, high-resolution radio continuum observations of cloud cores with the VLA (Ho, personal communication, 1984). Observations of star formation in cloud cores now favor the possibility that hot stars form at the earliest times. Because implosions driven by ionizing radiation can occur only after hot stars form, the numerous low-mass objects found by T. M. Adams et al. (1983) would necessarily be of recent origin. One reason that low-mass stars may be younger than is inferred from theoretical isochrones is the *reduction* in the contraction time scale due to dynamical (ram) effects of the implosion process (Stahler 1983*a*, J. H. Hunter 1979).

Finally, observational advances bring closer the possibility that low-mass star formation may be directly observed. Small-scale clumping exists in the molecular gas surrounding the compact H II region W 3(OH); the clump sizes ($\sim 2.5 \times 10^{16}$ cm) and densities ($n_{H_2} \cong 10^6$ cm^{-3}) were determined by Dickel et al. (1984). The clumps represent fragments in a dense molecular shell and each contains 2 to 5 $M_\odot$. These can be interpreted as resulting from radiation-driven implosions of density inhomogeneities in the cloud surrounding the central star. Secondary star formation may thus be observed to accompany the birth of a hot star if the surrounding cloud is sufficiently inhomogeneous.

III. DIRECTIONS FOR FUTURE RESEARCH

In this section we discuss some of the observational and theoretical research necessary for further understanding the secondary star formation process. For theoretical prospects, we view as essential the continuation and improvement of multidimensional numerical models. These provide the most detailed, fully nonlinear predictions of the observable signatures of secondary star formation. New, more detailed multidimensional calculations are crucially needed. In particular, it is important to delineate the differences in the interaction with clumpy molecular clouds of H II regions, stellar winds, and supernovae, and to quantify the relative efficiency of each. Our work strongly suggests that two-dimensional, radiation-hydrodynamics must include a variety of physical detail in order for us to understand the fundamental interaction of radiation-driven shocks in inhomogeneous clouds. A theoretical understanding of molecular cloud morphology is as important as numerical modeling.

The radiation-driven implosion model must be further studied with improved constitutive relations including the effects of additional molecular coolants, dust grains, and self-gravity. It is of particular importance to determine to what extent the masses of stars that can form depends on the degree of inhomogeneity and clumpiness in a molecular cloud. It will be important to combine the results of dynamical radiation implosion models with a code that calculates radio-line spectra. This will require the development of a new, two-dimensional moving-atmosphere, line-transport code. The resulting line spectra will provide the characteristic signatures for the different models.

The hydrodynamic interaction of colliding clouds requires further theoretical study in order to interpret observations of cloud cores. The process by which gas flows can lead to cloud inhomogeneities and to transonic, random motions can be understood by significantly increasing the resolution of observations made at the CO wavelengths. An analysis to determine the infrared signatures of cloud fragments would be useful to study the effects of gas-dust interactions.

The immediate future of star formation theory lies in the interpretation of newly acquired data. A complete far-infrared survey is needed to determine the extent of massive star formation within cloud interiors. Radio continuum studies with high dynamic range are needed of small-scale structures within ultracompact H II regions. These regions are potential sites of star formation due to radiation-driven implosions. Observations of continuum radiation surrounding these structures should provide important clues as to the relationship among new stars, their cloud environment, and the primary triggers. High-sensitivity studies of molecular lines in new secondary star-forming regions are needed to obtain detailed velocity maps. These are necessary to study the ionized gas ablated from cloud inhomogeneities as well as to define the velocity structure in shocked regions. Observations of shocked regions may enable us to distinguish between the various triggers responsible for secondary star formation.

Our understanding of secondary star formation will advance most rapidly if we can successfully combine new, numerical models with observations. Large-scale numerical calculations, similar to those described in this chapter, intelligently guided by analytic work, will play an increasingly essential role. New supercomputers, such as the Cray-XMP series and machines being designed in Japan, will stimulate the development of new numerical methods and will make possible calculations more amenable to observational interpretations. There can be little doubt that numerical calculations will continue to play an important role in astronomy, and it is possible that the next decade may bring forth a comprehensive theory for the formation of stars.

Acknowledgments. We acknowledge helpful discussions with many colleagues during the preparation of this review. Our collaboration in this work has been supported by the Livermore and Los Alamos National Laboratories, operated under contracts to the U. S. Dept. of Energy by the University of California.

PART III
Young Stellar Objects and Circumstellar Disks

YOUNG STELLAR OBJECTS AND THEIR CIRCUMSTELLAR DUST: AN OVERVIEW

A. ERIC RYDGREN
Rensselaer Polytechnic Institute

and

MARTIN COHEN
NASA Ames Research Center

The basic observational classes of young stellar objects are briefly reviewed and some of the theoretical and observational problems in pre-main sequence stellar evolution are discussed. We then summarize the evidence which indicates that the infrared excesses in recently formed stars are primarily due to thermal emission from circumstellar dust. Various indirect lines of evidence on the spatial distribution of the circumstellar dust are examined, and we conclude that the concentration of the dust in a large disk structure is the most viable model at this time.

Progress in the study of pre-main sequence stellar evolution has not been as rapid as in the study of post-main sequence evolution for several reasons. First, the initial conditions for pre-main sequence evolution are far less certain than those for post-main sequence evolution. Second, pre-main sequence stars are often complex objects, with erratic brightness variations, extreme chromospheric heating, strong mass outflows, and significant infrared excesses. It is difficult to advance rapidly in the understanding of these objects when one is still groping for the proper conceptual model. However, significant progress has been made in the study of young stars since the first Protostars and Planets meeting in 1978, and various lines of evidence seem to be converging toward a single unified picture. Some of this progress has come in unexpected ways, such as the emergence of the enigmatic Herbig-Haro objects as key actors in this unfolding drama.

Although there are a number of interesting astrophysical problems in the study of young stellar objects, one question is central to the theme of this book: do young stars show evidence of large dusty circumstellar disks from which planetary systems may form? In this chapter we will begin by reviewing the basic observational classes of young stellar objects and discussing the evolution of these stars in the HR diagram. We will then summarize the evidence which indicates that the infrared excesses associated with young stars are primarily due to thermal emission from circumstellar dust. Finally, we will look at some indirect evidence on the spatial distribution of the dust and note some outstanding problems.

I. CLASSES OF YOUNG STELLAR OBJECTS

From an observational standpoint, most young stellar objects fall into one of four natural groups as discussed in the following paragraphs.

1. T Tauri Stars. T Tauri stars are late-type (G, K, or M) pre-main sequence (PMS) stars with prominent line emission seen on classification spectrograms. The emission-line spectrum generally resembles the flash spectrum of the solar chromosphere, with the addition of a few low-excitation forbidden lines (Joy 1945; Herbig 1962*b*). The strongest emission lines at visible wavelengths are usually the Balmer series of hydrogen and the Ca II H and K lines. T Tauri spectra also generally show a "spectral veiling" which was described by Joy (1949) as resembling an overlying emission continuum which seemed to fill in the photospheric absorption lines. In the most extreme T Tauri stars, the numerous emission lines and spectral veiling are so strong that photospheric absorption features are not readily seen. However, these extreme T Tauri stars appear to form a continuous sequence in photometric colors and spectroscopic appearance with other T Tauri stars. There is also no clear-cut dividing line between the T Tauri stars and late-type stars with weaker line emission. Herbig and Rao (1972) and Bastian et al. (1983) suggest spectroscopic criteria for distinguishing T Tauri stars from emission-line M dwarfs, but the most reliable means of identifying young late-type stars is probably by the presence of the Li I absorption feature at 6707 Å. All legitimate T Tauri stars seem to possess an infrared excess due to thermal emission from circumstellar dust. It is now generally accepted that T Tauri stars are pre-main sequence stars with masses $\lesssim 3\ M_\odot$ and ages up to a few million years.

2. Herbig Ae/Be Stars. Herbig (1960) discusses a group of emission-line B- and A-type stars which are located in obscured regions and illuminate fairly bright reflection nebulosity. These stars seem to lie somewhat above the main sequence in the HR diagram (S. E. Strom et al. 1972; Finkenzeller and Mundt 1984) and are apparently destined to become main sequence B-type stars. The inferred ages of these stars are generally less than a million years. Like the T Tauri stars, the Herbig Ae/Be stars possess infrared excesses apparently due to thermal emission from circumstellar dust. Herbig Ae/Be stars are not nearly as common as T Tauri stars; a recent catalog by Finkenzeller and Mundt (1984) lists 57 confirmed or candidate members of this class.

3. Luminous Embedded Infrared Sources. Stars more massive than about 10 $M_\odot$ are apparently too obscured by dust to be readily observable at visible wavelengths during their pre-main sequence phase. The basic reason is that the Kelvin-Helmholtz contraction time scale for these massive stars becomes shorter than the protostellar accretion time scale, so that the star ignites its thermonuclear furnace while still accreting material. During the past decade, a number of extremely luminous infrared sources ($L > 10^3\ L_\odot$) have been discovered in dense molecular clouds. Some of these excite compact H II regions. It is difficult to place these objects in an evolutionary sequence, since only the luminosity is reasonably well known. However, there is little doubt that these represent recently formed stars of high mass. Habing and Israel (1979) review the observational and theoretical aspects of the formation of massive stars. These problems are considered further in the chapter by Thompson.

4. Pre-Main Sequence Stars with Weak Line Emission. Not all pre-main sequence stars are characterized by emission lines and infrared excesses. Observations indicate that stars with masses $\lesssim 3\ M_\odot$ lose their line emission and infrared excess prior to reaching the main sequence. These less-exotic PMS stars may be difficult to identify unless they lie in the PMS band of a young cluster or association and have membership confirmed by proper motion and radial velocity. Some candidate post-T Tauri K-type stars have recently been identified in X-ray studies (Feigelson and DeCampli 1981; Walter and Kuhi 1981) and spectroscopic surveys (Feigelson and Kriss 1983; Herbig unpublished). The Li I λ6707 absorption feature can also be used to distinguish recently formed late-type stars from field stars seen in the direction of a star-forming region. Much work remains to be done in the identification of additional weak-emission PMS stars and in following the evolution of solar-mass stars from the T Tauri phase to the main sequence.

II. EVOLUTION IN THE HR DIAGRAM

At least from a theoretical viewpoint, the star formation process involves two basic phases, a hydrodynamical collapse during which material is added by accretion (the protostellar phase) and a subsequent slow contraction to the main sequence on a Kelvin-Helmholtz time scale (the pre-main sequence

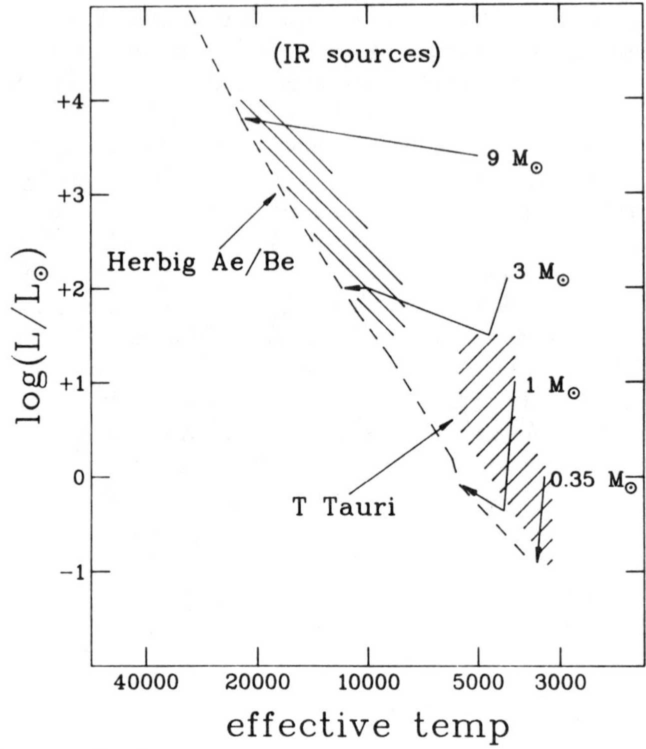

Fig. 1. Schematic HR diagram showing the main sequence (dashed line), pre-main sequence evolutionary tracks for stars of 9, 3, 1, and 0.35 $M_\odot$, and approximate loci of the Herbig Ae/Be stars and T Tauri stars (hatched regions).

phase). During the hydrodynamical collapse, the stellar core is expected to be obscured at visible wavelengths by the dust in the infalling material. The pre-main sequence star should become visible when the accretion phase terminates, perhaps because of an outflowing stellar wind. In the HR diagram the slow contraction phase consists in general of a nearly vertical "Hayashi track" characterized by a fully convective structure and a more horizontal "Henyey track" where the star has developed a radiative core. The least-massive stars evolve almost vertically downward in the HR diagram, retaining a fully convective structure.

These points are illustrated in Fig. 1, where we show a schematic HR diagram with main sequence (dashed line), PMS evolutionary tracks for stars of 9, 3, 1, and 0.35 $M_\odot$ (Iben 1965; M. Cohen and L. V. Kuhi 1979), and approximate loci of Herbig Ae/Be stars (Finkenzeller and Mundt 1984) and T Tauri stars (M. Cohen and L. V. Kuhi 1979). In general the duration of the pre-main sequence phase decreases with increasing stellar mass, until the accretion phase merges with the main sequence phase for masses $\gtrsim 10\ M_\odot$. The recent evidence for collimated mass outflows from some of the youngest identifiable low-mass stars suggests that the earliest phase of stellar evolution may be somewhat more complicated than first envisioned.

The pre-main sequence evolutionary tracks in the HR diagram are extremely important, because they provide the sole means by which we infer the masses and ages of recently formed stars. We should emphasize that *there are as yet no cases in which the mass of a T Tauri star or Herbig Ae/Be star has been fundamentally determined,* i.e., an eclipsing double-lined spectroscopic binary. Furthermore, there is no independent means for determining the ages of young stars. Specifically, the evidence for continuing star formation in various star-forming regions (see, e.g., M. Cohen and L. V. Kuhi 1979; M. T. Adams et al. 1983) prevents one from giving a unique age to a recently formed cluster. Thus, the assignment of masses and ages to pre-main sequence stars involves a "leap of faith" in the validity of the theoretical tracks.

To place a young star in the HR diagram, one must:

1. Estimate the spectral type;
2. Get the apparent bolometric luminosity from photometry at visible and infrared wavelengths;
3. Correct this bolometric luminosity for interstellar extinction;
4. Adopt a distance to the cluster or association containing the star.

Unfortunately, there are uncertainties associated with each of these steps. In some cases it is difficult to classify the spectrum accurately because of the line emission and spectral veiling. If the surfaces of young stars consist of regions of differing effective temperature (see, e.g., Herbig and Soderblom 1980; Appenzeller and Dearborn 1984), the spectral type could vary with wavelength and possibly also change with time as the star varies in brightness. When computing the bolometric luminosity of the star, one should include the flux from the circumstellar dust if the dust is heated directly or indirectly by the starlight. However, if the dust is heated by dissipational processes within a disk structure, it would seem inconsistent to include the dust emission in the bolometric luminosity, combine this luminosity with a photospheric effective temperature, and compare the resulting point in the HR diagram with theoretical evolutionary tracks for spherically symmetric diskless stars. The correction for interstellar extinction can also be uncertain, due to the influence of spectral veiling on stellar colors, the variability of stellar colors, and the possibility of anomalous interstellar extinction laws within some star-forming regions.

In the well-studied Taurus-Auriga dark cloud complex, most of the young stars appear to be in the convective phase of their pre-main sequence evolution. Since the convective tracks are nearly vertical, the deduced stellar masses will be insensitive to the derived bolometric luminosities. Although there will always be uncertainties with individual stars, the general mass range of the T Tauri stars in this region and the youth of these stars seems well established.

On the theoretical side, the fundamental question is whether most of the infalling material accretes directly onto the protostellar core in a nearly spherical collapse, or whether angular momentum forces much of the material to form a massive disk structure before accreting onto the stellar core. Even in the spherically symmetric case, there has been disagreement about the position of the star in the HR diagram at the end of the accretion phase. However, the work of Stahler et al. (1980*a*) seems to have clarified this situation. This question of protostellar and pre-main sequence evolution for low-mass stars is also discussed by Appenzeller (1983).

Recently Stahler (1983*a*) has extended the work of Stahler et al. to predict the "birthline" in the HR diagram for the youngest optically visible stars. He finds that the birthline derived from these spherically symmetric accretion models can be brought into agreement with the upper boundary of the observed T Tauri locus in the HR diagram through the adoption of a reasonable mass-accretion rate of 10^{-5} $M_{\odot}$ yr^{-1}. In the process of adjusting the isochrones in the HR diagram to common initial conditions, Stahler is also able to remove an apparent dip in the star formation rate for several well-studied regions of star formation. His work lends some credibility to the basic Hayashi-Henyey evolutionary scenario for low-mass young stars.

The YY Orionis stars are a subset of the T Tauri stars with strong ultraviolet continua and sporadic red-displaced absorption components in the stronger emission lines, indicating mass infall. Walker (1972) has suggested that the YY Ori stars are in the final stage of the accretion phase. However, further studies have not demonstrated that the YY Ori stars are systematically younger than other T Tauri stars, or that they have larger infrared excesses (M. Cohen and L. V. Kuhi 1979). One possibility is that the red-displaced absorption components arise from material which is falling back toward the star after being ejected at less than escape velocity (see, e.g., Mundt 1984).

A few low-luminosity pre-main sequence stars have been observed to brighten dramatically within a year, expelling shells of material and acquiring the spectrum of a low-surface-gravity A or F star (Herbig 1977a). This is apparently followed by a slow fading in brightness. In the case of V1057 Cygni, the progenitor star seems to have been a faint T Tauri star. Herbig (1977a) has suggested that these "FU Orionis" outbursts may occur in most young stars, perhaps many times. The mechanism currently favored for explaining these outbursts involves a temporary high rate of mass accretion from a circumstellar disk, which heats the outer layers of the star and causes them to swell up (see, e.g., R. B. Larson 1983; see also the chapter by Lin and Papaloizou).

III. THE NATURE OF INFRARED EXCESS

In 1978, there were differing opinions about the source of the infrared excess in the T Tauri stars and the Herbig Ae/Be stars. However, more recent

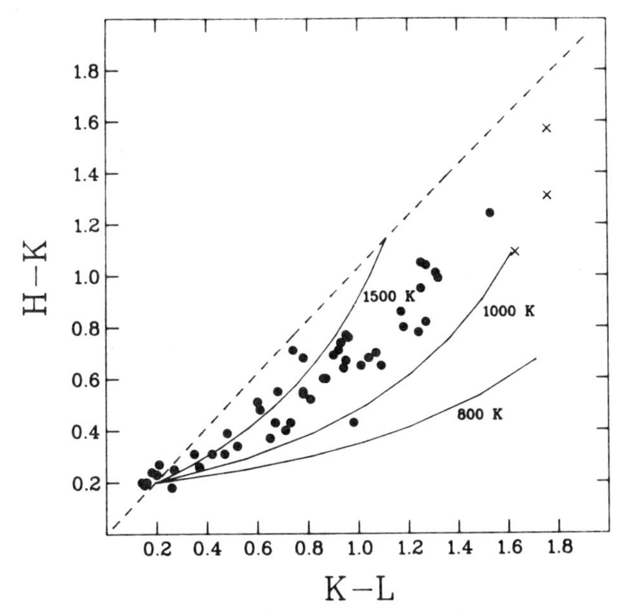

Fig. 2. Plot of *H-K* color against *K-L* color for a sample of T Tauri stars, based on photometry from Rydgren and Vrba (1981,1983) and Rydgren et al. (1982). The best-case disk stars HL Tau, DG Tau, and W90 (a Herbig Be star) are shown as X's. The solid lines are loci for the addition of blackbody infrared excess of temperature 1500, 1000, and 800 K to a late-type photosphere, while the dashed line is the locus of blackbodies in this diagram.

work has established that these excesses are primarily due to thermal emission from circumstellar dust.

The most compelling evidence for circumstellar dust in T Tauri systems comes from the analysis of *JHKL* (1 to 4 μm) photometric data. The *J-H, H-K,* and *K-L* colors can be accurately measured and are sensitive to emission from the hottest circumstellar dust. M. Cohen and L. V. Kuhi (1979) compare the observed T Tauri locus in the (*H-K, K-L*) diagram with the predictions of hot gaseous envelope and circumstellar dust models for infrared excess. They show that the *H-K* and *K-L* colors of T Tauri stars can be readily explained by thermal emission from hot (1000 to 1500 K) circumstellar dust, whereas the H-K and K-L colors of many T Tauri stars are much larger than expected from the hot gaseous envelope model. While some continuous emission from hot circumstellar gas cannot be precluded, it is clearly not a dominant effect and is not required to explain the infrared excesses in these stars.

Figure 2 shows the observed locus of T Tauri stars in the (*H-K, K-L*) diagram, based on the recent *JHKL* photometry of Rydgren and Vrba (1981,1983) and Rydgren et al. (1982). The dashed line is the locus of blackbodies, while the curved lines emanating from the lower left corner show the

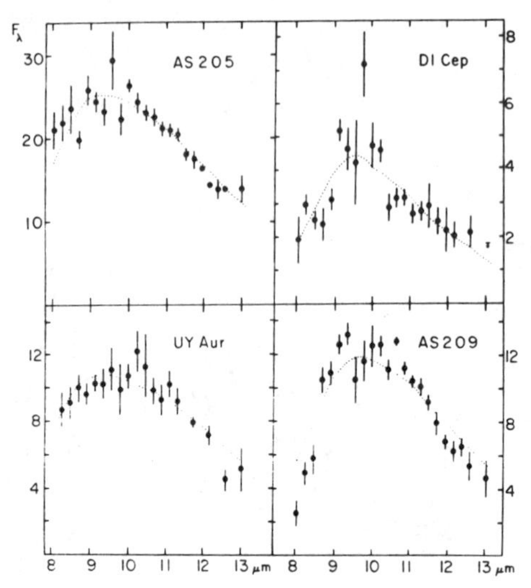

Fig. 3. Ten micron silicate emission features in four typical T Tauri stars, from M. Cohen and F. C. Witteborn (1985).

result of adding progressively more blackbody emission of temperature 1500, 1000, and 800 K to a late-type stellar photosphere. At larger *K-L* color where the temperature separation is better, the T Tauri infrared excesses clearly have a color temperature between 1000 and 1500 K. Rydgren et al. (1982) suggest that this uniformly high maximum dust temperature in T Tauri systems is evidence for a grain formation or destruction process.

The detection of 10 μm silicate emission or absorption features in a number of T Tauri stars provides further evidence that the infrared excess is due to circumstellar dust. M. Cohen and F. C. Witteborn (1985) have recently completed a 10-μm spectrophotometric survey of some 40 T Tauri stars. They find that silicate emission features are relatively common and that silicate absorption is rare. Some typical examples are shown in Figs. 3 and 4. The shapes of the silicate emission and absorption features in T Tauri stars are similar to the shape of the silicate emission feature observed in the Trapezium region of the Orion nebula.

The scarcity of silicate absorption features may provide a clue to the geometry of the circumstellar dust shells. HL Tau and DG Tau are the T Tauri stars with the deepest 10 μm absorption features and are also the T Tauri stars with the strongest linear polarization at visible wavelengths. M. Cohen (1983,1984) suggest that silicate absorption is seen in T Tauri stars only when a dusty disk is viewed nearly edge-on. The broadband spectral energy distributions of HL Tau and DG Tau are also quite unlike those of post-main sequence M stars with silicate absorption features.

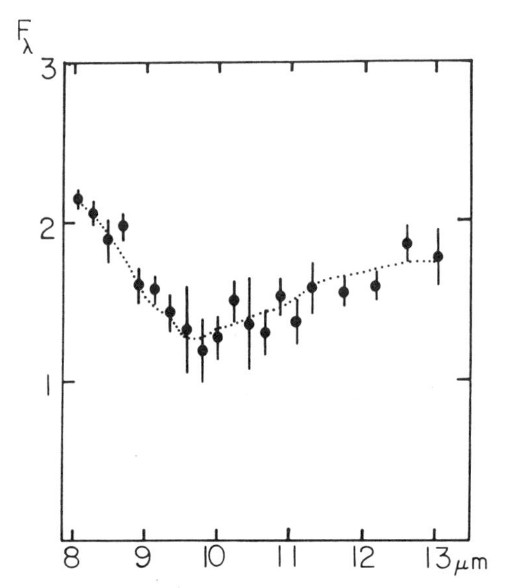

Fig. 4. Ten micron silicate absorption feature in the T Tauri star HL Tau, from M. Cohen and F. C. Witteborn (1985).

The time scale for the appearance and disappearance of the infrared excess in young solar-mass stars will also provide a useful constraint on models for the origin of circumstellar dust. The observations show that the infrared excess is clearly established in the youngest T Tauri stars which have inferred ages of order 10^5 yr since the beginning of protostellar collapse. Although the post-T Tauri stars are not yet well studied, it appears that prominent infrared excess, as measured by the *K-L* color, may be gone by an age of a few times 10^6 yr. This conclusion is based on *JHKL* photometry of a few weak-emission PMS K stars in the Taurus region (Rydgren and Vrba 1981,1983; Rydgren et al. 1982) and of a sample of G-type stars in the PMS band of the Orion Ic association (Rydgren and Vrba 1984).

The discovery of an infrared companion to T Tau (Dyck et al. 1982) is a useful reminder that stars cannot always be assumed to be single. However, the strong 20 μm excess in T Tau (due to the infrared companion) is not typical of the class. Furthermore, the invisible companion to T Tau was first noted at radio wavelengths (M. Cohen et al. 1982), yet the recent survey of Bieging et al. (1984) shows that radio-emitting companions to T Tauri stars are extremely rare. Some radio-quiet, very cool companions may be identified in the unbiased far-infrared sky survey from the Infrared Astronomy Satellite (IRAS). However, the existing 10 and 20 μm photometry of T Tauri stars shows that cool companions will not be the general rule, and in any case such companions will not significantly affect the *JHKL* colors which clearly indicate emission from hot circumstellar dust.

IV. CLUES TO THE SPATIAL DISTRIBUTION OF CIRCUMSTELLAR DUST

Observations of young stars have established beyond a reasonable doubt that the infrared excesses in recently formed stars are due to heated circumstellar dust. A key question is whether this dust is concentrated in a disk structure or is perhaps forming in the dense cooling winds from these stars. In this section we consider four indirect lines of evidence regarding the spatial distribution of the circumstellar dust associated with young stars.

A. T Tauri Variability

Both the T Tauri stars and the Herbig Ae/Be stars show irregular light variations. There is little doubt that erratic variations at shorter wavelengths in T Tauri stars are basically due to flaring and to changing chromospheric emission (see, e.g., Worden et al. 1981). However, it has been suggested (Gahm et al. 1974b; Walker 1978,1980) that changing circumstellar obscuration is responsible for the larger-amplitude variations (typically 1 to 2 mag) in these stars. If correct, this would imply that the dust is not confined to a thin disk structure, because we should not be seeing all these young stars exactly equator-on.

W. Herbst et al. (1983) have obtained extensive photometry of a sample of bright T Tauri stars and Herbig Ae/Be stars. Their results for the Herbig Ae star UX Ori and the T Tauri star CO Ori are especially interesting. The chromospheres in these stars are too weak to cause much variability, yet these stars have occasional aperiodic minima in their lightcurves which can be several magnitudes deep. Both of these stars become redder when fainter, but near minimum light the U-B and B-V colors "turn around" and start to become bluer again. The Hα emission equivalent width also becomes larger when these stars are fainter. This systematic dependence of colors and Hα emission on V magnitude is incompatible with variable circumstellar obscuration and implies that these large-amplitude brightness variations originate in the star itself.

A study of the color variations in typical T Tauri stars (Vrba et al. 1985) reveals that colors such as B-V and V-I usually correlate quite well with V magnitude and that these color slopes differ significantly between stars. Moreover, there is a clear trend for the larger color slopes to be associated with the stars of later spectral type, in accord with the view that the brightness variations are due to a changing mixture of hot plage regions (see, e.g., Herbig and Soderblom 1980) and cooler photospheric and dark spot regions on the stellar surface. The variable obscuration model for T Tauri variability does not readily explain the differences in color slope between stars or the dependence of color slope on spectral type.

It has recently been suggested that the development of strong magnetic flux near the surface of the star is responsible for the large-amplitude brightness variations in young stars. As demonstrated by Appenzeller and Dearborn (1984), this can lead to temporary changes in the structure of the outer part of the star which are sufficient to cause visual brightness changes as large as 3.5 mag. The observed blueing in *U-B* and *B-V* near minimum light in some stars is presumably due to the hotter chromospheric regions becoming more apparent. This theoretical work seems to provide a natural explanation for the deep lightcurve minima in both the Herbig Ae/Be stars and the T Tauri stars, without any need for variable circumstellar obscuration. Exceptionally strong surface magnetic fields could also be responsible for the extreme chromospheric heating in the T Tauri stars and are undoubtedly related to the ultraviolet and X-ray flares seen in T Tauri stars (see the chapter by Giampapa and Imhoff).

B. Limits on Circumstellar Reddening

The presence or absence of a general circumstellar reddening effect in T Tauri stars and Herbig Ae/Be stars could also provide a clue to the spatial distribution of the circumstellar dust. If the dust is in a spherical shell around the star, both reddening and infrared excess should increase with greater optical thickness in the dust shell. However, if the dust is confined to a thin disk structure with a central hole, the starlight should reach the observer unreddened by the dust except when the disk is viewed nearly edge-on. In this latter case there should be no general correlation between reddening and infrared excess.

Of the various nearby regions of star formation, the Taurus dark cloud complex has by far the best-observed sample of T Tauri stars. Figure 5 shows the *V-I* color excess (a reddening indicator) plotted against *H-K* color (a measure of infrared excess) for more than 40 T Tauri stars in the Taurus region. The interstellar reddening vector in this diagram is almost vertically upward, and many of these stars are expected to suffer significant intracluster reddening. Individual *V-I* color excesses should be uncertain by at least ± 0.10 mag. Looking at the lower envelope of the distribution, where the intracluster reddening should be smallest, one sees at best a slight correlation in the expected direction. These results are certainly consistent with no correlation between reddening and infrared excess in this sample.

If there is a spherical dust shell around typical T Tauri stars, the dust grains would have to be relatively large (at least several tenths of a micron in radius) to avoid reddening in excess of that observed. However, such grains would be larger than those found in the general interstellar medium (see, e.g., Mathis et al. 1977) and it remains to be seen if grains can grow quickly to this size in the cooling winds from these young stars.

M. Cohen and L. V. Kuhi (1979) have also looked for differences in reddening between pairs of T Tauri stars in the Taurus and Orion regions. They find similar reddening in most cases, as expected if the observed reddening were of interstellar rather than circumstellar origin.

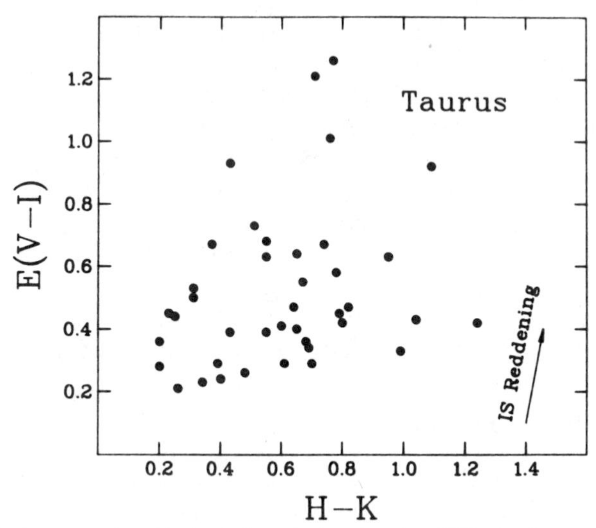

Fig. 5. *V-I* color excess (a reddening indicator) plotted against *H-K* color (a measure of infrared excess) for T Tauri stars in the Taurus region. The photometry is from Rydgren and Vrba (1981,1983) and Rydgren et al. (1982), while the spectral types are from M. Cohen and L. V. Kuhi (1979). The nearly vertical arrow indicates the effect of one magnitude of interstellar (IS) *V* extinction.

C. Comparison of Best-Case Disk Stars with Other Young Stars

Two of the T Tauri stars in the Taurus region (HL Tau and DG Tau) are strong candidates for possessing large-scale circumstellar disks. Both stars show well-collimated jets of glowing gas extending away from the star (Mundt and Fried 1983), 10 μm silicate absorption features as well as far-infrared excesses which appear to be elongated in the direction perpendicular to the jets (M. Cohen 1983,1984), and unusually strong linear polarization (Bastien 1982) with the electric vectors also perpendicular to the jet direction. W90 (LHα 25) is a Herbig Be star in the young cluster NGC 2264 and is also a good candidate for possessing a large circumstellar disk. This star falls several magnitudes below the main sequence in the (*V*, *B-V*) diagram for this cluster (Walker 1956) but has a spectroscopic luminosity which would place it about five magnitudes brighter (S. E. Strom et al. 1972). In all three cases it is suspected that the star is observed through a nearly edge-on disk structure.

Do these young stars with evidence for circumstellar disks differ from other recently formed stars in their infrared properties? In the (*H-K*, *K-L*) diagram, HL Tau, DG Tau and W90 all fall along the upper end of the observed T Tauri locus. This indicates that they have about the same maximum dust temperatures as other T Tauri stars and Herbig Ae/Be stars but that the ratio of thermal dust emission to photospheric light is especially large. A similar continuity between the best-case disk stars and other T Tauri stars is

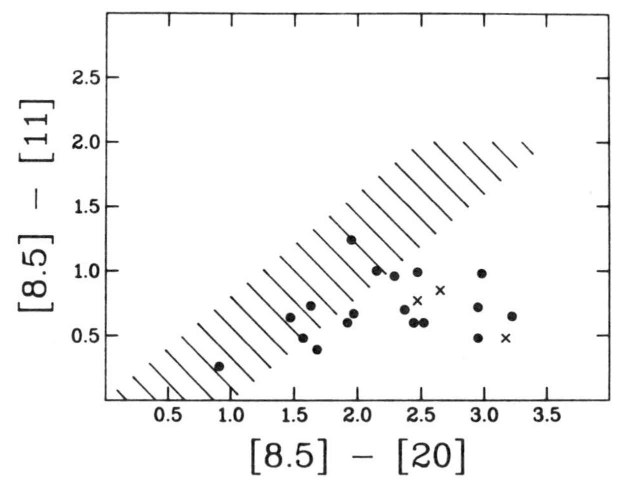

Fig. 6. [8.5 μm]–[11 μm] color plotted against [8.5 μm]–[20 μm] color for T Tauri stars, based on photometry from the literature as well as unpublished data from Grasdalen and from Vrba and Rydgren. The best-case disk stars HL Tau, DG Tau, and W90 (a Herbig Be star) are shown as X's. The hatched region is the locus of dusty M giants and supergiants.

seen at longer infrared wavelengths, where cooler dust is preferentially seen. Figure 6 shows the [8.5 μm]–[11 μm] color plotted against the [8.5 μm]–[20 μm] color for T Tauri stars in the Taurus region, based on photometry from the literature as well as unpublished data from Grasdalen and from Vrba and Rydgren. Again, the best-case disk stars HL Tau, DG Tau and W90 fall along the basic T Tauri locus. The locus of dusty M giants and supergiants in Fig. 6 is indicated by hatching. There is a significant difference between the dusty M stars and the T Tauri stars in this diagram, in the sense that the T Tauri stars tend to have larger 20 μm fluxes. It is not clear if this represents a fundamental difference in the dust distribution, or simply results from the heating of nearby dark cloud material by these young stars.

D. Radial Velocities of Forbidden Emission Lines

There is a growing body of evidence for well-collimated mass outflows from recently formed stars (see chapters by Schwartz and by Mundt). These flows are apparently responsible for shock-exciting the Herbig-Haro emission nebulosities. The forbidden emission lines seen in some T Tauri spectra (mainly [S II] and [O I]) are presumably also excited by a collimated flow or stellar wind at some distance from the star. Jankovics et al. (1983) find a strong tendency for the forbidden emission lines in T Tauri spectra to be blue-shifted relative to the star. As discussed by Appenzeller (1983), this is most easily understood if the forbidden emission lines are excited by a bipolar flow from the star but the redshifted lobe on the far side of the star is obscured by a large circumstellar disk.

To summarize this section, we find no compelling need for variable circumstellar obscuration to explain the large-amplitude variability of the T Tauri and Herbig Ae/Be stars and no convincing evidence that the reddening in young stars increases with infrared excess. Although the infrared colors of the best-case disk stars tend to be larger than those of other T Tauri stars, they seem to form a continuous sequence with the other stars and may result from these specific stars being observed with disks nearly edge-on. The systematic blueshift of the forbidden emission lines in T Tauri spectra is also most easily explained by the presence of a large circumstellar disk. All these lines of evidence are consistent with the circumstellar dust in T Tauri stars and Herbig Ae/Be stars being concentrated in large disk structures.

Several questions about the interpretation of the observed infrared excesses in young stars remain to be resolved. One problem is that the dynamical time scale of a disk formed during the star formation process is expected to be of order 10^5 yr, whereas the infrared excess in low-mass PMS stars seems to persist for a few times 10^6 yr. A second problem is that the mass of submicron grains needed to explain the infrared excesses in T Tauri stars is far smaller than the 10^{-2} to 10^{-1} $M_\odot$ associated with a minimum-mass solar nebula. Using the ice and silicate absorption features, M. Cohen (1983) estimates a mass of only about 5×10^{-10} $M_\odot$ in submicron particles for the disk around the young T Tauri star HL Tau. Models for the emitted flux from dust around T Tauri stars, based on reasonable density and temperature distributions, also suggest that $< 10^{-10}$ $M_\odot$ of submicron grains are needed to produce a typical T Tauri infrared excess of order 1 $L_\odot$. Thus if preplanetary disks are present around T Tauri stars, most of the mass is apparently either very cool or already in the form of larger particles. Although M. Cohen et al. (1984) estimate that the mass of cool gas and dust around several very young stars associated with Herbig-Haro objects may be comparable to a minimum-mass solar nebula, such large far-infrared excesses are not typical of older T Tauri stars. The observed infrared excesses in T Tauri and Herbig Ae/Be stars should be critically compared with the predicted infrared appearance of solar nebula models as a function of time to clarify the interpretation of the observations, especially with regard to possible planetary systems in formation.

V. PROSPECTUS

Significant progress has been made in the study of young stellar objects since 1978. Probably the most exciting results involve the discovery of collimated mass outflows from very young stars and the identification of Herbig-Haro objects as shock-excited clumps associated with these outflows (see chapters by Mundt and by Schwartz). Studies of luminous infrared sources embedded within dark clouds are also beginning to provide a clearer picture of the formation of massive stars. As a result of more thorough photometric and spectroscopic studies of young stars and more detailed theoretical modeling, it has been established that the infrared excesses are primarily due to circumstellar dust (see, e.g., M. Cohen and L. V. Kuhi 1979) and that most of the line emission seems to originate in a strong chromosphere (see, e.g., Calvet et al. 1984*b*). The sources of the brightness variations in these stars are also becoming better understood.

However, much more work remains to be done in the study of young stars. The spatial distribution of circumstellar dust needs to be studied further by both direct and indirect means. This should include attempts to spatially resolve the infrared emission regions around young stars with groundbased two-dimensional infrared arrays and eventually with high-resolution infrared telescopes in space. We also need to identify the energy source which powers the strong chromospheres in these stars. Is it passed up through the star via strong magnetic fields, or is it due to accretion onto the star? The nature of the flows and winds in these systems must be better understood and placed in a proper evolutionary sequence. Low-mass PMS stars with weak-line emission need to be identified in greater number and studied to determine if they are true post-T Tauri stars or stars which somehow are bypassing the T Tauri phase. Finally, there should be many pre-main sequence close-binary stars, but few have been found. A major point of contact between observation and theory involves the mass of stars; it is therefore important to search for binaries in order to obtain direct determinations of the masses of PMS stars.

Acknowledgment. We thank G. Grasdalen for providing unpublished photometry of T Tauri stars and D. Black and G. Herbig for thoughtful comments on the manuscript.

THE AMBIENT RADIATION FIELD OF YOUNG SOLAR SYSTEMS: ULTRAVIOLET AND X-RAY EMISSION FROM T TAURI STARS

MARK S. GIAMPAPA
National Solar Observatory

and

CATHERINE L. IMHOFF
Computer Sciences Corporation

We review the principal results that have emerged from ultraviolet and X-ray observations of T Tauri stars. We emphasize that, as seen in the ultraviolet, T Tauri stars are characterized by strongly enhanced emission. The X-ray emission and its variability, when detected, is also enhanced and indicative of the occurrence of violent flare activity near the stellar surface. We discuss possible effects of these enhanced emissions on conditions in early solar systems and the chemical evolution of early planetary atmospheres.

The introduction of new instrumentation and, therefore, new capabilities for viewing the universe inevitably yields information and insight that often revolutionize the course of research in astrophysics. The advent of ultraviolet, X-ray, and infrared satellite observatories has made available for analysis previously inaccessible wavelength regions. Consequently, new thermal regimes in stellar atmospheres and their immediate environs can be probed. In particular, the analyses and interpretation of the ultraviolet spectrum and soft X-ray emittance of the pre-main sequence T Tauri stars has resulted in significant advances in the understanding of the atmospheric properties of these objects and the relationship of these inferred properties to the origin and evolution of chromospheres and coronae in solar-type stars. The major results that have thus far emerged from ultraviolet and soft X-ray observations of T Tauri stars, as obtained with the *International Ultraviolet Explorer* (IUE) and the *Einstein* (HEAO-B) satellite observatories, respectively, have been recently reviewed by Giampapa (1984) and Feigelson (1984); see also forthcoming reviews by Imhoff (1984), Bertout (1984), and Montmerle (Preprint).

These review papers are primarily concerned with the definition of the kinds of compact and extended atmospheric structures (along with their variability) that characterize the T Tauri stars as deduced from ultraviolet and X-ray diagnostics. In this chapter we coalesce and summarize the principal results of these investigations that were discussed separately by Giampapa (1984) and by Feigelson (1984). Given that a T Tauri star was the source of the ambient radiation and particle fields for the young solar system, it is important to develop as realistic a model as possible for the emittance (and its variability) of this class of objects. We extend the previous reviews by describing in somewhat more detail the potential effects of ultraviolet and X-ray emission from T Tauri stars on the physical conditions of the prebiological atmosphere of the Earth and the young solar system in general. Implicit in this approach is the assumption that the Sun experienced a T Tauri phase in its pre-main sequence evolution and that currently observed T Tauri stars can be considered to be representative examples of the young Sun. Within this context, the investigation of the T Tauri stars becomes a fundamental aspect of solar-stellar physics.

I. ULTRAVIOLET EMISSION

In this section we describe the principal results that have been deduced from ultraviolet observations of T Tauri stars as obtained with the IUE satellite in the 1150 Å to 3200 Å wavelength range. We review the basic aspects of their ultraviolet line spectra, including atmospheric properties as inferred from line diagnostics and emission measures, observations of ultraviolet variability, and manifestations of multicomponent atmospheres.

A fundamental result of ultraviolet investigations of T Tauri stars is that the ultraviolet chromospheric ($T_e \sim 10^4$K) and transition region ($T_e \sim 10^5$K) emission lines are strongly enhanced, typically by factors of 10^2 to 10^4 relative to the quiet Sun and other late-type stars. The outer atmospheric line and continuous emission luminosity can, in total, be several percent of the stellar bolometric luminosity. Examples of the far-ultraviolet spectra of a small but representative sample of T Tauri stars are shown in Figs. 1 and 2. As in the case of many late-type stars, lines of transition-region ions such as C IV, Si IV, N V and He II (all characteristic of plasma at temperatures near $T \sim 10^5$K) can appear in the far-ultraviolet spectra of T Tauri stars. In addition, upper chromospheric lines of Si I, C II and O I (characteristic of plasma at temperatures T $\sim$ 10^4K) are present, while Fe II features can be identified throughout their spectra. Shell absorption or emission lines occur in the far-

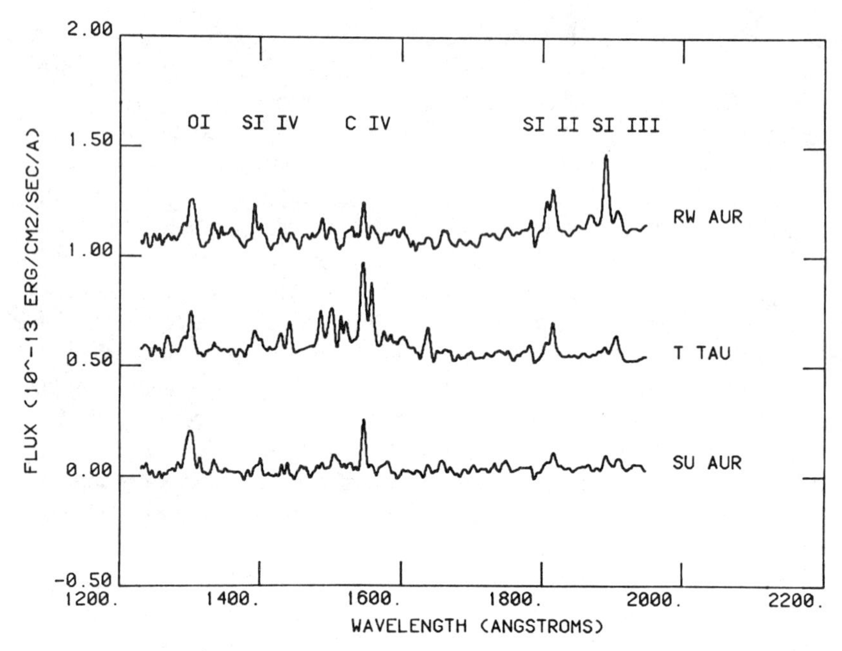

Fig. 1. Far-ultraviolet spectra of three T Tauri stars.

ultraviolet spectra of T Tauri stars with no or little photospheric contribution to the line spectrum in this wavelength range. Moreover, far-ultraviolet molecular H_2 emission has been identified by A. Brown et al. (1981) in the direction of the recently discovered infrared companion of the prototype T Tau. Furthermore, the ultraviolet continuum appears to be dominated, in many instances, by free-free and hydrogen-recombination radiation emanating from an extended region that is hotter than the underlying photosphere. A cool stellar photospheric energy distribution (corresponding to the known spectral type of a T Tauri star, as inferred from optical and infrared observations) often does not satisfy the observed ultraviolet spectral energy distribution. Clearly, the photospheric properties of T Tauri stars cannot be productively studied in the ultraviolet. Finally, Appenzeller et al. (1980) claim that there is a correlation between strong ultraviolet continuum emission and optical "blue veiling" (i.e., filling in of absorption lines) indicating that this so-called veiling can be the result of envelope emission seen against a cool photospheric background, as opposed to deep photospheric heating at the onset of a steep chromospheric temperature gradient. However, exceptions are noted, such as RU Lupi, which are strongly veiled in the optical but show a relatively weak ultraviolet continuum in near, simultaneously acquired, optical and ultraviolet spectra (Appenzeller et al. 1980). The blue veiling is likely the result of both an extended envelope contribution and deep photospheric heating giving rise to

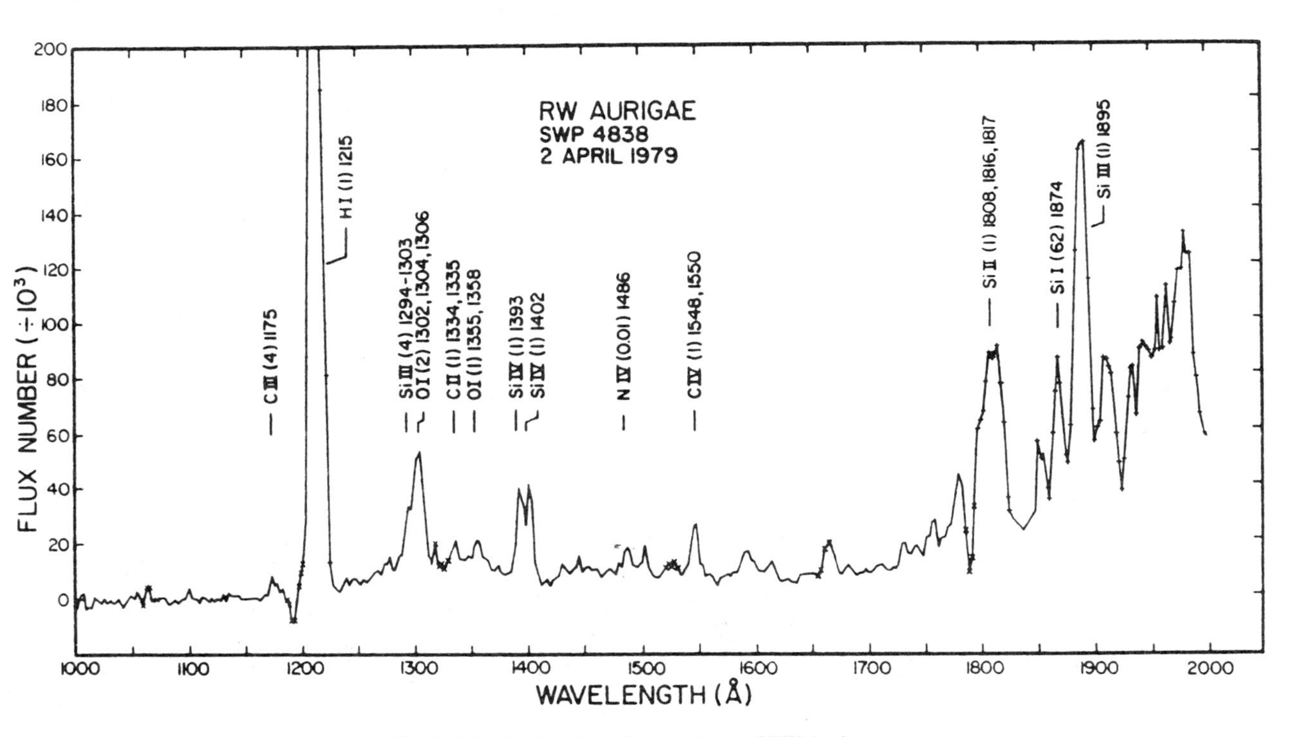

Fig. 2. The far-ultraviolet line spectrum of RW Aurigae.

chromospheric continuum emission (Calvet et al. 1984*b*). The relative importance of these processes in the veiling phenomenon is yet to be quantitatively assessed.

The pattern of enhanced far-ultraviolet line emission in the spectra of T Tauri stars is, in general, similar to that of other stars with active chromospheres, such as the dMe stars. However, some T Tauri stars exhibit a significant departure from this pattern in terms of the observed level of emission in the resonance lines of high-temperature ionic species relative to other, somewhat lower-temperature diagnostics. Imhoff and Giampapa (1980, 1982*b*,1984) originally noted the apparent weakening of far-ultraviolet lines of ions such as C IV, He II and even the absence of N V λ 1240 relative to Si IV, O I, and C II line emission within the spectra of some T Tauri stars. In fact, the absolute surface fluxes of the C IV, N V, and He II lines of these same T Tauri stars can be weakened relative to that of other T Tauri stars (although the fluxes in the detected features still remain considerably enhanced relative to the corresponding line fluxes in the quiet Sun). These investigators interpreted this observation as evidence that the maximum temperature T_{max} of plasma in these particular T Tauri stars is less than coronal temperatures ($T_{cor} \sim 10^{6-7}$K). The relative weakening, or even absence, of high-temperature transition-region line emission in some T Tauri stars must be due to a source of enhanced nonradiative cooling. More specifically, mass loss becomes the dominant component in the energy balance of the outer atmospheres of some T Tauri stars as opposed to radiative cooling via coronal X-ray emission. While the degree of mass loss is uncertain, estimates are generally in the range of 10^{4-7} times greater than the solar mass loss rate ($\sim 2 \times 10^{-14}$ $M_{\odot}$ yr^{-1}).

Imhoff and Giampapa (1980,1982*a,b,*1984) making use of ultraviolet and optical observations of T Tauri stars, proposed the above explanation. They noted that the optically "weak-emission" T Tauri stars (i.e., those stars with relatively less strong Balmer line and Ca II H and K line emission in the optical) show C IV, N V, Si IV and He II emission lines in normal relative strengths *and* were detected in the X-ray by *Einstein* (HEAO-B). The optically "strong-emission" T Tauri stars, which are presumably characterized by higher mass loss rates, displayed both a relative weakening of high-temperature lines in the ultraviolet and were *not* detected in the X-ray by *Einstein*. Theoretical support for this explanation has been advanced by L. Hartmann et al. (1982) in their models for Alfvén wave-driven mass loss from T Tauri stars (see also De Campli 1981). This model predicts that the local Alfvén wave heating rate increases more slowly with wind density than does the radiative cooling rate. Hence, there is a decline in the wind temperature T_{max} with increasing mass loss rate. In other words, those regions of highest Alfvén wave flux (and therefore highest heating rates) will also have the largest mass flux rates and will generally be cooler. This favors low-temperature emission over high-temperature emission. Thus, coronal temperatures are never attained in those T Tauri stars characterized by a sufficiently high Alfvén wave flux and correspondingly high mass loss rate. Furthermore, this fundamental result implies that the site of X-ray emission in T Tauri stars must be near the stellar surface and not in an "extended" or "wind" region. In Sec. II this conclusion is corroborated by X-ray observations.

In summary, the ultraviolet spectra of T Tauri stars are characterized by intense line and continuous emission indicative of a high degree of non-radiative heating occurring in the atmospheres of these stars. Some T Tauri stars exist that do not possess coronae, even though strong chromospheric line emission is present. In particular, the relative weakening or absence of high-temperature ultraviolet transition-region lines reveals that the maximum plasma temperature attained in these stars is only $T_{max} \sim 10^5$ K. Mass loss appears to become the dominant atmospheric cooling mechanism and this, in turn, constitutes corroborative evidence for the applicability of Alfvén wave-driven mass loss models to T Tauri stars. However, problems concerning the physical accuracy of Alfvén wind models still remain (MacGregor 1982).

Emission Measures and Densities

Emission measures ($\equiv \int n_e^2 dV$) (where n_e^2 is the electron density and dV is the volume element) combined with estimates of electron density can yield information on the depth (or geometric extent) of a line source region and the degree to which the stellar atmosphere can be regarded as plane-parallel and homogeneous. The emission measure is indicative of the power radiated by plasma. Cram et al. (1980) used density-sensitive ratios to obtain an estimate of the electron density in the source region of RU Lup. These investigators find a mean value of $\log n_e = 10.2 \pm 0.6$ but emphasize that the line ratios are highly sensitive to uncertain atomic parameters and atmospheric models. The propagation of uncertainties in estimates of emission measure and n_e lead to a range of 0.08 R_* to 4 R_* for the source depth of Si IV emission in RU Lup (Cram et al. 1980). This range is compatible with both a relatively compact emission region and a hot, extended region.

L. Hartmann et al. (1982) computed the flux ratio for Si IV/C III] where "]" denotes a semi-forbidden transition, within the context of an Alfvén wave model for mass loss in T Tauri stars. These investigators find Si IV/C III] = 0.94 which is to be compared with the observed mean value ⟨Si IV/C III]⟩ = 2.5 ± 1.5 and the range Si IV/C III] = 0.4 to 4.9 for seven T Tauri stars (Imhoff and Giampapa 1984). By contrast, the static, deep chromospheric, nonspecific model of T Tauri stars constructed by Cram (1979) yields Si IV/C III] = 2.61. While the observed mean value for this line flux ratio appears more consistent with the static, deep chromosphere model, the range of observed values is consistent with both models. We regard this as evidence for the presence of both compact regions and extended regions in the atmospheres of T Tauri stars. Interestingly, A. Brown et al. (1984) find, in their detailed examination of the spectrum of T Tau, discrepant values of the electron density as inferred from different line diagnostics formed in similar thermal regimes. They claim that this apparent discrepancy can only be reconciled by postulating a two-component atmosphere for T Tau (see also, Giampapa et al. 1981). In particular, A. Brown et al. (1984) state that the observed flux ratios and inferred electron densities can be understood if the permitted lines and X-ray emission are predominantly formed in a high-pressure, hydrostatic region while the semi-forbidden lines are formed in an extended, low-density component of the atmosphere. Moreover, the emission measure of the low-density region must be comparable, but not greater than, the emission measure of the high-density region.

The Mg II h and k Resonance Lines

The spectra of T Tauri stars in the 2000 Å to 3200 Å range are dominated by the resonance lines of Mg II. The results of an extensive, low-resolution study of the Mg II lines in a sample of T Tauri stars has been presented by Giampapa et al. (1981). According to these investigators, the Mg II h and k line fluxes indicate that the atmospheres of T Tauri stars are characterized by the most extreme degree of nonradiative heating among the class of single late-type stars (Giampapa et al. 1981, their Fig. 3). In fact, the Mg II lines, the Ca II lines, and the far-ultraviolet emission combined with any X-ray emission, can comprise 1% or more of the stellar bolometric luminosity. The total outer atmospheric radiation losses (including Balmer and Lyman line emission) in the specific case of T Tau have been estimated to be 8% of its bolometric luminosity (A. Brown et al. 1984).

High-resolution observations of both the Mg II h and k lines and the Ca II K line reveal P Cygni type profiles indicative of the presence of an expanding chromospheric region (Giampapa et al. 1984; see also, Lago and Penston 1982; Penston and Lago 1983; Lago 1982). Moreover, the Ca II K line profile and Mg II k line profile have similar shapes. Hence these lines arise from similar regions in a T Tauri atmosphere. This observation is consistent with the fact that the Mg II and Ca II resonance lines share similar formation characteristics. The Mg II h and k lines exhibit profile variability as well as variability in their total flux. We show in Fig. 3 two high-resolution observations of the Mg II h and k lines in SU Aur. The variability in the blue wings of these profiles is evident. Thus, the rate of mass loss is variable although a quantitative estimate of both the mass loss rate and the variation in this rate, as deduced from the profiles given in Fig. 3, would require the application of a hydrodynamic radiative transfer code. In summary, high-resolution observations of the Mg II h and k lines reveal the existence of an accelerating chromospheric component characterized by a variable mass loss rate.

Ultraviolet Variability

T Tauri stars are characterized by both rapid and long-term variability at practically all wavelengths. The ultraviolet variability of RW Aur has been described in detail by Imhoff and Giampapa (1981) and Imhoff (1984) and

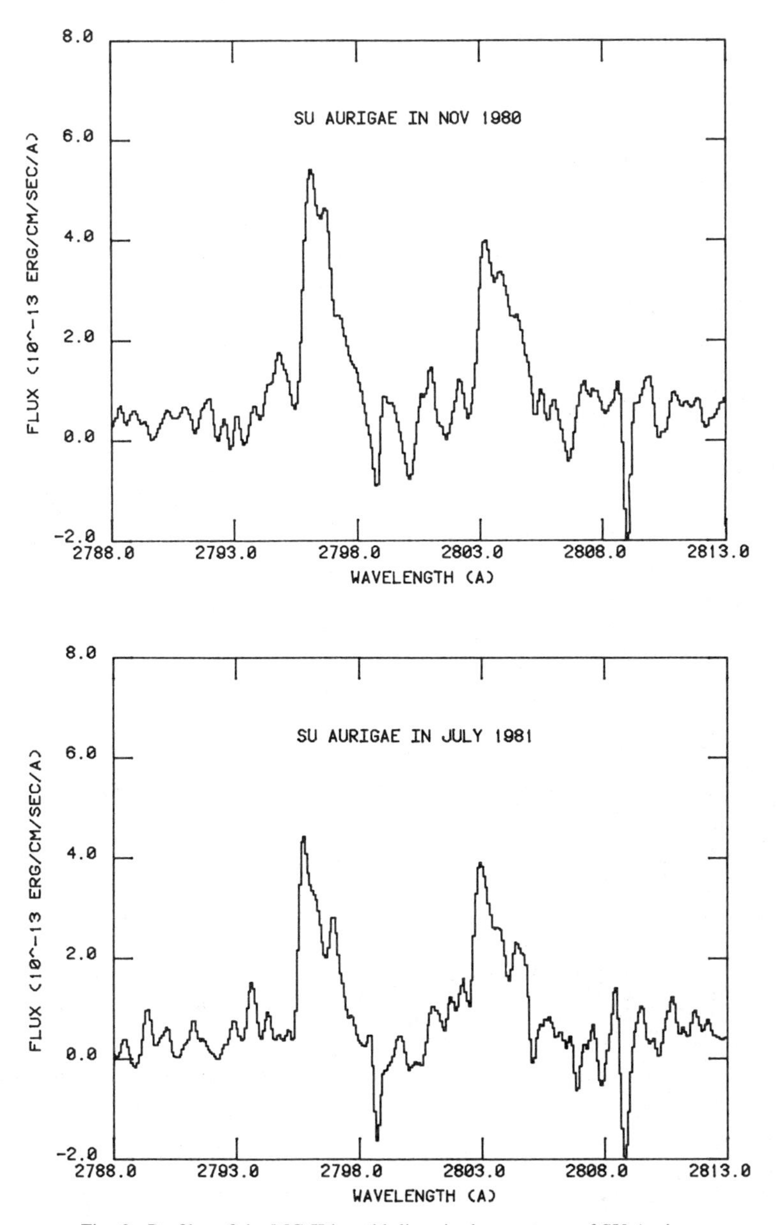

Fig. 3. Profiles of the MG II k and h lines in the spectrum of SU Aurigae.

those results are updated here (see also, Calvet et al. 1984*a*). In particular, the visual brightness of RW Aur increased by a factor of 2.5 between July 1978 and April 1979. During this time, the ultraviolet continuum at 2700 Å ± 100 Å and 2900 ± 100 Å increased by a factor of 5. The resonance lines of C IV and Si IV each increased by 4 to 5 times while the lower excitation lines of C II and Mg II (h and k) increased by factors of 2.5 and 1.5, respectively. In addition, the ultraviolet shell spectrum changed from emission to absorption. This latter observation implies that the enhanced ultraviolet continuum emission was interior to the shell. In subsequent observations, intermediate states were observed with behavior consistent with the earlier results.

The origin of the variability is unknown. However, the degree and pattern of the ultraviolet continuum and line-emission enhancement is reminiscent of violent flare activity. Violent outbursts and flare-like activity have been observed in the optical (Worden et al. 1981; Mundt 1984) and the X-ray (Feigelson and De Campli 1981; Feigelson 1984). Of course, temporal and spatial variability in the wind structure can presumably affect the far-ultraviolet high-temperature emission (see, e.g., Mundt and Giampapa 1982). The ultraviolet variability of T Tauri stars clearly merits further investigations with IUE and Space Telescope using improved temporal resolutions.

II. X-RAY EMISSION

In this section we discuss the principal results that have emerged from X-ray observations of T Tauri stars as obtained with the *Einstein* satellite observatory in the 0.2 to 4.0 keV passband.

A fundamental result of soft X-ray investigations of T Tauri stars is that: (1) not all observed T Tauri stars were detected in the X-ray; and (2) those T Tauri stars that were detected exhibited enhanced coronal emission as compared with the general class of late-type stars. The range in X-ray luminosity, L_x (erg s^{-1}), for T Tauri stars extends from nondetections, i.e., upper limits of $L_x \sim 10^{29}$, to several times 10^{31} or $\sim 10^3$ higher than the X-ray luminosities of local, main sequence stars. In particular, the local, low-mass disk stars are typically characterized by ages $\sim 10^9$ to 10^{10} yr and soft X-ray luminosities near that of the quiet Sun ($\sim 10^{27}$ erg s^{-1}) to an order of magnitude higher. The solar-type stars in the Hyades cluster, that has an estimated age of 0.75 byr, generally emit in the range $3 \times 10^{28} \leq L_x \leq 3 \times 10^{29}$. The level of X-ray emission among T Tauri stars, characterized by ages of a few times 10^6 yr, is generally in the range of $< 10^{29}$ to several times 10^{30}. In any event, the upper limit to observed T Tauri X-ray emission appears to be $L_x < 10^{34}$.

As a further illustration of the level of T Tauri X-ray emission, we give in Table I the median values of L_x for detected T Tauri stars in some well-known complexes in which star formation is occurring (Feigelson 1984).

TABLE I
X-Ray Emission from T Tauri Stars

Complex	Median L_x (erg s^{-1})
Orion	$\sim 8 \times 10^{30}$
Taurus-Auriga	$\sim 2 \times 10^{30}$
ρ Ophiuchi	$\sim 5 \times 10^{30}$
Chameleon[a]	$0.5\text{–}2 \times 10^{30}$

[a]Range in median values of L_x due to uncertainty in the distance of the Chameleon cloud.

As mentioned in the previous section, some T Tauri stars exist that were not detected in the X-ray and which exhibit a weakening of high-temperature, ultraviolet emission lines relative both to lower-temperature diagnostics within the spectrum and the same features in the spectra of X-ray-detected stars. Coronal temperatures are never attained in these X-ray quiet T Tauri stars (with the possible exception of transient X-ray flare outbursts). However, there may exist highly obscured T Tauri stars that appear to be X-ray quiet but which do possess coronae. In these cases, the outwardly directed coronal X-ray emission is presumably being extinguished by circumstellar material (Walter and Kuhi 1981). Future ultraviolet observations of X-ray quiet T Tauri stars characterized by high circumstellar extinction would prove decisive in determining the applicability of the so-called smothered coronae hypothesis.

Finally, we note that in those T Tauri stars that were detected in the X-ray, the level of X-ray emission is not enhanced to the degree that would be expected from an extrapolation of the far-ultraviolet fluxes observed in these stars combined with the ultraviolet and X-ray scaling relations given by Ayres et al. (1981).

Origin and Site of X-ray Emission

The observed rapid variability seen in the few available X-ray spectra of some T Tauri stars, the densities required to produce the observed levels of detected X-ray emission, along with the anticorrelation between high mass loss T Tauri stars and X-ray emission (discussed in the previous section), all indicate that X-ray emission in T Tauri stars arises from regions near the stellar surface and not in a wind or extended region (Feigelson and DeCampli 1981; Feigelson 1984). The observed X-ray emission is probably produced in large loops of magnetically confined plasma that are characterized by evolution on both rapid (10^2–10^4 s) and long time scales ($\geq 10^5$ s). The hard X-ray emission is a manifestation of high X-ray temperatures ($T \sim 10^7$K) that exceed those of the less-luminous main sequence coronae, although these kinds of hard components are seen in the active, more-evolved RS CVn systems

(Swank et al. 1981) with small filling factors (i.e., area coverages) on the stellar surface.

The observed high-amplitude X-ray variability is reminiscent of violent flare activity. This interpretation is indeed corroborated by high-speed optical photometry (Worden et al. 1981) and the ultraviolet observations discussed here. The rise times of X-ray events in T Tauri stars are a few times 10^{2-3}s while the decay times are 10^{3-4}s. The temperatures of these presumed flare events are 1 to 3 $\times$ 10^7 K and appear to occur over spatial scales of order 0.5 R_* (Feigelson 1984) where R_* is the stellar radius. The liberated flare energies in the X-ray are generally about 10^3 times higher than large solar events and dMe stellar flares. As illustrative examples, the luminosity of DG Tau varied from a 3 σ upper limit of $L_x < 6 \times 10^{30}$ to 4×10^{31} in $\sim 10^2$s; the X-ray luminosity of SU Aur was observed to decline by 50% in 16 hr.

In summary, ultraviolet and X-ray investigations of T Tauri stars combine to demonstrate that a realistic model of their atmospheres requires two components; a high-density, compact region analogous to closed magnetic field structures observed on the Sun, and open regions, analogous to solar coronal holes, that are characterized by a relatively strong mass loss rate and a $T_{max} \sim 10^5$ K. Hence the observed ultraviolet transition region line emission is composed of a wind, or extended-region contribution, and a compact-region contribution. The X-ray emission can only arise from the compact component since $T_{max} < T_{cor}$ in the wind. We suggest that this may explain why X-ray emitting T Tauri stars appear somewhat underluminous in X-rays than would be expected on the basis of the observed strong enhancement of their far-ultraviolet emission. The relative proportion of each contribution likely depends upon the extent and kind of magnetic field configurations existing on the stellar surface at the time of observation. In general, the X-ray luminosities of T Tauri stars are $\geq 10^3$ times that of the quiet Sun. The X-ray spectra are hard, indicating temperatures $T \sim 10^7$ K in the emitting regions. Moreover, violent flare-like variability is seen in the X-ray as well as in the ultraviolet and optical. We note that not all T Tauri stars are X-ray active. In particular, coronal temperatures are never attained in those T Tauri stars with apparently higher mass loss rates. The evolutionary relationship (if any) between X-ray active and X-ray quiet T Tauri stars is not yet clear nor has the temporal evolution of ultraviolet and X-ray emission during the pre-main sequence phase of stellar evolution been well defined.

III. AMBIENT EFFECTS: ULTRAVIOLET EMISSION

The recognition that the T Tauri stars are characterized by substantially enhanced ultraviolet and, in many cases, X-ray fluxes relative to that of the present-day Sun has significantly altered the research directions in related studies of the young solar system. These continuing investigations include (a) the origin and chemical evolution of the prebiological atmosphere of the early

Earth, and (b) the origin of abundance anomalies in the solar system. In the following, we will review briefly the impact of ultraviolet and X-ray investigations of T Tauri stars on these topics.

The understanding of biological evolution on Earth requires an accurate quantitative knowledge of the levels of free oxygen O_2 and ozone O_3 in the prebiological paleoatmosphere and the level of ultraviolet radiation ($\approx$ 1000 Å–3000 Å) incident on the surface of the Earth. In particular, ultraviolet radiation presumably initiated the photochemical processes responsible for the formation of O_2 and O_3 prior to the emergence of primitive organisms. For O_2 production and destruction the basic photochemical reactions are (Canuto et al. 1982,1983; see also, Zahnle and Walker 1982; Kasting et al. 1983):

$$H_2O + h\nu \rightarrow OH + H; \lambda \leq 2400 \text{ Å} \tag{1}$$

and

$$CO_2 + h\nu \rightarrow CO + O; \lambda \leq 2300 \text{ Å}. \tag{2}$$

From Reaction (1) we have

$$OH + OH \rightarrow O + H_2O \tag{3}$$

and Reactions (2) and (3) yield

$$O + O + M \rightarrow O_2 + M \tag{4}$$

$$O + OH \rightarrow O_2 + H \tag{5}$$

where M is a chemically inert species usually taken as the dominant constituent of the atmosphere. In the model computed by Canuto et al. (1983), M is identified with molecular nitrogen N_2. The destruction of O_2 occurs according to the following:

$$O_2 + h\nu \rightarrow O + O; \lambda \leq 2420 \text{ Å} \tag{6}$$

and

$$H + O_2 + M \rightarrow HO_2 + M \tag{7}$$

as well as

$$HCO + O_2 \rightarrow CO + HO_2. \tag{8}$$

The origin and evolution of ozone is, of course, coupled to that of oxygen. Ozone production (Reaction 9) and destruction (Reactions 10 and 11) occur according to the following:

$$O + O_2 + M \rightarrow O_3 + M \tag{9}$$

and

$$O_3 + h\nu \rightarrow O_2 + O; \lambda \leq 11{,}000 \text{ Å} \tag{10}$$

$$O_3 + O \rightarrow 2O_2 \tag{11}$$

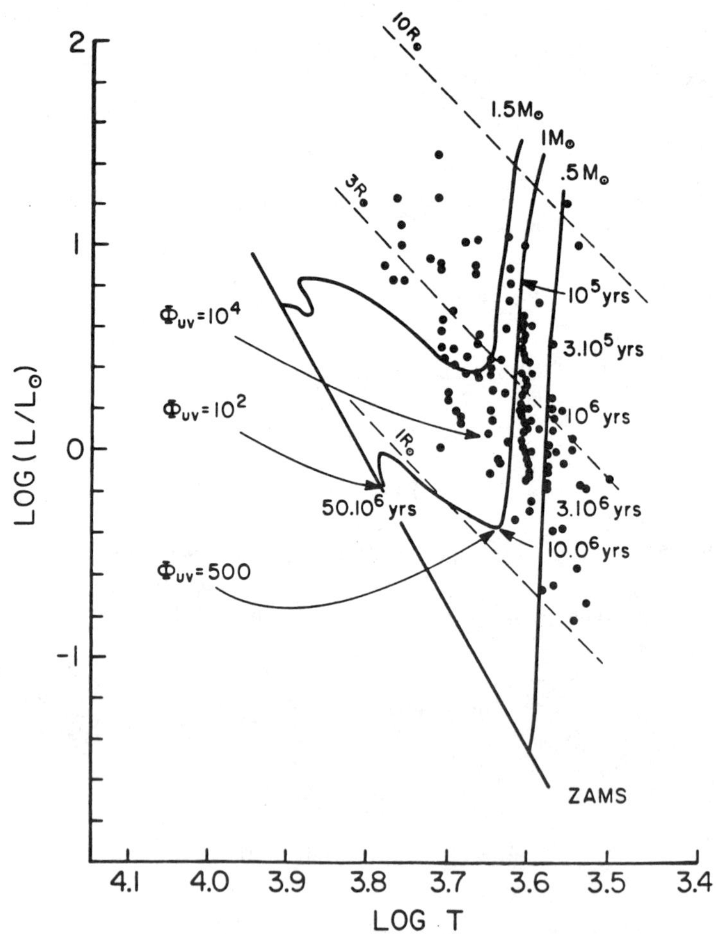

Fig. 4. The theoretical pre-main sequence evolutionary tracks of 0.5 $M_\odot$, 1 $M_\odot$, and 1.5 $M_\odot$ stars combined with observational data. The increase in the ultraviolet flux Φ_{uv}, relative to current levels, as well as the ages for a 1 $M_\odot$ star are shown.

and other reactions involving H, OH, HO_2, NO, Cl and so forth. Astrophysics enters through $h\nu$ in these photochemical processes. Previous investigations scaled the ultraviolet flux of the young Sun with its total luminosity, as inferred from theoretical stellar evolutionary tracks, prior to the availability of ultraviolet spectra of T Tauri stars (Fig. 4). Inspection of Fig. 4 reveals that during portions of the radiative phase of solar pre-main sequence evolution, the solar luminosity was less than it is today and so the ultraviolet flux was assumed to be correspondingly reduced. This approach was erroneous because stellar evolutionary models cannot predict detailed outer atmospheric (e.g., chromospheric, coronal, wind, extended-region, etc.) properties. We must rely on direct observations of stars that are examples of solar precursors.

Adopting this approach, Canuto et al. (1983) constructed a semi-empirical spectrum of the pre-main sequence Sun using ultraviolet and optical spectra of the three T Tauri stars SU Aur, RW Aur, and T Tau. These stars constitute a reasonable sample of this heterogeneous class, representing a range of T Tauri characteristics. The resulting mean spectrum, reduced to the flux intercepted at 1 AU and with a conservative correction for extinction by dust corresponding to $A_v = 0.5$ mag applied, is shown in Fig. 5, normalized to the present-day solar spectrum.

Inspection of this figure reveals that at the shortest wavelengths where the photochemical processes are most sensitive, a T Tauri star (i.e., the pre-main sequence Sun) is brighter than the present Sun by a factor $\sim 10^4$. Of course, this spectrum does not necessarily characterize the solar ultraviolet emittance during the subsequent evolution of the Sun on the radiative track (post-T Tauri phase). During the latter phase, the accretion of the Earth and atmospheric formation occurs. Hence, a necessary future refinement of this kind of empirical approach must involve the use of stellar input spectra that are more representative of the post-T Tauri stage. The difficult task of confidently identifying members of this class of stars combined with a quantitative study of their spectral properties is presently being pursued (see, e.g., Mundt et al. 1983*b*). In addition, the form of the decay law describing the decline of atmospheric emission during the pre-main sequence stage is a current topic of active investigation (see, e.g., Boesgaard and Simon 1982). Nevertheless, the utilization of the spectrum in Fig. 5 represents an important first step in the study of the photochemistry of the early atmosphere of the Earth in the presence of an enhanced ultraviolet radiation field.

Adopting the input spectrum given in Fig. 5, Canuto et al. (1983) computed a detailed one-dimensional model of the prebiological atmosphere of the Earth. These investigations discovered that the surface concentration of O_2 increases by 1 to 2 orders of magnitude relative to that which would be computed utilizing the present-day solar ultraviolet irradiance (however, Kasting, Pollack and Crisp [personal communication, 1984] came up with alternative results). The actual values of O_2 surface concentration depend critically on the adopted input values of CO_2 and H_2 concentration, respectively. The

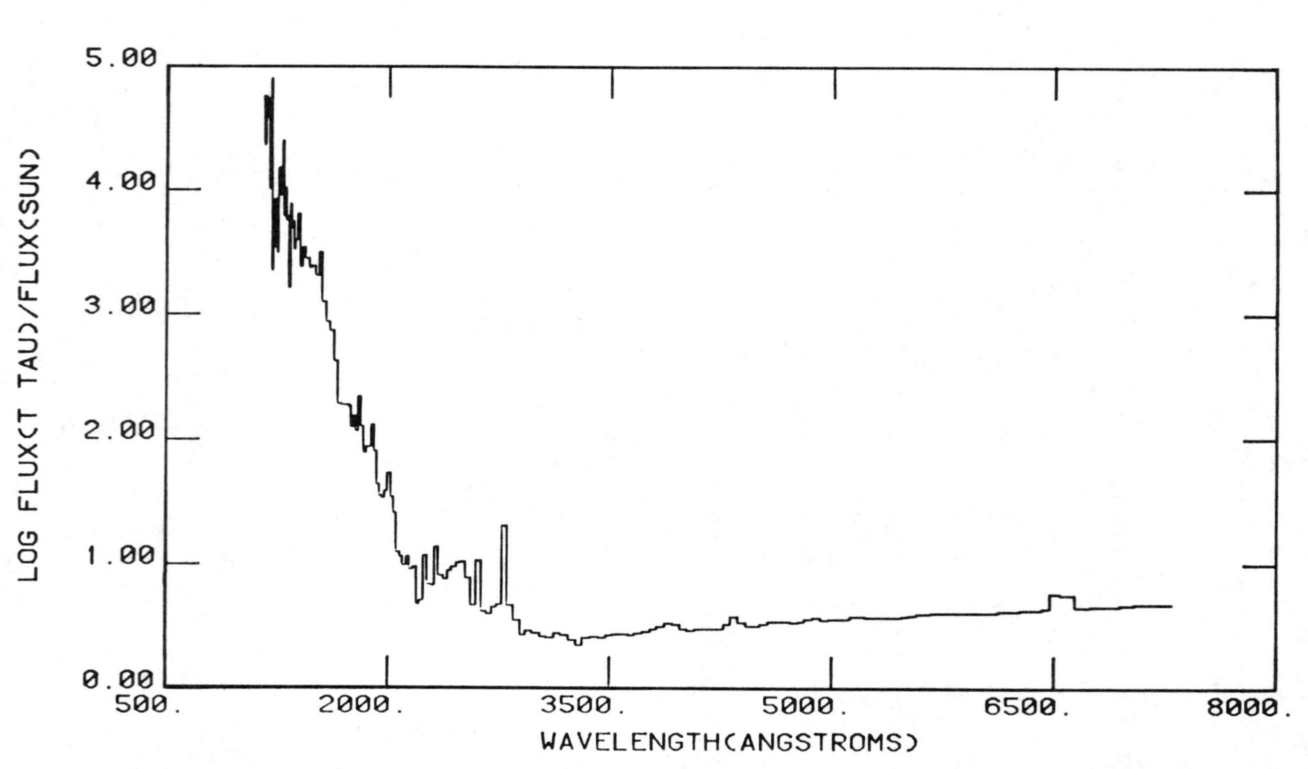

Fig. 5. The composite ultraviolet flux, normalized to the present solar values, as derived from spectra of three T Tauri stars.

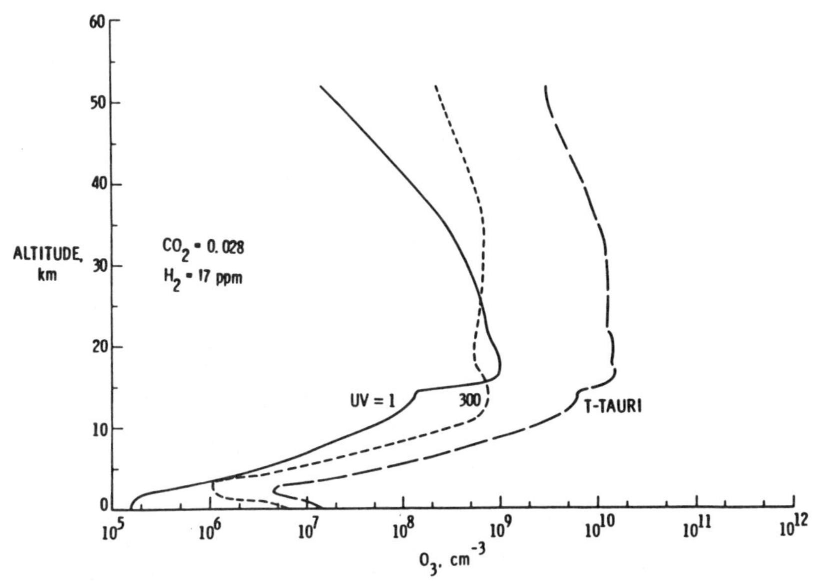

Fig. 6. The ozone profile for various enhancements of the ambient ultraviolet radiation field. UV = 1 corresponds to present-day levels, while the T Tauri curve is based on the spectrum in Fig. 5. UV = 300 is a wavelength independent enhancement.

computed column densities of ozone for various enhancements of the ultraviolet flux distribution are illustrated in Fig. 6. The atmospheric column density is the relevant physical parameter because of the important ability of ozone to absorb potentially harmful ultraviolet radiation. As can be seen from Fig. 6, the O_3 column density systematically increases with increasing ultraviolet flux. However, the highest O_3 column densities computed are still below the values required to provide a biologically effective ultraviolet shield (Kasting 1984, personal communication). In addition, the CO/CO_2 ratio is interesting because a recent laboratory experiment (Bar-Nun and Chang 1984) demonstrated that an ultraviolet irradiated mixture of water vapor (H_2O) and carbon monoxide (CO) yields an array of organic molecules relevant to the origin of life (in the absence of CO, irradiation of CO_2 cannot yield organics). Canuto et al. (1983) found that the CO/CO_2 ratio is of order unity at the altitude where these life reactions are believed to have occurred only in the cases of the enhanced ultraviolet models (see Fig. 7). Finally, we note the results concerning formaldehyde H_2CO. More specifically, H_2CO was a key molecule in the formation of complex organic molecules. Today, H_2CO is a trace species generated through a methane oxidation chain with the precursor, methane CH_4, produced by biogenic activity.

Interestingly, the lifetime of CH_4 in a prebiological atmosphere is very short because of photochemical destruction (Kasting et al. 1983). The ques-

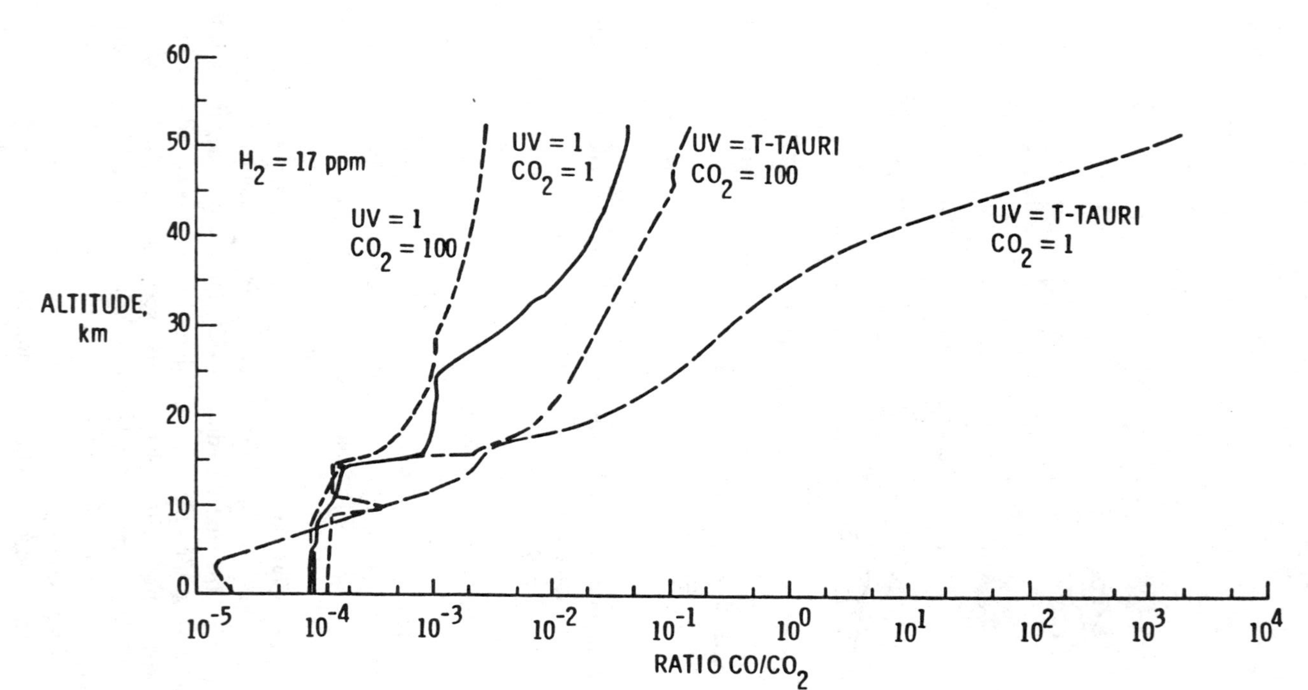

Fig. 7. The vertical profile of the CO/CO_2 ratio. The CO_2 concentration is given in units of the pre-industrial CO_2 level of 280 ppm by volume.

tion that arises is how can H_2CO be produced in the absence of a steady source of CH_4. A prebiological source for H_2CO has been suggested which is sensitive to ultraviolet flux and the abundances of CO_2, H_2, and H_2O (Pinto et al. 1980). The results obtained by Canuto et al. (1983) suggest that if H_2CO were indeed photochemically formed from CO_2 and H_2O in an ultraviolet-enhanced environment, then the level of CO_2 must have been greater in the early atmosphere than it is today.

IV. AMBIENT EFFECTS: X-RAY EMISSION

The effects of X-ray emission from T Tauri stars are either direct, mainly through influencing the local ionization equilibrium of surrounding material, or indirect, indicating violent, flare-like outbursts that give rise to high-energy particle fluxes. As reviewed by Feigelson (1984), T Tauri X-ray emission characterized by $T_e \sim 10^7$ K and $L_x \sim 10^{30}$ erg s^{-1} will produce a series of nested Stromgren spheres: a coronal region with an extent of order $\sim 10^{13}$ cm, an H II region ($T \sim 10^4$ K), a hot H I region, and a cool H_2-dominated region beginning at a distance $\sim 10^{14}$ cm (or about the orbit of Jupiter) from the star. We note, parenthetically, that the strong ultraviolet flux from the T Tauri stars will also influence the thermal state of surrounding material. For example, A. Brown et al. (1981) discovered ultraviolet molecular H_2 emission in the vicinity of T Tau and its infrared companion. The H_2 emission arises through fluorescence with the red wing of H Lyman α. In addition, photoabsorbed X rays are radiated in the near infrared by H_2 vibrational-rotational transitions and in the far infrared by dust grains. The integrated X-ray flux within complexes may be responsible for the ionized molecular species observed in their cores. Silk and Norman (1983) have further suggested that local X-ray ionization will couple the magnetic field to the gas and inhibit further collapse that could lead to star formation.

The appearance of high-amplitude, short-period outbursts in the X rays imply, as previously discussed, the occurrence of violent flare activity. As discussed by Worden et al. (1981), the proton flux from these powerful events could give rise to abundance anomalies in the early solar system without recourse to exotic hypotheses, such as nearby supernovae. Again assuming that the Sun experienced the same kind of T Tauri phase, the enhanced proton flux from flare activity could produce the isotopic and elemental anomalies found in lunar regolith and carbonaceous chondritic meteorites (see chapters by Kerridge and by Clayton et al.). The ^{26}Al anomaly could, in fact, result from enhanced proton irradiation associated with violent pre-main sequence flare activity (Feigelson 1984; Worden et al. 1981).

In summary, the ambient ultraviolent and X-ray radiation fields of T Tauri stars play significant roles in the origin and evolution of physical conditions in young solar systems and planetary atmospheres. In particular, assuming the Sun experienced a T Tauri-like phase in its pre-main sequence evolution, then violent flare activity and the associated high proton fluxes, as inferred from high X-ray variability, may account for various abundance anomalies we find in the solar system. Furthermore, current investigations in the prebiological chemical evolution of planetary atmospheres must now include the fact that the photochemical processes occurred within an enhanced ultraviolet environment. The evolution of the ultraviolet and X-ray radiation fields of solar-mass stars during their post-T Tauri phase is not well defined since actual examples of post-T Tauri stars are yet to be confidently identified. Furthermore, the origin of the profound differences in the chemical compositions of the atmospheres of Venus, Earth, and Mars, each of which presumably experienced the same enhanced ultraviolet environment of the young solar system, merits careful consideration. The advent of Space Telescope and the Advanced X-ray Astrophysics Facility (AXAF) will, undoubtedly, give us new light for these old problems.

Acknowledgments. We acknowledge valuable comments from G. Basri, D. Black, C. Jordan, and J. Kasting on an earlier version of this chapter. The authors also acknowledge support from the NASA/IUE guest observer program for their continuing investigations of the pre-main sequence stars. The National Solar Observatory is a division of the National Optical Astronomy Observatories, operated by the Association of Universities for Research in Astronomy, Inc., under contract with the National Science Foundation.

THE NATURE AND ORIGIN OF HERBIG-HARO OBJECTS

RICHARD D. SCHWARTZ
University of Missouri

A brief description of the nature of Herbig-Haro nebulae is given, and the shock-wave origin of the nebulae is discussed. Kinematical evidence suggests that Herbig-Haro objects are ejected in bipolar flows from young stars. Evidence from infrared observations of the stars that excite Herbig-Haro objects is summarized; these stars appear to be T Tauri stars. The origin of these nebulae is discussed emphasizing energy required to power them, and a number of questions are posed pertaining to outflow mechanisms associated with the exciting stars.

Since their discovery over three decades ago (Herbig 1951; Haro 1952), Herbig-Haro (HH) objects have been associated with star formation by virtue of their proximity to dark clouds which house young emission-line stars. The semistellar, emission-line HH nebulae possess forbidden-line spectra reminiscent of some T Tauri stars, and it was assumed that each HH nebula harbored a protostar which in time would reveal itself as a T Tauri star.

Within the past decade, observations spanning a wide spectral range have demonstrated that HH nebulae are not incipient stars, but rather the by-products of supersonic mass outflows from young stellar objects (YSOs). Observations by K. M. Strom et al. (1974*a*) and S. E. Strom et al. (1974) revealed the presence of embedded infrared sources which are proximate to, but not coincident with, some HH nebulae. Detection of broadband optical polarization (K. M. Strom et al. 1974*b*) in a few HH nebulae inspired the reflection model for HHs wherein the nebulae are the product of reflection from the circumstellar emission of a YSO (S. E. Strom et al. 1974). At the same time, observations of the emission nebula associated with T Tauri and the spectra of HH 1 and 2 led R. D. Schwartz (1974,1975) to conclude that the nebulae are the product of *in situ* shock-wave excitation created by supersonic mass outflows from YSOs. Spectropolarimetric observations of HH 24 by G. D. Schmidt and J. S. Miller (1979) revealed that its red continuum was highly polarized, indicative of reflection, whereas nebular emission lines from HH 24 were unpolarized, indicative of *in situ* excitation. Therefore, at least for those HHs possessing a detectable red continuum (indicative of reflection from a low-mass YSO), a compromise model has emerged in which *both* reflection and shock-wave excitation are at work.

In the early 1980s, a number of investigations involving ultraviolet, optical, infrared, and radio observations have revealed that many HH nebulae are the product of supersonic, bipolar mass outflows from YSOs. One such key study was that of Snell et al. (1980) who found bipolar CO flows associated with the L 1551 IRS 5 source which excites HH 28 and 29. Because the general topic of HH objects has been the subject of a number of reviews (Cantó 1981; R. D. Schwartz 1983*a,b*) and a major conference (Cantó and Mendoza 1983), this chapter will focus upon only the essential characteristics of HH nebulae and their exciting stars, with emphasis upon unanswered questions.

I. THE SHOCKING CIRCUMSTANCES OF HERBIG-HARO NEBULAE

Herbig-Haro Spectra

It is tempting to assert that we have a relatively cogent understanding of the spectra of HH nebulae based upon comparisons of synthesized spectra from shock-wave models with observed optical spectra (Dopita 1978; Raymond 1979; Dopita et al. 1982*a*). The nebulae, which are generally dominated by low-excitation emission lines, are characteristic of shocks with velocities in the range 50–140 km s^{-1}, preshock densities of 50–300 cm^{-3}, and solar abundances. It is now apparent that the red and blue continua of HH nebulae may have quite different origins. Only a small fraction of HHs exhibit prominent red continua, a probable product of reflection. Probably all HHs show a blue-ultraviolet continuum which increases in strength toward shorter wavelengths (Ortolani and D'Odorico 1980; Böhm et al. 1981; Dopita et al. 1982*a*). In combination with shock-wave models, Dopita et al. (1982*a*) have shown that this continuum likely originates from the hydrogen two-photon process which is collisionally enhanced in low-velocity shocks with low fractional preionization. Using the International Ultraviolet Explorer (IUE), Brugel et al. (1982) and R. D. Schwartz (1983*c*) have detected continuum peaks near 1500 Å in several HHs, indicative of the two-photon process. Böhm (1983) has reviewed details of optical and ultraviolet observations of HHs.

A remaining enigma in the interpretation of HH spectra is the presence of relatively intense ultraviolet emission lines of high-ionization species (e.g., C IV, C III]) in the higher-excitation HHs, which do not suggest the presence of such high ionization by their optical spectra (Böhm et al. 1981; Böhm-Vitense et al. 1982). The ultraviolet fluxes in HH 1 and 2H, corrected for extinction by a standard extinction curve, appear to be more than 10 times stronger than predicted by single velocity shock models which simulate the optical spectra. Moreover, the ultraviolet continuum appears to increase toward shorter wavelengths with a spectral index which is too large to be explained either by blackbody or two-photon emission.

It has become evident that part of this enigma may result from the use of an inappropriate extinction curve. Bohlin and Savage (1981) report a much flatter ultraviolet extinction for Θ Ori (in a dark cloud region) than is indicated by the standard curve (Savage and Mathis 1979), and Böhm and Böhm-Vitense (1982) indeed find that the Θ Ori curve gives more reasonable results for the ultraviolet emission from the region of the Cohen-Schwartz (C-S) star which excites HH 1 and HH 2. Second, we have had considerable success in combining the spectra of hot and cool shocks of differing areas to produce a match to the observed ultraviolet/optical spectra of HH 2H (Heuermann 1983). The higher ionization emission lines (C IV, etc.) arise from a hot ($\sim 3 \times 10^5$ K) component, whereas the lower ionization lines which dominate the optical spectrum originate from a larger area shock at $\sim 10^5$ K. The two-component shock model, although greatly oversimplified, is suggestive of the presence of a nonplanar shock which might possess one or more hot spots as a shock encounters dense inhomogeneities in the ambient medium. The two-component model can be considered as an approximation to the stellar wind bow shock around a dense cloudlet proposed by R. D. Schwartz (1978) as a model for HH nebulae. Figure 1 is a schematic cross-sectional view of such a shock. The high-temperature region of the shock is situated at the head of the bow wave where the shock is nearly planar, whereas the increasing obliquity of the bow wave at greater distances from the head of the shock produces a lower-temperature shock with a large area. If the dense cloudlet consists of molecular hydrogen with a density of 10^4–10^5 cm^{-3}, the secondary shock propagated into the cloudlet by the ram pressure of the post-bow shock gas on the front side of the cloudlet could excite the 2μm radition, due to transitions in H_2, that is seen from some HHs (see R. D. Schwartz 1981). In combination with the use of the Θ Ori extinction curve, the two-component shock model appears to remove the ultraviolet enigma. An important spin-off of this work in progress is the indirect confirmation that dark cloud ultraviolet extinction is indeed quite different from that of the general interstellar field. In a continuing IUE program we hope to obtain spectra of moderately reddened HHs and,

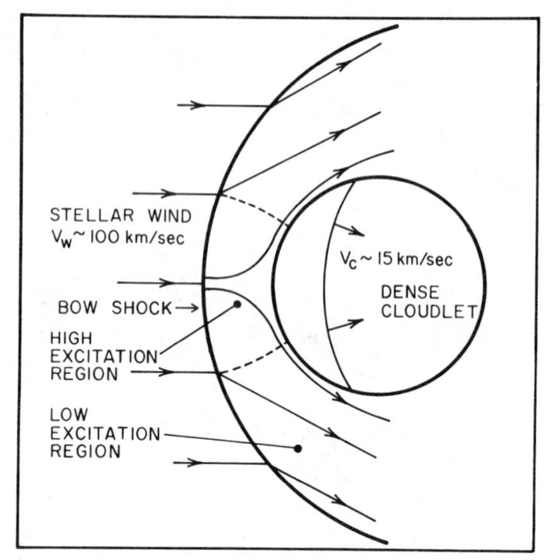

Fig. 1. A bow-shock model for HH nebulae. Only the component of the stellar wind velocity which is normal to the shock is thermalized, hence a high-temperature, subsonic regime exists at the head of the shock, and at greater distances along the bow shock the post-shock temperature decreases, yielding a large area low-excitation shock. The shock velocities noted are appropriate for the spectrum of HH 1 (see R. D. Schwartz 1978,1981). The cloudlets responsible for the emission knots in HH 1 and 2 would have diameters of about 10^3 AU.

with application of the theoretical two-photon distribution, to derive the extinction law for these objects.

Proper Motions and Radial Velocities

Cudworth and Herbig (1979) reported large proper motions for HH 28 and 29 in the L 1551 dark cloud, and subsequent studies have revealed that such motions are not uncommon among HHs (Herbig and Jones 1981; B. F. Jones and G. H. Herbig 1982). In general, the tangential velocities derived from these motions are consistent with the range of radial velocities observed for HHs, indicating a speed of up to 350 km s^{-1}, for example, for one of the knots in HH 1. As considerably lower shock velocities are required to produce their spectra, this implies that the shock waves are being produced by the encounter of a supersonic flow with material which already possesses a substantial velocity away from the star (the differential velocity will be the shock velocity).

In a few cases the geometrical alignment of HHs and a YSO is appropriate to allow observation of HHs moving in opposite directions from the YSO (e.g., HH 1–2, HH 46–47, HH 32). In a number of other cases HHs are aligned linearly on one side of a YSO, and the HH nebulae may be immersed in the blue-shifted lobe of a bipolar CO outflow from a YSO. Presumably, any

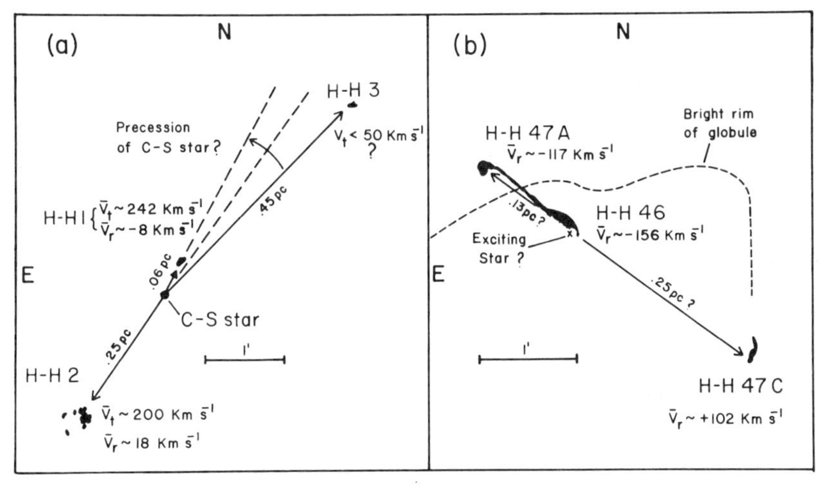

Fig. 2. The HH 1–3 system (a) and the HH 46–47 system (b), suggesting the effect of quasi-periodic, alternate north and south polar ejections. Recent proper motion measurements (R. D. Schwartz, Jones, and Sirk, unpublished) show that HH 47A and C are moving away from the exciting star with tangential velocities comparable to their radial velocities, suggesting an inclination of ~45° (figure from R. D. Schwartz 1983*b*).

HH nebulae in the red-shifted lobe would be too deeply embedded in the dark cloud to be visible (e.g., HH 7–11, HH 28/29). At the same time, some HHs exhibit diffuse structure with no obvious alignments (e.g., HH 24, HH 54); thus it is not certain that all HHs originate in highly collimated flows.

Because the properties of bipolar flows and jets associated with YSOs are discussed elsewhere in this book (see chapters by Rydgren and Cohen and by Mundt), discussion here is limited to a brief review of two particularly interesting systems in which HHs are observed to be rushing away in opposite directions from a YSO. Herbig and Jones (1981) measured high proper motions for HH 1 and 2, with oppositely-directed motions from a T Tauri star (the C-S star). A schematic of this system is shown in Fig. 2a where the combination of mean radial and tangential velocities for the objects suggest that the HH nebulae have been ejected nearly in the plane of the sky. Furthermore, there is a strong hint that the objects have been ejected at different times in alternate directions. If one assumes that the objects have moved at constant velocity since their ejection, HH 1 appears to have been ejected about 220 yr ago, and HH 2 about 1200 yr ago. Although it is not clear if HH 3 is a member of this system, its location is consistent with the geometry of the HH 1, 2, and C-S star alignment, and its tangential velocity would suggest that it has suffered deceleration since its ejection. Furthermore, the locations of HH 1, 2, and 3 with respect to the Cohen-Schwartz star are suggestive of a model in which the HHs have originated from a precessing source (see Fig. 2a).

Figure 2b exhibits the HH 46 and 47 system embedded in ESO 610-A, a cometary globule in the Gum Nebula (R. D. Schwartz 1977*a,b;* Dopita et al. 1982*b;* Graham and Elias 1983). HH 47A has been ejected into the foreground of the globule with a high negative relative velocity (i.e., toward the observer), whereas HH 47C is now just emerging in projection from behind the cloud with a high positive radial velocity. The line connecting these objects passes through an infrared source (Graham and Elias 1983; M. Cohen et al. 1984) embedded in the globule close to HH 46, an HH object with a sizeable reflection component from the embedded star. Particularly distinctive in this system is a bridge of material connecting HH 46 with HH 47A, apparently representing a highly collimated flow.

Models of Herbig-Haro Production

Although we are reasonably certain that HH nebulae are produced largely by shock-wave excitation, there is no single model capable of accounting for the details of all objects. R. D. Schwartz (1978) suggested that the HH nebulae are ambient cloudlets around which bow shock waves are formed by a supersonic stellar wind. In that model the stellar wind is the shocked agent. The inverse situation was suggested by C. A. Norman and J. Silk (1979) with an interstellar bullet model in which clumps of material are accelerated close to the star, and plow supersonically into a lower density, ambient medium. In their model the ambient medium is the shocked agent. The former model may have difficulty in accounting for the relatively rapid (~1000 yr) acceleration of a cloudlet to the velocities observed. The interstellar bullet model predicts that the higher-velocity clumps should exhibit the highest-excitation spectra, whereas the observations suggest that the lowest-excitation objects (e.g., HH 11, 43, 47) all possess high velocities. Models which achieve focusing of stellar winds have involved the elongated wind cavities proposed by Cantó (1980), and nozzle flow produced by expansion of a stellar wind in a circumstellar disk (Königl 1982). Tenorio-Tagle and Rozyczka (1984) have investigated the effects of a converging conical shock behind a cloudlet obstacle caught in a supersonic flow. Material overtaken by the shock condenses into fast, moving bullets which the authors identify with HH objects. Finally, the study of jets associated with YSOs (see the chapter by Mundt) suggests that in some cases HHs may be shock waves produced by instabilities at the interface of a supersonic, highly-collimated outflow with the ambient medium.

II. YOUNG STELLAR OBJECTS ASSOCIATED WITH HERBIG-HARO NEBULAE

The first HH objects shown to be clearly associated with a visible star were the emission nebulae associated with T Tauri (R. D. Schwartz 1975). We now recognize at least five other cases in which HHs are associated with visible YSOs, namely, HL Tau, DG Tau, the C-S star, AS 353A, and R Mon.

The majority of HH nebulae are not associated with visible stars. However, the pioneering infrared work of Strom and Strom and their colleagues demonstrated that embedded infrared sources could be identified in the vicinities of several HH nebulae. Furthermore, the infrared energy distributions ($\lambda \simeq 2$–20 μm) appeared to be typical of those found in T Tauri stars.

M. Cohen and R. D. Schwartz (1979,1980,1983) carried out systematic $\lambda = 2.2$ μm mapping of the vicinities of HH objects in search of their exciting stars. Photometry of probable exciting stars was obtained from both ground-based infrared observations and airborne (Kuiper Airborne Observatory) observations (M. Cohen et al. 1983, 1984), allowing estimates of the bolometric luminosities of the stars. Among 17 sources (including four of the visible stars noted above), the mean bolometric luminosity is about 17 $L_\odot$, with a range from 0.7 to 58 $L_\odot$. There are 8 sources with $L < 10$ $L_\odot$, and 3 sources with $L > 30$ $L_\odot$. R Mon, with $L \sim 1360$ $L_\odot$, has been excluded because it may represent a star of somewhat higher mass than the others. It should also be noted that some of the sources (e.g., Haro 6–10 and HH 100 IRS) have demonstrated marked variability, a feature common to many T Tauri stars. In summary, it appears that most of the YSOs identified as the probable exciting stars of HHs are relatively low-mass objects, probably of the T Tauri class.

III. CONCLUSIONS: ENERGY CONSTRAINTS IN THE PRODUCTION OF HERBIG-HARO OBJECTS

Apart from the problem of understanding the shocking details of HH nebulae looms the larger question: why are relatively low-mass YSOs compelled to eject material, perhaps in quasi-periodic and highly collimated streams? Are newborn stellar infants subject to a process by analogy to that of a newborn human infant who receives the mandatory slap by the attending physician to clear material from the infant's respiratory system? Who is the stellar doctor, and what receives the slap?

Analysis of the energetics of HHs suggests that the stellar doctor must be capable of delivering a sizeable amount of energy, much of which results in the kinetic energy of outflowing material. The most obvious candidate for the energy source is the gravitational potential energy of accreting material. That an accretion disk is present around a YSO is implied by the existence of bipolar flows. It is instructive to examine the energetics of a particular system to determine the required accretion rate.

Böhm et al. (1981) have estimated that about 17 Earth masses of material are responsible for emission in HH 1. A lower limit to the kinetic energy (E_k) of HH 1 can be computed by assuming that at least this amount of material is moving at $\sim$240 km s^{-1} (the mean velocity of four knots) as indicated by proper motion studies (Herbig and Jones 1981). For this case we find $E_k = 3.2 \times 10^{43}$ ergs. The radiative luminosity (including ultraviolet and optical emission) is rather difficult to estimate because it is critically dependent upon the interstellar extinction correction. For the standard extinction, one finds a luminosity of about 1 $L_\odot$, dominated by ultraviolet emission (the optical emission contributes only about 0.05 $L_\odot$; see Böhm et al. 1981). With the more reasonable Θ Ori extinction curve, HH 1 emits about 0.2 $L_\odot$ (ultraviolet and optical combined). If one assumes that the luminosity of HH 1 has been constant at this value since its ejection $\sim$ 220 yr ago, its radiative energy loss has been $E_r \simeq 5 \times 10^{42}$ erg. Thus the total energy involved is $E_t = E_k + E_r \sim 3.7 \times 10^{43}$ ergs.

Mundt and Hartmann (1983) have obtained detailed spectroscopic observations of the C-S star which reveal it to be a rather quiescent, weak-lined T Tauri star with no evidence of present mass loss. Their conclusion that HH 1 and 2 are the results of eruptive events from the star is given further impetus by the geometry of the system seen in Fig. 2a where it appears that ejections have occurred at intervals of $\sim 10^3$ yr. The energy source driving the ejections must thus have an equivalent mean luminosity of at least $E_t\ 10^{-3}$ yr $\sim 1.2 \times 10^{33}$ erg s$^{-1} \simeq 0.3$ $L_\odot$. The C-S star has a bolometric luminosity of about 27 $L_\odot$ (M. Cohen et al. 1983); thus the energy source driving the HHs is at least 1% of the stellar luminosity.

The luminosity due to accretion through a disk is given by $GM_*\dot{M}/2R_*$ where M_* and R_* are the mass and radius of the star, and $\dot{M}$ is the accretion rate. If 0.3 $L_\odot$ is a lower limit to the energy needed to drive the HH objects, one can compute a lower limit to the accretion rate required to produce this energy. For a 1 $M_\odot$ star with a radius of 2 $R_\odot$, one finds $\dot{M} \geq 4 \times 10^{-8}$ $M_\odot$ yr^{-1}. Due to viscous losses in the disk and to the ejection of material in addition to that seen in the HH objects, the accretion rate almost certainly is considerably greater than this value.

There are several unanswered questions about this scenario. First, how is the energy gained from accretion transformed into eruptive events? Is the problem purely one of gas dynamics, or can a global magnetic field help to serve as a reservoir for energy deposition, with accretion gradually filling the reservoir until the hydromagnetic dam bursts? Second, what kind of disk and/or magnetic field is required to produce highly collimated flows? The HH 1–2 system may be instructive in this regard because the proper motions suggest that we are viewing the star nearly in its equatorial plane. Hence any equatorial disk cannot have great vertical thickness because the star is clearly visible, suffering only about 6 mag of visual extinction (M. Cohen and R. D. Schwartz 1979). Perhaps HH 1 represents the results of the final outburst of this star which has now succeeded in shedding itself of placental material, settling down to a gentle journey to the main sequence. Alternatively, a dense but physically thin disk could exist around the star, a feature with obvious implications for planetary formation. But can a physically thin ($z \leq R_*$) disk effectively produce focused flows? Also, what mechanism is required to produce periodic ejections in alternate direction as suggested by Fig. 1?

Finally, in our search to understand mechanisms for the ejection of material from relatively cool YSOs, it may be instructive to deepen our understanding of similar phenomena on more normal stars. Brueckner and Bartoe (1983), for example, have found high-energy jets of material plowing through the solar corona with velocities of ~ 400 km s^{-1}. Giant stars, some of which occupy a region of the HR diagram common to YSOs, are known to possess stellar winds. Do the magnetohydrodynamical models for stellar winds in giants investigated by Mullan and Steinolfson (1983), for example, have any bearing on mass loss from YSOs?

Acknowledgment. This chapter has been prepared with support from the National Aeronautics and Space Administration, and from the National Science Foundation.

HIGHLY COLLIMATED MASS OUTFLOWS FROM YOUNG STARS

REINHARD MUNDT
Max-Planck-Institut für Astronomie

The observational evidence for highly collimated (bipolar) flows from young stars of low to medium luminosity ($0.1\ L_\odot \lesssim L \lesssim 100\ L_\odot$) is reviewed, with focus on optical studies of Herbig Haro (HH) objects and jets. These flows have typical lengths of 0.01 to 0.2 pc, opening angles of about 3 to 10 deg, and velocities of several hundred km s^{-1}. In at least two cases they are already collimated on length scales of < 70 AU. We discuss the possibility that these outflows are continuous and conclude that for several sources quasi-continuous outflow activity is indicated over time scales of up to several 10^4 yr. We consider that many HH objects may be powered by collimated flows. For some HH objects available kinematical data strongly suggest that they are powered by a jet emanating from a nearby T Tauri star or infrared source. Such an HH object model avoids some difficulties of other models, like the interstellar bullet model.

A large and growing body of observational material indicates that energetic, often bipolar, mass outflows are an important phase in early stellar evolution for probably all types of young stars. This refers to objects with a few solar luminosities (like T Tauri stars) as well as to objects with luminosities 10^5 times higher. Important observational evidence for such outflows has been provided by observations of the molecular gas components accelerated by the stellar winds of these objects (Rodriguez et al. 1982; Bally and Lada 1983; Edwards and Snell 1984). These investigations involved mostly the measurement of CO emission-line profiles.

Direct evidence for stellar winds from young stars is provided by P Cygni profiles observed in the optical spectra of some T Tauri and Herbig Ae/Be stars (see the chapter by Rydgren and Cohen for a review of these objects). For several decades, it has been known that some of them undergo mass loss (Kuhi 1964; Herbig 1960). However, within the last few years indications for mass loss has been found in many more of these objects by using high-quality optical spectra (Finkenzeller and Mundt 1984; Mundt 1984). These data generally indicate mass motions of up to several hundred km s^{-1} near the stellar surface and the terminal wind velocities are probably of the same order as indicated by velocities observed in associated Herbig-Haro (HH) objects (however, see Mundt 1984).

HH objects are an important probe for outflows from young stars. Their line emission originates in the cooling regions of high-velocity ($v \approx 100$ km s^{-1}) shockwaves, which probably arise from the dynamical interaction of winds from young stars with ambient medium (see, e.g., R. D. Schwartz 1975; chapter by Schwartz). HH emission is therefore well suited to probe the region close to these stars for the spatial distribution of high-velocity matter *directly* ejected by them. Recent proper motion and radial velocity measurements of HH objects have shown that this matter can be channeled into well-collimated, bipolar flows. In several cases, the high flow collimation is directly evident from the appearance of narrow high-velocity jets, extending away from T Tauri stars or from infrared sources of comparable luminosity. These jets and HH objects are strongly related phenomena, as is evident from their similar spectra and from their physical association observed in several cases (e.g., some catalogued HH objects form the brightest parts of these jets).

In this review we discuss highly-collimated, bipolar flows from young stars mainly from optical studies of HH objects and jets. These phenomena are in most cases associated with stars of low-to-moderate luminosity ($0.1\ L_\odot \lesssim L \lesssim 100\ L_\odot$) and therefore flows from high-luminosity objects are not part of this discussion. For a more general discussion of HH objects, the reader is referred to the chapter by Schwartz, to the reviews of Böhm (1983) or R. D. Schwartz (1983*a*), and to the Proceedings of the Symposium on HH Objects and T Tauri Stars (1983) honoring G. Haro. For recent work on molecular flows associated with HH-objects and young objects of low-to-moderate luminosity see Edwards and Snell (1984, and references therein).

I. OBSERVATIONAL EVIDENCE FOR HIGHLY-COLLIMATED, BIPOLAR MASS OUTFLOWS

A. Studies Prior to 1983

The first case that showed HH objects tracing a bipolar flow was found by proper motion measurements of HH 1 and HH 2 (Herbig and Jones 1981). These measurements established that the individual knots in HH 1 and HH 2 are moving in opposite directions with tangential velocities of 100 to 350 km s^{-1}, suggesting a highly anisotropic, bipolar flow being driven by a star lo-

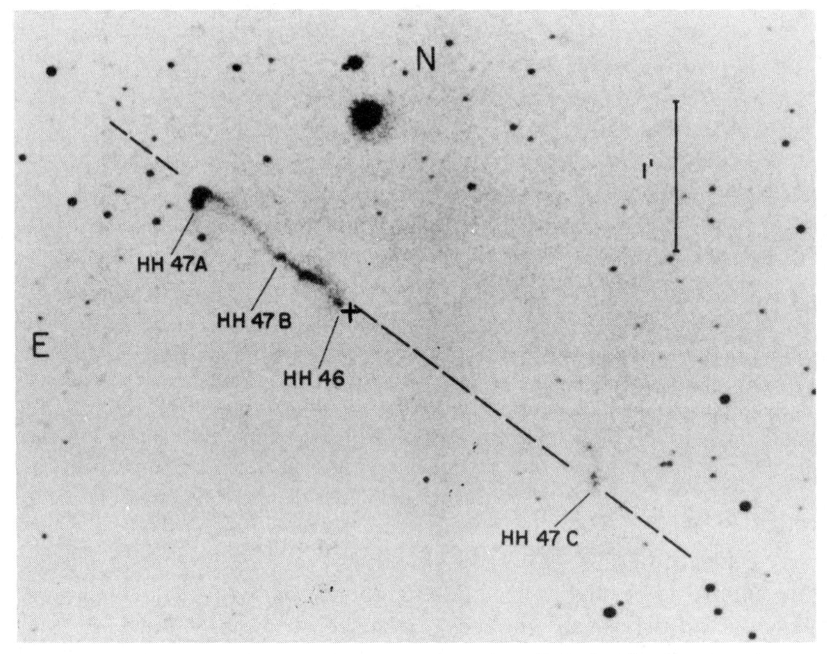

Fig. 1. [S II] 6716, 6731 line image of the HH 46/47 region. The broken line illustrates the close alignment of the HH objects tracing this highly-collimated, bipolar flow. The position of the infrared source driving this flow is indicated by a cross (figure from Graham and Elias 1983).

cated in between HH 1 and HH 2. A T Tauri star which could be the driving source of this flow was found by M. Cohen and R. D. Schwartz (1979) and has been designated as the C-S star.

One of the best examples for a highly-collimated, bipolar flow is provided by the HH 46/47 system (Graham and Elias 1983). An [S II] line image of the HH 46/47 region is reproduced in Fig. 1. It shows a 1 arcmin long jet with HH 46, HH 47B and in particular HH 47A being the brightest portions (hot spots) of this feature. Similar examples, for which previously known HH objects form the brightest portions of a jet, are given below. In this case, all the material in the jet is blue-shifted (v ≈ -100 to -190 km s^{-1}), while two tiny HH knots (HH 47C) located 1.5 arcmin southwest of HH 46 are red-shifted (v $\approx +100$ km s^{-1}). The jet and the two HH 47C knots lie on a straight line. The low-luminosity infrared source ($L \approx 12\ L_{\odot}$) driving this bipolar flow is located very close to this line, roughly midway between HH 47A and 47C (Emerson et al. 1984). Radial velocity measurements indicate higher velocities in the jet than in the bright knot HH 47A, suggesting a deceleration of the flow in this knot. Because HH 47A is located at the north-east end of the jet, the flow seems to terminate there.

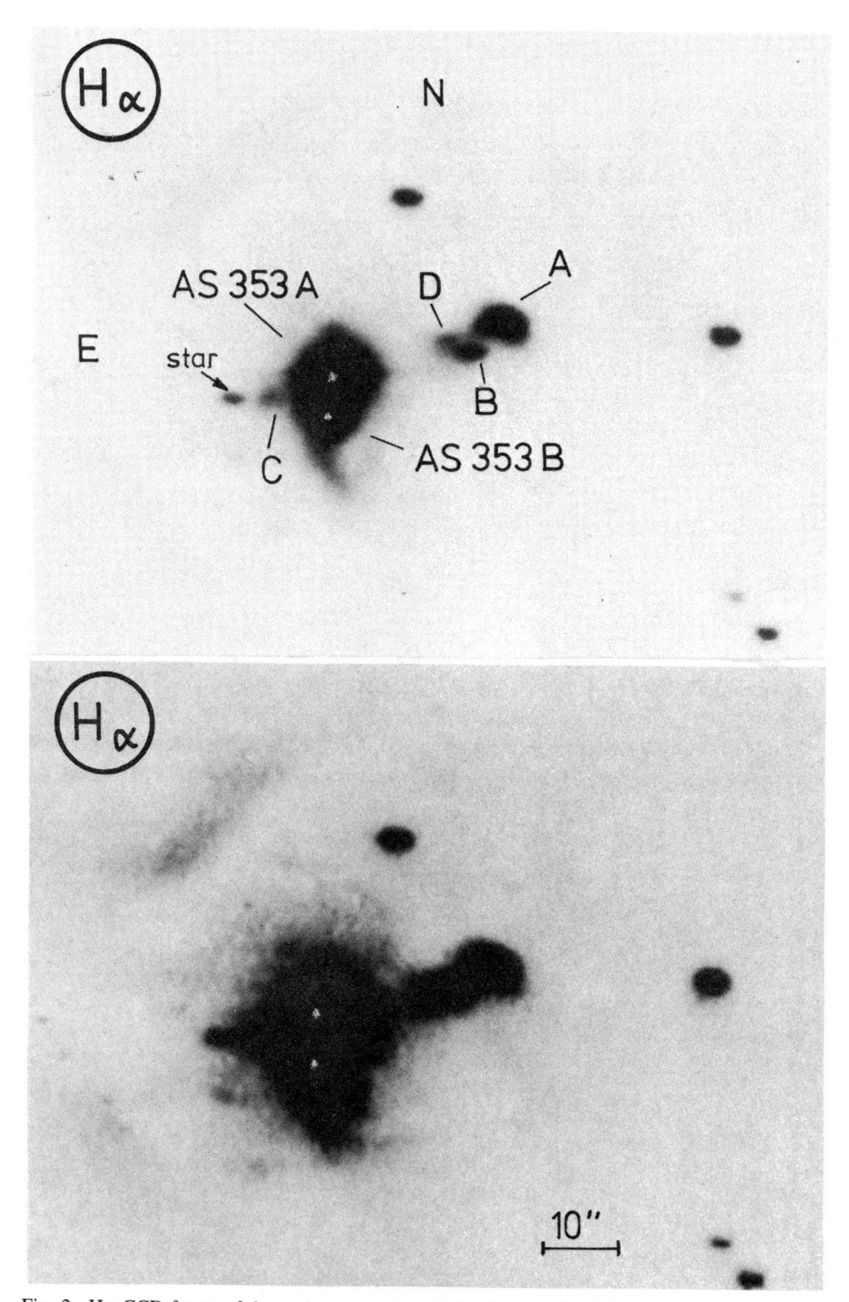

Fig. 2. Hα CCD frame of the region around the T Tauri star pair AS 353A and B shown at low (upper panel) and high contrast (lower panel). The position of these two stars is marked. Note the alignment of the HH objects HH 32A–D on the opposite sides of AS 353A which are moving in opposite directions (see Fig. 3). As evident from the high-contrast picture, the knots A, B and D are embedded in a jet-like nebula extending away from AS 353A (figure from Stocke et al., in preparation).

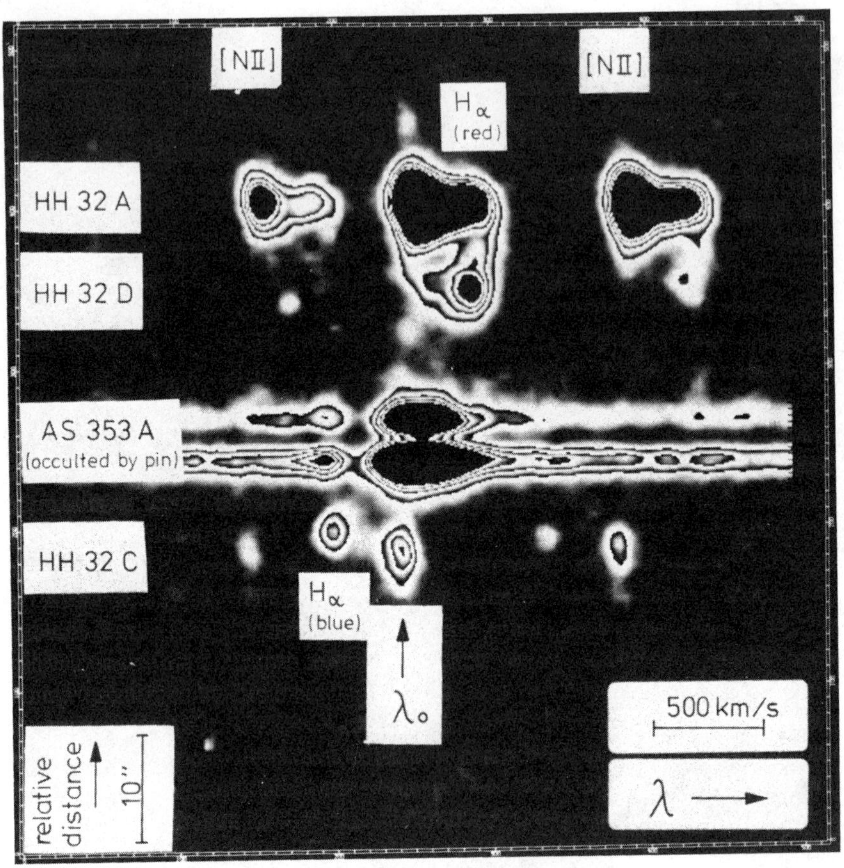

Fig. 3. A KPNO 4m long slit, medium resolution ($\Delta v = 65$ km s^{-1}) spectrogram of the HH 32 flow, with the slit oriented through HH 32C, AS 353A, and HH 32A and D, but not through HH 32B (see Fig. 2). The rest wavelength of the Hα line is indicated by an arrow. The HH objects on the opposite sides of AS 353A are moving in opposite directions with radial velocities of up to 350 km s^{-1} (figure from Stocke et al., in preparation).

Indications for bipolar mass motions similar to those observed for HH 1/2 and for HH 46/47 were found by Mundt et al. (1983*a*) for the T Tauri star AS 353A. On Hα and [S II] CCD images, they discovered a new HH object designated as HH 32C close to AS 353A. HH 32C is aligned, on the opposite side of AS 353A, with the two previously known HH objects HH 32A and B. A recently obtained deep Hα CCD frame of these objects is shown in Fig. 2 at low and high contrast. As evident from this picture, HH 32A, B, and D are the brightest portions of a jet extending away from AS 353A towards HH 32A where it apparently terminates (the same as for the HH 46/47 jet). Radial velocity measurements of the HH objects on the opposite sides of AS 353A have shown that the matter in these HH objects is moving in the opposite direction with velocities of up to $\approx$ 350 km s^{-1}. This is illustrated in Fig. 3, which shows a recent medium-resolution spectrogram (Δv = 65 km s^{-1}) around Hα and [N II] with the slit oriented through HH 32A and D, AS 353A and HH 32C. The line emission in HH 32A reaches velocities of up to +350 km s^{-1}, while in HH 32C the matter is moving into the opposite direction as evident from the high-velocity component of v $\approx$ -300 km s^{-1}. Very interesting is the predominantly high red-shifted material between HH 32 A and AS 353A (in the vicinity of HH 32 D). This is in contrast to HH 32A and HH 32C where a large fraction of the line emission originates in low-velocity material. These data suggest the following scenario: A collimated bipolar flow (or bipolar jet) of v $\approx$ 350 km s^{-1} is emanating from the vicinity of AS 353A and is moving into the direction of HH 32A and C. In HH 32A and C this flow becomes decelerated, as indicated by the strong low-velocity components observed in these two HH objects. The deceleration of this flow in the two HH objects is furthermore indicated by the *larger distance or extent of the low-velocity component* from the source, compared to the high-velocity component.

Another remarkable case are the objects HH 7–11. They form a chain of roughly equally spaced condensations which fall on a line approximately passing through an infrared-source (see also Fig. 11 in Sec. II.E). Deep Hα-CCD images by the author (unpublished data) show that HH 7–11 are the brightest portions of a narrow cone (jet) extending away from the position of HH 11 towards HH 7. HH 7, 8, 10 and 11 are all blue-shifted with the highest value (-150 km s^{-1}) measured for HH 11. For this object, Herbig and Jones (1983) have measured high proper motions (v $\approx$ 60 km s^{-1}) pointing away from the infrared source. All these data indicate that a well-collimated flow is emanating from this source. The CO observations by Snell and Edwards (1981) show this flow to be bipolar with HH 7–11 being located on the blue-shifted CO lobe. The one-sidedness of the flow in the optical can probably be explained by high reddening of the receeding (red-shifted) part of the flow which is moving into the molecular cloud and therefore affected by high extinction.

In addition to the examples discussed above a few more cases were known prior to 1983, for which HH object studies suggested collimated flows. These cases are HH 28/29 (Cudworth and Herbig 1979; see Fig. 7 in Sec. II.A), HH 39 (B. F. Jones and G. H. Herbig 1982), and HH 12 (Herbig and Jones 1983; K. M. Strom et al. 1983; see Fig. 10 in Sec. II.D).

B. Recent Observations at Calar Alto, Spain

Mundt and Fried (1983) and Mundt et al. (1984) carried out intensive CCD imaging of T Tauri stars and HH objects utilizing the 2.2 m telescope on Calar Alto, Spain. In the vicinity of these stars they searched for HH emission indicating collimated outflows from these stars or tried to find indications for such flows towards bright, catalogued HH objects (e.g., a jet). In all, they obtained deep CCD images of about 35 strong-emission-line T Tauri stars and

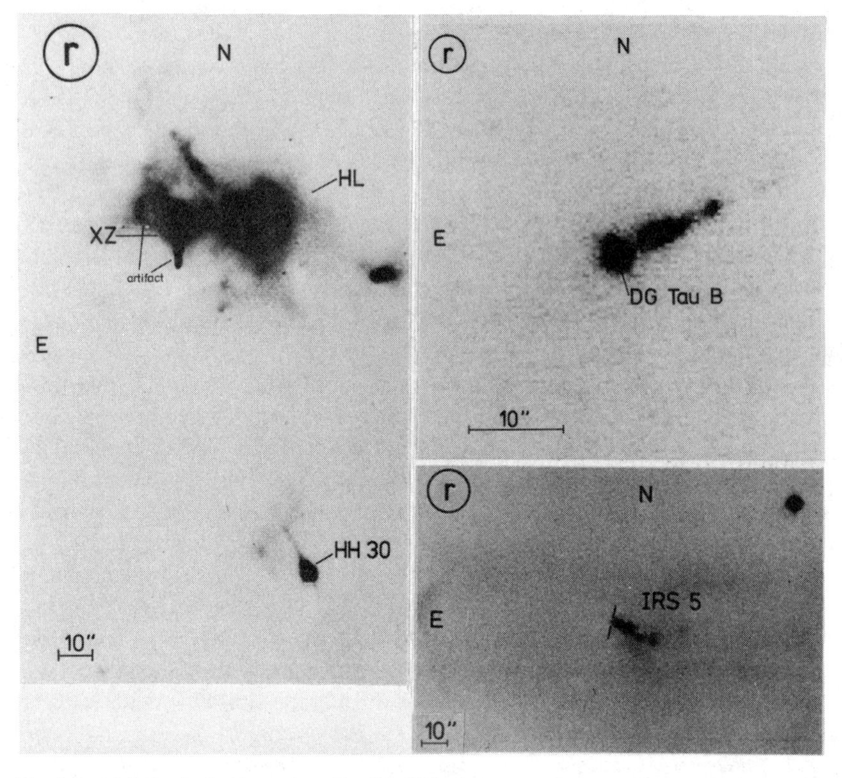

Fig. 4 (and 5 following). Examples of jet-like HH nebula extending away from T Tauri stars and low-luminosity IR-sources or being associated with bright, previously known HH objects (from Mundt and Fried 1983, and Mundt et al. 1984). See Plate 8 in the Color Section for color reproduction of L 1551-IRS 5.

about 15 HH objects. In order to discriminate HH nebulosities from continuum reflection nebulosities, they used filters *including* and *excluding* (the strongest) HH emission lines (like Hα, [N II], [S II]). They discovered 7 jet-like HH (or Hα) nebulosities similar to the features associated with HH 46/47 or HH 32. These jets either extend away from T Tauri stars and low-luminosity infrared sources or they point towards bright HH objects; they are associated with HL Tau, HH 30, DG Tau, DG Tau B, Haro 6-5 B, HH 33/40, HH 19. Furthermore, it was confirmed that a jet is extending away from L 1551-IRS 5 and another jet from the T Tauri-like star 1548C27 (M. Cohen et al. 1982; Craine et al. 1981). For 6 of these 9 jets, the radial velocities have been measured (see Table I). Most of the newly discovered jets are displayed in Figs. 4 and 5 (or in one of the following figures).

Mundt and Fried (1983) and Mundt et al. (1984) interpreted their data also in terms of well-collimated, high-velocity flows. This interpretation is strongly supported not only by the morphology of the newly discovered jets,

TABLE I

Properties of Highly Collimated Flows as Derived from Optical Studies of Jets and HH Objects

Associated Stars or HH Objects	Distance (pc)	Length (arcsec)	Length (pc)	Approximate Opening Angle (deg)	Radial Velocity (km s^{-1})	Proper Motion[b] (km s^{-1})	Luminosity of Driving Source ($L_{\odot}$)	References[c]
HH 12	350	120	0.2	5–10	−100	60–250 (12)	—	1,2
HH 7–11	350	65	0.1	10	−30 to −150	60 (11)	58	2,3,4
HL Tau	150	20	0.02	5	−170[a]	—	7.2	4,5,6
HH 30	150	13	0.01	5	—	—	—	5
L 1551-IRS 5	150	17	0.015	10	−80 to −190	150 (28), 170 (29)	32	4,5,6,7,8
DG Tau	150	8	0.006	10	−150 to −250	—	7.6	5,6
DG Tau B	150	17	0.015	5	—	—	—	5
Haro 6-5 B	150	60	0.05	5–10	−40[a] +50[a]	—	—	6,9
HH 1/2	460	55/110	0.12/0.25	10	−8[a] +17[a]	155–350 (1 A-F) 60–290 (2 A-I)	27	4,10
HH 33/40	460	75	0.17	5	+110 to +160	—		6,9
HH 19	460	30	0.07	—	+20	—	—	6,9
R Mon/HH 39	800	15/400	0.06/1.6	10	−75[a] +170[a]	60 (39 D) 275–310 (39 A,C,E)	1360	4,11,12
HH 46/47	400	75	0.15	5	−110 to −190[a] +120[a]	—	12	13,14
AS 353 A	300	29	0.035	10	−20 to −350[a] +60 to +380[a]	50 (32 A) 200 (32 B)	6.6	2,4,15,16
1548C27	?	45	$0.1\frac{d}{500\ \mathrm{pc}}$	3	−175	—	—	6,9

[a]Optically bipolar flow.

[b]Number in parenthesis gives HH object number.

[c]1: K. M. Strom et al. 1983; 2: Herbig and Jones 1983; 3: Snell and Edwards 1981; 4: M. Cohen et al. 1984; 5: Mundt and Fried 1983; 6: Mundt et al., in preparation; 7: Sarcander et al. 1984; 8: Cudworth and Herbig 1979; 9: Mundt et al. 1984; 10: Herbig and Jones 1981; 11: B. F. Jones and G. H. Herbig 1982; 12: Brugel et al. 1984; 13: Graham and Elias 1983; 14: Emerson et al. 1984; 15: Mundt et al. 1983*a*; 16: Stocke et al., in preparation.

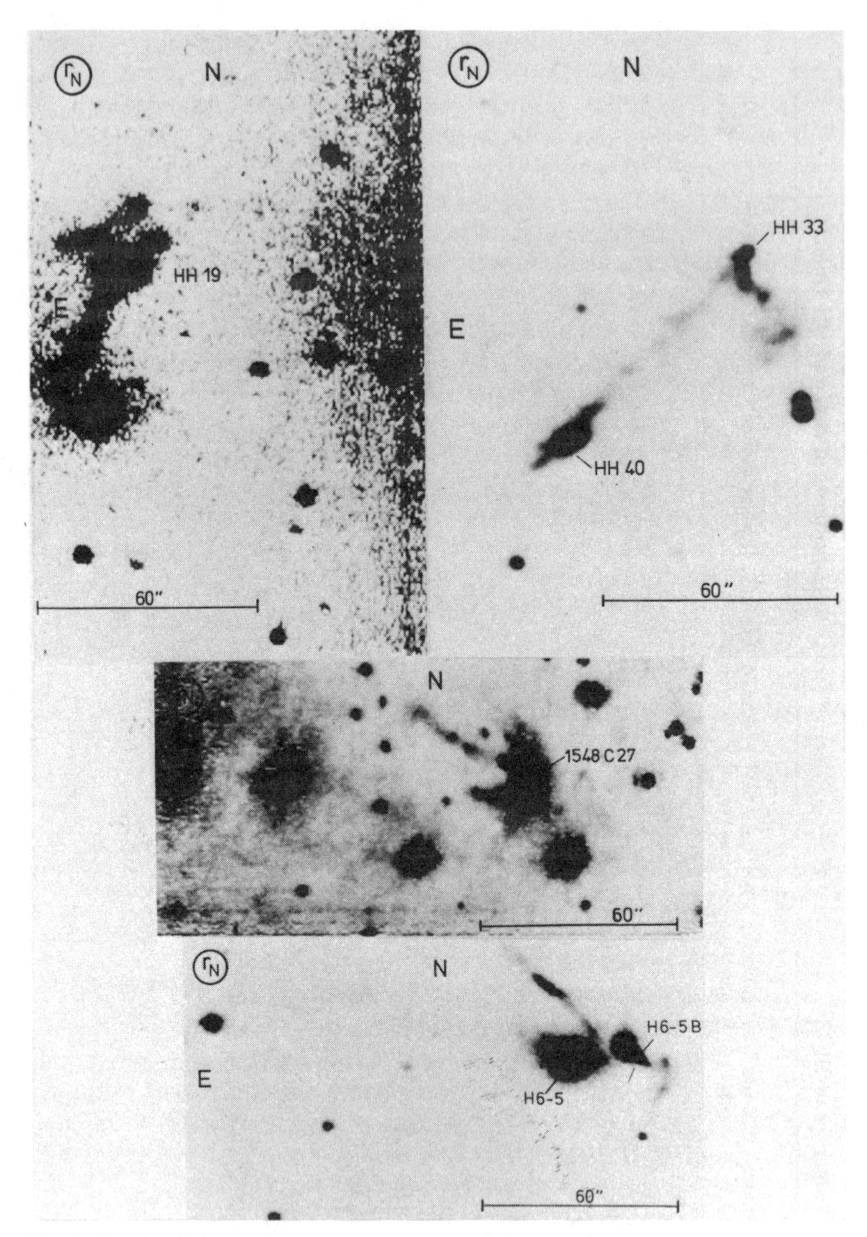

Fig. 5. See Fig. 4 above for text.

but also by their high velocities, with values of up to several hundred km s^{-1}. In addition, all jets studied spectroscopically so far have HH-like spectra, implying shock heating of the emitting material with velocities of about 100 km s^{-1}. Furthermore, this means that these jets can be regarded as highly elongated HH objects, but with much lower surface brightnesses than most classical HH objects. As mentioned above, the brightest parts of such jets or hot spots are in some cases previously known, stellar-like HH objects, which apparently mark the location of those shock fronts where larger amounts of the kinetic energy of the flow is dissipated. We note that the occurence of internal shock systems is a natural property of supersonic jets. Such shock systems can account for the heating of the jet material and also for the observed knotty structure (Norman et al. 1984; see also Sec. II.E.).

C. Jet Properties

Table I summarizes the properties of jets discovered at Calar Alto and those found by other investigators. The typical lengths of all these jets range from 0.01 to 0.2 pc (or 10″ to 100″ on the sky) and the opening angles have values between ≈ 3° and ≈ 10°. This indicates a much higher degree of collimation for the flows from young star than indicated by CO observations (at least close to the driving star). The measured radial velocities and proper

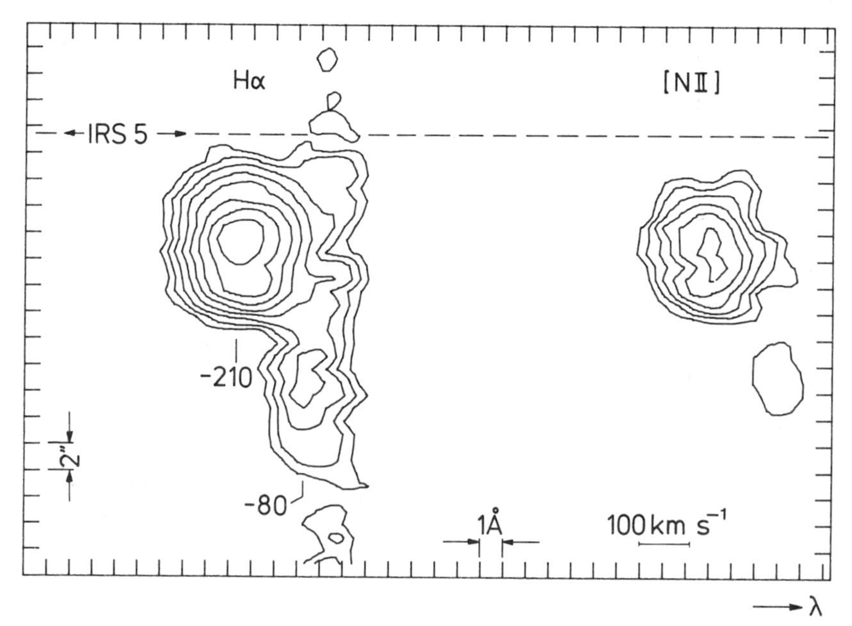

Fig. 6. Intensity contour plot of a long-slit image-tube spectrogram of the L 1551-IRS 5 jet in the region around Hα and [N II] λ 6583. The approximate position of IRS 5 is indicated. Higher radial velocities are observed towards the infrared source. The resolution of this spectrogram is about 100 km s^{-1} (figure from Sarcander et al. 1984). See Plate 8 in the color section for color reproduction of L 1551-IRS 5 jet.

motions indicate flow velocities of 100 to 400 km s^{-1} (without correcting for projection effects or for deceleration in shock fronts). As already discussed above, some of these flows show a relatively complex velocity structure. As a further example, we show in Fig. 6 an intensity contour plot of a long-slit image tube spectrogram of the L 1551-IRS 5 jet illustrating the strong changes in radial velocity along the jet (Sarcander et al. 1984). In many cases, the jet is spatially unresolved perpendicular to its axis ($\lesssim 2''$), providing upper limits for the jet diameter of $\approx 4 \times 10^{15}$ cm (for $r = 150$ pc). The stars driving these jets are either T Tauri stars or infrared sources with luminosities comparable to those of T Tauri stars ($L \approx 0.1$–$100\ L_{\odot}$). With the exception of R Mon (Brugel et al. 1984) no high-luminosity star (e.g., $L \gtrsim 1000\ L_{\odot}$) is known to be driving a jet. However, very little effort has been made to find jets around high-luminosity objects similar to those described here.

II. DISCUSSION

A. Collimation Length Scales

In the emission-line CCD frames of Mundt and Fried (1983), some of the jets (see e.g., those of L 1551-IRS 5, DG Tau B, and HH 30) could be traced to within 1 to 2″ from their respective stars, which within these limits appear unresolved in direction perpendicular to the jet. This observation implies collimation of these jets on length scales smaller than a few 10^{15} cm. For L 1551-IRS 5 and DG Tau, the Very Large Array (VLA) maps of Bieging et al. (1984) shown in Figs. 7 and 8 indicate even smaller collimation length scales. Already the innermost radio contours (resolution $0\overset{''}{.}5$) are elongated in a similar direction as optical jets of the two objects. The upper limit on the collimation length derived from these VLA data is about 10^{15} cm (or ≈ 70 AU).

Such small scales are in strong contrast to all models which channel these flows by large-scale (10^{17} to 10^{18} cm) density gradients (Torrelles et al. 1983). These models are based, however, on low-resolution radio observations with typical beam sizes on the order of 1 arcmin.

B. Are These Flows Continuous?

T Tauri stars are known to be highly variable objects. As many of the driving sources of optical jets or CO outflows are T Tauri stars (or infrared sources of similar luminosity), one must seriously consider whether their outflows are strongly variable as well. R. D. Schwartz (1983*b*) has even speculated that these flows are highly discontinuous and suggested the ejection of HH 1 and HH 2 from the C-S star during an eruptive event, like a FU Orionis outburst (see also Mundt and Hartmann 1983).

In Fig. 7 we compare, for the L 1551-IRS 5 flow, the information provided by various flow traces. The figure shows the bipolar CO flow together with the high proper motion HH objects HH 28/29, the optical jet, and the radio jet. It demonstrates the good alignment of the flow over a length scale

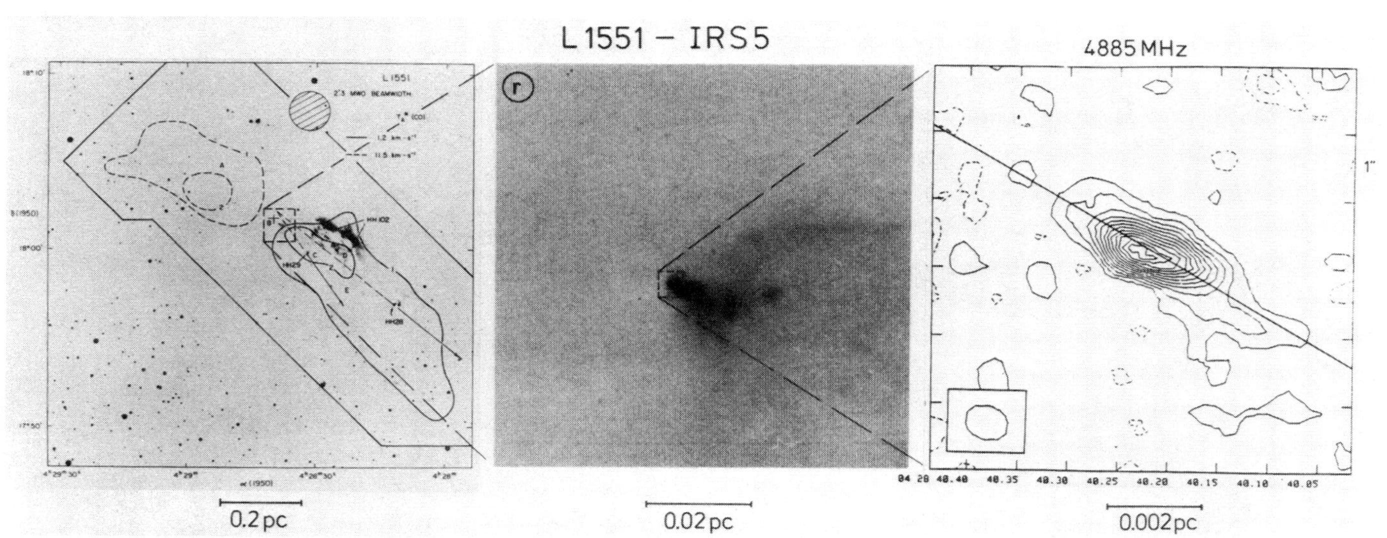

Fig. 7. L 1551-IRS 5 flow. Left panel: Optical image of the L 1551 dark cloud with the bipolar high-velocity CO gas lobes superposed (figure from Snell et al. 1980). Middle panel: CCD frame of the IRS 5 region showing the jet extending away from the position of the infrared source (figure from Mundt and Fried 1983). Right panel: 6 cm VLA radio continuum map of IRS 5 (figure from Bieging et al. 1984). Note the alignment of the flow in the same direction over a length scale range of 3 orders of magnitude.

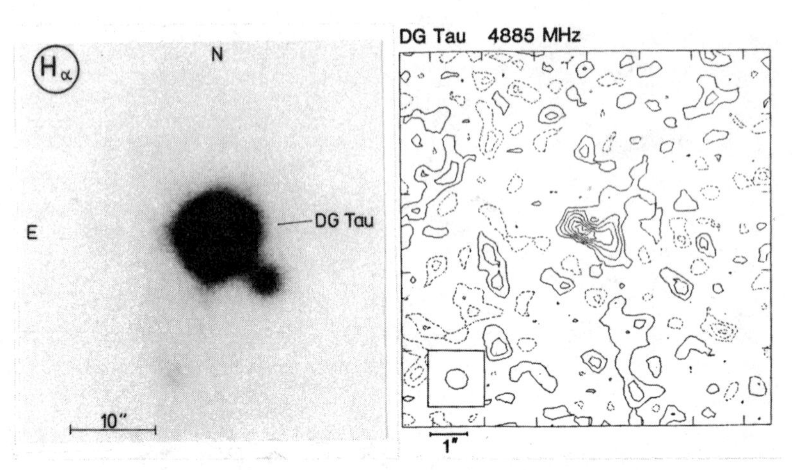

Fig. 8. Left panel: Hα CCD frame of DG Tau showing a 8″ long jet with a hot spot (HH knot) at the end (figure from Mundt and Fried 1983). Right panel: 6 cm VLA radio continuum map of DG Tau. An extended, 1.″5-long radio emission region is observed, oriented in nearly the same direction as the optical jet (figure from Bieging et al. 1984).

range of 3 orders of magnitude. Obviously, the various tracers of the bipolar flow from IRS 5 have very different dynamical time scales (e.g., 10 to 100 yr for the jet, 500 to 2000 yr for HH 29/28, and a few 10^4 yr for the CO gas). This strongly suggests a quasi-continuous outflow activity from IRS 5 over the last 10^4 yr. The data do not support a model for such a flow in which it is driven by a few single events.

For other objects various flow tracers also suggest quasi-continuous outflow activity over a broad range of time scales. In the case of AS 353A, DG Tau, HL Tau, and 1548C27, the P Cygni line profiles observed in Hα and/or Na D lines show that these objects are currently ejecting mass from their stellar surface, likely at relatively high rates (Mundt 1984; R. K. Ulrich, personal communication, 1983; Craine et al. 1983). The elongation of the jets emanating from these objects indicates that the ejection is essentially continuous over 10^2 to 10^3 yr (or longer, depending on the propagation velocity of the jet, which might be relatively low for a high-density ambient medium; see Eq. (1) in Sec. II.D below). Furthermore, around HL Tau and AS 353A, high-velocity CO gas has been found with a dynamical age of about 10^4 yr (Edwards and Snell 1982; Calvet et al. 1983).

It will be difficult to establish whether these flows are actually continuous (or the extent of their variability). To answer such questions it is necessary to know with some accuracy the mass flux and momentum of the various flow components like the stellar wind or the jet. In addition, one must know the physical mechanisms by which these flows are collimated and how the momentum from the stellar wind or the jet is transferred to the ambient molecular cloud medium.

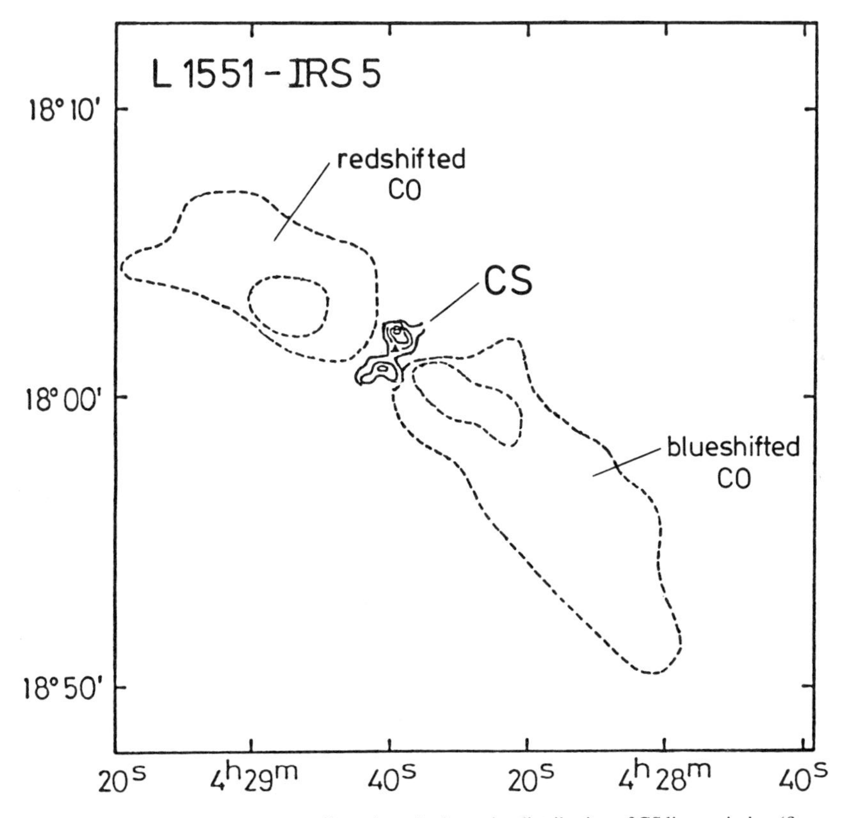

Fig. 9. The thick innermost contour lines show the intensity distribution of CS line emission (figure from Kaifu et al. 1984). These data suggest the presence of a toroid-like mass distribution around IRS 5 being oriented perpendicular to the CO flow (broken lines).

C. Indications for Circumstellar Disks

All models proposed to channel an originally isotropic stellar wind hydrodynamically into a bipolar jet require density gradients and some disk-like mass distribution around the star (Königl 1982). One may ask, is there any observational evidence for disks around the jet sources? The recent CS line observations of the L 1551-IRS 5 area by Kaifu et al. (1984) are of interest in this regard, because they have probed the dense regions ($N(H_2) \approx 10^5$ cm^{-3}) around this object. A map of the measured CS line emission is shown in Fig. 9, which suggests the presence of a toroid-shaped molecular cloud around IRS 5, oriented perpendicular to the optical jet and the CO flow. Further support for a disk-like mass distribution around IRS 5 is provided by its high polarization, $p = 21\%$ at 2.2 μm (Hodapp 1984), with the electric polarization vector *perpendicular* to the jet. Similar data exist for DG Tau and HL Tau (Mundt and Fried 1983). The high polarization (6 to 21%) in all these cases can, in principle, be explained by scattering of stellar light in the polar lobes of an optically thick disk (Elsässer and Staude 1978). In this model, the polarization vector is parallel to the disk and therefore disks perpendicular to the flow direction are suggested for these three objects. Polarization data similar to those of DG Tau or L 1551-IRS 5 have been obtained for several more luminous sources driving bipolar CO flows (Hodapp 1984). For a further discussion of circumstellar disks around young stellar objects see the chapter by Rydgren and Cohen.

D. Are Many HH Objects Powered by Jets? Difficulties of the Interstellar Bullet Model

An increasing number of bright HH objects are now known to be part of a jet from a young nearby star. In some cases, they are bright knots in the jet (e.g., HH 7–11, or HH 40), while in other cases the jet is apparently terminating at the position of an HH object. We discuss here the latter case; examples are HH 12, HH-DG Tau, HH 33, HH 19, HH 47A, and HH 32A and C. Because all these bright HH objects apparently represent the end of a jet (or jet head), it is suggested that they are powered by these jets. This means the kinetic energy of the jet (or part of it) is transformed there into heat in a shock front. In the context of the Blandford and Rees (1974) model of extragalactic radio jets, these bright HH objects trace the "working surface" of the jet. This model is strongly supported by velocity measurements of jets associated with DG Tau, HH 33/40, HH 46/47 and AS 353A, which indicate that all these flows, as required, are decelerated in the bright HH object located at the jet end (see Sec. I.A; Mundt and Fried 1983; Mundt et al. 1984). In this model the proper motion of the HH objects (e.g., HH 32A; see Table I) would be a natural consequence of the propagation of the working surface through the ambient medium. Its propagation velocity v_{ws} is approximately given by (Königl 1982):

$$v_{ws} \approx (n_j/n_a)^{1/2}\, v_j \qquad (1)$$

where v_j is the jet velocity and n_j and n_a are the particle density of the jet and the ambient medium, respectively. For n_j equal to the preshock densities derived for HH objects (≈ 100 cm^{-3}; see, e.g., R. D. Schwartz 1983*a*), for $n_a \approx 10^3$ cm^{-3}, and for $v_j \approx 300$ km s^{-1}, we derive $v_{ws} \approx 100$ km s^{-1}, a value which is consistent with the proper motion derived for HH 32A and other HH objects.

Is the appearance of an HH object at the jet end and the flow deceleration observed there consistent with the interstellar bullet model? In this model HH objects are dense clumps of gas moving with velocities of the order of 100 km s^{-1} through the ambient medium (C. A. Norman and J. Silk 1979). At the clump side facing the direction of motion, a bow shock is formed. If an HH

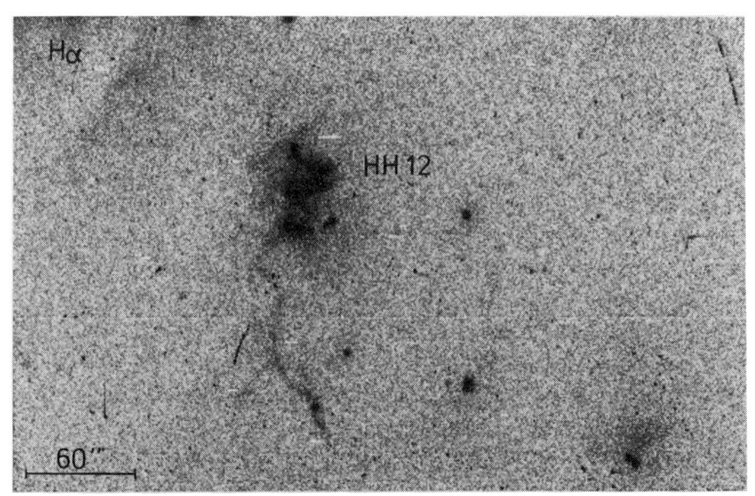

Fig. 10. Hα CCD frame of the HH 12 region. A quasi-sinuous jet seems to flow into HH 12 where this sinous-like pattern apparently continues (figure from K. M. Strom et al. 1983).

object at the jet end would represent such a "bullet," the jet must represent a glowing trail formed behind it, roughly comparable to that of a meteorite moving through the Earth's atmosphere. The emitting gas in such a glowing trail would be ambient gas that has been heated and ionized in the bow shock (see Choe et al. 1984). In order to form trails comparable in length to the observed jets, the recombination time and cooling time of the emitting gas must be of the order of the dynamical age of the jets, namely, 10^2 to 10^3 yr. This means that the emitting gas should have an electron density of 10^2 to 10^3 cm^{-3} (for $T_e \approx 10^4$ K), a rather low value considering that the bullet is expected to move through a molecular cloud medium with particle densities of 10^3 to 10^4 cm^{-3} and that this ambient medium will be, furthermore, highly compressed in the bow shock. Even if sufficiently large trails are formed behind a bullet, the observed flow deceleration near (or in) the HH object at the jet end would be difficult to explain. In this case, the bullet should have a velocity *higher* than the matter left behind it in the trail which is just opposite to what observations indicate. A further difficulty is encountered in trying to explain the wiggles shown by some jets (see Sec. II.E). Figure 10 shows one of the finest examples, the jet associated with HH 12. The quasi-sinusoidal appearance of this jet would be hard to understand as the glowing trail of a bullet, but could be explained by a "garden hose" instability (see, e.g., Woodward 1984).

The jets associated with HH 19 and HH 12 seem to terminate at these HH objects and, therefore, they might be powered by highly collimated flows as well. Both HH objects have a very knotty structure similar to other HH objects like HH 1, HH 2, and HH 39. For the latter three cases and for HH 12, the proper motions of the individual HH knots have been measured and in all cases the velocity dispersion of the knots is on the order of their average velocity. Therefore, these knots should spatially disperse in a fraction of their dynamical age if they have all been ejected from their driving star during a single event. This difficulty of the bullet picture can be avoided if energy is continuously deposited near (or in these knots) by a jet, resulting in the continuous formation and acceleration of new knots. In this picture, the individual HH knots could be due to various instabilities in the flow near the jet end, which according to numerical simulation (M. Norman 1984, personal communication) occur there more frequently than in the beam itself. Furthermore, the continuous deposition of energy in these HH objects by a jet may decrease the severe energetic problems of some HH models, like the bullet model (Mundt and Hartmann 1983).

In summary, there is evidence that some HH objects are powered by jets. This is evident in particular for those HH objects located at the jet end, because a flow deceleration is observed there. The extent to which this model may apply to typical HH objects must be tested by future observations. Besides more detailed kinematical studies of known jets and HH objects, further searches for jets connecting infrared sources and HH objects and detailed investigations of such new cases is a promising method to attack these questions.

E. Some Remarks on the Jet Morphology

Among the presently known jets, three morphological features (knots, wiggles, and cometary reflection nebulae) seem to be common to a significant number of them or to the nebulae around their driving stars. These are briefly discussed below.

Knots. Knots (or hot spots) are observed in nearly all presently known jets. As mentioned above, in some cases these hot spots are bright (catalogued) HH objects (e.g., HH 33, HH 40). Some jets (see Fig. 11) actually consist of a chain of roughly equally spaced knots. As evident from Fig. 11, this morphology is similar to some extragalactic jets. In other cases, like 1548C27 (see Fig. 5), only one or two knots are observed in the jet.

Wiggles. Several jets are not straight, but show changes in direction. One example is the HH 12 jet, discussed above (Sec. II.D and Fig. 10). Wiggles are also present in the H 6-5B jet (Fig. 5) and especially in the HH 46/47 jet (Fig. 1). Similar features seem to occur in the HL Tau jet and, as evident in Fig. 11, the knots in the L 1551-IRS 5 flow and the HH 7–11 flow are not on a straight line. A change in flow direction in the HL Tau jet near the star might also explain why HL Tau is not exactly aligned with this feature (Fig. 4).

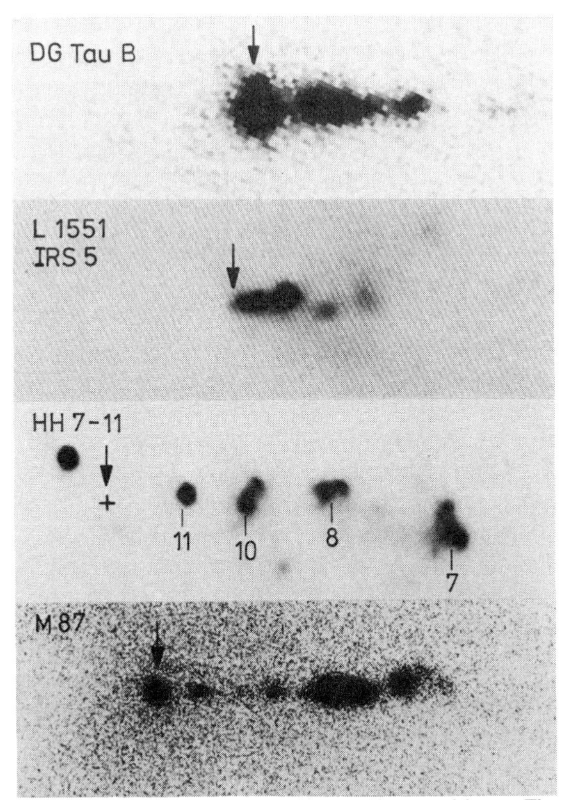

Fig. 11. Examples of jets showing strong, roughly equally spaced knots. The position of the driving stars of infrared sources is indicated by an arrow. A similar morphology is observed in the optical jet of the elliptical galaxy M 87.

Cometary Reflection Nebula. Some of the stars driving jets have cometary or cone-like reflection nebula with the jet being located close to the axis of the nebula. The prototype of this interesting morphological subgroup is 1548C27, shown in Fig. 5. Additional examples are L 1551-IRS 5, Haro 6-5 B (Figs. 4 and 5) and in particular R Mon (Brugel et al. 1984). Reflection nebula probably related to the above cases are observed near HH 46 (Graham and Elias 1983) and around DG Tau.

Possible formation mechanisms for knots in astrophysical jets have been discussed extensively by Norman et al. (1984). As noted above, these knots probably mark the location of increased energy dissipation in (strong) shock fronts. These shock fronts could be excited by a variety of fluid dynamical instabilities (e.g., pinching) or by external influences (e.g., pressure reconfinement). Shock systems caused by fluid instabilities are not stationary and the pattern velocity for different shocks (knots) can be quite different. Such effects might explain why in the case of HH 7–11 the individual HH knots have different proper motions (Herbig and Jones 1983). For the latter case, it has been suggested by Königl (1982) that this string of roughly equidistant HH knots represents a series of shock fronts (Mach disks) formed in an underexpanded jet which is excited to oscillations by the ambient medium. Königl (1982) considers a further possibility to form knots, namely, that they are due to obstacles in the flow (see also Mundt et al. 1983*a*). The observed wiggles could in principle be caused by precessing motions of the jet source. However, they can also be excited by flow instabilities (Woodward 1984; Norman et al. 1984).

The physical mechanisms causing the formation of cometary nebula around various jet sources are less clear. These nebula seem to be hollow cones (or hollow rotation paraboloids) with dense shells of gas and dust in which the light from the source is scattered. Apparently, matter from inside the cone has been pushed into the shell. Where did the necessary energy for this process come from? Has this energy been dissipated by the jet when it was moving through the ambient molecular cloud (Königl 1982), or could the cone simply be due to an expansion in a medium having a strong density gradient with the energy provided directly by the star (not by the jet)? Answers to these questions are also important for a better understanding of many other cometary nebula found around young stars.

IV. SUMMARY AND CONCLUSIONS

Young stars of low-to-medium luminosity ($1\ L_\odot \lesssim L \lesssim 100\ L_\odot$) are capable of producing well-collimated, high-velocity (100–400 km s^{-1}) flows. These jets are already collimated on length scales of $\approx$ 100 AU and close to their driving sources they are collimated to a much higher degree than could be inferred from the CO data. Due to their small collimation length scales, these jets can, in principle, provide important constraints on the circumstellar environment of young stars (e.g., disk properties).

Several catalogued HH objects (and perhaps many more) form the brightest part of a jet from a young star. For those HH objects located at the jet end, the existing kinematical data strongly suggest that they are powered by the jet. This means that these HH objects seem to represent the jet head (working surface). In other cases, they seem to be bright knots in a jet, tracing the location of shock fronts caused by flow instabilities.

Jets from young stars provide a unique opportunity to study cosmic jets in a much broader context. In contrast to extragalactic jets, one can measure important quantities like the electron density and radial velocity of the emitting gas or proper motion of hot spots. In some cases, the powering T Tauri star is sufficiently bright to allow detailed optical studies. Furthermore, in the near future millimeter and submillimeter observations at high resolution (5″ to 20″) will be possible and will allow the determination of the physical properties of ambient molecular gas in the immediate vicinity of the jets. Because many important parameters of these jets are (or will be) directly measurable, they may lead to a much better understanding of phenomena in supersonic jets than do extragalactic jets.

Acknowledgments. The author wishes to thank K.-H. Böhm, E. Brugel, Th. Bührke, S. Edwards, H. Elsässer, A. Königl, and G. Münch for comments and suggestions and for critically reading the manuscript.

HIGH MASS VERSUS LOW MASS STAR FORMATION

RODGER THOMPSON
University of Arizona

Advances in observational techniques, especially in the radio and infrared, have led to observations of high- and intermediate-mass young stellar objects (YSOs) while they are still embedded in their natal dust clouds. At the same time, these techniques, along with new optical observations, have been applied to the study of low-mass stars which are no longer completely obscured by dust. These observations have disrupted our previous view of star formation as a quiescent, spherically symmetric, gravitational contraction with subsequent hydrostatic evolution to the main sequence. Instead, most of the new observations have given evidence of very energetic asymmetric mass outflows and excess line emission. Our ideas about the sources of radiative and turbulent energy input to the surrounding natal molecular dark clouds also need reexamination in the light of the newly available data. Some of these new observations, especially for higher-mass stars, are examined in this chapter both to afford a better view of the stellar formation process and to illustrate the differences between the birth of high-mass and low-mass stars.

Some of the most exciting recent work on young stellar objects (YSOs) has concentrated on observations of intermediate and high-mass objects ($M > 3M_{\odot}$) in the luminosity range between 10^2 and 10^5 $L_{\odot}$. This chapter reviews the observations and considers the relationship between these stars and the better-studied low-mass YSOs which are treated by Rydgren, by Schwartz, and by Mundt in this book. Previous reviews of low-mass stars by Joy (1945,1949), Rydgren et al. (1976), Cohen and Kuhi (1979), Appenzeller (1983), Kuhi (1983), and Bertout (1984) are used in the following analysis. In particular, comparisons are made between high- and low-mass stars at roughly equal evolutionary stages.

Observations of high-mass stars have lagged those of low-mass stars due to dust obscuration still present around the faster evolving high-mass stars. This obscuration has prevented a direct view of the stellar photosphere, although some new far-red observations by McGregor et al. (1984) may have detected the photosphere in several cases. Most of our information has been obtained indirectly by observation of reemitted light, which bears as much similarity to the original light as good vodka bears to its natal grain. Considerable cleverness is therefore needed to extract useful information. The two main components of reprocessed radiation are thermal reradiation from dust, which gives information on the luminosity, and the radio continuum and atomic-line emission which give an indirect, sometimes uncertain, measure of the temperature.

I. THEORETICAL PROLOGUE

Although star formation is a period of rapidly changing temperature and luminosity, the mass of a pre-main sequence object should be obtainable by its location on a properly calculated evolutionary HR diagram. Numerical techniques have been used to model the initial collapse phase of a Jeans' mass protostellar cloud (cf. Larson 1973; Bodenheimer and Black 1978; Woodward 1978) to a point where viscous forces, nuclear burning, and convection become important. Stellar evolutionary codes (Iben 1965; Grossman and Graboske 1971) which treat these effects have been used to determine the radiative and convective tracks toward the zero age main sequence (ZAMS). Often the collapse tracks and the equilibrium tracks do not intersect. This mismatch has been the subject of recent work by Stahler et al. (1980*a,b*,1981). They have attempted to resolve the difficulty with analytical techniques which treat all of the stages of collapse and evolution to the ZAMS. Unfortunately, they have only calculated models of one-solar-mass stars in any detail with simplifying physical assumptions.

The main feature of all pre-main sequence evolutionary tracks is the development of a core-envelope structure which has been understood since the earliest calculations. This structure develops because the free-fall time for protostellar material is inversely proportional to the square root of the density:

$$t_{ff} = (3\pi/32G\rho)^{1/2}. \tag{1}$$

The dependence on density leads to a nonhomologus collapse; as the central density increases due to infall, the free-fall time decreases, which enhances the central density even faster. High-mass stars, therefore, evolve much more quickly than low-mass stars and develop a much more decoupled core-envelope structure due to the rapid collapse and higher luminosity of the core (R. B. Larson 1973). The core reaches a high-luminosity stage and even nuclear burning while still surrounded by an envelope or disk of natal material. Prediction of the emergent spectrum depends on a detailed knowledge of the radiative transfer and hydrodynamic interaction between the core and the surrounding material. Although several strides have been made in this direction (M. Simon et al. 1981*b;* Krolik and Smith 1981; M. Simon et al. 1983), a detailed and accurate calculation of the spectrum is not yet obtainable. In this case, it is the role of observations to help guide the theoretical development toward a better model of the evolving high-mass YSO.

II. OBSERVATIONS

This section discusses observations which give some insight into the nature of high- and intermediate-mass YSOs. Most observations are directed toward placing the objects in an evolutionary HR diagram which can be compared with and used to augment the theoretical picture. The presence of large-scale mass loss in most of the objects precludes a complete description of the evolution with a two-dimensional temperature-luminosity diagram.

Luminosity

Probably the most straightforward parameter to determine is the total luminosity of a YSO. Under the assumption that all of the stellar radiation, either directly or after reemission as line radiation, is absorbed by dust grains and reemitted as infrared radiation, the total infrared luminosity is equal to the total source luminosity. Because the majority of the reemitted power lies in the far-infrared telluric absorption regions, a true measurement of the luminosity requires observations from aircraft, balloons or spacecraft. To date most observations have been made with the Kuiper Airborne Observatory (KAO) (see, e.g., Harper 1974; Thronson and Harper 1979; Harvey et al. 1979; Werner et al. 1979). Public release of the infrared astronomical satellite (IRAS) data base will soon greatly increase the number of objects for which luminosities are available.

Several limitations exist on the luminosity measurements, most of which are shown in Fig. 1. The total luminosity will be underestimated if either photospheric photons escape without absorption by dust or are absorbed and reemitted by dust particles outside the telescope beam. On the other hand, multiple sources inside the telescope beam can cause confusion as to how the luminosity is divided among the several sources. In practice, the latter problem is the most vexing of the observational difficulties. Obscured YSOs often form in small clusters and the approximately $1'$ beams of the KAO and IRAS telescopes at far-infrared wavelengths are unable to spatially resolve such clusters. Assignment of total luminosity is usually based on extrapolation of mid- and near-infrared observations at higher spatial resolution. Despite these problems, luminosity is generally the most accurately known of the observational parameters.

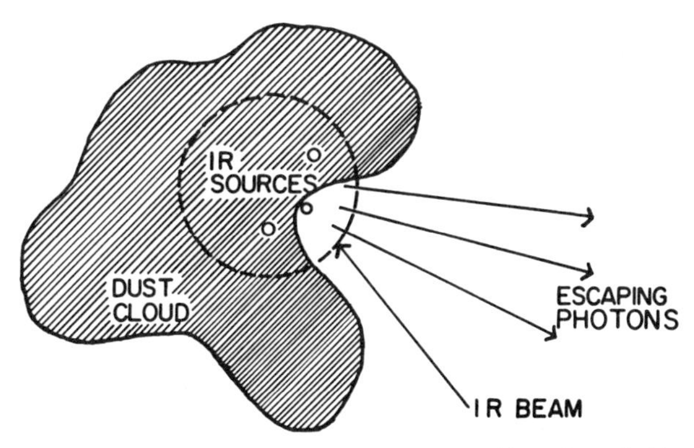

Fig. 1. The problems of luminosity determination are shown with a schematic diagram of a star formation region showing the far-infrared beam profile in a dust cloud with embedded sources and escaping photons.

Temperature

Since reradiation by dust destroys the spectral information contained in the photospheric radiation, determination of the temperature of the embedded source must be accomplished by other means. On the ZAMS, intermediate- and high-mass YSOs are hot enough to ionize the material around them. The degree of ionization provides a measure of the number of photons in the Lyman continuum through standard recombination theory. Comparison of the luminosity in the Lyman continuum with the total luminosity then fixes the temperature when compared to a blackbody or to computed stellar atmospheres. Panagia (1973) has computed the expected number of Lyman continuum photons and resultant ionization from various stellar atmospheres. These calculations have been redone with newer stellar atmospheres (Kurucz 1979) and extended to lower luminosities (Thompson 1984). The results are shown in Fig. 2, in which the parameter N_e^2V (electron density squared times the total volume) is used to measure the ionization. This facilitates comparison with the radio continuum measurements but assumes optically thin infrared recombination lines described by Menzel's Case B in which the Lyman α photons are trapped (Osterbrock 1974). As discussed below, this assumption is probably not accurate. Also shown in Fig. 2 is the ZAMS for Balmer continuum ionization. The appropriateness of this mechanism is discussed in (Sec. III).

A dominant characteristic of Fig. 2 is that most of the observations lie to the left or higher ionization side of the Lyman continuum ZAMS. This is not a prediction of any evolutionary model and is an indication that the physical processes used to construct the figure are not adequate to describe the nature of the objects observed. These objects, in fact, display a large number of common characteristics which are typical of YSOs in this mass range and

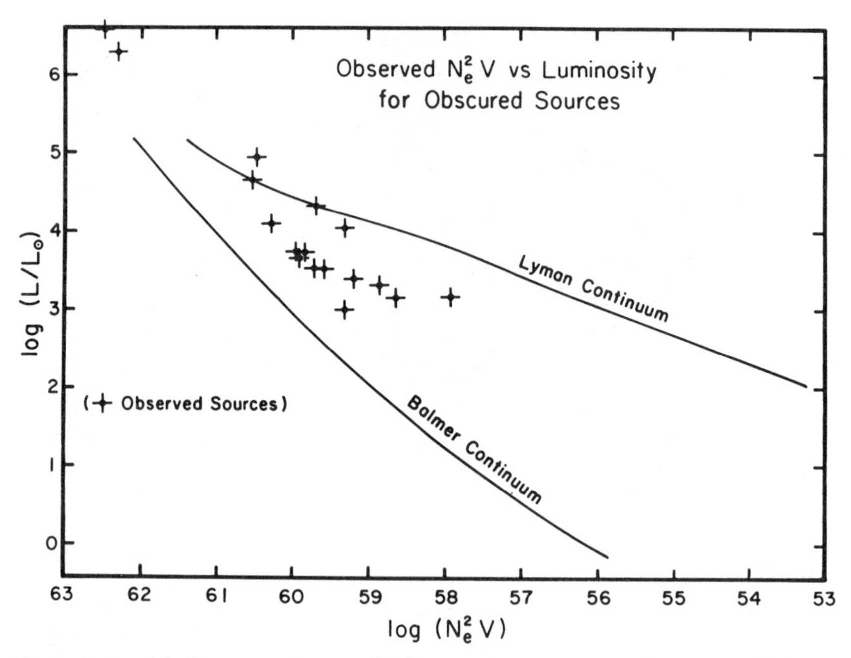

Fig. 2. A plot of the Lyman continuum and Balmer continuum zero age main sequence (ZAMS) in a N_e^2V vs. luminosity HR diagram. Measured sources are indicated with a cross which approximates the error estimates.

present one of the most fascinating set of problems encountered in star formation. Below is a list of the properties of these objects which has been taken from several sources (M. Simon et al. 1981*a,b;* Hall et al. 1978; Thompson 1982; Bally and Lada 1983; McGregor et al. 1984; M. Simon and L. Cassar 1984; Wynn-Williams 1984).

1. Most objects have an infrared line emission or N_e^2V values in excess of the value predicted by optically thin Case B recombination from a ZAMS star of the same luminosity.
2. All such line excess objects have luminosities $< 10^5$ $L_\odot$
3. The line ratios are not predicted by Case B recombination with dust extinction.
4. When detectable, the radio spectrum is optically thick and the flux is below that expected from an optically thin H II region with the N_e^2V value derived from either the infrared-line flux or from a ZAMS star of the same luminosity.
5. Lyman β pumped emission from O I 8446 and Mg II 2.1 μm lines is present in most of the objects.
6. Optically thick hydrogen Balmer, Paschen, and Brackett emission lines typify the spectra.

7. Broad emission line profiles $\geq$ 100 km s^{-1} FWHM with occasionally an associated narrow component are present in all high-resolution spectra.
8. Associated lower-temperature emission lines such as the Ca II far-red triplet emission appear in most of the spectra, with near-infrared CO emission also appearing in the Becklin-Neugebauer (BN) object in M42.
9. Many objects have associated high-velocity CO flows.
10. Many objects also have associated optical bipolar nebulae.

Several direct implications about the characteristics of YSOs can be derived from the above properties. The small radio flux cited in point 4 can be used to set an upper limit on the size of the ionized region at a few tens of AU assuming the radiation comes from a 10^4 K blackbody. Lyman β pumped O I 8446 emission (point 5) requires that Lyman α, β and Hα be optically thick (Strittmatter et al. 1977; Grandi 1980) which in turn requires a high $n = 2$ population in hydrogen. The N_e^2V values derived from the line emission, coupled with the small size, yield electron densities $> 10^8$ cm^{-3} and as high as 10^{12} cm^{-3}. At these densities and for a high $n = 2$ population, the assumptions of Menzel's Case B recombination are no longer valid; thus, the physical assumptions used to construct the ZAMS track in Fig. 2 are not appropriate. It is also true that if the infrared lines used to estimate the volume emission measure are optically thick, then this quantity has been underestimated, leading to an even higher discrepancy with the Lyman continuum ionization calculations.

Mass Loss

The high atomic line velocities (point 7) along with high-velocity CO flows and the presence of bipolar nebulae indicates that mass loss is occurring from the majority of YSOs. If the mass loss is spherically symmetric, the densities, sizes and velocities yield mass loss rates which vary from 10^{-7} to 10^{-5} $M_\odot$ yr^{-1}. At a velocity of 100 km s^{-1} the travel time through the emission region is on the order of a few months; therefore, material must be constantly replenished. The bipolar nature of the observed outflows indicates that the assumption of spherical symmetry is not valid. Except for the highly collimated jets associated with low-mass stars, most of the bipolar flows have opening angles of more than 90°, thus the mass loss rates and replenishment times estimated above are still valid. The existence of bipolar nebulae does, however, require collimation or restriction of the outflow at some point near the object. Recent Very Large Array (VLA) observations by Campbell (1984) of NGC 7538 IRS1 indicates that the collimation occurs at least as close as 50 AU to the central object.

High- and low-mass stars share many common properties with respect to mass loss, the most obvious being that YSOs in all mass ranges suffer mass loss. Bally and Lada (1983) have noted that most of the flows require forces which exceed the single scattering force available from the stellar radiation fields. Multiple scatterings are required, reinforcing the concept of high optical depths in the gas and dust surrounding the YSOs. In no case does the mechanical power required for the flow exceed the observed luminosity of the YSO. The mechanical power appears to be roughly proportional to luminosity (Bally and Lada 1983) which is consistent with radiatively driven outflow. Bally and Lada conclude that most of the mechanical energy injected into the ambient cloud by mass loss is contributed by stars in the mass range between 10 and 20 $M_\odot$. This conclusion will be examined in more detail in Sec. VI.

One significant difference between high- and low-mass stars is the degree of collimation of the mass outflow. Several low-mass stars have very collimated, essentially jet-like, flows which are observable in optical light. No such flows have been detected for intermediate- and high-mass stars although broad bipolar and fan-like nebulae (S106, NGC 2264IR) have been observed at optical wavelengths (Staude et al. 1982; Gehrz et al. 1982; D. A. Allen 1972). Observational details of the low-mass outflows are well described in the chapter by Mundt in this book.

It is tempting to attribute differences in collimation to different physical processes of mass outflow. The present state of observations, however, does not warrant such conclusions. A statistically more complete sampling of sources and an examination of the observational biases is needed to confirm the reality of such differences. An important difference in the observed emission is that the broad, fan-like structures are observed from photoionized regions whereas the emission from highly collimated jets arises from shocked material. In one case, the access of ionizing photons to a region is being restricted, in another case gas flow is being restricted. Emission from ionized gas could easily overwhelm the emission from collimated shocked gas in high- and intermediate-mass objects.

III. PROPOSED MODELS FOR LINE EXCESS

Several authors (Krolik and Smith 1981; M. Simon et al. 1981*a,b;* Thompson 1982; M. Simon et al. 1983; Scoville et al. 1983*a;* Thompson 1984) have considered the nature of the objects which exhibit the various characteristics discussed above. There is general concensus that the emission region is small (tens of AU) with high density (10^8–10^{12} cm^{-3}), and is undergoing outflow at velocities on the order of $\geq$ 100 km s^{-1}. There are differing views on the nature of the line-emission mechanism and very little speculation on the reasons for the mass loss.

A promising view has been presented by M. Simon et al. (1983) who suggest that ionization from the n = 2 level of hydrogen increases ionization in the emission region and gives rise to excess line emission. The Balmer continuum ionization curve in Fig. 2 shows that this mechanism can generate more ionization (Thompson 1984) than is required to explain the line fluxes. It is uncertain why the observations do not lie along the Balmer continuum

ZAMS if this mechanism is correct. Several answers are plausible. First, dust in the H II region may decrease the available ionizing flux, as is observed in very high-mass stars. Second, the region may be density- rather than ionization-bounded in the Balmer continuum. The third and most likely explanation, is that the Brackett lines are optically thick and are, therefore, decreased in intensity relative to their optically thin emission strength.

Support for the suggestion of ionization from $n = 2$ is provided by the Lyman β pumped O I 8446 (McGregor et al. 1984) and Mg II 2 μm (M. Simon and L. Cassar 1984) emission discussed above. To pump the O I and Mg II lines Lyman β photons must scatter enough times to have a high probability of encountering an oxygen or ionized magnesium atom. This requires a high optical depth in both Lyman β and Hα and, therefore, a high density of hydrogen in the $n = 2$ state. Hα must be optically thick to prevent Lyman β photons from being destroyed through the Hα, Lyman α cascade. A prime question is whether the $n = 2$ state is radiatively or collisionally populated.

Radiative excitation of the $n = 2$ state has been explored in the context of novae envelopes (Strittmatter et al. 1977) and for quasi stellar objects (QSO) clouds (cf. Kwan and Krolik 1981) with conditions similar to those of the emission regions around YSOs. These authors have found trapped Lyman α emission very effective in populating the $n = 2$ state and that collisional excitation does not compete at densities up to 10^{12} cm^{-3}. The O I 8446 emission appears to be further evidence of radiative excitation. If the hydrogen levels are collisionally populated then the lower levels of O I should also be collisionally controlled. The 8446 emission line would then not be enhanced over the other observable O I lines; therefore, the 8446 emission is radiatively controlled.

The 8446 emission-line region, however, should lie at the boundary of the hydrogen emission region since oxygen cannot remain neutral in an ionized hydrogen region. This should also be the lowest density part of the emission region for most reasonable outflow models. Collisional excitation could dominate in the inner part of the emission region where the densities are higher. This discussion has assumed a spherically symmetric emission region with a radial dependence for the density, ionization, and other parameters. The data of McGregor et al. (1984) show, within the signal-to-noise level, similar line profiles for the hydrogen emission lines, O I 8446, and the Ca II triplet which indicates similar velocity distributions for all emission regions. This implies that the emission region may be clumpy, with neutral regions intermixed with ionized regions. A clumpy region will greatly complicate the radiative transfer calculations.

Krolik and Smith (1981) suggest an alternative mechanism in which the hydrogen-level population is collisionally controlled in a hot outflowing wind. The wind is optically thick in the infrared lines which emit as blackbodies at a radius equal to the radius where they become optically thick. Exact predictions are difficult since they depend on the velocity and temperature structure of the outflow. The Krolik and Smith model appears consistent with most of the observations except for the Lyman β pumped emission lines. As discussed above, these lines occur at the outer boundaries of the emission region and may not reflect the conditions in the interior of the region.

All these models are disappointing in that they require the emission lines to be optically thick. If the emission lines are optically thick our view is again limited to the outer regions of the YSO. The only legitimate probes of the stellar photosphere appear to be the measurements by McGregor et al. (1984) shortward of 1 μm which reveal no observable absorption features indicative of photospheric temperature. These models also do not consider the mechanism that drives the outflow and may be the most important parameter of all.

A more sophisticated analysis of the radiative transfer and hydrodynamics in the emission region is certainly required. Many of the techniques now applied to the study of active galactic nuclei and quasars are directly applicable to the problem of YSOs. The YSOs are perhaps a more tractable problem which, when successfully addressed, will shed light on some of the more distant and energetic extragalactic phenomena. Another method of investigation is the two-dimensional finite difference calculation of the radiation and hydrodynamics employed by Sandford et al. (1982*a*) in their calculation of radiatively driven implosion of a molecular cloud.

IV. EVOLUTIONARY STATE

The strong-line emission from observed objects places their evolutionary state at least at the ZAMS because all pre-main sequence objects have a lower temperature than ZAMS stars of the same luminosity. Even so, there is difficulty accounting for the observed line emission. However, this is not the quiescent hydrostatic phase postulated for the ZAMS by stellar evolution calculations. We cannot clearly identify the high mass loss, hydrodynamic, state in which we observe most of these stars with a ZAMS location defined strictly in terms of temperature and luminosity. Initial contact with ZAMS appears to be made in an unstable state which requires adjustment of the star before normal hydrostatic evolution can take place.

Are there stars that can be identified at a stage of evolution earlier than the ZAMS? The possibility certainly exists: some stars (NGC 2024 IRS2 and W5 IRS1) in Fig. 2 have ionization lower than expected from Balmer continuum ionization and at least one (V645 Cyg) has less than that expected from Lyman continuum ionization. There are other objects which do not appear in the figure for which there is no detectable line flux, such as the sources in Mon R2 and S140 (Thompson and Tokunaga 1979). These sources have been excluded from Fig. 2 because there is evidence for multiplicity, making it difficult to assign accurate luminosities to them.

One of the objects lying on the Lyman ZAMS (NGC 2024 IRS2) is a well-studied infrared source. Its evolutionary state is ambiguous in that it may lie on the ZAMS by chance as it moves along a radiative equilibrium track toward the line excess region, or it may have passed through its instability period and has moved to its rightful place on the ZAMS. Most of the characteristics of this source are the same as those listed for the line excess sources described earlier (Sec. III), except for the line excess itself. Black and Wilner (1984) have found that the FWHM of the Bracket line is on the order of 100 km s^{-1}, so it appears that the object is still losing mass and is in the earlier stage of evolution.

The existence of Balmer continuum ionization raises the possibility of detecting stars at earlier stages of evolution and at lower masses than previously thought. Moorwood and Salinari (1984) discuss the evolution of several southern infrared objects with luminosities $> 10^5$ $L_\odot$. They identify several objects apparently at a pre-main sequence stage of evolution because they have ionization levels less than the Lyman continuum. M. Simon and L. Cassar (1984) have suggested that the evolutionary state of a pre-main sequence object can be measured by the ratio of Balmer to Lyman ionization. Their analysis suggests that the object Lk Hα 101 is very near or on the ZAMS. Unfortunately very few objects have both the high-resolution infrared spectra and the optical observations needed to determine this ratio.

V. COMPARISON WITH T TAURI STARS

Now that we have been able to spectroscopically observe intermediate- and high-mass YSOs it is appropriate to compare their characteristics with those of T Tauri stars for which an extensive set of optical observations exist. One can even ask if the T Tauri stars are simply low-mass extensions of the same phenomena which we have found in higher-mass stars. There are some basic similarities in that both classes of objects have line emission, broad-line profiles, and mass loss. The quantitative aspects of each of these properties, however, differs between the two classes of objects.

Line emission in T Tauri stars is best described as being produced by a chromosphere of mechanically heated gas around the star. The luminosity of the emission lines are a relatively small fraction of the total luminosity, with the bulk of the luminosity emerging as photospheric radiation. In the higher-mass stars, the luminosity in the lines is nearly equal to the total luminosity of the star which is characteristic of recombination radiation from a high-temperature object. Basically, the line radiation in T Tauri stars is consistent with a small amount of mechanically heated gas, while the line radiation from higher-mass objects is consistent with a radiatively excited gas.

Line profiles for the two classes of stars also differ. In the few cases of high-mass stars having resolved profiles, line widths are generally less than for T Tauri stars. T Tauri lines have widths on the order of 300–500 km s^{-1} as opposed to the 100 km s^{-1} profiles found for higher-mass stars, although some line-excess objects have broad-line wings on the order of the T Tauri widths (M. Simon and L. Cassar 1984; McGregor et al. 1984). P Cygni profiles are quite common in T Tauri stars and recently (cf. chapter by Mundt) a series of narrow-line absorption systems have been discovered. No equivalent systems have been found in higher-mass stars. Most profiles are symmetric with no absorption systems or have extended high-velocity plateaus or bases with relatively narrow lines. In most cases the signal-to-noise level is not high enough to detect absorption-line systems so it is difficult to rule out absorption systems for high-mass stars. It is also probable that the infrared continuum is due to dust emission rather than a photosphere. In this case, unless the high-velocity gas is outside the dust emission region, no absorption systems are expected.

Differences in the mass outflow properties have been discussed above and in more detail in other chapters (see chapters by Mundt, by Schwartz, and by Rydgren and Cohen). All these differences suggest that the quantitative aspects of the observed spectra are not similar between high- and low-mass stars. Outflow, however, is responsible for the spectra being different from hydrostatic stars of the same temperature and luminosity. In this sense, the only difference is the magnitude of the outflow and the temperature of the central object. The mechanism leading to the outflow may be the same in both low- and high-mass stars.

VI. IMPACT ON SURROUNDINGS AND SUBSEQUENT STAR FORMATION

Various papers (Zinnecker 1983; Stahler 1983*a;* Fleck 1983) have begun to discuss the effect of the physical conditions in a star formation region on the initial mass function (IMF). Since the IMF is one of the basic properties to be explained by star formation theories, one would like to believe that it can be derived and understood from first principles. The IMF, however, is determined by the physical conditions in the cloud from which the stars form, which are in turn influenced by the properties of the stars formed in the cloud. An extreme example of this is the formation of a very massive, early O star. The subsequent evolution of this star can, not only enhance the probability of star formation in a distant region (Elmegreen and Lada 1977), but can also completely stop star formation in its immediate vicinity by sweeping out all of the available star-forming mass. It is, therefore, apparent that stars of different mass will affect subsequent star formation in different ways.

It is well established that the observed velocity width in the radio observations of CO in molecular clouds is much broader than the Doppler width at the excitation temperature of the gas. This extra width is generally interpreted as turbulent motion (Myers 1983) on a relatively large scale. Myers has interpreted this turbulence with a Kolomogorov model in which the turbulent energy is supplied externally and greatest turbulent energy is present at the scale of the molecular cloud. There is some evidence, however, that the turbulent scale is not set by a Kolmogorov distribution but by the stars in the cloud itself (Fleck 1983) with the greatest turbulent energy at stellar size scales. Intermediate-mass stars with their high mass loss rates are prime candidates for this energy input. Bally and Lada (1983) have estimated that 65% of the energy input to a molecular cloud from newly formed stars is due to stars between 10 and 20 $M_\odot$ assuming that the energy input to the cloud is proportional to the main sequence nuclear luminosity.

The masses of stars which dominate the energy input to the ambient cloud may, however, depend on the nature of the energy input. Energy sources available to the forming star include gravitational, rotational, and eventually nuclear energy. It is evident that these energy sources depend on the mass of the star, and, therefore, the net contribution of each mass range to the total energy is proportional to the available energy source times the IMF. In this computation the nuclear energy source must be treated differently than the gravitational and rotational energy contributions since it is emitted over a time period much longer than the lifetime of the cloud. Bally and Lada take the available energy as the nuclear luminosity integrated over a time period of 10^5 yr.

The amount of energy available from gravity and rotation is determined by the change in potential energy of the accreted mass, either through direct infall, or disk accretion. The virial theorem requires that half of the potential energy be dissipated during the accretion process. The total potential energy change for a uniform density collapse is given by

$$E_{\rm pot} = \left(\frac{3}{5}\right) GM \left(\frac{M}{R}\right) \tag{2}$$

(where G, M, and R are gravitational constant, mass, and radius, respectively) which sets an upper limit on the energy that can be returned to the cloud. Since the mass to radius ratio is roughly constant on the main sequence (Allen 1976), the total energy returned to the cloud is proportional to the mass of the formed star. These considerations do not account for mechanisms such as coupling of the rotation of the collapsed core to the cloud via magnetic fields. They also assume that most of the matter accreted through the rotating disk reaches the boundary layer between the disk and the star. The full virial energy is then emitted to the cloud. The dissipation of rotational energy upon impact with the star (Lynden-Bell and Pringle 1974) has not been included. Inclusion of this energy would double the total emitted energy.

The same computation performed by Bally and Lada for nuclear energy can be performed for the gravitational and rotational energy. In the present case the total energy input is proportional to M rather than to $M^{3.3}$ which is the ZAMS luminosity function. It is also assumed that the gravitational contraction and accretion time is shorter than the cloud lifetime so that all of the energy can be utilized. The gravitational (E_g) and nuclear (E_n) energy inputs then are:

$$E_g = \left(\frac{3}{5}\right)\eta G\left(\frac{M_\odot}{R_\odot}\right)\int M(\mathrm{IMF})dM \tag{3}$$

$$E_n = \tau L_\odot \eta M_\odot^{-3.3}\int M^{3.3}(\mathrm{IMF})dM \tag{4}$$

where

$$(\mathrm{IMF}) = \begin{array}{ll} 18.2\,M^{-1.4} & 0.1\,M_\odot < M < 1\,M_\odot \\ 18.2\,M^{-2.5} & 1.0\,M_\odot < M < 10\,M_\odot \\ 115\,M^{-3.3} & 10\,M_\odot < M < 20\,M_\odot \end{array} \tag{5}$$

is the G. E. Miller and J. M. Scalo (1979) segmented IMF used by Bally and Lada, η is the efficiency of coupling the radiated energy to the cloud and τ is the lifetime for nuclear burning while the star is still in the cloud. We will assume with Bally and Lada, for the sake of comparison, that $\eta = 2 \times 10^{-2}$ and $\tau = 10^5$ yr. Integration of the above equations shows that the total gravitational and rotational energy available exceeds the nuclear energy by a factor of roughly 2. Because the nuclear energy input is directly proportional to the arbitrary interaction lifetime of 10^5 yr, it is reasonable to take the energy inputs as equal. Unlike the nuclear case, however, *45% of the gravitational energy input is from low-mass stars in the mass range between 0.1 and 1* $M_\odot$ *and 95% of the energy comes from stars with masses* $< 10\ M_\odot$.

Given the crude estimates on lifetimes, the form of the IMF, and the coupling constants, it is difficult to determine the exact ratio of energy input between high-mass and low-mass stars. The general conclusion is that the gravitational and rotational energy input is dominated by low-mass stars and is about equal to the nuclear energy input which is dominated by high-mass stars. Both high- and low-mass stars, therefore, contribute to the energetics of the cloud, with the low-mass stars making the initial contribution through gravitational contraction.

VII. FUTURE DIRECTIONS

Several directions for future research on YSOs are extremely important. More observations, especially spectroscopy in the radio, infrared, and far-red regions, are needed to form a statistically significant sample. From this sample, we can determine the percentage of objects showing mass loss and the duration of the mass loss phase. High spatial resolution radio continuum and line observations will also be extremely important in determining the spatial structure of the surrounding material. All of these observations should be extended to the low-luminosity objects available with present instrument sensitivities in order to study the contrasts and similarities in high-mass vs. low-mass star formation.

Theoretical studies of the radiative transfer and hydrodynamics in expanding, nonsymmetric, very optically-thick envelopes are needed to model line emission from observed objects. The escape probability method used in several QSO radiative transfer studies (see, e.g., Kwan and Krolik 1981) should be exploited in this study as well as the explicit calculation methods of Sandford et al. (1982*a*). Collapse and evolution models of high- and intermediate-mass stars which employ accurate hydrodynamics and radiative transfer are needed to assign an evolutionary state to the observed properties. Realistic modeling of disk evolution and structure is also required to understand the energetics and flows of the YSOs. This hoped for list of theoretical miracles will not appear soon; therefore, observations must lead the way by restricting and defining the parameter space of the acceptable models.

PROTOSTELLAR DISKS AND STAR FORMATION

PATRICK CASSEN
NASA Ames Research Center

FRANK H. SHU and SUSAN TEREBEY
University of California, Berkeley

The configuration of the solar system, as well as recent observations of young stellar objects, suggests that disks are a natural byproduct of the formation of stars from rotating interstellar clouds. We review current theories of angular momentum and mass transport in protostellar disks, present a model for their formation from the gravitational collapse of molecular cloud cores, and discuss some different evolutionary pathways possible for the combined protostar-disk systems that result. Although much of the theory of protostellar disks has been directed toward attempts to understand the formation of the solar system, we take a more general approach, concentrating on the connections between disks and the star formation process. However, we do not discuss the issue of planet building. We begin by stressing that, besides turbulent viscous mechanisms, gravitational and magnetic torques could play an important role in disk evolution. For modeling the primitive solar nebula, turbulent viscosity has the attraction that it concentrates the available angular momentum in a small fraction of the total mass, as has apparently happened in the solar system. However, stellar companions to stars may be even more prevalent than planetary systems, and it is possible that another mechanism predominates in many circumstances. Thus, a disk's ultimate fate may depend on factors not usually explored by disk modelers. An important consideration is undoubtedly the environment in which disk evolution is taking place. We try to synthesize from observational information a coherent account of processes which give rise to a molecular cloud core, control its rotation rate, lead to its eventual collapse, govern the appearance of the resulting protostar and nebular disk, and supply a bipolar stellar wind that eventually halts accretion. The model is idealized, and contains several gaps which need to be addressed by rigorous calculation, but the general picture is consistent with a wide variety of known astrophysical phenomena, and it makes specific and testable predictions.

The belief has long been held that the planets of the solar system originated from a disk of gas and dust surrounding the Sun at the time of its formation. The circumstantial evidence favoring the disk hypothesis remains noncontroversial: the near coplanarity and circularity of planetary orbits. Until recently, few new facts that bear on the question have been unearthed. For instance, it is still just beyond our capability to confirm the existence of other planetary systems, much less determine their characteristics. However, observations of young stellar objects, made with increasing spatial resolution, strongly suggest that the formation of disks around new stars is a common, if not ubiquitous phenomenon (see the chapter by Harvey). Furthermore, the implementation of new technology promises to provide a definitive test of the idea that planetary systems like our own are not uncommon (D. C. Black 1980). These developments, together with modern explorations of the solar system and advances in the theory of star formation, have provided impetus for new work on the theory of the formation and evolution of protostellar disks. In this chapter we review the status of theoretical work on disks, discuss the relation of that work to the theory of star formation, and suggest directions for future efforts.

We begin in Sec. I by discussing accretion disk theory and its application to models of the solar nebula and protostellar disks. In Sec. II we present a unified view of the process of star formation, starting from the evolution of molecular clouds, and leading naturally to the formation of protostellar disks. The models that are used to describe this process are idealized, and unlikely to be true in all details or under all circumstances; nevertheless, we believe that they provide good prototypes that represent well the essential hydromagnetic phenomena involved in star and disk formation. In Sec. III, we conclude with a qualitative discussion of several possible evolutionary paths and final configurations and how the outcomes depend on the relative efficiencies of various angular momentum transport processes. Each alternative raises specific theoretical problems that must be resolved before appreciable further progress can be made.

I. THEORY OF DISKS

At the outset, it should be stated that the theory of protostellar disks has largely been directed toward explaining the origin of the solar system, a natural consequence of our lack of detailed knowledge about other possibilities. Thus, some models presuppose conditions specially tailored to our planetary system; for example, they assume a disk containing only enough mass to account for the planetary masses reconstituted in volatiles to solar abundances, and having a radius of some tens of AU (the minimum-mass nebula). However, a theme of this chapter is the multiplicity of processes—many of them competing—that could conceivably determine the behavior of a disk. We should expect different evolutionary tracks leading to different end products, with a planetary system like ours being only one outcome among many.

The basic premises which almost all modern theories share are that stars form from the gravitational collapse of rotating dense molecular clouds (or portions thereof); that disks are natural byproducts of this collapse process; and that redistribution of angular momentum and mass is an important aspect of disk formation and evolution. In the case of the solar system, with nearly all of the angular momentum concentrated in less than one percent of the mass, it is highly unlikely that the original cloud fragment could have started with this distribution. Redistribution is necessary, and the key to describing disk evolution lies in understanding how this occurs.

It is also known that even the giant gas planets are significantly enriched in elements heavier than helium; therefore, considerable mass must have been lost from the outer solar system. The loss of this material has usually been attributed to the effects of a T Tauri wind, but whether the matter was swept to space or eroded from the nebula to become part of the Sun (B. G. Elmegreen 1978*b*) is not certain. The anisotropy often observed in mass flows from young stellar objects (see the chapters by Mundt and by Schwartz) may be rotationally controlled at some level, and therefore perhaps related to the existence of disks. A complete theory of protostellar disks should also treat the issue of how planets are built. Here, however, we restrict the discussion to the dynamical issues, from disk formation to disk dispersal. In this evolutionary story, the most uncertain part concerns the mechanisms which provide the requisite angular momentum redistribution. We begin our discussion, therefore, by outlining three candidates: viscous torques, gravitational torques, and magnetic torques.

A. Viscous Torques

The characteristics of viscous accretion disks are well understood, at least in a qualitative way. Early work by von Weizsacker (1948) and Lüst (1952), followed by Lynden-Bell and Pringle (1974) and others, described their essential behavior (see review by Pringle 1981). Viscous stresses acting on surfaces normal to the shear gradient cause the angular momentum in the outer parts of the disk to increase at the expense of that in the inner part. Thus, gas within a certain radius tends to spiral inwards, while that in the outer parts of the disk moves farther out. If the process continued indefinitely, a state would be approached in which all of the disk mass would be concentrated in the center, except for an infinitesimal amount orbiting infinitely far away, possessing all of the original angular momentum of the disk. In other words, viscous accretion disks tend to rearrange their mass and angular momentum in just the way that seems to have happened in the solar system. It is important to realize, however, that *any* dissipative process in which total angular momentum is conserved tends toward the same (minimum-energy) result (Lynden-Bell and Pringle 1974).

A quantitative description of the evolution of a viscous disk requires that the viscosity be specified, and therefore nebula modelers have expended some effort toward this end. If the kinematic viscosity arises because of turbulence, it will have a general form given in order of magnitude by

$$\nu = \mathrm{v}_t \, \ell \tag{1}$$

where v_t is the typical turbulent speed in the radial direction and ℓ is the effective horizontal mixing length.

To date, the only attempt to prescribe a turbulent viscous disk self-consistently is that of Lin and his coworkers (see the chapter by Lin and Papaloizou), who attributed the effective eddy viscosity to the convection which arises in the disk because of the energy released by the frictional accretion process itself. In particular, D. N. C. Lin and J. Papaloizou (1980) showed that even if overturning motions were absent to begin with, a thin disk, composed of well-mixed gas and dust in cosmic abundances, would become unstable to thermal convection as it cooled to space.

As the disk cooled, it would tend to contract in the vertical direction (parallel to the rotation axis); the heat flux consistent with this contraction would be sufficient to produce a superadiabatic temperature gradient in the vertical direction, thereby leading to instability. But with the initiation of convective eddies, a mechanism exists for the mixing of angular momentum of adjacent rings of gas, i.e., for producing turbulent viscosity. In the models of Lin and his coworkers, the disk does not continue indefinitely to contract vertically, but begins to evolve as a viscous accretion disk. The energy of viscous dissipation is ultimately drawn from the gravitational energy of the inwardly migrating gas and is transported to the faces of the disk by convective eddies. Lin and coworkers assume that the mean eddy speeds in the radial and vertical directions are comparable, and thereby calculate v_t by mixing-length theory applied to the vertical heat flux equation. But, realizing that rotation inhibits turbulent transport in the directions perpendicular to the rotation axis, they assume that the effective horizontal lifetime of a convective eddy is given, not by its value for vertical heat transport ($\sim H/\mathrm{v}_t$ where $H \sim c/\omega$; c and ω are the local sound speed and rotation rate), but approximately by ω^{-1}. As a consequence, their expression for the turbulent vicosity is given by Eq. (1) with $\ell = \mathrm{v}_t/\omega$. Since the epicyclic gyration radius (see Chandrasekhar 1942) of a free particle moving with radial velocity v_t is v_t/κ, where the epicyclic frequency κ equals ω in a Keplerian disk, this procedure is equivalent to assuming that eddies typically make one outward or inward epicyclic excursion before dissolving.

With the adoption of this prescription and with the application of the usual photospheric boundary conditions (see Eggleton 1967), D. N. C. Lin and J. Papaloizou (1980) were able to derive a closed set of equations for both the viscous evolution of the disk and its vertical structure. Furthermore, with the assumption that the grain opacity depends only on a power of temperature, the vertical and radial dependences of all variables can be separated, leading to a homogeneous nonlinear diffusion equation for the evolution of the viscous torque in the disk (D. N. C. Lin and P. Bodenheimer 1982):

$$\frac{\partial}{\partial \eta} f_g^{1/3} = j^{-8/3} \frac{\partial^2 f_g}{\partial j^2} . \qquad (2)$$

Here η is an appropriately defined dimensionless measure of time, j is the specific angular momentum, normalized to its value at some fiducial radius and therefore proportional to $r^{1/2}$ for a Keplerian disk (r is radius), and f_g is the normalized viscous torque (integrated over the vertical distance and multiplied by the circumference) which is exerted across a given radius by material inside that radius.

In order to solve Eq. (2), one must specify the initial conditions and boundary conditions. When homogeneous boundary conditions (zero torque at the center and at some large radius) are adopted, the solution of Eq. (2) depends on no explicit parameters other than those associated with the initial conditions; i.e., the only time scale is that for the relaxation of the initial distribution. Thereafter, the disks gradually wind down, with mass flowing in toward the center and the outer edge of the disk expanding at ever decreasing rates. This behavior is demonstrated explicitly by solutions obtained by D. N. C. Lin and P. Bodenheimer (1982) that approach a similarity form (see Fig. 1) in which the surface density diminishes as $t^{-5/14}$, the radial mass flux as $t^{-15/14}$, and the disk radius expands as $t^{1/7}$. It should be noticed that decay in time is generic to the process and not special to the detailed assumption about the form of the opacity. Basically, convection is needed to drive disk accretion, while (the energy released by) accretion is needed to drive convection. Thus, in the absence of torques exerted at the boundaries, the diffusive nature of the process leads to the decay of both accretion and convection.

The analysis of Lin and his coworkers is the only one that we are aware of that attempts to treat turbulent viscosity in a physically self-consistent way. The solution represents the gradual evolution of an optically thick disk as it cools to space, and thus may well represent a stage of nebula evolution after formation but before its dissipation or disruption. Lin and Bodenheimer conclude that such a stage would take 10^5 to 10^7 yr for a minimum-mass nebula.

In any part of the disk in which grains are not present, the opacity may be low enough for the disk to become optically thin. This will certainly be the case close to the protostar where both stellar heating and viscous dissipation are strong enough to evaporate grains. Lin (chapter by Lin and Papaloizou)

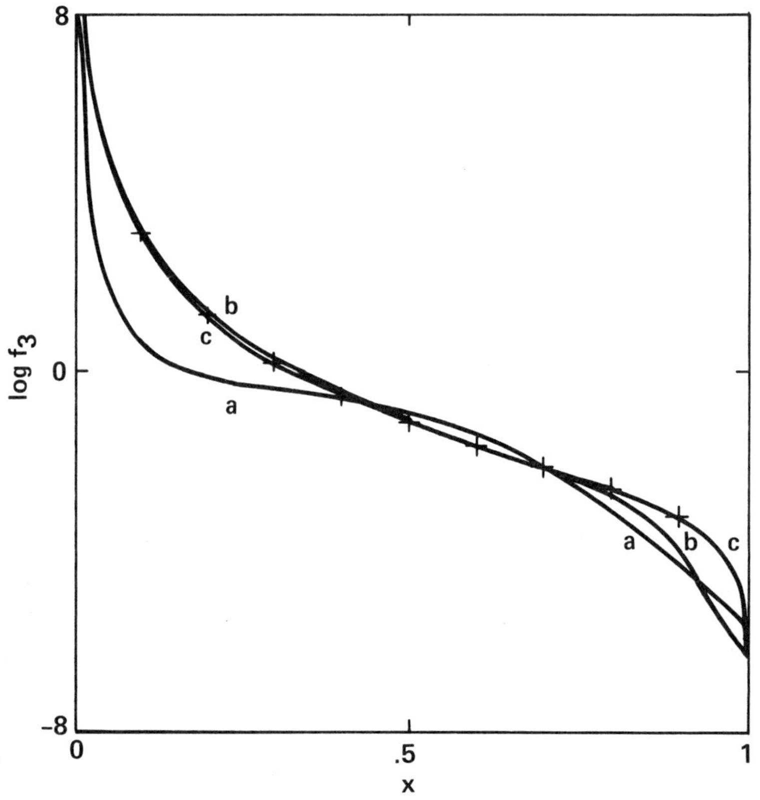

Fig. 1. Convecting accretion disks can be described by self-similar solutions of a nonlinear diffusion equation (Eq. 2). Shown above is the approach to the similarity form exhibited by a solution satisfying an arbitrary initial condition. The quantity $f_3 = j^{-7}\eta^{3/2}f_g$ and the similarity variable $x = j\eta^{-1/14}$, where η is a dimensionless measure of time, j is the specific angular momentum, and f_g is proportional to the normalized viscous couple. The crosses represent the similarity solution, while the solid lines are numerical solutions labeled in order of increasing time: $a = 2 \times 10^{-4}$; $b = 2.2 \times 10^{-3}$; $c = 2.222$ (figure from D. N. C. Lin and P. Bodenheimer 1982).

has argued that such a region may undergo cyclic states associated with short periods of rapid radial mass influx, identifiable as FU Orionis phenomena.

In C. Hayashi's (1981*b*) model of the solar nebula, it is assumed that grains rapidly accumulate near the midplane of the disk, thereby reducing the opacity to a point where thermal convection does not occur in the evolutionary stage. He considers the turbulent state to be a short episode ($\sim 10^3$ yr) following the collapse of the protostellar cloud (C. Hayashi 1981*b*). Thus, his models, like Lin's, treat viscous evolution, but they do not identify a specific turbulence-generating mechanism.

Both Lin and Hayashi presuppose the existence of a minimum-mass nebula; they do not attempt to model the formation process. Aside from the obvious bearing on initial conditions for their models, the question of how the nebula formed is important for other reasons.

First of all, protostellar formation times (and presumably those for disks) are estimated to be on the order of 10^5 yr, whereas grain coagulation and migration times in the nebula could be much less (Weidenschilling 1980). Therefore, as far as planet-building processes are concerned, all of the action might occur in the formation stage. In fact, it is possible that no relatively quiescent evolution stage such as that discussed above exists; the formation stage may end with the turning on of a wind which dissipates the nebula.

Second, whether a disk evolves into a planetary system, a multiple star system, or something else may depend on conditions attendant upon formation. Models that begin with a minimum-mass nebula (including Safronov's [1969; see also his chapter]) are designed to produce Earths, Jupiters, etc., and yield no information on other possibilities.

To model the formation of a disk, one must be concerned with the mass influx from the protostellar cloud, the exchange of angular momentum between the incoming gas and the disk, as well as other time-dependent quantities such as the gravitational potential of the protostar (and perhaps of the disk itself). Furthermore, thermal boundary conditions during formation are substantially different from those that apply during a quiescent state. Gas being accreted from a cloud is likely to arrive at the disk with supersonic velocity, as described in Sec. II, and must therefore pass through an accretion shock lying close to the disk. Thus, the disk may be heated primarily by radiation from the shocked gas. If this heating were the only source of energy for the disk, it would be stable, and the convective motions described by D. N. C. Lin and J. Papaloizou (1980) for the slowly evolving disk would not exist. The growing disk would be analogous to the growing protostars modeled by Stahler et al. (1980*a,b*), which remain nonconvective (before nuclear burning) due to the positive entropy gradient produced by the accretion process. Other instabilities may exist that would induce radial transport of angular momentum, and therefore allow the tapping of the gravitational energy of the disk. But in any event, shock heating should provide a stabilizing effect against convection, and we expect that the effective viscosity would be less than that calculated for the slowly evolving disk.

Models of the formation of the solar nebula have been discussed by Cameron (1978*a,c;* see also his chapter), by Morfill (1983*a,b*), and by Cassen and Summers (1983). These authors have taken rather different approaches and stressed different aspects of the problem, but it is possible to understand many of their results in terms of a few general principles. First, we briefly describe their premises and main conclusions regarding the dynamic evolution of the nebula.

Cameron's (1978*c*, 1983) models have been modified somewhat since they were first presented, the latest version being described in his chapter in this book. However, it is worth discussing the earlier models because of the important issues they have raised, and simply because they have provided much of the stimulation for subsequent work. Cameron (1978*c*) assumed that the total mass and angular momentum of the protosolar cloud were $2M_{\odot}$ and 8×10^{53} g cm^2 s^{-1}, respectively. This angular momentum estimate was based on his notion of the turbulent characteristics of interstellar clouds (Cameron 1973*a*), and is considerably higher than values estimated for a minimum-mass nebula, about 5×10^{51} g^2 cm^2 s^{-1}. He also assumed that mass was accreted onto the disk at a rate of 10^{-4} $M_{\odot}$ yr^{-1}, that the original protosolar cloud had uniform density and angular velocity, and that (for computational convenience) the infalling material hit the disk where its specific angular momentum matched that of the disk. The gravitational potential of the disk was approximated by an analytic function, which allowed explicit calculation of the angular velocity of the disk. For the viscosity, Cameron took $\nu = (2/9)cH$, where c and H are again the local sound speed and scale height, and he assumed the disk to be isothermal in the vertical direction. In other words, he explicitly assumed the existence and maintenance of turbulent radial motions of the order of the speed of sound. (This assumption is implicit to all so-called alpha disk models where the turbulent viscous stress is taken to be of order α times the gas pressure, and α is taken to be of order unity.) With these postulates, the disk evolution equation (Lynden-Bell and Pringle 1974) was solved numerically. The result was a disk that grew to several hundred AU and contained most of the accreted material by the time the accretion was terminated (and a mass loss phase was assumed to commence). Cameron noted that such a disk would be gravitationally unstable, and he argued that this would lead to the formation of giant gaseous protoplanets.

Cameron (see his chapter) no longer favors the gravitational instability mechanism for the formation of the outer planets, although he still considers such instabilities to be important for the formation of planetary cores in the inner solar system. He has calculated what he believes to be more realistic thermal structures based on an isentropic assumption (see Cameron and Fegley 1982). Further efforts have been made to refine the calculation of viscosity. The current models envision a rather violent formation stage, during which a hot, thick nebula spreads to become a convective disk, which feeds the growth of the protosun at the center and eventually becomes something similar in size and mass to a minimum-mass nebula. Attention is given to many details that could conceivably bear on the nature of presently observed solar system objects. In this sense, the models are inclusive, if not rigorous.

A rather different approach was pursued by Morfill (1983*a,b*), whose main interests were the thermal and chemical environments experienced by solid material in the nebula. The starting point for this model was the calculation of the collapse of a rotating gas cloud (again, with initially uniform density and angular velocity) by means of a numerical hydrodynamic code due to Tscharnuter (1980). The calculation shows the formation of a disk-like structure with dimensions comparable to the solar system. The gas from the collapsing cloud passes through an accretion shock surrounding the disk, within which it is assumed that turbulence produces an effective viscosity. Morfill (1983*a*) assumes the kinematic viscosity to be constant; in other calculations described in Morfill (1983*b*), an expression for the viscosity similar to Cameron's was used. It is reassuring that the nebula then evolves as expected for a viscous accretion disk, expanding with time as a central condensation builds up. The temperatures calculated are comparable to those found by D. N. C. Lin and J. Papaloizou (1980) and Cameron and Fegley (1982), although the mass of the disk is somewhat less than in those models.

Cassen and Moosman (1981) and Cassen and Summers (1983) were concerned mainly with the relation of disk properties to those of the protostellar cloud from which it formed. They constructed models of disk formation that relied on the analytic descriptions of accretion and collapse developed by Bondi (1952) and Shu (1977), modified to include rotational effects (Cassen and Pettibone 1976; Terebey et al. 1984*a*). They considered only a constant viscosity, or one with an *ad hoc* spatial dependence. The conditions under which their models are expected to apply are described in Sec. II, but some of their conclusions should be valid regardless of the particular collapse model used or the way in which the viscosity is specified. These conclusions are as follows.

The radius that the disk attains during formation will be approximately the larger of a centrifugal radius, $R_{CF} = J^2/k^2GM^3$, and a viscous radius $R_v = (\nu t)^{1/2}$. (Here, M and J are the total mass and angular momentum of the protostar plus disk, G is the gravitational constant, k is a constant of order unity whose precise value depends on the distributions of mass and angular momentum in the protostellar cloud—$k = 2/9$ in the model of Sec. II, t is the accretion time, and ν is the mean kinematic viscosity.) This simply means that the disk will have a size consistent with centrifugal support at its original distribution of angular momentum unless it spreads to even larger radii by viscous diffusion. An immediate consequence of this result is that the mass M_d of a disk for which $R_{CF} > R_v$ must be an appreciable fraction of the total mass accumulated in both protostar and disk; i.e., $M_d/M_* = O(1)$ (Cassen and Summers 1983). Such disks are subject to gravitational instability, as Cameron (1978*a*) has pointed out, and their subsequent evolution is likely to be determined by forces other than viscous stress. This situation will be discussed further below. If $R_v >> R_{CF}$, one can show that $M_d/M_* = O(R_{CF}/R_v)^{1/2}$; i.e., a low-mass disk will be produced.

Note that, as far as relative disk mass is concerned, it is the ratio (R_{CF}/R_v) that counts; high viscosity and low angular momentum both have the effect of allowing most of the mass to accumulate in the protostar, while the opposite conditions cause the mass to accumulate in the disk. Thus Cameron's (1978*a,c*) models have relatively massive disks, in spite of a high viscosity, because the initial cloud angular momentum was chosen to be large; for those models, $R_{CF}/R_v = O(1)$.

The success of viscous accretion disk models in representing the basic solar nebula configuration stems from the following two results: First, if the estimates of the viscosity, whether based on a simple dimensional argument or on a more detailed calculation such as Lin's, are believed, then they yield viscous lengths comparable to the size of the solar system, provided that the accretion time is within an order of magnitude of 10^5 yr. It follows from the arguments of Cassen and Summers (1983) that such viscous disks are consistent with low-mass nebulae.

Second, the temperature (at a given altitude) in a viscous accretion disk is a weak function of all parameters except radius, and typically falls from a few thousand degrees near the protosun to less than one hundred degrees several AU away. Lynden-Bell and Pringle (1974) showed that the photospheric temperature of an optically thick viscous disk is given approximately by

$$T \simeq \left(\frac{9GM\dot{M}}{8\sigma_s r^3}\right)^{1/4} \tag{3}$$

where σ_s is the Stefan-Boltzmann constant and $\dot{M}$ is the mass flux from disk to star. For a growing disk in which viscosity dominates the dynamics, the same formula holds with $\dot{M}$ being the total accretion rate from the cloud (Cassen and Moosman 1981). For $M = 1\ M_\odot$ and $\dot{M} = 10^{-5}\ M_\odot\ yr^{-1}$

$$T = 835\ r_{AU}{}^{-3/4}\ K. \tag{4}$$

Thus, viscous accretion disks satisfy the general requirement, inferred from the distribution of planetary masses, that the temperature in the inner solar system was high enough to evaporate refractory material, but that it diminished through the terrestrial planet zone to values for which ices could condense at Jupiter's distance and beyond (Fig. 2) (see, e.g., J. S. Lewis 1974). However, it should be noted that similar temperatures could also result simply from heating by the accretion shock created by the infalling matter from the protostellar cloud.

The two results cited above are characteristic of all nebula models for which viscosity provides the predominant angular momentum transport. They are substantive arguments for the viscous disk hypothesis, although it must be remembered that the methods that have been used to estimate the turbulent viscosity cannot be considered rigorous. Another note of caution is warranted with regard to the formation of minimum-mass nebulae. If the angular momentum of such a nebula was originally distributed smoothly in a protostellar cloud, the radius to which the gas would fall during collapse, R_{CF}, is much smaller than the size of the solar system. The values $J = 5 \times 10^{51}$ g-cm^2 s^{-1} and $M = 1\ M_\odot$ yield $R_{CF} \sim 10^{12}$ cm, the precise value depending on the distributions of density and angular momentum in the cloud. This means that much of the gas would fall directly onto the protostar, whose radius would

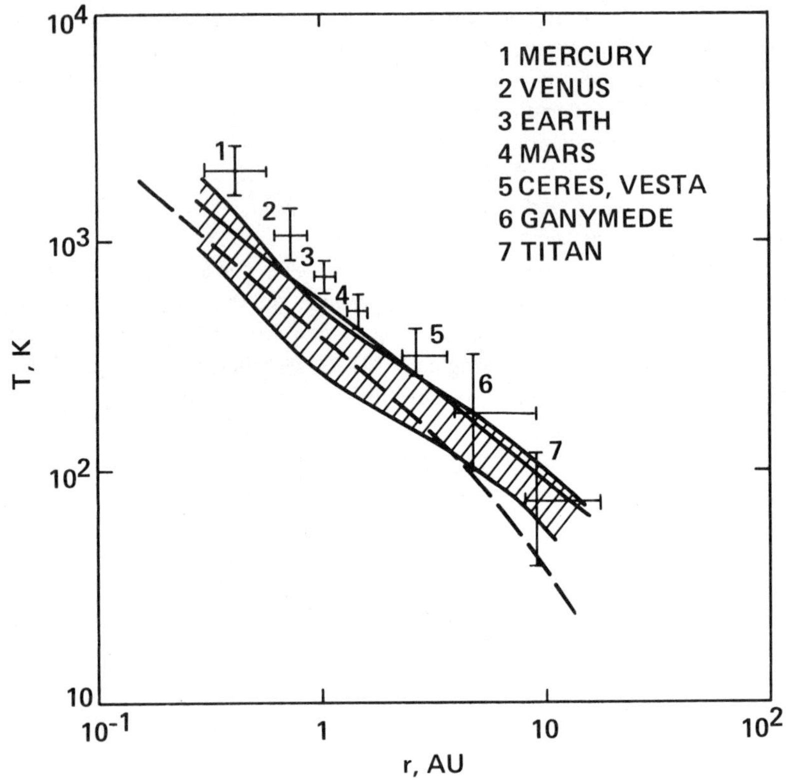

Fig. 2. The photospheric temperature of thin viscous disks follows an $r^{-3/4}$ law, regardless of the details of the viscous mechanism. This result is illustrated for three calculations, each based on a different formula for the viscosity. The solid line is the result of D. N. C. Lin's (1984) convective model; the dashed line is from Cassen and Summers (1983); and the shaded region represents values calculated by Morfill (1983*b*). The crosses show planetary accretion temperatures inferred by J. S. Lewis (1974) from the compositions of the planets.

likely be a few times 10^{11} cm (Stahler et al. 1980*a*). A viscous nebula of solar system size could then form only if material gained angular momentum by torques acting very close to the star (where there are no grains), and if turbulence could be maintained throughout the spreading disk (Cassen and Summers 1983). No adequate theory for either of these processes exists at present, although it has been suggested that gravitational or magnetic torques (Hoyle 1960) may be important.

B. Gravitational Torques

Any nonaxisymmetric distribution of gas in the cloud or disk will induce gravitational torques capable of affecting the angular momentum distribution. No doubt such inhomogeneities are inevitable; R. B. Larson (1984) and Boss

(1984*a*) have discussed some of their possible consequences. However, the relevant question for our purposes is whether systematic effects occur that contribute naturally to the formation and evolution of disks. There are several ways in which this might happen. First, as mentioned above, a gravitational instability is expected if $R_{CF} > R_v$, because disk mass then grows relative to protostar mass. The consequences of such an instability are not obvious; one might imagine the formation of rings, bars, spiral density waves, or fragments. The usual criterion for gravitational instability in a disk is that

$$Q \equiv \frac{b\kappa c}{\pi G \sigma} < 1 \tag{5}$$

where b is a constant of order unity, κ is the epicyclic frequency, and σ is the surface density (see, e.g., Toomre 1964; Goldreich and Lynden-Bell 1965*a*). This expression is derived from a local linear stability analysis, and is restricted to axisymmetric perturbations. Nevertheless, numerical calculations designed to test the stability of thin disks (see the review by Toomre [1981]; see also Cassen et al. [1981] for calculations made specifically in the context of protoplanetary nebulae) indicate that the parameter Q is one of the important indicators of stability. Another important discriminant measures essentially the fraction of the mass contained in the disk, weighted by a measure of its shear (see C. C. Lin and Y. Y. Lau 1979). If the disk has mass comparable to that contained in the protostar, so that this second discriminant plays a dynamical role, the numerical calculations suggest that disks that are axisymmetrically stable but not excessively so as defined by $Q > 1$, tend to develop spiral waves if the constraint of axial symmetry is removed. This result is in qualitative agreement with the linearized model analysis (see the summary of C. C. Lin and G. Bertin [1984]). Values of $Q < 1$ can lead to fragmentation.

The above results imply that a sufficiently massive nebular disk, in which Q starts out in the stable regime but is a decreasing function of time (as its surface density increases by infall of protostellar cloud matter, for instance), would generate spiral density waves before it would form rings and fragment. Waves of this type have been long thought to be responsible for the grand design of spiral galaxies, and have been studied in great detail, especially as their forced counterparts were predicted and then discovered in Saturn's rings (Goldreich and Tremaine 1978; Cuzzi et al. 1981; chapter by Lissauer and Cuzzi). The torques associated with them can transfer angular momentum outward, with a corresponding redistribution of mass, the details of which depend on the magnitude of the waves and the manner in which they are damped. On the other hand, if, despite this transport, the disk mass continued to grow and Q continued to drop as cloud collapse proceeded, large-scale fragmentation might result, leading possibly to the formation of giant gaseous protoplanets, or, more likely, to binary or multiple stars.

Another way in which gravitational torques could determine the course of disk evolution is through the action of density waves generated by a triaxial protostar. Under almost all circumstances, the protostar formed by the collapse of even a slowly rotating cloud would find itself in a state of rapid rotation. Studies of rotating, self-gravitating objects (Chandrasekhar 1969; Ostriker and Bodenheimer 1973; chapter by Durisen and Tohline) tell us that, as their angular momentum increases, such objects become unstable to modes that are characterized by ellipsoidal distortions. These distortions would excite spiral density waves in a surrounding disk, much as such a central bar can drive spiral structure in a barred galaxy (see, e.g., R. H. Sanders and J. M. Huntley 1976; W. W. Roberts and G. D. van Albada 1981; Yuan 1984). A preliminary analysis by Yuan and Cassen (1984) indicates that the rate at which such waves would transport angular momentum to the disk would be orders of magnitude greater than the gain of angular momentum by the growing protostar, if the protostar became even slightly distorted. This suggests that once the protostar attained the critical angular momentum for instability, all excess angular momentum brought with the subsequently accreted gas would be transferred to the disk, which would respond by expanding. It also indicates that there may be a limit to the amount of gas that can be accreted by the star from the disk, because all such gas brings with it the specific angular momentum corresponding to its orbital motion before it impacts the star.

Finally, protoplanets, once formed, would tidally interact with the nebular disk. D. N. C. Lin and J. Papaloizou (1979*a*) considered the transport of angular momentum due to the tidal friction associated with viscosity in a disk; in a more general context, Goldreich and Tremaine (1978,1979*a*) emphasized the transport of angular momentum by resonantly-forced, spiral-density waves. As mentioned previously, the latter phenomenon has been observed to occur between small satellites of Saturn and its rings. The dominant effect of either mechanism is the transfer of angular momentum *from* the more rapidly revolving disk material within the protoplanet's orbit, and *to* the more slowly revolving disk material outside the orbit. The result can be the clearing of a gap between protoplanet and disk, thereby inhibiting the growth of the protoplanet. Orbital evolution can also be driven by this kind of interaction. This type of gravitational interaction is discussed further in the chapter by Lissauer and Cuzzi.

It seems likely that gravitational torques will be important for disk evolution in many, if not all, circumstances. However, further research is required to define more precisely their role.

C. Magnetic Torques

In the recent literature, the only attempt to quantify magnetic effects in an accretion disk model of the solar nebula has been that of C. Hayashi (1981*b*). (Alfvén [1978] has proposed a solar system formation model in which electromagnetic forces dominate, but there is no disk in the usual sense. His theory has not attracted the interest enjoyed by accretion disk models.) Disk fields, perhaps generated by local dynamos acting on seed fields entrained during cloud collapse, would produce magnetic stresses on even slightly ionized disk gas. Sources of ionization exist due to heating associated with turbulent dissipation or shocks, protostellar radiation, cosmic rays, and radioactivity. C. Hayashi (1981*b*) showed that one can define a magnetic kinematic viscosity

$$\nu_B \equiv \frac{B_r^2 \tau_B}{4\pi\rho} \tag{6}$$

where B_r is the radial component of magnetic field, ρ is the density, and τ_B is the field decay time. In terms of ν_B, the torque-producing magnetic stress is given by $\rho\nu_B r\partial\omega/\partial r$, analogous to the expression for the ordinary viscous stress. The physical meaning of Eq. (6) is revealed by noting that it is equivalent to taking

$$\nu_B = V_A \, \ell_B \tag{7}$$

where V_A is the Alfvén speed and ℓ_B is the distance that such a disturbance travels in the field decay time. Thus, like all diffusion coefficients, ν_B is the product of a characteristic transport velocity and length.

C. Hayashi (1981*b*; see also the chapter by Hayashi et al.) estimated the degree of ionization at various positions in his nebula models, and compared the relative importance of magnetic and viscous stresses. He concluded that magnetic stresses could predominate in the outer regions of the nebula, where cosmic ray ionization is favored, and within Mercury's orbit, due to the effects of solar radiation.

Hoyle (1960) was also concerned with magnetic forces in the early solar system. Although his objective was different from those of recent investigators (he was attempting to explain the slow rotation of the Sun), and his concept of the star formation process was somewhat different from our present ideas, Hoyle recognized certain facts that are still important: (1) the radius of a minimum-mass nebula would initially be much smaller than the solar system; and (2) a protosolar magnetic field could produce a torque on the nebula that transferred angular momentum from star to disk, thereby slowing the star and expanding the disk. The key requirement is a magnetic link between the star and that part of the disk rotating more slowly than the star. Hoyle (1960) demonstrated that only modest magnetic field strengths need be invoked to produce a significant effect, but did not attempt to model the disk evolution. Apparently little work has been done on the problem since then.

II. UNIFIED VIEW OF STAR-DISK FORMATION

So far we have discussed a number of theoretical efforts to understand the evolution of protostellar disks, and have identified the processes that are thought to be important. We shall return to these processes in Sec. III, where we speculate on how they might lead to different nebula fates. What would be useful at the present stage of research on star and disk formation is a quantitative model that could guide both observations, which promise to reveal detail with increasing spatial resolution, and future theoretical work. It is desirable that such a model consider the issues of star formation in as broad a context as possible; be applicable to a wide range of conditions without the need to execute extensive numerical calculations for each new situation; be consistent with what we know about the interstellar medium and star-forming regions in particular; and most importantly, have consequences that can be tested. Thus, attention should focus on such quantitative issues as what masses and sizes of stars and disks can be expected from the collapse of interstellar clouds of given initial and boundary conditions.

The purpose of this section is to outline such a model. A full treatment should include all of the ingredients: self-gravity, thermal pressure, magnetic fields, rotation, and turbulence. Many numerical calculations in one, two, and three dimensions have been performed that address different parts of these issues (see reviews by Bodenheimer and Black 1978; Woodward 1978; and Tohline 1982). However, the simultaneous consideration of every effect is currently beyond our computational capacity; moreover, a piece-meal treatment which isolates the most important effects at each stage of the evolutionary process is probably more conducive to understanding. This is the approach we take on a first pass through the entire problem. After describing the model, we shall return to the question of disk formation and evolution to examine its consequences, and discuss what future work needs to be done for its development.

Briefly summarized, we envision a process in which stars are formed by the dynamical collapse of the dense cores of molecular clouds. The observational evidence suggests that the cores themselves are formed by a quasi-static process of contraction because (a) there are no known instances where dynamical collapse of a large portion of a molecular cloud is taking place; (b) the smooth and round isodensity contours of many molecular cloud cores hint at a balance of opposing forces; and (c) the velocity widths associated with cloud cores near the sites of low-mass star formation, at least, are only slightly more than thermal. A natural way to form a *quiet* core is by the diffusion of magnetic fields with respect to the neutral component of the gas of a molecular cloud clump. During the accompanying phase of quasi-static contraction, the rotation rate of the core will be set by magnetic coupling to its envelope. Upon exceeding a critical degree of central concentration, the core will suffer gravitational collapse, with the collapse proceeding nonhomologously, from inside-out. A protostar and, perhaps later, a nebular disk form at the center, with matter from the molecular cloud core continuing to rain down on top of them. During the main accretion phase, the central objects are visible only as infrared sources. The accretion is terminated upon the commencement of a strong stellar wind that draws its energy from the differential rotation of the protostar. In the rotating context, this wind is likely to take, at least in its initial stages, a bipolar form. The entire model is motivated by a combination of observational and theoretical considerations. Whether it is correct in all of its details is perhaps not so important for now as the fact that it is consistent with many known facts about present-day star formation and is rich in consequences that can be tested. We now begin a more detailed discussion of the proposed processes.

A. Support of Molecular Clouds

The bulk of the star formation which currently takes place in the Galaxy occurs in giant molecular cloud complexes of masses 10^5 to 10^6 $M_\odot$ (Burton 1976; Solomon and Sanders 1980; chapter by Elmegreen). The complexes themselves are made up of clumps having sizes of a few pc and masses of about 10^3 $M_\odot$ (see Fig. 3); they move with respect to each other at a speed of several km s^{-1}, roughly the virial speed associated with the complex; and they have cores which appear as hot spots in CO maps if massive star forma-

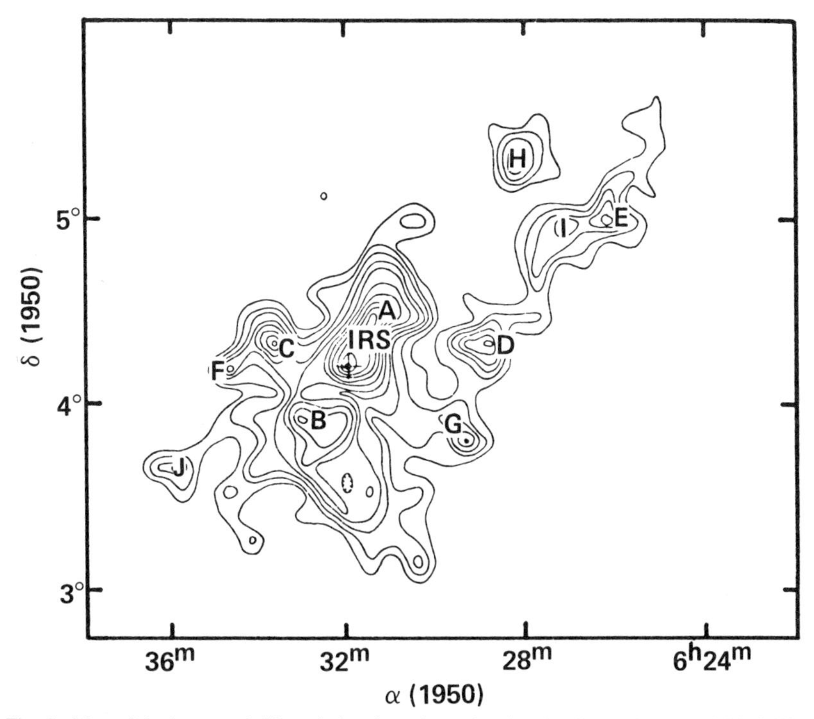

Fig. 3. Map of the integrated CO emission from the molecular cloud complex Mon OB2. Different clumps are indicated by different letters. A dense core can stand out in the CO contours if it contains an imbedded infrared source (IRS) of sufficient luminosity to heat up the surrounding gas and dust (figure from Blitz and Thaddeus 1980).

tion has taken place within the cores (Sargent 1977; Rowan-Robinson 1979; Blitz and Thaddeus 1980; chapter by Scalo).

With typical mean densities $\bar{\rho}$ corresponding to 10^2 to 10^3 H_2 cm^{-3} and temperatures $T \sim 10$ K, the Jeans mass associated with the clumps

$$M_J \sim (kT/G\mu)^{3/2} \bar{\rho}^{-3/2} \quad (8)$$

is still one or more orders of magnitude less than the mass of a clump. Because these molecular clumps cannot be gravitationally collapsing as a whole at anything like their free-fall rates, or there would be far too much star formation in the Galaxy (Zuckerman and Palmer 1974), it has been suggested that magnetic fields and/or rotation and/or turbulence provide the requisite support against the clump's self-gravity (see, e.g., Mouschovias 1976*b;* Field 1978; C. A. Norman and J. Silk 1980; R. B. Larson 1981). From this point of view, the emphasis lies not on finding a trigger for star formation, but on how to prevent it from happening even faster. The important trigger, if any, involves the origin of molecular cloud complexes (see, e.g., Shu 1984*b*); once these exist, they will inevitably find ways to form stars.

We adopt the view that most molecular cloud clumps (but not the complex as a whole) are supported against self-gravity primarily by magnetic fields. Direct and indirect estimates of the magnetic field strengths in molecular clouds (see, e.g., Chaisson and Vrba 1978; Draine 1980; Chernoff et al. 1982) are consistent with this suggestion. For instance, if magnetic fields provide the principal mechanism of support for a molecular cloud clump, the clump mass must not exceed the critical value

$$M_{cr} = 0.15 \frac{\Phi}{G^{1/2}} \quad (9)$$

where Φ is the total magnetic flux threading the clump (Strittmatter 1966; Mouschovias and Spitzer 1976; chapter by Mestel). For a cloud whose magnetic flux Φ is given by $B\pi R^2$, Eq. (9) may be expressed numerically as

$$M_{cr} = 10^3 M_\odot \left(\frac{B}{30\mu G}\right)\left(\frac{R}{2 \text{ pc}}\right)^2 \quad (10)$$

indicating that magnetic fields of strengths 30 μG or so would need little additional help to explain the equilibria of a typical molecular cloud clump (whose mass is 1500 $M_\odot$ according to Rowan-Robinson [1979]). It is also true that polarization studies show spatially well-ordered magnetic fields associated with some molecular clouds (see, e.g., Vrba et al. 1976; Vrba et al. 1981). The latter observations suggest that the field is strong enough to prevent the chaotic motions from tangling badly the lines of force.

The nature of interstellar turbulence is controversial, although many would attribute to it a prominent role in cloud dynamics (C. A. Norman and J. Silk 1980; R. B. Larson 1981; chapter by Dickman). However, the dynamical support which can be provided by turbulence (as contrasted with its transport of energy and angular momentum) is limited because it probably must remain submagnetosonic (cf. Goldreich and Kwan 1974; Field 1978; see, however, Scalo and Pumphrey 1982). Strong shock dissipation makes higher levels of turbulence difficult to sustain, even when sources, stellar winds, exist for the turbulence observed in molecular clouds (cf. Bally and Lada 1983). Thus, it is likely that turbulence can, at best, only give an amount of support comparable to magnetic fields or thermal pressure.

Rotation may help to support the envelope of a molecular clump, especially in the direction parallel to the magnetic field. Its influence on the cores of a molecular cloud, however, is small if the core rotates at nearly the same rate as the envelope (chapter by Goldsmith and Arquilla).

B. Formation of Molecular Cloud Cores

Magnetic support of a predominantly neutral medium inevitably leads to ambipolar diffusion (Mestel and Spitzer 1956; Langer 1978*a;* Mouschovias and Paleologou 1981; Nakano 1981; D. C. Black and E. H. Scott 1982), and the action of ambipolar diffusion in molecular clouds is to produce dense cores from density irregularities inside a clump (Fig. 4). Any initial enhancement (such as the nova disturbance considered by Cameron in his chapter) will do, as long as it is large enough. The reason is simple. Magnetic fields can exert forces only on charged particles; consequently, it is only the ions and electrons which are directly tied to the field. Under the pull of the self-gravity of a density enhancement in an otherwise uniform envelope, the neutral component of the molecular cloud gas will slip relative to the ionized component because the concentration of the latter is impeded by the presence of the magnetic field. It is the friction generated by this systematic slip that transfers the magnetic force to the neutrals; without the slip there would be no magnetic support. Thus, ambipolar diffusion (or plasma drift) is intrinsic to the support process; it is not a novel mechanism invented by theorists to confound observers.

As magnetic flux leaks out of a settling protocore, the Alfvén and magnetosonic velocities of the medium both drop, and therefore the level of turbulence that can be maintained (say, by a given background of stellar wind sources) also drops (see Myers 1983). Thus, magnetic fields and turbulence both become less effective relative to thermal pressure (and rotation) in helping to support the cloud core against its self-gravity. Eventually, we can expect the core to pass the margin of stability and to undergo dynamical collapse. If the magnetic field and turbulent velocity decay to small values, the precollapse core can be expected to have a mass comparable to the Jeans mass M_J given by Eq. (8), except that the average density that now applies would be

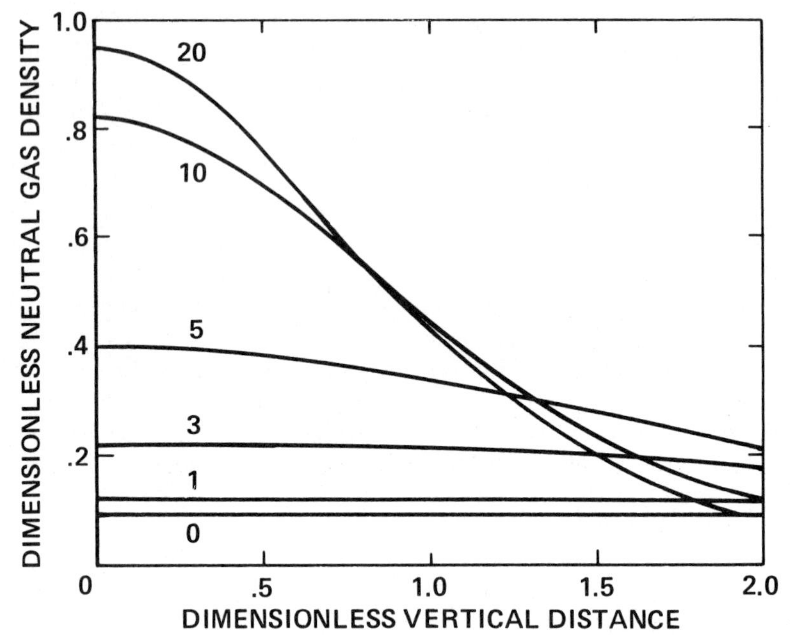

Fig. 4. The effect of ambipolar diffusion in molecular clouds is to produce dense, pressure-supported cores. Shown above is a one-dimensional analog of this process. A slab of gas initially supported against its own gravity predominantly by magnetic field relaxes asymptotically in time to hydrostatic equilibrium in which the magnetic field provides no support. The density profiles are labeled in order of increasing time (figure from Shu 1983).

that of the core, not that of the clump as a whole (the envelope). This set of circumstances seems to hold for the smallest cores which have been observed in molecular clouds (Myers and Benson 1983). Moreover, because the mass of the clump M greatly exceeds M_J, we see that the long-term trend in any molecular cloud clump would be *quasi-static* fragmentation into many cores.

The time scale required to form a molecular cloud core by this mechanism was estimated by Shu (1983; see also Shu and Terebey 1984) to be of the order 10^6 to 10^7 yr, and coincides roughly with the typical spread in ages of newly formed stars in a T association (see, e.g., Rydgren et al. 1976; M. Cohen and L. V. Kuhi 1979). In this interpretation, the individual stars of a T association form from individual molecular cloud cores, of which there are many in a given clump of molecular gas. The cores will usually look quiet because the lifetimes, 10^6 to 10^7 yr, in a quasi-static contraction stage exceed those in a collapse (protostellar) phase, 10^5 yr. There are some indications from the fraction of infrared sources found embedded in small molecular cloud cores (C. A. Beichman, personal communication) that the lifetimes of such cores are two or three times greater than those of the embedded sources. This would mean that the criterion used ($>$ 11 mag of visual extinction; Myers et al. 1983) selects preferentially the final stages of the quasi-static contraction process.

There is a theoretical bonus to the idea that magnetic flux loss occurs simultaneously with core formation: the asymptotic density distribution which the contraction produces becomes largely independent of initial and boundary conditions. In the one-dimensional isothermal problem, the asymptotic state is Spitzer's (1942) sech-squared density distribution (see Shu 1983). For the more realistic three-dimensional problem (Lizano and Shu, in preparation), we believe the state toward which the core tends (but can never exactly reach because it is in a state of unstable equilibrium) is the singular isothermal sphere with cloud sound speed a (Chandrasekhar 1957):

$$\rho = \frac{a^2}{2\pi G r^2}. \tag{11}$$

There are a number of reasons for pursuing this conjecture. First, the singular isothermal sphere is a natural candidate in any situation where all mechanisms of mechanical support other than thermal pressure are slowly drained from a very large gaseous mass. Under these circumstances, cores must form which have the properties that they smoothly join on to their envelopes and that any sphere of radius r centered on a density concentration tries to contain only one Jeans mass. The power-law density distribution (where the ratio of thermal energy to gravitational potential energy at every r is $3/4$) is the only one which has the latter behavior. Clearly, for it to join on smoothly to a (more or less) uniform density background (the envelope) requires us to specify a means of support for the envelope other than just thermal pressure. Fortunately, the precise means of this support is not crucial for the dynamics of the core because the entire mass M of the cloud clump is much greater than M_J (calculated on the basis of the conditions in the envelope), and the boundary conditions on the clump may, therefore, be regarded as effectively infinitely removed from the core.

Second, there is the observational suggestion that some resolved molecular cloud cores (which tend to be the larger ones) do have a power-law density structure like that of the singular isothermal sphere (see, e.g., Loren et al. 1983; Bally and Lada 1983), although the exponent cited may differ from 2 in some cases (chapter by Myers). Of course, the density profile near the center of a cloud core at the onset of dynamical collapse cannot be exactly r^{-2}; there will in fact be transient phases involving the collapse of a small amount of finite density material to form a central, condensed object (R. B. Larson 1969; Winkler and Newman 1980). However, as the collapse proceeds, we may expect the solution to look more and more like the collapse of the idealized r^{-2} model.

It so happens that the gravitational collapse of the singular isothermal sphere has an exact analytical solution (Shu 1977). The collapse proceeds in a

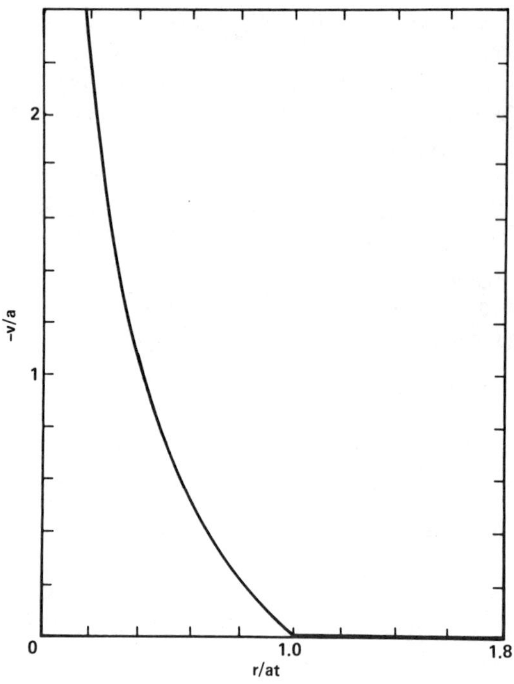

Fig. 5. Similarity solution for the velocity during the collapse of a nonrotating, singular isothermal sphere. An expansion wave whose front travels outward at the sound speed a produces inward motion in the initially static cloud. Within a radius $r = 0.4at$, the velocity becomes supersonic (figure from Shu 1977).

self-similar fashion (because the power-law [Eq. 11] has no intrinsic length or time scale), with the inner high-density regions falling in first to form a collapsed state (the protostar), on top of which material of initially lower density rains in steady accretion. The falling process steadily progresses outward as an expansion wave (Fig. 5). Because the dynamically collapsing portion of the cloud core is confined to its innermost parts, the infalling gas may only fill a small fraction of the beam of a radio telescope. In that case, it would be easy to understand why radio observations might yield relatively narrow molecular-line profiles even for cloud cores where the existence of an embedded mid-infrared source suggests that collapse is taking place to form a protostar (Myers and Benson 1983; Beichman et al. 1984; Baud et al. 1984). These observations of quiet cores with embedded protostars are much more difficult to understand in the context of proposals which envision dynamical implosions triggered by external shock waves or clump collisions, or in terms of gravitational fragmentation within the environment of a much larger collapsing cloud.

An important quantity for the accreting protostar is the mass accretion rate $\dot{M}$. For the self-similar collapse of the singular isothermal sphere, this rate is given by

$$\dot{M} = 0.975\frac{a^3}{G} \tag{12}$$

and equals 10^{-5} $M_\odot$ yr^{-1} for $a = 0.35$ km s^{-1}. The corresponding accretion luminosity (typically tens of $L_\odot$ for solar-type stars) should come out over a broad range of infrared wavelengths, with a characteristic color temperature of a few hundred degrees K. Such infrared protostars seem to be present in the Infrared Astronomical Satellite (IRAS) survey (Baud et al. 1984; Beichman et al. 1984). Buildup of the central protostar at rates comparable to 10^{-5} $M_\odot$ yr^{-1} seems to account for the observed positions of T Tauri stars in the HR diagram constructed by M. Cohen and L. V. Kuhi (1979; Stahler et al. 1980*a;* see also Stahler 1983*a*).

The purely spherical problem is, of course, highly idealized; worse, it cannot address the interesting problem of disk formation, the subject of this chapter. Fortunately, if a quasi-statically contracting molecular cloud core only rotates as fast as its envelope (because of magnetic braking), then its angular rate of rotation may be considered a small parameter at the onset of dynamical collapse. In this case, perturbation methods apply, and the rotating collapse flow is also amenable to an analytical treatment (Terebey et al. 1984*a*). It is to the discussion of this problem that we now turn.

C. Collapse of Rotating Cloud Cores

It is easy to show (see, e.g., Tassoul 1978) that the mass distributions of uniformly rotating, isothermal equilibria satisfy the partial differential equation

$$a^2 \nabla^2 \ln\rho + 4\pi G\rho = 2\Omega^2 \tag{13}$$

which is a generalization of the usual Lane-Emden equation. Terebey et al. (1984*a*) begin by finding an exact solution of Eq. (13) which, for simplicity, possesses the following properties:

1. It is unbounded (i.e., extends to infinity so we need not consider boundary effects),
2. It looks like a singular isothermal sphere (Eq. 11) for small r,
3. It approaches a constant-density state for large r with

$$\rho = \frac{\Omega^2}{2\pi G}. \tag{14}$$

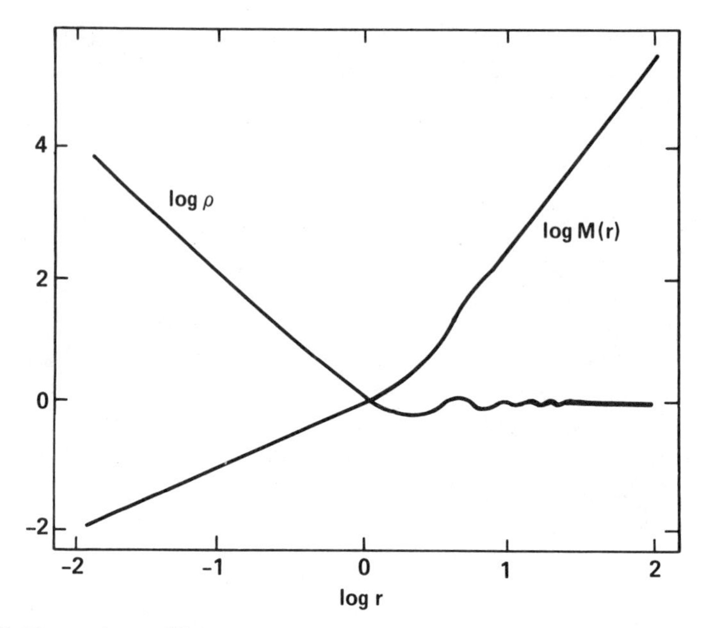

Fig. 6. The rotating equilibrium state satisfying Eq. (13), used as a model for the precollapse state of a cloud core and envelope. The radius r is normalized by a/Ω, and the density ρ by $\Omega^2/2\pi G$, and the mass within r by $2a^3/G\Omega$ (figure from Terebey et al. 1984*a*).

The formal equilibrium possesses spherical symmetry, which is possible even in the presence of rotation due to property (1) above. In any case, the effects of initial oblateness are small compared to the flattening produced by the dynamical collapse (see discussion in Terebey et al. 1984*a*).

Note that this initial state differs from those used in previous collapse calculations in two significant ways: first, instead of a uniform density distribution, equilibrium demands that there is a central concentration of matter; second, coupling the core's rotation to that of the envelope makes available more low angular momentum material from which to make configurations with dimensions comparable to the solar system or binary stars.

Figure 6 shows that the turnover from behavior given by Eq. (11), appropriate for the core of a molecular cloud, to behavior given by Eq. (14), appropriate for a (centrifugally-supported) envelope, occurs roughly at the radius $r_c = a/\Omega$. Inside r_c, rotation may be considered as a small addition to the support against self-gravity provided by an isotropic pressure. This ignores the effect of magnetic fields for helping to hold up the core, but this neglect helps to simplify a first study. We also neglect the slight additional braking which occurs once the core goes into dynamical collapse (cf. chapter by Mestel).

Not including the effect of magnetic fields is less justifiable for $r > r_c$, because the envelopes of molecular clouds are likely to be magnetically supported, as well as turbulent. Thus, the density of real molecular cloud envelopes (which nearly equals the mean density of clumps) can exceed the value given by Eq. (14). Nevertheless, the relation between the rotation rate and mean density found by Goldsmith and Arquilla (see their chapter, Eq. [10] and Fig. 6) is consistent with Eq. (14), at least for densities $< 10^3$ cm^{-3}, suggesting that the envelopes of some clumps have contracted to a rotationally supported equilibrium. In any case, the formal solution does illustrate the useful point that there will generally be a smooth transition between cloud cores and their envelopes.

To fix ideas, let us consider a specific numerical example. Suppose $a =$ 0.35 km s^{-1} and $\Omega = 10^{-14}$ s^{-1}. Then the turnover radius $r_c = a/\Omega = 3.5 \times 10^{18}$ cm and contains inside it an amount of mass $M_c = 2.18\, a^2 r_c/G = 70$ $M_\odot$. The mass M_c much exceeds what is needed to form, say, a 1 $M_\odot$ star. The radius R inside which resides $M(R) = 1$ $M_\odot$ corresponds to $R = 5.4 \times 10^{16}$ cm. If core collapse were to take place and if only the material inside R $<< r_c$ were to fall in before the accretion flow were reversed by a stellar wind, the total angular momentum

$$J = \frac{G^2 M^3 \Omega}{18 a^4} \tag{15}$$

possessed by the collapsed object would be 1.3×10^{52} g cm^2 s^{-1}, about three times the value for the augmented solar system (Cassen and Moosman 1981). If radial transport processes are slow compared to accretion time, the protostellar system will have a natural radial extent governed by centrifugal balance on the collapsed scale

$$R_{CF} = \frac{(\Omega R^2)^2}{GM}. \tag{16}$$

For our nominal choice of parameters, $R_{CF} = 6.4 \times 10^{12}$ cm, which is much smaller than the original radial extent R of the piece of cloud core which fell in to make the collapsed system. Thus, over a wide range of physical scales, $R_{CF} << r << r_c$, rotation may be considered a small effect, and a perturbational approach to the problem of core collapse is appropriate.

To be sure, other choices of the parameters a and Ω are possible; the ease with which we can vary our choice is, of course, one of the advantages of an analytical solution. The important thing to remember in order to apply the formulae given in this subsection is the need for a clean separation of the scale R_{CF} characterizing the dimension of the collapsed configuration, and the scale r_c characterizing the dimension of the original interstellar cloud core. Another way to state this constraint is that the perturbational approach applies if the cloud envelope rotates only a small fraction of a full turn in the time $t = M/\dot{M}$, where the mass accretion rate is given by Eq. (12). Since t is on the order of 10^5 yr, at least for the formation of stars on the lower main sequence (Stahler et al. 1980*a;* Myers and Benson 1983), and since the rotation periods of molecular clouds typically much exceed 10^6 yr (chapter by Goldsmith and Arquilla), the constraint seems likely to be satisfied in many realistic situations. For situations in which the core rotation rate is too high to satisfy this constraint, or where the core is initially not highly centrally condensed, the results of the collapse calculations reviewed by Tohline (1982) may be more relevant.

Whenever $R_{CF} << r_c$, an *intermediate region* exists, defined by

$$R_{CF} << r << r_c. \tag{17}$$

Exterior to the intermediate region the solution approaches the initial equilibrium state. In the intermediate region, the flow resembles the solution found for the nonrotating singular isothermal sphere except for the slight perturbations generated by the specific angular momentum being advected inwards on nearly radial streamlines. In particular, isodensity surfaces are almost spherical, corresponding to the equilibrium law (Eq. 11) outside the expansion wave, and to the free-fall form,

$$\rho = \frac{1}{4\pi G}\left(\frac{0.975}{2t}\right)^{1/2}\left(\frac{a}{r}\right)^{3/2} \tag{18}$$

as one approaches the collapsed configuration of total mass $M = 0.975a^3t/G$.

Interior to the intermediate region, the flow can be asymptotically matched to an appropriate inner solution; or more accurately, we may say that the inner limit of the intermediate solution provides outer boundary (matching) conditions for the inner problem. In this inner limit, the radial component of the velocity has the form,

$$u_r = \left(\frac{2G\dot{M}t}{r}\right)^{1/2} \tag{19}$$

Notice that u_r becomes much larger than a for $r << at$, the position of the head of the expansion wave, i.e., the flow becomes hypersonic. However, u_θ and u_ϕ both remain small in comparison with u_r as long as $r >> R_{CF}$. Thus, for the outer limit of the inner problem, the streamlines take the form of parabolic ballistic trajectories.

During the main accretion phase, the time required for the matter to cross the inner region and strike either the protostar(s) or the nebular disk(s) is short in comparison with the evolutionary time of the system as a whole. This has the consequence that the mass of the inflowing matter is negligible in comparison to that already stored in the protostar or disk, and that the simplifying assumption of quasi-steady state may be adopted for the solutions of the flows of matter and radiation (Stahler et al. 1980*a*, 1981; Cassen and Moosman 1981; Terebey et al. 1984*a*).

If at any time most of the total accreted mass resides in a nearly spherical protostar, then the gravitational field in the inner region may be taken as GM/r^2. In this case, parabolic trajectories remain a good approximation all the way in until the matter passes through an accretion shock on the surface of either the protostar or the accretion disk (Ulrich 1976; Cassen and Moosman 1981; Chevalier 1983). Then it can be shown that the equation describing the trajectory in a meridional plane is

$$\frac{(0.975)^3}{16}(\Omega t)^2\left(\frac{at}{r}\right) = \frac{\cos\theta_0 - \cos\theta}{\sin^2\theta_0 \cos\theta_0} \tag{20}$$

where θ_0 is the angle the asymptotically radial streamline makes with respect to the z-axis. Differentiating the above equation in time under the assumption that the accretion time t does not change appreciably on the short time scale of the variations of r and θ yields the following relation between $u_r = \dot{r}$ and $u_\theta = r\dot{\theta}$:

$$u_\theta = \frac{\cos\theta - \cos\theta_0}{\sin\theta} u_r. \tag{21}$$

Conservation of angular momentum yields the azimuthal component of the velocity as

$$u_\phi = \frac{\Omega(0.975\, at/2)^2 \sin\theta_0}{r \sin\theta} \tag{22}$$

whereas conservation of energy gives a third equation relating the three components of the velocity:

$$\frac{1}{2}(u_r^2 + u_\theta^2 + u_\phi^2) - \frac{G\dot{M}t}{r} = 0. \tag{23}$$

Equations (21)–(23) suffice to recover an Eulerian description of the axisymmetric velocity field if θ_0 as a function of (r,θ,t) is obtained from Eq. (20).

The density of the freely falling gas may be obtained by requiring conservation of mass flow through a streamtube:

$$\rho u_r r^2 \sin\theta \, d\theta \, d\phi = -\frac{\dot{M}}{4\pi}\sin\theta_0 \, d\theta_0 \, d\phi. \tag{24}$$

Terebey et al. (1984*a*) have explicitly shown that the outer limit of this inner solution (as $r \rightarrow \infty$) does indeed asymptotically match the inner limit of the intermediate solution described previously.

D. Gross Properties of Stars and Disks

An immediate result of the solution described above is the conclusion (Cassen and Summers 1983) that the maximum mass of a protostar that can result from direct accretion of infalling cloud material (rather than by accretion through a disk) is

$$M_* = \frac{1.5}{G}\left(\frac{16R_*a^8}{\Omega^2}\right)^{1/3} \tag{25}$$

where R_* is the radius of the protostar (nearly a constant in time according to the calculations of Stahler et al. [1980*a,b*]). This results because there is a limited amount of low angular momentum material that can fall within the protostellar radius. Indeed, once the disk forms, the star is able to acquire only about 50% more mass by direct accretion.

For the nominal parameter-space domain, $0.2 \text{ km s}^{-1} < a < 0.7 \text{ km s}^{-1}$ (corresponding to temperatures between 10 K and 100 K), $1 \text{ R}_\odot < R < 10 \text{ R}_\odot$, and $10^{-16} \text{ s}^{-1} < \Omega < 10^{-13} \text{ s}^{-1}$, Eq. (25) yields a range

$$0.01 \text{ M}_\odot < M_* < 100 \text{ M}_\odot \tag{26}$$

a realistic spread of stellar masses. This suggests that much of the matter of a high-mass star may need to be processed through an accretion disk, or that such stars only form from regions of high a or low Ω, or that their formation is triggered by processes very different from those described here (see, e.g., the chapter by Klein et al.).

What are the properties of disks produced by the process described here? If we ignore magnetic and gravitational torques, the disk properties are governed by the cloud parameters, a and Ω, and by the disk viscosity ν. In particular, the disk radius at time t, $r_d(t)$ will be the greater of R_v and R_{CF} (see Sec. I), where we now consider these quantities as functions of time. They may be written

$$R_\text{v} = (\nu t)^{1/2} \tag{27}$$

$$R_{CF} = (0.975)^3(\Omega t)^2 at/16. \tag{28}$$

Thus, whenever $\nu >> (\Omega t)^4 a^2 t/298$, the disk will spread viscously into the infalling material. In the formulation of Eq. (1), with $\text{v}_t < c$ and $\ell < H$, where c and H are the disk sound speed and vertical scale height, one can show that

$$r(t) < (H/r)^{4/3} at. \tag{29}$$

The above inequality, with $a < c$, assures that thin viscous disks spread subsonically, as expected. Little more can be said about this case without a detailed model for the frictional forces.

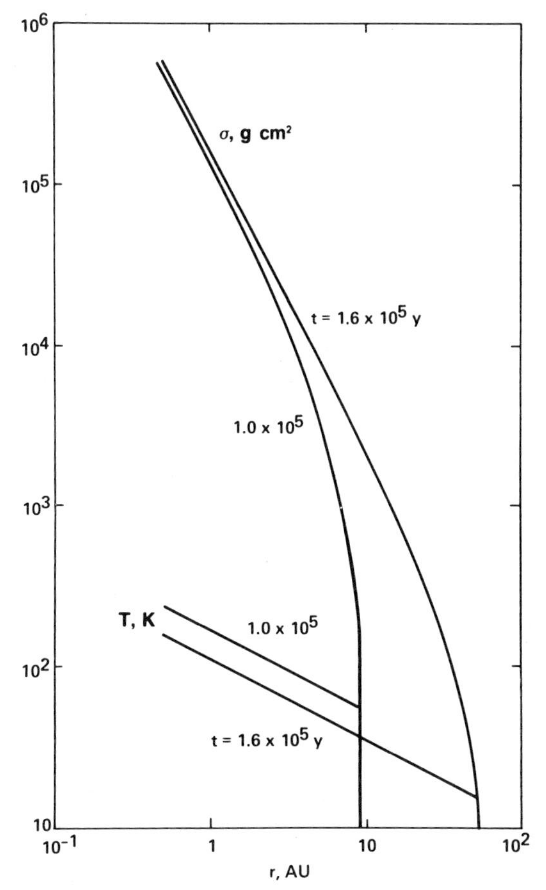

Fig. 7. Surface density and photospheric temperature at two times during the growth of a protostellar disk from the collapse of a rotating cloud core. The core is assumed to have a sound speed $a = 0.3$ km s^{-1} and rotation rate $\Omega = 4 \times 10^{-14}$ s^{-1}. The viscosity of the disk is taken to be 2.5×10^{15} cm^2 s^{-1}. The total mass accumulated after 1.6×10^5 yr is 1 $M_\odot$, of which approximately half resides in the disk. Since internal viscous torques are not strong enough to redistribute angular momentum over the disk radius, the photospheric temperature is not the $r^{-3/4}$ law characteristic of viscous disks, but nearly $r^{-1/2}$ (figure from Cassen and Summers 1983).

If the presence of an accretion shock suppresses convection, as suggested earlier, or whenever $\nu << (\Omega t)^4 a^2 t/298$, the disk radius increases as

$$r_d(t) = R_{CF}(t) \tag{30}$$

or

$$r_d(t) = 0.063 a^{-8} \, [GM(t)]^3 \tag{31}$$

when expressed in terms of the total mass $M(t)$ accreted in time t. Cassen and Moosman (1981) calculated disk surface density and photospheric temperature for this situation, in the approximation that the disk potential was negligible compared to that of the star (Fig. 7). This approximation of a $1/r^2$ gravitational field must fail at a time when the disk and star masses become comparable.

In this regard, it is well to remember that the particular intermediate solution described above could just as well have been matched to other inner solutions (ones containing, for example, binary stars). Conversely, the method used to construct the inner solution described above could also have been matched to other intermediate solutions (ones where the original rotation rate and accretion history gave a different distribution of specific angular momentum with time and polar angle than in Eq. (20). The technique and philosophy behind breaking up the calculations to outer, intermediate, and inner regions are much more general than the specific applications to which they have so far been applied.

For low-mass protostars, the inflowing matter in the inner region is heated by the outflowing radiation emerging from the accretion shock, but the gas pressure cannot rise fast enough to retard appreciably the infall. Accretion would continue until either radiation pressure acting on dust grains or an incipient stellar wind rises enough to reverse the inflow. For solar-type stars, radiation pressure is not enough, and we must appeal to strong stellar winds to prevent the entire mass of the molecular cloud core from falling into the central regions.

E. Origin of T Tauri Bipolar Flows

The observational and theoretical work done since 1970 allows the following interpretation. When protostars are accreting, they are not generally observable optically; they constitute infrared objects. When pre-main sequence stars are observable optically, they are not generally accreting; they possess powerful winds. Indeed, stars of all types seem to be born losing mass.

Ever since the discovery that giant molecular cloud complexes are the sites for most of the star formation which takes place in the Galaxy, it has been noted that the efficiency of star formation in them must be quite low (Zuckerman and Palmer 1974). Given that stars form out of molecular clouds having much more mass than needed to form those stars, it is tempting to conclude that the accretion process is halted, not because it runs out of gas, but because of the onset of a stellar wind which reverses the inflow. What is the mechanism which powers the outflow?

For T Tauri stars, Shu (1984*b;* see also, Shu and Terebey 1984) has proposed that the wind is powered by the release of energy stored as differential rotation when a low-mass protostar is driven completely convective by the onset of deuterium burning. The idea is that no matter what is the distribution of specific angular momentum in the core of the original molecular cloud, it can hardly know in advance that it is destined to be incorporated into a star. Thus, after the matter falls into the protostar, it will not necessarily correspond to a rotation law which the star can easily accomodate. In particular, the protostar will generally possess strong differential rotation. Now, as long as the protostar remains radiative and the distribution of specific angular momentum increases outward (the low angular momentum material always falls in first in the scenario described earlier), the differential rotation will not lead to any dynamical instabilities (Goldreich and Schubert 1967; Knobloch and Spruit 1982). Thus, a differentially rotating protostar of low mass can accrete passively as long as it remains radiative. The energy stored in this differential rotation will be a healthy fraction of the gravitational binding energy of the star—up to about 30% before the star becomes distortionally unstable (Ostriker and Bodenheimer 1973)—because any protostar that accretes from a rotating molecular cloud will attain equatorial speeds approaching Keplerian values if it acquires a nebular disk.

The ignition of deuterium, however, releases sufficient luminosity to drive a low-mass protostar completely convective (see Fig. 6 of Stahler et al. 1980*a*). The effect of convection must be to reduce the strong shear contained in the differential rotation (probably through the dynamo amplification of magnetic fields), thereby liberating energy. If a reasonable amount of this energy, brought to the surface by buoyant magnetic loops or in any other manner, is liberated as wind energy (and not as heat), and if the release occurs in a time short compared to the duration of accretion, then the wind will be able to reverse the inflow (cf. Chevalier 1983). At first, however, the accretion flow may smother any incipient stellar wind, much as the lid of a pressure cooker smothers the hot gases building up below. But the buildup of pressure will continue until the hot gases can eventually punch through at the weakest point. In the model of Sec. II.D, where the ram pressure of the swirling accretion flow acts as the lid on the pressure cooker, the lowest confining pressure is at the rotational poles. Thus, we may expect the outflow to be directed (at least initially) away from the star in two jets parallel and antiparallel to the rotation axis (cf. Konigl 1982).

This proposition is consistent with the measurements of Bally and Lada (1983) of the total energy of many bipolar flows. In bipolar flows, stellar winds of a few hundred km s^{-1} have typically been slowed down to a few tens of km s^{-1} by sweeping up ambient material. If we assume momentum conservation, the flow energy is then about one-tenth the total output of the original wind, and it is then interesting to note that bipolar flows, from B stars to T Tauri stars, have kinetic energies which are typically three orders of magnitude less than the gravitational binding energies of their stellar source, GM_*^2/R_*. Thus, the efficiency of conversion of stellar rotational energy to stellar wind energy need not be very high to account for the observed bipolar flows. Nevertheless, the observations stretch the viability of other mechanisms because, as Bally and Lada emphasize, the outward momentum in these flows far exceeds that carried by all the photons which have left the star. This indicates that the flow which swept up the bipolar lobes was *not* driven by photon luminosity. It is also worth noting that the velocities of the fastest Herbig-Haro objects (apparently associated with winds from T Tauri stars) are characteristic of escape velocities from stellar surfaces, rather than the much lower velocities that would be expected for winds originating from disks (see, e.g., C. A. Norman and R. E. Pudritz 1983).

Observations of the pre-main sequence stars themselves also lend support for the idea that winds terminate accretion. First, the youngest T Tauri stars apparently become visible at locations on the HR diagram consistent with the recent initiation of deuterium burning (c.f., Stahler 1983*a*). Second, observed pre-main sequence stars on convective tracks generally have low rotational velocities (Vogel and Kuhi 1981), suggesting that magnetic braking in an earlier phase of intense wind activity may have played a role before these stars could be revealed optically and develop the photospheric absorption lines that would allow their spectral classification.

F. Winds, Disks, and Massive Stars

If star formation is an accretion process from an essentially unlimited reservoir of matter, where the central protostar builds up from low masses to higher ones, why does not the onset of deuterium burning always occur at a low-mass stage where it can drive convection and produce a wind that will reverse the inflow? To produce a more massive star, note that the mass accretion rate $\dot{M}$, being proportional to a^3, is greater for hotter or more turbulent regions. This has a profound effect on protostellar development. If $\dot{M}_*$ is correspondingly high, then the radius R_* of the protostar will be larger at every mass than if $\dot{M}_*$ is low. The central temperature, proportional to M_*/R_*, will thus not achieve the critical value for deuterium ignition (about 10^6 K) until the central mass M_* is larger. If M_* reaches roughly 2 $M_\odot$ or more before deuterium burning commences, then the luminosity released will not exceed the value that a radiative envelope can carry stably. In the absence of an outer convection zone, there will not be the same dynamo mechanism to produce vigorous magnetic activity at the surface, and the limitation of the star's mass will have to rely on another process.

What might this process be? One possibility is that the amount of material which a high-mass star can accumulate is limited by that which can fall in directly, Eq. (25). This possibility would not solve the problem of its nebular disk becoming very massive. Another possibility (R. B. Larson and S. Starrfield 1971) is that radiation pressure acting on interstellar grains could be responsible for reversing the inflow, but the observational evidence (Bally and Lada 1983) suggests that the outflow associated with a newly born star of high mass has too much momentum to be due to this mechanism. A powerful stellar wind is apparently at work here too; perhaps weak instabilities act to bring the star from a strongly shearing state to one more nearly of uniform rotation, releasing energy in the process. For example, instabilities acting on a thermal time scale might suffice, since, for a high-mass star, the Kelvin-Helmholtz time is shorter than the accretion time. The latter fact is often invoked to explain why such stars already lie on the main sequence when they become optically visible (see, e.g., Appenzeller and Tscharnuter 1974). If strong magnetic fields are not involved in the redistribution of stellar angular momentum, efficient magnetic braking might not attend these winds, consistent with the rapid rotation of stars of early spectral type. It remains to be seen whether a strong enough wind could be powered this way.

If a strong wind does not form (until it is too late), the entire cloud core collapses, building up a very large and massive disk. It is interesting to speculate whether this is related to the origin of very large, rotating, flattened structures that have recently been discovered observationally, often in association with bipolar outflow sources. Examples of such sources and the sizes of their associated disks include NGC 2071, 4×10^{17} cm; S106, 2×10^{17} cm; Orion IRC2, 5×10^{16} cm; W30H, 2×10^{16} cm; L 1551, 7×10^{15} cm. These large structures, which typically do not contain much more mass than their central stars, cannot be understood on the basis of the scenario discussed so far, but we should keep in mind that the interpretation of the observations is complicated by the presence of vigorous winds and ionizing photons (in the case of early-type stars) which might ablate large quantities of mass from the so-called disk.

An issue of first importance is to determine whether the observed objects are true disks or merely gas distributions which happen to be flattened. If they are true disks, i.e., if they are rotationally supported with rotational velocities $V \sim (GM/r)^{1/2}$ much greater than the local sound speed c in the disk, then do they represent the original gas cloud, or have they formed from the collapse of an entire core? If they are merely flattened gas distributions, then are they supported laterally by magnetic tension, or have they arisen from the natural tendency of bipolar flows to evacuate cavities at the poles and leave matter behind on the equator? To decide among these possibilities, it is crucial for observers to establish the detailed mass distributions within such large disks and whether they are substantially centrifugally supported.

III. DISCUSSION

We begin this section with a few general comments on the implications of our views. First, in our development, the masses with which stars are born do not depend solely on processes that occur in the interstellar medium; they depend mainly on processes that occur in stars. In other words, the interstellar medium does not know about the masses that are required to produce the basic characteristics of stars (nuclear fusion, convective versus radiative envelopes, etc.); it is the stars that know these things. The mass scales about which the interstellar medium does seem to know (10^5 $M_\odot$ for complexes, 10^3 $M_\odot$ for clumps) are much too large to be typical of stars. If it is true that stars determine their own masses, then this puts the origin of the initial mass function in a considerably different light than currently envisaged by some workers; one cannot simply explain the initial mass function of stars as being due to the collapse of an appropriate assemblage of stellar-mass clouds (cf. Silk and Takahashi 1979). Fragmentation of molecular cloud cores dynamically into many pieces is a controversial possibility (see, e.g., Bodenheimer 1981; Silk 1982; Tohline 1982), but stellar mass objects can form from the collapse of clouds initially in equilibrium without invoking this hypothesis. Second, young stars are more active than evolved stars found in the same locations of the HR diagram (e.g., subgiants) because young stars have had much less time to adjust to the peculiarities of their rotational inheritance from the interstellar medium. The powerful winds that have long been invoked to sweep the primitive solar system clean may thus be an inevitable adjustment phase of every newly born star. Third, the process of the formation of high-mass stars may be distinguished from that of low-mass stars not only in the circumstances starting the collapse, but also in those which end it. Whether the mechanisms are fundamentally distinct, or merely are different manifestations of a continuous sequence remains to be established.

The initial conditions appropriate to the interstellar medium at the onset of collapse lead naturally to a variety of collapsed configurations. The parameters describing the cloud core, such as the effective sound speed a and the rotation rate Ω (and perhaps the left-over magnetic field as well), determine the accretion rate and the total amount of low angular momentum material that can fall directly onto the star. They may also affect the time at which the protostellar wind turns on, thereby controlling the total amount of material ultimately acquired by the protostar plus disk. This state of affairs leads to a multiplicity of outcomes, forcing us to take a wider view of protostars and disks than solely concentrating on the issue of the formation of the solar system.

From the basic model, disks of all masses and sizes can be produced. The next stage to be explored involves the evolutionary processes within nebular disks and the interactions of the newly formed protostars with their environment. Observations tell us that winds are dynamically important; hopefully, they will soon yield comparably useful information about the properties of disks near young stellar objects. Our theoretical understanding of the nebular evolution phase is quite incomplete; nevertheless, it is useful to discuss some of the scenarios possible within the context of our collapse model.

At the onset of gravitational instability, the cloud core begins to collapse at the center. As an expansion wave travels outward, more and more of the cloud is involved. The accretion rate into the center, $\dot{M} \sim a^3/G$, equals 10^{-5}

$M_\odot$ yr^{-1} for a = 0.35 km s^{-1}; thus, it takes typically 10^5 yr to form a 1 $M_\odot$ central object. At first, all of the matter falls onto the protostar, but, because the angular momentum of the original cloud increases outwards, material of progressively higher angular momentum rains down as the collapse proceeds. Soon, a disk forms, growing in radius and mass with time.

For low-mass stars, we have suggested that a stellar wind develops when deuterium burning drives a protostar completely convective. The mass of the protostar at which this occurs will usually not be the same as the mass of the protostar at the point when the disk begins to form. The period over which the disk is built up can be a significant fraction of the total accretion time, and therefore, either the disk may acquire an amount of mass comparable to or greater than that of the protostar, or it must process the equivalent amount through mechanisms that can transport angular momentum. A number of possible alternatives then present themselves depending on the efficiency of such tranport mechanisms. We discuss the disk evolution in terms of the three radii, the stellar radius R_*, the centrifugal radius R_{CF}, and the viscous radius R_v. We consider three cases, which do not exhaust the possibilities, but which provide useful examples.

1. $R_* > R_{CF}$ or R_v. All gas falls onto a single protostar and no disk forms at all. The mass accretion rate $\dot{M}$ is a function of the effective sound speed of the cloud core, and is greater for hotter clouds. The mass of the star is ultimately determined by when a stellar wind turns on.

2. $R_v >> R_{CF} > R_*$. A disk forms around the protostar and proceeds to spread due to the outward transport of angular momentum by viscous stresses. Material in the inner regions of the disk spirals inward to become part of the protostar. The mass of the disk is of order $(R_{CF}/R_v)^{1/2}\, M_* << M_*$, and the gaseous component therefore is not subject to gravitational instabilities. Examples of the formation of nebulae of this type have been calculated by Cassen and Summers (1983). The size of the disk is R_v unless other torques are also operative, in which case the disk may be larger. Two possibilities are magnetic torques due to, say, interaction with a protostellar field, and gravitational torques due to interaction with a triaxial protostar. In both cases, angular momentum must be extracted from the protostar itself, although it should be kept in mind that the protostar might continue to accrete mass and angular momentum from the cloud during the spreading of the disk. Because such a disk would possess relatively little mass, the masses of bodies formed by accumulation processes would be small compared to M_*. Thus given the example of our own solar system, one might suppose that disks like these eventually form planetary systems.

3. $R_{CF} >> R_v > R_*$. In this case, the disk grows by the accumulation of gas possessing ever greater specific angular momentum. Internal viscous torques are not strong enough to appreciably redistribute the angular momentum of the gas once it has become part of the disk, except perhaps that very close to the star. The mass of the protostar attains a limit given by Eq.

(25); the disk would grow to a comparable mass, at which time gravitational instabilities would come into play. If spiral density waves arise and can transport mass and angular momentum efficiently, then the disk could spread to radii appreciably greater than R_{CF}. The disk would be large, but its mass relative to the protostar's mass would be self-regulated by spiral instability. Its eventual fate might be to form many giant gaseous protoplanets (cf. Cameron 1978*c*) as the mass in the disk is spread over too large a range to be aggregated into another single body. On the other hand, if the transport by density waves cannot transport mass inwards and angular momentum outwards at a rate fast enough to keep up with the accretion flow from the interstellar cloud, a violent fragmentation instability may overtake the disk, resulting in the formation of a binary or a multiple star system. This instability, which would occur on a time scale of the rotation period of the disk, might well be the most common outcome, given the frequency of observed binary stars.

It seems that current theories of protostellar disks, although incomplete, have at least the potential for accounting for both planetary systems and binary or multiple star systems. (Of course, multiple star formation via triggering mechanisms, fission, or other schemes cannot be ruled out by disk theories.) It is clear that much work needs to be done to distinguish truly viable hypotheses. Some of the proposals listed above may appear to be possibilities only because we do not yet know enough to rule them out. We close this chapter by calling attention to several areas in which future efforts should be particularly rewarding.

1. A theory of protostellar structure that takes into account the accretion of rotating gas from both cloud and disk is needed to determine the timing of crucial events, such as the onset of nuclear burning or rotational instability, and the dependence of that timing on external parameters. Progress in this direction has been made by Mercer-Smith et al. (1984).
2. The stability of growing disks against thermal convection, in the presence of the accretion shock, needs to be investigated (as well as other hydrodynamic instabilities). If instability occurs, the effectiveness of the convectively driven turbulence for angular momentum transport must be determined, perhaps by methods such as those used by D. N. C. Lin and J. Papaloizou (1980).
3. The response of a growing disk to gravitational instability needs to be explored, both by the application of linear stability theory (see, e.g., C. C. Lin and Y. Y. Lau 1979), and the investigation of the nonlinear properties of density waves (see, e.g., Shu et al. 1984).
4. Hoyle's (1960) concept of magnetic protostar-disk coupling should be developed in the context of modern star formation theories.
5. A theory of mass outflow from young stellar objects that meets the growing body of observational constraints, and fits readily into an inclusive theory of star formation, needs to be developed.

6. The relationship between theoretical disk models and the very large flattened objects discussed in the observational literature should be examined carefully.

It is not hard to think of other important problems whose solution would enlighten our understanding of the formation of stars and planets. But we are led directly to those mentioned above by current models of the phenomenon, and, moreover, they seem tractable with the insights and analytic tools used to establish present theories.

OBSERVATIONAL EVIDENCE FOR DISKS AROUND YOUNG STARS

PAUL M. HARVEY
University of Texas at Austin

The detection of energetic bipolar outflows from a variety of young stellar objects has prompted a wealth of new observational data on the geometry of circumstellar clouds. The techniques used to probe the circumstellar environment include both observations that have resolved a disk-like structure or its effects and observations in a single spatial resolution element for which modeling suggests the presence of a small circumstellar disk. The best evidence for disks around young stars comes from the high angular resolution optical, infrared, and radio images of bipolar nebulosity requiring the presence of an axially symmetric circumstellar cloud. Observations in the next few years should make it possible to study the physical conditions and velocity structure in some of these clouds.

The idea that young stars at some stage in their formation should be surrounded by disks is standard in the folklore of astronomy. The effects of both angular momentum and magnetic fields should lead to an anisotropic collapse during the formation of a star. Current theories of formation of the solar system also require the presence of a large-scale disk in the protosolar nebula out of which the planetary system formed. In spite of these strong theoretical grounds for believing in the presence of circumstellar disks during star formation, it is interesting to note that there was essentially no discussion of observations of such structures in the first conference on "Protostars and Planets" (Gehrels 1978). It is the remarkable number of qualitatively new observations since that time which makes possible this review.

This work was triggered by the discovery of energetic bipolar outflows from young stellar objects (Snell et al. 1980). These observations and many others of similar objects at radio wavelengths (Bally and Lada 1983) and optical wavelengths (see, e.g., M. Cohen 1982) are generally taken as evidence for axial symmetry in the structure of circumstellar clouds. As discussed in Sec. II, such data constitute some of the best, albeit circumstantial, evidence for disks around young stars. But first, in Sec. I, I review a number of less direct observational methods for detecting disks. These methods are more controversial because they are often very model dependent; they may prove more valuable, in fact, as tools for the study of the detailed properties of disks which have been found by less controversial means.

In the following discussion it is assumed that the concept of circumstellar disk implies that the structure results from partial support against gravitational collapse in a plane, due to rotation or possibly to magnetic fields. Unfortunately, some geometries such as obscuring filaments or bars will mimic many of the observational properties of disks. In fact, one of the principal areas for further research will be to establish whether or not the disks which already have been proposed to explain many observations, are the result of rotational or magnetic support in an evolving protostellar cloud.

I. INDIRECT EVIDENCE FOR DISK STRUCTURES

High-resolution optical spectroscopy has been an effective tool for studying the temperature, density, and velocity structure of circumstellar material close to young stars. M. A. Smith et al. (1983) have argued that narrow emission and absorption line components in the spectra of pre-main sequence G stars can be best explained by circumstellar disks whose presence appears to be correlated with the rotational velocities of the stars. Because of the ages of these stars, however, they conclude that the disks are probably not primordial. Appenzeller (1983) has suggested that the asymmetry in the observability of blue- versus red-shifted forbidden lines in T Tauri stellar spectra can be explained by the presence of flattened clouds preferentially obscuring the red-shifted circumstellar regions. Both of these types of observations suffer from the serious handicap that the entire supposed disk structure is contained within one spatial resolution element of the observations. Therefore, it is not clear that geometries other than disks, when integrated over the line of sight and the plane of the sky, could not produce similar effects on line profiles and strengths. For instance, even the relatively simple problem of inflow or outflow implied by spectral line shapes of T Tauri stars is not clearly resolved (Ulrich 1976).

Continuum emission from a viscous accretion disk has been suggested as an explanation for excess emission in the optical spectral region for MWC 349 (Thompson et al. 1977). Thompson (1982) has also postulated such disk emission as a source of excess ionizing radiation to explain the presence of hydrogen recombination lines from stars otherwise not expected to produce observable H II regions. As in the discussion above of line shapes and intensities, the type of disk postulated to produce an optical or ultraviolet continuum is smaller than current angular resolution limits. Therefore, it is difficult to estimate how accurately the observations may be explained by such disk models rather than by some other geometry or unusual stellar photospheric property.

S. E. Strom et al. (1972) and H. A. Smith et al. (1982) have postulated disk-like circumstellar dust clouds to explain differences between the observed bolometric luminosity for young stars and that expected from de-reddening their optical photospheric emission. Both groups observed stars for which a stellar luminosity could be inferred from optical spectra and reddening corrections. In some cases these luminosities are a factor of ten greater than seen from ultraviolet, optical, and infrared photometry. They conclude that much of the luminosity is not being absorbed by local circumstellar dust clouds even though there is evidence that the substantial optical extinction is locally produced. Therefore, they argue that a roughly edge-on disk is producing the extinction but permitting much of the luminosity to escape perpendicular to the disk. Although persuasive, these arguments depend critically on three only partially substantiated assumptions: (1) the reddening is known; (2) the reddening is produced by dust close enough to the stars to be observable in emission in the infrared; and (3) the optical spectra give an accurate estimate of spectral type and class for stars whose photospheres are not completely understood.

Harvey et al. (1978) have suggested that nonspherically symmetric dust clouds are required to explain the properties of protostellar objects like AFGL 989. In particular, the simultaneous presence of high far-infrared optical depth together with visible/near-infrared emission can only be explained by a dense cloud that has relatively low optical depth along some directions, allowing the short wavelength radiation to escape. However, without other information on the shape of the cloud, this only argues against spherical symmetry, not necessarily for axial symmetry.

A final method which also involves deconvolution of processes within a single spatial resolution element is the inference of nonspherically symmetric dust cloud geometry from the observation of polarization in the infrared or optical spectral regions. Elsasser and Staude (1978) have proposed that scattering from compact, possibly disk-like dust clouds around protostellar objects may produce the relatively large degrees of polarization observed near many such objects. Although Dyck and Lonsdale (1981) have argued that these polarizations may also be produced by transmission through a cloud of aligned grains, there are several resolved objects in which the scattering mechanism is clearly evident (Tokunaga et al. 1981; Lacasse et al. 1981; Werner et al. 1983). Therefore, it seems likely that net polarization in some unresolved objects is produced by nonspherically symmetric dust distributions that may be disk shaped.

II. DIRECT OBSERVATION OF DISK-LIKE MORPHOLOGY

A. Low-Resolution Infrared and Radio Observations

The recent detection of high-velocity mass loss in collimated, bipolar flows from young stellar objects (Snell et al. 1980; R. D. Schwartz 1983*a;* Bally and Lada 1983) is one of the observations most suggestive of disk structure. In their initial analysis, Snell et al. proposed a model in which an accretion disk around a young star channeled a strong stellar wind into a collimated flow out along the axis of the disk. Disks of a variety of sizes and orientations have been invoked in all subsequent models for these flows. L. Hartmann and K. B. MacGregor (1982*a*) proposed that the flows are due to a centrifugally driven mass loss powered by the rotation of the central object coupled magnetically to the surrounding gas. This model requires the axis of rotation of the disk to be perpendicular to the plane of outflowing material; this prediction, however, may be inconsistent with some recent observations discussed below. Draine (1983) has proposed that a rotating protostar will produce magnetic stresses in the surrounding medium. These lead to the development of magnetic bubbles which expand most easily perpendicular to the plane of rotation of the star and circumstellar cloud, due both to the anisotropy of the magnetic stresses and the channeling effects of the presumed disk-like circumstellar cloud. Torbett (1984*a*) finds that bipolar outflow will result simply from anisotropic expansion from the accretion shock layer in a circumstellar accretion disk. In summary, all proposed scenarios for producing the spatially resolved bipolar outflows require an axially symmetric circumstellar cloud. Many of the models, in fact, require the effects of angular momentum and/or magnetic fields to power the outflows.

A variety of groups have attempted to observe these proposed circumstellar disk-like clouds rather than the effects of the supposed disks on the mass outflows. These observations have been motivated by the facts that: (1) disks with sizes comparable to the dimensions of the spatially resolved outflows (0.1–1 pc) may be required to explain the collimation of the outflows out to such distances (Torrelles et al. 1984); or (2) if there is a significant small-scale disk-like anisotropy due to rotation or magnetic fields, there would presumably be a larger-scale, but perhaps lower-contrast, anisotropy observable in the surrounding medium.

Observers of molecular lines have typically looked for disk-like structures in a line requiring reasonably high densities for excitation in order to avoid confusion from the extensive molecular cloud complexes surrounding most objects. Nonspherically symmetric clouds have been found associated with S 106 in NH_3 by Little et al. (1979*a*) and in ^{13}CO by Bally and Scoville (1982); with NGC 2071 by Bally (1982) in CS; with NGC 2071 and NGC 1333 by P. R. Schwartz et al. (1983*b*) in NH_3, CS, and $C^{18}O$; with HH 24–27 by N. Matthews and L. T. Little (1983) in NH_3; and with a variety of bipolar outflow sources in NH_3 by Torrelles et al. (1984). The circumstellar clouds found in these studies are typically elliptical in shape with the minor axis 50 to 75% of the major axis. In most (but not all) cases the disk plane is within 20° of perpendicular to the outflow direction. Individually, any one of these observations would be a relatively weak argument for the presence of axially symmetric structure. Taken together, however, they are quite suggestive of mildly flattened disk-like structures within the large-scale clouds surrounding many young stellar objects. Furthermore, observations of L 1551-IRS and the OMC-1 KL nebula (Kaifu et al. 1984; Hasegawa et al. 1984) present the most convincing single cases for such disks because of the relatively high, single-dish resolving power afforded by the Nobeyama Radio Observatory and because of the detection of significant rotational velocities.

Far-infrared ($\lambda \sim 100$ μm) observations have been used to study dust column densities and temperature distributions. Harvey et al. (1982) have reported on nonspherically symmetric dust structure in the cloud around S 106. The higher column density region of dust is in a disk or bar aligned with the elongated molecular cloud. A similar geometry was found in NGC 6334 IV by Harvey and Gatley (1983) although there are no other data for the object with which to compare the dust distribution. P. R. Schwartz et al. (1983*a*) find an elongated cloud in the far-infrared around S 140-IRS which they interpret as the result of asymmetric heating.

A major limitation of most of the above observations is the poor angular resolution of the telescopes together with the relatively low contrast between equator and poles of the supposed disks. Furthermore, clouds may be elliptically shaped for two entirely different reasons as illustrated by the different interpretations above for nonspherically symmetric far-infrared structure in S 140 and S 106. Clouds may be denser along an axis, leading to higher column densities and greater emission, or they may be less dense with less attenuation of heating radiation, higher temperatures, and greater emission. Therefore, temperature and column density information both are essential to interpret observations of elongated clouds, preferably with indicators sensitive to several different temperature and density regimes.

B. Well-Resolved Disk-Like Structure

The best observational evidence for disks around young stars comes from a variety of observations of bipolar, cometary, or jet-like nebulosity near young stellar objects at optical, infrared, and radio wavelengths. The prototype object of this class is R Mon (Zellner and Serkowski 1972). A number of other such optical nebulae have been discussed by Calvet and Cohen (1978), M. Cohen (1980,1982), Mundt (1983) and others. With a resolution of a few arcseconds and typical source distances of a few times 10^2 to 10^3 pc, the scale sizes of these disk-like clouds are on the order of 10^{-2} pc (2000 AU). Many are also associated with the larger-scale, elongated clouds discussed in the previous section.

In most of the well-studied biconical or cometary optical nebulae it has been established that part or all of the nebulosity is seen in reflection rather than emission and that the extinction to the nebulosity is much lower than that to the central illuminating object. This immediately implies a disk or, at least, a bar-like geometry for the circumstellar dust (see, e.g., M. Cohen and G. D. Schmidt 1981; Lacasse et al. 1981). This same phenomenology has been seen in more highly obscured objects in the near-infrared (1 to 4 μm). Werner et al. and Capps (1983) have observed the Orion KL nebula at 3.8 μm and find evidence for scattered near-infrared radiation from a highly obscured source of illumination not seen directly. Harvey and Wilking (1984) have observed a bipolar 2 to 3 μm nebula in NGC 6334 that is suggestive of an axially symmetric dust distribution around an obscured star; Harvey et al. (1984) find a possibly similar object near GGD 12-15.

Radio-continuum emission nebulosity has been observed in detail around the S 106 exciting star by Bally et al. (1983). A disk-like shadow on the emission region is clearly seen, implying a lack of ionized material. This is presumed to be due to the high density of circumstellar material in that region which prevents the ionization fronts from penetrating the disk.

High angular resolution observations of emission from, rather than extinction of a disk have proven extremely difficult. This is presumably due to the very high optical depths in the equatorial planes of most objects (Harvey and Wilking 1984) making the disk material too cool to emit at wavelengths (optical, near-infrared, or radio continuum) where high angular resolution has been available. However, Plambeck et al. (1982), using a millimeter wavelength interferometer with a resolution of 6″ have mapped an elongated cloud of SO emission in the Orion KL region. They interpret the major structure of this cloud as an "expanding doughnut" of high-density gas, rather than a rotating disk. Their velocity resolution, however, was $\Delta v \sim 4$ km s^{-1}, and the expected rotational velocity for a Keplerian disk of the observed mass and dimensions is only of order 2 km s^{-1}.

III. DISCUSSION

The best evidence for axially symmetric structures with a large contrast in density between equator and poles comes from the high angular resolution radio continuum, infrared, and optical observations discussed in Sec. II.B. On the other hand, the best evidence for rotational support of flattened disk-like clouds comes from the moderate angular resolution, high velocity resolution observations of molecular lines discussed in Sec. II.A, particularly results from the Nobeyama Telescope. Any one of these observations on its own constitutes marginal evidence for a circumstellar disk. Taken together, however, the quantity of circumstantial evidence and the fact that there is a great deal of overlap of lines of evidence for some objects suggests that circumstellar disks are almost certainly being observed in several cases. Table I lists some

TABLE I
Some Candidate Circumstellar Disk Objects

	References for Evidence[a]			Properties			
Name	Rotating, Flattened Molecular Cloud	Nonspherically Symmetric Extinction	Bipolar Outflow	Luminosity $L_\odot$	(Ref)[a]	Observed Scales Over Which Disk-like Structure Seen (pc)	Angular Velocity at Radius km s^{-1} pc^{-1} @ pc
L 1551-IRS 5	1	2	3	3.2×10^1	(4)	0.001–0.2	7 @ 0.05
OMC 1-KL Neb	5,6	7	8,9	10^5	(10)	0.01 –0.3	12 @ 0.1
NGC 2071-IRS	11		11	7.5×10^2	(12)	0.07 –0.5	1 @ 0.5
R Mon	13	14	13	8.6×10^2	(15)	0.01 –0.7	$\lesssim$3 @ 0.2
NGC 6334-V		16	17	6×10^4	(18)	0.02 –0.2	
S 106	19	20,21,22	23	4×10^4	(21)	0.006–1.0	~1 @ 0.5

[a]References: 1. Kaifu et al. 1984; 2. Strom et al. (unpublished); 3. Snell et al. 1980; 4. M. Cohen et al. 1984; 5. Plambeck et al. 1982; 6. Hasegawa et al. 1984; 7. Werner et al. 1983; 8. Erickson et al. 1982; 9. Olofsson et al. 1982; 10. Wynn-Williams et al. 1984; 11. Bally 1982; 12. Harvey et al. 1979*a*; 13. Cantó et al. 1981; 14. Zellner and Serkowski 1972; 15. Harvey et al. 1979*b*; 16. Harvey and Wilking 1984; 17. Fischer et al. 1982; 18. Harvey and Gatley 1983; 19. Bally and Scoville 1982; 20. Solf 1980; 21. Harvey et al. 1982; 22. Bally et al. 1983; 23. Solf and Carsenty 1982.

of these cases where there is an overlap of multiple lines of evidence. In a more speculative vein the facts that: (1) a large fraction of pre-main sequence stars may pass through a bipolar outflow stage (Bally and Lada 1983), and (2) all current theoretical models for these outflows require a circumstellar disk, suggest that pre-main sequence circumstellar disks may be ubiquitous.

Since theoretical arguments have long predicted that disks are a natural part of the stellar formation process, perhaps the most exciting aspect of the new observational results is the ability to provide quantitative measurements of disk properties. Table I also summarizes three of the most readily observable parameters of the listed objects. The range of luminosities shows that disks are seen over essentially as wide a range as that of observed, highly obscured young stars. The range of scale sizes is obviously limited on the low end by observational selection effects. The high end may indicate the scale at which significant anisotropy sets in during the collapse of a protostellar cloud. There is a tendency for the larger angular velocities to be associated with the clouds which have been observed with the spatially (rather than angularly) smallest beams. This is consistent, at least, with what would be expected if the range of initial angular velocities were rather small and the velocities increased as the clouds collapsed. In addition to these parameters, it is possible to make crude estimates of densities within and outside the disks based on molecular excitation at larger radii and extinction or H II properties at smaller radii. Although these estimates have uncertainties of at least a factor of 3 and perhaps as much as 10, they suggest that:

1. At radii of order 0.1 pc the molecular clouds have densities of 10^3 to 10^4 cm^{-3} and a density contrast between equator and poles of a factor of a few at most.
2. At radii of 10^{-3} to 10^{-2} pc the densities are probably of order 10^6 cm^{-3} in the equatorial planes and a factor of 10 to 100 less along the axes of the disks.

IV. FUTURE WORK

Clearly there is critical need for high angular resolution, high spectral resolution observations of circumstellar clouds to determine the density and velocity structure at the small radii where the disk structure seems most apparent. Such observations will not only confirm the presence of a disk but will provide measurements of the disk properties in a spatial regime where the axial symmetry of the cloud is strongly affecting both its appearance and, probably, the evolution of the circumstellar material. With the advent of high-sensitivity infrared spectrometers and millimeter-wavelength interferometers, the prospects are excellent for rapid progress in this area.

For objects which are not too deeply embedded in their own circumstellar material like many T Tauri stars, it may be more difficult to observe disks (if they exist) because of the smaller column densities. Careful observations and modeling of these kinds of objects, however, will aid in understanding the evolution of circumstellar disks because it is likely that some of these sources represent the late stages of evolution of previously obscured pre-main sequence stars surrounded by disks.

In general, much more sophisticated modeling of observed sources and of the bipolar outflow phenomenon is needed to understand and to take advantage of observations already existing. For instance, in the case of S 106 the morphology at all wavelengths gives the appearance of a circumstellar obscuring disk. Yet there are some details of the observations which are difficult to understand with a simple model. In particular, it is not clear why the radio continuum nebulosity is not visible close to the central star along the supposed disk axis where the density must be low enough to permit ionizing photons to escape to large radii (Bally et al. 1983). Likewise the 10 to 20 μm color temperature in S 106 shows a great deal of structure (Gehrz et al. 1982), little of which appears to result from a smooth, axially symmetric density distribution. In the case of the bipolar molecular outflow sources, there appears to be a wide range in the degree of collimation of the flows (Bally and Lada 1983) which may result from a range in some property of the supposed circumstellar disks such as the disk thickness-to-diameter ratio. However, without a better understanding of these objects, it is impossible to use the flow characteristics to determine the disk properties reliably.

In conclusion, disks and/or disk-like structures appear to have been observed around a number of young stellar objects in the last few years. Because this subject is so young observationally, progress in the study of the properties of the disks will be rapid over the next few years. Hopefully we may look forward to many more quantitative measurements which will permit improved constraints on models of star formation and perhaps even of planetary system formation.

DYNAMICAL AND CHEMICAL EVOLUTION OF THE PROTOPLANETARY NEBULA

G. E. MORFILL
Max-Planck-Institut für Physik und Astrophysik

W. TSCHARNUTER
Universität Wien

and

H. J. VÖLK
Max-Planck-Institut für Kernphysik

We review the currently available information on the dynamical collapse of rotating protostellar clouds, the formation and properties of viscous accretion disks obtained from such numerical experiments and the resulting constraints and consequences for the protoplanetary environment. Using this information we develop a transport theory for dust and gas phases and present results obtained from this theory. Possible modifications by intermittent turbulence are discussed, chemical fractionation effects are analyzed and the heterogeneity on small scales as well as the homogeneity on large scales of the primitive bodies in our solar system is examined within the framework of this theory. It is concluded that a turbulent protoplanetary nebula has a number of appealing consequences, especially from a dynamical (angular momentum transport) and from a chemical point of view. If the turbulence is intermittent, it may also provide the fastest means of growing planets from the solid dust component, an important point, because the giant planets at least must have formed before the gaseous component of the disk had dissipated.

I. INTRODUCTION

Ever since Nikolaus Kopernikus (1473–1543) removed the Earth from the center of the universe and Johannes Kepler (1571–1630) found the fundamental laws of the motion of the planets and their satellites, there has been a scientific case for a theory of solar system formation. However, efforts in this direction are hampered by two circumstances. First, no direct experiments can be made to analyze different aspects of the phenomenon as is normally done in physics; rather we can only collect all available pieces of information on the solar system and then try to assess their importance for the formation process. Second, the standard astrophysical approach would be to investigate an entire sample of different but similar objects observed in various regions of the sky as this usually allows classification according to common properties. These common properties would then be taken as characterizing the important physical processes whose range of consequences is bracketed by the observational variety of objects. In the end an evolutionary sequence might result.

Because our instruments are not yet advanced enough to observe other planetary systems (an IRAS-type satellite near VEGA would not have detected a trace of a planetary system around the Sun), this astronomical method is not directly available to us either. Much of the early thinking on planetary formation has therefore been concerned with *ad hoc* theories depending strongly on the personal bias of the author. This eclectic approach is visible up to the most recent times. In the mid-18th century Kant (1755) conceived his general picture, extended by Laplace (1796), which is commonly referred to as the Kant-Laplace cosmogony. Taken together, this theory considers planetary origin as a part of the formation of the Sun. Although this is a basic hypothesis, it has the advantage of reducing the number of *ad hoc* assumptions to a minimum. Such a scenario is, however, not very specific in its dynamic details. Yet it is surprising how modern the theory is in some aspects, for example in Kant's emphasis of the role of dust and of grain-grain coagulation (see Völk 1983*a*). In the form of a solar nebula the model has influenced much of cosmogonic thinking and is also the starting point of our considerations. Just because we believe that planetary systems around stars should be a rather common occurrence (without being able at this time to quantify this statement) and because stars are observed to form continuously in the Galaxy, it appears most reasonable to consider planetary formation as a part of star formation. The latter's rapidly growing observational basis then encompasses planet formation as well. Applying such a strategy specifically to the solar system requires the assumption that the interstellar medium has not changed drastically over the past 4.5×10^9 yr.

To be useful this concept must not assign too many special circumstances to stars with planets as opposed to those without. One special characteristic is rotation and angular momentum transport, whereby mass (in the central star) and angular momentum (in the planets) can be efficiently separated. In this context it is perhaps more a convenience rather than necessary to consider formation of single stars with planets as opposed to multiple star formation. In the solar case single star formation mainly avoids the awkward necessity to break up a multiple star system while keeping one member with an extended flat system of planets. The basic point is that it appears reasonable to assume that a significant redistribution of angular momentum can be provided by frictional transport as well as by a carefully tailored additional step of fragmentation. Whether this transport is due to turbulence in the gas, or due to magnetic fields is not clear. Presumably magnetic fields in protostars are maintained by turbulent dynamo effects in which case turbulent gas motions are always a part of the friction forces. Assuming then that the Sun originated as a single star, gravitational collapse together with turbulent viscosity is expected to lead to a protostar of the solar type.

Only a few preliminary collapse calculations have been performed, starting from gas densities $n_H = 10^4 \text{ cm}^{-3}$ characteristically observed in rotating diffuse interstellar clouds. None of these computations has as yet reached a stage where the central object is basically supported by static pressure balance rather than by centrifugal forces. Thus they serve more as a motivation for constructing rotating protostellar models rather than as a firm starting point for detailed studies of solar system origin. However, what makes these collapse studies very attractive from a practical point of view is their indication that protostellar evolution may, possibly over a wide range of parameters, proceed via an accretion disk. Although this conclusion entails the assumption of turbulent or magnetic viscosity as mentioned above, it allows the use of well-established hydrodynamic-disk models to extrapolate the numerical collapse calculations beyond their present status. This amounts to an approximate treatment which must be checked by a numerical calculation including global radiative transport.

Altogether then the concept of planetary formation in the framework of turbulent protostellar collapse, with angular momentum followed by a viscous accretion-disk period, appears very plausible. It corresponds to the present picture of star formation without making any *a priori* reference to planets and satellites. Very special events like the capture of a planet-type body by a star cannot be described in this context but whether such processes actually can operate is quite uncertain; the same is true for the probability of their occurrence. Therefore, we should be quite content if we find a way to form planets in protostellar accretion disks. This approach may lead to all sorts of planetary systems. In the absence of extended numerical parameter studies in protostellar collapse, we have to choose some of the model parameters such that our solar system can emerge. Whether we have made consistent choices will have to be investigated in future collapse studies.

Section II is devoted to discussion of available collapse calculations including detailed radiative energy transport. In Sec. III we discuss a simple accretion-disk model with parameters assumed appropriate for the solar system. In contrast to the Sun, at least the terrestrial planets and all the satellites contain a large complement of solid material; a central aspect, therefore, is the transport of trace constituents in the disk. Section IV discusses the processes, time scales, and approximations; it also covers a general transport theory. Constraints and limitations of the theory are discussed in Sec. V giving average characteristics and probabilities for the trace components which occur in the form of dust and gas. Assuming that large solid bodies form from this material, cosmochemical measurements constitute the empirical basis with which these material characteristics must be compared. Section VI contains some general results, and a few specific points are discussed in Sec. VII. Even though the accretion-disk is fairly successful in a general sense, significant questions remain unanswered. These are emphasized in Sec. VIII.

II. MODEL COLLAPSE CALCULATIONS INCLUDING TURBULENT VISCOSITY

Protostellar collapse with finite angular momentum is at least a two-dimensional phenomenon. Inclusion of realistic forms of radiative transport and frictional processes requires fully implicit numerical schemes which are at present restricted to one- and two-dimensional configurations. However, as far as radiative transfer is concerned, Boss (1984*b*) recently applied an implicit scheme for the Eddington approximation in the three-dimensional case, whereas the hydrodynamical part of the code is still an explicit formulation. Thus, as a first approximation, we discuss numerical models of rotating protostellar clouds with axial symmetry. Turbulent friction is included, which is expressed by assuming a simple form of kinematic viscosity

$$\nu = \frac{1}{3} \alpha c_s l \tag{1}$$

where c_s is the speed of sound, l the length scale of the largest turbulent eddies, e.g., the half-thickness of a turbulent disk, and α a free parameter < 1 in the case of subsonic turbulence. In all calculations discussed here α was assumed to be equal to 0.3. The set of equations that must be solved consists of the usual hydrodynamical equations (mass, momentum, energy) augmented by the moment equations (radiation energy, radiative flux) for radiative transport. The Eddington approximation has been used for closing the moment equations; no attempt has been made to determine the entire radiation field and the corresponding Eddington factors. Strictly speaking, the Eddington approximation applies only to optically thick layers where the radiation field is fairly isotropic (Tscharnuter and Winkler 1978; Tscharnuter 1985).

The first model sequence to be presented deals with a rapidly rotating fragment which, after a preceding collapse phase, achieved a stationary equilibrium without taking turbulent friction into account (Bodenheimer and

TABLE I
Initial Model Parameters

Parameter Case	A	B	C
Total mass, M ($M_\odot$)	3	3	1
Specific angular momentum, J/M ($cm^2\ s^{-1}$)	1.1×10^{21}	1.1×10^{20}	1.2×10^{21}
Density, ρ ($g\ cm^{-3}$)	10^{-20}	10^{-20}	1.4×10^{-18}
Radius, R (cm)	5.2×10^{17}	5.2×10^{17}	7×10^{16}
Temperature, T (K)	8.3	8.3	10
Angular velocity, Ω (s^{-1})	10^{-14}	10^{-15}	6.1×10^{-13}
E_{rot}/E_{grav}, β	1.2×10^{-2}	1.2×10^{-4}	0.32
Free-fall time t_{ff} (yr)	7×10^5	7×10^5	6×10^4
Mean molecular mass, μ (g)	4×10^{-24}	4×10^{-24}	4×10^{-24}

Tscharnuter 1979). The initial conditions chosen for this model sequence are listed in Table I (case C). The final equilibrium model is displayed in Fig. 1a. This axisymmetric configuration was shown to be unstable with respect to nonaxisymmetric perturbations; the outcome is a system of two fragments orbiting around each other (Rozyczka et al. 1980). Thus, studying effects of turbulent friction on such an unstable equilibrium model is a rather academic exercise. Nevertheless, it is an interesting question, whether frictional processes are always able to give rise to a single central body surrounded by a disklike nebula. Unexpectedly, the answer is no, as illlustrated in the three plots of Fig. 1.

Due to redistribution of angular momentum by turbulent friction, a velocity field directed towards the axis of rotation is built up. At first, the velocity vectors remain parallel to the equatorial (x,y) plane, because the equilibrium is pressure-supported in the z-direction (Fig. 1b). However, as a result of the increasing density in the vicinity of the axis of symmetry, gravitational forces there become dominant and collapse starts anew. At the same time, the shallow ring feature is more and more enhanced as it is shifted towards the center. The density of the ring increases due to geometrical engulfing of the center as well as to accumulation of material from the surrounding envelope. Eventually the ring itself becomes gravitationally unstable and collapses onto itself (Fig. 1c). Again, such a dynamical ring configuration will be highly unstable with respect to nonaxisymmetric perturbations. The final result will be a binary or multiple system of fragments. We may thus conclude that redistribution of angular momentum does not necessarily imply the formation of a single central object, unless frictional processes are already effectively operating in the collapsing cloud at a time when the hydrostatic core is going to form. If the removal of angular momentum from the central parts were to set in too late, the breakup of the core would be an almost unavoidable event.

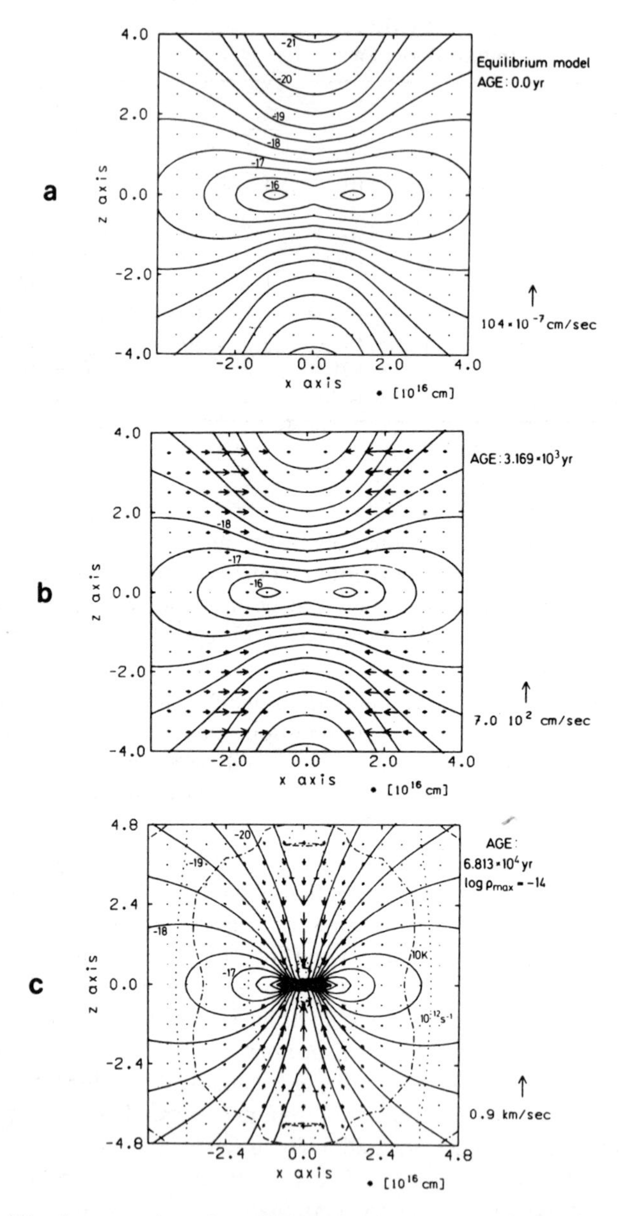

Fig. 1. Meridional cross sections of a rapidly rotating protostellar cloud as an illustration of the distortion of an equilibrium configuration, when turbulent friction is switched on (case C; see Table I). The vertical axis is the axis of rotation. Full lines represent the equidensity contours (numbers indicate logarithms of the density in g cm^{-3}). In plot c dotted lines are contours of constant angular velocity; the dash-dotted line displays the equitemperature contour of 10 K. The arrows denote the velocity field; the scale is given by the single arrow on the right-hand side of each drawing.

Another general feature of the collapse, which is not only pertinent to the case C we are dealing with, is indicated by Fig. 1c. The contour lines of constant angular velocity (dotted lines), which are parallel to the axis in the equilibrium model (Ω is constant on cylinders around the axis) are distorted in such a way that mass elements in the disk have an angular velocity different from the incoming material at the same distance from the axis. Physically, this situation is a consequence of both pressure effects in the disk and the time variation of the gravitational field generated by the changing density distribution in the disk. After having passed through the accretion shock, the incoming mass element will then find itself in a surrounding medium of different specific angular momentum. Consequently, exchange motions are present in the disk and a mixing process takes place. Because any mixing is more effective on smaller length scales, large-scale motions will decay to eventually set up a random velocity spectrum. It is thus conceivable that turbulence within a protostellar disk is triggered by the accretion rate.

We now turn to the discussion of two further model sequences (cases A and B; see Table I for the initial parameters) which differ only in the initial ratio $\beta = E_{rot}/E_{grav}$ (rotational energy/gravitational energy) due to a different initial rigid rotation rate; otherwise the initial configurations are identical. Although the initial β-values are quite low, 1.2×10^{-2} and 1.2×10^{-4} for A and B, respectively, it turned out (Tscharnuter 1978) that without assuming a sufficiently high transfer rate of angular momentum, the rotational energy of the quasi-hydrostatic core becomes far too large ($\beta_{core} \gtrsim 0.27$) to be able to survive as a single object. Due to our symmetry constraints, we only find a toroidal density wave to be excited which very much resembles the results of isothermal calculations (Bodenheimer and Tscharnuter 1979). In any case, it does not seem likely that a central object can accumulate. Test calculations with even smaller $\beta \simeq 10^{-5}$ exhibited the same tendency of ring formation; they also showed that the qualitative results are not changed if one starts out with a nonuniform initial density distribution as long as the density contrast between the outer parts of the fragment and its center does not exceed 1 to 2 orders of magnitude. Large variations of the initial angular velocity distribution, i.e., a large amount of differential rotation should not be present, because slowly rotating, dense fragments (as we consider them here) are expected to have already undergone considerable magnetic braking, making any noticeable differential rotation on the scale of a few tenths of a parsec quite unlikely. Thus, the idealized conditions of uniform density and uniform angular velocity as the simplest choice of the initial parameters are well justified (see, however, chapter by Cassen et al.).

From these results we draw the conclusion (if we want to model the origin of our solar system) that angular momentum should already be removed from the central regions of a protostellar cloud in the early phases of core formation. But, at this stage, the temperatures do not exceed 10^3 K and, accordingly, the degree of ionization should be too low for magnetic fields to become dynamically important. Likewise, high-amplitude density waves in flat disks that may exert gravitational torques on the disk material, which result in a net angular momentum transport, are presumably not dominant, because we are still dealing with a geometrically thick core. A thin disk in which density waves could be excited will therefore only take shape, if at all, much later in the evolution.

However, recent three-dimensional collapse calculations (Boss 1984*b*) indicate that quasi-static cores with densities up to 10^{-10} g cm^{-3} are surrounded by presolar nebulae which are barlike over the scale of the present solar system. It is argued that the amount of angular momentum transport in such a nebula is sufficiently large in order to avoid fission of the core. More conclusive calculations with much better numerical resolution of the crucial central parts of the protostellar cloud are desirable for a much more detailed investigation of this important problem.

From the above considerations, turbulent friction seems to be the most suitable mechanism for angular momentum transport in a collapsing protostellar cloud, although the source driving turbulence remains essentially unknown. On the basis of the simple assumption (Eq. 1) for the kinematic turbulent viscosity coefficient, the results of numerical calculations for case A are shown in the plots of Figs. 2 and 3. Qualitatively, case B exhibits the same evolutionary picture (Fig. 4). It is therefore convenient to discuss the common features of these results with sequence A.

As for case C, the viscosity parameter α was chosen to be 0.3. This amount of viscosity was sufficient to suppress the excitation of the toroidal density wave in the core. In addition, the excess angular momentum was quickly removed from the central parts of the cloud. An almost spherical innermost pressure-supported core region formed. As the temperature increased above 2000 K, H_2 dissociation set in and further collapse, analogous to the spherically symmetric models, led to almost stellar densities of about 10^{-4} g cm^{-3} in the center. The remainder of the core extending to about 4×10^{14} cm in the equatorial plane, and 1×10^{14} cm in the vertical z direction, can already be regarded as a precursor of a preplanetary nebula (Fig. 2a,b).

Contrary to what we know from spherical accretion which is a fairly stationary process, an excess pressure is built up resulting in an outward directed bulk motion of the material (Fig. 2c,d) due to viscous energy generation in the optically thick transition layer between the stellar core and the envelope (the length scale is several 10^{12} cm). Thus the accretion is inter-

Fig. 2. Time sequence of a collapsing, initially slowly rotating turbulent cloud (case A; see Table I for starting parameters). Displayed are meridional cross sections of selected models at various evolutionary phases during the first oscillation period (see text for further explanation). The surface of constant density = 10^{-13} g cm^{-3}, which roughly indicates the boundary of the optically thick core region, is separately plotted in addition to the respective cross section. Contour lines and symbols have the same meaning as Fig. 1.

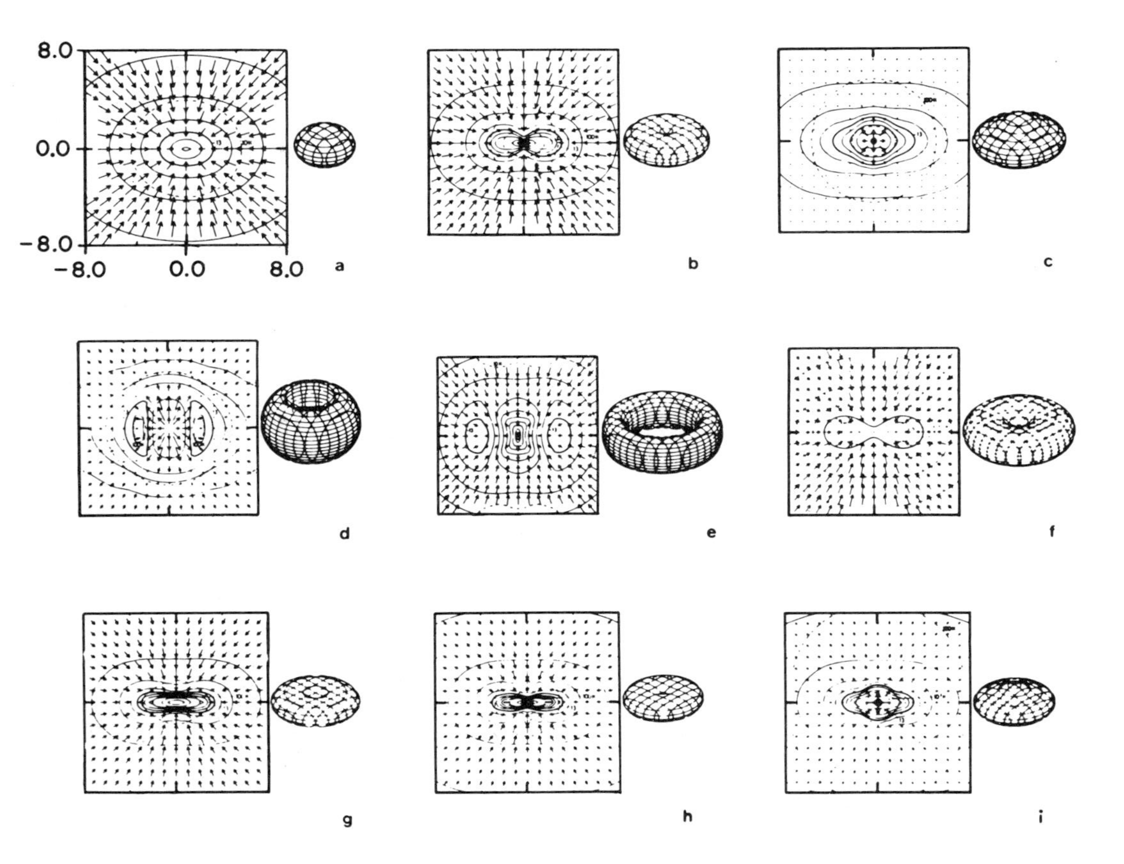
8.0
0.0
−8.0
−8.0
0.0
8.0
a
b
c
d
e
f
g
h
i

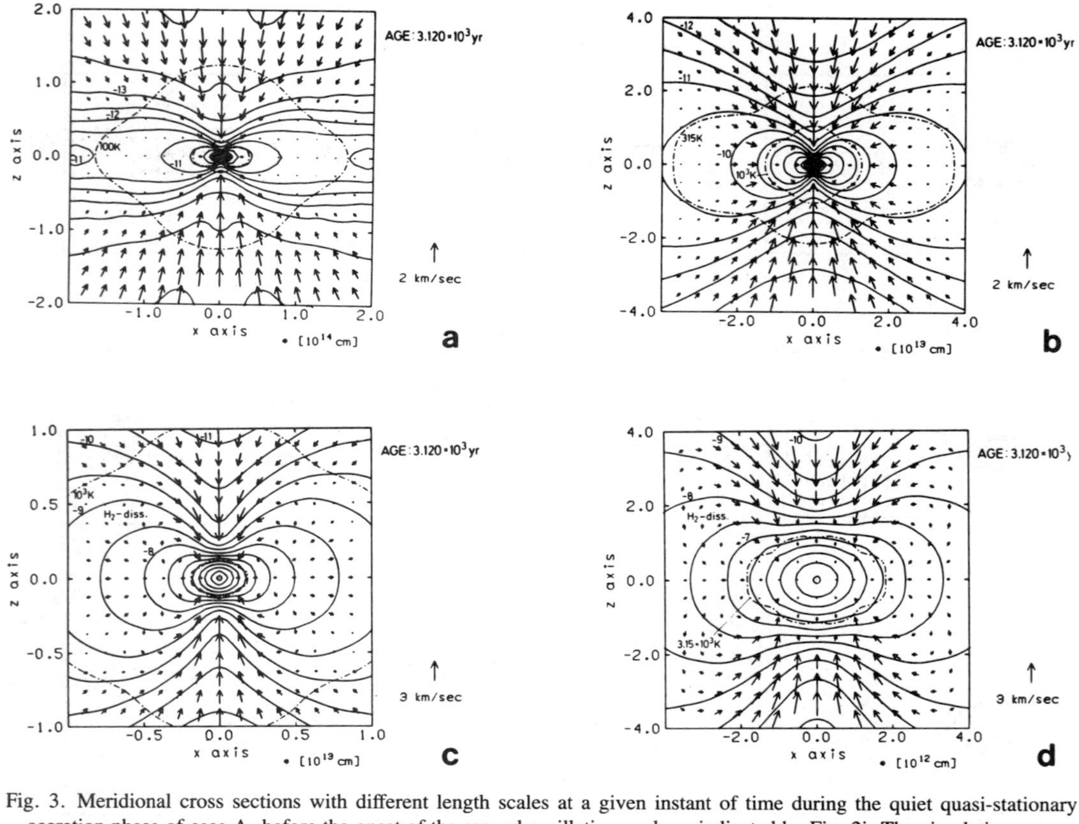

Fig. 3. Meridional cross sections with different length scales at a given instant of time during the quiet quasi-stationary accretion phase of case A, before the onset of the second oscillation cycle as indicated by Fig. 2i. The circulation pattern due to the dissociation process of H_2 molecules is well represented. Contour lines and symbols have the same meaning as in Fig. 1.

rupted and the evolution undergoes a nonstationary phase within which, according to the calculations, the central object is completely disrupted during the expansion phase. A thermal pulse moves outward, temporarily raising the temperature of the incoming material. Collapse starts afresh, when the excess heat has been radiated away (Fig. 2e–g), ending up with a second quasi-stationary accretion phase that looks quite similar to the first one (Fig. 2h). After a while, the next expansion phase begins (Fig. 2i); the time scale of one cycle is about 2000 yr.

Fig. 3a–d displays the structure of the inner parts of the disk, the transition region and the core, respectively, during a quiet accretion phase. The temperature and density contours near the axis (dash-dotted lines and full lines, respectively) indicate a polar hole from which radiation can escape more easily because of relatively low optical depths. Only in the innermost core downstream from the accretion shock, that forms polar caps, are the layers optically thick. There the temperature contours go along with the density contours (Fig. 3d). The basic physical reason for the low optical depths is the particular temperature variation of the opacity κ. When the temperature rises above 1700 K, most of the grains sublimate and κ drops several orders of magnitude from about unity to $10^{-4} - 10^{-5}$ cm^2 g^{-1}. At $1 - 2 \times 10^4$ K, depending on the density, κ is of the order of $10^4 - 10^5$ cm^2 g^{-1}. It is thus the opacity gap in the inner parts at temperatures, between 2000 and 6000 K, for example, and the density distribution near the axis that cause radiative energy to flow outward to produce the particular shape of the temperature contour lines.

In the transition region between core and disk on length scales of a few 10^{12} cm, a circulation pattern developed (Fig. 3c,d). This feature is a consequence of H_2 dissociation. In spherically symmetric protostars this process triggers a dynamical collapse, as the ratio γ of the specific heats becomes $<4/3$. However, nonzero angular momentum prevents matter in the equatorial plane from collapsing. Instead, the layers are convectively unstable. The circulation pattern, as obtained by the calculations, might be regarded as the lowest axisymmetric mode of a whole spectrum of convective patterns. In these transition layers, transport of angular momentum is very effective and, as a consequence, the viscous energy generation is enhanced. Because the layers are optically thick up to distances of a few 10^{13} cm, energy cannot be carried away by radiation at a sufficiently high rate. Thus, the thermal pressure steadily increases and drives the outward-directed bulk motions and oscillations mentioned above.

However, there are two main objections to these results. The first is based on some physical shortcomings of the numerical models. As emphasized above, the large-scale motions are triggered by the inefficiency of radiative energy transport, but small-scale convective motions will probably take over the energy redistribution. An excess pressure then is not likely to be built up. Unfortunately, a reliable theory of convection does not seem to be available. Another physical process that has not been included in the models is energy transfer associated with turbulence, i.e., turbulent heat conduction. It is hoped that this process can be incorporated into the energy balance equation in a rather straightforward fashion and will be sufficiently effective in lowering entropy gradients, wherever necessary, to such an extent that the importance of convection is drastically reduced.

The second objection addresses a technical point concerning the set up of the numerical scheme. With regard to numerical stability demands we were forced to choose an implicit difference scheme of a nonconservation type (Tscharnuter and Winkler 1979). Because it can be unclear whether such a code yields results that are not only globally but also locally accurate enough, one should be careful in discussing unexpected phenomena like nonstationary accretion phases in our case. It cannot be ruled out that small numerical inconsistencies between mass, momentum and energy fluxes across the discrete grid surfaces accumulate in such a way that spurious physical effects may result. This particular weakness of the models ought to be overcome by setting up a stable conservative two-dimensional scheme, which would basically be an appropriate extension of the one-dimensional scheme of Winkler et al. (preprint, 1983) designed for spherically symmetric protostellar collapse.

Sequence B starts out with smaller β (1.2×10^{-4}) but the same initial parameters as case A (see Table I). Fig. 4a-c illustrate again the same evolutionary picture as obtained for sequence A; collapse, formation of central condensation surrounded by a disk, the expansion and the outgoing thermal pulse. Again, a quasi-stationary accretion flow was not achieved. In order to proceed a step further by disregarding the dynamical oscillations which may be an artifact of missing physics or inadequate numerics, we introduced a semipermeable spherical inner boundary of radius 2.1×10^{12} cm. As a matter of fact, this boundary indicates roughly the region of the polar accretion shock, where the conversion of the infalling gas into heat and radiation takes place. If we are not interested in the detailed structure of the central stellar core, we now may allow the mass elements to cross this boundary only from the outside. There is then no other coupling than the increasing gravitational effects on the surrounding nebula due to continuous mass accumulation by the central object; in addition, the kinetic energy of the infalling material is assumed to be converted into radiation at that boundary analogous to the spherical models. These assumptions were sufficient to impede the oscillations and, after a period of very effective redistribution of angular momentum within the inner region of about 10^{14} cm (Fig. 5a), a quasi-stationary Keplerian, disk-like nebula around a rapidly rotating 0.5 $M_{\odot}$ protosun takes shape (Fig. 5b). The high temperatures indicated in Fig. 5a extending out to several 10^{13} cm do not imply that the region of planet formation is heated up in these phases. Mass elements that carry the specific angular momentum of the planets are still much farther outside, because the central mass contains only 0.1 $M_{\odot}$.

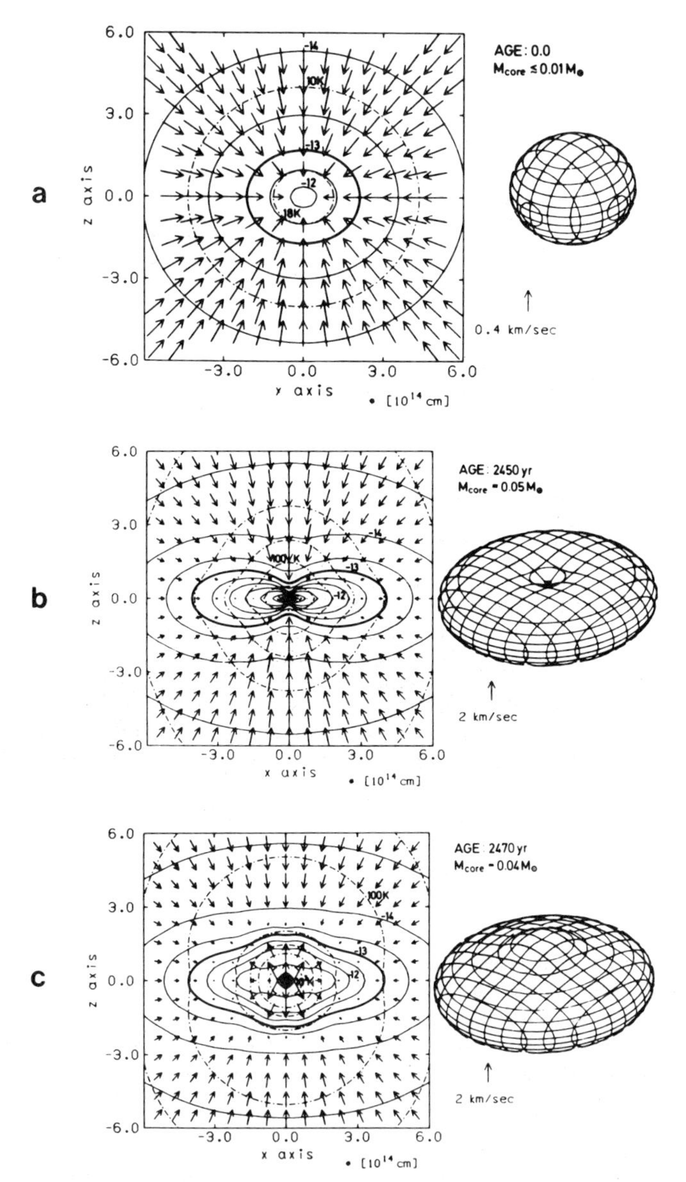

Fig. 4. Evolutionary sequence of an extremely slowly rotating protostellar cloud fragment (case B; see Table I for initial parameters). The close similarity with case A is obvious. Contour lines, the panoramic density surface and the symbols have the same meaning as in Figs. 2 and 1, respectively.

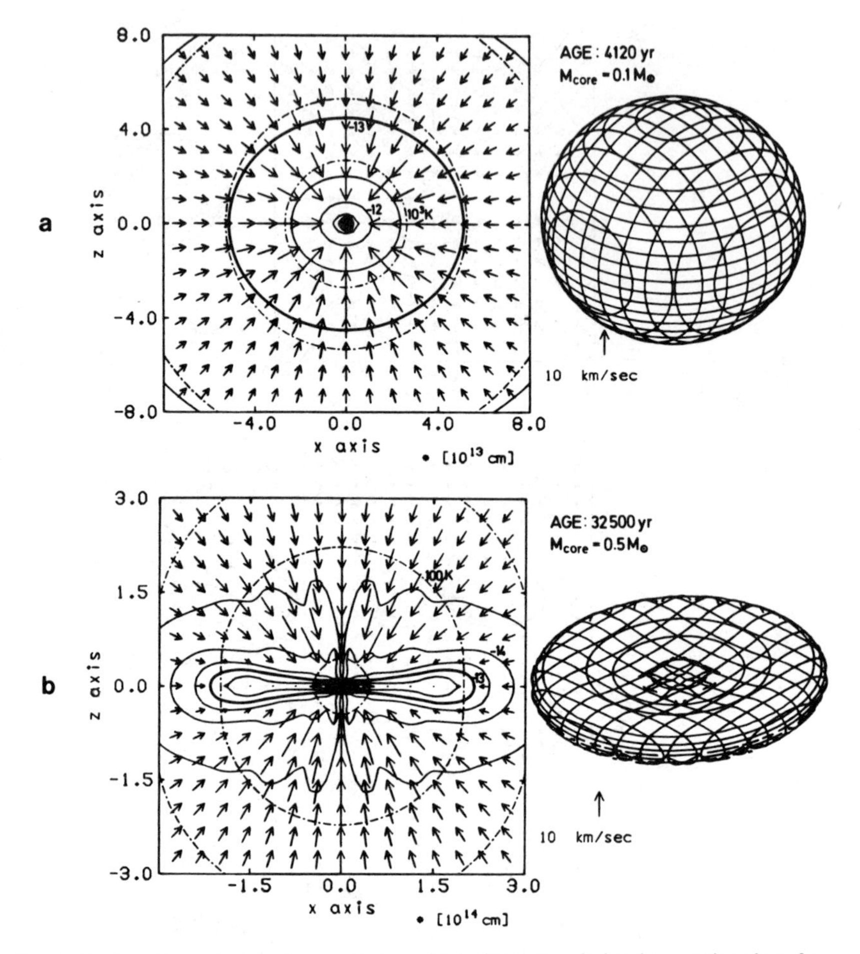

Fig. 5. Equidensity, equitemperature contours and velocity vectors during the accretion phase for model sequence B. Here the oscillations (cf. Fig. 2a-i) are artificially damped (see text for details). Plot (a) indicates a rapid redistribution phase of angular momentum, because the velocity field is almost radial and centrifugal forces thus have been weakened. Plot (b) is the final configuration; 0.5 $M_\odot$ have been accumulated in the central "sun" which is surrounded by a Keplerian nebula. Symbols have the same meaning as in Figs. 1 and 2, respectively.

Of course, the approximations introduced in order to make the problem at least partially tractable are indeed very crude. They should nevertheless provide a semi-quantitative evolutionary picture of the nebula. Now it may be regarded as a viscous accretion disk whose linear extension increases with time. In our case, this increase is due mainly to continuous angular momentum transfer in the collapsing region so that there are already mass elements of higher angular momentum which simply add to the disk. Only infall motions have been observed in the model.

A rough estimate for the minimum extension R_{min} of the disk may easily be obtained. If χ denotes the ratio of the centrifugal force and gravitation, we have for a mass element that is initially situated at the outer boundary in the equatorial plane.

$$\chi = \frac{R^3\Omega^2}{GM} \tag{2}$$

where R is the distance of the outer boundary, Ω the initial angular velocity, M the total mass and G the gravitational constant. The specific angular momentum J then is $R^2\Omega$. Assuming further that J is a conserved quantity for the mass element during the collapse, we have

$$\chi R = \frac{J^2}{GM} = \text{const.} \tag{3}$$

Thus the product $\chi\, R$ is an invariant and $\chi = 1$ at the point where centrifugal forces balance gravity. Then $R = R_{\mathrm{min}}$, i.e.

$$R_{\mathrm{min}} = \chi_{\mathrm{init}}\, R. \tag{4}$$

Putting in numbers for case B from Table I into Eq. (4), we get $R_{\mathrm{min}} = 1.9 \times 10^{14}$ cm for the minimum extension of the nebula. The amount of mass in the disk at a given instant in time depends on the mass flux from the freely falling envelope compared to the mass loss into the central star by viscous dissipation. The calculations had to be stopped, because the numerical resolution, being moderate anyway, became too poor as the nebula became more and more flattened. A further reason not to proceed was the fact that the core was estimated to have spun up again to the critical $\beta_{\mathrm{core}} \gtrsim 0.27$, thus indicating dynamical instability (see chapter by Durisen and Tohline). As Boss (1984*a*) pointed out, it could well be that angular momentum transfer by gravitational torques could keep β_{core} below 0.27 during the main accretion phase. Moreover, the total angular momentum of the central body became comparable to the angular momentum of the Keplerian disk that had formed around it. The structure of the disk could then change considerably if angular momentum were allowed to diffuse back into it. However, the total amount of angular momentum in the still almost freely falling extended envelope, from which the disk is continuously fed, is about three orders of magnitude larger than that of the core. From this, one should expect the disk to gain relatively more angular momentum during the accretion phase, when more material of higher angular momentum is added. A definite answer to this important problem can only be given if the interaction between the core and the nebula is treated in a consistent hydrodynamical way.

A fundamental question concerning the origin of our solar system is the temperature history of dust particles that represent the raw material for the planets and their satellites. Fig. 6 shows the temperature as a function of time

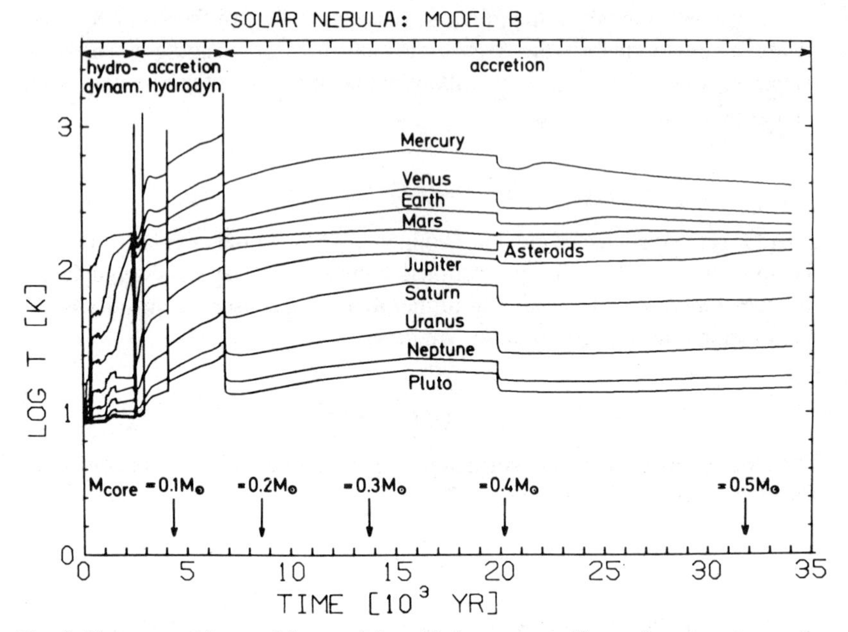

Fig. 6. Temperature history of dust particles with the same specific angular momentum as the various planets exhibit in our solar system. The accumulation time scale of the central "sun" is indicated by the sequence of increasing core masses M_{core}. The spikes in the temperature variations are probably artificial (see text for further discussion). The kinks occurring during the accretion phase (at about 2×10^4 yr) are due to a technical change in the numerical difference scheme for the radiative transport equation.

for particles carrying exactly the amount of specific angular momentum corresponding to the different planets in our solar system. The striking result is that the temperatures hardly rise beyond 10^3 K and that they remain fairly constant over long periods of time. The spikes are possibly due to spurious oscillations and accompanying release of heat. Because the temperature distribution depends mainly on the dust opacities which are reasonably well known and not so much on the initial conditions (at least if one starts out with masses of the order of one Jeans mass), the collapse models predict relatively low temperatures for the accumulation of the solids we observe in the form of meteorites today. A final point concerns the decelerated increase of the core mass, as demonstrated by Fig. 6. This result is in accordance with the basic nonhomology of the collapse flow.

III. THE PROTOPLANETARY DISK

In the previous section we showed that collapse calculations suggest the formation of a viscous protoplanetary disk as a possible evolutionary stage in the formation of single stars. The properties of this disk, as derived from these collapse calculations, are (see also Morfill 1985; Tscharnuter 1985) as follows:

1. Thin-disk approximations are valid;
2. The disk size is limited, and probably does not exceed by much the present size of the solar system (~40 AU);
3. The flow velocity is directed inward everywhere, even though angular momentum is transported outwards;
4. The mass accretion rate in the region is 10^{-6} $M_\odot$ yr^{-1}, and tends to decrease rather than increase as the system evolves;
5. The radial temperature gradient is typically $T \sim r^{-\delta}$, where $3/4 \leq \delta \leq 3/2$;
6. The temperature structure remains approximately constant in time;
7. The temperature at the outer edge of the disk is ~ 10 K;
8. The mass of the disk can be much less than the so-called minimum mass derived from augmented planetary masses;
9. The dominant heat source, during the early phases, is the conversion of kinetic energy of collapsing material into thermal energy;
10. In the simplest case (considered here) of the isolated collapse of one Jeans mass, collapse heating fades, once the system evolution has proceeded and the column of gas in the vicinity of the rotation axis is more and more depleted. The dominant heat source is then viscous dissipation of gravitational energy stored in the radial velocity shear of the gas;
11. The vertical disk structure during these later stages is not modified significantly by the residual ram pressure of the collapsing envelope;
12. The main supply of the matter during these stages comes from the radial outer edge of the disk. The radial outer edge is really a transition region between an inner, well defined, viscous disk and an outer collapsing envelope which has a large radial velocity component of O (km s^{-1}). For preliminary discussion see Tscharnuter (1985) and Morfill (1985);
13. Cooling is mainly by diffusive heat conduction and radiation perpendicular to the disk plane. In the planetary region, dust cooling is most important, and the opacities depend on the grain composition (e.g., ices, silicates, iron) as well as on the grain size distribution.

We shall base our description of the physics and chemistry in the early protoplanetary nebula on these results. As it is not possible to incorporate all the required physical and chemical processes into the collapse calculations, it is necessary to utilize a simpler method. For this purpose we construct a simplified accretion-disk model, which satisfies the necessary constraints of continuity of radial mass flow and perpendicular hydrostatic equilibrium, and takes its basic properties from those suggested by the collapse calculations.

The structure equations which are used to determine the properties of viscous accretion disks have been discussed and applied to various different physical situations for quite some time (see, e.g., Lüst 1952; Shakura and Sunyaev 1973; Lynden-Bell and Pringle 1974; D. N. C. Lin 1981*a;* Meyer and Meyer-Hofmeister 1982); for a recent review see Verbunt (1982). They are the basic hydrodynamic conservation laws, supplemented with formulations of essential physical processes, and are not repeated here. Instead, we restrict ourselves to a physical description and to model constraints derived from properties of our own solar system.

The physical process(es) leading to enhanced (relative to molecular viscosity induced) angular momentum transport in collapsing protostellar clouds are not yet clarified. One possibility, which has been discussed recently, is nonaxisymmetric perturbations of the rotating cloud core, leading to angular momentum transport through the excitation of density waves (see, e.g., chapter by Cassen et al.; Boss 1984*a;* R. B. Larson 1983; Michel 1984). Whether it works well in systems that are not strongly flattened is an open question as discussed in the previous section. The basic alternative, turbulence, is neither well described in its origin nor in its form as yet. A number of processes giving rise to turbulence have been suggested (see, e.g., Cameron 1976; Lin and Papaloizou 1980; Völk 1981; Sec. II above) for cool, molecular disks that result from gravitational collapse. The most attractive process in many ways is convective instability (D. N. C. Lin 1981*a*) The condition for this process to occur is the same as that discussed in stellar structure work. Effectively, it implies that the vertical optical depth of the disk should be larger than unity. Since the optical depth is determined by the grain chemistry and by the grain size distribution and number density, it is clear that the perpendicular optical depth in the disk is almost certainly modified by such processes as coagulation. As a result, we cannot exclude the possibility that certain regions of the protoplanetary nebula may at times not satisfy the condition for convective instability (Weidenschilling 1984). Apart from, e.g., nonuniform accretion (Völk 1983*b*) which can change the perpendicular temperature profile, this would be another factor that would make turbulence irregular. The possibility of intermittent turbulence as a general disk property was advocated originally by Völk (1982).

For the purposes of our discussion of trace constituent transport, we consider only the simplest viscous disk model, under the assumption that basic length scales, time scales and physical processes can be defined reasonably well in this simple model. It is important to establish a physical/chemical protoplanetary nebula model and to calculate observable properties. Only in this way can we hope to make comparisons with the wealth of cosmochemical data, and to utilize this important information to increase our knowledge regarding the early phases of solar system formation.

A. Simplified Disk Model

Following Morfill and Völk (1984), we present an internally consistent simplified accretion-disk model, which is in good accord with the convective-disk models discussed by, e.g., D. N. C. Lin (1981*a;* see also his chapter with J. Papaloizou). The basic ingredients are the continuity of radial mass flow, perpendicular hydrostatic equilibrium, and the assumption ν = constant, where ν is the kinematic coefficient of viscosity. For further information, see also Morfill (1985). We define ν in analogy with the molecular viscosity

$$\nu \equiv \tfrac{1}{3}\,\alpha\, c_s\, h = \text{const} \tag{5}$$

where $c_s(r)$ is the height-averaged sound speed, $h(r)$ is the disk scale height and α is a parameter ≤ 1 which we shall put equal to 1/3 for our numerical examples. This is a high value for α, corresponding to a turbulent energy density that is a fraction $\alpha^2 \approx 10^{-1}$ of the thermal energy density. Such α values appear to occur in the hot, fully-ionized disks around evolved compact stars. Adopting $\alpha = 1/3$ for the present case of neutral molecular protostellar disks constitutes an assumption that must be justified *a posteriori,* e.g., by comparing star formation time with observational limits.

Continuity of mass flow through the disk gives

$$\dot{M} = 4\,\pi\, rh\, \rho\, v_r = \text{const} \tag{6}$$

where $\rho\ (r)$ is the gas and dust density in the disk ($\rho \equiv \Sigma_k\, \rho_k$ for all components k) and v_r is the radial inward mean velocity resulting from the viscous angular momentum transport. All quantities are regarded as averaged over the disk thickness 2 h. This disregards the systematic convective interior structure of the disk as discussed by Urpin (1984). Solving the hydrodynamic structure equations for viscous disks yields approximately (see, e.g., Morfill 1985)

$$v_r \simeq -\frac{\nu}{r} \tag{7}$$

where the radial coordinate is considered positive outwards, as customary. The total mass of the disk, for a radial extent R is then

$$M = \dot{M}R^2/2\nu. \tag{8}$$

The disk thickness is determined from local vertical hydrostatic equilibrium, at a given radial distance r. Denoting the vertical axis as the z direction, we get

$$\frac{\mathrm{d}P}{\mathrm{d}z} = -\rho\frac{GM_c}{r^2}\frac{z}{r} \tag{9}$$

where we have assumed that the pressure gradient just balances the force due to gravity from the central protostar of mass M_c. The self gravity of the disk is neglected. In principle we should solve the energy equation also, in order to determine the pressure $P = \rho/\mu\, K_B T$, where μ is the mean molecular mass of the disk gas, K_B is Boltzmann's constant and T is the temperature. (P, ρ, T are functions of z, as well as r.) Assuming that the z dependence of T is much weaker than that of ρ (z), we can solve Eq. (9) and obtain the disk scale height

$$h^2 = \frac{3}{2}\sqrt{\frac{2R}{\gamma_G GM_c}}\frac{\dot{M}R^3}{\alpha M}\xi^{3/2} \tag{10}$$

where we have defined the dimensionless radial distance parameter

$$\xi \equiv \frac{r}{R}. \tag{11}$$

The ratio of specific heats γ_G enters, because the sound speed $c_S^2 \equiv \gamma_G K_B T/\mu$ is used. The disk properties derived from these simple considerations are summarized in Table II. They are consistent with the work by D. N. C. Lin (1981*a*).

B. Solar System Parameters

The input derived from the collapse calculations has already been discussed. It is difficult to infer further constraints about the Sun's protoplanetary disk from our contemporary solar system. A brief summary of the available additional information is given below; for a more thorough discussion, see Morfill and Völk (1984). The parameter values which we use are given in Table III. Constraints on the mass accretion rate may be derived from the time scales for star formation which are inferred from astronomical observations (Stahler 1983*a*) and from the spread in the formation ages inferred for meteorites (see, e.g., Kirsten 1978). The uncertainty in $\dot{M}$ is probably not greater than a factor 10.

The value for the size of the solar nebula is the extent of our present solar system out to the orbit of Neptune. This is only a minimum requirement; the nebula could have been larger, although the absence of planets beyond ~ 40 AU argues against this. Because obviously the protoplanetary disk must grow in size as the system evolves, and the protostellar mass M_c increases, we have

TABLE II
Disk Properties

Quantity	Radius Dependence
Disk thickness	$2h \propto r^{3/4}$
Gas and dust density	$\rho \propto r^{-3/4}$
Surface density	$\Sigma = \text{const}$
Gas temperature	$T \propto r^{-3/2}$
Coefficient of viscosity	$\nu = \text{const}$
Radial gas velocity	$v_r \propto r^{-1}$

TABLE III
Constraints for the Solar Nebula

Quantity	Magnitude
Mass accretion rate	$\dot{M} = 10^{-6}$ $M_\odot$ yr^{-1}
Radial size	$R(M_c = M_\odot) = 6 \times 10^{14}$ cm $R(M_c = 0.5\ M_\odot) = 3 \times 10^{14}$ cm
Temperature at outer edge	$T(r = R) = 10$ K
Mean molecular mass	$\mu = 4 \times 10^{-24}$ g
Ratio of specific heats	$\gamma_G = 1.4$

also specified an intermediate state. The uncertainties in these numbers are large, if judged only by our single observational input—the extent of the planetary region—but consistent with the collapse calculations (Sec. II).

The other constraints in Table III, we regard as *not* solar system specific; they should apply generally to accretion-disk systems formed in cool, dense interstellar clouds. The only problem here is the uncertainty in the temperature at the outer edge of the disk. A value of 10 K, adopted in Table III, should be regarded as a lower limit, corresponding to the precollapse temperature of the protostellar cloud. None of these considerations affect the principal results to be discussed here. They could be important, however, in questions of length scaling, boundary locations, time scales, etc. For this reason, results will be presented, wherever possible, in such a way that such scaling problems are minimized.

In Table IV we list some disk parameters which provide our best guess for a protosolar nebula (standard model) and some deviations from this if some of the key input parameters are changed. We see that increasing the temperature at the outer disk boundary results in a thicker, but less massive

TABLE IV
Typical Protoplanetary Disk Properties

Quantity	Standard model[a] (table III, $M_c = M_\odot$)	$T(r = R) = 50$ K	$R = 1.2 \times 10^{15}$ cm
$c_s(r = R)$	2.2×10^4 cm s^{-1}	4.9×10^4 cm s^{-1}	2.2×10^4 cm s^{-1}
H	2.4×10^{13} cm	5.3×10^{13} cm	6.7×10^{13} cm
$\rho(r = R)$	4.1×10^{-12} g cm^{-3}	3.6×10^{-13} g cm^{-3}	5.1×10^{-13} g cm^{-3}
Σ	192 g cm^{-2}	38 g cm^{-2}	68 g cm^{-2}
M	1.1×10^{-1} $M_\odot$	2.1×10^{-2} $M_\odot$	1.5×10^{-1} $M_\odot$
ν	5.2×10^{16} cm^2 s^{-1}	2.6×10^{17} cm^2 s^{-1}	1.5×10^{17} cm^2 s^{-1}

[a]The standard model differs from Morfill and Völk (1984) only through the choice of $\alpha = 1/3$ rather than 0.3.

disk. Because the vertical optical depth of the disk is proportional to Σ the vertical column density, this will ultimately make radiative cooling easier. In this way, we may expect the system to self-adjust, and there may be limits to the deviations which can be achieved. Increasing the disk size substantially becomes critical because the total disk mass then approaches 1 $M_{\odot}$. The collapse calculations indicate that low-mass disks are formed. Their structure is determined to a large extent by a dominant central mass M_c and not by self gravity; this provides a natural constraint.

C. Cooling Rates in Viscous Disks

Having fixed a disk geometry, by whatever solar system information is available, it is of course possible to extract information that has not been put in explicitly, and to see whether this is compatible with other available solar system information. One must ensure that independent measurements are used and, as usual, the decision whether or not the information derived is of cosmogonic importance is somewhat subjective. One such information is contained in the cooling rates of solid matter in the nebula, which we believe harbors an important clue about the early protoplanetary nebula. Measurements (Paque and Stolper 1983) indicate that refractory inclusions in the Allende meteorite have cooling rates in the range 1 to 20 K hr^{-1} at just below their condensation temperature. This temperature is ~ 1500 K, depending on surrounding gas pressure. Because the free space cooling rate is much faster, this immediately tells us something about the conditions in the early protoplanetary nebula at the indicated temperatures, assuming that these observations are cosmogonically significant.

Radiative cooling of a volume element of gas and dust at temperature T located inside the disk, with a residual column density Σ_R to the disk surface, gives a rate decrease in temperature (see Morfill 1983*a*)

$$\frac{\mathrm{d}T}{\mathrm{d}t}\bigg|_{\mathrm{rad}} \approx -\frac{c a_R T^5}{3 c_s^2 K_d \Sigma_R^2} \tag{12}$$

where $a_R = 7.56 \times 10^{-15}$ is the radiation constant, c is the speed of light and K_d the opacity. Taking an opacity law $K_d = 2 \times 10^{-6} T^2$, as appropriate for a mixture of silicates and metal grains (DeCampli and Cameron 1979) yields

$$\frac{\mathrm{d}T}{\mathrm{d}t} \approx -0.78 \times 10^{-6} \left(\frac{T}{\Sigma_R}\right)^2 \tag{13}$$

which for a temperature of 1500 K and our standard disk yields cooling rates larger than or equal to ~ 2 K hr^{-1}. The slowest cooling rate is for the largest value of $\Sigma_R = 1/2\Sigma$ in the equatorial plane. The radiative cooling rate assumes that a volume element of gas and dust, originally at temperature T can be transported rapidly into a cooler region of the nebula, and that the gas adjusts quickly to the new surroundings, leaving the hot grains to cool radiatively with the rate (Eq. 13), or else that at some point heating of the nebula is turned off rapidly.

One way in which this can be achieved (if we exclude rapid time variations of the whole disk) is by turbulent convection of a hot volume element into a cooler region. If this convection allows the volume element of gas to cool off faster than the rate (Eq. 13)—basically by adiabatic cooling as the volume element seeks pressure equilibrium in its new cooler surroundings, then Eq. (13) is the slower and hence the determining cooling rate for the embedded dust grains, that have been close to their condensation temperature.

Our disk model yields a temperature variation with radius, $T = T_o\, \xi^{-3/2}$, and the turbulent convection speed is αc_s. The rate of change of temperature of a given volume element of gas and dust, assuming instant adjustment to the height-averaged local values for outward, radial transport is

$$\frac{\mathrm{d}T}{\mathrm{d}t}\Big|_{1500\mathrm{K}} \approx -0.1 \text{ K hr}^{-1}. \tag{14}$$

For vertical transport, assuming a temperature gradient length scale $= h$, we have, of course, the same convection velocity αc_s, but the cooling rate is changed by the ratio $\frac{\mathrm{d}r}{\mathrm{d}h}\Big|_{1500\mathrm{K}}$. For our standard disk model this becomes

$$\frac{\mathrm{d}r}{\mathrm{d}h} = \frac{4R}{3H}\left(\frac{T(R)}{T}\right)^{1/6} \tag{14a}$$

and substituting values, we obtain for perpendicular transport away from the equatorial plane

$$\frac{\mathrm{d}T}{\mathrm{d}t}\Big|_{1500\mathrm{K}} \approx -1 \text{ K hr}^{-1}. \tag{15}$$

The convection pattern obtained in the relevant cloud regions as a consequence of H_2 dissociation (see Fig. 3) yields similar cooling rates as those given in Eq. (14). Other possibilities for obtaining cooling rates, at perhaps a higher rate, may be associated with the transition of dust particles through the accretion shock. For some information on the relevant physical processes see Wood (1984). In conclusion, viscous disk models do appear to give dust cooling rates compatible with observational requirements (Paque and Stolper 1983) although they may be a little on the low side.

IV. TRANSPORT OF DUST AND GAS IN TURBULENT DISKS

A. Processes Included, Time-Scales, Approximations

The turbulent disk serves as the substrate in which the transport of dust grains and, possibly, their gas must be considered. We discuss here the various physical processes that determine this transport. They will subsequently appear in simplified forms as part of a general transport theory for trace constituents. The physical processes we shall consider are:

1. Turbulent diffusion of dust grains, grain conglomerates, and grain gas in the disk;
2. Systematic frictional drifts of grains relative to the gas in the given gravity field of the central object, i.e., radial drifts and vertical sedimentation;
3. Evaporation and condensation of grains and gas, respectively, in the mean temperature field of the disk;
4. Coagulation of dust particles in inelastic collisions.

We disregard here the accretion of dust particles onto large planetesimal bodies formed at some earlier epoch. Coagulation is considered only in the framework of a monodispersive grain-size distribution and is treated independently of other transport processes. These are practical approximations which in addition involve slightly more subtle assumptions such as taking frictional forces on grains to be only due to the height-averaged density and velocity in the gas, or ignoring vertical temperature gradients for the condensation of gas in the disk. Possibly more basic is the assumption of a stationary level of turbulence in the disk and, generally, the disregard of slow evolutionary changes of disk properties. In this approximation, it appears as if the formation of planetesimals in a dense subdisk of grains cannot occur, because dust sedimentation is severely impeded by strong random gas motions. On the other hand, the gravitational fragmentation of the gaseous disk is not considered as it is quite implausible for the low-mass evolving disk configurations suggested by the protostellar collapse calculations regarded as appropriate for the solar system. As a consequence, the description stops at a stage of evolution where only the possible building blocks of planetesimals are determined in their properties. Formation of planetesimals themselves, their subsequent accumulation to large solid bodies like the terrestrial planets or the solid cores of giant gas planets, as well as the acquisition of gas envelopes by giant planets are therefore not discussed. This may in fact not be so critical because a rather plausible theory for each of these processes *per se* exists. We shall return to them in Sec. IX.

We believe that the most important aspect for comparison with cosmochemical evidence from meteorites and solid planets is the characterization of the building blocks of planetesimals and this is the basic aim of our accretion-disk analysis. Let us add a note of caution, however. It is quite clear that the following analysis is only a starting point. It shows that such an approach is feasible to begin with, provided that simplified descriptions of the relevant processes are used. At the same time, such simplifications may limit the scope of the theory. Therefore it is important to discuss in some detail approximate forms for the physical processes, listed above, which will appear in the model. In a certain sense they even define the model.

1. Turbulent Diffusion of Dust Grains. Turbulent gas motions induce random motions of embedded dust grains which have a frictional coupling with the gas. These random motions imply spatial diffusion (see, e.g., Taylor 1921) in a frame of reference moving with the average grain velocity relative to the center of the disk. The diffusion coefficient ν_k for the grain component k is $\nu_k = \nu \cdot \langle \delta v_k^2 \rangle^{\frac{1}{2}} / \langle \delta v^2 \rangle^{\frac{1}{2}}$, where ν is the turbulent viscosity of the gas (cf. Eq. 1) and $\langle \delta v_k^2 \rangle^{\frac{1}{2}}$ is the rms-induced grain velocity (see, e.g., Völk et al. 1980) in the gas with rms turbulent velocity $\langle \delta\, v^2 \rangle^{\frac{1}{2}} \equiv \alpha\, c_s$. For sufficiently small grains, ν_k approaches its maximum possible value ν, whereas very large (conglomerate) grains have $\nu_k \rightarrow 0$, effectively decoupling from the gas. For our standard disk model ν_k/ν decreases from unity with increasing grain size a as $(a/a_c)^{-1}$ beyond the critical size $a_c \approx 37$ cm. The value is independent of position in the disk. This critical size is quite large. Growth of the presolar submicron-sized grains by coagulation yields, even with optimistic assumptions, final radii much smaller than a_c. Thus, for the main arguments we shall assume $\nu_k/\nu \approx 1$, independent of grain size. Trace gases like grain gas follow exactly the gas. Therefore they diffuse with coefficient ν in a frame moving with the average gas velocity. For our standard disk model (see Table IV) a typical radial diffusion time scale measured in yr at radius $r = 1$ AU is then

$$t_{\text{diff}} \approx \frac{T^2}{\nu} \approx 91 r^2. \tag{16}$$

Thus, at least in the inner parts of the disk, the diffusion time is short compared to the mass accretion time of about 10^6 yr.

2. Radial Drifts and Vertical Sedimentation. In the absence of turbulence, dust grains have orbits that deviate from the streamlines of the gas around the disk center, because the grains are not pressure supported. On the other hand, they are subject to gravity from the central object and the disk itself, and they are coupled to the gas by friction. This leads to systematic motions relative to the gas which determine the average grain velocity. It is in this frame of reference that grain transport is purely diffusive.

For the low-mass disks considered here we neglect the gravity field of the disk. Therefore, to lowest order the grain orbits are Keplerian and, in projection onto the disk plane, circular due to gas friction. However, the particles tend both to sediment along the vertical z direction towards the disk midplane, and to drift radially into the center. The latter effect is due to the angular momentum loss to the more slowly rotating pressure-supported gas. In order to obtain an average transport equation of the dust component, we assume that both the sedimentation speed and the radial-drift velocity are given by the local terminal velocities in terms of the average disk parameters, irrespective of the history of the particle's trajectory. This presupposes that the frictional coupling time $1/\tau_f = 3\, \rho\, c_s/(2 \sqrt{\pi}\, \rho_s\, a)$ between dust and gas is short compared to all other time scales in the system. Basically it again implies small particles, similar in quantitative terms to the earlier condition for $\nu_k/\nu \approx 1$. In the expression for τ_f, ρ_s denotes the material density of a grain.

Under these simplifying circumstances the mean radial grain velocity v_{rs} is given by $v_{rs} = v_r + (\tau_f/\rho)\partial P/\partial r$, The mean sedimentation speed is $v_{zs} = g_z\tau_f = - GM_c\, z\, \tau_f/r^3$ (Whipple 1972; Adachi et al. 1976; Weidenschilling 1977*a*; Morfill 1985). In order to obtain simple moment equations in the size variable *a*, we will in addition assume that the grain radii *a* are rather narrowly distributed about a mean value $\langle a \rangle$ over the important regions of the disk.

3. Evaporation and Condensation of Grains. The simple disk model implies a considerable radial variation of the mean gas temperature from T $(r{=}R)$ = 10 K at the outer edge to much larger values in the planetary region. Because the central protosolar mass M_c grows with time as the nebula evolves, we must define "planetary region" more carefully. Assuming that a sufficiently large body will conserve its specific angular momentum once it has been formed, its orbital radius scales as $1/M_c$.This implies that our standard model, with M_c = 0.5 $M_\odot$, yields nebular temperatures of 169 K at Mars, 316 K at Earth, and 1313 K at the orbit of Mercury. The standard model, with M_c = $M_\odot$, yields nebular temperatures of 213 K at Jupiter and 1250 at Mars. A model that incorporates the opacity change in the grains at 150 K (where the ice mantle sublimes) merges the two standard models in this temperature range by introducing a radial band where the temperature gradient is small (see, e.g., D. N. C. Lin 1981*a*). In this sense, the 1 $M_\odot$ model appears more appropriate for the outer solar system (beyond Jupiter), whereas the 0.5 $M_\odot$ model should be more suitable for the terrestrial planet region (inside the orbit of Mars).

In any case, the above discussion shows clearly that cold grains entering the disk will be thermally modified long before they reach the central object, irrespective of where the latter's boundary is precisely located. During its random walk over the disk, a grain therefore will initially lose some of its volatiles, depending on how long it stayed in a region interior to the sublimation point of these constituents. Later it may acquire a layer of condensates, chemically and mineralogically transformed due to gas-phase reactions of the evaporated material and differential cooling. It may even go through such stages repeatedly because its radial propagation is of a stochastic nature. Interior to the mean evaporation point of the most refractory component, one expects the grain mass density to decrease sharply to zero because in fact the grains evaporate altogether. Just outside this zone, only refractory particles exist in a gaseous environment that contains the gas of almost all condensible molecules. As long as the dust is still well mixed with the gas, as assumed under point 2 (Sec. IV.A.2) this gas is entirely dominated in abundance by the cosmically dominant H and He. This dust evolutionary sequence is depicted schematically in Fig. 7.

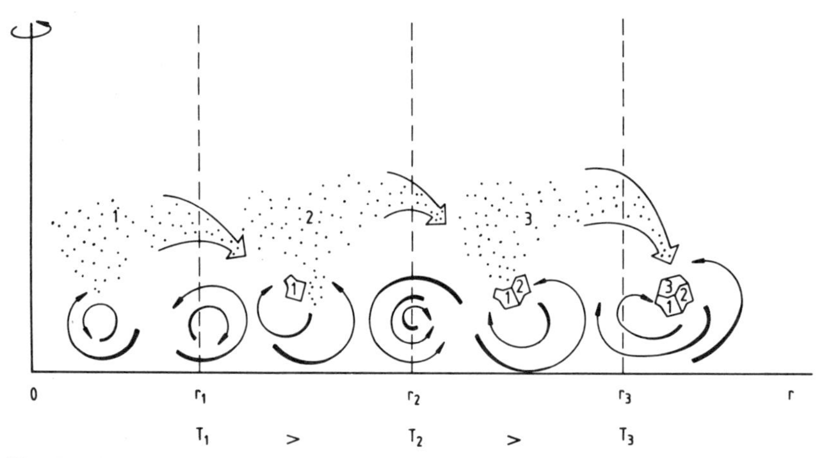

Fig. 7. Schematic diagram of the successive evaporation of a composite dust grain which is transported into the hot interior of the turbulent protoplanetary cloud. The evaporated gas disperses and may condense onto solid particles in the cooler regions.

For an analytic discussion, we make a number of technical simplifications none of which is essential or even necessary for a numerical computation. We assume instantaneous evaporation of certain grain constituents, independently of grain size, shape, or other constituents, as soon as the average ambient temperature exceeds the equilibrium evaporation point. Thus, for example, inside the innermost evaporation boundary, all grains are evaporated and only gas exists. Instead of essentially infinitely many, only one volatile and one refractory boundary are explicitly considered. Outside the boundaries the corresponding gas recondenses with a uniform rate per unit area and time. Condensation centers are the preexisting grains only; their size is uniform.

4. Coagulation of Dust Particles. Interstellar dust grains enter the accretion disk probably with an outer layer of condensed ices acquired during their residence in a supposedly cold, dense interstellar cloud from which the protostellar collapse originated. The sizes of these grains are still very small (in the micron range) simply because coagulating collisions are rare events even in dense clouds (Morfill et al. 1978; Völk et al. 1980). Within the protostellar disk, on the other hand, grain-grain collision times are short enough that rapid growth of sizeable conglomerates is possible. Thus coagulation is an essential process. It is a nonlinear effect and therefore it is complicated to evaluate it self-consistently within the framework of a transport theory. For simplicity we shall not only assume that particles coagulate independently of their thermodynamic time-history as mentioned above, but also that grain sizes are all equal and determined by the mean time of residence in the disk. Thus, only a mean type of particle is considered. Although this allows the calculation of mean drift velocities and sedimentation characteristics, we do not determine the fluctuations around the mean. However, work on these questions is in progress. A mean grain size over the main parts of the disk appears to be an acceptable first approximation (Morfill and Völk 1984). Using unit sticking efficiency (giving maximum growth) and a Kolmogorov-type turbulence spectrum with zero inner scale (see, however, Weidenschilling 1984), the mean grain size $\langle a \rangle$ turns out to be roughly 10 cm. This is an upper limit, and thus $\langle a \rangle \lesssim a_c$.

B. Transport Theory

For trace components k with a systematic velocity **v**k relative to the central mass, in a nonrotating frame, the general equation of continuity is given by

$$\frac{\partial}{\partial t}(\rho C_k) + \text{div}\,(\rho \mathbf{v}_k C_k) + \text{div}\,(\mathbf{j}_k) = -\left(\frac{\partial(\rho C_k)}{\partial t}\right)_{\text{st}} + Q - L. \quad (17)$$

Here $C_k \equiv \rho_k/\rho$ is the concentration based on the overall mass density ρ and

$$\mathbf{j}_k = -\rho\, \nu_k \,\text{grad}\, C_k \quad (18)$$

is the diffusive flux. Equation (17) expresses the fact, emphasized in the previous subsection, that in the absence of a coagulation loss rate $-[\partial(\rho C_k)]_{\text{st}}/\partial t$, and without gains Q, and losses L, e.g., through accretion, condensation and evaporation, the transport is diffusive in the drift frame. For the overall mass conservation we simply have

$$\frac{\partial}{\partial t}\rho + \text{div}\,(\rho \mathbf{v}) = 0 \quad (19)$$

where **v** is the mass velocity of the gas (see Sec. III). As long as v_{ks} and ν_k are identical for all solid grains, there is no net coagulation term for the respective solid mass density $\rho_{ks} \equiv \rho\, C_{ks}$, because coagulation is mass conserving:

$$\left(\frac{\partial(\rho C_{ks})}{\partial t}\right)_{st} = 0. \quad (20)$$

Correspondingly we have

$$\mathbf{j}_n = -\rho\, \nu_k \,\text{grad}\, C_n \quad (21)$$

as the diffusive number flux where $C_n \equiv n/\rho$ is the concentration for particle number density n. With the formal substitutions $C_k \rightarrow C_n$ and $\rho_k \rightarrow n$ in Eq. (17), we obtain the equation for the number density. As a result we have for solid concentrations C_{ks}, gas concentrations C_{kv}, and number concentrations C_n transport equations of the general form (Eq. 17) with appropriate source and loss terms as discussed in detail in Morfill and Völk (1984). The mean grain mass $\langle m \rangle$ is defined as $\langle m \rangle \equiv (1/n) \sum_k \rho_{ks}$. The assumptions on coagulation from the previous subsection amount to disregarding even the explicit coagulation term $(\partial n/\partial t)_{st}$ in the number density equation and to using instead a mean mass $\langle m \rangle$ for the particles as they are accreted at the outer boundary $r = R$. For analytic solutions of these transport equations, the disk may be considered to be thin with all quantities averaged in z direction over the gas scale height $h(r)$. Within the physical picture of Sec. III, it is convenient to introduce evaporation boundaries at radial distances that correspond to the critical mean temperature for evaporation (see, e.g., Morfill 1985). Inside the last boundary, all grains are evaporated and $n = 0$. All material is swallowed by the central object at its outer radius, without diffusing back.

V. LIMITATIONS OF THE TRANSPORT THEORY

The transport theory described in the previous section is a statistical theory. In order for it to apply in the first place, certain criteria have to be satisfied. In addition, the results obtainable in principle from such a theory are limited as well and have to be interpreted in terms of observable quantities. We discuss these limitations briefly below.

First, the stochastic forces acting on the dust particles and on the trace gas must be of such a nature that they describe a Markoff process. A Fokker-Planck description of the system is then applicable. There is essentially no way of proving that turbulence will evolve in a protoplanetary disk along these lines, where energy is transported in cascades to higher wave numbers, each successive hierarchy being independent of the previous one, until finally molecular viscosity takes over and the energy turns into heat. Nevertheless, it is the only plausible scenario developed so far, and is simply assumed.

Second, there are constraints imposed on the system itself. These manifest themselves in the necessary existence of a hierarchy of length or time scales. Some aspects have already been discussed in Sec. IV (see also Pringle 1976). The smallest relevant time scale is τ_f, the dust-gas frictional coupling time scale. Next we have the time it takes a volume element to adjust locally to vertical pressure equilibrium,

$$\tau_p \equiv h/c_s \tag{22}$$

which for our standard disk ($M_c = 0.5\ \mathrm{M}_\odot$) becomes $\tau_p \approx 5.5 \times 10^8\ \xi^{3/2}$ s. The thermal time scale obtained from radiative cooling is given in Eq. (12). For our standard disk we get $\tau_{\rm th} \approx 1.95 \times 10^{10}\ \xi^{3/2}$ s, where we have used an opacity law $K_d = 2 \times 10^{-6} T^2$, appropriate for a mixture of metal and silicate grains (De Campli and Cameron 1979).

The coherence time of the turbulence is

$$\tau_t \equiv h/\alpha c_s \approx 1.6 \times 10^9\ \xi^{3/2} \tag{23}$$

which for our values is simply three times the pressure time scale (Eq. 22). The values for τ_p, τ_t and τ_{th} should all be roughly similar, but significantly less than the turbulent mixing or diffusion time scale

$$\tau_{\text{diff}} \equiv r^2/\nu. \tag{24}$$

For our disk model we get $\tau_{\text{diff}} \approx 3.5 \times 10^{12}\ \xi^2$ s, and we see that in the planetary region, where the opacity is dominated by dust particles ($\xi \geq 0.034$, or $T \leq 1600$ K), $\tau_{\text{diff}} \gtrsim 10^3$ yr, and $\tau_{\text{diff}}/\tau_t \gtrsim 400$.

Another time scale of interest is the age of the disk τ_{disk}. Because the disk is fed from the outside by a large reservoir of gas and dust (i.e., the collapsing cloud fragment), and because this introduces another independent time scale τ_* (i.e., the exhaustion time scale for the whole cloud fragment), we define

$$\tau_{\text{disk}} \approx \frac{R}{R_*}\tau_*. \tag{25}$$

In other words, the disk grows with time, reaching a minimum radius R_* when the gas in the surrounding cloud is exhausted. From the collapse calculations we use $\tau_* = 10^6$ yr = the star formation time, and $R_* = 40$ AU = the size of our solar system. For our standard model ($M_c = 0.5\ \mathrm{M}_\odot$) we used $R/R_* = 0.5$, so that $\tau_{\text{disk}} \approx 1.5 \times 10^{13}$ s. This is roughly an order of magnitude larger than the maximum diffusion time scale, and implies that we may regard disk evolutionary structure changes as slow compared with turbulent transport effects; in other words, it suffices to consider a quasi-steady system. It appears, therefore, that the criteria for applying the transport equations described in Sec. IV are satisfied in protoplanetary disks, and that there are no principal limitations to the use of a stochastic transport description.

The results obtainable from a stochastic theory are limited, however, by the very nature of the description. Principal results that can be derived are: means, variances, two-point correlations and probabilities. In order to compare the results with solar system measurements, we have to make a number of assumptions. First, the theory describes a particular stage of the solar nebula evolution: the growth phase of the protosun. We assume that planet formation is concurrent with this stage of star formation. The fact that the outer giant planets contain large masses of gas (hydrogen, helium. . .) supports this view. Second, we must assume that the process of dust compaction into planetesimals does not change the properties of the nebula (e.g., opacity, thermal structure, level of turbulence, etc.) so that a given region in space always samples similar conditions. The hydrodynamic calculations indicate that the gas temperature at a given specific angular momentum location in the nebula is fairly constant with time, but of course they do not include grain coagulation and planetesimal formation as processes. On the other hand, planetesimal formation is not possible as long as $\nu_k/\nu \geq 1$ (see Sec. IV). In turbulent disks it seems very difficult to attain conditions favoring planetesimal formation, a problem which led to the suggestion that turbulence might be intermittent (Völk 1982). If turbulence is indeed intermittent, then we must assume that our calculations still apply as far as material properties are concerned. Any process which removes the driving cause for turbulence on a time scale shorter than the drift time of a typical grain and which lasts longer than a laminar sedimentation time is adequate. Such processes can be imagined to occur many times during the disk lifetime τ_{disk}, creating various generations of planetesimals. One such process, suggested by Weidenschilling (1984), involves lowering the perpendicular optical depth through a decrease of the (dust) opacity by grain coagulation. Then phases of turbulent accretion and coagulation of fresh small grains alternate with laminar phases of sedimentation and compaction of composite big grains. This assumes that convective instability is the main driving force of the turbulence.

Analytical derivations of time scales for the duration of turbulent periods and quiescent periods have been made (Völk et al., in preparation) and the fate of the dust particles has been discussed, i.e., their transport, sedimentation and coagulation. It appears that the periods of turbulence are sufficiently large that a statistical transport description for dust particles applies, but obviously in such a situation, the variances around the mean are somewhat larger. We can add the strong possibility that *on* and *off* phases may have a more local character in which case the solar nebula becomes even more complicated to describe. The region just inside the ice boundary, where the opacities drop significantly, may not be able to sustain turbulence as efficiently as other regions. This could lead to a temporary mass pile-up until grain drift populates the zone sufficiently well, so that turbulence is again possible. Clearly, a great deal of future work is necessary in order to understand and formulate such complications; however, if the diversity of small bodies, the heterogeneity on small scales and the homogeneity on large scales (mean planetary properties) are any indication, such work may be necessary. For the moment, it suffices to say that a transport theory for solids and associated gas phases in a turbulent disk has been formulated and some initial results are presented in the following sections. How these results relate to solar system measurements may not be straightforward to define; this depends to a large extent (especially in the case of variances) on the physical processes and on their time dependence. While the theory has been formulated quite generally, analytical results are only currently available for steady state solutions.

VI. GENERAL RESULTS FROM TRANSPORT THEORY

Results here are derived from steady state solutions, where time variations are assumed small. Details of the calculations may be found in Morfill (1985) and Morfill and Völk (1984); they are not repeated here.

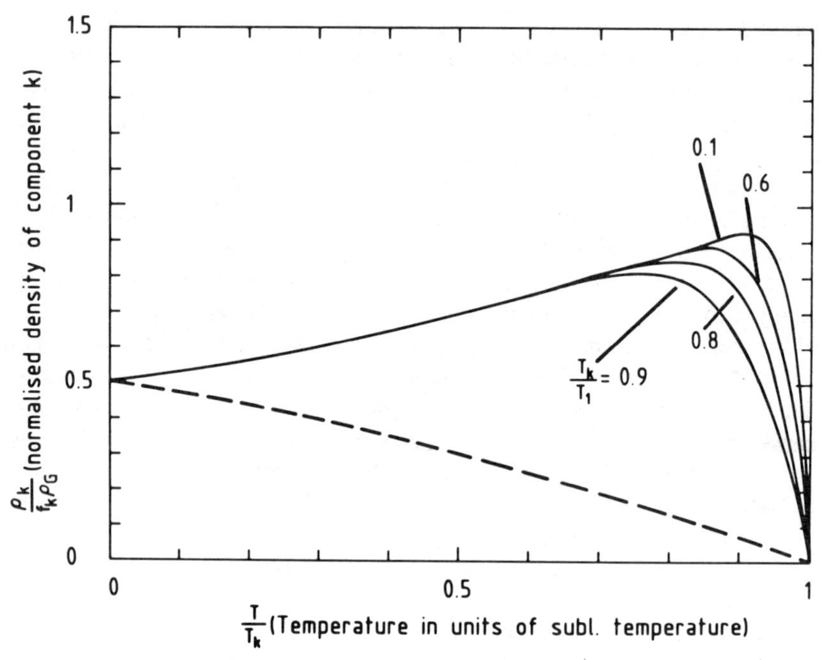

Fig. 8. The abundance of a chemical component k is plotted as a function of position in the protoplanetary accretion disk. The component is characterized by a sublimation temperature T_k. Abundances are normalized to the cosmic abundance; the position in the accretion disk is expressed in terms of the local gas temperature T, in units of T_k. Only the solid phase is considered, which cannot of course exist at temperatures $T/T_k > 1$. Different components are characterised by the ratios T/T_1, where T_1 defines the *total* sublimation boundary (of the most refractory component). The dashed line gives the normalized abundance variation if there were no sublimation and condensation. The curves shown correspond to particles of mean size such that the parameter $\lambda = 1$ (see text).

The first result, shown in Fig. 8, refers to the large-scale mean abundance of a given chemical component k in the solid phase. The abundance is normalized to the cosmic (or solar) value, and is plotted as a function of position in the nebula. Rather than use r, we have expressed position in terms of the local gas temperature $T(r)$, and we have normalized this to T_k, the sublimation/condensation temperature of the component k. Plotting the results in this way has the advantage of reducing the uncertainties of the disk model (and the built-in inadequacies) considerably. This may be seen when we compare the curves for different chemical components. All abundances go to zero at $T/T_k = 1$, as they must. Different components k characterized by the ratio

$$\frac{T_k}{T_1} \equiv \frac{\text{Sublimation temperature } (k)}{\text{Sublimation temperature } (1)} \tag{26}$$

where $k = 1$ corresponds to the most refractory component, show a similar behavior, except very close to their respective sublimation boundaries. The mean abundance of volatiles, which condense farther out (e.g., the $T_k/T_1 = 0.1$ curve) rises more sharply (in terms of T/T_k, but not in terms of T) than the mean abundance of refractories (e.g., the $T_k/T_1 = 0.8$ curve). The abundance of the solid phase far from the center of the nebula is equal to 0.5. This is due to the choice

$$\frac{\text{radial viscous transport velocity}}{\text{radial dust drift velocity (relative to the gas)}} \equiv 1 \tag{27}$$

which, as mentioned before, is in the appropriate region for the solar system type disks we have discussed. Also, it reflects the fact that we have simplified the problem by assuming rapid coagulation to raise almost instantly the particle mass to its global mean value. The implications of these simplifications have been discussed by Morfill and Völk (1984); they appear to be of little consequence in the inner solar system ($T/T_1 \gtrsim 0.1$).

The dashed line in Fig. 8 shows the results of the same calculation, performed without considering the transport and recondensation of the gas phase k. The dashed curve therefore represents an upper limit to the amount of unmodified presolar grain material.

Purely qualitatively, we see that at a given position in the nebula the probability of seeing unmodified presolar material increases as the sublimation temperature T_k increases; in other words, refractories should contain more information on presolar matter than should volatiles. Also, we see that in turbulent disks most of the matter has gone through at least one (possibly more) sublimation and condensation cycle. This is in agreement with cosmochemical analysis of primitive solar system material, which on average appears to follow typical condensation sequences. As we mentioned in Sec. V, evidence for microscale and macroscale inhomogeneities may be a result of incomplete sampling; so far the variances and two-point correlation functions have not been calculated.

In Fig. 9 we show some calculated, normalized abundance ratio (k/i) variations, plotted as a function of position within the nebula (again expressed in units T/T_k) where the chemical component k is taken to be the more volatile. Again the results apply to mean values. We see that the process of sublimation, transport and recondensation (ignoring gas-phase chemistry) leads to significant large-scale variations in abundance ratios in the disk. Obviously, with k being the more volatile component, the ratio (k/i) must go to zero at $T/T_k = 1$. Condensation of recycled gas then leads to a significant enhancement of (k/i) above the cosmic ratio $(k/i)_c$, and then the abundance ratio slowly approaches the cosmic value at the outer edge of the disk. The most apparent example of a large-scale abundance ratio variation is the Si/Fe content of the terrestrial planets. Our theory should be well suited to explain the mean properties of such large objects as planets, which have grown by the

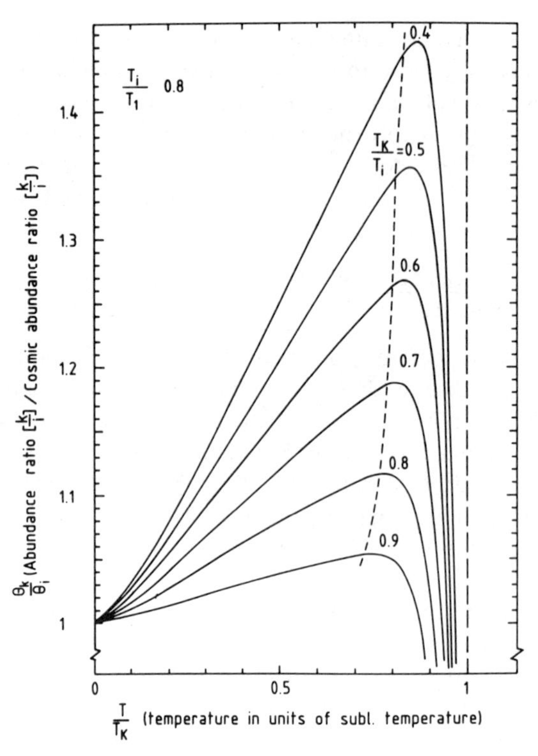

Fig. 9. The average abundance ratio of two chemical components k and i normalized to the cosmic ratio, is shown as a function of position in the protoplanetary accretion disk. (Position is expressed in terms of the local gas temperature, in units of T/T_k.) The component i is kept fixed (by choosing $T_i/T_I = 0.8$, i.e., i is a refractory component) and k is more volatile than i. To the left (low temperature side) of the dashed line, the curves are independent of the magnitude T_i/T_I.

accumulation of many smaller bodies and should therefore resemble mean values quite closely. In a high-pressure nebula (pressure $\gtrsim 50$ dyne cm^{-2}) the sublimation temperatures $T_{\text{Si}} < T_{\text{Fe}}$. Applying this to the solar system provides an easily understood explanation for the overabundance of Fe in Mercury. It implies that the temperature in the region where Mercury was formed should have been around 1250 to 1300 K. This is in reasonable accord with typical disk models that have a mass accretion rate of $\sim 10^{-6}$ $M_\odot$ yr^{-1}. The theoretical curve for the normalized abundance ratio, plotted against T/T_{Si} (silicates are the more volatile component), is shown in Fig. 10. Also shown, by their respective symbols, are the terrestrial planets. We see that the major puzzle, i.e., the large discrepancy between Mercury and the other planets, corresponds to the large-scale chemical trend. Although it may possibly be stretching the application of the theory too far, it is nevertheless interesting to

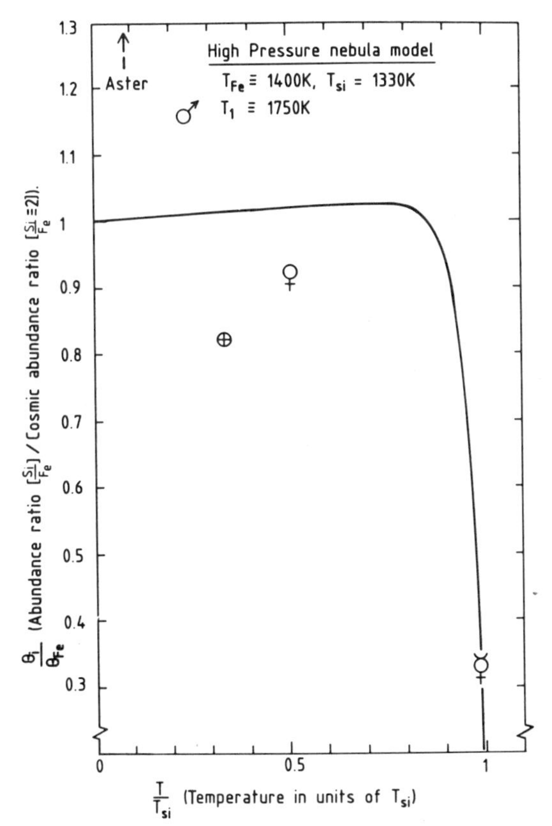

Fig. 10. In a high-pressure nebula model (see text) the Si/Fe ratio is plotted normalized to the cosmic abundance, as a function of local gas temperature, expressed in units of the sublimation temperature of silicates. The inner, terrestrial planets are shown by their respective symbols. The temperature assigned to each planet was taken from inferred accretion temperatures (Lewis 1974); and inferred abundance ratios Si/Fe were taken from Kaula (1976). The cosmic (solar) ratio Si/Fe = 2.

note that another abundant chemical component, FeS, condenses at about the position of Venus and yet another one, Fe_3O_4 (magnetite) at the position of Earth. This would lower the Si/Fe ratio for these planets. Similarly, hydrated silicates have a sublimation temperature which coincides roughly with the position of Mars, thus tending to increase the Si/Fe ratio again. All these chemical components have been found in meteorites, whereas no other significant chemical forms involving these elements have been detected. On the other hand, the uncertainties in the Si/Fe ratios determined for the terrestrial planets are still quite large (even more so for the asteroids) so that it may be premature to draw detailed conclusions at this time.

VII. SPECIFIC APPLICATIONS OF THE TRANSPORT THEORY

An important quantity to determine from the transport theory is the probability that a given volume element of dust particles, located at a position r_s at a given time (arbitrarily defined as t = o), may reach another location in the nebula r_M at any later time *before* it enters the hot central region where all the dust sublimes, i.e., before $r \lesssim r_1$. The interest centers here mainly on the observable consequences in unmodified primitive meteorites. In particular, we are interested in high-temperature residues, which are easily identified, their transport into cooler regions of the nebula, where they may become embedded in meteorite parent bodies, and the statistical and microscopic predictions of the transport theory. Detailed discussion of the consequences of the turbulent disk model for this application can be found in Morfill (1983*a*, 1985) and in Morfill and Völk (in preparation). We shall therefore describe only briefly the method for extracting the desired information and then discuss the results.

Consider the steady solution of the diffusion equation corresponding to a source located at r_s and an absorbing boundary at the position of the observer, at r_M. Dust particles sublime inside r_1, so we have to solve

$$\mathbf{div}\left(\mathrm{v}_s n - \nu\rho\,\mathbf{grad}\frac{n}{\rho}\right) = 0 \tag{28}$$

in each region ($r_1 \lesssim r \lesssim r_s$) and ($r_s \lesssim r \lesssim r_M$) (we always consider $r_1 < r_s < r_M$) subject to the boundary conditions $n(r_1) = n(r_M) = 0$ and the matching conditions in the limit $\epsilon \to 0$

$$[n]_{r_s-\epsilon} = [n]_{r_s+\epsilon} \tag{29}$$

$$\left[\mathrm{v}_s n - \nu\rho\,\mathbf{grad}\frac{n}{\rho}\right]_{r_s-\epsilon} = -S + \left[\mathrm{v}_s n - \nu\rho\,\mathbf{grad}\frac{n}{\rho}\right]_{r_s+\epsilon} \tag{30}$$

where S is the δ-function source at r_s. The flux into r_M can then be calculated, and the probability of seeing a dust particle at r_M, when it originally started off at r_s, is defined as

$$P(r_M, r_s) = \frac{\text{total particle current into } r_M}{\text{total source current}}. \tag{31}$$

Clearly, $P(r_M, r_1) = 0$ because the source is located at the total sublimation boundary, but for $r_s > r_l$, P is finite. Figure 11 shows the results of such a calculation. Distance is expressed in terms of temperature, normalized to the temperature of the source region, i.e., we have plotted $P\,(T_M, T_s)$ as a function of T_M/T_s. When $T_M/T_s = 1$, $P = 1$; when $T_M/T_s \to 0$, $P \to 0$, as expected. Calculations were performed for particles whose radial frictional drag-in-

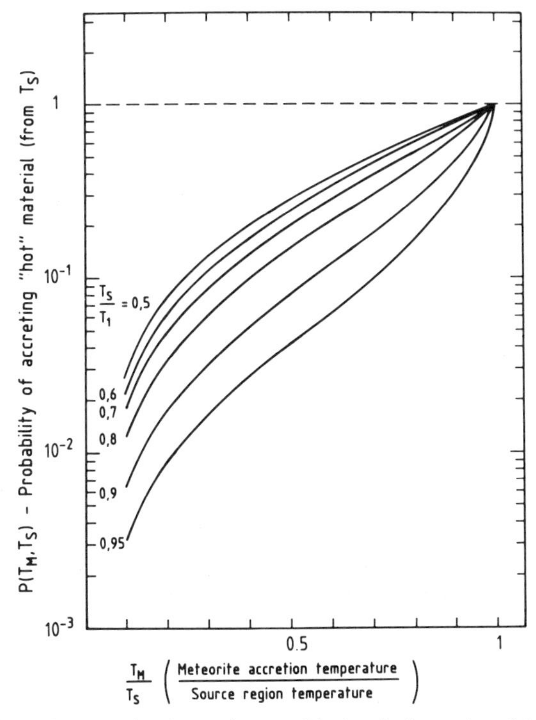

Fig. 11. Probability of transporting dust grains out of the interior hot region of the protoplanetary disk. The probability is calculated from our standard disk model; transport of grains from a hot region (temperature T_s) to a cooler region (temperature T_M) is expressed as a function T_m/T_s. For further details see text.

duced velocity equals the radial viscous contraction of the gas, and curves are shown for different source separations from the total sublimation boundary r_1 (or T_1), i.e., for different ratios T_s/T_1. Substituting some solar system values, we see that the probability of accreting high-temperature residues, where $T_s/T_I \simeq 0.9$ (e.g., $T_1 = 1700$ K, $T_s = 1530$ K), onto a body formed at a temperature corresponding to $T_M/T_s = 0.25$ ($T_M = 380$ K) is $\sim 2.5\%$. If the dust particles are smaller, and hence their systematic radial drift towards the center is less important, the probability increases, for the parameters used above, up to a maximum of $\sim 5\%$. There is a possibility that breakup due to selective sublimation of major volatiles (i.e., components with condensation temperatures below ~ 1500 K) may occur. This will lower the particle sizes considerably. The effect that the turbulent disk transport has on an ensemble of sizes is shown in Fig. 12. Basically the effect is one of selection in favor of small particles, the cutoff occurring at ~ 1 cm, if the material density of the residue is ~ 3 g cm^{-3}. This result, as the one shown in Fig. 11, is again a statistical result and takes no account of possible time variations, etc. The

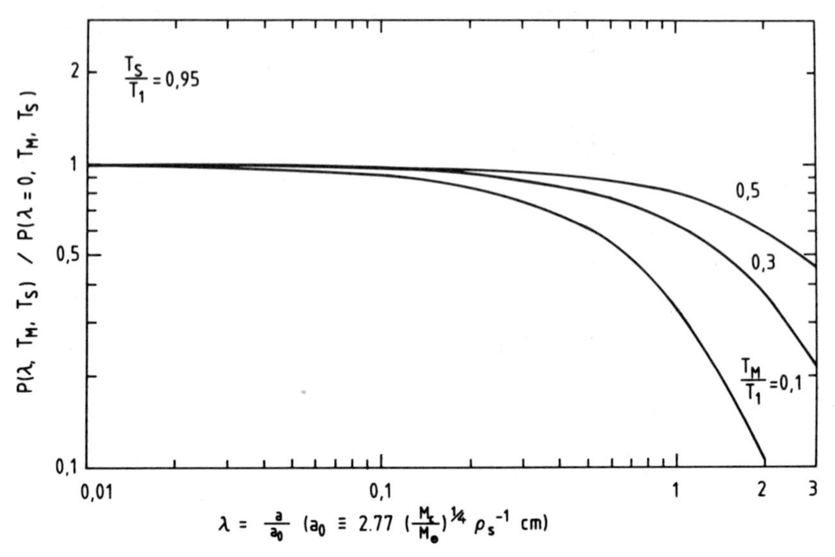

Fig. 12. Effects of dust particle size on their ability to be transported outwards into the cooler region. A size cutoff is expected due to transport effects, basically because the radial gas drag induced drift velocity increases with particle size, thus making it more difficult for large particles to diffuse very far out.

typical diffusive transport time scale from r_s to r_M is $t_{\mathrm{diff}}\ (r_s,\ r_M) = (r_M - r_s)^2/\nu$ which we can write in terms of temperature using the relationship $T = \mathrm{T_o}\,(r/R)^{-3/2}$. For our standard (0.5 $M_\odot$) model, this yields a transport time of ~ 300 yr, which is also the time scale for reaching steady state equilibrium. This is comparable to, possibly even larger than, the time scales which one can calculate for the duration of intermittent turbulence. We cannot exclude the possibility, therefore, that time dependent effects may play a role even on these relatively small scales. On the other hand, the most optimistic assumption would be that each consecutive turbulent eddy happens to convect the volume element outwards, with a mean velocity $0.5\ \alpha c_s$. The time taken to travel from r_s to r_M is then given from integrating

$$\frac{\mathrm{d}r}{\mathrm{d}t} = -\frac{\nu}{r} + 0.5\alpha c_s \tag{32}$$

where we have assumed small particles and hence a small gas-drag-induced radial drift. Integrating Eq. (32) from r_s to r_M yields $\Delta t \simeq 18$ yr for this most direct transport. The probability for this to happen is $P_d \approx 2^{-11}$ (there are at least 11 turbulent eddies, each of scale height $h(r)$ between r_s and r_M in our standard 0.5 $M_\odot$ disk model, and each new eddy has a probability of 0.5 to go outwards or to return inwards). This works out at roughly 5×10^{-4}. Although this probability is small, it is *not* negligible. Over the lieftime of the disk τ_*, this singular sequence of events could have happened $\sim N_d$ times, where

$$N_d \approx \frac{\tau_*}{\Delta t} P_d \frac{2\pi(r_s + r_M)}{h(r_s) + h(r_M)} \approx 2 \times 10^3. \tag{33}$$

These numbers, if nothing more, serve to illustrate that turbulence, far from automatically ensuring homogeneity in the chemical structure of the disk, does so only in a macroscopic sense, i.e., if the body from which the data is obtained has truly sampled many turbulent elements over a sufficiently long time. On a more microscopic level, corresponding to the mass contained in $\lesssim$ 1 turbulent eddy, individual aspects and heterogeneity may occur. In the region of interest $r_s < r < r_M$, this implies masses of $\sim 10^{22}$g, where we have assumed an abundance of solids of 10^{-4}. When compacted this corresponds to bodies of size $\lesssim$ 100 km.

VIII. UNSOLVED PROBLEMS

The description in Secs. VI and VII amounts to an extremely simplified picture of the early stages of a protoplanetary disk. Many aspects of it can be readily improved qualitatively as well as quantitatively, but a number of aspects remain which appear to require more basic modifications. Among these remaining problems are: (1) grain sedimentation and subsequent planetesimal formation, (2) the formation process for meteoritic chondrules, and (3) the sources for isotopic anomalies. We restrict ourselves here to a discussion of these three key questions.

(1) Within the main framework of the accretion-disk model, grains cannot sediment easily. It is uncertain whether a refined treatment of coagulation and statistical fluctuations in grain sizes can produce enough big grains which could sediment such that under steady state conditions a subdisk of particulates would form. This subdisk would necessarily be expected to fragment under its own gravity to form km-sized bodies in the manner described by Safronov (1969), and Goldreich and Ward (1973). A solution might be that planetesimals form only at the end of the disk evolution when turbulence fades away.

In order to preserve any effects of turbulence on the grain component, the turbulence must be turned off in such a way that particles can sediment into the equatorial plane, where they become compacted into planetesimals, before the radial transport due to the remaining viscous evolution and frictional drag forces remove the dust altogether. It can be shown by solving the radial transport and coagulation equations that the conditions for grain decoupling from the turbulence are more difficult to fulfill, the smaller the level of turbulence is. This means that turbulence must decay on a time scale faster than the radial drift time scale of the dust particles; otherwise the solids will be lost and no planets can form. This is a general requirement. Also the turbulence must decay faster than the rate with which mass accretion into the disk from the collapsing envelope slows down in the late stages, when the supply is exhausted. Mass accretion is expected to subside over times of the order of an initial free-fall time, i.e., in about 10^5 yr. The drift time is, therefore, the determining factor. A further requirement is that sedimentation, local gravitational instabilities in the dust subdisk, and in particular the collisional growth of planetesimals to bodies larger than one Earth mass and, in the case of the giant planets, the accumulation of gas, occur on a time scale faster than the dissipation time of the leftover gas disk. In addition, the left-over gas disk must have a mass greater than the minimum calculated from augmented planetary masses.

In the case of intermittent turbulence, our steady state calculations are still applicable, provided that, in addition to the rapid decay, the length of a turbulent cycle Δt (ξ) is comparable with or larger than the time scale to establish diffusional equilibrium, r^2/ν. During the quiescent periods, sedimentation and planetesimal formation takes place rapidly without changes in the disk material. Then, these large bodies are representative of the physicochemical state of the turbulent disk and can watch the subsequent evolution more or less as spectators from the sidelines of quasi-Keplerian orbits.

(2) Chondrules are significant constituents of large meteorite classes, quantitatively and qualitatively. Their sizes correspond to those of the larger conglomerate grains that can form in a turbulent disk. However, it is not clear whether the rather slow thermal processing in the disk (on a diffusion time scale) can transform such conglomerates into chondrules as they exist in meteorites. Some extra rapid heating and cooling may be required. Although we have not explicitly considered magnetic fields in the disk, they may nevertheless be amplified by turbulent dynamo action. Rapid flare-type magnetic reconnection processes might serve to flash-heat conglomerate grains. At least it is worth pointing out that such effects are not inconsistent with the general protoplanetary accretion-disk picture, as discussed in this chapter.

(3) Finally, we come to the question of the place of isotopic anomalies within the model. Turbulent diffusion, first of all, acts as a mixing process that would be expected to eradicate any isotopic nonuniformities rather than to produce them. One can readily demonstrate that within the present mathematical model of a turbulent disk, only mass-dependent isotope fractionation effects occur. Therefore any isotopic anomalies must be introduced into the disk from outside and then survive the interior transport processes in some form. Various possibilities for modification of the basic scenario are conceivable to account for isotopic nonuniformity, but they are not discussed here (see chapters by Clayton et al., by Wasserburg and by Kerridge and Chang).

IX. CONCLUSIONS

The concept of a turbulent protoplanetary accretion disk as the early and determining stage of solar system formation appears to be a viable model within which one can calculate a number of observable properties of solar system matter. Whether it is possible to account convincingly for planetesimal formation and isotopic anomalies by suitable modifications remains to be demonstrated but does not seem impossible.

By implication large planetary bodies as well as the cores of the giant gas planets are formed through gravitational accumulation (Safronov 1969) of planetesimals. However, the accumulation process as such is one of the various processes not addressed here. If it also forms the cores of the giant gas planets, then it must proceed in the massive presence of gas—a problem that has received much less attention in the past than gas-free accumulation. Generally, it should proceed much faster in a gaseous environment (see D. N. C. Lin 1981*a*) which is important because the subsequent acquisition of the gaseous envelope obviously requires the existence of a large gas reservoir at late times. We have not discussed the problem of how the gas is finally removed. The most widely assumed process is a protostellar wind. However, if the particulate disk apparently observed around Vega by the Infrared Astronomical Satellite (Aumann et al. 1984) has anything to do with a protoplanetary disk, then the removal of gas must not remove mm- and cm-sized grains as well. Here we have one example of the many important future astronomical observations which will transform planetary formation from a theoretician's toy to an empirical part of the evolution of the universe.

Acknowledgments. G. M. wishes to acknowledge the hospitality of the NASA Goddard Space Flight Center and the University of Maryland during part of the completion of this work.

FISSION OF RAPIDLY ROTATING FLUID SYSTEMS

RICHARD H. DURISEN
Indiana University

and

JOEL E. TOHLINE
Louisiana State University

Sufficiently rapid rotation can cause breakup of an equilibrium fluid system through the growth of dynamic distortional instabilities. This process is called fission and has been invoked to explain the formation of close binary and multiple systems of celestial bodies. Both classic and modern results of fission theory are reviewed in this chapter. Recent advances have been made in three areas: permitted equilibrium states for rotating self-gravitating fluids, stability limits for various distortional modes, and numerical hydrodynamic simulations of the nonlinear development of dynamically unstable modes—especially the bar modes. The hydrodynamic simulations have shown that, at least for n = *3/2 polytropes, the dynamic bar-mode instability leads to spiral-arm ejection of a ring or disk of material. Formation of a binary is not the direct outcome. Possible applications of these results to the theory of star and planet formation are discussed. The next decade promises even greater advances in both our understanding and computational capabilities.*

I. INTRODUCTION

A. Definition of the Problem

For over a century researchers have investigated whether and how rapid rotation causes breakup of self-gravitating fluid equilibrium states. The process of breakup is usually referred to as fission when it is initiated by the linear growth of an unstable mode of distortion in the shape of the equilibrium state. The fission problem, defined in this way, is distinguishable from the fragmentation problem which has recently monopolized the attention of star formation theorists. Fragmentation of a rotating system, such as a protostellar cloud, is usually envisioned as occurring during a dynamical collapse; fission, though a dynamical process, starts in states near equilibrium. Fragmentation is initiated by local or global compressional Jeans instabilities; fission is initiated by global distortional instabilities. These distinctions may become blurred in some limiting cases, but they are useful guidelines for most applications.

In this chapter, we will review progress in our understanding of the fission problem. To limit our scope, we will not discuss the fragmentation of collapsing protostellar clouds or local instabilities in circumstellar disks. Aspects of these subjects are addressed in companion chapters of this book (see for example chapters by Scalo, Cassen *et al.*, and Dickman) and a review of fragmentation may be found in Tohline (1982). Fission instabilities and related processes are of great potential importance in themselves. For instance, they have been invoked to explain close binary formation for stars (cf. Ostriker 1970), planets (cf. Lin 1981*b;* Durisen and Scott 1984), and even asteroids (Weidenschilling 1981*b;* Farinella et al. 1981). The possible effects of such instabilities must be considered when dealing with any rapidly rotating system. The recent developments in fission theory which will be of most interest to applications-oriented readers are probably contained in Sec. IV on numerical hydrodynamic simulations, where the actual outcome of fission instabilities is described. These readers should also study the summary and discussion in Sec. V. Although much of the remaining material is of a more technical nature, it does help to put the numerical results in perspective and will provide readers with a more comprehensive view of the subject.

B. Classic Results: Equilibrium States

Because fission was not discussed in the first *Protostars & Planets* book (Gehrels 1978), we feel that a somewhat detailed description of fission theory up to the mid 1970s will increase the usefulness of our chapter. This classic work dealt primarily with the existence and structure of equilibrium states and their stability to linear perturbations. Prior to the 1960s and going back to the 18th century, only incompressible fluid states were considered, because of their mathematical tractability by analytic methods (cf. Lyttleton 1953). In the 1960s and 1970s, these studies were extended by numerical methods to compressible fluids, principally polytropes, for which $P \sim \rho^{1 + 1/n}$, where P is the pressure, ρ is the mass density, and n is the polytropic index. The reader who wants more details and references than presented here will find Tassoul (1978) an excellent and thorough text. Other useful reviews with different emphases can be found in Lebovitz (1967, 1979), Ostriker (1970), Fricke and Kippenhahn (1972), and Moss and Smith (1981).

The earliest and most complete information on equilibrium states is available for incompressible ($n = 0$) fluids (cf. Chandrasekhar 1969). In particular, exact analytic results have long been available for uniform density, ellipsoidal equilibrium configurations that have a velocity field $\mathbf{v}$ which is linear in the Cartesian center of mass coordinates (x,y,z). The ellipsoidal surface figure is described by

$$\frac{x^2}{a^2} + \frac{y^2}{b^2} + \frac{z^2}{c^2} = 1 \tag{1}$$

where a, b, and c are the half lengths of the principal axes of the ellipse, usually chosen so that $a \geqslant b \geqslant c$. The simplest equilibrium states for uniform density, incompressible fluids are the Maclaurin spheroids. These uniformly rotating configurations are symmetrically flattened about the rotation axis so that $a = b \geqslant c$. With proper normalization in terms of mass M and density ρ, the Maclaurin spheroids become a one-parameter family of distinct objects. The most physically meaningful parameter to use is probably the total angular momentum J. Another choice of the parameter which will prove useful to us is the energy parameter

$$\beta = T/\ |W| \tag{2}$$

where T is the total rotational kinetic energy and W is the total gravitational energy. As J is increased, β increases monotonically and the spheroid becomes more flattened, i.e., c/a decreases.

A wide array of ellipsoidal equilibrium states are also available to incompressible self-gravitating fluids. For our purposes, the most important of these are the Jacobi, Dedekind, and Riemann S-type ellipsoids. All of these ellipsoids have the surface figure angular velocity $\boldsymbol{\Omega}_s$ and fluid vorticity $\boldsymbol{\zeta} = \boldsymbol{\nabla} \times \mathbf{v}$ aligned with the short axis of the ellipsoid. The Jacobi ellipsoids (see Fig. 3a in Sec. II. A below), like the Maclaurin spheroids, are rigid rotators ($\boldsymbol{\Omega}_s = \boldsymbol{\zeta}$). On the other hand, the Dedekind ellipsoids have a stationary surface figure in the inertial frame ($\Omega_s = 0$) with fluid streaming through it ($\zeta \neq 0$). The Riemann S-type ellipsoids exhibit the full range of other possible relationships between ζ and Ω_s. For each specified ζ/Ω_s, there is again a one-parameter family characterized by the total angular momentum J or by the energy parameter β. Each such sequence begins with $a = b$ at a particular Maclaurin spheroid as a limiting case. This behavior is referred to as bifurcation. At each β along the dynamically stable part of the Maclaurin sequence (see the next section), two sequences of equilibrium ellipsoids bifurcate or branch off. One sequence usually has Ω_s and ζ in much closer agreement (Jacobi-like) than the other (Dedekind-like). Some of the relationships between the equilibrium ellipsoids and the Maclaurin spheroids are illustrated in Fig. 1. In this figure, the Maclaurin spheriods lie along the right-hand axis ($a = b$), with $\beta = 0$ at the upper right ($a = b = c$) and β increasing as c/a decreases. The Jacobi and Dedekind ellipsoids bifurcate at the point S marked by the asterisk, which corresponds to $\beta \approx 0.1375$. With increasing J, the sequences for these ellipsoids follow the curve labeled JD from the point S to the origin. At all other points of the dynamically stable Maclaurin sequence, two sequences of Riemann S-type ellipsoids bifurcate and follow similar curves to the origin as J is increased.

It is also known classically that there are additional bifurcations of configurations with higher-order surfaces than spheroids or ellipsoids. These bifurcations occur both along the Maclaurin sequence and along the Riemann

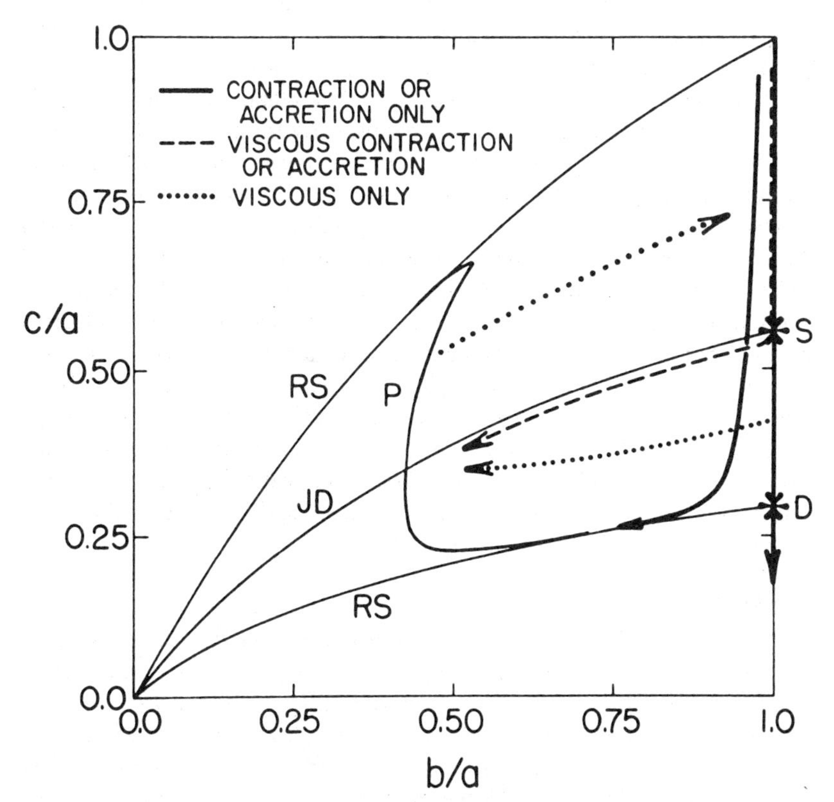

Fig. 1. The axis ratio plane for tumbling ellipsoidal equilibrium states. The Maclaurin sequence ($a = b$) lies along the right-hand edge. The Riemann S-type ellipsoids occur in the talon-shaped region bounded by the curves labeled RS. The Jacobi and Dedekind ellipsoids lie along the curve JD. The asterisks S and D denote the secular and dynamical bar-mode stability limits, respectively. Between S and D, Maclaurin spheroid bar modes are secularly unstable; below D, they are dynamically unstable. Dynamical pear-shaped instabilities occur at and to the left of the curve labeled P. The curves with arrowheads represent various possible evolutionary tracks. The dotted curves are adopted from Detweiler and Lindblom (1977); the heavy solid curves from Lebovitz (1972).

S-type ellipsoid sequences. The most important of these for fission theory is the bifurcation of pear-shaped figures along the Riemann sequences (see Fig. 3e in Sec. II.A below). The pear-shaped bifurcation point depends in a complex way on ζ/Ω_s and J and occurs either on or to the left of curve P in Fig. 1. Some of the other bifurcations are discussed later in Sec. II.A.

Accurate equilibrium models for rapidly rotating, self-gravitating, compressible ($n \neq 0$) fluids have been constructed by various numerical techniques involving expansions or finite differencing (see, e.g., Chandrasekhar and Lebovitz 1962; James 1964; Stoeckly 1965; Ostriker and Mark 1968; Clement 1974). All such models constructed before 1978 have symmetry about the rotation axis and so can be considered analogs of the Maclaurin spheroids. Both uniformly rotating (UR) and differentially rotating (DR) cases have been considered for polytropes (see, e.g., Tassoul and Ostriker 1970; Bodenheimer and Ostriker 1973) and degenerate dwarfs (see, e.g., Ostriker and Bodenheimer 1968). UR sequences for $n \geqslant 0.808$ truncate at a $\beta \leqslant 0.137$ due to critical rotation at the equator, but properly formulated DR sequences seem to be limited only by the virial equilibrium constraint $\beta \leqslant 0.5$. In general terms, polytropic models span the compressibility range of astrophysical interest. Roughly speaking, $n = 1/2$ corresponds to a relatively stiff celestial object, such as a terrestrial planet or a neutron star. Gas giant planets, like Jupiter and Saturn, have $n \approx 1$. Ordinary stars and degenerate dwarfs are well represented by the range $n = 3/2$ to 3. Interstellar clouds are nearly isothermal ($n \rightarrow \infty$). Attempts to model rotating main sequence stars or evolved stars with nuclear burning will not be discussed here, because they have generally not been subjected to bifurcation or stability analysis. Figure 2 gives an example of an $n = 3/2$ polytrope in rapid differential rotation. Almost nothing was known prior to the 1980s about the existence and nature of nonaxisymmetric equilibrium states for compressible fluids, except under the artificial assumption of uniform density (Lebovitz 1972,1974).

C. Classic Results: Stability

Several methods can be used for exact linear stability analyses of the uniform density, incompressible equilibrium states (cf. Lyttleton 1953; Chandrasekhar 1969). For instance, at each β along the Maclaurin sequence, there are two bar modes, also called the $1 = m = 2$ Kelvin or f modes, which distort the spheroids into ellipsoids ($a > b$) tumbling about the short axis. The linear eigenmodes of incompressible equilibrium states bear a direct and natural relationship to the bifurcation properties discussed in the preceeding section. In fact, the two bar modes at each β represent infinitesimal steps away from the Maclaurin sequence along the two sequences of Riemann S-type ellipsoids which branch off at the same β. In this sense, highly distorted Riemann S-type ellipsoids represent bar modes of nonlinear amplitude, and, correspondingly, the bar modes of the Maclaurin spheroids are just early members of the bifurcating Riemann sequences. For given β, each bar mode has a distinct pattern speed Ω_s and internal circulation ζ.

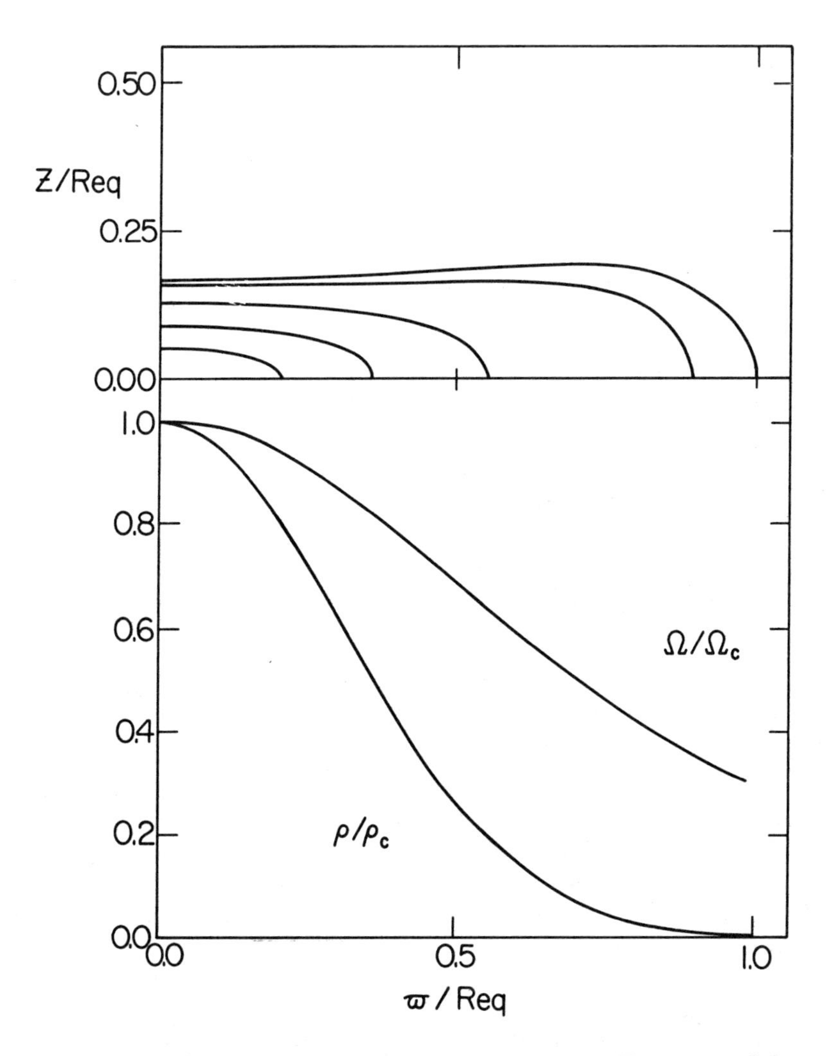

Fig. 2. The structure of a $\beta = 0.33$, $n = 3/2$ axisymmetric polytrope. The upper panel shows curves of constant density in a meridional plane for ratios of local to central density $\rho/\rho_c = 0.8$, 0.5, 0.1, 0.01, and 0.0. Here, ϖ and z are cylindrical radius from the rotation axis and height above the equatorial plane, respectively, and R_{eq} is the equatorial radius. The lower panel shows the distributions of density ρ and angular velocity in the equatorial plane, normalized to their central values (subscript c). This model was constructed with the self-consistent field code of Bodenheimer and Ostriker (1973).

The Maclaurin spheroid bar modes exhibit two limits beyond which different types of instability can occur:

$$\beta > \beta_s \approx 0.1375 \text{ secular bar instability} \tag{3a}$$

$$\beta > \beta_d \approx 0.2738 \text{ dynamical bar instability.} \tag{3b}$$

The secular instability occurs only in the presence of an appropriate dissipative mechanism. So far, it has been shown that viscosity alone or gravitational radiation alone are capable of driving this instability. For viscosity alone, the Jacobi-like bar mode will grow on a viscous time scale when $\beta_s < \beta < \beta_d$. For gravitational radiation alone, the Dedekind-like bar mode grows in the same β range on a time scale set by the rate at which gravitational radiation can carry away energy and angular momentum. The secular instability can be understood by means of simple energy arguments and conservation laws. Viscous evolution conserves J while evolution due to gravitational radiation conserves fluid circulation. For $\beta_s < \beta < \beta_d$, ellipsoidal configurations exist that have lower total energy but the same J or circulation as a Maclaurin spheroid, and so, as energy is dissipated, an ellipsoidal distortion grows. The end state for viscous evolution is the point on the Jacobi sequence with the same J. The end state for gravitational radiation evolution is the point on the Dedekind sequence with the same circulation. This is why the secular stability limit (S in Fig. 1) is also the point at which the Jacobi and Dedekind ellipsoids bifurcate. In the absence of dissipation, Maclaurin spheroids do not evolve for $\beta < \beta_d$ and are, in this sense, dynamically stable.

For $\beta > \beta_d$, S-type equilibrium ellipsoidal configurations do not exist. The two bar modes now have complex conjugate eigenfrequencies. The common real part is the pattern speed Ω_s. The imaginary parts lead to one mode which grows exponentially on a dynamical time scale and another which damps at the same rate. For $\beta > \beta_d$, then, the Maclaurin spheroids are dynamically unstable to growth of a bar mode. The time scale for growth is infinite at the marginal stability point $\beta = \beta_d$ but quickly becomes similar to a rotation period as β increases. The dynamical stability limit is point D in Fig. 1.

As J increases along the Riemann S-type ellipsoid sequences, dynamical instability is encountered at the same point where the pear-shaped figures bifurcate, i.e., on or to the left of curve P in Fig. 1. A small pear-shaped distortion will grow on a dynamical time scale beyond this point. For the Jacobi ellipsoids, in particular, the dynamical pear instability occurs for

$$\beta > \beta_p \approx 0.1628 \text{ dynamical pear instability.} \tag{4}$$

There is no corresponding secular instability for pear-shaped distortions. The first instability encountered for pear shapes as J is increased is the dynamical one. The Riemann S-type ellipsoids are, however, secularly unstable to changing their ellipsoidal axis ratios in the presence of dissipative mechanisms. Depending on the dissipative mechanism and on J, secular evolution of Riemann S-type ellipsoids leads to an end state on the Maclaurin, Jacobi, or Dedekind sequence. In some cases, this evolution will lead to crossing of curve P.

Secular and dynamical instabilities for higher-order modes of surface distortion than pears and bars also occur. Of the instabilities known classically, however, the secular bar mode instability is the one that occurs first along the Maclaurin sequence as J or β is increased. Similarly, the dynamical bar and pear-mode instabilities are the first dynamical instabilities that occur along the Maclaurin and Riemann sequences, respectively, as J is increased. As a result, these instabilities play a dominant role in most fission scenarios.

A proper linear eigenmode analysis for bar and pear distortions of rapidly rotating, compressible equilibrium states must be done numerically and is sufficiently difficult that it has still not been accomplished even for simple UR polytropes. Prior to 1978, the only results available were based on the linearized tensor virial equations (TVE's) of second order. The TVE method, as described in Chandrasekhar (1969) provides exact results for the bar and pear modes of the uniform, incompressible ellipsoids. Although not exact for compressible models, the linearized TVE's of second order as generalized by Tassoul and Ostriker (1968) have been used to provide approximate bar-mode pattern speeds, dynamical growth rates, and stability limits for rotating polytropes (Tassoul and Ostriker 1970; Ostriker and Bodenheimer 1973) and degenerate dwarfs (Ostriker and Tassoul 1969; Durisen 1975). Remarkably, the TVE bar-mode stability limits for all rotation laws (UR or DR) and equations of state (with n up to 3 tested) agree with the incompressible results of Eq. (3) to within the accuracy of the equilibrium models, namely about $\pm$ 0.001 in β for the secular instability point β_s and about $\pm$ 0.02 in β for the dynamical instability point β_d. At the time these results were obtained, it was unclear whether the close agreement was merely an artifact of the TVE method or a true regularity in the stability properties of rotating fluids. Because no analogs of Riemann ellipsoids were available for nonuniform, compressible fluids prior to 1978, nothing was known about pear stability limits.

D. Classic Routes to Fission

As the classic results became available over the years, the idea emerged that the formation of binary or multiple stars and of binary planets and satellite systems might be explained by some process of slow evolution along analogs of the Maclaurin and Riemann sequences and beyond. As J or β increases, successively higher-order surface distortions might occur, eventually leading to the splitting off of one or more discrete bodies. Historically, Poincaré, Darwin, Liapounoff, Jeans, and others envisioned this process as occurring via a succession of higher and higher-order secular instabilities. However, as we have seen, the pear-mode instability is dynamical not secular. More modern versions of the fission theory for binary formation invoke slow evolution up to and through a dynamical instability point, usually of a bar- or pear-shaped mode. The dynamical instability is then conjectured to produce the binary or multiple system directly.

As illustrated in Fig. 1, there are several possible ways for equilibrium objects to approach dynamical instability: (1) through viscosity or some other dissipative mechanism alone; (2) through increasing β by accretion or contraction in the presence of large viscosity; and (3) through accretion or contraction alone. Accretion can cause evolution toward states of increasing β if the accreted material has large specific angular momentum. Homologous quasi-static contraction, e.g., for a pre-main sequence star, can increase β when J and M are conserved because $T \sim J^2/MR^2$, $W \sim M^2/R$, and so $\beta \sim 1/R$.

Detailed evolutions of Riemann S-type ellipsoids have been computed with viscosity alone (see, e.g., Press and Teukolsky 1973), with gravitational radiation alone (see, e.g., Miller 1974), and with both at the same time (Detweiler and Lindblom 1977). Roughly speaking, above the curve JD in Fig. 1, any dissipation causes evolution toward the secularly stable Maclaurin spheroids which lie along the upper right-hand edge of the figure. For Maclaurin spheroids below the secular stability point S, viscosity alone causes the growth of a Jacobi-like bar mode, and the secular evolution ends either at the stable part of the Jacobi sequence or at pear-mode instability, depending on the initial β. Similarly, gravitational radiation causes growth and evolution of a Dedekind-like bar mode toward the stable Dedekind sequence or to pear-mode instability. These two dissipative mechanisms can actually compete with and stabilize each other (Lindblom and Detweiler 1977), when both are present, leading to relatively complicated trajectories not illustrated. Of course, for star and planet formation, gravitational radiation is unimportant.

The evolution of a contracting or accreting object will depend on how the contraction or accretion time scale, t_{con} or t_{acc}, compares with the characteristic viscous time scale t_{visc}. For a fluid of relatively high viscosity ($t_{visc} << t_{con}, t_{acc}$), the dashed trajectory is expected. Viscosity will keep the configuration axisymmetric up to point S and then will insure a configuration close to the Jacobi sequence until pear-mode instability is reached. Calculations of this sort have not yet actually been done. When $t_{visc} >> t_{con}, t_{acc}$, the heavy solid curves will result (Lebovitz 1972). Configurations with small initial deviations from axisymmetry will turn left near point D and head for a dynamical pear instability. Notice that only in the limit of inviscid, strictly axisymmetric contraction or accretion is a bar-mode instability the first opportunity for dynamical breakup.

Of the time scales involved, only t_{con} for stars and gas giant planets is reliably known. The time t_{acc} depends on initial conditions for stellar collapse and on planetesimal dynamics for terrestrial planets. The viscous time scale t_{visc} is the least well constrained in all cases, either because the appropriate material physics is not certain or because t_{visc} depends on assumptions about turbulence. Only a little progress has been made recently in elaborating any of these scenarios. Prior to the late 1970s, nothing was reliably known about the ultimate consequences of dynamical instabilities. The few early attempts to follow developing bar and pear-mode instabilities were hampered by extremely constrained geometry (see, e.g., Fujimoto 1968; Aubin 1973). As we have already mentioned, it has often been conjectured but never proven that the outcome of dynamical instability will be fission into a system of two or more coherent bodies. For the remainder of this chapter, we will refer to this idea as the binary fission hypothesis.

The summary here of work on the classic routes to fission has necessarily been brief. The interested reader will find a more detailed discussion and a complete reference list in Chapter 11 of Tassoul (1978).

Since the mid 1970s, significant progress has been made in three general areas:

1. Construction of equilibrium states for a wider range of conditions (Sec. II);
2. Improvements in linear stability theory, including the discovery of new types of instability (Sec. III);
3. Numerical hydrodynamic calculations which follow the nonlinear development of dynamical instabilities (Sec. IV).

These are sufficiently comprehensive categories that they provide a natural way to organize the remainder of this chapter. The emphasis will be rather theoretical, because, for the most part, studies have concentrated on relatively simple fluid systems with generalized characteristics, like polytropes, rather than on specific astrophysical applications with detailed physics. This is an appropriate state of development, at the present time, for a theoretical problem which is intrinsically three-dimenstional and, even in its simplest forms, already strains the limits of existing techniques. Recent success with the simpler systems has been encouraging. The three sections on the areas listed above are followed by a summary and discussion (Sec. V), where some comments are made about applications and about future prospects for calculations that include additional physical processes.

II. EQUILIBRIUM STATES

A. Uniformly Rotating Incompressible Fluids

The classic incompressible theory (as in Chandrasekhar 1969) dealt almost exclusively with ellipsoidal states and their infinitesimal higher-order distortions. For the most part, equilibrium sequences of the higher-order shapes were not constructed. In principle, this could be done using the TVE's of order higher than two, but these equations quickly become algebraically cumbersome. Although it could be argued that the axisymmetric and bar shapes and their secular and dynamical instability points provide all the essential features needed from incompressible theory, it is always necessary and often proves useful, in complex theoretical problems, to map out solution space thoroughly. In a continuing series of papers (referenced below), Eriguchi, Hachisu, and collaborators have been providing such a map by application of numerical techniques which permit them to calculate incompressible equilibrium states in rigid rotation (UR) with almost arbitrary surface shapes. The UR assumption restricts their work primarily to sequences which can trace their origin to the Maclaurin or Jacobi sequences by one or more bifurcations.

Along the Maclaurin sequence, it was known classically that for β values beyond the limits in Eq. (3), a series of additional secular and dynamical instability points occur for higher-order distortions. A particular subclass of these are neutral points, i.e., points where the eigenfrequency of a distortional mode becomes equal to zero in some preferred reference frame. Eriguchi and Sugimoto (1981) and Eriguchi and Hachisu (1982,1983*a*) have found that sequences of distorted, UR equilibrium sequences bifurcate from the Maclaurin sequence at all neutral points they investigated. For this application, the neutral points are determined in a frame rotating with the Maclaurin spheroid. Surprisingly, the sequences of nonaxisymmetric shapes all truncate at finite values of J due to critical rotation at the equator. The truncation by critical rotation is referred to as mass-shedding in these papers, but no physical demonstration of shedding is offered. Examples of these terminal configurations are given in Fig. 3f, g, and h for sequences which bifurcate at $\beta \cong 0.2035$, 0.2483, and 0.3244, respectively. This suggests that, of all the UR nonaxisymmetric equilibrium states bifurcating from the Maclaurin sequence, the Jacobi ellipsoids are the only ones which extend to infinite J. On the other hand, these authors have found axisymmetric sequences of one or two rings which extend to infinite J from bifurcations at $\beta \cong 0.3648$ and 0.4107. The first of these sequences connects the Maclaurin spheroids smoothly to the Dyson-Wong toroids known earlier.

Of particular interest for the binary fission hypothesis are the sequences of figures which bifurcate from the Jacobi sequence at $\beta \cong 0.1628$ and 0.1863 (Eriguchi et al. 1982; Hachisu and Eriguchi 1984*a*). The first of these is a pear-shaped sequence originating at the point of onset of dynamical pear instability, which also happens in this case to be a neutral point. This sequence truncates by mass-shedding at finite J with the configuration shown in Fig. 3e. At the higher β, higher J bifurcation, the sequence of configurations illustrated in Fig. 3a–d branches off. As one moves away from the Jacobi sequence, a pinch develops around the waist of the tumbling bar and deepens smoothly to create first a peanut shape (Fig. 3c), then a contact binary (Fig. 3d), and finally a sequence of detached binaries with increasing separation. Although the sequence of figures changes shape smoothly, the parameters β and even J are not monotonic, suggesting that, if the sequence were followed

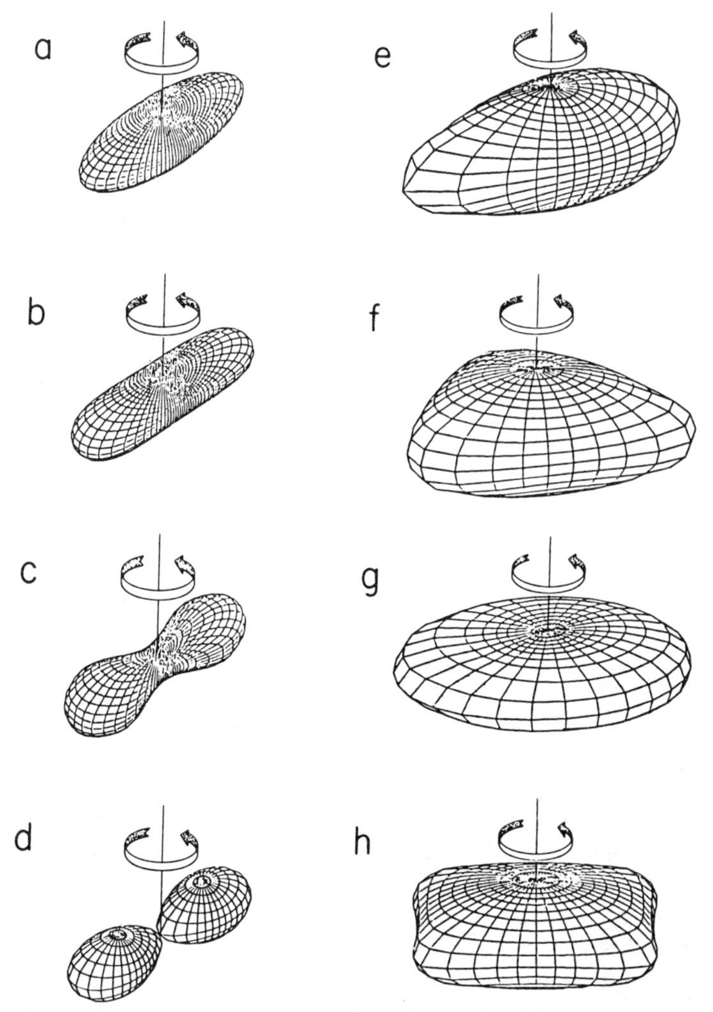

Fig. 3. Equilibrium configurations for uniformly rotating, incompressible fluids. Figures *a* through *d* show members of a sequence of equilibrium shapes beginning with a Jacobi ellipsoid of $\beta \cong 0.1863$ and progressing smoothly through peanut-shaped forms to a contact binary with $\beta \cong 0.1515$. The sequence can be continued further through detached binaries of increasing separation. Figure *e* is the terminal mass-shedding member of the pear-shaped sequence which bifurcates from the Jacobi sequence at $\beta \cong 0.1628$. Figures *f* to *h* are the terminal members of sequences which bifurcate from the Maclaurin sequence at $\beta \cong 0.2035$, 0.3244, and 0.2483, respectively.

in nature, some discontinuous jumps would occur where multiple solutions exist for the same J. Hachisu and Eriguchi (1984*b*) and Eriguchi and Hachisu (1983*b*) have extended some of this work to unequal mass ratios and multiple systems, but the sequences no longer connect to the Jacobi sequence.

The resemblence of Fig. 3a–d to classic binary fission and the smoothness of the transition are very seductive. However, a few caveats are in order concerning whether or not this mode of binary formation can ever be realized in nature. According to Fig. 1, the evolutionary scenario which follows the Jacobi sequence reaches the dynamical pear instability at $\beta = 0.1628$ before the point at $\beta = 0.1863$ where the peanut-like sequence begins. The instability that sets in at $\beta = 0.1628$ is dynamical, and it is difficult to imagine how it could be suppressed. Something dramatic must happen, and further evolution along the Jacobi sequence to $\beta = 0.1863$ will not occur. The fact that the pear-shaped sequence truncates without pinching off into a binary suggests, but does not prove, that the dynamical instability of the pear mode is more likely to eject mass in a disk or stream rather than a single lump. Even though the peanut equilibrium states do not seem to be directly accessible by quasi-static viscous evolution, it is important to know they exist. The nonlinear development of dynamical instabilities could lead to discontinuous jumps between distant sequences. Also, according to Lebovitz (1974), some inviscid, contracting, compressible ellipsoids may in fact circumvent pear-shaped instabilities.

Other difficulties with assessing this work are the restriction to strict UR by the numerical technique and incomplete information about the stability of the calculated configurations. For instance, it is known, in the axisymmetric case, that differential rotation actually permits toroids to exist with lower energy than the Maclaurin spheroids (Marcus et al. 1977). We also know, from the Riemann S-type ellipsoids, that, when internal motions are allowed, infinitely many new bifurcations become possible. Stability analyses, relaxation of UR, and introduction of internal motions are all extremely difficult.

In addition to the difference between truncated and untruncated sequences, as described above, the bifurcations studied by Hachisu et al. (1982) can be distinguished and classified by analogy to phase transitions (Hachisu and Eriguchi 1983). Bifurcations of shapes which preserve the symmetry of the original equilibrium state exhibit continuity of $d\Omega_s/dJ$ at the bifurcation and so represent second-order phase transitions, while bifurcations which change the symmetry show discontinuous $d\Omega_s/dJ$ and are first-order phase transitions. In this sense, the bifurcations from Maclaurin to Jacobi and from Jacobi to pear shape are first order, whereas the Jacobi to peanut shape is second order. The usefulness of this concept in future dynamical studies of fission remains to be seen.

B. Uniformly Rotating Polytropes

It has been known since the work of Jeans (1919), subsequently refined by James (1964), that, for $n \gtrsim 0.808$, UR axisymmetric polytrope sequences truncate due to critical equatorial rotation for $\beta \lesssim 0.137$ and never bifurcate into analogs of the Jacobi ellipsoids. For $n \lesssim 0.808$, James constructed equilibrium models of UR bars only for small degrees of distortion away from axisymmetry near $\beta = 0.137$. Three groups using different methods have now constructed complete sequences of rigidly tumbling polytropic bars for $n \lesssim 0.808$ (Vandervoort and Welty 1981; Ipser and Managan 1981; Hachisu and Eriguchi 1982,1983*a*). All these Jacobi ellipsoid analogs bifurcate from their corresponding UR axisymmetric counterparts at $\beta = 0.137$ to within ± 0.003, and, even for n as small as 0.1, they truncate due to critical rotation after only a small increase in J away from axisymmetry. Only for $n \lesssim 0.1$ can such sequences reach the dynamical pear instability. The terminal members of the sequences develop cuspy points at the ends of the bar. For $n \lesssim 0.02$, UR axisymmetric sequences even extend to high enough β (about 0.36) to exhibit bifurcating ring-like structures similar to those found for the Maclaurin sequence (Fukushina et al. 1980; Hachisu et al. 1982). It appears then that, until limited by the onset of critical rotation, the equilibrium configurations of UR polytropes show structures and bifurcations which are quantitatively similar to those of a UR incompressible fluid. Interesting effects are limited, however, to rather small degrees of compressibility.

C. Differentially Rotating Polytropes

As far as we know, except for nonrotational models in a Roche potential (Clement 1967), the UR polytropic bars discussed in the previous section are the only nonaxisymmetric single-body polytropic configurations which have been constructed directly by equilibrium techniques. Once both constraints of axisymmetry and rigid rotation are relaxed, the specific angular momentum j, i.e., the angular momentum per gram, of a fluid element will not in general be conserved during a rotation. Without some *a priori* algorithm to constrain this variation in j, it is impossible to construct a self-consistent equilibrium model directly. No one has yet been able to formulate such an algorithm for compressible fluids. For instance, Ipser and Managan (1981) have shown that the simplest attempt to generalize Dedekind ellipsoid configurations to polytropes fails. Nevertheless, considerable progress has been made by using various procedures to relax polytropic fluids into nonaxisymmetric steady state structures in hydrodynamic codes. In almost all cases, the configurations have been in differential rotation (DR) with internal motions.

In the first of these studies, which used smoothed particle hydrodynamics (see Sec. IV below), Gingold and Monaghan (1978) constructed initial states of nonrotating spherical polytropes. They then introduced uniform rotation. The resulting configurations were far from force balance and expanded. As they expanded, motions in the meridional planes were slowly damped. This introduced a dissipation which settled the configurations into low-energy equilibrium states while conserving angular momentum. For final $\beta \lesssim 0.14$, the polytropes relaxed into axisymmetric equilibrium states which were similar to Bodenheimer and Ostriker (1973) models. For final β's between about 0.14

and 0.17, the end state was roughly like an ellipsoid tumbling about its short axis. The axis ratios resembled those of Jacobi ellipsoids with comparable β, but the ends of the bar appeared to be slightly pointed. For $0.17 \lesssim \beta \lesssim 0.22$, final equilibrium states were found only after a pear-shaped figure appeared and shed mass to become a $\beta \cong 0.17$ tumbling ellipsoid. Similar results were obtained for both $n = 1/2$ and $n = 3/2$. The β-values bounding different behaviors (namely $\beta \lesssim 0.14$ axisymmetric, $0.14 \lesssim \beta \lesssim 0.17$ bar-like, $\beta \gtrsim 0.17$ mass-shedding pear shape) agree quantitatively with classic incompressible theory. The mass-shedding behavior of the pear shapes seems to presage Eriguchi et al.'s (1982) later discovery that the pear-shape sequence truncates due to critical rotation. However, the dumbbell shapes found by Eriguchi et al. do not seem to occur.

Boss (1980*c*, 1981*a*) has described the end states of two-dimensional and three-dimensional finite difference code calculations designed to test fragmentation during adiabatic collapse for polytropes of $n = 3/2$ and $n = 5/2$. When the geometry was confined to axisymmetry (two-dimensional), the final configurations had their maximum density at the center for $\beta \lesssim 0.43$ but became rings for $\beta \gtrsim 0.43$. This agrees well with the ring dynamical stability limit of $\beta \approx 0.437$ determined by Marcus et al. (1977). In similar three-dimensional collapses, roughly axisymmetric configurations resulted only for final $\beta \lesssim 0.27$. Whenever the final β would have been greater than 0.27, a binary formed before the collapse to a single equilibrium object could occur. Because the binary formation happened during collapse, it does not necessarily imply anything about the outcome of true fission instabilities. However, it is interesting that the collapses do seem to reflect some knowledge of the dynamical bar-mode instability point. Pear shapes were not seen, but this may be because the initial state was given a large amplitude bar-like perturbation only and no pear-like perturbation above machine noise.

A few three-dimensional dynamical calculations (Durisen and Tohline 1980; Durisen et al. 1984), to be described in more detail later (Sec. IV.B), have followed the growth of dynamical bar-mode instabilities in DR axisymmetric $n = 3/2$ equilibrium polytropes with $\beta > 0.27$. The end states after mass ejection were apparently stable, steady state bars with final $\beta \lesssim 0.2$. These bars were Dedekind-like in that the surface figure tumbled slowly while the fluid circulated quickly through it with the same sense of rotation. The axis ratios (1:0.65:0.3) resembled those of stable Dedekind-like Riemann *S*-type ellipsoids.

The evidence cited above suggests that, without the constraint of uniform rotation, compressible fluids exhibit the same set of ring-like, bar-like, and pear-shaped equilibrium figures which have been so thoroughly studied in incompressible theory. Furthermore, these configurations show quantitatively similar bifurcation properties in the parameter β. These conclusions are most firmly established for $n \lesssim 3/2$. It will be interesting to see if this regularity persists in more compressible cases.

D. Isothermal Equilibrium States

Our knowledge about extremely compressible rotating fluids has lagged behind results for $n \leqslant 3$ by a decade or so, even though isothermal conditions ($n \rightarrow \infty$) are directly relevant to star formation. The lag is due to the complication that a nonzero boundary pressure must be applied to obtain a finite model for $n \geqslant 5$. Only over the past few years has significant progress been made in understanding the equilibrium structure of rotating isothermal gas clouds and then only for axisymmetric cases.

Hayashi et al. (1982) have derived an analytic expression that describes exactly the equilibrium structure of an unbounded cloud that is rotating with a uniform rotational velocity (as opposed to a uniform angular velocity). Equilibrium structures having any value of β in the range $0 \leqslant \beta \leqslant 0.5$ can be constructed from their analytic model. Their model is unbounded in the sense that there is no external medium surrounding the cloud, and the equilibrium states are all infinite in extent. In addition, their equilibrium structures all have infinite central densities, and the density profile approaches an r^{-2} profile in the appropriate limiting cases. It can be shown that the equilibrium states found by Hayashi et al. are stable to global axisymmetric compressional modes only when they have large rotational energies; specifically, β must be greater than 0.2926 for dynamical stability.

Stahler (1983*b,c*) has used a self-consistent field method to numerically construct equilibrium structures in the presence of an external, bounding medium (see also Norman 1980). His models were chosen to have the same specific angular momentum distribution as a uniformly rotating, uniform density sphere. What Stahler found was that, for any given total cloud angular momentum (in his terminology, for a given "initial" β), he could construct an equilibrium structure that had a relatively small degree of central concentration and a relatively small equilibrium β ($\lesssim 0.3$) at a specific value of total cloud mass. In addition to this, for a certain range of total cloud angular momentum, a second equilibrium structure could be constructed that had a high degree of central concentration and a relatively high equilibrium β. In contrast to Hayashi et al., Stahler found that at any chosen value of total cloud angular momentum, his low β structures were dynamically stable against global axisymmetric compressional modes, while his high β structures were unstable. Stahler was unable to construct a model exactly like the one described by the analytic work of Hayashi et al. because of convergence difficulties with his numerical technique for structures with high degrees of central concentration.

Tohline (1984) and Hachisu and Eriguchi (1984*c*) have recently resolved the apparent discrepancy between the work of Hayashi et al. (1982) and Stahler (1983*b,c*). What they have shown is that, when the effects of an external, bounding medium are considered, a sequence of equilibrium figures can be constructed over the entire range of cloud masses $O < M < \infty$ for any given total cloud angular momentum. At small cloud mass, the equilibrium figure has a low value of β and the external medium plays a dominant role in defining the structure of the cloud. These low β structures appear to be stable against global axisymmetric compressional modes and are apparently only moderately centrally condensed (these are the structures found to be stable by Stahler [1983*c*]). At the largest cloud masses, the equilibrium figures have large values of β and the external medium plays a negligible role in defining the structure of the cloud. These high β structures also appear to be stable against global compressional modes and are apparently very highly centrally condensed. These structures are analogous to the ones found by Hayashi et al. At intermediate cloud masses, equilibrium figures can be constructed with intermediate values of equilibrium β, but they are often found to be dynamically unstable. The analysis by Tohline and by Hachisu and Eriguchi is consistent with both the results of Hayashi et al. and of Stahler. The former found only the high β equilibria to be stable because they did not include the effects of a bounding surface medium in their model; the latter found only the low β equilibria to be stable because his numerical technique did not allow convergence to the highly centrally condensed, high-β structures. Both groups had found that states at intermediate values of equilibrium β were dynamically unstable, and it is in this regime alone that their works directly overlap.

One final point of clarification should be made regarding the equilibrium structures of rotating, isothermal clouds. In the nonrotating Bonnor-Ebert sphere, there is a critical mass *above* which no equilibrium is attainable. Stahler (1983*c*) claimed that in rotating clouds, an analogous critical mass exists, but that the exact value of the critical mass is slightly different at different values of total cloud angular momentum. In the model described by Hayashi et al. (1982), for a given total cloud angular momentum, there is a critical mass *below* which no equilibrium is attainable. As Tohline (1984) points out, neither of these mass limits represents a true barrier to isothermal equilibrium states. When the effects of both surface pressure and rotation are considered, one finds that the Bonnor-Ebert type critical mass is always greater than or equal to the Hayashi et al. type critical mass—i.e., the two mass limits overlap. For a given total angular momentum, a cloud whose mass is above the Bonnor-Ebert mass can only reach equilibrium at high β, with a structure dominated by rotation. A cloud whose mass is below the Hayashi et al. mass can only reach equilibrium at low β, with a structure dominated by surface pressure. A cloud whose mass is in between the two critical values can actually attain an equilibrium at three different values of β, but the intermediate β structure will always be an unstable one.

The equilibrium structures studied by Hayashi et al. and Stahler, when looked at from this perspective, provide an excellent foundation upon which future studies of the nonaxisymmetric stability properties of rotating, isothermal clouds can be based. The results of these stability studies should indicate whether or not attempts to construct nonaxisymmetric equilibrium states will prove worthwhile. Nothing is known at present about where nonaxisymmetric or toroidal equilibrium structures may bifurcate from the axisymmetric forms. This will no doubt be complicated by the fact that the nonzero boundary pressure makes the isothermal axisymmetric structures a fully two-parameter family, rather than a one-parameter sequence, as for $n \lesssim 3$.

E. Nonisentropic Equilibrium States

So far, almost all studies of bifurcation for compressible equilibrium states have been done for isentropic fluids. For instance, the polytrope models which have been discussed so extensively are assumed to respond to adiabatic perturbations according to $P \sim \rho^{\gamma}$, where $\gamma = 1 + 1/n$ and n is the same as the polytropic index of the model. All the fluid in the configuration is thus constrained to start and remain on the same adiabat. This is realistic for some applications, such as highly degenerate or completely convective stars, but it is not realistic for pre-main sequence radiative tracks or subsequent phases of stellar evolution. Isentropic fluids are neutrally buoyant and so, in general, are indifferent to exchanges of fluid elements with the same j. For this reason, the internal energy of the configuration does not play a significant role in the growth of nonaxisymmetric distortions. Clement (1979) and Durisen and Imamura (1981) did consider secular bar-mode instability for cases where $\gamma \neq 1 + 1/n$ and found no significant differences in bifurcation properties, but these authors used the BFSS (Bardeen, Friedman, Schutz, Sorkin [1977]) criterion (see Sec. III.A below), where a trial eigenfunction for the bar mode is assumed to be similar to the Maclaurin bar mode. The insensitivity to choice of γ may just be an artifact of that assumption.

Lebovitz (1974) considered the bifurcation properties of uniform density, compressible ellipsoidal configurations. The latter are possible for an ideal gas, if an appropriate temperature structure is imposed, making the configurations nonisentropic. He found that pear-shaped figures are not then always the first to bifurcate along Riemann sequences but that fourth-harmonic distortions can be encountered first. He has also pointed out (Lebovitz 1984) that we do not know the stability properties of these compressible bifurcations. If they are dynamically stable and if fourth-harmonic distortions occur before pear-shaped (third-harmonic) ones, then analogs of the peanut-shaped sequences found by Eriguchi, Hachisu, and collaborators may lead to binary fission in the classical sense.

Lebovitz (1983,1984) is currently trying to work out a bifurcation and stability theory for compressible stars, but there are not yet any quantitative results. The possibility must be kept in mind that nonisentropic configurations may well behave very differently from strict polytropes. Most of the work on linear stability and numerical simulations discussed in this chapter deals only with isentropic fluids. This constraint clearly needs to be relaxed in future studies.

III. LINEAR STABILITY

A. The BFSS Criterion

The most important development of the past decade in linear stability theory for fission modes has been a critique of the TVE method leading to the formulation of a correct variational principle for Lagrangian perturbations. Although the quantitative consequences of this critique were negligible, they placed some results on a firmer theoretical footing and facilitated the discovery of secular instabilities previously unsuspected. Extensions of these results may prove useful in developing a more accurate and complete linear eigenmode theory for nonaxisymmetric disturbances in rotating celestial objects.

An infinitesimal perturbation of a hydrodynamic equilibrium state can be characterized by a Lagrangian displacement $\boldsymbol{\xi}(\mathbf{r},t)$. This vector prescribes the direction and magnitude that a fluid element is offset from its equilibrium position $\mathbf{r}$ at time t due to the perturbation. As emphasized by Hunter (1977), Bardeen et al. (1977), and Friedman and Schutz (1978*a*), there is an assumption implicit in the second-order linearized TVE method of Tassoul and Ostriker (1969). The bar-mode eigenfunction of the configuration being analyzed is assumed to be the same as the bar-mode eigenfunction for the Maclaurin spheroids, namely, in cylindrical coordinates (ϖ, ϕ, z),

$$\boldsymbol{\xi}_M = a\,(\varpi,\ -i\varpi,\ 0)\ e^{2i(\Omega_s t \pm \phi)} \tag{5}$$

where Ω_s is the pattern speed for tumbling of the bar-like surface figure (in stability theory parlance, $2\Omega_s$ is the eigenfrequency). In this sense, the TVE method is approximate and uses Eq. (5) as a trial eigenfunction in order to estimate Ω_s and determine where it goes to zero (neutral point) and where it becomes complex (dynamical instability).

Bardeen et al. (1977) and Friedman and Schutz (1978*a*) proved that in fact Eq. (5) is only a valid trial eigenfunction for uniformly rotating configurations, because, for differential rotation (DR), the perturbed motions induced by Eq. (5) do not satisfy the Kelvin circulation theorem. Many of the most interesting TVE applications have been for DR models with $n = 3/2$ and 3. Although the TVE results were known to be only approximate anyhow, the critique did raise one serious theoretical problem. These authors showed that, because of Eq. (5), the TVE method provides neither a necessary nor a sufficient condition for locating the point of secular instability when applied to DR models. By making slight modifications to Eq. (5), so that the Kelvin theorem was satisfied, they were able to develop a variational principle which *does* provide a sufficient condition for secular instability of bar modes in DR equilibrium states. We will refer to this as the BFSS (Bardeen, Friedman, Schutz, and Sorkin [1977]) criterion. (The reader is referred to sections 6.6 and 6.7 of Tassoul (1978) and the associated bibliography for background on the use of variational methods in the theory of rotating stars.)

The BFSS criterion has been applied to DR main sequence stars (Clement 1979) and to DR polytropes and degenerate dwarfs (Durisen and Imamura 1981). Durisen and Imamura found that, for $n = 3/2$ and $n = 3$ polytropic fluids and for a completely degenerate, noninteracting electron gas, the BFSS criterion gives a sufficient condition for secular instability which is only 1 to 7% higher in β than the classic Maclaurin spheroid value of Eq. (3). The deviation increases with degree of central concentration of mass and angular momentum over the range of compressibilities and angular momentum distributions studied. Clement's published results exhibited a much larger deviation but included a sign error in the stability analysis. In unpublished work, with the sign error corrected, Clement found good quantitative agreement between the BFSS stability limit for his main sequence models and the corresponding limit for the DR $n = 3$ polytropes of Durisen and Imamura.

Although the BFSS criterion did not produce a large quantitative difference from TVE results, it has improved our understanding of bar-mode instability in the following sense. For DR models, the TVE results were only approximate and of unknown accuracy, providing neither a necessary nor a sufficient condition for stability. There was really no way to assess the significance of agreement with Eq. (3). Now, we know certainly that DR configurations with compressibilities between $n = 3/2$ and 3 are secularly unstable when $\beta \gtrsim 0.14$ for a wide range of rotation laws. Combined with the reliable results on bifurcation for $n \lesssim 0.808$ UR configurations, this presents a remarkably consistent picture of bar-mode secular instability for all $n \lesssim 3$ (see Ostriker and Peebles [1973] and Vandervoort [1982] for similar considerations in the context of stellar systems). As a technical caveat, we should point out that the variational principle locates secular instabilities of only Dedekind-like ($\Omega_s << \zeta$) bar modes, whereas most evidence regarding bifurcation of bar configurations is for Jacobi ellipsoid analogs ($\Omega_s = \zeta$). As discussed in Sec. I.D, the different types of bar distortions are driven by different dissipative mechanisms. We still do not have a complete theory of bar-mode secular stability for all n which encompasses both types of dissipation. However, the results in hand exhibit suggestive regularity.

B. Generic Instability of Rotating Bodies

Another outcome of the recent work in Lagrangian perturbation theory is perhaps of even greater theoretical interest than the BFSS criterion. Using a variational principle, Friedman and Schutz (1978*a,b*) were able to demonstrate that *all* rotating fluid celestial objects are secularly unstable due to gravitational radiation, when modes of the form

$$\xi \sim e^{\pm im\phi} \tag{6}$$

are considered with m sufficiently large.

This is actually fairly easy to understand. Nonaxisymmetric Kelvin modes occur in pairs. As viewed in the rotating frame of the fluid, one mode will generally rotate faster than the fluid, the other more slowly. In an inertial reference frame, if the fluid rotates at just the right rate, the slow mode will be stationary. This represents a particular type of neutral point where a mode becomes neutral in an inertial frame (see Sec. II for a discussion of neutral points in a rotating frame; see also Chandrasekhar 1969, §36). For faster fluid rotation, the slow mode actually propagates in the same sense as the fluid when viewed by an inertial observer. The perturbation δJ to the fluid angular momentum J due to the slow mode is negative. Above the neutral point, the slow mode generates gravitational waves which carry off positive angular momentum. Above the neutral point, then, $d\delta J/dt < 0$, and the amplitude of the slow mode will grow as it radiates away fluid angular momentum. For large m, the slow modes are more easily dragged forward (pattern speed ~ eigenfrequency/m), and so, for any degree of fluid rotation, there will be modes of high enough m to be above their neutral points. Each particular slow mode grows and radiates until the fluid rotation rate approaches the neutral point for that mode.

The time scale τ_m for this instability was first estimated by Friedman and Schutz (1978*b*) and by Papaloizou and Pringle (1978) and was then determined exactly by Comins (1979*a,b*) for Maclaurin spheroids. Very roughly, for large m,

$$\tau_m \sim \frac{[(2m+1)!!]^2}{m}\left(\frac{P}{t_c}\right)^{2m+1}\frac{t_d{}^2}{P} \tag{7}$$

where P is the rotation period of the mode, $t_c = R_{eq}/c$ is the light-crossing time, and $t_d = (2\pi G\rho)^{-1/2}$ is the dynamical time. Because of the P/t_c factor, τ_m is only significantly shorter than the age of the Universe for the $m = 2$ modes of white dwarfs and the $m \leqslant 7$ modes of neutron stars. Even for neutron stars, not all these modes are effective. As shown quantitatively by Comins, viscosity will stabilize the gravitational radiation instability for all $m \geqslant 5$ in neutron stars. Lindblom and Detweiler (1977) and Lindblom and Hiscock (1983) have demonstrated similar viscous stabilization in more general contexts (see also Lindblom [1979] concerning effects of thermal conduction). For Maclaurin spheroids, neutral points in the inertial reference frame for $m =$ 2, 3, 4, and 5 occur at $\beta \cong 0.138$, 0.0991, 0.0771, and 0.0629, respectively. The possible relevance of instabilities at these neutral points for millisecond pulsars has been discussed in some detail by Friedman (1983). The latter reference is in fact an excellent mini-review on the generic instability. Imamura et al. (1984) have also used the BFSS variational principle to study high-m modes in polytropes and found that, unlike the $m = 2$ bar modes, the secular stability limits for $m \geqslant 3$ modes show notable sensitivity to fluid compressibility and angular momentum distribution.

In itself, an instability driven by gravitational radiation is clearly irrelevant to star and planet formation. It is described here in considerable detail, however, because it illustrates that there may be room for further surprises in the area of secular instabilities. For each m, the slow mode becomes unstable and grows above the neutral point because it radiates away angular momentum. A directly analogous case can be envisioned for a rotating protostar or protoplanet, if it is surrounded by material. Above its neutral point, one of the slow modes could grow if it could lose angular momentum to the surrounding material through surface stresses or gravitational torques (Durisen and Imamura 1979; Schutz 1983, §2.11).

C. Linear Stability of Fission Modes in Hydrodynamic Simulations

Although progress is being made in performing direct numerical linear eigenmode calculations for rapidly rotating stars (Clement 1981,1984) and breakthroughs should prove possible through further improvements in variational techniques (see, e.g., Schutz 1979), stability criteria (see, e.g., Lindblom 1983), and bifurcation theory (Hachisu and Eriguchi 1983; Lebovitz 1983,1984), much of the new useful information for compressible DR stars on linear stability, bifurcations, and even eigenmodes and eigenvalues has so far come from numerical hydrodynamic simulations. Detailed descriptions of these results are given in Sec. II.C earlier and also in Sec. IV.B below and so will not be repeated here. The paper by Tohline et al. (1984) is probably the best example to date of a numerical hydrodynamics code being used to study dynamical fission instabilities in the linear regime. In this technique, an unstable mode is allowed to organize itself and emerge from very low-level random initial noise (see Bardeen [1975] for a similar approach). Of course, this will only single out the fastest growing dynamically unstable mode, if any, and so is rather limited as a method for linear eigenmode analysis, because it can never be complete.

Overall, numerical hydrodynamics calculations have shown that, for n values between about 1/2 and 3, the bar-mode stability limits and, to a lesser extent, pear- or ring-mode dynamical stability limits seem quantitatively similar to the corresponding limits for Maclaurin spheroids. Tohline et al. have even found that the pattern speeds and growth rates for bar modes predicted by the TVE method are reasonably accurate for DR $n = 3/2$ models, despite the known inadequacies of TVE. Even the eigenmodes are fairly similar, except that for $\beta > 0.27$ they have a trailing spiral arm pattern. This pleasingly simple regularity, which has been emerging with accumulating data over the last two decades or so, is emphasized several times in this chapter because of its fundamental importance.

IV. NUMERICAL HYDRODYNAMIC SIMULATIONS

A. General Information

Computational facilities now have large enough storage capabilities and fast enough processing units to allow time-evolving, self-gravitating structures to be modeled in full three-dimensional generality. Many attempts have been made in the past few years to follow the growth of nonaxisymmetric structure in rapidly rotating, self-gravitating three-dimensional objects using the new computational tools. Five such studies have been specifically designed to address the fission problem, as defined in this review: Lucy (1977), Gingold and Monaghan (1978,1979), Durisen and Tohline (1980), Durisen et al. (1984); and Tohline et al. (1984). Although computational cost constraints have restricted these studies to models with fairly low spatial resolution, it appears as though certain large-scale nonaxisymmetric structures have been satisfactorily resolved. The distortional bar mode, at least beyond the dynamical instability point, appears to grow as a two-armed spiral pattern rather than as a coherent bar, and the nonlinear growth of this pattern can prevent classical binary fission from occurring. Confidence in these nontrivial numerical results stems, in part, from the fact that some direct overlap between numerical results and linear analyses has shown agreement and also, in part, from the fact that a number of quite different numerical techniques and codes have produced similar results. Before discussing the results, we will describe briefly the numerical codes that have been used to study the fission problem.

Four different computer programs were used in the five fission studies, as described in detail by Lucy (1977), Gingold and Monaghan (1977), Tohline (1980*a*), and Boss (1980*a*). All of the programs were designed to integrate the nonlinear equation of motion for a self-gravitating, compressible fluid (i.e., gas) in three dimensions, including the effects of gas pressure gradients. Of these four different computer programs, those written by Tohline and by Boss were based on a donor-cell finite-difference technique in which the Eulerian equation of motion was used to follow the motion of the gas through computational grid cells that were fixed in their spatial positions throughout an evolutionary calculation. This is sometimes called a fluid-in-cell technique. The programs written by Lucy and by Gingold and Monaghan used a smoothed-particle-hydrodynamic technique in which the gas was modeled as an ensemble of interacting particles whose positions changed in time as prescribed by the Lagrangian equation of motion. These two techniques are different in many respects. In studies of the free-fall collapse of isothermal gas clouds, they have apparently given, even at the qualitative level, quite different results (Gingold and Monaghan 1981,1982; Bodenheimer and Boss 1981). However, in the fission studies reported here, where the structures analyzed were initially in or near axisymmetric equilibrium and where the equations of state of the gas were much less compressible than the isothermal case, the two techniques have apparently produced very similar results. At this point in the development of the numerical techniques, we must be cautious about accepting, without careful scrutiny, the results of any single three-dimensional hydrodynamic simulation. An extensive discussion of the similarities and differences between the smoothed-particle-hydrodynamic and the finite-difference techniques is beyond the scope of this review. What should be emphasized here is that all four of the computer programs used in these studies were designed to integrate essentially the same set of coupled, partial differential equations and that results common to several different approaches are more likely to be physically correct.

In all cases, the time-dependent equations governing the evolutions were integrated forward in time using explicit integration techniques. Therefore, each evolution had its time steps restricted by the Courant condition (cf. Tohline [1982] for a discussion of various techniques) and, correspondingly, complete evolutions could only be followed for, at most, ten or so dynamical times. Deformations in the structures that were expected to arise on time scales much longer than the dynamical time scale (e.g., secular deformations) could not be adequately analyzed. Furthermore, as a result of low spatial resolution, small-scale deformations in the structures could not be accurately modeled. But as Lucy (1977) puts it, "Because the initial departure from axial symmetry is [believed to be] due to the onset of dynamical overstability for a mode of low order, we might reasonably hope that the subsequent evolution can be adequately followed with a low-resolution description of the protostar's structure."

The reader should keep in mind, when surveying the results of these numerical fission studies, that reflecting or periodic boundary conditions have often been imposed on the differential equations that describe the physical system in order to double or quadruple spatial resolution while using a fixed number of computational fluid elements. For example, when the northern and southern hemispheres of an object are expected to be mirror images of one another, reflection symmetry through the equatorial plane can be reasonably enforced and computational effort can be devoted to an analysis of one hemisphere only. Among the fission studies discussed here, the smoothed-particle-hydrodynamic models have imposed no symmetry constraints whatsoever, but the finite-difference models have consistently imposed reflection symmetry through both the equatorial plane and the rotation axis. This latter symmetry inherently prevents odd-mode nonaxisymmetric distortions from developing. Therefore, in the finite-difference models, bar-like distortions are accurately represented but the development of pear-shaped distortions, for example, is prohibited.

B. The Fission Studies

In order to study the fission problem with numerical codes like the ones just described, one must construct an initial model in or near axisymmetric equilibrium and yet one rotating fast enough to be dynamically unstable to nonaxisymmetric perturbations. Constructing such an initial model is not an easy task. One would ideally like to begin with a structure that has a slow rotation rate, then allow it to evolve slowly along one of the evolutionary paths in Fig. 1 until it surpasses a threshold for dynamical instability. Lucy has attempted to model such an evolution within the constraints of his explicit time-integration technique. All others have begun evolutions with models whose β values already exceeded that required for dynamical instability.

In all but Lucy's study, the same polytropic relation between the pressure and density was selected to specify not only the initial structure of the rotating fluid, but also to specify an adiabatic change in the gas pressure throughout the evolution. Polytropic indices of $n = 1/2$ and $n = 3/2$ were used. In the study by Lucy, a polytropic index of 1/2 was chosen to specify the initial structure, but the evolution included a rapid nonadiabatic, Kelvin-Helmholtz-like contraction phase and, therefore, a more complicated interplay between the equation of state for the $\gamma = 3$ perfect gas and the energy equation.

Among all five studies discussed here, the one reported in Durisen et al. (1984) is unique in that both the finite-difference and the smoothed-particle-hydrodynamic techniques were used to follow the evolution of identical, well-defined $n = 3/2$ polytropic models. In this respect, Durisen et al. should provide a useful standard of comparison for future work.

n = 3/2 Polytropes. Two of the five fission studies (Durisen et al. 1984; Tohline et al. 1984) focused specifically on the evolution of rapidly rotating $n = 3/2$ polytropes. These studies present results from both smoothed-particle-hydrodynamic and finite-difference-type codes at higher spatial resolutions than have been presented in all other published fission studies. They provide the most detailed and the most complete information currently available on the fission problem, at least as examined with time-dependent hydrodynamic techniques, and, hence, will be discussed in the greatest detail here.

For both studies, initial axisymmetric equilibrium models in differential rotation were constructed using the self-consistent field code of Bodenheimer and Ostriker (1973). The most extensive work has been performed on the model whose initial value of $\beta = 0.33$, but models with initial values of $\beta = 0.28$, 0.30, 0.35, and 0.38 have also been analyzed. In the remainder of this subsection, we will refer to the $\beta = 0.33$ model as the "standard" model. Its initial structure is shown in Fig. 2. After loading an initial model into one of the three-dimensional hydrodynamics codes, but before beginning the time-dependent calculation, the model was given a small perturbation to destroy the strict axisymmetric nature of its structure. In some studies, the initial perturbation was that prescribed by the Maclaurin spheroid bar eigenmode of Eq. (5) with $a \approx 0.06$; in others, random fluctuations in the density of $< 1\%$ amplitude were introduced throughout the model in an attempt not to bias any particular large-scale mode.

During its evolution, the standard model behaved in a manner typical of all $n = 3/2$ polytrope models of high initial β. Regardless of the type of initial perturbation or the type of numerical code used, as the model evolved, the amplitude of the bar mode steadily increased. The outer regions of the model developed into spiral arms trailing off the ends of the bar. With time, the bar transported angular momentum from its central regions out to the streaming arms and the arms separated from the bar. What ultimately developed was a central bar or bone-shaped object surrounded by a ring or disk of low mass, high angular momentum material. A small fraction of the debris from the

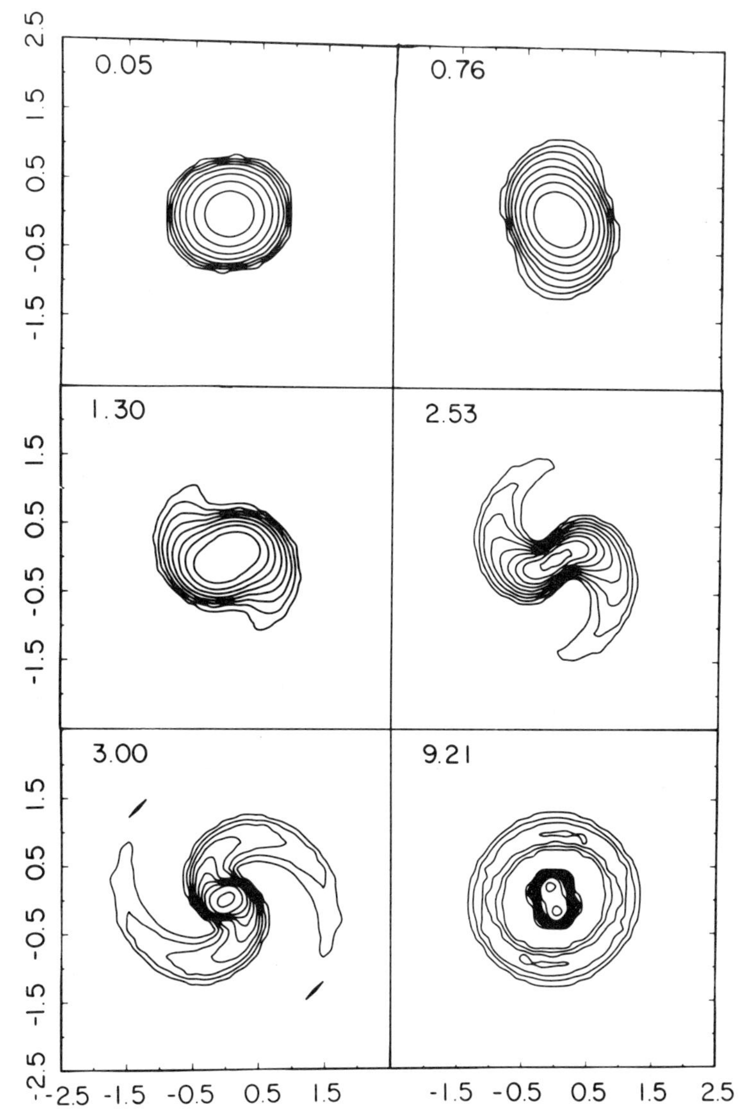

Fig. 4. Equatorial density contours as a function of time for a $\beta = 0.33$, $n = 3/2$ polytrope evolved by Tohline's code. The contouring interval is in factors of two down from maximum density. The length scales along the x and y axes are in units of the initial R_{eq}. The time for each frame in the upper left is given in units of the central initial rotation period (cirps) of the fluid. Note that, in the final frame at $t = 9.21$, the region between the ring and the central bone shape is empty. This diagram is adopted from Durisen et al. (1984).

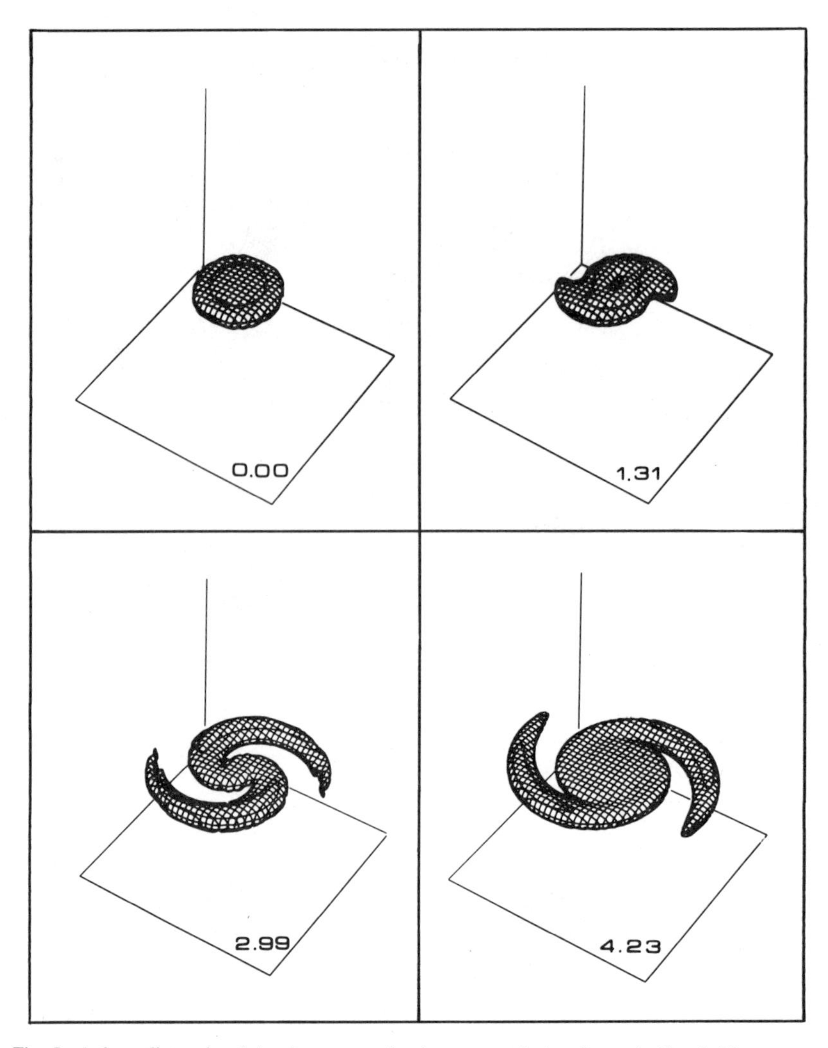

Fig. 5. A three-dimensional density contour for the same evolution shown in Fig. 4. The contour is chosen at a density level 8×10^{-4} times the initial central density of the polytrope. The time is given again in units of central initial rotation period (cirps).

arms was able to achieve escape velocity and leave the system. The central bar did not undergo binary fission. Figure 4 shows isodensity contours in the equatorial plane of the standard model at six different times in an extended evolution as calculated by Tohline's finite-difference code. Time zero, in this particular figure, marks the instant at which the Maclaurin spheroid bar eigenmode perturbation was introduced into the model, but it equally well represents the structure that occurs 9 central initial rotation periods (cirps) after a

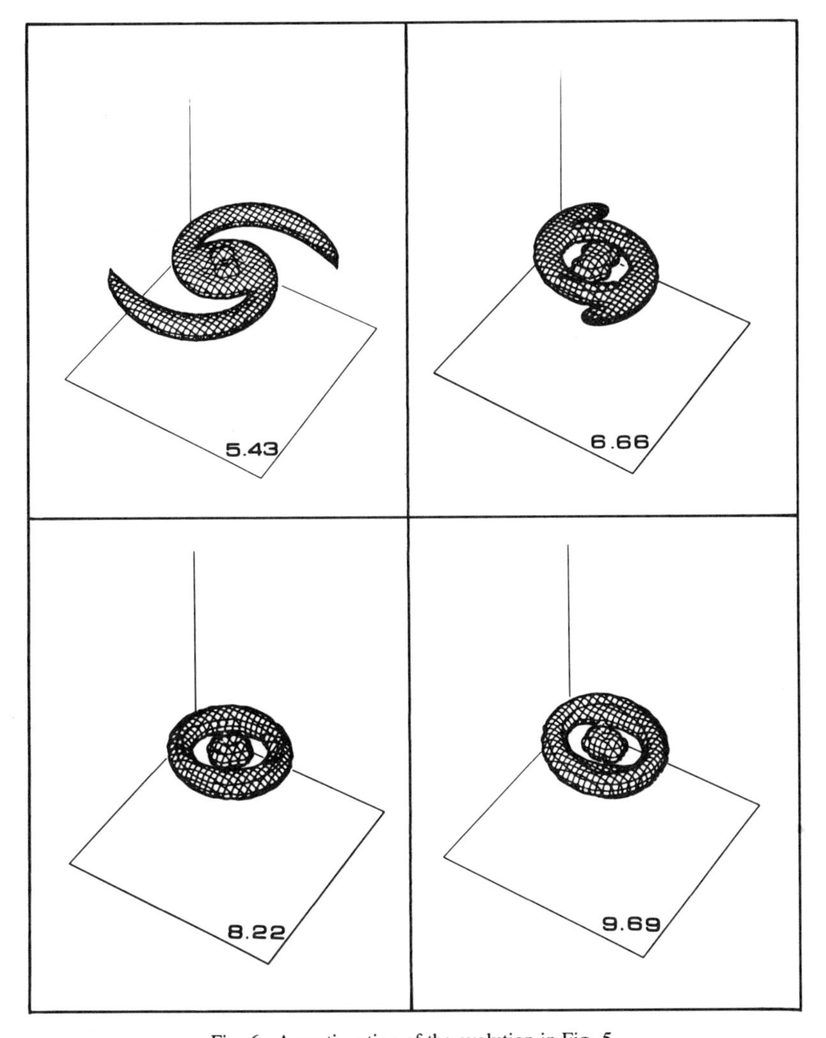

Fig. 6. A continuation of the evolution in Fig. 5.

very low amplitude random perturbation was introduced into the standard model, as reported by Tohline et al. (1984). As discussed further below, it is significant that the two evolutions converged to the same behavior from quite different starting conditions. The same evolution is displayed using three-dimensional density contours in Figs. 5, 6 and 7. The figures illustrate, better than words, the qualitative development of the dynamical instability in an $n = 3/2$ polytrope.

As reported in Durisen et al. (1984), Gingold and Monaghan's smoothed-particle-hydrodynamic code produced results that were very similar to the

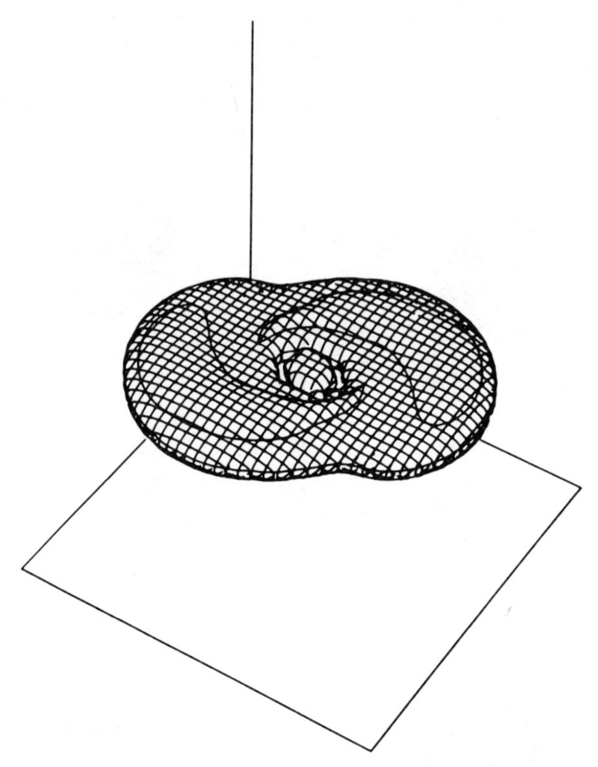

Fig. 7. Three-dimensional density contour for the end point of the evolution in Fig. 6. The density level for this contour is a hundred times lower than for Fig. 6. This illustrates that there is a low-mass disk of material extending outward from the ejected ring with a pronounced spiral pattern. Notice also that, even at this much lower density level, the gap between the central bar and ring is essentially empty.

results from Tohline's code (as just illustrated) for the standard model and for a model with an initial $\beta = 0.38$. Quantitatively, the codes agreed that, for the standard model, about 18% of the mass was ejected and that this mass contained more than half the original angular momentum. The central remnant bar was Dedekind-like with $\beta \lesssim 0.2$ and was apparently stable, because, by the end of the calculations, it had persisted virtually unchanged for several dynamical times. In terms of *a:b:c,* the final bars of both codes were about (1:0.65:0.3), which is to the right of the pear-mode instability line of Fig. 1. In fact, there was no marked deviation from reflection symmetry through the rotation axis in the smoothed-particle-hydrodynamic code. The peanut-like shape of the finite-difference bar in the last frame of Fig. 4 was probably just an artifact of axis problems in the cylindrical grid and probably does not represent any physically meaningful nonaxisymmetric distortion. Boss's finite-difference code was also used to study the $\beta = 0.38$ model. For this higher β model, all three numerical codes gave results that were similar to each other and qualitatively similar to the standard model. Quantitatively, somewhat greater mass and angular momentum fractions were ejected in the $\beta = 0.38$ model than in the standard model.

In each of these evolutions, gravitational torques were clearly important during both the initial phase of spiral arm development and a later phase of slow angular momentum transfer from the bar to the surrounding debris. In the finite-difference calculation of the standard model, material actually cleared away from the corotation radius where orbiting debris corotated with the central bar pattern. A two-armed spiral pattern, possibly driven by the bar, was actually visible in the extended disk of debris, as shown in Fig. 7. These gravitational interactions between the remnant bar and the surrounding disk or ring of ejecta would probably play an important role in the subsequent evolution of the system.

Tohline et al. (1984) have presented detailed results for the early phase of evolution of $n = 3/2$ models with only very low amplitude, random perturbations. The intent was to study whether the models would spontaneously select a unique unstable mode out of random noise and then to compare quantitatively the structure, growth rate, and pattern speed of the selected mode with the predictions of the TVE linear theory. To make this comparison, the TVE code of Ostriker and Bodenheimer (1973) was used to calculate expected pattern speeds and growth rates. With the exception of the $\beta = 0.28$ model, the lowest β model examined in this study, the evolutions all produced a dominating bar-like mode whose amplitude grew exponentially in time, as illustrated in Fig. 8. The bar was not perfectly coherent, but had a clearly defined trailing spiral pattern in the outer regions of the model, even in the initial stages of development when the pattern had a low amplitude. The bar-like mode selected a unique pattern speed which, in all models except $\beta = 0.28$, was within 20% of the pattern speed predicted by the TVE code. The growth rate of the mode was slower than the rate predicted by the linear theory, but most of this could be understood quantitatively as arising from the dissipative nature of the finite-difference technique at relatively low spatial resolution. The model with β initially equal to 0.33 was evolved for more than 10 cirps in order to allow its nonaxisymmetric structure to obtain a nonlinear amplitude. As mentioned above, by the time the bar mode was of large amplitude, the model looked nearly identical to the early stages of the evolutions discussed by Durisen et al. (1984). Hence, it seems clear that the results reported by that study are consistent with the results that would come from unbiased, low-amplitude nonaxisymmetric disturbances in dynamically unstable $n = 3/2$ polytropes.

Lower spatial resolution models of rapidly rotating, $n = 3/2$ polytropes have been described in the earlier work of Gingold and Monaghan (1978, 1979) and Durisen and Tohline (1980). It is interesting, and potentially useful to know, that many of the qualitative features seen in the high-resolution standard model just described were also observed in these earlier studies. Durisen

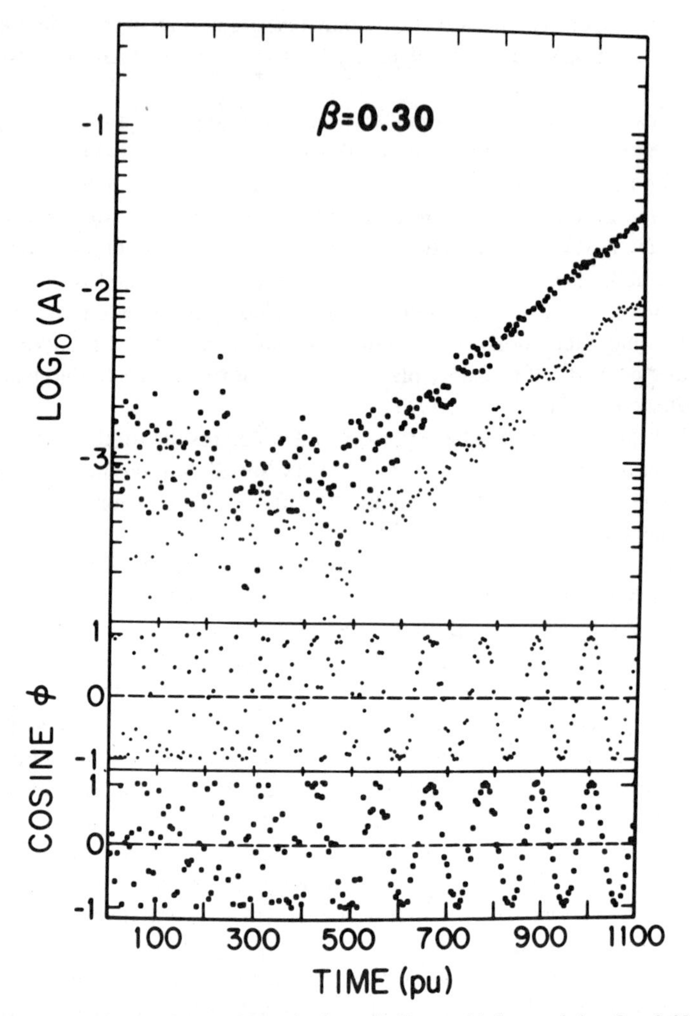

Fig. 8. The evolving amplitude and phase of a dynamically unstable bar mode in a $\beta = 0.30$, $n = 3/2$ polytrope. The amplitude A is defined here as the coefficient of the $m = 2$ azimuthal Fourier component of $\delta\rho/\rho$, where $\delta\rho/\rho$ is the fractional Eulerian change in density from its equilibrium value. Results are shown by the small and large symbols for two radii in the equatorial plane, $\varpi \approx R_{eq}/3$ and $2\,R_{eq}/3$, respectively. The phases of the same Fourier components are shown in the lower panels. Time is given in polytropic units defined so that 81.6 pu = 1 cirp (central initial rotation period). The data illustrate how a coherent two-armed pattern sorts itself out of low-level random initial noise and then grows exponentially on a dynamical time scale. This diagram is adopted from Tohline et al. (1984).

and Tohline followed the evolution of the standard model perturbed with the Maclaurin spheroid bar eigenmode using half the azimuthal resolution used in the later Durisen et al. (1984) work. Because the numerical diffusion was larger in this lower-resolution model, the rate of growth of the bar mode was artificially retarded. Qualitatively, however, the evolution proceeded in much the same manner as pictured here in Fig. 4 (see Durisen and Tohline's [1980] figure 1). Gingold and Monaghan did not use the standard initial model in their early work. Their starting models were uniformly rotating spheres with density profiles identical to the profile of a nonrotating polytrope of index $n = 3/2$. Because the models were spheres, they were not initially in hydrodynamic equilibrium, and they had β values that were initially larger than unity. A damping term was employed in the radial component of the equation of motion in the early part of each evolution in order to relax the meridional structure to its nonspherical equilibrium state. In the most rapidly rotating models, dynamical growth of nonaxisymmetric distortions began even during the early phases of radial damping. A bar-like (or bone-shaped) structure developed by the time the model relaxed to a β of 0.3. A diffuse stream of material was then ejected predominantly from one tip of the central object, carrying away about 25% of the total mass in one case and about 17% of the mass in another case. Some of the ejected debris seemed to reach escape velocity, and, as a whole, the debris housed a major fraction of the total angular momentum of the system. Even in this early study, Gingold and Monaghan concluded that, for the $n = 3/2$ polytropes, "Its ultimate state probably consists of a star surrounded by a diffuse disc."

n = *1/2 Polytropes.* The three earliest fission studies (Lucy 1977; Gingold and Monaghan 1978,1979; Durisen and Tohline 1980) examined, to some extent, the growth of nonaxisymmetric structure in rapidly rotating, $n = 1/2$ polytropes. All of these studies were performed using low spatial resolution, and the models examined in each individual study were distinctly different from the models examined in the other two studies. The results of these studies are mixed and, we feel, less trustworthy on the whole than the results for the $n = 3/2$ polytropes. Nevertheless, they deserve examination.

Lucy (1977) constructed initial models that were uniformly rotating, $n = 1/2$ polytropes with β values below the predicted dynamical stability limit of 0.27. In order to mimick the quasi-static contraction of a pre-main sequence star, Lucy allowed energy to radiate from the system in such a way as to cause the radius R to shrink via the relation $d\ln R/dt$ = constant. Because of the time constraints imposed by his explicit time-integration technique, the constant in this relation was chosen to produce an e-folding rate not too different from the rotation rate of the model. It has been pointed out that this can create difficulties in smoothed-particle-hydrodynamic simulations (Schussler and Schmitt 1981). Total angular momentum was conserved during contraction, so the rotational parameter β steadily increased. For one typical case reported, by the time $\beta = 0.26$, the model had developed rather slight nonaxisymmetry, $(a-b)/a \approx 0.08$. This suggests low numerical viscosity so that the evolutionary track for this calculation resembled the heavy solid curves in Fig. 1. By a comparable amount of time later, there was very rapid evolution of nonaxisymmetric structure. After completing a number of evolutions, Lucy identified the following trends: "(1) Following the appearance of departures from axial symmetry, a substantial fraction of mass is nearly always lost as debris. . . . The formation of debris is presumably due to the transfer of angular momentum from the protostar's interior to its surface by the action of gravitational torques. (2) Evolution into a bar-shaped structure is common. . . . (3) If fission [of the bar] does occur, it leads to a binary of small mass ratio, typically towards the lower end of the range 0.1–0.5." Thus, Lucy's work suggests that binary fission of an $n = 1/2$ polytrope can occur during a phase of contraction, but the shedding of debris is also important and may sometimes prevent binary fission from occurring.

As in their early study of $n = 3/2$ polytropes, Gingold and Monaghan (1979) did not use a model initially in axisymmetric equilibrium for their analysis of an $n = 1/2$ polytrope. Their starting model was a uniformly rotating sphere with a density profile identical to the profile of a nonrotating polytrope of index $n = 1/2$. A damping term was employed in the radial component of the equation of motion in the early part of the evolution in order to relax the meridional structure to its nonspherical equilibrium state. A "three-point star" formed by the time the model damped to a β of 0.48. Beyond this time, a more bar-like structure developed with a stream trailing predominantly off one of its ends. The bar ultimately underwent fission into a binary with a mass ratio of 0.28 comprising 91% of the initial mass surrounded by debris and a small stream with about 9% of the initial mass. The authors drew attention to the similarity between the immediate prefission phase in their evolution and a similar evolutionary stage in the detailed calculation reported by Lucy. Durisen and Tohline (1980) did not report details of any of their $n = 1/2$ polytrope evolutions, but they did state that one model with an initial $\beta = 0.33$ showed a behavior similar to the standard $n = 3/2$ model discussed above.

Related Studies. There are at least two other studies which are related to the fission work described above. Fujimoto and Sorensen (1977) used a two-dimensional hydrodynamics code to treat nonaxisymmetric breakup by modeling the initial equilibrium fluid system as an infinitesimally thin nonaxisymmetric disk with uniform pattern rotation and internal motions. The resultant disks could be considered flattened analogs of Riemann and Jacobi ellipsoids. A linear pressure surface density relation was adopted so that, at any instant of time, the disks were isothermal. The disks were evolved to higher β by cooling and were found to fragment into two or more pieces surrounded by considerable debris. A similar outcome was found by Rozyczka et al. (1980*a*) in fully three-dimensional evolutions starting with axisymmetric isothermal equilibrium states of high β. In both these investigations, the breakup process appeared to involve considerable compression of the fluid and so might more correctly be described as fragmentation rather than fission. In fact, it may be that, in the limit of extreme compressibility, the distinction between fission and fragmentation becomes blurred. This possibility needs to be explored further by considering larger n values in fission calculations of the type described in the preceeding subsections.

C. Summary of Results

One result stands out from the five fission studies: For $\beta > 0.27$ to 0.3, rotating polytropes are definitely unstable to the growth of nonaxisymmetric structure on a dynamical time scale. At least in the case of $n = 3/2$ polytropes, binary fission does not appear to be the ultimate outcome of the instability. Instead, the bar-like structure that develops from the instability is able to transport angular momentum from its central region out to its outer envelope via gravitational torques. The phase of rapid angular momentum transport is characterized by the development of material spiral arms. Upon gaining a significant amount of angular momentum, the material ejected in these arms (containing $\sim$ 10 to 20% of the total mass) forms a ring or circumstellar disk of debris. Some small portion of this material may achieve escape velocity from the system. Gingold and Monaghan (1978) appropriately describe the evolution by saying, "the $n = 1.5$ polytropes, in Eddington's (1922) phrase, 'are acting like a prudent balloonist, throwing out ballast to avoid the bump'." This result seems to hold independent of a wide variety of perturbations imposed on the initial starting model.

In the case of $n = 1/2$ polytropes, we cannot, with complete certainty, identify the outcome of the dynamical instability. Although Lucy reported the details of one evolution that did produce low mass ratio binary fission, he indicated in his closing remarks that he did observe other cases where the evolution was more like the $n = 3/2$ polytrope evolutions reported here. Gingold and Monaghan found that binary fission is always the outcome of their $n = 1/2$ evolutions, but this may be an artifact of the very high β values of their nonequilibrium starting conditions. In a sketchy account, Durisen and Tohline report that an $n = 1/2$, $\beta = 0.33$ evolution produced spiral arm ejection not binary fission. But if the development of odd-mode distortions is important in the $n = 1/2$ fission process, as suggested by the evolutions of both Lucy and of Gingold and Monaghan, then we must assume that by suppressing these modes, Durisen and Tohline have obtained an incorrect picture of the evolution.

We conclude that dynamical bar-mode distortional instability in rapidly rotating, $n = 3/2$ polytropes does not lead to binary fission. The situation for fluids of different degrees of compressibility and for pear-shaped distortional instabilities is much less certain and needs further study.

D. Future Expectations

It is clear that we have only scratched the surface with regard to following fission instabilities with hydrodynamic modeling techniques. Fission was originally envisioned as occurring as a result of secular evolution. The hydrodynamic calculations so far have only been able to treat the nonlinear growth of dynamical instabilities. Computer programs using implicit time-integration techniques will have to be written in three dimensions before secular evolutions can be completely studied. This type of work will demand large blocks of time on the largest and fastest computers presently available and is therefore probably the type of problem that will await solution for some time to come. In the meantime, however, models akin to the type explored by Lucy, where the secular evolution time scale is artificially sped up, should be used to test the behavior of secular instabilities with currently available (i.e., explicit) programs.

It is certainly worthwhile to continue exploring the nonlinear behavior of dynamical instabilities. For a while, it seems advisable to restrict investigations to relatively simple physical systems, such as polytropes. These systems can then be easily compared from one numerical technique to the next, and they have at least some contact with predictions of linear theory. Since different numerical techniques do not currently give identical results when working on identical models, it is clearly advisable to encourage substantial overlap between different studies. Ultimately, of course, the hope is that more complicated (and therefore more realistic) systems can be studied with these and similar techniques.

It is obvious that improvements in the current computational programs are desirable if we are to fully understand the outcome of even dynamical instabilities. The programs that rely on finite-difference techniques have, inherent in them, some unwanted numerical diffusion which tends to smear out nonaxisymmetric features. Either a substantial improvement in the spatial resolution of the computational grids will have to be made or higher-order finite-difference techniques (cf. Norman et al. 1980) will have to be employed before quantitative features of the calculations will be trustworthy. These programs will also need to relax the restriction that they have customarily employed of only allowing even-mode distortions to develop in the models. Pear-shaped instabilities, for example, cannot be studied if evolutions are restricted to even-mode distortions. This restriction is easy to relax but again demands a larger number of grid cells in each calculation and a correspondingly larger amount of computer storage and time. Those who use smoothed-particle-hydrodynamic techniques will also have to be continually aware of the fact that poor resolution in the outer envelopes of their models may inadvertantly alter the outcome of an evolution. For example, if some objects do avoid fission by transporting angular momentum from central regions to the outer envelope through gravitational torques, then the outer envelope must be accurately modeled as a fluid in order to receive the angular momentum at the necessary rate. The most straightforward way for both techniques to overcome their weakness is to improve the resolution (increase the number of particles) in the model.

V. SUMMARY AND DISCUSSION

A. Overall Status of Fission Theory in 1984

The following are the major results since the mid-1970s of research on the fission problem:

(1) At least for polytropes with indices $n \leqslant 3$, the linear stability limits and bifurcations, both secular and dynamical, for bar-, pear-, and ring-shaped distortions agree quantitatively in β with those for the incompressible Maclaurin spheroids. This result appears to be insensitive to rotation law, as long as a model sequence does not truncate at low β due to critical equatorial rotation. The bar modes in particular have been most thoroughly studied. The TVE method has proven to be quite accurate in practice, despite some serious theoretical drawbacks in principle. The extensive citation of Maclaurin spheroid stability limits over the years in astrophysical contexts appears to have been justified.

(2) The outcome of dynamical bar-mode instability for $n = 3/2$ polytropes is reliably known. The nonlinear growth of a bar mode leads to ejection of mass and angular momentum in two trailing spiral arms. The ejected material shows no tendency to condense into one or more discrete bodies but instead forms a ring or disk around the central remnant. For $\beta = 0.33$, the most well-studied case, 18% of the mass is ejected, containing more than half the angular momentum. The central remnant is a stable Dedekind-like bar with $\beta \leqslant 0.2$. Gravitational torques appear to play an important role in transferring angular momentum from the central regions to the ejecta, especially during the violent spiral arm ejection phase. Contrary to widespread expectations, binary fission is not the direct outcome of dynamical bar-mode instability for $n = 3/2$. For $n = 1/2$, the evidence is more ambivalent, with some calculations suggesting binary fission and others not. If result (1) above carries over to the nonlinear regime, $n = 1/2$ and $n = 3/2$ polytropes should behave similarly.

(3) Numerical three-dimensional hydrodynamic codes have proven their ability to treat a classic problem of astrophysical interest, both in the linear and nonlinear regimes. The fission problem is a better test case for these codes than the fragmentation problem they were originally designed for, because the fission problem has well-developed linear stability results to use as a basis for comparison. With the current state of the art, some caution is needed when attempting to identify physical as opposed to purely numerical features in results. It is useful, when possible, to apply several different codes to nearly identical tasks.

(4) Our knowledge about possible equilibrium states for rotating fluids is expanding rapidly. Progress has been especially notable in the construction of (a) nonaxisymmetric equilibrium states for uniformly rotating fluid with low compressibility ($n \lesssim 1$), and (b) axisymmetric equilibrium states of differentially rotating fluids with high compressibility ($n \rightarrow \infty$, i.e., isothermal) and finite boundary pressure.

B. Applications

Like the classic incompressible theory, modern fission calculations have been highly idealized, employing simple, generalized equations of state and rotation laws. Although detailed modeling will almost certainly prove possible in the coming decade, few calculations so far have attempted to tailor specific features to specific problems. In the following subsections, we will offer some general comments about potential applications in the hopes of stimulating further research and discussion.

1. Fission in Stellar Evolution. The formation of close binary stars has, of course, been one of the driving forces behind investigations of fission instabilities. On the basis of his own work with $n = 1/2$ polytropes, Lucy (1981) concluded that fission during quasi-static contraction was not an adequate explanation for close binary origin, because of the tendency he found to produce low mass ratios. For W UMa stars, the mass ratios are more nearly a half to one. The numerically improved results of Durisen et al. (1984) suggest that, for $n = 3/2$ polytropes, the situation is actually more severe. A binary system simply does not form directly by fission. The $n = 3/2$ equilibrium states should be realistic models for rapidly rotating Hayashi track stars. Neglect of energy transport effects in Durisen et al. is probably not a serious limitation, because, once the instability gets started, it occurs dynamically. Not enough time is available during the crucial spiral arm ejection phase for significant heat transfer. Thus, calculations for $n \lesssim 3/2$ make it clear that fission does not lead directly to a nearly equal mass close binary system in stars of relatively low compressibility.

It is tempting to conclude, once and for all, that close binaries do not form by fission during a Hayashi track stage, but some caution is needed. So far, only the bar-mode instability has been definitively resolved. The outcome of dynamical pear-mode instability is not as well known, although some of the work by Lucy (1977), Gingold and Monaghan (1979), and Eriguchi et al. (1982) again suggests debris shedding not binary fission. Even if ejection of debris should prove the rule for fission instabilities, we know little at present about the ultimate fate of the ejected debris or the central remnant. If the central remnant continues quasi-static contraction, it is conceivable that it could undergo a series of ring-ejection episodes reminiscent of the Kant-Laplace theory for formation of the solar system. On the basis of what we know now, this seems more likely than a later binary fission event. In this way, it is feasible that some part of the solar nebula was extruded in discrete rings. The processes in these rings would be similar to those hypothesized for a more traditional solar nebula and are more likely to produce a planetary system out of the debris rather than a single stellar companion.

As pointed out by Roxburgh (1966) for different reasons, the transition from convective to radiative equilibrium during pre-main sequence contraction may play an important role in fission theory. In some respects, a radiative star behaves like a more compressible gas, leading to an $n = 3$ centrally concentrated density structure. What little information is available concerning fission-like instabilities in highly compressible structures (Fujimoto and Sorensen 1977; Rozyczka et al. 1980*a*) indicates fragmentation into two or more comparable pieces. The actual situation is a bit more complicated, however, because a low-mass star in radiative equilibrium will respond to adiabatic perturbations with $P \sim \rho^{5/3}$, i.e., in this sense, as an $n = 3/2$ fluid. The simple polytrope models used to date are not directly applicable to such nonisentropic objects. If binary fission during the radiative phase proves possible hydrodynamically, certain constraints on the rotation state for pre-main sequence stars are required for it to occur. As discussed in companion chapters (see Rydgren and Cohen, and Giampapa and Imhoff), T Tauri stars have slower surface rotation than needed to achieve fission instability during subsequent contraction. For fission to occur, the interior must rotate several times faster than the surface. Future studies of solar oscillations and of the rotational histories of stars may help to resolve the issue of internal rotation rates.

The possibility also exists that fission instabilities occur in the cores of highly evolved stars during quasi-static evolution (Endal and Sofia 1978) or during a final phase of core collapse (Bodenheimer and Woosley 1983). The effects of fission instabilities in these cases could be profound but have not been explored in any detail.

2. *The Accretion Phase of Star Formation.* One free-fall time after the initiation of gravitational collapse in a protostellar cloud, a hydrodynamic equilibrium structure forms out of the centermost material and grows thereafter by accretion of the outlying material. As reviewed in Tohline (1982), agreement about the details of this process has only recently been achieved even for nonrotating collapse from a uniform Jeans mass initial condition. Such agreement is lacking for collapse from rotating initial conditions, and it is not at all clear what the proper initial conditions should be. A more detailed discussion of these issues can be found in other chapters of this book, especially in the chapter by Cassen et al. Researchers generally agree that clouds with large initial β fragment during collapse. The important consideration for us here is that, for some plausible parameter ranges, particularly low initial β's, it is possible to accrete a central rotating equilibrium body of steadily increasing β. For instance, Yuan and Cassen (1984) have shown that stars accreting from the self-similar isothermal collapse flows of Terebey et al.

(1984*b*) achieve final values of $J\ /\ (GM^3R_{eq})^{1/2}$ in excess of the bar-mode secular stability limits, as determined in this J parameter by Ostriker and Bodenheimer (1973).

The exact fate of such a secularly unstable equilibrium protostar depends on dissipation mechanisms and the presence of an external disk. With no external disk and moderately large viscosity, the object will follow the dashed trajectory in Fig. 1 to dynamical pear-mode instability. As discussed elsewhere, this will probably, but not certainly, lead to ejection of material. Because accretion will continue to drive the instability, the likely outcome is that the body will episodically extrude a disk of material. According to Yuan and Cassen (1984), the presence of a disk, whether formed by this extrusion process or formed directly by accretion into orbit around the central body, can drastically alter subsequent evolution, because the bar will lose angular momentum to the disk by stimulating a pair of spiral density waves at the outer Lindblad resonance. A similar process seems to occur during the later stages of the $\beta = 0.33$, $n = 3/2$ evolutions in Durisen et al. (1984) but is not quantitatively well resolved.

As suggested in Sec. III.B, gravitational torques exerted by external matter may act on bar modes in a manner analogous to gravitational radiation. If so, the work of Detweiler and Lindblom (1977) and Lindblom and Detweiler (1977) indicates that the evolution could become quite complex when both viscous and gravitational torques on a bar are strong. In addition, if the viscosity is small, an intriguing possibility arises. In cases where an external disk is always present due to direct accretion, the generic instability of modes with $m > 2$ may be triggered by gravitational torques even at low β and might severely limit the rotation rate of the central body. Clearly, a great deal of theoretical work remains to be done before we untangle this web of alternative evolutionary routes.

3. Planetary Satellite Systems. Fission instabilities may also have played a role in the formation of regular planetary satellites, both for the gas giant planets and for the Earth-Moon and Pluto-Charon systems.

As described in the chapters by Pollack and by Bodenheimer, there are basically two scenarios for gas giant planet formation. Gas giants may form all at once as bound units through a Jeans instability. Alternatively, the solid cores may accrete directly by aggregation of planetesimals, and the gaseous envelopes may then condense around the cores gravitationally. Evolutions of gas giant planets from solar nebula conditions toward their current states have been computed in detail under both sets of assumptions, but primarily for spherically symmetric, nonrotating cases. In the Jeans instability scenario, dissociation of H_2 causes an early hydrodynamic collapse phase down to radii about twice the current values for Jupiter and Saturn. With rotation included, this hydrodynamic collapse phase could lead to fragmentation or to disk formation, much as expected for protostellar cloud collapses. Many of the possibilities discussed in Sec. V.B.2 above are applicable. A central equilibrium object forming out of the collapse could undergo ejection episodes or could couple to a surrounding disk via gravitational torques.

The core accretion scenario is even more interesting from the viewpoint of fission instabilities. When the core mass exceeds a critical value, the gaseous envelope contracts down on the core rapidly but quasi-statically. The initial envelope extends out to the Hill sphere and so is likely to have large specific angular momentum. Extensive quasi-static contraction and rapid rotation provide ideal conditions for manifestation of fission instabilities. During the early phases, the envelope is very nearly a polytropic gas with $n = 3/2$, and so the results of existing hydrodynamic fission calculations for $n = 3/2$ polytropes are directly applicable. We would thus expect one or even several episodes of mass and angular momentum ejection by the contracting envelope. Portions of the planet's regular satellite system could form from each ring or disk ejection episode. It may prove useful in the future to perform detailed two- and three-dimensional simulations of the contraction using initial conditions as similar as possible to what would be expected from the core accretion picture.

In either scenario, a last ring ejection episode, as the central planet settles down to near current size, could lead to formation of a primordial ring system, such as Saturn's. One version of the primordial hypothesis supposes that the ring particles form by direct condensation from circumplanetary gas. As discussed in the chapter by Lissauer and Cuzzi, this could place tight constraints on time scales. The gaseous disk must last long enough for icy condensates to grow but not so long that gas drag will cause all the condensates to spiral into the planet. If the circumplanetary gas forms by fission, however, there will be a gap between the inner edge of the ejected gas and the planet (see Figs. 4, 5, and 6). If gravitational torques slowly expand the gaseous ring, condensates accumulated at the inner edge could be left behind to form the ring system.

One of the classic applications of fission theory has been as an explanation for the formation of the Earth-Moon system (Darwin 1879). This idea has remained one of the principal contending hypotheses for lunar origin (see O'Keefe and Sullivan [1978] for a modern version). As pointed out by Durisen and Scott (1984), the recent results from numerical hydrodynamic calculations may help to revitalize the hypothesis in the following ways. If a rapidly rotating proto-Earth fissioned, it may have ejected some 10 to 20% of its mass and more than half its angular momentum as a ring or disk of debris, rather than a single body. The Moon could then have aggregated in Earth orbit out of a small fraction of the debris. The rest of the mass and angular momentum would have been lost. This could explain why the current Earth-Moon system only has considerably less than the angular momentum needed to make the proto-Earth unstable. The accretion of the Moon in Earth orbit is a common feature of other nonfission hypotheses of lunar origin, and so the fission theory really encounters no greater difficulties than these others with regard to geochemical differences between the Earth and Moon. In fact, it has the distinct advantage of explaining the Moon's depletion in iron and siderophile elements, if one assumes that the Earth's core differentiated prior to fission. Fission does still suffer, however, from the possibly fatal flaw that it is difficult to imagine how the Earth could have been made to rotate so fast by accumulation from planetesimals. Also, fission calculations, so far, have used only polytropic fluids. Use of a more realistic rock-and-rubble equation of state may drastically alter the outcome (Boss and Mizuno, personal communication, 1984). Further consideration of this revised picture for a fissioned Moon certainly seems warranted. A related discussion of the Pluto-Charon system can be found in Lin (1981*b*).

4. Gravitational Torques. As discussed in the chapters by Cassen et al., by Lin and Papaloizou, and by Lissauer and Cuzzi, and in papers by Lin and Papaloizou (1979*b*), Larson (1984), and Boss (1984*a*) among others, there is a growing realization that gravitational torques play an important role in star and planet formation. Two major effects are expected: (1) net transfer of angular momentum (usually outwards) to or through a circumstellar disk and (2) tidal truncation of such a disk near regions of resonance between the disk and an imposed time-varying, nonaxisymmetric gravitational field. The gravitational field for the tidal truncation process could result from a central binary or nonaxisymmetric star or from an embedded planetary body. Gravitational torques may provide a mechanism for forming close binary stars. A wide binary might form by fragmentation during gravitational collapse and then spiral together due to gravitational torques exerted by surrounding material.

Both angular momentum transport and tidal truncation are manifest in the nonlinear calculations of dynamical fission instabilities. Angular momentum transport during the spiral arm ejection phase plays a role in surpressing binary formation. Truncation also occurs during later stages, as illustrated in Fig. 6 by the gap which opens near corotation between the central bar and the ejected mass. After truncation, tidal torques may continue to evolve the inner edge of the ejected ring or disk outward. Fission calculations may thus provide a useful laboratory for studies of gravitational torques under nontrivial but fully self-consistent conditions.

C. Prospects for the Next Six Years

Given our growing confidence in techniques and the increasing availability of powerful computational facilities, a continued or even accelerated pace of development can be expected in all aspects of fission theory which require numerical work. Reliable numerical simulations of dynamical instabilities for bar and pear modes should become available for a much wider range of compressibilities, and we will begin to see the inclusion of some additional physical effects (nonisentropic fluids, energy transport, more realistic equations of state). In this way, calculations will lend themselves more and more to realistic astrophysical applications. Numerical simulations will probably also provide many important new results on equilibrium states and stability. For instance, nonuniform, nonaxisymmetric, compressible equilibrium states with internal motions are almost impossible to construct directly by equilibrium techniques, but such objects seem to form naturally as end products of nonlinear hydrodynamic calculations. Accurate linear eigenmode analyses for rotating stars are also extremely difficult, but numerical codes have already proven their ability to provide useful results in the linear regime.

Some advances in equilibrium and linear stability theory will be made by direct methods as well. In particular, the nonaxisymmetric instabilities of the recently constructed isothermal equilibrium states will be tested. We will probably find that the seeming generality of stability properties for $n \leqslant 3$ does not carry over to highly compressible configurations with finite surface pressure. An effort is also needed to study fluids which are not strictly polytropic.

Although it is not hard to foresee what may be accomplished by numerical techniques, it is rather difficult to predict the outcome of more theoretical, speculative, or application-oriented efforts. In the realm of esoteric studies, we may continue to be jolted from time to time by surprising theorems obtained from variational principles and bifurcation theory. In the realm of more concrete theoretical studies, we expect that future research on gravitational torques will prove extremely profitable, both for fission theory itself and for the theory of star and planet formation generally. We have tried in this final section to outline possible ramifications of fission theory for a range of astrophysical problems. If we have piqued the interest of both theorists and observers to the extent that we had hoped, there should be a wide variety of fruitful applications over the coming years.

Acknowledgments. We would like to thank D. C. Black, A. P. Boss, P. Cassen, R. A. Gingold, I. Hachisu, N. R. Lebovitz, J. L. Tassoul, and C. Yuan for useful comments and correspondence. During preparation of this manuscript, the authors were supported by grants from the National Science Foundation.

PART IV
Chemistry and Grains

THE CHEMICAL STATE OF DENSE INTERSTELLAR CLOUDS: AN OVERVIEW

WILLIAM M. IRVINE, F. PETER SCHLOERB
University of Massachusetts

ÅKE HJALMARSON
Onsala Space Observatory

and

ERIC HERBST
Duke University

We list the currently known interstellar molecules and isotopes, discuss procedures for determining relative chemical abundances in molecular clouds, and present current best estimates for such abundances in regions of differing physical properties. Among the results are a general chemical similarity across a range of density and temperature for quiescent clouds, and some striking differences among regions which are not easily related to such physical parameters and may instead reflect cloud history and evolution. The possibility of constraining chemical models via measurements of relative abundances for the isomeric pairs HNC/HCN, CH_3NC/CH_3CN, and HOC^+/HCO^+ is discussed in detail.

Molecular clouds constitute the dominant mass phase of the interstellar medium and are of crucial significance to astronomy because they are the sites of star formation. The chemical composition of these objects is important both in its own right and as an aid to the study of cloud structure and evolution. Thus, observations of the excitation of molecular species probe the physical state of molecular clouds, including characteristics such as kinetic temperature, density, mass, and velocity structure. Moreover, the relative abundances of molecular species may also trace interstellar dynamical processes. As described in chapters by Glassgold and by Herbst, the chemical constitution of interstellar clouds naturally evolves along with their physical state, so we may hope that observed chemical abundances ultimately can serve to determine the state of evolution of collapsing molecular clouds. Processes such as the passage of shock waves or the outflow of material from both young protostellar objects and evolved stars will also leave traces through enhancements or depletions of particular molecules and isotopic species. The interpretation of such possible signposts of the evolutionary history of the cloud is likely to be complicated, because the chemical constituents of clouds serve not only as probes of star-forming regions but also directly affect their evolution. Together with C and C^+, abundant molecules such as H_2O and CO serve as cooling or heating agents and are thus fundamental to the evolution of these objects and to the efficiency of protostellar formation (P. F. Goldsmith and W. D. Langer 1978; Evans et al. 1982*a*; Takahashi et al. 1983). Similarly, the state of ionization, together with magnetic fields, certainly plays a crucial role in cloud dynamics and evolution (chapters by Langer and Mestel in this book; D. C. Black and E. H. Scott 1982; Paleologou and Mouschovias 1983). Thus, the challenge of interstellar molecular chemistry is to understand the origins of interstellar molecules within this very complex physical system. Once met, we may hope that this will lead to a better understanding of molecular clouds themselves.

Molecular clouds also provide laboratories where we can study both chemical processes and the properties of individual molecules in a way that cannot be duplicated on the Earth. It seems probable, for example, that radiative association plays a significant role in production schemes for complex molecules in the interstellar medium, although this process has only rarely been observed directly in the laboratory (chapter by Herbst). Moreover, a number of unstable molecules and radicals have been studied in more detail by astronomical means than by conventional methods, and some interstellar species have yet to be detected in the laboratory (see, e.g., Green 1981).

Finally, the chemistry of interstellar material may be directly linked to objects still surviving in our own solar system. Kerridge and Chang (see their chapter) discuss evidence for the preservation of interstellar matter in carbonaceous chondrites. In addition, the most primitive surviving matter in the solar system is generally believed to be that in comets (see Table I). The molecular species traditionally observed by optical means in comets consist almost entirely of unstable radicals and ions; these are generally believed to be dissociation products of the parent molecules that are the real constituents of cometary nuclei. Models of the chemistry in cometary comae suggest that the constitution of the nucleus is, in fact, closer to that of interstellar volatile material than to what one would expect as an equilibrium condensate from the solar nebula (L. Biermann et al. 1982; G. F. Mitchell et al. 1981). Although

TABLE I
Atomic and Molecular Species Observed in Comets[a]

Organic	C, C_2, C_3, CH, CN, CO, CS, (HCN)[b], (CH_3CN), (HC_3N), (HCO)
Inorganic	H, NH, NH_2, O, S, OH, Si, (H_2O), (NH_3), S_2
Metals	Na, Ca, Cr, Co, Mn, Fe, Ni, Cu, V, K
Ions	CO^+, CO_2^+, CH^+, CN^+, N_2^+, OH^+, H_2O^+, C^+, Ca^+, (H_2S^+)
Dust	Silicates, (Ices)

[a]For references, see Irvine (1983) and Delsemme (1982); and recent assignments by Altenhoff et al. (1983) for NH_3, by Hanner (1984) for H_2O ice, by A'Hearn et al. (1983) for S_2, by Irvine et al. (1984) for HC_3N, and by Cosmovici and Ortolami (1984) for HCO and H_2S^+.
[b]Parentheses indicate claimed but not yet confirmed.

this does not prove the survival of such material in comets, it does suggest the possibility that when our understanding of both interstellar chemistry and the nature of comets has increased sufficiently, we will have obtained important clues to the processes that formed our own solar system and may perhaps specify in some detail the type of interstellar cloud in which our Sun and planets were born.

I. CHEMICAL CONSTITUENTS OF INTERSTELLAR CLOUDS

A. Identified Interstellar Molecules

Our knowledge of the chemical constituents of the interstellar medium has considerably increased during the period since 1978. The number of molecular lines detected has roughly tripled relative to the ~400 quoted by Lovas et al. (1979), largely through the systematic spectral scans of Ori A, IRC + 10216, and Sgr B2 carried out with new, high-sensitivity instruments operated by the Onsala Space Observatory, Owens Valley Radio Observatory, and Bell Laboratories (Johansson et al. 1984; Sutton et al. 1984; Cummins et al. 1984). The majority of the new lines have been assigned to asymmetric rotors such as CH_3OH, CH_3OCHO, CH_3CH_2CN, $(CH_3)_2O$, and SO_2 and isotopic variants. About 15 new interstellar molecules, ions, or radicals have been discovered during this period (see Table II), as improved receivers at infrared, centimeter, and millimeter wavelengths have made possible more sensitive searches. Roughly 60 interstellar or circumstellar molecular species are now known. It is interesting to note that the galactic center source Sgr B2 no longer appears unique as it did a few years ago. Most of the truly interstellar molecules have now been detected in Ori A (Table III), and the nearby, cold, dark clouds in Taurus have yielded a peculiar chemistry of their own (Table IV), including twice the number of molecules known in these regions in 1978. References for the species in Tables II, III, and IV may be found in the spectral surveys cited above, and in reviews by Winnewisser et al. (1979), Irvine and Hjalmarson (1983), Mann and Williams (1980), Thaddeus (1981*a*), Lafont et al.

TABLE II
Interstellar Molecules[a]

Simple hydrides, oxides, sulfides, amides and related molecules				
H_2	CO	H_2CNH	CC	$NaOH$?
OH	OCS	H_3CNH_2	CS	
CH	NO	HN_2^+	SiS	
CH^+	HNO?		NS	
H_2O	SiO		SiC_2*	
H_2S	SO			
NH_3	SO_2			
CH_4*				
SiH_4*				
Nitriles, acetylene derivatives, and related molecules				
CN	HCN	H_3C-CN	$H_2C=CH_2$*	
$C\equiv CH$	$HC\equiv C-CN$	$H_3C-C\equiv C-CN$	H_3C-CH_2-CN	
$C\equiv C-CH$	$H(C\equiv C)_2-CN$	$H_3C-C\equiv CH$	$H_2C=CH-CN$	
$C\equiv C-C\equiv CH$	$H(C\equiv C)_3-CN$	$H_3C-(C\equiv C)_2-H$	$HN=C$	
$C\equiv C-CN$	$H(C\equiv C)_4-CN$		$HN=C=O$	
$C\equiv C-CO$	$H(C\equiv C)_5-CN$*		$HOCO^+$	
$HC\equiv CH$*			$HN=C=S$	
H_2NCN				
Aldehydes, alcohols, ethers, ketones, and related molecules				
$H_2C=O$	H_3COH	$HO-CH=O$	$HC=O^+$	
$H_2C=S$	H_3C-CH_2-OH	$H_3C-O-CH=O$	$HC=S^+$	
$H_3C-CH=O$	H_3CSH	$H_3C-O-CH_3$	$HC=O$	
$NH_2-CH=O$			HOC^+	
$H_2C=C=O$				

[a]Detections since 1978 are underlined. Previously reported lines of CO^+ and O_3 have now been assigned to CH_3OH and $^{13}CH_3OH$ by Sutton et al. (1984) and by G. A. Blake et al. (1984). For references to detections, see Sec. I and Bogey et al. 1984. (*) Indicates detected only in the envelope around the evolved star IRC+10216; (?) claimed but not yet confirmed.

(1982); see also recent assignments by Hollis and Rhodes (1982) for NaOH, Goldhaber and Betz (1984) for SiH_4, and Thaddeus et al. (1984) for SiCC, as well as the references for subsequent tables in this chapter.

Recently three new interstellar molecules have been detected. Three groups have independently reported the observation of several rotational transitions of $CH_3(C\equiv C)_2H$ in the cold cloud TMC-1 (Loren et al. 1984; Walmsley et al. 1984; MacLeod et al. 1984). This new polyacetylene thus adds to the complex mix of carbon chain species observed in the Taurus region. In contrast to CH_3C_4H, which has been studied in the laboratory for two decades, another new molecule, tricarbon monoxide ($C\equiv C-CO$), has only recently been identified on Earth (R. D. Brown et al. 1983). Detection is again reported for TMC-1 (Fig. 1), where the abundance appears to be substantially less than that of the radicals C_3N and C_4H. Finally, the suggested assignment to C_3H of two recently discovered doublets in the spectrum of the carbon star IRC + 10216 by Johansson et al. (1984) has been confirmed by subsequent detection of additional transitions at Bell Laboratories and at the Five College Radio Astronomy Observatory (FCRAO), by laboratory studies at the NASA Goddard Institute for Space Studies, and by the detection of the lowest-energy

TABLE III
Molecules Identified in the Orion Molecular Cloud[a]

Simple hydrides, oxides, sulfides, amides, and related molecules			
H_2	CO	H_3CNH_2	CS
OH	NO?	HN_2^+	
CH	OCS		
H_2O	SO		
H_2S	SO_2		
NH_3	SiO		
Nitriles, acetylene derivatives, and related molecules			
CN	HCN	$H_3C{-}CN$	$H_3C{-}CH_2{-}CN$
$C{\equiv}CH$	$HC{\equiv}C{-}CN$	$H_3C{-}C{\equiv}CH$	$\underline{H_2C{=}CH{=}CN}$
	$\underline{H(C{\equiv}C)_2{-}CN}$		$\underline{HN{=}C}$
			$\underline{HN{=}C{=}O}$
Aldehydes, alcohols, ethers, ketones, and related molecules			
$H_2C{=}O$	H_3COH	$\underline{H_3C{-}O{-}CH{=}O}$	$HC{=}O^+$
$\underline{H_2C{=}S}$	$\underline{H_3C{-}CH{=}O}$	$\underline{H_3C{-}O{-}CH_3}$	$\underline{HC{=}S^+}$
$\underline{H_2C{=}C{=}O}$		$\underline{HO{-}CH{=}O}$	$\underline{HCO?}$

[a]New detections since 1978 are underlined. Lines previously attributed to CH_4, CO^+, O_3, and CH_3CHO have been reassigned to species included here by Ellder et al. (1980), Johansson et al. (1984), G. A. Blake et al. (1984), and Sutton et al. (1984). See also Wootten et al. (1984*a*) for NO and unpublished Onsala data (B. Höglund, personal communication) for H_2CS and Sutton et al. (1984) for HCOOH; (?) claimed but not yet confirmed.

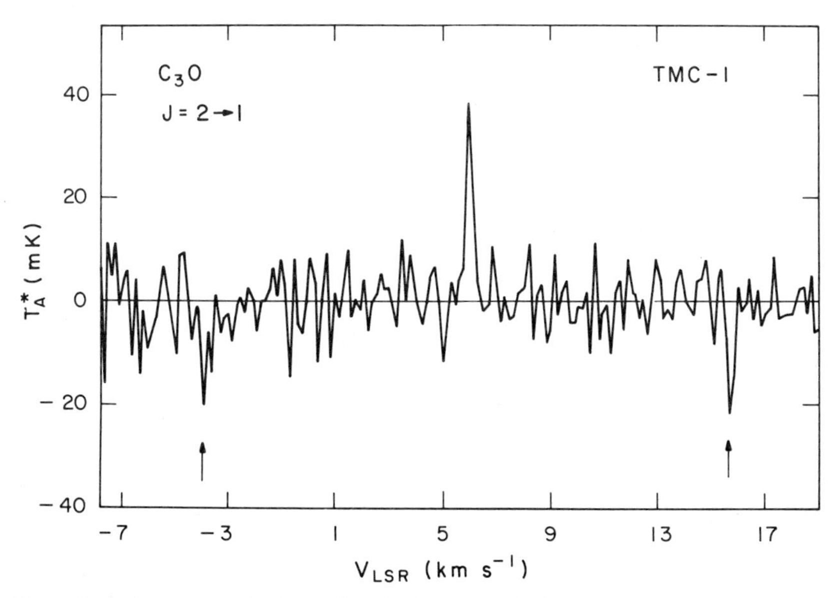

Fig. 1. Detection spectrum for interstellar tricarbon monoxide in the nearby dark cloud TMC-1. Arrows indicate "images" produced by frequency—switching technique (H. E. Matthews et al. 1984*b*).

TABLE IV
Molecules Identified in Dark Clouds[a]

Simple hydrides, oxides, sulfides, amides, and related molecules			
H_2[b]	CO	HN_2^+	CS
OH	$\underline{SO}$		$\underline{CC}$
CH	$\underline{SO_2}$		
NH_3			
Nitriles, acetylene derivatives, and related molecules			
CN	HCN	$\underline{H_3C{-}CN}$	$\underline{H_2C{=}CH{-}CN}$
$C{\equiv}CH$	$HC{\equiv}C{-}CN$	$\underline{H_3C{-}C{\equiv}C{-}CN}$	$\underline{HN{=}C}$
$\underline{C{\equiv}C{-}C{\equiv}CH}$	$H(C{\equiv}C)_2{-}CN$	$\underline{H_3C{-}C{\equiv}CH}$	$\underline{HN{=}C{=}O}$
$\underline{C{\equiv}C{-}CN}$	$H(C{\equiv}C)_3{-}CN$	$\underline{H_3C{-}(C{\equiv}C)_2{-}H}$	
$\underline{C{\equiv}C{-}CO}$	$\underline{H(C{\equiv}C)_4{-}CN}$		
$\underline{C{\equiv}C{-}CH}$			
Adlehydes, alcohols, ethers, ketones, and related molecules			
$H_2C{=}O$	$HC{=}O^+$		
$\underline{H_2C{=}S}$	$\underline{HC{=}S^+}$		
$\underline{H_2C{=}C{=}O?}$			
$\underline{H_3C{-}CH{=}O}$			

[a]New detections since 1978 are underlined. In addition to references in text (Sec. I), note for H_2CS (P. Vanden Bout, personal communication, 1984), for H_2CCO (H. Matthews, personal communication), and for CH_3CHO (H. E. Matthews et al. 1984*a*).
[b]Not observed directly, but surely the dominant constituent.

transition in TMC-1 at the Onsala Space Observatory (unpublished data). Inclusion of C_3H, C_3O, and CH_3C_4H in chemical schemes will provide new constraints on such models.

B. Interstellar Dust

Interstellar clouds include not only molecular gas but also small solid particles normally referred to as dust or grains. Canonically, the dust is estimated to include about 1% by mass of the material. Interstellar dust can be, and in fact has been, the subject of entire meetings and books (see, e.g., McDonnell 1978), and we shall leave its discussion in this book to the chapter by Draine (cf. also, Whittet 1981; Greenberg 1983; Zuckerman 1980; Willner et al. 1982). Here we discuss only the possible role of the dust in the gas-phase chemistry of the interstellar medium. We should remember, in this connection, that whether the dust is directly involved in the synthesis of molecules other than H_2, its role in shielding such species from the interstellar ultraviolet field is crucial to their survival (cf. Sandell 1978). Another important consideration for gas-phase chemistry is the percentage of elements heavier than helium that have been incorporated into the dust. Within an order of magnitude, it seems that the mass in dust and the mass in gas-phase heavy elements is comparable, although the situation for any particular element is often unclear. We shall return to this subject, in connection with our estimates of molecular abundances in interstellar clouds (Sec. III).

Several authors have speculated that grains may play a basic role in the formation of complex molecules, either as sites for surface catalysis or as source material which is returned to the gas phase upon grain fragmentation (see, e.g., Hayatsu and Anders 1981; Greenberg 1983; Tielens and Hagen 1982). The possible importance of dust may be shown rather easily; for instance, if every collison of a molecule with a dust grain in a dense cloud resulted in its adsorption, then molecules should be significantly depleted onto the dust in denser interstellar regions (see chapters by Herbst and by Draine). As we shall discuss below (Sec. III.B, conclusion 1), such depletion in dense clouds appears not to be a dominant effect, suggesting a mechanism for returning molecules to the gas phase. It is surprising, therefore, that there is no obvious observational evidence for the participation of the grains in the gas-phase chemistry (except for the presence of molecular hydrogen, that must be produced on grain surfaces). In fact, the relative abundances of many small polyatomic species seem to be reasonably well matched by purely gas-phase ion-molecule chemical models (see chapters by Herbst and by Glassgold; R. D. Brown and E. Rice 1981; Henning 1981; Millar and Freeman 1984*a,b;* Leung et al. 1984), although there appear to be some problems for nitrogen- and sulfur-containing species (Graedel et al. 1982; Prasad and Huntress 1982; Millar 1982). The abundances of positive ions and the presence of striking isotopic fractionation have also been emphasized as evidence for the predominance of gas-phase processes in interstellar chemistry (see review by W. D. Watson and C. M. Walmsley [1982], and earlier reference therein).

C. Isotopic Abundances

Relative isotopic abundances can be very diagnostic of the processes involved in molecular production and destruction, and the extremely large hydrogen/deuterium fractionation has traditionally been cited as evidence for the importance of gas-phase ion-molecule chemistry in the interstellar medium (see, e.g., W. D. Watson 1980). More recently, reported gradients in the $^{12}C/^{13}C$ ratio from the exterior towards the interior of clouds (and hence as a function of the ultraviolet field) have also been cited in support of gas-phase models (Langer et al. 1980; McCutcheon et al. 1980). The question of isotope abundances is also directly linked to efforts to understand the isotopic anomalies in meteorites that seem to imply a presolar (interstellar) origin for some of this material (chapter by Kerridge and Chang).

In view of existing reviews we shall not discuss these issues in detail (Penzias 1980; Wannier 1980; W. D. Watson and C. M. Walmsley 1982); we point out only some recent results. We also give a list of detected rare isotopic variants (see Table V).

TABLE V
Rare Interstellar Isotopic Variants[a]

Isotope	Detected in:
D	H_2, HCN, HCO^+, HN_2^+, H_2O, H_2CO, NH_3, HNC, HC_3N, HC_5N, CH_3OH
^{13}C	CO, CS, HCN, HCO^+, HNC, OCS, H_2CO, HC_3N, CH_3CN, CH_3OH, CN
^{15}N	HCN, HNC, NH_3, N_2H^+
^{17}O	CO, HCO^+
^{18}O	CO, OH, H_2CO, HCO^+, H_2O
^{29}Si	SiO, SiS
^{30}Si	SiO, SiS
^{33}S	CS
^{34}S	CS, SO, SO_2, OCS, SiS

[a]From Thaddeus (1981) and Winnewisser (1981), supplemented by recent detections from Johansson et al. (1984), Ziurys et al. (1984), Linke et al. (1983), and Gerin et al. (1984).

The D/H fractionation for several molecular species is reviewed from observational and theoretical aspects by R. D. Brown and E. Rice (1981) and W. D. Watson and C. M. Walmsley (1982). The studies of HC_3N and HC_5N show that substantial fractionation occurs even for these heavier species (Fig. 2). Present results indicate large variations between species, as well as for a given species among clouds (perhaps as a result of temperature). Unfortunately, for some molecules much of the basic observational data was obtained before the problems posed to abundance measurements by saturation and foreground absorption were fully appreciated (see Sec. III). For example, recent multiisotope studies of HNC by R. D. Brown and E. Rice (1981) and of HCO^+ by Guélin et al. (1982*b*) lowered previous estimates of the D/H fractionation in cold clouds by at least an order of magnitude. *We believe that further observations for at least HCN and NH_3 would produce important advances in this aspect of interstellar chemistry.*

Although there is general agreement that the $^{12}C/^{13}C$ ratio is significantly smaller in the galactic center (20 to 30) than in our local region, presumably because of nuclear processing of material, the precise value in the solar neighborhood (and its variability over the last 4.6 byr) remains uncertain. Compared with values discussed in the literature a few years ago, recent estimates of the total $^{12}C/^{13}C$ ratio are closer to the terrestrial value of 89. Thus, Scoville et al. (1983) find 96 ± 5 for $^{12}C/^{13}C$ as determined from CO in the Orion ambient molecular cloud; Henkel et al. (1982) give 80 ± 7 from measurements of H_2CO for the solar neighborhood; and R. W. Wilson et al. (1981) report 75 ± 8 for CO from nearby dark cloud data. In contrast, Wannier et al. (1982) give 55 ± 11 for CO in diffuse gas seen in absorption towards the star Zeta Oph. According to recent theoretical models, the total nuclear isotope abundance for $^{12}C/^{13}C$ should in most circumstances be bracketed by the values found for CO and for H_2CO (Langer et al. 1984).

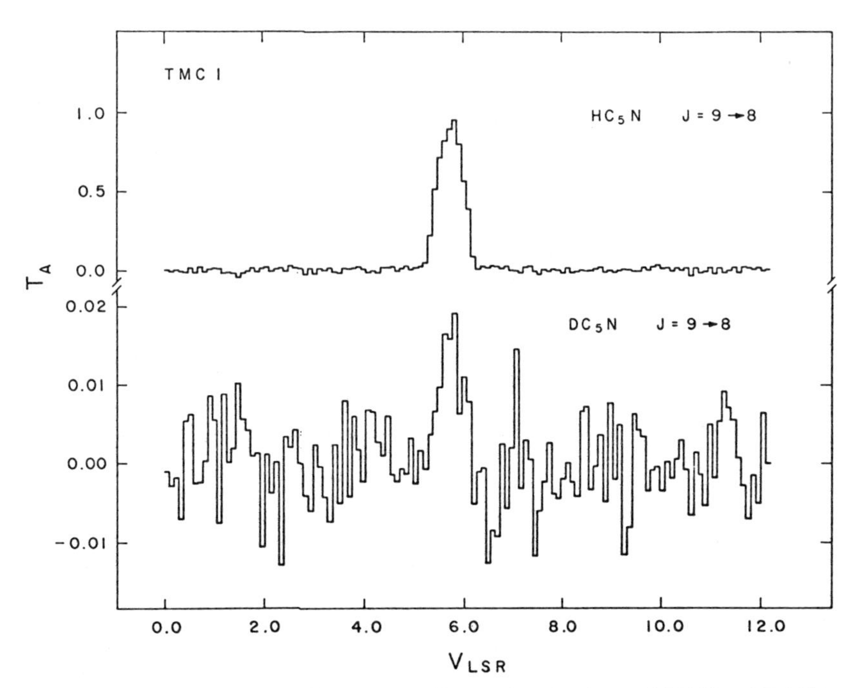

Fig. 2. An example of isotopic fractionation. The relative abundance of HC_5N and DC_5N is approximately given by the ratio of their observed line strengths and shows that $(DC_5N)/(HC_5N) \sim 10^{-2}$, which is much greater than the cosmic D/H ratio of $\sim 10^{-5}$ (Schloerb et al. 1981; see also MacLeod et al. 1981).

The relationship between observed interstellar fractionation and values preserved in presolar material incorporated into meteorites and/or comets depends on the chemical composition of such included material (cf. Vanýsek 1980; Vanýsek and Vanýsek, personal communication 1984). Because of the uncertainties in the interstellar values, particularly for HDO/H_2O (see, e.g., Olofsson 1984), quantitative comparisons are quite uncertain at present. Because, in addition to the total ratio of nuclei, the fractionation expected for a particular molecular species depends on physical conditions, predictions for the primitive solar nebula based on interstellar values cannot be precise at the present time, although qualitative links between solar and interstellar material seem to be established (chapter by Kerridge and Chang).

In this context the recently estimated $^{12}C/^{13}C$ ratio of 160 (+40, −55) in Jupiter, as inferred from Voyager 1 near-infrared observations of CH_4, is especially puzzling (Courtin et al. 1983; chapter by Gautier and Owen). After a careful analysis of the inherent uncertainties, these authors conclude that this surprisingly high value may represent the conditions in the primitive solar nebula. While the apparent decrease of the $^{12}C/^{13}C$ ratio from 160 to 80 in the solar neighborhood during the last 4.6 byr would be compatible with some models of galactic evolution (Vigroux et al. 1976), no plausable theory seems to predict such a difference between the inner solar system and Jupiter.

II. PHYSICAL CONDITIONS IN MOLECULAR CLOUDS

The determination not only of reasonably accurate relative abundances, but of *differences* in such abundances among molecular clouds with different physical conditions, is of critical importance to the evaluation of theoretical models and to the search for signposts of cloud evolution. Interstellar material in the Galaxy exists over a range of physical conditions, and classification of regions is somewhat subjective. Nonetheless, the categories listed below have proven useful (cf. Winnewisser 1981) and shall be used in our comparisons. The appropriate physical properties of those subregions for which we shall tabulate data are given in Table VI.

A. Giant Molecular Clouds

Giant molecular clouds are the most massive objects in the Galaxy and are often the sites of ongoing star formation, apparently including both the most massive O and B stars as well as less massive, solar-type stars. They are regions of great heterogeneity in temperature, density, and ionization state.

Although the giant molecular cloud near the galactic center, Sgr B2, remains a rich source of complex interstellar molecules, specification of abundances versus physical conditions is quite difficult at this distance. Striking inhomogeneities are expected to occur in such regions, difficult to spatially resolve in Sgr B2; the complex line shapes, that often include emission over a broad velocity range and superimposed absorption, support such inhomo-

TABLE VI
Physical Characteristics of Source Regions

Region	Mass ($M_\odot$)	Density (n) (H_2 molecule cm^{-3})	Temperature (T) (K)	References*
Diffuse	?	10^2–10^3	20–100	a,b
TMC-1/L 134N	10–10^2	10^3–10^5	10–20	c,d,e
Orion ridge**	10^2–10^3	10^4–10^6	50–100	f,g,h
Orion hot core	10	10^7–10^9	100–200	h,i,j
Orion plateau	3–30	10^6–10^7	90–150	h,k,l
IRC+10216***	1	10^3–10^{11}	10–1000	m,n

*References: (a) Nyman 1984; (b) Dickman et al. 1983; (c) Schloerb et al. 1983*b*; (d) Avery et al. 1982; (e) Snell et al. 1982; (f) Bastien et al. 1981; (g) Schloerb and Loren 1982; (h) Johansson et al. 1984; (i) Genzel et al. 1982; (j) P. F. Goldsmith et al. 1983; (k) Friberg 1984; (l) Plambeck et al. 1982; (m) Lafont et al. 1982; (n) Kwan and Linke 1982.
**Temperature may be higher (~150 K) in the southern condensation.
***Density and temperature decrease with radius.

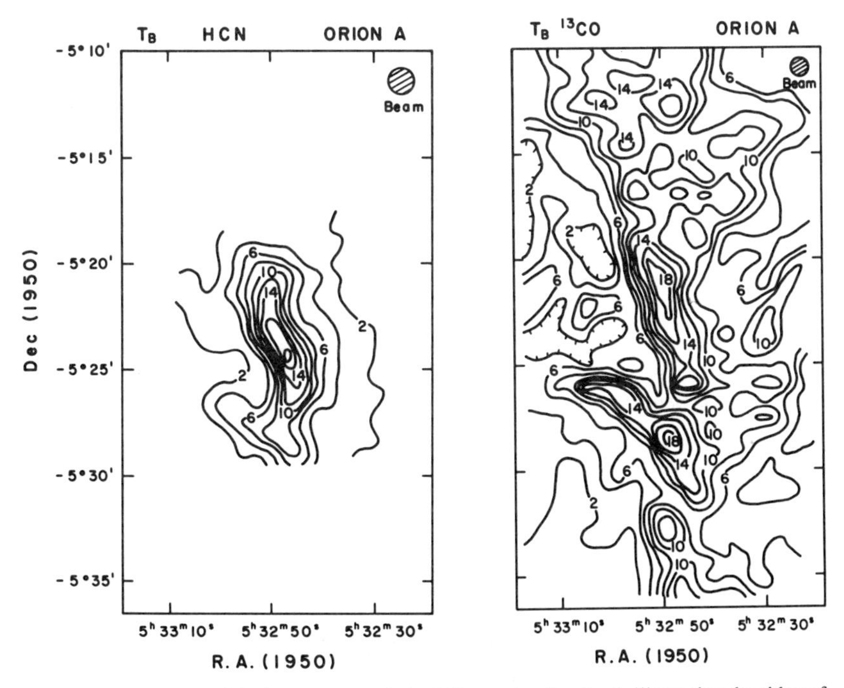

Fig. 3. Contour maps of the inner portion of the Orion molecular cloud, illustrating the ridge of emission apparent in emission from both low-excitation (^{13}CO) and high-excitation (HCN) molecules (in both cases J = 1 − 0) (Figure from Schloerb and Loren 1982.)

geneity. Careful examination of the spectra of isotopic species, as well as multiline excitation analyses, are probably required to interpret the data adequately, and detailed results await the publication of the Bell Laboratories spectral survey of this region (Cummins et al. 1984).

In this chapter we have chosen to discuss the nearest example of a giant molecular cloud to the solar system. This cloud is associated with the Orion Nebula and has been the subject of other extensive investigations (Glassgold et al. 1982). The inner portion of the region includes a dense *ridge* of material, illustrated in the emission from ^{13}CO and HCN in Fig. 3. High spatial and spectral resolution studies of the central portion of the ridge reveal an extremely complex structure, shown schematically in Fig. 4. Specifically, the ambient ridge is divided into a northern and southern component, with the latter containing a localized warmer clump which we shall refer to as the *southern condensation* (an interpretation of the inner ridge structure as a rotating, contracting disk is given by Hasegawa et al. [1984]). Certain complex molecules seem to be concentrated in this condensation (Johansson et al. 1984; Olofsson 1984). The *plateau,* named for the very broad shape of spec-

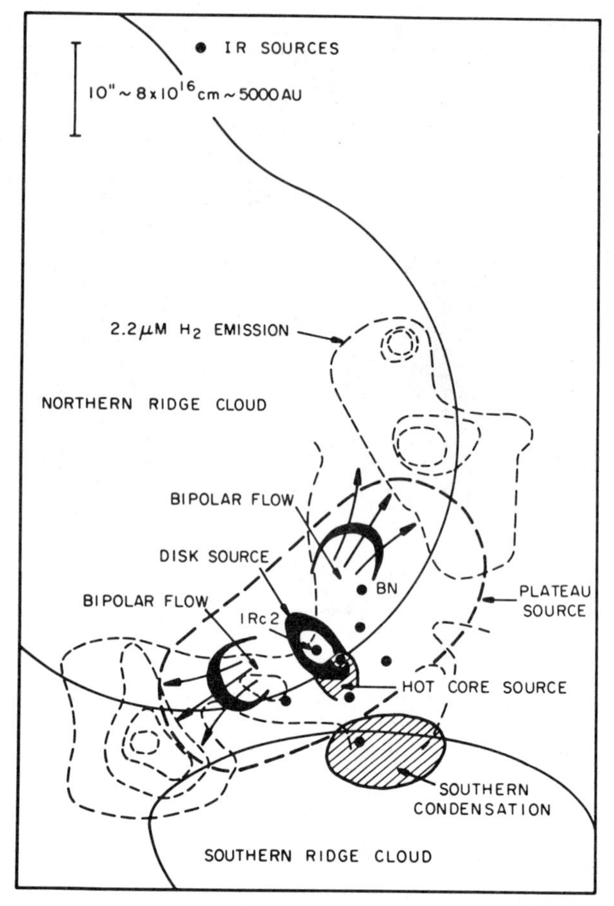

Fig. 4. Schematic of the core of the Orion molecular cloud illustrating the source regions discussed in the text (adapted from Olofsson 1983; see also Wynn-Williams et al. 1984; Friberg 1984).

tral lines and not for its spatial structure, consists of a spatially confined disk or toroid characterized by high-velocity gas flows and an approximately orthogonal bipolar outflow presumably responsible for shock-heated H_2 and lower-excitation CO and HCO^+ (Olofsson et al. 1982*a*; N. R. Erikson et al. 1982). The plateau contains several infrared sources, and the outflows are presumably related to mass loss from one or more of these pre-main sequence objects. The *hot core* is apparently a small, hot, dense region that may be associated with an individual compact infrared source (Genzel et al. 1982; P. F. Goldsmith et al. 1983). Even when these emissions from Orion subregions are not resolved by the beam of a radio telescope, they may be distinguished by their spectral signature (see Fig. 5).

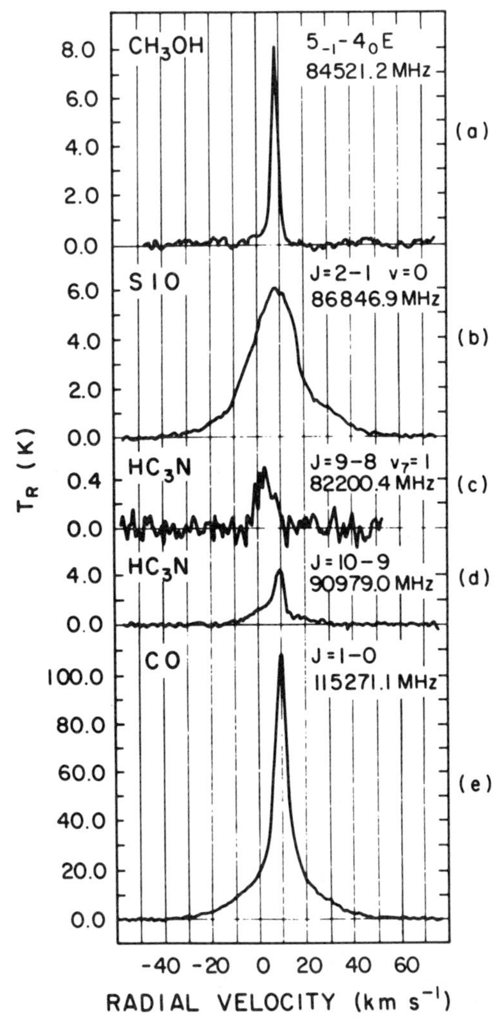

Fig. 5. Spectra of several molecular transitions towards the region shown in Fig. 4, illustrating characteristic differences in the emission from different subregions: (a) ridge lines are narrow and centered at velocity $V_{LSR} = 8 - 10$ km s^{-1}; (b) plateau lines are very wide but are also centered at $V_{LSR} \sim 9$ km s^{-1}; (c) intermediate-width lines from the hot core occur at $V_{LSR} \sim 5$ km s^{-1}; (d) and (e) emission from more than one region is evident in some transitions. Data from Johansson et al. 1984.

B. Dark Clouds

Nearby, lower-mass regions which appear dark against the stellar background are often called dark clouds (see, eg., chapters by Leung and by Myers). They lack embedded high-luminosity sources and are thus colder and typically more quiescent than the cores of giant molecular clouds. Nonetheless, an increasing number are being found to contain lower-luminosity sources, including T Tauri stars and possible low-mass protostellar objects (Elias 1978*c*; Keene et al. 1983; Frerking and Langer 1982). In the nearest relatively dense molecular clouds, some of these objects may be studied with high angular resolution.

The physical conditions in the two dark clouds considered here, TMC-1 and L 134N, appear very similar (Winnewisser 1981). The chemical differences between them (discussed below in Sec. III.B, conclusion 2) are, therefore, particularly interesting. Moreover, even though the available spatial resolution on these nearby objects is quite high, physical and chemical inhomogeneities are apparent (cf. Irvine et al. 1983), and considerable debate continues about their detailed structure (see, e.g., Schloerb et al. 1983*b*), including whether denser cores may be surrounded by lower-density (and temperature?) halos.

C. Outflows

Mass outflow is observed from evolved stars, from young stars, and from localized regions within larger clouds thought to contain protostars (Zuckerman 1980; Bally and Lada 1983; Edwards and Snell 1983). Such material may include both the results of high temperature and density, near-photospheric chemistry, as well as the products of photochemical processing during outflow and interaction between stellar material and the surrounding molecular cloud or protosolar nebula (W. D. Watson and C. M. Walmsley 1982; Lafont et al. 1982). As discussed above (Sec. II.A), the Orion plateau source appears to be an example of an outflow associated with a young star or stars. For an example of an outflow from an evolved star, one may consider the cloud of molecular and particulate material expelled by the carbon star IRC + 10216 (Olofsson et al. 1982*b*; Johansson et al. 1984). The chemistry in this latter case is expected to differ from that for outflows from young stars, because nuclear processing for IRC + 10216 has resulted in an envelope in which the abundance of carbon exceeds that of oxygen, contrary to the usual cosmic situation (cf. Iben and Renzini 1982).

D. Diffuse Clouds

For convenience less dense interstellar clouds, as well as the peripheries of denser regions, may be called diffuse. Our operational definition refers to regions where high dipole moment molecules (CS, HCO^+, HCN) are seen in absorption at millimeter wavelengths against background continuum sources, while CO and ^{13}CO are seen in emission. This seems to indicate densities in

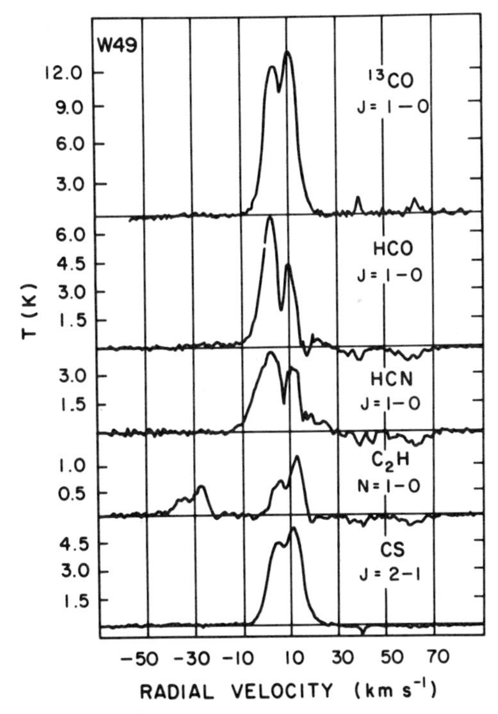

Fig. 6. The spectral features attributed here to diffuse clouds as seen at $\sim$ 40 km s^{-1} and $\sim$ 60 km s^{-1}, in emission for ^{13}CO and in absorption for HCO^+, HCN, C_2H, and CS. This material is contained in spiral arms between the Sun and the background continuum source W49, that produces the strong self-absorbed emission that is centered in the spectra (Nyman 1983,1984).

the range $100 \leq n(H_2) \leq 1000$ cm^{-3} (Dickman et al. 1983; Nyman 1983, 1984).

The diffuse clouds associated with galactic spiral arms in 21-cm line surveys give rise to such molecular features (Fig. 6). The arms are distinguished by their radial velocity with respect to the Sun. Each such absorption (emission) feature may sample more than one cloud in the line of sight; differences in excitation (or abundance) appear rather clearly between arms, and quite possibly within a given arm as a function of radial velocity (Nyman 1984). In addition, a considerable amount of data exists for local (Orion arm) gas from optical, ultraviolet, and radio studies (see, e.g., Dickman et al. 1983).

The density, temperature, and ultraviolet field certainly differ among such diffuse clouds. For the sake of a uniform data set, we cite below primarily results for the features towards W49, for which the most extensive observations currently exist, but we shall also include results for local gas for which fractional abundances can be reasonably well estimated by standard procedures. The reported Orion arm absorptions due to the polyacetylenic species HC_5N, HC_7N, and C_4H are interesting (Bell et al. 1981,1983); however, as the authors note, abundances remain rather uncertain, because the excitation state for these species is not well known.

III. DETERMINATION OF CHEMICAL ABUNDANCES

A. Procedure

A quantitative approach to the subject of chemistry in molecular clouds requires us to determine the abundance of various chemical species with respect to one another and to the dominant molecule in the cloud, H_2. Unfortunately, the determination of accurate abundances is one of the most difficult problems faced by radio astronomers. Thus, before we discuss the reasons for these difficulties and the method that forms the basis for the abundance determinations in this chapter, it seems appropriate to cite a recent poem by Erica Jong (1983):

> The poet fears failure
> & so she says
> "Hold on pen—
> what if the critics
> hate me?"
> & with that question
> she blots out more lines
> than any critic could. . . .
> [yet]
> It's in her nature
> to fear failure
> but not to let that fear
> blot out
>
> her lines.

It goes almost without saying that the chemistry of molecular clouds as studied by millimeter-wave radio astronomy has significant selection effects. First and foremost, rotational transitions in the millimeter and longer wavelength regions of the spectrum occur only for relatively heavy molecules with permanent electric dipole moments. Thus, potentially abundant species such as many simple hydrides, homonuclear molecules, and symmetric hydrocarbons cannot at present be directly probed. Hopefully, as high resolution spectroscopy pushes to higher frequencies and into space (to escape the telluric absorption at submillimeter wavelengths), at least those constraints regarding the lightest molecules may be lifted to allow a fuller sampling of the chemistry of molecular clouds. In some cases, it may be possible to estimate abundances of molecules with no electric dipole rotational transitions from observations of related species, such as unsymmetric isotopic variants possessing a very small dipole moment or protonated variants such as N_2H^+ or HCO_2^+, but data remains sparse (E. Herbst et al. 1977; Linke et al. 1983).

A second, more subtle, selection effect is that imposed by the energy level structure of a given molecule. Simple linear molecules, with only a small number of rotational levels populated at the densities and temperatures appropriate for molecular clouds, are far more readily detectable than are complex species, that may have an equal number of molecules distributed over a large number of states. Thus, consideration of the population distribution among the accessible energy levels (i.e., the partition function) is essential to a determination of molecular abundance.

We define the abundance $f(X)$ of a particular molecular species X to be the number fraction of molecules. Since molecular hydrogen accounts for virtually all of the gas in the cloud (apart from ~25% He by mass), we may write

$$f(x) = \frac{n(X)}{n(\mathrm{H}_2)} \tag{1}$$

where $n(X)$ denotes the number density of species X. In principle, $n(\mathrm{H}_2)$ can be determined indirectly in molecular clouds through analysis of the excitation of molecules by collisions, as H_2 is the dominant collision partner. In contrast, it is obvious that $n(X)$ will be difficult to determine directly, because, for optically thin transitions, the intensity of the emission depends upon the column density of molecules in the source, rather than on their number density. Thus, it is generally impossible to separate the effects of column density from number density without significant assumptions about the geometry and details of radiative transfer within the source (for examples of this approach, see Wootten et al. 1978, 1980*a,b*, 1982*b*).

An alternative method, which we advocate here, because it is applicable to a wide range of sources, is to compare the column densities of molecules as an indicator of their relative abundance. Therefore, we operationally define the abundance $f(X)$ to be

$$f(X) \approx \frac{N(X)}{N(\mathrm{H}_2)} \tag{2}$$

where $N(X)$ denotes the column density of species X (i.e., number of molecules per cm^2 in the telescope beam). Adoption of this method naturally means that the results so derived refer to some sort of average through the cloud. Moreover, even such average abundances may contain considerable uncertainties. We shall now discuss the use of this technique with particular attention to the assumptions made in each step of the analysis (see Table VII).

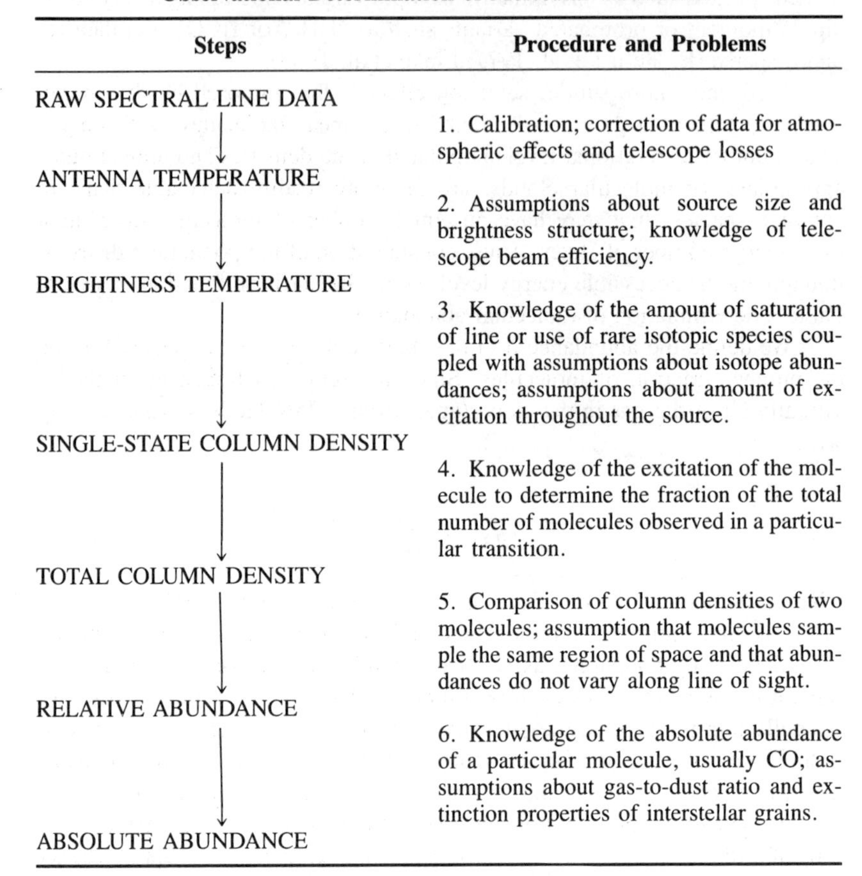

TABLE VII
Observational Determination of Molecular Abundance

Steps	Procedure and Problems
RAW SPECTRAL LINE DATA	
↓	1. Calibration; correction of data for atmospheric effects and telescope losses
ANTENNA TEMPERATURE	
↓	2. Assumptions about source size and brightness structure; knowledge of telescope beam efficiency.
BRIGHTNESS TEMPERATURE	
↓	3. Knowledge of the amount of saturation of line or use of rare isotopic species coupled with assumptions about isotope abundances; assumptions about amount of excitation throughout the source.
SINGLE-STATE COLUMN DENSITY	
↓	4. Knowledge of the excitation of the molecule to determine the fraction of the total number of molecules observed in a particular transition.
TOTAL COLUMN DENSITY	
↓	5. Comparison of column densities of two molecules; assumption that molecules sample the same region of space and that abundances do not vary along line of sight.
RELATIVE ABUNDANCE	
↓	6. Knowledge of the absolute abundance of a particular molecule, usually CO; assumptions about gas-to-dust ratio and extinction properties of interstellar grains.
ABSOLUTE ABUNDANCE	

1. Antenna Temperature. The first step is the determination of the antenna temperature (signal strength) of the spectral line from the raw, uncalibrated spectrum. Simple calibration of the data involves correction for the effects of atmospheric absorption and certain telescope losses. Currently, there is no general agreement on what method to use for these corrections nor even on which corrections to make at this stage (cf. Kutner and Ulich 1981). Thus, correct interpretation of the results reported in the literature requires a knowledge of the precise procedure used. Further, we note that the absolute calibration of millimeter data is probably not accurate to better than 10% under most circumstances.

2. Brightness Temperature. If the source observed were uniform and large compared to the angular resolution of the antenna beam, and if the antenna used were perfectly efficient, then the antenna temperature measured would simply be the brightness temperature of the emission from the cloud (i.e., the temperature of a blackbody filling the antenna beam and producing the observed monochromatic flux). It is this quantity which relates directly to the properties of the cloud itself. Unfortunately, the observed antenna temperature may deviate from the true brightness temperature by a wide margin. One of the greatest difficulties faced by radio astronomers is that real antenna patterns are such that there is often a nonnegligible amount of power received from the source outside of the main telescope beam (cf. P. F. Goldsmith et al. 1981). The ratio of the power received by the telescope in the main beam to the total amount of power received by the telescope is known as the main beam efficiency, which for telescopes currently in operation typically ranges between 20–80% for molecular-line observations. The remaining power not in the main beam may come from anywhere on the sky, and to the extent that it arises from elsewhere on the source, will cause difficulties in relating the antenna temperature to the brightness temperature of the emission. Thus, knowledge of the size and structure of the source is necessary to provide the possibility of a correction for these effects. Finally, we note that it might not be appropriate to characterize the observed emission with a single brightness temperature, because structure (clumping) on a size scale smaller than the beam could and often does exist (see, e.g., Pauls et al. 1983; Matsakis et al. 1981).

3. Single-State Column Density. The next step is to relate the brightness temperature to the population of the levels involved in the observed transition (see, e.g., Winnewisser et al. 1979; Penzias 1980). This procedure is most easily accomplished if the spectral line is optically thin (unsaturated, in spectroscopic terminology), because then the intensity is proportional to the column density of molecules in the upper (or lower) state of the transition. However, many molecular transitions in interstellar clouds are not optically thin and some allowance for saturation must then be made. Such effects are not always easily recognized and may be particularly troublesome if the cloud is clumpy. Under these circumstances it is possible that the emission arises from a number of small optically thick regions within the cloud, so that the intensity of the lines is proportional to the number of clumps rather than to the column density of molecules. A typical observational approach to the problem of saturation is to use rarer isotopic variants of the molecule, or weaker hyperfine components of the transition (if they exist), in the hope that these will be optically thin. This procedure, or use of higher-excitation transitions, also provides an estimate of possible self-absorption by low-excitation foreground material (Fig. 7). It may bring, however, the additional problem that the isotopic ratio must be known and must be assumed not to vary in the cloud. Finally, we expect that even if saturation of the lines is not a problem, the column density of molecules sensed by a given transition is determined by the

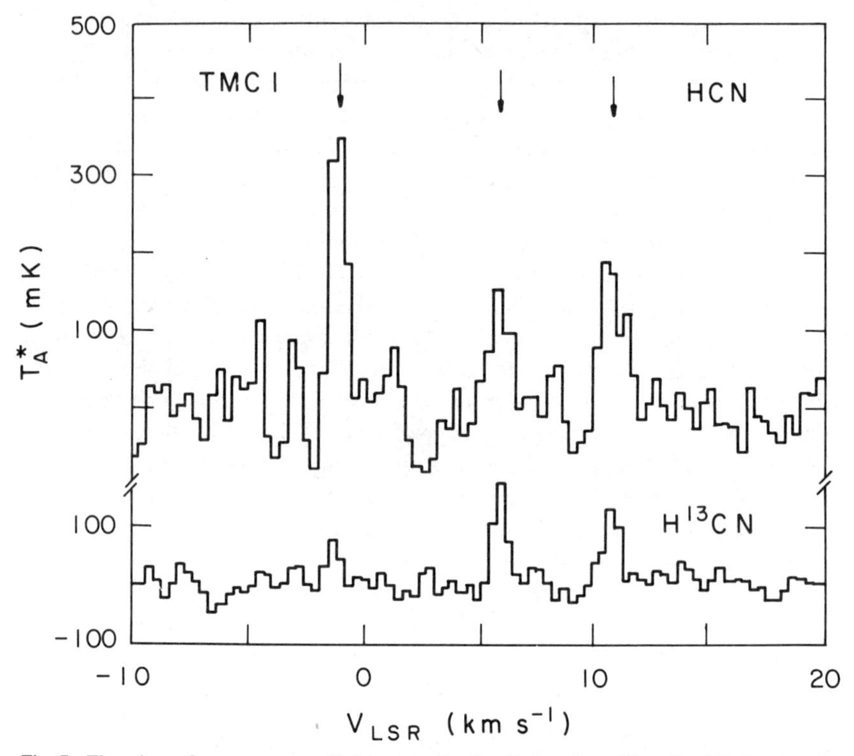

Fig. 7. Three hyperfine components (left to right F = 0 − 1, 2 − 1, and 1 − 1) of the J = 1 − 0 transitions of HCN and $H^{13}CN$ towards TMC-1. The line ratios for $H^{13}CN$ are as expected for optically-thin emission, while the anomalously large strength of the F = 0 − 1 component for HCN indicates absorption by low-excitation foreground material (Irvine and Schloerb 1984).

degree of excitation along the line of sight. For each individual molecular transition, a critical density is required for collisional excitation to balance radiative decay, and it is probable that some portions of the cloud do not achieve this critical density. Because different molecules have different excitation requirements, comparison of column densities for different species with the same absolute abundance could lead to an apparent difference in their relative abundance. Finally, we note that in very cold clouds in which the excitation temperature is close to the background temperature, the upper-state column density becomes very uncertain (see, e.g., Broten et al. 1984).

4. Total Column Density. The next step is the estimation of the total column density of molecules from the column density in one or more energy levels. In general, under the conditions in molecular clouds, the densities are not sufficient to thermalize all the levels, and assumptions of local thermodynamic equilibrium (LTE) to relate the column density of one transition to the total population are risky. A more reliable procedure, which has become widely practiced, is to use many transitions of a single molecule to better determine the population distribution. This has the added benefit of allowing an estimate both of the density and temperature of the region sampled and also of the validity of the assumptions intrinsic to the method (see, e.g., Schloerb et al. 1983*a*; Johansson et al. 1984; Snell et al. 1984). Unfortunately, interpretation is complicated if the source region sampled is not the same for each transition, as will be the case for data taken at widely differing frequencies with the same telescope. Often it is not possible to obtain multilevel observations of a particular molecular species, and as an alternative to assumptions of LTE, one might employ statistical equilibrium calculations to relate the observed column density in a single transition to the entire population. Even such elaborate schemes may be suspect, however, because the collisional cross sections are known accurately for only a very few molecules and the physical conditions are often poorly constrained (some recent examples of such calculations have been performed by Cummins et al. [1983]; Askne et al. [1984]; Andersson et al. [1984]).

5. *Relative Abundance.* Once the total column density has been estimated, one must compare values among species in order to calculate relative abundances. Given the cautionary tone of the preceeding discussion, however, it is obvious that direct comparison of the column densities for two molecules may suffer from significant systematic effects, with perhaps the greatest uncertainty resulting from the possibility that the two species sample different portions of the cloud owing to their differing excitation requirements. Because of this effect, one anticipates that molecules with easily excited transitions (small Einstein A-coefficient) are likely to be overestimated in abundance compared to molecules requiring high density for excitation. Relative abundance measurements are hence most reliable for molecules with nearly the same excitation requirements.

6. *From Relative to Absolute Abundance.* Finally, all that remains is to obtain an absolute abundance estimate from a comparison of the column density of a given molecular species to the column density of molecular hydrogen. As previously stated, however, this step normally requires comparing the data with a tracer of the molecular hydrogen column density, as the latter is not usually directly observable. Typically, this tracer is CO or, in fact, a rarer isotope like $C^{18}O$. The procedure may be summarized by the following equation

$$f(X) = \frac{N(X)}{N(H_2)} = \frac{N(X)}{N(CO)} \frac{N(CO)}{A_v} \frac{A_v}{N(H_2)} \tag{3}$$

where A_v is the visual extinction by dust. The relative abundance CO/H_2 has been derived not directly in cold molecular clouds, but indirectly through comparison of the CO column density and visual extinction by dust (Dickman 1978; Frerking et al. 1982), given the assumptions of a constant hydrogen gas-to-dust ratio and consistent optical properties of the dust within the clouds. Absolute abundance, then, is typically based upon the CO abundance *f*(CO) and the relative column densities of the molecule under consideration and CO. Since CO requires only relatively low densities for its excitation, we anticipate that such a comparison will tend to underestimate the abundances of most species. Direct observations of $N(H_2)$ are possible in cloud regions that are sufficiently hot to excite vibrational or electric quadrupole rotational transitions of H_2 (see, e.g., D. M. Watson 1984); or in principle in regions where H_2 absorption can be seen against a sufficiently luminous infrared source (J. H. Black and S. P. Willner 1984).

We may summarize the principal assumptions required to determine molecular abundances from a comparison of column densities as follows:

a. The source has a single, uniform set of chemical abundances; possible abundance gradients within the telescope beam are ignored (i.e., results are beam averages);
b. The source is uniform in spatial density structure; possible clumping and excitation gradients within the telescope beam are ignored (i.e., results may underestimate relative abundances of high excitation species);
c. CO (and its isotopic variants) is a valid tracer of molecular hydrogen column density; possible abundance gradient of CO with respect to H_2 density is ignored (cf. conclusion 1 in Sec. III.B; see also discussion of *f*(CO) below).

To the extent that the above assumptions are true for real clouds, the abundances derived by this method should be valid, provided that the excitation and hence the column densities are estimated accurately. However, each assumption has been questioned at one time or another, and it is an important goal for future work to better determine the interrelated physical and chemical internal structure of molecular clouds.

B. Results

In order to minimize problems associated with differing calibration and beam size, the results presented here are drawn, in so far as possible, from the uniformly calibrated spectral scan of the Orion cloud core and the envelope of the evolved star IRC + 10216 carried out at the Onsala Space Observatory (Johansson et al. 1984; Fig. 8) and from Onsala observations of what we have termed diffuse clouds (Nyman 1983,1984). The data sources for the dark cloud results are necessarily more diverse, but we have tried to select observations obtained at similar spatial resolution and for which saturation effects could be estimated (some results for which this was not possible are given as lower limits). In addition, we have when possible ratioed results for TMC-1 to those obtained for the physically similar dark cloud L 134N by the same

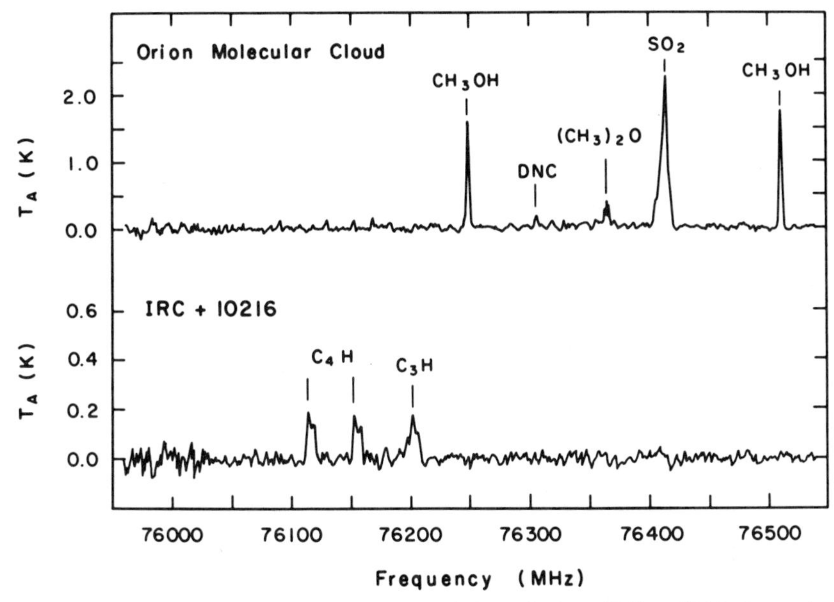

Fig. 8. Sample spectra of the Orion cloud core and the evolved star IRC + 10216 taken at the Onsala Space Observatory (Johansson et al. 1984).

investigator and telescope. Important new sources of data for Orion and Sgr B2 will be the surveys currently being completed at Bell Laboratories and the Owens Valley Radio Observatory (Cummins et al. 1984; Sutton et al. 1984).

To fix f(CO) we note that extensive studies for dark clouds using the theoretically based premise of a constant H_2/(visual extinction) ratio (Frerking et al. 1982; Dickman 1978), and observations of H_2 and CO emission from hotter Orion gas (D. M. Watson 1984), give very similar values, and we adopt in general for molecular clouds $f(CO) = 8(10)^{-5}$. This value is consistent with the lower bound on f(CO) found directly from infrared absorption measurements of H_2 and CO towards the molecular cloud NGC 2024 (J. H. Black and S. P. Willner 1984), and implies that ~12% of the carbon is in CO (however, note the discussion in conclusion 1 below). The CO column densities themselves are taken from $C^{18}O$ measurements (Guélin et al. 1982*b*) that give $N(CO) \approx 8 \times 10^{17}$ cm^{-2} for TMC-1 and L 134N, assuming a terrestrial $^{18}O/^{16}O$ ratio, and from Onsala ^{13}CO measurements, that yield $N(CO) = 1 \times 10^{19}$ cm^{-2} for the Orion ridge, assuming $^{12}CO/^{13}CO = 40$ and optically thin emission. An error in the CO column density or in f(CO) would translate into a systematic bias in all other abundances by the same factor, so we note that some estimates of N(CO) in Orion are up to four times larger than ours (Phillips and Huggins 1981), while a factor of 2 to 3 uncertainty in N(CO) for the dark clouds is certainly possible.

TABLE VIII
Molecular Abundances for Several Regions[a]

Species	Name	Abundance vs. H_2 ($\times 10^8$)			Abundance Ratio		Remarks and References	Dark Cloud References[b]
		Orion Ridge	TMC-1	IRC+10216	Plateau/ Ridge	L 134N/ TMC-1		
CO	Carbon monoxide	8000	8000	50000	1	1	adopted values, see text	1
CI	Atomic carbon	($\gtrsim$1000)			($\lesssim$0.5)		(2,3); L 134N $\gtrsim$ 400	2
C_2	Carbon dimer		5				20 arcmin from std. position	4
OH	Hydroxyl		30			0.25	beam size >> that for heavier species	53
CH	Methylidyne		2			0.5	beam size >> that for heavier species	5
C_2H	Ethynyl	2	$\gtrsim$1	80	$\lesssim$1	$\lesssim$1		6
C_4H	Butadiynyl	$\lesssim$0.03	2	100		0.05		7,8
CH_3C_2H	Methyl acetylene	0.5	0.6	$\lesssim$30		$\lesssim$0.2	(propyne)	8,9
CN	Cyanogen	(2)	3	(200)	$\lesssim$2	0.1	(10); IRC: increasing with radius	11,50
HCN	Hydrogen cyanide	2	2	1000	25			12
HNC	Hydrogen isocyanide	0.1	2	4	2	1	(see Table X)	12,13
CH_3CN	Methyl cyanide	0.1*	0.1	0.4		$\lesssim$1	(acetonitrile)	14
HC_3N	Cyanoacetylene	0.04	0.6	100	15	0.03		15–19
HC_5N	Cyanodiacetylene	0.006	0.3	300		0.03		19,20
CH_2CHCN	Vinylcyanide		0.02	$\lesssim$1		$\lesssim$0.5	(acrylonitile)	21
C_3N	Cyanoethynyl	$\lesssim$0.006	0.1	40		$\lesssim$0.2		7
CH_3C_3N	Methylcyanoacetylene		0.05					22
CH_3C_4H	Methyldiacetylene		0.2					47–49
HNCO	Isocyanic acid		0.02					23
N_2H^+		0.02	0.1			~1	unpublished Onsala data	24
NH_3	Ammonia	(20)	2	(2–10)	(10?)	10	(25–27)	19,28

HCO^+	Formyl ion	0.3	0.8	≲0.2	8	1		29
HDO	Deuterated water	0.3*			10		ridge position (30); H_2O, see text	
CH_3OH	Methanol	40*		≲80			ridge position (30)	
H_2CO	Formaldehyde	(≳8*)	2	≲1	(1)	1	(31–33,43)	34,35
H_2C_2O	Ketene	0.08						
$(CH_3)_2O$	Dimethyl ether	2*					ridge position (30,36)	
CH_3OCHO	Methyl formate	1*	≲0.1					19
CS	Carbon monosulfide	0.4	0.2	30	10	0.5	(36)	37
HCS^+	Thioformyl ion	0.03	0.06	≲1		≲1		38
H_2CS	Thioformaldehyde	0.2					unpublished Onsala data	
H_2S	Hydrogen sulfide	(≳0.2)			(~50)		(44); Plateau ~10 (39)	
OCS	Carbonyl sulfide	0.9*		≲20	8			
SO	Sulfur monoxide	0.5*	0.5	≲10	300	4	ridge structure (36)	40
SO_2	Sulfur dioxide	0.4*	≲0.1	≲30	300	≳4		38
SiO	Silicon monoxide	≲0.1	≲0.04	20	≳100			19
SiS	Silicon sulfide	≲0.001		60				
HC_7N	Cyanohexatriyne		0.1	(75)		≲0.02	IRC: HC_7N/HC_5N from (45)	19,55
HC_9N	Cyano-octatetra-yne		0.03				IRC detection (55)	19
$HC_{11}N$	Cyano-decapenta-yne			(50)			IRC: $HC_{11}N/HC_7N$ from (46)	
HOC^+	Isoformyl ion	≲0.001	≲0.002					41
HCO	Formyl	≲0.02						
CH_3CHO	Acetaldehyde	≲0.08	0.06			1		55
HC_2CHO	Propiolaldehyde		≲0.06				(propynal)	19
CH_2CHCHO	Acrolein	≲0.02						
CH_3CH_2OH	Ethanol	≲0.03						
HCOOH	Formic acid	≲0.08						
CH_3COOH	Acetic acid	≲1					from data in (42)	
CH_3NC	Methyl isocyanide	≲0.005	≲0.01				from data in (42)	12

TABLE VIII (*Continued*)

Species	Name	Abundance vs. H_2 ($\times 10^8$)			Abundance Ratio		Remarks and References	Dark Cloud References[b]
		Orion Ridge	TMC-1	IRC+10216	Plateau/ Ridge	L 134N/ TMC-1		
CH_3CH_2CN	Ethyl cyanide	$\lesssim 0.03$	$\lesssim 0.1$	$\lesssim 2$				14,19
NH_2CN	Cyanamide	$\lesssim 0.02$	$\lesssim 0.01$					12
NH_2CHO	Formamide	$\lesssim 0.03$	$\lesssim 0.2$					21
$(NH_2)_2CO$	Urea	$\lesssim 0.07$						
NH_2CH_2COOH	Glycine II	$\lesssim 0.05$	$\lesssim 0.04$				for conformer I see (51)	19
C_4H_4O	Furan	($\lesssim 0.07$)					**(52)	
C_4H_5N	Pyrrole	($\lesssim 0.03$)	$\lesssim 0.04$				**(52)	19
$C_3N_2H_4$	Imidazole	($\lesssim 0.1$)	$\lesssim 0.03$				**	19
CH_3SH	Methyl mercaptan	($\lesssim 0.06$)					**(54)	

[a]For Orion and IRC+10216 data is from Johansson et al. (1984) unless other citation given, in which case value is enclosed in parenthesis.

[b]Dark cloud references: (1) Frerking et al. 1982; (2) Phillips and Huggins 1981; (3) Beichman et al. 1982; (4) Hobbs et al. 1983; (5) Rydbeck et al. 1976; (6) Wootten et al. 1980*a*; (7) Guélin et al. 1982*a*; (8) Irvine et al. 1981; (9) Askne et al. 1984; (10) Wootten et al. 1982*a*; (11) Churchwell and Bieging 1983; (12) Irvine and Schloerb 1984; (13) Frerking et al. 1979; (14) H. E. Matthews and T. J. Sears 1983*a*; (15) Schloerb et al. 1983*b*; (16) Bujarrabal et al. 1981; (17) Snell et al. 1981; (18) Avery et al. 1982; (19) Irvine and Hjalmarson 1983; (20) Benson and Myers 1980; (21) H. E. Matthews and T. J. Sears 1983*b*; (22) Broten et al. 1984; (23) R. L. Brown 1981; (24) Linke et al. 1983; (25) Ziurys et al. 1981; (26) Betz and McLaren 1980; (27) Bell et al. 1982*b*; (28) Rydbeck et al. 1977; (29) Guélin et al. 1982*b*; (30) Olofsson 1984; (31) Bastien et al. 1981; (32) T. L. Wilson et al. 1980; (33) Myers and Buxton 1980; (34) Sherwood and Wilson 1981; (35) Evans and Kutner 1976; (36) Friberg 1984; (37) Snell et al. 1982; (38) Irvine et al. 1983; (39) T. B. H. Kuiper et al. 1981; (40) Rydbeck et al. 1980; (41) Woods et al. 1983; (42) Johansson et al. 1984; (43) Wootten et al. 1984*a*; (44) Thaddeus et al. 1972; (45) Winnewisser 1981; (46) Bell et al. 1982*a*; (47) MacLeod et al. 1984; (48) Loren et al. 1984; (49) Walmsley et al. 1984; (50) Crutcher et al. 1984; (51) Hollis et al. 1980; (52) Kutner et al. 1980; (53) Turner 1973; (54) Linke et al. 1979; (55) H. E. Matthews et al. 1984*a*.

(*)indicates dominant emission from the southern condensation (see text).

(**)beam size for Orion is considerably larger than for other molecular species; for confined emission (e.g., from *) the limit would be poorer by up to an order of magnitude.

TABLE IX
Relative Abundance for Varying Chemical Saturation

Region	CH_3CH_2CN/HC_3N	CH_2CHCN/HC_3N	References*
Orion hot core	7	1	a
Orion ridge	$\lesssim 1$	?	a
TMC-I	$\lesssim 0.2$	0.03	b,c
IRC+10216	$\lesssim 0.02$	$\lesssim 0.01$	a
Sgr B2	0.45	0.15	d

*(a) Johansson et al. 1984; (b) Irvine and Hjalmarson 1983; (c) H. E. Matthews and T. J. Sears 1983*b*; (d) Johnson et al. 1977.

There are also, of course, uncertainties in the column density determinations for other species, particularly those associated with estimating the partition function and optical depth (saturation). Values of column density quoted here are generally derived from observations of multiple transitions (perhaps of isotopes) that are consistent with a single excitation temperature and optically thin emission. We conclude that the abundances in Table VIII should not be relied on to better than an order of magnitude. Abundance ratios (other than to H_2) should be accurate to within a factor of a few, particularly if taken with the same equipment, and some individual ratios (e.g., Table IX and X) may be much better determined.

For the diffuse clouds, we note that in addition to using Eq. (3) with ^{13}CO, Nyman (1983,1984) has made use of measurements of $N(CH)$ and $N(H_2CO)$ vs. extinction and has compared $N(X)$ to these values and hence ultimately to $N(H_2)$ (Table XI). As for the denser clouds, $f(X)$ may be uncertain by an order of magnitude but ratios of abundances should be much more accurate.

Since the data are available, it is useful to compare these molecular cloud abundances with those found for the envelope of an evolved star. For the IRC

TABLE X
Abundance Ratios of Isomers

Region	HCN/HNC	CH_3CN/CH_3NC	HCO^+/HOC^+	References*
Orion ridge	20	$\gtrsim 20$	$\gtrsim 250$	a,b,c,e
Orion plateau	250			b
TMC-1	1	$\gtrsim 10$	$\gtrsim 400$	c,d
IRC+10216	250			b
Sgr B2			300	c

*(a) P. F. Goldsmith et al. 1981; (b) Johansson et al. 1984; (c) Woods et al. 1983; (d) Irvine and Schloerb 1984; (e) unpublished Onsala data.

TABLE XI
Abundances Relative to H_2 (× 10^8) in Quiescent Clouds[a] and Density[b] and Kinetic Temperature Information

Species	Diffuse	TMC-1/L 134N	Orion Ridge
CO	200/5000[c,d]	8000	8000
OH	8[e]	30	
CH^+	0.5[c]		
CH	2[c,e]	2	
C_2H	5	$\gtrsim$1	2
CN	1[c]	1	2
HCN	0.5	2	2
N_2H^+	$\lesssim$0.05	0.1	0.02
HCO^+	0.5	0.8	0.3
H_2CO	0.5, 0.2[f]	2	$\gtrsim$8
CS	0.5	0.2	0.4
NH_3	0.3[f]	2–20	20
n(cm^{-3})	10^2–10^3	10^3–10^5	10^4–10^6
T_k(K)	20–100	10–20	50–100

[a]Columns 3 and 4 from Table VIII; diffuse cloud data for spiral arm features towards W49 from Nyman (1983,1984), except as otherwise indicated.
[b]See Table VI.
[c]Observations towards 22 stellar lines of sight; CO is radio; CH^+, CH, and CN are optical data (Dickman et al. 1983).
[d]The first value is the average for low color-excess clouds; second value is for star BD+66°1674/75 where $E(B-V) \approx 1.4$, i.e. $A_v \sim 4$ mag, presumably representing the higher-density regime.
[e]Radio data (Rydbeck et al. 1976; Turner 1973).
[f]Spiral arm features towards Cas A (Batrla et al. 1984).

+ 10216 envelope both physical and chemical conditions certainly vary radially. Ratios of column densities can, therefore, be a misleading guide to *in situ* abundance ratios, particularly when column densities obtained by widely disparate techniques are compared. Our present results are undoubtedly averages over an extended region of the envelope, but they have been derived by using information from line shapes and spatial extent of the emission to deduce relative volume densities at an arbitrary reference radius (Olofsson et al. 1982*b*; Johansson et al. 1984); they, therefore, provide an internally consistent set of relative abundances. Values taken from similar modeling by other authors are given in parentheses in Table VIII.

The following seven general conclusions may be drawn from our tabulation:

Conclusion 1. A comparison of abundances in Tables VIII and XI (see also Fig. 9) reveals a surprising *uniformity* for the gas in quiescent regions over the rather wide range of temperature and density spanned by the diffuse clouds, the dark clouds TMC-1 and L 134N, and the Orion ridge. Although

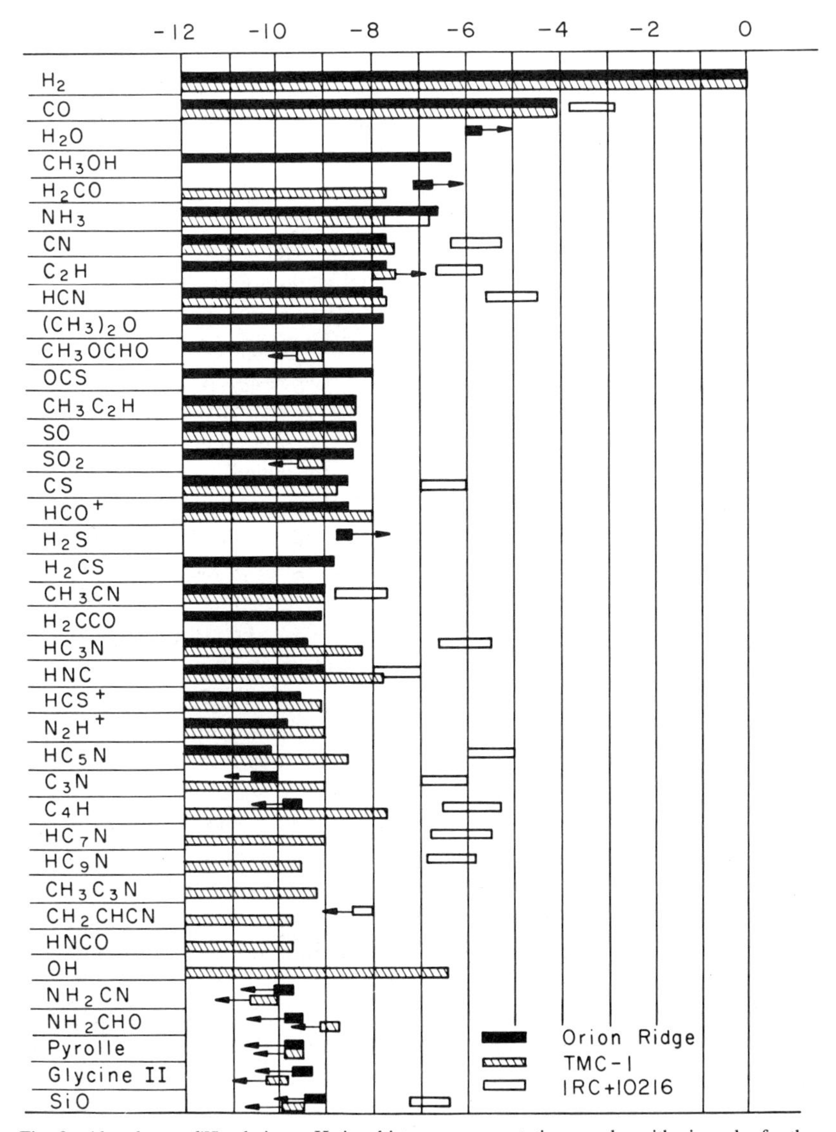

Fig. 9. Abundances $f(X)$ relative to H_2 in a histogram presentation on a logarithmic scale, for the molecular clouds TMC-1 and the Orion ridge. For IRC + 10216 the center of each rectangle gives $f(X)$, and its length is a rough estimate of the uncertainty (applicable also to molecular clouds). (Data from Table VIII in text.)

this may be compatible with gas-phase ion-molecule chemical models in which the major reactions are not very temperature-sensitive (W. D. Watson and C. M. Walmsley 1982), further calculations designed to treat consistently the full range of physical parameters represented by the present data would be desirable. Note that the observed uniformity is in direct contrast to some earlier studies that deduced a strong inverse dependence of abundance on cloud density for such species as HCO^+, HCN, CO, H_2CO, and C_2H (Wootten et al. 1978,1980*a,b*,1982*b*), an effect attributed to depletion of molecules onto grains in dense regions. Possibly such trends are obscured by other effects in the limited cloud sample discussed here. This seems to us to be unlikely, however. Although the Orion region was included only for C_2H in the earlier determinations, the previously derived $f(C_2H)$ was two orders of magnitude lower than our value in Table VIII. Also, recent results for the $H^{13}CO^+/^{13}CO$ ratio show great uniformity among Orion, L 134N, and many of the clouds included in the earlier studies (Wootten et al. 1984*b*). Strong support for the present conclusion is given by recent infrared absorption data which suggest even *less* CO depletion than we have assumed for the Orion ridge (~100% of carbon in gas-phase CO; Scoville et al. 1983). It seems most probable that the estimates from the previous pioneering studies suffer from serious systematic errors. Indeed, several basic assumptions/conclusions of the model have subsequently been shown to be seriously inaccurate (unsaturated emission lines, homogeneous source regions; cf. Guélin et al. 1982*b*; Frerking et al. 1979; R. D. Brown and E. Rice 1981; Turner and Thaddeus 1977; P. F. Goldsmith et al. 1981; P. F. Goldsmith 1984), and detailed theoretical analysis has shown that the simplifications employed in the radiative transfer calculations and the inverse correlation between errors in density and abundance can artificially produce just the reported variation of abundance with density (Stenholm 1983). Consistent with the present conclusions, the extensive dark cloud study by Frerking et al. (1982) found no evidence for CO depletion in regions with visual extinction as large as $A_v \sim 20$, and Crutcher et al. (1984) find no trend in f(CN) with cloud density. On the other hand, numerous authors have pointed out that the time scale theoretically expected for adsorption onto grains is short relative to estimated cloud ages (see chapter by Draine). Thus, the present results would seem to require some sort of desorption; such processes have been considered by Boland and de Jong (1982), Léger (1983), J. M. Greenberg (1983), and Draine (his chapter).

Conclusion 2. Some striking chemical *differences* are also evident among quiescent regions (Tables VIII and XII). Relative to both L 134N and the Orion ridge cloud, abundances in TMC-1 are *enhanced* for the acetylenic or polyacetylenic species C_3N, HC_3N, C_4H, HC_5N (and probably HC_7N, HC_9N, CH_3C_4H, and CH_3C_3N, but abundance limits for these species in Orion and L 134N are not well known). In contrast, abundances among these clouds are *similar* for CS, C_2H and (at least for Orion and TMC-1) for

TABLE XII
Relative Abundances in Two Dark Clouds[a]

TMC-1	L 134N
C_4H	NH_3
CH_3C_2H	SO
HC_3N	SO_2
HC_5N	
HC_7N	
C_3N	
HCS^+	
C_3O	
CH_3C_3N?	
CH_3C_4H?	

[a]Species listed under a particular source are $\gtrsim$5 times more abundant relative to CO than they are in the other source, at the specific positions probed. In contrast, abundances of CO, CS, HCO^+, N_2H^+, H_2CO, and CH_3CHO are quite similar in these two clouds. Data abstracted from Table VIII.

CH_3C_2H; while NH_3 is *depleted* in TMC-1 relative to the other clouds (note, however, that the tables refer to specific locations in these clouds, while the abundance of NH_3 and some sulfur-containing species may vary with position (Tölle et al. 1981; Ungerechts et al. 1980; Irvine et al. 1983). It may be important that the ratios of HC_3N, HC_5N, and C_4H within L 134N are similar to those within TMC-1. Because TMC-1 and L 134N have similar temperature and density, which in turn differ from those in Orion (Table VI), the results in Tables VIII and XII suggest that other factors may play a significant role in determining chemical composition (relative elemental abundances, cloud age or history, ultraviolet radiation field; Langer et al. 1984; Millar and Freeman 1984*a,b;* Stahler 1984*a*; chapters by Herbst and by Glassgold).

Conclusion 3. Several authors have previously pointed out that certain molecular abundances are *enhanced* in the Orion plateau relative to the ambient ridge cloud (see, e.g., T. B. H. Kuiper et al. 1981). This result is even more apparent in the present data, whose higher angular resolution makes the plateau (primarily the disk source) stand out more clearly (Table VIII and Fig. 10). Specifically, SO, SO_2, and SiO are *more than two orders of magnitude* more abundant than in the ridge, HCN and H_2S *more than ten times* as abundant, while the abundances of HDO, HC_3N, HCO^+, OCS, CS and possibly HNC are greater by at least a *factor of a few.* The $f(HCO^+)/f(CO)$ ratio increases still further in high-velocity gas over a spatially extended region that roughly coincides with the source of vibrationally excited H_2 emission

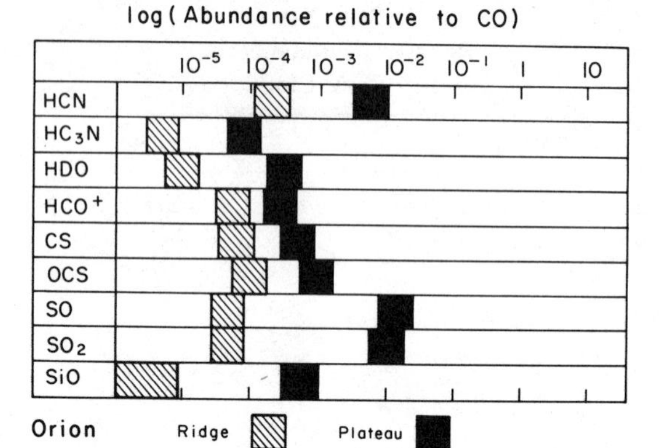

Fig. 10. Abundances on a logarithmic scale for the Orion ridge (ambient cloud) and the Orion plateau. (Data from Table VIII in text; figure from Irvine and Hjalmarson 1984.)

(Olofsson et al. 1982*a*), probably reflecting post-shock conditions (Wootten et al. 1984*b*). Since the plateau source seems to be associated (in an as yet poorly defined way) with star-formation activity, these chemical differences from the ambient ridge gas are particularly interesting. It has been suggested that the abundance enhancements may be due to shock heating, during which more endothermic reactions become possible, or, alternatively, to mass outflow from the envelope around an oxygen-rich star (Scalo and Slavsky 1980; W. D. Watson and C. M. Walmsley 1982; Dalgarno 1981; Hollenbach 1982; chapter by Herbst). On the other hand, the large abundances for SO, SO_2, and SiO can be produced under standard *in situ* cloud conditions provided that elemental S and Si are not strongly depleted (perhaps reflecting a warmer history for this gas; Prasad and Huntress 1982; Millar 1982). We also note that the observed HDO abundance implies that in the plateau $f(H_2O)/f(CO) \sim 0.1$, if the D/H fractionation is as large as for DCO^+/HCO^+ in the ridge, and would be even larger for lesser fractionation, as would be expected in warmer gas (Olofsson 1984).

Conclusion 4. Absolute abundances in the Orion hot core source are not known, because the CO column density is uncertain. If, however, we use the enhanced NH_3 abundances estimated from Very Large Array (VLA) observations of the hot core (Genzel et al. 1982; Pauls et al. 1983; similarly enhanced values have also been inferred for W31 and NGC 7538; T. L. Wilson et al. 1983), we would conclude that the hot core abundances for several species are similar to those in the plateau and thus substantially enhanced over the ambient Orion ridge cloud (Table XIII). Thus, the abundance of HDO in this region suggests an H_2O abundance that might approach or exceed that of CO (cf. Olofsson 1984) and hence conceivably might include a substantial fraction of the elemental oxygen. It is interesting to note that main-

TABLE XIII
Abundances in Orion Hot Core and Plateau[a]

Species	Hot Core Relative Abundance[b]	Plateau Fractional Abundance ($\times 10^8$)
HC_3N	1	0.6
CH_2CHCN	1	
CH_3CH_2CN	7	
CH_3CN	5	
HNCO	5	
HDO	10	3
HCN	200	50
NH_3	1000	200

[a]Johansson et al. (1984) and Ziurys et al. (1981).
[b]Same as fractional abundance if $X(NH_3) = 10^{-5}$ as indicated by Genzel et al. (1982) and Pauls et al. (1983); HC_3N column density is 10^{14} cm^{-2}.

ly nitrogen-containing molecules have thus far been observed in the hot core. Although this may be partially a matter of excitation, because typically these species have rather large dipole moments, other chemical differences relative to the ambient cloud are clearly indicated by Table IX, which shows that more chemically saturated species appear prominent in the hot core. Although SO has previously been reported in this source, its hot-core-like emission seems actually to be more extended and to have a different spatial origin (Friberg 1984).

Conclusion 5. Emission from several species (H_2CO, HDO, CH_3OH, $(CH_3)_2O$, CH_3OCHO, OCS, SO, SO_2, CH_3CN) observed in the Orion ridge appears to come predominantly from the southern condensation (Johansson et al. 1984). It is difficult to be certain that this reflects actual chemical differences along the ridge without more spatial mapping data, although this would seem to be the most likely explanation (there is no strong evidence for significant variations in density, total gas-column density, or temperature between the northern- and southern-ridge components within $\pm$ 1 arcmin of the Kleinmann-Low nebula, although the southern region seems somewhat warmer; Johansson et al. 1984; Olofsson 1984; Bastien et al. 1981; Friberg 1984). Many of these species seem chemically related, in terms of reactions involving the abundant species H_2O and/or atomic oxygen (chapter by Herbst; Millar 1982), perhaps suggesting a link with the plateau source.

Conclusion 6. There is a tendency for lower abundance of the positive ions in the Orion ridge compared to the dark clouds. This probably reflects a higher electron density in the case of Orion (Graedel et al. 1982).

Conclusion 7. There are many interesting examples of what might be called chemical selectivity apparent from the data, that may serve to constrain models of interstellar chemistry. For example:

a. Methyl formate, which may be considered as methylated formic acid, is in Orion at least 10 times as abundant as the simpler species; in contrast, formaldehyde (HCHO) is vastly more abundant than acetaldehyde (CH_3CHO), and the same holds for HCN relative to CH_3CN. Methanol is relatively abundant, but H_2O is even more so, although precise values remain highly uncertain because the deuterium fractionation is not well understood (W. D. Watson and C. M. Walmsley 1982).
b. Dimethyl ether in Orion is at least an order of magnitude more abundant than its isomer ethanol.
c. The isomeric pairs HCN/HNC, CH_3CN/CH_3NC, and HCO^+/HOC^+ all have roughly similar energy differences and barriers to rearrangement, yet the abundance ratios are vastly different (Table X), and in addition, the HCN/HNC ratio varies significantly among observed sources. The interstellar observations have thus stimulated both laboratory and theoretical study of these systems, as discussed in the next section.
d. Although neither cyclic molecules nor species with a branched heavy element backbone have been found in the interstellar medium, it is not clear that the failure to detect such molecules with current instrumental sensitivity has great significance. Acetic acid is such a branched chain, but the upper limit on its abundance is not below the measured abundance of its isomer methyl formate. Although glycine is less abundant than HC_5N in TMC-1, this is not necessarily so in Orion, and in neither source is it obviously less plentiful than other, lower molecular-weight H-C-N compounds. Similar remarks apply to those ring molecules for which abundance limits have been pushed to near the current sensitivity level (apart from the triangular SiCC, found in IRC + 10216; Thaddeus et al. 1984).
e. The differences in relative abundance of chemically saturated versus unsaturated species among interstellar and circumstellar sources (Table IX; Sec. III.B.4), apparently have not been commented upon previously. Conceivably they represent an effect of the thermal history of the gas.

Conclusion 8. As might naively be expected, the abundances in the IRC + 10216 circumstellar envelope are strikingly different from those in interstellar clouds (Table VIII, column 5, and Fig. 9). CO and particularly HCN are much more abundant in the case of IRC + 10216, presumably reflecting the equilibrium result that almost all oxygen and almost all nitrogen will be in these compounds (McCabe et al. 1979). Moreover, even though silicon is thought to be strongly depleted onto dust grains in IRC + 10216 (Lafont et al. 1982), SiO and SiS are vastly higher in abundance than in the Orion ridge and the dark clouds (although *f*(SiO) is similar to that in the Orion plateau), probably reflecting even greater depletion of Si in silicate grains in the interstellar medium. Although a qualitative similarity has sometimes been suggested for IRC + 10216 and TMC-1, because the cyanopolyynes appear prominently in the spectra of both objects, a quantitative comparison from Table VIII shows that this similarity is rather illusory; C_4H, HC_3N, C_3N and HC_5N and 10^2 to 10^3 times more abundant in IRC + 10216, and the contrast with the Orion ridge and L 134N is even greater. Nonetheless, the *relative* abundances of observed nitriles in TMC-1 and IRC + 10216 (e.g., normalized to HC_3N) differ significantly between these regions only in the case of the nonlinear species CH_3CN. One previously noted contrast between IRC + 10216 and interstellar abundances also proves to be perhaps misleading; although the failure to detect positive ions has been used as an argument against ion-molecule chemistry in the stellar envelope (Wannier and Linke 1978), even the present, more-stringent upper limit for HCO^+ is not significantly below the observed abundance in the Orion ridge. Likewise, the limits for a number of other oxygen-containing species are not stringent relative to interstellar values (a result of low-excitation conditions in the IRC + 10216 envelope, combined with typically large-partition function for these species).

IV. COMPARISONS WITH THEORY

Rather detailed calculations of the abundance of simple polyatomics now exist for both steady state and time-dependent chemical models including cloud evolution, embodying various boundary conditions on elemental abundances, thermal history, reaction rates, ambient radiation field, and the role of interstellar grains (see, e.g., the discussion and references in chapters by Glassgold and by Herbst). However, the large number of uncertain parameters makes a comprehensive comparison with observations very difficult. In our view, progress in constraining the models can best be achieved by carefully establishing the physical conditions in a variety of sources, and simultaneously obtaining accurate abundance ratios for chemically related species. Examples of such related species are chemical isomers, and we shall illustrate some of the chemical complexity that arises in practice for this approach by a discussion of three isomer pairs: HOC^+/HCO^+, HNC/HCN and CH_3NC/CH_3CN.

HOC^+/HCO^+. Woods et al. (1983) have identified a line in Sgr B2 that is probably due to the $J = 1 - 0$ transition of HOC^+. If the identification is correct, the HOC^+/HCO^+ abundance ratio in Sgr B2 is approximately 1/330. From negative HOC^+ observations in TMC-1 and Orion, Woods et al. (1983) deduce HOC^+/HCO^+ abundance ratios of $\leq 1/400$ and 1/200, respectively. Thus, it is clear that the metastable isomer HOC^+ is far less abundant than the stable isomer HCO^+ in interstellar sources (Table X).

HOC^+ is calculated to lie 33 kcal mol^{-1} (1.4 eV) above HCO^+ in energy (DeFrees et al. 1984), and to be stable against isomerization because of a large activation energy barrier of 40 kcal mol^{-1} (1.7 eV) which prevents migration of the hydrogen atom from the oxygen to the carbon. Since interstellar collision time scales are much longer than those in the laboratory, it might still be conceivable that interstellar HOC^+ is depleted relatively rapidly by tunneling under this barrier. However, DeFrees et al. (1984) have calculated a tunneling time scale at interstellar cloud temperatures of 10^{14} to 10^{15} yr. Even though tunneling calculations are not very reliable, it is difficult to believe that the calculation is sufficiently in error to be wrong in the conclusion that HOC^+-HCO^+ tunneling is not the cause of the low HOC^+/HCO^+ abundance ratio. What then causes the relatively small abundance of HOC^+?

There are two possibilities: either HOC^+ is formed at a less rapid rate than HCO^+, or it is depleted more rapidly by reactions with some constituents of the interstellar gas. These possibilities are, of course, not mutually exclusive. Woods et al. (1983) suggested that HOC^+ and HCO^+ are formed at approximately the same rate via the reactions

$$CO + H_3^+ \rightarrow HCO^+ + H_2 \quad (4a)$$

$$\rightarrow HOC^+ + H_2 \quad (4b)$$

and that HOC^+ is preferentially depleted via the exothermic "catalysis" reaction

$$HOC^+ + H \rightarrow H + HCO^+ . \quad (5)$$

Assuming a typical ion-molecule rate coefficient for this latter reaction, it is still necessary to posit a fractional hydrogen atom abundance of 5×10^{-3} to explain the observed HOC^+/HCO^+ abundance ratio. This fractional abundance is higher than believed by many astronomers to pertain to the inner sections of interstellar clouds.

Green (1983) and DeFrees et al. (1984) have both calculated that reaction (5) possesses a small but nonzero activation energy of 1 to 4 kcal mol^{-1}. The latter set of authors has estimated the rate coefficient for surmounting and/or tunneling under this barrier to be about 10 orders of magnitude *below* the standard ion-molecule rate at 10 K. Defrees et al. (1984) have also investigated reactions between HOC^+ and H_2; even a relatively slow reaction with molecular hydrogen could be rate-determining under interstellar conditions. However, all reaction pathways leading to "catalysis"

$$HOC^+ + H_2 \rightarrow H_2 + HCO^+ \quad (6)$$

are blocked by activation energy barriers of at least 15 kcal mol^{-1} and are much too slow to be important. Even the very exothermic and potentially rapid radiative association reaction

$$HOC^+ + H_2 \rightarrow HOCH_2^+ + h\nu \tag{7}$$

is blocked by a large barrier. Thus, it would appear that differential depletion rates cannot by themselves explain the HOC^+/HCO^+ abundance ratio.

What about the suggestion that HOC^+ and HCO^+ are formed at the same rate via reactions (4a) and (4b)? Illies et al. (1982,1983) have now measured these rates in the laboratory via an indirect method and have determined that only (6 ± 5)% of the product is HOC^+. The large uncertainty in this result, however, makes it unclear whether the interstellar HOC^+/HCO^+ abundance ratio can be explained in this way. If we accept the 6% figure, we are left with an HOC^+/HCO^+ ratio of 1/16, not 1/200 or 1/400. More definitive experiments should be undertaken. There is one theoretical result for the relative rates of reactions (4a) and (4b); DeFrees et al. (1984) have used the approximate phase space theory to find that ~1% of the reaction product is HOC^+. This result is in better agreement with interstellar HOC^+ and HCO^+ abundances but cannot be regarded as definitive. The problem is made more complex by the probability that reactions (4a) and (4b) are not the only source of HOC^+ and HCO^+. Another likely reaction is the one between C^+ and H_2O

$$C^+ + H_2O \rightarrow HCO^+ + H \tag{8a}$$

$$\rightarrow HOC^+ + H. \tag{8b}$$

Reaction (8) has been studied in the laboratory but the branching ratio has not yet been determined. Phase space calculations indicate that HOC^+ is slightly preferred. Although reaction (4a) is a much more important source of interstellar HCO^+ than is reaction (8a), it is not at all clear whether reaction (4b) is a more important source of HOC^+ than is reaction (8b). *More experimental work is needed.*

Finally a detailed experimental study of the exothermic reaction

$$HOC^+ + CO \rightarrow HCO^+ + CO \tag{9}$$

should be undertaken to determine if it contributes to the low relative HOC^+ abundance. Present indirect indications (Illies et al. 1982) are that it is slow.

From all of the above information, we can only conclude that ion-molecule theory is qualitatively correct in calculating less HOC^+ than HCO^+; a quantitative determination of the HOC^+/HCO^+ abundance ratio based on accurate rate coefficient data still eludes us.

HNC/HCN. The calculation of the HNC/HCN abundance ratio is plagued even more severely by a lack of laboratory data. The variation of this abundance ratio from one interstellar source to another is striking; it is unity in TMC-1 (Fig. 11) and 1/20 in the warmer Orion (ridge) source (P. F. Gold-

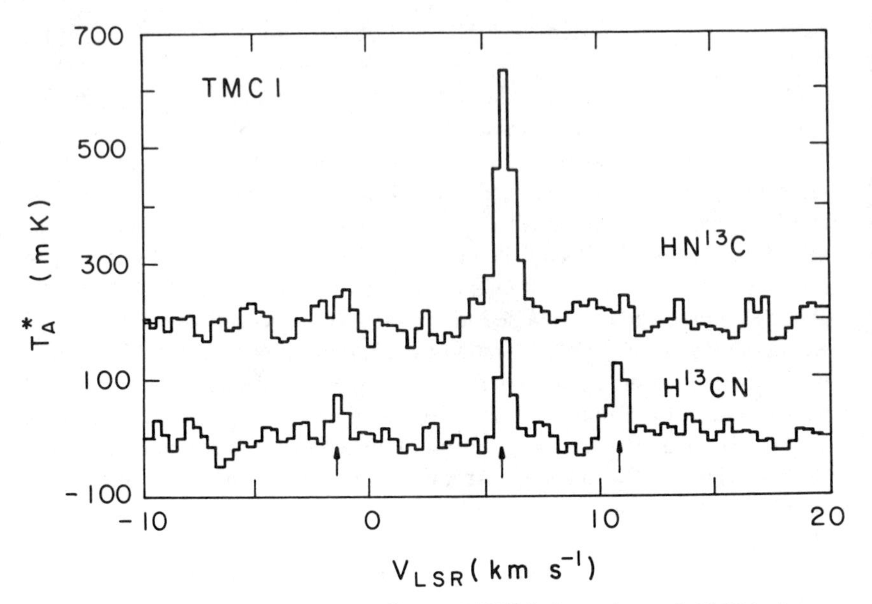

Fig. 11. The J = 1 − 0 transitions of H^{13}CN and HN^{13}C observed towards TMC-1. Arrows indicate three hyperfine components for H^{13}CN (see Fig. 7). Comparison of the integrated intensities shows an abundance ratio HNC/HCN ~ 1.5 (Irvine and Schloerb 1984), which may be compared with values of 0.3 to 0.02 in some giant molecular clouds (P. F. Goldsmith et al. 1981).

smith et al. 1981; Table X). Basic ion-molecule theory predicts a constant abundance ratio of order unity, but this prediction is not based on solid experimental evidence. The reaction that is supposed to produce both HCN and HNC is the ion-electron dissociative recombination between H_2CN^+ and electrons

$$HCNH^+ + e \rightarrow HCN + H, HNC + H. \tag{10}$$

The HNC/HCN product branching ratio for this reaction has never been measured in the laboratory. The structure of the ground state $HCNH^+$ ion is a linear one, and it seems plausible to suggest that HCN and HNC formation from ground state $HCNH^+$ is equally probable. A phase space calculation of the branching ratio (E. Herbst 1978*b*) also suggests this possibility. However, no detailed theoretical treatment of this reaction has yet been undertaken. It is unlikely on energetic grounds that the branching ratio will show much of a temperature dependence.

If reaction (10) is the dominant source of HNC and HCN, and if these species are depleted at the same rate, then a temperature-independent prediction of unity for HNC/HCN is obtained. (A ratio even higher than unity, as seen in some dark clouds, can be rationalized by preferential HNC formation from an excited state of $HCNH^+$ that has the structure H_2NC^+ [see Allen et al. 1980; R. D. Brown 1977].) Are both species depleted at the same rate? The ion-molecule depletion rates are unlikely to be substantially different, although small activation energy barriers for selected HNC and/or HCN reactions cannot be ruled out. As with the HOC^+ case, one must also investigate the possibility of the metastable isomer tunneling under an activation energy barrier to form the stable isomer. However, the time scale calculated for this process by both DeFrees et al. (1984) and S. K. Gray et al. (1980) is 10^{22} to 10^{23} yr. What about reaction of the "radical" HNC with reactive neutral species such as atoms? It is conceivable that HNC might prove to be more reactive in this sense than HCN; detailed quantum chemical investigations of the potential surfaces of systems such as HNC + O and HCN + O would be most useful. The chemical rule-of-thumb for neutral molecules, however, is that they are only reactive if in nonsinglet electronic ground states (the metastable isomer HNC is in a singlet electronic ground state). An additional depletion mechanism for HNC would be most useful in explaining the present results, which show that the HCN fractional abundance is the same in Orion (ridge) and TMC-1 whereas the HNC fractional abundance is diminished in the warmer source. Perhaps there is an additional depletion reaction with a small activation energy that is more important in the warmer Orion (ridge) source than in TMC-1.

P. F. Goldsmith et al. (1981) have argued that instead of a preferential depletion of HNC, there is a preferential formation of HCN. These authors suggest that HCN but not HNC can be produced via the neutral-neutral reaction

$$CH_2 + N \rightarrow HCN + H. \tag{11}$$

They argue that under the condition of relatively high electron abundance, this reaction can play a significant role in HCN production. DeFrees (personal communication, 1983) is currently investigating its quantum mechanical potential surface to determine whether or not it leads preferentially to HCN. If reaction (10) is a preferential producer of HCN, then the reaction

$$NH_2 + C \rightarrow HNC + H \tag{12}$$

might be a preferential HNC production mechanism. DeFrees plans to investigate this system as well.

If reaction (11) does indeed produce HCN preferentially, P. F. Goldsmith et al. (1981) calculate that reaction (11) is relatively more important in warmer clouds like the Orion (ridge) source, and that an HNC/HCN abundance ratio significantly less than unity results in these sources. In order to reconcile this opinion with the result that HCN has the same fractional abundance in

TMC-1 and Orion (ridge), it is necessary that the common $HCNH^+ + e$ route to HNC and HCN be diminished in importance in the warmer source. This could certainly be the case if the precursor ions that lead to $HCNH^+$ production (CH_3^+, C^+) are less abundant (in a fractional sense) in Orion.

As in the HOC^+/HCO^+ situation, the net result is that a lack of detailed theoretical and experimental information on rate coefficients and reaction pathways precludes a definitive gas-phase prediction for the HNC/HCN abundance ratio under differing physical conditions.

CH_3NC/CH_3CN. The metastable isomer CH_3NC exists as a laboratory species but has not been seen in dense interstellar clouds. The data in Table X indicate an abundance ratio CH_3NC/CN_3CN of $\leq 1/10$. The lack of detection of CH_3NC appears to be in agreement with the limited amount of experimental and theoretical data available concerning relevant gas-phase reactions. It is thought that the synthesis of CH_3CN proceeds through the rapid radiative association reaction

$$CH_3^+ + HCN \rightarrow CH_3CNH^+ + h\nu \tag{13}$$

which has been studied in the laboratory. The structure of the product ion has been investigated by D. Bohme and coworkers (personal communications, 1983) and found to be as written above; that is, protonated cyanomethane (CH_3CN) and *not* protonated iso-cyanomethane (CH_3NC). This is an important result, because it means that as the CH_3^+ and HCN approach one another, they do *not* form much CH_3NCH^+, which is the product ion one might naively expect if the reaction proceeded directly. Rather, the CH_3NCH^+ structure, even if formed initially at long range, isomerizes easily to CH_3CNH^+. From the precursor ion CH_3CNH^+, only CH_3CN can then be formed without significant activation energy

$$CH_3CHN^+ + e \rightarrow CH_3CN + H \tag{14}$$

$$CH_3CNH^+ + e \not\rightarrow CH_3NC + H. \tag{15}$$

Given the evidence that the CH_3NCH^+ ion is not strongly metastable at the energy of reaction (13), it would appear most unlikely that this ion could be formed in the more exothermic radiative association reaction between CH_3^+ and HNC

$$CH_3^+ + HNC \rightarrow CH_3CNH^+ + h\nu. \tag{16}$$

Thus, the available evidence indicates that much CH_3NC cannot be formed by the same pathway thought to be dominant in CH_3CN production. Additional quantum chemical studies that probe for the barrier to rearrangement between

CH_3NCH^+ and CH_3CNH^+ would be useful. With such information, a phase space treatment can be undertaken to determine the small amount of CH_3NCH^+ produced in reaction (13).

Additional Metastable Isomers. In their study of possible metastable isomers in dense interstellar clouds, Green and Herbst (1979) considered a wide variety of species. One of these species, H_2CC or vinylidine, was thought to be important because it possesses a dipole moment and its detection could indicate the amount of small hydrocarbon formation in clouds. These authors predicted vinylidene to form from the same ion-election recombination reaction as acetylene (HCCH)

$$C_2H_3^+ + e \rightarrow H_2CC + H \tag{17}$$

$$C_2H_3^+ + e \rightarrow HCCH + H. \tag{18}$$

However, recent tunneling calculations by Osamura et al. (1981) show the barrier against tunneling of vinylidene into acetylene to be sufficiently low that the calculated lifetime against tunneling is at most 10^{-9} s. Therefore, this species will have a negligible abundance in dense clouds. Tunneling calculations need to be performed for a number of other metastable isomers proposed by Green and Herbst (1979); before such calculations can be undertaken, detailed quantum chemical potential surfaces must be available. If the tunneling rates of metastable isomers are rapid, then their interstellar chemistry will be simple; they will possess vanishingly small abundances and will not be detected.

V. CONCLUSIONS

The results described in previous sections lead us to some general questions, that imply certain directions for future research.

1. Why is depletion of molecules onto grains in dense regions not more apparent (e.g., the similar abundances for many species in spiral arm diffuse clouds, dark clouds, and in the Orion ridge)? Which desorption processes may occur?
2. Why, in spite of question (1), do some striking abundance differences exist among and within rather quiescent cloud regions? Are these evolutionary effects?
3. How can one distinguish among explanations for "anomalous" chemistry in apparently "active" regions, such as the Orion plateau and hot core? Do they imply that there are real chemical tracers of protostellar activity?
4. What can interstellar isotopic ratios tell us about the formation of the solar system and about the chemical evolution of galaxies?

5. Can we anticipate substantial increments in the number of known interstellar molecules through the next decade or so?

To answer questions such as 1, 2, and 3, clearly we must first determine relative abundances both for a variety of clouds and within individual clouds. From our discussion of the necessary procedures, it should be apparent that this requires knowledge of the physical state of a cloud and its variation with position. Thus, physics and chemistry are inextricably linked, and elucidation would seem to require multitransition, multiisotope, and multispecies mapping of a number of regions (cf. Guélin et al. 1982*b*).

In relation to the origin of the solar system, better isotopic ratios are needed, both for solar system objects like comets and the giant planets, and for a variety of potentially condensible interstellar species, particularly H_2O, NH_3, and (if it proves to be abundant) CH_4. The apparent variations in $^{13}CO/^{12}CO$ within and among molecular clouds also need further study.

The high degree of success ($\sim$90%) in the assignment of newly detected interstellar lines to known molecular species by the broad frequency range surveys of giant molecular clouds and IRC + 10216 (Johansson et al. 1984; Sutton et al. 1984; Cummins et al. 1984) suggests that detections of new molecules might most profitably be pursued in new frequency ranges for these sources such as the submillimeter region, and in other types of sources such as the cold, dark clouds.

Acknowledgments. We are grateful to many colleagues across both oceans for communicating unpublished results and for helpful discussions. This research was supported in part by grants from NASA, from NSF and from the Swedish Natural Science Research Council (NFR).

GRAIN EVOLUTION IN DARK CLOUDS

B. T. DRAINE
Princeton University

Observations pertaining to the sizes and composition of grains in diffuse clouds and dark clouds are briefly reviewed. The physical processes acting to change the composition and size distribution of grains in dark clouds are examined. Accretion of ice mantles on grains can deplete molecules from the gas phase on time scales short compared to cloud lifetimes. Temperature fluctuations can prevent mantle formation only on very small (radii $\lesssim$ 10 Å) grains. Turbulence-driven coagulation may act to remove the smaller grains from the grain size distribution, resulting in grains of a conglomerate character; however, whether or not grains colliding at 0.01–0.1 km s^{-1} will actually stick together is uncertain. Fragmentation following high-velocity grain-grain collisions (e.g., induced in shock waves) may replenish the small particles, and sputtering of grains in shock waves can erode ice mantles. The FU Orionis phenomenon in active star-forming clouds may act to sublime ice mantles and return molecules to the gas phase, but is probably of secondary importance.

Although dust grains are an important component of dark (molecular) clouds, our knowledge of their properties is limited. It is clear from observations that the dust grains in molecular clouds differ from dust in diffuse clouds and the evolutionary processes responsible for these differences are undoubtedly complex. This chapter surveys the physical processes which may act to change the composition and size distribution of grains located within dark clouds, at densities ranging up to $\sim 10^6 cm^{-3}$. While an assessment of the relative importance of the different mechanisms can be made, there is no attempt to put together a complete model for grain evolution in dark clouds. Evolution of the grain population in collapsing protostars or accretion disks (see the chapter by Morfill et al.) will not be addressed here.

The properties of grains in diffuse clouds (the precursors of molecular clouds) are reviewed in Sec. I. Observational information concerning the size and composition of grains in dark clouds is reviewed in Sec. II. In Sec. III, the evolution of grains in quiescent molecular clouds is discussed, with attention given to a number of distinct physical processes. The extent to which active star formation in a cloud may be expected to affect the grain population is considered in Sec. IV. Section V contains a brief summary.

I. DUST IN DIFFUSE CLOUDS

It is plausible to assume that the matter (gas and grains) now in dark clouds was, at some earlier time, in a less dense state characteristic of diffuse interstellar clouds; this must be true if the lifetimes of molecular clouds are appreciably shorter than the age of the Galaxy. As a starting point for a discussion of grain evolution in dark clouds, it is therefore useful to outline our present understanding of the nature of dust grains in diffuse clouds, because these are, by assumption, the precursors of molecular clouds. A review of the observed properties of interstellar dust may be found in Savage and Mathis (1979).

There is as yet no universal agreement as to the nature of the interstellar dust population in diffuse clouds. Although considerable variations are seen in the wavelength dependence of the extinction, particularly in the ultraviolet (cf. Witt et al. 1984), it is useful to discuss an average extinction law. Several properties of interstellar grains in diffuse clouds deduced from observations are listed below, in order of decreasing certainty:

1. The observed extinction curve requires that a substantial range of grain sizes be present, with grains of radii $a \gtrsim 0.1$ μm to produce the bulk of the visual extinction, and $a \lesssim 0.02$ μm grains required by the steep rise in extinction at ultraviolet wavelengths. [Observations of polarization of starlight by aligned dust grains of course require that a substantial fraction of the grains producing the visual extinction be nonspherical; here a loosely refers to radii of equivalent spherical grains of equal volume.]
2. The strong infrared emission/absorption features at 10 and 20 μm are strong evidence that silicate material (probably amorphous) constitutes approximately 50% of the interstellar grain volume, and is the repository of nearly all of the Fe, Mg, and Si present in the interstellar medium.
3. The strong bump in the extinction curve at $\lambda \approx 2200$ Å is probably due to graphite particles with radii $a \approx 100$–200 Å (although other materials have been proposed, including polycyclic carbon macromolecules [Donn 1968], small MgO particles [MacLean et al. 1982], and nongraphitic carbonaceous material [Sakata et al. 1983]).

Unfortunately, even if one accepts that the 2200 Å band is due to graphite, these three constraints do not uniquely determine the nature of the grain population. For example, Hong and Greenberg (1980) proposed a grain model consisting of $a \approx 0.15$ μm core-mantle grains (with $a \approx 0.05$ μm silicate cores and a mantle of unspecified molecular material), plus a population of $a \approx 0.02$ μm graphite grains (to explain the 2200 Å feature), plus $a \approx 0.01$ μm silicate grains. In this model, the bulk of the visual extinction is due to the core-mantle grains. While the mantle material in core/mantle grain models was originally assumed to be predominantly H_2O ice, failure to detect the 3.1 μm ice band in diffuse clouds (see Sec. II.B) has rendered this untenable. Advocates of the core/mantle model for grains in diffuse clouds now argue that photolysis by interstellar ultraviolet will convert a frozen mixture of initially simple molecules (H_2O, NH_3, CO, H_2CO, CH_4, etc.) into a complex organic refractory residue. This hypothesis is supported in part by laboratory experiments (see, e.g., J. M. Greenberg et al. 1983), and in part by observation of an interstellar absorption band at 3.4 μm which is attributed to the C–H stretch mode (see Sec. II.B below).

Mathis, Rumpl, and Nordsieck (1977; hereafter MRN) have proposed an alternative model for grains in diffuse clouds. In the MRN graphite-silicate grain model only bare silicate and bare graphite particles are present, each having a simple power-law size distribution

$$dn = An_{\mathrm{H}}a^{-3.5}da \qquad a_{\min} < a < a_{\max} \tag{1}$$

extending between upper and lower size cutoffs $a_{\min} \approx 0.005$ μm and $a_{\max} \approx 0.25$ μm. The graphite and silicate grains are present in approximately equal numbers: $A(\text{graphite}) = 10^{-25.16}\text{cm}^{2.5}\ \text{H}^{-1}$, $A(\text{silicate}) = 10^{-25.11}\text{cm}^{2.5}\ \text{H}^{-1}$ (Draine and Lee 1984). Subsequent papers by Mathis (1979) and Mathis and Wallenhorst (1981) have concluded that this simple model is consistent with observations of interstellar extinction and polarization, and that observations of anomalous extinction and polarization can be accomodated by simple adjustment of the upper and lower size cutoffs. Recent studies have concluded that the simple MRN graphite-silicate model is apparently consistent with much available evidence relating to infrared and far-infrared properties of the dust population (Mezger et al. 1982; Mathis et al. 1983; Draine and Lee 1984). However, recent observations of an absorption band at 3.4 μm (see Sec. II.B below) suggest that grains in diffuse clouds may include a hydrocarbon component, a feature not present in the unadorned MRN model.

In view of the simplicity and relative success of the MRN model, it will be adopted as the basis for discussion below, but it should be remembered that its validity is not firmly established. One interesting quantity is the total grain geometric cross section per H atom:

$$\Sigma \equiv \int \frac{n(a)}{n_{\mathrm{H}}} \pi a^2 da = 2\pi A(a_{\min}^{-0.5} - a_{\max}^{-0.5}) \ . \tag{2}$$

For the MRN distribution one finds $\Sigma_{21} = 1.1$, where $\Sigma_{21} \equiv \Sigma/(10^{-21}\text{cm}^2\text{H}^{-1})$; approximately 80% of Σ is due to particles with $a < 0.05$ μm, which are required to explain the far-ultraviolet extinction observed in diffuse clouds. Conversely, most of the mass is in the larger particles, about 2/3 of the mass residing in grains with $a > 0.05$ μm.

II. OBSERVED PROPERTIES OF GRAINS IN DARK CLOUDS

Grains in dark clouds are known to differ from those in diffuse clouds in at least two respects: (1) they show wavelength-dependent extinction and polarization in the 0.5–3 μm region which differs from that found on lines of sight passing only through diffuse clouds; (2) they sometimes manifest a number of infrared absorption bands, of which the 3.1 μm ice band is the best example, which are not observed in diffuse clouds.

A. Variations in the Wavelength Dependence of Reddening and Polarization

It has been customary to use the ratio $R \equiv A_V/(A_B - A_V)$ as a measure of the slope of the extinction law between the V (5500 Å) and B (4400 Å) bands (A_λ is the extinction at wavelength λ). In the diffuse interstellar medium, it now appears that $R \approx 3.1$ is a representative average value (Savage and Mathis 1979); this value is consistent with the MRN size distribution. Values of R smaller than ~ 3 are not observed. In dark clouds, values of R appreciably larger than 3.1 are sometimes found. A study by Vrba and Rydgren (1984) of background stars seen through the Cha T1 and R CrA dark clouds obtained normal values of R in the outer parts of the Cha T1 Cloud, but $R > 4$ for several stars near the cloud core where $A_V \gtrsim 2$ mag; a similar increase in R toward the cloud center was found for the R CrA Cloud. Anomalous values of R have also been observed in other clouds by a number of investigators (e.g., Carrasco et al. 1973; Whittet et al. 1976).

Independent evidence for anomalous extinction comes from observations of the wavelength dependence of the linear polarization produced by aligned grains. The wavelength λ_{max} at which the polarization peaks has a mean value in diffuse clouds $< \lambda_{max} > \approx 0.545$ μm (Serkowski et al. 1975). In dark clouds, however, larger values of λ_{max} are often found; this is generally taken to reflect an increase in the size of the grains which provide the polarization. For example, Vrba et al. (1981) found a mean value $< \lambda_{max} > = 0.75 \pm 0.09$ μm for stars in the R CrA Cloud. An empirical relationship between R and λ_{max} has been proposed: $R \approx 5.7\ (\lambda_{max}/\mu\text{m})$ (Serkowski et al. 1975; Vrba et al. 1981), which is consistent with the notion that an increase in the mean grain size will produce increases in both λ_{max} and R; however, Chini and Krugel (1983) find that R and λ_{max} toward M17 do not satisfy the proposed empirical relationship. Chini and Krugel have in fact argued that anomalously large R values need *not* indicate the presence of larger-than-average grains. It is, however, the case that their grain models, which reproduce anomalous extinction curves, include silicate grains which are larger than average, even though the average size of the graphite grains is reduced. Thus it seems safe to conclude that grain growth is required for at least the silicate grains.

The above data alone do not appear sufficient to identify the mechanism responsible for increasing the grain size. Accretion of H_2O ice mantles, if restricted only to grains with pre-accretion radii $\gtrsim 0.05$ μm, could increase the radius of the larger grains by a factor as large as ~1.75 (assuming all of the available oxygen to be converted to solid H_2O), which might then account for λ_{max} values as large as ~0.95 μm or so (of course, strong ice bands would then be expected for lines of sight with these large λ_{max} values). Alternatively, the increase in the typical particle size could also be the result of coagulation.

An important datum bearing on this question comes from measurements of the extinction relative to the hydrogen column density. Accretion of material onto a grain almost inevitably results in an increase in the extinction cross section of the grain at all wavelengths (ignoring small oscillations in the extinction cross section with wavelength due to interference effects; these oscillations would be averaged out for any reasonable size distribution). The line of sight toward ρ Oph has $R \approx 4$ (Carrasco et al. 1973; R. L. Brown and B. Zuckerman 1975), but Jura (1980) has pointed out that this anomalous value of R cannot be the result of accretion from the gas phase because this line of sight has A_V/N_H approximately a factor of two below the average interstellar value. Jura concludes that coagulation is apparently required if the grains in this cloud originally evolved from a size distribution typical of diffuse clouds. However, Mathis and Wallenhorst (1981) note that small ($a \approx 0.015$ μm) graphite particles are required toward ρ Oph to account for the ultraviolet extinction, and have suggested that the measured N_H may be in error.

B. Molecular Bands

The strongest band observed in dark clouds is the silicate band near 10 μm, which is also observed in diffuse clouds and in the circumstellar dust shells around stars with $C/O < 1$. To within observational error, the band is found to show the same shape in both emission from hot grains in the Trapezium H II region and in absorption in dark clouds (Gillett et al. 1975*a*). Many lines of sight through dense clouds show an absorption feature at 3.1 μm, which is usually attributed to solid H_2O. The explanation of the 3.1 μm band profile is controversial; Léger et al. (1979, 1983) have attributed it to pure, amorphous H_2O ice grains of rather large size ($a \approx 0.5$ μm), while Hagen et al. (1983a) have shown that it can be reproduced by more normal-sized ($a \approx 0.15$ μm) grains composed of an amorphous mixture of H_2O and other molecules.

The 3.1 μm band is found only on lines of sight that pass through dense, dark regions. D. H. Harris et al. (1978) have concluded that in the ρ Oph region the 3.1 μm band appears only on lines of sight with $A_V \gtrsim 25$ mag. However, Whittet et al. (1983) have detected the 3.1 μm band in stars in Taurus with A_V as small as 4.5 mag., and Goebel (1983) has recently detected the 3.1 μm ice band toward the B8V star HD29647 in the Taurus Cloud complex, with $A_V = 3.6$ mag. For HD29647, Goebel determined the ice band strength relative to A_V to be $\Delta A_{3.1\mu m}/A_V = 0.028$, which is within the range 0.021 to 0.058 found by Harris et al. for the lines of sight in the ρ Oph Cloud with $A_V \gtrsim 25$ mag. It is noteworthy that IUE (International Ultraviolet Explorer) observations of HD29647 (Snow and Seab 1980) reveal an anomalous ultraviolet extinction law, lacking the 2200 Å extinction feature. Goebel has pointed out that this may be the result of ice mantles coating the small graphite particles which are usually invoked to explain the 2200 Å extinction feature. There is also evidence indicating a rapid rise in the extinction toward HD29647 shortward of 1300 Å, which Goebel suggests may be due to ice coatings on small particles. If this is so, it indicates that coagulation has not greatly reduced the abundance of small ($\lesssim 200$ Å) core particles in this cloud, even though conditions were apparently conducive to ice mantle formation.

When the 3.1 μm ice band is detected, its strength $\Delta A_{3.1\mu m}$ relative to A_V varies by at least a factor of ten. The lines of sight where Harris et al. detected the ice band had $\Delta A_{3.1\mu m}/A_V$ ranging from 0.021 to 0.058. It should be emphasized, however, that some lines of sight have upper limits to $\Delta A_{3.1\mu m}/A_V$ as small as 0.002 (toward VI Cyg. No. 12; Gillett et al. 1975*b*) so that failure to detect the ice band on at least some lines of sight is not due only to the generally smaller amounts of dust present on lines of sight where $\Delta A_{3.1\mu m}$ falls below the detection threshold.

In very dark clouds, estimation of A_V is uncertain, and it is more convenient to compare the strength of the ice band to that of the 9.7μm silicate feature. The line of sight to VI Cyg. No. 12 has $\Delta A_{9.7\mu m}/A_V \approx 0.07$ (Gillett et al. 1975*b*), but this ratio may differ in very dense clouds. A study of the spectra of 13 compact infrared sources in dark clouds (Willner et al. 1982) found $\Delta A_{3.1\mu m}/\Delta A_{9.7\mu m}$ ranging from 0.07 (for AFGL 2059) to 0.77 (for AFGL 961); all of the sources considered had $A_V \gtrsim 20$ mag. Toward W33 A the ice feature is unusually strong: $\Delta A_{3.1\mu m}/\Delta A_{9.7\mu m} > 1.1$ (Capps et al. 1978). It is evident that the grain populations vary considerably from one dark, star-forming cloud to another.

Other molecular bands have been detected. The identification of the 3.1 μm band with ice has been strengthened by the detection of a second H_2O ice band at 6.0 μm (Willner et al. 1982), and tentative detection of a third band at 45 μm (E. F. Erickson et al. 1981). There is still some question as to why the H_2O band at 12 μm is not present at its expected strength (Capps et al. 1978). The presence of other molecules (e.g., CO, NH_3, CH_4) mixed together with H_2O in the ice mixture may shift the wavelength of the band to $\lambda > 13$ μm (Hagen et al. 1983*b*) where it cannot be observed because of atmospheric absorption. It is interesting to note that there is one source (OH 0739-14 ≡

OH 231.8 + 4.2; Soifer et al. 1981) where the 3.1 μm and 12 μm ice bands are present with band strengths and profiles consistent with *pure* amorphous H_2O ice; in this source the extinction is dominated by dust which has formed in a stellar wind from this evolved giant star and so has not undergone the evolutionary history of dust in molecular clouds (Hagen et al. 1983*a*).

Knacke et al. (1982*b*) have detected an absorption band at 2.97 μm which they identify as due to a frozen NH_3/H_2O mixture. Other spectral features include a band at 3.4 μm, probably due to C-H stretching modes in hydrocarbons (Wickramasinghe and Allen 1980; T. J. Jones et al. 1983); a 4.60 μm band, probably due to a CN stretching mode in an unidentified molecule (Lacy et al. 1984); a 4.67 μm band due to solid phase CO (Lacy et al. 1984); and a band at 6.8 μm for which a number of identifications have been suggested (cf., Evans et al. 1983).

The 3.4 μm band is of particular interest, as it has been observed on lines of sight which lack the 3.1 μm ice feature—the line of sight to the Galactic center is one example (D. A. Allen and D. T. Wickramasinghe 1981)—thus suggesting that this feature is associated with grains in relatively diffuse clouds. Identification of the 3.4 μm feature as due to the C–H stretching mode in solid hydrocarbons seems reasonable. The observed band is quite weak ($\Delta A_{3.4}/A_V \approx 0.003$–$0.006$; Wickramasinghe and Allen 1980), but it is difficult to infer the amount of hydrocarbon required. For an assumed band strength of 1500 $cm^2\ g^{-1}$, the observed 3.4 μm feature would require about 15% of interstellar carbon to be present in hydrocarbons, but this number is very uncertain because the 3.4 μm band strength varies considerably among hydrocarbons: from 340 $cm^2\ g^{-1}$ for C_6H_6 to 8400 $cm^2\ g^{-1}$ for $C_{42}H_{88}$ (Duley and Williams 1979). Ultraviolet photolysis has been proposed as one mechanism for producing such hydrocarbons (see Sec. III.D below). Such hydrocarbons are of course not present in the MRN graphite-silicate model for grains in diffuse clouds. If hydrocarbons do in fact contribute an appreciable fraction of the grain volume in diffuse clouds, the MRN model for grains in diffuse clouds may have to be rejected in favor of core-mantle grain models.

III. PHYSICAL PROCESSES IN QUIESCENT DARK CLOUDS

A. Accretion of Atoms or Molecules

If grain velocities are small compared to thermal velocities of gas atoms, as is normally the case, the time scale for depletion of a species from the gas phase is

$$\tau_{\rm depl} = \left(\frac{\pi\mu}{8kT}\right)^{1/2} \frac{1}{n_{\rm H}\alpha\Sigma}$$

$$= \frac{3 \times 10^5\ {\rm yr}}{\alpha n_4 \Sigma_{21}} \left(\frac{\mu}{20 m_{\rm H}}\right)^{1/2} \left(\frac{20\ {\rm K}}{T}\right)^{1/2} \tag{3}$$

where α is the sticking coefficient, which is expected to be of order unity (Burke and Hollenbach 1983), and $n_4 \equiv n_H/10^4$ cm^{-3}. We see that this time scale is short enough that appreciable depletion of condensible species may occur during the lifetime of a cloud if no mechanism acts to return material to the gas phase.

To what extent can this accretion process change the typical grain size? It will be seen that this depends sensitively on the lower size cutoff a_{min}. As noted above, the grains in diffuse clouds already incorporate nearly all of the Mg, Si, and Fe, probably a substantial fraction (~2/3) of the C, and ~15–20% of the O. The grain volume per H atom in the MRN model is 5.7×10^{-27} cm^3 H^{-1}. Depletion of the remaining 80% of the cosmic abundance of O, in the form of H_2O accreted onto grains, would add ~2.0×10^{-26} cm^3 H^{-1} to the grain volume, increasing the grain volume by a factor ~5.

Suppose that the grains prior to accretion have a MRN size distribution (Eq. 1), with $\xi \equiv a_{min}/a_{max}$. If the sticking coefficient is independent of grain size, then accretion will increase the radius of every grain by the same amount $\Delta a \equiv \beta a_{max}$, and the total grain volume will increase by a factor

$$\frac{V_{new}}{V_{old}} = 1 + \frac{3\beta}{\xi^{1/2}} + \frac{\beta^2}{\xi^{3/2}} \frac{1 - \xi^{3/2}}{1 - \xi^{1/2}} + \frac{\beta^3}{5\xi^{5/2}} \frac{1 - \xi^{5/2}}{1 - \xi^{1/2}}. \quad (4)$$

Observations of ultraviolet extinction require $a_{min} \lesssim 100$ Å in diffuse clouds, and $a_{min} = 50$ Å is often assumed. With $a_{min} = 50$ Å and $a_{max} = 2500$ Å one has $\xi = 0.02$, and one finds from Eq. (2) that a fivefold increase in grain volume requires $\beta = 0.063$. Thus uniform accretion of ice mantles onto the MRN model will increase the grain radii by only $\Delta a = \beta a_{max} = 160$ Å if the lower limit on the initial size distribution is $a_{min} = 50$ Å. Observations of infrared emission from reflection nebulae suggest the presence of even smaller grains undergoing temperature fluctuations due to absorption of individual ultraviolet photons. Sellgren (1984) concludes that the MRN size distribution extends down to $a_{min} = 10$ Å and if this is the case, a size-independent sticking coefficient implies that a volume increase of ~5 will occur for $\beta = 0.02$, or $\Delta a = 50$ Å. Even if a_{min} is assumed to be as large as $a_{min} = 200$ Å, one finds that ice mantles can increase the grain size by at most $\Delta a = 460$ Å, an increase of radius of only 18% for the largest grains. Thus we see that the MRN distribution, modified only by accretion with size-independent sticking coefficient α, results in a maximum grain size only slightly larger than in the initial distribution. Therefore, acc etion alone (with a size-independent sticking coefficient) cannot explain observations of significantly increased grain size in some clouds.

B. Denudation of Very Small Grains by Thermal Spikes

The discussion above has assumed that grain growth was limited only by the accretion rate, resulting in a mantle thickness Δa independent of the initial radius. For very small grains, however, absorption of a visible or ultraviolet photon may heat the grain sufficiently to desorb a number of adsorbed ice molecules, as has been discussed by Purcell (1976). Let u be the photon energy density and $h\nu$ the typical energy, and let Q_{abs} be the absorption efficiency. Suppose that for grain radius a, absorption of a photon heats the grain to a sufficiently high temperature that $\varepsilon h\nu/L$ molecules are desorbed from the surface, where L_{sub} is the heat of sublimation per molecule and $\varepsilon < 1$ is an "efficiency." If the number density of condensible species is fn_H, then one may easily show that the ratio of the desorption rate to the adsorption rate is

$$\frac{N_{desorp}}{N_{adsorp}} = \frac{6 \times 10^3 \varepsilon Q_{abs}}{n_4}\left(\frac{u}{0.001 \text{ eV cm}^{-3}}\right)\left(\frac{0.5 \text{ eV}}{L_{sub}}\right)\left(\frac{10^{-4}}{f}\right) \quad (5)$$

where we have assumed 0.1 km s^{-1} as a typical thermal speed for a condensible species in the gas. In the diffuse interstellar medium the energy density of starlight is $u \approx 0.45$ eV cm^{-3}. A silicate grain of radius $a = 10$Å has $Q_{abs} \approx 0.001$ at $\lambda = 3000$ Å (Draine and Lee 1984). Graphite grains have larger values of Q_{abs}. Thus if ε were of order unity then, for a typical cloud density $n_H = 10^4$ cm^{-3} and condensible fraction $f \approx 10^{-5}$, thermal desorption during temperature spikes would keep a 10 Å silicate or graphite grain bare for u as small as $\sim 5 \times 10^{-5}$ eV cm^{-3}, four orders of magnitude below the radiation energy density outside the cloud (and corresponding to an extinction optical depth ~10 into the cloud at the wavelengths in question; cf. Flannery et al. 1980). At larger optical depths into the cloud the radiation field may be dominated by Lyman and Werner band photons from H_2 excited by cosmic rays, with an energy density $\sim 5 \times 10^{-7}$ eV cm^{-3} (Prasad and Tarafdar 1983); even though Q_{abs} for silicate grains is about an order of magnitude larger at 1200 Å compared to 3000 Å, this energy density is probably not large enough to be important in clouds with $n_H \gtrsim 10^4$ cm^{-3}.

How small must a grain be for these thermal spikes to be important; i.e., how does ε vary with a? The probability per unit time ν that a given molecule at the grain surface will be thermally desorbed may be approximated $\nu \approx \nu_0 \exp(-\theta/T_{gr})$, where T_{gr} is the grain temperature, ν_0 is a vibration frequency, and $k\theta \approx L_{sub}$ is the adsorption energy per molecule. Low-temperature vapor pressure data have been used to obtain ν_0 and θ for various candidate mantle constituents in Table I (ν_0 and θ of course apply to the pure solid in each case).

Now consider desorption following a thermal spike which instantaneously raises the grain temperature to a peak value T_p. If energy loss to desorption is ignored, then the temperature decays as the grain emits infrared photons, with a time scale $\tau_{rad} \equiv |d\ln T/dt|^{-1}$ given by

$$\tau_{rad}(T) = \frac{Ca}{3 \langle Q \rangle \sigma T^3} \quad (6)$$

where C is the heat capacity per volume, $\langle Q \rangle$ the Planck-averaged absorption efficiency at temperature T, and σ the Stefan-Boltzmann constant. If

TABLE I
Desorption Parameters[a]

Substance	θ(K)	$k\theta$(eV)	ν_0(s^{-1})	n(10^{22} cm^{-3})
CO	951	0.082	1.6×10^{16}	1.7
CH_4	1120	0.097	2.9×10^{13}	1.6
NH_3	3894	0.34	1.6×10^{16}	2.4
HCN	4531	0.39	4.1×10^{15}	1.5
H_2O	6113	0.53	1.3×10^{16}	3.3

[a] ν_0 and θ calculated from low-temperature vapor pressure data (Weast 1978) assuming a sticking coefficient of unity: $n^{2/3}\nu = p/(2\pi mkT)^{1/2}$.

$T_p/\theta << 1$, as will be the case, the expected number of desorbed molecules is approximately

$$N_{\mathrm{desorb}} \approx 4\pi a^2 n_a^{2/3} \int \nu \mathrm{d}t \approx 4\pi a^2 n_a^{2/3} \nu_0 \tau_{\mathrm{rad}}(T_p) \frac{T_p}{\theta} \exp(-\theta/T_p) \qquad (7)$$

where n_a is the number density (molecules per volume) of the adsorbed species, assumed to totally cover the grain surface. We will generally be interested in peak temperatures of order ~100 K, in which case the heat capacity $C \approx 3Nk$, where $N \approx 10^{23}\mathrm{cm}^{-3}$ is the number density of atoms in the grain. For $100 \lesssim T \lesssim 200$ K, Planck-averaged emissivities are approximately (Draine and Lee 1984)

$$<Q> \approx 0.013\left(\frac{a}{0.1\mu\mathrm{m}}\right) T_2^{5/4} \text{ for silicate} \qquad (8)$$

$$<Q> \approx 0.006\left(\frac{a}{0.1\mu\mathrm{m}}\right) T_2^{3/5} \text{ for graphite} \qquad (9)$$

where $T_2 \equiv T/100\mathrm{K}$. These emissivities imply $\tau_{\mathrm{rad}} \approx 200\ T_2^{-4.25}$ s for graphite, and $\tau_{\mathrm{rad}} \approx 400\ T_2^{-3.6}$ s for silicate.

Observations of the 3.1 μm band point to H_2O, with a binding energy k θ = 0.53 eV, as a primary constituent of the grain mantles formed in dense clouds. For a pure H_2O mantle, one finds from Eq. (7) that ~1 H_2O molecule will be desorbed from an $a = 10^{-7}$ cm grain if $T_p = \theta/42 = 146$ K. We now must estimate the energy required to heat the grain to this temperature, using the Debye model for the heat capacity. The Debye temperature for graphite is $\theta_D = 420$ K; for SiO_2 it is $\theta_D = 470$ K. For $n_{\mathrm{gr}} = 1 \times 10^{23}$ cm^{-3} and $\theta_D = 450$ K, one finds that the energy required to heat the grain to 146 K is ~$4.3(a/10\ \text{Å})^3$ eV. From this we conclude that thermal spikes induced by photons may be able to prevent water molecules from accreting on the surfaces of grains smaller than ~10 Å, although it is clear that the available photon energies preclude this process being effective for grain radii $\gtrsim$ 20 Å. (More volatile species could be removed from larger grains, because lower peak temperatures are required, but not H_2O).

Other heating mechanisms are possible. Chemical reactions at the grain surface have been proposed by M. Allen and G. W. Robinson (1975), but the available energies are again <10 eV, so that chemical reactions can only be effective for desorbing H_2O from $\lesssim$ 10 Å grains. Cosmic rays may also produce thermal spikes. Let H be the heat deposited in the grain by a cosmic ray, and suppose the number of desorbed molecules to be $\epsilon H/L_{sub}$, where $\epsilon < 1$ is the efficiency. If ζ is the primary ionization rate for a hydrogen atom, one can show that the ratio of the desorption rate to the accretion rate is approximately

$$\frac{\dot{N}_{desorb}}{\dot{N}_{adsorb}} = \frac{0.01\varepsilon}{n_4}\left(\frac{a}{10\text{Å}}\right)\left(\frac{0.5\text{ eV}}{L_{sub}}\right)\left(\frac{\zeta}{10^{-17}\text{s}^{-1}}\right)\left(\frac{10^{-4}}{f}\right). \tag{10}$$

Cross sections appropriate to 10 MeV protons have been assumed, but this result is nearly independent of the cosmic ray energy E since both the H ionization cross section and the energy loss rate in the solid vary as E^{-1}. In $a \lesssim 10$ Å grains, the efficiency ϵ may approach unity, but ϵ is expected to be small for $a \gtrsim 100$ Å, as the energy deposited by the cosmic ray will not raise the grain temperature sufficiently. Because $\zeta < 10^{-16}\text{s}^{-1}$ in dense clouds (Langer 1984), thermal desorption by cosmic rays is not expected to be important. Thus there does not appear to be any mechanism to inhibit accretion onto grains as small as 10 Å. Thus accretion, even in the presence of thermal spikes, can have only a small effect on the sizes of the $a \gtrsim 0.1$ μm grains which are responsible for extinction and polarization at visible wavelengths.

C. Coagulation and Fragmentation

Grain-grain collisions can occur in dense clouds, and may have important consequences for the grain size distribution (Scalo 1977*a*). If the collision velocity is small, the grains may adhere to one another, a process known as coagulation.

The important quantity determining the grain-grain collision rate is essentially just the velocity dispersion of the grains at a given position: frequent grain-grain collisions are possible only if the velocity dispersion is large. The time scale for a given grain, of radius a_1, to collide with some other grain, with radius in the interval $(a_2, a_2 + da_2)$, is

$$d\tau_{coll}^{-1} \approx (3 \times 10^5\text{ yr})^{-1}\left(\frac{a_1 + a_2}{a_2}\right)^2 n_4\left(\frac{\delta v(a_1,a_2)}{0.1\text{ km s}^{-1}}\right)d\Sigma_{21} \tag{11}$$

where $d\Sigma_{21}$ is the contribution to Σ_{21} by grains with radii in the interval $[a_2, a_2 + da_2]$, and $\delta v(a_1,a_2)$ is the mean relative speed between grains of radii a_1 and a_2.

Brownian motion will give the grains a velocity dispersion on the order of $\sim 10^{-3}(a/100\ \text{Å})^{-3/2}$km s^{-1}, too small to be of consequence for $a \gtrsim 50$ Å. [Recall that for the MRN size distribution (Eq. 1), grains with $a < 100$ Å contain less than 20% of the total grain mass, so that accretion of all $a < 100$ Å grains onto $a > 0.05$ μm grains can increase the radii of the latter by only $\sim$10%.] Variations in drift velocities of grains due to radiation pressure, ambipolar diffusion, or gravitational sedimentation can also contribute to $\delta\nu_{gr}$ but, except under unusual circumstances (e.g., near a luminous young star, or in a shock wave), the resulting random velocities are probably too small to lead to substantial modifications in the grain population. Suprathermal rotational and translational velocities can result from a number of effects which may act if the gas, grains, and radiation field depart from thermodynamic equilibrium, as discussed by Purcell (1979). In a dark cloud where little atomic H or ultraviolet radiation is present, the main effect may be due to variation in the accomodation coefficient over the grain surface, which can lead to suprathermal motions if the gas and grain temperatures differ. The expected velocities are, however, not large.

Turbulence appears to be the mechanism most likely to produce significant grain-grain collision rates, which may be estimated by the following heuristic argument. Because the turbulent velocity field of the gas fluctuates in both time and space, the grain population will acquire velocities which vary between grains of different sizes, and even among grains of a single size. Consider for the moment a grain of size a. If it acquires a velocity relative to the gas, that velocity will tend to decay as $\exp(-t/t_{drag})$, when the drag time scale (for subsonic motion) is $t_{drag} = (a\rho_{gr}/n)\,(\pi/8\mu kT)^{1/2}$. Now suppose we have a turbulent velocity field with a random velocity v_{max} on a length scale l_{max}. On a length scale l, where $l_{min} < l < l_{max}$, the turbulent velocity field is estimated to have a Kolmogorov spectrum: $v_l \approx v_{max}\,(l/l_{max})^{1/3}$, with a turnover time $\tau_l = l/v_l \approx l^{2/3}\,l_{max}^{1/3}\,v_{max}^{-1}$. In terms of the turnover time τ, we have $v_\tau \approx v_{max}\,(\tau/\tau_{max})^{1/2}$, where $\tau_{max} \equiv l_{max}/v_{max}$.

Now consider grains of a single radius a. It seems clear that these grains will have a velocity dispersion that is primarily determined by the component of the turbulent velocity field fluctuating on a time scale comparable to t_{drag}; the grain cannot effectively respond to more rapidly fluctuating components, while more slowly fluctuating components will simply result in an overall advection of the grain population, with little internal velocity dispersion at a given point. Thus we may estimate that the characteristic velocity of a grain relative to the gas will be approximately equal to the amplitude of the turbulent eddies having a characteristic turnover time comparable to the grain inertial time scale $\tau \approx t_{drag}$; this in turn means that grains of radius a cross from one such eddy to another, and one expects that $\delta v\,(a,a)$ will be of the same order as the grain velocity relative to the gas. Furthermore, since the induced velocity $\propto \tau_{drag}^{1/2}$, and $\tau_{drag} \propto a$, the largest grains will have the largest velocities relative to the gas. Thus

$$\delta v(a_1,a_2) \approx \frac{v_{max}^{3/2}}{l_{max}^{1/2}}\left(\frac{\rho_{gr}}{4n}\right)^{1/2}\left(\frac{2\pi}{\mu kT}\right)^{1/4}[\max(a_1,a_2)]^{1/2}$$

$$\approx \frac{0.07 \text{ km s}^{-1}}{n_4}\left(\frac{\max(a_1,a_2)}{10^{-5}\text{ cm}}\right)^{1/2}\left(\frac{v_{max}}{\text{km s}^{-1}}\right)^{3/2}\left(\frac{10^{18}\text{ cm}}{l_{max}}\right)^{1/2}. \quad (12)$$

This rough estimate, in fact, agrees fairly well with the results of a more detailed analysis by Völk et al. (1980).

We now evaluate the collision rate for three cases: $a_2 << a_1$, $a_2 \approx a_1$, and $a_2 >> a_1$. Assuming the MRN distribution: $d\Sigma \propto a^{-0.5} d\ln a$, one obtains

$$d\tau_{coll}^{-1} \approx B\left(\frac{a_1}{a_2}\right)^{5/2} d\ln a_2 \quad \text{for } a_2 << a_1 \quad (13a)$$

$$\approx 4B\, d\ln a_2 \quad \text{for } a_2 \approx a_1 \quad (13b)$$

$$\approx B\, d\ln a_2 \quad \text{for } a_2 >> a_1 \quad (13c)$$

where

$$B \equiv \frac{1}{2} A \pi n_H^{1/2} \rho_{gr}^{1/2}\left(\frac{2\pi}{\mu kT}\right)^{1/4}\frac{v_{max}^{3/2}}{l_{max}^{1/2}}$$

$$= (4 \times 10^6 \text{ yr})^{-1} n_4^{1/2}\left(\frac{v_{max}}{\text{km s}^{-1}}\right)^{3/2}\left(\frac{10^{18}\text{ cm}}{l_{max}}\right)^{1/2} \quad (14)$$

where $A = 10^{-24.8}$ cm$^{2.5}$ H^{-1} has been taken for the MRN distribution. It is therefore clear that: (1) a given grain is more likely to collide with a grain much smaller than itself, if such grains are present; (2) the small grains near the lower cutoff of the size distribution are as likely to collide with a large grain as with another small one; and (3) the time scale for depleting the small grains from the size distribution is of order 10^6 yr for a typical cloud, if the grains stick when they collide.

The outcome of a grain-grain collision will be a function of the collision speed Δv. Collisions with $\Delta v \gtrsim 20$ km s^{-1} have sufficient center-of-mass energy to vaporize at least the smaller of the two colliding grains, and this is presumably the outcome. Collisions with $\Delta v \gtrsim 5$ km s^{-1} probably result in vaporization of any ice mantle material present, at least on the smaller grain. Shattering of grains may be important at lower velocities (or a large grain might be shattered in a high Δv impact with a smaller grain). The minimum relative velocity required for fracturing the refractory grain cores is of order $v_{cr} = (6\sigma/a\rho)^{1/2} \approx 0.15\,(0.1\,\mu\text{m}/a)^{1/2}$ km s^{-1}, where $\sigma \approx 10^3$erg cm^{-2} is the surface free energy, and $\rho \approx 3$ g cm^{-3} the density of the grain material.

For $\Delta v >> v_{cr}$ the collision may result in a large number of fragments, with the total surface area of the fragments being much larger than that of the original collision partners (Burke and Silk 1976). The MRN size distribution may itself be the result of such fragmentation (P. Biermann and M. Harwit 1980).

For the collision velocities $\lesssim 0.1$ km s^{-1} indicated by Eq. (11), it thus appears that the refractory cores will remain intact. In a totally inelastic collision between two identical grains, the grain temperature would rise to $T \approx$ 200 K $(\Delta v/\text{km s}^{-1})^2$, so that thermal effects should be unimportant for $\Delta v \lesssim$ 0.1 km s^{-1}. Some mantle material may perhaps be removed in the form of fragments. Grain mantle explosions may be triggered by low-velocity collisions; ultraviolet-irradiated ices can store chemical energy in the form of frozen free radicals in sufficient quantity that when the temperature is raised above ~30 K (e.g., by $\Delta v > 0.04$ km s^{-1} collisions), runaway chemical reactions lead to substantial ejection of mantle material (J. M. Greenberg and A. S. Yencha 1973; d'Hendecourt et al. 1982).

What is most uncertain is whether or not the colliding grains will stick together. For $\Delta v \lesssim 10^{-3}$ $(0.1\ \mu\text{m}/a)$ km s^{-1} particles appear likely to stick, as the potential energy due to surface forces exceeds the initial center of mass kinetic energy. For Δv in the important range 0.001 to 0.1 km s^{-1} the outcome is less clear. Scalo (1977*a*) has concluded that sticking will occur for Δv $< v_{cr}$, the threshold for shattering. One consequence of coagulation will be the formation of conglomerate grains consisting of an aggregation of graphite and silicate particles, possibly cemented together by ice. This makes calculation of grain optical properties in dark clouds especially uncertain.

D. Processing by Ultraviolet or Cosmic Rays

We have seen in Sec. II.B that the thermal spikes resulting from absorption of individual photons are able to thermally desorb molecules like H_2O only from extremely small ($a \lesssim 10$ Å) grains. It is also possible for desorption to occur nonthermally, as a quantum process; this is known as photodesorption. Photodesorption appears most likely to occur when the candidate molecule is first electronically excited as the result of absorption of an ultraviolet photon; if the electronically excited molecule has a sufficiently repulsive interaction with the substrate, it may be ejected. Let ϵ be the probability of ejection following photoexcitation. From the limited experimental evidence available, Draine and Salpeter (1979*b*) estimated $\epsilon \approx 0.005$, although with considerable uncertainty, corresponding to a photodesorption cross section $\sigma_{pd} \approx 10^{-20}$ cm^2. The desorption probability per unit time for a molecule at the surface is then approximately (Draine and Salpeter 1979*b*)

$$\nu_{des} \approx (5 \times 10^4\ \text{yr})^{-1}\,(\sigma_{pd}/10^{-20}\ \text{cm}^2). \tag{15}$$

In a dark cloud this rate would be reduced by a factor ~$0.5 \exp(-1.8 A_V)$ (Boland and de Jong 1982). We see that photodesorption may be able to appreciably erode the mantles of grains near the surface of dark clouds, particularly if the photodesorption cross section is larger than the estimated value $\sim 10^{-20}$ cm^2. If turbulence in dark clouds is effective at recirculating gas and dust between the surface layers and the dark interior, as has been suggested by Boland and de Jong (1982), then photodesorption could have important effects on the grains throughout the cloud. Ultraviolet irradiation can also alter the chemical composition of the grain mantle by photolysis (J. M. Greenberg et al. 1983); this has been suggested as the mechanism for producing (from an ice containing H_2O, CO, NH_3, etc.) species containing $C \equiv N$ bonds, the existence of which is inferred from the 4.60 μm absorption feature observed by Lacy et al. (1984), and C—H bonds, to account for the observed 3.4 μm absorption feature (see Sec. II.B above).

Cosmic rays can sputter molecules from the grain. However, the erosion rates are totally negligible for realistic cosmic ray fluxes (Draine and Salpeter 1979*a*). Cosmic rays (and X rays) may also cause polymerization in carbon-rich molecular ices (Strazzulla et al. 1983). The time scale for polymerization of, for example, CH_4 ice is estimated to be $\tau \approx 10^9\,(10^{-17}\,\mathrm{s}^{-1}/\zeta)$ yr, where ζ is the ionization rate per H atom.

E. Effects of Galactic Supernovae

During the $\sim 10^6$ yr time scale of grain evolution in a dark cloud, what is the probability that a supernova may occur sufficiently near the cloud for the optical luminosity to have an effect on the grain? The effects of supernovae on dust in the diffuse interstellar medium have been discussed by Draine and Salpeter (1979*b*), and this discussion may be readily extended to grains in dark clouds.

Suppose we wish to assess the effects of supernova flashes on grains located an optical depth τ_{cl} inside a cloud. The peak temperature attained by a grain heated only by direct photons from a supernova of peak luminosity L_{SN} a distance r away is simply

$$T_p = \left(\frac{L_{SN} e^{-\tau_{cl}} Q_*}{16\pi r^2 < Q(T_p) > \sigma}\right)^{1/4} \tag{16}$$

where σ is the Stefan-Boltzmann constant, and Q_* is the absorption efficiency factor averaged over the supernova spectrum. The grain will undergo a thermal spike due to the supernova flash, but with the temperature decay time τ determined not by the radiative cooling time of the grain but rather by the decay time $\tau \approx 10^6$ s of the supernova lightcurve. To desorb 100 monolayers, we require $\nu_0 \tau\,(T_p/\theta) \exp(-\theta/T_p) = 100$; for H_2O this has the solution $T_p = \theta/42 = 150$ K.

With a color temperature of $\sim 10^4$ K, the supernova spectrum will have a mean photon energy $< h\nu > \approx 2.3$ eV. Since grain absorption cross sections increase rapidly with energy, we take $\lambda \approx 3500$ Å ($h\nu \approx 3.5$ eV) as representative for estimating the heating effects of supernovae. At $\lambda \approx 3500$ Å the absorption efficiency factors are approximately (Draine and Lee 1984)

$$Q_* \approx 0.3\,(a/0.1\ \mu\text{m}) \qquad \text{for silicate} \tag{17}$$

$$Q_* \approx 1.5\,(a/0.1\ \mu\text{m}) \qquad \text{for graphite, } a \lesssim 0.1\ \mu\text{m}. \tag{18}$$

From Eq. (16) for the grain temperature we now find that the maximum distance from the supernova is

$$R = 6 \times 10^{18} \exp(-\tau_{cl}/2)\ \text{cm} \qquad \text{for silicate} \tag{19}$$

$$R = 2 \times 10^{19} \exp(-\tau_{cl}/2)\ \text{cm} \qquad \text{for graphite} \tag{20}$$

where a peak luminosity $L_{SN} = 10^9\ L_\odot$ (Tammann 1977) has been assumed for the supernova. Even with $\tau_{cl} = 0$ this distance is $\lesssim 6$ pc, and the galactic supernova rate of $\sim 10^{-13}$ pc^{-3} yr^{-1} implies a negligible probability that a supernova will occur within this distance of the molecular cloud during the $\lesssim 10^7$ yr period of interest.

Boland and de Jong (1982) have concluded that thermal spikes due to supernovae can have an important effect on depletions in clouds. Their conclusion is based on the assumption that $\theta \approx 1000$ K is a reasonable value for grain mantles. As seen from Table I, however, only the most weakly bound molecular materials (e.g., pure, solid CO) have such small values of θ. If grain mantles are in large part composed of H_2O, with NH_3 also present in appreciable amounts, it would be remarkable if the representative value of θ were not at least $\sim$3000 K; while a weakly bound CO molecule might be desorbed from the surface monolayer, a CO molecule trapped in the H_2O/NH_3 matrix could not be removed unless the matrix can first be desorbed, and this desorption will be characterized by fairly large values of θ. We conclude that supernova flashes are probably not likely to have an appreciable effect on depletions or grain mantles in dense clouds.

IV. GRAIN EVOLUTION IN STAR-FORMING CLOUDS

A. Shock Waves

It is becoming increasingly clear that activities associated with star formation can drive shock waves into the surrounding gas. Theoretical models of shock waves in molecular gas have taken on a new aspect as it has been realized that magnetic fields can play a major role in the shock structure as a result of the dissipation associated with ion-neutral streaming in a gas of low fractional ionization. An extensive set of shock models has been published by Draine et al. (1983).

The effects of these shock waves on dust have only begun to be assessed. Dust grains are in general charged, and hence are affected by motions of the magnetized ion-electron plasma. Streaming of the ions (and magnetic field) through the neutral gas in a shock wave causes the charged grain to be driven through the neutral gas at a velocity that is intermediate between the streaming velocity of the ions and the flow velocity of the neutral gas. Collisions of He atoms and H_2 molecules with the grain can cause sputtering of volatile grain mantles; Draine et al. (1983) found that a shock velocity of ~25 km s^{-1} was sufficient to fully erode a grain mantle composed of H_2O or any other molecular ice. However, in this case only a small fraction of the gas in a star-forming cloud is processed by such fast shocks. Grain-grain collisions may be important in lower-velocity shocks, but this has not yet been investigated for MHD shocks in molecular clouds.

B. Radiation from Newly Formed Stars

Consider a star of luminosity L. Within what distance R_{cr} will the temperature of dust grains be raised above a critical temperature $T_{cr} \approx 10^2$ K sufficient for mantle sublimation? Ignoring extinction and other heat sources, R_{cr} is given by

$$R_{cr} = \left[\frac{L}{16\pi\sigma T_{cr}^4}\frac{<Q(T_*)>}{<Q(T_{cr})>}\right]^{1/2}$$

$$= 1.2 \times 10^{15}\ \mathrm{cm}\left(\frac{L}{L_\odot}\right)^{1/2}\left(\frac{<Q(T_*)>/<Q(T_{cr})>}{10^2}\right)^{1/2}\left(\frac{T_{cr}}{10^2\ \mathrm{K}}\right)^{-2}. \quad (21)$$

Note that the cross section πR_{cr}^2 is proportional to L. Now suppose a cloud of volume V contains a number of stellar sources, of total luminosity L. Suppose that the gas has a large-scale turbulent velocity v_{turb}, and the stars have random velocities v_*. Then the average time before a grain enters into one of the critical spheres surrounding a stellar source is simply

$$\tau = \left(\frac{V}{L}\right)\frac{16\sigma T_{cr}^4}{(v_* + v_{turb})}\frac{<Q(T_{cr})>}{<Q(T_*)>}$$

$$= 4 \times 10^8\ \mathrm{yr}\left(\frac{10^4\ L_\odot\ \mathrm{pc}^{-3}}{L/V}\right)\left(\frac{\mathrm{km\ s}^{-1}}{v_* + v_{turb}}\right)$$

$$\left(\frac{T_{cr}}{100\ \mathrm{K}}\right)^4\left(\frac{<Q(T_{cr})>/<Q(T_*)>}{0.01}\right). \quad (22)$$

Consider, for example, the core ($A_V > 50$ mag) of the ρ Oph Cloud, with a volume $V \approx 0.5$ pc^3 and mean density $n_H \approx 1.3 \times 10^4$ cm^{-3}, and containing 30 identified stellar sources, with a total luminosity $L \approx 1230$ $L_\odot$ (Wilking and Lada 1983). Taking $v_{turb} \approx v_* \approx 1$ km s^{-1}, and emissivities for Eqs. (9) and (17), we find $\tau \approx 3 \times 10^8$ $(T_{cr}/100\ \mathrm{K})^{5.25}$ yr for the typical time before a graphite grain has its temperature raised above T_{cr} as the result of approaching too closely to a stellar source. Silicate grains, because of their higher albedo in the near infrared and visual bands, tend to remain cooler than graphite grains. A luminosity density $L/V \gtrsim 10^5\ L_\odot\ pc^{-3}$ would appear to be required for this mechanism to be important on the $\lesssim 10^7$ yr time scale during which the cloud may contain a young star cluster.

We note, however, that grain denudation by thermal desorption is so extremely temperature sensitive that even rare excursions to higher temperatures could be very important. Hence, if temporary luminosity increases occur due to flare phenomena associated with star formation, each star might affect a much larger volume of the cloud than the above estimate based on observed (i.e., average) luminosity densities. Such flare phenomena have been observed: FU Ori stars undergo sudden luminosity increases by a factor $\sim 10^2$, returning to their original luminosity on a time scale of decades. Herbig (1977*a*) has estimated a repetition time scale $t_{rep} \approx 10^4$ yr for such eruptions in an average T Tauri star, so that only one out of 10^3 T Tauri stars is in such a phase at any given time. Taking a luminosity increase of a factor $\alpha \approx 10^2$, one can show that the factor by which the processing rate τ^{-1} is increased is

$$\sim \frac{4\alpha^{3/2} R_{cr}}{3(v_* + v_{turb}) t_{rep}} \quad \text{if } \alpha^{1/2} R_{cr} < (v_* + v_{turb}) t_{rep} \tag{23a}$$

$$\sim \alpha \quad \text{if } \alpha^{1/2} R_{cr} > (v_* + v_{turb}) t_{rep} \tag{23b}$$

where R_{cr} is evaluated using the average luminosity of the star. Taking $10 L_\odot$ for the average T Tauri star (probably an overestimate) and $\alpha = 10^2$, we find from Eq. (21) $\alpha^{1/2} R_{cr} \approx 6 \times 10^{16}$ cm for graphite. Since $(v_* + v_{turb})\ t_{rep} \approx 6 \times 10^{16}$ cm (for $v_* = v_{turb} = 1$ km s^{-1} and $t_{rep} = 10^4$ yr), we find that the FU Ori phenomenon increases the denudation rate due to T Tauri stars by a factor $\alpha \approx 10^2$. Hence the FU Ori phenomenon would appear to be important only in clouds containing an unusually high density ($\gtrsim 10^2$ pc^{-3}) of T Tauri stars. Given our present limited knowledge, however, the possibility remains open that flare phenomena could play an important role in grain modification in star-forming clouds.

C. Grain Processing by Protostellar Activity

Protostars may expel a significant amount of material which almost became part of the star, but was driven away by radiation pressure and a protostellar wind. The grains in this material may be significantly modified, and this process may be important in determining the characteristics of interstellar grains (Burke and Silk 1976). Here we note that graphite grains are not thermodynamically stable in a gas of cosmic composition (i.e., O/C > 1). If these grains are allowed to reach chemical equilibrium at temperatures <1000 K, the graphite would be chemically converted to CO, or molecules containing CO. Silicate grains, on the other hand, are thermodynamically stable. This conversion of graphite into CO does not occur in diffuse interstellar gas because of kinetic barriers (Draine 1979). In the dense circumprotostellar environment, however, it is possible that chemical attack by O, H_2O, etc. may succeed in destroying graphite grains. If each protostar processes and ejects a mass of gas comparable to the mass eventually incorporated into the star, then clouds in which star formation has proceeded with high efficiency might show an underabundance of graphite relative to silicate grains. Chemical destruction of graphite grains can also occur in compact H II regions.

V. SUMMARY AND QUESTIONS

Observational evidence and theoretical studies lead to a number of conclusions regarding grain evolution in dark clouds, including the following points:

1. Grains in diffuse clouds (the presumed precursors to grains in molecular clouds) have a size distribution extending to radii $a \lesssim 0.01$ μm (=100 Å) and $a \gtrsim 0.1$ μm, and may consist of graphite and silicate particles.
2. Observational evidence strongly suggests an increase in grain sizes in at least some dark clouds.
3. Observed infrared absorption bands indicate that grain mantles in dark clouds consist of a frozen mixture of H_2O, NH_3, CO, and probably molecules containing CH and CN bonds.
4. A recently observed 3.4 μm absorption band, which has been observed on lines of sight not showing the 3.1 μm H_2O ice feature, suggests the existence of a hydrocarbon component which is more stable than H_2O ice. This component may be present in diffuse clouds.
5. Accretion of gas-phase atoms and molecules will initially act on the diffuse cloud grain distribution (with $\Sigma_{21} \approx 1$) on a time scale $\sim 3 \times 10^5$ n_4^{-1} yr.
6. Thermal spikes can prevent mantle formation only on very small ($a \lesssim 10$ Å) grains.
7. Turbulence-driven coagulation appears likely to remove the smaller grains from the size distribution, provided they stick to larger grains in collisions with Δv in the range 0.01 to 0.1 km s^{-1}. This will act to increase the mean grain size, and to reduce the grain surface area (to Σ_{21} ~ 0.1?).
8. Even if the initial grain distribution consisted of pure graphite and silicate grains, composite (conglomerate) grains are likely to result from coagulation.
9. In star-forming clouds, the grain-grain collision velocities induced by

turbulence or shock waves may be large enough to result in grain shattering, thereby replenishing the small grains and increasing Σ_{21}.

10. Grain mantles can be removed by sputtering in $v_s \gtrsim 25$ km s^{-1} shock waves.
11. The FU Orionis phenomenon or other flare activity in star-forming clouds may contribute to sublimation of volatile mantles.

Our understanding of grain evolution in dark clouds obviously remains quite incomplete. A number of important questions do not yet have adequate answers, for example:

a. What is the actual composition and structure of grains in dark clouds? We may hope to move toward answering this question by means of more careful spectroscopy of embedded infrared sources, and by more laboratory work on the optical properties of candidate grain materials.
b. Laboratory measurements of photodesorption cross sections (see, e.g., Bourdon et al. 1982) need to be extended into the ultraviolet to determine the possible importance of this process.
c. What happens when two dust grains collide with relative velocities in the range 0.001 to 0.1 km s^{-1}? While difficult, it may be possible to study this experimentally.
d. We need a better description of the turbulent velocity field in clouds to improve estimates of turbulence-driven grain-grain collision rates. The Kolmogorov turbulence spectrum is likely to be a poor description of the actual velocity field, especially if the magnetic field is dynamically important.

Acknowledgments. I am grateful to J. M. Greenberg, J. P. Ostriker, and L. Spitzer for helpful comments. I am particularly indebted to L. J. Allamandola and A. G. G. M. Tielens for numerous suggestions which have improved this chapter. This research was supported in part by the National Science Foundation, and in part by the Alfred P. Sloan Foundation.

TIME-DEPENDENT INTERSTELLAR CHEMISTRY

A. E. GLASSGOLD
New York University

Time-dependent calculations have now become almost standard in interstellar chemistry, but time-dependent effects have not been definitely established in interstellar clouds except in shocked regions and circumstellar envelopes. The relatively large amounts of C and C_2H which have been observed can be understood in this way, but other explanations are also possible. In a similar vein, chemical dating of interstellar clouds has still not been convincingly demonstrated. Further studies of the chemical evolution of collapsing clouds would be of interest, particularly with different initial conditions than used in the past. A basic obstacle to further progress in interstellar chemistry is that the current theory is basically gas-phase, and it may not apply to densities greater than $\sim 10^5$ cm^{-3}. We are hindered by our incomplete understanding of dust-surface chemistry. Integrating gas and dust chemistry is the main challenge facing interstellar chemistry today.

It became clear by 1978 that time-dependent considerations are important for interstellar chemistry (see reviews by Watson 1978 and by E. Herbst 1978*a*). Shu (1973) had suggested earlier that chemical dating of interstellar clouds might be possible using species whose chemical time scales are of the same order or longer than dynamical time scales. He considered the particular case of the conversion of atomic to molecular hydrogen in isolated dark clouds, and concluded that such clouds are transient objects with lifetimes $< 10^7$ yr. The time to develop the main carbon species, i.e. CO, in a typical intermediate-density cloud was estimated by Langer and Glassgold (1976) to be in the range 10^5–10^6 yr. More elaborate time-dependent chemical calculations have been published (see, e.g., Iglesias 1977; Prasad and Huntress

1980*b*; Graedel et al. 1982) which confirm and extend these earlier conclusions.

The current theoretical emphasis on time-dependent phenomena is motivated by several factors. One is that steady-state abundance calculations may be inaccurate unless cloud dynamical time scales are considerably longer than chemical time scales. Alternatively, measurements of molecular abundances offer possibilities for estimating the ages of the clouds, following Shu's (1973) suggestion. Furthermore, simplistic ideas that interstellar clouds are relatively quiescent and homogeneous have been challenged as more sensitive and higher-resolution observations (both spatial and spectral) have been made. If we estimate the dynamical time scale in terms of a characteristic length l and a characteristic velocity v,

$$t_{dy} = \frac{l}{\mathrm{v}} = 10^5 \ \mathrm{yr} \left(\frac{1/\mathrm{pc}}{\mathrm{v}/10 \ \mathrm{km}} \right) \tag{1}$$

we see that relatively short dynamical time scales are likely when we consider phenomena on small spatial scales. Good examples are winds around both young and evolved objects where $l < 0.1$ pc and/or $\mathrm{v} > 10 \ \mathrm{km \ s^{-1}}$. If some part of a cloud is in gravitational collapse, then the characteristic free-fall time for (a sphere with) hydrogen density n is

$$t_{ff} = \left(\frac{3\pi}{32 \ G \ n} \right)^{1/2} = 4.44 \times 10^7 \ \mathrm{yr} \ n^{-1/2}. \tag{2}$$

At intermediate densities (100–1000 $\mathrm{cm^{-3}}$), the chemical time scales for H_2 and CO are at least this long, i.e., $t_{ch} \gtrsim t_{ff}$.

Once steady state is abandoned, the theory becomes considerably more complicated, physically as well as mathematically. One important point is that initial conditions must be specified, and these involve physical conditions (density, velocity, and temperature) as well as abundances. It has been assumed frequently that the appropriate initial condition for time-dependent calculations of interstellar chemistry is a diffuse (semi-transparent) cloud without molecules, usually at rest and homogeneous. The tacit assumption is that chemical evolution involves the formation of the complex from the simple. One must regard this simplistic point of view with considerable skepticism, however, especially if the calculations are intended to describe real star-forming, molecular clouds. The observations would suggest that initial conditions for a collapse calculation, for example, should describe a thick, dense molecular cloud.

I. LIMITATIONS OF STEADY-STATE MODELS

Conventional gas-phase chemistry is severely inhibited in cool interstellar clouds shielded from stellar ultraviolet radiation because of activation energy barriers in reactions between neutral species. This problem was overcome by invoking ion-molecule reactions of various types (for further background, see J. H. Black and A. Dalgarno 1976 and W. D. Watson 1976). Ions are produced primarily by cosmic rays or, in transparent regions, by interstellar ultraviolet radiation. Considerable confidence in this idea has been provided by the detection of the molecular ions HCO^+ and N_2H^+ (surrogates for their progenitor $H_3{}^+$, which has not yet been found in interstellar clouds) and their deuterated versions. Additional support for the ion-molecule, gas-phase interstellar chemistry comes from qualitative agreement between measured abundances and theoretical estimates based on steady-state formulations of the theory. It must be recognized, however, that most abundances are rather uncertain at this stage, whether they are based on observations or on theory.

A major difficulty for the theory has arisen in the early 1980s from the detection of the 610 μm fine-structure line of neutral carbon in strong emission from a fair number of molecular clouds with hot centers (Phillips et al. 1980; Phillips and Huggins 1981). The position and velocity of the neutral carbon correspond well with earlier observations of carbon monoxide, and the abundance ratio is surprisingly large, $C/CO \sim 1$. The steady-state theory had predicted much smaller ratios, $C/CO << 1$, except for transition zones which are partially transparent to ultraviolet radiation. The transition zones of clouds with only one simple boundary appear to be insufficient to account for the observed column densities.

This difficulty with the steady-state formulation of interstellar gas-phase chemistry has been compounded by observations of C_2H by Huggins et al. (1984). Analogous to the C I result, Huggins et al. find surprisingly large amounts of this radical associated with CO, i.e., $C_2H/^{13}CO \sim 0.01$.

Phillips and Huggins (1981) offered several possible explanations for the large observed C/CO ratio (which would also apply to C_2H/CO):

1. The clouds may have higher ultraviolet levels than expected for a uniform cloud exposed only to external radiation.
2. The interior regions of clouds may be continually rejuvenated by turbulent mixing out to regions where CO can be photodissociated.
3. The clouds may not have completed their chemical evolution from a more diffuse state, and are at stages where substantial amounts of C remain to be processed into molecules.

Each of these possibilities can be supported in various ways: e.g. (1) by the generation of internal ultraviolet radiation by cosmic rays (Prasad and Tarafdar 1983), (2) by estimates of turbulence and grain processes (Boland and de Jong 1982), and (3) by simply invoking the time scale for conversion of C to CO. However, it has not yet been possible to decide which of these explanations is to be preferred, or whether some completely new idea is required.

Calculations by Langer et al. (1984) offer one possibility for retaining the steady-state chemistry by modifying the usual assumption that the gas in interstellar clouds is oxygen rich, following the suggestion of Tarafdar et al. (1983). Their results for C/O = 1.28 give good agreement for the C/CO and $C_2H/^{13}CO$ ratios, ~ 1 and ~ 0.01 respectively, at densities ~ 10^4 cm^{-3}. Langer et al. point out that a possible mechanism for altering the C/O ratio is differential depletion onto grains of O and C^+, discussed originally by W. D. Watson and E. E. Salpeter (1972). Measurements of the total gaseous abundance of carbon and oxygen are quite uncertain, even in diffuse clouds, so consideration of C/O ratios somewhat greater as well as less than one (Tarafdar et al. 1983) seems permisable. Observational checks of this idea involve difficult measurements of the abundances of oxygen-bearing molecules like O_2 and H_2O, which should be underabundant in a carbon-rich cloud, and the detection of ice bands in the infrared. Another way to preserve the steady state chemistry is to find mechanisms which could extend surface transition regions (Tielens and Hollenbach 1984).

The above discussion suggests several lines of research. It is important to investigate, theoretically, various proposals to explain the large C/CO and C_2H/CO ratios in quantitative detail, in order to determine whether time-dependent chemistry is required to explain the available data. It would be highly desirable to have more measurements of neutral (and the accompanying ionized) carbon, particularly at higher spatial resolution; however, such measurements will probably require new technical developments in far-infrared spectroscopy. Additional measurements of radical intermediates, like C_2H, would also be useful in this connection.

II. CHEMICAL DATING OF INTERSTELLAR CLOUDS

Until the discovery of C in dense clouds, it seemed clear from the observation of strong CO emission alone that interstellar clouds were at least several million years old. If the presence of C has to be explained as a time-dependent phenomenon, then this number becomes an upper rather than a lower limit. It is worth noting that a chemical age is the time since the relevant species were *not* associated, and that the actual age of the cloud as a dynamical entity could be longer. This is important in the context of explanation (2) (cf. Sec. I) involving the periodic rejuvenation of CO by mixing (Phillips and Huggins 1981; Boland and de Jong 1982). In this case, the chemical and mixing times are about the same (~ 10^6 yr), but the clouds could be considerably older.

Relatively few chemical determinations of cloud ages have been attempted so far. Shu's (1973) estimates for dark clouds, based on H I abundances, were followed up by M. Allen and G. W. Robinson (1976), who used different ideas about the formation of molecular hydrogen on grains, but obtained similar ages, i.e., 10^6–10^7 yr. These conclusions have had to be revised, however, in the light of more definitive measurements of the H I abundance in dark clouds (McCutcheon et al. 1978). Its relatively small abundance, about 10^{-3}, seems consistent with near steady conditions where the H I is produced by cosmic ray destruction of molecular hydrogen (Solomon and Werner 1971).

Although there appears to be relatively little H I in isolated dark clouds, molecular cloud complexes may contain much more. Blitz and Shu (1980) reviewed the evidence for this conclusion, and Wannier et al. (1983) have detected peripheral H I in several clouds. The observed column densities are on the order of 10^{20} cm^{-2}, characteristic of the transition regions derived from steady-state theory (Federman et al. 1979). The presence of a substantial fraction of atomic hydrogen in an interstellar cloud does not necessarily mean that it is young. A real cloud is inhomogeneous, and most of the H I will probably be in diffuse regions (including the periphery, as observed). Such regions have a relatively small molecular component which comes into chemical equilibrium fairly rapidly. In any case, chemical dating can only be based on observations of cloud material with the same physical history. This requirement may not be met when a simple comparison is made of the total H I mass with the total molecular hydrogen mass.

Blitz and Shu (1980) also suggested that the presence of chain molecules in dark clouds implies that the clouds are not older than 3×10^7 yr; otherwise the molecules would have been adsorbed onto dust grains (M. Allen and G. W. Robinson 1976). Stahler (1984*b*) has reconsidered this problem, and derives lower limits of 2, 1, 0.3, and 0.33×10^6 yr for B 335, TMC 1, TMC 2, and L 134N clouds, respectively. His analysis is based on cyanoacetylene chain abundance ratios, and involves certain assumptions about the chemistry, such as sequential formation of the chain members and time-independent destruction on dust grains. In practice, his age estimates are based on the time scale for destruction by adsorption on dust, as in previous discussions by Shu (1973) and by M. Allen and G. W. Robinson (1976).

The most important of Stahler's assumptions to discuss here is whether or not molecules are actually adsorbed on dust grains with high efficiency—one of the most fundamental issues in interstellar chemistry. If the effective dust area per proton is taken from measurements of extinction to be about 10^{-21} cm^2, and if the sticking probability is assumed to be unity, then the time scale for sticking of an atom or molecule of mass m at density n and temperature T is

$$t_s = 2 \times 10^9 \text{ yr}\left[\frac{(m/m_{\rm H})^{1/2}}{n\,T^{1/2}}\right] \tag{3}$$

The fact that this becomes small for moderate densities (e.g. $< 10^4$ yr for $n > 10^5$ cm^{-3}), has of course been known for a long time, but the implications have been avoided by most interstellar chemists who tend to regard grain physical chemistry as being nearly intractable. A few investigators have taken Eq. (3) at face value, and have depleted neutral species in time-dependent calculations using sticking coefficients close to unity (see, e.g., M. Allen and G. W. Robinson 1976; Iglesias 1977; Stahler 1984*b*; Tarafdar et al. 1984). Although theory does say that the sticking probabilities for neutrals are close to unity (W. D. Watson and E. E. Salpeter 1972; Burke and Hollenbach 1983), processes probably exist which remove neutrals from grain surfaces. These processes may operate in conjunction with the rejuvenation of the gas by dynamic mixing, as discussed by Phillips and Huggins (1981) and by Boland and de Jong (1982). Until we have a better understanding of the chemical interaction between gas and dust, it will be difficult to make progress on important chemical questions concerning interstellar clouds, such as their ages. It is noteworthy that a new initiative to unify gas-phase and grain chemistry has been made by d'Hendecourt et al. (personal communication, 1984).

Despite the difficulty, serious scientific investigation of the physical and chemical properties of interstellar dust must be considered the single most important challenge facing interstellar chemists today. Several years ago, Greenberg and his collaborators initiated a program of laboratory and theoretical investigations of surface chemistry (Hagen et al. 1980). His group condensed simple molecules onto cold surfaces, and found that the infrared spectra were similar to those observed toward compact H II regions (Lacy et al. 1984).

Some clouds are found to have substantial solid CO, but the amount seems to be sensitive to past as well as present physical conditions. Leger (1983) demonstrated theoretically that the saturation pressure for CO is ~ 17 K for typical interstellar clouds. From the fact that substantial CO is observed in emission, he concluded that an efficient desorption mechanism must be operative. Léger and his collaborators have suggested that molecular desorption from small dust grains may occur during the thermal spikes induced by cosmic ray collisions.

III. THE CHEMISTRY OF COLLAPSING CLOUDS

Observations suggest that theoretical models should incorporate the essential aspects of the inhomogeneous structure and time-dependent nature of interstellar clouds. Relatively little definitive work of this kind has been done so far, partly because of the technical difficulties in solving multidimensional hydrodynamic problems while simultaneously including realistic treatments of chemistry along with energy and radiative transfer. One area where some model calculations have been carried out is the evolution of gravitationally unstable, spherically symmetric clouds, reviewed below. Considerable progress has been made in applying one-dimensional shock theory to the gas in the BN-KL region of Orion (Draine and Roberge 1982; Chernoff et al. 1982), and in the chemistry of the circumstellar envelopes around evolved stars (reviewed by Glassgold and Huggins 1984).

A series of studies of collapsing clouds was completed in 1969 before the molecular character of interstellar clouds was established (J. H. Hunter 1969; Penston 1969; Disney et al. 1969). No chemical transformations were considered, and the main thermal elements were cosmic ray heating and cooling by fine-structure lines, in addition to mechanical work. Usually started as diffuse, uniform, spheres at rest, the unstable model clouds evolved within one free-fall time to cool, dense configurations, rather similar to the initial condition employed in Larson's protostar calculations (but without molecular hydrogen). The low temperatures were achieved in part by too little heating, a long-standing difficulty for diffuse regions when metal coolants with solar abundances are used (D. W. Goldsmith et al. 1969).

A model collapsing cloud in which chemical and thermal evolution were treated consistently was developed by Gerola and Glassgold (1978). The calculation modeled a spherically symmetric cloud, but the treatments of the chemistry and energy transfer were reasonably complete for the regime of interest, and would still be acceptable today. The gas-phase ion-molecule chemistry was simplified by treating species with very short chemical time scales in steady state. The evolution was followed until the density in the central core reached a value such that the grain sticking time became comparable to the free-fall time. For the case reported in detail, a cloud with mass 2.7×10^4 $M_\odot$ and initial radius $R = 11$ pc, this density was 5×10^4 cm^{-3}. The effects of rotation, magnetic fields, and turbulence were ignored.

In these calculations, the center of the cloud cooled down quickly to a temperature of about 15 K, while the surface region remained at about 25 K. Somewhat warmer surface temperatures would have been obtained with an improved treatment of grain photoelectron heating (Federman and Glassgold 1980). The chemical evolution of the cloud could be understood in terms of the relevant time scales and a few simple ideas on the physical basis of the chemistry. For example, the dense cloud ion-molecule chemistry cannot operate until the electron fraction (ratio of electron to total hydrogen density) becomes very small. This means that the surface regions will remain relatively free of molecules, and will resemble a diffuse cloud, at least until they evolve to higher densities. The chemistry in the dense inner regions or core can develop fully, but only after certain prerequisites are fulfilled; this requires some time. First, there must be sufficient molecular hydrogen, and second, C^+ and other heavy ions must recombine. The final result was that the transition from C to CO does not take place in the center of this cloud until close to a free-fall time, i.e., $\sim 4 \times 10^6$ yr in this case.

As a cloud collapses, the chemistry in the core changes gradually from that characteristic of a diffuse cloud to that characteristic of a dense cloud. After significant evolution, there will be a similar spatial variation from the outside of a cloud into its core. The chemical and thermal properties do indeed resemble those observed in the cores of molecular clouds.

The simultaneous treatment of thermal, chemical, and radiative transfer phenomena together with the hydrodynamics is very demanding of computing power, especially for calculations in more than one dimension. The particular calculations discussed by Gerola and Glassgold did not show strong thermal-chemical effects on the dynamical evolution. Rather, it was the dynamics which was the main driving force for the chemical and thermal evolution. This suggests that an approximate understanding of the chemical evolution of a grossly unstable cloud can be achieved by decoupling the hydodynamics from the calculation of the chemistry, radiation transfer, and the heating and cooling. Tarafdar et al. (1984) have done this by specifying the temperature profile as a function of density and distance to the outside of the cloud. Their preliminary results are similar to those of Gerola and Glassgold but, by reducing the computational burden considerably, they are able to study the chemical evolution of collapsing clouds for a large range of masses and initial densities. They are investigating the hypothesis that essentially all interstellar clouds are related by thermal-chemical evolution during collapse, along the lines of stellar evolution. The fact that clouds are observed to have diverse properties may then be explained as a result of evolution, or of their different masses and possibly different initial densities. A quantitative test of this idea is difficult to make because of limitations in the data; for example, the variation of observed molecular abundances with cloud density is uncertain (see the chapter by Irvine et al.). In addition, many of the clouds for which abundances have been estimated may have been influenced by embedded or nearby stars, and thus do not conform to the model assumption of isolated clouds collapsing under gravity.

IV. SUMMARY

I have attempted to deal with some current problems in interstellar chemistry in the context of time-dependent calculations. Indeed most large calculations are now done as a function of time, usually at constant density. Moreover, time-dependent theories have been invoked recently to explain certain abundance ratios which are not readily understood with steady-state theory. The most important example is the observation of neutral carbon species in large abundance relative to and in association with CO. This fact is consistent with time-dependent chemistry, but the explanation is not unique.

The second main point to be made is that further progress in the interstellar chemistry of dense clouds requires a much better understanding of the physical chemistry of the gas-dust interaction. A direct consequence of our present ignorance of this subject is that molecular abundances (other than molecular hydrogen) cannot be used to date interstellar clouds.

The chemical evolution of collapsing clouds has been studied for spherical clouds only. The results of the model calculations indicate that the interiors of moderately dense clouds cool down, and that the resulting negative pressure gradient generally assists rather than opposes gravity. The interiors of collapsing clouds evolve into dense phases with molecular abundances similar to those observed in real interstellar clouds. This does not prove, however, that interstellar clouds are collapsing. (In fact, interstellar clouds have not yet been observed directly to be in a collapse phase, but the mere fact that stars are observed to have been formed recently inside interstellar clouds, strongly suggests that gravitational collapse does occur.) The collapse calculations are limited, as is all of gas-phase interstellar chemistry, to moderate densities, e.g., $< 10^5$ cm^{-3}, such that the time scale for sticking of molecules to dust grains is no less than the dynamical time scale. Until this restriction is removed, the really interesting calculation, the chemical evolution of a dense core to protostellar cloud densities, cannot be made.

Several fruitful areas for the application of time-dependent chemistry not discussed here, e.g. shocks and circumstellar envelopes. Together with the issues raised above on interstellar clouds, these problems demonstrate that time-dependent considerations have become an important part of interstellar chemistry.

Acknowledgment. This work has been supported in part by a grant from NASA.

ELECTRON ABUNDANCES AND IONIZATION RATES IN INTERSTELLAR CLOUDS

WILLIAM D. LANGER
AT&T Bell Laboratories

Observations of the isotopes of H_2 and HCO^+ make it possible to estimate bounds on the electron abundance and cosmic ray and X-ray ionization rates in interstellar clouds. The theory behind these methods is reviewed and the current state of the observations summarized.

Electrons and ions are important constituents in interstellar chemistry and the dynamics of the interstellar gas because the ions mediate the coupling of the neutral gas to magnetic fields. Their abundances are a function of the sources which ionize and heat the gas. Until recently, the determination of electron abundances and cosmic ray and X-ray ionization rates in the dense clouds has been poorly known. The detection of deuterated molecules and polyatomic ions now makes it possible to measure, or at least place useful limits on, these properties in interstellar clouds. All of these determinations are based on the abundance relationships given by gas-phase ion-molecule chemistry. In this chapter I review the theory of ionization balance in clouds, the ion-molecule chemistry relevant to determining the electron abundance and cosmic ray ionization rate, and the results of recent observational studies of these properties.

I. IONIZATION BALANCE

The ionization balance in interstellar clouds is determined by the production, transfer, and destruction of atomic and molecular ions, shown schemat-

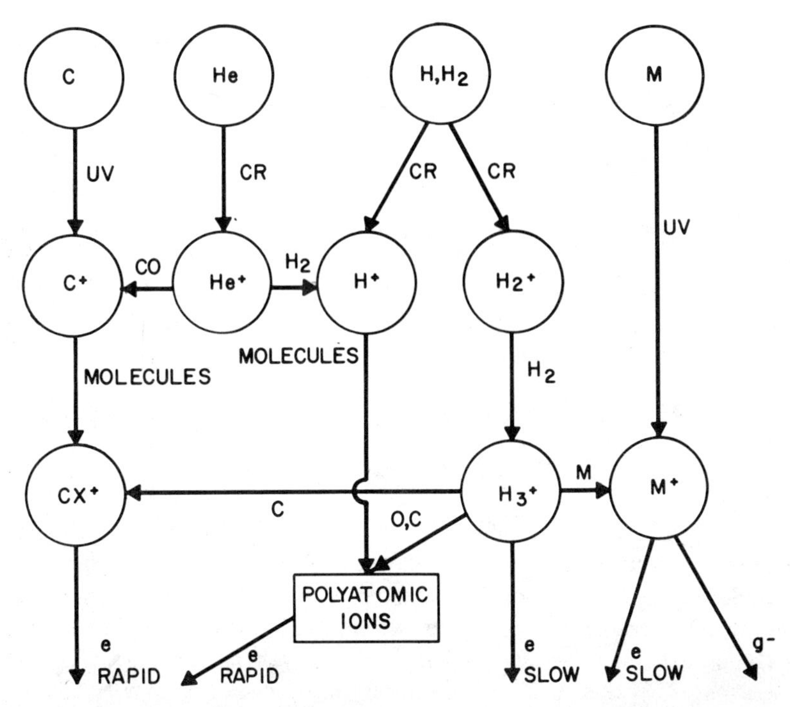

Fig. 1. A schematic picture of the ionization balance in interstellar clouds with extinctions $A_v \gtrsim 0.5$ mag. The clouds are primarily H_2 surrounded by a thin layer of atomic hydrogen. Species with ionization potentials < 13.6 eV (S, Si, C, etc.) are almost completely ionized on the outside; however, as the ultraviolet radiation is attenuated (by absorption on dust grains) these trace elements are mostly in neutral atomic and molecular form. Cosmic rays maintain some degree of ionization throughout the cloud and produce directly and indirectly the molecular ions that drive much of the gas-phase ion-molecule chemistry. The initially ionized species transfer their charge to other constituents by charge-exchange or ion-molecule reactions. This sequence is terminated by recombination with electrons in the gas phase (radiative and dissociative) or on grains. The ratio $n(H_3{}^+)/n(He^+)$ is governed by the electron abundance.

ically in Fig. 1. Photoionization is dominant both in the exterior unshielded parts of clouds and close to energetic embedded stellar sources. Species with ionization potentials < 13.6 eV (e.g., C, S, Si, Mg, CH, H_2O, and OH) are readily photoionized while those with greater ionization potentials remain predominantly neutral (e.g., H_2, He, CO, O, and N). Ionization of hydrogen and helium, which compose the bulk of the cloud's gas, is by cosmic rays (*cr*) through the following reactions:

$$\begin{aligned} cr\ &+ H \rightarrow H^+ + e + cr \\ &+ H_2 \rightarrow H^+ + H + e + cr \\ &+ H_2 \rightarrow H_2{}^+ + e + cr \\ &+ He \rightarrow He^+ + e + cr. \end{aligned}$$

X rays also ionize the hydrogen and helium gas in clouds via reactions such as

$$\nu + H_2 \rightarrow H_2^+ + e$$

$$\nu + He \rightarrow He^+ + e$$

where the bulk of the photon energy goes into the photoelectron. In addition to the ionization rate due to the primary ionization of hydrogen, the energetic electron produces secondary ionization of the gas in much the same manner as a cosmic ray (cf. Glassgold and Langer 1973). Furthermore, X rays produce ionized species of C, N, and O through *K*- and *L*-shell photoionization

$$\nu + C \rightarrow C^+ + e, L \text{ shell}$$

$$\text{v} + C \rightarrow C^{++} + e, K \text{ shell.}$$

X rays are produced by hot diffuse structures or discrete sources; these can be in close proximity to a cloud or, in the case of discrete sources, embedded. The role of X rays in ionizing the gas and producing molecules is discussed in some detail by Langer (1978*b*) and by Krolik and Kallman (1983). This X-ray ionization in clouds is similar to that of cosmic rays because most of the net ionization of H_2 due to X rays results from the photoelectrons (secondary ionization).

The atomic and molecular ions produced by these ionization sources can transfer their charge to other species, or form new ionic species, through a variety of ion-molecule reactions and charge-exchange processes. Some examples of these are given below along with the type of process in parenthesis:

$$C^+ + H_2 \rightarrow CH_2^+ + h\nu \text{ (radiative association)}$$

$$C^{++} + H_2 \rightarrow C^+ + H^+ + H \text{ (dissociative charge transfer)}$$

$$C^+ + H_2O \rightarrow HCO^+ + H$$

$$H_2^+ + H_2 \rightarrow H_3^+ + H$$

$$H_3^+ + O \rightarrow OH^+ + H_2 \text{ (proton transfer)}$$

$$H_3^+ + CO \rightarrow HCO^+ + H_2$$

$$H_3^+ + (Mg, Fe) \rightarrow (Mg^+, Fe^+) + H_2 + H \text{ (charge transfer)}$$

$$He^+ + H_2 \rightarrow He + H + H^+ \text{ (dissociative charge transfer)}$$

$$He^+ + CO \rightarrow He + C^+ + O.$$

The eventual neutralization of the ions is by recombination in the gas phase or on grains,

$$HCO^+ + e \rightarrow CO + H \text{ (dissociative)}$$

$$H_3{}^+ + e \rightarrow H_2 + H \text{ (dissociative)}$$

$$\rightarrow H + H + H$$

$$(C^+, Mg^+, Fe^+) + e \rightarrow (C, Mg, Fe) + h\nu \text{ (radiative)}$$

$$(C^+, HCO^+) + (\text{grain})^- \rightarrow (C, H + CO) + h\nu + (\text{grain})^0$$

$$(Mg^+, Fe^+) + (\text{grain})^- \rightarrow (Mg, Fe) + h\nu \text{ (grain)}^0.$$

A more extensive discussion of the ion-molecule chemistry including tables of reactions and reaction-rate coefficients can be found in Prasad and Huntress (1980*a*) and Graedel et al. (1982). Table I gives the reactions used for this discussion along with their reaction-rate coefficients k_i; the cosmic ray ionization rates are denoted by ζ and are usually calculated in terms of the ionization rate per hydrogen. The following nomenclature is used throughout for specie $X{:}n(x) \equiv$ density in cm^{-3}; $[X] = n(x)/n(H_2) \equiv$ fractional abundance (useful only where the hydrogen gas is almost entirely molecular, otherwise $[X] = n(x)/n$, where $n = 2n(H_2) + n(HI)$, as in diffuse regions).

It is clear from the outline above that the molecular abundances in the interstellar ion-molecule chemistry are functions of the ionization rate and

TABLE I
Reactions and Rate Coefficients

Reaction	Reaction Rate Coefficient ($cm^3\ s^{-1}$)[a]	Reference
$H_3^+ + HD \rightarrow H_2D^+ + H_2$	1.7×10^{-9}	b
$H_3^+ + e \rightarrow H_2 + H$ and $H + H + H$	$7.2 \times 10^{-6} T^{-1/2}$	c
	$<2 \times 10^{-8}$	d
$H_2D^+ + e \rightarrow HD + H$, $H_2 + D$, and $2H + D$	$6.5 \times 10^{-6} T^{-1/2}$	c
	$<2 \times 10^{-8}$	d
$HCO^+ + e \rightarrow CO + H$	$1.4 \times 10^{-5} T^{-0.85}$	d
$H_3^+ + CO \rightarrow HCO^+ + H_2$	1.8×10^{-9}	b
$H_2D^+ + CO \rightarrow HCO^+ + HD$	$2/3 \times 1.6\ 10^{-9}$	b
$\rightarrow DCO^+ + HD$	$1/3 \times 1.6 \times 10^{-9}$	b
$H_3^+ + O \rightarrow OH^+ + H_2$	8.0×10^{-10}	b
$H_3^+ + N_2 \rightarrow N_2H^+ + H_2$	1.7×10^{-9}	b
$H_3^+ + H_2O \rightarrow H_3O^+ + H_2$	5.9×10^{-9}	b
$H_3^+ + O_2 \rightarrow O_2H^+ + H_2$	6.4×10^{-10}	b
$H_3^+ + Mg,Fe \rightarrow (Mg^+,Fe^+) + H + H_2$	10^{-9}	b
$(Mg^+,Fe^+) + (\text{grain})^- \rightarrow (Mg,Fe) + (\text{grain})^\circ$	$3.2 \times 10^{-18} T^{0.5}$	b

[a]Temperature T is in K.
[b]These are taken from Table B1 of Guélin et al. (1982*b*), and Graedel et al. (1982).
[c]J. B. A. Mitchell et al. (1983); McGowan et al. (1979).
[d]N. G. Adams et al. (1984).

electron abundance. One can, in principle, determine ζ and $[e]$ by observing the appropriate species and using the abundance relationships dictated by ion-molecule chemistry. The abundance relationships depend quantitatively on the reaction rate coefficients for the gas phase reactions. Most of the reactions basic to the interstellar chemistry have been studied in the laboratory at room temperature or down to $\sim$ 90 K, but only a very few have been measured at the temperatures prevailing in the cold clouds (10–30 K). Reaction-rate coefficients have had to be extrapolated to low temperatures, a procedure that appears reasonable in most cases. For example, measurements of ion-molecule or ion-atom reactions that are large, near the Langevin (or collision) rate, at room temperature are large at very low temperatures (according to measurements at 20 to 80 K). The same is not true for reactions which are small at room temperature (e.g., He^+ or N^+ with H_2).

The electron recombination rate coefficients are, obviously, very important to the state of ionization in the clouds. The dissociative recombination rate coefficients $k_e(X)$ are generally large at room temperature, $\sim 5 \times 10^{-7}$ cm^3 s^{-1}, and increase at low temperatures as $T^{-1/2}$ (cf., Auerbach et al. 1977). There are exceptions to these results, however, and H_2^+ and possibly H_3^+ are among these. The value of k_e for H_3^+, $k_e(H_3^+)$ has profound implications for the chemistry and interpretations of the state of ionization; consequently it is important to review the status of measurements of k_e (H_3^+).

A large value for k_e (H_3^+) is indicated by the room temperature measurements of Leu et al. (1973) and also down to 100 K by those of Auerbach et al. (1977) and J. B. A. Mitchell et al. (1983), resulting in k_e $(H_3^+) \simeq 7 \times 10^{-6}$ $T^{-1/2}$ cm^3 s^{-1}. N. G. Adams et al. (1984), however, find a very small rate coefficient, at least two orders of magnitude smaller, for this process at 95 and 300 K, therefore setting an upper limit at 95 K, k_e $(H_3^+) < 2 \times 10^{-8}$ cm^3 s^{-1} (the same limit was also measured for k_e (D_3^+). Explanations for the different laboratory values are reviewed by D. Smith and N. G. Adams (1984) and summarized below. They suggest that the difference is perhaps due to contamination of the laboratory plasmas in the earlier experiments by vibrationally excited H_3^+ and/or H_5^+, both of which recombine rapidly. [The effects of vibrational state on k_e were found to be important for H_2^+ by McGowan et al. (1979) and k_e $(H_2^+ (v = 0))$ is probably small.] On the other hand, they find that ground state H_3^+ recombines very slowly because the laboratory plasma used by N. G. Adams et al. (1984) was composed primarily of H_3^+ $(v = 0)$. At the present time, these explanations must be considered possible but not certain, because in the earlier experiments the presence of H_3^+ $(v > 0)$ and H_5^+ was either excluded or not found to affect the measurements.

Until the existing differences regarding the laboratory measurements of the H_3^+ recombination rate coefficient can be clarified (and measurements made at 10–20 K), both possibilities should be considered in the interstellar chemistry and in evaluating the electron abundance in dense clouds. I shall adopt just such an approach in this review, presenting the results for both the large and small value of k_e (H_3^+), leaving the appropriate choice to the reader when the correct value is finally known.

In the dense shielded regions, the ions can be separated into three categories: polyatomic ions (H_3^+, HCO^+, CH_3^+, . . .), reactive atomic ions (C^+, O^+, . . .), and nonreactive metal ions (e.g., Mg^+, Fe^+, Si^+, Ca^+, Na^+). The last group are destroyed very slowly (in general) compared with the other two, and can comprise a significant fraction of the ion abundance even though their elemental abundance is small. The degree to which these metals are depleted determines, in part, whether they contribute significantly to the ion abundance. Two limiting possibilities are considered to illustrate these different contributions: (1) only polyatomic ions are present, $[e] = \Sigma$ [polyatomic ions]; and (2) metal ions dominate, $[e] = [M^+] >> $ [polyatomic ions], where $[M^+] = \Sigma$ [all nonreactive metal ions]. These limits will also be discussed for the two different experimental values of the dissociative recombination rate coefficient for H_3^+.

A. k_e (H_3^+) Large

In the first case where the polyatomic ions dominate

$$[e] = [H_3^+] + \Sigma[X_iH^+] \tag{1}$$

where Σ is over all polyatomic ions produced by proton transfer (except H_3^+). The following relationship can be derived from the chemical balance equations

$$\Sigma[X_iH^+] \simeq \frac{[H_3^+]\Sigma k_i[X_i]}{k_e(X_iH^+)[e]} \tag{2}$$

and

$$[H_3^+] \simeq \frac{\zeta/n(H_2)}{k_e(H_3^+)[e] + \Sigma k_i[X_i]} . \tag{3}$$

The expression for the ion abundance reduces to

$$[H_3^+] + \Sigma[X_iH^+] \simeq \frac{\zeta/n(H_2)}{k_e[e]} \tag{4}$$

using $k_e(H_3^+) \simeq k_e(X_iH^+)$. This expression assumes that all polyatomic ions are generated from cosmic rays directly through formation of H_3^+ and indirectly by proton transfer of H_3^+ to atoms and molecules (cf. Langer 1976). Solving Eq. (4) yields the electron abundance

$$[e] \simeq \left(\frac{\zeta}{k_e n(H_2)}\right)^{1/2} \tag{5}$$

For dense clouds with $\zeta = 4 \times 10^{-17}$ s^{-1}, and $n(H_2) = 10^4$ cm^{-3}, Eq. (5) yields $[e] \simeq 4 \times 10^{-8}$ where $k_e \simeq 2.3 \times 10^{-6}$ cm^3 s^{-1}. (Recombination on grains can be neglected for values of $[e] \gtrsim 10^{-10}$.)

In the second case, the metal ions are produced from $H_3{}^+$ and destroyed by radiative association and grain recombination, with the latter dominating for $[e] \lesssim 10^{-6}$. Therefore

$$[e] \simeq [M^+] = \frac{k[H_3^+][M]}{k_g} \tag{6}$$

$$[M^+] = \frac{k[M]\zeta/n(H_2)}{k_e(H_3^+)k_g[e]}. \tag{7}$$

Assuming $[M^+]/[M] << 1$ and $[e] = [M^+]$ leads to the solution

$$[e] \cong \left(\frac{k[M]\zeta}{k_e k_g n(H_2)}\right)^{1/2}. \tag{8}$$

This expression yields $[e] \simeq 4 \times 10^{-7}$ for $\zeta = 4 \times 10^{-17}$ s^{-1}, $n(H_2) = 10^4$ cm^{-3}, and $[M] = 10^{-6}$, where $k_e \simeq 2.3 \times 10^{-6}$ cm^3 s^{-1}, $k \simeq 10^{-9}$ cm^3 s^{-1}, and $k_g \simeq 10^{-17}$ cm^3 s^{-1} (see Table I; Graedel et al. 1982).

B. $k_e(H_3^+)$ Small

If $k_e(H_3^+) < 2 \times 10^{-8}$ cm^3 s^{-1}, then the destruction of H_3^+ is by proton transfer to atoms and molecules rather than recombination at low electron concentration ($[e] < 5 \times 10^{-6}$) and

$$[H_3^+] \simeq \frac{\zeta/n(H_2)}{\Sigma k_i[X_i]}. \tag{9}$$

On the other hand, the remaining polyatomic ions (HCO^+, N_2H^+, . . .) are still primarily destroyed by recombination

$$\Sigma[XH^+] \simeq [H_3^+]\frac{\Sigma k_i[X_i]}{k_e[e]}. \tag{10}$$

Under interstellar cloud conditions $\Sigma[X_i]$ is typically $\sim 10^{-4}$, so that $[H_3^+] > \Sigma[X_iH^+]$ for $[e] > 10^{-7}$, and conversely for $[e] < 10^{-7}$. The electron abundance is again considered for the two limiting cases regarding the metal abundance.

First, consider the case where the metal abundance is too small to contribute to the ion balance

$$[e] \simeq [H_3^+] + \Sigma[X_iH^+]. \tag{11}$$

Combining Eqs. (9), (10), and (11) yields

$$[e] \simeq \frac{\zeta}{n(\mathrm{H_2})}\left[\frac{1}{\Sigma k_i[X_i]} + \frac{1}{k_e[\mathrm{XH^+}][e]}\right] \tag{12}$$

which reduces to the following approximate solutions

$$[e] \simeq \frac{\zeta/n(\mathrm{H_2})}{k\Sigma[X_i]},\ [e] > 10^{-7}$$

$$[e] \simeq \left(\frac{\zeta/n(\mathrm{H_2})}{k_e}\right)^{1/2},\ [e] < 10^{-7}. \tag{13}$$

For $\zeta = 4 \times 10^{-17}$ s^{-1} and $n(\mathrm{H_2}) \simeq 10^3$ cm^{-3}, the electron concentration is $[e] \sim 4 \times 10^{-7}$.

In the limit where the gas phase metal abundance, $[M_t] = [M] + [M^+]$, is large enough to dominate the ion abundance, we have

$$[e] = [M^+] \simeq \frac{[M]}{k_g}(k[\mathrm{H_3^+}] + \Sigma k_i[\mathrm{XH^+}]) \tag{14}$$

where the M^+ is produced by charge transfer and neutralized on grains. Substituting into Eq. (14) and rewriting $[M^+]$ in convenient units yields

$$[M^+] \simeq \frac{\zeta/10^{-17}}{n(\mathrm{H_2})/10^4}\left(1 + \frac{10^{-7}}{[e]}\right)[M]. \tag{15}$$

For $[e] > 10^{-7}$ and $n(\mathrm{H_2}) < 10^4$ cm^{-3} nearly all the metals are ionized and

$$[e] \simeq [M^+] \simeq [M_t] \tag{16}$$

where obviously $[M_t] \lesssim$ few $\times 10^{-6}$ for the above to be valid. On the other hand, if $[e] < 10^{-7}$, then

$$[e] \lesssim [M^+] = [M_t] \quad \text{for } n(\mathrm{H_2}) < 10^4 \text{ cm}^{-3} \tag{17}$$

and

$$[e] \gtrsim 10^{-7}\left(\frac{[M_t]}{10^{-7}}\frac{\zeta/10^{-17}}{n(\mathrm{H_2})/10^4}\right)^{1/2} \quad \text{for } n(\mathrm{H_2}) >> 10^4 \text{ cm}^{-3} \tag{18}$$

where $[M_t]$ must be $<10^{-7}$.

II. DETERMINATION OF ELECTRON ABUNDANCE AND IONIZATION RATE

The abundance relationship between the ions and molecules is a function of the electron abundance, as is evident in Figs. 2 and 3, and, therefore from the discussion above, a function of the cosmic ray ionization rate, hydrogen density, and metal abundance. In diffuse regions with very little shielding, $A_v \lesssim 1$ mag, the HD abundance is a function of the cosmic ray ionization rate and observations of HD and H_2 can be used to determine ζ. In shielded regions, the best determinations of [*e*] and ζ come from observations of the isotopes of HCO^+.

In diffuse clouds the HD abundance is coupled to the ionization rate through the following chain of reactions (Fig. 2) as discussed by Dalgarno et al. (1973), J. H. Black and A. Dalgarno (1973), and Watson (1973*a,b*):

$$H^+ + D \rightleftarrows H + D^+$$

$$D^+ + H_2 \rightarrow HD + H^+.$$

The charge transfer reaction is exothermic to the right and the second reaction is rapid (O'Donnell and Watson 1974). The HD molecule is destroyed by line photodissociation similar to H_2

$$\nu + HD \rightarrow H + D.$$

In contrast to H_2, however, the HD lines do not self-shield as much because of their lower abundance, and the dissociating lines of HD remain optically thin until very large depths into the cloud. The abundance of HD is a function of ζ because H^+ is produced by cosmic ray ionization of H and H_2. By contrast, catalytic reactions on grain surfaces produce molecular hydrogen. Therefore the ratio $n(HD)/(n(H) + 2n(H_2))$ is a function of ζ in diffuse clouds, and the latter can be estimated by modeling this ratio (O'Donnell and Watson 1974).

In shielded regions, the ion-molecule chemistry is driven by the cosmic ray ionization of H_2 and He. The $H_2{}^+$ reacts to form $H_3{}^+$ (see Sec. I) which converts atoms into molecular ions, while the He^+ destroys molecules. This cycle is shown schematically in Fig. 3, which includes additional basic aspects of the chemistry. The relative efficiency of production versus destruction of molecules depends on the ratio of $H_3{}^+$ to He^+ which is a function of the

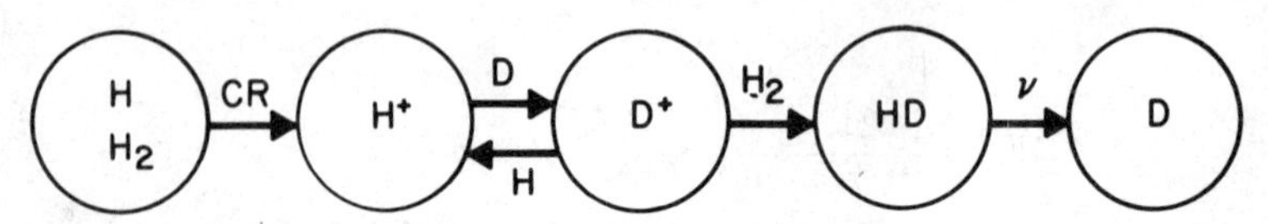

Fig. 2. The reaction chain for producing HD in interstellar clouds. The production of HD is coupled to the cosmic ray ionization rate through the production of H^+, while H_2 is produced on grains. The HD/H_2 ratio, therefore, can be used to determine ζ.

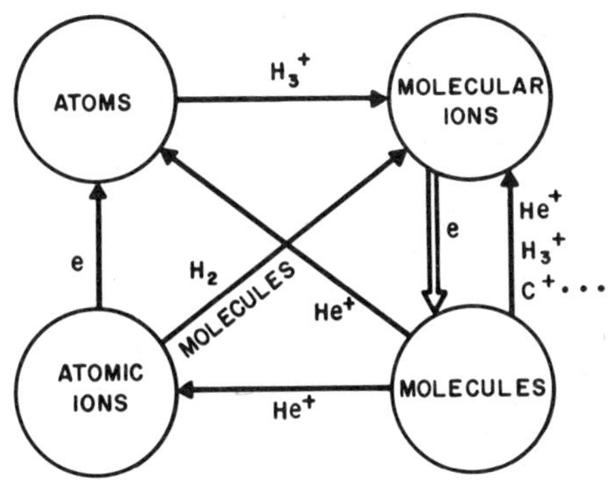

Fig. 3. The basic ion-molecule cycle in dark interstellar clouds. Atoms are converted by $H_3{}^+$ into molecular ions which then react with electrons (and other atoms and molecules) to form molecules. The molecules are destroyed by interactions with He^+, sometimes after conversion to other forms by neutral-neutral reactions. The conversion from atoms into molecules becomes efficient when the $H_3{}^+$ abundance is the order of or greater than the He^+ abundance.

electron abundance (cf. Graedel et al. 1982). If $H_3{}^+/He^+ \gtrsim 1$, a substantial fraction of the carbon and oxygen is converted to molecules.

In dense shielded regions, the deuterium abundance in other molecules is suggested to be a sensitive function of the electron abundance rather than an indicator of the deuterium to hydrogen ratio. Near the Sun the elemental abundance ratio of deuterium was determined from ultraviolet absorption in diffuse clouds to be $[D/H] \sim 10^{-5}$. Early radio surveys attempted to measure this ratio throughout the interstellar medium by studying deuterated molecules, but first observations found that $[DCN/HCN] \sim 10^{-2}$, which is too large to be consistent with cosmological and nucleosynthesis models. Watson (1974 *a,b*) explained this large ratio as due to chemical fractionation of deuterium in one of the progenitor molecules in the gas-phase chemistry, most likely $H_3{}^+$. In this case the deuterated molecules (DCN, N_2D^+, DCO^+, etc.) are produced from H_2D^+. The deuterium enhancement in this ion results from the isotope exchange

$$H_3{}^+ + HD \rightleftarrows H_2D^+ + H_2 + \Delta E \tag{19}$$

which is exothermic by an amount $\Delta E \simeq 200$ K (due to a difference in zero point vibrational energies). In thermodynamic equilibrium, the constituents in reaction (19) would be in the ratio

$$\frac{H_2D^+}{H_3^+} = \frac{HD}{H_2} \exp\,(\Delta E/kT) \tag{20}$$

and this ion ratio is of order 1 at $T \sim 10$ to 20 K for $HD/H_2 \sim 10^{-5}$ (the D to H ratio measured in the local interstellar medium).

In interstellar clouds the enhancement of deuterium in steady state is limited by the destruction of H_2D^+ due to recombination with electrons and

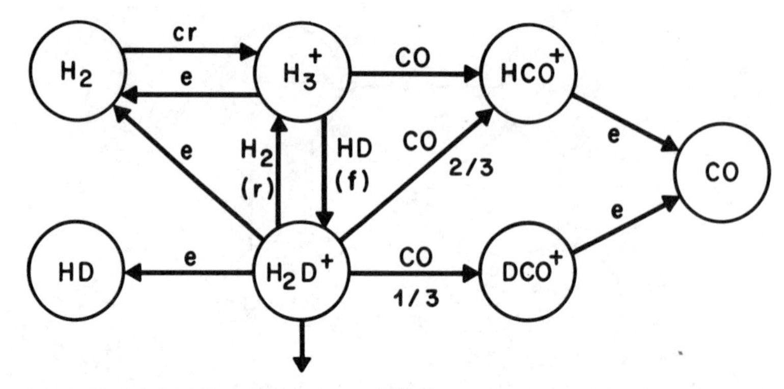

Fig. 4. The chemical reaction network for deuterium enhancement of H_2D^+ is shown along with the transfer of this enhancement to DCO^+. Also shown are the reactions balancing $H_3{}^+$, HCO^+, and CO, which are functions of $[e]$ and the cosmic ray ionization rate.

reactions with atoms and molecules. The chemical network which determines the deuterium enhancement is shown schematically in Fig. 4. In steady state the abundance ratio is

$$\frac{[H_2D^+]}{[H_3^+]} = \frac{k_f[HD]}{k_r + k_e[e] + \sum_i k_i[X_i]} \tag{21}$$

where k_f and k_r ($= k_f \exp(-\Delta E/T)$) are the forward and reverse reaction rate coefficients for reaction (19); k_e (H_2D^+) is for dissociative recombination (similar to $H_3{}^+$); and the sum is over all reactions with trace atoms and molecules (C, O, CO, N_2, H_2O, . . .). Upper limits on $[e]$, [CO], and T can be derived if the ratio $[H_2D^+]/[H_3{}^+]$ can be measured, given [HD]. Unfortunately this ratio cannot be measured directly because these ions have not been observed, and one must instead use other deuterated species that are produced from H_2D^+ and $H_3{}^+$.

The formyl ion isotopes are perhaps the best species for measuring the deuterium enhancement because: (1) HCO^+ is formed only by gas phase reactions and its formation is well understood; (2) many reactions involving this ion have been measured in the laboratory and the reaction rate coefficients are well known; and (3) HCO^+ is abundant and widespread in interstellar clouds. In addition, a lower limit to the electron density can be inferred from the HCO^+ abundance.

DCO^+ in interstellar clouds is produced mainly through the reaction, $H_2D^+ + CO \rightarrow DCO^+ + H_2$, so that the enhancement of H_2D^+ is reflected in DCO^+. Therefore, $[DCO^+]/[HCO^+] \simeq ⅓\,[H_2D^+]/[H_3{}^+]$, where the factor of ⅓ takes account of the fact that in only one out of three collisions will the deuterium ion be transferred (otherwise a proton is transferred and HCO^+ is formed). The deuterium abundance ratio, denoted R, is the abundance ratio of the deuterated to nondeuterated form. For HCO^+, for example, R is given by

$$R = \gamma \frac{k_f[\mathrm{HD}]}{k_r + k_e[e] + \Sigma k_i[X_i]} \tag{22}$$

where $\gamma = 1/3$ (for further details see Guélin et al. 1982*b*). Therefore, if dissociative recombination of H_2D^+ is rapid ($k_e \sim 7 \times 10^{-6}\, T^{-1/2}$ cm^3 s^{-1}), observations of DCO^+ and HCO^+ can provide the following limits on the electron abundance, $6 \times 10^{-9}/R > [e] > [HCO^+]$, and for the trace species $[CO + N_2 + 3H_2O + \ldots] \lesssim 10^{-5}/R$, where we have used $HD/H_2 = 2.8 \times 10^{-5}$. Note that for $[CO + N_2 + \ldots] \simeq$ few $\times 10^{-4}$, the value usually found from direct observations or inferred from chemical models, the ratio R provides a good value for $[e]$ only down to $[e] \sim (1 - 2) \times 10^{-7}$. Below this value, R yields only an upper limit on $[e]$. However, if k_e $(H_2D^+) < 2 \times 10^{-8}$ cm^3 s^{-1}, then the limits on the electron abundance become $3 \times 10^{-7}/R > [e] > [HCO^+]$ and the upper limit is considerably larger in this case.

Another technique for measuring either the electron abundance or the ionization rate exploits the relationship between HCO^+ and CO in the ion-molecule chemistry (Wootten et al. 1979) and is illustrated in Fig. 4. In steady state the HCO^+ and H_3^+ abundances are

$$[\mathrm{HCO}^+] \simeq \frac{k[\mathrm{H}_3^+][\mathrm{CO}]}{k_e'[e] + \Sigma k_i'[X_i']} \tag{23}$$

$$[\mathrm{H}_3^+] = \frac{\zeta/n(\mathrm{H}_2)}{k_e[e] + \Sigma k_i[X_i]} \tag{24}$$

where the sum is over all reacting atoms and molecules, and the prime refers to HCO^+ reactions. These relationships can be combined to form the quantity Z defined by Wootten et al. (1979)

$$Z = n(\mathrm{H}_2)\frac{[\mathrm{HCO}^+]}{[\mathrm{CO}]} = \frac{k\zeta}{(k_e'[e] + \Sigma k_i'[X_i'])(k_e[e] + \Sigma k_i[X_i])}. \tag{25}$$

For a large value of $k_e(H_3^+)$ the following relationships hold: if ζ is known (or at least an upper limit known), then an upper limit on $[e]$ can be found from $[e] < (k_e k_e' Z/k\zeta)^{1/2}$; on the other hand, if a value or an upper limit to $[e]$ is available from other measurements, an upper limit to the cosmic ray ionization rate can be determined from $\zeta \lesssim (k_e k_e' Z[e]^2/k)$.

Instead, if $k_e = k_e\,(H_3{}^+) < 2 \times 10^{-8}$ cm^3 s^{-1} and $[e] <$ few $\times 10^{-6}$, then, $[e] \lesssim \zeta/(k_e'\, Z\, \Sigma\, [X_i]) \leq \zeta/(k_e'\, n\,(H_2)[HCO^+])$. To within a factor of two, $\Sigma[X_i] \gtrsim [CO]$ and the relationships (24) or (25) above can be expressed in terms of HCO^+, and the following limits can be established for the electrons, $[HCO^+] < [e] < \zeta/(k_e'\,(HCO^+)\, n\,(H_2)[HCO^+])$. If $[e]$ is known from other methods, then a limit on the ionization rate can be found $\zeta \lesssim k_e'\,(HCO^+)\, n\,(H_2)[HCO^+]/[e]$.

This discussion has stressed the determination of the cosmic ray ionization rate. As noted above, however, X-ray ionization of the bulk of the gas is nearly indistinguishable from cosmic ray ionization. The determination of ζ is basically the sum of these components ζ_{cr} and ζ_{XR}. To separate out the X-ray component requires a detailed analysis of their effects on the trace constituents of the gas. Model calculations of the effects of X rays on the abundances of trace molecules have been made by Langer (1978*b*) and Krolik and Kallman (1983). These authors found that the absolute abundances of some molecular species depend on the X-ray ionization rate. For example, in dense clouds with embedded X-ray sources, Krolik and Kallman (1983) find that a few species are particularly sensitive to the X-ray ionization rate; of these only CN and HCN are readily observable. In principle then ζ_{XR} can be determined; in practice, however, it is difficult to do so because current chemical models have many uncertainties surrounding the calculation of absolute abundances (cf. Graedel et al. 1982).

III. OBSERVATIONAL RESULTS

The DCO^+ molecule was first detected in interstellar clouds by Hollis et al. (1976) in the $J = 1 \rightarrow 0$ rotational state; subsequently the $J = 2 \rightarrow 1$ emission was also observed in a number of sources by Guélin et al. (1977). These early detections gave ratios $[DCO^+/HCO^+] \sim 1$ in a few dark clouds and yielded very small limits on the electron fraction, $[e] < 10^{-8}$. These limits, however, depended on four critical assumptions:

1. The line intensities were proportional to the abundance of the molecular species;
2. The rates of deuterium isotope exchange at low temperatures (10–20 K) could be derived from those at 300 K;
3. The dissociative recombination rate coefficient was known at low temperatures for H_2D^+ (v = 0);
4. The [D/H] elemental ratio was $\sim$ few $\times 10^{-5}$.

Subsequent studies have shown that the first two assumptions, and possibly the third, were incorrect. The detection of $H^{13}CO^+$ (W. D. Watson et al. 1978; Langer et al. 1978) and $HC^{18}O^+$ in dark clouds (Langer et al.) demonstrated that the lines of the abundant isotopic species HCO^+ were optically very thick. Furthermore, recent laboratory measurements of deuterium exchange have revised upwards by about a factor of five the reaction-rate coefficients at 10 to 20 K (N. G. Adams and D. Smith 1981). The net effect of these corrections was to revise upwards the limits on $[e]$ in cold regions ($\sim$10 K) to $\sim$ few $\times 10^{-7}$. Finally, if $k_e\,(H_2D^+) < 2 \times 10^{-8}$ cm^3 s^{-1}, then the upper limits on $[e]$ must be further revised upwards to $\sim 10^{-5}$ (in this case, the method employing the HCO^+ abundance can be used to provide a better estimate of $[e]$ if the ionization rate is known; see below).

To date only two extensive surveys have been made to derive the electron abundance in clouds using the weak isotopes. Guélin, Langer, and Wilson

(1982*b*), hereafter (GLW), observed five isotopes of the formyl ion (HCO^+, DCO^+, $H^{13}CO^+$, $HC^{18}O^+$, and $D^{13}CO^+$) in five interstellar clouds (TMC 1, TMC 2, L 183, NGC 1333, and NGC 2264) and made extensive maps using the first three isotopes. In addition, some DCO^+ ($J = 2 \rightarrow 1$) emission was observed to provide excitation information for DCO^+. Wootten, Loren, and Snell (1982*b*), hereafter (WLS), surveyed more than thirty clouds for HCO^+, $H^{13}CO^+$, and DCO^+ emission and detected DCO^+ in twelve of these. The DCO^+ and $H^{13}CO^+$ were extensively mapped in only one source, NGC 1333. Observations of DCO^+ $J = 1 \rightarrow 0$ and $J = 2 \rightarrow 1$ emission were made using telescopes of similar beam size at each wavelength to provide accurate modeling of the DCO^+ excitation temperature.

Figure 5 shows the spectra towards one position in TMC 1 from (GLW). The anomalous isotope ratio in the emission intensity, T_{peak} ($H^{13}CO^+$) > T_{peak} ($H^{12}CO^+$), is a consequence of the absorption of HCO^+ photons coming from the core (high-excitation region) by a low-excitation foreground layer, shown schematically in Fig. 6 (Langer et al. 1978). It is clear that the HCO^+ cannot be used to derive [DCO^+/HCO^+] or [HCO^+] because of

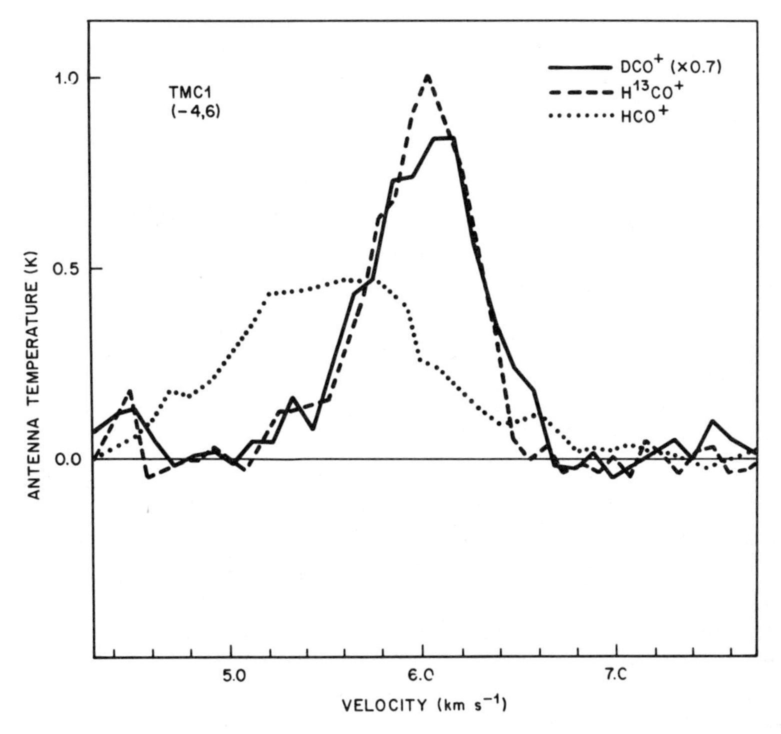

Fig. 5. The $J = 1 \rightarrow 0$ transition of the three main HCO^+ isotopic species and the $J = 2 \rightarrow 1$ transition of DCO^+ in TMC 1 at the offset position (−4′, 6′) as shown in Guélin et al. (1982). The absorption of HCO^+ is clearly shown by the low intensity of its emission compared to that of $H^{13}CO^+$ and the different line shapes. Note that DCO^+ has a line shape similar to that of $H^{13}CO^+$ indicating that they arise from the same region. The weakest isotopes, $HC^{18}O^+$ and $D^{13}CO^+$, are shown in Guélin et al. (1982*b*, their Fig. 1c).

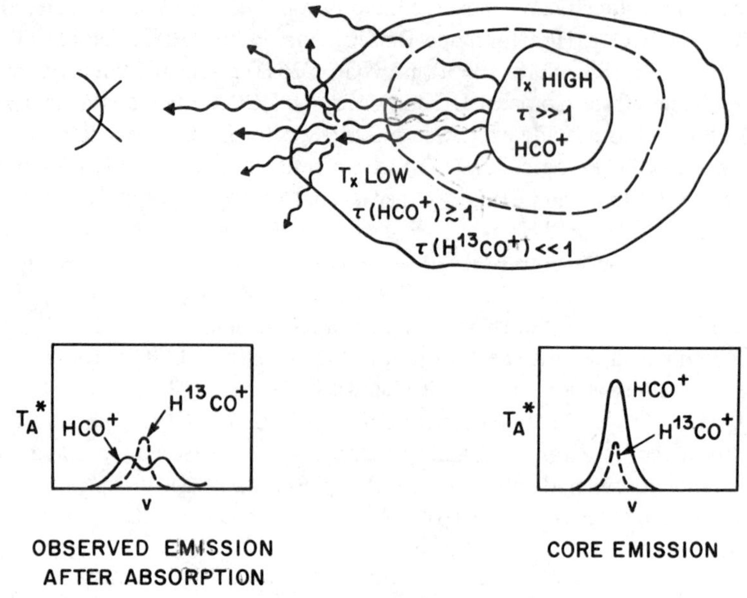

Fig. 6. A schematic picture showing how HCO^+ emission from a high-excitation core can be absorbed by foreground material and produce an anomalous intensity compared to $H^{13}CO^+$. The latter's emission is hardly modified by the foreground because its abundance is low and therefore the absorption opacity is small.

saturation and/or self-absorption and instead $H^{13}CO^+$ (or ideally $HC^{18}O^+$) should be used. For these isotopes, the deuterium ratio or HCO^+ abundance is determined by scaling by the appropriate carbon and oxygen isotope ratios, for example

$$\frac{DCO^+}{HCO^+} = \frac{DCO^+}{H^{13}CO^+} \times \frac{H^{13}CO^+}{H^{12}CO^+} \quad . \tag{26}$$

The HCO^+ isotopic ratios are reasonably well known (within a factor of two) and show only moderate variations in carbon and hardly any variation in oxygen due to isotopic fractionation (cf. Langer et al. 1984). The similarity of the DCO^+ and $H^{13}CO^+$ line shapes means that their emission arises from the same spatial region and that they have similar excitation conditions. Therefore, the abundance ratio $[DCO^+/H^{13}CO^+]$ and $[H^{13}CO^+]$ are reliable measures of the deuterium fractionation and HCO^+ fractional abundance, respectively.

Contour maps of the integrated intensity of DCO^+, $H^{13}CO^+$, and HCO^+ in TMC 1 are shown in Fig. 7. The DCO^+ emission is extended over a large region and $[DCO^+/H^{13}CO^+]$ does not vary much in the densest region. The $H^{13}CO^+$ map shows evidence of fragments in this cloud and the

HCO^+ emission appears to be confined to a region with large visual extinction, $A_v \gtrsim 3$ mag. The constraints on [*e*] derived by (GLW) and (WLS) are summarized in Sec. IV below. In addition, (WLS) compared the temperature dependence of the observed DCO^+/HCO^+ to the dependence in Eq. (3) in order to evaluate ΔE. The value of ΔE is uncertain and its determinations range from 140 K (N. G. Adams and D. Smith 1981) to 230 K (Porter and Solomon, personal communication, 1977; see also Wootten et al. 1979).

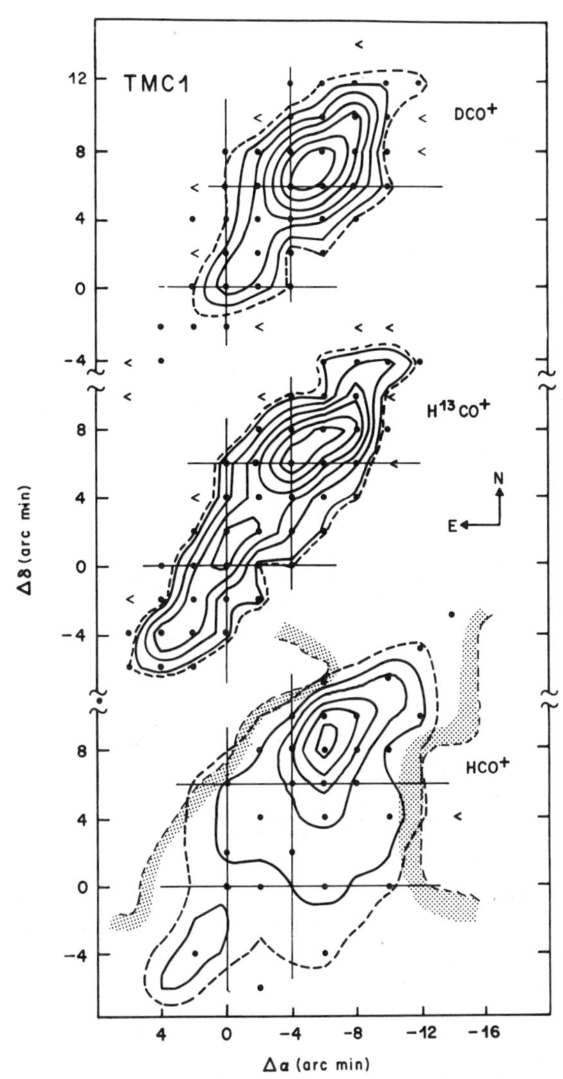

Fig. 7. Contour map of the integrated antenna temperature $\int T_A^* dv$, for TMC 1 as displayed by Guélin et al. (1982*b*). The central position in 1950 coordinates is $\alpha = 4^h38^m39^s$, $\delta = 25° 25'45''$. The shaded line delineates the borders of the high optical obscuration regions, $A_v \gtrsim 3$ mag.

IV. SUMMARY AND CONCLUSIONS

Listed below are some conclusions of (GLW) and (WLS) regarding the electron and ion abundances, cosmic ray ionization rates, and deuterium chemistry in interstellar clouds. First the conclusions based on the laboratory measurements yielding $k_e\,(H_3^+) \simeq 7 \times 10^{-6}\,T^{-1/2}$ cm^3 s^{-1} are summarized (1–9); afterwards, the changes and conclusions appropriate for the laboratory measurements indicating $k_e\,(H_3^+) < 2 \times 10^{-8}$ cm^3 s^{-1} are given (I–VII). In cases where only one of these two sets of authors is responsible for a conclusion, the appropriate reference follows:

1. $[DCO^+/HCO^+] \sim 10^3\,[D/H]_{\rm interstellar}$ and somewhat uniform where $A_v \gtrsim 3$–5 mag providing that $T_{\rm kin} \simeq 10$–15 K.
2. Scaling the $HC^{18}O^+$ abundance by 500 yields $[HCO^+] \sim 10^{-8}$ and therefore $[e] \gtrsim 10^{-8}$ (GLW).
3. The upper limit on $[e]$ is typically a few $\times\,10^{-7}$ to 10^{-6} in regions where DCO^+ is detected and $\lesssim 10^{-7}$ at a few places.
4. The limits on the trace molecules and atoms implied by the $[DCO^+/HCO^+]$ observations is $[CO + N_2 + 3H_2O + C + O] \lesssim 3 \times 10^{-4}$, which is consistent with direct abundance determinations of CO, C, and N_2 (using N_2H^+). If these trace species limit the deuterium enhancement, then the value of $[e]$ may be smaller than the upper limits given above.
5. The temperature dependence of the deuterium-fractionation chemistry is confirmed and $\Delta E/k \simeq 180$ K for reaction (19) (WLS).
6. The fractional metal abundance $[Mg + Fe + Ca + Na] \lesssim 10^{-6}$, a factor ~40 times smaller than the total cosmic fractional abundance, implying substantial depletion onto grains.
7. In the five sources studied by (GLW), $\zeta < 4 \times 10^{-16}$ s^{-1}, about ten times larger than the rate expected for these clouds. Values of 10^{-17} to 10^{-16} s^{-1} have been derived from HD studies in diffuse clouds (O'Donnell and Watson 1974) and $(0.4$–$1.6) \times 10^{-16}$ s^{-1} from recombination line studies towards 3C273 (Payne et al. 1984). The limit derived from the $[HCO^+/CO]$ ratio is very sensitive to the value of $[e]$ and $n(H_2)$, and ζ could be a factor of ten times smaller.
8. Assuming an upper limit of 10^{-16} s^{-1} for ζ (from the HD study), (WLS) derived limits on $[e]$ using the $[HCO^+/CO]$ ratio and found $[e] < 10^{-7}$ in all but two sources.
9. The observed range for $[DCO^+/HCO^+]$ in nearby clouds (WLS) is about that observed at different galactocentric radii by Penzias (1979). These variations in $[DCO^+/HCO^+]$ are probably dominated more strongly by local conditions than by [D/H] variations in the Galaxy (WLS).

If the dissociative recombination rate coefficients for H_3^+ and H_2D^+ are as small as is indicated by the measurements of Smith and Adams (1984), $k_e(H_3^+) < 2 \times 10^{-8}$ cm^3 s^{-1}, and if this limit is valid near 10 K, then the following conclusions can be drawn from the HCO^+, DCO^+, and CO surveys of (GLW) and (WLS):

I. The lower limit on $[e]$, as determined from $[HC^{18}O^+]$ is still 10^{-8}.

II. The upper limit on $[e]$, using $[\mathrm{DCO^+/HCO^+}]$, is typically $\sim 10^{-5}$ in dense regions; this limit is much greater than the value of $[e]$ suggested by theoretical models in these regions and is, therefore, too large to be useful in analyzing cloud conditions.

III. The limit on trace molecules and atoms is still, $[\mathrm{CO} + \mathrm{N_2} + \mathrm{C} + \mathrm{O} + \ldots] \lesssim 3 \times 10^{-4}$, close to the direct abundance determinations of CO, C, and N_2; thus, the maximum deuterium enhancement in cold clouds ($\lesssim 20$ K) is limited by these trace species.

IV. The near constancy of $[\mathrm{DCO^+/HCO^+}]$ in cold regions is a consequence of the nearly uniform abundance of trace species which limit the deuterium enhancement.

V. The temperature dependence of the deuterium-fractionation chemistry is still present in the data of (WLS), but their analysis has to be repeated to solve for the value of ΔE in reaction (19).

VI. The electron abundance can be determined, or at least a realistic upper bound set, from observations of HCO^+ if ζ is known,

$$[e] \lesssim \frac{\zeta / n(\mathrm{H_2})}{k_e(\mathrm{HCO^+})[\mathrm{HCO^+}]}. \tag{27}$$

A value of $\zeta \simeq 5 \times 10^{-17}\ \mathrm{s^{-1}}$ seems reasonable from the studies of O'Donnell and Watson (1974) and Payne et al. (1984). Adopting this value yields $10^{-8} < [e] \lesssim 3 \times 10^{-7}$ in the dense regions where $[\mathrm{HCO^+}] \simeq 10^{-8}$ and $n(\mathrm{H_2}) \simeq 10^4\ \mathrm{cm^{-3}}$. These low values of $[e]$ in the dense, shielded regions are similar to those usually quoted in the literature using the deuterium enhancement method and based on a large value of $k_e\,(\mathrm{H_2D^+})$, though the deuterium method no longer yields such low values if $k_e\,(\mathrm{H_2D^+})$ is small (see conclusion II above). Therefore, by fortuitous circumstances, many conclusions in the literature based on the earlier evaluations need not be changed. The uncertainty in $[e]$, however, is now directly dependent on the uncertainty in ζ.

VII. The ion abundance is probably dominated by the metals in dense regions with $[e] \simeq [M^+] \simeq [M_t]$. The metals are depleted by factors of about 30 to 50 because $[e] \lesssim 3 \times 10^{-7}$.

Many of these conclusions are different from those determined from the earliest observations of the isotopes of HCO^+. Sensitive observations of weak isotopes and new laboratory data on the deuterium reactions and (possibly) the dissociative recombination reactions have changed substantially the interpretation of electron and ion abundances, ionization rates, and metal depletion since 1978. Still problems remain. Better lower limits on $[e]$ are needed and more observations of the very weak isotopes (e.g., $\mathrm{HC^{18}O^+}$) must be made. Further mapping in clouds is needed to study variations of $[e]$, perhaps as the ultraviolet ionization rate varies with extinction. Finally, the different laboratory measurements of the dissociative recombination rate coefficients for H_3^+ and H_2D^+ must be reconciled and a correct value established, especially at low temperatures (10–20 K).

GAS-PHASE CARBON CHEMISTRY IN DENSE CLOUDS

ERIC HERBST
Duke University

The literature concerning the chemistry of complex molecules in dense interstellar clouds is reviewed and recent gas-phase studies are emphasized. Detailed ion-molecule pathways leading to large interstellar molecules are presented. These pathways include radiative association reactions; therefore these reactions are discussed in some detail. The few existing models of dense clouds that include complex molecules are compared. Finally, the chemistry of collapsing protostellar regions is briefly explored.

In 1978, the chemistry of dense interstellar clouds was reviewed (E. Herbst 1978*a*) in *Protostars and Planets* (Univ. of Arizona Press, Space Science Series). Since that time, much work has been accomplished in this field. This review will concentrate on developments concerning the gas-phase synthesis of complex, carbon-containing molecules in dense clouds. Before proceeding to this topic however, it is necessary to consider other salient developments in interstellar chemistry.

In 1980, Prasad and Huntress (1980*a,b*) published a large, detailed model of dense cloud chemistry that is time dependent in the sense that chemical abundances are allowed to evolve from initial values under constant physical conditions (gas density, temperature, visual extinction). The model consists of over 1400 gas-phase reactions involving 137 species. Grains are utilized only as a passive site for the production of molecular hydrogen. Elements included in the calculation are H, He, C, N, O, Si, S, Mg, Fe, and Na at the so-called depleted cosmic abundance ratios. In general, neutral molecules with more than four atoms (called "complex" throughout this work) are not included or are included with little detail. The steady-state abundances (reached by $\approx 10^7$ yr) of many species were found to be in reasonable agreement with observation; some serious disagreements were also found. Prasad and Huntress (1982) later made improvements in the sulfur chemistry. Many of the reactions used by Prasad and Huntress (1980*a,b*) had been studied in the laboratory; in particular, a large percentage of the laboratory effort came from Huntress and co-workers (Huntress 1977).

Graedel et al. (1982) have undertaken a time-dependent chemical model of dense cloud chemistry similar to that of Prasad and Huntress (1980*a,b*) but somewhat reduced in size. These authors utilized two sets of gaseous elemental abundance ratios in their calculations: a normal set of depleted abundances and a set of abundances in which metals are depleted additionally by two orders of magnitude. The latter set of abundances normally leads to better agreement with observation at steady state for sulfur-containing molecules, molecular ions, and the overall fractional ionization. In their manuscript, Graedel et al. (1982) present results in the form of three-dimensional diagrams in which abundances of species are plotted vs. time and gas density. One can see from these diagrams (see Fig. 1) that significant molecular abundances are developed at times (10^5–10^6 yr) long before steady state is realized. This is an important point for several reasons. Gas-phase models tend to ignore grain adsorption, a process which many investigators believe to occur on a time scale of $\sim 3 \times 10^9 / n(H_2)$ yr, where $n(H_2)$ is the H_2 gas density in cm^{-3}, assuming sticking rates of unity. If these models are to include grain adsorption, rapid molecular production times are needed in order for gas-phase syntheses to be effective. In addition, as discussed in Sec. III, complex molecule production may be only efficient at short cloud lifetimes and it would be disconcerting if smaller species were not also present at such times.

In addition to improvements in modeling of clouds under ambient conditions, progress in shock chemistry has also occurred during the earlier 1980s. G. F. Mitchell and T. S. Deveau (1983) examine the consequences of a shock on a diffuse cloud of initial density 100 cm^{-3} with a significant abundance of C^+ and find that small hydrocarbons like CH_4 are formed at high abundance and remain so for $\sim 10^5$ yr. This work and its extension by G. F. Mitchell (1983) to larger hydrocarbons is discussed more fully in Sec. II of this chapter. Finally, G. F. Mitchell (1984) has also considered the effects of shocks on dense clouds.

Progress in dust chemistry has not been as rapid as that in gas-phase reactions. As stated by W. D. Watson (1983), "Except for molecular hydrogen, there is in my opinion still no convincing mechanism for returning molecules from grain surfaces back to the gas." However, a detailed model calculation of the chemistry of grain mantles has been made by Tielens and Hagen (1982) and work by Greenberg and coworkers continues to show the potential for complex molecule production on and in grains (see, e.g., J. M. Greenberg 1983). The most detailed model of grain formation of gas-phase

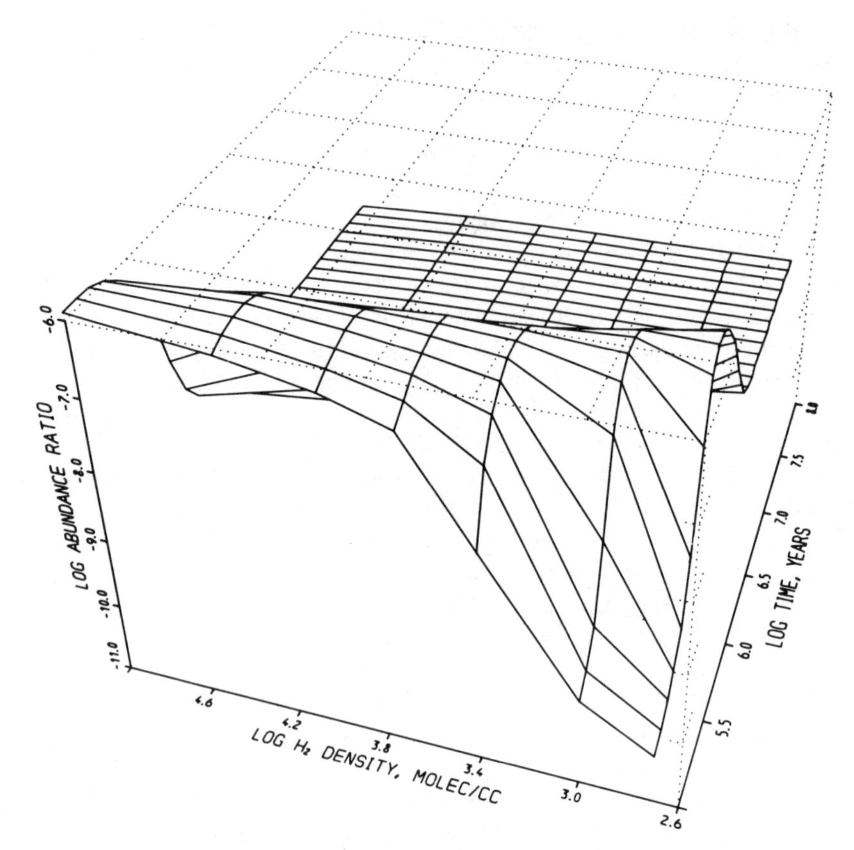

Fig. 1. Log molecular abundance (H_2CO) plotted against both evolutionary time and molecular hydrogen density.

molecules remains that of M. Allen and G. W. Robinson (1977). The problem of returning molecules to the gas once they are on a cold grain surface is still a vexing concern even if surface chemistry does not occur.

Neither of the ambient gas-phase models discussed above treats complex molecules in any detail. The difficulty in including complex molecules rests in the need for larger and larger sequences of rapid exothermic ion-molecule and/or neutral-neutral reactions. Sooner or later, known roadblocks in the form of endothermic reactions or exothermic reactions with activation energy appear and curtail syntheses under ambient interstellar conditions. To overcome these roadblocks, I and others have advocated the inclusion of radiative association reactions, in which two species come together and literally stick, forming a more complex species. These processes are discussed in Sec. I and their inclusion in ion-molecule syntheses of selected complex molecules is discussed in Sec. II. Section III contains a discussion of those models which include complex, carbon-containing species. Finally, the fate of carbon species and all other molecules during protostellar collapse is speculatively explored in Sec. IV.

I. RADIATIVE ASSOCIATION

A. Theory and Experiment

Ion-molecule radiative association reactions are processes in which an atomic or molecular ionic species A^+ collides with a neutral species B to produce a product molecule AB^+. The process is best visualized as a two-step one in which the species A^+ and B first form a "reaction complex" AB^{+*} which either redissociates or is stabilized by emission of a photon:

$$A^+ + B \rightleftarrows AB^{+*} \tag{1}$$

$$AB^{+*} \rightarrow AB^+ + h\nu. \tag{2}$$

The need for radiative association reactions in ion-molecule reaction schemes arises from the fact that normal ion-molecule reactions leading to more than one product molecule, *viz.*,

$$A^+ + B \rightarrow C^+ + D \tag{3}$$

are often endothermic and, therefore, cannot occur under interstellar cloud conditions. In addition, the process would seem to be an efficient way of synthesizing more complex species from smaller ones.

Radiative association is a difficult process to study in the laboratory where it is often dominated by three-body association, in which the complex AB^{+*} is collisionally stabilized by a third body C:

$$AB^{+*} + C \rightarrow AB^+ + C. \tag{4}$$

Very low laboratory pressures are needed for radiative association to become more rapid than three-body association. Several radiative association rates have been measured in the laboratory but most of the information on rate coefficients of reactions of interstellar importance comes from theory or analogy based on measured three-body rates.

Radiative association has provoked a great deal of theoretical interest. There are two theoretical problems involved in a determination of a radiative association rate coefficient: the formation and destruction rates of the intermediate complex AB^{+*}, and the radiative stabilization rate of the complex. In an early treatment (E. Herbst 1976) of AB^{+*} formation and dissociation, I used independent approximate theories to describe the rates of these processes. Such an approach has since been followed by Freed et al. (1982). However, the formation and dissociation rates are related by microscopic reversibility. Theories that incorporate this principle explicitly have been formulated by E. Herbst (1980,1981,1982*a*), Bates (1979,1983*a,b*), and Bass et al. (1979). A detailed discussion of the differences among these theories is not appropriate here; it is sufficient to mention that their results for specific rate coefficients can differ by an order of magnitude or more at low temperatures. All of these calculations show that the redissociation rates of complexes decrease as temperature decreases and as their size and the stabilization energy of AB^+ increase. Thus, favorable radiative association reactions tend to involve larger reactants that can form species with higher dissociation energies. Low temperatures will enhance all such reactions.

The radiative stabilization rate of the complex depends strongly on whether or not it is in its ground or in an excited electronic state. For the former case, stabilization must proceed via infrared photon emission. A somewhat detailed theoretical treatment of this process (E. Herbst 1982*b*) confirms the rate of $\approx 10^2$ s^{-1} used by most investigators; however, more theoretical work on this rate is planned. If the complex can stabilize itself via relaxation from an excited to a ground electronic state with concomitant emission of a visible or ultraviolet photon, the stabilization rate can be enhanced dramatically. An example is the CH_2^{+*} complex, formed via the collision of C^+ and H_2, which is calculated to stabilize itself at a rate of $\approx 10^5$ s^{-1} (E. Herbst 1982*a*).

The theoretical approaches to the formation and dissociation rates of the complex can be tested in calculations of three-body association rates. Comparison with experiments (N. G. Adams and D. Smith 1981*a*) shows that the so-called modified thermal level of treatment (Bates 1979; E. Herbst 1981; N. G. Adams and D. Smith 1981*a*), which involves only a modest amount of computational effort, is quite successful in reproducing many measured rates at temperatures down to 80 K. However, certain three-body reactions appear not to have any measurable rate coefficients in the laboratory (see, e.g., E. Herbst et al. 1983). These systems are thought to possess activation energy barriers in the complex formation channel. Therefore, the analogous radiative association reactions will also be vanishingly slow and must not be used in interstellar models.

Assuming the theoretical approach to radiative stabilization of the complex to be valid, comparison of calculated and measured three-body results indicates that calculated radiative association rate coefficients should be accurate to perhaps an order of magnitude if activation energy is not present. This statement is bolstered by two as yet unpublished experimental measurements of Dunn and coworkers (personal communication, G. Dunn) on radiative association rates at 10 K in a low pressure ion trap. They measured an *upper limit* of 2×10^{-15} cm^3 s^{-1} for the rate coefficient of the process

$$C^+ + H_2 \rightarrow CH_2^+ + h\nu \tag{5}$$

and a value of 1.8×10^{-13} cm^3 s^{-1} for the rate coefficient of the process

$$CH_3^+ + H_2 \rightarrow CH_5^+ + h\nu. \qquad (6)$$

The latest calculated result for $C^+ + H_2$ is 8×10^{-16} cm^3 s^{-1} (E. Herbst 1982*a*), obtained via both the modified-thermal and more detailed phase-space approaches, whereas the modified-thermal value for the $CH_3^+ + H_2$ is 1.0×10^{-14} cm^3 s^{-1} (E. Herbst 1983) or 3.3×10^{-14} cm^3 s^{-1} (Bates 1983*a*).

For some radiative association reactions suggested as being important in the synthesis of complex interstellar molecules (Huntress and Mitchell 1979), there are competing normal exothermic channels. Bates (1983*b*) has argued that the existence of a competing channel will dramatically diminish complex lifetimes and reduce radiative association reaction rates severely. There are observed reaction systems in the laboratory, however, where three-body association reactions compete surprisingly effectively with normal two-body reactions (D. Smith and N. G. Adams 1978). This problem is currently being examined in some detail. Preliminary calculations show that the rate of radiative association can be severely affected by the shape of the potential energy surface in the vicinity of the complex geometry.

B. Important Interstellar Radiative Association Reactions

Two classes of radiative association reactions have proved particularly useful to investigators of complex molecule syntheses. These classes are discussed below.

Hydrogenation. These reactions, of the general class

$$A^+ + H_2 \rightarrow AH_2^+ + h\nu \qquad (7)$$

serve to hydrogenate ions when normal hydrogenating ion-molecule reactions of the type

$$A^+ + H_2 \rightarrow AH^+ + H \qquad (8)$$

are endothermic. Key reactions include $A^+ = C^+, CH_3^+$, and $C_2H_2^+$. E. Herbst et al. (1983) have shown, however, that many of these systems involving hydrocarbon ions possess activation energy.

CH_3^+ Reactions. The ion CH_3^+ is thought to lead to a variety of complex species via reactions such as

$$CH_3^+ + H_2O \rightarrow CH_3OH_2^+ + h\nu \qquad (9)$$

$$CH_3OH_2^+ + e \rightarrow CH_3OH + H \qquad (10)$$

$$CH_3^+ + HCN \rightarrow CH_3CNH^+ + h\nu \qquad (11)$$

$$CH_3CNH^+ + e \rightarrow CH_3CN + H \quad (12)$$

$$CH_3^+ + CH_3OH \rightarrow CH_3OCH_4^+ + h\nu \quad (13)$$

$$CH_3OCH_4^+ + e \rightarrow CH_3OCH_3 + H. \quad (14)$$

Among the investigators who have analyzed the importance of these reactions are D. Smith and N. G. Adams (1978), Huntress and Mitchell (1979), and Schiff and Bohme (1979). The relative importance of CH_3^+ derives from the fact that it is depleted only slowly by radiative association with H_2 and is more abundant than many similar ions.

II. ION-MOLECULE SYNTHESES OF HYDROCARBONS AND CYANOACETYLENES

In the early 1980s, a variety of investigators have discussed possible ion-molecule reaction networks leading to complex molecule production. A representative but not exhaustive list of these investigators' work should include D. Smith and N. G. Adams (1978), Huntress and Mitchell (1979), Suzuki (1979), G. F. Mitchell et al. (1979), Schiff and Bohme (1979), Freed et al. (1982), Suzuki (1983), E. Herbst et al. (1983), E. Herbst (1983), G. F. Mitchell (1983), and Millar and Freeman (1984*a,b*). The major focus of most of the studies has been on the synthesis of hydrocarbons and cyanoacetylenes although other complex species have also been considered. In this section, the major current ideas on gas-phase molecular synthesis are discussed. In all of the proposed reaction networks leading to complex molecules, radiative association processes of the types mentioned in Sec. I.B play an important role.

A. Hydrocarbons

The synthesis of hydrocarbons proceeds from the precursor species C^+ (C II) and C(C I) via the reactions

$$C^+ + H_2 \rightarrow CH_2^+ + h\nu \quad (15)$$

$$C + H_3^+ \rightarrow CH^+ + H_2. \quad (16)$$

From CH^+ and CH_2^+, ion-molecule and ion-electron reactions produce one-carbon hydrocarbons such as methane (CH_4):

$$CH^+ + H_2 \rightarrow CH_2^+ + H \quad (17)^*$$

$$CH_2^+ + H_2 \rightarrow CH_3^+ + H \quad (18)^*$$

$$CH_3^+ + H_2 \rightarrow CH_5^+ + h\nu \quad (19)^*$$

$$CH_3^+ + e \rightarrow CH_2 + H; CH + H_2 \quad (20)$$

$$CH_5^+ + e \rightarrow CH_4 + H; CH_3 + H_2. \quad (21)$$

[An asterisk following a reaction number indicates that both the reaction rate coefficient and reaction products have been studied in the laboratory.] For ion-electron dissociative recombination reactions, rate coefficients have been measured for many reactions, but product branching ratios have been determined for only one system, that of $H_3^+ + e$. Consequently, branching ratios must be estimated from theory.

Once elementary hydrocarbons are formed, two-carbon species can be synthesized via insertion reactions involving C^+ and/or C with one-carbon species, *viz.*,

$$C^+ + CH_4 \rightarrow C_2H_3^+ + H; C_2H_2^+ + H_2 \quad (22)^*$$

$$C + CH_3^+ \rightarrow C_2H_2^+ + H \quad (23)$$

$$C + CH_2 \rightarrow C_2H + H \quad (24)$$

followed by reactions such as

$$C_2H_2^+ + H_2 \rightarrow C_2H_4^+ + h\nu \quad (25)$$

$$C_2H_2^+ + e \rightarrow C_2H + H; C_2 + H_2 \quad (26)$$

$$C_2H_3^+ + e \rightarrow C_2H_2 + H; C_2H + H_2 \quad (27)$$

$$C_2H_4^+ + e \rightarrow C_2H_3 + H; C_2H_2 + H_2. \quad (28)$$

In addition, condensation reactions between hydrocarbon ions and neutrals can occur. Consider the reaction

$$CH_3^+ + CH_4 \rightarrow C_2H_5^+ + H_2 \quad (29)^*$$

followed by

$$C_2H_5^+ + e \rightarrow C_2H_4 + H; C_2H_3 + H_2. \quad (30)$$

An important consideration in hydrocarbon synthetic schemes is whether hydrocarbon ions can be hydrogenated via normal reaction with molecular H_2 (e.g., reaction 17) or, if this is impossible, via radiative association with H_2 (e.g., reactions 19 and 25). Some hydrogenation is necessary to maintain the synthetic process because reactions like (22), (23), and (29) have as a common feature the replacement of hydrogen atoms with carbon atoms. Reactions involving C^+ or C can eject one or two H atoms from the ion product (see, e.g., reaction 22) whereas condensation reactions produce ion products with fewer total hydrogen atoms than those of the combined reactants. As synthesis proceeds, only skeletal carbon species will be produced in the absence of hydrogenation.

E. Herbst et al. (1983) have investigated normal and three-body association reactions between an assortment of hydrocarbon ions through C_4H^+ in complexity with H_2. They conclude that some hydrogenation reactions can proceed via normal exothermic channels, some via three-body channels in the laboratory (and, by inference, via radiative association channels in interstellar clouds), and some do not proceed at all due to large activation energy barriers. Unfortunately, for those interested in modeling interstellar clouds, no general pattern can be seen in these results except that hydrogenation is by no means certain. For two-carbon ions, E. Herbst et al. (1983) found that $C_2H_2{}^+$ can be hydrogenated (via association) but that $C_2H_3{}^+$, $C_2H_4{}^+$ and $C_2H_5{}^+$ cannot by either normal or associative mechanisms. Thus, a synthesis of the ion $C_2H_5{}^+$ must occur via a condensation reaction such as reaction (29) and not via hydrogenation of $C_2H_3{}^+$ or $C_2H_4{}^+$.

Difficulty in hydrogenation via reaction types (7) and (8) manifests itself clearly in the synthesis of three-carbon and four-carbon species. Consider the reaction sequence

$$C^+ + C_2H_2 \rightarrow C_3H^+ + H \qquad (31)^*$$

$$C_3H^+ + H_2 \rightarrow C_3H_3{}^+ + h\nu \qquad (32)$$

$$C_3H_3{}^+ + H_2 \not\rightarrow \text{no products} \qquad (33)^*$$

$$C_3H_3{}^+ + e \rightarrow C_3H_2 + H; C_3H + H_2 \qquad (34)$$

$$C^+ + C_3H_2 \rightarrow C_4H^+ + H \qquad (35)$$

$$C_4H^+ + H_2 \not\rightarrow \text{no products} \qquad (36)^*$$

$$C_4H^+ + e \rightarrow C_4 + H. \qquad (37)$$

This sequence cannot produce the well-known interstellar species C_4H, which can however be produced via

$$C + C_3H_3{}^+ \rightarrow C_4H_2{}^+ + H \qquad (38)$$

$$C_4H_2{}^+ + e \rightarrow C_4H + H \qquad (39)$$

or via the condensation reaction

$$C_2H_2{}^+ + C_2H_2 \rightarrow C_4H_3{}^+ + H; C_4H_2{}^+ + H_2 \qquad (40)^*$$

followed by dissociative recombination. The hydrogen-rich species C_3H_4 (methyl acetylene) involves an even more hydrogenated precursor ion ($C_3H_5{}^+$ or $C_3H_6{}^+$). Such an ion cannot be formed via hydrogenation of $C_3H_3{}^+$ or $C_3H_4{}^+$ but must be formed via a condensation reaction of the type

$$C_2H_3{}^+ + CH_4 \rightarrow C_3H_5{}^+ + H_2. \qquad (41)^*$$

Millar and Freeman (1984*a,b*) have extended the above reaction networks to include five-carbon species in an attempt to explain the observed abundance of the complex hydrocarbon CH_3C_4H. As these authors note, measurements of hydrogenation reaction rates for five-carbon ionic species are necessary before accurate predictions can be made.

Although the above types of reactions represent perhaps a majority view of important ion-molecule pathways to hydrocarbon formation, other related ideas have been expressed. Freed et al. (1982) have argued that radiative association reactions of the type

$$C^+ + C_n \rightarrow C_{n+1}^+ + h\nu \tag{42}$$

followed by hydrogenation can efficiently synthesize long chain hydrocarbons. The success of this idea rests on rapid radiative association rates, large C^+ abundances, and successful partial hydrogenation of bare carbon ions C_{n+1}^+ to lead to neutral C_{n+1} via dissociative recombination. Contrast

$$C_{n+1}^+ + e \rightarrow C_n + C; \text{ etc.} \tag{43}$$

which results in no net synthesis with

$$C_{n+1}^+ + H_2 \rightarrow C_{n+1}H^+ + H \tag{44}$$

$$C_{n+1}H^+ + e \rightarrow C_{n+1} + H; \text{ etc.} \tag{45}$$

which leads to production of C_{n+1} from C_n. The efficacy of the network depends on unmeasured rate coefficients and has been questioned (Bates 1983*b*).

G. F. Mitchell (1983) has considered the synthesis of interstellar hydrocarbons as complex as C_6H_2 following the passage of a shock through a cloud of density 10^2 cm^{-3}. He utilizes ion-molecule condensation reactions, C^+ reactions, and hydrogenation reactions of the kind discussed above. However, the high temperature of the post-shock gas allows selected reactions of significant activation energy to turn on. One result is that hydrogenation is always rapid. In addition to ion-molecule reactions, G. F. Mitchell (1983) also invokes neutral-neutral reactions like

$$C_2H_2 + C_2H_2 \rightarrow C_4H_3 + H. \tag{46}^*$$

All of the reaction networks discussed here are comprised of reactions that have been studied in the laboratory as well as those that have not. More laboratory work is clearly needed and must be used in models wherever possible. A word of caution is in order here, however. Laboratory data are not always correct and, even if correct, may not be directly applicable. Consider the vexing case of $C_3H^+ + H_2$. E. Herbst et al. (1983) originally reported that the reaction products at 80 K were $C_3H_2^+$ + H. However, as noted by Bohme and co-workers (D. Bohme, personal communication), the correct product for purely thermal reactants in a high pressure laboratory reactor is $C_3H_3^+$, formed via three-body association. Addition of a small amount of extra kinetic energy to the C_3H^+ reactant leads to the opening of the $C_3H_2^+$ + H channel. In interstellar clouds, the product branching ratio between $C_3H_2^+$ and $C_3H_3^+$ is not well determined from these experiments. If the C_3H^+ ion is formed with a small amount of kinetic energy via an exothermic ion-molecule reaction, it can either react with H_2 to form $C_3H_2^+$ and $C_3H_3^+$ (via radiative association) or collide with H_2 elastically and inelastically to lose this energy. Once the energy is lost, C_3H^+ can only radiatively associate with H_2. We are currently working on estimating the correct branching ratio for interstellar conditions.

The unusual reaction between C_3H^+ and H_2 brings up the question of how important nonthermal effects are in gas-phase ion-molecule syntheses. In general, collision times are long enough so that excess vibrational excitation of molecular species will radiate away before reaction. Rotational excitation is not normally associated with much of a change in reaction rates. Thus, only translational excitation of reactants need be seriously considered. Even translational excitation is cooled relatively rapidly via collisions with H_2 unless the excess energy can be channeled into reaction with H_2 as in the above case. A detailed study of these effects has not yet been undertaken.

B. Cyanoacetylenes

Several ion-molecule pathways have been proposed for the production of simple cyanoacetylenes such as C_3N, HC_3N, C_5N, and HC_5N. These synthetic routes are based on precursor species such as HCN, H_2CN^+ and N. In addition to the cyanoacetylenes, several other organo-nitrogen compounds can be synthesized via similar networks.

The first synthesis of HC_3N and C_3N to be published in the literature was the sequence (Huntress 1977; Schiff and Bohme 1979)

$$C_2H_2^+ + HCN \rightarrow H_2C_3N^+ + H \qquad (47)^*$$

$$H_2C_3N^+ + e \rightarrow HC_3N + H; C_3N + H_2 \qquad (48)$$

but it was considered too slow by G. F. Mitchell et al. (1979) who proposed, in its place, the radiative association reaction

$$HCNH^+ + C_2H_2 \rightarrow H_4C_3N^+ + h\nu \qquad (49)$$

followed by

$$H_4C_3N^+ + e \rightarrow HC_3N + H_2 + H; C_3N + 2H_2. \qquad (50)$$

E. Herbst (1983), Suzuki (1983), and Millar and Freeman (1984*a,b*) prefer reactions of hydrocarbons and atomic nitrogen for the synthesis of C_3N and HC_3N. E. Herbst (1983) showed that the calculated rate coefficient of the radiative association process (reaction 49) is too slow to explain the observed HC_3N abundance in the TMC-1 cloud. Reactions with N atoms seem appropriate because of the large calculated abundance of N. However, it must be mentioned that only one hydrocarbon ion-N reaction has been studied in the laboratory, namely the $C_2H_2^+$-N system. Possible reactions with N atoms leading to HC_3N and C_3N production are

$$C_3H_2^+ + N \rightarrow HC_3N^+ + H \quad (51)$$

$$HC_3N^+ + H_2 \rightarrow H_2C_3N^+ + H \quad (52)$$

$$C_3H_3^+ + N \rightarrow H_2C_3N^+ + H \quad (53)$$

followed by reaction (48). In addition to ion-molecule reactions, possible neutral-neutral reactions include

$$C_3H + N \rightarrow C_3N + H \quad (54)$$

$$C_3H_2 + N \rightarrow HC_3N + H. \quad (55)$$

Care must be exercised in the use of neutral-neutral reactions, however. Except for radical-radical reactions, activation energies remain a real possibility in the absence of laboratory evidence to the contrary. Another problem with both hydrocarbon neutral and hydrocarbon ion reactions with N is the possibility of lack of conservation of electronic spin between ground electronic state reactants and products. A change of spin often entails activation energy. Note that atomic nitrogen is in a ^4S electronic ground state (spin $= 3/2$) and that a reaction with a singlet species can only produce species with individual spins capable of coupling to form a resultant state with spin $= 3/2$ in the absence of a spin change. Based on this analysis, it is possible that reactions (53) and (55) possess considerable activation energy. Laboratory studies and/or theoretical calculations of relevant potential surfaces are needed.

More complex cyanoacetylenes can be produced via extensions of the above networks to larger molecules. Consider first the analog of the radiative association reaction (49) (G. F. Mitchell et al. 1979).

$$H_2C_3N^+ + C_2H_2 \rightarrow H_4C_5N^+ + h\nu \quad (56)$$

followed by

$$H_4C_5N^+ + e \rightarrow HC_5N + H_2 + H; C_5N + 2H_2. \quad (57)$$

Leung et al. (1984) calculated the radiative association between $H_2C_3N^+$ and C_2H_2 to occur on virtually every collision, if competing exothermic pathways can be ignored in the calculation. The N atom reaction sequence can also lead to C_5N and HC_5N production (Millar and Freeman 1984*a,b*) via reactions such as

$$C_5H^+ + N \rightarrow C_5N^+ + H \quad (58)$$

$$C_5N^+ + H_2 \rightarrow HC_5N^+ + H \quad (59)$$

$$HC_5N^+ + H_2 \rightarrow H_2C_5N^+ + H \quad (60)$$

$$H_2C_5N^+ + e \rightarrow HC_5N + H; C_5N + H_2 \quad (61)$$

although it is unclear whether or not the hydrogenation reactions can occur or whether C_5H^+ itself is efficiently formed. Ion-molecule syntheses of even larger cyanoacetylenes are even more speculative and will not be discussed here.

C. Other Complex Molecules

Several organo-nitrogen compounds that are not cyanoacetylenes can be synthesized via N atom reactions of the type discussed in Sec. II.B. For example, Millar and Freeman (1984*a*) pointed out that vinyl cyanide (C_2H_3CN) can be formed via

$$C_3H_5^+ + N \rightarrow H_4C_3N^+ + H \quad (62)$$

followed by

$$H_4C_3N^+ + e \rightarrow C_2H_3CN + H. \quad (63)$$

Note that the precursor ion $H_4C_3N^+$ is also formed via the radiative association reaction (49).

Although the possibility exists that some organic compounds involving oxygen (alcohols, ethers, aldehydes, esters) can be formed via reactions between hydrocarbon ions and/or neutrals with atomic oxygen which lead to precursor ions or the species, it is more likely, based on sparse ion-molecule laboratory data (D. Smith and N. G. Adams 1978; Huntress and Mitchell 1979), that these species derive from radiative association reactions involving the relatively abundant ion CH_3^+ or similar species. A glance at the reaction list shown in Sec. I.B demonstrates some of the species formed in this manner. For more reactions, the reader is referred to D. Smith and N. G. Adams (1978) and Huntress and Mitchell (1979). We note that all of the utilized CH_3^+ associations have been studied in the laboratory under three-body conditions (high pressure) and that two have been studied under low pressure, radiative conditions (CH_3^+ + HCN, H_2). All studied reactions are at least as rapid as indicated by theory.

III. MODELS INVOLVING COMPLEX MOLECULES

In 1983–84, several models of gas-phase chemistry in dense clouds have been constructed that include some complex molecules. The majority of these models are concerned with dark clouds and the discussion here will emphasize these regions. Prior to this spate of work, Suzuki (1979) had included some complex molecules in a giant calculation of ion-molecule chemistry similar to the work of Prasad and Huntress (1980*a,b*).

E. Herbst (1983) has constructed a "semi-detailed" model in which he fixes the abundances of selected atomic and smaller molecular species and then solves for the steady-state abundances of more complex molecules, in particular hydrocarbons and cyanoacetylenes. The model consists of 305 gas-phase reactions, which are comprised of synthetic reactions of the types discussed in Sec. II and depletion reactions such as reactions between stable neutrals and the ions H^+, He^+, C^+, H_3^+, and HCO^+, and reactions between unstable neutrals and the above ions as well as the atoms O and N. Herbst's model is able to reproduce the observed abundances of species such as CH, C_2H, C_3H_4, C_4H, and C_3N in the TMC 1 cloud if and only if he assumes a high fractional abundance of atomic carbon ($\sim 10^{-5}$). Although C has not been detected in TMC 1, its observed upper limit (Phillips and Huggins 1981) does not preclude such a high fractional abundance and, of course, high C abundances have been observed in other sources. Herbst's original model did not reproduce the observed HC_3N abundance, but a newer version with new experimental reaction rates does.

The necessity for a high abundance of atomic C in this model derives from the fact that hydrocarbon synthesis starts from C and C^+. A high fractional abundance of C^+ would also aid hydrocarbon synthesis, as advocated by Suzuki (1983). Observations of TMC 1, however, indicate a fractional ionization of $< 10^{-7}$. Most detailed models of the gas-phase chemistry of smaller molecular species (see, e.g., Prasad and Huntress 1980*a,b*) show that at steady state CO is the dominant form of carbon by several orders of magnitude and that C is quite low in abundance. However, Graedel et al. (1982) calculated that high C abundances at steady state are possible under some unusual conditions. The normal ion-molecule time dependence does show that the dominant forms of carbon in the evolution of a diffuse cloud into a dense one are first C^+, then C, and finally CO. Several authors have thought of ways of enhancing calculated C abundances at steady state. High internal ultraviolet fields, a C/O abundance ratio greater than unity, and a complex interaction among turbulence, adsorption of C onto grains before conversion into CO, and grain mantle evaporation have been advocated.

A more detailed model of TMC 1 in the steady-state limit has been undertaken by Millar and Freeman (1984*a*) using similar reactions to those considered by E. Herbst (1983). In this model, all abundances are calculated ones and a finite visual extinction (A_v = 5) is utilized so that photodissociation processes come into play. [Herbst's model does not include photodissociation explicitly.] The model contains 500 reactions and 140 species. Cosmic abundance ratios of C, N, and O are depleted by only a factor of 3. Calculated fractional abundances of C and C^+ are 1.1×10^{-7} and 7.3×10^{-9}, respectively. With a small amount of C, calculated abundances of some hydrocarbons might be expected to be and are low; C_3H_4 is calculated to have a fractional abundance of 4.2×10^{-11}, and C_4H a fractional abundance of 7.6×10^{-11}. The observed values are $(2–4) \times 10^{-9}$ and $(0.8–1.5) \times 10^{-8}$, respectively. The model is much more satisfactory for species like C_3N, HC_3N, and C_2H. A similar model is available for cloud L 183 (Millar and Freeman 1984*b*) where more severe cosmic abundance ratio depletions are utilized to determine different results.

A time-dependent model including complex molecules has recently been completed by Leung et al. (1983; see also chapter by Glassgold). The model contains 200 chemical species and over 1800 reactions. Some complex neutral molecules in the model are HC_3N, C_4N, C_4H, CH_3CN, C_5N, CH_3OH, HC_5N, CH_3NH_2, CH_3CHO, C_2H_5OH, and CH_3OCH_3. The model starts with a set of representative diffuse cloud abundances and follows the chemistry until steady state is reached. Two sets of elemental gaseous abundances are utilized: the normal depleted ones, and a set in which metals are depleted by an additional 2 orders of magnitude (Graedel et al. 1982). Physical time dependence is excluded; the gas density is assumed constant as is the temperature. An extinction A_v sufficiently large to exclude photon-induced processes is assumed. The reaction set includes virtually all the reactions discussed in Secs. I and II.

Leung et al. (1983) found that calculated abundances of complex molecules are in reasonable agreement with observed values in the clouds TMC 1, Orion (ridge) and Sgr B2 at times *before* steady state is reached, somewhat after the neutral carbon atom abundance has peaked and is declining. Because the calculated steady-state abundances of most complex molecules are lower than observed values, these authors concluded, as did E. Herbst (1983), that a large C I abundance is required for significant production of complex species. The time scales for complex molecule productions are quite short ($10^5–10^6$ yr) and might even survive the inclusion of gas depletion via adsorption onto grains in the model. Inclusion of grain adsorption and physical time dependence are planned by these authors but before such inclusion, it is probably not profitable to speculate on what their results signify regarding cloud lifetimes. After all, the depletion of C I in steady state can be mitigated by processes involving ultraviolet internal radiation or grain adsorption and evaporation.

Finally, G. F. Mitchell (1983) has extended the shock model of G. F. Mitchell and T. J. Deveau (1983) to include hydrocarbons up to six carbon atoms in size. The chemical system consists of 108 species and 1252 reactions. The types of hydrocarbon reactions included have been discussed in Sec. II. The pre-shock gas is at a density of 100 cm^{-3}. A 10 km s^{-1} shock produces a temperature of 3680 K and a density of 480 cm^{-3}. As cooling proceeds, the density eventually reaches 3×10^4 cm^{-3}. Hydrocarbon concentrations are enhanced by large abundances of C^+ and/or C which are present up through times of 10^5 yr after the shock. In general, large fractional abundances of complex hydrocarbons are attained. The key to hydrocarbon production in this and other models appears to be to maintain a large abundance of C^+ and/or C and to prevent near total conversion of the cosmic abundance of carbon into CO. Note that in the dense cloud shock model of G. F. Mitchell (1984), complex hydrocarbon abundances are generally small.

To conclude this section, we note that despite the dissimilarities of the models discussed, a major conclusion is that under certain conditions ion-molecule syntheses can produce sufficient abundances of complex molecules to explain observational results. The lack of experimental data on numerous important reactions and the complex interplay of chemical and physical effects make it difficult to pose major observational tests for the competing models. Perhaps one such test is the presence of deuterated cyanoacetylenes in TMC 1. As noted by Millar and Freeman (1984*a*), the existence of these compounds appears to require more H_2D^+ than can exist in the unusual model of Suzuki (1983). In general, however, normal ion-molecule chemistry is best tested by observations that do not require full models but only small subsets of reactions. An excellent example, discussed in the chapter by Irvine et al., is metastable isomer/stable isomer abundance ratios.

IV. CHEMISTRY AND PROTOSTARS

What happens to the abundances of interstellar molecules during the evolutionary track of a piece of interstellar cloud toward the main sequence? Are there chemical signatures of protostellar phases? These intriguing questions have not been addressed in the previously discussed models. The importance of these questions, however, merits a serious attempt to answer them. In the near future, Leung, Herbst, and Huebner plan to extend their detailed time-dependent chemical model to include gravitational collapse. Hydrodynamic collapse calculations with some chemistry have been undertaken by Gerola and Glassgold (1978) and by Tarafdar et al. (1984), but have not been pushed past the dense cloud stage. In this section, we speculate briefly on what might happen to the chemistry of a rapidly collapsing object.

The chemistry of a collapsing region will be affected by the increase in density and the subsequent increase in temperature. The increase in density will probably result in the screening of cosmic rays from the collapsing object;

an object of gas density 10^{10} cm^{-3} with extent 100 AU (10^{15} cm) possesses a total gas column density of 10^{25} cm^{-2} which is 10 times that estimated to effectively block cosmic rays. The exclusion of cosmic rays will eventually turn off normal ion-molecule chemistry. Even if some cosmic rays do penetrate, the time scale of ion-molecule processes will gradually fall behind that of the collapsing region and become less relevant unless another ionization source exists. Similar arguments have been made against ion-molecule synthesis in the expanding shell IRC + 10216.

What will happen under these conditions? Fractional abundances of ions will decrease due to recombination of ions and electrons. Fractional abundances of major neutral species will become frozen. Reactive neutral species will eventually recombine via three-body association reactions at densities $> 10^{11}$ cm^{-3}. Significant adsorption of species onto grains is also a likely outcome unless temperatures increase too rapidly.

Eventually, increasing densities will, of course, lead to increasing temperatures. The chemistry of protostellar regions will be significantly affected by the extent of the time lag between density increases and temperature increases. A large temperature increase will permit desorption of grain mantles, turn-on endothermic reactions and exothermic reactions with activation energy barriers and, sooner or later, drive the material towards chemical equilibrium. Although the initial result of an increase in temperature may well be an increase in molecular complexity, in close analogy with that calculated for certain shocks by G. F. Mitchell (1983), the steady increase in temperature will eventually lead to the destruction of all but the hardiest molecular species in the immediate vicinity of the newly forming star. Significant molecular complexity may persist, however, at some reasonable distance from the star.

These brief qualitative statements need to be reinforced or disproved by quantitative treatments. Given the importance of the topics of star and planetary formation and advances in modeling techniques, it is reasonable to expect that quantitative approaches to the chemistry of protostellar regions will be forthcoming in the near future.

Acknowledgments. I would like to acknowledge the support of my research program by grants from the National Science Foundation. Stimulating conversations on the topic of the chemistry of protostars, held at the Aspen Center for Physics during August 1983, were of significant benefit to me.

PART V

Chemistry and Grains in the Solar Nebula

METEORITIC CONSTRAINTS ON PROCESSES IN THE SOLAR NEBULA

JOHN A. WOOD
Harvard-Smithsonian Center for Astrophysics

Chondritic meteorites contain abundant components—chondrules, Ca,Al-rich inclusions (CAIs)—that clearly were formed as dispersed objects in the solar nebula by high-temperature processes. There is no consensus as to the nature of these processes. In spite of this, several broad conclusions can be drawn about chondrite-forming processes in the nebula: (1) the chondrules and CAIs were formed in regions where the dust/gas ratio was orders of magnitude greater than the cosmic value; (2) after these objects were formed, they began to accrete very promptly (a time scale of years or maybe even hours); (3) formation and aggregation happened at the same radial distance in the nebula where chondrites reside today (the asteroid belt, ~3 AU); (4) the ambient temperature of the nebula at that time and place was relatively low (<700 K), and was not responsible for the thermal processing of the chondrules and CAIs. Some other, transient, high-energy events or processes formed them.

The substance of the terrestrial planets existed as interstellar dust grains before the solar system was formed. In the solar nebula, a series of events and processes transformed these grains into planetesimals. We would have little hope of ever understanding how this happened were it not for the remarkable fact that samples of only partly transformed planetary material have survived from that time, in the form of chondritic meteorites. A wealth of information is preserved in the fine structure of these meteorites, and in principle it should be possible to infer much about the properties and evolution of the nebula from a perceptive study of this fine structure. In practice, however, workers have placed a variety of interpretations on the data, and no consensus has been reached about the required nebular conditions.

TABLE I
Comparison of Selected Elemental Abundances (Atomic Ratios) in Principal Subtypes of Chondritic Meteorite[a]

	Fe/Si	Ca/Si	In/(10^7Si)	Tl/(10^7Si)	Bi/(10^7Si)	C/Si
C1	0.9	0.07	1	2	1.5	0.8
C2	0.8	0.065	0.5	0.9	0.7	0.4
CV3	0.75	0.095	0.3	0.4	0.5	0.15
CO3	0.85	0.07	0.2	0.5	0.4	0.05
H3-H6	0.8	0.05	0.002	0.007	0.01	~0
L3-L6	0.6	0.05	0.1	0.04	0.1	~0
LL3-LL6	0.55	0.05	0.04	0.007	0.06	~0
EH4-EH5	0.95	0.035	0.5	0.9	0.9	0.05
EL6	0.65	0.035	0.03	0.07	0.1	0.05
Solar	0.90	0.072	1.9	1.9	1.4	15

[a]Data from Anders et al. (1976), Ebihara et al. (1982), Holweger (1979), Keays et al. (1971), Laul et al. (1973), Sears et al. (1982), Wasson (1978), and Wolf et al. (1980).

I. CHONDRITE SUBTYPES

Chondritic material from roughly 2000 discrete falls (witnessed or unwitnessed) is preserved in the meteorite collections of the world. These samples are not uniform in character: nine major chondrite subtypes that differ in chemical composition, mineralogy, and rock texture have been recognized (Table I). Presumably they represent nine variations on the type of nebular processing that affected protoplanetary material.

Differences in the proportions of abundant, nonvolatile elements among the subtypes are small but real (Fe/Si, Ca/Si in Table I). Abundances of the more volatile elements (e.g. In, Tl, Bi) vary more widely. The depletion of these elements relative to the nebular (solar) abundance is attributed to high temperatures in the nebula, which prevented the elements from condensing onto (or which distilled them out of) the solid particles that later accreted into chondritic planetesimals. Differences in the depletion factors of volatile elements among chondrite subtypes stem from differences in the degree of thermal processing and details of the accretional history of the various subtypes. It is tempting to dwell on these chemical differences among chondrite subtypes, but it is much more important to appreciate how small they are. Compared with the diversity of compositions found among terrestrial and lunar rocks, the chondrites are remarkably uniform in composition and similar to the solar atmosphere in their abundances of condensable elements (Fig. 1).

Letter-number designations are used for the chondrite subtypes (Table I). The numbers in these indicate the degree of planetary metamorphism the chondrites experienced in planetesimals, after final accretion. 1 and 2 denote low-temperature aqueous metamorphism, while 3 through 6 indicate in-

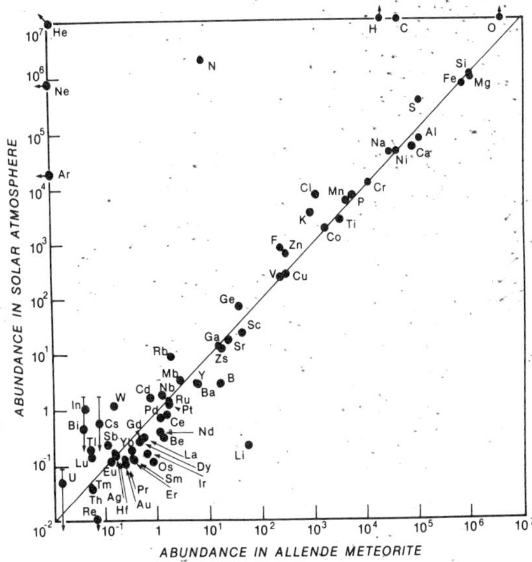

Fig. 1. A comparison of element abundances in the Allende CV3 chondrite and the solar atmosphere, both normalized to Si = 10^6. Only gaseous elements and molecules are underrepresented in Allende. Li has been depleted in the Sun by thermonuclear reactions (figure from Wood 1981*b*).

creasingly severe levels of high-temperature anhydrous metamorphism. This chapter is not concerned with the effects of planetary metamorphism, so the discussion will be confined to the least-metamorphosed categories of chondrites (types 1, 2, and 3). The subtypes with C in their titles are carbonaceous chondrites. Most of the carbon they contain is in the form of complex, high-molecular-weight, organic compounds.

II. FINE STRUCTURE OF CHONDRITES

Chondrites are not homogeneous rocks, but aggregations of smaller objects. These are clearly visible in thin sections of chondrites (Fig. 2). (Thin sections are rock slices ~30 μm thick, prepared by a process of sawing, grinding, and polishing, and mounted on glass slides. They can be illuminated by transmitted light and studied under a microscope.) Three principal categories of aggregated objects are present in the chondrites: chondrules, Ca,Al-rich inclusions (CAIs), and matrix. There is widespread agreement that these objects formed separately, dispersed in the nebula, then accreted into chondritic aggregations.

Chondrules

Chondrules (Figs. 2–4) are rounded to irregular masses of igneous rock. Temperatures of 1500 to 1900 K were required to melt them. Their range of

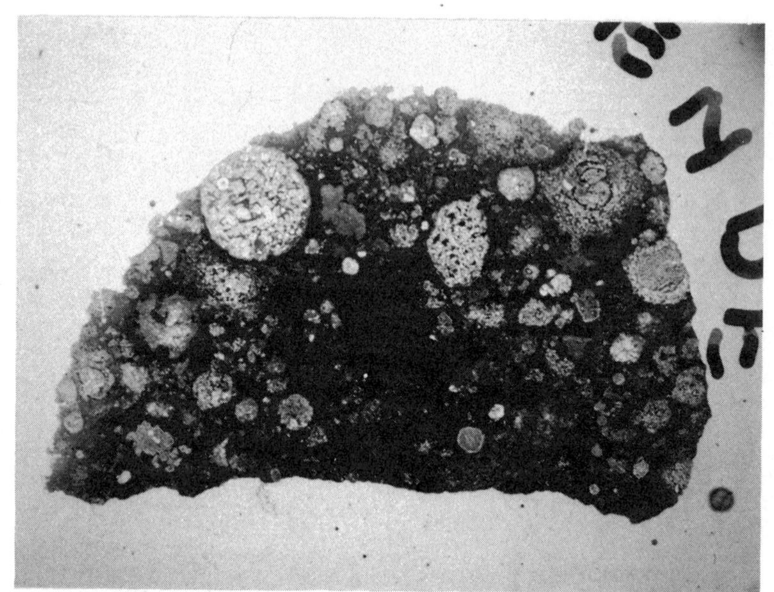

Fig. 2. Thin section of a 1-cm sample of the Allende C3V chondrite. Illumination is by transmitted light; bright areas are semitransparent, dark areas opaque. Chondrules (more or less rounded, transparent bodies) and matrix (the opaque material between them) are easily recognized. CAIs (dark-gray, irregular zones in the matrix) are less conspicuous.

dimensions is sharply limited; most chondrules have diameters between 0.3 and 1.5 mm. They are very abundant (up to 70% by volume) in some chondrite subtypes (H3, L3, LL3, CV3, CO3): these chondrites are essentially close-packed arrays of chondrules. C1 chondrites, on the other hand, the subtype that most closely reproduces the solar elemental abundance levels (Table I), contain no chondrules. Since they make up the bulk of many chondrites, chondrules must contain approximately solar proportions of the major, nonvolatile elements. An exception to this is Fe, which is present in approximately solar porportions in some chondrules but not in others. The principal minerals in chondrules are olivine, $(Fe,Mg)_2SiO_4$, and orthopyroxene, $(Fe,Mg)SiO_3$. Chondrules that occur side by side in the same chondrite display significant compositional differences, especially in their Fe/Mg ratios. There are systematic differences in the morphologies and mean dimensions of chondrules in the various chondrite subtypes (Figs. 3 and 4).

Ca,Al-rich Inclusions

CAIs have a specialized chemical composition: they are much enriched in the least volatile elements (Ca, Al, Ti, Ba, Sr, Sc, Y, rare earth elements, Zr, Hf, Th, V, Nb, Ta, Mo, W, U, Re, Ru, Os, Rh, Ir, Pt) and depleted in the more volatile elements, including the most abundant condensable elements

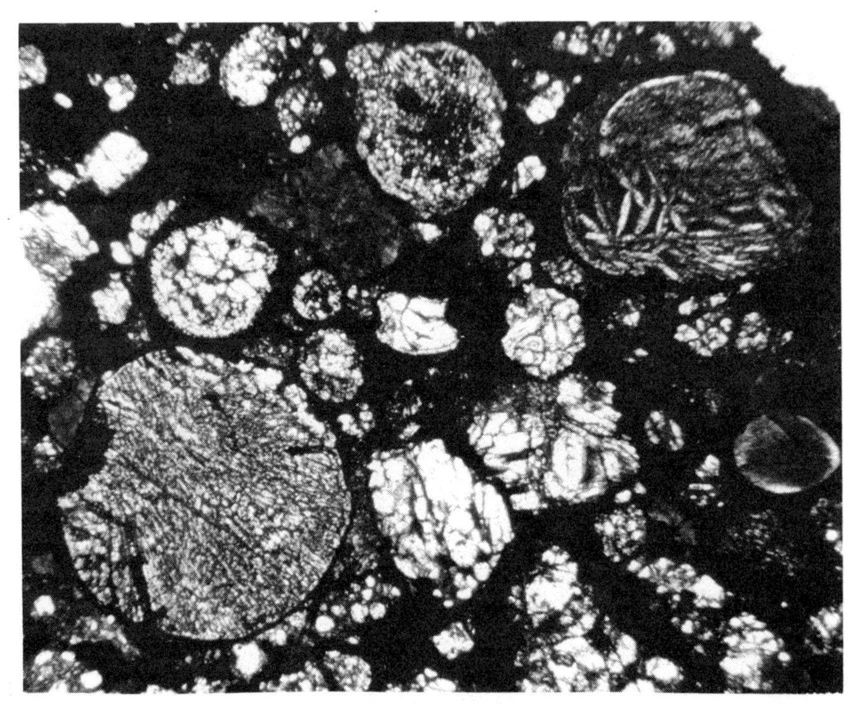

Fig. 3. Thin section of the H3 chondrite Tieschitz. Chondrules in the H3, L3, and LL3 chondrites are especially abundant and obviously droplet-shaped. Width of field is 4 mm.

(Si, Mg, Fe, S). The minerals contained in CAIs are an expression of their peculiar chemistry: examples are hibonite, $CaAl_{12}O_{19}$; perovskite, $CaTiO_3$; spinel, $MgAl_2O_4$; and melilite, $Ca_2(Al_2,MgSi)SiO_7$. The abundance of CAIs varies among chondrite subtypes. CV3 chondrites (Allende is the most notable example) contain ~10% by volume; C2 chondrites, ~1%; H3, L3, LL3, and C1 chondrites contain essentially none.

Two types of CAIs are named in the literature, coarse-grained CAIs and fine-grained CAIs, though actually these objects display a continuum of grain sizes. Coarse-grained CAIs range in diameter from 0 to ~2.5 cm, but most (by volume percent) such objects are near the large end of this range. Many coarse-grained CAIs have igneous textures and rounded shapes (Fig. 5), and were clearly melted globules (~1600 K) at one time. All coarse-grained CAIs may have had an igneous history, but this point is disputed.

Fine-grained CAIs make up a smaller volume fraction of their host chondrites than coarse-grained CAIs. Fine-grained CAIs range in dimension from 0 to ~1 cm; their abundances (volume percent) are fairly evenly distributed across this range. Fine-grained CAIs have a complex superfine structure. Many are themselves agglomerations of tiny (10–100 μm) bodies, each of which may have a concentrically banded pattern of minerals (Fig. 6). These

Fig. 4. A comparison of thin sections of CV3 (Vigarano, left) and CO3 (Warrenton, right) chondrites. The glass disks are one inch in diameter. Most of the aggregated objects are chondrules; some are CAIs. The character of chondrules and CAIs differs greatly between the two subtypes, yet the latter have almost identical overall chemical compositions. The chondrules also differ markedly from those in Tieschitz (Fig. 3), most being more tortuous in shape.

structures record a complicated history of distillation and/or condensation of elements, followed by aggregation into fine-grained CAIs and then, ultimately, aggregation of the CAIs, along with other material, into chondritic planetesimals.

Matrix

Matrix is the dark (opaque) material that fills in the space between chondrules and CAIs in little-metamorphosed chondrites (Figs. 2–4). It has approximately the same cosmic composition as bulk chondritic material. The carbonaceous chondrites' complement of organic carbon is sited in their matrices. Matrix makes up as little as 30% of the volume of the chondrites in which chondrules and CAIs are abundant; on the other hand, virtually 100% of the substance of C1 chondrites can be regarded as matrix.

Matrix consists of a more or less porous aggregation of very fine (<10 μm) mineral grains, the character of which differs among chondrite subtypes. In most cases it is unclear to what degree the matrix grains have been trans-

Fig. 5. A large, coarse-grained CAI from the Allende CV3 chondrite (diameter, 2.4 cm). Most of the crystals visible are melilite; anorthite, fassaite, and spinel are also present. The igneous texture and spheroidal shape make it clear that this object was once a molten globule (figure from R. S. Clarke et al. 1970).

formed by the mild metamorphism that has affected all the chondrite subtypes. (Because of their small grain size and large exposed surface area, matrix minerals would be more vulnerable to metamorphism than chondrule or CAI minerals.) There is wide agreement that either the matrix mineral grains or their pre-metamorphic precursor grains were once dispersed dust in the solar nebula.

CV3 matrix (Fig. 7) consists mostly of plates of Fe-rich olivine, along with smaller grains of olivine and other minerals. The matrices of H3, L3, and LL3 chondrites are basically similar, except that the plate morphology for olivine is missing. C1 and C2 matrices consist largely of hydrated, layer-lattice Mg-silicate minerals. In terrestrial rocks, such minerals are formed by the hydrous, low-temperature metamorphism of anhydrous precursor Mg-silicate minerals, such as olivine. In the C1 and C2 chondrites it is clear from textural evidence that some of the hydrated matrix minerals, at least, were also formed *in situ* by the alteration of an earlier generation of minerals (Bunch and Chang 1980). It is not clear that this is universally the case, however; it is possible that some of the hydrated mineral grains in these meteorites are primary, and existed in that form while dispersed in the nebula.

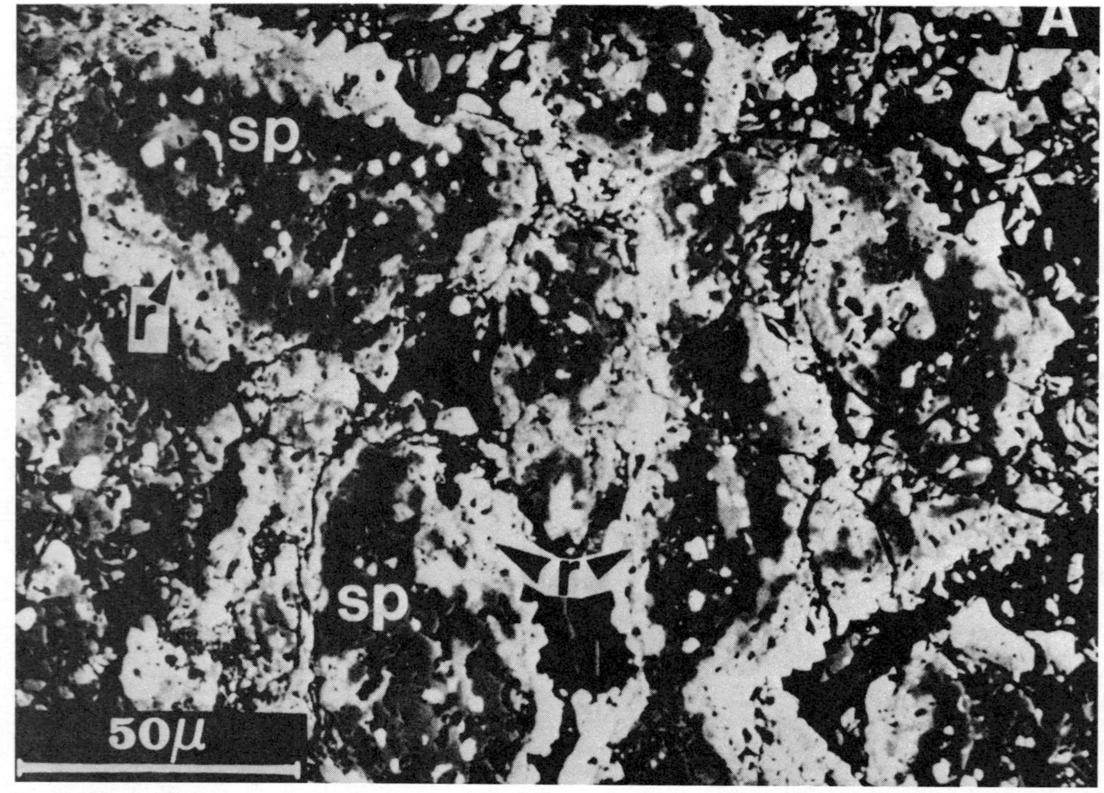

Fig. 6. Superfine structure in a fine-grained CAI. S E M backscattered electron (BSE) image of a polished section; bright areas correspond to high-molecular-weight minerals, which reflect impinging electrons more efficiently. The CAI is an aggregation of ~50 μm concentric objects, having spinel (*sp*) cores and banded rims (*r*) of pyroxenes and garnet. Bright spots in spinel cores are perovskite crystals. Figure courtesy of A. S. Kornacki.

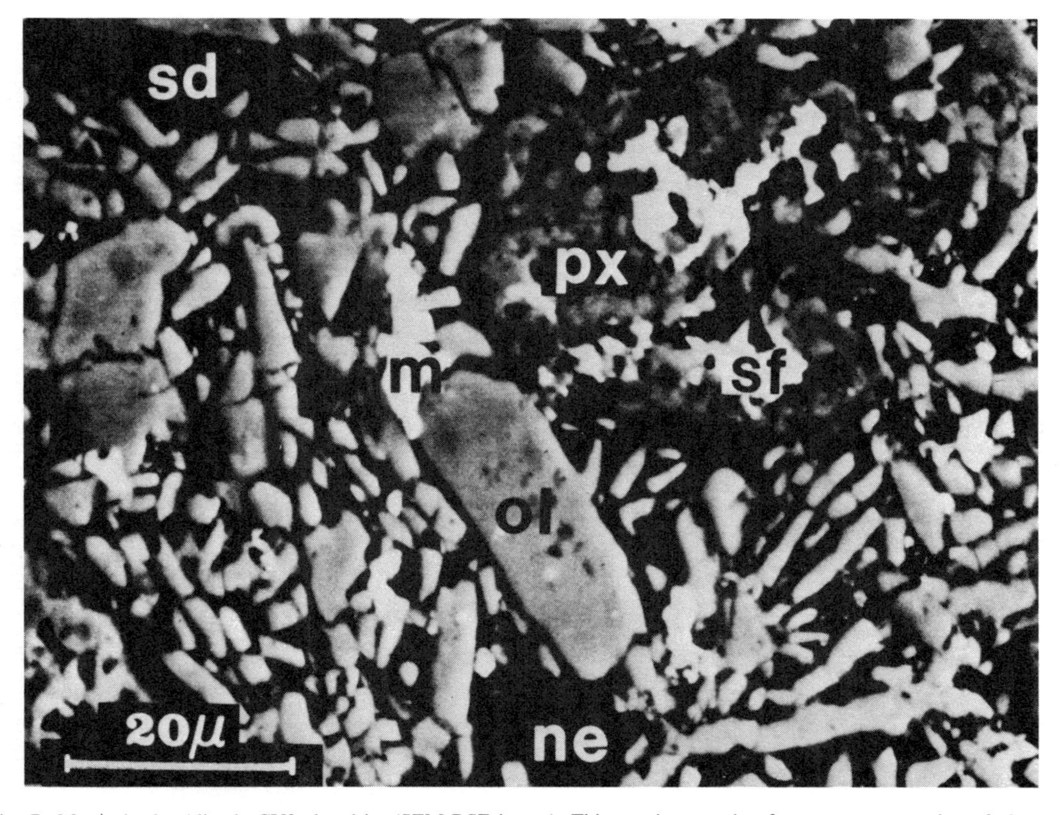

Fig. 7. Matrix in the Allende CV3 chondrite (SEM BSE image). This consists mostly of a porous aggregation of plates of Fe_2SiO_4-rich olivine (*ol*), but minor amounts of pyroxene px: (Fe, Mg, Ca)SiO_3; sulfide sf: (Fe,Ni)S; *m*: Fe, Ni metal; nepheline ne: $NaAlSiO_4$; and sodalite sd: $3NaAlSiO_4 \cdot NaCl$ are also present (figure from Kornacki and Wood 1984).

In all chondrite subtypes, the dust that was to become matrix appears to have formed at lower temperatures than the chondrules and CAIs that now accompany it. This is inferred from the matrices' high content of ferrous iron, sulfur, organic compounds, and (possibly primary) bound water, all of which are stable as condensed species in the nebula only at temperatures well below the melting range of chondrules and the stability range of CAI minerals.

III. ORIGIN OF THE STRUCTURAL COMPONENTS OF CHONDRITES

Chondrules and CAIs clearly were formed at high temperatures. These objects, at least, were affected by high-energy events or processes of some sort. We can imagine four plausible settings in which the chondrules, the CAIs, and the matrix grains might have developed the properties they now display.

1. Any or all of the chondrite components might represent surviving presolar interstellar grains;
2. They might be interstellar grains that have been more or less thermally processed in the nebula;
3. They might be condensates from the solar nebula (i.e., interstellar grains that were totally vaporized in the nebula, then recondensed);
4. They might be material that has been processed on or in early planetesimals (e.g., melted and dispersed by high-velocity impacts).

Table II shows a matrix of these four possibilities vs. the three structural components of chondrites. Almost every possible combination of origin and component has been proposed at one time or another. (Only selected references are given. These are by no means comprehensive bibliographies, nor do they always include the earliest reference to the concept.)

All four interpretations may be correct for matrix material. The preservation in carbonaceous chondrite matrix of noble gas components having presolar isotopic anomalies (D. C. Black 1972*b*; R. S. Lewis et al. 1979) shows that some intact presolar interstellar grains are present in the matrix. The severe thermal processing that affected the chondrules and CAIs undoubtedly would have produced thermally processed dust grains as well, which are unlikely to have been excluded from the chondritic aggregations. The high-temperature processing that produced CAIs must have involved the vaporization of metallic elements; these would have condensed as dust in the aftermath, and the evidence is clear that matrix grains were to some extent transformed by planetary hydrothermal metamorphism in the C1 and C2 chondrites.

The chondrule and CAI columns of Table II, however, are a battlefield. The modes of origin of these objects are vigorously debated. If a poll were taken of workers active in the field, a bare majority of them probably would be found to favor the formation of CAIs by condensation from a hot nebula,

TABLE II
Four Different Interpretations
Applied to the Three Principal Structural Components of Chondrites

	Chondrules	CAIs	Matrix
Interstellar material, more or less well-preserved	—	Wark (1979)	Cameron (1973*c*)
Interstellar material, processed (melted, distilled)	D. D. Clayton (1980), Wood (1984)	Kurat (1970), Chou et al. (1976), D. D. Clayton (1977)	Kornacki and Wood (1984)
Nebular condensate	Suess (1963), Wood (1963)	L. Grossman (1972, 1975)	Wood (1963), Anders (1964)
Planetary material, more or less processed	Fredriksson (1963), Urey (1967), Dodd (1971)	Meeker et al. (1983)	DuFresne and Anders (1962), Bunch and Chang (1980)

and the formation of chondrules by some form of nebular thermal processing of an earlier generation of accreted matrix material. (Let me stress that these are majority opinions, not necessarily correct interpretations, nor are they my preferred interpretations.) Curiously, most workers see the chondrules and CAIs as having been formed by two different, almost unrelated, processes rather than by a single process acting with different degrees of intensity.

IV. CONSTRAINTS ON THE FORMATION OF CHONDRITES

Although it may appear from the lack of consensus just noted that not much can be learned from chondrites about nebular processes, this is not the case. Four important constraints can be placed on the chondrite-forming processes, without knowing exactly what they were. These provide information about the nebula which until now has not been factored into our thinking about it.

A. *Gas/dust fractionation occurred in the presolar interstellar medium and/or the nebula and chondrules were formed in the zones of dust enrichment.* This can be concluded from the compositions of minerals in the chondrules. The equilibrium partitioning of Fe between olivine or pyroxene and coexisting metal alloy is controlled by the O/H ratio of the gaseous environment surrounding the minerals. In a gas with solar O/H, at 1500 K (which approximates the temperature range in which the chondrules were molten), olivine should contain only ~0.001 mole % Fe_2SiO_4 (Larimer

1968). The rest of the olivine should be Mg_2SiO_4, and any other Fe present in the condensed system should appear in a metallic alloy. In more oxidizing systems (higher O/H), correspondingly more of the Fe appears as Fe_2SiO_4 in the olivine and less as metallic allow. At O/H $\sim 10^6$ times the solar value, essentially all the system's Fe can occur in the olivine if the system has enough Si to make this much Fe_2SiO_4. At even high O/H ratios, a new and more highly oxidized Fe phase, Fe_3O_4, appears in the system. The higher the O/H is in this oxidation regime, the more Fe appears as Fe_3O_4 and the less is present in olivine.

None of the chondrite subtypes have chondrules that contain olivine with as little as 0.001 mole % Fe_2SiO_4, as they should if they equilibrated with a gas of solar composition. In most subtypes this parameter lies between 0.05 and 50. This could mean that the chondrules formed in environments with O/H greater than the solar value, or it could mean that the chondrules were formed by melting a highly oxidized, condensed, precursor material, and the melted droplets cooled and solidified before they had time to equilibrate completely with the reducing environment surrounding them. The latter interpretation may well be the main reason for the high Fe_2SiO_4 content of some chondrule olivines. The wide range of Fe_2SiO_4 mole percent observed among chondrules in a given chondrite can then be ascribed to differing degrees of reaction and equilibration with the gas phase.

Even in this disequilibrium situation, however, there is a way of inferring the gas composition. Figure 8 shows histograms of the Fe_2SiO_4 content of olivines found in random surveys of several C2 chondrites. Continued reaction of initially oxidized droplets with a reducing (low O/H) environment would have the effect of moving the olivine compositions to the left in the diagram. There are major peaks in the histograms at very low mole percent Fe_2SiO_4 values, which must represent the limit corresponding to essentially complete equilibrium between the droplets and their environment. Fredriksson and Keil (1964) place this limit at $\sim$0.5 mole percent Fe_2SiO_4 for one particular C2 chondrite; Wood (1967) reports a value of $\sim$1 mole %.

These olivine compositions would be at equilibrium with gases having O/H $\sim$100 times and $\sim$1000 times the solar value, respectively. The only feasible way in which O/H could have been enhanced in the nebula would have been a gas/dust fractionation followed by partial or total vaporization of the dust, which added its content of oxygen to the gas phase. The gas/dust fractionation factors invoked are not small. An even more extreme fractionation may be required to rationalize the mineralogy of chondrules in Allende and other CV3 chondrites, which contain Fe_3O_4 coexisting with olivine. Snellenberg (1978) and Haggerty and McMahon (1979) have noted that, taken at face value, this mineralogy requires a nine order of magnitude enhancement of O/H over the solar value.

B. *Chondrules, CAIs, and matrix began to accrete very promptly (probably less than an orbital period) after the chondrules and CAIs were formed.*

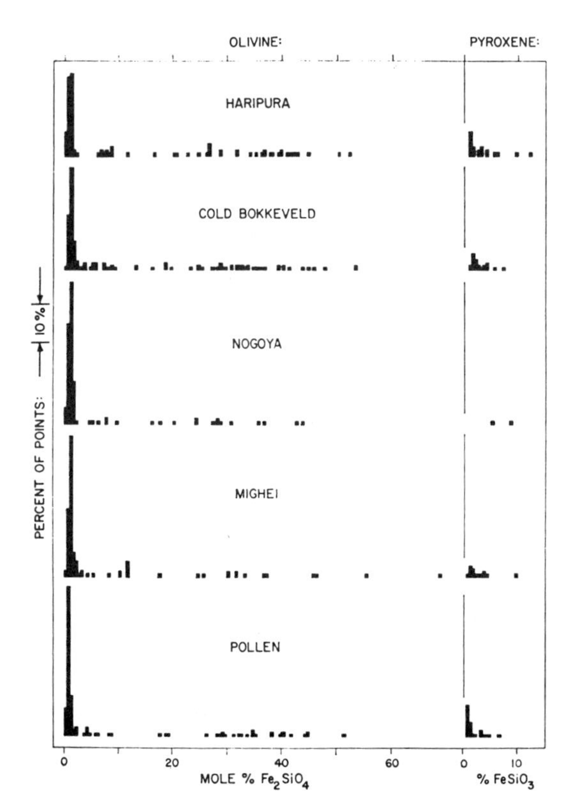

Fig. 8. Histograms of Fe_2SiO_4 content of olivine and pyroxene crystals, from random surveys of chondrules and dispersed mineral grains in five C2 chondrites, by Wood (1967).

This follows from the distinct morphological differences between chondrite subtypes (note especially the difference between CV3 and CO3 chondrites [Fig. 4], which have almost identical major-element chemical compositions). These must reflect minor variations, temporal or spatial, in the chondrule-forming mechanism. If the chondrules continued to orbit independently for thousands of years after they were formed, these morphologic populations would have been thoroughly mixed and the textural identities of chondrite subtypes would have been lost. The distinction between subtypes cannot have been created by physical fractionation processes in the nebula (e.g., an aerodynamic sorting effect that left large chondrules in one region, where they eventually became CV3 chondrites, and let smaller ones go elsewhere to join CO3 planetesimals), because such a fractionation would also create chemical differences between the products that are not observed, notably between the CV3 and CO3 subtypes.

This point can be made even more strongly by observing that there is a compositional complementarity between the structural components of some chondrites. Consider the C2 chondrite Murchison: Fe/Si (atomic ratio) in its overall composition is 0.81, close to the solar ratio (0.90), but Fe/Si in its matrix is 1.23 (McSween and Richardson 1977). Fuchs et al.'s (1973) survey of mineral compositions in Murchison indicates that Fe/Si in its nonmatrix material (chondrules, CAIs, isolated crystals) is ~0.20. Mass balance in this situation requires that Murchison contains ~41% nonmatrix material. The actual value is not known. A survey of one thin section made by the author yielded 31 ± 4 volume percent nonmatrix material. Conversion to weight percentages would bring this number closer to 41%, as the matrix is porous and underdense.

Thus, two dissimilar components, each with nonsolar Fe/Si, have accreted in the right proportions to bring the overall value of Fe/Si (as well as other element ratios) very nearly to the solar value. This cannot be accidental: the matrix and nonmatrix must have formed in a complementary way from a batch of material of solar composition; then both components were *comprehensively* accreted. Presumably Fe lost from chondrules by preferential volatilization during the chondrule-forming event quantitatively recondensed at lower temperatures into matrix dust. The point is that these two mechanically dissimilar components, ~1-mm chondrules and ~10-μm dust particles, could not remain dispersed in the nebula for any appreciable time, or experience any significant amount of spatial transport, without being aerodynamically sorted so that the net Fe/Si in any given batch of material was no longer solar. Accretion must have begun, at least, *very promptly* after the high-energy event or process that distributed Fe unevenly between matrix and nonmatrix. To stipulate that it began within an orbital period is very conservative. Within hours or minutes is probably closer to the mark. (This argument cannot be applied to most other chondrite subtypes. The chondrules of CV3, CO3, H3, L3, and LL3 chondrites contain much more Fe than do C2 chondrules, roughly as much Fe as their matrices do. Thus a physical fractionation of matrix from chondrules and CAIs would not produce a conspicuous change in Fe/Si, or any other major element ratios.)

C. *The high-energy events or processes that created chondrules and CAIs occurred at the radial distance of the present repository of meteorites, the asteroid belt (~3 AU).* Alternatively, if the mass of the protosolar system at that time was significantly different from its present value, these events occurred wherever the specific orbital angular momentum had the value it currently has in the asteroid belt. This follows from the previous constraint. The chondrite components could not be moved to the asteroid belt from elsewhere in the nebula prior to accretion without homogenizing textures and losing the solar elemental abundance pattern, and masses of accreted chondritic material larger than a few cm could not be moved by gas-dynamic effects in the thin gas near 3 AU.

D. *Midplane temperatures at 3 AU were relatively cool (< ~700 K), and since chondrules and CAIs were produced locally, the ambient gas tem-*

perature cannot have been responsible for their processing. This can be said because chondritic material, once it began to aggregate, would have gravitated to the midplane. The midplane must have been cool because some chondritic material, specifically layer-lattice silicates in the matrices of C1 and C2 chondrites, are unstable at high temperatures, tending to decompose into anhydrous silicates plus water vapor at more than 700 K. These meteorites, once formed, had to be stored in a relatively cool place.

An alternate explanation would seem to be that the chondritic aggregations grew so large (hundreds of meters) before they reached the hot nebular midplane that they insulated their interiors from high ambient gas temperatures for thousands of years. This would raise the intriguing possibility that the post-accretional high-temperature effects seen in most meteorites (metamorphism, melting and differentiation) resulted from the ambient gas temperature, not ^{26}Al decay or electrical heating induced by a T Tauri solar wind, and the planetesimals had inverted thermal gradients. However, this interpretation is very unlikely; it requires unrealistically swift and efficient accretion to build chondritic planetesimals to ~100 m dimensions in the year or so before their orbital motions cause them to cross the midplane.

Ambient gas temperature as a heat source for chondrule formation can also be rejected on other grounds. The nature of the crystalline textures in chondrules is controlled by, among other things, the rates at which the chondrules cooled. Laboratory experimentation has established that the meteoritic chondrules cooled through the temperature range of crystallization at very roughly 1 K s^{-1} (Hewins 1984). The scale of the nebula is too large by orders of magnitude for objects to be able to move from hot regions to cooler regions at this relatively rapid rate, even at sonic velocity.

V. CONCLUSIONS

Chondrules and CAIs were formed in the solar nebula at a radial distance of ~3 AU. The nebular gas was relatively cool at this time and place. Some type of transient high-energy event or process, operating in zones of dust enrichment, furnished the high temperatures needed to process the chondrules and CAIs. These objects cooled on a time scale of minutes, then on a similarly short time scale began to aggregate, together with dust (and possibly ice), into chondritic masses having approximately the solar proportions of major condensable elements.

In my opinion our difficulty in understanding the nature of the energetic chondrule- and CAI-forming processes stems not from a shortage of data or observational opportunities, but from the fact that the problem lies squarely on the boundary between two dissimilar disciplines, mineralogy and astrophysics. Mineralogists lack the background to understand the forces and processes in the nebular regime where their meteorites were formed; astrophysicists lack the background to evaluate the conflicting interpretations they hear from mineralogists, or to appreciate the reality of the constraints that meteoritic data place on nebular processes. Solution of the chondrite problem will require a degree of collaboration between the two disciplines that has not yet been attempted.

Acknowledgment. This work was supported in part by a grant from the National Aeronautics and Space Administration.

SHORT-LIVED NUCLEI IN THE EARLY SOLAR SYSTEM

G. J. WASSERBURG
California Institute of Technology

The status of the ongoing search for presently extinct or exceedingly rare radioactive nuclei in the early solar system is reviewed. In selected materials from meteorites, the nuclides ^{26}Mg, ^{107}Ag, ^{129}Xe and unshielded Xe isotopes in many meteorites are found to be in excess over typical values found in terrestrial, lunar or bulk meteoritic samples. The excesses of these nuclides indicated by asterisks are shown to be correlated with particular elements as follows: $^{26}Mg^*$:Al; $^{107}Ag^*$:Pd; $^{129}Xe^*$:I; Xe:U-Th-Sm-Nd and fission tracks. It is shown for many samples that $^{26}Mg^*/^{27}Al \approx 5 \times 10^{-5}$; $^{107}Ag^*/^{108}Pd \approx 2 \times 10^{-5}$; $^{129}Xe^*/^{127}I \approx 10 \times 10^{-5}$. The high degree of correlation of excess nuclides in coexisting minerals of variable chemical composition in meteorites indicates that the excess nuclides are due to the decay of a parent nuclide whose chemical characteristics are of the respective chemical species indicated. It is concluded that these isotopic enhancements are produced by the decay of ^{26}Al ($\bar{\tau} = 1.1 \times 10^6$ yr); ^{107}Pd ($\bar{\tau} = 9.4 \times 10^6$ yr) and ^{129}I ($\bar{\tau} = 23 \times 10^6$ yr). The source of the excess unshielded Xe isotopes in some samples is due to spontaneous fission of ^{244}Pu ($\bar{\tau} = 120 \times 10^6$ yr). These observations establish the presence of rather short-lived nuclear species in the matter from which the solar system formed. Assuming that these nuclei were produced by stellar nucleosynthesis, it is possible to establish a time scale for formation of the solar system from the interstellar medium and the time required to form small planets. The last time of major r-process contribution to solar matter is between 10^8 and 4×10^8 yr prior to formation of the solar system. A small late-stage injection of $\sim 10^{-4} M_\odot$ of freshly synthesized material is required to explain the short-lived nuclei. This late-stage nucleosynthesis must have taken place $\lesssim \sim 3 \times 10^6$ yr before solar system formation. A direct and rather immediate connection between the precursor interstellar medium and the early solar system is implied.*

I. INTRODUCTION

This chapter presents an overview of the existence and abundance of short-lived nuclei that were present in the early solar system. It summarizes the status of research in this area in a succinct tutorial fashion and does not represent an extensive scholarly survey of the field. The literature references are not meant to be complete but should be adequate to give the reader a selection of primary sources. As a complement to this review, the reader is urged to read the chapter by Cameron on star formation and extinct radioactivities for important views.

It is now known that the atomic weights (i.e., isotopic abundances) of many of the chemical elements in the solar system are not constant but somewhat variable. In most cases, the isotopic abundances of each of the chemical elements are only subject to minor variations in their proportions, but in some cases these shifts are very large. This is particularly evident in comparing meteoritic with lunar and terrestrial materials. Many of these variations are not attributable to small mass dependent fractionation. For small mass dependent fractionation, we would expect the shifts in isotopic abundance in a sample to be linear $(N_i/N_j) = (N_i/N_j)_o\,[1 + \alpha(i\text{-}j)]$; that is, the ratio of the number of isotopic nuclei of mass number i to those of mass number j would (relative to typical terrestrial values o) vary linearly for a sample in proportion to the difference in mass numbers $(i\text{-}j)$ by a fractionation factor α. Deviations from this rule are called nonlinear isotopic anomalies. These deviations would indicate the operation of other effects: either very large chemical fractionations, selective mass independent fractionation, or nuclear effects. Nuclear effects may include the modification of normal materials by particle bombardment or by the result of radioactive decay of a parent nucleus, or by varying proportions of nucleosynthetic components which go to make up the element.

It has been well known for three-quarters of a century that some large shifts are due to the decay of long-lived radioactive nuclei (e.g., Pb from U, Th). In the present discussion, we will pursue those variations that cannot be related to long-lived nuclei but may be due to rather short-lived nuclei. We will show that there are excesses of certain isotopes that may be assigned to the decay of short-lived radioactive nuclei corresponding to an abundance of about 10^{-4} relative to a neighboring stable isotope. This abundance is roughly similar to the level of general nonlinear isotopic anomalies present in meteoritic material (excluding oxygen) which are not attributable to short-lived nuclei (see the chapter by Clayton et al.). If these results are representative of the bulk solar system, they indicate that $\sim 10^{-4} M_{\odot}$ of freshly synthesized material of peculiar composition was added to the protosolar nebula as it was formed from the interstellar medium (ISM). This chapter will not address the problem of general isotopic anomalies (see Begemann 1980; Wasserburg et al. 1980).

A fully self-consistent model of the local abundances of the nuclear species should: (1) present the rates of nucleosynthesis over the history of the universe; (2) designate the particular processes and sites (nuclear and astrophysical) which produced the nuclide; and, (3) trace the participating matter from the initial state to include all parts that would contribute to our local sample.

In a particular parcel of matter we may write

$$\frac{dN_i(\mathbf{r},\tau)}{d\tau} = -\lambda_i N_i(\mathbf{r},\tau) + P_i \tag{1}$$

where N_i is the number of nuclei (i) at the location $\mathbf{r}$, $1/\lambda_i$ is the mean life and P_i is the local birth rate, destruction rate, injection etc.

Figure 1 is a cartoon showing P as a function of time. In this figure, the total duration of net production is T. This is followed by an interval Δ over which P is zero for the matter parcel which is to become the solar system. At the end of the interval Δ, objects of the solar system start forming. Subsequent to T, it is assumed that there is no further nucleosynthetic activity that affects material in the solar system except for radioactive decay. This assumes effects of galactic and solar cosmic rays and post-formation injection of extrasolar

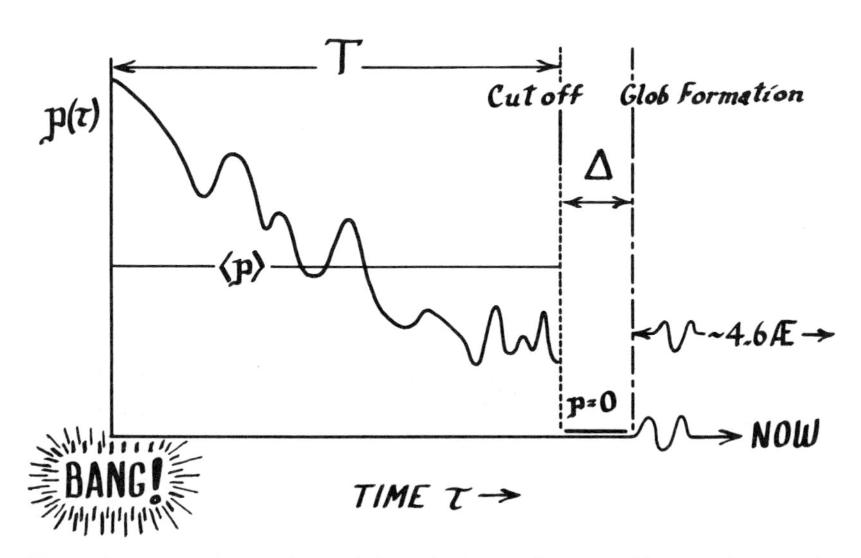

Fig. 1. Cartoon showing the characteristic production rate for some arbitrary nucleus over the age of the universe from Bang (at $\tau = 0$) to $\tau = T$ at which time the last synthesized material to be added to the solar system is made. Subsequent to this cutoff the production contributing to the solar system is taken as zero. The time interval of free decay prior to the formation of globs of identifiable matter making up the solar system is Δ. The total time scale is $T + \Delta + 4.6$ AE.

system material to be negligible. The number of i nuclei present at $T + \Delta$ may be written

$$N_i(T + \Delta) = \exp[-\lambda_i(T + \Delta)] \int_0^T \exp(\lambda_i\xi)\, P_i(\xi)\, d\xi \tag{2}$$

which for a stable nucleus (k) is $N_k(T + \Delta) = N_k(T) = \langle P_k \rangle\, T$ where the brackets refer to the average value. For a short-lived nuclide

$$N_j(T + \Delta) \approx P_j(T) \exp(-\lambda_j\Delta)/\lambda_j \tag{3}$$

As the various nuclear species are produced by different processes, it is not in general possible to relate the different production functions ($P_i(\tau)/P_j(\tau) \neq$ const).

Let us first consider the classical case of the two most abundant uranium isotopes which must be made in the same process. The mean life of ^{235}U is 1.0 AE and of ^{238}U is 6.5 AE (1 AE $= 10^9$ yr). The presennt abundance $N(^{235}U)/N(^{238}U) = 1/137.8$. At 4.6 AE ago $N(^{235}U)/N(^{238}U) = 1/3$. If these isotopes were produced in a single event with $P(^{235}U)/P(^{238}U) = 1.5$, then the time scale for T is 1.8 AE with a total time of $1.8 + 4.6 = 6.4$ AE. The essence of this calculation was first done by Rutherford (although he did not know the 4.6 AE number). If the production is uniform, then T becomes 8 AE with a total time of 12.6 AE. These rough calculations indicate the long galactic time scale for the production of the actinides (cf., Burbidge et al. 1957; Fowler and Hoyle 1960).

Our considerations here are to be directed to the short time scale Δ. If a pair of short-lived and stable nuclei were all formed in one event then for an observed ratio $N_j(T + \Delta)P_k/N_k(T+\Delta)\, P_j \equiv R_{jk}$, we have $\Delta_{jk} = -\,1/\lambda_j \ln[R_{jk}]$. For $R_{jk} = 10^{-4}$, $\Delta_{jk} \sim 9.2/\lambda_j$. This would give $\Delta \simeq 0.23$ AE for $1/\lambda = 24.5 \times 10^6$ yr and $\Delta = 0.01$ AE for $1/\lambda = 10^6$ yr. It follows that the existence of a short-lived nuclide in matter participating in the formation of the solar system will require a rather short time interval Δ between the termination of nucleosynthetic activity and solar system formation.

A more reasonable treatment of this problem is to consider production over a long time scale so that $N_j(T)/N_k(T) \cong P_j(T)/\langle P_k(T) \rangle\, \lambda_j T$. This will yield a lower value for Δ_{jk} than calculated above for sudden synthesis. The problem of establishing the ratio of the short-term production rate $P_j(T)$ for the short-lived nuclide j in the neighborhood of T relative to the long-time average $\langle P_k(T) \rangle$ for the stable nuclide k must be resolved in each case, as the processes responsible for production may involve very different astrophysical sites and nuclear mechanisms. In either case, *the strictest time limit comes from the nuclide with the shortest mean life* assuming that its presence can be established. This time scale is related to the particular nucleosynthetic process associated with the particular nuclide. [For a more complete discussion see Schramm and Wasserburg (1970) and Schramm (1982) and references therein.

TABLE I
Radioactive Isotopes with $\bar{\tau} > 10^4$ yr

Score[a]	Parent[b]	Daughter[c]	$\bar{\tau}$ (in 10^6 yr)	Score[a]	Parent[b]	Daughter[c]	$\bar{\tau}$ (in 10^6 yr)
	^{10}Be	^{10}B	2.3	+	^{107}Pd	^{107}Ag	9.4
+	^{26}Al	^{26}Mg	1.1	+	^{129}I	^{129}Xe	23
	^{36}Cl	^{36}Ar	0.4	?	^{135}Cs	^{135}Ba	3.3
−	^{41}Ca	^{41}K	0.19		^{137}La	^{137}Ba	~0.09
	^{53}Mn	^{53}Cr	5.5	?	^{146}Sm	^{142}Nd	149
	^{60}Fe[d]	^{60}Ni	~1		^{150}Gd	(^{146}Sm)	2.6
	^{59}Ni[d]	^{59}Co	0.1		^{182}Hf	^{182}W	13
	^{79}Se	^{79}Br	≤0.09		^{202}Pb[d]	^{202}Hg	0.8
	^{81}Kr	^{81}Br	0.3		^{205}Pb	^{205}Tl	20
	^{93}Zr	^{93}Nb	2.2		^{239}Pu	α,SF	0.03
	^{92}Nb	^{92}Zr	231		^{242}Pu	α,SF	0.5
	^{99}Tc	^{99}Ru	0.3	+	^{244}Pu	α,SF	120
	^{98}Tc	^{98}Ru	6.1	−	^{247}Cm	(^{235}U)	22.5
	^{97}Tc	^{97}Mo	3.8		^{248}Cm	(^{244}Pu)	0.5

[a](+) Positive identification; (?) positive identification or hint; (−) strict upper limit.
[b]Parents which produce a daughter element with only one isotope, or part of the ^{235}U, ^{238}U, ^{232}Th chain, are not listed.
[c]( ) Radioactive intermediate daughter.
[d]^{60}Fe revised after Kutschera et al. (1984); see also search by Hinton et al. (1984); revised mean lives of ^{59}Ni and ^{202}Pb from Nishiizumi et al. (1981) and Nagai et al. (1981).

A presentation of some models which consider aspects of the chemical evolution of the Galaxy may be found in Clayton (1983,1984).]

Certainly all nuclei originally present 4.6 AE ago with mean lives < 35 × 10^6 yr will be extinct today. (This corresponds to the decay of all particles in 1 $M_\odot$ over a time of 4.6 AE.) In order to obtain the strictest limit on Δ, attention must be devoted to nuclei with rather short life times. Table I is a list of radioactive nuclei of pertinence, their mean life and the stable daughter nucleus to which they decay. The current status of searches for each species is given. The only cases that may be considered to be clear cut in terms of their existence in the early solar system are ^{26}Al, ^{107}Pd, ^{129}I and ^{244}Pu and are indicated by (+). Species for which there is a positive indication or even a strong hint are indicated by (?). Those for which a strict and meaningful upper limit is established are indicated by (−).

In order to demonstrate the presence in the early solar system of a nuclide which is presently extinct, the following steps are necessary:

1. Identify objects which were produced in the solar system and of sufficient antiquity;
2. Establish that subsequent to their formation, the objects have not been subject to significant late modification which would seriously redistribute the constituent elements and nuclei;

3. Establish that there exist excesses of the daughter nuclide;
4. Show that the excesses of the daughter nuclide correlate with the chemical behavior of the chemical species of the parent. If the chemically different phases in an object were crystallized contemporaneously from an isotopically homogeneous mass, then the correlation of the daughter nuclide with the chemistry of the parent element implies the *in situ* decay of the parent nucleus in the object. If this crystallization took place within the solar system, then the parent nuclide had to be present in the solar system.

To study very old solar system objects that are relatively well preserved focuses most studies on meteorites. It should be remembered that most stony meteorites are aggregates of many distinct smaller rock fragments, spherules (chondrules), crystals and of microscopic and submicroscopic particles. Each of these objects may possibly have different histories, some of which may precede the formation of planetary bodies. Many of the fragments and chondrules show the results of crystallization from silicate liquids on a small scale. The iron meteorites are different in that they were produced by partial melting of small planetary objects. Because the original abundance of the presently extinct nuclide may be very small, it is almost always necessary to find objects which contain very high concentrations of the chemical element associated with the parent and with very low concentrations of the element associated with the daughter. Otherwise the enrichment in the daughter isotopic abundance above the "normal" value will be below detection. It must also be shown that the total exposure of the object under study to cosmic ray particles is insufficient to produce the effects observed. The age of the solar system is $\sim 1.4 \times 10^{17}$ s, so the total cosmic ray proton fluence for a small object is $(\sigma\phi\tau) \sim 3 \times 10^{-7}$ for a one barn cross section. As a result, some substantial isotopic shifts are in fact produced by cosmic rays.

The importance of an old undisturbed object and of adequate enrichment of parent relative to daughter element is illustrated by the following example. Consider two crystals each of 1 g, one made of normal magnesium and one of normal aluminum but also containing ^{26}Al in the abundance ^{26}Al/^{27}Al $=$ 10^{-4}. Then the total Mg produced in the aluminum cube after the decay of the ^{26}Al is $\sim 10^{-4}$g of pure ^{26}Mg but the total ^{26}Mg present in the magnesium cube is $\sim 10^{-1}$g. For the bulk system, the total fractional increase in ^{26}Mg is only 10^{-3} which is just measurable. Note that the Al/Mg chosen for this bulk system is ten times the average solar system value (see Cameron 1982). Thus if the Mg isotopes were caused to mix or equilibrate in a high Mg environment, there would not be a detectable effect.

The demonstration that the daughter nuclide behaves as if it had the chemistry of the parent element is usually shown as follows. Consider a set of objects formed at the same time with a uniform initial isotopic composition o but each object different from one another in having different relative abundances of the elements. Again using the ^{26}Al example, if ^{26}Mg* is the excess of ^{26}Mg, the isotopic composition in object m subsequent to the decay of ^{26}Al is given by

$$\begin{aligned}(^{26}\mathrm{Mg}/^{24}\mathrm{Mg})_m &= (^{26}\mathrm{Mg}/^{24}\mathrm{Mg})_o + (^{26}\mathrm{Mg}^*/^{24}\mathrm{Mg})_m \\ &= (^{26}\mathrm{Mg}/^{24}\mathrm{Mg})_o + (^{26}\mathrm{Mg}^*/^{27}\mathrm{Al})_m\,(^{27}\mathrm{Al}/^{24}\mathrm{Mg})_m \\ &= (^{26}\mathrm{Mg}/^{24}\mathrm{Mg})_o + (^{26}\mathrm{Al}/^{27}\mathrm{Al})_o\,(^{27}\mathrm{Al}/^{24}\mathrm{Mg})_m. \end{aligned} \quad (4)$$

This relationship shows that the Mg isotopic ratio would be a linear function of the ^{27}Al/^{24}Mg ratio in the array of systems m described above with a slope of $(^{26}\mathrm{Al}/^{27}\mathrm{Al})_o$ and an initial Mg isotopic composition given by the intercept when $^{27}\mathrm{Al}/^{24}\mathrm{Mg} = 0$ (see Fig. 2). The initial value of $(^{26}\mathrm{Mg}/^{24}\mathrm{Mg})_o$ is of great importance in inferring the evolutionary history. For the case of ^{26}Al, the percentage enrichment p in ^{26}Mg/^{24}Mg is $p = 908(\mathrm{Al/Mg})\,(^{26}\mathrm{Al}/^{27}\mathrm{Al})$. For the solar ratio of $(\mathrm{Al/Mg})_o \sim 0.1$ and $^{26}\mathrm{Al}/^{27}\mathrm{Al} \sim 10^{-4}$, we obtain $p \sim 10^{-2}\%$. It is evident that substantial enrichments of Al/Mg relative to solar, must be obtained in order to observe effects of even 1%. The same considerations outlined above pertain to all the extinct nuclides.

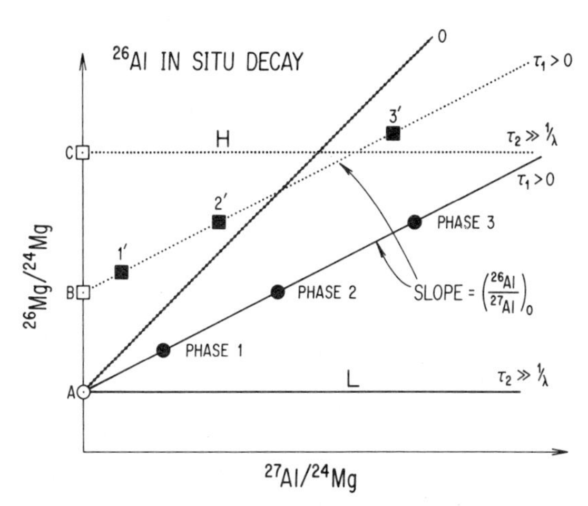

Fig. 2. Isotopic systematics of extinct parent-daughter systems for a sample crystallized from an isotopically homogeneous initial state. The example shown is for ^{26}Al but applies to all other systems. A series of phases (1,2,3) which formed at time τ_1 with the same initial Mg isotopic composition A and with $(^{26}\mathrm{Al}/^{27}\mathrm{Al})_o$ would, after the complete decay of ^{26}Al, yield the linear array passing through A and show a direct correlation of ^{26}Mg/^{24}Mg with ^{27}Al/^{24}Mg. A similar system formed at $\tau = 0$ (τ_1 yr earlier) from the same material would define the line marked 0 and have a higher slope $[(^{26}\mathrm{Al}/^{27}\mathrm{Al})_o \exp(\lambda\tau_1)]$. Material formed at τ_1 from a reservoir with a high ratio H of ^{27}Al/^{24}Mg and originating at $\tau = 0$ would have a high initial value of $(^{26}\mathrm{Mg}/^{24}\mathrm{Mg})_o$ due to ^{26}Al decay in the interval 0 to τ_1 (point B). If an isolated reservoir contains ^{26}Al and has a sufficiently high ratio of Al/Mg, then the isotopic composition of the Mg will be shifted due to ^{26}Al decay (horizontal line H passing through point C). For crystals formed at a time τ_2 from such a reservoir ($\tau_2 >> 1/\lambda$). All these effects (and more complex ones) have been observed for the ^{26}Al-^{26}Mg and ^{107}Pd-^{107}Ag systems (figure after Lee et al. 1977).

As pointed out by D. D. Clayton (1975, 1979) the existence of isotopic enrichments in phases rich in the parent element is not sufficient to demonstrate *in situ* decay within the solar system as this could result from decay of the parent nuclide in dust grains in the interstellar medium well before any stages of solar system formation occurred. In the case that the decay took place in interstellar dust grains, there is no necessary temporal connection with the solar system and the interval Δ has a very different meaning. In order to eliminate this possibility, it is necessary to establish that the correlation discussed above is in material that was formed in the solar system. More broadly, two questions arise: What materials or objects in the solar system contain preserved grains or small aggregates of grains of interstellar origin? How can these grains be identified and analyzed? This is an area of intensive and difficult research. Indeed, D. D. Clayton (1982) has presented interpretations which attribute many of the observations discussed here to result from admixtures of interstellar nuggets and grains in larger solar system objects. With regard to the general isotopic anomalies (not due to radioactive decay), Consolmagno and Cameron (1980) explain them as resulting from incomplete mixing of the "standard" nucleosynthetic components which make up the solar system. In this chapter, we discuss radioactive species and their daughter products. For most of the cases presented in this review, we consider the crystals containing the peculiar nuclei to be the product of processing and crystallization within the solar system. The fractionation processes are considered to have separated the elements into gas, liquid and solid phases from macroscopic parcels of preexisting heterogeneous matter within the solar system. For these objects, we consider that the interstellar gas and dust which went to make them up had been locally homogenized at least on a local scale ($\gtrsim 10^{-3}$ g of condensed matter). As a result, preexisting correlations between parent and daughter nuclei in interstellar grains would be destroyed and new ones produced by chemical and physical fractionation within the solar system. The bulk object may of course still display an anomalous isotopic pattern for many chemical species. The individual phases within the object will be isotopically uniform and not vary isotopically from phase to phase except for changes resulting from subsequent radioactive decay. There is still the real possibility that interstellar dust grains are preserved in meteorites and will someday be clearly identified. However, we do not consider the larger-scale, well-crystallized objects (0.1 cm to 100 cm) discussed herein to be preserved presolar bodies. The original proposal by D. D. Clayton (1975*a*) which sought to assign the observed correlated daughter products of ^{129}I and ^{244}Pu to decay in the interstellar medium and not in the solar system was not and is not in accord with the observations according to this author's view. The extent to which this viewpoint can plausibly explain the more complete observations on the radioactive-daughter systems reviewed herein should be judged by the reader.

II. THE POSITIVE CASES

^{129}I

The nuclide ^{129}I ($\bar{\tau} = 23 \times 10^6$ yr) is inferred to have been present in the early solar system from a correlation of ^{129}Xe with ^{127}I (see Fig. 3). This correlation was shown for macroscopic meteorite samples by Jeffery and Reynolds (1961). These meteorites are aggregates of many small grains possibly of very different origin. The actual sites or phases within these meteorites which contain the ^{129}Xe atoms are not established but they are obviously

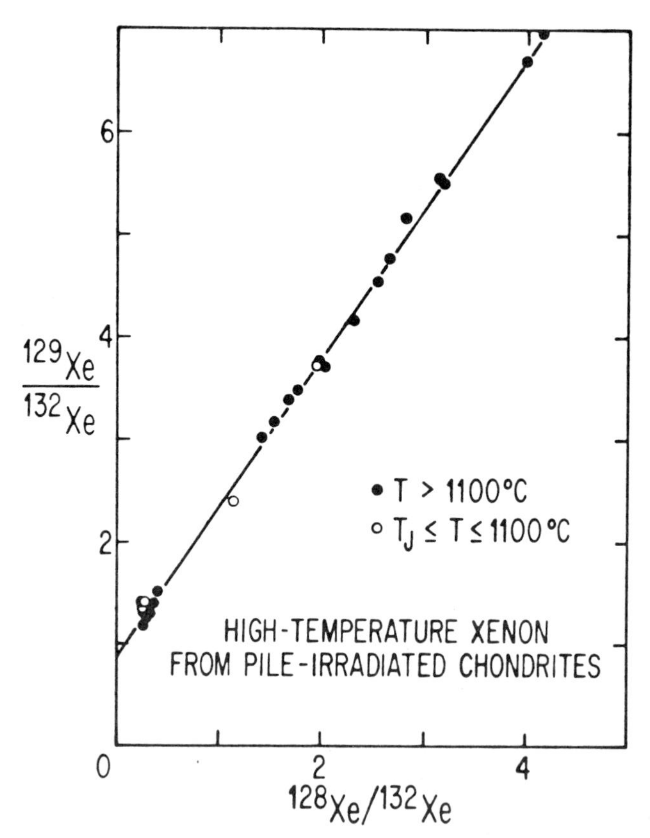

Fig. 3. Data on some chondritic meteorites that have been irradiated with neutrons in the laboratory to produce ^{128}Xe from the ^{127}I naturally present in the sample. The other Xe isotopes are almost completely unaffected by the irradiation. $(^{129}\text{Xe}/^{132}\text{Xe})_T = (^{129}\text{Xe}/^{132}\text{Xe})_o + (^{129}\text{I}/^{132}\text{Xe})_T = (^{129}\text{Xe}/^{132}\text{Xe})_o + K(^{129}\text{I}/^{127}\text{I})_o[(^{128}\text{Xe}/^{132}\text{Xe})_T - (^{128}\text{Xe}/^{132}\text{Xe})_o]$; K is a constant of the irradiation. The observed correlation shows that the excess ^{129}Xe released is associated with ^{128}Xe (produced by neutron capture on ^{127}I in the pile irradiating) during the laboratory heating at temperature T. This method does not require physical separation of microscopic I-rich and I-poor phases. The value of $^{129}\text{Xe}^*/^{127}\text{I} = ^{129}\text{I}/^{127}\text{I} \cong 10^{-4}$ (figure after Jeffery and Reynolds 1961 and Hohenberg et al. 1967).

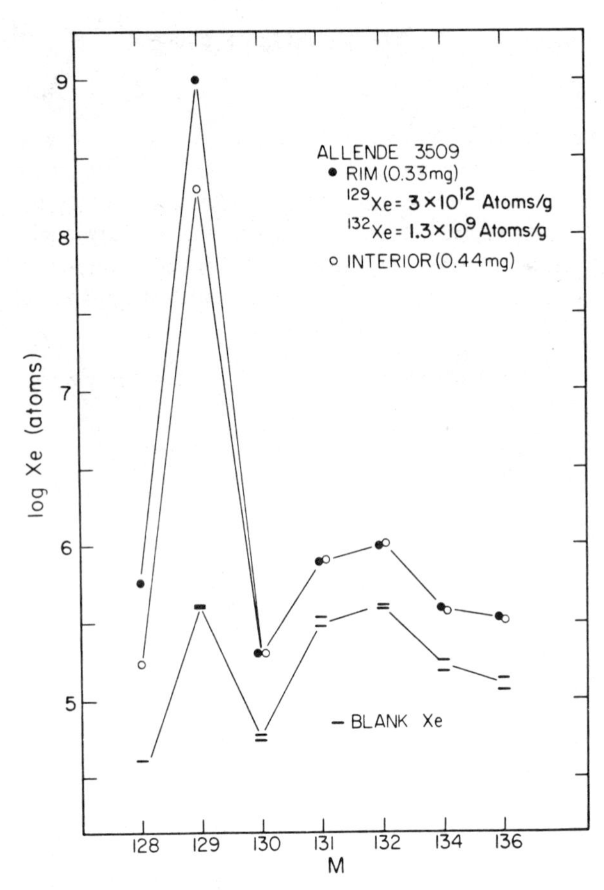

Fig. 4. Isotopic composition of Xe from the rim and interior of Allende inclusion 3509. Note that the ordinate is logarithmic. The sample was selected because it was already known to contain halogen-rich material (see Fig. 5). It contains almost pure ^{129}Xe; the other six isotopes are present at levels just somewhat above the procedural blank (Wasserburg and Huneke 1979).

correlated with I. It is known that there are halogen-rich phases present in these meteorites. A clear example of xenon consisting of essentially pure ^{129}Xe has been found in halogen-rich crystals distributed through a 1 cm spheroid in the Allende meteorite (see Figs. 4 and 5). The textures exhibited by the different phases throughout this spheroid indicate crystallization from a melt. This sample was selected for study because it was evidently rich in halogens based on the mineralogy and chemistry (Mason and Taylor 1982). Table II shows a chemical analysis of this inclusion. As can be seen from the table, the inclusion is enriched in Cl far above the typical matrix of the meteorite in which it was found. The Cl/Br ratio measured (not shown) is approximately typical for carbonaceous meteorites (it is in solar proportions) and

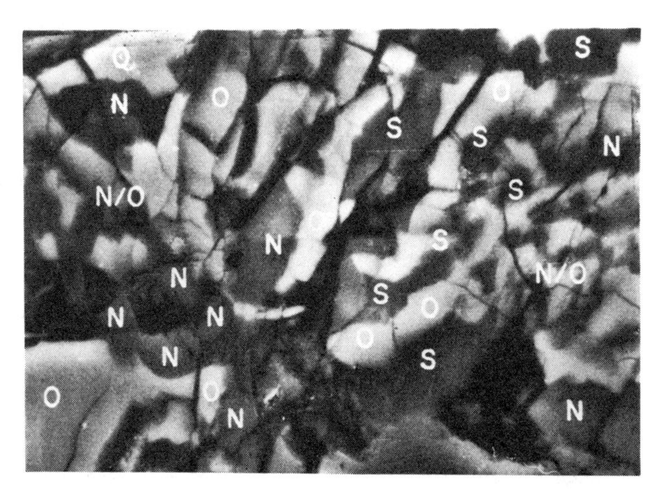

Fig. 5. Backscattered electron image of a 1-cm spherical inclusion 3509 from Allende (white bar is a 10-μm scale). Labels are: O–olivine $(Fe,Mg)_2SiO_4$; N–nepheline $NaAlSiO_4$; and S–sodalite $(Na_8(AlSiO_4)_6Cl_2$. The intimately intergrown phases appear to have crystallized from a melt. This halogen-rich material is not uniformly distributed throughout the inclusion (see Table II) but shows $^{129}Xe^*/^{127}I = 1.0 \times 10^{-4}$ for samples from both the rim and interior (see Fig. 4).

implies high I content. This is confirmed by an independent iodine analysis. The sample was then analyzed for Xe and found to contain nearly pure ^{129}Xe and gave $^{129}Xe/^{127}I \approx 1.0 \times 10^{-4}$. This simple experiment demonstrates that for ancient objects, if you select ones rich in halogens, they will contain almost pure $^{129}Xe^*$ (the excess ^{129}Xe is represented by $^{129}Xe^*$). This parent-daughter scheme is optimal because of the enormous chemical fractionation between Xe and I that takes place upon formation of solids. This correlation of $^{129}Xe^*$ with I has been found for a diverse variety of meteorites (see Podosek 1970; Niemeyer 1980; Hohenberg et al. 1981). ^{129}I was therefore widely distributed throughout all condensed solar system matter. [Some $^{129}Xe^*$ is even found in relatively young terrestrial and lunar materials. The

TABLE II
Major Chemical Abundances (%) in Allende Inclusion 3509[a]

Sample	Al_2O_3	FeO	MgO	CaO	Na_2O	K_2O	Cl
Interior	17.3	8.10	13.8	6.8	10.1	0.545	0.895
Rim	18.2	7.00	16.2	4.1	10.5	1.20	1.02
Meteorite Matrix	2.82	36.0	22.4	2.1	0.35	0.027	0.031

[a]After J. C. Laul, personal communication.

latter $^{129}Xe^*$ is naturally not correlated with I in the young material but presumably is trapped after its production early in the history of these planets or the debris accumulated to make the planets.] The ratio of $^{129}Xe^*$ to ^{127}I in many meteorites is $^{129}Xe^*/^{127}I \approx 1.0 \times 10^{-4}$ although some variations exist. These results lead to a firm value of $^{129}I/^{127}I \approx 10^{-4}$ in the early solar system. Until the past decade, this was the only truly short-lived nucleus that was known to have been present. Taking a production ratio $P(^{129}I)/P(^{127}I) \sim 1$, the above ratio gave $\Delta \sim 1.6 \times 10^8$ yr and indicated a long interval between the last nucleosynthetic event and the formation of the solar system. However, remember that the value of 10^{-4} may result from a long interval Δ or the addition of only a small amount of freshly synthesized material to old material.

^{26}Al

The nuclide ^{26}Al ($\bar{\tau} = 1.1 \times 10^6$ yr) is inferred to have been present in the early solar system from a correlation of ^{26}Mg with ^{27}Al (see Figs. 6, 7 and 8). All of these samples have $^{25}Mg/^{24}Mg$ the same as the terrestrial value to within a few parts per thousand, thus showing that the high $^{26}Mg/^{24}Mg$ ratios are actually due to the addition of $^{26}Mg^*$ to ambient normal magnesium. Samples of Al-rich/Mg-poor material (with Al/Mg $\sim 10^3$ times solar) have been found to contain magnesium with $^{26}Mg/^{24}Mg = 0.1710$ as compared

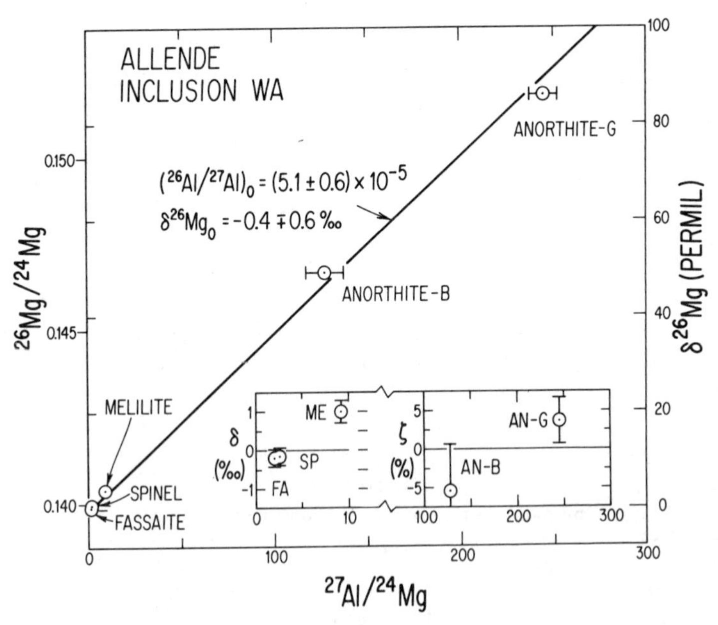

Fig. 6. Correlation of $^{26}Mg/^{24}Mg$ with $^{27}Al/^{24}Mg$ for Ca,Al-rich inclusion WA from Allende. Data for three different phases with distinctive chemical compositions are shown. Inset shows the quality of the fit relative to the line. This shows an excellent correlation of $^{26}Mg/^{24}Mg$ with Al/Mg implying that $^{26}Mg^*/^{27}Al$ is constant for the different phases (figure after Lee et al. 1977).

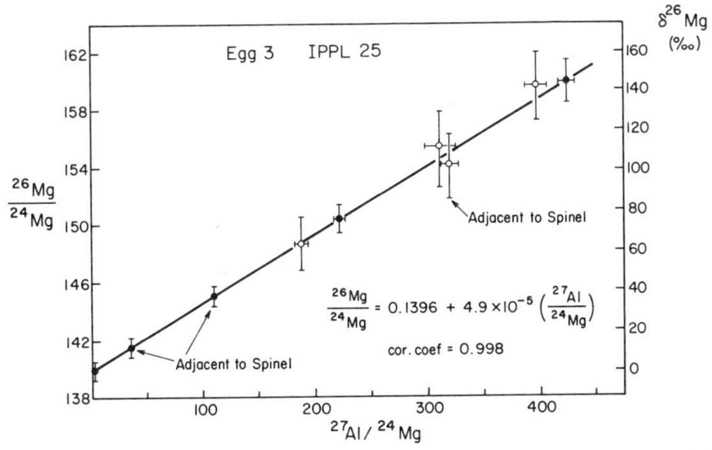

Fig. 7. Correlation of the ^{26}Mg isotopic abundance with $^{27}Al/^{24}Mg$ for the Ca,Al-rich inclusion Egg 3 from Allende. This spheroidal inclusion is about 1 cm in radius. Analyses are shown for plagioclase crystals ($CaAl_2Si_2O_8$) and spinel crystals ($MgAl_2O_4$) in different geometrical relationships. Shifts in $^{26}Mg/^{24}Mg$ in parts per thousand are indicated on the right-hand scale (figure after Armstrong et al. 1984).

with the terrestrial or normal meteoritic value of 0.139805. This corresponds to a 22% increase in the abundance of ^{26}Mg. The slope of the correlation line in Fig. 6, Fig. 7 and the upper line in Fig. 8 is $^{26}Mg^*/^{27}Al = 5 \times 10^{-5}$. These data are on crystals of coexisting mineral phases (up to $\sim$ 1 mm) from an individual fragment (2 cm in diameter). [The inclusions containing crystals with $^{26}Mg^*$ are often microscopic (< 100 μm) and are not always as large as in the examples given in the figures.] The densely interlocked crystals of these inclusions [WA, Egg-3, and USNM 3529-26] are considered to have crystallized from a silicate liquid at elevated temperatures. With the exception of the oxygen isotopic composition, which is unusual and locally heterogeneous, the other chemical elements in themselves have essentially normal (or uniform) isotopic compositions and are almost indistinguishable from terrestrial values. From these observations, it is concluded that the $^{26}Mg^*$ is correlated with Al in different phases and is thus produced from the *in situ* decay of ^{26}Al in the individual associated crystals with a uniform initial abundance $^{26}Al/^{27}Al = 5 \times 10^{-5}$.

A search has been made to establish the level of homogeneity of $^{26}Mg^*$ in crystals of $CaAl_2Si_2O_8$ from inclusion WA of Allende. While variations of ^{24}Mg concentration were observed, the ratio of $^{26}Mg^*/Al$ was constant for volumes down to 10^{-13} cm^3 (Huneke et al. 1982). It should be noted that while the abundance of $^{26}Al/^{27}Al \approx 5 \times 10^{-5}$ is typical for many inclusions, there are several other meteorite samples that show lower ratios, others that do not have well-defined correlation lines and some with initial values $(^{26}Mg/^{24}Mg)_o$ above or below the terrestrial value (see Wasserburg and Papanastassiou 1982;

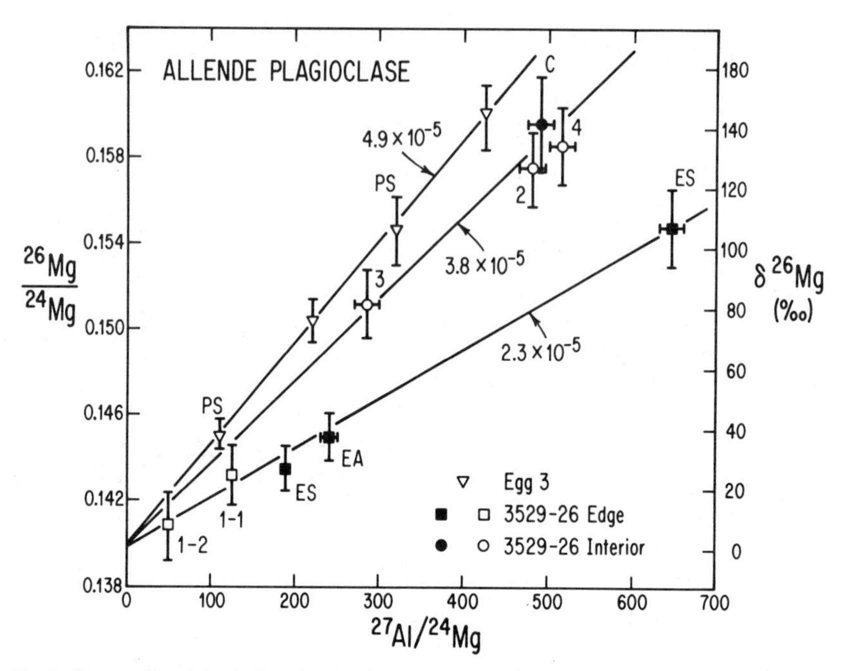

Fig. 8. Data on Egg-3 (as in Fig. 7) (triangles) and on plagioclase crystals from another inclusion (3529-26) (circles and squares) in different geometrical relationships. Note the lack of precise correlation in the latter inclusion. This is plausibly due to migration of Mg and other elements caused by metamorphism (Armstrong et al. 1984).

Hutcheon 1982; Papanastassiou et al. 1977). In addition, there are other samples that are also believed to be very ancient which contain Al-rich/Mg-poor phases which do not exhibit any $^{26}Mg^*$ (Lee et al. 1979). It follows that there are some complexities in the ^{26}Al-^{26}Mg system. Some of these are reasonably explainable in terms of later recrystallization of the phases over their history (Wasserburg and Papanastassiou 1982). Indeed, it is quite plausible (but not necessarily true) that intense heating of small planets by decay of ^{26}Al could contribute to the migration and redistribution of the elements which is observed in Fig. 8.

The number of different meteorites containing crystals or inclusions that exhibit $^{26}Mg^*$ are only a few (the C2 stony meteorites Allende, Leoville, and Murchison). Most meteorites typically show $^{26}Mg/^{24}Mg$ ratios that are indistinguishable from the terrestrial value. While the above observations appear to require the presence of ^{26}Al in the early solar system, it is not yet possible to demonstrate that it was ubiquitous. A recent study has reported positive evidence for ^{26}Al in a microscopic refractory inclusion contained within a normal chondritic meteorite (Hinton and Bischoff 1984). These workers found a hibonite crystal which shows an excellent correlation of $^{26}Mg^*$ with Al corresponding to $^{26}Al/^{27}Al \sim 0.8 \times 10^{-5}$, a factor of six less than the "canonical" value of 5×10^{-5}. This result is quite important in that it shows that fragments of "old" refractory material containing ^{26}Al were included in normal chondrites. However, no evidence of ^{26}Al has been found in Al-rich phases which comprise the typical material of chondrites, achondrites or mesosiderites. A gap still exists for ^{26}Al between early (?) Ca, Al-rich materials and typical meteoritic materials. Demonstration of the extent to which ^{26}Al was widespread and uniform throughout the solar system remains a key issue. From the existing data, it is very possible that ^{26}Al was heterogeneously distributed. The ^{26}Al abundance and distribution at the time of planet formation is critical to all models of early planetary heating. The reader is referred to an earlier extensive review on short-lived nuclei with emphasis on ^{26}Al (Wasserburg and Papanastassiou 1982).

^{107}Pd

The nuclide ^{107}Pd ($\bar{\tau} = 9.4 \times 10^6$ yr) is the most recently identified short-lived nuclide (Kelly and Wasserburg 1978). The existence of ^{107}Pd is of great importance as this suggests that both ^{129}I and ^{107}Pd (which are associated with only slightly rapid, or rapid neutron capture processes) were produced and injected at the same time. Because the mean life of ^{107}Pd is a factor of 2.4 less than that of ^{129}I, it would not have survived with a long time ($\sim 10^8$ yr) of free decay. Evidence for ^{107}Pd has been established by showing a correlation of ^{107}Ag* with Pd. As distinct from ^{26}Al, whose presence has been established in small inclusions or fragments, the only objects with a sufficiently high ratio of parent to daughter element are iron meteorites which are produced by planetary melting processes. The presence of ^{107}Pd in planets provides a direct connection between the time scales of nucleosynthetic processes and planet formation and development. In some classes of iron meteorites that have high Pd and are extremely depleted in Ag, it has been found that $^{107}Ag/^{109}Ag = 9$ as compared to the normal value of 1.09. There is a correlation of ^{107}Ag* with Pd in several different classes of iron meteorites. Examples are shown in Figs. 9–12. The first set of results shown are on iron meteorites of class IV B. Bulk samples of metal show an excellent correlation with $^{107}Ag/^{109}Ag$ increasing regularly with the $^{108}Pd/^{109}Ag$ ratio. If it is assumed that these different meteorites are all of the same age, then the slope determined is $^{107}Ag^*/^{108}Pd = 2 \times 10^{-5}$. However, the correlation line extrapolates to $(^{107}Ag/^{109}Ag)_o \sim 0$. A recent study of the IV A iron meteorites by Chen and Wasserburg (1984) is exhibited in Fig. 10. These data show an excellent correlation for different metal fragments of Gibeon (a twenty-ton object with coarse grained FeNi crystals). In addition, other IV A irons appear to plot approximately along the same trend. The correlation line extrapolates to $(^{107}Ag/^{109}Ag)_o \sim 1.2$. The data on sulfide inclusions from IV A meteorites are shown in Fig. 11. Studies of two other irons of the class III AB show small enrichments of $^{107}Ag/^{109}Ag$ and a correlation line with a slope of $^{107}Ag^*/^{108}Pd = 1.8 \times 10^{-5}$ for data from metal and sulfide phases which passes through normal silver at Pd = 0 (Fig. 12). The III AB meteorites are usually associated with melting and core formation of small planetary bodies.

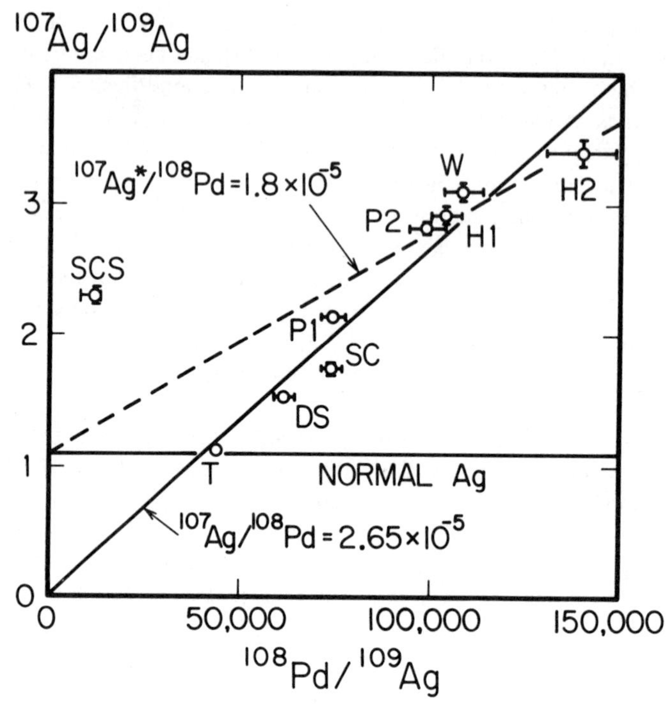

Fig. 9. Ag-Pd data on different iron meteorites of the class IV B (after Kaiser and Wasserburg 1983). The typical $^{107}Ag^*$ concentrations are around 10^{11} atoms g^{-1}. Note that the correlation line passes near the origin and not through the value for normal silver (dashed line). Note also that the two points labeled SC (Santa Clara metal) and SCS (Santa Clara sulfide) indicate some discrepant behavior. If these different iron meteorites are of the same age, this implies the presence of ^{109}Ag without any ^{107}Ag as distinct from the normal value $^{107}Ag/^{109}Ag = 1.09$.

The widespread occurrence of $^{107}Ag^*$ correlated with Pd in many different iron meteorites attests to the presence of ^{107}Pd in small planetary bodies at the time they melted and segregated.

There are some problems that are puzzling. A study of FeS nodules included in Gibeon (which have low Pd and high Ag) show that they lie in the neighborhood of the extrapolated correlation line (Fig. 11). However, there are obviously wide variations of $^{107}Ag/^{109}Ag$ within the Pd-poor sulfides. These data do not lie on the correlation line. This effect appears similar to the high $^{107}Ag/^{109}Ag$ ratio observed in the sulfide of Santa Clara by Kaiser and Wasserburg (1983) (see Fig. 9). The group IV meteorites have all been subject to shock heating and sometimes heat treatment by humans. In addition, there is still the matter of a peculiar low intersection of the correlation line in Fig. 9 corresponding to an initial $^{107}Ag/^{109}Ag \approx 0$ for the IV B meteorites mentioned above. This may be explained if the correlation line is not an isochron, which could result from (small) differences in the ages for meteorites within

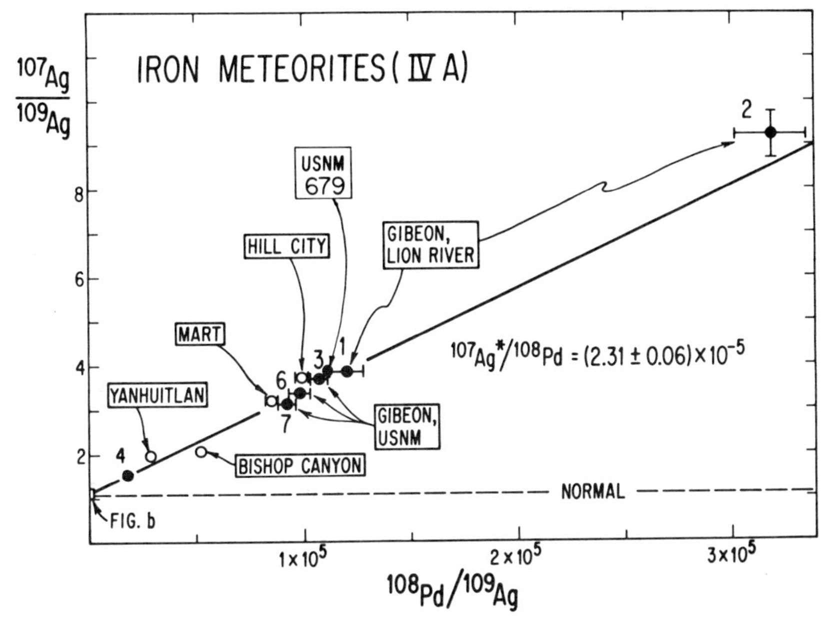

Fig. 10. Correlation of the silver isotopic composition with the $^{108}Pd/^{109}Ag$ ratio for IV A iron meteorites (after Chen and Wasserburg 1984). The filled circles are analyses of seven different samples of the metal phase of the Gibeon meteorite. Note the extremely high ratios of $^{107}Ag/^{109}Ag$. Data on four other meteorites of this class are also shown. Note that Yanhuitlan and Bishop Canyon are somewhat off the line, indicating a different age. Error bars are shown if larger than the points. The inset near Pd/Ag = 0 is shown in Fig. 11.

this class. However, the constancy of $^{107}Ag/^{108}Pd$ is most remarkable for this class of meteorites as T. Kaiser (personal communication) has pointed out. While my colleagues and I are confident of the existence of ^{107}Pd in planetary bodies, we are certainly not comfortable about the details.

^{244}Pu

Extensive experimental data have shown that a variety of meteoritic materials contain clear evidence for the presence of ^{244}Pu ($\bar{\tau} = 120 \times 10^6$ yr) through its characteristic decay (fission products and excess fission tracks). Excesses of unshielded Xe isotopes were first found in a meteorite rich in U and Th by Rowe and Kuroda (1965). The distinctive spontaneous fission spectrum is shown in Fig. 13. In fact, the fission yield of ^{244}Pu was measured on man-made Pu after it was firmly established that a fissionable short-lived transuranic nuclide R_x with a unique fission spectrum was present in the early solar system. The spontaneous fission Xe spectrum of ^{244}Pu was measured in the laboratory on a separated ^{244}Pu sample. The results are the same as for the spectrum observed in several different types of meteorites within the range of uncertainties shown in Fig. 13. Xenon with this isotopic composition corre-

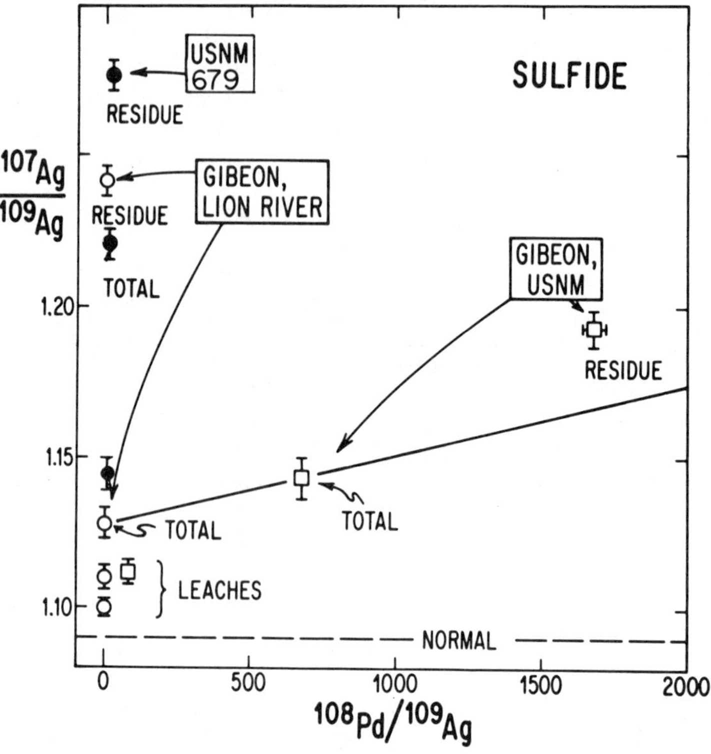

Fig. 11. Analyses of Gibeon FeS nodules with low Pd/Ag from three different samples (Chen and Wasserburg 1984). Each sample was subjected to a series of leaches with acids and analyzed. The remaining material (residue) was dissolved and analyzed. All data points for leaches and residue are shown. The coordinates of the total sulfide were calculated from the components (totals). These data show gross isotopic heterogeneity within the sulfide phases with large excesses of ^{107}Ag which are not yet explained (see points SC and SCS in Fig. 9).

lates with fission tracks, and with U and Nd concentration. Among the meteorites in which ^{244}Pu fission Xe and excess fission tracks were found is Angra Dos Reis (Hohenberg 1970; Storzer and Pellas 1977; Lugmair and Marti 1977; Wasserburg et al. 1977*b*). This meteorite is well recognized to be the product of planetary differentiation processes. It is a clear example of melting and crystal segregation thus showing that the ^{244}Pu decay took place in planetary bodies. Because of its long life, ^{244}Pu must still be present in nature today at a very low level. So far no intensive effort has been made to search for this nuclide on Earth. [A preliminary report by D. C. Hoffman et al. (1971) indicated that ^{244}Pu had been found in a terrestrial sample. This work was not adequate to establish the claim and further efforts by the same group have not yet presented evidence in support of it.] For a more extensive discussion of ^{244}Pu in meteorites, the reader is referred to recent articles concerning the

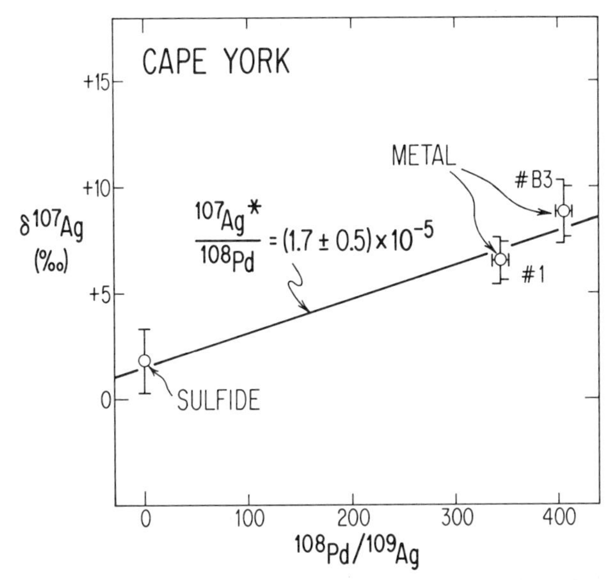

Fig. 12. Shifts in $^{107}Ag/^{109}Ag$ in parts per thousand correlated with $^{108}Pd/^{109}Ag$ in the Cape York meteorite (III AB) (after Chen and Wasserburg 1983). Data are for two samples of metal and a coexisting nodule of FeS with low Pd and high Ag content. The point 0 corresponds to the isotopic composition of normal terrestrial silver. The isotopic shifts are small because of the low value of Pd/Ag (compare the scale with Figs. 9 and 10).

subject (Alexander et al. 1971; Burnett et al. 1982; Wasserburg and Papanastassiou 1982; Hudson et al. 1984; Pellas and Storzer 1981).

III. INDICATIONS AND HINTS

^{146}Sm

This nuclide has a rather long mean life of $\bar{\tau} = 149 \times 10^6$ yr. It lies on the neutron deficient side of the stability field and is associated with low-abundance nuclei produced by the *p*-process or γ-n-process (Woosley and Howard 1978). It is most reasonable to expect it to be present in the early solar system (and also present today) if it were produced in sufficient abundance. The possible use of ^{146}Sm as a *p*-process chronometer was proposed by Audouze and Schramm (1972). Several workers have claimed the existence of ^{146}Sm in the early solar system. Some of the results are manifestly in error and others are inconsistent. We concentrate here on the most recent results.

There is evidence of excesses of ^{142}Nd in fine-grained residues from the Allende meteorite (Lugmair et al. 1983). Table III shows an example of the results. It has recently been demonstrated that this ^{142}Nd* is associated with recoil phenomena as confirmed by the presence of ^{234}U excesses from ^{238}U decay and recoil (Shimamura and Lugmair 1984). Lugmair et al. (1983) interpreted the data to result from the recoil of the residual nuclei produced by α

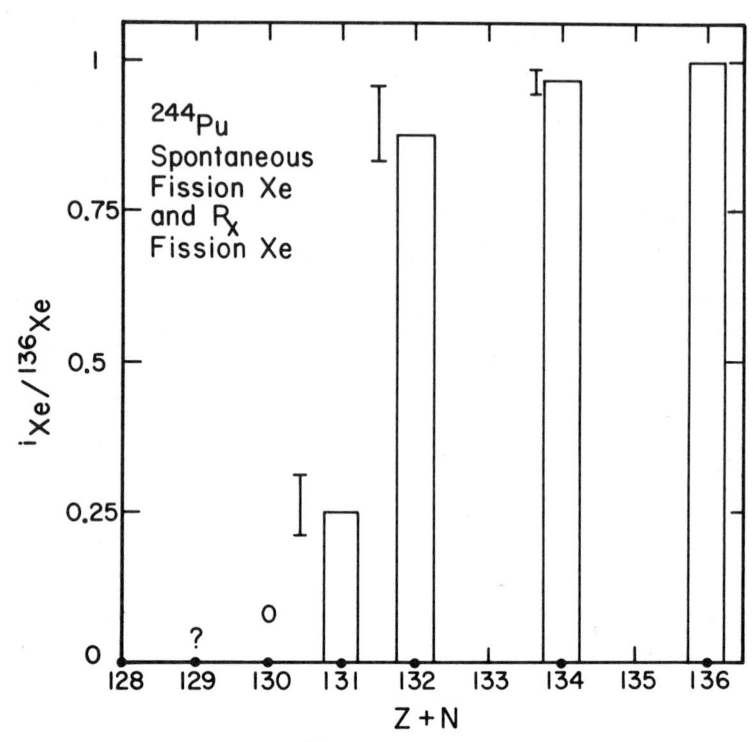

Figure 13. Xenon fission spectrum of ^{244}Pu as determined on separated Pu which was stored for several months and the R_x fission spectrum determined on whitlockite from the St. Severin meteorite and on a U-rich meteorite. The meteorite data has some small uncertainties as shown due to cosmic ray spallation on the rare earth elements (figure after Alexander et al. 1971; Wasserburg et al. 1969; Hohenberg 1970).

decay. The fine-grained material surrounding larger grains thus acted like a catcher foil. Previous data indicated hints of small ^{142}Nd excesses but these were not self-consistent results which correlated with Sm. A report by Jacobsen and Wasserburg (1984) has shown self-consistent correlations of ^{142}Nd* with Sm in two meteorites. The inferred ratio ^{146}Sm/^{144}Sm is different for these two meteorites but appears to be a consequence of their different ages. The results are shown in Figs. 14 and 15. These data all yield no detectable ^{142}Nd excess for samples with average chondritic (e.g., solar) relative abundances of Nd/Sm. This is compatible with the model described in Sec. I. It should be noted that the results by Jacobsen and Wasserburg (1984) do not support previous observations by other workers on the same samples. The effects reported are quite small and the error bars are large so that current observations, while strongly suggestive, do not constitute a reasonable proof.

Using the isotopic data on acid residues, Lugmair et al. (1983) obtain $^{146}\text{Sm}/^{144}\text{Sm} = 4.5 \times 10^{-3}$. The results obtained by Jacobsen and Wasser-

TABLE III
Neodymium in Residues from the Allende Meteorite[a,b]

	δ_{142}	δ_{143}	δ_{145}	δ_{146}	δ_{150}
Allende residue CF-1	4.69 ± 0.18	277.59 ± 0.14	-1.04 ± 0.26	1.04 ± 0.21	-0.3 ± 0.4
Terrestrial standard	-0.06 ± 0.16	0.05 ± 0.14	-0.02 ± 0.15	-0.03 ± 0.07	0.1 ± 0.4

[a]Table after Lugmair et al. 1983.
[b]$\delta_j = 10^3$ [${}^iR$ sample/R_c standard $-$ 1] ${}^iR = {}^i\mathrm{Nd}/{}^{144}\mathrm{Nd}$ corrected for mass fractionation. Note the small but clear excess of ^{142}Nd and the enormous excess of ^{143}Nd. These are attributed to α recoil from the decay of ^{146}Sm and ^{147}Sm ($\bar{\tau}_{^{147}\mathrm{Sm}} = 1.6 \times 10^{11}$ yr), respectively. Assuming the shifts are due to recoil over the history of the solar system $^{146}\mathrm{Sm}/^{144}\mathrm{Sm} = (^{142}\mathrm{Nd}/^{143}\mathrm{Nd})_\odot(^{147}\mathrm{Sm}/^{144}\mathrm{Sm})_\odot(\delta 142/\delta 143)(\bar{\tau}^{142}\mathrm{Sm}/4.6 \times 10^9)$. This gives $^{146}\mathrm{Sm}/^{144}\mathrm{Sm} \approx 5 \times 10^{-3}$.

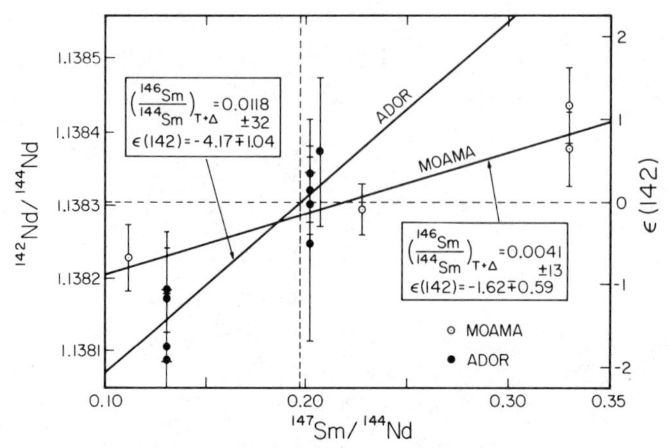

Fig. 14. Isotopic ratio of ^{142}Nd/^{144}Nd versus ^{147}Sm/^{144}Nd for different phases from the Angra Dos Reis (ADOR) and Moama meteorites. The deviations of ^{142}Nd/^{144}Nd relative to terrestrial Nd in parts in 10^4 ($\epsilon(142)$) is indicated on the right-hand scale. The intersection of the dashed lines corresponds to the bulk solar system value of today. The error bars are $\pm 2\ \sigma$. Samples with Sm/Nd = $(\mathrm{Sm/Nd})_\odot$ show $\epsilon(142) = 0$; samples with higher or lower Sm/Nd values appear to show excesses or shortages, respectively, of ^{142}Nd, although the effects are quite small and not precisely defined (Jacobsen and Wasserburg 1984).

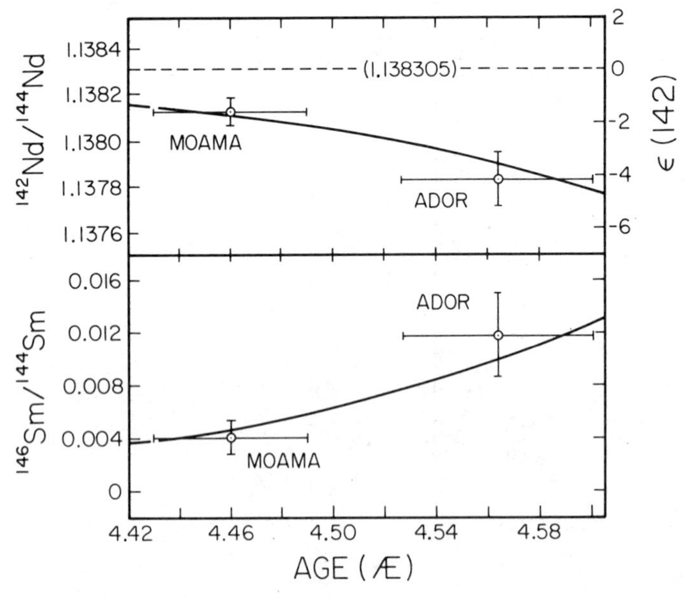

Fig. 15. Graphs showing the change in ^{142}Nd/^{144}Nd and ^{146}Sm/^{144}Sm over early solar system history due to the decay of ^{146}Sm, assuming the reliability of the effects shown in Fig. 14. The ages of the meteorites were independently determined in the same experiment using the decay of $^{147}\mathrm{Sm} \rightarrow {}^{143}\mathrm{Nd}$. The differences in ^{146}Sm abundance for the two meteorites appear to reflect their difference in age (Jacobsen and Wasserburg 1984).

burg are a factor of ~ 2 greater. Either there is an age difference (10^8 yr) or else there is a discrepancy. In addition, as pointed out by Lugmair et al. (1983) and Jacobsen and Wasserburg (1984), there is no clear independent guide from the theoretical production rates as to what the production rate of ^{146}Sm should be. Indeed, the values of (5–10) $\times 10^{-3}$ quoted above are far higher than calculated by Woosley and Howard (1978) for the γ-n-process. More careful and imaginative experimental and theoretical work needs to be done here.

^{135}Cs

There is only one observation that suggests the presence of this nuclide. This is found in the existence of a shortage of ^{135}Ba (by 2 parts in 10^4) in only one case from the Allende sample C-1 (McCulloch and Wasserburg 1978). This sample is one of the rare ones that show large *fractionation* α (see Sec. I) and nonlinear anomalies in several elements attributed to unknown nuclear effects (called FUN samples). Although the lifetime of 3.3×10^6 yr for ^{135}Cs and the magnitude of the effect are compatible with the ^{26}Al data, strong evidence is lacking. If an ancient Cs-rich/Ba-poor meteorite sample could be found, this would aid in clarifying the matter, but so far none has been discovered.

^{41}Ca

This nuclide is crucial because of its very short mean life of 0.19×10^6 yr. As argued in Sec. I, positive identification of a short-lived nucleus gives the most stringent control on Δ for the related nucleosynthetic process. D. D. Clayton (1975*b*) suggested that excesses of ^{41}K might be found because most ^{41}K is synthesized as ^{41}Ca. The first attempt to search for this effect was made by Begemann and Stegmann (1976) who found $^{41}Ca/^{40}Ca \leq 5 \times 10^{-5}$. A positive identification of $^{41}K^*$ in some Allende samples was reported by Huneke et al. (1981). These workers searched for $^{41}K^*$ in high Ca and very low K crystals in Allende and found $^{41}K^*/^{40}Ca \sim 10^{-7}$ to 5×10^{-8}. A more intensive study of similar inclusions was recently reported by Hutcheon et al. (1984) who found that the presence of $(^{40}Ca^{42}Ca)^{++}$ in the mass 41^+ ion beam during analysis was a more serious interference than Huneke et al. thought. $CaCa^{++}$ correlates perfectly with Ca^+ and is not distinguished from $^{41}K^+$ by its mass ($\Delta M \simeq 1.2 \times 10^{-3}$ amu). Upon careful investigation of the yield of $(^{40}Ca^{41}Ca)^{++}$ relative to $^{42}Ca^+$, it was found that this yield was very sensitive to the composition and nature of the sample. Upon establishing that $(^{40}Ca^{41}Ca)^{++}/^{42}Ca^+ \sim 2 \times 10^{-5}$ for appropriate silicates, Hutcheon et al. (1984) were able to place a strict upper limit of $^{41}K^*/^{40}Ca < (8 \pm 3) \times 10^{-9}$. This would require $\Delta \geq 1.8 \times 10^6$ yr if the initial ratio of $^{41}Ca/^{40}Ca$ were $\sim 10^{-4}$. It now appears that the hint of ^{41}Ca reported earlier must be replaced by the upper limit given by Hutcheon et al. (1984). It will not be an easy task to establish the presence of ^{41}Ca at these very low abundance levels.

^{247}Cm

This nuclide is of great importance because it is a transuranic which must have been produced in the same process and site that made ^{244}Pu. By comparison, it has a relatively short lifetime so its abundance places a strong limit on Δ for the last *r*-process contribution. ^{247}Cm decays to make ^{235}U, thus the segregation of Cm relative to U should produce excesses (or shortfalls) of ^{235}U relative to ^{238}U as proposed by Blake and Schramm (1973). There are numerous papers claiming to find excesses of ^{235}U in meteoritic materials in *Nature*, the most recent of which is an article by Tatsumoto and Shimamura (1980). Intensive studies of the $^{235}U/^{238}U$ ratio using much improved experimental procedures including a matched ^{233}U–^{236}U tracer for internal calibration has shown the previous results claiming effects to be in error. Chen and Wasserburg (1981*a,b*) have shown for several samples of Allende material (known to contain $^{26}Mg^{*}$ and large fractionations of U, Th and Nd relative to each other) that any variations in $^{235}U/^{238}U$ must be $< \pm 0.4\%$. Early variations in Cm/U mean that the fractional shifts in ^{235}U abundance observed today are given by $\Delta^{235}U/^{235}U = N(^{247}Cm)/N(^{235}U) f_{Cm/U}$. Taking a low value of 1 for the enrichment factor ($f_{Cm/U}$) they obtained $^{247}Cm/^{235}U \leq 4 \times 10^{-3}$.

IV. SUMMARY AND DISCUSSION

^{244}Pu was alive and well in the early solar system. The abundance of this nuclide at 4.6 AE is not precisely fixed but is in the neighborhood of $^{244}Pu/^{238}U \approx 10^{-2}$. I have chosen the value of 7×10^{-3} from Hudson et al. (1984) (see Table IV). The presence of ^{247}Cm has not been detected but an upper limit to its abundance is $^{247}Cm/^{235}U \leq 4 \times 10^{-3}$. Both ^{244}Pu and ^{247}Cm are only produced by *r*-process nucleosynthesis. The stellar site where this takes place is not established but must involve very high temperatures and neutron densities. From the limits on ^{247}Cm and the presence of ^{244}Pu, it follows that the last substantial addition of real r-process nuclei took place somewhere between about 10^8 yr and 5×10^8 yr prior to solar system formation. Obviously, this could have taken place in either discrete events or continuously. The time scale

TABLE IV

Nuclide	Solar System Abundance at ≈ 4.6 AE
^{41}Ca	$^{40}Ca/^{41}Ca \leq (8 \pm 4) \times 10^{-9}$
^{26}Al	$^{26}Al/^{27}Al \approx 5.0 \times 10^{-5}$
^{107}Pd	$^{107}Pd/^{108}Pd \approx 2.0 \times 10^{-5}$
^{129}I	$^{129}I/^{127}I \approx 1.0 \times 10^{-4}$
^{146}Sm ?	$^{146}Sm/^{144}Sm \sim (0.012; 0.005)$
^{247}Cm	$^{247}Cm/^{235}U \leq 5 \times 10^{-3}$
^{244}Pu	$^{244}Pu/^{238}U \approx 7 \times 10^{-3}$
	$^{235}U/^{238}U = 0.33$

for the last substantial r-process contribution to the solar system appears fixed by these data. (We note that no results on the presence of superheavy elements has been found to stand either experimental testing or theoretical scrutiny.) The preponderence of the U and Th in the solar system must have been produced much before this time of late r-process addition ($\sim$ 4.7 to 5.1 AE). From the precisely established ages of some solar system objects at 4.555 AE, the $^{235}U/^{238}U$ ratio at that time and the estimated relative production rate of 1.5, a firm lower limit to the age of the universe, is 6.4 AE. This yields an upper limit for the Hubble constant of 100 km s^{-1} Mpc^{-1} using the standard $\tau H = 2/3$ relation. Various models of galactic nuclesynthesis can fit a $\sim$ 10 AE time scale quite well but there are no nuclear chronometers yet indentified which *require* this or greater times. The extent to which the astronomical observations and theoretical considerations of stellar and galactic evolution and of globular clusters may require a much greater time is not decided (Flannery and Johnson 1982).

The nuclei ^{26}Al, ^{107}Pd and ^{129}I were all present in the early solar system. The abundance of these species relative to a nearby isotope of the same element are all in the neighborhood of 2×10^{-5} to 10^{-4} (see Table IV). The fact that they are in roughly comparable abundance in spite of the wide range of mean lives (1.1 to 23×10^6 yr) must mean that they were produced in events closely related temporally (i.e., over a very short time scale). Their low abundance is thus not due to a large interval of decay without production which would decrease the abundance of each species i by $e^{-\lambda_i \Delta}$. Rather, they are the result of adding freshly synthesized matter (about 10^{-4} of the nuclear species) to old material. The time scale for this last-gasp addition to solar system matter is between 10^6 to 9×10^6 yr, with the higher value being a maximum. The usual explanation for the presence of ^{26}Al is shown in the cartoon in Fig. 16. The general view is that an association of rapidly evolving stars in a cloud injects freshly synthesized material from either supernova or nova explosions into the surrounding cloud medium. This material (a mixture of old and freshly formed nuclei) then condenses to form the solar system on a $\sim 1 \times 10^6$ yr time scale. This time scale is about a factor of 20 to 180 less than was obtained by just considering ^{129}I alone. With regard to ^{129}I, it is quite reasonable to consider that some of the ^{129}I represents matter produced over a longer time scale and some was a late addition with the other shorter-lived nuclei. The proportion of the ^{129}I that was old is difficult to estimate but could be from 10 to 90% of the total. The mechanism of production of ^{26}Al is distinct from ^{107}Pd and ^{129}I. Production of the former is from proton reactions while the production of the latter must involve (single) neutron capture by Pd and Te or else multiple neutron capture in a mini-r-process. From the abundances measured and the time scale indicated, all of these nuclei must have been produced in a completely different and smaller event than the pure r-process nucleus ^{244}Pu. The older ^{244}Pu contribution to the then ambient medium (a few %) was far greater than this late injection of ^{26}Al, ^{107}Pd and ^{129}I, etc. (0.01%).

LATE ADDITION

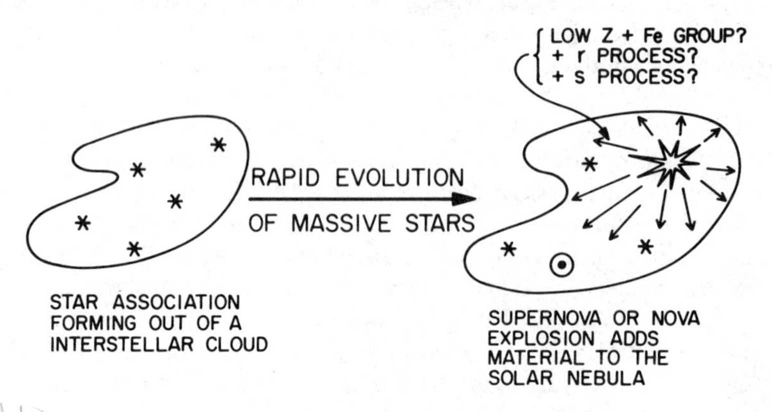

$^{26}Al/^{27}Al = 5 \times 10^{-5} \Rightarrow$ CONDENSATION WITHIN 3×10^{6} yrs OF EXPLOSION

Fig. 16. One general conceptual framework for explaining the presence of freshly synthesized material in the solar nebula from injection of supernova or nova debris into the vicinity of the protosolar nebula environment (figure from Lee et al. 1976*a,b*).

The upper limit recently obtained for ^{41}Ca shows that, if this nuclide was produced, it must have been produced earlier than 1.8×10^{6} yr prior to the formation of the solar system. If we assume all of the short-lived nuclei were produced at essentially the same time, then the last gasp injection took place sometime between 1.8×10^{6} yr and a maximum of 9×10^{6} yr prior to solar system formation.

The most direct inference from all of the preceding considerations is that the Sun condensed from a medium that contained recently ejected nucleosynthetic debris at a modest level of concentration. This debris came from distinctive sites of stellar nucleosynthesis and were reasonably (but not perfectly) mixed on a 10^{6} yr time scale. As the above nuclei have been identified in different solar system objects, they appear to tie certain solar system events to late injection. The ^{26}Al effects have so far only been found in material *associated* with early solar nebular condensation (see L. Grossman [1980] for the proposed origins and formation of Ca,Al-rich inclusions). ^{129}I was present both in these materials and possibly in small planets. The ^{107}Pd is present in iron meteorites which must have formed in differentiating planets. A general scheme outlining the above chronology is shown in Fig. 17. If we assume that all of the short-lived nuclei were originally present prior to solar system formation, then relative to the time of last nucleosynthesis we obtain the chronology shown. If the precursor medium had $^{26}Al/^{27}Al = 10^{-3}$, then the total time to form the Ca,Al-rich inclusions which contained ^{26}Al was 3×10^{6} yr. To produce and melt small planets which contained ^{107}Pd as a radioactive nuclide would require planet accumulation on a time scale not much greater than the mean life. From the measurements of the ages of meteorites and the ages determined for the earliest differentiation of the Earth and Moon, a time gap of 5×10^7 to 10^8 yr is known to exist. These time scales are commensurate with the accumulation times of cm to 100 m, km and 10^3 km planetary bodies as calculated by Safronov (1969) and Ward and Goldreich (1973). If the time scales for coalescing the protosolar system from the precursor medium is estimated from the free-fall time ($\tau_{ff} = 4 \times 10^7/\sqrt{n_H}$) and taking the time scale to be given by the lifetime of ^{26}Al ($\bar{\tau}_{^{26}Al} \sim 1.0 \times 10^6$ yr), this implies a hydrogen density n_H of $\sim 2 \times 10^3$ H_2 cm^{-3} for the initial state. This is quite compatible with the densities in a molecular cloud. The whole process of collapse and aggregation appears to be compatible within the mythologies shared by cosmochemistry and astrophysics. There seems to be a most immediate connection between the solar system and the interstellar medium at the time of formation.

The presence of ^{26}Al in the abundance found above applied throughout the whole solar system would provide an ample heat source for melting small planets formed within a few million years. This problem was first identified by Urey (1955). If ^{26}Al was the heat source for melting small planets, some evidence should be preserved in these bodies if they were cooled on a time scale comparable with ~ 1 to 2×10^6 yr. To find this evidence we must have very old planetary objects. To date, *no clear evidence of ^{26}Al has been found in planetary objects to prove that this was the source of heating*. The objects which are produced by planetary differentiation (basaltic achondrites) which caused Urey to propose the ^{26}Al heat source so far show no evidence of ^{26}Al (Schramm et al. 1970). The oldest object yet dated directly is the meteorite Angra Dos Reis whose age is 4.555 ± 0.005 AE (see Jacobsen and Wasserburg [1984] and Wasserburg et al. [1977*b*] and references therein). This is the best precision (2σ) in age determination so far achieved using long-lived radioactivities. Angra Dos Reis is a cumulate igneous rock which must have formed on a planet and thus should be somewhat younger than the early-formed objects we would want. In principle, it is possible to use one of the

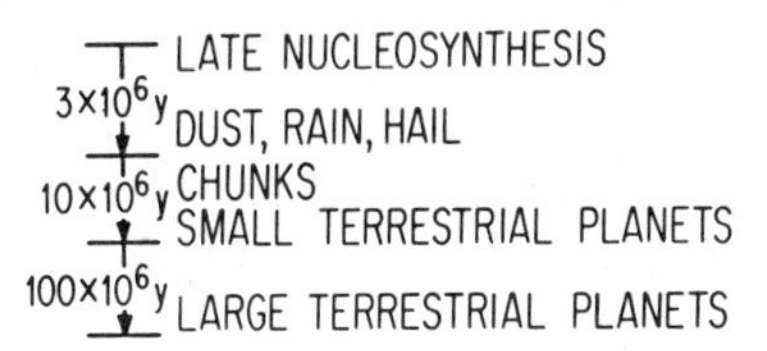

Fig. 17. A time schedule for producing the solar system relative to a late nucleosynthetic event. Note that this presupposes that the nucleosynthetic event preceeded small planet formation and was not subsequent to or contemporaneous with it.

short-lived nuclei themselves to provide a very precise chronometer as one can measure $^{26}Mg^{*}/^{27}Al$ or $^{129}Xe^{*}/^{127}I$ to within a few percent. This would allow us to place objects in a self-consistent time scale which is very precise (10^5 yr for ^{26}Al and 10^6 yr for ^{129}I). However, this requires that the freshly synthesized nuclei be well homogenized in solar system matter in order to have an initial reference state, as pointed out by Cameron (1973*d*). From the available data for both systems, this does not appear to be the case. Some efforts to establish such chronology have been made for ^{129}I but they remain subject to this problem (Podosek 1978). A sensible experiment might be to obtain both the absolute age and evidence for ^{26}Al on various meteorites. One problem is that of finding planetary materials with high enough Al/Mg (10^3–10^4) in order to observe any effect. There is also enormous difficulty in establishing absolute ages with the precision required (better than 5 million yr in 4600 million years) in order *a priori* to select the correct old sample. It is more likely that the problem may be solved by a search for plausible candidate meteorites.

Having laid out the basic observations and canonical inferences, it is of merit (i.e., provocative) to consider some problems and alternatives. The model of a single (or closely connected) nucleosynthetic event(s) to explain the abundances of the short-lived nuclei is most plausible. That this happened before solar system formation is however not required. We cannot yet place the meteorites and their constituent parts in an absolute chronometric scale measured back from today with adequate precision. The arguments associating the refractory inclusions in Allende with early solar condensates are reasonable (see L. Grossman 1980) but are not unequivocal as a chronometer or as proof of the actual processes. Indeed, there are many conflicts between a condensation model for a hot part of the solar nebula and the mineral phases observed in so-called refractory (and sometimes volatile-rich) inclusions. Material has certainly been subject to complex reprocessing and it is not possible to relate a given object to a well-defined evolution. It is not clear from purely theoretical considerations as to how much of the solar nebula was hot. The heating was probably very restricted in space and time and it is not obvious how to relate the meteorites and planets which we study to the medium from which we think they came. Some recent attempts to tie the complex chemical evolution together with infall and circulation dynamics around the protosun has been made by Morfill (1983*b*). This is an important area requiring further study and closer ties between mineral chemistry and "cooking" either by the Sun or in planets (cf. Meeker et al. 1983). Secondly, while the ratio of short-lived/stable and exotic/stable nuclei appears to be $\sim 10^{-4}$ (excluding oxygen), this observation only pertains to condensed materials directly associated with the terrestrial planets. This may or may not be attributable to the whole solar system including the Sun.

If the results are representative of the solar system, this requires the injection of $M_{\text{FRESH}} \sim 10^{-4} M_{\odot}$. This presumably occurred at the same time as the general isotopic anomalies for many elements not associated with short-lived nuclei were introduced. Whether the injecting sources (e.g., supernova, nova) are a trigger or simply a coincidence is a matter of debate (see Truran and Cameron 1978; Cameron and Truran 1977). It is difficult to assign causal phenomenon for our particular star in a stellar nursery. In terms of the precursor state prior to condensation of the Sun, ^{26}Al (if it were in high abundance) would yield a major source of ionizing radiation in the interstellar medium (Consolmagno and Jokipii 1978). If the isotopic results are not representative of the bulk solar system (and thus $<< 10^{-4}M_{\odot}$), then they could indicate either injection into the placental interstellar medium (ISM) or production within the proto or early solar system (see Fig. 18). The basic question is whether these effects are due to pre- or post-solar system processes (see Fig. 19). The time scale implied by ^{26}Al is substantially less than the time for 1 $M_{\odot}$ to evolve onto the main sequence. As a result it is of primary importance to assess the effects of early solar evolution on the residual matter surrounding the Sun.

While local irradiation of a solar mass to produce ^{26}Al is excluded due to ionization loss (Lee 1978), it is possible that the radioactive nuclei could have been produced by irradiation of a dust cloud or of planetary surfaces. A schematic diagram showing a local irradiation by the Sun at sufficient proton energy to produce ^{26}Al is presented in Fig. 18. If the Sun went through a T-Tauri phase

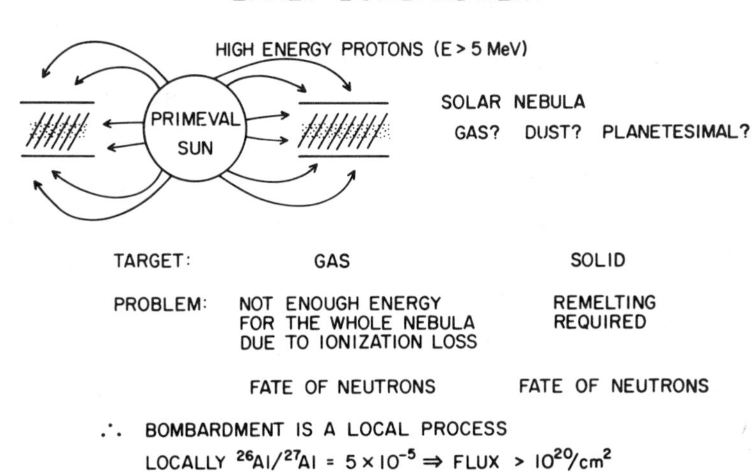

Fig. 18. Cartoon illustrating the problems associated with proton bombardment of the early solar nebula with energetic protons to produce ^{26}Al throughout the gas and dust. The possibility of a local process possibly involving just dust and planetesimals and the fate of secondary neutrons have not been intensively studied (figure from Lee et al. 1976*a,b*).

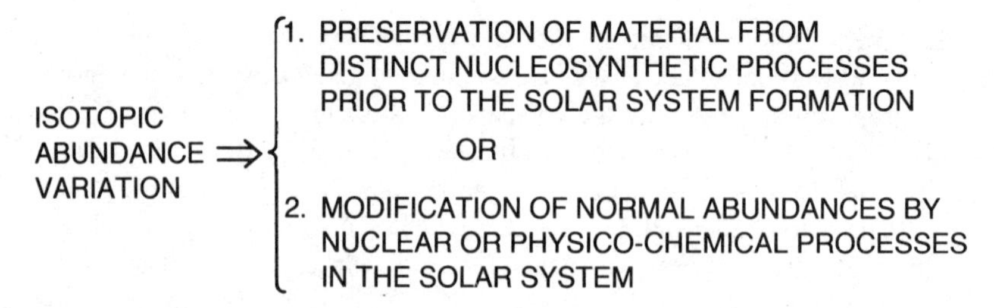

Fig. 19. One of the basic alternatives required to explain general isotopic variations in the solar system (figure from Lee et al. 1976 *a,b*). The question of how to store or preserve isotopic heterogeneities (both radioactive and stable) without complete mixing or gross heterogeneity is still a puzzle as the density changes by a factor of 10^{12} in going from a dense cloud to the solar nebula (see Fig. 16).

with a luminosity of $10^4 L_\odot$, it is plausible to expect extreme temperatures within 1 AU (causing surface heating and evaporation) and an energetic solar cosmic ray fluence of 10^{21} to $10^{22}\,p\,\mathrm{cm}^{-2}$ ($E \gtrsim 30$ MeV). This corresponds to a flux about 10^5 to 10^6 times greater than the current solar cosmic ray flux for a period of $\sim 3 \times 10^6$ yr. Although some workers do not believe the Sun could have gone through a T-Tauri phase (A. G. W. Cameron, personal communication), there is a very high probability that it could have been an X-ray star. While this would not provide much flash heating at 1 AU, the high energetic particle flux could produce many nuclear reactions. This type of model has not been intensively investigated. While ^{26}Al would certainly be produced in such a scenario, it is not evident that sufficient production of ^{107}Pd and ^{129}I from secondary neutrons would take place. However, this has been suggested to explain the extrapolated initial value of $(^{107}\mathrm{Ag}/^{109}\mathrm{Ag}) = 0$ for IV B meteorites. The problems of isotopic heterogeneity of both ^{26}Al and of the stable nuclei of several elements might be more easily explained by local solar system processes. The absence of ^{26}Al in some samples that have been interpreted as early solar condensates or supercooled droplets must in any case require gross heterogeneity in ^{26}Al distribution (Esat et al. 1978; Hutcheon 1982; Hinton and Bischoff 1984). A reconsideration of the solar irradiation scenario by Fowler, Greenstein and Hoyle (1962) with less extreme conditions and more modest expectations is warranted. A local irradiation model must include the irradiation of dust and small planetary bodies. It is not clear that icy bodies are required for neutron scattering (suggested by Fowler et al.)for the nuclei under consideration.

The preceding comments are not meant to deny the plausible interpretation of a direct and immediate connection with the ISM and stellar nucleosynthesis. However, since the first clear indications of short-lived nuclei were found, it has always been evident that in principle there were two alternative scenarios (see Figs. 16 and 20). This must be kept in mind while we are searching for the true solution to the problem. No self-consistent theory has as yet been proposed that adequately explains the key isotopic, chemical and mineralogic observations.

While alternatives require study, the recent exciting report of ^{26}Al as a source in the Galaxy certainly appears to support the view that the sources for the solar system came from the ISM and not local production. The study by Mahoney et al. (1982) has reported the possible presence of a 1.808 MeV γ line as a diffuse (?) source in the Galaxy. A more extensive analysis of this complex data set by Mahoney et al. (1984) shows a well-defined signal for ^{26}Al. The signal is there but it is not strong compared to spacecraft background. The most recent analyses of the HEAO 3 (High Energy Astronomical Observatory) data are shown in Fig. 21. These workers have identified this γ-ray line as due to the decay of ^{26}Al, resulting from the transition of ^{26}Mg from the first excited state to the ground state. Further measurements of this γ flux will be required to improve the counting statistics and to more firmly define the nature and distribution of the sources. These workers have estimated a $^{26}\text{Al}/^{27}\text{Al} \sim 10^{-5}$ for the Galaxy which is remarkably close to the abundance found for the early solar system. A recent discussion of the HEAO results is given by D. D. Clayton (1984). The result by Mahoney et al. (1984) is most exciting and would appear to eliminate supernova sources (which are relatively rare) because of the high ^{26}Al abundance (about 4 $M_{\odot}$ of ^{26}Al in the Galaxy). From the observations on meteorites, it should be noted that supernovae (SN) as a source of ^{26}Al, were never very plausible because of the low $^{26}\text{Al}/^{27}\text{Al}$ ratio for the average ejected matter. In order to provide the ^{26}Al observed in meteorites from a supernova source it was necessary to assume that the ^{26}Al came from the matter in a special zone in the supernova with very high ^{26}Al yield. With no dilution from the ISM and for a standard supernova production ratio of $(^{26}\text{Al}/^{27}\text{Al})_{SN} \sim 10^{-3}$ (see Table 2 of Woosley and Weaver [1982] and see Rodney and Rolfs [1982] for a discussion of the Al-Mg cycle), this would require that $\sim$ 5% of all the ^{27}Al in the solar system was

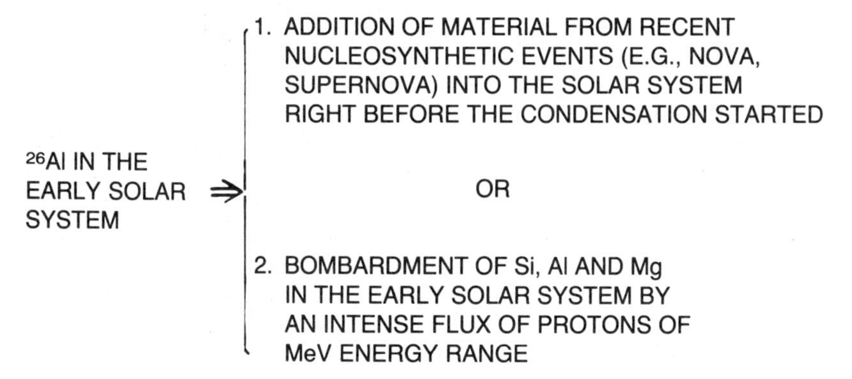

Fig. 20. The obvious alternatives for explaining the presence of ^{26}Al in the early solar system (figure from Lee et al. 1976*a,b*).

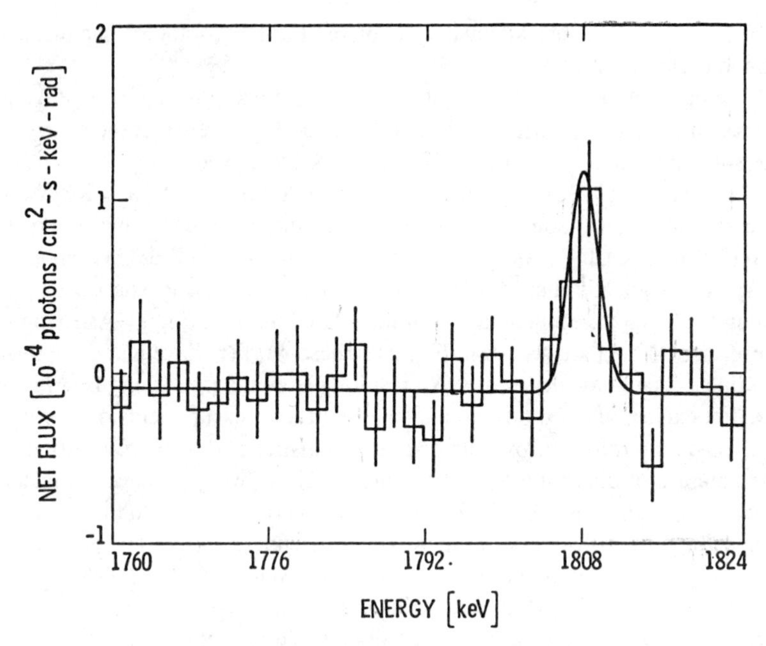

Fig. 21. Gamma ray spectrum of galactic emission in the region of the 1.808 MeV ^{26}Al line (after Mahoney et al. 1982 and Mahoney et al. 1984). These data were taken with the HEAO-3 spacecraft during a 2-week period in 1979 and identify the presence of ^{26}Al as a diffuse source throughout the galactic plane. The line represents the transition of the first excited state of ^{26}Mg (produced by decay of ^{26}Al in the interstellar medium) to the ground state of ^{26}Mg.

from a single late event, a view which my colleagues and I have always felt was most unreasonable. Other proposals and recent experimental studies showed that red giants could produce ^{26}Al/^{27}Al $\sim$ 1 which then eliminates the problem of adding too much fresh material at the last event (Arnould et al. 1980; Nørgaard 1980; Champagne et al. 1983*a,b,c*). Also, other studies have concluded that novae would readily produce ^{26}Al. However, none of the mechanisms listed above would produce ^{107}Pd in an obvious fashion as this requires neutrons ($\langle \sigma\phi\tau \rangle \sim 10^{-4}$).

In a steady state within a dense cloud in which fresh injection is taking place uniformly, we would expect the abundance of a short-lived nuclide to be proportional to the mean life. The abundance of ^{107}Pd and ^{26}Al should also be quite different if they were in steady state in a molecular cloud. This does not appear to be the case for the three nuclides under discussion. As a result, it seems that ^{107}Pd (and possibly ^{26}Al) must have been produced rather locally within the intersellar medium or that the production rates need serious reconsideration. It is possible that the value of $\sim 10^{-4}$ for the abundance of ^{26}Al, ^{107}Pd and ^{129}I is a coincidence.

Insofar as a dense molecular cloud was the placental site from which the matter of the Sun was derived, it is necessary that this medium contain rapidly evolving stars ejecting material. From present considerations, it is believed that molecular clouds were isolated from the general ISM and galactic nucleosynthesis for $\sim 10^8$ yr. This is compatible with the ^{244}Pu and ^{247}Cm results indicating that these nuclei are produced during galactic nucleosynthesis long before the formation of the molecular cloud that appears to have been the medium from which the Sun formed. Some of the ^{129}I could be from this precloud source. However, the short-lived nuclei would have to be produced within the cloud and would require rapidly evolving stars dispersing fresh products into the medium which are rapidly separated and condensed on a time scale of $\sim 10^6$ yr.

With regard to further searches for other extinct nuclides, it appears that ^{53}Mn is a prime target as its lifetime is in the correct range. A very tempting problem would be to look for the decay products of technetium isotopes. This element is known to be present in stars and may provide the best identification of the nature of one of the stellar sources as the production mechanisms have been studied. The recent revision of the mean life of ^{60}Fe (Kutschera et al. 1984) suggests a search for ^{60}Ni*. The absence of γ lines associated with ^{60}Fe and its daughters (Mahoney et al. 1982,1984) and the presence of ^{26}Al as a diffuse source may also have a larger astrophysical significance with the longer mean life for ^{60}Fe. The lines associated with ^{60}Fe would then surely have been observed if they were produced along with ^{26}Al. The absence of γ lines from ^{60}Fe might imply that supernova injections on a galactic scale were much more frequent in early times than at present or that Fe production is completely separated from ^{26}Al production and comes from some stages preceding molecular cloud formation.

In restricting our attention to relatively long-lived nuclides ($\gtrsim 10^6$ yr), it must not be forgotten that some isotopic anomalies present in solar system material suggest radioactive progenitors with much shorter lifetimes. Demonstration of the presence of almost pure ^{22}Ne (Ne-E) in solar system material (see Figs. 22 and 23) by D. C. Black (1972*b*) suggested the presence of dust grains of interstellar origin containing trapped presolar system materials. D. D. Clayton (1975*b*) suggested that freshly synthesized ^{22}Na ($\bar{\tau}$ = 3.6 yr) trapped in dust grains was the source of this anomaly. This explanation would require the separation of stellar debris on an almost instantaneous time scale and the preservation (or trapping) of the noble gas daughter for the trip from another star to make meteorites in the solar system (D. C. Black 1972*b;* Eberhardt 1974; Eberhardt et al. 1979,1981). Some theoretical analyses of possible stellar production mechanisms and the trapping of Ne-E have been carried out (cf., D. D. Clayton, 1975*a,b,*1979; Arnould and Nørgaard 1978). It could mean preservation of rapidly condensed grains around some precursor stars or local production by the early Sun. In either case, the trapping mecha-

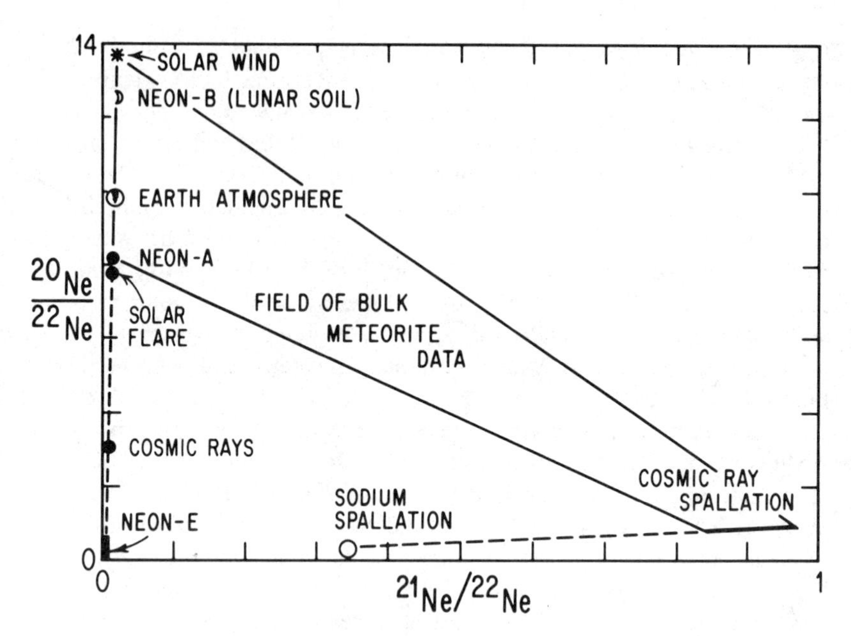

Fig. 22. The isotopic ratios of Ne for different reservoirs represented on a three-isotope correlation diagram. Mixtures of two end members lie on a line. The location of the Ne-E is close to the origin and appears to have closer isotopic affinity with cosmic rays than any other source but is even more enriched in ^{22}Ne (see Fig. 23). The gross nature of isotopic variability of Ne in the solar system is evident. The contrast with the composition of galactic cosmic rays should also be noted. *Note added in proof:* Mewaldt et al. (1984) have obtained a more precise value of $^{20}Ne/^{22}Ne = 9.2$ for solar flare nuclei which is substantially above their previous value (but within the original errors). This new datum for flares plots about half way between Ne-A and Ne-B.

nism for noble gases is quite obscure at the present time. The puzzle of short-lived nuclei and the formation of stars is certainly not quite put together.

Acknowledgments. This chapter is dedicated to Naomi and Jesse Greenstein in honor of their 1.573×10^9 s anniversary of rotating together as a beautiful binary system. The joy of their company and the wisdom of their ways prompted me to leave a pleasant conference a bit early to bask in their noneclipsing sunshine. I wish to thank my colleagues D. A. Papanastassiou, J. Chen, J. Armstrong and I. Hutcheon for permitting me to use their most recent results in preparing this summary. Figures 16, 18, 19 and 20 in this chapter are copies of the original figures used by Lee et al. (1976*a,b*) in their reports at the American Geophysical Union Meeting and the American Astronomical Society meetings in the spring of 1976. They do not yet appear to require change; the questions and alternatives are still there; we are still obliged to either answer the questions or else rephrase them in a clearer way so that we can get answers. This review was much enhanced by insightful and

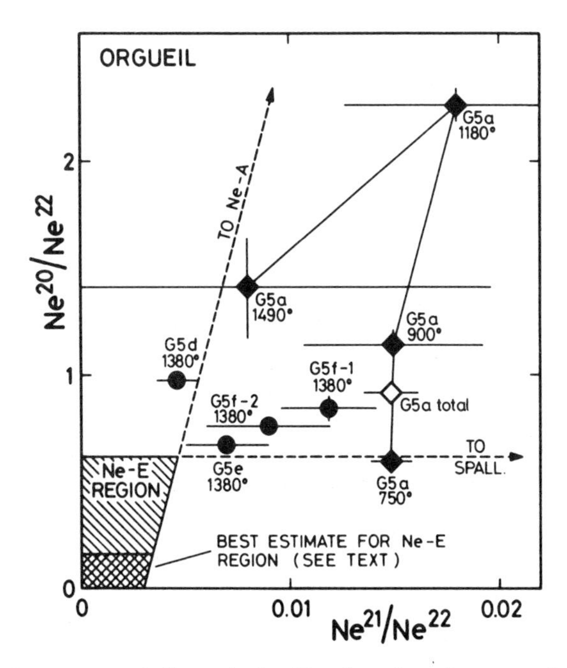

Fig. 23. New data on extremely fine-grained residues from the carbonaceous chondrite Orgueil. These data show the measured isotopic composition of Ne obtained by heating the residues to the temperatures indicated. The Ne-E component is highly enriched in ^{22}Ne as compared with terrestrial atmosphere ^{20}Ne (90%), ^{21}Ne (0.3%), ^{22}Ne (9%) (figure after Eberhardt et al. 1979).

critical comments by D. C. Black, T. Kirsten and an ongoing education by A. G. W. Cameron. The intensive and thorough review by D. D. Clayton was particularly helpful. Some of the stronger and clearer statements were made in response to his comments. This work is supported by a grant from the National Aeronautics and Space Administration.

SURVIVAL OF INTERSTELLAR MATTER IN METEORITES: EVIDENCE FROM CARBONACEOUS MATERIAL

JOHN F. KERRIDGE
University of California at Los Angeles

and

SHERWOOD CHANG
NASA Ames Research Center

An interstellar origin has been attributed to certain organic and apparently elemental forms of C which occur in carbonaceous meteorites. Evidence for such an origin comes from anomalous isotopic composition either of the C itself or of elements, such as N, H, or the noble gasses, which are combined with or trapped within the carbonaceous material. Suggested sources for these anomalous compositions include atmospheres of red giant stars, novae and supernovae, and interstellar molecular clouds.

For many years, it was believed by cosmochemists that the early solar system was isotopically homogeneous. Analyses of meteorites revealed isotopic ratios, for all elements, which differed from normal terrestrial values of those ratios only in ways which could be ascribed to familiar processes such as mass-dependent fractionation, spallation by cosmic rays, or decay of radionuclides. Between 1969 and 1973; however, this view began to change as evidence appeared for isotopic compositions, in certain meteorites, which could not be explained in terms of processes indigenous to the solar system (D. C. Black and R. O. Pepin 1969; R. N. Clayton et al. 1973). Those meteorites were mainly carbonaceous chondrites, a category making up some 5% of stone meteorites, and vigorous study of such meteorites during the past decade has revealed evidence for anomalous isotopic compositions in most elements analysed thus far, although for most elements such anomalies are restricted to minor or trace components within the meteorites (see, e.g., Begemann 1980; chapter by R. N. Clayton et al. in this book). To the extent that those anomalies cannot be understood in the context of solar system processes, one must appeal to an extrasolar origin of the anomaly; by association, its host solid is, therefore, interpreted as surviving interstellar material. Such components may be either crystalline or amorphous solids, and the isotopic anomaly may be present in the composition of either a major element making up the interstellar particle itself (e.g., oxygen), or a trace constituent within that particle (e.g., neon). In either case, the solid is generally described as the host phase of the anomaly. In principle, interstellar matter may also be preserved in substances that do not form discrete directly observable entities but rather occur highly dispersed within meteorites as molecular species either adsorbed or deposited on solid constituents.

Anomalous isotopic composition remains our only key to the identification of surviving interstellar material. In this chapter we shall focus on anomalies contained within host phases composed predominantly of carbon, either in largely elemental form or combined with other elements into organic matter. (Anomalies in noncarbonaceous host phases are discussed in the chapter by R. N. Clayton et al.) We shall consider what sort of information about interstellar material is available from the study of carbonaceous solids, how that information may be obtained, what we have learned so far, and what we might hope to learn in the future. We begin, however, by discussing briefly the nature of the environment in which much of the evidence currently resides, i.e., those meteorites known as carbonaceous chondrites. Some of the same anomalies occur also in ordinary chondrites (see, e.g., Niederer and Eberhardt 1977; Alaerts et al. 1979; Moniot 1980; McNaughton et al. 1981; Yang and Epstein 1983), but attention is focused here on occurrences in carbonaceous chondrites as they have so far been the most informative and contain examples of all the anomalies considered here.

I. CARBONACEOUS CHONDRITES

Not all carbonaceous chondrites are alike, and classification of (and hence understanding of the relationships between) them is far from complete. For present purposes, however, the most important properties may be described by generalizations which embrace most, if not all, types. Note, incidentally, that although there are some 38 carbonaceous chondrites, only half a dozen have been studied in reasonable detail for the elements discussed here.

Like every other meteorite studied so far, each carbonaceous chondrite was once part of a planetesimal-sized object located probably in the asteroid belt. The number of such parent bodies responsible for the presently sampled population of carbonaceous chondrites is not known but probably lies between one and ten (Dodd 1981). While part of its parent body, presumably near its surface, each carbonaceous chondrite was made by the mixing together of materials formed previously in a variety of environments. Accumulation must have been at low-to-moderate temperatures because many of the components are still strikingly out of chemical equilibrium with each other; thus, oxidized and reduced species coexist in close proximity, and minerals of high-temperature, often igneous, origin are embedded in a low-temperature, hydrated matrix (Kerridge and Bunch 1979; Bunch and Chang 1980). By analogy with lunar breccias, of admittedly very different mineralogy, it seems likely that fragmentation of preexisting lithologies, transport and hence mixing of lithic components, and induration into the final meteorite were impact-driven processes. It is not clear, however, whether they took place in a regolith environment, similar to that on the present moon, or during accretion of the parent planetesimal from earlier generations of, presumably smaller, planetesimals. Observational evidence may be found in support of either view (Price et al. 1975; Goswami and Lal 1979).

The lithologies sampled by a carbonaceous chondrite show evidence for a variety of processes including melting, and even partial vaporization, of silicates, thermal metamorphism and loss of volatiles, and aqueous alteration and secondary mineralization (Bunch and Chang 1980; McSween 1979; chapter by Boynton). The spatial and temporal relationships of these processes to each other and to final accumulation are largely unknown. In most cases, the last such process apparently obscured the record of earlier processing, although there is petrographic evidence for partial aqueous alteration of an earlier generation of high-temperature, possibly igneous, silicates (Bunch and Chang 1980). Radiochronological information, though currently sparse, suggests that both high- and low-temperature mineralizations, including the aqueous activity, were among the earliest events in the primitive solar system, although there is some evidence for relatively recent processing as well (C. M. Gray et al. 1973; R. S. Lewis and E. Anders 1975; Macdougall and Kothari 1976; Macdougall et al. 1984).

In summary, the material of carbonaceous chondrites is not pristine but is likely to have experienced significant thermal and/or aqueous processing on the meteorite parent body. Any attempt to infer the nebular or presolar history of a constituent of such a meteorite must, therefore, take such secondary alteration into consideration. Similarly, processing in the solar nebula may have affected the record of presolar events, so that disentangling the effects of several very different environments is a fundamental problem in meteorite research.

The carbon in carbonaceous chondrites occurs in inorganic, organic and apparently elemental forms. The inorganic forms include carbonate, CO_2 and CO (Yuen et al. 1984); these will not be discussed further for lack of evidence for anomalous isotopic composition. For the organic C, i.e., C bonded to hydrogen or nitrogen as well as to itself, a useful distinction can be made between compounds that can be solubilized and extracted with solvents and acids, on the one hand, and a complex macromolecular material, on the other, that is virtually insoluble and makes up a major portion of the C in most carbonaceous chondrites. The soluble organic fraction is prone to terrestrial contamination, though careful work has shown that it is characterized by real, if modest, isotopic variability (Chang et al. 1978; Becker and Epstein 1982). Although most analyses have been performed on bulk extracts, a recent study of individual carboxylic acids and hydrocarbons did not reveal unusual C isotopic compositions (Yuen et al. 1984). The available evidence, therefore, indicates a solar system origin for this material and it will not be further considered here. Instead, this discussion focuses on the insoluble organic matter and apparently elemental forms of C, which appear to play host to material of interstellar origin.

II. SEPARATION AND ANALYSIS OF THE PRESOLAR COMPONENT

All of the carbonaceous host phases considered here exist as relatively minor constituents dispersed throughout a meteorite with essentially no known petrographic relationship either to each other or to the lithic fabric of the meteorite. Optimization of the analysis conditions requires such a host phase to be prepared in as pure a state as possible. This is generally achieved by utilizing a variety of chemical and physical properties of the host phase to separate it from the bulk of the meteorite. A typical purification procedure might involve the following steps (see, e.g., R. S. Lewis and E. Anders 1983).

After grinding to a fine powder, the meteorite is treated with acids like HF and HCl to dissolve the major inorganic minerals, which are mainly silicates. Unwanted by-products, such as fluorides and elemental sulfur, are removed by means of appropriate solvents, and other reagents may be used to dissolve possibly interfering species. The residue is then divided into a number of fractions based upon grain size and/or density, and the fraction containing the phase of interest is heated to progressively higher temperatures, either in a vacuum or in an atmosphere of pure oxygen. Gaseous products are collected and analyzed isotopically; those generated in the interval within which the host phase breaks down are identified as due to that phase. Several features of this approach require further comment.

First, the petrographic context of the host phase is lost; any potentially meaningful associations between the host phase and other minerals in the original meteorite invariably fail to survive the dissolution procedure. Consequently, in view of its brecciated nature, it is difficult to use the petrology of the meteorite to infer anything about the history of the host phase, which may, indeed, have been mixed into the meteorite at the last minute. Elucidation of the extent to which different meteoritic constituents have shared a common history should be a major goal of work in this area.

Second, not only does the purification procedure deprive the host phase of mineral associations which it enjoyed in the meteorite but it can create apparent associations which are probably spurious. Any mineral with the same physical/chemical properties as the host phase, but genetically unrelated to it, will end up in the same fraction and be analyzed with it. This not only causes dilution of interesting material but, if both minerals release isotopically distinguishable components over the same temperature range during analysis, this can confuse interpretation of the data.

Finally, although resolution of the previous problem can be aided, if not accomplished, by microscopic characterization of each fraction prior to and/or following analysis, such studies have been of only limited value so far. Not only is it quite difficult to correlate isotopic data with specific particles imaged in a microscope, but, given the very fine-grain size and frequently amorphous or partially crystalline nature of many apparent host phases, it is not always possible to make precise identifications of such particles. This, too, is an area in which much more work is needed (Lumpkin 1981). Note that noncarbonaceous host phases generally require much less harsh separation procedures so that many of the comments given above do not apply to them.

In summary, although the procedures described above have led to several important recent discoveries, they are not without their problems and caution is still required in interpretation of their results. We turn next to the nature of such results and their interpretations.

III. ISOTOPIC SYSTEMATICS OF CARBONACEOUS MATERIAL

Interpretation of variations in isotopic composition is greatly aided if the effects of nuclear processes, such as nucleosynthesis or cosmic ray spallation, can be distinguished from those due to mass-dependent fractionations, such as those associated with diffusive loss or chemical reactions. For elements with three or more stable isotopes, the latter case can generally be recognized by an observed dependence of the degree of fractionation upon the fractional mass difference of the relevant isotopes (see, e.g., R. N. Clayton et al. 1973; also see, Thiemens and Heidenreich 1983). This approach is applicable to oxygen, one of the four major elements making up carbonaceous material; however, C, H, and N are characterized by only two stable isotopes each (see Table I), so that this approach cannot be used. Instead, it is necessary to adopt a less direct approach; the one most commonly employed may be termed "guilt by association." If a C-, H-, or N-component with a particular isotopic composition is released over the same temperature range as an independently well characterized component in a multiisotope element, such as one of the noble gases, or if the abundances of the two components correlate with each other in different samples, a similar origin for the two components may be inferred. Bearing

TABLE I
Stable Isotopes of Hydrogen, Carbon, and Nitrogen with Abundances and Principal Modes of Nucleosynthetic Production

	Principal Mode of Nucleosynthesis	Isotopic Ratio: Heavy/Light	
		Terrestrial Standard	Protosolar Value
^{1}H ^{2}D	Big bang Big bang Spallation ?	1.55×10^{-4}	2×10^{-5}
^{12}C ^{13}C	Helium burning CNO cycle	1.1×10^{-2}	1.1×10^{-2}
^{14}N ^{15}N	CNO cycle Explosive CNO	3.67×10^{-3}	?

in mind the difficulty of preparing pure specimens of different host phases, the possibilities for ambiguity in this approach are obvious.

Despite the problems outlined above, the search for interstellar carbonaceous matter in meteorites is worthwhile for a number of reasons.

1. Carbon is an abundant element in the cosmos so that in most regions of the universe it constitutes a significant fraction of the potentially condensible elements.
2. The formation of C-containing compounds and their conversion to solid phases can take place in a wide variety of astrophysical environments, from cold interstellar molecular clouds to hot ejecta from red giant stars.
3. Condensed carbonaceous material appears to be relatively durable in the various environments which it might experience between its formation and emplacement in the meteorite. These include the interstellar medium, the primordial solar nebula and the meteorite parent body. Nonetheless, as stressed earlier, the possibility of alteration must always be borne in mind.

Paralleling the three factors described above, we may define three types of information which we might obtain from successful identification of interstellar material in a meteorite.

1. We can recognize at least some of the nucleosynthetic sources which supplied material to the primitive solar system.
2. We can learn something about the processes which led to production of the dust which entered the early solar system.
3. Survival of interstellar dust can enable us to place constraints on conditions in the solar nebula and on the meteorite parent bodies.

In this context, it is interesting to note that carbonaceous host phases occur in various types of primitive meteorites that differ in chemical and mineralogical composition, redox state, and inferred thermal history. Theories of protosolar collapse, nebular thermochemical evolution, and parent body formation should take into account the distribution and preservation of these components in meteorites.

The concepts outlined above are best illustrated by specific examples drawn from recent studies of apparently interstellar carbonaceous material preserved in meteorites. These are described in the next section (see also R. S. Lewis and E. Anders 1983).

IV. INTERSTELLAR CARBONACEOUS COMPONENTS

Of the numerous forms in which C occurs in carbonaceous chondrites, four are supposed to have an interstellar origin. The evidence leading to such interpretations and the apparent nature of the components are discussed below, using the nomenclature of R. S. Lewis and E. Anders (1983). This nomenclature is by no means universally accepted but it seems to us to provide a useful taxonomic approach. Note that, for the first three carbonaceous components, each is defined on the basis of the presence within it of a specific isotopic anomaly in a noble gas.

Carbon-Alpha

This designation is given to the carrier of a trapped component of neon, termed Ne-E, which is essentially pure ^{22}Ne (D. C. Black and R. O. Pepin 1969; D. C. Black 1972*b*; Eberhardt et al. 1981), an isotope which constitutes only about 10% of "normal" Ne (Fig. 1). The evidence favoring a carbonaceous carrier is indirect: low cosmogenic ^{21}Ne content implying low *Z*; susceptibility to oxidative decomposition; density less than 2.3 g cm^{-3}; and release of Ne-E at about 600°C in stepwise pyrolysis. A direct correlation of C with Ne-E remains to be established. Nonetheless, taken together, all of these properties are difficult to reconcile with a non-carbonaceous carrier.

The grain size of C-alpha is speculated to be in the range 1 to 10 μm, but its degree of crystallinity and chemical structure are unknown. It is roughly estimated to comprise less than about 5 ppm (parts per million) of a typical carbonaceous chondrite (R. S. Lewis and E. Anders 1983).

Although it is possible for ^{22}Ne to be made preferentially to the other isotopes of Ne during nucleosynthesis under certain conditions (Arnould and Norgaard 1978), it is generally believed that Ne-E resulted from the decay of extinct ^{22}Na (D. C. Black and R. O. Pepin 1969). The shortness of its half-life, 2.6 yr, requires either that the ^{22}Na was produced in the early solar system by charged particle irradiation (Heymann and Dziczkaniec 1976; Lee 1978), for which supporting evidence in other isotopic systems is lacking, or, more probably, that decay occurred in an environment close to the site of its stellar nucleosynthesis (D. D. Clayton 1975).

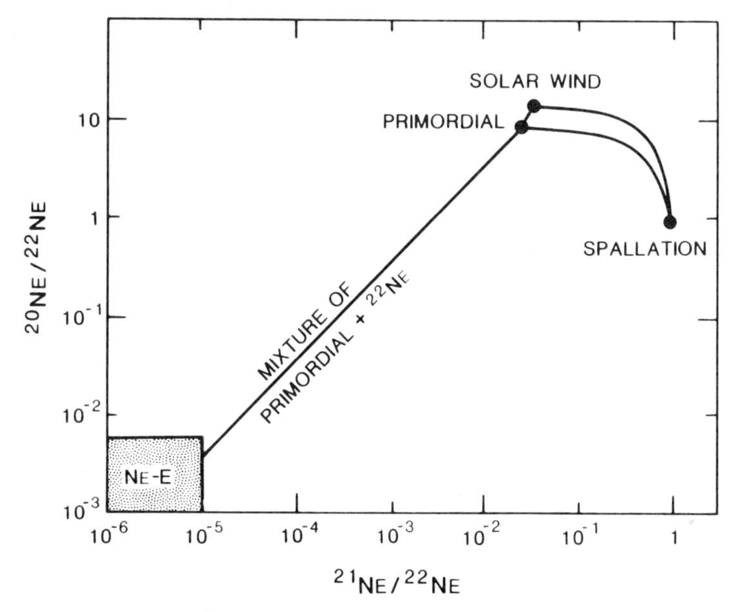

Fig. 1. A logarithmic three-isotope plot for the element neon. The anomalous neon component known as Ne-E, consisting of essentially pure ^{22}Ne, is dramatically different in isotopic composition from the common components of Ne in the solar system. Figure after Eberhardt et al. (1981).

One possible site is the envelope of an exploding nova and, since elemental C is believed to be the first solid to condense from such a gaseous envelope (D. D. Clayton and F. Hoyle 1976), C-alpha may be interpreted as a nova-condensate. Carbon isotopic measurements have not yet been obtained in analyses in which the C released is directly correlatable with Ne-E released by the same extraction technique. Therefore, a firm estimate of the $^{13}C/^{12}C$ ratio of C-alpha is not yet available, though a value in excess of 0.0125 has been suggested (R. S. Lewis and E. Anders 1983), which would be consistent with origin in a nova, though not diagnostic. Carbon condensed from a nova should contain ^{14}N from decay of freshly synthesized ^{14}C, but any N gas trapped by the C grains would be rich in ^{15}N so that any prediction of the isotopic composition of N associated with Ne-E is problematic (D. D. Clayton and F. Hoyle 1976). Such N has not yet, in fact, been detected (R. S. Lewis et al. 1983*b*).

A further ambiguity for the nova hypothesis should be acknowledged; in the same meteorites, Eberhardt et al. (1981) discovered additional Ne-E trapped in the mineral apatite [$Ca_2PO_4(OH,F)$] which cannot plausibly have condensed in a nova, or in any other stellar environment. Further work is obviously needed.

Carbon-Beta

Apparently another elemental form, C-beta is defined as the carbonaceous host phase of an anomalous Xe component, termed s-Xe. This component has a unique isotopic spectrum (Srinivasan and Anders 1978) which is an excellent match with that predicted by D. D. Clayton and R. A. Ward (1978) for Xe produced nucleosynthetically by the slow-neutron-capture, or s-process (Fig. 2). That process is believed to take place in intermediate zones

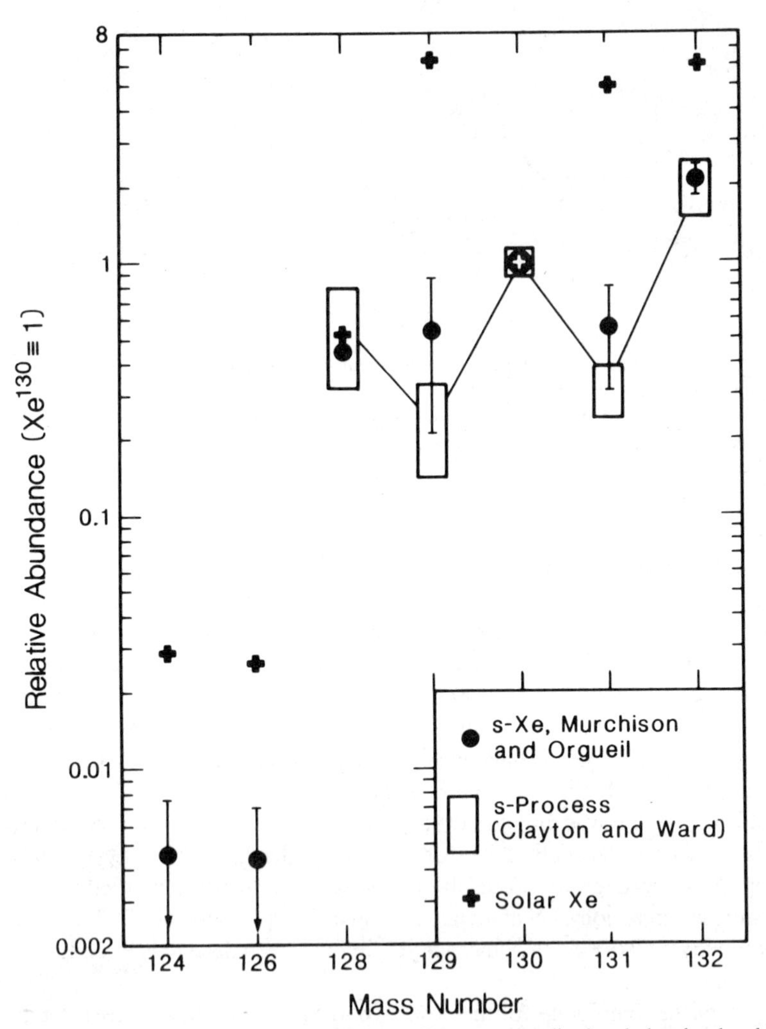

Fig. 2. The relative abundance of xenon isotopes. Data are normalized such that the abundance of ^{130}Xe is set at unity. The isotopic composition of Xe associated with the carbonaceous component termed C-beta, corresponds very closely with that calculated by D. D. Clayton and R. A. Ward (1978) for Xe produced during *s*-process nucleosynthesis. Figure from Anders (1981).

in red giant stars. An elemental form of C is inferred from the observation that the carrier retains s-Xe during treatment with NaOCl and Na_2O_2, but loses s-Xe and is apparently destroyed upon further oxidation by $HClO_4$.

Little is known about the physical characteristics of C-beta, which is thought to account for $\lesssim 5$ ppm of a typical carbonaceous chondrite. Its grain diameter is suspected to be about 1 μm, but whether it is crystalline or amorphous is not known nor is its prevailing chemical structure. It begins to break down and release its trapped Xe at temperatures above about 1100°C during vacuum pyrolysis, suggesting that it is quite refractory.

When a sample of an acid-resistant residue containing s-Xe was combusted in stepwise fashion, C released as CO_2 above 900°C exhibited a $^{13}C/^{12}C$ ratio greater than 0.024 (Swart et al. 1983), i.e., more than twice the average solar system value (Table 1). The occurrence of this C in the same samples as s-Xe is consistent with them possessing a carrier-trapped gas relationship. The release of s-Xe has been accomplished, however, by pyrolysis at 1100–1600°C rather than by combustion, and the association between the ^{13}C-enriched C and s-Xe may be coincidental.

Carbon isotopic ratios as high as 0.5 have been associated with red giant and carbon stars (see, e.g., Lambert and Sneden 1977; Scalo 1977*b*), and these types of stars are known to have C-containing molecules and carbonaceous grains, possibly including silicon carbide, in circumstellar shells. Material can also be ejected during subsequent evolution of red giants as novae or supernovae. Elemental C and SiC are plausible condensates to form from such material and the isotopic composition inferred for C-beta seems to be consistent with such origins, if not strictly diagnostic. Note that the observed enrichment in ^{13}C is not on its own uniquely characteristic of a nuclear anomaly as fractionation of such a magnitude could in principle be produced by ion-molecule reactions in interstellar clouds at temperatures below about 30 K (W. D. Watson 1977). However, consistent nucleosynthetic origins for both s-Xe and ^{13}C-enriched C lend support to a nuclear interpretation of the anomalous C and to its identification with C-beta.

If formed in a red giant, C-beta could have trapped N with an anomalously low value of $^{15}N/^{14}N$, but analyses of C-beta have not yet revealed evidence of such isotopically light N (R. S. Lewis et al. 1983*b*).

If C-beta is the ^{13}C-enriched component, as supposed, it is unusually resistant to oxidation. Its characteristic isotopic signature emerges after carbonaceous acid residues have first been subjected to wet oxidation and then only during subsequent stepwise combustion in O_2, whereupon it appears most strongly at temperatures >900°C. It has been suggested (Swart et al. 1983) that the high-combustion temperatures required for C-beta may reflect a relatively high degree of crystallinity, but this has not been established. By comparison, combustion of highly crystalline graphite and diamond is virtually complete at 900°C (Swart et al. 1982). If the ^{13}C-enriched C is the carrier of s-Xe, it is broken down over the same temperature range by either pyrolysis or combustion. In any case, it exhibits chemical properties unexpected for a form of elemental C.

Although the association between ^{13}C-enriched C and s-Xe is not rigorously established, it is nonetheless a reasonable one, so that for the present, C-beta may be interpreted as having condensed from the atmosphere of a red giant star during either a quiescent or an explosive stage (R. S. Lewis and E. Anders 1983). Because of its refractory nature, its survival tells us less about temperatures in the early solar system than may be inferred from survival of some of the less refractory species discussed in subsequent sections.

Carbon-Delta

Yet another apparently elemental form, C-delta is also characterized by the presence of isotopically unusual trapped Xe (see, e.g., Anders 1981). As illustrated in Fig. 3, this Xe is characterized by enrichments, relative to average solar system Xe, in both light and heavy isotopes. For that reason, it is convenient to refer to it as Xe-HL, although numerous other terms have been used in the past.

The carbonaceous composition of the Xe-HL host phase is more firmly established than that of any of the other gas components described above. This follows from its low density (<2.3 g cm^{-3}), partial susceptibility to wet oxidation, complete combustibility in O_2 at about 600°C, and direct correla-

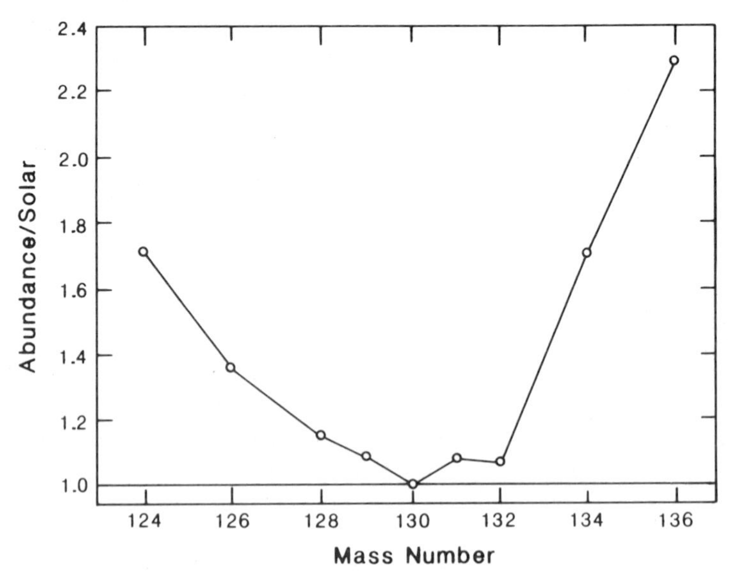

Fig. 3. Same as for Fig. 2, except abundances are plotted on a linear rather than a logarithmic scale. Xenon associated with the carbonaceous component known as C-delta, is strikingly enriched in heavy and light isotopes compared with normal solar system Xe. Figure from R. S. Lewis and E. Anders (1983).

tion of C abundance with that of the noble gases (Ott et al. 1981; Frick et al. 1983). Carbon-delta appears to be much less resistant to oxidation than is C-beta.

The two principal hypotheses for the origin of Xe-HL invoked fission and supernova production, respectively. The fission hypothesis (see, e.g., Anders 1981) identified the "H" part of the spectrum with Xe produced by fission of a transuranic element. Although the observed meteoritic spectrum grossly resembled that characteristic of fission, it failed to match in detail that of any known transuranic element. Consequently, the parent nuclide was identified as a now extinct superheavy element with a postulated proton number around 114. The "L" part of the spectrum could not be explained by fission, but its resemblance to a mass-fractionation trend led to the suggestion that HL-Xe consisted of average solar system Xe, heavily mass-fractionated in favor of the light isotopes, to which the superheavy-fission-Xe was added. The mechanism by which that marriage took place was conjectural.

The supernova hypothesis, on the other hand (see, e.g., Manuel et al. 1972; Ott et al. 1981), invoked production of the "H" part of the Xe spectrum in a neutron-rich zone within a supernova (i.e., a site of rapid neutron capture, or *r*-process, nucleosynthesis), whereas the "L" part originated in a proton-rich zone within the same supernova (i.e., a site of proton capture, or *p*-process, nucleosynthesis). Mixture of these two zones somewhat later in the supernova explosion would have produced the observed Xe-HL and trapped it in condensing grains.

Two lines of permissive evidence have recently suggested that the supernova hypothesis is more likely to be correct. In the first place, a search within samples of C-delta for evidence of fissiogenic isotopes, predicted by the fission hypothesis to be present in the elements barium, neodymium, and samarium, proved to be negative (R. S. Lewis et al. 1983*a*; Lugmair et al. 1983). In both hypotheses, Xe-H would be associated with isotopically anomalous Ba, Nd, and Sm, but *in situ* fission, as hypothesized, would freeze in that association in the host phase, whereas in the supernova hypothesis, a plausible, though *ad hoc,* suggestion is that the relatively refractory elements Ba, Nd, and Sm condensed in a different region of the expanding supernova from that in which Xe was trapped (R. S. Lewis and E. Anders 1983).

The second line of evidence involves the isotopic composition of N which is apparently trapped or combined in C-delta, in association with Xe-HL. Again, the association is not rigorously established but seems reasonable, in light of Xe-HL and N-isotopic determinations made simultaneously on the same sample (Frick et al. 1983; see also R. S. Lewis and E. Anders 1983). Such N is relatively highly enriched in ^{14}N, with a $^{15}N/^{14}N$ ratio less than 2.5×10^{-3} (Frick et al. 1983; R. S. Lewis et al. 1983*b*). This value is substantially below the terrestrial value, given in Table I, although the protosolar value is currently unknown. Because the fission hypothesis postulates a local, i.e., solar system, origin for C-delta, isotopically anomalous N in such material is less likely to be consistent with that hypothesis, whereas the presence of ^{14}N-enriched N requires an alien, i.e. extrasolar, source for C-delta and hence for Xe-HL. This is regarded as loosely consistent with the supernova hypothesis (R. S. Lewis et al. 1983*b*), although an actual nucleosynthetic connection between Xe-HL and the ^{14}N-enriched N has not been established.

The isotopic composition inferred for C-delta itself seems to be mundane, with a $^{13}C/^{12}C$ value of about 0.011 (Ott et al. 1981; Frick et al. 1983; R. S. Lewis and E. Anders 1983). Superficially, this would appear to favor a local, i.e. fission, origin rather than an exotic, i.e. supernova, origin for Xe-HL, but the possibility of dilution by normal solar system C reduces the force of this argument. Indeed, there is growing evidence that much, if not most, of the acid-insoluble carbonaceous matter in meteorites contains little noble gas of either local or exotic origin, which underscores the dilution problem (Ott et al. 1984), and the need for caution in assigning C and N isotopic compositions to host phases of anomalous noble gas components.

The grain size of C-delta is quite uncertain (Lumpkin 1981; Swart et al. 1983), as is its degree of crystallinity, whether amorphous (Lumpkin 1981; R. S. Lewis and E. Anders 1983) or turbostratic (Lumpkin 1981). Its abundance in a typical carbonaceous meteorite may be as much as 200 ppm. Unlike s-Xe in C-beta, Xe-HL is released at relatively low temperatures during pyrolysis, beginning at 500 to 600°C with peak release at about 800°C. This suggests that it could not have experienced such temperatures in the early solar system.

Organic Matter

The preceding carbonaceous host phases were all prepared by treating demineralized residues with chemical oxidizing agents in order to remove the abundant insoluble organic matter present in these meteorites. Unfortunately, no procedure has yet been devised for performing the inverse process, i.e., preparing a sample of the insoluble organic matter free from elemental C. Isotopic analyses of nominally organic residues are, therefore, contaminated with C and N derived from forms of C such as those described above. It is possible to correct such data for contributions due to nonorganic material, yielding small but significant variations in $^{13}C/^{12}C$ and $^{15}N/^{14}N$, but the most striking isotopic feature characteristic of organic matter is found in the H present in such material.

Substantial enrichment of D is apparent in the organic material as a whole (Kolodny et al. 1980), but the most striking D contents are observed in the insoluble organic fraction (Robert and Epstein 1982), which may constitute up to about 3% by weight of a carbonaceous chondrite (Hayes 1967). Bulk samples of the insoluble organic matter, which, like terrestrial kerogen, apparently consists of highly condensed aromatic moieties attached to short aliphatic bridges and side chains (see, e.g., Hayatsu et al. 1977), give D/H ratios up to 5.4×10^{-4}, while individual fractions in a stepwise release can reach values of 6.9×10^{-4} (Robert and Epstein 1982). It seems likely that

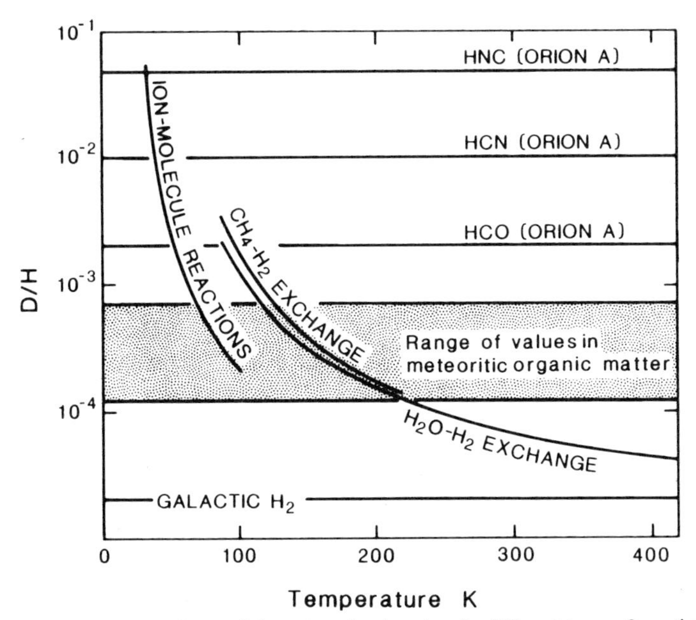

Fig. 4. Temperature dependence of deuterium fractionation for different types of reaction compared with deuterium content of some representative astrophysical and meteoritic materials. Meteoritic organic matter and interstellar molecules are enormously enriched in D compared with the galactic H_2 from which the solar system formed. This enrichment is too great to have been achieved during reactions involving neutral species (right-hand curves), but is consistent with the fractionations associated with ion-molecule reactions at interstellar cloud temperatures (left-hand curve). Figure after Geiss and Reeves (1980).

protosolar H had a D/H ratio of about 2×10^{-5} (D. C. Black 1973; Geiss and Reeves 1980), leading to enrichments of D in meteoritic material by nearly a factor of 35. Fractionations of such magnitude could only be achieved in the early solar system by recourse to extreme conditions which seem inconsistent with our current understanding (D. C. Black 1973; Geiss and Reeves 1980). Thus, as illustrated in Fig. 4, isotopic exchange between H_2 and an organic molecule such as CH_4 could theoretically concentrate D to that extent in the latter molecule at temperatures below about 130 K (Reeves and Bottinga 1972). However, at such low temperatures, the reaction would take many times the age of the universe (R. S. Lewis and E. Anders 1983).

The only known mechanism which seems capable of explaining the D enrichment observed in meteoritic organic matter is the fractionation associated with ion-molecule reactions at temperatures below about 100 K as illustrated in Fig. 4. Such reactions are thought to be responsible for observed enrichments of D by factors up to 2.5×10^3 in dense interstellar clouds which are characterized by appropriately low temperatures (see, e.g., W. D. Watson 1977; Snell and Wootten 1979). The possibility of similar reactions having taken place in the primitive solar nebula cannot be ruled out but the lack of evidence for suitably high levels of ionization and low temperatures makes this seem less likely. Consequently, even though it is strictly an argument by default, the conclusion that D-enriched organic molecules synthesized by ion-molecule reactions in interstellar clouds entered the early solar system and were incorporated into carbonaceous meteorites (Kerridge 1980,1983; Geiss and Reeves 1980) has been widely accepted (see, e.g., Hayatsu and Anders 1981; Robert and Epstein 1982; Yang and Epstein 1983).

Even within the context of an interstellar origin for the meteoritic D enrichment, several issues remain unresolved. It is not clear whether meteoritic D/H values reflect mixing of material possessing the extreme enrichments observed, for example, in Orion A, with less fractionated, or even unfractionated, material, or whether they correspond to ion-molecule reactions taking place at somewhat higher temperatures than those in Orion. Put another way, are there two kinds of kerogen-like material in the meteorites, one exotic and one local, or is it all of interstellar origin? The mundane $^{13}C/^{12}C$ values measured for insoluble organic matter (Smith and Kaplan 1970; Robert and Epstein 1982; Kerridge 1983) suggest that the former is more likely, although the issue of dilution by ordinary C remains unresolved. Another question concerns whether or not the apparently exotic kerogen-like material was itself made in an interstellar cloud; an alternative is that it was made within the solar system by polymerization of simpler organic molecules synthesized earlier in the interstellar environment. Evidence is currently lacking on this issue; the actual D/H ratio of interstellar carbonaceous grains, as distinct from gaseous molecules, is unknown, as is the involvement of ion-molecule reactions in their formation. A question alluded to above involves the isotopic composition of C and N in the exotic organic matter. Both elements are significantly fractionated isotopically during ion-molecule reactions at low temperatures (W. D. Watson 1977; N. G. Adams and D. Smith 1981*b*). Consequently, it would be anticipated that D-enriched H would be associated with isotopically fractionated C and N, though the magnitude, and even the sign, of such fractionation is not yet predictable. As noted above, however, organic matter has so far yielded only "normal" isotopic compositions for C and N. The potential problem of isotopically anomalous exotic C or N being masked by such abundant "normal" material needs to be assessed.

Several structural studies have been made of the insoluble organic fraction in meteorites and a number of features of this material have been revealed (Bandurski and Nagy 1976; Hayatsu et al. 1977,1980). However, if, as seems likely, the exotic material is mixed with abundant material produced locally, it would be unwise to attribute the observed structural features to the interstellar material. The one physical property of the exotic material which is reasonably well defined is its thermal stability. From its behavior during vacuum pyrolysis, it seems unlikely that it could have experienced temperatures above about 300°C (Robert and Epstein 1982) so that, if it entered the solar system in this form, a constraint is imposed on the nebular temperature at some time and location, as yet unknown. If, as suggested above, the kerogen-like material were polymerized locally, some of its exotic precursor molecules could have had even lower thermal stability.

V. CONCLUSIONS AND FUTURE DIRECTIONS

Taking the interpretations given above at face value, the conclusion may be drawn that grains formed in the atmospheres of red giants, novae and supernovae, and molecules, and possibly grains, synthesized in interstellar clouds, survived entry into the primitive solar system and incorporation into meteorite parent bodies. Future work will attempt to consolidate those conclusions, as well as to define more closely both the conditions in which the exotic material was formed and the environments which it has experienced since formation. Some possible lines of inquiry are discussed briefly below.

In the case of the apparently elemental forms of C, it is important to characterize those forms more precisely using microscopic and crystallographic techniques. Included in that study would be search for genetic links between different forms of carbonaceous material. For example, there is a suggestion that meteorites which show increased evidence for thermal metamorphism also contain a greater proportion of relatively well-ordered C. This raises the possibility that some forms of C, now apparently distinct, are actually related genetically. Such a study should also lead to a recognition of criteria indicative of secondary alteration of interstellar material.

An important issue to be addressed by microscopic investigation of putative interstellar material in meteorites is the possible relationship between such material and the core-mantle structures proposed by J. M. Greenberg (1982*a*) to explain the spectral properties of interstellar grains. As noted above, experiments to date have destroyed any petrographic associations which the carbonaceous matter may have had in the meteorite and future studies should include a search for possibly surviving core-mantle structures. It may turn out that the silicate cores have failed to survive but that characteristic morphology may be recognizable in some of the carbonaceous material, such as hollow spheres or sheaths.

Further purification of the various exotic forms of C is desirable, most particularly with a view to obtaining isotopic data which might be more precisely diagnostic of their mode of origin. Examples here are the $^{15}N/^{14}N$ ratio of C-beta and the $^{13}C/^{12}C$ ratio of C-delta. Similarly, more precise evaluation of the isotopic compositions of C and N associated with D-enriched organic matter is necessary, and this leads to an important contribution which is needed from the astrophysical community. As noted earlier, it is not yet clear what isotopic fractionations in C and N would be expected to accompany incorporation of D-enriched H into interstellar molecules. Construction of a model capable of making such predictions is obviously not a trivial task, but even a semi-quantitative estimate would be of value at this time.

Furthermore, except for carbonate/CO_2, few O isotopic analyses have been reported on carbonaceous substances in meteorites. The 3-isotope O system may reveal anomalies in components studied so far that exhibit only mundane compositions for C, H, and N. Along these same lines, analyses of C, H, N, and O in discrete compounds of the extractable organic fraction may yet uncover possibly anomalous features.

Discrimination between exotic and local organic matter is obviously desirable and to this end the isotopic fractionations associated with production of organic matter by plausible mechanisms in the solar system need further exploration. In addition, it is by no means certain that the meteorites analyzed so far have sampled the full range of isotopic variability characteristic of the early solar system; as noted earlier, less than one sixth of the carbonaceous meteorites have been analyzed in any detail, and there is increasing evidence for small quantities of organic matter in noncarbonaceous meteorites. Clearly, there is plenty still to do.

Acknowledgments. Support by the National Aeronautics and Space Administration through the Planetary Materials and Exobiology Programs is gratefully acknowledged.

ISOTOPIC VARIATIONS IN SOLAR SYSTEM MATERIAL: EVAPORATION AND CONDENSATION OF SILICATES

ROBERT N. CLAYTON, TOSHIKO K. MAYEDA,
and CAROL A. MOLINI-VELSKO
University of Chicago

Simultaneous determination of stable isotope abundances of oxygen and silicon in meteoritic chondrules and inclusions shows that these two abundant elements undergo fundamentally different cosmochemical processes. Silicon isotopic variations reflect processes of evaporation and condensation at high temperatures, probably during brief melting events. Oxygen isotopic variations reflect interaction between isotopically distinct reservoirs, probably involving a gaseous component of the solar nebula. There remains uncertainty as to the origin of the various reservoirs.

It was once believed that all of the matter comprising our solar system had passed through a stage of thorough homogenization, probably in a high-temperature gaseous state, so that any preexisting isotopic heterogeneities were erased. For example, Suess (1965) wrote: "Among the very few assumptions which, in the opinion of the writer, can be considered well justified and firmly established, is the notion that the planetary objects, i.e., planets, their satellites, asteroids, meteorites, and other objects of our solar system, were formed from a well-mixed primordial nebula of chemically and isotopically uniform composition. At some time between the time of formation of the elements and the beginning of condensation of the less volatile material, this nebula must have been in the state of a homogeneous gas mass of a temperature so high that no solids were present." Isotopic heterogeneities are to be expected in presolar matter since several different astrophysical sites and processes are required for nucleosynthesis of all of the known nuclides (Burbidge et al. 1957). In particular, different isotopes of an element are likely to have been produced, at least in part, in different processes, so that lack of subsequent homogenization should lead to isotopic variability inherited from nucleosynthesis. More recent evidence (D. C. Black 1972; R. N. Clayton et al. 1973) showed that isotopic homogenization of the matter in the solar system was not complete, so that the residual isotopic variability can, in principle, be used both to provide experimental information related to nucleosynthesis and to provide tracers of chemical and physical processes in the early solar system.

During the last decade, studies of meteorites have shown that several elements have nonstandard abundances of their stable isotopes; see reviews by R. N. Clayton (1978), Lee (1979), and Begemann (1980). Those isotopic variations that are not understood in terms of known solar system processes have been called isotopic anomalies. Despite the large amount of experimental information that has accumulated, we still lack even the most rudimentary understanding of the origin of these anomalies. Appeal has been made to nucleosynthetic origins, both near and remote in time (R. N. Clayton et al. 1973; Cameron and Truran 1977; D. D. Clayton 1982), to energetic particle irradiation in the T Tauri stage of solar evolution (Lee 1978), and to unusual photochemical isotope effects (Thiemens and Heidenreich 1983). A part of the difficulty in interpreting the evidence no doubt stems from our inadequate understanding of the histories of the individual small bodies (chondrules, crystals, dust aggregates, etc.) that are the bearers of the observed signals. A goal of this chapter is to use some of the isotopic information available in small components of meteorites to follow the chemical and thermal histories of these small particles before they were incorporated into their larger parent bodies, in order to see how nebular processes may have modified the primary raw materials.

Partitioning of elements between a gaseous phase and condensed phases played a major role in determining the chemical compositions of the solid bodies in the solar system; see the reviews by Anders (1971), and by L. Grossman and J. W. Larimer (1974). The partitioning processes are combinations of evaporation of solids or liquids, condensation of vapors, and exchange between condensed and gaseous reservoirs. While a great deal has been written on the effect of these processes in producing the observed elemental abundances in planets and in meteorites, the location of the processes in space and time in the early solar system remains poorly understood. We shall explore the application of light-element isotope abundance variations to the study of the volatility related processes. The emphasis will be on abundant rock-forming elements: oxygen, silicon, magnesium, calcium, and titanium, for which the isotopic fractionations in intraplanetary processes are either negligibly small or are well understood, in order to study the effects of cosmochemical processes without confusion due to planetary processes. Those elements which are very volatile in cosmochemical processes (the noble

TABLE I
Solar-System Abundances

O	18.4×10^6
Mg	1.06×10^6
Si	1.00×10^6
Fe	0.90×10^6
Na	6.4×10^4
Al	8.5×10^4
Ca	6.3×10^4

gases, hydrogen, carbon, and nitrogen) also exhibit a large and mystifying variety of isotopic variations which are not obviously related to any of the effects discussed here; see, for example, Podosek (1978) and Yang and Epstein (1983).

Oxygen, magnesium, and silicon are the three most abundant rock-forming elements in the inner solar system. (We shall not be concerned here with the ices that are abundant farther out.) Each of these elements has three stable isotopes, so that mass-dependent processes can be recognized by their isotopic fractionation patterns. The information available in studying all three elements is not redundant, due to the peculiar cosmochemical nature of oxygen. Wood (1981*a*) has emphasized the unique property of oxygen in that it must occupy two major reservoirs, one gaseous and one solid, over a wide range of nebular temperatures. This is simply a consequence of the solar system elemental abundances. Table I gives the relative atomic abundances of the seven most abundant rock-forming elements. It can be seen from simple stoichiometry that at temperatures below about 1300 K (L. Grossman and J. W. Larimer 1974), all of the major metallic elements should be in the solid state, whereas only about 20% of the oxygen is combined as solid oxides or silicates, the remainder being bound as gaseous carbon monoxide and water vapor. The existence of isotopic anomalies in oxygen on scales ranging from micrometers to hundreds of kilometers is most easily understood in terms of interactions of isotopically distinct reservoirs. The most obvious way to maintain isotopic heterogeneity on the scales observed is to have different isotopic abundances in the gaseous and solid reservoirs. This difference has been interpreted to be a consequence of different nucleosynthetic histories of the oxygen in the two reservoirs (R. N. Clayton 1981).

Interaction between gaseous and condensed reservoirs will be discussed for three sets of meteoritic components: chondrules, calcium-aluminum-rich inclusions (CAIs), and Allende inclusions with fractionation and unknown nuclear effects (FUN). Chondrules are once-molten spherules, typically 0.1 to 1.0 mm in diameter, that are the principal constituent of the most common stony meteorites, the ordinary chondrites. CAIs are rounded to irregular objects, typically 1 to 10 mm in size, strongly enriched in refractory elements, and found most abundantly in Type 3 carbonaceous chondrites. Allende FUN inclusions are rare CAIs with isotopic anomalies in many elements. Objects in each of these categories have experienced an event or a period of high-temperature chemical processing which has left a record in the abundances of the isotopes of the major elements. It will be seen that this series from chondrules to CAIs to FUN inclusions is one of progressively larger isotopic variations.

I. ANALYTICAL METHODS

Isotopic analysis of oxygen and silicon in a single sample can readily be carried out, since the bromine pentafluoride preparative procedure gives quantitative yields of O_2 and SiF_4, both of which are suitable for gas-source mass spectrometry with precisions of 0.1‰. Measurement of small mass-dependent fractionation in magnesium is much more difficult, since both thermal ionization and ion microprobe techniques have inherent fractionations. These techniques are suitable for measurement of natural fractionation effects of a few parts per thousand per mass unit. Much greater precision is, of course, achieved in the measurement of excesses or deficiencies of a specific isotope such as radiogenic ^{26}Mg, after normalization for instrumental fractionation effects. For elements with four or more stable isotopes, such as calcium and titanium, double-spiking techniques permit the resolution of natural fractionation from instrumental effects and thus make precise fractionation measurements possible by thermal ionization techniques.

II. CHONDRULES

Oxygen and silicon exhibit fundamentally different isotopic behavior in chondrules from both ordinary chondrites and carbonaceous chondrites. On a three-isotope plot of δ^{29}Si versus δ^{30}Si (^{29}Si/^{28}Si and ^{30}Si/^{28}Si variations, respectively, in parts per thousand relative to a terrestrial silicon standard), data from individual chondrules or groups of chondrules fall along a slope-½ line, as expected for mass-dependent fractionation effects. The slope-½ is a consequence of a one mass-unit difference in the ratio on the ordinate, and a two mass-unit difference in the ratio on the abscissa. On the corresponding oxygen isotope graph of δ^{17}O versus δ^{18}O (^{17}O/^{16}O and ^{18}O/^{16}O variations) for the same samples, the data fall along a line of slope = 1. These contrasting patterns are illustrated in Figs. 1 and 2 for individual ferromagnesian chondrules from Allende (CV3 carbonaceous chondrite) (R. N. Clayton et al. 1983*b*) and in Figs. 3 and 4 for composite samples of size-sorted chondrules from Dhajala (H3 unequilibrated ordinary chondrite) (R. N. Clayton et al. 1983*a*). Both types of behavior can be rationalized in terms of isotopic modification of a more or less homogeneous precursor material during the chondrule-forming process. The mean value of δ^{30}Si for solar system matter is

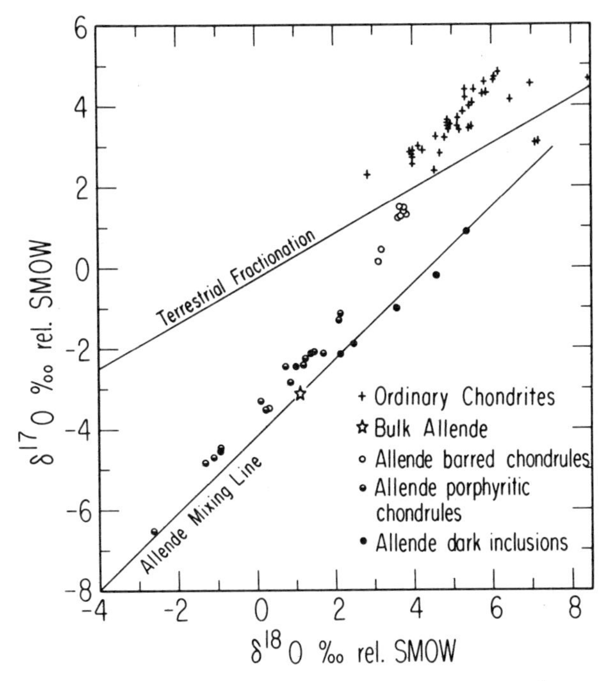

Fig. 1. Oxygen isotopic compositions of individual chondrules from the Allende carbonaceous chondrite and from several ordinary chondrites. The ordinate shows variations of $^{17}O/^{16}O$ ratios in parts per thousand (‰) relative to the terrestrial mean ocean water standard (SMOW); the abscissa shows variations in $^{18}O/^{16}O$ ratios. The Terrestrial Fractionation line is the locus of all terrestrial oxygen samples, and has a slope of 0.52, determined by mass-dependent isotopic fractionation. The Allende Mixing Line is determined from measurements on calcium-aluminum-rich inclusions (CAI) for which most of the data points lie off the diagram to the lower left. The variations in composition among the Allende chondrules probably result from an exchange process between a solid precursor with isotopic composition similar to Bulk Allende and a gaseous reservoir with a composition near the cluster of Allende barred chondrules. The variation in composition of chondrules from ordinary chondrites is also probably due to exchange with an external gaseous reservoir. Textural evidence indicates that the most ^{16}O-rich chondrules have been most intensely heated and are the most exchanged. Thus, the precursor solids were ^{16}O-poor, and the gaseous reservoir had a composition very similar to that which exchanged with the Allende chondrules. This interpretation of the data requires a minimum of three isotopically distinct oxygen reservoirs: two solids and one gas.

−0.5‰ (i.e., the $^{30}Si/^{28}Si$ ratio of solar system matter is 0.5 parts per thousand less than the ratio of the terrestrial standard silicon). The Dhajala data show that the large chondrules have silicon isotopic composition very close to that of mean solar system matter, whereas the small chondrules are depleted in the light isotopes by about 0.5‰ per mass unit, as is known to occur on evaporation of silicon from melts (Molini-Velsko et al. 1982). The slope-1 trend in the oxygen isotope data on the same samples cannot be the result of

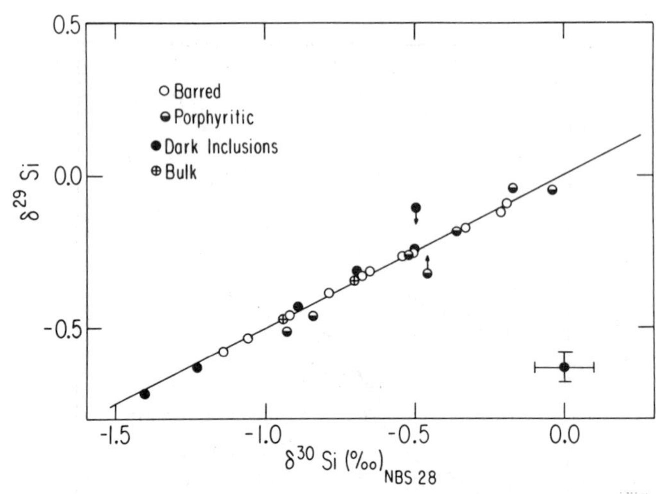

Fig. 2. Silicon isotopic compositions of the same Allende chondrules as shown in Fig. 1. The same δ-notation is used to represent variations of $^{29}Si/^{28}Si$ (ordinate) and $^{30}Si/^{28}Si$ (abscissa) in parts per thousand relative to NBS 28 quartz standard. Within analytical uncertainty, all points fall on a normal mass-dependent fractionation line. Based on analyses of a large number of stony meteorites, average solar system matter has $\delta^{30}Si = -0.5‰$.

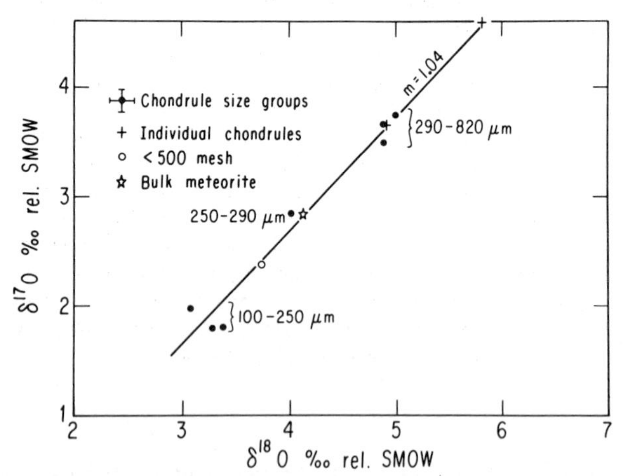

Fig. 3. Oxygen isotopic compositions of Dhajala (H3) chondrules. Each data point represents a composite sample of about 50 chondrules within a narrow size range. The data define a line of unit slope, resulting from mixture of two end-members differing in ^{16}O content. The large chondrules have predominantly porphyritic textures indicating incomplete melting; about half of the small chondrules have cryptocrystalline, dendritic or radial textures indicating complete melting. The data are compatible with gas-liquid isotope exchange upon melting in the chondrule-forming event, with more extensive exchange in the smaller, hotter particles.

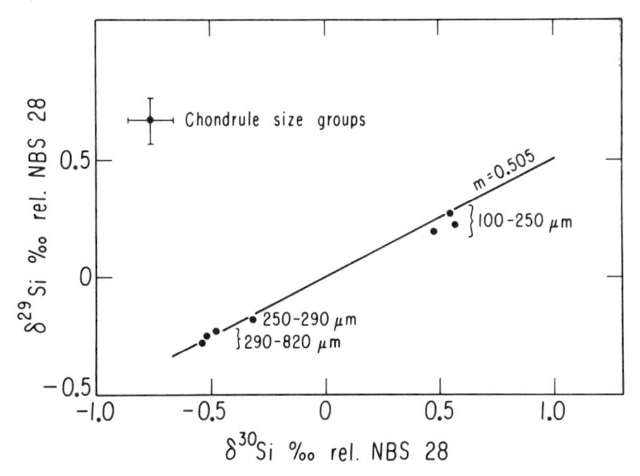

Fig. 4. Silicon isotopic compositions of the same Dhajala chondrules as in Fig. 3. The large chondrules have average solar system compositions. The small chondrules are enriched in heavy isotopes along a slope-½ mass-dependent fractionation line. Similar enrichments have been observed in laboratory experiments as a consequence of partial evaporation of silicon from a melt (Molini-Velsko et al. 1982).

an evaporation process, because that would lead to a slope-½ line with light-isotope depletion in the smaller chondrules, just as is seen for silicon. The observed light-isotope enrichment, specifically an ^{16}O enrichment, must have occurred by exchange of the melt with the ambient oxygen-containing gas.

The Allende data show somewhat similar, but by no means identical, behavior, in that the silicon data define a mass-dependent fractionation line, with a range of about 0.7‰ per mass unit, but now showing both heavy isotope enrichments and depletions relative to the bulk meteorite or to mean solar abundances. The oxygen isotopes in Allende chondrules form an array which departs from the slope-1 line determined by the Allende CAIs, in having a substantially steeper slope. The direction of action of the modification process appears to be from the Allende bulk composition toward a more ^{18}O- and ^{17}O-rich composition. Evidence for this conclusion is that the chondrules most enriched in heavy isotopes are all of the texture types indicative of complete melting, whereas the other chondrules have porphyritic textures, as are found in laboratory experiments in which chondrules were incompletely melted (Nagahara 1983), and hence less intensely heated. Thus the direction of apparent gas-chondrule exchange is different for the chondrules from ordinary chondrites and chondrules from carbonaceous chondrites. However, the two groups appear to converge toward a common oxygen isotopic composition in the most extensively exchanged chondrules, implying the possibility of a similar composition for the gaseous reservoir for the two groups of chondrules. Note that this interpretation requires that two isotopically distinct solid

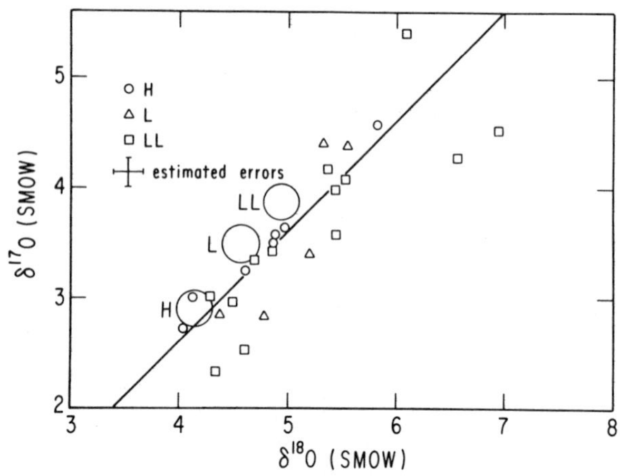

Fig. 5. Oxygen isotopic compositions of individual chondrules from unequilibrated ordinary chondrites. The circles labelled H, L, and LL represent the isotopic compositions of whole-rock samples of equilibrated chondrites from the three iron groups. (The radius of the circle is equal to the standard deviation of the population for each group.) It is clear that the isotopic compositions of the chondrules scatter much more widely than those of the bulk meteorites, and that the compositions bear no relationship to the iron group of the bulk meteorite. The spread is probably due to gas/chondrule exchange at high temperature, which implies chondrule formation in a nebular process before the existence of separate parent bodies with characteristic iron groupings.

precursor reservoirs be prepared before chondrule formation: one ^{16}O-rich for carbonaceous chondrites, one ^{16}O-poor for ordinary chondrites.

An important characteristic of the oxygen isotopic composition of individual chondrules from ordinary chondrites is that they form a single isotopic mixing array for all ordinary chondrites, independent of the iron group (H, L, or LL) of the parent meteorite (Gooding et al. 1983). This is in contrast to the isotopic compositions of the bulk meteorites, which have distinctive values for each iron group (Fig. 5). This shows that the chondrules acquired their compositions before the parent bodies were formed. We can therefore conclude that the chondrules acquired their oxygen isotopic compositions in nebular processes when the gaseous reservoir was still present.

III. CALCIUM-ALUMINUM-RICH INCLUSIONS

Isotopic variations in Allende CAIs are dominated by mass-dependent fractionation effects for silicon, magnesium, and calcium (Molini-Velsko et al. 1983; Niederer and Papanastassiou 1984), but by reservoir exchange processes for oxygen. Superimposed on these general effects are specific anomalies of nuclear origin, such as ^{26}Mg excesses due to decay of extinct ^{26}Al (C. M. Gray and C. W. Compston 1974; Lee and Papanastassiou 1974) and ^{50}Ti

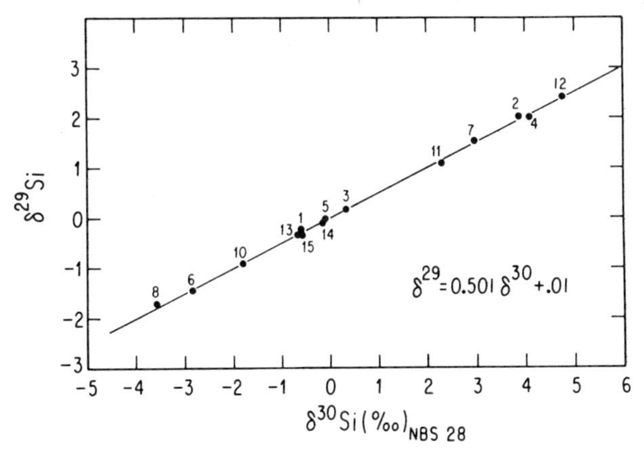

Fig. 6. Silicon isotopic compositions of individual CAI from Allende F series. Both coarse-grained and fine-grained inclusions were analyzed. Several points cluster near the mean solar value (δ^{30}Si = −0.5‰); others are both enriched and depleted in heavy isotopes, following a slope-½ fractionation line. The range of compositions is about six times greater than that seen in ferromagnesian chondrules. There is a tendency for coarse-grained inclusions to be enriched in heavy isotopes, as expected for evaporation residues, and for fine-grained inclusions to be depleted in heavy isotopes, as expected for condensates. However, more complex processes are required (see text).

excesses (Heydegger et al. 1979; Niederer et al. 1980; Niemeyer and Lugmair 1981). These specific anomalies form major separate topics related to solar system chronology and to nucleosynthesis, and are beyond the scope of this chapter; for a recent review, see Begemann (1980).

Silicon isotopic compositions for a series of coarse- and fine-grained Allende inclusions (F series) are shown in Fig. 6 (Molini-Velsko 1983). The range of observed fractionations is about six times as great as that of the Allende ferromagnesian chondrules, being about 5‰ per mass unit. A similar range is observed for calcium isotope fractionation, while the magnesium fractionation range is about 20‰ per mass unit (Niederer and Papanastassiou 1984).

The magnitudes of the isotope effects in silicon, magnesium, and calcium are very much larger than the effects occurring in igneous processes, as exemplified by terrestrial rocks. Only cosmochemical processes of evaporation and condensation (and possibly sputtering) can be expected to produce such large effects. The equilibrium isotope exchange of silicon between olivine and SiO gas should produce < 1‰ per mass unit enrichment in the heavy isotopes in olivine at temperatures above 1500 K (R. N. Clayton et al. 1978). Therefore, production of the observed range requires a multi-stage process, such as a Rayleigh distillation. Condensation from an isotopically homogeneous gas would produce a small and almost constant heavy-isotope enrichment for the first few percent of silicon condensed, which is the fraction

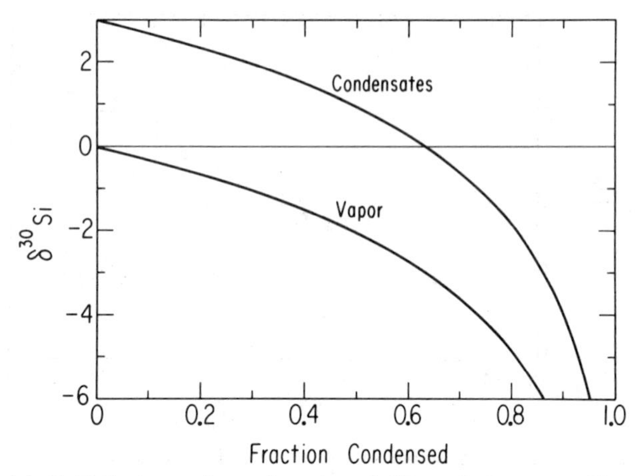

Fig. 7. Rayleigh distillation curve for condensation of silicates from a vapor. The equilibrium fractionation factor was taken as a 3‰ enrichment of ^{30}Si in condensates relative to vapor. During condensation of the first 5% of the gas (as would be appropriate for hot nebular condensation of CAIs), $\delta^{30}Si$ remains almost constant at +2.8 to +3.0‰. This is not consistent with the observations shown in Fig. 6.

of the solar abundance that is represented in the CAIs (Fig. 7). On the other hand, a Rayleigh evaporation of an initially homogeneous melt would produce both light-isotope enrichment in the vapor and large and variable heavy-isotope enrichment in the final few percent of the residual liquid (Fig. 8). Thus, the qualitative conclusion derived from the isotopic data on silicon, magnesium, and calcium is that evaporation and/or condensation involving major fractions of these abundant elements have been an important part of the solar

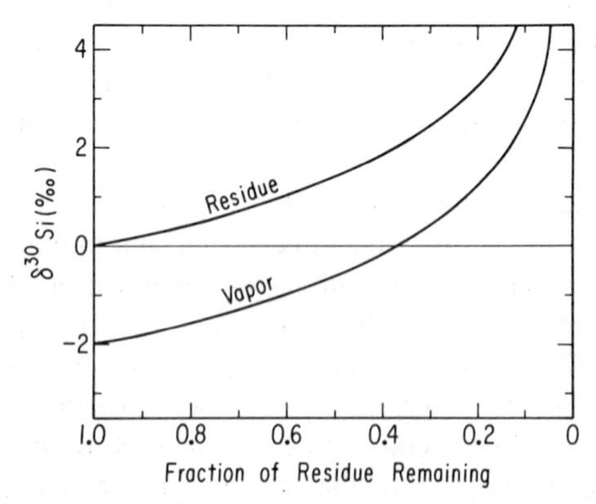

Fig. 8. Rayleigh evaporation curve for silicate liquids. The equilibrium fractionation factor was taken as a 2‰ enrichment of ^{30}Si in liquid relative to vapor. The process produces both isotopically light vapors and isotopically heavy and highly variable residual liquids, with effects of several parts per thousand in the final few percent of residual liquid. This type of process is capable of giving the variations illustrated in Fig. 6.

system history of the CAIs (Molini-Velsko 1983; Niederer and Papanastassiou 1984).

For all three elements, both heavy-isotope enrichments and light-isotope enrichments are observed in CAIs. In magnesium and silicon, there is some tendency for heavy-isotope enrichment in coarse-grained inclusions and light-isotope enrichment in fine-grained inclusions. However, the correlation between the isotope effects in different elements shows that multi-stage processes are required. For example, Niederer and Papanastassiou (1984) show that the directions of the isotope effects in calcium and magnesium are usually opposite, i.e., magnesium is usually enriched in heavy isotopes, and calcium in light isotopes. This requires a series of evaporation/condensation events which affect the two elements differently due to their difference in volatility. Magnesium and silicon, which are of similar volatility, correlate reasonably well in their isotope effects, although the magnesium effects are considerably larger, perhaps because silicon vaporizes as the more massive SiO molecules. The relationship between volatility and the direction of isotopic enrichment is still more complicated, in that titanium effects correlate positively with magnesium effects (G. J. Wasserburg, personal communication), even though titanium is more refractory than calcium in the vaporization of material of Allende composition (Notsu et al. 1978). At present, there are too few CAIs for which isotopic fractionation has been measured on three or four major elements to reveal all of the systematic relationships. In any case, *the existing data are totally incompatible with a simple history of a single stage of condensation during monotonic cooling from an initially hot gas, the first-order framework on which many cosmochemical models have been built* (see, e.g., Suess 1965; Anders 1971; J. S. Lewis 1972*a*; L. Grossman and J. W. Larimer 1974; Boynton 1975).

Since the isotopic variations in major cationic elements indicate large-scale evaporation and recondensation processes in the formation of CAIs, stoichiometry requires comparably large-scale processes for oxygen, with concomitant large mass-dependent isotopic fractionation. This is not, however, what is observed. Bulk oxygen isotopic compositions of individual CAIs from Allende and other carbonaceous chondrites fall on a line on the three-isotope graph with a slope near one. This is illustrated for the F-series in Fig. 9. The total absence of correlation between oxygen and silicon isotope effects is shown in Fig. 10, indicating that different processes control the isotopic compositions of these two elements. It is also found that each CAI has internal oxygen isotopic heterogeneities on a scale of micrometers, with different minerals having greatly different isotopic compositions. In fact, separated minerals from a single CAI can generate the same line as shown in Fig. 9, but with even greater range in isotopic composition (R. N. Clayton et al. 1977). The internal variations have been interpreted as resulting from a secondary exchange process, occurring after crystallization of the major minerals, and possibly associated with alteration processes in which more volatile elements, such as alkalis, iron, and halogens, have been introduced. The most ^{16}O-rich

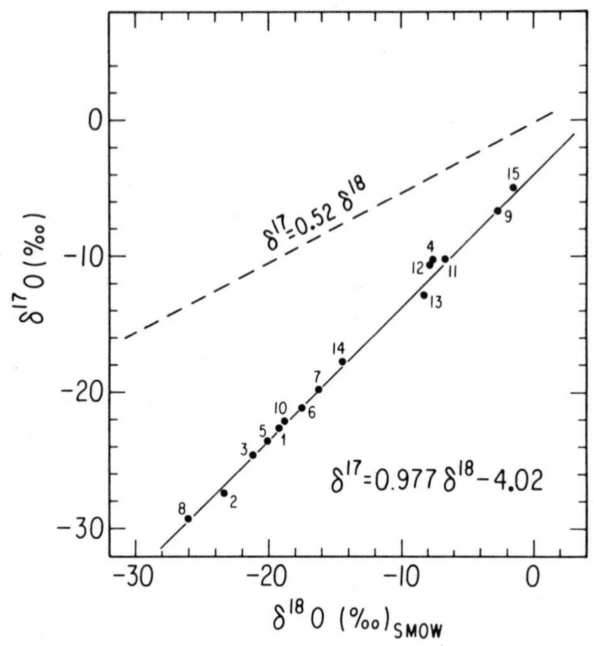

Fig. 9. Oxygen isotopic compositions of the same Allende F series CAIs as in Fig. 6. The dashed line is the terrestrial fractionation line. The Allende data show the same slope-1 mixing line as reported previously from other samples (R. N. Clayton et al. 1977). Once again, mixing of ^{16}O-rich and ^{16}O-poor reservoirs is implied. Arguments by R. N. Clayton et al. (1977) identify these as a ^{16}O-rich solid, with initial $\delta^{18}O \sim -40‰$, $\delta^{17}O \sim -42‰$, and a gas with a composition in the vicinity of the terrestrial line.

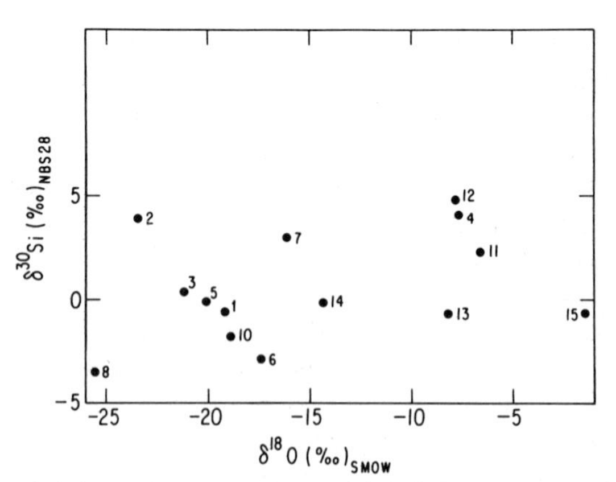

Fig. 10. Graph of $\delta^{30}Si$ versus $\delta^{18}O$ for the Allende F series CAIs. The total absence of correlation indicates that the isotopic compositions of these most abundant elements are controlled by different processes. The silicon variations result from combinations of evaporation and condensation, whereas the oxygen variations are a consequence of a later mineralogically controlled gas/solid isotope exchange (R. N. Clayton et al. 1977).

minerals, typically spinel, forsterite, and pyroxene, are least disturbed by this secondary exchange, whereas the least ^{16}O-rich minerals, melilite and anorthite, appear to have undergone more extensive exchange, due to greater diffusion rates in these minerals (Muehlenbachs and Kushiro 1974; T. Hayashi and K. Muehlenbachs 1984).

In contrast to oxygen, magnesium, silicon, and calcium do not have large mineral-to-mineral isotopic variations within inclusions. No measurable internal variations are observed in magnesium and calcium (Lee 1979; Niederer and Papanastassiou 1984) and the silicon effects are a few tenths of 1‰, as expected for normal mass-dependent mineral/mineral fractionation. Thus, the secondary processes which greatly modified the oxygen composition had no effect on the compositions of other elements, once again attesting to the importance of an external oxygen reservoir.

If the secondary alteration process is a correct interpretation of the oxygen isotopic variations within CAIs, then their preexchange compositions must have lain at or beyond the most ^{16}O-rich spinel compositions (R. N. Clayton et al. 1977), corresponding to the point marked "spinel" in Fig. 11. Furthermore, the extensive exchange of oxygen which must have accompanied the evaporation/condensation of the major cationic elements should have resulted in a close approach to oxygen isotopic equilibrium between the gaseous and condensed reservoirs in this region of the nebula. Since the equi-

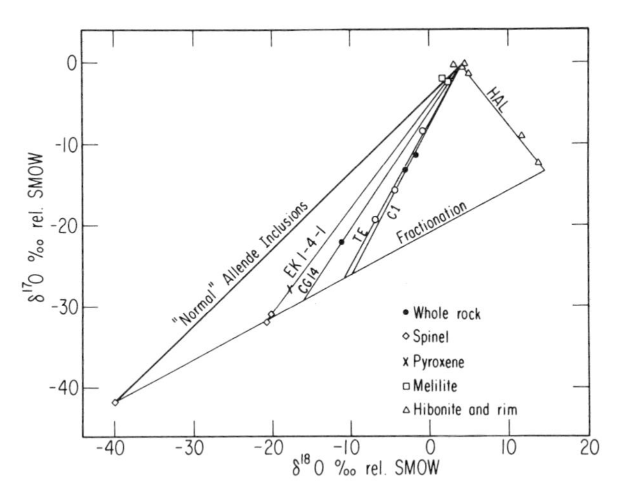

Fig. 11. Oxygen isotopes in FUN inclusions. The line labeled "Normal" Allende Inclusions is the same as the line in Fig. 9, determined by Allende CAIs. In the two-stage model it is postulated that all inclusions initially had isotopic compositions near the most ^{16}O-rich point on the line, at $\delta^{18}O = -40‰$, $\delta^{17}O = -42‰$. "Normal" CAIs then exchanged with an external reservoir poorer in ^{16}O; FUN inclusions underwent an intermediate process, moving along the line marked Fractionation, to varying degrees for each inclusion. Then each inclusion exchanged with the same ^{16}O-poor reservoir.

librium isotopic fractionations between gas and solids or liquids are only 1–2‰/mass unit at the temperatures of evaporation and condensation of silicates (Onuma et al. 1972), it follows that the isotopic composition of this gaseous reservoir was close to that now found in spinel and pyroxene in Allende.

In the Sec. II on ferromagnesian chondrules, it was argued that three components with initially different oxygen isotopic compositions were required to account for the observations. Allende chondrules required a ^{16}O-enriched solid precursor; ordinary chondrite chondrules required a ^{16}O-depleted solid precursor; both required a gaseous reservoir with a composition near the terrestrial value. The arguments based on Allende CAIs imply the existence of a ^{16}O-enriched gaseous reservoir as well. However, this does not necessarily require a fourth separate reservoir, since the ^{16}O-rich solid reservoir could have been made by high-temperature equilibration with the ^{16}O-rich gas. Is it feasible to have two isotopically different gas reservoirs in the solar nebula? They may have existed either at different times or in different regions, because evidence for the large ^{16}O enrichments have been observed only in the carbonaceous chondrites.

There are two possibilities for the origin of isotopically distinct reservoirs: (1) they may be inherited from heterogeneities in the interstellar medium; or (2) they may be produced within the solar nebula. The first of these possibilities has been considered the more likely ever since the first observation of the oxygen anomalies (R. N. Clayton et al. 1973). Solid and gaseous components may have different proportions of materials with different nucleosynthetic histories, which are reflected in their chemical state. The second category, production of anomalies within the solar system, can be subdivided into nuclear and nonnuclear processes. Production of a ^{16}O-rich component by bombardment by energetic particles from the Sun, resulting in preferential destruction of ^{17}O and ^{18}O, has been discussed by Lee (1978). His calculations show that to produce the ^{16}O excess in the Earth alone requires an enormous proton fluence and a special distribution of matter in the nebula during irradiation. This scenario predicts certain large isotope effects in other elements, which have thus far not been observed. The possibility of producing a ^{16}O-rich reservoir and a ^{16}O-poor reservoir simultaneously by purely chemical means has been demonstrated experimentally (Thiemens and Heidenreich 1983). The processes involved are not well understood, but may be related to self-shielding by the abundant ^{16}O-bearing molecules in a photodissociation reaction. Self-shielding effects in carbon monoxide in molecular clouds have been proposed to account for enhancements in ^{13}CO in their outer regions (Bally and Langer 1982; Frerking et al. 1982). From both theoretical and observational considerations, a substantial isotopic enrichment begins at CO column densities of 10^{14} cm^{-2} and probably extends to a density of about 10^{18} cm^{-2}, by analogy with calculations for self-shielding in O_2 (Navon and Wasserburg 1984). For a gas of solar composition at a total pressure of 10^{-5} atm, this column density corresponds to distances of only a few kilometers.

TABLE II
Properties of FUN Inclusions

Property	C1	EK 1-4-1	CG14	TE	HAL	EGG 3
Fractionated O	Yes	Yes	Yes	Yes	Yes	No
〃 Mg	Yes	Yes	Yes	Yes	No	Yes
〃 Si	Yes	Yes	Yes	Yes	No	No
Radiogenic ^{26}Mg	No	No	?	?	No	Yes
Heavy element anomalies	Yes	Yes	No	No	?	?
Excess ^{29}Si	Yes	Yes	Yes	No	No	Yes
Excess ^{25}Mg	Yes	Yes	Yes	Yes	No	Yes
Cerium deficiency	Yes	No	No	No	Yes	?
References[a]	1, 2	1, 2, 7	3	3	4, 5	6

[a](1) Wasserburg et al. 1977*a*; (2) R. N. Clayton and T. K. Mayeda 1977; (3) R. N. Clayton et al. 1984; (4) Lee et al. 1979; (5) Lee et al. 1980; (6) Esat et al. 1980; (7) Yeh and Epstein 1978.

Hence, ultraviolet radiation originating in the protosun could not affect a significant fraction of the nebular gas. It may be possible to develop models with energy sources distributed throughout the nebula, or models in which the nebular disk is irradiated on its surfaces by the Sun and neighboring stars. Furthermore, it is not sufficient to have an isotope-selective photodissociation: efficient chemical trapping of the products is required if the process is to produce two isotopically distinct reservoirs with separate chemical identity. These problems remain to be worked out in order to determine whether the self-shielding phenomenon was important in the early solar system.

IV. FUN INCLUSIONS

Two CAIs in Allende (Cl and EK 1-4-1) have a collection of isotopic properties that are so remarkable that they have been separatedly categorized. These, labeled FUN, have strongly fractionated isotopic abundances of oxygen, magnesium, and silicon; they lack radiogenic ^{26}Mg; they have nuclear anomalies in all elements studied: oxygen, magnesium, silicon, calcium, titanium, strontium, barium, samarium, and neodymium. Searches for additional FUN inclusions have shown that there is no single set of properties that sets this group off from other Allende CAIs. This is illustrated in Table II, which shows properties of several inclusions with varying degrees of FUN-ness. It has further been noted that no chemical, mineralogical, or textural property has been recognized that would serve to identify FUN inclusions before isotopic analysis (R. N. Clayton et al. 1984). Nevertheless, several of these inclusions have a common set of characteristics in the fractionation patterns of oxygen, magnesium, and silicon. Oxygen isotope data are shown in Fig. 11. The oxygen in all of these inclusions appears to have had an initial isotopic composition like that of all other Allende CAIs: $\delta^{18}O \sim -40‰$, $\delta^{17}O \sim -42‰$. The FUN inclusions, like ordinary CAIs, later underwent a

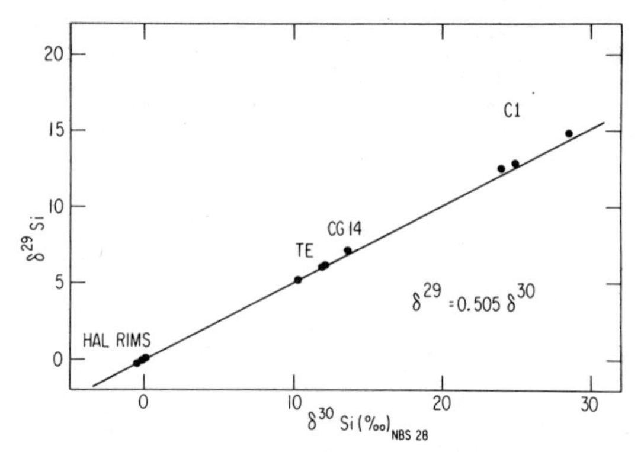

Fig. 12. Silicon isotopes in Allende FUN inclusions. Except for HAL, which has normal silicon, the FUN inclusions are very strongly fractionated, and some have small but significant excesses of ^{29}Si.

final oxygen exchange with an external (probably gaseous) reservoir with a near-terrestrial isotopic composition. However, the inclusions Cl, EK 1-4-1, HAL, CG 14, and TE, all passed through an intermediate stage, in which the major elements, oxygen, magnesium, and silicon underwent large isotopic fractionations, enriching the heavy isotopes. The silicon isotope data are shown in Fig. 12. The magnitudes of the effects are similar to the inverse square root of the mass ratios (Figs. 13 and 14) and are thus about an order of

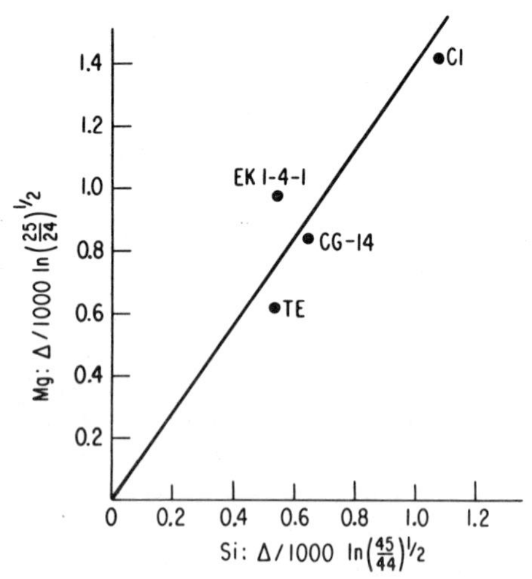

Fig. 13. Correlation between magnesium isotope fractionation and silicon isotope fractionation in FUN inclusions. The measured fractionations have been normalized to the inverse square root of the masses, using the most abundant vapor phase species: atomic Mg for magnesium, and SiO molecules for silicon. If both elements were fractionated by a factor equal to the inverse square root of the masses, the data point would plot at 1.0 on both axes.

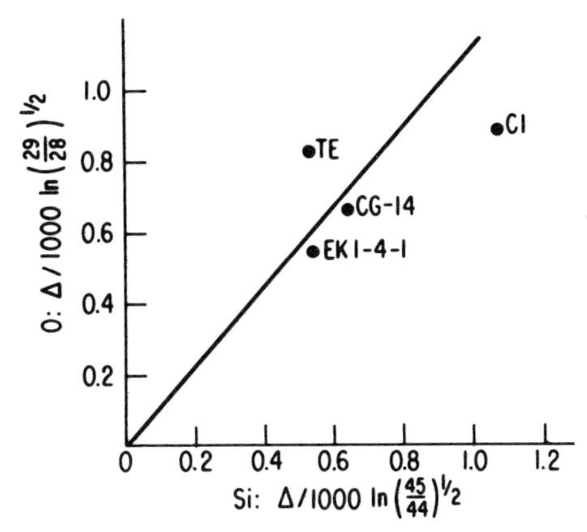

Fig. 14. Correlation between oxygen isotope fractionation and silicon isotope fractionation, normalized as in Fig. 13, with oxygen taken as CO molecules and silicon as SiO molecules.

magnitude greater than usual equilibrium or kinetic isotope effects at temperatures appropriate for evaporation and condensation. Extreme Rayleigh evaporation processes are unlikely, since CG 14 and TE, which are relatively rich in magnesium and silicon, have strong isotopic fractionations for these elements, whereas HAL, which has very little magnesium, also has very little magnesium isotope enrichment. Furthermore, the oxygen isotope behavior in the intermediate step in formation of FUN inclusions is unlike that seen in any chondrules: mass-dependent fractionation without evidence of exchange with an external reservoir. Thus, it would appear either that oxygen fractionation occurred in the absence of an external reservoir or that a unidirectional process was involved. Wood (1981*a*) chose the second alternative, claiming that slow evaporation of silicates could lead to the large fractionations found in FUN inclusions. However, in order to suppress back-reaction with the surroundings, a gas-free environment would also be required. Another unidirectional process that might produce large isotopic fractionation is sputtering in interstellar space (D. D. Clayton 1981). In this cold environment, no back-reaction would be expected. Although there have been no experimental tests on the efficacy of this process for the elements considered, a theoretical model (Haff et al. 1981) predicts effects that are too small, and in the case of oxygen, even in the wrong direction. It is our conclusion that no convincing mechanism has been proposed to explain the very large fractionation effects found in these unusual inclusions. Until this question is answered, it remains very difficult to interpret, in any detailed way, the significance of the many nuclear anomalies found in inclusions Cl and EK 1-4-1.

Acknowledgments. We have benefited from discussion with L. Grossman, A. M. Davis, and G. J. MacPherson. This research was supported by a grant from the National Science Foundation.

METEORITIC EVIDENCE CONCERNING CONDITIONS IN THE SOLAR NEBULA

WILLIAM V. BOYNTON
University of Arizona

Chondritic meteorites contain evidence of many fractionation processes, which separate groups of chemical elements with similar properties. These fractionations provide strong evidence that high temperatures (1400–1500 K) occurred at least for short periods (hours) and that moderately high temperatures (1200 K) occurred for longer times. These temperatures obtained as far from the Sun as the asteroid belt. Many episodes of condensation and isolation of grains from the gas are recorded and provide evidence of a chaotic environment. Nevertheless, the absence of Ca,Al-rich inclusions in ordinary chondrites suggests that mixing of solid materials over significant heliocentric distances did not occur.

I. INTRODUCTION

In the past, the solar nebula was generally considered, or at least assumed, to be a simple environment. It was thought to have been well mixed and to have cooled monotonically from peak temperatures which decreased smoothly as a function of heliocentric distance (Cameron 1969; Larimer 1967; Grossman 1972; Lewis 1972*a*; Grossman and Larimer 1974). We now know, however, that the solar nebula was far from homogeneous; different meteorite parent bodies have different oxygen isotopic composition and different components of a single meteorite (in particular, the Allende carbonaceous chondrite) have different isotopic composition in a variety of trace elements as well as oxygen (see the chapter by Clayton in this book). Boynton (1978*a*) suggested that the various components in Allende could not have formed under a simple scenario of monotonic cooling and isolation from the nebula.

This chapter attempts to explain to the non-meteoriticist the evidence that meteorites provide concerning the nature of the solar nebula. Because other chapters deal with petrographic (Wood) and isotopic (Clayton; Kerridge; Wasserburg) studies on meteorites, this chapter will focus on information learned from chemical analyses of meteorites and their components.

In order to understand these data, it is useful first to consider the diagrams used by cosmochemists. Elemental abundances determined from analyses of materials relating to early nebular processes are generally normalized to CI carbonaceous chondrites, which provide our best estimate of solar abundances for the condensible elements. On the other hand, theoretical calculations of the abundance of elements present in a solid at a given temperature following condensation in the solar nebula are usually presented as the fraction condensed. Because the theoretical calculations generally consider a system with solar abundances of the elements, a plot of the fractions condensed of a suite of elements should look identical to the CI-normalized abundances determined in a nebular condensate found in a meteorite.

This can best be understood from Fig. 1, in which a series of hypothetical elements are plotted in order of increasing volatility. At a certain temperature, elements A-C are totally condensed, elements D-F are partially condensed and elements G-I remain entirely in the gas phase. If a sample of this hypothetical condensate were analyzed, its elemental abundances relative to CI chondrites would plot with an identical shape except that the ordinate scale would differ by some factor. This factor is the fraction of total condensible mass that is in the sample. For example, if the hypothetical solid of Fig. 1

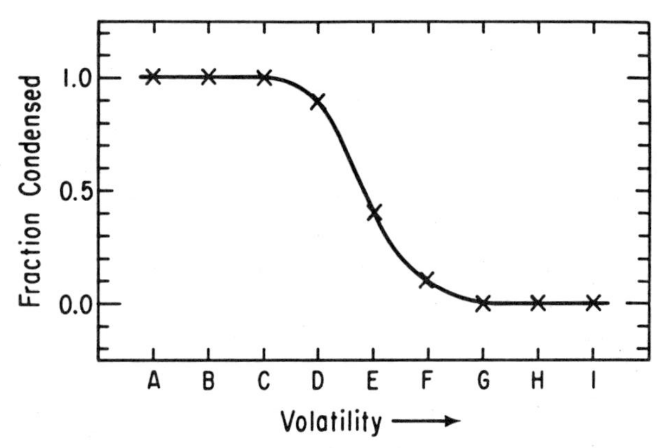

Fig. 1. Fraction of hypothetical elements in an early condensate isolated from the solar nebula. The elements A, B, and C, being the most refractory, are totally condensed while G, H, and I, the most volatile, are totally in the gas phase; elements, D, E, and F are only partially condensed.

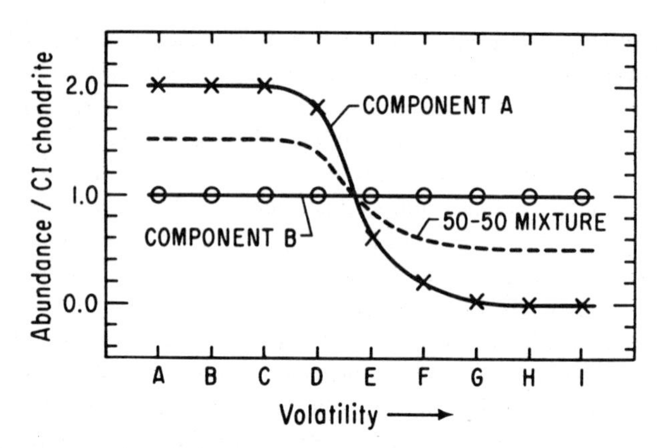

Fig. 2. Abundances in hypothetical meteorite components normalized to CI chondrites, our best measure of solar abundances. Component A was isolated at a high temperature when only 50% of the condensible mass of the nebula had condensed, yielding a factor of 2 enrichment in the elements which are totally condensed. Several meteorite groups have abundance patterns similar to the 50-50 mixture.

were isolated when only one-tenth of the total condensible mass had in fact condensed, then the totally condensed elements, elements A to C, would have abundances, in units of grams/gram, 10 times those found in CI chondrites because they would not have the effect of dilution by the remaining 90% of the condensible mass.

In the solar nebula, of course, the situation was not so simple. Samples may be a mixture of components isolated at two different temperatures, or the sample may have only been partially isolated from the gas while condensation continued. In Fig. 2, abundance patterns of two hypothetical solids are plotted. Component A was isolated at a high temperature where only 50% of the mass had condensed, and component B was isolated at a low temperature such that all elements were totally condensed. The abundance pattern of the mixture of the two components has a shape similar to that of component A except that the amplitude is decreased. If these data were plotted with a log scale for the ordinate, as is often done, it is obvious that the pattern of the mixture would look very different from that of component A.

Unfortunately from an interpretive standpoint, there is another scenario by which a pattern identical to that of the mixture can arise. If condensation progressed to a point similar to that of Fig. 1, but, if instead of being isolated from the gas, the solid condensates were concentrated in a region where they were enriched relative to the remaining gas, and condensation continued, then the resultant pattern would be identical to the mixture in Fig. 2. This concentration could arise if, for example, the grains settled to the median plane of the nebula, and later mixing of gases above the plane with the grains was small.

Figures 1 and 2 also illustrate an important concept in cosmochemistry, namely that most elements tend to condense over a fairly narrow temperature range. At any given temperature, most elements will tend to be either entirely in the condensed phase or entirely in the gas phase; very few will be partitioned with comparable amounts in both phases. For example, La is calculated to change from 10% in the solid to 90% in the solid over a temperature range of only 67 K.

It is also useful in this section to discuss two broad types of data that are available to us. The first comes from the study of bulk properties of large objects, i.e., the composition of the planets, asteroids or meteorite parent bodies. The second type of data comes from a study of components of primitive meteorites which have preserved objects that remain relatively unaltered since their accretion to the meteorite parent body. The former can be referred to as macroscopic properties, the latter as microscopic. Macroscopic properties tend to provide a broad picture of conditions in the solar nebula because the process of accretion combines material formed under a variety of conditions. One of the advantages of these data is that, at least with respect to the planets, we know the part of the solar system in which they formed. Microscopic properties allow one to examine processes that are obliterated by the averaging process of accretion. For example, we shall see that isolation of a wide variety of components from the nebula occurred at a high temperature in a large part of the nebula, but because the complementary material ultimately condensed in another form and accreted to the same parent body, a bulk analysis of the parent body has little record of the high-temperature isolation process.

II. EVIDENCE FOR LARGE-SCALE NEBULAR FRACTIONATIONS

In this section, we discuss the evidence from studies of macroscopic properties of meteorites and planets that have a bearing on nebular-wide fractionations. Most of these studies have been discussed previously in other reviews (see, e.g., Wasson 1974; Grossman and Larimer 1974; Dodd 1981) and will be mentioned only briefly here. The work is important, nevertheless, because it provides evidence that the processes inferred from microscopic properties may have been important in large parts of the solar system.

A. Metal-Silicate Fractionation

The fact that fractionation of metal from silicate occurred within the various classes of chondrites has been known for some time (Urey and Craig 1953). The three groups of ordinary chondrites have different metal and/or total iron content; the H-chondrites are high in both metal and total iron content, the L-chondrites have low total iron content and the LL chondrites are low in both total iron and metal. (The letter designations of these groups are, in fact, based on these properties.) The terrestrial planets also show evidence of an iron-silicate fractionation process with Mercury having approximately twice the iron content of the other terrestrial planets. Lewis (1972*a*) suggested that this observation could be explained by high nebular temperatures near Mercury which would permit most of the iron, but little of the silicate, to condense. Weidenschilling (1978), on the other hand, suggested that gas drag between solids and the non-Keplerian-rotating gas would cause the less dense silicate-rich meter-sized bodies to be preferentially swept from the Mercury formation region.

Neither of these explanations, however, can explain the metal-silicate fractionation seen in chondrites. As noted by Larimer and Anders (1970), the metal-silicate fractionation also removed volatile elements such as Au, Ga, and Ge. These volatile elements condense at temperatures far below the condensation of the silicates and would thus not be subject to the Lewis (1972*a*) separation mechanism whereby the iron is fractionated due to its high condensation temperature. The Weidenschilling (1978) mechanism requires large (meter-sized) bodies and is a very strong function of heliocentric distance. Because the ordinary chondrites show the iron-silicate fractionation on a centimeter scale (and they almost certainly did not form near the orbit of Mercury), some other mechanism must be responsible.

Based on condensation temperatures of the volatile siderophile elements, Larimer and Anders (1970) calculated that the temperature of the metal-silicate fractionation was in the range of 680 to 1050 K. As the Curie temperature for Fe-Ni alloys is in the upper end of this range, Larimer and Anders suggested that the metal may have been separated from the silicates by a magnetic field.

B. Refractory-Element Fractionation

It is also well known that the Mg/Si ratio in chondrites varies between the various groups (Urey 1961; Ahrens 1964). This fractionation was shown by Larimer and Anders (1970) to include other refractory elements and has been most recently discussed by Wasson (1978) and Kerridge (1979). This fractionation can be seen from Fig. 3, in which various refractory-element/Si ratios are plotted vs. Ca/Si. The fractionation spans nearly a factor of three among the groups and the trends are parallel with slopes of approximately unity, indicating that the refractory elements were removed (or added) in constant proportions. Wasson (1978) argued that this fractionation of a large number of elements, all of which have high condensation temperatures, argues that high temperatures in the solar nebula were wide spread; they did not just occur close to the Sun. Additional evidence for high temperatures will be discussed in Sec. III dealing with microscopic properties. Although these data provide evidence for a high-temperature fractionation event, Kerridge (1979) noted that the amount of Mg in the high-temperature component seemed inconsistent with a simple condensate formed under chemical equilibrium. He

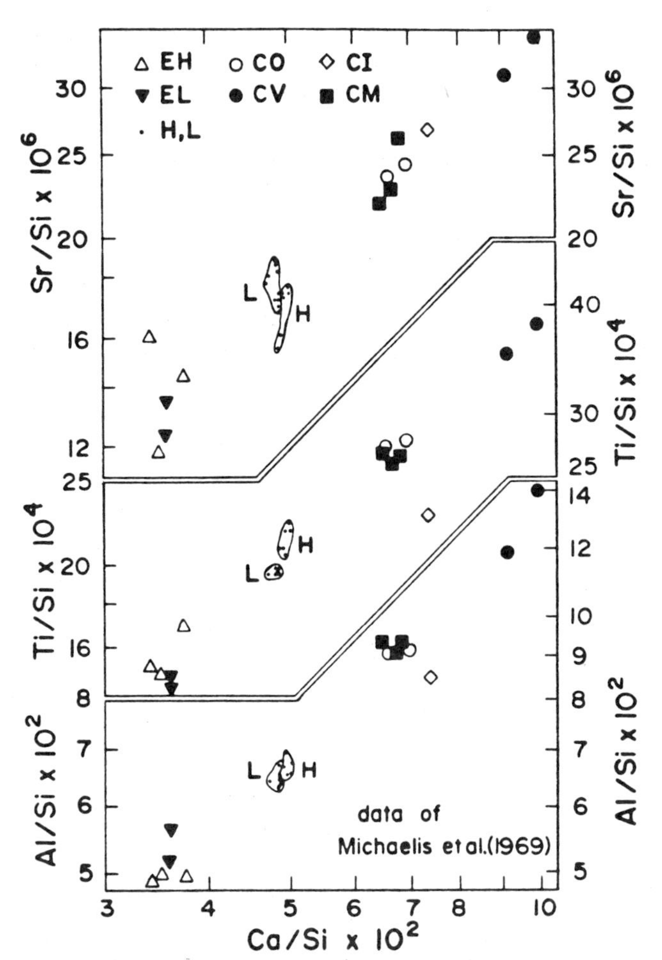

Fig. 3. Plot showing that refractory-element-to-Si ratios vary among the chondrite groups (Wasson 1978). This refractory-element fractionation requires the separation of Si (and other volatile elements) from the refractory elements. The only plausible way to do this is with Si in the gas phase which requires temperatures in excess of 1200 K.

suggested that the fractionated component may have contained an additional Mg-rich, Si-poor phase. As will be shown later (Sec. III), there is much evidence based on small inclusions in the meteorites that suggests several episodes of high-temperature fractionation; each of these episodes could generate one or more components.

C. Volatile-Element Fractionation

Larimer and Anders (1967) showed that volatile elements in chondrites are also fractionated. They showed that 28 volatile elements in CI (=C1), CM

(=C2), CO and CV (=C3), and E3, 4 groups of chondrites, are present with group/CI ratios of approximately 1.0, 0.6, 0.3 and 0.7, respectively. They interpreted these data as requiring a mixture of two components, a low-temperature component with the full complement of both volatile and nonvolatile elements, and a high-temperature component with only nonvolatile elements. This situation is similar to that described with respect to the hypothetical mixture in Fig. 2. As mentioned then, one cannot distinguish the case of a mixture from the case where part of the volatile-rich gas is lost at some temperature in the middle of the condensation sequence. However, in this case, Larimer and Anders (1967) noted that the proportion of high-temperature minerals, mainly chondrules, were present in these meteorites in about the same proportions as the amount of the high-temperature component inferred from the volatile-element data. Thus, the identification of a plausible high-temperature component argues in favor of the two-component model rather than a gas-loss model.

This volatile-element fractionation is not related to the refractory-element fractionation discussed in Sec. II.B. If the groups most depleted in volatile elements were the ones most enriched in refractory elements, it could be said to be a continuous fractionation. It can be seen from Fig. 3 that the abundances of refractory elements decrease in the order CV $>$ CI $\approx$ CM $\approx$ CO $>$ H $\approx$ L $>$ E. The volatile elements, however, decrease in the order CI $>$ E $>$ CM $>$ CV $\approx$ CO $>$ H $\approx$ L. Clearly, two different temperature-related events were responsible.

In the case of the ordinary chondrites, however, the fractionation pattern is not so simple as with the carbonaceous chondrites. Wai and Wasson (1977) showed that the CI-normalized abundances of moderately volatile elements (i.e., those with condensation temperatures in the range of approximately 1100 K to 600 K at 10^{-6}atm total nebular pressure) decrease continuously with decreasing condensation temperature. They suggested a continuous volatile-loss model in which either the volatile-rich nebular gas was lost continuously as temperatures decreased, or small volatile-rich grains, formed increasingly by homogeneous nucleation as temperatures fell, failed to accrete to the meteorite parent body. Anders (1977), however, claimed that most of the volatile elements had abundances that clustered around 0.25 times the CI values and suggested that those with values in the range of 0.9 to 0.25 were partially lost from chondrules in the heating event responsible for chondrule formation. At the time of this Wasson-Anders debate, not enough was known about the volatile-element composition of chondrules to test the Anders hypothesis that chondrules are devoid of volatile elements and constitute the high-temperature component. Recently, however, Grossman and Wasson (1982) and Wilkening et al. (1984) have analyzed individual chondrules from ordinary chondrites and found that the volatile content of the chondrules are not significantly different from the bulk meteorite. Although this evidence does not directly argue for the Wai and Wasson (1977) model, it does remove the strong association between volatile depletion and abundance of chondrules in meteorites that argued for the two-component model.

The discussion of this section demonstrates that at least three chemical fractionations occurred within the solar nebula. At least two of these, the refractory-element and volatile-element fractionations, were caused by changes in conditions as a function of temperature. The cause of the third, the metal-silicate fractionation, is a little less clear but may involve interactions with a nebular magnetic field.

These fractionations must have occurred over some substantial portion of the solar nebula; exactly how large is difficult to determine. Because the asteroid belt is not well mixed (Gradie and Tedesco [1982] showed that different classes of asteroids cluster at different heliocentric distances), the fractionation must have at least occurred throughout the asteroid belt. The regions of the solar nebula which accreted to form the meteorite parent bodies must have remained unmixed until the planetesimals were several meters in size, because meteorites which are mixtures of several different groups are relatively rare. This subject of mixing will be discussed in more detail later (Sec. IV.A).

It will be shown in the next section that a variety of very efficient fractionation events occurred to form the different inclusions found in meteorites. Although the process of accretion to form meteorite parent bodies is efficient at mixing material, it will be shown that for each class of meteorite this mixing occurred over a relatively limited region of the solar system.

III. EVIDENCE FOR SMALL-SCALE FRACTIONATIONS

As discussed by Wood (see his chapter), the carbonaceous chondrites contain a variety of inclusions and chondrules. Several of the various types of Ca,Al-rich inclusions (CAI's) will be described in this section because different types record different kinds of high-temperature fractionation events. Most of the CAI's which have been studied are from the Allende CV3 carbonaceous chondrite, although similar objects from other meteorites have been studied. The CAI's are generally divided into groups, such as coarse-grained and fine-grained, based on the size of the minerals that comprise them, or into about six groups based on trace element content.

A. Ca,Al-Rich Inclusions From Allende

Grossman and Ganapathy (1976) showed that 15 refractory elements in the coarse-grained CAI's from Allende were enriched fairly uniformly to abundances of about 18 times the CI chondrites. Other workers have extended the list to about 30 elements (see Grossman [1980] for a review of the properties of Allende coarse-grained CAI's). The enrichment of 18 times CI chondrites suggests that these inclusions represent the most refractory 5.5% ($=1/18$) of the mass of the condensible elements. This material could be a high-temperature condensate from the solar nebula or a refractory residue from a high-temperature vaporization event. Clearly, in either case high temperatures, around 1400 K or above, are required to make these objects. The nature of the high-temperature event(s) is not understood nor is the mechanism by which the objects became isolated from the gas at high temperature.

Evidence for several other fractionation events can be found in the fine-grained CAI's. These objects, often referred to as fine-grained aggregates because they are an assemblage of many μm-sized grains, are classified mainly on the basis of their abundances of trace elements, particularly the rare-earth elements (REE). The REE, or lanthanides, consist of the fifteen elements from La to Lu in the periodic table which differ from each other only in the number of 4f-electrons. Because the 4f-electrons do not readily participate in chemical bonding, the REE generally have very similar chemical properties. It is well known, for example, that in geological fractionation processes, the REE behave as a coherent group. Only the size of the REE, which decreases monotonically with increasing atomic number, serves to fractionate the members of this group, and the fractionations are always a smooth function of size. (An exception occurs for Eu, which under highly reducing conditions can exist in a two-plus oxidation state and behaves differently from the other three-plus REE.)

The volatility of the REE is one property, of considerable interest to cosmochemists, which does not vary as a smooth function of size. The volatility of the different REE under conditions expected in the solar nebula spans a factor of 10^7 (Boynton 1975), and because their volatilities vary in a way which is not a smooth function of size, it is easy for us to distinguish a nebular gas/solid fractionation from some later fractionation caused by a planetary process.

Examples of typical CAI REE patterns, i.e., chondrite-normalized abundances arranged in order of increasing atomic number—and decreasing ionic size—are shown in Fig. 4 (Boynton 1983). The Group I pattern is quite flat (unfractionated) and is generally typical of the coarse-grained inclusions mentioned above. It shows the characteristic enrichment of about 18 times CI abundances. The abundance of Eu is slightly greater than that of the other REE, and is almost certainly caused by the two-plus valence state. (Because REE patterns studied in terrestrial geochemical systems are generally smooth, such an excess of Eu is referred to as a positive Eu anomaly. Abundance anomalies for REE other than Eu are unknown in terrestrial, lunar or meteoritic igneous processes, but are quite common in CAI's.) The Group III pattern is also quite flat except for deficiencies in Eu and Yb. These elements are the two most volatile REE (Boynton 1975) and their low abundance suggests that the Group III material was isolated from the nebula at somewhat higher temperatures than the Group I material. As in the case of the Group I inclusions, the Group III inclusions could have been isolated following either condensation or partial vaporization. Although Eu can exhibit anomalous behavior due to its two-plus valence, it is very difficult to reduce Yb to the two-

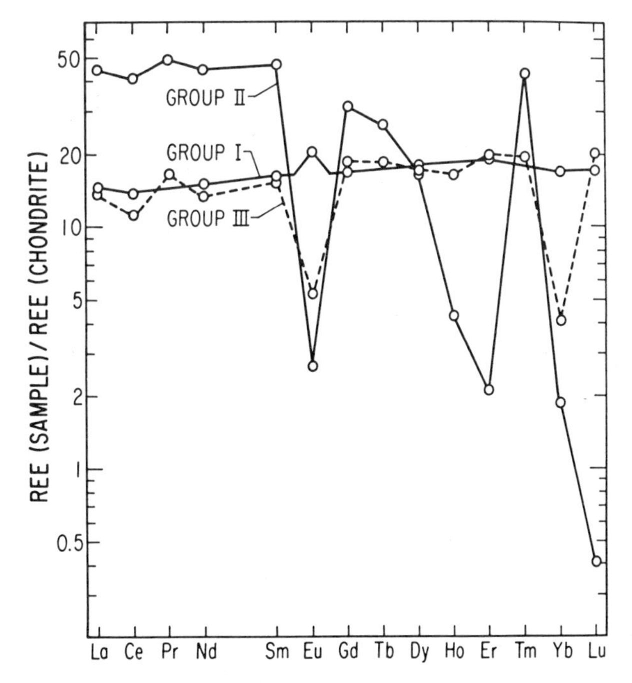

Fig. 4. Rare-earth element (REE) patterns for three different groups of Allende Ca,Al-rich inclusions (from Boynton 1983). Group I inclusions have nearly unfractionated REE patterns at about 18 times CI chondrites, suggesting that the inclusion is the most refractory 5.5% of the condensible matter of the solar system. The Group III material was apparently isolated at higher temperatures when the most volatile REE, Eu and Yb, were still in the gas phase. In addition to Eu and Yb, Group II inclusions also lost the most refractory REE to an ultrarefractory component. These fractionations require temperatures of 1400 K; they also require a very chaotic environment capable of multiple episodes of condensation and isolation of components.

plus state even under the highly reducing conditions expected in the nebula (Boynton 1975); the depletion of Yb is almost certainly due to its higher volatility.

The Group II REE pattern, although now quite common in fine-grained CAI's, was quite remarkable when it was discovered (Tanaka and Masuda 1973) because it was unlike any known REE pattern. It has very high abundances of the light REE, sharply decreasing abundances going to the heavier REE, and significant anomalies in Eu, Tm and Yb. Clearly, this pattern cannot be due to an igneous fractionation process.

Boynton (1975) showed that these REE patterns can be explained if the Group II CAI's are condensates from the solar nebula. Figure 5 shows a Group II REE pattern with values calculated based on the REE volatilities. Considering that there is an uncertainty of a factor of 2 to 3 in the volatilities, the agreement between theory and observation is quite good. These peculiar

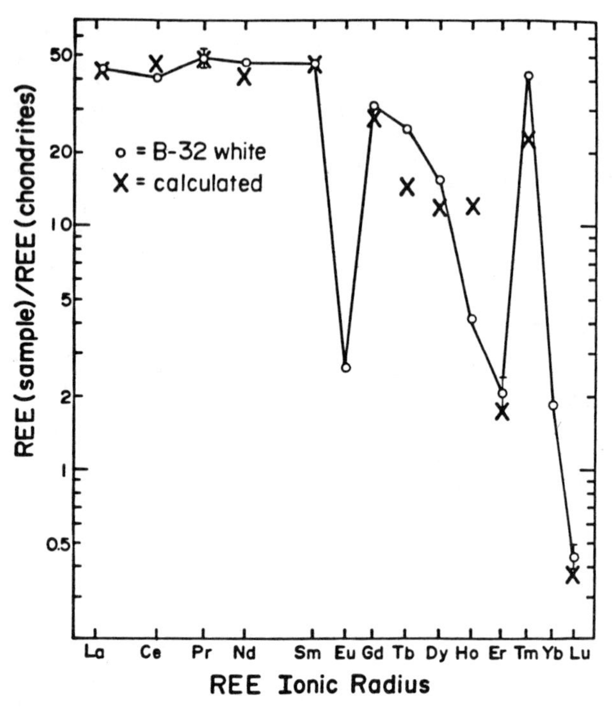

Fig. 5. Plot showing that the peculiar REE patterns of Group II inclusions can be duplicated with calculations based on REE vapor pressure. To achieve this fractionated pattern under conditions approaching equilibrium requires time on the order of an hour or more, too long for some transient heating events such as lightning.

REE patterns are now known to be quite common in Allende CAI's. Davis and Grossman (1979) plotted REE patterns of an additional 19 Group II CAI's and showed that they are all consistent with origins as nebular condensates.

However, the material of the Group II inclusions cannot be the first solids to condense out of the solar nebula. The elements that are the most refractory, Lu and Ho, are the most depleted, but they should be the most enriched if this were the first condensate from a gas of solar composition. The calculated values in Fig. 5 were, in fact, generated assuming that the REE in the Group II material remained in the gaseous form following removal of the more refractory REE from the gas into a hypothetical initial ultrarefractory component (either a refractory residue or a condensate). Thus, the REE pattern in the Group II CAI's was established *when the REE were in the gas phase* in equilibrium with an ultrarefractory component. This gas then condensed to form the Group II material. Note that whereas the Group I and Group III material (Fig. 4) could have an origin as either a condensate or a refractory residue, the REE pattern in the Group II CAI's could not be generated as a refractory residue of a partial vaporization event.

It can also be seen from Figs. 4 and 5 that the two most volatile REE, Eu and Yb, are also absent from the Group II material. Apparently, as for the Group III material, the grains making up the Group II CAI's were isolated from the gas at a sufficiently high temperature so that these more volatile REE did not condense. Thus the Group II CAI's are a sample of the middle of the refractory-element condensation sequence.

The fractionation involved with the fine-grained aggregates, the Group II and Group III CAI's, are even more complicated when lower temperatures are considered. For example, the volatile elements Na and Cl (Grossman and Ganapathy 1975) and Zn (Grossman and Ganapathy 1976) are also substantially enriched in these CAI's relative to abundances in CI chondrites. Thus, the Group II aggregates are missing the ultrarefractory elements, as well as the more volatile elements in the middle of the condensation sequence (where most of the mass would condense), but managed to collect the most volatile elements. The Group III inclusions experienced a similar history, except that they retained their full complement of ultrarefractory elements.

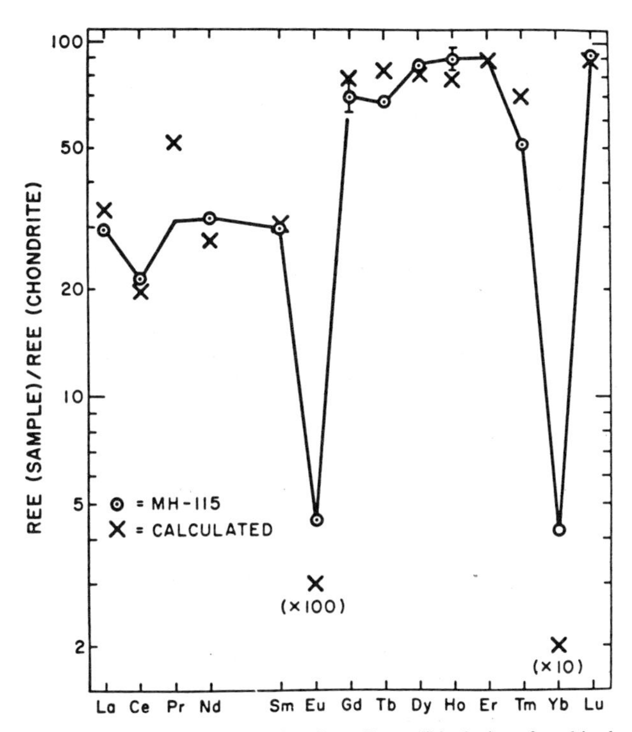

Fig. 6. The ultrarefractory component missing from Group II inclusions found in the Murchison meteorite (Boynton et al. 1980). The enrichment of 100 times CI chondrites suggests that this material is the first 1% of the condensible matter to form in the solar system.

B. The Ultrarefractory Component

The hypothetical ultrarefractory component has never been found in Allende, but it was found in Murchison, a CM carbonaceous chondrite (Boynton et al. 1980). The REE pattern of this ultrarefractory CAI is plotted in Fig. 6 along with calculated REE abundances. The most refractory REE, Lu and Ho, are totally condensed, and their abundances, at about 100 times CI chondrites, suggest that this CAI represents the first 1% of the condensible mass of the solar nebula. More recently, Palme et al. (1982) found an even more refractory inclusion in the Ornans CO3 chondrite, which has Lu enriched to 10,000 times CI chondrites.

It seems clear that in the regions where carbonaceous chondrites formed, many episodes of condensation and isolation from the gas occurred at high temperatures. Some of these isolation processes were very efficient. For example, the Group II fine-grained aggregates must have formed in a reservoir completely isolated from grains having the REE pattern of the ultrarefractory component (Fig. 6). Because Group II aggregates typically have a Lu abundance around 0.5 times chondrites, at most they could contain 0.5% of grains with the 100-times chondrite Lu abundance of the ultrarefractory component. Considering that the Group II aggregates are cm-sized bodies composed of μm-sized grains, it is hard to imagine the environment in which the small grains could have agglomerated without trapping any ultrarefractory grains. It seems clear though, that the ultrarefractory grains must have been in a different part of the solar nebula for the entire time that these aggregates were forming. This requirement further supports the suggestion made by Wood (see his chapter) that many of these processes were very fast.

IV. WHAT METEORITE DATA TELL US ABOUT THE SOLAR NEBULA

Although the elemental abundance patterns seen in bulk meteorites and in components of meteorites are not fully understood, they do provide constraints on the conditions in the solar nebula. These constraints are discussed below.

A. Evidence Against Widespread Mixing in the Solar Nebula

The chondrites are mixtures of a diverse set of components. For example, all but the CI group of carbonaceous chondrites contain chondrules and a wide variety of CAI's, in addition to a volatile-rich matrix. However, these mixtures differ considerably between the groups of chondrites. The CI carbonaceous chondrites are essentially devoid of chondrules and CAI's, while the ordinary chondrites contain chondrules and a different kind of matrix, but no CAI's.

Morfill et al. (see their chapter) have suggested that the high-temperature processing of objects found in meteorites occurred at distances relatively close to the Sun, and that these objects were then transported outward to the meteorite formation location by eddy diffusion in a turbulent nebula. By this mechanism, they proposed to provide for the high temperatures required by the meteorites without the embarrassment (at least to some astrophysical models) of having high temperatures out as far as the asteroid belt. However, because diffusion will only provide a net transport of material from a region of higher concentration to one of lower concentration, the high-temperature CAI's should be found in all chondrite classes which formed closer to the Sun than the carbonaceous chondrites which have these inclusions.

Gradie and Tedesco (1982) have shown that the asteroid belt contains a compositional structure. This compositional sorting must preserve a record of the relative distances from the Sun where the various groups formed because there is no plausible mechanism to sort an initially random collection of asteroids. The carbonaceous chondrites are most similar to the C asteroids (Chapman et al. 1975; Gaffey and McCord 1979), which are found at the outermost parts of the asteroid belt. Therefore, one would expect CAI's to be even more abundant in the ordinary chondrites, which is clearly not the case. It appears then that the high-temperature components could not have been formed close to the Sun and been transported to the meteorite formation locations.

Quite apart from the question of where the CAI's formed, their presence in high concentration in some chondrite groups and absence in others puts strong constraints on mixing during the early stages of accretion. Apparently, each of the various chondrite groups sampled only a very limited region of the solar nebula (Wasson and Wetherill 1979). Wood, in his chapter, suggests that this observation requires that accretion began soon after the formation of chondrules and inclusions, certainly less than a few orbital periods. It is possible though, that if orbits were very circular at this stage, zones could persist for many orbital periods with little or no mixing between them. Such quiescent conditions, however, hardly seem capable of generating the various episodes of condensation and isolation from the gas that are recorded in the CAI's. Solving this conundrum in a way that satisfies both astrophysical and cosmochemical constraints remains a challenge.

B. High Temperatures in the Region of the Asteroid Belt

The Group II REE patterns (Fig. 5), the ones with a deficiency of ultrarefractory REE, provide clear evidence that the REE which they contain were at one time in the gas phase. The temperatures required to vaporize these elements are a function of total pressure in the nebula, but for reasonable pressures in the range of 10^{-6} to 10^{-4} atm, temperatures on the order of 1400–1500 K are required. It seems clear that these objects formed in our solar system because they have an isotopic composition typical of solar system material (see Clayton's chapter). Although they may contain a mixture of some grains of material which were not processed in our solar system, it is not likely that the REE pattern was established outside our solar system. Fegley and Kornacki (1984) have suggested that the Group II inclusions could form entirely from interstellar grains, but they require a physical separation of the ultrarefractory grains from the Group II grains. Considering that the separation was at least 99.5% complete as discussed earlier (Sec. III.B), such a separation of μm-sized grains seems unlikely.

The duration of the heating events is not known, but some constraints can be provided. The REE pattern of the Group II inclusions (Fig. 5) must have been established under conditions approaching equilibrium. Such conditions may only require times on the order of hours, which may not seem like much of a constraint, but they probably cannot be generated in seconds that may be expected from certain transient events such as electrostatic discharge. Very rapid heating would vaporize the REE by a mechanism controlled by diffusion of the REE from the interior to the surface of the grains. This process would produce a fractionation which is a smooth function of diffusion coefficients and hence size. The time required to establish equilibrium is thus a function of the grain size of the ultrarefractory component. This component must not be too small because it must be separated from the gas before the Group II grains form. This separation requires that the grains be large enough to avoid being carried with the gas. Although it does not seem possible to quantify without generating a numerical model, the refractory-element fractionation observed among the chondrite groups (Fig. 3) suggests that longer times were required. For example, some reasonable length of time was probably required to add refractory elements (or remove Si) from the zone where the CV chondrites were forming. Unless the refractory component was grossly different in grain size from the Si-rich component, it seems necessary to require that the more volatile, Si-rich component was in the gas phase while material was being gained or lost from the zones. In addition, the observation discussed above, that there appears to be little transport of objects such as CAI's between regions of different heliocentric distance, suggests that the material being lost or gained was in the gas phase. Temperatures on the order of 1200 K are required to keep 50% of the Si in the gas phase (Grossman 1972).

C. Most of the High-Temperature Fractionations Occurred in a Region of Near-Solar H/O and Gas/Dust Ratios

Normally this conclusion would not be worth discussing because solar ratios are generally assumed to be present. However, in his chapter, Wood (1984) suggests that the oxidized iron content of chondrules implies an O/H ratio, and hence a dust/gas ratio, enhanced by factors of 10^2 to 10^9. Most of the CAI's which show a fractionated REE pattern, however, must have undergone the fractionation event in a region of near-solar H/O ratio. The evidence for this statement comes from the fact that the relative volatilities of several of the REE are a strong function of the oxygen pressure, which in the solar nebula is directly related to the H/O ratio. For example, Boynton and Cunningham

(1981) showed that under oxidizing conditions, Ce becomes much more volatile and Yb much more refractory. Under oxidizing conditions the Group III and Group II REE patterns should show Ce and Eu depletions rather than the Yb and Eu depletions that are actually observed. A limit of about 10 to 100 increase in the O/H ratio is the maximum that can be tolerated. Exceptions have been observed for two CAI's, however. These CAI's have large negative Ce anomalies and clearly formed under highly oxidizing conditions (Boynton 1978*b*; Davis et al. 1982). These two CAI's are also exceptional in having the fractionation and unknown nuclear (FUN) isotopic anomalies (see Clayton's chapter).

V. CONCLUSIONS

Meteorites have preserved a record of extensive fractionations on both the macroscopic and microscopic scale. These fractionations suggest that moderately high nebular temperatures (~1200 K) persisted for some time in the region of the asteroid belt and that higher temperatures (1400–1500 K) were present for times on the order of hours or more. The microscale fractionations show further that the solar nebula was very chaotic with many episodes of condensation and isolation of grains from gas. The lack of mixing of CAI's between various meteorite classes, however, suggests that only limited mixing and transport of grains occurred between regions of different heliocentric distances.

Acknowledgments. This work benefited by constructive reviews by J. T. Wasson and J. A. Wood.

PART VI
Formation of Giant Planets

FORMATION OF THE GIANT PLANETS AND THEIR SATELLITE–RING SYSTEMS: AN OVERVIEW

JAMES B. POLLACK
NASA Ames Research Center

This chapter presents an overview of the nature of the outer giant planets and their satellite and ring systems, theories of their origins, and models of the planets' evolution. The giant planets may have formed either as a result of gravitational instabilities in the solar nebula that gave rise to their envelopes, with their cores being formed later (the gas-instability hypothesis); or as a result of the formation of massive solid cores, chiefly by binary accretion of smaller planetesimals, which permanently captured massive amounts of gas from the surrounding solar nebula (the core-instability hypothesis). Neither model fits all the relevant observational constraints in an obvious fashion. However, the core-instability hypothesis is currently favored because it leads to a "first principles" prediction of the value and near constancy of the core masses. During and after their formation, the giant planets experienced three major evolutionary phases: an early quasi-hydrostatic phase (phase 1) in which their dimensions (10^2–10^3 R_P) changed slowly on time scales of $\geqq 10^5$ yr, where R_P is the planet's current radius; a rapid contraction phase (phase 2) during which they shrank to a size of several R_P on a much shorter time scale than that of phase 1; and a quasi-hydrostatic phase (phase 3) during which they very slowly contracted to their present dimensions. The observed excess luminosities of Jupiter, Saturn, and Neptune (they each radiate to space about twice the solar energy they absorb) are due to a release of gravitational energy, either in the past or present. Jupiter's excess luminosity is currently generated at the expense of internal energy, which was built up during its formation and early contraction history. However, Saturn's excess luminosity may be due chiefly to the segregation of He from H in its envelope due to the partial immiscibility of He in metallic H at sufficiently low temperatures ($\leqq 10^4$ K). Neptune's excess luminosity, like Jupiter's, appears to be due to a cooling of its interior. The regular satellites and ring material of the giant planets formed within viscous accretion disks that *came into existence around their parent planets during the latter's rapid contraction phase. The evolution of their mass, dimensions, and perhaps the radial distribution of temperature within them were controlled by turbulent viscosity, although the early high luminosity of the planets (several orders of magnitude above their current luminosities) may have also helped to heat them. Pressures in these planetary disks were far higher than in the solar nebula, their dimensions relative to their primary were far smaller, and they contained much less of the system's angular momentum. Satellites and ring material formed within these viscous accretion disks chiefly by binary accretion of grains, both preexisting and formed by the condensation of nebula gases. Thus, the composition of the regular satellites and rings of the outer planets reflect the temperature/pressure conditions in the region of their parent planet's nebula where they formed. The irregular satellites of the outer planets probably were captured during the early history of the solar system, due to gas drag they experienced in passing through the highly extended outer reaches of the planet's envelopes, just prior to the rapid contraction phase. Alternatively, capture may have been effected through the collision of two stray bodies within the planet's gravitational sphere of influence.*

I. INTRODUCTION

Our solar system is rich in the diversity of the objects it contains, including the Sun, planets, satellites, asteroids and comets. The Sun is an ordinary G2 main sequence star that is composed principally of H and He and that is presently powered by nuclear fusion in its deep, hot interior. It is thought to have formed within a giant molecular cloud 4.6×10^9 yr ago, with the cloud experiencing a series (perhaps partly hierarchial in nature) of gravitational instabilities that ultimately produced a multitude of stars, including the Sun. Unlike some or most of its siblings, the protosun did not solve its angular momentum problem—the inhibition of continued contraction due to a buildup of rotational velocity—by forming a multiple star system. Rather, a viscous accretion disk is thought to have formed around a growing central object, the early Sun. It was within this disk, the primitive solar nebula, that all the other objects of the solar system formed.

The objects that formed within the solar nebula vary enormously in composition. Objects that formed within the inner solar system (the terrestrial planets, their moons, and the asteroids) are made almost entirely of compounds that were capable of existing in the solid phase at the ambient temperatures of the solar nebula, i.e., they contain virtually no H or He.

Even the atmospheres of the largest of these objects (i.e., the Earth, Venus, and Mars) reflect the more volatile components of their planet's interior and hence are rich in gases made of C, N, and O rather than H and He. Furthermore, temperatures in the inner region of the solar nebula were not low enough ($\geqq$ 160 K) for gases such as water vapor to condense into water ice. Thus, water is not a major constituent of any object of the inner solar system. The water contained in these objects is derived from the hydration of silicate minerals, which occurs at temperatures below about 400 K. For convenience, I will refer to the bulk composition of the objects of the inner solar system as "rock," a variety of compounds dominated by Mg, Fe, Si, and O.

In contrast to objects in the inner solar system, many of the smaller objects of the outer solar system (Pluto, satellites, and comets) contain a significant ice component as well as a rock component. By "ice," I mean to denote a compositional class including water as well as condensed materials that form under even lower temperature conditions in the solar nebula, such as ammonia ice, ammonia monohydrate ($NH_3 \cdot H_2O$), and a number of clathrates, such as methane clathrate ($CH_4 \cdot 6H_2O$). The term ice is used regardless of whether the material is currently in a liquid, solid, or even gas phase. The inner satellites of Jupiter and perhaps Neptune's Triton represent the only notable exceptions to the above compositional makeup of small objects in the outer solar system. Rather than being made of comparable proportions of ice and rock, they are constituted mostly of rock.

A third compositional class is made up of the outer giant planets—Jupiter, Saturn, Uranus, and Neptune. Gaseous envelopes composed chiefly of H_2 and He represent a significant fraction of these planets' masses. However, they also contain central cores made of rock and ice. Thus, they are composites of relatively little chemically differentiated pieces of the solar nebula (their gaseous envelopes) and the highly differentiated, high-Z material (their central cores), the dominant material of all the other objects in the solar system, except, of course, the Sun. Objects belonging to this third class exhibit a wide variance in the relative proportion of envelope and core; most of Jupiter's and Saturn's mass is contained in their envelopes, whereas the reverse is true for Uranus and Neptune.

There appears to be two ways in which matter in the universe organizes itself into condensed objects. On the one hand, large objects, such as the Sun, form from gravitational instabilities that occur within a very rarefied gaseous medium. The end result is a high density, compact gaseous body, whose initial composition is essentially identical to that of the gaseous medium within which it formed. On the other hand, very tiny objects, such as comets and asteroids, form by accretion of only the solid component of the medium of their birth. Thus, their composition reflects that of material that was able to remain in the solid phase or condense from gases at the ambient temperatures of the region where they formed. Furthermore, although gravitational instabilities may have been important at some stage in the growth of these bodies (Goldreich and Ward 1973), successive collisions between pairs of smaller-sized predecessors of the final object, binary accretion, is thought to play a major role in the formation of this class of objects.

One fundamental question that concerns the outer planets is their mode of formation. Did they form by analogy to the Sun and other stars as a result of a gas instability that took place in the solar nebula and at a later time accrete their high-Z cores? Or did they form by analogy to the comets and asteroids through the accretionary growth of their high-Z cores that became massive enough to permanently capture gas from the solar nebula? In trying to answer this question, we are attempting to find the dividing line that separates objects formed from the two basic condensation mechanisms: How big can an object be and yet form by binary accretion? How small can an object be and yet form by gas instability?

A second fundamental question involving the outer giant planets has to do with the mode of formation of their satellites and, in particular, the likely occurrence of a viscous accretion disk within which many of them originated. Most of the moons of the outer giant planets have low eccentricity, low inclination orbits and travel in a prograde direction (same direction as the planet rotates) about their primary. These orbital characteristics strongly suggest that these "regular" satellites were not formed somewhere else in the solar system and later captured by their primary. Rather, they probably formed by binary accretion within a viscous accretion disk that developed about its primary as the primary contracted toward its present size. If so, we have several additional examples of such disks besides the solar nebula that can test our understanding of the nature of these fundamental objects. How similar were the physical processes that operated in the accretion disk of the Sun, (the solar nebula), on the one hand and those of the outer giant planets, on the other hand? What factors determined the mass and composition of objects that formed in these disks?

It is worth noting in the context of the above discussion that there are several tens of regular satellites in the outer solar system and only three in the inner solar system. In part, the relative deficiency of satellites in the inner solar system may be due to tidal friction, which could have caused large, close moons to have spiraled into Mercury and Venus and have led to a sweep-up of smaller objects by the Moon as its orbit evolved outward (see, e.g., Ward and Reed 1973). But this difference may also reflect the absence of a large gaseous envelope around the inner planets, which, therefore, made them incapable of forming accretion disks that could efficiently produce moons. But are all three moons of the inner planets captured objects? Or could terrestrial planets somehow, perhaps through fission, produce their own satellites? While these questions are outside the scope of this chapter, I, nevertheless, wished to raise them.

The outermost satellites of Jupiter, Saturn, and Neptune have highly inclined, highly eccentric orbits and some of them travel in retrograde directions. Thus, these "irregular" satellites may be captured objects. If so, their composition may provide constraints on the temperature conditions within the outer part of the solar nebula at about the time the giant planets formed. But were they captured? If so, how?

In this chapter, I discuss the questions raised above regarding the origin of the outer planets and their satellite systems. In so doing, I attempt to keep the discussion within the framework of astrophysical theories of the origin of stars and the nature of viscous accretion disks. However, I also point out unique aspects of these problems. In Sec. II, additional background material is presented on the nature of the outer planets and their satellite systems. Subsequent sections deal with the origin of the outer planets (Sec. III), their subsequent evolution (Sec. IV), and the origin of their regular (Sec. V) and irregular (Sec. VI) satellites. The discussion of the origin of the regular satellites is focused around viscous accretion disks and includes a short discussion of the origins of the rings of the outer planets.

This chapter is intended to provide an overview of the formation of the outer planets. More detailed discussion on the origin and evolution of the outer planets, observational constraints on their origin, physical processes characterizing their ring and satellite systems, and the origin of comets are given in the chapters by Bodenheimer; Gautier and Owen as well as Podolak and Reynolds; Lissauer and Cuzzi; and Weissman, respectively.

II. NATURE OF THE OUTER PLANET SYSTEMS

In this section, I summarize properties of the outer planets and their associated satellites and rings that are relevant to the formation of these systems and their subsequent evolution.

Table I summarizes some of the basic properties of the four giant outer planets, which range in size from about 4 times to about 10 times the size of the Earth and have masses varying from about 15 to about 300 times the Earth's mass ($M_{\oplus}$). Information about the bulk composition of these planets can be obtained by fitting hydrostatic equilibrium models to their observed masses and radii. Such fits are illustrated in Fig. 1 for four different chemically homogeneous models: pure H, a solar mixture of elements (H-He), ice, and rock (Stevenson 1982*b*). The solid curves show the theoretical mass-radius relationship for 0 K isothermal models, while the dashed curves denote the relationship for models having adiabatic interiors. The latter is a good approximation for the interiors of at least Jupiter, Saturn, and Neptune, which have strong internal heat sources, as discussed more fully below (cf. Hubbard 1980).

Several important qualitative conclusions can be drawn about the bulk composition of the four giant planets from Fig. 1. First, Jupiter and Saturn are made mostly of a solar mixture of elements (Cameron 1973*b*), while Uranus and Neptune are composed chiefly of a mixture of ice and rock. Second, both Jupiter and Saturn require a minor fraction of high-Z material, in excess of that expected for solar abundances, to match their observed mass and radius. Third, Saturn contains a larger fractional amount of high-Z material than does Jupiter. This latter point can be readily seen by noting that the radius of the adiabatic, solar composition model is almost independent of mass. Thus, the observation that Saturn has a smaller radius than Jupiter leads to this third conclusion.

TABLE I
Properties of the Outer Planets

Property	Jupiter	Saturn	Uranus	Neptune
Radius, km	6.98×10^4	5.83×10^4	2.55×10^4	2.45×10^4
Mean density, g cc^{-1}	1.334	0.69	1.26	1.67
Total mass[a]	318.1	95.1	14.6	17.2
Envelope mass[a,b]	288.1–303.1	72.1–79.1	1.3–3.6	0.7–3.2
Core mass[a,b,c]	15–30	16–23	11–13.3	14–16.5
Atmospheric composition[d]:				
He/H	0.117 ± 0.025	0.062 ± 0.011	—	—
C/H[e]	$(1.09 \pm 0.08) \times 10^{-3}$	$(7.5–17.4) \times 10^{-4}$	$(1.9–9.4) \times 10^{-3}$	$(9.4–118) \times 10^{-4}$
N/H[f]	$(1–2) \times 10^{-4}$	2×10^{-4}	$< 1 \times 10^{-4}$	$< 1 \times 10^{-4}$
D/H	$(2.0–4.1) \times 10^{-5}$	$(1.1–4.8) \times 10^{-5}$	$(3.3–6.3) \times 10^{-5}$	—
Excess luminosity[g]:				
Luminosity, W	$(3.35 \pm 0.26) \times 10^{17}$	$(8.63 \pm 0.6) \times 10^{16}$	$< 1.5 \times 10^{15}$	$(2 \pm 1) \times 10^{15}$
Luminosity/mass, W g^{-1}	$(1.76 \pm 0.14) \times 10^{-13}$	$(1.52 \pm 0.11) \times 10^{-13}$	$< 2 \times 10^{-14}$	$(2 \pm 1) \times 10^{-14}$

[a]In units of $M_\oplus = 5.98 \times 10^{27}$ g.
[b]Based on the models of Hubbard and MacFarlane (1980) and Podolak and Reynolds (1981).
[c]Includes all high-*Z* material in excess of solar elemental values of Cameron (1973*b*), including high-*Z* material in the gas envelope.
[d]Based on data summarized by Gautier and Owen (1983*b*) and Hanel et al. (1981*a*). In all cases ratios refer to mole fractions.
[e]Lambert's (1978) value for the Sun is 4.7×10^{-4}.
[f]Lambert's (1978) value for the Sun is 9.8×10^{-5}. The possible depletion of NH_3 in the atmospheres of Uranus and Neptune based on radio thermal emission has been called into question by a sizable variation in this emission for Uranus with orientation of the axis of rotation (Gulkis et al. 1983).
[g]Based on data summarized by Hanel et al. (1983) and Hubbard (1980).

More detailed information about the relative proportions of solar gas and high-*Z* material, and their radial distribution, can be obtained by constructing hydrostatic models having several compositional zones and by fitting such models not only to the observed mass and radius, but also to their observed J_2 and J_4 gravitational moments (see, e.g., Hubbard and MacFarlane 1980; Podolak and Reynolds 1981). The estimates of envelope and core masses (solar gas and high-*Z* material) given in Table I were obtained in this way. The range of values reflects uncertainties in both the equation of state (see Stevenson 1982*b*) and observed quantities, especially J_2 and J_4. A key fact emerges from this analysis; while the envelope masses of the outer planets range over

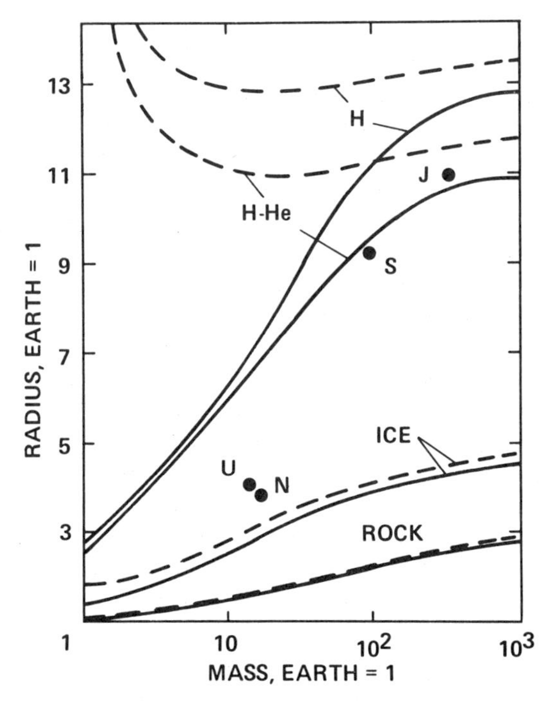

Fig. 1. Radius of a giant planet, in units of the Earth's radius, as a function of its mass, in units of $M_\oplus$. The dots with J, S, U, and N next to them show the observed values for Jupiter, Saturn, Uranus, and Neptune, respectively. The solid and dashed curves show theoretical relationships for isothermal, 0 K interiors and adiabatic interiors, respectively, for planets made of various materials.

about two orders of magnitude, the core masses are the same to within a factor of 2.

Unfortunately, it is not possible to determine the relative proportion of ice and rock in the cores of Jupiter and Saturn because their cores are embedded deep in the planet (and hence only influence J_2) and because they constitute only a minor fraction of their planet's mass. However, it is possible to bound crudely the ice/rock ratio for Uranus and Neptune because their cores represent most of their planet's mass. According to the models of Podolak and Reynolds (1981, 1984), the mass ratio of ice to rock in these planets (including possibly high-Z enriched envelopes) lies between 1 and 3.6. The upper bound corresponds to the solar elemental value and is based on ice being more volatile than rock.

While H_2 and He are the most abundant gases in the atmospheres of the giant planets, there is now good evidence for significant departures from solar elemental abundances, as summarized in Table I (see the chapter by Gautier and Owen for more detail). For example, the He/H ratio is approximately solar in the observable atmosphere of Jupiter but is about a factor of 2 smaller than solar in Saturn's atmosphere. Because it is very difficult to conceive of a way of segregating H from He in the solar nebula, this result for Saturn probably reflects an inhomogeneity in the radial distribution of He in that planet's interior. A possible cause for such an inhomogeneity will be discussed in Sec. IV. According to Table I, the C/H ratio is enhanced over solar by a factor of 2 in the atmospheres of Jupiter and Saturn and by as much as a factor of several tens in the atmospheres of Uranus and Neptune. Thus, not all excess high-Z material is contained in the cores. Also, N/H is enhanced over solar by perhaps a factor of 2 in the atmospheres of Jupiter and Saturn, while it appears to be underabundant in the atmospheres of Uranus and Neptune. Finally, the D/H ratio in the atmospheres of all the giant planets is approximately the same (the quoted uncertainties are large), with the nominal value exceeding that for the present interstellar medium by a factor of several (Geiss and Reeves 1981) and that of the protosun by a factor of ~ 2 (D. C. Black 1973).

Figure 2 presents one current set of models of the interior structure of the giant planets (Stevenson 1982*b*). Pressures become large enough ($\geqq 2$ Mbar) in the fluid envelopes of Jupiter and Saturn for H to undergo a phase transition from molecular hydrogen, a good electrical insulator, to metallic hydrogen, a good electrical conductor. However, the pressures in the envelopes of Uranus and Neptune never reach high enough values to produce metallic hydrogen. If ice and rock are segregated in the cores of Uranus and Neptune, an ionic water ocean may overlie a hot rock central region. Ammonia is expected to dissolve readily in such an ocean (perhaps causing the underabundant N/H ratio in these atmospheres), while methane is expected to be far less soluble (perhaps contributing to the enhanced C/H ratio in these atmospheres).

Measurements of the radiative energy budgets of the outer planets shows that Jupiter, Saturn, and Neptune radiate to space about twice as much energy as the amount of sunlight they absorb. However, only an upper bound exists on Uranus' excess. As quantitatively summarized in Table I, these results imply the occurrences of a sizable internal heat source for at least three of the four giant planets. By contrast, the internal heat flux of the Earth, derived chiefly from the decay of long-lived radioisotopes, is only about 10^{-4} of the flux of absorbed sunlight.

There are 16, 18, 5, and 2 known satellites in orbit about Jupiter, Saturn, Uranus, and Neptune, respectively. Table II summarizes several basic properties of the regular and irregular satellites of these systems. Only Uranus lacks a known irregular satellite. At least one (Nereid) and perhaps both (Triton as well) of Neptune's satellites fall into the irregular category. The high inclination of Triton's orbit places it in the irregular category, while its orbital distance, and large mass imply that it belongs to the class of regular satellites.

As shown in Table II, the regular satellites have sizes that range from values comparable to a small asteroid or very large cometary nucleus (several tens of kilometers) to ones somewhat larger than that of the planet Mercury

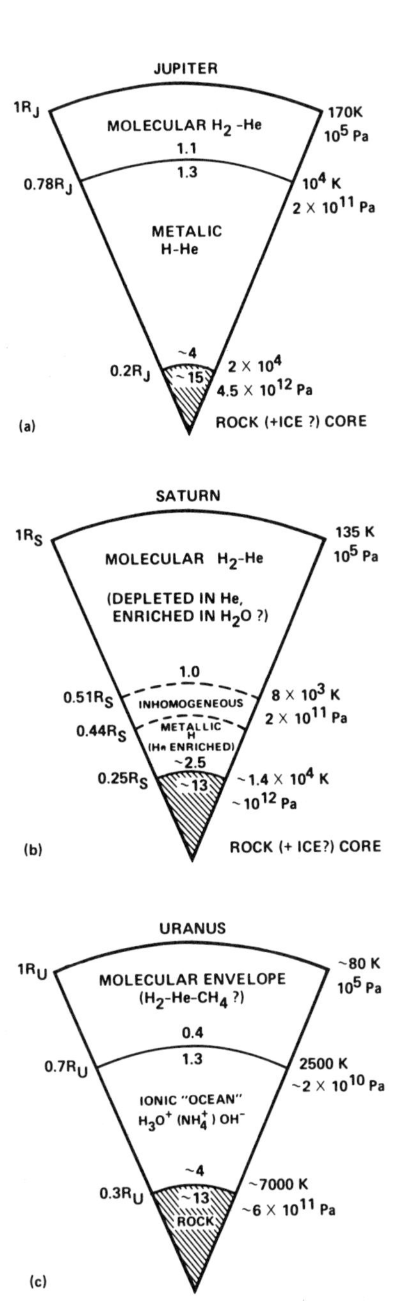

Fig. 2. Schematic representation of typical interior models of Jupiter (a), Saturn (b), and Uranus (c). The numbers on the left side show distance from the center in units of the planet's radius, the numbers on the right side show temperature and pressure, and the numbers in the center show densities in g cc^{-1}. The corresponding interior model for Neptune is very similar to that of Uranus.

TABLE II
Properties of the Satellites of the Outer Planets[a]

	Jupiter	Saturn	Uranus	Neptune
Regular satellites:				
Number	8	17	5	1[c]
Distance[b]	1.8–26.6	2.3–59.0	5.1–23	14.6
Radius, km	20–2630	15–2575	150–400	1600
Irregular satellites:				
Number	8	1	0	1[c]
Distance[b]	156–333	216	—	227
Radius, km	5–85	110	—	150

[a]Based on data given by E. C. Stone and E. D. Miner (1982), Morrison (1982), and Morrison et al. (1977).
[b]In units of planetary radii.
[c]Triton is assumed to be a regular satellite in view of its large mass and proximity to Neptune and despite its high orbital inclination and retrograde motion.

(several thousands of kilometers). In all cases, they are situated from several to several tens of planetary radii from the center of their primary, with the smaller ones tending to be located at closer distances.

As mentioned in Sec. I, the satellites of the outer planets are made of a mixture of ice and rock. As far as is known from the mean densities of the larger moons and the albedoes and infrared spectra of the smaller ones, ice and rock occur in comparable and hence approximately (within a factor of several) solar proportions in all the regular satellites of Saturn and Uranus and in the two outermost large Galilean satellites of Jupiter, i.e. Callisto and Ganymede. However, rock is certainly the dominant component of the two inner Galilean satellites, Io and Europa, and perhaps the still closer, smaller regular satellites of Jupiter. Indeed, there is a monotonic decrease of mean density and hence rock/ice ratio with increasing distance from Jupiter among the four Galilean satellites. Finally, currently available data suggest that Triton may be made mostly of rock (R. Greenberg 1984).

On the basis of cosmic abundances (O > C > N) and infrared spectral observations, water ice is believed to be the dominant ice species. However, there is evidence for the occurrence of other ice species in the makeup of some of the ice-rich satellites. The best such evidence is provided by the occurrence of N_2 and CH_4 in Titan's massive atmosphere (the atmospheric pressure at the surface is 1.6 bar) (Tyler et al. 1981; Samuelson et al. 1981). These gases were very likely derived from material that made up this satellite (e.g., outgassing of the interior) rather than from gases captured from the primordial nebula of Saturn (Owen 1982; Pollack and Consolmagno 1984). Similarly N_2 and CH_4 have been detected on the surface and/or in the atmosphere of Neptune's moon Triton (Cruikshank et al. 1983). More indirect evidence for the occurrence of ices other than water ice is provided by the longevity of geological activity (e.g., tectonism, resurfacing events) on several of the larger moons of Saturn (Ellsworth and Schubert 1983) and the low albedo of a portion of Saturn's moon Iapetus (Pollack and Consolmagno 1984).

The currently known irregular satellites of the outer solar system include the outer eight satellites of Jupiter, the outermost moons of Saturn (Phoebe) and Neptune (Nereid) and perhaps Neptune's other moon, Triton. If we exclude Triton from this group for reasons discussed above, the irregular satellites are characterized by sizes in the range of tens to hundreds of kilometers and distances from their primaries of several hundred planetary radii (cf. Table II). The eight irregular satellites of Jupiter cluster into two families, with members of each family having very similar orbital inclinations and semimajor axes.

The visible albedo and visible and near-infrared spectra of Phoebe and several of the largest irregular satellites of Jupiter imply that they are made of carbonaceous chondritic-like material, i.e., hydrous silicates containing complex carbon compounds (Degewij et al. 1980). This material is prevalent in the outer part of the asteroid belt. Little information currently exists on the composition of the other irregular satellites. It should be noted that excavation of the subsurface by large impact events prevents the intrinsic surface from being masked by meteoroid dust. As illustrated perhaps by the surface of Callisto, undifferentiated bodies containing mixtures of ice and rock are expected to have higher visible albedoes than those characterizing the irregular satellites. Thus, there appears to be a real bulk compositional difference between the irregular satellites and the outer regular satellites; the irregular satellites, whose composition has been evaluated, lack water ice, although they may contain water of hydration.

Jupiter, Saturn, and Uranus are known to have rings that lie interior to their satellite systems. In all three cases, the main mass of the rings lies within the classical Roche limit for tidal disruption of a gravitationally bound body, although the role of tidal forces in creating the present population of particles within the rings remains elusive (see, e.g., Pollack and Cuzzi 1981). Also there are several rings of Saturn that lie outside this classical limit. In all cases, the rings consist of a multitude of small solid objects in independent orbit about their primary. Collisions between ring particles in almost all cases (except for the halo component of Jupiter's ring and Saturn's E ring) result in very flat rings.

Many of the particles in Saturn's rings have sizes that lie between 1 mm and 10 m (Cuzzi et al. 1980; Tyler et al. 1981), although micron-sized particles are also present in some of Saturn's rings (Pollack 1981) and several kilometer-sized bodies also appear to occur within a few of the ring gaps (Cuzzi et al. 1983). Micron-sized particles are the dominant optical scatterers in the rings of Jupiter (Jewitt 1982). But, in view of their short lifetimes ($<<10^9$ years), they may be derived from the micrometeoroid erosion of much larger-sized bodies in the region of these rings (J. A. Burns et al. 1980). The particles in Uranus' rings have a characteristic size of a few centimeters (Elliot and Nicholson 1984).

At present, there exists good compositional information only for the rings of Saturn. Water ice is a dominant component of the surfaces and interiors of the millimeter-to-meter-sized particles of the main rings of Saturn (Pilcher et al. 1970; Pollack et al. 1973; Cuzzi et al. 1980). Although rock may be present as a minor coloring agent for these particles (Lebofsky et al. 1970), the bulk ice-to-rock ratio seems to be significantly in excess of its solar value (Cuzzi et al. 1980). If so, these are the only objects in the outer solar system known to have a super-solar ice-to-rock ratio.

The three ring systems differ significantly in their overall architecture and visible optical depth (Jewitt 1982; Cuzzi et al. 1984; Elliot and Nicholson 1984). Most of the rings of Saturn and the rings of Jupiter are quite broad (greater than thousands of kilometers in radial extent), while the rings of Uranus and Saturn's F ring are quite narrow ($<$ 100 km). Jupiter's rings have a very low optical depth ($\sim 10^{-5}$), while the rings of Saturn and Uranus have optical depths on the order of unity. Small moons (1–100 km) may play important roles in the ring systems by confining narrow rings between pairs of them (Goldreich and Tremaine 1979*a*) and by serving as a source of smaller ring particles (Burns et al. 1980). The mass of Saturn's rings is comparable to that of the intermediate-sized (200 km radius) Saturnian satellite Mimas (Holberg et al. 1982), while the masses of the rings of Uranus and Jupiter are $\sim 10^{-4}$ and $\sim 10^{-6}$ that of Saturn's rings (Elliot and Nicholson 1984; J. A. Burns et al. 1984). A much more detailed discussion of the properties of the rings of the outer solar system is given in the chapter by Lissauer and Cuzzi.

III. ORIGIN OF THE OUTER PLANETS

As discussed in the introduction, there are two classes of theories for the origin of the giant planets. On the one hand, they may have resulted from gravitational instabilities within the solar nebula that initially generated a compact gaseous object having solar elemental abundances. Subsequently, this giant gaseous protoplanet obtained high-Z material that collected into a core. I will refer to this model as the "gas-instability" model (Cameron 1978*a*). On the other hand, the core may have formed first, mainly through the binary accretion of small solid planetesimals. Ultimately the core may have grown to be sufficiently massive to effectively concentrate and hence capture a massive gaseous envelope from the solar nebula. I will refer to this model as the "core-instability" model (Perri and Cameron 1974; Mizuno et al. 1978; Harris 1978*a;* Mizuno 1980; Bodenheimer and Pollack 1984). It is interesting to note that these models require the presence of the solar nebula during most (gas-instability model) or all (core-instability model) of the period of planet formation. Below, we discuss in greater detail the nature of these two models and then compare their predictions against various constraints on the origin of the giant planets (e.g., Table I). A further discussion of this topic is given in the chapter by Bodenheimer.

A. Gas-Instability Model

According to the gas-instability model, the solar nebula underwent a series of gravitational instabilities that resulted in giant gaseous protoplanets (Cameron 1978*a*). The first of these may have been a global scale instability involving global azimuthal perturbations that resulted in rings of enhanced gas density. Such instabilities occur when the ratio of the mass of the solar nebula to that of the protosun (the central condensation of the nebula) exceeds a certain critical value, with this value depending on the radial wavelength of the perturbation. For wavelengths comparable to the mean spacing of the outer planets, this ratio is on the order of 10^{-1}. According to Cameron (1978*a*), as the gas density in these rings increases, they become locally unstable in a manner analogous to a Jeans-type instability in which the gravitational energy of a gas ball exceeds in absolute magnitude its thermal energy. Coalescence of gas balls in a given ring and/or the elimination of some by gravitational scattering ultimately produces a single giant gaseous protoplanet at a given radial distance, the predecessor of the gaseous envelope of the giant planets. While the above scenario is physically plausible, it has not been studied in sufficient detail to determine whether the postulated instabilities are the fastest growing ones. For example, spiral density waves rather than rings might result when the solar nebula becomes unstable to global perturbations (Cassen, personal communication).

Giant gaseous protoplanets could develop high-Z cores in several ways. For the moment, we will ignore the potentially critical issue of the solubility of high-Z material in the high-pressure zone of the envelope, which may prevent giant gaseous protoplanets from ever developing cores. This topic will be examined in Sec. III.C. Grains in an envelope may coalesce and grow to a sufficient size to fall rapidly to the center. The conversion of some grains from a solid phase to a liquid phase, which can occur under relevant pressure and temperature conditions in the envelopes, may greatly abet coalescence (Slattery et al. 1980). However, there must be an efficient resupply of grains to the envelopes from the solar nebula in order for the resultant cores to represent true excesses of solar elemental abundances for the planet as a whole, rather than a mere redistribution of elements; and in order to produce cores of the correct fractional mass range of 3–90% vs. $\sim$ 1% for solar abundances. Alternatively, most or almost all of the envelope gases were lost to space after the cores formed, due either to tidal stripping by the forming Sun (Cameron 1978*a*) or thermal evaporation as the solar nebula heated up (Cameron et al. 1982).

Another mechanism by which the postulated giant gaseous protoplanets may have gained cores is through the gas drag capture of planetesimals from the surrounding solar nebula (Pollack et al. 1979). Such capture may have been particularly efficient at times soon after the protoplanets formed, when they were much larger than the current dimensions of the giant planets (10^2–10^3 times larger). Furthermore, as discussed in Sec. VI, the irregular satellites may have been captured during a very short interval of the early history of the giant planets. If so, one might expect that satellites other than those that now remain were forced by gas drag to spiral into the giant planets, adding to the mass of solid interior material.

B. Core-Instability Model

According to the core-instability model, the giant planets formed by the binary accretion of a massive core, which was then able to capture permanently a massive envelope from the surrounding solar nebula. Much of the work that has been done on this class of models has centered on determining the critical core mass at which a massive envelope is captured and, to a lesser degree, in determining the physical process that permits such an efficient capture to occur. The initial core instability calculations were carried out by Perri and Cameron (1974), who constructed a series of static, hydrostatic, core-envelope models with completely adiabatic envelopes, i.e., the envelopes were assumed to be in convective equilibrium throughout their extent. The pressure, density, and temperature of these models were matched with those of the ambient solar nebula at the tidal radius of the planet, i.e., that distance at which the Sun's tidal force equaled the planet's gravitational force. They found that once the core mass exceeded a critical value (on the order of 100 $M_{\oplus}$), it was not possible to construct a static model. Stability tests carried out by Perri and Cameron for models having core masses close to the critical value suggested that they experienced a hydrodynamical instability once the critical value was exceeded. Thus, for core masses in excess of the critical value, gas is added very rapidly to the developing planet. Unfortunately, the value of the critical core mass found by Perri and Cameron is almost an order of magnitude larger than that inferred for the giant planets. Its value was found to be sensitive to the boundary conditions with the solar nebula. Both aspects of these results are inconsistent with the properties of the actual core masses (cf. Table I).

Much more satisfactory results were obtained by Mizuno et al. (1978), Harris (1978*a*), and Mizuno (1980), who calculated the temperature structure of the envelopes by first finding the radiative equilibrium temperature gradient and then testing to see if it was convectively unstable. If so, the gradient was set equal to the adiabatic value, a very good assumption according to mixing length theory for the densities characterizing these envelopes. The dominant source of opacity in these envelopes was ice and rock grains. According to Mizuno's calculations, a substantial zone of the outermost portion of the envelope was in radiative equilibrium and therefore these zones had subadiabatic temperature gradients. Due chiefly to this difference between these models and the earlier ones of Perri and Cameron (1974), the former were characterized by critical core masses of about 10 $M_{\oplus}$, with the critical mass being very insensitive to the exterior boundary conditions and to the amount of grain opacity. Critical core mass was once again operationally defined as the value above which it was not possible to construct a hydrostatic model, with the presumption being that larger core masses had envelopes undergoing a hydrodynamical collapse.

A substantially lower critical core mass $\sim 0.1\ M_{\oplus}$ has been obtained by Stevenson (1984), who investigated the properties of grain-free envelopes. He assumes that the cores were built from planetesimals that were constructed in a very efficient fashion from grains in the solar nebula. Thus, virtually no grains were left in the solar nebula at the times when the cores began to accrete. A key consequence of this type of model is that the opacity in the envelope is many orders of magnitude smaller (being derived solely from gases such as hydrogen and water vapor) than for envelopes containing a solar abundance complement of grains. The great reduction in opacity leads to a much smaller critical core mass. In such a scenario, the giant planets are produced by the subsequent accretion of "super-ganymedean puffballs," the objects resulting after the critical core mass is reached.

Such a scenario has the potential advantage of forming the giant planets much more rapidly than in the more conventional core instability picture. But, it does not readily explain why all four giant planets have a core mass of $\sim 15\ M_{\oplus}$. Also, the accreting puffballs should be mostly envelope since they will add much more gas than core after they reach a critical core mass. Thus, all the giant planets would be expected to have the predominant fraction of their mass residing in their envelopes, contrary to the deduced structure of Uranus and Neptune. Below, we discuss only the more conventional core-instability model, but Stevenson's model deserves further study.

Evolutionary sequences of accreting core/envelope models have recently been constructed by Bodenheimer and Pollack (1985). These calculations provide a means for defining the physical process that enables the envelope mass ultimately to grow more rapidly than the core mass and hence leads to a more well-defined criterion for the critical core mass. In addition to being actual evolutionary calculations, these models also differ from their predecessors by introducing a second criterion, besides the tidal one, for defining the exterior boundary of the planet. The accretion radius is also determined, where this radius refers to the location at which the thermal energy of an element of gas equals, in absolute value, its gravitational binding energy to the planet. When the accretion radius was less than the tidal radius, as it was during at least the early phases of evolution, it was used to define the exterior boundary of the model. Grains, in solar elemental proportions, provide the dominant source of opacity at temperatures $\lesssim 2000$ K, as in the models of Mizuno (1980).

As illustrated in Fig. 3, the evolutionary calculations of Bodenheimer and Pollack (1985) show that initially the core grows much more rapidly than the envelope. However, when the envelope mass becomes comparable to the core mass, the envelope undergoes a rapid, but nonhydrodynamic, contrac-

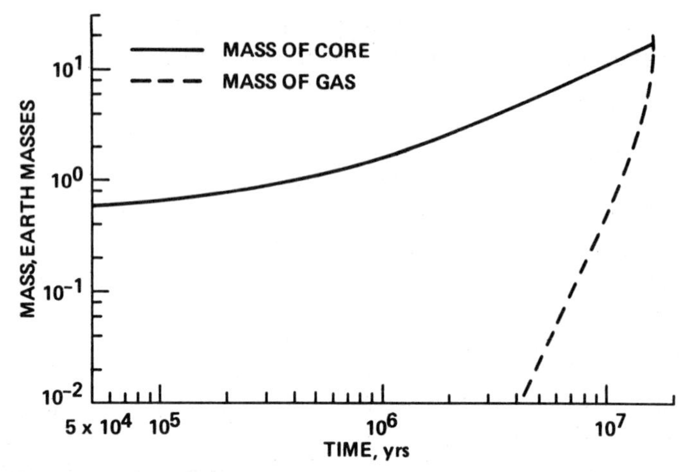

Fig. 3. Core mass (solid line) and envelope mass (dashed line) as a function of time for an evolutionary sequence having a constant core accretion rate of 1 $M_\oplus$ 10^{-6} yr and exterior boundary temperature and density of 150 K and 5.8×10^{-11} g cc^{-1}, respectively.

tion in which the energy of contraction rather than the energy of core accretion becomes the dominant source of energy for the planet. At this stage, the rate of envelope accretion increases greatly and exceeds by a large factor the growth rate of the core (cf. Fig. 3). The transition between the situation where the growth rate of the core exceeds that of the envelope and its converse takes place continuously over a small but finite range of core masses. If the critical core mass is defined as the value at which the two growth rates are equal, these calculations show that it has a value on the order of 15 $M_\oplus$, i.e., comparable to Mizuno's result, for a constant core accretion rate of 10^{-6} $M_\oplus$ yr^{-1}. This value was found to be insensitive to the nebula boundary conditions, but to vary somewhat with the core accretion rate. For example, the critical core mass increased by about 60% when the accretion rate was increased by a factor of 10. Harris (1978*a*) also pointed out the dependence of the critical core mass on the core accretion rate. The critical core mass shows less sensitivity to opacity, as determined by the abundance of small-sized grains.

If an assumption is made about the abundance of small-sized grains during giant planet formation, and if it is assumed that the giant planets formed in a one step process (as contrasted to Stevenson's [1984] scenario described above), it is possible to estimate crudely the core accretion rate from the observed core mass of the giant planets. If small-sized grains were present with solar elemental abundances, then a core accretion rate of 10^{-6} $M_\oplus$ yr^{-1} is implied in the region of the giant planets. Larger accretion rates would be required if much of the solid material had already accreted into planetesimals.

TABLE III
Truth Table for Alternative Models for the Origin of the Giant Planets

Constraint	Pass or Fail	
	Gas-instability model	**Core-instability model**
Value of core mass: 15 $M_\oplus$ and nearly the same for all four giant planets.	Fail (?)	Pass
Value of envelope mass: ~1–300 $M_\oplus$; large variance among four giants.	Pass	Pass
Ratio of envelope to core mass: ~0.1–20.	Fail	?
Time scale for formation of the giant planets:		
(a) $<4.6 \times 10^9$ yr.	Pass	Fail (?)
(b) Less than time to form Mars and a single giant asteroid.	Pass (?)	Fail
Separation of core and envelope.	Fail (?)	Pass
Enrichment of C/H in envelope: 2–20 times solar.	Pass	Pass
Depletion of N/H in envelopes of Uranus and Neptune.	Pass	Pass
Properties of solar nebula.	?	?

C. Critical Tests of the Models

Table III provides a summary of the successes and failures of the two models for the origin of the giant planets in explaining a variety of properties of these planets and relevant aspects of other parts of the solar system. In the remainder of this section, I discuss each of these constraints and the basis for assigning a pass or fail grade to the origin models.

As discussed in Sec. II, a remarkable property of the four giant planets is that they have high-Z core masses that are the same within a factor of 2. A persuasive argument in favor of the core-instability model is the ability of such a model to consistently predict both the absolute value of the core mass and its relative insensitivity to conditions in the solar nebula and other factors. By way of contrast, the gas-instability model does not lead in any obvious way to a particular value for the core mass and, furthermore, it provides no obvious expectation for a nearly constant core mass.

The final envelope masses of the giant planets are predicted to have a large variance in both models. In the case of the gas-instability model, the envelope mass of the initial giant gaseous protoplanets can be expected to decrease significantly with distance from the center of the solar nebula due to a strong gradient in the nebula's gas density with distance. In the case of the core-instability model, the final envelope mass depends on the amount of solar nebula gas added after the critical core mass is reached. The latter depends, among other factors, on the time it takes a given planet to reach its critical core mass relative to the time at which the solar nebula was largely dissipated (the more remote giant planets took longer to reach their critical core masses and so had less time to add gas from the solar nebula); the density of the solar nebula at the location of the giant planets (it was less for the more distant ones); and on the total mass of a giant planet at which it tidally truncates the solar nebula by gravitational torques (Lin and Papaloizou 1979*a*), thereby creating an empty zone in the solar nebula about itself.

While either model has no obvious difficulty in generating planets in which most of the mass resides in the envelope (i.e., Jupiter and Saturn), the gas-instability model is pressed to generate planets in which the core constitutes the bulk of the planet's mass (i.e., Uranus and Neptune). For example, if the cores of the giant planets formed principally by grain coagulation and sedimentation in this model, then either almost the entire envelope had to be subsequently lost (the present envelopes of Uranus and Neptune would represent only 10^{-3} of the initial envelope mass), or there had to be an incredibly massive influx of grains from the solar nebula. If, for this model, the cores formed principally by capture of planetesimals from the surrounding solar nebula, it seems strange that the envelopes of proto-Uranus and Neptune had so much less mass than that contained in the gas of the surrounding solar nebula; the latter would have had to be at least 10^3 times more massive than the former for a solar abundance high-Z to H mass ratio of 10^{-2}. This is a strange result because proto-Uranus and Neptune were presumed to be produced by a series of gravitational instabilities in the solar nebula, including a global instability. The envelope mass had a value that was about 70% of the critical core mass in Mizuno's calculations. If so, the core-instability model would not readily explain why the envelopes of Uranus and Neptune represent only $\sim$ 10 to 20% of the corresponding core masses. But, the evolutionary calculations of Bodenheimer and Pollack (1985) show that the rate of growth of the envelope equals that of the core when the envelope mass is $\sim$ 15% of that of the core. Thus, within the context of the core-instability model, Uranus and Neptune just reached the critical core mass value when accretion ceased, whereas Jupiter and Saturn continued to accrete far beyond this point.

The above discussion gives rise to some speculations concerning the factors determining the zone within which giant planets form. Interior to the region of the giant planets, i.e., in the zone of the terrestrial planets, the solid cores of the developing planets may not have become massive enough (i.e., $\sim$ 15 $M_\oplus$) for them to capture efficiently a massive envelope from the surrounding solar nebula before it dissipated. If envelope accretion is halted by tidal truncation in the case of the outer planets and if the minimum mass of a planet required for tidal truncation decreases with increasing distance (e.g., due to decreasing gas density), then Uranus and Neptune may have been the most distant planets capable of reaching a critical core mass before truncating the solar nebula. More distant planets (e.g., Pluto?) may have reached the minimum mass required for tidal truncation before achieving a critical core mass and hence would be essentially solid planets with very little envelope (cf. Fig. 3).

There are several constraints on the time scale over which the giant planets formed. Quite obviously, they had to form in a time less than the age of the solar system or 4.6×10^9 yr. Because the gas instability model involves giant planet formation as a result of several gravitational instabilities, the time to form the gaseous envelope of a giant planet is probably not much larger than a free-fall time for this type of model. This time is very much less than the age of the solar system. It is more difficult to quantify the time required for a giant gaseous protoplanet to obtain a core of 15 $M_\oplus$ because this depends on the core formation mechanism. However, it seems likely that core formation could take place in less than the age of the solar system. Thus, although a number of aspects of giant planet formation within the context of the gas-instability model need to be worked out, it seems likely that giant planets can be formed in less than the age of the solar system for such a model.

The above, apparently very modest upper bound on the time scale for giant planet formation, provides a real challenge to the core-instability model. The rate limiting step in giant planet formation for such a model is the construction of a 15 $M_\oplus$ core through binary accretion. Models that have been widely used to study the formation of the terrestrial planets by binary accretion in a gas-free environment, i.e., subsequent to the dissipation of the solar nebula, suggest that $\sim 10^7$ to 10^8 yr would be required to construct the cores of Jupiter and Saturn, but that time scales comparable to or greater than the age of the solar system would be required for Uranus and Neptune (Safronov and Ruskol 1982). While the cores of the giant planets had to form before the solar nebula was dissipated in the core-instability model, it is not clear that core formation in the presence of a gas medium will lead to a substantial reduction in the time required to produce the cores of Uranus and Neptune. This absolute time scale problem for the core-instability model is made even more difficult by the results of Bodenheimer and Pollack (1985). In order to obtain the observed core masses, the core accretion rate had to be $\sim 10^{-6}$ $M_\oplus$ yr^{-1} throughout the giant planet zone.

Possible solutions to this absolute time scale problem include (A. Harris, personal communication):

1. The mass of solids initially present in the solar nebula may have been much greater than the amount of solid material incorporated into the planets. In this case, the volume density of planetesimals would be much higher than in the converse situation and the time scales for binary accretion would be reduced proportionately.
2. The critical core mass may have been much smaller than the observed core masses of the giant planets (cf. Stevenson 1984, and the discussion above).
3. Accreting matter in a gaseous medium may have had much greater mobility than in a gas-free environment. Small planetesimals experience a rapid radial drift due to gas drag, while larger ones may show a significant radial drift due to tidal torques exerted on them by the surrounding solar nebula (see, e.g., W. R. Ward 1984).

Another constraint on the time of giant planet formation is a relative one. The small mass of Mars compared to the mass of Earth and Venus and the occurrence of many small bodies in the asteroid belt rather than a single large object could be attributed to Jupiter forming before planet formation was completed in the inner solar system and its gravity perturbing the orbits of nearby objects (Weidenschilling 1975). As a result, nearby objects having similar semimajor axes had high relative velocities and were more likely to disrupt one another when they collided rather than forming a larger composite body. If the giant planets formed by gas instabilities and the terrestrial planets by binary accretion, it seems possible to meet the above constraint on their relative times of formation, especially as it is only necessary to have formed Jupiter's massive envelope prior to the completion of planet formation in the inner solar system. However, if binary accretion was responsible for building both the terrestrial planets and the cores of the giant planets, as is the case for the core-instability model, this time constraint is much more difficult to achieve. Both the larger mass of Jupiter's core than that of a terrestrial planet (15 $M_\oplus$ vs. 0.1–1 $M_\oplus$) and a lower density of planetesimals in Jupiter's zone of formation than that in the inner solar system, reflecting the lower gas density in the former region, imply that it should have taken longer to form Jupiter's core, a prerequisite for capturing a massive envelope, than to form a terrestrial planet or an asteroidal parent body. This problem is not overcome in a direct way by postulating that ice condensed in Jupiter's zone but not farther inward in the solar nebula, because this increases the fractional mass available for core formation by only a factor of several. But, as pointed out by Pollack et al. (1985), thermal convection in the solar nebula may have been much more vigorous in regions containing ice grains than in ones lacking them, so that it is not inconceivable that Jupiter reached a critical core mass before planet formation was over in the inner solar system.

The J_2 gravitational moments of the giant planets imply that much of the excess high-Z material is segregated toward their centers in cores. Recent thermodynamic calculations suggest that rock and ice are soluble in hydrogen at high pressures (D. Stevenson and W. Hubbard, personal communication). If so, it may be very difficult to create the high-Z cores of Jupiter and Saturn if their envelopes formed first, but this would not be a problem if the cores largely formed first. Note that once a core is in place, it takes a very long time to dissolve it into the envelope. This constraint, therefore, may provide a key discriminant among the two origin models. However, further work needs to be done on the solubility of rock and ice in hydrogen and in showing that a model in which rock and ice are well mixed in the high pressure hydrogen zone of Jupiter or Saturn is inconsistent with the planet's J_2 moment.

The observed enrichment of C/H above its solar abundance value in the observable atmospheres of the giant planets can probably be accommodated in both origin theories. In the case of the core-instability model, planetesimals experienced increasing ablation on their way to the core as the envelope's mass increased. In the case of the gas-instability model, it was even more difficult for high-Z material to go to the core region because the envelope formed first. Additional enrichments of C/H in the envelopes of all the giant planets may have taken place after the completion of planetary formation by physical collisions with comets and other stray bodies and in the envelopes of Uranus and Neptune by the expulsion of some methane from the core region due to its low solubility in a possible ionic water ocean (cf. Fig. 1c). Conversely, N may be depleted in the atmospheres of Uranus and Neptune due to the high solubility of NH_3 in putative ionic water oceans (Stevenson 1982*b*).

Finally, the two origin models could be further tested if the properties of the solar nebula were well constrained. In this case, one could determine, for example, whether the solar nebula was unstable against the global scale perturbation envisioned by the gas-instability model. Unfortunately, there is much too large a variance among modern models of the solar nebula for this to be a fruitful avenue of attack at present (e.g., cf. the models of Cameron [1978*a*] and Lin and Papaloizou [1980]; chapters by Lin, by Cameron, by Hayashi et al., and by Safronov and Ruzmaikina).

In summary, neither model of the origin of the giant planets is without serious problems (cf. Table III). However, I am impressed by the ability of the core-instability model to predict correctly the core masses of the giant planets. Also, the difficulty of forming central cores for Jupiter and Saturn within the context of the gas-instability model favors the core-instability hypothesis.

IV. EVOLUTION OF THE GIANT PLANETS

In this section we first outline the main phases of evolution that the giant planets underwent during and after their formation and then discuss in detail the source of their current excess luminosities (cf. Table I). The phases of evolution form a starting point for theories of the origin of the satellites and rings of the outer planets, which are presented in Secs. V and VI. A much more complete discussion of the evolution of the giant planets is given in the chapter by Bodenheimer and additional comments on the origin of satellites and rings is contained in the chapter by Lissauer and Cuzzi.

Jupiter, Saturn, and presumably Uranus and Neptune passed through three major phases from their inception to the present. These are an early quasi-hydrostatic stage (stage 1) in which their dimensions changed comparatively slowly on time scales on the order of 10^6 yr; an intermediate stage (stage 2) of rapid, possibly hydrodynamic contraction; and a final stage (stage 3) of slow quasi-hydrostatic contraction that has been occurring up until the present. These stages apply equally to the two theories of the origin of the giant planets, although, in detail, the early stages are different for the two theories.

In the case of the gas-instability theory, the envelopes of the giant planets formed relatively rapidly by a series of gas instabilities and binary accretion and had initial dimensions on the order of a tidal radius of $\sim 10^3$ times their current dimensions R_P. At this point, stage 1 commences. While core material was added to the protoplanets, their dimensions changed slowly and their interiors experienced a steady increase in temperature due to a combination of the gravitational energy gained by the change in the planet's dimensions and the gravitational energy released by core formation. In the case of Jupiter and Saturn, the former energy source was probably always the dominant one and these planets contracted on a Kelvin-Helmholtz time scale, i.e., the time scale determined by their net luminosity and gravitational energy. Stage 1 lasted for $\sim 4 \times 10^5$ yr for Jupiter and $\sim 5 \times 10^6$ yr for Saturn (Bodenheimer et al. 1980). In the case of Uranus and Neptune, accretional energy may have been the dominant energy source during at least part of stage 1, as implied by the large fraction of their current mass represented by their cores. If so, these planets may have experienced expansion during part of stage 1, with the duration of stage 1 being determined in part by the time scale required to acquire a 15 $M_{\oplus}$ core.

In the case of the core-instability hypothesis, stage 1 encompasses the interval in which their cores grew to the critical value. Consequently, the giant planets grew in size during this stage as both the accretionary and tidal radii increased in value as the planet's mass increased. The duration of stage 1 was set by the time required for the planets to accrete a critical core mass. As for the gas-instability model, the interiors became progressively hotter during stage 1.

Stage 1 ended and stage 2 began when the planet underwent a rapid contraction. This could have occurred for two reasons. Molecular hydrogen became dissociated over a large fraction of the deeper portions of the envelope so that the planet became hydrodynamically unstable and collapsed to a much smaller size on a free-fall time scale of 1 yr; or, the envelope became sufficiently massive so that its rapid, but nonhydrodynamic contraction was needed to meet the luminosity requirements of the protoplanet. In the case of the gas-instability theory, evolutionary calculations for Jupiter and Saturn for nonrotating models show that stage 1 ended when they had radii of about 200 R_P and 60 R_P, respectively, at which point their deep interiors were hot enough (2500 K) to initiate a hydrodynamical collapse (Bodenheimer et al. 1980). Figure 4 illustrates the time history of a model Jupiter during stages 1 and 2 for the gas-instability theory. No comparable calculations exist for Uranus and Neptune.

As discussed in Sec. III, evolutionary calculations performed for the core-instability theory show that stage 1 ended when the envelope mass became comparable to the core mass and therefore mass began to be added to the envelope at a large enough rate from the surrounding solar nebula to power a rapid, but nonhydrodynamical contraction (Bodenheimer and Pollack 1984). Thus, the planet's mass, mainly its envelope, continued to increase to its present value during the early part of stage 2. Growth ceased as a result of:

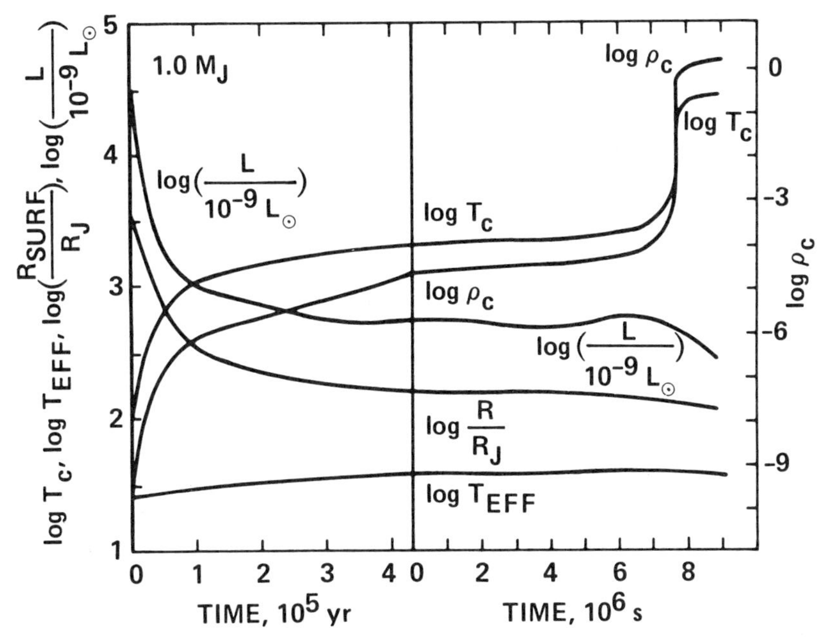

Fig. 4. Properties of models of Jupiter during phases 1 and 2 for origins involving gas instabilities. The quantities T_c, ρ_c, L, R, and T_{eff} refer to the central temperature, central density, internal luminosity, radius, and effective temperature, respectively, in units of degrees K g cc^{-1}, 10^{-9} solar luminosities, present radius, and degrees K, respectively.

(1) the planet became massive enough to tidally truncate the surrounding solar nebula (i.e., the planet's gravitational torque exceeded the viscous torque of the nebula) (Lin and Papaloizou 1979*a*); (2) the solar nebula was dissipated; or (3) the planet swept up all the available gas. The third possibility seems unlikely as the bulk composition of the planet should have been approximately solar in that case.

At the end of the rapid contraction, the deep interior became dense enough to be almost incompressible and hence rapid contraction ceased. This occurred when the planet had a size of several times R_P. Calculations performed for the gas-instability model by Bodenheimer et al. (1980) suggest that Jupiter and Saturn had dimensions of 1.3 and 3.4 R_P, respectively, at the end of their hydrodynamical collapse, although the former value is not well determined and might be somewhat larger ($\sim 3\, R_P$ according to A. Grossman, personal communication). At this point, essentially all theories for the origin of the giant planets follow a common path: stage 3. The planets continued to contract toward their present dimensions, but at progressively slower rates as their interiors became more and more incompressible (Graboske et al. 1975*b*; Pollack et al. 1977; Grossman et al. 1980). The interiors may or may not have warmed somewhat during the earliest portions of stage 3, depending on the

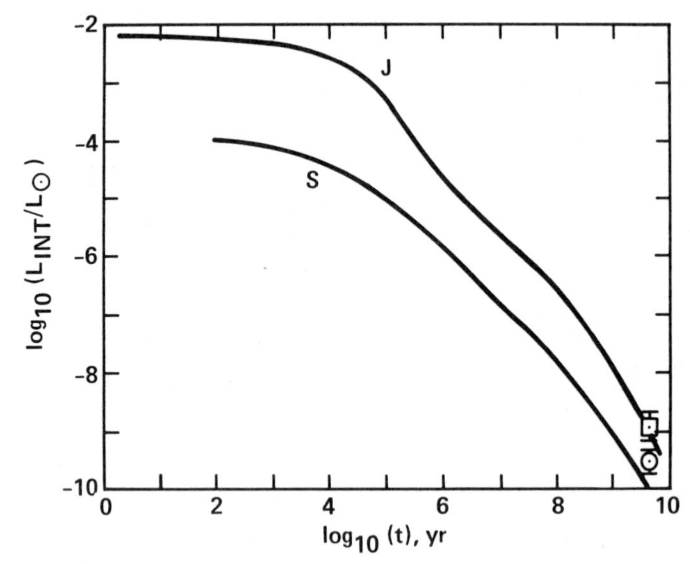

Fig. 5. Intrinsic luminosity of Jupiter (J) and Saturn (S), in units of the Sun's luminosity, as a function of time during phase 3.

planet's dimensions at the start of this stage. But during almost all or all of stage 3, the interior cooled as contraction was not rapid enough to supply gravitational energy at the rate needed to balance their net loss of energy arising from radiation to space.

Figure 5 illustrates the temporal variation of excess (or internal) luminosity for models of Jupiter and Saturn during phase 3 (Pollack et al. 1977). According to this figure, both planets had excess luminosities that were several orders of magnitude larger than the current values during the earliest portions of stage 3, at times when their satellite systems were forming. The potential influence of these high luminosities on the composition of these planets' satellites and rings will be discussed in the next section. Note that at these times, total and excess luminosities were virtually identical as they differ by about a factor of 2 at present and the excess was much larger during the early part of phase 3 than it is today.

The square and circle in Fig. 5 show the observed excess luminosities of Jupiter and Saturn, respectively, at the present epoch. While the evolutionary tracks of Fig. 5 refer to models having only solar elemental abundance envelopes, the relative agreement between theory and observation is not significantly changed when allowance is made for the presence of a high-Z core (Grossman et al. 1980) and the Voyager determinations of the excess luminosity is used (Hanel et al. 1981*a*, 1983). According to this figure, the current excess luminosity of Jupiter can be entirely accounted for by the loss of internal energy from a hot interior. The initially high temperatures of Jupiter's interior (50,000 K at the base of the envelope at the start of stage 3 vs. about 25,000 K today) were produced by the release of gravitational energy at times when the planet rapidly contracted. However, the predicted excess luminosity of Saturn at the current epoch is about a factor of 3 smaller than the observed value.

Because the cooling, contraction model of Fig. 5 gave a good fit to Jupiter's excess luminosity, its inability to fit the data for Saturn is probably not due to an inaccuracy in the calculation, but rather to the presence of an additional energy source in the case of Saturn. The only type of energy source that is quantitatively capable of supplying Saturn's excess luminosity over a nontrivial fraction of the age of the solar system is gravitational energy release. For example, the decay of long-lived radioactive nuclides in Saturn's core and the infall of interplanetary material fail by orders of magnitude to generate enough energy (Hubbard 1980). Also, temperatures in the deep portions of the envelopes of the giant planets are much too low (by at least an order of magnitude) for any type of fusion to occur (Graboske et al. 1975*b*). In terms of gravitational energy sources, the results of Fig. 5 illustrate the ability of planetwide contraction to supply the needed internal luminosity. A second source of gravitational energy release is that due to chemical differentiation, wherein a compound having a high density sinks toward the planet's center and material having a low density rises toward its surface.

In order for differentiation to provide the needed internal energy source for Saturn, a major component of the interior must be involved (Pollack et al. 1977). There are two such components: helium and high-Z material. Because the excess high-Z material has been largely situated in the central core since the planet's formation, attention has focused on the segregation of He from H in the envelope as the major source of Saturn's current excess luminosity (Pollack et al. 1977; Stevenson 1980). In particular, thermodynamic calculations suggest that at sufficiently low temperatures, helium becomes partially immiscible in hydrogen (Smoluchowski 1967; Salpeter 1973; Stevenson and Salpeter 1977). If this occurs, helium-rich fluid elements form and grow to a sufficiently large size to sink rapidly toward the planet's interior. Heat is generated by viscous drag on these droplets. In this way, gravitational energy is converted to thermal energy that supplies the excess luminosity. The interior of the planet continues to cool with time, as for a chemically homogeneous cooling situation, but at a slower rate and one determined by the excess luminosity budget. Thus, a servo system exists in which the interior temperature must decrease at a rate such that the attendant release of gravitational energy due to helium precipitation plus the rate of loss of internal energy balances the excess radiation emitted to space.

As illustrated in Fig. 6, helium is expected to become immiscible first in the outer part of the envelope's metallic hydrogen zone (Stevenson and Salpeter 1977; Stevenson 1982). According to the thermodynamic calculations of Stevenson and Salpeter (1977), helium begins to become immiscible when the

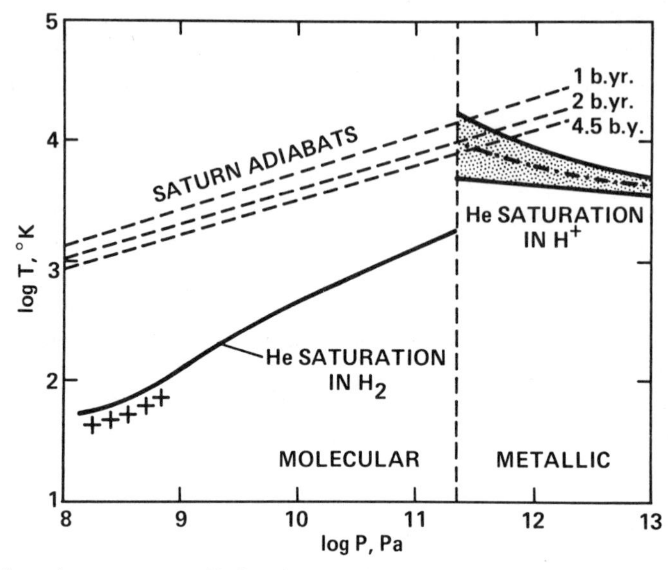

Fig. 6. Saturation temperature of helium in a cosmic mixture of elements as a function of pressure. At temperatures below the saturation temperature, helium partially separates from hydrogen. The dashed curves labeled "Saturn adiabats" show the run of temperatures through its interior at various times from the beginning of phase 3.

temperature in this region falls below about 10,000 K. Such a temperature is achieved after chemically homogeneous models of Saturn have evolved for 1–2 $\times 10^9$ yr of stage 3 (Pollack et al. 1977). By way of contrast, even today the interior temperatures in the outer part of Jupiter's metallic hydrogen zone are larger than the value at which helium segregation begins, according to Stevenson and Salpeter (1977). Thus, it is possible for helium differentiation to be occurring in Saturn's interior but not Jupiter's.

Because the metallic hydrogen zone constitutes only about 20% of the mass of Saturn's fluid envelope, there needs to be an efficient exchange of helium between the outer portion of the metallic hydrogen zone and the molecular hydrogen region of the envelope for helium differentiation in the metallic H zone to power the excess luminosity for several billion years. If the phase transition from molecular to metallic hydrogen is a zero order phase transition and there is no density discontinuity at their boundary, such an exchange should occur, given the presence of vigorous internal thermal convection. However, it might not occur if the phase transition is a first order one. It is not currently known what type of phase transition occurs between molecular and metallic hydrogen.

Good evidence supporting the above model of helium segregation in the metallic H zone of Saturn is given by analyses of Voyager infrared spectra (Hanel et al. 1981*a*). These results indicate that the He/H ratio for Saturn's atmosphere is about a factor of 2 smaller than the ratio for Jupiter's atmosphere and for the Sun (cf. Table I). However, the thermodynamic basis for helium immiscibility in Saturn's interior has been called into question by recent calculations of MacFarlane and Hubbard (1984). They find a temperature for the onset of helium immiscibility in metallic hydrogen that is about an order of magnitude smaller than those of Stevenson and Salpeter (1977). If so, helium segregation would not now be taking place in Saturn's metallic H zone. The large discrepancy in these two thermodynamic calculations reflect the great accuracy needed in determining the Gibbs free energy (to $\sim 0.1\%$) in order to obtain good immiscibility temperatures. Both approaches, of necessity, involved approximations in determining the free energy. In summary, I still consider the helium immiscibility mechanism to be the leading candidate (maybe the only one) for explaining most of Saturn's excess luminosity. But much more accurate thermodynamic calculations of H-He immiscibility need to be carried out to show that the temperatures within Saturn's metallic H zone are cool enough for partial immiscibility to occur but that they are still too warm in Jupiter's analogous zone.

Simulations of the evolutionary histories of Uranus and Neptune during stage 3 have focused on attempts to understand the occurrence of a detectable excess luminosity for Neptune and the lack of one for Uranus (cf. Table I). Let us first consider Neptune. As for Jupiter and Saturn, the decay of long-lived radionuclides in Neptune's core is insufficient by a large factor to account for its excess luminosity (Hubbard 1980). Another possible energy source for Neptune is that provided by the tidal dissipation associated with the decay of Triton's (Neptune's largest satellite) orbit (Trafton 1974). But, for this source to be quantitatively adequate, Triton's orbit would have to be evolving so rapidly that it will reach Neptune's surface in a time much less than the age of the solar system. Thus, once again, gravitational energy appears to be the ultimate source of Neptune's excess luminosity. Much gravitational energy release occurred during the planet's early history in the form of gravitational energy associated with core formation and an early rapid contraction of its envelope. As for Jupiter and Saturn, Neptune's interior has probably been cooling off since the earliest portion of stage 3. In this case, the current excess luminosity of Neptune could be attributed directly to the current rate of loss of internal thermal energy, which, in turn, was built up by very early gravitational energy release.

The first law of thermodynamics can be used to quantitatively study the relationship between the excess luminosities of Uranus and Neptune and the cooling of their interiors (Hubbard 1980). Thus, there is a basic energy balance between the rate at which heat is radiated to space in excess of the rate at which sunlight is being absorbed and the rate of loss of internal energy. The former term is proportioned to $(T_e^4 - T_o^4)$, when T_e and T_o are the actual effective photospheric temperature and its value when only solar energy absorption is considered. The latter factor is proportional to the planet's mass M, average heat capacity C_v, and average rate of temperature decrease, $d\bar{T}/dt$; $\bar{T}$ and T_e are connected by an adiabat. Thus, the rate of interior cooling, $d\bar{T}/dt$, is proportional to:

$$\frac{d\bar{T}}{dt} \propto \frac{(T_e^4 - T_o^4)}{\bar{C}_v M} \propto \frac{(\bar{T}^4 - \bar{T}_o^4)}{\bar{C}_v M} \tag{1}$$

where the bar denotes a planet-wide average.

The difference between the internal luminosities of Uranus and Neptune can be understood qualitatively in terms of the difference in the amount of sunlight absorbed by the two planets (due mainly to their differing distance from the Sun). Because $d\bar{T}/dt \propto (\bar{T}^4 - \bar{T}_o^4)$, it greatly diminishes as T_e approaches T_o. Thus, as T_e approaches T_o, interior convection becomes progressively less efficient in transporting internal energy to the surface and the excess luminosity tends toward a small value. Figure 7 illustrates quantitative simulations of the cooling histories of Uranus and Neptune, in which the initial temperatures, due to early, rapid contraction, were assumed to be significantly larger than current values. To within the considerable current uncertainties in the values of the excess luminosity and interior specific heats (C_v), these calculations are in approximate accord with the observed effective temperatures of Uranus and Neptune at the current epoch.

However, it is possible that Uranus and Neptune cooled somewhat more rapidly than is the case for time invariant compositional models (e.g., Fig. 7) due to the partial dissolution of the planets' cores and the mixing of high-Z material into their envelopes (Stevenson 1982*b*). If core material is soluble in the envelopes, as was indicated in the discussion of Sec. III, this process could well be operating, although its quantitative importance has not yet been evaluated.

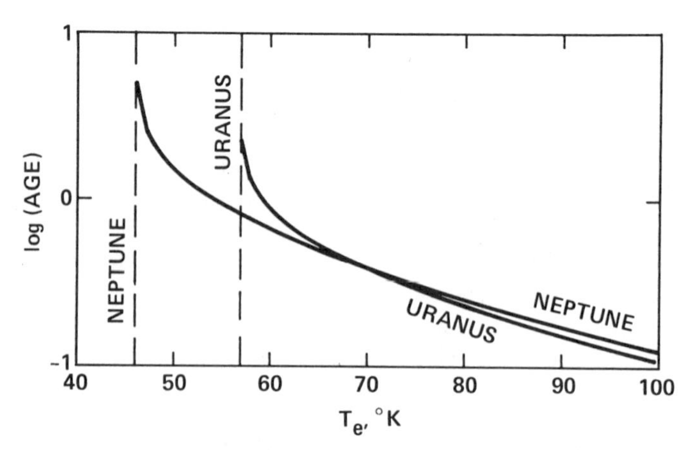

Fig. 7. Cooling curves for Uranus and Neptune. Log (age) = 0 and −1 refer to the present epoch and start of phase 3, respectively. The vertical dashed lines show the planet's effective temperature T_e due solely to the absorption of sunlight.

In summary, Jupiter's excess luminosity is due to the loss of internal heat; Saturn's is probably due mostly to the segregation of He from H in the metallic H zone; and that of Neptune and Uranus are due to the loss of internal heat, perhaps accelerated by the partial dissolution of their cores. Uranus' internal luminosity is less than Neptune's, perhaps due to the capping effect of adsorbed sunlight on convection within Uranus' interior.

V. ORIGIN OF THE REGULAR SATELLITES

As indicated in Sec. I, the orbital characteristics of many of the satellites of the outer solar system imply that these regular satellites were formed within a disk that developed soon after their parent planet formed. In this section, I discuss the nature of these disks, compare them with the solar nebula, and investigate the relationship between the composition and other properties of the regular satellites and rings of the outer planets and the environmental conditions in their nebulae of birth.

According to Table II, the regular satellites of all four outer planets are located between several R_P and several tens R_P from their primary. Thus, a reasonable first approximation for the radius of the satellite-forming accretion disk is several tens R_P. This size may be contrasted with a radius of several 10^3 R_s for the solar nebula based on the current location of the planets, where R_s is the radius of the Sun. As we will see shortly, consequences of the much smaller dimensions of the planetary accretion disks include much higher pressures and smaller angular momentum transfer from their primary than was the case for the solar nebula.

The above radii of the planetary accretion disks suggest that they began to form toward the end of stage 2 when the size of the protoplanet became less than several tens of R_p. This might have occurred in the following way. Specific angular momentum probably increased outward in a protoplanet during stage 1 due to the operation of viscous transfer processes (see, e.g., Lynden-Bell and Pringle 1974). If the protoplanet underwent a very rapid contraction or collapse during stage 2, specific angular momentum would have been conserved in each shell of matter during this time. Therefore, the outermost layers were most likely to be halted in their contraction at some distance from the main mass of the planet due to a steadily increasing centrifugal force (centrifugal force is proportional to j^2/r^3, where j is specific angular momentum and r is distance from the planet's center). In addition, the radius of the nebula disk probably increased during the earliest portion of stage 3 due to an outward transfer of angular momentum from the planet and inner parts of the disk by viscous torques.

The above concepts of the formation of the satellite accretion disks are consistent with the present-day orbital characteristics of the regular satellites and rotation rates of their planets. In particular, the specific angular momentum (angular momentum per unit mass) of the orbital motion of the larger regular satellites is 1 to 2 orders of magnitude greater than that associated with their planet's rotation. However, the angular velocity of these satellites' orbital motion is less than that of their planet's rotation. Note that viscous transfer processes drive rotating systems toward a state of solid body rotation. Thus, the outer parts of a cloud of gas have angular velocities that are at most equal to that of the inner parts and more typically less than that of the inner parts.

Because of the smaller dimensions of the accretion disks of the outer planets in comparison with that of the solar nebula relative to their respective primaries, the satellites of the outer solar system contain a much smaller fraction of their system's total angular momentum than do the planets. Thus, while the orbital motions of the planets contain, collectively, much more total angular momentum than that associated with the Sun's rotation, the reverse is true for the satellites' orbital motions and their planet's rotation.

Although the mass and size of the satellite viscous accretion disks evolved with time, it is useful to obtain a ballpark estimate of pressure conditions in these disks at a representative time during satellite formation on the basis of the current masses of the satellites. In this spirit, a minimum mass nebula can be derived by adding a cosmic complement of H_2 and He to the masses of the regular satellites (Pollack and Consolmagno 1984). In this way, minimum masses for the nebula of $10^{-2}\ M_P$ are derived for the Jupiter, Saturn, and Uranus systems, where M_P is the mass of the corresponding planet. If Triton is considered to be a regular satellite, the corresponding minimum mass for Neptune's nebula is $10^{-1}\ M_P$. It seems reasonable to take M_P as an extreme upper bound on the mass of these nebulae. Using the above bounds on the mass of these nebulae and assuming the nebular mass was uniformly distributed throughout the current dimensions of the satellite system with a vertical dimension of 10^{-1} of its radial dimension (cf. Lin 1981*a*) and a temperature of 100 K, I derive typical nebula densities of 10^{-5} to 10^{-3} g cc^{-1} and pressures of 0.1 to 10 bar. These densities and pressures are several orders of magnitude larger than their counterparts for the solar nebula.

One implication of the high pressures of the nebulae of the outer planets is that local thermodynamic equilibrium was approached more closely within them than it was in the solar nebula. This matter is of particular interest with regard to the dominant C- and N-containing gas species in these nebulae as ices made of these species may have been part of the building materials for some of the satellites and precursors of the gases contained in Titan's atmosphere. Under strict thermodynamic equilibrium, CO and N_2 are the chief C- and N-containing gases at high temperatures in a solar composition nebula, while CH_4 and NH_3 are the corresponding chief species at low temperatures ($\leqq$ 500–1000 K). The temperature marking the transition from the high- to low-temperature regime increases with nebula pressure and is higher for CO/CH_4 than for N_2/NH_3. However, if the rate of conversion of the high-temperature species to the low-temperature ones is too slow in the colder regions of the nebula and if there is mixing of material between the two zones, the high-temperature forms may be the dominant species in the low-temperature region. Lewis and Prinn (1980) have shown that there may have been strong kinetic inhibitions to the conversion of CO into CH_4 and N_2 into NH_3 in the outer portions of the solar nebula. Hence CO and N_2 may have been the chief C and N gas species throughout the solar nebula. If so, CO- and N_2-containing ices, rather than CH_4- and NH_3-containing ices, may have been part of the material that formed comets in the early history of the solar system. However, as a result of the much higher pressures in the nebulae of the outer planets, kinetic inhibitions were far less effective in preventing the conversion of the high-temperature forms of C and N gases into their low-temperature forms (Prinn and Fegley 1981). Thus, NH_3- and CH_4-containing ices were the chief N- and C-containing ices incorporated into some of the satellites of the outer planets, although minor but nontrivial amounts of N_2- and CO-containing ices may have also been incorporated into some of the satellites.

The composition of the particulate matter of the nebulae that served as the raw material for making satellites and rings depends on the ambient temperature and pressure. At high temperatures ($\sim$ 500–1500 K), only refractory rocky grains are present, namely ones made of metallic iron and unhydrated silicates. Such material may be a major ingredient of the innermost, small satellites of Jupiter. At intermediate temperatures ($\sim$ 150–500 K), low-temperature silicates, containing water, oxidized iron, and perhaps carbonaceous matter, are the dominant components of grains. This material may be the major component of the two inner large moons of Jupiter and Neptune's Triton and one of the two major components (ice being the other one) of virtually all the other regular moons of the outer planets. Finally, at very low temperatures ($<$ 150 K), grains containing ices, as well as low-temperature silicates, are present.

Figure 8 illustrates the thermodynamic equilibrium phase diagram for low-temperature rocky material and various types of ices as a function of temperature and pressure. The left-hand and right-hand dashed curves in this figure show adiabatic temperature tracks through minimum and maximum mass nebulae for the Saturn system (Pollack and Consolmagno 1984). Also shown in this figure are possible conditions at the orbits of some of the major satellites of Saturn. For all pressures of interest, ices containing H_2O are the highest-temperature ice condensates. At pressures comparable to the minimum mass nebula, this condensate is water ice. However, at pressures comparable to that of the maximum mass nebula, this condensate is a liquid solution of water and ammonia. The highest-temperature ice condensates containing NH_3 for low-pressure nebulae and CH_4 for all pressure conditions are ones that also contain H_2O (in particular, ammonia hydrate, $NH_3{\cdot}H_2O$, and methane clathrate, $CH_4{\cdot}6H_2O$). Thus, the incorporation of NH_3 and CH_4 gas into these ice species is dependent upon there being a large surface area of previously condensed H_2O that is in contact with the gases of the nebula.

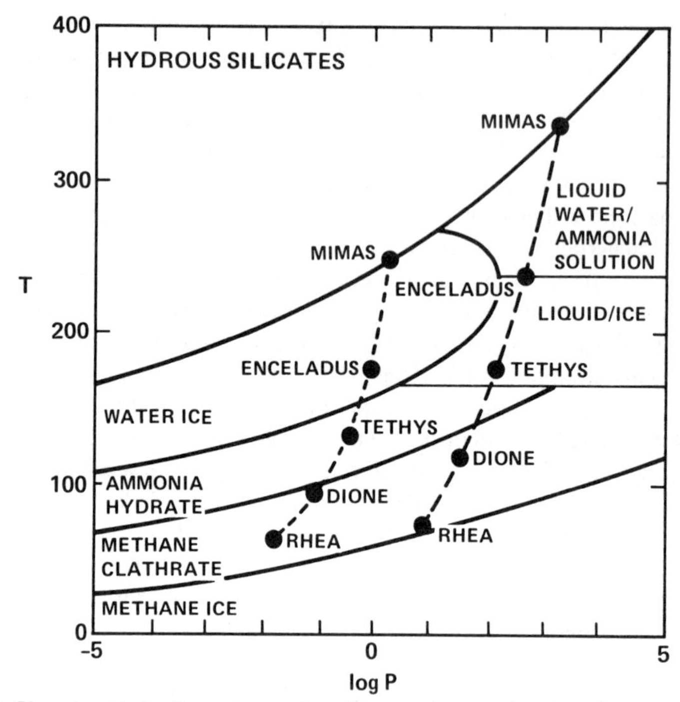

Fig. 8. Phase boundaries for various rock and ice species as a function of temperature T in degrees K and pressure P in bars. The left and right dashed lines show hypothetical adiabats for minimum and maximum mass nebulae of Saturn.

Therefore, the amount of NH_3 and especially CH_4 incorporated into ices may be limited. At temperatures comparable to those at which methane clathrate forms, clathrates containing CO, N_2, and Ar can also occur.

In order to relate phase diagrams such as Fig. 8 to the composition of the material that formed satellites around a given planet, it is necessary to determine the variation of pressure and temperature as a function of space (especially radial and vertical distance) and time within a model nebula. This can be accomplished by using three principal equations: a momentum equation in which the chief balance is between gravity and pressure gradient in the vertical direction (i.e., hydrostatic equilibrium) and between gravity and centrifugal force in the radial direction; an energy equation involving radiative and convective heat transport, with viscous dissipation and the planet's luminosity being the two major energy sources; and a mass equation, in which viscous torques cause mass in the inner part of the nebula to move inward and eventually be incorporated into the planet and mass in the outer part to move farther outward (see, e.g., Lin and Papaloizou 1980).

There are two principal uncertainties in constructing such models of the planetary accretion disks: specifying the viscosity and determining the key energy source. With regard to viscosity, molecular viscosity would have caused a negligible evolution of these disks and the solar nebula over the age of the solar system and thus turbulent processes generated almost all the viscosity. A variety of processes including thermal convection, accretion of material from the solar nebula onto the disk, and meridional circulation currents could be key sources of turbulent viscosity in these disks, as for the solar nebula. Lin and Papaloizou (1980) have suggested that thermal convection in the vertical direction is the chief source of viscous dissipation in the solar nebula, although the importance of other sources still needs to be assessed. Vigorous thermal convection in the vertical direction can be expected in the planetary accretion disks for essentially the same reason it occurs in the solar nebula; grains produce a high opacity in the vertical direction and the steep temperature dependence of their opacity causes the radiative equilibrium temperature gradient to be convectively unstable. In addition, even in the absence of grain opacity, as may occur in the late evolution of the disk due to accretionary processes, the pressure induced transitions of H_2 may generate optical depths in excess of unity in the vertical direction as a result of the high pressures in the planetary accretion disks (Lunine and Stevenson 1982). By way of contrast, once grains have accreted into planetesimals in the solar nebula (or even meter-sized particles), pressures would have been too low in this nebula for gases to supply much opacity in the vertical direction.

Even if the chief source of turbulent viscosity is known, there is still a considerable problem of evaluating the kinematic viscosity. In the case of the solar nebula and astronomical accretion disks, the kinematic viscosity is usually assumed to be proportional to the speed of sound and scale height, with the constant of proportionality being assumed to be on the order of 10^{-1} (Cameron 1978*a*). A new approximate theory linking turbulence and viscosity developed by Canuto et al. (1984) may provide a less arbitrary basis for evaluating the kinematic viscosity.

There are two principal energy sources for heating the accretion disks of the outer planets: viscous dissipation and the luminosity of the planet. In modern models of the solar nebula, viscous dissipation is assumed to be the dominant energy source (Cameron 1978*a*; Lin and Papaloizou 1980). In this case, the ultimate energy source is the release of gravitational energy due to mass inflow in the disk toward the central object and mass flow onto the disk from a surrounding cloud. Models of the solar nebula for which thermal convection in the vertical direction generates turbulent viscosity yield the following dependence of temperature T on radial distance r: $T \propto r^{-3/2}$ (Lin 1981*a*). Such a temperature gradient is super-adiabatic; whether it would be unstable to convection in the radial direction or preserved by rotation is unclear.

In almost all models of the nebulae of the outer planets (except for that of Lunine and Stevenson 1982), it has been assumed that the high luminosity of the central object was the chief heating source for its nebula (see, e.g., Pollack and Reynolds 1974; Pollack et al. 1976). In this case, an adiabatic temperature gradient $T \propto r^{-1}$ might apply throughout the inner part of the nebula due to its high opacity, with the gradient becoming subadiabatic and merging with the temperature of the solar nebula in the outer part of the nebula. Although studies need to be conducted to ascertain the conditions under which turbulent viscosity or planetary luminosity is the dominant energy source for these accretion disks, in either case, there is a steep temperature gradient in the inner part of the disk and a shallower one farther out.

We next relate the temperature structure of the planetary accretion disks to the composition of the regular satellites and rings that formed in them. Due to the monotonic decrease of nebula temperature with increasing radial distance, satellites made in the inner regions should be made of more refractory material than ones formed farther out. A classical example of such a compositional gradient is provided by the Galilean satellites of Jupiter (Pollack and Reynolds 1974). As mentioned in Sec. II, the monotonic decrease of these satellites' mean density with increasing distance from Jupiter implies a progressive increase in their bulk water ice content with distance, with the inner two moons lacking any large ice content and the outer two moons having an approximately equal mixture of rock and water ice, i.e., a factor of 2–3 less water ice than expected from solar elemental abundances. Presumably, the innermost small satellites of Jupiter and its rings are composed chiefly of high-temperature rock.

There is no obvious systematic trend of satellite density with distance in the Saturn system (Smith et al. 1982). This lack of a trend may be due, in part, to the presence of a significant ice component throughout the Saturn system (Pollack and Consolmagno 1984). Thus, the chief compositional variation expected among the regular satellites of Saturn is in the variety of ices present in them (cf. Fig. 8). The innermost ones might contain only water ice (or water/ammonia ice) and rock, while the ones farther out might also have ammonia and methane in their interiors (as well as perhaps a little CO, N_2, and Ar). Because the elements that form these lower-temperature ices (e.g., C and N) are less abundant than O (that forms H_2O) and because there is less of a density contrast among the various ices than between them and rock, no large systematic gradient in satellite density with distance is expected in the Saturn system.

The occurrence of a significant ice component throughout the Saturn system, including the rings, implies that temperature conditions were much colder in Saturn's nebula than in Jupiter's at the times of satellite formation. This difference could be attributed to a lower luminosity L for Saturn during the earliest part of stage 3 ($L \sim M_P^2$), or to a lower viscous dissipation rate in the less massive Saturn nebula. Extrapolation of this difference to the systems of Uranus and Neptune (if Triton is considered to be a regular satellite) implies that ices other than water ice are prevalent throughout the moons and rings of these systems.

Viscous dissipation promotes a steady loss of material from the disk to the planet. Thus, it may be one key factor in determining the lifetime of a nebula and, by implication, provide a bound on the time scale over which satellites and ring material were created. Once accretion of material from outside the disk/central object system has ceased, such dissipation of the nebula can be quite rapid. Models of the solar nebula indicate that half of it was lost to the Sun in only about 10^5 yr, with almost all of it being removed in 10^6–10^7 yr (Lin 1981*a*). Scaling these results to the nebula of the outer planets results in viscous dissipation time scales that are much shorter (Stevenson et al. 1984).

However, there may be reasons for believing that the nebulae of the outer planets had longer lifetimes than those characterizing viscous dissipation. A lower bound on the lifetime of Saturn's nebula may be derived from a comparison of the dimensions of the main rings with that of the planet (Pollack and Consolmagno 1984). At the beginning of stage 3, Saturn's radius was about 3.4 R_p (Bodenheimer et al. 1980), i.e., well outside the region of the main rings. It was not until about 5×10^5 yr after the start of stage 3 that Saturn contracted to within the inner edge of the B ring. Furthermore, another $\sim 5 \times 10^6$ yr needed to pass before Saturn's luminosity decreased sufficiently so that water vapor could begin condensing in this region. If the ring material formed within Saturn's nebula, as seems to be very likely, and, if this material formed at its present location, then 5×10^5 yr is certainly a lower bound on the lifetime of Saturn's nebula and perhaps 5×10^6 yr is a more appropriate lower bound.

A lower bound on the lifetime of Saturn's nebula can also be derived from the composition of Titan's atmosphere. The occurrence of methane in this atmosphere implies that temperatures were cold enough to permit the formation of methane clathrate at Titan's distance from Saturn. If Saturn's luminosity were the chief source of nebula heating at this distance, an assumption obviously open to question, then a lower bound of 10^6 yr is derived from evolutionary models of the planet's luminosity during the earliest part of stage 3 (Pollack et al. 1977). Note that He separation would not have been occurring at this time so that the planet's luminosity can be calculated with more confidence during the early part of stage 3 than during the later part of stage 3.

The above bounds on nebula lifetime appear to exceed those expected from dissipation. If these bounds are meaningful they may imply that accretion of gas from the solar nebula helped resupply gas to the nebulae of the outer planets for 5×10^5 to 5×10^6 yr or more and that after the elimination of this outside source of gas, these nebulae quickly disappeared. Whether such a resupply of gas was possible is not clear as the giant planets may have tidally truncated the solar nebula near them. Another possibility is that the turbulent viscosity coefficient, a notoriously difficult quantity to estimate, was significantly smaller than is conventionally assumed. However, it is conceivable, perhaps even likely, that the nebula lifetime was much shorter than 10^6 yr, in which case there are flaws in the arguments given above for bounding the nebula lifetime.

In summary, the regular satellites and ring material of the outer planets formed within viscous accretion disks that were similar to the solar nebula in terms of the physical processes governing the structure and evolution of these nebulae. The composition of the satellites and rings reflect temperature and pressure conditions within their nebulae of birth. The disks of the outer planets quantitatively differed from the solar nebula in that the size of the former, and hence its specific and total angular momenta were smaller than those of the latter, and pressures and densities were much higher in the former; however, the fraction of the primary's mass contained in these two types of nebulae may have been similar (10^{-2} to 1 M_P).

VI. ORIGIN OF THE IRREGULAR SATELLITES

As discussed in Sec. II, the orbital and compositional properties of the irregular satellites of the outer solar system are quite different from those of the regular satellites. Hence, they did not form in the accretion disks of their planets with their current orbital properties. Either they formed elsewhere in the solar system and were later captured or they are indeed native to their current system, but experienced major changes in their orbital characteristics. The compositional dissimilarities between the irregular satellites of Jupiter and Saturn and the outer regular satellites of these systems (the former lack ice) provides a powerful argument in favor of the former hypothesis. Recall that the irregular satellites are located farther out in these systems than are the outer regular moons. Thus, if the irregular moons formed in planetary accretion disks, they accreted in colder places than the regular moons. If the outer regular moons contain ice, so should the irregular ones in this case. However, an open mind should be maintained on this matter pending additional information on the composition of the irregular satellites. If the irregular satellites were in fact captured, where did they form and when were they captured? Below, we discuss three hypotheses for the origin and subsequent history of the irregular satellites that attempt to answer these questions.

A. Lagrange Point Capture Hypothesis

According to the first of these theories, the Lagrange point capture hypothesis, the irregular satellites formed in the solar nebula and were temporarily captured when they passed at a low relative velocity through the planet's interior Lagrange point (Heppenheimer 1975). However, the laws of motion are symmetric with respect to time. Therefore, within about a hundred orbital periods or less, the captured object exits through the Lagrange point back into orbit about the Sun unless something happens during its time of capture. Permanent capture could be achieved if the Sun lost several tens of percent of its mass during the capture period or if the planet's mass increased by a comparable amount (Heppenheimer and Porco 1977). Such a permanent capture could therefore occur only during the early history of the solar system. However, there are no reasons (e.g., astrophysical analogies to T Tauri stars) to suspect that the Sun suffered such a great loss of mass in so short a period ($<$ 100 yr). Furthermore, while a giant planet's mass might well have increased by this amount in so short a period (e.g., after reaching a critical core mass in the core-instability theory), the initial capture would then have taken place within the gaseous envelope of the planet, in which case drag forces would have been very important. In this event, this theory is closely equivalent to the gas drag capture hypothesis which will be discussed in Sec. VI.C. Also, because the satellites have a much larger cross section per unit mass than does their planet, the drag they experienced by isotropic accretion was much more effective at transporting them inwards than the inward transport due to the resulting gain in mass by the planet (Harris 1978*b*).

B. Collision Hypothesis

The collision hypothesis for the origin of the irregular satellites postulates that an outer regular satellite suffered a catastrophic collision with a large stray body (Columbo and Franklin 1971). The end result is a family of inner prograde satellites (the largest fragments of the satellite) and a family of outer retrograde satellites, the largest fragments of the stray body. This hypothesis was developed for the irregular satellites of Jupiter that, in fact, show such a clustering of orbital properties into two families. Furthermore, this theory does predict approximately the correct dispersion velocity among the family members, which is comparable to the escape velocity of the satellites. However, its applicability to the single known retrograde irregular satellite of Saturn, Phoebe, is unclear. It also does not readily explain the compositional differences between the larger prograde irregular satellites and the outer two regular moons of Jupiter.

An interesting and perhaps relevant variation of the collision hypothesis discussed above involves collisions between two stray bodies within a planet's gravitational sphere of influence (Ruskol 1982; Stevenson et al. 1984). Some or all of the collisional fragments may be permanently captured. If the larger fragments do not reaccrete with one another, a family of irregular satellites, such as occurs in Jupiter's system, could result. If they do reaccrete, a large quasi-irregular object, such as Triton, could be produced. Finally, if they do reaccrete and continue to gain mass by colliding with other stray bodies, inner satellites might be produced.

In my opinion, the above collision hypothesis is least relevant for explaining the origin of the regular satellites. It is not clear how such a theory could readily account for the systematic variation of ice/rock ratio with distance among the Galilean satellites of Jupiter. Also, a circumplanetary nebula would probably need to be postulated to explain the low orbital inclination of the regular satellites to the planet's equatorial plane, in which case the nebula should have been capable of generating its own system of satellites.

The collision hypothesis can, however, account for some of the properties of the irregular satellites. The occurrence of highly inclined, eccentric orbits, with some satellites traveling in a retrograde direction, is expected for collisions between stray bodies initially in orbit about the Sun. Also, one could readily understand the compositional differences between the irregular satellites of Jupiter and Saturn and the outer regular ones of these systems as the two classes of bodies would have originated in different places.

Nevertheless, there are a few possible problems that this origin theory may encounter. First, it is not obvious that the velocity dispersion among the largest fragments would be ~ 2 orders of magnitude smaller than the heliocentric velocities of the two stray bodies (~ 100 m s^{-1} vs. ~ 10 km s^{-1}), as occurs for the two families of Jupiter's irregular satellites. Second, if the largest collision fragments of Jupiter's irregular satellites did not reaccumulate into single objects, is it reasonable that such a reaccumulation would have occurred for Phoebe? As far as is known, Phoebe does not belong to a family of irregular satellites and so, within the context of this origin theory, one must postulate that it formed through the reaccumulation of the larger fragments.

C. Gas Drag Hypothesis

Finally, the gas drag hypothesis invokes gas drag capture of stray planetesimals of the solar nebula, which passed through the distended outer envelope of the protoplanets during phase 1 (Pollack et al. 1979). Earlier versions of this theory, not cast into the framework of modern theories of the origin of the outer planets, were given by See (1910) and Kuiper (1951*b*). Such capture occurs when the object passes through a nebula gas mass comparable to its own mass (more exactly $\sim 10\%$ of its own mass). Thus, for conditions appropriate to the outer envelopes of the protoplanets during phase 1, bodies having a size comparable to or smaller than approximately several to several hundred kilometers could be so captured. Even bigger objects, perhaps of the size of Triton, could be captured if they passed through a more interior point of the protoplanet envelope. Normally, once capture occurs, gas drag would continue to operate on the captured body and it would quickly spiral into the deep interior of the planet, perhaps adding to its core. But, if capture occurred just before the start of hydrodynamical collapse or very rapid contraction (within ~ 10 yr), gas might be removed quickly from the orbit of the satellite and hence it could remain a satellite. This scenario depends, of course, on the absence of further accretion of solar nebula gas by the protoplanet during this time period or at least on this continued accretion being small.

The gas drag hypothesis successfully, but very crudely, predicts the observed size of the irregular satellites (cf. Table II). It also predicts approximately correctly their orbital positions. They should have semimajor axes comparable to or somewhat less than the radius of the protoplanet at the start of hydrodynamical collapse or very rapid contraction. For the gas-instability model, evolutionary calculations for spherically symmetric (i.e., nonrotating) models yield planetary radii of 200 R_P and 60 R_P for Jupiter and Saturn, respectively (Bodenheimer et al. 1980). Somewhat larger values might be realized for models in which rotation is included. Even larger values (factors of several) may characterize core-instability models, with the exact value depending on the mass of the Sun at the time of very rapid contraction. Recall that the outer boundary of these models at the time when they reach a critical core mass may be set by the tidal radius. This radius varies inversely with the cube root of the Sun's mass. These predictions are consistent with the semimajor axes of the irregular satellites shown in Table II. It must be noted, however, that the presently favored core-instability models may not necessarily undergo a fast enough contraction to prevent continued spiraling inward of the captured bodies and they might continue to accrete solar nebula gas during and after the period of very rapid contraction.

If the gas drag model is correct, it supplies some useful insights into conditions in the early solar system. First, much more mass would have spiraled into the deep interior of the planet than was left behind as irregular satellites. An irregular satellite is the end result of gas drag capture occurring within a very restricted time interval and for a certain size range; very small bodies get dragged along with the collapsing envelope. The material that ended up in the planet could have helped build the high-Z core and/or enrich the envelope in high-Z material. This hypothesis also implies that a high volume density of planetesimals were present in the solar nebula at the times the outer planets formed, a conclusion which is consistent with but does not prove the core-instability theory of their origin. Finally, the composition of the irregular satellites may provide constraints on temperature conditions in the nearby solar nebula at the time they formed. If so, these temperatures were appropriate for forming low-temperature silicates (150 to 500 K). But did the irregular satellites form close to the orbital position of the planets that captured them? If so, was the solar nebula warmer at the time the irregular satellites were constructed than later prevailed in the nebulae of the outer planets so that ices could condense?

In summary, it is likely that the irregular satellites formed within the solar nebula and were captured by the giant planets during the early history of the solar system. Capture by gas drag close to the end of phase 1 and by collision between two stray bodies within the planet's sphere of influence are the most likely capture mechanisms.

VII. CONCLUSIONS

Much progress has been made in understanding the origin and evolution of the giant planets through the application of modern astrophysical concepts and the derivation of a host of observational constraints. The former include the extensive use of stellar evolution codes and the incipient employment of models of viscous accretion disks. The latter have been greatly strengthened by a variety of groundbased, airborne, and spacecraft measurements taken over the last decade, including the Pioneer and Voyager flybys of Jupiter and Saturn, and hopefully including, in the future, Voyager flybys of Uranus and Neptune.

Two rival hypotheses for the origin of the giant planets have been formulated and developed to the point where their predictions can be tested in a variety of ways, as summarized in Table III. One is certainly impressed by the ability of the core-instability model to predict successfully, *ab initio,* the core masses of the giant planets. In addition, the apparent solubility of rock and ice in high pressure H may make it very difficult to create a central core if the envelope is generated first. For these reasons, almost everyone, including the gas-instability theory's champion of the last decade, A. G. W. Cameron, now favor the core-instability theory. However, this hypothesis is not without serious problems, the most important of which are concerned with relative and absolute time scales of formation. There is still much profit to be found in carrying out hard numerical calculations of the predictions of the rival models and in the search for ways of obviating their problems.

Many fundamental problems remain to be attacked with the advent of viscous accretion disk models to study the formation of the regular satellites and ring material of the giant planets. How did these disks form? How similar and dissimilar were they to the solar nebula? What controlled the temperature structure of these disks? What was the source of their viscosity? How long did they last? How did satellites accrete in such disks? We note that the latter question involves a broad and fundamental question that concerns not only satellites but also the terrestrial planets and the cores of the giant planets. Traditionally, the terrestrial planets have been considered to have largely formed in a gas-free environment, after the dissipation of the solar nebula. This has been motivated, in part, by the problems associated with understanding the rare gas abundances in the atmospheres of Venus, Earth, and Mars if they were entirely assembled before the dissipation of the solar nebula. However, recently, efforts have been made to study the consequences of such a formation scenario. Clearly, the favored status of the core instability model implies that objects more massive than the terrestrial planets, i.e., the cores of the giant planets, were assembled before the dissipation of the solar nebula. We must further ask whether the nebula disks of the giant planets were present throughout the satellite formation period and what was the nature and rate of accretion during this period.

The gas drag hypothesis for the capture of the irregular satellites is the present front runner among hypotheses of their origin. However, for this model to be applicable within the context of the core-instability hypothesis, the rapid accretion of envelope gas had to cease or greatly slow down before the onset of the hydrodynamical collapse or the end of the very rapid contraction phase; otherwise, the captured object would have continued to experience gas drag and its orbit would have rapidly decayed. But which comes first: tidal truncation of the solar nebula by the giant planets and/or the dissipation of the solar nebula or the onset of hydrodynamical collapse/very rapid contraction? If the gas drag hypothesis is correct, much can be learned about the nature of planetesimals in the surrounding solar nebula and environmental conditions there from the compositional properties of the irregular satellites. Clearly a good start has been made in the topic of this chapter. However, not surprisingly, this field is ripe with opportunity for future studies.

OBSERVATIONAL CONSTRAINTS ON MODELS FOR GIANT PLANET FORMATION

DANIEL GAUTIER
Observatoire de Paris

and

TOBIAS OWEN
State University of New York at Stony Brook

We review current information about element abundances and isotope ratios in the atmospheres of Jupiter, Saturn, Uranus, and Neptune. The observed enhancement of C/H compared with the solar value favors models for the origin of these bodies that invoke the accretion and degassing of an ice-rock core followed by the accumulation of a solar composition envelope. Titan may represent an example of a core-forming planetesimal. Observations of D/H and other isotope ratios must be accommodated by these models in ways that are not yet completely clear. We suggest some additional tests.

A large amount of new observational evidence about the composition of the atmospheres of the outer planets has become available during the last decade. In particular, the recent observations of Jupiter and Saturn with the infrared spectrometer (IRIS) on Voyager Spacecraft permit more precise derivations of some isotopic ratios and some abundances of minor constituents than was possible from previous studies (D. Gautier et al. 1981; Kunde et al. 1982; D. Gautier et al. 1982; Courtin et al. 1983, preprint 1984; Conrath et al. 1984). When these new Voyager results are combined with groundbased observations that are available for all of the outer planets, it is possible to derive useful constraints on possible modes of planet formation.

In this chapter, we first briefly review the two principal theories for outer planet origin that have been in vogue in recent years, and then see how the new data can be used to discriminate between them. In the process, we explore the implications of this new material for other problems involving comets and outer planet satellites, and suggest some additional observations that will further constrain the problem. This chapter is an updated synthesis and elaboration of two previous papers (D. Gautier and T. Owen 1983*a,b*). The reader is also referred to the chapter in this book by Podolak and Reynolds for a complementary discussion.

I. MODELS FOR PLANETARY FORMATION

In the case of Jupiter and Saturn, one general approach has been to assume that these planets were formed by local gravitational condensation of matter in the primordial solar nebula. These condensations initially encompassed volumes several thousand times greater than the present radii of the planets, in accord with the low density envisaged for the nebula (see, e.g., Bodenheimer 1974). Subsequent hydrodynamical collapse of these giant protoplanets, followed by slow contraction in hydrostatic equilibrium, led to the present conditions. Throughout this process the composition remained solar and homogeneous. If this is really how these planets formed, the noncondensible components of their atmospheres (H_2, He, CH_4, Ne, etc.) should be in solar proportions. At warm enough levels in the lower atmospheres, even condensible species such as NH_3 and H_2O should be present in amounts consistent with solar elemental abundances.

A second general approach, known as the nucleation model, assumes that the cores of the giant planets formed first. Grains of refractory materials (silicates, iron, etc.) and ices of H_2O, NH_3, CH_4, and mixed clathrate hydrates of CH_4 and other highly volatile compounds (N_2, CO, etc.) accreted to form these cores. The exact proportions of the potentially available constituents that would actually make up such cores are difficult to predict *a priori*. However, the model does predict that, when the mass of such a core reaches a critical value, surrounding gaseous envelopes mainly made of the abundant gases still more difficult to condense (H_2, He, Ne, etc.) will collapse and become gravitationally attached to the core. In such a case, it is likely that during the formation of the core, accretional heating vaporizes the ices, which would have subsequently enriched the gaseous envelopes in CH_4 and possibly NH_3, H_2O, etc., depending on the composition of the core, its early thermal history, and on the mixing in the resulting atmosphere. Models of this type have been developed by Perri and Cameron (1974), Mizuno et al. (1978), and Mizuno (1980).

Hubbard and MacFarlane (1980*a*) and Stevenson (1982*a,b*) have discussed the advantages and weaknesses of both scenarios in the framework of theories of the internal structures of the giant planets. Recent models of

Jupiter and Saturn (Slattery 1977; Hubbard et al. 1980) and of Uranus and Neptune (Hubbard and MacFarlane 1980*a,b*) favor the existence of massive cores in all four of these planets. At least for Uranus and Neptune, the models require a large proportion of ices in the composition of the cores; this would favor the nucleation model and the assumption that the temperature was low enough to condense a substantial amount of various ices from the gas phase at the distances from the Sun at which these outermost giant planets formed.

A model for the origin of Saturn's satellite Titan, and its atmosphere, is emerging from analyses of Voyager and groundbased data. This model also favors the accretion of clathrate hydrates as a very early step (Owen 1982; Strobel 1982). Because of Titan's ability to retain the heavier gases released during its accretion, the volatile and nonvolatile inventories of this satellite may offer us a good example of the kind of material that was available to form the cores of the larger planets. On the other hand, since Titan probably formed in the proto-Saturnian nebula where the pressure, temperature, and composition must have differed significantly from conditions in the adjacent solar nebula, one must be cautious in adopting Titan as a model. The problem is that we do not know under what conditions the core-producing planetesimals accreted. Comet nuclei represent another possible analogue, but the time and place of origin of these bodies is also unknown. We shall see later that in both cases there are critical observations that can help to remove these uncertainties.

II. CONSTRAINTS IMPOSED BY ATMOSPHERIC ABUNDANCES

We can make a direct test of the condensation model by comparing the chemical composition of the atmospheres of the outer planets to the composition of the Sun. The most recent determinations of elemental abundance ratios for all of these planets are given in Table I, which also contains the solar values. Information on isotope abundances is given in Table II. The spread in these numbers requires some discussion.

A. Helium

Both modes of planet formation predict a solar ratio for He/H. Jupiter may satisfy this condition considering present uncertainties in solar helium abundances, while Saturn clearly does not. The best current values are 0.44 to 1.00 × solar for Jupiter and 0.03 to 0.45 × solar for Saturn (D. Gautier et al. 1981; D. Gautier 1983; Gough 1983; Conrath et al. 1984). The apparent depletion of helium in Saturn's outer envelope is commonly interpreted to be a result of helium condensation in the planet's interior, arising from the immiscibility of helium in hydrogen, as predicted by Stevenson and Salpeter (1977*a*). Saturn's smaller mass has allowed it to cool more rapidly than Jupiter, so that in fact an additional source of energy beyond that provided by the primordial heat is required to explain the observed excess over reradiated

TABLE I
Element Abundance Ratios in Giant Planet Atmospheres

	Jupiter/Sun	Ref.[a]	Saturn/Sun	Ref.[a]	Uranus/Sun	Ref.[a]	Neptune/Sun	Ref.[a]	Sun	Ref.[a]
He/H	1.0–0.44	(1)	0.45–0.03	(1)					$7–9 \times 10^{-2}$	(1)
C/H	2.32 ± 0.18	(2)	$4.8^{+2.6}_{-2.0}$	(3)	~20	(4)	~25	(4)	4.7×10^{-4}	(7)
					10–100	(5)	~2	(6)		
					~4	(6)				
N/H	1 ± 0.5 (1 bar)	(8)	0.25–1.0 (above 2 bar)	(3)	<1	(11)	<1	(11)	9.8×10^{-5}	(7)
	~2 (12 bar)	(9,10)	2.4 (3 bar)	(9,10)						
O/H	1/30 (see text)	(8)							6.8×10^{-4}	(12)
P/H	1 ± 0.3	(8)	2.8 ± 1.6	(3)					2.4×10^{-7}	(12)

[a]References: (1): Conrath et al. 1984; (2): D. Gautier et al. 1982 and D. Gautier and T. Owen 1983*b*; (3): Courtin et al. preprint 1984; (4): Lutz et al. 1976; (5): Wallace 1980; (6): Fink and Larson 1979; (7): Lambert 1978; (8): Kunde et al. 1982; (9): Courtin 1982; (10): Marten et al. 1980; (11): Gulkis et al. 1977; (12): Cameron 1982.

TABLE II
Deuterium/Hydrogen in Giant Planet Atmospheres

Planet	D/H	Molecule	References
Jupiter	$3.3 \pm 1.1 \times 10^{-5}$	CH_3D	Courtin et al. 1984 preprint[a]
	5.1 ± 0.7	HD	Trauger et al. 1977
	$1.2 - 3.1$	HD	Encrenaz and Combes 1982
Saturn	$1.6^{+1.3}_{-1.2}$	CH_3D	Courtin et al. 1984 preprint[a]
	5.5 ± 2.9	HD	Macy and Smith 1978
	2	CH_3D	Fink and Larson 1978
	<2.9	CH_3D	de Bergh et al. 1985
Uranus	$9^{+9}_{-4.5}$	CH_3D	de Bergh et al. (in preparation)[b]
	>5	HD	Trafton and Ramsay 1980; modified by Cochran and Smith 1983
Neptune	(Not yet measured as of June, 1984)		

[a]Assumes the C/H values given in Table I and a fractionation factor $f = 1.37$.
[b]Assumes a fractionation factor of 1.

solar energy. The difference in received versus emitted energy is made up by frictional dissipation of gravitational energy provided by droplets of helium migrating toward the center of the planet (Stevenson 1982*a*). This internal differentiation leads to a helium deficiency in the atmosphere. However, the most recent value of He/H (Table I) indicates more depletion than originally predicted. There are also uncertainties about the degree of immiscibility of helium in metallic hydrogen (W. B. Hubbard, personal communication, 1984). In any case, it appears at present that we should attribute the departure of He/H on Saturn from the solar value to evolutionary effects rather than to original conditions.

B. Carbon

Prior to Voyager infrared measurements, a large uncertainty existed in the value of the CH_4 mole fraction in Jupiter and Saturn. Most of the determinations of the CH_4 abundance from groundbased measurements were made in the near infrared where the effect of scattering of sunlight by aerosols makes the interpretation of the results strongly model-dependent. In the middle infrared, large uncertainties come from telluric absorption, from the difficulty to calibrate spectra properly, and from the uncertainty in the temperature profile of the atmosphere being studied. Recent investigations led to values of the CH_4 abundance varying from the solar value to 5 times the solar value (Wallace and Hunten 1978; Combes and Encrenaz 1979; Buriez and de Bergh 1981). The excellent precision and the high spatial resolution of infrared spectra recorded by Voyager (permitting selection of data from areas clear of clouds) lead to the conclusion that the carbon in Jupiter is enhanced over the solar value of Lambert (1978) by a factor of 2 (D. Gautier et al. 1982). The factor of 2.3 ± 0.2 given in Table I includes a correction required by the most recent determination of the intensity of the ν_4 band of CH_4 (L. Brown 1982) used in the methane abundance determination. Voyager IRIS results for Saturn indicate an even greater enrichment of carbon in the atmosphere of that planet (Courtin et al. preprint 1984), *viz.*, by a factor of $4.8\pm^{2.6}_{2.0}$. This substantial enrichment has been confirmed by a different approach to the analysis of Voyager data carried out by Bezard and Gautier (1984).

The above is a very important conclusion, because methane is the only detected minor constituent whose abundance in these atmospheres is free from the effects of condensation (as clouds or hazes) or major chemical reactions with other atmospheric gases. Thus, the methane/hydrogen ratio should correspond to a solar value of C/H, if Jupiter and Saturn formed by direct condensation with no subsequent accretion of large masses of methane-rich planetesimals. The fact that carbon is enriched is therefore a strong argument against a simple condensation model for these planets.

Voyager II has not yet visited Uranus and Neptune, and at present there is a considerable spread in the C/H determinations for these objects from groundbased observations. Nevertheless, the existing data all require an enrichment of carbon on both planets compared with the solar value (Table I). A recent revision of the broadening coefficient for the hydrogen quadrupole lines will lead to even higher values of C/H in these atmospheres (Cochran and Smith 1983). The result is consistent, in turn, with conclusions reached simply on the basis of the relatively high mean densities of these planets; Uranus and Neptune must be made of matter distinctly deficient in light elements compared with the solar composition (see, e.g., Hubbard and MacFarlane 1980*a*).

C. Nitrogen

Determination of the N/H ratio is more difficult, even on Jupiter, for two reasons. First, ammonia condenses in the atmospheres of all of these planets at temperatures lower than about 140 K so that the abundances measured in the upper tropospheres (where we obtain most of our data) do not provide the bulk composition. Second, even at deeper levels, it is possible that some of the NH_3 has been trapped to form clouds of NH_4SH or of $NH_3 \cdot H_2O$. These clouds could exist on Jupiter at pressures of 3 to 5 bar. No information at levels deeper than 5 bar can be inferred from visible and infrared spectra, although it can be obtained from radio emission in the centimeter range. Marten et al. (1980) have inferred the NH_3 vertical distribution of Jupiter and Saturn from radio measurements. For Jupiter, they derive an altitude-dependent distribution of NH_3 in the deep atmosphere and find a significant increase of the NH_3 mixing ratio at 3.5 bar, reaching about twice the solar abundance at 12 bar. This result is confirmed and its accuracy improved by the use of the thermal profile retrieved from Voyager data and the use of the Voyager H_2/He ratio (D. Gautier et al. 1981) in an adiabatic extrapolation to levels below those actually probed by Voyager instruments (Courtin 1982).

Similarly, the use of information on these parameters derived from Voyager data on Saturn permits an improvement in the previous results of M. J. Klein et al. (1978) and Marten et al. (1980) for this planet. The conclusion is that the N/H ratio on Saturn is at least twice the solar ratio at atmospheric levels deeper than about 4 bar (Courtin 1982). These results would be modified, however, if some additional opacity, for example from H_2O droplets, occurred at centimeter wavelengths (M. J. Klein et al. 1978). In view of this difficulty, we agree with Courtin et al. (1984 preprint) that it is premature to conclude that nitrogen is enriched on Saturn.

Interpretation of the 1–21 cm spectrum of Uranus had led Gulkis et al. (1977) to conclude that NH_3 is *depleted* compared to the solar value in the deep atmosphere of Uranus. The explanation of such a depletion is not clear. The most plausible interpretations are the chemical removal of NH_3 at the 150 K level by a superabundance of sulfur (forming NH_4SH clouds) as originally suggested by Gulkis et al. (1977), or a solution of NH_3 in an ionic "ocean" of water as proposed by Stevenson (1982*a*). Further studies of the latter possibility indicate the necessity of considering a binary mixture of H_2 and H_2O to determine under what conditions such an ocean would form. A preliminary analysis suggests that temperatures of 605 K and pressures of a few kilobars might be reasonable, but the ocean may in fact be a thick layer of water clouds (D. J. Stevenson, personal communication, 1984). Depletion of ammonia and enrichment of helium are both consequences of this model. Under these conditions, atmospheric ammonia is no longer useful to establish constraints on bulk composition. Similar considerations apply to Neptune.

D. Oxygen

H_2O is detectable in the 5-μm window of Jupiter (H. P. Larson et al. 1975). From Voyager IRIS spectra, Kunde et al. (1982) have derived a vertical distribution of H_2O in clear regions of the planet where it is possible to detect radiation emerging from levels well below the visible clouds. At the 4-bar level, the H_2O mixing ratio is still only 1/30 of the solar value. M. Podolak (personal communication, 1984) has suggested that a wet adiabat would allow a low enough temperature at 4 bar to keep water on the vapor pressure saturation curve. If this is indeed the case, the observed mixing ratio would be expected and would not indicate an underabundance of oxygen.

Alternatively, H. P. Larson et al. (1975) have suggested that H_2O may be depleted in these clear regions as a result of dynamical processes. If the clear areas in the Jovian cloud layer are caused by local subsidence—as seems likely from cloud morphology in those places where it can be studied (Owen and Terrile 1981)—then these are just the regions where one would expect a low H_2O mixing ratio as the dry cold air of the upper troposphere moves downward toward warmer levels. Yet another possibility is offered by models for the interior of Jupiter in which a large core is enriched in ices, which could lead to a depletion of H_2O in the gaseous envelope (Hubbard et al. 1980). The resolution of this controversy requires an accurate determination of O/H below the region where water clouds could form. Such a measurement will be carried out by the Galileo probe in 1988.

H_2O has not yet been detected on Saturn, Uranus, or Neptune. Upper limits are well above the observed mixing ratio on Jupiter, since the measurement requires detection of 5-μm thermal radiation which is simply too weak to permit spectroscopy with appropriate resolution using currently available instrumentation.

E. Phosphorus

Another piece of information comes from the measurement of the mixing ratio of phosphine. On the basis of chemical equilibrium models, PH_3 should not be present in the upper atmospheres of these planets (J. S. Lewis 1976). Yet it is observed in significant quantities on Jupiter and Saturn, probably as a result of vigorous vertical mixing in the lower troposphere. In view of the expected interaction between PH_3 and H_2O, it is not obvious that PH_3 abundances measured in the upper atmospheres can be used to infer the P/H ratio for these planets. The observational evidence tends to support this concern, since the values of P/H found for Jupiter and Saturn are very different (Table I). Nevertheless, the apparent enrichment of P/H on Saturn compared with the solar value seems hard to explain if the true value is simply solar, because the chemical reaction with H_2O would tend to reduce the amount of PH_3 actually reaching observable levels in the upper troposphere.

It is presently impossible to detect phosphine in the atmospheres of Uranus and Neptune. Voyager 2 might be able to carry out the necessary observations of Neptune during the August 1989 encounter, if the sensitivity of the IRIS instrument remains sufficiently high. In the case of Uranus, the search for PH_3 is hindered by the low temperature of the upper troposphere. A brightness temperature of 74 ± 1 K at 10.3 μm was reported for the first time in 1983; the planet has still not been detected with groundbased telescopes at 8.9 μm (Orton et al. 1983). These are the two wavelengths at which phosphine absorbs in this region. Even if one could carry out a sensitive search, it is unlikely, however, that PH_3 would be detected on Uranus because there is no strong internal energy source in this planet to drive the necessary vertical convection. Finally, if a dense water cloud is responsible for the observed depletion of ammonia on both Uranus and Neptune, it seems unlikely that PH_3 will be present above this level on either planet.

III. CONCLUSIONS FROM ELEMENTAL ABUNDANCES

Taken together, these results clearly favor the nucleation model. There is definitely an enhancement of carbon relative to solar abundances in all four giant planets. There is also a strong indication that Jupiter and Saturn are enriched in nitrogen. In cases where a depletion is observed (O/H on Jupiter, He/H on Saturn, N/H on Uranus and Neptune), there are reasonable explanations involving atmospheric dynamics or models for the planets' internal structures. As these explanations could obviously account for departures from a strictly solar composition as well, we must ask if there are other ways of enriching carbon in the outer envelopes of these objects.

Stevenson (1983) has argued that convection could not have been sufficiently strong to move core constituents upward into the observable regions of the atmospheres of Jupiter and Saturn. If so, the global enrichment in less volatile elements caused by the formation of a large core should not be noticeable. He proposed instead that the observed enrichment in methane is the result of an accretion of several Earth masses of icy planetesimals by Jupiter and Saturn *after* their formation. A difficulty with this suggestion is that it would lead to a comparable enhancement of H_2O, which is not observed (although it may exist—see discussion above). Other problems are posed by the source(s) and trajectories of this enormous mass of late accreting material.

Perhaps a solution can be found in terms of a temporary, secondary atmosphere developed by the accreting core. Such an atmosphere could be deficient in H_2O if accretion were slow enough to keep the temperature of the lower atmosphere below the value corresponding to the saturation vapor pressure of water. If this temporary atmosphere (consisting predominantly of CH_4, NH_3, CO, and N_2) then mixed with the envelopes of gas from the nebula during their collapse, rather than subsequently, the slowness of normal convective processes could be avoided. This scenario needs to be tested by calculations that model the chemistry and dynamics of the collapse.

The apparent enhancement of carbon in the atmospheres of the giant planets seems to require condensation or trapping of CH_4 and/or CO in icy planetesimals, probably in the form of clathrate hydrates (e.g., $CH_4 \cdot 7\ H_2O$), as proposed by several authors, starting with S. L. Miller (1961). However, it is not obvious where these planetesimals formed. Temperatures below 40 K are required for clathrate formation at plausible pressures in the solar nebula, and such low temperatures suggest distances well beyond Jupiter. Alternatively, one could imagine the formation of these planetesimals (or at least their continued growth) in protoplanetary disks. At the higher pressures in such locales, clathrate formation at higher temperatures is feasible, as has been pointed out for the specific case of Titan (Owen 1982; Strobel 1982). If the planetesimals did form at larger distances and then migrated inward to form the Jovian core, they must have been rather large objects to avoid loss of their most volatile constituents in times shorter than the core-forming process; just how massive will depend on the model for the solar nebula which is used to characterize the temperature and radiation field through which the accreting planetesimals must pass. One must also ponder the effect of such a massive late accumulation on the satellites. Cratering records on these satellites have not yet produced independent support for this scenario.

IV. CONSTRAINTS FROM ISOTOPE RATIOS

We have discussed the implications of the values of D/H determined in the atmospheres of Jupiter and Saturn in detail elsewhere (D. Gautier and T. Owen 1983*a*). The most recent determinations of these ratios (Table II) support our previous contention that they represent the local intersteller value at the time these two planets formed. The best determination seems to be for Jupiter in spite of remaining discrepancies between derivations of D/H from CH_3D and from HD, and in spite of some difficulties in the interpretation of CH_3D measurements at 5 μm (Courtin et al. preprint 1984). The result given in Table II is D/H = $3.3 \pm 1.1 \times 10^{-5}$, based on the combination of Voyager measurements at 8.6 μm (Kunde et al. 1982) and the high-resolution ground-based observations of Knacke et al. (1982*a*) at the same wavelength. The value of $1.6\pm^{1.3}_{1.2} \times 10 - {}^{5}$ derived for Saturn (Courtin et al. preprint 1984) is distinctly lower, but the uncertainties overlap. We conclude from these two planetary determinations that the primordial solar nebula value of D/H was near 2.5×10^{-5}.

This result supports the recent calculation by Geiss and Bochsler (1979) who derived a value of $2.0^{+1.5}_{-1.0} \times 10^{-5}$ for D/H in the primordial nebula from the ^{3}He abundance in the solar wind. (The first attempts using ^{3}He led to a value of $1.5^{+1.5}_{-0.7} \times 10^{-5}$ [D. C. Black 1972*a*] and an upper limit of $< 4.0 \times 10^{-5}$ [Geiss and Reeves 1972]). Evidently the relative amount of deuterium in the primordial solar nebula was distinctly higher than the present local interstellar value, that is now thought to be near 3×10^{-6} (Vidal-Madjar et al. 1983).

As D. Gautier and T. Owen (1983*a*) have pointed out, the observed abundances of helium and deuterium in the solar system can be used to constrain models of big-bang nucleosynthesis. In the framework of these cosmological theories, both helium and deuterium are mainly produced during the big bang. Subsequently, some helium is generated within stars and enriches the interstellar medium as a result of nova and supernova explosions and stellar winds. On the other hand, deuterium is steadily burned in stars so its interstellar abundance decreases with time. (A comparison of the solar system value of D/H, corresponding to the interstellar abundance of 4.6×10^9 yr ago, with the present interstellar value indicates that this decrease indeed occurs.) Thus, the solar system abundance of helium represents an upper limit on the primordial helium abundance, while the deuterium abundance gives a lower limit on the primordial abundance of this isotope. A comparison of these limits with the theoretical predictions of the standard model of the big bang suggests that either deuterium is more efficiently destroyed in galactic nucleosynthesis than previously estimated, or that the standard model of big-bang nucleosynthesis should be revised to produce less primordial helium (D. Gautier and T. Owen 1983*a*). In either case, we are not learning about planetary origins.

The situation could be rather different on Uranus and Neptune, as Hubbard and MacFarlane (1980*b*) and MacFarlane and Hubbard (1982) have pointed out. Both of these planets appear to have large cores with relatively thin gaseous envelopes. Hence the composition of the atmospheres of these bodies should more closely reflect the composition of their cores. In the extreme case, one could assume that the atmospheres are nothing more than the result of degassing by the ices in the core, rather like the situation assumed for the atmosphere of Titan but with the ability to retain hydrogen. Hubbard and MacFarlane (1980*b*) suggested that the ices in the core represent low-temperature equilibration of D and H in the outer solar nebula. If so, this same equilibration should leave its mark in the hydrogen found in atmospheric gases.

As an example, for $T < 150$ K, Beer and Taylor (1973) in their pioneering work on this topic showed that the fractionation factor f is > 10 in the relation

$$[CH_3D]/[CH_4] = 4\ ([D]/[H])f. \qquad (1)$$

In other words, a large enrichment in the deuterium in methane should result. This methane is trapped in the ices that form the cores of Uranus and Neptune and is subsequently released to the atmospheres, where it can exchange deuterium with the atmospheric hydrogen. Because the methane mixing ratio on these two planets is higher than that for Jupiter and Saturn, one would expect a noticeable enhancement in D/H as well. Hubbard and MacFarlane (1980) predict an enhancement of $\gtrsim 5$ times the nebular value.

However, Beer and Taylor (1973) pointed out that low-temperature equilibration of isotope exchange reactions may not occur because of insufficient time. In particular, the time to reach equilibrium in reaction (1) for $T < 400$ K is greater than the age of the universe. This means that there is insufficient time for the large enrichment postulated by Hubbard and MacFarlane to occur by the equilibrium process. Conversely, if for some reason the temperatures at which Uranus and Neptune formed were well above 400 K, equilibrium would be reached but the enrichment would be low. Thus, if a large enhancement of D/H is discovered, catalysis or other nonequilibrium processes must be invoked (Podolak 1982). Deuterium enrichment in the ices prior to formation of the solar nebula is yet another possibility. Ion-molecule reactions in the interstellar medium lead to an enhancement of D in many molecules (Solomon and Woolf 1973), and if these molecules are frozen into ices, one has another reservoir for species enriched in D (V. Vanýsek and P. Vanýsek, personal communication, 1984). There is good evidence for the enrichment of D in interstellar water, which may again be expected to lead to large values of D/H in primordial solar system ices (Ip 1984).

It is difficult at the present time to conclude whether or not this is the case on Uranus, on the basis of published measurements. Previous determinations from HD (Macy and Smith 1978; Trafton and Ramsay 1980) and preliminary results from CH_3D/CH_4 (de Bergh et al. 1981) suggested a D/H value similar to the Jovian value. However, a revision of the CH_3D analysis by de Bergh (1984) leads to substantially higher values, possibly by a factor 3 to 5. Moreover, the reanalysis of H_2 quadrupole lines in the Uranus spectrum by Cochran and Smith (1983) showing that the previous determinations of the H_2 abundance should be revised downwards, would lead to an increase of the HD/H_2 ratio. Thus, the apparent disagreement between D/H determined from HD and from CH_3D may be spurious. Evidently the situation is still uncertain, and we must wait for further study before attempting to derive firm conclusions.

Note that a large enhancement in deuterium is found on both Earth and Venus. D/H in terrestrial oceans is 1.5×10^{-4}, whereas on Venus it is approximately 100 times this value, as a result of massive escape of hydrogen (Donahue et al. 1982). The enhanced value on Earth may result from equilibration of deuterium between H_2O and H_2 in the inner solar nebula or it may have been set in the meteoroids and/or comets that brought volatiles to the inner planets (Anders and Owen 1977). Determination of D/H in a comet would be extremely helpful both for its possible relevance to the ratio in terrestrial waters and for the index it would provide on the abundances of these important isotopes in primordial ices in the solar nebula.

Evidence for another isotope anomaly in the atmosphere of Jupiter, has been reported by Courtin et al. (1983). The evaluation of $^{12}CH_4/^{13}CH_4$ from Voyager measurements indicates that $^{12}C/^{13}C = 160 \pm ^{40}_{50}$, i.e., $1.8^{+0.4}_{-0.6}$ times the terrestrial-solar value of 89. Previous attempts to determine this ratio from studies of the $3\nu_3$ band of CH_4 near 1.1 μm gave values that were compatible with the terrestrial value for both Jupiter and Saturn (Fox et al. 1972; Combes et al. 1977). Refinements of these determinations are in progress and seem again to lead to values of $^{12}C/^{13}C$ near 90 (C. de Bergh, personal communication, 1984).

If the Voyager IRIS result should, nevertheless, prove to be correct (the present conflict will certainly be resolved by the Galileo probe in 1988), a plausible explanation of the variation of the $^{12}C/^{13}C$ ratio between the inner part of the solar system and its periphery must be found. The most familiar physical processes which could have occurred in Jupiter (chemical fractionation or neutron irradiation) would have resulted in a decrease of the $^{12}C/^{13}C$ ratio compared to the one in the primitive nebula. That is at odds with the observed value if the solar nebular value is, as generally assumed, equal to the terrestrial ratio. Furthermore, neutron irradiation would have affected the carbon isotopes in the inner planets as well, whereas Venus, Mars, and Earth all exhibit the solar value to within 5 to 10% (Owen et al. 1977; Nier and McElroy 1977; J. H. Hoffman et al. 1980). (The value of $^{12}C/^{13}C = 185 \pm 69$ recently derived by Clancy and Muhleman [1983] from observations of microwave CO lines in the spectrum of Venus may indicate fractionation of ^{13}C between ^{13}CO and $^{13}CO_2$ in the planet's mesosphere. It does not apply to the total atmospheric carbon, which is mainly found as CO_2.) Other problems and implications of the Voyager result are discussed by Courtin et al. (1983).

The only other isotopic information we have is a tentative determination of $^{15}N/^{14}N = 0.006 \pm 0.001$ on Jupiter by Tokunaga et al. (1979). This is 1.6 times the solar ratio, but the authors stress that despite the low value of their quoted error, uncertainties in the model atmosphere used to derive this number allow it to be consistent with the solar ratio. Once again, there is a clear need for additional observations.

V. CONCLUSIONS

A large improvement in the determinations of elemental abundances of helium, deuterium, and carbon has been provided by the Voyager encounters with Jupiter and Saturn. In this review, we have addressed the problem of the origin of the giant planets using all of the additional abundance data currently available. While some of these determinations are still less certain than one would like, the following conclusions are consistent with what we know at this time.

1. The atmospheres of Jupiter and Saturn are enriched in carbon compared to the solar atmosphere by about a factor 2 and 5, respectively. The enrichment in Uranus and Neptune seems to be still greater. These results favor inhomogeneous models of formation of the giant planets through accretion of planetesimals from the primordial nebula.

2. In such a scenario, the outer planets should also be enriched in other heavy elements. There is an indication that such an enrichment in nitrogen exists in the deep tropospheres of both Jupiter and Saturn, but an unambiguous answer for this element and for oxygen will have to wait *in situ* mass spectrometer measurements made from probes sent into the atmospheres of these planets. The available data suggest that oxygen may, in fact, be depleted in these atmospheres, but enriched in the cores.

3. The manner in which an enrichment of heavy elements took place is not clear. One approach to this problem is to examine the composition of plausible candidates for core-forming planetesimals. Titan may offer us an example of the type of object that accreted to form the giant planet cores, if these planetesimals grew in protoplanetary disks. We cannot assess the total volatile inventory that Titan contains because we do not know to what extent it has degassed. The icy nuclei of "new" comets from the Oort cloud provide another possible model for these planetesimals.

4. A rock-ice core of solar composition with the ice consisting entirely of methane and CO clathrate would be able to produce the observed carbon enrichment on Jupiter, if $M_{core} \approx 13\ M_{\oplus}$. Following this idea, one would assume that the accreting cores developed methane (plus CO, N_2, ^{36}Ar, ^{38}Ar) atmospheres as they formed, so that these gases would easily mix with the later collapsing envelope dominated by H_2 and He. In such a case, one pre-

dicts a solar value for Ne/H, but ^{36}Ar, ^{38}Ar should be enhanced. The reason for this difference is that neon will not form a clathrate hydrate at temperatures and pressures in any current models for the solar nebula, or in protoplanetary disks. Hence any neon in the atmosphere will be captured with the hydrogen and helium during the collapsing envelope phase of the formation process. The neon should therefore be in cosmic proportions to these two most abundant elements. In contrast, methane, argon, and nitrogen all form clathrates at the higher pressures expected in protoplanetary disks. Ammonia will easily condense as a hydrate, while the clathrate-forming properties of CO are presently under study.

5. Water might have remained frozen in the core under these conditions, leading to a depletion of H_2O in the planetary tropospheres. Conversely, late accretion of the several Earth masses of icy, clathrate-containing planetesimals required to enhance the Jovian methane abundance would cause an enrichment of water in the outer envelopes even greater than that of methane.

6. Accurate determinations of D/H in the atmospheres of Uranus and Neptune would test models for the formation of these planets. An improvement in the available observations can be made from Earth. The abundance of deuterium on Titan does not help us, as this isotope is easily enriched as hydrogen escapes. On the other hand, the value of D/H in comets will furnish an important test of the low-temperature fractionation of deuterium in icy bodies in the solar nebula. This determination may be made during the 1986 apparition of Comet Halley.

7. Probe missions into the atmospheres of all of these outer planets and Titan can be expected to reveal a wealth of additional information on element abundances and isotope ratios. This is the only way to determine abundances of the isotopes of noble gases and constituents that condense in the upper layers of these atmospheres. H/He and H/Ne will provide valuable tracers for the amount of nebular gas accreted by planetary cores, and for its subsequent fractionation in planets. On Uranus, for example, a solar value of Ne/He with supersolar He/H would indicate the presence of the hydrogen-water clouds suggested by D. J. Stevenson (personal communication, 1984). The absence of He and Ne would indicate a purely secondary atmosphere, etc. Intercomparisons of isotope ratios will enable considerable refinement of present ideas about the homogeneity of the solar nebula as well as better definitions of models for the origins and internal structures of these massive planets and their satellites.

8. Noble gas isotopes offer a useful test of physical fractionation processes owing to their chemical inertness. Evaluation of these species in the atmospheres of Saturn, Uranus, and Neptune compared with the values to be measured in comets might offer significant clues toward establishing points of origin for the latter. As another example, the presence of several percent of primordial argon on Titan would indicate the importance of clathrate hydrates

for the origin of that satellite's atmosphere, and for the emplacement of gases in comets and core-forming planetesimals.

At present, we have the prospect of a probe into Jupiter's atmosphere in 1988 as part of the NASA Galileo Project. This will be preceded by ESA and USSR missions to Halley's Comet in 1986, which will provide another pathway toward knowledge about the nature of volatiles incorporated by ices in the outer solar system. We can anticipate further progress in determinations of isotope ratios from groundbased and Space Telescope observations, but we will need a family of probes into the atmospheres of Titan, Saturn, Uranus, and Neptune to provide the additional constraints on planetary formation that only such data can supply. The mission to the Saturn system being studied as a possible joint mission by ESA and NASA would provide another step toward this goal.

WHAT HAVE WE LEARNED FROM MODELING GIANT PLANET INTERIORS?

MORRIS PODOLAK
Tel Aviv University

and

RAY T. REYNOLDS
NASA Ames Research Center

Models of the giant planets are reviewed. The theoretical techniques used in computing the models are described, and the observational and experimental inputs are summarized. Special emphasis is placed on uncertainties in these input data. The models are then examined and the results of various authors presented. It is demonstrated that all the planets have heavy-element enhancements of between 10 and 40 $M_{\oplus}$, *with a large fraction of this material residing in the core. It is also shown that the ratio of ice to rock in Uranus and Neptune is on the order of three. The implications of these results for theories of the origin of the solar system are discussed.*

In dealing with so nebulous a subject as the origin of the solar system, it is essential to tie speculations as firmly as possible to observations. At present these observations must be limited largely to our own solar system, and since we are some 4.5×10^9 yr from that origin, the observations must be indirect. For this reason, it is important to find objects that have been little altered, in some important feature, since the time of their formation. In this respect, the giant planets Jupiter, Saturn, Uranus, and Neptune provide an excellent set of objects for study. In their present state, mass loss from these planets is very low. The standard form of Jeans' escape formula (Hunten 1982) shows that thermal escape of even the lightest constituents is entirely negligible, even when integrated over the age of the solar system. Thus, for Jupiter, assuming an exospheric temperature of 200 K (which is, if anything, too high), the loss rate of H^+, which because of its charge has an effective mass of 8.3×10^{-25}g, is such that only about 10^5 g would be lost over the age of the solar system. For the other giant planets the loss rate is even lower. The rate of accretion of mass under present-day conditions is also small. Let us take the maximum gravitational radius, b_{max}, for adjacent Keplerian orbits to be (Weidenshilling 1974)

$$\pi b_{max}^2 = \frac{\pi}{2}\left[R^2 + \left(R^4 + \frac{4v_e R^2}{K^2}\right)^{1/2}\right] \quad (1)$$

where R is the planetary radius, v_e is the escape velocity from the planet, and

$$K^2 = \frac{GM}{4r^3} \quad (2)$$

where G is Newton's constant, M is the mass of the Sun, and r is the planetary orbital radius. For a mean mass density in the vicinity of Jupiter's orbit of 5×10^{-24} g cm^{-3} (Dubin and McCracken 1962), the amount of mass accreted in this way in the past 4.5×10^9 yr would be only 1.5×10^{25} g, or about 0.003 Earth masses ($M_\oplus$). Humes (1976) estimates a mass accretion on Jupiter a factor of two higher, based on Pioneer 10 and 11 data.

Equation (1) gives similar values for the other giant planets. Thus, the overall elemental composition of these planets has remained unchanged since they acquired their present form. A comparison of these planetary compositions with those expected from cosmogonic theories should provide important clues as to how these planets were formed. Alternatively, if one were convinced that certain elements were accumulated by the planets in their solar ratios, the compositions of these planets would provide those ratios. Thus, Peebles (1964) studied models of Jupiter and Saturn with a view to determining the cosmic helium-to-hydrogen ratio, while Beer (1976) has suggested that Jupiter might be an excellent place to determine the cosmic boron abundance.

The obvious difficulty with using the giant planets as sources for the determination of cosmic abundances is that the composition of these planets is not uniform with radius. Even the earliest models allowed for shells of differing composition (Wildt 1938; Miles and Ramsey 1952; DeMarcus 1958; DeMarcus and Reynolds 1962), and more recent work (see below) has shown that dense cores are, almost certainly, features of all the giant planets. Theoretical models are, therefore, needed to tie the bulk composition of a planet to that observed for the outermost layers. Such models offer the additional advantage of supplying a pressure, density, and temperature profile for the planetary interior, which allows one to assess the possibility of magnetic fields being generated in the interior (Smoluchowski 1975, 1979*a;* Torbett and Smoluchowski 1979, 1980*a*). Alternatively, the measurements of the magnetic field can be used as an additional check on the validity of the models (Hide 1981; Russell 1980). Finally, the luminosity of the planet can be estimated. This too can serve as a check on the validity of the model. Indeed, evolutionary models provide, in addition, an insight into the effect of the evolving planet on its surroundings (Graboske et al. 1975*b;* Pollack and Reynolds 1974; Pollack et al. 1977). Because of the importance of interior models for cosmogonic theories, as well as the intrinsic interest of the models themselves, there have been a number of attempts at computing such models. In Sec. I we present the theoretical and observational basis for the modeling effort; in Sec. II we present the models and assess them; and in Sec. III we summarize what has been learned, and suggest where future research should be directed.

I. THE MODELING PROCEDURE

The problem of modeling the planetary interior can be divided into several parts. The first is essentially mathematical in nature, and consists of merely stating that mass is conserved, and that forces balance in hydrostatic equilibrium. If we assume, for the moment, that the planet is not rotating, and is, therefore, spherical, then if $\rho(r)$ is the density at some point r from the center of the planet, and $M(r)$ is the mass contained in a sphere of radius r, then

$$\frac{\mathrm{d}M(r)}{\mathrm{d}r} = 4\pi r^2 \rho(r). \tag{3}$$

If $P(r)$ is the pressure, then force balance requires

$$\frac{\mathrm{d}P(r)}{\mathrm{d}r} = -\frac{GM(r)\rho(r)}{r^2}. \tag{4}$$

If the planet is rotating, however, then it will not be spherical, and these equations must be amended. The classical procedure in this case is to use the level surfaces approach, which dates back to the work of Airy (1826). Here one assumes that surfaces of equal density are also surfaces of equal potential, and that they can be described by ellipsoids of revolution whose equations are

$$r(s,\theta) = s\left[1 + \sum_{n=1}^{\infty} e_{2n} P_{2n}(\cos\theta)\right] \tag{5}$$

where s is the mean radius of the ellipsoid, defined as the radius of a sphere with an equivalent volume, P_n is the nth Legendre polynomial, θ is the colatitude, and the e_n are shape parameters. Because the planet is expected to be symmetric about the equator, only even terms are expected. In addition, there is symmetry about the rotation axis, so no azimuthal terms appear. For a slowly rotating planet, the centrifugal force can be treated as a perturbation, where the small parameter of the problem is q, the ratio between the centrifugal and gravitational forces

$$q = \frac{\omega^2 R^3}{GM(R)} \tag{6}$$

where ω is the angular velocity, R the equatorial radius, and $M(R)$ the planetary mass. For the giant planets we find that e_{2n} is typically of the order of q^n. Equations 3 and 4 can then be rewritten to second order in q simply by replacing r with s, and adding the term $2\,\omega^2 s\,\rho(s)/3$ to the right-hand side of Eq. 4 (DeMarcus 1958). This centrifugal term is small relative to the gravitational term, and only for Saturn does it approach 10% near the surface. To this extent, second-order theory is quite adequate. A consequence of the ellipsoidal shape of the planet is the addition of higher-order terms to the external gravitational potential, which can then be written as

$$V(r,\theta) = -\frac{GM}{r}\left[1 - \sum_{n=1}^{\infty}\left(\frac{R_e}{r}\right)^{2n} J_{2n}P_{2n}(\cos\theta)\right] \tag{7}$$

where R_e is the planet's equatorial radius and the J_n's are the higher moments of the field, and can in principle be determined observationally. These moments can also be computed from the density distribution, and thus provide an important check on the models. Here, however, the second-order theory is inadequate, because for Jupiter and Saturn third-order terms affect J_2 by about 4% and J_4 by about 6% (Zharkov et al. 1972). Models using second-order theory (those before about 1975) will, thus, have overestimated J_2 and the absolute value of J_4, resulting in planets that are too centrally condensed. Uranus and Neptune, which have q's less than half those of Jupiter and Saturn, are not much affected by the addition of third-order terms.

A second method for dealing with rotating planets was developed by Hubbard and his colleagues (Hubbard 1974; Hubbard et al. 1975) based on the self-consistent field method (James 1964; Ostriker and Mark 1968). Here the θ-dependence of the density is retained, ρ being expanded in Legendre polynomials, and the potential is computed in a self-consistent manner either by solving Poisson's equation, or evaluating the appropriate integrals. This method has several advantages. In the first place, it allows for a direct determination of the shape of the surface; secondly, it is less cumbersome to use in carrying out the expansions to fourth order and higher. For the purpose of checking the derived density profile of a model planet with the observed gravitational field, either method is adequate, and both agree well for a given density profile.

In addition to the two mathematical relations (Eqs. 3 and 4), one also has to determine a relation between the interior temperature and the temperature near the surface. For the case of Jupiter, this is relatively straightforward. Its luminosity is considerably larger than the value expected in equilibrium with solar heating (Hanel et al. 1981*b*), and by using the computed thermal conductivity of metallic hydrogen, which comprises the bulk of the planet, one can show that conduction is too slow to provide the observed flux. Most of the planet must, therefore, be convecting (Hubbard 1968; Bishop and DeMarcus 1970). This implies that the temperature gradient throughout most of Jupiter's interior is very close to adiabatic, provided that there are no major phase transitions. A similar argument can be applied to Saturn, since its luminosity is about one third that of Jupiter's, and a greater fraction of its hydrogen is in the low-conductivity molecular state. Thus, it too must be largely convecting, and have a nearly adiabatic temperature gradient in its interior.

The case for Uranus is less clear. It has a very small internal heat source, for which at present only an upper limit has been measured; thus the above argument fails. However, Zharkov and Trubitsyn (1972) have argued that the cooling time is long enough so that the initial profile, due to an adiabatic compression of the planet during its formation, should be maintained over the lifetime of the solar system. In addition, the observation of radio bursts (L. W. Brown 1976) and Lyman α hydrogen emission similar to the Jovian polar aurora (Durrance and Moos 1982) imply the presence of a magnetic field on Uranus. This in turn implies a magnetic dynamo, and hence convection in the interior. Thus, although the argument is not as convincing as for Jupiter, there is good evidence that Uranus too is convecting throughout a large part of its interior. The argument of Zharkov and Trubitsyn applies to Neptune as well, and here we have the additional point of a measurable internal heat flux. No evidence has been found of a Neptunian magnetic field, but that may be a consequence of higher internal pressures and densities which cause the innermost portion of the core to solidify, and so reduce the amount of material participating in the magnetic dynamo, even though the temperatures throughout the planet may still be close to adiabatic (see Torbett and Smoluchowski 1980*a*).

We conclude, then, that for all four planets an adiabatic temperature gradient should adequately represent the internal temperature profile. This assumption, that has been made in all current models of the giant planets, gives us the relation between surface temperature and the interior temperature profile.

To complete the equation set for the construction of model planets, we need an equation of state for the planetary material. This is generally broken up into two parts, the problem of computing the equations of state for the individual species, and the problem of combining these to form an equation of state for the material as a whole. The most important material in Jupiter and Saturn is hydrogen. It exists in two forms: a low-pressure molecular phase, and a high-pressure metallic phase. Experimental data on the metallic phase is essentially nonexistant (Ross et al. 1981), but theoretically the problem is in good shape. There are several ways of computing the energy of the metal, and for zero temperature they agree to within $\leq 5\%$ over the range of interest (Ross and Shishkevich 1977; Zharkov et al. 1978). Even the older work of DeMarcus (1958) does not differ, in this range, by $> 7\%$ from the more modern calculations.

Of more concern is the equation of state at finite temperature. Here again there are several approaches. One is to treat the pressure P as a sum of a zero temperature part P_o and a thermal perturbation P_T which is generally computed from Debye theory (Hubbard 1969; Podolak and Cameron 1974). This method has the advantage that it enables one to see directly the relative contributions of the zero temperature and thermal parts. In addition, there is a convenient way to compute an adiabat in the Debye formalism, and combinations of species can be handled easily. The disadvantages consist of the difficulty of estimating the thermodynamic quantities necessary for the computation, such as the Debye temperature and its logarithmic derivative (Gruniesen parameter) and the fact that the method works only when P_T is a small perturbation to P_o. When the temperatures become sufficiently high, this approximation breaks down.

A second approach has been to use the dielectric function method, which computes the self-consistent perturbations to the free electrons in the plane wave states due to the proton potential. This method is very flexible, since it does not require any particular lattice structure for the protons, and is thus applicable at high temperatures. A disadvantage of the method is that the required computations are lengthy, involving Monte Carlo simulations (Hubbard and Slattery 1971). Stevenson (1975) has presented a scheme based on choosing an appropriate trial solution for the ionic distribution, that leads to considerably fewer computations, although the number is still sufficiently formidable to make it an unpleasant task to compute a series of models with different compositions. The difference between the Debye-theory method and the dielectric-function method is fairly small provided one uses the correct value of the Gruniesen parameter (Kopyshev 1965) and multiplies by the Debye function. This latter function is nearly one when the temperature is much higher than the Debye temperature, and is therefore often omitted, but its presence in the expression for P_T brings the Debye theory into much closer agreement with Monte Carlo results. For densities and temperatures like those found in Jupiter, the two methods agree to within 10%. For Saturn, where the thermal perturbation is smaller, the agreement will be better, while for Uranus and Neptune the metallic hydrogen region is small, if it exists at all, and so will play only a small role in the model.

At lower pressures hydrogen exists as a molecule, and because it is not spherical, its equation of state is much more difficult to compute. On the other hand, there are experimental points available with which to support the theories. A review of the progress made in computing the equation of state of molecular hydrogen can found in Ross and Shishkevitch (1977; see also, Ross et al. 1981). Here again, the theoretically more rigorous method is based on Monte Carlo or molecular dynamics computations using intermolecular potentials fitted to experimental data (Neese et al. 1971; Slattery and Hubbard 1973; Graboske et al. 1975*a*). These computations are lengthy, and here again it is convenient to use an appropriate zero-temperature equation of state with the thermal perturbation computed from Debye theory (Trubitsyn 1965, 1971; Zharkov et al. 1978; Podolak and Cameron 1974). Since the molecular fluid must gradually become an ideal gas as the atmosphere is approached, the Debye solid approximation must break down, and some sort of interpolation must be applied to bridge the gap between the dense molecular fluid and the ideal gas (see Podolak and Cameron 1974). Here, as in the metallic hydrogen case, the method lacks rigor, but it enables one to treat various mixtures easily, thus allowing great flexibility in the computation of models. In addition, the procedure seems to give rather accurate results. Thus, an equation of state based on a fit to Monte Carlo calculations (Hubbard et al. 1980) fixes a point on Jupiter's adiabat (P_o = 1 bar, T_o = 165 K) at P = 36.1 Mbar, ρ = 3.67 g cm^{-3}, and T = 19,300 K. Using the Debye theory approximation with the same initial pressure-temperature point, the procedure described in Podolak and Cameron (1974) finds ρ = 3.53 g cm^{-3} and T = 18,500 K at P = 36.1 Mbar. Thus, adiabats computed by these two different methods agree rather closely, and either procedure can be relied upon to give good values for the pressure, density, and temperature at some interior point.

There is one additional uncertainty in the hydrogen equation of state, and that is the pressure at which the molecular fluid transforms into the metallic state. Again, there is no reliable experimental data. Several groups have reported seeing a phase transition at high pressures, but all of the experiments are subject to large uncertainties (see Ross and Shishkevich 1977). The transition pressure is generally believed to be in the 2 to 3 Mbar range. Because the density difference at these pressures should be about 10%, and the pressure uncertainty falls in a rather narrow range, the effect on the models is small.

A second major component of planetary interiors is helium. As in the case of hydrogen, there have been two approaches. The Debye solid approach computes the zero temperature pressure, and adds the thermal component as a perturbation. For mixtures of hydrogen and helium the density of the mixture is computed from the additive volume law

$$\frac{1}{\rho} = \frac{X}{\rho_1} + \frac{Y}{\rho_2} \tag{8}$$

where X and Y are the mass fractions of hydrogen and helium in the mixture, and ρ_1 and ρ_2 are their respective densities. Equation (8) can be obviously generalized to accomodate any number of components in the mixture. Again, the advantage of the Debye method is in its flexibility, but more detailed methods have shown interesting effects not evident in the Debye approximation, such as departures from the additive volume law (Stevenson 1975; Stevenson and Salpeter 1977*a*), and immiscibility of helium in hydrogen (Stevenson and Salpeter 1977*b*). For hydrogen-helium mixtures under conditions of interest for model planets, the departure from volume additivity is small (Hubbard 1972).

A class of substances which are of interest from a cosmogonic point of view are the ices (H_2O, NH_3, CH_4 etc.), because these are solid under some conditions and gaseous under others. These substances are more difficult to treat theoretically, although progress is being made as a result of a combination of high-pressure experiments and more sophisticated theoretical approaches (Liu 1982; A. S. Mitchell and W. J. Nellis 1982; Ree 1982). In Fig. 1 we have plotted the experimental points of Liu determined from static compression of ice VII (the stable polymorph of ice at room temperature and pressures $\gtrsim$ 20 Kbar) at 20°C and the reflected shock points of Mitchell and Nellis. Since the shock points correspond to temperatures of about 10^4 K, the zero-temperature values will have a pressure smaller by some 10–12% for a given density. Also shown is the equation of state given by Zharkov et al. (1978). As can be seen, the agreement is rather good, although the highest pressure point indicates that the potential may get somewhat softer at Mbar pressures. Other ices are currently being studied by similar techniques, and it is hoped that in the near future good equations of state will be available. At present, uncertainties in the behavior of ices at high pressures is a major source of uncertainty in models of Uranus and Neptune.

The last class of substances that we wish to consider is the rocky-type materials (SiO_2, MgO, etc.) which will remain solid under almost all conditions in the primitive solar nebula. The equation of state for rock is more difficult to compare among the various authors, because one must compute the equation of state for a mixture of rock-forming substances, and there are different choices among modelers as to what that mixture should be. The usual method of computing the equation of state is to combine experimental data, which now extends to several Mbar, with a theoretical high-pressure equation of state such as that of Thomas-Fermi-Dirac (Salpeter and Zapolsky 1967) or the quantum statistical model (Kirzhnits 1967; Zharkov et al. 1978). Figure 2 shows a comparison between these two methods at zero temperature for a mixture of 39% SiO_2, 27% MgO, and 34% Fe-Ni. As can be seen, the Salpeter-Zapolsky equation of state is harder, although agreement improves at the highest pressures. Different choices of composition will typically lead to differences of this magnitude among the various rock equations of state.

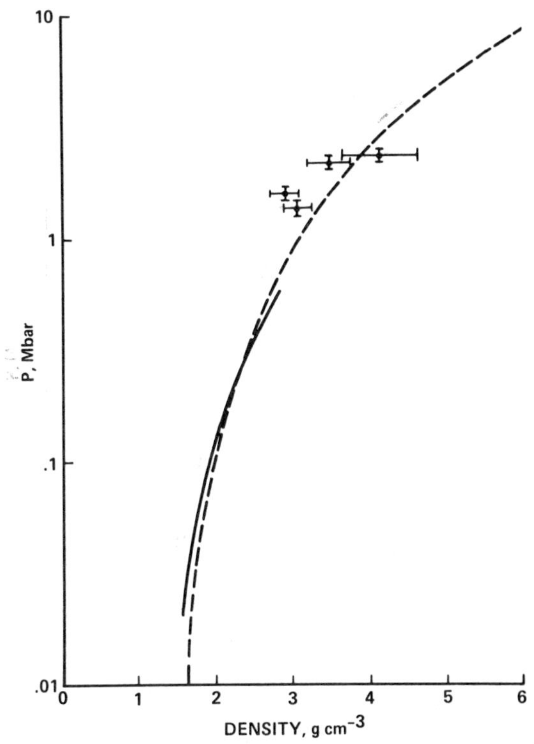

Fig. 1. H_2O equation of state. Solid curve from static compression experiments (Liu 1982), dashed curve from theory (Zharkov et al. 1978), points with error bars from shock experiments (A. S. Mitchell and W. J. Nellis 1982).

We see from the preceding discussion that the various approaches to computing an equation of state and adiabat lead to differences that are generally under 10% in the pressure, density, and temperature within a planet. This in itself is not very disconcerting, but there is a subtle effect that is of importance for making deductions about planetary composition. For the Jovian envelope, the mass fractions of hydrogen, helium, and ices in solar composition are roughly 0.77, 0.22, and 0.01, respectively. At a pressure of 0.5 Mbar, for example, the densities of the individual substances are about 0.54 g cm^{-3} for hydrogen, 1.5 g cm^{-3} for helium, and about 2 g cm^{-3} for an ice mixture, giving a total density, by Eq. (8) of 0.63 g cm^{-3}. Note that by far the most important term both because of its large mass fraction and because of its low density is hydrogen. Thus, a 5.6% increase in the density of hydrogen (which is within the range of uncertainty) will cause a 5% increase in the density of the mixture. A similar increase can be achieved by raising the mass fraction of ices by a factor of 8. Thus, a 5% error in the hydrogen equation of state can

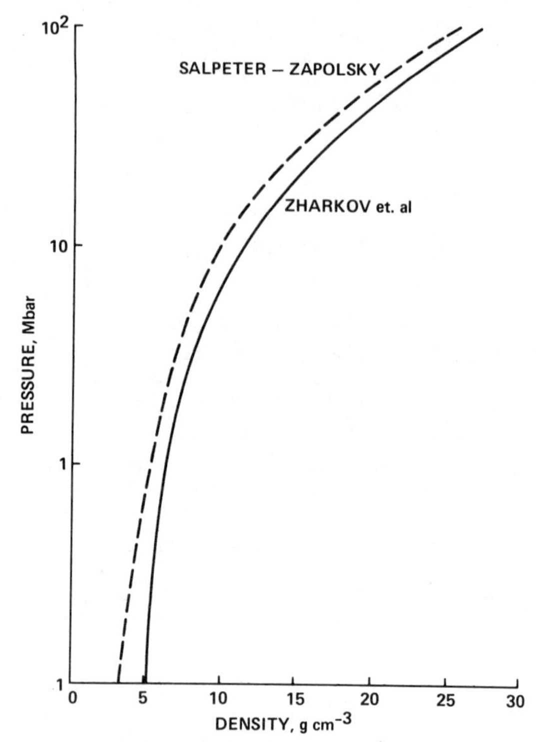

Fig. 2. Comparison of two equations of state for rock (39% SiO_2, 27% MgO, 34% Fe-Ni). Solid curve is the quantum statistical model (Zharkov et al. 1978), dashed curve is from Thomas-Fermi-Dirac theory (Salpeter and Zapolsky 1967).

cause a much larger error in the estimate of the ice abundance. This point must be borne in mind when assessing the abundances determined from model planet computations.

One final issue that must be mentioned in this context, is the question of the miscibility of various mixtures at high pressure and temperature. Thus, several authors have suggested that the imiscibility of helium in hydrogen might be responsible for some of the excess heat flow from Jupiter and Saturn (Salpeter 1973; Smoluchowski 1973; Stevenson and Salpeter 1977*b;* Pollack et al. 1977). Recent work (MacFarlane and Hubbard 1983) has, however, questioned whether this imiscibility, in fact, exists under the conditions found in Jupiter and Saturn. The related question of the miscibility of ices in hydrogen has relevance to the structure of Uranus and Neptune; this issue is discussed in Secs. II.C and D.

In addition to the theoretical and observational input required by the models, there is still enough freedom in choosing the structure that some sort of additional constraint must be placed on the problem. We can call this the philosophical input. For most current models the philosophy is based on some sort of scenario for planet formation. Thus, one would expect that hydrogen and helium were acquired in the same way, and would, therefore, be in solar ratio to each other. Ices which are sometimes gaseous and sometimes solid, would have been acquired possibly in a different way, while rock, which would have always been solid could have been acquired in still a third way. It is, thus, the arrangement and relative abundance of the members of these three classes of substances that reduces the number of free parameters of the model to something reasonable.

II. THE MODELS

A. Jupiter

The mass of Jupiter is readily determined from the orbits of its satellites, and, more recently, from the trajectories of spacecraft. The best current value is 317.735 $M_\oplus$ (Null 1976), or 1.897×10^{30} g. The equatorial radius, determined from Voyager measurements is 71,541 $\pm$ 4 km at the 100 mbar level (Lindal et al. 1981). This becomes 71,492 $\pm$ 4 km when reduced to a standard pressure of one bar (Hubbard and Horedt 1983). The first few gravitational moments have been determined from the trajectories of Pioneers 10 and 11 to be $J_2 = 0.014733 \pm 0.000004$, $J_4 = -0.000587 \pm 0.000007$, and $J_6 = (34 \pm 50) \times 10^{-6}$, all referred to a normalizing radius of 71,398 km (Null 1976). The absence of odd terms and second tesseral harmonics indicates that most of the planet is indeed in a state of hydrostatic equilibrium. There are several rotation periods associated with various features in the Jovian atmosphere, but for the purpose of modeling, the most relevant is the period associated with the rotation of the magnetic field, which is most directly related to the rotation of the bulk of Jupiter's interior. This period has been measured to be $9^h\ 55^m\ 29\overset{s}{.}7$ (Donivan and Carr 1969). Using this period, Anderson (1978) computed a dynamical oblateness of 0.0651 for Jupiter that is in excellent agreement with the optical oblateness of 0.0650 determined from the β Sco occultation (Hubbard 1977). Once again, we see strong evidence for a state of hydrostatic equilibrium. The temperature at the one-bar level must be determined from model fits to thermal flux measurements. The consensus of the most recent efforts is that the one-bar temperature is 165 $\pm$ 5 K (Lindal et al. 1981). The most recent estimate of Jupiter's internal heat flux is $(5.444 \pm 0.425) \times 10^{-4}$ W cm^{-2}, giving a ratio of total energy emitted to absorbed solar energy of 1.668 $\pm$ 0.085 (Hanel et al. 1981*b*). The helium abundance has been measured in two independent ways. An inversion of the infrared Voyager measurements implies a mass fraction of helium Y of 0.19 $\pm$ 0.05, while the radio occultation experiment gives $Y = 0.21 \pm 0.06$ (D. Gautier et al. 1981).

The first question one can ask of the models is whether Jupiter has a heavy-element core. Although the current trend is to answer in the affirmative (Podolak and Cameron 1974, 1975; Zharkov and Trubitsyn 1976; Stevenson

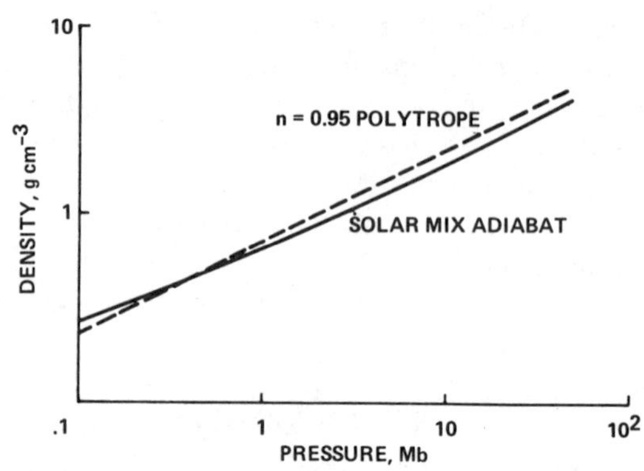

Fig. 3. Density vs. pressure relation for a $x = 0.95$ polytrope which matches Jupiter's $n = 0.95$ J_2 and J_4 (dashed curve), compared with a solar-mix adiabat (solid curve).

and Salpeter 1976; Podolak 1977; Slattery 1977; Hubbard et al. 1980), this answer is not necessarily the final one. Hubbard and Slattery (1976) have shown that a polytrope of index 0.95 can fit the observed values of J_2 and J_4 to within the observed uncertainties. This model does not require a core of heavy elements, although it does imply a milder inhomogeneity, that of an increase of helium, or some other heavy material, with decreasing radius. This can be seen from Fig. 3, where we have plotted the density as a function of pressure for a solar-mix adiabat (solid curve) and for an $n = 0.95$ polytrope (dashed curve). The difference in density at the higher pressures can be taken to indicate the presence of heavier elements. If one insists on a solar composition envelope, this extra mass must be collected and put in the form of a core. A second possibility for a coreless model is the one computed by Slattery (1977), who finds that if he takes a solar composition adiabat, and enhances the metallic hydrogen region with heavy elements, a model of Jupiter is produced that fits the observations when the enhancement is about 15 $M_\oplus$. A rather similar model is presented by Zharkov and Trubitsyn (1976). They found, however, that the envelope must be enriched by some 30 to 40 $M_\oplus$. It is difficult, from the data presented in Zharkov and Trubitsyn, to evaluate the source of the discrepancy, but it is likely that the difference is due to small differences in the equation of state of hydrogen. As illustrated above, this has a large effect on the estimated heavy-element abundance. Thus, in their earlier work, using a somewhat different equation of state for hydrogen, Hubbard and Slattery (1976) found that about 30 $M_\oplus$ of heavy elements were needed to bring the model into agreement with the observations.

It does not seem that the abundance of heavy elements can go down much farther, however. In a fairly thorough study by Stevenson and Salpeter (1976) of the effects of various uncertainties in the equation of state of hydrogen-helium mixtures, they showed that if a solar envelope is assumed, a core of some 9 to 11 $M_\oplus$ is required. Other models featuring a solar-mix envelope are those of Zharkov and Trubitsyn (1976) which have cores of 10 to 13 $M_\oplus$, and those of Slattery (1977) with cores of 14 and 16 $M_\oplus$. These last models assumed a core density of 22 and 13 g cm^{-3}, respectively. The first of these densities is appropriate for a core of rock, the second for a core of ice. Note that the size of the core is quite insensitive to the choice of material. The models of A. S. Grossman et al. (1980) were chosen primarily on the basis of their calculated luminosity after 4.5×10^9 yr of cooling. They find that a core of the order of 20 $M_\oplus$ gives a good fit as well to the gravitational moments. Finally, there are the models of Podolak (1977) which have both an enhancement in the envelope and a core. For the models which most closely satisfy the observational constraints, the core is about 18 $M_\oplus$, with an additional 30 $M_\oplus$ of water in the envelope. If the material in the envelope is rock, then the core mass rises to 28 $M_\oplus$, with an additional 14 $M_\oplus$ of rock in the envelope; these two possibilities give a total of heavy-element content of 42 to 48 $M_\oplus$. It should be noted that in these models J_4 is about 12% too high, which indicates that the heavy-element content should be reduced somewhat in the envelope. Although this model is quite different from the others listed, it is in rather good agreement with the findings of Hubbard and Horedt (1983). By inverting the gravity data, they found that a good fit could be obtained for a model with 31 $M_\oplus$ of heavy elements in the envelope, and 11 $M_\oplus$ in the core. As a measure of the uncertainties inherent in such a calculation, they note that 15 $M_\oplus$ of heavy elements in the envelope and 15 $M_\oplus$ in the core would give an equally good fit.

In summary, there is little doubt that Jupiter is enhanced with respect to solar composition. There must be an enhancement of heavy elements either in the form of a core or dispersed throughout the planet with the concentration increasing towards the center. A core together with an enhanced envelope is also possible. The total enhancement is between 10 and 40 $M_\oplus$, and it is currently beyond the power of the models to distinguish between ice and rock, or, for that matter, between helium and rock. It is almost certainly possible to construct a model of Jupiter which would fit all of the gravitational data, and be composed of hydrogen and helium alone, provided that one allowed the helium fraction to increase towards the center. The total helium content would be considerably higher than the solar value, and it is difficult to see how such a fractionation could be maintained in view of the miscibility of helium in hydrogen at these temperatures, and the convective mixing expected in Jupiter's interior; however, such a model cannot as yet be ruled out. A further refinement of the measured helium abundance for Jupiter, as well as measurements of the water content in the deeper layers of the atmosphere are vital to help us distinguish between the various theoretical models. In addition, more work is needed in refining the equation of state of hydrogen, since, as we have seen, it is a major cause of uncertainty in abundance determinations. Current models are summarized in Table I.

TABLE I
Jupiter Models

Authors	Core Size	Heavy-Element Enhancement of the Envelope	Comments
Zharkov and Trubitsyn (1976)	no core 10–13 $M_\oplus$	30–40 $M_\oplus$ none	—
Hubbard and Slattery (1976)	no core	Increased enhancement towards the center	Polytropic model with $n = 0.95$ fit to J_2 and J_4.
Stevenson and Salpeter (1976)	9–11 $M_\oplus$	none	—
Podolak (1977)	~18 $M_\oplus$ ~28 $M_\oplus$	~30 $M_\oplus$ ~14 $M_\oplus$	Ices enhanced in the envelope. Rock enhanced in envelope (J_4 ~12% high for both cases).
Slattery (1977)	no core 14–16 $M_\oplus$	15 $M_\oplus$ none	—
Hubbard et al. (1980)	30 $M_\oplus$	none	Treats the least certain part of the hydrogen equation of state as a free parameter.
A. S. Grossman et al. (1980)	~20 $M_\oplus$	none	Essentially an evolutionary model with the core size fixed to give the proper moments.
Hubbard and Horedt (1983)	11 $M_\oplus$ 15 $M_\oplus$	31 $M_\oplus$ 15 $M_\oplus$	Inversion of gravity data using parameterized equation of state.

B. Saturn

The mass of Saturn is well determined from the orbits of its satellites and from the tracking of spacecraft. The best current value is 95.142 $M_\oplus$, or 5.68×10^{29} g (Null et al. 1981). The equatorial radius is 6.00×10^9 cm, and the first two multipole moments of the gravitational field are $J_2 = 0.016479 \pm 0.000018$, $J_4 = -0.000937 \pm 0.000038$ for an assumed value of $J_6 = 0.000084$ (Null et al. 1981). Detection of radio bursts from Saturn has allowed the determination of the rotation period of the magnetic field. The value was found to be $10^h\,39^m\!.9 \pm 0^m\!.3$ (M. L. Kaiser et al. 1980, 1984). As with Jupiter this period is supposed to refer to the rotation of the bulk of the planet's interior. The temperature at the one-bar level is again model dependent, but both the Voyager infrared and radio measurements give a value of about 135 ± 5 K at this pressure level (Hanel et al. 1981*b*; Tyler et al. 1981). There is some uncertainty regarding the value of the helium mass fraction in Saturn. Preliminary results indicate that it is significantly lower than the Jovian value, being about 0.13 to 0.14 (D. Gautier and T. Owen 1983*a*); this may have some implications towards the miscibility of helium at higher pressures.

Models of Saturn have been traditionally more difficult to construct than models of Jupiter. At least as far back as in the models of DeMarcus (1958), Saturn displayed its idiosyncrasies. DeMarcus found that he could not get the models to fit J_4 well, and that he required a rather large fraction of the outer envelope to be helium-free (i.e., of low density). Nevertheless, he required a rapid increase in the helium abundance in the inner 35% of the planetary radius, giving a helium core of some 20 $M_\oplus$. The total helium content, including the helium mixed with hydrogen in the layers just above the core, gave $Y = 0.32$, much higher than the value found for Jupiter. These models were at near-zero temperature, and used a second-order theory in computing the gravitational moments. In addition, the equation of state for molecular hydrogen gave densities some 10% too low at the higher pressures; nonetheless, the qualitative differences between Jupiter and Saturn models have persisted in even the most recent investigations. Peebles (1964) was also unable to find a model with a sufficiently low J_4, and he too required a 20 $M_\oplus$ core and $Y = 0.48$. Hubbard (1969) was able to construct models with no core, but his J_4 was also some 30% too high, and he required $Y = 0.73$. Podolak and Cameron (1974) computed a core of 20 $M_\oplus$ with an additional 10 $M_\oplus$ of ices in the envelope. They too were unable to get a J_4 sufficiently low (in absolute value) to match the observations. All of the above models used second-order theory to compute the moments, and all were ultimately unsuccessful at fitting the observations, but all had several similarities. In every case, Saturn was relatively more enhanced in heavy elements than Jupiter, and in each case, with the exception of the models of Hubbard (1969), most of this enhancement was concentrated in a core of some 20 $M_\oplus$. Even Hubbard's best Saturn model is a poor fit to the observations and cannot be considered as a significant exception to the trend described above. The fact that both J_2 and J_4 in his models are too high, indicates insufficient central condensation.

More recent models have succeeded in matching J_4, but the differences from Jupiter still remain. Slattery (1977) found that a solar envelope ($Y = 0.22$) and a core of 15 $M_\oplus$ of rock ($\rho = 15$ g cm^{-3}) or a core of 17 $M_\oplus$ of ice ($\rho = 10$ g cm^{-3}) fit the observations provided that the adiabat started at a temperature of 160 K at one bar. This requirement of high temperature implies that the density of a solar mixture of material is probably too high, and this must be offset by increased thermal expansion. This may also be related to the fact that Slattery used $Y = 0.22$ (as did all of the other models to be discussed) instead of the observed value. Podolak (1978) found that an envelope with 4 $M_\oplus$ of ice surrounding a core of 22 $M_\oplus$ gave a good fit to the observations.

Here again, the envelope was much less enhanced in heavy elements than the corresponding Jupiter model. It should be noted that although in that paper it is stated that the J_4's of the models were consistently too high, the code used in computing those moments had difficulty in dealing with discontinuous density distributions, and this led to an overestimate in the (absolute) value of J_4. In fact, the J_4 computed for the models with an improved code is in very good agreement with the observations. This model is similar to one by Zharkov and Trubitsyn (1976) which has a 23 $M_\oplus$ core with an additional 5 to 10 $M_\oplus$ of ices in the envelope. A second model of theirs has a 25 $M_\oplus$ core with a solar envelope. The models of Hubbard et al. (1980) have a solar envelope ($Y = 0.19$) with an 18.6 $M_\oplus$ core of rock and ice in solar proportions, and A. S. Grossman et al. (1980) in trying to obtain a better fit to Saturn's luminosity with their evolutionary code, also found a core of 19 to 20 $M_\oplus$. Finally, Hubbard and Horedt (1983), using their inversion technique, found that the constraints on J_4 are still too weak to decide between models with Y between 0.11 and 0.26.

Thus, models of Saturn seem to require a stronger concentration of heavy elements towards the center than do those of Jupiter. All recent models show this core, and give its mass as between 15 and 25 $M_\oplus$ regardless of the equation of state chosen for the core material itself (rock or ice), and regardless of composition (within reason) of the envelope. In fact, however, most of these models have envelopes of nearly solar composition which would imply a total abundance of $\sim$1 $M_\oplus$. The largest calculated enhancements, in the models of Zharkov and Trubitsyn (1976), are some 5 to 10 $M_\oplus$ in the envelope of 72 $M_\oplus$. This, of course, assumes that $Y = 0.2$, or thereabouts, as in solar com-

TABLE II
Saturn Models

Authors	Core Size	Heavy-Element Enhancement of the Envelope	Comments
Zharkov and Trubitsyn (1976)	23 $M_\oplus$ 25 $M_\oplus$	5–10 $M_\oplus$ none	—
Slattery (1977)	15–17 $M_\oplus$	none	Temperature at one bar taken to be 160 K.
Podolak (1978)	22 $M_\oplus$	4 $M_\oplus$	—
Hubbard et al. (1980)	18.6 $M_\oplus$	none	Treats the least certain part of the hydrogen equation of state as a free parameter.
A. S. Grossman et al. (1980)	19–20 $M_\oplus$	none	Essentially an evolutionary model with the core size fixed to give the proper moments.

position. If $Y = 0.14$ as the Voyager measurements seem to indicate, then the heavy-element enhancement would be somewhat larger. Current Saturn models are summarized in Table II. These differences between Jupiter and Saturn are intriguing, and will be speculated upon in the concluding section.

C. Uranus

The mass of Uranus, determined from the orbits of its rings is 14.5 $M_\oplus$, or 8.668×10^{28} g (Nicholson et al. 1982; Elliot et al. 1981). The equatorial radius, derived from occultation measurements is 26,228 $\pm$ 30 km (Elliot et al. 1980). This refers to the upper atmosphere, and the radius at the one-bar level is about 25,800 km. This is considerably larger than the value of 23,700 km used in model planet calculations before 1970. Thus, these early models fit to a density 30% too high, and this makes them of little value for the determination of abundances. The gravitational moments have been measured from the precession of the rings to be $J_2 = 0.003349 \pm 0.000005$, and $J_4 = -0.000038 \pm 0.000009$ (Nicholson, personal communication; Eliott and Nicholson 1984). The value of J_4 has been determined to useful accuracy only recently, and has not yet been fully utilized in existing models. The rotation period is poorly known, the measured periods differing from each other by as much as a factor of two (see Podolak and Reynolds [1984] for a summary of these measurements). Curiously, most of the determinations seem to center around two numbers: 16 and 24 hr. Both of these are much higher than the value of 10.8 hr (Moore and Menzel 1930) that was used in models up until 1976. Because this number is vital for fixing the interior through J_2 (which varies as the square of the period) and J_4 (which varies as the fourth power), a broad class of models is admissible by the current observational constraints. It is possible to relate the oblateness ϵ of a slowly rotating planet to J_2 to second order by

$$\varepsilon = \left(1 + \frac{3}{2}J_2\right)\left(\frac{3}{2}J_2 + \frac{\omega^2 R^3}{2GM}\right) \tag{9}$$

where G is Newton's constant, M is the mass of the planet, R its equatorial radius, and ω its angular velocity. Measurements of the oblateness by two independent methods give 0.022 ± 0.001 (Franklin et al. 1980*a*) and 0.024 ± 0.003 (Elliot et al. 1981). From Eq. 9 one can then deduce rotation periods of 16.6 ± 0.5 and 15.5 ± 1 hr, respectively, giving strong support to the 16 hr value, although the 24 hr period may still be regarded as a (weak) possibility. The temperature at one bar, as derived from the model atmosphere calculations is about 75 ± 5 K (D. Gautier and R. Courtin 1979). An upper limit to the internal heat source is 122 erg cm^{-2}s^{-1} (Lockwood et al. 1983), and the lower limit appears to be nonzero but $< 5 \times 10^{-4}$ of the total flux (Wallace 1980).

Although there has been a considerable effort in modeling Uranus, many of the early models must be discarded due to significant changes in the observed values of the radius J_2, the oblateness, and the rotation period. In addition to the difficulty of obtaining good observational data, there is uncertainty in which materials to represent in the models. Unlike Jupiter and Saturn, where the heavy elements were merely perturbations to the hydrogen, in both Uranus and Neptune, they are a major component. A number of suggestions have been examined, such as metallic NH_4 (Porter 1961), "mud" (DeMarcus and Reynolds 1962), "CHONNE" (Ramsay 1967), and methane (Podolak 1976) but the most recent models follow the suggestion of Reynolds and Summers (1965) and use mixtures of rock, ice, and hydrogen and helium. These materials can be arranged in three distinct layers (Hubbard and MacFarlane 1980; MacFarlane and Hubbard 1982; Podolak and Reynolds 1981, 1984), or in two layers, with the ice mixed into the hydrogen-helium envelope (Podolak and Reynolds 1981,1984). It is now generally agreed that, if a three-shell model is to be acceptable, some of the ices in the middle shell must be mixed with the hydrogen and helium in the envelope (Podolak and Reynolds 1981; MacFarlane and Hubbard 1982). Briefly the results of the models of Podolak and Reynolds (1984) can be summarized as follows: for two-shell models, the constraints provided by J_2 and J_4 require a rotation period of more than 18 hr, and the envelope must be enhanced in ices to the extent that the ratio of ice in the envelope to rock in the core $I/R > 1$. Thus, if one accepts the observational results that imply a period < 18 hr, one must discard the two-shell model. On the other hand, if model planet calculations show that only a two-shell model is stable under the Uranian conditions of pressure and temperature, one must question the observations that imply a period < 18 hr. The three-shell models can be made to fit the observations with a 16 hr period, provided that some of the ice from the middle shell is mixed into the atmosphere. In this case I/R for the whole planet (including any ice in the middle shell) must be $\gtrsim 3$. If longer rotation periods are allowed, more ice must be shifted into the envelope to provide a fit to J_2. If I/R is taken to be one, periods between 18.5 and 19.5 hr can be fit by judicious shifting of the ice between envelope and shell. For $I/R = 2$, periods between 17.5 and 22 hr work well, while for $I/R = 3.6$, periods between 14.5 and 24 hr give a good fit. Although there is some uncertainty as to what ratio would be expected in solar composition, a value of about 3 is generally used (Zharkov and Trubitsyn 1978; Cameron 1982). The three-shell models, then, require I/R to be close to the solar value. Table III summarizes the models discussed above. We see from the table that the mass of the rock core is low (< 4 $M_\oplus$) unless we are willing to accept a rotation period of 19 to 20 hr, and even in this case the core mass rises to something over 5 $M_\oplus$. The total mass of ice plus rock is between 10 and 13 $M_\oplus$.

All the recent models have assumed that Uranus' interior is convecting and that the thermal gradient is, therefore, close to adiabatic. The luminosity of Uranus is low, however, and on this basis it is impossible to argue that convection is needed to supply the necessary heat flux. Nevertheless, as we have

TABLE III
Summary of Uranus Models

Atmospheric[a] Enhancement (× Solar)	Central Pressure (Mbar)	Core Mass ($M_\oplus$)	Shell[b] Mass ($M_\oplus$)	Central Temperature (K)	Required[c] Period (hr)	$-J_4$[d] ($\times 10^{-5}$)	I/R
20	23.4	10.8	—	4400	11.1	11.9	0.07
20	19.5	6.19	5.61	6000	13.0	6.62	1.0
20	16.2	4.22	7.98	6300	13.8	5.37	2.0
20	13.4	2.81	9.69	6500	14.8	4.46	3.6
75	18.1	7.89	—	3900	15.6	5.58	0.5
75	16.8	5.78	2.94	4800	16.2	5.38	1.0
75	14.5	3.98	5.48	5200	16.6	4.92	2.0
75	12.2	2.65	7.28	5300	17.1	4.48	3.6
130	13.4	5.36	—	3700	19.2	4.29	$\lesssim$1.0
130	13.4	5.22	0.12	3700	19.5	4.28	1.0
130	12.5	3.81	2.41	4300	19.7	4.20	2.0
130	10.9	2.55	4.52	4500	19.8	4.05	3.6
170	10.2	3.64	—	3700	21.9	3.69	2.0
170	9.82	2.49	2.20	4000	22.1	3.49	3.6
210	7.53	2.15	—	3500	24.2	3.22	4.2

[a]The solar ratio of ice to hydrogen-helium is taken to be 0.0125.
[b]Two-shell models are denoted by a dash.
[c]Period required to match J_2.
[d]The observed value is $J_4 = -(3.8 \pm 0.9) \times 10^{-5}$.

pointed out above, the evidence for a magnetic field and crude estimates of the cooling time of the planet seem to indicate that the interior should be convecting. To explain the absence of a sizable heat source on Uranus, Hubbard (1978) showed that the cooling time for Uranus is short enough so that it could cool to a temperature approximately equal to the temperature in equilibrium with the solar radiation field in a time comparable to the age of the solar system. For the case of Neptune, solar insolation is less, so that its heat source is still visible. More recent detailed computations by Hubbard and MacFarlane (1980) show that the cooling times for both Uranus and Neptune are longer than the age of the solar system and that they must have accreted at rather low temperatures in order to have the effective temperatures that we observe today. With regard to this, Stevenson (1982*a*) has raised the interesting point that as Uranus accreted a core of ice and rock, and a hydrogen-helium atmosphere fell onto it, the energy input involved in lifting the ices in order to mix them into the envelope would act to cool the planet considerably. As a final point, we would like to note that our own computations of cooling times for Uranus and Neptune (Podolak and Reynolds, unpublished) show that both Uranus and Neptune can cool from high initial temperatures within the age of the solar system. The difference between our results and those of Hubbard and MacFarlane stems from slight differences in the computation of the adiabat through the envelope. This uncertainty in the details of the physics, combined with the uncertainty in the effective temperatures of the present-day Uranus and Neptune is sufficient to cause significant differences in computed evolution times.

D. Neptune

The mass of Neptune is 17.13 $M_\oplus$, or $(1.022 \pm 0.002) \times 10^{29}$ g (A. W. Harris 1983), as derived from the orbit of Triton. The equatorial radius, determined from a stellar occultation is 25,265 $\pm$ 36 km at the occultation level, which corresponds to 24,753 $\pm$ 59 km at one atmosphere (Freeman and Lynga 1970). This is fully 15% larger than the radius used in models prior to this date, so that these models have a density more than 50% too high. J_2 has been determined from the orbit of Triton to be 0.0043 $\pm$ 0.0003 if Neptune's rotation is prograde as seems likely and 0.00325 $\pm$ 0.00017 if it is retrograde (A. W. Harris 1983). J_4 is still unmeasured. While there was a controversy over the rotation period of Neptune several years ago, the most recent determinations seem to be in good agreement, giving a value of about 18 hr (Belton et al. 1981; Slavsky and Smith 1981; R. A. Brown et al. 1981). It should be noted that the optical oblateness, derived from a stellar occultation and given as 0.026 $\pm$ 0.005 (Freeman and Lynga 1970) and 0.021 $\pm$ 0.004 (Kovalevsky and Link 1969) implies, through Eq. (9), a significantly shorter period, between 11.6 and 14.7 hr in the first case, and between 13.2 and 17 hr in the second. It must be stressed, however, that these measurements are difficult to interpret, and the photometric value is considered more reliable (Belton et al.

1980). The temperature at one bar is rather close to the value found for Uranus, 75 ± 5K (D. Gautier and R. Courtin 1979). Finally, Neptune does have a significant internal heat source, amounting to $2.6^{+2.4}_{-0.9}$ times the solar input (D. Gautier and R. Courtin 1979).

Models of Neptune face difficulties similar to those of Uranus. In the first place, the observational quantities are still not sufficiently well determined for the purpose of producing unambiguous models; in the second place, since the heavy elements are again a major component, their nature and distributions must be more carefully determined. Essentially the only models that fit the latest data with the J_2 for prograde rotation are those of Podolak and Reynolds (1981,1984). Again these models can be divided into two- and three-shell models. For the case of two-shell models, a fit to J_2 can be obtained only if the atmosphere is so enhanced in ice that the value of I/R is very close to the solar value. Thus, one model that fits well has a core of 3.47 $M_\oplus$ and 11.1 $M_\oplus$ of ice in the envelope, leading to a total mass of condensibles of 14.6 $M_\oplus$ or 85% of the planet: The three-shell models are not very different, because a large enhancement in the envelope is still needed. Thus, if the value of I/R is taken to be 3.6, the ice shell has a mass of only 1.01 $M_\oplus$, while the rock core has a mass of 3.16 $M_\oplus$, with 11.3 $M_\oplus$ of ice in the envelope. Thus, the total mass of heavy elements is 14.5 $M_\oplus$, again nearly all the mass of the planet. These numbers can be modified somewhat if the rotation period is shorter, say 17 hr, or if the sense of rotation is actually retrograde, but the conclusion would remain that considerable enhancements of ices are required, and that the ratio of ice to rock, even in the case of retrograde rotation with a 17 hr period, must be > 1.

III. CONCLUSIONS

In order to assess the meaning of the models for cosmogonic theories, it is useful to give a brief review of some of the scenarios for planet formation. More details can be found in papers by Podolak (1982), Podolak and Reynolds (1984), and Pollack (1984). Typically these scenarios fall into two classes, those dealing with giant protoplanets and those dealing with accretion. In the first case, gravitational instabilities result in the formation of giant gaseous protoplanets of the order of a Jupiter mass, which are initially of solar composition. As the protoplanet contracts, it must lose some of its mass, and evolve into the object we see today. Examples of this type of theory can be found in papers by Cameron (1978*b,c*), McCrea (1978), and Woolfson (1978*a,b*). If the original composition was indeed solar, one can estimate the mass of the original object from the mass of heavy elements that remained. Thus, in the case of Jupiter, with a heavy-element content of 20 to 30 $M_\oplus$, a protoplanet mass of 1100 to 1600 $M_\oplus$ is implied. Interestingly enough, the mass of heavy elements in Saturn is almost identical, implying that proto-Saturn had nearly the same mass as proto-Jupiter. The study of the evolution of such objects has been rather limited (a review can be found in a paper by Bodenheimer [1982]), and it is still unclear exactly how such objects would interact with the surrounding nebula. We can see, however, that Jupiter should have lost some 75 to 80% of its initial mass, while Saturn would have lost 90 to 95%. This is rather drastic, although probably not impossible, through tidal stripping (DeCampli and Cameron 1970) or thermal evaporation (Cameron et al. 1982). The situation for Uranus and Neptune is even more striking. They have a heavy-element content of about 14 $M_{\oplus}$, implying a protoplanetary mass of 750 $M_{\oplus}$. They must have lost 98% of their original mass. It will be an interesting task to devise a means of losing so much mass from so large an object in a reasonable time.

A second set of theories centers around the concept of accretion. Here solids in a given region accumulate into protoplanetary embryos (Safronov 1969). Near the Sun only refractory materials are solid, while farther out ices become available. With this additional mass, larger embryos can form, and an instability can be induced in the surrounding gas, causing some of it to fall onto the embryo (Mizuno 1980; Nakagawa et al. 1983). The ratio of ice to rock will depend on the ambient temperature of the surrounding nebula, and I/R will increase as one goes farther from the Sun. J. S. Lewis and R. G. Prinn (1980) have pointed out that at high temperatures, like those expected in the inner part of the nebula, the equilibrium composition of solar-mix gas will be such that nitrogen, carbon, and oxygen will be in the form of N_2, CO, H_2O, and small amounts of CO_2. In the colder regions these substances will be reduced by hydrogen to NH_3, CH_4, and H_2O. As Lewis and Prinn show, the kinetics for hydrogen reduction is extremely slow at these low temperatures, and does not go to completion. Thus, in the vicinity of Uranus and Neptune, where the temperatures should be about 50 K, the solids, in addition to rock, are such minor species as CO_2, NH_4COONH_2 and NH_4NCO_3, in addition to H_2O. Under these conditions, since most of the oxygen is tied up as CO gas, the I/R expected for Uranus and Neptune should be < 0.5. As we have seen, it must be $>> 1$ according to the models, thus putting the idea of kinetic inhibition of hydrogen reduction into question. This does not exclude the possibility that other mechanisms, such as Fischer-Tropsch reactions (Hayatsu and Anders 1981) contributed to the ice content of these planets. This is a difficulty of accretional theories that will have to be overcome by future work. In addition, such theories must explain how cores of similar mass (differing by no more than a factor of 4) gave rise to such different planets as Jupiter and Uranus.

It is quite possible that both accretion mechanism and gravitational instability were at work in different parts of the nebula, or in the same area at different times. Thus, one explanation for the massive cores of the giant planets is that the original protoplanet did, indeed, have a solar composition, but after the initial instability and collapse, additional solids were captured through gas drag (see Pollack 1984). Such a scenario, however, is difficult to apply to Uranus and Neptune, where the total hydrogen and helium content is about 2 $M_\oplus$. This is rather small for a gravitational instability to have formed it, and it seems rather unlikely that so small an object would have enough of a subnebula to form a regular satellite system. For these planets to have been formed, either accretion played a major role, or nearly all of the original protoplanet mass was somehow lost.

Having discussed the bulk features of giant planet composition, we would like to say a few words about observations of some specific elements (a detailed discussion of this point can be found in the chapter by Gautier and Owen in this book). In the first place, there is the apparent difference in the helium content between Jupiter and Saturn. As we have pointed out above, helium and hydrogen may not be miscible at the lower temperatures found in Saturn, and we may be seeing evidence of helium precipitation in that planet. Indeed, this mechanism may help explain the relatively short cooling times derived from evolutionary models for Saturn (Pollack et al. 1977). The recent measurement of water in Jupiter's lower atmosphere (Kunde et al. 1982) which puts the abundance at 30 times less than the solar value at the 4-bar level, is no cause for concern, as that region is probably still on the saturated part of the vapor pressure curve, due to the lower temperatures along a wet adiabat. Pressure levels of the order of ten bars will have to be probed before we can have definitive values of the water content of Jupiter's atmosphere. The NH_3 abundance has been measured for all of the giant planets, and the most striking result is the apparent absence of NH_3 in the upper atmosphere of Uranus and Neptune (Gulkis et al. 1978; Gulkis 1981, personal communication). The suggestion that the ammonia is trapped in the form of NH_4SH clouds (Prinn and Lewis 1973) has the drawback that the sulfur abundance must be at least equal to the nitrogen abundance, while in solar composition it is only 20% of that value. Possibly the NH_3 is hidden deeper in the planet; recent work suggests that the reduced NH_3 abundance is simply due to the reduction of the NH_3 vapor pressure due to NH_3-H_2O cloud formation (D. Stevenson, personal communication). If the nitrogen depletion were, in fact, planet-wide, it would provide a major difficulty for accretion theories. In giant protoplanet theories, the nitrogen remains in the form of N_2, and may be removed together with the H_2 and He.

A final species that has important consequences for cosmogony is deuterium. Although its abundance on Jupiter and Saturn is not inconsistent with the protosolar value, of $D/H = 1.7 \times 10^{-5}$ (Cameron 1982), it seems to be considerably above the protosolar value on Uranus (Cochran and Smith 1983). A possible explanation, in terms of an accretion model could be based on the argument of Hubbard and MacFarlane (1980*b*) who noted that for a gas of solar composition, the relative amount of deuterium that will be confined to ices rather than to gaseous hydrogen depends on the temperature. At low temperatures, the ices will contain between 10 and 10^3 times more deuterium than at higher temperatures. Since Uranus contains a large mass of ices accreted at low temperatures, the deuterium excess could have been redistributed partly into the atmosphere, explaining the observed excess. Indeed the

TABLE IV
Important Issues

Issue	Giant Protoplanet Scenarios	Accretion Scenarios
I/*R* for Uranus and Neptune ~3.	*I*/*R* expected = 2 to 3.6.	*I*/*R* expected (with kinetic inhibition of hydrogen reduction) = 0.5; otherwise ~3.
Low NH_3 abundance observed on Uranus and Neptune.	N_2 lost with H_2 and He.	NH_3—H_2O cloud system.
High deuterium abundance observed on Uranus.	D/H expected to be near the solar value.	D/H expected to be > solar value due to accretion of deuterium-rich ices.
Problems of planet formation.	Why do giant-planet masses vary by a factor of 20 while their complement of heavy elements varies by no more than a factor of 4?	Why do giant-planet masses vary by a factor of 20 while their complement of heavy elements varies by no more than a factor of 4?
	Can giant gaseous protoplanets lose enough mass to form the planets as we see them?	Can Uranus and Neptune be accreted in a time less than the age of the solar system?

observed excess is still too low to be the result of complete retainment of deuterium in the ices, but for this reaction as well, the kinetics is extremely slow, and so the accreted ices probably did not have their full complement of deuterium at the time of their accretion. In the giant protoplanet scenario, however, it is hard to see how deuterium could be at all enriched over its protosolar value. The arguments for and against the two scenarios for planet formation are given in Table IV.

The theory and the observations relevant to giant planet models has considerable room for improvement. Clearly there is much work still to be done. Such quantities as the rotation rate of Uranus, the deuterium abundance of Neptune, the helium-to-hydrogen ratio on Uranus and Neptune, the water abundance on Jupiter, and the luminosity of Uranus are of obvious importance in refining the models. More varied interiors for Uranus and Neptune are also interesting topics for study. In this respect the ices play an interesting role. There is still some uncertainty in the behaviour of ices in the interiors of Uranus and Neptune. Ross (1981) suggested, on the basis of shock-wave experiments, that CH_4 might be pyrolized to metallic carbon or even diamond under the high temperatures and pressures characteristic of the interiors of these planets. The notion that the interiors of Uranus and Neptune contain a substantial mass of diamonds is intriguing but, if there is a large percentage of hydrogen mixed with the ices, equilibrium may still favor the presence of CH_4. Shock experiments have also shown that the conductivity of ices increases at high pressure and temperature, indicating that significant molecular ionization occurs (A. S. Mitchell and W. J. Nellis 1982). On the basis of studies like these, Stevenson (1982*a*) suggested that a model of Uranus or Neptune should consist of an outer envelope of H_2, He, and CH_4, a shell consisting of an ionic ocean of H_3O^+, $NH_4{}^+$ and OH^-, and a rock core. Although these models differ from the three-shell models of Podolak and Reynolds (1984) with regard to the exact distribution of the ices throughout the envelope and shell, they are qualitatively similar in that they have an enhanced envelope, an ice shell, and a rock core. In solar abundance, the mass ratio of CH_4 to $H_2O + NH_3$ is about 0.6. Thus, the enhancement in the envelope should be related to the mass of the ice shell by this value. For our models there are two cases: those that fit a 24-hr period have a small ice shell, and the enhancement in the envelope required to fit the gravitational moments implies a CH_4 to $H_2O + NH_3$ well in excess of 0.6. For models that fit a 16-hr period, however, the ice shell is substantially larger, and the enhancement of the envelope deduced from our models implies a CH_4 to $H_2O + NH_3$ ratio significantly less than 0.6 if CH_4 is confined to the envelope. This does not take into account the fact that the density of CH_4 is considerably less than the density of the ice mixture used in the Podolak-Reynolds models. Allowance for this difference would, in fact, increase the computed ratio, though it is not clear by how much. At present, we can only say that the model suggested by Stevenson (1982*a*) is inconsistent with a 24-hr period for Uranus, but may be consistent with a 16-hr period.

One final example of the types of questions models of Uranus and Neptune can answer is given by the suggestion of Belton (1982). He pointed out that the low deuterium and nitrogen abundances, and the shape of Uranus's microwave spectrum between 1 and 21 cm might be explained if there were a "surface" near the 280 K level (plus a depletion of NH_3). We have examined this possibility with our models and found that, if one assumed that Uranus consisted of a core of 2×10^{-3} M_{Uranus} (necessary for the stability of the computer program), surrounded by a shell of CH_4 (the least dense of the common ices) and a thin H_2-He atmosphere, the mean density would still be a bit high. It is possible, though unlikely, that further improvements in our knowledge of the equation of state of CH_4 will bring its density down sufficiently so that a fit could still be achieved. A second argument against this model comes from the gravitational moments. As the "surface" moves closer to the one-bar level, J_2 increases for a given rotation period. To match the observed J_2, periods in excess of 24 hr are required (probably $\sim$ 30 hr). Again, there are still enough uncertainties in the parameters, both observational and theoretical, so that it is impossible to state conclusively that such a model is untenable; nevertheless, the evidence is strongly against it. The future of model planet interiors holds promise for reducing the uncertainties in the structure and composition of these objects, and efforts should be made to bring cosmogonic theories to the stage where they can make unambiguous statements about what these structures and compositions should be like. Then the giant planets will provide useful constants for such theories.

Acknowledgments. This work was supported by the NASA Planetary Geology and Geophysics Program.

EVOLUTION OF THE GIANT PLANETS

PETER BODENHEIMER
Lick Observatory

The theory of the evolution of the giant planets is discussed with emphasis on detailed numerical calculations in the spherical approximation. Initial conditions are taken to be those provided by the two main hypotheses for the origin of the giant planets. If the planets formed by gravitational instability in the solar nebula, the initial mass is comparable to the present mass or larger. The evolution then goes through the following phases: (1) an initial contraction phase in hydrostatic equilibrium; (2) a hydrodynamic collapse induced by molecular dissociation; and (3) a second equilibrium phase involving contraction and cooling to the present state. During phase (1) a rock-ice core must form by precipitation or accretion. If, on the other hand, the giant planets formed by first accreting a solid core and then capturing gas from the surrounding nebula, then the evolutionary phases are as follows: (1) a period during which planetesimals accrete to form a core of about one earth mass, composed of rock and ice; (2) a gas accretion phase, during which a relatively low-mass gaseous envelope in hydrostatic equilibrium exists around the core, which itself continues to grow to 10 to 20 Earth masses; (3) the point of arrival at the "critical" core mass at which point the accretion of gas is much faster than the accretion of the core, and the envelope contracts rapidly; (4) continuation of accretion of gas from the nebula and buildup of the envelope mass to its present value (for the case of Jupiter or Saturn); and (5) a final phase, after termination of accretion, during which the protoplanet contracts and cools to its present state. Some observational constraints are described, and some problems with the two principal hypotheses are discussed.

The epoch of formation of the giant planets represents an interesting and critical phase of the evolution of the solar system. A wide variety of physical processes are involved including hydrodynamics, radiation transport, convective transport, condensation of solid nuclei, growth of the nuclei by accretion, gas accretion onto solid cores, tidal, viscous, rotational, and magnetic effects. In its most general form, the problem must be attacked by the methods of magnetohydrodynamics in three space dimensions; in practice, the research on the problem up to the present time has involved treatment of a limited number of these physical processes, generally with the further restrictive assumption of spherical symmetry. Furthermore, a protoplanet in its formation stage and during its entire evolution cannot be treated in isolation; tidal and radiative effects of the Sun as well as the pressure, temperature, and gas flow in the neighboring solar nebula all must be taken into account. In turn the giant planets, once they have accumulated sufficient mass, may exert a decisive influence on the further evolution of the nebula itself (D. N. C. Lin and J. Papaloizou 1979*a*,1980). The nature of these interactions is one of the fundamental questions that must be investigated. The study of the giant planets is also of interest with regard to the question of detection of extrasolar planets. In their earliest evolutionary stages, the giant planets have luminosities that are much higher than present values, and therefore such objects near young stars are prime candidates for discovery. It is important for this purpose to determine the luminosity and surface temperature as a function of time.

This discussion deals primarily with studies of the evolution of the giant planets, given well-defined initial and boundary conditions. These conditions, of course, must be provided by a theory of formation, and we discuss calculations based on the two major hypotheses. According to the *first hypothesis,* the giant planets formed as a consequence of an instability in the gas of the solar nebula, which developed subcondensations of planetary mass that began to evolve as a unit from a very early stage. In the *second hypothesis,* the giant planets formed by gradual accumulation of solid particles into a rock-ice core of a few Earth masses ($M_{\oplus}$), and then at a later stage the nebular gas accreted onto the core. An important goal of evolutionary calculations is to distinguish between these two possibilities, a difficult task because most of the available observational material describes a late stage of planetary evolution (the present) rather than the formation stage. However, certain constraints must be satisfied by the evolutionary theories.

1. The composition of the outer envelopes of the giant planets is near solar, although Saturn is somewhat deficient in helium (see the chapter by Gautier and Owen). Since this gas must have come from the primordial solar nebula, the giant planets must have acquired their gaseous envelopes at a stage before the nebular mass in their vicinity fell below some minimum value. Since the evolution time of the solar nebula at this stage is roughly 10^6 yr (Cameron 1978*a;* D. N. C. Lin and P. Bodenheimer 1982), this argument puts a strong constraint on the formation time of the giant planets.
2. The masses, internal luminosities, effective temperatures, gravitational

moments, temperature at 1 bar pressure in the atmosphere, and radii for the giant planets are well known at the present epoch, particularly for Jupiter and Saturn. Evolutionary models must fit these observations after a time of 4.6×10^9 yr has elapsed since formation.

3. Observations (Stevenson 1982*a*) of the current mean densities and the dipole and quadrupole moments of the gravitational field indicate the presence of central cores composed of elements heavier than helium, including both rocky and icy constituents. The core masses for all the giant planets are similar, probably 15 to 30 $M_\oplus$, although the total masses differ by a factor of more than 20. The similarity of the core masses is one of the fundamental properties that the evolutionary theory should be able to explain.
4. The temperatures during the formation phase, as deduced from the equilibrium condensation model for grains published by J. S. Lewis (1974) range from roughly 125 K at Jupiter to 25 K in the outer part of the solar nebula. The corresponding densities range from about 10^{-10} to 10^{-12} g cm^{-3}, respectively. Although these values for temperature and density are not based on direct measurements and are model dependent, they provide a useful guide to initial conditions.
5. All of the giant planets except Neptune have satellite systems with regularly spaced prograde, nearly circular orbits that lie in the equatorial plane of the planet. The angular momentum of the material out of which the planets formed must be distributed in such a way as to account for these orbits.

This chapter discusses theoretical calculations that have been guided by the above constraints. The discussion emphasizes the planets Jupiter and Saturn, since extensive data are available for them, and it also emphasizes the earlier phases of the evolution. Discussions of topics not treated in detail here, such as the evolution of Uranus and Neptune and the physical state of the present-day giant planets, may be found in related review articles such as those by Stevenson (1982*a,b*), Hubbard (1980), Podolak (1982), and Pollack (1984) as well as in chapters by Podolak and Reynolds and by Pollack in this book. In the following section the physics involved in evolutionary calculations is reviewed. In Secs. II and III numerical calculations are described which are based on initial conditions derived, respectively, from the two main hypotheses of giant planet formation. Concluding remarks are presented in Sec. IV.

I. PHYSICAL ASSUMPTIONS

Most calculations of the evolution of protoplanets have been carried out under the assumption of spherical symmetry, although a few approximate calculations that included rotational effects were reported by Bodenheimer (1977). The composition of the gaseous regions is assumed to be homogeneous and solar, with hydrogen mass fraction $X = 0.74$ and helium mass fraction $Y = 0.24$. Intermixed with the gas is dust with mass fraction 0.01, provided that the temperature is less than the evaporation temperature of 1000 to 2000 K, depending on density. The core region is assumed to be rock with a constant density of about 3 g cm^{-3}. The evolution of the gas can be well approximated by the standard stellar structure equations. If M_r is the mass interior to radius r, P the pressure, T the temperature, L_r the energy in erg s^{-1} crossing a sphere at radius r, and ρ the density, the equations to be solved include the equation of mass conservation

$$\frac{\partial M_r}{\partial r} = 4\pi r^2 \rho \tag{1}$$

the equation of motion

$$\frac{1}{\rho}\frac{\partial P}{\partial r} + \frac{GM_r}{r^2} + \frac{\partial^2 r}{\partial t^2} = 0 \tag{2}$$

the equation of energy conservation

$$\frac{\partial L_r}{\partial r} = 4\pi r^2 \rho \left(\epsilon - P\frac{\partial V}{\partial t} - \frac{\partial E}{\partial t} \right) \tag{3}$$

where ϵ is the energy source in erg g^{-1} s^{-1}, $V = 1/\rho$, and E is the internal energy per unit mass; and the diffusion equation for radiative transfer

$$\frac{\partial T}{\partial r} = -\frac{3}{4ac}\frac{\kappa\rho}{T^3}\frac{L_r}{4\pi r^2} \tag{4}$$

where c is the velocity of light, a is the radiation density constant, and κ is the opacity in cm^2 g^{-1}.

During almost the entire evolution the gas is quite optically thick, local thermodynamic equilibrium holds, and Eq. (4) is sufficiently accurate. If the protoplanet is unstable to convection in a region, the temperature gradient there is taken to be the adiabatic gradient (Schwarzschild 1958), rather than that given by Eq. (4). These equations are supplemented by relations for the equation of state of the gas, the opacity, and the energy generation rate. During the earliest stage of evolution, where the densities are $< 10^{-6}$ g cm^{-3} and temperatures are < 2000 K, the equation of state can be taken to be that of an ideal gas of molecular hydrogen mixed with atomic helium. Above 2000 K dissociation of the H_2 must be taken into account. At later stages, when densities in the interior of protoplanets fall in the range 10^{-2} to 1 g cm^{-3} and temperatures approach 2-3 $\times$ 10^4 K, nonideal effects must be included, such as pressure dissociation, Coulomb effects, and excluded volume corrections (Graboske et al. 1975*b*; A. S. Grossman et al. 1980). In the region of molecular dissociation, a dynamical instability can set in if the adiabatic exponent Γ, defined as d(log P)/d(log ρ) at constant entropy, falls below 4/3 in a sufficiently large region. In this phase, the full equation of motion (Eq. 2) must be employed; however, during most of the protoplanetary evolution, both before and after the dissociation phase, the configuration contracts in quasi-hydrostatic equilibrium and the acceleration term in this equation may be set to zero. The behavior of the radiative opacity is important in determining the location of convection zones. At the lowest temperatures the major contribution comes from water-ice grains. At temperatures above the evaporation point of the ice (160 K), the opacity is dominated by grains of magnetite, metallic iron, and hydrated or nonhydrated silicates with or without iron. Above the evaporation point of these mineral grains, the opacity drops considerably and the main source becomes molecules of water and TiO. Above around 3000 K, these molecules dissociate and the negative hydrogen ion begins to be important. Information regarding the low-temperature opacities is provided by Alexander (1975), Bodenheimer et al. (1980*a*), DeCampli and Cameron (1979), and Alexander et al. (1983). Nuclear energy generation is of negligible importance in the mass range considered here; thus the term ϵ in Eq. (3) may be set to zero. Four boundary conditions must be specified to complete this set of equations, two of which may be applied at the center, where $r = 0$ and $L_r = 0$ at $M_r = 0$. The two surface boundary conditions differ according to the various cases considered, and they will be discussed in the following sections.

II. EVOLUTION OF GIANT GASEOUS PROTOPLANETS

Although the idea of a giant gaseous protoplanet, formed by gravitational instability in the solar nebula, was proposed by G. P. Kuiper (1951*a*), the first detailed numerical calculations of the evolution of such objects were performed by Bodenheimer (1974) for the earliest stages and by A. S. Grossman et al. (1972) and Graboske et al. (1975*b*) for the later stages. The initial condition is dictated by the Jeans criterion, namely the requirement that a subcondensation of approximately Jovian mass be self-gravitating. The gravitational energy, given for a uniform-density object by $-0.6GM^2/R$, where M is the total mass and R the total radius, must be greater in absolute value than the thermal energy, given by $1.5\, R_g M\, T/\mu$, where T is the mean temperature, R_g the gas constant, and μ the mean atomic weight per free particle, which is about 2.3 for a solar-composition gas with the hydrogen in molecular form. In the case of Jupiter, a nebular temperature of 150 K results in a maximum radius of about 140 $R_\odot$ and a minimum mean density of about 5×10^{-10} g cm^{-3}. This density is similar to the estimated density in the solar nebula at Jupiter's position at the time of its formation (J. S. Lewis 1974; Cameron 1978 *a*). If in fact the Sun were already a condensed object at this time, the tidal radius at Jupiter's present position [R (tidal) $= D(m_p/3M_\odot)^{1/3}$, where D is the distance to the Sun and m_p the planetary mass] would be about 75 $R_\odot$ or about 0.35 AU, and the condensation would be disrupted unless its mean density were $> 3 \times 10^{-9}$ g cm^{-3}. However it is presumed that during the formation of the subcondensations most of the mass of the solar system was in the nebula rather than in the Sun, in which case the tidal radius is larger and the required mean density smaller. DeCampli and Cameron (1979) have shown, based on the evolving nebula model of Cameron (1978*a,b*), that contracting protoplanets at either the Jupiter or Saturn formation positions fall inside their tidal radii during practically their entire evolution, even if they start near the Jeans limit. The time scale for decrease of the tidal radius is the nebular evolution time, which is comparable to the contraction time for Jovian-mass protoplanets in their earliest phases.

For the evolutionary calculations, two different surface boundary conditions have been employed, corresponding to (1) an isolated protoplanet (Bodenheimer et al. 1980*a*; DeCampli and Cameron 1979), and (2) a protoplanet embedded in the solar nebula (Cameron et al. 1982). In the first case, the protoplanet is assumed to have essentially zero pressure at the boundary and a surface luminosity L (erg s^{-1}) given by the blackbody relation

$$L = 4\pi R^2 \sigma T_e^4 \tag{5}$$

where σ is the Stefan-Boltzmann constant, R the total radius, and T_e the temperature at optical depth 1. The results of calculations based on these conditions show that protoplanets contract in quasi-hydrostatic equilibrium during their early evolution, as long as the central temperature is < 2000 K. The time scale of evolution in this phase is 4×10^5 yr for the case of Jupiter and 4×10^6 yr for the case of Saturn, during which time the radius decreases by a factor of 25. The models contain extensive convection zones in the regions where the opacity is dominated by grains. When the temperature in the central regions exceeds 1700 K, the grains evaporate, the opacity drops, and the regions become radiative. The mean luminosity during this phase is 5×10^{-7} $L_\odot$ for Jupiter and about a factor of 30 lower for Saturn, while surface temperatures are 30 to 60 K.

DeCampli and Cameron (1979) and Slattery et al. (1980) have considered the possibility of the formation of a rocky core during this evolutionary phase by precipitation of grain material to the center. The pressure-temperature relation at the center of protoplanets of one Jovian mass or less was found to pass through the region where liquid metallic iron could exist; the temperatures at this point fall in the range 1500 to 2000 K. Calculations show that liquid droplets can form, grow, and rain to the center in a time short compared with the protoplanetary evolution time. Also, convective currents can circulate the grains from the outer part of the protoplanet into the liquid region, so that in a protoplanet of one Jovian mass a core of roughly 1 $M_\oplus$ could form. Further accumulation of core material must then result from continued inflow of solid material from the nebula, for example, by capture of planetesimals (Pollack et al. 1979). Alternatively, one could argue that the initial mass of a giant planet was much larger than a Jovian mass so that a larger supply of condensable material was available, and that the excess envelope mass was evaporated after core formation.

A number of difficulties occur when one considers core formation by these processes. If the initial mass of a giant planet were much larger than a Jovian mass, then the central regions of the protoplanet would never pass into the regime where liquid iron could exist. One should of course investigate the uncertainty in the critical mass for the occurrence of precipitation. Slattery et al. (1980) indicate that two major effects could increase it: first, an increase in the iron abundance, and second, the existence of a liquid phase for another grain constituent at temperatures of 1600 to 1800 K and pressures below 1 bar. However, these effects result in an uncertainty in the critical mass by probably a factor 2. Another difficulty arises from the fact that the grain material is probably soluble in metallic hydrogen (Stevenson 1982*b*). Thus the entire core-formation process, including the capture of planetesimals, must have been completed during the earliest stages of the evolution, before the central pressure increases to the point where metallic hydrogen forms. For a homogeneous model of one Jovian mass such a condition is reached shortly after completion of the dynamic collapse phase. A final difficulty with this core-formation process is that the formation of relatively large (1 cm) droplets in the liquid zone sharply reduces the opacity, causing the region to become radiative and, therefore, limiting the amount of solid material that can be convected into the central regions.

The occurrence of convection zones during a significant portion of the early contraction phase for Jupiter and Saturn is important when the effects of rotation are considered. The approximate (but fully two dimensional) rotating models of Bodenheimer (1977) showed that if angular momentum were conserved for each mass element during this phase, then models with sufficient angular momentum to account for the orbital angular momentum of the Galilean satellites reach a point of dynamical instability before the central temperature reaches 2000 K. This instability, which occurs when the ratio of rotational kinetic energy to gravitational potential energy exceeds 0.26, could possibly result in fission of the protoplanet into two orbiting fragments. However, it was further pointed out that, if angular momentum transfer toward the surface of the model occurred during the contraction, a distribution of angular momentum could be approached which would not lead the protoplanet into instability but would favor the formation of a central object and a surrounding stable nebular disk, out of which satellites could form. The turbulent viscosity induced by convective motions could well be the physical mechanism by which the required angular momentum transfer took place. A modified version of this argument would be expected to apply also in the case that the planets formed by the core-accretion, gas-capture scenario.

When the central temperature of the protoplanet reaches 2000 K, a dynamical instability sets in, caused by the dissociation of the hydrogen molecules. The time scale of evolution is suddenly reduced to the order of days as gravitational collapse occurs. This collapse is practically adiabatic, since the collapse time is short compared with the time for radiation to be transported out from the collapsing region. The relatively long radiation diffusion time is in part caused by the increase in opacity at temperatures $>$ 2000 K where molecular sources and the negative hydrogen ion become important. The collapse stops at the center when dissociation becomes nearly complete there and the pressure increases because of nonideal effects in the equation of state. In the case of Jupiter the temperature has increased to 2×10^4 K and the density by four orders of magnitude to about 0.2 g cm^{-3}. A region in hydrostatic equilibrium forms at the center and quickly grows in mass as the outer layers of the collapsing protoplanet are accreted. A shock forms at the outer boundary of the core; it maintains a fairly constant radius of about 2×10^{10} cm. At the conclusion of the accretion phase a Jupiter-mass protoplanet has a central density of about 3 g cm^{-3}, a central temperature of 35,000 K, $\log L/L_\odot = -5.65$, $T_e = 600$ K, and $R = 10^{10}$ cm, not much larger than the present radius (7×10^9 cm) of Jupiter. The corresponding quantities for Saturn are 0.3 g cm^{-3}, 18,000 K, -4.78, 660 K, and 2×10^{10} cm, about 3.4 times the present radius of Saturn. The evolution of Jupiter and Saturn is illustrated in the ($\log \rho_c$, $\log T_c$) plane in Fig. 1, where the solid dots mark the onset of collapse and the asterisks the completion of collapse for the entire mass. The evolution in a HR diagram is shown in Fig. 2.

The later phases of evolution have been calculated for Jupiter by Graboske et al. (1975*b*) and for Saturn by Pollack et al. (1977), under the assumption of a uniform (gaseous) composition. Calculations including the effects of a rocky core of 20 to 25 $M_\oplus$ for both planets were presented by A. S. Grossman et al. (1980) with similar results. During this phase the planets are in hydrostatic equilibrium, and because of the very high opacity to radiation in the interior, the gas is convectively unstable. The energy loss rate of the protoplanet is controlled by the opacity in the very outermost layers near optical depth unity. These layers, therefore, must be treated in detail with nongray model atmospheres, where the principal opacity source is pressure-induced transitions of molecular hydrogen (Pollack and Ohring 1973). The equation of state is complicated in the deep interior at late stages of the evolution when liquid metallic hydrogen is present. The interior equations solved are then Eqs. (1) and (2) with the acceleration term set to zero, Eq. (3) with ϵ set to zero, and the adiabatic temperature-pressure relation.

If, during the evolution of a planetary-mass object, the equation of state is close to that of an ideal gas, then, as the virial theorem predicts, the gravitational energy released from contraction goes about equally into heating and into radiation from the surface. Since the objects are convective, the contraction proceeds at nearly constant surface temperature and with decreasing luminosity, as was shown by C. Hayashi (1961) for stars in the analogous phase of

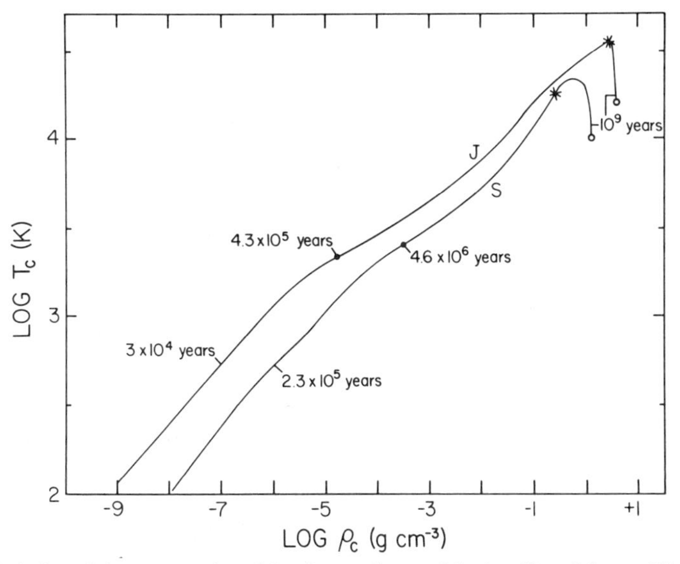

Fig. 1. Evolution of the centers of models of protoplanets of Jupiter (J) and Saturn (S) mass, under the assumption that they formed by gravitational instability. Filled circles: the point of onset of hydrodynamic collapse; Asterisks; the beginning of the final phase of contraction and cooling; Open circles: models of present planets (figure after Bodenheimer et al. 1980*a*).

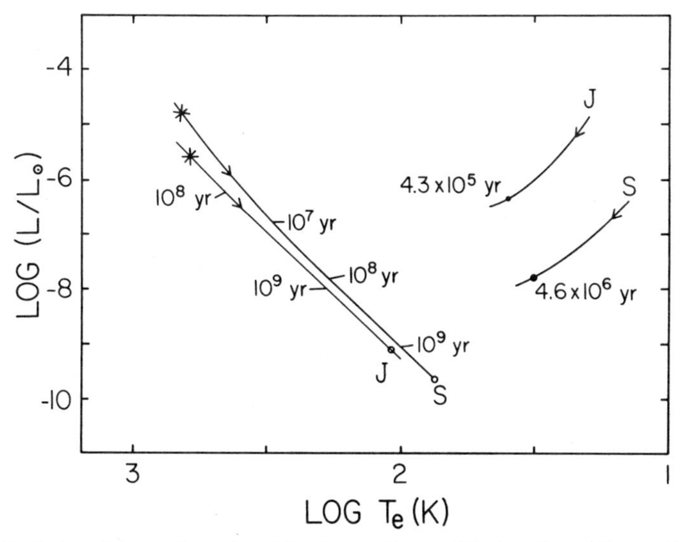

Fig. 2. Evolution of the surface of models of protoplanets of Jupiter (J) and Saturn (S) mass, under the assumption that they formed by gravitational instability. Symbols have the same meaning as in Fig. 1, except that the open circles give the approximate observed values.

their evolution. Planetary contraction takes place at considerably smaller values of T_e and L than for the case of stars. As the contraction proceeds, the equation of state becomes nonideal and the interior material becomes less and less compressible as it undergoes a gradual transition to the liquid phase. Eventually further gravitational compression is unable to supply the energy lost at the surface, some of which then must be provided by internal energy. The following phase of cooling and decreasing luminosity at nearly constant radius is analogous to the white-dwarf phase of stellar evolution. It turns out that for both Jupiter and Saturn a large fraction of the mass of the models is already in the nonideal state by the time the hydrodynamical collapse has been completed. In the case of Saturn, a temperature maximum of 21,000 K occurs at the center shortly after completion of collapse. For the case of Jupiter, the central temperature maximum is 35,000 K, the value obtained at the end of collapse at which point the object is already in the cooling phase. The evolutionary tracks are shown in Figs. 1 and 2. The evolutionary time scale for Jupiter increases from about 10^7 yr at the beginning of this phase to more than 10^9 yr toward the end.

The fit of these models to the present observed properties of Jupiter and Saturn at an assumed age of 4.6×10^9 yr has been discussed by Graboske et al. (1975*b*), Pollack et al. (1977), and A. S. Grossman et al. (1980). The latter study included rotational effects and a central rocky core, and the observed mean rotation rate as well as the gravitational moments J_2 and J_4 were considered in the fit. The evolution of a fully convective structure is practically independent of initial conditions; thus, this procedure is unable to distinguish between the two possibilities: (1) that the core formed first and then collected the gas, and (2) that the core was formed by precipitation within or accretion by a giant gaseous protoplanet. In the case of Jupiter, the calculated and observed radii agree to within 0.1% and the theoretical luminosity agrees with the observed value to within the observational error. The best fit was obtained for the following values of parameters: $X = 0.77$, $Y = 0.21$, and core mass of 20 $M_{\oplus}$, where X and Y are, respectively, the hydrogen and helium mass fractions in the envelope. The results show that at the present time, only about 25% of Jupiter's energy is being produced by contraction and the other 75% by release of heat generated by contraction in the past.

In the case of Saturn, the same fitting procedure results in a notable discrepancy. The values of radius, luminosity, and temperature at 1 bar pressure match well if the composition is taken to be the same as that for Jupiter and if the core mass is also taken to be 20 $M_{\oplus}$. However, the age at this point is calculated to be only 2.6×10^9 yr, and the gravitational moments J_2 and J_4 do not agree well with observations. Another way of stating the discrepancy is to say that for models at Saturn's expected age (4.6×10^9 yr), all reasonable values of core mass and chemical composition in the envelope give radii which are slightly too small and luminosities which are a factor 3 too low. An additional energy source must therefore be found. The evolutionary models have physical conditions which are favorable for the precipitation or settling of the helium toward the center because of its limited solubility in the metallic hydrogen zone, a suggestion first made by Smoluchowski (1967,1973) and Salpeter (1973) and discussed in detail by Stevenson and Salpeter (1977*a,b*) and Stevenson (1980). This process would release additional gravitational energy and would predict an inhomogeneous envelope for Saturn, with the helium concentration increasing inward. The Voyager observations (Conrath et al. 1984) do indicate that the helium abundance in Saturn's atmosphere ($Y = 0.06 \pm 0.05$) is significantly less than in Jupiter's ($Y = 0.18 \pm 0.04$) and thereby provide support for the settling hypothesis; note that Jupiter's helium abundance is similar to that of the Sun ($Y = 0.22 \pm 0.03$). However, recent theoretical calculations by MacFarlane and Hubbard (1983) indicate that the solubility of helium in metallic hydrogen is strongly model-dependent and may decrease significantly only at much lower temperatures than those expected in Saturn's interior.

The evolution of protoplanets during their earliest phases under a modified set of surface boundary conditions has been discussed by Cameron et al. (1982). The protoplanet is assumed to be imbedded in the solar nebula, which evolves in time according to the models of Cameron (1978*a,b*). The temperature-density variation in these models, both as a function of distance from the center and as a function of time, was found to fall close to a given adiabatic curve whose entropy depended on the assumed viscous dissipation rate in the nebula. The adiabat corresponding to Cameron's published models is here referred to as the "high" adiabat, calculated under the assumption that the turbulent viscosity is given by one-third the sound speed times the scale height in the nebula. For example, along this adiabat the maximum temperature attained at the Jupiter position in the nebula is about 165 K, and the maximum density about 3×10^{-9} g cm^{-3}. At the Earth's position in the nebula the corresponding quantities are 680 K and 10^{-8} g cm^{-3}, respectively. Further study has shown that these dissipation rates are probably too high, and that a lower adiabat would be more realistic. Some of the calculations reported here are based on the "low" adiabat, which is still uncertain and which has the same density variation with time as the "high" adiabat but with temperatures reduced by a factor of 3.9. Generally speaking, the nebular models show that the temperatures and pressures increase gradually over a period of about 5×10^4 yr, as a consequence of continued infall of material into the nebula from the surrounding cloud. Beyond that time, temperatures and pressures gradually decrease over a somewhat longer time scale (2×10^5 yr) as nebular material escapes via a presumed coronal wind and also evolves as a consequence of the action of turbulent viscous stress.

The modified surface boundary condition on the protoplanet becomes

$$aT_b^4 = L/(2\pi cR^2) + aT_N^4 \tag{6}$$

where T_N is the nebular temperature (a function of time), L and R are, respectively, the luminosity and radius of the protoplanet, and T_b is the boundary temperature. The first term on the right of Eq. (6) represents the contribution from the thermal contraction of the protoplanet, which is generally small compared with the second term. A critical parameter in this problem is evidently the ratio N of the time scale of increase of the nebular temperature to the contraction time of the protoplanet.

In the first calculation of this type, the nebular temperature is assumed to increase by 10^{-3} deg yr^{-1}, corresponding to the variation found by Cameron (1978*a,b*) in his high adiabat calculation at the Jupiter position in the solar nebula. The protoplanetary mass is taken to be one Jovian mass, and thus N is somewhat less than one, meaning specifically, that the time scale for an increase by a factor of 10 in the internal temperature (10^5 yr) is longer than the corresponding time for the external temperature (5×10^4 yr). The surface density is given the small and constant value of 4.6×10^{-12} g cm^{-3}, an assumption which can be justified on the basis of calculations by D. N. C. Lin and J. Papaloizou (1979*a*) who show that the tidal effects of a Jovian-mass protoplanet can truncate the nebula at the position of the protoplanet's orbit and prevent further mass flow onto or into the vicinity of the object. The evolution is illustrated in Fig. 3. The protoplanet contracts by about a factor 5 during the first 3×10^4 yr, then expands. The central density and temperature increase at first, then level off and sharply decrease at 1.4×10^5 yr. Clearly,

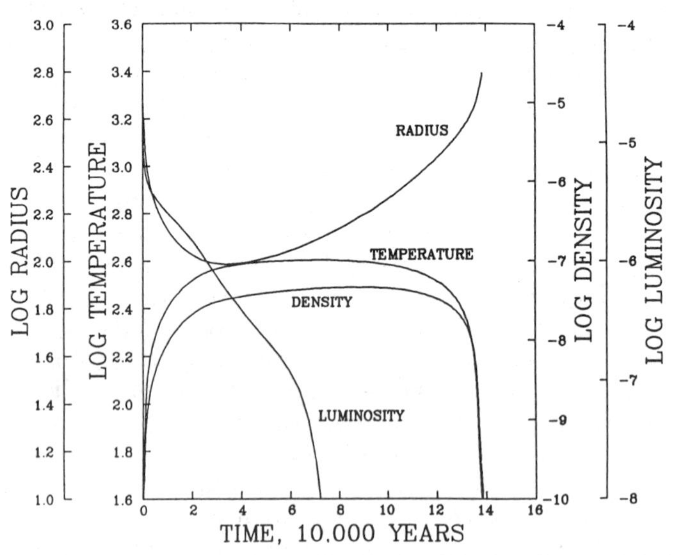

Fig. 3. The evolution of the radius and luminosity (in solar units), the central density (in g cm^{-3}), and the central temperature (K) of a giant gaseous protoplanet immersed in a thermal bath with the external temperature increasing at a constant rate of 10^{-3} deg yr^{-1} (from Cameron et al. 1982).

the end result is complete evaporation of the object, as a consequence of the fact that, after a time of 7×10^4 yr, the nebula is hot enough so that the direction of energy flow reverses. Energy is transferred from the nebula to the protoplanet, which has its internal energy increased to the point where it becomes unbound.

Two more calculations were performed in which the nebular temperature variation was assumed to follow Cameron's low-adiabat nebula. For a Jovian mass, the maximum nebular temperature is a factor 3.9 less than in the previous case and the corresponding time scale therefore a factor 3.9 longer; thus, the ratio N is > 1. The protoplanet survives, because it can contract and radiate sufficient energy to remain bound. The evolution in this case resembles that of an isolated protoplanet, except that the contraction time is slightly longer (8×10^5 yr). The situation changes when the Jovian-mass protoplanet is placed at the position of the asteroid belt, where the maximum T_N is 107 K. Here the rate of increase in T_N is rapid enough so that N is < 1, and the object evaporates. Thus, in the low-adiabat case Jovian-mass objects would be expected to evaporate during their very early evolution if their formation position were at or interior to the asteroid belt. Note carefully, however, that these conclusions depend on the assumed values of T_N, which depend on the uncertain energy dissipation rate caused by turbulent viscosity in the solar nebula.

It is of interest to perform similar calculations for a higher-mass protoplanet. The fragmentation of the solar nebula is likely to produce objects more massive than a Jovian mass, in fact closer to 100 Jovian masses (Cameron 1978*a,b*). Furthermore, if the present rocky core in Jupiter is to be explained largely by the precipitation of the heavy elements within a giant protoplanet, the original mass would have to be at least 2000 $M_\oplus$, or about 6 Jovian masses. The theory seems to require fairly massive original protoplanets, which subsequently precipitate a substantial fraction of their heavy elements (leaving enough to explain present abundances), after which most of the envelope is evaporated either by tidal interaction with the Sun or by heating of the nebula. Some of the difficulties with these ideas have been discussed above. The further question of evaporation by heating is addressed in a recent calculation by Korycansky et al. (1984) for the case of 10 Jovian masses. The time scale for changes in the solar nebula is the same as above; however, because the luminosity of an isolated protoplanet during the early contraction goes roughly as M^3, the evolutionary time to a given radius $t = GM^2/(RL)$ is proportional to M^{-1}. Thus the ratio N tends to be large for high masses. The calculations show that for the low adiabat the protoplanet contracts rapidly enough to reach the point of molecular dissociation before the heating effect of the nebula becomes significant. This conclusion holds for the Jupiter position in the nebula, the asteroid position, or even the Earth position. Evaporation of a protoplanet of this mass was found to occur only for the high adiabat and at the Earth position or farther in. The questions of tidal evaporation and of prior heavy-element precipitation must still be addressed.

If the protoplanet is not able to open up a gap in the nebula, then the above boundary condition, known as the "thermal bath," does not apply, since both the temperature variation and the pressure variation in the nebula must be considered. Calculations were performed by Cameron et al. (1982) for a Jovian-mass protoplanet at three different positions in the solar nebula and for both high and low adiabats. For the case of the high adiabat and the Jupiter position, the first part of the evolution is similar to that in the thermal bath case. The object contracts by about a factor of 5 in about 6×10^4 yr, at which time T_N has reached its maximum of 165 K, and the nebular density has reached 3×10^{-9} g cm^{-3}. For comparison, the central quantities at this time are 250 K and 7.5×10^{-9} g cm^{-3}. Expansion begins and the central temperature and central density drop. The reason for the expansion in this case is not so much the rapid input of energy, however, but the release of the external pressure, which, during the phase of rapid nebular temperature increase, is able to prevent significant expansion. The radius increases by a factor 1.7, then contraction resumes at 1.6×10^5 yr when both pressure and temperature in the nebula are low enough so that they have negligible influence on the protoplanet. The evolution then reverts to that in the isolated case, except that the evolution time to the point of dissociation is increased by roughly a factor 1.5, because of the additional energy input to the object in its earliest phases.

In the low-adiabat case, the rate of temperature increase at the surface is much slower and there is no expansion phase. The object contracts rapidly enough so that an outward temperature gradient is always maintained. The high surface pressure in this case is more important than the temperature, and it in fact results in higher temperature gradients at the edge of the object and a more rapid rate of energy release. Thus, the luminosity is higher and the evolution time shorter in this case than in the high-adiabat case. The total contraction time for a Jovian-mass protoplanet at the Jupiter position to the point of onset of molecular dissociation at the center is 1.4×10^5 yr in this case, as compared to 6.1×10^5 yr in the high-adiabat case and 4.1×10^5 yr in the isolated case. At the other positions in the solar nebula, the results are qualitatively similar. In all cases the surface pressure on the protoplanet prevents its disruption.

The evolution of the high-mass protoplanet discussed above is of interest in another context. The young star T Tauri is found to have a companion, detectable in the infrared and radio regions of the spectrum (M. Cohen et al. 1982; Dyck et al. 1982; T. Simon et al. 1983; P. R. Schwartz, personal communication, 1984). The infrared luminosity is found to be 1.5 $L_\odot$, and the effective temperature to be about 700 to 800 K. It was suggested by Hanson et al. (1983) that the companion is a protoplanet of perhaps 10 Jovian masses, consisting of a condensed core plus a steady-state accreting envelope nearly in free fall and being supplied by material in a circumstellar disk. In this situation, the term "core" refers to the entire portion of the protoplanetary mass (gas plus rock) which has reached hydrostatic equilibrium with a radius comparable to present planetary radii. This suggestion has been examined more carefully by Korycansky et al. (1984). The evolution of the 10 Jovian mass protoplanet (with isolated boundary conditions) was carried through the dynamical collapse phase to the point where most of the material had come into equilibrium in the central core. The location of the accretion shock at the outer edge of the core could be fairly well determined to be 5×10^{10} cm. This radius as well as an assumed core mass (the core radius is only weakly dependent on the mass) were then used as central conditions in a calculation of a steady-state accretion flow with material assumed to be continuously supplied by the disk. The equations of steady-state flow, including the continuity equation, the equation of motion with pressure effects included, the energy equation, and the diffusion equation for radiative transport were solved, starting at a point just outside the shock and extending out to the infrared photosphere. At the shock, it was assumed that all of the kinetic energy of inflow was converted into radiated luminosity, which is then given by

$$L_{\rm acc} = GM_c\dot{M}/R_c \tag{7}$$

where $\dot{M}$ is the accretion rate and M_c and R_c are, respectively, the mass and radius of the core. This approximation has been found to hold true during the analogous phase of protostellar evolution (Winkler and Newman 1980). Detailed grain and molecular opacities were used in the envelope (Bodenheimer et al. 1980*a*). It was found that both the infrared luminosity and the surface temperature could be matched with an inflow rate of about 10^{-6} $M_\odot$ yr^{-1} and a core mass of about 20 Jovian masses. The envelope was found to be practically in free fall, to have a photospheric radius of 10^{13} cm, and to have nearly constant luminosity as a function of radius (equal to $L_{\rm acc}$ in Eq. 7). This interpretation has been criticized by Bertout (1983) who remarks that it leaves unexplained the properties of the radio spectrum. He explains the object as a protostar with a core of 2 or 3 $M_\odot$, having an extensive, high-temperature, largely ionized surrounding infalling envelope in which the thermal radio emission is produced. This model, however, predicts a much higher infrared luminosity than is observed; the problem of the inconsistency of the deduced radio and infrared luminosities in this object has not been resolved.

III. CORE ACCRETION-GAS CAPTURE MODEL

Some of the difficulties involved in forming protoplanets by gravitational instability in the solar nebula include: (1) the physical conditions may not be favorable for precipitation of the heavy elements to form the core; (2) there is no natural explanation for the fact that the core masses of the giant planets are all about the same; and (3) the question of tidal or evaporative stripping of a large fraction of the envelope of the original protoplanet has not been satisfactorily resolved. However, a reasonable alternative assumption involves the formation of the core first, by gravitational instability in the dust layer of the solar nebula followed by gradual accretion of planetesimals, in the same manner as the presumed formation of the terrestrial planets. Once the core mass becomes large enough, the gas from the surrounding solar nebula can be captured as a result of the gravitational influence of the core. Although the entire evolution of a protoplanet under this hypothesis has not yet been calculated, several interesting model calculations of the early phases have been made. The late phase of contraction and cooling is essentially the same as under the gravitational condensation theory and has already been discussed in Sec. II.

The set of equations to be solved is essentially the same as described above (Eqs. 1–4). However, since gas accretion begins while core buildup is still taking place, an additional energy source for the gaseous envelope is considered, namely the accretion energy of the solid planetesimals as they arrive at the core boundary. This energy is assumed to be liberated at or near the core surface and can therefore be calculated according to the expression

$$L_{\rm acc} = \dot{M}\int_{R_c}^{R_{\rm out}} \frac{GM_r}{r^2}\,{\rm d}r \tag{8}$$

where $\dot{M}$ is the mass accretion rate and $R_{\rm out}$ is the outer boundary of the protoplanet's gaseous envelope. The boundary condition at the inner edge of the envelope is simply that $M_r = M_c$ and $L_r = L_{\rm acc}$ at $r = R_c$. At the outer edge the boundary conditions are dependent on the properties of the solar nebula. The accretion radius of the protoplanet is given by $R_A = {\rm G}\ m_p/c^2$, where c^2 is the sound speed in the solar nebula. Beyond this radius, a particle has greater thermal energy than binding energy to the protoplanet and could not be accreted onto the object. The tidal radius of the protoplanet is given by $R_T = D\ (m_p/3\ {\rm M}_\odot)^{1/3}$ where D is the distance to the Sun. This value for R_T assumes that the mass of the Sun has already built up to its present value, but in the present context the initial phases of core accretion take so long that by the time an appreciable amount of gas has been collected, the assumption is reasonable. The boundary condition is then $\rho = \rho_o$, $T = T_o$, where ρ_o and T_o have so far been taken to be constant in time and are obtained from a model of the solar nebula. This condition is applied at $R_{\rm out} = R_A$ or $R_{\rm out} = R_T$, whichever is smaller.

The calculations are generally started at the time when the core mass is about 0.1 $M_\oplus$; at that point the stellar structure equations for the envelope become valid; that is, it is possible to construct an optically thick gaseous envelope in hydrostatic equilibrium. The envelope mass at this time turns out to be about 10^{-6} $M_\oplus$ and the radius about 2×10^{10} cm. The first studies of this problem were those of Perri and Cameron (1974), who assumed an adiabatic envelope, and Mizuno et al. (1978), who calculated a model consisting of an inner adiabatic layer and an outer isothermal layer. In neither of these models was the energy transport, which turns out to be both by radiation

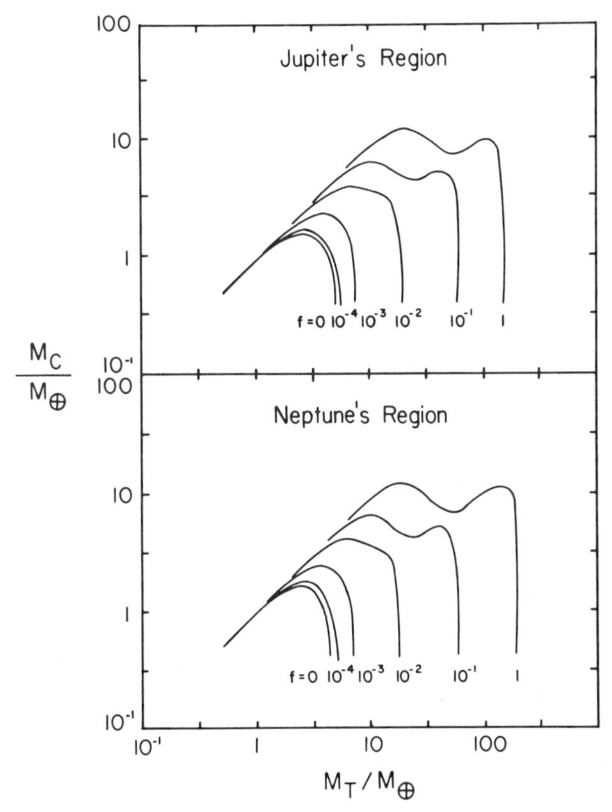

Fig. 4. Mizuno's (1980) static core-envelope models. The core mass in units of the Earth mass ($M_\oplus$) is plotted against the total mass in the same units. The first maximum of each curve corresponds to the critical core mass. f is the ratio of the assumed (constant) grain opacity to the estimated value for interstellar clouds, which Mizuno takes to be 1 cm^2 g^{-1}.

and convection, correctly treated. The first complete solution of Eqs. (1)–(4), including the accretion energy source but neglecting the time-dependent terms in the energy equation, was obtained by Mizuno (1980). He assumed a constant grain opacity, a constant value of $\dot{M}$ with time, a constant L_r as a function of radius in a given model, an ideal-gas equation of state, and an outer boundary condition applied at $R_{\rm out} = R_T$. At the Jupiter position, for example, $\rho_o = 1.5 \times 10^{-10}$g cm^{-3} and $T_o = 97$ K. By varying the assumed core mass, he obtained a sequence of purely static models with increasing total mass. The results are shown in Fig. 4. A typical model consists of a nearly isothermal outer layer near the photosphere, a radiative region, and an inner convective zone. For relatively low values of the core mass (a few Earth masses or less), the envelope mass is small compared with the core mass. A critical value of the core mass is obtained, above which there is no solution in strict hydrostatic equilibrium. For standard interstellar grain abundance, solar chemical composition in the gaseous envelope, and accretion rate $\dot{M} = 10^{-6}$ $M_{\oplus}$ yr^{-1}, $M_{crit} = 12$ $M_{\oplus}$ and the corresponding envelope mass is comparable, about 8 $M_{\oplus}$. The tidal radius for Jupiter at this point is about 2×10^{12} cm. The value of M_{crit} does depend to some extent on parameters: if the grain opacity is reduced, M_{crit} is reduced (see Fig. 4); if the mean molecular weight is increased, M_{crit} is reduced; and if $\dot{M}$ is increased (by a factor 10) then M_{crit} is increased (by a factor 2). Of considerable interest is the fact that M_{crit} is insensitive to the outer boundary condition. Thus, if the assumed position of the model in the solar nebula is varied, the value of M_{crit} remains about the same as long as $\dot{M}$ is held constant. The effects of the various parameters are discussed by Stevenson (1982*b*) who constructs a simple analytic model that approximately reproduces Mizuno's results. The analytical model shows that as long as the outer parts of the protoplanet are radiative, as they are in Mizuno's case, the temperature and pressure as a function of radius are independent of ρ_o and T_o once sufficient depth below the surface has been reached. This so-called radiative-zero solution has been described by Schwarzschild (1958). Stevenson is also able to show, using the stellar structure equations, that the critical core mass is only weakly dependent on the assumed value of the outer radius. It has been assumed that rapid dynamical collapse occurs once the critical mass has been reached, to a radius of a few times 10^{10} cm. However, later results, to be described below, indicate that the instability is not a dynamical one.

These results of Mizuno are significant in that (1) they predict a critical core mass which is comparable to the core masses deduced from observations of the giant planets, and (2) they predict that the core masses of all the giant planets should be similar, in agreement with observational data. They do not, however, explain the dissimilarity of the envelope masses, because at the critical point the envelope mass is comparable to the core mass. In the case of Jupiter and Saturn, additional material from the solar nebula must be accreted to build up the envelope mass; the time scale and nature of the subsequent accretion for these planets, up to their present masses, has been discussed by Safronov and Ruskol (1982). Factors which they include are the structure and angular momentum of the surrounding solar nebula and the time for material of the nebula to diffuse into the tidal radius of the protoplanet. The accretion is terminated in their model when the nebular gas is dissipated from the solar system. An alternative mechanism for termination of accretion has, however, been proposed by D. N. C. Lin and J. Papaloizou (1979*a*), which would operate even if the accretion time scale for the protoplanet were shorter than the nebular lifetime. Once the protoplanet reaches a critical mass, proportional to the inverse of the local Reynolds number in the nebula, its tidal influence on the surrounding gas opens up a gap in the nebula and prevents further material from entering the region around the protoplanet. The critical mass is, in fact, estimated to be comparable to Jupiter's mass at Jupiter's position in the nebula. These considerations, however, do not appear to apply to the case of Uranus and Neptune, whose core accretion times are longer than the lifetime of the solar nebula, whose envelope masses are too small to be explained by the critical mass argument, and whose masses are probably too small to tidally truncate the nebula.

The first time-dependent evolutionary calculations under the core-accretion gas-capture hypothesis have been performed by Bodenheimer and Pollack (1984). These calculations involve the complete solution to Eqs. (1)–(4), including the time-dependent terms in the energy equation but assuming hydrostatic equilibrium. The luminosity supplied by accreting planetesimals is included through the ϵ term in the energy equation, and this energy source is assumed to be concentrated near the bottom of the envelope. Detailed tables of the opacity and of the equation of state, including nonideal effects, are used (Bodenheimer et al. 1980*a*). At the outer boundary the assumed values of T_o and ρ_o are applied at $R_{out} = R_A$ or R_T, whichever is smaller. Mass is added to the core at a fixed rate $\dot{M}$; thus, with a fixed core density (3 g cm^{-3}), its radius increases slowly with time. At the outer boundary sufficient mass is added to the envelope at every time step so that the outer radius is in fact equal to R_{out}, under the assumption that mass transport times in the solar nebula are always short enough so that this condition can be satisfied.

To provide a comparison with the Mizuno model, the evolutionary calculation was made with similar parameters, that is, with $T_o = 100$ K, $\rho_o = 1.2 \times 10^{-10}$ g cm^{-3}, and with $R_{out} = R_T$, which was evaluated at the Jupiter position. The equation of state and opacity are somewhat but not fundamentally different from those used by Mizuno. The evolution started with a core mass of 0.1 $M_{\oplus}$ and continued for 1.66×10^7 yr, at which time the core had built up to a mass of 16.6 $M_{\oplus}$. The evolution in the (log ρ_c, log T_c) plane is shown in Fig. 5, where it is compared with that for a giant gaseous protoplanet of one Jovian mass. The central position here refers to the core edge. The luminosity is mainly provided by accretion, and it increases from 10^{-8} to 10^{-5} $L_{\odot}$ during the evolution. At a core mass of 16 $M_{\oplus}$, the structure of the model is quite similar to that of the Mizuno model at the critical core mass, with an inner convection zone extending out to 2×10^{10} cm at which point the temperature is 2.1×10^3 K.

The evolution continues beyond this point without undergoing dynamical collapse. However, a noticeable change occurs, in that the envelope begins to contract rapidly as more mass is added. Gravitational energy begins to dominate over accretion energy. The contraction is quasi-static, however, with no noticeable departure from hydrostatic equilibrium. In fact, there is no mechanism available to cause hydrodynamic collapse; the dissociation zone of molecular hydrogen, in a model just beyond the point where the critical core mass is reached, includes only 9% of the total mass, which is insufficient to result in dynamical instability. A large portion of the inner envelope has contracted to densities on the order of 0.1 g cm^{-3}, where the equation of state has

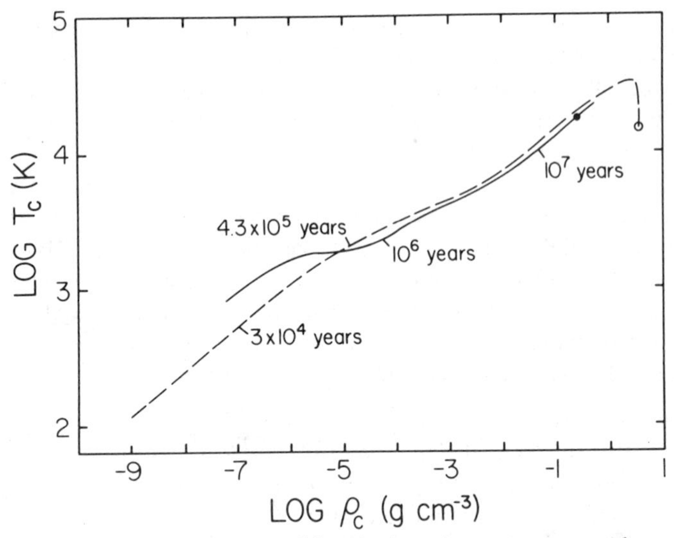

Fig. 5. The evolution of the central conditions in the rock-core gaseous envelope model of Bodenheimer and Pollack (1984) is indicated by the solid line. The outer boundary condition is similar to that used by Mizuno (1980). For comparison purposes, the corresponding evolution of a Jovian-mass homogeneous gaseous protoplanet (Bodenheimer et al. 1980*a*) with "isolated" surface boundary conditions is given by the dashed line. The core-envelope model has a final mass of about 60 $M_\oplus$. The point where the critical mass is reached is indicated by a solid dot.

become nonideal and $\Gamma_1 > 4/3$. Further increase in the envelope mass occurs rapidly, in part because the mass accretion time is determined by the contraction time, and in part because R_T itself is increasing rapidly because of the increasing total mass. Thus a runaway increase in envelope mass occurs, and on this time scale of 5×10^4 yr, the core mass hardly changes. The phenomenon of a "critical core mass" is found to be not so much an effect of a dynamical instability but rather a point where the envelope accretion time scale becomes drastically shorter than the core accretion time scale.

The evolution was continued until the envelope mass reached 44 $M_\oplus$. At this point the luminosity was 1×10^{-5} $L_\odot$ most of which (95%) was being supplied by gravitational contraction rather than accretion. At this point, core and envelope accretion were terminated and the model was allowed to contract at constant mass, still with constant values of ρ_o and T_o at the boundary. This contraction occurred very rapidly at first, taking 1600 yr for the radius to decrease from 2.7×10^{12} cm to 5×10^{11} cm at a luminosity of about 2×10^{-5} $L_\odot$. The rate of contraction then slowed down considerably, with a radius of 5.6×10^{10} cm reached after a further time interval of 7.2×10^4 yr, after which time the luminosity had declined to 1.6×10^{-6} $L_\odot$. The evolution time then continued to lengthen, and further contraction to 2.6×10^{10} cm took 6×10^7 yr. At this ending point the model had clearly entered the final phase of contraction, with $\rho_c = 0.53$ g cm^{-3}, $T_c = 2.26 \times 10^4$ K, and a fully convective structure. Mizuno (personal communication, 1983) has made a similar calculation for a constant core mass slightly larger than his "critical" 12 $M_\oplus$ and for a constant envelope mass of 8 $M_\oplus$. He also finds contraction without dynamical instability to a size on the order of present giant planetary radii.

A more extensive set of evolutionary calculations has been performed by Bodenheimer and Pollack (1984), which takes into account the accretion radius R_A and which considers the effects of the various parameters such as position in the solar nebula, boundary temperature and density, and accretion rate. The results of these calculations will appear elsewhere, but the following general conclusions can be reached.

1. When the "critical" core mass is reached, the models are dynamically stable, but rapid gravitational contraction occurs. As the envelope continues to accrete mass, it is possible but not certain that in particular cases dynamic collapse occurs at a later stage.
2. If mass flow into the tidal or accretion radius continues beyond the point where the critical core mass is reached, the envelope mass increases much more rapidly than the core mass.
3. The critical core mass, which depends on parameters, falls into the range of 10 to 40 $M_\oplus$. However, note that core masses this large require the use of the full interstellar grain opacity.
4. The critical core mass is determined by the point at which the gravitational influence of the core becomes large enough to overcome nebular thermal and tidal effects and to attract a comparable mass of gas.

IV. CONCLUSIONS

The main issues to be considered in calculations of the evolution of the giant planets are as follows:

1. What is the formation mechanism?
2. How can the similarity of the core masses and the wide range of envelope masses be explained?
3. How is the formation time scale related to that for the inner planets?
4. How is the formation process related to the properties of the solar nebula?

With regard to questions (1) and (2), neither of the two major hypotheses can explain the low-mass envelopes of Uranus and Neptune. In the case of Jupiter and Saturn, the observations require the presence of a high-density core, which both theories can provide. However, the theory of gravitational instability runs into some difficulty in forming the core, because it requires a large-mass envelope, which later must be removed, in order to provide sufficient material. There is also a problem of precipitating heavy materials to the center, because in the presence of convective motions they must be insoluble in the gas in order to precipitate out. In fact, materials such as iron and silicates are reported to be soluble in the deeper layers of the protoplanetary envelope, where metallic hydrogen has formed (Stevenson 1982*b*). These questions require further investigation, as does the question of the solubility of helium in metallic hydrogen during the late phases of the evolution of Saturn. The core-envelope model provides a natural explanation for the similarity of the core masses, but even for the planets Jupiter and Saturn the processes which determine the size of the final envelope remain to be worked out.

With regard to the formation time scale, the theory of gravitational condensation clearly has the advantage, because core accretion times up to 10 $M_\oplus$ are estimated at 10^8 yr for Jupiter, much longer than the presumed lifetime of the solar nebula and inconsistent with the available evidence that the terrestrial planets formed after the giant planets. Two suggestions have been made recently which indicate the possibility of drastically reduced core formation times. Stevenson (1984) discusses the formation of a dense adiabatic H_2O-H_2 shell around the core and the consequent reduction of the critical core mass and its formation time scale. D. N. C. Lin (1983) has given several arguments to indicate that the core formation time at Jupiter can be reduced to the order of 10^6 yr if the presence of the surrounding gas is taken into account in the process of accretion of solid material. A still unresolved problem, however, is the long accretion time for Uranus and Neptune. One of the important items for future work is the examination of the possibility that the core-accretion, gas-capture model can form giant planets fast.

Finally, the question of the relation between the properties of the solar nebula and the formation and evolution of the giant planets requires extensive further investigation. Only very simple boundary conditions representing average conditions in the solar nebula have been employed in calculations of giant-planet evolution. In the earlier stages, the two-dimensional structure of the nebula as well as its evolution must be taken into account in the specification of boundary conditions. Various nebular models should be considered; the properties of the giant planets could possibly thereby provide some clues as to the nature of the solar nebula. The question of mass transport of nebular material into the zones of gravitational influence of the protoplanets should be considered in more detail. Rotating models under the core-envelope hypothesis should be constructed, including considerations of angular momentum transport from orbital motion in the nebula into the spin of the protoplanet. The evolution of the giant planets not only is influenced by the solar nebula but may represent a closely analogous process to the evolution of the nebula itself, only on a smaller scale. Further investigations of the similarities and differences in the evolution of these systems could well provide further insight on the basic question of the origin of the solar system.

Acknowledgment. This work was supported in part by grants from the National Science Foundation.

THE ORIGIN OF COMETS: IMPLICATIONS FOR PLANETARY FORMATION

PAUL R. WEISSMAN
Jet Propulsion Laboratory

Theories of cometary origin fall into two groups: primordial theories in which the comets formed concurrently with the Sun and planets and have been stored in the Oort cloud over the history of the solar system, and episodic theories in which comets either formed in distant interstellar clouds and were subsequently captured by the solar system or comets formed as a result of catastrophic, recent, and possibly repeated events in the planetary region. The primordial theories have, in general, achieved far wider acceptance than the episodic theories. However, there is still considerable debate as to just where in the primordial solar nebula the comets formed. Candidate formation sites are the Uranus-Neptune zone, the outer primitive solar accretion disk, subfragments of the solar nebula in orbit about it, or neighboring nebular fragments from other protostars forming in the same star cluster as the Sun. The present dynamical evidence is not sufficient to discriminate between these different formation sites. Because all the formation sites provide for cometary accretion at relatively low temperatures, it is expected that comets preserve a valuable cosmochemical record of conditions in the primordial solar nebula, or possibly even the pre-solar interstellar cloud. But because the overlap between the cometary and planetary formation zones is small at best, and nonexistent at worst, it is not clear that comets can yield a great deal of information on accretionary processes in the primordial solar nebula. The possible existence of a massive inner Oort cloud is also discussed.

In studying the origin of the Sun and planets it is desirable to know the composition of the primordial solar nebula out of which they formed. For this reason much of the past two decades of planetary exploration has been concerned with a search for the most pristine samples of original solar nebula material. That search has moved, in turn, from the major planets to their satellite systems to the smaller bodies in the solar system, i.e., meteors, asteroids, and comets.

Meteorite samples have already provided a wealth of information about the composition and processes in the primordial nebula. But even the most pristine meteorites, the carbonaceous chondrites, are greatly depleted relative to solar abundances in the more volatile elements. Because of their location relatively close to the Sun, the asteroids can similarly be expected not to contain a significant fraction of volatiles. However, Whipple's (1950) icy-conglomerate model of the cometary nucleus predicts a roughly 50–50 mixture of ice and dust, mostly water ice but also CO_2, NH_3, CH_4, and other volatiles either freely mixed or bound as clathrates in the water ice lattice.

The study of comets is thus important for the information it can provide on the formation of the planetary system. That is true if, in fact, the comets did form concurrently with the Sun and planets. As will be discussed below (Sec. III), some researchers believe that comets may have formed in interstellar clouds and were subsequently captured by the solar system, or that comets may be the result of relatively recent processes within the planetary system. Though these theories do not all enjoy wide support, they do need to be considered.

TABLE I
Theories of Cometary Origin

Primordial Theories	
Oort (1950); G. P. Kuiper (1951*a*); Safronov (1969)	Icy planetesimals ejected from the planetary region, primarily the Uranus-Neptune zone.
Cameron (1973*a*); Donn (1976); L. Biermann and K. W. Michel (1978)	Formation *in situ* in distant subfragments of the solar nebula or related star-forming clouds.
Cameron (1978*b*)	Formation in outer region of protosolar accretion disk, and ejection to the Oort cloud due to solar and nebula mass loss.
Episodic Theories	
Lyttleton (1948)	Formation in solar wake due to passage through interstellar clouds.
Vsekhsvyatskii (1967)	Eruption from giant planets and/or satellites.
McCrea (1975)	Formation in compressed interstellar clouds at galactic spiral-arm shocks.
Van Flandern (1978)	Exploded planet in the asteroid belt.
Clube and Napier (1982)	Repeated Oort cloud disruption and recapture from giant molecular clouds.

In general then, we can group the theories of cometary origin into one of two categories: primordial, in which the origin of comets is associated with the origin of the Sun and planets, and episodic, in which comets are formed independently of the planets and usually much more recently. The major theories and their principal proponents are summarized in Table I. In general, the primordial origin theories all recognize the existence of the Oort cloud, while differing over where the comets formed and how they were brought to their current orbits. The episodic origin theories are in less agreement and several discount Oort's interpretation of the inverse semimajor axes $1/a_o$ distribution as proof of a massive cloud of comets surrounding the solar system. All the origin theories discussed accept the Whipple (1950) icy-congolmerate model for the cometary nucleus, with the exception of Lyttleton (1948) who prefers his own "sandbank" model. O'Dell (1973) suggested a gravitationally bound "rubble pile" model for cometary nuclei, similar to some current asteroid models, but the idea has received little subsequent attention.

The later sections of this chapter will examine both the primordial (Sec. II) and episodic (Sec. III) theories of cometary origin. But first, to better prepare the reader, a brief description of the Oort cloud based on observational evidence from long-period comets and dynamical modeling through computer simulation techniques will be presented.

I. THE OORT CLOUD

The present-day picture of the cometary nucleus is Whipple's (1950) icy-conglomerate model, an irregularly shaped, heterogeneous mixture of ices and dust, several km in diameter. A conceptual illustration of a cometary nucleus and some of the sublimation-related processes which have been suggested on the nucleus surface is shown in Fig. 1. As the nucleus approaches the Sun in a highly eccentric orbit the ices (primarily water ice) sublimate and the evolving gas carries off ice and dust particles to form a neutral atmosphere or coma around the nucleus. Gas molecules are ionized by charge exchange with the solar wind and are carried off to form plasma tails, called Type I tails, while dust particles blown back by solar radiation pressure form dust tails, called Type II tails. Some basic attributes of cometary nuclei, orbits, and observations are given in Tables II, III and IV. A more complete summary of cometary properties is given by Marsden and Roemer (1982).

Most comets enter the planetary region on long-period, near-parabolic orbits. When corrected for planetary perturbations, it is found that the energy of the cometary orbits, expressed by the inverse of the orbits' semimajor axes $1/a_o$, tends to cluster at very small, gravitationally bound values. The distribution of original inverse semimajor axes for 190 long-period comets, as found by Marsden et al. (1978), is shown in Fig. 2. These orbits are weakly bound to the Sun and extend out roughly to interstellar distances. Marsden et al. showed that the most distant comets had mean aphelion distances of about

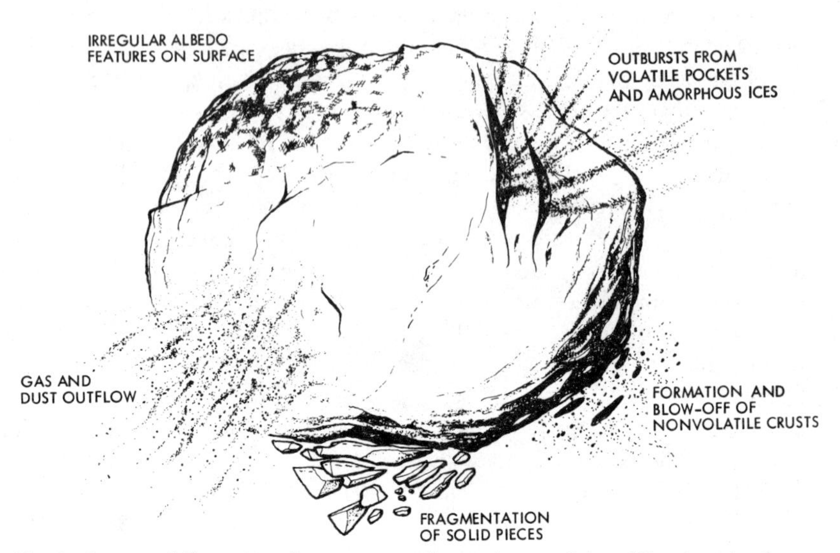

Fig. 1. Conceptual illustration of a cometary nucleus and some of the sublimation-related processes suggested as acting on the nucleus surface.

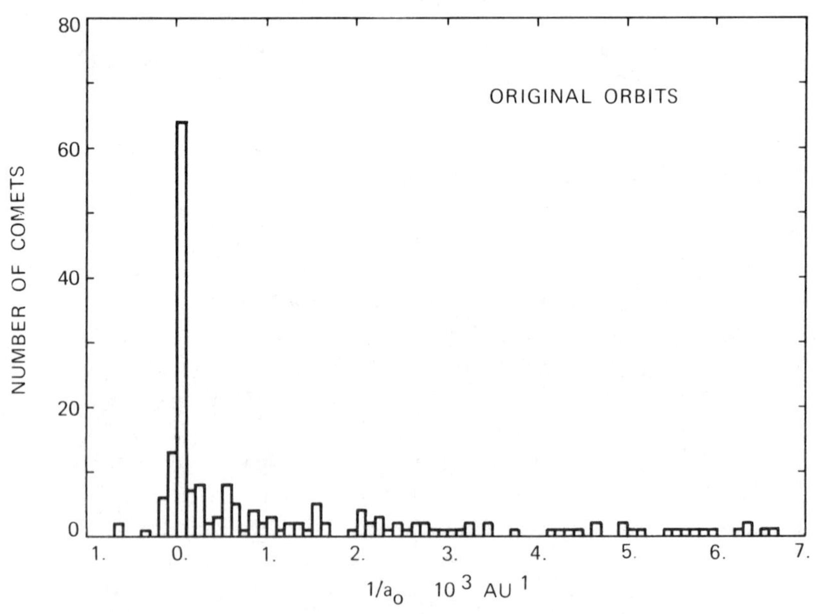

Fig. 2. Distribution of original inverse semimajor axes for 190 observed long-period comets as found by Marsden et al. (1978). The large spike of comets at the left represents new comets from the Oort cloud.

TABLE II
Basic Data on Cometary Nuclei

Diameter	1–10 km
Density	1.0–1.4 g cm^{-3}
Mass	10^{15}–10^{19} g
Ice:dust ratio	0.5–2.0 (by mass)
Rotation period	10 hr–5 days
Composition	H_2O, CH_4, CO_2, NH_3, HCN Silicate and absorbing (carbon?) dust
Mass loss per orbit	0.1–1.0%

TABLE III
Distribution of Cometary Orbit Types

Comets	Semimajor Axis (AU)	Period (yr)
"New" (from Oort cloud)	$10^4 < a < 5 \times 10^4$	$10^6 < p < 10^7$
Long-period	$a > 35$	$p > 200$
(randomly oriented over celestial sphere)		
Short-period	$a < 35$	$p < 200$
"Jupiter family"	$3 < a < 8$	$5 < p < 20$
mostly direct, low inclination		
(<30°) orbits		

TABLE IV
Statistics of Comet Orbits and Observations

	Long-period	Short-period
Average returns	5	100–1000
Total number observed	600	125
Total returns observed	600	540
Observed per year (1972–1982 time base)	4.6	9.7

4.3×10^4 AU, or about 0.22 parsecs. A small number of apparently hyperbolic original orbits in Fig. 2 are presumed to be due to errors in the orbit determination.

Oort (1950) interpreted the distribution of $1/a_o$ for long-period comets as being indicative of a huge spherical cloud of comets at these large distances, slowly feeding comets into the planetary region as a result of random perturbations from passing stars. The low, flat distribution of comets to the right of the spike in Fig. 2 are comets perturbed inward to smaller semimajor axes by planetary perturbations on previous perihelion passages. To account for the observed flux of long-period comets, Oort estimated that there were 1.8×10^{11} comets in the cloud with a total mass of 0.01 to 0.1 Earth masses. Oort theorized that the cloud was formed from small bodies ejected from the asteroid belt at the time of solar system formation.

The idea of Oort's cloud of comets surrounding the solar system and gravitationally bound to it has received wide acceptance over the past thirty years. Dynamical studies by Kendall (1961), Whipple (1962), Weissman (1979), and Fernandez (1980*a*) have shown that the cloud hypothesis does predict the observed distribution of cometary orbits. Once new comets enter the planetary region from the cloud, their motion tends to be dominated by planetary perturbations, primarily by Jupiter and Saturn. The planetary perturbations cause the comets to diffuse in $1/a$, either being dynamically ejected from the solar system or captured to shorter period orbits. Other factors such as random disruption, collisions with the Sun, planets, and asteroids, and loss of volatiles through sublimation also contribute to the removal of comets from the solar system.

The effect of stellar perturbations on the orbits of comets in the Oort cloud and the subsequent dynamical evolution of the cloud has been studied by a number of authors. Faintich (1971), Yabushita (1972), Rickman (1976), and Weissman (1980) found the total stellar perturbation on comets in the cloud over the history of the solar system, expressed as the sum of many small impulsive velocity perturbations, was between 110 and 150 m s^{-1}. Equating these estimates to the escape velocity sets a dynamical limit on the Oort cloud of about 10^5 AU. The fact that this is about twice the mean aphelion distance found by Marsden et al. (1978) suggests that there may be other perturbers in addition to passing stars not included in the calculations. Recent work by Clube and Napier (1982) suggests that those additional perturbers are likely the giant molecular clouds which populate the galaxy.

The dynamical evolution of cometary orbits in the Oort cloud has been studied by Weissman (1982) and Fernandez (1982), both using Monte Carlo simulation techniques. These authors have investigated the primordial theories of cometary origin listed in Table I: formation of comets as icy planetesimals in the Uranus-Neptune zone, 15–30 AU from the Sun; formation in the outer regions of the primitive solar accretion disk, 10^2 to 10^3 AU from the Sun; or formation in subfragments of the primordial solar nebula in distant orbits at $\geq 10^4$ AU. They found that planetary and stellar perturbations over the history of the solar system would randomize the distribution of orbits in the Oort cloud and make it impossible to discern the original cometary orbits from study of currently observed orbits.

In addition, both found that the original Oort cloud had been severely depleted over the history of the solar system by the combined planetary and stellar perturbations. For comet origin in the Uranus-Neptune zone they found that only 10 to 16% of the original cloud population was still bound to the Sun at the current time. For comets originating in subfragments of the primordial solar nebula, the surviving fraction at present is 50 to 70% of the original population.

Comets are lost from the Oort cloud by a variety of dynamical mechanisms. Comets which reenter the planetary region are either dynamically ejected on hyperbolic orbits or captured to short-period orbits as a result of perturbations by Jupiter and Saturn. Stars passing through the Oort cloud will dynamically eject all comets within a particular distance of their trajectory, typically about 500 AU. More gentle, distant stellar perturbations will pump up the semimajor axes of cometary orbits to the point where they exceed the Sun's sphere of influence and will slowly diffuse away. Comets surviving these various end-states constitute the present-day Oort cloud.

The relative fraction of comets in each of these end-states (planetary, ejected, stellar loss, and survivor as defined above) as a function of initial perihelion distance as found by Weissman (1982) is shown in Fig. 3. The number of directly ejected comets is too small to be shown; Weissman's model emphasized distant stellar perturbations and not the more violent but far

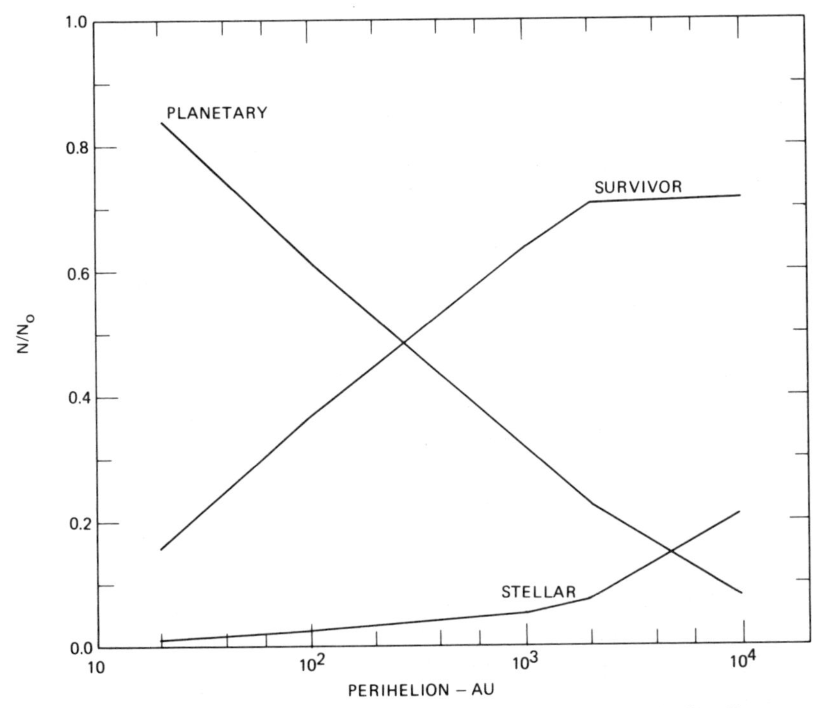

Fig. 3. Dynamical end-states for Oort cloud comets as a function of initial perihelion distance as found by Weissman (1982). The fraction of the Oort cloud comets that survives increases with increasing perihelion distance. The number of comets directly ejected from the Oort cloud by stellar perturbations was too small to show on this diagram. For the definition of end-states, see text.

less frequent close encounters which would eject comets. Weissman (1980) estimated that about 10% of the comets in the Oort cloud would be ejected by close stellar encounters.

Weissman (1983*a*), using an improved version of his dynamical simulation model, found the present population of comets in the Oort cloud to be between 1.8 and 2.2 $\times$ 10^{12} comets, independent of the assumed formation site for the comets. The population of the cloud is found by comparing the relative number of hypothetical comets entering the planetary region in the computer simulation with the estimated observed flux of long-period comets. On the other hand, Fernandez (1982) found a probable population of 6 $\times$ 10^{10} comets for an assumed origin in the Uranus-Neptune zone, and 3 $\times$ 10^{12} comets for formation in orbiting nebula subfragments. Two factors contribute to these differences. First, Fernandez used an observed flux of comets through the planetary system roughly an order of magnitude less than that used by Weissman. Fernandez only considered relatively bright comets for which observational statistics were reasonably complete, while Weissman used the flux of both bright and faint long-period comets corrected for observational selection effects as found by Everhart (1967).

The second effect involves the different initial orbits assumed by the two authors for comets formed in distant nebular fragments. Weissman assumed a relatively eccentric initial orbit with perihelion of 10^4 AU and aphelion of 4 $\times$ 10^4 AU. Fernandez assumed circular orbits at 10^4, 2.5 $\times$ 10^4, and 5 $\times$ 10^4 AU. Such circular initial orbits are weakly bound to the solar system and are more likely to escape before their perihelia can be evolved into the planetary region, a change which requires a near total loss of angular momentum. Fragments of the primordial solar nebula in which comets might have formed would be expected to be in highly eccentric orbits; if the initial angular momentum was high enough to give circular orbits, then the nebula would never collapse.

The total mass of comets in the Oort cloud is estimated by Weissman to be 2.5 to 3.0 Earth masses, regardless of where the comets formed; Fernandez estimated 2.6 or 130 Earth masses for comets formed in the Uranus-Neptune zone or in nebula subfragments, respectively. Weissman (1983*b*) estimated the mass of the average cometary nucleus to be 7.3 $\times$ 10^{15} g; Fernandez (1982) found 2.6 $\times$ 10^{17} g. The difference is due, again, to whether or not one includes the faint comets in the calculation. A secondary factor is the assumed albedo of the surfaces of observed cometary nuclei: Weissman used a figure of 0.6; Fernandez used 0.4. Lower albedos result in greater estimates of the nucleus mass. Current thinking is that both these albedos are high; typical current estimates for the albedo of dirty ice surfaces range from 0.1 to 0.3 (R. N. Clark and P. G. Lucy 1983).

If one adopts a revised value of 0.3 for the average surface albedo of cometary nuclei then Weissman's estimate of the mass of the current Oort cloud increases to between 7.0 and 8.5 Earth masses. Assuming the depletion factors given above, the original mass of the Oort cloud could have been as great as 42–50 Earth masses for a Uranus-Neptune zone origin, or 10–12 Earth masses for origin in distant fragments of the solar nebula.

The cometary mass estimates which go into the above calculations are highly uncertain. Most mass estimates are based on measurements of the brightness of cometary nuclei at large solar distances where water ice sublimation is believed to be negligible. But the large solar distance is not in itself proof that there is not some residual, unresolved coma still present around the nuclei, contaminating the measurements and making nuclei appear brighter than they actually are. Beyond that, there is no direct data on nucleus surface albedo, as discussed above. Albedo is also expected to vary with cometary age as sublimation from repeated perihelion passages leaves behind an increasingly thick layer of nonvolatile materials on the nucleus surface. The mass estimates given in this chapter should be considered uncertain by about one order of magnitude.

Hills (1981) showed that stars passing through the Oort cloud at distances $< 2 \times 10^4$ AU from the Sun could cause sudden showers of comets into the planetary region. These shower comets would come from a massive inner Oort cloud where the typical distant stellar perturbations had relatively little effect on the orbits. Hills suggested that this inner cloud would have formed from material in the outer protosolar nebula and that it might contain up to 10^2 times the number of comets in the "active" Oort cloud beyond 2×10^4 AU from the Sun.

Whitmire and Jackson (1984) and M. Davis et al. (1984) have suggested that an unseen solar companion with a distant, eccentric orbit passing through the Oort cloud could induce periodic comet showers and thus help explain possible periodicities in the terrestrial extinction record. Though an intriguing idea, there are many problems with this hypothesis. The additional flux of cometary showers would raise the cratering rates on the Earth and Moon by a factor of 5 to 18; yet, there is no evidence for such a high cratering rate in the recent past. The period of the alleged companion star, 28×10^6 yr, implies an aphelion distance near the limits of the Sun's sphere of influence; the stability of such an orbit is highly questionable. Lastly, to have escaped detection, the mass and luminosity of the companion must be fairly low, < 0.08 solar masses at the very least, and no evidence for such an object has been found.

Recent studies of the perturbation of the Oort cloud by giant molecular clouds (Clube and Napier 1982) have suggested that the Oort cloud can not survive over the history of the solar system, and that it must be repeatedly replenished. Suggested sources for replenishment are capture of new comets from the interstellar medium, or pumping up of orbits from an unseen inner Oort cloud. The details of the Clube and Napier hypothesis will be discussed in Sec. III on episodic theories of cometary origin.

There is no evidence for any observed comets having an interstellar origin. Although some comets shown in Fig. 2 have slightly hyperbolic orbits, these are likely due to errors in the orbit determinations. For example, nongravitational forces resulting from jetting of volatiles on the sunward surfaces of cometary nuclei act to make the orbits appear more hyperbolic. The most eccentric original orbit observed is 1.00064 for comet 1955V, corresponding to an original hyperbolic excess velocity of only 0.8 km s^{-1}. Typical interstellar comets would be expected to have excess velocities of at least 20 km s^{-1} and orbital eccentricities of ≥ 1.45. Estimates have been made that perhaps one in a thousand observed long-period comets would be of interstellar origin. At present there are well-determined orbits for about 600 long-period comets (Table IV).

II. THEORIES OF PRIMORDIAL COMETARY ORIGIN

In his original paper on the subject Oort (1950) suggested that comets were material dynamically ejected from the asteroid belt due to gravitational interactions with Jupiter and other planets. G. P. Kuiper (1951*a*) pointed out that ice would not be a major constituent of bodies formed that close to the Sun and suggested that the formation zone be moved farther from the Sun to coincide with the major planets. Safronov (1972*a*) showed that close encounters with Jupiter and Saturn tended to eject material from the solar system on hyperbolic orbits, rather than to the distant elliptical orbits characteristic of the Oort cloud. He showed that far greater efficiency could be obtained by assuming that the majority of comets were icy planetesimals ejected from the Uranus-Neptune zone. This has become the most widely accepted and widely studied theory of cometary origin.

The present picture of planetary formation in the outer solar nebula is reviewed in R. Greenberg et al. (1984); it can be summarized as follows. Solid condensates in the rotating solar nebula settle towards the equatorial plane of the nebula forming an accretion disk. When the density is high enough the disk breaks into uniformly sized clumps, due to gravitational instabilities as described by Goldreich and Ward (1973). These clumps accrete to form planetesimals on the order of 3 to 5 km in diameter in the region of the terrestrial planets, and on the order of 100 to 300 km in the outer planets zone. Inelastic collisions between planetesimals begin building larger bodies. Once a sufficiently large embryo planet begins, it tends to sweep up quickly all the remaining planetesimals in its orbital zone. In the latter stages of this sweeping, the primary process changes from accretion to ejection as the random velocities of the planetesimals are pumped up by encounters with the largest bodies, causing the gravitational capture cross section of the protoplanet to shrink.

One immediate problem with the scenario above is the size of the protocomets. The planetesimals created by Goldreich-Ward instabilities in the Uranus-Neptune zone are roughly 20 to 100 times larger than the typical sizes currently estimated for cometary nuclei. This implies that the presently observed comet population is collisional debris from impacting planetesimals. This problem is discussed further below.

Safronov (1969) studied the conditions for accretion in different regions of the solar nebula. For the Uranus-Neptune zone he found accretionary times were on the order of 10^{11} yr, far longer than the 4.5×10^9 yr age of the solar system. To get around this he assumed that the initial surface density of planetesimals in the zone was enough to total 10 times the current mass of Uranus and Neptune. According to Safronov (1972*a*) most of this material was hyperbolically ejected from the solar system but about 1 to 2% of it found its way to distant elliptical orbits to form the Oort cloud.

However, the ejection of so much mass from the Uranus-Neptune zone had to have extracted a great deal of angular momentum from the growing protoplanets, causing them to spiral in towards the Sun. This would have put the actual formation zone farther out in the system, on the order of ≥ 100 AU, further worsening the problem of accretionary times because of the longer orbital period and requiring yet a higher density of planetesimals to speed it up. In fact, it will be shown below that this did not happen.

R. Greenberg et al. (1984) have attempted to resolve some of these problems. They created dynamical models of accretion in the Uranus-Neptune zone and experimented with different initial distributions of planetesimals. Starting with 100-km-sized planetesimals they found that they could never get sufficient numbers of small bodies through collisional breakup to account for the currently estimated number of comets in the Oort cloud. To solve this problem they suggested that planetesimal formation in the accretion disk may have started before all the nebula condensates had settled to the equatorial plane. They envisioned clumping of large grains in the nebula causing them to settle to the disk more rapidly than small particles. In this manner the accretion disk could reach the critical density for onset of gravitational instability when $< 1\%$ of the nebula condensates had settled to the central plane. According to R. Greenberg et al. the resulting planetesimals would be on the order of 6 to 8 km in diameter.

Inserting these smaller planetesimals into their dynamical model, Greenberg et al. found that they could grow Neptune to its present size in a reasonable time, on the order of 10^6 yr (though the actual integrations were terminated when the planetary embryo had reached 10% of its current size), while also generating a size distribution and number of small bodies close to that for the Oort cloud comets. Further modeling led Greenberg and his colleagues to the conclusion that the only way to get the current size distribution for cometary nuclei was to start with planetesimals with the same distribution.

The shape of the size distribution, shown in Fig. 4, is particularly interesting since it is not a uniform power-law size distribution, but rather has a knee in the curve at about 8 km diameter; collisional breakup produces lots of smaller protocomets but accretionary processes produce few large bodies. Everhart (1967) found that the intrinsic distribution of cometary brightness for

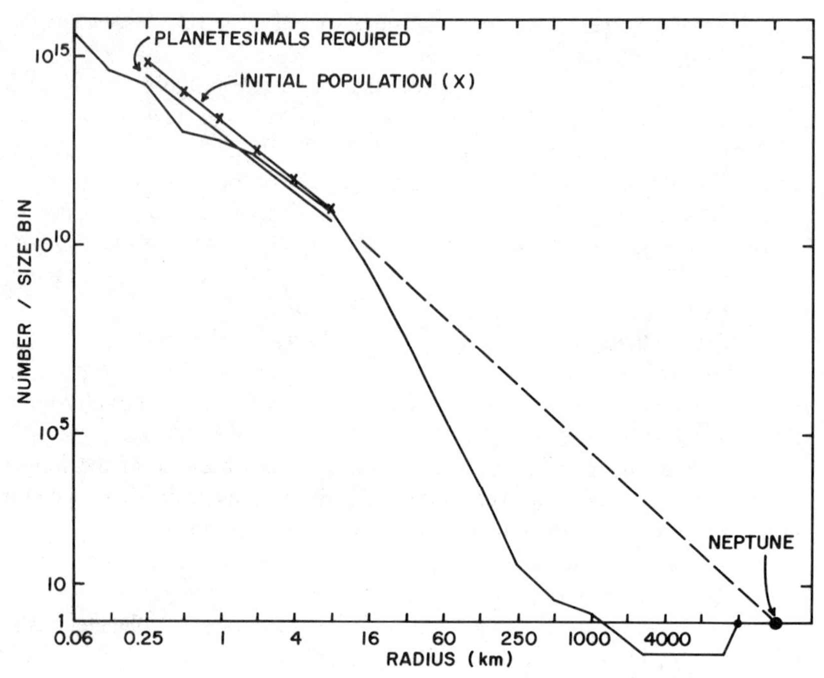

Fig. 4. Size distribution of planetesimals in the Uranus-Neptune zone after 1.4×10^5 yr as modeled by R. Greenberg et al. (1984). The initial planetesimal's size distribution is chosen to match the size distribution of observed comets.

long-period comets also had a knee in the curve, at an absolute magnitude H_{10} of about 5.0. Everhart's derived brightness distribution is shown in Fig. 5. Weissman (1983*b*) converted Everhart's curve to a mass distribution and found that the knee occurs for a mass of 10^{17} g or a diameter of about 5.8 km. But Weissman used an albedo for the surfaces of the cometary nuclei of 0.6, so adjusting his mass estimates to a more reasonable value for dirty ice, say 0.3, gives a diameter for the comets at the knee in the curve of 8.2 km, in good agreement with Greenberg et al. The mass of such comets would be 2.9×10^{17} g.

Fernandez and Ip (1981,1983) investigated the problem of ejecting material from the different outer planet zones. They found that Neptune was the most efficient in ejecting material to the Oort cloud rather than to escape, 72% of all ejecta winding up in the cloud. The corresponding figures are 57%, 14%, and 3% for Uranus, Saturn, and Jupiter, respectively. These efficiencies are considerably greater that the values of 2%, 1.2%, 0.5%, and 0.2% estimated by Safronov (1972*a*) for Neptune, Uranus, Saturn and Jupiter respectively; the reasons for this disagreement are not immediately obvious. Safronov did define a more narrow Oort cloud zone and correcting for this difference would increase his capture efficiencies by a factor of 2.7. Also,

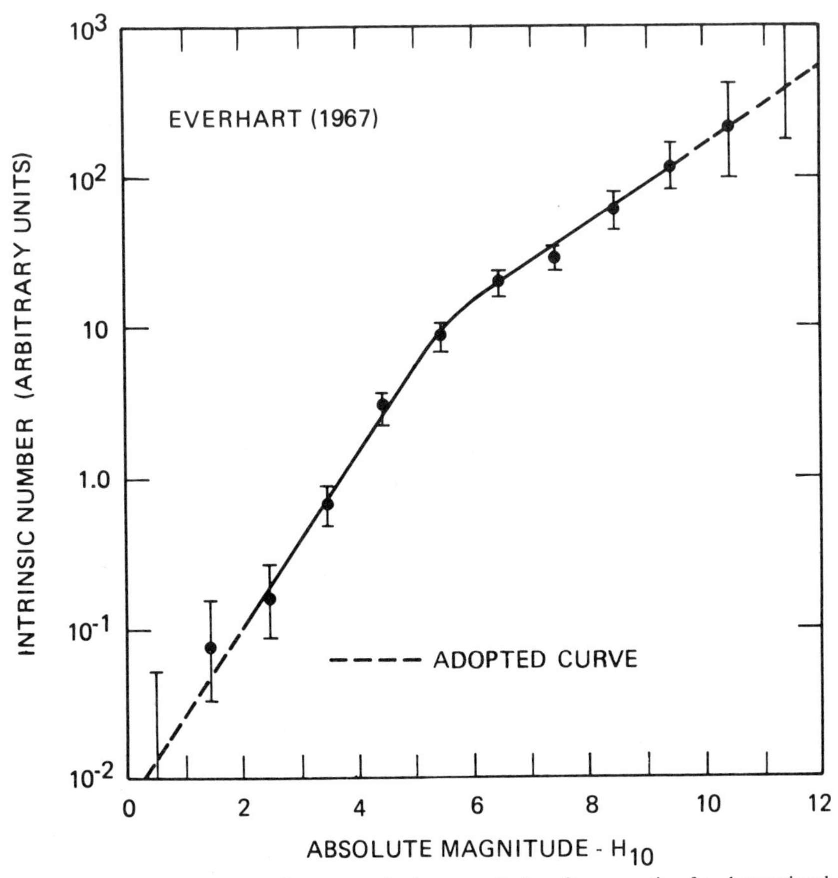

Fig. 5. Intrinsic distribution of cometary absolute magnitudes after correction for observational selection effects, as found by Everhart (1967). The knee in the curve at $H_{10} = 5$ corresponds to a nucleus mass of 2.9×10^{17} g and a diameter of 8.2 km.

Fernandez and Ip considered capture of Uranus-Neptune zone planetesimals to short-period comet orbits as an end-state, whereas it is really only a transient state; most of those objects would be ejected by Jupiter in $\leq 10^6$ yr. Other differences can likely be traced to the different methods used: an analytical approach by Safronov, and a Monte Carlo integration by Fernandez and Ip.

Wetherill (1977) showed that the Uranus-Neptune zone planetesimals would provide the major source of impactors in the region of the terrestrial planets during the first 1.5×10^9 yr of the solar system's history. After that, the Oort cloud would be the primary source. At present, Fernandez and Ip calculate that $< 1\%$ of the comets entering the terrestrial planet region are from the Uranus-Neptune zone.

More recently, Fernandez and Ip (1984) have considered the exchange of angular momentum between the outer planets and the planetesimals in their zones. They found that Uranus and Neptune did initially move inward as they lost angular momenta to the planetesimals ejected or placed in the Oort cloud. But they also sent a substantial flux of protocomets inward to Saturn and Jupiter crossing orbits, as previously pointed out by Safronov (1972*a*) and Wetherill (1977). The larger gravity fields of the inner giant planets ejected the comets from the system before Uranus or Neptune could. The net effect was that Uranus and Neptune both moved back outward, to final semimajor axes somewhat beyond their initial starting points. Saturn also experienced a net movement outward. Jupiter spiraled inward due to the loss of angular momentum but the change in semimajor axis was < 0.2 AU due to the planet's large mass. The evolution of the planets' semimajor axes with time is shown in Fig. 6.

One problem which should be noted with Fernandez and Ip's work is that they always start their integrations with a proto-Uranus or proto-Neptune in the accretion zone which is already 0.1 or 0.2 times the final mass of the planet to be created. Such a large initial protoplanet easily leads to runaway accretion and planetesimal ejection. What remains unclear is what is the time scale for creation of the initial planetary embryo, or why are there not multiple large embryos growing at the same time? However, if one merges Fernandez and Ip's model with the initial planetesimal accretion calculations of Greenberg et al., then a somewhat consistent scenario is achieved.

Shoemaker and Wolfe (1984) have studied the evolution of planetesimals in the Uranus-Neptune zone as a source for the late heavy bombardment on the Jovian and Saturnian satellites. As with Fernandez and Ip, they found a residual population of about 0.5% of the initial planetesimals in the zone surviving to the present time. Their initial model only considered planetary perturbations; subsequently, they added stellar perturbations to see if some of the longer-period planetesimal orbits, though not large enough to be in the Oort cloud, might still be affected by random passing stars. This resulted in the number of surviving planetesimals increasing to 9% of the original population, because the stellar perturbations raised the perihelia of the planetesimal orbits out of the Uranus-Neptune zone, greatly increasing their dynamical lifetimes against ejection.

Shoemaker and Wolfe estimate that the total mass of residual planetesimals is on the order of 100 to 200 Earth masses. This massive inner cloud can then serve as a replenishment source for the larger (in linear dimension though not in mass) Oort cloud which is actively being stirred by passing stars. As the outer shell of the Oort cloud is stripped away the inner cloud orbits are pumped up to fill the void. At present, Shoemaker and Wolfe estimate that the active Oort cloud population is being kept in a steady state, but eventually the inner cloud will begin to be depleted and the Oort cloud population will decrease.

A problem with all these models of planetesimal evolution is that they assume that these events occurred in a gas-free medium. Models of the forma-

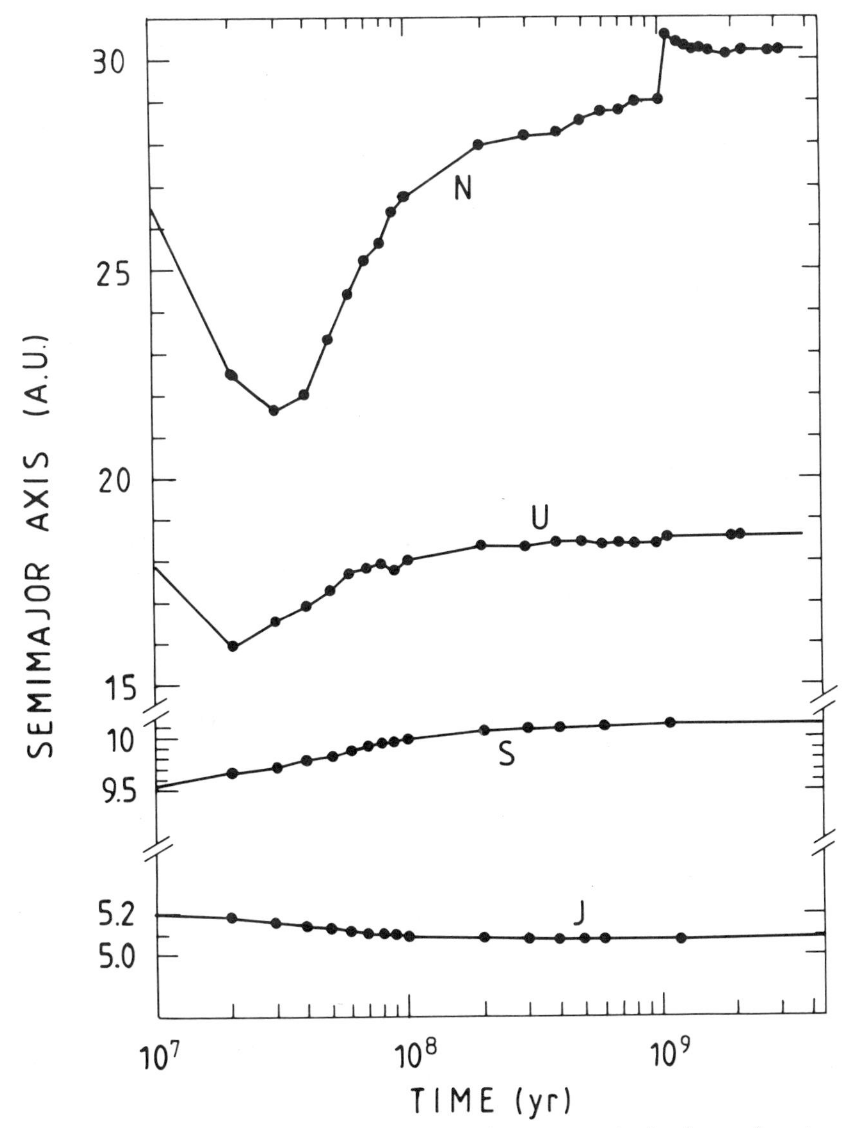

Fig. 6. Variation of the semimajor axes of the Jovian planets as a result of exchange of angular momentum with planetesimals as found by Fernandez and Ip (1984). The initial semimajor axes are 5.203, 9.54, 20, and 30 AU for Jupiter, Saturn, Uranus, and Neptune, respectively.

tion of the outer planets such as the gaseous protoplanet or core-accretion models both assume far more complex and involved processes than the simple gravitational interactions of a swarm of small planetesimals described here (see Bodenheimer's chapter in this book). Further modeling should include the effects of the gas on accretionary time scales, radiative transfer and energy dissipation in the nebula, collisional mechanisms between planetesimals, etc.

The presumed pristine nature of comets and their apparent compositional similarities with the interstellar material has suggested to J. M. Greenberg (1982*b*) and others that comets were formed outside the planetary region. Cameron (1973*a*) suggested that comets formed in subfragments of the primordial solar nebula. Though not large enough to form stars, these fragments, with masses of 0.1 to 0.2 solar masses, would still contract to form accretion disks and then small bodies through Goldreich-Ward gravitational instabilities in the disk. In Cameron's model several such nebula fragments in distant elliptical orbits about the Sun would be sufficient to produce the Oort cloud in its current position, with no need for a dynamical mechanism to place comets in such distant orbits.

One variation on this idea was provided by L. Biermann and K. W. Michel (1978) who suggested that comets formed in a similar manner but at about 10^4 AU from the Sun on the edge of the protosolar accretion disk. The orbits of the protocomets were enlarged to their current values due to subsequent mass loss from the Sun during its presumed T Tauri stage and the dispersal of the remnant solar nebula. Another variation on this idea by Donn (1976) suggests that the Sun formed in a cluster of protostars and that comets formed in neighboring cloud fragments moving at small velocities relative to the protosun. Some fraction of these comets were captured prior to dispersal of the star cluster.

The important similarity in these three ideas is that the comets would form at very low temperatures, on the order of 10 to 30 K, from largely unprocessed interstellar grains and gases. Modest compositional differences between comets and other solar system bodies could be explained by inhomogeneities in the makeup of the various cloud fragments and/or neighbors. Again, by forming the comets *in situ* there is no need for a dynamical mechanism to transport them to Oort-cloud distances.

The major problem with all these theories is the lack of detailed development of cometary accretion models for distant nebular fragments. The physics of planetary accretion disks around forming protostars, based in part on observations of accretion disks around young stellar objects, is a well-developed field. But the formation of disks and gravitational instabilities in low-mass nebular fragments where star formation may or may not be taking place is a far more poorly studied and understood phenomenon.

Hills (1982) proposed a mechanism where small clouds of absorbing dust in the distant nebula (10^3–10^4 AU from the protosun) felt a net external radiation pressure forcing them together, to the point where self-gravitation took over and the cloud agglomerated into a protocometary nucleus. To produce nuclei with typical masses of 3×10^{15} g, Hills calculates that the initial dust cloud or "clump" radius had to be ~0.02 AU. This may provide one mechanism for growing protocomets without the need for an accretionary disk in the nebula and/or nebular fragments. A similar mechanism was proposed by Whipple and Lecar (1976) involving an early enhanced solar wind acting on the protosolar nebula.

There are some dynamical differences between the three proposed theories discussed here. Biermann and Michels' suggestion of forming the comets at 10^4 AU and then expanding the nebula due to solar and nebula mass loss is reasonable but the fraction and time scale for the mass loss must be within certain limits so as to pump up the orbits sufficiently, yet not lose an overwhelming fraction of the protocomets to interstellar space. Cameron's comet formation in distant nebula fragments is also reasonable; if the fragments are in highly elliptical orbits they will spend most of their time at aphelion, moving very slowly, and will not be greatly perturbed or lost due to the Sun's presumed T Tauri stage and nebula dispersal. Donn's idea of capture of comets from neighboring nebula fragments is a bit more troubled. As will be discussed in Sec. III, the solar system's cross section for cometary capture is extremely low except for hyperbolic encounter velocities of < 1 km s^{-1}. Random velocities in collapsing protostar clusters might be expected to be somewhat higher, on the order of 3 to 5 km s^{-1}. Still, the capture process might be aided by the presence of perturbing third bodies in the form of other protostars, or by energy dissipation processes such as gas drag in the nebulae.

Cameron (1978*b*) modified and improved his earlier solar nebula model and suggested that comets formed on the edge of a massive primitive solar accretion disk at distances of 10^2 to 10^3 AU from the Sun. Similar to the Biermann and Michel model, the cometary orbits were enlarged due to the dispersal of the massive primitive nebula. The nebula dispersal process must have occurred in a time span short relative to the orbital period of the protocomets, so as to change the orbits from near circular to near parabolic. At the same time, the nebular dispersal time must have been long relative to the orbital periods of the major planets so that they would slowly spiral out from the Sun, rather than being thrown into highly eccentric orbits.

One problem with Cameron's model is that the Sun-nebula combination must lose almost exactly half its mass to convert the circular protocometary orbits to near-parabolic. Dermott and Gold (1978) solved this problem by suggesting that a succession of mass loss events occurred. In this manner, it is possible to postulate a wide range of mass losses, each pumping up the protocometary orbits in a succession of small steps.

Again, the detailed development of comets forming in the primitive accretion disk at large solar distances has not been performed. Since the formation site in this case is much closer to the planetary region, it is not unreasonable to accept an accretion disk extending out to these distances. The limit on the dimensions of the observed planetary region seems to have been set by the dynamical accretion time for planetary-sized bodies, and not by any physical cut-off of the accretion disk.

Cameron's model also suggests the existence of a massive inner Oort cloud of comets whose orbits were not pumped up as much by the central mass loss. This inner reservoir of comets would be difficult to sense since the orbits have perihelia outside the planetary region and aphelia on the order of $\leq 10^4$ AU, too small to be significantly perturbed by random passing stars. As demonstrated below, there are increasing reasons to believe that such a massive inner Oort cloud does indeed exist.

III. THEORIES OF EPISODIC COMETARY ORIGIN

Lyttleton (1948) hypothesized that comets were formed by gravitational focusing of interstellar material in the Sun's wake as the Sun and solar system passed through interstellar clouds. The focus point would be approximately 10 to 10^3 AU behind the Sun and random motions in the interstellar gas and dust, as well as planetary perturbations, would cause the orbit aphelion directions to be dispersed over a wide area of the sky. Lyttleton linked this theory of cometary origin to the sandbank model for comets in which there is no nucleus but rather a loosely bound cloud of dust particles with adsorbed volatiles.

The gravitational focusing hypothesis is unable to explain the $1/a_o$ distribution for the long-period comets since it predicts a cometary source much closer to the planetary region than shown in Fig. 1. There are asymmetries in the distribution of aphelion directions of cometary orbits on the sky (Bogart and Noerdlinger 1982) and they do appear to be associated with the solar antapex direction. However, the asymmetries are generally not very large and alternate explanations are possible. For example, L. Biermann et al. (1983) have proposed that an excess of comets orbiting in any direction might be the result of the perturbations by a recent star passage through the Oort cloud in that direction. Also, it is possible that the Sun's motion relative to the local group of stars introduces preferential perturbations which bring more comets in from the Oort cloud along the Sun's apex and antapex directions, or conversely, the greatest perturbations may be perpendicular to the Sun's motion and the cloud has been depleted in that direction over the history of the solar system. The problem is further complicated by observational selection effects such as a seasonal dependence of cometary discovery rates and the unequal number of northern and southern hemisphere observers.

Vsekhsvyatskii (1967) believed that comets were the result of eruptions from the giant planets, an idea originally proposed by Lagrange in 1814. Because short-period comets are almost always in Jupiter-crossing orbits, and because many are shown to have made a close encounter to Jupiter shortly before their discovery, Vsekhsvyatskii reasoned that Jupiter was indeed the source of these comets. This idea has received little support because of the problems associated with ejecting cometary-sized bodies intact from planets with such strong gravitational fields. The escape velocity from Jupiter is 57.5 km s^{-1}; for Saturn it is 35.4 km s^{-1}.

The discovery of active volcanism on Io, the innermost of the Galilean satellites of Jupiter, momentarily resurrected the eruption hypothesis because of the lower-escape velocities of the satellites, on the order of 2 to 5 km s^{-1}. However, even these lower ejection velocities seem beyond any reasonable physical explanation. In addition, ejected comets from satellites would still have to escape the central planet's gravity field. Fesenkov (1963) and Radzievskii (1981) showed that the observed distribution of cometary orbits is consistent with comets evolving from near-parabolic long-period orbits to elliptical short-period orbits, and not the reverse.

McCrea (1975) proposed that comets formed in interstellar clouds which had been compressed at galactic spiral-arm shocks. Icy interstellar grains would be drawn together by their mutual gravity in compressed regions with H_2 densities of 10^{10} to 10^{12} cm^{-3}. The solar system could be expected to pass through the spiral arms about once every 10^8 yr, or about 50 times in the history of the planetary system. During each of these passages the Sun might be expected to capture on the order of 10^8 comets in highly elliptical orbits, giving the appearance of an Oort cloud without actually requiring such a massive reservoir of comets.

The weakness in McCrea's theory, as in other interstellar capture hypotheses, is the low probability of capture of interstellar comets by the solar system. Valtonen and Innanen (1982) showed that the typical capture probability, given the Sun's random velocity of 20 km s^{-1} relative to the local group of stars, was on the order of 10^{-13} per comet. Only for encounter velocities on the order of ≤ 1 km s^{-1} did Valtonen and Innanen find any reasonable probability of capture. Valtonen (1983) showed that the existence of an unseen solar companion of 0.1 solar mass at 10^4 AU from the Sun could enhance the capture probability but not enough to capture the number of comets required by McCrea or other capture proponents. In any case, there is no evidence for such a solar companion.

Another, specific problem with McCrea's theory is the relatively high value, and very narrow range of densities required to form cometary-sized bodies. Such high densities are not generally observed in interstellar clouds, even among some of the denser globules. McCrea's conditions for cometary formation may be too restrictive to expect significant numbers of cometary nuclei actually to form.

Van Flandern (1978) proposed that the comets, asteroids, and meteorites all originated from a 70 Earth mass planet in the asteroid belt which exploded between 5 and 15 $\times$ 10^6 yr ago. He explained the apparent Oort cloud comets as fragments ejected from the explosion on very long-period ellipses making their first return to the planetary region at the present time. The problems with Van Flandern's idea are so extensive that it has received little serious consideration. It does not explain the observed distribution of cometary orbits. It cannot be reconciled with either the primordial ages or primitive, undifferentiated composition of many meteorite types. It cannot explain the distribution of asteroid orbits or the distribution of compositional types within the asteroid belt. It predicts a recent heavy bombardment of planetary surfaces for which no evidence can be found. Lastly, it suggests no mechanism for actually causing the hypothetical planet to explode.

Considerable attention has focused recently on a theory by Clube and Napier (1982) that the Oort cloud would be disrupted by encounters between the solar system and giant molecular clouds (GMCs), and that the solar system might subsequently capture a new cloud of interstellar comets from the encountered GMC. They calculate that there may have been 10 to 15 encounters with GMCs of mass $\geq 2 \times 10^5$ solar masses over the history of the solar system, each encounter capable of stripping away the Oort cloud beyond about 10^4 AU from the Sun. At the same time, they expect that the solar system will capture on the order of 10^{11} new comets formed in the GMC by processes similar to that proposed by McCrea. Clube and Napier associate these repeated disruption-capture events with periods of terrestrial bombardment and resulting catastrophes, such as the Cretacious-Tertiary extinction event 65×10^6 yr ago.

Bailey (1983*a*) has examined the dynamical arguments of Clube and Napier and found that they overestimated the magnitude of the GMC perturbations by a factor of between 2 and 10. If the reduction factor is only 2 then the GMCs may still pose a significant disruption threat to the Oort cloud; for a reduction factor of 10 the GMC perturbations are not catastrophic. At the heart of the question is the Sun's past dynamical history in the galaxy. The Sun's current random velocity relative to the local standard of rest at this radius in the galaxy is only 16 km s^{-1}, as compared with a typical rms value of 60 km s^{-1} for G-type stars. The normal means for accelerating stars up to these velocities is encounters with GMCs. The Sun's low velocity implies that either it has encountered very few GMCs in its history, or its current random velocity is anomalously low and has been much higher in the past. Statistically, the second explanation is the more probable.

Another problem with Clube and Napier's hypothesis is the same as that pointed out for McCrea's theory, namely the difficulty of capture of comets from interstellar clouds. Very low encounter velocities between the solar system and the GMCs, and some reasonable dispersion of cometary velocities within the GMCs, are required. Clube and Napier (1984) have attempted to show that such a combination of conditions is possible but require a density of comets in the GMC of about 0.1 AU^{-3}. The density of comets in the Oort cloud, assuming a population of 2×10^{12} comets and a spherical volume with radius of 6×10^4 AU, is 0.002 AU^{-3}.

Clube and Napier (1984) admit that the problem of the stripping away of the outer Oort cloud can be solved if there is a massive inner cloud, on the order of 10^{13} comets, available to replenish it. This is consistent with arguments by Bailey (1983*a*) that the observed orbits of long-period comets are best explained by a "centrally condensed" Oort cloud with much of the mass located closer to the Sun than the new comets typically observed. As noted previously and discussed further below, there is increasing reason to believe that such a massive inner cloud actually does exist.

IV. DISCUSSION

Support for the various primordial theories of cometary origin is generally divided between dynamicists who want to form the comets as close to the planetary region as possible, and cosmochemists who prefer to form comets in the most distant regions of the protosolar nebula. Dynamical arguments favoring the outer planetary zone are the higher expected accretion disk density and the shorter dynamical accretion times. Arguments favoring the outer nebula are the similarities in observed properties of cometary and interstellar dust, the identification of known interstellar molecules in comets, and the dynamical efficiency of forming comets *in situ* at Oort-cloud distances.

The similarities between cometary and interstellar dust as discussed by J. M. Greenberg (1982*b*) are interesting but not convincing. Both exhibit a 10 μm emission feature, but that is only an indicator of amorphous rather than crystalline surface structure. The same amorphous solids can be found in carbonaceous chondrites which appear to be the major surface material among the outer asteroid belt objects. The size range of interstellar dust particles is typically an order of magnitude smaller than cometary dust particles: interstellar dust is typically < 0.2 μm in diameter; the bulk of cometary dust particles are micron size or larger. Though cometary particles are often aggregates of smaller particles, this still requires some accretionary mechanism to form the much larger cometary particles. Lastly, the presence of apparently presolar grains in meteorites demonstrates that interstellar material may survive nebular processes relatively close to the protosun.

The same arguments may be applied to interstellar molecules in comets. All the primordial theories of cometary origin have comets forming far from the Sun in relatively cool regions of the solar nebula. The temperature distribution in the primitive nebula as a function of time as calculated by Cameron (1978*b*) is shown in Fig. 7. The maximum derived temperature in the Uranus zone is ~80 K; for Neptune it is ~60 K. Though these temperatures are certainly warmer than in cold molecular clouds where $T \sim 10$ K, they are not at all unusual for temperatures observed in some dense cloud cores. Thus, cometary material in the outer-planets zone may never be heated to temperatures that would dissociate interstellar molecules or melt amorphous solids into crystalline rocks.

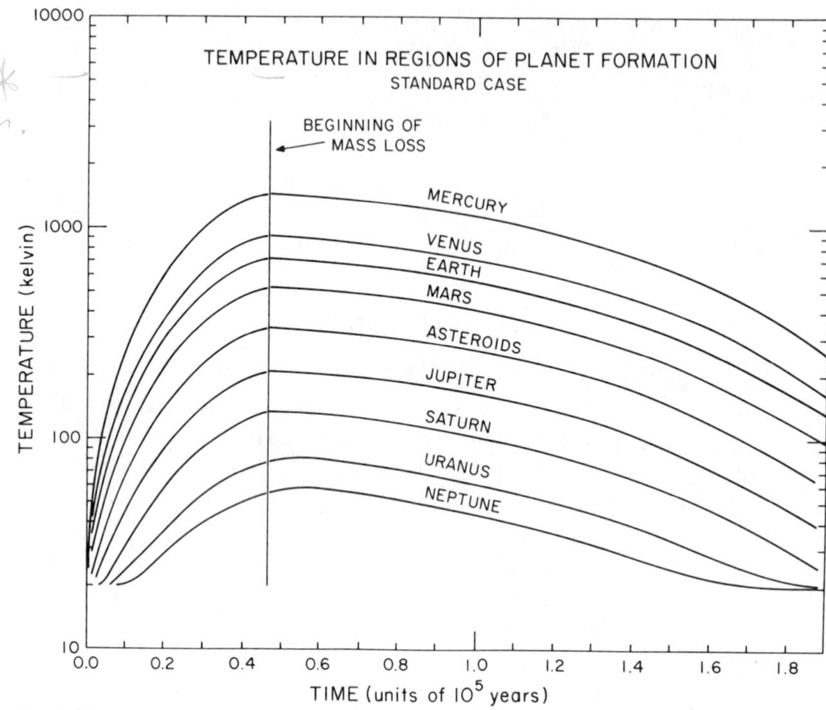

Fig. 7. Temperature in the regions of planet formation as a function of time, as found by Cameron (1978*b*). The regions each start out very cold, possibly as cold as the background radiation temperature in the cloud fragment, 20 K. Note the relatively low maximum temperatures of 80 K for Uranus and 60 K for Neptune.

Isotopic differences may be a clue as to how processed or unprocessed cometary materials may be. Unfortunately the available data is fairly limited. Vanýsek and Rahe (1978) have pointed out that the measured $^{12}C/^{13}C$ ratio in comets is on the order of ≥ 100, as compared with 89 for the Earth, approximately 75 for the local interstellar medium, and 160 for Jupiter. Though this is certainly not conclusive, it does seem to suggest some processing of cometary materials.

Essentially, it is most important to recognize that the entire solar system formed out of interstellar material and that outer solar system conditions likely remained very similar to those in interstellar clouds. Thus, it may be difficult to discriminate between materials formed at 20 to 30 AU in the primordial nebula, and those formed at 10^4 AU. Both are likely relatively unprocessed interstellar materials.

Dynamical arguments do not allow us to discriminate between the three primordial theories. The current Oort cloud has been so randomized by stellar perturbations that it is impossible to discern any dynamical record of how it formed. Uranus-Neptune zone planetesimals certainly seem capable of providing the mass of the Oort cloud within a factor of two. The low state of development of the theories invoking more distant formation sites is a weakness but does not allow us to discard them; further work on the details of these distant formation theories would be very desirable.

The possibility of a massive inner Oort cloud has already been mentioned several times. An inner Oort cloud could be the replenishment source if the outer cloud is stripped away by encounters with GMCs. It could be the source of Hills' comet showers and of the impacting flux on the outer planet satellites as suggested by Shoemaker and Wolfe. Cameron (1962,1978*b*) suggested, from solar nebular model considerations, the existence of a massive inner cloud. Whipple (1964*a*) and Bailey (1983*b*) have suggested that such a comet cloud or belt could explain the perturbations on Neptune's orbit. Whipple estimated the belt mass to be 10 to 20 Earth masses at 50 AU; Bailey suggested a far more massive inner Oort cloud, possibly as large as Hills' (1981) value of 10^{-2} solar masses. Bailey (1983*c*) suggested that this inner comet cloud may be observable by its infrared radiation and may be within the detection limits of the Infrared Astronomical Satellite (IRAS). Low et al. (1984) have reported that IRAS did, in fact, detect considerable structure in the infrared background which they interpreted as cold material in the outer solar system. Further analysis of IRAS results is required to better pin down the nature and distance of this cold material.

Fernandez (1980*b*) has shown that a comet belt beyond Neptune may be the source of the short-period comets. Hamid et al. (1968) investigated perturbations on known short-period comets with large semimajor axes and set an upper limit of 1.3 Earth masses for the total mass of comets in a possible belt at 40 or 50 AU. This result does not preclude a more massive comet belt farther out.

Chiron, the only known minor planet in the outer solar system, is likely a Uranus-Neptune planetesimal that has survived over the history of the solar system, possibly stored in the inner Oort cloud. Scholl (1979) has shown that Chiron is evolving inwards from the Uranus-Neptune zone and will likely be ejected eventually due to a close encounter with Saturn or Jupiter. The ever decreasing estimates of the mass of Pluto and its satellite Charon have brought the size of these objects down to values (1500 km radius for Pluto, 500 km for Charon) not much greater than the size of the largest asteroids. Thus, they too may be surviving Uranus-Neptune planetesimals, their dynamical lifetimes prolonged by the well-known 2:3 orbital resonance with Neptune (J. G. Williams and G. S. Benson 1971) that keeps them from making close encounters with that planet. The high-inclination, retrograde orbit of Neptune's large satellite Triton also suggests that it may be a captured planetesimal, though the rotation pole of Neptune is undetermined and may be similarly inclined, thus making Triton a regular satellite.

Weissman (1984) has suggested that the ring of material around the star Vega recently discovered by IRAS (Aumann et al. 1984) may be a cometary belt around that star. At a distance of 85 AU from Vega and with an observed temperature of 85 K, the shell or ring of particles is in a region where icy planetesimals may form. The long orbital period of ~550 yr at that distance would probably preclude planets having formed that far from Vega in its estimated 10^8 yr lifetime. Weissman estimates the mass of the Vega ring at 15 Earth masses based on the limited data publicly available from IRAS. Aumann (1984) reports that IRAS has found Vega-like dust shells around 6% of the stars in the solar neighborhood.

Wood (1979) suggested that asteroids from the inner solar system may also be placed in the Oort cloud due to Jupiter perturbations, though this would be a very inefficient means of supplying material to the cloud. Some fraction of these asteroids would evolve back into the planetary region, eventually to orbits similar to short-period comet orbits where they would serve as a source of meteorites striking the Earth. Wood estimated that this contamination of the Oort cloud by nonvolatile objects may be on the order of 10^{-4} the total cometary population. A problem with this scenario is how the nonvolatile asteroids would have reduced their aphelia down to values within the orbit of Jupiter, without nongravitational forces from jetting of surface ices.

If comets formed in the Uranus-Neptune zone then they may be able to yield considerable information on accretionary processes in the primitive solar nebula at that distance. This would be a key piece of data for extrapolating to regions of the nebula and accretion disk closer in toward the Sun. Comets formed in the Uranus-Neptune zone set a dynamical time limit for the clearing of planetesimals out of that region so as to explain the observed time scales for the late heavy bombardment on the Moon's surface. Unfortunately, the Uranus-Neptune zone is the region of the solar system which we presently know the least about. Hopefully the Voyager 2 spacecraft will correct that situation. It is expected to encounter Uranus in January, 1986 and Neptune in 1989.

If comets formed farther out in the primordial solar nebula then it is more difficult to relate them to planetary formation processes. The comets should preserve a cosmochemical record of conditions in the primordial nebula at large solar distances and that in itself is a very valuable piece of information. But any constraints the comets could place on planetary formation would be vague at best.

The coming spacecraft missions to Halley's comet in 1986 and the plans for a comet rendezvous mission in the early 1990's will go a long way towards resolving many of our questions about where comets formed. *In situ* measurements of cometary compositions and isotopic ratios will tell whether these objects are indeed unprocessed interstellar materials, or if they have gone through considerable differentiation in the primordial solar nebula. Detailed study of cometary dust will establish its relationship to interstellar dust particles. We can only await these results with patient anticipation.

Acknowledgments. The author is grateful to R. Greenberg, J. A. Fernandez, W. H. Ip, and E. Shoemaker for providing preprints or other results of their current work in progress, and for many useful technical discussions. The author also thanks G. Wetherill, M. Hanner, R. Carlson, A. Harris, and A. Graps for additional helpful discussions. This work was supported by the NASA Planetary Geophysics and Geochemistry Program and was performed at the Jet Propulsion Laboratory under contract with the National Aeronautics and Space Administration.

RINGS AND MOONS: CLUES TO UNDERSTANDING THE SOLAR NEBULA

JACK J. LISSAUER and JEFFREY N. CUZZI
NASA Ames Research Center

Planetary satellites, ranging in size from ring particles to the Galilean moons of Jupiter, constitute a small but diverse component of the solar system. In some respects, the satellite systems of the giant planets can be thought of as miniature solar systems. The chemical and physical properties of these systems can yield clues to the conditions under which they may have formed, and thus can constrain models of the circumplanetary nebulae and the solar nebula. Planetary rings contain a great deal of structure, most strikingly displayed in Voyager spacecraft images. Understanding the causes of this structure can yield important insights into the dynamics of the protoplanetary disk. Taken together, rings and moons offer clues as to the nature of the processes which led to the formation of the planets themselves.

Planetary ring and satellite systems constitute families of bodies with unique interrelationships. The contents and distribution of these systems provide important clues to the physical conditions at the place and time of their formation: the vicinity of newly formed gas giant planets. However, before we may interpret these clues, we must first understand the processes, both active and extinct, that have influenced the observed properties of these systems.

Many satellites are planetary bodies in all but name; Ganymede, the largest, is larger than Mercury; Titan, the second largest, has an atmosphere more massive than that of the Earth. Satellite systems record the bulk composition of nebular solids; however, most of this material is hidden in their interiors, and we must infer its makeup from remote measurements such as mass and radius. The surfaces of most of the satellites are undegraded by atmospheric weathering and record a rich and varied history. This record may date back, in many cases, to the final stages of accretion. However, it has not yet been possible to fully distinguish accretional history from the subsequent surficial evolution, which results from a combination of internal metamorphosis and external bombardment. In this chapter, we review the significant evolutionary processes and attempt to highlight open questions of relevance to planetary origin.

In planetary rings, the collective action of the ensemble often obscures the individuality of the particles. These collective effects, which are subtle and often counter-intuitive, have acted over time to shape and redistribute the ring material. One of the goals of ring studies is to understand how conceptually similar collective effects have acted to produce three quite dissimilar appearing ring systems. Indications are that at least Saturn's ring system is in a highly evolved state. Thus, evolutionary processes must be better understood before inferences may be drawn about local properties at the time of planet formation. In this chapter, we review current knowledge of ring properties and evolutionary processes, and note areas in need of attention.

The satellite systems of the giant planets mimic in many ways the planetary system of the Sun. It is generally believed that the circumplanetary disks, of which planetary ring/moon systems are the dregs, mimic the larger circumsolar particle disk in which the planets themselves formed. Thus, the circumstances in, and formation processes of rings and moons are surely of interest to the larger solar system problem. Furthermore, many processes of importance in the evolution of particle disks are currently active and on display in the three known ring systems. Here they are prevented from proceeding to completion by strong planetary tides which thwart the accumulation of ring/moon systems. Their dynamical interrelationships illustrate intriguing parallels between ring/moon systems and the solar system.

In Sec. I, we summarize the physical characteristics and orbital properties of the known planetary satellites, including rings. Section II contains a discussion of several aspects of the physics of flat, orbiting disk systems. We concentrate on the topics of accretion and gravitational torques, two areas in which planetary rings may give us clues to processes once operative in the protoplanetary disk. In Sec. III, we concentrate on the origin and evolution of planetary satellites. We emphasize the properties of the moons that depend most heavily on the environment in which they accreted, and subsequent processes which may have blurred these clues to satellite origin. One of these processes is impact cratering, which provides a history of stray debris in the solar system. We conclude, in Sec. IV, with a discussion of the relationship between rings, small ringmoons and the larger moons of the giant planets, and with a comparison between the solar system and the satellite and ring systems of the giant planets.

TABLE I
Properties of Planetary Satellites[a]

Satellite	Orbital Radius (10^3 km)	Orbital Period (days)	Eccentricity[b] (degrees)	Inclination[b] (degrees)	Radius[c] (km)	Mass[c] (10^{23} g)	Density[d] (g cm^{-3})
Earth							
Moon	384.4	27.3217	0.05490	5.1	1,738	734.9	3.34
Mars							
M1 Phobos	9.378	0.319	0.015	1.02	14 11 9	(9.6×10^{-5})	(2.0)
M2 Deimos	23.46	1.262	0.00052	1.82	(8) 6 (5)	(2×10^{-5})	(1.9)
Jupiter							
Main Ring	123–129	0.28–0.30	0	0	—	—	—
J16 Metis	127.96	0.2948	0	0	(20)	—	—
J14 Adrastea	128.98	0.2983	0	0	13 10 8	—	—
J5 Amalthea	181.3	0.498	0.003	0.45	135 85 75	—	—
J15 Thebe	221.9	0.675	0.013	0.9	55 — 45	—	—
J1 Io	421.6	1.769	0.000	0.027	1,815	892	3.55
J2 Europa	670.9	3.551	0.000	0.468	1,569	487	3.04
J3 Ganymede	1070	7.155	0.001	0.183	2,631	1490	1.93
J4 Callisto	1880	16.689	0.007	0.253	2,400	1075	1.83
J13 Leda	11,110	240	0.147	26.7	(5)	—	—
J6 Himalia	11,470	251	0.158	27.6	(90)	—	—
J10 Lysithea	11,710	260	0.130	29.0	(10)	—	—
J7 Elara	11,740	260	0.207	24.8	(40)	—	—
J12 Ananke	20,700	617	0.17	147	(10)	—	—
J11 Carme	22,350	692	0.21	164	(15)	—	—
J8 Pasiphae	23,300	735	0.38	145	(20)	—	—
J9 Sinope	23,700	758	0.28	153	(15)	—	—
Saturn							
Main Rings	75–137	0.24–0.60	0	0	—	(0.3)	—
S17 Atlas	137.67	0.602	0.002	0.3	(19) — (13)	—	—
S16 1980S27	139.35	0.613	0.004	0.0	70 (50) (37)	—	—

S15 1980S26	141.70	0.629	0.004	0.1	(55) (42) (33)	—	—
S11 Epimetheus	151.42	0.694	0.009	0.3	(70) (57) (50)	—	—
S10 Janus	151.47	0.695	0.007	0.1	110 95 80	—	—
S1 Mimas	185.54	0.942	0.020	1.52	197	0.375[d]	1.18[d]
S2 Enceladus	238.04	1.370	0.004	0.02	251	(0.84)	(1.12)
S3 Tethys	294.67	1.888	0.000	1.09	530	(7.55)	(1.20)
S13 Telesto	294.67	1.888	—	—	(15) (12) (8)	—	—
S14 Calypso	294.67	1.888	—	—	— (12) (11)	—	—
S4 Dione	377.42	2.737	0.002	0.02	560	10.5	1.43
S12 1980S6	377.42	2.737	0.005	0.2	(18) — —	—	—
S5 Rhea	527.10	4.518	0.001	0.35	765	24.9	1.33
S6 Titan	1221.86	15.945	0.029	0.33	2,575	1346	1.88
S7 Hyperion	1481	21.277	0.104	0.4	175 117 100	—	—
S8 Iapetus	3561	79.331	0.028	7.52	730	18.8	1.16
S9 Phoebe	12,954	550.4	0.163	174	110	—	—
Uranus[e]							
Rings	41–50	0.25–0.34	—	—	—	—	—
U5 Miranda	130	1.413	0.0027	4.2	(200)	—	—
U1 Ariel	192	2.520	0.0034	0.3	665	(16)	(1.3)
U2 Umbriel	267	4.144	0.0050	0.4	555	(10)	(1.4)
U3 Titania	438	8.706	0.0022	0.14	800	(59)	(2.7)
U4 Oberon	586	13.46	0.0008	0.1	815	(60)	(2.6)
Neptune							
N1 Triton	354.3	5.877	<0.0005	158.5	(1750)	—	—
N2 Nereid	5510	365.21	0.75	27.6	(200)	—	—
Pluto							
P1 Charon	19.7	6.387	—	—	(500)	—	—

[a]Most of the data in this table are taken from Burns (1986).
[b]We have given free eccentricities and inclinations. Inclinations are given with reference to the location Laplace plane. Roughly, this means with reference to the planet's equator for near satellites and planet's orbit for distant satellites.
[c]Values given in parentheses are uncertain by > 10%. Parenthesized values of physical properties of the satellites of Uranus, Neptune, and Pluto have much higher uncertainties.
[d]Voyager results yield 20% higher values for Mimas' mass and density, in conflict with listed values from perturbations of Tethys (see Tyler et al. 1982).
[e]Most of the data presented for the Uranian satellites are from Veillet (1983).

I. DISTRIBUTION AND COMPOSITION OF SATELLITES AND RINGS

The satellite systems of the outer planets (see Table I) have often been likened to the planetary system itself. In each case (excepting Neptune), multiple secondaries orbit their primary, with most being in nearly circular, coplanar, prograde orbits having a certain regularity to their spacings. Unlike the planets, most satellite rotations have been synchronously locked by strong planetary tidal forces (see Secs. III.C and III.E); thus, their initial rotation states are unknown. Exterior to their families of regular satellites, outer planet satellite systems include members with highly irregular orbits; also, close in (and even intermingled with) their ring systems, families of small ringmoons are seen in prograde, coplanar and roughly circular orbits. Most of these small ringmoons have been discovered by spacecraft; in the case of Uranus, they have only been hypothesized as ring shepherds (see Sec. II). Confirmation of their existence by Voyager in 1986, although difficult, will be sought.

Attention has often been focused on the geometrical progression seen between the semimajor axes of successive secondaries in the solar and outer planet systems (D. C. Black 1971; Nieto 1972). The semimajor axes in all four regular secondary systems show a similar, nonrandom, progression (Fig. 1). The fundamental significance of this was probably first grasped by Kirkwood (Nieto 1972); the orbital spacing intervals Δr_n are roughly proportional to the distance from the primary

$$\Delta r_n \equiv r_{n+1} - \mathrm{r}_n \approx \mathrm{K}_r r_n \qquad (1)$$

with K_r ranging from ~0.3 (Saturn's satellites) to nearly 1.0 (the planets). This relationship is probably telling us something about sweep-up mechanisms in planetesimal disks. Many different processes could account for this distribution of orbits. For example, planetesimals in crossing orbits of eccentricity e will sweep out regions of width $\sim 2er$ (Vityazev et al. 1978). A problem with this approach is in maintaining large enough values of e ($\sim \mathrm{K}_r/2$) for accretion to continue (Wetherill, 1980*a*; see, however, Wetherill and Cox 1984). Alternatively, the regularities in spacing between successive satellites (and planets) can be expressed as relationships between periods (or orbital frequencies, Ω) instead of radii (Dermott 1973):

$$\Omega_{n+1} \approx K_\Omega \Omega_n \qquad (2)$$

with K_Ω varying from ~0.4 (the planets) to ~0.8 (Saturn's moons). Equations (1) and (2) are essentially equivalent for orbits in a Keplerian gravitational potential; however, Eq. (2) emphasizes the possibility that resonant interactions between protosatellites (protoplanets) may be responsible for the observed spacing.

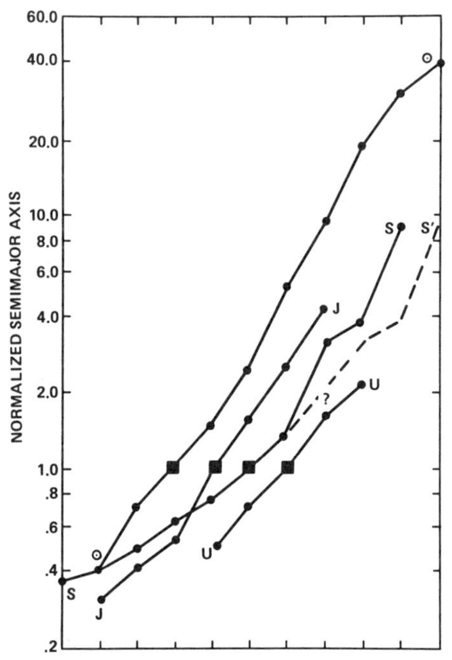

Fig. 1. The Titius-Bode-type relationship shown by the normalized orbital radii of the planets (curve denoted ⊙) and the regular satellite systems of Jupiter (J), Saturn (S), and Uranus (U). The choice of the normalization point (■) is arbitrary, and subsequent satellites are sequentially positioned. The asteroid belt is included in the curve for the planets. If one were to treat the Saturn system similarly (leaving a gap at the location denoted between Rhea and Titan, due to Titan's dominant mass in that system), the dashed curve S′ is obtained (see also D. C. Black 1971). These curves are considerably smoother and of a different slope than would result from a random selection of radii over the observed range, and probably reflect some physics of the accretion process.

In addition to the spacing regularities, other noticeable systematic trends are seen, such as the huge gap between the large regular satellites, which tend to reside between 3 and 30 planetary radii, and the smaller irregular satellites, which lie out at a few hundred planetary radii. A similar discrepancy is seen in the spectral appearance of these outer irregulars: they are much darker and more primitive (i.e., similar in appearance to undifferentiated meteorites) than the regular satellites. Most of the regular satellites apparently have icy surfaces; however, their bulk densities imply a significant rocky component, suggesting that they have undergone at least some differentiation (B. A. Smith et al. 1979a,b,1981,1982). These differences may tell us something important about the nature of protosatellite disks. The highly eccentric, often retrograde, orbits of the irregular satellites have led many to believe that they are captured objects, possibly losing energy at encounter due to gas drag in the outer regions of a primitive, greatly distended, gaseous protoplanet (See 1910; Pollack et al. 1979; T.-Y. Huang and K. A. Innanen 1983). The same calculations also demonstrate a rapid orbital decay following capture; orbits will circularize and decay into the planet on a time scale of tens of orbital periods (Pollack et al. 1979). The lack of a continuous inward progression of increasingly regular objects lying between 20 and 150 planetary radii indicates that, (a) the observed objects are only those captured very close to the end of the distended protoplanet era, and (b) the distended protoplanet era must have ceased abruptly, on a time scale shorter than 10 to 10^2 yr. This seems to suggest a rapid collapse phase, during which the protoplanet shrank from $\sim 10^3\ R_p$ to $\sim 10^2\ R_p$ (see chapters by Pollack and by Bodenheimer).

Furthermore, the limited radial extent of the families of regular satellites (tens of planetary radii; see Table I), and their coplanar, prograde, "Titius-Bode Law" radial progressions (Fig. 1) argue for formation from a flattened disk. However, the extent of this disk seems to have been no more than several tens of planetary radii. Again, this may tell us something useful about the planet/satellite accretion process. By way of contrast, the solar disk, that produced the planets, extended for $\sim 10^4$ solar radii. We note that although the Sun probably formed from a molecular cloud of $> 10^5$ times the Sun's final radius, the protoplanets (or protoenvelopes of the giant planet cores) were truncated by protosolar tidal influences at $\lesssim 10^3$ times their current radii; thus, the relatively small radii of the circumplanetary nebula which would have resulted from protoplanetary collapse, might be interpreted in terms of simple angular momentum scaling arguments. It is also of interest that regular satellite orbits lie well within the three-body regions of stability (Pendleton and Black 1983) and the analogous "Hill spheres" of the outer planets ($\sim 10^3$ planetary radii), which strengthens hypotheses of a nebular-related satellite origin. Other crude comparisons between the protodisks of the Sun and the giant planets are shown in Table II.

These aspects point the way for future study. The formation of a disk around a collapsing protoplanet may have been significantly different from the formation of the protoplanetary disk about the protosun, despite the fact that both resulted from retention of angular momentum (chapters by Cassen et al., by Bodenheimer and by Durisen and Tohline). Thus, the collapse must be investigated in more detail if we are to make the fullest use of the satellite systems of the giant planets as analogies to the solar system.

The Saturn system is replete with 10 to 30 km-sized chunks, several of which occupy stable Lagrangian positions along the orbits of Tethys and Dione (B. A. Smith et al. 1982; see Table I). There is a systematic trend toward lower mass among the regular satellites as one moves inwards in the regular satellite systems (Fig. 2), although closer satellite spacing suggests that the surface mass densities of circumplanetary nebulae increased towards the planets (Fig. 3). In the Saturn and Jupiter systems, in the limit as the rings are approached, there may be a continuous range of sizes between, say, Atlas or

TABLE II

Typical Minimum-Mass[a], Viscous[b], Nebula Characteristics

	Circumplanetary	Circumsolar
Radius	$\sim 30\ R_p \approx 2 \times 10^{11}$ cm	$\sim 10^4\ R_\odot \approx 5 \times 10^{14}$ cm $\approx 2 \times 10^{-4}$ pc
Surface density (gas & dust)	10^4–10^5 g cm^{-2}	10^2–10^3 g cm^{-2}
Luminosity[c]	$\sim 10^{-3}\ L_\odot$	$\sim 10^{-2}\ L_\odot$
Viscous dissipation time[d]	$\sim 10^2$ yr	$\sim 10^5$ yr
Main condensable volatiles	H_2O, NH_3, CH_4 (?)	H_2O, N_2, CO (?)
Vertical scale height	$H \sim R_p \sim 10^{10}$ cm (at 15 R_p)	$H \sim 10^{13}$ cm (at 10 AU)

[a] Assuming solar composition and that all of the refractory elements were retained in the satellites (planets). If the circumplanetary disks were enhanced in refractories, as are the giant planets, their masses may have been less than this "minimum."

[b] Assuming the viscosity to be $\nu = \alpha\, cH$ with $\alpha \approx 0.1$.

[c] $L = \pi\, r^2 \sigma \nu \left(r \frac{d\Omega}{dr} \right)^2$.

[d] $t \sim r^2/\nu$.

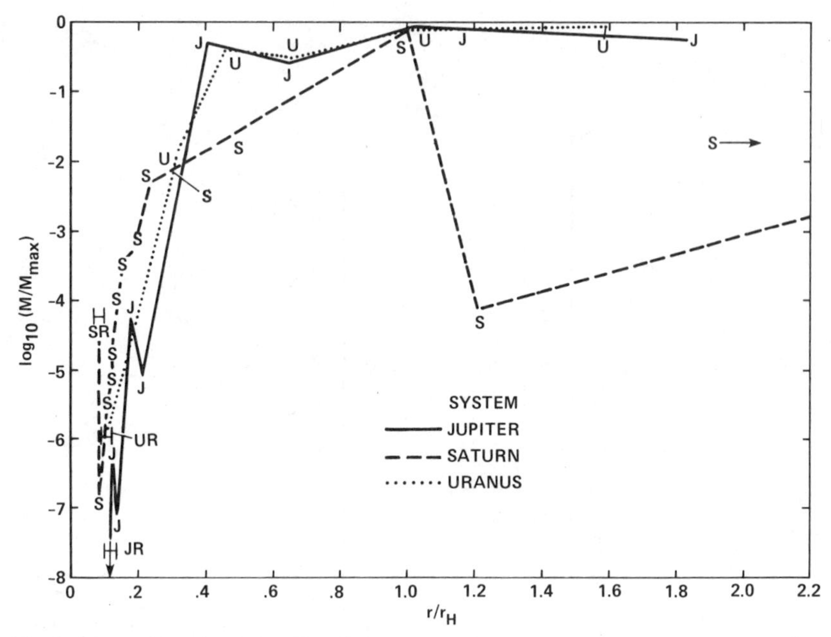

Fig. 2. Masses M and orbital radii r of planetary satellites for the Jupiter, Saturn, and Uranus systems. The masses are given as fractions of the most massive satellite in the system, and the radii as fractions of the radius r_H where an orbiting body would have had the same specific angular momentum as the satellite system (see, e.g., D. C. Black [1971] for a similar treatment). Note that satellite masses decrease as the primary is approached.

Metis (~20 km radius), several 10 km Encke division moonlets (Cuzzi and Scargle 1984), a variety of km-sized F ring clumps, and possibly even a large number of 100-m-sized objects within the rings (see, e.g., Sec. III.D of Cuzzi et al. 1984). The origin and redistribution of these small chunks of material are unsolved problems of potentially great importance.

The systematic increase in density of the Galilean satellites with decreasing distance from Jupiter (Table I) has been used in attempts to constrain the thermal environment in which they formed, and thus, models of proto-Jupiter and its circumplanetary disk (Pollack and Reynolds 1974; Lunine and Stevenson 1982). Recent data from Saturn (B. A. Smith et al. 1982; Table I) are difficult to interpret in the same manner; we return to this subject and discuss other aspects of circumplanetary nebulae in Sec. III.A.

The properties of planetary rings are at least as diverse as those of moons (Pollack and Cuzzi 1981). Saturn's main ring system, the only one known before 1977, is very broad, comparable to Saturn's radius in width, and has an average optical depth near unity. The particles which make up Saturn's rings are made of dirty water ice; their diameters range from microns to tens of meters, with an area-weighted average of about one meter (Cuzzi et al. 1984).

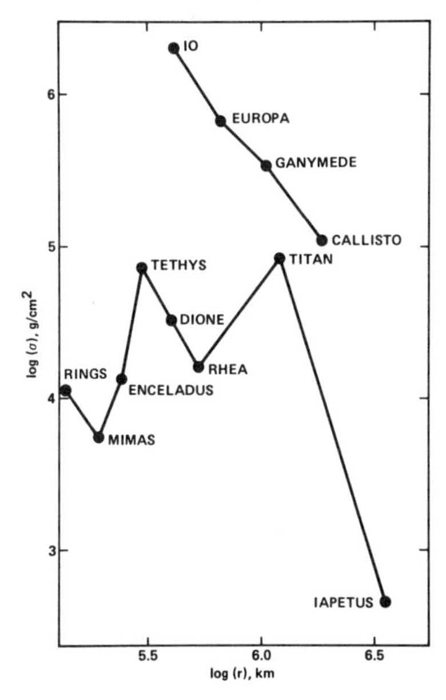

Fig. 3. Surface densities of minimum-mass circumplanetary nebulae of Jupiter and Saturn. These densities were calculated by spreading the mass of each satellite over an annulus extending halfway to each of its neighbors, and augmenting this observed solid component with sufficient gas to produce nebulae of solar composition. If the circumplanetary nebulae were enriched in condensable material relative to a solar mixture, densities would have been less than this minimum-mass model. Figure adapted from Pollack and Consolmagno (1984).

A great deal of structure exists in Saturn's rings; we will discuss some of the mechanisms which produce this structure in Sec. II. The Uranian rings consist of narrow bands of very dark particles, presumably centimeters in size (Elliot and Nicholson 1984). The visible Jovian ring has an extremely low optical depth (10^{-4}). It contains micro-sized particles, probably composed of silicates, which have very short lifetimes before being destroyed by erosive sputtering and other mechanisms; presumably they are constantly being replenished by erosion of larger bodies ("mooms"; J. A. Burns et al. 1984). Saturn also possesses narrow ringlets similar to the Uranian rings both within and outside its main ring system. Broad, tenuous rings of small particles (the E and G rings) exist outside Saturn's main ring system.

The above discussion has focused on the extensive ring and satellite systems of Jupiter, Saturn, and Uranus. The systematic, regular characteristics of these satellite systems argue for formation in circumplanetary disks, analogous to the protoplanetary disk out of which the planets are believed to have accreted (chapter by Lin and Papaloizou). The less extensive and less regular satellite systems of the terrestrial planets and Pluto are not suggestive of circumplanetary disks surrounding these solid bodies. According to many theories of planet formation (see, e.g., Safronov 1969; chapter by Safronov and Ruzmaikina; Wetherill 1980*a*), no gaseous circumplanetary disks ever existed around these bodies. The moons of Earth, Mars, Pluto and possibly some asteroids, may have formed by fission from the primary (Ringwood 1979; Durisen and Scott 1984), or may be captured bodies (Singer 1968,1971; Lambeck 1979; Yoder 1982; A. W. Harris and W. R. Ward 1982). Satellite formation may be much more difficult and unpredictable under such circumstances. Alternatively, evolutionary processes may have depleted formerly extensive satellite systems of terrestrial planets and other solid bodies (Sec. III.C).

Neptune's irregular satellite system (it has no known ring) may somehow be related to its low gas content, which may imply a surrounding nebular disk, if any, quite different from those of the other gas giants. Models of the ice giant planets (Podolak and Reynolds 1981) show that Neptune may have less than one half as much H_2 and He as are present in Uranus. Neptune's long accretion time scale (Wetherill 1980*a*; Fernandez and Ip 1981) also suggests that Neptune may have never been surrounded by a massive disk.

We shall continue to emphasize the regular satellite systems of Jupiter, Saturn and Uranus because they are more likely to have been formed in a manner analogous to the planetary system. However, it should be remembered that not all satellite systems are so regular, and that other satellite formation scenarios are possible.

II. DISK DYNAMICS

Saturn's rings, as the best observed flat rotating astrophysical system in the universe, can serve as a testing ground for many of the theories developed for that mythical object from which our planetary system supposedly formed—the protoplanetary disk. Both systems consist of interacting particles orbiting a large central mass in a centrifugally supported disk. In both cases, interactions are dominated by gravity and physical collisions (or pressure); large-scale electromagnetic interactions probably play a smaller role, although they are important in some solar nebula models (Hoyle 1960; Alfvén and Arrhenius 1976). Also, many aspects of the behavior of both systems can be treated as fluid phenomena.

There are, of course, many differences between Saturn's rings and the protoplanetary disk, the most obvious one being scale. Saturn's rings are a remarkably thin system; although they cover a radial extent of 10^{10} cm, their vertical thickness is only $\sim 10^3$ cm. The protoplanetary disk, in addition to being larger in radial extent, 10^{15} cm, had a much greater relative thickness, of order 10^{13} cm for the gas, and thus, pressure was more important. A much more flattened disk of condensed particles, present during at least one phase of most nebular models, would be more analogous to Saturn's rings in this respect. The dynamical ages of the two systems are also quite distinct. Whereas Saturn's rings have probably persisted for $\sim 10^{12}$ revolutions, the protoplanetary disk only lasted for 10^4 to 10^8 orbits, depending on which solar nebular model is assumed and on whether we are concerned with the region in which the terrestrial planets or the giant planets formed.

A. Roche's Limit

Other differences between planetary rings and the protoplanetary disk can also be illuminating. The planets were able to accrete in the solar nebula, or from its residue; however, the rings around Saturn and Uranus have been unable to accrete, or coalesce into single objects, due to their proximity to these planets. (The visible dust ring surrounding Jupiter and Saturn's E and G rings are composed of ephemeral particles, which have little mutual interaction before being destroyed by external influences, and thus are not significant for our discussion here; see J. A. Burns et al. [1984] for a useful review.) Tidal forces on a satellite vary as the inverse cube of its distance from the primary; thus, it is possible that if a moon wandered too close to a planet, it could be tidally disrupted. Roche (1847) was the first person to analyze this problem in detail. He showed that a perfectly fluid satellite (held together only by its self-gravity), in a circular orbit and synchronously rotating, would have no stable equilibrium configuration, i.e., would be disrupted, if its distance from the center of the primary was less than:

$$r_R = 2.45\left(\frac{\rho_p}{\rho_s}\right)^{1/3} R_p \qquad (3)$$

where R_p is the primary's radius and ρ_p and ρ_s are the densities of the primary and the secondary. The radius r_R is known as Roche's limit. Real moons are not, however, completely fluid. H. H. Aggarwal and V. R. Oberbeck (1974) suggested that rock and ice moons up to ~200 km in size may be able to survive well inside Roche's limit. On the other hand, one may ask whether moons could have accreted in such a harsh tidal environment. Various Roche-limit-like criteria can be derived for accretion. All are of the same form as Eq. (3); however, the coefficient can differ by as much as a factor of two depending on such factors as the shapes, deformabilities, rotation rates, and relative sizes of the two bodies trying to accrete (Weidenschilling et al. 1984). Electromagnetic interactions, either between particles with net charge or upon physical contact of particle surfaces, can lead to accretion criteria not in the form of Eq. (3). These are likely to be most important for very small particles. A detailed discussion of the Roche limit is beyond the scope of this chapter (see J. A. Burns [1982]; Smoluchowski [1979*b*]; Weidenschilling et al. [1984] for further details). We do, however, note that although the Roche zone is a fuzzy boundary, it is nonetheless very important to the long-term existence of major planetary rings. Major ring systems are located within or near the Roche zones of their planets, whereas major moons are well outside this region (Fig. 4). An intermediate class of bodies, including moonlets and shep-

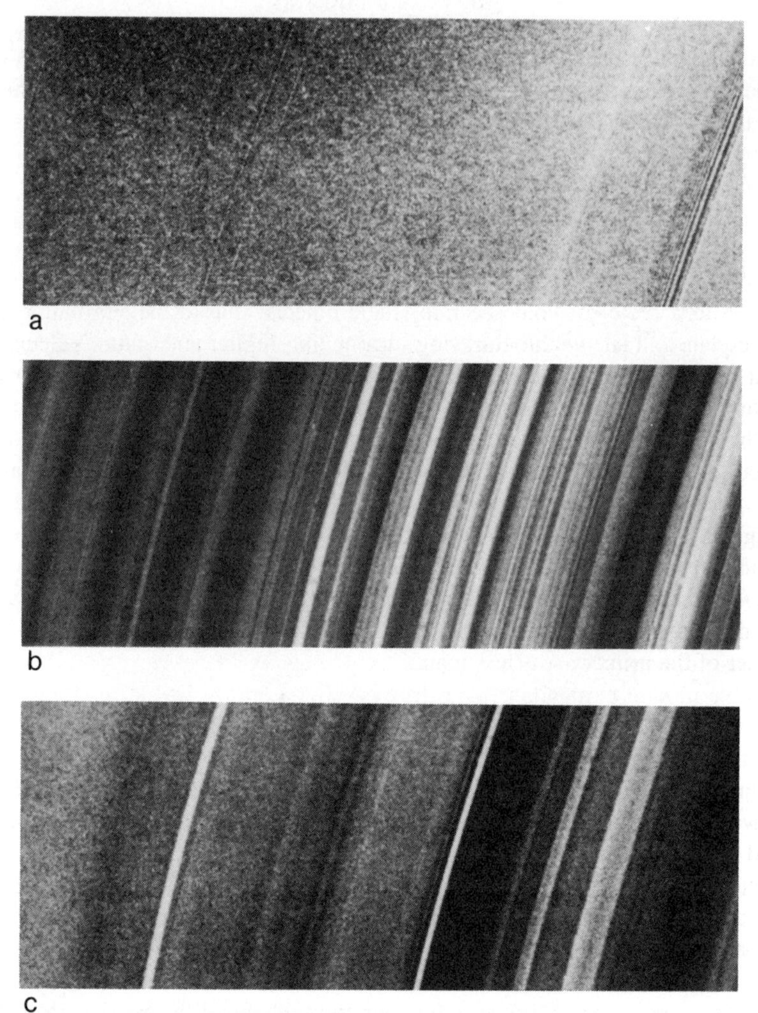

Fig. 4. Three Voyager images representing typical regions in each of Saturn's three major rings. Each image has been processed separately. The C ring is actually significantly darker than the A ring, whereas the B ring is slightly brighter. The image of the A ring has been stretched (contrast enhanced) much more than those of the other rings, due to the A ring's relative uniformity. The images each display an area roughly 4000 kilometers in radial extent; Saturn is off to the right of each image. Image a: density and bending waves excited at resonances with Saturn's many moons dominate the structure of the A ring. Most of the waves in this image are unresolved and appear as bright bands. Image b: Saturn's B ring is dominated by an irregular structure, with variations in optical depth existing on many length scales. The cause of this structure is unknown; however, viscous instabilities have been suggested (W. R. Ward 1981; D. N. C. Lin and P. Bodenheimer 1981; Lukkari 1981). Image c: Saturn's C ring is a low optical depth region which contains several gaps and optically thick ringlets. It resembles Cassini's Division in many respects. The cause of most of this structure is unknown; however, moonlets (Lissauer et al. 1981) and resonances (Holberg et al. 1982) have been implicated for some features.

herding moons, can exist in both regions. We return to this point in Sec. IV.A, where we discuss the relationships between these various types of planetary satellites of all sizes. However, we wish to point out that tidal effects are much more important in planetary rings than they are thought to have been in the protoplanetary disk (although, e.g., giant gaseous protoplanets would not have been able to accrete inside 1 AU after the Sun had formed).

B. Resonances

Many other dynamical processes, unrelated to Roche-type tidal influences, also affect the structure of Saturn's rings. The most clearly observed and understood of these structures are caused by resonant interactions with Saturn's many moons. As similar processes may have been quite important in both the protoplanetary and protosatellite disks, we discuss them here in detail.

Angular momentum can be secularly transferred between a moon and ring particles at a resonance, where the ratio of their orbital periods is roughly of the form $(j + k)/j$, where j and k are integers and k is small. Two main types of features were observed at resonances in Saturn's rings by the Voyager spacecraft: spiral waves and gaps or ring boundaries. Spiral waves, which consist of density waves (horizontal disturbances) and bending waves (which involve motion perpendicular to the ring plane) are excited at resonances in regions of moderate optical depth (see Fig. 5). These waves propagate away from resonance carrying with them angular momentum received from the exciting moon. The angular momentum is deposited in the ring particles as the waves are damped. Bending waves are excited only at vertical resonances, and typically transport much less angular momentum than do density waves (Shu et al. 1983); thus, their importance to the evolution of the protoplanetary disk is likely to be far less than that of density waves. We therefore restrict our discussion to density waves.

The theory of density waves was developed by C. C. Lin and F. H. Shu (1964) to explain the spiral structure observed in most disk galaxies. In 1978, Goldreich and Tremaine suggested that Saturn's moons would excite spiral density waves at resonant locations within the rings of Saturn. These waves have been observed in Voyager data by Cuzzi et al. (1981), Lane et al. (1982), Holberg et al. (1982), Holberg (1982), and Esposito et al. (1983).

A moon or protoplanet exerts a coherent and repetitive force on disk material at resonance. This forcing perturbs the trajectories of near-resonant material, causing clumping. The pattern of clumping remains fixed in the frame rotating with the perturbing satellite (or, for some resonances of eccentric and inclined satellites, the frame rotating with the frequency of the component of the satellite's potential responsible for the resonance). The satellite exerts a torque on the nonaxisymmetric clumping that it has induced at resonance. As both the amplitude of the clumps and the force per unit mass on the clumps are proportional to the perturbing satellite's mass, the torque exerted at

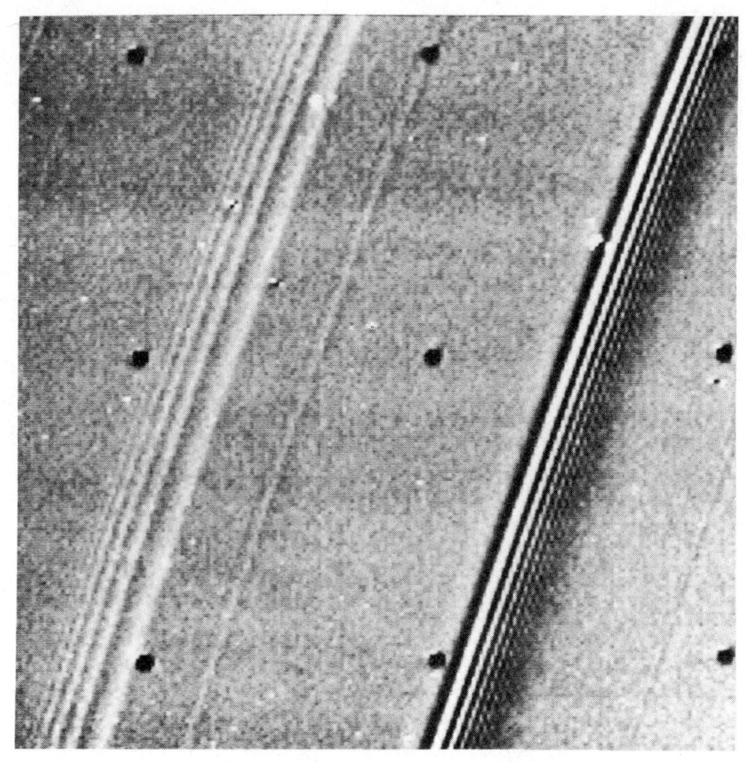

Fig. 5. Spiral density waves (left) and spiral bending waves excited by Saturn's moon Mimas at its 5:3 horizontal and vertical resonances within Saturn's rings. Density waves are variations of the quantity of particles at a given position within the rings. Bending waves are vertical corrugations of the ring plane, and are observed due to the low illumination angle of the Sun. The locations of the two resonances are separated due to the oblateness of Saturn, that causes orbital apsides to advance and orbital nodes to regress (see Shu et al. [1983] for further details).

a resonance (in the linear approximation of small fractional mass clumping) varies as the square of the perturber's mass. The angular momentum input at resonances is transported within the disk due to the disk's self-gravity or pressure. The compressions and rarefactions of material induced by the satellite exert forces on nearby disk material, causing further clumping. This process continues, resulting in a spiral density wave propagating away from resonance (Fig. 6; Shu 1984*a*). As the wave damps, the angular momentum which it carries is deposited in the disk. In the Saturn system, the moons exciting observed density waves are all exterior to the rings, and set up waves with pattern speed (angular velocity of the frame in which the wave pattern is fixed) lower than the orbital motions of the ring particles through which they propagate. Such waves carry negative angular momentum, and, as they are damped, ring material drifts inward. The moons, in turn, are pushed outward

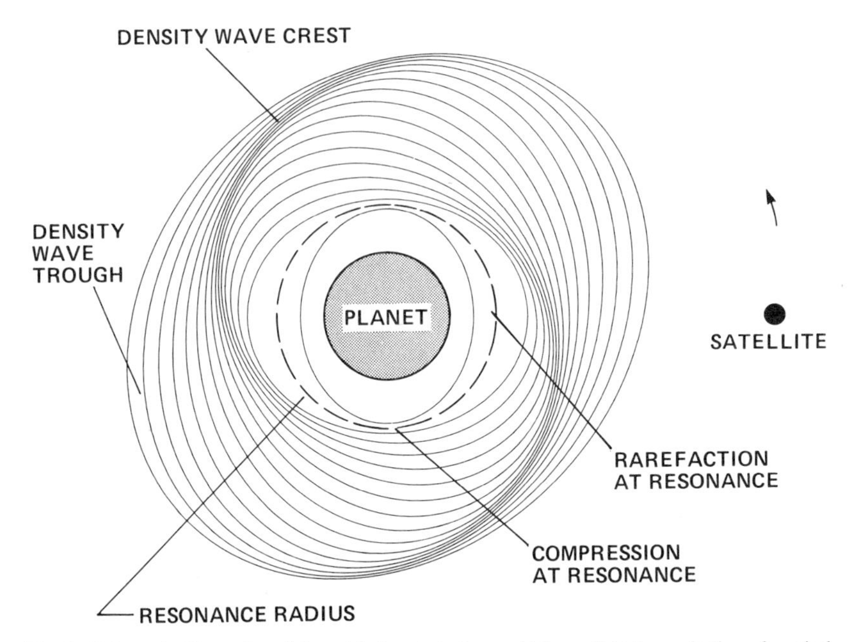

Fig. 6. Schematic illustration of the orbital perturbations which result in the excitation of a spiral density wave. The case shown is for the region near the 2:1 resonance with an external moon; a two-armed spiral wave is excited at this resonance. The ovals represent particle streamlines as seen in the frame rotating with the moon. The Keplerian ellipses of the instantaneous particle orbits map into closed, planet-centered ovals (2-lobed figures) in the rotating frame, because the particles undergo two epicycles between each passage of the moon. Streamlines straddling the exact resonance (dashed line) have periapse and apoapse longitudes reversed; where they bunch together, particle density enhancements occur. These enhancements are fixed in the rotating frame, and serve to induce coherent oscillations, or epicycles, in neighboring particles which drift past due to differential rotation. In the absence of self-gravity, streamlines at successively larger radii drift at increasing rates in the rotating frame, and no stable pattern can ensue. However, ring self-gravity slows the horizontal oscillations of the particles, and successive streamlines exterior to resonance may be made to remain fixed in the rotating frame. For this to occur, a small phase lag between successive streamlines is required. As seen in the figure, the resulting locus of streamline compression maps out the crests of a spiral density wave. A small mass-density is less able to counteract the differential rotation, and a larger phase shift is required, which yields a more tightly wrapped wave; thus the tightness of the spiral pattern is a measure of the surface mass-density near resonance (cf. Cuzzi et al. 1981).

as they excite the waves. For resonances located exterior to the satellite's orbit, as would often have been the case in the protoplanetary disk, the satellite (protoplanet) would yield angular momentum to the disk. In both cases, net angular momentum is transported to material farther away from the primary. (If gas pressure was sufficient to cause material interior to the protoplanet's orbit to orbit more slowly than the protoplanet, angular momentum may have been transferred inwards; we return to this point when we discuss gap clearing in Sec.II.C.)

The ability of spiral density waves to transport angular momentum outwards (and mass inwards) is what makes them of such great potential importance to models of the solar nebula. In our solar system today, over 99.8% of the mass is concentrated in the Sun, whereas 98% of the angular momentum resides in the orbital motions of the planets. A massive T-Tauri stage solar wind may have carried away much of the Sun's momentum (E. J. Weber and L. Davis 1967); however, even if the Sun were rotating near breakup (chapter by Durisen and Tohline) before such a wind began, its specific angular momentum would still have been $< 1\%$ that of the planets. As mass and angular momentum were undoubtedly more evenly distributed in the protosolar material, some type of redistribution must have taken place. Viscous spreading also yields the correct flow directions (Lynden-Bell and Pringle 1974); however, the value of viscosity required for this process to be quantitatively meaningful is many orders of magnitude greater than the molecular viscosity. Thus, unless the protoplanetary disk was turbulent, viscous spreading and related angular momentum transport may have been minimal (see chapters by Cassen et al. and by Lin and Papaloizou).

The torque T exerted by an Earth-sized planetesimal located in Earth's current orbit on material orbiting twice as rapidly (i.e., located at Earth's 2:1 resonance) can be calculated from formulae derived by Goldreich and Tremaine (1979*a*), and would have been

$$T = 2 \times 10^{30} \, \sigma \tag{4}$$

g cm^2 s^{-2} where σ is the surface density in g cm^{-2} of the disk in the resonant region. Integrated for one year, this torque is equivalent to that required to remove the entire angular momentum of a 30-m annulus located at the resonant region, or to move a 2500-km annulus inward by 2500 km. Damping of the waves due to viscosity and/or nonlinear effects would have spread angular momentum loss over a larger region; however, the induced drift may still have been significant. The back reaction on the planetesimal would have depended on the surface density of the disk in the resonant region. If $\sigma = 10^4$ g cm^{-2}, the Earth-sized planetesimal would have moved outwards at the rate of 60 km yr^{-1}. Rates would have been much smaller for small planetesimals, but much larger for the Jovian planets and their cores. Significantly more torque would have been exerted at the closer 3:2, 4:3, etc. resonances; torques at exterior resonances with more slowly orbiting material would drive protoplanets inwards. Goldreich and Tremaine (1980) showed that the time scale for motion of Jupiter within the protoplanetary disk may have been as rapid as 10^4 yr. Of course, in the real situation, torques due to inner and outer Lindblad resonances will partially cancel out, making predictions strongly dependent on the assumed radial mass distribution. We now examine torques in regions near a protoplanet, using observations of Saturn's rings as a guide.

C. Gap Clearing

Satellites can clear gaps around themselves in a manner analogous to the resonant effects discussed above. Resonant torques increase as one approaches a satellite (Goldreich and Tremaine 1979*a;* Lissauer and Cuzzi 1982). Near the orbit of a satellite, resonances pile up, overlap, and their influences blend together (Goldreich and Tremaine 1979*a*). The combination of these two effects allows even small satellites to exert a significant torque on nearby disk material. As the flow of angular momentum is always outwards, the satellite removes angular momentum from matter inside its orbit and yields angular momentum to matter orbiting farther out. Material both interior and exterior to the satellite is thus repelled and a gap can be created. Goldreich and Tremaine (1979*b*) proposed that diffusive spreading of the sharp, narrow rings of Uranus was prevented by shepherding moons pushing on them from both sides. Although the proposed Uranian ring shepherds have not yet been detected, the F ring of Saturn was found to possess a shepherding moon on either side (B. A. Smith et al. 1981). These shepherds are fairly large (> 50 km radius) and apparently confine most of the F ring material to a few narrow strands. Unobserved small moons have been proposed as explanations for much of the radial structure seen in Saturn's rings, especially nonresonant clear gaps (Lissauer et al. 1981; Hénon 1981,1984). No such moonlet has yet been observed (B. A. Smith et al. 1982); however, strong evidence exists for their presence, especially within Encke's Division. This mostly empty annulus in the outer portion of Saturn's A ring has scalloped edges in some Voyager images (Fig. 7), indication of the recent passage of a moonlet at nearby azimuths (Cuzzi and Scargle 1985). The important question for the solar nebular problem is, how large must a protoplanet have been in order to clear a gap around itself, as this would presumably have halted its accretion.

The physics of shepherding is still poorly understood (R. Greenberg 1983; Borderies et al. 1984). Perturbations due to a shepherd's passages may be damped out before individual resonances can be set up. If perturbations are damped very rapidly, on a time scale of less than the synodic period of the shepherding moon and the nearby ring material, the eccentricities and, thus, the streamlines of ring particles will vary in azimuth. This accounts for the azimuthal variations of the scallops observed on the edges of Encke's Division. The torque exerted by such moonlets will be similar to that calculated by summing isolated resonances, provided that the particle motions are not damped significantly during their brief encounter with the moonlet, and are damped to circular orbits (or have their eccentricity vectors randomized) before their next encounter. In this case, the particulars of this damping process only affect the detailed shape of the ring edge; they do not change the net torque (R. Greenberg 1983).

The torque exerted by a moonlet (or protoplanet) must balance the viscous torque due to differential motion of the surrounding material in order for a gap to be created. In the linear, axisymmetric theory (small deviation of

Fig. 7. Voyager 2 image of the Encke Division reprojected and horizontally magnified. Running along the right-hand edge are seen roughly sinusoidal edge waves or radial oscillations, which are not noticeable on the left side. These edge waves may be the visible trace of particle streamlines that have been perturbed by the recent passage of a moonlet embedded with the Encke Division. Such waves have been observed on both inner and outer edges of the Encke Division, at different longitudes. One of several kinky ringlets may be seen in the center of the division. Figure from Cuzzi and Scargle (1985).

streamlines), the specific torque (torque per unit disk surface density) increases as the inverse fourth power of distance from the moonlet (Goldreich and Tremaine 1979*b*). The torque levels off within one disk scale height of the moonlet, where the thickness of the disk makes an important contribution to the distance between moonlet and ring particle (Goldreich and Tremaine 1980). An edge sharper than the scale height of the disk is extremely difficult to maintain against diffusion, so this may be viewed as a lower limit to the size of a gap created (Papaloizou and Lin 1984). However, if solid material were concentrated in a thin layer near the midplane of the disk, as proposed in some nebular models (Safronov 1969; Goldreich and Ward 1973; C. Hayashi 1981*a*) and were massive enough to move independently of the gas, it could have been acted upon separately. A gap within this layer need only have been larger than the scale height of the condensed material, which is given by the particle random velocities. A moonlet (protoplanet) must, of course, also be able to create a gap larger than its own diameter, or it will collide with disk material. The size of the gap that a moonlet of a given density can create is proportional to the square of the moonlet's radius (Lissauer et al. 1981); thus, only bodies larger than some critical radius are able to create gaps wider than their own size; smaller bodies continue to have physical collisions with surrounding material, enabling them to accrete or to be eroded away, depending on the circumstances. Protoplanets which migrated rapidly through the disk, due to asymmetric gravitational torques or other forces, would have been required to exert a greater torque on nearby material in order to clear a gap before moving onward (W. R. Ward 1984). If they were able to clear gaps, they would have pushed disk material through the nebula ahead of themselves.

Nonlinearities may limit the rate of increase in torque as one approaches a moonlet or protoplanet (Borderies et al. 1984; see, however, Shu et al. 1985). In addition, the energy conveyed to the disk along with the torque can increase the local viscosity and make gap clearing more difficult (Borderies et al. 1984). Future detection of moonlets in Encke's Division and other locations within the rings of Saturn (and Uranus) will provide useful tests for theories of shepherding.

The mass at which a protoplanet may have been able to clear a gap in the solar nebula depends on the conditions characterizing its location in the protoplanetary disk. Viscosity, an important parameter for these calculations, is uncertain by many orders of magnitude; thus, all numerical estimates are quite uncertain. D. N. C. Lin and J. Papaloizou (1979*a*) argue that the ratio of the mass of the smallest protoplanet which could have cleared a gap to the mass of the protosun must have been greater than the inverse Reynolds number, $\mathcal{R}^{-1} \equiv \nu / r^2\Omega$, of the disk; in other words, that the minimum mass for gap clearing is proportional to the disk's viscosity.

Gas pressure is also important. As gas in the solar nebula was partially pressure-supported (see Sec. III.A), its orbital velocity was less than that of planetesimals and protoplanets located at the same position within the nebula. Resonances, which depend on orbital rates, would thus have been displaced from being centered around the protoplanet inward to a center where the gas corotated with (i.e., had the same orbital velocity as) the protoplanet. This would have weakened the ability of protoplanets to create gaps around themselves, but may have increased their ability to migrate within the nebula due to density-wave torques. However, a self-consistent analysis, which includes pressure gradients induced by clearing of disk material in the process of gap formation, is needed in order to resolve this point (D. Lin 1984, personal communication).

Papaloizou and Lin (1984) estimate that a protoplanet would have needed a mass of 10^{-4} to 10^{-3} $M_\odot$, in other words, at least twice the mass of Neptune, in order to have cleared a gap in a reasonably viscous (e.g., turbulent) protoplanetary disk of moderate scale height ($H \approx 0.05\ r$–$0.1\ r$, where r is the distance to the protosun). A gap in the midplane dust layer could have been cleared much more easily, and could have slowed the accretion rate of planetesimals much smaller than the Earth, if most of the mass of the condensed matter was in small bodies (R. Greenberg 1982).

D. Gravitational Instabilities

Spiral density waves excited by instabilities in the protoplanetary disk may have had an even more significant effect on its evolution than waves excited by protoplanets. These waves are produced when a disk is sufficiently massive and flat that the gravitational potential energy released by clumping is sufficient to overcome the stabilizing effects of shear due to rotation. Mathematically, the required flatness can be stated as the fluid analog of Toomre's (1964) criterion for stability of a stellar disk. A fluid disk is unstable to axisymmetric perturbations if

$$Q \equiv \frac{\kappa c}{\pi G \sigma} < 1. \tag{5}$$

In Eq. (5), κ is the epicyclic (radial) frequency, c is the dispersion velocity or sound speed (which varies as disk thickness), and σ is the (unperturbed) surface density of the disk. Nonaxisymmetric perturbations will also begin to grow when $Q \lesssim 1$. Waves excited in this manner are short spiral density waves, in which pressure plays a larger role than does disk self-gravity (see Shu [1984*a*] for a more complete discussion).

Self-excited density waves have the potential to transport large amounts of angular momentum; in fact, they are probably the major cause of angular momentum redistribution in spiral galaxies. Unlike resonantly excited density waves, which in the solar nebula would have depended on the (relatively small) masses of the protoplanets, self-excited waves could have involved the whole mass of the protoplanetary disk. Unfortunately, there are still zeroth-order questions which must be answered before applying this theory to solar nebular models. Although the criterion for instability (Eq. 5) is fairly well understood, the instability may lead to local clumping rather than production of spiral density waves. Such clumping has been suggested to have resulted in kilometer-sized planetesimals, if the instability occurred in a flattened sheet of condensed material in the midplane of the protoplanetary disk (Safronov 1969; Goldreich and Ward 1973), or giant gaseous protoplanets, if uncondensed material was also involved in the instability (Cameron 1978). (The reason for such a vast difference in the masses of the resulting clumps, much larger than the gas-to-dust ratio of 100, is that the volume of material engulfed in these clumps varies as the cube of the scale height of the material involved in the instability.) A linear perturbation analysis of the type used to derive Toomre's criterion cannot tell us what ultimately forms as a result of instabilities in the protoplanetary disk, but only the conditions under which the instabilities are likely to occur. To estimate the mass required for a nebula to become unstable, we make the Keplerian orbit approximation:

$$\Omega = \left(\frac{GM_{\odot}}{r^3}\right)^{1/2} = \kappa. \tag{6}$$

We note that the scale height of the nebula is $H = c/\Omega$; thus, from Eq. (5) we have

$$Q = \frac{M_{\odot}}{\pi r^2 \sigma} \frac{H}{r}. \tag{7}$$

Equation (7) roughly states that the ratio of the mass of the disk to that of the protosun must be greater than the reciprocal of the disk's aspect ratio for an instability to occur. The disk's mass must have been at least 0.1 $M_{\odot}$ to 0.5 $M_{\odot}$ for gravitational instabilities in the gas to have been important in the region in which the planets formed ($M_{\odot}$ here refers to the mass of the protosun if accretion was still occurring). For all nebula models in which a significant fraction of the heavy elements condense out and drift to the midplane, this very thin sheet of solids is unstable to clumping. In the region where the Earth formed, the dust layer of a minimum mass nebula, with 10 g cm^{-2} of condensed material, would have become unstable once its thickness dropped below ~1000 km. The same region of this nebula, with a gas surface density of 1000 g cm^{-2}, would have to have had a thickness $\lesssim 10^5$ km in order for gaseous gravitational instability to occur. In other words, the temperature of such a nebula would have to have been $< 10^{-2}$ K at 1 AU.

Numerical simulations have confirmed Toomre's criterion for instability, and can be used to illustrate some possible consequences for the protoplanetary disk (Cassen et al. 1981). Saturn's rings may be thin enough for gravitational instabilities to be important (Shu 1984*a*); however, the density waves that could be produced by this mechanism would be too short to have been observed by Voyager. We note, however, that the largest common particles in Saturn's rings (5–10 m radius) are approximately the size of the clumps which appear due to these instabilities. These particles may currently stabilize the rings (see Shu [1984*a*] for a more detailed discussion). The mechanism which determines the size distribution of Saturn's ring particles may also help us understand the size spectrum of planetesimals, although Roche-type tidal effects, important for planetary rings but negligible for planetesimals, argue for cautious use of such comparisons.

Viscous and diffusional instabilities have been suggested as the cause of much of the irregular structure observed in regions of high optical depth in Saturn's rings (W. R. Ward 1981; D. N. C. Lin and P. Bodenheimer 1981;

Lukkari 1981; Stewart et al. 1984; Fig, 4b). Such instabilities probably could not have occurred in the gaseous material in the protoplanetary disk because pressure effects in a gas have a stabilizing effect. Gravitational instabilities would have been dominant in a nonturbulent disk of condensed matter, causing clumping which would have reduced the optical depth to a value too low for viscous instabilities to operate. If the differential velocities of planetesimals, due to gas drag, produced sufficient turbulence to prevent a gravitational instability within the layer of condensed matter (Weidenschilling 1980), this turbulence would also have inhibited viscous instabilities.

III. ORIGIN AND EVOLUTION OF PLANETARY SATELLITES

A. Satellite Formation

Questions regarding the origin of planetary satellites have traditionally focused on the chemical composition of the satellites, and its interpretation in terms of the thermal and chemical properties of the surrounding local nebula as functions of position and time during the accumulation process. Following the approach of J. S. Lewis (1974), who showed that the inwardly increasing temperature in the protosolar nebula may account for the inward increase in uncompressed planetary density characterizing higher-temperature condensates, Pollack and Reynolds (1974) presented a scenario in which the inward increase in density shown by the Galilean satellites (Fig. 8) could be due to the high luminosity of proto-Jupiter. The two inner Galilean satellites are known to be silicate-rich, whereas the two outer ones are of lower density and contain roughly equal quantities of silicates and water ice. Application of the same approach to the Saturn system (Pollack et al. 1976; see also, Consolmagno and Lewis 1976) has been less satisfactory. No dramatic density gradient is seen there as the satellites are all fairly rich in icy material (B. A. Smith et al. 1982; Pollack and Consolmagno 1984; Fig. 9). Another possible difficulty in the Saturn system is the evidence for albedo and, therefore, compositional inhomogeneity within Saturn's rings (Cuzzi et al. 1984). Such inhomogeneities would not be expected to arise in a quiescent, slowly cooling nebula in which the earliest (rocky) condensates are filtered out and removed by gas drag long before the condensation of ices. These observations may, however, be explicable as minor surface modifications of the ring particles. Alternatively, the rings may have been changed substantially from their form in the accretion era (see Sec. IV.A). The situation is even more confused in the case of Uranus. Because Uranus is farther from the Sun and less massive than Saturn, even lower-density, more icy material would be expected; however, although water ice is seen on the surfaces of all four large regular satellites, no evidence for water ice (or any other ice) has been found in the rings, and the ring particles are quite black (Elliot and Nicholson 1984). In addition, there is a preliminary suggestion that the outer two large Uranian satellites may have significantly higher densities than the inner two (Table I; Veillet 1983).

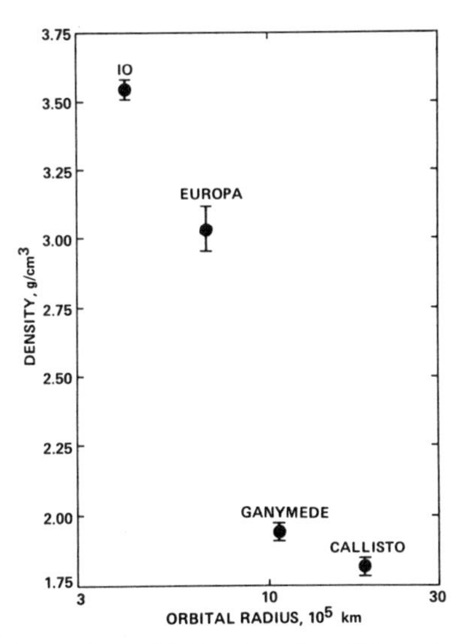

Fig. 8. Densities of the major satellites of Jupiter plotted vs. distance from the planet. Note that a systematic trend of increasing density approaching the planet exists among the Jovian moons, probably due to a temperature gradient in the circum-Jovian nebula.

Thus, one is led to suggest that the accretion picture may not have been as simple, calm, and static as presumed in the slowly cooling models thus far explored. This complexity is mirrored in the abundant meteorite evidence for intermingling of high- and low-temperature condensates (see chapter by Boynton).

While planetesimals were accreting, a variety of mixing processes may have occurred in circumplanetary nebulae (CPN). Presumably, the CPN was

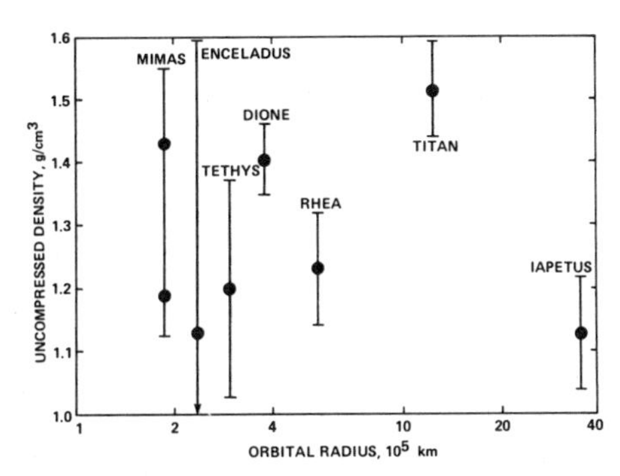

Fig. 9. Bulk densities, adjusted to remove the effects of compression, of the major satellites of Saturn plotted against distance from the planet. No clear systematic trend, as seen for the Jovian moons (Fig. 8) exists in the Saturn system. Note that the scales in Figs. 8 and 9 are quite different. The two data points given for Mimas were derived using two differing values of that satellite's mass (see Tyler et al. 1982).

cooling slowly; if all other properties were constant, the ratio of the cooling time t_c to the accretion time t_a determined the particular sequence of condensates. For $t_c >> t_a$, the early condensates would have been removed from the gas phase before other compounds, which might have involved them in part (e.g., $NH_3 \cdot H_2O$, $CH_4 \cdot H_2O$) could have formed (disequilibrium condensation). For $t_c << t_a$, the condensate at any time would have been a thermodynamic-equilibrium species (see J. S. Lewis 1972*b*). Of course, the processes which formed and heated the CPN are not fully understood; hydrodynamic collapse and formation of a disk, as well as continual infall onto and flow through a possibly viscously heated disk are important scenarios which have not had their thermal and chemical histories explored at all. It is quite possible that the energy generated in viscous dissipation may have produced more heating in the disk than that generated by the protoplanet (see chapters by Cassen et al. and by Lin and Papaloizou). Furthermore, such a disk would have had a vertical thermal gradient, and objects moving vertically may have accreted high- and low-temperature condensates at the same radial location.

The situation becomes even more complex when the potential for radial motion of accreting satellitesimals is realized. Several processes may have produced significant radial migration on the time scale of accretion. The first to be recognized were aerodynamic processes (gas drag). These processes were quantified for the solar nebula by Adachi et al. (1976) and by Weidenshilling (1977*a*), following earlier suggestions by Whipple (1964*b*); they have also been discussed by J. A. Burns (1977), Pollack et al. (1979) and Torbett (1984*b*). The gaseous component of the nebula experienced a radial pressure gradient which modified its net inward acceleration by a term Δg

$$g' = g + \Delta g = -\frac{GM}{r^2} + \frac{1}{\rho}\frac{dP}{dr} \tag{8}$$

such that the gas was partially pressure-supported and orbited more slowly ($g' < g$) than massive solid particles at r, which responded only to gravity. This relative velocity created an apparent headwind which removed angular momentum from particles' orbits, causing them to decay. The rate of decay was size-dependent, and very rapid (Weidenshilling 1977*a*). From Table II we see that typical CPNs were denser than the protoplanetary disk, and because CPN rotation rates were quite rapid, gas drag decay times t_g were typically as short as 10 to 10^2 yr for 10 to 10^3 m-sized objects. Medium-sized particles moved inwards fastest because they had a significant velocity relative to the gas, but were still small enough to respond to the strong headwind. Gas drag could, however, also have aided the accretion process. The smallest solid particles moved along with the gas, colliding with more rapidly moving larger particles. If collisions were gentle enough (relative velocities small enough), accretion would have resulted. The differing inward migration rates due to gas drag also could have been important in rapidly bringing together particles from neighboring regions of the CPN.

A planetesimal must have attained a size of > 100 km before it could have remained radially fixed for a time on the order of an accretion time (Weidenshilling 1981). However, for objects of this size and larger, gravitational torques may have come into play, as discussed in Sec. II. These torques may have produced rapid radial motion, enhancing accretion, or may have truncated the disk, thereby halting the accretion process. Both gravitational-torque effects must be modeled to better understand the accretion process. Furthermore, there are ongoing and/or extinct evolutionary processes, occurring after the satellites formed, which must be better understood before primordial conditions may be inferred from currently observed properties. Below, we discuss several of the most important evolutionary processes and their implications for origin hypotheses.

B. Orbital Evolution

Tidal evolution can alter the orbits of moons both before and after the nebular disk is removed; it is, in some cases, still important today (Peale and Burns 1985). This process involves the outward transfer of angular momentum to a secondary from a displaced tidal bulge which it has raised on the primary (Fig. 10; see also, Kaula 1964). Similar effects will tend to cause the axial rotation of the satellite to approach the synchronous rate (Goldreich and Peale 1968; Peale 1976*a*). The orbits of all the inner satellites of the Jovian planets, except for the very small ones, may have been significantly expanded over the age of the solar system. This process may be responsible for the existence of orbit-orbit resonances, as we discuss below.

The current satellite orbits display a large number of commensurabilities in which orbital periods of different satellites are in very nearly whole-number ratios (Peale 1976*b*; R. Greenberg 1977). Some of these commensurabilities are quite precise; the Laplace relation between the mean motions of the three inner Galilean satellites has been verified to ten significant figures. Additionally, their mean motions are, respectively, in nearly a 4:2:1 ratio. Other commensurabilities and near-commensurabilities are found in the Saturn and Uranus systems (R. Greenberg 1975,1976,1977,1979*b*).

These commensurabilities are too numerous to be explained by chance (Goldreich 1965; Dermot 1973). It has been suggested that satellites preferentially formed in such resonant orbits; this suggestion has received fresh support (cf. Goldreich and Tremaine 1978*b*) from recent understanding of the properties of spiral density waves (Sec. II). However, the prevailing belief is that tidal evolution, as discussed above, may have driven initially nonresonant satellites into such stable commensurabilities (Goldreich 1965; Yoder 1979). This is because the orbits of inner satellites evolve faster than more remote ones of the same mass. Once a commensurability is reached, angular momentum may be transferred to the outer member so that the commensurate pair evolves as a unit (Yoder 1979).

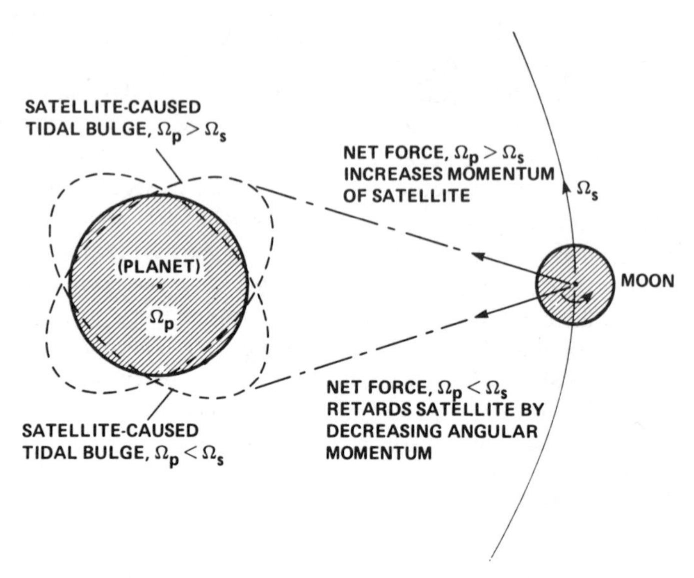

Fig. 10. An illustration of the tidal bulge raised on a planet by one of its satellites. If the mean motion of the satellite Ω_s is smaller than the rotational frequency of the planet Ω_p, the near lobe of the tidal bulge is carried ahead of the satellite and produces a net torque which increases the satellite's angular momentum and energy, causing it to recede from the planet. Simultaneously, the rotational angular momentum of the planet is decreased by the satellite's pull on the asymmetric mass distribution of the tidal bulge. If $\Omega_p < \Omega_s$, the satellite falls inwards toward the planet.

Significant orbital evolution may also be caused by density-wave interactions with planetary rings. Such resonant torques are especially important even today in the Saturn system, as discussed in Sec. II, and could have caused Saturn's inner ringmoons to evolve all the way from the edge of the ring to their current locations in only $\sim 10^7$ to 10^8 yr (Goldreich and Tremaine 1982). We note that density-wave torques can also move ring material over large radial distances (Sec. II; see also, Lissauer et al. 1984). Additionally, ring material can be moved due to bombardment by meteoroids (Durisen 1984; Ip 1983; Lissauer 1984).

With regard to the properties of circumplanetary nebulae, it appears that the various possibilities for radial motion and orbital evolution preclude more detailed inference of nebular thermal structure from simple assumptions as to current composition and location of the satellites. Furthermore, several of these processes suggest ways of transporting material over the large distances required by the observed separation between satellites.

C. Why Moons Do Not Have Moons

Although most of the planets in our solar system have satellites in orbit about them, no natural satellite has ever been observed in orbit about a moon.

One would not suspect the sequence of sun, planet, moon . . . to continue indefinitely; however, the abrupt break in the chain, where none of the nearly four dozen known moons in the solar system (Table I) possesses an observed secondary, requires an explanation. There is no shortage of small bodies. Several Saturnian moons share the same orbits (B. A. Smith et al. 1982; Chenette and Stone 1983; Table I); most reside near the stable L_4 or L_5 Lagrange points as do Trojan asteroids of Jupiter, with one pair being in a very loose coorbital lock which encloses three Lagrange points (Dermott and Murray 1981*a,b*). Two differences between the planets and the moons, one relating to formation and one to evolution, may contribute to an understanding of this situation.

The first difference relates to formation. Most planetary satellites are observed to orbit the gas giant planets. These planets presumably possessed circumplanetary nebulae in which their regular satellites accreted; their irregular satellites may have been captured in an early distended protoplanet phase (Sec. I). The terrestrial planets presumably did not possess CPNs. They have few satellites, whose formation is poorly understood, and may have depended on special circumstances (Sec. I). Moons, which are also likely to have lacked orbiting disks, would similarly have had difficulty in acquiring secondaries.

Alternatively, some moons may have had secondaries which have since escaped or been destroyed. Again, an analogy between moons and the terrestrial planets can be made. Mercury and Venus may have, at one time, had satellites. Subsequently, the satellite orbits could have tidally decayed into the planet (J. A. Burns 1973; W. R. Ward and M. J. Reid 1973). (This process will lead to Mars' consumption of its inner moon Phobos within the next few hundred million years [J. A. Burns 1978].) Moons of the inner planets may also have escaped due to tidal and solar perturbations (A. W. Harris and W. R. Ward 1982).

The dynamical environment of a satellite orbiting a moon would have been more hostile than that of an inner planet satellite. For long-term stability, a secondary must orbit outside its primary's Roche limit (unless it is small; see Sec. II.A), but well inside the primary's sphere of influence. Although there is room for stable orbits about many moons, the amount of available room is much smaller, even relative to the size of the system, than is available for satellites of the inner planets. For example, using the criterion developed by Pendleton and Black (1983), it can be shown that orbits about the Earth are stable out to $\sim$100 Earth radii, whereas bodies in stable orbits about Callisto and Titan must be within $\sim$10 radii of these moons. Given the small sizes of their stable regions, even if these moons had formed with satellites, loss mechanisms proposed for satellites of terrestrial planets would have been even more efficient in removing satellites of moons. Specifically, loosely bound satellites could have been freed by perturbations from the central planet, possibly in conjunction with resonant effects of other moons. Relatively massive satellites of moons could have evolved tidally due to their proximity to the moon. The spins of all major moons in the solar system are synchronous with their orbital motions (Sec. IV.B). An object within a moon's sphere of influence would have, therefore, orbited that moon more rapidly than the moon rotated. Tidal torques would thus have caused orbital decay (Fig. 10), and tides would have increased in magnitude as the object approached the moon. Reid (1973) estimated that no object with radius larger than ~100 km could have remained in orbit around any major moon over the age of the solar system, regardless of the object's initial orbit. On the other hand, small satellites could have been disrupted or scattered during the final stage of accretional bombardment. One or more of the above reasons may explain the lack of satellites of moons; although such a satellite may yet be discovered, we know that such objects are, if they exist, uncommon. They also help us to illustrate some of the differences between solar and planetary systems (Sec. IV.B).

D. Internal Evolution: Endogenic Processes

Although most of our knowledge of the satellites comes from observations of their surfaces (see Sec. III.E), the surface material may not truly represent the bulk properties of the satellite. For instance, the surface of Europa is nearly pure ice; however, its bulk composition, as inferred from its density, is dominated (>80%) by silicates. Regarding ice-rock satellites in general, there have been discussions of the possibility for initial accretion of a rocky core subsequently overlaid by an icy mantle (see, e.g., Pollack and Reynolds 1974). Alternatively, a more heterogeneous ice and rock accretion may have been followed by radioisotope heating, melting, and differentiation (see, e.g., J. S. Lewis 1971; Consolmagno and Lewis 1976,1977,1978; Schubert et al. 1981; Friedson and Stevenson 1983). There is a likelihood that the mere energy of accretion itself was sufficient to thoroughly melt the ice and differentiate at least the outer layers of the largest satellites, regardless of the order in which the material accumulated (see, e.g., Cassen et al. 1982). If the growing satellites were surrounded by their own primordial atmospheres, as suggested by Lunine and Stevenson (1982), much of the energy of accretion may have been dissipated in the atmospheres. Also, the early thermal history of the protomoons may have been affected by heat transport properties of their atmospheric blankets. These processes must be studied in greater detail.

The surfaces, at least, of the satellites must have cooled and solidified relatively quickly; theoretical calculations for the Galilean satellites (Reynolds and Cassen 1979) imply upper limits of $\sim 5 \times 10^8$ yr for solidification of their mantles. Actual cooling times from an initial, partially molten state could have been as short as $\sim 10^5$ yr. The smaller Saturnian satellites would have frozen even more rapidly.

In spite of the above conclusions, it was predicted by Peale et al. (1979) just prior to Voyager Jupiter encounter that Io should show evidence for "widespread and recurrent surface volcanism." These authors had noted that the eccentricity of Io's orbit was continually reinforced by the Laplace commensurability with Europa (Sec. III.B) and that the ensuing energy dissipation due to varying tidal deformation would provide a potent, ongoing heat source. The success of this model (B. A. Smith et al. 1979*a,b;* see also, Yoder 1979) led to its application to Europa (Cassen et al. 1979; Squyres et al. 1983*a*). The observation of apparently recent surface flows on Saturn's small satellite Enceladus (B. A. Smith et al. 1981,1982) supported a similar tidal explanation (Yoder 1979); however, the magnitude of the current eccentricity, forced by Dione, is too small to drive a significant tidal heating effect unless Enceladus contains substantial quantities of ammonia and has other internal properties favorable to tidal heating (Squyres et al. 1983*b;* Yoder 1981). Alternatively, Lissauer et al. (1984) suggest that the eccentricity of Enceladus could have been substantially amplified while in a recently disrupted 2:1 commensurability with Saturn's ringmoon Janus. In this case, the eccentricity forcing would have resulted from energy and angular momentum transferred via ring density waves driven by Janus.

E. Cratering: Exogenic Processes

The satellites' surfaces may also be modified by external agents; in fact, impact craters are the dominant landform on all visible satellite surfaces except for Io, Europa, and restricted regions of Enceladus, which are ostensibly resurfaced more rapidly than they are impacted. The spatial distribution, size distribution, number density, and morphology of these craters provide potentially valuable constraints on the surface/subsurface nature (viscosity) of moons, as well as on the final sweep-up phase of the accretion process itself. Most workers agree that the current cratering flux, integrated over the age of the solar system, is far too small to explain the size and quantity of craters (Shoemaker and Wolfe 1982; Shaeffer et al. 1981); therefore, the observed crater distribution probably represents a final late heavy-bombardment phase of the accretion process which occurred between the formation of the present-day crust and $\sim 4 \times 10^9$ yr ago, analogous to that seen on the lunar surface (W. K. Hartmann 1972,1975). Unfortunately, many of the details of the cratering record, and inferences therefrom, remain controversial. Below, we summarize our perceptions of the current state of affairs, and of possible implications of these data for planetary formation.

We first discuss the nature of the impacting population; e.g., its size distribution and source region. There is ongoing debate (W. K. Hartmann 1984) as to whether even the most heavily cratered surfaces observed have achieved a sufficiently high crater density to be saturated (the situation where the crater density distribution is unchanged by further impacts). Saturation would imply that the crater size distribution does not reflect a unique meteoroidal size-frequency distribution (Woronow et al. 1982). If the surfaces are not saturated, and if surface processes (e.g., viscous relaxation) have not selectively erased large craters while leaving only somewhat smaller ones unaffected (or vice versa), then the observed differences in crater size distribution imply different meteoroidal size distributions for the inner and outer solar system, and possibly differences between the size distributions of meteoroids impacting the Jovian and Saturnian satellites as well (Woronow et al. 1982; see, however, W. K. Hartmann 1984). In a similar way, the distribution of crater density with respect to the satellite's apex of motion (center of leading hemisphere in its orbital motion) contains information on the relative velocity vector of the meteoroids; if a substantial apex asymmetry exists, a population of extra-system objects, probably in moderate-velocity heliocentric orbits, is implied. This is because the satellites' axial rotations are synchronously locked to their orbital motion about the planet, and thus, a symmetric population of meteoroids, when viewed from the satellites, appears to be emanating from the direction in front of the satellite (Shoemaker and Wolfe 1982). A distribution of objects impacting from adjacent planetocentric orbits will not create such a crater density asymmetry. Neither will an assymetry be visible even for heliocentric impactors if the satellites are not continually synchronously locked with the same hemispheres always leading and trailing in their orbital motion.

Although several of Saturn's moons qualitatively display the predicted apex asymmetry in unresolved craters, as represented by variations in albedo and underlying structural patterns (Plescia 1983), the predicted apex asymmetry is not observed systematically in the densities of the ancient, large craters under discussion (Passey and Shoemaker 1982; Shoemaker et al. 1982; Plescia and Boyce 1983; Plescia 1983; S. Squyres 1984, personal communication). Although this has been ascribed to multiple reorientation of satellite spin axes and/or apex locations by large impacts, this explanation may be untenable due to the large sizes of the required impactors (see, e.g., Peale 1975, 1976*a;* Lissauer 1985; see however, McKinnon 1981.)

If the impacting population can be inferred to have roughly the same size distribution for all cratered surfaces in the inner and outer solar system, and if it can be shown to have been in heliocentric orbits, then a common, universal, late heavy-bombardment population may be postulated, possibly coincident with the formation of the Oort cloud by ejection of Uranus-Neptune region planetesimals (Wetherill 1975). Aside from its attractive simplicity, such a scenario would allow the Apollo absolute-age time scale of lunar cratering (W. K. Hartmann 1972) to be applied to date the ages of surfaces on the outer planet satellites (Shoemaker and Wolfe 1982; Plescia and Boyce 1983). However, the surface relaxation processes which would be needed to reconcile the observed crater-size populations with a single meteoroidal-size population are, as mentioned above, not at all universally accepted (Woronow et al. 1982; Chapman and McKinnon 1986).

To summarize, the characteristics of observed crater size distributions and regional crater density variations, taken together, do not conclusively support a simultaneous, solar-system-wide epoch of late heavy bombardment by objects in heliocentric orbit with a well-defined particle size distribution. Although there is evidence for a far higher flux of large impacts at some time in the past, we suggest that at least some of the objects may have originated in planetocentric orbits. One obvious difficulty with the concept is the very short lifetime for the sweep-up of planetocentric material. Clearly, however, this important question deserves more study.

F. Atmospheric Evolution

Both Titan and Triton are known to have outgassed significant amounts of nitrogen, which currently exists as N_2; in the case of Titan, it resides in a dense atmosphere (Hunten et al. 1984); in the case of Triton it may reside in a liquid ocean (Cruikshank et al. 1984). Titan's atmosphere, and a possible ethane ocean (Lunine et al. 1983), contains large amounts of CH_4, and possibly as much as 10% Ar. The details of the elemental, isotopic, and molecular makeup of these atmospheres, when measured by atmospheric probes, will provide many useful constraints on the mode of accretion (Owen 1982; chapter by Gautier and Owen).

For instance, the relative abundances of Ar, N, Ne, and C, and their isotopes, will allow us to determine whether these important volatiles came in as reduced or oxidized condensates (Prinn and Fegley 1981), or as clathrates of water ice (see chapter by Gautier and Owen). If N was originally in NH_3, as currently suspected based on estimates of nebular densities (Table II), then the surface temperature of Titan must have been much warmer in the past (>150 K) than it is presently (93 K), in order for sufficient NH_3 to have resided in the atmosphere where it could have been photodissociated to produce the observed N_2 (Atreya et al. 1978). Of course, the composition of Titan's organic molecular constituents (already known to include ethane, acetylene, hydrogen cyanide, and other biogenic precursor molecules) will also be of great interest.

IV. CONCLUDING REMARKS

A. Planetary Satellites: Moons vs. Rings

Planetary satellites have classically been divided into two categories. Bodies large enough to be identified as individual entities have been referred to as moons, whereas smaller bodies, detectable only collectively, have been called ring particles. As our detection techniques continue to improve, we are resolving individually smaller and smaller objects. New methods are yielding more information on the size distribution and properties of ring particles (Marouf et al. 1983). The interactions between moons and rings are very complex (Sec. II; Cuzzi et al. 1984). For example, the visible Jovian ring and Saturn's faint E ring are both composed of ephemeral small particles that must be recently eroded from, or otherwise produced by, larger satellites (J. A. Burns et al. 1984). The Uranian rings probably assume their current narrow forms due to gravitational torques exerted by small shepherding moons located between them (Goldreich and Tremaine 1979*b;* Elliot and Nicholson 1984). Especially in the Saturn system, there is a systematic inward trend of decreasing satellite mass (Fig. 2; Table I). Continuing this trend, small ringmoons exist both just outside, and embedded within, the rings. These moons are responsible for much of the structure observed in the rings (Sec. II); thus, one cannot separate rings and moons into categories of interacting and noninteracting particles. The rapid rates of orbital evolution of rings and ringmoons due to the density-wave torques exerted between them (Sec. II) support the concept of a high degree of interaction between the small and very small inner satellites of Saturn. Physical collisions may even occur between the F ring and its inner shepherding moon, 1980S27 (Borderies et al. 1984). Earlier in the history of the solar system even stronger connections may have existed; rings may be fragments of a collisionally and/or tidally disrupted moon (B. A. Smith et al. 1982), or possibly the ringmoons accreted from the rings (N. Borderies and P. Goldreich 1982, personal communication). We are thus motivated to examine planetary ring particles and regular moons together as a single system, presumed to have formed in a flattened nebula in the equatorial plane of a young planet.

B. Planetary Satellites vs. Solar Satellites

We now address the question of how planetary satellites, as a group, compare with the major and minor planets in orbit about the Sun. One important difference is that of scale, with distances and masses being hundreds to thousands of times larger in the solar system than in circumplanetary systems. Dynamically, this means that satellite systems are older, i.e., moons have undergone more orbits than have planets. One major physical consequence of this difference in scale is that the Sun is a star shining brightly at a temperature of 5700 K, whereas the giant planets, at ~100 K, are quite cool. However, the difference between the temperature of the protosun and, at least, proto-Jupiter and proto-Saturn was significantly less than at present because each of these bodies was deriving energy from gravitational collapse (see chapter by Bodenheimer). Because the satellites of the giant planets are much closer to their central body than the planets are to the Sun, some of the same general consequences of central body luminosity on the condensation of matter, and thus, on the current compositions of the planets and satellites, are observed (Sec. III).

Another consequence of the scale differences between the solar and planetary systems is the differing spin states of the secondaries. Tidal influences of the planets have synchronized the spin states of almost all of their regular satellites (Secs. III.C and III.E) whereas, due to the greater distances involved (which more than offset the larger mass of the Sun), only the inner two planets have had their spin significantly altered due to tides raised by the Sun. The differences in relative angular momentum between the protosun and protoplanets and their respective nebulae have been mentioned earlier (Sec. I).

Most planetary satellites are observed in orbit about their gas giant planets. These planets presumably formed in a manner somewhat analogous with the Sun, i.e., they were surrounded by a disk, which resulted from conservation of angular momentum when the protoplanets collapsed from an early distended phase (chapters by Pollack and by Bodenheimer). Regular satellites could have formed in these disks as the planets did in the solar nebula; the irregular outer satellites may have been captured just before the envelopes collapsed (Sec. I; chapter by Pollack). The presumed lack of corresponding circumsatellite disks may be the reason that no moon possesses an observed secondary (Sec. I). Alternatively, the lack of such secondaries may be another consequence of the differences in scale between satellite systems and the planetary system (Sec. III.C).

Although the major satellites of the Sun and the planets are in many ways comparable, the asteroid belt is not a solar ring. The asteroid belt is far outside the Sun's Roche limit. The asteroids are well separated; collisions are uncommon and relative velocities are so high that they usually result in excavation or disruption. The source of these high random velocities is unknown; however, Jupiter is implicated in most theories, due to the location of strong Jovian resonances in and near the asteroid belt (see, e.g., Torbett and Smoluchowski 1980*b;* R. Greenberg and H. Scholl 1979). Particles comprising the major ring systems of Saturn and Uranus have very low random velocities, but are unable to accrete due to the strong tidal influence of their primary (Sec. II.A). No such ring exists within the Sun's Roche limit. (The recently discovered solar dust ring [Isobe et al. 1983] is almost certainly composed of ephemeral particles which evolve into the region due to Poynting-Robertson drag and vaporize there.) It is unlikely that any solid material inside the solar Roche limit could have survived over the age of the solar system. Thus, rings that are in stable collisional equilibrium (at least over reasonably long time intervals [see Secs. II and IV.A]) are probably a phenomenon unique to planets and other cool bodies.

C. Planetary Rings vs. the Protoplanetary Disk

As discussed in Sec. II, planetary rings can serve as a model for the protoplanetary disk. Many of the physical processes observed in Saturn's rings are thought to have been important to the dynamics of the protoplanetary and protosatellite disks; however, these processes must have operated in differing forms due to dissimilarities between these systems.

There is ample evidence for disk-satellite interactions in planetary rings. Moons excite spiral density waves or truncate rings at many resonance locations, receiving angular momentum from distant ring material in the process. In a similar way, resonant interactions between planetesimals/protoplanets and the protoplanetary disk may have been an important mechanism for angular momentum transport within the solar nebula (Sec. II.B). Protoplanets embedded within a disk could have excited density waves at both inner and outer resonances. The net transfer of angular momentum would have depended on the surface mass-density profile of the disk. Although even the direction of protoplanet motion is uncertain, Goldreich and Tremaine (1980) have shown that these torques could have moved Jupiter through the protoplanetary disk on a time scale of 10^4 yr; smaller protoplanets would have moved less rapidly.

Small moonlets in Saturn's (and probably Uranus') rings clear gaps around themselves via gravitational torques and shepherd ring particles into narrow ringlets. Similarly, protoplanets may have been able to clear gaps around themselves, thereby halting accretion (Sec. II.C). Because local (shepherding) torques are stronger than remote (Lindblad resonance/density-wave) torques, even a slight imbalance in the surface mass-density from one side of the planetesimal/protoplanet to the other could have offset any tendency for the protoplanet to drift through the nebula. Even if it did move, the protoplanet may have "snowplowed" the nebula along with it (Hourigan and Ward 1984). Due to the fact that the nebula's mass greatly exceeded that of the protoplanets, it may be that no long-range movement of protoplanets occurred. These topics are ripe for further study.

Collective gravitational instabilities may influence the particle size distribution in Saturn's rings (Shu 1984*a*). Similar instabilities in a particulate layer in the midplane of the protoplanetary disk could have created ~1-km-sized planetesimals (Safronov 1969; Goldreich and Ward 1973; Sec. II.D); however, gas drag in protoplanetary and protosatellite disks may have increased relative velocities in the dust layer sufficiently to have removed this instability (Torbett 1984*b*). If the protoplanetary disk was sufficiently massive ($\gtrsim 0.5\ M_{\odot}$), gravitational instabilities may have created giant gaseous protoplanets (Cameron 1978). Unfortunately, little is currently known as to the actual form these instabilities may take. For instance, we do not know whether a gravitational instability forms a protoplanet or merely a free spiral density that propagates away through the disk. Also, the extent to which such free waves can contribute to redistribution of mass and angular momentum is unknown. Perhaps they account for what most models now attribute to an *ad hoc* turbulent viscosity. Again, much work needs to be done.

Planetary rings remain segregations of countless small satellites, whereas portions of the protoplanetary disk were able to accrete into the planets. The classic explanation, namely, that rings exist where the strong gravitational field of a nearby planet prevents accretion or can tidally disrupt a fluid moon (Roche 1847), remains generally accepted; however, the situation is not as simple as Roche had envisioned. Rings and moons are seen interspersed in a region near the Roche limit for tidal disruption. The cause of this distribution of bodies in the Roche zone is poorly understood. Possible explanations include inward motion of moons due to gas drag, breakup or erosion of moons near Roche's limit, and limits on the size to which weakly aggregated moons can grow in this region. Understanding the interactions responsible for this distribution may yield information on the general dynamics of the accretion process (Sec. II.A).

D. Summary

We have discussed several ways in which observations of planetary moons and rings can help us understand the origin of the solar system. Regular planetary satellites can be used as tracers; their compositions, locations and sizes yield clues to the environments of the circumplanetary nebulae in which they supposedly formed (Secs. I and III). Irregular satellites also may help us understand more about the formation of the giant planets (Sec. I). Planetary rings offer several valuable clues towards understanding dynamical processes operative in the solar nebula. Their very existence, as many small bodies instead of one large moon, tells us that tidal influences can greatly affect the accretion process (Sec. II.A). The mechanisms which cause the observed structure in Saturn's rings may have been important in the disks out of which the moons and planets formed. Density waves, driven by moons, are observed at many locations within Saturn's rings; density waves excited in the same manner by protoplanets may have transported large amounts of material and angular momentum within the solar nebula (Sec. II.B). Moonlets orbiting within Saturn's (and probably Uranus') rings are able to shepherd the paths of nearby ring particles, thereby clearing gaps around themselves; similar processes, operating in the solar nebula, may have halted the accretion of protoplanets (Sec. II.C).

All of the topics mentioned above are active research areas. On the observational front, we can look forward to the Voyager encounters of Uranus and Neptune in 1986 and 1989, respectively. We will then have a third regular satellite system to compare with those of Jupiter and Saturn, and we may learn more about why Neptune's satellite system is so irregular. The Galileo mission will give us a much more detailed look at Jupiter's moons and rings; eventually an orbiter should also be sent to the Saturn system. Groundbased observations and those made from the Hubble Space Telescope will also be very helpful.

On the theoretical front, a better understanding of the processes creating structure in Saturn's rings is needed. Nonlinear spiral density waves and shepherding may have been very important in the protoplanetary disk, and theoretical studies which can be confirmed by observations of Saturn's rings will be useful. Theoretical studies should also be extended to include the effects of both gas and particulates in the solar nebula. These processes may then be incorporated into models of circumplanetary and circumstellar disks (see, e.g., chapters by Cassen et al., by Cameron, by Lin and Papaloizou, and by Safronov and Ruzmaikina).

Acknowledgments. We wish to thank our colleagues for many helpful conversations and comments on preliminary versions of this manuscript, especially A. Boss, J. Burns, P. Cassen, C. Chapman, P. Goldreich, W. Kaula, D. Lin, J. Pollack, R. Reynolds, F. Shu, S. Squyres, G. Stewart, S. Tremaine, W. Ward and G. Wetherill. We are very grateful to M. Gomes for her help in preparing the manuscript. J. J. L. was supported by an NAS-NRC Resident Research Associateship.

PART VII

Solar Nebula Models and Other Planetary Systems

FORMATION OF THE SOLAR NEBULA AND THE PLANETS

V. S. SAFRONOV AND T. V. RUZMAIKINA
USSR Academy of Sciences

Any scenario for the formation of the Sun and solar nebula is dependent on the angular momentum of the presolar nebula. In this chapter, a slowly rotating presolar nebula with angular momentum J $\sim 10^{52}$ $(M/M_{\odot})^{5/3}$ *g* $cm^2 s^{-1}$ *and mass* M $\simeq M_{\odot}$ *is considered. The transfer of angular momentum within the core from its central part to the surface by magnetic forces prevents the core from fragmenting and establishes differential quasi-Keplerian rotation in the surface equatorial layer of the core. Continuous outflow of angular momentum results in an increasing equatorial radius of the core and formation of an embryo of the solar nebula. Accretion of the gas from the surrounding presolar cloud onto the forming solar nebula hinders the disk expansion and at the same time is a probable source of turbulance and therefore of viscous outward angular momentum transfer.*

The problem of the accumulation of planets from many embryos in the zone of each planet is considered. It is shown that at an intermediate stage there is a moderate runaway growth of planet embryos, and that the parameter θ *determining the velocities of bodies increases significantly. However, at the final stages* θ, *is of order unity. The total time scale of growth of the terrestrial planets remains* 10^8 *yr as it was in earlier investigations.*

Two main ideas in solar system cosmogony are now widely recognized: (1) a common origin of the Sun and the protoplanetary cloud from the same nebula; (2) formation of the planets by accumulation of solid material. The first idea is in some sense a return to classical hypotheses by Kant and Laplace which began about two decades ago (Hoyle 1960; Cameron 1962; Schatzman 1967). However, it is only now that ways to overcome difficulties in explaining the main properties of the solar system have been found. The second idea was revived by Schmidt (1944,1957) on the basis of geophysical and geochemical considerations, and by Urey (1952) on the basis of physico-chemical investigations of meteorites. Workers in many countries are presently actively involved on this idea. The main stages of evolution of the preplanetary cloud have been studied and a theory for the accumulation of the planets has been developed. At the same time the interaction between solar system cosmogony and other related sciences has been widened so that the subject is now closely connected with astrophysics, cosmochemistry, meteoritics, space research, and Earth sciences. The development of cosmogony now depends strongly on the successes in these sciences.

In this chapter we turn our attention to questions which are important for the correct choice of the main mechanisms responsible for the formation of the solar system. The following terminology will be used. A fragment of about one solar mass which is isolated from a molecular cloud and collapses under self-gravitation we refer to as a protostellar cloud or, more specifically, the presolar nebula (PSN). A quasi-Keplerian gaseous disk developing around a dense central core (the forming protosun) is referred to as the solar nebula (SN) or preplanetary cloud.

I. FORMATION OF THE SOLAR NEBULA

There is growing evidence for the existence of circumstellar disks around some pre-main sequence stars (see, e.g., the chapter by Harvey; Cohen 1983), and for a dust ring around Vega and probably some other stars (Walgate 1983). This evidence gives support to the idea of a common origin for stars and planets. It reveals also that the origin of the solar nebula is closely connected with the problem of star formation.

An attractive mechanism for the formation of a single star with a disk is associated with the angular momentum transfer from the center outward during the collapse of a protostellar cloud (Tscharnuter 1978,1980; Safronov and Ruzmaikina 1978; Ruzmaikina 1981*a*,1982; Cassen and Moosman 1981; Cassen and Summers 1983). According to Lynden-Bell and Pringle (1974) the energy of a rotating system with constant mass and total angular momentum has a minimum when most of the mass is concentrated at the center while most of angular momentum is contained in distant negligibly small mass. The present solar system where almost all the mass is concentrated in the Sun and most of the angular momentum is in the planets, represents an example of such an energetically preferable system.

On the other hand, about a half of all stars are members of binaries or multiple systems of stars. Their angular momenta are on the average much higher than the angular momentum of the solar system. So it can be expected that single stars with disks and, particularly, the solar system are mainly formed as a result of collapse of protostellar clouds rotating more slowly than clouds from which binaries are formed (Fig. 1).

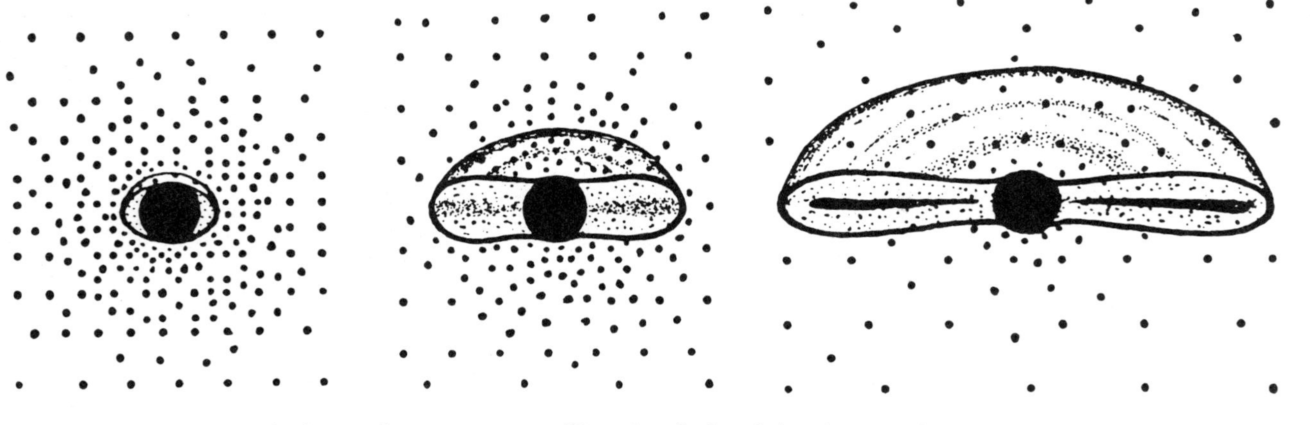

Fig. 1. Schematic pictures of consequent stages of formation of solar nebula and protosun from slowly rotating PSN.

According to observations, interstellar molecular clouds have angular velocities with range $\Omega \sim 10^{-13}$ rad s^{-1} to probably $\Omega \lesssim 10^{-15}$ rad s^{-1} (chapter by Goldsmith and Arquilla; Field 1978; Bodenheimer 1980). The specific angular momentum of a spherical mass M with uniform number density n is

$$J/M \sim 10^{20} \frac{\Omega}{10^{-15}} \left(\frac{M}{M_{\odot}} \right)^{2/3} \left(\frac{n}{10^3} \right)^{-2/3} \tag{1}$$

Clouds with $\Omega \sim 10^{-13}$ to 10^{-15} rad s^{-1}, $M = 1$ M$_{\odot}$ and $n \sim 10^3$ cm^{-3} have $J/M \sim 10^{20}$ to 10^{22} cm^{2}s^{-1} and hence $J \sim 10^{53}$ to 10^{55} g cm^2 s^{-1}. The collapse of clouds rotating with so large a value of J/M has been studied by many authors. The results were reviewed by Bodenheimer (1980), Tscharnuter (1980) and Tohline (1982) (see also Boss and Black [1982] and Boss [1983]). During hydrodynamic contraction of the cloud the ratio, β, of the rotational energy to the absolute value of gravitational energy increases rapidly (up to $\gtrsim 0.3$) in the central part of the cloud and the cloud breaks up on a dynamic time scale. Neither magnetic fields nor turbulent viscosity could prevent the fragmentation at this stage of collapse. The process seems to finally terminate in the formation of double or multiple stellar systems (Regev and Shaviv 1981).

The minimum mass M_{SN} and angular momentum J_{SN} of the primitive solar nebula have been estimated by several authors by adding the volatiles to the present mass of the planets in order to restore solar composition (Hoyle 1960; Kusaka et al. 1970; Weidenschilling 1977*b*). The estimated values lie in the range $M_{SN} \simeq 0.01$–0.02 M$_{\odot}$, $J_{SN} \simeq 3 - 5 \times 10^{51}$ g cm^2 s^{-1}. It should be noted that gravitational perturbations produced by the giant planets in the process of their accumulation could result in the ejection of a substantial mass of solid material from the solar system, perhaps comparable with the total mass of the outer planets. Hence, the mass of the solar nebula must have been at least 0.02 to 0.03 M$_{\odot}$.

On the other hand, the possibility of diffusion of solid bodies due to their mutual gravitational pertubations should also be taken into account. The most effective perturbations were probably due to large planets. Jupiter and Saturn ejected bodies not only into hyperbolic and parabolic orbits, but also into eccentric elliptical orbits with large semimajor axes. Most of the bodies returned as a rule to the inner zone. However, the bodies had a nonnegligible probability of colliding with each other and so remain in the region of the outer planets ($R > 10$ AU) through their orbits becoming more circular. This should be included in further quantitative considerations. The angular momentum of the solar nebula with an initial radius about equal to the radius of Saturn's orbit ($\simeq 10$ AU) and mass $M_{SN} = 0.02$–0.03 M$_{\odot}$ is $J_{SN} = 5.10^{51}$–10^{52} g cm^{2}s^{-1}.

It has been recognized in recent years that a slowly rotating cloud can evolve into a single protostar with a disk of low mass (Safronov and Ruzmaikina 1978; Tscharnuter 1980; Ruzmaikina 1981*a,b;* Cassen and

Moosman 1981). When the cloud is barely Jeans unsteable, it contracts highly nonhomogeneously. A quasi-equilibrium protostellar core with a mass of a few percent of the total mass $M_c \sim 10^{-2}\, M_\odot$, with radius $R_c \sim 10^{12}$ cm and central density $\rho_c \sim 10^{-2}$ g cm^{-3} can be formed in an initially uniform and uniformly rotating cloud of mass $M_o \simeq 1 M_\odot$ with an angular momentum (g cm^{2}s^{-1})

$$J_o \lesssim 10^{52}\,(M/M_o)^{5/3} \tag{2}$$

and correspondingly with initial value of $\beta_o \lesssim 4 \times 10^{-6}$ (Ruzmaikina 1981*b*). According to the test calculations of Tscharnuter (1980), the initial values β_o should be $< 10^{-5}$ to avoid the ring mode of instability setting in before the core forms.

Development of a slowly rotating protostellar cloud with angular momentum given by Eq. (2) seems to be quite feasible. If a fragment of the solar mass corotates with the Galaxy ($\Omega \sim 10^{-15}$ s^{-1}) at the onset of collapse ($n = 10^5$ cm^{-3}, $T = 10$ K), it has an angular momentum $J \simeq 10^{52}$ g cm^{2}s^{-1}. Magnetic braking of molecular clouds is efficient at early stages of contraction (Mouschovias 1978; Mestel and Paris 1979; Dorfi 1982). The braking can maintain a nearly constant angular velocity comparable to that of galactic rotation up to the same densities $n = 10^5$ cm^{-3} resulting in $J \sim 10^{52}$g cm^{2}s^{-1} for $M = 1\, M_\odot$. Protostellar clouds with such an angular momentum can also be produced in a repeated process of collapse and fragmentation of a molecular cloud (Bodenheimer 1980).

Observational evidence shows that molecular clouds are clumpy down to the smallest resolvable scales and have chaotic supersonic internal motions with a velocity difference Δ in km s^{-1} over length scales L (measured in parsecs) given by the relation

$$\Delta = 1.1 \left(\begin{matrix} +0.4 \\ -0.3 \end{matrix} \right) \cdot L^{0.38} \tag{3}$$

for $0.1 < L < 100$ pc (see Larson 1981, and references therein). The clumping can be produced by a thermal instability in the shocked gas (Burduzha and Ruzmaikina 1974; Smith 1980). The chaotic motions can produce an effective viscosity strong enough to ensure the uniform rotation of a molecular cloud at a relatively prolonged stage of evolution when hydrodynamic collapse is hindered by the chaotic internal motion. Then, the angular momentum per unit mass in the developing central condensation decreases to a value lower than that given by the condition (Eq. 2) for the formation of a single core in the PSN.

The core evolves on the accretion time scale $\tau_a = M_c/\dot{M}$ which is much longer than the free-fall time so that a redistribution of the angular momentum within the core can be effective enough to prevent its fission and allow the development of a disk. Transfer of angular momentum within the core can be associated with a torque produced by a magnetic field enhanced in the collapse (chapter by Mestel; Scott and Black 1980; Black and Scott 1982; Ruzmaikina 1981,1984). At the early stages of collapse the behavior of the magnetic field is determined by the drift of ions through the neutrals (ambipolar diffusion). The strength B of the magnetic field can be estimated by equating a characteristic diffusion time with the time scale of contraction. It depends on the degree of ionization, which is determined by a balance between the rate of ionization ξ by cosmic rays (or by inner radioactive sources) and the rate of recombination on dust grains. Ruzmaikina (1984) finds for spherical collapse that

$$B \sim 2\pi^{3/5} \cdot \mu_i^{3/5} \left(\frac{GM_o}{c_s^2 R_o} \Phi_o \right)^{1/5} \left(\frac{\rho_g r_g \langle \sigma_i u_i \rangle}{f} \xi n \right)^{2/5} \tag{4}$$

where M_o and R_o are the mass and initial radius of the cloud, $\Phi_o = \pi B_o R_o^2$ is the initial magnetic flux through the equatorial plane, ρ_g and r_g are the density and typical radius of grains, f is a fraction of the cloud mass contained in the dust, the quantities μ_i, σ_i, u_i are, respectively, the reduced ion-neutral mass, the cross section for collisions between neutrals and ions, and their relative velocities. A similar relation $B \propto n^{0.44}$ was obtained by Black and Scott (1982) in two-dimensional numerical computations of collapse of magnetized clouds with ambipolar diffusion.

When the density (cm^{-3}) increases to

$$n \simeq 5 \times 10^{13} B \tag{5}$$

in Gauss, Ohmic dissipation becomes important and any further increase in B is stopped. At density $\rho = 2nm_H \sim 10^{-8}$–$10^{-7}$ g cm^{-3} the temperature reaches around 1600 K, the dust evaporates, the alkaline metals are thermally ionized, and the magnetic field becomes frozen in. Up to the end of contraction the field intensifies 10^3 to 10^4 times.

Consider a protostellar cloud of $M_o = 1\ M_\odot$, $c_s = 3 \times 10^4$ cm s^{-1}, $R_o = 10^{17}$ cm, $n_o = 10^5$ cm^{-3}, $B_o = 10^{-5}$ Gauss, $< \sigma_i u_i > = 10^{-9}$ cm^3 s^{-1}, and the dust with the usually taken parameters $f = 2 \times 10^{-2}$, $\rho_g = 1$ g cm^{-3}, $r_g = 10^{-5}$ cm. In the center of the protostellar core just formed ($M_c \sim 10^{-2} M_o$, $R_c \sim 10^{12}$ cm, $\rho_c \sim 10^{-2}$ g cm^{-3}) a poloidal component B of the magnetic field becomes of order 10^3 Gauss, provided that the gas is ionized by cosmic rays with $\xi \sim 10^{-18}$ s^{-1} at the initial stage of the collapse.

The core is rotating differentially with $\Omega \sim 10^{-4}$ s^{-1} in the center ($\rho_c \sim 10^{-2}$g cm^{-3}) and $\Omega \sim 10^{-6}$ s^{-1} near the surface ($\rho \sim 10^{-5}$ g cm^{-3}), provided that initially PSN was uniform in density and angular velocity ($\Omega_o = 10^{-15}$ s^{-1}), and its local angular momentum was conserved during the collapse. The differential rotation generates an azimuthal component B_φ from the poloidal one. When the dissipation of the field is unimportant, B_φ increases in accordance with the equation

$$\frac{\partial B_\varphi}{\partial t} = R(\mathbf{B} \nabla \Omega) \tag{6}$$

where R is a distance from the axis of rotation (Ω is assumed to be directed along the z axis). The intensification proceeds for some period of time which is dependent on the magnetic Reynolds number

$$\mathrm{Re}_m = \frac{4\pi\sigma_e}{c^2} L_B^2 \Omega \tag{7}$$

where σ_e is an electrical conductivity and L_B is a length scale of the magnetic field.

The components of the magnetic field with the axis of symmetry initially parallel ($B_\parallel$) and perpendicular ($B_\perp$) to the axis of rotation evolve to the different final states (Parker 1979). The component $B_\parallel$ generates the azimuthal field B_φ which increases over a time scale

$$t_\parallel \sim \mathrm{Re}_m \, \Omega^{-1} \tag{8}$$

according to Eq. (6), and then attains the steady state

$$B_\varphi \sim \mathrm{Re}_m \, B_\parallel \, . \tag{9}$$

The component $B_\perp$ generates B_φ temporarily up to the time

$$t_\perp \sim \mathrm{Re}_m{}^{1/3} \, \Omega^{-1} \tag{10}$$

when B_φ reaches maximum and afterwards is excluded from the region of differential rotation.

For a fully ionized gas in the core $4\pi\sigma_e/c^2 = 10^{-13} T^{3/2} \mathrm{cm}^{-2}\mathrm{s}$. Taking for the core $R \sim L_B \sim 10^{12}$ cm, $\Omega = 10^{-6}\mathrm{s}^{-1}$ and $T = 2 \times 10^4$ K, we obtain $\mathrm{Re}_m \simeq 3 \times 10^{11}$ and the time scales $t_\perp \sim 10^3$ yr, $t_\parallel \sim 10^{10}$ yr.

Thus, the differential rotation, if supported in some way in the nonturbulent core, could increase B by wrapping the lines of force of B for the total time of accretion (10^5–10^6 yr).

A magnetic torque, produced by combined action of poloidal $\mathbf{B}$ and toroidal (B_φ) magnetic field, transports the angular momentum outwards to lower Ω. The change of the angular momentum per unit mass $j = \Omega R^2$ is described by equation

$$\frac{\partial j}{\partial t} = \frac{1}{4\pi\rho} \mathbf{B} \cdot \nabla(R_\varphi). \tag{11}$$

If B_φ is described by Eq. (6), then Eq. (11) transforms into the wave equation, for Ω, and a time scale for the transport of j over a distance R is

$$\tau \sim R/V_A \tag{12}$$

where $V_A = [B \cdot B_{\parallel}/(4\ \pi\ \rho)]^{1/2}$ is the Alfvén velocity determined by the poloidal component of $B = (B_{\parallel}^2 + B_{\perp}^2)^{1/2}$.

Angular momentum is transported to the edge of the core in about 10^2 yr if $B \sim B_{\parallel} \sim 10^3$ Gauss in the center. The time scale of the core contraction caused by accretion of new mass becomes longer than 10^2 yr when the core mass increases to a few percent of a solar mass. After that, the magnetic torque effectively transports the angular momentum from the center to the surface. On acquiring the angular momentum, an equatorial region of the core moves outward and transforms into a differentially rotating embryo of the solar nebula. During further evolution, the magnetic field links the disk with the core and removes the excess of the angular momentum from the core brought by infalling gas.

Note that the angular momentum of the core flows out in a plane inclined to the equatorial plane of the PSN, if $B_{\perp} \neq 0$, i.e., if the principal direction of the magnetic field in the PSN does not coincide with the axis of rotation. It is likely that the presence of $B_{\perp}$ could result in the observed 7° inclination of the solar equatorial plane to that of the ecliptic. Hoyle (1960) was the first to recognize that a magnetic field can transport angular momentum from the protosun to the solar nebula. Recent observational evidence indicates that 3/4 of pre-main sequence stars of solar type in the association Ori Ic are already slow rotators, i.e., $< (\vartheta_{\varphi} \sin i)^2 >^{1/2} = 12$ km s^{-1} (Smith et al. 1983) suggesting that braking mechanism is efficient.

The magnetic field can effectively also transport angular momentum in the inner dust-free region of the solar nebula where the temperature $T \gtrsim 1600$ K, and alkine metals are ionized. The radius R_1 of this zone can be estimated assuming that the innermost part of the PSN, including the growing disk, is in quasi-thermal equilibrium due to the high opacity of the dust-containing envelope. The envelope is opaque to thermal radiation with $T \gtrsim 2000$ K at least for $M_c \lesssim 0.5$–$0.8\ M_{\odot}$ assuming the law of opacity provided by Decampli and Cameron (1979). Within the central dust-free region the temperature (K) then is

$$T \gtrsim \left(\frac{L}{\sigma_{ST}} \right)^{1/4} R_1^{-1/2} = 394 \left(\frac{L}{L_{\odot}} \right)^{1/4} R_1^{-1/2} \tag{13}$$

where σ_{ST} is the Stefan-Boltzman constant, and L is the luminosity of the core (Allen 1973). According to numerical calculations by Tscharnuter and Winkler (1979), L is $\gtrsim 30\ L_{\odot}$ for core masses $0.1 \lesssim \frac{M}{M_{\odot}} \lesssim 0.8$, and the dust-free zone has a radius $R_1 \gtrsim 0.3$ AU. Stahler et al. (1981) found $R_1 \gtrsim 0.58$ AU for $\dot{M}/M_{\odot} = 10^{-5}$yr^{-1}.

In the region $R < R_1$ the degree of ionization is $x_i \gtrsim 10^{-8}$. The conductivity for a slightly ionized gas is $\sigma_e = 1.1 \times 10^{16}\ x_i$ s^{-1} (Hayashi 1981). Then the magnetic Reynolds number is $\mathrm{Re}_m = 10^6$ near the edge of the region

$R \simeq R_1$, where $T \simeq 1600$ K and the scale height $h = c_s\,\Omega^{-1} \simeq 0.4\,R_1$ and increases inward. The magnetic field dragged out from the core can be highly amplified by differential rotation in this region and so it can effectively transport angular momentum outwards to $R \sim R_1$ over $t \sim 10^3$ yr.

The role of a magnetic field in the rest of the solar nebula is questionable. Hayashi (1981) found that the degree of ionization in the main part of the SN is negligible and the field decays over a time scale smaller than $2\,\pi\,\Omega^{-1}$. The component B_φ could be temporarily intensified for high z coordinates and in the region of the outer planets which would be transparent to cosmic rays. However, the intensification is of short duration and so it is not essential in the presence of the dust. It is only after depletion of the dust that the field can intensify to a dynamically important value.

If helical turbulence developes in the core or in the disk, field amplification by a hydromagnetic dynamo effect is possible in the region with $R_m >> 1$ (Zeldovich et al. 1984). The maximum strength of a large-scale magnetic field which can be generated by turbulence with typical velocities v_t and mixing length l is of order

$$\frac{B_r B}{4\pi\rho} \sim v_t h\Omega. \tag{14}$$

The torque associated with this field is a factor h/l of that produced by turbulent viscosity itself. Hence, in the case $l < h$ the torque associated with this field can dominate in the central region $R \lesssim R_1$ and, possibly, in a low-density subsurface layer as well as at the edge of the disk.

A magnetic torque can cause the expansion of the embryo solar nebula to a radius $R_d \sim R_1$. However, another mechanism of redistribution of the angular momentum is needed for further expansion. If the solar nebula is turbulent, it can continue to expand due to turbulent viscosity. The radius R_d of the viscous disk increases with time as

$$R_d \simeq 3(\nu\, t)^{1/2}. \tag{15}$$

(Lynden-Bell and Pringle [1974]; and with accretion, Kolychalov and Syunyaev [1980]). According to this relation, a turbulent viscosity $\nu \sim 3 \times 10^{15}$ cm^2 s^{-1} is needed to form a solar nebula of the size $R_d \sim 10^{15}$ cm during the time of accretion $\sim 10^6$ yr (Ruzmaikina 1981*a*). Such viscosity could be produced by a rather moderate turbulence with $l \sim h$ and $\nu_t \sim 10^{-2}\,c_s$, where c_s is the sound speed.

Evolution of a thin viscous Keplerian disk around the core at the stage of accretion of PSN gas is described by the equations of conservation of mass and angular momentum

$$\frac{\partial\sigma}{\partial t} + \frac{1}{R}\frac{\partial}{\partial R}(\sigma\, R\, V_r) = \dot{\sigma}_a \tag{16}$$

$$\frac{\partial (j\sigma)}{\partial t} + \frac{1}{R}\frac{\partial}{\partial R}(j\,\sigma\,R\,V_r) = j_a\dot{\sigma}_a - \frac{1}{R}\frac{\partial}{\partial R}\left(\sigma\,\nu\,R^3\,\frac{\partial\Omega}{\partial R}\right) \tag{17}$$

(Cassen and Moosman 1981; Safronov 1982), where σ is the surface density, σ_a is the mass infalling onto unit area of the disk in the plane $z = 0$ per unit time, V_r is the radial velocity component, within the disk, j and j_a are the specific angular momenta of the disk material and infalling gas, respectively. The last term on the right side of Eq. (17) describes the radial angular momentum transfer by viscous torque. Elimination of $\partial\sigma/\partial t$ from Eqs. (16) and (17) gives

$$V_r = -\frac{3}{\sigma R^{1/2}}\frac{\partial}{\partial R}(\sigma\,\nu\,R^{1/2}) - \frac{2R\dot{\sigma}_a}{\sigma}\left(1 - \frac{j_a}{j}\right) - \frac{R\dot{M}_o}{M_o}. \tag{18}$$

Inserting this expression for V_r into Eq. (16), we obtain the following basic equation for σ (R, t)

$$\frac{\partial\sigma}{\partial t} = \frac{3}{R}\frac{\partial}{\partial R}\left[R^{1/2}\frac{\partial}{\partial R}(\nu\,\sigma\,R^{1/2})\right] +$$

$$\frac{1}{R}\frac{\partial}{\partial R}\left[2R^2\,\dot{\sigma}_a\left(1 - \frac{j_a}{j}\right) + \frac{\dot{M}_c R^2}{M_c}\right] + \dot{\sigma}_a. \tag{19}$$

Cassen and Moosman (1981), assuming that the PSN gas falls only onto the central part of the disk, have shown that the relative importance of viscosity in the disk is determined by a dimensionless parameter

$$P = (R_V/R_{JM})^2 \tag{20}$$

where $R_V = [\nu\,\tau_a\,(M_o)]^{1/2}$ is the diffusion length associated with the collapse time, $R_{JM} = J^2/k^2\,GM_o^3$ is the radius at which the gas of the outermost equatorial region of the cloud would be in centrifugal balance with gravity (k is a number which depends on the initial distribution of angular momentum in the cloud; $k = 2/5$ for uniform and uniformly rotating clouds). When $P << 1$, the radius of the disk R_d is about R_{JM}. When P is very large, an approximate solution of Eq. (19) shows that the disk expands radially almost at the same rate (Eq. 15) as in the absence of accretion. Cassen and Summers (1983) obtained an analytical similarity solution of the evolution equation for the viscous disk with ν = const and $\nu \propto R$, assuming that all material falls onto the core. The length scale of the disk with uniform viscosity is found to be $R_d = 2\,(\nu\,t)^{1/2}$, i.e., 1.5 times less than that given by Eq. (15) because of increase of the core mass.

At the end of accretion the mass of the disk is

$$M_{SN} = 0.193\,P^{-1/4}\,M_\odot. \tag{21}$$

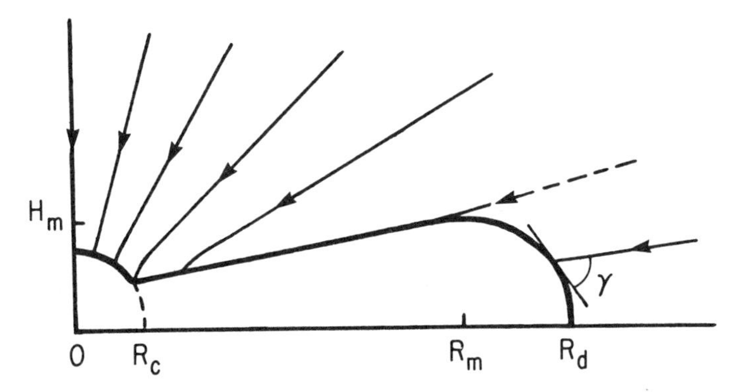

Fig. 2. Disk geometry. Gas near the equatorial plane falls onto the outer part of the disk.

We must take $P \sim 10^4$ and R_{JM} 10^{-2} $R_{\rm SN}$, respectively, in order to obtain a reasonable value for the solar nebula mass $M_{\rm SN} \simeq 0.02$ $M_{\odot}$. Then we obtain $R_{JM} \simeq 0.1$ AU and the angular momentum of PSN

$$J = k\,(GM^3\,R_{JM})^{1/2} \sim 1.10^{52} \tag{22}$$

for the final value of $R_{\rm SN} \simeq 3 \times 10^{14}$ cm and $k = 2/5$.

It is evident from geometrical considerations (Fig. 2) that an appreciable fraction of the PSN material moving near the equatorial plane in the solid angle $2\,\pi\,h_m/R_m$ falls onto the outer part of the disk, namely onto a ring with the width $\delta\,R_d = R_d–R_m$, where h_m is the maximum thickness of the disk for $R = R_m$. The total mass infalling onto the outer part of the disk per unit time is about

$$\dot{M}_d = \dot{M}\,\frac{h_m}{2R_m} \sim 10^{-1}\,\dot{M}. \tag{23}$$

The accretion of gas with low angular momentum onto the disk edge both hinders the radial spreading of the disk and at the same time gives rise to turbulence in the outer region making possible viscous expansion.

It follows from Eqs. (16) and (17) that the mass and angular momentum transport in the disk is characterized by three time scales: the viscous time scale

$$\tau_\nu = R_d^2/\nu \tag{24}$$

the time scale of accretion of PSN material onto the disk

$$\tau_\sigma = \frac{\sigma}{\dot{\sigma}_a} \tag{25}$$

(this material, possessing a low angular momentum, impels the disk to contract), and the time scale of a general contraction because of increase of the core mass

$$\tau_a = \frac{M_c}{\dot{M}_c} \tag{26}$$

For the outer part of the disk we have

$$\frac{\sigma}{\dot{\sigma}_a} \lesssim \frac{2\delta R_d}{R_d}\frac{M_d}{\dot{M}_d} \simeq \frac{4\delta R_d}{h_m}\frac{M_d}{\dot{M}} \tag{27}$$

and

$$\frac{\tau_\sigma}{\tau_a} = \frac{4\delta R_d}{h_m}\frac{M_d}{M_c} < < 1. \tag{28}$$

Thus, the influx of the gas at the perephery is the dominant effect opposing the viscous expansion of the disk. The expansion dominates ($\tau_\nu < \tau_\sigma$) when

$$\nu > \nu_e = \frac{\dot{\sigma}_a}{\sigma} R_d^2 \sim \frac{R_d^2 M_c}{\tau_a M_d}\frac{h_m}{R_d} \tag{29}$$

Using Eq. (29) and conditions for the conservation of mass and angular momentum, one can estimate the radius and mass of the viscous disk with $\nu = \nu_e$ at the end of accretion, when most of the angular momentum is concentrated in the disk (Ruzmaikina 1982).

$$R_d = \left(\frac{J^2}{GM_c^3}\right)^{1/5} \left(\tau_a \nu\right)^{2/5} \left(\frac{h_m}{\delta R_d}\right)^{-2/5} \tag{30}$$

$$M_d = \left(\frac{J^2}{GM_c^3}\right)^{2/5} \left(\tau_a \nu\right)^{-1/5} \left(\frac{h_m}{\delta R_d}\right)^{1/5} \tag{31}$$

For $J = 10^{52}$ g cm^{2}s^{-1} and $M = 1\ M_\odot$, the viscosity $\nu \simeq 10^{17}$ cm^{2}s^{-1} provides the growth of the disk to a radius $R_d = 3 \times 10^{14}$ cm over the time τ_a $(M_o) \sim 10^6$ yr.

Numerical solution of Eq. (19) was undertaken together with S. V. Maeva (personal communication). The results are in agreement with these estimates and will be published elsewhere.

We prefer to relate the solar nebula formation to the accretion stage because an infall of the PSN gas gives a definite source of the disk turbulence. This turbulence is due to the Rayleigh and shear instabilities in the mixing layer between the gas of the disk and the accreted gas, and also to the exciting of inertial waves with frequencies less than 2 Ω penetrating to a larger depth than the thickness of the boundary layer. Development of turbulence in the SN after the end of accretion remains problematical. The tangential velocity of the infalling gas relative to the disk material is of the order of the Keplerian velocity $(GM/R)^{1/2}$. Hence, it is reasonable to assume that the energy which feeds the turbulence is a fraction α' of the accretion energy. Then for the turbulent energy dissipation per unit mass and unit time we obtain

$$\epsilon_t = \frac{v_t^3}{l} \simeq \alpha' \frac{GM}{R} \frac{\dot{\sigma}_a}{\sigma} \tag{32}$$

and for the turbulent viscosity

$$\nu = \frac{1}{3} v_t l \simeq \frac{l^2}{3v_t^2} \alpha' \frac{\dot{\sigma}_a}{\sigma} \frac{GM}{R} \tag{33}$$

where v_t and l are the turbulent velocity and the mixing length. A general structure of motions in the disk can be understood for the region of the disk where $\dot{\sigma}_a$ can be approximated by a power-law function of R. From Eqs. (18) and (19) we find for a low-mass Keplerian disk with $\dot{\sigma}_a \propto R^{-na}$ and $v_t = \Omega\, l$ (see below) that

$$V_r = \frac{R\dot{\sigma}_a}{\sigma} \left(n_a - \frac{5}{2} \right) \alpha' - 2 - \zeta \tag{34}$$

and

$$\frac{\partial \sigma}{\partial t} = \dot{\sigma}_a(5-2n_a)\,[1 + (2 - n_a)\frac{\alpha'}{2} + 0(\zeta)] \tag{35}$$

where $\zeta = \dfrac{\sigma \dot{M}_c}{\dot{\sigma}_a M_c} \sim 10^{-1}$ (see Eq. 27). Thus, when $n_a < 5/2$ the material in the disk moves inward ($V_r < 0$) but its surface density increases ($\partial\sigma/\partial t > 0$) because of $\dot{\sigma}_a$.

The power law approximation fails near the edge of the disk. In this region σ depends on $\dot{\sigma}_a$ which in turn is dependent on the shape of disk. Another simplification in this model is in the assumption of uniform turbulence over the whole thickness of the disk. A "sandwich" structure of the disk cannot be excluded (Safronov and Vitjazev 1984). The upper layer moves inward and causes turbulence in an underlayer without mixing. The latter moves outward due to the viscous force. A deeper region is not turbulent and slowly contracts because of the increasing central mass.

The magnitude of the turbulent viscosity is a major mystery in the study of the SN formation. von Weizsäcker (1948) has suggested that turbulence in the solar nebula can be induced by a convection in radial direction due to a decrease in the temperature with distance R from the Sun. Later it was shown

(Safronov and Ruskol 1957; Safronov 1958) that the solar nebula should be stable. Due to the increase of the angular momentum outwards ($j \propto R^{1/2}$), too high a temperature gradient would be needed to exite convection. Shakura and Sunyaev (1973) have assumed that the turbulent velocity v_t and the mixing length l are approximately equal to the sound velocity c_s and the thickness h of the disk, respectively. The turbulent viscosity has been taken in the form

$$\nu = \alpha\, c_s h \tag{36}$$

where $\alpha \lesssim 1$. The value $\alpha \simeq 1$ can be considered as a reasonable upper limit if there is sufficient energy for turbulence. Sometimes Eq. (36) is applied for weak turbulence with $\alpha << 1$. In this case, real values of v_t and/or l should be considerably smaller than c_s and h, respectively.

In another approach, a linear dependence

$$l \sim \frac{v_t}{\Omega} \tag{37}$$

has been assumed (Safronov 1969,1982; Lin and Papaloisou 1980). Relation (37) is similar to that between amplitudes of radial displacements and velocities of a body moving on a slightly elliptical orbit. The relation satisfies the condition that dissipation of the turbulent energy per unit time $\epsilon_t \sim v_t^3/l$ is approximately equal to the rate of a viscous dissipation of energy in the differentially rotating Keplerian disk due to turbulent viscosity ($\nu \sim 1/3 l v_t$),

$$\epsilon_\nu = \nu R^2 \left(\frac{\partial \Omega}{\partial R} \right)^2 = \frac{3}{4}\, l v_t\, \Omega = \frac{3 v_t^{\,3}}{4 l}\ . \tag{38}$$

Lynden-Bell and Pringle (1974) have assumed that, due to various instabilities, turbulence is established in a disk with the effective viscosity ν_{eff} which being inserted in the expression for the Reynolds number $\text{Re} = R^2/\nu_{\text{eff}}$ gives the critical value $\text{Re} \sim 10^3$ for the onset of turbulence.

Only a few attempts to specify possible sources of turbulence in the disk have been undertaken. Zeldovich (1981) studied conditions of turbulence onset due to the shear of a fluid between two cylinders (the Couette flow when the outer cylinder rotates $\partial j/\partial t > 0$) and gave arguments in favor of a turbulence in the Keplerian disk.

Lin and Papaloizou (1980) have suggested convective motion in the z direction (along the axis of rotation) as a mechanism for maintaining the turbulence in the disk. They have found that in the case of a power law dependence of the opacity $\tilde{\kappa}$ on temperature, $\tilde{\kappa} \propto T^{\xi}$, the necessary condition for the convection is $\xi > 1$. The convection cools the disk and without turbulence it contracts towards the equatorial plane. However, turbulent dissipation of energy heats the disk. The authors find that there exists some stationary state of the disk when the gain of energy balances the loss. Thus, in this model, the turbulence is self-maintained and the turbulent viscosity redistributes the disk material in the radial direction. This mechanism is interesting but, in order to have a more definite judgement about its efficiency, further investigations are needed.

Vitjazev and Pechernikova (1982) using Berlage's condition of the minimum energy dissipation have found a quasi-equilibrium density distribution in the solar nebula. They believe that the turbulence could exist until establishment of the equilibrium structure.

At the final stage of accretion, the gas of PSN falling nearly radially to the center runs across the disk and forms the shock which becomes a physical boundary of the disk. The shock is strong at the central part and edge of the disk. The cooling rate behind the shock front is high due to gas-grain collisions (Cassen and Moosman 1981). Therefore, an equation of continuity of the normal component of momentum flux at the front reduces to

$$c_s^2 \rho_s = \rho_n v_n^2 \sin^2\gamma \tag{39}$$

where γ is the angle between the direction of the flow and the local tangent to the surface in the meridional plane, ρ_s is the density in the disk near the surface, ρ_n and v_n are the density and velocity of the accretion flow

$$\rho_n = \frac{\dot{M}}{4\pi R^2 v_n}\ ,\ v_n^2 = \frac{2GM}{R} = 2V_k^2\ . \tag{40}$$

The conditions in the outermost part $R > R_m = R\,(h_m)$ of the disk are different from the conditions in the remainder of the disk. The gas from the equatorial region of the PSN falls onto the outermost part of the disk and begins to slide from the equator along the surface. The angular momentum decreases outward across the boundary layer from $j = (GMR)^{1/2}$ to $j_a << j$ and Rayleigh instability sets in. It results in the mixing of accreted gas with that of the disk (the disk gas with higher angular momentum tends to float up and gas with smaller tends to sink), followed by strong turbulence in that region. A possible clumping of the infalling gas (due to the thermal instability in the cooling layer behind the shock) promotes the development of the instability.

In the region $R < R_m$, the gas falls onto the disk at small angles and continues to flow to the center along the disk surface. A shear layer developes between the disk and outer flow because of the differences $\sim V_k$ between both their radial and their azimutal velocities. In the layer the turbulence developes with typical velocity fluctuations $v_t \sim V_k/30$ (Morkovin 1972; Tennekes and Lumley 1978). A thickness of the layer is $D \sim v_t/\Omega$. It is determined by Eq. (36) for the turbulent length scale $l \sim D$. Thus, we find $D \sim R/30$ and viscosity $\nu \sim D\, v_t/3 \sim 10^{17}$ cm^{2}s^{-1}. The boundary layer contains a relatively small fraction of the disk mass. However, inhomogeneities in the accreted gas and fluctuations of velocity, density, and pressure in the layer can exite inertial waves with frequencies < 2 Ω (Greenspan 1968). The waves penetrate into deeper layers and increases their viscosity.

II. FORMATION OF SOLID BODIES IN THE DISK

The low-mass solar nebula considered above is stable against fragmentation and formation of gaseous protoplanets. A theory for planetary accumulation from small solids meets with no major difficulty, though some points in the process are not yet clear and deserve discussion.

Grains carried by infalling gas can survive passage through the shock at the surface of almost the whole of disk excluding the innermost part (see Hollenbach and McKee 1979). Transport of dust particles in a turbulent disk and chemical processes associated with repeated evaporations and recondensations of grains have been considered by Morfill (1983; see also his chapter). The idea of intermittent turbulence in the disk has been suggested by Völk (1982). In the regions where the turbulence disappears for some time, or becomes weak enough, a rapid sedimentation of mm-sized particles occurs and gravitational instability sets in. As a result seed bodies form which survive the resuming of turbulence and can grow into larger bodies. When the turbulence reduces, particles precipitate through the gas toward the central plane of the disk. Larger particles settle faster than the smaller ones and sticking causes growth. Their settling rate increases and the growth accelerates. In the nonturbulent gas the time scale of precipitation of particles coagulating at every collision is a few 10^3 revolutions around the Sun and their final radius is about 1 cm (Safronov 1969; Weidenshilling 1980; Nakagawa et al. 1981).

However, there are two complications in the process which make its study difficult. First, the structure and physical properties of small particles at this stage and their ability to coagulate are poorly known. Compact spherical silicate particles can stick together due to the action of van der Waals and electrostatic forces only at very low relative velocities (Coradini et al. 1981). The better sticking properties of iron particles have invoked an idea of possible chemical fractionation of particles during their sedimentation (Makalkin 1980). At the same time, suggestions in favor of the formation of porous particles with a complex structure and a mixed composition have been offered. It has been proposed that these particles coagulate much more efficiently without an appreciable fractionation (Weidenshilling 1980). Fractionation relative to solar composition is not shown in the composition of Cl carbonaceous meteorites which are usually considered as the most primitive solid material in the solar system. The growth of small particles in the protoplanetary cloud should be studied much more carefully. The question is important also for understanding of the evolution of planetary rings.

One way or another a dust subdisk is formed in the central plane. At some critical density it becomes gravitationally unstable and breaks up into separate condensations (Safronov 1960,1969; Goldreich and Ward 1973; Genkin and Safronov 1975). For a thin disk of thickness h, the dispersion equation is found to be

$$\omega^2 = \kappa^2 + k^2 c_p^2 - 2\,\pi\, G\, K\,(1 + kh/2)^{-1} \tag{41}$$

where κ is epicyclic frequency, $\kappa^2 = d\,(\Omega\, R^2)^2/R^3\, \mathrm{d}R$ and c_p is the random velocities of the particles. The same equation, only with $h = 0$, has been derived by Goldreich and Ward for an infinitely thin disk. In the dust disk with decreasing thickness h, the instability begins at $kh/2 \simeq 0.4$.

When the thermal velocities of particles are equal in the R and z directions (as for the gas), then, in a disk of finite thickness, the instability begins at a surface density σ which is 40% higher than in the infinitely thin disk. Weidenshilling (1980) points out that turbulence arising at the boundary of the dust layer, due to its faster rotation with respect to the gas, prevents the contraction of the layer and delays the onset of instability until the particles grow to meter-sized bodies. Probably a more important process for hindering the instability can be produced by differences in radial velocities of particles because of the dependence of gas-drag forces on the size of solid bodies (Torbett 1984).

Dust condensations formed by the instability cannot contract immediately because of their rotation and have initial densities only a few times the Roche density ρ_R. However, the contraction occurs when they coalesce at collisions. In the Earth's zone they have transformed into solid bodies of about 10-km size in 10^4 yr after the increase of their initial masses by a factor of 10^2. In Jupiter's zone the condensations have transformed into solid bodies of 10^3 km in 10^6 yr. The stage of condensations has been much more lengthy in Uranus' and Neptune's zones owing to their much smaller initial densities $\rho_o \sim 3 \times 10^{-10}$ g cm^{-3} as $\rho_R \propto R^{-3}$. However, even there it was substantially shorter than the accumulation time of the planets. Remnants of diffuse matter around the planets can mix with the inner parts of the circomplanetary satellite swarms and form planetary rings.

Assuming $\omega^2 = 0$ one can find from Eq. (41) that the critical density is a minimum at $\lambda = 2\,\pi/k \simeq 8\,h$ and $\rho_{cr} \simeq 2\,\rho^*$, where $\rho^* = 3\,M_\odot/4\,\pi\,R^3$. In the dust layer this density can be attained due to growth of particles to the size when their relative velocities become $\lesssim 10$ cm s^{-1} in the Earth's zone, and 3×10^2 cm s^{-1} in the zone of Jupiter. Velocities of molecules are much higher ($\sim 10^5$ cm s^{-1} and the density of the gas is much lower than ρ_{cr}. It was shown long ago that in the gaseous protoplanetary cloud gravitational instability and the formation of massive gaseous protoplanets would be possible only if the mass of the cloud is comparable with that of the Sun (Ruskol 1960). Nevertheless, models of gaseous protoplanets continue to be suggested (see, e.g., Cameron 1978*c*). To clarify the situation, Cassen et al. (1981) have numerically investigated the gravitational instability of the circumsolar gaseous disk, modeling it by 60,000 gravitating points. The earlier conclusion has been confirmed: a disk with mass 1 $M_\odot$ is unstable at temperatures below 300 K, and a disk with the mass $M_\odot/3$, at temperatures below 100 K. Therefore, in a low-mass protoplanetary cloud ($M_d < 0.1\ M_\odot$), formation of gaseous protoplanets is impossible by gravitational instabilities. Attempts to relate the formation of such protoplanets with a fragmentation of the PSN during its very early evolution do not remove the difficulties mentioned above.

III. ACCUMULATION OF THE PLANETS

Study of the evolution of a swarm of protoplanetary bodies began with a determination of its most important characteristics—the distribution of masses and relative velocities of the bodies. Velocities increase due to gravitational perturbations of bodies and depend on their masses. At high velocities, smaller bodies disintegrate and the mass distribution changes. In other words, the coupled evolution of the two distributions should be investigated. Strict analytical treatment of the problem seems to be impossible. Numerical solutions also meet with serious difficulties—inevitable simplifications qualify the results. In analytical considerations, the evolution of the mass distribution of bodies was decoupled from the evolution of their random velocities (Safronov 1969). The velocities of bodies were calculated with the assumed inverse power law of mass distribution

$$n\,(m) = C\,(t)\,(m')^{-q},\ \ m' \leq m(t),\ \ q < 2 \tag{42}$$

where $c(t)$ is a time-dependent coefficient, and the mass distribution function has been obtained with an assumed velocity of bodies. It has been found that in the system of bodies (planetesimals) with the above distribution law (Eq. 42), the average random velocities depend only slightly on masses and rather quickly tend to an equilibrium value v which has been determined from a balance of energy of random motions gained due to gravitational stirring by encounters and lost due to collisions and gas drag. The stirring is ultimately driven by the Keplerian shear of the system. The equilibrium value is proportional to but a few times lower than the escape velocity on the surface of the largest body m in its zone of feeding

$$\mathrm{v} = \left(\frac{Gm}{\theta r}\right)^{1/2}. \tag{43}$$

The nondimensional parameter θ depends on q and the gas drag. At the stage of gas-free accumulation, $\theta \simeq 3$ to 5 for $q = 1.8$. At the earlier stage, the gas reduced chaotic motions and θ was several times higher. Evolution of velocities with assumed values of q was studied analytically by Kaula (1979).

The mass distribution of bodies was found with the aid of the coagulation equation. As one can see from Eq. (18), the velocities of bodies increase with time proportionally to the radius r of the largest body. At the early stage of accumulation they are low (a few tens m/s for $r = 100$ km) and bodies can collisionally combine without fragmentation. The asymptotic solution of the coagulation equation found for this case is satisfactorily approximated by a power law (Eq. 18) with the exponent $q \simeq 1.6$. At later stages, velocities become higher and smaller bodies disintegrate at collisions. In this case, the asymptotic solution is approximated by the same law with $q \simeq 1.8$ (Zvjagina et al. 1973). In both cases $q < 2$. This means that most of the mass is concentrated in the largest bodies and that the assumption underlying Eq. (18) is fulfilled. Thus, the equilibrium solutions found for a decoupled evolution of velocities and mass distributions are self-consistent.

In the range of a few largest bodies, asymptotics and statistics do not work. Gravitational cross sections of these bodies are proportional to the fourth power of their radii when the velocities of other bodies are described by Eq. (43). As a result a so-called runaway growth of the largest body in its feeding zone takes place, which separates the body from the continuous mass distribution given by Eq. (42). Then in Eq. (43) for v, one should take m and r of some effective body intermediate between the first largest body and the second one. The runaway accretion entails important consequences; it leads to formation of planet embryos and shortens the time scale of accumulation. Under such conditions, the equilibria mass and velocity distributions cannot be attained. Their coupled evolution at an early stage beginning from a system of equal kilometer-sized bodies has been numerically calculated by R. Greenberg et al. (1978*b*). It was found that a few bodies grow rapidly to hundreds of kilometers in size and the remainder do not grow significantly. In some sense, the mass distribution is like that given by Eq. (42) with $q > 2$. The velocities of bodies also increase very slowly, not in agreement with Eq. (43). However, the calculations cannot be extended into the later stages; in that case the gravitational cross sections taken from the two-body approximation is an overestimate. From numerical simulation of the final stage of accumulation of the terrestrial planets, Wetherill (1978,1980*a*) has found the time scale for growth $\sim 10^8$ yr and $\theta \simeq 2$.

A model of many-embryo accumulation can be briefly described in the following way. A runaway growth of the largest bodies leads to formation of a more or less regular system of embryos with their own ring-shaped zones of feeding. The eccentricities e of the embryos' orbits are considerably smaller than those ($\bar{e}$) of other bodies, and the half-width of the feeding zone R_f is determined mainly by the latter ones

$$\begin{aligned} R_f = (\bar{e}+e)R = \frac{\sqrt{2}\,(1+e/\bar{e})\,\mathrm{v}}{\Omega} = \frac{1+e/\bar{e}}{\Omega}\left(\frac{8\pi G\delta}{3\,\theta}\right)^{1/2} r \\ = \Delta R_p \left(\frac{m}{m_p}\right)^{1/3} \cdot \left(\frac{\theta_p}{\theta}\right)^{1/2} \end{aligned} \tag{44}$$

where m, r and δ are the mass, the radius, and the density of the embryo, $2\,\Delta R_p$ is the width of the feeding zone of the planet and θ_p is the value of θ at the end of accumulation. The total number of embryos in the zone of the planet is not less than

$$N_m = \frac{\Delta R_p}{\Delta R_f}\left(\frac{m_p}{m}\right)^{1/3}\left(\frac{\theta}{\theta_p}\right)^{1/2} \tag{45}$$

and they contain a fraction μ_e of the whole mass of solids in the zone

$$\mu_e = \frac{Nm}{m_p} \simeq \left(\frac{m}{m_p}\right)^{2/3}\left(\frac{\theta}{\theta_p}\right)^{1/2}. \tag{46}$$

While the embryos are in different feeding zones, there is no runaway growth of one of them relative to the others. However, with the increase of their masses, their zones widen and overlap more and more. The distance of effective gravitational interaction between neighboring embryos m and m_1

$$\Delta R_G \simeq [(e + e_1) + 3(m/M_\odot)^{1/3}]R \tag{47}$$

is about two times smaller than ΔR_f. Thus, in the feeding zone of the embryo there are also at least two other, neighboring embryos and the total number of embryos in the zone of the planet could be about twice greater than N_m. For $m << m_p$, $\mu_e << 1$, and embryos grow mainly due to sweeping out smaller bodies in their zones and not to the neighboring embryos. When ΔR_G becomes greater than the radial distance ΔR between the embryos, the smaller one m_1 experiences strong perturbation of m and leaves its quasi-circular orbit. Before colliding with a larger embryo, it has several times higher probability of disintegrating at close passage inside its Roche limit (Ziglina 1978). The number of embryos decreases until they consume all solid material between them and until the distances between their orbits become large enough to secure orbital stability of the system for billions of years.

At an earlier stage, gravitational stirring of bodies by the embryos is unimportant because they contain only a small fraction μ_e of the total mass of the system. But this fraction increases with time. On the other hand, one massive planet embryo moving in nearly circular orbit would not produce a systematic increase of velocities of bodies because their encounters with it always take place at the same distance from the Sun. However, this is not the case when there are several embryos in the zone. Thus, at the later stage of accumulation they could appreciably increase the velocities of bodies. The runaway growth and evolution of velocities have been estimated in a model of gas-free accumulation of many embryos with the following simplifying assumptions (Safronov 1982):

1. The system consists of embryos with about the same mass m and of other bodies m' which have the power mass distribution given by Eq. (42) with the largest body m_1;

2. The feeding zones of the embryos overlap and in every zone $2\,\Delta R_f$ there are n embryos;
3. The runaway growth of embryos m relative to m_1 begins at initial values r_{10} and $(m/m_1)_0$; the mass distribution is conserved for all bodies $m' < m_1$.

It was shown that the parameter θ characterizing relative velocities of bodies according to Eq. (43) can be determined from the expression

$$\theta = \theta_q \left(\frac{m}{m_1} \right)^{2/3} \left[1 + \frac{nfm^2}{(1-\mu_e)m_1 m_p} \right]^{-1/2} \tag{48}$$

where θ_q is the value of θ for the system with the mass distribution given by Eq. (42) and containing no embryos; $f \simeq 3$ to 4 for $q = 1.8$. The ratio $m/m_1 = (r/r_1)^3$ is determined from

$$\frac{dr}{dr_1} = \frac{1 + 2\theta}{1 + 2\theta r_1^2/r^2} . \tag{49}$$

From Eqs. (45), (46), (48), and (49), the quantities N, θ_e, m/m_1 and θ can be found as functions of m/m_p. Preliminary estimates have shown that to the end of accumulation θ tends to be about unity. Calculations were performed for $q = 11/6$, $\theta_q = 3$, $\theta_p = 1$, $r_{10} = 100$ km, $(m/m_1)_o = 1.5$. It has been found that the ratio m/m_1 increases to 30 by the end of accumulation. At first, the parameter θ increases. It reaches a maximum $\theta_m \simeq 10$ for $m/m_p \sim 10^{-2}$ and $m/m_1 \sim 10$, then it decreases to a value of 3 for $m/m_p \simeq 0.16$, $\mu_e \simeq 0.5$, and finally $\theta_p \simeq 1$ as $m \rightarrow m_p$. The number of embryos decreases from the initial value of about 10^2 to a value of 4 at $m/m_p \simeq 10^{-1}$. Using an initial value $(m/m_1)_o = 2$ gives a five-times higher final ratio m/m_1 and only one-and-a-half-times higher θ_m.

These results show that at the intermediate stage of accumulation there is a moderate runaway growth of planet embryos. The velocities of bodies increases more slowly than the radii of the embryos, and θ is several times higher than in the case of a simple power law mass distribution. Gas drag additionally increases θ, but at the final stage for $m > 0.1\ m_p$, the embryos control the velocities of bodies, θ diminishes and feeding zones widen. The radial diffusion (spreading) of bodies additionally enlarge the zones by a factor of two or three. Finally, the number of embryos in the terrestrial zone is reduced to the four present planets. So, the assistance of massive Jupiter is not necessary to account for the small number of terrestrial planets. A similar conclusion has been obtained from a study of evolution of eccentricities of preplanetary embryos (Pechernikova and Vitjazev 1980).

The rate of growth of a planet depends on the parameter θ in a simple way

$$\frac{dm}{dt} = \pi r_{\rm eff}^2\, \rho v = 2(1 + 2\,\theta)\, r^2 \Omega \sigma. \tag{50}$$

Some authors have argued that the increase of θ due to a runaway growth of the embryos leads to a decrease of the total duration of formation of the planets. In fact, only an intermediate stage of the planetary growth is shortened. The long final stage of accumulation is even more prolonged. Therefore, the whole time scale of accumulation remains the same as found earlier (10^8 yr for the Earth).

IV. CONCLUSION

A systematic quantitative study of the whole process of the formation of the Sun and planets has begun only recently. A model of their formation as the final result of the evolution of a slowly rotating presolar nebula seems to be the most promising. A solar nebula can probably grow to the size of the planetary system at the accretion stage of collapse of the PSN due to an outward angular momentum transport within the core and the nebula. Magnetic torque is likely to dominate such transport in the central part of the PSN, and the turbulent viscosity is important in the outer parts of the nebula. Sources and intensity of turbulence have not been well studied yet. Clarification of these questions would promote considerable progress in understanding the origin of the solar system. A detailed study of the model of planetary accumulation of solid bodies shows that there are no major difficulties in the theory; nevertheless, a number of questions remain to be answered.

ON THE DYNAMICAL ORIGIN OF THE SOLAR SYSTEM

D. N. C. LIN
Lick Observatory

and

J. PAPALOIZOU
Queen Mary College

It is generally believed that the Sun and its planetary system were formed about 4.6×10^9 yr ago from a rotating cloud of interstellar gas and dust with the protosun at its center. Revolving around the protosun was a primordial solar nebula out of which the planetary system was formed. In general, the basic physical properties of the primordial solar nebula are very similar to those of an accretion disk. The evolution of the nebula can be characterized by three stages: infall, diffusive, and clearing stages. During each stage, the structure, stability and evolution of the nebula are primarily determined by the efficiency of the angular momentum and heat transport processes. The dominant transport mechanism may be turbulence and convection. In this chapter, we outline a quantitative model with which the distribution of temperature, density, pressure, thickness, and mass transfer rate of the nebula can be determined at any given epoch. Observational data, such as the spatial and mass distribution of the planets, as well as meteorite data are utilized to provide constraints on these models. The most favorable model is one in which the protogiant planets as well as the prototerrestrial planets formed through several stages of coagulation, accumulation, and accretion. Owing to constraints on (1) grain condensation temperature, (2) the minimum nebular mass required to allow accretion of the protoplanets, and (3) the magnitude of nebular viscosity in the vicinity of protogiant planets' orbits around the protosun, the final epoch, when the protoplanets acquired most of their masses, must have lasted $\lesssim 10^6$ yr. Furthermore, during this epoch, the solar nebula was considerably less massive than the present Sun. Quantitative analysis can also be carried out for the inner *regions of the nebula where most of the energy is dissipated. The emerging photons from these inner regions are primarily in the optical frequency range. These nebular models are directly relevant to T Tauri stars. According to our model, the inner regions of the nebula may be thermally unstable. Such an instability may be the underlying origin of the FU Ori phenomenon. In the outer regions, where the nebula is sufficiently cool for grains to condense, both the rate of energy release and the spectral characteristics can be calculated and applied to T Tauri stars to check for consistency.*

I. INTRODUCTION

Studies of the origin of the primordial solar nebula date back to the work of Descartes (1644). Amongst widely different theories prior to this century, the model most qualitatively compatible with observation is that due to Kant (1755) and Laplace (1796). In the Kant-Laplace model the Sun and its planets were formed from a rotationally flattened gaseous nebula. The Sun was formed at the center and the planets at various positions in the nebula.

Today, most models of solar system formation still adopt the basic Kant-Laplacian concept that the Sun and its planetary system had a common origin. However, today's theories are constrained by the vast amount of data accumulated over the past three decades from meteorite analyses, groundbased observations, and interplanetary probes. Consequently, modern theories necessarily pay more attention to the detailed processes at each stage of solar system formation. It is hoped that a close compatibility between theory and observation on many minor issues may eventually provide an incontrovertible body of evidence to support a grand unified theory of the origin of the solar system. With this guiding philosophy, we shall describe some of the recent developments in the nebular hypothesis, particularly in the area of how the initial angular momentum was distributed and rearranged through a process of dynamical evolution.

To initiate our discussion, we review briefly the standard Kant-Laplacian scenario in which the nebula contains a small fraction of a solar mass and its dynamics is dominated by the gravity of the protosun. The main motion of the nebula is nearly circular rotation around the protosun, with local Keplerian velocities, so that the gravity of the protosun is approximately balanced everywhere by centrifugal force. According to Laplace's original model, the nebula may have broken up into several discrete rings, due to rotational instabilities, out of which the planets formed. This aspect of the Laplacian model was later challenged by Jeffreys (1924) who compared the nebular flow pattern with typical instances of laboratory shear flows in which molecular collisions induce viscous stress. The general effect of viscous stress is to reduce the velocity gradient in the flow. In the solar nebula, the Keplerian velocity law implies that the rotation velocity U_ϕ decreases with nebular radius r. In this case, the viscous stress acts to reduce the difference in the angular velocity between gases in adjacent radial zones. In order to speed up the slower-moving gas in the outer region of the nebula, angular momentum must be transferred from the inner to the outer region. Upon acquiring more angular momentum, however, the nebular gas in the outer region experiences a stronger centrifugal force and expands to larger radii until a new force balance can be achieved. Similarly, upon losing its angular momentum, gas in the inner part of the nebula drifts inward. Consequently, the viscous stress induces a spread of the nebular material instead of a reduction in the velocity gradient. The accumulated effect of the viscous stress over the entire nebula is to transport angular momentum continuously to the outer regions of the nebula. Once sufficient time has elapsed, a small fraction of the nebular material at the outer edge absorbs most of the angular momentum and expands considerably. The bulk of the nebular material loses angular momentum and thereby drifts inward.

Although Jeffreys indicated a general tendency for the nebula to spread out, the efficiency of the viscous transport process remains to be investigated. The rate of angular momentum transfer is essentially determined by the magnitude of the viscous stress W which is characterized by the product of a kinematic viscosity coefficient ν and the local rate of strain. If molecular collisions provide the only source of friction, we find from standard kinetic theory of gases (Jeans 1967) that $\nu_{mol} \approx c_s l$ where c_s is the local sound speed, $l \approx 1/(nA)$ is the collisional mean free path of the molecules, where n is the local number density and A is the molecular collision cross section. In the solar nebula, the necessary condition for grain condensation is $T \lesssim 10^3$K so that $c_s \lesssim 10^5$cm s^{-1}. In order for there to be sufficient material to build up the planets in the solar system, $n \gtrsim 10^{15}$ cm^{-3}. Adopting a typical value for an atomic cross section A ($\approx 10^{-16}$ cm^2), we find $\nu_{mol} \approx 10^6$ cm^2 s^{-1}. The typical time scale for significant mass redistribution over a radius $r \approx 10^{13}$ cm is $\tau_\nu \approx r^2/\nu_{mol} \approx 10^{12}$ yr, which is much longer than the present age of the solar system. Thus, if molecular viscosity provides the only momentum transport mechanism, the nebula's mass distribution would not be significantly changed, during the course of its evolution, from its initial conditions. However, there are other processes which can induce efficient momentum transport causing the nebula to evolve on a relatively short time scale (von Weizsäcker 1943,1948).

Quantitatively, the effectiveness of momentum transfer is normally measured in terms of a dimensionless Reynolds number $\mathcal{R}_{eff} (= UL/\nu)$, where U and L are the characteristic velocity and length scales. In typical laboratory shear flows, when molecular interaction provides a molecular Reynolds number $\mathcal{R}_{mol} \gtrsim 10^3$, the velocity gradient of the flow is too large for the molecular viscous momentum transfer process to smooth it out within a characteristic time scale L/U, it is found that small perturbations, either intrinsic to the flow or due to the physical conditions at the flow boundaries, have a general tendency to grow vigorously into large-scale turbulence. In turbulent flows, gas or liquid from different regions of the flow are mixed together mechanically so that momentum transfer is considerably more effective than that provided by molecular collisions and the effective Reynolds number, $\mathcal{R}_{\text{eff}} \approx 10^{2-3}$ (Townsend 1976). If we extrapolate this result to the solar nebula, the fact that $\mathcal{R}_{\text{mol}} \gtrsim 10^{13}$ suggests that the nebula may be unstable and turbulent (Lynden-Bell and Pringle 1974). In fact, there are several qualitative reasons to speculate that $\mathcal{R}_{\text{eff}} << \mathcal{R}_{\text{mol}}$ in stellar and protostellar accretion disks which are generically similar to the solar nebula. (1) The lack of remanent nebulae around stars in the post T Tauri phase suggests that the characteristic viscous diffusion time scale $\tau_\nu \lesssim 10^6$yr (or $\mathcal{R}_{\text{eff}} \approx \tau_\nu \Omega \lesssim 10^6$). (2) The characteristic time scale for lightcurve variations in dwarf novae and X-ray binaries is on the order of days to years which suggests that typically $\mathcal{R}_{\text{eff}} \lesssim 10^{3-5}$ in accretion disks around compact degenerate stars (Pringle 1981). (3) We shall show in Sec. V that unless $\mathcal{R}_{\text{eff}} \lesssim 10^{3-5}$, any protoplanet with a mass less than the present mass of Jupiter would tidally truncate the nebula and thereby impede its own growth.

If the nebula is turbulent, the energy associated with chaotic motion is continually transferred to small eddies and dissipated on some viscous length scale (Tennekes and Lumley 1972). Thus, in order for the nebula to sustain a turbulent flow pattern, it must be driven to do so by some intrinsic instabilities. There are many potential instabilities which may cause the gas flow to become turbulent in the nebula. For example, for some time after the formation of the nebula, infall of material may continue. Because the infalling material may have velocities and angular momentum substantially different from those in the nebula at the arrival locations, considerable shear in the interface regions can be induced. Such an intense shear may cause the flow near the interface to become Kelvin-Helmholtz unstable. The mixing of angular momentum between the infalling and nebular material can provide additional stirring (Cameron 1976). Another possible origin of turbulence is magnetohydrodynamic effects. Typical T Tauri stars radiate an intense ultraviolet flux which may ionize at least a small fraction of the nebular material. These ionized particles may be coupled to a seed magnetic field in the nebula. Through collisions between the ionized and the neutral particles in the nebula, the ionized particles are forced to move together with the neutral particles. Consequently, the field may be dragged along with the flow and stretched by differential rotation of the nebula. The field strength, as a result, would grow until field lines break and reconnect. If the amplified random field can be effectively coupled to the nebular material, gas from different regions of the nebula would be connected and angular momentum would be transferred between these regions (Hoyle 1960; Hayashi 1981*b*). Today it is still an open question whether a significant fraction of gas may be ionized to allow this process to operate because, for typical nebular temperatures, most of the nebular material is in a neutral state. There are still other possible instability mechanisms, but for most cases rigorous stability analysis has not yet been carried out. Therefore, it is often difficult to assess just how much confidence should be attached to each of these qualitative arguments. There is, however, one instability mechanism which has been examined rigorously. This mechanism is convection. We shall show in Sec. III that most regions of the nebula are convectively unstable.

If the nebula is intrinsically unstable, von Weizsäker argues that turbulence must necessarily lead to dissipation of energy. Because the accompanying viscous stress also induces inward drift of the nebular material, there is a continuous release of gravitational energy from the nebula. This energy is fed into turbulent eddies to make up the loss of energy by dissipation so that chaotic motion is maintained. Consequently, the thermal properties of the nebula may also be determined by the action of viscous stress.

The mathematical foundation of viscous disk accretion flow was first established by Lüst (1952). However, his pioneering work was poorly known to most astronomers and astrophysicists. In the past two decades, the discovery of quasars, cataclysmic variables, and powerful X-ray sources in binary systems have generated considerable interests in accretion-disk theories (Prendergast and Burbidge 1968; Lynden-Bell 1969). In an attempt to understand the underlying physical processes governing mass transfer flows in binary systems and around massive black holes, important theoretical understanding has been gained (Lynden-Bell 1969; Shakura and Sunyaev 1973; Lynden-Bell and Pringle 1974). The general basis of these accretion-disk theories is to consider the effect of internal friction on the evolution of a differentially rotating disk of gas in which the centrifugal force acting on the gas is everywhere balanced by gravitational force. The primary conclusion of these investigations is that the rate of evolution of an accretion disk is determined by the efficiency of angular momentum transport which itself is determined by the magnitude of the internal frictional force in the flow. This new information and knowledge has been applied to models of solar system formation (Cameron 1976,1978). Today accretion-disk theory has become an integral part of quantitative models for the origin and evolution of the solar nebula (Lin and Papaloizou 1980; Lin 1981; Lin and Bodenheimer 1982). The physical nature and the mathematical foundation of accretion-disk theory is extensively reviewed by Pringle (1981) for a wide range of astrophysical contexts. In Sec. II we briefly recapitulate the salient features of accretion-disk theory in the context of the primordial solar nebula.

With appropriate mathematical formulation, rigorous analysis of the dynamics of the nebula can be carried out. In the absence of turbulence, there is no significant angular momentum transfer or radial migration of the nebular material. Nonetheless, the nebular material will contract towards the midplane as thermal energy is lost from the surface of the nebula. At any given radius, the vertical structure of the nebula can be derived. In Sec. III we show that these solutions indicate that the nebula is intrinsically unstable against convection in the vertical direction, in which case large-scale turbulence may develop. Turbulence in the nebula has three major effects:

1. It induces effective angular momentum transfer;
2. It induces effective thermal energy transport;
3. It generates thermal energy by viscous dissipation.

A region of the nebula is in local thermodynamic equilibrium if the thermal energy generated by viscous dissipation is balanced by the local heat flux. With a given prescription for turbulent viscosity, a model that is locally in thermal equilibrium can be constructed such that all the local thermodynamical variables can be expressed in terms of the surface density of the nebula. In Sec. III we utilize three *ad hoc* prescriptions for turbulent viscosity to determine the vertical structure of the nebula.

Once a prescription for viscosity is specified, the global viscous evolution of the nebula can be studied from an arbitrary intitial mass distribution. Although there is no reliable information on the initial mass distributions, there are phenomenological models with which we may estimate the mass distribution during the stage of protoplanetary formation under certain assumptions. Early quantitative investigations of the structure of the solar nebula were based on the hypothesis that the planets and satellites collected all the rocky and icy material within the solar nebula. By adopting the assumption that the chemical composition of the original solar nebula was similar to that of the Sun today, and that most of the volatile materials in the nebula were not condensed into the planets and satellites, a lower limit of the nebular mass may be determined by augmenting the mass of the planets to account for the missing constituents, namely hydrogen and helium gas (Cameron 1962). The mass obtained in this way is about 2% of the present mass of the Sun. For the remainder of this chapter, we shall refer to models with a few percent of the solar mass as the minimum-mass nebula model. If the efficiency of collection of rocky or icy material by the protoplanets is less than unity, the mass of the nebula must have been larger than this amount.

The simplest phenomenological model for the initial mass distribution of the solid material is based on the assumption that the solid material in the solar system has preserved its original spatial distribution. Consequently, the mass distribution of all the nebular material may be constructed, provided that limited spatial mixing is allowed (Cameron 1973*a,b*; Kusaka et al. 1970). In addition, the temperature distribution of certain regions in the nebula may be estimated through analysis of the condensation temperature required for various chemical elements which make up the bulk of the mass of the terrestrial planets and the giant planets' satellites (Lewis 1972,1974). If the terrestrial planets are formed at approximately the same time and near their present orbital radii, the cosmochemical data would imply that the temperature of the nebula was roughly inversely proportional to distance from the Sun. However, this approximation is extremely inaccurate because (1) the efficiency of collection of solid material may be different at different locations, (2) the protoplanets may not have formed at the same time, and (3) the condensed rocky cores which make up the protoplanets may have undergone considerable post-formation orbital evolution. Although these results do not provide a firm constraint on the initial state of the nebula, they do indicate that the primary source of thermal energy of the nebula is probably not due to the absorption of the solar radiation; otherwise the temperature would be inversely proportional to the square root of the distance from the Sun throughout the evolution of the nebula. Another possible observational constraint may come from future observation of disks around T Tauri stars. The distribution of the disk surface density may be strongly constrained by the spectrum of the emerging photons. In Sec. IV we discuss the evolution of the solar nebula in detail. The above observational data are used to constrain various evolutionary models.

Despite the adoption of the above observational constraints, the detailed dynamics of the solar nebula are still subject to considerable uncertainty. Due to the lack of an accurate, non *ad hoc* prescription for turbulent viscosity, we cannot assess with confidence the effectiveness of angular momentum transport in the nebula. There is an indirect argument with which we can set limits on the magnitude of viscosity in the nebula. This argument is based on the protoplanet-nebula tidal interaction. When a protoplanet has gained sufficient mass, it induces tidal transfer of angular momentum with the nebula in a manner similar to tidal interaction between Saturn's satellites and rings. Unless the magnitude of kinematic viscosity is sufficiently high, the protoplanet can tidally truncate the nebula by opening up gaps in the vicinity of its orbit. Thereafter, the nebular material can no longer be accreted by the protoplanet and the protoplanet's growth is terminated. Thus, from the present mass of the giant planets, we may estimate the magnitude of viscosity at the time of protoplanetary formation. In Sec. V we discuss protoplanet-nebula tidal interaction in detail.

The primary purpose of constructing a solar nebula model is to evaluate the conditions under which protoplanets were formed. Although there are still considerably divergent opinions on the details of protoplanetary formation, we can already utilize our results to cast constraints on different scenarios. For example, there are two major scenarios for the formation of the solar system: (1) Cameron's gravitational instability scenario, and (2) the Safronov-Wetherill gas-free coagulation scenario (see chapter by Safronov and Ruzmaikina).

During the first conference on Protostar and Planets, Cameron (1978) proposed that protoplanets formed out of the fragments of a gravitationally unstable gaseous nebula (see also chapter by Cameron in this book). Cameron further speculated that refractory materials may condense and segregate towards the center of protoplanets with mass comparable to or less than that of Jupiter. In order to explain the chemical composition of the terrestrial planets, Cameron proposed that the outer gaseous envelope may have been removed by the tidal action of the protosun. Cameron's scenario is attractive on philosophical grounds because it provides a working hypothesis for the formation of giant as well as terrestrial planets. However, it is based on numerous *ad hoc* assumptions. The implication of some of these assumptions can be deduced from solar nebula models. In Sec. VI we use the nebula models to deduce the condition required for gravitational instability in the gaseous component of the nebula. In order for the nebula to be gravitationally unstable, the force associated with the nebula's self gravity must be comparable to or larger than that associated with the pressure gradient in the direction normal to the plane of the nebula. The nebula's self gravity increases linearly with its surface density. However, because the nebula is convectively unstable and turbulent, it is heated by viscous dissipation. According to the accretion-disk models to be presented in Sec. III, the heating rate and nebular temperature increase with surface density. Consequently, at any given radius, the nebula is gravitationally unstable only when the characteristic nebular mass inside this radius is comparable to the mass of the protosun. This quantitative result has several important implications.

1. A gravitationally unstable nebula evolves on a dynamical time scale. The nebula can only evolve into a gravitationally unstable state during an initial infall phase in which the nebular mass is gradually accumulated. The results in Sec. V indicate that at any radius the surface density in the nebula can increase if the local mass infall rate exceeds the local viscous diffusion rate. In order to sustain mass accumulation in the nebula until it becomes unstable, the infall rate must be sufficiently high so that this phase cannot continue for longer $\approx 10^5$yr.
2. As the infall material impinges onto the nebula, its kinetic energy is shock-dissipated at the interface between it and the nebula. With such a large infall flux, the nebula is predominantly heated by this shock dissipation. The temperature of the nebula inside 1 AU may well exceed the condensation temperature of typical refractory grains. Furthermore, the heat generated by the shock dissipation provides yet another stabilizing effect against gravitational instability.
3. Because the nebula can only evolve into a gravitationally unstable state during the infall phase, we do not expect the regions of the nebula at different radii to become gravitationally unstable at the same time unless the infall flux onto the nebula is of a particular form. If the infall flux decreases rapidly with radius, the inner regions of the nebula may become self gravitating first. In this case, we may have global bar instability rather than local gravitational instability (see, e.g., Cassen et al. 1981). The formation of a strong central bar induces strong tidal transport of angular momentum and mass redistribution on a dynamical time scale. Consequently, further accumulation in the nebula may be terminated and the nebula may be stabilized against local gravitational instability.
4. If the infall flux does not decrease rapidly with radius, the outer regions of the nebula may become self gravitating and locally gravitationally unstable. In this case, the outer regions of the nebula may break up into many

fragments. The characteristic mass of the fragments, estimated from a standard stability analysis, is much larger than the mass of giant as well as terrestrial planets (Lin 1981). Thus, the fragments must not only lose more than 99% of their original mass but also must adjust appropriately their chemical abundances to become protogiant planets. In the standard gravitational stability analysis, the tidal action of the protosun is already taken into account. The fragments are more likely to contract gravitationally along some low-mass protostellar evolutionary tracks on the HR diagram rather than lose their envelope by tidal stripping process.

5. Even in the unlikely event that the fragments can indeed lose an appropriate amount of mass from their envelope, the stripped fragments can only evolve into protoplanets if the mass loss occurred on a time scale shorter than the dynamical relaxation time scale of the fragments in order to avoid the introduction of large orbital eccentricities among the fragments. It would seem unlikely that condensation and sedimentation of the grains can occur in the fragments' envelope while such rapid and efficient mass loss from the bulk of the fragments is taking place. These implications provide major difficulties for Cameron's gravitational instability scenario.

There is an alternative formation scenario for the terrestrial planets (see chapters by Hayashi et al. and by Safronov and Ruzmaikina). Terrestrial planets contain refractory solid materials. The crater record on their surface is indicative that solid-particle accretion may have played an important role in the formation process of terrestrial planets. In the past two decades considerable effort has been devoted to the study of the coagulation of planetesimals (see, e.g., Safronov 1969, 1972*b*; Kusaka et al. 1970; Wetherill 1980). Most of these investigations address the growth processes of planetesimals in a gas-free environment. Although the principal reason for the gas-free assumption is for computational convenience, it also provides a natural explanation for why the chemical abundance of heavy inert gases on Earth is considerably less than that found in the Sun (Wetherill 1980). In a gas-free environment, solid-particle planetesimals grow through cohesive collisions. The growth time scale is determined by the collision frequency which in turn depends on the velocity dispersion of the planetesimals. However, in order to prevent destructive collisions, the velocity dispersion is limited to the escape velocity of the largest planetesimals. These two requirements constrain the growth time scale for an Earth-mass object to be $\approx 10^{7-8}$ yr. Although the gas-free coagulation scenario provides a reasonable working hypothesis for the formation of the terrestrial planets, can we utilize part of this scenario to examine the formation of the protogiant planets? Giant planets are composed primarily of hydrogen and helium and must have been formed in a gas-rich environment. Because the nebula is heated by viscous dissipation, the protogiant planet must first form a core with a mass comparable to or larger than that of the Earth before it can accrete gas efficiently (see chapters by Bodenheimer and by Pollack). Suppose we adopt the simplest and most naive assumption that the gaseous background does not affect the growth of the planetesimals until the stage when the escape velocity of the planetesimals is comparable to the nebula sound speed. In order for the nebula to retain a sufficient supply of gas, the growth time scale of the solid core of the protogiant planet must be comparable to or shorter than the viscous diffusion time scale for the minimum-mass nebula. From the nebula model presented in Secs. III and IV, we find that this time scale must be $\lesssim 10^6$ yr. We can also use the protoplanet-nebula tidal interaction argument presented in Sec. V to reach a similar conclusion. This required growth time scale is much shorter than that found in the gas-free coagulation scenario. Thus, it is generally assumed that the prototerrestrial and protogiant planets formed at different epochs from different processes in different environments (Wetherill 1980).

Ideally, we would like to provide an unified scenario to address the formation processes for both prototerrestrial as well as protogiant planets. One possible scenario is that all protoplanets formed in a gas-rich environment. The growth of solid material is strongly influenced by the presence of the gas in several ways.

1. The presence of a turbulent gaseous background can induce large-velocity dispersion among grains and small particles which enhances the efficiency of coagulation (Weidenschilling 1984).
2. Viscous drag between intermediate-mass planetesimals and the nebular gas can induce orbital migration which also increases collision frequency (Adachi et al. 1976).
3. Tidal interaction between Jupiter-size protoplanets and the nebula can cause orbital migration and enhances the accretion process (Lin and Papaloizou 1980; Goldreich and Tremaine 1980).
4. As the nebula is viscously diffused, the nebula temperature everywhere decreases. The nucleation and condensation process in such an environment may be continuous. The orbital evolution of the newly formed small particles are essentially determined by the viscous drag of the nebular gas. These particles are brought to the neighborhood of the most massive planetesimals and accreted by them. Under some circumstances, this process may be more efficient at feeding the large planetesimals than coagulation of similar size planetesimals (Lin 1982).

In the inner regions of the solar nebula, the temperature of the nebula may be sufficiently high even after the planetesimals have consumed most of the heavy elements, so that the planetesimals still do not have sufficient mass to accrete gas efficiently. The temperature of the gas in the outer regions of the nebula is relatively low so that the critical mass required for efficient gas accretion is smaller. Furthermore, the mass distribution in the nebula provides a favorable condition for the formation of more massive planetesimals at larger radii. Once gas accretion proceeds, its efficiency increases with the protoplanet's mass so that the gas accretion phase is relatively short-lived. Ultimately, the tidal interaction between the protogiant planets and the nebula may be sufficiently strong to open up a gap in the vicinity of the protoplanet's orbit so that the nebular gas can no longer reach the protoplanet and the growth of the protoplanet is essentially terminated. This qualitative scenario provides an attractive scheme for the formation of the solar system because it is based on relatively few *ad hoc* assumptions. However, the basic physical nature of such growth processes is still rather uncertain and considerable effort must be devoted to establish a quantitative basis for it in the future. For the purposes of providing the reader an introduction to this scenario, we discuss briefly some qualitative aspects of protoplanetary formation in a gas-rich environment in Sec. VI.

T Tauri stars are among the youngest optically detected stars in the Galaxy. These stars probably have ages ranging from 10^{5-6} yr (Herbig 1962; Cohen and Kuhi 1979). This age range is comparable to the viscous diffusion time scale in the nebula so that there may still be protoplanetary nebulae around this class of stars. In Sec. VII we present a model for the inner region of the nebula and compare it with the observational properties of T Tauri stars. These dynamical solutions indicate that the inner region of the nebula is probably thermally unstable and may undergo a thermal relaxation cycle. Such a cycle may be responsible for the FU Ori phenomenon. Finally, in Sec. VIII we summarize these results.

II. THE DYNAMICS OF ACCRETION DISKS

The evolution of the mass distribution in the nebula can be derived directly from the fundamental equations of fluid dynamics. The rate of change in density ρ is described by the continuity equation

$$\frac{\partial \rho}{\partial t} + \nabla \cdot \rho \mathbf{U} = 0 \tag{1}$$

where $\mathbf{U}$ is the flow velocity. If the nebula is relatively thin, i.e., its characteristic density scale height in the vertical direction $\mathbf{H}$ is small compared with r, the continuity equation may be integrated over the z direction. In an axisymmetric nebula, Eq. (1) can be expressed simply in cylindrical coordinates such that

$$\frac{\partial \Sigma}{\partial t} + \frac{1}{r}\frac{\partial}{\partial r}\Sigma U_r r - S_\Sigma(r,t) = 0 \tag{2}$$

where $S_\Sigma(r,t) = -\rho(r,z,t)U_z(r,z,t)\,\big|_{z=-\infty}^{z=\infty}$ is the infall flux onto the nebula per unit area, U_r and U_z are the velocities in the radial and vertical directions and $\Sigma = \int_{-\infty}^{\infty} \rho dz$ is the surface density. The total mass infall rate onto the nebula between the inner boundary r_{in}, and the outer boundary r_{out}, is $\dot{M}_* = 2\pi \int_{r_{in}}^{r_{out}} S_\Sigma(r,t) r dr$. In the absence of infall, $S_\Sigma = 0$, so that Eq. (2) is reduced to

$$\frac{\partial \Sigma}{\partial t} + \frac{1}{r}\frac{\partial}{\partial r}\Sigma U_r r = 0. \tag{3}$$

The characteristic velocity in the nebula can be determined from the momentum equation

$$\frac{\mathrm{D}\mathbf{U}}{\mathrm{D}t} = -\frac{1}{\rho}\nabla p + \frac{1}{\rho}\nabla\boldsymbol{\sigma} - \nabla\Psi \tag{4}$$

where p is the pressure. The material derivative $\mathrm{D}/\mathrm{D}t$ is composed of both the local time derivative $\partial/\partial t$ and the advective contributions $\mathbf{U}\cdot\nabla$. The stress tensor $\sigma_{ij} = 2\eta(e_{ij} - \Delta\delta_{ij}/3) + \rho\zeta\Delta\delta_{ij}$, where η $(= \rho\nu)$ and ζ are the shear and bulk viscosity, the rate of strain tensor $e_{ij} = (\partial U_i/\partial x_j + \partial U_j/\partial x_i)/2$ and $\Delta = e_{ii}$. If self gravity of the nebula is not important, the gravitational potential $\Psi = -GM/r$ where M is the mass of the Sun. Equation (4) can be decomposed into three components. Although each component contains a large number of terms, some contributions are relatively unimportant. For example, in the absence of infall, the dominant component of $\mathbf{U}$ in a relatively thin, rotationally flattened nebula is U_ϕ such that $|U_\phi| >> |U_r| >> |U_z|$. We now justify this inequality. In order for a nebula to be geometrically thin, the pressure scale height in the vertical direction H and the sound speed c_s must satisfy $H << r$ and $c_s << (GM/r)^{1/2}$. Provided Σ and T are smoothly distributed in the nebula, i.e., $|\partial \ln\Sigma/\partial \ln r| \approx 1$ and $|\partial \ln T/\partial \ln r| \approx 1$, $|(1/\rho)(\partial p/\partial r)| \approx c_s^2/r << GM/r^2$ so that the r component of the equation of motion is reduced to

$$\frac{U_\phi^2}{r} = \frac{GM}{r^2} \tag{5}$$

which is the condition for Keplerian flow. Note that contributions from U_r terms are ignored in the above equation because the nebula is approximately rotationally supported. The contribution from the pressure gradient in the radial direction is only important if either the nebular temperature is high or there is a sharp thermal transition in the radial direction.

From Eq. (5), we find that unless the mass of the protosun changes rapidly, U_ϕ is independent of time. Thus, after integrating over z, the ϕ component of equation of motion becomes

$$\Sigma U_r r\frac{\mathrm{d}\Omega r^2}{\mathrm{d}r} = \frac{\partial}{\partial r}\Sigma\nu r^3\frac{\mathrm{d}\Omega}{\mathrm{d}r} + S_\Sigma(r,t)J(r,t)r \tag{6}$$

where $J(r,t)$ is the excess specific angular momentum carried by the infalling material and we follow standard conventions by defining $\Omega r = U_\phi$. In the absence of infall, we can deduce U_r from Eq. (6) such that

$$U_r = -\frac{3\nu}{r}\left(\frac{\partial \ln \Sigma}{\partial \ln r} + \frac{\partial \ln \nu}{\partial \ln r} + \frac{1}{2}\right). \tag{7}$$

If Σ and ν are smoothly distributed in r, $|U_r| \approx \nu/r \approx \Omega r/\mathcal{R}_{\rm eff}$ where $\mathcal{R}_{\rm eff}$ is the Reynolds number. In typical accretion disk models, $\mathcal{R}_{\rm eff}$ is much larger than unity so that $|U_r| << \Omega r$ which justifies, in a self consistent manner, the assumption made in deriving Eq. (5). Note that U_r can change with time if either Σ or ν evolve with time.

Combining Eqs. (3) and (5), we deduce the standard mass diffusion equation

$$\frac{\partial \Sigma}{\partial t} - \frac{1}{r}\frac{\partial}{\partial r}\left[3r^{1/2}\frac{\partial}{\partial r}(\Sigma \nu r^{1/2}) - \frac{2S_\Sigma(r,t)J(r,t)}{\Omega}\right] - S_\Sigma(r,t) = 0. \tag{8}$$

In the absence of infall, Eq. (8) becomes

$$\frac{\partial \Sigma}{\partial t} - \frac{3}{r}\frac{\partial}{\partial r} r^{1/2}\frac{\partial}{\partial r}(\Sigma \nu r^{1/2}) = 0. \tag{9}$$

The evolution of any initial Σ distribution can be calculated provided a prescription for the coefficient of kinematic viscosity ν is specified. The most uncertain and important aspect of any accretion-disk theory is the determination of the magnitude and dependence of ν. For example, if ν is a function of r only, both Eqs. (8) and (9) are linear in Σ and may be solved in terms of modified Bessel functions. In this case, the evolution of any initial Σ distribution can be followed by using a Green's function approach (Lynden-Bell and Pringle 1974). However, ν may be more generally determined by the local conditions, i.e., it may be a function of p, ρ, r, and t. In that case, both Eqs. (8) and (9) are nonlinear and the evolution of the disk can only be found by numerical solutions. In Sec. IV we shall discuss, in more detail, the evolution of the nebula with or without infall.

If the ν is indeed a function of local thermodynamic properties, we must utilize additional equations to determine the thermal structure of the nebula. In the absence of infall, the equation of motion in the vertical direction of a geometrically thin, non-self-gravitating accretion disk is reduced to

$$\frac{1}{\rho}\frac{\partial p}{\partial z} = -\Omega^2 z. \tag{10}$$

The isothermal sound speed is given by $c_s^2 = (\partial p/\partial \rho)_T$. For an isothermal perfect gas temperature T, Eq. (10) implies $\rho \propto \exp(-z^2/2H^2)$ where $H =$

c_{s0}/Ω is the vertical density scale height and c_{s0} is the sound speed evaluated at the midplane of the nebula. For a general polytropic equation of state, $p \propto \rho^\gamma$,

$$\rho \propto [1 - (\gamma - 1)\Omega^2 z^2/2c_{s0}^2]^{1/(\gamma-1)}$$

so that the characteristic length scale for density change in the vertical direction is again $\approx c_{s0}/\Omega$. Thus, at the orbital radii of typical planets, $H << r$ unless the disk temperature is greater than 10^4K. It is also interesting to note that if T is smoothly distributed, $|U_z/U_r| \approx H/r << 1$ so that our initial approximation is again justified in a self-consistent manner. The rate of change of thermal energy can be derived from the energy equation such that

$$C_v\rho\left(\frac{DT}{Dt} - (\Gamma_3 - 1)\frac{T}{\rho}\frac{D\rho}{Dt}\right) = D - \nabla \cdot \mathcal{F} \quad (11)$$

where C_v and Γ_3 are the specific heat and adiabatic index and $\mathcal{F}$ is the total heat flux. If the disk is radiative, heat is entirely transported by radiation at a rate

$$\mathcal{F} = \mathbf{F_{rad}} = -\frac{4acT^3}{3\kappa\rho}\nabla T. \quad (12)$$

If the disk is convectively unstable, a portion of the heat may be transported by convection so that

$$\mathcal{F} = \mathbf{F_{rad}} + \mathbf{F_{con}} \quad (13)$$

where the convective flux $\mathbf{F_{con}}$ can be derived from a simple mixing-length prescription (Cox and Giuli 1968). The local rate of dissipation is given by

$$D = 2\rho\nu(e_{ij} - \Delta\delta_{ij}/3)(e_{ij} - \Delta\delta_{ij}/3) + \rho\zeta\Delta^2. \quad (14)$$

In the absence of infall, $|U_z| << |U_r| << \Omega r$, so that D is reduced to

$$D = 9\nu\rho\Omega^2/4. \quad (15)$$

For an axisymmetric thin nebula, Eq. (11) can be integrated with respect to z to obtain

$$C_v\left[\frac{\partial T}{\partial t} + U_r\frac{\partial T}{\partial r} - (\Gamma_3 - 1)\frac{T}{\Sigma}\left(\frac{\partial \Sigma}{\partial t} + U_r\frac{\partial \Sigma}{\partial r}\right)\right]$$

$$= \frac{9}{4}\nu\Omega^2 - 2\frac{\mathcal{F}_s}{\Sigma} - \frac{2H}{r\Sigma}\frac{\partial r\mathcal{F}_r}{\partial r} + \frac{S_\Sigma(r,t)S_T(r,t)}{\Sigma} \quad (16)$$

where $\mathcal{F}_r$ is the total flux in the r direction, $\mathcal{F}_s(= \sigma T_e^4)$ is the surface flux in the z direction, T_e is the effective temperature at the nebula surface and $S_T(r,t)$ is the excess specific thermal energy carried by the infalling material. It is of interest to note that if the radial distribution of Σ and ν is relatively smooth, the magnitude of the advective transport terms satisfy $|C_\nu U_r \partial T/\partial r| \approx \mathcal{R}T\nu/r^2 \approx c_s^2\nu/r^2 \approx (H/r)^2 D/\rho << D/\rho$. If the nebula were out of local thermal equilibrium, for example if $DH >> |\mathcal{F}_z|$, the characteristic time scale for T to change would be $\tau_{\rm th} \approx |T/(\partial T/\partial t)| \approx (c_{s0}H/\nu)(1/\Omega)$. This time scale is considerably shorter than the characteristic time scale for significant viscous diffusion in the radial direction, $\tau_\nu \approx |r/U_r| \approx \mathcal{R}_{\rm eff}/\Omega \approx (r/H)^2(c_{s0}H/\nu)(1/\Omega) \approx (r/H)^2\,\tau_{\rm th}$ for a relatively thin nebula. Thus, if the nebula is thermally stable, radiative transport and viscous dissipation would balance, forming a local thermal equilibrium during most stages of nebula evolution.

While $\mathcal{F}_s$ in Eq. (16) is determined by T_e, $D/\rho(= 9\nu\Omega^2/4)$ is determined by the magnitude of ν which may be determined by the temperature on the midplane of the nebula, T_c. Therefore, it is important for us to establish a relationship between T_e and T_c. In a thin thermally stable nebula, where $H/r << 1$, most contributions in Eq. (16) are negligibly small compared with the local dissipation rate and the heat flux in the vertical direction so that Eq. (16) reduces to

$$2\sigma T_e^4 = 9\Sigma\nu\Omega^2/4 \tag{17}$$

in the absence of infall. In those regions where the nebular flow is in a steady state, Eq. (9) becomes

$$3\pi\Sigma\nu = \dot{M} \tag{18}$$

where $\dot{M}$ is the mass flux in the nebula which is independent of r. From Eqs. (17) and (18) we find $T_e \propto r^{-3/4}$.

Detailed solutions for the vertical structure of the nebula can also be constructed. For example, if the nebula is in thermal equilibrium, Eq. (11) reduces to

$$\frac{\partial \mathcal{F}_z}{\partial z} = \frac{9}{4}\rho\nu\Omega^2 \tag{19}$$

in the absence of infall. In this case, the vertical structure of the nebula is determined from the solutions of Eqs. (10), (13), and (19) provided that the equation of state, the opacity law, and the viscosity prescription are specified. From these solutions, the viscous evolution of the nebula can be deduced using Eq. (9). If the viscosity is actually a function of local thermodynamic variables, these equations provide self-consistent solutions.

III. LOCAL ANALYSIS AND VERTICAL STRUCTURE

The above analysis indicates that the evolution of the nebula is basically determined by the magnitude of ν. If ν is a function of local thermodynamic variables such as T and ρ, we must solve the energy transport equation in order to evaluate the magnitude of ν. In principle, the evolution of an axisymmetric nebula should be analyzed in a two-dimensional time-dependent framework. However, if the nebula is relatively thin, i.e., $H/r << 1$, the r and z dependences in the momentum and energy transport equations can be essentially decoupled. In this section we shall utilize various approximations to examine a variety of simplified and idealized problems which address several important issues.

A. Convective Stability of the Nebula

Here we first consider the convective stability of the nebula. The nebula is assumed to be nonturbulent initially so that $\nu = \nu_{\rm mol} \approx 0$, $D = 0$, $U_r = 0$, and the only motion other than Keplerian rotation in the nebula is contraction in the vertical direction. The rate of internal energy generation by PdV work is balanced by the heat flux in the vertical direction. We wish to show here that such a flow pattern is intrinsically unstable against convection.

Convective Stability without Infall. Consider the vertical structure of a turbulent-free geometrically thin nebula in the absence of infall. The basic equations which determine the structure and evolution of the disk can be simplified considerably by ignoring contributions due to viscous stress and advective transport in the radial direction, so that the continuity Eq. (1) becomes

$$\frac{\partial \rho}{\partial t} + \frac{\partial(\rho U_z)}{\partial z} = 0. \tag{20}$$

The momentum Eq. (4) becomes

$$\frac{\partial U_z}{\partial t} + U_z \frac{\partial U_z}{\partial z} = -\frac{1}{\rho}\frac{\partial p}{\partial z} - z\Omega^2. \tag{21}$$

Similarly, the energy Eq. (11) becomes

$$C_{\rm v}\left(\frac{\partial T}{\partial t} + U_z \frac{\partial T}{\partial z}\right) - \frac{p}{\rho^2}\left(\frac{\partial \rho}{\partial t} + U_z \frac{\partial \rho}{\partial z}\right) = -\frac{1}{\rho}\frac{\partial \mathcal{F}_z}{\partial z}. \tag{22}$$

For a geometrically thin nebula,

$$|\partial \mathcal{F}_z/\partial z| \approx |(r/H)^2(1/r)(\partial r\mathcal{F}_r/\partial r)| >> |(1/r)(\partial r\mathcal{F}_r/\partial r)|.$$

For the initial equilibrium laminar solution, only the radiative flux contributes to the heat-transfer process, i.e.

$$\mathcal{F}_z = F_{\mathrm{rad},z} = \frac{-4acT^3}{3\kappa\rho}\frac{\partial T}{\partial z} \tag{23}$$

where κ is the opacity.

For arbitrary initial conditions, the nebula would not be in a state of hydrostatic equilibrium. It evolves in two stages. First, the nebula undergoes a phase of rapid readjustment. Through shock dissipation, a state of quasi-hydrostatic equilibrium may be achieved on a dynamical time scale $\tau_d \approx 1/\Omega$. Thereafter, the nebula contracts towards the midplane subsonically in a quasi-stationary manner. The rapid evolutionary phase can be analyzed by solving Eqs. (20) to (23) with a standard one-dimensional numerical hydrodynamic scheme with appropriate boundary conditions, equation of state, opacity table, and treatment of shock waves. Such a numerical analysis is currently being carried out by S. Ruden (personal communication). His results indicate that the thermal energy released from gravitational contraction towards the midplane can induce a sufficiently large temperature gradient so that the nebula is convectively unstable.

In the subsequent quasi-stationary evolution, the nebula loses thermal energy due to surface heat flux. In the absence of viscous dissipation, the heat content in the nebula decreases with time and the nebula contracts towards the midplane. The contraction provides a release of gravitational energy through $P\mathrm{d}V$ work. During this stage of evolution, the vertical structure of the nebula is continually adjusted. We now examine the convective stability of the nebula by considering the adiabatic gradient of a purely radiative solution. The flow pattern during this stage is best solved in terms of a Lagrangean variable z_l, such that the continuity equation implies $\rho \mathrm{d}z_l (= \mathrm{d}\Sigma_l)$ is independent of time (Landau and Lifshitz 1959) provided that $S_\Sigma = 0$. In this case, Eqs. (21) and (22) are reduced to

$$\left(\frac{\partial p}{\partial \Sigma}\right)_t = -\Omega^2 z_l \tag{24}$$

and

$$T\left(\frac{\partial S}{\partial t}\right)_{z_l} = -\frac{1}{\rho}\left(\frac{\partial \mathcal{F}_z}{\partial z_l}\right)_t \tag{25}$$

where S is the entropy and $T\mathrm{d}S = C_v \mathrm{d}T - (p/\rho^2)\mathrm{d}\rho$. In Eq. (24), we ignore the contribution due to $(\partial U_z/\partial t)_{z_l}$, because the quasi-stationary contraction is subsonic. If, in addition, the opacity is a power-law function of T only, say

$\kappa = \kappa_o T^\beta$, then it is possible to construct a set of homologously contracting solutions with $z = z_o z_l(t)$, $\rho = \rho_o(z_o)/z_l(t)$, $T = T_o(z_o)z_l(t)^2$, and $F_{\mathrm{rad},z} = F_o(z_o)z_l(t)^{8-2\beta}$. With these homologously contracting solutions, the t and z_o dependences in Eq. (25) can be separated, where z_l is a time-independent variable (Lin and Papaloizou 1980), such that

$$z_l(t)^{2\beta-7}\frac{\mathrm{d}z_l(t)}{\mathrm{d}t} = -\frac{1}{\rho_o(2C_\mathrm{v}T_o + \mathcal{R}T_o/\mu)}\frac{\partial F_o}{\partial z_o}. \tag{26}$$

For a gas with a constant specific heat ratio γ, the solution of Eq. (26) is

$$z_l(t) = [1 - k(t_o - t)]^{-1/(6-2\beta)} \tag{27}$$

where t_o is an integration constant, and

$$\frac{1}{\rho_o}\frac{\partial F_o}{\partial z_o} = \frac{k(\gamma + 1)\mathcal{R}T_o}{\mu(6 - 2\beta)(\gamma - 1)} = \epsilon(T_o) \tag{28}$$

where k is an arbitrary constant that can be chosen to provide any value of the surface flux required. The variables F_o, ρ_o, p_o, and T_o are functions of z_o only. Thus, despite the lack of viscous dissipation, internal energy is generated, through $P\mathrm{d}V$ work, at a rate $z_l(t)^{8-2\beta}\epsilon(T_o)$. The time-dependent components in Eqs. (24) and (25) can also be factored out with the result

$$\frac{\partial p_o}{\partial z_o} = -\rho_o\Omega^2 z_o \tag{29}$$

and

$$F_o = -\frac{4acT_o^{3-\beta}}{3\kappa_o\rho_o}\frac{\partial T_o}{\partial z_o}. \tag{30}$$

The ordinary differential Eqs. (28), (29), and (30) can then be integrated simply to provide the structure of a quasi-stationary homologously contracting nebula. Before presenting the results obtained from numerical integrations, we present an analytic criterion for the sufficient condition for convective instability.

The condition for convective instability for a gas with constant second adiabatic exponent γ_2 (see, e.g., Schwarzschild 1958) is

$$\nabla = \frac{\mathrm{dlog}T}{\mathrm{dlog}p} \geq \nabla_\mathrm{ad} = \frac{(\gamma_2 - 1)}{\gamma_2}. \tag{31}$$

This criterion can be generalized to include rotational effects in the disk. If the disk is relatively thin, the rotational effects cannot significantly modify the above stability criterion (Fricke and Kippenhan 1972).

Utilizing the homologous relationships, we find ∇ to be a function of z_o only. From Eqs. (29) to (31), we find

$$\nabla = \frac{p_o \partial T_o / \partial z_o}{T_o \partial p_o / \partial z_o} = \frac{3\kappa_o T_o^{\beta - 4} p_o F_o}{4ac\Omega^2 z_o}. \tag{32}$$

Note that since ∇ is assumed to be independent of t, stable models stay stable and unstable models stay unstable throughout the contraction phase. We now derive the condition which separates the two different types of models. It is convenient to introduce a variable

$$X = \Omega^2 z_o / F_o \tag{33}$$

so that

$$\nabla = \frac{3\kappa_o T_o^{\beta - 4} p_o}{4acX}. \tag{34}$$

Because

$$\frac{dX}{dz_o} = \frac{\Omega^2}{F_o}\left(1 - \frac{\rho_o \epsilon z_o}{\int_0^{z_o} \rho_o \epsilon dz_o}\right) \tag{35}$$

if ϵ is a monotonically decreasing function of z_o, $dX/dz_o > 0$ and the minimum and maximum values of X are achieved at the midplane and the surface, respectively. Furthermore, because $dT_o/dz_o \leq 0$, X decreases monotonically with increasing T_o. Using this fact we can generate the inequality

$$p_o - p_s = \int_{T_s}^{T_o} \frac{4acT^{3-\beta} X dT}{3\kappa_o} \geq \frac{4Xac(T_o^{4-\beta} - T_s^{4-\beta})}{3(4 - \beta)\kappa_o} \tag{36}$$

where p_s and T_s are the pressure and temperature at the nebula's surface. If $p_o >> p_s$ and $T_o >> T_s$, we find from Eqs. (34) and (36)

$$\nabla \geq 1/(4 - \beta). \tag{37}$$

Thus, a sufficient criterion for the onset of convection instability is

$$1/(4 - \beta) \geq (\gamma - 1)/\gamma. \tag{38}$$

In most regions of the nebula, its temperature is well below 2000 K so that the hydrogen gas is primarily in the form of molecules and the metals condense into micron-sized dust grains. In this case, $\gamma \approx 1.4$ so that the critical β for convective instability is ≈ 0.6. At these temperatures, if the dust grains are well mixed with the gas, they provide the dominant sources of opacity. The magnitude of the grain opacity roughly varies as the square of the temperature (Kellman and Gaustad 1969; DeCampli and Cameron 1979). For example, if $T \lesssim 160$ K

$$\kappa \approx \kappa_{\rm ice} = 2 \times 10^{-4}\, T^2 \tag{39}$$

in units of cm^2 g^{-1}. Above $\approx$ 160 K, ice may be evaporated so that the opacity can decrease rapidly over narrow ranges of temperature. However, other grains such as SiO_2 and Fe grains, which can remain in a condensed form for $T > 1000$ K, may also contribute to the opacity so that

$$\kappa_{\rm Fe} = 6 \times 10^{-6}\, T^2 \tag{40}$$

and

$$\kappa_{\rm SiO_2.} = 10^{\circ}\, T^2\,. \tag{41}$$

In all cases, $\beta > 0.6$ so that *the nebula is convectively unstable.*

The above results may be interpreted in the following physical manner. Because $\kappa \propto T^2$, a cooler region of the nebula has a lower opacity and therefore can cool off more quickly than the hotter regions in the nebula. The top regions of the nebula, being the most outlying, cool off first. As their temperature decreases, these regions become easier to cool. Eventually the thermal structure of the nebula adjusts to a new equilibrium in which the top surface layers of the nebula are considerably cooler than the midplane. Owing to the large temperature and pressure gradients in the vertical direction, the nebula may become superadiabatic and convectively unstable in that direction. In some parts of the inner region, the grain population changes rapidly with temperature due to evaporation and condensation processes. In this temperature range, the opacity also undergoes rapid variations. The stability analysis is more complicated but similar argument can still be applied to show that the nebula is convectively unstable.

In the above analytic discussion, we adopt the idealized opacity law for three representative grain species. Standard opacity tables are available which incorporate additional grains found in the interstellar medium. In this case, the above analytic results may be simply verified by standard numerical integration of Eqs. (28), (29) and (30) together with tabulated values for κ. (We are indebted to J. Pollack and P. Bodenheimer for the use of their opacity table.) The appropriate boundary conditions are (a) $F = 0$ at $z_l = 0$; (b) $F = \sigma T_e^4$ at $z_l = H$; (c) $T = T_e$ at $z_l = H$; and (d) $\tau \approx \rho\kappa(T_e)\, H = 2/3$ at $z_l = H$.

In the above analysis, the functional form of the radiative flux in the energy transport equation is only relevant for an optically thick nebula. When the optical depth becomes of order unity, we can no longer assume $p_0 >> p_s$ and $T_0 >> T_s$ in Eq. (36). The temperature gradient may become so small that the vertical structure is essentially isothermal. An isothermal nebula is convectively stable because $\nabla = 0$. When the optical depth becomes less than unity, the thermal structure of the nebula may be analyzed by detailed radiative transfer calculations in a manner analogous to stellar atmosphere calculations. Our numerical solutions indicate that the nebula is convectively unstable even in the limit that τ is of order unity. For $\tau \lesssim 1$, the nebula is convectively stable. The optical depth of the nebula decreases significantly when the surface density and temperature of the nebular gas are reduced during the course of viscous evolution and when the micron-sized grains, that are the primary sources of opacity, are depleted as they coagulate into large grains and plantesimals (Weidenschilling 1984). However, the time scale for optical-depth variation is sensitively determined by:

1. Collision properties of the grains;
2. The detailed nature of gas-grain interaction in a turbulent medium;
3. The feed back response of the nebula, such as the convective velocity and the thickness of the convective zone;
4. The dependence of the opacity law on the grains' size-distribution function.

All of these physical processes are poorly known at present and any detailed computation must necessarily invoke a large number of assumptions. Furthermore, the results generated from first attempts indicate that the outcome of grain coagulation may be very sensitive to the actual assumptions adopted (Weidenschilling 1984). Therefore, it is still too early to speculate whether or not the nebula may be stabilized against convection. Nonetheless, investigation in this area should be and will be carried out.

Convective Stability with Infall. In the absence of infall, optically thick regions of the nebula are convectively unstable because there is a sufficiently large temperature gradient in the vertical direction. This temperature gradient is induced by the *PdV* work which generates thermal energy primarily near the midplane where most of the disk material is concentrated. For the infall case, as external material arrives onto the nebula, the infall velocity is reduced from supersonic to subsonic flow through a shock wave. In the shock front, intense energy dissipation acts to reduce the temperature gradient in the vertical direction. If the infall flux is sufficiently large, ∇ in most regions of the nebula may be reduced below $(\gamma - 1)/\gamma$ so that the nebula is stabilized against convection. In addition, the infalling material may carry very different angular momentum than the nebular material. Near the shock interface, large shear may cause the nebula to become Kelvin-Helmholtz unstable and thereby induce large-scale turbulence. Under the idealized assumptions that the infalling material arrives at the nebula without any excess specific angular momentum so that the interface is Kelvin-Helmholtz stable and convection is the *only* mechanism for generating turbulence in the solar nebula, the efficiency of viscous transport of angular momentum would vanish if the infall rate were sufficiently high to stabilize against convection. In this case, the infalling material would accumulate in the vicinity of the infall radius and Σ there would increase. Below the shock surface near the midplane, the nebular material would adjust into a quasi-hydrostatic equilibrium. Due to the absence of viscous dissipation of energy, any heat loss would lead to a temperature decrease and a contraction in the vertical direction. If there is sufficiently large Σ, the PdV work resulting from such contraction would become comparable to or larger than the shock dissipation rate and the nebula could once more become convectively unstable. Thus, for any given mass infall rate S_Σ, there is a critical Σ_c above which the nebula is convectively unstable. We now construct quantitative models which can be used to deduce a $\Sigma_c - S_\Sigma$ relation.

Qualitatively, the infall case, like the infall-free case, can be examined in two stages.

1. In the *hydrodynamic stage,* the nebular structure departs considerably from thermal equilibrium. In this case, the nebula would evolve on a shock heating time scale $\tau_{sh} \approx \Sigma \mathcal{R} T / \mu S_\Sigma S_T$ which may be as short as a few years for $\dot{M}_* \approx 10^{-6}$ $M_\odot$ yr^{-1}. Strong shock dissipation would induce intense heating and thereby quickly adjust the thermal structure of the nebula until it evolves into a quasi-stationary stage.
2. The nebula enters into a *quasi-stationary stage* when τ_{sh} becomes comparable to the cooling time scale which is given by $\tau_c \approx \Sigma \mathcal{R} T / \mu \sigma T_s^4$. If the nebula contains 0.1 $M_\odot$ which is uniformly spread out in the present size of the solar system and has a temperature of a few hundred degrees, τ_c is of the order a few hundred years. In the quasi-stationary stage, there is a standing shock, at some distance H_s, above the nebula. In the absence of diffusion in the radial direction, the location of the shock front evolves in the vertical direction on a mass buildup time scale $\tau_b \approx \Sigma / S_\Sigma$ which is considerably longer than τ_{sh}, τ_c, and the dynamical time scale Ω^{-1}. Below the shock front, the flow is very subsonic and the nebula is essentially in a hydrostatic state, i.e., the vertical pressure gradient approximately provides the support against the gravitational pull of the Sun's gravity towards the midplane. Because the quasi-stationary stage lasts much longer than the hydrodynamic stage, we shall primarily consider the evolution of this stage in this subsection.

The quasi-stationary stage can be rigorously analyzed with similar mathematical methods as those adopted for star formation calculations (Larson 1969; Stahler et al. 1980*a,b*). Above the shock front, the flow is supersonic and is characterized by a density ρ_1, a vertical velocity $U_z = V_1$, and a temperature T_1. Across the shock front, the standard Rankine-Hugoniot relationships imply

$$\rho_1(V_1 - V_s) = \rho_2(V_2 - V_s) \quad (42)$$

$$p_1 + \rho_1(V_1 - V_s)^2 = p_2 + \rho_2(V_2 - V_s)^2 \quad (43)$$

and

$$\rho_1(V_1 - V_s)[2h_1 + (V_1 - V_s)^2] + 2F_1$$
$$= \rho_2(V_2 - V_s)[2h_2 + (V_2 - V_s)^2] + 2F_2 \quad (44)$$

where the subscript 2 denotes the quantities in the subsonic region below the shock front, h ($= p/\rho + E$) is the enthalpy, E is the internal energy, F is the heat flux, and V_s is the propagation velocity of the shock front. Because the flow above the shock front is highly supersonic, p_1, h_1, and V_s on the left hand side of Eqs. (42) and (44) can be ignored. Below the shock front, h, E, and p can be evaluated from ρ and T. Considerable care must be taken to evaluate F_1 and F_2. If the flow above the shock front is optically thin and that below the shock front is optically thick, $F_2 = \sigma T_2^4$ and $F_1 = fF_2$ where f is a dimensionless parameter to be determined (Larson 1969; Stahler et al. 1980*a*,*b*). In the limit that infall is highly supersonic, Eqs. (42) to (44) reduce to

$$\rho_1 V_1^2 = p_2 + \rho_1^2 V_1^2/\rho_2 \quad (45)$$

and

$$\rho_1 V_1^3 = \rho_1^3 V_1^3/\rho_2^2 + 2(1 - f)\sigma T_2^4 + 2\rho_1 V_1 h_2. \quad (46)$$

From Eqs. (45) and (46) and the equation of state, T_2 and ρ_2 as well as p_2, h_2, E_2, and V_2 can be uniquely determined from V_1 and ρ_1 once the dimensionless parameter f is specified. Note that in the above expressions V_1 is negative for infall.

Between the shock front and the midplane, the flow pattern is best analyzed with a Lagrangian variable $\Sigma_l(z) = \int_0^z \rho(z')dz'$ at any arbitrary value of z. Before solving these equations, it is important to incorporate the effect of infall. Infall implies that the boundary value of Σ, $\Sigma_B = \Sigma_l(z = H_s)$, where H_s is the distance between the shock front and the midplane, increases with time such that

$$\partial \Sigma_B / \partial t = - \rho_1 V_1 = S_\Sigma. \quad (47)$$

For arbitrary values of z_l, the continuity equation becomes

$$\left(\frac{\partial \rho}{\partial t}\right)_{z_l} + \frac{\rho}{\partial z/\partial z_l}\frac{\partial}{\partial z_l}\left(\frac{\partial z}{\partial t}\right)_{z_l} = S_\rho(z_l) \tag{48}$$

which has the solution

$$\frac{\rho \partial z}{\partial z_l} = \rho_o \exp\left(\int_0^t \frac{S_\rho}{\rho} dt\right) \tag{49}$$

or

$$d\Sigma_l(t) = d\Sigma_{lo} \exp\left(\int_0^t \frac{S_\rho}{\rho} dt\right). \tag{50}$$

If the rate of density increase due to infall at any location is proportional to the density already there, i.e., $S_\rho(z_l) \propto \rho(z_l)$, it would be computationally convenient to introduce a parameter $q = \Sigma_{lo}/\Sigma_l(t)$ which is independent of position, where Σ_{lo} is evaluated at some initial time t_o, so that $0 \leq q \leq 1$. Then, the important structure equations become

$$\left(\frac{\partial z_l}{\partial \Sigma_l}\right)_t = \frac{1}{\rho} \tag{51}$$

$$\left(\frac{\partial p}{\partial \Sigma_l}\right)_t = -\Omega^2 z_l \tag{52}$$

and

$$\left(\frac{\partial \mathcal{F}_z}{\partial \Sigma_l}\right)_t = -\left(\frac{T \partial S}{\partial t}\right)_{\Sigma_{lo}}. \tag{53}$$

For radiative solutions

$$\mathcal{F}_z = \frac{-4acT^3}{3\kappa}\left(\frac{\partial T}{\partial \Sigma_l}\right)_t. \tag{54}$$

Equations (47) to (54) from a complete set of equation for quasi-stationary evolution and they reduce to the standard vertical structure equations in the absence of infall when $\partial\Sigma/\partial t = 0$. The four variables are z_l, T, ρ, and $\mathcal{F}_z$. The boundary conditions are (1) $\mathcal{F}_z = 0$ and (2) $z_l = 0$ at $\Sigma_l = 0$, (3) $T = T_2$ and (4) $\rho = \rho_2$ at $\Sigma = \Sigma_i$, where Σ_i is the total surface density of the nebula at a given time t. The value of z_l at $\Sigma = \Sigma_i$ is the height of the shock front H_s. For arbitrary values of ρ_1 and V_1, detailed solutions can be obtained by first constructing an initial equilibrium model with $(T\partial S/\partial t)_{\Sigma_{lo}} = 0$. For a given value of Σ_i, only a unique value of H_s for a given f can provide a solution which satisfies all the boundary conditions. The quasi-stationary model can be advanced to the next step by calculating the new value of Σ from $q(t)$, evaluated from Eq. (47), into Eqs. (51) to (54). From the evolution of the thermodynamical variables, the time derivative contribution in Eq. (53) can be calculated.

This approach has recently been adopted by S. Ruden (personal communication), and his preliminary results indicate that as Σ increases with time, the contribution due to PdV work indeed increases such that the temperature gradient in the vertical direction has a tendency to become superadiabatic.

B. Vertical Structure of a Convectively Unstable Nebula

If the nebula is convectively unstable, turbulent motion would be induced. Convectively driven turbulence has three major consequences for the evolution of the nebula.

1. Convection increases the efficiency of thermal energy transfer from the midplane to the nebula surface layers.
2. Turbulent eddies induce mechanical mixing of material over a radial extent comparable to the size of the largest convective eddies and thereby increase the efficiency of angular momentum transport.
3. Turbulent motion induces a continual transfer of kinetic energy from the differential rotation in the main flow to the largest eddies.

The energy associated with the chaotic motion is then transferred to smaller eddies as the largest eddies break up on an eddy turnover time scale. Eventually, the eddies decay to a scale such that the local viscous stress, which is the product of the local velocity gradient and molecular viscosity, is sufficiently large that the random kinetic energy of the small eddies is dissipated into heat on an eddy turnover time scale. The heat released from this energy dissipation provides the primary source of thermal energy for the nebula. Throughout this chain of events, the drain of kinetic energy from the main flow is eventually compensated by the release of the gravitational energy as material drifts toward the protosun.

The relative importance of these three effects is very different. For example, while convection may only modestly supplement the heat transport efficiency, turbulence induced by it can fundamentally change the nature of momentum transport and viscous dissipation processes. Let us first examine the effect of turbulence-induced viscous dissipation on the vertical structure of the nebula. Equation (15) indicates that the rate of energy dissipation is $D = 2.25\rho\nu_{\rm eff}\Omega^2$. There are numerous *ad hoc* prescriptions for $\nu_{\rm eff}$. For example, although the fluctuating velocity components due to turbulence $\mathbf{u}_t$ vanish when averaged over a long period of time, the Reynolds stress, $< (\mathbf{u}_t \cdot \nabla)\mathbf{u}_t >$ $= (1/T_a)\int_0^{T_a} (\mathbf{u}_t \cdot \nabla)\mathbf{u}_t dt$ for arbitrarily large T_a, does not generally vanish (Tennekes and Lumley 1972). In laboratory flow, the Reynolds stress provides an effective Reynolds number $\mathcal{R}_{\rm eff} = U_0 L_0/\nu_{\rm eff} \approx 10$ to 10^3 where U_0 and L_0

are the characteristic main-stream velocity and length scale (Townsend 1976). By analogy with laboratory turbulent flow, we could assume that $\mathcal{R}_{\rm eff}$ is constant throughout the nebula at all times so that

$$\nu_{\rm eff} = \mathcal{R}_{\rm eff}^{-1}\,\Omega r^2 \tag{55}$$

is a function of r only. In this case, the diffusion Eq. (9) becomes linear and can be solved in terms of modified Bessel functions (Lynden-Bell and Pringle 1974).

However, the solar nebula, unlike laboratory flows, has considerable temperature and density range. It is entirely possible that the magnitude of $\mathcal{R}_{\rm eff}$ may be determined by local variables such as T and ρ. An alternative, widely adopted, *ad hoc* prescription for $\nu_{\rm eff}$ in accretion disk theories is the α model (Shakura and Sunyaev 1973). In standard α models,

$$\nu_{\rm eff} = \frac{2}{3}\alpha_s c_s H = \frac{2}{3}\alpha_s c_s^2/\Omega \tag{56}$$

where α_s is a dimensionless parameter which is comparable to or less than unity for transonic or subsonic turbulence. Although the α prescription can be regarded as having a basis similar to that for molecular viscosity which can be calculated from the kinetic theory of gases (Jeans 1967), there is little rigorous physical justification for it because fluid elements cannot be treated in the same manner as molecules. In addition, although the prescription is for a turbulent viscosity, it does not explicitly contain a stability condition, i.e., the origin of the turbulence is not stated. Nonetheless, the α prescription is very useful because it provides a $\nu_{\rm eff}$ which is determined by the local thermal and dynamical conditions.

In the context of solar nebula models, the α prescription can be substantiated with a slightly more rigorous treatment by incorporating an assumption that convection is the *only* mechanism which can generate turbulence. In this case, we set $\nu_{\rm eff} = 0$ in the radiative region and

$$\nu_{\rm eff} = \alpha_c {\rm v}_c l_c = \alpha_c {\rm v}_c^2/\Omega \tag{57}$$

in the convective region, where ${\rm v}_c$ is the convective velocity and α_c is a dimensionless parameter comparable to or less than unity. Because typical convective eddies break up on a time scale comparable to $\tau_{\rm eddy} \approx \Lambda/c_s \approx \Omega^{-1}$, where Λ is the scale height or the mixing length, we estimate the mixing length in the radial direction $l_c \approx {\rm v}_c \tau_{\rm eddy} \approx {\rm v}_c \Omega^{-1}$ (Lin and Papaloizou 1980). In this prescription, the rotation of the nebula is effectively taken into account. The effective turbulent stress, $1/\Sigma r^2\, \partial/\partial r\, (\Sigma \nu_{\rm eff} r^3\, \partial\Omega/\partial r) \approx \nu_{\rm eff}\Omega/r \approx \alpha_c {\rm v}_c^2/r$ which adequately represents the Reynolds stress estimated from the standard mixing-length approximation that $< ({\bf u_t} \cdot \nabla){\bf u_t} > \approx {\rm v}_c^2/r$. If the convective velocity is comparable to c_s, Eqs. (56) and (57) provide very similar estimates for $\nu_{\rm eff}$.

This prescription for $\nu_{\rm eff}$ can be applied not only to the energy equation, to evaluate the magnitude of D, but also to the mass-diffusion equation, to determine the evolution of the nebula. From the discussion in Sec. II, we showed that in a nebula evolving due to viscous effects, a relatively short time is required for the nebula to enter a state of local thermal equilibrium. In such a state, the nebula is not only in vertical pressure equilibrium but also has local dissipation balanced by radiative losses in the vertical direction. In the absence of infall and efficient advective heat transport in the radial direction, the vertical structure can be determined from a set of ordinary Eqs. (10), (12), (13), (19), and a prescription for F_c. The effect of infall can be included to modify these equations. In principle, these equations can be solved exactly, once the boundary conditions are specified. In order to reveal the most important contributions to the vertical structure of the nebula, we first proceed with a conventional thin-disk approximation, in which we adopt Eq. (56) for $\nu_{\rm eff}$ and ignore the convective contribution in the total vertical heat flux, $\mathcal{F}_z$. With this approach, approximate analytic solutions can be obtained and the results can be verified by exact numerical integrations. We then include the convective contribution in $\mathcal{F}_z$ but still adopt the standard α prescription for $\nu_{\rm eff}$ to determine the relative importance of convective heat transport. Finally, we consider a self-consistent treatment in which convection is not only included in the heat transport but also in the heat generation process.

1. Conventional Local Thermal Equilibrium Thin-Disk Approximation. Here, we adopt a conventional radiative thin-disk model by analytically integrating the energy equation over the vertical direction (Shakura and Sunyaev 1973). In order to calculate $\mathcal{F}_z$, we must first specify the functional form of κ. Although κ can be obtained from standard opacity tables, for the purpose of obtaining analytic solutions, we approximate κ in the relevant temperature and density range by an analytic functional fit such that

$$\kappa = \kappa_0 \rho^{m-1} T^{3-n} \tag{58}$$

where m and n are constants. For ice grain opacity, $m = 1$ and $n = 1$. If we ignore convective heat transport, the z integral of Eq. (12) is

$$\int_0^{\Sigma/2} \frac{3F_{\rm rad,z}}{4ac} {\rm d}\Sigma = \int_{T_e}^{T_c} \frac{T^3}{\kappa} {\rm d}T. \tag{59}$$

If $F_{\rm rad,z}$ is a slowly varying function of Σ except near the midplane where it vanishes, the left-hand side of Eq. (59) would be $\approx 3F_s\Sigma/8ac$. Applying the standard Eddington approximation for plane-parallel atmospheres, the surface heat flux, $F_s = \sigma T_e^4$ and T_e is the effective temperature at $\tau = 2/3$. If κ is primarily determined by T (as in the case of grain opacity), the right-hand side of Eq. (59) would be $\approx T_c^4/(1+n)\kappa_c$ provided that $n > -1$, where T_c and κ_c are the temperature and opacity at midplane. Thus, Eq. (59) would imply $T_e^4 \approx 32\, T_c^4/3(1+n)\kappa_c\Sigma$. For more general opacity laws, we adopt the standard thin-disk approximation in which

$$T_e^4 = \frac{4T_c^4}{3\tau_c} = \frac{8T_c^4}{3\kappa_c\Sigma} \tag{60}$$

which differs from the above approximate analysis by a factor of order unity. The integral of Eq. (19) implies

$$2.25\nu_c\Sigma\Omega^2 = 2F_s = 0.5acT_e^4. \tag{61}$$

Using the standard α prescription for $\nu_{\rm eff}$ (Eq. 56), Eqs. (60) and (61) give

$$\frac{3}{2}\alpha_s\Sigma\frac{\mathcal{R}T_c}{\mu}\Omega = \frac{4\mathrm{acT}_c^4}{3\kappa_c\Sigma} \tag{62}$$

which can be solved once the opacity law is specified. In general κ_c is a function of T_c and ρ_c. In a thin nebula, the solution of the pressure balance Eq. (10) implies

$$H = \left(\frac{\mathcal{R}T_c}{\mu}\right)^{1/2}\frac{1}{\Omega} \tag{63}$$

$$\rho_c = \Sigma/2H \tag{64}$$

so that ρ_c can be expressed in terms of T_c and Σ. Consequently, κ_c becomes a function of T_c and Σ. At this point, we can apply standard opacity tables to Eq. (62) to determine T_c and ν (or equivalently T_e) as functions of α, Σ and r (Fig. 1). Note that there is a slight transition at T_e between 100 K to 200 K. At these temperatures, ice grain condensation is marginal. The presence of ice grains can change κ considerably.

The best available opacity table can be approximated with the analytic fits

$$\kappa_1 = \kappa_{01}\, T^2 \tag{65a}$$

for $T < T_{12} = 170$ K where $\kappa_{01} = 2 \times 10^{-4}$

$$\kappa = \kappa_{02}T^{-7} \tag{65b}$$

for $T_{12} < T < T_{23} = 210$ K where $\kappa_{02} = 2 \times 10^{16}$

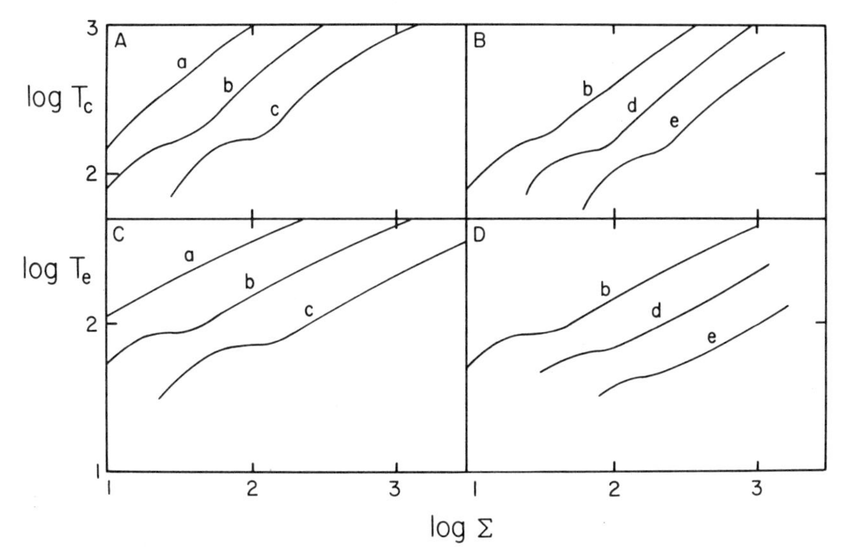

Fig. 1. The $T_c - \Sigma$ and $T_e - \Sigma$ relationships deduced from the local thermal equilibrium conditions, i.e., Eq. (62). The surface density Σ is in units of g cm^{-2} whereas both the midplane temperature T_c and the surface temperature T_e are in units of K. The value of α_s for cases *a, b,* and *c* are 1, 0.1 and 0.01, respectively. These three cases are for $r = 1$ AU. The value of α_s for both cases *d* and *e* are 0.1 whereas the value of $r = 4$ AU is for cases *d* and 16 AU is for case *e*.

$$\kappa_3 = \kappa_{03} T \tag{65c}$$

for $T_{23} < T < T_{34} \approx 2000\ K$ where $\kappa_{03} = 5 \times 10^{-3}$. The temperatures separating different opacity laws are derived by equating the different power-law expression of κ. Such a fit can be extended continuously over all $T <$ 2000 K by the fit

$$\kappa_f = [(\kappa_1^{-2} + \kappa_2^{-2})^{-2} + (\kappa_3/(1 + 10^{22}T^{-10}))^4]^{1/4}. \tag{66}$$

From the general analytic fit (Eq. 58), we find

$$\kappa_c = \kappa_0 \left[\frac{1}{2}\left(\frac{\mu}{\mathcal{R}}\right)^{1/2}\right]^{m-1} (\Sigma\Omega)^{m-1} T_c^{(7-2n-m)/2}. \tag{67}$$

Omitting some algebra, we find from Eqs. (62) and (67) that

$$T_c^{2n+m-1} = \frac{3^4}{2^{2m+4}}\left(\frac{\alpha_s \kappa_0}{ac}\right)^2 \left(\frac{\mathcal{R}}{\mu}\right)^{3-m} \Omega^{2m} \Sigma^{2m+2} \tag{68a}$$

$$\nu^{2n+m-1} = 1.5^{5-2n-m} 4^{-m} \alpha_s^{2n+m+1} \left(\frac{\kappa_0}{ac}\right)^2 \left(\frac{\mathcal{R}}{\mu}\right)^{2n+2} \Omega^{1+m-2n} \Sigma^{2m+2} \tag{68b}$$

and

$$\rho_c^{m+1} = \left(\frac{2ac}{9\alpha_s\kappa_0}\right)\left(\frac{\mu}{\mathcal{R}}\right)^2 \Omega T_c^{n-1}. \tag{68c}$$

Thus, once Σ is specified at a given radius, all other physical quantities can be determined in terms of it. In the case of the solar nebula, $m = 1$ and $n = 1$ in the regions where grain opacity dominates so that from the Eqs. (68a) to (68c) we find

$$T_c = \frac{9\alpha_s\kappa_{01}\mathcal{R}\Omega\Sigma^2}{8ac\mu} \tag{69a}$$

$$\Sigma\nu = \left(\frac{3\kappa_{01}}{4ac}\right)\left(\frac{\alpha_s\mathcal{R}}{\mu}\right)^2\Sigma^3 \tag{69b}$$

and

$$\rho_c = \left(\frac{2ac\Omega}{9\alpha_s\kappa_{01}}\right)^{1/2}\left(\frac{\mu}{\mathcal{R}}\right). \tag{69c}$$

Similarly, the transition regions where $m = 1$ and $n = 10$ we find

$$T_c = \left(\frac{9\alpha_s\kappa_{02}\mathcal{R}\Omega\Sigma^2}{8ac\mu}\right)^{1/10} \tag{70a}$$

$$\Sigma\nu = 3^{-8/10}2^{7/10}\left(\frac{\kappa_{02}}{4ac}\right)^{1/10}\left(\frac{\alpha_s\mathcal{R}}{\mu}\right)^{11/10}\Omega^{9/10}\Sigma^{12/10} \tag{70b}$$

and

$$\rho_c = \frac{1}{2}\left(\frac{\mu}{\mathcal{R}}\right)^{0.1}\left(\frac{9\alpha_s\kappa_{02}}{8ac\Omega}\right)^{0.4}\Omega^{1.4}\Sigma^{1.4}. \tag{70c}$$

In the relative warm regions

$$T_c = \left(\frac{9\alpha_s\kappa_{03}\mathcal{R}\Omega\Sigma^2}{8ac\mu}\right)^{1/2} \tag{71a}$$

$$\Sigma\nu = 2^{-1/2}\left(\frac{\kappa_{03}}{ac}\right)^{1/2}\left(\frac{\alpha_s\mathcal{R}}{\mu}\right)^{3/2}\Omega^{-1/2}\Sigma^2 \tag{71b}$$

and

$$\rho_c = \left(\frac{\mu\Omega}{\mathcal{R}}\right)^{3/4}\left(\frac{ac}{18\alpha_s\kappa_{03}}\right)^{1/4}\Sigma^{1/2}. \tag{71c}$$

The analytic solutions Eqs. (69) to (71) are in general agreement with the numerical solution in Fig. 1.

2. *Role of Convective Heat Transport.* Analytic or semi-analytic solutions are particularly useful for solving the diffusion Eq. (9) to determine the evolution of the nebula. However, before we utilize them extensively, it is worthwhile to examine the limitation and validity of various adopted assumptions. To begin with, we examine the Eddington approximation and the effect of convection. First, we neglect the effect of convection in both D and $\mathcal{F}_z$ and construct numerical solutions for the vertical structure Eqs. (10), (12) and (19) with an α prescription (Eq. 56). In a thermal equilibrium state, all the thermal energy generated by viscous dissipation at a given radius is transported to the surface layers and radiated away. Therefore, the relevant boundary conditions are the same as those stated in Sec. III.A.1. The pressure balance condition must also be established at the surface which implies that $p = 2\Omega^2 H/3\kappa$ at $z = H$. The value of Σ at H is the integrated surface density for the nebula. The results obtained from the numerical integration of Eqs. (10), (12) and (19) are presented in Fig. 2A. It is of interest to note that the numerically integrated solutions are in excellent agreement with those obtained from simple thin-disk approximations.

We now modify the treatment of heat transport by including the convective flux while retaining the standard α prescription, i.e., Eq. (56), for $\nu_{\rm eff}$. To do this, we impose the condition that wherever the convective stability condition is violated, i.e., wherever, $\nabla = \dfrac{p}{T}\dfrac{\partial T/\partial z}{\partial p/\partial z} \geq \dfrac{(\gamma - 1)}{\gamma}$, the total flux becomes

$$\mathcal{F}_z = F_{{\rm rad},z} + F_{{\rm con},z} \tag{72}$$

where $F_{{\rm con},z}$ is the vertical component of the convective flux $\mathbf{F}_{\rm con}$ which can be estimated from a simple mixing-length model (Böhm-Vitense 1958; Cox and Giuli 1968) such that

$$F_{\rm con} = \frac{C_p \Lambda_1 {\rm v}_c}{2}\left(-\nabla T + \nabla'_{\rm ad}\frac{T}{p}\nabla p\right). \tag{73}$$

The convective velocity of the eddies, ${\rm v}_c$ can be derived according to the mixing-length prescription

$${\rm v}_c = \left(\frac{\Omega^2 z Q}{8T}\right)^{1/2}\left|-\nabla T + \nabla'_{\rm ad}\frac{T}{p}\nabla p\right|^{1/2}\Lambda_2 \tag{74}$$

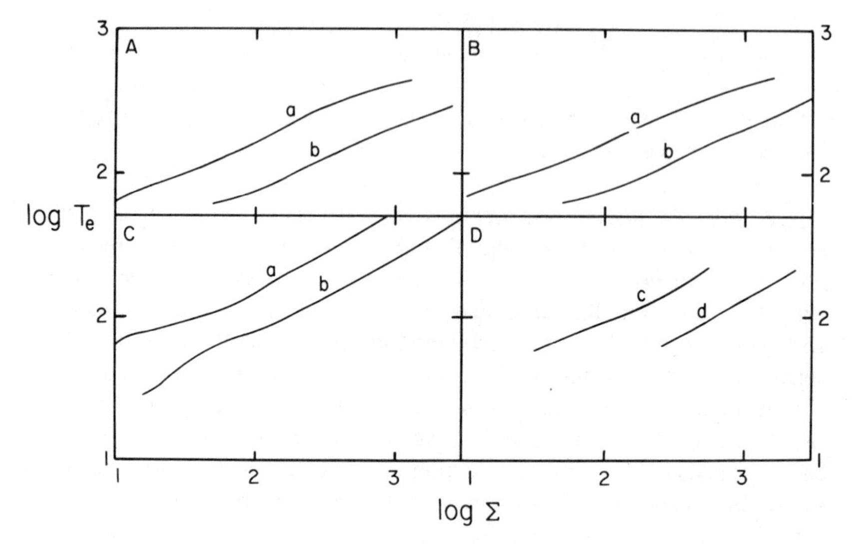

Fig. 2. The local thermal equilibrium solution deduced from detailed numerical solutions of the vertical structure. In plot A only radiative flux is included in $\mathcal{F}_z$ whereas in plot B both radiative and convective flux are included. Plot C represents solutions obtained from conventional thin-disk approximation. In A, B, and C the standard α model is used for viscosity prescription. In plot D, a self-consistent viscosity prescription is adopted. All of these computations are for $r = 1$ AU. In cases *a* and *b*, α_s equals 0.15 and 0.015, respectively. In cases *c* and *d*, α_c equals 1.5 and 0.15, respectively. The physical parameters Σ, T_e and T_c are in identical units as those in Fig. 1.

and

$$-Q = \left(\frac{\partial \ln \rho}{\partial \ln \mu}\right)_{p,T} \left(\frac{\partial \ln \mu}{\partial \ln T}\right)_{p,\rho} + \left(\frac{\partial \ln \rho}{\partial \ln T}\right)_{p,\mu}. \tag{75}$$

The mixing lengths, Λ_1 and Λ_2 are normally determined by the pressure scale height, $p/g\rho$. However, near the midplane where the scale height is large compared with z, Λ_1 is set to be z so that the heat generated from one side of the nebula will not be transported by convective eddies from the other side.

Using the same boundary conditions as for the pure radiative case, the convective solutions are determined numerically and presented in Fig. 2B. These results are again in good agreement with the results obtained from the Eddington approximation and the pure radiative case (Fig. 2C). At a given Σ, convection increases the total flux by less than one-third which may be interpreted to indicate that convection never carries a dominant fraction of the flux. This behavior is also found for the partially ionized regions, where H^- opacity is important in a disk around dwarf novae (Faulkner et al. 1983) and for the hot fully ionized, electron scattering regions of a disk surrounding a neutron star or a black hole (Shakura et al. 1978; Roberson and Tayler 1981).

3. *Self-Consistent Convective Model.* In the above analysis, it is evident that the α prescription is very useful for constructing thermal equilibrium models for the nebula. Perhaps the weakest assumption in the α prescription is that the origin of turbulence is not specified. In the solar nebula, we have already demonstrated that there is *at least one* source of turbulence, i.e., convection. If we assume that convection is the *only* source of turbulence in the solar nebula, we can construct a self-consistent model in which all three major effects of convection are taken into account. Quantitatively we can utilize Eq. (57) for ν_{eff} so that $D = 2.25\ \alpha_c \rho v_c^2 \Omega$ in the convective zone and we set $\nu = D = 0$ in the radiative zone.

Adopting the same physical conditions at the midplane and the surface layer of the nebula as above, the numerical solution for a given surface heat flux or equivalently for surface temperature can be obtained (Lin and Papaloizou 1980; Lin 1981). In Fig. 3, we plot T, ρ, τ, and $\mathcal{F}_z$ as a function of z. These results indicate that convection only sets in at the point where the nebula becomes opaque so that a sufficiently large temperature gradient can be established to induce the convective instability. These results can be generalized for different surface conditions (Fig. 4). The convection zone is surrounded by a radiative surface layer where $\nu = 0$ and $\mathcal{F}_z$ remains constant. Because the radiated energy emerging from the surface layer ultimately arises from the gravitational energy of the inwardly diffusing gas in the nebula, a relatively large heat flux can only result from a relatively high surface density. Thus, there is a one-to-one correspondence between T_e and Σ and the general results are plotted as functions of Σ. These results indicate that T_c, ρ_c, p_c, ν,

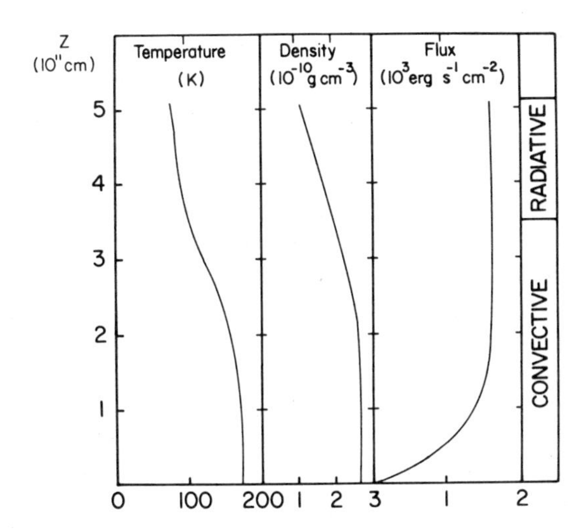

Fig. 3. The vertical structure deduced from a particular solution of the self-consistent local thermal equilibrium treatments of nebular structure. The relevant values are $r = 10^{13}$ cm, $T_e = 75$ K and $\alpha_c = 1$.

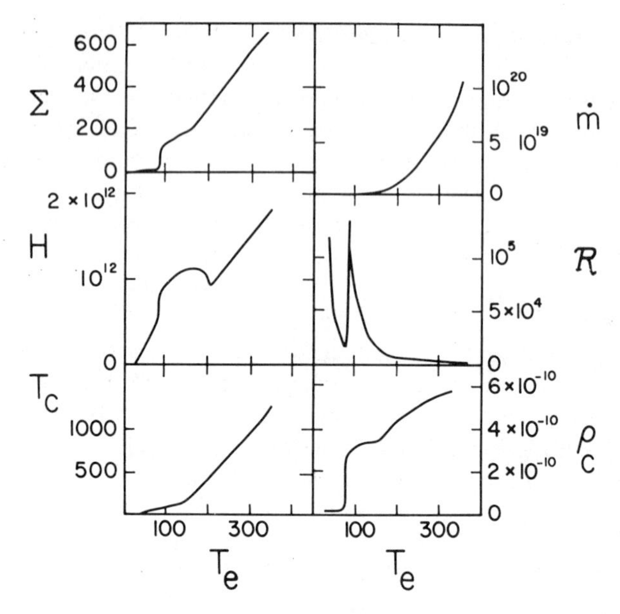

Fig. 4. Numerical solutions deduced from self-consistent local thermal equilibrium treatments of nebular structure for various surface temperatures. The physical parameter for Σ, H, T_c, $\dot{m}$ and ρ_c are in cgs units. The values of r and α_c are 10^{13} cm and 1, respectively.

and H all increase with Σ. The $\nu - \Sigma$ relation derived here is particularly useful for solving the diffusion Eq. (9) (see discussions in Sec. IV).

These results are also in good agreement with the solutions deduced from the conventional, radiative, α-prescription, thin-disk approximation (Fig. 2D). The only difference is that α_c is approximately equivalent to 0.02 α_s. This difference results from the fact that the convection zone only occupies a fraction of the disk, i.e., $l < H$, and that convective motion is subsonic, i.e., $v_c \lesssim 0.2\, c_s$. Nonetheless, the general functional form of the solutions is essentially reproduced from the simple standard thin-disk approximations for most values of Σ. At large Σ, because H/r approaches unity, the thin-disk approximation is no longer appropriate. Note that $Q_T \equiv c_s\Omega/\pi G\Sigma)$ is a very weak function of Σ and that a very large Σ is needed for the self gravity of the nebula to become important ($Q_T \lesssim 1$). A large Σ also implies a large ν and therefore a large mass transfer rate. Thus, a large Σ cannot be sustained over a long period of time. Note that the ice grain transition effect is more pronounced in these solutions than in the thin-disk approximation. Numerical results indicate that the convection zone is much thinner if the ice grains cannot condense, and this would greatly reduce ν. In order to maintain the nebular temperature above 160 K, however, there must be a corresponding heat flux which is ultimately determined by the rate of gravitational energy release by inwardly drifting nebular matter. In order to maintain a relatively high mass flux with a low radial velocity, the surface density must be relatively high. Thus, a minor change in T_e implies a large change in Σ.

In the limit that the nebula is geometrically thin, the r and z dependences can be separated such that the solution obtained at one radius may be scaled to other radii (Lin and Papaloizou 1980; Lin 1981). For example, if we assume

$$T = T_0 g_T(z_1) f_T(r),\ \rho = \rho_0 g_\rho(z_1) f_\rho(r)$$

$$\mathcal{F} = \mathcal{F}_0 g_{\mathcal{F}}(z_1) f_{\mathcal{F}}(r),\ p = p_0 g_p(z_1) f_p(r) \tag{76}$$

where $z_1 = z/H$ and $H = H_0 f_H(r)$, we find from Eq. (10) that

$$\frac{1}{z_1 gp(z_1)} \frac{\mathrm{d}g_p(z_1)}{\mathrm{d}z_1} = -\frac{\rho_0 H_o^2}{p_o} \frac{GMf_\rho(r) f_H(r)^2}{f_p(r) r^3}. \tag{77}$$

Because the z and r dependences are totally separated, the two sides of Eq. (77) must be equal to a constant. Thus $\mathrm{d}g_p/\mathrm{d}z_1 = a g_\rho z_1$ and $f_p r^3 = b f_H^2 f_\rho$. Other functional relationships can be derived from Eqs. (19) and (56). Finally, if we assume that either convection does not carry a significant amount of the total flux or the fraction of total flux carried by convection is a weak function of r, we can also utilize Eqs. (72) and (73). From these functionals we find that

$$f_T = f_\Sigma^2 r^{-3/2},\ f_\rho = r^{-3/4},\ f_\nu = f_\Sigma^2$$

$$f_H = f_\Sigma r^{3/4},\ f_F = f_\Sigma^3 r^{-3}. \tag{78}$$

These functional dependences, which are identical to those in Eqs. (69) to (71), are particularly useful for evolution calculations. In Sec. IV below we see that these analytic scaling laws are verified for the results of full-scale numerical integrations for the steady state case.

C. A Summary of Local Analysis and Vertical Structure

The results presented in this subsection indicate the following:

1. The solar nebula adjusts to a quasi-stationary, local hydrostatic equilibrium on dynamical time scales.
2. From the conditions required for such a local hydrostatic equilibrium, we can deduce the vertical structure of the nebula in the absence of turbulence with or without infall.
3. In the absence of infall, the nebula undergoes homologous contraction. Due to the temperature dependent of the grain opacity, the central portion of the nebula is superadiabatic so that the gas flow in the nebula is convectively unstable.

4. Infall can provide an additional heat source near the surface of the nebula, thereby reducing the temperature gradient in the vertical direction, and temporarily stabilizing against convection.
5. Preliminary solutions for the infall case indicate that there is always a critical Σ_c for which the nebula is convectively unstable. If convection is the only source of turbulence, infall would induce Σ to increase to Σ_c before the nebula becomes unstable.
6. The solution for local thermodynamic equilibrium can be rigorously determined from the vertical structure equations with any prescription for viscosity with or without infall.
7. With the standard α prescription, a set of analytic solutions can be obtained. These analytic solutions adequately describe the overall properties of the nebula.
8. In detailed numerical solutions for the α model, we find that although the midplane region of the nebula is convective, convection does not strongly modify the efficiency of heat transport.
9. Convection qualitatively changes the rate of viscous dissipation and reduces the magnitude of $\mathcal{R}_{\rm eff}$ to values typically on the order of a few thousand.
10. The vertical structure equations can also be solved with a self-consistent approach and the solutions obtained therefrom are in good agreement with those found with the α prescription for viscosity.

IV. EVOLUTION OF THE NEBULA

Due to the action of viscous stress, there is a general evolutionary tendency for an accretion disk to continuously transport angular momentum outward and therefore to allow the nebular material to fall inward. The evolution of the nebula can be deduced in a quantitative manner. Under certain circumstances, the flow pattern is three dimensional. For example, (1) during its formation stage, the nebula may be relatively thick and the flow in the r and z direction is strongly coupled if the infall rate is extremely large or if the protosun is considerably less massive. (2) The infalling material may carry very different amounts of specific angular momentum from the nebular material, and nonaxisymmetric shearing instability may be triggered. (3) If the self gravity of the nebula is sufficiently strong, nonaxisymmetric density waves may be induced (Cassen et al. 1981). (4) If the protosun is a rapid rotator, the nonaxisymmetric tidal interaction between a rotationally distorted protosun and the inner regions of the nebula may also induce nonaxisymmetric flow patterns. (5) Tidal interaction between protogiant planets and the outer regions of the nebula may also play a dominant role in the late stages of solar nebula evolution (Lin and Papaloizou 1980).

Some of these circumstances may be best investigated with an ultimate three-dimensional numerical analysis (Tscharnuter 1980; Boss 1983). However, because a three-dimensional numerical analysis is complicated and time consuming, it is impractical to use it to examine the evolution of the nebula over many orbital periods. For most stages of the evolution, the nebula may be relatively thin and the flow pattern is approximately axisymmetric so that a thin-disk treatment may be sufficient. In this section, we utilize the results obtained in Secs. II and III and evaluate quantitatively the evolution of the nebula.

For the convenience of general discussion, we divide the evolution of the nebula into three major stages:

1. The infall stage, when the evolution of the nebula is strongly influenced by the influx of external material;
2. The viscous stage, when the external supply of material is exhausted and the evolution of the nebula is primarily determined by its internal viscous frictional force;
3. The clearing stage, when most of the nebular material is accreted by either the protosun or the protoplanets, and through various physical processes the remaining nebular material is driven away.

Also for the convenience of general discussion, we divide the radial distribution into five regions.

1. The innermost region extends approximately from the inner boundary near the protosun to the present radius of Mercury. In this radial range, the nebula is relatively hot, i.e., $T \gtrsim 2000$ K, and contains no grains. We discuss the dynamics of this region in Sec. VII.
2. The inner region covers the present orbital radii of the terrestrial planets. This region is dominated by metal grains during the protoplanetary formation epoch.
3. The intermediate region ranges from Jupiter's to Uranus' radius where most of the nebula's mass is contained. In this region, ice grains provide the dominant sources of opacity.
4. The outer region extends between the orbital radii of Uranus and Neptune where the nebular material may have a tendency to diffuse outwards rather than inwards.
5. The outermost region is beyond the orbit of Neptune where the nebula may become optically thin.

We now show that as the nebula evolves viscously, most of the angular momentum in the nebula is deposited in the outermost region while the bulk of the nebula's mass is accreted by the protosun.

A. The Infall Stage

In the early stages of nebular formation, matter from the collapsing cloud continuously arrives at the nebula at a rate faster than that of the redistribution of matter induced by the viscous frictional force, so that the mass of the nebula gradually builds up. The reservoir of cloud material is depleted on a characteristic collapse time scale τ_{ff}, which for a 10 K, 1 $M_\odot$ molecular cloud, may be as short as 10^5 yr (Cameron 1978) or indefinitely long (see chapter by Cassen et al.). As the infall flux decreases and the viscous frictional stress increases with the mass of the nebula, the viscous diffusion rate eventually becomes comparable to the infall rate and the infall stage evolves into a viscous stage (Cassen and Moosman 1982; Lin and Bodenheimer 1982).

Quantitatively, the effects of infall can be incorporated into the evolution calculation for the infall stage. In Sec. III we showed that the nebula can establish a thermal equilibrium on a dynamical time scale such that Eq. (16) is reduced to

$$\frac{9\Sigma\nu\Omega^2}{4} + S_\Sigma S_T = 2\sigma T_e^4 = \frac{16\sigma T_c^4}{3\kappa_c\Sigma}. \tag{79}$$

Thus, the magnitude of T_c can be determined from Σ once prescriptions for ν, S_Σ, S_T are specified. The evolution of Σ is determined from Eq. (8) which can be solved with a simple one-dimensional numerical hydrodynamic scheme (see Richtmyer and Morton 1967) once the initial Σ distribution and the boundary conditions at small and large r are specified.

In order to elucidate some aspects of nebular evolution during the infall stage, we further divide it up into two substages: (1) the initial buildup stage, and (2) the quasi-stationary stage. In the initial buildup stage, shock dissipation reduces the temperature gradient in the vertical direction and may stabilize convection when Σ is relatively small. If convection is the only source of turbulence, viscous diffusion is essentially absent until Σ becomes larger than Σ_c. In this case, Eq. (8) reduces to

$$\frac{\partial\Sigma}{\partial t} + \frac{2}{r}\frac{\partial}{\partial r}\left(\frac{S_\Sigma(r,t)J(r,t)}{\Omega}\right) - S_\Sigma(r,t) = 0. \tag{80}$$

The above approximation is particularly relevant if the nebula is indeed turbulent-free; e.g., for $\Sigma < \Sigma_c(S_\Sigma)$.

Although Eq. (80) can be analytically or numerically integrated for arbitrary functions of $S_\Sigma(r,t)$ and $J(r,t)$, we adopt some particular prescriptions for them for illustrative purposes. Let us consider the case that the infalling material joins the nebula with a fixed excess fraction of specific angular momentum, i.e., $J(r,t) = \beta_J\Omega r^2$, and carries with it a fixed excess fraction of the orbital kinetic energy, i.e., $S_T(r,t) = \beta_T\Omega^2 r^2$, where $\beta_J \lesssim 1$ and $\beta_T \lesssim 1$ are constants. For the infall flux, we adopt a more general functional form $S_\Sigma(r,t) = S_o r^a e^{-\beta_\Sigma t}$, where $S_o = |\dot{M}_*|/2\pi \int_{r_{in}}^{r_{out}} r^{a+1} dr$ and it is positive for infall. If the infall rate is uniform, $a = 0$; and if it is constant in time, $\beta_\Sigma = 0$. With these prescriptions

$$\Sigma(r,t) = (2 - 4a\beta_J - 8\beta_J)S_o r^a f_\Sigma(t)/2 \tag{81}$$

where $f_\Sigma(t) = t$ for $\beta_\Sigma = 0$, otherwise $f_\Sigma(t) = (1 - e^{-\beta_\Sigma t})/\beta_\Sigma$.

The temperature distribution is determined by Eq. (79) which reduces to

$$\frac{16\sigma T_c^4}{3\kappa_c\Sigma} = S_\Sigma(r)S_T(r). \tag{82}$$

If κ_c is due to grain opacity (i.e., $\kappa = \kappa_0 T^2$), we find from Eqs. (81) and (82) that

$$T_c^2(r,t) = \frac{3(2 - 4a\beta_J - 8\beta_J)\kappa_0}{32\sigma} GM\beta_T S_o^2 r^{2a-1} f_\Sigma(t)e^{-\beta_\Sigma t}. \tag{83}$$

Thus, T_c increases with time as material accumulates in the nebula. The radial distribution of T_c is completely determined by the r dependence of S_Σ, S_T, and J. For the constant and uniform S_Σ case where $a = \beta_\Sigma = 0$, $\Sigma = \Sigma_o t$ and $T_c = T_{co}(t/r)^{1/2}$ where Σ_o and T_{co} are constants.

As Σ increases to sufficiently large values, shock dissipation can no longer stabilize the convective instability. Thereafter, viscous stress increases with Σ until the viscous diffusion rate in the nebula becomes comparable to the infall rate so that Σ changes slowly with time. Generally, the outer edge of the nebula can expand freely and the entire nebula cannot establish a steady state. If, however, the boundary conditions J, S_Σ, and S_T change slowly with time, at least the inner regions of the nebula may evolve towards a quasi-steady state. In such a state $\partial\Sigma/\partial t$ is relatively small so that Eq. (8) is reduced to

$$\Sigma\nu = \frac{(2 - 4a\beta_J - 8\beta_J)S_o}{3(a + 2)(2a + 5)} r^{a+2} + A + Br^{-1/2}. \tag{84}$$

The values of the integration constants A and B are determined by the boundary conditions. If $a > -2$, contributions associated with A and B are negligible for inner and intermediate radii. As is always the case, we must adopt a ν prescription in order to determine the radial distribution in Σ. For example, if $\mathcal{R}_{\rm eff}$ is constant, Eq. (84) implies $\Sigma \propto S_o r^{a+3/2}$ for inner and intermediate radii. Furthermore, from Eq. (79) we find $T_c \propto S_o r^{a+1/4}$ when grains provide the dominant source of opacity. However, if ν is a function of local physical parameters, we must determine T_c and Σ simultaneously from Eqs. (79) and (84). With an α prescription (Eq. 56) for $\nu_{\rm eff}$, $\Sigma \propto S_o^{1/3} r^{(a+2)/3}$, and $T_c \propto S_o^{2/3} r^{(4a-1)/6}$ for inner and intermediate radii.

The above results have several implications worth noting.

1. During the initial buildup phase the typical time scale for the nebula to increase its surface density to a particular value Σ_1 is $\tau_1 \approx \Sigma_1/S_\Sigma$. In order for infall to dominate over the viscous diffusion process, $\tau_1 \lesssim \tau_\nu(\Sigma_1) \approx$

$r^2/\nu(\Sigma_1) = \mathcal{R}_{\rm eff}/\Omega$. This initial buildup stage is short-lived for relatively small $\mathcal{R}_{\rm eff}$.

2. During the quasi-equilibrium stage, $\Sigma\nu \approx S_\Sigma r^2$. If, during this stage, the mass of the nebula is comparable to the minimum amount needed to make up for the "missing" gaseous constituents, the infall rate needed to sustain such a quasi-steady flow is $\dot{M} \approx 10^{-2}\mathcal{R}_{\rm eff}^{-1}\,{\rm M}_\odot\,{\rm yr}^{-1}$. If $\mathcal{R}_{\rm eff} \approx 10^3$, $\dot{M} \approx 10^{-5}\,{\rm M}_\odot\,{\rm yr}^{-1}$ which can hardly last for more than 10^5 yr.
3. The equilibrium surface density in the $\nu = \nu(r)$ case has a stronger dependence on S_o than that in the α prescription case. In the latter case, because the nebula temperature and thickness increase with Σ and S_o, ν also increases with Σ and S_o (Lin and Bodenheimer 1982). Consequently, the time scale for the establishment of such an equilibrium decreases with S_o and may be considerably shorter than the $\nu = \nu(r)$ case for large S_o.
4. In both the initial buildup stage and the quasi-steady stage, the radial distribution of T_c is determined by the radial dependence of S_Σ. Rather restricted functional forms are required for $S_\Sigma(r,t)$ to reproduce the temperature distribution ($T \propto r^{-1}$) traced out from the condensation temperature of the dominant rocky material in terrestrial planets and satellites.
5. Because both Σ and T_c increase with the infall rate, the self-gravitating effect of the nebula is a weak function of S_o. For example, when the infall flux is uniform and constant, the stability factor ($Q_T = c_s\Omega/\pi G\Sigma$) varies as $Q_T \propto S_o^{-1/2} r^{-a/2-23/8}$ for the constant Reynolds number case and $Q_T \propto$ $r^{-9/4}$ for the α prescription. Thus, a large infall flux does not necessarily imply that the nebula is more prone to gravitational instability.

B. The Viscous Stage

Although infall may continue throughout the evolution of the nebula, infall flux probably decreases on a time scale comparable to the free-fall time for the original cloud. When the infall flux is reduced to a relatively low level, the solar nebula evolves under the action of its own internal friction and enters a viscous stage. During this stage, angular momentum is transported outward to allow most of the matter to fall inward, so that nebular mass decreases and nebular size increases on a characteristic diffusion time scale $\tau_\nu \approx r/U_r \approx r^2/\nu$. One important issue to be addressed is the evolutionary time scale for a minimum-mass nebula during the viscous stage. Quantitatively, the evolutionary time scale is model dependent. In the $\nu = \nu(r)$ case, the characteristic diffusion time scale in the nebula is $\approx \mathcal{R}_{\rm eff}/\Omega$. If, however, ν is a function of local thermal variables, the magnitude of ν can change during the evolution of the nebula. In this case, a self-consistent global analysis is needed to evaluate the evolution.

1. The $\nu = \nu(r)$ Prescription. These physical effects can be examined rigorously with diffusion Eq. (9) and a ν prescription. Let us first consider the case where $\nu = \nu(r) = \nu_o r^\delta$ so that Eq. (9) is linear. For a constant Reynold's number model, $\delta = 1/2$. The general solution for Eq. (9) in this case is $\Sigma(r,t) = \Sigma_1(r)e^{-s_o t}$ where $\Sigma_1(r)$ is the solution of

$$\Sigma_1 + \frac{3\nu_o}{s_o r}\frac{\partial}{\partial r}r^{1/2}\frac{\partial}{\partial r}(\Sigma_1 r^{1/2+\delta}) = 0. \tag{85}$$

Under the coordinate transformation $g = \Sigma_1(r)r^{\delta + 1/2}$, Eq. (85) becomes

$$g + \frac{3\nu_o r^\delta}{s_o}\left(\frac{\partial^2 g}{\partial r^2} + \frac{1}{2r}\frac{\partial g}{\partial r}\right) = 0. \tag{86}$$

With further coordinate transformation $x = (s_o/3\nu_o)^{1/2} r^{1-\delta/2}/(1 - \delta/2)$ and $G = g/x^q$, we find

$$\frac{\partial^2}{\partial x^2}G + \frac{1}{x}\frac{\partial G}{\partial x} + \left(1 - \frac{q^2}{y^2}\right)G = 0 \tag{87}$$

where $q = 1/2(2 - \delta)$. The solutions to Eq. (87) are modified Bessel's functions (Lynden-Bell and Pringle 1974) such that

$$\Sigma = \int_0^\infty r^{-\delta - 1/4} e^{-3\nu_o k^2 t}[A(k)J_q(ky) + B(k)J_{-q}(ky)]kdk \tag{88}$$

where $y = r^{1-\delta/2}/(1 - \delta/2)$. Regularity conditions at the inner boundary imply $B(k) = 0$ and $A(k)$ can be obtained from the initial Σ distribution such that

$$A(k) = \int_0^\infty \Sigma r^{\delta + 1/4} J_q(ky)ydy. \tag{89}$$

With these solutions, Lynden-Bell and Pringle (1974) found, for various initial Σ distributions, that there is a critical point outside which a small amount of disk material drifts outwards to absorb most of the angular momentum and inside which the bulk of the disk matter diffuses inwards and is accreted by the central object. From Eq. (7), we find that such a radius R_W occurs at $\partial \ln \Sigma \nu r^{1/2}/\partial \ln r = 0$. (1) Because $\Sigma \nu r^{1/2}$ is just the viscous stress W, the critical radius is the location where W attains a maximum value in the nebula. In the present context, R_W occurs when $\partial \ln\Sigma/\partial \ln r = -(\delta + ½)$. The viscous stress is positive definite throughout the nebula. If the outer edge of the nebula can expand freely and if there is not significant viscous coupling between the accreting star and the nebula, $W = 0$ at both boundaries. In this case, W must attain a maximum value somewhere in the nebula. Lynden-Bell and Pringle's results indicate that R_W is located in the outer regions of the nebula and that

R_W actually increases with time. (2) The surface density in all, except the outermost, regions of the nebula decreases with time. Material near the outer edge of the nebula absorbs angular momentum from the bulk of nebular material and expands outwards. Due to this outward diffusion of nebular material, Σ just interior to the outer edge increases with time. In addition, the outer-boundary radius increases as $\approx t^{1/2}$. (3) The characteristic evolutionary time scale is $\approx R_h^2/\nu \approx R_h^{2-\delta}/\nu_o$ where R_h is the radius where half of the initial disk mass is initially contained. Thus, in the $\nu = \nu(t)$ case, the surface density essentially vanishes on a time scale $\approx \mathcal{R}_{\rm eff}/\Omega$ which is $\lesssim 10^5$ yr if $\mathcal{R}_{\rm eff} \lesssim 10^3$. (4) If the nebula is in a local thermal equilibrium everywhere and the dominant opacity is due to grains, we find from Eqs. (17) and (60) that

$$T_c = \frac{3}{8}\left(\frac{GM\nu_0\kappa_0\Sigma^2}{\sigma}\right)^{1/2} r^{(\delta-3)/2}. \tag{90}$$

Because Σ generally decreases with r, T_c decreases with r more rapidly that r^{-1} unless δ is larger than unity.

2. *The α and Mixing-Length Prescriptions.* The magnitude of $\nu_{\rm eff}$ may be a function of local thermodynamic variables such as in Eqs. (56) and (57). If $\nu_{\rm eff}$ decreases with T_c (and therefore with Σ), the evolutionary time scale lengthens as material is depleted in the nebula. This effect is best demonstrated with quasi-steady solutions in which $\partial\Sigma/\partial t$ is relatively small compared with the viscous stress contributions in Eq. (8). Quasi-steady state solutions are particularly relevant in the inner and intermediate regions of the nebula because there is a general evolutionary tendency for the nebula to establish a quasi-steady state flow pattern there (Lin and Bodenheimer 1982). These quasi-steady regions can extend to about a few times 10^{14} cm, and are particularly interesting because this is the region where most of the protoplanets may be formed.

a. Quasi-Steady State Solutions. In a quasi-steady state, the continuity Eq. (3) implies that $\Sigma U_r r$ is independent of radius or, alternatively, from Eq. (9)

$$\Sigma\nu = A + Br^{-1/2} \tag{91}$$

where A and B are integration constants. The mass and net angular momentum transfer rates in the nebula are independent of radius such that $\dot{M}_s = -3\pi A$ and $\dot{J}_s = 3\pi B(GM)^{1/2}$. These are also the accretion rates into the protosun so that the values of A and B are determined by conditions at the inner boundary. Thus, for most intermediate radii well outside the inner boundary, the contribution from the second term in Eq. (91) is small. If there is no viscous stress between the protosun and the nebula, $B = 0$. If we now apply the $\nu - \Sigma$

relationship derived for the simple α model, we find from Eqs. (91) and (69) that

$$\begin{aligned} \nu &= A_1(A + Br^{-1/2})^{2/3} \\ \Sigma &= A_2(A + Br^{-1/2})^{1/3} \\ T_c &= A_3 r^{-3/2}(A + Br^{-1/2})^{2/3} \\ T_e &= A_4 r^{-3/4}(A + Br^{-1/2})^{1/4} \\ \rho_c &= A_5 r^{-3/4} \end{aligned} \tag{92}$$

where A_1, A_2, A_3, A_4 and A_5 are constants. Similar functional forms can also be obtained from analytic approximation to the self-consistent convectively driven turbulent models, i.e., Eq. (78). These numerical results indicate that:

1. Σ and ν are essentially independent of r in the grain dominated regions. These quantities undergo a near-discontinuous change at the ice-grain condensation point.
2. $H \propto r^{3/4}$ so that the outer regions of the nebula are always shielded by the inner regions from the radiation released by the protosun.
3. $T_c \propto r^{-3/2}$ and $T_e \propto r^{-3/4}$. Typically, T_c is hotter than T_e by a factor between 1 and 10.
4. The spectrum has a characteristic form in which the intensity per unit frequency interval is given by $F_\nu \propto \nu^{1/3}$.

In Sec. III we showed that in the limit that the nebula is geometrically thin, an analytic approximation can be obtained for the self-consistent convective model by separating the r and z dependences in various physical variables (Lin 1981). From these relationships, we can scale the thermal properties of the nebula at different locations from a set of numerical solutions obtained at one particular radius. These results indicate that

$$\begin{aligned} T_c &= 200\alpha_c^{-1/3}\dot{m}^{2/3}m^{1/2}r^{-3/2}\mathrm{K} \\ T_e &= 180\dot{m}^{1/4}m^{1/4}r^{-3/4}\mathrm{K} \\ \rho_c &= 2 \times 10^{-10}\alpha_c^{-1/2}m^{1/4}r^{-3/4}\mathrm{g\ cm^{-3}} \\ \mathcal{R}_{\mathrm{eff}} &= 5 \times 10^4\alpha_c^{-2/3}\dot{m}^{-2/3}m^{1/2}r^{1/2} \\ \Sigma &= 100\alpha_c^{-2/3}\dot{m}^{1/3}\mathrm{g\ cm^{-2}} \end{aligned} \tag{93}$$

in the relatively warm regions where SiO_2, Fe, and other metal grains dominate the opacity. The dimensionless mass of the protosun m, radius of the nebula r and the mass accretion rate $\dot{m}$, are in units of 1 $M_\odot$, 10^{13} cm and 10^{18} g s^{-1}, respectively. In the outer, cooler ice-grain dominated regions,

$$T_c = 400\ \alpha_c^{-1/3} \dot{m}^{2/3} m^{1/2} r^{-3/2} \mathrm{K}$$
$$\rho_c = 2 \times 10^{-11}\ \alpha_c^{-1/2} m^{1/4} r^{-3/4} \mathrm{g\ cm^{-3}}$$
$$\mathcal{R}_{\mathrm{eff}} = 7 \times 10^3\ \alpha_c^{-2/3} \dot{m}^{-2/3} m^{1/2} r^{1/2}$$
$$\Sigma = 15\ \alpha_c^{-2/3} \dot{m}^{1/3} \mathrm{g\ cm^{-2}}. \tag{94}$$

These two regions are separated at $R_c \approx 10^{13}\ \alpha_c^{-2/3}\ \dot{m}^{4/9} m^{1/3}$cm.

Alternatively, the thermal structure of the nebula may be constructed from the numerical solutions of the vertical structure equations. When the nebula is in a steady state, $T_e^4 = 3\ GM\dot{M}/8\pi\sigma r^3$. Thus, we can follow the numerical procedures outlined in Sec. III to construct numerical models with $T_e \propto r^{-3/4}$ as the surface boundary condition (Lin and Papaloizou 1980; Lin 1981). The numerical solutions are presented in Fig. 5. These solutions are in general agreement with the expressions in Eqs. (93) and (94). Slight depar-

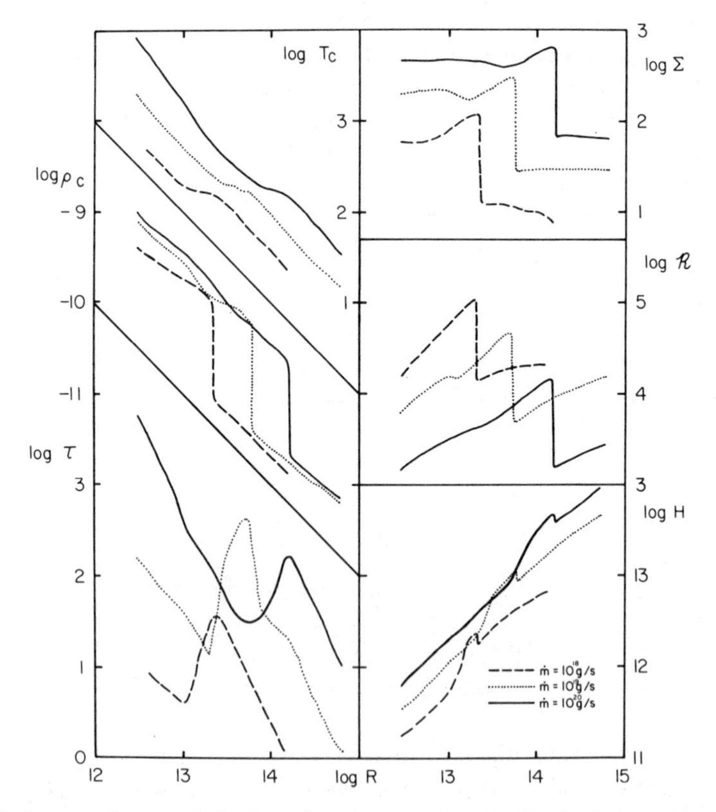

Fig. 5. Quasi-steady state solution for various mass transfer rates. The physical quantities are in cgs units and the value of α_c is taken to be unity.

tures are due to the difference between the approximate opacity law and the opacity table, and the analytic solution is a one-zone approximation which is likely to be inadequate for those regions where the opacity varies rapidly due to grain evaporations.

These results may be compared with cosmochemical data (Lewis 1972,1974; Boynton 1978) to provide useful interpretations. A detailed fit between the quasi-steady state model of the nebula (Lin 1981) and the cosmochemical data reveals that: (1) the different composition of the satellites of giant planets and that of the terrestrial planets is consistent with the temperature distribution in a quasi-steady state model with an appropriately chosen mass flux; and (2) the particular model with this appropriate temperature distribution has a mass flux around 10^{-6} $M_\odot$ yr^{-1} and $\Sigma \approx 10^3$g cm^{-3}, which corresponds to a few percent $M_\odot$ out to the orbit of Saturn.

Although the above quasi-steady state results are useful for deducing the general physical conditions of the inner and intermediate regions of the nebula, we still need to justify the basic assumption that there is an evolutionary tendency towards a quasi-steady flow pattern in these regions. Such a justification can only be obtained from a global evolutionary calculation. Another important question to be addressed by a global evolutionary calculation is how long a minimum-mass nebula can be sustained during the viscous stage. This time scale could result in strong constraints on the formation time of protoplanets. It is also of interest to note the general evolution of the thermal structure of the nebula because it may have profound effects on the evolution of planetesimals. Finally, a global treatment is necessary for determining the structure and evolution of the outer regions of the nebula. Therefore, a more appropriate comparison between observation and theory is a detailed fit of the data to a fully evolutionary rather than a quasi-steady state model.

b. Global Evolution Solutions. As the mass of the nebula is gradually redistributed, its vertical structure adjusts to new thermal equilibria on a characteristic thermal time scale, $\tau_{\rm th} \approx \Sigma c_s^2/F_z$. For a thin disk, $\tau_\nu >> \tau_{\rm th}$ so that local quasi-stationary thermal equilibrium is always maintained. The results in Sec. III (Eqs. 69 to 71) indicate that in these thermal equilibria, T_c, H, and ν decrease as Σ decreases. Thus, the evolution time scale increases as the nebular material is depleted.

We now apply the thermal equilibrium solutions to the diffusion Eq. (9) and determine the quantitative evolutionary trend of the nebula numerically. Let us first consider the physical conditions of those regions of the nebula where $T_c \lesssim 2000$ K so that grain opacity dominates the radiation transfer process, convection drives turbulence, and the nebula is both thermally and viscously stable. When the nebula is in local thermal equilibrium, Eq. (70) or Fig. 2 indicates that $\Sigma\nu = \nu_o \Sigma^3/\Sigma_o^2$, where $\nu_o = \nu(r_o)$, $\Sigma_o = \Sigma(r_o)$, and r_o is an arbitrary radius. If this scaling law applies throughout the nebula with an unique value of A_Σ, the diffusion Eq. (9) can be simplified to

$$\frac{\partial \Sigma}{\partial t} = \frac{3\nu_o}{\Sigma_o^2 r} \frac{\partial}{\partial r} r^{1/2} \frac{\partial}{\partial r} \Sigma^3 r^{1/2}. \tag{95}$$

Under a suitable coordinate transformation, $j = (r/r_o)^{1/2}$ and $\eta = 3\nu_o t/4r_o^2 = 3\Omega_o t/4\mathcal{R}_{\mathrm{eff},o}$, Eq. (95) can be reduced to a nondimensional form such that

$$\frac{\partial \Sigma}{\partial \eta} = \frac{1}{j^3 \Sigma_o^2} \frac{\partial^2 (\Sigma^3 j)}{\partial j^2} \tag{96}$$

which is clearly nonlinear. Equation (96) can be solved with standard numerical procedures for studying heat conduction equations (Richtmayer and Morton 1967; Lin and Bodenheimer 1982). For boundary conditions, one may set the viscous stress, $W = 3\Sigma\nu\Omega r^2/2$, equal to zero at both the inner boundary, which corresponds to no viscous coupling between the nebula and the protosun, and the outer boundary, which corresponds to free expansion.

The mathematically idealized diffusion Eq. (96) is useful because, for arbitrary initial conditions, the solution evolves towards a similarity form (Lin and Bodenheimer 1982). In order to demonstrate this property, we substitute for Σ a nondimensionalized viscous stress, $f_g = W(r)/W(r_o) = (\Sigma/\Sigma_o)^3 j$, into Eq. (96) so that

$$\frac{\partial}{\partial \eta} f_g^{1/3} = j^{-8/3} \frac{\partial^2}{\partial j^2} f_g. \tag{97}$$

With the thermal equilibrium solutions, the radial distribution of f_g can be used to determine the radial distribution of other physical quantities with the scaling law in Eq. (78): $f_\Sigma = j^{-1/3} f_g^{1/3}$, $f_T = j^{-11/3} f_g^{2/3}$, $f_\rho = j^{-3/2}$, $f_{\dot{m}} = \partial f_g^{1/3}/\partial j$, $f_p = j^{-31/6} f_g^{2/3}$, $f_h = j^{7/6} f_g^{1/3}$, $f_\nu = j^{-2/3} f_g^{2/3}$, and $f_{\mathcal{R}} = j^{5/3} f_g^{-2/3}$. We plot the distribution of f_g for four different initial conditions (Fig. 6). After an initial transient phase, the evolution becomes remarkably similar in all four cases. The outer edge nebula gradually spreads outward. The radius where f_g attains a maximum value, R_W also moves outwards. Material outside R_W diffuses outward whereas material interior to R_W drifts inward. The solutions in Fig. 6 indicate the existence of a set of similarity solutions. We now try to determine the similarity variable in terms of j and η.

If there is a similarity variable with a power-law dependence on η and j such that $x = j\eta^\gamma$, the most general functional transformation is $f_g = j^\lambda f_1(x)$ where γ and λ are two dimensionless indices to be determined. One constraint on the values of γ and λ can be established from the conservation of the total angular momentum in the nebula, $J_{\mathrm{tot}} = 4\pi \int_{r_{\mathrm{in}}}^{r_{\mathrm{out}}} \Sigma r^2 \mathrm{d}J = 4\pi \Sigma_o r_o^2 J_o \int_{r_{\mathrm{in}}}^{r_{\mathrm{out}}} f_g^{1/3} j^{11/3} \mathrm{d}j$. As the nebula evolves, the outer radius of the nebula expands such that $r_o(\eta) = r_o(\eta_o)(\eta_o/\eta)^{2\gamma}$. In terms of x, $J_{\mathrm{tot}} = 4\pi\Sigma_o r_o^2 J_o \eta^{-\gamma(\lambda+14)/3} \int_{x_{\mathrm{in}}}^{x_{out}} f_1^{1/3}(x) \times x^{(11+\lambda)/3} \mathrm{d}x$. Thus conservation of J_{tot} implies that $\lambda = -14$ and Eq. (97) reduces to

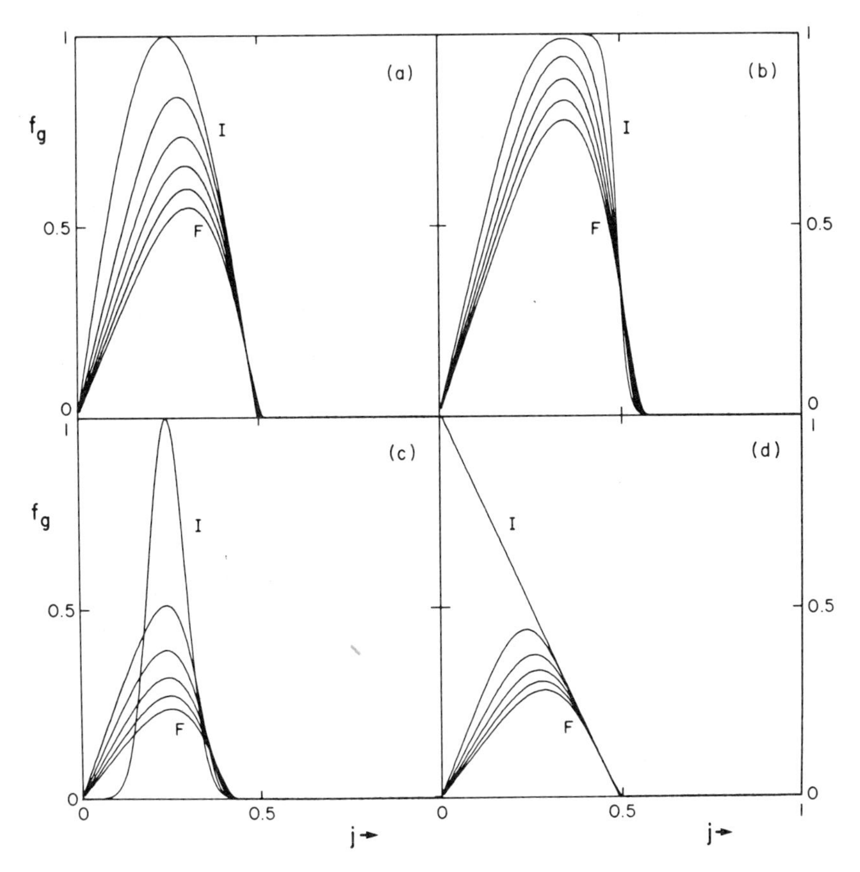

Fig. 6. Numerical solution of diffusion Eq. (97) for four different initial conditions. The dimensionless viscous couple f_g is plotted against the dimensionless specific angular momentum j. The six curves in each diagram are separated by equally spaced dimensionless time intervals, $\Delta\eta = 4 \times 10^{-5}$, with f_g decreasing with η. Plots (a), (b), (c), and (d) refer, respectively, to a sinusoidal, step function, Gaussian, and linear initial distribution of f_g with respect to j. The vertical scale is normalized to the maximum initial value of f_g.

$$\frac{\gamma}{j\eta^{1+15\gamma}} \frac{\partial f_1^{1/3}}{\partial x} = \frac{\partial^2}{\partial x^2} x^{-14} f_1 . \tag{98}$$

Similarity solutions for Eq. (98) exist only if $\gamma = -1/14$; consequently, $f_g = j^{-14} f_1(j\eta^{-1/14})$. Although Eq. (97) is reduced to an ordinary differential equation, it is still nonlinear and its global solutions can only be obtained numerically. However, in the limit of small x, Eq. (98) has an approximate solution, $f_1 = bx^{15} + ax^{14}$ so that $f_g = bj\eta^{-15/14} + a\eta^{-1}$, where a and b are arbitrary constants. For most regions outside the inner boundary, the contribution from $a\eta^{-1}$ is small so that $f_g = bj\eta^{-15/14}$, $f_\Sigma = \eta^{-5/14}$, $f_T = j^{-3}\eta^{-5/7}$, $f_\rho = j^{-3/2}$, $f_H = j^{3/2}\eta^{-5/14}$, $f_\nu = \eta^{-5/7}$, $f_{\mathcal{R}} = j\eta^{5/7}$, and $f_{\dot{M}} = \eta^{-15/14}$. These approximate solutions are in good agreement with the global evolution solution presented in Fig. 7 over most inner and intermediate radii. Because the functional form of $\dot{M}$ is approximately independent of radius, the flow pattern is in a quasi-steady state and at any given time Eqs. (93) and (94) provide a reasonable approximation to the structure of the nebula in the inner and intermediate regions.

The above solutions serve another useful purpose for demonstrating the evolutionary properties of the nebula. Because all the thermal quantities have a power-law dependence on $\eta = 3\Omega_o t/4\mathcal{R}_{\mathrm{eff},o}$, the viscous evolution process in the nebula indeed slows down. It is also of interest to note that Σ decreases with time more slowly than $\dot{M}$. As $\dot{M}$ decreases by an order of magnitude, Σ

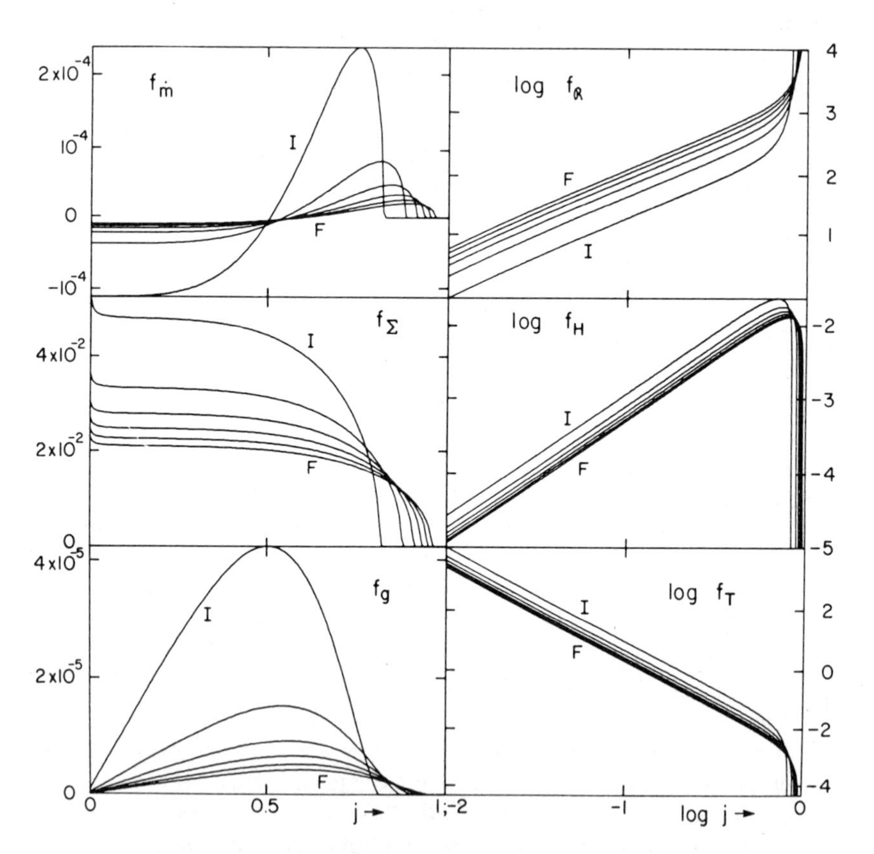

Fig. 7. The solution of diffusion equations with an idealized viscosity law. It is the continuation of case (c) in Fig. 6 over a time interval 10^6 times longer. The dimensionless functional dependence of mass flux $f_{\dot{m}}$, surface density f_Σ, viscous stress f_g, Reynolds number $f_{\mathcal{R}}$, scale height f_H and temperature f_T are plotted against dimensionless specific angular momentum j. The initial and final masses for each quantity are labeled by I and F, respectively.

only decreases by a factor of two. At any given time, $T_c \propto r^{-3/2}$ and $T_e \propto r^{-3/4}$ in the inner region. However, T_c everywhere evolves with time. In the inner and intermediate regions it decreases more rapidly than Σ. If the condensation temperature deduced from the cosmochemical data (Lewis 1972) indeed describes the T_c distribution at the time of grain formation, the above results would indicate that grains in most inner regions of the solar nebula are condensed over a period Δt comparable to several viscous diffusion time scales, i.e., $\approx \eta \mathcal{R}_{\mathrm{eff},o} \Omega_o^{-1}$. For a minimum-mass nebula with a temperature distribution $T \propto 500(r/1\ \mathrm{AU})^{-1}$, $\Delta t \approx 10^5 \alpha_c^{-1}$ yr. Note that as the nebula's mass is depleted with increasing η, Δt actually increases so that the evolutionary process is slowed down. Of further interest is the rate of increase in the outer radius, $r_{\mathrm{out}} \propto t^{1/7}$. Both of these evolutionary properties are in strong contrast to the $\nu = \nu(r)$ case where the evolutionary time scale does not change with time and $r_{\mathrm{out}} \propto t^{1/2}$.

The above analytic approximations break down in the outer regions of the nebula. There we have to use numerical techniques to analyze the nebula's dynamics. In this case, there is little advantage in adopting the idealized scaling relationships resulting from assuming $\kappa = \kappa_0 T^2$. Instead, we can utilize the equilibrium solution obtained from the analytic fit of the general opacity table (see Sec. III.B.1) to solve Eq. (9). The results of these numerical computations are presented in Fig. 8.

The main results of the numerical computations are:

1. The evolution of a minimum-mass nebula with a physical dimension comparable to that of the present solar system takes place on a time scale of $\approx 10^6$ yr for $\alpha_c \approx 1$. In principle, the evolutionary time scale can be indefinitely long if $\alpha_c << 1$.
2. After an initial transition phase, the evolution of the nebula, regardless of how the nebular mass was distributed at the termination of the infall stage, follows a unique pattern. Thus there is a self-adjusting process in the overall nebular structure.
3. The nebular mass, surface density, density, temperature and thickness at all but the outermost regions of the nebula decrease with time. Because the scale height also decreases with time, angular momentum transport gradually becomes less efficient and the evolutionary time scale gradually lengthens.
4. A minimum-mass nebula contains sufficient surface density such that it is optically thick throughout the nebula between the orbital radius of Mercury and that of Neptune, even in the case that α_c is of order unity. For smaller α_c, the surface density may be higher and the nebula has a larger optical depth unless the primary sources of opacity, namely grains, are significantly depleted.
5. In these optically thick regions, T_c may be larger than T_e by a factor of several. It is of interest to note that convective eddies may exert a viscous

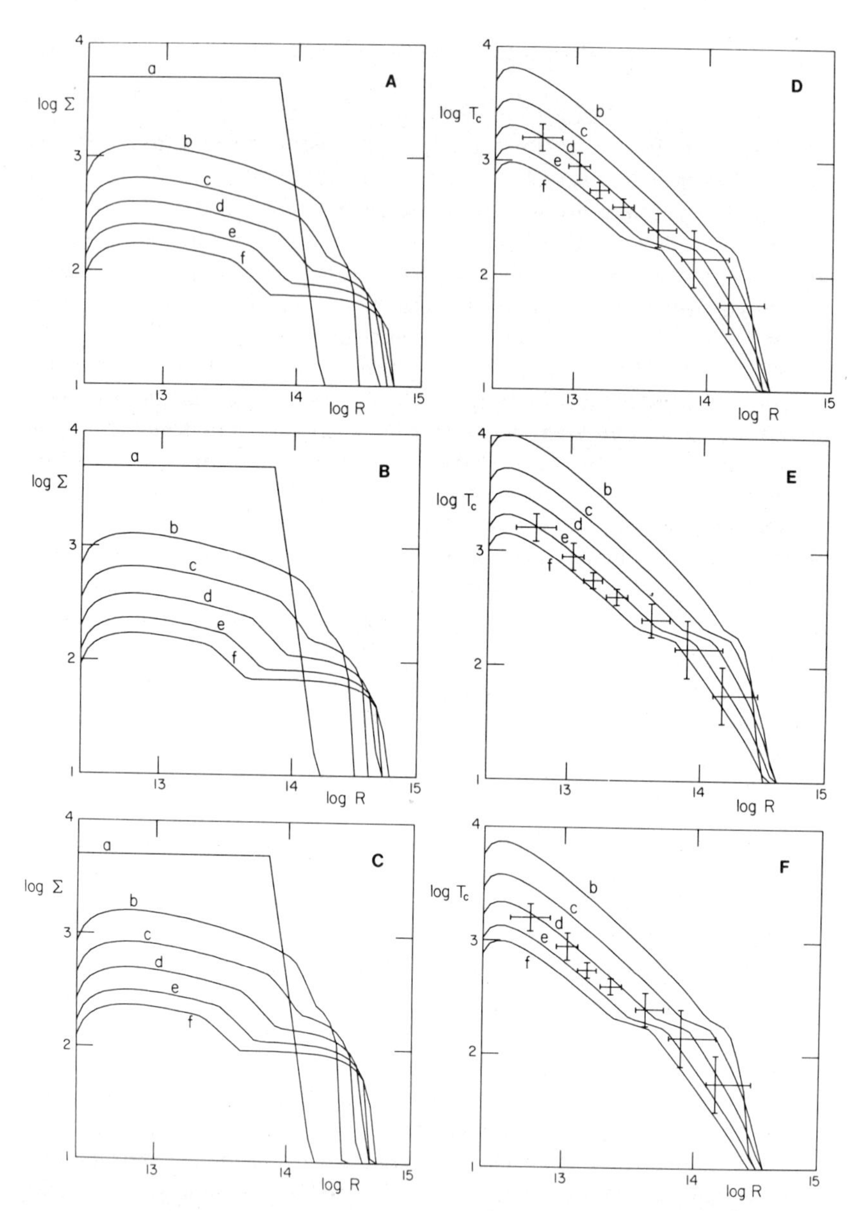

Fig. 8. The evolution of viscous diffusion equation with an α-model viscosity prescription. The surface density distribution at various times is plotted in cgs units in plots A, B, and C whereas the midplane temperature distribution at various times is plotted in D, E, and F. Three values of α are used. In plots A and D, $\alpha_s = 2 \times 10^{-2}$ (or eqivalently $\alpha_c = 1$). The initial Σ distribution in all cases is indentical and the curves labeled by b, c, d,e and f represent, respectively, $t = 2 \times 10^4$, 7×10^4, 1.6×10^5, 3×10^5 and 5×10^5 yr. In plots B and E, $\alpha_s = 10^{-2}$ (or equivalently $\alpha_c = 0.5$) and the curves labeled by b, c, d, e and f represent, respectively, $t = 5 \times 10^4$, 2×10^5, 4×10^5, 8×10^5 and 1.3×10^6 yr. In C and F, $\alpha_s = 5 \times 10^{-3}$ (or equivalently $\alpha_c = 0.25$) and the curves labeled by b, c, d, e and f represent, respectively, $t = 10^5$, 3.5×10^5, 8×10^5, 1.4×10^6, 2×10^6 yr. The condensation temperature derived by Lewis (1974) for various planets and satellites are marked, with ranges of uncertainties in the temperature distribution plots.

drag on small-sized grains, inducing them to circulate between the relatively cool nebular surface region and the relatively hot nebular interior. For certain types of grains, such a circulation provides a mechanism for repeated condensation and evaporation (Morfill 1983; Cameron and Fegley 1982). Fine-structure data obtained from the Allende meteorite also indicates that the nebula may indeed have undergone repeated condensation and evaporation (see Boynton's chapter).

6. The inner and intermediate regions of the nebula always evolve into a quasi-steady state in which the mass transfer rate is independent of the distance from the protosun.
7. At any given time, $T_c \propto r^{-3/2}$ and $T_e \propto r^{-3/4}$ near the inner and intermediate regions where the flow is quasi-steady and both decrease more slowly with r at the outer regions of the nebula. Provided $0.1 \lesssim \alpha_c \lesssim 1$, the temperature distribution in a minimum-mass nebula model approximately resembles that deduced from cosmochemical data. For smaller values of α_c, larger nebular surface density, or equivalently nebular mass, is needed to provide sufficient viscous dissipation. These temperature distributions can be maintained over several viscous diffusion time scales at that epoch.
8. In the intermediate region of the nebula, there is a first-order phase-transition zone for water molecules. Inside this zone the temperature of the nebula is sufficiently hot that most of the water molecules are in a gaseous state. At larger radii, most of the water molecules can undergo a first-order phase transition and condense into ice grains. Similar transition zones exist for other molecules. Because iron and silicon grains have a much higher phase-transition (condensation) temperature than ice grains, their transition zones are at much smaller radii. Consequently, metal grains are probably found throughout the nebula whereas ice grains are only found at large radii.
9. Inside the ice-grain transition zone, the opacity is somewhat lower which causes the convective zone to be smaller. If we adopt a self-consistent prescription for viscosity, (Eq. 57), the nebular viscosity would also be smaller. When a steady state flow is established in this region, the nebular surface density is higher inside the transition zone by an order of magnitude than outside the zone.
10. The ice-grain transition zone evolves inward in time as the nebula gradually cools off and exhausts most of its material. Melting and recondensation of ice grains are possible as a consequence of convective circulation in the transition zone.
11. At the outer regions of the nebula, there is a critical radius R_W at which the viscous stress W reaches a maximum. Inside this radius, the inward diffusion velocity decreases with radius whereas the nebular material diffuses outward outside R_W. As the nebular material is depleted, R_W evolves outwards.

12. In the outer region of the nebula near R_W, typical fluid elements undergo orbital migration. However, since they cannot diffuse rapidly towards small radii, these fluid elements continually evolve towards a cooler environment. As a result, the gas elements may become supersaturated, creating new nucleation sites which enhance grain condensation. In these regions, grain condensation is a continuous process throughout the viscous stage. These continuous condensation regions generally spread inward in time.

In the absence of a comprehensive model for turbulence in the solar nebula, the evolutionary time scale is uncertain to a factor of α_c. Some circumstantial evidence for the time scale or early nebular processes can be obtained from the studies of isotopic composition of meteorites. It has been discovered that the Allende meteorite contains an excess of the isotope ^{26}Mg, a decay product of the short-lived isotope ^{26}Al. The implication of this important discovery is that ^{26}Al must have been present at the time of grain formation (Lee et al. 1976*c*). As ^{26}Al can only be produced through the thermonuclear reactions in a stellar interior or in a supernova explosion and has a half-life of $\approx 10^6$ yr, its presence in the grains is extremely intriguing. If we assume that the relatively large-size grains, in which this ^{26}Al is found, condensed in the solar nebula, we would deduce from the inclusion of ^{26}Al that the grain condensation time is $< 10^6$ yr (Cameron and Truran 1976). The grain-condensation time scale deduced from these meteoritic data is consistent with the diffusion time scale $\approx 10^6$ yr if $\alpha_c \gtrsim 0.5$. In Sec. V, we shall present somewhat stronger dynamical constraints on the values of α_c.

Very near the protosun, the nebula may be too hot for grains to condense so that the dominant sources of opacity are due to molecular transitions in, e.g., H_2O and CO molecules, and atomic hydrogen ionization processes. Because grains cannot condense in this region, protoplanets cannot form there. Nonetheless, most of the gravitational energy of the inwardly drifting material is released in this region so that to an observer it is an interesting region to study. This region, too, is convectively unstable. However, unlike elsewhere in the solar nebula, most of the energy generated through viscous dissipation is not released locally as heat. Instead it is absorbed by the atoms, the molecules and the electrons in dissociation and ionization processes. Consequently, the radiation released from this region lacks the characteristic spectrum we discussed earlier even if the flow is in the form of a quasi-steady state. This region, furthermore, is likely to be thermally unstable as a consequence of the opacity law. We shall discuss the dynamics of this region in more detail in Sec. VI.

C. The Clearing Stage

Today, there is very little gas left between the planets. At some stage of the solar nebula evolution, the gas must have been cleared away. There are four basic models to account for the clearing of the nebula. First, sometime after the formation of the protosun, there may have been a very strong stellar wind which swept out the bulk of the nebular material (Horedt 1978; Elmegreen 1978). Although there is no satisfactory model to account for the origin of such a wind, mass outflow has been observed in a class of newly forming stars, T Tauri stars (Herbig 1962; Cohen and Kuhi 1979). With the typical mass outflux and wind velocity in T Tauri stars, a minimum-mass nebula may be cleared in a few thousand to a few million years. However, there is observational evidence for anisotropy in the mass outflow from some T Tauri stars (Cohen 1981; Cohen et al. 1982). In these cases, the most likely direction of the T Tauri wind is perpendicular to the plane of the nebula. Consequently, the wind-induced clearing process is unlikely to be efficient in removing gas from most outer regions of the nebula. A second model is based on the general expectation that there may be a hot coronal layer above the surface layer of the nebula which is established by the dissipation of mechanical waves generated by turbulence (Cameron 1978). Such a model is analogous to one of the two models that account for the solar corona. Just as the presence of a hot corona in the Sun causes a hydrodynamic flow of mass away from the Sun, coronal layers high in the atmosphere of the nebula may induce mass flow away from the nebula (DeCampli 1980). The quantitative verification of these arguments is still to be carried out.

A third possible mechanism for clearing is associated with the photoionization effect due to solar ultraviolet radiation. Although the Sun's present ultraviolet flux is so low that it hardly has any significant effect on the upper atmosphere of the Earth, its ultraviolet flux may have been considerably higher during its T Tauri phase. If a small portion of the nebular gas is ionized by ultraviolet photons of the early Sun, it may have sufficient kinetic energy to leave the solar system. Consequently, a radiatively driven ionization wind may be induced. Such a process could be particularly efficient for the outer regions of the nebula, say beyond the orbit of Jupiter, where the escape velocity is less than about 10 km s^{-1}. In fact, we would expect a considerable amount of material to accumulate in the outer regions of the nebula for two reasons. First, a fraction of the nebular material must move outwards to carry the angular momentum viscously removed from the infalling material. If most of the mass in the nebula is initially located near the present orbit of Jupiter, about 30% of the infalling material must expand to the present orbit of Neptune to absorb the redistributed angular momentum. Second, near the outer regions of the nebula, the temperature is sufficiently cool that the nebula is likely to become optically thin. In this case, if convection is switched off and there is no other source of turbulence, nebular material would accumulate at these large radii. Note that this process is only effective for removing the gaseous component of the nebula. Small grains may undergo orbital decay under the action of the Poynting-Robertson effect, whereas larger grains or planetestimals may be left behind. The disk of large-sized grains may have observational properties similar to those which are being attributed to a planetesimal disk around Vega by the Infrared Astronomical Satellite (IRAS) team.

A fourth model is based on protoplanet-gas interaction. Protoplanets with very small mass have little effect on the structure and evolution of the nebula. When they have grown to sufficiently large masses, the protoplanets can induce a tidal torque on the nebula and eventually open up gaps in the vicinity of the protoplanets' orbits (Lin and Papaloizou 1980; Cameron 1979). Subsequently, further mass growth by accretion from the nebula is terminated and the nebular flow between the protoplanets becomes unstable so that the remaining nebular gas trapped between the prototplanets may be ejected by a wind or, perhaps more likely, be driven toward the protosun. The advantage of this tidal model is that it allows most of the protoplanets to acquire their mass before the remaining nebular gas is cleared away. It also provides a strong dynamical constraint on the mass of the nebula as we show in Sec. V below.

V. PROTOPLANETARY-NEBULAR TIDAL INTERACTION: CONSTRAINTS ON NEBULAR EVOLUTION

We have shown in the last section that the radial structure and evolution of the nebula are basically determined by the magnitude of ν_{eff}. In the absence of a deterministic theory for turbulence, the magntidue of ν_{eff} for the solar nebula cannot be accurately derived from theoretical models. However, there are indirect dynamical constraints which can be utilized to estimate the magnitude of ν_{eff}. In this section, we examine in detail one particular dynamical process, the protoplanetary-nebular tidal interaction.

Giant planets contain mainly hydrogen and must have been formed in a gas-rich environment. If the gas they contain was accreted from the nebula, their present properties, such as mass, chemical composition and orbital position may have been determined by the details of the dynamical interaction with the nebula. For example, strong tidal interaction may have occurred between a protoplanet and the nebula. Because a protoplanet revolves around the Sun faster than does the nebula outside its orbit, angular momentum is transferred from the protoplanet's orbit to the nebula in order to achieve a lower-energy state, namely uniform rotation. Similarly, angular momentum is transferred from the nebula inside the protoplanets' orbits to the protoplanet. The rate of this tidally induced angular momentum transfer can be calculated in various ways (Lin and Papaloizou 1979*a,b*; Goldreich and Tremaine 1980). This rate is proportional to the square of the mass of the protoplanet and to the nebular surface density. As the nebula evolves, the mass of the protoplanet grows, so that the rate of tidal transfer of angular momentum is increased. At the same time, the nebular material is exhausted, and therefore the rate of the viscous transfer of angular momentum in the nebula is decreased. Eventually, the tidal effect can redistribute angular momentum faster than can the viscous effect. When angular momentum is being transported tidally to the protoplanet from the nebular gas interior to its orbit, the gas moves inward and away from the protoplanet. Similarly, the nebular gas exterior to the orbit of the protoplanet moves outward and away from the protoplanet. Consequently, the tidal effect begins to open up gaps in the nebula in the vicinity of the protoplanets' orbits and further mass growth by accretion will be inhibited (Lin and Papaloizou 1980). Furthermore, depending on the total mass and structure of the nebula, the net angular momentum exchange rate may cause secular orbital evolution of the protoplanet (Goldreich and Tremaine 1980). It is clear that the present positions and masses of the giant planets may enable us to put constraints on some of the nebular properties during their formation.

The basic purpose of the above qualitative discussion is to indicate the importance of a detailed analysis of the protoplanet-disk tidal interaction. We now turn to a discussion of previous analyses of this problem and their limitations and the motivation for future detailed mathematical analyses and calculations. The analysis to be presented here is closely analogous to that for tidal interaction between binary stars and accretion disks (Lin and Pringle 1976; Papaloizou and Pringle 1977; Lin and Papaloizou 1979*a,b*), and that of planetary rings and satellites (Goldreich and Tremaine 1980,1982, and references therein). In these contexts the angular momentum exchange rate has been calculated by a variety of methods. In this section, we first introduce a simple impulse approximation to provide an heuristic estimate for the efficiency of tidal transport of angular momentum. We then consider the effect of gas pressure. Finally, we examine the orbital evolution of protogiant planets in the solar nebula.

A. Impulse Approximation

When nebular material orbits at a radius close to that of a protoplanet, because the mass ratio q between typical protoplanets and the Sun is much less than unity, the tidal force of the protoplanet can only be felt during the orbital phase when particles have a close approach with the protoplanet. For simplicity, let us first neglect the curvature of the unperturbed orbits, and the influence of the Sun during such a close approach. Then, as a fluid element, interior to the protoplanet's orbit, flows past the protoplanet, the tidal force deflects the orbit by a small angle δ such that

$$\cot^2(\delta/2) = u^4a^2/G^2M_p^2 \tag{99}$$

where u is the relative velocity, a is the impact parameter, and M_p is the mass of the protoplanet. The corresponding angular momentum per unit mass transferred is

$$\Delta h = ur_p(1 - \cos\delta) \tag{100}$$

where r_p is the orbital radius of the protoplanet in cylindrical coordinates. To be more definite, we consider the protoplanet to orbit externally to the nebula. However, a similar analysis applies when the protoplanet is interior to the nebula. The only difference is that the protoplanet loses rather than gains angular momentum. Thus all formulae apply apart from appropriate sign changes. When the motion of the protoplanet and the unperturbed nebular material is nearly circular, $u = (\Omega - \omega)r_p$, where Ω and ω are the orbital angular frequency of the nebula and the protoplanet, respectively. Because the time between successive encounters is $2\pi/(\Omega - \omega)$, the rate of specific angular momentum transfer is

$$\dot{h} = G^2M_p^2/[\pi r_p^2 a^2(\Omega - \omega)^2]. \tag{101}$$

Using Taylor's expansion of Ω, and performing an integration over the nebula, we can deduce that the rate of total angular momentum transfer rate is given by

$$\dot{H}_T = \frac{2G^2M_p^2 r_p \Sigma}{r_p^2(\mathrm{d}\Omega/\mathrm{d}r)^2}\int_{\Delta_o}^{\infty}\frac{\mathrm{d}a}{a^4} = \frac{8}{27}\left(\frac{M_p}{M}\right)^2\Sigma\omega^2 r_p^4\left(\frac{r_p}{\Delta_o}\right)^3 \tag{102}$$

where Σ is the surface density, and Δ_o is the distance between the protoplanet and the nebula's edge. In the limit that the protoplanet influences the nebula strongly, Δ_o can be taken to be the Roche radius of the protoplanet, i.e., $\Delta_o = R_L = r_p(q/3)^{1/3}$ so that

$$\dot{H}_T = 8q\Sigma r_p^4\omega^2/9. \tag{103}$$

The above analysis can be modified to include the influence of the Sun on the close encounter (see appendix A in Lin and Papaloizou [1979*a*]). In this case, the equations of motion for the fluid element are

$$\frac{\mathrm{d}^2R}{\mathrm{d}t^2} - R\left(\frac{\mathrm{d}\phi}{\mathrm{d}t}\right)^2 = -\frac{\partial\Psi}{\partial R} \tag{104}$$

and

$$R^2\frac{\mathrm{d}^2\phi}{\mathrm{d}t^2} + 2R\frac{\mathrm{d}R}{\mathrm{d}t}\frac{\mathrm{d}\phi}{\mathrm{d}t} = -\frac{\partial\Psi}{\partial\phi} \tag{105}$$

where R and ϕ are the radial and azimuthal position of the fluid element. The total gravitational potential is due to the combined contribution from the Sun and the protoplanet and is given by

$$\Psi = \Psi_s + \Psi_p \tag{106}$$

where

$$\Psi_s = -GM/r \tag{107}$$

and in the inertial frame

$$\Psi_p = \frac{Gm_pR\cos(\phi - \omega t)}{r_p^2} - \frac{Gm_p}{[r_p^2 + R^2 - 2Rr_p\cos(\phi - \omega t)]^{1/2}}. \tag{108}$$

In Eq. (108), we define the line $\phi = 0$ as the line joining the Sun and the protoplanet at the instant $t = 0$. We consider a fluid element which is also on this line at $t = 0$. The perturbation due the protoplanet's gravitational field causes angular momentum transfer between the protoplanet and the fluid element. Because a typical protoplanet's mass is much smaller than that of the Sun, energy and angular momentum can only be effectively transferred between the fluid element and the protoplanet when a close encounter is occurring. In order to estimate the amount of angular momentum transferred per encounter Δh, Lin and Papaloizou (1979*a*) consider the tidal interaction for $-\pi/2(\Omega - \omega) < t < \pi/2(\Omega - \omega)$. Then

$$\Delta h = -\int_{-\pi/2(\Omega-\omega)}^{\pi/2(\Omega-\omega)} \left(\frac{\partial \Psi_p}{\partial \phi}\right) \mathrm{d}t. \tag{109}$$

At the earlier time, $\phi - \omega t$ for the unperturbed orbit is $-\pi/2$, while at the later time it is $\pi/2$. In the impulse approximation, we assume the unperturbed orbit at the earlier time to be circular. In the limit of $M_p/M << 1$, the perturbation potential due to the protoplanet can be expressed in terms of Fourier series. After some algebra, it can be shown (see appendix A in Lin and Papaloizou [1979*a*]) that to within a multiplicative factor of order unity,

$$\Delta h = \frac{\pi^2}{4(\omega - \Omega)^3} \frac{(GM_p)^2}{r_p^2 R_L^2}. \tag{110}$$

for the case $a = R_L$. The results in Eqs. (100) and (110) are in essential agreement. Actually, under the action of a tidal torque, the Jacobi constant $E - \omega h$, where E is the orbital energy, cannot be altered. Consequently, fluid elements interior to the orbit of the protoplanet lose angular momentum at too fast a rate to remain in circular orbits; thus their orbits become eccentric. After the close encounter, tidal effects become very weak. If the viscous force is sufficiently strong, it would tend to dissipate any radial motion and thereby circularize the orbits. If we assume that the fluid element loss $|\Delta h|$ per close encounter, the rate of tidal transfer of angular momentum in Eq. (103) can also be derived from this approach.

In both impulse approaches above, we take the equation of motion for a fluid element to be that of a noninteracting particle. The fluid nature of the nebula flow is described by continuity, momentum and energy equations. In the limit $M_p/M << 1$, the protoplanet's tide has little effect on the nebular flow except near orbital resonances where Ω is commensurable with ω. The rate of resonant tidal transfer of angular momentum can be calculated in many different ways (for reviews, see Goldreich and Tremaine [1982] and chapter by Lissauer and Cuzzi). The rate of tidal transfer of angular momentum can also be calculated by summing up a series of point-like contributions from low-order Lindblad resonances ranging from the inner 2:1 resonance to those near the edge of the disk (Goldreich and Tremaine 1980). The basic functional dependence of dH_T/dt, derived from the two different approaches, is identical. The numerical coefficient differs from that in Eq. (102) by a factor of 2.83.

The results of the impulse approximation can also be obtained from numerical simulation (Lin and Papaloizou 1979*a*). If we ignore the gas pressure effect, the flow pattern can be essentially simulated with a Lagrangian particle scheme (Lin and Pringle 1976). For illustration purposes, we present in Fig. 9, some results from such simulations.

What do these results imply in the context of providing constrants for the dynamical evolution of the nebula? These results were applied by Lin and Papaloizou (1980) to determine the protoplanetary tidal interaction for their model of the solar nebula. This model has a turbulent viscosity as a result of

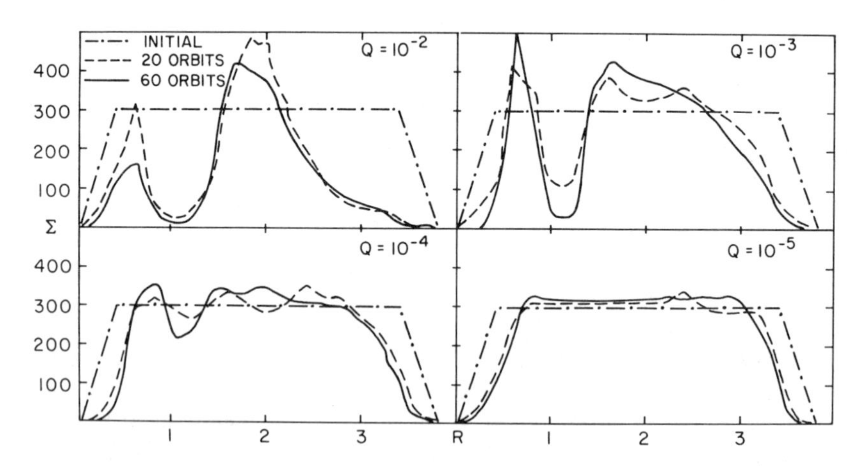

Fig. 9. The evolution of surface density in the nebula as a function of the mass ratio, M_p/M between the protoplanet and protosun. The Reynolds number of the nebula is 625 $\mathcal{R}^{-1}$. When the mass ratio is relatively large, say $M_p/M = 10^{-2}$, i.e., when the protoplanet is relatively massive, the nebula is clearly truncated tidally. However, the presence of the protoplanet has very little tidal effect on the nebula when the mass of the protoplanet is relatively small, e.g., say in the $Q = 10^{-5}$ case.

convection. As in accretion-disk theory (Lynden-Bell and Pringle 1974), matter flows in towards the Sun on a viscous diffusion time scale. Lin and Papaloizou (1980) concluded that the viscous transport of angular momentum would be overwhelmed once M_p/M exceeds 10^{-4} to 10^{-3}. For larger M_p a gap would form in the disk and further mass accretion would be inhibited. Similar applications have been made to study the formation of satellites around giant planets (Coradini et al. 1981).

However, caution in applying the results of these calculations to a gaseous disk has been urged by Lunine and Stevenson (1983). They have stressed the need to consider the detailed nature of the dissipative process, and have noted that the tidally induced disturbance may propagate throughout the disk, delocalizing the tidal torque, when the dissipation is very weak. They also noted that the formation of a gap changes the surface-density profile and in turn that could change the position of the Lindblad resonances in the disk. The angular momentum transfer rates could then differ from those given by the existing calculations which assume uniform surface densities. Greenberg (1983) has also emphasized the need for proper consideration of the dissipative process. He suggests that the tidal torque may be significantly reduced if the dissipation is weak.

B. Nebular Response to Tidal Perturbations

From the discussion in Secs. III and IV we find that for typical models of the solar nebula, the vertical scale height of the nebula is 0.05 to 0.1 times the distance to the Sun. This scale height is comparable to the Roche lobe radius for protoplanets with mass in the range 10^{-4} to $10^{-3} M_\odot$. When a gap is first formed, it would be expected to be about one disk scale height wide so that the disk surface density changes on that scale length. In this situation the Lindblad resonance conditions are changed and the conventional tidal torque calculations need to be reexamined. This effect is best demonstrated by analyzing the radial component of the equation of motion for a moderately thick nebula. After vertical averaging, the radial component of Eq. (4), with both tidal and viscous effects neglected, becomes

$$\frac{1}{\Sigma_o}\frac{dP_o}{dr} + \frac{d\Psi}{dr} - \Omega^2 r = 0 \tag{111}$$

where the gravitational potential Ψ is due to the protosun and P_o is the vertically integrated pressure, which only acts in the horizontal direction. We assume that there is a functional relation between P_o and Σ_o. Using this we find it convenient to define the sound speed c_s through $dP_o/d\Sigma_o = c_s^2$ and the thickness or scale height H_D by $H_D = c_s\Omega^{-1}$. If the surface density varies on a scale length comparable to the radius, $\Omega^2 = GM/r^3\{1 + O(H_D^2/r^2)\}$ which is essentially Keplerian for small H_D/r. Another important parameter is the epicyclic frequency κ_f which is given by

$$\kappa_f^2 = \frac{2\Omega}{r}\frac{\mathrm{d}(\Omega r^2)}{\mathrm{d}r} = \frac{1}{r^3}\frac{\mathrm{d}}{\mathrm{d}r}[(\Omega r^2)^2] = \frac{GM}{r^3} + \frac{1}{r^3}\frac{\mathrm{d}}{\mathrm{d}r}\left[\frac{r^3 c_s{}^2}{\Sigma_o}\frac{\mathrm{d}\Sigma_o}{\mathrm{d}r}\right]. \quad (112)$$

The behavior of κ determines not only the location of the resonances but also the gradient of angular momentum in the nebula. From Eq. (112) we find that if Σ_o varies on a scale comparable to H_D, significant departures of order unity from the Keplerian equation, $\kappa_f = \Omega$ can occur. Moreover, a disk edge, at which Σ decreases rapidly to zero on a scale length shorter than H_D in general implies $\kappa_f^2 < 0$, so that specific angular momentum decreases with radius. A differentially rotating flow with a negative angular momentum gradient is unstable against axisymmetric perturbations according to the Rayleigh criterion (Landau and Lifshitz 1959). When the Rayleigh criterion is violated, turbulence induced by Taylor vortices would act to smooth out the density gradient (Rayleigh 1917; Greenspan 1969). Thus the shortest radial scale length one should consider in the disk is H_D, and if variations on that scale do occur, κ would deviate significantly from its Keplerian value.

These considerations are important for estimating the positions of the Lindblad resonances. These are given by $\omega = \Omega \pm \kappa_f/m$ for interger m, (see Goldreich and Tremaine 1980). If short-scale variations of Σ_o occur, these resonances would be affected, and so also is the calculation of the tidal interaction. This would be the case for a disk edge with scale length H_D which has the orbiting body a distance comparable to H_D from the edge.

What is needed is a full nonlinear solution of the hydrodynamic equations governing the tidal interaction, i.e., the solution of the two-dimensional (r and ϕ) momentum equation including a nonaxisymmetric perturbing potential ψ

$$\frac{\partial U_r}{\partial t} + U_r\frac{\partial U_r}{\partial r} + \frac{U_\phi}{r}\frac{\partial U_r}{\partial \phi} - \frac{U_\phi^2}{r} = -\frac{1}{\Sigma}\frac{\partial P}{\partial r} - \frac{GM}{r^2} - \frac{\partial \Psi_p}{\partial r} + F_r \quad (113)$$

$$\frac{\partial U_\phi}{\partial t} + U_r\frac{\partial U_\phi}{\partial r} + \frac{U_\phi}{r}\frac{\partial U_\phi}{\partial \phi} + \frac{U_r U_\phi}{r} = -\frac{1}{\Sigma r}\frac{\partial P}{\partial \phi} - \frac{1}{r}\frac{\partial \Psi_p}{\partial \phi} + F_\phi \quad (114)$$

where

$$F_r = 2\nu\frac{\partial}{\partial r}\left(\frac{U_r}{r}\right) + \frac{2}{\Sigma}\frac{\partial}{\partial r}\left(\Sigma\nu\frac{\partial U_r}{\partial r}\right) + \frac{1}{\Sigma}\frac{\partial}{\partial r}\left[\frac{\Sigma(\zeta - \frac{2}{3}\nu)}{r}\left(\frac{\partial r U_r}{\partial r} + \frac{\partial U_\phi}{\partial \phi}\right)\right]$$

$$+ \nu\left[\frac{1}{r^2}\frac{\partial^2 U_r}{\partial \phi^2} + r^2\frac{\partial}{\partial r}\left(\frac{1}{r^3}\frac{\partial U_\phi}{\partial \phi}\right)\right] + \frac{2}{\Sigma r^2}\frac{\partial \Sigma\nu}{\partial \phi}\left(U_r + \frac{\partial U_\phi}{\partial \phi}\right) \quad (115)$$

and

$$F_\phi = \frac{1}{\Sigma r^2}\frac{\partial}{\partial r}\left(\Sigma \nu r^3 \frac{\partial}{\partial r}\left(\frac{U_\phi}{r}\right)\right) + \frac{1}{r^2\Sigma}\frac{\partial}{\partial \phi}\left[\Sigma\left(\zeta - \frac{2}{3}\nu\right)\left(\frac{\partial r U_r}{\partial r} + \frac{\partial U_\phi}{\partial \phi}\right)\right]$$

$$+ \frac{2U_r}{\Sigma r}\frac{\partial \Sigma \nu}{\partial \phi} + \frac{2}{r^2}\left[\frac{2}{\Sigma}\frac{\partial}{\partial \phi}\left(\Sigma \nu \frac{\partial U_\phi}{\partial \phi}\right) - \nu \frac{\partial^2 U_\phi}{\partial \phi^2}\right]$$

$$+ \frac{1}{\Sigma r^2}\frac{\partial}{\partial \phi}\left(\Sigma \nu \frac{\partial (U_r r)}{\partial r}\right) + \frac{1}{\Sigma r}\left(\frac{\partial U_r}{\partial \phi}\right)\left(\frac{\partial \Sigma \nu}{\partial r}\right) \tag{116}$$

and the continuity equation

$$\frac{\partial \Sigma}{\partial t} + \frac{1}{r}\frac{\partial}{\partial r}\Sigma U_r r + \frac{1}{r}\frac{\partial \Sigma U_\phi}{\partial \phi} = 0. \tag{117}$$

The solutions of Eqs. (113) to (117) are needed whether or not the perturbations of the nebula by the protoplanet can be treated using linear theory because the surface density profile has to be determined self-consistently as a result of the interplay between tidal and viscous processes. When the protoplanet is sufficiently massive, it can induce spiral shock waves in the nebula. Recently, Lin and Papaloizou have initiated a numerical analysis of this problem. The first results of this analysis are presented in Fig. 10 below. Because the results of the fully nonlinear calculations are very complex, we first set up a simple, linear, analytic framework. In this framework, we approximate the surface density profile near the orbit to be fixed and assume that the nonlinear effects, if they occur, act to produce enhanced dissipation in the disk so as to restore the validity of the linear approximation (Papaloizou and Lin 1984). With this approach, it is possible to demonstrate that, although angular momentum waves excited by the protoplanet on a relatively thick nebula can propagate to large distances and suffer reflections even when the effective viscosity is moderately large, the angular momentum flux is essentially independent of the dissipation, no matter how weak it may be, provided this flux is calculated by averaging over a range of relevant orbital frequencies. This approach corresponds to acknowledging that the tidal evolution of the nebula and the orbiting protoplanet are sufficiently slow that the viscous effect can dissipate the angular momentum wave excited by the tides and that the full tidal response in the nebula can be established. This is true if either linear theory is strictly valid or if the effect of nonlinearity is to cause enhanced dissipation in such a way as to restore the validity of linear theory. Finally, Papaloizou and Lin adopt specific nebula models with a polytropic relationship between the vertically integrated pressure and the surface density to calculate the tidal torque. These models provide a set of general formulae for the torque density when the edge of the nebula is about one vertical scale height H_∞ from the protoplanet. For example, in one model Papaloizou and Lin set the polytropic index to be 1.5 and the surface density to have a profile

$\Sigma(\Delta) = \Sigma_\infty\{\tanh[(\Delta - \Delta_o)/H]\}^{1.5}$ where Δ is the distance from the protoplanet; the location of the edge of the nebula is given by $r = r_o$, or $\Delta = \Delta_o$; H determines the radial scale of the density and is given by $1.5^{1/2}H_\infty$. Σ_∞ and H_∞ are, respectively, the values of Σ and H_D at large distance from the protoplanet. In this case,

$$\dot{H}_T = 0.23q^2\Sigma_\infty r_o^4\omega^2\,(r_o/H_\infty)^3 \tag{118}$$

for the maximum value of applied torque. The result in Eq. (118) is in remarkably good agreement with that in Eq. (103) if H_∞ is set equal to R_L, which implies that the modification of the Lindblad resonances has little effect on the torque density and the rate of angular momentum transport.

How do the above results modify the conditions for gap formation? The central idea of gap formation is that there is a balance between the rate of angular momentum transport due to the action of viscosity on Keplerian shear and that due to tidal effects. To be specific we concentrate on the case when the body orbits outside the disk and express the total loss rate of angular momentum from the disk in the form

$$\frac{\mathrm{d}H_T}{\mathrm{d}t} = 2\pi\int_{r_1}^{r_o}\Sigma r\dot{h}\mathrm{d}r \tag{119}$$

where the functional form of the rate of loss of specific angular momentum $\dot{h}$ reflects the manner of tidal dissipation in the disk, and it is assumed to vanish at some characteristic radius r_1. If this loss is balanced by the action of viscosity on shear, we must have

$$\frac{\mathrm{d}}{\mathrm{d}r}\left(\nu\Sigma r^3\frac{\mathrm{d}\Omega}{\mathrm{d}r}\right) = \dot{h}\Sigma r \tag{120}$$

(see Sec. IV and Lynden-Bell and Pringle [1974]). Given a form for $\dot{h}$, Eq. (120) can be solved for the structure of the gap assuming that $\Sigma = 0$ at $r = r_o$, some small distance from the perturber. Our preliminary nonlinear calculations show that for values of ν and c_s appropriate to the solar nebula, Ω does not deviate greatly from the Keplerian value. Therefore, we use the Keplerian values of Ω in the perturbed region as well as for $r < r_1$. Integrating Eq. (120) between r_o and r_1, we find

$$\frac{\mathrm{d}H_T}{\mathrm{d}t} = [3\pi\nu\Sigma r^2\Omega]_{r=r_1}. \tag{121}$$

Using the fact that $\mathrm{d}H_T/\mathrm{d}t$ must be less than the value given by Eq. (118), taking $\Sigma(r_1) = \Sigma_\infty$, and assuming r_1 to be the same order as r_o, we find that in order to produce a gap, we require

$$\frac{\nu}{r_o^2\omega} \lesssim 0.025q^2\left(\frac{r_o}{H_\infty}\right)^3 \tag{122}$$

or equivalently

$$q \gtrsim \left(\frac{40\nu}{r_o^2\omega}\right)^{1/2}\left(\frac{H_\infty}{r_o}\right)^{3/2}. \tag{123}$$

This condition does not depend very much on the functional form of $\dot{h}$ which determines the scale of the gap. However, the time scale to form the gap is determined by the rate at which tides can significantly redistribute the angular momentum of the material originally in the gap region. This clearly lengthens if the gap region is enlarged.

Setting $H_\infty = q^{1/3}r_o$ in Eq. (123), the condition to form a gap becomes

$$q \gtrsim \frac{40\nu}{r_o^2\omega} = \frac{40}{\mathcal{R}} = 40\alpha_s\left(\frac{H_\infty}{r_o}\right)^2. \tag{124}$$

This is to within a numerical factor, the same condition as given by Lin and Papaloizou (1979*a*), that the mass ratio should be greater than the inverse Reynolds number. The results of this linear approximation are also in general agreement with that deduced from the fully nonlinear calculations despite the

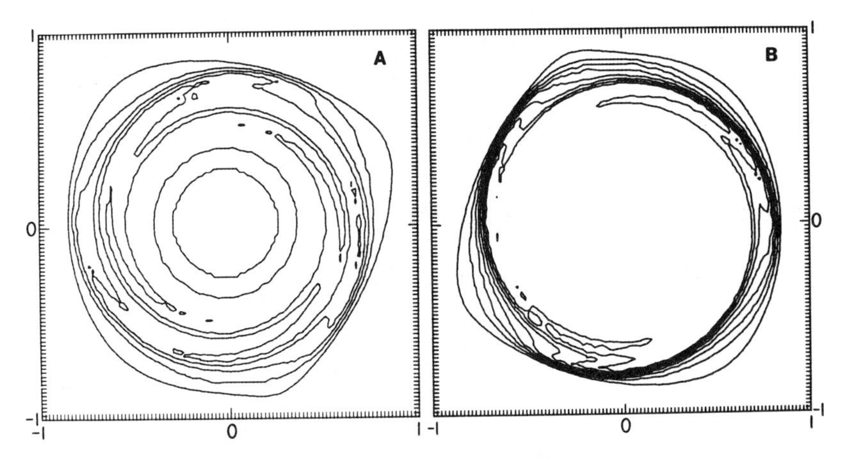

Fig. 10. The dimensionless surface-density distribution of the nebula when it is strongly influenced by the tidal force of a protoplanet with a $M_P = 10^{-3}$ $M_\odot$. The dimensionless sound speed is 0.03 ωr_0 and the dimensionless viscosity is 10^{-5} ωr_0^2. In order to demonstrate the global pattern, we plot the surface density contour with $\Delta\Sigma = 0.1$ between each level in plot A. The spiral pattern of the density maxima is associated with shock dissipation. Two arm spirals at the 2:1 resonance and three spiral arms at the 3:2 resonance are visible. The length of each reference scale is $2r_0$. The structure of the outer edge is more clearly revealed from the fine surface density contours with $\Delta\Sigma = 0.05$ between each level as seen in plot B. The Sun and the protoplanet are located at (0,0) and (1,0), respectively.

fact that we relax the assumptions on the surface density profile, the role of nonlinear shock dissipation, and the variation of Ω. In this case, we adopt a viscosity law in which $\nu = \nu_o(\Sigma/\Sigma_o)^2$ where ν_o and Σ_o are scaling constant. For the equation of state, we use a polytropic gas model. The shock dissipation, when it occurs, is computed with von Neumann's method. In Fig. 10, we plot the surface density contours and dissipation pattern. The condition for truncation is slightly modified due to a small departure of Ω from the Keplerian rotation law near the disk edge.

C. Orbital Evolution

The tidal effect of protoplanet-nebula interaction can be incorporated into the evolution calculation of the nebula. In this case, the ϕ component of the equation of motion (Eq. 6) becomes

$$\Sigma U_r \frac{dr^2\Omega}{dr} = \frac{1}{r}\frac{\partial}{\partial r}\nu\Sigma r^3 \frac{d\Omega}{dr} + \Sigma\Lambda \tag{125}$$

where U_r is the radial velocity and Λ is the injection rate of angular momentum per unit mass by tidal interaction with the protoplanet. In Sec. II we showed that in most regions of the nebula Ω is closely approximated by the Keplerian value which is given by $\Omega = (GM/r^3)^{1/2}$. We also showed in the previous subsection that near the orbit of the protoplanet, the small departure from Keplerian rotation does not significantly affect the local angular momentum transfer rate provided the Rayleigh stability criterion is satisfied. Recall that the Rayleigh criterion is satisfied if the Roche radius of the protoplanet is larger than H_∞. Because we are primarily interested in the orbital migrations of protoplanets which can truncate the nebula, we only consider the case where the Rayleigh stability criterion is satisfied and adopt a Keplerian approximation for Ω. Then Eq. (125) becomes

$$\Sigma r U_r = 2\Sigma\Lambda \frac{r^{3/2}}{(GM)^{1/2}} - 3r^{1/2}\frac{\partial \nu\Sigma r^{1/2}}{\partial r}. \tag{126}$$

Using Eq. (126) to eliminate $\Sigma U_r r$, the continuity Eq. (3) gives

$$\frac{\partial\Sigma}{\partial t} = \frac{1}{r}\frac{\partial}{\partial r}\left(3r^{1/2}\frac{\partial \nu\Sigma r^{1/2}}{\partial r} - \frac{2\Lambda\Sigma r^{3/2}}{(GM)^{1/2}}\right). \tag{127}$$

To calculate the evolution of a nebula with an embedded protoplanet, we need to specify the functional form of Λ. As we have seen above, the functional dependence of Λ on r depends on the distribution of tidal dissipation in the nebula. If ν is relatively large, e.g., $\alpha_c \gtrsim 0.1$, this dissipation is concentrated near the edge of the nebula in the vicinity of the protoplanet's orbit (Papaloizou and Lin 1984). If ν is relatively small, nonlinear effects may cause the dissipation to be concentrated near the edge. In this case, the impulse approximation, Eq. (101), or

$$\Lambda = \text{sign}(r - r_p)\frac{fq^2GM}{2r}\left(\frac{r}{\Delta_p}\right)^4 \tag{128}$$

probably provides an adequate estimate for Λ. In Eq. (128), f is a constant of order unity and Δ_p is the maximum of $|r - r_p|$, the nebula thickness H_∞ and the protoplanet's Roche radius $(q/3)^{1/3}r_p$. The rate of tidal angular momentum transfer between the nebula and the protoplanet is $2\pi \int_{R_i}^{R_o} \Lambda r \Sigma dr$. By total angular momentum conservation, the protoplanet's orbit must also evolve such that

$$\frac{M_p}{2}\left(\frac{GM}{r_p}\right)^{1/2}\frac{dr_p}{dt} = -2\pi\int_{r_{\text{in}}}^{r_{\text{out}}} \Lambda r \Sigma dr. \tag{129}$$

Equations (127) – (129) together describe the nebula and protoplanetary orbit evolution. They can be solved for arbitrary viscosity laws. For example, in the α prescription, $\nu = \nu_o(\Sigma/\Sigma_o)^2$ (see Eq. 69). For computational convenience, we can introduce a set of dimensionless variables such that $\tau = t\nu_o/r_o^2$, $\Sigma_1 = \Sigma/\Sigma_o$, and $y = r/r_o$ where r_o is some suitably chosen radial scaling length, e.g., the initial radius of the outermost radius of the nebula. With these variables, Eqs. (127) and (129) become

$$\frac{\partial \Sigma_1}{\partial \tau} = \frac{3}{y}\frac{\partial}{\partial y}\left[y^{1/2}\frac{\partial(y^{1/2}\Sigma_1^3)}{\partial y} - \text{sign}(y - y_p)\Sigma_1 y^{1/2} A\left(\frac{r}{\Delta_p}\right)^4\right] \tag{130}$$

and

$$\frac{\partial y_p}{\partial \tau} = -2BAy_p^{1/2}\int_{y_i}^{y_o}\Sigma_1\left(\frac{r}{\Delta_p}\right)^4 \text{sign}(y - y_p)dy \tag{131}$$

where $A \equiv (GMr_o)^{1/2}q^2f/(3\nu_o)$ measures the strength of the tidal effects in comparison to viscosity and $B \equiv 3\pi\Sigma_o r_o^2/M_p$ measures the mass ratio between the nebula and the protoplanet. In order to open up a gap in the nebula, the two terms in the bracket in Eq. (130) must become comparable, i.e., for a gap with a scale H_∞, and $\Sigma_1 \approx 1$, we require $r_o/H_\infty \approx |A|(r_o/H_\infty)^4$, or $|A| \approx (H_\infty/r_o)^3$. If $|A|$ becomes larger, the gap becomes wider than H_∞. These conditions on $|A|$ are easily seen to be equivalent to that given by Eq. (122) above. The dimensionless parameter $B/|A|$ measures the nebular mass to the mass of the protoplanet. When $B = 0$, the mass of the nebula vanishes.

D. Preliminary Numerical Calculation

Equations (130) and (131) can be solved numerically with standard explicit finite-difference techniques (Richtmeyer and Morton 1967). As an example, we run several cases with $H_\infty/r_o = 0.1$, and $|A| = 10^{-3}$. The value of B is chosen to be 5, 20, and 80 so that a wide range of nebular mass is represented. In the B = 80 case, the mass of the nebula is considerably larger than the protoplanet's mass initially but is depleted to a value comparable to M_p after a few diffusion time scales. We start the computation with $\Sigma_1 = 1$ for $0 < y < y_{\rm max}$, and $\Sigma_1 = 0$ for $y > y_{\rm max}$, where $y_{\rm max}$ is chosen to be unity. The outer boundary of the nebula is allowed to expand freely while at the inner boundary we specify $\partial\Sigma_1/\partial y = 0$, as required for regularity.

In Fig. (11) we present the evolution of a protoplanetary orbit and the nebular surface density with two different initial orbital radii for the protoplanet and two different values of B. In the first case, $y_{\rm max} = 1, B = 20$, and $y_p = 0.6$ initially. In the absence of the protoplanet, $u_r < 0$ at $y \leq 0.6$, i.e.,

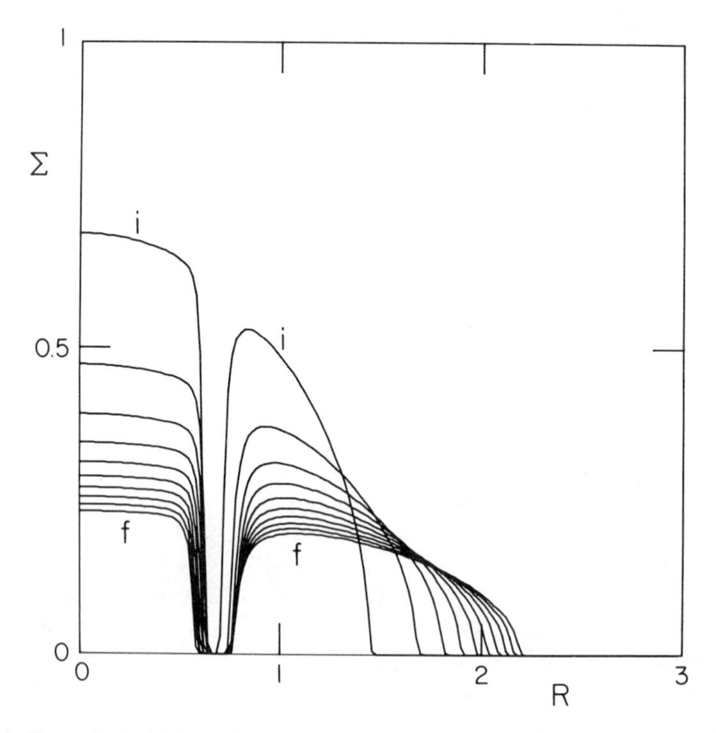

Fig. 11. The evolution of the surface-density distribution of the nebula when it is strongly influenced by the tidal force of a protoplanet with $A = 10^{-3}$ and $B = 20$. The value of $H_\alpha/R = 0.05$. The initial value of y_p is 0.6. The nebula has a uniform unit surface density out to $R \simeq 1$ initially. The ten curves represent the distribution of Σ at ten instants of time which are separated by a uniform interval, $\Delta\tau = 0.13$.

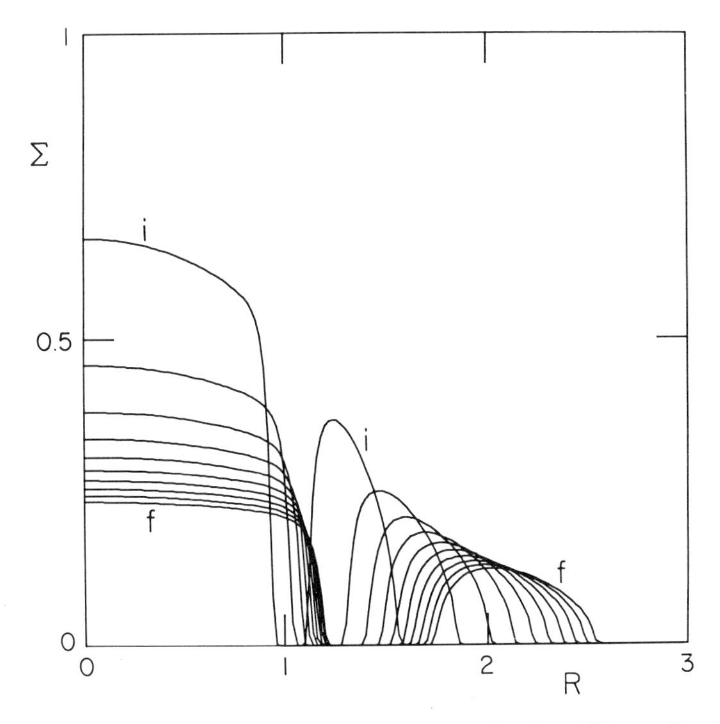

Fig. 12. The evolution of nebular surface density. The values of A, B, and $H_\propto$ as well as initial values of Σ are identical to those in Fig. 11. The initial value of y_p is 0.8 and the time interval between different curves is 0.18.

the nebular material near y_p would undergo inward diffusion. In the present case, $|A|$ is sufficiently large to ensure eventual gap formation in the nebula so that the protoplanet and the nebula are strongly coupled together. At least initially, the mass of the nebula is somewhat larger than that of the protoplanet so that the protoplanet simply passes angular momentum between the two regions of the nebula separated by the protoplanet. The results in Fig. 11 indicate that a gap is well established. Because the magnitude of ν in the nebula decreases with Σ_1, the condition for gap formation becomes increasingly favorable as the nebular material is depleted. Prior to the gap formation, the orbit of the protoplanet would hardly change because Σ_1 on either side of the protoplanet is sufficiently similar that the angular momentum transfer rate on the two sides is delicately balanced to cancel out. Even after the gap formation, the inward migration of the protoplanet is exceedingly slow so that after Σ_1 has reduced by a factor of 4, the radial position of the protoplanet has hardly changed. Although the mass of the nebular material beyond the orbit of the protoplanet is comparable to or larger than M_p, the mutual tidal interaction is ineffective in causing y_p to change significantly over a diffusion time scale. In fact, the surface density decreases everywhere while

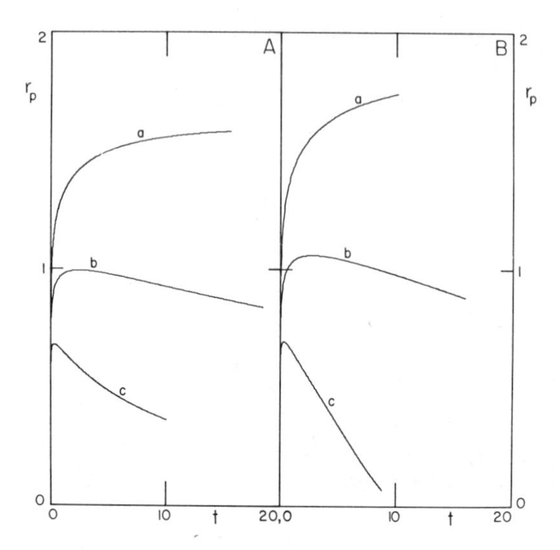

Fig. 13. The evolution of the orbital radius R_p ($= r_p$) of a protoplanet in a nebula with $H_\infty/r_0 = 0.05$ and $A = 10^{-3}$. The curves a, b, and c represent, respectively, initial $y_p = 0.6, 0.7$, and 0.8. In plot A, $B = 5$ whereas in plot B, $B = 80$. The label for the abscissa r_p represents y_p whereas the label for the ordinate t represents η in this figure.

the outer boundary of the nebula expands with time. Subsequently, the evolution of the system comes to a halt as ν decreases with Σ_1. In the second case (Fig. 12), $y_p = 0.8$ and $y_{max} = 1$. Initially the nebular material at $y = 0.8$ diffuses outwards. Due to strong tidal coupling, the protoplanet moves outwards with the outwardly diffusing material. In this case too, the protoplanet's orbital radius only increases by a factor of two while Σ_1 is depleted by a factor of 4. In the case where $B = 5$, because the initial mass of the nebula is much smaller than the above cases, less orbital migration is found (Fig. 13A). Similarly when in the $B = 80$ case, the orbital migration of the protoplanet is completely determined by the motion of the nebular gas (Fig. 13B). Even in this case, the orbital migration proceeds on the viscous diffusion time scale of the nebula.

In summary, these calculations indicate in general that the orbital evolution of a protoplanet always occurs on a time scale comparable to or longer than the viscous time scale of the nebular gas after some initial transients. If the protoplanet starts at some intermediate position in the disk, a feedback mechanism operates to balance inward and outward angular momentum transfer so that little orbital evolution occurs. Even if the protoplanet is placed on the outside of the disk, after rapidly reaching an equilibrium from the disk edge, further evolution occurs on the slow viscous time scale.

VI. PROTOPLANET FORMATION

From the nebula models outlined in Secs. III and IV, the temperature, mass, density and pressure distribution can be deduced at any given epoch of solar nebula evolution. From these results we are now in a position to examine the physical conditions in the solar nebula at the time the protoplanets acquired most of their mass and find constraints on and evidence for or against protoplanetary-formation scenarios. For example, the tidal constraint in Sec. V implies that the viscous evolution time scale in the nebula must be $\lesssim 10^6$ yr in order to avoid tidal truncation and gap formation. Therefore, Jupiter and Saturn must have been formed within a million years after the initial formation of the Sun. Although this argument provides a strong limit on the duration of the protoplanetary formation epoch, there is still considerable uncertainty on just when and how the protoplanets did form. The diversity of opinion may be divided into three major scenarios: (1) All protoplanets formed within 10^5 to 10^6 yr of the infall stage by gravitational instability which occurred in a very massive and extended nebula (Cameron 1978); (2) The protogiant and terrestrial planets formed at different epochs through different processes. While protogiant planets may have formed through massive gravitational instability, as in the first scenario, the terrestrial planets formed through a series of gradual accretion processes in an essentially gas-free environment (Safronov 1969, 1972*b*; Kusaka et al. 1970; Wetherill 1980); (3) The protogiant as well as prototerrestrial planets formed through a process of gradual accretion, first of dust, then of gases in a gas-rich environment (Hayashi et al. 1977; Hayashi 1981*a*; Lin 1982).

A. Gravitational Instability Scenario

Quantitative investigations indicate that a nebula may become gravitationally unstable if Q_T $(= \Omega c_s / \pi G \Sigma)$ is comparable to or less than unity (Safronov 1969, 1972*b*; Goldreich and Ward 1972; Cameron 1978). In Cameron's model a small value for Q_T can be attained if the nebula is massive and extended, say one solar mass spreads beyond the orbit of Neptune, or if the central protosun has a low mass, say $<< 1 M_\odot$. If the nebula is convective or turbulent such that the standard α model is applicable, T_c can be determined in terms of Σ and Ω as shown in Sec. III. Adopting a grain opacity law, $\kappa = 2 \times 10^{-4} T^2$, we find that the nebula is in local thermal equilibrium if $T = \kappa_o \mathcal{R} \alpha_s \Omega \Sigma^2 / (\sigma \mu)$ so that $Q_T = (\mathcal{R} / \pi G \mu)(\kappa_o \alpha_s \Omega^3 / \sigma)^{1/2} \approx 10^{15} \Omega^{3/2}$ for $\alpha_s \approx 0.1$. In this case, $Q_T \lesssim 1$ only if $\Omega \lesssim 10^{-10} s^{-1}$ which is possible only well beyond the present orbit of Neptune or if the protosun is much less massive than $1 M_\odot$ (Lin and Papaloizou 1980; Lin 1981). If the nebula is gravitationally unstable, the mass of a typical fragment can be calculated from the characteristically most unstable wavelength $\lambda \approx 2\pi^2 \Sigma r^3 / M$ (see, e.g., Safronov 1969). According to the gravitational instability scenario, the fragments typically have masses $\approx 0.1 M_\odot$ (Cameron 1978; Lin 1981). Thus, in order for fragments to evolve into the protoplanets of our solar system, the following sequence of events must take place:

1. The protosun must gain the other 90% of its present mass after the protoplanets have already formed.

2. The massive fragments must lose 99% of their mass to evolve into protogiant planets and 99.99% of their mass to evolve into prototerrestrial planets.
3. Gas must be lost preferentially in order for the fragments to evolve into prototerrestrial planets.
4. The fragments' orbital radii must be reduced by at least an order of magnitude.
5. These evolutionary processes must occur on a time scale of $\lesssim 10^5$ yr, even in the most optimistic case, in order to prevent orbital instability from randomizing the protoplanets' orbits and inducing the ejection of low-mass fragments and a very large orbital eccentricity in the protoplanetary system that is not seen today.

Although it may not be impossible for this chain of events to occur and ultimately to lead to the formation of the present solar system, it is difficult to stabilize the gravitationally unstable nebula unless its initial gas content is substantially depleted as a consequence of fragmentation. Therefore, it may be impossible for the nebula to retain sufficient gaseous material, in the post-fragmentation era, to feed the protosun in order for it to grow substantially. In addition, both Jupiter and Saturn probably have solid cores today. It has been pointed out by D. Stevenson (personal communication) that it may be impossible for a solid core to form inside these fragments.

B. The Gas-Free Accumulation Scenario

In this scenario, the formation epoch of prototerrestrial planets continues for about 10^7 to 10^8 yr (Safronov 1972*b*; Wetherill 1980). The mass of the nebula, from which the protoplanets formed, is a small fraction, say about 10 to 20%, of the present mass of the Sun. Because of the relatively low nebular mass and surface density, the nebula is stable against its own gravity, i.e., the pressure-gradient effect is always sufficient to balance the gravity in the vertical direction. In the first stage of planetary formation, grains in the solar nebula coagulate through inelastic collisions. In the absence of turbulence, the dispersion velocity of the grains is dissipated through inelastic collisions so that they settle towards the midplane (Safronov 1972*b*; Goldreich and Ward 1973). Although the solid grains may only constitute a small fraction of the nebula's mass, their self gravity is much more important for them than for the gas after their velocity dispersion is damped well below the sound speed of the nebular gas. Eventually, when a sufficiently dense layer has been formed near the midplane, the grains become gravitationally unstable and collapse into kilometer-sized planetesimals. If the nebula is turbulent, turbulent eddies may exert a viscous drag on the grains in random directions so that the dispersion velocity is maintained at a relatively large value to stabilize against gravitational instability (Weidenschilling 1977). However, a moderately large velocity dispersion also induces more frequent collisions between the grains.

Through a series of inelastic and cohesive collisions, the grains may also grow into kilometer-sized planetesimals before they settle down to the midplane of the nebula (Weidenschilling 1980).

If a large solid body forms through cohesive grain collisions, its radius R_s would grow at a rate $dR_s/dt = \rho_{gr} <v> /\rho_p$ where ρ_{gr} is the grains distribution density, $<v>$ is the grains' dispersion velocity and ρ_p is the internal density of individual grains. If the distribution of grains is stable against gravitational collapse, we must have $<v> \gtrsim \pi G \Sigma_{gr}/\Omega$, where $\Sigma_{gr} = \rho_{gr} H_{gr}$ is the surface density of the grains and H_{gr} is their scale height. We see that for a stable distribution

$$\frac{da}{dt} \geqslant \frac{\rho_{gr}}{\rho_p} \frac{\pi G \Sigma_{gr}}{\Omega} \approx \frac{M_{gr}}{M_\odot} \frac{r}{H_{gr}} \frac{\Sigma_{gr}\Omega}{\rho_p} \tag{132}$$

where M_{gr} is the mass of the nebula residing in grains. Taking the lower limits, $M_{gr}/M_\odot \approx 10^{-5}$, $r/H_{gr} \approx 10$, $\Sigma_{gr} \approx 1$ g cm^{-2}, and $\rho_p \approx 1$ g cm^{-3}, we find $da/dt \gtrsim 10^{-4}\,\Omega$. This result implies that even in a turbulent nebula, with no gravitational instability, grains with radius $\approx 10^2$cm can form within about 10^6 yr. It is easy to show that bodies of such a size must decouple from the turbulence so that H_{gr}/r must decrease and that growth to a size of ≈ 1 km can then occur rapidly. In this case, the gas drag due to the turbulence can no longer introduce a significant velocity dispersion in the vertical direction among the kilometer-sized planetesimals so that they settle to the mid-plane of the nebula.

If a significant fraction of the heavy elements in the solar nebula is assembled into a two-dimensional distribution of kilometer-sized planetesimals, their orbits overlap such that direct or glancing collisions occur frequently. These collisions can lead both to coalescence and disruption of the planetesimals and a size distribution quickly develops (Zvygina et al. 1973; Greenberg et al. 1978). After $\approx 10^3$ yr, planetesimals with diameters of the order of 1000 km are produced. At this stage, the nearly circular orbits of the planetesimals are separated by about one Roche radius, and so their mass M_{pl} is given approximately by $M_{pl}/[2\pi r^2 (M_{pl}/M_\odot)^{1/3}] = \Sigma_{gr}$, or $M_{pl}/M_\odot \approx (M_{gr}/M_\odot)^{3/2}$. Thereafter, the collisional frequency is relatively low unless the planetesimals develop considerable random velocities. Random velocities may be induced by the gravitational influence planetesimals exert on each other when they undergo close but noncollisional encounters. Random velocities may also be damped by dissipative collisions (Safronov 1969; Kaula 1979; Wetherill 1978,1980; Stewart and Kaula 1980). Despite the lack of knowledge of the elastic, erosive and coagulational properties of the direct physical collisions and the complicated nature of the mutual gravitational interactions between planetesimals, the dynamics of planetesimals in the coagulation stage can be analyzed in detail under simplified and somewhat *ad hoc* assumptions such as neglecting the presence of nebular gas. Nonetheless, these computations can provide useful insights. For example, the critical random velocity, at which the maximum growth rate of the planetesimals is achieved, may be estimated from such an approach (Wetherill 1980). Towards the final stages of accumulation, when most of the mass is contained in several massive protoplanetary embryos, it is essential for the embryos to maintain relatively large random velocities until they have coagulated into four bodies corresponding to the progenitors of the four terrestrial planets (Ziglina 1976; Ziglina and Safronov 1976; Cox et al. 1978; Cox and Lewis 1980; Wetherill 1980). Owing to the low collisional frequency, the typical growth time scale during the final stage is around 10^7 to 10^8 yr.

The gas-free scenario provides a more plausible mechanism for prototerrestrial planet formation than does the massive gravitational instability scenario, principally because the mass and the composition of the prototerrestrial planets may be deduced without a great number of *ad hoc* assumptions. However, it cannot provide a unified scheme for the formation of all protoplanets. Because the giant planets are gas-rich, they must have been formed in a gas-rich environment. Thus, if prototerrestrial planets formed through coagulation in a gas-free environment, we must invoke a scenario in which the prototerrestrial and protogiant planets formed not only at a different epoch and at different distances from the Sun, but also through different processes (Wetherill 1980).

C. Gas-Rich Scenario

Alternatively, we can hypothesize that both the prototerrestrial planets and the solid core of protogiant planets formed through coagulation processes in a gas-rich environment. This scenario naturally satisfies the requirement that protogiant planets must have had a solid core with a mass comparable to that of the terrestrial planets before they could have efficiently accreted from the relatively hot nebula which was continually heated by viscous dissipation. The presence of a turbulent gaseous background can induce a relatively large velocity dispersion among the grains and thereby enhance the coagulation process (Weidenschilling 1984). The viscous drag between the planetesimals and the nebular gas can also induce orbital migration of the planetesimals in the radial direction (Whipple 1971; Adachi et al. 1976). Most important, an evolving nebular background may provide a continuous supply of solid material for the most massive planetesimals to accrete and thereby shorten the characteristic growth time for the final-stage embryo coagulation from 10^{7-8} yr down to 10^{5-6} yr as required by the tidal truncation constraint. Continuous supply of grains is possible during the viscous evolution stage when the nebular mass, surface density and temperature decrease with time (see Sec. IV). In the inner and intermediate regions of the nebula, the radial distribution of the condensation temperature, deduced from the cosmochemical data (Lewis 1972), can be adequately represented if most of the solid material in the nebula condensed over a few viscous diffusion time scales. In the outer regions, the time scale for nebular temperature to decrease is faster than that for the inward viscous diffusion of the gas. In the outermost regions, the nebular gas actually diffuses outwards to carry away excess angular momentum. Consequently, in these regions, the gas cools off with viscous evolution. As a result, the gas may become supersaturated, creating new nucleation sites which enhance grain condensation. In these regions, grain condensation is a continuous process throughout the viscous evolution stage. These continuous condensation regions generally spread towards smaller radii with time. Quantitatively, the evolution of the solid particle's mass and phase space distribution function, $f(m_g, v_g, r)$, can be calculated from a set of coagulation equations (Safronov 1969). The continuous condensation processes can be incorporated into the coagulation equations with an additional source term. It should be noted that the particles' growth rate may be affected by motion in the vertical direction when there is a large vertical temperature gradient.

Fresh solid particles may be continually brought to the vicinity of massive planetesimals by the nebular gas-drag effect. We now estimate the accretion cross section of the largest planetesimals. Due to frequent collisional encounters, the orbital eccentricities of the largest planetesimals are probably relatively small (Greenberg 1979). As the solid particles migrate through the nebula, their orbits may come within 1 Roche radius, $R_L = (M_p/M_\odot)^{1/3} r$, of a massive planetesimal with mass M_p. In this case, the particle's motion is strongly influenced by the tidal force of the planetesimal provided the eccentricity of its orbit around the protosun is smaller than $\approx (M_p/M_\odot)^{1/3}$. Because we are primarily interested in the accretion cross section of the massive planetesimal, which may have a geometric radius $R_g \approx 10^{-2} R_L$ we search for those perturbed orbits which come within R_g of the perturbing planetesimal.

The trajectories of the particles that undergo close encounters with the planetesimals can be calculated from the restricted three-body problem. Orbits of this type have been considered previously in the case of the Earth by Dole (1962) and Giuli (1968). We now provide a rediscussion for M_p ranging from 10^{-14} to $10^{-6} M_\odot$ primarily to evaluate the cross section for particle accretion by massive planetesimals. For initial conditions, we primarily consider the impacting trajectory of particles whose initial orbits around the protosun are circular. These computations are particularly relevant because the particle's dispersive motion is continually dissipated by the viscous-drag effect associated with the nebular gas. Particles larger than a few centimeters may be segregated from the turbulent flow and their dispersion velocity and orbital eccentricities may be very small.

The main results of these trajectory calculations indicate that for particles whose initial circular orbital radius r_g is to within R_L of the orbital radius r_p of the planetesimal, the orbit reverses direction before getting within R_L of the planetesimal. Particles on these orbits cannot be accreted by the planetesimal. Particles with $1.2 \lesssim x \lesssim 1.32$, where $x = |r_g - r_p|/R_L$ follow nearly hyperbolic orbits around the planetesimal. Their close approaches to the planetesimal are still larger than R_g but smaller than R_t. Particles with $1.32 \lesssim x \lesssim 1.39$ and $1.55 \lesssim x \lesssim 1.58$ actually approach to within R_p of the planetesimal. In these cases, the particles would have direct impacts with the planetesimal. For sufficiently small $M_p/M_\odot$, these trajectories apply to particles whose initial orbits are either inside or outside r_p. With these trajectories, it is a simple matter to estimate the accretion rate, dM_p/dt, from an ambient background with a surface density Σ_a of particles. Taking accretion from both inside and outside r_p into account, we find that $dM_p/dt \approx 0.14\,\Sigma_a(M_p/M_\odot)^{2/3}\Omega r_p^2$. This formula is somewhat similar to that proposed by Lyttleton (1972).

According to the vertical structure models we presented in Sec. III, most regions of the nebula are optically thick owing to grain opacity. In a particle disk, the collision frequency is $\approx \tau\Omega$ (Goldreich and Tremaine 1978) where $\tau \approx \Sigma_a a^2$, so that particles with radius $a \lesssim 1$ cm collide with each other many times per orbit. Near the Roche lobe of the planetesimal, the collision frequency may be higher because the dispersion velocity is increased by the tidal effect of the planetesimal. If collisions among particles are inelastic, their relative motion would be damped. In this case, the accretion cross section may be strongly affected by collisions. The effect of collisions can be studied numerically with a Lagrangian particle interaction method (Lin and Pringle 1976; Franklin et al. 1980). In this method, a large number of particles follow their orbits as determined by the gravitational potential of the protosun and the planetesimal. These particles are forced to interact with all neighboring particles within some mixing length l, usually taken to be $10^{-2}R_L$ at time intervals short compared with Ω^{-1}. With this approach, Lin and Papaloizou calculated the flow pattern of a group of frequently colliding particles in the neighborhood of a planetesimal with $M_p = 10^{-6}M_\odot$. Their results indicate that the flow pattern is remarkably smooth (see Fig. 14). The resulting accretion rate is similar to but about four times larger than the collisionless case discussed above. Note that if the particles are not fully settled to the midplane and their scale height H_p is larger than R_L then the accretion rate should be reduced by a factor R_L/H_p, so that

$$\frac{dM_p}{dt} \approx 0.6\Sigma_a r_p^2\Omega\left(\frac{M_p}{M_\odot}\right)\left(\frac{r_p}{H_p}\right). \tag{133}$$

From these detailed numerical computations, we find that it may be possible to reduce the time scale required for a planetesimal to grow to a mass about that of the Earth, at the Earth's orbital radius, to about 10^{5-6} yr, provided that the nebula is somewhat more massive than the minimum-mass nebula. In fact, such a nebula may be required if a significant fraction of the solid particles is coupled to the gas flow in the nebula and is accreted by the Sun so that the heavy elements in the nebula are not entirely retained by the planetesimals.

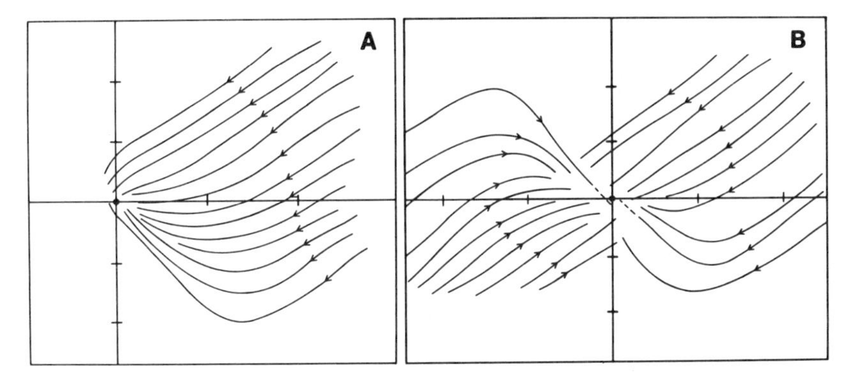

Fig. 14. The trajectories of particles on neighboring circular orbit around a low-mass protoplanet. Axes are marked at intervals of 0.4 $q^{1/3}D$. The mass ratio q is taken to be 10^{-6}. Collision effects are included. The characteristic Reynolds number of the flow is 10^3. In plot A, only orbits interior to the protoplanet are considered whereas in plot B, particles inside and outside the protoplanet's orbit are injected symmetrically. The protoplanet is represented by a circle with a radius equal to 0.01 $a^{1/3}D$.

In order for a planetesimal to be able to accrete gas at all, its surface escape velocity must be larger than the sound speed of the nebular gas. This condition implies a minimum mass, $M_{\rm min} \approx 10^{10}\rho_p^{-1/2}c_s^3$ where ρ_p is the density of the planetesimal which may be of order 1 g cm^{-3}. But note that if an optically thick atmosphere can be formed, a larger mass than $M_{\rm min}$ is required for the onset of gas accretion (see chapters by Bodenheimer and by Pollack). This mass $M_{\rm min}$ is comparable to the mass of the Moon. If the density scale height of the nebular gas is larger than the Roche radius of the planetesimal, the maximum possible gas-accretion rate would be determined by the Bondi formula in which $dM_p/dt \approx 4\pi\rho(GM_p)^2c_s{}^{-3}$ which depends on different powers of the nebular temperature and of the planetesimal's mass than does the particle accretion rate. There is a critical mass at every orbital radius such that the two accretion rates are equal. For example, in a minimum-mass nebula, this mass is approximately 1 Earth mass ($M_\oplus$) at the present orbit of Jupiter. At the present orbit of Earth, this mass is approximately 10 $M_\oplus$. Because the particle accretion rate depends on a lower power of the accretor's mass, particle accretion dominates gas accretion for masses $<< 1\ M_\oplus$. The gas-accretion rate rapidly exceeds the dust-accretion rate as the accretors grow beyond the critical mass. The gaseous envelope surrounding the solid core mass becomes optically thick, forming a protogiant planet. At the present orbital radius of Jupiter, the fastest time scale for doubling the mass of an object of 1 $M_\oplus$ is about 10^5 yr, and that for doubling Jupiter's mass is about 10^3 yr.

What processes determine the final mass of protogiant planets? The growth of protogiant planets may be terminated when the nebula is sufficiently depleted or when their tidal effects cause the truncation of the nebula in the vicinity of their orbits (see Sec. V). The present mass ratio of the giant planets to the Sun is between 10^{-4} and 10^{-3}. If the mass of the planets and the Sun has not changed significantly since their formation, the Reynolds number $\mathcal{R}_{\rm eff}$ at their formation radius must have been relatively low. Although $\mathcal{R}_{\rm eff}$ is relatively low near the protosun (down to $\approx 10^4$), protogiant planets could not have been formed there because, in this case, the nebular temperature interior to its orbital radius would be too hot to allow grain condensation which is essential for the formation of solid cores of giant planets and the prototerrestrial planets. Just outside the ice-grain condensation zone, however, $\mathcal{R}_{\rm eff} \lesssim 10^4$ for a minimum-mass nebula so that the tidal truncation constraint can be satisfied. Farther outside this transition zone, $\mathcal{R}_{\rm eff}$ increases with radius. Thus, if all the protogiant planets were formed at the same epoch, the more distant protogiant planets would terminate their growth at lower masses. If they were not formed at the same epoch, the termination condition must be modified accordingly. As for the prototerrestrial planets, tidal truncation probably did not play a role; instead their final mass may have been determined by the availability of solid material and the efficiency of particle accretion in the nebula.

The basic attractiveness of the above scenario of planetary formation is that it provides a natural explanation for the mass distribution, chemical abundance, and spatial distribution of the present planetary system. While some detailed quantitative analysis still needs to be carried out, the model provides a possible scheme for protoplanet formation with minimum *ad hoc* assumptions.

VII. THERMAL STRUCTURE OF THE INNERMOST REGIONS OF THE NEBULA

Up to now, we have limited our discussion primarily to the outer regions of the nebula where grain opacity is dominant. These solutions are particularly relevant if we want to study the environment out of which the protoplanets are formed. As in any other branches of science, a hypothesis remains as such until it is verified by observational or experimental data. A possible test of our models is to utilize the cosmochemical data obtained from meteorites and interplanetary probes. Presumably, the chemical composition and physical structure of celestial bodies is strongly influenced by the nebular environment out of which they were formed. However, the discussion in Sec. VI indicates that the chemical and physical structure of the protoplanets might also be strongly affected by the formation process. Thus, the implications of these tests are not clear.

Alternatively, our models of the solar nebula may be tested with the observations of protoplanetary disks around T Tauri stars which are thought to be the youngest stars in the Milky Way Galaxy. In some cases, they may have a circumstellar disk around them (Hanson et al. 1983; Cohen 1983). In these cases, we may wish to concentrate on the observable properties of the disk in those wavelength ranges that contribute to most of the disk's emerging radiation. Most of the radiation of the disk actually emerges as optical photons from the innermost regions of the disk, where the rate of viscous dissipation of differential rotation in the disk is at a maximum rate. Therefore, if we wish to search amongst T Tauri stars for evidence of disks, it would be appropriate to construct quantitative models for the inner regions where molecular and atomic opacity is more relevant.

A. Structure of the Inner Regions of the Disk

With an opacity table, the thin-disk nebula structure Eq. (62) can be solved for an arbitrary viscosity prescription. In the inner regions of the disk where $T \gtrsim 1500$ K, a Cox-Stewart-Alexander opacity table is relevant; we are indebted to P. Bodenheimer for providing us with such a table. The tabulated value of κ in these tables can be fitted in the form of a power-law relation (Eq. 58) such that for 4000 K $\lesssim T \lesssim$ 30,000 K

$$\kappa_7 = \kappa_{o7} \rho T^{-2.5} \mathrm{g}^{-1}\ \mathrm{cm}^2 \tag{134}$$

for bound-free and free-free opacity in the fully ionized limit, where $\kappa_{07} = 1.5 \times 10^{20}$, and

$$\kappa_6 = \kappa_{o6} \rho^{1/3} T^{10} \mathrm{g}^{-1}\ \mathrm{cm}^2 \tag{135}$$

for H^- and molecular opacity in the partially ionized limit, where $\kappa_{06} = 10^{-36}$. These analytic expressions produce peak opacities along a ridge line given by $T_{67} = 3 \times 10^4 \rho^{4/75}$ K (e.g., $T \approx T_{67} = 1.17 \times 10^4$ K for $\rho = 10^{-8}$g cm^{-3}), which is obtained by equating expressions (134) and (135). In the temperature range 1500 K $\lesssim T \lesssim$ 4000 K, there is a major transition in the opacity law due to the evaporation of grains and dissociation of molecules. For the relevant density and temperature range, we find that the opacity table can be fitted by the following analytic approximations:

$$\kappa_5 = \kappa_{05} \rho^{2/3}\, T^3 \mathrm{g}^{-1}\ \mathrm{cm}^2 \tag{136}$$

where $\kappa_{05} = 2 \times 10^{-8}$ and

$$\kappa_4 = \kappa_{04} \rho^{2/3} T^{-9} \mathrm{g}^{-1}\ \mathrm{cm}^2 \tag{137}$$

where $\kappa_{04} = 2 \times 10^{34}$. The critical temperature which separates κ_5 and κ_6 is $T_{56} = 1.1 \times 10^4 \rho^{1/21}$ K. Similarly, $T_{45} = 3000$ K separates κ_4 and κ_5 and $T_{34} = 4.6 \times 10^3 \rho^{1/15}$ separates κ_4 from the grain opacity κ_3 in Eq. (65). Note that the subscript for different values κ denotes the relevant opacity law in decreasing order of gas temperature. These analytic approximations are supplemented by the opacity laws for different grains (Eq. 65) and together they provide a complete opacity law for the present purpose.

Substituting the above results into Eqs. (68a) to (68c), we find

$$T_c = 4 \times 10^4 \alpha_s{}^{1/6} \Omega^{1/3} \Sigma^{1/2} \text{ K} \tag{138a}$$

$$\nu = 4.5 \times 10^{12} \alpha_s{}^{7/6} \Omega^{-2/3} \Sigma^{1/2} \text{ cm}^2 \text{s}^{-1} \tag{138b}$$

and

$$\rho = 2 \times 10^{-7} \alpha_s{}^{-1/12} \Omega^{5/6} \Sigma^{3/4} \text{g cm}^{-3} \tag{138c}$$

for the hot region where $T > T_{67}$ and Eq. (134) is relevant. In this case, μ is taken to be unity. The vertical structure solutions in Sec. III indicate that the nebular temperature decreases with z.

The standard thin-disk approximation for radiative flux, i.e., Eq. (60), must be modified if a significant portion of the nebula has a temperature $T_{56} \lesssim T \lesssim T_{67}$ so that Eq. (135) is relevant. In this case, the surface boundary condition in the nebula becomes important (Faulkner et al. 1983). This effect is best demonstrated by integrating the radiative flux in the vertical direction through a nebula with $T_e < T_i$. For simplicity, we ignore the density dependence in the opacity law such that $\kappa = \kappa_6 \approx \kappa_{10} T^{10}$ for $T < T_{67}$ and $\kappa = \kappa_{07} T^{-2}$ for $T > T_{67}$. From Eq. (59) and the atmospheric boundary conditions, we find

$$T_e^4 \approx \frac{T_e^4}{\tau_e} + \frac{T_c^4}{\tau_c} \tag{139}$$

for $T_e < T_{67} < T_c$, and

$$T_e^4 \approx \frac{T_e^4}{\tau_e} - \frac{T_c^4}{\tau_c} \tag{140}$$

for $T_e < T_c < T_{67}$ where $\tau_e = \kappa_e \Sigma_e / 2$, $\tau_c = \kappa_c \Sigma_c / 2$, $\kappa_e = \kappa(T_e)$, $\kappa_c = \kappa(T_c)$, Σ_e is the column density above the "photosphere" (where $T = T_e$) and $\Sigma_c = \Sigma$ (see Faulkner et al. 1983). Now, consider the dependence of T_e on T_c as the latter decreases. For large $T_c (> T_{67})$ and small $\tau_c (<< \tau_e)$ Eq. (139) implies the standard $T_e^4 \approx T_c^4 / \tau_c$ relation. As T_c becomes smaller than T_{67}, and τ_c much larger than τ_e, all that survives is $T_e^4 \approx T_e^4 / \tau_e$ or equivalently

$$\tau_e = \frac{1}{2} 10^{-36} \rho_e^{1/3} T_e^{10} \Sigma_e \approx 1. \tag{141}$$

If we assume that Σ_e is a significant fraction of Σ, $T_e \approx 8 \times 10^3 \Sigma^{-0.1}$ K for a wide range of density, i.e., 10^{-10}g cm$^{-1} \lesssim \rho \lesssim 10^{-6}$g cm^{-3}. From the condition for local energy equilibria (Eq. 61) we find

$$\nu\Sigma \approx 2 \times 10^{11}\Omega^{-2}\Sigma^{-2/5}\text{g s}^{-1}. \tag{142}$$

The above result should be contrasted with the $(\nu\Sigma)$ relationship in the hot zones, where, from Eqs. (61), (137) and (138), $\nu\Sigma \propto \Sigma^{3/2}$. Ignoring the density dependence in κ_6, we find that the critical values Σ_B and ν_B separating these two solutions are

$$\Sigma_B = 0.2\alpha_s^{-35/57}\Omega^{-40/57}\text{g cm}^{-2} \tag{143a}$$

and

$$\dot{M}_B = 3\pi\Sigma_B\nu_B = 3.6 \times 10^{12}\alpha_s^{14/57}\Omega^{-98/57}\text{g s}^{-1}. \tag{143b}$$

When the density dependence in κ_6 is included, we find (Faulkner et al. 1983) that

$$\Sigma_B = 0.2\alpha_s^{-26/45}\Omega^{-13/18}\text{g cm}^{-2} \tag{144a}$$

and

$$\dot{M}_B = 4.4 \times 10^{12}\alpha_s^{3/10}\Omega^{-7/4}\text{g s}^{-1}. \tag{144b}$$

Similarly, the results in Eq. (142) should be contrasted with the $(\nu\Sigma)$ relation for temperature $< T_{56}$. From Eqs. (68) and (136), we find

$$\Sigma\nu = 10^{11}\alpha_s^4\Omega^4\Sigma^9 \tag{145}$$

for $T_{45} \lesssim T_c \lesssim T_{56}$. In fact, for all temperatures $< T_{56}$, $\nu\Sigma$ monotonically increases with Σ. The critical values of Σ and $\dot{M}$ separating solutions (142) and (145) are

$$\Sigma_A = \alpha_s^{-20/47}\Omega^{-30/47}\text{g cm}^{-2} \tag{146a}$$

and

$$\dot{M}_A = 10^{11}\alpha_s^{8/47}\Omega^{-82/47}\text{g s}^{-1}. \tag{146b}$$

If the density dependence in κ_6 is included, we find

$$\Sigma_A = \alpha_s^{-19/45}\Omega^{-29/45}\text{g cm}^{-2} \tag{147a}$$

and

$$\dot{M}_A = 10^{11}\alpha_s^{1/5}\Omega^{-9/5}\text{g s}^{-1}. \tag{147b}$$

For $T_{34} \lesssim T_c \lesssim T_{45}$, we deduce from the standard thin-disk approximation, i.e., Eq. (68), that $\Sigma\nu \propto \alpha_s^{40/37}\Omega^{32/37}\Sigma^{45/37}$. In principle, the standard thin-disk approximation should also be modified in this case because the opacity in the cooler regions above the midplane may be considerably larger than that near the midplane. However, the opacity varies rapidly over a small temperature range such that even the modified treatment, in which the surface condition is emphasized, may be inadequate. In addition, the opacity reaches a local minimum in this temperature range and the nebula may become optically thin for certain values of $\dot{M}$ and nebular radius. It is not at all clear how the efficiency of local viscous heating may be affected in the optically thin region of the nebula. For example, turbulence may decay on a thermal time scale so that α_s may be reduced well below unity. On the other hand, the optical depth in certain absorption lines may be sufficiently large so that a significant amount of heat may be trapped throughout to maintain large-scale turbulence. Detailed analyses on radiation transfer and hydrodynamics relevant to this temperature range should be carried out in the future.

When the mass transfer rate is relatively large, the nebula is optically thick even when κ attains a minimum value at $T_c \approx T_{45} \approx 3000$ K. In this case, the above analytic approximation can be verified by detailed numerical integration of the vertical structure equations by neglecting radial and time derivatives. The boundary conditions in this case are identical to those outlined in Sec. III. We present the results, which are obtained for a purely radiative and a fully convective treatment, in the form of a $\nu\Sigma - \Sigma$ relationship (Fig. 15A and B). When appropriate comparisons are made, the numerical results of the convective computation (curve 1 in plot B) are in general agreement with those obtained by previous investigators (Smak 1982; Meyer and Meyer-Hofmeister 1981; Cannizzo et al. 1982). Once again, the radiative and convective curves in Fig. 15 are remarkably similar. In addition, the general functional form of the $\nu\Sigma - \Sigma$ relationship, obtained from the full integration, is in good agreement with the analytic approximation above.

The primary reason for the above discussion is to indicate how to evaluate the flux of energy $\mathcal{F}_z$ in the energy transport equation so that evolution of ν and Σ can be subsequently determined from the energy and mass transport Eqs. (6) and (16). In these evolution calculations to be presented below, analytic expressions for $\mathcal{F}_z$ in terms of T_c and Σ are very useful. From the above discussions, we find that $\mathcal{F}_z$ can be characterized by either $\mathcal{F}_z = \sigma T_{e1}^4 = 4\sigma T_c{}^4/3\tau_c$, or $\mathcal{F}_z = \sigma T_{e2}^4$ with $\tau(T_{e2}) \approx 10^{-36}(\Sigma/c_s(T_c))^{1/3}T_{e2}^{10}\,\Sigma = E$, depending on the physical conditions. The constant $E = (\rho_c/\rho_e)^{1/3}\Sigma/\Sigma_e$, where ρ_e and Σ_e are evaluated at the surface of the nebula (i.e., where $T = T_e$), is a factor of order unity and may be chosen to give a good fit to the vertical integration. A fairly simple algorithm may be constructed to satisfy the above constraints in the appropriate regions such that

$$\frac{1}{T_e^n} = \frac{1 - f_a}{(T_{e1}^m + T_{e2}^m)^{n/m}} + \frac{f_a}{T_e^n} \tag{148}$$

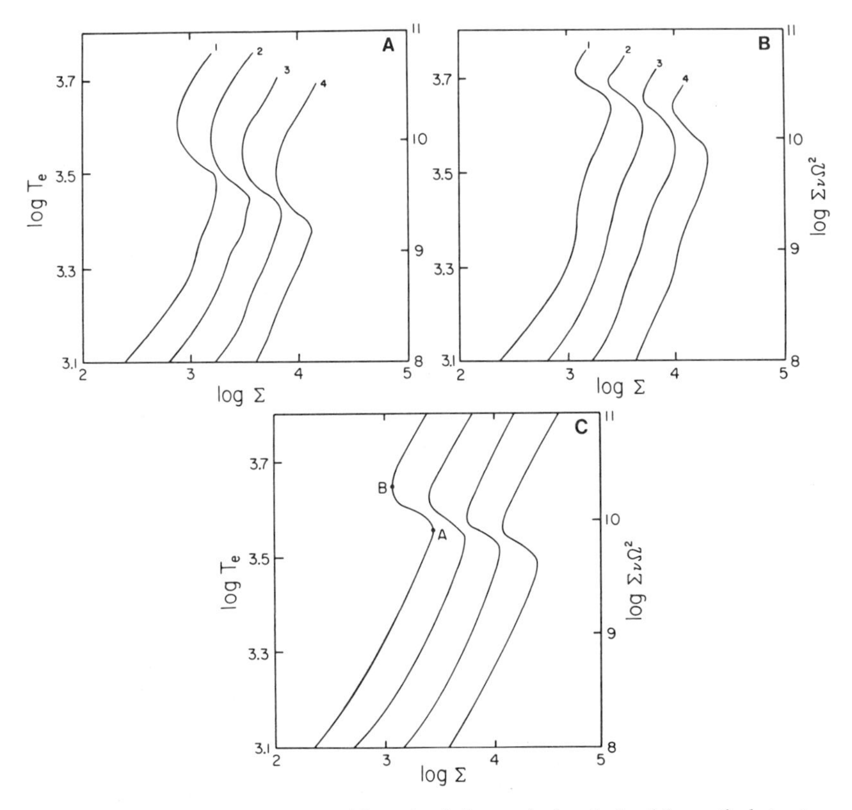

Fig. 15. The $\Sigma - \Sigma\nu$ relation deduced from detailed numerical analysis of the vertical structure of the nebula in the neighborhood of the protosun. The mass of the protosun is taken to be 1 $M_\odot$. The standard α model is used for the viscosity prescription with $\alpha_s = 0.1$. All symbols are in cgs units. Curves 1, 2, 3 and 4 represent results obtained for 5 $R_\odot$, 10 $R_\odot$, 20 $R_\odot$ and 40 $R_\odot$. In plot A, only radiative flux is used for $\mathcal{F}_z$ and in B, both convective flux and radiative flux are included. In plot C, the analytic approximation of these thin-disk local-thermal-equilibrium solutions is plotted. The symbols *A* and *B* represent the likely location for thermal transition.

where $f_a = 1/[1 + (T_c/4000 \text{ K})^8]$. A reasonable analytic fit can be obtained for $m = 4$, $n = 2$ and $E = 5.6$. The equilibrium solution derived from the analytic fit is plotted in Fig. 15C and it is in excellent agreement with the results of the full integration for these regions.

B. Thermal and Viscous Stability

The above results can be used to study the stability of nebular flow. The simplest stability analysis is local short wavelength analysis. Accretion-disk flow is locally unstable against viscous diffusion if the local equilibrium viscous stress $\nu\Sigma$ decreases with Σ (Lightman and Eardley 1974). Such an instability may be responsible for producing surface density variation in the B ring around Saturn (Ward 1981; Lin and Bodenheimer 1981). Disk flow is locally thermally unstable if the heating rate, due to viscous dissipation D, increases more rapidly with T_c than the local cooling rate $\mathcal{F}_z$ (Pringle et al. 1973). Thermal time scales are generally much shorter than viscous time scales in thin disks so that thermal instabilities, if present, dominate over viscous instabilities (Shakura and Sunyaev 1976).

The $(\nu\Sigma,\Sigma)$ relation presented in Fig. 15 is useful for a preliminary discussion of stability. The thermal equilibrium solution clearly divides the $(\nu\Sigma,\Sigma)$ plots into two domains. In domain H, the local heating rate, due to viscous dissipation, exceeds the local cooling rate. In this case, the disk will continue to heat up until a thermal equilibrium is achieved. In domain C, the local heating rate is insufficient to balance the local cooling rate. The disk will continue to cool towards a thermal equilibrium of lower temperature. Thus, if domain H is above the equilibrium curve, the solution is thermally unstable whereas if domain C is above the equilibrium, the solution is thermally stable. Between points A and B lies a range of $\dot{M}(\approx 3\pi\Sigma\nu)$ for which the equilibrium solution is thermally unstable.

C. Prospects for a Thermal Limit Cycle

Let us consider, with the aid of Fig. 16, a situation in which the disk is built up from low Σ. At any given region, a distinction should be drawn between $\dot{M}$, the rate at which the disk can viscously diffuse disk material away from that region, and $\dot{M}_{\rm in}$, the rate at which mass is transfered into that region. The magnitude of $\dot{M} = 2\pi\Sigma U_r r$ is determined by local conditions whereas $\dot{M}_{\rm in}$ is determined by conditions outside the region. The mass transfer into this region may be viscously processed through the disk at neighboring radii or be associated with infall. If $\dot{M}_{\rm in} > \dot{M}$, the local Σ will increase. The representative point will move along the lower branches of the thermal equilibrium curve. In this way, local thermal properties adjust to different mass input rates. If, however, $\dot{M} > \dot{M}_A$, there is no corresponding equilibrium Σ on the lower branch of the thermal equilibrium solution. In attempting to adjust $\dot{M}$ to $\dot{M}_{\rm in}$, the representative point encounters thermal instability at point A and moves thereafter in the general direction of the upper thermal equilibrium branch. This transition occurs on fairly rapid thermal time scales. Once on the upper branch, much larger rates of local mass diffusion occur. If $\dot{M}$ now exceeds $\dot{M}_{\rm in}$, reduction of local Σ ensues and the representative point moves towards lower values of Σ and thus $\nu\Sigma$. If $\dot{M}_{\rm in}$ lies between $\dot{M}_B$ and $\dot{M}_A$, the representative point, in seeking an equilibrium, encounters the same problem at B which it met at A. It is forced to make another transition, this time to the lower branch. Thus, if $\dot{M}_{\rm in}$ lies between $\dot{M}_A$ and $\dot{M}_B$, this part of the disk will spend its time in a fruitless search for an unattainable steady state. The above schematic local analysis may be applied to different radial zones. The radial dependence of $\dot{M}_A$ and $\dot{M}_B$ is given in Eqs. (144) and (147). These results indicate that $\Sigma_A > \Sigma_B$ and $\dot{M}_A < \dot{M}_B$ for all r and therefore a radially dependent

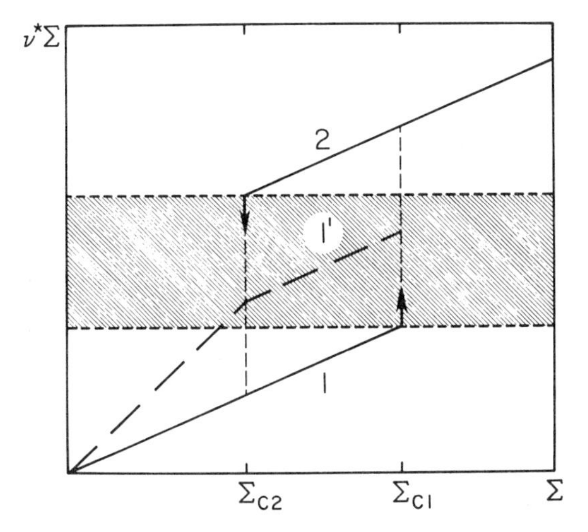

Fig. 16. A schematic $\nu^* \Sigma - \Sigma$ relationship for the analog problem. Curves 1 and 2 correspond to the uniform prescription whereas 1′ corresponds to the nonuniform prescription. The shaded area is the excluded range for steady state solution. Σ_{c1} and Σ_{c2} correspond to the critical Σ for the two branches of steady state solutions. The general relationship need not be linear, but could still possess analogous hysteresis features.

range of unstable $\dot{M}$ exists throughout the disk. Note that at relatively large radii, say $\gtrsim$50 $R_\odot$, $\dot{M}_A \gtrsim 10^{20}$g s^{-1}, is larger than the mass transfer rates found in typical solar nebula model. Thus, the solutions at large radii are primarily on the lower thermal equilibrium branch.

D. Global Evolution

The above heuristic arguments suggest the possibility of a thermal relaxation limit cycle. However, this local stability analysis has two important shortcomings: (1) without some form of linkage (hydrodynamic and/or thermodynamic) between adjacent radial zones, there is no apparent reason for the outburst to operate collectively, and (2) both ionization and radial thermal contact may provide stabilizing effects. These shortcomings can only be overcome in a self-consistent global analysis.

The problem is approached in two stages. In the first, a much simplified analog treatment can be used to mimic the hydrodynamic features of a relaxation cycle. The character of the flow can be explored with an *ad hoc* viscosity prescription in which there are two uniform critical values of Σ denoted by Σ_{c1} and Σ_{c2} which satisfy $\Sigma_{c1} > \Sigma_{c2}$. Material remains in a low-viscosity state, state 1, until $\Sigma > \Sigma_{c1}$, makes a transition to state 2 and remains in that state until $\Sigma < \Sigma_{c2}$. Mathematically, we denote $3\nu\Sigma = \Sigma$ for $\Sigma < \Sigma_{c1} = 0.1$ and $3\nu\Sigma = \Sigma + C$ for $\Sigma > \Sigma_{c2} = 0.07$ where C is a constant. A discontinuous transition of $C/3$ occurs in the prescribed form of $\nu\Sigma$ when Σ passes through Σ_{c1} and Σ_{c2}. In order to conserve both mass and angular momentum, the actual transition may be approximated with $d\nu/dt = -(\nu - \nu_o)/\tau_\nu$, where ν_o is the initial value of ν and τ_ν is some characteristic transition time scale. With this prescription, the evolution of the nebula can be deduced from Eq. (79).

With the above prescription for equilibrium viscosity, for certain ranges of mass input rate, no steady state is possible; instead the accretion-disk flow establishes a limit cycle (see Fig. 17). It is of interest to note that with the above purely local viscosity prescription, the transport properties throughout the nebula are decoupled in the radial direction. Nevertheless, phase transitions can propagate throughout the disk collectively. This coherent phenomenon can be attributed to an avalanche effect. As soon as a small region has acquired a larger viscosity after having an upward transition, it behaves somewhat like an isolated viscous annulus. The inner part diffuses inwards, the outer part outwards. In so doing, adjacent low-viscosity zones have their surface density increased beyond Σ_{c1}, and themselves acquire a large viscosity. Consequently, two transition waves propagate away from the initiating transition point to enlarge the high-viscosity zone until the entire disk is in a high state.

In the second, more realistic approach, appropriate thermodynamic effects are included in a modified thin-disk α-model approximation. In this case, the mass diffusion Eq. (8) and the energy transport Eq. (16) are solved

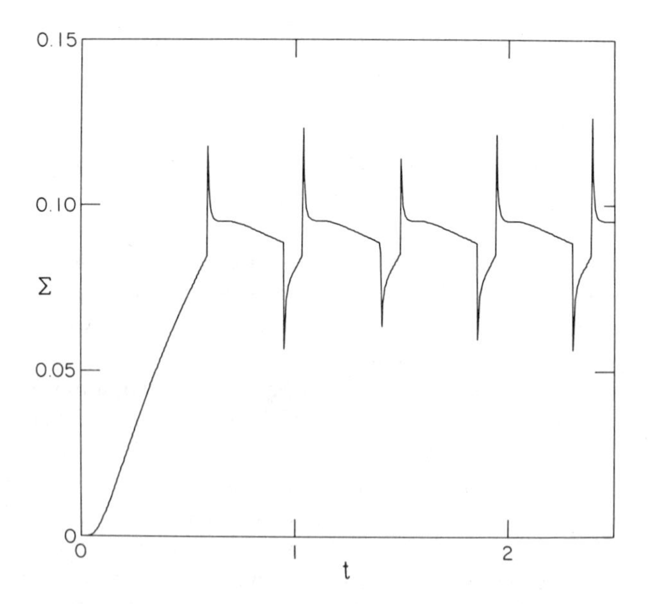

Fig. 17. Evolution of surface density at the inner boundary. Both Σ and t are in terms of non-dimensional units. The mass-input rate $S_0 = 1$ and $C = 0.13$. A uniform ν^* prescription is used. A limit cycle is clearly established after $t = 0.75$.

simultaneously with an analytic approximate (Eq. 148) for the local flux. In such an approach, the radial contact is established by advective transport and radiative flux in the radial direction (Papaloizou et al. 1983). When appropriate boundary conditions are applied to the solar nebula, we find that the disk is thermally unstable under certain circumstances. Nonlocal effects and viscosity limitations are important in promoting the collective nature of the instability and its growth. The instability may be touched off at one radial position, but have its most noticeable effects at another.

The results of these global analyses indicate that the stability criterion deduced from the local analyses is reasonably accurate. However, when thermal instability is first triggered somewhere in the disk, the thermodynamic properties, including the state of ionization, undergo a rapid phase transition at that location. This local phase transition induces a sharp temperature gradient such that $r/H \gtrsim |\partial \ln T/\partial \ln r| >> 1$. The upper limit in the magnitude of the temperature gradient in the radial direction is constrained by the Rayleigh dynamical stability criterion for axisymmetric, differentially rotating flows. In addition, the transition front is partially ionized such that $C_v >> \mathcal{R}$. Consequently, when one radial zone has undergone a thermal phase transition, its adjacent regions are strongly affected by the advective and radiative transfer processes in the radial direction. These global influences are strongly emphasized in the energy transport equation where the magnitude of terms such as $C_v \dot{U}_r \partial T_c/\partial r$ becomes comparable to or larger than both the local viscous dissipation $(9\nu\Omega^2/4)$ and the local radiative term (F_z/Σ) near the transition front. Through these advective and radiative processes, adjacent regions are physically connected, in the radial direction, such that thermal instability may coherently propagate away from its initial radial position.

In order to demonstrate this behavior, we calculated a standard case, in which $\dot{M}_* = 10^{-5} M_\odot \text{ yr}^{-1}$, $M_1 = 1 \text{ M}_\odot$, $R_d = 100 \text{ R}_\odot$ so that the outer region of the nebula is always stable. However, the inner boundary of the nebula is set to be 5 $\text{R}_\odot$ so that the inner region of the nebula is thermally unstable according to the local analysis. The onset of the thermal phase transition in the inner region induces an outwardly propagating upward transition wave. This property is well illustrated in Fig. 18a in which we plot the Σ, T_c, and $\dot{M}$ distribution at ten successive instances of time separated by unequal time intervals of the order one month. Near the transition front where the temperature gradient is very large, the radial velocity of the disk material U_r, $[= -3(\nu/r)(\partial \ln\Sigma/\partial \ln r + \partial \ln T/\partial \ln r - \partial \ln\mu/\partial \ln r + 2)]$, becomes comparable to $\approx \nu/H \approx \alpha_s c_s$. This magnitude of U_r is considerably larger than those derived under steady state conditions where the temperature gradient is relatively small such that $U_r \approx \alpha_s c_s^2/(\Omega r) << c_s$. Furthermore, the sign of U_r implies that the nebula material is spreading outward from the hot fully ionized region. As this fast-moving material is intercepted by the disk material in the cool, mostly neutral regions, the surface density near the transition front increases beyond Σ_A so that an upward thermal transition is triggered in the

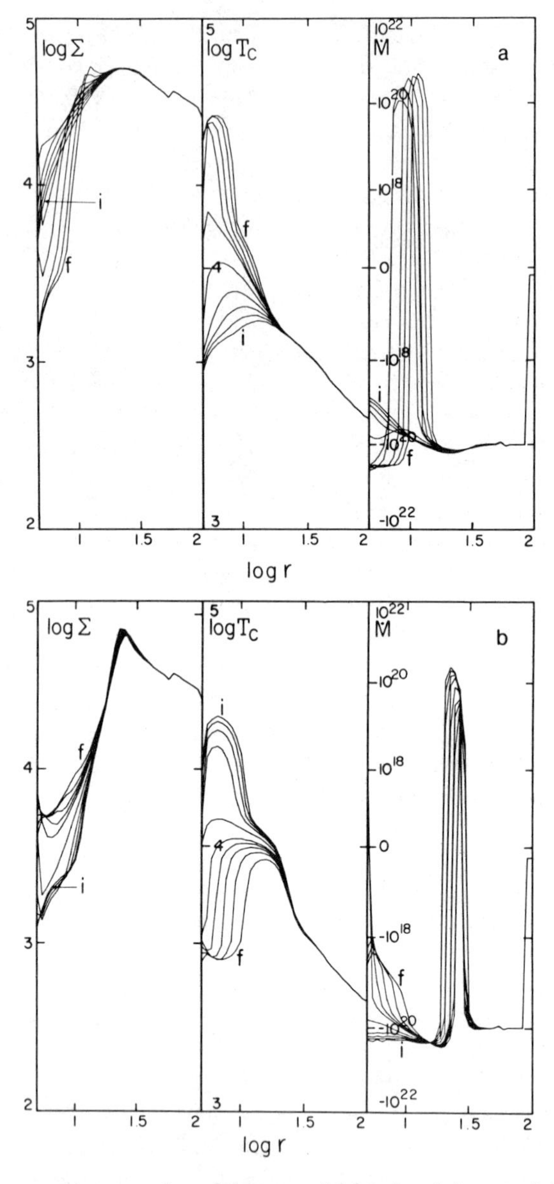

Fig. 18. The onset and propagation of the upward (plot a) and downward (plot b) transition waves. The mass input rate $M_* = 10^{-5}$ $M_\odot$ yr^{-1}, and $\alpha_h = 10^{-1}$ and $\alpha_q = 10^{-2}$. The surface densities, temperature of the midplane *T*, and mass diffusion $\dot{M}$ (= $\Sigma U_r r$ which differs from that defined in the text by 2π) distribution are plotted at ten different instants of time in cgs units. The radius *r* is measured in solar radius. The interval between the time represented for each curve is not equal. In plot a, the ten curves represent 0.65, 1.41, 2.18, 2.95, 3.71, 4.21, 4.29, 4.43, 4.61 and 4.78 × 10^7s after the onset of instability. In plot b the time interval is ~2.2 × 10^6s.

mostly neutral zone immediately adjacent to the transition front. This sequence of events induces the transition wave to propagate at a speed $\approx U_r \approx \alpha_s c_s$. According to local analyses, the dominant cause for the thermal transition is the steep T-dependence in the H^- opacity which is dominant only over a narrow range of temperature. Thus, the temperature and the sound speed, c_s ($\approx 10^6$ cm s^{-1}), near the transition front are practically independent of the radius. This behavior promotes the upward transition wave to propagate at an approximately uniform speed. The characteristic propagation time scale, $\tau_r \approx r/(\alpha_s c_{s,f})$, is considerably less than the viscous diffusion time scale at a similar temperature range.

As the upward transition wave propagates to the outer regions of the nebula, Σ_A becomes larger. At a sufficiently large radius r_{cr}, the runaway ceases when the wave cannot increase the column density beyond $\Sigma_A(r_{cr})$. Thus, the mass transfer rate into the hot fully-ionized interior is limited to be less than $\dot{M}_{cr} = \dot{M}_A(r_{cr})$. Consequently, the transfer rate in the disk, $\dot{M} > \dot{M}_A > \dot{M}_{cr}$ interior to r_{cr} and the nebula is gradually depleted and Σ decreases on a characteristic viscous time scale relevant for the high state at r_{cr}, i.e., $\tau_{d,h} \approx (GMr_{cr})^{1/2}/(\alpha_s \mathcal{R}T)_h$. The subscript h is adopted to denote the value of α_s and T, during the fully ionized high state. These quantities may have different magnitudes during the mostly neutral low state when they are denoted by subscript q. When somewhere in the outer regions of the disk, Σ decreases below the locally appropriate values of Σ_B, a downward transition is triggered. In this transition, the disk material evolves from a fully or highly ionized state to a mostly neutral state. The optical depth of the disk becomes comparable to or less than unity. The propagation of the downward transition wave is illustrated in Fig. 18b where we plot the Σ, T_c, and $\dot{M}$ distribution separated by ten approximately equal time intervals with $\Delta t \approx 2.2 \times 10^6$s. During the decline stage, the global depletion of the disk material proceeds in an approximately quasi-steady manner. This behavior implies that the outer regions cool below the ionization temperatures first so that there is a general tendency for the onset of the downward transition to first occur in the outer regions of the disk. As the hot, fully ionized regions continue to be depleted, the downward transition front propagates inwards.

The results in Eqs. (144), (146), and (147) indicate that the critical values of Σ are weakly dependent on the dimensionless parameter α_s. Although the stability conditions are not strongly affected by the value of α, the propagation of the transition waves and the duration of the mostly neutral low state are essentially determined by the values of α_h and α_q respectively. In Sec. III, we indicate that the α prescription is *ad hoc* and the value of α_h and α_q is rather uncertain. It could be argued that the value of α_h is probably close to unity because during the high state, the nebula is fully ionized and optically thick so that (1) random magnetic field may be amplified (Shakura and Sunyaev 1973), and (2) the loss of kinetic energy associated with the turbulent motion may be compensated by the gain of gravitational energy-associated accretion on an

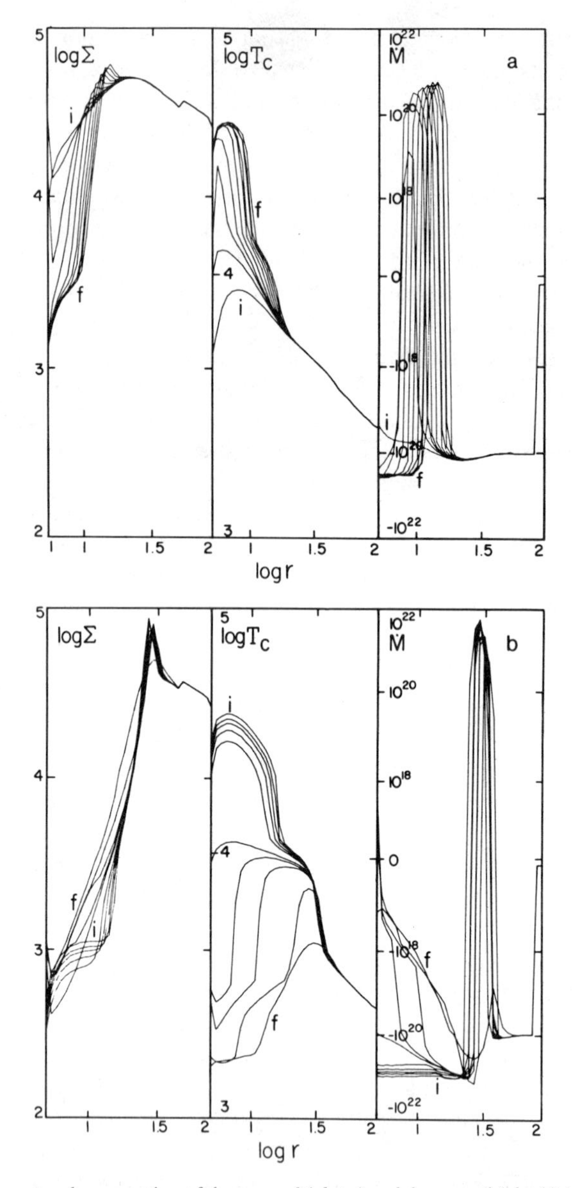

Fig. 19. The onset and propagation of the upward (plot a) and downward (plot b) transition wave for the $\alpha_h = 1$ and $\alpha_q = 10^{-2}$ case. The symbols used and the value of $\dot{M}_*$ are identical to those in Fig. 18. The ten curves in plot a represent 1.265, 2.275, 2.301, 2.325, 2.347, 2.369, 2.391, 2.413, 2.436 and 2.462 $\times$ 10^8s after we started from some initial surface-density distribution. In plot b, the ten curves represent 0.6, 1.3, 1.9, 2.7, 3.5, 4.6, 4.9, 5.3, 7.2 and 35 $\times$ 10^6s after maximum light.

appropriate time scale (Shakura et al. 1978). When the nebula is in the cool and mostly neutral state, if convection is the only process which can generate turbulence, the results in Sec. III indicate that $\alpha_q \approx 10^{-2}$. For present illustration purposes, we adopt a set of values $\alpha_h = 0.1$ and $\alpha_q = 0.01$ to produce the results in Fig. 18. We also run a second case in which $\alpha_h = 1$ and $\alpha_q = 0.01$ (Fig. 19). Note that although the $\dot{M}_*$ and other nebula parameters are identical in the two cases, the latter case produces a transition wave with considerably larger amplitude and propagation speed than the first case. Obviously, a more thorough computational study is needed in the future.

As the nebula evolves through these different stages, $\dot{M}$ is neither steady nor uniform. However, the variation in Σ, T_c and $\dot{M}$ in the disk is periodic and the evolution of the disk takes on the form of a thermal relaxation cycle. In order to make a comparison with observational data, we plot the most directly observable quantity, the visual magnitude M_v, in Fig. 20. These results indicate that the basic oscillation period for the limit cycle in both cases is $\approx$ 10 yr although the duration of the outbursts ranges from several years in the first case and several months in the second case. For the $\alpha_h = 0.1$ case, the characteristic rise time scale of the M_v lightcurve, near 1 mag below outburst maximum, ranges between $\approx 1 - 10 \times 10^6$s/mag and the decline time scale of the M_v lightcurve near the same magnitude somewhat longer. About 30% of the cycle is spent in a state where M_v is within about 1 mag from its maximum value. In the $\alpha_h = 1$ case, less than one tenth of the cycle is spent in the high state.

The above analysis indicates that the total energy output from the nebula may be regulated by the thermal instability. However, over half of the energy released due to the accretion process must be generated in the boundary regions separating the nebula and the central protostar. As a star evolves through

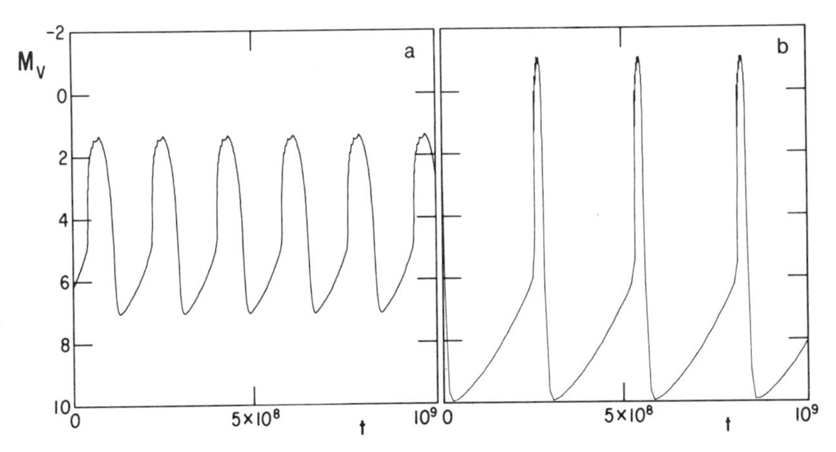

Fig. 20. The limit cycle in the M_v lightcurves. The same values of $\dot{M}_*$, α_h and α_q are used in plot a as in Fig. 18, whereas in plot b, the same values of $\dot{M}_*$, α_h and α_q are used as in Fig. 19.

its T Tauri phase, part of the energy released from its photosphere is generated from gravitational contraction of the stellar envelope and core. Typically an isolated 1 $M_\odot$ T Tauri star has a luminosity about several times $L_\odot$. But during the thermal relaxation cycle, $\dot{M}$ in the inner regions of the nebula can temporarily increase beyond $10^{-5} M_\odot$ yr^{-1} which implies an energy generation rate comparable to or larger than $\approx 10^{35}$erg^{-1}. If most of this energy is generated in the surface boundary regions between the protostar and the nebula, the protostellar envelope must expand and adjust to accommodate this additional source of energy. Rigorous analysis in this area has not been carried out.

E. Observable Properties of Nebular Thermal Instability and the FU Ori Phenomenon

With the above quantitative model for the inner region of the nebula, we can now calculate the observable properties associated with these thermal instabilities. First, the bolometric luminosity of the entire nebula is given by

$$L_{\rm Bol} = \int_{r_{\rm in}}^{r_{\rm out}} 4\pi F_s(r) r dr \tag{149}$$

which may be "integrated" numerically by summing the surface flux over various radial zones. Similarly, the visual magnitude may be obtained in terms of the standard detector response functions, V_λ (Allen 1981), such that

$$M_{\rm v} = -2.5\log\left\{\frac{\rm cm^2 s}{(10{\rm pc})^2{\rm erg}}\int_0^\infty\int_{r_{\rm in}}^{r_{\rm out}} V_\lambda F_\lambda(r)[1 - \exp(-\tau_\lambda(r))] r dr d\lambda\right\} - 14.08 \tag{150}$$

where

$$F_\lambda(r) = \frac{4\pi hc^2\lambda^{-5}}{[\exp(hc/k\lambda T_e(r)) - 1]} \tag{151}$$

is the local Planck function at radius r and τ_λ is the optical depth of the nebula in the vertical direction. Although the U and B magnitude can also be constructed in the same manner, these frequency ranges may be contaminated by strong associated emission lines.

The continuum spectrum is given by

$$S_\lambda = \int_{r_{\rm in}}^{r_{\rm out}} F_\lambda(r)[1 - \exp(-\tau_\lambda(r))] r dr. \tag{152}$$

If the continuum spectrum is dominated by radiation from those optically thick regions of the nebula which are in a steady state, such that $T_e \propto r^{-3/4}$, we find from Eqs. (151) and (152) that $S_\lambda \propto \lambda^{-7/3}$ or equivalently, $S_\nu \propto \nu^{1/3}$. However, the above analysis indicates that the inner regions of the nebula are never in a steady state. As most of the optical photons emerge from the inner regions of the nebula, the continuum spectrum would not generally resemble that deduced for a steady state disk model. Instead, we must numerically integrate the spectrum from a global disk model.

In the above discussion, we have indicated that some inner regions of the nebula can become marginally optically thin to continuum radiation. In this case, line radiation may become prominent. Provided that the nebula is optically thick to these line radiations so that the condition for local thermodynamic equilibrium is satisfied, the integrated flux of a line is given by

$$S_l \propto \int_{r_{\rm in}}^{r_{\rm out}} F_\lambda \exp(-\tau_c) \Delta\nu_D \int_{-\infty}^{\infty} [1 - \exp(-\tau_{\nu_o} \exp(-x^2))] {\rm d}x r {\rm d}r \quad (153)$$

where τ_c and τ_{ν_o} are the continuum and line center optical depths, and $\Delta\nu_D$ is the assumed Doppler width of the line (Williams 1980). Using this method, we can evaluate the spectral characteristics during quiescent or outburst phases of nebular evolution.

Another interesting spectral characteristic is continuum and line polarization. If a significant fraction of the radiation released from the surface of the central star or in the central regions of the nebula can illuminate the surface of the outer regions of the nebula, grain scattering would cause the reflected photons to be polarized. The most favorable condition for such a polarization process to operate is when the inner region of the nebula is in a quiescent state such that it is relatively optically and geometrically thin. We are currently constructing these radiation transfer models for the inner regions of the nebula in order to provide a more quantitative observational prediction. These results may be applied to T Tauri stars still containing a remnant disk and to dwarf nova binaries which undergo mass transfer between them.

FU Ori stars are a class of T Tauri stars that are undergoing eruptive outbursts (Herbig 1977). Some of the observational characteristics can be fitted with the above model. However, the results presented here are still too preliminary to provide overwhelming evidence to support the scenario that the FU Ori phenomenon is associated with thermal instabilities in accretion disks. For example, the period of the cycle in both cases in Fig. 20 is considerably shorter than those thought to occur in FU Ori stars. Considerable theoretical investigation is still needed to determine under what conditions these long-period oscillations can be achieved. In any case, future theoretical investigation on the evolution of the inner regions of the nebula may provide observable clues to identify the best observational candidates from which the riddle of solar system formation may be unveiled.

VIII. SUMMARY

In the past few years, theoretical investigations on the origin and evolution of the solar nebula have progressed rapidly. Through these investigations, a new grand unified scenario for the formation of the solar nebula has emerged. In this scenario, the nebular phase of the solar system is turbulent, of relatively low mass, and short-lived. Convectively driven turbulence not only provides an energy transport process in the vertical direction but also induces viscous energy dissipation and mass transfer in the radial direction. Consequently, the presence of turbulence prevents a buildup of the gaseous mass of the nebula. Various dynamical constraints suggest that prototerrestrial and protogiant planetary formation proceeds through a series of gradual accumulation processes. During the final epoch of protoplanetary formation, the mass of the nebula was around 0.1 $M_{\odot}$. Most of the protoplanets probably acquired the bulk of their mass within about 10^6 yr after the formation of the protosun.

Acknowledgments. We thank D. Black, P. Bodenheimer, S. J. Aarseth, and S. Ruden for useful conversations and helpful comments. This work was supported in part by a grant from the National Science Foundation. We are also grateful to the staff, especially A. Karp, at the IBM Palo Alto Scientific Center for generously providing assistance and valuable computing time on the IBM 3081 computer to enable us to carry out some parts of the numerical computation presented in this chapter.

FORMATION AND EVOLUTION OF THE PRIMITIVE SOLAR NEBULA

A. G. W. CAMERON
Harvard-Smithsonian Center for Astrophysics

A disturbance mechanism is suggested to initiate the development of cores in dense molecular clouds and to bring into them the shorter-lived extinct radioactivities that were present in the early solar system. This mechanism involves a red-giant star of about 1 $M_{\odot}$ in its asymptotic giant branch which sheds its envelope and forms a planetary nebula. The collapse of a molecular cloud core to form the primitive solar nebula is discussed. Steady mass flow models of the primitive solar nebula are constructed having a range of masses and dissipation rates. With these as a guide, a history of the solar nebula is proposed into which a variety of phenomena are fitted. This history has three main stages, consisting of a buildup of mass in the nebula, a main dissipation phase in which the Sun is formed, and a thermalization stage in which dissipation effectively ends.

I. RISE AND FALL OF THE SUPERNOVA TRIGGER

A few years ago Truran and I proposed the formation of the primitive solar nebula as a consequence of the explosion of a supernova (of Type II) near an interstellar cloud, which would thereby be compressed to the threshold of gravitational collapse (Cameron and Truran 1976). This theory enjoyed a brief period of popularity but now seems unlikely to provide a model for any significant amount of star formation.

This nearby supernova seemed needed to inject live ^{26}Al into the material that would form the solar system. Much primitive meteoritic material, rich in aluminum and poor in magnesium, shows excesses of ^{26}Mg which are interpreted as the decay product of ^{26}Al (see the review by Wasserburg in this book). However, D. D. Clayton (1984) has recently interpreted HEAO-3 measurements of the gamma-ray lines of ^{26}Al from the interstellar medium (Mahoney et al. 1982) to show that there is ten times as much ^{26}Al in the interstellar medium as can be produced by supernovae. Clayton estimated that the observed amount of ^{26}Al could easily be produced by nova production throughout the galactic disk.

The nearby supernova also seemed needed to produce those isotopic anomalies in a few rare samples of solar system material characterized by Wasserburg et al. as FUN (fractionated and with unknown nuclear abundance contributions) anomalies (see, e.g., Wasserburg et al. 1980). However, Consolmagno and Cameron (1980) showed that, within the errors of measurement or estimation of the relevant neutron capture cross sections, FUN anomalies can be produced by local renormalization of normal s-, r-, and p-process abundances in the sample material. Thus, there no longer is a need for the creation of peculiar abundance anomalies in an individual nearby nucleosynthesis event in order to produce these FUN anomalies.

The proximity of the supernova to the interstellar cloud seemed necessary so that the cloud could be enveloped in a hot, higher-pressure, lower-density gas; the cloud would then be compressed to the point of gravitational collapse and Rayleigh-Taylor instabilities would promote fragmentation. Although this physical mechanism still seems possible, there is no convincing observational evidence for this kind of star formation. Probably the cloud would recoil from the site of the coming supernova explosion due to the strong ultraviolet flux from the presupernova star, which would not irradiate the cloud isotropically.

As made abundantly clear in this book, there is now a different observational basis for understanding star formation than was the case when the concept of a supernova trigger was introduced. While this theory has thus outlived its usefulness, the problem remains that if the formation of the solar system is to be investigated as a postulated typical star-forming event, then it is necessary to find one or more plausible sources of short-lived extinct radioactivities which were apparently live when the solar system was formed.

WHY?

II. CHARACTERISTICS OF STAR-FORMING REGIONS

Dense molecular clouds have many "cores" distributed throughout their interiors. The following characteristics, taken from the study by Myers and Benson (1983), appear to be most pertinent to the phenomena that we wish to discuss.

1. Clustering. These cores and young emission line stars (T Tauri stars) tend to be clumped together at several spots within a molecular cloud.

2. Dimensions. The diameter is typically about 0.1 pc, containing about 1 to 2 $M_{\odot}$ (the average may be a little less than this estimate if there is an observational bias toward discovering larger cores).
3. Thermodynamic quantities. The densities within the cores are about 10^4 to 3×10^4 cm^{-3} (above a background density of about 10^3 cm^{-3}), and the temperature of 10 K is rather accurately established.
4. Random velocities. Very few velocity profiles in the cores are dominated by dynamic collapse or rotation. Most profiles show a random velocity distribution characterized by a Mach number of about 0.7.
5. Ages. The T Tauri stars associated with the cores have ages varying from 10^5 to several times 10^6 yr (Mercer-Smith et al. 1984).
6. Collapse thresholds. All cores seem to be close to the threshold for gravitational collapse, with a significant fraction of the cores having crossed this threshold. It appears that those cores that have not crossed the collapse threshold are likely to do so on a time scale of the order of 10^5 yr.

These are the environmental conditions which appear observationally relevant to the bulk of new star formation in the Galaxy at this time. It is in this set of conditions that one must formulate a theory of star formation, and it is therefore also attractive to postulate that the formation of the solar system should be connected to this environment.

III. FORMATION OF CLOUD CORES

A theory of star formation must start with a theory of formation of the cores in dense molecular clouds. If the theory is also to account for the formation of the solar system, then it must also provide newly formed cores with a supply of fresh radioactivities like ^{26}Al.

Shu (see Shu and Terebey 1984), starting from the observation that motions in cloud cores are subsonic, has discussed the formation of these cores with the assumption that the environment is quiet. In his view, a cloud core is a part of the cloud in which there is a slow growth of a density perturbation. During this growth the magnetic field maintains control of the density perturbation, and so Shu assumes that the core rotates at the cloud normal rotation rate estimated to be about 10^{-14} radians s^{-1}. The field gradually escapes from the core by ambipolar diffusion (Shu 1983). A one solar mass core collapsing from these initial conditions puts about half the mass directly into a star and the rest into a relatively small disk surrounding it. I, however, do not regard this as a promising model for the formation of the primitive solar nebula.

On the other hand, I am impressed by the fact that the internal motions in the cloud cores are only just subsonic. A region with mean velocities of Mach 0.7 is a rather violent place, and the rate of dissipation is very high (and varying as the fifth power of the Mach number). This is suggestive of a violent dynamical compression of gas into the core followed by a rapid dissipation of the resulting shocks. What is required appears to be a disturbance in the cloud which accelerates the gas leading to the core formation. Such a disturbance should result in the formation of a bubble in the cloud due to local deposition of heat and momentum, causing local expansion of the gas. This will spread out the magnetic lines of force passing near the center of the bubble, crowding them near the periphery, and thus applying a pinch to the material tied to the flux lines so that it is accelerated away from the pinch in each of the two available directions. Such a disturbance would then normally result in the formation of two cloud cores. I have considered several candidates for the primary disturbance which can initiate the formation of cloud cores, and a secondary disturbance which can propagate the formation of cloud cores (Cameron 1984). The postulated primary disturbance is the late evolutionary stage of a star of about 1 $M_{\odot}$ when it passes through its asymptotic giant branch (AGB) stage and shortly thereafter sheds its envelope as a planetary nebula. The postulated secondary disturbance is from the T Tauri stellar winds or bipolar outflows of the stars newly formed from the collapse of cloud cores.

IV. EXTINCT RADIOACTIVITIES IN THE EARLY SOLAR SYSTEM

We are particularly interested in those extinct radioactivities which have half-lives much shorter than the expected lifetimes of the dense molecular clouds. There are three isotopes of particular interest: (1) ^{26}Al, half-life 0.74 $\times$ 10^6 yr, giving ^{26}Mg enrichments in aluminum-rich and magnesium-poor minerals; (2) ^{107}Pd, which decays with a half-life of 6.5 $\times$ 10^6 yr into ^{107}Ag and yields ^{107}Ag excesses in silver-poor iron meteorites; and (3) ^{129}I, which decays with a half-life of 16 $\times$ 10^6 yr into ^{129}Xe which is widely found as a rare gas component in meteorites.

^{26}Al is produced by hydrogen burning at the somewhat elevated temperatures associated with hydrogen shell-sources in red-giant stars and by hydrogen thermonuclear runaways occurring in classical nova eruptions. The reaction which produces ^{26}Al is ^{25}Mg(p,γ)^{26}Al. Hillebrandt and Thielemann (1982) showed that high ^{26}Al/^{27}Al ratios should occur in nova ejecta. Although an equally detailed investigation of ^{26}Al production in red-giant envelopes has not yet been carried out, I estimate on the basis of the recently revised rate for the ^{25}Mg(p,γ)^{26}Al reaction (Champagne et al. 1983*c*) that a major portion of the ^{25}Mg in the envelope will be transformed to ^{26}Al.

^{107}Ag and ^{129}Xe are not normally considered to be products of s-process nucleosynthesis. ^{107}Pd does lie on the s-process capture path; it is unstable against beta decay; but the r-process contribution dominates. ^{129}I is usually considered to be an r-process product, but it can also be formed in an s-process if the neutron flux becomes very large in a pulse and heavy nuclei capture several neutrons during the pulse. The pulsed s-process may take place during the shell-flashing stages of a red-giant star. It is likely that the neutron source responsible for brief r-process-like flashes of element production in these stars is ^{13}C, produced when hydrogen is mixed into the helium zone, following a helium shell flash and the development of an extensive convection layer, in an asymptotic giant branch (AGB) star of roughly 1 $M_{\odot}$ (Iben and Renzini 1984).

The time from the cessation of active nucleosynthesis in the interstellar gas to the formation of the solar system can be estimated by taking the ratio of the amount of the radioactivity present in the early solar system to that estimated to be present in the interstellar gas. In order to obtain consistency between the abundances of ^{244}Pu and ^{129}I in the early solar system, this isolation time must be about 2×10^8 yr. This represents some 12 half-lives of ^{129}I, during which its abundance falls by a factor of 4000. ^{244}Pu has a half-life of 82×10^6 yr and so is quite likely to be a product of ordinary galactic nucleosynthesis. The abundance of ^{244}Pu drops by a factor of five in 2×10^8 yr. However, an isolation time of 2×10^8 yr for ^{107}Pd would decrease its abundance by a factor of 10^9, much too large to be consistent with its observed level in the early solar system. Kelly and Wasserburg (1978) originally reached this conclusion by similar arguments. Hence this radioactivity, and perhaps much of the ^{129}I as well, have been derived from a source within the molecular cloud. In that case there cannot be a valid estimate of an isolation time.

V. DISTURBANCES IN THE CLOUD INTERIOR

The most effective type of disturbance in a cloud interior that can initiate large-scale mass flows appears to be a stellar source which emits an intense stellar wind. Such stars can eject a few percent of a solar mass of material during the time that they move 0.1 pc through the cloud (the dimension of a cloud core); if the emission velocity is several tens of km s^{-1}, then the emitted energy will be roughly 10^{45} ergs during this time. Newly formed stars within the cloud should be especially effective, because they would move rather slowly relative to the gas, and they eject quite a lot of gas in a short interval after formation, both as stellar wind and as bipolar outflows. Each such disturbance may blow a bubble in the cloud interior and thus initiate mass flow in both directions along the local field lines. Hence this type of star formation in dense molecular clouds may be a multiplicative process, increasing the local rate of star formation progressively until a significant depletion of the gas has occurred.

Stars that may be effective both in forming interesting nucleosynthesis products and in causing significant disturbances in the interiors of dense clouds are red giants, having large stellar winds. In particular, AGB stars, which undergo frequent helium shell flashes followed quickly by the loss of the envelope in a time ranging from $< 2 \times 10^6$ yr for stars 1 $M_{\odot}$ to 10^5 yr for stars of 8 $M_\odot$ (Iben 1984), are likely to be important initiators of star formation. Neutron production takes place during the shell flashes. Especially important is the shedding of a planetary nebula shortly after the AGB phase, when at least several tenths of a solar mass of material is quickly thrown off with a velocity of at least 30 km s^{-1}. A disturbance of that nature should be very effective in initiating core formation.

VI. NUCLEAR PROCESSES IN AN ASYMPTOTIC GIANT BRANCH STAR

Recent studies of stellar evolution indicate that red-giant stars of 1 to 8 $M_\odot$ undergo dredge-up during the AGB phase in which the products of helium reactions in the interior are brought to the surface (Iben 1984). These stars may be divided into two classes: (1) those stars with 2 to 8 $M_\odot$ appear primarily responsible for producing the s-process abundances, made by repeated neutron production by the $^{22}Ne(\alpha,n)^{25}Mg$ reaction during helium shell flashes. The ^{129}I extinct radioactivity is not made in this s-process, and so I do not consider this range of stars to be candidate precursors of the solar system, although they should contribute to the disturbances that can produce star formation; (2) those stars with masses near 1 $M_\odot$ appear to produce quite high fluxes of neutrons during shell flashes through the $^{13}C(\alpha,n)^{16}O$ reaction. In the following discussion we assume that one of these stars played a key role in producing the extinct radioactivities of the early solar system.

The rate at which AGB activity occurs in the interiors of dense molecular clouds can be estimated in the following way. Let us assume that differences in the scale heights of these stars and of gas above the galactic plane can be ignored. About 4×10^9 $M_\odot$ of gas exist in the Galaxy (Salpeter 1977). Approximately one star of 1 to 2 $M_\odot$ enters the AGB phase per year. About 10^6 $M_\odot$ of material form a dense cloud complex; an average mass element in the Galaxy should therefore experience local AGB contamination once every 4000 yr. In a dense cloud complex the density is three orders of magnitude higher than the mean, and hence a star of roughly 1 $M_\odot$ will go through the AGB in the interior of the complex about once in several million years.

According to Myers and Benson (1983) roughly 20 stars form in the Taurus-Auriga complex in an interval of 10^5 yr, roughly the same as the number of cores that they estimate will be formed in that time (about two-thirds of which have so far not been detected). This star formation rate is nearly 1000 times the rate of AGB disturbances.

In the AGB phase of 1 $M_\odot$ stars, the repeated helium shell flashes create extensive convection zones that nearly reach the overlying hydrogen layer. Between these episodes, the outer convection zone extends far enough down to mix material into that part of the mass zone previously reached by the inner convection zones. Hence there is a two-way flow of material. Material with enhanced heavy-element abundances derived from neutron-capture zones passes into the upper convection zone at the time of its deepest penetration, increasing the abundances of these elements at the surface of the star. Small amounts of hydrogen are mixed downward to react with ^{12}C and form ^{13}C. During the next helium shell flash when the helium convection zone reaches this material, it is mixed inward to provide a neutron source. Iben does not give an estimate of the neutron number density in the burning region, but it would be about 10^{10} neutrons cm^{-3}. Heavy nuclei have neutron capture cross sections of 0.1 to 1 barn (except near closed neutron shells) at helium-burning temperatures. The lifetimes of such nuclei against neutron capture are thus typically of the order of 1 to 7 days. Hence ^{129}I should be a product of the neutron capture in the AGB phase. The neutron capture path leading to this nucleus is rather spread out since most of the relevant isotopes with mass numbers 125 to 127, of $Z = 50$ to 52, have half-lives of several days or longer, which are comparable to or shorter than the expected neutron capture lifetimes. ^{107}Pd can be formed by neutron capture on any time scale.

In a pulsed s-process the abundances of heavier nuclei will be built up by factors (relative to solar system material) of the order of 100 to 1000. From the 30 keV Maxwell-averaged measurements of the neutron capture cross sections (M. J. Harris 1981; Macklin 1984) the production ratio of $^{107}Pd/^{108}Pd$ is found to be 0.194. If the neutron capture path went only through the iodine isotopes, then from similar measurements of their neutron capture cross sections (Macklin 1983), the production ratio of $^{129}I/^{127}I$ would be 1.4. Since neutron capture cross sections generally decrease with increasing mass number, this ratio is probably a reasonably good number to take for the actual capture path.

Norgaard (1980) has pointed out that significant amounts of ^{26}Al will be made in the outer convection zone in a red giant, and that the ejection of this may be of cosmogonic significance. This view has been strengthened by the discovery of a low-lying resonance in the $^{25}Mg(p,\gamma)^{26}Al$ reaction, that probably converts most of the ^{25}Mg to ^{26}Al near the base of the convection zone during the AGB phase of stars over a wide mass range (Champagne et al. 1983*c*). If we only consider the reactions on ^{25}Mg, this may give a $^{26}Al/^{27}Al$ ratio of unity in the outer envelope, as the abundances of ^{25}Mg and ^{27}Al are comparable (Cameron 1982). However, it appears that at hydrogen shell-burning temperatures of 4 to 6×10^7 K, other hydrogen reactions on intermediate mass nuclei can enhance the ^{26}Al yield (Caughlan et al. 1984). The abundance of ^{24}Mg is ten times that of ^{27}Al, so that much of this may be converted into ^{26}Al at temperatures in excess of 6×10^7 K. Also, some of the initial ^{26}Mg may be converted into ^{27}Al. The hydrogen-burning shell temperature is probably not high enough to destroy aluminum nuclei in the evolution time of the AGB stage. Hence the AGB star may become enriched to a $^{26}Al/^{27}Al$ ratio of from a few to as much as 10, which is the asymptotic value in the MgAl cycle (R. A. Ward and W. A. Fowler 1980).

S giant stars typically show overabundance factors of a few to a few tens of the heavier elements. Thus dredged-up material has been diluted by a factor of about 10 to 100. Hence the $^{107}Pd/^{108}Pd$ and $^{129}I/^{127}I$ ratios will be little changed after mixing into the surface layers.

According to D. D. Clayton (1984), if the ^{26}Al in the interstellar medium was formed in supernova explosions with a $^{26}Al/^{27}Al$ ratio of 10^{-3}, then over the course of galactic history ten times too much ^{27}Al would have been injected into the interstellar medium. The much higher ratios of these isotopes which can be ejected from red-giant stars suggests the possibility that the observed ^{26}Al may be principally contributed by these objects.

VII. DYNAMICS OF THE DISTURBANCES

We have seen that an AGB star in the 1 $M_\odot$ range should shed its envelope in the interior of a 10^6 $M_\odot$ molecular cloud complex once every few 10^6 yr. These stars typically have a velocity relative to such a cloud of the order of 10 to 30 km s^{-1} and will cross the interior of the cloud in a few 10^6 yr. The most intensive mass-shedding part of the AGB phase of the star takes much less than a crossing time and results in the ejection of a planetary nebula. Suppose that this ejection has occurred, that a bubble has been blown into the interior of the cloud, and that the magnetic field has redirected the velocity of the expanding gas largely along the direction of the field lines. We can make a crude estimate of the time required to sweep up enough gas to form a pair of cloud cores, each nominally of about 2 $M_\odot$. Assume that in one direction along the magnetic flux tube 0.1 $M_\odot$ of material move at 30 km s^{-1}. Assume conservation of momentum in the subsequent flow. Let ρ be the cloud gas density, A be the cross-sectional area of the magnetic flux tube, x be the distance along the tube, m_o the initial mass of driving gas, and v_o the initial velocity of the driving gas. It follows that the velocity of the gas a distance x along the tube is

$$v = \frac{m_0 v_0}{\rho A x}. \tag{1}$$

The time required for the sweep-up to move this distance is

$$t = \frac{\rho A x^2}{2 m_0 v_0}. \tag{2}$$

If the density of the cloud is 10^3 molecules cm^{-3} and the radius of the tube is 0.1 pc, then when 2 $M_\odot$ of material have been swept up, the distance traveled along the magnetic tube is 1 pc and the time from Eq. (2) is 3×10^5 yr. This forms a cloud core with a density of the order of 3×10^4 cm^{-3}, approximately at the threshold for gravitational collapse. The dynamical collapse time of 466 $\rho^{-1/2}$ (where ρ is density and time is in seconds) is, for this density, some 5×10^4 yr. The time required for the dissipation of the resulting accretion disk may take significantly longer than the free-fall time, as we discuss below; we take 10^5 yr as the combined infall and dissipation time up to the onset of the T Tauri stage with the ejection of a strong stellar wind and with bipolar outflows.

This star-formation process probably initiates a chain of additional star formation. The T Tauri stellar wind can put as much or more energy and momentum into the surrounding molecular cloud than can the ejection of material from an active red giant (C. Norman and J. Silk 1980). The wind will blow a new bubble in the cloud and initiate the formation of two new cores due to the pinch effect. In this way, a chain of star formation is propagated. This is probably the underlying cause of the clumping together of cloud cores and T Tauri stars within a cloud. In the early T Tauri phase probably a few 0.1 $M_{\odot}$ will be ejected in the stellar wind at a velocity of several tens of km s^{-1}. This event is thus comparable in energy and momentum emission to the planetary nebula shedding event which initiated the chain. Thus, again we take the time for the formation of a new cloud core to be 3×10^5 yr, and the time for the infall and disk dissipation to be a further 10^5 yr. I take 4×10^5 yr to be a standard generation time in the star-formation chain.

VIII. ABUNDANCES OF EXTINCT RADIOACTIVITIES

We now evaluate the meaning of the extinct radioactivity abundances in the early solar system. The values of these abundances quoted below are taken from a summary of abundances obtained by G. J. Wasserburg and his colleagues (see Chen 1984). The various production ratios of extinct radioactivities relative to their neighboring isotopes will have been decreased due to two effects: mass dilution and radioactive decay. The following formalism models these two effects.

Let R_p be the production ratio of a radioactive isotope relative to a neighboring reference isotope, x be the abundance ratio of the reference isotope in the stellar envelope relative to the original amount there, D be the dilution factor (ratio of the mass of material into which the envelope material is mixed relative to that ejected from the envelope), t_{ss} be the time in years between the start of the mixing and the time of formation of the primitive solar nebula, τ be the mean life of the radioactivity, and R_f be the final isotope ratio in primitive solar system material. In terms of these definitions, the initial envelope isotope ratio is $x\,R_p\,/\,(x + 1)$ after the source has mixed into the envelope.

We must make a crude model of the mixing in order to obtain a relation between the dilution factor D and the time t_{ss}. We have estimated that 0.1 $M_{\odot}$ of material in the initial star-formation generation are mixed into an amount of mass greater by a factor of 20. Subsequently, we assume that half the available extinct radioactive material is mixed into the newly formed star and that the remainder is divided in half and mixed into an equal amount of material, giving a dilution factor of 4 per stellar generation. With the assumed 4×10^5 yr per stellar generation, we convert these discrete estimates to a continuous expression for the dilution time

$$t_{ss} = 2.89 \times 10^5 \ln D - 4.64 \times 10^5 \text{ yr}. \tag{3}$$

Following dilution, the final isotope radio is

$$R_f = x \frac{R_p \exp(-t_{ss}/\tau)}{(x + 1 + D)}. \tag{4}$$

From Eqs. (3) and (4), if we know R_p, x, and R_f, D and t_{ss} can be computed by successive iterations.

From the available data it is not possible to compute definitive values for these parameters since we are ignorant of x for the radioactivities in the AGB source star. There is about a factor of four spread in the values of x that produce a consistent set of results for the three extinct radioactivities. We give a solution within the acceptable range. Let $x_{108} = x_{127} = 2.0$ and $x_{27} = 0.3$ (meaning that ^{27}Al has been enhanced by a factor of 1.3 over that originally present in the envelope). The isotope least affected by decay is ^{129}I. We assumed its production ratio relative to ^{127}I to be 1.4. The ^{129}I/^{127}I ratio is found in some early solar system minerals to be 1.7×10^{-4}. Hence D = 14898 and $t_{ss} = 2.31 \times 10^6$ yr, and the radioactive decay factor is 0.905. ^{107}Pd is also little affected by decay. We took the ^{107}Pd/^{108}Pd production ratio to be 0.194; the ratio commonly observed in early solar system materials is about 2.3×10^{-5}. Hence $D = 13228$ and $t_{ss} = 2.28 \times 10^6$ yr, with a corresponding radioactive decay factor of 0.784.

The above determinations of the dilution factor are made independently and are well within agreement considering their errors. We adopt an average $D = 14000$. Hence $t_{ss} = 2.30 \times 10^6$ yr, implying that there were 5 to 6 generations of star formation in the cloud at the time of formation of the solar nebula. The abundance enhancement of ^{26}Al over ^{26}Mg in the AGB envelope is uncertain, but is probably in the range 1 to 10. Within the formalism presented above, the ^{26}Al/^{27}Al production ratio lies in the range 3 to 33. The observed ratio in early solar system material is 5×10^{-5}. For the adopted value of t_{ss}, the radioactive decay factor is 0.110. Corresponding to the adopted value of D, the production ratio $R_p = 21$. This is within the expected range.

Hence, one can put together a consistent picture based on a 1 $M_\odot$ AGB star as an initiator of star formation in a molecular cloud, which can produce extinct radioactivities for the early solar system in the observed abundance ratios. There remain uncertainties in the numbers but these are essentially constrained within a range of about a factor of 4.

IX. COLLAPSE OF CLOUD CORES

We have discussed the formation of a cloud core by dynamical compression of gas along magnetic field lines. Self-gravitation has pulled the matter together to a mean density much above the cloud mean, and although some ambipolar diffusion will have taken place, there has been much less time available than in Shu's quiet hypothesis (Sec. III), and the magnetic field does not fully control the motions. In accordance with observation, we assume the core interior to be turbulent with random motions having Mach 0.7. The mean motion of the cloud core relative to the cloud itself may be somewhat larger. Thus a core with a radius of about 0.05 pc may be moving at roughly 1 km s^{-1} with smaller random internal velocities. The motion of matter in the core about halfway out the radius vector, relative to the cloud background, would correspond to a rotation rate of 10^{-12} radians s^{-1} relative to a nonmoving core center. The net actual rotation rate about the center of mass of the core will be the dispersion in the rotation rate computed in this way after correcting for the motion of the center of mass. It may well be about 10^{-13} radian s^{-1} or more. This net angular velocity results from fluctuations in the sweeping motion of the gas along the field lines, and the net angular momentum vector may have any orientation except along the field lines.

The angular momentum of the cloud core may be much higher than the value 10^{-14} radian s^{-1} assumed by Shu (see Sec. III). When the core collapses, most of the mass in the collapsing gas is likely to form a disk rather than to fall into a central protostar. In fact, because the infalling gas will be highly turbulent, the initial distribution of the gas, rather than forming a small central core, is likely to form a roughly planar but highly chaotic distribution.

The effective disk-formation time is the time between the first permanent halting of the infalling gas at the center of the disk and the time that most of the infalling matter has impacted the disk. We have noted above that the free-fall collapse time of a cloud core is about 5×10^4 yr. The dispersion in this time, which is the effective disk-formation time, is likely to be 1 or 2×10^4 yr. This does not count the last straggling infall of the gas from the periphery of the core, which may not succeed in reaching the disk.

X. DISK DISSIPATION

The basic theory of disk dissipation has been developed by Lynden-Bell and Pringle (1974). If the main source of kinematic viscosity in the disk can be specified as a function of position and thermodynamic variables, then the theory allows the computation of the energy, angular momentum, and mass fluxes within the disk as a function of time. It also specifies the rate at which the frictional process will dissipate energy locally. I mention one simple relationship here which follows for a disk with steady inward mass flow (Cameron 1978*a*, his equation 45)

$$F = -3 \pi \nu \sigma \tag{5}$$

Here F is the outward mass flux, ν is the viscosity, and σ is the surface density of the disk. This approximation has been used in the computation of the segments of the primitive solar nebula presented below.

Due to lack of knowledge of the physics of the viscosity, it has been customary to parameterize it with a constant α:

$$\nu = \alpha c_s H \tag{6}$$

where c_s is the sound speed at midplane and H is the scale height above midplane. This formulation is due to Shakura and Sunyaev (1973). A theory relating α to the growth rates of the largest modes in a thermally convecting region has been developed by Canuto, Goldman, and Hubickyj (CGH) (1984) and some extensions have been developed by Cabot (in preparation). The viscosity involved in disk dissipation is thus identified to be the turbulent viscosity of the thermally convecting medium.

In fully developed turbulence the energy fed into the medium at a wavenumber k is partly dissipated by viscous effects and partly fed into the disturbances with higher wavenumbers, and this leads to an infinite set of connected equations for the turbulent velocity components. It has been traditional in turbulence theory to deal with an approximation for those eddy sizes that are much smaller than the preferred eddy sizes at which the energy is predominantly input to the system; for such a range of eddy sizes the flow of energy is constant from one scale of eddy sizes to the next, and power-law relationships emerge that are commonly known as the Kolmogoroff spectrum. However, in the problem of a turbulent disk, it is evident that the turbulent viscosity depends on the largest eddies present which are also those at which the main input to the turbulent energy occurs.

The theory of CGH is an application of a procedure outlined by Ledoux et al. (1961). These authors expressed the energy flow in the system in terms of the growth rates of the unstable modes that feed energy into the system. CGH make the point that for the limited purpose of calculating α, a full theory of convection is not needed, but only a theory applicable to the scale of the largest eddies effective in the determination of the turbulent viscosity. They thus use the formalism of Ledoux et al. to set up integral expressions for α in terms of growth rates for the largest and fastest-growing modes in the system. The application of the theory then requires that a sufficiently representative sample of the modes present must have their growth rates evaluated for a particular disk geometry.

In the initial formulation of the theory, rotational terms were omitted from the dispersion relation, and for that case one obtains a value of about 0.24 for α. In the next step CGH added axisymmetric rotational terms to the dispersion relation, and upon evaluating representative growth rates they obtained a value of about 0.14 for α. In an independent evaluation of the axisymmetric case, Cabot found a value of about 0.16 for α, in good agreement. Cabot has gone on to add nonaxisymmetric terms to the dispersion relation. The evaluation of the growth rates is considerably more complicated for this case, but Cabot in current work is typically finding values for α approaching the 0.24 of the nonrotational case.

I have adopted $\alpha = 0.24$ for the numerical work reported below, and have also done some calculations with $\alpha = 0.10$ to show the degree of sensitivity of the results to the choice of this parameter. It is necessary to emphasize that these developments are new and all numerical results are subject to improvement.

XI. COMPUTATION OF DISK STRUCTURE

Some calculations are reported below of the structure of the inner solar nebula for a range of rates for steady inward flow of mass and for two values of α. These calculations are intended to be exploratory rather than precise, and so a variety of approximations have been made to lessen the burden of calculation. The calculations are intended to determine the structure of the disk during the terminal stage of the formation of the Sun. Hence a mass of one Sun is assumed at the center of the disk (although the fully accreted Sun probably had significantly more mass which was lost in the T Tauri stage).

A basic assumption made in calculating α is that the environment of the disk is thermally convective. This means that everywhere where there is rapid dissipation, the temperature gradient must be superadiabatic. In general, we can expect that the interior of a disk is dense enough so that the superadiabaticity is small, as in the outer envelope of a star. In constructing disk models I achieve a considerable simplification by assuming that the interior is isentropic, which should be a reasonable first approximation. This means that the interior of the disk must be assumed both radially and vertically isentropic. Normally the vertical component of the adiabatic temperature gradient is zero at midplane and becomes quite steep near the surface. The real vertical structure would gradually change from near adiabatic to a radiative temperature gradient near the surface; in practice I have continued an upward integration on an adiabatic temperature gradient until the radiative flux that would be carried by this adiabatic gradient becomes equal to σT^4, where T is the local temperature. At this point the vertical column is topped with an isothermal atmosphere of optical depth 2/3. I also insist that the optical depth between photosphere and midplane must be large enough to support the radiation of the energy; this prevents regions of very low opacity from having their photospheres located near midplane. This procedure should slightly underestimate the photospheric temperature, but that should not be important for the present exploratory purposes.

The equation for hydrostatic equilibrium in the vertical direction is

$$\frac{dP}{dz} = \frac{GM\rho z}{(R^2 + z^2)^{3/2}}. \tag{7}$$

The equation for the surface density σ is

$$\frac{d\sigma}{dz} = 2\rho. \tag{8}$$

Here P is the pressure, M the mass at the center of the disk, z the height above midplane, R the radial distance from the center of the disk, and ρ is the density. Note that we do not make a thin-disk approximation here, but we do ignore the self-gravity of the disk, which is a reasonable approximation. This structure must be solved by numerical integration. The thermodynamic variables are connected by the imposed isentropic relationships.

The simple surface condition involves the comparison of two measures of luminosity (remember that there are two surfaces in a vertical column):

$$L_1 = 2\sigma T^4 \tag{9}$$

$$L_2 = \frac{16 A_\tau A_\Gamma L_1}{3\chi P_\rho} \cdot \left(-\frac{dP}{dz}\right) \tag{10}$$

where $A_\tau = 1 - e^{-\tau}$, and $A_\Gamma = (\Gamma_2 - 1)/\Gamma_2$. Here τ is the optical depth integrated outward from midplane

$$\frac{d\tau}{dz} = \rho\kappa. \tag{11}$$

In these expressions κ is the opacity and Γ_2 is the second adiabatic exponent. The first luminosity is the rate at which the column would radiate into space if uncovered, and the second luminosity is the radiative energy transport by the adiabatic gradient. The factor including the exponential term is the empirical correction inserted to assure that there is sufficient optical depth beneath the photosphere to radiate energy at the required rate. These two luminosities are equal at the assumed photosphere.

The numerical evaluations were facilitated by my use of a large table of the equation of state with associated thermodynamic quantities and opacities which is stored on the computer for our general use. This is calculated for solar composition and uses the high opacities discussed by DeCampli and Cameron (1979). For any particular choice of internal adiabat in the disk construction, a preliminary program calculated a detailed path through this table and stored the quantities needed in a convenient form.

The above procedures enable a vertical section of the disk to be computed, provided a thermodynamic starting condition (such as the pressure) is specified at midplane. This condition is chosen in such a way as to assure the validity of Eq. (5) for a constant inward mass flux F in the inner regions of the disk to which the procedures are intended to apply. The viscosity ν in Eq. (5) is obtained from Eq. (6), and the sound speed in Eq. (6) is taken to be

$$c_s = \left(\frac{\Gamma_1 P}{\rho}\right)^{1/2} \tag{12}$$

which is evaluated at midplane. The scale height H in Eq. (6) is taken to be the height above midplane at which the density has fallen by a factor e following the adiabat. This definition was strictly enforced in order to prevent sudden radial discontinuities in the thermodynamic quantities which would be unphysical, as the general procedure used does not allow the model to adjust for changes in the radial energy transport.

A trial model of a disk can now be computed. At this stage we do not know the actual value of the mass flux, but we can specify a beginning midplane pressure at a minimum radial distance and integrate outwards keeping the quantity F/α constant at its initial value through iteration on the midplane pressure for each vertical profile. The models reported here were started at 10^{12} cm and integrated to 10^{14} cm (or slightly beyond in a few cases).

Most disk model builders have assumed that the energy locally deposited due to dissipation is also locally radiated away. In general, this will be only approximately true. The vertical adiabatic gradient is very small near midplane and only steepens as the surface is approached; the radial adiabatic gradient at midplane is also rather small but can be larger than the vertical superadiabatic gradient there. It is natural to assume that the superadiabatic gradient components are proportional to the gradients themselves, but this is by no means assured. For example, if radiation in the vertical direction were to be hindered by a locally high opacity, then the superadiabatic gradient in the radial direction could steepen so that there is an enhanced radial transport of energy through thermal convection. Rotational effects tend to make the large convective eddies smaller in the radial direction than perpendicular to midplane, but nevertheless, they are quite capable of transporting substantial quantities of heat in the radial direction. For this reason I have not made the traditional assumption that the heat locally deposited by viscous friction must be locally radiated from the disk surface; instead I have assumed that there must be a balance between energy deposition and radiation over a reasonable radial distance. The radial distance so chosen starts with the large jump in the opacity accompanying the condensation of iron and silicates in matter of solar composition, and ends with the somewhat smaller jump in opacity that accompanies the condensation of water ice.

The procedure for determining the vertical structure gives the photospheric temperature at a given radial distance in the disk; the total energy radiated over the chosen interval is thus simply the quantity L_1 integrated over the designated surface area. The energy deposition D is given by the viscous dissipation in the theory of Lynden-Bell and Pringle (1974) as

$$D = g \frac{\left(-\frac{\partial \Omega}{\partial R}\right)}{2\pi R} \tag{13}$$

where g is the viscous couple and Ω is the angular velocity

$$\Omega = \left(\frac{GM}{R^3}\right)^{1/2}. \tag{14}$$

For the steady inward flow case (Cameron (1978*a*)

$$g = -F(GMR)^{1/2}. \tag{15}$$

Note that all of the quantities in Eqs. (13) to (15) are known for the calculated structure except the constant F. These can therefore be integrated over the disk within the chosen radial interval, remaining uncertain by the factor F. Equating the result to the energy emitted, it is possible to obtain the value of F for the structure. Since the structure had been calculated with a trial value of F/α, it thus follows that α is determined. This whole procedure is then iterated with new values of F/α until the desired value of α is obtained.

XII. STAR FORMATION FROM A DISK

Before describing the details of the new solar nebula models that have been calculated, it is desirable to pause to consider some recent calculations of star formation assumed to occur as the result of the dissipation of a disk (Mercer-Smith et al. 1984). In these calculations, disk dissipation was simulated in an otherwise ordinary mass-accreting stellar evolution study by imposing a surface accretion luminosity in addition to the internal luminosity and by varying the rate of mass accretion onto the star.

We may express the accretion luminosity in the form

$$L_{\rm acc} = \frac{\xi G M \dot{M}}{R}. \tag{16}$$

Here $\dot{M}$ is the mass-accretion rate from the disk onto the protostar and ξ is an efficiency factor determined by the accretion mechanism. As accretion takes place onto a protostar from a disk, much of the incoming angular momentum must be removed from the protostar or else it will not be able to contract during the subsequent evolution. If we do not know the removal mechanism, we do not know in detail how much of the incoming rotational energy will be removed with the angular momentum in nonradiative form and how much is dissipated in the surface layers of the protosun and lost by radiation. For most of the cases studied, it was assumed that the incoming gas accretes to the equator, and then spreads out over the surface of the protostar, and the rotational energy accompanies the outward flux of angular momentum. For this case, $\xi = 1/4$. The accretional luminosity is added to the internal luminosity of the star when determining the outer boundary conditions. The mass-accretion rate determines how fast mass must be added into the surface layers; in doing this it was assumed that the incoming mass had the same specific entropy as the mass in the surface layers. Otherwise this is a standard stellar evolution calculation.

The early history of a 1 $M_\odot$ star is shown in Fig. 1 for the case of a uniform accretion rate of 1×10^{-5} $M_\odot$ yr^{-1}. The accretion phase is shown plotted on a HR diagram as a series of open circles, ending with an open star plotted where the mass becomes 1 $M_\odot$. At this point in the evolution, the accretion was suddenly turned off, and the star quickly readjusted to the right in the diagram, thereafter evolving downward and somewhat to the left until reaching the main sequence. The points appearing in the figure are the positions of T Tauri stars in 4 young clusters. As may be seen from Fig. 1, for an accretion time of 10^5 yr, a uniform rate of accretion does not produce stars with the high observed luminosities of the youngest T Tauri stars. The initial protosolar envelope is not large enough. It starts at about 3.0 solar radii, whereas what is needed is closer to 4.5 solar radii. In fact, Stahler (1983*a*) reports the results of a radial infall calculation to form new stars, for which $\xi = 1$, and finds that the initial stellar radii are large enough to place the post-accretion tracks initially among the positions of the most luminous T Tauri stars.

The observed higher luminosities are reproduced when we take the accretion rate from the disk to be proportional to the elapsed time and the total elapsed time to be 5×10^4 yr. Both of these effects enhance the thermal energy content of the protostar. This nonuniform rate of accretion should actually be a better representation of true disk dissipation for reasons discussed below. It may be recalled that the disk formation times are probably more like 2×10^4 yr. If the disk dissipation time is comparable to the disk formation time, then there should be no problem in reproducing the luminosities of the youngest T Tauri stars. But we can firmly state that the mean mass inflow rate within the solar nebula should be more than 10^{-5} $M_\odot$ yr^{-1} if the Sun behaved like a typical T Tauri star.

Most of the accretion takes place with a luminosity 20 to 30 times the present solar luminosity. Early in the course of the accretion a fairly extensive

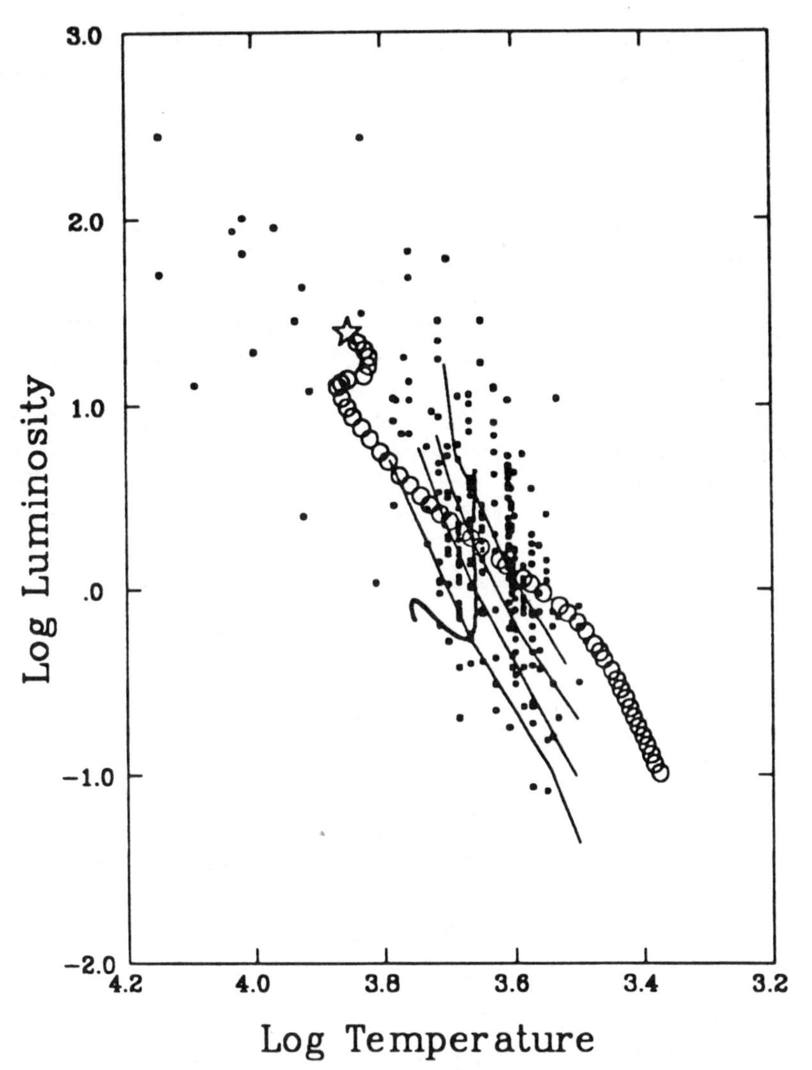

Fig. 1. HR diagram showing the accretion and early evolution of a star of 1 $M_\odot$ with a steady accretion rate of 1×10^{-5} $M_\odot$ yr^{-1}. The small dots are the positions of T Tauri stars in the Orion, Taurus, NGC2264, and NGC700 young open clusters (M. Cohen and L. V. Kuhi 1979). The track defined by the mostly overlapping open circles is the accretion track, ending in a star when the mass is 1 $M_\odot$. The line which descends vertically and then curves up to the left is the pre-main sequence evolutionary track for 1 $M_\odot$. The four thin lines descending from upper left to lower right are isochrones of 3×10^5, 1×10^6, 3×10^6, and 1×10^7 yr, respectively, in descending order.

convection zone develops in the deep interior of the protosun, except in a small central region, initially too cold to participate in the convection. Deuterium-burning in the interior helps to raise the entropy there, so that the convection zone can gradually approach the surface (the accretion leaves the envelope with outward increasing entropy, so the convection zone can approach the surface only as the interior entropy is raised). For models having an increasing accretion rate, only when the accretion stops can the outer convection zone break through to the high-entropy surface. Thus, the onset of the T Tauri solar wind should occur at the approximate cessation of accretion.

It is thus clear that times in the range 10^4 to 10^5 yr appear to be associated with the star-formation process. It is important to know whether the intrinsic dissipation time in a disk like the primitive solar nebula is long or short compared to this time. If the dissipation time were long, then the formation of the solar nebula could be taken as essentially instantaneous to be followed by a long dissipation process, whereas if the dissipation time were short, the solar nebula would adjust its structure to the details of the mass infall. It turns out that neither extreme is a good description of reality.

Prior to the onset of the T Tauri solar wind, the star and the inner solar nebula should be viewed as a continuous fluid mass, in which there is no clear boundary at which one describes the star as ending and the disk as beginning. Rather, there is a smooth transition from mostly hydrostatic pressure support of the matter to mostly centripetal support.

XIII. SOLAR NEBULA MODELS

One of the parameters that enters into the determination of solar nebula models is the entropy, which I have taken to be constant throughout the interior. Four adiabats were chosen to form the basis for four models of the solar nebula for each of the two values of α; these were labeled A, B, C, and D, from low to high entropy. In each case, it was assumed that the rate of mass flow inwards through the model was constant. The models were started at an inner radius of 10^{12} cm and extended to 10^{14} cm unless it was necessary to integrate further to reach the point of ice condensation (which was true in case A). The midplane pressure was varied at the inner radius to bring about equality between energy deposited in the iron-to-ice condensation range and the radiation from that interval. When this adjustment was made, it was possible to determine the value of the inward mass flux, and hence the dissipation time for the model, defined as the time required for the inward mass flux to transport 1 $M_\odot$ to the center of the disk.

The results for the four different models with $\alpha = 0.24$ are shown in Table I. The values of the mass shown are those out to 10^{14} cm. In the steady flow region the mass increases outward approximately in proportion to the radius. Thus, the mass values shown are likely to be much less than the real mass that the solar nebula would have; the total mass may well be roughly an order of magnitude more than the values shown.

TABLE I
Solar Nebula Models with $\alpha = 0.24$

Model	Mass[a] ($M_{\odot}$)	Flux (g s^{-1})	Dissipation Time[b] (yr)
A	0.073	2.6×10^{21}	2.48×10^{4}
B	0.024	5.4×10^{20}	1.17×10^{5}
C	0.0074	9.6×10^{19}	6.6×10^{5}
D	0.0023	1.7×10^{19}	3.8×10^{6}

[a]Mass contained within radius of 10^{14} cm.
[b]Dissipation time is 1 $M_{\odot}$ divided by the mass flux.

Looking first at model D, we note that the mass shown is $<< 1\%\ M_{\odot}$. For this case, the dissipation time is much longer than the characteristic infall time of $< 10^5$ yr. Thus, it appears that for such a small amount of mass, little dissipation would take place during the infall time; model D would principally just accrete mass from the infall during that time. In fact, the situation at the start of infall would be more severe than that shown, since model D is computed for a central stellar mass of 1 $M_{\odot}$, whereas at the start of infall there would be much less mass at the center and the mass in the disk would be much more spread out. This is part of the justification for preferring the star-formation prescription in which the accretion rate varies as the elapsed time; the actual relationship may be even more sensitive than this. Model C contains substantially more mass than D, but the same general conclusions apply. It is apparent that the mass flux rises very rapidly with the mass in the disk.

Models B and A straddle the expected mass flux for the terminal stage of the accretion of the Sun, although the situation probably lies close to or even beyond A. Thus, it may be expected that the disk mass at that time was an order of magnitude greater than the 7% $M_{\odot}$ that one can interpolate between A and B. This also means that model A contains the physical characteristics that are most likely to be appropriate to this terminal stage, and the A curves showing these characteristics in the figures should be viewed with this in mind.

The models in Table II were all prepared with $\alpha = 0.10$. The model numbers A,B,C, and D have the same adiabats as the corresponding models in Table I. It may be seen that for a given adiabat, the Table II characteristics involve a considerably higher mass, a slightly greater mass flux, and a slightly smaller dissipation time. However, the mass has increased much more rapidly than the mass flux, so that for a given mass disk the dissipation time is increased compared to Table I. Thus, decreasing the α parameter makes the disk dissipation less efficient, as is to be expected.

Figures 2 through 6 show the radial dependence of several parameters of the models in Table I. Shown in Fig. 2 is the variation of the midplane tem-

TABLE II
Solar Nebula Models with α = 0.10

Model	Mass[a] ($M_\odot$)	Flux ($g\ s^{-1}$)	Dissipation Time[b] (yr)
A	0.176	3.0×10^{21}	2.1×10^{4}
B	0.057	6.9×10^{20}	9.1×10^{4}
C	0.018	1.3×10^{20}	4.8×10^{5}
D	0.0060	2.4×10^{19}	2.6×10^{6}

[a]Mass contained within radius of 10^{14} cm.
[b]Dissipation time is 1 $M_\odot$ divided by the mass flux.

perature with radial distance for each of the models. There is a particularly striking feature of the region between models A and B in this figure. Iron does not condense until the temperature falls through about 1300 K on these adiabats. For the interesting region, this occurs out in the asteroid belt. There does not appear to be any easy way to avoid the conclusion that, in the terminal stages of solar accretion, the entire inner region of the solar nebula was so hot that solids would not condense there. The photospheric temperatures, shown in Fig. 3, are slightly lower, with iron condensing between the orbits of Earth and Mars, but the surface condensates will be continually vaporized when the thermal convection sweeps them down toward midplane. In principle, other condensates would form from material moving upwards, but such condensation faces a nucleation barrier and tends to make quite large particles (Cameron and Fegley 1982), that contribute only inefficiently to the opacity.

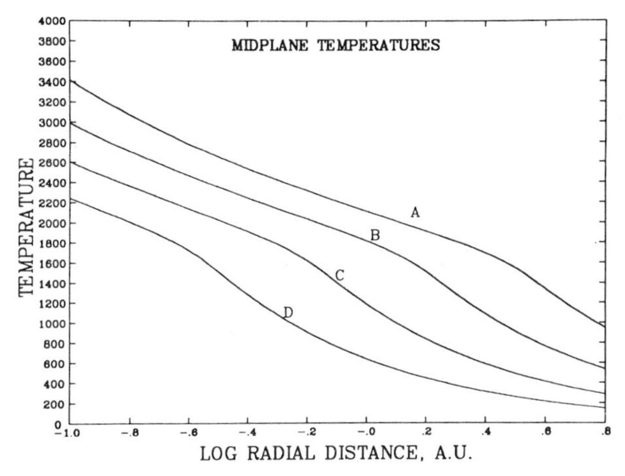

Fig. 2. Midplane temperatures as a function of radial distance for the four nebula models whose characteristics are listed in Table I.

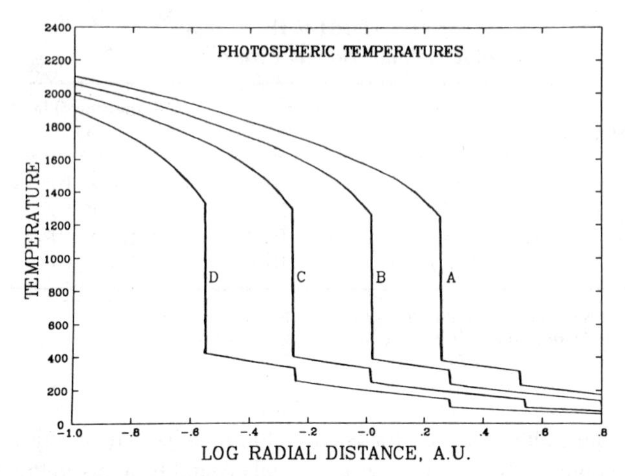

Fig. 3. Photospheric temperatures as a function of radial distance for the four nebula models whose characteristics are listed in Table I.

Figure 4 shows the radial dependence of the midplane pressure. Note that under iron condensation conditions the pressure is only around 10^{-5} bar. This is a better value to adopt in chemical condensation calculations than the conventional 10^{-3} bar. Figure 5 shows the height of the photosphere for the models. The largest step function in the curves shows where iron condenses and the opacity becomes very much higher. Figure 6 shows the variation of the surface density of the models with radial distance. The masses of the models are obtained from the radial integration of this quantity.

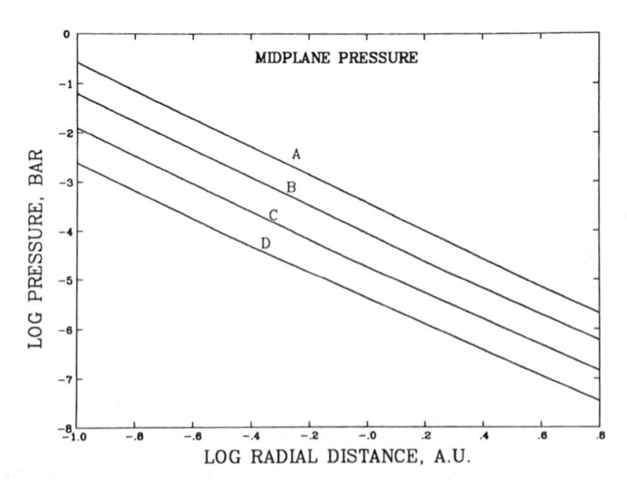

Fig. 4. The pressure at midplane as a function of radial distance for the four nebula models whose characteristics are listed in Table I.

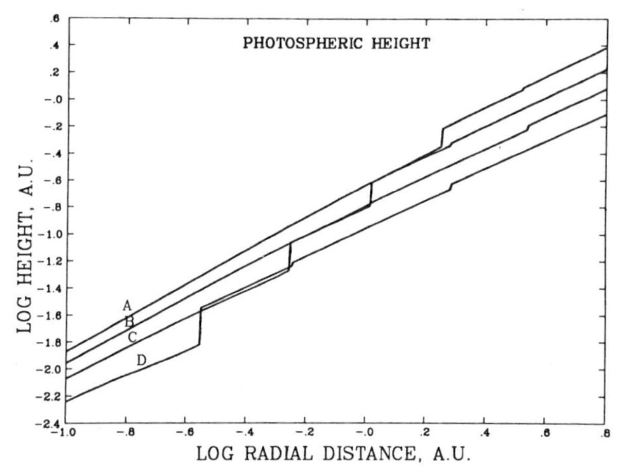

Fig. 5. The height above midplane of the photosphere as a function of radial distance for the four nebula models whose characteristics are listed in Table I.

XIV. EVOLUTION OF THE SOLAR NEBULA

In this section is presented my attempt to construct an evolutionary picture of the solar nebula that responds to the various calculations discussed above. This picture divides the evolution of the nebula into three stages.

Stage 1

Within just over two million years of the AGB mass-shedding event, a cloud core forms and becomes dense enough to collapse under the influence of

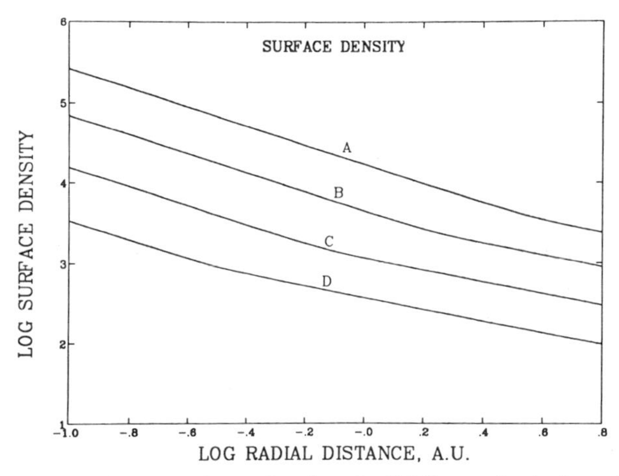

Fig. 6. The surface density (g cm^{-2}) as a function of radial distance for the four nebula models whose characteristics are listed in Table I.

its own gravitational field. Because it is formed by a process of dynamic compression, the gas in the core is fully turbulent. When the infall of this interstellar material is first halted, the turbulence in the gas will lead to the formation of a roughly planar chaotic mass of gas which is the initial phase of the primitive solar nebula. Computer simulations of this infall frequently show the formation of rings, but these simulations frequently have more angular momentum than is expected for the solar nebula and they are constrained by the imposition of axial symmetry. It seems to me more likely that the imposition of turbulent motions on the earliest infalling gas velocities will produce an initial density distribution that is very lumpy.

Gravitational instabilities in this gas are thus likely to be common. Probably most of these instabilities are intrinsically three-dimensional in character. Any two-dimensional (ring-like) instabilities are likely to break up quickly into isolated fragments. I call the products of these instabilities and fragmentations giant gaseous protoplanets; they are self-gravitating spheres with low internal temperatures and densities and large dimensions (DeCampli and Cameron 1979). They probably have masses comparable to that of Jupiter, but with a spread in mass of a factor ten or more. Those protoplanets formed near the center of the nebula will interact among themselves both gravitationally and tidally, leading to many collisions and mergers. Those near the center of the nebula which have merged together will form the initial core of the Sun. Those farther away from the center are likely to last longer.

As gas accumulates, the temperature in the nebula will rise. Cameron, et al. (1982) have found that this leads to the thermal expansion and evaporation of the envelopes in the giant gaseous protoplanets on rather short time scales, provided the protoplanets are not situated farther out than the region of formation of the asteroids. The solid material originally present as interstellar grains in the protoplanet envelopes may, at least in part, grow by collisions into dust balls of significant size (millimeters), precipitate through the envelopes, and collect near the centers of the protoplanets. When the envelopes are thermally stripped away, these planetary cores composed of refractory materials may thus remain behind, later to form much of the interiors of the present planets.

Local superadiabatic gradients within accumulating gas will induce thermal convection, and hence local dissipation; it can be expected that such regions will gradually spread to form a fully convective disk, whose global dissipation can then begin. Because of the large initial distances and the small initial amounts of mass in the nebula, the initial rates of mass flow will be small and not very evolutionarily significant, as can be extrapolated from Table I.

Thus, the principal events to be associated with Stage 1 are the accumulation of gas in the nebula, the formation of the gas into a convective disk, and the formation and thermal evaporation of giant gaseous protoplanets, leaving behind in the gas a number of bodies composed of condensed matter with masses of the order of those of the present inner planets.

Stage 2

As the Sun accretes at the spin axis of the nebula, the gravitational binding of the gas to the Sun increases. Conservation of angular momentum will cause all of the gas to spiral inward during the accretion of the Sun. This increases the surface density of the nebula and hence (from Table I) increases even more the inward mass flow and the rate of dissipation. This is why it is better to approximate the growth of a star from the nebula by an accretion rate proportional to the elapsed time than to use a constant accretion rate, but the true relationship may be even more sensitive than this.

As the Sun approaches its final mass, the inner part of the nebula will be at a high enough temperature (about 1000 K) to sustain a significant amount of thermal ionization, and hence to trap magnetic lines of force. In fact, the thermal convection may sustain a dynamo. Hence magnetic fields in the upper layers of the nebula will exhibit many of the same phenomena as do solar magnetic fields today. In particular, the magnetic field will become twisted with storage of magnetic energy, and this should be relieved at high altitudes by large nebula flares, similar to large solar flares (Levy 1982).

The solar nebula is initially filled with interstellar grains that to a large extent may collect together as dust balls, and the thermal convection will stir these thoroughly throughout the gas, provided that the temperature does not become too high and evaporate them. I expect that the dustballs at high altitudes will be exposed to the effects of the nebula flares, being heated and melted by ultraviolet radiation and possibly particle bombardment, and forming chondrules and calcium-aluminum inclusions, depending upon their location and degree of evaporation. Convection will spread these throughout the nebula.

The ultraviolet radiation from these flares may cause a separation of ^{16}O from ^{17}O and ^{18}O due to different self-shielding effects in oxygen-containing molecules in the gas, as has been found to occur in laboratory experiments (Thiemens et al. 1983). This has been suggested as a method for producing the ^{16}O excesses that are commonly found in extraterrestrial materials. It seems likely that differential photodissociation can occur in the upper layers of the nebula at this time, but whether the subsequent chemical steps needed to preserve the isotopic effect in solid materials can occur is a matter of current controversy.

When the surface density of the nebula becomes large, tidal interactions of the gas with the largest giant gaseous protoplanet cores (perhaps about 10 $M_\oplus$ of condensed materials) may be able to move these cores into the outer part of the nebula, where they are likely to become locked into position with respect to the gas (Goldreich and Tremaine 1980; Ward 1983). They would then be in a position to capture the surrounding gas to form quite massive envelopes (Mizuno 1980, and references therein). I believe this to be the most promising hypothesis for the mechanism by which the giant outer planets were formed; they were nucleated in the inner solar nebula and later completed in the outer nebula in a relatively short time.

Once the surface density of the nebula reaches values typical of the cases A and B of Table I, mass inflow to form the Sun becomes rapid and accelerates, forming Stage 2 of solar nebula history. It is possible that the formation of chondrules and inclusions, the generation of oxygen isotopic anomalies, and the formation of giant planets are relics of this stage of nebula history.

Stage 3

With the diminishing rate of interstellar infall, the dissipation of the nebula will lower the surface density and decrease the rate of accumulation of the Sun. The solar convection zone can then break through to the surface of the Sun and the T Tauri wind can start. The luminosity of the accreting Sun is then about 20 to 30 times the present luminosity, as can be seen from Fig. 1. The high solar luminosity leads, in the presence of a significant amount of surrounding scattering matter, to a heating of the nebula surfaces at large distances, leading to a decrease in the rate at which the nebula radiates energy from the interior. In turn, this decreases the rate of mass inflow and the rate of energy deposition, and lowers the normal surface temperatures, thus progressively shutting down the thermal convection and effectively terminating disk dissipation at large distances. The effect should then propagate inwards. I call this the thermal stabilization phase. The density distribution in the nebula at this time is probably close to model B of Table I. As we do not know the effectiveness of several possible mechanisms for removing the solar nebula, the configuration of the thermal stabilization phase may last for a considerable time.

With the cessation of thermal convection, solid matter can now settle to midplane and collect through gravitational instabilities into cometary and asteroidal bodies with radii of a few kilometers (Goldreich and Ward 1973). The giant gaseous protoplanet cores will collect many of these to complete their growth as planets, acquiring their more volatile elements from these small bodies which are in chemical equilibrium with much cooler gases than were the planetary cores at the time of their formation.

The cometary bodies in the outer nebula can be moved out to the Oort cloud only if there is a relatively rapid loss of solar and nebula mass (Cameron 1978*a*). I do not see how such a rapid mass removal from the solar nebula could occur within the context of the cosmogonical picture presented here. Instead, it seems more likely that the comets were formed by a variant of a mechanism suggested by Hills (1982). In Hills' mechanism there is a radial infall of material to form the Sun, and toward the end of that process the radiation streaming outward from the protosun enhances fluctuations in the distribution of interstellar grains in the infalling matter. This causes the regions with a density enhancement of the grains to suffer a slight imbalance in the radiation pressure owing to absorption and scattering of the radiation passing through the regions. The net radiation pressure is inward toward the center of the fluctuations. Hills shows that these can develop into comets in a matter of centuries. His theory requires that the sunlight should not have suffered reddening due to absorption and reradiation; he also would welcome the contribution from the light from other nearby newly formed stars.

In the present cosmogony, this picture would undergo some changes. The presence of turbulence contributes angular momentum to the infalling gas, so that the infall is no longer strictly radial. This means that the time available for differential radiation pressure on grain fluctuations is greatly spread out, which is a positive change. The radiation from the Sun is undoubtedly greatly reddened by absorption and reradiation within the gas, so that its scattering cross section on interstellar grains is much reduced: a negative change. On the other hand, the presence of turbulence should promote the gentle collisions among interstellar grains, building up clumps of them which will have a larger cross section for light in the farther infrared: a positive change. There will be no other nearby newly-formed stars in this scenario: a negative change. Hills pointed out that his theory applies to regions of grain fluctuations of all sizes beyond some minimum, but in the present context we have the requirement that the cometary bodies must be able to clump together faster than the material is sheared apart by turbulence, thus limiting the size scale. In fact, the preferred size scale should be the smallest eddy size in a turbulent medium (Cameron 1973*a*). It happens that for the numerical example given by Hills, the smallest eddy size is approximately equal to the size of the region from which his typical comet is formed, which is very encouraging for this modified picture. The theory needs to be redone in detail; it would form very large numbers of comets at a few thousand astronomical units from the Sun, from which stellar encounters can perturb them into the Oort cloud.

This thermal stabilization stage of the nebula is Stage 3. The amount of mass in the nebula at that time is a few percent of that of the Sun, about what the traditional views of the solar nebula predict, but for totally different reasons. This is when the T Tauri stage of the Sun turns on, when the dissipation of the nebula ceases, when small bodies are formed at midplane in the nebula, and when the planets finish growing with the incorporation of volatiles contained in the small bodies that collide with them. The comets should also form at this time, but not as a solar nebula process.

FORMATION OF THE SOLAR SYSTEM

CHUSHIRO HAYASHI
Kyoto University

KIYOSHI NAKAZAWA, YOSHITSUGU NAKAGAWA
University of Tokyo

We review the over-all evolution of the solar system, which has been studied by our group in Kyoto. Starting from a disk-like solar nebula with small mass composed of H_2 and He gases and dust grains, we pursue multistep evolutionary processes through which the present planets, satellites, and asteroids have been formed. It is shown that interactions between solid bodies (i.e., small grains as well as large protoplanets) and the gas of the solar nebula are important in determining the various stages of the evolution. The basic philosophy of our study and remaining problems are briefly described.

I. INTRODUCTION

The study of the origin of the solar system entered a very scientific stage in about 1960, when observations and theories of pre-main sequence stars (especially T Tauri stars) began to be developed, the solar chemical composition became clear, and a large amount of chemical and mineralogical data of meteorites became available. In accordance with this situation, we have investigated since 1970 an evolutionary sequence of multistep processes which start from the formation of the solar nebula and end with the formation of the present planets (terrestrial, minor, and giant), satellites, and meteorites. Our results are called, for simplicity, a Kyoto model of over-all evolution.

There are now three representative models of the formation of the solar system, i.e., Cameron's (1978*a*), Safronov's (1969), and the Kyoto models. Cameron's model is quite different from the other two: it assumes a massive solar nebula, from which so-called giant gaseous protoplanets are first born. On the contrary, Safronov's and the Kyoto models assume a low-mass nebula in which the planetary growth proceeds through accumulation processes. These two models follow a common evolution until the formation of planetesimals stage. The difference between them is that Safronov assumes gas-free planetary accretion, whereas in the Kyoto model the accumulation proceeds in the nebular gas except for distant regions far from the Sun, as described in the following sections.

The outline of the Kyoto model is most concisely shown in Table I (see also Fig. 24 in Sec. XI for a more detailed space-time description). In our investigation, the evolution starts from a preplanetary solar nebula with small mass, 0.01 to 0.04 $M_\odot$, which has a disk-like structure where the three forces (i.e., solar gravity, gas pressure, and centrifugal force) are all balanced. In constructing this solar nebula model, we assumed that the gas and dust grains of interstellar origin are well mixed and that the radial distribution of surface mass density of the disk has such a form that the present terrestrial planets as well as the cores of the giant planets have been formed through minimum migration of dust materials in radial direction. As described below, we have pursued the evolution of this solar nebula by taking fully into account the effects of the gas component. Through a series of our studies (see, e.g., Hayashi 1981*a*), we have constructed a theory which describes the whole evolution of the solar system. Important epochs and stages of this evolution are summarized in Table I, each of which is described in detail. Sections of the text are arranged nearly in the chronological order of evolution. As seen from the table, our theory may explain, in a consistent way, the characteristics of the present solar system.

Nevertheless, many quantitative problems still remain to be solved. One of the greatest problems, i.e., a remaining gap in the Kyoto model, is the formation of the solar nebula itself. Although progress has been made in this decade in the theory and observation of star formation in molecular clouds, our knowledge on star formation is not yet complete enough to understand, for example, the reason why a single star is formed instead of a binary system and why stars have been born with the mass spectrum as observed. This is the present situation in the study of the formation of a bulk star. The formation of the preplanetary solar nebula is more complicated because it may be existing initially in the very outermost region of a collapsing cloud and its mass is very small compared with that of the Sun. Furthermore, we have to clarify the mechanism of angular momentum transport by magnetic and/or turbulent viscosity. The above is the reason why our study started from the stage where the above-mentioned equilibrium solar nebula has once been formed.

The second problem is the dissipation of the gas component of the solar

TABLE I
Chronological Table of Planetary Formation

Time (yr)	
-10^{6-5}	Collapse & fragmentation of a giant molecular cloud
-10^{5-4}	Collapse of a rotating presolar cloud
0	Formation of protosun and solar nebula
	⟨Growth & sedimentation of dust-grains⟩
10^4	Fragmentation of dust layer into planetesimals
10^5	⟨Accumulation of planetesimals⟩
10^6	
	Formation of the Earth
10^7	Formation of Jupiter's core & accretion of gas onto it
	Dissipation of nebular gas
10^8	Formation of Saturn's core & capture of remaining gas
	Formation of Asteroid Belt
10^9	Formation of Uranus(?)
	Formation of Neptune(?)
10^{10}	

nebula. This is very important for our theory because, according to our results as described in Sec. VI, the existence of the nebular gas reduces considerably the growth time of the terrestrial planets as well as the cores of the giant planets. Furthermore, Jupiter, Saturn, and even the Moon can hardly be formed without the gas, according to our detailed calculations. The time of dissipation or of the duration of the gas has been roughly estimated to be 10^6 to 10^7 yr, nearly in agreement with the lifetime of T Tauri stars, but a more precise evaluation is necessary.

The third problem is the origin of meteorites, on which we have not yet developed our theory to such an extent that we can compare it directly with observations. For this, we have to clarify a series of physical, chemical, and mineralogical processes. According to our results on planetary formation in the asteroid region, we are now considering that the above processes were initiated by high-velocity collisions of asteroids which were 10^3 times as numerous as at present and which gained high random velocities through Jovian perturbation in stages after the gas was dissipated (see Sec. X).

II. METHOD OF APPROACH

A brief discussion is given here on the philosophy underlying our theoretical study. Our philosophy may be common to the philosophies of all investigations in natural science, except that we must deal with long-term evolution which is composed of multistep processes. As described in Sec. I, we start from the construction of the solar nebula model because of the incompleteness of present theories of star formation. We have investigated each of the multistep processes as shown below, e.g., growth and sedimentation of dust grains in the nebula, formation of planetesimals due to fragmentation of a dust layer, radial migration and accumulation of planetesimals to form large bodies, etc.

Each of the above processes should, of course, be studied by making calculations as precise as possible. Furthermore, the sequence of all the above-mentioned processes should be consistent as a whole and, in this sense, we should try to find clear causal relations among various processes of evolution. Finally, the results obtained for the last stage of evolution should be compared with observations of the present solar system, because at present we have no observations of other planetary systems. If there would be some discrepancies between theory and observation, we should, of course, examine whether we have overlooked some important processes or made oversimplified assumptions. With corrections introduced through this examination, we must reconstruct the whole sequence of processes or even modify the initial structure of the solar nebula assumed.

We have performed the above-mentioned recurrence procedure several times in constructing the present Kyoto model, mainly because many macroscopic physical laws, which should be used as key-stones of our study, had not been fully established. Contrary to the case of atomic or nuclear physics, we have to deal with a system having a great number of degrees of freedom (e.g., gases plus grains or gases plus planetesimals) and also we must pursue irreversible processes continuing for a very long run of time. For such a system, we have to disentangle complicated processes and pick up a few basic processes, each of which will be described by a relatively simple macroscopic law.

There are many examples of such laws for which we have already found or are trying to find by computer simulations. Some examples are the sticking cross section of planetesimals moving in the solar gravity field (not in free space), the gas drag coefficient of a protoplanet attracting gases by its own gravity and the cross section of tidal disruption of planetesimals during their close encounters with a protoplanet (these are all described in Sec. VI). Further, in order to understand the origin of meteorites, we need to find several new laws which characterize thermal and chemical processes expected to occur during the stage of high-velocity collision of asteroids as mentioned in Sec. I. As to the microscopic data and laws, from which the macroscopic laws are to be derived, we consider that they are almost fully available at present, contrary to the case before 1960.

III. FORMATION OF PROTOSUN AND SOLAR NEBULA

Our present knowledge about star formation is not complete enough to construct a precise model of the solar nebula, from which the long-term evolution of the solar system starts. Hence, in this section we only comment on several problems of star formation, which are closely related to our main theme of planetary formation.

Observationally, molecular clouds contain many dense regions which may be called prestellar clouds. These clouds have typical temperatures ~ 10 K and densities $\sim 10^{-20}$g cm^{-3}. For a gas with solar composition, the sound velocity in cm s^{-1} is given by

$$c_s = \left(\frac{kT}{\mu m_{\mathrm{H}}}\right)^{1/2} = 1.88 \times 10^4 \left(\frac{T}{10\mathrm{K}}\right)^{1/2} \left(\frac{2.34}{\mu}\right)^{1/2} \tag{1}$$

where μ (= 2.34) is the mean molecular weight of the gas composed mainly of H_2 and He.

For a cloud of about 1 $M_\odot$ to begin collapsing to form the Sun, its gravity should be greater than the pressure force and, accordingly, its radius should be smaller than the Jeans' length (or the well-known sonic point radius in a stellar wind problem), which is given by

$$a = GM_\odot/2c_s^2 = 1.26 \times 10^4 (10\mathrm{K}/T)\ \mathrm{AU}\ . \tag{2}$$

Furthermore, until the cloud condenses to a radius of about 10^2 AU (corresponding to the mean density of about 10^{-13} g cm^{-3}), it is transparent to infrared radiation and its temperature is kept nearly constant ($T \sim 10$ K). After the cloud becomes opaque, the temperature rises and the gas pressure increases as $P \propto \rho^{7/5}$ until H_2 begins to dissociate at $T \sim 1600$ K. The collapse is finally stopped when the temperature rises well above 10^4 K where hydrogen is almost completely ionized.

On the collapse of a rotating cloud in the isothermal stages mentioned above, a large number of numerical computations have so far been conducted by many authors (see Bodenheimer [1981] for a review). However, many problems still remain, e.g., the fragmentation problem and the central runaway or centrifugal bounce problem, which indicate that the collapse is not as simple a process as first imagined. Recently, in order to solve the above problems, Miyama et al. (1984) and Narita et al. (1984) have performed a great variety of two- and three-dimensional computations. They used both Lagrangian and Eulerian methods of computation in order to check the accuracy of their results.

Miyama et al. (1984) performed computations for the usual and simple initial condition that a homogeneous sphere is rotating uniformly. In this case, the initial condition is specified by the two parameters, $\alpha = E_{th}/|E_{grav}|$ and $\beta = E_{rot}/|E_{grav}|$, i.e., the ratios of thermal, gravitational, and rotational energies. They computed various cases of α and β where $\beta > 0.1$ and found that the product $\alpha\beta$ determines whether the collapsing cloud fragments or not, that is:

Case 1. The cloud collapses but does not fragment for $0.20 > \alpha\beta > 0.12$;
Case 2. The cloud collapses and fragments into three or more pieces for $0.12 > \alpha\beta$.

The above fragmentation is due to the formation of a very flattened disk-like core which is known to be gravitationally unstable (Goldreich and Lynden-Bell 1965*a,b*). Case 2 indicates the formation of a binary or multiple stellar system. In case 1, a single star like the Sun may be formed but the initial total angular momentum J (which is proportional to $\alpha\beta$) is very large, i.e., 2×10^{54} gcm^2 s^{-1} compared with 3×10^{50} gcm^2 s^{-1} of the present solar system. Accordingly, if this is the case of solar formation, a large amount of angular momentum would have to be lost outward by some unknown friction mechanism.

It seems more likely that the initial cloud has a much smaller angular momentum, i.e., $\alpha\beta << 1$. Tscharnuter (1981) computed the contraction of a cloud of 3 $M_\odot$ with the initial values, $\alpha = 1$ and $\beta = 1.2 \times 10^{-4}$ (corresponding to $J = 6 \times 10^{53}$ gcm^2 s^{-1} for 3 $M_\odot$), assuming the existence of turbulences with a viscosity coefficient as large as $\eta = 0.1 c_s z$, where z is the thickness of the cloud. His result shows that after 3×10^4 yr the central condensate becomes a dense core with 0.5 $M_\odot$ and, at the same time, a rotating disk-like nebula with $\rho = 10^{-11} - 10^{-13}$ g cm^{-3}, $T = 10^3 - 10^2$ K and $a = 0.3 - 15$ AU is formed. The total life-time of this nebula was estimated by Tscharnuter to be about 10^7 yr. The mechanism for excitation of strong turbulences that he assumed is not known, but his result indicates one possible way of the formation of the solar nebula in a case where some kind of shear viscosity is present in the collapsing cloud.

Recently, Narita et al. (1984) computed a model for the three-dimensional collapse of a nonrotating invicid isothermal cloud (i.e., the case with $\beta = 0$), whose initial state is a homogeneous oblate spheroid with an axis ratio of 4 (note that most molecular clouds observed are not spherical). Their result shows that, if α is as small as 0.2, the cloud becomes highly flattened and finally fragments into pieces, whereas in the case of $\alpha = 0.4$, the cloud does not fragment even after the central condensation proceeds to a considerable extent. The above result indicates that fragmentation depends on the initial condition of a cloud, and in the problem of star formation we have to know, in turn, how the initial condition itself is set up. In connection with the above fragmentation problem, we pose a question to the giant protoplanet hypothesis on the origin of the solar system as developed by Cameron and his co-workers (Cameron 1978*a;* DeCampli and Cameron 1979), i.e., a question as to how the massive cloud they assumed leads to the formation of a single star, instead of forming a system of double or triple stars.

Recent radio observations have discovered about 20 bipolar flow objects, each of which is associated with a rotating disk-like cloud. An infrared source is situated at the center of this disk. These objects are thought to be on the way to star formation. One of them, a disk with the smallest mass (probably 2 $M_\odot$ or so) has been found by Kaifu et al. (1984) around the central star IRS 5 in a dark cloud L 1551. This disk has a diameter of about 0.1 pc in agreement with Eq. (2). The bipolar flow ejected from the central star has a velocity of the order of 10^2 km s^{-1} and we consider it very likely that this stellar wind originates from a circumstellar hot region (say, a very active stellar corona) where rotational energy is being effectively converted into magnetic energy. The existence of the strong stellar wind indicates that, on the formation of the protosun and also the solar nebula, we must take into account the effect of magnetic viscosity on angular momentum transport through the disk as well as on the ejection of angular momentum occuring from the circumstellar region in all directions (note that the disk has a shadow effect for this ejection).

On the above magnetism problem, a few comments are made here. Hayashi (1981*b*) investigated the rate of amplification of magnetic fields in the solar nebula due to its differential rotation and also the rate of their decay due to Joule loss. As a result, he showed that magnetic viscosity depends strongly on the ionization degree of the gas and that, in the Kyoto model of the solar nebula, magnetic viscosity is very high in hot regions ($T > 10^3$ K) inside, say, Mercury's orbit, while it is very low in most of the cool and dense regions of the solar nebula. This means that the preplanetary solar nebula can be in a stationary rotating state for a relatively long time. Furthermore, in low-density regions lying outside, say, the Uranus or Neptune orbit, cosmic rays can penetrate to the equatorial plane and the degree of ionization is much higher than in the inner regions. This means that magnetic viscosity is also high in these outer regions.

In the above, we have pointed out many processes that have to be taken into account in the study of the formation of the solar nebula. Computer simulations of the formation may not be simple but are not prohibitive presently in view of the recent rapid development of high-speed computers.

IV. MODEL OF THE INITIAL SOLAR NEBULA

According to the amount of total mass, current models of the solar nebula fall into two types, i.e., a low-mass nebula with mass of the order of 0.02 $M_{\odot}$ and a massive nebula with mass of 1 $M_{\odot}$. First, we briefly describe the latter model developed by Cameron and his co-workers (Cameron and Pine 1973; Cameron 1978*a;* DeCampli and Cameron 1979; chapter by Cameron). Their scenario is as follows.

At first, the massive nebula gravitationally fragments into so-called giant gaseous protoplanets with mass of the order of that of Jupiter. In the interior of the protoplanet, dust grains sink toward the center to form a core. Afterwards, the protoplanet becomes a Jovian-type planet if the surrounding gaseous envelope contracts and accretes onto the core. On the other hand, it becomes a terrestrial planet if the envelope is ultimately stripped off by solar tidal action and also by mutual collisions. In this model, the Jovian-type planets might be formed rather easily, but the formation of the terrestrial planets is not clear because the above dissipation of the envelope has not been studied quantitatively. Furthermore, it is necessary to find a mechanism through which a large amount of the disused nebular gas, left after planetary formation, can be removed from the solar system. This model contains many unsolved problems besides verification of the assumption that such a massive nebula does not break up to form a system of binary stars, as mentioned in Sec. III.

Now we turn to the low-mass solar nebula studied so far by many workers including us (see, e.g., Safronov 1969; Kusaka et al. 1970; Weidenschilling 1977*b;* Wetherill 1980*b;* Hayashi 1981*a,b*). A total mass as small as 0.01 $M_{\odot}$ has been derived from the mass of the present planetary system by adding a certain amount of H_2 and He gases required to reproduce the solar chemical abundance. Here, we should consider that 0.01 $M_{\odot}$ is a minimum mass of the nebula because a comparable amount of mass may fall onto the Sun during various stages of planetary formation.

As mentioned in the above, the radial distribution of the nebula mass is derived from the masses and orbits of the present planets. The structure of the solar nebula, which has already settled into equilibrium, is determined by a force balance as well as by a heat balance. The equilibrium nebula is a flattened disk circulating around the Sun with a velocity slightly lower than that of circular Keplerian motion, owing to the presence of a small gradient of gas pressure in the radial direction (see Fig. 1). The temperature of the nebula is fixed by the balance between heating due to solar radiation and cooling due to emission of thermal radiation.

Contrary to Cameron's massive nebula, the low-mass nebula itself never fragments because of its small self-gravity, and at first, dust grains begin to sink toward the equatorial plane of the nebula (described in Sec. V). When most of the dust grains have sunk to the equatorial plane, the nebula becomes transparent to solar visible radiation and as a result, the temperature of the

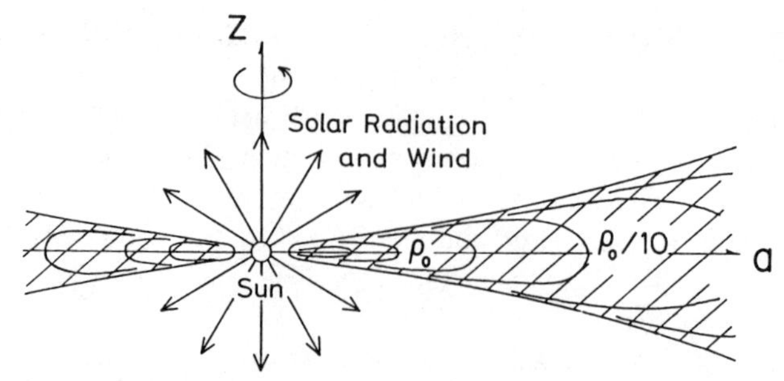

Fig. 1. The Sun and the solar nebula. The solid curves represent equidensity contours in the nebula.

nebular gas is determined mainly by this radiation. In this case, the temperature T at the distance a from the Sun[a] is given by

$$T = 280\left(\frac{L}{L_\odot}\right)^{1/4}\left(\frac{a}{1\ \mathrm{AU}}\right)^{-1/2} \tag{3}$$

where L and $L_\odot$ are the solar luminosity at this stage and at present, respectively. Hereafter we put $L = 1\ L_\odot$ for simplicity, in view of the weak dependence of T on L, although we are now considering the T Tauri stage of the protosun with higher luminosities.

Hayashi (1981*b*) presented a disk model of the equilibrium nebula at the above transparent stage. He first determined the surface density distribution of dust materials in the disk σ_{dust} from the masses and orbital radii of the present planets (envelope masses of giant planets being excluded) under the assumption that the planets have later been formed through minimum radial displacement of these dust materials. As a result, he obtained

$$\sigma_{\mathrm{dust}} = \begin{cases} 7.1 \times \left(\dfrac{a}{1\ \mathrm{AU}}\right)^{-3/2} \mathrm{gcm}^{-2} & \text{for } 0.35\ \mathrm{AU} < a < 2.7\ \mathrm{AU} \\ 30 \times \left(\dfrac{a}{1\ \mathrm{AU}}\right)^{-3/2} \mathrm{gcm}^{-2} & \text{for } 2.7\ \mathrm{AU} < a < 36\ \mathrm{AU} \end{cases} \tag{4}$$

where the difference in the coefficients comes from the condition that, in addition to rocky materials, icy materials can condense in the outer region where $T < 170$ K (the condensation temperature of H_2O ice), i.e., $a > 2.7$ AU according to Eq. (3). Next he showed that the surface gas density in the whole region σ_{gas} required to reproduce the solar chemical composition, can be expressed in the simple form,

$$\sigma_{\mathrm{gas}} = 1.7 \times 10^3\left(\frac{a}{1\ \mathrm{AU}}\right)^{-3/2} \mathrm{gcm}^{-2} \quad \text{for } 0.35\ \mathrm{AU} < a < 36\ \mathrm{AU}. \tag{5}$$

[a]Throughout the text, a and z denote the distance from the Sun (or the semimajor axis in the case of planetary motion) and the height from the equatorial plane, respectively.

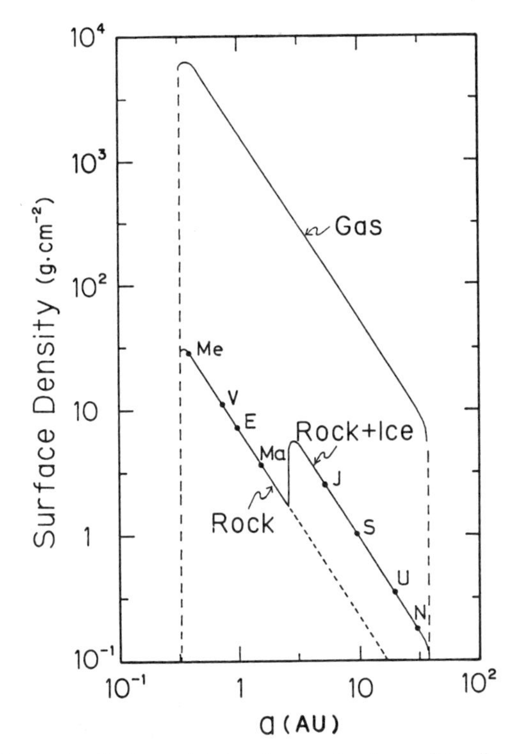

Fig. 2. Surface density distributions of dust and gas as functions of the distance from the Sun. Positions of the planets are denoted by Me, V, E, Ma, J, S, U and N (Hayashi 1981*b*).

The above distributions of σ_{dust} and σ_{gas} are shown in Fig. 2. The total mass of this nebula is 0.013 $M_\odot$. Hayashi (1981*b*) further obtained, from the equilibrium condition in the *z*-direction, the gas density ρ_{gas} in the equatorial plane and the half-thickness of the nebula z_o, i.e.,

$$\rho_{gas} = 1.4 \times 10^{-9}\left(\frac{a}{1\ \mathrm{AU}}\right)^{-11/4} \mathrm{gcm}^{-3} \tag{6}$$

and

$$z_o = \left(\frac{2kT}{\mu m_H}\frac{a^3}{GM_\odot}\right)^{1/2} = 0.047\left(\frac{a}{1\ \mathrm{AU}}\right)^{5/4} \mathrm{AU}. \tag{7}$$

Equations (3) through (7) are, of course, approximate equations, but they are all expressed in very simple forms convenient for our study of the later evolution of the solar nebula. Hereafter, we adopt the above formulae whenever we need the evaluation of physical quantities in various stages of evolution.

V. FORMATION OF PLANETESIMALS

The first process occurring in the solar nebula, after it has settled into equilibrium, is the separation of the gas and solid components of the nebula. It is likely that at early stages there would remain some turbulent motion in the nebula. However, such turbulence soon decays, if there is no source of its excitation (ter Haar 1950). Some convection may also occur in an early phase of the solar nebula (Lin and Papaloizou 1980; Lin 1981). After the turbulent motion and convection have decayed below a certain level, dust grains sink toward the equatorial plane of the nebula, owing to the z-component of the solar gravity, and at the same time they grow by coalescence. As a result, a very thin dust layer is formed around the equatorial plane, which finally fragments into a number of planetesimals, as described below.

A. Growth and Sedimentation of Dust Grains

First we consider the sedimentation process. Since the gas drag is sufficiently large, dust grains sink in the z-direction with terminal velocity (see, e.g., Nakagawa et al. 1981)

$$v_z = -\frac{\rho_{mat}}{\rho_{gas}} \frac{r}{v_{th}} \frac{GM_{\odot}}{O^3} z \tag{8}$$

where ρ_{mat} and r are the material density and the radius of a dust grain, respectively, ρ_{gas} is the gas density, and v_{th} is the mean thermal velocity of gas molecules, i.e., $(8\, kT/\pi \mu m_H)^{1/2}$. For example, the velocity v_z has a value as small as 10^{-2}cm s^{-1} for $r = 1\ \mu m$ and at $a = 1$ AU.

In sinking toward the equatorial plane, dust grains mutually collide and grow by sticking. We now consider vertical sweeping of mass by a grain and estimate the final radius of the grain. The growth of the dust grain is described by the sweep equation,

$$\frac{dm}{dz} = -p_s \pi r^2 \rho_{dust} \tag{9}$$

where m $(= 4\pi\rho_{mat} r^3/3)$, p_s, and ρ_{dust} $(= \sigma_{dust}/2z_o)$ are the mass of the dust grain, the sticking probability, and the spatial density of an ensemble of dust grains, respectively. Let us consider that a grain with an initial radius $r_o < 1$ μm begins to sink from a height z_o, where z_o is the half-thickness of the nebula given by Eq. (7). Integrating Eq. (9) where p_s is assumed to be constant, we find that the radius r of the grain, when it has just reached a height z, is given by

$$r = r_o + (1 - z/z_o) r_m \tag{10}$$

with

$$r_m = \frac{p_s \sigma_{\text{dust}}}{8\rho_{\text{mat}}} \tag{11}$$

where σ_{dust} is the surface density of dust grains given by Eq. (4). The above r_m is much greater than r_o and hence, gives a maximum radius attained by the vertical sweeping (Safronov 1969). Putting $p_s = 1$ for simplicity, we obtain $r_m = 4.4$, 3.2, and 0.2 mm at the orbits of the Earth, Jupiter, and Neptune, respectively, where we have taken $\rho_{\text{mat}} = 2$ g cm^{-3} for the first and 1 g cm^{-3} for the latter two orbits.

Next, we estimate the sedimentation time. As a grain grows, its sedimentation velocity v_z increases as $v_z \propto r$. As a result, its collision with other small grains becomes more and more frequent and hence, its growth is further accelerated; that is, the growth and the sedimentation are cooperative. By integration of the equation, $dt = dz/v_z$, using Eqs. (7), (8), and (10), we find that the sedimentation time from z_o to z is given by

$$t_{\text{sed}} = \frac{4}{\pi^{3/2}} \frac{t_K}{p_s \zeta} \frac{1}{1 + r_o/r_{\text{m}}} \ln\left(\frac{z_o r}{z r_o}\right) \tag{12}$$

where t_K $(= 2\pi(a^3/GM_\odot)^{1/2})$ is the Keplerian period for a planetary region considered and ζ is the mass fraction of dust grains contained originally in the nebula (i.e., $\zeta = 1/240$ for $a < 2.7$ AU and $1/56$ for $a > 2.7$ AU in our model described in Sec. IV). Now putting, for example, $z = z_o/10$, $p_s = 1$, and $r_o = 1$ μm, we find that t_{sed} is 2×10^3, 4×10^3, and 4×10^4 yr at the orbits of the Earth, Jupiter, and Neptune, respectively.

It should be noticed that the above values of t_{sed} are larger than t_K by a factor of about 10^3 and that t_{sed} neither depends on the gas density nor on the temperature; that is, it is determined essentially by the mass fraction of dust grains and by the Keplerian period in a region considered. The sedimentation may occur in an early contracting phase of the solar nebula before it becomes in equilibrium. Even in such a phase, the sedimentation time is given by Eq. (12) so long as dust grains sink with the terminal velocity given by Eq. (8). Then, the sedimentation in that phase may not be appreciable, because t_{sed} is about 10^3 times larger than t_K (i.e., a free-fall time) as mentioned above.

Nakagawa et al. (1981) and Weidenschilling (1980) numerically simulated the above growth and sedimentation in an equilibrium solar nebula by computing the temporal and spatial variation of the mass spectrum of dust grains. These studies obtained similar results, which are essentially in good agreement with the above simple estimation. For example, Nakagawa et al. (1981) showed that in the Earth zone, a thin dust layer composed of cm-sized grains is formed in about 3×10^3 yr after the beginning of sedimentation (see Fig. 3).

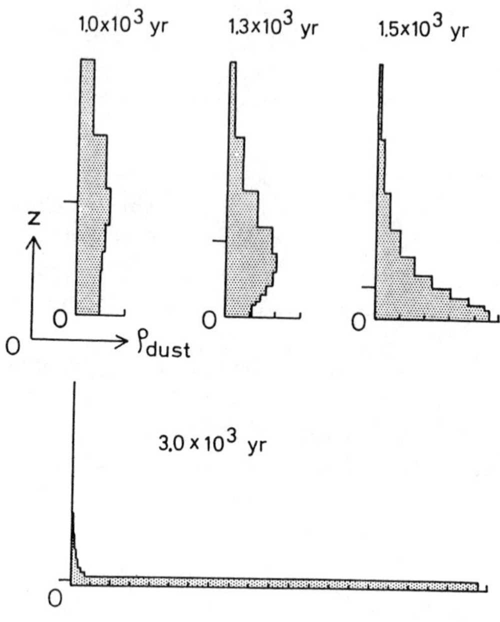

Fig. 3. Density distribution of sinking dust grains in the *z*-direction in the Earth zone. Abscissa: the density $\rho_{dust}(z)$ in units of the initial value. Ordinate: the *z*-coordinate (Nakagawa et al. 1981).

In addition to the vertical velocity v_z considered above, dust grains have also radial velocity v_r due to drag of the nebula gas which is circulating with an angular velocity slightly less than the Keplerian. In early stages of sedimentation, v_r is negligibly small compared with v_z but it becomes larger than v_z when the dust grains come close to the equatorial plane, i.e., when z becomes smaller than about 0.1 z_o. In this case, dust grains may further grow beyond the maximum radius r_m given by Eq. (11), owing to accretion in the radial direction. Nakagawa et al. (1985) have shown that, if this effect is taken into account and if the sticking probability is unity, dust grains can grow to radii of 18, 5, and 0.5 cm in the Earth, Jupiter, and Neptune zones, respectively. As to these maximum sizes of dust grains, there still remains a problem. For example, Weidenschilling (1980) pointed out that the sticking probability for such large bodies may be much smaller than unity so that dust grains, for example, in the Earth zone may not grow above 1 cm.

B. Gravitational Fragmentation of a Dust Layer

As the sedimentation proceeds, the dust layer becomes thinner and thinner with a considerable increase of density. When the density exceeds the Roche density 3.5 $M_\odot/a^3$ (Jeans 1929), the dust layer is expected to be gravitationally unstable to fragmentation. At 1 AU, for example, the density of the dust layer reaches the Roche density when the thickness of the layer becomes about $1/10^5$ of the nebula. Safronov (1969), Hayashi (1972), and Goldreich and Ward (1973) investigated the gravitational instability of such a thin dust layer and reached the same conclusion that numerous planetary embryos, so-called planetesimals, are born as a result of gravitational fragmentation of the dust layer.

Next, we consider the formation of planetesimals due to gravitational instability. In the case of a thin layer circulating with Keplerian angular velocity Ω_K [$= GM_\odot/a^3)^{1/2}$], the dispersion relation for a ring-mode perturbation of the form $e^{i(\omega t + kr)}$ is given by

$$\omega^2 = \Omega_K^2 - 2\pi G\sigma k + c_s^2 k^2 \tag{13}$$

where σ is the surface density of the layer and c_s is the sound velocity (Toomre 1964; Goldreich and Lynden-Bell 1965*a,b*). If $\omega^2 < 0$, the layer is unstable to perturbation and, if $\omega^2 > 0$, it is stable. As ω^2 is quadratic in k, it is easily found that the instability occurs when the sound velocity c_s is smaller than the critical value $c_s^* = \pi G\sigma/\Omega_K$ or, equivalently, when the surface density σ is larger than the critical value $\sigma^* = \Omega_K c_s/\pi G$. Applying this criterion to our gaseous solar nebula model, we find that it is always stable because the surface density σ_{gas} given by Eq. (5) is smaller than σ^*. Therefore, our low-mass nebula never breaks into gaseous fragments, whereas, Cameron's model nebula is found to be unstable because $\sigma > \sigma^*$ (Cameron and Pine 1973; Cameron 1978*a*). It may fragment into a number of gaseous protoplanets or into a relatively small number of larger bodies, i.e., double or triple stars.

Next, we apply the above dispersion relation to the dust layer. In this case, we see that the dust layer contains also gas molecules whose pressure can never be neglected although mass density is negligible. In general, the sound velocity in such a medium is given by $c_s^2 = \gamma(p_{gas} + p_{dust})/(\rho_{gas} + \rho_{dust})$, where γ is the ratio of specific heats which has a value of the order of unity, and p_{gas}, ρ_{gas}, p_{dust}, and ρ_{dust} are the pressure and the mass density of the gas component of the dust layer and those of the dust component, respectively. As the sedimentation of dust grains proceeds, the sound velocity c_s becomes small, as we have approximately $c_s^2 \simeq \gamma p_{gas}/\rho_{dust}$ where p_{gas} is constant while ρ_{dust} increases with the sedimentation. When c_s becomes smaller than the critical value c_s^*, ω^2 becomes negative for a wave number near the critical value $k^* = \Omega_K^2/\pi G\sigma$ (see Fig. 4). Accordingly, the dust layer begins to fragment into a number of rings, and further, these rings break into many pieces. This is a mechanism for formation of planetesimals. Now, using the above expression for k^*, we can estimate the mean mass of planetesimals as

$$m = \pi\sigma_{dust}\left(\frac{2\pi}{k^*}\right)^2 = (2\pi)^2\left(\frac{\sigma_{dust}\pi a^2}{M_\odot}\right)^3 M_\odot \tag{14}$$

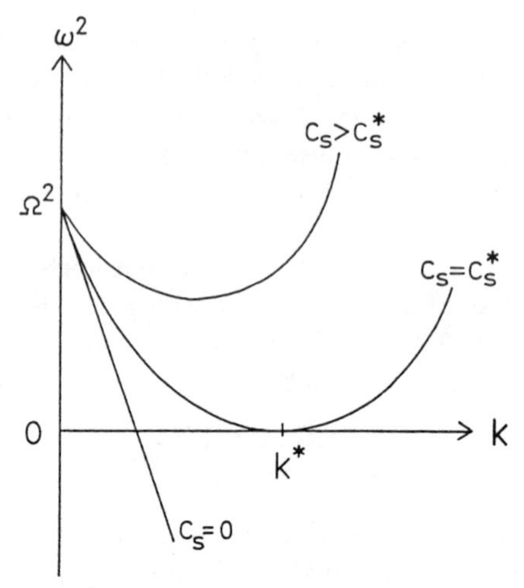

Fig. 4. Dispersion relation for a thin rotating disk. When the sound velocity c_s becomes smaller than the critical value c_s*, instability occurs around the critical wave number k*.

which gives 1.2×10^{18}, 1.1×10^{21}, and 1.6×10^{22} g for the Earth, Jupiter, and Neptune zones, respectively. It should be noticed that Eq. (14) gives only an order of magnitude of the mean mass because the fragmentation of a contracting ring has not yet been well studied.

From the condition $c_s = c_s$*, we find that the half-thickness of the dust layer at the onset of instability is as small as 1 m at 1 AU. The density ρ_{dust} at this time is 0.1 g cm^{-3} at 1 AU, and this is much larger than the Roche limit, 2×10^{-6} g cm^{-3}. These extreme values are partly due to the above treatment of the dust layer as a one-component fluid and partly due to the neglect of the external pressure acting on the outer boundary of the dust layer. In fact, Hayashi (1977) investigated the gravitational instabilities of a two-component fluid composed of gas and dust, taking into account the frictional interaction between them, and showed that the instability begins when the density slightly exceeds the Roche limit. Coradini et al. (1981) also obtained similar results based on the two-component theory and, furthermore, pointed out that the contraction of a fragment should be subject to a gas-drag force. Recently, Sekiya (1983) studied the problem in more detail and derived a dispersion relation in which the external pressure as well as the perturbed motion in the z-direction are fully taken into account. He found that the instability begins earlier and grows faster than that considered in the previous one-component theory but the mass of a planetesimal so formed is only slightly smaller than that given by Eq. (14). Summarizing the results of this reexamination of the instability, we can now consider that the fragmentation of the dust layer begins when the density becomes somewhat larger than the Roche limit and that the mean mass of planetesimals is still given by Eq. (14).

Finally, we should keep in mind that there may exist some turbulence in the boundary layers which are formed on both sides of the dust layer by the velocity difference between the nebular gas and dust layer, as pointed out by Goldreich and Ward (1973) and Weidenschilling (1980). Such a turbulence may stir up the dust grains and prevent the dust layer from becoming thin. Weidenschilling (1980) shows that the dust grains, if they have grown to 1 cm size at 1 AU and to a smaller size at outer regions, are not subject to turbulence. As seen above, the dust grains in the dust layer are considered to have grown to such sizes. Hence, we can consider that such a turbulence does not inhibit the gravitational instability of the dust layer.

VI. FORMATION OF TERRESTRIAL PLANETS AND CORES OF GIANT PLANETS

The planetesimals, formed as fragments of the dust layer as described in Sec. V.B, accumulate through mutual collisions and grow to form finally terrestrial-type planets and also cores of giant planets. The accumulation of planetesimals has been studied by many authors. Most have studied the accumulation occurring in a gas-free condition, under the implicit assumption that the gaseous nebula can be easily dissipated in very early stages of accumulation (see e.g., Safronov 1969; R. Greenberg et al. 1978*b*; Wetherill 1980*a,b*). On the other hand, we have developed our theory, as described below, from a standpoint that the gas component of the solar nebula still remains in the accumulation stage, at least until the formation of Jupiter and Saturn (the dissipation time of the gas is estimated in Sec. VII.C). In this section we first consider coalescence cross section and relative velocity between planetesimals. Next, in terms of these quantities, we estimate the growth time in a certain simplified case. Then, we present the results of our detailed numerical study of planetary growth. Finally, we show that, when massive bodies are formed, their growth is accelerated because their gravity attracts the surrounding gas and hence, gas drag is greatly increased.

A. Coalescence Cross Section

The gravitational collision cross section for an isolated body, i.e., in the absence of the solar gravity, is given by

$$\sigma_{col} = \pi (r + r')^2(1 + 2\theta) \qquad (15)$$

with

$$\theta = G(m + m')/\mathrm{v}^2(r + r') \qquad (16)$$

where r, r', m, and m' are the radii and masses of two colliding planetesimals, respectively, v is their relative velocity, and θ is the Safronov number (as named by Wetherill [1980*b*]) which represents the enhancement of the cross section due to gravitational focusing. However, it is not yet clear whether the above formula (15) is a good approximation to planetesimals circulating around the Sun, except for the case of high-velocity encounters, i.e., the case where $\theta < 1$.

Nishida (1983) has investigated the rate of collision occurring in the solar gravity field by means of direct numerical computations of a great number of orbits in the framework of a coplanar restricted three-body problem. His results show that, in the case of low-velocity encounter, the collision rate $\sigma_{col}v$ is enhanced by a factor of about 2 compared with that in free space. As Nishida has not yet computed the case of off-plane collisions, the enhancement factor in general cases is not well known at present but we should keep in mind that there is a general tendency for enhancement.

On the other hand, it becomes obvious that we cannot ignore tidal effects on the enhancement. Nakazawa and Hayashi (1985) found an effective increase of the coalescence cross section due to tidal disruption in their numerical study of binary encounter; that is, they assumed a gas-free condition for simplicity and simulated close encounters of planetesimals, composed of N gravitating bodies (N = 560), with a planet which has a radius R_p. Their results show that, if the distance at their closest approach is less than 1.8 R_p in the case of a rocky planetesimal (with material density ρ_{mat} = 3.34 g cm^{-3}) and less than 2.8 R_p for an icy planetesimal (with ρ_{mat} = 1.67 g cm^{-3}), the planetesimal is almost totally disrupted; furthermore, almost no fragments escape from the planetary Hill sphere, but are captured in bound orbits around the planet (Fig. 5). Here, we should recall the existence of the solar nebula; during revolutions around the planet, the fragments will soon accrete onto the

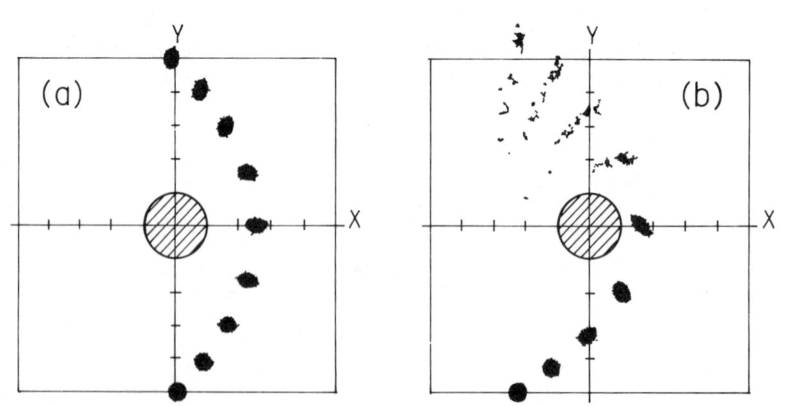

Fig. 5. Tidal disruption of a planetesimal in a close encounter with a planet. Panel (a): If the pericenter distance $> f \cdot R_P$ (f and R_P being the factor shown in the text and the radius of the central planet, respectively), the incoming planetesimal is not disrupted but only deformed. Panel (b): If the pericenter distance $< f \cdot R_P$, the incoming planetesimal is disrupted and almost all of the fragments are gravitationally trapped by the central planet.

planet owing to the drag of the surrounding gas. This is equivalent to coalescence and we must consider that coalescence occurs if the distance at the closest approach is less than $f(r + r')$, where $f = 1.8$ and 2.8 for the inner and outer regions of the solar system, respectively. Accordingly, we should rewrite Eq. (15) as

$$\sigma_{col} = \pi(r + r')^2 f^2 (1 + 2\theta/f). \tag{17}$$

Thus, the coalescence cross section, σ_{col}, is enlarged by a factor of f^2 in the case where $2\theta/f < 1$. Hereafter, we adopt Eq. (17) for the coalescence cross section, keeping in mind that there still remains some uncertainty due to the free-space approximation as mentioned above.

B. Random Velocity

We now consider the magnitude of relative velocity between planetesimals which are circulating around the Sun. As a result of gravitational scattering due to their encounters with each other, their random motion is excited. By the term random motion we mean here a deviation from Keplerian motion which has the same semimajor axis but has null eccentricity and inclination. On the other hand, the random motion is dissipated by the drag of the nebular gas if the random velocity is too large. Consequently, planetesimals have equilibrium random velocities which are determined by a balance between the above excitation and dissipation. Relative velocities between planetesimals are of the same order of magnitude as random velocities.

First of all, we consider the excitation of random motion due to scattering. In the case of gravitational scattering in free space, the excitation time t_{exc} is given by Chandrasekhar's relaxation time (Chandrasekhar 1942, pp. 48–79), i.e.,

$$t_{exc} = \frac{1}{n\, \sigma_{ext} \mathrm{v}} \tag{18}$$

with

$$\sigma_{exc} = \pi \left(\frac{2Gm}{\mathrm{v}^2} \right)^2 \ln \Lambda \tag{19}$$

and

$$\Lambda = \left(\frac{\mathrm{v}^2}{2Gmn} \right)^{1/3} \tag{20}$$

where n is the number density of planetesimals, v is the equilibrium random velocity, and we have assumed for simplicity that all the planetesimals have the same mass m. Equation (20) indicates that Λ is the ratio of the mean separation distance $n^{-1/3}$ to the impact parameter for 90° deflection, $2Gm/\mathrm{v}^2$. Equation (19) shows that the excitation cross section is larger than that of 90° deflection by a factor $\ln \Lambda$. Note that Eq. (18) has been derived under the assumption of successive two-body scatterings in free space. We adopt this equation hereafter, although it is not yet clear whether it is always a good approximation to the case of planetesimals circulating around the Sun.

Next, we consider the gas drag effect on the decay of random motion. The gas drag force F_D acting on a planetesimal can be expressed as

$$F_D = \tfrac{1}{2} C_D \pi r^2 \rho_{\mathrm{gas}} (\Delta u)^2 \tag{21}$$

where C_D is a dimensionless drag coefficient which depends on the Reynolds number Re as well as on the Mach number $\mathfrak{M}$, and Δu is a relative velocity between the planetesimal and the nebular gas. For simplicity we put $C_D = 1$ because, according to Adachi et al. (1976) and Weidenschilling (1977*a*), it takes a value between 0.5 and 1.5 for a planetesimal with mass larger than 10^{18} g, if its gravity effect on the surrounding gas is negligible (see Sec. VI.E for this gravity effect). Now the dissipation time t_{dis} of random motion is given by

$$t_{\mathrm{dis}} = \frac{m\mathrm{v}}{F_D} \simeq \frac{2m}{\pi r^2 \rho_{\mathrm{gas}} \mathrm{v}} \tag{22}$$

where we have put $\Delta u \simeq \mathrm{v}$ for simplicity.

Putting $t_{\mathrm{exc}} = t_{\mathrm{dis}}$, we find that the equilibrium random velocity v as well as the corresponding eccentricity e and inclination i are given by

$$\frac{\mathrm{v}}{\mathrm{v}_K} = e = \sqrt{2}\, i = \left[\left(\frac{a\sigma_{\mathrm{dust}}}{r^2 \rho_{\mathrm{gas}}}\right) \ln\Lambda\right]^{1/5} \left(\frac{m}{M_\odot}\right)^{2/5} \tag{23}$$

where a is the heliocentric distance and v_K $(=(GM_\odot/a)^{1/2})$ is the Keplerian velocity (for derivation, see Hayashi et al. [1977] and Nakagawa [1978]). Note that there exists an approximate equipartition between eccentricity e and inclination i. Expressions for σ_{dust} and ρ_{gas} of our solar nebula model are given by Eqs. (4) and (6), respectively. Substituting these into Eq. (23), we obtain the following results.

The values of e at the orbits of the Earth, Jupiter, and Neptune are plotted as a function of mass m in Fig. 6. This figure shows that e is of the order of 10^{-4} to 10^{-3} for $m = 10^{18}$ g and becomes gradually larger with increasing mass. This reflects the fact that, with the increase of mass, the dissipation of random motion by gas drag becomes less effective while the excitation due to scattering becomes more effective. Furthermore, using the value of v, we can calculate the values of 2θ and t_{exc}. In Fig. 6, the enhancement factor $2\theta/f$ appearing in Eq. (17) is shown by the dashed curves. We see that $2\theta/f$ takes a

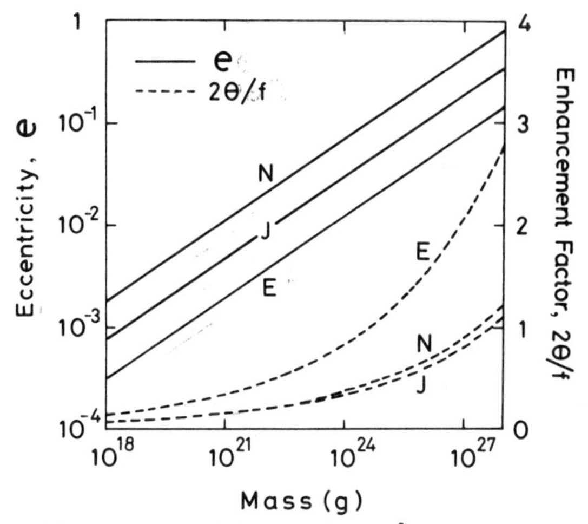

Fig. 6. The eccentricity e and the enhancement factor $2\tilde{\theta}/f$ as functions of the mass of a planetesimal in the equal mass approximation. The zones of the Earth, Jupiter, and Neptune are denoted by E, J, and N, respectively. We have put $f = 1.8$ for E and 2.8 for J and N (Nakagawa et al. 1983).

value of the order of unity and hence, the gravitational enhancement of a collision cross section is not large so long as the planetesimals collide with the equilibrium velocity as given by Eq. (23). The excitation time t_{exc} ($= t_{\text{dis}} \propto m^{1/15}$) has a very weak dependence on mass m and takes values of about 10^4, 10^6, and 10^8 yr in the Earth, Jupiter, and Neptune zones, respectively. It should be noticed that this excitation time is much smaller than the growth time of planets as described in Secs. VI.C–VI.F.

C. Growth Time in the Case of No Migration

Here, we also assume as in Sec. VI.B that planetesimals have the same mass. Furthermore, we assume for simplicity that there exists no radial migration of planetesimals and hence, the surface mass density distribution is given by Eq. (4) (the above two assumptions will be removed in Sec. VI.E). Then, the number density n of planetesimals is given by

$$n = \frac{\sigma_{\text{dust}}\Omega_K}{2m\text{v}} \tag{24}$$

where Ω_K ($= 2\pi/t_K$) is the Keplerian angular frequency; note that v/Ω_K ($\simeq \sqrt{2}ai$) represents the half-thickness of the planetesimal layer (i.e., a disk-like region in which the planetesimals are orbiting). We can now estimate the growth time of planets t_{grow} using the coalescence cross section σ_{col} given by Eq. (17) and the relative velocity v given by Eq. (23), i.e., we obtain

$$t_{\text{grow}} = \frac{1}{n\sigma_{\text{col}}\text{v}} = \frac{\rho_{\text{mat}}r}{3\pi\sigma_{\text{dust}}} \frac{t_K}{f^2(1 + 2\theta/f)} \tag{25}$$

where ρ_{mat} ($= 3m/4\pi r^3$) is the material density of planetesimals. Then, using Eq. (4) for σ_{dust}, we finally obtain

$$t_{\text{grow}} = \begin{pmatrix} 1.5 \\ 0.22 \end{pmatrix} \frac{10^4}{f^2(1 + 2\theta/f)} \left(\frac{m}{10^{18} g}\right)^{1/3} \left(\frac{a}{1 \text{ AU}}\right)^3 \text{yr} \tag{26}$$

where the coefficients 1.5 and 0.22 in parentheses correspond to the regions $a < 2.7$ AU and $a > 2.7$ AU, respectively.

It should be noted that the growth time t_{grow} depends on the random velocity v only through the Safronov number θ. Hence, in the case where $2\theta/f < 1$ as shown in Fig. 6, it is proportional to ra^3. Namely, with increasing distance from the Sun, t_{grow} increases rapidly as a^3, because both t_K and $1/\sigma_{\text{dust}}$ are proportional to $a^{3/2}$. Now, from Eq. (26), we obtain 3×10^7 yr for the growth time of the Earth (with $m = 1\ M_\oplus$, $M_\oplus$ being the present Earth mass) and 1×10^9 yr for that of Jupiter's core (with $m = 10\ M_\oplus$), whereas, for Neptune's core (with $m = 10\ M_\oplus$), we obtain 2×10^{11} yr which is far beyond the age of the solar system. Such an unrealistic growth time does suggest that we have overlooked some important process which accelerates the planetary growth, as we show in the next section.

D. Effect of Radial Migration on the Growth Time

In the above we have not considered the migration of planetesimals in the radial direction, i.e., the change in semimajor axes of their orbits. In reality, mutual gravitational scattering gives rise to a radial diffusion of planetesimals (Nakagawa 1978; Nakagawa et al. 1983). Furthermore, the gas drag causes a gradual flow of planetesimals toward the Sun (Adachi et al. 1976; Weidenschilling 1977*a*). Such migration from a region where σ_{dust} is large may accelerate planetary growth. First of all we consider radial diffusion. In terms of eccentricity e and excitation time t_{exc}, the diffusion coefficient D_r is written as $D_r = e^2a^2/t_{\text{exc}}$. Using this expression for D_r, we now define a characteristic time of radial diffusion t_{dif} as a time required for planetesimals to diffuse over a radial length $a/10$, i.e.,

$$t_{\text{dif}} = (a/10)^2/D_r = t_{\text{exc}}/100\, e^2. \tag{27}$$

This diffusion time t_{dif} is much larger than the growth time t_{grow} in early stages of accumulation. However, t_{dif} becomes smaller than t_{grow} when the mass of planetesimals grows above 10^{25} g because the random velocity increases with mass. Accordingly, the effects of diffusion on planetary growth can no longer be neglected and hence, the growth time is no longer given by

Eq. (25). In fact, Nakagawa et al. (1983) found by numerical simulation that the diffusion considerably accelerates planetary growth, as described below in Sec. VI.E.

Next we consider the inward flow of planetesimals due to gas drag. The radial flow velocity v_r is given by

$$v_r = -\left(\frac{C_D \pi r^2}{m}\right)\rho_{gas} a(e + i + \eta)\eta v_K \tag{28}$$

where η is one-half the ratio of the gas-pressure gradient to the solar gravity, i.e.,

$$\eta = -\frac{1}{2}\left(\frac{d\rho_{gas}}{da}\right) \Big/ \left(\frac{GM_{\odot}\rho_{gas}}{a^2}\right) = 1.8 \times 10^{-3}\left(\frac{a}{1\ \mathrm{AU}}\right)^{1/2} \tag{29}$$

(Adachi et al. 1976; Weidenschilling 1977*a*). The characteristic time of the inward flow t_{flow} is given by a time required for planetesimals to travel over a radial distance $a/10$, i.e.,

$$t_{flow} = \frac{a/10}{|v_r|} = \frac{t_{dis}}{10\eta}(1 + \eta/e). \tag{30}$$

The flow velocity v_r is quite small and so the flow time t_{flow} is larger than the growth time t_{grow}. Hence, we can expect that radial flow is not essential to planetary growth.

Besides the above migration, we should also take into account the size distribution of planetesimals in order to estimate the growth time more precisely, as the migration itself brings about a mixing of planetesimals with different masses. For this, we need a numerical simulation as described below.

E. Numerical Simulation of Accumulation

There are two alternative approaches to the study of the accumulation process: one is to solve the *N*-body problem by direct computation of orbits and another is to simulate it by means of a statistical treatment using a distribution function. The former has an advantage that, in principle, no knowledge of scattering and coalescence cross section is needed if the orbits of all planetesimals are computed directly. However, if the particle number *N* is very large, practical difficulties arise. The number of planetesimals at the initial stage is on the order of 10^{12}. Therefore, the latter approach using a distribution function is now preferable at least for the early stages of planetary growth.

Nakagawa et al. (1983) have derived a growth equation for a distribution of planetesimals as a function of *m* and *a*. The equation contains terms representing the coalescence as well as the inward flow and the radial diffusion mentioned in Sec. VI.D. The whole equation is so complicated that here we write it only schematically as (see Nakagawa et al. 1983, for details)

$$\frac{\partial}{\partial t}\sigma(m,a) = \text{(inward flow)} + \text{(diffusion)} + \text{(coalescence)} \qquad (31)$$

where $\sigma(m,a)\mathrm{d}m$ is a surface mass density of planetesimals at heliocentric distance a and with mass between m and $m + \mathrm{d}m$. In Eq. (31), the diffusion term has a form such that angular momentum is strictly conserved. Also, the effect of tidal disruption has been taken into account in the coalescence term through the factor f mentioned in Sec. VI.A. However, a term representing collisional fragmentation has not been included in Eq. (31) for the following reason. (Here we should recall the effect of gas drag.) Even if collisional fragmentation occurs, the fragments may be easily decelerated owing to gas drag and may possibly reaccrete owing to gravitation. Fujiwara and Tsukamoto (1980) studied experimentally the motion of fragments produced at the time of a catastrophic destruction and found that, except for fine-grained particles ejected from an impact site, most fragments have relatively small velocities ($\lesssim 10^2$m s^{-1}). From these results, we can expect that in the gaseous nebula collisional fragmentation may have no essential effect on the accumulation.

Nakagawa et al. (1983) numerically integrated Eq. (31) for the four representative planetary zones, those of Earth, Jupiter, Saturn, and Neptune and found that proto-Earth, proto-Jupiter, proto-Saturn, and proto-Neptune with masses of 1×10^{27} g are formed in a period of 5×10^6, 1×10^7, 1.6×10^8, and 4.6×10^9 yr, respectively (Table II). They also found that after the largest body has grown to about 10^{25} g, radial diffusion effectively accelerates the accumulation. In other words, planetesimals lying in the regions where a is smaller grow faster as seen from Eq. (26), and the resulting larger planetesimals, which also have larger mobility, diffuse to regions where a is larger; i.e., planetary growth is initially slower. As a result, the accumulation in these outer regions is accelerated. In fact, the computed growth times are

TABLE II
Growth Time of Planets (yr)

Mass	Planet		
	Earth	Jupiter	Saturn
1×10^{27} g[a]	5×10^6	1×10^7	2×10^8
Present mass or critical core mass[b]	1×10^7	4×10^7	6×10^8

[a]Growth times for this mass are the results of numerical simulation by Nakagawa et al. (1983).
[b]The critical core mass is assumed to be 10 $M_\oplus$. These values of growth time, obtained from the scaling law, i.e., Eq. (26), are further reduced by a factor 2 or 3 owing to the gas drag effect found by Takeda et al. (1985).

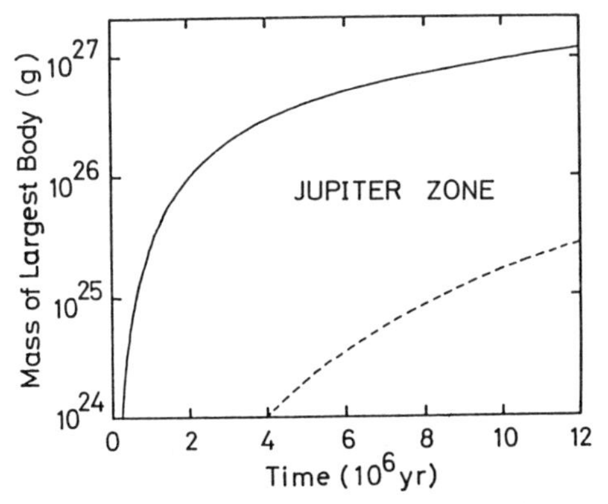

Fig. 7. Mass of the largest body growing in the Jupiter zone (the solid curve). For comparison, the growth of mass given by Eq. (26) for the case of no migration is also shown by the dashed curve (Nakagawa et al. 1983).

considerably smaller than those obtained from Eq. (26) (see Fig. 7), although the enhancement factor $2\theta/f$ still remains of the order of unity. The shape of the mass spectrum obtained is shown in Fig. 8, where $m\sigma(m,a)$ is plotted as a function of mass m.

Nakagawa et al. (1983) terminated the computation at a stage where the most massive planetesimal had grown to 1×10^{27} g because, in the later stages, their treatment using the distribution function $\sigma(m,a)$ became meaningless, i.e., there existed only a very small fraction of a massive body in every numerical zone. They estimated the final growth time of the Earth to be 1×10^7 yr by extrapolation, using the scaling law, $t_{\mathrm{grow}} \propto m^{1/3}a^3$, as given by

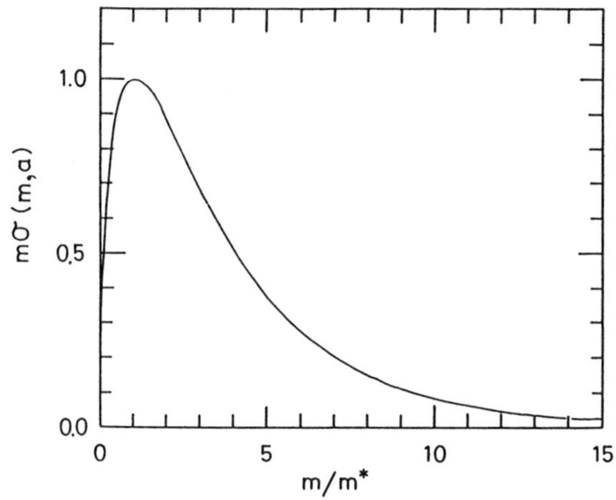

Fig. 8. Typical form of $m\sigma(m,a)$ as a function of m/m^*, where m^* denotes the peak mass of the curve.

Eq. (26). This growth time is about one-tenth of those obtained by Safronov (1969) and Wetherill (1980*a,b*) who did not take into account the effects of tidal coalescence and radial diffusion. Both proto-Jupiter and proto-Saturn with mass of 1×10^{27} g must grow further to bodies with mass of about 10 $M_\oplus$ before the surrounding atmospheres become gravitationally unstable and hence, accrete onto the bodies (described in Sec. VII). Using the above scaling law, we can estimate that the masses of proto-Jupiter and proto-Saturn grow up to 10 $M_\oplus$ in 4×10^7 and 6×10^8 yr, respectively (Table II).

The above result indicates that in the Neptune zone, the most massive body has grown up only to 1×10^{27} g in the whole age of the solar system. This suggests that the above study is still incomplete, as we point out again in Sec. VI.F. Here, we have calculated a mean value of the growth time by means of a statistical treatment of a number of interacting bodies. However, this treatment will break down if the planetary growth is a stochastic process where a fluctuation is greater than the mean value, e.g., if a large scattering of two Keplerian bodies is predominant over successive small-angle scatterings.

F. Acceleration of Growth: Gas Drag Effect on Gravitating Massive Bodies

Accumulation of solids is accelerated if the surface mass density σ_{dust} is higher or if the Safronov number θ is larger, as shown by Eq. (25). However, we cannot increase the surface mass density by a significant factor, for it seems difficult to remove from the solar system a large amount of disused mass remaining after the completion of planetary formation. Hence, in the following, we consider the latter possibility.

To obtain a larger value of θ, a smaller collision velocity is required. In their numerical computation, Nakagawa et al. (1983) used the equilibrium random velocity which was based on the gas drag law for a nongravitating body. We should reconsider this gas drag law. Protoplanets larger than 10^{26} g attract the surrounding nebular gas to form their own atmospheres (see Sec. VII.A). The gas drag acting on such massive bodies will be much larger than that expected from the law for nongravitating bodies. Recently, Takeda et al. (1985) numerically simulated steady gas flow around a spherical gravitating body and found that the drag coefficient C_D defined by Eq. (21) is significantly larger than unity in the supersonic case and also in the subsonic case where $\theta >> 1$. Some of their results are shown in Fig. 9, where C_D is plotted as a function of the Mach number $\mathfrak{M}$ and the parameter $Gm/r\text{v}^2$, where v is the flow velocity at infinity, i.e., the same as the relative velocity Δu appearing in Eq. (21). Note that the parameter $Gm/r\text{v}^2$ has a value nearly equal to the Safronov number. We see that the drag coefficient C_D increases rapidly with increases of $Gm/r\text{v}^2$ and the Mach number. From these results, Takeda et al. estimate that the growth times of planets given in Table II will be reduced by a factor of 2 or 3, at least.

We cannot expect the above reduction of the growth time for the accumulation in the Neptune zone because, as described in Sec. VII.D, Neptune

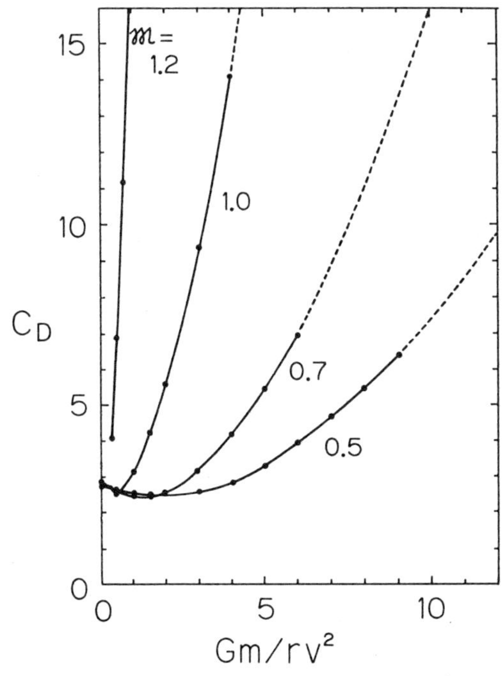

Fig. 9. Drag coefficient C_D as a function of Gm/rv^2 and the Mach number $\mathfrak{M}$. The ratio of specific heats and the Reynolds number are 1.4 and 20, respectively. The solid and dashed curves represent the results of numerical simulation and theoretically extrapolated values, respectively (Takeda et al. 1985). When $\mathfrak{M}$ is large, C_D increases rapidly with Gm/rv^2. The initial decrease of C_D with increase of Gm/rv^2, as seen in the curves of $\mathfrak{M} = 1$, 0.7, and 0.5, is due to the fact that the gas density near the stagnation point is higher than that in the other regions and hence, the sphere is gravitationally attracted by the gas in a direction opposite to the flow. On the other hand, when Gm/rv^2 is large, the gas density in the rear regions becomes much higher and the gravity of the gas acts in a direction of the flow to enhance the gas drag.

(and Uranus also) is to be considered as growing mostly in stages when the solar nebula has already disappeared. Therefore, as to the dissipation of random motion of proto-Uranus and proto-Neptune, we may have to consider the effects of so-called dynamical friction and various inelastic processes (e.g., coalescence, disruption, and tidal action), instead of gas drag.

VII. FORMATION OF GIANT PLANETS AND DISSIPATION OF THE SOLAR NEBULA

In the previous section, we have shown that the terrestrial planets grow through accumulation of planetesimals in the presence of the nebular gas. The growing protoplanet with mass $> 10^{26}$ g attracts the gas by its gravity and it is surrounded by a relatively dense gas, which we will hereafter call the primordial atmosphere. The atmospheric mass increases with the growth of the protoplanet, and in the case of a very massive protoplanet, it becomes comparable with or greater than the protoplanetary mass (the thermal effect of the primordial atmosphere is described in Sec. VIII.A). In this case, the self-gravity of the atmosphere can no longer be neglected and it is expected that the atmosphere becomes gravitationally unstable and collapses onto the surface of the protoplanet.

The above stability of the primordial atmosphere was first studied by Perri and Cameron (1974) under the assumption that the atmosphere is wholly convective. Mizuno et al. (1978) investigated the same problem for a different boundary condition (which corresponds to our Kyoto model of the solar nebula) and also under the improved assumption that the atmosphere consists of an outer isothermal region and an inner adiabatic region. Further, Mizuno (1980) calculated the atmospheric structure taking fully into account the effect of heat transport. In this section, we describe first the above-mentioned instability of the primordial atmosphere, and then the subsequent gas capture process which leads to the formation of the present giant planets. Finally, we discuss the dissipation of the solar nebula, which explains the reason why the masses of the outer giant planets are smaller.

A. Stability of the Primordial Atmosphere

The existence of the primordial atmosphere with solar chemical composition as well as its importance to planetary formation was first pointed out by Hayashi et al. (1979). They assume that the atmosphere is spherically symmetric and in hydrostatic equilibrium. Generally, it consists of an outer radiative region and an inner convective region. Throughout these regions, the outflow of energy L is assumed to be constant and equal to the rate of release of gravitational energy of accreting planetesimals, i.e.,

$$L = GM_p\dot{M}_p/R_p \tag{32}$$

where M_p and R_p are the mass and radius of a protoplanet, respectively, and $\dot{M}_p$ is the growth rate of planetary mass. The opacity of the gas κ which is important in determining the temperature distribution in the atmosphere, is given by the sum $\kappa_{mol} + \kappa_{dust}$, where κ_{mol} and κ_{dust} are the contributions from molecules and dust grains, respectively. The value of κ_{mol} was calculated in detail by Mizuno (1980), but κ_{dust} is now regarded as a parameter because it depends on the size and number density of dust grains existing in the atmosphere, both of which are not well known at present.

As an example, let us now consider the case of proto-Jupiter. In this case, we can put $\dot{M}_p = 10\ M_\oplus/10^7$ yr in accordance with the result obtained in Sec. VI. Then, by integration of equations of hydrostatic equilibrium and heat transfer, we can calculate the atmospheric structure if the values of M_p, R_p, and κ_{dust} are given. Figure 10 shows the M_{total}-M_p relation (M_{total} being the

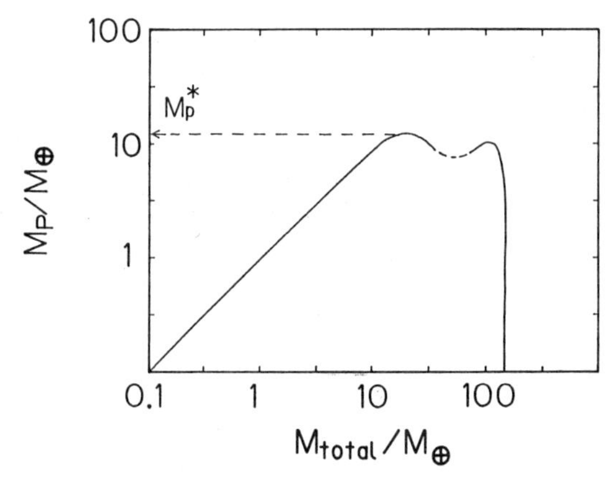

Fig. 10. Relation between the mass of a protoplanet M_P and the total mass M_{total}. Parameters are set as follows: $a = 5.2$ AU (i.e., the Jupiter region), $\dot{M}_P = 10\ M_\oplus/10^7$ yr and $\kappa_{\text{dust}} = 1\ \text{cm}^2\ \text{g}^{-1}$. The dashed curve shows unstable atmospheric models and $M_P{}^*$ denotes a critical mass for instability (Mizuno 1980).

sum of planetary and atmospheric masses) obtained by Mizuno (1980) for the case of $\kappa_{\text{dust}} = 1\ \text{cm}^2\ \text{g}^{-1}$. When M_p is small, the mass attracted into the Hill sphere is very small and hence, we have $M_p \simeq M_{\text{total}}$. For the other extreme case, $M_{\text{total}} >> M_p$, we have a purely gaseous protoplanet with no core. One of the remarkable results is that there exists no equilibrium atmosphere when the protoplanet has a mass greater than a certain critical value $M_p{}^*$. This is due to a kind of gravitational instability similar to that of Jeans instability.

The critical mass $M_p{}^*$ depends on the value of κ_{dust}. As seen from Fig. 11, $M_p{}^*$ increases monotonically with κ_{dust}. This can be easily understood: if

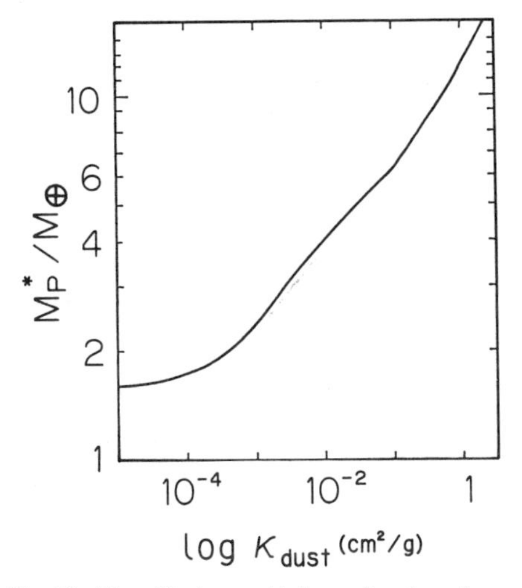

Fig. 11. The critical mass $M_P{}^*$ as a function of κ_{dust}.

the opacity is high, we have a steep temperature gradient and the resulting steep pressure gradient can sustain a massive atmosphere against gravity. In the limiting case where $\kappa_{dust} >> 1$ cm^2 g^{-1}, M_p* becomes as large as 70 $M_\oplus$ and this mass corresponds to that obtained by Perri and Cameron (1974). An optimum value of κ_{dust} is of the order of 1 cm^2 g^{-1} (Mizuno 1980; Mizuno and Wetherill 1984). Furthermore, M_p* was found to be almost independent of the planetary distance from the Sun, if κ_{dust} does not change appreciably throughout the solar nebula. Thus, we have for M_p* a universal value of about 10 $M_\oplus$. [We consider that most of the mass of accreting planetesimals is of ice beyond the asteroidal region. Ice will presumably be vaporized and mixed in the innermost convective region of the atmosphere after the accretion onto the massive protoplanet. Even in this case, however, the critical mass M_p* is not changed much if the core mass M_p is regarded as the sum of metallic/rocky and icy components.]

The above result is important to the formation of giant planets. For instance, when a protoplanet grows above the critical mass M_p*, then the surrounding atmosphere cannot be in hydrostatic equilibrium but has to contract toward the surface of the protoplanet to form a compact proto-giant-planet. One should note that the total mass M_{total} of this proto-giant-planet is about 2 M_p*, i.e., about 20 $M_\oplus$. This is much smaller than the masses of the present Jupiter and Saturn. Accordingly, the nebular gas in the outer regions has to continue to accrete onto the proto-giant-planet until Jupiter or Saturn is formed.

B. Capture of the Nebula Gas

When the above proto-giant-planet is formed, the Hill sphere of the protoplanet becomes almost vacant. As a result, the nebular gas outside the Hill sphere will flow into it. How and in what time does the nebular gas flow into the vacant Hill sphere? This problem has been investigated recently by Sekiya et al. (1984). They studied hydrodynamics of this gas inflow by means of three-dimensional computations with a so-called smoothed particle method (using 4000 particles). They assumed that proto-Jupiter has null eccentricity and that, when the inflowing gas reaches the planetary surface, it accretes immediately onto the surface. As an example of their results, Fig. 12 shows the streamlines of fluid elements at a stage where the mass of proto-Jupiter is one-tenth of its present size. As seen from this figure, the nebular gas existing originally in a ring-like belt around the Sun with a width of about $4r_H$ (r_H being the radius of the Hill sphere given by Eq. (33), can easily flow into the vacant Hill sphere of proto-Jupiter (see also Fig. 13). They also found that the inflow rate of the nebular gas is of the order of 1 $M_\oplus$/100 yr. Considering that the Helmholz-Kelvin time (i.e., the time of contraction due to radiative energy loss) of a gaseous envelope of proto-Jupiter is longer than the above-mentioned accretion time, they estimated that Jupiter is formed in a period of 10^5 to 10^6 yr, i.e., in a typical time of the Helmholz-Kelvin contraction. This

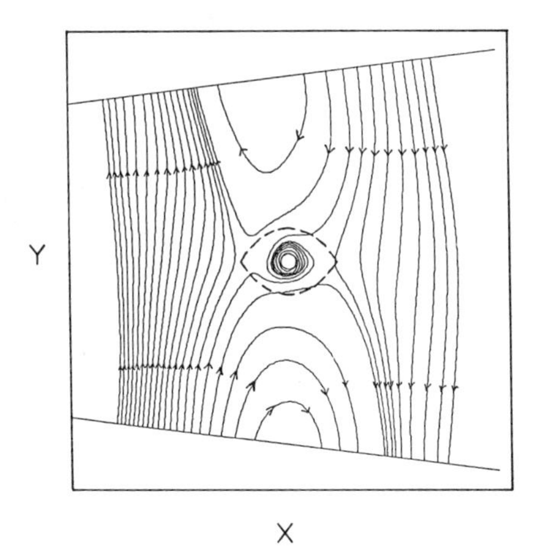

Fig. 12. Equatorial streamlines of nebular gas elements around proto-Jupiter with one tenth of the present mass of Jupiter. The dashed curves denote the Hill sphere of proto-Jupiter, and x and y are the heliocentric coordinates in a rotating frame where both the Sun and proto-Jupiter are at rest.

formation time is short compared with the dissipation time of the solar nebula, which is of the order of 10^7 yr, as shown below in Sec. VII.C.

Using the above results, we now estimate the final masses of the giant planets. Owing to the gas accretion, the total planetary mass $M_{\rm total}$ increases with time and hence, the Hill radius r_H (and also the width of the ring $4r_H$) increases according to the equation

$$r_H = a\left(\frac{M_{\rm total}}{3M_\odot}\right)^{1/3} \tag{33}$$

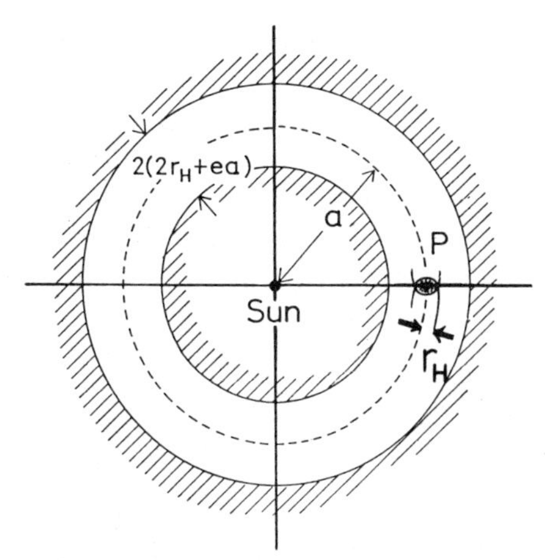

Fig. 13. Solar nebula with a vacant ring-like belt formed as a result of accretion of the nebular gas onto a proto-giant-planet. Here, P denotes the proto-giant-planet and a and e are the semimajor axis and the eccentricity of the proto-giant-planet, respectively.

where a is the distance of the protoplanet from the Sun. The expansions of the Hill sphere and the ring enable a new gas to flow into the Hill sphere. These processes continue successively and the mass of the protoplanet becomes larger and larger. The growth of mass cannot continue infinitely but stops at a certain final mass for the following reason. The mass of the solar nebula available for the accretion is proportional to the width of the ring $4r_H$, whereas r_H itself increases only as $r_H \propto (M_{\text{total}})^{1/3}$, according to Eq. (33).

More precisely speaking, if we consider the eccentric motion of the planet, the available mass is given by (see Fig. 13)

$$M = 2\pi a \cdot 2(2r_H + ea)\sigma_{\text{gas}} \tag{34}$$

where a and e are the semimajor axis and the eccentricity of the orbit, respectively. Now, putting $M = M_{\text{total}}$ in Eqs. (33) and (34), we can find the final masses of the giant planets if σ_{gas} is given. If we assume that the solar nebula has not yet dissipated and adopt the value of σ_{gas} given by Eq. (5), then we find that the final masses are $3.5 \times 10^2 M_\oplus$, $5.2 \times 10^2 M_\oplus$, $8.4 \times 10^2 M_\oplus$ and $1.1 \times 10^3 M_\oplus$ for the regions of Jupiter, Saturn, Uranus and Neptune, respectively. These masses do not agree with those of the present planets except for Jupiter. The discrepancy is considered as due to the above assumption that the nebular gas is still existing at stages where the cores of Saturn, Uranus and Neptune have grown to the critical mass of gravitational instability. This discrepancy can be removed if we consider the dissipation of the nebular gas (see Sec. VII.D).

C. Escape of the Solar Nebula

At present, the solar nebular gas, which has played important roles in various stages of planetary formation, no longer exists; the nebular gas must have escaped at some stage of evolution of the solar system. Elmegreen (1978) and Horedt (1978) first studied the escape process considering the effect of a strong solar wind existing in the T Tauri stage of the early-Sun. Besides the wind particles, the effect of ultraviolet radiation and also magnetic field accompanied by the wind must be studied. At present, however, we have no precise theory in which all of these effects have been taken into account.

Following the analysis by Sekiya et al (1980*a*), let us estimate the order of magnitude of the escape time of the solar nebula. For our model of the solar nebula described in Sec. III, the gravitational binding energy E_b is given by

$$E_b = -\int \frac{GM_\odot}{2r}\sigma_{\text{gas}} 2\pi r dr = -3.6 \times 10^{43} \text{ erg.} \tag{35}$$

If we consider the T Tauri stellar wind and ultraviolet radiation, the energy flow available for escape of the nebula is written as

$$L_{\text{av}} = \frac{\Omega}{4\pi}(\zeta_{\text{sw}} L_{\text{sw}} + \zeta_{\text{uv}} L_{\text{uv}}) \tag{36}$$

where the ζ's and L's are the efficiency of energy conversion and the luminosity, respectively, and subscripts sw and uv denote the solar wind and ultraviolet radiation, respectively. Both ζ_{sw} and ζ_{uv} are known to be of the order of 0.1 (Sekiya et al. 1981). Furthermore, Ω is the heliocentric solid angle subtended by the disk of the solar nebula and is nearly equal to π for our nebula model. The characteristic escape time is then given by

$$t_{es} = \frac{|E_b|}{L_{av}} = 1.3 \times 10^6 \frac{10^{-3} L_\odot}{\zeta_{sw} L_{sw} + \zeta_{uv} L_{uv}} \frac{\pi}{\Omega} \text{yr.} \tag{37}$$

The mean values of L_{sw} and L_{uv} of T Tauri stars are not well known at present. Observations of one T Tauri star show that the far-ultraviolet luminosity is about 10^3 times that of the present Sun, i.e., we have nearly $L_{uv} = 10^{-2} L_\odot$ (Gahm et al. 1979). The wind-particle luminosity is also believed to be as large as 10^5 to 10^6 times the present solar value. Hence, $L_{sw} + L_{uv}$ is probably of the order of $1 \times 10^{-2} L_\odot$, or smaller if we take into account its violent time variation. If we put $L_{sw} + L_{uv} = 3 \times 10^{-3} L_\odot$, we obtain several $\times\ 10^6$ yr for the escape time of the solar nebula. Note here that this time is of the same order of magnitude as that of the T Tauri stage of stellar evolution, as calculated by Ezer and Cameron (1965). Thus, the solar nebula may be dissipated almost completely in a period of about 10^7 yr.

D. Final Masses of the Giant Planets

Four giant planets, namely, Jupiter, Saturn, Uranus, and Neptune are known to have a structure quite different from that of the terrestrial type planets; namely, a core composed of mineral and ice is surrounded by a more or less massive envelope composed mainly of H_2 and He. Slattery (1977) and Hubbard and MacFarlane (1980) studied extensively the structure of the giant planets and obtained a remarkable result that all the giant planets have almost the same core mass of about 15 $M_\oplus$ in spite of large differences in total mass and in distance from the Sun. In other words, the envelope masses of giant planets decrease with the increase of distance from the Sun.

In Sec. VII.B, we obtained tentative values for the final masses of the giant planets. Now, we will revise these values by comparing the growth time of the planetary core with the escape time of the solar nebula. In Jupiter's region, the core can grow to the critical mass of instability in a period of the order of 10^7 yr, i.e., it grew before the dissipation of the solar nebula. Thus, Jupiter has the final mass estimated in the above.

As seen in Sec. VI, however, the growth time of the core in Saturn's region is of the order of 10^8 yr and this is longer than that in the Jupiter region. Therefore, a part of the nebula has already escaped, before the core of Saturn grows to the critical mass by accumulating solid planetesimals. Accordingly, in the estimation of the final mass, we have to use a value of σ_{gas} which is diminished by a certain factor from the value given by Eq. (5). At present, the theory of nebula dissipation is not precise enough to predict the above diminution factor. Inversely, if we adopt 1/3.5 for the factor, we obtain 100 $M_\oplus$ for Saturn's final mass; that is, the core of Saturn would have grown to the critical mass when the density of the solar nebula has decreased by a factor of, e.g., 3.5. This estimate may not be so far off.

The growth time of the cores of the most distant planets, Uranus and Neptune, is estimated to be of the order of 10^8 to 10^9 yr, which is far greater than the lifetime of T Tauri stars. Hence, it is likely that both of these planets grew mostly in a gas-free condition and the final mass obtained earlier has no meaning, i.e., the nebular gas has already dissipated when their cores grow to the critical mass. This provides a reason for the masses of their envelopes (i.e., about 1 $M_\oplus$) to be very small compared with their cores. Nevertheless, there remains a question as to how and when they could capture the gas with mass as large as 1 $M_\oplus$. This problem has not yet been well studied, but the following points should be noted: a solid protoplanet with mass of about 1 $M_\oplus$ can possess its own primordial atmosphere with mass of about 0.1 $M_\oplus$ (see Sec. VIII.A below); furthermore, as seen in Sec. VI.D, solid protoplanets can migrate radially over a large distance in planetary space when they are growing. Hence, we may suppose that protoplanets, which grew in the gaseous nebula, traveled to the outer distant regions and formed Uranus or Neptune by accumulating small planetesimals existing in those regions.

VIII. EVOLUTION OF PROTO-EARTH

In Sec. VI, we have already seen that the Earth grew to its present mass within 10^6 to 10^7 yr, i.e., before the nebula gas began to escape. This means that proto-Earth was surrounded by the primordial atmosphere and the existence of this atmosphere gives us a new scenario for the evolution of the early Earth, which is quite different from that obtained under the assumption of gas-free accumulation (see, e.g., Safronov 1969). Here we describe first the atmospheric blanketing effect, which is one of the most striking effects of the primordial atmosphere. Next, we discuss briefly the thermal history of the early Earth and the formation of its core and mantle structure. Finally, we examine a consistency between the existences of the primordial atmosphere and that of the present atmosphere.

A. Blanketing Effect of the Primordial Atmosphere

The primordial atmosphere has a blanketing effect such that the planetary surface (i.e., the bottom of the atmosphere) is kept very hot. In order to understand this, we should first estimate the magnitude of the blanketing effect in a qualitative manner.

First, we consider a hypothetical case where the primordial atmosphere does not exist. Then, the temperature at the planetary surface T_f is determined by a heat balance, i.e., a balance between the loss of energy due to emission of blackbody radiation and the release of gravitational energy due to accretion of planetesimals. The former is given by $4\pi R_p{}^2\sigma T_f{}^4$, where σ is the Stefan-Boltzmann constant and R_p is the planetary radius. The latter is already given by Eq. (32). Equating these, we have for the surface temperature

$$T_f^4 = \frac{L}{4\pi R_p^2 \sigma}. \tag{38}$$

Next, we consider the case where a planet is surrounded by a primordial atmosphere. For simplicity, we assume that the atmosphere is wholly in radiative equilibrium. Then, in terms of the optical depth τ $(= \int \kappa\rho dr)$, the temperature distribution is expressed as

$$\frac{dT^4}{d\tau} = \frac{3}{4\sigma}\frac{L}{4\pi r^2}. \tag{39}$$

From this equation, the temperature at the bottom of the atmosphere T_b can be estimated, in order of magnitude, as

$$T_b^4 = \frac{L\tau_b}{4\pi R_p^2 \sigma} = \tau_b T_f^4 \tag{40}$$

where τ_b is the optical depth at the planetary surface. For a dense atmosphere where $\tau_b >> 1$, we find that T_b is greater than T_f by the factor $\tau_b{}^{1/4}$. This is the blanketing effect of the atmosphere.

The details of the blanketing effect on the proto-Earth have already been calculated by Hayashi et al. (1979) and Nakazawa et al. (1985*b*), by taking into account the existence of a convective region. We summarize their results here. As mentioned in Sec. VII.A, the atmospheric structure depends on the growth rate of the proto-Earth $\dot{M}_p$, its mass M_p, and the opacity due to dust grains κ_{dust}. Of these three parameters, we put $\dot{M}_p = 1\ M_\oplus/10^6$ yr in accordance with the result obtained in Sec. VI.F and, further, put $\kappa_{dust} = 1$ cm^2 g^{-1} (Mizuno 1980; Mizuno and Wetherill 1984). The planetary mass M_p is regarded here as a free parameter. In Fig. 14, the calculated temperature T_b is plotted as a function of M_p. For comparison, we also show the hypothetical surface temperature T_f given by Eq. (38). This figure shows that the blanketing effect becomes important when $M_p > 0.1\ M_\oplus$, and the effect increases with increasing M_p, because a massive body attracts a large amount of the nebular gas which forms a dense atmosphere with large τ_b. Furthermore, for $M_p > 0.2\ M_\oplus$, T_b is higher than the melting temperature of planetary materials. This surface melting causes an important effect on the structure and the thermal history of the growing Earth; that is, the melting gives rise to a chemical differentiation of planetary materials (see Sec. VIII.B) and, as a result, a large amount of gravitational energy is released inside the proto-Earth.

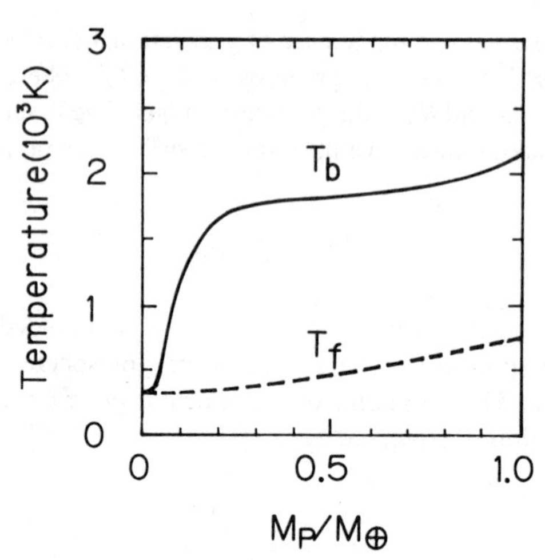

Fig. 14. The bottom temperature T_b of the primordial atmosphere of the proto-Earth as a function of the planetary mass M_P (Nakazawa et al. 1985*b*). A hypothetical temperature T_f is also shown by the dashed curve.

From a quite different viewpoint, Safronov (1978) and Kaula (1979*b*) examined the melting of the growing Earth. They considered the blanketing effect due to planetary materials themselves such that impacts of infalling planetesimals turn up the surface planetary materials and, as a result, a part of kinetic energy of planetesimals is embedded, as thermal energy, in a deep region. This self-blanketing effect depends strongly on the size of infalling planetesimals; the effect is appreciable in the case of large planetesimals but negligible for small planetesimals. At present, we have no precise knowledge about the mass spectrum of planetesimals infalling onto a growing planet and hence, there still remains an uncertainty in their evaluation of the temperature of the growing planet.

B. Formation of the Core-Mantle Structure of the Earth

As shown above, the temperature at the bottom of the primordial atmosphere exceeds the melting temperature of planetary materials when the proto-Earth grows to a mass greater than 0.2 $M_\oplus$. As a result, the surface of the proto-Earth is melted. Planetesimals newly accreting onto this molten surface can be completely melted within a period of about 10^3 yr even if their sizes are as large as 100 km (Sasaki et al. 1983). Such melting must be immediately followed by chemical differentiation between low-density silicate and high-density metal: heavy metal sediments through the molten region. As the melting time of newly accreting planetesimals and the sedimentation time of heavy metal are both very short compared with the growth time of the Earth, it is likely that the growing proto-Earth greater than 0.2 $M_\oplus$ is composed of three layers as illustrated in Fig. 15, i.e., the innermost unmelted protocore composed of a mixture of silicate and metal, the intermediate metallic layer and

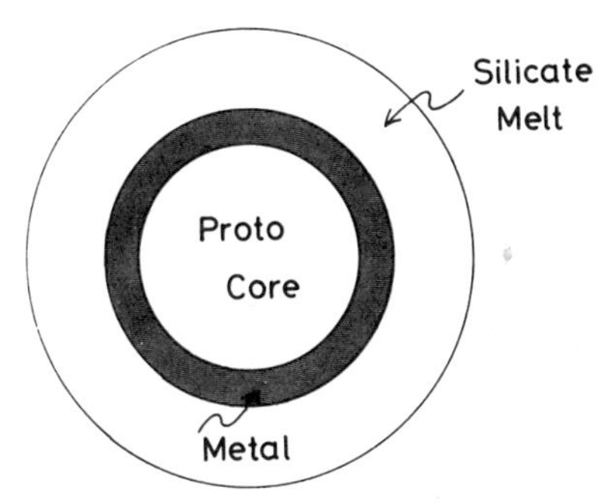

Fig. 15. Internal structure of the growing Earth, which consists of three layers: the innermost protocore, metallic layer, and outermost silicate melt.

the outermost molten silicate layer. The evolution of the proto-Earth with such a structure was first discussed qualitatively by Stevenson (1981) and recently studied in detail by Sasaki and Nakazawa (1985) as described below.

In the above-mentioned chemical differentiation and subsequent sedimentation of heavy metal, an enormous amount of gravitational energy is released in the molten silicate layer. Adibatic compression due to the increase in mass of the proto-Earth also contributes to the rise of temperature inside the proto-Earth. These two are the most important heating sources, whereas the decay of radioactive elements is negligible because the time span of evolution concerned is as short as 10^7 yr or so. On the other hand, we have outward energy flow due to convection which is developed in the silicate melt (conductive heat transport is negligible). Taking into account the above heat sources and heat transport, Sasaki and Nakazawa (1985) computed numerically the thermal evolution of the growing Earth until it has grown to the present mass. They assumed that the surface temperature of the growing Earth is equal to T_b shown in Fig. 14. Their result indicates that, after the proto-Earth has grown to the mass of 0.2 $M_\oplus$, the temperature of the outermost silicate layer is always slightly higher than the melting temperature[a] (see Fig. 16), because of the following heat transfer mechanism. If the temperature of the silicate melt is high enough, the viscosity of the melt is so small that heat is transferred efficiently outward by fully developed convection. On the other hand, if the temperature and hence, the degree of partial melting are low, then convection is suppressed strongly. As a result, the temperature cannot decrease further.

In the intermediate metallic layer, however, the temperature is not so high because there exists no energy source except for the adiabatic compression. One should note that a part of the metal can react chemically during the sedimentation and takes the form of metallic compounds such as FeO and FeS. Hence, the metallic layer must also be molten, in spite of its relatively

[a]Note that the melting temperature in deep regions of the silicate layer is much higher than that in surface regions owing to the pressure effect.

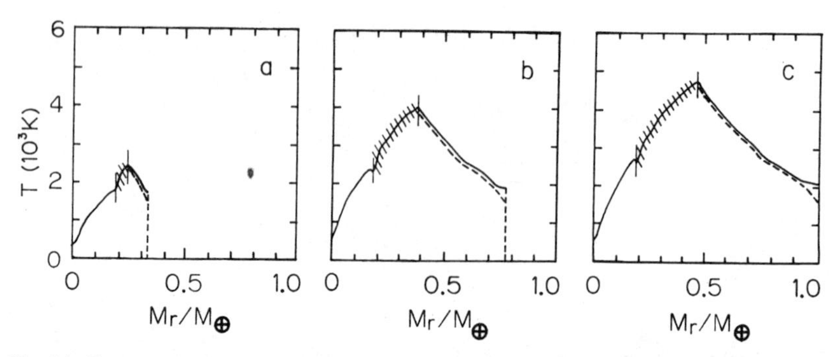

Fig. 16. Temperature distribution in the growing Earth at stages when the mass is 0.32 $M_\oplus$ (panel a), 0.77 $M_\oplus$ (panel b), and 1.0 $M_\oplus$ (panel c). Dashed curves show the melting temperature of silicate material and hatched region denotes the intermediate metallic layer.

low temperature, because the melting temperatures of such compounds are very low.

It is clear that such a molten layered structure as shown in Fig. 15 is gravitationally unstable, because the density of the intermediate metallic layer is sufficiently higher than that of the innermost protocore which consists of an unmelted mixture of silicate and metal. The instability will occur with some complicated three-dimensional modes of motion and we have not yet solved this difficult problem of gravitational instability. Hence, at present, we do not know at what stage of evolution and in what time scale the gravitational instability occurs. However, it is certain that the instability occurs at some stage and, as a result, the unmelted silicate-metal protocore is replaced by a metallic core; that is, the present core of the Earth is formed (Stevenson 1981). At the time of this replacement, gravitational energy is released and hence, the unmelted silicate-metal mixture, which is originally composed of volatile rich primordial materials, will be at least partially melted. This may also explain the origin of the present atmosphere as well as that of the lower mantle.

C. Dissipation of the Primordial Atmosphere

The mass of the primordial atmosphere is as large as 1×10^{26} g and the pressure at the planetary surface is about 50 atm, at the time when the Earth grows to its present mass (Nakazawa et al. 1985*b*). This massive atmosphere has a chemical composition completely different from that of our present atmosphere; it has a solar composition and hence, the amount of rare gases contained in the mass 1×10^{26} g is much larger than that contained in today's atmosphere (e.g., 2.6×10^6 times for Ne and 8.5×10^2 times for Xe). Hence, it is important to examine whether or not the existence of this atmosphere is in conflict with that of our present atmosphere.

It is generally accepted that the present terrestrial atmosphere was formed by the outgassing from the Earth's interior (Brown 1949), which oc-

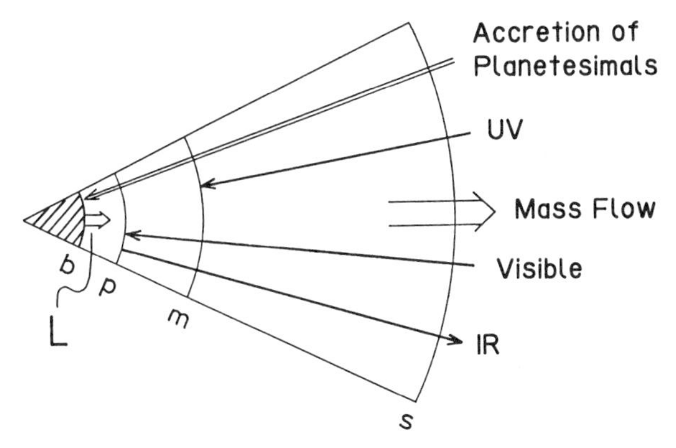

Fig. 17. Schematic picture of an escaping primordial atmosphere. The bottom of the atmosphere is denoted by *b*, the photosphere by *p*, the minimum temperature point by *m* and the sonic point by *s*.

curred catastrophically within 5×10^8 yr after the formation of the Earth (Hamano and Ozima 1978). This gives the following two constraints on the existence of the primordial atmosphere. One is that it must be dissipated within the above period 5×10^8 yr (note that the solar nebula is dissipated within 10^7 yr, as mentioned in Sec. VII.C). The second is that, besides the main components H_2 and He, heavy components such as Xe and Kr, mentioned above, must also be dissipated. These two conditions were examined by Sekiya et al. (1980*a,b*) and Sekiya et al. (1981), as described below.

We suppose that the nebular gas has already dissipated and hence, the primordial atmosphere is directly irradiated by solar radiation. The gas of the atmosphere is heated by this radiation and begins to flow outward. For simplicity, we assume that the outflow is spherical and steady. A schematic picture of the escaping atmosphere is illustrated in Fig. 17. Now, the energy equation for the outflowing gas is written as

$$\rho u \frac{dH}{d\gamma} = \Gamma \tag{41}$$

where Γ is the rate of heating by solar radiation, u is the velocity of the outflowing gas, and H is the sum of the specific enthalpy and the gravitational energy, i.e.,

$$H = \frac{1}{2}u^2 + \frac{\gamma}{\gamma - 1}\frac{P}{\rho} - \frac{GM_\oplus}{r}. \tag{42}$$

Using the relation, $\dot{M}_{\rm atm} = -4\pi\rho u r^2 = $ constant, we can rewrite Eq. (41) in the form

$$|\dot{M}_{\rm atm}| = \int_{r_m}^{r_s} 4\pi r^2 \Gamma dr \Big/ \left[\frac{GM_\oplus}{r_m} + \frac{5 - 3\gamma}{4(\gamma - 1)}\frac{GM_\oplus}{r_s} + \frac{\gamma + 1}{4}\left(\frac{\Gamma r}{\rho u}\right)_s\right] \tag{43}$$

where we have used the sonic point condition for the outer boundary (see Sekiya et al. 1980*a*), i.e.,

$$c_s^2 = u_s^2 = \frac{GM_\oplus}{2r_s} + \frac{\gamma - 1}{2}\left(\frac{\Gamma r}{\rho u}\right)_s \tag{44}$$

and neglected the terms, u_m^2 and P_m/ρ_m, which are smaller than $GM_\oplus/r_m$. In the above, the subscripts m and s denote the minimum temperature point and the sonic point, respectively (see Fig. 17).

First, in order to estimate roughly the mass loss rate $\dot{M}_{atm}$, we assume that the dominant heating source is solar ultraviolet radiation. Then, Γ is approximately given by

$$\Gamma = F_{uv}\zeta_{uv}\kappa_{uv}\rho \tag{45}$$

where F_{uv}, ζ_{uv}, and κ_{uv} are the energy flux of ultraviolet radiation, the heating efficiency and the opacity for ultraviolet radiation, respectively. Further, for simplicity, we neglect terms other than the first term in the bracket in Eq. (43). Since the optical depth for ultraviolet radiation is about unity at r_m, the mass loss rate is approximately given by

$$|\dot{M}_{atm}| = \frac{4\pi r_m^3 F_{uv}\zeta_{uv}}{GM_\oplus}. \tag{46}$$

If we take $r_m = 5\ R_\oplus$, $\zeta_{uv} = 0.4$ (see below) and $F_{uv} = 4 \times 10^3 \text{erg cm}^{-2}\text{s}^{-1}$ (which corresponds to $L_{uv} = 3 \times 10^{-3} L_\odot$, i.e., the value adopted in Sec. VII.C), then we obtain a mass loss rate as large as $5 \times 10^{19}\text{g yr}^{-1}$. Although this is a very rough value, it is found that the primordial atmosphere with mass of 1×10^{26} g can be dissipated within a period of the order of 10^6 yr.

An exact calculation of the mass loss rate was made by Sekiya et al. (1980*a*, 1981) by numerical integration of hydrodynamic equations. They considered heating processes due to extreme-ultraviolet, far-ultraviolet and visible radiation, as well as cooling due to emission of infrared radiation and adiabatic expansion of the gas. The mass loss rate that they obtained is illustrated in Fig. 18 as a function of the energy flux of ultraviolet radiation at 1 AU. This figure shows that $|\dot{M}_{atm}|$ is not always proportional to F_{uv} because, in the real case, the terms other than $GM_\oplus/r_m$ in Eq. (43) cannot be neglected, contrary to the assumption made in the above rough estimation. They also found that the dominant heat source is the far-ultraviolet radiation, which can dissociate H_2O molecules and has a heating efficiency ζ_{uv} as large as 0.4. As mentioned earlier, the far-ultraviolet flux of T Tauri stars is of the order of $10^4 \text{erg cm}^{-2}\ \text{s}^{-1}$ at a distance of 1 AU from the star (Gahm et al. 1979). If we use $F_{uv} = 4 \times 10^3 \text{erg cm}^{-2}\text{s}^{-1}$, then $|\dot{M}_{atm}|$ is about 1×10^{19} g yr^{-1}. Thus, they reached the following conclusion. The primordial atmosphere began to dissi-

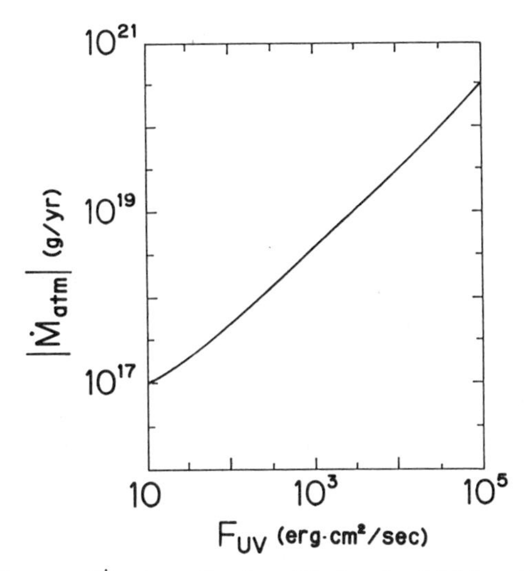

Fig. 18. The mass loss rate, $|\dot{M}_{atm}|$, as a function of the far-ultraviolet flux at 1 AU (Sekiya et al. 1981).

pate after the nebular gas had dissipated, and was completely lost from the Earth's Hill sphere within a period of 10^6 to 10^7 yr, i.e., before the formation of the present atmosphere.

Sekiya et al. (1980*b*) also examined the dissipation of the heavy rare gases such as Xe and Kr, by using the two-component gas kinetic theory. According to these authors, heavy atoms can also escape owing to the drag of outflowing H_2 molecules, and their contents are reduced below the levels observed in the present atmosphere.

The above results show that there is no inconsistency between the existence of the primordial atmosphere and that of the present atmosphere.

IX. ORIGIN OF SATELLITES AND RINGS

Since Darwin first discussed the lunar origin in 1879, many efforts have been made to solve this problem from the viewpoint of fission origin (see e.g., Ringwood and Kesson 1977; Binder 1977; O'Keefe and Sullivan 1978) as well as from other viewpoints of capture origin. Although fission origin may be able to provide an explanation for the chemical similarity between lunar regolith and the terrestrial mantle, it must solve the so-called angular momentum difficulty. Capture origin, on the other hand, also has difficulties; in order that an initially unbound body may be trapped within the Hill sphere of the Earth, its kinetic energy must be dissipated by some mechanism. Almost all the mechanisms, which have so far been proposed are not satisfactory because they are so speculative and do not fit with recent theories of planetary formation. Hence, here we will describe a theory of capture origin of the

Moon which has recently been constructed according to our Kyoto model. Furthermore, we will try to explain the origin of the other satellites together with the planetary rings.

A. Origin of the Moon

In studying the capture process of our Moon, one of the most important problems is to find a mechanism of energy dissipation for a planetesimal which has once entered the terrestrial Hill sphere. For this purpose, it is worthwhile first to examine its motion in the Hill sphere for a case where energy dissipation does not exist at all, i.e., a case where we neglect all forces other than the gravity of the Earth and the Sun. Furthermore, for simplicity, we assume that the mass of the incoming planetesimal is negligibly small compared with that of the Earth, and that the Earth is moving around the Sun along a circular orbit. In this case, the orbital motion of a planetesimal is given by a solution to the restricted three-body problem. If we adopt a corotating frame where both the Sun and the Earth are at rest, then we have the Jacobi integral, which is given by

$$E = \frac{1}{2}(\dot{x}^2 + \dot{y}^2 + \dot{z}^2) - \frac{\Omega^2}{2}(x^2 + y^2) - \frac{GM_\odot}{r_1} - \frac{GM_\oplus}{r_2} + \text{const.} \quad (47)$$

The first, second, third, and fourth terms on the right-hand side of the above equation show kinetic energy, centrifugal potential, gravitational potential due to the Sun, and that due to the Earth, respectively. The total energy E of a planetesimal is conserved in this frame.

We choose the zero-point of E so that E is negative for planetesimals which are in bound orbits around the Earth. Now, we suppose that an unbound planetesimal with some positive energy E comes from a relatively distant region and once enters the Hill sphere of the Earth. From a number of orbital computations, we find that the characteristics of its subsequent motion are quite different for the two cases: a low-energy case ($E < 17GM_\oplus/a$) and a high-energy case ($E > 17GM_\oplus/a$), where a is the distance between the Earth and the Sun, i.e., $a = 1$ AU.

In the low-energy case, a planetesimal moves around the Earth in a complicated manner as seen in Fig. 19a and after a large but limited number of revolutions, it escapes out of the Hill sphere. The number of revolutions varies from case to case, depending on the energy E as well as on the other initial conditions of motion; the number spreads over a wide range from several to several hundred and, on the average, it increases as E decreases (Nakazawa et al. 1983). One of the remarkable features of the orbital motion is that the planetesimal rotates always counterclockwise around the Earth, i.e., in the same direction as that of the orbital motion of the Earth around the Sun (Hayashi et al. 1977; Heppenheimer and Porco 1977). Furthermore, the planetesimal never comes very close to the Earth, that is, its perigee distance (the

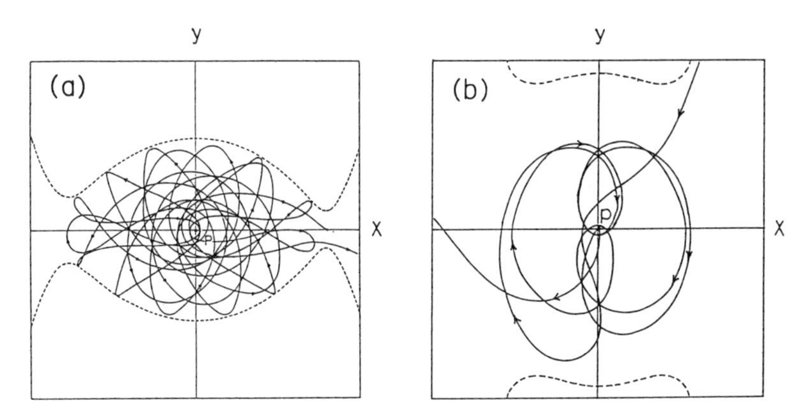

Fig. 19. Examples of equatorial orbits around the Earth (denoted by *P*) in the low-energy case (plot a) ($E = 0.33\ GM_{\oplus}/a$) and the high-energy case (plot b) ($E = 104\ GM_{\oplus}/a$). Dashed curves show the zero-velocity curves.

minimum distance from the Earth) is always greater than several times the Earth's radius.

In the high-energy case, on the other hand, a planetesimal rotates irregularly around the Earth and, in most cases, it escapes from the Hill sphere after only a few revolutions as seen in Fig. 19b (Nishida 1983). The orbital motion around the Earth is both clockwise and counterclockwise, these two directions occurring with nearly equal frequencies. Furthermore, the perigee distance becomes frequently as small as a few Earth radii and, occasionally, it collides directly with the Earth.

The above features of orbital motion of low-energy and high-energy planetesimals are of great use when we examine the mechanism of dissipation of kinetic energy. Until now, three different kinds of energy dissipation have been suggested: gas drag of the primordial atmosphere, tidal action (Mignard 1980,1981) and collision with other planetesimals moving in the Earth's Hill sphere (Ruskol 1972). One should note here that the tidal effect depends strongly on the perigee distance and hence, it becomes important only when the incoming planetesimal comes very close to the Earth. Such an encounter occurs in the above-mentioned high-energy case, but the possibility of trapping inside the Hill sphere is rather small because a large amount of energy must be dissipated in this case and, further, the planetesimal suffers from tidal disruption if it comes close to the Earth. On the other hand, collision between unbound planetesimals in the Hill sphere is a rare event because they can stay in the Hill sphere for only less than 10^2 yr.

Thus, the gas drag effect of the primordial atmosphere is the most probable mechanism for trapping the Moon. This mechanism is important only for the low-energy case because a high-energy planetesimal escapes from the Hill sphere easily before its energy is dissipated by gas drag. Note also that, if a low-energy planetesimal enters the Hill sphere at some stage before the primordial atmosphere begins to escape, the gas drag is too strong and it will eventually fall onto the Earth's surface after a large number of revolutions. On the other hand, if a planetesimal encounters the Earth at a stage after the complete dissipation of the atmosphere, it cannot be trapped but easily goes out of the Hill sphere. For this reason, Nakazawa et al. (1983) considered a case such that a low-energy proto-Moon came into the Hill sphere at a stage when the primordial atmosphere was just beginning to escape. According to Nakazawa et al. (1983), the density distribution of the escaping atmosphere at time t is written in the form factorized with respect to r and t, i.e.,

$$\rho(r,t) = \rho_o(r)\cdot \mathrm{f}(t), \;\; \mathrm{f}(t) = \exp(-\, t/t_{\mathrm{atm}}) \tag{48}$$

where $\rho_o(r)$ is the initial density distribution of the atmosphere and t_{atm} is the characteristic time of atmospheric dissipation, which is of the order of 10^6 to 10^7 yr (see Sec. VIII.C). We now consider the motion of the proto-Moon in the Hill sphere filled by the gas with the above density distribution. In this case, the energy E is no longer conserved but changes with time. Let ΔE be the change of energy during one period of revolution. Clearly, ΔE will be proportional to $\rho(r,t)$ as well as to a constant A which measures the magnitude of the gas drag effect and is given by

$$A = \frac{C_D \pi \mathrm{R}_{\mathrm{M}}^2}{2\mathrm{M}_{\mathrm{M}}} = 3.2 \times 10^{-10}\mathrm{cm}^2\mathrm{g}^{-1} \tag{49}$$

where R_{M} and M_{M} are the radius and the mass of the present Moon, respectively, and C_D is the gas drag coefficient (in this section we put $C_D = ½$). Using the ordinary perturbation method applied to the equation of motion containing a gas drag term, Nakazawa et al. (1983) found that an exact expression for ΔE is given by

$$\Delta E = -2\pi A G \mathrm{M}_{\oplus} \rho_o(r_p) \mathrm{f}(t)\, \mathcal{J} \tag{50}$$

where r_p is the perigee distance of the orbit and $\mathcal{J}$ is a correction factor which has a value of about 0.5. This factor comes from the fact that the orbit is not circular but eccentric and, further, the atmospheric density is not uniform.

Let us first make a rough estimation of ΔE. Putting $\rho_o = 1 \times 10^{-5}$ g cm^{-3} (which is the atmospheric density at $r = 6\ \mathrm{R}_{\oplus}$), we obtain $\Delta \mathrm{E} = -0.17\mathrm{f}(t)G\mathrm{M}_{\oplus}/a$ from Eq. (50). If the incoming proto-Moon has an initial energy of about 0.33 $G\mathrm{M}_{\oplus}/a$, it can revolve around the Earth more than 50 times before it escapes. The amount of energy dissipated during 50 revolutions is about $8.5\mathrm{f}(t)G\mathrm{M}_{\oplus}/a$. Thus, if $\mathrm{f}(t) > 1/25$, then the proto-Moon can be trapped within the Hill sphere.

The above estimate is very crude and, in fact, the evaluation of the trapping condition is more complicated. This is due to the fact that $\rho_o(r)$ is a rapidly decreasing function of r and the perigee distance r_p changes appreciably from revolution to revolution around the Earth (see Fig. 19a). Considering these effects, Nakazawa et al. (1983) calculated the trapping probability by means of Monte Carlo procedures. The result shows that the Moon can be trapped with an appreciable probability if $f(t) > 1/50$, i.e., if the proto-Moon comes into the terrestrial Hill sphere at some stage before the density of the escaping atmosphere is decreased to 1/50 of the initial.

After the proto-Moon has been trapped into a bound orbit, it continues to lose energy owing to gas drag. As a result, the orbital radius decreases gradually and, if the gas density of the atmosphere is too high, the proto-Moon will fragment into pieces owing to a tidal effect when the orbital radius becomes less than the Roche limit. Nakazawa et al. (1985*a*) studied this problem by calculating the effects of both gas drag and tides on the long-term orbital motion of the proto-Moon around the Earth. Their result shows that the trapped proto-Moon can survive as a satellite without falling onto the Earth and, furthermore, can reach the present lunar position, if it enters the Hill sphere at some stage where $f(t) < 1/5$. This condition is consistent with the trapping condition (i.e., $f(t) > 1/50$) and, therefore, we can conclude that the entrance of the low-energy proto-Moon into the Hill sphere at some stage, where the condition $1/5 > f(t) > 1/50$ is satisfied, is one of the possible ways of the lunar origin. One should note that the trapping in this low-energy case explains the counterclockwise revolution of the Moon around the Earth.

B. Other Satellites and Planetary Rings

Except for Mercury and Venus, the planets have one or more satellites around them; more than 30 satellites have been discovered in our solar system until now. As is well known, the satellites have a wide variety of masses as well as modes of orbital motion. Some of them are as massive as the Moon, and others are very small. In most cases, they revolve around the parent planets along nearly circular orbits and in the same direction as that of the planetary spin (i.e., counterclockwise). On the other hand, some of them have large inclinations to the equator and rotate in a direction opposite to the planetary spin.

For the Galilean satellites, Pollack et al. (1979) examined the capture process due to gas drag, and Lunine and Stevenson (1982) proposed a nebula theory in which a system of Jupiter and Galilean satellites is regarded as a miniature of the solar system. Hunten (1979) attempted to explain the origin of Martian satellites due to gas drag of the primordial atmosphere. However, at present, we have no complete theory which explains the origin of all the satellites. Hence, here we only give a few comments on this problem.

First, it should be noted that the tidal effect is important to the orbital evolution of trapped planetesimals. In a case where the spin of the parent planet is very slow or in a direction opposite to the revolution of a trapped planetesimal, the tidal effect acts in such a way that the planetesimal is decelerated and finally falls onto the surface of the planet. This may be the reason why Venus and Mercury have no satellite.

At present, we have no quantitative theory about the origin of the big satellites of Jupiter and Saturn. However, it is likely that they were formed through capture processes due, mainly, to gas drag. As mentioned in Sec. VII.D, proto-Jupiter as well as proto-Saturn was surrounded by a gas which finally accreted to form the giant planet. This is a situation convenient for satellites to be captured by gas drag. Furthermore, this capture process explains the fact that all of these big satellites rotate counterclockwise and circularly around the parent planet. For the case of Triton, which is the largest satellite of Neptune and rotates clockwise around it, the capture process is not clear at present but the clockwise rotation indicates that the capture was probably due to tidal dissipation.

Neither gas drag nor tidal action can explain the origin of the small satellites for the following reason. Unless the gas density is very low, the gas drag is too strong for these small bodies; that is, they cannot survive as satellites but fall easily onto the planetary surface. On the other hand, tidal dissipation is too weak because of their small masses. One probable way of their origin is that they were formed by tidal disruption of a planetesimal which had a close encounter with a planet (see Fig. 5).

Planetary rings found around Jupiter, Saturn and Uranus are considered to be an ensemble of fragments of a satellite. As mentioned above, the orbit of a satellite is unstable when it is revolving in a direction opposite to the planetary spin. In this case, the orbital radius will decrease until it reaches the Roche limit and, finally, the satellite fragments into a number of pieces owing to the tidal effect. This tidal disruption has been simulated by us numerically by means of the particle method. According to our preliminary result, soon after the orbital radius of a satellite becomes smaller than the Roche limit, the satellite is broken into a large number of pieces and, afterward, they are gradually scattered through their gravitational interaction. As a result, fragments rotate differentially with each other around the planet, forming a ring with a certain width (see Fig. 20).

X. ASTEROIDS AND METEORITES

Now, let us consider the evolution of an ensemble of planetesimals in the asteroidal zone. As seen in Fig. 2, the surface density σ_{dust} has a local minimum in this zone. Because of this low surface density, the growth of planetesimals in the asteroidal zone is very slow. Using the growth time given by Eq. (25) and also the mass spectrum shown in Fig. 8, we estimate that, at the stage when the nebular gas has just disappeared, the mass of the largest body in the zone is only of the order of 10^{26} g, their mean mass is as small as 4

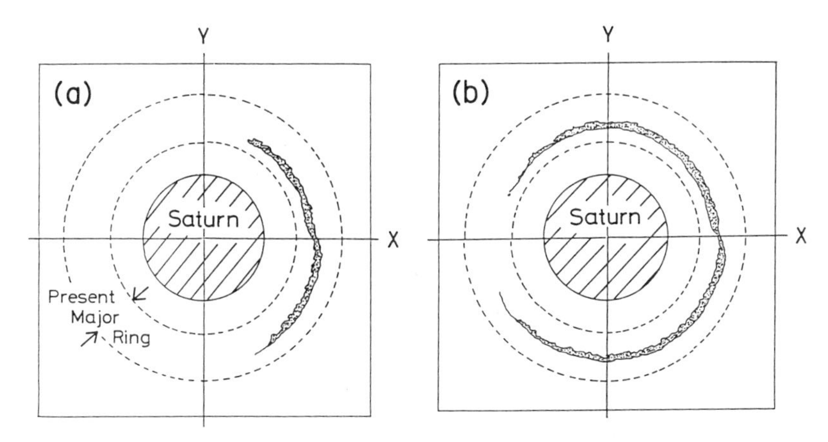

Fig. 20. Tidal disruption of a small satellite which is put initially inside the Roche limit around Saturn. Fragments spread gradually around Saturn owing to mutual gravity. Dashed circles show the edges of the present major ring of Saturn.

$\times\ 10^{24}$ g and the total mass is 10^3 times greater than at present. Hereafter, these small bodies will be called protoasteroids, for simplicity.

At the stage mentioned above, the Earth as well as Jupiter have already grown to their present masses, as mentioned in Sec. VI and VII. Hence, in the later gas-free stages, the orbital motion of protoasteroids will be greatly perturbed by the gravity of these planets, and especially, by that of Jupiter. Their random motion is excited largely because of the absence of gas drag which suppresses the excitation. In this section, we first study how Jupiter's gravity excites the random motion and, next, describe the evolution of the protoasteroids which is initiated by their high-velocity collisions with each other. We expect that this evolution will explain the origin of the present asteroid belt as well as the origin of meteorites.

A. Perturbation due to Jupiter's Gravity and Origin of the Asteroid Belt

Recently, Nakagawa and Hayashi (1985) have studied Jupiter's perturbation by direct numerical computations of a number of orbits of a protoasteroid. They used the 8th-order Runge-Kutta method and checked the accuracy of results by performing time-reversal computations of the orbits. According to these authors, the equations of motion of a protoasteroid (with a negligible mass) and Jupiter are written

$$\frac{d\mathbf{v}}{dt} = -GM_{\odot}\frac{\mathbf{r}}{|\mathbf{r}|^3} - GM_J\left(\frac{r - r_J}{|\mathbf{r} - \mathbf{r}_J|^3} + \frac{r_J}{|r_J|^3}\right) \quad (51)$$

and

$$\frac{d\mathbf{v}_J}{dt} = -G(M_\odot + M_J)\frac{r_J}{|r_J|^3} \tag{52}$$

where $\mathbf{r}$, $\mathbf{r}_J$, $\mathbf{v}$, and $\mathbf{v}_J$ are the heliocentric position and velocity vectors of the protoasteroid and Jupiter, respectively, and M_J is Jupiter's mass. In order to see a long-term behavior of the orbital elements, they pursued the orbital motion of the protoasteroid as well as that of Jupiter for 10^4 to 10^5 yr in the coplanar case. They adopted the initial condition such that the eccentricity of the protoasteroid is zero and that of Jupiter is the present value, i.e., 0.048.

Their results show that, if the protoasteroid starts from a position near the 3/1, 2/1, or 3/2 resonance with an appropriate phase angle relative to Jupiter, its eccentricity can easily increase up to 0.3 or larger in a period of 10^3 to 10^4 yr, owing to the so-called long-period perturbation (see Figs. 21 and 22). Even if it starts from an off-resonant position, it can still gain an eccentricity as large as 0.05 to 0.15 in a period of 10^4 yr, owing to secular perturbation (see Figs. 21 and 22). Furthermore, they found that, if it starts from a position

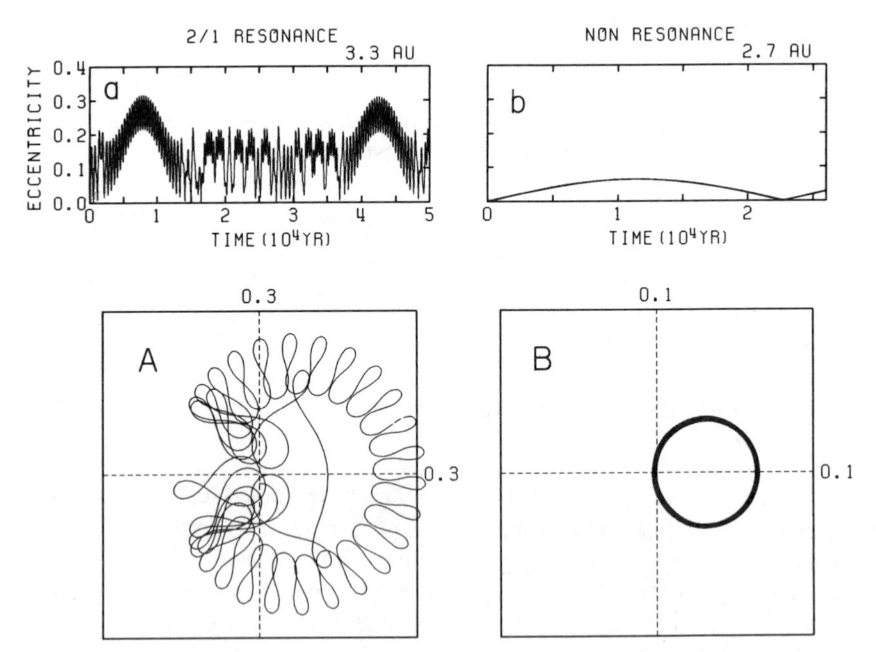

Fig. 21. Time variations of the eccentricity e and the perihelion longitude $\tilde{\omega}$ of a particle perturbed by Jupiter. For the two typical cases, a 2/1-resonance and a nonresonance, the time variation of e is shown in plots (a) and (b), respectively and the corresponding locus in the ($x = e\cos\tilde{\omega}$, $y = e\sin\tilde{\omega}$) plane is shown in plots (A) and (B). In the nonresonant case (see (b) and (B)) the time variations of e and $\tilde{\omega}$ show a typical behavior of the secular perturbation.

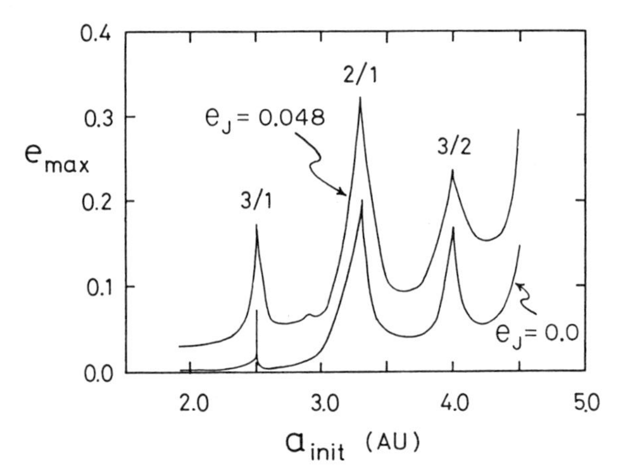

Fig. 22. Relation between the maximum eccentricity, e_{max}, attained in 10^4 to 10^5 yr and the initial semimajor axis, a_{init}. The initial eccentricity of a particle is assumed to be zero. The two curves represent the cases, $e_J = 0$ and $e_J = 0.048$ (i.e., the present value of Jupiter), where e_J is Jupiter's eccentricity assumed to be constant in the computation.

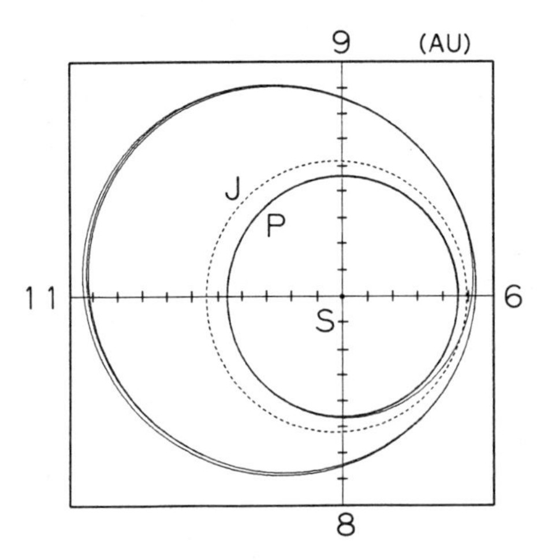

Fig. 23. An example of orbits of a particle making a close encounter with Jupiter. The solid curve denoted by P is the orbit of the particle and the dashed curve denoted by J is that of Jupiter. The position of the Sun is denoted by S. The initial heliocentric distance of the particle is 4.6 AU in this example.

outside the 3/2 resonance, it enters easily the Hill sphere of Jupiter. Afterwards, in some cases, it is scattered largely by Jupiter and comes into a very eccentric orbit which extends greatly beyond Jupiter (see Fig. 23). The above results are important for understanding the origin and evolution of the asteriod belt. In addition, the examples of highly eccentric orbits are interesting also in connection with the origin of comets.

Observationally, the present asteroids have many well-known features (see a review by Chapman et al. 1978). The total amount of their masses is approximately 6×10^{24} g, much less than the lunar mass. Their orbits are considerably eccentric and inclined, compared with those of the ordinary planets. Moreover, most of their semimajor axes are distributed in a zone between 2.2 and 3.2 AU with several gaps located at positions commensurable with Jupiter.

These features of the asteroids can be understood from the above results of orbital computations. According to the surface density σ_{dust} given by Eq. (4), the total mass of the protoasteroids distributed in the asteroidal zone is of the order of 1 $M_{\oplus}$, i.e., above 10^3 times larger than at present. As the mean mass is about 4×10^{24} g at the stage where the nebular gas has dissipated, as mentioned before, the total number of such mean bodies is about 10^3. After dissipation of the gas, they soon gain high random velocities, i.e., large eccentricities and inclinations through Jupiter's perturbation as found by the above orbital computations.

Now let us consider collisions between protoasteroids which have very high random velocities (see Sec. X.B, below, for collision frequencies). In most cases, a collision does not result in coalescence but leads to fragmentation; if we put $e = 0.3$ for the eccentricity, the mean random velocity ev_K is about several km s^{-1} and is larger than the sound velocity in solid materials and also larger than the escape velocity from the largest body existing at that time. Therefore, the protoasteroids are considered to stop growing at the stage when the nebula has disappeared. Together with the low value of the surface density σ_{dust} in the asteroidal zone, this is a reason why large planets could not be formed in that zone.

Moreover, most of the orbits of the fragments are not stable; that is, fragments existing in regions near the resonances or outside the 3/2 resonance are easily scattered by Jupiter within a period of 10^4 yr or so. The scattered fragments escape from the asteroidal zone and, finally, most of them will be captured by Jupiter as well as by the Earth and Mars. On the other hand, off-resonant fragments diffuse gradually in the radial direction until they come into a resonant position and, hence, escape from the asteroidal zone. Through the above processes, the number of fragments and their total mass decrease gradually with time. In a period of 4.5×10^9 yr, the total mass will decrease to about 10^{-3} of the initial, i.e., the present value. This is a scenario for the formation of the asteroid belt deduced from our theory of planetary formation together with the above results of orbital computations.

B. Origin of Meteorites

It is very likely that the asteroids are parent bodies of meteorites. In fact, there exist some similarities between the reflection spectra of asteroids and meteorites (McCord and Gaffey 1974; Chapman et al. 1978). Furthermore, the Pribram and the Lost City meteorites are considered from the analysis of their infalling orbits to have come from the asteroid belt. On the other hand, meteorites are known to have a wide variety of isotopic, chemical, and mineralogical features. This indicates that they have been formed through a series of multistep processes occurring in various physical (i.e., thermal as well as mechanical) and chemical conditions. We have not yet studied these processes quantitatively. However, on the basis of the above-mentioned formation process of the asteroids, we can outline the history of meteorites, as described below.

We again consider the gas-free stages where the protoasteroids are colliding frequently with each other. In the equal mass approximation, we obtain the mean collision time t_{col} by putting simply $f = 1$ and $\theta = 0$ in Eq. (25), i.e.,

$$t_{col} = \frac{1}{n\sigma_{col}V} = \left(\frac{\rho_{mat}r}{3\pi\sigma_{dust}}\right) t_K \tag{53}$$

where r is the radius of a protoasteroid. Considering that the radius corresponding to the mean mass 4×10^{24}g is 700 km and that we have $\sigma_{dust} = 2$ gcm^{-2} at $a = 2.7$ AU according to Eq. (4), we rewrite Eq. (53) as

$$t_{col} = 2 \times 10^7 \left(\frac{\sigma_{dust}}{2 \text{ gcm}^{-2}}\right)^{-1} \left(\frac{r}{700 \text{ km}}\right) \text{yr}. \tag{54}$$

One should notice that collisions become infrequent with time because of the decrease of σ_{dust} and that the collision time for the present asteroids is of the order of 10^{10} yr.

The relative collision velocity is of the order of several km s^{-1}, as mentioned above, and hence, collisions give rise to cratering and surface melting as well as to fragmentation of protoasteroids. Equation (54) indicates that, in early stages, such high-velocity collisions must occur with a time interval on the order of 10^7 yr. We should recall here that the so-called relative formation interval of meteorites, which is known from relative excess of ^{129}Xe produced by the *in situ* decay of ^{129}I, is about 3×10^7 yr; that is, the condensation epoch of most meteorites has a spread of about 3×10^7 yr. The agreement of this value with the above collision time t_{col} may not be pure coincidence but possibly suggests that the condensation is a result of surface melting on the protoasteroids or their fragments, which occurred in the active phase of high-velocity collisions.

It should be noted that the wide variety of chemical features and mineralogical patterns seen in meteorites were formed through multistep processes rather than through one event only. Materials forming the meteorites have experienced, at least, three stages (or three locations) where physical and chemical circumstances are very different:

1. Interstellar space in which original dust grains were first formed;
2. Early solar nebula where some of the dust grains have experienced heating and chemical mixing;
3. Stage of high-velocity collisions of protoasteroids, as mentioned above.

Of these three stages, probably the last one is the most important and essential in forming the variety of chemical and mineralogical features. In this stage the collision velocity has a mean value of about several km s^{-1}. Strictly speaking, however, the relative velocity spreads widely from a very low value (say, 0.1 km s^{-1}) to 10 km s^{-1} or higher. Accordingly, solid materials experience a number of successive collisions that occur with various relative velocities.

Concerning these collision processes, it is important to note that there is the following evidence. Some of materials, of which meteorites are composed, have experienced very rapid cooling (e.g., Tsuchiyama and Nagahara [1981] found that the cooling rate required for the formation of chondrules is of the order of 1 K hr^{-1}), while others experienced very slow cooling. These may be explained easily if we consider the above collisional process; e.g., when protoasteroids collide violently with each other, they fragment into small pieces or droplets, which are sometimes hot enough to be melted. In some circumstances these droplets are cooled down rapidly by emitting radiation almost freely, while in other cases the cooling is very slow owing to the blanketing effect of the surrounding droplets themselves. On the other hand, if we consider heating and cooling processes in the early solar nebula, we can never expect a rapid cooling because the surrounding gas is optically thick.

Mixing of planetesimals due to radial migration is also important for the origin of meteorites. Planetesimals formed in regions far distant from the Sun (i.e., Jupiter's regions or more distant regions) contain a large amount of volatile elements such as H_2O and carbon compounds, whereas planetesimals formed in the Earth's region are composed of volatile-poor but refractory-rich materials. Through gravitational scattering between them, some of them can migrate to the asteroidal zone, as mentioned in Sec. VI, and mix with planetesimals formed in that zone. This mixing process may lead to the variety of chemical and mineralogical features seen in meteorites. For example, it may be possible that some of the planetesimals, which have traveled from a distant region to the asteroidal zone without any metamorphism, gather a large number of tiny droplets (i.e., chondrules) which have been ejected at the time of high-speed collisions of protoasteroids. These may be regarded as the formation of parent bodies of carbonaceous chondrites. In any case, we should note that migration of planetesimals as well as successive collisions are the most important physical processes, through which parent bodies of meteorites acquired a wide variety of isotopic, chemical, and mineralogical features.

After the above-mentioned physico-chemical metamorphism, a part of fragments, that have been produced by high-velocity collision, may also gain large random velocities (i.e., large eccentricities and inclinations) through the Jovian perturbation. In a long period of time (say, 10^9 yr), these fragments would further break very gradually into tiny pieces to the sizes of meteorites, owing to erosion due to solar irradiation (i.e., breakup due to thermal stress) and, further, some of them would travel to the Earth as meteorites. In view of the above long erosion time, this scenario of meteorite formation may well explain the fact that their cosmic ray exposure ages are rather young (e.g., $\leqq 3 \times 10^8$ yr) even though their formation ages are very old.

Thus, we may explain the origin of the meteorites in the framework of our theory of planetary formation. However, many quantitative problems concerning the origin are left unsolved. Especially, we must study the above-mentioned thermal and mechanical processes which follow high-velocity as well as low-velocity collisions, i.e., the fragmentation and reaccretion processes in which both collisional heating and radiative cooling of bodies with a variety of sizes are taken into account.

XI. SUMMARY

In the preceding sections, we have described the long-term evolution of the whole solar system, which starts from an equilibrium solar nebula composed of both solid and gaseous components. Here, we summarize the results of our study on the overall evolution. In Fig. 24, a space-time diagram of evolution is illustrated schematically, in which important events and epochs are denoted. Numerical values in this figure may contain some uncertainties which should be diminished by future study.

As seen in Fig. 24, the nebular gas era continued for about a few ten-millions of years almost independently of distance from the Sun, whereas the formation time of each planet increases with increasing distance. This difference between the time of nebula escape and that of planetary formation gives rise to a diversity of planets. All of the terrestrial-type planets grew in the nebular gas era and, except for asteroids, were surrounded by dense primordial atmospheres when they were born. High temperatures at the planetary surface due to blanketing effect of this atmosphere would give rise to the heterogeneity of the interiors, i.e., the core and mantle structure of the Earth and Venus. On the other hand, Mars was born with a small mass because of the shortage of planetary materials existing in its zone. Therefore, the mass of the Martian primordial atmosphere should be smaller and hence, the blanketing effect should be minor. This may be a reason why the Martian mean density is lower than the densities of the Earth and Venus.

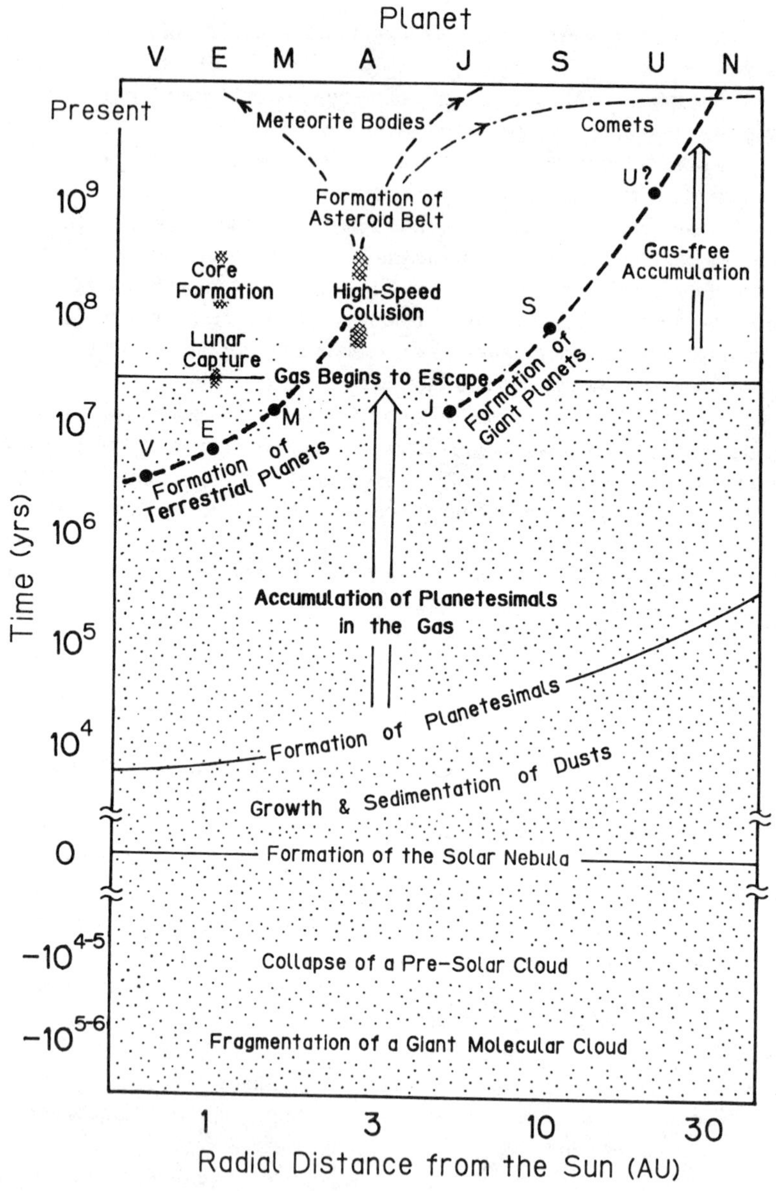

Fig. 24. A space-time diagram of evolution of the solar system. The dotted region shows the gaseous era. The present positions of the planets are shown at the top of the figure.

In the asteroidal zone, the planetary growth was very slow because of the low surface density of planetary materials and hence, protoplanets could not grow to large bodies before the nebular gas escaped. In the gas-free condition after the escape, random motions of protoasteroids were excited easily by the perturbation of Jovian gravity; they collided with each other and fragmented into pieces. The subsequent successive reaccumulation and fragmentation probably account for the origin of the asteroids. Meteorites may be a part of these fragments which traveled to the Earth. A wide variety of chemical and mineralogical metamorphisms observed in meteorites must be the results of the above-mentioned successive collisions.

Jupiter and Saturn grew in the nebular-gas era (although in the case of Saturn, the gas had partially escaped when it was born). After their cores grew by accretion of planetesimals to the critical mass for gravitational instability of the surrounding atmospheres, they could gather large amounts of the nebular gas required to form the massive giant planets. On the other hand, the growth time of Uranus and Neptune is very long compared with the escape time of the gas which is nearly equal to the period of the T Tauri stage and, hence, they grew mostly in a gas-free era. It is likely that the small masses of their gaseous envelopes are due to this gas-free condition.

As to the growth time of Neptune, we again emphasize that, according to our theory, it is comparable to or even greater than the age of the solar system. This is a problem to be solved in the future; we must examine whether we have ignored or overlooked some important process.

Besides the above Neptune problem, there also remain other important problems to be solved, i.e., the formation as well as the dissipation of the solar nebula itself. We hope that these theoretical gaps existing in the evolutionary sequence of our Kyoto model will be filled by means of quantitative calculations in the near future. In any case, our Kyoto model of over-all evolution seems to explain, in a consistent way, the outline of origin of all the hierarchy in the solar system, even though the above-mentioned problems remain to be solved.

Acknowledgments. We would like to thank D. C. Black and M. S. Matthews for encouraging us to write this chapter. We also wish to thank H. Takeda and M. Sekiya for preparing figures and for their helpful comments.

Glossary

GLOSSARY*

Compiled by Melanie Magisos

accretion — the agglomeration of matter together to form larger bodies such as stars, planets, and moons

activation energy — the minimum amount of energy required to produce a chemical reaction (an activation energy may be required to surmount an energy barrier even for exothermic reactions)

adiabatic lapse rate — temperature decrease with altitude for an atmospheric parcel that does not exchange heat with its surroundings

AE — aeon, 10^9yr

albedo — geometric albedo: ratio of planet brightness at zero phase angle to the brightness of a perfectly diffusing disk with the same position and apparent size as the planet; Bond albedo: fraction of the total incident light reflected by a spherical body

Alfvén speed — the speed at which hydromagnetic waves are propagated in a magnetic field: $\nu_A = B/(4\pi\rho)^{1/2}$ where B is the strength of the magnetic field and ρ is the inverse specific volume of a plasma

Alfvén waves — a transverse magnetohydrodynamic wave generated by tensions in lines of magnetic force. These waves are believed to be associated with the heating of stellar coronae, and possibly stellar chromospheres

aliphatic hydrocarbon — a straight or branched open-chain hydrocarbon having the empirical formula C_nH_{2n+2}, such as methane or phytane

*We have used various definitions from *Glossary of Astronomy and Astrophysics* by J. Hopkins (by permission of the University of Chicago Press, copyright 1980 by the University of Chicago) and from *Astrophysical Quantities* by W. W. Allen (London: Athlone Press, 1973).

AMU	atomic mass unit; amu is also used
aphelion	distance of greatest separation between two bodies orbiting in an eccentric orbit
association	a low-density stellar aggregate consisting of stars of similar physical type
AU	the mean distance between Earth and Sun (1.496×10^{13} cm $\simeq$ 500 light seconds)
Balmer emission lines	the spectral lines of hydrogen corresponding to electron energy transitions involving the second energy level
barn	a measure of cross section in particle interactions (1 barn $= 10^{-24}$ cm^2)
Barnard objects	dark nebulae in the plane of our Galaxy which were catalogued by Barnard in 1927
Becklin-Neugabauer object (BN object)	a compact (diameter $<$ 200 AU) infrared source (color temperature about 600 K) in the Orion Nebula. It is one of the brightest infrared objects known at $\lambda \leqslant 10$ μm ($L = 10^4$ $L_\odot$), and is not coincident with any distinctive optical or radio continuum feature. It was discovered in 1966
bending waves	wave motion in which particles in a ring oscillate normal to the ring plane. The wave propagates due to collective gravitational effect of particles on neighboring regions. The wave can be driven by inclination resonance
blackbody	an idealized body which absorbs all radiation of all wavelengths incident on it. The radiation emitted by a blackbody is a function of temperature only. Because it is a perfect absorber, it is also a perfect emitter. The luminosity L of a blackbody is given in terms of its area A and temperature T by $L = \sigma AT^4$ where σ is the Stefan-Boltzman constant
blister model	model for the structure of an H II region at the edge of a cloud. The ionizing star is near the edge but outside of the neutral cloud, and the ionized gas flows off of the cloud in one direction, expanding into the low-density environment

bolometric luminosity	the total energy radiated by an object per unit time in all wavelengths
Bonner-Ebert sphere	an isothermal equilibrium state of a self-gravitating sphere of gas. Such a sphere has a finite radius and non-zero surface pressure
bound cluster	a grouping of stars which is gravitationally bound such that their velocity dispersion is less than the cluster escape velocity
branching ratio	for a chemical or physical reaction with more than one possible set of products, this ratio describes the relative statistical likelihood of such sets
breccias	rocks characterized by a broken, fragmented internal texture
Brownian motion	random motion of microscopic particles suspended in a gas due to collisions with gas molecules
Ca-, Al-rich inclusions	mm- to cm-size components of chondritic meteorites. Compositions are greatly depleted from cosmic in the relatively volatile elements, hence enriched in Ca, Al, and other involatile elements. They are thought to have formed as dispersed objects in the solar nebula, at high temperatures
CAI	Ca-, Al-rich inclusions
carbon stars	in the HD stellar classification system, a rather loose category of peculiar red giant stars, usually of spectral types R and N, whose spectra show strong bands of C_2, CN, or other carbon compounds and unusually high abundances of lithium. Carbon stars resemble S stars in the relative proportion of heavy and light metals, but they contain so much carbon that these bands dominate their spectra. They are believed to be in an advanced stage of evolution, as evidenced by the surface enhancement of ^{13}C and Tc
champagne flow	the flow of bubbles and champagne out of a newly opened champagne bottle is, in some respects, similar to the initial flow of ionized gas from the cavity surrounding an O-type star embedded in a molecular cloud. The

	flow goes through a narrow channel at the edge of the cloud that is opened up by high-pressure gas
chondrites	most abundant and compositionally primitive class of meteorites. They consist of aggregations of small objects that formed dispersed in the solar nebula, in different temperature regimes
chondrules	mm-size, more or less spherical igneous objects that make up the bulk of most chondrites. Most chondrules contain roughly cosmic proportions of Mg, Fe, Si, Ca, and Al
chromosphere	the region in a late-type stellar atmosphere that is between the stellar photosphere and corona. It is characterized by a positive temperature gradient believed to be due to the presence of mechanical heating, such as by magneto-acoustic waves. The temperature generally ranges from a few times 10^3K to about 2×10^4K
CI problem	inability of standard ion-molecule models to reproduce the high CI (neutral atomic carbon) abundance observed in some dense clouds
circumplanetary nebulae (disks)	flattened disks of gas and solids in orbit about (proto-) planets. Most planetary satellites are presumed to have formed in these disks
column density	number of atoms or molecules per cm^2 in the "line of sight"
cometary globule	a small dark cloud embedded in an H II region. The side of the globule facing the source of ionizing radiation often has a bright rim, whereas the opposite side of the globule has a more diffuse taillike appearance
cometesimal	an icy planetesimal
comets	small icy solar system bodies in highly eccentric orbits
commensurabilities	orbital frequencies that are in a ratio of small whole numbers
complex molecules	species containing four or more atoms

condensation reaction	reaction between two small molecules to form two products, one of which is significantly larger than the reactants
Coriolis effect	the acceleration which a body in motion experiences when observed in a rotating frame. This force acts at right angles to the direction of the angular velocity
corona	the outermost layer of late-type stars like the Sun. The corona is characterized by relatively low-density plasma at temperatures in excess of 10^6K
coronal loops	X-ray emitting, coronal plasma confined by magnetic fields that emerge through the stellar photosphere
Courant condition	a limitation on the largest time step that can be used in numerical hydrodynamic calculations. The time step must be short enough that a sound wave cannot travel across a spatial cell in one time step
cyanoacetylene	organic molecule composed of very unsaturated hydrocarbon and CN group
dark cloud	a relatively dense (10^3 to 10^6 particles per cm^3), cool (about 10 to 20 K) cloud of interstellar matter (primarily molecular) whose dust particles obscure the light from stars beyond it and give the cloud the appearance of a region devoid of stars
Debye length	the maximum distance that a given electron can be from a given positive ion and still be influenced by the electric field of that ion in a plasma. Although according to Coulomb's law oppositely charged particles continue to attract each other at infinite distances, Debye showed that there is a cutoff of this force owing to shielding by other charged particles in between. This critical separation decreases for increased density $l_D = (kT/4\pi\, n_e e^2)^{1/2}$
Debye temperature	parameter appearing in the Debye theory which is given by $\theta = h\omega/k$ where h is Planck's constant, ω is the maximum vibrational frequency of the thermal vibration of the lattice, and k is Boltzmann's constant
dense interstellar cloud	interstellar region of gas (99% by mass) and dust (1%) with gas density exceeding $10^3\ cm^{-3}$

density waves in galaxies	moving waves in the distribution of the density of stars and gas in a galactic disk
diffuse nebula	an irregularly shaped cloud of interstellar gas or dust whose spectrum may contain emission lines (emission nebula) or absorption lines characteristic of the spectrum of nearby illuminating stars (reflection nebula)
dust layer (subdisk)	a dust subdisk formed in the solar nebula because of sedimentation of dust particles to the central plane. When the density in the layer reaches a critical value $\rho_{cr} = 2\rho^* = 3M_{\odot}/2\pi R^3$, the layer becomes gravitationally unstable and breaks up into numerous dust condensations evolving further into solid bodies
Eddington approximation	method developed by Eddington to treat radiative transfer in stars. The crucial aspect of the approximation is that the radiation pressure K at some point can be expressed in terms of J, the mean intensity of the radiation at that point, as follows: $K = J/3c$, where c is the speed of light
Einstein coefficient	an emission (or absorption) coefficient. A_{ji} is the coefficient of spontaneous emission; B_{ji} is the coefficient of stimulated emission; B_{ij} is the coefficient of absorption, where i is the lower level and j is the upper level
emission measure	a quantity, inferred from line diagnostics, that is a measure of the total radiated power within a given plasma in an energy interval
endothermic reaction	one which requires energy
exothermic reaction	one which releases energy
extinction	the combined effect of scattering and true absorption giving rise to loss of light along a given line of sight
fractionation	separation of chemical elements or isotopes by chemical and physical processes
FTS	Fourier transform spectroscope

FUN	fractionation and unknown nuclear effects
FU Orionis phenomenon	a flare phenomenon associated with T Tauri stars
FWHM	full width at half maximum, the width of a spectral line, of a ring feature, or of any other feature on a quantitative trace where the intensity equals one-half of the maximum intensity
GMC	giant molecular cloud
Goldreich-Ward gravitational stability criteria	in a flat differentially rotating disk there is a critical surface density above which the self-gravity of the disk would overwhelm the effect of pressure as well as differential rotation. The Goldreich-Ward gravitational stability criteria refers to this critical surface density for the grains which can settle to the mid-plane of the solar nebula. The proposed outcome of such an instability is the formation of planetesimals with sizes comparable to or larger than a few kilometers
Gould belt	the local system of stars and gas within about 300 pc of the Sun. It is an expanding belt $\leqslant 1$ kpc in diameter and inclined about 20° to the galactic plane in which OB stars, young clusters and associations, and dark clouds are concentrated. It is a source of high-energy γ-rays
grain mantles	coating believed to be present on some dust grains
Gruniesen parameter	parameter appearing in the Debye theory which is given by $\gamma = d(\ln \theta)/d(\ln V)$ where θ is the Debye temperature and V is the volume
HEAO	High Energy Astronomical Observatory
Herbig-Haro nebulae	semi-stellar, emission-line nebulae which are produced by shock waves in the supersonic outflow of material from young stars; also referred to as Herbig-Haro objects
Hill sphere	a spherical region within which a planet's gravity dominates the motion of solid objects

H-K color	measure of infrared excess obtained by comparison of flux at two standard wavelengths, the so-called H(λ = 1.65 μm) and K(λ = 2.2 μm) bands
HR diagram (Hertzsprung-Russell diagram)	in present usage, a plot of bolometric magnitude against effective temperature for a group of stars. Related plots are the color-magnitude plot (absolute or apparent visual magnitude against color index) and the spectrum-magnitude plot (visual magnitude versus spectral type, the original form of the HR diagram)
Hubble time	(H_o^{-1}) the characteristic age of the Universe since the big bang (19.4 + 1.6 × 10^9 yr for a value of the Hubble constant H_o = 50 km $s^{-1}Mpc^{-1}$ and a constant expansion rate)
HVF	high velocity flow
Hyades cluster	a young (5 × 10^8 yr) moving cluster (radial velocity, + 36 km s^{-1}) of more than 200 stars (mainly FO-KO) visible to the naked eye in Taurus, about 44 to 48 pc distant. Main-sequence turnoff mass is about 1.9 $M_\odot$. There are several white dwarfs in the cluster
hydrocarbon	organic molecule composed solely of hydrogen and carbon atoms
hydrogenation reaction	ion-molecule reaction in which a hydrogen atom (or atoms) is added to a molecule
hyperfine components	the result of the splitting of a spectral line due to interaction between a nonzero nuclear spin and other atomic or molecular angular momenta
igneous material	material formed by solidification from a molten state
initial mass function (IMF)	the distribution of stellar masses at birth: relative number of stars formed per unit mass
insertion reactions	ion-molecule reactions in which a C^+ ion or a CI neutral inserts itself into a larger species thereby increasing the number of carbon atoms in the species
interstellar extinction	the dimming of starlight due to absorption and scattering by interstellar dust

ion-molecule reaction	gas-phase chemical reaction between a charged and a neutral species. Unlike a neutral-neutral reaction, it can normally occur at temperatures under 50 K
IRAS	Infrared Astronomical Satellite
IRS	infrared source
isomers	molecules with the same elemental composition but a different structure (e.g., HCN and HNC)
IUE	International Ultraviolet Explorer, an Earth-orbiting observatory
Jeans escape formula	Jeans showed that if the velocity of the molecules of a particular gas exceeds one-fifth the escape velocity of the system they are in (e.g., a planet's gravity field), then about 50% of these molecules will escape in about 1000 yr
Jeans instability	gravitational instability in an idealized, infinite, homogeneous medium
Jeans length	a critical length for infinitesmal perturbations in a self-gravitating gas. Perturbations larger than this length will have self-gravitational forces that exceed the internal pressure forces. Such perturbations must collapse. Smaller perturbations can be supported by internal pressure and will not collapse
Jeans mass	the mass necessary for protostellar collapse $M_j \geqslant 1.795 (RT/G)^{3/2}\rho^{-1/2}$, where R is the gas constant, T the cloud temperature, G the gravitational constant, and ρ the cloud density
Kelvin-Helmholtz instability	the tendency of waves to grow on a shear boundary between two regions in relative motion parallel to the boundary
kerogen	insoluble organic material found in rocks
Kolmogorov spectrum	in a homogeneous and isotropic turbulent medium, energy is continually transferred between turbulent eddies of different sizes. The distribution of energy cascading

	down the different scales is referred to as the Kolmogorov spectrum
kpc	kiloparsec = 10^3 pc = 3.086×10^{21} cm
K stars	stars of spectral type K are cool, orange to red stars with surface temperatures of about 3600 to 5000 K. Their spectra resemble those of sunspots, in which the hydrogen lines have been greatly weakened. The H and K lines reach their greatest intensity. Strongest lines are Ca I (4227 Å) and the G band (4303 Å)
$L_{\odot}$	solar luminosity = 3.826×10^{33} erg s^{-1}
Lagrange points (L^4, L^5)	the triangular points, 60° ahead of and behind the secondary in a three-body problem. Orbits about these points are stable if the primary-to-secondary mass ratio is greater than 26. The Trojan asteroids are bodies in such orbits
Laplace relation	the relationship between the orbital frequencies of the three inner Galilean moons of Jupiter, Io (1), Europa (2), and Ganymede (3): $\Omega_1 - 3\Omega_2 + 2\Omega_3 = 0$
late-type stars	stars of spectral classes K, M, S, and C
LDR	Large Deployable Reflector
lightcurve	brightness values plotted as a function of time
Lindblad resonance	a resonance of first order in the eccentricity of a perturbed particle. This terminology comes from galactic dynamics; in the 1920s Lindblad invoked such resonances to explain spiral arms
lithology	the physical character of a rock or rocks
Local Group	the cluster of galaxies to which the Milky Way galaxy belongs. It is an irregular cluster of probably more than 30 members, including three spirals (our Galaxy, M31 = NGC 224, and M33 = NGC 598), four irregulars (LMC = A0524, SMC = A0051, IC 1613 = DDO 8, and NGC 6822 = DDO 209), and the intermediate or dwarf ellipticals NGC 147, NGC 185, NGC 205, NGC 221 =

M32, Fornax, Sculptor, Leo I and II, Ursa Minor, Draco, and three dE companions to M31. Probable members include IC 10, WLM (q.v.), Leo A, the Pegasus dwarf, IC 5152, and DDO 210. Several other unnamed members have been discovered at radio and infrared wavelengths. Possible additions include Sextans A and B, NGC 3109, a galaxy in Carina (170 kpc distant), and one in Phoenix (1.8 Mpc distant). The diameter of the Local Group is at least 3 Mpc; crossing time 10^{10} yr

Lorentz force	force on a charged particle due to the presence of an electric field and motion across a magnetic field. The latter component is the cross product of the particle's velocity with the magnetic flux density
LTE	local thermodynamic equilibrium, the approximation that a radiation field can be computed assuming that all atomic and molecular energy states are in thermal equilibrium
luminosity function	number distribution of stars or galaxies with respect to their absolute magnitudes. The luminosity function shows the number of stars of a given luminosity (or the number of galaxies per integrated magnitude band) in a given volume of space
Lyman and Werner band photons	hydrogen molecule bands
Lyman continuum	continuum radiation at wavelengths < 912 angstroms
$M_{\odot}$	solar mass $= 1.989 \times 10^{33}$g
$M_{\oplus}$	Earth mass $= 5.976 \times 10^{27}$g
Mach number	the ratio of speed of a flow to the local speed of sound
Maclaurin spheroids	a form which a homogeneous self-gravitating mass can take when in a state of uniform rotation. Its eccentricity varies from zero (when it is not rotating) to 1 in the limit of infinite angular momentum

mag	logarithmic measure of the brightness of astronomical objects. Five magnitudes correspond to a factor of 100 in intensity or brightness
magnetic Reynolds number	$R_m = lv/\nu_m$, where l is the length scale of motion, v is a velocity and ν_m is a magnetic viscosity, ($\nu_m = c^2/4\pi\sigma_e$, where σ_e is an electrical conductivity, c is a speed of light)
Magnetohydro-dynamics (MHD)	the study of the collective motions of charged particles in a magnetic field
Markoff approximation	an approximation used in the study of probabilistic processes. It assumes that the future development of a system is completely determined by the *present* state of that system and not by the way in which that system achieved its present state
matrix	fine-grained (1 to 10 μm) mineral material that fills the spaces between chondrules and CAIs in unmetamorphosed chondrites, thought to have existed as dispersed dust in the solar nebula prior to chondrite accretion. Matrix material contains approximately cosmic proportions of the major elements
megaparsec	Mpc $= 10^6$ pc $= 3.086 \times 10^{24}$ cm
metastable isomer	a form of a polyatomic molecule that is structurally distinct and of higher energy than the lower energy, or stable, isomer. Interstellar examples are HOC^+ and HNC
metastable state	a state characterized by a potential energy minimum, but which is not the ground state of a system
Mie scattering	the scattering of light by particles with size comparable to the wavelength of the light. Exact albedo and phase functions are highly oscillatory with direction and wavelength; however, slight irregularities and distributions of size smooth out the features, leaving predominantly forward scattering behavior
"modified thermal" theory	theory of association reactions in which there is a modified assumption of thermal equilibrium

molecular self-shielding	molecules near the edge of a cloud will reach an equilibrium between photodissociation by the ambient ultraviolet radiation field and reformation on grain surfaces or by gas-phase chemical reactions. The radiation that these molecules absorb when they photodissociate is prevented from reaching the interior of the cloud. Thus molecules in the interior can exist for a long time without photodissociation by external sources because they are shielded from such sources by similar molecules near the cloud edge
moonlet (shepherding moon)	a moon orbiting within a planetary ring system which is massive enough to clear a gap around itself via gravitational torques. Two massive moonlets may be able to shepherd ring material into a narrow band
NBS	National Bureau of Standards
nonthermal effect	effect on reaction rates of reactants that are not in thermal equilibrium
nova	a star that exhibits a sudden surge of energy, temporarily increasing its luminosity by as much as 17 mag or more (although 12 to 14 mag is typical). Novae are old disk-population stars, and are all close binaries with one component a main sequence star filling its Roche lobe, and the other component a white dwarf. Unlike supernovae, novae retain their stellar form and most of their substance after the outburst. Since 1925, novae have been given variable star designations
OB associations	associations of stars of spectral types OB-2. In general they are recognized only within about 3 kpc of the Sun
oblateness	parameter describing the shape of the planet, and equal to the difference between the equatorial and polar radii, divided by the equatorial radius
olivine	$(Mg, Fe)_2\ SiO_4$, the most abundant mineral in chondritic meteorites
Oort cloud	cloud of around 10^{12} comets surrounding the solar system at distances up to 10^5 AU

optically thin	unsaturated, in spectroscopic nomenclature, so that the intensity of an optically-thin transition is proportional to the column density of molecules
parent molecules	the hypothesized nucleus constituents which are the precursors to observed cometary radicals and ions
Parker instability	instability caused by energetic particles in the magnetic field interacting with the interstellar gas
Paschen emission lines	atomic hydrogen series in the infrared
pc	parsec: 1 parsec = the distance where 1 AU subtends 1 arcsec = 206,265 AU = 3.26 lightyear = 3.086×10^{18} cm
P Cygni profiles	blue-shifted absorption line superposed on a broad emission line indicative of mass outflow
perihelion	the point in a solar system orbit closest to the Sun
petrography	the branch of geology dealing with the description and systematic classification of rocks, especially igneous and metamorphic rocks, usually by means of microscopic examination of thin sections
Pfund lines of H I	a spectral series of hydrogen lines in the far-infrared, representing transitions between the fifth energy level and higher levels
phase-space theory	rigorous statistical theory of all ion-molecule reactions
photodesorption	ejection of a molecule from a surface following absorption of a photon
Planck function	the energy distribution of blackbody radiation under conditions of thermal equilibrium at a temperature T: $B_\nu = (2h_\nu{}^3/c^2)\,[\exp\,(h_\nu/kT)-1]^{-1}$, where h is Planck's constant and $_\nu$ is the frequency
planetesimal	small rocky or icy body formed from the primordial solar nebula, and 1 to 10 km diameter

PMS	pre-main sequence
polarization	the property of a beam of electromagnetic radiation in which the direction of the electric vector is nonrandom. It may be confined to a plane (linear polarization), or it may rotate systematically with time (circular polarization)
polyacetylene	a molecule whose primary structural elements are linked, triply-bonded pairs of carbon atoms [e.g., $H(C \equiv C)_2 CN$]
Poynting-Robertson drag (PR drag)	a loss of orbital angular momentum by orbiting particles associated with their absorption and reemission of solar radiation
ppm	parts per million
p-process	the name of the hypothetical nucleosynthetic process thought to be responsible for the synthesis of the rare heavy proton-rich nuclei which are bypassed by the *r*- and *s*-processes. It is manifestly less efficient (and therefore rarer) than the *s*- or *r*-process because the protons must overcome the Coulomb barrier, and may in fact work as a secondary process on the *r*- and *s*-process nuclei. It seems to involve primarily (p, γ) reactions above cerium (where neutron separation energies are low). The *p*-process is assumed to occur in supernova envelopes at a temperature $\gtrsim 10^9$ K and at densities $\lesssim 10^4$ g cm^{-3}
protoplanetary disk	the flattened disk surrounding the protosun out of which the planets are believed to have formed
protoplanetary swarm	a disklike swarm of solid bodies (planetesimals) formed in the dust layer and evolving as a dynamical system of colliding and gravitationally interacting bodies with a characteristic distribution of mass and velocities transforming finally into a system of planets
radiative association	a reaction in which two atoms/molecules combine, and energy is conserved by the emission of a photon
radical	an atom or molecule possessing an unpaired electron

R associations — spatially coincident groups of stars in reflection nebulae. They typically consist of BO-A2 stars with occasional red giants or supergiants

Rayleigh-Taylor instability — a type of hydrodynamic instability for static fluids (e.g., cold dense gas above hot rarefied gas)

red giant — a late-type (K or M) high-luminosity (brighter than $M_v = 0$) star of very large radius that occupies the upper-right portion of the HR diagram. Red giants are post-main sequence stars that have exhausted the nuclear fuel in their cores, and whose luminosity is supported by energy production in a hydrogen-burning shell. Within the lifetime of the Galaxy, only main sequence stars of type F and earlier have had time to evolve to the red giant phase (or beyond). The red giant phase corresponds to the establishment of a deep convective envelope. Red giants in a globular cluster are about 3 times more luminous than RR Lyrae stars in the same cluster. Red supergiants have a maximum luminosity of $M_v \sim -8$

regolith — a layer of fragmentary debris produced by meteoritic impact on the surface of the Moon or a minor planet

remnant diffuse cloud — predicted substructure in a young molecular cloud which forms by conglomeration of smaller clouds that were formerly diffuse clouds

Reynolds number — a dimensionless number ($R = L\mathrm{v}/\nu$, where L is a typical dimension of the system, v is a measure of the velocities that prevail, and ν is the kinematic viscosity) that governs the conditions for the occurrence of turbulence in fluids. See also magnetic Reynolds number

Roche limit — the minimum distance at which a satellite influenced by its own gravitation and that of a central mass can be in mechanical equilibrium. For a satellite of negligible mass, zero tensile strength, and the same mean density as its primary, in a circular orbit around its primary, the critical distance at which the satellite will break up is 2.44 times the radius of the primary

r-process — the capture of neutrons on a very rapid time scale (i.e., one in which a nucleus can absorb neutrons in rapid succession, so that regions of great nuclear instability are bridged), a theory advanced to account for the existence of all elements heavier than bismuth as well as the neutron-rich isotopes heavier than iron. The essential feature of the *r*-process is the release of great numbers of neutrons in a very short time (less than 100 s). The presumed source for such a large flux of neutrons is a supernova, at the boundary between the neutron star and the ejected material

Runge-Kutta method — a specialized method of numerical integration

Safronov number — a parameter θ relating the random velocity in a swarm of particles to the escape velocity of a characteristic size particle (usually the largest in the swarm). θ = (escape velocity/random velocity)^2/2

shock wave — a sharp change in the pressure, temperature, and density of a fluid which develops when the velocity of the fluid begins to exceed the velocity of sound

SIRTF — Space Infrared Telescope Facility

SMOW — standard mean ocean water

solar nebula — a gas-dust disk developing around the protosun. Mass of the solar nebula is usually assumed to be in the range from 0.02 to 0.05 $M_\odot$. The term protoplanetary cloud is also sometimes used as a synonym of the solar nebula

spiral waves — density or bending waves that have characteristic spiral patterns in a rotating disk of particles

s-process — a process in which heavy, stable, neutron-rich nuclei are synthesized from iron-peak elements by successive captures of free neutrons in a weak neutron flux, so there is time for β decay before another neutron is captured (cf., *r*-process). This is a slow process of nucleosynthesis which is assumed to take place in the intershell regions during the red-giant phase of evolution, at densities up to 10^5 g cm^{-3} and temperatures of about 3×10^8 K (neutron densities assumed are 10^{10} cm^{-3})

sputtering — expulsion of atoms from a solid, caused by impact of energetic particles

star complexes	regions of former or current star formation extending for several hundred parsecs, and including one or more star clusters and OB associations. These are probably the largest scales for star formation in spiral galaxies. They form in the largest molecular clouds, which may contain $10^7\ M_\odot$ of gas
star formation efficiency	a time dependent quantity which measures the fraction of total stellar mass relative to the initial cloud mass from which the stars have formed
steady state abundance	chemical abundance with no time dependence
Stefan-Boltzman constant	constant of proportionality relating the radiant flux per unit area from a blackbody to the fourth power of its absolute temperature: $\sigma = 5.67 \times 10^{-8}\ \mathrm{Wm^{-2}\,K^{-4}}$
stellar wind	a steady particle flux from a star resulting, in part, from the expansion of the outer atmosphere under its own pressure gradient
Strömgren radius	the radius of an emission nebula within which the hydrogen is nearly all H II. H is a function of the temperature of the central star and the density of the nebular gas. If the star is not hot enough to ionize the entire cloud, the nebula is said to be radiation bounded. If the star is hot enough (or if the density of the cloud is small enough) that the entire cloud is ionized, the nebula is said to be gas bounded
supernova	a gigantic stellar explosion in which the star's luminosity suddenly increases by as much as 20 mag. Most of the star's mass is blown off, leaving behind, at least in some cases, an extremely dense core which (as in the Crab Nebula) may be a neutron star and a pulsar
supernovae Type I	supernovae which have a spectrum characterized by detailed structure without a well-defined continuum, and appear to be deficient in hydrogen. They occur in the halo and in old disk populations (Type II stellar population) in galaxies
supernovae Type II	supernovae which have a spectrum characterized by a continuum with superimposed hydrogen lines and some

metal lines. They occur in the arms of spiral galaxies, implying that their progenitors are massive stars (Type I stellar population)

synchronous rotation — the state in which a body rotates at its orbital frequency. For example, one side of the Moon is never visible from Earth because the Moon rotates synchronously

T association — stellar association containing many T Tauri stars

thermal clumps — predicted clumps in a molecular cloud that are the size of a Jeans length, resulting from the local condensation of gas to a density where thermal pressure gradients can resist the local self-gravitational force

Titius-Bode law — a mnemonic device discovered by Titius in 1766 and advanced by Bode in 1772, used for remembering the distances of the planets from the Sun. Take the series 0, 3, 6, 12, . . . ; add 4 to each member of the series, and divide by 10. The resulting sequence 0.4, 0.7, 1.0, 1.6, . . . gives the approximate distance from the Sun (in AU) of Mercury, Venus, Earth, Mars, . . . , out to Uranus. The law fails for Neptune and beyond

transition region — the interface between a stellar chromosphere and corona, it is characterized by the onset of a very steep temperature gradient that coincides with the nearly complete ionization of neutral hydrogen. Far-ultraviolet line emission indicative of plasma temperatures of order 10^5 K arise in this region

T Tauri stars — young, late-type stars that are precursors to solar-mass stars characterized by emission line spectra, infrared excesses, and irregular variability. The prototype for this class of stars is T Tau

turbostratic — partially ordered polymer-like structure

two-photon process — the decay of a metastable state of an atom via the emission of two photons whose energies sum to the total energy of the transition. For neutral hydrogen, this is the 2s − 1s transition

Van der Waals forces — the relatively weak attractive forces operative between neutral atoms and molecules

veiling	the filling in of stellar absorption line cores as a result of chromospheric and/or extended region continuous emission processes. This phenomenon is especially evident in the spectra of T Tauri stars
V-I color-reddening indicator	index of rather low interstellar extinction of electromagnetic waves
virialized	state of statistical equilibrium of the velocities of stars in a cluster attained through the randomizing effect of numerous stellar encounters
virial theorem	for a bound gravitational system the long-term average of the kinetic energy is one half of the potential energy
VLA (Very Large Array)	an array of radio telescopes sited in the southwestern United States
X	hydrogen mass fraction
Y	helium mass fraction
YSO	young stellar object
Z	mass fraction of all elements heavier than helium
ZAMS	zero-age main sequence
Zeeman effect	spectral line broadening due to the influence of magnetic fields. A multiplet of lines is produced, with distinct polarization characteristics. The Zeeman effect is measured by measuring the difference between right-hand and left-hand polarization across a spectral line

Color Section

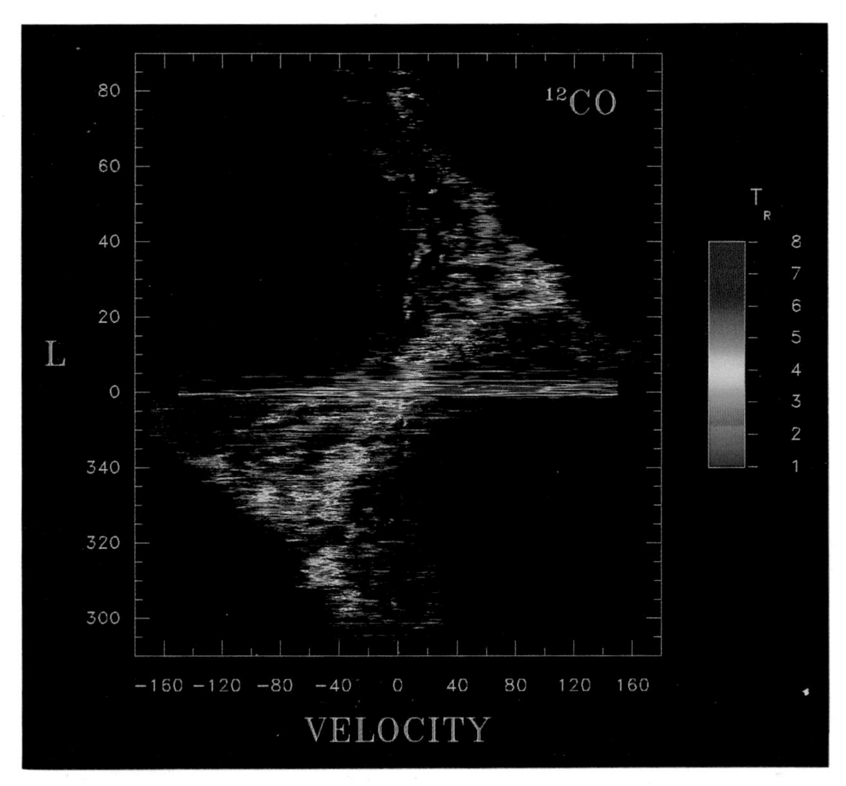

Color Plate 1. Color image of Fig. 3 in the chapter by Solomon and Sanders, with a lower cutoff at 1.0 K. The intensity scale is in units of T_A^*. Saturation is at 6 K with a lower cutoff at 1.0 K. The strongest emission arising from giant molecular clouds (GMC's) and clusters of GMC's is apparent in the yellow and red structures. For example, at longitude = 15°, velocity = 15 km s^{-1} (M17) and longitude 30°.5, velocity = 90 km s^{-1} (W43), two well-known GMC's associated with active star formation and ionized gas (from hot stars) are very prominent. The strong emission near longitude 0° is from the galactic center region. The processing of the color image in this Plate (and for Plates 3, 4, and 5) was done at the Univ. of Massachusetts Remote Sensing Center with the assistance of D. Chelsey.

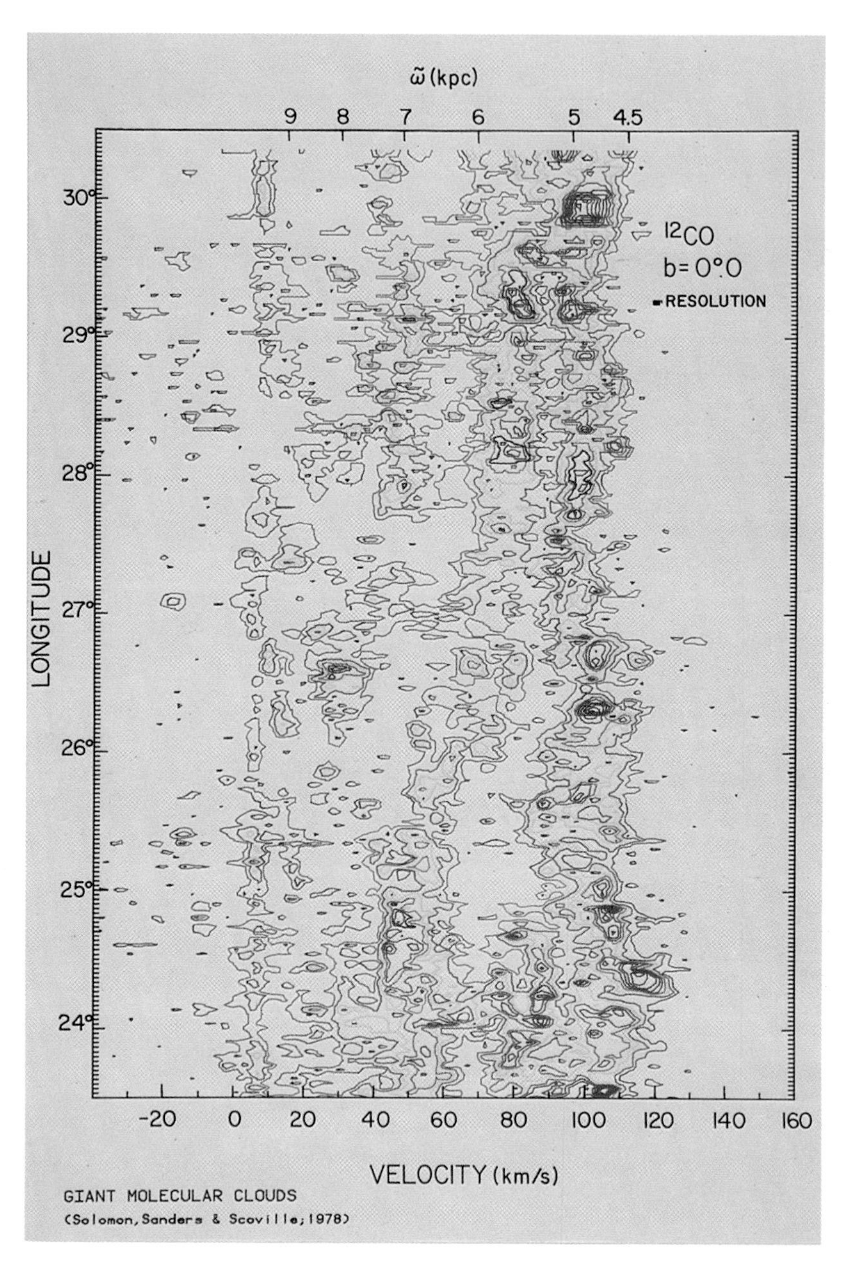

Color Plate 2. CO emission from data taken every 2′ in the galactic plane displayed in the longitude-velocity plane. Contour intervals are in steps of $T_A^* = 1$ K ($T_A^*/\eta = 1.6$ K). The diagram is a synthesis of 215 separate observations. The resolution is indicated by the black rectangle. The breakup into giant molecular clouds is very striking above the third contour. The data extend from $\ell = 23°.533$ to $\ell = 30°.333$ (see the chapter by Solomon and Sanders).

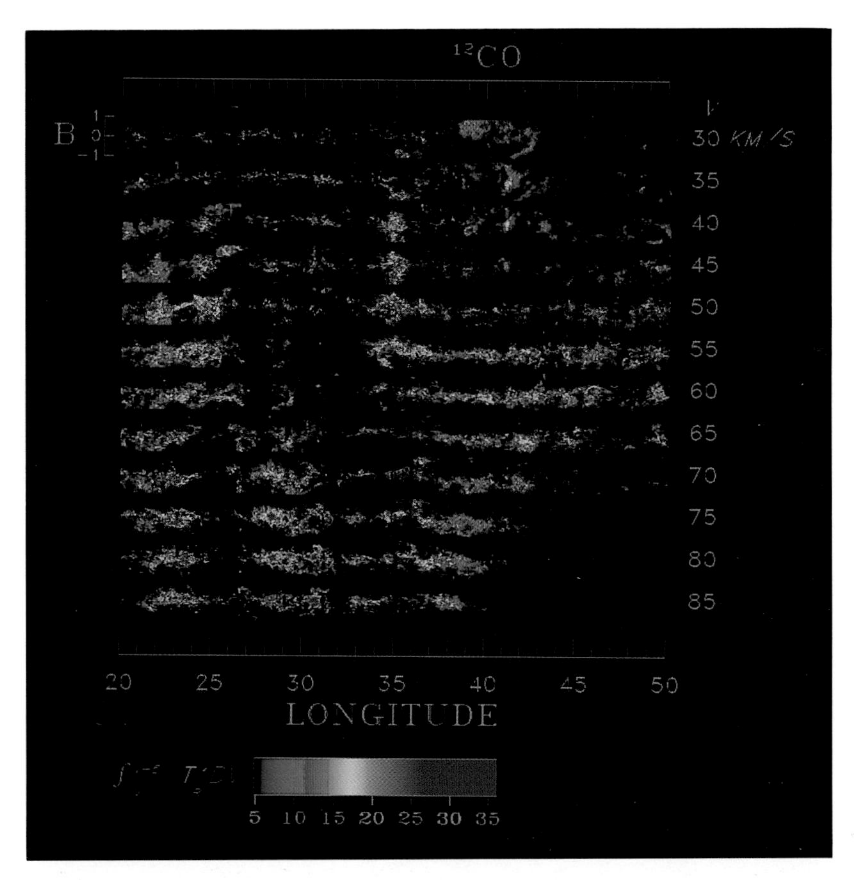

Color Plate 3. A map showing the location and intensity of CO emission from interstellar molecular clouds. Each strip is a view of the galactic plane from longitude 20° to 50° and latitude − 1° to + 1° at a fixed velocity (Doppler shift); the map consists of a composite of 24,600 observations at 0°.05 intervals. Individual GMC's, as well as clusters of molecular clouds, which are the most massive objects in the Galaxy, dominate the emission. The clouds in this picture are at a distance of from 6000 to 36,000 lightyears from the Sun. These data are from the Massachusetts–Stony Brook survey of the Galaxy (see the chapter by Solomon and Sanders).

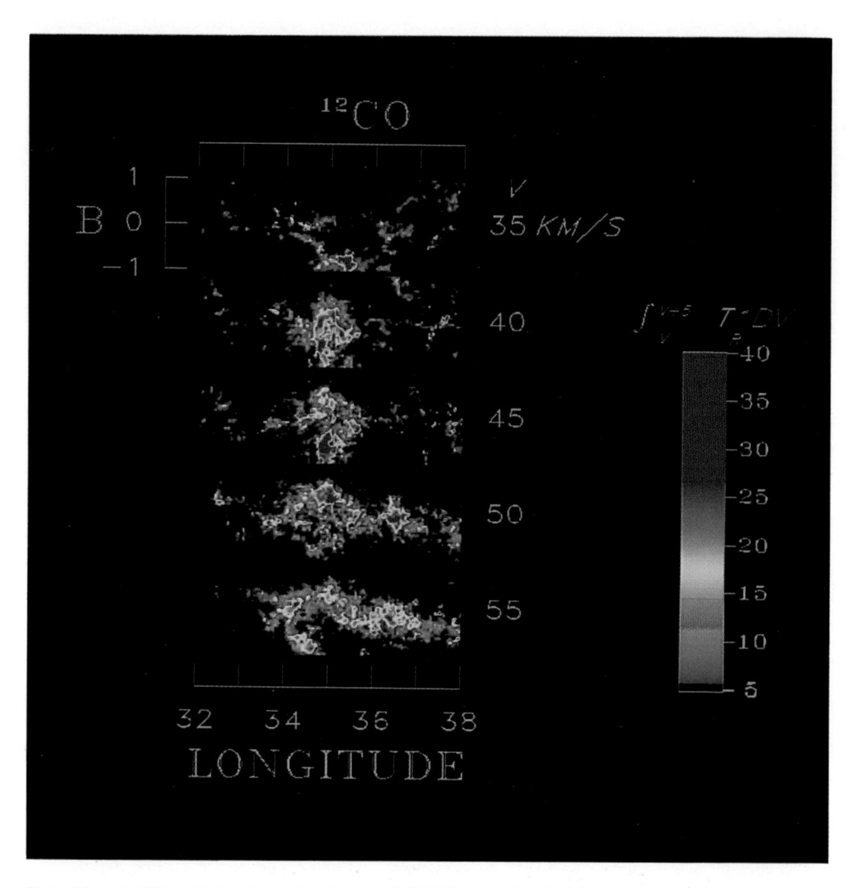

Color Plate 4. CO emission from the cluster of GMC's associated with the supernova remnant W44 at a distance of 3 kpc from the Sun. The cluster contains at least eight clouds with diameters >20 pc and has a total diameter of approximately 110 pc. The total velocity width is 25 km s^{-1}. The virial mass and CO luminosity mass estimates both give $M_T = 4 \times 10^6\ M_\odot$ (see the chapter by Solomon and Sanders).

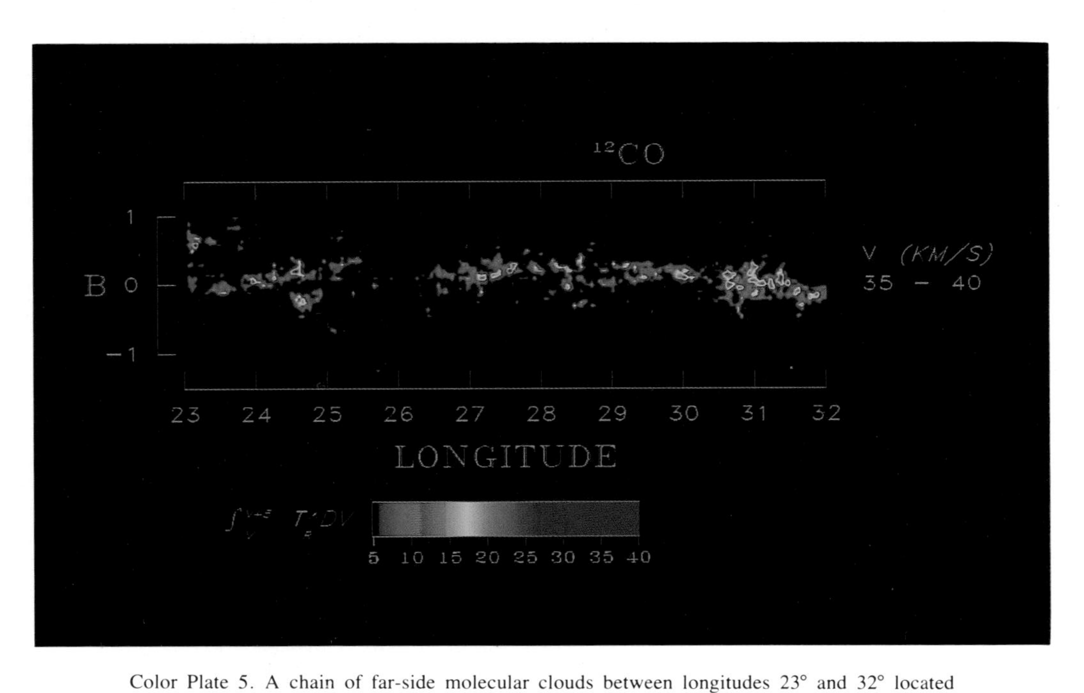

Color Plate 5. A chain of far-side molecular clouds between longitudes 23° and 32° located approximately 14 kpc from the Sun. The color scale is in units of CO-integrated intensity, $\int T_A^*$ (CO) dv, in 5 km s^{-1} bins (see the chapter by Solomon and Sanders).

Color Plate 7. A false color image of the Scorpius-Ophiuchus star-forming region produced with data from the Infrared Astronomical Satellite. The intensity at infrared wavelengths is converted into colors with blue depicting 12 μm, green 60 μm, and red 100 μm. The image covers a region of approximately 30° by 30° centered on the Rho Ophiuchi dark cloud. The bright band in the lower left corner is the plane of the Galaxy. (See the chapter by Wilking and Lada.)

◀ Color Plate 6. A false color image of the region of sky around the constellation Orion, produced from observations made with the Infrared Astronomical Satellite (IRAS). The intensity at infrared wavelengths is converted into colors with blue depicting 12 μm, green 60 μm, and red 100 μm. The bright area in the lower center of the picture is the star-forming region in the Orion molecular cloud while the circular structure in the upper center is heated dust associated with the star Lambda Orionis. IRAS, responsible for this image as well as that in Plate 7, was developed and operated by the Netherlands Agency for Aerospace Programs (NIVR), the US National Aeronautics and Space Administration (NASA), and the UK Science and Engineering Council (SERC). (See the chapter by Wilking and Lada.)

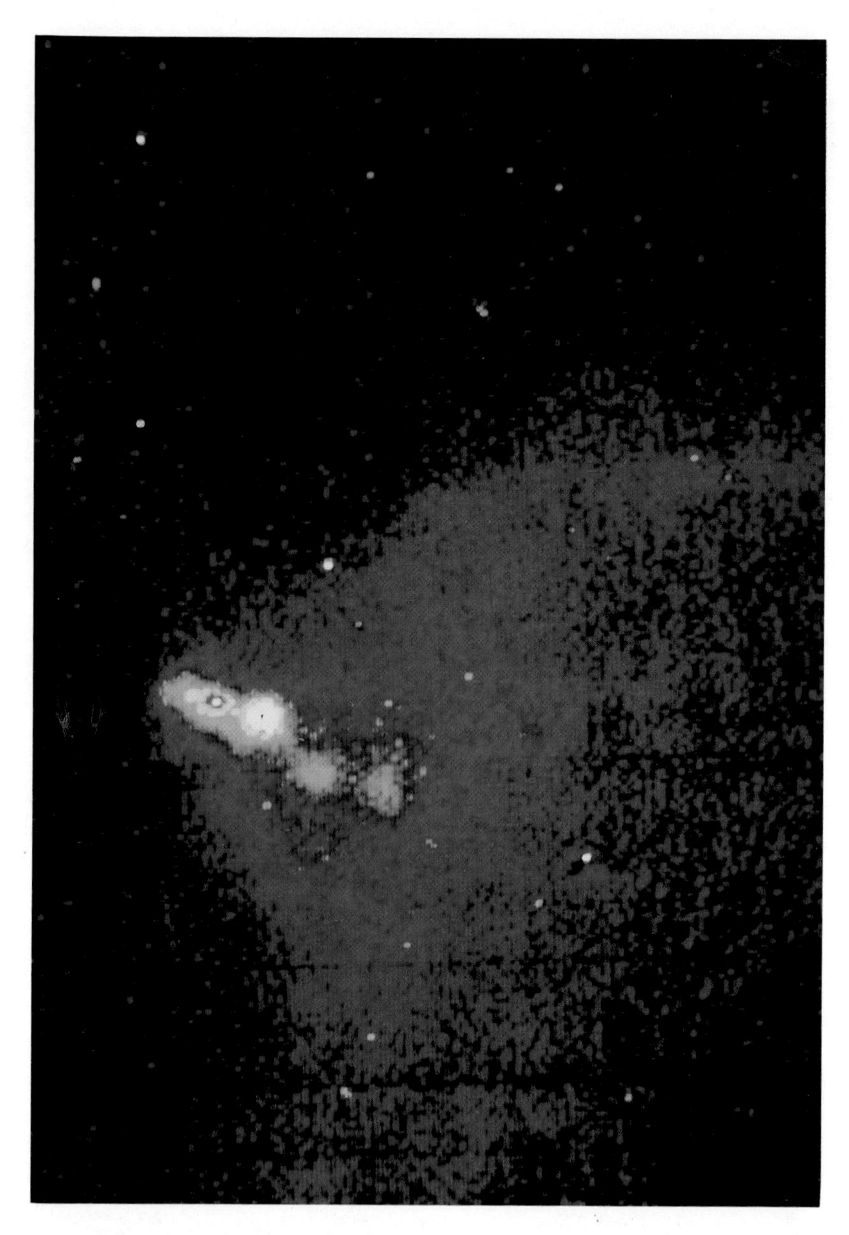

Color Plate 8. Color image of the L 1151-IRS 5 jet. North is up and east is to the left in this CCD image obtained with the 2.2-m telescope on Calar Alto, Spain. It shows a 20 arcsec-long, knotty jet which is emanating from a weak JR source ($L \approx 30\ L_\odot$) located at the very northeast (left) end of the jet. In the two brightest knots radial velocities of -200 km s^{-1} have been measured, while the two weaker knots show velocities of -100 km s^{-1}. The jet shows an emission-line spectrum typical of Herbig-Haro objects and is located near the axis of a cone-like nebula, which is probably due to scattered light from the HR source (see the chapter by Mundt).

References

REFERENCES

Compiled by Terry S. Mullin and Melanie Magisos

Abt, H. A., Chaffee, F. H., and Suffolk, G. 1972. Rotational velocities of Ap stars. *Astrophys. J.* 175:779–785.

Adachi, I., Hayashi, C., and Nakazawa, K. 1976. The gas drag effect on the elliptical motion of a solid body in the primordial solar nebula. *Prog. Theor. Phys.* 56:1756–1771.

Adams, M. T., Strom, K. M., and Strom, S. E. 1983. The star-forming history of the young cluster NGC 2264. *Astrophys. J. Suppl.* 53:893–936.

Adams, N. G., and Smith, D. 1981*a*. The rate coefficients for several ternary association reactions at thermal energies. *Chem. Phys. Lett.* 79:563–567.

Adams, N. G., and Smith, D. 1981*b*. $^{15}N/^{14}N$ isotope fractionation in the reaction $N_2H^+ + N_2$: Interstellar significance. *Astrophys. J.* 247:L123–L125.

Adams, N. G., and Smith, D. 1981*c*. A laboratory study of the reaction $H^+3 + HD \rightleftarrows H_2D^+ +$ H. *Astrophys. J.* 248:373–379.

Adams, N. G., Smith, D., and Alge, E. 1984. Measurements of dissociative recombination coefficients of $H^+{}_3$, HCO^+, and $CH^+{}_5$. *J. Chem. Phys.* 81:1778–1784.

Adams, W. S. 1941. Some results with the Coudé spectrogram of the Mount Wilson Observatory. *Astrophys. J.* 93:11–23.

Adams, W. S. 1943. The structure of interstellar H and K lines in fifty stars. *Astrophys. J.* 97:105–111.

Adams, W. S. 1949. Observations of interstellar H and K, molecular lines, and radial velocities in the spectra of 300 O and B stars. *Astrophys. J.* 109:354–379.

Aggarwal, H. A., and Oberbeck, V. R. 1974. Roche limit of a solid body. *Astrophys. J.* 191:577–588.

Aggarwal, S. S., and Kalra, G. L. 1984. Wave-wave interaction in a magneto-gravitating plasma. *Astrophys J.* 280:792–796.

A'Hearn, M. F., Feldman, P. D., and Schleicher, D. G. 1983. The discovery of S_2 in comet IRAS-Araki-Alcock 1983d. *Astrophys. J.* 274:L99–L103.

Ahrens, L. H. 1964. Si-Mg fractionation in chondrites. *Geochim. Cosmochim. Acta* 28:1869–1870.

Airy, G. B. 1826. *Phil. Trans. Roy. Soc. Lond.* Ser. A, part III, p. 548.

Alaerts, L., Lewis, R. S., and Anders, E. 1979. Isotopic anomalies of noble gases in meteorites and their origins. III. LL chondrites. *Geochim. Cosmochim. Acta* 43:1399–1416.

Alexander, D. R. 1975. Low temperature Rosseland opacity tables. *Astrophys. J. Suppl.* 29:363–374.

Alexander, D. R., Johnson, H. R., Rypma, R. L. 1983. Effect of molecules and grains on Rosseland mean opacities. *Astrophys. J.* 272:773–780.

Alexander, E. C. Jr., Lewis, R. S., Reynolds, J. H., and Michel, M. C. 1971. Plutonium-244-Confirmation as an extinct radioactivity. *Science* 172:837–846.

Alfvén, H. 1978. From dark interstellar clouds to planet and satellites. In *Protostars & Planets,* ed. T. Gehrels (Tucson: Univ. of Arizona Press), pp. 533–544.

Alfvén, H., and Arrhenius, G. 1976. *Evolution of the Solar System* (Washington: NASA SP-345).

Allen, C. W. 1973. *Astrophysical Quantities,* 1st ed. (London: Athlone Press).

Allen, C. W. 1976. *Astrophysical Quantities,* 3rd ed. (London: Athlone Press), p. 209.

Allen, D. A. 1972. Infrared objects in HII regions. *Astrophys. J.* 172:L55–L58.

Allen, D. A., and Wickramasinghe, D. T. 1981. Diffuse interstellar absorption bands between 2.9 and 4.0 μm. *Nature* 294:239–240.

Allen, M., and Robinson, G. W. 1975. Formation of molecules on small interstellar grains. *Astrophys. J.* 195:81–90.
Allen, M., and Robinson, G. W. 1976. Molecular hydrogen in interstellar dark clouds. *Astrophys. J.* 207:745–757.
Allen, M., and Robinson, G. W. 1977. The molecular composition of dense interstellar clouds. *Astrophys. J.* 212:396–415.
Allen, R. J., and Goss, W, M. 1979. The giant spiral galaxy M101:V. A complete synthesis of the distribution and motions of the neutral hydrogen. *Astron. Astrophys. Suppl.* 36:135–162.
Allen, T. L., Goddard, J. D., and Schaefer, H. F. III. 1980. A possible role for triplet H_2CN^+ isomers in the formation of HCN and HNC in interstellar clouds. *J. Chem. Phys.* 73:3255–3263.
Altenhoff, W. J., Downes, D., Pauls, T., and Schraml, J. 1978. Survey of the galactic plane at 4.875 GHz. *Astron. Astrophys. Suppl.* 35:23–54.
Altenhoff, W. J., Batrla, W., Huchtmeier, W. K., Schmidt, S., Stumpff, P., and Walmsley, M. 1983. Radio observations of comet 1983d. *Astron. Astrophys.* 125:L19–L22.
Ambartsumian, V. A. 1955. Stellar systems of positive total energy. *Observatory* 75:72–78.
Ames, S., and Heiles, C. 1970. Observations of the spatial structure of interstellar hydrogen. II. Optical determination of distances in a small region. *Astrophys. J.* 160:59–64.
Anders, E. 1964. Origin, age, and composition of meteorites. *Space Sci. Rev.* 3:583–714.
Anders, E. 1971. Meteorites and the early solar system. *Ann. Rev. Astron. Astrophys.* 9:1–34.
Anders, E. 1977. Critique of "Nebular condensation of moderately volatile elements and their abundances in ordinary chondrites" by Chien M. Wai and John T. Wasson. *Earth Planet. Sci. Lett.* 36:14–20.
Anders, E. 1981. Noble gases in meteorites: Evidence for presolar matter and superheavy elements. *Proc. Roy. Soc. London* A374:207–238.
Anders, E., and Owen, T. 1977. Mars and Earth: Origin and abundance of volatiles. *Science* 198:453–465.
Anders, E., Higuchi, H., Ganapathy, R., and Morgan, J. W. 1976. Chemical fractionations in meteorites. IX. C3 chondrites. *Geochim. Cosmochim. Acta* 40:1131–1139.
Anderson, J. D. 1978. Fourth order equilibrium figure for the atmospheres of Jupiter and Saturn. *Lunar Planet. Sci.* IX:15–16 (abstract).
Anderson, R. C., Bruce, W., Henry, R. C., Feldman, P. D., and Fastie, W. G. 1977. Rocket observations of the far-ultraviolet radiation field at high galactic latitudes. *Bull. Amer. Astron. Soc.* 9:629–630 (abstract).
Andersson, M., Askne, J., and Hjalmarson, Å. 1984. Temperatures in the Orion molecular cloud as estimated from methyl cyanide observations. *Astron. Astrophys.* 136:243–249.
Appenzeller, I. 1983. Recent advances in the theoretical interpretation of T-Tauri stars. *Rev. Mexicana Astron. Astrophys.* 7:151–168.
Appenzeller, I. 1974. Interaction between the Barnard Loop nebula and the interstellar magnetic field. *Astron. Astrophys.* 36:99–105.
Appenzeller, I., and Tscharnuter, W. 1974. The evolution of a massive protostar. *Astron. Astrophys.* 30:423–430.
Appenzeller, I., and Dearborn, D. S. P. 1984. Brightness variations caused by surface magnetic fields in pre-main sequence stars. *Astrophys. J.* 278:689–694.
Appenzeller, I., Chavarria, C., Krautter, J., Mundt, R., and Wolf, B. 1980. UV spectrograms of T Tauri stars. *Astron. Astrophys.* 90:184–191.
Armstrong, J. T., Hutcheon, I. D., and Wasserburg, G. J. 1984. Disturbed Mg isotopic systematics in Allende CAI. *Lunar Planet. Sci.* XV:15 (abstract).
Armstrong, J. W., and Rickett, B. J. 1981. Power spectrum of small-scale density irregularities in the interstellar medium. *Mon. Not. Roy. Astron. Soc.* 194:623–638.
Arnal, E. M., and Gergely, T. E. 1977. Search for neutral hydrogen in Bok globules. *Astron. Astrophys.* 54:961–964.
Arnal, E. M., Goss, W. M., Dickel, H. R., and Forster, J. R. 1982. Westerbork observations of 6-cm H_2CO in W3. *Mon. Not. Roy. Astron. Soc.* 201:317–330.
Arnould, M., and Nørgaard, H. 1978. Thermonuclear origin of Ne-E. *Astron. Astrophys.* 64:195–213.
Arnould, M., Nørgaard, H., Thielemann, F. K., and Hillebrandt, W. 1980. Synthesis of ^{26}A1 in explosive hydrogen burning. *Astrophys. J.* 237:931–950.

Arny, T. 1971. Importance of the internal velocity field in star formation. *Astrophys. J.* 169:289–292.

Arny, T., and Weissman, P. 1973. Interaction of proto-stars in a collapsing cluster. *Astron. J.* 78:309–315.

Arons, J., and Max, C. E. 1975. Hydromagnetic waves in molecular clouds. *Astrophys. J.* 196:L77–L82.

Arquilla, R. 1984. The Structure and Angular Momentum Content of Dark Clouds. Ph.D. thesis, Univ. of Massachusetts.

Arquilla, R., and Goldsmith, P. F. 1984. The molecular cloud associated with the infrared source GL437. *Astrophys. J.* 279:664–672.

Arshutkin, L. N., and Kolesnik, I. G. 1981. Structure and characteristics of turbulent interstellar clouds. *Astrophys. J.* 17:199–203.

Askne, J., Höglund, B., Hjalmarson, Å., and Irvine, W. M., 1984. Methyl acetylene as a temperature probe in molecular clouds. *Astron. Astrophys.* 130:311–318.

Atreya, S. K., Donahue, T. M., and Kulin, W. R. 1978. Evolution of a nitrogen atmosphere on Titan. *Science* 201:611–613.

Aubin, G. 1973. Evolution Dynamique des Figures Piriformes de Poincare. Ph.D. thesis, Univ. of Montreal.

Audouze, J., and Schramm, D. N. 1972. ^{146}Sm: A chronometer for p-process nucleosynthesis. *Nature* 237:447–449.

Auerbach, D., Cacak, R., Caudano, R., Gaily, T. D., Keyser, C. J., McGowan, J. W., Mitchell, J. B. A., and Wilk, S. F. J. 1977. Merged electron ion beam experiments. I. *J. Phys.* 10B:3797–3820.

Aumann, H. H. 1984. IRAS observations of nearby main-sequence dwarfs. *Bull. Amer. Astron. Soc.* 16:483 (abstract).

Aumann, H. H., Gillett, F. C., Beichman, C. A., de Jong, T., Houck, J. R., Low, F., Neugebauer, G., Walker, R. G., and Wesselius, P. 1984. Discovery of a shell around Alpha Lyrae. *Astrophys. J.* 278:L23–L27.

Avery, L. W. 1980. Long chain carbon molecules in the interstellar medium. In *Interstellar Molecules,* ed. B. Andrew (Dordrecht: D. Reidel), pp. 47–58.

Avery, L. W., MacLeod, J. M., and Broten, N. W. 1982. A model of Taurus molecular cloud 1 based on HC_3N observations. *Astrophys. J.* 254:116–125.

Ayres, T. R., Marstad, N. C., and Linsky, J. L. 1981. Outer atmospheres of cool stars. IX. A survey of ultraviolet emission from F-K dwarfs and giants with IUE. *Astrophys. J.* 247:545–559.

Baars, J. W. M. 1970. Dual Beam Parabolic Antennae in Radio Astronomy. Ph.D. thesis, Delft, Holland.

Bahcall, N. A., and Soneira, R. A. 1984. A supercluster catalog. *Astrophys. J.* 277:27–37.

Baierlein, R. 1983. Stellar clustering as induced by a supernova: The non-linear evolution. *Mon. Not. Roy. Astron. Soc.* 205:669–673.

Baierlein, R., Schwing, E., and Herbst, W. 1981. Stellar clustering as induced by supernova. *Icarus* 48:49–58.

Bailey, M. E. 1983*a*. The structure and evolution of the solar system comet cloud. *Mon. Not. Roy. Astron. Soc.* 204:603–633.

Bailey, M. E. 1983*b*. Comets, Planet X and the orbit of Neptune. *Nature* 302:399–400.

Bailey, M. E. 1983*c*. Theories of cometary origin and the brightness of the infrared sky. *Mon. Not. Roy. Astron. Soc.* 205:47–52.

Baker, P. L. 1973. Statistical investigation of neutral hydrogen profiles. *Astron. Astrophys.* 23:81–92.

Bally, J. 1982. Energetic activity in a star-forming molecular cloud core: A disk-constrained bipolar outflow in NGC 2071. *Astrophys. J.* 261:558–568.

Bally, J., and Langer, W. D. 1982. Isotope-selective photodestruction of carbon monoxide. *Astrophys. J.* 255:143–148.

Bally, J., and Scoville, N. Z. 1982. Structure and evolution of molecular clouds near H II regions. II. The disk constrained H II region S106. *Astrophys. J.* 255:497–509.

Bally, J., and Lada, C. J. 1983. The high-velocity molecular flows near young stellar objects. *Astrophys. J.* 265:824–847.

Bally, J., Snell, R. L., and Predmore, R. 1983. Radio images of the bipolar H II region S106. *Astrophys. J.* 272:154–162.

Bandurski, E. L., and Nagy, B. 1976. The polymer-like organic material in the Orgueil meteorite. *Geochim. Cosmochim. Acta* 40:1397–1406.

Bania, T. M. 1981. Numerical simulations of the interstellar medium. In *Phases of the Interstellar Medium,* ed. J. M. Dickey (Green Bank: N.R.A.O.), pp. 7–14.

Bania, T. M., and Lyon, J. G. 1980. OB stars and the structure of the interstellar medium: Cloud formation and the effects of different equations of state. *Astrophys. J.* 239:173–192.

Bania, T. M., and Lockman, F. J. 1984. A survey of the latitude structure of galactic HI on small angular scales. *Astrophys. J. Suppl.* 54:513–546.

Baran, G. 1975. Quoted in Cohen, M., and Kuhi, L. V. 1979. Observational studies of pre-main-sequence evolution. *Astrophys. J. Suppl.* 41:743–843.

Baran, G. P., and Thaddeus, P. 1977. A CO survey of the dust clouds in Taurus and Perseus. *Bull. Amer. Astron. Soc.* 9:353 (abstract).

Bardeen, J. M. 1975. Global instabilities of disks. In *Dynamics of Stellar Systems, IAU Symposium No. 69,* ed. A. Hayli (Dordrecht: D. Reidel), pp. 297–320.

Bardeen, J. M., Friedman, J. L., Schutz, B. F., and Sorkin, R. 1977. A new criterion for secular instability of rapidly rotating stars. *Astrophys. J.* 217:L49–L53.

Barnard, E. E. 1913. Photographs of the Milky Way and of comets. *Publ. Lick Obs.* XI.

Barnard, E. E. 1919. On the dark markings of the sky with a catalogue of 182 objects. *Astrophys. J.* 49:1–23.

Barnard, E. E. 1927. *A Photographic Atlas of Selected Regions of the Milky Way,* eds. E. B. Frost and M. R. Calvert (Washington: Carnegie Inst. of Washington).

Bar-Nun, A., and Chang, S. 1984. *J. Geophys. Res.* In press.

Barrett, A. H., Ho, P. T. P., and Myers, P. C. 1977. Ammonia in the Kleinmann-Low nebula. *Astrophys. J.* 211:L39–L43.

Bash, F. N. 1979. Density wave induced star formation: The optical surface brightness of galaxies. *Astrophys. J.* 233:534–538.

Bash, F. N., and Peters, W. L. 1976. Dynamics of CO molecular clouds in the galaxy. *Astrophys. J.* 205:786–797.

Bash, F. N., Green, E. M., and Peters, W. L. III. 1977. The galactic density wave, molecular clouds, and star formation. *Astrophys. J.* 217:464–472.

Bash, F. N., Hausman, M., and Papaloizou, J. 1981. A turbulent model for giant molecular clouds. *Astrophys. J.* 245:92–98.

Bass, L. M., Chesnavich, W. J., and Bowers, M. T. 1979. Gas phase ion-molecule association reactions. A statistical phase space theory approach. *J. Amer. Chem. Soc.* 101:5493–5502.

Bastian, U., Finkenzeller, U., Jaschek, C., and Jaschek, M. 1983. The definition of T Tauri and Herbig Ae/Be stars. *Astron. Astrophys.* 126:438–439.

Bastien, P. 1981. Fragmentation, fragment interactions, and the stellar mass spectrum. *Astron. Astrophys.* 93:160–162.

Bastien, P. 1982. A linear polarization survey of T Tauri stars. *Astron. Astrophys. Suppl.* 48:153–164.

Bastien, P. 1983. Gravitational collapse and fragmentation of isothermal, non-rotating, cylindrical clouds. *Astron. Astrophys.* 119:109–116.

Bastien, P., Bieging, J., Henkel, C., Martin, R. N., Pauls, T., Walmsley, C. M., Wilson, T. L., and Ziurys, L. M. 1981. Small scale clumping in the Orion molecular cloud. *Astron. Astrophys.* 98:L4–L7.

Batchelor, G. K. 1956. *The Theory of Homogeneous Turbulence* (Cambridge: Cambridge Univ. Press).

Bates, D. R. 1979. Ion-molecule association. *J. Phys.* B12:4135–4146.

Bates, D. R. 1983*a*. Theory of molecular formation by radiative association in interstellar clouds. *Astrophys. J.* 270:564–577.

Bates, D. R. 1983*b*. Radiative association and the synthesis of long carbon chain molecules in interstellar clouds. *Astrophys. J.* 267:L121–L124.

Batrla, W., Wilson, T. L., and Rahe, J. 1981. HI and dust in Kutner's Cloud. *Astron. Astrophys.* 96:202–206.

Batrla, W., Wilson, T., Bastien, P., and Ruf, K. 1983*a*. Clumping in molecular clouds. The region between OMC1 and 2. *Astron. Astrophys.* 128:279–290.

Batrla, W., Wilson, T., and Martin-Pintado, J. 1983*b*. Formaldehyde toward Cas A: L Cloud sizes and H_2 densities. *Astron. Astrophys.* 119:139–144.

Batrla, W., Walmsley, C. M., and Wilson, T. L. 1984. Ammonia clouds in absorption against Cas A. *Astron. Astrophys.* 136:127–132.

Baud, B., Young, E., Beichman, C. A., Beintema, D. A., Emerson, J. P., Habing, H., Harris, S., Jennings, R. E., Marsden, P. L., and Wesselius, P. 1984. High sensitivity IRAS observations of the Chamaeleon I dark cloud. *Astrophys. J.* 278:L53–L55.

Baudry, A., Cernicharo, J., Perault, M., de la Noe, J., and Despois, D. 1981. Observations of HCO^+, $H^{13}CO^+$, ^{13}CO, and $C^{18}O$ in Taurus cloudlets. *Astron. Astrophys.* 104:101–115.

Beck, R. 1973. Abundance of CH_3D and the D/H ratio in Jupiter. *Astrophys. J.* 179:309–327.

Beck, R. 1982. The magnetic field in M31. *Astron. Astrophys.* 106:121–132.

Beck R. 1983. Magnetic fields and spiral structure. In *Internal Kinematics and Dynamics of Galaxies,* ed. E. Athanassoula (Dordrecht: D. Reidel), pp. 159–160.

Becker, R. H., and Epstein, S. 1982. Carbon, hydrogen and nitrogen isotopes in solvent-extractable organic matter from carbonaceous chondrites. *Geochim. Cosmochim. Acta* 46:97–103.

Beckwith, S. 1981. The implications of molecular hydrogen emission. In *Infrared Astronomy, IAU Symposium No. 96,* eds. C. G. Wynn-Williams and D. Cruikshank (Dordrecht: D. Reidel), pp. 169–178.

Beckwith, S., Evans, N. J. II, Gatley, I., Gull, G., and Russell, R. W. 1983*a*. Observations of the extinction and excitation of the molecular hydrogen emission in Orion. *Astrophys. J.* 264:152–160.

Beckwith, S., Natta, A., and Salpeter, E. E. 1983*b*. High-velocity outflow sources in molecular clouds: The case for low-mass stars. *Astrophys. J.* 267:596–602.

Beckwith, S., Zuckerman, B., Skrutskie, M. F., and Dyck, H. M. 1984. Discovery of solar system-size halos around young stars. *Astrophys. J.* 287:793–800.

Bedijn, P. J., and Tenorio-Tagle, G. 1980. On the sequential formation of subgroups in OB associations. *Astron. Astrophys.* 88:58–60.

Beer, R. 1976. Jupiter and the boron problem. *Icarus* 29:193–200.

Beer, R., and Taylor, F. W. 1973. The abundance of CH_3D and the D/H ratio in Jupiter. *Astrophys. J.* 179:309–327.

Begemann, F. 1980. Isotopic anomalies in meteorites. *Reports Prog. Phys.* 43:1309–1356.

Begemann, F., and Stegmann, W. 1976. Implications from the absence of a ^{41}K anomaly in an Allende inclusion. *Nature* 259:549–550.

Beichman, C. 1984. IRAS observations of solar type stars. Presented at Protostars and Planets II conference, Tucson, Arizona, January.

Beichman, C. A., and Harris, S. 1981. The formation of a T Tauri star: Observations of the infrared source in L1551. *Astrophys. J.* 245:589–592.

Beichman, C. A., Becklin, E. E., and Wynn-Williams, C. G. 1979. New multiple systems in molecular clouds. *Astrophys. J.* 232:L47–L52.

Beichman, C. A., Phillips, T. G., Wootten, A., and Frerking, M. A. 1982. An upper limit to the atomic carbon abundance in the Orion plateau. In *Regions of Recent Star Formation,* eds. R. S. Rogers and P. E. Dewdney (Dordrecht: D. Reidel), pp. 445–452.

Beichman, C. A., Jennings, R. E., Emerson, J. P., Baud, B., Harris, S., Rowan-Robinson, M., Aumann, H. H., Gautier, T. N., Gillet, F. C., Habing, H. J., Marsden, P. L., Neugebauer, G., and Young, E. 1984. The formation of solar type stars: IRAS observations of the dark cloud Barnard 5. *Astrophys. J.* 278:L45–L48.

Bell, M. B., Feldman, P. A., and Matthews, H. E. 1981. Cyanopolyne absorption in the direction of Cassiopeia A. *Astron. Astrophys.* 101:L13–L16.

Bell, M. B., Feldman, P. A., Kwok, S., and Matthews, H. E. 1982*a*. Detection of $HC_{11}N$ in IRC+10216. *Nature* 295:389–391.

Bell, M. B., Kwok, S., Matthews, H. E., and Feldman, P. A. 1982*b*. Further observations of microwave inversion lines of ammonia in IRC10216. *Astron. J.* 87:404–407.

Bell, M. B., Feldman, P. A., and Matthews, H. E. 1983. Detection of butadiynyl (C_4H) in absorption against Cassiopeia A. *Astrophys. J.* 273:L35–L39.

Belton, M. J. S. 1982. An introductory review of our present understanding of the structure and composition of Uranus' atmosphere. In *Uranus and the Outer Planets,* ed. G. Hunt (Cambridge: Cambridge Univ. Press), pp. 155–172.

Belton, M. J. S., Wallace, L., Hayes, S. H., and Price, M. J. 1980. Neptune's rotation period. A correction and speculation on the differences between photometric and spectroscopic results. *Icarus* 42:71–78.

Belton, M. J. S., Wallace, L., and Howard, S. 1981. The periods of Neptune: Evidence for atmospheric motions. *Icarus* 46:263–274.

Bennetts, D. A., Bader, M. J., and Marles, R. H. 1982. Convective cloud merging and its effect on rainfall. *Nature* 300:42–45.

Benson, P. J. 1983. Microwave Observations of Dense Cores in Local Interstellar Clouds. Ph.D. thesis, Massachusetts Institute of Technology, Cambridge.

Benson, P. J., and Myers, P. C. 1980. Detection of HC_5N in four dark clouds. *Astrophys. J.* 242:L87–L91.

Benson, P. J., and Myers, P. C. 1983. Dense cores in dark clouds. IV. HC_5N observations. *Astrophys J.* 270:589–604.

Benson, P. J., Myers, P. C., and Wright, E. L. 1984. Dense cores in dark clouds: Young embedded stars at 2 micrometers. *Astrophys. J.* 279:L27–L30.

Benz, W. 1985. On fragmentation and turbulence in collapsing gas clouds. *Bull. Amer. Astron. Soc.* 16:878 (abstract).

Beran, M. J. 1968. *Statistical Continuum Theories*. (New York: Interscience).

Bernes, C., and Sandqvist, A. 1977. Anisotropic scattering in dark clouds and formaldehyde lifetimes. *Astrophys. J.* 217:71–77.

Bertin, G. 1980. On the density wave theory for normal spiral galaxies. *Phys. Rep.* 61:1–69.

Bertout, C. 1983. T Tauri south: A protostar? *Astron. Astrophys.* 126:L1–L4.

Bertout, C. 1984. T Tauri stars: An overview. *Reports on Progress in Modern Physics*. Preprint.

Betz, A. L., and McLaren, R. A. 1980. Infrared heterodyne spectroscopy of circumstellar molecules. In *Interstellar Molecules,* ed. B. H. Andrew (Dordrecht: D. Reidel), pp. 503–508.

Bevington, P. R. 1969. *Data Reduction and Error Analysis for the Physical Sciences* (New York: McGraw-Hill), chap. 7.

Bezard, B., and Gautier, D. 1984. A seasonal model of Saturnian upper troposphere: Comparison with Voyager infrared measurements. *Icarus*. In press.

Bhatia, V. B., and Kalra, G. L. 1980. Wave-wave interaction in a rotating gravitating gas cloud. *Astrophys. J.* 239:146–150.

Bhatt, H. C., Rowse, D. P., and Williams, I. P. 1984. The mass distribution of dark clouds. *Mon. Not. Roy. Astron. Soc.* 209:69–75.

Bieging, J., Cohen, M., and Schwartz, P. R. 1984. VLA observations of T Tauri stars. II. A luminosity-limited survey of Taurus-Auriga. *Astrophys. J.* 282:699–708.

Biermann, L., and Michel, K. W. 1978. On the origin of cometary nuclei in the presolar nebula. *Moon and Planets* 18:447–464.

Biermann, L., Giguere, P. T., and Huebner, W. F. 1982. A model of a comet coma with interstellar molecules in the nucleus. *Astron. Astrophys.* 108:221–226.

Biermann, L., Huebner, W. F., and Lust, R. 1983. Aphelion clustering of "new" comets: Star tracks through Oort's cloud. *Proc. Nat. Acad. Sci. U.S.* 80:5151–5155.

Biermann, P., and Harwit, M. 1980. On the origin of the grain-size spectrum of interstellar dust. *Astrophys. J.* 241:L105–L107.

Binder, A. B. 1977. Fission origin of the Moon: Accumulating evidence. *Proc. Lunar Sci. Conf.* 8:118–120.

Bishop, E. V., and DeMarcus, W. C. 1970. Thermal histories of Jupiter models. *Icarus* 12:317–330.

Blaauw, A. A. 1964. The O associations in the solar neighborhood. *Ann. Rev. Astron. Astrophys.* 2:213–247.

Black, D. C. 1971. On the equivalence of the planet-satellite formation processes. *Icarus* 15:115–119.

Black, D. C. 1972*a*. On the origins of trapped helium, neon and argon isotopic variations in meteorites. I. Gas-rich meteorites, lunar soil and breccia. *Geochim. Cosmochim. Acta* 36:347–375.

Black, D. C. 1972*b*. On the origins of trapped helium, neon, and argon isotopic variations in meteorites. II. Carbonaceous meteorites. *Geochim. Comsmochim. Acta* 36:377–394.

Black, D. C. 1973. Deuterium in the early solar system. *Icarus* 19:154–159.

Black, D. C. 1980. In search of other planetary systems. *Space Sci. Rev.* 25:35–81.

Black, D. C., and Pepin, R. O. 1969. Trapped neon in meteroites, II. *Earth Planet. Sci. Lett.* 6:395–405.
Black, D. C., and Scott, E. H. 1982. A numerical study of the effects of ambipolar diffusion on the collapse of magnetic gas clouds. *Astrophys. J.* 263:696–715.
Black, J. H., and Dalgarno, A. 1973. The cosmic abundance of deuterium. *Astrophys. J.* 184:L101–L104.
Black, J. H., and Dalgarno, A. 1976. Molecule formation in the interstellar gas. *Rep. Prog. Phys.* 39:573–612.
Black, J. H., and Willner, S. P. 1984. Interstellar absorption lines in the infrared spectrum of NGC 2024 IRS2. *Astrophys. J.* 279:673–678.
Blake, G. A., Sutton, E. C., Masson, C. R., Phillips, T. G., Herbst, E., Plummer, G. M., and De Lucia, F. 1984. ^{13} CH_3OH in OMC-1. *Astrophys. J.* 286:586–590.
Blake, J. B., and Schramm, D. N. 1973. ^{247}Cm as a short-lived r-process chronometer. *Nature* 243:138–140.
Blandford, R. D., and Rees, M. M. 1974. A "twin-exhaust" model for double radio sources. *Mon. Not. Roy. Astron. Soc.* 169:395–415.
Blitz, L. 1978. A Study of the Molecular Complexes Accompanying Mon OB1, Mon OB2, and CMa OB1. Ph.D. thesis, Columbia University, New York.
Blitz, L. 1980. Star forming molecular clouds towards the galactic anticentre. In *Giant Molecular Clouds in the Galaxy,* eds. P. M. Solomon and M. G. Edmunds (Oxford: Pergamon), pp. 211–229.
Blitz, L., and Shu, F. H. 1980. The origin and lifetime of molecular cloud complexes. *Astrophys. J.* 238:148–157.
Blitz, L., and Thaddeus, P. 1980. Giant molecular complexes and OB associations. I. The Rosette molecular complex. *Astrophys. J.* 241:676–696.
Blitz, L., and Stark, A. A. 1982. A CO survey of 372 optical HII regions. In *Regions of Recent Star Formation,* eds. R. S. Roger and P. E. Dewdney (Dordrecht: D. Reidel), pp. 209–212.
Bodenheimer, P. 1974. Calculations of the early evolution of Jupiter. *Icarus* 23:319–325.
Bodenheimer, P. 1977. Calculations of the effects of angular momentum on the early evolution of Jupiter. *Icarus* 31:356–368.
Bodenheimer, P. 1978. Evolution of rotating interstellar clouds. II. On the formation of multiple star systems. *Astrophys. J.* 224:488–496.
Bodenheimer, P. 1981. The effects of rotation during star formation. In *Fundamental Problems in the Theory of Stellar Evolution, IAU Symposium No. 93,* eds. D. Sugimoto, D. Q. Lamb, and D. N. Schramm (Dordrecht: D. Reidel), pp. 5–26.
Bodenheimer, P. 1982. Origin and evolution of the giant planets. In *The Comparative Study of the Planets,* eds. A. Coradini and M. Fulchignoni (Dordrecht: D. Reidel), pp. 25–48.
Bodenheimer, P., and Ostriker, J. P. 1973. Rapidly rotating stars. VIII. Zero-viscosity polytropic sequences. *Astrophys. J.* 180:159–169.
Bodenheimer, P., and Black, D. C. 1978. Numerical calculations of protostellar hydrodynamic collapse. In *Protostars & Planets,* ed. T. Gehrels (Tucson: Univ. of Arizona Press), pp. 288–322.
Bodenheimer, P., and Tscharnuter, W. M. 1979. A comparison of two independent calculations of the axisymmetric collapse of a rotating protostar. *Astron. Astrophys.* 74:288–293.
Bodenheimer, P., and Boss, A. P. 1981. Fragmentation in rotating protostar: A re-examination of comparison calculations. *Mon. Not. Roy. Astron. Soc.* 197:477–485.
Bodenheimer, P., and Woosley, S. E. 1983. A two-dimensional supernova model with rotation and nuclear burning. *Astrophys. J.* 269:281–291.
Bodenheimer, P., and Pollack, J. B. 1984. Evolution of core/envelope model of the outer planets. In preparation.
Bodenheimer, P., Grossman, A. S., DeCampli, W. M., Marcy, G., and Pollack, J. B. 1980*a*. Calculations of the evolution of the giant planets. *Icarus* 41:293–308.
Bodenheimer, P., Tohline, J. E., and Black, D. C. 1980*b*. Criteria for fragmentation in a collapsing rotating cloud. *Astrophys. J.* 242:209–218.
Boesgaard, A. M., and Simon, T. 1982. Ultraviolet observations of young field stars. In *Second Cambridge Workshop on Cool Stars, Stellar Systems and the Sun,* eds. M. S. Giampapa and L. Golub, SAO Spec. Rept. 392, vol. 1, pp. 161–169.

Bogart, R. S., and Noerdlinger, P. D. 1982. On the distribution of orbits among long-period comets. *Astron. J.* 87:911–917.

Bogey, M., Demuynck, C., and Destombes, J. L. 1984. Laboratory detection of the protonated carbon dioxide by submillimeter wave spectroscopy. *Astron. Astrophys.* In press.

Bohlin, R. C., and Savage, B. D. 1981. Ultraviolet interstellar extinction toward stars in the Orion nebula and toward HD 147889. *Astrophys. J.* 249:109–117.

Böhm, K.-H. 1983. Optical and UV observations of HH-objects. *Rev. Mexicana Astron. Astrophys.* 7:55–70.

Böhm, K.-H., and Böhm-Vitense, E. 1982. Ultraviolet radiation from the environment of the Cohen-Schwartz star. *Astrophys. J.* 263:L35–L38.

Böhm, K.-H., Böhm-Vitense, E., and Brugel, E. W. 1981. The ultraviolet spectrum of Herbig-Haro object 1. *Astrophys. J.* 245:L113–L117.

Böhm-Vitense, E. 1958. Über die wasserstoffkonvektionszone in sternen versciedener effktivtemperaturen und leuchtkräfte *Z. Astrophys.* 46:108–143.

Böhm-Vitense, E., Böhm, K.-H., Cardelli, J. A., and Nemec, J. M. 1982. The ultraviolet continuous and emission-line spectra of the Herbig-Haro objects HH2 and HH1. *Astrophys. J.* 262:224–233.

Bok, B. J. 1934. The stability of moving clusters. *Harvard Circ.* No. 384.

Bok, B. J. 1955. Gas and dust in interstellar clouds. *Astron. J.* 60:146–148.

Bok, B. J. 1977. Dark nebulae, globules, and protostars. *Publ. Astron. Soc. Pacific* 89:597–611.

Bok, B. J. 1978. Star formation in or very close to a southern globule. *Publ. Astron. Soc. Pacific* 90:489–490.

Bok, B. J., and Reilly, E. F. 1947. Small dark nebulae. *Astrophys. J.* 105:255–257.

Bok, B. J., and Cordwell, C. S. 1973. A study of dark nebulae. In *Molecules in the Galactic Environment,* eds. M. A. Gordon and L. E. Snyder (New York: Wiley-Interscience), pp. 53–92.

Bok, B. J., and McCarthy, C. C. 1974. Optical data for selected Barnard objects. *Astron. J.* 79:42–44.

Bok, B. J., Cordwell, C. S., and Cromwell, R. H. 1971. Globules. In *Dark Nebulae, Globules, and Protostars,* ed. B. T. Lynds (Tucson: Univ. of Arizona Press), pp. 33–55.

Boland, W., and de Jong, T. 1982. Carbon depletion in turbulent molecular cloud cores. *Astrophys. J.* 261:110–114.

Boland, W., and de Jong, T. 1984. Hydrostatic models of molecular clouds. II. Steady state models of spherical clouds. *Astron. Astrophys.* 134:87–98.

Bondi, H. 1952. On spherically symmetrical accretion. *Mon. Not. Roy. Astron. Soc.* 112:195–204.

Bonnor, W. B. 1956. Boyle's law and gravitational instability. *Mon. Not. Roy. Astron. Soc.* 116:351–359.

Borderies, N., Goldreich, P., and Tremaine, S. 1984. Unsolved problems in planetary ring dynamics. In *Planetary Rings,* eds. R. Greenberg and A. Brahic (Tucson: Univ. of Arizona Press), pp. 713–734.

Bosma, A. 1978. 21-cm Line Studies of Spiral Galaxies. Ph.D. thesis, Gröningen, Netherlands.

Boss, A. P. 1980*a*. Protostellar formation in rotating interstellar clouds. I. Numerical methods and tests. *Astrophys. J.* 236:619–627.

Boss, A. P. 1980*b*. Protostellar formation in rotating interstellar clouds. III. Nonaxisymmetric collapse. *Astrophys. J.* 237:866–876.

Boss, A. P. 1980*c*. Collapse and equilibrium of rotating, adiabatic clouds. *Astrophys. J.* 242:699–709.

Boss, A. P. 1981*a*. Collapse and fragmentation of rotating, adiabatic clouds. *Astrophys. J.* 250:636–644.

Boss, A. P. 1981*b*. Numerical three dimensional calculations of tidally induced binary protostar formation. *Astrophys. J.* 246:866–878.

Boss, A. P. 1982. Hydrodynamical models of presolar nebula formation. *Icarus* 51:623–632.

Boss, A. P. 1983. Fragmentation of a nonisothermal protostellar cloud. *Icarus* 55:181–184.

Boss, A. P. 1984*a*. Angular momentum transfer by gravitational torques and the evolution of binary protostars. *Mon. Not. Roy. Astron. Soc.* 209:543–567.

Boss, A. P. 1984*b*. Protostellar formation in rotating interstellar clouds. IV. Nonisothermal collapse. *Astrophys. J.* 277:768–782.

Boss, A. P. 1985. Angular momentum transfer by gravitational torques and the evolution of binary protostars. Preprint.

Boss, A. P., and Bodenheimer, P. 1979. Fragmentation in a rotating protostar: A comparison of two three-dimensional computer codes. *Astrophys. J.* 234:289–295.

Boss, A. P., and Black, D. C. 1982. Collapse of accreting, rotating, isothermal, interstellar clouds. *Astrophys. J.* 258:270–279.

Bothun, G. D. 1984. Searching for the optimal means of comparative H I analysis. *Astrophys. J.* 277:532–541.

Boulanger, F., Stark, A. A., and Combes, F. 1981. A complete CO map of a spiral arm region in M31. *Astron. Astrophys.* 93:L1–L4.

Bourdon, E. B., Prince, R. H., and Duley, W. W. 1982. An experimental determination of the cross section for photodesorption. *Astrophys. J.* 260:909–913.

Bowers, P. F., Kerr, F. J., and Hawarden, T. G. 1980. HI self-absorption in the southern coalsack dust complex. *Astrophys. J.* 241:183–196.

Boynton, W. V. 1975. Fractionation in the solar nebula: Condensation of yttrium and the rare earth elements. *Geochim. Cosmochim. Acta* 39:569–584.

Boynton, W. V. 1978*a*. The chaotic solar nebula: Evidence for episodic condensation in several distinct zones. In *Protostars & Planets,* ed. T. Gehrels, (Tucson: Univ. Arizona Press), pp. 427–438.

Boynton, W. V. 1978*b*. Rare-earth elements as indicators of supernova condensation. *Lunar Planet. Sci.* IX:120–122 (abstract).

Boynton, W. V. 1983. Cosmochemistry of the rare earth elements: Meteorite studies. In *Rare Earth Element Geochemistry,* ed. P. Henderson (Amsterdam: Elsevier Science Publishers), pp. 63–114.

Boynton, W. V., and Cunningham, C. C. 1981. Condensation of refractory lithophile trace elements in the solar nebula and in supernovae. *Lunar Planet. Sci.* XII:106–108 (abstract).

Boynton, W. V., Frazier, R. M., and Macdougall, J. D. 1980. Identification of an ultra-refractory component in the Murchison meteorite. *Lunar Planet. Sci.* XI:103–105 (abstract).

Brand, P. W. J. L. 1981. Ionization front interactions and the formation of globules. *Mon. Not. Roy. Astron. Soc.* 197:217–233.

Brand, P. W. J. L., Hawarden, T. G., Longmore, A. J., Williams, P. M., and Caldwell, J. A. R. 1983. Cometary globule 1. *Mon. Not. Roy. Astron. Soc.* 203:215–222.

Braunsfurth, E., and Feitzinger, J. V. 1983. Young stars and bubbles in the Large Magellanic Cloud. *Astron. Astrophys.* 127:113–131.

Brenner, N. M. 1976. The fast Fourier transform. In *Methods of Experimental Physics, Vol. 12, Part C,* ed. M. L. Meeks (New York: Academic), pp. 286–295.

Bridle, A. H., and Kesteven, M. J. 1970. A massive H I cloud surrounding some compact H II regions. *Astron. J.* 75:902–909.

Brinks, E. 1981. NGC 206, a hole in M31. *Astron. Astrophys.* 95:L1–L4.

Brinks, E., and Bajaja, E. 1983. H I-shells in M31. In *Internal Kinematics and Dynamics of Galaxies,* ed. E. Athanassoula (Dordrecht: D. Reidel), pp. 139–140.

Broten, N. W., MacLeod, J. M., Avery, L. W., Irvine, W. M., Höglund, B., Friberg, P., and Hjalmarson, Å. 1984. The detection of interstellar methylcyano-acetylene. *Astrophys. J.* 276:L25–L29.

Brown, A., Jordan, C., Millar, T. J., Gondhalekar, P., and Wilson, R. 1981. H α emission in the EUV spectrum of T Tauri and Burnham's nebula. *Nature* 290:34–36.

Brown, A., Ferraz, M. C. de M., and Jordan, C. 1984. The chromosphere and corona of T Tauri. *Mon. Not. Roy. Astron. Soc.* 207:831–859.

Brown, H. 1949. Rare gases and the formation of the Earth's atmosphere. In *The Atmospheres of the Earth and Planets,* ed. G. P. Kuiper (Chicago: Univ. Chicago Press), pp. 258–266.

Brown, L. W. 1976. Possible radio emission from Uranus at 0.5 MHz. *Astrophys. J.* 207:L209–L212.

Brown, L. 1982. Experimental line strengths for the ν_4 and ν_2 bands of methane in the 7 μm region. *Bull. Amer. Astron. Soc.* 14:733–734 (abstract).

Brown, R. A., Cruikshank, D. P., and Tokunaga, A. T. 1981. The rotation period of Neptune's upper atmosphere. *Icarus* 47:159–165.

Brown, R. D. 1977. Deuterium enrichment in interstellar HCN and HNC. *Nature* 270:39–41.

Brown, R. D., and Rice, E. 1981. Interstellar deuterium chemistry. *Phil. Trans. R. Soc. London* A303:523–533.

Brown, R. D., Eastwood, F. W., Elmes, P. S., and Godfrey, P. D. 1983. Tricarbon monoxide. *J. Amer. Chem. Soc.* 105:6496–6497.

Brown, R. L. 1981. Isocyanic acid in the Taurus molecular cloud 1. *Astrophys. J.* 248:L119–L122.

Brown, R. L., and Zuckerman, B. 1975. Compact HII regions in the Ophiuchus and R Coronae Austrinae dark clouds. *Astrophys. J.* 202:L125–L128.

Brown, R. L., and Chang, C.-A. 1983. The relation between magnetic field and gas density in interstellar clouds. *Astrophys. J.* 264:134–138.

Brueckner, G. E., and Bartoe, J.-D. F. 1983. Observations of high energy jets in the corona above the quiet Sun, the heating of the corona, and the acceleration of the solar wind. *Astrophys. J.* 272:329–348.

Brugel, E. W., Shull, J. M., and Seab, C. G. 1982. The ultraviolet spectrum of Herbig-Haro object 2H. *Astrophys. J.* 262:L35–L39.

Brugel, E. W., Mundt, R., and Bührke, T. 1984. Jets from young stars. IV. The case of R Monocerotis. Submitted to *Astrophys. J.* (Lett.).

Bruhweiler, F. C., Gull, T. R., Kafatos, M., and Soffia, S. 1980. Stellar winds, supernovae, and the origin of the H I supershells. *Astrophys. J.* 238:L27–L30.

Buff, J., Gerola, H., and Stellingwerf, R. F. 1979. A predictability limit for collapsing isothermal spheres. *Astrophys. J.* 230:839–841.

Bujarrabal, V., Guélin, M., Morris, M., and Thaddeus, P. 1981. The abundance and excitation of the carbon chains in interstellar molecular clouds. *Astron. Astrophys.* 99:239–247.

Bunch, T. E., and Chang, S. 1980. Carbonaceous chondrites. II. Carbonaceous chondrite phyllosilicates and light element geochemistry as indicators of parent body processes and surface conditions. *Geochim. Cosmochim. Acta* 44:1543–1577.

Burbidge, E. M., Burbidge, G. R., Fowler, W. A., and Hoyle, F. 1957. Synthesis of the elements in stars. *Rev. Mod. Phys.* 29:547–650.

Burdjuzha, V. V., and Ruzmaikina, T. V. 1974. Possible mechanism for the formation of OH and H_2O maser sources associated with compact HII region. *Astron. Zh.* 51:346–356.

Buriez, J. C., and de Bergh, C. 1981. A study of the atmosphere of Saturn based on methane line profiles near 1.1 μm. *Astron. Astrophys.* 94:382–390.

Burke, J. R., and Silk, J. 1976. The dynamical interaction of a newly formed protostar with infalling matter: The origin of interstellar grains. *Astrophys. J.* 210:341–364.

Burke, J. R., and Hollenbach, D. J. 1983. The gas-grain interaction in the interstellar medium: Thermal accommodation and trapping. *Astrophys. J.* 265:223–234.

Burnett, D. S., Stapanion, M. L., and Jones, J. H. 1982. Meteorite actinide chemistry and cosmochronology. In *Essays in Nuclear Astrophysics,* eds. C. A. Barnes, D. D. Clayton, and D. N. Schramm (Cambridge: Cambridge Univ. Press), pp. 242–285.

Burns, J. A. 1973. Where are the satellites of the inner planets? *Nature* 242:23–25.

Burns, J. A., ed. 1977. Orbital evolution. In *Planetary Satellites* (Tucson: Univ. of Arizona Press), pp. 113–156.

Burns, J. A. 1982. The dynamical evolution of the solar system. In *Formation of Planetary Systems,* ed. A. Brahic (Toulouse, France: Cepadues), pp. 403–501.

Burns, J. A. 1986. Introduction. In *Satellites,* eds. J. A. Burns and M. S. Matthews (Tucson: Univ. of Arizona Press). In press.

Burns, J. A. 1982. The dynamical evolution of the solar system. In *Formation of Planetary Systems,* ed. A. Brahic (Toulouse, France: Cepadues), pp. 403–501.

Burns, J. A., Showalter, M. R., Cuzzi, J. N., and Pollack, J. B. 1980. Physical processes in Jupiter's ring: Clues to its origin by jove. *Icarus* 44:339–360.

Burns, J. A., Showalter, M. R., and Morfill, G. E. 1984. The ephemeral rings of Jupiter and Saturn. In *Planetary Rings,* eds. R. Greenberg and A. Brahic (Tucson: Univ. of Arizona Press), pp. 200–272.

Burton, W. B. 1976. The morphology of hydrogen and of other tracers in the galaxy. *Ann. Rev. Astron. Astrophys.* 14:275–306.

Burton, W. B., and Gordon, M. A. 1976. CO in the galaxy: The thickness of the galactic CO layer. *Astrophys. J.* 207:L189–L193.

Burton, W. B., and Gordon, M. A. 1978. Carbon monoxide in the galaxy. III. The overall nature of its distribution in the equatorial plane. *Astron. Astrophys.* 63:7–27.

Burton, W. B., Gordon, M. A., Bania, T. M., and Lockman, F. J. 1975. The overall distribution of CO in the plane of the galaxy. *Astrophys. J.* 202:30–49.
Byrd, G. G., and Valtonen, M. J. 1985. Origin of redshift asymmetry in galaxy groups. *Bull. Amer. Astron. Soc.* 16:962 (abstract).
Calamai, G., Felli, M., and Giardinelli, S. 1982. Star formation in the NH_3 cloud of the NGC 2071 region. *Astron. Astrophys.* 109:123–130.
Calvet, N., and Cohen, M. 1978. Studies of bipolar nebulae. V. The general phenomenon. *Mon. Not. Roy. Astron. Soc.* 182:687–704.
Calvet, N., Canto, J., and Rodriguez, L. F. 1983. Stellar winds and molecular clouds: T Tauri stars. *Astrophys. J.* 268:739–752.
Calvet, N., Basri, G., and Kuhi, L. V. 1984. The chromospheric hypothesis for the T Tauri phenomenon. *Astrophys. J.* 277:725–737.
Calvet, N., Basri, G. S., Imhoff, C. L., and Giampapa, M. S. 1985. Simultaneous observations of Ca II K and Mg II k in T Tauri stars. *Astrophys. J.* In press.
Cameron, A. G. W. 1962. The formation of the sun and the planets. *Icarus* 1:13–69.
Cameron, A. G. W. 1969. Physical conditions in the primitive solar nebula, Chapter 2. In *Meteorite Research,* ed. M. Millman (Dordrecht: D. Reidel), pp. 7–15.
Cameron, A. G. W. 1973*a*. Accumulation processes in the primitive solar nebula. *Icarus* 18:407–450.
Cameron, A. G. W. 1973*b*. Abundances of the elements in the solar system. *Space Sci. Rev.* 15:121–146.
Cameron, A. G. W. 1973*c*. Interstellar grains in museums? In *Interstellar Dust and Related Topics,* eds. J. M. Greenberg, and H. C. van de Hulst (Dordrecht: D. Reidel), pp. 545–547.
Cameron, A. G. W. 1973*d*. Are large time differences in meteorite formation real? *Nature* 246:30–32.
Cameron, A. G. W. 1978*a*. Physics of the primitive solar accretion disk. *Moon and Planets* 18:5–40.
Cameron, A. G. W. 1978*b*. The primitive solar accretion disk and the formation of the planets. In *The Origin of the Solar System,* ed. S. F. Dermott (New York: J. Wiley & Sons), pp. 49–75.
Cameron, A. G. W. 1978*c*. Physics of primitive solar nebula and of giant gaseous protoplanets. In *Protostars & Planets,* ed. T. Gehrels (Tucson: Univ. of Arizona Press), pp. 453–487.
Cameron, A. G. W. 1979. The interaction between giant gaseous protoplanets and the primitive solar nebula. *Moon and Planets* 21:173–183.
Cameron, A. G. W. 1982. Elementary and nuclidic abundances in the solar system. In *Essays in Nuclear Astrophysics,* eds. C. A. Barnes, D. D. Clayton, and D. N. Schramm (Cambridge: Cambridge Univ. Press), pp. 23–43.
Cameron, A. G. W. 1983. Dissipation of thick accretion disks. *Astrophys. Space Sci.* 93:295–303.
Cameron, A. G. W. 1984. Star formation and extinct radioactivities. *Icarus* 60:416–427.
Cameron, A. G. W., and Pine, M. R. 1973. Numerical models of the primitive solar nebula. *Icarus* 18:377–406.
Cameron, A. G. W., and Truran, J. W. 1977. The supernova trigger for formation of the solar system. *Icarus* 30:447–461.
Cameron, A. G. W., and Fegley, M. B. 1982. Nucleation and condensation in the primitive solar nebula. *Icarus* 52:1–13.
Cameron, A. G. W., DeCampli, W. M., and Bodenheimer, P. H. 1982. Evolution of giant gaseous protoplanets embedded in the primitive solar nebula. *Icarus* 49:298–312.
Campbell, B. 1984. VLA observations of collimated outflow at NGC 7538 IRS 1. Preprint.
Cannizzo, J. K., Ghosh, P., and Wheeler, J. C. 1982. Convective accretion disk and the onset of dwarf nova outbursts. *Astrophys. J.* 260:L83–L86.
Cantó, J. 1980. A stellar wind model for Herbig-Haro objects. *Astron. Astrophys.* 86:327–338.
Cantó, J. 1981. Herbig-Haro objects. In *Investigating the Universe,* ed. F. D. Kahn (Dordrecht: D. Reidel), pp. 95–124.
Cantó, J., and Mendoza, E. E., eds. 1983. Proceedings of the Symposium on Herbig-Haro Objects, T Tauri Stars and Related Phenomena, to Honor Guillermo Haro. *Rev. Mexicana Astron. Astrof.* 7.
Cantó, J., Rodriguez, L. F., Barral, J. F., and Carrall, P. 1981. Carbon monoxide observations of R Monocerotis, NGC 2261, and Herbig-Haro 39: The interstellar nozzle. *Astrophys. J.* 244:102–114.

Cantó, J., Franco, J., Rodriguez, L. F., and Torrelles, J. M. 1984. On the correlation between size and turbulent velocity in molecular clouds. Preprint.

Canuto, V. M., Levine, J. S., Augustsson, R. R., and Imhoff, C. L. 1982. UV radiation from the young sun and oxygen and ozone levels in the prebiological paleoatmosphere. *Nature* 296:816–820.

Canuto, V. M., Levine, J. S., Augustsson, R. R., Imhoff, C. L., and Giampapa, M. S. 1983. The young sun and the atmosphere and photochemistry of the early Earth. *Nature* 305:281–286.

Canuto, V. M., Goldman, I., and Hubickyj, O. 1984. A formula for the Shakura-Sunyaev turbulent viscosity parameter. *Astrophys. J.* 238:L55–L58.

Capps, R. W., Gillett, F. C., and Knacke, R. F. 1978. Infrared observations of the OH source W33 A. *Astrophys. J.* 226:863–868.

Carr, J. 1984. A CO study of the Cep Ak molecular cloud. Preprint.

Carrasco, L., Strom, S. E., and Strom, K. M. 1973. Interstellar dust in the Rho Ophiuchi dark cloud. *Astrophys. J.* 182:95–109.

Carruthers, G. R. 1970. Rocket observations of interstellar molecular hydrogen. *Astrophys. J.* 161:L81–L85.

Casoli, F., and Combes, F. 1982. Can giant molecular clouds form in spiral arms? *Astron. Astrophys.* 110:287–294.

Casoli, F., Combes, F., and Gerin, M. 1984. Observations of molecular clouds in the second galactic quadrant. *Astron. Astrophys.* 113:99–109.

Cassen, P., and Pettibone, D. 1976. Steady accretion of a rotating fluid. *Astrophys. J.* 208:500–511.

Cassen, P. M., and Moosman, A. 1981. On the formation of protostellar disks. *Icarus* 48:353–376.

Cassen, P. M., and Summers, A. L. 1983. Models of the formation of the solar nebula. *Icarus* 53:26–40.

Cassen, P. M., Reynolds, R. T., and Peale, S. J. 1979. Is there liquid on Europa? *Geophys. Res. Lett.* 6:731–734.

Cassen, P. M., Smith, B. F., Miller, R. H., and Reynolds, R. T. 1981. Numerical experiments on the stability of preplanetary disks. *Icarus* 48:377–392.

Cassen, P. M., Peale, S. J., and Reynolds, R. T. 1982. Structure and thermal evolution of the Galilean satellites. In *Satellites of Jupiter,* ed. D. Morrison (Tucson: Univ. of Arizona Press), pp. 93–128.

Castor, J. I., Abbott, D. C., and Klein, R. I. 1975*a*. Radiation-driven winds in Of stars. *Astrophys. J.* 195:157–174.

Castor, J. I., McCray, R., and Weaver, R. 1975*b*. Interstellar bubbles. *Astrophys. J.* 200:L107–L110.

Caughlan, G. R., Fowler, W. A., Harris, M. J., and Zimmerman, B. A. 1984. Tables of thermononuclear reaction rates for low-mass nuclei ($1 \leq Z \leq 14$). *Atmos. Data Nucl. Tables.* In Press.

Cernicharo, J., Guélin, M., and Askne, J. 1984. TMCI-like cloudlets in HCL2. *Astron. Astrophys.* 138:371–379.

Chaisson, E. J., and Vrba, F. J. 1978. Magnetic field structures and strengths in dark clouds. In *Protostars & Planets,* ed. T. Gehrels (Tucson: Univ. of Arizona Press), pp. 189–208.

Champagne, A. E., Howard, A. J., and Parker, P. D. 1983*a*. Threshold states in ^{26}Al: (I) Experimental investigations. *Nuclear Physics* A402:159–178.

Champagne, A. E., Howard, A. J., and Parker, P. D. 1983*b*. Threshold states in ^{26}Al: (II) Extraction of resonance strengths. *Nuclear Physics* A402:179–188.

Champagne, A. E., Howard, A. J., and Parker, P. D. 1983*c*. Nucleosynthesis of ^{26}Al at low stellar temperatures. *Astrophys. J.* 269:686–688.

Chandrasekhar, S. 1942. *Principles of Stellar Dynamics* (Chicago: Univ. of Chicago Press).

Chandrasekhar, S. 1951*a*. The invariant theory of isotropic turbulence in magneto-hydrodynamics. *Proc. Roy. Soc.* A204:435–449.

Chandrasekhar, S. 1951*b*. The fluctuations of density in isotropic turbulence. *Proc. Roy. Soc.* A210:18–25.

Chandrasekhar, S. 1951*c*. The gravitational instability of an infinite homogeneous turbulent medium. *Proc. Roy. Soc.* A210:26–29.

Chandrasekhar, S. 1954. The gravitational instability of an infinite homogeneous medium when Coriolis force is acting and a magnetic field is present. *Astrophys. J.* 119:7–9.
Chandrasekhar, S. 1955. A theory of turbulence. *Proc. Roy. Soc.* A229:1–19.
Chandrasekhar, S. 1956. Theory of turbulence. *Phys. Rev.* 102:941–952.
Chandrasekhar, S. 1957. *Stellar Structures* (New York: Dover).
Chandrasekhar, S. 1969. *Ellipsoidal Figures of Equilibrium* (New Haven: Yale Univ. Press).
Chandrasekhar, S., and Münch, G. 1952. The theory of the fluctuations in brightness of the Milky Way. V. *Astrophys. J.* 115:103–123.
Chandrasekhar, S., and Fermi, E. 1953. Problems of gravitational instability in the presence of a magnetic field. *Astrophys. J.* 118:116–141.
Chandrasekhar, S., and Lebovitz, N. R. 1962. On the oscillations and the stability of rotating gaseous masses. III. The distorted polytropes. *Astrophys. J.* 136:1082–1104.
Chang, S., Mack, R., and Lennon, K. 1978. Carbon chemistry of separated phases of Murchison and Allende meteorites. *Lunar Planet. Sci.* IX:157–159 (abstract).
Chapman, C. R., and McKinnon, W. B. 1986. Cratering. In *Satellites,* eds. J. A. Burns and M. S. Matthews (Tucson: Univ. of Arizona Press). In press.
Chapman, C. R., Morrison, D., and Zellner, B. 1975. Surface properties of asteroids: A synthesis of polarimetry, radiometry and spectrophotometry. *Icarus* 25:104–130.
Chapman, C. R., Williams, J. G., and Hartmann, W. K. 1978. The Asteroids. *Ann. Rev. Astron. Astrophys.* 16:33–75.
Chen, J. H. 1984. Talk at "Fifteenth Lunar and Planetary Science Conference," Lunar and Planetary Institute, Houston, TX.
Chen, J. H., and Wasserburg, G. J. 1981*a*. The isotopic composition of uranium and lead in Allende inclusions and meteorite phosphates. *Earth Planet. Sci. Lett.* 52:1–15.
Chen, J. H., and Wasserburg, G. J. 1981*b*. Isotopic determination of uranium in picomole and subpicomole quantities. *Analy. Chem.* 53:2060–2067.
Chen, J. H., and Wasserburg, G. J. 1983. The isotropic composition of silver and lead in two meteorites: Cape York and Grant. *Geochim. Cosmochim. Acta* 47:1725–1737.
Chen, J. H., and Wasserburg, G. J. 1984. The origin of excess ^{107}Ag in Gibeon (IVA) and other iron meteorites. In *Lunar Planet. Sci.* XV:144 (abstract).
Chenette, D. L., and Stone, E. C. 1983. The Mimas ghost revisited: An analysis of the electron flux and electron microsignatures observed in the vicinity of Mimas at Saturn. *J. Geophys. Res.* 88:8755–8764.
Chernoff, D. F., McKee, C. F., and Hollenbach, D. J. 1982. Molecular shock waves in the BN-KL region of Orion. *Astrophys. J.* 259:L97–L102.
Chesterman, J. F., Warren-Smith, R. F., and Scarrott, S. M. 1982. Optical polarization in M17. *Mon. Not. Roy. Astron. Soc.* 200:965–970.
Cheung, A. C., Rank, D. M., Townes, C. H., Thornton, D. D., and Welch, W. J. 1968. Detection of NH_3 molecules in the interstellar medium by their microwave emission. *Phys. Rev. Lett.* 21:1701.
Cheung, L. H., Forgel, J. A., Gezari, D. Y., and Hauser, M. G. 1980. 1.0 mm maps and radial density distributions of southern HII/molecular cloud complexes. *Astrophys. J.* 240:74–83.
Chevalier, R. A. 1983. The environments of T Tauri stars. *Astrophys. J.* 268:753–765.
Chieze, J. 1985, quoted in Silk, J. 1985*b*. Molecular cloud evolution and star formation. In *Birth and Infancy of Stars* (1983 Les Houches Lectures), eds. R. Lucas, A. Omont, and L. R. Stora (Amsterdam: North-Holland). In press.
Chieze, J., and Lazareff, B. 1980. A model for the H I cloud spectrum in the solar neighborhood. *Astron. Astrophys.* 91:290–301.
Chini, R., and Krugel, E. 1983. Abnormal extinction and dust properties in M16, M17, NGC6357 and the Ophiuchus dark cloud. *Astron. Astrophys.* 117:289–296.
Choe, S.-U. 1984. Bipolar Molecular Outflows: T Tauri Stars and Herbig-Haro Objects. Ph.D. thesis, Univ. of Minnesota.
Choe, S.-U., Böhm, K.-H., and Solf, J. 1984. The line profiles generated in the bow shocks of a Herbig-Haro object. Submitted to *Astrophys. J.* (Lett.).
Chou, C.-L., Baedecker, P. A., and Wasson, J. T. 1976. Allende inclusions: Volatile-element distribution and evidence for incomplete volatilization of presolar solids. *Geochim. Cosmochim. Acta* 40:85–94.
Churchwell, E., and Bieging, J. H. 1983. CN in dark clouds. *Astrophys. J.* 265:216–222.

Clancy, R. T., and Muhleman, D. O. 1983. A measurement of $^{12}CO/^{13}CO$ ratio in the mesosphere of Venus. *Astrophys. J.* 273:829–836.

Clark, F. O., and Johnson, D. R. 1978. Rotation and velocity structure in the core of the optical condensation B213NW and its relation to the parent gas. *Astrophys. J.* 220:550–609.

Clark, F. O., and Johnson, D. R. 1981. The L134-483-L1778 system of interstellar clouds. *Astrophys. J.* 247:104–111.

Clark, F. O., Giguere, P. T., and Crutcher, R. M. 1977. Radio observations of fragmentation and localized multiple velocity components in a group of dust clouds in Taurus. *Astrophys. J.* 215:511–516.

Clark, R. N. and Lucey, P. G. 1983. Spectral properties of ice-particulate mixtures: Implications for remote sensing. I. Intimate mixtures. *J. Geophys. Res.* 89:6341–6348.

Clarke, R. S., Jarosewich, E., Mason, B., Nelen, J., Gómez, M., and Hyde, J. R. 1970. The Allende, Mexico meteorite shower. *Smith. Contrib. Earth Sci.* No. 5, 53 pp.

Clavel, J., Viala, Y. P., and Bel, N. 1978. Chemical and thermal equilibrium in dark clouds. *Astron. Astrophys.* 65:435–448.

Clayton, D. D. 1975*a*. Extinct radioactivities: Trapped residuals of presolar grains. *Astrophys. J.* 199:765–769.

Clayton, D. D. 1975*b* Na^{22}, Ne-E, extinct radioactive anomalies and unsupported Ar^{40}. *Nature* 257:36–37.

Clayton, D. D. 1977. Solar system isotopic anomalies: Supernova neighbor or presolar carriers? *Icarus* 32:255–269.

Clayton, D. D. 1979. Supernovae and the origin of the solar system. *Space Sci. Rev.* 24:147–226.

Clayton, D. D. 1980. Chemical energy in cold-cloud aggregates: The origin of meteoritic chondrules. *Astrophys. J.* 239:L37–L41.

Clayton, D. D. 1981. Origins of Ca-Al-rich inclusions. II. Sputtering and collisions in the three-phase interstellar medium. *Astrophys. J.* 251:374–386.

Clayton, D. D. 1982. Cosmic chemical memory: A new astronomy. *Quart. Jour. Roy. Astron. Soc.* 23:174–212.

Clayton, D. D. 1983. Discovery of s-process Nd in Allende residue. *Astrophys. J.* 271:L107–L109.

Clayton, D. D. 1984. ^{26}Al in the interstellar medium. *Astrophys. J.* 280:144–149.

Clayton, D. D., and Hoyle, F. 1976. Grains of anomalous isotopic composition from novae. *Astrophys. J.* 203:490–496.

Clayton, D. D., and Ward, R. A. 1978. S-process studies: Xenon isotopic anomalies. *Astrophys. J.* 244:1000–1006.

Clayton, R. N. 1978. Isotopic anomalies in the early solar system. *Ann. Rev. Nucl. Part. Sci.* 28:501–522.

Clayton, R. N. 1981. Isotopic variations in primitive meteorites. *Phil. Trans. Roy. Soc. London* A303:339–349.

Clayton, R. N., and Mayeda, T. K. 1977. Correlated oxygen and magnesium isotope anomalies in Allende inclusions. I. Oxygen. *Geophys. Res. Lett.* 4:295–298.

Clayton, R. N., Mayeda, T. K., and Grossman, L. 1973. A component of primitive nuclear composition in carbonaceous chondrites. *Science* 182:485–488.

Clayton, R. N., Onuma, N., Grossman, L., and Mayeda, T. K. 1977. Distribution of the presolar component in Allende and other carbonaceous chondrites. *Earth Planet. Sci. Lett.* 34:209–224.

Clayton, R. N., Mayeda, T. K., and Epstein, S. 1978. Isotopic fractionation of silicon in Allende inclusions. *Proc. Lunar Planet. Sci. Conf.* 9:1267–1278.

Clayton, R. N., Mayeda, T. K., Molini-Velsko, C. A., and Goswami, J. N. 1983*a*. Oxygen and silicon isotopic compositions of Dhajala chondrules. *Meteoritics* 18:282–283.

Clayton, R. N., Onuma, N., Ikeda, Y., Mayeda, T. K., Hutcheon, I. D., Olsen, E. J., Molini-Velsko, C. A. 1983*b*. Oxygen isotopic compositions of chondrules in Allende and ordinary chondrites. In *Chondrules and Their Origins,* ed. E. A. King (Houston: Lunar & Planetary Inst.), pp. 37–43.

Clayton, R. N., MacPherson, G. J., Hutcheon, I. D., Davis, A. M., Grossman, L., Mayeda, T. K., Molini-Velsko, C. A., Allen, J. M., and El Goresy, A. 1984. Two forsterite-bearing FUN inclusions in the Allende meteorite. *Geochim. Cosmochim. Acta* 48:535–548.

Clemens, D. P., Sanders, D. B., Scoville, N. Z., and Solomon, P. M. 1985. Massachusetts-Stony Brook galactic plane CO survey: 1-v maps of the first quadrant. *Astrophys. J. Suppl.* In press.

Clement, M. J. 1967. Triaxial figures of equilibrium for centrally condensed masses. *Astrophys. J.* 148:159–174.

Clement, M. J. 1974. On the solution of Poisson's equation for rapidly rotating stars. *Astrophys. J.* 194:709–714.

Clement, M. J. 1979. On the equilibrium and secular instability of rapidly rotating stars. *Astrophys. J.* 230:230–242.

Clement, M. J. 1981. Normal modes of oscillation for rotating stars. I. The effect of rigid rotation on four low-order pulsations. *Astrophys. J.* 249:746–760.

Clement, M. J. 1984. Normal modes of oscillation for rotating stars. II. Variational solutions. *Astrophys. J.* 276:724–736.

Clifford, P., and Elmegreen, B. G. 1983. A collision cross section for interactions between magnetic diffuse clouds. *Mon. Not. Roy. Astron. Soc.* 202:629–646.

Clube, S. V. M., and Napier, W. M. 1982. Spiral arms, comets, and terrestrial catastrophism. *Quart. J. Roy. Astron. Soc.* 23:45–66.

Clube, S. V. M., and Napier, W. M. 1984. Comet capture from molecular clouds: A dynamical constraint on star and planet formation. *Mon. Not. Roy. Astron. Soc.* 208:575–588.

Cochran, W. D., and Smith, W. H. 1983. Desaturation of H_2 quadrupole lines of the outer planets. *Astrophys. J.* 271:850–864.

Cohen, J. G. 1976. Emission from the disks of spiral galaxies. *Astrophys. J.* 203:587–592.

Cohen, M. 1980. Red and nebulous objects in dark clouds: A survey. *Astron. J.* 85:29–35.

Cohen, M. 1981. Are we beginning to understand T Tauri stars? *Sky and Telescope* 62:300–303.

Cohen, M. 1982. The case for anisotropic mass loss from T-Tauri stars. *Publ. Astron. Soc. Pacific* 94:266–270.

Cohen, M. 1983. HL Tau and its circumstellar disk. *Astrophys. J.* 270:L69–L71.

Cohen, M. 1984. Circumstellar disks around HL Tau and DG Tau. Presented at Protostars and Planets II conference, Tucson, Arizona, January (abstract).

Cohen, M., and Kuhi, L. V. 1979. Observational studies of pre-main-sequence evolution. *Astrophys. J. Suppl.* 41:743–843.

Cohen, M., and Schwartz, R. D. 1979. The exciting star of Herbig-Haro object 1. *Astrophys. J.* 233:L77–L80.

Cohen, M., and Schwartz, R. D. 1980. A search for the exciting stars of Herbig-Haro objects. *Mon. Not. Roy. Astron. Soc.* 191:165–168.

Cohen, M., and Schmidt, G. D. 1981. Spectropolarimetry of Herbig-Haro objects and the exciting star of HH30. *Astron. J.* 86:1228–1231.

Cohen, M., and Schwartz, R. D. 1983. The exciting stars of Herbig-Haro objects. *Astrophys. J.* 265:877–900.

Cohen, M., and Witteborn, F. C. 1985. Ten micron spectroscopy of T Tauri stars. *Astrophys. J.* In press.

Cohen, M., Bieging, J. H., and Schwartz, P. R. 1982. VLA observations of mass loss from T Tauri stars. *Astrophys. J.* 253:707–715.

Cohen, M., Harvey, P. M., Schwartz, R. D., and Wilking, B. A. 1984. Far-infrared studies of Herbig-Haro objects and their exciting stars. *Astrophys. J.* 278:671–678.

Cohen, M., Harvey, P. M., Schwartz, R. D., and Wilking, B. A. 1984. Far-infrared studies of Herbig-Haro objects and their exciting stars. *Astrophys. J.* 278:671–678.

Cohen, R. S., and Thaddeus, P. 1977. An out of plane galactic CO survey. *Astrophys. J.* 217:L155–L159.

Cohen, R. S., Cong, H., Dame, T. M., and Thaddeus, P. 1980. Molecular clouds and galactic spiral structure. *Astrophys. J.* 239:L53–L56.

Colomb, F. R., Poppel, W. G. L., and Heiles, C. 1980. Galactic H I at /b/ greater than or equal to 10°. II. Photographic presentation of the combined southern and northern data. *Astron. Astrophys. Suppl.* 40:47–55.

Columbo, G., and Franklin, F. A. 1971. On the formation of the outer satellite groups of Jupiter. *Icarus* 15:186–191.

Combes, M., and Encrenaz, T. 1979. A method for the determination of abundance ratios in the outer planets: Application to Jupiter. *Icarus* 39:1–27.
Combes, M., Maillard, J. P., and de Bergh, C. 1977. Evidence for a telluric value of the $^{12}C/^{13}C$ ratio in the atmosphere of Jupiter and Saturn. *Astron. Astrophys.* 61:531–537.
Comins, N. 1979*a*. On secular instabilities of rigidly rotating stars in general relativity. I. Theoretical formalism. *Mon. Not. Roy. Astron. Soc.* 189:233–253.
Comins, N. 1979*b*. On secular instabilities of rigidly rotating stars in general relativity. II. Numerical results. *Mon. Not. Roy. Astron. Soc.* 189:255–272.
Condon, E. U., and Odishaw, H. 1967. *Handbook of Physics,* 2nd ed. (New York: McGraw Hill).
Conrath, B. J., Gautier, D., Hanel, R. A., and Hornstein, J. S. 1984. The helium abundance of Saturn from Voyager measurements. *Astrophys. J.* In press.
Consolmagno, G. J., and Lewis, J. S. 1976. Structural and thermal models of icy Galilean satellites. In *Jupiter,* ed. T. Gehrels (Tucson: Univ. of Arizona Press), pp. 1035–1051.
Consolmagno, G. J., and Lewis, J. S. 1977. Preliminary thermal history models of icy satellites. In *Planetary Satellites,* ed. J. A. Burns (Tucson: Univ. of Arizona Press), pp. 492–500.
Consolmagno, G. J., and Jokipii, J. R. 1978. ^{26}Al and the partial ionization of the solar nebula. *Moon and Planets* 19:253–259.
Consolmagno, G. J., and Lewis, J. S. 1978. The evolution of icy satellite interiors and surfaces. *Icarus* 34:280–293.
Consolmagno, G. J., and Cameron, A. G. W. 1980. The origin of the "FUN" anomalies and the high temperature inclusions in the Allende meteorite. *Moon and Planets* 23:3–25.
Cook, T. L. 1977. Three-Dimensional Dynamics of Protostellar Evolution. Ph.D. thesis, Rice University (also report LA-6841T, Los Alamos National Laboratory).
Coradini, A., Magni, G., and Federico, C. 1977. Grain accretion processes in a protoplanetary nebula. II. Accretion time and mass limit. *Astrophys. Space Sci.* 48:79–87.
Coradini, A., Federico, C., and Magni, G. 1981. Formation of planetesimals in an evolving protoplanetary disk. *Astron. Astrophys.* 98:173–185.
Cosmovici, C. B., and Ortolani, S. 1984. Detection of new molecules in the visible spectrum of Comet IRAS-Araki-Alcock (1983d). *Nature* 310:122–124.
Courtin, R. 1982. La Structure Thermique et le Composition des Atmospheres des Planetes Geantes à artir de leur Spectre Infrarouge Lointain. Ph.D. thesis, Université de Paris.
Courtin, R., Gautier, D., Marten, A., and Kunde, V. 1983. The $^{12}C/^{13}C$ ratio in Jupiter from the Voyager infrared investigations. *Icarus* 53:121–132.
Cowie, L. L. 1980. On the dynamics of agglomerating ensembles of clouds. *Astrophys. J.* 236:868–879.
Cowie, L. L. 1981. Cloud fluid compression and softening in spiral arms and the formation of giant molecular cloud complexes. *Astrophys. J.* 245:66–71.
Cowie, L. L., Hu, E. M., Taylor, W., and York, D. G. 1981. A search for expanding supershells of gas around OB associations. *Astrophys. J.* 250:L24–L29.
Cox, D. P. 1979. Mechanical heating of the interstellar medium. I. The source and rate. *Astrophys. J.* 234:863–875.
Cox, D. P. 1983. Self-regulating star formation: The rate limit set by ionizing photons. *Astrophys. J.* 265:L61–L62.
Craine, E. C., Boeshaar, G. O., and Byard, P. L. 1981. 1548C27: An interesting new cometary nebula. *Astron. J.* 86:751–754.
Craine, E. C., Turnshek, D. E., Turnshek, D. A., Lynds, B. T., and O'Neil, E. J. 1983. The spectrum of 1548C27. Steward Obs. Preprint 413.
Cram, L. E. 1979. Atmospheres of T Tauri stars: The photosphere and low chromosphere. *Astrophys. J.* 234:949–957.
Cram, L. E., Giampapa, M. S., and Imhoff, C. L. 1980. Emission measures derived from far ultraviolet spectra of T Tauri stars. *Astrophys. J.* 238:905–908.
Crovisier, J. 1978. Kinematics of neutral hydrogen clouds in the solar vicinity from the Nancay 21-cm absorption survey. *Astron. Astrophys.* 70:43–50.
Crovisier, J. 1981. Statistical properties of interstellar neutral hydrogen from 21-cm absorption surveys. *Astron. Astrophys.* 94:162–174.
Crovisier, J., and Dickey, J. M. 1983. The spatial power spectrum of galactic neutral hydrogen from observations of the 21-cm emission line. *Astron. Astrophys.* 123:282–296.

Cruikshank, D. P., Brown, R. H., and Clark, R. N. 1983. Nitrogen on Triton. *Bull. Amer. Astron. Soc.* 15:857 (abstract).
Cruikshank, D. P., Brown, R. H., and Clark, R. N. 1984. Nitrogen on Titan. *Icarus* 58:293–305.
Crutcher, R. M. 1983. Interstellar magnetic fields: Detection of OH Zeeman splitting. *Bull. Amer. Astron. Soc.* 15:931 (abstract).
Crutcher, R. M., Troland, T. H., and Heiles, C. 1981. Magnetic fields in molecular clouds: OH Zeeman observations. *Astrophys. J.* 249:134–137.
Crutcher, R. M., Churchwell, E., and Ziurys, L. M. 1984. CN in dark interstellar clouds. *Astrophys. J.* 283:668–674.
Cudworth, K. M., and Herbig, G. H. 1979. Two large proper-motion Herbig-Haro objects. *Astron. J.* 84:548–551.
Cummins, S. E., Green, S., Thaddeus, P., and Linke, R. A. 1983. The kinetic temperature and density of the Sagittarius B2 molecular cloud from observations of methyl cyanide. *Astrophys. J.* 266:331–338.
Cummins, S. E., Linke, R. A., and Thaddeus, P. 1984. A survey of the mm-wave spectrum of Sgr B2 with column densities for 21 molecular species. In preparation.
Cuzzi, J. N., and Scargle, J. 1985. Encke's division has wavy edges. *Astrophys. J.* In press.
Cuzzi, J. N., Pollack, J. B., and Summers, A. L. 1980. Saturn's rings: Particle composition and size distribution as constrained by observations at microwave wavelengths. II. Radio interferometric observations. *Icarus* 44:683–705.
Cuzzi, J. N., Lissauer, J. J., and Shu, F. H. 1981. Density waves in Saturn's rings. *Nature* 292:703–707.
Cuzzi, J. N., Scargle, J. D., Showalter, M., and Esposito, L. W. 1983. Saturn's rings: Indirect evidence for moonlets embedded within Encke's division. *Bull. Amer. Astron. Soc.* 15:813 (abstract).
Cuzzi, J. N., Lissauer, J. J., Esposito, L. W., Holberg, J. B., Marouf, E. A., Tyler, G. L., and Boischot, A. 1984. Saturn's rings: Properties and processes. In *Planetary Rings*, eds. R. Greenberg and A. Brahic (Tucson: Univ. of Arizona Press), pp. 73–199.
Dalgarno, A. 1981. Chemical processes in the shocked interstellar gas. *Phil. Trans. Roy. Soc. London* A303:513–522.
Dalgarno, A., and McCray, R. A. 1972. Heating and ionization of HI regions. *Ann. Rev. Astron. Astrophys.* 10:375–426.
Dalgarno, A., Black, J. H., and Weisheit, J. 1973. Ortho-para transitions in H_2 and the fractionation of HD. *Astrophys. Lett.* 114:77–79.
Dame, T. 1983. Molecular Clouds and Galactic Spiral Structure. Ph.D. thesis, Columbia University, New York.
Dame, T., Elmegreen, B. G., Cohen, R. S., and Thaddeus, P. 1984. The sizes and positions of the largest molecular cloud complexes in the galaxy. *Astrophys. J.* In press.
Darwin, G. H. 1879. On the precession of a viscous spheroid and on the remote history of the Earth. *Phil. Trans. Roy. Soc. Part II* 170:447–530.
Davidson, J., and Jaffe, D. 1984. Far-infrared and submillimeter observations of the low-luminosity protostars L1455 FIR and L1551 IRS5: The confinement of bipolar outflows. *Astrophys. J.* 277:L13–L16.
Davidson, K., and Harwit, M. 1967. Infrared and radio appearance of cocoon stars. *Astrophys. J.* 148:443–448.
Davies, R. D., Elliott, K. H., and Meaburn, J. 1976. The nebular complexes of the Large and Small Magellanic Clouds. *Mem. Roy. Astron. Soc.* 81:89–128.
Davis, A. M., and Grossman, L. 1979. Condensation and fractionation of rare earths in the solar nebula. *Geochim. Cosmochim. Acta* 43:1611–1632.
Davis, A. M., Tanaka, T., and Grossman, L. 1982. Chemical composition of HAL, an isotopically-unusual Allende inclusion. *Geochim. Cosmochim. Acta* 46:1627–1651.
Davis, M., Hut, P., and Muller, R. A. 1984. Extinction of species by periodic comet showers. *Nature* 308:715–717.
de Bergh, C., Lutz, B. L., Lockwood, G. W., Owen, T., and Buriez, J. C. 1981. Detection of CH_3CH_4 ratio on Titan and Uranus. *Bull. Amer. Astron. Soc.* 13:703 (abstract).
deBoer, K. S., and Morton, D. C. 1974. Interstellar CI lines in ρ Ophiuchus. *Astron. Astrophys.* 37:305–311.

DeCampli, W. M. 1981. T Tauri winds. *Astrophys. J.* 244:124–146.

DeCampli, W. M., and Cameron, A. G. W. 1979. Structure and evolution of isolated giant gaseous protoplanets. *Icarus* 38:367–391.

DeFrees, D. J., McLean, A. D., and Herbst, E. 1984. Calculations concerning the HCO^+ /HOC^+ abundance ratio in dense interstellar clouds. *Astrophys. J.* 279:322–334.

Degewij, J., Cruikshank, D. P., and Hartmann, W. K. 1980. Near-infrared colorimetry of J6 Himalia and S9 Phoebe: A summary of 0.3 to 2.2 μm reflectances. *Icarus* 44:541–547.

DeGioia-Eastwood, K., Grasdalen, G. L., Strom, S. E., and Strom, K. M. 1984. Massive star formation in NCG 6946. *Astrophys. J.* 278:564–574.

deGraauw, T., Lidholm, S., Fitton, B., Beckman, J., Israel, F. P., Nieuwenhuijzzen, H., and Vermue, J. 1981. CO ($J=2\rightarrow1$) observations of southern H II regions. *Astron. Astrophys.* 102:257–264.

Deissler, R. G. 1976. Gravitational collapse of a turbulent vortex with application to star formation. *Astrophys. J.* 209:190–204.

Deissler, R. G. 1984. Turbulent solutions of the equations of fluid motion. *Rev. Mod. Phys.* 56:223–254.

de Jong, T., Dalgarno, A., and Boland, W. 1980. Hydrostatic models of molecular clouds. *Astron. Astrophys.* 91:68–84.

Delsemme, A. H. 1982. Chemical composition of cometary nuclei. In *Comets,* ed. L. L. Wilkening (Tucson: Univ. of Arizona Press), pp. 85–130.

DeMarcus, W. C. 1958. The constitution of Jupiter and Saturn. *Astron. J.* 63:1–28.

DeMarcus, W. C., and Reynolds, R. T. 1962. The constitution of Uranus and Neptune. *Mem. Soc. Roy. Sci. Liège* 7:51–64.

Dermott, S. F. 1973. Bode's law and the resonant structure of the solar system. *Nature Physical Sci.* 244:18–21.

Dermott, S. F., and Gold, T. 1978. On the origin of the Oort cloud. *Astron. J.* 83:449–450.

Dermott, S. F., and Murray, C. D. 1981*a*. The dynamics of tadpole and horseshoe orbits. I. Theory. *Icarus* 48:1–11.

Dermott, S. F., and Murray, C. D. 1981*b*. The dynamics of tadpole and horseshoe orbits. II. The co-orbital satellites of Saturn. *Icarus* 48:12–22.

Descartes, R. 1644. *Principia Philosophia* (Amsterdam).

Detweiler, S. L., and Lindblom, L. 1977. On the evolution of the homogeneous ellipsoidal figures. *Astrophys. J.* 213:193–199.

DeYoung, D. S. 1980. Turbulent generation of magnetic fields in extended extragalactic radio sources. *Astrophys. J.* 241:81–97.

Dibai, E. A. 1958. Evolution of globules in the vicinity of hot stars. *Sov. Astron. J.* 2:429–432.

Dibai, E. A. 1960. The origin of cometary nebulae. *Sov. Astron. J.* 4:13–18.

Dibai, E. A., and Kaplan, S. A. 1965. Cumulative shock waves in interstellar space. *Sov. Astron.* 8:520–523.

Dickel, H. R., Lubenow, A. F., Goss, W. M., Forster, J. R., and Rots, A. H. 1983. VLA observations of H_2CO in DR21. *Astron. Astrophys.* 120:74–84.

Dickel, H. R., Lubenow, A. F., Goss, W. M., and Rots, A. H. 1984. VLA observations of H_2CO in W 3(OH). *Astron. Astrophys.* 135:107–115.

Dickey, J. M. 1984. Small scale structure and motions in the interstellar gas. In *The Milky Way,* ed. H. van Woerden (Dordrecht: D. Reidel). In press.

Dickey, J. M., Crovisier, J., and Kazes, I. 1984. The smallest sizes of diffuse interstellar clouds. Preprint.

Dickman, R. L. 1975. A survey of carbon monoxide emission in dark clouds. *Astrophys. J.* 202:50–57.

Dickman, R. L. 1977. Bok globules. *Sci. Amer.* 236:66–81.

Dickman, R. L. 1978. The ratio of carbon monoxide to molecular hydrogen in interstellar dark clouds. *Astrophys. J. Suppl.* 37:407–427.

Dickman, R. L., and Kleiner, S. C. 1981. Mesoturbulence in molecular clouds: A numerical approach. *Bull. Amer. Astron. Soc.* 13:864 (abstract).

Dickman, R. L., and Clemens, D. C. 1983. A gravitationally stable Bok globule. *Astrophys. J.* 271:143–160.

Dickman, R. L., and Kleiner, S. C. 1985. Large-scale structure of the Taurus molecular cloud complex. III. Methods for turbulence. *Astrophys. J.* In press.

Dickman, R. L., McCutcheon, W., and Shuter, W. 1979. Carbon monoxide isotope fractionation in the dust cloud Lynds 134. *Astrophys. J.* 234:100–110.
Dickman, R. L., Kutner, M. L., Pasachoff, J. M., and Tucker, K. D. 1980. Turbulence in the dust cloud L134: High-resolution of observations of 6 centimeter formaldehyde absorption. *Astrophys. J.* 238:853–859.
Dickman, R. L., Somerville, W. B., Whittet, D. C. B., McNally, D., and Blades, J. C. 1983. Abundances of carbon-bearing diatomic molecules in diffuse interstellar clouds. *Astrophys. J. Suppl.* 53:55–72.
Dieter, N. H. 1976. Interstellar formaldehyde near stars of the Orion population. *Astrophys. J.* 199:289–296.
Disney, M. J., and Hopper, P. B. 1975. The alignment of interstellar dust clouds and the differential z-field of the galaxy. *Mon. Not. Roy. Astron. Soc.* 170:177–184.
Disney, M. J., McNally, D., and Wright, A. E. 1969. The collapse of interstellar clouds. IV. Models of collapse and a theory of star formation. *Mon. Not. Roy. Astron. Soc.* 146:123–160.
Dodd, R. T. 1971. The petrology of chondrules in the Sharps meteorite. *Contrib. Mineral. Petrol.* 31:201–227.
Dodd, R. T. 1981. *Meteorites: A Petrologic-Chemical Synthesis* (Cambridge: Cambridge Univ. Press).
Dole, S. H. 1962. Gravitational concentration of particles in space near the Earth. *Planet. Space Sci.* 9:541–553.
Donahue, T. M., Hoffman, J. H., Hodges, R. R. Jr., and Watson, A. J. 1982. Venus was wet: A measurement of the ratio of deuterium to hydrogen. *Science* 216:630–633.
Donivan, F. F., and Carr, T. D. 1969. Jupiter's decametric rotation period. *Astrophys. J.* 157:L65–L68.
Donn, B. 1968. Polycyclic hydrocarbons, platt particles, and interstellar extinction. *Astrophys. J.* 152:L129–L133.
Donn, B. 1976. Comets, interstellar clouds, and star clusters. In *The Study of Comets* (Washington: NASA SP-393), pp. 663–672.
Dopita, M. A. 1978. Optical emission from shocks. IV. The Herbig-Haro objects. *Astrophys. J. Suppl.* 37:117–144.
Dopita, M. A., Binette, L., and Schwartz, R. D. 1982*a*. The two-proton continuum in Herbig-Haro objects. *Astrophys. J.* 261:183–194.
Dopita, M. A., Schwartz, R. D., and Evans, I. 1982*b*. Herbig-Haro objects 46 and 47: Evidence for bipolar ejection from a young star. *Astrophys. J.* 263:L73–L77.
Dorfi, E. 1982. 3D models for self-gravitating, rotating, magnetic interstellar clouds. *Astron. Astrophys.* 114:151–164.
Douglas, A. E., and Herzberg, G. 1941. CH^+ in interstellar space and in the laboratory. *Astrophys. J.* 94:381.
Downes, D., Wilson, T. L., Bieging, J., Wink, J. 1980. H110 alpha, and H_2CO survey of galactic radio sources. *Astron. Astrophys. Suppl.* 40:379–394.
Downes, D., Genzel, R., Becklin, E. E., and Wynn-Williams, C. G. 1981. Outflow of matter in the KL nebula: The role of IRc2. *Astrophys. J.* 244:869–883.
Draine, B. T. 1979. On the chemisputtering of interstellar graphite grains. *Astrophys. J.* 230:106–115.
Draine, B. T. 1980. Interstellar shock waves with magnetic precursors. *Astrophys. J.* 241:1021–1038.
Draine, B. T. 1983. Magnetic bubbles and high-velocity outflows in molecular clouds. *Astrophys. J.* 270:519–536.
Draine, B. T., and Salpeter, E. E. 1979*a*. On the physics of dust grains in hot gas. *Astrophys. J.* 231:77–94.
Draine, B. T., and Salpeter, E. E. 1979*b*. Destruction mechanisms for interstellar dust. *Astrophys. J.* 231:438–445.
Draine, B. T., and Roberge, W. G. 1982. A model for the intense molecular line emission from OMC-1. *Astrophys. J.* 259:L91–L96.
Draine, B. T., and Lee, H. M. 1984. Optical properties of interstellar graphite and silicate grains. *Astrophys. J.* 285. In press.
Draine, B. T., Roberge, W. G., and Dalgarno, A. 1983. Magneto-hydrodynamic shock waves in molecular clouds. *Astrophys. J.* 264:485–507.

Drapatz, S., and Zinnecker, H. 1984. The size and mass distribution of galactic molecular clouds. *Mon. Not. Roy. Astron. Soc.* 210:11P–14P.

Dreher, J. W., Johnston, K. J., Welsh, W. J., and Walker, R. C. 1983. Ultracompact structure in the HII region W49N. Submitted to *Astrophys. J.*

Dubin, M., and McCracken, C. W. 1962. Measurements of distributions of interplanetary dust. *Astron. J.* 67:248.

Duerr, R., Imhoff, C. L., and Lada, C. J. 1982. Star formation in the λ Orionis region. I. The distribution of young objects. *Astrophys. J.* 261:135–150.

DuFresne, E. R., and Anders, E. 1962. On the chemical evolution of the carbonaceous chondrites. *Geochim. Cosmochim. Acta* 26:1085–1114.

Duley, W. W., and Williams, D. A. 1979. Are there organic grains in the interstellar medium. *Nature* 277:40–41.

Dunham, T. Jr. 1937. Interstellar neutral potassium and neutral calcium. *Publ. Astron. Soc. Pacific* 49:26–28.

Dunham, T. Jr. 1939. *Proc. Amer. Phil. Soc.* 81:227.

Dunham, T. Jr. 1941. The concentration of interstellar molecules. *Publ. Amer. Astron. Soc.* 10:123–124 (abstract).

Dunham, T. Jr., and Adams, W. S. 1937. New interstellar lines in the ultra-violet spectrum. *Publ. Amer. Astron. Soc.* 9:5–6 (abstract).

Durisen, R. H. 1975. Upper mass limits for stable rotating white dwarfs. *Astrophys. J.* 199:179–183.

Durisen, R. H. 1984. Transport effects due to particle erosion mechanisms. In *Planetary Rings,* eds. R. Greenberg and A. Brahic (Tucson: Univ. of Arizona Press), pp. 416–446.

Durisen, R. H., and Imamura, J. N. 1979. Improved secular stability limits for rotating white dwarfs. In *White Dwarfs and Variable Degenerate Stars,* eds. H. M. Van Horn and V. Weidemann (Rochester: Univ. of Rochester Press), pp. 43–47.

Durisen, R. H., and Tohline, J. E. 1980. A numerical study of the fission hypothesis for rotating polytropes. *Space Sci. Rev.* 27:267–273.

Durisen, R. H., and Imamura, J. N. 1981. Improved secular stability limits for differentially rotating polytropes and degenerate dwarfs. *Astrophys. J.* 243:612–616.

Durisen, R. H., and Scott, E. H. 1984. Implications of recent numerical calculations for the fission theory of the origin of the Moon. *Icarus.* 58:153–158.

Durisen, R. H., Gingold, R. A., Tohline, J. E., and Boss, A. P. 1984. The binary fission hypothesis: A comparison of results from different numerical codes. In preparation.

Durrance, S. T., and Moos, H. W. 1982. Intense Lyman alpha emission from Uranus. *Nature* 299:428–429.

Dyck, H. M., and Lonsdale, C. J. 1981. Polarimetry of infrared sources. In *Infrared Astronomy,* eds. C. G. Wynn-Williams and D. P. Cruikshank (Dordrecht: D. Reidel), pp. 223–236.

Dyck, H. M., and Howell, R. R. 1982. Speckle interferometry of molecular cloud sources at 4.8 μm. *Astron. J.* 87:400–403.

Dyck, H. M., Simon, T., and Zuckerman, B. 1982. Discovery of an infrared companion to T Tauri. *Astrophys. J.* 255:L103–L106.

Dyck, H. M., Beckwith, S., and Zuckerman, B. 1983. Speckle interferometry of IRC+10216 in the fundamental vibration-rotation lines of CO. *Astrophys. J.* 271:L79–L84.

Dyson, J. E. 1968. The dynamics of the Orion Nebula. I. Neutral condensations in an H II region. *Astrophys. Space Sci.* 1:388.

Eberhardt, P. 1974. A neon-E-rich phase in the Orgueil carbonaceous chondrite. *Earth Planet. Sci. Lett.* 24:182–187.

Eberhardt, P., Jungck, M. H. A., Meier, F. O., and Niederer, F. R. 1979. Presolar grains in Orgueil: Evidence from Neon-E. *Astrophys. J.* 234:L169–L171.

Eberhardt, P., Jungck, M. H. A., Meier, F. O., and Niederer, F. R. 1981. A neon-E rich phase in Orgueil: Results obtained on density separates. *Geochim. Cosmochim. Acta* 45:1515–1528.

Ebert, R. 1955. Temperatur des interstellaren Gases bei großen Dichten. *Z. Astrophys.* 37:222–229.

Ebert, R., and Zinnecker, H. 1981. Effect of mass gain on stellar evolution. In *Effects of Mass Loss in Stellar Evolution,* eds. C. Chiosi and R. Stalio (Dordrecht: D. Reidel), pp. 361–371.

Ebert, R., von Hoerner, S., and Temesvary, S. 1960. *Die Entstehung von Sternen durch Kondensation Diffuser Materie* (Berlin: Springer-Verlag), pp. 184–324.

Ebihara, M., Wolf, R., and Anders, E. 1982. Are Cl chondrites chemically fractionated? A trace element study. *Geochim. Cosmochim. Acta* 46:1849–1862.
Eckmann, J. P. 1981. Roads to turbulence in dissipative dynamical systems. *Rev. Mod. Phys.* 53:643–654.
Edwards, S., and Snell, R. L. 1982. A search for high-velocity molecular gas around T Tauri stars. *Astrophys. J.* 261:151–160.
Edwards, S., and Snell, R. L. 1983. A survey of high-velocity molecular gas in the vicinity of Herbig-Haro objects. I. *Astrophys. J.* 270:605–619.
Edwards, S., and Snell, R. L. 1984. A survey of high-velocity molecular gas near Herbig-Haro objects. II. *Astrophys. J.* 281:237–249.
Efremov, Yu.M. 1979. Star complexes. *Sov. Astron. Lett.* 4:66–69.
Eggen, O. J. 1982. The Hyades main sequence. *Astrophys. J. Suppl.* 50:221–240.
Eggleton, D. P. 1967. The structure of narrow shells in red giants. *Mon. Not. Roy. Astron. Soc.* 135:243.
Eilik, J. A., and Henriksen, R. N. 1984. The electron energy spectrum produced in radio sources by turbulent, resonant acceleration. *Astrophys. J.* 277:820–831.
Elias, J. H. 1978*a*. A study of the IC 5146 dark cloud complex. *Astrophys. J.* 223:859–875.
Elias, J. H. 1978*b*. An infrared study of the Ophiuchus dark cloud. *Astrophys. J.* 224:453–472.
Elias, J. H. 1978*c*. A study of the Taurus dark cloud complex. *Astrophys. J.* 224:857–872.
Ellder, J., Friberg, P., Hjalmarson, Å., Höglund, B., Irvine, W. M., Johansson, L. E. B., Olofsson, H., Rydbeck, G., Rydbeck, O. E. H., and Guélin, M. 1980. On methyl formate, methane, and deuterated ammonia in Orion A. *Astrophys. J.* 242:L93–L97.
Elliot, J. L., and Nicholson, P. D. 1984. The rings of Uranus. In *Planetary Rings,* eds. R. Greenberg and A. Brahic (Tucson: Univ. of Arizona Press), pp. 25–72.
Elliot, J. L., Dunham, E., Mink, D. J., and Churms, J. 1980. The radius and ellipticity of Uranus from its occultation of SAO 158687. *Astrophys. J.* 236:1026–1030.
Elliot, J. L., French, R. G., Frogel, J. A., Elias, J. H., Mink, D. J., and Liller, W. 1981. Orbits of nine Uranian rings. *Astron. J.* 86:444–455.
Ellsworth, K., and Schubert, G. 1983. Saturn's icy satellites: Thermal and structural models. *Icarus* 54:490–510.
Elmegreen, B. G. 1978*a*. On the determination of magnetic fields in dense cloud complexes by the observation of Zeeman splitting. *Astrophys. J.* 225:L85–L88.
Elmegreen, B. G. 1978*b*. On the interaction between a strong stellar wind and a surrounding disk nebula. *Moon and Planets* 19:261–277.
Elmegreen, B. G. 1979*a*. Gravitational collapse in dust lanes and the appearance of spiral structure in galaxies. *Astrophys. J.* 231:372–383.
Elmegreen, B. G. 1979*b*. Magnetic diffusion and ionization fractions in dense molecular clouds: The role of charged grains. *Astrophys. J.* 232:729–739.
Elmegreen, B. G. 1980. Star formation behind shocks. In *Giant Molecular Clouds in the Galaxy,* eds. P. M. Solomon and M. G. Edmunds (Oxford: Pergamon), pp. 255–264.
Elmegreen, B. G. 1981*a*. The role of magnetic fields in constraining the translational motions of giant cloud complexes. *Astrophys. J.* 243:512–525.
Elmegreen, B. G. 1981*b*. From clouds to stars. In *The Formation of Planetary Systems,* ed. A Brahic (Toulouse: Cepadues Editions), pp. 61–174.
Elmegreen, B. G. 1982*a*. The Parker instability in a self-gravitating gas layer. *Astrophys. J.* 253:634–654.
Elmegreen, B. G. 1982*b*. The formation of giant cloud complexes. In *Submillimeter Wave Astronomy,* eds. J. E. Beckman and J. P. Phillips (Cambridge: Cambridge Univ. Press), pp. 3–14.
Elmegreen, B. G. 1982*c*. The formation of giant cloud complexes by the Parker-Jeans instability. *Astrophys. J.* 253:655–665.
Elmegreen, B. G. 1983*a*. The initial stellar mass function as a statistical ensemble and implications for the formation of bound clusters. In *The Nearby Stars and the Stellar Luminosity Function, IAU Colloquium No. 76,* eds. A. G. D. Phillip, and A. R. Upgren (Schenectady, NY: L. Davis Press), pp. 235–240.
Elmegreen, B. G. 1983*b*. Long-range propagating star formation. *Bull. Amer. Astron. Soc.* 15:990 (abstract).
Elmegreen, B. G. 1983*c*. Quiescent formation of bound galactic clusters. *Mon. Not. Roy. Astron. Soc.* 203:1011–1020.

Elmegreen, B. G. 1985. Primary and secondary mechanisms of giant cloud formation. In *Star Formation, Lectures Delivered at Les Houches Summer School,* eds. A. Omont and R. Lucas (Amsterdam: North-Holland). In press.

Elmegreen, B. G., and Lada, C. J. 1976. Discovery of an extended (85 pc) molecular cloud associated with the M17 star-forming complex. *Astron. J.* 81:1089–1094.

Elmegreen, B. G., and Lada, C. J. 1977. Sequential formation of subgroups in OB associations. *Astrophys. J.* 214:725–741.

Elmegreen, B. G., and Elmegreen, D. M. 1978. Star formation in shock-compressed layers. *Astrophys. J.* 220:1051–1062.

Elmegreen, B. G., and Chiang, W.-H. 1982. Runaway expansion of giant shells driven by radiation pressure from field stars. *Astrophys. J.* 253:666–678.

Elmegreen, B. G., and Elmegreen, D. M. 1983. Regular strings of HII regions and superclouds in spiral galaxies: Clues to the origin of cloudy structure. *Mon. Not. Roy. Astron. Soc.* 203:31–45.

Elmegreen, B. G., Lada, C. J., and Dickinson, D. F. 1979. The structure and extent of the giant molecular cloud near M17. *Astrophys. J.* 230:415–427.

Elmegreen, B. G., Elmegreen, D. M., and Morris, M. 1980. On the abundance of CO in galaxies: A comparison of spiral and Magellanic irregular galaxies. *Astrophys. J.* 240:455–463.

Elmegreen, D. M., and Elmegreen, B. G. 1978. CO observations of a high latitude molecular cloud associated with NGC 7023. *Astrophys. J.* 220:510–515.

Elmegreen, D. M., and Elmegreen, B. G. 1982. CO observations of the SAB galaxies NGC 157, 2903, 4321, and 5248, and the Seyfert Galaxy NCG 1068. *Astron. J.* 87:626–634.

Elmegreen, D. M., and Elmegreen, B. G. 1984. Blue and near-infrared surface photometry of spiral structure in 34 non-barred grand design and flocculent galaxies. *Astrophys. J. Suppl.* 54:127–149.

Elsässer, H., and Staude, H. J. 1978. On the polarization of young stellar objects. *Astron. Astrophys.* 70:L3–L6.

Emerson, J. P., Harris, S., Jennings, R. E., Beichman, C. A., Baud, B., Beintema, D. A., Marsden, P. L., and Wesselius, P. R. 1984. IRAS observations near young objects with bipolar outflows: L 1551 and HH 46–47. *Astrophys. J.* 278:L49–L52.

Encrenaz, T., and Combes, M. 1982. On the C/H and D/H ratios in the atmospheres of Jupiter and Saturn. *Icarus* 52:54–61.

Endal, A. S., and Sofia, S. 1978. The evolution of rotating stars. II. Calculations with time-dependent redistribution of angular momentum for 7 $M_\odot$ and 10 $M_\odot$ stars. *Astrophys. J.* 220:279–290.

Erickson, E. F., Knacke, R. F., Tokunaga, A. T., and Haas, M. R. 1981. The 45 micron H_2O ice band in the Kleinmann-Low nebula. *Astrophys. J.* 245:148–153.

Erickson, N. R., Goldsmith, P. F., Snell, R. L., Berson, R. L., Huguenin, G. R., Ulich, B. L., and Lada, C. J. 1982. Detection of bi-polar CO outflow in Orion. *Astrophys. J.* 261:L103–L107.

Eriguchi, Y., and Sugimoto, D. 1981. Another equilibrium sequence of self-gravitating and rotating incompressible fluid. *Prog. Theor. Phys.* 65:1870–1875.

Eriguchi, Y., and Hachisu, I. 1982. New equilibrium sequences bifurcating from Maclaurin sequence. *Prog. Theor. Phys.* 67:844–851.

Eriguchi, Y., and Hachisu, I. 1983*a*. Two kinds of axially symmetric equilibrium sequences of self-gravitating and rotating incompressible fluid. Two-ring sequence and core-ring sequence. *Prog. Theor. Phys.* 69:1131–1136.

Eriguchi, Y., and Hachisu, I. 1983*b*. Gravitational equilibrium of a multi-body fluid system. *Prog. Theor. Phys.* 70:1534–1541.

Eriguchi, Y., Hachisu, I., and Sugimoto, D. 1982. Dumbbell shape equilibria and mass-shedding pear-shape of self-gravitating incompressible fluid. *Prog. Theor. Phys.* 67:1068–1075.

Esat, T., Lee, T., Papanastassiou, D. A., and Wasserburg, G. J. 1978. Search for ^{26}Al effects in the Allende FUN inclusion Cl. *Geophys. Res. Lett.* 5:807–810.

Esat, T. M., Papanastassiou, D. A., and Wasserburg, G. J. 1980. This initial state of ^{26}Al and ^{26}Mg/^{24}Mg in the early solar system. *Lunar Planet. Sci.* XI:262–264 (abstract).

Esposito, L. W., O'Callaghan, M., and West, R. A. 1983. The structure of Saturn's rings: Implications from the Voyager stellar occultation. *Icarus* 56:439–452.

Evans, A., Bode, M. F., Whittet, D. C. B., Davies, J. K., Kilkenny, D., and Baines, D. W. T. 1982*b*. The variability of Ry Lupi. *Mon. Not. Roy. Astron. Soc.* 199:37P–43P.
Evans, N. J. 1978. Star formation in molecular clouds. In *Protostars & Planets,* ed. T. Gehrels (Tucson: Univ. of Arizona Press), pp. 152–164.
Evans, N. J. II, and Kutner, M. L. 1976. H_2CO emission at 2 mm in dark clouds. *Astrophys. J.* 204:L131–L134.
Evans, N. J. II, Plambeck, R. L., and Davis, J. H. 1979. Detection of the $3_{12} \rightarrow 2_{11}$ transition of interstellar formaldehyde at 1.3 mm. *Astrophys. J.* 227:L25–L28.
Evans, N. J. II, Blair, G., Harvey, P., Israel, F., Peters, W., Scholtes, M., deGraauw, T., and Vanden Bout, P. 1981. The energetics of molecular clouds. IV. The S88 molecular cloud. *Astrophys. J.* 250:200–212.
Evans, N. J. II, Blair, G. N., Nadeau, D., and Vanden Bout, P. 1982. The energetics of molecular clouds. V. The S37 molecular cloud. *Astrophys. J.* 253:115–130.
Evans, N. J., Carr, J., Beckwith, S., Skrutskie, M., and Wyant, J. 1983. Spectroscopy of molecular cloud sources at 6–7 microns. *Publ. Astron. Soc. Pacific* 95:648–652.
Everhart, E. 1967. Intrinsic distributions of cometary perihelia and magnitudes. *Astron. J.* 72:1002–1011.
Ezer, D., and Cameron, A. G. W. 1965. A study of solar evolution. *Can. J. Phys.* 43:1497–1517.
Faintich, M. B. 1971. Interstellar Gravitational Perturbations of Cometary Orbits. Ph.D. thesis, Univ. of Illinois, Urbana.
Falgarone, E., and Gilmore, W. 1981. Partial aperture synthesis of five dark clouds at 1.4 GHz. *Astron. Astrophys.* 95:32–38.
Faltonen, M. J., and Innanen, K. 1982. The capture of interstellar comets. *Astrophys. J.* 255:307–315.
Farinella, P., Paolicchi, P., Tedesco, E. F., and Zappala, V. 1981. Triaxial equilibrium ellipsoids among the asteroids? *Icarus* 46:114–123.
Faulkner, J., Lin, D. N. C., and Papaloizou, J. 1983. On the evolution of accretion disc flow in cataclysmic variables. I. The prospect of a limit cycle in dwarf nova systems. *Mon. Not. Roy. Astron. Soc.* 205:359–375.
Fazio, G. G., Wright, E. L., Zeilik, M., and Low, F. J. 1976. A far-infrared map of the Ophiuchus dark cloud region. *Astrophys. J.* 206:L165–L169.
Federman, S. R., and Glassgold, A. E. 1980. Modeling of diffuse interstellar clouds: The case of gamma arae. *Astron. Astrophys.* 89:113–117.
Federman, S. R., Glassgold, A. E., and Kwan, J. 1979. Atomic to molecular hydrogen transition in interstellar clouds. *Astrophys. J.* 227:446–473.
Fegley, B., and Kornacki, A. S. 1984. The origin and mineral chemistry of Group II inclusions in carbonaceous chondrites. *Lunar Planet. Sci.* XV:262–264 (abstract).
Feigelson, E. D. 1984. X-ray emission from pre-main-sequence stars. In *Third Cambridge Workshop on Cool Stars, Stellar Systems and the Sun,* eds. S. L. Baliunas and L. Hartmann (Berlin-Heidelberg: Springer-Verlag), pp. 27–42.
Feigelson, E. D., and DeCampli, W. M. 1981. Observations of x-ray emission from T Tauri stars. *Astrophys. J.* 243:L89–L93.
Feigelson, E. D., and Kriss, G. A. 1983. A search for weak Hα emission line pre-main-sequence stars. *Astron. J.* 88:431–438.
Feitzinger, J. V., and Stuwe, J. A. 1984. *A Dark Nebula Catalogue for Southern Skies* (Bochum, West Germany: Astronomical Institute of Ruhr Univ.).
Felli, M., Johnston, K. J., and Churchwell, E. 1980. An unusual radio point source in M17. *Astrophys. J.* 242:L57–L161.
Felli, M., Churchwell, E., and Massi, M. 1984. A high resolution study of M17 at 1.3, 2, 6, and 21 cm. *Astron. Astrophys.* 136:53–64.
Fernandez, J. A. 1980*a*. Evolution of comet orbits under the perturbing influence of the giant planets and nearby stars. *Icarus* 42:406–421.
Fernandez, J. A. 1980*b*. On the existence of a comet belt beyond Neptune. *Mon. Not. Roy. Astron. Soc.* 192:481–491.
Fernandez, J. A. 1982. Dynamical aspects of the origin of comets. *Astron. J.* 87:1318–1332.
Fernandez, J. A., and Ip, W.-H. 1981. Dynamical evolution of a cometary swarm in the outer planetary region. *Icarus* 47:470–479.

Fernandez, J. A., and Ip, W.-H. 1983. On the time evolution of the cometary influx in the region of the terrestrial planets. *Icarus* 54:377–387.
Fernandez, J. A., and Ip, W.-H. 1984. Some dynamical aspects of the accretion of Uranus and Neptune: The exchange of angular momentum with planetesimals. *Icarus* 58:109–120.
Ferrini, F., Machesoni, F., and Vulpiani, A. 1982. A turbulent model for molecular clouds. *Phys. Lett.* 92A:47–50.
Ferrini, F., Marchesoni, F., and Vulpiani, A. 1983*a*. A hierarchical model for gravitational compressible turbulence. *Astrophys. Space Sci.* 96:83–93.
Ferrini, F., Marchesoni, F., and Vulpiani, A. 1983*b*. On the initial mass function and the fragmentation of molecular clouds. *Mon. Not. Roy. Astron. Soc.* 202:1071–1086.
Ferrini, F., Marchesoni, F., and Vulpiani, A. 1984. A hierarchical model for gravitational compressible turbulence. *Mon. Not. Roy. Astron. Soc.* In press.
Fesenkov, V. G. 1963. On the nature and origin of comets. *Soviet Astron. J.* 6:459–464.
Field, G. B. 1970. Theory of star formation. In *Evolution stellaire avant la séquence principale* (Liège: Mem. Soc. Roy. Liège), 19:29–45.
Field, G. B. 1978. Conditions in collapsing clouds. In *Protostars and Planets,* ed. T. Gehrels (Tucson: Univ. of Arizona Press), pp. 243–264.
Fink, U., and Larson, H. P. 1978. Deuterated ethane observed on Saturn. *Science* 201:343–345.
Fink, U., and Larson, H. P. 1979. The infrared spectra of Uranus, Neptune, and Titan from 0.8 to 2.5 microns. *Astrophys. J.* 233:1021–1040.
Finkenzeller, U., and Mundt, R. 1984. The Herbig Ae/Be stars associated with nebulosity. *Astron. Astrophys. Suppl.* 55:109–141.
Fischer, J., Joyce, R. R., Simon, M., and Simon, T. 1982. Near-infrared observations of the far-infrared source V region in NGC 6334. *Astrophys. J.* 258:165–169.
FitzGerald, M. P. 1968. The distribution of reddening material. *Astron. J.* 73:983–994.
FitzGerald, M. P., Stephens, T. C., and Witt, A. N. 1976. Surface brightness profiles of dark nebulae: The thumbprint nebula in Chamaeleon. *Astrophys. J.* 208:709–717.
Flannery, B. P., and Johnson, B. 1982. A statistical method for determining ages of globular clusters by fitting isochrones. *Astrophys. J.* 263:166–186.
Flannery, B. P., and Press, W. H. 1979. An ionization-coupled acoustic instability of the interstellar medium. *Astrophys. J.* 231:688–696.
Flannery, B. P., Roberge, W., and Rybicki, G. B. 1980. The penetration of diffuse ultraviolet radiation into interstellar clouds. *Astrophys. J.* 236:598–608.
Fleck, R. C. Jr. 1980. Turbulence and the stability of molecular clouds. *Astrophys. J.* 242:1019–1022.
Fleck, R. C. Jr. 1981. On the generation and maintenance of turbulence in the interstellar medium. *Astrophys. J.* 246:L151–L154.
Fleck, R. C. Jr. 1982*a*. Cosmic turbulence and the angular momenta of astronomical systems. *Astrophys. J.* 261:631–635.
Fleck, R. C. Jr. 1982*b*. Star formation in turbulent molecular clouds: The initial stellar mass function. *Mon. Not. Roy. Astron. Soc.* 201:551–559.
Fleck, R. C. Jr. 1983*a*. A note on compressibility and energy cascade in turbulent molecular clouds. *Astrophys. J.* 272:L45–L48.
Fleck, R. C. Jr. 1983*b*. On scaling the magnetic field strength in interstellar clouds: Resolution of the "B versus n dilemma." *Astrophys. J.* 264:139–140.
Fleck, R. C. Jr. 1984. The Kelvin-Helmholtz instability in interstellar environments. I. Morphology of the Corona Australis complex. *Astron. J.* 89:506–508.
Fleck, R. C. Jr., and Clark, F. O. 1981. A turbulent origin for the rotation of molecular clouds. *Astrophys. J.* 245:898–902.
Ford, J. 1983. How random is a coin toss? *Phys. Today* 36(4):40–47.
Forster, J. R., Goss, W. M., Dickel, H. R., and Habing, H. J. 1981. H_2CO mapping toward DR21 and W58. (K3-50). *Mon. Not. Roy. Astron. Soc.* 197:513–527.
Fowler, W. A., and Hoyle, F. 1960. Nuclear cosmochronology. *Ann. Phys.* 10:280–302.
Fowler, W. A., Greenstein, J. L., and Hoyle, F. 1962. Nucleosynthesis during the early history of the solar system. *Geophys. J.* 6:148–220.
Fox, K., Owen, T., Mantz, A. W., and Rao, K. N. 1972. A tentative identification of $^{13}CH_4$ and an estimate of $^{12}C/^{13}C$ in the atmosphere of Jupiter. *Astrophys. J.* 176:L81–L84.

Franco, J. 1983. Protostellar rotation: Turbulence and heating of molecular clouds. *Astrophys. J.* 264:508–516.

Franco, J. 1984. Winds from low-mass stars. Self-regulated star formation in Taurus, Ophiuchus, NGC2264, and Orion. *Astron. Astrophys.* 137:85–91.

Franco, J., and Cox, D. P. 1983. Self-regulated star formation in the galaxy. *Astrophys. J.* 273:243–248.

Franklin, F. A., Avis, C. C., Colombo, G., and Shapiro, I. I. 1980*a*. The geometric oblateness of Uranus. *Astrophys. J.* 236:1021–1034.

Franklin, F. A., Lecar, M., Lin, D. N. C., and Papaloizou, J. 1980. Tidal torque on infrequently colliding particle disks in binary systems and the truncation of the asteroid belt. *Icarus* 42:272–280.

Fredriksson, K. 1963. Chondrules and the meteorite parent bodies. *Trans. N.Y. Acad. Sci.* 25:756–769.

Fredriksson, K., and Keil, K. 1964. The iron, magnesium, calcium, and nickel distribution in the Murray carbonaceous chondrite. *Meteoritics* 2:201–217.

Freed, K., Oka, T., and Suzuki, H. 1982. On the n-dependence of the reaction rate for $C^+ + C_n \rightarrow C^+{}_{n+1}$ in interstellar space. *Astrophys. J.* 263:718–722.

Freeman, K. C. 1977. Star formation and the gas content of galaxies.In *The Evolution of Galaxies and Stellar Populations,* eds. B. M. Tinsley and R. B. Larson (New Haven: Yale Univ. Obs.), pp. 133–156.

Freeman, K. C., and Lynga, G. 1970. Data for Neptune from occultation observations. *Astrophys. J.* 160:767–780.

Frerking, M. A., and Langer, W. D. 1982. Detection of pedestal features in dark clouds: Evidence for formation of low mass stars. *Astrophys. J.* 256:523–529.

Frerking, M. A., Langer, W. D., and Wilson, R. W. 1979. Determination of the hyperfine structure of $HN^{13}C$ and HNC. *Astrophys. J.* 232:L65–L68.

Frerking, M. A., Langer, W. D., and Wilson, R. W. 1982. The relationship between carbon monoxide abundance and visual extinction in interstellar clouds. *Astrophys. J.* 262:590–605.

Friberg, P. 1984. $SO(3_2\text{-}2_1)$ mapping of the Orion KL cloud components. *Astron. Astrophys.* 132:265–277.

Frick, U., Becker, R. H., and Pepin, R. O. 1983. Carbon, nitrogen and xenon components in the Allende carbonaceous meteorite. *Lunar Planet. Sci.* XIV:217–218 (abstract).

Fricke, K., and Kippenhahn, R. 1972. Evolution of rotating stars. *Ann. Rev. Astron. Astrophys.* 10:45–72.

Friedlander, S. K., and Topper, L. 1961. *Turbulence: Classic Papers on Statistical Theory* (New York: Interscience).

Friedman, J. L. 1983. An upper limit on the frequency of pulsars. *Phys. Rev. Lett.* 51:11–14.

Friedman, J. L., and Schutz, B. F. 1978*a*. Lagrangian perturbation theory of nonrelativistic fluids. *Astrophys. J.* 221:937–957.

Friedman, J. L., and Schutz, B. F. 1978*b*. Secular instability of rotating Newtonian stars. *Astrophys. J.* 222:281–296.

Friedson, A. J., and Stevenson, D. J. 1983. Viscosity of ice-rock mixtures and applications to the evolution of icy satellites. *Icarus* 56:1–14.

Frisch, U., Sulem, P. U., and Nelkin, M. 1978. A simple dynamical model of intermittent fully developed turbulence. *J. Fluid Mech.* 87:719–736.

Fuchs, L. H., Olsen, E., and Jensen, K. J. 1973. Mineralogy, mineral-chemistry, and composition of the Murchison (C2) meteorite. *Smith. Contrib. Earth Sci.* No. 10, 39 pp.

Fujimoto, M. 1968. Gravitational collapse of rotating gaseous ellipsoids. *Astrophys. J.* 152:523–536.

Fujimoto, M., and Sorensen, S.-A. 1977. A computation study of the fission of self-gravitating, rotating, and elongated gaseous disks. *Astron. Astrophys.* 60:251–257.

Fujiwara, A., and Tsukamoto, A. 1980. Experimental study on the velocity of fragments in collisional breakup. *Icarus* 44:142–153.

Fukushima, T., Eriguchi, Y., Sugimoto, D., and Bisnovatyi-Kogan, G. S. 1980. Concave hamburger equilibrium of rotating bodies. *Prog. Theor. Phys.* 63:1957–1970.

Fulkerson, S. A., and Clark, F. O. 1984. The gas density gradient for three dark interstellar clouds. *Astrophys. J.* 287:723–727.

Gaffey, M. J., and McCord, T. B. 1982. Mineralogical and petrological characterizations of asteroid surface materials. In *Asteroids,* ed. T. Gehrels (Tucson: Univ. of Arizona Press), pp. 688–723.

Gahm, G. F., Nordh, H. L., Olofsson, S. G., and Carlborg, N. C. J. 1974. Simultaneous spectroscopic and photoelectric observations of the T Tauri star RU Lupi. *Astron. Astrophys.* 33:399–411.

Gahm, G. F., Fredga, K., Liseau, R., and Dravins, D. 1979. The far-UV spectrum of the T Tauri star RU Lupi. *Astron. Astrophys.* 73:L4–L6.

Gaida, M., Ungerechts, H., and Winnewisser, G. 1984. Ammonia observations and star counts in the Taurus dark cloud complex. *Astron. Astrophys.* 137:17–25.

Gail, H. P., and Sedlmayer, E. 1979. Dynamical evolution of spherical gas-dust nebulae, including diffusion effects. *Astron. Astrophys.* 76:158–167.

Garay, G., Reid, M., and Moran, J. 1984. Compact H II regions: Hydrogen recombination and OH maser lines. *Astrophys. J.* In press.

Garrison, R. F. 1967. Some characteristics of the B and A stars in the upper Scorpius complex. *Astrophys. J.* 147:1003–1016.

Gatewood, G., Breakiron, L. A., Goebel, R., Kipp, F., Russell, J. L., and Stern, J. W. 1980. On the astrometric detection of neighboring planetary systems. II. *Icarus* 41:205–231.

Gautier, D. 1983. Helium and deuterium in the outer solar system. In *Primordial Helium, ESO Workshop,* eds. P. Shaver, D. Kunth, and K. Kjar (Garching: Federal Republic of Germany), pp. 139–161.

Gautier, D., and Courtin, R. 1979. Atmospheric thermal structures of the giant planets. *Icarus* 39:28–45.

Gautier, D., and Owen, T. 1983*a*. Cosmogonical implications of helium and deuterium abundances on Jupiter and Saturn. *Nature* 302:215–218.

Gautier, D., and Owen, T. 1983*b*. Cosmogonical implications of elemental isotopic abundances in the atmospheres of the giant planets. *Nature* 304:691–694.

Gautier, D., Conrath, B., Flasar, M., Hanel, R., Kunde, V., Chedin, A., and Scott, N. 1981. The helium abundance of Jupiter from Voyager. *J. Geophys. Res.* 86:8713–8720.

Gautier, D., Bezard, B., Marten, A., Balutear, J. P., Scott, N., Chedin, A., Kunde, V., and Hanel, R. 1982. The C/H ratio in Jupiter from the Voyager infrared investigation. *Astrophys. J.* 257:901–912.

Gautier, T., Fink, U., Treffers, R., and Larson, H. 1976. Detection of molecular hydrogen quadrupole emission in the Orion nebula. *Astrophys. J.* 207:L129–L134.

Gehrels, T., ed. 1978. *Protostars & Planets* (Tucson: Univ. of Arizona Press).

Gehrz, R. D., Grasdalen, G. L., Castelaz, M., Gullixson, C., Mozurkewich, D., and Hackwell, J. A. 1982. Anatomy of a region of star formation: Infrared images of S106 (AFGL 2584). *Astrophys. J.* 254:550–561.

Geiss, J., and Reeves, H. 1972. Cosmic and solar system abundances of deuterium and helium-3. *Astron. Astrophys.* 18:126–132.

Geiss, J., and Bochsler, P. 1979. On the abundances of rare ions in the solar wind. In *Proceedings of Fourth Solar Wind Conference,* Burghausen (Berlin: Springer-Verlag).

Geiss, J., and Reeves, H. 1981. Deuterium in the solar system. *Astron. Astrophys.* 93:189–199.

Geller, M. J., and Beers, T. C. 1982. Substructure within clusters of galaxies. *Publ. Astron. Soc. Pacific* 94:421–439.

Genkin, I. L., and Safronov, V. S. 1975. Instability of rotating systems with radial perturbations. *Astr. Zh. USSR* 52:306–315 (in Russian).

Genzel, R., and Downes, D. 1983. Mass outflow in molecular clouds: A new phase in the evolution of newly-formed stars? In *Highlights of Astronomy,* ed. West (Dordrecht: D. Reidel), pp. 6, 689–706.

Genzel, R., Downes, D., Ho, P. T. P., and Bieging, J. 1982. NH_3 in Orion KL: A new interpretation. *Astrophys. J.* 259:L103–L107.

Georgelin, Y. M., and Georgelin, Y. P. 1976. The spiral structure of our galaxy determined from H II regions. *Astron. Astrophys.* 49:57–79.

Georgelin, Y. M., Georgelin, Y. P., Laval, A., Monnet, G., and Rosado, M. 1983. Observations of giant bubbles in the Large Magellanic Cloud. *Astron. Astrophys. Suppl.* 54:459–469.

Gerin, M., Combes, F., Encrenaz, P., Linke, R., Destombes, J. L., and Demuynck, C. 1984. Detection of ^{13}CN in three galactic sources. *Astron. Astrophys.* 136:L17–L20.

Gerola, H., and Glassgold, A. E. 1978. Molecular evolution of contracting clouds: Basic methods and initial results. *Astrophys. J. Suppl.* 37:1–25.

Giampapa, M. S. 1984. Results from ultraviolet observations of T Tauri stars. In *Third Cambridge Workshop on Cool Stars, Stellar Systems and the Sun,* eds. S. L. Baliunas and L. Hartmann (Berlin-Heidelberg: Springer-Verlag), pp. 14–26.

Giampapa, M. S., Calvet, N., Imhoff, C. L., and Kuhi, L. V. 1981. IUE observations of the pre-main-sequence stars. I. Mg II and Ca resonance line fluxes for T Tauri stars. *Astrophys. J.* 251:113–125.

Giampapa, M. S., Imhoff, C. L., Morossi, C., and Ramella, M. 1984. IUE observations of the pre-main-sequence stars. III. Mg II and Ca II line profiles for T Tauri stars. Submitted to *Astrophys. J.*

Giaretta, D. L. 1979. Instabilities in molecular hydrogen clouds. *Astron. Astrophys.* 78:328–334.

Giaretta, D. L. 1984. On absolute and convective instabilities. *Astron. Astrophys.* 88:113–116.

Gilden, D. L. 1984. Clump collisions in molecular clouds: Gravitational instability and coalescence. *Astrophys. J.* 279:335–349.

Gillett, F. C., Forrest, W. J., Merrill, K. M., Capps, R. W., and Soifer, B. T. 1975*a*. The 8–13 μm spectra of compact HII regions. *Astrophys. J.* 200:609–620.

Gillett, F. C., Jones, T. W., Merrill, K. M., and Stein, W. A. 1975*b*. Anisotropy of constituents of interstellar grains. *Astron. Astrophys.* 45:77–81.

Gillis, J., Mestel, L., and Paris, R. B. 1974. Magnetic braking during star formation. I. *Astrophys. Space Sci.* 27:167–194.

Gillis, J., Mestel, L., and Paris, R. B. 1979. Magnetic braking during star formation. II. *Mon. Not. Roy. Astron. Soc.* 187:311–335.

Gilmore, W. 1980. Radio continuum interferometry of dark clouds. I. A search for newly formed HII regions; and II. A study of physical properties of local newly formed HII regions. *Astron. J.* 85:894–944.

Gingold, R. A., and Monaghan, J. J. 1977. Smoothed particle hydrodynamics: Theory and application to nonspherical stars. *Mon. Not. Roy. Astron. Soc.* 181:375–389.

Gingold, R. A., and Monaghan, J. J. 1978. Binary fission in damped rotating polytropes. *Mon. Not. Roy. Astron. Soc.* 184:481–499.

Gingold, R. A., and Monaghan, J. J. 1979. Binary fission in damped rotating polytropes II. *Mon. Not. Roy. Astron. Soc.* 188:39–44.

Gingold, R. A., and Monaghan, J. J. 1981. The collapse of a rotating, nonaxisymmetric isothermal cloud. *Mon. Not. Roy. Astron. Soc.* 197:461–475.

Gingold, R. A., and Monaghan, J. J. 1982. The reliability of finite difference and particle methods for fragmentation problems. *Mon. Not. Roy. Astron. Soc.* 199:115–119.

Gingold, R. A., and Monaghan, J. J. 1983. On the fragmentation of differentially rotating clouds. *Mon. Not. Roy. Astron. Soc.* 204:715–733.

Giuli, R. T. 1968. On the rotation of the Earth produced by gravitational accretion of particles. *Icarus* 8:301–323.

Giuliani, J. L. 1979. The hydrodynamic stability of ionization-shock fronts. Linear theory. *Astrophys. J.* 233:280–293.

Giuliani, J. L. 1980. Instabilities and star formation within ionization-shock fronts. *Astrophys. J.* 242:219–225.

Glassgold, A. E., and Langer, W. D. 1973. Heating of molecular-hydrogen clouds by cosmic rays and x-rays. *Astrophys. J.* 186:859–888.

Glassgold, A. E., and Langer, W. D. 1976. Thermal-chemical instabilities in CO clouds. *Astrophys. J.* 204:403–407.

Glassgold, A. E., and Huggins, P. J. 1984. Circumstellar chemistry. In *M, S, and C Stars,* eds. H. Johnson and F. Querci (NASA, CRNS). In press.

Glassgold, A. E., Huggins, P. J., and Schucking, E. L., eds. 1982. *Symposium on the Orion Nebula to Honor Henry Draper.* In *Ann. New York Acad. Sci.* 395.

Goebel, J. H. 1983. Observation of ice mantles toward HD 29647. *Astrophys. J.* 268:L41–L45.

Goldhaber, D. M., and Betz, A. L. 1984. Silane in IRC+10216. *Astrophys. J.* 279:L55–L58.

Goldreich, P. 1965. An explanation for the frequent occurrence of commensurable mean motions in the solar system. *Mon. Not. Roy. Astron. Soc.* 130:159–182.

Goldreich, P., and Lynden-Bell, D. 1965*a*. I. Gravitational instability of uniformly rotating disks. *Mon. Not. Roy. Astron. Soc.* 130:97–124.

Goldreich, P. and Lynden-Bell, D. 1965*b*. II. Spiral arms as sheared gravitational instabilities. *Mon. Not. Roy. Astron. Soc.* 130:125–158.

Goldreich, P., and Schubert, G. 1967. Differential rotation in stars. *Astrophys. J.* 150:571–587.

Goldreich, P., and Peale, S. 1968. Dynamics of planetary rotations. *Ann. Rev. Astron. Astrophys.* 6:287–320.

Goldreich, P., and Ward, W. R. 1973. The formation of planetesimals. *Astrophys. J.* 183:1051–1061.

Goldreich, P., and Kwan, J. 1974. Molecular clouds. *Astrophys. J.* 189:441–453.

Goldreich, P., and Tremaine, S. 1978*a*. The velocity dispersion in Saturn's rings. *Icarus* 34:227–239.

Goldreich, P., and Tremaine, S. 1978*b*. The formation of the Cassini Division in Saturn's rings. *Icarus* 34:240–253.

Goldreich, P., and Tremaine, S. 1979*a*. The excitation of density waves at the Lindblad and corotation resonances of an external potential. *Astrophys. J.* 233:857–871.

Goldreich, P., and Tremaine, S. 1979*b*. Towards a theory for the Uranian rings. *Nature* 277:97–99.

Goldreich, P., and Tremaine, S. 1980. Disk-satellite interactions. *Astrophys. J.* 241:425–441.

Goldreich, P., and Kylafis, N. 1981. On mapping the magnetic field direction in molecular clouds by polarization measurements. *Astrophys. J.* 243:L75–L78.

Goldreich, P., and Kylafis, N. 1982. Linear polarization of radio frequency lines in molecular clouds and circumstellar envelopes. *Astrophys. J.* 253:606–621.

Goldreich, P., and Tremaine, S. 1982. The dynamics of planetary rings. *Ann. Rev. Astron. Astrophys.* 20:249–283.

Goldsmith, D. W., Habing, H. J., and Field, G. B. 1969. Thermal properties of interstellar gas heated by cosmic rays. *Astrophys. J.* 158:173–183.

Goldsmith, P. F. 1984. Submillimeter observations of molecules and the structure of giant molecular clouds. In *Galactic and Extragalactic Infrared Spectroscopy,* eds. M. F. Kessler and J. P. Phillips (Dordrecht: D. Reidel), pp. 233–250.

Goldsmith, P. F., and Langer, W. D. 1978. Molecular cooling and thermal balance in dark interstellar clouds. *Astrophys. J.* 222:881–895.

Goldsmith, P. F., and Sernyak, M. Jr. 1984. Structure of the L1535 dark cloud and the velocity field in the Taurus molecular cloud complex. *Astrophys. J.* In press.

Goldsmith, P. F., Langer, W. D., Schloerb, F. P., and Scoville, N. Z. 1980. High angular resolution observations of CS in the Orion Nebula. *Astrophys. J.* 240:524–531.

Goldsmith, P. F., Langer, W. D., Ellder, J., Irvine, W. M., and Kollberg, E. 1981. Determination of the HNC to HCN abundance ratio in giant molecular clouds. *Astrophys. J.* 249:524–531.

Goldsmith, P. F., Krotkov, R., Snell, R. L., Brown, R. D., and Godfrey, P. 1983. Vibrationally excited CH_3CN and HC_3N in Orion. *Astrophys. J.* 274:184–194.

Goldsmith, P. F., Snell, R. L., Hemeon-Heyer, M., and Langer, W. D. 1984. Bipolar outflows in dark clouds. *Astrophys. J.* 286:599–608.

Gondhalekar, P. M., and Wilson, R. 1975. The interstellar radiation field between 912 and 2740 Ångstroms. *Astron. Astrophys.* 38:329–333.

Gondhalekar, P. M., Phillips, A. P., and Wilson, R. 1980. Observations of the interstellar ultraviolet radiation field from the S2/68 sky-survey telescope. *Astron. Astrophys.* 85:272–280.

Gooding, J. L., Mayeda, T. K., Clayton, R. N., and Fukuoka, T. 1983. Oxygen isotopic heterogeneities, their petrological correlations, and implications for melt origin of chondrules in unequilibrated ordinary chondrites. *Earth Planet. Sci. Lett.* 65:209–224.

Gordon, M. A., and Burton, W. B. 1976. Carbon monoxide in the galaxy. I. The radial distribution of CO, H_2 and nuclei. *Astrophys. J.* 208:346–353.

Gordon, M. A., Heidmann, J., and Epstein, E. E. 1982. A search for CO in clumpy irregular galaxies. *Publ. Astron. Soc. Pacific* 94:415–420.

Goswami, J. N., and Lal, D. 1979. Formation of the parent bodies of the carbonaceous chondrites. *Icarus* 40:510–521.

Gough, D. O. 1983. The proto-solar helium abundance. In *Primordial Helium, ESO Workshop,* eds. P. Shaver, D. Kunth, and K. Kjar (Garching: Federal Republic of Germany), pp. 117–136.

Graboske, H. C. Jr., Olness, R. J., and Grossman, A. S. 1975*a*. Thermodynamics of dense hydrogen-helium fluids. *Astrophys. J.* 199:255–264.

Graboske, H. C. Jr., Pollack, J. B., Grossman, A. S., and Olness, R. J. 1975*b*. The structure and evolution of Jupiter: The fluid contraction stage. *Astrophys. J.* 199:265–281.

Gradie, J., and Tedesco, E. 1982. Compositional structure of the asteroid belt. *Science* 216:1405–1407.

Graedel, T. E., Langer, W. D., and Frerking, M. A. 1982. The kinetic chemistry of dense interstellar clouds. *Astrophys. J. Suppl.* 48:321–368.

Graham, J. A., and Elias, J. H. 1983. Herbig-Haro objects in the dust globule ESO 210-6A. *Astrophys. J.* 272:615–626.

Grandi, S. A. 1980. OI 8446 emission in Seyfert 1 galaxies. *Astrophys. J.* 238:10–16.

Grappin, R., Frisch, U., Leorat, J., and Poquet, A. 1982. Alfvénic fluctuations as asymptotic states of MHD turbulence. *Astron. Astrophys.* 105:6–14.

Grasdalen, G. L. 1976. Brackett-alpha emission in the Becklin-Neugebauer object. *Astrophys. J.* 205:L83–L86.

Grasdalen, G. L., Strom, S. E., Strom, K. M., Capps, R. W., Thomspon, D., and Castelaz, M. 1984. High spatial resolution IR observations of young steller objects: A possible disk surrounding HL Tau. *Astrophys. J.* 283:L57–L62.

Gray, C. M., and Compston, W. 1974. Excess ^{26}Mg in the Allende meteorite. *Nature* 251:495–497.

Gray, C. M., Papanastassiou, D. A., and Wasserburg, G. J. 1973. The identification of early condensates from the solar nebula. *Icarus* 20:213–239.

Gray, S. K., Miller, W. H., Yamaguchi, Y., and Schaefer, H. F. III. 1980. Reaction path Hamiltonian: Tunneling effects in the unimolecular isomerization HNC-HCN. *J. Chem. Phys.* 73:2733–2739.

Green, S. 1981. Interstellar chemistry: Exotic molecules in space. *Ann. Rev. Phys. Chem.* 32:103–138.

Green, S. 1983. Metastability of isoformyl ions in collisions with helium and hydrogen. *Astrophys. J.* 277:900–906.

Green, S., and Herbst, E. 1979. Metastable isomers: A new class of interstellar molecules. *Astrophys. J.* 229:121–131.

Greenberg, J. M. 1982*a*. Dust in dense clouds: One stage in a cycle. In *Submillimetre Wave Astronomy,* eds. J. P. Phillips and J. E. Beckman (Cambridge: Cambridge Univ. Press), pp. 261–306.

Greenberg, J. M. 1982*b*. What are comets made of? A model based on interstellar dust. In *Comets,* ed. L. L. Wilkening (Tucson: Univ. of Arizona Press), pp. 131–163.

Greenberg, J. M. 1983. The largest molecules in space: Interstellar dust. In *Cosmochemistry and the Origin of Life,* ed. C. Ponnamperuma (Dordrecht: D. Reidel), pp. 71–112.

Greenberg, J. M., and Yencha, A. J. 1973. Exploding interstellar grains and complex molecules. In *Interstellar Dust and Related Topics, IAU Symposium No. 52,* eds. J. M. Greenberg and H. C. van de Hulst (Dordrecht: D. Reidel), pp. 369–373.

Greenberg, J. M., van de Bult, C. E. P. M., and Allamandola, L. J. 1983. Ices in Space. *J. Phys. Chem.* 87:4243–4260.

Greenberg, R. 1975. The dynamics of Uranus' satellites. *Icarus* 24:325–332.

Greenberg, R. 1976. The Laplace relation and the masses of Uranus' satellites. *Icarus* 29:427–433.

Greenberg, R. 1977. Orbit-orbit resonances in the solar system: Varieties and similarities. In *Vistas in Astronomy* (New York: Pergamon), pp. 209–239.

Greenberg, R. 1979*a*. Growth of large, late stage planetesimals. *Icarus* 39:140–151.

Greenberg, R. 1979*b*. The motions of the Uranian satellites: Theory and applications. In *Dynamics of the Solar System,* ed. R. L. Duncombe (Dordrecht: D. Reidel), pp. 177–180.

Greenberg, R. 1982. Planetesimals to planets. In *Formation of Planetary Systems,* ed. A. Brahic (Toulouse, France: Cepedues), pp. 516–569.

Greenberg, R. 1983. The role of dissipation in shepherding of ring particles. *Icarus* 53:207–218.

Greenberg, R. 1984. Satellite masses in the Uranus and Neptune systems. In *Uranus and Neptune,* ed. J. T. Bergstralh (Pasadena: JPL Conference Proceedings), pp. 463–480.

Greenberg, R., and Scholl, H. 1979. Resonances in the asteroid belt. In *Asteroids,* ed. T. Gehrels (Tucson: Univ. of Arizona Press), pp. 310–333.

Greenberg, R., Hartmann, W. K., Chapman, C. R., and Wacker, J. F. 1978*a*. The accretion of planets from planetesimals. In *Protostars and Planets,* ed. T. Gehrels (Tucson: Univ. of Arizona Press), pp. 599–624.

Greenberg, R., Wacker, J. F., Hartmann, W. L., and Chapman, C. R. 1978*b*. Planetesimals to planets: Numerical simulation of collisional evolution. *Icarus* 35:1–26.

Greenberg, R., Weidenschilling, S. J., Chapman, C. R., and Daivs, D. R. 1984. From icy planetesimals to outer planets and comets. *Icarus* 59:87–113.

Greenspan, H. P. 1968. *The Theory of Rotating Fluids* (Cambridge: Cambridge Univ. Press).

Grossman, A. S., and Graboske, H. C. Jr. 1971. Evolution of low-mass stars. III. Effects of nonideal thermodynamic properties during the pre-main-sequence contraction. *Astrophys. J.* 164:475–490.

Grossman, A. S., Graboske, H. C., Pollack, J. B., Reynolds, R. T., and Summers, A. L. 1972. An evolutionary calculation of Jupiter. *Phys. Earth Planet Interiors* 6:91–98.

Grossman, A. S., Pollack, J. B., Reynolds, R. T., Summers, A. L., and Graboske, H. C. Jr. 1980. The effect of dense cores on the structure and evolution of Jupiter and Saturn. *Icarus* 42:358–372.

Grossman, J. N., and Wasson, J. T. 1982. Evidence for primitive nebular components in chondrules from the Chainpur chondrite. *Geochim. Cosmochim. Acta* 46:1081–1099.

Grossman, L. 1972. Condensation in the primitive solar nebula. *Geochim. Cosmochim. Acta.* 36:597–619.

Grossman, L. 1975. Petrography and mineral chemistry of Ca, Al-rich inclusions in the Allende meteorite. *Geochim. Cosmochim. Acta* 39:433–454.

Grossman, L. 1980. Refractory inclusions in the Allende meteorite. *Ann. Rev. Earth Planet. Sci.* 8:559–608.

Grossman, L., and Larimer, J. W. 1974. Early chemical history of the solar system. *Rev. Geophys. Space Phys.* 12:71–101.

Grossman, L., and Ganapathy, R. 1975. Volatile elements in Allende inclusions. *Proc. Lunar Planet. Sci. Conf.* 6:1729–1736.

Grossman, L., and Ganapathy, R. 1976. Trace elements in the Allende meteorite. II. Fine-grained, Ca-rich inclusions. *Geochim. Cosmochim. Acta* 40:967–977.

Guélin, M., Langer, W. D., Snell, R. L., and Wootten, A. H. 1977. Observations of DCO^+: The electron abundance in dark clouds. *Astrophys. J.* 217:L165–L168.

Guélin, M., Friberg, P., and Mezaoui, A. 1982*a*. Astronomical study of the C_3N and C_4H radicals: Hyperfine interactions and rho-type doubling. *Astron. Astrophys.* 109:23–31.

Guélin, M., Langer, W. D., and Wilson, R. W. 1982*b*. The state of ionization in dense molecular clouds. *Astron. Astrophys.* 107:107–127.

Gulkis, S., Janssen, M. A., and Olsen, E. T. 1978. Evidence for the depletion of ammonia in the Uranus atmosphere. *Icarus* 34:10–19.

Gulkis, S., Olsen, E. T., Klein, M. J., and Thompson, T. J. 1983. Uranus: Variability of the microwave spectrum. *Science* 221:453–455.

Güsten, R., Chini, R., and Neckel, T. 1984. A circumstellar disk in Cep A. Submitted to *Astron. Astrophys.*

Habing, H. J. 1968. *Bull. Astron. Inst. Netherlands* 19:421.

Habing, H. J., and Israel, F. P. 1979. Compact H II regions and OB star formation. *Ann. Rev. Astron. Astrophys.* 17:345–385.

Hachisu, I., and Eriguchi, Y. 1982. Bifurcation and fission of three dimensional, rigidly rotating and self-gravitating polytropes. *Prog. Theor. Phys.* 68:206–221.

Hachisu, I., and Eriguchi, Y. 1983. Bifurcations and phase transitions of self-gravitating and uniformly rotating fluid. *Mon. Not. Roy. Astron. Soc.* 204:583–589.

Hachisu, I., and Eriguchi, Y. 1984*a*. Fission of dumb-bell equilibrium and binary state of rapidly rotating polytropes. *Publ. Astron. Soc. Japan.* In press.

Hachisu, I., and Eriguchi, Y. 1984*b*. Binary fluid star. *Publ. Astron. Soc. Japan.* In press.

Hachisu, I., and Eriguchi, Y. 1984*c*. A criterion for the fragmentation of a rotating and collapsing gas cloud. *Astron. Astrophys.* 140:259–264.

Hachisu, I., Eriguchi, Y., and Sugimoto, D. 1982. Rapidly rotating polytropes and concave hamburger equilibrium. *Prog. Theor. Phys.* 68:191–205.

Haff, P. K., Watson, O. C., and Tombrello, T. A. 1981. Possible isotopic fractionation effects in material sputtered from minerals. *J. Geophys. Res.* 86:9553–9561.

Hagen, W., Allamandola, L. J., and Greenberg, J. M. 1980. Infrared absorption lines by molecules in grain mantles. *Astron. Astrophys.* 86:L1.

Hagen, W., Tielens, A. G. G. M., and Greenberg, J. M. 1983*a*. The three micron "ice" band in grain mantles. *Astron. Astrophys.* 117:132–140.

Hagen, W., Tielens, A. G. G. M., and Greenberg, J. M. 1983*b*. A laboratory study of the infrared spectra of interstellar ices. *Astron. Astrophys. Suppl.* 51:389–416.

Haggerty, S. E., and McMahon, B. M. 1979. Magnetite-sulfide-metal complexes in the Allende meteorite. *Proc. Lunar Planet. Sci. Conf.* 10:851–870.

Hall, D. N. B., Kleinmann, S. G., Ridgway, S. T., and Gillett, F. C. 1978. High-resolution 1.5-5 micron spectroscopy of the Becklin-Neugebauer source in Orion. *Astrophys. J.* 223:L47–L50.

Hamano, Y., and Ozima, M. 1978. Earth-atmosphere evolution model based on Ar isotopic data. In *Abundance in Earth and Planetary Science,* eds. E. C. Alexander, and M. Ozima (Tokyo: Center for Academic Publ.), vol. 3, pp. 155–171.

Hamid, S. E., Marsden, B. G., and Whipple, F. L. 1968. Influence of a comet belt beyond Neptune on the motions of periodic comets. *Astron. J.* 73:727–729.

Hanel, R. A., Conrath, B. J., Flasar, F. M., Kunde, V. G., Maguire, W., Pearl, J. C., Pirraglia, J. A., Samuelson, R., Hearth, L. W., Allison, M., Cruikshank, D. P., Gautier, D., Gierasch, P., Horn, L., Koppany, R., and Ponnamperuma, C. 1981*a*. Infrared observations of the Saturnian system from Voyager 1. *Science* 212:192–200.

Hanel, R. A., Conrath, B. J., Hearth, L. W., Kunde, V. G., and Pirraglia, J. A. 1981*b*. Albedo, internal heat, and energy balance of Jupiter: Preliminary results of the Voyager infrared investigation. *J. Geophys. Res.* 86:8705–8712

Hanel, R. A., Conrath, B. J., Kunde, V. G., Pearl, J. C., and Pirraglia, J. A. 1983. Albedo, internal heat source, and energy balance of Saturn. *Icarus* 53:262–285.

Hanner, M. S. 1984. Comet Cernis: Icy grains at last? *Astrophys. J.* 277:L75–L78.

Hansen, S. S. 1982. The magnetic fields in the Orion Kleinmann-Low nebula as derived from hydroxyl maser radiation. *Astrophys. J.* 260:599–603.

Hanson, R. B., Jones, B. F., and Lin, D. N. C. 1983. The astrometric position of T Tauri and the nature of its companion. *Astrophys. J.* 270:L27–L30.

Harlow, F. H., and Amsden, A. A. 1975. Numerical calculation of multiphase fluid flow. *J. Comp. Phys.* 17:19–52.

Haro, G. 1952. Herbig's nebulous objects near NGC 1999. *Astrophys. J.* 115:572–573.

Harper, D. A. 1974. Far-infrared emission from HII regions. II. Multicolor photometry of selected sources and $2'.2$ resolution maps of M42 and NGC 2024. *Astrophys. J.* 192:557–571.

Harris, A. W. 1978*a*. The formation of the outer planets. *Lunar Planet. Sci.* IX:459–461 (abstract).

Harris, A. W. 1978*b*. Satellite formation II. *Icarus* 34:128–145.

Harris, A. W. 1983. Physical characteristics of Neptune and Triton inferred from the orbital motion of Triton. Presented at "Natural Satellites Conference," Ithaca, NY, July (abstract).

Harris, A. W., and Ward, W. R. 1982. Dynamical constraints on the formation and evolution of planetary bodies. *Ann. Rev. Earth Planet. Sci.* 10:61–108.

Harris, A. W., Townes, C., Matsakis, D., and Palmer, P. 1983. Small rotating clouds of stellar mass in Orion molecular cloud I. *Astrophys. J.* 265:L63–L66.

Harris, D. H., Woolf, N. J., and Rieke, G. H. 1978. Ice mantles and abnormal extinction in the Rho Opiuchi cloud. *Astrophys. J.* 226:829–838.

Harris, M. J. 1981. 30 keV (n,γ) cross sections from the nuclear statistical model. *Astrophys. Space Sci.* 77:357–367.

Hartmann, L., and MacGregor, K. B. 1982*a*. Protostellar mass and angular momentum loss. *Astrophys. J.* 259:180–192.

Hartmann, L., and MacGregor, K. B. 1982*b*. Wave-driven winds from cool stars. I. Some effects of magnetic field geometry. *Astrophys. J.* 257:264–268.

Hartmann, L., Edwards, S., and Avrett, E. H. 1982. Wave driven winds from cool stars. I. Models for T Tauri stars. *Astrophys. J.* 261:279–292.

Hartmann, W. K. 1972. Paleocratering of the moon: Review of post-Apollo data. *Astrophys. Space Sci.* 17:48–64.

Hartmann, W. K. 1975. Lunar "cataclysm": A misconception? *Icarus* 24:181–187.

Hartmann, W. K. 1984. Does crater "saturation equilibrium" occur in the solar system? *Icarus* 60:56–74.

Hartmann, W. K., and Davis, D. R. 1975. Satellite-sized planetesimals and lunar origin. *Icarus* 24:504–515.

Harvey, P. M., and Gatley, I. 1983. Infrared observations of OB star formation in NGC 6334. *Astrophys. J.* 269:613–624.

Harvey, P. M., and Wilking, B. A. 1984. NGC 6334-V. An infrared bi-polar nebula. *Astrophys. J.* (Lett.). In press.

Harvey, P. M., Campbell, M. F., and Hoffmann, W. F. 1978. Far infrared emission from compact sources in NGC 2264 and the Rosette nebula. *Astrophys. J.* 215:151–154.

Harvey, P. M., Campbell, M. F., Hoffman, W. F., Thronson, H. A. Jr., and Gatley, I. 1979*a*. Infrared observations of NCG 2071 (IRS) and AFGL 490: Two low-luminosity young stars. *Astrophys. J.* 229:990–993.

Harvey, P. M., Thronson, H. A. Jr., and Gatley, I. 1979*b*. Far-infrared observations of optical emission-line stars: Evidence for extensive cool dust clouds. *Astrophys. J.* 231:115–123.

Harvey, P. M., Thronson, H. A. Jr., Gatley, I., and Werner, M. W. 1982. Far-infrared mapping of the double-lobed H II region S106. *Astrophys. J.* 258:568–571.

Harvey, P. M., Wilking, B. A., and Joy, M. 1984. An infrared study of the bi-polar outflow region GGD 12-15. Submitted to *Astrophys. J.*

Haschick, A. D., and Ho, P. T. 1983. Formation of OB clusters: W33 complex. *Astrophys. J.* 267:638–646.

Hasegawa, T., Kaifu, N., Inatani, J., Morimoto, M., Chikada, Y., Hirabashi, H., Iwashita, H., Morita, K., Tojo, A., and Akabane, K. 1984. CS around Orion-KL: A large rotating disk. *Astrophys. J.* 283:117–122.

Hausman, M. A. 1981. Collisional mergers and fragmentation of interstellar clouds. *Astrophys. J.* 245:72–91.

Hausman, M. A. 1982. Theoretical models of the mass spectrum of interstellar clouds. *Astrophys. J.* 261:532–542.

Hausman, M. A., and Roberts, W. W. 1984. Spiral structure and star formation. II. Stellar lifetimes and cloud kinematics. *Astrophys. J.* 282:106–117.

Hawarden, T. G., and Brand, P. W. J. L. 1976. Cometary globules and the structure of the Gum nebula. *Mon. Not. Roy. Astron. Soc.* 175:19P–22P.

Hayakawa, S., Yamashita, K., and Yoshioka, S. 1969. Diffuse component of the cosmic far uv radiation and interstellar dust grains. *Astrophys. Space Sci.* 5:493–502.

Hayashi, C. 1961. Stellar evolution in early phases of gravitational contraction. *Publ. Astron. Soc. Japan* 13:450–458.

Hayashi, C. 1972. Origin of the solar system. In *Report of 5th Lunar Planet. Symp. ISAS,* Tokyo, pp. 13–18.

Hayashi, C. 1977. The gravitational instability of a rotating disk of two-component fluid. Unpublished.

Hayashi, C. 1981*a*. Formation of planets. In *Fundamental Problems in the Theory of Stellar Evolution, IAU Symposium No. 93,* eds. D. Sugimoto, D. Q. Lamb, and D. N. Schramm (Dordrecht: D. Reidel), pp. 113–128.

Hayashi, C. 1981*b*. Structure of the solar nebula, growth and decay of magnetic fields and effects of magnetic and turbulent viscosities on the nebula. *Prog. Theor. Phys. Suppl.* 70:35–53.

Hayashi, C., Nakazawa, K., and Adachi, I. 1977. Long-term behavior of planetesimals and the formation of the planets. *Publ. Astron. Soc. Japan* 29:163–196.

Hayashi, C., Nakazawa, K., and Mizuno, H. 1979. Earth's melting due to the blanketing effect of the primordial dense atmosphere. *Earth Planet. Sci. Lett.* 43:22–28.

Hayashi, C., Narita, S., and Miyama, S. M. 1982. Analytic solutions for equilibrium of rotating isothermal clouds. *Prog. Theor. Phys.* 68:1949–1966.

Hayashi, T., and Muehlenbachs, K. 1984. Rapid oxygen diffusion in melilite and its relevance to meteorites. *EOS Trans. Amer. Geophys. Union* 65:308 (abstract).

Hayatsu, R., and Anders, E. 1981. Organic compounds in meteorites and their origins. *Topics Curr. Chem.* 99:1–37.

Hayatsu, R., Matsuoka, S., Scott, R. G., Studier, M. H., and Anders, E. 1977. Origin of organic matter in the early solar system. VII. The organic polymer in carbonaceous chondrites. *Geochim. Cosmochim. Acta* 41:1325–1339.

Hayatsu, R., Winans, R. E., Scott, R. G., McBeth, R. L., Moore, L. P., and Studier, M. H. 1980. Phenolic ethers in the organic polymer of the Murchison meteorite. *Science* 207:1202–1204.

Hayden, F. A. 1952. *Photographic Atlas of the Southern Milky Way* (Washington: Carnegie Inst. of Washington).

Hayes, J. M. 1967. Organic constituents of meteorites: A review. *Geochim. Cosmochim. Acta* 31:1395–1440.

Heathcote, S. R., and Brank, P. W. J. L. 1983. The state of clouds in violent interstellar medium. *Mon. Not. Roy. Astron. Soc.* 203:67–86.
Heiles, C. 1967. Observations of the spatial structure of interstellar hydrogen. I. High-resolution observations of a small region. *Astrophys. J. Suppl.* 15:97–130.
Heiles, C. 1968. Normal OH emission and interstellar dark clouds. *Astrophys. J.* 151:919–934.
Heiles, C. 1969. Neutral hydrogen in dark dust clouds. *Astrophys. J.* 156:493–499.
Heiles, C. 1971. Physical conditions and chemical constitution of dark clouds. *Ann. Rev. Astron. Astrophys.* 9:293–322.
Heiles, C. 1974. A modern look at interstellar clouds. In *Galactic Radio Astronomy,* eds. F. J. Kerr and S. C. Simonson (Dordrecht: D. Reidel), pp. 13–44.
Heiles, C. 1979. H I shells and supershells. *Astrophys. J.* 229:533–544.
Heiles, C. 1980. Is the intercloud medium pervasive? *Astrophys. J.* 235:833–839.
Heiles, C. 1984. H I shells, supershells, shell-like objects, and "worms." *Astrophys. J.* 55:585–597.
Heiles, C., and Jenkins, E. B. 1976. An almost complete survey of 21-cm line radiation for $|b| \geq 10°$. V. Photographic presentation and qualitative comparison with other data. *Astron. Astrophys.* 46:333–360.
Heiles, C., and Troland, T. 1982. Measurements of magnetic field strengths in the vicinity of Orion. *Astrophys. J.* 260:L23–L26.
Heinze, K. G., and Mendoza, V. 1973. Emission-line stars in the Chamaeleon T association. *Astrophys. J.* 180:115–119.
Heisenberg, W. 1948. On the theory of statistical and isotropic turbulence. *Proc. Roy. Soc.* A195:402–406.
d'Hendecourt, L. B., Allamandola, L. J., Baas, F., and Greenberg, J. M. 1982. Interstellar grain explosions: Molecule cycling between gas and dust. *Astron. Astrophys.* 109:L12–L14.
Henkel, C., Wilson, T. L., and Bieging, J. 1982, Further ($^{12}C/^{13}C$) ratios from formaldehyde: A variation with distance from the galactic center. *Astron. Astrophys.* 109:344–351.
Henning, K. 1981. Molecule formation in interstellar clouds by gas phase reactions. *Astron. Astrophys. Suppl.* 44:405–435.
Hénon, M. 1981. A simple model of Saturn's rings. *Nature* 293:33–35.
Hénon, M. 1984. A simple model of Saturn's rings, revisited. In *Planetary Rings, IAU Colloquium No. 75,* ed. A. Brahic (Toulouse, France: Cepadues), pp. 363–384.
Henriksen, R. N., and Turner, B. E. 1984. Star cloud turbulence. *Astrophys. J.* 287:200–207.
Henry, R. C. 1973. Ultraviolet background radiation. *Astrophys. J.* 179:97–102.
Henry, R. C. 1977. Far-ultraviolet studies. I. Predicted far ultraviolet interstellar radiation field. *Astrophys. J.* 33:451–458.
Henry, R. C., Feldman, P. D., Fastie, W. G., and Weinstein, A. 1974. Far ultraviolet brightness of the north and south galactic pole regions from Apollo 17. *Bull. Amer. Astron. Soc.* 6:461 (abstract).
Henry, R. C., Swandic, J. R., Shulman, S. D., and Fritsz, G. 1977. Far-ultraviolet studies. II. Galactic latitude dependence of the 1530 Å interstellar radiation field. *Astrophys. J.* 212:707–713.
Heppenheimer, T. A. 1975. On the presumed capture origin of Jupiter's outer satellites. *Icarus* 24:172–180.
Heppenheimer, T. A., and Porco, C. 1977. New contributions to the problem of capture. *Icarus* 30:385–401.
Herbig, G. H. 1951. The spectra of two nebulous objects near NGC 1999. *Astrophys. J.* 113:697–698.
Herbig, G. H. 1954. Emission-line stars associated with the nebulous cluster NGC 2264. *Astrophys. J.* 119:483–495.
Herbig, G. H. 1960. The spectra of Be- and Ae-type stars associated with nebulosity. *Astrophys. J. Suppl.* 4:337–368.
Herbig, G. H. 1962*a*. Spectral classification of faint members of the Hyades and Pleiades and the dating problem in galactic clusters. *Astrophys. J.* 135:736–747.
Herbig, G. H. 1962*b*. The properties and problems of T Tauri stars and related objects. *Adv. Astron. Astrophys.* 1:47–103.
Herbig, G. H. 1974. On the nature of the small dark globules in the Rosette nebula. *Publ. Astron. Soc. Pacific* 86:604–608.
Herbig, G. H. 1977*a*. Eruptive phenomena in early stellar evolution. *Astrophys. J.* 217:693–715.

Herbig, G. H. 1977*b*. Radial velocities and spectral types of T Tauri stars. *Astrophys. J.* 214:747–758.

Herbig, G. H. 1978. Some aspects of early stellar evolution that may be relevant to the origin of the solar system. In *The Origin of the Solar System,* ed. S. F. Dermott (New York: Wiley), pp. 219–235.

Herbig, G. H., and Rao, N. K. 1972. Catalog of emission-line stars of the Orion population. *Astrophys. J.* 174:401–423.

Herbig, G. H., and Soderblom, D. R. 1980. Observations and interpretation of the near-infrared line spectra of T Tauri stars. *Astrophys. J.* 242:628–637.

Herbig, G. H., and Jones, B. F. 1981. Large proper motions of the Herbig-Haro objects HH1 and HH2. *Astron. J.* 86:1232–1244.

Herbig, G. H., and Jones, B. F. 1983. Proper motions of Herbig Haro objects. III. HH-7 through -11, HH-12, and HH-32. *Astron. J.* 88:1040–1052.

Herbst, E. 1976. Radiative associations in dense, H_2-containing interstellar clouds. *Astrophys. J.* 205:94–102.

Herbst, E. 1978*a*. The current state of interstellar chemistry of dense clouds. In *Protostars and Planets,* ed. T. Gehrels (Tucson: Univ. of Arizona Press), pp. 88–99.

Herbst, E. 1978*b*. What are the products of polyatomic ion-electron dissociative recombination reactions? *Astrophys. J.* 222:508–516.

Herbst, E. 1980. An additional uncertainty in calculated radiative association rates of molecular formation at low temperature. *Astrophys. J.* 241:197–199.

Herbst, E. 1981. Theories of ion-molecule association reactions compared with new experimental data on $C^+ + H_2 + H_2 \rightarrow CH_3^+ + H$. *J. Chem. Phys.* 75:4413–4416.

Herbst, E. 1982*a*. A reinvestigation of the rate of the $C^+ + H_2$ radiative association reaction. *Astrophys. J.* 252:810–813.

Herbst, E. 1982*b*. An approach to the estimation of polyatomic vibrational radiative relaxation rates. *Chem. Phys.* 65:185–195.

Herbst, E. 1983. Ion-molecule syntheses of interstellar molecular hydrocarbons through C_4H: Towards molecular complexity. *Astrophys. J. Suppl.* 53:41–53.

Herbst, E., Green, S., Thaddeus, P., and Klemperer, W. 1977. Indirect observation of unobservable interstellar molecules. *Astrophys. J.* 215:503–510.

Herbst, E., Adams, N. G., and Smith, D. 1983. Laboratory measurements of ion-molecule reactions pertaining to interstellar hydrocarbon synthesis. *Astrophys. J.* 269:329–333.

Herbst, W., and Assousa, G. E. 1977. Observational evidence for supernovae-induced star formation: Canis Major R1. *Astrophys. J.* 217:473–487.

Herbst, W., and Assousa, G. E. 1978. The role of supernovae in star formation and spiral structure. In *Protostars & Planets,* ed. T. Gehrels (Tucson: Univ. of Arizona Press), pp. 368–383.

Herbst, W., and Miller, D. P. 1982. The age spread and initial mass function of NGC 3293: Implications for the formation of clusters. *Astron. J.* 87:1478–1490.

Herbst, W., Holtzman, J. A., and Klasky, R. S. 1983. Photometric variations of Orion population stars. II. Ae-irregular variables and T Tauri stars. *Astron. J.* 88:1648–1664.

Heuermann, R. W. 1983. Theoretical Models of Shock Wave Excitation for Herbig-Haro Objects. MS thesis, Univ. of Missouri.

Hewins, R. H. 1984. Dynamic crystallization experiments as constraints on chondrule genesis. In *Chondrules and Their Origins,* ed. E. A. King (Houston: Lunar & Planetary Inst.), pp. 122–133.

Heydegger, H. R., Foster, J. J., and Compston, W. 1979. Evidence of a new isotopic anomaly from titanium isotope ratios in meteoritic materials. *Nature* 278:704–707.

Heymann, D., and Dziczkaniec, M. 1976. Early irradiation of matter in the solar system: Magnesium (proton, neutron) scheme. *Science* 191:79–81.

Hide, R. 1981. On the rotation of Jupiter. *Geophys. J. Roy. Astron. Soc.* 64:283–289.

Higdon, J. C. 1985. Density fluctuations in the interstellar medium: Evidence for anisotropic magnetogasdynamic turbulence. *Astrophys. J.* In press.

Hildebrand, R. H. 1983. The determination of cloud masses and dust characteristics from submillimeter thermal emission. *Quar. J. Roy. Astron. Soc.* 24:267–282.

Hillebrandt, W., and Thielemann, F.-K. 1982. Nucleosynthesis in novae: A source of Ne-E and ^{26}Al? *Astrophys. J.* 255:617–623.

Hills, J. G. 1980. The effect of mass loss on the dynamical evolution of a stellar system: Analytic approximations. *Astrophys. J.* 225:986–991.

Hills, J. G. 1981. Comet showers and the steady-state infall of comets from the Oort cloud. *Astron. J.* 86:1730–1740.

Hills, J. G. 1982. The formation of comets by radiation pressure in the outer protosun. *Astron. J.* 87:906–910.

Hinton, R. W., and Bischoff, A. 1984. Ion microprobe magnesium isotope analyses of plagioclase and hibonite from ordinary chondrites. *Nature* 308:169–172.

Hinton, R. W., Scatema-Wachel, D. E., and Davis, A. M. 1984. A search for ^{60}Fe in meteorites using the ion probe microanalyzer. *Lunar Planet. Sci.* XV:365 (abstract).

Ho, P. T. P., and Haschick, A. D. 1981. Formation of OB clusters: VLA observations. *Astrophys. J.* 248:622–637.

Ho, P. T. P., and Townes, C. H. 1983. Interstellar ammonia. In *Ann. Rev. Astron. Astrophys.*, ed. G. Burbidge, D. Layzer, and J. Phillips (Palo Alto: Annual Reviews, Inc.), vol. 21, pp. 239–270.

Ho, P., Barrett, A., Myers, P., Matsakis, D., Cheung, A., Chui, M., Townes, C., and Yngvesson, K. 1979. Ammonia observations of the Orion molecular cloud. *Astrophys. J.* 234:912–921.

Ho, P., Martin, R., and Barrett, A. 1981. Molecular clouds associated with compact H II regions. I. General properties. *Astrophys. J.* 246:761–787.

Ho, P., Genzel, R., and Das, A. 1983. VLA observations of warm NH_3 associated with mass outflows in W51. *Astrophys. J.* 266:596–601.

Hobbs, L. M. 1974. Statistical properties of interstellar clouds. *Astrophys. J.* 191:395–399.

Hobbs, L. M., Black, J. H., and van Dishoeck, E. F. 1983. Interstellar C_2 molecules in a Taurus dark cloud. *Astrophys. J.* 271:L95–L99.

Hodapp, K.-W. 1984. Infrared polarization of sources with bipolar mass outflow. Submitted to *Astron. Astrophys.*

Hoffman, D. C., Lawrence, F. O., Mewherter, J. L., and Rourke, F. M. 1971. Detection of Plutonium-244 in nature. *Nature* 234:132–134.

Hoffman, J. H., Hodges, R. R., Donahue, T. M., and McElroy, M. B. 1980. Composition of the Venus lower atmosphere from Pioneer Venus mass spectrometer. *J. Geophys. Res.* 85:7882–7890.

Hohenberg, C. M., Podosek, F. A., and Reynolds, J. H. 1967. Xenon-iodine dating: sharp isochronism in chondrites. *Science* 156:202–206.

Hohenberg, C. M. 1970. Xe from the Angra dos Reis meteorite. *Geochim. Cosmochim. Acta* 34:185–191.

Hohenberg, C. M., Hudson, G. B., Kennedy, B. M., and Podosek, F. A. 1981. Noble gas retention chronologies for the St. Severin meteorite. *Geochim. Cosmochim. Acta* 45:535–546.

Holberg, J. B. 1982. Identification of 1980S27 and 1980S26 resonances in Saturn's A ring. *Astron. J.* 87:1416–1422.

Holberg, J. B., Forrester, W. T., and Lissauer, J. J. 1982. Identification of resonance features within the rings of Saturn. *Nature* 297:115–120.

Hollenbach, D. J. 1982. Shock waves in Orion. In *Symposium on the Orion Nebula to Honor Henry Draper,* eds. A. E. Glassgold, P. J. Huggins, and E. L. Schucking (New York: Ann. N.Y. Acad. Sci.), 395:242–256.

Hollenbach, D., and McKee, C. F. 1979. Molecule formation and infrared emission in fast interstellar shocks. I. Physical processes. *Astrophys. J. Suppl.* 41:555–592.

Hollenbach, D. J., Werner, M. W., and Salpeter, E. E. 1971. Molecular hydrogen in HI regions. *Astrophys. J.* 163:165–180.

Hollis, J. M., and Rhodes, P. J. 1982. Detection of interstellar sodium hydroxide in self-absorption toward the galactic center. *Astrophys. J.* 262:L1–L5.

Hollis, J. M., Snyder, L. E., Lovas, F. J., and Buhl, D. 1976. Radio detection of interstellar DCO^+. *Astrophys. J.* 209:L83–L95.

Hollis, J. M., Snyder, L. E., Suenram, R. D., and Lovas, F. J. 1980. A search for the lowest-energy conformer of interstellar glycine. *Astrophys. J.* 241:1001–1006.

Holweger, H. 1979. Abundances of the elements in the sun. In *Les Éléments et Leurs Isotopes dans l'Universe,* 22nd Coll. Int. d'Astrophys. Liège, June, pp. 117–139.

Hong, S. S., and Greenberg, J. M. 1980. A unified model of interstellar grains: A connection between alignment efficiency, grain model size, and cosmic abundance. *Astron. Astrophys.* 88:194–202.

Hopper, P. B., and Disney, M. J. 1974. The alignment of interstellar dust clouds. *Mon. Not. Roy. Astron. Soc.* 168:639–650.
Horedt, G. P. 1978. Blow-off of the protoplanetary cloud by a T Tauri-like solar wind. *Astron. Astrophys.* 64:173–178.
Horedt, G. P. 1982. On the angular momentum of colliding interstellar clouds. *Astron. Astrophys.* 106:29–33.
Houlahan, P., and Scalo, J. M. 1985. Statistical characterization of interstellar structures. In preparation.
Hourigan, K., and Ward, W. R. 1984. Radical migration of preplanetary material: Implications for the accretion time scale problem. *Icarus* 60:29–39.
Howell, R. R., McCarthy, D. W., and Low, F. J. 1981. One-dimensional infrared speckle interferometry. *Astrophys. J.* 251:L21–L26.
Hoyle, F. 1953. On the fragmentation of gas clouds into galaxies and stars. *Astrophys. J.* 118:513–528.
Hoyle, F. 1960. On the origin of the solar system. *Quart. J. Roy. Astron. Soc.* 1:28–55.
Hsu, J.-C. 1984. A Polarimetric Study of the Interstellar Medium in the Taurus Dark Clouds. Ph.D. thesis, University of Texas, Austin.
Hu, E. 1981. High latitude HI shells in the galaxy. *Astrophys. J.* 248:119–127.
Huang, S. S., and Struve, O. 1954. Stellar rotation. *Ann. d'Astrophys.* 17:85–93.
Huang, T.-Y., and Innanen, K. A. 1983. The gravitational escape/capture of planetary satellites. *Astron. J.* 88:1537–1548.
Hubbard, W. B. 1968. Thermal structure of Jupiter. *Astrophys. J.* 152:745–754.
Hubbard, W. B. 1969. Thermal models of Jupiter and Saturn. *Astrophys. J.* 155:333–344.
Hubbard, W. B. 1972. Statistical mechanics of light elements at high pressure. II. Hydrogen and helium alloys. *Astrophys. J.* 176:525–531.
Hubbard, W. B. 1974. Gravitational field of a rotating polytrope of index one. *Astron. Zh.* 51:1052.
Hubbard, W. B. 1977. de Sitter's theory flattens Jupiter. *Icarus* 30:311–313.
Hubbard, W. B. 1978. Comparative thermal evolution of Uranus and Neptune. *Icarus* 35:177–181.
Hubbard, W. B. 1980. Intrinsic luminosities of the Jovian planets. *Rev. Geophys. Space Phys.* 18:1–9.
Hubbard, W. B., and Slattery, W. L. 1971. Statistical mechanics of light elements at high pressure I. *Astrophys. J.* 168:131–139.
Hubbard, W. B., and Slattery, W. L. 1976. Interior structure of Jupiter: Theory of gravity sounding. In *Jupiter,* ed. T. Gehrels (Tucson: Univ. of Arizona Press), pp. 176–194.
Hubbard, W. B., and MacFarlane, J. J. 1980*a*. Structure and evolution of Uranus and Neptune. *J. Geophys. Res.* 85:225–234.
Hubbard, W. B., and MacFarlane, J. J. 1980*b*. Theoretical predictions of deuterium abundances in the Jovian planets. *Icarus* 44:676–681.
Hubbard, W. B., and Horedt, G. P. 1983. Computation of Jupiter interior models from gravitational inversion theory. *Icarus* 54:456–465.
Hubbard, W. B., Slattery, W. L., and DeVito, C. L. 1975. High zonal harmonics of rapidly rotating planets. *Astrophys. J.* 199:504–516.
Hubbard, W. B., MacFarlane, J. J., Anderson, J. D., Null, G. W., and Biller, E. D. 1980. Interior structure of Saturn inferred from Pioneer 11 gravity data. *J. Geophys. Res.* 85:5909–5916.
Hudson, G. B., Kennedy, B. M., Podosek, F. A., and Hohenberg, C. M. 1984. The early solar system abundance of ^{244}Pu as inferred from the St. Severin Chondrite. *J. Geophys. Res.* In press.
Huggins, P. J., Carlson, W. C., and Kinney, A. L. 1984. The abundance and distribution of interstellar CH. *Astron. Astrophys.* 133:347–356.
Hughes, V. A. 1982. Formation of BO.5 star due to the interactions of a shock wave with a molecular cloud in IC1805. In *Regions of Recent Star Formation,* eds. R. S. Roger and P. E. Dewdney (Dordrecht: D. Reidel), pp. 349–355.
Hughes, V. A., and Wouterloot, J. G. A. 1984. The star-forming regions in Cepheus A. *Astrophys. J.* 276:204–210.

Humes, D. H. 1976. The Jovian meteoroid environment. In *Jupiter,* ed. T. Gehrels (Tucson: Univ. of Arizona Press), pp. 1054–1067.

Humphreys, R. M., and McElroy, D. B. 1984. The IMF for massive stars in the galaxy and in the Magellanic clouds. *Astrophys. J.* 284:565–577.

Huneke, J. D., Armstrong, J. T., and Wasserburg, G. J. 1981. ^{41}K and ^{26}Mg in Allende inclusions and a hint of ^{41}Ca in the early solar system. *Lunar Planet. Sci.* XII:482–484 (abstract).

Huneke, J. D., Armstrong, J. T., Shaw, H. F., and Wasserburg, G. J. 1982. High resolution ion microprobe measurements of Mg in Allende plagioclase and standard glass. *Lunar Planet. Sci.* XIII:348–349 (abstract).

Hunten, D. M. 1979. Capture of Phobos and Deimos by protoatmospheric drag. *Icarus* 37:113–123.

Hunten, D. M. 1982. Thermal and nonthermal escape mechanisms for terrestrial bodies. *Planet. Space Sci.* 30:773–783.

Hunten, D. M., Tomasko, M. T., Flasar F. M., Samuelson, R. E., Strobel, D. F., and Stevenson, D. J. 1984. Titan. In *Saturn,* eds. T. Gehrels and M. S. Matthews (Tucson: Univ. of Arizona Press), pp. 671–759.

Hunter, C. 1962. The instability of the collapse of a self-gravitating gas cloud. *Astrophys. J.* 136:594–608.

Hunter, C. 1977. On secular stability, secular instability, and points of bifurcation of rotating gaseous masses. *Astrophys. J.* 213:497–517.

Hunter, J. H. 1969. The collapse of interstellar gas clouds and the formation of stars. *Mon. Not. Roy. Astron. Soc.* 142:473–498.

Hunter, J. H. Jr. 1979. The influence of initial velocity fields upon star formation. *Astrophys. J.* 233:946–949.

Hunter, J. H., and Schweiker, K. S. 1981. On the development of vorticity and waves in shearing media with preliminary application to the solar nebula. *Astrophys. J.* 243:1030–1039.

Hunter, J. H. Jr., and Fleck, R. C. Jr. 1982. Star formation: The influence of velocity fields and turbulence. *Astrophys. J.* 256:550–558.

Hunter, J. H., and Horak, T. 1983. The development of structure in shearing, viscous media. II. *Astrophys. J.* 265:402–416.

Hunter, J. H. Jr., Sandford, M. T. II, Whitaker, R. W., and Klein, R. I. 1984. Star formation in colliding gas flows. In preparation.

Huntress, W. T. Jr. 1977. Laboratory studies of bimolecular reactions of positive ions in interstellar clouds, in comets, and in planetary atmospheres of reducing composition. *Astrophys. J. Suppl.* 33:495–514.

Huntress, W. T. Jr., and Mitchell, G. F. 1979. The synthesis of complex molecules in interstellar clouds. *Astrophys. J.* 231:456–467.

Hutcheon, I. D. 1982. Ion probe magnesium measurements of Allende inclusions. In *Nuclear and Chemical Dating Techniques,* Amer. Chem. Soc. Symp. Ser., No. 176:95–128.

Hutcheon, I. D., Armstrong, J. T., and Wasserburg, G. J. 1984. Excess ^{41}K in Allende CAZ: Confirmation of a hint. *Lunar Planet. Sci.* XV:387 (abstract).

Hyland, A. R. 1981. Globules, dark clouds, and low mass pre-main sequence stars. In *Infrared Astronomy,* eds. C. G. Wynn-Williams and D. Cruikshank (Dordrecht: D. Reidel), pp. 125–151.

Hyland, A. R., Jones, T. J., and Mitchell, R. M. 1982. A study of the Chamaeleon dark cloud complex: Survey, structure, and embedded sources. *Mon. Not. Roy. Astron. Soc.* 201:1095–1117.

Ibanez, M. H. 1981. The onset of fragmentation of prestellar clouds by H_2 formation. *Mon. Not. Roy. Astron. Soc.* 196:13–22.

Iben, I. Jr. 1965. Stellar evolution. I. The approach to the main sequence. *Astrophys. J.* 141:993–1018.

Iben, I. 1984. Nucleosynthesis in low and intermediate mass stars on the asymptotic giant branch. In *Proceedings of the William Alfred Fowler Conference on Nucleosynthesis.* To be published.

Iben, I. Jr., and Talbot, R. J. 1966. Stellar formation rates in young clusters. *Astrophys. J.* 144:968–977.

Iben, I., and Renzini, A. 1982. On the formation of carbon star characteristics and the production of neutron-rich isotopes in asymptotic giant branch stars of small core mass. *Astrophys. J.* 263:L23–L27.

Iben, I., and Renzini, A. 1984. Single star evolution. I. Massive stars and early evolution of low and intermediate mass stars. *Phys. Rep.* In press.

Ichimaru, S. 1976. Magnetohydrodynamic turbulence in disk plasmas and magnetic field fluctuations in the galaxy. *Astrophys. J.* 208:701–705.

Icke, V. 1982. Transitions between epicyclic stellar orbits induced by massive gas clouds. *Astrophys. J.* 254:517–537.

Iglesias, E. 1977. The chemical evolution of molecular clouds. *Astrophys. J.* 218:697–715.

Illies, A. J., Jarrold, M. F., and Bowers, M. T. 1982. On the formation of HCO^+ and HOC^+ from the reaction between $H_3{}^+$ and CO. *J. Chem. Phys.* 77:5847–5848.

Illies, A. J., Jarrold, M. F., and Bowers, M. T. 1983. Experimental investigation of gaseous ionic structures of interstellar importance: HCO^+ and HOC^+. *J. Amer. Chem. Soc.* 105:2562–2565.

Imamura, J. N., Friedman, J. L., and Durisen, R. H. 1984. On the secular stability of rotating stars. In *The Birth and Evolution of Neutron Stars: Issues Raised by the Millisecond Pulsars,* eds. S. T. Reynolds and D. R. Stinebring (Greenbank: NRAO), pp. 191–199.

Imhoff, C. L. 1984. UV observations and results on pre-main-sequence stars. In *Future of Ultraviolet Astronomy Based on Six Years of IUE Research,* ed. J. M. Mead, R. D. Chapman, and Y. Kondo (Washington: NASA, Scientific and Technical Information Branch), pp. 81–92.

Imhoff, C. L., and Giampapa, M. S. 1980. The ultraviolet spectrum of the T Tauri star RW Aurigae. *Astrophys. J.* 239:L115–L119.

Imhoff, C. L., and Giampapa, M. S. 1981. The ultraviolet variability of the T Tauri star RW Aurigae. In *The Universe at Ultraviolet Wavelengths: The First Two Years of IUE,* ed. R. D. Chapman (NASA Conf. Publ. 2171), pp. 185–191.

Imhoff, C. L., and Giampapa, M. S. 1982*a*. Far ultraviolet and x-ray evidence concerning the chromospheres and coronae of the T Tauri stars. In *The Second Cambridge Workshop on Cool Stars, Stellar Systems and the Sun,* eds. M. Giampapa and L. Bolub (SAO Special Rep. 392), pp. 175–179.

Imhoff, C. L., and Giampapa, M. S. 1982*b*. Chromospheres and coronae in the T Tauri stars. In *Advances in Ultraviolet Astronomy: Four Years of IUE Research,* eds. Y. Kondo, J. M. Mead, and R. D. Chapman (NASA Conf. Publ. 2238), pp. 456–459.

Imhoff, C. L., and Giampapa, M. S. 1984. IUE observations of the pre-main-sequence stars. II. Far ultraviolet line fluxes for T Tauri stars. To be submitted to *Astrophys. J.*

Inoue, M., and Tabara, H. 1981. Structure of the galactic magnetic field in the solar neighborhood. *Publ. Astron. Soc. Japan* 33:603–615.

Ip, W.-H. 1983. Collisional interactions of ring particles: The ballistic transport. *Icarus* 54:253–262.

Ip, W.-H. 1984. Condensation and agglomeration of cometary ice: The $HDO{:}H_2$ ratio as tracer. Preprint.

Ipser, J. R., and Managan, R. A. 1981. On the existence and structure of inhomogeneous analogs of the Dedekind and Jacobi ellipsoids. *Astrophys. J.* 250:362–372.

Irvine, W. M. 1983. The chemical composition of the pre-solar nebula. In *Cometary Exploration I,* ed. T. I. Gombosi (Budapest: Central Res. Inst. Physics, Hungarian Acad. Sci.), pp. 3–10.

Irvine, W. M., and Hjalmarson, Å. 1983. Comets, interstellar molecules, and the origin of life. In *Cosmochemistry and the Origin of Life,* ed. C. Ponnamperuma (Dordrecht: D. Reidel), pp. 113–142.

Irvine, W. M., and Hjalmarson, Å. 1984. The chemical composition of interstellar molecular clouds. *Origins Life* 14:15–25.

Irvine, W. M., and Schloerb, F. P. 1984. Cyanide and isocyanide abundances in the cold dark cloud TMC-1. *Astrophys. J.* 288:516–521.

Irvine, W. M., Höglund, B., Friberg, P., Askne, J., and Ellder, J. 1981. The increasing chemical complexity of the Taurus dark clouds: Detection of CH_3CCH and C_4H. *Astrophys. J.* 248:L113–L117.

Irvine, W. M., Good, J. C., and Schloerb, F. P. 1983. Observations of SO_2 and HCS^+ in cold molecular clouds. *Astron. Astrophys.* 127:L10–L13.

Irvine, W. M., Abraham, Z., A'Hearn, M., Altenhoff, W., Andersson, C., Bally, J., Batrla, W., Baudry, A., Bockelee-Morvan, D., Crovisier, J., de Pater, I., Despois, D., Ekelund, L., Gerard, E., Heiles, C., Hollis, J. M., Huchtmeier, W., Levreault, R., Masson, C. R., Palmer, P., Perault, M., Rickard, L. J., Sargent, A. I., Scalise, E., Schloerb, F. P., Schmidt, S.,

Stark, A. A., Stumpff, P., Sutton, E., Swade, D., Sykes, M., Turner, B., Wade, C., Walmsley, M., Webber, J., Winnberg, A., and Wootten, A. 1984. Radioastronomical observations of comets IRAS-Araki-Alcock (1973d) and Sugano-Saigusa-Fujikawa (1983e). *Icarus* 60:215–220.

Isobe, S., and Sasaki, G. 1982. Globules in the Orion nebula. III. Age range of the stars in the Orion nebula. *Publ. Astron. Soc. Japan* 34:241–247.

Isobe, S., Tanabe, H., Maihara, T., Mizutani, H., and Koma, Y. 1983. Solar dust ring observed by balloon-borne optical and infrared polarimeters during solar eclipse on June 11, 1983. *Bull. Amer. Astron. Soc.* 15:959 (abstract).

Israel, F. P. 1978. HII regions and CO clouds: The blister model. *Astron. Astrophys.* 70:769–775.

Israel, F. P., Gatley, I., Matthews, K., and Neugebauer, G. 1982. Observations of NGC 604 over six decades in frequency. *Astron. Astrophys.* 105:229–235.

Jacobsen, S. B., and Wasserburg, G. J. 1984. Sm-Nd isotopic evolution of chondrites and achondrites. *Earth Planet. Sci. Lett.* 67:137–150.

Jaffe, D. T., and Fazio, G. G. 1982. Star formation in the M 17 giant molecular cloud. *Astrophys. J.* 257:L77–L82.

Jaffe, D. T., Hildebrand, R. H., Keene, J., Harper, D. A., Lowenstein, R. F., and Moran, J. M. 1984. Far-IR selected star formation regions. *Astrophys. J.* In press.

James, R. A. 1964. The structure and stability of rotating gas masses. *Astrophys. J.* 140:552–582.

Jankovics, I., Appenzeller, I., and Krauter, J. 1983. Blueshifted forbidden lines in T Tauri stars. *Publ. Astron. Soc. Pacific* 45:883–885.

Jarvis, J. F., and Tyson, J. A. 1981. FOCAS: Faint object classification and analysis system. *Astron. J.* 86:476–495.

Jeans, J. H. 1919. *Problems of Cosmogony and Stellar Dynamics* (Cambridge: Cambridge University Press).

Jeans, J. H. 1929. The configurations of rotating liquid masses. In *Astronomy and Cosmogony* (Cambridge: Cambridge Univ. Press).

Jeans, J. H. 1967. *An introduction to the kinetic theory of gases* (Cambridge: Cambridge Univ. Press).

Jeffrey, P. M., and Reynolds, J. H. 1961. Origin of excess ^{129}Xe in stone meteorites. *J. Geophys. Res.* 66:3582–3583.

Jeffreys, H. 1924. *The Earth* (Cambridge: Cambridge Univ. Press).

Jenkins, E. B., and Savage, B. D. 1974. Ultraviolet photometry from the orbiting ultraviolet observatory XIV: An extension of the survey of Lyman alpha absorption from interstellar hydrogen. *Astrophys. J.* 187:243–255.

Jensen, E. B., Talbot, R. J., and Dufour, R. J. 1981. M83. III. Age and brightness of young and old stellar populations. *Astrophys. J.* 243:716–735.

Jewitt, D. C. 1982. The rings of Jupiter. In *Satellites of Jupiter,* ed. D. Morrison (Tucson: Univ. of Arizona Press), pp. 44–64.

Jog, C. J., and Solomon, P. M. 1984*a*. A galactic disk as a two-fluid system: Consequences for the critical stellar velocity dispersion and the formation of condensations in the gas. *Astrophys. J.* 276:127–134.

Jog, C. J., and Solomon, P. M. 1984*b*. Two-fluid gravitational instabilities in a galactic disk. *Astrophys. J.* 276:114–126.

Johansson, L. E. B., Andersson, C., Ellder, J., Friberg, P., Hjalmarson, Å., Höglund, B., Irvine, W. M., Olofsson, H., and Rydbeck, G. 1984. Spectral scan of Orion A and IRC+10216 from 72 to 91 GHz. *Astron. Astrophys.* 130:227–256.

Johnson, D. R., Lovas, F. S., Gottlieb, C. A., Gottlieb, E. W., Litvak, M. M., Guélin, M., Thaddeus, P. 1977. Detection of interstellar ethyl cyanide. *Astrophys. J.* 218:370–376.

Jones, B. F. 1970. Internal motions in the Pleiades. *Astron. J.* 75:563–574.

Jones, B. 1976. The origin of galaxies: A review of recent theoretical developments and their confrontation with observation. *Rev. Mod. Phys.* 48:107–150.

Jones, B. F., and Herbig, G. H. 1979. Proper motions of T Tauri variables and other stars associated with the Taurus-Auriga dark clouds. *Astron. J.* 84:1872–1889.

Jones, B. F., and Herbig, G. H. 1982. Proper motions of Herbig-Haro objects. II. The relationship of HH-39 to R Monocerotis and NGC 2261. *Astron. J.* 87:1223–1232.

Jones, T. J., Hyland, A. R., Robinson, G., Smith, R., and Thomas, J. 1980. Infrared observations of a Bok globule in the Southern Coalsack. *Astrophys. J.* 242:132–140.

Jones, T. J., Ashley, M., Hyland, A. R., and Ruelas-Mayorga, A. 1981. A search for the infrared counterpart of type II OH masers. I. A model for the IR background source confusion. *Mon. Not. Roy. Astron. Soc.* 197:413–428.

Jones, T. J., Hyland, A. R., and Allen, D. A. 1983. 3 μm spectroscopy of IRS7 towards the galactic centre. *Mon. Not. Roy. Astron. Soc.* 205:187–190.

Jones, T. J., Hyland, A. R., and Bailey, J. 1984*a*. The inner core of a Bok globule. *Astrophys. J.* 282:675–682.

Jones, T. J., Hyland, A. R., Harvey, P. M., Wilking, B. A., and Joy, M. 1984*b*. The Chamaeleon dark cloud complex. II. A deep survey around HD 97300. In preparation.

Jong, E. 1983. The poet fears failure. In *Ordinary Miracles* (New York: New Amer. Lib.), pp. 38.

Joy, A. H. 1945. T-Tauri variable stars. *Astrophys. J.* 102:168–195.

Joy, A. H. 1949. Bright-line stars among the Taurus dark clouds. *Astrophys. J.* 110:424–437.

Jura, M. 1974. Formation and destruction rates of interstellar H_2. *Astrophys. J.* 191:375–379.

Jura, M. 1979. The mean intensity of radiation at 2μm in the solar neighborhood. *Astrophys. Lett.* 20:89–91.

Jura, M. 1980. Origin of large interstellar grains toward ρ Ophiuchi. *Astrophys. J.* 253:63–65.

Kadanoff, L. P. 1984. Roads to chaos. *Phys. Today* 36:46–53.

Kahn, F. D. 1969. The flow of ionized gas from a globule in interstellar space. *Physica* 41:172–189.

Kaifu, N., Suzuki, S., Hasegawa, T., Morimoto, M., Inatani, J., Nagane, K., Miyazawa, K., Chikada, Y., Kanzawa, T., and Akabane, K. 1984. Rotating gas disk around L1551 IRS-5. *Astron. Astrophys.* 134:7–12.

Kaiser, M. L., Desch, M. D., Warwick, J. W., and Pearce, J. B. 1980. Voyager detection of nonthermal radio emission from Saturn. *Science* 209:1238–1240.

Kaiser, M. L., Desch, M. D., Kurth, W. S., Lecacheux, A., Genova, F., Pedersen, B. M., and Evans, D. R. 1984. Saturn as a radio source. In *Saturn,* eds. T. Gehrels and M. S. Matthews (Tucson: Univ. of Arizona Press), pp. 378–415.

Kaiser, T., and Wasserburg, G. J. 1983. The isotopic composition and concentration of Ag in iron meteorites and the origin of exotic silver. *Geochim. Cosmochim. Acta* 47:43–58.

Kampé de Fériet, J. 1955. Discussion of the measurement of turbulence in nebulae. In *Gas Dynamics of Cosmic Clouds* (Amsterdam: North Holland), pp. 134–136.

Kant, I. 1755. *Allgemeine Naturgeschichte und Theorie des Himmels.*

Kaplan, S. A. 1966. *Interstellar Gas Dynamics* (Oxford: Pergamon).

Kaplan, S. A., and Klimishin, I. A. 1964. Methods for analyzing interstellar turbulence. *Sov. Astron. J.* 8:210–216.

Kaplan, S. A., and Pikel'ner, S. B. 1970. *The Interstellar Medium* (Cambridge: Harvard Univ. Press).

Kasting, J. F., Zahnle, K. J., and Walker, J. C. G. 1983. Photochemistry of methane in the Earth's early atmosphere. *Precambrian Res.* 20:121–148.

Kaula, W. M. 1964. Tidal dissipation by solid friction and the resulting orbital evolution. *Rev. Geophys.* 2:661–685.

Kaula, W. M. 1979. Equilibrium velocities of planetesimal populations. *Icarus* 40:262–275.

Kaula, W. M. 1979*b*. Thermal evolution of Earth and Moon growing by planetesimal impacts. *J. Geophys. Res.* 84:999–1008.

Keays, R. R., Ganapathy, R., and Anders, E. 1971. Chemical fractionations in meteorites. IV. Abundances of fourteen trace elements in L-chondrites: Implications for cosmothermometry. *Geochim. Cosmochim. Acta* 35:337–363.

Keene, J. 1981. Far-infrared observations of globules. *Astrophys. J.* 245:115–123.

Keene, J., Harper, D. A., Hildebrand, R. H., and Whitcomb, S. E. 1980. Far infrared observations of the globule B335. *Astrophys. J.* 240:L43–L46.

Keene, J., Davidson, J. A., Harper, D. A., Hildebrand, R. H., Jaffe, D. T., Lowenstein, R. F., Low, F. J., and Pernic, R. 1983. Far-infrared detection of low-luminosity star formation in the Bok globule B335. *Astrophys. J.* 274:L43–L47.

Kegel, W. H., and Traving, G. 1976. On the Jeans instability in the steady state interstellar gas. *Astron. Astrophys.* 50:137–139.

Kegel, W. H., and Völk, H. J. 1983. Velocity fluctuations in the interstellar medium due to the gravitational interaction with the system of stars. *Astron. Astrophys.* 119:101–108.

Kellman, S. A., and Gaustad, J. E. 1969. Roseeland and Planck mean absorption coefficients for particles of ice, graphite, and silicon dioxide. *Astrophys. J.* 157:1465–1467.

Kelly, W. R., and Wasserburg, G. J. 1978. Evidence for the existence of ^{107}Pd in the early solar system. *Geophys. Res. Lett.* 5:1079–1082.

Kendall, D. G. 1961. Some problems in the theory of comets, I and II. In *Berkeley Symposium on Mathematical Statistics and Probability* (Berkeley: Univ. of California Press), 3:99–148.

Kennicutt, R. C. Jr. 1983. The rate of star formation in normal disk galaxies. *Astrophys. J.* 272:54–67.

Kennicutt, R. C. Jr., and Kent, S. M. 1983. A survey of Hα emission in normal galaxies. *Astron. J.* 88:1094–1107.

Kenyon, S., and Starrfield, S. 1979. On the structure of Bok globules. *Publ. Astron. Soc. Pacific* 91:271–275.

Kerridge, J. F. 1979. Fractionation of refractory lithophile elements among chondritic meteorites. *Proc. Lunar Planet. Sci. Conf.* 10:989–996.

Kerridge, J. F. 1980. Isotopic clues to organic synthesis in the early solar system. *Lunar Planet. Sci.* XI:538–540 (abstract).

Kerridge, J. F. 1983. Isotopic composition of carbonaceous-chondrite-kerogen: Evidence for an interstellar origin of organic matter in meteorites. *Earth Planet. Sci. Lett.* 64:186–200.

Kerridge, J. F., and Bunch, T. E. 1979. Aqueous activity on asteroids: Evidence from carbonaceous meteorites. In *Asteroids,* ed. T. Gehrels (Tucson: Univ. of Arizona Press), pp. 745–764.

Khavtassi, D. Sh. 1955. A statistical study of dark nebulae. *Bull. Abastumani Obs.* 18:29–114.

Khavtassi, D. Sh. 1960. *Atlas of Galactic Dark Nebulae* (Abastumani, USSR: Abastumani Astrophys. Obs.).

Kirsten, T. 1978. Time and the solar system. In *The Origin of the Solar System,* ed. S. F. Dermott (Chichester, UK: J. Wiley), p. 267–346.

Kirzhnits, D. A. 1967. *Field Theoretical Methods of Many Body Systems* (Oxford: Pergamon Press).

Klein, M. J., Janssen, M. A., Gulkis, S., and Olsen, E. T. 1978. Saturn's microwave spectrum: Implications for the atmosphere and rings. In *The Saturn System,* eds. D. Hunten and D. Morrison (Washington: NASA Rep. CP 2068), pp. 195–216.

Klein, R. I., and Chevalier, R. A. 1978. X-ray bursts from type II supernovae. *Astrophys. J.* 223:L109–L112.

Klein, R. I., Sandford, M. T. II., and Whitaker, R. W. 1980. Two-dimensional radiation hydrodynamics calculations of the formation of O-B associations in dense molecular clouds. *Space Sci. Rev.* 27:275–281.

Klein, R. I., Sandford, M. T. II, and Whitaker, R. W. 1983. Star formation within OB subgroups: Implosion by multiple sources. *Astrophys. J.* 271:L69–L73.

Kleiner, S. C. 1985. Correlation Analysis of the Taurus Molecular Cloud Complex. Ph.D. thesis, Univ. of Massachusetts.

Kleiner, S. C., and Dickman, R. L. 1984. Large-scale structure of the Taurus molecular cloud complex. I. Density fluctuations - a fossil Jeans length? *Astrophys. J.* 286:255–262.

Kleiner, S. C., and Dickman, R. L. 1985*a*. Large-scale structure of the Taurus molecular cloud complex. II. Velocity fluctuations and turbulence. *Astrophys. J.* In press.

Kleiner, S. C., and Dickman, R. L. 1985*b*. The turbulent velocity field of TMC-1. Submitted to *Astrophys. J.*

Kleinmann, S. G., Hall, D. N. B., and Scoville, N. Z. 1984. Infrared spectra of young stellar objects. In preparation.

Knacke, R. F., Kim, S. J., Ridgway, S. T., and Tokunaga, A. T. 1982*a*. The abundances of CH_4, CH_3D, NH_3, and PH_3 in the troposphere of Jupiter derived high resolution 1100-1200 cm^{-1} spectra. *Astrophys. J.* 262:388–395.

Knacke, R. F., McCorkle, S., Puetter, R. C., Erickson, E. F., and Kratschmer, W. 1982*b*. Observations of interstellar ammonia ice. *Astrophys. J.* 260:141–146.

Knapp, G. R. 1972. HI Observations of Dark Clouds. Ph.D. thesis, Univ. of Maryland.

Knapp, G. R. 1974. Observations of HI in dense interstellar dust clouds. I. A survey of 88 clouds. *Astrophys. J.* 79:527–540.

Knobloch, E., and Spruit, H. C. 1982. Stability of differential rotation in stars. *Astron. Astrophys.* 113:261–268.

Knude, J. 1979. Interstellar reddening in clouds in the solar vicinity. *Astron. Astrophys.* 71:344–351.

Kolmogorov, A. N. 1941*a*. The local structure of turbulence in an incompressible viscous fluid for very large Reynolds numbers. *Comptes Rendus de l'Academie des Sciences de l'URSS* 30:301–305. Reprinted in Friedlander, S. K., and Topper, L., eds. 1961. *Turbulence: Classic Papers on Statistical Theory* (New York: Interscience).

Kolmogorov, A. N. 1941*b*. On degeneration of isotropic turbulence in an incompressible viscous liquid. *Comptes Rendus de l'Academie des Sciences de l'URSS* 31:538. Reprinted in Friedlander and Topper, 1961.

Kolmogorov, A. N. 1941*c*. Dissipation of energy in locally isotropic turbulence. *Comptes Rendus de l'Academie des Sciences de l'URSS* 32:16. Reprinted in Friedlander and Topper, 1961.

Kolodny, Y., Kerridge, J. F., and Kaplan, I. R. 1980. Deuterium in carbonaceous chondrites. *Earth Planet. Sci. Lett.* 46:149–158.

Kolychalov, P. I., and Sunjaev, R. A. 1980. Outer parts of accreting disks around super massive black holes. *Pisma Astron. Zh.* 6:680–686.

Königl, A. 1982. On the nature of bipolar sources in dense molecular clouds. *Astrophys. J.* 261:115–134.

Kopyshev, V. P. 1965. Gruneisen constant in the Thomas Fermi approximation. *Sov. Phys. Doklady* 10:338–339.

Kornacki, A. S., and Wood, J. A. 1984. The mineralogy, chemistry, and origin of inclusion matrix and meteorite matrix in the Allende CV3 chondrite. *Geochim. Cosmochim. Acta* 48:1663–1676.

Korycansky, D., Bodenheimer, P., Ruden, S., and Lin, D. N. C. 1984. Steady-state models for accreting protoplanets. In preparation.

Kossacki, K. 1968. On the possibility of pre-stellar bodies formation in clouds compressed by shock waves. *Acta Astron.* 18:221–253.

Kovalevsky, J., and Link, F. 1969. Diameter, flattening and optical properties of the upper atmosphere of Neptune as derived from the occulation of the star BD-17°4388. *Astron. Astrophys.* 2:398–412.

Kraichnan, R. H., and Montgomery, D. 1980. Two-dimensional turbulence. *Rep. Prog. Phys.* 43:537–619.

Krebs, J., and Hillebrandt, W. 1984. The interaction of supernova shock fronts and nearby interstellar clouds. *Astron. Astrophys.* In press.

Krolik, J. H., and Smith, H. A. 1981. Infrared atomic hydrogen line formation in luminous stars. *Astrophys. J.* 249:628–636.

Krolik, J. H., and Kallman, T. R. 1983. X-ray ionization and the Orion molecular cloud. *Astrophys. J.* 267:610–624.

Krügel, E., Stenholm, L. G., Steppe, H., and Sherwood, W. A. 1983. The physical structure of globule B335. *Astron. Astrophys.* 127:195–200.

Kuhi, L. V. 1964. Mass loss from T Tauri stars. *Astrophys. J.* 140:1409–1433.

Kuhi, L. V. 1983. Optical and X-ray observations of T Tauri stars. *Rev. Mexicana Astron. Astrophys.* 7:127–139.

Kuiper, G. P. 1951*a*. On the origin of the solar system. In *Astrophysics,* ed. J. A. Hynek (New York: McGraw Hill), pp. 357–424.

Kuiper, G. P. 1951*b*. On the origin of the irregular satellites. *Proc. Nat. Acad. Sci. U.S.* 37:717–721.

Kuiper, T. B. H., Zuckerman, B., and Rodriquez-Kuiper, E. N. 1981. High velocity molecular emission in Orion: A case for stellar winds. *Astrophys. J.* 251:88–102.

Kunde, V. G., Hanel, R. A., Maguire, W., Gautier, D., Paluteau, J. P., Marten, A., Chedin, A., Husson, N., and Scott, N. 1982. The tropospheric gas composition of Jupiter's north equatorial belt (NH_3, PH_3, CH_3D, GeH_4, H_2O) and the Jovian D/H isotopic ratio. *Astrophys. J.* 263:443–467.

Kurat, G. 1970. Zur genese der Ca-Al-reichen einschlusse in chondriten von Lancé. *Earth Planet. Sci. Lett.* 9:225–231.

Kurt, V. G., and Sunyaev, R. A. 1968. Observations and interpretations of the ultraviolet radiation of the galaxy. *Sov. Astron.* 11:928–931.

Kurucz, R. L. 1979. Model atmospheres for G, F, A, B, and O stars. *Astrophys. J. Suppl.* 40:1–340.
Kusaka, T., Nakano, T., and Hayashi, C. 1970. Growth of solid particles in the primodial solar nebula. *Prog. Theor. Phys.* 44:1580–1596.
Kutner, M. 1973. A study of formaldehyde 6-cm anomalous absorption. In *Molecules and the Galactic Environment,* eds. M. Gordon and L. Snyder (New York: Wiley), pp. 199–206.
Kutner, M., and Meade, K. 1981. Molecular clouds outside the solar circle in the first quadrant of our galaxy. *Astrophys. J.* 249:L15–L18.
Kutner, M. L., and Ulich, B. L. 1981. Recommendations for calibration of millimeter-wavelength spectral line data. *Astrophys. J.* 250:341–348.
Kutner, M. L., Evans, N. J., and Tucker, K. D. 1976. A dense molecular cloud in the OMC-1/OMC-2 region. *Astrophys. J.* 209:452–461.
Kutner, M. L., Tucker, K. D., Chin, G., and Thaddeus, P. 1977. The molecular complexes in Orion. *Astrophys. J.* 215:521–528.
Kutner, M. L., Machnik, D. E., Tucker, K. D., and Dickman, R. L. 1980. Search for interstellar pyrrole and furan. *Astrophys. J.* 242:541–544.
Kutschera, W., Billquist, P. J., Frekers, D., Henning, W., Jensen, K. J., Xiuzeng, M. A., Pardo, R., Paul, M., Rehm, K. E., Smither, R. K., and Yntema, J. L. 1984. Half-Life of ^{60}Fe. Proceedings of the 3rd Int'l Symp. on Accelerator Mass Spectrometry, April 10–13, to be published in *Nucl. Instr. and Meth. B.*
Kwan, J. 1978. Radiation transport and the kinematics of molecular clouds. *Astrophys. J.* 223:147–160.
Kwan, J., and Scoville, N. 1976. The nature of the broad molecular line emission at the Kleinmann-Low nebula. *Astrophys. J.* 210:L39–L44.
Kwan, J., and Krolik, J. H. 1981. The formation of emission lines in quasars and Seyfert nuclei. *Astrophys. J.* 250:478–507.
Kwan, J., and Linke, R. A. 1982. Circumstellar molecular emission of evolved stars and mass loss: IRC+10216. *Astrophys. J.* 254:587–593.
Lacasse, M. G., Boyle, D., Levreault, R., Pipher, J. L., and Sharpless, S. 1981. Polarimetric observations of S106. *Astron. Astrophys.* 104:57–64.
Lacy, J. H., Beck, S. C., and Geballe, T. R. 1982. Infrared emission line studies of the structure and excitation of H II regions. *Astrophys. J.* 255:510–523.
Lacy, J. H., Bass, F., Allamandola, L. J., Persson, S. E., McGregor, P. J., Lonsdale, C. J., Geballe, T. R., and van de Bult, C. E. P. 1984. 4.6 μm absorption features due to solid phase CO and Cyano group molecules toward compact infrared sources. *Astrophys. J.* 276:533–543.
Lada, C. J. 1980. Formation of massive stars in OB associations and giant molecular clouds. In *Giant Molecular Clouds in the Galaxy,* eds. P. M. Solomon and M. G. Edmunds (Oxford: Pergamon), pp. 239–253.
Lada, C. J. 1985. Cold outflows, energetic winds, and enigmatic jets around young stellar objects. *Ann. Rev. Astron. Astrophys.* 23. In press.
Lada, C. J., and Black, J. H. 1976. CO observations of the bright-rimmed cloud B35. *Astrophys. J.* 203:L75–L79.
Lada, C., and Wooden, D. 1979. Molecular-line observations of the S 252 (NGC 2175) star-forming complex. *Astrophys. J.* 232:158–168.
Lada, C. J., and Wilking, B. A. 1980. ^{13}CO self-absorption in the ρ Oph dark cloud. *Astrophys. J.* 238:620–626.
Lada, C. J., and Wilking, B. A. 1984. The nature of the embedded population in the Rho Ophiuchi dark cloud: Mid-infrared observations. *Astrophys. J.* 287:610–621.
Lada, C. J., Blitz, L., and Elmegreen, B. G. 1978. Star formation in OB associations. In *Protostars & Planets,* ed. T. Gehrels (Tucson: Univ. of Arizona Press), pp. 341–367.
Lada, C. J., Margulis, M., and Dearborn, D. 1984. The formation and early dynamical evolution of open clusters. *Astrophys. J.* 285:141–152.
Lafont, S., Lucas, R., and Omont, A. 1982. Molecular abundances in IRC+10216. *Astron. Astrophys.* 106:201–213.
Lago, M. T. V. T. 1982. A new investigation of the T Tauri star RU Lupi. II. The physical conditions of the line emitting region. *Mon. Not. Roy. Astron. Soc.* 198:445–456.
Lago, M. T. V. T., and Penston, M. V. 1982. A new investigation of the T Tauri star RU Lupi. I. Observation and immediate analysis. *Mon. Not. Roy. Astron. Soc.* 198:429–443.

Lambeck, K. 1979. On the orbital evolution of the Martian satellites. *J. Geophys. Res.* 84:5651–5658.

Lambert, D. L. 1978. The abundance of the elements in the solar photosphere. VIII. Revised abundances of carbon, nitrogen and oxygen. *Mon. Not. Roy. Astron. Soc.* 182:249–272.

Lambert, D. L., and Sneden, C. 1977. The $^{12}C/^{13}C$ ratio in stellar atmospheres. VII. The very metal-deficient giant HD122563. *Astrophys. J.* 215:597–602.

Landau, L. D., and Lifshitz, E. M. 1959. *Fluid Mechanics* (London: Pergamon).

Landolt, A. V. 1979. Contracting stars in the Pleiades. *Astrophys. J.* 231:468–476.

Lane, A. L., Hord, C. W., West, R. A., Esposito, L. W., Coffeen, D. L., Sato, M., Simmons, K. E., Pomphry, R. B., and Morris, R. B. 1982. Photopolarimetry from Voyager 2: Preliminary results on Saturn, Titan and the rings. *Science* 215:537–543.

Langer, W. D. 1976. The carbon monoxide abundance in interstellar clouds. *Astrophys. J.* 206:699–712.

Langer, W. D. 1978*a*. The stability of interstellar clouds containing magnetic fields. *Astrophys. J.* 225:95–106.

Langer, W. D. 1978*b*. The formation of molecules from singly and multiply ionized atoms. *Astrophys. J.* 225:860–868.

Langer, W. D. 1984. Physical and chemical processes of molecular clouds. In *Star Formation, Lectures Delivered at Les Houches Summer School,* eds. A. Omont and R. Lucas. (Amsterdam: North-Holland). In press.

Langer, W. D., and Glassgold, A. E. 1976. Time scales for molecule formation by ion-molecule reactions. *Astron. Astrophys.* 48:395–403.

Langer, W. D., Wilson, R. W., Henry, P. S., and Guélin, M. 1978. Observations of anomalous intensities in the lines of the HCO^+ isotopes. *Astrophys. J.* 225:L139–L142.

Langer, W. D., Goldsmith, P. F., Carlson, E. R., and Wilson, R. W. 1980. Evidence for isotopic fractionation of carbon monoxide in dark clouds. *Astrophys. J.* 235:L39–L44.

Langer, W. D., Graedel, T. E., Frerking, M. A., and Armentrout, P. B. 1984. Carbon and oxygen isotope fractionation in dense interstellar clouds. *Astrophys. J.* 277:581–604.

Laplace, P. S. de. 1796. *Exposition du système du monde* (Paris).

Larimer, J. W. 1967. Chemical fractionations in meteorites-I. Condenstation of the elements. *Geochim. Cosmochim. Acta* 31:1215–1238.

Larimer, J. W. 1968. Experimental studies on the system Fe-MgO-SiO_2-O_2 and their bearing on the petrology of chondritic meteorites. *Geochim. Cosmochim. Acta* 32:1187–1207.

Larimer, J. W., and Anders, E. 1967. Chemical fractionations in meteorites-II. Abundance patterns and their interpretation. *Geochim. Cosmochim. Acta* 31:1239–1270.

Larimer, J. W., and Anders, E. 1970. Chemical fractionations in meteorites-III. Major element fractionations in chondrites. *Geochim. Cosmochim. Acta* 34:367–387.

LaRosa, T. N. 1983. Radiatively induced star formation. *Astrophys. J.* 274:815–821.

Larson, H. P., Fink, U., Treffers, R. R., and Gautier, T. N. 1975. Detection of water vapor on Jupiter. *Astrophys. J.* 197:L137–L140.

Larson, R. B. 1969. Numerical calculations of the dynamics of a collapsing protostar. *Mon. Not. Roy. Astron. Soc.* 145:271–295.

Larson, R. B. 1973. The evolution of protostars. Theory. *Fund. Cosmic Phys.* 1:1–70.

Larson, R. B. 1978*a*. Calculations of three-dimensional collapse and fragmentation. *Mon. Not. Roy. Astron. Soc.* 184:69–85.

Larson, R. B. 1978*b*. The stellar state: Formation of solar-type stars. In *Protostars & Planets,* ed. T. Gehrels (Tucson: Univ. of Arizona Press), pp. 43–57.

Larson, R. B. 1979. Stellar kinematics and interstellar turbulence. *Mon. Not. Roy. Astron. Soc.* 186:479–490.

Larson, R. B. 1981. Turbulence and star formation in molecular clouds. *Mon. Not. Roy. Astron. Soc.* 194:809–826.

Larson, R. B. 1982. Mass spectra of young stars. *Mon. Not. Roy. Astron. Soc.* 200:159–174.

Larson, R. B. 1983. Angular momentum and protostellar disks. *Rev. Mexicana Astron. Astrof.* 7:219–227.

Larson, R. B. 1984. Gravitational torques and star formation. *Mon. Not. Roy. Astron. Soc.* 206:197–207.

Larson, R. B. 1985. Cloud fragmentation and stellar masses. Preprint.

Larson, R. B., and Starrfield, S. 1971. On the formation of massive stars and the upper limit of stellar masses. *Astron. Astrophys.* 13:190–197.
Lathrop, K. D., and Brinkley, F. W. 1973. TWOTRAN-II: An interfaced, exportable version of the TWOTRAN code for two-dimensional transport. Report LA-4848-MS (Los Alamos National Lab.).
Lattanzio, J. C., Monaghan, J. J., Pongracic, H., and Schwarz, M. P. 1984. Interstellar cloud collisions. Preprint.
Laul, J. C., Ganapathy, R., Anders, E., and Morgan, J. W. 1973. Chemical fractionations in meteorites. VI. Accretion temperatures of H-, LL-, and E-chondrites, from abundance of volatile trace elements. *Geochim. Cosmochim. Acta* 37:329–357.
Layzer, D. 1963. On the fragmentation of self-gravitating gas clouds. *Astrophys. J.* 137:351–362.
Lebofsky, L. A., Johnson, T. V., and McCord, T. B. 1970. Saturn's rings: Spectral reflectivity and compositional implications. *Icarus* 13:226–230.
Lebovitz, N. R. 1967. Rotating fluid masses. *Ann. Rev. Astron. Astrophys.* 5:465–480.
Lebovitz, N. R. 1972. On the fission theory of binary stars. *Astrophys. J.* 175:171–183.
Lebovitz, N. R. 1974. The fission theory of binary stars. II. Stability to third-harmonic disturbances. *Astrophys. J.* 190:121–130.
Lebovitz, N. R. 1979. Rotating self-gravitating masses. *Ann. Rev. Fluid Mech.* 11:229–246.
Lebovitz, N. R. 1983. On the fission theory of binary stars. III. The formulation of the bifurcation problem. *Astrophys. J.* 275:316–329.
Lebovitz, N. R. 1984. On the fission theory of binary stars. IV. Exact solutions in polynomial spaces. *Astrophys. J.* 284:364–380.
Lebrun, F., and Huang, Y.-L. 1984. Nearby molecular clouds. I. Ophiuchus-Sagittarius, $b > 10°$. *Astrophys. J.* 281:634–638.
Ledoux, P., Schwarzschild, M., and Spiegel, E. A. 1961. On the spectrum of turbulent convection. *Astrophys. J.* 133:184–197.
Lee, T. 1978. A local proton irradiation model for isotopic anomalies in the solar system. *Astrophys. J.* 224:217–226.
Lee, T. 1979. New isotopic clues to solar system formation. *Rev. Geophys. Space Phys.* 17:1591–1611.
Lee, T., and Papanastassiou, D. A. 1974. Mg isotopic anomalies in the Allende meteorite and correlation with O and Sr effects. *Geophys. Res. Lett.* 1:225–228.
Lee, T., Papanastassiou, D. A., and Wasserburg, G. J. 1976*a*. The presence of ^{26}Al in the early solar nebula. *Bull. Amer. Astron. Soc.* 8:457 (abstract).
Lee, T., Papanastassiou, D. A., and Wasserburg, G. J. 1976*b*. Internal isochrons of Allende inclusions by the ^{26}Al-^{26}Mg method. *Trans. Amer. Geophys. Union* 57:278.
Lee, T., Papanastassiou, D. A., and Wasserburg, G. J. 1976*c*. Demonstration of ^{26}Mg excess in Allende and evidence for ^{26}Al. *Geophys. Res. Lett.* 3:109–112.
Lee, T., Papanastassiou, D. A., and Wasserburg, G. J. 1977. Aluminum-26 in the early solar system: Fossil or fuel? *Astrophys. J.* 211:L107–L110.
Lee, T., Russell, W. A., and Wasserburg, G. J. 1979. Calcium isotopic anomalies and the lack of aluminum-26 in an unusual Allende inclusion. *Astrophys. J.* 228:L93–L98.
Lee, T., Mayeda, T. K., and Clayton, R. N. 1980. Oxygen isotopic anomalies in Allende inclusion HAL. *Geophys. Res. Lett.* 7:493–496.
Léger, A. 1983. Does CO condense on dust in molecular clouds? *Astron. Astrophys.* 123:271–278.
Léger, A., Klein, J., de Cheveigne, S., Guinet, C., Defourneau, D., and Belin, M. 1979. The 3.1 μm absorption in molecular clouds is probably due to amorphous H_2O ice. *Astron. Astrophys.* 79:256–259.
Léger, A., Gauthier, S., Defourneau, D., and Rouan, D. 1983. Properties of amorphous H_2O ice and origin of the 3.1 μm absorption. *Astron. Astrophys.* 117:164–169.
Leisawitz, D. 1985. A study of the association of molecular gas with young open clusters. In preparation.
Lequeux, J. 1977. Kinematics and dynamics of dense clouds. In *Star Formation,* eds. T. de Jong and A. Maeder (Dordrecht: D. Reidel), pp. 69–94.
Lequeux, J., Peimbert, M., Rayo, J. F., Serrano, A., and Torres-Peimbert, S. 1979. Chemical

composition and evolution of irregular and blue compact galaxies. *Astron. Astrophys.* 80:155–166.

Leslie, D. C. 1973. *Developments in the Theory of Turbulence* (Oxford: Clarendon).

Leu, M. T., Biondi, M. A., and Johnsen, R. 1973. Measurements of recombination of electrons with H^+_3 and H^+_5 ions. *Phys. Rev.* 8A:413–422.

Leung, C. M. 1975*a*. Radiation transport in dense interstellar dust clouds. I. Grain temperature. *Astrophys. J.* 199:340–358.

Leung, C. M. 1975*b*. Radiation Transport in Dense Interstellar Dust Clouds. Ph.D. thesis, Univ. of California, Berkeley.

Leung, C. M. 1978. Radiative transfer effects and the interpretation of interstellar molecular cloud observations. I. Basic physics of line formation. *Astrophys. J.* 225:427–441.

Leung, C. M., and Liszt, H. S. 1976. Radiation transport and non-LTE analysis of interstellar molecular lines. *Astrophys. J.* 208:732–746.

Leung, C. M., and Brown, R. L. 1977. On the interpretation of CO self-absorption profiles seen toward embedded stars in dense interstellar clouds. *Astrophys. J.* 214:L73–L78.

Leung, C. M., Kutner, M. L., and Mead, K. N. 1982. On the origin and structure of isolated dark globules. *Astrophys. J.* 262:583–589.

Leung, C. M., Herbst, E., and Huebner, W. F. 1984. Synthesis of complex molecules in dense interstellar clouds via gas-phase chemistry: A pseudo time-dependent calculation. *Astrophys. J. Suppl.* 56:231–256.

Levinson, F. H., and Brown, R. L. 1980. Analysis and interpretation of HI self-absorption lines. I. *Astrophys. J.* 242:416–423.

Levreault, R. M. 1983. A survey of mass loss among pre-main-sequence stars: Preliminary results. *Bull. Amer. Astron. Soc.* 15:679–680 (abstract).

Levy, E. H. 1978. Magnetic field in the primitive solar nebula. *Nature* 276:481.

Levy, E. H. 1982. Comments at Chondrules and Their Origins Conference, Houston, TX.

Levy, E. H., and Sonett, C. P. 1978. Meteorite magnetism and early solar-system magnetic fields. In *Protostars & Planets,* ed. T. Gehrels (Tucson: Univ. of Arizona Press), pp. 516–532.

Lewis, J. S. 1971. Satellites of the outer planets: Their physical and chemical nature. *Icarus* 15:174–185.

Lewis, J. S. 1972*a* Metal/silicate fractionation in the solar system. *Earth Planet. Sci. Lett.* 15:286–290.

Lewis, J. S. 1972*b*. Low-temperature condensation from the solar nebula. *Icarus* 16:241–252.

Lewis, J. S. 1974. The temperature gradient in the solar nebula. *Science* 186:440–443.

Lewis, J. S. 1976. Equilibrium and disequilibrium chemistry of adiabatic, solar-composition planetary atmospheres. In *Chemical Evolution of the Giant Planets,* ed. C. Ponnamperuma (New York: Academic Press), pp. 13–26.

Lewis, J. S., and Prinn, R. G. 1980. Kinetic inhibition of CO and N_2 reduction in the solar nebula. *Astrophys. J.* 238:357–364.

Lewis, R. S., and Anders, E. 1975. Condensation time of the solar nebula from extinct ^{129}I in primitive meteorites. *Proc. Nat. Acad. Sci. U.S.* 72:268–273.

Lewis, R. S., and Anders, E. 1983. Interstellar matter in meteorites. *Sci. Amer.* 249:66–77.

Lewis, R. S., Alaerts, L., Matsuda, J.-I., and Anders, E. 1979. Stellar condensates in meteorites: Isotopic evidence from noble gases. *Astrophys. J.* 234:L165–L168.

Lewis, R. S., Anders, E., Shimamura, T., and Lugmair, G. W. 1983*a*. Barium isotopes in Allende meteorite: Evidence against an extinct superheavy element. *Science* 222:1013–1015.

Lewis, R. S., Anders, E., Wright, I. P., Norris, S. J., and Phillinger, C. T. 1983*b*. Isotopically anomalous nitrogen in primitive meteorites. *Nature* 305:767–771.

Lighthill, M. J. 1955. The effect of compressibility on turbulence. In *Gas Dynamics of Cosmic Clouds* (Amsterdam: North Holland), p. 121–129.

Lightman, A. P., and Eardley, D. M. 1974. Black holes in binary systems: Instability of disk accretion. *Astrophys. J.* 187:L1–L3.

Lillie, C. F., and Witt, A. N. 1976. Ultraviolet photometry from the orbiting astronomical observatory. XXV. Diffuse galactic light in the 1500-4200 Å region and the scattering properties of interstellar dust grains. *Astrophys. J.* 208:64–74.

Lin, C. C., and Shu, F. H. 1964. On the spiral structure of disk galaxies. *Astrophys. J.* 140:646–655.
Lin, C. C., and Shu, F. H. 1968. Theory of spiral structure. In *Galactic Astronomy,* vol. 2, eds. H. Y. Chiu and A. Muriel (New York: Gordon & Breach), pp. 1–93.
Lin, C. C., and Lau, Y. Y. 1979. Density wave theory of spiral structure of galaxies. *Stud. Appl. Math.* 60:97–163.
Lin, C. C., and Bertin, G. 1984. Formation and maintenance of sprial structure in galaxies. In *The Milky Way Galaxy, IAU Symposium No. 106,* ed. H. van Woerden (Dordrecht: D. Reidel). In press.
Lin, D. N. C. 1981*a*. Convective accretion disk model for the primitive solar nebula. *Astrophys. J.* 246:972–984.
Lin, D. N. C. 1981*b*. On the origin of the Pluto-Charon system. *Mon. Not. Roy. Astron. Soc.* 197:1081–1085.
Lin, D. N. C. 1982. The nebular origin of the solar system. In *Rubey Colloquium, Vol. 3, The Solar System,* ed. M. Kivelson (Englewood, NJ: Prentice-Hall). In press.
Lin, D. N. C. 1984. The nebular origin of the solar system. Lick Observatory Contribution No. 426. Preprint.
Lin, D. N. C., and Pringle, J. E. P. 1976. Numerical simulation of mass transfer and accretion disc flow in binary systems. In *Structure and Evolution of Close Binary Systems, IAU Symposium No. 73,* eds. P. Eggleton, S. Mitton, and J. Whelan (Dordrecht: D. Reidel), pp. 237–252.
Lin, D. N. C., and Papaloizou, J. 1979*a*. Tidal torques on accretion disks in binary systems with extreme mass ratios. *Mon. Not. Roy. Astron. Soc.* 186:799–812.
Lin, D. N. C., and Papaloizou, J. 1979*b*. On the evolution of a circumbinary accretion disk and the tidal evolution of commensurable satellites. *Mon. Not. Roy. Astron. Soc.* 188:191–201.
Lin, D. N. C., and Papaloizou, J. 1980. On the structure and evolution of the primordial solar nebula. *Mon. Not. Roy. Astron. Soc.* 191:37–48.
Lin, D. N. C., and Bodenheimer, P. 1981. On the stability of Saturn's rings. *Astrophys. J.* 248:L83–L86.
Lin, D. N. C., and Bodenheimer, P. 1982. On the evolution of convective accretion disk models of the primordial solar nebula. *Astrophys. J.* 262:768–779.
Lindal, G. F., Wood, G. E., Levy, G. S., Anderson, J. D., Sweetnam, D. N., Hotz, H. B., Buckles, B. J., Holmes, D. P., Doms, P. E., Eshleman, V. R., Tyler, G. L., and Croft, T. A. 1981. The atmosphere of Jupiter: An analysis of the Voyager occultation measurements. *J. Geophys. Res.* 86:8721–8727.
Lindblad, P. O. 1967. 21-cm observations in the region of the galactic anti-centre. *Bull. Astron. Inst. Netherlands* 19:34–73.
Lindblom, L. 1979. On the secular instability from thermal conductivity in rotating stars. *Astrophys. J.* 233:974–980.
Lindblom, L. 1983. Necessary conditions for the stability of rotating Newtonian stellar models. *Astrophys. J.* 267:402–408.
Lindblom, L., and Detweiler, S. L. 1977. On the secular instabilities of the Maclaurin spheroids. *Astrophys. J.* 211:565–567.
Lindblom, L., and Hiscock, W. A. 1983. On the stability of rotating stellar models in general relativity theory. *Astrophys. J.* 267:384–401.
Linke, R. A., and Goldsmith, P. F. 1980. Observations of interstellar carbon monosulfide. Evidence for turbulent cores in giant molecular clouds. *Astrophys. J.* 235:437–451.
Linke, R. A., Frerking, M. A., and Thaddeus, P. 1979. Interstellar methyl mercaptan. *Astrophys. J.* 234:L139–L142.
Linke, R. A., Guélin, M., and Langer, W. D. 1983. Detection of $H^{15}NN^+$ and $HN^{15}N^+$ in interstellar clouds. *Astrophys. J.* 271:L85–L88.
Lissauer, J. J. 1984. Ballistic transport in Saturn's rings: An analytic theory. *Icarus* 57:63–71.
Lissauer, J. J. 1985. Can cometary bombardment disrupt synchronous rotation of planetary satellites? Submitted to *J. Geophys. Res.*
Lissauer, J. J., and Cuzzi, J. N. 1982. Resonances in Saturn's rings. *Astron. J.* 87:1051–1058.
Lissauer, J. J., Shu, F. H., and Cuzzi, J. N. 1981. Moonlets in Saturn's rings? *Nature* 292:707–711.

Lissauer, J. J., Peale, S. J., and Cuzzi, J. N. 1984. Ring torque on Janus and the melting of Enceladus. *Icarus* 58:159–168.

Liszt, H. S., and Leung, C. M. 1977. Radiative transport and non-LTE analysis of interstellar molecular lines. II. Carbon monosulfide. *Astrophys. J.* 218:396–405.

Liszt, H. S., Wilson, R. W., Penzias, A. A., Jefferts, K. B., Wannier, P. G., and Solomon, P. M. 1974. CO and CS in the Orion nebula. *Astrophys. J.* 190:557–564.

Liszt, H. S., Xiang, D., and Burton, W. B. 1981. Properties of the galactic molecular cloud ensemble from observations of ^{13}CO. *Astrophys. J.* 249:532–549.

Little, L. T., Macdonald, G. H., Riley, P. W., and Matheson, D. N. 1979*a*. Ammonia observations of the molecular cloud near S106. *Mon. Not. Roy. Astron. Soc.* 188:429–435.

Little, L. T., MacDonald, G. H., Riley, P. W., and Matheson, D. N. 1979*b*. The relative distribution of ammonia and cyanobutadiyne emission in Heiles' 2 dust cloud. *Mon. Not. Roy. Astron. Soc.* 189:539–550.

Little, L. T., Brown, A. T., MacDonald, G. H., Riley, P. W., and Matheson, D. N. 1980. Ammonia observations of the molecular clouds near S68, S140, OMC2 and S106. *Mon. Not. Roy. Astron. Soc.* 193:115–128.

Liu, L. 1982. Compression of ice VII to 500 kbar. *Earth Planet. Sci. Lett.* 61:359–364.

Lockman, F. J. 1979. The distribution of dense HII regions in the inner galaxy. *Astrophys. J.* 232:761–781.

Lockwood, G. W., Lutz, B. L., Thompson, D. T., and Warnock, A. III. 1983. The albedo of Uranus. *Astrophys. J.* 266:402–414.

Loren, R. B. 1976. Colliding clouds and star formation in NGC 1333. *Astrophys. J.* 209:466–488.

Loren, R. B. 1977*a*. The Monoceros R2 cloud: Near infrared and molecular observations of a rotating collapsing cloud. *Astrophys. J.* 215:129–150.

Loren, R. B. 1977*b*. The star-formation process in molecular clouds associated with Herbig Be/Ae stars. I. LkHα 198, BD + 40°4124, and NGC 7129. *Astrophys. J.* 218:716–735.

Loren, R. B. 1981. The densities of the molecular clouds associated with Herbig Ae/Be and other young stars. *Astron. J.* 86:69–83.

Loren, R. B., Plambeck, R. L., Davis, J. H., and Snell, R. L. 1981. High resolution J=2-1 and J=1-0 CO self-reversed line profiles toward molecular clouds. *Astrophys. J.* 245:495–511.

Loren, R. B., Sandqvist, A. A., and Wootten, H. A. 1983. Molecular clouds on the threshold of star formation: The radial density profile of the cores of the ρ Oph and R CrA clouds. *Astrophys. J.* 270:620–640.

Loren, R. B., Wootten, A., and Mundy, L. G. 1984. The detection of interstellar methyldiacetylene. *Astrophys. J.* 286:L23–L26.

Lovas, F. J., Snyder, L. E., and Johnson, D. R. 1979. Recommended rest frequencies for observed interstellar molecular transitions. *Astrophys. J. Suppl.* 41:451–480.

Low, F. J., Beintema, D. A., Gautier, T. N., Gillett, F. C., Beichman, C. A., Neugebauer, G., Young, E., Aumann, H. H., Boggess, N., Emerson, J. P., Habing, H. J., Hauser, M. G., Houck, J. R., Rowan-Robinson, M., Soifer, B. T., Walker, R. G., Wesselius, P. R. 1984. Infrared cirrus: New components of the extended IR emission. *Astrophys. J.* 278:L19–L22.

Lowe, C., and Lynden-Bell, D. 1976. The minimum Jeans mass or when fragmentation must stop. *Mon. Not. Roy. Astron. Soc.* 176:367–390.

Lucke, P. B. 1978. The distribution of color excesses and reddening material in the solar neighborhood. *Astron. Astrophys.* 64:367–377.

Lucy, L. 1977. A numerical approach to the testing of the fission hypothesis. *Astron. J.* 82:1013–1024.

Lucy, L. 1981. The formation of binary stars. In *Fundamental Problems in the Theory of Stellar Evolution,* eds. D. Sugimoto, D. Q. Lamb, and D. N. Schramm (Dordrecht: D. Reidel), pp. 75–83.

Lugmair, G. W., and Marti, K. 1977. Sm-Nd-Pu timepieces in the Angra Dos Reis meteorite. *Earth Planet. Sci. Lett.* 35:273–284.

Lugmair, G. W., Shimamura, T., Lewis, R. S., and Anders, E. 1983. Samarium-146 in the early solar system: Evidence from neodymium in the Allende meteorite. *Science* 222:1015–1018.

Lukkari, J. 1981. Collisional amplification of density fluctuations in Saturn's rings. *Nature* 292:433–435.

Lumpkin, G. R. 1981. Electron microscopy of carbonaceous matter in Allende acid residues. *Proc. Lunar Planet. Sci. Conf.* 12:1153–1166.

Lundmark, K. 1926. The distribution of the dark nebulae. *Uppsala Obs. Medd.* No. 12 (in Swedish with English abstract).

Lunine, J. I., and Stevenson, D. J. 1982. Formation of the Galilean satellites in a gaseous nebula. *Icarus* 52:14–39.

Lunine, J. I., Stevenson, D. J., and Yung, Y. L. 1983. Ethane ocean on Titan. *Science* 222:1229–1230.

Lüst, R. 1952. Die entwicklung einer um einen zeutralkorper rotierenden gasmasse. I. Loesungen de hydrodynamischen gleichungen mit turbulenter reibung. *Z. Naturforschung* 7a:87–98.

Lutz, B. L., Owen, T., and Cess, R. D. 1976. Laboratory band strengths of methane and their application to the atmospheres of Jupiter, Saturn, Neptune, and Titan. *Astrophys. J.* 203:541–551.

Lynden-Bell, D. 1966. The role of magnetism in spiral structure. *Observatory* 86:57–60.

Lynden-Bell, D. 1969. Galactic nuclei as collapsed old quasars. *Nature* 223:690–694.

Lynden-Bell, D., and Pringle, J. E. 1974. The evolution of viscous disks and the origin of nebular variables. *Mon. Not. Roy. Astron. Soc.* 168:603–637.

Lynds, B. T. 1962. Catalogue of dark nebulae. *Astrophys. J. Suppl.* 7:1–52.

Lynds, B. T. 1967. The surface brightness of dark nebulae. In *Modern Astrophysics: A Memorial to Otto Struve,* ed. M. Hack (Paris: Gauthier-Villars), pp. 67–81.

Lynds, B. T. 1968. Dark nebulae. In *Nebulae and Interstellar Matter,* eds. B. M. Middlehurst and L. H. Aller (Chicago: Univ. of Chicago Press), pp. 119–140.

Lyttleton, R. A. 1948. On the origin of comets. *Mon. Not. Roy. Astron. Soc.* 108:465–475.

Lyttleton, R. A. 1953. *The Stability of Rotating Liquid Masses* (Cambridge: Cambridge Univ. Press).

Lyttleton, R. A. 1972. On the formation of the planets from the solar nebula. *Mon. Not. Roy. Astron. Soc.* 158:463–483.

Macdougall, J. D., and Kothari, B. 1976. Formation chronology for C2 meteorites. *Earth Planet. Sci. Lett.* 33:36–44.

Macdougall, J. D., Lugmair, G. W., and Kerridge, J. F. 1984. Early solar system aqueous activity: Sr isotope evidence from the Orgueil CI meteorite. *Nature* 307:249–251.

MacFarlane, J. J., and Hubbard, W. B. 1982. Internal structure of Uranus. In *Uranus and the Outer Planets,* ed. G. Hunt (Cambridge: Cambridge Univ. Press), pp. 111–124.

MacFarlane, J. J., and Hubbard, W. B. 1983. Statistical mechanics of light elements at high pressure. V. Three-dimensional Thomas-Fermi-Dirac theory. *Astrophys. J.* 272:301–310.

MacGregor, K. B. 1982. Future prospects for the theory of solar-stellar winds. *Bull. Amer. Astron. Soc.* 14:946–947 (abstract).

Macklin, R. L. 1983. Neutron capture cross sections and resonances of Iodine-127 and Iodine-129. *Nucl. Sci. Engin.* 85:350–361.

Macklin, R. L. 1984. Neutron capture measurements on fission-product Palladium-107. Submitted to *Nucl. Sci. Engin.*

MacLean, S., Duley, W. W., and Millar, T. J. 1982. A laboratory simulation of the interstellar 220 nanometer feature. *Astrophys. J.* 256:L61–L64.

MacLeod, J. M., Avery, L. W., and Broten, N. W. 1981. Detection of deuterated cyanodiacetylene (DC_5N) in Taurus molecular cloud 1. *Astrophys. J.* 251:L33–L37.

MacLeod, J. M., Avery, L. W., and Broten, N. W. 1984. The detection of interstellar methyldiacetylene (CH_3C_4H). *Astrophys. J.* 282:L89–L92.

Macy, W. Jr., and Smith, W. M. 1978. Detection of HD on Saturn and Uranus, and the D/H ratio. *Astrophys. J.* 222:L73–L75.

Mahoney, W. A., Ling, J. C., Jacobson, A. S., and Lingenfelter, R. E. 1982. Diffuse galactic gamma-ray line emission from nucleosynthetic ^{60}Fe, ^{26}Al, and ^{22}Na: Preliminary limits from HEAO 3. *Astrophys. J.* 262:742–748.

Mahoney, W. A., Ling, J. C., Wheaton, W. A., and Jacobsen, A. S. 1984. HEAO-3 discovery of ^{26}Al in the interstellar medium. *Astrophys. J.* 286:578–585.

Makalkin, A. B. 1980. Possibility of formation of an originally inhomogeneous Earth. *Phys. Earth Planet. Inter.* 22:302–312.

Mann, A. P. C., and Williams, D. A. 1980. A list of interstellar molecules. *Nature* 283:721–725.

Manuel, O. K., Hennecke, E. W., and Sabu, D. D. 1972. Xenon in carbonaceous chondrites. *Nature* 240:99–101.

Marcus, P. S., Press, W. H., and Teukolsky, S. A. 1977. Stablest shapes for an axisymmetric body of gravitating, incompressible fluid. *Astrophys. J.* 214:584–597.

Marouf, E. A., Tyler, G. L., Zebker, H. A., and Eshleman, V. R. 1983. Particle size distribution in Saturn's rings from Voyager 1 radio occultation. *Icarus* 54:189–211.

Marsden, B. G., and Roemer, E. 1982. Basic information and references. In *Comets,* ed. L. L. Wilkening (Tucson: Univ. of Arizona Press), pp. 707–733.

Marsden, B. G., Sekanina, Z., and Everhart, E. 1978. New osculating orbits for 110 comets and analysis of original orbits for 200 comets. *Astron. J.* 83:64–71.

Marten, A., Courtin, R., Gautier, D., and Lacombe, A. 1980. Ammonia vertical density profiles in Jupiter and Saturn from their radioelectric and infrared emissivities. *Icarus* 41:410–422.

Martin, H. M., Sanders, D. B., and Hills, R. E. 1984. CO emission from fragmentary molecular clouds: A model applied to observations of M17SW. *Mon. Not. Roy. Astron. Soc.* 208:35–55.

Martin, R. N., and Barrett, A. H. 1975. CS in dense clouds and globules. *Astrophys. J.* 202:L83–L86.

Martin, R. N., and Barrett, A. H. 1978. Microwave spectral lines in galactic dust globules. *Astrophys. J. Suppl.* 36:1–51.

Mason, B., and Taylor, S. R. 1982. Inclusions in the Allende meteorite. *Smithsonian Contributions to Earth Sciences 25* (Washington, D.C.: Smithsonian Institution Press).

Mathewson, D. S., van der Kruit, P. C., and Brouw, W. N. 1972. A high resolution radio continuum survey of M51 and NGC 5195 at 1415 MHz. *Astron. Astrophys.* 17:468–486.

Mathieu, R. D. 1983. Dynamical constraints on star formation efficiency. *Astrophys. J.* 267:L97–L101.

Mathis, J. S. 1979. The size distribution of interstellar particles. II. Polarization. *Astrophys. J.* 232:747–753.

Mathis, J. S., and Wallenhorst, S. G. 1981. The size distribution of interstellar particles. III. Peculiar extinctions and normal infrared extinction. *Astrophys. J.* 244:483–492.

Mathis, J. S., Rumpl, W., and Nordsieck, K. H. 1977. The size distribution of interstellar grains. *Astrophys. J.* 217:425–433.

Mathis, J. S., Mezger, P. G., and Panagia, N. 1983. Interstellar radiation field and dust temperatures in the diffuse interstellar matter and in giant molecular clouds. *Astron. Astrophys.* 128:212–229.

Matsakis, D. N., Brandshaft, D., Chui, M. F., Cheung, A. C., Ynguesson, K. S., Cardiasmenos, A. G., Shanley, J. F., and Ho, P. T. P. 1977. Anomalous ammonia absorption in DR 21. *Astrophys. J.* 214:L67–L71.

Matsakis, D. N., Hjalmarson, Å., Palmer, P., Cheung, A. C., and Townes, C. H. 1981. VLA observations of DR21 $NH_3(1,1)$ absorption: Direct evidence for clumping. *Astrophys. J.* 250:L85–L89.

Matthews, H. E., and Sears, T. J. 1983*a*. Detection of the J=1-0 transition of CH_3CN. *Astrophys. J.* 267:L53–L57.

Matthews, H. E., and Sears, T. J. 1983*b*. The detection of vinyl cyanide in TMC-1. *Astrophys. J.* 272:149–153.

Matthews, H. E., Friberg, P., and Irvine, W. M. 1984*a*. The detection of acetaldehyde in cold dust clouds. *Astrophys. J.* In press.

Matthews, H. E., Irvine, W. M., Friberg, P., Brown, R. D., and Godfrey, P. D. 1984*b*. A new interstellar molecule: Tricarbon monoxide. *Nature* 310:125–128.

Matthews, N., and Little, L. T. 1983. Ammonia observation of the Herbig-Haro objects HH24-27. *Mon. Not. Roy. Astron. Soc.* 205:123–130.

Mattila, K. 1970. Interpretation of the surface brightness of dark nebulae. *Astron. Astrophys.* 9:53–63.

Mattila, K., Winnberg, A., and Grasshof, M. 1979. OH observations of the dark nebula Lynds 134. *Astron. Astrophys.* 78:275–286.

Mavrokoukoulakis, N. D., Ho, K. L., Cole, R. S. 1978. Temporal spectra of atmospheric amplitude scintillations at 110 GHz and 36 GHz. *IEEE Trans. Antennas and Propagation* AP26:875–877.

May, R. M. 1976. Simple mathematical models with very complicated dynamics. *Nature* 261: 459–467.

Mayer-Hasselwander, H. A., Bennett, K., Bignami, G. F., Buccheri, R., Carvero, P. A., Hermsen, W., Kanbach, G., Lebrun, F., Lichti, G. G., Masnou, J. L., Paul, J. A., Pinkau, K., Sacco, B., Scansi, L., Swanenberg, B. N., and Wills, R. D. 1984. Large-scale distribution of galactic gamma radiation observed by COS-B. *Astron. Astrophys.* 105:164–175.

Mazurek, T. J. 1980. H II bubbles and disruption of molecular clouds. *Astron. Astrophys.* 90:65–69.

McCabe, E. M., Smith, R. C., and Clegg, R. E. S. 1979. Molecular abundances in IRC+10216. *Nature* 281:263–266.

McCarthy, D. W. 1982. Triple structure of infrared source 3 in the Monoceros R2 molecular cloud. *Astrophys. J.* 257:L93–L98.

McCord, T. B., and Gaffey, M. T. 1974. Asteroid: Surface composition from reflection spectroscopy. *Science* 186:352–355.

McCrea, W. H. 1955. Colloquium on the formation of stars. *Observatory* 75:206–211.

McCrea, W. H. 1957. The formation of population I stars. I. *Mon. Not. Roy. Astron. Soc.* 117:562–578.

McCrea, W. H. 1975. Solar system as space probe. *Observatory* 95:239–255.

McCrea, W. H. 1978. The formation of the solar system: A protoplanet theory. In *The Origin of the Solar System,* ed. S. F. Dermott (New York: J. Wiley & Sons), pp. 75–110.

McCulloch, M. T., and Wasserburg, G. J. 1978. Ba and Nd isotopic anomalies in the Allende meteorite. *Astrophys. J.* 220:L15–L19.

McCuskey, S. W. 1938. The galactic structure in Taurus. I. Surface distribution of stars. *Astrophys. J.* 88:209–227.

McCutcheon, W. H., Shuter, W. L. H., and Booth, W. S. 1978. Observations of the 21-cm line in dark clouds. *Mon. Not. Roy. Astron. Soc.* 185:755–769.

McCutcheon, W. H., Dickman, R. L., Shuter, W. L. H., and Roger, R. S. 1980. The $^{13}CO/C^{18}O$ ratio in interstellar dark clouds: Evidence for isotopic fractionation. *Astrophys. J.* 237:9–18.

McCutcheon, W. H., Roger, R. S., and Dickman, R. L. 1982. The molecular cloud complex in the vicinity of IC 5146. *Astrophys. J.* 256:139–150.

McDavid, D. 1984. The role of the galactic magnetic field in the evolution of a dark globular filament in Cygnus. *Astrophys. J.* 284:141–143.

McDonnell, J. A. M., ed. 1978. *Cosmic Dust* (New York: J. Wiley).

McGee, R. X., and Murray, 1961. A sky survey of neutral hydrogen at λ21 cm. I. The general distribution and motions of the local gas. *Aust. J. Phys.* 14:260–278.

McGee, R. X., and Milton, J. A. 1964. A sky survey of neutral hydrogen at 21 cm. III. Gas at higher radial velocities. *Aust. J. Phys.* 17:125–157.

McGee, R. X., and Milton, J. A. 1966. 21 cm hydrogen-line survey of the Large Magellanic Cloud. II. Distribution and motions of neutral hydrogen. *Aust. J. Phys.* 19:343–374.

McGee, R. X., Murray, J. D., and Milton, J. A. 1963. A sky survey of neutral hydrogen at λ21 cm. II. The detailed distribution of low-velocity gas. *Aust. J. Phys.* 16:136–170.

McGlynn, T. A. 1984. Dissipationless collapse of galaxies and initial conditions. *Astrophys. J.* 281:13–30.

McGowan, J. W., Mul, P. M., D'Angelo, V. S., Mitchell, J. B. A., Defrance, P., and Froelich, H. R. 1979. Energy dependence of dissociative recombination below 0.08 eV. *Phys. Rev. Lett.* 42:373–375.

McGregor, P. J., Persson, S. E., and Cohen, J. G. 1984. Spectrophotometry of compact embedded infrared sources in the 0.6–1.0 micron wavelength region. Preprint.

McKee, C. F., and Ostriker, J. P. 1977. A theory of the interstellar medium: Three components regulated by supernova explosions in an inhomogeneous substrate. *Astrophys. J.* 218:148–169.

McKellar, A. 1940. Evidence for the molecular origin of some hitherto unidentified interstellar lines. *Publ. Astron. Soc. Pacific* 52:187–192.

McKinnon, W. B. 1981. Reorientation of Ganymede and Callisto by impact and interpretation of the cratering record. *EOS* 62:318 (abstract).

McMillan, R. S., Smith, P. H., Frecker, J. E., Merline, W. J., and Perry, M. L. 1984. The LPL radial accelerometer. In *Stellar Radial Velocities, IAU Colloquium No. 88* (Schenectady: L. Davis Press). In press.

McMillan, S. L. W., Flannery, B. P., and Press, W. H. 1980. Nonlinear hydrodynamics of acoustic instabilities in diffuse clouds. *Astrophys. J.* 240:488–498.

McNaughton, N. J., Borthwick, J., Fallick, A. E., and Pillinger, C. T. 1981. Deuterium/hydrogen ratios in unequilibrated ordinary chondrites. *Nature* 294:639–641.

McSween, H. Y. 1979. Are carbonaceous chondrites primitive or processed? A review. *Rev. Geophys. Space Phys.* 17:1059–1078.

McSween, H. Y., and Richardson, S. M. 1977. The composition of carbonaceous chondrite matrix. *Geochim. Cosmochim. Acta* 41:1145–1161.

Meaburn, J. 1980. The giant and supergiant shells of the Magellanic Clouds. *Mon. Not. Roy. Astron. Soc.* 192:365–375.

Meeker, G. P., Wasserburg, G. J., and Armstrong, J. T. 1983. Replacement textures in CAI and implications regarding planetary metamorphism. *Geochim. Cosmochim. Acta* 47:707–721.

Menon, T. K. 1958. Interstellar structure of the Orion region. I. *Astrophys. J.* 127:28–47.

Menten, K., and Walmsley, C. 1984. Ammonia observations of L1551. Preprint.

Menton, K. M., Walmsley, C. M., Krugel, E., and Ungerechts, H. 1984. Ammonia in B335. *Astron. Astrophys.* 137:108–112.

Mercer-Smith, J. A., Cameron, A. G. W., and Epstein, R. I. 1984. On the formation of stars from disk accretion. *Astrophys. J.* 279:363–366.

Mestel, L. 1963. On the galactic law of rotation. *Mon. Not. Roy. Astron. Soc.* 126:553–575.

Mestel, L. 1965*a*. Problems of star formation. I. *Quart. J. Roy. Astron. Soc.* 6:161–198.

Mestel, L. 1965*b*. Problems of star formation. II. *Quart. J. Roy. Astron. Soc.* 6:265–298.

Mestel, L. 1966*a*. The magnetic field of contracting gas cloud. I. *Mon. Not. Roy. Astron. Soc.* 133:265–284.

Mestel, L. 1966*b*. A note on the spin of sub-condensations forming in a differentially rotating medium. *Mon. Not. Roy. Astron. Soc.* 131:307–310.

Mestel, L. 1969. The role of the magnetic field in star formation. In *Plasma Instabilities in Astrophysics,* eds. D. A. Tidman and D. G. Wentzel (New York: Gordon & Breach), pp. 329–352.

Mestel, L. 1977. Theoretical processes in star formation. In *Star Formation,* eds. T. de Jong and A. Maeder (Dordrecht: D. Reidel), pp. 213–232.

Mestel, L., and Spitzer, L. Jr. 1956. Star formation in magnetic dust clouds. *Mon. Not. Roy. Astron. Soc.* 116:505–514.

Mestel, L., and Strittmatter, P. A. 1967. The magnetic field of a contracting gas cloud. II. *Mon. Not. Roy. Astron. Soc.* 137:95–105.

Mestel, L., and Paris, R. B. 1979. Magnetic braking during star formation. III. *Mon. Not. Roy. Astron. Soc.* 187:337–356.

Mestel, L., and Paris, R. B. 1984. Star formation and the galactic magnetic field. *Astron. Astrophys.* 136:98–120.

Mestel, L., and Ray, T. P. 1985. Disk-like magneto-gravitational equilibria. *Mon. Not. Roy. Astron. Soc.* 212:275–300.

Meusinger, H. 1983. On the past star formation rate in the solar neighborhood. *Astron. Nachr.* 304:285–298.

Mewaldt, R. A., Spalding, J. D., and Stone, E. C. 1984. A high-resolution study of the isotopes of solar flare nuclei. *Astrophys. J.* 280:892–901.

Meyer, F., and Meyer-Hofmeister, E. 1981. On the elusive case of cataclysmic variable outbursts. *Astron. Astrophys.* 104:L10–L12.

Mézáros, P. 1968. Dust and atomic hydrogen near rho Ophiuchi. *Astrophys. Space Sci.* 2:510–519.

Mezger, P. G., Mathis, J. S., and Panagia, N. 1982. The origin of the diffuse galactic far infrared and sub-millimeter emission. *Astron. Astrophys.* 105:372–388.

Michel, F. C. 1984. Hydraulic jumps in "viscous" accretion disks. *Astrophys. J.* 279:807–813.

Mignard, F. 1980. The evolution of the lunar orbit revised. II. *Moon and Planets* 23:185–201.

Mignard, F. 1981. The evolution of the lunar orbit revised. III. *Moon and Planets* 24:189–207.

Miles, B., and Ramsey, W. H. 1952. On the internal structure of Jupiter and Saturn. *Mon. Not. Roy. Astron. Soc.* 112:234–243.

Millar, T. J. 1982. Dense cloud chemistry. *Astrophys. Space Sci.* 87:435–453.

Millar, T. J., and Freeman, A. 1984*a*. Chemical modeling of molecular sources. I. TMC-1. *Mon. Not. Roy. Astron. Soc.* 207:405–424.

Millar, T. J., and Freeman, A. 1984*b*. Chemical modeling of molecular sources. II. L183. *Mon. Not. Roy. Astron. Soc.* 207:425–432.
Miller, B. D. 1974. The effect of gravitational radiation-reaction on the evolution of the Riemann S-type ellipsoids. *Astrophys. J.* 187:609–620.
Miller, G. E., and Scalo, J. M. 1978. On the birthplaces of stars. *Publ. Astron. Soc. Pacific* 90:506–513.
Miller, G. E., and Scalo, J. M. 1979. The initial mass function and stellar birthrate in the solar neighborhood. *Astrophys. J. Suppl.* 41:513–547.
Miller, S. L. 1961. The occurrence of gas hydrates in the solar system. *Proc. Nat. Acad. Sci. U.S.* 47:1798–1808.
Milman, A. S. 1977. Carbon monoxide observations of a rotating dust globule. *Astrophys. J.* 211:128–134.
Milman, A. S., Knapp, G. R., Kerr, F. J., Knapp, S. L., and Wilson, W. J. 1975. Carbon monoxide observations of a dust cloud in the Orion region: L1630. *Astron. J.* 80:93–110.
Mineur, H. 1939. Equilibré des nuages galactiques et des amas ouverts dans la voie lactée évolution des amas. *Ann. d'Astrophys.* 2:1–244.
Mitchell, A. S., and Nellis, W. J. 1982. Equation of state and electrical conductivity of water and ammonia shocked to 100 GPa (10 Mbar) pressure range. *J. Chem. Phys.* 76:6273–6281.
Mitchell, G. F. 1983. The synthesis of hydrocarbon molecules in a shocked interstellar cloud. *Mon. Not. Roy. Astron. Soc.* 205:765–772.
Mitchell, G. F. 1984. Effects of shocks on the molecular composition of a dense interstellar cloud. *Astrophys. J. Suppl.* 54:81–101.
Mitchell, G. F., and Deveau, T. J. 1983. Effects of a shock on the molecular composition of a diffuse interstellar cloud. *Astrophys. J.* 266:646–661.
Mitchell, G. F., Huntress, W. T. Jr., and Prasad, S. S. 1979. Interstellar synthesis of the cyanopolyynes and related molecules. *Astrophys. J.* 233:102–108.
Mitchell, G. F., Prasad, S. S., and Huntress, W. T. 1981. Chemical model calculations of C_2, C_3, CH, CN, OH, and NH_2 abundances in cometary comae. *Astrophys. J.* 244:1087–1093. 1093.
Mitchell, J. B. A., Forand, J. L., Ng, C. T., Levac, D. P., Mitchell, R. E., Mul, P. M., Claeys, W., Sen, A., and McGowan, J. W. 1983. Measurement of the branching ratio for dissociative recombination of $H^+{}_3$ + e. *Phys. Rev. Lett.* 51:885–888.
Miyama, S. M., Hayashi, C., and Narita, S. 1984. Criteria for collapse and fragmentation of rotating interstellar clouds. *Astrophys. J.* 279:621–652.
Mizuno, H. 1980. Formation of the giant planets. *Prog. Theor. Phys.* 64:544–557.
Mizuno, H., and Wetherill, G. W. 1984. Grain abundance in the primordial atmosphere of the Earth. *Icarus* 59:74–86.
Mizuno, H., Nakazawa, K., and Hayashi, C. 1978. Instability of gaseous envelope surrounding planetary core and formation of giant planets. *Prog. Theor. Phys.* 60:699–710.
Molini-Velsko, C. A. 1983. Isotopic Composition of Silicon in Meteorites. Ph.D. thesis, Univ. of Chicago.
Molini-Velsko, C. A., Clayton, R. N., and Mayeda, T. K. 1982. Silicon isotopes: Experimental vapor fractionation and tektites. *Meteoritics* 17:255–256 (abstract).
Molini-Velsko, C. A., Mayeda, T. K., and Clayton, R. N. 1983. Silicon isotopes in components of the Allende meteorite. *Lunar Planet. Sci.* XIV:509–510 (abstract).
Monetti, A., Pipher, J. L., Helfer, H. L., McMillan, R. S., and Perry, M. L. 1984. Magnetic field structure in the Taurus dark cloud. *Astrophys. J.* 282:508–515.
Monin, A. S., and Yaglom, A. M. 1975. *Statistical Fluid Mechanics* (Cambridge: MIT Press), 2 vols.
Moniot, R. K. 1980. Noble-gas-rich separates from ordinary chondrites. *Geochim. Cosmochim. Acta* 44:253–271.
Montmerle, T., Koch-Miramond, L., Falgarone, E., and Grindlay, J. E. 1983. Einstein observations of the Rho Ophiuchi dark cloud: An X-ray Christmas tree. *Astrophys. J.* 269:182–201.
Moore, J. H., and Menzel, D. H. 1930. The rotation of Uranus. *Publ. Astron. Soc. Pacific* 42:330–335.
Moorwood, A. F. M., and Salinari, P. 1984. Infrared objects near to H_2O masers in regions of active star formation. III. Evolution phases deduced from IR recombination line and other data. *Astron. Astrophys.* In press.

Moran, J. M. 1983. High resolution observations of star formation. Colloquium presented at the University of Massachusetts, Oct. 1983.

Moran, J., Reid, M., Lada, C., Yen, J., Johnston, K., and Spencer, J. 1978. Evidence for the Zeeman effect in the OH maser emission from W3(OH). *Astrophys. J.* 224:L67–L72.

Moran, J. M., Garay, G., Reid, M. J., Genzel, R., Wright, M. C. H., and Plambeck, R. L. 1983. Detection of radio emission from the Becklin-Neugebauer object. *Astrophys. J.* 271:L31–L34.

Morfill, G. E. 1983*a*. Some cosmochemical consequences of a turbulent protoplanetary cloud. *Icarus* 53:41–54.

Morfill, G. E. 1983*b*. Physics and chemistry in the primitive solar nebula. Summary of Les Houches Summer School course, MPE Preprint 21. Also Preprint 21 at MPI für Physik und Astrophys.

Morfill, G. E. 1985. Physics and chemistry in the primitive solar nebula. In *The Birth and Infancy of Stars,* ed. R. Lucas, A. Omont and L. R. Stora. (Amsterdam: North-Holland). In press.

Morfill, G. E., and Völk, H. J. 1984. Transport of dust and vapor and chemical fractionation in the early protosolar cloud. *Astrophys. J.* 287:371–395.

Morfill, G. E., Röser, S., Tscharnuter, W., and Völk, H. 1979. The dynamics of dust in a collapsing protostellar cloud and its possible role in planet formation. *Moon and Planets* 19:211–220.

Morgan, D. H., Nandy, K., and Thompson, G. I. 1978. Ultraviolet observations of the diffuse galactic light from the S2/68 sky-survey telescope. *Mon. Not. Roy. Astron. Soc.* 185:371–380.

Morgan, J., Wolff, S., Strom, S. E., and Strom, K. M. 1984. Narrowband imaging and velocity maps of young stellar objects: Initial results. *Astrophys. J.* 285:L71–L73.

Morkovin, M. V. 1972. Effects of compressibility on turbulent flows. In *Mécanique de la Turbulence, No. 108, Colloques Internatoux du Centre National de la Récherch Scientifique,* Paris.

Morris, M., Snell R. L., and Vanden Bout, P. 1977. Emission from highly excited rotational states of HC_3N in dense clouds. *Astrophys. J.* 216:738–746.

Morrison, D., ed. 1982. Introduction to the satellites of Jupiter. In *Satellites of Jupiter* (Tucson: Univ. of Arizona Press), pp. 3–43.

Morrison, D., Cruikshank, D. P., and Burns, J. A. 1977. Introducing the satellites. In *Planetary Satellites,* ed. J. A. Burns (Tucson: Univ. of Arizona Press), pp. 3–17.

Moss, D., and Smith, R. C. 1981. Stellar rotation and magnetic stars. *Rep. Prog. Phys.* 44:831–891.

Mould, J. R., and Wallis, R. E. 1977. Band strengths of M stars in the Orion population. *Mon. Not. Roy. Astron. Soc.* 181:625–635.

Mouschovias, T. Ch. 1976*a*. Nonhomologous contraction and equilibrium of self-gravitating, magnetic interstellar clouds embedded in an intercloud medium. Star formation. I. *Astrophys. J.* 206:753–767.

Mouschovias, T. Ch. 1976*b*. Nonhomologous contraction and equilibrium of self-gravitating, magnetic interstellar clouds embedded in an intercloud medium. Star formation. II. *Astrophys. J.* 207:141–158.

Mouschovias, T. Ch. 1977. A connection between the rate of rotation of interstellar clouds, magnetic fields, ambipolar diffusion, and the periods of binary stars. *Astrophys. J.* 211:147–151.

Mouschovias, T. Ch. 1978. Formation of stars and planetary systems in magnetic interstellar clouds. In *Protostars & Planets,* ed. T. Gehrels (Tucson: Univ. of Arizona Press), pp. 209–242.

Mouschovias, T. Ch. 1981. The role of magnetic fields in the formation of stars. In *Fundamental Problems in the Theory of Stellar Evolution, IAU Symposium No. 93,* eds. D. Sugimoto, D. Q. Lamb, and D. N. Schramm (Dordrecht: D. Reidel), pp. 27–62.

Mouschovias, T. Ch., and Spitzer, L. 1976. Note on the collapse of magnetic interstellar clouds. *Astrophys. J.* 210:326–327.

Mouschovias, T. Ch., and Paleologou, E. V. 1979. The angular momentum problem and magnetic braking: An exact, time-dependent solution. *Astrophys. J.* 230:204–222.

Mouschovias, T. Ch., and Paleologou, E. V. 1980. The angular momentum problem and magnetic breaking during star formation: Exact solutions for an aligned and perpendicular rotator. *Moon and Planets* 22:31–45.

Mouschovias, T. Ch., and Paleologou, E. V. 1981. Ambipolar diffusion in interstellar clouds, time-dependent solutions in one spatial dimension. *Astrophys. J.* 246:48–64.

Mouschovias, T. Ch., Shu, F. H., and Woodward, P. R. 1974. On the formation of interstellar cloud complexes, OB associations and giant HII regions. *Astron. Astrophys.* 33:73–77.

Muehlenbachs, K., and Kushiro, I. 1974. Oxygen isotope exchange and equilibrium of silicates with CO_2 or O_2. *Carnegie Inst. Of Washington Year Book* 73:232–236.

Mullan, D. J., and Steinolfson, R. S. 1983. Closed and open magnetic fields in stellar winds. *Astrophys. J.* 266:823–830.

Münch, G. 1958. Internal motions in the Orion nebula. *Rev. Mod. Phys.* 30:1035–1041.

Mundt, R. 1983. Jets from young stars. *Rev. Mexicana Astron. Astrof.* 7:234 (abstract).

Mundt, R. 1984. Mass loss in T Tauri stars: Observational studies of the cool parts of their stellar winds and expanding shells. *Astrophys. J.* 280:749–770.

Mundt, R., and Giampapa, M. S. 1982. Observations of rapid line profile variability in the spectra of T Tauri stars. *Astrophys. J.* 256:156–167.

Mundt, R., and Fried, J. W. 1983. Jets from young stars. *Astrophys. J.* 274:L83–L86.

Mundt, R., and Hartmann, L. 1983. HH1 and HH2: The results of an eruptive event in the Cohen-Schwartz star? *Astrophys. J.* 268:766–777.

Mundt, R., Stocke, J., and Stockman, H. S. 1983*a*. Jets from pre-main-sequence stars: AS 353A and its associated Herbig-Haro objects. *Astrophys. J.* 265:L71–L75.

Mundt, R., Walter, F. M., Feigelson, E. D., Finkenzeller, V., Herbig, G. H., and Odell, A. P. 1983*b*. Observations of suspected low mass post-T Tauri stars and their evolutionary status. *Astrophys. J.* 269:229–238.

Mundt, R., Bührke, T., Fried, J. W., Neckel, T., Sarcander, M., and Stocke, J. 1984. Jets from young stars. III. The case of Haro 6-5 B, HH 33/40, HH 19, and 1548C27. *Astron. Astrophys.* In press.

Mundy, L. 1984. The Density and Molecular Column Density Structure of Three Molecular Cloud Cores. Ph.D. Thesis, Univ. of Texas, Austin.

Myers, P. C. 1980. Asymmetric ^{13}CO lines in dark clouds: Evidence for contraction. *Astrophys. J.* 242:1013–1018.

Myers, P. C. 1982. Low-mass star formation in the dense interior of Barnard 18. *Astrophys. J.* 257:620–632.

Myers, P. C. 1983. Dense cores in dark clouds. III. Subsonic turbulence. *Astrophys. J.* 270:105–118.

Myers, P. C., and Buxton, R. B. 1980. Observations of H_2CO in the Orion nebula at 1 centimeter wavelength. *Astrophys. J.* 239:515–518.

Myers, P. C., and Benson, P. J. 1983. Dense cores in dark clouds. II. NH_3 observations and star formation. *Astrophys. J.* 266:309–320.

Myers, P. C., Ho, P. T. P., Schneps, M. H., Chin, G., Pankonia, V., and Winfoerg, A. 1978. Atomic and molecular observations of the ρ Ophiuchi dark cloud. *Astrophys. J.* 220:864–882.

Myers, P. C., Linke, R. A., and Benson, P. J. 1983. Dense cores in dark clouds. I. CO observations and column densities of high extinction regions. *Astrophys. J.* 264:517–537.

Nachman, P. 1979. Molecular line studies of dark clouds with associated young stellar objects. *Astrophys. J. Suppl.* 39:103–133.

Nagahara, H. 1983. Chondrules formed through incomplete melting of the preexisting mineral clusters and the origin of chondrules. In *Chondrules and Their Origins,* ed. E. A. King (Houston: Lunar & Planetary Inst.), pp. 211–222.

Nagai, H., Nitoh, O., and Honda, M. 1981. Half-life of ^{202}Pb. *Radiochim. Acta* 29:169–172.

Nakagawa, Y. 1978. Statistical behavior of planetesimals in the primitive solar system. *Prog. Theor. Phys.* 59:1834–1851.

Nakagawa, Y., and Hayashi, C. 1984. Formation of asteroids due to Jupiter's perturbation. In preparation.

Nakagawa, Y., Nakazawa, K., and Hayashi, C. 1981. Growth and sedimentation of dust grains in the primordial solar nebula. *Icarus* 45:517–528.

Nakagawa, Y., Hayashi, C., and Nakazawa, K. 1983. Accumulation of planetesimals in the solar nebula. *Icarus* 54:361–376.

Nakagawa, Y., Sekiya, M., and Hayashi, C. 1985. Sedimentary growth of dust grains in a non-turbulent solar nebula. Submitted to *Icarus*.

Nakano, T. C. 1976. Fragmentation of magnetic interstellar clouds by ambipolar diffusion. I. *Publ. Astron. Soc. Japan* 28:355–369.

Nakano, T. C. 1981. Fundamental processes in star formation. *Prog. Theor. Phys. Suppl.* 70:54–76.

Nakano, T. C. 1983*a*. Non-conservation of magnetic flux in star formation. *Publ. Astron. Soc. Japan* 35:87–90.

Nakano, T. C. 1983*b*. Quasistatic contraction of magnetic clouds due to plasma drift. III. Formation of massive protostars. *Publ. Astron. Soc. Japan* 35:209–223.

Nakano, T. C. 1984. The role of magnetic fields in star formation. *Fund. Cosmic Phys.* In press.

Nakano, T. C., and Umebayashi, T. 1980. Behavior of grains in the drift of plasma and magnetic field in dense interstellar clouds. *Publ. Astron. Soc. Japan* 32:613–621.

Nakazawa, K., and Hayashi, C. 1985. Tidal disruption in binary encounter of protoplanetary bodies. In preparation.

Nakazawa, K., Komuro, T., and Hayashi, C. 1983. Origin of the Moon: Capture by gas drag of the Earth's primordial atmosphere. *Moon and Planets* 28:311–327.

Nakazawa, K., Hayashi, C., and Komuro, T. 1985*a*. Origin of the Moon. II: Condition for survival of the trapped proto-moon as a satellite. Submitted to *Moon and Planets*.

Nakazawa, K., Mizuno, H., Sekiya, M., and Hayashi, C. 1985*b*. Structure of the primordial atmosphere surrounding the early-Earth. *J. Geomag. Geoelectr.* 37. In press.

Narita, S., McNally, D., Pearce, G. L., and Sørensen, S. A. 1983. The collapse of a fast rotating interstellar gas cloud. *Mon. Not. Roy. Astron. Soc.* 203:491–515.

Narita, S., Hayashi, C., and Miyama, S. M. 1984. Characteristics of collapse of rotating cloud. *Prog. Theor. Phys.* In press.

Navon, O., and Wasserburg, G. J. 1984. Self-shielding in O_2—A possible explanation for oxygen isotopic anomalies in meteorites? *Lunar Planet. Sci.* XV:589–590 (abstract).

Neckel, Th., and Klare, G. 1980. The spatial distribution of the interstellar extinction. *Astron. Astrophys. Suppl.* 42:251–281.

Neece, G. A., Rogers, F. J., and Hoover, W. G. 1971. Thermodynamic properties of compressed solid hydrogen. *J. Comp. Phys.* 7:621–636.

Neugebauer, G., Becklin, E. E., and Matthews, K. 1982. The double structure of W3-IRS 5 as determined from high-resolution spatial scans. *Astron. J.* 87:395–399.

Newton, K. 1980. Neutral hydrogen in IC 342. I. The large scale structure. *Mon. Not. Roy. Astron. Soc.* 191:169–184.

Nicholson, P. D., Matthews, K., and Goldreich, P. 1982. Radial widths, optical depths and eccentricities of the Uranian rings. *Astron. J.* 87:433–447.

Niederer, F. R., and Eberhardt, P. 1977. A neon-E-rich phase in Dimmitt. *Meteoritics* 12:327–331.

Niederer, F. R., and Papanastassiou, D. A. 1984. Ca isotopes in refractory inclusion. *Geochim. Cosmochim. Acta.* 48:1279–1294.

Niederer, F. R. Papanastassiou, D. A., and Wasserburg, G. J. 1980. Endemic isotopic anomalies in titanium. *Astrophys. J.* 240:L73–L77.

Niemeyer, S. 1980. I-Xe and ^{40}Ar-^{39}Ar dating of silicate from Weekeroo Station and Netschaevo IIE iron meteorites. *Geochim. Cosmochim. Acta* 44:33–44.

Niemeyer, S., and Lugmair, G. W. 1981. Ubiquitous isotopic anomalies in Ti from normal Allende inclusions. *Earth Planet. Sci. Lett.* 53:211–225.

Nier, A. O., and McElroy, M. B. 1972. Composition and structure of Mars' upper atmosphere: Results from the neutral mass spectrometers on Viking 1 and 2. *J. Geophys. Res.* 82:4341–4350.

Nieto, M. M. 1972. *The Titius-Bode Law of Planetary Distances: Its History and Theory* (New York: Pergamon).

Niimi, H. 1970. Stability of the system of stars, gas and magnetic fields. *Astrophys. Space Sci.* 6:297–314.

Nishida, S. 1983. Collisional processes of planetesimals with a protoplanet under the gravity of the proto-Sun. *Prog. Theor. Phys.* 70:93–105.

Nishiizumi, K., Gensho, R., and Honda, M. 1981. Half-life of ^{59}Ni. *Radiochim. Acta* 29:113–116.

Nittmann, J., Falle, S. A. E. G., and Gaskell, P. H. 1982. The dynamical destruction of shocked gas clouds. *Mon. Not. Roy. Astron. Soc.* 201:833–847.

Nørgaard, H. 1980. ^{26}Al from red giants. *Astrophys. J.* 236:895–898.

Norman, C. A., and Silk, J. 1979. Interstellar bullets: H_2O masers and Herbig-Haro objects. *Astrophys. J.* 228:197–205.

Norman, C. A., and Silk, J. 1980. Clumpy molecular clouds: A dynamic model self-consistently regulated by T Tauri star formation. *Astrophys. J.* 238:158–174.

Norman, C. A., and Pudritz, R. E. 1983. Centrifugally driven winds from contracting molecular disks. *Astrophys. J.* 274:677–697.

Norman, M. L. 1980. A Numerical Study of Rotating Interstellar Clouds: Equilibrium and Collapse. Ph. D. thesis, Univ. of California, Davis.

Norman, M. L., Wilson, J. R., and Barton, R. T. 1980. A new calculation on rotating protostar collapse. *Astrophys. J.* 239:968–981.

Norman, M. L., Smarr, L., and Winkler, K.-H. 1984. Fluid dynamical mechanisms for knots in astrophysical jets. In *Numerical Astrophysics,* ed. J. Cantrella (Boston: Jones and Bartlett).

Notsu, K., Onuma, N., Nishida, N., and Nagasawa, H. 1978. High temperature heating of the Allende meteorite. *Geochim. Cosmochim. Acta* 42:903–908.

Null, G. W. 1976. Gravity field of Jupiter and its satellites from Pioneer 10 and Pioneer 11 tracking data. *Astron. J.* 81:1153–1161.

Null, G. W., Lau, E. L., Biller, E. D., and Anderson, J. D. 1981. Saturn gravity results obtained from Pioneer 11 tracking data and Earth-based Saturn satellite data. *Astron. J.* 86:456–468.

Nyman, L. A. 1983. Detection of HCO^+ and HCN absorption towards three galactic H II regions. *Astron. Astrophys.* 120:307–312.

Nyman, L. A. 1984. Detection of CS and C_2H in absorption. Submitted to *Astron. Astrophys.*

Obukhov, A. M. 1941. On the distribution of energy in the spectrum of turbulent flow. *Izv. Akad. Nauk SSSR, ser Geogr. i Geofiz.* 5:453 (in Russian).

O'Dell, C. R. 1973. A new model for cometary nuclei. *Icarus* 19:137–146.

O'Donnell, E. J., and Watson, W. D. 1974. Upper limit to the flux of cosmic rays and x-rays in interstellar clouds. *Astrophys. J.* 191:89–92.

Ogura, K., and Hasegawa, T. 1983. A survey of northern Bok globules for H-alpha emission stars. *Publ. Astron. Soc. Japan* 35:299–315.

O'Keefe, J. A., and Sullivan, E. C. 1978. Fission origin of the Moon: Causing and timing. *Icarus* 35:272–283.

Olofsson, H. 1983. Galactic and Extragalactic Molecular Clouds. Ph.D. thesis, Chalmers University of Technology (Tech. Rept. 139, School of Electrical Engineering).

Olofsson, H. 1984. Deuterated water in Orion KL and W51M. *Astron. Astrophys.* 134:36–44.

Olofsson, H., Ellder, J., Hjalmarson, Å., and Rydbeck, G. 1982*a*. Extended and anisotropic high-velocity gas flows in the Orion-KL region. *Astron. Astrophys.* 113:L18–L21.

Olofsson, H., Johansson, L. E. B., Hjalmarson, Å., and Nguyen-Quang-Rieu. 1982*b*. High sensitivity molecular line observations of IRC+10216. *Astron. Astrophys.* 107:128–144.

Olson, D. W., and Sachs, R. K. 1973. The production of verticity in an expanding, self-gravitating fluid. *Astrophys. J.* 185:91–104.

Onuma, N., Clayton, R. N., and Mayeda, T.K. 1972. Oxygen isotopic cosmothermometer. *Geochim. Cosmochim. Acta* 36:169–188.

Oort, J. H. 1950. The structure of the cometary cloud surrounding the solar system and a hypothesis concerning its origin. *Bull. Astron. Inst. Netherlands* 11:91–110.

Oort, J. H. 1954. Outline of a theory on the origin and acceleration of interstellar clouds and O associations. *Bull. Astron. Inst. Netherlands* 12:177–186.

Oort, J. H. 1983. Superclusters. *Ann. Rev. Astron. Astrophys.* 21:373–428.

Öpik, E. J. 1953. Stellar associations and supernovae. *Irish Astron. J.* 2:219–233.

Oppenheimer, M. 1977. Isentropic instabilities in the interstellar gas. *Astrophys. J.* 211:400–403.

Ortolani, S., and D'Odorico, S. 1980. A discussion on the nature of the Herbig-Haro object no. 1 from its far UV spectrum. *Astron. Astrophys.* 83:L8–L9.

Orton, G. S., Tokunaga, A. T., and Caldwell, J. 1983. Observational constraints of the atmospheres of Uranus and Neptune from new measurements near 10 μm. *Icarus* 56:147–164.

Osamura, Y., Schaefer, H. F. III, Gray, H. F., and Miller, W. H. 1981. Vinylidene: A very shallow minimum on the C_2H_2 potential energy surface. Static and dynamical considerations. *J. Amer. Chem. Soc.* 103:1904–1907.

Osterbrock, D. E. 1961. On ambipolar diffusion in HI regions. *Astrophys. J.* 134:270–272.

Osterbrock, D. E. 1974. *Astrophysics of Gaseous Nebulae* (San Francisco: W. H. Freeman & Co.).

Ostriker, J. P. 1970. Fission and the origin of binary stars. In *Stellar Rotation,* ed. A. Slettebak (Dordrecht: D. Reidel), pp. 147–156.

Ostriker, J. P., and Bodenheimer, P. 1968. Rapidly rotating stars. II. Massive white dwarfs. *Astrophys. J.* 151:1089–1098.

Ostriker, J. P., and Mark, J. W.-K. 1968. Rapidly rotating stars. I. The self-consistent-field method. *Astrophys. J.* 151:1075–1088.

Ostriker, J. P. and Tassoul, J.-L. 1969. On the oscillations and stability of rotating stellar models. II. Rapidly rotating white dwarfs. *Astrophys. J.* 155:987–997.

Ostriker, J. P., and Bodenheimer, P. 1973. On the oscillations and stability of rapidly rotating stellar models. III. Zero-viscosity polytropic sequences. *Astrophys. J.* 180:171–180.

Ostriker, J. P., and Peebles, P. J. E. 1973. A numerical study of the stability of flattened galaxies: Or, can cold galaxies survive? *Astrophys. J.* 186:467–480.

Ostriker, J. P., and Cowie, L. L. 1981. Galaxy formation in an intergalactic medium dominated by explosions. *Astrophys. J.* 243:L127–L131.

Ott, E. 1981. Strange attractors and chaotic motions of dynamical systems. *Rev. Mod. Phys.* 53:655–672.

Ott, H. A., and Sanders, W. L. 1980. On the evolution of Barnard globules. *Astron. Astrophys.* 85:365–366.

Ott, U., Mack, R., and Chang, S. 1981. Noble-gas-rich separates from the Allende meteorite. *Geochim. Cosmochim. Acta* 45:1751–1788.

Ott, U., Kronenbitter, J., Flores, J., and Chang, S. 1984. Colloidally separated samples from Allende residues: Noble gases, carbon and an ESCA study. *Geochim. Cosmochim. Acta* 48:267–280.

Owen, T. 1982. The composition and origin of Titan's atmosphere. *Planet. Space Sci.* 30:833–838.

Owen, T., and Terrile, R. 1981. Colors on Jupiter. *J. Geophys. Res.* 86:8797–8814.

Owen, T., Biermann, K., Rushneck, D. R., Biller, J. E., Howarth, D. W., and Lafleur, A. L. 1977. The composition of the atmosphere at the surface of Mars. *J. Geophys. Res.* 82:4635–4640.

Paleologou, E. V., and Mouschovias, T. Ch. 1983. The magnetic flux problem and ambipolar diffusion during star formation: One-dimensional collapse. I. Formulation of the problem and method of solution. *Astrophys. J.* 275:838–857.

Palla, F., Salpeter, E. E., and Stahler, S. W. 1983. Primordial star formation: The role of molecular hydrogen. *Astrophys. J.* 271:632–641.

Palme, H., Wlotzka, F., Nagel, K., and El Goresy, A. 1982. An ultra-refractory inclusion from the Ornans carbonaceous chondrite. *Earth Planet Sci. Lett.* 61:1–12.

Palmer, P., Zuckerman, B., Buhl, D., and Snyder, L. E. 1969. Formaldehyde absorption in dark nebulae. *Astrophys. J.* 156:L147–L150.

Panagia, N. 1973. Some physical parameters of early-type stars. *Astron. J.* 78:929–934.

Panchev, S. 1971. *Random Functions and Turbulence* (Oxford: Pergamon).

Papaloizou, J., and Pringle, J. E. P. 1977. Tidal torque on acretion discs in close binary systems. *Mon. Not. Roy. Astron. Soc.* 181:441–454.

Papaloizou, J., and Pringle, J. E. 1978. Gravitational radiation and the stability of rotating stars. *Mon. Not. Roy. Astron. Soc.* 184:501–508.

Papaloizou, J., and Lin, D. N. C. 1984. On the tidal interaction between protoplanets and the primordial solar nebula. I. Linear calculation of the role of angular momentum exchange. *Astrophys. J.* 285:818–834.

Papaloizou, J., Faulkner, J., and Lin, D. N. C. 1983. On the evolution of accretion disc flow in cataclysmic variables. II. The existence and nature of collective relaxation oscillations in dwarf nova systems. *Mon. Not. Roy. Astron. Soc.* 205:487–513.

Papanastassiou, D. A., Lee, T., and Wasserburg, G. J. 1977. Evidence for ^{26}Al in the solar system. In *Comets, Asteroids and Meteorites,* ed. A. H. Delsemme (Toledo: Univ. of Toledo Press), pp. 343–349.

Papoulis, A. 1965. *Probability, Random Variables, and Stochastic Processes* (New York: McGraw Hill).

Paque, J. M., and Stolper, E. 1983. Experimental evidence for slow cooling of Type-B CAIs from a partially molten state. *Lunar and Planet. Sci.* XIV:596–597 (abstract).

Paresce, F., Morgan, B., Bowyer, S., and Lampton, M. 1979. Observations of the cosmic background radiation at 1440 angstroms. *Astrophys. J.* 230:304–310.

Paris, R. B. 1971. Magnetism and Cosmogony. Ph.D. thesis, University of Manchester, England.

Parker, D. A. 1973. The equilibrium of an interstellar magnetic gas cloud. *Mon. Not. Roy. Astron. Soc.* 163:41–65.

Parker, D. A. 1974. The equilibrium of an interstellar magnetic disk. *Mon. Not. Roy. Astron. Soc.* 168:331–344.

Parker, E. N. 1966. The dynamical state of the interstellar gas and field. *Astrophys. J.* 145:811–833.

Parker, E. N. 1979. *Cosmical Magnetic Fields* (Oxford: Clarendon Press).

Passey, Q. R., and Shoemaker, E. M. 1982. Craters and basins on Ganymede and Callisto: Morphological indicators of crustal evolution. In *Satellites of Jupiter,* ed. D. Morrison (Tucson: Univ. of Arizona Press), pp. 379–434.

Pauls, T. A., Wilson, T. L., Bieging, J. H., and Martin, R. N. 1983. Clumping in Orion KL: 2-arcsecond maps of ammonia. *Astron. Astrophys.* 124:23–38.

Payne, H. E., Salpeter, E. E., and Terzian, Y. 1984. An upper limit to hydrogen ionization rates from radio recombination line observations toward 3C123. Astron. J. 89:668–672.

Peale, S. J. 1975. Dynamical consequences of meteorite impacts on the moon. *J. Geophys. Res.* 80:4939–4946.

Peale, S. J., 1976*a*. Excitation and relaxation of the wobble, precession, and libration of the moon. *J. Geophys. Res.* 81:1813–1827.

Peale, S. J. 1976*b*. Orbital resonances in the solar system. *Ann. Rev. Astron. Astrophys.* 14:215–246.

Peale, S. J., and Burns, J. A. 1985. Dynamic evolution. In *Satellites of the Solar System,* eds. J. A. Burns. D. Morrison, and M. S. Matthews (Tucson: Univ. of Arizona Press). In press.

Peale, S. J., Cassen, P. M., and Reynolds, R. T. 1979. Melting of Io by tidal dissipation. *Science* 203:892–894.

Pechernikova, G. V., and Vityazev, A. V. 1980. The evolution of eccentricities of the orbits of the planets in the process of their accumulation. *Astron. Zh. USSR* 57:799–811 (in Russian).

Peebles, P. J. E. 1964. The structure and composition of Jupiter and Saturn. *Astrophys. J.* 140:328–347.

Peebles, P. J. E. 1980. *The Large-Scale Structure of the Universe* (Princeton: Princeton Univ. Press).

Peebles, P. J. E., and Dickel, R. H. 1968. Origin of the globular star clusters. *Astrophys. J.* 154:891–908.

Pellas, P., and Storzer, D. 1981. ^{244}Pu fission track thermometry and its application to stony meteorites. *Proc. Roy. Soc. London* A374:253–270.

Pels, G., Oort, J. H., and Pels-Kluyver, H. A. 1975. New members of the Hyades cluster and a discussion of its structure. *Astron. Astrophys.* 43:423–441.

Pendleton, Y. J., and Black, D. C. 1983. Further studies on criteria for the onset of dynamical instability in general three-body systems. *Astron. J.* 88:1415–1419.

Penston, M. V. 1969. Dynamics of self-gravitating gaseous spheres. *Mon. Not. Roy. Astron. Soc.* 145:457–485.

Penston, W. V., and Lago, M. T. V. T. 1983. Optical and ultraviolet line profiles in the T Tauri star LH alpha 332–21. *Mon. Not. Roy. Astron. Soc.* 2077–2084.

Penston, W. V., Munday, V. A., Sickland, D. J., and Penston, M. J. 1969. Interstellar clouds. *Mon. Not. Roy. Astron. Soc.* 142:355–386.

Penzias, A. A. 1975. Observations and physics of dense neutral clouds. In *Les Houches XXVI. Atomic and Molecular Physics and Interstellar Matter,* ed. R. Balian, P. Encrenaz, and J. Lequeux (Amsterdam: North Holland), pp. 373–408.

Penzias, A. A. 1979. Interstellar HCN, HCO^+ and the galactic deuterium gradient. *Astrophys. J.* 228:430–434.

Penzias, A. A. 1980. Nuclear processing and isotopes in the galaxy. *Science* 208:663–669.

Penzias, A. A., Solomon, P. M., Wilson, R. W., and Jefferts, K. B. 1971. Interstellar carbon monosulfide. *Astrophys. J.* 168:L53–L58.

Penzias, A. A., Solomon, P. M., Jefferts, K. B., and Wilson, R. W. 1972. Carbon monoxide observations of dense interstellar clouds. *Astrophys. J.* 174:L43–L48.

Perault, M., Falgarone, E., and Puget, J. L. 1985*a*. ^{13}CO observations of cold giant molecular clouds. *Astron. Astrophys.* In press.

Perault, M., Falgarone, E., and Puget, J. L. 1985*b*. Fragmented molecular clouds: Statistical analysis of the ^{13}CO (δ=1-0) emission distribution. *Astron. Astrophys.* In press.

Perri, F., and Cameron, A. G. W. 1974. Hydrodynamic instability of the solar nebula in the presence of a planetary core. *Icarus* 22:416–425.

Perry, C. L., and Johnston, L. 1982. A photometric map of interstellar reddening within 300 parsecs. *Astrophys. J. Suppl.* 50:451–516.

Phillipps, S., Kersey, S., Osborne, J. L., Haslam, C. G. T., and Stoffel, H. 1981. Distribution of galactic synchrotron emission I. *Astron. Astrophys.* 98:286–294.

Phillips, T. G., and Beckman, J. 1980. The nature of the Kleinmann Low nebula. *Mon. Not. Roy. Astron. Soc.* 193:245–260.

Phillips, T. G., and Huggins, P. J. 1981. Abundance of atomic carbon (CI) in dense interstellar clouds. *Astrophys. J.* 251:533–540.

Phillips, T. G., Huggins, P. J., Kuiper, T. B. H., and Miller, R. E. 1980. Detection of the 610 micron (294 Ghz) line of interstellar atomic carbon. *Astrophys. J.* 238:L103–L106.

Pikelner, S. B., and Sorochenko, R. L. 1974. Fluctuations of density and velocity in young Orion-type nebulae. *Soviet Astron.* 17:443–450.

Pilcher, C. B., Chapman, C. R., Lebofsky, L. A., and Kieffer, H. H. 1970. Saturn's rings: Identification of water frost. *Science* 167:1372–1373.

Pinto, J. P., Gladstone, G.R., and Yung, Y. L. 1980. Photochemical production of formaldehyde in Earth's primitive atmosphere. *Science* 210:183–184.

Pitz, E., Leinert, C., Schulz, A., and Link, H. 1979. Rocket photometry of ultraviolet galactic light. *Astron. Astrophys.* 72:92–96.

Plambeck, R. 1984. Millimeter interferometry of star-forming regions. Presented at URSI International Symposium on Millimeter and Submillimeter Wave Radio Astronomy, Granada, Spain.

Plambeck, R. L., Wright, M. C. H., Welch, W. J., Bieging, J. H., Baud, B., Ho, P. T. P., and Vogel, S. N. 1982. Kinematics of Orion-KL: Aperture synthesis maps of 86 GHz SO emission. *Astrophys. J.* 259:617–624.

Plescia, J. B. 1983. The geology of Dione. *Icarus* 56:255–277.

Plescia, J. B., and Boyce, J. M. 1983. Crater numbers and geological histories of Iapetus, Enceladus, Tethys, and Hyperion. *Nature* 301:666–670.

Podolak, M. 1976. Methane rich models of Uranus. *Icarus* 27:473–477.

Podolak, M. 1977. The abundance of water and rock in Jupiter as derived from interior models. *Icarus* 30:155–162.

Podolak, M. 1978. Models of Saturn's interior: Evidence for phase separation. *Icarus* 33:342–348.

Podolak, M. 1982. The origin of Uranus: Compositional considerations. In *Uranus and the Outer Planets, IAU Colloquium No. 81,* ed. G. E. Hunt (Cambridge: Cambridge Univ. Press), pp. 93–109.

Podolak, M., and Cameron, A. G. W. 1974. Models of the giant planets. *Icarus* 22:123–148.

Podolak, M., and Cameron, A. G. W. 1975. Further investigations of Jupiter models. *Icarus* 25:627–634.

Podolak, M., and Reynolds, R. T. 1981. On the structure and composition of Uranus and Neptune. *Icarus* 46:40–50.

Podolak, M., and Reynolds, R. T. 1984. Consistency tests of cosmogonic theories from models of Uranus and Neptune. *Icarus* 57:102–111.

Podosek, F. A. 1970. Dating of meteorites by the high-temperature release of iodine-correlated Xe^{129}. *Geochim. Cosmochim. Acta* 34:341–365.

Podosek, F. A. 1978. Isotopic structures in solar system materials. *Ann. Rev. Astron. Astrophys.* 16:293–334.

Pollack, J. B. 1981. Phase curve and particle properties of Saturn's F ring. *Bull. Amer. Astron. Soc.* 13:727 (abstract).

Pollack, J. B. 1984. Origin and history of the outer planets: Theoretical models and observational constraints. *Ann. Rev. Astron. Astrophys.* 22:389–424.

Pollack, J. B., and Ohring, G. 1973. A numerical method for determining the temperature structure of planetary atmospheres. *Icarus* 19:34–42.
Pollack, J. B., and Reynolds, R. T. 1974. Implications of Jupiter's early contraction history for the composition of the Galilean satellites. *Icarus* 21:248–253.
Pollack, J. B., and Cuzzi, J. N. 1981. Rings in the solar system. *Sci. Amer.* 245:105–129.
Pollack, J. B., and Consolmagno, G. 1984. Origin and evolution of the Saturn system. In *Saturn*, eds. T. Gehrels and M. S. Matthews (Tucson: Univ. of Arizona Press), pp. 811–866.
Pollack, J. B., Summers, A. L., and Baldwin, B. 1973. Estimates of the size of the particles in the rings of Saturn and their cosmogonic implications. *Icarus* 20:263–278.
Pollack, J. B., Grossman, A. S., Moore, R., and Graboske, H. C. Jr. 1976. The formation of Saturn's satellites and rings as influenced by Saturn's contraction history. *Icarus* 29: 35–48.
Pollack, J. B., Grossman, A. S., Moore, R., and Graboske, H. C. 1977. A calculation of Saturn's gravitational contraction history. *Icarus* 30:111–128.
Pollack, J. B., Burns, J. A., and Tauber, M. E. 1979. Gas drag in primordial circumplanetary envelopes: A mechanism for satellite capture. *Icarus* 37:587–611.
Pollack, J. B., McKay, C., and Christofferson, B. 1985. A calculation of the Rosseland mean opacity of dust grains in primordial solar system nebulae. Submitted to *Icarus*.
Pöppel, W. G. I., Rohlfs, K., and Celnik, W. 1983. Formaldehyde, cold neutral hydrogen and dust distribution in a globular filament in Taurus. *Astron. Astrophys.* 126:152–160.
Poquet, A. 1979. Fully developed magnetohydrodynamic turbulence: Numerical simulation and closure techniques. *Intern. J. Fusion Energy* 2:39–66.
Porter, W. S. 1961. The constitutions of Uranus and Neptune. *Astron. J.* 66:243–245.
Prasad, S. S., and Huntress, W. T. Jr. 1980*a*. A model for gas phase chemistry in interstellar clouds. I. The basic model, library of chemical reactions, and chemistry among C, N, and O compounds. *Astrophys. J. Suppl.* 43:1–35.
Prasad, S. S., and Huntress, W. T. Jr. 1980*b*. A model for gas phase chemistry in interstellar clouds. II. Nonequilibrium effects and effects of temperature and activation energy. *Astrophys. J.* 239:151–165.
Prasad, S. S., and Huntress, W. T. Jr. 1982. Sulfur chemistry in dense interstellar clouds. *Astrophys. J.* 260:590–598.
Prasad, S. S., and Tarafdar, S. P. 1983. UV radiation field inside dense clouds: lts possible existence and chemical implications. *Astrophys. J.* 267:603–609.
Prendergast, K. H., and Burbidge, G. R. 1968. On the nature of some Galactic x-ray sources. *Astrophys. J.* 151:L83–L88.
Press, W. H., and Teukolsky, S. A. 1973. On the evolution of the secularly unstable viscous Maclaurin spheroids. *Astrophys. J.* 181:513–517.
Price, P. B., Hutcheon, I. R., Braddy, D., and Macdougall, J. D. 1975. Track studies bearing on solar system regoliths. *Proc. Lunar Sci. Conf.* 6:3449–3469.
Prigogine, I. 1980. *From Being to Becoming: Time and Complexity in the Physical Sciences*. (San Francisco: W. H. Freeman & Co.).
Pringle, J. 1976. Thermal instabilities in accretion discs. *Mon. Not. Roy. Astron. Soc.* 177:65–71.
Pringle, J. E. 1981. Accretion disks in astrophysics. *Ann. Rev. Astron. Astrophys.* 19:137–162.
Pringle, J. E. P., Rees, M. J., and Pacholzyk, A. G. 1973. Accretion onto massive black holes. *Astron. Astrophys.* 29:179–184.
Prinn, R. G., and Lewis, J. S. 1973. Uranus atmosphere: Structure and composition. *Astrophys. J.* 179:333–342.
Prinn, R. G., and Fegley, B. Jr. 1981. Kinetic inhibition of CO and N_2 reduction in circumplanetary nebulae: Implications for satellite composition. *Astrophys. J.* 249:308–317.
Proceedings of the Symposium on HH Objects and T Tauri Stars 1983. *Rev. Mexicana Astron. Astrophys.* vol. 7.
Pudritz, R. E., and Norman, C. A. 1983. Centrifugally driven winds from contracting molecular disks. *Astrophys. J.* 274:677–697.
Pumphrey, W. A., and Scalo, J. M. 1983. Simulation models for the evolution of cloud systems. I. Introduction and preliminary simulations. *Astrophys. J.* 269:531–559.
Pumphrey, W. A., and Scalo, J. M. 1984. Simulation models for the evolution of cloud systems. II. Collisions, drag, and accretion in self-gravitating systems of supported fragments. Submitted to *Astrophys. J.*

Purcell, E. M. 1976. Temperature fluctuations in very small interstellar grains. *Astrophys. J.* 206:685–690.

Quinn, P. J. 1983. On the formation and dynamics of shells around elliptical galaxies. In *External Kinematics and Dynamics of Galaxies, IAU Symposium No. 100,* ed. E. Athanassoula (Dordrecht: D. Reidel), pp. 347–348.

Quiroga, R. J. 1983. Hydrodynamic and turbulent motions in the galactic disk. *Astrophys. Space Sci.* 93:37–52.

Quiroga, R. J., and Varsavsky, C.M. 1970. The ratio of atomic hydrogen to dust in the omega nebula. *Astrophys. J.* 160:83–88.

Radzievskii, V. V. 1981. A qualitative analysis of the problem of comet migration. *Soviet Astron. A. J.* 15:32–35.

Ramsey, W. H. 1967. On the constitutions of Uranus and Neptune. *Planet. Space Sci.* 15:1609–1623.

Rayleigh, Lord. 1917. On the dynamics of revolving fluids. *Proc. Roy. Soc.* A93:148–154.

Raymond, J. C. 1979. Shock waves in the interstellar medium. *Astrophys. J. Suppl.* 39:1–27.

Ree, F. H. 1982. Molecular interaction of dense water at high temperature. *J. Chem. Phys.* 76:6287–6300.

Rees, M. J., and Ostriker, J. 1977. Cooling, dynamics and fragmentation of massive gas clouds: Clues to the masses and radii of galaxies and clusters. *Mon. Not. Roy. Astron. Soc.* 179:541–559.

Reeves, H., ed. 1972. *On the Origin of the Solar System* (Paris: CNRS)

Reeves, H., and Bottinga, Y. 1972. The D/H ratio in Jupiter's atmosphere. *Nature* 238:326–327.

Regev, O., and Shaviv, G. 1981. Formation of protostars in collapsing, rotating, turbulent clouds. *Astrophys. J.* 245:934–959.

Reid, M. J. 1973. The tidal loss of satellite-orbiting objects and its implications for the lunar surface. *Icarus* 20:240–248.

Reid, M., Haschick, A., Burke, B., Moran, J., Johnson, K., and Swenson, G. 1980. The structure of interstellar hydroxyl masers: VLBI synthesis observations of W3 (OH). *Astrophys. J.* 239:89–111.

Reipurth, B. 1983. Star formation in Bok globules and low-mass clouds. I. The cometary globules in the Gum nebula. *Astron. Astrophys.* 117:183–198.

Reipurth, B. 1984. Bok globules. *Mercury* 13:50–56.

Reipurth, B., and Bouchet, P. 1984. Star formation in Bok globules and low-mass clouds. II. A collimated flow in the Horsehead. *Astron. Astrophys.* 137:L1–L4.

Reynolds, R. T., and Summers, A. L. 1965. Models of Uranus and Neptune. *J. Geophys. Res.* 70:199–208.

Reynolds, R. T., and Cassen, P. M. 1979. On the internal structure of the major satellites of the outer planets. *Geophys. Res. Lett.* 6:121–124.

Richtmyer, R. D., and Morton, K. W. 1967. *Difference Method for Initial-Value Problems,* 2nd ed. (New York: Interscience-Wiley).

Rickard, L. J., Palmer, P., Buhl, D., and Zuckerman, B. 1977. Observations of formaldehyde absorption in the region of NGC 2264 and other Bok globules. *Astrophys. J.* 213:654–672.

Rickman, H. 1976. Stellar perturbations of orbits of long-period comets and their significance for cometary capture. *Bull. Astron. Czech.* 27:92–105.

Riegel, K. W. 1967. 21-cm line observations of galactic HII regions. *Astrophys. J.* 148:87–103.

Ringwood, A. E. 1979. *Origin of the Earth and Moon* (New York: Springer-Verlag).

Ringwood, A. E., and Kesson, S. E. 1977. Composition and origin of the Moon. *Proc. Lunar. Sci. Conf.* 8:371–398.

Rivolo, A.R., Solomon, P. M., and Sanders, D. B. 1985. Statistical clustering properties of giant molecular cloud cores. *Bull. Amer. Astron. Soc.* 16:937 (abstract).

Roberson, R. A., and Crowe, C. T. 1980. *Engineering Fluid Mechanics* (Boston: Houghton Mifflin).

Robert, F., and Epstein, S. 1982. The concentration and isotopic composition of hydrogen, carbon and nitrogen in carbonaceous meteorites. *Geochim. Cosmochim. Acta* 46:81–95.

Roberts, M. S. 1957. The numbers of early-type stars in the galaxy and their relation to galactic clusters and associations. *Publ. Astron. Soc. Pacific* 69:59–64.

Roberts, W. W. 1969. Large scale shock formation in spiral galaxies and its implications on star formation. *Astrophys. J.* 158:123–143.

Roberts, W. W., and Yuan, C. 1970. Applications of the density wave theory to the spiral structure of the Milky Way system. III. Magnetic fields and large scale shock formation. *Astrophys. J.* 161:887–902.

Roberts, W. W., and van Albada, G. D. 1981. A high resolution study of gas flows in barred spirals. *Astrophys. J.* 246:740–750.

Roberts, W. W., and Hausman, M. A. 1984. Spiral structure and star formation. I. Formation mechanisms and mean free paths. *Astrophys. J.* 277:744–767.

Roberts, W. W., Roberts, M. S., and Shu, F. H. 1975. Density wave theory and the classification of spiral galaxies. *Astrophys. J.* 196:381–405.

Robertson, J. A., and Tayler, R. J. 1981. Thermal stability of optically thick accretion discs. *Mon. Not. Roy. Astron. Soc.* 196:185–195.

Robinson, B. J., Manchester, R. N,, Whiteoak, J. B., Sanders, D. B., Scoville, N. Z., Clemens, D. P., McCutcheon, W.H., and Solomon, P. M. 1984. The distribution of CO in the galaxy for longitudes 294° to 86°. *Astrophys. J.* 283:L31–L36.

Roche, E. 1847. *Acad. Sci. Lett. Montpellier* 243–262.

Rodney, W. S., and Rolfs, C. 1982. Hydrogen burning in massive stars. *Essays in Nuclear Astrophysics,* eds. C. A. Barnes, D. D. Clayton, and D. N. Schramm (Cambridge: Cambridge Univ. Press), pp. 193–232.

Rodriguez, L. F., Carral, P., Ho, P. T. P., and Moran, J. M. 1982. Anisotropic mass outflows in regions of star formation. *Astrophys. J.* 260:635–646.

Ross, F. E., and Calvert, M. R. 1934. *Atlas of the Northern Milky Way* (Chicago: Univ. of Chicago Press).

Ross, M. 1981. The ice layer in Uranus and Neptune—Diamonds in the sky? *Nature* 292:435–436.

Ross, M., and Shishkevich, C. 1977. Molecular and metallic hydrogen. *Rand Report* R-2056-ARPA.

Ross, M., Graboske, H. C. Jr., and Nellis, W. J. 1981. Equation of state experiments and theory relevant to planetary modeling. *Phil. Trans. Proc. Roy. Soc. London* A303:303–313.

Rossano, G. S. 1978*a*. Distribution of extinction in the Corona Australis dark cloud complex. *Astron. J.* 83:234–240.

Rossano, G. S. 1978*b*. Distribution of extinction in several dark cloud complexes in Scorpius and Ophiuchus. *Astrophys. J.* 83:241–243.

Rots, A. H. 1975. Distribution and kinematics of neutral hydrogen in the spiral galaxy M81. *Astron. Astrophys.* 45:43–55.

Rowan-Robinson, M. 1979. Clouds of dust and molecules in the galaxy. *Astrophys. J.* 234:111–128.

Rowan-Robinson, M., Clegg, P. E., Beichmann, C. A., Neugebauer, G., Soifer, B. T., Aumann, H. H., Beintema, D. A., Boggess, N., Emerson, J. P., Gautier, T. N., Gillett, F. C., Hauser, M. G., Houck, J. R., Low, F. J., and Walker, R. G. 1984. The IRAS minisurvey. *Astrophys. J.* 278:L7–L10.

Rowe, M. W., and Kuroda, P. K. 1965. Fissiogenic Xe from the Pasamonte meteorite. *J. Geophys. Res.* 70:709–714.

Roxburgh, I. W. 1966. On the fission theory of the origin of binary stars. *Astrophys. J.* 143:111–120.

Rozyczka, M. 1983. 3-D simulations of the collapse of nonspherical interstellar clouds. *Astron. Astrophys.* 125:45–51.

Rozyczka, M., Tscharnuter, W. M., Winkler, K.-H., and Yorke, H. W. 1980*a*. Fragmentation of interstellar clouds: Three-dimensional hydrodynamical calculations. *Astron. Astrophys.* 83:118–128.

Rozyczka, M., Tscharnuter, W. M., and Yorke, H. W. 1980*b*. Three-dimensional numerical models of the collapse of turbulent interstellar clouds. *Astron. Astrophys.* 81:347–350.

Rumstay, K. S., and Kaufman, M. 1983. H II regions and star formation in M83 and M33. *Astrophys. J.* 274:611–631.

Ruskol, E. L. 1960. To the problem of formation of protoplanets. *Probl. of Cosmogony* 7:8–14 (in Russian).

Ruskol, E. L. 1972. The origin of the Moon. III. Some aspects of the dynamics of the circumterrestrial swarm. *Soviet Astron. A.J.* 15:646–654.

Ruskol, E. L. 1982. Origin of planetary satellites. *Izvestiya Earth Physics* 18:425–433.

Russell, C. T. 1980. Planetary magnetism. *Rev. Geophys. Space Sci.* 18:77–106.
Ruzmaikina, T. V. 1981*a*. On the role of the magnetic field and turbulence in the evolution of the presolar nebula. *23rd COSPAR Meeting, Budapest, June 1980. Adv. Space Res.* 1:49–53.
Ruzmaikina, T. V. 1981*b*. Angular momentum of protostars giving the birth to preplanetary disks. *Pizma Astron. Zh.* 7:188–192 (in Russian).
Ruzmaikina, T. V. 1982. In *Diskussions forum Ursprung des Connen-systems,* ed. H. Völk. *Mitt. Astron. Ges.* 57:49–53.
Ruzmaikina, T. V. 1984. The role of magnetic field in star formation. In *Magnetic Field in Astrophysics,* eds. Ya. B. Zeldovich, A. A. Ruzmaikin, D. D. Sokoloff, and P. H. Roberts (London: Gordon & Breach Sci. Publ.), pp. 267–291.
Rybicki, G., and Hummer, J. 1978. A generalization of the Sobolev method for flows with nonlocal radiative coupling. *Astrophys. J.* 219:654–675.
Rydbeck, O. E. H., Kollberg, E., Hjalmarson, Å., Sume, A., Ellder, J., and Irvine, W. M. 1976. Radio observations of interstellar CH. I. *Astrophys. J. Suppl.* 31:333–415.
Rydbeck, O. E. H., Sume, A., Hjalmarson, Å., Ellder, J., Rönnäng, B. O., and Kollberg, E. 1977. Hyperfine structure of interstellar ammonia in dark clouds. *Astrophys. J.* 215:L35–L40.
Rydbeck, O. E. H., Irvine, W. M., Hjalmarson, Å., Rydbeck, G., Ellder, J., and Kollberg, E. 1980. Observations of SO in dark and molecular clouds. *Astrophys. J.* 235:L171–L175.
Rydgren, A. E., and Vrba, F. J. 1981. Nearly simultaneous optical and infrared photometry of T Tauri stars. *Astron. J.* 86:1069–1075.
Rydgren, A. E., and Vrba, F. J. 1983. Additional UBVRI and JHKL photometry of T Tauri stars in the Taurus region. *Astron. J.* 88:1017–1026.
Rydgren, A. E., and Vrba, F. J. 1984. The incidence of infrared excesses among G-type stars in the direction of the Orion Ic association. *Astron. J.* 89:399–405.
Rydgren, A. E., Strom, S. E., and Strom, K. M. 1976. The nature of the objects of Joy: A study of the T Tauri phenomenon. *Astrophys. J. Suppl.* 30:307–336.
Rydgren, A. E., Schmelz, J. T., and Vrba, F. J. 1982. Evidence for a characteristic maximum temperature in the circumstellar dust associated with T Tauri stars. *Astrophys. J.* 256:168–176.
Safronov, V. S. 1958. On the turbulence in the protoplanetary cloud. *Rev. Mod. Physics* 30:1023–1024.
Safronov, V. S. 1960. On the gravitational instability in flattened systems with axial symmetry and non-uniform rotation. *Ann. d'Astrophys.* 23:901–904.
Safronov, V. S. 1969. *Evolution of the Protoplanetary Cloud and Formation of the Earth and Planets* (Moscow: Nauka Press); also NASA TTF-677, 1972.
Safronov, V. S. 1972*a*. Ejection of bodies from the solar system in the course of the accumulation of the giant planets and the formation of the cometary cloud. In *Motion, Evolution of Orbits, and Origin of Comets,* eds. C. A. Chebotarev and E. I. Kazimirchak-Polonskaya (Dordrecht: D. Reidel), pp. 329–334.
Safronov, V. S. 1972*b*. Accumulation of the planets. In *On the Origin of the Solar System,* ed. H. Reeves (Paris: CNRS), pp. 89–113.
Safronov, V. S. 1978. The heating of the Earth during its formation. *Icarus* 33:1–12.
Safronov, V. S. 1982. Present-day state of the theory of the Earth's origin. *Izvestia Acad. Sci. USSR, Fizika Zemli* 6:5–24 (in Russian).
Safronov, V. S., and Ruskol, E. L. 1957. On the hypothesis of turbulence in the protoplanetary cloud. *Probl. of Cosmogony* 5:22–46.
Safronov, V. S., and Ruzmaikina, T. V. 1978. On the angular momentum transfer and accumulation of solid bodies in the solar nebula. In *Protostars & Planets,* ed. T. Gehrels (Tucson: Univ. of Arizona Press), pp. 545–564.
Safronov, V. S., and Ruskol, E. L. 1982. On the origin and initial temperature of Jupiter and Saturn. *Icarus* 49:284–296.
Safronov, V. S., and Vitjazev, A. V. 1983. Origin of the solar system. In *Astrophysics and Space Physics: Soviet Scientific Reviews,* ed. R. A. Synjaev. In press.
Saito, T., Ohtani, H., and Tomita, Y. 1981. The extinction and the H I content of the dark cloud complex Khavtasi 141. *Publ. Astron. Soc. Japan* 33:327–340.
Sakata, A., Wada, S., Okutsu, Y., Shintani, H., and Nakada, Y. 1983. Does a 2200 Å hump observed in an artificial carbonaceous composite account for UV interstellar extinction? *Nature* 301:493–494.

Salpeter, E. E.1973. On convection and gravitational layering in Jupiter and in stars of low mass. *Astrophys. J.* 181:L83–L86.
Salpeter, E. E.1977. Formation and destruction of dust grains. *Ann. Rev. Astron. Astrophys.* 15:267–293.
Salpeter, E. E., and Zapolsky, H. S. 1967. Theoretical high pressure equations of state including correlation energy. *Phys. Rev.* 158:876–886.
Samuelson, R. E., Hanel, R. A., Kunde, V. G., and Maguire, W. C. 1981. Mean molecular weight and hydrogen abundance of Titan's atmosphere. *Nature* 292:688–693.
Sancisi, P. R., and Wesselius, P. R. 1970. Cold neutral hydrogen in the Taurus dust clouds. *Astron. Astrophys.* 7:341–348.
Sandell, G. 1978. Lifetime of molecules in a dark cloud model. *Astron. Astrophys.* 69:85–101.
Sandell, G., and Mattila, K. 1975. Radiation density and lifetimes of molecules in interstellar dust clouds. *Astron. Astrophys.* 42:357–364.
Sanders, D. B., Clemens, D. P., Scoville, N. Z., and Solomon, P. M. 1984*a*. Molecular cloud clusters and chains. *The Milky Way Galaxy, IAU Symposium No. 106*. In press.
Sanders, D. B., Solomon, P. M., and Scoville, N. Z. 1984*b*. Giant molecular clouds in the galaxy. I. The axisymmetric distribution of H_2. *Astrophys. J.* 276:182–203.
Sanders, D. B., Clemens, D. P., Scoville, N. Z., and Solomon, P. M. 1984*c*. Massachusetts-Stony Brook galactic plane CO survey: b-v maps of the first quadrant. *Astrophys. J. Suppl.* In press.
Sanders, D. B., Scoville, N. Z., and Solomon, P. M. 1985*b*. Giant molecular clouds in the galaxy. II. Characteristics of discrete features. *Astrophys. J.* In press.
Sanders, D. B., Solomon, P. M., and Yahil, A. 1985*c*. Clusters of giant molecular clouds. To be submitted to *Astrophys. J.*
Sanders, R. H., and Huntley, J. M. 1976. Gas response to oval distortions in disk galaxies. *Astrophys. J.* 209:53–65.
Sanford, M. T. II, and Whitaker, R. W. 1983. Hydrodynamic models of Herbig-Haro objects. *Mon. Not. Roy. Astron. Soc.* 205:105–121.
Sanford, M. T. II, Whitaker, R. W., and Klein, R. I. 1982*a*. Radiation-driven implosions in molecular clouds. *Astrophys. J.* 260:183–201.
Sandford, M. T. II, Whitaker, R. W., and Klein, R. I. 1982*b*. Radiation-hydrodynamics of HII regions and molecular clouds. In *Regions of Recent Star Formation,* eds. R. S. Roger and P. E. Dewdney (Dordrecht: D. Reidel), pp. 129–132.
Sandford, M. T. II, Whitaker, R. W., Klein, R. I. 1984. Radiatively driven dust-bounded implosions: Formation and stability of dense globules. *Astrophys. J.* 282:178–190.
Sarcander, M., Neckel, T., and Elsässer, H. 1984. Spectroscopic observations in L1551. Submitted to *Astrophys. J.* (Lett.).
Sargent, A. I. 1977. Molecular clouds and star formation. I. Observations of the Cepheus OB3 molecular cloud. *Astrophys. J.* 218:736–748.
Sargent, A. I. 1979. Molecular clouds and star formation. II. Star formation in the Cepheus OB3 and Perseus OB2 molecular clouds. *Astrophys. J.* 233:163–181.
Sasaki, S., and Nakazawa, K. 1985. Internal evolution of the growing proto-Earth: The effect of mass accretion and metal-silicate differentiation. In preparation.
Sasaki, S., Nakazawa, K., and Mizuno, H. 1983. Melting of materials forming the proto-Earth and the early evolution of the Earth. *Proc. 16th Lunar Planet. Symp. ISAS,* Tokyo, pp. 139–142.
Sasao, T. 1973. On the generation of density fluctuations due to turbulence in self-gravitating media. *Publ. Astron. Soc. Japan* 25:1–33.
Savage, B. D., and Mathis, J. S. 1979. Observed properties of interstellar dust. *Ann. Rev. Astron. Astrophys.* 17:73–111.
Scalo, J. M. 1977*a*. Grain size control in dense interstellar clouds. *Astron. Astrophys.* 55:253–260.
Scalo, J. M. 1977*b*. On the frequency distribution of $^{12}C/^{13}C$ ratios in G-K giants and carbon stars. *Astrophys. J.* 215:194–199.
Scalo, J. M. 1978. The stellar mass spectrum. In *Protostars & Planets,* ed. T. Gehrels (Tucson: Univ. of Arizona Press), pp. 265–287.
Scalo, J. M. 1984. Turbulent velocity structure in interstellar clouds. *Astrophys. J.* 277:556–561.
Scalo, J. M. 1985*a*. The stellar initial mass function. *Fund. Cos. Phys.* In press.

Scalo, J. M. 1985*b*. Evidence for universality of the stellar initial mass function. Submitted to *Astrophys. J.*

Scalo, J. M. 1985*c*. Theoretical constraints on hierarchical fragmentation of the interstellar medium. In preparation.

Scalo, J. M., and Slavsky, D. B. 1980. Chemical structure of circumstellar shells. *Astrophys. J.* 239:L73–L77.

Scalo, J. M., and Pumphrey, W. A. 1982. Dissipation of supersonic turbulence in interstellar clouds. *Astrophys. J.* 258:L29–L33.

Scalo, J. M., and Struck-Marcell, C. 1985. Nonlinear behavior of a cloud fluid model: Time delays, limit cycles, and star formation bursts in galaxies. Submitted to *Astrophys. J.* (Lett.).

Schaeffer, O. A., Nagel, K., Fechtig, H., and Neukum, G. 1981. Space erosion of meteorites and the secular variation of cosmic rays (over 10^9 years). *Planet. Space Sci.* 29:1109–1118.

Schatzman, E. 1967. Cosmogony of the solar system and origin of the deuterium. *Ann. d'Astrophys.* 30:963–973.

Scheffler, H. 1967. Über das massenspektrum der interstellaren wolken. *Z. f. Astrophys.* 66:33–44.

Schiff, H. I., and Bohme, D. K. 1979. An ion-molecule scheme for the synthesis of hydrocarbon-chain and organonitrogen molecules in dense interstellar clouds. *Astrophys. J.* 232:740–746.

Schloerb, F. P., and Loren, R. B. 1982. CO and shocks related to the evolution of the Orion Nebula. In *Symposium on the Orion Nebula to Honor Henry Draper,* eds. A. E. Glassgold, P. J. Huggins, and E. L. Schucking (New York: Ann. N. Y. Acad. Sci.), 395:32–48.

Schloerb, F. P., and Snell, R. L. 1982. Hierarchical fragmentation and star formation in Heiles cloud 2. *Bull. Amer. Astron. Soc.* 14:969 (abstract).

Schloerb, F. P., and Snell, R. L. 1984. Large scale structure of molecular gas in Heiles cloud 2: A remarkable rotating ring. *Astrophys. J.* 283:129–139.

Schloerb, F. P., Snell, R. L., Langer, W. D., and Young, J. S. 1981. Detection of deuteriocyanobutadiyne (DC_5N) in the interstellar cloud TMC-1. *Astrophys. J.* 251:L37–L41.

Schloerb, F. P., Friberg, P., Hjalmarson, Å., Höglund, B., and Irvine, W. M. 1983*a*. Observations of sulfur dioxide in the Kleinmann-Low nebula. *Astrophys. J.* 264:161–171.

Schloerb, F. P., Snell, R. L., and Young, J. S. 1983*b*. Structure of dense molecular gas in TMC-1 from observations of three transitions of HC_3N. *Astrophys. J.* 267:163–173.

Schmidt, E. G. 1975. The structure of the Bok globule B361. *Mon. Not. Roy. Astron. Soc.* 172:401–409.

Schmidt, G. D., and Miller, J. S. 1979. The emission/reflection nature of HH 24. *Astrophys. J.* 234:L191–L194.

Schmidt, O. Yu. 1944. Meteoritic theory of the origin of the earth and planets. *Kokl. AN SSSR* 45:245–249 (in Russian).

Schmidt, O. Yu. 1957. *Chetyre Lektsii o Teorii Proiskhozhdeniya Zemli (Four Lectures on the Theory of the Earth's Origin),* 3rd ed. (Modkva: Izdatel'stvo AN SSSR, Russian; London: Lawrence and Wishart, 1959, English).

Schneider, S., and Elmegreen, B. G. 1979. A catalogue of dark globular filaments. *Astrophys. J. Suppl.* 41:87–95.

Schneps, M. H., Martin, R. N., Ho, P. T. P., and Barrett, A. H. 1978. The effects of rotation on microwave spectral line profiles: A study of CRL 437. *Astrophys. J.* 221:124–136.

Schneps, M. H., Ho, P. T. P., and Barrett, A. H. 1980. The formation of elephant-trunk globules in the Rosette nebula: CO observations. *Astrophys. J.* 240:84–98.

Schoenberg, E. 1964. *Veroffentl. Sternwerte. Munchen* 5, No. 21.

Scholl, H. 1979. History and evolution of Chiron's orbit. *Bull. Amer. Astron. Soc.* 11:801 (abstract).

Schramm, D. N. 1982. The r-process and nucleocosmochronology. In *Essays in Nuclear Astrophysics,* eds. C. A. Barnes, D. D. Clayton, and D. N. Schramm (Cambridge: Cambridge Univ. Press), pp. 325–353.

Schramm, D. N., and Wasserburg, G. J. 1970. Nucleochronologies and the mean age of the elements. *Astrophys. J.* 162:57–69.

Schramm, D. N., Tera, F., and Wasserburg, G. J. 1970. The isotopic abundance of ^{26}Mg and limits of ^{26}Al in the early solar system. *Earth Planet. Sci. Lett.* 10:44–59.

Schubert, G., Stevenson, D. J., and Ellsworth, K. 1981. Internal structures of the Galilean satellites. *Icarus* 47:46–59.

Schussler, M., and Schmitt, D. 1981. Comments on smoothed particle hydrodynamics. *Astron. Astrophys.* 97:373–379.
Schutz, B. F. 1979. General variational principle for normal modes of rotating stars. *Astrophys. J.* 232:874–877.
Schutz, B. F. 1983. Problems in astrophysical fluid dynamics. *Lectures Appl. Math.* 20:99–140.
Schwartz, P. R. 1982. New observations of the T Tauri radio source. Paper presented at Protostars and Planets conference, Tucson, AZ, January.
Schwartz, P. R., Cheung, A. C., Bologna, J. M., Chiu, M. F., Wrak, J. A., and Matsakis, D. 1977. Observations of ammonia in selected galactic regions. *Astrophys. J.* 218:671–676.
Schwartz, P. R., Thronson, H. A. Jr., Lada, C. J., Smith, H. A., Glaccum, W., Harper, D. A., and Knowles, S. H. 1983*a*. Far-infrared and submillimeter observations of stellar radiative and wind heating in S140-IRS. *Astrophys. J.* 271:625–631.
Schwartz, P. R., Waak, J. A., and Smith, H. A. 1983*b*. High-density gas associated with "molecular jets": NGC 1333 and NGC 2071. *Astrophys. J.* 267:L109–L114.
Schwartz, R. D. 1974. The T Tauri emission nebula. *Astrophys. J.* 191:419–432.
Schwartz, R. D. 1975. T Tauri nebulae and Herbig-Haro nebulae: Evidence for excitation by a strong stellar wind. *Astrophys. J.* 195:631–642.
Schwartz, R. D. 1977*a*. A survey of southern dark clouds for Herbig-Haro objects and H α emission stars. *Astrophys. J. Suppl.* 35:161–170.
Schwartz, R. D. 1977*b*. Evidence of star formation triggered by expansion of the Gum nebula. *Astrophys. J.* 212:L25–L26.
Schwartz, R. D. 1978. A shocked cloudlet model for Herbig-Haro objects. *Astrophys. J.* 223:884–900.
Schwartz, R. D. 1981. High dispersion spectra of Herbig-Haro objects: Evidence for shock-wave dynamics. *Astrophys. J.* 243:197–203.
Schwartz, R. D. 1983*a*. Herbig-Haro objects. *Ann. Rev. Astron. Astrophys.* 21:209–237.
Schwartz, R. D. 1983*b*. Herbig-Haro objects: An overview. *Rev. Mexicana Astron. Astrof.* 7:27–54.
Schwartz, R. D. 1983*c*. Ultraviolet continuum and H_2 fluorescent emission in Herbig-Haro objects 43 and 47. *Astrophys. J.* 268:L37–L40.
Schwarzschild, M. 1958. *Structure and Evolution of the Stars* (Princeton: Princeton Univ. Press).
Schweizer, F. 1976. Photometric studies of spiral structure: The disks and arms of six SbI and ScI galaxies. *Astrophys. J. Suppl.* 31:313–332.
Schweizer, F. 1983. Observational evidence for mergers. In *Internal Kinematics and Dynamics of Galaxies, IAU Symposium No. 100,* ed. E. Athanassoula (Dordrecht: D. Reidel), pp. 319–329.
Scott, E. H. 1984. Ambipolar diffusion in equilibrium self-gravitating gaseous configurations. I. *Astrophys. J.* 278:396–408.
Scott, E. H., and Black, D. C. 1980. Numerical calculations of the collapse of nonrotating, magnetic gas clouds. *Astrophys. J.* 239:166–172.
Scoville, N. Z. 1984. Astrophysical interpretation of molecular spectra. In *Galactic and Extragalactic Infrared Spectroscopy, XVIth Eslab Symposium,* eds. M. F. Kessler and J. P. Phillips (Dordrecht: D. Reidel), pp. 165–174.
Scoville, N. Z., and Solomon, P. M. 1975. Molecular clouds in the galaxy. *Astrophys. J.* 199:L105–L109.
Scoville, N. Z., and Hersh, K. 1979. Collisional growth of giant molecular clouds. *Astrophys. J.* 229:578–582.
Scoville, N. Z., and Wannier, P. G. 1979. Molecular cloud structure from 2-cm formaldehyde and 2.6-mm carbon monoxide lines. *Astron. Astrophys.* 76:140–149.
Scoville, N. Z., and Young, J. S. 1983. The molecular gas distribution in M51. *Astrophys. J.* 265:148–165.
Scoville, N. Z., Solomon, P. M., and Sanders, D. B. 1976. The galactic distribution (in radius and Z) of interstellar molecular hydrogen. In *The Structure and Content of the Galaxy and Galactic Gamma Rays,* eds. C. E. Fichtel and F. W. Stecker (Greenbelt: Goddard Space Flight Center), pp. 151–162.
Scoville, N. Z., Solomon, P. M., and Sanders, D. B. 1979. CO observations of spiral structure and the lifetime of giant molecular clouds. In *The Large Scale Characteristics of the Galaxy, IAU Symposium No. 84,* ed. W. B. Burton (Dordrecht: D. Reidel), pp. 277–283.

Scoville, N. Z., Krotkov, N. Z., and Wang, D. 1980. Collisional and infrared radiative pumping of molecular vibrational states: The CO infrared bands. *Astrophys. J.* 199:L105–L109.

Scoville, N. Z., Kleinmann, S. G., Hall, D. N. B., and Ridgway, S. T. 1983*a*. The circumstellar and nebular environment of the Becklin-Neugebauer object: λ = 2-5 micron spectroscopy. *Astrophys. J.* 275:201–224.

Scoville, N. Z., Young, J. S., and Lucy, L. B. 1983*b*. The distribution of molecular clouds in the nuclear region of NGC 1068. *Astrophys. J.* 270:443–464.

Searle, L., Sargent, W. L. W., and Bagnuolo, W. G. 1973. The history of star formation and the colors of late-type galaxies. *Astrophys. J.* 179:427–438.

Sears, D. W., Kallemeyn, G. W., and Wasson, J. T. 1982. The compositional classification of chondrites. II. The enstatite chondrite groups. *Geochim. Cosmochim. Acta* 46:597–608.

See, T. J. J. 1910. *Researches on the Evolution of the Stellar Systems, Vol. II, The Capture Theory of Cosmical Evolution* (Lynn, MA: R. P. Nichols & Sons), chapters 10 and 11.

Seiden, P. E., and Gerola, H. 1982. Propagating star formation. *Fund. Cosmic Phys.* 7:241–311.

Sekiya, M. 1983. Gravitational instabilities in a dust-gas layer and formation of planetesimals in the solar nebula. *Prog. Theor. Phys.* 69:1116–1130.

Sekiya, M., Nakazawa, K., and Hayashi, C. 1980*a*. Dissipation of the primordial terrestrial atmosphere due to irradiation of the solar EUV. *Prog. Theor. Phys.* 64:1968–1985.

Sekiya, M., Nakazawa, K., and Hayashi, C. 1980*b*. Dissipation of the rare gases contained in the primordial Earth's atmosphere. *Earth Planet. Sci. Lett.* 50:197–201.

Sekiya, M., Hayashi, C., and Nakazawa, K. 1981. Dissipation of the primordial terrestrial atmosphere due to irradiation of the solar far-UV during T Tauri stage. *Prog. Theor. Phys.* 66:1301–1316.

Sekiya, M., Miyama, S. M., and Hayashi, C. 1984. Accretion of the nebula gas onto the surface of the proto-Jupiter. In preparation.

Sellgren, K. 1983. Properties of young clusters near reflection nebulae. *Astron. J.* 88:985–997.

Sellgren, K. 1984. The near-infrared continuum emission of visual reflection nebulae. *Astrophys. J.* 277:623–633.

Serkowski, K., Mathewson, D. S., and Ford, V. L. 1975. Wavelength dependence of interstellar polarization and ratio of total to selective extinction. *Astrophys. J.* 196:261–290.

Shakura, N. J., and Sunyaev, R. A. 1973. Black holes in binary systems. Observational appearance. *Astron. Astrophys.* 24:337–355.

Shakura, N. J., and Sunyaev, R. A. 1977. A theory of the instability of disc accretion onto black holes and variability of binary x-ray source, galactic nuclei, and quasars. *Mon. Not. Roy. Astron. Soc.* 175:613–632.

Shakura, N. J., Sunyaev, R. A., and Zilitinkevich, S. S. 1978. On the turbulent energy transport in accretion disks. *Astron. Astrophys.* 62:179–187.

Sherwood, W. A., and Wilson, T. L. 1981. A comparison of visual extinction with H_2CO and H I absorption in Heiles cloud 2. *Astron. Astrophys.* 101:72–78.

Shields, G. A., and Tinsley, B. M. 1976. Composition gradients across spiral galaxies. II. The stellar mass limit. *Astrophys. J.* 203:66–71.

Shimamura, T., and Lugmair, G. W. 1984. Uranium isotopic abundance in Allende residue. *Lunar Planet. Sci.* XV:776–778 (abstract).

Shoemaker, E. M., and Wolfe, R. F. 1982. Cratering time scales for the Galilean satellites. In *Satellites of Jupiter,* ed. D. Morrison (Tucson: Univ. of Arizona Press), pp. 277–339.

Shoemaker, E. M., and Wolfe, R. F. 1984. Evolution of the Uranus-Neptune planetesimal swarm. *Lunar Planet. Sci.* XV:780–781 (abstract).

Shoemaker, E. M., Lucchitta, B. K., Plescia, J. B., Squyres, S. W., and Wilhelms, D. E. 1982. The geology of Ganymede. In *Satellites of Jupiter,* ed. D. Morrison (Tucson: Univ. of Arizona Press), pp. 435–520.

Shu, F. H. 1973. On the genetic relation between interstellar clouds and dust clouds. In *Interstellar Dust and Related Topics, IAU Sympoisum No. 52,* eds. J. M. Greenberg and H. D. van de Hulst (Dordrecht: D. Reidel), pp. 257–262.

Shu, F. H. 1977. Self-similar collapse of isothermal spheres and star formation. *Astrophys. J.* 214:488–497.

Shu, F. H. 1983. Ambipolar diffusion in self-gravitating isothermal layers. *Astrophys. J.* 273:202–213.

Shu, F. H. 1984*a*. Waves in planetary rings. In *Planetary Rings,* eds. R. Greenberg and A. Brahic (Tucson: Univ. of Arizona Press), pp. 513–561.
Shu, F. H. 1984*b*. Star formation in molecular clouds. In *The Milky Way Galaxy, IAU Symposium No. 106,* ed. H. van Woerden (Dordrecht: D. Reidel). In press.
Shu, F. H., and Terebey, S. 1984. The formation of cool stars from cloud cores. In *Third Cambridge Meeting on Cool Stars* (Cambridge: Cambridge Univ. Press). In press.
Shu, F. H., Yuan, C., and Lissauer, J. J. 1985. Nonlinear spiral density waves: An inviscid theory. *Astrophys. J.* In press.
Shull, J. M. 1980. Stellar winds in molecular clouds. *Astrophys. J.* 238:860–866.
Silk, J. 1977*a*. On the fragmentation of cosmic gas clouds. I. The formation of galaxies and the first generation of stars. *Astrophys. J.* 211:638–648.
Silk, J. 1977*b*. On the fragmentation of cosmic gas clouds. III. The initial stellar mass function. *Astrophys. J.* 214:718–724.
Silk, J. 1978. Fragmentation of molecular clouds. In *Protostars & Planets,* ed. T. Gehrels (Tucson: Univ. of Arizona Press), pp. 172–188.
Silk, J. 1981. Molecular clouds and star formation. In *Star Formation* (10th Advanced Course of the Swiss Society of Astronomy and Astrophysics, Saas-Fee).
Silk, J. 1982. Does fragmentation occur on protostellar mass scales during the dynamic collapse phase? *Astrophys. J.* 256:514–522.
Silk, J. 1983. The first stars. *Mon. Not. Roy. Astron. Soc.* 205:705–718.
Silk, J. 1985*a*. Macroscopic turbulence in molecular clouds. Preprint.
Silk, J. 1985*b*. Molecular cloud evolution and star formation. In *Birth and Infancy of Stars* (1983 Les Houches Lectures), eds. R. Lucas, A. Omont, and R. Stora (Amsterdam: North-Holland). In press.
Silk, J., and Takahashi, T. 1979. A statistical model for the initial mass function. *Astrophys. J.* 229:242–256.
Silk, J., and Norman, C. A. 1983. X-ray emission from pre-main-sequence stars, molecular clouds, and star formation. *Astrophys. J.* 272:L49–L53.
Sim, M. E. 1968. *Royal Obs. Edinburgh Publ.* 6, No. 8.
Simard-Normandin, M., and Kronberg, P. P. 1980. Rotation measures and the galactic magnetic field. *Astrophys. J.* 242:74–94.
Simon, M., and Cassar, L. 1984. Velocity resolved IR spectroscopy of LkH alpha-101. Preprint.
Simon, M., Righini-Cohen, G., Felli, M., and Fischer, J. 1981*a*. VLA observations of the Becklin-Neugebauer object, CRL 490, Monoceros R2 IRS 3, M8 E, and CRL 2591. *Astrophys. J.* 245:552–559.
Simon, M., Righini-Cohen, G., Fischer, J., and Cassar, L. 1981*b*. Velocity resolved spectroscopy of the Brackett-gamma line emission of CRL 490 and M17 IRS 1. *Astrophys. J.* 251:552–556.
Simon, M., Felli, M., Cassar, L., Fischer, J., and Massi, M. 1983. Infrared line and radio continuum emission of circumstellar ionized regions. *Astrophys. J.* 266:623–645.
Simon, T., Simon, M., and Joyce, R. R. 1979. Bα line survey of compact infrared sources. *Astrophys. J.* 230:127–132.
Simon, T., Schwartz, P. R., Dyck, H. M., and Zuckerman, B. 1983. New results on the binary companion of T Tauri. In *Activity in Red-Dwarf Stars, IAU Colloquium No. 71,* eds. P. B. Byrne and M. Rodono (Dordrecht: D. Reidel). In press.
Simonetti, J. H., Cordes, J. M., and Spangler, S. R. 1984. Small-scale variations in the galactic magnetic field: The rotation measure structure function and birefringence in interstellar scintillations. *Astrophys. J.* 284:126–134.
Singer, S. F. 1968. The origin of the moon and geophysical consequences. *Geophys. J. Roy. Astron. Soc.* 15:205–226.
Singer, S. F. 1971. The Martian satellites. In *Physical Studies of the Minor Planets,* ed. T. Gehrels (Washington: NASA SP-267), pp. 399–402.
Slattery, W. L. 1977. The structure of the planets Jupiter and Saturn. *Icarus* 32:58–72.
Slattery, W. L., and Hubbard, W. B. 1973. Statistical mechanics of light elements at high pressure. III. Molecular hydrogen. *Astrophys. J.* 181:1031–1038.
Slattery, W. L., DeCampli, W. M., and Cameron, A. G.W. 1980. Protoplanetary core formation by rain-out of minerals. *Moon and Planets* 23:381–390.

Slavsky, D., and Smith, H. J. 1981. Further evidence for rotation periods of Uranus and Neptune. *Bull. Amer. Astron. Soc.* 13:733 (abstract).

Smak, J. 1982. Accretion in close binaries. I. Modified α-discs with convection. *Acta Astron.* 32:199–224.

Smith, B. A., Soderblom, L. A., Johnson, T. V., Ingersoll, A. P., Collins, S. A., Shoemaker, E. M., Hunt, G. E., Masursky, H., Carr, M. H., Davies, M. E., Cook, A. F. II. Boyce, J., Danielson, G. E., Owen, T., Sagan, C., Beebe, R. F., Veverka, J., Strom, R. G., McCauley, J. F., Morrison, D., Briggs, G. A., and Suomi, V. E. 1979*a*. The Jupiter system through the eyes of Voyager 1. *Science* 204:13–31.

Smith, B. A., Soderblom, L. A., Beebe, R. F., Boyce, J., Briggs, G. A., Carr, M. H., Collins, S. A., Cook, A.F. II, Danielson, G. E., Davies, M. E., Hunt, G. E., Ingersoll, A. P., Johnson, T. V., Masursky, H., McCauley, J., Morrison, D., Owen, T., Sagan, C., Shoemaker, E. M., Strom, R. G., Suomi, V. E., and Veverka, J. 1979*b*. The Galilean satellites and Jupiter: Voyager 2 imaging science results. *Science* 206:927–950.

Smith, B. A., Soderblom, L. A., Beebe, R. F., Boyce, J., Briggs, G. A., Bunker, A., Collins, S. A., Hansen, C., Johnson, T. V., Mitchell, J., Terrile, R., Carr, M. H., Cook, A. F. II, Cuzzi, J. N., Pollack, J. B., Danielson, G. E., Ingersoll, A. P., Davies, M. E., Hunt, G. E., Owen, T., Sagan, C., Veverka, J., Strom, R. G., and Suomi, V. E. 1981. Encounter with Saturn: Voyager 1 imaging results. *Science* 212:163–191.

Smith, B. A., Soderblom, L. A., Batson, R., Bridges, P., Inge, J., Masursky, H., Shoemaker, E. M., Beebe, R. F., Boyce, J., Briggs, G. A., Bunker, A., Collins, S. A., Hansen, C., Johnson, T. V., Mitchell, J., Terrile, R., Cook, A. F. II, Cuzzi, J. N., Pollack, J. B., Danielson, G. E., Ingersoll, A. P., Davies, M. E., Hunt, G. E., Morrison, D., Owen, T., Sagan, C., Veverka, J., Strom, R. G., and Suomi, V. E. 1982. A new look at the Saturn system. *Science* 215:504–537.

Smith, D., and Adams, N. G. 1978. Molecular synthesis in interstellar clouds. Radiative association reactions of $CH_3{}^+$ ions. *Astrophys. J.* 220:L87–L92.

Smith, D., and Adams, N. G. 1984. Dissociative recombination coefficients for $H^+{}_3$ HCO^+, N_2H^+, and $CH^+{}_5$ at low temperature: Interstellar implications. *Astrophys J.* 284:L13–L16.

Smith, G. H. 1982. The role of HII regions during star formation and chemical enrichment in globular clusters. *Astrophys. J.* 259:607–616.

Smith, G. H. 1985. Possible consequences of gas accretion on the initial mass function of star clusters. *Astrophys. J.* In press.

Smith, H. A., Thronson, H. A. Jr., Lada, C. J., Harper, D. A., Loewenstein, R. F., and Smith, J. 1982. Far-infrared observations of FU Orionis. *Astrophys. J.* 258:170–176.

Smith, J. A. 1980. Cloud-cloud collisions in the interstellar medium. *Astrophys. J.* 238:842–852.

Smith, J. W., and Kaplan, I. R. 1970. Endogenous carbon in carbonaceous meteorites. *Science* 167:1367–1370.

Smith, L. F., Biermann, P., and Mezger, P. 1978. Star formation rates in the galaxy. *Astron. Astrophys.* 66:65–76.

Smith, M. A., Beckers, J. M., and Barden, S. C. 1983. Rotation among Orion in G stars: Angular momentum loss considerations in pre-main-sequence stars. *Astrophys. J.* 271:237–254.

Smoluchowski, R. 1967. Internal structure and energy emission of Jupiter. *Nature* 215:691–695.

Smoluchowski, R. 1973. Dynamics of the Jovian interior. *Astrophys. J.* 185:L95–L99.

Smoluchowski, R. 1975. Jupiter's molecular hydrogen layer and the magnetic field. *Astrophys. J.* 200:L119.

Smoluchowski, R. 1979*a*. Origin of the magnetic fields in the giant planets. *Phys. Earth Planet. Int.* 20:247–254.

Smoluchowski, R. 1979*b*. The ring systems of Jupiter, Saturn, and Uranus. *Nature* 280:377–378.

Snell, R. L. 1981. A study of nine interstellar dark clouds. *Astrophys. J. Suppl.* 45:121–175.

Snell, R. L. 1983. Bipolar molecular outflows near Herbig-Haro objects. *Rev. Mexicana Astron. Astrof.* 7:79–94.

Snell, R. L., and Loren, R. B. 1977. Self-reversed CO profiles in collapsing molecular clouds. *Astrophys. J.* 211:122–127.

Snell, R. L., and Wootten, A. 1979. Observations of interstellar HNC, DNC and $HN^{13}C$: Temperature effects on deuterium fractionation. *Astrophys. J.* 228:748–758.

Snell, R. L., and Edwards, S. 1981. High velocity molecular gas near Herbig-Haro objects HH 7-11. *Astrophys. J.* 251:103–107.
Snell, R. L., and Edwards, S. 1983. A survey of high velocity molecular gas in the vicinity of Herbig-Haro objects. I. *Astrophys. J.* 270:605–619.
Snell, R. L., and Edwards, S. 1984. A survey of high velocity molcular gas near Herbig-Haro objects. II. *Astrophys. J.* 281:237–249.
Snell, R. L., Loren, R. B., and Plambeck, R. L. 1980. Observations of CO in L1551: Evidence for stellar wind driven shocks. *Astrophys. J.* 239:L17–L22.
Snell, R. L., Schloerb, F. P., Young, J. S., Hjalmarson, Å., and Friberg, P., 1981. Observations of HC_3N, HC_5N., and HN_7N in molecular clouds. *Astrophys. J.* 244:45–53.
Snell, R. L., Langer, W. D., and Frerking, M. A. 1982. Determination of density structure in dark clouds from CS observations. *Astrophys. J.* 255:149–159.
Snell, R. L., Mundy, L. G., Goldsmith, P. F., Evans, N. J. II, and Erickson, N. R. 1984. Models of molecular clouds: I. Multitransition studies of CS. *Astrophys. J.* 276:625–645.
Snellenberg, J. W. 1978. A petrogenetic grid applicable to chondrule formation. *Lunar Planet. Sci.* IX:1080–1081 (abstract).
Snow, T. P., and Seab, C. G. 1980. An anomalous ultraviolet extinction curve in the Taurus dark cloud. *Astrophys. J.* 242:L83–L86.
Soifer, B. T., Willner, S. P., Capps, R. W., and Rudy, R. J. 1981. 4-8 micron spectrophotometry of OH 0739-14. *Astrophys. J.* 250:631–635.
Solf, J. 1980. The spectrum and structure of the bipolar nebula S106. *Astron. Astrophys.* 92:51–56.
Solf, J., and Carsenty, U. 1982. The kinematical structure of the bipolar nebula S106. *Astron. Astrophys.* 113:142–149.
Solomon, P. 1978. Physics of molecular clouds from millimeter wave line observations. In *Infrared Astronomy,* ed. G. Fazio and G. Setti (Dordrecht: D. Reidel), pp. 97–115.
Solomon, P. M. 1981. The molecular cloud distribution in N981 between 3 and 15 kpc. In *NRAO Symposium on Extragalactic Molecules,* eds. L. Blitz, and M. Kutner (Green Bank: NRAO Publications Division), pp. 41–50.
Solomon, P. M., and Wickramasinghe, N. C. 1969. Molecular hydrogen in dense interstellar clouds. *Astrophys. J.* 158:449–460.
Solomon, P. M., and Werner, W. M. 1971. Low-energy cosmic rays and the abundance of atomic hydrogen in dark clouds. *Astrophys. J.* 165:41–49.
Solomon, P. M., and Woolf, N. J. 1973. Interstellar deuterium: Chemical fractionation. *Astrophys. J.* 180:L89–L92.
Solomon, P. M., and Sanders, D. B. 1980. Giant molecular clouds as the dominant component of interstellar matter in the galaxy. In *Giant Molecular Clouds in the Galaxy,* eds. P. M. Solomon and M. G. Edmunds (New York: Pergamon), pp. 41–73.
Solomon, P. M., Sanders, D. B., and Scoville, N. Z. 1979*a*. Giant molecular clouds in the galaxy: Distribution, mass, size, and age. In *The Large Scale Characteristics of the Galaxy, IAU Symposium No. 84,* ed. W. B. Burton (Dordrecht: D. Reidel), pp. 35–54.
Solomon, P. M., Scoville, N. Z., and Sanders, D. B. 1979*b*. Giant molecular clouds in the galaxy: The distribution of ^{12}CO emission in the galactic plane. *Astrophys. J.* 232:L89–L94.
Solomon, P. M., Barrett, J. M., Sanders, D. B., and deZafra, R. 1983*a*. CO emission and the optical disk in the giant Sc galaxy M101. *Astrophys. J.* 266:L103–L106.
Solomon, P. M., Stark, A. A., and Sanders, D. B. 1983*b*. CO emission in the outer galaxy between longitudes 50° and 72°. *Astrophys. J.* 267:L29–L36.
Solomon, P. M., Sanders, D. B., Scoville, N. Z., and Clemens, D. P. 1985. Massachusetts-Stony Brook galactic plane CO survey: l-b maps of the first quadrant. *Astrophys. J. Suppl.* In press.
Sonett, C. P., Colburn, D. S., and Schwartz, K. 1968. Electrical heating of meteorite parent bodies and planets by dynamo induction from a pre-main sequence T-Tauri "solar wind." *Nature* 219:924–926.
Spencer, R. G., and Leung, C. M. 1978. Infrared radiation from dark globules. *Astrophys. J.* 222:140–152.
Spitzer, L. 1942. The dynamics of the interstellar medium. III. Galactic distribution. *Astrophys. J.* 95:329–394.

Spitzer, L. 1958. Disruption of galactic clusters. *Astrophys. J.* 127:17–27.
Spitzer, L. Jr. 1978. *Physical Processes in the Interstellar Medium* (New York: Wiley).
Spitzer, L. 1982. Acoustic waves in supernova remnants. *Astrophys. J.* 262:315–321.
Spitzer, L., and Härm, R. 1958. Evaporation of stars from isolated clusters. *Astrophys. J.* 127:544–550.
Spitzer, L., and Jenkins, E. B. 1975. Ultraviolet studies of the interstellar gas. *Ann. Rev. Astron. Astrophys.* 13:133–164.
Spitzer, L. Jr., Drake, J. F., Jenkins, E. B., Morton, D. C., Rogerson, J. B., and York, D. G. 1973. Spectrophotometric results from the Copernicus satellite. *Astrophys. J.* 181:L116–L121.
Spitzer, L. Jr., Cochran, W. D., and Hirshfield, A. 1974. Column densities of interstellar molecular hydrogen. *Astrophys. J. Suppl.* 28:373–389.
Squyres, S. W., Reynolds, R. T., Cassen, P. M., and Peale, S. J. 1983*a*. Liquid water and active resurfacing of Europa. *Nature* 301:225–226.
Squyres, S. W., Reynolds, R. T., Cassen, P. M., and Peale, S. J. 1983*b*. The evolution of Enceladus. *Icarus* 53:319–331.
Srinivasan, B., and Anders, E. 1978. Noble gases in the Murchison meteorite: Possible relics of s-process nucleosynthesis. *Science* 201:51–56.
Stahler, S. W. 1983*a*. The birthline for low mass stars. *Astrophys. J.* 274:822–829.
Stahler, S. W. 1983*b*. The equilibria of rotating, isothermal clouds. I. Method of solution. *Astrophys. J.* 268:155–164.
Stahler, S. W. 1983*c*. The equilibria of rotating, isothermal clouds. II. Structure and dynamical stability. *Astrophys. J.* 268:165–184.
Stahler, S. W. 1984*a*. The cyanopolynes as a chemical clock for molecular clouds. *Astrophys. J.* 281:209–218.
Stahler, S. W. 1984*b*. A chemical clock for molecular clouds. Submitted to *Astrophys. J.*
Stahler, S. W., Shu, F. H., and Taam, R. E. 1980*a*. The evolution of protostars. I. Global formulation and results. *Astrophys. J.* 241:637–654.
Stahler, S. W., Shu, F. H., and Taam, R. E. 1980*b*. The evolution of protostars. II. The hydrostatic core. *Astrophys. J.* 242:226–241.
Stahler, S. W., Shu, F. H., and Taam, R. E. 1981. The evolution of protostars. III. The accretion envelope. *Astrophys. J.* 248:727–737.
Stal'bovski, O. I., and Schevchenco, V. S. 1981. The structure of star formation regions. III. Individual regions: Spatial extent, mass, and age of SFR Sagittarius I. *Soviet Astron.* 25:25–32.
Stark, A. A. 1979. Ph.D. thesis, Princeton University.
Stark, A. A. 1984. Kinematics of molecular clouds. I. Velocity dispersion in the solar neighborhood. *Astrophys. J.* 281:624–633.
Staude, H. J., Lenzen, R., Dyck, H. M., and Schmidt, G. D. 1982. The bipolar nebula S106: Photometric, polarimetric, and spectropolarimetric observations. *Astrophys. J.* 255:95–102.
Stauffer, J. R. 1980. Observations of pre-main-sequence stars in the Pleiades. *Astron. J.* 85:1341–1353.
Stauffer, J. R. 1984. Optical and infrared photometry of late type stars in the Pleiades. *Astrophys. J.* 280:189–201.
Stenholm, L. G. 1983. Structure of molecular clouds. *Astron. Astrophys.* 117:41–45.
Stenholm, L. G. 1984. The fluctuation spectrum of molecular clouds. *Astron. Astrophys.* 137:133–137.
Stephenson, G. 1961. The gravitational instability of an infinite homogeneous rotating viscous medium in the presence of a magnetic field. *Mon. Not. Roy. Astron. Soc.* 122:455–459.
Stevenson, D. J. 1975. Thermodynamics and phase separation of dense fully ionized hydrogen-helium mixtures. *Phys. Rev.* 12B:3999–4007.
Stevenson, D. J. 1980. Saturn's luminosity and magnetism. *Science* 208:746–748.
Stevenson, D. J. 1981. Models of the Earth's core. *Science* 165:611–619.
Stevenson, D. J. 1982*a*. Interiors of the giant planets. *Ann. Rev. Earth Planet. Sci.* 10:257–295.
Stevenson, D. J. 1982*b*. Formation of the giant planets. *Planet. Space Sci.* 30:755–764.
Stevenson, D. J. 1983. Structure of the giant planets: Evidence for nucleated instabilities and post-formational accretion. *Proc. Lunar Planet. Sci. Conf.* (abstract).
Stevenson, D. J. 1984. On forming the giant planets quickly (superganymedean puff balls). *Lunar Planet. Sci.* XV:822–823 (abstract).

Stevenson, D. J., and Salpeter, E. E. 1976. Interior models of Jupiter. In *Jupiter,* ed. T. Gehrels (Tucson: Univ. of Arizona Press), pp. 85–112.
Stevenson, D. J., and Salpeter, E. E. 1977*a*. The phase diagram and transport properties of hydrogen-helium fluid planets. *Astrophys. J. Suppl.* 35:221–237.
Stevenson, D. J., and Salpeter, E. E. 1977*b*. The dynamics and helium distribution in hydrogen-helium fluid planets. *Astrophys. J. Suppl.* 35:239–261.
Stevenson, D. J., Harris, A. W., and Lunine, J. I. 1984. Origins of satellites. In *Satellites of the Solar System,* eds. J. A. Burns, D. Morrison, and M. S. Matthews (Tucson: Univ. of Arizona Press).
Stewart, G. R., and Kaula, W. M. 1980. A gravitational kinetic theory for planetesimals. *Icarus* 44:154–171.
Stewart, G. R., Lin, D. N. C., and Bodenheimer, P. 1984. Collision-induced transport processes in planetary rings. In *Planetary Rings,* eds. R. Greenberg and A. Brahic (Tucson: Univ. of Arizona Press), pp. 447–512.
Stoeckly, R. 1965. Polytropic models with fast, nonuniform rotation. *Astrophys. J.* 142:208–228.
Stone, E. C., and Miner, E. D. 1982. Voyager 2 encounter with the Saturnian system. *Science* 215:499–504.
Stone, M. E. 1970. Collisions between H I clouds. II. Two-dimensional model. *Astrophys. J.* 159:293–307.
Storzer, D., and Pellas, P. 1977. Angra dos Reis: Plutonium distribution and cooling history. *Earth Planet. Sci. Lett.* 35:285–293.
Straizys, V., and Meistas, E. 1980. Interstellar extinction in the dark Taurus clouds. I. *Acta Astron.* 30:541–552.
Strazzulla, G., Calcagno, L., and Foti, G. 1983. Polymerization induced on interstellar grains by low energy cosmic rays. *Mon. Not. Roy. Astron. Soc.* 204:59–62.
Strittmatter, P. A. 1966. Gravitational collapse in the presence of a magnetic field. *Mon. Not. Roy. Astron. Soc.* 132:359–378.
Strittmatter, P. A., Woolf, N. J., Thompson, R. I., Wilderson, S., Angel, J. R. P., Stockman, H. S., Gilbert, G., Grandi, S. A., Larson, H., and Fink, U. 1977. The spectral development of nova Cygni 1975. *Astrophys. J.* 216:23–32.
Strobel, D. J. 1982. Chemistry and evolution of Titan's atmosphere. *J. Planet. Space Sci.* 30:833–838.
Strom, K. M., Strom, S. E., and Grasdalen, G. L. 1974*a*. An infrared source associated with a Herbig-Haro object. *Astrophys. J.* 187:83–86.
Strom, K. M., Strom, S.E., and Kinman, T. 1974*b*. Optical polarization of selected Herbig-Haro objects. *Astrophys. J.* 191:L93–L94.
Strom, K. M., Strom, S. E., and Vrba, F. J. 1976. Infrared surveys of dark cloud complexes. I. The Lynds 1630 dark cloud. *Astron. J.* 81:308–313.
Strom, K. M., Strom, S. E., and Stocke, J. 1983. Optical study of a possible bipolar flow associated with HH 12. *Astrophys. J.* 271:L23–L26.
Strom, S. E. 1983. Recent progress in the study of young stellar objects. *Rev. Mexicana Astron. Astrof.* 7:201–218.
Strom, S. E., Strom, K. M., Yost, J., Carrasco, L., and Grasdalen, G. 1972. The nature of the Herbig Ae and Be-type stars associated with nebulosity. *Astrophys. J.* 173:353–366.
Strom, S. E., Grasdalen, G. L., and Strom, K. M. 1974. Infrared and optical observations of Herbig-Haro objects. *Astrophys. J.* 191:111–142.
Strom, S. E., Jensen, E. B., and Strom, K. 1976. Density waves in the disks of two spiral galaxies. *Astrophys. J.* 206:L11–L14.
Struck-Marcell, C., and Scalo, J. M. 1984. Continuum models for gas in disturbed galaxies. II. Stability of simplified model systems. *Astrophys. J.* 277:132–148.
Struve, O. 1937. On the interpretation of the surface brightness of diffuse galactic nebulae. *Astrophys. J.* 85:194–212.
Struve, O., and Elvey, C. T. 1936. Photometric observations of some of Barnard's dark nebulae. *Astrophys. J.* 83:162–172.
Suess, H. E. 1963. Properties of chondrules-I. In *Origin of the Solar System,* eds. R. Jastrow and A. G. W. Cameron (New York: Academic Press), pp. 143–146.
Suess, H. E. 1965. Chemical evidence bearing on the origin of the solar system. *Ann. Rev. Astron. Astrophys.* 3:217–234.

Sutton, E. C., Blake, G., Masson, C., and Phillips, T. G. 1984. Molecular line survey of Orion A from 215 to 247 GHz. Submitted to *Astrophys. J.*

Suzuki, H. 1979. Molecular evolution in interstellar clouds. I. *Prog. Theor. Phys.* 62:936–956.

Suzuki, H. 1983. Synthesis of chain molecules in partially ionized carbon regions. *Astrophys. J.* 272:579–590.

Swank, J. H., White, N. E., Holt, S. S., and Becker, R. H. 1981. Two component x-ray emission from RS Canum Venaticorum binaries. *Astrophys. J.* 246:208–214.

Swart, P. K., Grady, M. M., Wright, I. P., and Pillinger, C. T. 1982. Carbon components and their isotopic composition in the Allende meteorite. *J. Geophys. Res. Suppl.* 87:A283–A289.

Swart, P. K., Grady, M. M., Pillinger, C. T., Lewis, R. S., and Anders, E. 1983. Interstellar carbon in meteorites. *Science* 220:406–410.

Sweet, P. A. 1950. The effect of turbulence on a magnetic field. *Mon. Not. Roy. Astron. Soc.* 110:69–83.

Swings, P., and Rosenfeld, L. 1937. Considerations regarding interstellar molecules. *Astrophys. J.* 86:483–485.

Swinney, H. L., and Gollub, H. P. 1978. The transition to turbulence. *Phys. Today* 31(8):41–49.

Tacconi, L. J., and Young, J. S. 1984. CO abundances and star formation in the three irregular galaxies NGC 4449, 4214 and 3738. *Bull. Amer. Astron. Soc.* 16:538 (abstract).

Tajima, T., and Leboeuf, J. N. 1980. Kelvin-Helmholtz instability in supersonic and super-Alfvénic fluids. *Phys. Fluids* 23:884–888.

Takahashi, T., Hollenbach, D. J., and Silk, J. 1983. H_2O heating in molecular clouds: Line transfer and thermal balance in a warm dusty medium. *Astrophys. J.* 275:145–162.

Takano, T., Stutzki, J., Winnewisser, G., and Fukui, Y. 1984. Role of the rotating molecular disk around the bipolar flow source near NGC 2071. Preprint.

Takeda, H., Matsuda, T., Sawada, K., and Hayashi, C. 1985. Drag on a gravitating sphere moving through a gas body. *Prog. Theor. Phys.* In press.

Tammann, G. A. 1977. A progress report on supernova statistics. In *Supernovae,* ed. D. N. Schramm (Dordrecht: D. Reidel), pp. 95–116.

Tanaka, T., and Masuda, A. 1973. Rare-earth elements in matrix, inclusions and chondrules of the Allende meteorite. *Icarus* 4:523–530.

Tarafdar, S. P., Prasad, S. S., and Huntress, W. T. 1983. Dependence of interstellar depletion on hydrogen column density: Possible implications. *Astrophys. J.* 267:156–162.

Tarafdar, S. P., Prasad, S. S., Huntress, W. T., Villere, K. R., and Black, D. C. 1985. Chemistry in dynamically evolving clouds. *Astrophys. J.* In press.

Tassoul, J.-L. 1978. *Theory of Rotating Stars* (Princeton: Princeton Univ. Press).

Tassoul, J.-L., and Ostriker, J. P. 1968. On the oscillations and stability of rotating stellar models. I. Mathematical techniques. *Astrophys. J.* 154:613–626.

Tassoul, J.-L., and Ostriker, J. P. 1970. Sur la stabilité séculaire des polytrope en equilibre relatif. *Astron. Astrophys.* 4:423–427.

Tatarskii, V. I. 1961. *Wave Propagation in a Turbulent Medium* (New York: McGraw Hill).

Tatsumoto, M., and Shimamura, T. 1980. Evidence for live ^{247}Cm in the early solar system. *Nature* 286:118–122.

Taylor, G. I. 1921. Diffusion by continuous movements. *Proc. London Math Soc.,* Ser. 2, 20:196–211.

Taylor, G. I. 1935. Statistical theory of turbulence. Parts I–IV. *Proc. Roy. Soc.* A151:421–478.

Tennekes, H., and Lumley, J. L. 1972. *A First Course in Turbulence* (Cambridge: MIT Press).

Tenorio-Tagle, G. 1977. The time evolution of a globule immersed in an HII region. *Astron. Astrophys.* 54:517–524.

Tenorio-Tagle, G. 1979. The gas dynamics of HII regions. *Astron. Astrophys.* 71:59–65.

Tenorio-Tagle, G. 1980. The formation of super-rings. *Astron. Astrophys.* 88:61–65.

Tenorio-Tagle, G. 1981. The collision of clouds with a galactic disk. *Astron. Astrophys.* 94:338–344.

Tenorio-Tagle, G., and Rozyczka, M. 1984. Bullets, interstellar plops and plunks. Preprint.

Terebey, S., Shu, F. H., and Cassen, P. 1984*a*. The collapse of the cores of slowly rotating isothermal clouds. *Astrophys. J.* 286:529–551.

Terebey, S., Shu, F. H., and Cassen, P. M. 1984*b*. Rotating cloud collapse: 10^{18} cm to 10^{11} cm. Presented at Protostars and Planets II conference, Tucson, Arizona, January.

ter Haar, D. 1950. Further studies on the origin of the solar system. *Astrophys. J.* 111:179–190.
Thaddeus, P. 1977. Molecular clouds. In *Star Formation,* eds. T. de Jong and A. Maeder (Dordrecht: D. Reidel), p. 37–54.
Thaddeus, P. 1981. Radio observations of molecules in the interstellar gas. In *Molecules in Interstellar Space,* eds. A. Carrington, and D. A. Ramsay (London: The Royal Society), pp. 5–15.
Thaddeus, P., and Dame, T. M. 1984. In *Edinburgh Workshop in Star Formation,* ed. R. D. Wolstancroft (Edinburgh: Royal Observatory), pp. 15–26.
Thaddeus, P., Kutner, M. L., Penzias, A. A., Wilson, R. W., and Jefferts, K. B. 1972. Interstellar hydrogen sulfide. *Astrophys. J.* 176:L73–L76.
Thaddeus, P., Cummins, S. E., and Linke, R. A. 1984. Detection of the SiCC radical towards IRC+10216: The first molecular ring in an astronomical source. *Astrophys. J.* 283:L45–L48.
Thiemens, M. H., and Heidenreich, J. E. 1983. The mass-independent fractionation of oxygen: A novel isotope effect and its possible cosmochemical implications. *Science* 219:1073–1075.
Thiemens, M. H., Heidenreich, J. E., and Lundberg, L. 1983. Photochemical production of isotopic anomalies in the early solar system. *Lunar Planet. Sci.* XIV: 785–786 (abstract).
Thompson, R. I. 1982. Excess line emission in protostellar objects. *Astrophys. J.* 257:171–178.
Thompson, R. I. 1984. Lyman and Balmer continuum ionization in ZAMS stars: Applications to the line excess phenomenon. *Astrophys. J.* In press.
Thompson, R. I., and Tokunaga, A. T. 1978. Analysis of obscured infrared objects. II. Allen's infrared source NGC 2264. *Astrophys. J.* 226:119–123.
Thompson, R. I., and Tokunaga, A. T. 1979. Infrared spectroscopy of lineless objects associated with star formation regions. *Astrophys. J.* 229:153–157.
Thompson, R. I., and Tokunaga, A. T. 1980. Analysis of obscured infrared objects. VI. H and He lines in W51 and K3-50. *Astrophys. J.* 235:889–893.
Thompson, R. I., Strittmatter, P. A., Erickson, E. F., Witteborn, F. C., and Strecker, D. W. 1977. Observation of pre-planetary disks around MWC 349 and LKH α 101. *Astrophys. J.* 218:170–180.
Thronson, H. A. Jr., and Harper, D. A. 1979. Compact HII regions in the far-infrared. *Astrophys. J.* 230:133–148.
Tielens, A. G. G. M., and Hagen, W. 1982. Model calculations of the molecular composition of interstellar grain mantles. *Astron. Astrophys.* 114:245–260.
Tielens, A. G. G. M., and Hollenbach, D. J. 1984. CI and the lifetime of molecular clouds. Submitted to *Astrophys. J.*
Toelle, F., Ungerechts, H., Walmsley, C. M., Winnewisser, G., and Churchwell, E. 1981. A molecular line study of the elongated dark dust cloud TMC 1. *Astron. Astrophys.* 95:143–155.
Tohline, J. E. 1980*a*. Fragmentation of rotating protostellar clouds. *Astrophys. J.* 235:866–881.
Tohline, J. E. 1980*b*. The gravitational fragmentation of primordial gas clouds. *Astrophys. J.* 239:417–427.
Tohline, J. E. 1982. Hydrodynamic collapse. *Fund. Cos. Phys.* 8:1–82.
Tohline, J. E. 1984. Do we understand rotating isothermal collapses yet? Submitted to *Icarus.*
Tohline, J. E. 1985. Star formation: Phase transition, not Jeans instability. Preprint.
Tohline, J. E., Durisen, R. H., and McCollough, M. 1984. The linear and nonlinear dynamic stability of rapidly rotating $n = 3/2$ polytropes. In preparation.
Tokunaga, A. T., Knacke, R. F., Ridgway, S. T., and Wallace, L. 1979. High resolution spectra of Jupiter in the 744-980 inverse centimeter spectral range. *Astrophys. J.* 232:603–615.
Tokunaga, A. T., Lebofsky, M. J., and Rieke, G. H. 1981. Infrared reflection nebulae in S106 and NGC 7538 E. *Astron. Astrophys.* 99:108–110.
Tölle, F., Ungerechts, H., Walmsley, C. M., Winnewisser, G., and Churchwell, E. 1981. A molecular line study of the elongated dark dust cloud TMC-1. *Astron. Astrophys.* 95:143–155.
Tomisaka, K. 1984. Coagulation of interstellar clouds in spiral gravitational potential and formation of giant molecular clouds. Preprint.
Tomita, Y., Saito, T., and Ohtani, H. 1979. The structure and dynamics of large globules. *Publ. Astron. Soc. Japan* 31:407–416.
Toomre, A. 1964. On the gravitational stability of a disk of stars. *Astrophys. J.* 139:1217–1238.

Toomre, A. 1981. What amplifies the spirals? In *The Structure and Evolution of Normal Galaxies,* eds. S. M. Fall and D. Lynden-Bell (London/New York: Cambridge Univ. Press), pp. 111–136.

Torbett, M. V. 1983. Stellar formation through disk accretion bipolar nebulae and the T Tauri phase. *Bull. Amer. Astron. Soc.* 14:957 (abstract).

Torbett, M. V. 1984*a*. Hydrodynamic ejection of bipolar flows from objects undergoing disk accretion: T Tauri stars, massive pre-main sequence objects, and cataclysmic variables. *Astrophys. J.* 278:318–325.

Torbett, M. V. 1984*b*. Gas drag and planetesimal accumulation. Submitted to *Icarus.*

Torbett, M. V., and Smoluchowski, R. 1979. The structure and the magnetic field of Uranus. *Geophys. Res. Lett.* 6:675–676.

Torbett, M. V., and Smoluchowski, R. 1980*a*. Hydromagnetic dynamo in the cores of Uranus and Neptune. *Nature* 286:237–239.

Torbett, M. V., and Smoluchowski, R. 1980*b*. Sweeping of Jovian resonances and the evolution of the asteroids. *Icarus* 44:722–729.

Torrelles, J. M., Rodriguez, L. F., Cantó, J., Carral, P., Marcaide, J., Moran, J. M., and Ho, P. T. P. 1983. Are interstellar toroids the focusing agent of the bipolar molecular outflows? *Astrophys. J.* 274:214–230.

Torrelles, J. M., Rodriguez, L. F., Cantó, J., Carral, P., Marcaide, J., Moran, J. M., and Ho, P. T. P. 1984. Interstellar toroids as the focusing agent of the bipolar molecular outflows. *Astrophys. J.* In press.

Townes, C. 1976. Interstellar molecules. *Mem. Soc. Roy. Liège* 9:453–474.

Townes, C. H., Genzel, R., Watson, D. M., and Storey, J. W. V. 1983. Detection of interstellar NH_3 in the far-infrared: Warm and dense gas in Orion-KL. *Astrophys. J.* 269:L11–L16.

Townsend, A. A. 1976. *The Structure of Turbulent Shear Flow* (Cambridge: Cambridge Univ. Press).

Townsend, A. A., and Stewart, R. W. 1951. *Phil. Trans. Roy. Soc. London* A243:48–50.

Trafton, L. 1974. The source of Neptune's internal heat and the value of Neptune's tidal dissipation factor. *Astrophys. J.* 193:477–480.

Trafton, L., and Ramsay, D. A. 1980. The D/H ratio in the atmosphere of Uranus: Detection of the R_5 (.1) line of HD. *Icarus* 41:423–429.

Tranger, J. T., Roesler, F. L., and Mickelson, M. E. 1977. The D/H ratio on Jupiter, Saturn, and Uranus based on new HD and H_2 data. *Bull. Amer. Astron. Soc.* 91:516 (abstract).

Troland, T. H., and Heiles, C. 1982*a*. The Zeeman effect in 21 centimeter line radiation: Methods and initial results. *Astrophys. J.* 252:179–192.

Troland, T. H., and Heiles, C. 1982*b*. Magnetic field measurements in two expanding H I shells. *Astrophys. J.* 260:L19–L22.

Trubitsyn, V. P. 1965. Equation of state of solid hydrogen. *Sov. Phys. Solid State* 7:2708–2714.

Trubitsyn, V. P. 1971. Phase diagrams of hydrogen and helium. *Sov. Astron. A. J.* 15:303–309.

Truran, J. W., and Cameron, A. G. W. 1978. ^{26}Al production in explosive carbon burning. *Astrophys. J.* 219:226–229.

Tscharnuter, W. M. 1978. Collapse of the presolar nebula. *Moon and Planets* 19:229–236.

Tscharnuter, W. M. 1980. 1D, 2D and 3D collapse of interstellar clouds. In *Stellar Hydrodynamics,* ed. A. N. Cox and D. S. King (Los Alamos, NM: Los Alamos National Lab.); also *Space Sci. Rev.* 27:235–246.

Tscharnuter, W. H. 1981. Accumulation of a rapidly rotating protostar and formation of an associated nebula. In *Fundamental Problems in the Theory of Stellar Evolution, IAU Symposium No. 93,* eds. D. Sugimoto, D. Q. Lamb, and D. N. Schramm (Dordrecht: D. Reidel), pp. 105–106.

Tscharnuter, W. M. 1985. In *The Birth and Infancy of Stars,* ed. R. Lucas, A. Omont, and L. R. Stora (Amsterdam: North-Holland). In press.

Tscharnuter, W. M., and Winkler, K.-H. 1979. A method for computing self gravitating gas flows with radiation. *Computer Phys. Comm.* 18:171–199.

Tsuchiyama, A., and Nagahara, H. 1981. Effects of pre-cooling thermal history and cooling rate on the texture of chondrites. *Mem. Nat. Inst. Polar Res.* 20:175–192.

Tucker, K. D., Kutner, M. L., and Thaddeus, P. 1973. A large carbon monoxide cloud in Orion. *Astrophys. J.* 186:L13–L17.

Turner, B. E., 1973. Nonthermal OH emission in interstellar dust clouds. II. *Astrophys. J.* 186:357–395.

Turner, B. E., and Thaddeus, P. 1977. On the relationship of interstellar N_2H^+, HCO^+, HCN, and CN. *Astrophys. J.* 211:755–771.

Turon, P., and Mennesier, M. O. 1975. The fine structure of the galactic dust cloud distribution near the sun. *Astron. Astrophys.* 44:209–213.

Tyler, G. L., Eshleman, V. R., Anderson, J. D., Levy, G. S., Lindal, C. F., Wood, G. E., and Croft, T. A. 1981. Radio science investigations of the Saturn system with Voyager 1: Preliminary results. *Science* 212:201–206.

Tyler, G. L., Eshleman, V. R., Anderson, J. D., Levy, G. S., Lindal, G. F., Wood, G. E., and Croft, T. A. 1982. Radio science with Voyager 2 at Saturn: Atmosphere and ionosphere and the masses of Mimas, Tethys and Iapetus. *Science* 215:553–558.

Ulrich, R. K. 1976. An infall model for the T Tauri phenomenon. *Astrophys. J.* 210:377–391.

Ungerechts, H., Walmsley, C. M., and Winnewisser, G. 1980. Ammonia and cyanoacetylene observations of the high density core of L183 (L134N). *Astron. Astrophys.* 88:259–266.

Ungerechts, H., Walmsley, C. M., and Winnewisser, G. 1982. Ammonia observations of cold cloud cores. *Astron. Astrophys.* 111:339–345.

Urey, H. C. 1952. *The Planets, Their Origin and Development* (New Haven: Yale Univ. Press).

Urey, H. C. 1955. The cosmic abundances of potassium, uranium, and thorium and the heat balances of the Earth, the Moon, and Mars. *Proc. Nat. Acad. Sci. U.S.* 41:127–144.

Urey, H. C. 1961. Criticism of Dr. B. Mason's paper on "The origin of meteorites." *J. Geophys. Res.* 66:1988–1991.

Urey, H. C. 1967. Parent bodies of the meteorites and the origin of chondrules. *Icarus* 7:350–359.

Urey, H. C., and Craig, H. 1953. The composition of stone meteorites and the origin of the meteorites. *Geochim. Cosmochim. Acta* 4:36–82.

Urpin, G. 1984. Hydrodynamic flows in accretion disks. *Sov. Astron. A. J.* 28:50–55.

Vallée, J. P. 1983. Large-scale magnetic field in the Perseus spiral arm. *Astron. Astrophys.* 124:147–150.

Valtonen, M. J. 1983. On the capture of comets into the solar system. *Observatory* 103:1–4.

Valtonen, M. J., and Innanen, K. 1982. The capture of interstellar comets. *Astrophys. J.* 255:307–315.

van den Bergh, S. 1972. A preliminary classification scheme for interstellar absorbing clouds. In *Vistas in Astronomy,* ed. A. Beer (New York: Pergamon), pp. 265–277.

Vanden Bout, P. A., and Evans, J. J. II 1982. Energetics of molecular clouds—Summary of Sharpless molecular cloud observations. *Bull. Amer. Astron. Soc.* 14:604 (abstract).

Vanden Bout, P. A., Loren, R. B., Snell, R. L., and Wootten, A. 1983. Cyanoacetylene as a density probe of molecular clouds. *Astrophys. J.* 271:161–169.

Vandervoort, P. O. 1982. The dynamical instability of a rotating, axisymmetric galaxy with respect to a deformation into a bar. *Astrophys. J.* 256:L41–L44.

Vandervoort, P. O., and Welty, D. E. 1981. On the construction of models of rotating stars and stellar systems. *Astrophys. J.* 248:504–515.

Van Flandern, T. C. 1978. A former asteroidal planet as the origin of comets. *Icarus* 36:51–74.

Vanýsek, V. 1980. Isotopic fractionation in interstellar carbon-bearing molecules unrelated to carbon monoxide. In *Interstellar Molecules,* ed. B. H. Andrew (Dordrecht: D. Reidel), pp. 423–426.

Vanýsek, V., and Rahe, J. 1978. The $^{12}C/^{13}C$ isotope ratio in comets, stars, and interstellar matter. *Moon and Planets* 18:441–446.

Veillet, C. 1983. De l'observation et du mouvement des satellites d'Uranus. Ph.D. thesis, Univ. of Paris, France.

Verbunt, F. 1982. Accretion disks in stellar x-ray sources *Space Sci. Rev.* 32:379–404.

Verschuur, G. L. 1974*a*. Studies of neutral-hydrogen cloud structure. *Astrophys. J. Suppl.* 27:65–112.

Verschuur, G. L. 1974*b*. Studies of neutral-hydrogen cloud structure in the vicinity of the north polar spur. *Astrophys. J. Suppl.* 27:283–306.

Viallefond, F., Allen, R. J., and Goss, W. M. 1981. The giant spiral galaxy M101. VII. Associations of H I concentrations and H II complexes. *Astron. Astrophys.* 104:127–141.

Vidal-Madjar, A., Laurent, C., Gry, C., Bruston, P., Ferlet, R., and York, D. G. 1983. The ratio of deuterium to hydrogen in interstellar space. *Astron. Astrophys.* 120:58–62.

Vigroux, L., Audouze, J., and Lequeux, J. 1976. Isotopes of C, N and O and chemical evolution of galaxies (II). *Astron. Astrophys.* 52:1–9.

Villere, K. R., and Black, D. C. 1980. Collapsing cloud models for Bok globules. *Astrophys. J.* 236:192–200.

Villere, K. R., and Black, D. C. 1982. Collapse models for dark interstellar clouds. *Astrophys. J.* 252:524–528.

Villumsen, J. V., 1984. Violent relaxation and dissipationless collapse. *Astrophys. J.* 284:75–89.

Vishniac, E. T. 1983. The dynamic and gravitational instabilities of spherical shocks. *Astrophys. J.* 274:152–167.

Visser, H. D. C. 1980. The dynamics of the spiral galaxy M81. *Astron. Astrophys.* 88:149–158.

Vithal, K. L., and Vats, R. P. 1983*a*. Cascading in nonlinear interactions of random modes. *Astrophys. Space Sci.* 89:301–311.

Vithal, K. L., and Vats, R. P. 1983*b*. Unified approach to weak turbulence. *Astrophys. Space Sci.* 91:245–272.

Vityazev, A. V., and Pechernikova, G. V. 1982. Models of protoplanetary disks about stars of the F-G type. *Pis'ma Lett. Astron. Zh.* 8:371–377 (in Russian).

Vityazev, A. V., Pechernikova, A. V., and Safronov, V. S. 1978. Limiting masses, distances and times for accumulation of the planets of the terrestrial group. *Sov. Astron. A. J.* 22:60–63.

Vogel, S. N., and Kuhi, L. V. 1981. Rotational velocities of pre-main-sequence stars. *Astrophys. J.* 245:960–976.

Völk, H. 1982. Physical processes of relevance to the formation of the planetary system. In *Sun and Planetary System 6th Eur. Reg. Meet. Astron. Dubrovnik 19–23 Okt. 1981*, eds. W. Fricke and G. Teleki (Dordrecht: D. Reidel), pp. 233–242.

Völk, H. 1983*a*. Accretion disks: Stars and planets. In *Topics in Plasma-, Astro-, and Space Physics,* ed. G. Haerendel, pp. 25.

Völk, H. 1983*b*. Formation of planetesimals in turbulent protoplanetary accretion disks. *Meteoritics* 18:412–413.

Völk, H. J., Jones, F. C., Morfill, G. E., and Röser, S. 1980. Collisions between grains in a turbulent gas. *Astron. Astrophys.* 85:316–325.

von Hoerner, S. 1962. Strong shock fronts. In *The Distribution and Motion of Interstellar Matter in Galaxies,* ed. L. Woltjer (New York: Benjamin), pp. 193–204.

von Hoerner, S. 1968. The formation of stars. In *Interstellar Ionized Hydrogen,* ed. V. Terzian (New York: Benjamin), pp. 101–170.

von Neumann, J. 1949. Recent theories of turbulence (unpublished). In *Collected Works: Volume VI,* ed. A. H. Taub (New York: MacMillan, 1963), pp. 437–472.

von Weizsäcker, C. F. 1943. Über die entstenlung des planetensystems. *Z. Astrophys.* 22:319–355.

von Weizsäcker, C. F. 1948. Die rotation kosmischyer Gasmassen. *Z. Naturforschung* 3a:524–539.

von Weizsäcker, C. F. 1951. The evolution of galaxies and stars. *Astrophys. J.* 114:165–186.

Vrba, F. J. 1977. The role of magnetic fields in the evolution of five dark cloud complexes. *Astron. J.* 82:198–208.

Vrba, F. J., and Rydgren, A. E. 1984. The ratio of total-to-selective extinction in the Chameleon T1 and R Coronae Australis dark cloud. *Astrophys. J.* 283:123–128.

Vrba, F. J., Strom, K. M., Strom, S. E., and Grasdalen, G. L. 1975. Further study of the stellar cluster embedded in the Ophiuchus dark cloud complex. *Astrophys. J.* 197:77–84.

Vrba, J. F., Strom, K. M., and Strom, S. E. 1976. Magnetic field structure in the vicinity of five dark cloud complexes. *Astron. J.* 81:958–969.

Vrba, F. J., Coyne, G. V., and Tapia, S. 1981. Observation of grain and magnetic field properties of the R Coronae Australis dark cloud. *Astrophys. J.* 243:489–511.

Vrba, F. J., Rydgren, A. E., Zak, D. S., and Schmelz, J. T. 1985. Some systematic trends in the color variations of T Tauri stars at visible wavelengths. *Astron. J.* 90:326–332.

Vsekhsvyatskii, S. K. 1967. *The Nature and Origin of Comets and Meteors* (Moscow: Prosveschcheniye Press).

Wai, C. M., and Wasson, J. T. 1977. Nebular condensation of moderately volatile elements and their abundances in ordinary chondrites. *Earth. Planet. Sci. Lett.* 36:1–13.

Walgate, R. 1983. Emerging solar system in view. *Nature* 304:681.

Walker, M. F. 1956. Studies of extremely young clusters. I. NGC 2264. *Astrophys. J. Suppl.* 2:365–388.

Walker, M. F. 1972. Studies of extremely young clusters. VI. Spectroscopic observations of the ultraviolet-excess stars in the Orion nebula cluster and NGC 2264. *Astrophys. J.* 175:89–116.

Walker, M. F. 1978. Spectroscopic and photometric observations of YY Orionis. *Astrophys. J.* 224:546–557.

Walker, M. F. 1980. Simultaneous spectroscopic and photometric observations of BM Andromedae. *Publ. Astron. Soc. Pacific* 92:66–71.

Walker, M. F. 1983. Studies of extremely young clusters. VII. Spectroscopic observations of faint stars in the Orion nebula. *Astrophys. J.* 271:642–662.

Wallace, L. 1980. The structure of the Uranus atmosphere. *Icarus* 43:231–259.

Wallace, L., and Hunten, D. M. 1978. The Jovian spectrum in the region 0.4–1.1 μm: The C/H ratio. *Rev. Geophys. Space Phys.* 16:289–319.

Waller, W. 1984. Molecular birthsites of massive stars in the Milky Way. Paper presented at 164th meeting of the AAS.

Walmsley, C. M., Winnewisser, G., and Toelle, F. 1980. Cyanoacetylene and cyanodiacetylene in interstellar clouds. *Astron. Astrophys.* 81:245–250.

Walmsley, C. M., Jewell, P. R., Snyder, L. E., and Winnewisser, G. 1984. Detection of interstellar methyldiacetylene (CH_3C_4H) in the dark dust cloud TMC-1. *Astron. Astrophys.* 134:L1–L14.

Walter, F. M., and Kuhi, L. V. 1981. The smothered coronae of T Tauri stars. *Astrophys. J.* 250:254–261.

Wannier, P. G. 1980. Nuclear abundances and evolution of interstellar medium. *Ann. Rev. Astron. Astrophys.* 18:399–437.

Wannier, P. G., and Linke, R. A. 1978. Isotope abundance anomalies in IRC+10216. *Astrophys. J.* 225:130–137.

Wannier, P. G., Penzias, A. P., and Jenkins, E. B. 1982. The $^{12}CO/^{13}CO$ abundance ratio toward ρ Ophiuchi. *Astrophys. J.* 254:100–107.

Wannier, P. G., Lichten, S., and Morris, M. 1983. Warm HI halos around molecular clouds. *Astrophys. J.* 268:727–738.

Ward, R. A., and Fowler, W. A. 1980. Thermalization of long-lived nuclear isomeric states under stellar conditions. *Astrophys. J.* 238:266–286.

Ward, W. R. 1981. On the radial structure of Saturn's rings. *Geophys. Res. Lett.* 8:641–643.

Ward, W. R. 1983. Presented at Solar Nebula Working Group, NASA Ames Research Center.

Ward, W. R. 1984. The solar nebula and the planetesimal disk. In *Planetary Rings,* eds. R. Greenberg and A. Brahic (Tucson: Univ. of Arizona Press), pp. 660–684.

Ward, W., and Goldreich, P. 1973. Formation of planetesimals. *Astrophys. J.* 183:1051–1061.

Ward, W. R., and Reid, M. J. 1973. Solar tidal friction and satellite loss. *Mon. Not. Roy. Astron. Soc.* 164:21–32.

Ward, W., and Cameron, A. G. W. 1978. Disk evolution within the Roche limit. *Lunar Planet. Sci.* IX:1205 (abstract).

Wark, D. A. 1979. Birth of the presolar nebula: The sequence of condensation revealed in the Allende meteorite. *Astrophys. Space Sci.* 65:275–295.

Wasserburg, G. J., and Huneke, J. C. 1979. I-Xe dating of I bearing phases in Allende. *Lunar Planet. Sci.* X:1307–1309 (abstract).

Wasserburg, G. J., and Papanastassiou, D. A. 1982. Some short-lived nuclides in the early solar system—a connection with the placental ISM. In *Essays in Nuclear Astrophysics,* eds. C. A. Barnes, D. D. Clayton and D. N. Schramm (Cambridge: Cambridge Univ. Press), pp. 77–140.

Wasserburg, G. J., Huneke, J. C., and Burnett, D. S. 1969. Correlation between fission tracks and fission-type xenon from an extinct radioactivity. *Phys. Rev. Lett.* 22:1198–1201.

Wasserburg, G. J., Lee, T., and Papanastassiou, D. A. 1977*a*. Correlated O and Mg isotopic anomalies in Allende inclusions: II. Magnesium. *Geophys. Res. Lett.* 4:299–302.

Wasserburg, G. J., Tera, F., Papanastassiou, D. A., and Huneke, J. C. 1977*b*. Isotopic and chemical investigations on Angra dos Reis. *Earth Planet. Sci. Lett.* 35:294–316.

Wasserburg, G. J., Papanastassiou, D. A., and Lee, T. 1980. Isotopic heterogeneities in the solar system. In *Early Solar System Processes and the Present Solar System* (Bologna: Soc. Italiana di Fisica), pp. 144–191.

Wasson, J. R. 1974. *Meteorites* (New York: Springer-Verlag).

Wasson, J. T. 1978. Maximum temperatures during the formation of the solar nebula. In *Protostars & Planets,* ed. T. Gehrels (Tucson: Univ. of Arizona Press), pp. 488–501.

Wasson, J. T., and Wetherill, G. W. 1979. Dynamic, chemical and isotopic evidence regarding the formation locations of asteroids and meteorites. In *Asteroids,* ed. T. Gehrels (Tucson: Univ. of Arizona Press), pp. 926–974.

Watson, D. M. 1982. Far-infrared CO line emission from Orion-KL. In *Symposium on the Orion Nebula to Honor Henry Draper,* eds. A. E. Glassgold, P. J. Huggins, and E. L. Schudking (New York: New York Academy of Sciences), pp. 136–141.

Watson, D. M. 1984. Far infrared spectroscopy of molecular clouds. In *Galactic and Extragalactic Infrared Spectroscopy, XVIth Eslab Symposium,* eds. M. F. Kessler and J. P. Phillips (Dordrecht: D. Reidel), pp. 195–219.

Watson, W. D. 1973*a*. Deuterium in interstellar molecules. *Astrophys. J.* 181:L129–L133.

Watson, W. D. 1973*b*. Formation of the HD molecule in the interstellar medium. *Astrophys. J.* 182:L73–L76.

Watson, W. D. 1974*a*. Ion-molecule reactions, molecule formation, and hydrogen-isotope exchange in dense interstellar clouds. *Astrophys. J.* 188:35–42.

Watson, W. D. 1974*b*. Ion-molecule reactions, molecule formation, and hydrogen-isotope exchange in dense interstellar clouds. *Astrophys. J.* 189:221–225.

Watson, W. D. 1976. Interstellar molecule reactions. *Rev. Mod. Phys.* 48:513–522.

Watson, W. D. 1977. Isotopic fractionation in interstellar molecules. In *CNO Isotopes in Astrophysics,* ed. J. Audouze (Dordrecht: D. Reidel), pp. 105–114.

Watson, W. D. 1978. Current problems in interstellar chemistry. In *Protostars & Planets,* ed. T. Gehrels (Tucson: Univ. of Arizona Press), pp. 77–87.

Watson, W. D. 1980. Molecule formation in cool, dense interstellar clouds. In *Interstellar Molecules,* ed. B. H. Andrew (Dordrecht: D. Reidel), pp. 341–353.

Watson, W. D. 1983. Gas phase chemistry in the interstellar medium. Paper presented at "XVIth Eslab Symposium," Univ. of Illinois preprint ILL-(AST)-83-26.

Watson, W. D., and Salpeter, E. E. 1972. Molecule formation on interstellar grains. *Astrophys. J.* 174:321–340.

Watson, W. D., and Walmsley, C. M. 1982. Chemistry relevant to molecular clouds near H II regions. In *Regions of Recent Star Formation,* eds. R. S. Roger and P. E. Dewdney (Dordrecht: D, Reidel), pp. 367–377.

Watson, W. D., Anicich, V., and Huntress, W. 1976. Measurement and significance of the reaction $^{13}C^+ + ^{12}CO \rightleftarrows ^{12}C^+ + ^{13}CO$ for alteration of the $^{13}CO/^{12}CO$ ratio in interstellar molecules. *Astrophys. J.* 205:L165–168.

Watson, W. D., Snyder, L. E., and Hollis, J. M. 1978. The DCO^+/HCO^+ abundance ratio and the electron density in cool interstellar clouds. *Astrophys. J.* 222:L145–L147.

Weast, R. C. 1978. *CRC Handbook,* 59th edition, (W. Palm Beach: CRC Press), pp. 231–255.

Weaver, H. 1974. Some aspects of galactic structure derived from the Berkeley low latitude survey of neutral hydrogen. In *Galactic Radio Astronomy,* eds. F. J. Kerr and S. C. Simonson (Dordrecht: D. Reidel), pp. 573–586.

Weaver, H. F. 1979. Large supernova remnants as common features of the disk. In *The Large Scale Characteristics of the Galaxy, IAU Symposium No. 84,* ed. W. B. Burton (Dordrecht: D. Reidel), pp. 295–300.

Weaver, H., and Williams, D. R. W. 1973. The Berkeley low-latitude survey of neutral hydrogen. Part I. Profiles. *Astron. Astrophys. Suppl.* 8:1–503.

Weber, E. J., and Davis, L. 1967. The angular momentum of the solar wind. *Astrophys. J.* 148:217–227.

Weber, S. V. 1976. Oscillation and collapse of interstellar clouds. *Astrophys. J.* 208:113–126.

Weidenschilling, S. J. 1974. A model for the accretion of terrestrial planets. *Icarus* 22:426–435.

Weidenschilling, S. J. 1975. Mass loss from the region of Mars and the asteroid belt. *Icarus* 26:361–366.

Weidenschilling, S. J. 1977*a*. Aerodynamics of solid bodies in the solar nebula. *Mon. Not. Roy. Astron. Soc.* 180:57–70.

Weidenschilling, S. J. 1977*b*. The distribution of mass in the planetary system and solar nebula. *Astrophys. Space Sci.* 51:153–158.
Weidenschilling, S. J. 1978. Iron/silicate fractionation and the origin of Mercury. *Icarus* 35:99–111.
Weidenschilling, S. J. 1980. Dust to planetesimals: Settling and coagulation in the solar nebula. *Icarus* 44:172–189.
Weidenschilling, S. J. 1981*a*. Aspects of accretion in circumplanetary nebula. *Lunar Planet. Sci.* XII:1170–1171 (abstract).
Weidenschilling, S. J. 1981*b*. How fast can an asteroid spin? *Icarus* 46:124–126.
Weidenschilling, S. J. 1984. Evolution of grains in a turbulent solar nebula. *Icarus* 60:555–567.
Weidenschilling, S. J., Chapman, C. R., Davis, D. R., and Greenberg, R. 1984. Ring particles: Collisional interactions and physical nature. In *Planetary Rings,* eds. R. Greenberg and A. Brahic (Tucson: Univ. of Arizona Press), pp. 367–415.
Weissman, P. R. 1979. Physical and dynamical evolution of long-period comets. In *Dynamics of the Solar System,* ed. R. L. Duncombe (Dordrecht: D. Reidel), pp. 277–282.
Weissman, P. R. 1980. Stellar perturbations of the cometary cloud. *Nature* 288:242–243.
Weissman, P. R. 1982. Dynamical history of the Oort cloud. In *Comets,* ed. L. L. Wilkening (Tucson: Univ. of Arizona Press), pp. 637–658.
Weissman, P. R. 1983*a*. Diffusion of Oort cloud comets into the planetary region. *Bull. Amer. Astron. Soc.* 15:869 (abstract).
Weissman, P. R. 1983*b*. The mass of the Oort cloud. *Astron. Astrophys.* 118:90–94.
Weissman, P. R. 1984. The Vega particulate shell: Comets or asteroids? *Science* 224:987–989.
Welter, G. L. 1982. Gravitationally driven stabilities in shock compressed gas layers. *Astron. Astrophys.* 105:237–241.
Welter, G. L., and Schmid-Burgk, J. 1981. On the possibility of star formation behind interstellar shocks. *Astrophys. J.* 245:927–933.
Welter, G. L., and Nepveu, M. 1982. Shock induced star formation: The effects of magnetic fields and turbulence. *Astron. Astrophys.* 113:277–284.
Werner, M. W. 1982. Infrared studies of star formation in Orion. In *Symposium on the Orion Nebula to Honor Henry Draper,* eds. A. E. Glassgold, P. J. Huggins, and E. L. Schucking (New York: New York Academy of Sciences), pp. 79–99.
Werner, M. W., Becklin, E. E., Gatley, I., Matthews, K., Neugebauer, G., and Wynn-Williams, C. G. 1979. An infrared study of the NGC 7538 region. *Mon. Not. Roy. Astron. Soc.* 188:463–479.
Werner, M. W., Dinerstein, H. L., and Capps, R. W. 1983. The polarization of the infrared cluster in Orion: The spatial distribution of the 3.8 micron polarization. *Astrophys. J.* 265:L13–L17.
Wesselius, P. R., Beintema, D. A., and Olnon, F. M. 1984. IRAS observations of two early-type pre-main sequence stars in the association Cha I. *Astrophys. J.* 278:L37–L39.
Westbrook, W. E., Werner, M. W., Elias, J. H., Gezari, D. Y., Hauser, M. G., Lo, K. Y., and Neugebauer, G. 1976. One-mm continuum emission studies of four molecular clouds. *Astrophys. J.* 209:94–101.
Wetherill, G. W. 1975. Late heavy bombardment of the moon and terrestrial planets. *Proc. Lunar Sci. Conf.* 6:1539–1561.
Wetherill, G. W. 1977. Pre-mare cratering and early solar system history. In *The Soviet-American Conference on Cosmochemistry of the Moon and Planets* (Washington, D.C.: NASA SP-370), pp. 553–567.
Wetherill, G. W. 1978. Accumulation of the terrestrial planets. In *Protostars & Planets,* ed. T. Gehrels (Tucson: Univ. of Arizona Press), pp. 565–598.
Wetherill, G. W. 1980*a*. Formation of the terrestrial planets. *Ann. Rev. Astron. Astrophys.* 18:77–113.
Wetherill, G. W. 1980*b*. Numerical calculations relevant to the accumulation of the terrestrial planets. In *The Continental Crust and Its Mineral Deposits,* ed. W. Strangway (Waterloo, Ontario: Geological Assoc. of Canada), pp. 3–24.
Wetherill, G. W., and Cox, L. P. 1984. The range of validity of the two-body approximation in models of terrestrial planet accumulation. I. Gravitational perturbations. *Icarus* 60:40–55.
Wheeler, J. C., Mazurek, T. J., and Sivarzmakrishnan, A. 1980. Supernovae in molecular clouds. *Astrophys. J.* 237:781–792.

Whipple, F. L. 1950. A comet model. I. Acceleration of comet Encke. *Astrophys. J.* 111:374–394.

Whipple, F. L. 1962. On the distribution of semimajor axes among comet orbits. *Astron. J.* 67:1–9.

Whipple, F. L. 1964*a*. Evidence for a comet belt beyond Neptune. *Proc. Nat. Acad. Sci. U.S.* 51:711–718.

Whipple, F. L. 1964*b*. The history of the solar system. *Proc. Nat. Acad. Sci. U.S.* 52:565–594.

Whipple, F. L. 1972. On certain aerodynamic processes for asteroids and comets. *From Plasma to Planet: Proc. Nobel Symposium No. 21,* ed. A. Elvius (New York: Wiley), pp. 211–232.

Whipple, F. L., and Lecar, M. 1976. Comet formation induced by solar wind. In *The Study of Comets* (Washington, D.C.: NASA SP-393), pp. 660–662.

White, R. E. 1977. Microturbulence, systematic motions and line formation in molecular clouds. *Astrophys. J.* 211:744–753.

Whitmire, D. P., and Jackson, A. A. IV. 1984. Are periodic mass extinctions driven by a distant solar companion? *Nature* 308:713–715.

Whittet, D. C. B. 1981. The composition of interstellar grains. *Quart. J. Roy. Astron. Soc.* 22:3–21.

Whittet, D. C. B., van Breda, I. G., and Glass, I. S. 1976. Infrared photometry, extinction curves, and *R* values for stars in the Southern Milky Way. *Mon. Not. Roy. Astron. Soc.* 177:625–643.

Whittet, D. C. B., Bode, M. F., Longmore, A. J., Baines, D. W. T., and Evans, A. 1983. Interstellar ice grains in the Taurus molecular clouds. *Nature* 303:218–220.

Whitworth, A. P. 1976. UV radiation fields in dark clouds. In *Solid State Astrophysics,* eds. N. C. Wickramasinghe and D. J. Morgan (Dordrecht: D. Reidel), pp. 207–225.

Whitworth, A. 1979. The erosion and dispersal of massive molecular clouds by young stars. *Mon. Not. Roy. Astron. Soc.* 186:59–68.

Whitworth, A. P. 1981. Globule gravitational stability for one-dimensional polytropes. *Mon. Not. Roy. Astron. Soc.* 195:967–977.

Wickramasinghe, D. T., and Allen, D. A. 1980. The 3.4 μm interstellar absorption feature. *Nature* 287:518–519.

Wielen, R. 1971. On the lifetimes of galactic clusters. *Astrophys. Space Sci.* 13:300–308.

Wildt, R. 1938. On the state of matter in the interior of the planets. *Astrophys. J.* 87:508–516.

Wilkening, L. L., Boynton, W. V., and Hill, D. H. 1984. Trace elements in rims and interiors of Chainpur chondrules. *Geochim. Cosmochim. Acta.* In press.

Wilking, B. A., and Lada, C. J. 1983. The discovery of new embedded sources in the centrally condensed core of the Rho Ophiuchi dark cloud: The formation of a bound cluster? *Astrophys. J.* 274:698–716.

Wilking, B. A., Lebofsky, M. J., Rieke, G. H., and Kemp, J. C. 1979. Infrared polarimetry in the ρ Ophiuchi dark cloud. *Astron. J.* 84:199–203.

Wilking, B. A., Harvey, P. M., Lada, C. J., Joy, M., and Doering, C. R. 1984. The formation of massive stars along the W5 ionization front. *Astrophys. J.* 279:291–303.

Wilking, B. A., Harvey, P. M., Joy, M., Hyland, A. R., and Jones, T. J. 1985. Far-infrared observations of young clusters embedded in the R Coronae Australis and Rho Ophiuchi dark clouds. *Astrophys. J.* In press.

Williams, I. P., and Bhatt, H. C. 1982. The dust distribution in Bok globules. *Mon. Not. Roy. Astron. Soc.* 199:465–470.

Williams, J. G., and Benson, G. S. 1971. Resonances in the Neptune-Pluto system. *Astron. J.* 76:167–177.

Williams, R. E. 1980. Emission lines from accretion disk in cataclysmic variables. *Astrophys. J.* 235:939–944.

Willner, S. P., Gillett, F. C., Herter, T. L., Jones, B., Krassner, J., Merrill, K. M., Pipher, J. L., Puetter, R. C., Rudy, R. J., Russell, R. W., and Soifer, B. T. 1982. Infrared spectra of protostars: Composition of the dust shells. *Astrophys. J.* 253:174–187.

Wilson, R. W., Jefferts, K. B., and Penzias, A. A. 1970. Carbon monoxide in the Orion nebula. *Astrophys. J.* 161:L43–L44.

Wilson, R. W., Langer, W. D., and Goldsmith, P. F. 1981. A determination of the carbon and oxygen isotope ratios in the local interstellar medium. *Astrophys. J.* 243:L47–L52.

Wilson, T. L., and Jaffe, D. 1981. Observations of $2_{12} \rightarrow 1_{11}$ line of H_2CO. *Astrophys. J.* 245:866–870.
Wilson, T. L., Walmsley, C. M., Henkel, C., Pauls, T., and Mattes, H. 1980. Observations of the 3_{12}-3_{13} line of H_2CO. *Astron. Astrophys.* 91:36–40.
Wilson, T. L., Mauersberger, R., Walmsley, C. M., and Batrla, W. 1983. Non-metastable ammonia absorption toward compact H II regions. *Astron. Astrophys.* 127:L19–L22.
Winkler, K.-H., and Newman, M. J. 1980. Formation of solar-type stars in spherical symmetry. I. The key role of the accretion shock. *Astrophys. J.* 236:201–211.
Winnewisser, G. 1981. The chemistry of interstellar molecules. *Topics Cur. Chem.* 99:39–71.
Winnewisser, G., Churchwell, E., and Walmsley, C. M. 1979. Astrophysics of interstellar molecules. In *Modern Aspects of Microwave Spectroscopy,* ed. G. W. Chantry (New York: Academic Press), pp. 313–499.
Witt, A. N., and Johnson, M. W. 1973. The interstellar radiation density between 1250 and 4250 angstroms. *Astrophys. J.* 181:363–368.
Witt, A. N., and Stephens, T. C. 1974. Monte Carlo calculations of the surface brightness profiles of spherical dark nebulae. *Astron. J.* 79:948–953.
Witt, A. N., Bohlin, R. C., and Stecher, T. P. 1984. The variation of galactic interstellar extinction in the ultraviolet. Preprint.
Wolf, R., Richter, G. R., Woodrow, A. B., and Anders, E. 1980. Chemical fractionations in meteorites. XI. C2 chondrites. *Geochim. Cosmochim. Acta* 44:711–717.
Wolff, S. C., Edwards, S., and Preston, G. W. 1982. The origin of stellar angular momentum. *Astrophys. J.* 252:322–336.
Wood, D. 1982. Fragmentation in rotating interstellar gas clouds. *Mon. Not. Roy. Astron. Soc.* 199:331–343.
Wood, J. A. 1963. On the origin of chondrules and chondrites. *Icarus* 2:152–180.
Wood, J. A. 1967. Olivine and pyroxene compositions in type II carbonaceous chondrites. *Geochim. Cosmochim. Acta* 31:2095–2108.
Wood, J. A. 1979. The Oort cloud as a source of Apollo/Amor asteroids. In *Reports of Planetary Geology Program, 1978–79* (Washington, D.C.: NASA TM 80339), pp. 11–12.
Wood, J. A. 1981*a*. The interstellar dust as a precursor of Ca, Al-rich inclusions in carbonaceous chondrites. *Earth Planet. Sci. Lett.* 56:32–44.
Wood, J. A. 1981*b*. Meteorites. In *The New Solar System,* eds. J. K. Beatty, B. O'Leary, and A. Chaikin (Cambridge: Sky Publ. Co.), pp. 187–196.
Wood, J. A. 1984. On the formation of meteoritic chondrules by aerodynamic drag heating in the solar nebula. *Earth Planet. Sci. Lett.* 70:11–26.
Woods, R. C., Gudeman, C. S., Dickman, R. L., Goldsmith, P. F., Huguenin, G. R., Irvine, W. M., Hjalmarson, Å., Nyman, L. A., and Olofsson, H. 1983. The $(HCO^+)/(HOC^+)$ abundance ratio in molecular clouds. *Astrophys. J.* 270:583–588.
Woodward, P. R. 1976. Shock-driven implosion of interstellar clouds and star formation. *Astrophys. J.* 207:484–501.
Woodward, P. R. 1978. Theoretical models of star formation. *Ann. Rev. of Astron.* 16:555–584.
Woodward, P. R. 1984. Piecewise-parabolic methods for astrophysical fluid dynamics. In *Astrophysical Radiation Hydrodynamics,* ed. K. H. A. Winkler and M. L. Norman (Dordrecht: D. Reidel).
Woolfson, M. M. 1978*a*. The capture theory and the origin of the solar system. In *The Origin of the Solar System,* ed. S. F. Dermott (New York: J. Wiley & Sons), pp. 179–198.
Woolfson, M. M. 1978*b*. The evolution of the solar system. In *The Origin of the Solar System,* ed. S. F. Dermott (New York: J. Wiley & Sons), pp. 199–217.
Woosley, S. E., and Howard, W. M. 1978. The p-process in supernovae. *Astrophys. J. Suppl.* 36:285–304.
Woosley, S. E., and Weaver, T. A. 1982. Nucleosynthesis in two 25M stars of different population. In *Essays in Nuclear Astrophysics,* eds. C. A. Barnes, D. D. Clayton, and D. N. Schramm (Cambridge: Cambridge Univ. Press), pp. 401–426.
Wootten, A. 1981. A dense molecular cloud impacted by the W28 supernova remnant. *Astrophys. J.* 245:105–114.
Wootten, A., Evans, N. J., Snell, R., and Vanden Bout, P. 1978. Molecular abundance variations in interstellar clouds. *Astrophys. J.* 225:L143–L148.

Wootten, A., Snell, R., and Glassgold, A. E. 1979. The determination of electron abundances in interstellar clouds. *Astrophys. J.* 234:876–880.

Wootten, A., Bozyan, E. P., Garrett, D. B., Loren, R. B., and Snell, R. L. 1980*a*. Detection of C_2H in cold dark clouds. *Astrophys. J.* 239:844–854.

Wootten, A., Snell, R., and Evans, N. J. II. 1980*b*. Models of molecular clouds and the abundance of H_2CO and HCO^+. *Astrophys. J.* 240:532–546.

Wootten, A., Lichten, S. M., Sahai, R., and Wannier, P. G. 1982*a*. CN abundance variations in the shell of IRC+10216. *Astrophys. J.* 257:151–160.

Wootten, A., Loren, R. B., and Snell, R. L. 1982*b*. A study of DCO^+ emission regions in interstellar clouds. *Astrophys. J.* 255:160–175.

Wootten, A., Loren, R. B., and Bally, J. 1984*a*. Formaldehyde in the Orion molecular flow: Evidence for a gentle acceleration. *Astrophys. J.* 277:189–195.

Wootten, A., Loren, R. B., Sandquist, A., Friberg, P., and Hjalmarson, Å. 1984*b*. The evolution of star-bearing molecular clouds: The high-velocity HCO^+ flow in NGC 2071. *Astrophys. J.* 279:633–649.

Worden, S. P., Schneeberger, T. J., Kuhn, J. R., and Africano, J. L. 1981. Flare activity on T Tauri stars. *Astrophys. J.* 244:520–527.

Woronow, A., Strom, R. G., and Gurnis, M. 1982. Interpreting the cratering record: Mercury to Ganymede and Callisto. In *Satellites of Jupiter,* ed. D. Morrison (Tucson: Univ. of Arizona Press), pp. 237–276.

Wouterloot, J. G. A. 1981. The Large-Scale Structure of Molecular Clouds. Ph.D. dissertation. Sterrewacht Leiden.

Wouterloot, J. G. A. 1984. OH observations of the Ophiuchus complex. *Astron. Astrophys.* 135:32–38.

Wouterloot, J. G. A., Habing, H. J., and Herman, J. 1980. A new main line OH maser with a probable Zeeman pattern. *Astron. Astrophys.* 81:L11–L12.

Wray, J. D., and deVaucouleurs, G. 1980. The brightest superassociations in spiral and irregular galaxies as extragalactic distance indicators. *Astron. J.* 85:1–8.

Wynn-Williams, C. G. 1982. The search for infrared protostars. *Ann. Rev. Astron. Astrophys.* 20:587–618.

Wynn-Williams, C. G. 1984. Infrared hydrogen emission lines from HII regions and "protostars." Preprint for publication of XVIth Eslab Symposium.

Wynn-Williams, C. G., and Becklin, E. E. 1974. Infrared emission from H II regions. *Publ. Astron. Soc. Pacific* 86:5–25.

Wynn-Williams, C. G., Genzel, R., Becklin, E. E., and Downes, D. 1984. The Kleinmann-Low nebula: An infrared cavity. *Astrophys. J.* 281:172–183.

Yabushita, S. 1972. Stellar perturbations of orbits of long-period comets. *Astron. Astrophys.* 16:395–403.

Yang, J., and Epstein, S. 1983. Interstellar organic matter in meteorites. *Geochim. Cosmochim. Acta* 47:2199–2216.

Yeh, H.-W., and Epstein, S. 1978. $^{29}Si/^{28}Si$ and $^{30}Si/^{28}Si$ of meteorites and Allende inclusions. *Lunar Planet. Sci.* IX:1289–1291 (abstract).

Yoder, C. F. 1979. How tidal heating in Io drives the Galilean orbital resonance locks. *Nature* 279:767–770.

Yoder, C. F. 1981. Tidal friction and Enceladus' anomalous surface. *EOS* 62:939 (abstract).

Yoder, C. F. 1982. Tidal rigidity of Phobos. *Icarus* 49:327–346.

Yoneyama, T. 1973. Thermal instability in reacting gas. *Publ. Astron. Soc. Japan* 25:349–373.

Yorke, H. W., and Shustov, B. M. 1981. The spectral appearance of dusty protostellar envelopes. *Astron. Astrophys.* 98:125–132.

Young, J. S. 1983. Molecular clouds and star formation in spiral galaxies. In *Surveys of the Southern Galaxy,* eds. W. B. Burton and F. P. Israel (Dordrecht: D. Reidel), pp. 253–264.

Young, J. S., and Scoville, N. Z. 1982*a*. Extragalactic CO: Gas distributions which follow the light in IC 342 and NGC 6946. *Astrophys. J.* 258:467–489.

Young, J. S., and Scoville, N. 1982*b*. The dependence of CO emission on luminosity and the rate of star formation in Sc galaxies. *Astrophys. J.* 260:L11–L18.

Young, J. S., Langer, W. D., Goldsmith, P. F., and Wilson, R. W. 1981. Coupling of the magnetic field and rotation in the dark cloud B5. *Astrophys. J.* 251:L81–L84.

Young, J. S., Goldsmith, P. F., Langer, W. D., Wilson, R. W., and Carlson, E. R. 1982.

Physical conditions and carbon monoxide abundance in the dark cloud B5. *Astrophys. J.* 261:513–531.

Young, J. S., Gallagher, J. S., and Hunter, D. A. 1984. CO emission from the star burst irregular galaxy NGC 1569. *Astrophys. J.* 276:476–479.

Yuan, C. 1984. On the 3-kpc arm—Resonance excitation of linear and nonlinear waves by an oval distortion in the central region. *Astrophys. J.* In press.

Yuan, C., and Cassen, P. 1984. How much angular momentum can a central star accrete during the collapse of a rotating interstellar cloud? Presented at Protostars and Planets II conference, Tucson, AZ, January.

Yuen, G., Blair, N., DesMarais, D. J., and Chang, S. 1984. Carbon isotope composition of low molecular weight hydrocarbons and carboxylic acids from Murchison meteorite. *Nature* 307:252–254.

Zahnle, K. J., and Walker, J. C. G. 1982. The evolution of solar ultraviolet luminosity. *Rev. Geophys. Space Phys.* 20:280–292.

Zealy, W. J., Ninkov, Z., Rice, E., Hartley, M., and Tritton, S. B. 1983. Cometary globules in the Gum-Vela complex. *Astrophys. Lett.* 23:119–131.

Zeldovich, Ya. B. 1981. On the friction of fluids between rotating cylinders. *Proc. Roy. Soc. London* A374:299–312.

Zeldovich, Ya. B., Ruzmaikin, A. A., and Sokoloff, D. D. 1984. *Magnetic Field in Astrophysics* (London: Gordon & Breach Sci. Publ.).

Zellner, B. H., and Serkowski, K. 1972. Polarization of light by circumstellar material. *Publ. Astron. Soc. Pacific* 84:619–626.

Zharkov, V. N., and Trubitsyn, V. P. 1972. Adiabatic temperatures in Uranus and Neptune. *Izv. Earth Phys.* 7:120–127.

Zharkov, V. N., and Trubitsyn, V. P. 1976. Structure, composition, and gravitational field of Jupiter. In *Jupiter,* ed. T. Gehrels (Tucson: Univ. of Arizona Press), pp. 133–175.

Zharkov, V. N., and Trubitsyn, V. P. 1978. *Physics of Planetary Interiors,* ed. W. B. Hubbard (Tucson: Pachart).

Zharkov, V. N., Trubitsyn, V. P., and Makalkin, A. B. 1972. The high gravitational moments of Jupiter and Saturn. *Astrophys. J.* 10:L159–L161.

Zharkov, V. N., Tsarevsky, I. A., and Trubitsyn, V. P. 1978. Equations of state of hydrogen, hydrogen compounds, crystals of inert gases, oxides, iron and FeS. NASA TM 75311.

Zheng, X., Ho, P., Reid, M., and Schneps, M. 1984. Molecular clouds associated with compact H II regions. II. The rapidly rotating condensation associated with ON1. Submitted to *Astrophys. J.*

Ziglina, I. N. 1976. Effect on eccentricity of a planet's orbit of its encounter with bodies of the swarm. *Sov. Astron. A. J.* 20:730–733.

Ziglina, I. N. 1978. Tidal disintegration of bodies near the planet. *Izvestija Acad. Sci. SSSR, Ser. Physics of Earth* 7:3–10 (in Russian).

Ziglina, I. N., and Safronov, V. S. 1976. Averaging of the orbital eccentricity of bodies which are accumulating into a planet. *Sov. Astron. A. J.* 20:244–248.

Zimmerman, B. A., Fowler, W. A., and Caughlan, G. R. 1975. Tables of thermonuclear reaction rates. Orange Aide Preprint No. 399, California Institute of Technology.

Zinnecker, H. 1981. The Mass Spectrum of Star Formation. Ph.D. thesis, Techn. Univ., Munchen: Report MPI-PAE/Extraterr. No. 167.

Zinnecker, H. 1982. Prediction of the protostellar mass spectrum in the Orion near-infrared cluster. In *Symposium on the Orion Nebula to Honor Henry Draper,* eds. A. E. Glassgold, P. J. Huggins, and E. L. Schucking (New York: New York Acad. of Sciences), pp. 226–235.

Zinnecker, H. 1983. A Random Hierarchical Fragmentation Theory for the Log-Normal Initial Mass Function of Stars. Ph.D. thesis, Max-Planck Institut für Physik und Astrophysik, Institut für Extraterrestrische Physik, Garching, West Germany.

Ziurys, L. M., Martin, R. N., Pauls, T. A., and Wilson, T. L. 1981. Ammonia in Orion. *Astron. Astrophys.* 104:288–295.

Ziurys, L. M., Clemens, D. P., Saykelly, R. J., Colvin, M., and Schaefer, H. F. 1984. A search for interstellar silicon nitride. *Astrophys. J.* 281:219–224.

Zuckerman, B. 1973. A model of the Orion nebula. *Astrophys. J.* 183:863–869.

Zuckerman, B. 1980. Envelopes around late-type giant stars. *Ann. Rev. Astron. Astrophys.* 18:263–288.

Zuckerman, B., and Evans, N. J. 1974. Models of massive molecular clouds. *Astrophys. J.* 192:L149–L152.

Zuckerman, B., and Palmer, P. 1974. Radio radiation from interstellar molecules. *Ann. Rev. Astron. Astrophys.* 12:279–313.

Zuckerman, B., Kuiper, T. B. H., and Rodriguez-Kuiper, E. N. 1976. High-velocity gas in the Orion infrared nebula. *Astrophys. J.* 209:L137–L142.

Zuckerman, B., Morris, M., and Palmer, P. 1981. The location of the hot molecular core in Orion. *Astrophys. J.* 250:L39–L42.

Zvygina, E. V., Pechernikova, G. V., and Safronov, V. S. 1973. A qualitative solution of the coagulation equation taking into account the fragmentation of bodies. *Sov. Astron. A. J.* 17:793–800.

Zweibel, E. G., and Josafatsson, K. 1983. Hydromagnetic wave dissipation in molecular clouds. *Astrophys. J.* 270:511–518.

LIST OF CONTRIBUTORS WITH ACKNOWLEDGMENTS TO FUNDING AGENCIES

The following people helped to make this book possible, by organizing, writing, refereeing, or otherwise.

M. A'Hearn, Astronomy Program, University of Maryland, College Park, MD.
S. Aiello, Instituto de Fisica Superiore, Università di Firenze, Florence, Italy.
L. Allamandola, NASA Ames Research Center, Moffett Field, CA.
T. Armstrong, National Radio Astronomical Observatory, Charlottesville, VA.
R. Arquilla, Radio Astronomy, University of Massachusetts, Amherst, MA.
G. Arrhenius, Scripps Institute of Oceanography, University of California, La Jolla, CA.
G. Augason, NASA Ames Research Center, Moffett Field, CA.
D. E. Backman, Institute for Astronomy, University of Hawaii, Honolulu, HI.
G. Basri, Astronomy Department, University of California, Berkeley, CA.
F. N. Bash, Astronomy Department, University of Texas, Austin, TX.
P. Bastien, Department of Physics, University of Montreal, Montreal, Canada.
E. E. Becklin, Institute for Astronomy, University of Hawaii, Honolulu, HI.
C. A. Beichman, Jet Propulsion Laboratory, Pasadena, CA.
W. Benz, Observatoire de Geneve, Sauverny, Switzerland.
C. Bertout, Landessternwarte, Heidelberg, West Germany.
D. C. Black, NASA Ames Research Center, Moffett Field, CA.
J. H. Black, Steward Observatory, University of Arizona, Tucson, AZ.
M. Blain, The Composing Room, Grand Rapids, MI.
L. Blitz, Astronomy Program, University of Maryland, College Park, MD.
P. H. Bodenheimer, Lick Observatory, University of California, Santa Cruz, CA.
N. W. Boggess, NASA Headquarters, Washington, D.C.
A. P. Boss, Carnegie Institution of Washington, Washington, D.C.
M. Bossi, Osservatorio Astronomico, Merate, Italy.
P. Bouwens, The Composing Room, Grand Rapids, Mi.
W. V. Boynton, Department of Planetary Sciences, University of Arizona, Tucson, AZ.
T. Brooke, Department of Earth and Space Sciences, State University of New York, Stony Brook, NY.
J. D. Burke, Jet Propulsion Laboratory, Pasadena, CA.
J. A. Burns, Center for Radiophysics and Space Research, Cornell University, Ithaca, NY.
A. G. W. Cameron, Harvard College Observatory, Cambridge, MA.
B. Campbell, Steward Observatory, University of Arizona, Tucson, AZ.
H. Campins, Planetary Science Institute, Tucson, AZ.
V. Canuto, Institute for Space Studies, New York, NY.
E. D. Carlson, Adler Planetarium, Chicago, IL.
P. Cassen, NASA Ames Research Center, Moffett Field, CA.
S. Chang, NASA Ames Research Center, Moffett Field, CA.
C. R. Chapman, Planetary Science Institute, Tucson, AZ.
J. Christy, Hughes Aircraft, Tucson, AZ.
D. D. Clayton, Department of Physics and Astronomy, Rice University, Houston, TX.
R. N. Clayton, Enrico Fermi Institute, University of Chicago, Chicago, IL.
D. Clemens, University of Massachusetts, Amherst, MA.
M. Cohen, NASA Ames Research Center, Moffett Field, CA.
C. Cook, University of Arizona Press, Tucson, AZ.
S. F. Cox, University of Arizona Press, Tucson, AZ.
D. Davis, Planetary Science Institute, Tucson, AZ.

D. den Boer, The Composing Room, Grand Rapids, MI.
K. Denomy, Lunar and Planetary Laboratory, University of Arizona, Tucson, AZ.
L. d'Hendencourt, Université de Paris, France.
R. L. Dickman, Five College Radio Astronomy Observatory, University of Massachusetts, Amherst, MA.
H. L. Dinnerstein, Department of Astronomy, University of Texas, Austin, TX.
B. T. Draine, Princeton University Observatory, Princeton University, Princeton, NJ.
R. H. Durisen, Astronomy Department, Indiana University, Bloomington, IN.
D. Eardley, Institute of Theoretical Physics, University of California, Santa Barbara, CA.
P. Eberhardt, Physics Institute, Universitat Bern, Bern, Switzerland.
S. Edwards, Clark Science Center, Northampton, MA.
B. G. Elmegreen, IBM T. J. Watson Research Center, Yorktown Heights, NY.
J. P. Emerson, Department of Physics, Queen Mary College, London, England.
N. Epchtein, Observatoire de Paris, Meudon, France.
N. J. Evans II, Department of Astronomy, University of Texas, Austin, TX.
M. Feierberg, NASA Ames Research Center, Moffett Field, CA.
F. Ferrini, Instituto di Astronomia, Pisa, Italy.
U. Finkenzeller, Landessternwarte, Heidelberg, West Germany.
R. C. Fleck, Physics Science Laboratory, Embry-Riddle Aeronautical University, Daytona Beach, FL.
M. Frerking, Jet Propulsion Laboratory, Pasadena, CA.
H. Fuhrmann, Jet Propulsion Laboratory, Pasadena, CA.
Y. Fukui, Department of Physics, Nagoya University, Nagoya, Japan.
G. Fuller, Department of Astronomy, University of California, Berkeley, CA.
B. Gaastra, The Composing Room, Grand Rapids, MI.
T. N. Gautier, Jet Propulsion Laboratory, Pasadena, CA.
T. Gehrels, Lunar and Planetary Laboratory, University of Arizona, Tucson, AZ.
R. Gehrz, Department of Physics and Astronomy, University of Wyoming, Laramie, WY.
R. Genzel, Department of Physics, University of California, Berkeley, CA.
D. Gilden, Institute for Advanced Study, Princeton, NJ.
F. C. Gillett, Astronomical Society of the Pacific, San Francisco, CA.
A. Glassgold, Physics Department, New York University, New York, NY.
I. Goldman, Goddard Institute for Space Studies, New York, NY 10025.
P. Goldsmith, Five College Radio Astronomy Observatory, University of Massachusetts, Amherst, MA.
J. Graham, Cerro Tololo, Chile.
J. M. Greenberg, Huygens Laboratory, Universiteit Leiden, Leiden, Netherlands.
M. R. Haas, NASA Ames Research Center, Moffett Field, CA.
P. Hacking, Jet Propulsion Laboratory, Pasadena, CA.
B. W. Hapke, Department of Geology and Planetary Science, University of Pittsburgh, PA.
A. W. Harris, Jet Propulsion Laboratory, Pasadena, CA.
W. Hartmann, Planetary Science Institute, Tucson, AZ.
P. Harvey, Department of Astronomy, University of Texas, Austin, TX.
T. Hasegawa, Nobeyama Radio Observatory, Nagano, Japan.
M. G. Hauser, NASA Goddard Space Flight Center, Greenbelt, MD.
C. E. Heiles, Astronomy Department, University of California, Berkeley, CA.
M. Hemcon-Heyer, Five College Radio Astronomy Observatory, University of Massachusetts, Amherst, MA.
A. Hendry, Flandrau Planetarium, University of Arizona, Tucson, AZ.
C. Henkel, Max-Planck-Institut für Radiophysik, Bonn, West Germany.
F. L. Herbert, Lunar and Planetary Laboratory, University of Arizona, Tucson, AZ.
G. H. Herbig, Lick Observatory, University of California, Santa Cruz, CA.
E. Herbst, Department of Physics, Duke University, Durham, NC.
Å. Hjalmarson, Onsala Space Observatory, Onsala, Sweden.
D. J. Hollenbach, NASA Ames Research Center, Moffett Field, CA.
W. Hubbard, Lunar and Planetary Laboratory, University of Arizona, Tucson, AZ.
O. Hubicjyj, Institute of Space Studies, New York, NY.
C. L. Imhoff, NASA Goddard Space Flight Center, Greenbelt, MD.

W.-H. Ip, Max-Planck-Institut für Radiophysik, Bonn, West Germany.
S. Isobe, Tokyo Astronomical Observatory, Tokyo, Japan.
T. Jones, Physics and Astronomy Department, University of Minnesota, Minneapolis, MN.
C. Jordan, Department of Theoretical Physics, Oxford University, Oxford, England.
W. M. Kaula, Department of Earth and Space Science, University of California, Los Angeles, CA.
J. Keene, Downs Laboratory of Physics, California Institute of Technology, Pasadena, CA.
J. F. Kerridge, Institute of Geophysics, University of California, Los Angeles, CA.
R. I. Klein, Lawrence Livermore National Laboratory, University of California, Livermore, CA.
S. Kleiner, Radio Astronomy, University of Massachusetts, Amherst, MA.
R. F. Knacke, Earth and Space Science Department, State University of New York, Stony Brook, NY.
A. Konigl, Astronomy and Astrophysics Center, University of Chicago, Chicago, IL.
L. V. Kuhi, Astronomy Department, University of California, Berkeley, CA.
A. Kyrala, Department of Physics, Arizona State University, Tempe, AZ.
J. Lacy, Department of Astronomy, University of California, Berkeley, CA.
C. Lada, Steward Observatory, University of Arizona, Tucson, AZ.
D. Lal, Institute of Geophysics and Planetary Physics, University of California, Los Angeles, CA.
A. P. Lane, National Radio Astronomy Observatory, Charlottesville, VA.
W. Langer, Plasma Physics Laboratory, Princeton University, Princeton, NJ.
R. B. Larson, Astronomy Department, Yale University, New Haven, CT.
N. R. Lebovitz, Department of Mathematics, University of Chicago, Chicago, IL.
D. Lester, Department of Astronomy, University of Texas, Austin, TX.
C. M. Leung, Department of Physics, Rensselaer Polytechnic Institute, Troy, NY.
R. Levreault, College of Natural Sciences, University of Texas, Austin, TX.
E. H. Levy, Lunar and Planetary Laboratory, University of Arizona, Tucson, AZ.
D. N. C. Lin, Lick Observatory, University of California, Santa Cruz, CA.
J. J. Lissauer, NASA Ames Research Center, Moffett Field, CA.
F. J. Low, Steward Observatory, University of Arizona, Tucson, AZ.
I. MacKinnon, NASA Johnson Space Center, Houston, TX.
M. Magisos, Lunar and Planetary Laboratory, University of Arizona, Tucson, AZ.
G. Magni, Instituto de Astrofisica Spaziale, Rome, Italy.
M. Margulis, Steward Observatory, University of Arizona, Tucson, AZ.
S. Marinus, Lunar and Planetary Laboratory, University of Arizona, Tucson, AZ.
H. Martin, National Radio Astronomy Observatory, Charlottesville, VA.
K. Mason, Tucson, AZ.
C. Masson, Downs Laboratory of Physics, California Institute of Technology, Pasadena, CA.
R. D. Mathieu, Center for Astrophysics, Cambridge, MA.
C. Matthews, Department of Chemistry, University of Illinois, Chicago, IL.
M. A. Matthews, Jurisprudence and Social Policy Program, University of California, Berkeley, CA.
M. S. Matthews, Lunar and Planetary Laboratory, University of Arizona, Tucson, AZ.
D. W. McCarthy, Steward Observatory, University of Arizona, Tucson, AZ.
C. F. McKee, Physics Department, University of California, Berkeley, CA.
G. McLaughlin, Lunar and Planetary Laboratory, University of Arizona, Tucson, AZ.
R. S. McMillan, Lunar and Planetary Laboratory, University of Arizona, Tucson, AZ.
D. McNally, University of London Observatory, London, England.
L. Mestel, Astronomy Center, University of Sussex, Brighton, England.
G. F. Mitchell, Department of Astronomy, St. Mary's University, Halifax, Nova Scotia, Canada.
H. Mizuno, Department of Terrestrial Magnetism, Carnegie Institution of Washington, Washington, D.C.
A Moneti, Department of Physics and Astronomy, University of Rochester, Rochester, NY.
J. Moran, Center for Astrophysics, Cambridge, MA.
G. Morfill, Max-Planck-Institut für Extraterrestrische Physik, Garching, West Germany.
T. S. Mullen, Lunar and Planetary Laboratory, University of Arizona, Tucson, AZ.

R. Mundt, Max-Planck-Institut für Astronomie, Heidelberg, West Germany.
P. Myers, Center for Astrophysics, Cambridge, MA.
A. Nadda, Osservatorio Astrofisica di Arcetri, Florence, Italy.
T. Nakano, Department of Physics, Kyoto University, Kyoto, Japan.
H. E. Newson, Max-Planck-Institut für Chemie, Mainz, West Germany.
M. L. Norman, Max-Planck-Institut für Astrophysik, Garching, West Germany.
J. Nuth, NASA Goddard Space Flight Center, Greenbelt, MD.
T. Owen, Department of Earth and Space Science, State University of New York, Stony Brook, NY.
R. Padman, Astronomy Department, University of California, Berkeley, CA.
S. J. Peale, Physics Department, University of California, Santa Barbara, CA.
S. Pearce, Department of Geophysics and Astronomy, University of British Columbia, Vancouver, B.C., Canada.
E. Persson, Las Campanas Observatory, Pasadena, CA.
M. Podolak, NASA Ames Research Center, Moffet Field, CA.
F. Podosek, McDonnell Center for the Space Sciences, Washington University, St. Louis, MO.
J. B. Pollack, NASA Ames Research Center, Moffett Field, CA.
S. Pope, Lunar and Planetary Laboratory, University of Arizona, Tucson, AZ.
A. Poveda, Kitt Peak National Observatory, Tucson, AZ.
S. S. Prasad, Jet Propulsion Laboratory, Pasadena, CA.
R. Probst, Kitt Peak National Observatory, Tucson, AZ.
R. Pudritz, Astronomy Department, University of California, Berkeley, CA.
D. M. Rank, Lick Observatory, University of California, Santa Cruz, CA.
H. Reeves, Centre d'Etudes Nucléaires de Saclay, Gif-sur-Yvette, France.
J. Reynolds, Department of Physics, University of California, Berkeley, CA.
R. Reynolds, NASA Ames Research Center, Moffett Field, CA.
F. J. M. Reitmeyer, NASA Johnson Space Center, Houston, TX.
S. Ruden, Lick Observatory, University of California, Santa Cruz, CA.
E. Rydgren, Space Telescope Institute, Baltimore, MD.
D. B. Sanders, Downs Laboratory, California Institute of Technology, Pasadena, CA.
K. Sanders, The Composing Room, Grand Rapids, MI.
M. T. Sanford II, Los Alamos National Laboratory, Los Alamos, NM.
A. Sargent, Downs Lab for Physics, California Institute of Technology, Pasadena, CA.
J. M. Scalo, Department of Astronomy, University of Texas, Austin, TX.
P. Schloerb, Astronomy Department, University of Massachusetts, Amherst, MA.
A. Schulz, Max-Plank-Institut für Radioastronomie, Bonn, West Germany.
P. R. Schwartz, Naval Research Laboratory, Washington, D.C.
R. D. Schwartz, Department of Physics, University of Missouri, St. Louis, MO.
E. H. Scott, NASA Goddard Space Flight Center, Greenbelt, MD.
N. Z. Scoville, Division of Math, Physics, and Astronomy, California Institute of Technology, Pasadena, CA.
H. Shaffer III, University of Arizona Press, Tucson, AZ.
W. Shelton, The Composing Room, Grand Rapids, MI.
D. Shirley, University of California, Los Angeles, CA.
F. H. Shu, Astronomy Department, University of California, Berkeley, CA.
J. I. Silk, Department of Astronomy, University of California, Berkeley, CA.
M. Simon, Department of Earth and Space Science, State University of New York, Stony Brook, NY.
W. L. Slattery, Los Alamos Scientific Laboratory, Los Alamos, NM.
D. H. Smith, *Sky and Telescope*, Cambridge, MA.
P. H. Smith, Lunar and Planetary Laboratory, University of Arizona, Tucson, AZ.
R. L. Snell, Five College Radio Astronomy Observatory, University of Massachusetts, Amherst, MA.
P. Solomon, Astronomy Program, State University of New York, Stony Brook, NY.
G. F. Spagna, Department of Physics, Rensselaer Polytechnic Institute, Troy, NY.
S. W. Stahler, Center for Astrophysics, Cambridge, MA.
S. Staley, Indianapolis, IN.
D. Stevenson, Geological and Planetary Sciences, California Institute of Technology, Pasadena, CA.

G. Stewart, NASA Ames Research Center, Moffett Field, CA.
G. Strazzulla, Osservatorio Astrofisico di Catania, Catania, Italy.
G. Streetfellow, Lick Observatory, University of California, Santa Cruz, CA.
S. Strom, Astronomy Program, University of Massachusetts, Amherst, MA.
J. Tarter, NASA Ames Research Center, Moffett Field, CA.
S. R. Taylor, Research School of Earth Science, Australia National University, Canberra Act, Australia.
S. Terebey, Astronomy Department, University of California, Berkeley, CA.
P. Thaddeus, Goddard Institute for Space Studies, New York, NY.
R. I. Thompson, Steward Observatory, University of Arizona, Tucson, AZ.
A. G. G. M. Tielens, NASA Ames Research Center, Moffett Field, CA.
J. Tohline, Department of Physics and Astronomy, Louisiana State University, Baton Rouge, LA.
M. V. Torbett, Department of Physics and Astronomy, Murray State University, Murray, KY.
J. L. Turner, Department of Astronomy, University of California, Berkeley, CA.
R. K. Ulrich, Astronomy Department, University of California, Los Angeles, CA.
P. vanden Bout, Astronomy Department, University of Texas, Austin, TX.
K. R. Villere, NASA Ames Research Center, Moffett Field, CA.
S. Vogel, Department of Astronomy, University of California, Berkeley, CA.
H. Völk, Max-Planck-Institut für Kernphysik, Heidelberg, West Germany.
F. Vrba, U. S. Naval Observatory, Flagstaff, AZ.
R. Walker, Jamieson Science Engineering, Bethesda, MD.
P. G. Wannier, Jet Propulsion Laboratory, Pasadena, CA.
D. W. Wark, Lunar and Planetary Laboratory, University of Arizona, Tucson, AZ.
G. J. Wasserburg, Geological and Planetary Sciences, California Institute of Technology, Pasadena, CA.
W. D. Watson, Department of Physics, University of Illinois, Urbana, IL.
S. Weidenschilling, Planetary Science Institute, Tucson, AZ.
P. Weissman, Jet Propulsion Laboratory, Pasadena, CA.
W. J. Welch, Department of Radio Astronomy, University of California, Berkeley, CA.
M. Werner, NASA Ames Research Center, Moffett Field, CA.
G. W. Wetherill, Department of Terrestrial Magnetism, Carnegie Institution of Washington, Washington, D.C.
R. W. Whitaker, Los Alamos National Laboratory, Los Alamos, NM.
G. White, Department of Physics, Queen Mary College, London, England.
B. A. Wilking, Department of Astronomy, University of Texas, Austin, TX.
I. P. Williams, Applied Math Department, Queen Mary College, London, England.
F. E. Witteborn, NASA Ames Research Center, Moffett Field, CA.
J. A. Wood, Center for Astrophysics, Cambridge, MA.
D. Wooden, NASA Ames Research Center, Moffett Field, CA.
P. R. Woodward, Lawrence Livermore Laboratory, University of California, Livermore, CA.
A. Wootten, National Radio Astronomy Observatory, Charlottesville, VA.
E. Young, Steward Observatory, University of Arizona, Tucson, AZ.
C. Yuan, NASA Ames Research Center, Moffett Field, CA.
D. S. Zak, Physics Department, Rensselaer Polytechnic Institute, Troy, NY.
H. Zinnecker, Royal Observatory, Edinburgh, Scotland.
M. Zolensky, NASA Johnson Space Center, Houston, TX.

The editors acknowledge the support of NASA Consortium Interchange NCA–1A040–402 for the preparation of the book. The following authors wish to acknowledge specific funds involved in supporting the preparation of their chapters.

Arquilla, R.; NSF Grant AST-82-12252.
Bodenheimer, P.; NSF Grants AST-81-00163 and AST-83-01229.
Boynton, W. V.; NASA Grant NAG-9-37.
Cameron, A. G. W.; NASA Grant NGR-22-007-269 and NGR-22-007-272; and NSF Grant AST-81-19545.

Clayton, R. N.; NSF Grant EAR-78-23680.
Dickman, R. L.; NSF Grant AST-82-12252.
Draine, B. T.; NSF Grant AST-82-09569.
Durisen, R. H.; NSF Grant AST-81-20367.
Elmegreen, B. G.; NSF Grant AST-82-19711.
Evans, N. J. II; NASA Grants NAG-2-199 and NAG-2-253; NSF Grant AST-83-12332.
Glassgold, A. E.; NASA Grant NGR-33-016-196.
Goldsmith, P. F.; NSF Grant AST-82-12252.
Harvey, P. M.; NASA Grant NAG-2-67.
Herbst, E.; NSF Grants AST-80-20321 and AST-83-12270.
Imhoff, C. L.; NASA Contract NAS-5-25774.
Irvine, W. M.; NASA Grant NAGW-436; NSF Grants AST-80-20321, AST-83-12270, AST-82-12252, and INT-82-05661; and Swedish Natural Science Research Council Grants F-FU 4619-101 and F-FU 4967-115.
Klein, R. I.; Livermore National Laboratory Contract W-7405-ENG-48 to the U. S. Dept. of Energy by the Univ. of California.
Lada, C. J.; NSF Grant AST-82-10643.
Levy, E. H.; NASA Grant NSG-7419.
Lin, D. N. C.; NSF Grant AST-83-01223.
Mayeda, T. K.; NSF Grant EAR-78-23680.
Molini-Velsko, C. A.; NSF Grant EAR-78-23680.
Owen, T.; NASA Grant NGR-33-015-141; and NASA Contracts 953614 and 955290.
Papaloizou, J.; NSF Grant AST-83-01223.
Podolak, M.; NASA Grant NCA2-OR340-002.
Rydgren, A. E.; NSF Grant AST-82-17851.
Sandford, M. T. II; Los Alamos National Laboratory Contract W-7405-ENG-3 to the U. S. Dept. of Energy by the Univ. of California.
Scalo, J. M.; NSF Grant AST-83-04032.
Schwartz, R. D.; NASA Grant NAG-5-243 and NSF Grant AST-82-01430.
Shu, F. H.; NSF Grant AST-83-14682.
Tohline, J. E.; NSF Grant AST-82-17744.
Wasserburg, G. J.; NASA Grant NAG-9-43.
Whitaker, R. W.; Los Alamos National Laboratory Contract W-7405-ENG-3 to the U. S. Dept. of Energy by the Univ. of California.
Wilking, B. A.; NSF Grants AST-81-16403 and AST-82-10643.
Wood, J. A.; NASA Grant NAG-9-28.

INDEX

INDEX